Atomic Numbers and Atomic Weights[a]

Actinium	Ac	89	(227)	Mercury	Hg	80	200.59
Aluminum	Al	13	26.981538	Molybdenum	Mo	42	95.94
Americium	Am	95	(243)	Neodymium	Nd	60	144.24
Antimony	Sb	51	121.760	Neon	Ne	10	20.1797
Argon	Ar	18	39.948	Neptunium	Np	93	(237)
Arsenic	As	33	74.92160	Nickel	Ni	28	58.6934
Astatine	At	85	(210)	Niobium	Nb	41	92.90638
Barium	Ba	56	137.327	Nitrogen	N	7	14.00674
Berkelium	Bk	97	(247)	Nobelium	No	102	(259)
Beryllium	Be	4	9.012182	Osmium	Os	76	190.23
Bismuth	Bi	83	208.98038	Oxygen	O	8	15.9994
Boron	B	5	10.811	Palladium	Pd	46	106.42
Bromine	Br	35	79.904	Phosphorus	P	15	30.97376
Cadmium	Cd	48	112.41	Platinum	Pt	78	195.07
Calcium	Ca	20	40.078	Plutonium	Pu	94	(244)
Californium	Cf	98	(251)	Polonium	Po	84	(209)
Carbon	C	6	12.011	Potassium	K	19	39.0983
Cerium	Ce	58	140.116	Praseodymium	Pr	59	140.90765
Cesium	Cs	55	132.90545	Promethium	Pm	61	(145)
Chlorine	Cl	17	35.4527	Protactinium	Pa	91	231.03588
Chromium	Cr	24	51.9961	Radium	Ra	88	(226)
Cobalt	Co	27	58.93320	Radon	Rn	86	(222)
Copper	Cu	29	63.546	Rhenium	Re	75	186.207
Curium	Cm	96	(247)	Rhodium	Rh	45	102.90550
Dysprosium	Dy	66	162.50	Rubidium	Rb	37	85.4678
Einsteinium	Es	99	(252)	Ruthenium	Ru	44	101.07
Erbium	Er	68	167.26	Samarium	Sm	62	150.36
Europium	Eu	63	151.965	Scandium	Sc	21	44.95591
Fermium	Fm	100	(257)	Selenium	Se	34	78.96
Fluorine	F	9	18.998403	Silicon	Si	14	28.0855
Francium	Fr	87	(223)	Silver	Ag	47	107.8682
Gadolinium	Gd	64	157.25	Sodium	Na	11	22.989770
Gallium	Ga	31	69.723	Strontium	Sr	38	87.62
Germanium	Ge	32	72.61	Sulfur	S	16	32.066
Gold	Au	79	196.96655	Tantalum	Ta	73	180.9479
Hafnium	Hf	72	178.49	Technetium	Tc	43	(98)
Helium	He	2	4.002602	Tellurium	Te	52	127.60
Holmium	Ho	67	164.93032	Terbium	Tb	65	158.92534
Hydrogen	H	1	1.0079_4	Thallium	Tl	81	204.3833
Indium	In	49	114.818	Thorium	Th	90	232.0381
Iodine	I	53	126.90447	Thulium	Tm	69	168.93421
Iridium	Ir	77	192.22	Tin	Sn	50	118.710
Iron	Fe	26	55.845	Titanium	Ti	22	47.867
Krypton	Kr	36	83.80	Tungsten	W	74	183.84
Lanthanum	La	57	138.9055	Uranium	U	92	238.0289
Lawrencium	Lr	103	(262)	Vanadium	V	23	50.9415
Lead	Pb	82	207.2	Xenon	Xe	54	131.29
Lithium	Li	3	6.941	Ytterbium	Yb	70	173.04
Lutetium	Lu	71	174.967	Yttrium	Y	39	88.90585
Magnesium	Mg	12	24.3050	Zinc	Zn	30	65.39
Manganese	Mn	25	54.93805	Zirconium	Zr	40	91.224
Mendelevium	Md	101	(258)				

[a]From *Pure Appl. Chem.,* **71,** 1593 (1999). A value in parentheses is the mass number of the longest-lived isotope.

PHYSICAL CHEMISTRY

PHYSICAL CHEMISTRY

Fifth Edition

Ira N. Levine

Chemistry Department
Brooklyn College
City University of New York
Brooklyn, New York

Boston   Burr Ridge, IL   Dubuque, IA   Madison, WI   New York   San Francisco   St. Louis
Bangkok   Bogotá   Caracas   Kuala Lumpur   Lisbon   London   Madrid   Mexico City
Milan   Montreal   New Delhi   Santiago   Seoul   Singapore   Sydney   Taipei   Toronto

McGraw-Hill Higher Education

*A Division of The **McGraw-Hill** Companies*

PHYSICAL CHEMISTRY, FIFTH EDITION

Published by McGraw-Hill, a business unit of The McGraw-Hill Companies, Inc., 1221 Avenue of the Americas, New York, NY 10020. Copyright © 2002, 1995, 1988, 1983, 1978 by The McGraw-Hill Companies, Inc. All rights reserved. No part of this publication may be reproduced or distributed in any form or by any means, or stored in a database or retrieval system, without the prior written consent of The McGraw-Hill Companies, Inc., including, but not limited to, in any network or other electronic storage or transmission, or broadcast for distance learning.

Some ancillaries, including electronic and print components, may not be available to customers outside the United States.

This book is printed on acid-free paper.

1 2 3 4 5 6 7 8 9 0 DOW/DOW 0 9 8 7 6 5 4 3 2 1

ISBN 0-07-253495-8
ISBN 0-07-112242-7 (ISE)

Executive editor: *Kent A. Peterson*
Developmental editor: *Shirley R. Oberbroeckling*
Marketing manager: *Thomas Timp*
Marketing coordinator: *Tami Petsche*
Senior project manager: *Kay J. Brimeyer*
Production supervisor: *Sherry L. Kane*
Coordinator of freelance design: *David W. Hash*
Cover designer: *Rokusek Design*
Cover image: © *PhotoDisc, Inc.*
Supplement producer: *Jodi K. Banowetz*
Compositor: *Lachina Publishing Services*
Typeface: *10/12 Times Roman*
Printer: *R. R. Donnelley & Sons Company/Willard, OH*

Cover: Fourier-transform infrared spectrum of gas-phase methanol courtesy of Dr. David L. Sullivan and Dr. John Ekerdt, Department of Chemical Engineering, University of Texas at Austin.

Library of Congress Cataloging-in-Publication Data

Levine, Ira N. (date)
    Physical chemistry / Ira N. Levine.—5th ed.
      p. cm.
    Includes bibliographical references and index.
    ISBN 0-07-253495-8
    1. Chemistry, Physical and theoretical. I. Title.

QD453.3 .L48   2002
541.3—dc21   2001024009
                                                 CIP

INTERNATIONAL EDITION ISBN 0-07-112242-7
Copyright © 2002. Exclusive rights by The McGraw-Hill Companies, Inc., for manufacture and export. This book cannot be re-exported from the country to which it is sold by McGraw-Hill. The International Edition is not available in North America.

www.mhhe.com

To my mother and to the memory of my father

Table of Contents

Preface

This textbook is for the standard undergraduate course in physical chemistry.

In writing this book, I have kept in mind the goals of clarity, accuracy, and depth. To make the presentation easy to follow, the book gives careful definitions and explanations of concepts, full details of most derivations, and reviews of relevant topics in mathematics and physics. I have avoided a superficial treatment, which would leave the student with little real understanding of physical chemistry. Instead, I have aimed at a treatment that is as accurate, as fundamental, and as up-to-date as can readily be presented at the undergraduate level.

IMPROVEMENTS IN THE FIFTH EDITION

Problems. A key element of any physical chemistry course is problem solving. The following improvements in the problems were made in the new edition:

*   Sections 1.9 on study suggestions and 2.12 on problem solving were expanded.
*   Many true/false problems requiring thought rather than calculation were added to allow students to quickly assess how well they understand a given section.
*   Several new worked-out examples were added at places where I found students had difficulty solving problems.
*   The classification of problems by section was changed from a table at the beginning of the problems to headings for individual sections, making it easier to find the problems for a section.
*   Scientist trivia questions that add human interest were added (Probs. 18.71, 19.72, and 20.63).

Spreadsheet Applications. Spreadsheets provide an easy way for students to solve complicated physical chemistry problems without the need for programming. Detailed instructions are given for using a spreadsheet to

*   fit a C_P polynomial (Sec. 5.6)
*   solve a nonlinear equation (Sec. 6.4)
*   solve the simultaneous nonlinear equations occurring in simultaneous-equilibria problems (Sec. 6.5)
*   do linear and nonlinear least-squares fits of data (Sec. 7.3)
*   use an equation of state to calculate vapor pressure and molar volumes of liquid and vapor in equilibrium (Sec. 8.5)
*   compute a liquid–liquid phase diagram by minimization of G using the simple-solution model (Sec. 12.11)

Additional spreadsheet applications are

*   numerical integration of rate equations using the modified Euler method (Sec. 17.7 and Prob. 17.57)
*   fitting spectroscopic data to determine molecular constants (Prob. 21.32)

Quantum Chemistry. Advances in the power of computers and the development of new quantum-mechanical calculation methods such as density-functional theory are making molecular quantum-mechanical calculations an essential tool in many areas of chemistry. The treatment of quantum chemistry was strengthened by adding the following new sections:

- Section 20.10 on density-functional theory (DFT) gives students an understanding of the most computationally efficient and widely used computational method in molecular quantum chemistry.
- Section 20.12 on performing quantum chemistry calculations familiarizes students with the use of quantum chemistry programs. Several homework problems ask students to carry out quantum-mechanical and molecular-mechanics calculations (Probs. 20.55 to 20.59).
- Section 20.13 on molecular mechanics explains the basic ideas of this method widely used for treating very large molecules.
- Section 18.16 on Hermitian operators discusses some advanced topics in quantum mechanics. This section is optional and can be omitted without affecting the understanding of later sections.
- Section 21.17 on group theory introduces symmetry point groups and representations. This section can be omitted if desired.

Biological Applications. To reflect the ever-increasing importance of biological applications in chemistry, the following biochemical topics were added:

- negative pressures and ascent of sap (Sec. 8.4)
- freezing-point depression and antifreeze proteins in organisms (Sec. 12.3)
- mass-spectrometric methods to determine protein molecular weights (Sec. 12.3)
- estimation of bioconcentration factors from octanol–water partition coefficients (Sec. 12.7)
- use of a printed electrode as a glucose biosensor (Sec. 14.5)
- gel- and capillary-electrophoresis separation of biomolecules (Sec. 16.6)
- fluorescence in DNA sequencing (Sec. 21.11)
- CD spectra in studying conformations of biomolecules (Sec. 21.14)
- the atomic force microscope (Sec. 24.10)
- molecular-dynamics computer simulation of protein folding (Sec. 24.14)

Other New Topics. Other added or expanded topics are

- black-hole entropy (Sec. 3.8)
- failures of Le Châtelier's principle (Sec. 6.6)
- estimating thermodynamic properties using group additivity (Sec. 5.10)
- simultaneous reaction equilibria (Secs. 6.5 and 6.6)
- metallic liquid hydrogen (Sec. 7.4)
- computer programs for estimating thermodynamic properties (Sec. 5.10), calculating phase diagrams (Sec. 12.11), doing multiple-equilibria calculations (Sec. 11.6), integrating rate equations (Sec. 17.7), and performing quantum-chemistry calculations (Sec. 20.12).
- supercritical fluids (Sec. 8.3)
- models for nonelectrolyte activity coefficients (Sec. 10.5)
- treatments of electrolyte activity coefficients at high concentrations (Sec. 10.8)
- the surface microlayer (Sec. 13.3)
- use of aerogels to capture stardust (Sec. 13.6)
- magnetization and relaxation in NMR (Sec. 21.12)
- room-temperature ionic liquids (Sec. 24.5)

Simplifications. Some discussions have been simplified or shortened:

- Section 10.6 on electrolyte solutions was simplified by moving the thermodynamics of ion pairing to Sec. 10.9.
- Chapters 14 (Electrochemical Systems) and 16 (Transport Processes) were shortened by dropping a few minor topics.
- The Sec. 9.8 discussion of partial molar properties and equilibrium in ideally dilute solutions was shortened.
- The Sec. 15.10 equipartition-of-energy discussion was simplified and shortened.

Units and Symbols. The usage of units and symbols was revised:

- Discussion of gaussian units was dropped and only SI units are used for electricity and magnetism. However, some commonly used non-SI units such as atmospheres, torrs, calories, angstroms, debyes, and poises are still used occasionally to familiarize students with them.
- Symbols were changed to more fully conform with IUPAC recommendations.

FEATURES

Mathematics Level. Although the treatment is an in-depth one, the mathematics has been kept at a reasonable level and advanced mathematics unfamiliar to students is avoided. Since mathematics is a stumbling block for many students trying to master physical chemistry, I have included reviews of calculus (Secs. 1.6, 1.8, and 8.9).

Organization. The book is organized so that students can see the broad structure and logic of physical chemistry rather than feel they are being bombarded with a hodgepodge of formulas and ideas presented in random order. In line with this, the thermodynamics chapters are grouped together, as are those on quantum chemistry. Statistical mechanics is taken up after thermodynamics and quantum chemistry.

Equations. The equation numbers of equations important enough to memorize are starred. In developing theories and equations, I have clearly stated the assumptions and approximations made, so that students will be aware of when the results apply and when they do not apply. The conditions of applicability of important thermodynamic equations are explicitly stated alongside the equations.

Quantum Chemistry. The presentation of quantum chemistry steers a middle course between an excessively mathematical treatment that would obscure the physical ideas for most undergraduates and a purely qualitative treatment that does little beyond repeating what the student has learned in previous courses. The book discusses modern ab initio, density-functional, semiempirical, and molecular-mechanics calculations of molecular properties so that students can appreciate the value of such calculations to nontheoretical chemists.

Problems. Each chapter has a wide variety of problems (both calculational and conceptual), and answers to many of the numerical problems are given. The class time available for going over problems is usually limited, so a manual of solutions to the problems has been prepared and can be purchased by students on authorization of the instructor. Sections 2.12 and 1.9 give practical suggestions for problem solving and studying.

ACKNOWLEDGMENTS

The following professors provided reviews for the preparation of the fifth edition:

Linda Casson	Rutgers University
Lisa Chirlian	Bryn Mawr College
James Diamond	Linfield College
Jon Draeger	University of Pittsburgh, Bradford
Michael Eastman	Northern Arizona University
Drannan Hamby	Linfield College
James F. Harrison	Michigan State University
Robert Howard	University of Tulsa
Darrell Iler	Eastern Mennonite University
Robert A. Jacobson	Iowa State University
Raj Khanna	University of Maryland
Arthur Low	Tarleton State University
Jennifer Mihalick	University of Wisconsin, Oshkosh
Brian Moores	Randolph-Macon College
Thomas Murphy	University of Maryland
Stephan Prager	University of Minnesota
James Riehl	Michigan Technological University
Sanford Safron	Florida State University, Tallahassee
Donald Sands	University of Kentucky
Paul Siders	University of Minnesota, Duluth
Agnes Tenney	University of Portland
Michael Tubergen	Kent State University
Gary Washington	U.S. Military Academy
Michael Wedlock	Gettysburg College
John C. Wheeler	University of California, San Diego
Robb Wilson	Louisiana State University, Shreveport
Nancy Wu	Florida Memorial College
Gregory Zimmerman	Tennessee State University

Reviewers of previous editions were Professors Alexander R. Amell, S. M. Blinder, C. Allen Bush, Thomas Bydalek, Paul E. Cade, Donald Campbell, Gene B. Carpenter, Jefferson C. Davis, Jr., Allen Denio, Luis Echegoyen, Eric Findsen, L. Peter Gold, George D. Halsey, David O. Harris, Denis Kohl, Leonard Kotin, Willem R. Leenstra, John P. Lowe, Jack McKenna, Howard D. Mettee, George Miller, Alfred Mills, Mary Ondrechen, Laura Philips, Peter Politzer, Frank Prochaska, John L. Ragle, Roland R. Roskos, Theodore Sakano, George Schatz, Richard W. Schwenz, Robert Scott, Charles Trapp, George H. Wahl, Thomas H. Walnut, Grace Wieder, Robert Wiener, Richard E. Wilde, John R. Wilson, and Peter E. Yankwich.

Helpful suggestions for this and previous editions were provided by Professors Thomas Allen, Fitzgerald Bramwell, Dewey Carpenter, Norman C. Craig, John N. Cooper, Thomas G. Dunne, Hugo Franzen, Darryl Howery, Daniel J. Jacob, Bruno Linder, Madan S. Pathania, Jay Rasaiah, J. L. Schreiber, Fritz Steinhardt, Vicki Steinhardt, John C. Wheeler, and Grace Wieder and by my students.

I thank all these people for the considerable help they provided.

The help I received from the developmental editor Shirley Oberbroeckling, the project manager Kay Brimeyer, the designer David Hash, and their coworkers at McGraw-Hill is gratefully acknowledged.

I welcome any suggestions for improving the book that readers might have.

Ira N. Levine
INLevine@brooklyn.cuny.edu

Learning Aids

Mathematics and Physics Reviews

- Sections 1.6, 1.8, and 8.9 review calculus.
- Sections 2.1, 14.1, and 21.12 review classical mechanics, electrostatics, and magnetism.

Suggestions for Study and Problem Solving

- Section 1.9 gives practical advice on how to study physical chemistry.
- Section 2.12 gives detailed strategies for solving physical chemistry problems.

Summaries

- Each chapter concludes with a summary of the key points.
- Each chapter summary ends with a list of the specific kinds of calculations the student is expected to be able to do.
- Many longer sections end with a section summary.
- Systematic lists are given showing how to calculate q, w, ΔU, ΔH (Sec. 2.9), and ΔS (Sec. 3.4) for various kinds of processes.

72

2.8 A nonideal gas is heated slowly and expands reversibly at a constant pressure of 275 torr from a volume of 385 to 875 cm³. Find w in joules.

2.9 Using the P_1, V_1, P_2, and V_2 values of Example 2.2, find w for a reversible process that goes from state 1 to state 2 in Fig. 2.3 via a straight line (*a*) by calculating the area under the curve; (*b*) by using $w_{rev} = -\int_1^2 P\,dV$. [*Hint:* The equation of the straight line that goes through points x_1, y_1 and x_2, y_2 is $(y - y_1)/(x - x_1) = (y_2 - y_1)/(x_2 - x_1)$.]

2.10 It was stated in Sec. 2.2 that for a given change of state, w_{rev} can have any positive or negative value. Consider a change of state for which $P_2 = P_1$ and $V_2 > V_1$. For this change of state, use a *P-V* diagram to (*a*) sketch a process with $w_{rev} < 0$; (*b*) sketch a process with $w_{rev} > 0$. Remember that neither P nor V can be negative.

Section 2.3

2.11 Specific heats can be measured in a *drop calorimeter*; here, a heated sample is dropped into the calorimeter and the final temperature is measured. When 45.0 g of a certain metal at 70.0°C is added to 24.0 g of water (with $c_P = 1.00$ cal/g-°C) at 10.0°C in an insulated container, the final temperature is 20.0°C. (*a*) Find the specific heat capacity of the metal. (*b*) How much heat flowed from the metal to the water? *Note:* In (*a*), we are finding the average c_P over the temperature range of the experiment. To determine c_P as a function of T, one repeats the experiment many times, using different initial temperatures for the metal.

Section 2.4

2.12 True or false? (*a*) For every process, $\Delta E_{syst} = -\Delta E_{surr}$. (*b*) For every cyclic process, the final state of the system is the same as the initial state. (*c*) For every cyclic process, the final state of the surroundings is the same as the initial state of the surroundings. (*d*) For a closed system at rest with no fields present, the sum $q + w$ has the same value for every process that goes from a given state 1 to a given state 2. (*e*) If systems A and B each consist of pure liquid water at 1 bar pressure and if $T_A > T_B$, then the internal energy of system A must be greater than that of B.

2.13 For which of these systems is the system's energy conserved in every process: (*a*) a closed system; (*b*) an open system; (*c*) an isolated system; (*d*) a system enclosed in adiabatic walls?

2.14 One food calorie $= 10^3$ cal $= 1$ kcal. A typical adult ingests 2200 kcal/day. (*a*) Show that an adult uses energy at about the same rate as a 100-W lightbulb. (*b*) Calculate the total annual metabolic-energy expenditure of the 6×10^9 people on earth and compare it with the 4×10^{20} J per year energy used by the world economy. (Neglect the fact that children use less metabolic energy than adults.)

2.15 A mole of water vapor initially at 200°C and 1 bar undergoes a cyclic process for which $w = 145$ J. Find q for this process.

2.16 William Thomson tells of running into Joule in 1847 at Mont Blanc; Joule had with him his bride and a long ther-mometer with which he was going to "try for elevation of temperature in waterfalls." The Horseshoe Falls at Niagara Falls is 167 ft high and has a summer daytime flow rate of 2.55×10^6 L/s. (*a*) Calculate the maximum possible temperature difference between the water at the top and at the bottom of the falls. (The maximum possible increase occurs if no energy is transferred to such parts of the surroundings as the rocks at the base of the falls.) (*b*) Calculate the maximum possible internal-energy increase of the 2.55×10^6 L that falls each second. (Before it reaches the falls, more than half the water of the Niagara River is diverted to a canal or underground tunnels for use in hydroelectric power plants beyond the falls. These plants generate 4.4×10^9 W. A power surge at one of these plants led to the great blackout of November 9, 1965, which left 30 million people in the northeast United States and Ontario, Canada, without power for many hours.)

2.17 Imagine an isolated system divided into two parts, 1 and 2, by a rigid, impermeable, thermally conducting wall. Let heat q_1 flow into part 1. Use the first law to show that the heat flow for part 2 must be $q_2 = -q_1$.

2.18 Sometimes one sees the notation Δq and Δw for the heat flow into a system and the work done during a process. Explain why this notation is misleading.

2.19 Explain how liquid water can go from 25°C and 1 atm to 30°C and 1 atm in a process for which $q < 0$.

2.20 The potential energy stored in a spring is $\frac{1}{2} kx^2$, where k is the force constant of the spring and x is the distance the spring is stretched from equilibrium. Suppose a spring with force constant 125 N/m is stretched by 10.0 cm, placed in 112 g of water in an adiabatic container, and released. The mass of the spring is 20 g, and its specific heat capacity is 0.30 cal/(g °C). The initial temperature of the water and the spring is 18.000°C. The water's specific heat capacity is 1.00 cal/(g °C). Find the final temperature of the water.

2.21 Consider a system enclosed in a vertical cylinder fitted with a frictionless piston. The piston is a plate of negligible mass, on which is glued a mass m whose cross-sectional area is the same as that of the plate. Above the piston is a vacuum. (*a*) Use conservation of energy in the form $dE_{syst} + dE_{surr} = 0$ to show that for an adiabatic volume change $dE_{syst} = -mg\,dh - dK_{pist}$, where dh is the infinitesimal change in piston height, g is the gravitational acceleration, and dK_{pist} is the infinitesimal change in kinetic energy of the mass m. (*b*) Show that the equation in part (*a*) gives $dw_{irrev} = -P_{ext}\,dV - dK_{pist}$ for the irreversible work done on the system, where P_{ext} is the pressure exerted by the mass m on the piston plate.

2.22 Suppose the system of Prob. 2.21 is initially in equilibrium with $P = 1.000$ bar and $V = 2.00$ dm³. The external mass m is instantaneously reduced by 50% and held fixed thereafter, so that P_{ext} remains at 0.500 bar during the expansion. After undergoing oscillations, the piston eventually comes to rest. The final system volume is 6.00 dm³. Calculate w_{irrev}.

Section 2.5

2.23 True or false? (*a*) The quantities H, U, PV, ΔH, and $P\,\Delta V$ all have the same dimensions. (*b*) ΔH is defined only for

678

19.70 Is there a gravitational attraction between the electron and the proton in the H atom? If there is, why is this not taken into account in the Hamiltonian? Do a calculation to support your answer.

19.71 (a) Show that the maximum value of ψ_{2p_z} for $Z = 1$ is $\psi_{max} = 1/(2a)^{3/2}\pi^{1/2}e$. (b) Write a computer program that will vary z/a from 0.01 to 10 in steps of 0.01 and for each value of z/a will calculate values of y/a for which $|\psi_{2p_z}/\psi_{max}|$ is equal to a certain constant k, where the value of k is input at the start of the program. Note that for some values of z/a, there are no values of y/a that satisfy the condition. Be careful that spurious values of y/a are eliminated. (The output of this program can be used as input to a graphing program to graph contours of the $2p_z$ orbital.)

19.72 *Physicist trivia question.* Name the physicist referred to in each of the following descriptions. All names appear in Chapter 19. Two of these physicists have elements named after them. (a) This experimental physicist (rated the 10th greatest physicist of all times in a 1999 poll) was weak in mathematics. Norman Ramsey took a course given by him in the 1930s and found that when this physicist tried to derive in class the formula for Rutherford scattering of alpha particles, "he got completely fouled up in the math, and he finally ended up telling us to go home and work it out for ourselves." Later, Ramsey came to recognize the great physical insight this physicist had and Ramsey concluded that "an ability to make a formal mathematical derivation was not the criterion of being a good physicist." (b) He was friends with the Swiss psychoanalyst Carl Jung and contributed a chapter to a book written by Jung. Jung published analyses of many of the dreams of this physicist; the number 4 often occurred in these dreams. (c) He was one of the few twentieth-century physicists who did outstanding work in both experiments and theory. In the mid-1930s, he and coworkers bombarded many elements with neutrons and produced radioactive products. He found that uranium irradiated with neutrons gave products whose atomic numbers did not lie in the range 86 to 92 and concluded that he had produced new elements with atomic numbers of 93 and 94, which he called ausenium and hasperium, respectively. He received a Nobel Prize in physics "for his demonstration of the existence of new radioactive elements produced by neutron irradiation, and for his related discovery of nuclear reactions brought about by slow neutrons." In fact, he had not prepared elements with $Z > 92$. One month after he received his Nobel Prize, Hahn and Strassmann published work showing that neutron irradiation of uranium gave barium as one product. Meitner and Frisch used Bohr's liquid-drop model of the nucleus to interpret the Hahn–Strassmann results as the fission of a uranium nucleus to produce two lighter nuclei. On December 2, 1942, the first human-produced self-sustaining nuclear-fission chain reaction was achieved on a squash court at the University of Chicago in a uranium pile constructed under the direction of the subject of this question. The success of the experiment was reported in a coded telephone conversation with the words "The Italian navigator has just landed in the New World."

19.73 True or false? (a) In this chapter, e stands for the charge on an electron. (b) The exact helium-atom ground-state wave function is a product of wave functions for each electron. (c) The wave function of every system of fermions must be antisymmetric with respect to interchange of all coordinates of any two particles. (d) The spin quantum number s of an electron has the possible values $\pm\frac{1}{2}$. (e) The shape of a $2p_z$ orbital is two tangent spheres. (f) All states belonging to the same electron configuration of a given atom must have the same energy. (g) Every solution of the time-independent Schrödinger equation is a possible stationary state. (h) The ground-state wave function of a lithium atom cannot be expressed as a spatial factor times a spin factor. (i) Every linear combination of two solutions of the time-independent Schrödinger equation is a solution of this equation.

Examples

- Numerous worked-out examples show students how to solve problems.
- Examples are generally followed by an exercise (with answer given) to allow students to test their ability to work a similar problem.

Equations

- Fundamental equations that students should memorize have their equation numbers marked with an asterisk.
- The conditions of applicability of thermodynamics equations are stated alongside the equations.

Problems

- A great variety of end-of-chapter problems, arranged by section, are provided.
- Many true/false questions requiring thought rather than calculation allow students to quickly assess their understanding.

Figure 7.8

Spreadsheet for finding $\Delta_{vap}H_m$. The lower view shows the formulas.

	A	B	C	D	E	F	G
1	vapor pressure H2O CC eqn lst sqs						
2	t₉₀/C	T/K	P/torr	ln P/torr	1/T	P(exp fit)	res-sqs
3	39.99	313.14	55.364	4.01393	0.003193	55.58574	0.04917
4	49.987	323.137	92.592	4.5282	0.003095	92.37595	0.046677
5	59.984	333.134	149.51	5.00736	0.003002	148.9069	0.363762
6	69.982	343.132	233.847	5.45467	0.002914	233.4571	0.152036
7	79.979	353.129	355.343	5.87308	0.002832	356.7985	2.118484
8						sumsqres	2.73013
9						b	m
10						20.43627	-5141.24

	A	B	C	D	E	F	G
1	vapor press						
2	t₉₀/C	T/K	P/torr	ln P/torr	1/T	P(exp fit)	res-sqs
3	=40-0.01	=A3+273.15	55.364	=LN(C3)	=1/B3	=EXP($F$10+$G$10/B3)	=(F3-C3)^2
4	=50-0.013	=A4+273.15	92.592	=LN(C4)	=1/B4	=EXP($F$10+$G$10/B4)	=(F4-C4)^2
5	=60-0.016	=A5+273.15	149.51	=LN(C5)	=1/B5	=EXP($F$10+$G$10/B5)	=(F5-C5)^2
6	=70-0.018	=A6+273.15	233.847	=LN(C6)	=1/B6	=EXP($F$10+$G$10/B6)	=(F6-C6)^2
7	=80-0.021	=A7+273.15	355.343	=LN(C7)	=1/B7	=EXP($F$10+$G$10/B7)	=(F7-C7)^2
8						sumsqres	=SUM(G3:G7)
9						b	m
10						20.43627	-5141.24

use Excel to get the coefficients *m* and *b* that give the best least-squares fit to the straight line *y* = *mx* + *b*, choose Data Analysis from the Tools menu and then choose Regression in the scroll-down list. (If Data Analysis is not on the Tools menu, choose Add-Ins on the Tools menu, check Analysis ToolPak and click OK.) In the dialog box that opens, enter D3 : D7 a[...] Input X range. Click the Output range butto[...] upper left cell of the least-squares-fit output[...] Residual Plots box, and click on OK. You g[...] The desired constants *b* and *m* are the tw[...] Coefficients. One finds 20.4363 as the inter[...] (the coefficient of the X Variable). (You are[...] bility that the slope lies in the range −509[...] [the deviations of the experimental ln (*P*/to[...] using the straight-line fit] are small, note tha[...] parabolic with positive residuals for the mic[...] als for the first and last points. This indicat[...] better than a straight line. (We know that [...] other approximations have been made.) For [...] of a straight-line fit, the residuals are rand[...] way in Excel to get the coefficients of a str[...] =SLOPE(D3:D7, E3:E7) and =INTEF[...] empty cells. A third way is to graph the data[...]

Spreadsheet and Computer Programs

- Detailed instructions (Secs. 5.6, 6.4, 6.5, 7.3, 8.5, and 12.11) teach students how to use a spreadsheet to easily solve complex physical chemistry problems such as simultaneous equilibria.
- Instructions are given for performing molecular electronic quantum chemistry calculations (Sec. 20.12).
- Internet addresses are given for obtaining software (some of it free) for doing equilibrium, kinetics, and quantum chemistry calculations (Secs. 5.10, 11.6, 12.11, 17.7, and 20.12).

EXAMPLE 16.4   Ionic radii in solution

Estimate the radii of $Li^+(aq)$ and $Na^+(aq)$, given that the viscosity of water at 25°C is 0.89 cP.

Equation (16.70), the u^∞ value of $Li^+(aq)$ in the table earlier in this section, and the relation 1 P = 0.1 N s m^{-2} [Eq. (16.14)] give

$$r(Li^+) \approx \frac{1(1.6 \times 10^{-19}\,C)}{6\pi(0.89 \times 10^{-3}\,N\,s\,m^{-2})[40 \times 10^{-5}(10^{-2}\,m)^2\,V^{-1}\,s^{-1}]}$$

$$\approx 2.4 \times 10^{-10}\,m = 2.4\,Å$$

Na^+ and Li^+ have the same charges, and (16.70) gives the radii as inversely proportional to the mobilities. Therefore, $r(Na^+) \approx (40/52)(2.4\,Å) = 1.8\,Å$. The larger size of $Li^+(aq)$ (despite the smaller atomic number of Li) is due to the larger n_h value of Li.

Exercise

In CH_3OH at 25°C and 1 atm, $u^\infty(Li^+) = 4.13 \times 10^{-4}$ cm^2/V-s, $u^\infty(Na^+) = 4.69 \times 10^{-4}$ cm^2/V-s, and $\eta = 0.55$ cP. Estimate the radii of Li^+ and Na^+ ions in methanol and compare with the values in water. (*Answers:* 3.7 Å and 3.3 Å.)

Stokes' law can be used to estimate how long it takes an ion to reach its terminal speed after the electric field is applied. From (16.70), the terminal speed equals $|z|eE/6\pi\eta r$. The force due to the electric field is $|z|eE$. If we neglect the frictional resistance, Newton's second law $F = ma = m\,dv/dt$ gives $|z|eE \approx m\,dv/dt$, which integrates to $v \approx |z|eEt/m$. Setting v equal to the terminal speed, we get $|z|eEt/m \approx |z|eE/6\pi\eta r$. The time needed to reach the terminal speed is then $t \approx m/6\pi\eta r$. For $m = 10^{-22}$ g, $\eta = 10^{-2}$ g s^{-1} cm^{-1}, and $r = 10^{-8}$ cm, we get $t \approx 10^{-13}$ s. Since we neglected F_{fr}, the actual time required is somewhat longer.

Electrophoresis

The migration of charged polymeric molecules (*polyelectrolytes*) and charged colloidal particles in an electric field is called **electrophoresis**. Electrophoresis can separate different proteins and different nucleic acids and is commonly done with a polymer gel (Sec. 13.6) as the medium. Electrophoresis "is the most important physical technique available" in biochemistry and molecular biology (K. E. van Holde et al., *Principles of Physical Biochemistry*, Prentice-Hall, 1998, sec. 5.3).

When electrophoresis is done in a free solvent, heating of the solvent by the electric current will produce convectional flow, which destroys the desired separation. Use of a gel eliminates the undesirable effects of convection. One commonly used gel is an agarose gel, which contains an aqueous medium dispersed in the pores of a three-dimensional network formed by a polysaccharide obtained from agar.

In a DNA (deoxyribonucleic acid) molecule, each phosphate group that links two deoxyriboses has one acidic hydrogen (hence the A in DNA). Ionization of these hydrogens gives DNA a negative charge in aqueous solution. The R side chains of 3 of the 20 amino acids $NH_2CHRCOOH$ that occur in proteins contain an amine group and the R chains of two contain a COOH group. In a buffered highly basic (high pH) solution of a protein, neutralization of the COOH groups produces COO^- groups that give the protein a negative charge. In buffered low-pH solutions, protonation of the amine groups gives the protein a positive charge. At a certain intermediate pH (the *isoelectric point*), the protein is uncharged.

Biochemical Topics

- Biochemical applications such as gel- and capillary electrophoresis separation of biomolecules (Sec. 16.6), the use of fluorescence in DNA sequencing (Sec. 21.11), and estimation of bioconcentration factors from octanol–water partition coefficients (Sec. 12.7) have been added.

Thermodynamics

1.1   PHYSICAL CHEMISTRY

Physical chemistry is the study of the underlying physical principles that govern the properties and behavior of chemical systems.

A chemical system can be studied from either a microscopic or a macroscopic viewpoint. The **microscopic** viewpoint is based on the concept of molecules. The **macroscopic** viewpoint studies large-scale properties of matter without explicit use of the molecule concept. The first half of this book uses mainly a macroscopic viewpoint; the second half uses mainly a microscopic viewpoint. The term *chemical physics* denotes those aspects of physical chemistry that study phenomena at the molecular level.

We can divide physical chemistry into four areas: thermodynamics, quantum chemistry, statistical mechanics, and kinetics (Fig. 1.1). **Thermodynamics** is a macroscopic science that studies the interrelationships of the various equilibrium properties of a system and the changes in equilibrium properties in processes. Thermodynamics is treated in Chapters 1 to 14.

Molecules and the electrons and nuclei that compose them do not obey classical mechanics. Instead, their motions are governed by the laws of quantum mechanics (Chapter 18). Application of quantum mechanics to atomic structure, molecular bonding, and spectroscopy gives us **quantum chemistry** (Chapters 19 to 21).

The macroscopic science of thermodynamics is a consequence of what is happening at a molecular (microscopic) level. The molecular and macroscopic levels are related to each other by the branch of science called **statistical mechanics.** Statistical mechanics gives insight into why the laws of thermodynamics hold and allows calculation of macroscopic thermodynamic properties from molecular properties. We shall study statistical mechanics in Chapters 15, 16, 22, 23, and 24.

Kinetics is the study of rate processes such as chemical reactions, diffusion, and the flow of charge in an electrochemical cell. The theory of rate processes is not as well developed as the theories of thermodynamics, quantum mechanics, and statistical mechanics. Kinetics uses relevant portions of thermodynamics, quantum chemistry, and statistical mechanics. Chapters 16, 17, and 23 deal with kinetics.

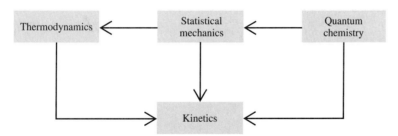

Figure 1.1

The four branches of physical chemistry. Statistical mechanics is the bridge from the microscopic approach of quantum chemistry to the macroscopic approach of thermodynamics. Kinetics uses portions of the other three branches.

The principles of physical chemistry provide a framework for all branches of chemistry.

Organic chemists use kinetics studies to figure out the mechanisms of reactions, use quantum-chemistry calculations to study the structures and stabilities of reaction intermediates, use symmetry rules deduced from quantum chemistry to predict the course of many reactions, and use nuclear-magnetic-resonance (NMR) and infrared spectroscopy to help determine the structure of compounds. Inorganic chemists use quantum chemistry and spectroscopy to study bonding. Analytical chemists use spectroscopy to analyze samples. Biochemists use kinetics to study rates of enzyme-catalyzed reactions; use thermodynamics to study biological energy transformations, osmosis, and membrane equilibrium, and to determine molecular weights of biological molecules; use spectroscopy to study processes at the molecular level (for example, intramolecular motions in proteins are studied using NMR); and use x-ray diffraction to determine the structures of proteins and nucleic acids.

Chemical engineers use thermodynamics to predict the equilibrium composition of reaction mixtures, use kinetics to calculate how fast products will be formed, and use principles of thermodynamic phase equilibria to design separation procedures such as fractional distillation. Geochemists use thermodynamic phase diagrams to understand processes in the earth. Polymer chemists use thermodynamics, kinetics, and statistical mechanics to investigate the kinetics of polymerization, the molecular weights of polymers, the flow of polymer solutions, and the distribution of conformations of a polymer molecule.

The term "physical chemistry" was used occasionally in the 1700s. For example, the Russian poet, physicist, and chemist Mikhail Lomonosov outlined a course in physical chemistry in 1752. Widespread recognition of physical chemistry as a discipline began in 1887 with the founding of the journal *Zeitschrift für Physikalische Chemie* by Wilhelm Ostwald with J. H. van't Hoff as coeditor. Ostwald investigated chemical equilibrium, chemical kinetics, and solutions and wrote the first textbook of physical chemistry. He was instrumental in drawing attention to Gibbs' pioneering work in chemical thermodynamics and was the first to nominate Einstein for a Nobel prize. Surprisingly, Ostwald argued against the atomic theory of matter and did not accept the reality of atoms and molecules until 1908. Ostwald, van't Hoff, and Arrhenius are generally regarded as the founders of physical chemistry. (In Sinclair Lewis's 1925 novel *Arrowsmith,* the character Max Gottlieb, a medical school professor, proclaims that "Physical chemistry is power, it is exactness, it is life.")

In its early years, physical chemistry research was done mainly at the macroscopic level. With the discovery of the laws of quantum mechanics in 1925–1926, emphasis began to shift to the molecular level. (The *Journal of Chemical Physics* was founded in 1933 in reaction to the refusal of the editors of the *Journal of Physical Chemistry* to publish theoretical papers.) Nowadays, the power of physical chemistry has been greatly increased by experimental techniques that study properties and processes at the molecular level and by fast computers that (*a*) process and analyze data of spectroscopy and x-ray crystallography experiments, (*b*) accurately calculate properties of molecules that are not too large, and (*c*) perform simulations of collections of hundreds of molecules.

1.2   THERMODYNAMICS

Thermodynamics

We begin our study of physical chemistry with thermodynamics. **Thermodynamics** (from the Greek words for "heat" and "power") is the study of heat, work, energy, and the changes they produce in the states of systems. In a broader sense, thermodynamics studies the relationships between the macroscopic properties of a system. A key property in thermodynamics is temperature, and thermodynamics is sometimes defined as the study of the relation of temperature to the macroscopic properties of matter.

We shall be studying **equilibrium thermodynamics,** which deals with systems in equilibrium. (**Irreversible thermodynamics** deals with nonequilibrium systems and rate processes.) Equilibrium thermodynamics is a macroscopic science and is independent of any theories of molecular structure. Strictly speaking, the word "molecule" is not part of the vocabulary of thermodynamics. However, we won't adopt a purist attitude but will often use molecular concepts to help us understand thermodynamics. Thermodynamics does not apply to systems that contain only a few molecules; a system must contain a great many molecules for it to be treated thermodynamically. The term "thermodynamics" in this book will always mean equilibrium thermodynamics.

Thermodynamic Systems

The macroscopic part of the universe under study in thermodynamics is called the **system.** The parts of the universe that can interact with the system are called the **surroundings.**

For example, to study the vapor pressure of water as a function of temperature, we might put a sealed container of water (with any air evacuated) in a constant-temperature bath and connect a manometer to the container to measure the pressure (Fig. 1.2). Here, the system consists of the liquid water and the water vapor in the container, and the surroundings are the constant-temperature bath and the mercury in the manometer.

An **open system** is one where transfer of matter between system and surroundings can occur. A **closed system** is one where no transfer of matter can occur between system and surroundings. An **isolated system** is one that does not interact in any way with its surroundings. An isolated system is obviously a closed system, but not every closed system is isolated. For example, in Fig. 1.2, the system of liquid water plus water vapor in the sealed container is closed (since no matter can enter or leave) but not isolated (since it can be warmed or cooled by the surrounding bath and can be compressed or expanded by the mercury). For an isolated system, neither matter nor energy can be transferred between system and surroundings. For a closed system, energy but not matter can be transferred between system and surroundings. For an open system, both matter and energy can be transferred between system and surroundings.

A thermodynamic system is either open or closed and is either isolated or nonisolated. Most commonly, we shall deal with closed systems.

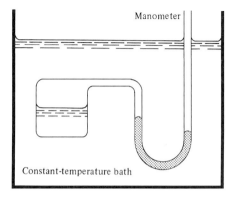

Manometer

Constant-temperature bath

Figure 1.2

A thermodynamic system and its surroundings.

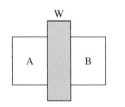

Figure 1.3

Systems A and B are separated by a wall W.

Walls

A system may be separated from its surroundings by various kinds of walls. (In Fig. 1.2, the system is separated from the bath by the container walls.) A wall can be either **rigid** or **nonrigid** (movable). A wall may be **permeable** or **impermeable,** where by "impermeable" we mean that it allows no matter to pass through it. Finally, a wall may be **adiabatic** or **nonadiabatic.** In plain language, an adiabatic wall is one that does not conduct heat at all, whereas a nonadiabatic wall does conduct heat. However, we have not yet defined heat, and hence to have a logically correct development of thermodynamics, adiabatic and nonadiabatic walls must be defined without reference to heat. This is done as follows.

Suppose we have two separate systems A and B, each of whose properties are observed to be constant with time. We then bring A and B into contact via a rigid, impermeable wall (Fig. 1.3). If, no matter what the initial values of the properties of A and B are, we observe no change in the values of these properties (for example, pressures, volumes) with time, then the wall separating A and B is said to be **adiabatic.** If we generally observe changes in the properties of A and B with time when they are brought in contact via a rigid, impermeable wall, then this wall is called **nonadiabatic** or **thermally conducting.** (As an aside, when two systems at different temperatures are brought in contact through a thermally conducting wall, heat flows from the hotter to the colder system, thereby changing the temperatures and other properties of the two systems; with an adiabatic wall, any temperature difference is maintained. Since heat and temperature are still undefined, these remarks are logically out of place, but they have been included to clarify the definitions of adiabatic and thermally conducting walls.) An adiabatic wall is an idealization, but it can be approximated, for example, by the double walls of a Dewar flask or thermos bottle, which are separated by a near vacuum.

In Fig. 1.2, the container walls are impermeable (to keep the system closed) and are thermally conducting (to allow the system's temperature to be adjusted to that of the surrounding bath). The container walls are essentially rigid, but if the interface between the water vapor and the mercury in the manometer is considered to be a "wall," then this wall is movable. We shall often deal with a system separated from its surroundings by a piston, which acts as a movable wall.

A system surrounded by a rigid, impermeable, adiabatic wall cannot interact with the surroundings and is isolated.

Equilibrium

Equilibrium thermodynamics deals with systems in **equilibrium.** An isolated system is in equilibrium when its macroscopic properties remain constant with time. A nonisolated system is in equilibrium when the following two conditions hold: (*a*) The system's macroscopic properties remain constant with time; (*b*) removal of the system from contact with its surroundings causes no change in the properties of the system. If condition (*a*) holds but (*b*) does not hold, the system is in a *steady state.* An example of a steady state is a metal rod in contact at one end with a large body at 50°C and in contact at the other end with a large body at 40°C. After enough time has elapsed, the metal rod satisfies condition (*a*); a uniform temperature gradient is set up along the rod. However, if we remove the rod from contact with its surroundings, the temperatures of its parts change until the whole rod is at 45°C.

The equilibrium concept can be divided into the following three kinds of equilibrium. For **mechanical equilibrium,** no unbalanced forces act on or within the system; hence the system undergoes no acceleration, and there is no turbulence within the system. For **material equilibrium,** no net chemical reactions are occurring in the system, nor is there any net transfer of matter from one part of the system to another or between the system and its surroundings; the concentrations of the chemical species in the various parts of the system are constant in time. For **thermal equilibrium** between a system and its surroundings, there must be no change in the properties of the system or surroundings when they are separated by a thermally conducting wall. Likewise, we can insert a thermally conducting wall between two parts of a system to test whether the parts are in thermal equilibrium with each other. For thermodynamic equilibrium, all three kinds of equilibrium must be present.

Thermodynamic Properties

What properties does thermodynamics use to characterize a system in equilibrium? Clearly, the **composition** must be specified. This can be done by stating the mass of each chemical species that is present in each phase. The **volume** V is a property of the system. The pressure P is another thermodynamic variable. **Pressure** is defined as the magnitude of the perpendicular force per unit area exerted by the system on its surroundings:

$$P \equiv F/A \qquad (1.1)*$$

where F is the magnitude of the perpendicular force exerted on a boundary wall of area A. The symbol $\equiv$ indicates a definition. *An equation with a star after its number should be memorized.* Pressure is a scalar, not a vector. For a system in mechanical equilibrium, the pressure throughout the system is uniform and equal to the pressure of the surroundings. (We are ignoring the effect of the earth's gravitational field, which causes a slight increase in pressure as one goes from the top to the bottom of the system.) If external electric or magnetic fields act on the system, the field strengths are thermodynamic variables; we won't consider systems with such fields. Later, further thermodynamic properties (for example, temperature, internal energy, entropy) will be defined.

An **extensive** thermodynamic property is one whose value is equal to the sum of its values for the parts of the system. Thus, if we divide a system into parts, the mass of the system is the sum of the masses of the parts; mass is an extensive property. So is volume. Properties that do not depend on the amount of matter in the system are called **intensive.** Density and pressure are examples of intensive properties. We can take a drop of water or a swimming pool full of water, and both systems will have the same density.

If each intensive macroscopic property is constant throughout a system, the system is **homogeneous.** If a system is not homogeneous, it may consist of a number of homogeneous parts. A homogeneous part of a system is called a **phase.** For example, if the system consists of a crystal of AgBr in equilibrium with an aqueous solution of AgBr, the system has two phases: the solid AgBr and the solution. A phase can consist of several disconnected pieces. For example, in a system composed of several AgBr crystals in equilibrium with an aqueous solution, all the crystals are part of the same phase. Note that the definition of a phase does not mention solids, liquids, or gases. A system can be entirely liquid (or entirely solid) and still have more than one phase. For example, a system composed of the nearly immiscible liquids H_2O and CCl_4 has two phases. A system composed of the solids diamond and graphite has two phases.

A system composed of two or more phases is **heterogeneous.**

The **density** ρ (rho) of a phase of mass m and volume V is

$$\rho \equiv m/V \qquad (1.2)*$$

Figure 1.4 plots some densities at room temperature and pressure. The symbols s, l, and g stand for solid, liquid, and gas.

Suppose that the value of every thermodynamic property in a certain thermodynamic system equals the value of the corresponding property in a second system. The

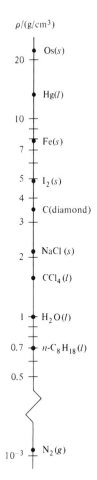

Figure 1.4

Densities at 25°C and 1 atm. The scale is logarithmic.

systems are then said to be in the same **thermodynamic state.** The state of a thermo-dynamic system is defined by specifying the values of its thermodynamic properties. However, it is not necessary to specify all the properties to define the state. Specification of a certain minimum number of properties will fix the values of all other properties. For example, suppose we take 8.66 g of pure H_2O at 1 atm (atmosphere) pressure and 24°C. It is found that in the absence of external fields all the remaining properties (volume, heat capacity, index of refraction, etc.) are fixed. (This statement ignores the possibility of surface effects, which are considered in Chapter 13.) Two thermodynamic systems each consisting of 8.66 g of H_2O at 24°C and 1 atm are in the same thermodynamic state. Experiments show that, for a single-phase system con-taining specified fixed amounts of nonreacting substances, specification of two addi-tional thermodynamic properties is generally sufficient to determine the thermody-namic state, provided external fields are absent and surface effects are negligible.

A thermodynamic system in a given equilibrium state has a particular value for each thermodynamic property. These properties are therefore also called **state func-tions,** since their values are functions of the system's state. The value of a state func-tion depends only on the present state of a system and not on its past history. It doesn't matter whether we got the 8.66 g of water at 1 atm and 24°C by melting ice and warm-ing the water or by condensing steam and cooling the water.

1.3 TEMPERATURE

Suppose two systems separated by a movable wall are in mechanical equilibrium with each other. Because we have mechanical equilibrium, no unbalanced forces act and each system exerts an equal and opposite force on the separating wall. Therefore each system exerts an equal pressure on this wall. Systems in mechanical equilibrium with each other have the same pressure. What about systems that are in thermal equilibrium (Sec. 1.2) with each other?

Just as systems in *mechanical* equilibrium have a common *pressure,* it seems plau-sible that there is some thermodynamic property common to systems in *thermal* equi-librium. This property is what we *define* as the **temperature,** symbolized by θ (theta). By definition, *two systems in thermal equilibrium with each other have the same tem-perature; two systems not in thermal equilibrium have different temperatures.*

Although we have asserted the existence of temperature as a thermodynamic state function that determines whether or not thermal equilibrium exists between systems, we need experimental evidence that there really is such a state function. Suppose that we find systems A and B to be in thermal equilibrium with each other when brought in con-tact via a thermally conducting wall. Further suppose that we find systems B and C to be in thermal equilibrium with each other. By our definition of temperature, we would assign the same temperature to A and B ($\theta_A = \theta_B$) and the same temperature to B and C ($\theta_B = \theta_C$). Therefore, systems A and C would have the same temperature ($\theta_A = \theta_C$), and we would expect to find A and C in thermal equilibrium when they are brought in con-tact via a thermally conducting wall. If A and C were not found to be in thermal equi-librium with each other, then our definition of temperature would be invalid. It is an experimental fact that:

Two systems that are each found to be in thermal equilibrium with a third sys-tem will be found to be in thermal equilibrium with each other.

This generalization from experience is *the zeroth law of thermodynamics.* It is so called because only after the first, second, and third laws of thermodynamics had been for-mulated was it realized that the zeroth law is needed for the development of thermody-namics. Moreover, a statement of the zeroth law logically precedes the other three. The zeroth law allows us to assert the existence of temperature as a state function.

Having defined temperature, how do we measure it? Of course, you are familiar with the process of putting a liquid-mercury thermometer in contact with a system, waiting until the volume change of the mercury has ceased (indicating that thermal equilibrium between the thermometer and the system has been reached), and reading the thermometer scale. Let us analyze what is being done here.

To set up a temperature scale, we pick a reference system r, which we call the **thermometer.** For simplicity, we choose r to be homogeneous with a fixed composition and a fixed pressure. Furthermore, we require that the substance of the thermometer must always expand when heated. This requirement ensures that at fixed pressure the volume of the thermometer r will define the state of system r uniquely— two states of r with different volumes at fixed pressure will not be in thermal equilibrium and must be assigned different temperatures. Liquid water is unsuitable for a thermometer since when heated at 1 atm, it contracts at temperatures below 4°C and expands above 4°C (Fig. 1.5). Water at 1 atm and 3°C has the same volume as water at 1 atm and 5°C, so the volume of water cannot be used to measure temperature. Liquid mercury always expands when heated, so let us choose a fixed amount of liquid mercury at 1 atm pressure as our thermometer.

We now assign a different numerical value of the temperature θ to each different volume V_r of the thermometer r. The way we do this is arbitrary. The simplest approach is to take θ as a linear function of V_r. We therefore *define* the temperature to be $\theta \equiv aV_r + b$, where V_r is the volume of a fixed amount of liquid mercury at 1 atm pressure and a and b are constants, with a being positive (so that states which are experienced physiologically as being hotter will have larger θ values). Once a and b are specified, a measurement of the thermometer's volume V_r gives its temperature θ.

The mercury for our thermometer is placed in a glass container that consists of a bulb connected to a narrow tube. Let the cross-sectional area of the tube be A, and let the mercury rise to a length l in the tube. The mercury volume equals the sum of the mercury volumes in the bulb and the tube, so

$$\theta \equiv aV_r + b = a(V_{\text{bulb}} + Al) + b = aAl + (aV_{\text{bulb}} + b) \equiv cl + d \qquad (1.3)$$

where c and d are constants defined as $c \equiv aA$ and $d \equiv aV_{\text{bulb}} + b$.

To fix c and d, we define the temperature of equilibrium between pure ice and liquid water saturated with dissolved air at 1 atm pressure as 0°C (for centigrade), and we define the temperature of equilibrium between pure liquid water and water vapor at 1 atm pressure (the normal boiling point of water) as 100°C. These points are called the *ice point* and the *steam point.* Since our scale is linear with the length of the mercury column, we mark off 100 equal intervals between 0°C and 100°C and extend the marks above and below these temperatures.

Having armed ourselves with a thermometer, we can now find the temperature of any system B. To do so, we put system B in contact with the thermometer, wait until thermal equilibrium is achieved, and then read the thermometer's temperature from the graduated scale. Since B is in thermal equilibrium with the thermometer, B's temperature equals that of the thermometer.

Note the arbitrary way we defined our scale. This scale depends on the expansion properties of a particular substance, liquid mercury. If we had chosen ethanol instead of mercury as the thermometric fluid, temperatures on the ethanol scale would differ slightly from those on the mercury scale. Moreover, there is at this point no reason, apart from simplicity, for choosing a linear relation between temperature and mercury volume. We could just as well have chosen θ to vary as $aV_r^2 + b$. Temperature is a fundamental concept of thermodynamics, and one naturally feels that it should be formulated less arbitrarily. Some of the arbitrariness will be removed in Sec. 1.5, where the ideal-gas temperature scale is defined. Finally, in Sec. 3.6 we shall define the most fundamental temperature scale, the thermodynamic scale. The mercury centigrade

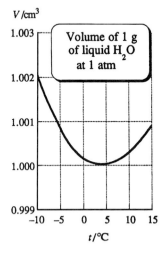

Figure 1.5

Volume of 1 g of water at 1 atm versus temperature. Below 0°C, the water is supercooled (Sec. 7.4).

scale defined in this section is not in current scientific use, but we shall use it until we define a better scale in Sec. 1.5.

Let systems A and B have the same temperature ($\theta_A = \theta_B$), and let systems B and C have different temperatures ($\theta_B \neq \theta_C$). Suppose we set up a second temperature scale using a different fluid for our thermometer and assigning temperature values in a different manner. Although the numerical values of the temperatures of systems A, B, and C on the second scale will differ from those on the first temperature scale, it follows from the zeroth law that on the second scale systems A and B will still have the same temperature, and systems B and C will have different temperatures. Thus, although numerical values on any temperature scale are arbitrary, the zeroth law assures us that the temperature scale will fulfill its function of telling whether or not two systems are in thermal equilibrium.

Since virtually all physical properties change with temperature, properties other than volume can be used to measure temperature. With a *resistance thermometer,* one measures the electrical resistance of a metal wire. A *thermistor* (which is used in a digital fever thermometer) is based on the temperature-dependent electrical resistance of a semiconducting metal oxide. A *thermocouple* involves the temperature dependence of the electric potential difference between two different metals in contact (Fig. 14.4). Very high temperatures can be measured with an *optical pyrometer,* which examines the light emitted by a hot solid. The intensity and frequency distribution of this light depend on the temperature (Fig. 18.1*b*), and this allows the solid's temperature to be found (see *Quinn,* chap. 7; references with the author's name italicized are listed in the Bibliography).

Temperature is an abstract property that is not measured directly. Instead, we measure some other property (for example, volume, electrical resistance, emitted radiation) whose value depends on temperature and (using the definition of the temperature scale and calibration of the measured property to that scale) we deduce a temperature value from the measured property.

Thermodynamics is a macroscopic science and does not explain the molecular meaning of temperature. We shall see in Sec. 15.3 that increasing temperature corresponds to increasing average molecular kinetic energy, provided the temperature scale is chosen to give higher temperatures to hotter states.

1.4 THE MOLE

We now review the concept of the mole, which is used in chemical thermodynamics.

The ratio of the average mass of an atom of an element to the mass of some chosen standard is called the **atomic weight** or **relative atomic mass** A_r of that element (the r stands for "relative"). The standard used since 1961 is $\frac{1}{12}$ times the mass of the isotope ^{12}C. The atomic weight of ^{12}C is thus exactly 12, by definition. The ratio of the average mass of a molecule of a substance to $\frac{1}{12}$ times the mass of a ^{12}C atom is called the **molecular weight** or **relative molecular mass** M_r of that substance. The statement that the molecular weight of H_2O is 18.015 means that a water molecule has on the average a mass that is 18.015/12 times the mass of a ^{12}C atom. We say "on the average" to acknowledge the existence of naturally occurring isotopes of H and O. Since atomic and molecular weights are *relative* masses, these "weights" are dimensionless numbers. For an ionic compound, the mass of one formula unit replaces the mass of one molecule in the definition of the molecular weight. Thus, we say that the molecular weight of NaCl is 58.443, even though there are no individual NaCl molecules in an NaCl crystal.

The number of ^{12}C atoms in exactly 12 g of ^{12}C is called **Avogadro's number.** Experiment (Sec. 19.2) gives 6.02×10^{23} as the value of Avogadro's number.

Avogadro's number of ^{12}C atoms has a mass of 12 g, exactly. What is the mass of Avogadro's number of hydrogen atoms? The atomic weight of hydrogen is 1.0079, so each H atom has a mass 1.0079/12 times the mass of a ^{12}C atom. Since we have equal numbers of H and ^{12}C atoms, the total mass of hydrogen is 1.0079/12 times the total mass of the ^{12}C atoms, which is (1.0079/12) (12 g) = 1.0079 g; this mass in grams is numerically equal to the atomic weight of hydrogen. The same reasoning shows that Avogadro's number of atoms of any element has a mass of A_r grams, where A_r is the atomic weight of the element. Similarly, Avogadro's number of molecules of a substance whose molecular weight is M_r will have a mass of M_r grams.

The average mass of an atom or molecule is called the **atomic mass** or the **molecular mass.** Molecular masses are commonly expressed in units of **atomic mass units** (amu), where 1 amu is one-twelfth the mass of a ^{12}C atom. With this definition, the atomic mass of C is 12.011 amu and the molecular mass of H_2O is 18.015 amu. Since 12 g of ^{12}C contains 6.02×10^{23} atoms, the mass of a ^{12}C atom is (12 g)/(6.02×10^{23}) and 1 amu = (1 g)/(6.02×10^{23}) = 1.66×10^{-24} g. The quantity 1 amu is called 1 dalton by biochemists, who express molecular masses in units of daltons.

A **mole** of some substance is defined as an amount of that substance which contains Avogadro's number of elementary entities. For example, a mole of hydrogen atoms contains 6.02×10^{23} H atoms; a mole of water molecules contains 6.02×10^{23} H_2O molecules. We showed earlier in this section that, if $M_{r,i}$ is the molecular weight of species i, then the mass of 1 mole of species i equals $M_{r,i}$ grams. The mass per mole of a pure substance is called its **molar mass** M. For example, for H_2O, $M = 18.015$ g/mole. The molar mass of substance i is

$$M_i \equiv \frac{m_i}{n_i} \qquad (1.4)*$$

where m_i is the mass of substance i in a sample and n_i is the number of moles of i in the sample. The molar mass M_i and the molecular weight $M_{r,i}$ of i are related by $M_i = M_{r,i} \times 1$ g/mole, where $M_{r,i}$ is a dimensionless number.

Following Eq. (1.4), n_i was called "the number of moles" of species i. Strictly speaking, this is incorrect. In the officially recommended SI units (Sec. 2.1), the **amount of substance** (also called the **chemical amount**) is taken as one of the fundamental physical quantities (along with mass, length, time, etc.), and the unit of this physical quantity is the mole, abbreviated mol. Just as the SI unit of mass is the kilogram, the SI unit of amount of substance is the mole. Just as the symbol m_i stands for the mass of substance i, the symbol n_i stands for the amount of substance i. The quantity m_i is not a pure number but is a number times a unit of mass; for example, m_i might be 4.18 kg (4.18 kilograms). Likewise, n_i is not a pure number but is a number times a unit of amount of substance; for example, n_i might be 1.26 mol (1.26 moles). Thus the correct statement is that n_i is the amount of substance i. The number of moles of i is a pure number and equals n_i/mol, since n_i has a factor of 1 mol included in itself.

Since Avogadro's number is the number of molecules in one mole, the number of molecules N_i of species i in a system is

$$N_i = (n_i/\text{mol}) \cdot (\text{Avogadro's number})$$

where n_i/mol is the number of moles of species i in the system. The quantity (Avogadro's number)/mol is called the **Avogadro constant** N_A. We have

$$N_i = n_i N_A \qquad \text{where } N_A = 6.02 \times 10^{23} \text{ mol}^{-1} \qquad (1.5)*$$

Avogadro's number is a pure number, whereas the Avogadro constant N_A has units of mole^{-1}.

Equation (1.5) applies to any collection of elementary entities, whether they are atoms, molecules, ions, radicals, electrons, photons, etc. Written in the form $n_i = N_i/N_A$,

Eq. (1.5) gives the definition of the amount of substance n_i of species i. In this equation, N_i is the number of elementary entities of species i.

If a system contains n_i moles of chemical species i and if n_{tot} is the total number of moles of all species present, then the **mole fraction** x_i of species i is

$$x_i \equiv n_i/n_{tot} \tag{1.6}*$$

The sum of the mole fractions of all species equals 1; $x_1 + x_2 + \cdots = n_1/n_{tot} + n_2/n_{tot} + \cdots = (n_1 + n_2 + \cdots)/n_{tot} = n_{tot}/n_{tot} = 1$.

1.5 IDEAL GASES

The laws of thermodynamics are general and do not refer to the specific nature of the system under study. Before studying these laws, we shall describe the properties of a particular kind of system, namely, an ideal gas. We shall then be able to illustrate the application of thermodynamic laws to an ideal-gas system. Ideal gases also provide the basis for a more fundamental temperature scale than the liquid-mercury scale of Sec. 1.3.

Boyle's Law
Boyle investigated the relation between the pressure and volume of gases in 1662 and found that, for a fixed amount of gas kept at a fixed temperature, P and V are inversely proportional:

$$PV = k \qquad \text{constant } \theta, m \tag{1.7}$$

where k is a constant and m is the gas mass. Careful investigation shows that Boyle's law holds only approximately for real gases, with deviations from the law approaching zero in the limit of zero pressure. Figure 1.6a shows some observed P-versus-V curves for 28 g of N_2 at two temperatures. Figure 1.6b shows plots of PV versus P for 28 g of N_2. Note the near constancy of PV at low pressures (below 10 atm) and the significant deviations from Boyle's law at high pressures.

Note how the axes in Fig. 1.6 are labeled. The quantity P equals a pure number times a unit; for example, P might be 4.0 atm = 4.0 × 1 atm. Therefore, P/atm (where the slash means "divided by") is a pure number, and the scales on the axes are marked with pure numbers. If $P = 4.0$ atm, then P/atm = 4.0. (If a column in a table is labeled $10^3 P$/atm, then an entry of 5.65 in this column would mean that $10^3 P$/atm = 5.65 and simple algebra gives $P = 5.65 \times 10^{-3}$ atm.)

Boyle's law is understandable from the picture of a gas as consisting of a huge number of molecules moving essentially independently of one another. The pressure exerted by the gas is due to the impacts of the molecules on the walls. A decrease in volume causes the molecules to hit the walls more often, thereby increasing the pressure. We shall derive Boyle's law from the molecular picture in Chapter 15, starting from a model of the gas as composed of noninteracting point particles. In actuality, the molecules of a gas exert forces on one another, so Boyle's law does not hold exactly. In the limit of zero density (reached as the pressure goes to zero or as the temperature goes to infinity), the gas molecules are infinitely far apart from one another, forces between molecules become zero, and Boyle's law is obeyed exactly. We say the gas becomes **ideal** in the zero-density limit.

Pressure and Volume Units
From the definition $P \equiv F/A$ [Eq. (1.1)], pressure has dimensions of force divided by area. In the SI system (Sec. 2.1), its units are newtons per square meter (N/m²), also called **pascals** (Pa):

$$1 \text{ Pa} \equiv 1 \text{ N/m}^2 \tag{1.8}*$$

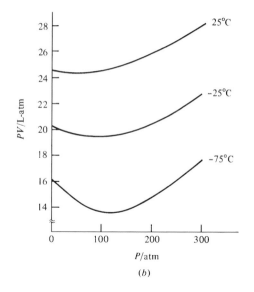
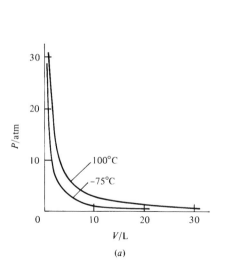

Figure 1.6

Plots of (a) P versus V and (b) PV versus P for 1 mole of N_2 gas at constant temperature.

Because 1 m^2 is a large area, the pascal is an inconveniently small unit of pressure, and its multiples the kilopascal (kPa) and megapascal (MPa) are often used: 1 kPa $\equiv 10^3$ Pa and 1 MPa $= 10^6$ Pa.

Chemists customarily use other units. One **torr** (or 1 mmHg) is the pressure exerted at 0°C by a column of mercury one millimeter high when the gravitational acceleration has the standard value $g = 980.665$ cm/s^2. The downward force exerted by the mercury equals its mass m times g. Thus a mercury column of height h, mass m, cross-sectional area A, volume V, and density ρ exerts a pressure P given by

$$P = F/A = mg/A = \rho V g/A = \rho A h g/A = \rho g h \tag{1.9}$$

The density of mercury at 0°C and 1 atm is 13.5951 g/cm^3. Converting this density to kg/m^3 and using (1.9) with $h = 1$ mm, we have

$$1 \text{ torr} = \left(13.5951 \frac{\text{g}}{\text{cm}^3}\right)\left(\frac{1 \text{ kg}}{10^3 \text{ g}}\right)\left(\frac{10^2 \text{ cm}}{1 \text{ m}}\right)^3 (9.80665 \text{ m/s}^2)(10^{-3} \text{ m})$$

$$1 \text{ torr} = 133.322 \text{ kg m}^{-1} \text{ s}^{-2} = 133.322 \text{ N/m}^2 = 133.322 \text{ Pa}$$

since 1 N = 1 kg m s^{-2} [Eq. (2.7)]. One **atmosphere** (atm) is defined as exactly 760 torr:

$$1 \text{ atm} \equiv 760 \text{ torr} = 1.01325 \times 10^5 \text{ Pa} \tag{1.10}$$

Another widely used pressure unit is the **bar**:

$$1 \text{ bar} \equiv 10^5 \text{ Pa} = 0.986923 \text{ atm} = 750.062 \text{ torr} \tag{1.11}$$

The bar is slightly less than 1 atm. The approximation

$$1 \text{ bar} \approx 750 \text{ torr} \tag{1.12}*$$

will usually be accurate enough for our purposes. See Fig. 1.7.

Common units of volume are cubic centimeters (cm^3), cubic decimeters (dm^3), cubic meters (m^3), and liters (L or l). The liter was originally defined as the volume of 1000 g of water at 3.98°C and 1 atm pressure. This made the liter equal to 1000.028 cm^3. However, in 1964 the **liter** was redefined as exactly 1000 cm^3. One liter equals 10^3 cm$^3 = 10^3(10^{-2}$ m$)^3 = 10^{-3}$ m$^3 = (10^{-1}$ m$)^3 = 1$ dm^3, where one decimeter (dm) equals 0.1 m.

$$1 \text{ liter} = 1 \text{ dm}^3 = 1000 \text{ cm}^3 \tag{1.13}*$$

Figure 1.7

Units of pressure. The scale is logarithmic.

Pressure Measurement.   Moderate pressures are measured with a *manometer,* a U-tube filled with mercury. Various gauges are used to measure low pressures. A *thermocouple gauge* uses a thermocouple to measure the temperature of a heated wire filament placed in the gas whose pressure is to be determined. At low pressures, the rate at which the gas conducts heat away from the filament is proportional to the gas pressure (see Sec. 16.2), so the wire's temperature depends on the gas pressure. A thermocouple gauge is calibrated with a *McLeod gauge.* Here, a large volume of low-pressure gas is compressed to a much smaller volume, where its pressure is measured. Boyle's law is then used to find the original pressure. High pressures can be measured with a *Bourdon gauge,* which contains a C-shaped or spiral hollow metal tube closed at one end. Fluid pressure within the tube tends to straighten it and thereby moves an indicator on a dial. The Bourdon gauge is calibrated using a fluid confined by a piston on which known weights are placed. Bourdon gauges are used on the outlet regulator valves of cylinders of compressed gas.

Charles' Law

Charles (1787) and Gay-Lussac (1802) measured the thermal expansion of gases and found a linear increase in volume with temperature (measured on the mercury centigrade scale) at constant pressure and fixed amount of gas:

$$V = a_1 + a_2\theta \qquad \text{const. } P, m \tag{1.14}$$

where a_1 and a_2 are constants. For example, Fig. 1.8 shows the observed relation between V and θ for 28 g of N_2 at a few pressures. Note the near linearity of the curves, which are at low pressures. The content of Charles' law is simply that the thermal expansions of gases and of liquid mercury are quite similar. The molecular explanation for Charles' law lies in the fact that an increase in temperature means the molecules are moving faster and hitting the walls harder and more often. Therefore, the volume must increase if the pressure is to remain constant.

The Ideal-Gas Absolute Temperature Scale

Charles' law (1.14) is obeyed most accurately in the limit of zero pressure; but even in this limit, gases still show small deviations from Eq. (1.14). These deviations are due to small differences between the thermal-expansion behavior of ideal gases and that of liquid mercury, which is the basis for the θ temperature scale. However, in the zero-pressure limit, the deviations from Charles' law are the *same* for different gases. In the limit of zero pressure, all gases show the same temperature-versus-volume behavior at constant pressure.

Extrapolation of the N_2 low-pressure V-versus-θ curves in Fig. 1.8 to low temperatures shows that they all intersect the θ axis at the same point, approximately $-273°$ on the mercury centigrade scale. Moreover, extrapolation of such curves for any gas, not just N_2, shows they intersect the θ axis at $-273°$. At this temperature, any ideal

Figure 1.8

Plots of volume versus centigrade temperature for 1 mole of N_2 gas at constant pressure.

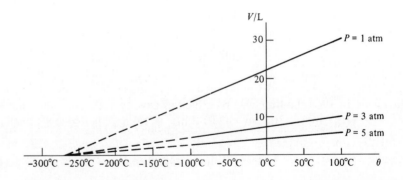

gas is predicted to have zero volume. (Of course, the gas will liquefy before this temperature is reached, and Charles' law will no longer be obeyed.)

As noted, all gases have the same temperature-versus-volume behavior in the zero-pressure limit. Therefore, to get a temperature scale that is independent of the properties of any one substance, we shall define an ideal-gas temperature scale T by the requirement that the T-versus-V behavior of a gas be exactly linear (that is, obey Charles' law exactly) in the limit of zero pressure. Moreover, because it seems likely that the temperature at which an ideal gas is predicted to have zero volume might well have fundamental significance, we shall take the zero of our ideal-gas temperature scale to coincide with the zero-volume temperature. We therefore define the **absolute ideal-gas temperature** T by the requirement that the relation $T \equiv BV$ shall hold exactly in the zero-pressure limit, where B is a constant for a fixed amount of gas at constant P, and where V is the gas volume. Any gas can be used.

To complete the definition, we specify B by picking a fixed reference point and assigning its temperature. In 1954 it was internationally agreed to use the triple point (tr) of water as the reference point and to define the absolute temperature T_{tr} at this triple point as exactly 273.16 K. The K stands for the unit of absolute temperature, the **kelvin,** formerly called the degree Kelvin (°K). (The water *triple point* is the temperature at which pure liquid water, ice, and water vapor are in mutual equilibrium.) At the water triple point, we have 273.16 K $\equiv T_{tr} = BV_{tr}$, and $B = (273.16 \text{ K})/V_{tr}$, where V_{tr} is the gas volume at T_{tr}. Therefore the equation $T \equiv BV$ defining the absolute ideal-gas temperature scale becomes

$$T \equiv (273.16 \text{ K}) \lim_{P \to 0} \frac{V}{V_{tr}} \qquad \text{const. } P, m \qquad (1.15)$$

How is the limit $P \to 0$ taken in (1.15)? One takes a fixed quantity of gas at some pressure P, say 200 torr. This gas is put in thermal equilibrium with the body whose temperature T is to be measured, keeping P constant at 200 torr and measuring the volume V of the gas. The gas thermometer is then put in thermal equilibrium with a water triple-point cell at 273.16 K, keeping P of the gas at 200 torr and measuring V_{tr}. The ratio V/V_{tr} is then calculated for $P = 200$ torr. Next, the gas pressure is reduced to, say, 150 torr, and the gas volume at this pressure is measured at temperature T and at 273.16 K; this gives the ratio V/V_{tr} at $P = 150$ torr. The operations are repeated at successively lower pressures to give further ratios V/V_{tr}. These ratios are then plotted against P, and the curve is extrapolated to $P = 0$ to give the limit of V/V_{tr} (see Fig. 1.9). Multiplication of this limit by 273.16 K then gives the ideal-gas absolute temperature T of the body. In practice, a constant-volume gas thermometer is easier to use than a constant-pressure one; here, V/V_{tr} at constant P in (1.15) is replaced by P/P_{tr} at constant V.

Accurate measurement of a body's temperature with an ideal-gas thermometer is tedious, and this thermometer is not useful for day-to-day laboratory work. What is done instead is to use an ideal-gas thermometer to determine accurate values for several fixed points that cover a wide temperature range. The fixed points are triple points and normal melting points of certain pure substances (for example, O_2, Ar, Zn, Ag). The specified values for these fixed points, together with specified interpolation formulas that use platinum resistance thermometers for temperatures between the fixed points, constitute the International Temperature Scale of 1990 (ITS-90). The ITS-90 scale is designed to reproduce the ideal-gas absolute scale within experimental error and is used to calibrate laboratory thermometers. Details of ITS-90 are given in B. W. Mangum, *J. Res. Natl. Inst. Stand. Technol.,* **95,** 69 (1990); *Quinn,* sec. 2-12 and appendix II.

Since the ideal-gas temperature scale is independent of the properties of any one substance, it is superior to the mercury centigrade scale defined in Sec. 1.3. However, the ideal-gas scale still depends on the limiting properties of *gases.* The thermodynamic

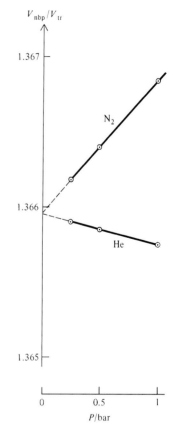

Figure 1.9

Constant-pressure gas thermometer plots to measure the normal boiling point (nbp) of H_2O. Extrapolation gives $V_{nbp}/V_{tr} = 1.36595_5$, so $T_{nbp} = 1.36595_5(273.16 \text{ K}) = 373.124$ K $= 99.974°C$.

temperature scale, defined in Sec. 3.6, is independent of the properties of any particular kind of matter. For now we shall use the ideal-gas scale.

The present definition of the **Celsius** (centigrade) **scale** t is in terms of the ideal-gas absolute temperature scale T as follows:

$$t \equiv T - 273.15° \qquad\qquad (1.16)*$$

The triple point of water is at $273.16 \text{ K} - 273.15° = 0.01°\text{C}$, exactly. On the present Celsius and Kelvin scales, the ice and steam points (Sec. 1.3) are not fixed but are determined by experiment, and there is no guarantee that these points will be at $0°\text{C}$ and $100°\text{C}$. However, the value 273.16 K for the water triple point and the number 273.15 in (1.16) were chosen to give good agreement with the old centigrade scale, so we expect the ice and steam points to be little changed from their old values. Experiment gives 0.0000 ± 0.0001 and $99.974°\text{C}$ for the ice and steam points.

Since the absolute ideal-gas temperature scale is based on the properties of a general class of substances (gases in the zero-pressure limit, where intermolecular forces vanish), one might suspect that this scale has fundamental significance. This is true, and we shall see in Eqs. (15.14) and (15.15) that the average kinetic energy of motion of molecules through space in a gas is directly proportional to the absolute temperature T. Moreover, the absolute temperature T appears in a simple way in the law that governs the distribution of molecules among energy levels; see Eq. (22.69), the Boltzmann distribution law.

From Eq. (1.15), at constant P and m we have $V/T = V_{tr}/T_{tr}$. This equation holds exactly only in the limit of zero pressure but is pretty accurate provided the pressure is not too high. Since V_{tr}/T_{tr} is a constant for a fixed amount of gas at fixed P, we have

$$V/T = K \qquad \text{const. } P, m$$

where K is a constant. This is Charles' law. However, logically speaking, this equation is not a law of nature but simply embodies the *definition* of the ideal-gas absolute temperature scale T. After defining the thermodynamic temperature scale, we can once again view $V/T = K$ as a law of nature.

The General Ideal-Gas Equation

Boyle's and Charles' laws apply when T and m or P and m are held fixed. Now consider a more general change in state of an ideal gas, in which the pressure, volume, and temperature all change, going from P_1, V_1, T_1 to P_2, V_2, T_2, with m unchanged. To apply Boyle's and Charles' laws, we imagine this process to be carried out in two steps:

$$P_1, V_1, T_1 \xrightarrow{(a)} P_2, V_a, T_1 \xrightarrow{(b)} P_2, V_2, T_2$$

Since T and m are constant in step (a), Boyle's law applies and $P_1V_1 = k = P_2V_a$; hence $V_a = P_1V_1/P_2$. Use of Charles' law for step (b) gives $V_a/T_1 = V_2/T_2$. Substitution of $V_a = P_1V_1/P_2$ into this equation gives $P_1V_1/P_2T_1 = V_2/T_2$, and

$$P_1V_1/T_1 = P_2V_2/T_2 \qquad \text{const. } m, \text{ ideal gas} \qquad (1.17)$$

What happens if we vary the mass m of ideal gas while keeping P and T constant? Volume is an extensive quantity, so V is directly proportional to m for any one-phase, one-component system at constant T and P. Thus V/m is constant at constant T and P. Combining this fact with the constancy of PV/T at constant m, we readily find (Prob. 1.23) that PV/mT remains constant for any variation in P, V, T, and m of any pure ideal gas: $PV/mT = c$, where c is a constant. There is no reason for c to be the same for different ideal gases, and in fact it is not. To obtain a form of the ideal-gas law that has the same constant for every ideal gas, we need another experimental observation.

In 1808 Gay-Lussac noted that the ratios of volumes of gases that react with one another involve small whole numbers when these volumes are measured at the same temperature and pressure. For example, one finds that two liters of hydrogen gas react with one liter of oxygen gas to form water. This reaction is $2H_2 + O_2 \rightarrow 2H_2O$, so the number of hydrogen molecules reacting is twice the number of oxygen molecules reacting. The two liters of hydrogen must then contain twice the number of molecules as does the one liter of oxygen, and therefore one liter of hydrogen will have the same number of molecules as one liter of oxygen at the same temperature and pressure. The same result is obtained for other gas-phase reactions. We conclude that equal volumes of different gases at the same temperature and pressure contain equal numbers of molecules. This idea was first recognized by Avogadro in 1811. (Gay-Lussac's law of combining volumes and Avogadro's hypothesis are strictly true for real gases only in the limit $P \rightarrow 0$.) Since the number of molecules is proportional to the number of moles, Avogadro's hypothesis states that equal volumes of different gases at the same T and P have equal numbers of moles.

Since the mass of a pure gas is proportional to the number of moles, the ideal-gas law $PV/mT = c$ can be rewritten as $PV/nT = R$ or $n = PV/RT$, where n is the number of moles of gas and R is some other constant. Avogadro's hypothesis says that, if P, V, and T are the same for two different gases, then n must be the same. But this can hold true only if R has the same value for every gas. R is therefore a universal constant, called the **gas constant.** The final form of the ideal-gas law is

$$PV = nRT \qquad \text{ideal gas} \qquad \textbf{(1.18)}*$$

Equation (1.18) incorporates Boyle's law, Charles' law (more accurately, the definition of T), and Avogadro's hypothesis.

An **ideal gas** is a gas that obeys $PV = nRT$. Real gases obey this law only in the limit of zero density, where intermolecular forces are negligible.

Using $M \equiv m/n$ [Eq. (1.4)] to introduce the molar mass M of the gas, we can write the ideal-gas law as

$$PV = mRT/M \qquad \text{ideal gas}$$

This form enables us to find the molecular weight of a gas by measuring the volume occupied by a known mass at a known T and P. For accurate results, one does a series of measurements at different pressures and extrapolates the results to zero pressure (see Prob. 1.20). We can also write the ideal-gas law in terms of the density $\rho = m/V$ as

$$P = \rho RT/M \qquad \text{ideal gas}$$

The only form worth remembering is $PV = nRT$, since all other forms are easily derived from this one.

The gas constant R can be evaluated by taking a known number of moles of some gas held at a known temperature and carrying out a series of pressure–volume measurements at successively lower pressures. Evaluation of the zero-pressure limit of PV/nT then gives R (Prob. 1.19). The experimental result is

$$R = 82.06 \, (\text{cm}^3 \, \text{atm})/(\text{mol K}) \qquad \textbf{(1.19)}*$$

Since 1 atm $= 101325$ N/m^2 [Eq. (1.10)], we have 1 cm^3 atm $= (10^{-2}$ m$)^3 \times 101325$ N/m$^2 = 0.101325$ m^3 N/m$^2 = 0.101325$ J. [One newton-meter $=$ one joule (J); see Sec. 2.1.] Hence $R = 82.06 \times 0.101325$ J/(mol K), or

$$R = 8.314_5 \, \text{J}/(\text{mol K}) = 8.314_5 \, (\text{m}^3 \, \text{Pa})/(\text{mol K}) \qquad \textbf{(1.20)}*$$

Using 1 atm $= 760$ torr and 1 bar ≈ 750 torr, we find from (1.19) that $R = 83.14_5$ (cm^3 bar)/(mol K). Using 1 calorie (cal) $= 4.184$ J [Eq. (2.44)], we find

$$R = 1.987 \, \text{cal}/(\text{mol K}) \qquad \textbf{(1.21)}*$$

Accurate values of physical constants are listed inside the back cover.

Ideal Gas Mixtures

So far, we have considered only a pure ideal gas. In 1810 Dalton found that the pressure of a mixture of gases equals the sum of the pressures each gas would exert if placed alone in the container. (This law is exact only in the limit of zero pressure.) If n_1 moles of gas 1 is placed alone in the container, it would exert a pressure n_1RT/V (where we assume the pressure low enough for the gas to behave essentially ideally). Dalton's law asserts that the pressure in the gas mixture is $P = n_1RT/V + n_2RT/V + \cdots = (n_1 + n_2 + \cdots)RT/V = n_{tot}RT/V$, so

$$PV = n_{tot}RT \qquad \text{ideal gas mixture} \qquad (1.22)*$$

Dalton's law makes sense from the molecular picture of gases. Ideal-gas molecules do not interact with one another, so the presence of gases 2, 3, . . . has no effect on gas 1, and its contribution to the pressure is the same as if it alone were present. Each gas acts independently, and the pressure is the sum of the individual contributions. For real gases, the intermolecular interactions in a mixture differ from those in a pure gas, and Dalton's law does not hold accurately.

The **partial pressure** P_i of gas i in a gas mixture (ideal or nonideal) is defined as

$$P_i \equiv x_iP \qquad \text{any gas mixture} \qquad (1.23)*$$

where $x_i = n_i/n_{tot}$ is the mole fraction of i in the mixture and P is the mixture's pressure. For an ideal gas mixture, $P_i = x_iP = (n_i/n_{tot})(n_{tot}RT/V)$ and

$$P_i = n_iRT/V \qquad \text{ideal gas mixture} \qquad (1.24)*$$

The quantity n_iRT/V is the pressure that gas i of the mixture would exert if it alone were present in the container. However, for a nonideal gas mixture, the partial pressure P_i as defined by (1.23) is not necessarily equal to the pressure that gas i would exert if it alone were present.

EXAMPLE 1.1 Density of an ideal gas

Find the density of F_2 gas at 20.0°C and 188 torr.

The unknown is the density ρ, and it is often a good idea to start by writing the definition of what we want to find: $\rho \equiv m/V$. Neither m nor V is given, so we seek to relate these quantities to the given information. The system is a gas at a relatively low pressure, and it is a good approximation to treat it as an ideal gas. For an ideal gas, we know that $V = nRT/P$. Substitution of $V = nRT/P$ into $\rho = m/V$ gives $\rho = mP/nRT$. In this expression for ρ, we know P and T but not m or n. However, we recognize that the ratio m/n is the mass per mole, that is, the molar mass M. Thus $\rho = MP/RT$. This expression contains only known quantities, so we are ready to substitute in numbers. The molecular weight of F_2 is 38.0, and its molar mass is $M = 38.0$ g/mol. The absolute temperature is $T = 20.0° + 273.15° = 293.2$ K. Since we know a value of R involving atmospheres, we convert P to atmospheres: $P = (188 \text{ torr})(1 \text{ atm}/760 \text{ torr}) = 0.247$ atm. Then

$$\rho = \frac{MP}{RT} = \frac{(38.0 \text{ g mol}^{-1})(0.247 \text{ atm})}{(82.06 \text{ cm}^3 \text{ atm mol}^{-1} \text{ K}^{-1})(293.2 \text{ K})} = 3.90 \times 10^{-4} \text{ g/cm}^3$$

Note that the units of temperature, pressure, and amount of substance (moles) canceled. The fact that we ended up with units of grams per cubic centimeter, which is a correct unit for density, provides a check on our work. *It is strongly recommended that the units of every physical quantity be written down when doing calculations.*

Exercise

Find the molar mass of a gas whose density is 1.80 g/L at 25.0°C and 880 torr. (*Answer:* 38.0 g/mol.)

1.6   DIFFERENTIAL CALCULUS

Physical chemistry uses calculus extensively. We therefore review some ideas of differential calculus. (In the novel *Arrowsmith,* Max Gottlieb asks Martin Arrowsmith, "How can you know physical chemistry without much mathematics?")

Functions and Limits

To say that the variable y is a **function** of the variable x means that for any given value of x there is specified a value of y; we write $y = f(x)$. For example, the area of a circle is a function of its radius r, since the area can be calculated from r by the expression πr^2. The variable x is called the *independent variable* or the *argument* of the function f, and y is the *dependent variable*. Since we can solve for x in terms of y to get $x = g(y)$, it is a matter of convenience which variable is considered to be the independent one. Instead of $y = f(x)$, one often writes $y = y(x)$.

To say that the **limit** of the function $f(x)$ as x approaches the value a is equal to c [which is written as $\lim_{x \to a} f(x) = c$] means that for all values of x sufficiently close to a (but *not* necessarily equal to a) the difference between $f(x)$ and c can be made as small as we please. For example, suppose we want the limit of $(\sin x)/x$ as x goes to zero. Note that $(\sin x)/x$ is undefined at $x = 0$, since 0/0 is undefined. However, this fact is irrelevant to determining the limit. To find the limit, we calculate the following values of $(\sin x)/x$, where x is in radians: 0.99833 for $x = \pm 0.1$, 0.99958 for $x = \pm 0.05$, 0.99998 for $x = \pm 0.01$, etc. Therefore

$$\lim_{x \to 0} \frac{\sin x}{x} = 1$$

Of course, this isn't a rigorous proof. Note the resemblance to taking the limit as $P \to 0$ in Eq. (1.15); in this limit both V and V_{tr} become infinite as P goes to zero, but the limit has a well-defined value even though ∞/∞ is undefined.

Slope

The **slope** of a straight-line graph, where y is plotted on the vertical axis and x on the horizontal axis, is defined as $(y_2 - y_1)/(x_2 - x_1) = \Delta y/\Delta x$, where (x_1, y_1) and (x_2, y_2) are the coordinates of any two points on the graph, and Δ (capital delta) denotes the change in a variable. If we write the equation of the straight line in the form $y = mx + b$, it follows from this definition that the line's slope equals m. The **intercept** of the line on the y axis equals b, since $y = b$ when $x = 0$.

The **slope** of any curve at some point P is defined to be the slope of the straight line tangent to the curve at P. For an example of finding a slope, see Fig. 9.3. Students sometimes err in finding a slope by trying to evaluate $\Delta y/\Delta x$ by counting boxes on the graph paper, forgetting that the scale of the y axis usually differs from that of the x axis in physical applications.

In physical chemistry, one often wants to define new variables to convert an equation to the form of a straight line. One then plots the experimental data using the new variables and uses the slope or intercept of the line to determine some quantity.

EXAMPLE 1.2 Converting an equation to linear form

According to the Arrhenius equation (17.66), the rate coefficient k of a chemical reaction varies with absolute temperature according to the equation $k = Ae^{-E_a/RT}$, where A and E_a are constants and R is the gas constant. Suppose we have measured values of k at several temperatures. Transform the Arrhenius equation to the form of a straight-line equation whose slope and intercept will enable A and E_a to be found.

The variable T appears as part of an exponent. By taking the logs of both sides, we eliminate the exponential. Taking the natural logarithm of each side of $k = Ae^{-E_a/RT}$, we get $\ln k = \ln(Ae^{-E_a/RT}) = \ln A + \ln(e^{-E_a/RT}) = \ln A - E_a/RT$, where Eq. (1.65) was used. To convert the equation $\ln k = \ln A - E_a/RT$ to a straight-line form, we define new variables in terms of the original variables k and T as follows: $y \equiv \ln k$ and $x \equiv 1/T$. This gives $y = (-E_a/R)x + \ln A$. Comparison with $y = mx + b$ shows that a plot of $\ln k$ on the y axis versus $1/T$ on the x axis will have slope $-E_a/R$ and intercept $\ln A$. From the slope and intercept of such a graph, E_a and A can be calculated.

Exercise

The moles n of a gas adsorbed divided by the mass m of a solid adsorbent often varies with gas pressure P according to $n/m = aP/(1 + bP)$, where a and b are constants. Convert this equation to a straight-line form, state what should be plotted versus what, and state how the slope and intercept are related to a and b. (*Hint:* Take the reciprocal of each side.)

Derivatives

Let $y = f(x)$. Let the independent variable change its value from x to $x + h$; this will change y from $f(x)$ to $f(x + h)$. The average rate of change of y with x over this interval equals the change in y divided by the change in x and is

$$\frac{\Delta y}{\Delta x} = \frac{f(x + h) - f(x)}{(x + h) - x} = \frac{f(x + h) - f(x)}{h}$$

The *instantaneous* rate of change of y with x is the limit of this average rate of change taken as the change in x goes to zero. The instantaneous rate of change is called the **derivative** of the function f and is symbolized by f':

$$f'(x) \equiv \lim_{h \to 0} \frac{f(x + h) - f(x)}{h} = \lim_{\Delta x \to 0} \frac{\Delta y}{\Delta x} \qquad (1.25)*$$

Figure 1.10 shows that *the derivative of the function $y = f(x)$ at a given point is equal to the slope of the curve of y versus x at that point.*

As a simple example, let $y = x^2$. Then

$$f'(x) = \lim_{h \to 0} \frac{(x + h)^2 - x^2}{h} = \lim_{h \to 0} \frac{2xh + h^2}{h} = \lim_{h \to 0} (2x + h) = 2x$$

The derivative of x^2 is $2x$.

A function that has a sudden jump in value at a certain point is said to be **discontinuous** at that point. An example is shown in Fig. 1.11a. Consider the function $y = |x|$, whose graph is shown in Fig. 1.11b. This function has no jumps in value anywhere and so is everywhere **continuous.** However, the slope of the curve changes suddenly at $x = 0$. Therefore, the derivative y' is discontinuous at this point; for negative x the

Figure 1.10

As point 2 approaches point 1, the quantity $\Delta y/\Delta x = \tan \theta$ approaches the slope of the tangent to the curve at point 1.

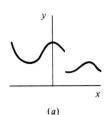

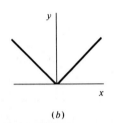

Figure 1.11

(a) A discontinuous function.
(b) The function $y = |x|$.

function y equals $-x$ and y' equals -1, whereas for positive x the function y equals x and y' equals $+1$.

Since $f'(x)$ is defined as the limit of $\Delta y/\Delta x$ as Δx goes to zero, we know that, for small changes in x and y, the derivative $f'(x)$ will be approximately equal to $\Delta y/\Delta x$. Thus $\Delta y \approx f'(x)\,\Delta x$ for Δx small. This equation becomes more and more accurate as Δx gets smaller. We can conceive of an infinitesimally small change in x, which we symbolize by dx. Denoting the corresponding infinitesimally small change in y by dy, we have $dy = f'(x)\,dx$, or

$$dy = y'(x)\,dx \qquad (1.26)*$$

The quantities dy and dx are called **differentials.** Equation (1.26) gives the alternative notation dy/dx for a derivative. Actually, the rigorous mathematical definition of dx and dy does not require these quantities to be infinitesimally small; instead they can be of any magnitude. (See any calculus text.) However, in our applications of calculus to thermodynamics, we shall always conceive of dy and dx as infinitesimal changes.

Let a and n be constants, and let u and v be functions of x; $u = u(x)$ and $v = v(x)$. Using the definition (1.25), one finds the following derivatives:

$$\frac{da}{dx} = 0, \qquad \frac{d(au)}{dx} = a\,\frac{du}{dx}, \qquad \frac{d(x^n)}{dx} = nx^{n-1}, \qquad \frac{d(e^{ax})}{dx} = ae^{ax}$$

$$\frac{d \ln ax}{dx} = \frac{1}{x}, \qquad \frac{d \sin ax}{dx} = a \cos ax, \qquad \frac{d \cos ax}{dx} = -a \sin ax$$

$$(1.27)*$$

$$\frac{d(u + v)}{dx} = \frac{du}{dx} + \frac{dv}{dx}, \qquad \frac{d(uv)}{dx} = u\,\frac{dv}{dx} + v\,\frac{du}{dx}$$

$$\frac{d(u/v)}{dx} = \frac{d(uv^{-1})}{dx} = -uv^{-2}\frac{dv}{dx} + v^{-1}\frac{du}{dx}$$

The chain rule is often used to find derivatives. Let z be a function of x, where x is a function of r; $z = z(x)$, where $x = x(r)$. Then z can be expressed as a function of r; $z = z(x) = z[x(r)] = g(r)$, where g is some function. The *chain rule* states that $dz/dr = (dz/dx)(dx/dr)$. For example, suppose we want $(d/dr) \sin 3r^2$. Let $z = \sin x$ and $x = 3r^2$. Then $z = \sin 3r^2$, and the chain rule gives $dz/dr = (\cos x)(6r) = 6r \cos 3r^2$.

Equations (1.26) and (1.27) give the following formulas for differentials:

$$d(x^n) = nx^{n-1}\,dx, \qquad d(e^{ax}) = ae^{ax}\,dx$$

$$d(au) = a\,du, \qquad d(u + v) = du + dv, \qquad d(uv) = u\,dv + v\,du \qquad (1.28)*$$

We often want to find a maximum or minimum of some function $y(x)$. For a function with a continuous derivative, the slope of the curve is zero at a maximum or minimum point (Fig. 1.12). Hence to locate an extremum, we look for the points where $dy/dx = 0$.

The function dy/dx is the first derivative of y. The **second derivative** d^2y/dx^2 is defined as the derivative of the first derivative: $d^2y/dx^2 \equiv d(dy/dx)/dx$.

Partial Derivatives

In thermodynamics we usually deal with functions of two or more variables. Let z be a function of x and y; $z = f(x, y)$. We define the **partial derivative** of z with respect to x as

$$\left(\frac{\partial z}{\partial x}\right)_y \equiv \lim_{\Delta x \to 0} \frac{f(x + \Delta x, y) - f(x, y)}{\Delta x} \qquad (1.29)$$

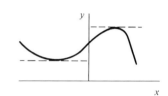

Figure 1.12

Horizontal tangent at maximum and minimum points.

This definition is analogous to the definition (1.25) of the ordinary derivative, in that if y were a constant instead of a variable, the partial derivative $(\partial z/\partial x)_y$ would become just the ordinary derivative dz/dx. The variable being held constant in a partial derivative is often omitted and $(\partial z/\partial x)_y$ written simply as $\partial z/\partial x$. In thermodynamics there are many possible variables, and to avoid confusion it is essential to show which variables are being held constant in a partial derivative. The partial derivative of z with respect to y at constant x is defined similarly to (1.29):

$$\left(\frac{\partial z}{\partial y}\right)_x \equiv \lim_{\Delta y \to 0} \frac{f(x, y + \Delta y) - f(x, y)}{\Delta y}$$

There may be more than two independent variables. For example, let $z = g(w, x, y)$. The partial derivative of z with respect to x at constant w and y is

$$\left(\frac{\partial z}{\partial x}\right)_{w, y} \equiv \lim_{\Delta x \to 0} \frac{g(w, x + \Delta x, y) - g(w, x, y)}{\Delta x}$$

How are partial derivatives found? To find $(\partial z/\partial x)_y$ we take the ordinary derivative of z with respect to x while regarding y as a constant. For example, if $z = x^2 y^3 + e^{yx}$, then $(\partial z/\partial x)_y = 2xy^3 + ye^{yx}$; also, $(\partial z/\partial y)_x = 3x^2y^2 + xe^{yx}$.

Let $z = f(x, y)$. Suppose x changes by an infinitesimal amount dx while y remains constant. What is the infinitesimal change dz in z brought about by the infinitesimal change in x? If z were a function of x only, then [Eq. (1.26)] we would have $dz = (dz/dx)\, dx$. Because z depends on y also, the infinitesimal change in z at constant y is given by the analogous equation $dz = (\partial z/\partial x)_y\, dx$. Similarly, if y were to undergo an infinitesimal change dy while x were held constant, we would have $dz = (\partial z/\partial y)_x\, dy$. If now both x and y undergo infinitesimal changes, the infinitesimal change in z is the sum of the infinitesimal changes due to dx and dy:

$$dz = \left(\frac{\partial z}{\partial x}\right)_y dx + \left(\frac{\partial z}{\partial y}\right)_x dy \qquad \textbf{(1.30)}*$$

In this equation, dz is called the **total differential** of $z(x, y)$. Equation (1.30) is often used in thermodynamics. An analogous equation holds for the total differential of a function of more than two variables. For example, if $z = z(r, s, t)$, then

$$dz = \left(\frac{\partial z}{\partial r}\right)_{s, t} dr + \left(\frac{\partial z}{\partial s}\right)_{r, t} ds + \left(\frac{\partial z}{\partial t}\right)_{r, s} dt$$

Three useful partial-derivative identities can be derived from (1.30). For an infinitesimal process in which y does not change, the infinitesimal change dy is 0, and (1.30) becomes

$$dz_y = \left(\frac{\partial z}{\partial x}\right)_y dx_y \qquad (1.31)$$

where the y subscripts on dz and dx indicate that these infinitesimal changes occur at constant y. Division by dz_y gives

$$1 = \left(\frac{\partial z}{\partial x}\right)_y \frac{dx_y}{dz_y} = \left(\frac{\partial z}{\partial x}\right)_y \left(\frac{\partial x}{\partial z}\right)_y$$

since from the definition of the partial derivative, the ratio of infinitesimals dx_y/dz_y equals $(\partial x/\partial z)_y$. Therefore

$$\left(\frac{\partial z}{\partial x}\right)_y = \frac{1}{(\partial x/\partial z)_y} \qquad \textbf{(1.32)}*$$

Note that the same variable, y, is being held constant in both partial derivatives in (1.32). When y is held constant, there are only two variables, x and z, and you will probably recall that $dz/dx = 1/(dx/dz)$.

For an infinitesimal process in which z stays constant, Eq. (1.30) becomes

$$0 = \left(\frac{\partial z}{\partial x}\right)_y dx_z + \left(\frac{\partial z}{\partial y}\right)_x dy_z \qquad (1.33)$$

Dividing by dy_z and recognizing that dx_z/dy_z equals $(\partial x/\partial y)_z$, we get

$$0 = \left(\frac{\partial z}{\partial x}\right)_y\left(\frac{\partial x}{\partial y}\right)_z + \left(\frac{\partial z}{\partial y}\right)_x \quad \text{and} \quad \left(\frac{\partial z}{\partial x}\right)_y\left(\frac{\partial x}{\partial y}\right)_z = -\left(\frac{\partial z}{\partial y}\right)_x = -\frac{1}{(\partial y/\partial z)_x}$$

where (1.32) with x and y interchanged was used. Multiplication by $(\partial y/\partial z)_x$ gives

$$\left(\frac{\partial x}{\partial y}\right)_z\left(\frac{\partial y}{\partial z}\right)_x\left(\frac{\partial z}{\partial x}\right)_y = -1 \qquad \textbf{(1.34)*}$$

Equation (1.34) looks intimidating but is actually easy to remember because of the simple pattern of variables: $\partial x/\partial y$, $\partial y/\partial z$, $\partial z/\partial x$; the variable held constant in each partial derivative is the one that doesn't appear in that derivative.

Sometimes students wonder why the ∂y's, ∂z's, and ∂x's in (1.34) don't cancel to give $+1$ instead of -1. One can cancel ∂y's etc. only when the same variable is held constant in each partial derivative. The infinitesimal change dy_z in y with z held constant while x varies is not the same as the infinitesimal change dy_x in y with x held constant while z varies. [Note that (1.32) can be written as $(\partial z/\partial x)_y(\partial x/\partial z)_y = 1$; here, cancellation occurs.]

Finally, let dy in (1.30) be zero so that (1.31) holds. Let u be some other variable. Division of (1.31) by du_y gives

$$\frac{dz_y}{du_y} = \left(\frac{\partial z}{\partial x}\right)_y\frac{dx_y}{du_y}$$

$$\left(\frac{\partial z}{\partial u}\right)_y = \left(\frac{\partial z}{\partial x}\right)_y\left(\frac{\partial x}{\partial u}\right)_y \qquad \textbf{(1.35)*}$$

The ∂x's in (1.35) can be canceled because the same variable is held constant in each partial derivative.

A function of two independent variables $z(x, y)$ has the following four second partial derivatives:

$$\left(\frac{\partial^2 z}{\partial x^2}\right)_y \equiv \left[\frac{\partial}{\partial x}\left(\frac{\partial z}{\partial x}\right)_y\right]_y, \qquad \left(\frac{\partial^2 z}{\partial y^2}\right)_x \equiv \left[\frac{\partial}{\partial y}\left(\frac{\partial z}{\partial y}\right)_x\right]_x$$

$$\frac{\partial^2 z}{\partial x\, \partial y} \equiv \left[\frac{\partial}{\partial x}\left(\frac{\partial z}{\partial y}\right)_x\right]_y, \qquad \frac{\partial^2 z}{\partial y\, \partial x} \equiv \left[\frac{\partial}{\partial y}\left(\frac{\partial z}{\partial x}\right)_y\right]_x$$

Provided $\partial^2 z/(\partial x\, \partial y)$ and $\partial^2 z/(\partial y\, \partial x)$ are continuous, as is generally true in physical applications, one can show that they are equal (see any calculus text):

$$\frac{\partial^2 z}{\partial x\, \partial y} = \frac{\partial^2 z}{\partial y\, \partial x} \qquad \textbf{(1.36)*}$$

The order of partial differentiation is immaterial.

Fractions are sometimes written with a slant line. The convention is that

$$a/bc + d \equiv \frac{a}{bc} + d$$

1.7   EQUATIONS OF STATE

Experiment generally shows the thermodynamic state of a homogeneous system with a fixed composition to be specified when the two variables P and T are specified. If the thermodynamic state is specified, this means the volume V of the system is specified. Given values of P and T of a fixed-composition system, the value of V is determined. But this is exactly what is meant by the statement that V is a function of P and T. Therefore, $V = u(P, T)$, where u is a function that depends on the nature of the system. If the restriction of fixed composition is dropped, the state of the system will depend on its composition as well as on P and T. We then have

$$V = f(P, T, n_1, n_2, \ldots) \tag{1.37}$$

where $n_1, n_2, \ldots$ are the numbers of moles of substances 1, 2, ... in the homogeneous system and f is some function. This relation between P, T, $n_1, n_2, \ldots$, and V is called a **volumetric equation of state,** or, more simply, an **equation of state.** If the system is heterogeneous, each phase will have its own equation of state.

For a one-phase system composed of n moles of a single pure substance, the equation of state (1.37) becomes $V = f(P, T, n)$, where the function f depends on the nature of the system; f for liquid water differs from f for ice and from f for liquid benzene. Of course, we can solve the equation of state for P or for T to get the alternative form $P = g(V, T, n)$ or $T = h(P, V, n)$, where g and h are certain functions. The laws of thermodynamics are general and cannot be used to deduce equations of state for particular systems. Equations of state must be determined experimentally. One can also use statistical mechanics to deduce an approximate equation of state starting from some assumed form for the intermolecular interactions in the system.

An example of an equation of state is $PV = nRT$, the equation of state of an ideal gas. In reality, no gas obeys this equation of state.

The volume of a one-phase, one-component system is clearly proportional to the number of moles n present at any given T and P. Therefore the equation of state for any pure one-phase system can be written in the form

$$V = nk(T, P)$$

where the function k depends on what substance is being considered. Since the functional dependence of V on n is the same for any pure substance, and since we usually deal with closed systems (n fixed), it is convenient to eliminate n and write the equation of state using only intensive variables. To this end, we define the **molar volume** V_m of any pure, one-phase system as the volume per mole:

$$V_m \equiv V/n \tag{1.38}*$$

V_m is a function of T and P; $V_m = k(T, P)$. For an ideal gas, $V_m = RT/P$. The m subscript in V_m is often omitted when it is clear that a molar volume is meant. (Commonly used alternative symbols for V_m are $\bar{V}$ and $\tilde{V}$.)

For any extensive property of a pure one-phase system, we can define a corresponding molar quantity. For example, the molar mass of a substance is m/n [Eq. (1.4)].

What about equations of state for real gases? We shall see in Chapter 15 that ignoring forces between the molecules leads to the ideal-gas equation of state $PV = nRT$. Actually, molecules initially attract each other as they approach and then repel each other when they collide. To allow for intermolecular forces, van der Waals in 1873 modified the ideal-gas equation to give the **van der Waals equation**

$$\left(P + \frac{an^2}{V^2}\right)(V - nb) = nRT \tag{1.39}$$

Each gas has its own a and b values. Determination of a and b from experimental data is discussed in Sec. 8.4, which lists some a and b values. Subtraction of nb from V corrects for intermolecular repulsion. Because of this repulsion, the volume available to the gas molecules is less than the volume V of the container. The constant b is approximately the volume of one mole of the gas molecules themselves. (In a liquid, the molecules are quite close together, so b is roughly the same as the molar volume of the liquid.) The term an^2/V^2 allows for intermolecular attraction. These attractions tend to make the pressure exerted by the gas [given by the van der Waals equation as $P = nRT/(V - nb) - an^2/V^2$] less than that predicted by the ideal-gas equation. The parameter a is a measure of the strength of the intermolecular attraction; b is a measure of molecular size.

For most liquids and solids at ordinary temperatures and pressures, an approximate equation of state is

$$V_m = c_1 + c_2T + c_3T^2 - c_4P - c_5PT \qquad (1.40)$$

where $c_1, \ldots, c_5$ are positive constants that must be evaluated by fitting observed V_m versus T and P data. The term c_1 is much larger than each of the other terms, so V_m of the liquid or solid changes only slowly with T and P. In most work with solids or liquids, the pressure remains close to 1 atm. In this case, the terms involving P can be neglected to give $V_m = c_1 + c_2T + c_3T^2$. This equation is often written in the form $V_m = V_{m,0}(1 + At + Bt^2)$, where $V_{m,0}$ is the molar volume at $0°C$ and t is the Celsius temperature. Values of the constants A and B are tabulated in handbooks. The terms $c_2T + c_3T^2$ in (1.40) indicate that V_m usually increases as T increases. The terms $-c_4P - c_5PT$ indicate that V_m decreases as P increases.

For a single-phase, pure, closed system, the equation of state of the system can be written in the form $V_m = k(T, P)$. One can make a three-dimensional plot of the equation of state by plotting P, T, and V_m on the x, y, and z axes. Each possible state of the system gives a point in space, and the locus of all such points gives a surface whose equation is the equation of state. Figure 1.13 shows the equation-of-state surface for an ideal gas.

If we hold one of the three variables constant, we can make a two-dimensional plot. For example, holding T constant at the value T_1, we have $PV_m = RT_1$ as the equation of state of an ideal gas. An equation of the form $xy = $ constant gives a hyperbola when plotted. Choosing other values of T, we get a series of hyperbolas (Fig. 1.6a). The lines of constant temperature are called **isotherms,** and a constant-temperature process is called an **isothermal process.** We can also hold either P or V_m constant and plot **isobars** (P constant) or *isochores* (V_m constant).

Figure 1.14 shows some isotherms and isobars of liquid water.

We shall find that thermodynamics enables us to relate many thermodynamic properties of substances to partial derivatives of P, V_m, and T with respect to one another. This is useful because these partial derivatives can be readily measured. There are six such partial derivatives:

$$\left(\frac{\partial V_m}{\partial T}\right)_P, \quad \left(\frac{\partial V_m}{\partial P}\right)_T, \quad \left(\frac{\partial P}{\partial V_m}\right)_T, \quad \left(\frac{\partial P}{\partial T}\right)_{V_m}, \quad \left(\frac{\partial T}{\partial V_m}\right)_P, \quad \left(\frac{\partial T}{\partial P}\right)_{V_m}$$

The relation $(\partial z/\partial x)_y = 1/(\partial x/\partial z)_y$ [Eq. (1.32)] shows that three of these six are the reciprocals of the other three:

$$\left(\frac{\partial T}{\partial P}\right)_{V_m} = \frac{1}{(\partial P/\partial T)_{V_m}}, \qquad \left(\frac{\partial T}{\partial V_m}\right)_P = \frac{1}{(\partial V_m/\partial T)_P}, \qquad \left(\frac{\partial P}{\partial V_m}\right)_T = \frac{1}{(\partial V_m/\partial P)_T}$$

$$(1.41)$$

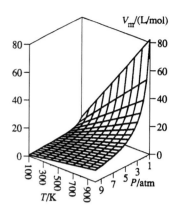

Figure 1.13

Equation-of-state surface for an ideal gas.

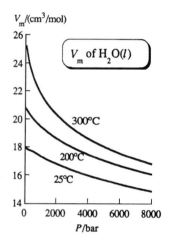

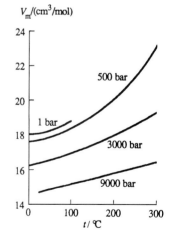

Figure 1.14

Molar volume of $H_2O(l)$ plotted versus P and versus T.

Furthermore, the relation $(\partial x/\partial y)_z(\partial y/\partial z)_x(\partial z/\partial x)_y = -1$ [Eq. (1.34)] with x, y, and z replaced by P, V_m, and T, respectively, gives

$$\left(\frac{\partial P}{\partial V_m}\right)_T\left(\frac{\partial V_m}{\partial T}\right)_P\left(\frac{\partial T}{\partial P}\right)_{V_m} = -1$$

$$\left(\frac{\partial P}{\partial T}\right)_{V_m} = -\left(\frac{\partial P}{\partial V_m}\right)_T\left(\frac{\partial V_m}{\partial T}\right)_P = -\frac{(\partial V_m/\partial T)_P}{(\partial V_m/\partial P)_T} \tag{1.42}$$

where $(\partial z/\partial x)_y = 1/(\partial x/\partial z)_y$ was used twice.

Hence there are only two independent partial derivatives: $(\partial V_m/\partial T)_P$ and $(\partial V_m/\partial P)_T$. The other four can be calculated from these two and need not be measured. We define the **thermal expansivity** (or **cubic expansion coefficient**) α (alpha) and the **isothermal compressibility** κ (kappa) of a substance by

$$\alpha(T, P) \equiv \frac{1}{V}\left(\frac{\partial V}{\partial T}\right)_{P,n} \equiv \frac{1}{V_m}\left(\frac{\partial V_m}{\partial T}\right)_P \tag{1.43}*$$

$$\kappa(T, P) \equiv -\frac{1}{V}\left(\frac{\partial V}{\partial P}\right)_{T,n} \equiv -\frac{1}{V_m}\left(\frac{\partial V_m}{\partial P}\right)_T \tag{1.44}*$$

α and κ tell how fast the volume of a substance increases with temperature and decreases with pressure. The purpose of the $1/V$ factor in their definitions is to make them intensive properties. Usually, α is positive; however, liquid water decreases in volume with increasing temperature between 0°C and 4°C at 1 atm. One can prove from the laws of thermodynamics that κ must always be positive (see *Zemansky and Dittman,* sec. 14-9, for the proof). Equation (1.42) can be written as

$$\left(\frac{\partial P}{\partial T}\right)_{V_m} = \frac{\alpha}{\kappa} \tag{1.45}$$

EXAMPLE 1.3 α and κ of an ideal gas

For an ideal gas, find expressions for α and κ and verify that Eq. (1.45) holds.

To find α and κ from the definitions (1.43) and (1.44), we need the partial derivatives of V_m. We therefore solve the ideal-gas equation of state $PV_m = RT$ for V_m and then differentiate the result. We have $V_m = RT/P$. Differentiation with respect to T gives $(\partial V_m/\partial T)_P = R/P$. Thus

$$\alpha = \frac{1}{V_m}\left(\frac{\partial V_m}{\partial T}\right)_P = \frac{1}{V_m}\left(\frac{R}{P}\right) = \frac{P}{RT}\frac{R}{P} = \frac{1}{T} \tag{1.46}$$

$$\kappa = -\frac{1}{V_m}\left(\frac{\partial V_m}{\partial P}\right)_T = -\frac{1}{V_m}\left[\frac{\partial}{\partial P}\left(\frac{RT}{P}\right)\right]_T = \frac{1}{V_m}\left(\frac{RT}{P^2}\right) = \frac{1}{P} \tag{1.47}$$

$$\left(\frac{\partial P}{\partial T}\right)_{V_m} = \left[\frac{\partial}{\partial T}\left(\frac{RT}{V_m}\right)\right]_{V_m} = \frac{R}{V_m} \tag{1.48}$$

But from (1.45), we have $(\partial P/\partial T)_{V_m} = \alpha/\kappa = T^{-1}/P^{-1} = P/T = nRTV^{-1}/T = R/V_m$, which agrees with (1.48).

Exercise

For a gas obeying the equation of state $V_m = RT/P + B(T)$, where $B(T)$ is a certain function of T, (a) find α and κ; (b) find $(\partial P/\partial T)_{V_m}$ in two different ways. [*Answer:* $\alpha = (R/P + dB/dT)/V_m$; $\kappa = RT/V_mP^2$; $(\partial P/\partial T)_{V_m} = P/T + P^2(dB/dT)/RT$.]

For solids, α is typically 10^{-5} to 10^{-4} K^{-1}. For liquids, α is typically $10^{-3.5}$ to 10^{-3} K^{-1}. For gases, α can be estimated from the ideal-gas α, which is $1/T$; for temperatures of 100 to 1000 K, α for gases thus lies in the range 10^{-2} to 10^{-3} K^{-1}.

For solids, κ is typically 10^{-6} to 10^{-5} atm^{-1}. For liquids, κ is typically 10^{-4} atm^{-1}. Equation (1.47) for ideal gases gives κ as 1 and 0.1 atm^{-1} at P equal to 1 and 10 atm, respectively. Solids and liquids are far less compressible than gases because there isn't much space between molecules in liquids and solids.

The quantities α and κ can be used to find the volume change produced by a change in T or P.

EXAMPLE 1.4 Expansion due to a temperature increase

Estimate the percentage increase in volume produced by a 10°C temperature increase in a liquid with the typical α value 0.001 K^{-1}, approximately independent of temperature.

Equation (1.43) gives $dV_P = \alpha V \, dT_P$. Since we require only an approximate answer and since the changes in T and V are small (α is small), we can approximate the ratio dV_P/dT_P by the ratio $\Delta V_P/\Delta T_P$ of finite changes to get $\Delta V_P/V \approx \alpha \, \Delta T_P = (0.001 \ K^{-1}) \ (10 \ K) = 0.01 = 1\%$.

Exercise

For water at 80°C and 1 atm, $\alpha = 6.412_7 \times 10^{-4} \ K^{-1}$ and $\rho = 0.971792$ g/cm³. Using the approximation $dV_P/dT_P \approx \Delta V_P/\Delta T_P$ for ΔT_P small, find the density of water at 81°C and 1 atm and compare with the true value 0.971166 g/cm³. (*Answer:* 0.971169 g/cm³.)

The single most comprehensive collection of physical and chemical data is Landolt-Börnstein: *Zahlenwerte und Funktionen* (*Numerical Data and Functional Relationships*), published by Springer-Verlag. The sixth edition of Landolt-Börnstein consists of 28 books published from 1950 to 1980. A "New Series" of volumes was begun in 1961 and so far contains over 250 books, each with text in both English and German. A useful source of experimental data is the *Handbook of Chemistry and Physics* (published annually by CRC Press); the most recent editions of the *Handbook* are the most reliable.

1.8 INTEGRAL CALCULUS

Differential calculus was reviewed in Sec. 1.6. Before reviewing integral calculus, we recall some facts about sums.

Sums
The definition of the summation notation is

$$\sum_{i=1}^{n} a_i \equiv a_1 + a_2 + \cdots + a_n \qquad \textbf{(1.49)*}$$

For example, $\Sigma_{i=1}^{3} i^2 = 1^2 + 2^2 + 3^2 = 14$. When the limits of a sum are clear, they are often omitted. Some identities that follow from (1.49) are (Prob. 1.58)

$$\sum_{i=1}^{n} ca_i = c \sum_{i=1}^{n} a_i, \qquad \sum_{i=1}^{n} (a_i + b_i) = \sum_{i=1}^{n} a_i + \sum_{i=1}^{n} b_i \qquad \textbf{(1.50)*}$$

$$\sum_{i=1}^{n} \sum_{j=1}^{m} a_i b_j = \sum_{i=1}^{n} a_i \sum_{j=1}^{m} b_j \qquad (1.51)$$

Integral Calculus

Frequently one wants to find a function $y(x)$ whose derivative is known to be a certain function $f(x)$; $dy/dx = f(x)$. The most general function y that satisfies this equation is called the **indefinite integral** (or *antiderivative*) of $f(x)$ and is denoted by $\int f(x)\,dx$.

$$\text{If} \quad dy/dx = f(x) \qquad \text{then} \quad y = \int f(x)\,dx \qquad \textbf{(1.52)*}$$

The function $f(x)$ being integrated in (1.52) is called the **integrand.**

Since the derivative of a constant is zero, the indefinite integral of any function contains an arbitrary additive constant. For example, if $f(x) = x$, its indefinite integral $y(x)$ is $\frac{1}{2}x^2 + C$, where C is an arbitrary constant. This result is readily verified by showing that y satisfies (1.52), that is, by showing that $(d/dx)(\frac{1}{2}x^2 + C) = x$. To save space, tables of indefinite integrals usually omit the arbitrary constant C.

From the derivatives given in Sec. 1.6, it follows that

$$\int af(x)\,dx = a\int f(x)\,dx, \qquad \int [f(x) + g(x)]\,dx = \int f(x)\,dx + \int g(x)\,dx$$

$$\textbf{(1.53)*}$$

$$\int dx = x + C, \qquad \int x^n\,dx = \frac{x^{n+1}}{n+1} + C \qquad \text{where } n \neq -1 \qquad \textbf{(1.54)*}$$

$$\int \frac{1}{x}\,dx = \ln x + C, \qquad \int e^{ax}\,dx = \frac{e^{ax}}{a} + C \qquad \textbf{(1.55)*}$$

$$\int \sin ax\,dx = -\frac{\cos ax}{a} + C, \qquad \int \cos ax\,dx = \frac{\sin ax}{a} + C \qquad \textbf{(1.56)*}$$

where a and n are nonzero constants and C is an arbitrary constant. For more complicated integrals than those in Eqs. (1.53) through (1.56), use a table of integrals or a computer program that does indefinite integration.

A second important concept in integral calculus is the definite integral. Let $f(x)$ be a continuous function, and let a and b be any two values of x. The **definite integral** of f between the limits a and b is denoted by the symbol

$$\int_a^b f(x)\,dx \qquad (1.57)$$

The reason for the resemblance to the notation for an indefinite integral will become clear shortly. The definite integral (1.57) is a number whose value is found from the following definition. We divide the interval from a to b into n subintervals, each of width Δx, where $\Delta x = (b - a)/n$ (see Fig. 1.15). In each subinterval, we pick any point we please, denoting the chosen points by $x_1, x_2, \ldots, x_n$. We evaluate $f(x)$ at each of the n chosen points and form the sum

$$\sum_{i=1}^n f(x_i)\Delta x = f(x_1)\Delta x + f(x_2)\Delta x + \cdots + f(x_n)\Delta x \qquad (1.58)$$

We now take the limit of the sum (1.58) as the number of subintervals n goes to infinity, and hence as the width Δx of each subinterval goes to zero. This limit is, by definition, the definite integral (1.57):

$$\int_a^b f(x)\,dx \equiv \lim_{\Delta x \to 0} \sum_{i=1}^n f(x_i)\Delta x \qquad (1.59)$$

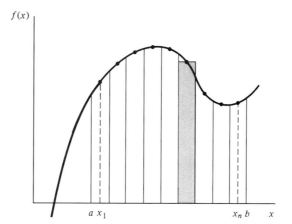

Figure 1.15

Definition of the definite integral.

The motivation for this definition is that the quantity on the right side of (1.59) occurs very frequently in physical problems.

Each term in the sum (1.58) is the area of a rectangle of width Δx and height $f(x_i)$. A typical rectangle is indicated by the shading in Fig. 1.15. As the limit $\Delta x \to 0$ is taken, the total area of these n rectangles becomes equal to the area under the curve $f(x)$ between a and b. Thus we can interpret the definite integral as an area. Areas lying below the x axis, where $f(x)$ is negative, make negative contributions to the definite integral.

Use of the definition (1.59) to evaluate a definite integral would be tedious. The fundamental theorem of integral calculus (proved in any calculus text) enables us to evaluate a definite integral of $f(x)$ in terms of an indefinite integral $y(x)$ of $f(x)$, as

$$\int_a^b f(x)\, dx = y(b) - y(a) \qquad \text{where } y(x) = \int f(x)\, dx \qquad \textbf{(1.60)*}$$

For example, if $f(x) = x$, $a = 2$, $b = 6$, we can take $y = \frac{1}{2}x^2$ (or $\frac{1}{2}x^2$ plus some constant) and (1.60) gives $\int_2^6 x\, dx = \frac{1}{2}(6^2) - \frac{1}{2}(2^2) = 16$.

The integration variable x in the definite integral on the left side of (1.60) does not appear in the final result (the right side of this equation). It thus does not matter what symbol we use for this variable. If we evaluate $\int_2^6 z\, dz$, we still get 16. In general, $\int_a^b f(x)\, dx = \int_a^b f(z)\, dz$. For this reason the integration variable in a definite integral is called a **dummy variable.** (The integration variable in an indefinite integral is not a dummy variable.) Similarly it doesn't matter what symbol we use for the summation index in (1.49). Replacement of i by j gives exactly the same sum on the right side, and i in (1.49) is a dummy index.

Two identities that readily follow from (1.60) are $\int_a^b f(x)\, dx = -\int_b^a f(x)\, dx$ and $\int_a^b f(x)\, dx + \int_b^c f(x)\, dx = \int_a^c f(x)\, dx$.

An important method for evaluating integrals is a change in variables. For example, suppose we want $\int_2^3 x \exp(x^2)\, dx$. Let $z \equiv x^2$; then $dz = 2x\, dx$, and

$$\int_2^3 xe^{x^2}\, dx = \frac{1}{2}\int_4^9 e^z\, dz = \frac{1}{2}e^z\Big|_4^9 = \frac{1}{2}(e^9 - e^4) = 4024.2$$

Note that the limits were changed in accord with the substitution $z = x^2$.

From (1.52), it follows that the derivative of an indefinite integral equals the integrand: $(d/dx)\int f(x)\, dx = f(x)$. Note, however, that a definite integral is simply a number and not a function; therefore $(d/dx)\int_a^b f(x)\, dx = 0$.

Integration with respect to x for a function of two variables is defined similarly to (1.52) and (1.59). If $y(x, z)$ is the most general function that satisfies

$$\left[\frac{\partial y(x, z)}{\partial x}\right]_z = f(x, z) \tag{1.61}$$

then the indefinite integral of $f(x, z)$ with respect to x is

$$\int f(x, z)\, dx = y(x, z) \tag{1.62}$$

For example, if $f(x, z) = xz^3$, then $y(x, z) = \frac{1}{2}x^2z^3 + g(z)$, where g is an arbitrary function of z. If y satisfies (1.61), one can show [in analogy with (1.60)] that a definite integral of $f(x, z)$ is given by

$$\int_a^b f(x, z)\, dx = y(b, z) - y(a, z) \tag{1.63}$$

For example, $\int_2^6 xz^3\, dx = \frac{1}{2}(6^2)z^3 + g(z) - \frac{1}{2}(2^2)z^3 - g(z) = 16z^3$.

The integrals (1.62) and (1.63) are similar to ordinary integrals of a function $f(x)$ of a single variable in that we regard the second independent variable z in these integrals as constant during the integration process; z acts as a parameter rather than as a variable. (A **parameter** is a quantity that is constant in a particular circumstance but whose value can change from one circumstance to another. For example, in Newton's second law $F = ma$, the mass m is a parameter. For any one particular body, m is constant, but its value can vary from one body to another.) In contrast to the integrals (1.62) and (1.63), in thermodynamics we shall often integrate a function of two or more variables in which all the variables are changing during the integration. Such integrals are called line integrals and will be discussed in Chapter 2.

EXAMPLE 1.5 Change of volume with change in temperature

A liquid with thermal expansivity α is initially at temperature and volume T_1 and V_1. If the liquid is heated from T_1 to T_2 at constant pressure, find an expression for V_2 using the approximation that α is independent of T.

We have $\alpha \equiv (1/V)(\partial V/\partial T)_P = V^{-1}\, dV_P/dT_P$, where dV_P and dT_P are infinitesimal changes at constant P. Thus $V^{-1}\, dV_P = \alpha\, dT_P$. Integration at constant P gives

$$\int_1^2 V^{-1}\, dV = \int_1^2 \alpha\, dT \quad \text{so} \quad \ln V \big|_1^2 \approx \alpha T \big|_1^2 \quad \text{and} \quad \ln V_2 - \ln V_1 \approx \alpha(T_2 - T_1)$$

Exercise

If a liquid is isothermally compressed from the state P_1, V_1 to P_2, V_2, express V_2 in terms of P_1, V_1, P_2, and κ; neglect the pressure dependence of κ. [*Answer:* $\ln V_2 \approx \ln V_1 - \kappa(P_2 - P_1)$.]

Logarithms

Integration of $1/x$ gives the natural logarithm $\ln x$. Because logarithms are used so often in physical chemistry derivations and calculations, we now review their properties. If $x = a^s$, then the exponent s is said to be the **logarithm** (log) of x to the base a: if $a^s = x$, then $\log_a x = s$. The most important base is the irrational number $e = 2.71828\ldots$, defined as the limit of $(1 + b)^{1/b}$ as $b \to 0$. Logs to the base e are called

natural logarithms and are written as ln x. For practical calculations, one often uses logs to the base 10, called **common logarithms** and written as log x, $\log_{10} x$, or lg x. We have

$$\ln x \equiv \log_e x, \qquad \log x \equiv \log_{10} x \qquad \textbf{(1.64)*}$$

$$\text{If } 10^t = x, \quad \text{then } \log x = t. \qquad \text{If } e^s = x, \quad \text{then } \ln x = s. \qquad (1.65)$$

From (1.65), we have

$$e^{\ln x} = x \qquad \text{and} \qquad 10^{\log x} = x \qquad (1.66)$$

From (1.65), it follows that $\ln e^s = s$. Since $e^{\ln x} = x = \ln e^x$, the exponential and natural logarithmic functions are inverses of each other. The function e^x is often written as exp x. Thus, exp $x \equiv e^x$. Since $e^1 = e$, $e^0 = 1$, and $e^{-\infty} = 0$, we have $\ln e = 1$, $\ln 1 = 0$, and $\ln 0 = -\infty$. One can take the logarithm or the exponential of a dimensionless quantity only.

Some identities that follow from the definition (1.65) are

$$\ln xy = \ln x + \ln y, \qquad \ln (x/y) = \ln x - \ln y \qquad \textbf{(1.67)*}$$

$$\ln x^k = k \ln x \qquad \textbf{(1.68)*}$$

$$\ln x = (\log_{10} x)/(\log_{10} e) = \log_{10} x \ln 10 = 2.3026 \log_{10} x \qquad (1.69)$$

To find the log of a number greater than 10^{100} or less than 10^{-100}, which cannot be entered on most calculators, we use $\log(ab) = \log a + \log b$ and $\log 10^b = b$. For example,

$$\log_{10} (2.75 \times 10^{-150}) = \log_{10} 2.75 + \log_{10} 10^{-150} = 0.439 - 150 = -149.561$$

To find the antilog of a number greater than 100 or less than -100, we proceed as follows. If we know that $\log_{10} x = -184.585$, then

$$x = 10^{-184.585} = 10^{-0.585} 10^{-184} = 10^{-184}/10^{0.585}$$

$$= 10^{-184}/3.85 = 2.60 \times 10^{-185}$$

These procedures can also be used to find logs and antilogs using tables.

1.9   STUDY SUGGESTIONS

A common reaction to a physical chemistry course is for a student to think, "This looks like a tough course, so I'd better memorize all the equations, or I won't do well." Such a reaction is understandable, especially since many of us have had teachers who emphasized rote memory, rather than understanding, as the method of instruction.

Actually, comparatively few equations need to be remembered (they have been marked with an asterisk), and most of these are simple enough to require little effort at conscious memorization. Being able to reproduce an equation is no guarantee of being able to apply that equation to solving problems. To use an equation properly, one must understand it. Understanding involves not only knowing what the symbols stand for but also knowing when the equation applies and when it does not apply. Everyone knows the ideal-gas equation $PV = nRT$, but it's amazing how often students will use this equation in problems involving liquids or solids. Another part of understanding an equation is knowing where the equation comes from. Is it simply a definition? Or is it a law that represents a generalization of experimental observations? Or is it a rough empirical rule with only approximate validity? Or is it a deduction from the basic laws of thermodynamics made without approximations? Or is it a deduction from the laws of thermodynamics made using approximations and therefore of limited validity?

As well as understanding the important equations, you should also know the meanings of the various defined terms (closed system, ideal gas, etc.). Boldface type (for example, **isotherm**) is used to mark very important terms when they are first defined. Terms of lesser importance are printed in italic type (for example, *isobar*). If you come across a term whose meaning you have forgotten, consult the index; the page number where a term is defined is printed in boldface type.

Working problems is essential to learning physical chemistry. Suggestions for solving problems are given in Sec. 2.12. It's a good idea to test your understanding of a section by working on some relevant problems as soon as you finish each section. Do not wait until you feel you have mastered a section before working some problems. The problems in this book are classified by section.

Keep up to date in assignments. Cramming does not work in physical chemistry because of the many concepts to learn and the large amount of practice in working problems that is needed to master these concepts. Most students find that physical chemistry requires a lot more study and problem-solving time than the typical college course, so be sure you allot enough time to this course.

Make studying an active process. Read with a pencil at hand and use it to verify equations, to underline key ideas, to make notes in the margin, and to write down questions you want to ask your instructor. Sort out the basic principles from what is simply illustrative detail and digression. In this book, small print is used for historical material, for more advanced material, and for minor points.

After reading a section, make a written summary of the important points. This is a far more effective way of learning than to keep rereading the material. You might think it a waste of time to make summaries, since chapter summaries are provided. However, preparing your own summary will make the material much more meaningful to you than if you simply read the one at the end of the chapter.

A psychologist carried out a project on improving student study habits that raised student grades dramatically. A key technique used was to have students close the textbook at the end of each section and spend a few minutes outlining the material; the outline was then checked against the section in the book. [L. Fox in R. Ulrich et al. (eds.), *Control of Human Behavior,* Scott, Foresman, 1966, pp. 85–90.]

Before reading a chapter in detail, browse through it first, reading only the section headings, the first paragraph of each section, the summary, and some of the problems at the end of the chapter. This gives an idea of the structure of the chapter and makes the reading of each section more meaningful. Reading the problems first lets you know what you are expected to learn from the chapter.

You might try studying occasionally with another person. Discussing problems with someone else can help clarify the material in your mind.

Some suggestions to help you prepare for exams are

1. Learn the meanings of all terms in boldface type.
2. Memorize all starred equations *and* their conditions of applicability. (Do not memorize unstarred equations.)
3. Make sure you *understand* all starred equations.
4. Review your class notes.
5. Rework homework problems you had difficulty with.
6. Work some unassigned problems for additional practice.
7. Make summaries if you have not already done so.
8. Check that you understand all the concepts mentioned in the end-of-chapter summaries and can do each type of calculation listed in these summaries.
9. Prepare a practice exam by choosing some relevant homework problems and work them in the time allotted for the exam.

Since, as with all of us, your capabilities for learning and understanding are finite and the time available to you is limited, it is best to accept the fact that there will probably be some material you may never fully understand. No one understands everything fully.

1.10   SUMMARY

The four branches of physical chemistry are thermodynamics, quantum chemistry, statistical mechanics, and kinetics.

Thermodynamics deals with the relationships between the macroscopic equilibrium properties of a system. Some important concepts in thermodynamics are *system* (*open* versus *closed*; *isolated* versus *nonisolated*; *homogeneous* versus *heterogeneous*); *surroundings*; *walls* (*rigid* versus *nonrigid*; *permeable* versus *impermeable*; *adiabatic* versus *thermally conducting*); *equilibrium* (*mechanical, material, thermal*); *state functions* (*extensive* versus *intensive*); *phase*; and *equation of state.*

Temperature was defined as an intensive state function that has the same value for two systems in thermal equilibrium and a different value for two systems not in thermal equilibrium. The setting up of a temperature scale is arbitrary, but we chose to use the ideal-gas absolute scale defined by Eq. (1.15).

An ideal gas is one that obeys the equation of state $PV = nRT$. Real gases obey this equation only in the limit of zero density. At ordinary temperatures and pressures, the ideal-gas approximation will usually be adequate for our purposes. For an ideal gas mixture, $PV = n_{tot}RT$. The partial pressure of gas i in any mixture is $P_i \equiv x_i P$, where the mole fraction of i is $x_i \equiv n_i / n_{tot}$.

Differential and integral calculus were reviewed, and some useful partial-derivative relations were given [Eqs. (1.30), (1.32), (1.34), and (1.36)].

The thermodynamic properties α (thermal expansivity) and κ (isothermal compressibility) are defined by $\alpha \equiv (1/V)(\partial V/\partial T)_P$ and $\kappa \equiv -(1/V)(\partial V/\partial P)_T$ for a system of fixed composition.

Understanding, rather than mindless memorization, is the key to learning physical chemistry.

Important kinds of calculations dealt with in this chapter include

- Calculation of P (or V or T) of an ideal gas or ideal gas mixture using $PV = nRT$.
- Calculation of the molar mass of an ideal gas using $PV = nRT$ and $n = m/M$.
- Calculation of the density of an ideal gas.
- Calculations involving partial pressures.
- Use of α or κ to find volume changes produced by changes in T or P.
- Differentiation and partial differentiation of functions.
- Indefinite and definite integration of functions.

FURTHER READING AND DATA SOURCES

Temperature: *Quinn; Shoemaker, Garland, and Nibler,* chap. XVIII; *McGlashan,* chap. 3; *Zemansky and Dittman,* chap. 1. Pressure measurement: *Rossiter, Hamilton, and Baetzold,* vol. VI, chap. 2. Calculus: C. E. Swartz, *Used Math for the First Two Years of College Science,* Prentice-Hall, 1973.

ρ, α, and κ values: *Landolt-Börnstein,* 6th ed., vol. II, part 1, pp. 378–731.

PROBLEMS

Section 1.2

1.1 True or false? (*a*) A closed system cannot interact with its surroundings. (*b*) Density is an intensive property. (*c*) The Atlantic Ocean is an open system. (*d*) A homogeneous system must be a pure substance. (*e*) A system containing only one substance must be homogeneous.

1.2 State whether each of the following systems is closed or open and whether it is isolated or nonisolated: (*a*) a system enclosed in rigid, impermeable, thermally conducting walls; (*b*) a human being; (*c*) the planet earth.

1.3 How many phases are there in a system that consists of (*a*) $CaCO_3(s)$, $CaO(s)$, and $CO_2(g)$; (*b*) three pieces of solid AgBr, one piece of solid AgCl, and a saturated aqueous solution of these salts.

1.4 Explain why the definition of an adiabatic wall in Sec. 1.2 specifies that the wall be rigid and impermeable.

1.5 The density of Au is 19.3 g/cm³ at room temperature and 1 atm. (*a*) Express this density in kg/m³. (*b*) If gold is selling for $300 per troy ounce, what would a cubic meter of it sell for? One troy ounce = 480 grains, 1 grain = $\frac{1}{7000}$ pound, 1 pound = 453.59 g.

Section 1.4

1.6 True or false? (*a*) One gram is Avogadro's number of times as heavy as 1 amu. (*b*) The Avogadro constant N_A has no units. (*c*) Mole fractions are intensive properties. (*d*) One mole of water contains Avogadro's number of water molecules.

1.7 For O_2, give (*a*) the molecular weight; (*b*) the molecular mass; (*c*) the relative molecular mass; (*d*) the molar mass.

1.8 A solution of HCl in water is 12.0% HCl by mass. Find the mole fractions of HCl and H_2O in this solution.

1.9 Calculate the mass in grams of (*a*) one atom of carbon; (*b*) one molecule of water.

Section 1.5

1.10 True or false? (*a*) On the Celsius scale, the boiling point of water is slightly less than 100.00°C. (*b*) Doubling the absolute temperature of an ideal gas at fixed volume and amount of gas will double the pressure. (*c*) The ratio PV/mT is the same for all gases in the limit of zero pressure. (*d*) The ratio PV/nT is the same for all gases in the limit of zero pressure. (*e*) All ideal gases have the same density at 25°C and 1 bar. (*f*) All ideal gases have the same number of molecules per unit volume at 25°C and 10 bar.

1.11 Do these conversions: (*a*) 5.5 m³ to cm³; (*b*) 1.0 GPa to bar (where 1 GPa ≡ 10^9 Pa); (*c*) 1.0 GPa to atm; (*d*) 1.5 g/cm³ to kg/m³.

1.12 In Fig. 1.2, if the mercury levels in the left and right arms of the manometer are 30.43 and 20.21 cm, respectively, above the bottom of the manometer, and if the barometric pres-

sure is 754.6 torr, find the pressure in the system. Neglect temperature corrections to the manometer and barometer readings.

1.13 (*a*) A seventeenth-century physicist built a water barometer that projected through a hole in the roof of his house so that his neighbors could predict the weather by the height of the water. Suppose that at 25°C a mercury barometer reads 30.0 in. What would be the corresponding height of the column in a water barometer? The densities of mercury and water at 25°C are 13.53 and 0.997 g/cm³, respectively. (*b*) What pressure in atmospheres corresponds to a 30.0-in. mercury-barometer reading at 25°C at a location where g = 978 cm/s²?

1.14 Derive Eq. (1.17) from Eq. (1.18).

1.15 (*a*) What is the pressure exerted by 24.0 g of carbon dioxide in a 5.00-L vessel at 0°C? (*b*) A rough rule of thumb is that 1 mole of gas occupies 1 ft³ at room temperature and pressure (25°C and 1 atm). Calculate the percent error in this rule. One inch = 2.54 cm.

1.16 A sample of 87 mg of an ideal gas at 0.600 bar pressure has its volume doubled and its absolute temperature tripled. Find the final pressure.

1.17 For a certain hydrocarbon gas, 20.0 mg exerts a pressure of 24.7 torr in a 500-cm³ vessel at 25°C. Find the molar mass and the molecular weight and identify the gas.

1.18 Find the density of N_2 at 20°C and 0.967 bar.

1.19 For 1.0000 mol of N_2 gas at 0.00°C, the following volumes are observed as a function of pressure:

P/atm	1.0000	3.0000	5.0000
V/cm³	22405	7461.4	4473.1

Calculate and plot PV/nT versus P for these three points and extrapolate to $P = 0$ to evaluate R.

1.20 The measured density of a certain gaseous amine at 0°C as a function of pressure is

P/atm	0.2000	0.5000	0.8000
ρ/(g/L)	0.2796	0.7080	1.1476

Plot P/ρ versus P and extrapolate to $P = 0$ to find an accurate molecular weight. Identify the gas.

1.21 After 1.60 mol of NH_3 gas is placed in a 1600-cm³ box at 25°C, the box is heated to 500 K. At this temperature, the ammonia is partially decomposed to N_2 and H_2, and a pressure measurement gives 4.85 MPa. Find the number of moles of each component present at 500 K.

1.22 A student attempts to combine Boyle's law and Charles' law as follows. "We have $PV = K_1$ and $V/T = K_2$. Equals multiplied by equals are equal; multiplication of one equation by the other gives $PV^2/T = K_1K_2$. The product K_1K_2 of two constants is a constant, so PV^2/T is a constant for a fixed amount of ideal gas." What is the fallacy in this reasoning?

1.23 Prove that the equations $PV/T = C_1$ for m constant and $V/m = C_2$ for T and P constant lead to $PV/mT = $ a constant.

1.24 A certain gas mixture is at 3450 kPa pressure and is composed of 20.0 g of O_2 and 30.0 g of CO_2. Find the CO_2 partial pressure.

1.25 A 1.00-L bulb of methane at a pressure of 10.0 kPa is connected to a 3.00-L bulb of hydrogen at 20.0 kPa; both bulbs are at the same temperature. (*a*) After the gases mix, what is the total pressure? (*b*) What is the mole fraction of each component in the mixture?

1.26 A student decomposes $KClO_3$ and collects 36.5 cm^3 of O_2 over water at 23°C. The laboratory barometer reads 751 torr. The vapor pressure of water at 23°C is 21.1 torr. Find the volume the dry oxygen would occupy at 0°C and 1.000 atm.

1.27 Two evacuated bulbs of equal volume are connected by a tube of negligible volume. One bulb is placed in a 200-K constant-temperature bath and the other in a 300-K bath, and then 1.00 mol of an ideal gas is injected into the system. Find the final number of moles of gas in each bulb.

1.28 An oil-diffusion pump aided by a mechanical fore-pump can readily produce a "vacuum" with pressure 10^{-6} torr. Various special vacuum pumps can reduce P to 10^{-11} torr. At 25°C, calculate the number of molecules per cm^3 in a gas at (*a*) 1 atm; (*b*) 10^{-6} torr; (*c*) 10^{-11} torr.

1.29 A certain mixture of He and Ne in a 356-cm^3 bulb weighs 0.1480 g and is at 20.0°C and 748 torr. Find the mass and mole fraction of He present.

1.30 The earth's radius is 6.37×10^6 m. Find the mass of the earth's atmosphere. (Neglect the dependence of g on altitude.)

1.31 (*a*) If $10^5 P/bar = 6.4$, what is P? (*b*) If $10^{-2}T/K = 4.60$, what is T? (*c*) If $P/(10^3 \text{ bar}) = 1.2$, what is P? (*d*) If $10^3(K/T) = 3.20$, what is T?

1.32 A certain mixture of N_2 and O_2 has a density of 1.185 g/L at 25°C and 101.3 kPa. Find the mole fraction of O_2 in the mixture. (*Hint:* The given data and the unknown are all intensive properties, so the problem can be solved by considering any convenient fixed amount of mixture.)

1.33 The mole fractions of the main components of dry air at sea level are $x_{N_2} = 0.78$, $x_{O_2} = 0.21$, $x_{Ar} = 0.0093$, $x_{CO_2} = 0.0003$. (*a*) Find the partial pressure of each of these gases in dry air at 1.00 atm and 20°C. (*b*) Find the mass of each of these gases in a 15 ft $\times$ 20 ft $\times$ 10 ft room at 20°C if the barometer reads 740 torr and the relative humidity is zero. Also, find the density of the air in the room. Which has a greater mass, you or the air in the room of this problem?

Section 1.6
1.34 On Fig. 1.15, mark all points where df/dx is zero and circle each portion of the curve where df/dx is negative.

1.35 Let $y = x^2 + x - 1$. Find the slope of the y-versus-x curve at $x = 1$ by drawing the tangent line to the graph at $x = 1$ and finding its slope. Compare your result with the exact slope found by calculus.

1.36 Find d/dx of (*a*) $2x^3 e^{-3x}$; (*b*) $5e^{-3x^2}$; (*c*) $\ln 2x$; (*d*) $1/(1 - x)$; (*e*) $x/(x + 1)$; (*f*) $\ln (1 - e^{-2x})$; (*g*) $\sin^2 3x$.

1.37 (*a*) Find dy/dx if $xy = y - 2$. (*b*) Find $d^2(x^2 e^{3x})/dx^2$. (*c*) Find dy if $y = 5x^2 - 3x + 2/x - 1$.

1.38 Use a calculator to find: (*a*) $\lim_{x \to 0} x^x$ for $x > 0$; (*b*) $\lim_{x \to 0} (1 + x)^{1/x}$.

1.39 (*a*) Evaluate the first derivative of the function $y = e^{x^2}$ at $x = 2$ by using a calculator to evaluate $\Delta y/\Delta x$ for $\Delta x = 0.1$, 0.01, 0.001, etc. Note the loss of significant figures in Δy as Δx decreases. If you have a programmable calculator, you might try programming this problem. (*b*) Compare your result in (*a*) with the exact answer.

1.40 Find $\partial/\partial y$ of: (*a*) $\sin (axy)$; (*b*) $\cos (by^2 z)$; (*c*) $xe^{x/y}$; (*d*) $\tan (3x + 1)$; (*e*) $1/(e^{-a/y} + 1)$; (*f*) $f(x)g(y)h(z)$.

1.41 Take $(\partial/\partial T)_{P,n}$ of (*a*) nRT/P; (*b*) P/nRT^2 (where R is a constant).

1.42 (*a*) If $y = 4x^3 + 6x$, find dy. (*b*) If $z = 3x^2 y^3$, find dz. (*c*) If $P = nRT/V$, where R is a constant and all other quantities are variables, find dP.

1.43 Let $z = x^5/y^3$. Evaluate the four second partial derivatives of z; check that $\partial^2 z/(\partial x \, \partial y) = \partial^2 z/(\partial y \, \partial x)$.

1.44 (*a*) For an ideal gas, use an equation like (1.30) to show that $dP = P(n^{-1} dn + T^{-1} dT - V^{-1} dV)$ (which can be written as $d \ln P = d \ln n + d \ln T - d \ln V$). (*b*) Suppose 1.0000 mol of ideal gas at 300.00 K in a 30.000-L vessel has its temperature increased by 1.00 K and its volume increased by 0.050 L. Use the result of (*a*) to estimate the change in pressure, ΔP. (*c*) Calculate ΔP exactly for the change in (*b*) and compare with the estimate given by dP.

Section 1.7
1.45 Find the molar volume of an ideal gas at 20.0°C and 1.000 bar.

1.46 (*a*) Write the van der Waals equation (1.39) using the molar volume instead of V and n. (*b*) If one uses bars, cubic centimeters, moles, and kelvins as the units of P, V, n, and T, give the units of a and of b in the van der Waals equation.

1.47 For a liquid obeying the equation of state (1.40), find expressions for α and κ.

1.48 For H_2O at 50°C and 1 atm, $\rho = 0.98804$ g/cm^3 and $\kappa = 4.4 \times 10^{-10}$ Pa^{-1}. (*a*) Find the molar volume of water at 50°C and 1 atm. (*b*) Find the molar volume of water at 50°C and 100 atm. Neglect the pressure dependence of κ.

1.49 For an ideal gas: (*a*) sketch some isobars on a V_m-T diagram; (*b*) sketch some isochores on a P-T diagram.

1.50 A hypothetical gas obeys the equation of state $PV = nRT(1 + aP)$, where a is a constant. For this gas: (a) show that $\alpha = 1/T$ and $\kappa = 1/P(1 + aP)$; (b) verify that $(\partial P/\partial T)_V = \alpha/\kappa$.

1.51 Use the following densities of water as a function of T and P to estimate α, κ, and $(\partial P/\partial T)_{V_m}$ for water at 25°C and 1 atm: 0.997044 g/cm^3 at 25°C and 1 atm; 0.996783 g/cm^3 at 26°C and 1 atm; 0.997092 g/cm^3 at 25°C and 2 atm.

1.52 For H$_2$O(l) at 50°C and 1 atm, $\alpha = 4.576 \times 10^{-4}$ K^{-1}, $\kappa = 44.17 \times 10^{-6}$ bar^{-1}, and $V_m = 18.2334$ cm^3/mol. (a) Estimate V_{m,H_2O} at 52°C and 1 atm and compare the result with the experimental value 18.2504 cm^3/mol. Neglect the temperature dependence of α. (b) Estimate V_{m,H_2O} at 50°C and 200 bar and compare with the experimental value 18.078 cm^3/mol.

1.53 By drawing tangent lines and measuring their slopes, use Fig. 1.14 to estimate for water: (a) α at 100°C and 500 bar; (b) κ at 300°C and 2000 bar.

1.54 For H$_2$O at 17°C and 1 atm, $\alpha = 1.7 \times 10^{-4}$ K^{-1} and $\kappa = 4.7 \times 10^{-5}$ atm^{-1}. A closed, rigid container is completely filled with liquid water at 14°C and 1 atm. If the temperature is raised to 20°C, estimate the pressure in the container. Neglect the pressure and temperature dependences of α and κ.

1.55 Give a molecular explanation for each of the following facts. (a) For solids and liquids, κ usually decreases with increasing pressure. (b) For solids and liquids, $(\partial \kappa/\partial T)_P$ is usually positive.

1.56 Estimate the pressure increase needed to decrease isothermally by 1% the 1-atm volume of (a) a typical solid with $\kappa = 5 \times 10^{-6}$ atm^{-1}; (b) a typical liquid with $\kappa = 1 \times 10^{-4}$ atm^{-1}.

Section 1.8

1.57 (a) Evaluate $\sum_{J=0}^{4} (2J + 1)$. (b) Write the expression $x_1V_1 + x_2V_2 + \cdots + x_sV_s$ using summation notation. (c) Write out the individual terms of the double sum $\sum_{i=1}^{2} \sum_{j=4}^{6} b_{ij}$.

1.58 Prove the sum identities in (1.50) and (1.51). (*Hint:* Write out the individual terms of the sums.)

1.59 Evaluate: (a) $\int_3^{-2} (2V + 5V^2)\, dV$; (b) $\int_2^4 V^{-1}\, dV$; (c) $\int_1^\infty V^{-3}\, dV$; (d) $\int_0^{\pi/2} x^2 \cos x^3\, dx$.

1.60 Find (a) $\int \sin ax\, dx$; (b) $\int_0^\pi \sin ax\, dx$; (c) $(d/da) \int_0^\pi \sin ax\, dx$; (d) $\int (a/T^2)\, dT$.

1.61 State whether each of the following is a number or is a function of x: (a) $\int e^{x^2}\, dx$; (b) $\int_1^2 e^{x^2}\, dx$; (c) $\sum_{x=1}^{203} e^{x^2}$.

1.62 In which of the following is t a dummy variable? (a) $\int e^{t^2}\, dt$; (b) $\int_0^3 e^{t^2}\, dt$; (c) $\sum_{t=1}^{100} t^5$.

1.63 (a) If $df(x)/dx = 2x^3 + 3e^{5x}$, find $f(x)$. (b) If $\int f(x)\, dx = 3x^8 + C$, where C is a constant, find $f(x)$.

1.64 (a) Use a programmable calculator or a computer to obtain approximations to the integral $\int_2^3 x^2\, dx$ by evaluating the sum (1.58) for $\Delta x = 0.1, 0.01,$ and 0.001; take the x_i values at the left end of each subinterval. Compare your results with the exact value. (b) Use (1.58) with $\Delta x = 0.01$ to obtain an approximate value of $\int_0^1 e^{-x^2}\, dx$.

1.65 (a) Find $\log_{10} (4.2 \times 10^{1750})$. (b) Find $\ln (6.0 \times 10^{-200})$. (c) If $\log_{10} y = -138.265$, find y. (d) If $\ln z = 260.433$, find z.

1.66 Find (a) $\log_2 32$; (b) $\log_{43} 1$; (c) $\log_{26} 8$.

General

1.67 Classify each property as intensive or extensive; (a) temperature; (b) mass; (c) density; (d) electric field strength; (e) α; (f) mole fraction of a component.

1.68 For O$_2$ gas in thermal equilibrium with boiling sulfur, the following values of PV_m versus P are found:

P/torr	1000	500	250
PV_m/(L atm mol^{-1})	59.03	58.97	58.93

(Since P has units of pressure, P/torr is dimensionless.) From a plot of these data, find the boiling point of sulfur.

1.69 True or false? (a) Every isolated system is closed. (b) Every closed system is isolated. (c) For a fixed amount of an ideal gas, the product PV remains constant during any process. (d) The pressure of a nonideal gas mixture is equal to the sum of the partial pressures defined by $P_i \equiv x_iP$. (e) dy/dx is equal to $\Delta y/\Delta x$ for all functions y. (f) dy/dx is equal to $\Delta y/\Delta x$ only for functions that vary linearly with x according to $y = mx + b$. (g) $\ln (b/a) = -\ln (a/b)$. (h) If $\ln x$ is negative, then x lies between 0 and 1. (i) Ideal-gas isotherms further away from the axes of a P-versus-V plot correspond to higher temperatures. (j) $\sum_{i=1}^{n} a_i b_i = \sum_{i=1}^{n} a_i \sum_{i=1}^{n} b_i$.

The First Law of Thermodynamics

Chapter 1 introduced some of the vocabulary of thermodynamics and defined the important state function temperature. Another key state function in thermodynamics is the internal energy U, whose existence is postulated by the first law of thermodynamics; this law is the main topic of Chapter 2. The first law states that the total energy of system plus surroundings remains constant (is conserved). Closely related to the internal energy is the state function enthalpy H, defined in Sec. 2.5. Other important state functions introduced in this chapter are the heat capacities at constant volume and at constant pressure, C_V and C_P (Sec. 2.6), which give the rates of change of the internal energy and enthalpy with temperature [Eq. (2.53)]. As a preliminary to the main work of this chapter, Sec. 2.1 reviews classical mechanics.

The internal energy of a thermodynamic system is the sum of the molecular energies, as will be discussed in detail in Sec. 2.11. *Energy is a key concept in all areas of physical chemistry.* In quantum chemistry, a key step to calculating molecular properties is solving the Schrödinger equation, which is an equation that gives the allowed energy levels of a molecule. In statistical mechanics, the key to evaluating thermodynamic properties from molecular properties is to find something called the partition function, which is a certain sum over energy levels of the system. The rate of a chemical reaction depends strongly on the activation energy of the reaction. More generally, the kinetics of a reaction is determined by something called the potential-energy surface of the reaction.

The importance of energy in the economy is obvious. World consumption of energy in the year 2000 was 4×10^{20} J, mainly from fossil fuels (oil, coal, and natural gas).

2.1   CLASSICAL MECHANICS

Two important concepts in thermodynamics are work and energy. Since these concepts originated in classical mechanics, we review this subject before continuing with thermodynamics.

Classical mechanics (first formulated by the alchemist, theologian, physicist, and mathematician Isaac Newton) deals with the laws of motion of macroscopic bodies whose speeds are small compared with the speed of light c. For objects with speeds not small compared with c, one must use Einstein's **relativistic mechanics.** Since the thermodynamic systems we consider will not be moving at high speeds, we need not worry about relativistic effects. For nonmacroscopic objects (for example, electrons), one must use **quantum mechanics.** Thermodynamic systems are of macroscopic size, so we shall not need quantum mechanics at this point.

Newton's Second Law
The fundamental equation of classical mechanics is **Newton's second law of motion:**

$$\mathbf{F} = m\mathbf{a} \qquad\qquad (2.1)*$$

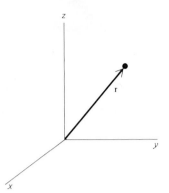

Figure 2.1

The displacement vector **r** from the origin to a particle.

where m is the mass of a body, **F** is the vector sum of all forces acting on it at some instant of time, and **a** is the acceleration the body undergoes at that instant. **F** and **a** are vectors, as indicated by the boldface type. **Vectors** have both magnitude and direction. **Scalars** (for example, m) have only a magnitude. To define acceleration, we set up a coordinate system with three mutually perpendicular axes x, y, and z. Let **r** be the vector from the coordinate origin to the particle (Fig. 2.1). The particle's **velocity v** is the instantaneous rate of change of its position vector **r** with respect to time:

$$\mathbf{v} \equiv d\mathbf{r}/dt \qquad (2.2)*$$

The magnitude (length) of the vector **v** is the particle's **speed** v. The particle's **acceleration a** is the instantaneous rate of change of its velocity:

$$\mathbf{a} \equiv d\mathbf{v}/dt = d^2\mathbf{r}/dt^2 \qquad (2.3)*$$

A vector in three-dimensional space has three components, one along each of the coordinate axes. Equality of vectors means equality of their corresponding components, so a vector equation is equivalent to three scalar equations. Thus Newton's second law **F** = m**a** is equivalent to the three equations

$$F_x = ma_x, \qquad F_y = ma_y, \qquad F_z = ma_z \qquad (2.4)$$

where F_x and a_x are the x components of the force and the acceleration. The x component of the position vector **r** is simply x, the value of the particle's x coordinate. Therefore (2.3) gives $a_x = d^2x/dt^2$, and (2.4) becomes

$$F_x = m\frac{d^2x}{dt^2}, \qquad F_y = m\frac{d^2y}{dt^2}, \qquad F_z = m\frac{d^2z}{dt^2} \qquad (2.5)$$

The *weight W* of a body is the gravitational force exerted on it by the earth. If g is the acceleration due to gravity, Newton's second law gives

$$W = mg \qquad (2.6)$$

Units

Two systems of units are commonly used in mechanics. In the mks system, the units of length, time, and mass are meters (m), seconds (s), and kilograms (kg), respectively. A force that produces an acceleration of one meter per second^2 when applied to a one-kilogram mass is defined as one **newton** (N):

$$1 \text{ N} \equiv 1 \text{ kg m/s}^2 \qquad (2.7)$$

In the cgs system, the units of length, time, and mass are centimeters (cm), seconds, and grams (g). The cgs unit of force is the *dyne* (dyn):

$$1 \text{ dyn} \equiv 1 \text{ g cm/s}^2 \qquad (2.8)$$

The reader can verify that

$$1 \text{ N} = 10^5 \text{ dyn} \qquad (2.9)$$

In 1960, the General Conference on Weights and Measures recommended a single system of units for worldwide scientific use. This system is called the **International System of Units** (*Système International d'Unités*), abbreviated **SI.** In mechanics, the SI uses mks units. If one were to adhere to SI units, pressures would always be given in newtons/meter^2 (pascals). However, it seems clear that many scientists will continue to use such units as atmospheres and torrs for many years to come. The current scientific literature shows a trend toward increased use of SI units, but since much of the literature continues to be in non-SI units, it is necessary to be familiar with both SI units and commonly used non-SI units. SI units for some quantities introduced previously are meters for length, kilograms for mass, seconds for time, cubic meters (m^3) for volume, kg/m^3 for density, newtons for force, pascals for pressure, kelvins for temperature, moles for amount of substance, and kg/mol for molar mass.

Work

Suppose a force **F** acts on a body while the body undergoes an infinitesimal displacement dx in the x direction. The infinitesimal amount of **work** dw done on the body by the force **F** is defined as

$$dw \equiv F_x \, dx \qquad\qquad (2.10)*$$

where F_x is the component of the force in the direction of the displacement. If the infinitesimal displacement has components in all three directions, then

$$dw \equiv F_x \, dx + F_y \, dy + F_z \, dz \qquad\qquad (2.11)$$

Consider now a noninfinitesimal displacement. For simplicity, let the particle be moving in one dimension. The particle is acted on by a force $F(x)$ whose magnitude depends on the particle's position. Since we are using one dimension, F has only one component and need not be considered a vector. The work w done by F during displacement of the particle from x_1 to x_2 is the sum of the infinitesimal amounts of work (2.10) done during the displacement: $w = \Sigma \, F(x) \, dx$. But this sum of infinitesimal quantities is the definition of the definite integral [Eq. (1.59)], so

$$w = \int_{x_1}^{x_2} F(x) \, dx \qquad\qquad (2.12)$$

In the special case that F is constant during the displacement, (2.12) becomes

$$w = F(x_2 - x_1) \qquad \text{for } F \text{ constant} \qquad\qquad (2.13)$$

From (2.10), the units of work are those of force times length. The SI and cgs units of work are the **joule** (J) and the *erg*, respectively:

$$1 \text{ J} \equiv 1 \text{ N m} = 1 \text{ kg m}^2/\text{s}^2 \qquad\qquad (2.14)$$

$$1 \text{ erg} \equiv 1 \text{ dyn cm} \qquad\qquad (2.15)$$

The use of (2.9) gives

$$1 \text{ J} = 10^7 \text{ ergs} \qquad\qquad (2.16)$$

Power P is defined as the rate at which work is done. If an agent does work dw in time dt, then $P \equiv dw/dt$. The SI unit of power is the *watt* (W): $1 \text{ W} \equiv 1 \text{ J/s}$.

Mechanical Energy

We now prove the *work–energy theorem*. Let **F** be the total force acting on a particle, and let the particle move from point 1 to point 2. Integration of (2.11) gives as the total work done on the particle:

$$w = \int_1^2 F_x \, dx + \int_1^2 F_y \, dy + \int_1^2 F_z \, dz \qquad\qquad (2.17)$$

Newton's second law gives $F_x = ma_x = m(dv_x/dt)$. Also, $dv_x/dt = (dv_x/dx)(dx/dt) = (dv_x/dx)v_x$. Therefore $F_x = mv_x(dv_x/dx)$, with similar equations for F_y and F_z. We have $F_x \, dx = mv_x \, dv_x$, and (2.17) becomes

$$w = \int_1^2 mv_x \, dv_x + \int_1^2 mv_y \, dv_y + \int_1^2 mv_z \, dv_z$$

$$w = \tfrac{1}{2}m(v_{x2}^2 + v_{y2}^2 + v_{z2}^2) - \tfrac{1}{2}m(v_{x1}^2 + v_{y1}^2 + v_{z1}^2) \qquad\qquad (2.18)$$

We now define the **kinetic energy** K of the particle as

$$K \equiv \tfrac{1}{2}mv^2 = \tfrac{1}{2}m(v_x^2 + v_y^2 + v_z^2) \qquad \textbf{(2.19)*}$$

The right side of (2.18) is the final kinetic energy K_2 minus the initial kinetic energy K_1:

$$w = K_2 - K_1 = \Delta K \qquad \text{one-particle syst.} \qquad (2.20)$$

where ΔK is the change in kinetic energy. The *work–energy theorem* (2.20) states that the work done on the particle by the force acting on it equals the change in kinetic energy of the particle. It is valid because we defined kinetic energy in such a manner as to make it valid.

Besides kinetic energy, there is another kind of energy in classical mechanics. Suppose we throw a body up into the air. As it rises, its kinetic energy decreases, reaching zero at the high point. What happens to the kinetic energy the body loses as it rises? It proves convenient to introduce the notion of a *field* (in this case, a gravitational field) and to say that the decrease in kinetic energy of the body is accompanied by a corresponding increase in the *potential energy* of the field. Likewise, as the body falls back to earth, it gains kinetic energy and the gravitational field loses a corresponding amount of potential energy. Usually, we don't refer explicitly to the field but simply ascribe a certain amount of potential energy to the body itself, the amount depending on the location of the body in the field.

To put the concept of potential energy on a quantitative basis, we proceed as follows. Let the forces acting on the particle depend only on the particle's position and not on its velocity, or the time, or any other variable. Such a force $\mathbf{F}$ with $F_x = F_x(x, y, z)$, $F_y = F_y(x, y, z)$, $F_z = F_z(x, y, z)$ is called a *conservative force*, for a reason to be seen shortly. Examples of conservative forces are gravitational forces, electrical forces, and the Hooke's law force of a spring. Some nonconservative forces are air resistance, friction, and the force you exert when you kick a football. For a conservative force, we define the **potential energy** $V(x, y, z)$ as a function of x, y, and z whose partial derivatives satisfy

$$\frac{\partial V}{\partial x} \equiv -F_x, \qquad \frac{\partial V}{\partial y} \equiv -F_y, \qquad \frac{\partial V}{\partial z} \equiv -F_z \qquad (2.21)$$

Since only the partial derivatives of V are defined, V itself has an arbitrary additive constant. We can set the zero level of potential energy anywhere we please.

From (2.17) and (2.21), it follows that

$$w = -\int_1^2 \frac{\partial V}{\partial x}\, dx - \int_1^2 \frac{\partial V}{\partial y}\, dy - \int_1^2 \frac{\partial V}{\partial z}\, dz \qquad (2.22)$$

Since $dV = (\partial V/\partial x)\, dx + (\partial V/\partial y)\, dy + (\partial V/\partial z)\, dz$ [Eq. (1.30)], we have

$$w = -\int_1^2 dV = -(V_2 - V_1) = V_1 - V_2 \qquad (2.23)$$

But Eq. (2.20) gives $w = K_2 - K_1$. Hence $K_2 - K_1 = V_1 - V_2$, or

$$K_1 + V_1 = K_2 + V_2 \qquad (2.24)$$

When only conservative forces act, the sum of the particle's kinetic energy and potential energy remains constant during the motion. This is the law of conservation of mechanical energy. Using E_{mech} for the total **mechanical energy,** we have

$$E_{\text{mech}} = K + V \qquad (2.25)$$

If only conservative forces act, E_{mech} remains constant.

What is the potential energy of an object in the earth's gravitational field? Let the x axis point outward from the earth with the origin at the earth's surface. We have $F_x = -mg$, $F_y = F_z = 0$. Equation (2.21) gives $\partial V/\partial x = mg$, $\partial V/\partial y = 0 = \partial V/\partial z$. Integration gives $V = mgx + C$, where C is a constant. (In doing the integration, we assumed the object's distance above the earth's surface was small enough for g to be considered constant.) Choosing the arbitrary constant as zero, we get

$$V = mgh \tag{2.26}$$

where h is the object's altitude above the earth's surface. As an object falls to earth, its potential energy mgh decreases and its kinetic energy $\frac{1}{2}mv^2$ increases. Provided the effect of air friction is negligible, the total mechanical energy $K + V$ remains constant as the object falls.

We have considered a one-particle system. Similar results hold for a many-particle system. (See H. Goldstein, *Classical Mechanics,* 2d ed., Addison-Wesley, 1980, sec. 1-2, for derivations.) The kinetic energy of an n-particle system is the sum of the kinetic energies of the individual particles:

$$K = K_1 + K_2 + \cdots + K_n = \frac{1}{2} \sum_{i=1}^{n} m_i v_i^2 \tag{2.27}$$

Let the particles exert conservative forces on one another. The potential energy V of the system is not the sum of the potential energies of the individual particles. Instead, V is a property of the system as a whole. V turns out to be the sum of contributions due to pairwise interactions between particles. Let V_{ij} be the contribution to V due to the forces acting between particles i and j. One finds

$$V = \sum_i \sum_{j>i} V_{ij} \tag{2.28}$$

The double sum indicates that we sum over all pairs of i and j values except those with i equal to or greater than j. Terms with $i = j$ are omitted because a particle does not exert a force on itself. Also, only one of the terms V_{12} and V_{21} is included, to avoid counting the interaction between particles 1 and 2 twice. For example, in a system of three particles, $V = V_{12} + V_{13} + V_{23}$. If external forces act on the particles of the system, their contributions to V must also be included. [V_{ij} is defined by equations similar to (2.21).]

One finds that $K + V = E_{\text{mech}}$ is constant for a many-particle system with only conservative forces acting.

The mechanical energy $K + V$ is a measure of the work the system can do. When a particle's kinetic energy decreases, the work–energy theorem $w = \Delta K$ [Eq. (2.20)] says that w, the work done on it, is negative; that is, the particle does work on the surroundings equal to its loss of kinetic energy. Since potential energy is convertible to kinetic energy, potential energy can also be converted ultimately to work done on the surroundings. Kinetic energy is due to motion. Potential energy is due to the positions of the particles.

EXAMPLE 2.1 Work

A woman slowly lifts a 30.0-kg object to a height of 2.00 m above its initial position. Find the work done on the object by the woman, and the work done by the earth.

The force exerted by the woman equals the weight of the object, which from Eq. (2.6) is $F = mg = (30.0 \text{ kg}) (9.81 \text{ m/s}^2) = 294 \text{ N}$. From (2.12) and (2.13), the work she does on the object is

$$w = \int_{x_1}^{x_2} F(x)\,dx = F\,\Delta x = (294 \text{ N})(2.00 \text{ m}) = 588 \text{ J}$$

The earth exerts an equal and opposite force on the object compared with the lifter, so the earth does -588 J of work on the object. This work is negative because the force and the displacement are in opposite directions. The total work done on the object by all forces is zero. The work–energy theorem (2.20) gives $w = \Delta K = 0$, in agreement with the fact that the object started at rest and ended at rest. (We derived the work–energy theorem for a single particle, but it also applies to a perfectly rigid body.)

Exercise

A sphere of mass m is attached to a spring, which exerts a force $F = -kx$ on the sphere, where k (called the force constant) is a constant characteristic of the spring and x is the displacement of the sphere from its equilibrium position (the position where the spring exerts no force on the sphere). The sphere is initially at rest at its equilibrium position. Find the expression for the work w done by some-one who slowly displaces the sphere to a final distance d from its equilibrium position. Calculate w if $k = 10$ N/m and $d = 6.0$ cm. (*Answer:* $\frac{1}{2}kd^2$, 0.018 J.)

2.2　*P-V* WORK

Work in thermodynamics is defined as in classical mechanics. When part of the surroundings exerts a macroscopically measurable force $\mathbf{F}$ on matter in the system while this matter moves a distance dx at the point of application of $\mathbf{F}$, then the surroundings has done **work** $dw = F_x\,dx$ [Eq. (2.10)] on the system, where F_x is the component of $\mathbf{F}$ in the direction of the displacement. $\mathbf{F}$ may be a mechanical, electrical, or magnetic force and may act on and displace the entire system or only a part of the system. When F_x and the displacement dx are in the same direction, positive work is done on the system: $dw > 0$. When F_x and dx are in opposite directions, dw is negative.

Reversible *P-V* Work

The most common way work is done on a thermodynamic system is by a change in the system's volume. Consider the system of Fig. 2.2. The system consists of the matter contained within the piston and cylinder walls and has pressure P. Let the external pressure on the frictionless piston also be P. Equal opposing forces act on the piston, and it is in mechanical equilibrium. Let x denote the piston's location. If the external pressure on the piston is now increased by an infinitesimal amount, this increase will produce an infinitesimal imbalance in forces on the piston. The piston will move inward by an infinitesimal distance dx, thereby decreasing the system's volume and increasing its pressure until the system pressure again balances the external pressure. During this infinitesimal process, which occurs at an infinitesimal rate, the system will be infinitesimally close to equilibrium.

The piston, which is part of the surroundings, exerted a force, which we denote by F_x, on matter in the system at the system–piston boundary while this matter moved a distance dx. The surroundings therefore did work $dw = F_x\,dx$ on the system. Let F be the magnitude of the force exerted by the system on the piston. Newton's third law (action = reaction) gives $F = F_x$. The definition $P = F/A$ of the system's pressure P

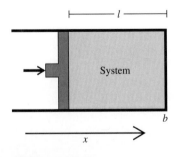

Figure 2.2

A system confined by a piston.

gives $F_x = F = PA$, where A is the piston's cross-sectional area. Therefore the work $dw = F_x\,dx$ done on the system in Fig. 2.2 is

$$dw = PA\,dx \qquad (2.29)$$

The system has cross-sectional area A and length $l = b - x$ (Fig. 2.2), where x is the piston's position and b is the position of the fixed end of the system. The volume of this cylindrical system is $V = Al = Ab - Ax$. The change in system volume when the piston moves by dx is $dV = d(Ab - Ax) = -A\,dx$. Equation (2.29) becomes

$$dw_{\text{rev}} = -P\,dV \qquad \text{closed system, reversible process} \qquad \textbf{(2.30)*}$$

The subscript rev stands for reversible. The meaning of "reversible" will be discussed shortly. We implicitly assumed a closed system in deriving (2.30). When matter is transported between system and surroundings, the meaning of work becomes ambiguous; we shall not consider this case. We derived (2.30) for a particular shape of system, but it can be shown to be valid for every system shape (see *Kirkwood and Oppenheim,* sec. 3-1).

We derived (2.30) by considering a contraction of the system's volume ($dV < 0$). For an expansion ($dV > 0$), the piston moves outward (in the negative x direction), and the displacement dx of the matter at the system–piston boundary is negative ($dx < 0$). Since F_x is positive (the force exerted by the piston on the system is in the positive x direction), the work $dw = F_x\,dx$ done on the system by the surroundings is negative when the system expands. For an expansion, the system's volume change is still given by $dV = -A\,dx$ (where $dx < 0$ and $dV > 0$), and (2.30) still holds.

In a contraction, the work done on the system is positive ($dw > 0$). In an expansion, the work done on the system is negative ($dw < 0$). (In an expansion, the work done on the surroundings is positive.)

So far we have considered only an infinitesimal volume change. Suppose we carry out an infinite number of successive infinitesimal changes in the external pressure. At each such change, the system's volume changes by dV and work $-P\,dV$ is done on the system, where P is the current value of the system's pressure. The total work w done on the system is the sum of the infinitesimal amounts of work, and this sum of infinitesimal quantities is the following definite integral:

$$w_{\text{rev}} = -\int_1^2 P\,dV \qquad \text{closed syst., rev. proc.} \qquad (2.31)$$

where 1 and 2 are the initial and final states of the system, respectively.

The finite volume change to which (2.31) applies consists of an infinite number of infinitesimal steps and takes an infinite amount of time to carry out. In this process, the difference between the pressures on the two sides of the piston is always infinitesimally small, so finite unbalanced forces never act and the system remains infinitesimally close to equilibrium throughout the process. Moreover, the process can be reversed at any stage by an infinitesimal change in conditions, namely, by infinitesimally changing the external pressure. Reversal of the process will restore both system and surroundings to their initial conditions.

A **reversible process** is one where the system is always infinitesimally close to equilibrium, and an infinitesimal change in conditions can reverse the process to restore both system and surroundings to their initial states. A reversible process is obviously an idealization.

Equations (2.30) and (2.31) apply only to reversible expansions and contractions. More precisely, they apply to mechanically reversible volume changes. There could be a chemically irreversible process, such as a chemical reaction, occurring in the system during the expansion, but so long as the mechanical forces are only infinitesimally unbalanced, (2.30) and (2.31) apply.

The work (2.31) done in a volume change is called **P-V work.** Later on, we shall deal with electrical work and work of changing the system's surface area, but for now, only systems with *P-V* work will be considered.

We have defined the symbol w to stand for work done *on* the system by the surroundings. Some texts use w to mean work done *by* the system on its surroundings. Their w is the negative of ours.

Line Integrals

The integral $\int_1^2 P\,dV$ in (2.31) is not an ordinary integral. For a closed system of fixed composition, the system's pressure P is a function of its temperature and volume: $P = P(T, V)$. To calculate w_{rev}, we must evaluate the negative of

$$\int_1^2 P(T, V)\,dV \tag{2.32}$$

The integrand $P(T, V)$ is a function of *two* independent variables T and V. In an ordinary definite integral, the integrand is a function of *one* variable, and the value of the ordinary definite integral $\int_a^b f(x)\,dx$ is determined once the function f and the limits a and b are specified. For example, $\int_1^3 x^2\,dx = 3^3/3 - 1^3/3 = 26/3$. In contrast, in $\int_1^2 P(T, V)\,dV$, both of the independent variables T and V may change during the volume-change process, and the value of the integral depends on how T and V vary. For example, if the system is an ideal gas, then $P = nRT/V$ and $\int_1^2 P(T, V)\,dV = nR \int_1^2 (T/V)\,dV$. Before we can evaluate $\int_1^2 (T/V)\,dV$, we must know how both T and V change during the process.

The integral (2.32) is called a **line integral.** Sometimes the letter L is put under the integral sign of a line integral. The value of the line integral (2.32) is defined as the sum of the infinitesimal quantities $P(T, V)\,dV$ for the particular process used to go from state 1 to state 2. This sum equals the area under the curve that plots P versus V. Figure 2.3 shows three of the many possible ways in which we might carry out a reversible volume change starting at the same initial state (state 1 with pressure P_1 and volume V_1) and ending at the same final state (state 2).

In process (*a*), we first hold the volume constant at V_1 and reduce the pressure from P_1 to P_2 by cooling the gas. We then hold the pressure constant at P_2 and heat the gas to expand it from V_1 to V_2. In process (*b*), we first hold P constant at P_1 and heat the gas until its volume reaches V_2. Then we hold V constant at V_2 and cool the gas until its pressure drops to P_2. In process (*c*), the independent variables V and T vary in an irregular way, as does the dependent variable P.

For each process, the integral $\int_1^2 P\,dV$ equals the shaded area under the *P*-versus-*V* curve. These areas clearly differ, and the integral $\int_1^2 P\,dV$ has different values for processes (*a*), (*b*), and (*c*). The reversible work $w_{rev} = -\int_1^2 P\,dV$ thus has different values for each of the processes (*a*), (*b*), and (*c*). We say that w_{rev} (which equals minus the shaded area under the *P*-versus-*V* curve) depends on the *path* used to go from state 1 to 2, meaning that it depends on the specific process used. There are an infinite number of ways of going from state 1 to state 2, and w_{rev} can have any positive or negative value for a given change of state.

The plots of Fig. 2.3 imply pressure equilibrium within the system during the process. In an irreversible expansion (see after Example 2.2), the system may have no single well-defined pressure, and we cannot plot such a process on a *P-V* diagram.

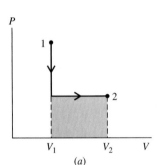

(a)

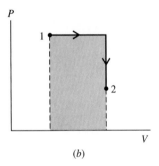

(b)

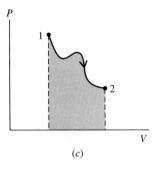

(c)

Figure 2.3

The work w done on the system in a reversible process (the heavy lines) equals minus the shaded area under the *P*-versus-*V* curve. The work depends on the process used to go from state 1 to state 2.

EXAMPLE 2.2 *P-V* work

Find the work w_{rev} for processes (*a*) and (*b*) of Fig. 2.3 if $P_1 = 3.00$ atm, $V_1 = 500$ cm^3, $P_2 = 1.00$ atm, and $V_2 = 2000$ cm^3. Also, find w_{rev} for the reverse of process (*a*).

We have $w_{rev} = -\int_1^2 P\,dV$. The line integral $\int_1^2 P\,dV$ equals the area under the *P*-versus-*V* curve. In Fig. 2.3*a*, this area is rectangular and equals

$$(V_2 - V_1)P_2 = (2000\ cm^3 - 500\ cm^3)(1.00\ atm) = 1500\ cm^3\ atm$$

Hence $w_{rev} = -1500\ cm^3$ atm. The units cm^3 atm are not customarily used for work, so we shall convert to joules by multiplying and dividing by the values of the gas constant $R = 8.314$ J/(mol K) and $R = 82.06\ cm^3$ atm/(mol K) [Eqs. (1.19) and (1.20)]:

$$w_{rev} = -1500\ cm^3\ atm\ \frac{8.314\ J\ mol^{-1}\ K^{-1}}{82.06\ cm^3\ atm\ mol^{-1}\ K^{-1}} = -152\ J$$

An alternative procedure is to note that no work is done during the constant-volume part of process (*a*); all the work is done during the second step of the process, in which *P* is held constant at P_2. Therefore

$$w_{rev} = -\int_1^2 P\,dV = -\int_{V_1}^{V_2} P_2\,dV = -P_2\int_{V_1}^{V_2} dV = -P_2 V\Big|_{V_1}^{V_2}$$

$$= -P_2(V_2 - V_1) = -(1.00\ atm)(1500\ cm^3) = -152\ J$$

Similarly, we find for process (*b*) that $w = -4500\ cm^3$ atm $= -456$ J (see the exercise in this example).

Processes (*a*) and (*b*) are expansions. Hence the system does positive work on its surroundings, and the work *w* done on the system is negative in these processes.

For the reverse of process (*a*), all the work is done during the first step, during which *P* is constant at 1.00 atm and *V* starts at 2000 cm^3 and ends at 500 cm^3. Hence $w = -\int_{2000\ cm^3}^{500\ cm^3} (1.00\ atm)\,dV = -(1.00\ atm)(500\ cm^3 - 2000\ cm^3) = 152$ J.

Exercise

Find w_{rev} for process (*b*) of Fig. 2.3 using the P_1, V_1, P_2, V_2 values given for process (*a*). (*Answer:* $-4500\ cm^3$ atm $= -456$ J.)

Irreversible *P-V* Work

The work *w* in a mechanically irreversible volume change sometimes cannot be calculated with thermodynamics.

For example, suppose the external pressure on the piston in Fig. 2.2 is suddenly reduced by a finite amount and is held fixed thereafter. The inner pressure on the piston is then greater than the outer pressure by a finite amount, and the piston is accelerated outward. This initial acceleration of the piston away from the system will destroy the uniform pressure in the enclosed gas. The system's pressure will be lower near the piston than farther away from it. Moreover, the piston's acceleration produces turbulence in the gas. Thus we cannot give a thermodynamic description of the state of the system.

We have $dw = F_x\,dx$. For *P-V* work, F_x is the force at the system–surroundings boundary, which is where the displacement dx is occurring. This boundary is the inner face of the piston, so $dw_{irrev} = -P_{surf}\,dV$, where P_{surf} is the pressure the system exerts on the inner face of the piston. (By Newton's third law, P_{surf} is also the pressure the piston's inner face exerts on the system.) Because we cannot use thermodynamics to calculate P_{surf} during the turbulent, irreversible expansion, we cannot find dw_{irrev} from thermodynamics.

The law of conservation of energy can be used to show that, for a frictionless piston (Prob. 2.21),

$$dw_{irrev} = -P_{ext}\,dV - dK_{pist} \tag{2.33}$$

where P_{ext} is the external pressure on the outer face of the piston and dK_{pist} is the infinitesimal change in piston kinetic energy. The integrated form of (2.33) is $w_{irrev} = -\int_1^2 P_{ext}\,dV - \Delta K_{pist}$. If we wait long enough, the piston's kinetic energy will be dissipated by the internal friction (viscosity—see Sec. 16.3) in the gas. The gas will be heated, and the piston will eventually come to rest (perhaps after undergoing oscillations). Once the piston has come to rest, we have $\Delta K_{pist} = 0 - 0 = 0$, since the piston started and ended at rest. We then have $w_{irrev} = -\int_1^2 P_{ext}\,dV$. Hence we can find w_{irrev} after the piston has come to rest. If, however, part of the piston's kinetic energy is transferred to some other body in the surroundings before the piston comes to rest, then thermodynamics cannot calculate the work exchanged between system and surroundings. For further discussion, see D. Kivelson and I. Oppenheim, *J. Chem. Educ.*, **43**, 233 (1966), and *de Heer*, sec. 4.4.

Summary

For now, we shall deal only with work done due to a volume change. The work done *on* a closed system in an infinitesimal mechanically reversible process is $dw_{rev} = -P\,dV$. The work $w_{rev} = -\int_1^2 P\,dV$ depends on the path (the process) used to go from the initial state 1 to the final state 2.

2.3   HEAT

When two bodies at unequal temperatures are placed in contact, they eventually reach thermal equilibrium at a common intermediate temperature. We say that heat has flowed from the hotter body to the colder one. Let bodies 1 and 2 have masses m_1 and m_2 and initial temperatures T_1 and T_2, with $T_2 > T_1$; let T_f be the final equilibrium temperature. Provided the two bodies are isolated from the rest of the universe and no phase change or chemical reaction occurs, one experimentally observes the following equation to be satisfied for all values of T_1 and T_2:

$$m_2 c_2 (T_2 - T_f) = m_1 c_1 (T_f - T_1) \equiv q \tag{2.34}$$

where c_1 and c_2 are constants (evaluated experimentally) that depend on the composition of bodies 1 and 2. We call c_1 the **specific heat capacity** (or **specific heat**) of body 1. We define q, the amount of **heat** that flowed from body 2 to body 1, as equal to $m_2 c_2 (T_2 - T_f)$.

The unit of heat commonly used in the nineteenth and early twentieth centuries was the **calorie** (cal), defined as the quantity of heat needed to raise one gram of water from 14.5°C to 15.5°C at 1 atm pressure. (This definition is no longer used, as we shall see in Sec. 2.4.) By definition, $c_{H_2O} = 1.00$ cal/(g °C) at 15°C and 1 atm. Once the specific heat capacity of water has been defined, the specific heat capacity c_2 of any other substance can be found from (2.34) by using water as substance 1. When specific heats are known, the heat q transferred in a process can then be calculated from (2.34).

Actually, (2.34) does not hold exactly, because the specific heat capacities of substances are functions of temperature and pressure. When an infinitesimal amount of heat dq_P flows at constant pressure P into a body of mass m and specific heat capacity at constant pressure c_P, the body's temperature is raised by dT and

$$dq_P \equiv mc_P\,dT \tag{2.35}$$

where c_P is a function of T and P. Summing up the infinitesimal flows of heat, we get the total heat that flowed as a definite integral:

$$q_P = m \int_{T_1}^{T_2} c_P(T)\,dT \qquad \text{closed syst., } P \text{ const.} \tag{2.36}$$

The pressure dependence of c_P was omitted because P is held fixed for the process. The quantity mc_P is the *heat capacity at constant pressure C_P* of the body: $C_P \equiv mc_P$. From (2.35) we have

$$C_P = dq_P/dT \tag{2.37}$$

Equation (2.34) is more accurately written as

$$m_2 \int_{T_f}^{T_2} c_{P2}(T) \, dT = m_1 \int_{T_1}^{T_f} c_{P1}(T) \, dT = q_P \tag{2.38}$$

If the dependence of c_{P2} and c_{P1} on T is negligible, (2.38) reduces to (2.34).

We gave examples in Sec. 2.2 of reversible and irreversible ways of doing work on a system. Likewise, heat can be transferred reversibly or irreversibly. A reversible transfer of heat requires that the temperature difference between the two bodies be infinitesimal. When there is a finite temperature difference between the bodies, the heat flow is irreversible.

Two bodies need not be in direct physical contact for heat to flow from one to the other. Radiation transfers heat between two bodies at different temperatures (for example, the sun and the earth). The transfer occurs by emission of electromagnetic waves by one body and absorption of these waves by the second body. An adiabatic wall must be able to block radiation.

Equation (2.36) was written with the implicit assumption that the system is closed (m fixed). As is true for work, the meaning of heat is ambiguous for open systems. (See R. Haase, *Thermodynamics of Irreversible Processes,* Addison-Wesley, 1969, pp. 17–21, for a discussion of open systems.)

2.4 THE FIRST LAW OF THERMODYNAMICS

As a rock falls toward the earth, its potential energy is transformed into kinetic energy. When it hits the earth and comes to rest, what has happened to its energy of motion? Or consider a billiard ball rolling on a billiard table. Eventually it comes to rest. Again, what happened to its energy of motion? Or imagine that we stir some water in a beaker. Eventually the water comes to rest, and we again ask: What happened to its energy of motion? Careful measurement will show very slight increases in the temperatures of the rock, the billiard ball, and the water (and in their immediate surroundings). Knowing that matter is composed of molecules, we find it easy to believe that the macroscopic kinetic energies of motion of the rock, the ball, and the water were converted into energy at the molecular level. The average molecular translational, rotational, and vibrational energies in the bodies were increased slightly, and these increases were reflected in the temperature rises.

We therefore ascribe an **internal energy** U to a body, in addition to its macroscopic kinetic energy K and potential energy V, discussed in Sec. 2.1. This internal energy consists of: molecular translational, rotational, vibrational, and electronic energies; the relativistic rest-mass energy $m_{rest}c^2$ of the electrons and the nuclei; and potential energy of interaction between the molecules. These energies are discussed in Sec. 2.11.

The total energy E of a body is therefore

$$E = K + V + U \tag{2.39}$$

where K and V are the macroscopic (not molecular) kinetic and potential energies of the body (due to motion of the body through space and the presence of fields that act on the body) and U is the internal energy of the body (due to molecular motions and intermolecular interactions). Since thermodynamics is a macroscopic science, the development of thermodynamics requires no knowledge of the nature of U. All that is needed is some means of measuring the change in U for a process. This will be provided by the first law of thermodynamics.

In most applications of thermodynamics that we shall consider, the system will be at rest and external fields will not be present. Therefore, K and V will be zero, and the total energy E will be equal to the internal energy U. (The effect of the earth's gravitational field on thermodynamic systems is usually negligible, and gravity will usually be ignored; see, however, Sec. 15.8.) Chemical engineers often deal with systems of flowing fluids; here, $K \neq 0$.

With our present knowledge of the molecular structure of matter, we take it for granted that a flow of heat between two bodies involves a transfer of internal energy between them. However, in the eighteenth and nineteenth centuries the molecular theory of matter was controversial. The nature of heat was not well understood until about 1850. In the late 1700s, most scientists accepted the caloric theory of heat. (Some students still do, unhappily.) Caloric was a hypothetical fluid substance present in matter and supposed to flow from a hot body to a cold one. The amount of caloric lost by the hot body equaled the amount gained by the cold body. The total amount of caloric was believed to be conserved in all processes.

Strong evidence against the caloric theory was provided by Count Rumford in 1798. In charge of the army of Bavaria, he observed that, in boring a cannon, a virtually unlimited amount of heating was produced by friction, in contradiction to the caloric-theory notion of conservation of heat. Rumford found that a cannon borer driven by one horse for 2.5 hr heated 27 lb of ice-cold water to its boiling point. Addressing the Royal Society of London, Rumford argued that his experiments had proved the incorrectness of the caloric theory.

> Rumford began life as Benjamin Thompson of Woburn, Massachusetts. At 19 he married a wealthy widow of 30. He served the British during the American Revolution and settled in Europe after the war. He became Minister of War for Bavaria, where he earned extra money by spying for the British. In 1798 he traveled to London, where he founded the Royal Institution, which became one of Britain's leading scientific laboratories. In 1805 he married Lavoisier's widow, adding further to his wealth. His will left money to Harvard to establish the Rumford chair of physics, which still exists.

Despite Rumford's work, the caloric theory held sway until the 1840s. In 1842 Julius Mayer, a German physician, noted that the food that organisms consume goes partly to produce heat to maintain body temperature and partly to produce mechanical work performed by the organism. He then speculated that work and heat were both forms of energy and that the total amount of energy was conserved. Mayer's arguments were not found convincing, and it remained for James Joule to deal the death blow to the caloric theory.

Joule was the son of a wealthy English brewer. Working in a laboratory adjacent to the brewery, Joule did experiments in the 1840s showing that the same changes produced by heating a substance could also be produced by doing mechanical work on the substance, without transfer of heat. His most famous experiment used descending weights to turn paddle wheels in liquids. The potential energy of the weights was converted to kinetic energy of the liquid. The viscosity (internal friction) of the liquid then converted the liquid's kinetic energy to internal energy, increasing the temperature. Joule found that to increase the temperature of one pound of water by one degree Fahrenheit requires the expenditure of 772 foot-pounds of mechanical energy. Based on Joule's work, the first clear convincing statement of the law of conservation of energy was published by the German surgeon, physiologist, and physicist Helmholtz in 1847.

The internal energy of a system can be changed in several ways. Internal energy is an extensive property and thus depends on the amount of matter in the system. The internal energy of 20 g of H_2O at a given T and P is twice the internal energy of 10 g of H_2O at that T and P. For a pure substance, the **molar internal energy** U_m is defined as

$$U_m \equiv U/n$$

where n is the number of moles of the pure substance. U_m is an intensive property that depends on P and T.

We usually deal with closed systems. Here, the system's mass is held fixed.

Besides changing the mass of a system by adding or removing matter, we can change the energy of a system by doing work on it or by heating it. The **first law of thermodynamics** asserts that there exists an extensive state function E (called the **total energy** of the system) such that for any process in a closed system

$$\Delta E = q + w \qquad \text{closed syst.} \tag{2.40}$$

where ΔE is the energy change undergone by the system in the process, q is the heat flow into the system during the process, and w is the work done on the system during the process. The first law also asserts that a change in energy ΔE of the system is accompanied by a change in energy of the surroundings equal to $-\Delta E$, so *the total energy of system plus surroundings remains constant* (is conserved). For any process,

$$\Delta E_{syst} + \Delta E_{surr} = 0$$

We shall restrict ourselves to systems at rest in the absence of external fields. Here $K = 0 = V$, and from (2.39) we have $E = U$. Equation (2.40) becomes

$$\Delta U = q + w \qquad \text{closed syst. at rest, no fields} \qquad \textbf{(2.41)*}$$

where ΔU is the change in internal energy of the system. U is an extensive state function.

Note that, when we write ΔU, we mean ΔU_{syst}. We always focus attention on the system, and *all thermodynamic state functions refer to the system* unless otherwise specified. The conventions for the signs of q and w are set from the system's viewpoint. When heat flows into the system from the surroundings during a process, q is positive ($q > 0$); an outflow of heat from the system to the surroundings means q is negative. When work is done on the system by the surroundings (for example, in a compression of the system), w is positive; when the system does work on its surroundings, w is negative. A positive q and a positive w each increase the internal energy of the system.

For an infinitesimal process, Eq. (2.41) becomes

$$dU = dq + dw \qquad \text{closed syst.} \tag{2.42}$$

where the other two conditions of (2.41) are implicitly understood. dU is the infinitesimal change in system energy in a process with infinitesimal heat dq flowing into the system and infinitesimal work dw done on the system.

The internal energy U is (just like P or V or T) a function of the state of the system. For any process, ΔU thus depends only on the final and initial states of the system and is independent of the path used to bring the system from the initial state to the final state. If the system goes from state 1 to state 2 by any process, then

$$\Delta U = U_2 - U_1 = U_{final} - U_{initial} \qquad \textbf{(2.43)*}$$

The symbol Δ always means the final value minus the initial value.

A process in which the final state of the system is the same as the initial state is called a **cyclic process;** here $U_2 = U_1$, and

$$\Delta U = 0 \qquad \text{cyclic proc.}$$

which must obviously be true for the change in any state function in a cyclic process.

In contrast to U, the quantities q and w are not state functions. Given only the initial and final states of the system, we cannot find q or w. The heat q and the work w depend on the path used to go from state 1 to state 2.

Suppose, for example, that we take 1.00 mole of liquid H_2O at 25.0°C and 1.00 atm and raise its temperature to 30.0°C, the final pressure being 1.00 atm. What is q? The answer is that we cannot calculate q because the process is not specified. We could, if we like, increase the temperature by heating at 1 atm. In this case, $q = mc_P \Delta T = 18.0 \text{ g} \times 1.00 \text{ cal}/(\text{g °C}) \times 5.0°C = 90$ cal. However, we could instead emulate James Joule and increase T solely by doing work on the water, stirring it with a paddle (made of an adiabatic substance) until the water reached 30.0°C. In this case, $q = 0$. Or we could heat the water to some temperature between 25°C and 30°C and then do enough stirring to bring it up to 30°C. In this case, q is between 0 and 90 cal. Each of these processes also has a different value of w. However, no matter how we bring the water from 25°C and 1.00 atm to 30.0°C and 1.00 atm, ΔU is always the same, since the final and initial states are the same in each process.

EXAMPLE 2.3 Calculation of ΔU

Calculate ΔU when 1.00 mol of H_2O goes from 25.0°C and 1.00 atm to 30.0°C and 1.00 atm.

Since U is a state function, we can use any process we like to calculate ΔU. A convenient choice is a reversible heating from 25°C to 30°C at a fixed pressure of 1 atm. For this process, $q = 90$ cal, as calculated above. During the heating, the water expands slightly, doing work on the surrounding atmosphere. At constant P, we have

$$w = w_{rev} = -\int_1^2 P \, dV = -P \int_1^2 dV = -P(V_2 - V_1)$$

where (2.31) was used. Because P is constant, it can be taken outside the integral. The volume change is $\Delta V = V_2 - V_1 = m/\rho_2 - m/\rho_1$, where ρ_2 and ρ_1 are the final and initial densities of the water and $m = 18.0$ g. A handbook gives $\rho_2 = 0.9956 \text{ g/cm}^3$ and $\rho_1 = 0.9970 \text{ g/cm}^3$. We find $\Delta V = 0.025 \text{ cm}^3$ and

$$w = -0.025 \text{ cm}^3 \text{ atm} = -0.025 \text{ cm}^3 \text{ atm} \frac{1.987 \text{ cal mol}^{-1} \text{ K}^{-1}}{82.06 \text{ cm}^3 \text{ atm mol}^{-1} \text{ K}^{-1}}$$

$$= -0.0006 \text{ cal}$$

where two values of R were used to convert w to calories. Thus, w is completely negligible compared with q, and $\Delta U = q + w = 90$ cal. Because volume changes of liquids and solids are small, usually P-V work is significant only for gases.

Exercise

Calculate q, w, and ΔU when 1.00 mol of water is heated from 0°C to 100°C at a fixed pressure of 1 atm. Densities of water are 0.9998 g/cm³ at 0°C and 0.9854 g/cm³ at 100°C. (*Answer:* 1800 cal, −0.006 cal, 1800 cal.)

Although the values of q and w for a change from state 1 to state 2 depend on the process used, the value of $q + w$, which equals ΔU, is the same for every process that goes from state 1 to state 2. This is the experimental content of the first law.

Since q and w are not state functions, it is meaningless to ask how much heat a system contains (or how much work it contains). Although one often says that "heat and work are forms of energy," this language, unless properly understood, can mislead one into the error of regarding heat and work as state functions. Heat and work are defined only in terms of processes. Before and after the process of energy transfer between system and surroundings, heat and work do not exist. *Heat is an energy transfer between system and surroundings due to a temperature difference. Work is an energy transfer between system and surroundings due to a macroscopic force acting through a distance.* Heat and work are forms of energy *transfer* rather than forms of energy. Work is energy transfer due to the action of macroscopically observable forces. Heat is energy transfer due to the action of forces at a molecular level. When bodies at different temperatures are placed in contact, collisions between molecules of the two bodies produce a net transfer of energy to the colder body from the hotter body, whose molecules have a greater average kinetic energy than in the colder body. Heat is work done at the molecular level.

Much of the terminology of heat is misleading because it is a relic of the erroneous caloric theory of heat. Thus, one often refers to "heat flow" between system and surroundings. In reality, the so-called heat flow is really an energy flow due to a temperature difference. Likewise, the term "heat capacity" for C_P is misleading, since it implies that bodies store heat, whereas heat refers only to energy transferred in a process; bodies contain internal energy but do not contain heat.

Heat and work are measures of energy transfer, and both have the same units as energy. The unit of heat can therefore be defined in terms of the joule. Thus the definition of the calorie given in Sec. 2.3 is no longer used. The present definition is

$$1 \text{ cal} \equiv 4.184 \text{ J} \qquad \text{exactly} \qquad \textbf{(2.44)}*$$

where the value 4.184 was chosen to give good agreement with the old definition of the calorie. The calorie defined by (2.44) is called the *thermochemical calorie,* often designated cal_{th}. (Over the years, several slightly different calories have been used.)

It is not necessary to express heat in calories. The joule can be used as the unit of heat. This is what is done in the officially recommended SI units (Sec. 2.1), but since some of the available thermochemical tables use calories, we shall use both joules and calories as the units of heat, work, and internal energy.

Although we won't be considering systems with mechanical energy, it is worthwhile to consider a possible source of confusion that can arise when dealing with such systems. Consider a rock falling in vacuum toward the earth's surface. Its total energy is $E = K + V + U$. Since the gravitational potential energy V is included as part of the system's energy, the gravitational field (in which the potential energy resides) must be considered part of the system. In the first-law equation $\Delta E = q + w$, we do not include work that one part of the system does on another part of the system. Hence w in the first law does not include the work done by the gravitational field on the falling body. Thus for the falling rock, w is zero; also, q is zero. Therefore $\Delta E = q + w$ is zero, and E remains constant as the body falls (although both K and V vary). In general, w in $\Delta E = q + w$ does not include the work done by conservative forces (forces related to the potential energy V in $E = K + V + U$).

Sometimes people get the idea that Einstein's special relativity equation $E = mc^2$ invalidates the conservation of energy, the first law of thermodynamics. This is not so. All $E = mc^2$ says is that a mass m always has an energy mc^2 associated with it and an energy E always has a mass $m = E/c^2$ associated with it. The total energy of system plus surroundings is still conserved in special relativity; likewise, the total relativistic mass of system plus surroundings is conserved in special relativity. Energy cannot disappear; mass cannot disappear. The equation $\Delta E = q + w$ is still valid in special relativity. Consider, for example, nuclear fission. Although it is true that the sum of the *rest* masses of the nuclear fragments is less than the rest mass of the original nucleus, the fragments are moving at high speed. The relativistic mass of a body increases with increasing speed, and the total relativistic mass of the fragments exactly equals the relativistic mass of the original nucleus.

2.5 ENTHALPY

The **enthalpy** H of a thermodynamic system whose internal energy, pressure, and volume are U, P, and V is defined as

$$H \equiv U + PV \tag{2.45}*$$

Since U, P, and V are state functions, H is a state function. Note from $dw_{rev} = -P \, dV$ that the product of P and V has the dimensions of work and hence of energy. Therefore it is legitimate to add U and PV. Naturally, H has units of energy.

Of course, we could take any dimensionally correct combination of state functions to define a new state function. Thus, we might define $(3U - 5PV)/T^3$ as the state function "enwhoopee." The motivation for giving a special name to the state function $U + PV$ is that this combination of U, P, and V occurs often in thermodynamics. For example, let q_P be the heat absorbed in a constant-pressure process in a closed system. The first law $\Delta U = q + w$ [Eq. (2.41)] gives

$$U_2 - U_1 = q + w = q - \int_{V_1}^{V_2} P \, dV = q_P - P \int_{V_1}^{V_2} dV = q_P - P(V_2 - V_1)$$

$$q_P = U_2 + PV_2 - U_1 - PV_1 = (U_2 + P_2V_2) - (U_1 + P_1V_1) = H_2 - H_1$$

$$\Delta H = q_P \qquad \text{const. } P \text{, closed syst., } P\text{-}V \text{ work only} \tag{2.46}*$$

since $P_1 = P_2 = P$. In the derivation of (2.46), we used (2.31) ($w_{rev} = -\int_1^2 P \, dV$) for the work w. Equation (2.31) gives the work associated with a volume change of the system. Besides a volume change, there are other ways that system and surroundings can exchange work, but we won't consider these possibilities until Chapters 13 and 14. Thus (2.46) is valid only when no kind of work other than volume-change work is done. Note also that (2.31) is for a mechanically reversible process. A constant-pressure process is mechanically reversible since, if there were unbalanced mechanical forces acting, the system's pressure P would not remain constant. Equation (2.46) says that for a closed system that can do only P-V work, the heat q_P absorbed in a constant-pressure process equals the system's enthalpy change.

For any change of state, the enthalpy change is

$$\Delta H = H_2 - H_1 = U_2 + P_2V_2 - (U_1 + P_1V_1) = \Delta U + \Delta(PV) \tag{2.47}$$

where $\Delta(PV) \equiv (PV)_2 - (PV)_1 = P_2V_2 - P_1V_1$. For a constant-pressure process, $P_2 = P_1 = P$ and $\Delta(PV) = PV_2 - PV_1 = P \, \Delta V$. Therefore

$$\Delta H = \Delta U + P \, \Delta V \qquad \text{const. } P \tag{2.48}$$

An error students sometimes make is to equate $\Delta(PV)$ with $P \, \Delta V + V \, \Delta P$. We have

$$\Delta(PV) = P_2V_2 - P_1V_1 = (P_1 + \Delta P)(V_1 + \Delta V) - P_1V_1$$

$$= P_1 \Delta V + V_1 \Delta P + \Delta P \Delta V$$

Because of the $\Delta P \, \Delta V$ term, $\Delta(PV) \neq P \, \Delta V + V \, \Delta P$. For infinitesimal changes, we have $d(PV) = P \, dV + V \, dP$, since $d(uv) = u \, dv + v \, du$ [Eq. (1.28)], but the corresponding equation is not true for finite changes. [For an infinitesimal change, the equation after (2.48) becomes $d(PV) = P \, dV + V \, dP + dP \, dV = P \, dV + V \, dP$, since the product of two infinitesimals can be neglected.]

Since U and V are extensive, H is extensive. The molar enthalpy of a pure substance is $H_m = H/n = (U + PV)/n = U_m + PV_m$.

Consider now a constant-volume process. If the closed system can do only P-V work, then $dw = -P \, dV = 0$, since $dV = 0$. The sum of the dw's is zero, so $w = 0$. The first law $\Delta U = q + w$ then becomes for a constant-volume process

$$\Delta U = q_V \qquad \text{closed syst., } P\text{-}V \text{ work only, } V \text{ const.} \tag{2.49}$$

where q_V is the heat absorbed at constant volume. Comparison of (2.49) and (2.46) shows that in a constant-pressure process H plays a role analogous to that played by U in a constant-volume process.

From Eq. (2.47), we have $\Delta H = \Delta U + \Delta(PV)$. Because solids and liquids have comparatively small volumes and undergo only small changes in volume, in nearly all processes that involve only solids or liquids (*condensed* phases) at low or moderate pressures, the $\Delta(PV)$ term is negligible compared with the ΔU term. (For example, recall the example in Sec. 2.4 of heating liquid water, where we found $\Delta U = q_P$.) For condensed phases not at high pressures, the enthalpy change in a process is essentially the same as the internal-energy change: $\Delta H \approx \Delta U$.

2.6 HEAT CAPACITIES

The **heat capacity** C_{pr} of a closed system for an infinitesimal process pr is defined as

$$C_{pr} \equiv dq_{pr}/dT \qquad (2.50)^*$$

where dq_{pr} and dT are the heat flowing into the system and the temperature change of the system in the process. The subscript on C indicates that the heat capacity depends on the nature of the process. For example, for a constant-pressure process we get C_P, the **heat capacity at constant pressure** (or *isobaric heat capacity*):

$$C_P \equiv \frac{dq_P}{dT} \qquad (2.51)^*$$

Similarly, the **heat capacity at constant volume** (or *isochoric heat capacity*) C_V of a closed system is

$$C_V \equiv \frac{dq_V}{dT} \qquad (2.52)^*$$

where dq_V and dT are the heat added to the system and the system's temperature change in an infinitesimal constant-volume process. Strictly speaking, Eqs. (2.50) to (2.52) apply only to reversible processes. In an irreversible heating, the system may develop temperature gradients, and there will then be no single temperature assignable to the system. If T is undefined, the infinitesimal change in temperature dT is undefined.

Equations (2.46) and (2.49) written for an infinitesimal process give $dq_P = dH$ at constant pressure and $dq_V = dU$ at constant volume. Therefore (2.51) and (2.52) can be written as

$$C_P = \left(\frac{\partial H}{\partial T}\right)_P, \qquad C_V = \left(\frac{\partial U}{\partial T}\right)_V \qquad \text{closed syst. in equilib., } P\text{-}V \text{ work only} \quad (2.53)^*$$

C_P and C_V give the rates of change of H and U with temperature.

To measure C_P of a solid or liquid, one holds it at constant pressure in an adiabatically enclosed container and heats it with an electrical heating coil. For a current I flowing for a time t through a wire with a voltage drop V across the wire, the heat generated by the coil is VIt. If the measured temperature increase ΔT in the substance is small, Eq. (2.51) gives $C_P = VIt/\Delta T$, where C_P is the value at the average temperature of the experiment and at the pressure of the experiment. C_P of a gas is found from the temperature increase produced by electrically heating the gas flowing at a known rate.

The thermodynamic state of an equilibrium system at rest in the absence of applied fields is specified by its composition (the number of moles of each component present in each phase) and by any two of the three variables P, V, and T. Commonly, P and T are used. For a closed system of fixed composition, the state is specified by P and T. Any state function has a definite value once the system's state is specified. Therefore any state function of a closed equilibrium system of fixed composition is a

$C_{P,m}$/(J/mol-K)

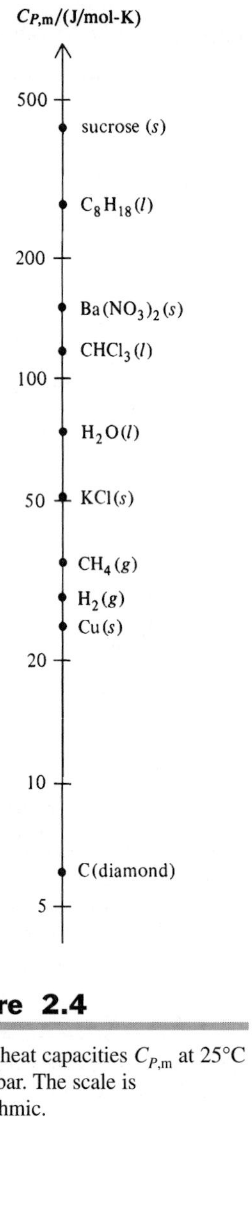

Figure 2.4

Molar heat capacities $C_{P,m}$ at 25°C and 1 bar. The scale is logarithmic.

function of T and P. For example, for such a system, $H = H(T, P)$. The partial derivative $(\partial H(T, P)/\partial T)_P$ is also a function of T and P. Hence C_P is a function of T and P and is therefore a state function. Similarly, U can be taken as a function of T and V, and C_V is a state function.

For a pure substance, the **molar heat capacities** at constant P and at constant V are $C_{P,m} = C_P/n$ and $C_{V,m} = C_V/n$. Some $C_{P,m}$ values at 25°C and 1 atm are plotted in Fig. 2.4. The Appendix gives further values. Clearly, $C_{P,m}$ increases with increasing size of the molecules. See Sec. 2.11 for discussion of $C_{P,m}$ values.

For a one-phase system of mass m, the **specific heat capacity** c_P is $c_P \equiv C_P/m$. The adjective **specific** means "divided by mass." Thus, the **specific volume** v and **specific enthalpy** h of a phase of mass m are $v \equiv V/m = 1/\rho$ and $h \equiv H/m$.

Do not confuse the heat capacity C_P (which is an extensive property) with the molar heat capacity $C_{P,m}$ or the specific heat capacity c_P (which are intensive properties). We have

$$C_{P,m} \equiv C_P/n \qquad \text{pure substance} \qquad (2.54)*$$

$$c_P \equiv C_P/m \qquad \text{one-phase system} \qquad (2.55)*$$

$C_{P,m}$ and c_P are functions of T and P. Figure 2.5 plots some data for $H_2O(g)$. These curves are discussed in Sec. 8.6.

One can prove from the laws of thermodynamics that C_P and C_V must both be positive. (See *Münster*, sec. 40.)

$$C_P > 0, \qquad C_V > 0 \qquad (2.56)$$

Exceptions to (2.56) are systems where gravitational effects are important. Such systems (for example, black holes, stars, and star clusters) can have negative heat capacities [D. Lynden-Bell, *Physica A*, **263**, 293 (1999)].

What is the relation between C_P and C_V? We have

$$C_P - C_V = \left(\frac{\partial H}{\partial T}\right)_P - \left(\frac{\partial U}{\partial T}\right)_V = \left(\frac{\partial(U + PV)}{\partial T}\right)_P - \left(\frac{\partial U}{\partial T}\right)_V$$

$$C_P - C_V = \left(\frac{\partial U}{\partial T}\right)_P + P\left(\frac{\partial V}{\partial T}\right)_P - \left(\frac{\partial U}{\partial T}\right)_V \qquad (2.57)$$

We expect $(\partial U/\partial T)_P$ and $(\partial U/\partial T)_V$ in (2.57) to be related to each other. In $(\partial U/\partial T)_V$, the internal energy is taken as a function of T and V; $U = U(T, V)$. The total differential of $U(T, V)$ is [Eq. (1.30)]

$$dU = \left(\frac{\partial U}{\partial T}\right)_V dT + \left(\frac{\partial U}{\partial V}\right)_T dV \qquad (2.58)$$

Equation (2.58) is valid for any infinitesimal process, but since we want to relate $(\partial U/\partial T)_V$ to $(\partial U/\partial T)_P$, we impose the restriction of constant P on (2.58) to give

$$dU_P = \left(\frac{\partial U}{\partial T}\right)_V dT_P + \left(\frac{\partial U}{\partial V}\right)_T dV_P$$

where the P subscripts indicate that the infinitesimal changes dU, dT, and dV occur at constant P. Division by dT_P gives

$$\frac{dU_P}{dT_P} = \left(\frac{\partial U}{\partial T}\right)_V + \left(\frac{\partial U}{\partial V}\right)_T \frac{dV_P}{dT_P}$$

The ratio of infinitesimals dU_P/dT_P is the partial derivative $(\partial U/\partial T)_P$, so

$$\left(\frac{\partial U}{\partial T}\right)_P = \left(\frac{\partial U}{\partial T}\right)_V + \left(\frac{\partial U}{\partial V}\right)_T \left(\frac{\partial V}{\partial T}\right)_P \qquad (2.59)$$

Substitution of (2.59) into (2.57) gives the desired relation:

$$C_P - C_V = \left[\left(\frac{\partial U}{\partial V} \right)_T + P \right] \left(\frac{\partial V}{\partial T} \right)_P \qquad (2.60)$$

What is the physical reason for the fact that $C_P \neq C_V$? The definitions $C_P = dq_P/dT$ and $C_V = dq_V/dT$ show that the origin of the difference lies in the difference between dq_P and dq_V, the heats added at constant pressure and at constant volume to produce the same temperature change dT. The first law $dU = dq + dw$ gives $dq = dU - dw = dU + P\,dV$ for a closed system with only P-V work. It follows that $dq_P = dU_P + P\,dV_P$ and $dq_V = dU_V$, where the subscripts indicate constant P or V. Therefore,

$$dq_P - dq_V = dU_P - dU_V + P\,dV_P \qquad (2.61)$$

There are thus two reasons for the difference between dq_P and dq_V. In a constant-pressure process, part of the added heat goes into the work of expansion (the $P\,dV_P$ term), whereas in a constant-volume process the system does no work on the surroundings. Also, the change in internal energy that occurs at constant pressure differs from the change that occurs at constant volume: $dU_P \neq dU_V$.

The state function $(\partial U/\partial V)_T$ in (2.60) has dimensions of pressure and is sometimes called the *internal pressure*. Clearly, $(\partial U/\partial V)_T$ is related to that part of the internal energy U that is due to intermolecular potential energy. A change in the system's volume V will change the average intermolecular distance and hence the average intermolecular potential energy. For gases not at high pressure, the smallness of intermolecular forces makes $(\partial U/\partial V)_T$ in (2.60) and the related quantity $dU_P - dU_V$ in (2.61) small. Here, the main contribution to $C_P - C_V$ comes from the $P\,dV_P$ term in (2.61). For liquids and solids, where molecules are close to one another, the large intermolecular forces make $(\partial U/\partial V)_T$ and $dU_P - dU_V$ large. Here, the main contribution to $C_P - C_V$ comes from the difference $dU_P - dU_V$. Measurement of $(\partial U/\partial V)_T$ in gases is discussed in Sec. 2.7.

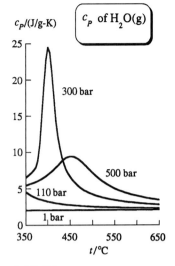

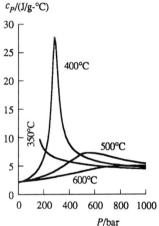

Figure 2.5

Specific heat of $H_2O(g)$ plotted versus T and versus P.

2.7 THE JOULE AND JOULE–THOMSON EXPERIMENTS

In 1843 Joule tried to determine $(\partial U/\partial V)_T$ for a gas by measuring the temperature change after free expansion of the gas into a vacuum. This experiment was repeated by Keyes and Sears in 1924 with an improved setup (Fig. 2.6).

Initially, chamber A is filled with a gas, and chamber B is evacuated. The valve between the chambers is then opened. After equilibrium is reached, the temperature change in the system is measured by the thermometer. Because the system is surrounded by adiabatic walls, q is 0; no heat flows into or out of the system. The expansion into a vacuum is highly irreversible. Finite unbalanced forces act within the system, and as the gas rushes into B, there is turbulence and lack of pressure equilibrium. Therefore $dw = -P\,dV$ does not apply. However, we can readily calculate the work $-w$ done by the system. The only motion that occurs is within the system itself. Therefore the gas does no work on its surroundings, and vice versa. Hence $w = 0$ for expansion into a vacuum. Since $\Delta U = q + w$ for a closed system, we have $\Delta U = 0 + 0 = 0$. This is a constant-energy process. The experiment measures the temperature change with change in volume at constant internal energy, $(\partial T/\partial V)_U$. More precisely, the experiment measures $\Delta T/\Delta V$ at constant U. The method used to get $(\partial T/\partial V)_U$ from $\Delta T/\Delta V$ measurements is similar to that described later in this section for $(\partial T/\partial P)_H$.

We define the *Joule coefficient* μ_J (mu jay) as

$$\mu_J \equiv (\partial T/\partial V)_U \qquad (2.62)$$

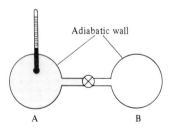

Figure 2.6

Sketch of the Keyes–Sears modification of the Joule experiment.

How is the measured quantity $(\partial T/\partial V)_U = \mu_J$ related to $(\partial U/\partial V)_T$? The variables in these two partial derivatives are the same (namely, T, U, and V). Hence we can use $(\partial x/\partial y)_z(\partial y/\partial z)_x(\partial z/\partial x)_y = -1$ [Eq. (1.34)] to relate these partial derivatives. Replacement of x, y, and z with T, U, and V gives

$$\left(\frac{\partial T}{\partial U}\right)_V \left(\frac{\partial U}{\partial V}\right)_T \left(\frac{\partial V}{\partial T}\right)_U = -1$$

$$\left(\frac{\partial U}{\partial V}\right)_T = -\left[\left(\frac{\partial T}{\partial U}\right)_V\right]^{-1}\left[\left(\frac{\partial V}{\partial T}\right)_U\right]^{-1} = -\left(\frac{\partial U}{\partial T}\right)_V\left(\frac{\partial T}{\partial V}\right)_U$$

$$\left(\frac{\partial U}{\partial V}\right)_T = -C_V\mu_J \tag{2.63}$$

where $(\partial z/\partial x)_y = 1/(\partial x/\partial z)_y$, $(\partial U/\partial T)_V = C_V$, and $\mu_J = (\partial T/\partial V)_U$ [Eqs. (1.32), (2.53), and (2.62)] were used.

Joule's 1843 experiment gave zero for μ_J and hence zero for $(\partial U/\partial V)_T$. However, his setup was so poor that his result was meaningless. The 1924 Keyes–Sears experiment showed that $(\partial U/\partial V)_T$ is small but definitely nonzero for gases. Because of experimental difficulties, only a few rough measurements were made.

In 1853 Joule and William Thomson (in later life Lord Kelvin) did an experiment similar to the Joule experiment but allowing far more accurate results to be obtained. The **Joule–Thomson experiment** involves the slow throttling of a gas through a rigid, porous plug. An idealized sketch of the experiment is shown in Fig. 2.7. The system is enclosed in adiabatic walls. The left piston is held at a fixed pressure P_1. The right piston is held at a fixed pressure $P_2 < P_1$. The partition B is porous but not greatly so. This allows the gas to be slowly forced from one chamber to the other. Because the throttling process is slow, pressure equilibrium is maintained in each chamber. Essentially all the pressure drop from P_1 to P_2 occurs in the porous plug.

We want to calculate w, the work done on the gas in throttling it through the plug. The overall process is irreversible since P_1 exceeds P_2 by a finite amount, and an infinitesimal change in pressures cannot reverse the process. However, the pressure drop occurs almost completely in the plug. The plug is rigid, and the gas does no work on the plug, or vice versa. The exchange of work between system and surroundings occurs solely at the two pistons. Since pressure equilibrium is maintained at each piston, we can use $dw_{rev} = -P\,dV$ to calculate the work at each piston. The left piston does work w_L on the gas. We have $dw_L = -P_L\,dV = -P_1\,dV$, where we use subscripts L and R for left and right. Let all the gas be throttled through. The initial and final volumes of the left chamber are V_1 and 0, so

$$w_L = -\int_{V_1}^{0} P_1\,dV = -P_1\int_{V_1}^{0} dV = -P_1(0 - V_1) = P_1V_1$$

The right piston does work dw_R on the gas. (w_R is negative, since the gas in the right chamber does positive work on the piston.) We have $w_R = -\int_0^{V_2} P_2\,dV = -P_2V_2$. The work done on the gas is $w = w_L + w_R = P_1V_1 - P_2V_2$.

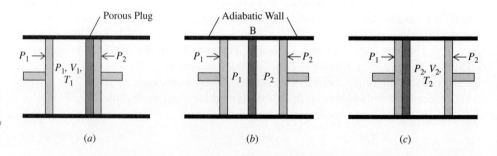

Figure 2.7

The Joule–Thomson experiment.

(a)   (b)   (c)

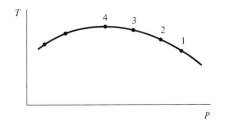

Figure 2.8

An isenthalpic curve obtained
from a series of Joule–Thomson
experiments.

The first law for this adiabatic process ($q = 0$) gives $U_2 - U_1 = q + w = w$, so
$U_2 - U_1 = P_1 V_1 - P_2 V_2$ or $U_2 + P_2 V_2 = U_1 + P_1 V_1$. Since $H \equiv U + PV$, we have

$$H_2 = H_1 \qquad \text{or} \qquad \Delta H = 0$$

The initial and final enthalpies are equal in a Joule–Thomson expansion.

Measurement of the temperature change $\Delta T = T_2 - T_1$ in the Joule–Thomson
experiment gives $\Delta T/\Delta P$ at constant H. This may be compared with the Joule experiment, which measures $\Delta T/\Delta V$ at constant U.

We define the **Joule–Thomson coefficient** μ_{JT} by

$$\mu_{JT} \equiv \left(\frac{\partial T}{\partial P} \right)_H \qquad (2.64)*$$

μ_{JT} is the ratio of infinitesimal changes in two intensive properties and therefore is an
intensive property. Like any intensive property, it is a function of T and P (and the
nature of the gas).

A single Joule–Thomson experiment yields only $(\Delta T/\Delta P)_H$. To find $(\partial T/\partial P)_H$ values, we proceed as follows. Starting with some initial P_1 and T_1, we pick a value of P_2
less than P_1 and do the throttling experiment, measuring T_2. We then plot the two
points (T_1, P_1) and (T_2, P_2) on a T-P diagram; these are points 1 and 2 in Fig. 2.8. Since
$\Delta H = 0$ for a Joule–Thomson expansion, states 1 and 2 have equal enthalpies. A rep-
etition of the experiment with the same initial P_1 and T_1 but with the pressure on the
right piston set at a new value P_3 gives point 3 on the diagram. Several repetitions,
each with a different final pressure, yield several points that correspond to states of
equal enthalpy. We join these points with a smooth curve (called an *isenthalpic curve*).
The slope of this curve at any point gives $(\partial T/\partial P)_H$ for the temperature and pressure
at that point. Values of T and P for which μ_{JT} is negative (points to the right of point
4) correspond to warming on Joule–Thomson expansion. At point 4, μ_{JT} is zero. To
the left of point 4, μ_{JT} is positive, and the gas is cooled by throttling. To generate further isenthalpic curves and get more values of $\mu_{JT}(T, P)$, we use different initial temperatures T_1.

Values of μ_{JT} for gases range from $+3$ to $-0.1°C/atm$, depending on the gas and
on its temperature and pressure. Figure 2.9 plots some μ_{JT} data for N_2 gas.

Joule–Thomson throttling is used to liquefy gases. For a gas to be cooled by a
Joule–Thomson expansion ($\Delta P < 0$), its μ_{JT} must be positive over the range of T and
P involved. In Joule–Thomson liquefaction of gases, the porous plug is replaced by a
narrow opening (a needle valve). Another method of gas liquefaction is an approximately reversible adiabatic expansion against a piston.

A procedure similar to that used to derive (2.63) yields (Prob. 2.34a)

$$\left(\frac{\partial H}{\partial P} \right)_T = -C_P \mu_{JT} \qquad (2.65)$$

We can use thermodynamic identities to relate the Joule and Joule–Thomson coefficients; see Prob. 2.34b.

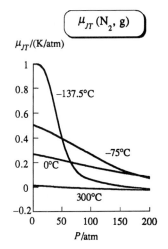

$\mu_{JT} (N_2, g)$

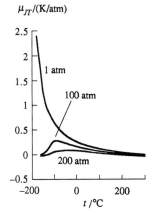

Figure 2.9

The Joule–Thomson coefficient
of $N_2(g)$ plotted versus P and
versus T.

PERFECT GASES AND THE FIRST LAW

Perfect Gases

An ideal gas was defined in Chapter 1 as a gas that obeys the equation of state $PV = nRT$. The molecular picture of an ideal gas is one with no intermolecular forces. If we change the volume of an ideal gas while holding T constant, we change the average distance between the molecules, but since intermolecular forces are zero, this distance change will not affect the internal energy U. Also, the average translational kinetic energy of the gas molecules is a function of T only (as is also true of the molecular rotational and vibrational energies—see Sec. 2.11) and will not change with volume. We therefore expect that, for an ideal gas, U will not change with V at constant T and $(\partial U/\partial V)_T$ will be zero. However, we are not yet in a position to prove this thermodynamically. To maintain the logical development of thermodynamics, we therefore now define a **perfect gas** as one that obeys both the following equations:

$$PV = nRT \quad \text{and} \quad (\partial U/\partial V)_T = 0 \qquad \text{perfect gas} \qquad (2.66)^*$$

An ideal gas is required to obey only $PV = nRT$. Once we have postulated the second law of thermodynamics, we shall prove that $(\partial U/\partial V)_T = 0$ follows from $PV = nRT$, so there is in fact no distinction between an ideal gas and a perfect gas. Until then, we shall maintain the distinction between the two.

For a closed system in equilibrium, the internal energy (and any other state function) can be expressed as a function of temperature and volume: $U = U(T, V)$. However, (2.66) states that for a perfect gas U is independent of volume. Therefore U of a perfect gas depends only on temperature:

$$U = U(T) \qquad \text{perf. gas} \qquad (2.67)^*$$

Since U is independent of V for a perfect gas, the partial derivative $(\partial U/\partial T)_V$ in Eq. (2.53) for C_V becomes an ordinary derivative: $C_V = dU/dT$ and

$$dU = C_V \, dT \qquad \text{perf. gas} \qquad (2.68)^*$$

It follows from (2.67) and $C_V = dU/dT$ that C_V of a perfect gas depends only on T:

$$C_V = C_V(T) \qquad \text{perf. gas} \qquad (2.69)^*$$

For a perfect gas, $H \equiv U + PV = U + nRT$. Hence (2.67) shows that H depends only on T for a perfect gas. Using $C_P = (\partial H/\partial T)_P$ [Eq. (2.53)], we then have

$$H = H(T), \qquad C_P = dH/dT, \qquad C_P = C_P(T) \qquad \text{perf. gas} \qquad (2.70)^*$$

Use of $(\partial U/\partial V)_T = 0$ [Eq. (2.66)] in $C_P - C_V = [(\partial U/\partial V)_T + P](\partial V/\partial T)_P$ [Eq. (2.60)] gives

$$C_P - C_V = P(\partial V/\partial T)_P \qquad \text{perf. gas} \qquad (2.71)$$

From $PV = nRT$, we get $(\partial V/\partial T)_P = nR/P$. Hence for a perfect gas $C_P - C_V = nR$ or

$$C_{P,m} - C_{V,m} = R \qquad \text{perf. gas} \qquad (2.72)^*$$

We have $\mu_J C_V = -(\partial U/\partial V)_T$ [Eq. (2.63)]. Since $(\partial U/\partial V)_T = 0$ for a perfect gas, it follows that $\mu_J = 0$ for a perfect gas. Also, $\mu_{JT} C_P = -(\partial H/\partial P)_T$ [Eq. (2.65)]. Since H depends only on T for a perfect gas, we have $(\partial H/\partial P)_T = 0$ for such a gas, and $\mu_{JT} = 0$ for a perfect gas. Surprisingly, as Fig. 2.9 shows, μ_{JT} for a real gas does not go to zero as P goes to zero. (See Prob. 8.37 for analysis of this fact.)

We now apply the first law to a perfect gas. For a reversible volume change, $dw = -P \, dV$ [Eq. (2.30)]. Also, (2.68) gives $dU = C_V \, dT$ for a perfect gas. For a fixed amount of a perfect gas, the first law $dU = dq + dw$ (closed system) becomes

$$dU = C_V \, dT = dq - P \, dV \qquad \text{perf. gas, rev. proc., } P\text{-}V \text{ work only} \quad (2.73)$$

EXAMPLE 2.4 Calculation of q, w, and ΔU

Suppose 0.100 mol of a perfect gas having $C_{V,m} = 1.50R$ independent of temperature undergoes the reversible cyclic process $1 \to 2 \to 3 \to 4 \to 1$ shown in Fig. 2.10, where either P or V is held constant in each step. Calculate q, w, and ΔU for each step and for the complete cycle.

Since we know how P varies in each step and since the steps are reversible, we can readily find w for each step by integrating $dw_{\text{rev}} = -P \, dV$. Since either V or P is constant in each step, we can integrate $dq_V = C_V \, dT$ and $dq_P = C_P \, dT$ [Eqs. (2.51) and (2.52)] to find the heat in each step. The first law $\Delta U = q + w$ then allows calculation of ΔU.

To evaluate integrals like $\int_1^2 C_V \, dT$, we will need to know the temperatures of states 1, 2, 3, and 4. We therefore begin by using $PV = nRT$ to find these temperatures. For example, $T_1 = P_1 V_1 / nR = 122$ K. Similarly, $T_2 = 366$ K, $T_3 = 732$ K, $T_4 = 244$ K.

Step $1 \to 2$ is at constant volume, no work is done, and $w_{1 \to 2} = 0$. Step $2 \to 3$ is at constant pressure and

$$w_{2 \to 3} = -\int_2^3 P \, dV = -P(V_3 - V_2) = -(3.00 \text{ atm})(2000 \text{ cm}^3 - 1000 \text{ cm}^3)$$

$$= -3000 \text{ cm}^3 \text{ atm } (8.314 \text{ J})/(82.06 \text{ cm}^3 \text{ atm}) = -304 \text{ J}$$

where two values of R were used to convert to joules. Similarly, $w_{3 \to 4} = 0$ and $w_{4 \to 1} = 101$ J. The work w for the complete cycle is the sum of the works for the four steps, so $w = -304$ J $+ 0 + 101$ J $+ 0 = -203$ J.

Step $1 \to 2$ is at constant volume, and

$$q_{1 \to 2} = \int_1^2 C_V \, dT = nC_{V,m}\int_1^2 dT = n(1.50R)(T_2 - T_1)$$

$$= (0.100 \text{ mol})1.50[8.314 \text{ J}/(\text{mol K})](366 \text{ K} - 122 \text{ K}) = 304 \text{ J}$$

Step $2 \to 3$ is at constant pressure, and $q_{2 \to 3} = \int_2^3 C_P \, dT$. Equation (2.72) gives $C_{P,m} = C_{V,m} + R = 2.50R$, and we find $q_{2 \to 3} = 761$ J. Similarly, $q_{3 \to 4} = -608\frac{1}{2}$ J and $q_{4 \to 1} = -253\frac{1}{2}$ J. The total heat for the cycle is $q = 304$ J $+ 761$ J $- 608\frac{1}{2}$ J $- 253\frac{1}{2}$ J $= 203$ J.

We have $\Delta U_{1 \to 2} = q_{1 \to 2} + w_{1 \to 2} = 304$ J $+ 0 = 304$ J. Similarly, we find $\Delta U_{2 \to 3} = 457$ J, $\Delta U_{3 \to 4} = -608\frac{1}{2}$ J, $\Delta U_{4 \to 1} = -152\frac{1}{2}$ J. For the complete cycle, $\Delta U = 304$ J $+ 457$ J $- 608\frac{1}{2}$ J $- 152\frac{1}{2}$ J $= 0$, which can also be found from $q + w$ as 203 J $-$ 203 J $= 0$. An alternative procedure is to use the perfect-gas equation $dU = C_V \, dT$ to find ΔU for each step.

For this cyclic process, we found $\Delta U = 0$, $q \neq 0$, and $w \neq 0$. These results are consistent with the fact that U is a state function but q and w are not.

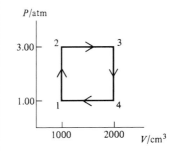

Figure 2.10

A reversible cyclic process.

Exercise

Use the perfect-gas equation $dU = C_V \, dT$ to find ΔU for each step in the cycle of Fig. 2.10. (*Answer:* 304 J, 456 J, -609 J, -152 J.)

Exercise

Verify that w for the reversible cyclic process in this example equals minus the area enclosed by the lines in Fig. 2.10.

Reversible Isothermal Process in a Perfect Gas

Consider the special case of a reversible isothermal (constant-T) process in a perfect gas. (Throughout this section, the system is assumed closed.) For a fixed amount of a perfect gas, U depends only on T [Eq. (2.67)]. Therefore $\Delta U = 0$ for an isothermal change of state in a perfect gas. This also follows from $dU = C_V dT$ for a perfect gas. The first law $\Delta U = q + w$ becomes $0 = q + w$ and $q = -w$. Integration of $dw_{rev} = -P \, dV$ and use of $PV = nRT$ give

$$w = -\int_1^2 P \, dV = -\int_1^2 \frac{nRT}{V} \, dV = -nRT \int_1^2 \frac{dV}{V} = -nRT(\ln V_2 - \ln V_1)$$

$$w = -q = nRT \ln \frac{V_1}{V_2} = nRT \ln \frac{P_2}{P_1} \qquad \text{rev. isothermal proc., perf. gas} \qquad (2.74)$$

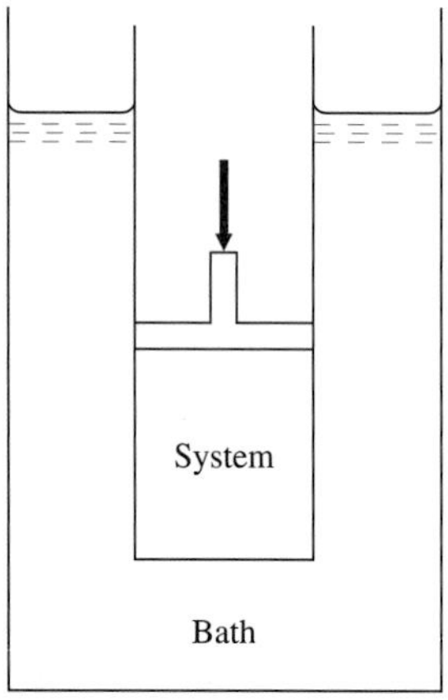

where Boyle's law was used. If the process is an expansion ($V_2 > V_1$), then w (the work done *on* the gas) is negative and q (the heat added to the gas) is positive. All the added heat appears as work done by the gas, maintaining U as constant for the perfect gas. It is best *not* to memorize an equation like (2.74), since it can be quickly derived from $dw = -P \, dV$.

To carry out a reversible isothermal volume change in a gas, we imagine the gas to be in a cylinder fitted with a frictionless piston. We place the cylinder in a very large constant-temperature bath (Fig. 2.11) and change the external pressure on the piston at an infinitesimal rate. If we increase the pressure, the gas is slowly compressed. The work done on it will transfer energy to the gas and will tend to increase its temperature at an infinitesimal rate. This infinitesimal temperature increase will cause heat to flow out of the gas to the surrounding bath, thereby maintaining the gas at an essentially constant temperature. If we decrease the pressure, the gas slowly expands, thereby doing work on its surroundings, and the resulting infinitesimal drop in gas temperature will cause heat to flow into the gas from the bath, maintaining constant temperature in the gas.

Figure 2.11

Setup for an isothermal volume change.

EXAMPLE 2.5 Calculation of q, w, and ΔU

A cylinder fitted with a frictionless piston contains 3.00 mol of He gas at $P = 1.00$ atm and is in a large constant-temperature bath at 400 K. The pressure is reversibly increased to 5.00 atm. Find w, q, and ΔU for this process.

It is an excellent approximation to consider the helium as a perfect gas. Since T is constant, ΔU is zero [Eq. (2.68)]. Equation (2.74) gives

$$w = (3.00 \text{ mol})(8.314 \text{ J mol}^{-1} \text{ K}^{-1})(400 \text{ K}) \ln(5.00/1.00) = (9980 \text{ J}) \ln 5.00$$

$$w = (9980 \text{ J})(1.609) = 1.61 \times 10^4 \text{ J}$$

Also, $q = -w = -1.61 \times 10^4$ J. Of course, w (the work done on the gas) is positive for the compression. The heat q is negative because heat must flow from the gas to the surrounding constant-temperature bath to maintain the gas at 400 K as it is compressed.

Exercise

0.100 mol of a perfect gas with $C_{V,m} = 1.50R$ expands reversibly and isothermally at 300 K from 1.00 to 3.00 L. Find q, w, and ΔU for this process. (*Answer:* 274 J, −274 J, 0.)

Reversible Constant-P (or Constant-V) Process in a Perfect Gas

The calculations of q, w, and ΔU for these processes were shown in Example 2.4.

Reversible Adiabatic Process in a Perfect Gas

For an adiabatic process, $dq = 0$. For a reversible process in a system with only P-V work, $dw = -P \, dV$. For a perfect gas, $dU = C_V \, dT$ [Eq. (2.68)]. Therefore, for a reversible adiabatic process in a perfect gas, the first law $dU = dq + dw$ becomes

$$C_V \, dT = -P \, dV = -(nRT/V) \, dV$$

$$C_{V,m} \, dT = -(RT/V) \, dV$$

where $PV = nRT$ and $C_{V,m} = C_V/n$ were used. To integrate this equation, we separate the variables, putting all functions of T on one side and all functions of V on the other side. We get $(C_{V,m}/T) \, dT = -(R/V) \, dV$. Integration gives

$$\int_1^2 \frac{C_{V,m}}{T} \, dT = -\int_1^2 \frac{R}{V} \, dV = -R(\ln V_2 - \ln V_1) = R \ln \frac{V_1}{V_2} \qquad (2.75)$$

For a perfect gas, $C_{V,m}$ is a function of T [Eq. (2.69)]. If the temperature change in the process is small, $C_{V,m}$ will not change greatly and can be taken as approximately constant. Another case where $C_{V,m}$ is nearly constant is for monatomic gases, where $C_{V,m}$ is essentially independent of T over a very wide temperature range (Sec. 2.11 and Fig. 2.15). The approximation that $C_{V,m}$ is constant gives $\int_1^2 (C_{V,m}/T) \, dT = C_{V,m} \int_1^2 T^{-1} \, dT = C_{V,m} \ln (T_2/T_1)$, and Eq. (2.75) becomes $C_{V,m} \ln (T_2/T_1) = R \ln (V_1/V_2)$ or

$$\ln (T_2/T_1) = \ln (V_1/V_2)^{R/C_{V,m}}$$

where $k \ln x = \ln x^k$ [Eq. (1.68)] was used. If $\ln a = \ln b$, then $a = b$. Therefore

$$\frac{T_2}{T_1} = \left(\frac{V_1}{V_2}\right)^{R/C_{V,m}} \qquad \text{perf. gas, rev. adiabatic proc., } C_V \text{ const.} \qquad (2.76)$$

Since C_V is always positive [Eq. (2.56)], Eq. (2.76) says that, when $V_2 > V_1$, we will have $T_2 < T_1$. A perfect gas is cooled by a reversible adiabatic expansion. In expanding adiabatically, the gas does work on its surroundings, and since q is zero, U must decrease; therefore T decreases. A near-reversible, near-adiabatic expansion is one method used in refrigeration.

An alternative equation is obtained by using $P_1 V_1/T_1 = P_2 V_2/T_2$. Equation (2.76) becomes

$$P_2 V_2/P_1 V_1 = (V_1/V_2)^{R/C_{V,m}} \quad \text{and} \quad P_1 V_1^{1+R/C_{V,m}} = P_2 V_2^{1+R/C_{V,m}}$$

The exponent is $1 + R/C_{V,m} = (C_{V,m} + R)/C_{V,m} = C_{P,m}/C_{V,m}$, since $C_{P,m} - C_{V,m} = R$ for a perfect gas [Eq. (2.72)]. Defining the *heat-capacity ratio* γ (gamma) as

$$\gamma \equiv C_P/C_V$$

we have

$$P_1 V_1^{\gamma} = P_2 V_2^{\gamma} \qquad \text{perf. gas, rev. ad. proc., } C_V \text{ const.} \qquad (2.77)$$

For an adiabatic process, $\Delta U = q + w = w$. For a perfect gas, $dU = C_V \, dT$. With the approximation of constant C_V, we have

$$\Delta U = C_V(T_2 - T_1) = w \qquad \text{perf. gas, ad. proc., } C_V \text{ const.} \qquad (2.78)$$

To carry out a reversible adiabatic process in a gas, the surrounding constant-temperature bath in Fig. 2.11 is replaced by adiabatic walls, and the external pressure is slowly changed.

We might compare a reversible isothermal expansion of a perfect gas with a reversible adiabatic expansion of the gas. Let the gas start from the same initial P_1 and V_1 and go to the same V_2. For the isothermal process, $T_2 = T_1$. For the adiabatic expansion, we showed that $T_2 < T_1$. Hence the final pressure P_2 for the adiabatic expansion must be less than P_2 for the isothermal expansion (Fig. 2.12).

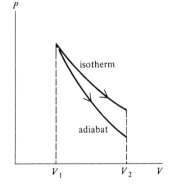

Figure 2.12

Ideal-gas reversible isothermal and adiabatic expansions that start from the same state.

Summary

A perfect gas obeys $PV = nRT$, has $(\partial U/\partial V)_T = 0 = (\partial H/\partial P)_T$, has U, H, C_V, and C_P depending on T only, has $C_P - C_V = nR$, and has $dU = C_V\,dT$ and $dH = C_P\,dT$. These equations are valid only for a perfect gas. A common error students make is to use one of these equations where it does not apply.

2.9 CALCULATION OF FIRST-LAW QUANTITIES

This section reviews thermodynamic processes and then summarizes the available methods for the calculation of q, w, ΔU, and ΔH in a process.

Thermodynamic Processes

When a thermodynamic system undergoes a change of state, we say it has undergone a **process.** The **path** of a process consists of the series of thermodynamic states through which the system passes on its way from the initial state to the final state. Two processes that start at the same initial state and end at the same final state but go through different paths (for example, a and b in Fig. 2.3) are different processes. (The term "change of state" should not be confused with the term "phase change." In thermodynamics, a system undergoes a **change of state** whenever one or more of the thermodynamic properties defining the system's state change their values.)

In a **cyclic** process, the system's final state is the same as the initial state. In a cyclic process, the change in each state function is zero: $0 = \Delta T = \Delta P = \Delta V = \Delta U = \Delta H$, etc. However, q and w need not be zero for a cyclic process (recall Example 2.4 in Sec. 2.8).

In a **reversible** process, the system is always infinitesimally close to equilibrium, and an infinitesimal change in conditions can restore both system and surroundings to their initial states. To perform a process reversibly, one must have only infinitesimal differences in pressures and temperatures, so that work and heat will flow slowly. Any changes in chemical composition must occur slowly and reversibly; moreover, there must be no friction. We found that the work in a mechanically reversible process is given by $dw_{\text{rev}} = -P\,dV$. In Chapter 3, we shall relate the heat dq_{rev} in a reversible process to state functions [see Eq. (3.20)].

In an **isothermal** process, T is constant throughout the process. To achieve this, one encloses the system in thermally conducting walls and places it in a large constant-temperature bath. For a perfect gas, U is a function of T only, so U is constant in an isothermal process; this is not necessarily true for systems other than perfect gases.

In an **adiabatic** process, $dq = 0$ and $q = 0$. This can be achieved by surrounding the system with adiabatic walls.

In a **constant-volume** (isochoric) process, V is held constant throughout the process. Here, the system is enclosed in rigid walls. Provided the system is capable of only P-V work, the work w is zero in an isochoric process.

In a **constant-pressure** (isobaric) process, P is held constant throughout the process. Experiments with solids and liquids are often performed with the system open to the atmosphere; here P is constant at the atmospheric pressure. To perform a constant-P process in a gas, one encloses the gas in a cylinder with a movable piston, holds the external pressure on the piston fixed at the initial pressure of the gas, and slowly warms or cools the gas, thereby changing its volume and temperature at constant P. For a constant-pressure process, we found that $\Delta H = q_P$.

Students are often confused in thermodynamics because they do not understand whether a quantity refers to a *property* of a system in some particular thermodynamic state or whether it refers to a *process* a system undergoes. For example, H is a property of a system and has a definite value once the system's state is defined; in contrast, $\Delta H \equiv H_2 - H_1$ is the *change* in enthalpy for a process in which the system goes from state 1 to state 2. Each state of a thermodynamic system has a definite value of H. Each change of state has a definite value of ΔH.

There are two kinds of quantities for a process. The value of a quantity such as ΔH, which is the change in a state function, is independent of the path of the process and depends only on the final and the initial states: $\Delta H = H_2 - H_1$. The value of a quantity such as q or w, which are not changes in state functions, depends on the path of the process and cannot be found from the final and initial states alone.

We now review calculation of q, w, ΔU, and ΔH for various processes. In this review, we assume that the system is closed and that only P-V work is done.

1. **Reversible phase change at constant T and P.** A **phase change** or **phase transition** is a process in which at least one new phase appears in a system without the occurrence of a chemical reaction. Examples include the melting of ice to liquid water, the transformation from orthorhombic solid sulfur to monoclinic solid sulfur (Sec. 7.4), and the freezing out of ice from an aqueous solution (Sec. 12.3). For now, we shall be concerned only with phase transitions involving pure substances.

   The heat q is found from the measured latent heat (Sec. 7.2) of the phase change. The work w is found from $w = -\int_1^2 P \, dV = -P \, \Delta V$, where ΔV is calculated from the densities of the two phases. If one phase is a gas, we can use $PV = nRT$ to find its volume (unless the gas is at high density). ΔH for this constant-pressure process is found from $\Delta H = q_P = q$. Finally, ΔU is found from $\Delta U = q + w$. As an example, the measured (latent) heat of fusion (melting) of H_2O at 0°C and 1 atm is 333 J/g. For the fusion of 1 mol (18.0 g) of ice at this T and P, $q = \Delta H = 6.01$ kJ. Thermodynamics cannot furnish us with the values of the latent heats of phase changes or with heat capacities. These quantities must be measured. (One can use statistical mechanics to calculate theoretically the heat capacities of certain systems, as we shall later see.)

2. **Constant-pressure heating with no phase change.** A constant-pressure process is mechanically reversible, so

$$w = w_{\text{rev}} = -\int_1^2 P \, dV = -P \, \Delta V \qquad \text{const. } P$$

where ΔV is found from the densities at the initial and final temperatures or from $PV = nRT$ if the substance is a perfect gas. If the heating (or cooling) is reversible, then T of the system is well defined and $C_P = dq_P/dT$ applies. Integration of this equation and use of $\Delta H = q_P$ give

$$\Delta H = q_P = \int_{T_1}^{T_2} C_P(T) \, dT \qquad \text{const. } P \qquad (2.79)$$

Since P is constant, we didn't bother to indicate that C_P depends on P as well as on T. The dependence of C_P and C_V on pressure is rather weak. Unless one deals with high pressures, a value of C_P measured at 1 atm can be used at other pressures. ΔU is found from $\Delta U = q + w = q_P + w$.

If the constant-pressure heating is irreversible (for example, if during the heating there is a finite temperature difference between system and surroundings or if temperature gradients exist in the system), the relation $\Delta H = \int_1^2 C_P \, dT$ still applies, so long as the initial and final states are equilibrium states. This is so because H is a state function and the value of ΔH is independent of the path (process) used to connect states 1 and 2. If ΔH equals $\int_1^2 C_P \, dT$ for a reversible path between states 1 and 2, then ΔH must equal $\int_1^2 C_P \, dT$ for any irreversible path between states 1 and 2. Also, in deriving $\Delta H = q_P$ [Eq. (2.46)], we did not assume the heating was reversible, only that P was constant. Thus, Eq. (2.79) holds for any constant-pressure temperature change in a closed system with P-V work only.

Since H is a state function, we can use the integral in (2.79) to find ΔH for any process whose initial and final states have the same pressure, whether or not the entire process occurs at constant pressure.

3. **Constant-volume heating with no phase change.** Since V is constant, $w = 0$. Integration of $C_V = dq_V/dT$ and use of $\Delta U = q + w = q_V$ give

$$\Delta U = \int_1^2 C_V \, dT = q_V \qquad V \text{ const.} \tag{2.80}$$

As with (2.79), Eq. (2.80) holds whether or not the heating is reversible. ΔH is found from $\Delta H = \Delta U + \Delta(PV) = \Delta U + V \Delta P$.

4. **Perfect-gas change of state.** Since U and H of a perfect gas depend on T only, we integrate $dU = C_V \, dT$ and $dH = C_P \, dT$ [(2.68) and (2.70)] to give

$$\Delta U = \int_{T_1}^{T_2} C_V(T) \, dT, \qquad \Delta H = \int_{T_1}^{T_2} C_P(T) \, dT \qquad \text{perf. gas} \tag{2.81}$$

If $C_V(T)$ or $C_P(T)$ is known, we can use $C_P - C_V = nR$ and integrate to find ΔU and ΔH. The equations of (2.81) apply to any perfect-gas change of state including irreversible changes and changes in which P and V change. The values of q and w depend on the path. If the process is reversible, then $w = -\int_1^2 P \, dV = -nR \int_1^2 (T/V) \, dV$, and we can find w if we know how T varies as a function of V. Having found w, we use $\Delta U = q + w$ to find q.

5. **Reversible isothermal process in a perfect gas.** Since U and H of the perfect gas are functions of T only, we have $\Delta U = 0$ and $\Delta H = 0$. Also, $w = -\int_1^2 P \, dV = -nRT \ln(V_2/V_1)$ [Eq. (2.74)] and $q = -w$, since $q + w = \Delta U = 0$.

6. **Reversible adiabatic process in a perfect gas.** The process is adiabatic, so $q = 0$. We find ΔU and ΔH from Eq. (2.81). The first law gives $w = \Delta U$. If C_V is essentially constant, the final state of the gas can be found from $P_1 V_1^\gamma = P_2 V_2^\gamma$, where $\gamma \equiv C_P/C_V$.

7. **Adiabatic expansion of a perfect gas into vacuum.** Here (Sec. 2.7) $q = 0$, $w = 0$, $\Delta U = q + w = 0$, and $\Delta H = \Delta U + \Delta(PV) = \Delta U + nR \Delta T = 0$.

Equations (2.79) and (2.80) tell us how a temperature change at constant P or at constant V affects H and U. At this point, we are not yet able to find the effects of a change in P or V on H and U. This will be dealt with in Chapter 4.

A word about units. Heat-capacity and latent-heat data are often tabulated in calories, so q is often calculated in calories. Pressures are often given in atmospheres, so P-V work is often calculated in cm^3 atm. The SI unit for q, w, ΔU, and ΔH is the joule. Hence we frequently want to convert between joules, calories, and cm^3 atm. We do this by using the values of R in (1.19) to (1.21). See Example 2.2 in Sec. 2.2.

A useful strategy to find a quantity such as ΔU or q for a process is to write the expression for the corresponding infinitesimal quantity and then integrate this expression from the initial state to the final state. For example, to find ΔU in an ideal-gas change of state, we write $dU = C_V \, dT$ and $\Delta U = \int_1^2 C_V(T) \, dT$; to find q in a constant-pressure process, we write $dq_P = C_P \, dT$ and $q_P = \int_1^2 C_P \, dT$. The infinitesimal change in a state function under the condition of constant P or T or V can often be found from the appropriate partial derivative. For example, if we want dU in a constant-volume process, we use $(\partial U/\partial T)_V = C_V$ to write $dU = C_V \, dT$ for V constant, and $\Delta U = \int_1^2 C_V \, dT$, where the integration is at constant V.

When evaluating an integral from state 1 to 2, you can take quantities that are constant outside the integral, but anything that varies during the process must remain inside the integral. Thus, for a constant-pressure process, $\int_1^2 P \, dV = P \int_1^2 dV = P(V_2 - V_1)$, and for an isothermal process, $\int_1^2 (nRT/V) \, dV = nRT \int_1^2 (1/V) \, dV$

$= nRT \ln (V_2/V_1)$. However, in evaluating $\int_1^2 C_P(T) \, dT$, we cannot take C_P outside the integral, unless we know that it is constant in the temperature range from T_1 to T_2.

EXAMPLE 2.6 Calculation of ΔH

$C_{P,m}$ of a certain substance in the temperature range 250 to 500 K at 1 bar pressure is given by $C_{P,m} = b + kT$, where b and k are certain known constants. If n moles of this substance is heated from T_1 to T_2 at 1 bar (where T_1 and T_2 are in the range 250 to 500 K), find the expression for ΔH.

Use of (2.79) gives

$$\Delta H = q_P = \int_1^2 nC_{P,m} \, dT = n \int_{T_1}^{T_2} (b + kT) \, dT = n(bT + \tfrac{1}{2} kT^2)\Big|_{T_1}^{T_2}$$

$$\Delta H = n[b(T_2 - T_1) + \tfrac{1}{2}k(T_2^2 - T_1^2)]$$

Exercise

Find the ΔH expression when n moles of a substance with $C_{P,m} = r + sT^{1/2}$, where r and s are constants, is heated at constant pressure from T_1 to T_2. [*Answer:* $nr(T_2 - T_1) + \tfrac{2}{3}ns(T_2^{3/2} - T_1^{3/2})$.]

2.10 STATE FUNCTIONS AND LINE INTEGRALS

We now discuss ways to test whether some quantity is a state function. Let the system go from state 1 to state 2 by some process. We subdivide the process into infinitesimal steps. Let db be some infinitesimal quantity associated with each infinitesimal step. For example, db might be the infinitesimal amount of heat that flows into the system in an infinitesimal step ($db = dq$), or it might be the infinitesimal change in system pressure ($db = dP$), or it might be the infinitesimal heat flow divided by the system's temperature ($db = dq/T$), etc. To determine whether db is the differential of a state function, we consider the line integral ${}_L\!\int_1^2 db$, where the L indicates that the integral's value depends in general on the process (path) used to go from state 1 to state 2.

The line integral ${}_L\!\int_1^2 db$ equals the sum of the infinitesimal quantities db for the infinitesimal steps into which we have divided the process. If b is a state function, then the sum of the infinitesimal changes in b is equal to the overall change $\Delta b \equiv b_2 - b_1$ in b from the initial state to the final state. For example, if b is the temperature, then ${}_L\!\int_1^2 dT = \Delta T = T_2 - T_1$; similarly, ${}_L\!\int_1^2 dU = U_2 - U_1$. We have

$$\int_{\substack{1 \\ L}}^{2} db = b_2 - b_1 \qquad \text{if } b \text{ is a state function} \qquad (2.82)$$

Since $b_2 - b_1$ is independent of the path used to go from state 1 to state 2 and depends only on the initial and final states 1 and 2, the value of the line integral ${}_L\!\int_1^2 db$ is independent of the path when b is a state function.

Suppose b is not a state function. For example, let $db = dq$, the infinitesimal heat flowing into a system. The sum of the infinitesimal amounts of heat is equal to the total heat q flowing into the system in the process of going from state 1 to state 2; we have ${}_L\!\int_1^2 dq = q$; similarly, ${}_L\!\int_1^2 dw = w$, where w is the work in the process. We have seen that q and w are not state functions but depend on the path from state 1 to state

2. The values of the integrals $_L\int_1^2 dq$ and $_L\int_1^2 dw$ depend on the path from 1 to 2. In general, if b is not a state function, then $_L\int_1^2 db$ depends on the path. Differentials of a state function, for example, dU, are called *exact differentials* in mathematics; the differentials dq and dw are *inexact*. Some texts use a special symbol to denote inexact differentials and write đq and đw (or Dq and Dw) instead of dq and dw.

From (2.82), it follows that, if the value of the line integral $_L\int_1^2 db$ depends on the path from state 1 to state 2, then b cannot be a state function.

Conversely, if $_L\int_1^2 db$ has the same value for every possible path from state 1 to state 2, b is a state function whose value for any state of the system can be defined as follows. We pick a reference state r and assign it some value of b, which we denote by b_r. The b value of an arbitrary state 2 is then defined by

$$b_2 - b_r = \int_r^2 db \qquad (2.83)$$

Since, by hypothesis, the integral in (2.83) is independent of the path, the value of b_2 depends only on state 2; $b_2 = b_2(T_2, P_2)$, and b is thus a state function.

If A is any state function, ΔA must be zero for any cyclic process. To indicate a cyclic process, one adds a circle to the line-integral symbol. If b is a state function, then (2.82) gives $\oint db = 0$ for any cyclic process. For example, $\oint dU = 0$. But note that $\oint dq = q$ and $\oint dw = w$, where the heat q and work w are not necessarily zero for a cyclic process.

We now show that, if

$$\oint db = 0$$

for every cyclic process, then the value of $_L\int_1^2 db$ is independent of the path and hence b is a state function. Figure 2.13 shows three processes connecting states 1 and 2. Processes I and II constitute a cycle. Hence the equation $\oint db = 0$ gives

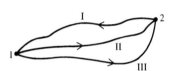

Figure 2.13

Three processes connecting states 1 and 2.

$$\int_2^1 db + \int_1^2 db = 0 \qquad (2.84)$$
$$_{\text{I}} \qquad _{\text{II}}$$

Likewise, processes I and III constitute a cycle, and

$$\int_2^1 db + \int_1^2 db = 0 \qquad (2.85)$$
$$_{\text{I}} \qquad _{\text{III}}$$

Subtraction of (2.85) from (2.84) gives

$$\int_1^2 db = \int_1^2 db \qquad (2.86)$$
$$_{\text{II}} \qquad _{\text{III}}$$

Since processes II and III are arbitrary processes connecting states 1 and 2, Eq. (2.86) shows that the line integral $_L\int_1^2 db$ has the same value for every process between states 1 and 2. Therefore b must be a state function.

Summary

If b is a state function, then $_L\int_1^2 db$ equals $b_2 - b_1$ and is independent of the path from state 1 to state 2. If b is a state function, then $\oint db = 0$.

If the value of $_L\int_1^2 db$ is independent of the path from 1 to 2, then b is a state function. If $\oint db = 0$ for every cyclic process, then b is a state function.

2.11   THE MOLECULAR NATURE OF INTERNAL ENERGY

Internal energy is energy at the molecular level. The molecular description of internal energy is outside the scope of thermodynamics, but a qualitative understanding of molecular energies is helpful.

Consider first a gas. The molecules are moving through space. A molecule has a translational kinetic energy $\frac{1}{2}mv^2$, where m and v are the mass and speed of the molecule. A **translation** is a motion in which every point of the body moves the same distance in the same direction. We shall later use statistical mechanics to show that the total molecular translational kinetic energy $U_{tr,m}$ of one mole of a gas is directly proportional to the absolute temperature and is given by [Eq. (15.14)] $U_{tr,m} = \frac{3}{2}RT$, where R is the gas constant.

If each gas molecule has more than one atom, then the molecules undergo rotational and vibrational motions in addition to translation. A **rotation** is a motion in which the spatial orientation of the body changes, but the distances between all points in the body remain fixed and the center of mass of the body does not move (so that there is no translational motion). In Chapter 22, we shall use statistical mechanics to show that except at very low temperatures the energy of molecular rotation $U_{rot,m}$ in one mole of gas is RT for linear molecules and $\frac{3}{2}RT$ for nonlinear molecules [Eq. (22.112)]: $U_{rot,lin,m} = RT$; $U_{rot,nonlin,m} = \frac{3}{2}RT$.

Besides translational and rotational energies, the atoms in a molecule have vibrational energy. In a molecular **vibration,** the atoms oscillate about their equilibrium positions in the molecule. A molecule has various characteristic ways of vibrating, each way being called a vibrational normal mode (see, for example, Figs. 21.26 and 21.27). Quantum mechanics shows that the lowest possible vibrational energy is not zero but is equal to a certain quantity called the molecular zero-point vibrational energy (so-called because it is present even at absolute zero temperature). The vibrational energy contribution U_{vib} to the internal energy of a gas is a complicated function of temperature [Eq. (22.113)]. For most light diatomic (two-atom) molecules (for example, H_2, N_2, HF, CO) at low and moderate temperatures (up to several hundred kelvins), the average molecular vibrational energy remains nearly fixed at the zero-point energy as the temperature increases. For polyatomic molecules (especially those with five or more atoms) and for heavy diatomic molecules (for example, I_2) at room temperature, the molecules usually have significant amounts of vibrational energy above the zero-point energy.

Figure 2.14 shows translational, rotational, and vibrational motions in CO_2.

In classical mechanics, energy has a continuous range of possible values. Quantum mechanics (Chapter 18) shows that the possible energies of a molecule are restricted to certain values called the **energy levels.** For example, the possible rotational-energy values of a diatomic molecule are $J(J + 1)b$ [Eq. (18.81)], where b is a constant for a given molecule and J can have the values 0, 1, 2, etc. One finds (Sec. 22.5) that there is a **distribution** of molecules over the possible energy levels. For example (Prob. 22.54), for CO gas at 298 K, 0.93% of the molecules are in the $J = 0$ level, 2.7% are in the $J = 1$ level, 4.4% are in the $J = 2$ level, . . . , 3.1% are in the $J = 15$ level, ... As the temperature increases, more molecules are found in higher energy levels, the average molecular energy increases, and the thermodynamic internal energy and enthalpy increase (Fig. 5.11).

Besides translational, rotational, and vibrational energies, a molecule possesses **electronic energy** ε_{el} (epsilon el). We define this energy as $\varepsilon_{el} \equiv \varepsilon_{eq} - \varepsilon_{\infty}$, where ε_{eq} is the energy of the molecule with the nuclei at rest (no translation, rotation, or vibration) at positions corresponding to the equilibrium molecular bond lengths and angles, and ε_{∞} is the energy when all the nuclei and electrons are at rest at positions infinitely far apart from one another, so as to make the electrical interactions between all the

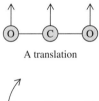

A translation

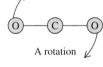

A rotation

A vibration

Figure 2.14

Kinds of motions in the CO_2 molecule.

charged particles vanish. (The quantity ε_∞ is given by the special theory of relativity as the sum of the rest-mass energies $m_{rest}c^2$ for the electrons and nuclei.) For a stable molecule, ε_{eq} is less than ε_∞.

The electronic energy ε_{el} can be changed by exciting a molecule to a higher electronic energy level. Nearly all common molecules have a very large gap between the lowest electronic energy level and higher electronic levels, so at temperatures below, say, 5000 K, virtually all the molecules are in the lowest electronic level and the contribution of electronic energy to the internal energy remains constant as the temperature increases (provided no chemical reactions occur).

In a chemical reaction, the electronic energies of the product molecules differ from those of the reactant molecules, and a chemical reaction changes the thermodynamic internal energy U primarily by changing the electronic energy. Although the other kinds of molecular energy generally also change in a reaction, the electronic energy undergoes the greatest change.

Besides translational, rotational, vibrational, and electronic energies, the gas molecules possess energy due to attractions and repulsions between them (**intermolecular forces**); intermolecular attractions cause gases to liquefy. The nature of intermolecular forces will be discussed in Sec. 22.10. Here, we shall just quote some key results for forces between neutral molecules.

The force between two molecules depends on the orientation of one molecule relative to the other. For simplicity, one often ignores this orientation effect and uses a force averaged over different orientations so that it is a function solely of the distance r between the centers of the interacting molecules. Figure 22.21a shows the typical behavior of the potential energy v of interaction between two molecules as a function of r; the quantity σ (sigma) is the average diameter of the two molecules. Note that, when the intermolecular distance r is greater than $2\frac{1}{2}$ or 3 times the molecular diameter σ, the intermolecular potential energy v is negligible. *Intermolecular forces are generally short-range.* When r decreases below 3σ, the potential energy decreases at first, indicating an attraction between the molecules, and then rapidly increases when r becomes close to σ, indicating a strong repulsion. Molecules initially attract each other as they approach and then repel each other when they collide. The magnitude of intermolecular attractions increases as the size of the molecules increases, and it increases as the molecular dipole moments increase.

The average distance between centers of molecules in a gas at 1 atm and 25°C is about 35 Å (Prob. 2.54), where the **angstrom** (Å) is

$$1 \text{ Å} \equiv 10^{-8} \text{ cm} \equiv 10^{-10} \text{ m} \qquad (2.87)*$$

Typical diameters of reasonably small molecules are 3 to 6 Å [see (16.26)]. The average distance between gas molecules at 1 atm and 25°C is 6 to 12 times the molecular diameter. Since intermolecular forces are negligible for separations beyond 3 times the molecular diameter, the intermolecular forces in a gas at 1 atm and 25°C are quite small and make very little contribution to the internal energy U. Of course, the spatial distribution of gas molecules is not actually uniform, and even at 1 atm significant numbers of molecules are quite close together, so intermolecular forces contribute slightly to U. At 40 atm and 25°C, the average distance between gas molecules is only 10 Å, and intermolecular forces contribute substantially to U.

Let $U_{intermol,m}$ be the contribution of intermolecular interactions to U_m. $U_{intermol,m}$ differs for different gases, depending on the strength of the intermolecular forces. Problem 4.22 shows that, for a gas, $U_{intermol,m}$ is typically -1 to -10 cal/mol at 1 atm and 25°C, and -40 to -400 cal/mol at 40 atm and 25°C. ($U_{intermol}$ is negative because intermolecular attractions lower the internal energy.) These numbers may be compared with the 25°C value $U_{tr,m} = \frac{3}{2}RT = 900$ cal/mol.

The fact that it is very hard to compress liquids and solids tells us that in condensed phases the molecules are quite close to one another, with the average distance between molecular centers being only slightly greater than the molecular diameter. Here, intermolecular forces contribute very substantially to U. In a liquid, the molecular translational, rotational, and vibrational energies are, to a good approximation (Sec. 22.11), the same as in a gas at the same temperature. We can therefore find $U_{intermol}$ in a liquid by measuring ΔU when the liquid vaporizes to a low-pressure gas. For common liquids, ΔU_m for vaporization typically lies in the range 3 to 15 kcal/mol, indicating $U_{intermol,m}$ values of -3000 to -15000 cal/mol, far greater in magnitude than $U_{intermol,m}$ in gases and $U_{tr,m}$ in room-temperature liquids and gases.

Discussion of U in solids is complicated by the fact that there are several kinds of solids (see Sec. 24.3). Here, we consider only molecular solids, those in which the structural units are individual molecules, these molecules being held together by intermolecular forces. In solids, the molecules generally don't undergo translation or rotation, and the translational and rotational energies found in gases and liquids are absent. Vibrations within the individual molecules contribute to the internal energy. In addition, there is the contribution $U_{intermol}$ of intermolecular interactions to the internal energy. Intermolecular interactions produce a potential-energy well (similar to that in Fig. 22.21a) within which each entire molecule as a unit undergoes a vibrationlike motion that involves both kinetic and potential energies. Estimates of $U_{intermol,m}$ from heats of sublimation of solids to vapors indicate that for molecular crystals, $U_{intermol,m}$ is in the same range as for liquids.

For a gas or liquid, the molar internal energy is

$$U_m = U_{tr,m} + U_{rot,m} + U_{vib,m} + U_{el,m} + U_{intermol,m} + U_{rest,m}$$

where $U_{rest,m}$ is the molar rest-mass energy of the electrons and nuclei, and is a constant. Provided no chemical reactions occur and the temperature is not extremely high, $U_{el,m}$ is a constant. $U_{intermol,m}$ is a function of T and P. $U_{tr,m}$, $U_{rot,m}$, and $U_{vib,m}$ are functions of T.

For a perfect gas, $U_{intermol,m} = 0$. The use of $U_{tr,m} = \frac{3}{2}RT$, $U_{rot,nonlin,m} = \frac{3}{2}RT$, and $U_{rot,lin,m} = RT$ gives

$$U_m = \frac{3}{2}RT + \frac{3}{2}RT \text{ (or } RT) + U_{vib,m}(T) + \text{const.} \qquad \text{perf. gas} \qquad (2.88)$$

For monatomic gases (for example, He, Ne, Ar), $U_{rot,m} = 0 = U_{vib,m}$, so

$$U_m = \frac{3}{2}RT + \text{const.} \qquad \text{perf. monatomic gas} \qquad (2.89)$$

The use of $C_{V,m} = (\partial U_m / \partial T)_V$ and $C_{P,m} - C_{V,m} = R$ gives

$$C_{V,m} = \frac{3}{2}R, \qquad C_{P,m} = \frac{5}{2}R \qquad \text{perf. monatomic gas} \qquad (2.90)$$

provided T is not extremely high.

For polyatomic gases, the translational contribution to $C_{V,m}$ is $C_{V,tr,m} = \frac{3}{2}R$; the rotational contribution is $C_{V,rot,lin,m} = R$, $C_{V,rot,nonlin,m} = \frac{3}{2}R$ (provided T is not extremely low); $C_{V,vib,m}$ is a complicated function of T—for light diatomic molecules, $C_{V,vib,m}$ is negligible at room temperature.

Figure 2.15 plots $C_{P,m}$ at 1 atm versus T for several substances. Note that $C_{P,m} = \frac{5}{2}R$ = 5 cal/(mol K) for He gas between 50 and 1000 K. For H_2O gas, $C_{P,m}$ starts at $4R =$ 8 cal/(mol K) at 373 K and increases as T increases. $C_{P,m} = 4R$ means $C_{V,m} = 3R$. The value $3R$ for this nonlinear molecule comes from $C_{V,tr,m} + C_{V,rot,m} = \frac{3}{2}R + \frac{3}{2}R$. The increase above $3R$ as T increases is due to the contribution from $C_{V,vib,m}$ as excited vibrational levels become populated.

The high value of $C_{P,m}$ of liquid water compared with that for water vapor results from the contribution of intermolecular interactions to U. Usually C_P for a liquid is substantially greater than that for the corresponding vapor.

The theory of heat capacities of solids will be discussed in Sec. 24.12. For all solids, $C_{P,m}$ goes to zero as T goes to zero.

Figure 2.15

$C_{P,m}$ at 1 atm versus T for several substances; s, l, and g stand for solid, liquid, and gas.

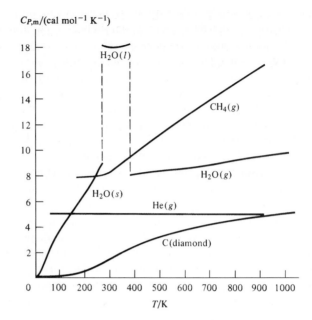

The heat capacities $C_{P,m} = (\partial H_m/\partial T)_P$ and $C_{V,m} = (\partial U_m/\partial T)_V$ are measures of how much energy must be added to a substance to produce a given temperature increase. The more ways (translation, rotation, vibration, intermolecular interactions) a substance has of absorbing added energy, the greater will be its $C_{P,m}$ and $C_{V,m}$ values.

2.12 PROBLEM SOLVING

Trying to learn physical chemistry solely by reading a textbook without working problems is about as effective as trying to improve your physique by reading a book on body conditioning without doing the recommended physical exercises.

If you don't see how to work a problem, it often helps to carry out these steps:

1. List all the relevant information that is given.
2. List the quantities to be calculated.
3. Ask yourself what equations, laws, or theorems connect what is known to what is unknown.
4. Apply the relevant equations to calculate what is unknown from what is given.

Although these steps are just common sense, they can be quite useful. The point is that problem solving is an active process. Listing the given information and the unknown quantities and actively searching for relationships that connect them gets your mind working on the problem, whereas simply reading the problem over and over may not get you anywhere. In listing the given information, it is helpful to *translate the words in the problem into equations*. For example, the phrase "adiabatic process" is translated into $dq = 0$ and $q = 0$; "isothermal process" is translated into $dT = 0$ and $T = $ constant.

In steps 1 and 2, sketches of the system and the process may be helpful. In working a problem in thermodynamics, one must have clearly in mind which portion of the universe is the system and which is the surroundings. The nature of the system should be noted—whether it is a perfect gas (for which many special relations hold), a nonideal gas, a liquid, a solid, a heterogeneous system, etc. Likewise, be aware of the kind of process involved—whether it is adiabatic, isothermal (T constant), isobaric (P constant), isochoric (V constant), reversible, etc.

Of course, the main hurdle is step 3. Because of the many equations in physical chemistry, it might seem a complex task to find the right equation to use in a problem. However, there are relatively few equations that are best committed to memory. These are usually the most fundamental equations, and usually they have fairly simple forms. For example, we have several equations for mechanically reversible P-V work in a closed system: $dw_{rev} = -P\,dV$ gives the work in an infinitesimal reversible process; $w_{rev} = -\int_1^2 P\,dV$ gives the work in a finite reversible process; the work in a constant-pressure process is $-P\,\Delta V$; the work in an isothermal reversible process in a perfect gas is $w = nRT \ln (V_1/V_2)$. The only one of these equations worth memorizing is $dw_{rev} = -P\,dV$, since the others can be quickly derived from it. Moreover, rederiving an equation from a fundamental equation reminds you of the conditions under which the equation is valid. *Do not memorize unstarred equations.* Readers who have invested their time mainly in achieving an understanding of the ideas and equations of physical chemistry will do better than those who have spent their time memorizing formulas.

Many of the errors students make in thermodynamics arise from using an equation where it does not apply. To help prevent this, many of the equations have the conditions of validity stated next to them. *Be sure the equations you are using are applicable to the system and process involved.* For example, students asked to calculate q in a reversible isothermal expansion of a perfect gas sometimes write "$dq = C_P\,dT$ and since $dT = 0$, we have $dq = 0$ and $q = 0$." This conclusion is erroneous. Why? (See Prob. 2.62.)

If you are baffled by a problem, the following suggestions may help you. (*a*) Ask yourself what given information you have not yet used, and see how this information might help solve the problem. (*b*) Instead of working forward from the known quantities to the unknown, try working backward from the unknown to the known. To do this, ask yourself what quantities you must know to find the unknown; then ask yourself what you must know to find these quantities; etc. (*c*) Write down the definition of the desired quantity. For example, if a density is wanted, write $\rho \equiv m/V$ and ask yourself how to find m and V. If an enthalpy change is wanted, write $H \equiv U + PV$ and $\Delta H = \Delta U + \Delta(PV)$ and see if you can find ΔU and $\Delta(PV)$. (*d*) In analyzing a thermodynamic process, ask yourself which state functions stay constant and which change. Then ask what conclusions can be drawn from the fact that certain state functions stay constant. For example, if V is constant in a process, then the P-V work must be zero. (*e*) Stop working on the problem and go on to something else. The solution method might occur to you when you are not consciously thinking about the problem. A lot of mental activity occurs outside of our conscious awareness.

When dealing with abstract quantities, it often helps to take specific numerical values. For example, suppose we want the relation between the rates of change dn_A/dt and dn_B/dt for the chemical reaction $A + 2B \rightarrow$ products, where n_A and n_B are the moles of A and B and t is time. Typically, students will say either that $dn_A/dt = 2\,dn_B/dt$ or that $dn_A/dt = \frac{1}{2}\,dn_B/dt$. (Before reading further, which do you think is right?) To help decide, suppose that in a tiny time interval $dt = 10^{-3}$ s, 0.001 mol of A reacts, so that $dn_A = -0.001$ mol. For the reaction $A + 2B \rightarrow$ products, find the corresponding value of dn_B and then find dn_A/dt and dn_B/dt and compare them.

In writing equations, a useful check is provided by the fact that *each term in an equation must have the same dimensions.* Thus, an equation that contains the expression $U + TV$ cannot be correct, because U has dimensions of energy = mass $\times$ length^2/time^2, whereas TV has dimensions of temperature $\times$ volume = temperature $\times$ length^3. From the definitions (1.25) and (1.29) of a derivative and a partial derivative, it follows that $(\partial z/\partial x)_y$ has the same dimensions as z/x. The definitions (1.52) and (1.59) of indefinite and definite integrals show that $\int f\,dx$ and $\int_a^b f\,dx$ have the same dimensions as fx.

When writing equations, do not mix finite and infinitesimal changes in the same equation. Thus, an equation that contains the expression $P\,dV + V\,\Delta P$ must be wrong because dV is an infinitesimal change and ΔP is a finite change. If one term in an equation contains a single change in a state function, then another term that contains only state functions must contain a change. Thus, an equation cannot contain the expression $PV + V\,\Delta P$ or the expression $PV + V\,dP$.

As to step 4, performing the calculations, errors can be minimized by carrying units of all quantities as part of the calculation. *Make sure you are using a self-consistent set of units.* Do not mix joules and kilojoules or joules and calories or joules and cm^3 atm in the same equation. If you are confused about what units to use, a strategy that avoids errors is to express all quantities in SI units. *Inconsistent use of units is one of the most common student errors in physical chemistry.*

Express your answer with the proper units. A numerical answer with no units is meaningless.

In September 1999, the $125 million U.S. Mars Climate Orbiter spacecraft was lost. It turned out that the engineers at Lockheed Martin sent data on the thrust of the spacecraft's thrusters to scientists at the Jet Propulsion Laboratory in units of pounds, but the JPL scientists assumed the thrust was in units of newtons, and so their programming of rocket firings to correct the trajectory produced an erroneous path that did not achieve orbit (*New York Times,* Oct. 1, 1999, p. A1). You don't have to be a rocket scientist to mess up on units.

Express the answer to the proper number of significant figures. Use an electronic calculator with keys for exponentials and logarithms for calculations. After the calculation is completed, it is a good idea to check the entire solution. If you are like most of us, you are probably too lazy to do a complete check, but it takes only a few seconds to check that the sign and the magnitude of the answer are physically reasonable. Sign errors are especially common in thermodynamics, since most quantities can be either positive or negative.

A solutions manual for problems in this textbook is available from the publisher on authorization of your instructor.

2.13 SUMMARY

The work done on a closed system when it undergoes a mechanically reversible infinitesimal volume change is $dw_{\text{rev}} = -P\,dV$.

The line integral $\int_1^2 P(T, V)\,dV$ (which equals $-w_{\text{rev}}$) is defined to be the sum of the infinitesimal quantities $P(T, V)\,dV$ for the process from state 1 to state 2. In general, the value of a line integral depends on the path from state 1 to state 2.

The heat transferred to a body of constant composition when it undergoes a temperature change dT at constant pressure is $dq_P = C_P\,dT$, where C_P is the body's heat capacity at constant pressure.

The first law of thermodynamics expresses the conservation of the total energy of system plus surroundings. For a closed system at rest in the absence of fields, the total energy equals the internal energy U, and the change in U in a process is $\Delta U = q + w$, where q and w are the heat flowing into and the work done on the system in the process. U is a state function, but q and w are not state functions. The internal energy U is energy that exists at the molecular level and includes molecular kinetic and potential energies.

The state function enthalpy H is defined by $H \equiv U + PV$. For a constant-pressure process, $\Delta H = q_P$ in a closed system with P-V work only.

The heat capacities at constant pressure and constant volume are $C_P = dq_P/dT = (\partial H/\partial T)_P$ and $C_V = dq_V/dT = (\partial U/\partial T)_V$.

The Joule and Joule–Thomson experiments measure $(\partial T/\partial V)_U$ and $(\partial T/\partial P)_H$; these derivatives are closely related to $(\partial U/\partial V)_T$ and $(\partial H/\partial P)_T$.

A perfect gas obeys $PV = nRT$ and $(\partial U/\partial V)_T = 0$. The changes in thermodynamic properties for a perfect gas are readily calculated for reversible isothermal and reversible adiabatic processes.

The methods used to calculate q, w, ΔU, and ΔH for various kinds of thermodynamic processes were summarized in Sec. 2.9.

The line integral $_L\int_1^2 db$ is independent of the path from state 1 to state 2 if and only if b is a state function. The line integral $\oint db$ is zero for every cyclic process if and only if b is a state function.

The molecular interpretation of internal energy in terms of intramolecular and intermolecular energies was discussed in Sec. 2.11.

Important kinds of calculations dealt with in this chapter include calculations of q, w, ΔU, and ΔH for

- Phase changes (for example, melting).
- Heating a substance at constant pressure.
- Heating at constant volume.
- An isothermal reversible process in a perfect gas.
- An adiabatic reversible process in a perfect gas with C_V constant.
- An adiabatic expansion of a perfect gas into vacuum.
- A constant-pressure reversible process in a perfect gas.
- A constant-volume reversible process in a perfect gas.

FURTHER READING

Zemansky and Dittman, chaps. 3, 4, 5; *Andrews* (1971), chaps. 5, 6, 7; *de Heer,* chaps. 3, 9; *Kestin,* chap. 5; *Reynolds and Perkins,* chaps. 1, 2; *Van Wylen and Sonntag,* chaps. 4, 5.

PROBLEMS

Section 2.1

2.1 True or false? (*a*) The kinetic energy of a system of several particles equals the sum of the kinetic energies of the individual particles. (*b*) The potential energy of a system of interacting particles equals the sum of the potential energies of the individual particles.

2.2 Give the SI units of (*a*) energy; (*b*) work; (*c*) volume; (*d*) force; (*e*) speed; (*f*) mass.

2.3 Express each of these units as a combination of meters, kilograms, and seconds: (*a*) joule; (*b*) pascal; (*c*) liter; (*d*) newton; (*e*) watt.

2.4 An apple of mass 155 g falls from a tree and is caught by a small boy. If the apple fell a distance of 10.0 m, find (*a*) the work done on the apple by the earth's gravitational field; (*b*) the kinetic energy of the apple just before it was caught; (*c*) the apple's speed just before it was caught.

2.5 An apple of mass 102 g is ground up into applesauce (with no added sugar) and spread evenly over an area of 1.00

m^2 on the earth's surface. What is the pressure exerted by the applesauce?

Section 2.2

2.6 True or false? (*a*) The P-V work in a mechanically reversible process in a closed system always equals $-P\,\Delta V$. (*b*) The symbol w in this book means work done on the system by the surroundings. (*c*) The infinitesimal P-V work in a mechanically reversible process in a closed system always equals $-P\,dV$. (*d*) The value of the work w in a reversible process in a closed system can be found so long as we know the initial state and the final state of the system. (*e*) The value of the integral $\int_1^2 P\,dV$ is fixed once the initial and final states 1 and 2 and the equation of state $P = P(T, V)$ are known. (*f*) The equation $w_{\text{rev}} = -\int_1^2 P\,dV$ applies only to constant-pressure processes. (*g*) $\int_1^2 P\,dV = \int_1^2 nR\,dT$ for every reversible process in an ideal gas.

2.7 If $P_1 = 175$ torr, $V_1 = 2.00$ L, $P_2 = 122$ torr, $V_2 = 5.00$ L, find w_{rev} for process (*b*) of Fig. 2.3 by (*a*) finding the area under the curve; (*b*) using $w_{\text{rev}} = -\int_1^2 P\,dV$.

2.8 A nonideal gas is heated slowly and expands reversibly at a constant pressure of 275 torr from a volume of 385 to 875 cm^3. Find w in joules.

2.9 Using the P_1, V_1, P_2, and V_2 values of Example 2.2, find w for a reversible process that goes from state 1 to state 2 in Fig. 2.3 via a straight line (a) by calculating the area under the curve; (b) by using $w_{rev} = -\int_1^2 P\, dV$. [Hint: The equation of the straight line that goes through points x_1, y_1 and x_2, y_2 is $(y - y_1)/(x - x_1) = (y_2 - y_1)/(x_2 - x_1)$.]

2.10 It was stated in Sec. 2.2 that for a given change of state, w_{rev} can have any positive or negative value. Consider a change of state for which $P_2 = P_1$ and $V_2 > V_1$. For this change of state, use a P-V diagram to (a) sketch a process with $w_{rev} < 0$; (b) sketch a process with $w_{rev} > 0$. Remember that neither P nor V can be negative.

Section 2.3

2.11 Specific heats can be measured in a *drop calorimeter*; here, a heated sample is dropped into the calorimeter and the final temperature is measured. When 45.0 g of a certain metal at 70.0°C is added to 24.0 g of water (with $c_P = 1.00$ cal/g-°C) at 10.0°C in an insulated container, the final temperature is 20.0°C. (a) Find the specific heat capacity of the metal. (b) How much heat flowed from the metal to the water? *Note:* In (a), we are finding the average c_P over the temperature range of the experiment. To determine c_P as a function of T, one repeats the experiment many times, using different initial temperatures for the metal.

Section 2.4

2.12 True or false? (a) For every process, $\Delta E_{syst} = -\Delta E_{surr}$. (b) For every cyclic process, the final state of the system is the same as the initial state. (c) For every cyclic process, the final state of the surroundings is the same as the initial state of the surroundings. (d) For a closed system at rest with no fields present, the sum $q + w$ has the same value for every process that goes from a given state 1 to a given state 2. (e) If systems A and B each consist of pure liquid water at 1 bar pressure and if $T_A > T_B$, then the internal energy of system A must be greater than that of B.

2.13 For which of these systems is the system's energy conserved in every process: (a) a closed system; (b) an open system; (c) an isolated system; (d) a system enclosed in adiabatic walls?

2.14 One food calorie = 10^3 cal = 1 kcal. A typical adult ingests 2200 kcal/day. (a) Show that an adult uses energy at about the same rate as a 100-W lightbulb. (b) Calculate the total annual metabolic-energy expenditure of the 6×10^9 people on earth and compare it with the 4×10^{20} J per year energy used by the world economy. (Neglect the fact that children use less metabolic energy than adults.)

2.15 A mole of water vapor initially at 200°C and 1 bar undergoes a cyclic process for which $w = 145$ J. Find q for this process.

2.16 William Thomson tells of running into Joule in 1847 at Mont Blanc; Joule had with him his bride and a long ther-

mometer with which he was going to "try for elevation of temperature in waterfalls." The Horseshoe Falls at Niagara Falls is 167 ft high and has a summer daytime flow rate of 2.55×10^6 L/s. (a) Calculate the maximum possible temperature difference between the water at the top and at the bottom of the falls. (The maximum possible increase occurs if no energy is transferred to such parts of the surroundings as the rocks at the base of the falls.) (b) Calculate the maximum possible internal-energy increase of the 2.55×10^6 L that falls each second. (Before it reaches the falls, more than half the water of the Niagara River is diverted to a canal or underground tunnels for use in hydroelectric power plants beyond the falls. These plants generate 4.4×10^9 W. A power surge at one of these plants led to the great blackout of November 9, 1965, which left 30 million people in the northeast United States and Ontario, Canada, without power for many hours.)

2.17 Imagine an isolated system divided into two parts, 1 and 2, by a rigid, impermeable, thermally conducting wall. Let heat q_1 flow into part 1. Use the first law to show that the heat flow for part 2 must be $q_2 = -q_1$.

2.18 Sometimes one sees the notation Δq and Δw for the heat flow into a system and the work done during a process. Explain why this notation is misleading.

2.19 Explain how liquid water can go from 25°C and 1 atm to 30°C and 1 atm in a process for which $q < 0$.

2.20 The potential energy stored in a spring is $\frac{1}{2}kx^2$, where k is the force constant of the spring and x is the distance the spring is stretched from equilibrium. Suppose a spring with force constant 125 N/m is stretched by 10.0 cm, placed in 112 g of water in an adiabatic container, and released. The mass of the spring is 20 g, and its specific heat capacity is 0.30 cal/(g °C). The initial temperature of the water and the spring is 18.000°C. The water's specific heat capacity is 1.00 cal/(g °C). Find the final temperature of the water.

2.21 Consider a system enclosed in a vertical cylinder fitted with a frictionless piston. The piston is a plate of negligible mass, on which is glued a mass m whose cross-sectional area is the same as that of the plate. Above the piston is a vacuum. (a) Use conservation of energy in the form $dE_{syst} + dE_{surr} = 0$ to show that for an adiabatic volume change $dE_{syst} = -mg\, dh - dK_{pist}$, where dh is the infinitesimal change in piston height, g is the gravitational acceleration, and dK_{pist} is the infinitesimal change in kinetic energy of the mass m. (b) Show that the equation in part (a) gives $dw_{irrev} = -P_{ext}\, dV - dK_{pist}$ for the irreversible work done on the system, where P_{ext} is the pressure exerted by the mass m on the piston plate.

2.22 Suppose the system of Prob. 2.21 is initially in equilibrium with $P = 1.000$ bar and $V = 2.00$ dm^3. The external mass m is instantaneously reduced by 50% and held fixed thereafter, so that P_{ext} remains at 0.500 bar during the expansion. After undergoing oscillations, the piston eventually comes to rest. The final system volume is 6.00 dm^3. Calculate w_{irrev}.

Section 2.5

2.23 True or false? (a) The quantities H, U, PV, ΔH, and $P\,\Delta V$ all have the same dimensions. (b) ΔH is defined only for a constant-pressure process. (*c*) For a constant-volume process in a closed system, $\Delta H = \Delta U$.

2.24 Which of the following have the dimensions of energy: force, work, mass, heat, pressure, pressure times volume, enthalpy, change in enthalpy, internal energy, force times length?

2.25 The state function H used to be called "the heat content." (*a*) Explain the origin of this name. (*b*) Why is this name misleading?

2.26 We showed $\Delta H = q$ for a constant-pressure process. Consider a process in which P is not constant throughout the entire process, but for which the final and initial pressures are equal. Need ΔH be equal to q here? (*Hint:* One way to answer this is to consider a cyclic process.)

2.27 A certain system is surrounded by adiabatic walls. The system consists of two parts, 1 and 2. Each part is closed, is held at constant P, and is capable of P-V work only. Apply $\Delta H = q_P$ to the entire system and to each part to show that $q_1 + q_2 = 0$ for heat flow between the parts.

Section 2.6

2.28 True or false? (*a*) C_P is a state function. (*b*) C_P is an extensive property.

2.29 (*a*) For $CH_4(g)$ at 2000 K and 1 bar, $C_{P,m} = 94.4$ J mol^{-1} K^{-1}. Find C_P of 586 g of $CH_4(g)$ at 2000 K and 1 bar. (*b*) For C(diamond), $C_{P,m} = 6.115$ J mol^{-1} K^{-1} at 25°C and 1 bar. For a 10.0-carat diamond, find c_P and C_P. One carat = 200 mg.

2.30 For $H_2O(l)$ at 100°C and 1 atm, $\rho = 0.958$ g/cm^3. Find the specific volume of $H_2O(l)$ at 100°C and 1 atm.

Section 2.7

2.31 (*a*) What state function must remain constant in the Joule experiment? (*b*) What state function must remain constant in the Joule–Thomson experiment?

2.32 For air at temperatures near 25°C and pressures in the range 0 to 50 bar, the μ_{JT} values are all reasonably close to 0.2°C/bar. Estimate the final temperature of the gas if 58 g of air at 25°C and 50 bar undergoes a Joule–Thomson throttling to a final pressure of 1 bar.

2.33 Rossini and Frandsen found that, for air at 28°C and pressures in the range 1 to 40 atm, $(\partial U_m/\partial P)_T = -6.08$ J mol^{-1} atm^{-1}. Calculate $(\partial U_m/\partial V_m)_T$ for air at (*a*) 28°C and 1.00 atm; (*b*) 28°C and 2.00 atm. [*Hint:* Use (1.35).]

2.34 (*a*) Derive Eq. (2.65). (*b*) Show that

$$\mu_{JT} = -(V/C_P)(\kappa C_V \mu_J - \kappa P + 1)$$

where κ is defined by (1.44). [*Hint:* Start by taking $(\partial/\partial P)_T$ of $H = U + PV$.]

2.35 Is μ_J an intensive property? Is μ_J an extensive property?

Section 2.8

2.36 For a fixed amount of a perfect gas, which of these statements must be true? (*a*) U and H each depend only on T. (*b*) C_P is a constant. (*c*) $P \, dV = nR \, dT$ for every infinitesimal process. (*d*) $C_{P,m} - C_{V,m} = R$. (*e*) $dU = C_V \, dT$ for a reversible process.

2.37 (*a*) Calculate q, w, ΔU, and ΔH for the reversible isothermal expansion at 300 K of 5.00 mol of a perfect gas from 500 cm^3 to 1500 cm^3. (*b*) What would ΔU and w be if the expansion connects the same initial and final states as in (*a*) but is done by having the perfect gas expand into vacuum?

2.38 One mole of He gas with $C_{V,m} = 3R/2$ essentially independent of temperature expands reversibly from 24.6 L and 300 K to 49.2 L. Calculate the final pressure and temperature if the expansion is (*a*) isothermal; (*b*) adiabatic. (*c*) Sketch these two processes on a P-V diagram.

2.39 For $N_2(g)$, $C_{P,m}$ is nearly constant at $3.5R = 29.1$ J/(mol K) for temperatures in the range 100 to 400 K and low or moderate pressures. (*a*) Calculate q, w, ΔU, and ΔH for the reversible adiabatic compression of 1.12 g of $N_2(g)$ from 400 torr and 1000 cm^3 to a final volume of 250 cm^3. Assume perfect-gas behavior. (*b*) Suppose we want to cool a sample of $N_2(g)$ at room T and P (25°C and 101 kPa) to 100 K using a reversible adiabatic expansion. What should the final pressure be?

2.40 Find q, w, ΔU, and ΔH if 2.00 g of He(g) with $C_{V,m} = \frac{3}{2}R$ essentially independent of temperature undergoes (*a*) a reversible constant-pressure expansion from 20.0 to 40.0 dm^3 at 0.800 bar; (*b*) a reversible heating with P going from 0.600 to 0.900 bar while V remains fixed at 15.0 dm^3.

Section 2.9

2.41 True or false? (*a*) A thermodynamic process is defined by the final state and the initial state. (*b*) $\Delta T = 0$ for every isothermal process. (*c*) Every process that has $\Delta T = 0$ is an isothermal process. (*d*) $\Delta U = 0$ for a reversible phase change at constant T and P. (*e*) q must be zero for an isothermal process.

2.42 State whether each of the following is a property of a thermodynamic system or refers to a noninfinitesimal process: (*a*) q; (*b*) U; (*c*) ΔH; (*d*) w; (*e*) C_V; (*f*) μ_{JT}; (*g*) H.

2.43 Give the value of C_{pr} [Eq. (2.50)] for (*a*) the melting of ice at 0°C and 1 atm; (*b*) the freezing of water at 0°C and 1 atm; (*c*) the reversible isothermal expansion of a perfect gas; (*d*) the reversible adiabatic expansion of a perfect gas.

2.44 (*This problem is especially instructive.*) For each of the following processes deduce whether each of the quantities q, w, ΔU, and ΔH is positive, zero, or negative. (*a*) Reversible melting of solid benzene at 1 atm and the normal melting point. (*b*) Reversible melting of ice at 1 atm and 0°C. (*c*) Reversible adiabatic expansion of a perfect gas. (*d*) Reversible isothermal expansion of a perfect gas. (*e*) Adiabatic expansion of a perfect gas into a vacuum (Joule experiment). (*f*) Joule–Thomson adiabatic throttling of a perfect gas. (*g*) Reversible heating of a perfect gas at constant P. (*h*) Reversible cooling of a perfect gas at constant V.

2.45 For each process state whether each of q, w, and ΔU is positive, zero, or negative. (*a*) Combustion of benzene in a sealed container with rigid, adiabatic walls. (*b*) Combustion of benzene in a sealed container that is immersed in a water bath at 25°C and has rigid, thermally conducting walls. (*c*) Adiabatic expansion of a nonideal gas into vacuum.

2.46 One mole of liquid water at 30°C is adiabatically compressed, *P* increasing from 1.00 to 10.00 atm. Since liquids and solids are rather incompressible, it is a fairly good approximation to take *V* as unchanged for this process. With this approximation, calculate q, ΔU, and ΔH for this process.

2.47 The molar heat capacity of oxygen at constant pressure for temperatures in the range 300 to 400 K and for low or moderate pressures can be approximated as $C_{P,m} = a + bT$, where $a = 6.15$ cal mol^{-1} K^{-1} and $b = 0.00310$ cal mol^{-1} K^{-2}. (*a*) Calculate q, w, ΔU, and ΔH when 2.00 mol of O_2 is reversibly heated from 27°C to 127°C with *P* held fixed at 1.00 atm. Assume perfect-gas behavior. (*b*) Calculate q, w, ΔU, and ΔH when 2.00 mol of O_2 initially at 1.00 atm is reversibly heated from 27°C to 127°C with *V* held fixed.

2.48 For this problem use 79.7 and 539.4 cal/g as the latent heats of fusion and vaporization of water at the normal melting and boiling points, $c_P = 1.00$ cal/(g K) for liquid water, $\rho = 0.917$ g/cm^3 for ice at 0°C and 1 atm, $\rho = 1.000$ g/cm^3 and 0.958 g/cm^3 for water at 1 atm and 0°C and 100°C, respectively. Calculate q, w, ΔU, and ΔH for (*a*) the melting of 1 mol of ice at 0°C and 1 atm; (*b*) the reversible constant-pressure heating of 1 mol of liquid water from 0°C to 100°C at 1 atm; (*c*) the vaporization of 1 mol of water at 100°C and 1 atm.

2.49 Calculate ΔU and ΔH for each of the following changes in state of 2.50 mol of a perfect monatomic gas with $C_{V,m} = 1.5R$ for all temperatures: (*a*) (1.50 atm, 400 K) → (3.00 atm, 600 K); (*b*) (2.50 atm, 20.0 L) → (2.00 atm, 30.0 L); (*c*) (28.5 L, 400 K) → (42.0 L, 400 K).

2.50 Can q and w be calculated for the processes of Prob. 2.49? If the answer is yes, calculate them for each process.

2.51 For a certain perfect gas, $C_{V,m} = 2.5R$ at all temperatures. Calculate q, w, ΔU, and ΔH when 2.00 mol of this gas undergoes each of the following processes: (*a*) a reversible isobaric expansion from (1.00 atm, 20.0 dm^3) to (1.00 atm, 40.0 dm^3); (*b*) a reversible isochoric change of state from (1.00 atm, 40.0 dm^3) to (0.500 atm, 40.0 dm^3); (*c*) a reversible isothermal compression from (0.500 atm, 40.0 dm^3) to (1.00 atm, 20.0 dm^3). Sketch each process on the same *P-V* diagram and calculate q, w, ΔU, and ΔH for a cycle that consists of steps (*a*), (*b*), and (*c*).

Section 2.11

2.52 Classify each of the following as kinetic energy, potential energy, or both: (*a*) translational energy; (*b*) rotational energy; (*c*) vibrational energy; (*d*) electronic energy.

2.53 Explain why $C_{P,m}$ of He gas at 10 K and 1 atm is larger than $\frac{5}{2}R$.

2.54 (*a*) Calculate the volume of 1 mole of ideal gas at 25°C and 1 atm. Let the gas be in a cubic container. If the gas molecules are distributed uniformly in space with equal spacing between adjacent molecules (of course, this really isn't so), the

gas volume can be divided into Avogadro's number of imaginary equal-sized cubes, each cube containing a molecule at its center. Calculate the edge length of each such cube. (*b*) What is the distance between the centers of the uniformly distributed gas molecules at 25°C and 1 atm? (*c*) Answer (*b*) for a gas at 25°C and 40 atm.

2.55 Estimate $C_{V,m}$ and $C_{P,m}$ at 300 K and 1 atm for (*a*) Ne(*g*); (*b*) CO(*g*).

2.56 Use Fig. 2.15 to decide whether U_{intermol} of liquid water increases or decreases as *T* increases.

General

2.57 (*a*) Use Rumford's data given in Sec. 2.4 to estimate the relation between the "old" calorie (as defined in Sec. 2.3) and the joule. Use 1 horsepower = 746 W. (*b*) The same as (*a*) using Joule's data given in Sec. 2.4.

2.58 Students often make significant-figure errors in taking reciprocals, in taking logs and antilogs, and in taking the difference of nearly equal numbers. (*a*) For a temperature of 1.8°C, calculate T^{-1} (where *T* is the absolute temperature) to the proper number of significant figures. (*b*) Find the common logs of the following numbers: 4.83 and 4.84; 4.83×10^{20} and 4.84×10^{20}. From the results, formulate a rule as to the proper number of significant figures in the log of a number known to *n* significant figures. (*c*) Calculate $(210.6 \text{ K})^{-1} - (211.5 \text{ K})^{-1}$ to the proper number of significant figures.

2.59 (*a*) A gas obeying the van der Waals equation of state (1.39) undergoes a reversible isothermal volume change from V_1 to V_2. Obtain the expression for the work w. Check that your result reduces to (2.74) for $a = 0 = b$. (*b*) Use the result of (*a*) to find w for 0.500 mol of N_2 expanding reversibly from 0.400 L to 0.800 L at 300 K. See Sec. 8.4 for the *a* and *b* values of N_2. Compare the result with that found if N_2 is assumed to be a perfect gas.

2.60 (*a*) If the temperature of a system decreases by 8.0°C, what is ΔT in kelvins? (*b*) A certain system has $C_P = 5.00$ J/°C. What is its C_P in joules per kelvin?

2.61 Explain why Boyle's law $PV = $ constant for an ideal gas does not contradict the equation $PV^\gamma = $ constant for a reversible adiabatic process in a perfect gas with C_V constant.

2.62 Point out the error in the Sec. 2.12 reasoning that gave $q = 0$ for a reversible isothermal process in a perfect gas.

2.63 A perfect gas with $C_{V,m} = 3R$ independent of *T* expands adiabatically into a vacuum, thereby doubling its volume. Two students present the following conflicting analyses. Genevieve uses Eq. (2.76) to write $T_2/T_1 = (V_1/2V_1)^{R/3R}$ and $T_2 = T_1/2^{1/3}$. Wendy writes $\Delta U = q + w = 0 + 0 = 0$ and $\Delta U = C_V \Delta T$, so $\Delta T = 0$ and $T_2 = T_1$. Which student is correct? What error did the other student make?

2.64 A perfect gas undergoes an expansion process at constant pressure. Does its internal energy increase or decrease? Justify your answer.

2.65 Classify each of the following properties as intensive or extensive and give the SI units of each: (a) density; (b) U; (c) H_m; (d) C_P; (e) c_P; (f) $C_{P,m}$; (g) P; (h) molar mass; (i) T.

2.66 A student attempting to remember a certain formula comes up with $C_P - C_V = TV\alpha^m/\kappa^n$, where m and n are certain integers whose values the student has forgotten and where the remaining symbols have their usual meanings. Use dimensional considerations to find m and n.

2.67 Because the heat capacities per unit volume of gases are small, accurate measurement of C_P or C_V for gases is not easy. Accurate measurement of the heat-capacity ratio γ of a gas (for example, by measurement of the speed of sound in the gas) is easy. For gaseous CCl_4 at 0.1 bar and 20°C, experiment gives $\gamma = 1.13$. Find $C_{P,m}$ and $C_{V,m}$ for $CCl_4(g)$ at 20°C and 1 bar.

2.68 Give the SI units of each of the following properties and state whether each is extensive or intensive. (a) $(\partial V/\partial T)_P$; (b) $V^{-1}(\partial V/\partial T)_P$; (c) $(\partial V_m/\partial P)_T$; (d) $(\partial U/\partial V)_T$.

2.69 True or false? (a) ΔH is a state function. (b) C_V is independent of T for every perfect gas. (c) $\Delta U = q + w$ for every thermodynamic system at rest in the absence of external fields. (d) A process in which the final temperature equals the initial temperature must be an isothermal process. (e) For a closed system at rest in the absence of external fields, $U = q + w$. (f) U remains constant in every isothermal process in a closed system. (g) $q = 0$ for every cyclic process. (h) $\Delta U = 0$ for every cyclic process. (i) $\Delta T = 0$ for every adiabatic process in a closed system. (j) A thermodynamic process is specified by specifying the initial state and the final state of the system. (k) If a closed system at rest in the absence of external fields undergoes an adiabatic process that has $w = 0$, then the system's temperature must remain constant. (l) P-V work is usually negligible for solids and liquids. (m) If neither heat nor matter can enter or leave a system, that system must be isolated. (n) For a closed system with P-V work only, a constant-pressure process that has $q > 0$ must have $\Delta T > 0$. (o) $\int_1^2 (1/V)\,dV = \ln(V_2 - V_1)$. (p) The value of ΔU is independent of the path (process) used to go from state 1 to state 2.

CHAPTER 3

The Second Law of Thermodynamics

A major application of thermodynamics to chemistry is to provide information about equilibrium in chemical systems. If we mix nitrogen and hydrogen gases together with a catalyst, portions of each gas react to form ammonia. The first law assures us that the total energy of system plus surroundings remains constant during the reaction, but the first law cannot say what the final equilibrium concentrations will be. We shall see that the second law provides such information. The second law leads to the existence of the state function entropy S, which possesses the property that for an isolated system the equilibrium position corresponds to maximum entropy. The second law of thermodynamics is stated in Sec. 3.1. The deduction of the existence of the state function S from the second law is carried out in Secs. 3.2 and 3.3. The rest of this chapter shows how to calculate entropy changes in processes (Sec. 3.4), shows the relation between entropy and equilibrium (Sec. 3.5), defines the thermodynamic temperature scale (Sec. 3.6), and discusses the molecular interpretation of entropy (Sec. 3.7).

Energy is both a molecular and a macroscopic property and plays a key role in both quantum chemistry and thermodynamics. Entropy is a macroscopic property but is not a molecular property. A single molecule does not have an entropy. Only a collection of a large number of molecules can be assigned an entropy. Entropy is a less intuitively obvious property than energy. The concept of entropy has been applied and perhaps misapplied in many fields outside the physical sciences as evidenced by books with titles such as *Entropy and Art*, *Social Entropy Theory*, *Entropy in Urban and Regional Modeling*, and *Economics, Entropy and the Environment*.

3.1 THE SECOND LAW OF THERMODYNAMICS

In 1824 a French engineer named Sadi Carnot published a study on the theoretical efficiency of steam engines. This book (*Reflections on the Motive Power of Fire*) pointed out that, for a heat engine to produce continuous mechanical work, it must exchange heat with two bodies at different temperatures, absorbing heat from the hot body and discarding heat to the cold body. Without a cold body for the discard of heat, the engine cannot function continuously. This is the essential idea of one form of the second law of thermodynamics. Carnot's work had little influence at the time of its publication. Carnot worked when the caloric theory of heat held sway, and his book used this theory, incorrectly setting the heat discarded to the cold body equal to the heat absorbed from the hot body. When Carnot's book was rediscovered in the 1840s, it caused confusion for a while, since Joule's work had overthrown the caloric theory. Finally, about 1850, Rudolph Clausius and William Thomson (Lord Kelvin) corrected Carnot's work to conform with the first law of thermodynamics.

Carnot died of cholera in 1832 at age 36. His unpublished notes showed that he believed the caloric theory to be false and planned experiments to demonstrate this. These planned experiments included the vigorous agitation of liquids and measurement of "the motive power consumed and the heat produced." Carnot's notes stated: "Heat is simply motive power, or rather motion, which has changed its form. . . . [Motive] power is, in quantity, invariable in nature; it is . . . never either produced or destroyed. . . ."

There are several equivalent ways of stating the second law. We shall use the following statement, the **Kelvin–Planck statement of the second law of thermodynamics,** due originally to William Thomson and later rephrased by Planck:

It is impossible for a system to undergo a cyclic process whose sole effects are the flow of heat into the system from a heat reservoir and the performance of an equivalent amount of work by the system on the surroundings.

By a *heat reservoir* or *heat bath* we mean a body that is in internal equilibrium at a constant temperature and that is large enough for flow of heat between it and the system to cause no significant change in the temperature of the reservoir.

The second law says that it is impossible to build a cyclic machine that converts heat into work with 100% efficiency (Fig. 3.1). Note that the existence of such a machine would not violate the first law, since energy is conserved in the operation of the machine.

Like the first law, the second law is a generalization from experience. There are three kinds of evidence for the second law. First is the failure of anyone to construct a machine like that shown in Fig. 3.1. If such a machine were available, it could use the atmosphere as a heat reservoir, continuously withdrawing energy from the atmosphere and converting it completely to useful work. It would be nice to have such a machine, but no one has been able to build one. Second, and more convincing, is the fact that the second law leads to many conclusions about equilibrium in chemical systems, and these conclusions have been verified. For example, we shall see that the second law shows that the vapor pressure of a pure substance varies with temperature according to $dP/dT = \Delta H/(T \, \Delta V)$, where ΔH and ΔV are the heat of vaporization and the volume change in vaporization, and this equation has been experimentally verified. Third, statistical mechanics shows that the second law follows as a consequence of certain assumptions about the molecular level.

The first law tells us that work output cannot be produced by a cyclic machine without an equivalent amount of energy input. The second law tells us that it is impossible to have a cyclic machine that completely converts the random molecular energy of heat flow into the ordered motion of mechanical work. As some wit has put it: The first law says you can't win; the second law says you can't break even.

Note that the second law does not forbid the complete conversion of heat to work in a *noncyclic* process. Thus, if we reversibly and isothermally heat a perfect gas, the gas expands and, since $\Delta U = 0$, the work done by the gas equals the heat input [Eq. (2.74)]. Such an expansion, however, cannot be made the basis of a continuously operating machine. Eventually, the piston will fall out of the cylinder. A continuously operating machine must use a cyclic process.

An alternative statement of the second law is the *Clausius statement:*

It is impossible for a system to undergo a cyclic process whose sole effects are the flow of heat into the system from a cold reservoir and the flow of an equal amount of heat out of the system into a hot reservoir.

Proof of the equivalence of the Clausius and Kelvin–Planck statements is outlined in Prob. 3.7.

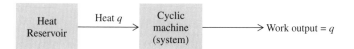

Figure 3.1

A system that violates the second law of thermodynamics.

3.2   HEAT ENGINES

We shall use the second law to deduce theorems about the efficiency of heat engines. Chemists have little interest in heat engines, but our study of them is part of a chain of reasoning that will lead to the criterion for determining the position of chemical equilibrium in a system. Moreover, study of the efficiency of heat engines is related to the basic question of what limitations exist on the conversion of heat to work.

Heat Engines

A **heat engine** converts some of the random molecular energy of heat flow into macroscopic mechanical energy (work). The working substance (for example, steam in a steam engine) is heated in a cylinder, and its expansion moves a piston, thereby doing mechanical work. If the engine is to operate continuously, the working substance has to be cooled back to its original state and the piston has to return to its original position before we can heat the working substance again and get another work-producing expansion. Hence the working substance undergoes a cyclic process. The essentials of the cycle are the absorption of heat q_H by the working substance from a hot body (for example, the boiler), the performance of work $-w$ by the working substance on the surroundings, and the emission of heat $-q_C$ by the working substance to a cold body (for example, the condenser), with the working substance returning to its original state at the end of the cycle. The system is the working substance.

Our convention is that w is work done *on* the system. Work done *by* the system is $-w$. Likewise, q means the heat flowing into the system, and $-q_C$ is the heat that flows from the system to the cold body in one cycle. For a heat engine, $q_H > 0$, $-w > 0$, and $-q_C > 0$, so $w < 0$ and $q_C < 0$. The quantity w is negative for a heat engine because the engine does positive work on its surroundings; q_C is negative for a heat engine because positive heat flows out of the system to the cold body.

Although this discussion is an idealization of how real heat engines work, it contains the essential features of a real heat engine.

The *efficiency e* of a heat engine is the fraction of energy input that appears as useful energy output, that is, that appears as work. The energy input per cycle is the heat input q_H to the engine. (The source of this energy might be the burning of oil or coal to heat the boiler.) We have

$$e = \frac{\text{work output per cycle}}{\text{energy input per cycle}} = \frac{-w}{q_H} = \frac{|w|}{q_H} \tag{3.1}$$

For a cycle of operation, the first law gives $\Delta U = 0 = q + w = q_H + q_C + w$, and

$$-w = q_H + q_C \tag{3.2}$$

where the quantities in (3.2) are for one cycle. Equation (3.2) can be written as $q_H = -w + (-q_C)$; the energy input per cycle, q_H, is divided between the work output $-w$ and the heat $-q_C$ discarded to the cold body. Use of (3.2) in (3.1) gives

$$e = \frac{q_H + q_C}{q_H} = 1 + \frac{q_C}{q_H} \tag{3.3}$$

Since q_C is negative and q_H is positive, the efficiency is less than 1.

To further simplify the analysis, we assume that the heat q_H is absorbed from a hot reservoir and that $-q_C$ is emitted to a cold reservoir, each reservoir being large enough to ensure that its temperature is unchanged by interaction with the engine. Figure 3.2 is a schematic diagram of the heat engine.

Since our analysis at this point will not require specification of the temperature scale, instead of denoting temperatures with the symbol T (which indicates use of the ideal-gas scale; Sec. 1.5), we shall use τ (tau). We call the temperatures of the hot and

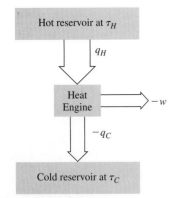

Figure 3.2

A heat engine operating between two temperatures. The heat and work quantities are for one cycle. The widths of the arrows indicate that $q_H = -w - q_C$.

cold reservoirs τ_H and τ_C. The τ scale might be the ideal-gas scale, or it might be based on the expansion of liquid mercury, or it might be some other scale. The only restriction we set is that the τ scale always give readings such that the temperature of the hot reservoir is greater than that of the cold reservoir: $\tau_H > \tau_C$. The motivation for leaving the temperature scale unspecified will become clear in Sec. 3.6.

Carnot's Principle

We now use the second law to prove *Carnot's principle: No heat engine can be more efficient than a reversible heat engine when both engines work between the same pair of temperatures τ_H and τ_C.* Equivalently, the maximum amount of work from a given supply of heat is obtained with a reversible engine.

To prove Carnot's principle, we assume it to be false and show that this assumption leads to a violation of the second law. Thus, let there exist a superengine whose efficiency e_{super} exceeds the efficiency e_{rev} of some reversible engine working between the same two temperatures as the superengine:

$$e_{super} > e_{rev} \qquad (3.4)$$

where, from (3.1),

$$e_{super} = \frac{-w_{super}}{q_{H,super}}, \qquad e_{rev} = \frac{-w_{rev}}{q_{H,rev}} \qquad (3.5)$$

Let us run the reversible engine in reverse, doing positive work w_{rev} *on* it, thereby causing it to absorb heat $q_{C,rev}$ from the cold reservoir and emit heat $-q_{H,rev}$ to the hot reservoir, where these quantities are for one cycle. It thereby functions as a heat pump, or refrigerator. Because this engine is reversible, the magnitudes of the two heats and the work are the *same* for a cycle of operation as a heat pump as for a cycle of operation as a heat engine, except that all signs are changed. We couple the reversible heat pump with the superengine, so that the two systems use the same pair of reservoirs (Fig. 3.3).

We shall run the superengine at such a rate that it withdraws heat from the hot reservoir at the same rate that the reversible heat pump deposits heat into this reservoir. Thus, suppose 1 cycle of the superengine absorbs 1.3 times as much heat from the hot reservoir as 1 cycle of the reversible heat pump deposits into the hot reservoir. The superengine would then complete 10 cycles in the time that the heat pump completes 13 cycles. After each 10 cycles of the superengine, both devices are back in their original states, so the combined device is cyclic.

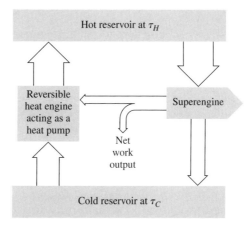

Figure 3.3

Reversible heat pump coupled with a superengine.

Since the magnitude of the heat exchange with the hot reservoir is the same for the two engines, and since the superengine is by assumption more efficient than the reversible engine, the equations of (3.5) show that the superengine will deliver more work output than the work put into the reversible heat pump. We can therefore use *part* of the mechanical work output of the superengine to supply *all* the work needed to run the reversible heat pump and still have some net work output from the superengine left over. This work output must by the first law have come from some input of energy to the system of reversible heat pump plus superengine. Since there is no net absorption or emission of heat to the hot reservoir, this energy input must have come from a net absorption of heat from the cold reservoir. The net result is an absorption of heat from the cold reservoir and its complete conversion to work by a cyclic process.

However, this cyclic process violates the second law of thermodynamics (Sec. 3.1) and is therefore impossible. We were led to this impossible conclusion by our initial assumption of the existence of a superengine with $e_{super} > e_{rev}$. We therefore conclude that this assumption is false. We have proved that

$$e(\text{any engine}) \leq e(\text{a reversible engine}) \tag{3.6}$$

for heat engines that operate between the same two temperatures. (To increase the efficiency of a real engine, one can reduce the amount of irreversibility by, for example, using lubrication to reduce friction.)

Now consider two reversible heat engines, A and B, that work between the same two temperatures with efficiencies $e_{rev,A}$ and $e_{rev,B}$. If we replace the superengine in the above reasoning with engine A running forward and use engine B running backward as the heat pump, the same reasoning that led to (3.6) gives $e_{rev,A} \leq e_{rev,B}$. If we now interchange A and B, running B forward and A backward, the same reasoning gives $e_{rev,B} \leq e_{rev,A}$. These two relations can hold only if $e_{rev,A} = e_{rev,B}$.

We have shown that (1) all reversible heat engines operating between the same two temperatures have the same efficiency e_{rev}, and (2) this reversible efficiency is the maximum possible for any heat engine that operates between these temperatures, so

$$e_{irrev} \leq e_{rev} \tag{3.7}$$

These conclusions are independent of the nature of the working substance used in the engines and of the kind of work, holding also for non-P-V work. The only assumption made was the validity of the second law of thermodynamics.

Calculation of e_{rev}

Since the efficiency of any reversible engine working between the temperatures τ_H and τ_C is the same, this efficiency e_{rev} can depend only on τ_H and τ_C:

$$e_{rev} = f(\tau_H, \tau_C) \tag{3.8}$$

The function f depends on the temperature scale used. We now find f for the ideal-gas temperature scale, taking $\tau = T$. Since e_{rev} is independent of the nature of the working substance, we can use any working substance to find f. We know the most about a perfect gas, so we choose this as the working substance.

Consider first the nature of the cycle we used to derive (3.8). The first step involves absorption of heat q_H from a reservoir whose temperature remains at T_H. Since we are considering a reversible engine, the gas also must remain at temperature T_H throughout the heat absorption from the reservoir. (Heat flow between two bodies with a finite difference in temperature is an irreversible process.) Thus the first step of the cycle is an *isothermal* process. Moreover, since $\Delta U = 0$ for an isothermal process in a perfect gas [Eq. (2.67)], it follows that, to maintain U as constant, the gas must expand and do work on the surroundings equal to the heat absorbed in the first step.

The first step of the cycle is thus a reversible isothermal expansion, as shown by the line from state 1 to state 2 in Fig 3.4a. Similarly, when the gas gives up heat at T_C, we have a reversible isothermal compression at temperature T_C. The T_C isotherm lies below the T_H isotherm and is the line from state 3 to state 4 in Fig. 3.4a. To have a complete cycle, we must have steps that connect states 2 and 3 and states 4 and 1. We assumed that heat is transferred only at T_H and T_C. Therefore the two isotherms in Fig. 3.4a must be connected by two steps with no heat transfer, that is, by two reversible **adiabats.**

This reversible cycle is called a Carnot cycle (Fig. 3.4b). The working substance need not be a perfect gas. A **Carnot cycle** is defined as a *reversible* cycle that consists of two isothermal steps at different temperatures and two adiabatic steps.

We now calculate the Carnot-cycle efficiency e_{rev} on the ideal-gas temperature scale T. We use a perfect gas as the working substance and restrict ourselves to P-V work. The first law gives $dU = dq + dw = dq - P\,dV$ for a reversible volume change. For a perfect gas, $P = nRT/V$ and $dU = C_V(T)\,dT$. The first law becomes

$$C_V\,dT = dq - nRT\,dV/V$$

for a perfect gas. Dividing by T and integrating over the Carnot cycle, we get

$$\oint C_V(T)\,\frac{dT}{T} = \oint \frac{dq}{T} - nR \oint \frac{dV}{V} \qquad (3.9)$$

Each integral in (3.9) is the sum of four line integrals, one for each step of the Carnot cycle in Fig. 3.4b. We have

$$\oint C_V(T)\,\frac{dT}{T} = \int_{T_1}^{T_2} \frac{C_V(T)}{T}\,dT + \int_{T_2}^{T_3} \frac{C_V(T)}{T}\,dT + \int_{T_3}^{T_4} \frac{C_V(T)}{T}\,dT$$

$$+ \int_{T_4}^{T_1} \frac{C_V(T)}{T}\,dT \qquad (3.10)$$

Each integral on the right side of (3.10) has an integrand that is a function of T only, and hence each such integral is an ordinary definite integral. Use of the identity $\int_a^b f(T)\,dT + \int_b^c f(T)\,dT = \int_a^c f(T)\,dT$ (Sec. 1.8) shows that the sum of the first two integrals on the right side of (3.10) is $\int_{T_1}^{T_3} (C_V/T)\,dT$ and the sum of the last two integrals on the right side of (3.10) is $\int_{T_3}^{T_1} (C_V/T)\,dT$. Hence the right side of (3.10) equals $\int_{T_1}^{T_3} (C_V/T)\,dT + \int_{T_3}^{T_1} (C_V/T)\,dT = \int_{T_1}^{T_1} (C_V/T)\,dT = 0$. Therefore (3.10) becomes

$$\oint C_V(T)\,\frac{dT}{T} = 0 \qquad (3.11)$$

The cyclic integral in (3.11) must vanish because $[C_V(T)/T]\,dT$ is the differential of a state function, namely, a certain function of T whose derivative is $C_V(T)/T$. (Recall Sec. 2.10.) Note, however, that the integral of $P\,dV$ does not vanish for a cycle, since $P\,dV$ is not the differential of a state function.

The second integral on the right side of (3.9) must also vanish. This is because dV/V is the differential of a state function (namely, $\ln V$), and its line integral is therefore zero for a cyclic process.

Hence (3.9) becomes

$$\oint \frac{dq}{T} = 0 \qquad \text{Carnot cycle, perf. gas} \qquad (3.12)$$

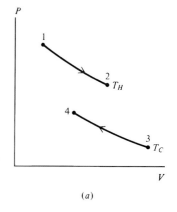

(a)

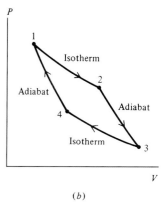

(b)

Figure 3.4

(a) Isothermal steps of the reversible heat-engine cycle. (b) The complete Carnot cycle. (Not to scale.)

We have

$$\oint \frac{dq}{T} = \int_1^2 \frac{dq}{T} + \int_2^3 \frac{dq}{T} + \int_3^4 \frac{dq}{T} + \int_4^1 \frac{dq}{T} \qquad (3.13)$$

Since processes $2 \to 3$ and $4 \to 1$ are adiabatic with $dq = 0$, the second and fourth integrals on the right side of (3.13) are zero. For the isothermal process $1 \to 2$, we have $T = T_H$. Since T is constant, it can be taken outside the integral: $\int_1^2 T^{-1} dq = T_H^{-1} \int_1^2 dq = q_H/T_H$. Similarly, $\int_3^4 T^{-1} dq = q_C/T_C$. Equation (3.12) becomes

$$\oint \frac{dq}{T} = \frac{q_H}{T_H} + \frac{q_C}{T_C} = 0 \qquad \text{Carnot cycle, perf. gas} \qquad (3.14)$$

We now find e_{rev}, the maximum possible efficiency for the conversion of heat to work. Equations (3.3) and (3.14) give $e = 1 + q_C/q_H$ and $q_C/q_H = -T_C/T_H$. Hence

$$e_{\text{rev}} = 1 - \frac{T_C}{T_H} = \frac{T_H - T_C}{T_H} \qquad \text{Carnot cycle} \qquad (3.15)$$

We derived (3.15) using a perfect gas as the working substance, but since we earlier proved that e_{rev} is independent of the working substance, Eq. (3.15) must hold for any working substance undergoing a Carnot cycle. Moreover, since the equations $e_{\text{rev}} = 1 + q_C/q_H$ and $e_{\text{rev}} = 1 - T_C/T_H$ hold for any working substance, we must have $q_C/q_H = -T_C/T_H$ or $q_C/T_C + q_H/T_H = 0$ for any working substance. Therefore

$$\oint \frac{dq}{T} = \frac{q_C}{T_C} + \frac{q_H}{T_H} = 0 \qquad \text{Carnot cycle} \qquad (3.16)$$

Equation (3.16) holds for any closed system undergoing a Carnot cycle. We shall use (3.16) to derive the state function entropy in Sec. 3.3.

Note from (3.15) that the smaller T_C is and the larger T_H is, the closer e_{rev} approaches 1, which represents complete conversion of the heat input into work output. Of course, a reversible heat engine is an idealization of real heat engines, which involve some irreversibility in their operation. The efficiency (3.15) is an upper limit to the efficiency of real heat engines [Eq. (3.7)].

Most of our electric power is produced by steam engines (more accurately, steam turbines) that drive conducting wires through magnetic fields, thereby generating electric currents. A modern steam power plant might have the boiler at 550°C (with the pressure correspondingly high) and the condenser at 40°C. If it operates on a Carnot cycle, then $e_{\text{rev}} = 1 - (313 \text{ K})/(823 \text{ K}) = 62\%$. The actual cycle of a steam engine is not a Carnot cycle because of irreversibility and because heat is transferred at temperatures between T_H and T_C, as well as at T_H and T_C. These factors make the actual efficiency less than 62%. The efficiency of a modern steam power plant is typically about 40%. (For comparison, James Watt's steam engines of the late 1700s had an efficiency of roughly 15%.) River water is commonly used as the cold reservoir for power plants. A 1000-MW power plant uses roughly 2 million L of cooling water per minute (Prob. 3.30). About 10% of the river flow in the United States is used by power plants for cooling. A cogeneration plant uses some of the waste heat of electricity generation for purposes such as space heating, thereby increasing the overall efficiency.

The analysis of this section applies only to heat engines, which are engines that convert heat to work. Not all engines are heat engines. For example, in an engine that uses a battery to drive a motor, the energy of a chemical reaction is converted in the battery to electrical energy, which in turn is converted to mechanical energy. Thus, chemical energy is converted to work, and this is not a heat engine. The human body converts chemical energy to work and is not a heat engine.

3.3   ENTROPY

For any closed system that undergoes a Carnot cycle, Eq. (3.16) shows that the integral of dq_{rev}/T around the cycle is zero. The subscript rev reminds us of the reversible nature of a Carnot cycle.

We now extend this result to an *arbitrary* reversible cycle, removing the constraint that heat be exchanged with the surroundings only at T_H and T_C. This will then show that dq_{rev}/T is the differential of a state function (Sec. 2.10).

The curve in Fig. 3.5a depicts an arbitrary reversible cyclic process. We draw reversible adiabats (shown as dashed lines) that divide the cycle into adjacent strips (Fig. 3.5b). Consider one such strip, bounded by curves ab and cd at the top and bottom. We draw the reversible isotherm mn such that the area under the zigzag curve $amnb$ equals the area under the smooth curve ab. Since these areas give the negative of the reversible work w done on the system in each process, we have $w_{amnb} = w_{ab}$, where ab is the process along the smooth curve and $amnb$ is the zigzag process along the two adiabats and the isotherm. ΔU is independent of the path from a to b, so $\Delta U_{amnb} = \Delta U_{ab}$. From $\Delta U = q + w$, it follows that $q_{amnb} = q_{ab}$. Since am and nb are adiabats, we have $q_{amnb} = q_{mn}$. Hence $q_{mn} = q_{ab}$. Similarly, we draw the reversible isotherm rs such that $q_{rs} = q_{cd}$. Since mn and rs are reversible isotherms and ns and rm are reversible adiabats, we could use these four curves to carry out a Carnot cycle; Eq. (3.16) then gives $q_{mn}/T_{mn} + q_{sr}/T_{sr} = 0$, and

$$\frac{q_{ab}}{T_{mn}} + \frac{q_{dc}}{T_{sr}} = 0 \qquad (3.17)$$

We can do exactly the same with every other strip in Fig. 3.5b to get an equation similar to (3.17) for each strip.

Now consider the limit as we draw the adiabats closer and closer together in Fig. 3.5b, ultimately dividing the cycle into an infinite number of infinitesimally narrow strips, in each of which we draw the zigzags at the top and bottom. As the adiabat bd comes closer to the adiabat ac, point b on the smooth curve comes closer to point a and, in the limit, the temperature T_b at b differs only infinitesimally from that at a. Let T_{ab} denote this essentially constant temperature. Moreover, T_{mn} in (3.17) (which lies between T_a and T_b) becomes essentially the same as T_{ab}. The same thing happens at the bottom of the strip. Also, the heats transferred become infinitesimal quantities in the limit. Thus (3.17) becomes in this limit

$$\frac{dq_{ab}}{T_{ab}} + \frac{dq_{dc}}{T_{dc}} = 0 \qquad (3.18)$$

The same thing happens in every other strip when we take the limit, and an equation similar to (3.18) holds for each infinitesimal strip. We now add all the equations like (3.18) for each strip. Each term in the sum will be infinitesimal and of the form dq/T, where dq is the heat transfer along an infinitesimal portion of the arbitrary reversible cycle and T is the temperature at which this heat transfer occurs. The sum of the infinitesimals is a line integral around the cycle, and we get

$$\oint \frac{dq_{rev}}{T} = 0 \qquad (3.19)$$

The subscript rev reminds us that the cycle under consideration is reversible. If it is irreversible, we can't relate it to Carnot cycles and (3.19) need not hold. Apart from the reversibility requirement, the cycle in (3.19) is arbitrary, and (3.19) is the desired generalization of (3.16).

Since the integral of dq_{rev}/T around any reversible cycle is zero, it follows (Sec. 2.10) that the value of the line integral $\int_1^2 dq_{rev}/T$ is independent of the path between

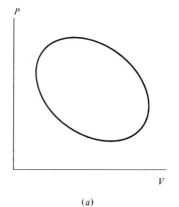

(a)

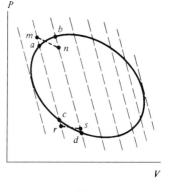

(b)

Figure 3.5

An arbitrary reversible cycle and its relation to Carnot cycles.

states 1 and 2 and depends only on the initial and final states. Hence dq_{rev}/T is the differential of a state function. This state function is called the **entropy** S:

$$dS \equiv \frac{dq_{rev}}{T} \qquad \text{closed syst., rev. proc.} \qquad \textbf{(3.20)}*$$

The entropy change on going from state 1 to state 2 equals the integral of (3.20):

$$\Delta S = S_2 - S_1 = \int_1^2 \frac{dq_{rev}}{T} \qquad \text{closed syst., rev. proc.} \qquad \textbf{(3.21)}*$$

Throughout this chapter we have been considering only closed systems; q is undefined for an open system.

If a system goes from state 1 to state 2 by an irreversible process, the intermediate states it passes through may not be states of thermodynamic equilibrium and the entropies, temperatures, etc., of intermediate states may be undefined. However, since S is a state function, it doesn't matter how the system went from state 1 to state 2; ΔS is the same for any process (reversible or irreversible) that connects states 1 and 2. But it is only for a reversible process that the integral of dq/T gives the entropy change. Calculation of ΔS in irreversible processes is considered in the next section.

Clausius discovered the state function S in 1854 and called it the transformation content (*Verwandlungsinhalt*). Later, he renamed it entropy, from the Greek word *trope*, meaning "transformation," since S is related to the transformation of heat to work.

Entropy is an extensive state function. To see this, imagine a system in equilibrium to be divided into two parts. Each part, of course, is at the same temperature T. Let parts 1 and 2 receive heats dq_1 and dq_2, respectively, in a reversible process. From (3.20), the entropy changes for the parts are $dS_1 = dq_1/T$ and $dS_2 = dq_2/T$. But the entropy change dS for the whole system is

$$dS = dq/T = (dq_1 + dq_2)/T = dq_1/T + dq_2/T = dS_1 + dS_2 \qquad (3.22)$$

Integration gives $\Delta S = \Delta S_1 + \Delta S_2$. Therefore $S = S_1 + S_2$, and S is extensive.

For a pure substance, the **molar entropy** is $S_m = S/n$.

The commonly used units of S in (3.20) are J/K or cal/K. The corresponding units of S_m are J/(mol K) or cal/(mol K).

The path from the postulation of the second law to the existence of S has been a long one, so let us review the chain of reasoning that led to entropy.

1. Experience shows that complete conversion of heat to work in a cyclic process is impossible. This assertion is the Kelvin–Planck statement of the second law.
2. From statement 1, we proved that the efficiency of any heat engine that operates on a (reversible) Carnot cycle is independent of the nature of the working substance but depends only on the temperatures of the reservoirs: $e_{rev} = -w/q_H = 1 + q_C/q_H = f(\tau_C, \tau_H)$.
3. We used a perfect gas as the working substance in a Carnot cycle and used the ideal-gas temperature scale to find that $e_{rev} = 1 - T_C/T_H$. From statement 2, this equation holds for any system as the working substance. Equating this expression for e_{rev} to that in statement 2, we get $q_C/T_C + q_H/T_H = 0$ for any system that undergoes a Carnot cycle.
4. We showed that an arbitrary reversible cycle can be divided into an infinite number of infinitesimal strips, each strip being a Carnot cycle. Hence for each strip, $dq_C/T_C + dq_H/T_H = 0$. Summing the dq/T's from each strip, we proved that $\oint dq_{rev}/T = 0$ for any reversible cycle that any system undergoes. It follows that the integral of dq_{rev}/T is independent of the path. Therefore dq_{rev}/T is the differential of a state function, which we call the entropy S; $dS \equiv dq_{rev}/T$.

Don't be discouraged by the long derivation of $dS = dq_{rev}/T$ from the Kelvin–Planck statement of the second law. You are not expected to memorize this derivation. What you are expected to do is be able to apply the relation $dS = dq_{rev}/T$ to calculate ΔS for various processes. How this is done is the subject of the next section.

3.4 CALCULATION OF ENTROPY CHANGES

The entropy change on going from state 1 to state 2 is given by Eq. (3.21) as $\Delta S = S_2 - S_1 = \int_1^2 dq_{rev}/T$, where T is the absolute temperature. For a reversible process, we can apply (3.21) directly to calculate ΔS. For an irreversible process pr, we cannot integrate dq_{pr}/T to obtain ΔS because dS equals dq/T only for reversible processes. For an irreversible process, dS is not necessarily equal to dq_{irrev}/T. However, S is a state function, and therefore ΔS depends only on the initial and final states. We can therefore find ΔS for an irreversible process that goes from state 1 to state 2 if we can conceive of a reversible process that goes from 1 to 2. We then calculate ΔS for this reversible change from 1 to 2, and this is the same as ΔS for the irreversible change from 1 to 2 (Fig. 3.6).

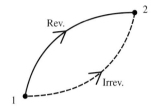

In summary, to calculate ΔS for any process; (a) Identify the initial and final states 1 and 2. (b) Devise a convenient *reversible* path from 1 to 2. (c) Calculate ΔS from $\Delta S = \int_1^2 dq_{rev}/T$.

Figure 3.6

Reversible and irreversible paths from state 1 to state 2. Since S is a state function, ΔS is the same for each path.

Let us calculate ΔS for some processes. Note that, as before, *all state functions refer to the system,* and ΔS means ΔS_{syst}. Equation (3.21) gives ΔS_{syst} and does not include any entropy changes that may occur in the surroundings.

1. **Cyclic process.** Since S is a state function, $\Delta S = 0$ for every cyclic process.
2. **Reversible adiabatic process.** Here $dq_{rev} = 0$; therefore

$$\Delta S = 0 \qquad \text{rev. ad. proc.} \qquad (3.23)$$

Two of the four steps of a Carnot cycle are reversible adiabatic processes.

3. **Reversible phase change at constant T and P.** At constant T, (3.21) gives

$$\Delta S = \int_1^2 \frac{dq_{rev}}{T} = \frac{1}{T} \int_1^2 dq_{rev} = \frac{q_{rev}}{T} \qquad (3.24)$$

q_{rev} is the latent heat of the transition. Since P is constant, $q_{rev} = q_P = \Delta H$ [Eq. (2.46)]. Therefore

$$\Delta S = \frac{\Delta H}{T} \qquad \text{rev. phase change at const. } T \text{ and } P \qquad (3.25)$$

Since $\Delta H = q_P$ is positive for reversible melting of solids and vaporization of liquids, ΔS is positive for these processes.

EXAMPLE 3.1 ΔS for a phase change

Find ΔS for the melting of 5.0 g of ice (heat of fusion = 79.7 cal/g) at 0°C and 1 atm. Find ΔS for the reverse process.

The melting is reversible and Eq. (3.25) gives

$$\Delta S = \frac{\Delta H}{T} = \frac{(79.7 \text{ cal/g})(5.0 \text{ g})}{273 \text{ K}} = 1.46 \text{ cal/K} = 6.1 \text{ J/K} \qquad (3.26)$$

For the freezing of 5.0 g of liquid water at 0°C and 1 atm, q_{rev} is negative, and $\Delta S = -6.1 \text{ J/K}$.

Exercise

The heat of vaporization of water at 100°C is 40.66 kJ/mol. Find ΔS when 5.00 g of water vapor condenses to liquid at 100°C and 1 atm. (*Answer:* -30.2 J/K.)

4. **Reversible isothermal process.** Here T is constant, and $\Delta S = \int_1^2 T^{-1} \, dq_{rev} = T^{-1} \int_1^2 dq_{rev} = q_{rev}/T$. Thus

$$\Delta S = q_{rev}/T \qquad \text{rev. isotherm. proc.} \qquad (3.27)$$

Examples include a reversible phase change (case 3 in this list) and two of the four steps of a Carnot cycle.

5. **Reversible change of state of a perfect gas.** From the first law and Sec. 2.8, we have for a reversible process in a perfect gas

$$dq_{rev} = dU - dw_{rev} = C_V \, dT + P \, dV = C_V \, dT + nRT \, dV/V \qquad (3.28)$$
$$dS = dq_{rev}/T = C_V \, dT/T + nR \, dV/V$$

$$\Delta S = \int_1^2 C_V(T) \, \frac{dT}{T} + nR \int_1^2 \frac{dV}{V}$$

$$\Delta S = \int_{T_1}^{T_2} \frac{C_V(T)}{T} \, dT + nR \ln \frac{V_2}{V_1} \qquad \text{perf. gas} \qquad (3.29)$$

If $T_2 > T_1$, the first integral is positive, so increasing the temperature of a perfect gas increases its entropy. If $V_2 > V_1$, the second term is positive, so increasing the volume of a perfect gas increases its entropy. If the temperature change is not large, it may be a good approximation to take C_V constant, in which case $\Delta S \approx C_V \ln (T_2/T_1) + nR \ln (V_2/V_1)$. A mistake students sometimes make in using (3.29) is to write $\ln (V_2/V_1) = \ln (P_1/P_2)$, forgetting that T is changing. The correct expression is $\ln (V_2/V_1) = \ln (P_1 T_2/P_2 T_1)$.

6. **Irreversible change of state of a perfect gas.** Let n moles of a perfect gas at P_1, V_1, T_1 irreversibly change its state to P_2, V_2, T_2. We can readily conceive of a reversible process to carry out this same change in state. For example, we might (*a*) put the gas (enclosed in a cylinder fitted with a frictionless piston) in a large constant-temperature bath at temperature T_1 and infinitely slowly change the pressure on the piston until the gas reaches volume V_2; (*b*) then remove the gas from contact with the bath, hold the volume fixed at V_2, and reversibly heat or cool the gas until its temperature reaches T_2. Since S is a state function, ΔS for this reversible change from state 1 to state 2 is the same as ΔS for the irreversible change from state 1 to state 2, even though q is not necessarily the same for the two processes. Therefore Eq. (3.29) gives ΔS for the irreversible change. Note that the value of the right side of (3.29) depends only on T_2, V_2, and T_1, V_1, the state functions of the final and initial states.

EXAMPLE 3.2 ΔS for expansion into a vacuum

Let n moles of a perfect gas undergo an adiabatic free expansion into a vacuum (the Joule experiment). (*a*) Express ΔS in terms of the initial and final temperatures and volumes. (*b*) Calculate ΔS_m if $V_2 = 2V_1$.

(*a*) The initial state is T_1, V_1, and the final state is T_1, V_2, where $V_2 > V_1$. T is constant because $\mu_J = (\partial T/\partial V)_U$ is zero for a perfect gas. Although the process is adiabatic ($q = 0$), ΔS is not zero because the process is irreversible. Equation (3.29) gives $\Delta S = nR \ln(V_2/V_1)$, since the temperature integral in (3.29) is zero when $T_2 = T_1$. (*b*) If the original container and the evacuated container are of equal volume, then $V_2 = 2V_1$ and $\Delta S = nR \ln 2$. We have

$$\Delta S/n = \Delta S_m = R \ln 2 = [8.314 \text{ J/(mol K)}](0.693) = 5.76 \text{ J/(mol K)}$$

Exercise

Find ΔS when 24 mg of $N_2(g)$ at 89 torr and 22°C expands adiabatically into vacuum to a final pressure of 34 torr. Assume perfect-gas behavior. (*Answer:* 6.9 mJ/K.)

---

7. **Constant-pressure heating.** First, suppose the heating is done *reversibly*. At constant pressure (provided no phase change occurs), $dq_{rev} = dq_P = C_P \, dT$ [Eq. (2.51)]. The relation $\Delta S = \int_1^2 dq_{rev}/T$ [Eq. (3.21)] becomes

$$\Delta S = \int_{T_1}^{T_2} \frac{C_P}{T} \, dT \qquad \text{const. } P \text{, no phase change} \tag{3.30}$$

If C_P is essentially constant over the temperature range, then $\Delta S = C_P \ln (T_2/T_1)$.

EXAMPLE 3.3 ΔS for heating at constant P

The specific heat capacity c_P of water is nearly constant at 1.00 cal/(g °C) in the temperature range 25°C to 75°C at 1 atm (Fig. 2.15). (*a*) Find ΔS when 100 g of water is reversibly heated from 25°C to 50°C at 1 atm. (*b*) Without doing the calculation, state whether ΔS for heating 100 g of water from 50°C to 75°C at 1 atm will be greater than, equal to, or less than ΔS for the 25°C to 50°C heating.

(*a*) The system's heat capacity is $C_P = mc_P = (100 \text{ g})[1.00 \text{ cal/(g °C)}] = 100$ cal/K. (A temperature change of one degree Celsius equals a change of one kelvin.) For the heating process, (3.30) with C_P constant gives

$$\Delta S = \int_{T_1}^{T_2} \frac{dq_{rev}}{T} = \int_{T_1}^{T_2} \frac{C_P}{T} \, dT = C_P \ln \frac{T_2}{T_1}$$

$$= (100 \text{ cal/K}) \ln \frac{323 \text{ K}}{298 \text{ K}} = 8.06 \text{ cal/K} = 33.7 \text{ J/K}$$

(*b*) Since C_P is constant, the reversible heat required for each of the processes with $\Delta T = 25$°C is the same. For the 50°C to 75°C change, each infinitesimal element of heat dq_{rev} flows in at a higher temperature than for the 25°C to 50°C change. Because of the $1/T$ factor in $dS = dq_{rev}/T$, each dq_{rev} produces a smaller increase in entropy for the higher-temperature process, and ΔS is smaller for the 50°C to 75°C heating. The higher the temperature, the smaller the entropy change produced by a given amount of reversible heat.

Exercise

Find ΔS when 100 g of water is reversibly heated from 50°C to 75°C at 1 atm. (*Answer:* 31.2 J/K.)

Now suppose we heat the water *irreversibly* from 25°C to 50°C at 1 atm (say, by using a bunsen-burner flame). The initial and final states are the same as for the

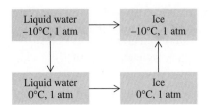

Figure 3.7

Irreversible and reversible paths
from liquid water to ice at $-10°C$
and 1 atm.

reversible heating. Hence the integral on the right side of (3.30) gives ΔS for the irreversible heating. Note that ΔS in (3.30) depends only on T_1, T_2, and the value of P (since C_P will depend somewhat on P); that is, ΔS depends only on the initial and final states. Thus ΔS for heating 100 g of water from 25°C to 50°C at 1 atm is 33.7 J/K, whether the heating is done reversibly or irreversibly. For irreversible heating with a bunsen burner, portions of the system nearer the burner will be at higher temperatures than portions farther from the burner, and no single value of T can be assigned to the system during the heating. Despite this, we can imagine doing the heating reversibly and apply (3.30) to find ΔS, provided the initial and final states are equilibrium states. Likewise, if we carry out the change of state by stirring at constant pressure as Joule did, rather than by heating, we can still use (3.30).

To heat a system reversibly, we surround it with a large constant-temperature bath that is at the same temperature as the system, and we heat the bath infinitely slowly. Since the temperature of the system and the temperature of its surroundings differ only infinitesimally during the process, the process is reversible.

8. **General change of state from (P_1, T_1) to (P_2, T_2).** In paragraph 7 we considered ΔS for a change in temperature at constant pressure. Here we also need to know how S varies with pressure. This will be discussed in Sec. 4.6.

9. **Irreversible phase change.** Consider the transformation of 1 mole of supercooled liquid water at $-10°C$ and 1 atm to 1 mole of ice at $-10°C$ and 1 atm. This transformation is irreversible. Intermediate states consist of mixtures of water and ice at $-10°C$, and these are not equilibrium states. Moreover, withdrawal of an infinitesimal amount of heat from the ice at $-10°C$ will not cause any of the ice to go back to supercooled water at $-10°C$. To find ΔS, we use the following reversible path. We first reversibly warm the supercooled liquid to 0°C and 1 atm (paragraph 7). We then reversibly freeze it at 0°C and 1 atm (paragraph 3). Finally, we reversibly cool the ice to $-10°C$ and 1 atm (paragraph 7). ΔS for the irreversible transformation at $-10°C$ equals the sum of the entropy changes for the three reversible steps, since the irreversible process and the reversible process each connect the same two states. Figure 3.7 shows the two paths. Numerical calculations are left as a problem (Prob. 3.15).

10. **Mixing of different inert perfect gases at constant P and T.** Let n_a and n_b moles of the inert perfect gases a and b, each at the same initial P and T, mix (Fig. 3.8). By inert gases, we mean that no chemical reaction occurs on mixing. Since the

Figure 3.8

Mixing of perfect gases at
constant T and P.

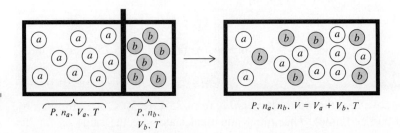

gases are perfect, there are no intermolecular interactions either before or after the partition is removed. Therefore the total internal energy is unchanged on mixing, and T is unchanged on mixing.

The mixing is irreversible. To find ΔS, we must find a way to carry out this change of state reversibly. This can be done in two steps. In step 1, we put each gas in a constant-temperature bath and reversibly and isothermally expand each gas separately to a volume equal to the final volume V. Note that step 1 is not adiabatic. Instead, heat flows into each gas to balance the work done by each gas. Since S is extensive, ΔS for step 1 is the sum of ΔS for each gas, and Eq. (3.29) gives

$$\Delta S_1 = \Delta S_a + \Delta S_b = n_a R \ln (V/V_a) + n_b R \ln (V/V_b) \qquad (3.31)$$

Step 2 is a reversible isothermal mixing of the expanded gases. This can be done as follows. We suppose it possible to obtain two *semipermeable* membranes, one permeable to gas a only and one permeable to gas b only. For example, heated palladium is permeable to hydrogen but not to oxygen or nitrogen. We set up the unmixed state of the two gases as shown in Fig. 3.9a. We assume the absence of friction. We then move the two coupled membranes slowly to the left. Figure 3.9b shows an intermediate state of the system.

Since the membranes move slowly, membrane equilibrium exists, meaning that the partial pressures of gas a on each side of the membrane permeable to a are equal, and similarly for gas b. The gas pressure in region I of Fig. 3.9b is P_a and in region III is P_b. Because of membrane equilibrium at each semipermeable membrane, the partial pressure of gas a in region II is P_a, and that of gas b in region II is P_b. The total pressure in region II is thus $P_a + P_b$. The total force to the right on the two movable coupled membranes is due to gas pressure in regions I and III and equals $(P_a + P_b)A$, where A is the area of each membrane. The total force to the left on these membranes is due to gas pressure in region II and equals $(P_a + P_b)A$. These two forces are equal. Hence any intermediate state is an equilibrium state, and only an infinitesimal force is needed to move the membranes. Since we pass through equilibrium states and exert only infinitesimal forces, step 2 is reversible. The final state (Fig. 3.9c) is the desired mixture.

The internal energy of a perfect gas depends only on T. Since T is constant for step 2, ΔU is zero for step 2. Since only an infinitesimal force was exerted on the membranes, $w = 0$ for step 2. Therefore $q = \Delta U - w = 0$ for step 2. Step 2 is adiabatic as well as reversible. Hence Eq. (3.23) gives $\Delta S_2 = 0$ for the reversible mixing of two perfect gases.

ΔS for the irreversible mixing of Fig. 3.8 equals $\Delta S_1 + \Delta S_2$, so

$$\Delta S = n_a R \ln (V/V_a) + n_b R \ln (V/V_b) \qquad (3.32)$$

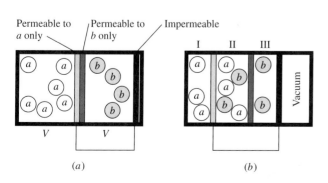

(a)

(b)

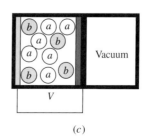

(c)

Figure 3.9

Reversible isothermal mixing of perfect gases. The system is in a constant-temperature bath (not shown).

The ideal-gas law $PV = nRT$ gives $V = (n_a + n_b)RT/P$ and $V_a = n_aRT/P$, so $V/V_a = (n_a + n_b)/n_a = 1/x_a$. Similarly, $V/V_b = 1/x_b$. Substituting into (3.32) and using $\ln(1/x_a) = \ln 1 - \ln x_a = -\ln x_a$, we get

$$\Delta_{mix}S = -n_aR \ln x_a - n_bR \ln x_b \qquad \text{perf. gases, const. } T, P \qquad (3.33)$$

where mix stands for mixing and x_a and x_b are the mole fractions of the gases in the mixture. Note that $\Delta_{mix}S$ is positive for perfect gases.

The term "entropy of mixing" for $\Delta_{mix}S$ in (3.33) is perhaps misleading, in that the entropy change comes entirely from the volume change of each gas (step 1) and is zero for the reversible mixing (step 2). Because ΔS is zero for step 2, the entropy of the mixture in Fig. 3.9c equals the entropy of the system in Fig. 3.9a. In other words, *the entropy of a perfect gas mixture is equal to the sum of the entropies each pure gas would have if it alone occupied the volume of the mixture at the temperature of the mixture.*

Equation (3.33) applies only when a and b are different gases. If they are identical, then the "mixing" at constant T and P corresponds to no change in state and $\Delta S = 0$.

The preceding examples show that the following processes increase the entropy of a substance: heating, melting a solid, vaporizing a liquid or solid, increasing the volume of a gas (including the case of mixing of gases).

Summary

To calculate $\Delta S \equiv S_2 - S_1$, we devise a reversible path from state 1 to state 2 and we use $\Delta S = \int_1^2 (1/T)\, dq_{rev}$. If T is constant, then $\Delta S = q_{rev}/T$. If T is not constant, we use an expression for dq_{rev} to evaluate the integral; for example, $dq_{rev} = C_P\, dT$ for a constant-pressure process or $dq_{rev} = dU - dw_{rev} = C_V\, dT + (nRT/V)\, dV$ for a perfect gas.

3.5   ENTROPY, REVERSIBILITY, AND IRREVERSIBILITY

In Sec. 3.4, we calculated ΔS for the *system* in various processes. In this section we shall consider the total entropy change that occurs in a process; that is, we shall examine the sum of the entropy changes in the system and the surroundings: $\Delta S_{syst} + \Delta S_{surr}$. We call this sum the entropy change of the universe:

$$\Delta S_{univ} = \Delta S_{syst} + \Delta S_{surr} \qquad (3.34)*$$

where the subscript univ stands for universe. Here, "universe" refers to the system plus those parts of the world that can interact with the system. Whether the conclusions of this section about ΔS_{univ} apply to the entire universe in a cosmic sense will be considered in Sec. 3.8. We shall examine separately ΔS_{univ} for reversible processes and irreversible processes.

Reversible Processes

In a reversible process, any heat flow between system and surroundings must occur with no finite temperature difference; otherwise the heat flow would be irreversible. Let dq_{rev} be the heat flow into the system from the surroundings during an infinitesimal part of the reversible process. The corresponding heat flow into the surroundings is $-dq_{rev}$. We have

$$dS_{univ} = dS_{syst} + dS_{surr} = \frac{dq_{rev}}{T_{syst}} + \frac{-dq_{rev}}{T_{surr}} = \frac{dq_{rev}}{T_{syst}} - \frac{dq_{rev}}{T_{syst}} = 0$$

Integration gives

$$\Delta S_{univ} = 0 \qquad \text{rev. proc.} \tag{3.35}$$

Although S_{syst} and S_{surr} may both change in a reversible process, $S_{syst} + S_{surr} = S_{univ}$ *is unchanged in a reversible process.*

Irreversible Processes

We first consider the special case of an *adiabatic* irreversible process in a closed system. This special case will lead to the desired general result. Let the system go from state 1 to state 2 in an irreversible adiabatic process. The disconnected arrowheads from 1 to 2 in Fig. 3.10 indicate the irreversibility and the fact that an irreversible process cannot in general be plotted on a *P-V* diagram since it usually involves non-equilibrium states.

To evaluate $S_2 - S_1 = \Delta S_{syst}$, we connect states 1 and 2 by the following reversible path. From state 2, we do work adiabatically and reversibly on the system to increase its temperature to T_{hr}, the temperature of a certain heat reservoir. This brings the system to state 3. From Eq. (3.23), ΔS is zero for a reversible adiabatic process. Hence $S_3 = S_2$. (As always, state functions refer to the system unless otherwise specified. Thus S_3 and S_2 are the system's entropies in states 3 and 2.) We next either add or withdraw enough heat $q_{3\to4}$ isothermally and reversibly at temperature T_{hr} to make the entropy of the system equal to S_1. This brings the system to state 4 with $S_4 = S_1$. ($q_{3\to4}$ is positive if heat flows into the system from the reservoir during the process $3 \to 4$ and negative if heat flows out of the system into the reservoir during $3 \to 4$.) We have

$$S_4 - S_3 = \int_3^4 \frac{dq_{rev}}{T} = \frac{1}{T_{hr}} \int_3^4 dq_{rev} = \frac{q_{3\to4}}{T_{hr}}$$

Since states 4 and 1 have the same entropy, they lie on a line of constant S, an *isentrop*. What is an isentrop? For an isentrop, $dS = 0 = dq_{rev}/T$, so $dq_{rev} = 0$; an isentrop is a reversible adiabat. Hence to go from 4 to 1, we carry out a reversible adiabatic process (with the system doing work on the surroundings). Since S is a state function, we have for the cycle $1 \to 2 \to 3 \to 4 \to 1$

$$0 = \oint dS_{syst} = (S_2 - S_1) + (S_3 - S_2) + (S_4 - S_3) + (S_1 - S_4)$$

$$\oint dS_{syst} = (S_2 - S_1) + 0 + q_{3\to4}/T_{hr} + 0 = 0$$

$$S_2 - S_1 = -q_{3\to4}/T_{hr}$$

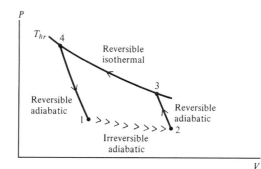

Figure 3.10

Irreversible and reversible paths between states 1 and 2.

The sign of $S_2 - S_1$ is thus the same as the sign of $-q_{3 \to 4}$. We have for the cycle

$$\oint dU = 0 = \oint (dq + dw) = q_{3 \to 4} + w$$

The work done on the system in the cycle is thus $w = -q_{3 \to 4}$. The work done by the system on the surroundings is $-w = q_{3 \to 4}$. Suppose $q_{3 \to 4}$ were positive. Then the work $-w$ done on the surroundings would be positive, and we would have a cycle ($1 \to 2 \to 3 \to 4 \to 1$) whose sole effect is extraction of heat $q_{3 \to 4}$ from a reservoir and its complete conversion to work $-w = q_{3 \to 4} > 0$. Such a cycle is impossible, since it violates the second law. Hence $q_{3 \to 4}$ cannot be positive: $q_{3 \to 4} \leq 0$. Therefore

$$S_2 - S_1 = -q_{3 \to 4}/T_{hr} \geq 0 \qquad (3.36)$$

We now strengthen this result by showing that $S_2 - S_1 = 0$ can be ruled out. To do this, consider the nature of reversible and irreversible processes. In a reversible process, we can make things go the other way by an infinitesimal change in circumstances. When the process is reversed, both system *and* surroundings are restored to their original states; that is, the universe is restored to its original state. In an irreversible process, the universe cannot be restored to its original state. Now suppose that $S_2 - S_1 = 0$. Then $q_{3 \to 4}$, which equals $-T_{hr}(S_2 - S_1)$, would be zero. Also, w, which equals $-q_{3 \to 4}$, would be zero. (Points 3 and 4 would coincide.) After the irreversible process $1 \to 2$, the path $2 \to 3 \to 4 \to 1$ restores the system to state 1. Moreover, since $q = 0 = w$ for the cycle $1 \to 2 \to 3 \to 4 \to 1$, this cycle would have no net effect on the surroundings, and at the end of the cycle, the surroundings would be restored to their original state. Thus we would be able to restore the universe (system + surroundings) to its original state. But by hypothesis, the process $1 \to 2$ is irreversible, and so the universe cannot be restored to its original state after this process has occurred. Therefore $S_2 - S_1$ cannot be zero. Equation (3.36) now tells us that $S_2 - S_1$ must be positive.

We have proved that *the entropy of a closed system must increase in an irreversible adiabatic process:*

$$\Delta S_{syst} > 0 \qquad \text{irrev. ad. proc., closed syst.} \qquad (3.37)$$

A special case of this result is important. An isolated system is necessarily closed, and any process in an isolated system must be adiabatic (since no heat can flow between the isolated system and its surroundings). Therefore (3.37) applies, and *the entropy of an isolated system must increase in any irreversible process:*

$$\Delta S_{syst} > 0 \qquad \text{irrev. proc., isolated syst.} \qquad (3.38)$$

Now consider $\Delta S_{univ} = \Delta S_{syst} + \Delta S_{surr}$ for an irreversible process. Since we want to examine the effect on S_{univ} of only the interaction between the system and its surroundings, we must consider that during the irreversible process the surroundings interact only with the system and not with any other part of the world. Hence, for the duration of the irreversible process, we can regard the system plus its surroundings (syst + surr) as forming an isolated system. Equation (3.38) then gives $\Delta S_{syst+surr} \equiv \Delta S_{univ} > 0$ for an irreversible process. We have shown that S_{univ} *increases in an irreversible process:*

$$\Delta S_{univ} > 0 \qquad \text{irrev. proc.} \qquad (3.39)$$

where ΔS_{univ} is the sum of the entropy changes for the system and surroundings.

We previously showed $\Delta S_{univ} = 0$ for a reversible process. Therefore

$$\Delta S_{univ} \geq 0 \qquad \qquad \mathbf{(3.40)}^*$$

depending on whether the process is reversible or irreversible. Energy cannot be created or destroyed. Entropy can be created but not destroyed.

The statement that

dq_{rev}/T **is the differential of a state function S that has the property $\Delta S_{univ} \geq 0$ for any process**

can be taken as a third formulation of the second law of thermodynamics, equivalent to the Kelvin–Planck and the Clausius statements. (See Prob. 3.24.)

We have shown (as a deduction from the Kelvin–Planck statement of the second law) that S_{univ} *increases for an irreversible process and remains the same for a reversible process.* A reversible process is an idealization that generally cannot be precisely attained in real processes. Virtually all real processes are irreversible because of phenomena such as friction, lack of precise thermal equilibrium, small amounts of turbulence, and irreversible mixing; see *Zemansky and Dittman,* chap. 7, for a full discussion. Since virtually all real processes are irreversible, we can say as a deduction from the second law that S_{univ} is continually increasing with time. See Sec. 3.8 for comment on this statement.

Entropy and Equilibrium

Equation (3.38) shows that, for any irreversible process that occurs in an isolated system, ΔS is positive. Since all real processes are irreversible, when processes are occurring in an isolated system, its entropy is increasing. Irreversible processes (mixing, chemical reaction, flow of heat from hot to cold bodies, etc.) accompanied by an increase in S will continue to occur in the isolated system until S has reached its maximum possible value subject to the constraints imposed on the system. For example, Prob. 3.20 shows that heat flow from a hot body to a cold body is accompanied by an increase in entropy. Hence, if two parts of an isolated system are at different temperatures, heat will flow from the hot part to the cold part until the temperatures of the parts are equalized, and this equalization of temperatures maximizes the system's entropy. When the entropy of the isolated system is maximized, things cease happening on a macroscopic scale, because any further processes can only decrease S, which would violate the second law. By definition, the isolated system has reached equilibrium when processes cease occurring. Therefore (Fig. 3.11):

Thermodynamic equilibrium in an isolated system is reached when the system's entropy is maximized.

Thermodynamic equilibrium in nonisolated systems is discussed in Chapter 4.

Thermodynamics says nothing about the *rate* at which equilibrium is attained. An isolated mixture of H_2 and O_2 at room temperature will remain unchanged in the absence of a catalyst. However, the system is not in a state of true thermodynamic equilibrium. When a catalyst is introduced, the gases react to produce H_2O, with an increase in entropy. Likewise, diamond is thermodynamically unstable with respect to conversion to graphite at room temperature, but the rate of conversion is zero, so no one need worry about loss of her engagement ring. ("Diamonds are forever.") It can even be said that pure hydrogen is in a sense thermodynamically unstable at room temperature, since fusion of the hydrogen nuclei to helium nuclei is accompanied by an increase in S_{univ}. Of course, the rate of nuclear fusion is zero at room temperature, and we can completely ignore the possibility of this process.

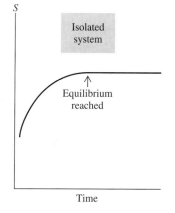

Figure 3.11

The entropy of an isolated system is maximized at equilibrium.

3.6   THE THERMODYNAMIC TEMPERATURE SCALE

In developing thermodynamics, we have so far used the ideal-gas temperature scale, which is based on the properties of a particular kind of substance, an ideal gas. The state functions P, V, U, and H are not defined in terms of any particular kind of substance, and it is desirable that a fundamental property like temperature be defined in a more general way than in terms of ideal gases. Lord Kelvin pointed out that the second law of thermodynamics can be used to define a *thermodynamic temperature scale* that is independent of the properties of any kind of substance.

We showed in Sec. 3.2 that, for a Carnot cycle between temperatures τ_C and τ_H, the efficiency e_{rev} is independent of the nature of the system (the working substance) and depends only on the temperatures: $e_{rev} = 1 + q_C/q_H = f(\tau_C, \tau_H)$, where τ symbolizes any temperature scale whatever. It follows that the heat ratio $- q_C/q_H$ (which equals $1 - e_{rev}$) is independent of the nature of the system that undergoes the Carnot cycle. We have

$$-q_C/q_H = 1 - f(\tau_C, \tau_H) \equiv g(\tau_C, \tau_H) \tag{3.41}$$

where the function g (defined as $1 - f$) depends on the choice of temperature scale but is independent of the nature of the system. We shall use the equation $-q_C/q_H = g(\tau_C, \tau_H)$ to define the thermodynamic temperature scale.

Before doing so, we examine the requirements the function g must satisfy. By considering two Carnot engines working with one reservoir in common, one can show that Carnot's principle (3.6) (which is a consequence of the second law) requires that the function g have the form

$$g(\tau_C, \tau_H) = \phi(\tau_C)/\phi(\tau_H) \tag{3.42}$$

where ϕ (phi) is some function. The proof of (3.42) is outlined in Prob. 3.26. Equation (3.41) becomes

$$-q_C/q_H = \phi(\tau_C)/\phi(\tau_H) \tag{3.43}$$

For convenience we shall also require that the temperature scale be such that the temperature of the hotter reservoir will always be greater than the temperature of the colder reservoir.

Within the framework of these two requirements, we are free to choose a scale defined in terms of the ratio $-q_C/q_H$. The simplest choice for the function ϕ in (3.43) is "take the first power." This choice gives the *thermodynamic temperature scale* Θ (capital theta). Temperature ratios on the thermodynamic scale are thus defined by

$$\frac{\Theta_C}{\Theta_H} \equiv \frac{-q_C}{q_H} \tag{3.44}$$

Equation (3.44) fixes only the ratio Θ_C/Θ_H. We complete the definition of the Θ scale by choosing the temperature of the triple point of water as $\Theta_{tr} = 273.16°$.

To measure the thermodynamic temperature Θ of an arbitrary body, we use it as one of the heat reservoirs in a Carnot cycle and use a body composed of water at its triple point as the second reservoir. We then put any system through a Carnot cycle between these two reservoirs and measure the heat q exchanged with the reservoir at Θ and the heat q_{tr} exchanged with the reservoir at $273.16°$. The thermodynamic temperature Θ is then calculated from (3.44) as

$$\Theta = 273.16° \frac{|q|}{|q_{tr}|} \tag{3.45}$$

Since the heat ratio in (3.45) is independent of the nature of the system put through the Carnot cycle, the Θ scale does not depend on the properties of any kind of substance.

How is the thermodynamic scale Θ related to the ideal-gas scale T? We proved in Sec. 3.2 that, on the ideal-gas temperature scale, $T_C/T_H = -q_C/q_H$ for any system that undergoes a Carnot cycle; see Eq. (3.16). Moreover, we chose the ideal-gas temperature at the water triple point as 273.16 K. Hence for a Carnot cycle between an arbitrary temperature T and the triple-point temperature, we have

$$T = 273.16 \text{ K} \frac{|q|}{|q_{tr}|} \tag{3.46}$$

where q is the heat exchanged with the reservoir at T. Comparison of (3.45) and (3.46) shows that *the ideal-gas temperature scale and the thermodynamic temperature scale are numerically identical.* We need not distinguish between them and will henceforth use the same symbol T for each scale. The thermodynamic scale is the fundamental scale of science, but as a matter of practical convenience, extrapolated measurements on gases, rather than Carnot-cycle measurements, are used to measure temperatures accurately.

3.7 WHAT IS ENTROPY?

Each of the first three laws of thermodynamics leads to the existence of a state function. The zeroth law leads to temperature. The first law leads to internal energy. The second law leads to entropy. It is not the business of thermodynamics, which is a macroscopic science, to explain the microscopic nature of these state functions. Thermodynamics need only tell us how to measure T, ΔU, and ΔS. Nevertheless it is nice to have a molecular picture of the macroscopic thermodynamic state functions.

Temperature is readily interpreted as some sort of measure of the average molecular energy. Internal energy is interpreted as the total molecular energy. Although we have shown how ΔS can be calculated for various processes, the reader may feel frustrated at not having any clear picture of the physical nature of entropy. Although entropy is not as easy a concept to grasp as temperature or internal energy, we can get some understanding of its physical nature.

Molecular Interpretation of Entropy

We saw in Sec. 3.5 that the entropy S of an isolated system is maximized at equilibrium. We therefore now ask: What else is maximized at equilibrium? In other words, what really determines the equilibrium position of an isolated thermodynamic system? To answer this, consider a simple example, the mixing at constant temperature and pressure of equal volumes of two different inert perfect gases d and e in an isolated system (Fig. 3.12). The motion of the gas molecules is completely random, and the molecules do not interact with one another. What then makes 2 in Fig. 3.12 the equilibrium state and 1 a nonequilibrium state? Why is the passage from the unmixed state 1 to the mixed state 2 irreversible? (From 2, an isolated system will never go back to 1.)

Clearly the answer is *probability*. If the molecules move at random, any d molecule has a 50% chance of being in the left half of the container. The probability that all the d molecules will be in the left half and all the e molecules in the right half (state 1) is extremely small. The most probable distribution has d and e molecules each equally distributed between the two halves of the container (state 2). An analogy to the

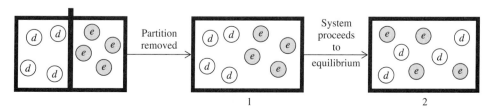

Figure 3.12

Irreversible mixing of perfect gases at constant T and P.

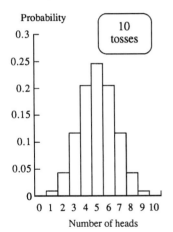

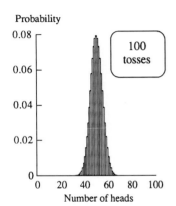

Figure 3.13

Probabilities for various numbers of heads when a coin is tossed 10 times and 100 times.

spatial distribution of 1 mole of d molecules would be tossing a coin 6×10^{23} times. The chance of getting 6×10^{23} heads is extremely tiny. The most probable outcome is 3×10^{23} heads and 3×10^{23} tails, and only outcomes with a very nearly equal ratio of heads to tails have significant probabilities. The probability maximum is extremely sharply peaked at 50% heads. (For example, Fig. 3.13 shows the probabilities for obtaining various numbers of heads for 10 tosses of a coin and for 100 tosses. As the number of tosses increases, the probability of significant deviations from 50% heads diminishes.) Similarly, any spatial distribution of the d molecules that differs significantly from 50% d in each container has an extremely small probability because of the large number of d molecules; similarly for the e molecules.

It seems clear that *the equilibrium thermodynamic state of an isolated system is the most probable state.* The increase in S as an isolated system proceeds toward equilibrium is directly related to the system's going from a state of low probability to one of high probability. We therefore postulate that the entropy S of a system is a function of the probability p of the system's thermodynamic state:

$$S = f(p) \tag{3.47}$$

Amazingly, use of the single fact that entropy is an extensive state function allows us to find the function f in our postulate (3.47). To do this, we consider a system composed of two independent, noninteracting parts, 1 and 2, separated by a rigid, impermeable, adiabatic wall that prevents flow of heat, work, and matter between them. Entropy is an extensive property, so the entropy of the composite system $1 + 2$ is $S_{1+2} = S_1 + S_2$, where S_1 and S_2 are the entropies of parts 1 and 2. Substitution of (3.47) into this equation gives

$$h(p_{1+2}) = f(p_1) + g(p_2) \tag{3.48}$$

where f, g, and h are three functions. Since systems 1, 2, and $1 + 2$ are not identical, the functions f, g, and h are not necessarily identical. What is the relation between the probability p_{1+2} of the composite system's thermodynamic state and the probabilities p_1 and p_2 of the states of parts 1 and 2? The probability that two independent events will both happen is shown in probability theory to be the product of the probabilities for each event. For example, the probability of getting two heads when two coins are tossed is $\frac{1}{2} \times \frac{1}{2} = \frac{1}{4}$. Since parts 1 and 2 behave independently of each other, we have $p_{1+2} = p_1 p_2$. Equation (3.48) becomes

$$h(p_1 p_2) = f(p_1) + g(p_2) \tag{3.49}$$

Our task is to find the functions that satisfy

$$h(xy) = f(x) + g(y) \tag{3.50}$$

Before reading ahead, you might try to guess a solution for h.

It isn't hard to prove that the only way to satisfy (3.50) is with logarithmic functions. Problem 15.52 shows that the functions in (3.50) must be

$$f(x) = k \ln x + a, \qquad g(y) = k \ln y + b, \qquad h(xy) = k \ln (xy) + c \tag{3.51}$$

where k is a constant and a, b, and c are constants such that $c = a + b$. The constant k must be the same for all systems [otherwise, (3.50) would not be satisfied], but the additive constant (a, b, or c) differs for different systems.

Since we postulated $S = f(p)$ in Eq. (3.47), we have from (3.51) that

$$S = k \ln p + a \tag{3.52}$$

where k and a are constants and p is the probability of the system's thermodynamic state. Since the second law allows us to calculate only *changes* in entropy, we cannot use thermodynamics to find a. We can, however, evaluate k as follows.

Consider again the spontaneous mixing of equal volumes of two different perfect gases (Fig. 3.12). State 1 is the unmixed state of the middle drawing of Fig. 3.12, and state 2 is the mixed state. Equation (3.52) gives for the process $1 \rightarrow 2$:

$$\Delta S = S_2 - S_1 = k \ln p_2 + a - k \ln p_1 - a$$
$$S_2 - S_1 = k \ln (p_2/p_1) \qquad (3.53)$$

(Don't confuse the probabilities p_1 and p_2 with pressures.) We want p_2/p_1. The probability that any particular d molecule is in the left half of the container is $\frac{1}{2}$. Since the perfect-gas molecules move independently of one another, the probability that every d molecule is in the left half of the container is the product of the independent probabilities for each d molecule, namely, $(\frac{1}{2})^{N_d}$, where N_d is the number of d molecules. Likewise, the probability that all the e molecules are in the right half of the container is $(\frac{1}{2})^{N_e}$. Since d and e molecules move independently, the simultaneous probability that all d molecules are in the left half of the box and all e molecules are in the right half is the product of the two separate probabilities, namely,

$$p_1 = (\tfrac{1}{2})^{N_d}(\tfrac{1}{2})^{N_e} = (\tfrac{1}{2})^{N_d+N_e} = (\tfrac{1}{2})^{2N_d} \qquad (3.54)$$

since $N_d = N_e$. (We took equal volumes of d and e at the same T and P.)

State 2 is the thermodynamic state in which, to within the limits of macroscopic measurement, the gases d and e are uniformly distributed through the container. As noted, the probability of any departure from a uniform distribution that is large enough to be directly detectable is vanishingly small because of the large number of molecules composing the system. Hence the probability of the final state 2 is only "infinitesimally" less than one and can be taken as one: $p_2 = 1$. Therefore, for the mixing, (3.53) and (3.54) give

$$S_2 - S_1 = k \ln (1/p_1) = k \ln 2^{2N_d} = 2N_d k \ln 2 \qquad (3.55)$$

However, in Sec. 3.4 we used thermodynamics to calculate ΔS for the constant-T-and-P irreversible mixing of two perfect gases; Eq. (3.33) with mole fractions set equal to one-half gives

$$S_2 - S_1 = 2n_d R \ln 2 \qquad (3.56)$$

Equating the thermodynamic ΔS of (3.56) to the statistical-mechanical ΔS of (3.55), we get $n_d R = N_d k$ and $k = Rn_d/N_d = R/N_A$, where $N_A = N_d/n_d$ [Eq. (1.5)] is the Avogadro constant. Thus

$$k = \frac{R}{N_A} = \frac{8.314 \text{ J mol}^{-1}\text{ K}^{-1}}{6.022 \times 10^{23} \text{ mol}^{-1}} = 1.38 \times 10^{-23} \text{ J/K} \qquad (3.57)$$

We have evaluated k in the statistical-mechanical formula $S = k \ln p + a$. The fundamental physical constant k, called **Boltzmann's constant,** plays a key role in statistical mechanics. The connection between entropy and probability was first recognized in the 1870s by the physicist Ludwig Boltzmann. The application of $S = k \ln p + a$ to situations more complicated than the mixing of perfect gases requires knowledge of quantum and statistical mechanics. In Chapter 22 we shall obtain an equation that expresses the entropy of a system in terms of its quantum-mechanical energy levels. Our main conclusion for now is that *entropy is a measure of the probability of a state.* Apart from an additive constant, the entropy is proportional to the log of the probability of the thermodynamic state.

Equation (3.52) reads $S = (R/N_A) \ln p + a$. This relation is valid for any system, not just an ideal gas. The occurrence of R in this general equation shows that the constant R is more universal and fundamental than one might suspect from its initial occurrence in the ideal-gas law. (The same is true of the ideal-gas absolute temperature T.)

We shall see in Chapter 22 that R/N_A, the gas constant per molecule (Boltzmann's constant), occurs in the fundamental equations governing the distribution of molecules among energy levels and thermodynamic systems among quantum states.

Disordered states generally have higher probabilities than ordered states. For example, in the mixing of two gases, the disordered, mixed state is far more probable than the ordered, unmixed state. Hence it is often said that entropy is a measure of the molecular *disorder* of a state. Increasing entropy means increasing molecular disorder. However, order and disorder are subjective concepts, whereas probability is a precise quantitative concept. It is therefore preferable to relate S to probability rather than to disorder.

For mixing two different gases, the connection between probability and entropy is clear. Let us examine some other processes. If two parts of a system are at different temperatures, heat flows spontaneously and irreversibly between the parts, accompanied by an increase in entropy. How is probability involved here? The heat flow occurs via collisions between molecules of the hot part with molecules of the cold part. In such collisions, it is more probable for the high-energy molecules of the hot part to lose some of their energy to the low-energy molecules of the cold part than for the reverse to happen. Thus, internal energy is transferred from the hot body to the cold until thermal equilibrium is attained, at which point it is equally probable for molecular collisions to transfer energy from one part to the second part as to do the opposite. It is therefore more probable for the internal molecular translational, vibrational, and rotational energies to be spread out among the parts of the system than for there to be an excess of such energy in one part.

Now consider an isolated reaction mixture of H_2, Br_2, and HBr gases. During molecular collisions, energy transfers can occur that break bonds and allow the formation of new chemical species. There will be a probability for each possible outcome of each possible kind of collision, and these probabilities, together with the numbers of molecules of each species present, determine whether there is a net reaction to give more HBr or more H_2 and Br_2. When equilibrium is reached, the system has attained the most probable distribution of the species present over the available energy levels of H_2, Br_2, and HBr.

These last two examples indicate that entropy is related to the distribution or spread of energy among the available molecular energy levels. The total energy of an isolated system is conserved, and it is the *distribution* of energy (which is related to the entropy) that determines the direction of spontaneity. The equilibrium position corresponds to the most probable distribution of energy.

We shall see in Sec. 22.6 that the greater the number of energy levels that have significant occupation, the larger the entropy is. Increasing a system's energy (for example by heating it) will increase its entropy because this allows higher energy levels to be significantly occupied, thereby increasing the number of occupied levels. It turns out that increasing the volume of a system at constant energy also allows more energy levels to be occupied, since it lowers the energies of many of the energy levels. (In the preceding discussion, the term "energy levels" should be replaced by "quantum states" but we won't worry about this point now.)

Fluctuations

What light does this discussion throw on the second law of thermodynamics, which can be formulated as $\Delta S \geq 0$ for an isolated system (where $dS = dq_{rev}/T$)? The reason S increases is because an isolated system tends to go to a state of higher probability. *However*, it is not absolutely impossible for a macroscopic isolated system to go spontaneously to a state of lower probability, but such an occurrence is highly unlikely. Hence the second law is only a law of probability. There is an extremely small, but nonzero, chance that it might be violated. For example, there is a possibility of observing the spontaneous unmixing of two mixed gases, but because of the huge numbers of molecules present, the probability that this will happen is fantastically small. There is an extremely tiny probability that the random motions of oxygen molecules in the air around you might carry them all to one corner of the room, causing you to die for lack of oxygen, but this possibility is nothing to lose any sleep over. The mixing of gases is irreversible because the mixed state is far, far more probable than any state with significant unmixing.

To show the extremely small probability of significant macroscopic deviations from the second law, consider the mixed state of Fig. 3.12. Let there be $N_d = 0.6 \times 10^{24}$ molecules of the perfect gas d distributed between the two equal volumes. The most likely distribution is one with 0.3×10^{24} molecules of d in each half of the container, and similarly for the e molecules. (For simplicity we shall consider only the distribution of the d molecules, but the same considerations apply to the e molecules.) The probability that each d molecule will be in the left half of the container is $\frac{1}{2}$.

Probability theory (*Sokolnikoff and Redheffer,* p. 645) shows that the standard deviation of the number of d molecules in the left volume equals $\frac{1}{2}N_d^{1/2} = 0.4 \times 10^{12}$. The *standard deviation* is a measure of the typical deviation that is observed from the most probable value, 0.3×10^{24} in this case. Probability theory shows that, when many observations are made, 68% of them will lie within 1 standard deviation from the most probable value. (This statement applies whenever the distribution of probabilities is a normal, or gaussian, distribution. The gaussian distribution is the familiar bell-shaped curve at the upper left in Fig. 18.18.)

In our example, we can expect that 68% of the time the number of d molecules in the left volume will lie in the range $0.3 \times 10^{24} \pm 0.4 \times 10^{12}$. Although the standard deviation 0.4×10^{12} molecules is a very large number of molecules, it is negligible compared with the total number of d molecules in the left volume, 0.3×10^{24}. A deviation of 0.4×10^{12} out of 0.3×10^{24} would mean a fluctuation in gas density of 1 part in 10^{12}, which is much too small to be directly detectable experimentally. A directly detectable density fluctuation might be 1 part in 10^6, or 0.3×10^{18} molecules out of 0.3×10^{24}. This is a fluctuation of about 10^6 standard deviations. The probability of a fluctuation this large or larger is found (Prob. 3.27) to be approximately $1/10^{200,000,000,000}$. The age of the universe is about 10^{10} years. If we measured the density of the gas sample once every second, it would take (Prob. 3.28) about

$$\frac{0.7 \times 10^{200,000,000,000}}{3 \times 10^7} \approx 10^{200,000,000,000} \tag{3.58}$$

years of measurements for the probability of finding a detectable density fluctuation of 1 part in 10^6 to reach 50%. For all practical purposes, such a fluctuation in a macroscopic system is "impossible."

Probability theory shows that we can expect fluctuations about the equilibrium number density to be on the order of $\sqrt{N}$, where N is the number of molecules per unit volume. These fluctuations correspond to continual fluctuations of the entropy about its equilibrium value. Such fluctuations are generally unobservable for systems of macroscopic size but can be detected in special situations (see below). If a system had 100 molecules, we would get fluctuations of about 10 molecules, which is an easily detectable 10% fluctuation. A system of 10^6 molecules would show fluctuations of 0.1%, which is still significant. For 10^{12} molecules ($\approx 10^{-12}$ mole), fluctuations are 1 part per million, which is perhaps the borderline of detectability. The validity of the second law is limited to systems where N is large enough to make fluctuations essentially undetectable.

In certain situations, fluctuations about equilibrium are experimentally observable. For example, tiny (but still macroscopic) dust particles or colloidal particles suspended

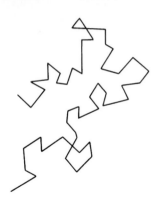

Figure 3.14

A particle undergoing Brownian motion.

in a fluid and observed through a microscope show a ceaseless random motion (Fig. 3.14), called **Brownian motion** (after its discoverer, the botanist Robert Brown). These motions are due to collisions with the molecules of the fluid. If the fluid pressure on all parts of the colloidal particle were always the same, the particle would remain at rest. (More accurately, it would sink to the bottom of the container because of gravity.) However, tiny fluctuations in fluid pressures on the colloidal particle cause the random motion. Such motion can be regarded as a small-scale violation of the second law.

Similarly, random fluctuations in electron densities in an electrical resistor produce tiny internal currents, which, when amplified, give the "noise" that is always present in an electronic circuit. This noise limits the size of a detectable electronic signal, since amplification of the signal also amplifies the noise.

The realization that the second law is not an absolute law but only one for which observation of macroscopic violations is in general overwhelmingly improbable need not disconcert us. Most laws dealing with the macroscopic behavior of matter are really statistical laws whose validity follows from the random behavior of huge numbers of molecules. For example, in thermodynamics, we refer to the pressure P of a system. The pressure a gas exerts on the container walls results from the collisions of molecules with the walls. There is a possibility that, at some instant, the gas molecules might all be moving inward toward the interior of the container, so that the gas would exert zero pressure on the container. Likewise, the molecular motion at a given instant might make the pressure on some walls differ significantly from that on other walls. However, such situations are so overwhelmingly improbable that we can with complete confidence ascribe a single uniform pressure to the gas.

3.8   ENTROPY, TIME, AND COSMOLOGY

In the spontaneous mixing of two different gases, the molecules move according to Newton's second law, $\mathbf{F} = m\, d^2\mathbf{r}/dt^2 = m\, d\mathbf{v}/dt$. This law is symmetric with respect to time, meaning that, if t is replaced by $-t$ and $\mathbf{v}$ by $-\mathbf{v}$, the law is unchanged. Thus, a reversal of all particle motions gives a set of motions that is also a valid solution of Newton's equation. It is therefore possible for the molecules to become spontaneously unmixed, and this unmixing does not violate the law of motion $\mathbf{F} = m\mathbf{a}$. However, as noted in the previous section, motions that correspond to a detectable degree of unmixing are extremely improbable (even though not absolutely impossible). Although Newton's laws of motion (which govern the motion of individual molecules) do not single out a direction of time, when the behavior of a very large number of molecules is considered, the second law of thermodynamics (which is a statistical law) tells us that states of an isolated system with lower entropy must precede in time states with higher entropy. The second law is not time-symmetric but singles out the direction of increasing time; we have $dS/dt > 0$ for an isolated system, so the signs of dS and dt are the same.

If someone showed us a film of two gases mixing spontaneously and then ran the film backward, we would not see any violations of $\mathbf{F} = m\mathbf{a}$ in the unmixing process, but the second law would tell us which showing of the film corresponded to how things actually happened. Likewise, if we saw a film of someone being spontaneously propelled out of a swimming pool, with the concurrent subsidence of waves in the pool, we would know that we were watching a film run backward. Although tiny pressure fluctuations in a fluid can propel colloidal particles about, Brownian motion of an object the size of a person is too improbable to occur.

The second law of thermodynamics singles out the direction of increasing time. The astrophysicist Eddington stated that "entropy is time's arrow." The fact that $dS/dt > 0$ for an isolated system gives us the *thermodynamic arrow* of time. Besides the thermodynamic arrow, there is a *cosmological arrow* of time. Spectral lines in light reaching us from other galaxies show wavelengths that are longer than the corresponding wavelengths of light from objects at rest (the famous red shift). This red shift indicates that all galaxies are moving away from us. (The frequency shift results from the Doppler effect.) Thus the universe is expanding with increasing time, and this expansion gives the cosmological arrow. Many physicists believe that the thermodynamic and the cosmological arrows are directly related, but this question is still undecided. [See T. Gold, *Am. J. Phys.,* **30,** 403 (1962); S. F. Savitt (ed.), *Time's Arrows Today,* Cambridge University Press, 1997.]

The second law of thermodynamics shows that S increases with time for an isolated system. Does this statement apply to the entire physical universe? In Sec. 3.5 we used "universe" to mean the system plus those parts of the world that interact with the system. In this section, "universe" shall mean everything that exists—the entire cosmos of galaxies, intergalactic matter, electromagnetic radiation, etc. Physicists in the late nineteenth century believed the second law to be valid for the entire universe, but nowadays people are not so sure. Most of our thermodynamic observations are on systems that are not of astronomic size, and we must be cautious about extrapolating thermodynamic results to encompass the entire universe. There is no guarantee that laws that hold on a terrestrial scale must also hold on a cosmic scale. Although there is no evidence for a cosmic violation of the second law, our experience is insufficient to rule out such a violation. Entropy changes in cosmic processes such as nuclear fusion in stars and formation and decay of black holes are discussed in S. Frautschi, *Science,* **217,** 593 (1982).

> The general theory of relativity predicts that when a star whose mass is greater than about three times the sun's mass runs out of nuclear fuel and ends nuclear fusion reactions, the elimination of the outward radiation pressure allows the gravitational attraction between the atoms of the star to collapse the star to a black hole. A *black hole* is a region of space whose gravitational field is so strong that no matter or radiation can escape from the hole. It was originally thought that if one throws a system into a black hole, this represents a violation of the second law of thermodynamics since the system's entropy is no longer observable. However, work by Bekenstein and by Hawking showed that one could attribute an entropy to a black hole and the value of the entropy could be calculated from an observable property of the hole (namely, the area of its boundary). Bekenstein formulated a *generalized second law of thermodynamics,* which states that in all interactions with a black hole, the sum of the entropies of the hole and of everything outside the hole does not decrease [J. D. Bekenstein, *Phys. Rev. D,* **7,** 2333 (1973); S. Hawking, *A Brief History of Time,* Bantam, 1988, chap. 7].

The currently accepted cosmological model is the big bang model, in which it is hypothesized that, about 12 or 15 billion (12 or 15×10^9) years ago, all the matter and radiation of the universe was present in a highly condensed state in a tiny volume. Explosion of this ball of matter and radiation produced the presently observed expanding universe. Astronomical observations suggest that this expansion will continue forever but are not definitive enough to completely eliminate the possibility that the universe contains enough matter to allow gravitational attractions in time to overcome the force of the initial explosion. If the latter is true, the universe will eventually reach a maximum expansion, from which it will begin to contract, ultimately bringing all matter together again. Perhaps a new big bang would then initiate a new cycle of expansion and contraction. Thus we could have a cyclic or oscillatory cosmological model.

If the cyclic expansion–contraction cosmological model is correct, what will happen in the contraction phase of the universe? If the universe returns to a state essentially the same as the initial state that preceded the big bang (and if we accept the applicability of the concept of entropy to the entire universe), then the entropy of the universe would decrease during the contraction phase. This expectation is further supported by the arguments for a direct connection between the thermodynamic and cosmological arrows of time. But what would a universe with decreasing entropy be like? Would time run backward in a contracting universe? What is the meaning of the statement that "time runs backward"?

A concept that follows from the second law of thermodynamics is the *principle of the degradation of energy.* One can prove (see *Zemansky and Dittman,* sec. 8-10) that any process that occurs makes an amount of energy $T_C \Delta S_{univ}$ unavailable for conversion to work, where T_C is the temperature of the coldest reservoir at hand and ΔS_{univ} is the entropy change in the process. Since S_{univ} is continually increasing, energy is continually being made unavailable for conversion to work. If the second law applies to the entire universe, ultimately no energy will be available for doing work. The universe will have reached a uniform temperature with its entropy maximized, and all processes, including life, will cease. This gloomy prospect has been christened the *heat death* of the universe. In the past, there has been much philosophical speculation based on this supposed heat death. It seems clear at present that we need to know a lot more about cosmology before we can be certain what the fate of the universe will be. (Entropy and heat death are favorite themes of science-fiction writers; see, for example, Isaac Asimov's story "The Last Question," in *Nine Tomorrows,* Doubleday, 1959, Fawcett, 1969.)

An article on the long-term fate of the universe states that "a continually expanding universe never reaches true thermodynamic equilibrium and hence never reaches a constant temperature. Classical heat death is thus manifestly avoided. However, the expansion can . . . become purely adiabatic so that the entropy in a given comoving volume . . . approaches (or attains) a constant value. In this case, the universe can still become a . . . lifeless place with no ability to do physical work. We denote this latter possibility as cosmological heat death" [F. C. Adams and G. Laughlin, *Rev. Mod. Phys.,* **69,** 337 (1997)]. This article speculates that cosmological heat death won't occur before 10^{100} years from now but is possible after that time; after 10^{100} years it is speculated that protons will have decayed, black holes evaporated, and the universe will consist of the waste products of these processes, namely, electrons, positrons, neutrinos (Sec. 17.19), and very low-energy photons.

3.9   SUMMARY

We assumed the truth of the Kelvin–Planck statement of the second law of thermodynamics, which asserts the impossibility of the complete conversion of heat to work in a cyclic process. From the second law, we proved that dq_{rev}/T is the differential of a state function, which we called the entropy S. The entropy change in a process from state 1 to state 2 is $\Delta S = \int_1^2 dq_{rev}/T$, where the integral must be evaluated using a reversible path from 1 to 2. Methods for calculating ΔS were discussed in Sec. 3.4.

We used the second law to prove that the entropy of an isolated system must increase in an irreversible process. It follows that thermodynamic equilibrium in an isolated system is reached when the system's entropy is maximized. Since isolated systems spontaneously change to more probable states, increasing entropy corresponds to increasing probability p. We found that $S = k \ln p + a$, where the Boltzmann constant k is $k = R/N_A$ and a is a constant.

Important kinds of calculations dealt with in this chapter include:

- Calculation of ΔS for a reversible process using $dS = dq_{rev}/T$.
- Calculation of ΔS for an irreversible process by finding a reversible path between the initial and final states (Sec. 3.4, paragraphs 6, 7, and 9).
- Calculation of ΔS for a reversible phase change using $\Delta S = \Delta H/T$.
- Calculation of ΔS for constant-pressure heating using $dS = dq_{rev}/T = (C_P/T) \, dT$.
- Calculation of ΔS for a change of state of a perfect gas using Eq. (3.29).
- Calculation of ΔS for mixing perfect gases at constant T and P using Eq. (3.33).

FURTHER READING

Denbigh, pp. 21–42, 48–60; *Kestin*, chap. 9; *Zemansky and Dittman*, chaps. 6, 7, 8.

PROBLEMS

Section 3.2

3.1 True or false? (*a*) Increasing the temperature of the hot reservoir of a Carnot-cycle engine must increase the efficiency of the engine. (*b*) Decreasing the temperature of the cold reservoir of a Carnot-cycle engine must increase the efficiency of the engine. (*c*) A Carnot cycle is by definition a reversible cycle. (*d*) Since a Carnot cycle is a cyclic process, the work done in a Carnot cycle is zero.

3.2 Consider a heat engine that uses reservoirs at 800°C and 0°C. (*a*) Calculate the maximum possible efficiency. (*b*) If q_H is 1000 J, find the maximum value of $-w$ and the minimum value of $-q_C$.

3.3 Suppose the coldest reservoir we have at hand is at 10°C. If we want a heat engine that is at least 90% efficient, what is the minimum temperature of the required hot reservoir?

3.4 A Carnot-cycle heat engine does 2.50 kJ of work per cycle and has an efficiency of 45.0%. Find w, q_H, and q_C for one cycle.

3.5 Heat pumps and refrigerators are heat engines running in reverse; a work input w causes the system to absorb heat q_C from a cold reservoir at T_C and emit heat $-q_H$ into a hot reservoir at T_H. The coefficient of performance K of a refrigerator is q_C/w, and the coefficient of performance ε of a heat pump is $-q_H/w$. (*a*) For reversible Carnot-cycle refrigerators and heat pumps, express K and ε in terms of T_C and T_H. (*b*) Show that ε_{rev} is always greater than 1. (*c*) Suppose a reversible heat pump transfers heat from the outdoors at 0°C to a room at 20°C. For each joule of work input to the heat pump, how much heat will be deposited in the room? (*d*) What happens to K_{rev} as T_C goes to 0 K?

3.6 Use sketches of the work w_{by} done *by* the system for each step of a Carnot cycle to show that w_{by} for the cycle equals the area enclosed by the curve of the cycle on a *P-V* plot.

3.7 Prove that the Clausius statement of the second law is equivalent to the Kelvin–Planck statement. To prove that statements A and B are logically equivalent, we must show that (*a*) if we assume A to be true, then B must be true; (*b*) if we assume B to be true, then A must be true. Thus we must be able to deduce the Clausius statement from the Kelvin–Planck statement, and vice versa. Here are some hints on how to proceed. First, assume the Kelvin–Planck statement to be true. Temporarily, suppose the Clausius statement to be false. Let an anti-Clausius device (a cyclic device that absorbs heat from a cold reservoir and delivers an equal amount of heat to a hot reservoir with no other effects) be coupled with a heat engine that uses the same pair of reservoirs. Show that, if the heat engine is run so

that it discards heat to the cold reservoir at the same rate the anti-Clausius device removes heat from this reservoir, we have a device that violates the Kelvin–Planck statement. Hence the existence of an anti-Clausius device is incompatible with the truth of the Kelvin–Planck statement, and the Clausius statement has been deduced from the Kelvin–Planck statement. To deduce the Kelvin–Planck statement from the Clausius statement, assume the Clausius statement to be true and couple an anti-Kelvin–Planck heat engine with a heat pump.

Section 3.4

3.8 True or false? (*a*) A change of state from state 1 to state 2 produces a greater increase in entropy when carried out irreversibly than when done reversibly. (*b*) The heat q for an irreversible change of state from state 1 to 2 might differ from the heat for the same change of state carried out reversibly. (*c*) The higher the absolute temperature of a system, the smaller the increase in its entropy produced by a given positive amount dq_{rev} of reversible heat flow. (*d*) The entropy of 20 g of $H_2O(l)$ at 300 K and 1 bar is twice the entropy of 10 g of $H_2O(l)$ at 300 K and 1 bar. (*e*) The molar entropy of 20 g of $H_2O(l)$ at 300 K and 1 bar is equal to the molar entropy of 10 g of $H_2O(l)$ at 300 K and 1 bar. (*f*) For a reversible isothermal process in a closed system, ΔS must be zero. (*g*) The integral $\int_1^2 T^{-1}C_V\,dT$ in Eq. (3.29) is always equal to $C_V \ln (T_2/T_1)$. (*h*) The system's entropy change for an adiabatic process in a closed system must be zero. (*i*) Thermodynamics cannot calculate ΔS for an irreversible process. (*j*) For a reversible process in a closed system, dq is equal to $T\,dS$. (*k*) The formulas of Sec. 3.4 enable us to calculate ΔS for various processes but do not enable us to find the value of S of a thermodynamic state.

3.9 The molar heat of vaporization of Ar at its normal boiling point 87.3 K is 1.56 kcal/mol. (*a*) Calculate ΔS for the vaporization of 1.00 mol of Ar at 87.3 K and 1 atm. (*b*) Calculate ΔS when 5.00 g of Ar gas condenses to liquid at 87.3 K and 1 atm.

3.10 Find ΔS when 2.00 mol of O_2 is heated from 27°C to 127°C with P held fixed at 1.00 atm. Use $C_{P,\text{m}}$ from Prob. 2.47.

3.11 Find ΔS for the conversion of 1.00 mol of ice at 0°C and 1.00 atm to 1.00 mol of water vapor at 100°C and 0.50 atm. Use data from Prob. 2.48.

3.12 Find ΔS when 1.00 mol of water vapor initially at 200°C and 1.00 bar undergoes a cyclic process for which $q = -145$ J.

3.13 Calculate ΔS for each of the following changes in state of 2.50 mol of a perfect monatomic gas with $C_{V,\text{m}} = 1.5R$ for all temperatures: (*a*) (1.50 atm, 400 K) → (3.00 atm, 600 K); (*b*) (2.50 atm, 20.0 L) → (2.00 atm, 30.0 L); (*c*) (28.5 L, 400 K) → (42.0 L, 400 K).

3.14 For $N_2(g)$, $C_{P,m}$ is nearly constant at 29.1 J/(mol K) for temperatures in the range 100 K to 400 K and low or moderate pressures. Find ΔS for the reversible adiabatic compression of 1.12 g of $N_2(g)$ from 400 torr and 1000 cm³ to a final volume of 250 cm³. Assume perfect-gas behavior.

3.15 Find ΔS for the conversion of 10.0 g of supercooled water at $-10°C$ and 1.00 atm to ice at $-10°C$ and 1.00 atm. Average c_P values for ice and supercooled water in the range 0°C to $-10°C$ are 0.50 and 1.01 cal/(g °C), respectively. See also Prob. 2.48.

3.16 State whether each of q, w, ΔU, and ΔS is negative, zero, or positive for each step of a Carnot cycle of a perfect gas.

3.17 After 200 g of gold [$c_P = 0.0313$ cal/(g °C)] at 120.0°C is dropped into 25.0 g of water at 10.0°C, the system is allowed to reach equilibrium in an adiabatic container. Find (a) the final temperature; (b) ΔS_{Au}; (c) ΔS_{H_2O}; (d) $\Delta S_{Au} + \Delta S_{H_2O}$.

3.18 Calculate ΔS for the mixing of 10.0 g of He at 120°C and 1.50 bar with 10.0 g of O_2 at 120°C and 1.50 bar.

3.19 A system consists of 1.00 mg of ClF gas. A mass spectrometer separates the gas into the species ^{35}ClF and ^{37}ClF. Calculate ΔS. Isotopic abundances: $^{19}F = 100\%$; $^{35}Cl = 75.8\%$; $^{37}Cl = 24.2\%$.

3.20 Let an isolated system be composed of one part at T_1 and a second part at T_2, with $T_2 > T_1$; let the parts be separated by a wall that allows heat flow at only an infinitesimal rate. Show that, when heat dq flows irreversibly from T_2 to T_1, we have $dS = dq/T_1 - dq/T_2$ (which is positive). (*Hint:* Use two heat reservoirs to carry out the change of state reversibly.)

Section 3.5

3.21 True or false? (a) For a closed system, ΔS can never be negative. (b) For a reversible process in a closed system, ΔS must be zero. (c) For a reversible process in a closed system, ΔS_{univ} must be zero. (d) For an adiabatic process in a closed system, ΔS cannot be negative. (e) For a process in an isolated system, ΔS cannot be negative. (f) For an adiabatic process in a closed system, ΔS must be zero. (g) An adiabatic process cannot decrease the entropy of a closed system. (h) For a closed system, equilibrium has been reached when S has been maximized.

3.22 For each of the following processes deduce whether each of the quantities ΔS and ΔS_{univ} is positive, zero, or negative. (a) Reversible melting of solid benzene at 1 atm and the normal melting point. (b) Reversible melting of ice at 1 atm and 0°C. (c) Reversible adiabatic expansion of a perfect gas. (d) Reversible isothermal expansion of a perfect gas. (e) Adiabatic expansion of a perfect gas into a vacuum (Joule experiment). (f) Joule–Thomson adiabatic throttling of a perfect gas. (g) Reversible heating of a perfect gas at constant P. (h) Reversible cooling of a perfect gas at constant V. (i) Combustion of benzene in a sealed container with rigid, adiabatic walls. (j) Adiabatic expansion of a nonideal gas into vacuum.

3.23 (a) What is ΔS for each step of a Carnot cycle? (b) What is ΔS_{univ} for each step of a Carnot cycle?

3.24 Prove the equivalence of the Kelvin–Planck statement and the entropy statement [the set-off statement after Eq. (3.40)] of the second law. [*Hint:* Since the entropy statement was derived from the Kelvin–Planck statement, all we need do to show the equivalence is to assume the truth of the entropy statement and derive the Kelvin–Planck statement (or the Clausius statement, which we proved equivalent to the Kelvin–Planck statement in Prob. 3.7) from the entropy statement.]

Section 3.6

3.25 Willard Rumpson (in later life Baron Melvin, K.C.B.) defined a temperature scale with the function ϕ in (3.43) as "take the square root" and with the water triple-point temperature defined as 200.00°M. (a) What is the temperature of the steam point on the Melvin scale? (b) What is the temperature of the ice point on the Melvin scale?

3.26 Let the Carnot-cycle reversible heat engine A absorb heat q_3 per cycle from a reservoir at τ_3 and discard heat $-q_{2A}$ per cycle to a reservoir at τ_2. Let Carnot engine B absorb heat q_{2B} per cycle from the reservoir at τ_2 and discard heat $-q_1$ per cycle to a reservoir at τ_1. Further, let $-q_{2A} = q_{2B}$, so that engine B absorbs an amount of heat from the τ_2 reservoir equal to the heat deposited in this reservoir by engine A. Show that

$$g(\tau_2, \tau_3)g(\tau_1, \tau_2) = -q_1/q_3$$

where the function g is defined as $1 - e_{rev}$. The heat reservoir at τ_2 can be omitted, and the combination of engines A and B can be viewed as a single Carnot engine operating between τ_3 and τ_1; hence $g(\tau_1, \tau_3) = -q_1/q_3$. Therefore

$$g(\tau_1, \tau_2) = \frac{g(\tau_1, \tau_3)}{g(\tau_2, \tau_3)} \tag{3.59}$$

Since τ_3 does not appear on the left side of (3.59), it must cancel out of the numerator and denominator on the right side. After τ_3 is canceled, the numerator takes the form $\phi(\tau_1)$ and the denominator takes the form $\phi(\tau_2)$, where ϕ is some function; we then have

$$g(\tau_1, \tau_2) = \frac{\phi(\tau_1)}{\phi(\tau_2)} \tag{3.60}$$

which is the desired result, Eq. (3.42). [A more rigorous derivation of (3.60) from (3.59) is given in *Denbigh*, p. 30.]

3.27 For the gaussian probability distribution, the probability of observing a value that deviates from the mean value by at least x standard deviations is given by the following infinite series (M. L. Abramowitz and I. A. Stegun, *Handbook of Mathematical Functions*, Natl. Bur. Stand. Appl. Math. Ser. 55, 1964, pp. 931–932):

$$\frac{2}{\sqrt{2\pi}} e^{-x^2/2} \left(\frac{1}{x} - \frac{1}{x^3} + \frac{3}{x^5} - \cdots \right)$$

where the series is useful for reasonably large values of x. (a) Show that 99.7% of observations lie within ± 3 standard deviations from the mean. (b) Calculate the probability of a deviation $\geq 10^6$ standard deviations.

3.28 If the probability of observing a certain event in a single trial is p, then clearly the probability of not observing it in one trial is $1 - p$. The probability of not observing it in n independent trials is then $(1 - p)^n$; the probability of observing it at least once in n independent trials is $1 - (1 - p)^n$. (a) Use these ideas to verify the calculation of Eq. (3.58). (b) How many times must a coin be tossed to reach a 99% probability of observing at least one head?

General

3.29 For each of the following sets of quantities, all the quantities except one have something in common. State what they have in common and state which quantity does not belong with the others. (In some cases, more than one answer for the property in common might be possible.) (a) H, U, q, S, T; (b) T, ΔS, q, w, ΔH; (c) q, w, U, ΔU, V, H; (d) ρ, S_m, M, V; (e) ΔH, ΔS, dV, ΔP; (f) U, V, ΔH, S, T.

3.30 Estimate the volume of cooling water used per minute by a 1000-MW power plant whose efficiency is 40%. Assume that the cooling water undergoes a 10°C temperature rise (a typical value) when it cools the steam.

3.31 A certain perfect gas has $C_{V,m} = a + bT$, where $a = 25.0$ J/(mol K) and $b = 0.0300$ J/(mol K²). Let 4.00 mol of this gas go from 300 K and 2.00 atm to 500 K and 3.00 atm. Calculate each of the following quantities for this change of state. If it is impossible to calculate a quantity from the given information, state this. (a) q; (b) w; (c) ΔU; (d) ΔH; (e) ΔS.

3.32 Classify each of these processes as reversible or irreversible: (a) freezing of water at 0°C and 1 atm; (b) freezing of supercooled water at -10°C and 1 atm; (c) burning of carbon in O_2 to give CO_2 at 800 K and 1 atm; (d) rolling a ball on a floor with friction; (e) the Joule–Thomson experiment; (f) adiabatic expansion of a gas into vacuum (the Joule experiment); (g) use of a frictionless piston to infinitely slowly increase the pressure on an equilibrium mixture of N_2, H_2, and NH_3, thereby shifting the equilibrium.

3.33 For each of the following pairs of systems, state which system (if either) has the greater U and which has the greater S. (a) 5 g of Fe at 20°C and 1 atm or 10 g of Fe at 20°C and 1 atm; (b) 2 g of liquid water at 25°C and 1 atm or 2 g of water vapor at 25°C and 20 torr; (c) 2 g of benzene at 25°C and 1 bar or 2 g of benzene at 40°C and 1 bar; (d) a system consisting of 2 g of metal M at 300 K and 1 bar and 2 g of M at 310 K and 1 bar or a system consisting of 4 g of M at 305 K and 1 bar. Assume the specific heat of M is constant over the 300 to 310 K range and the volume change of M is negligible over this range; (e) 1 mol of a perfect gas at 0°C and 1 atm or 1 mol of the same perfect gas at 0°C and 5 atm.

3.34 Which of these cyclic integrals must vanish for a closed system with P-V work only? (a) $\oint P\, dV$; (b) $\oint (P\, dV + V\, dP)$; (c) $\oint V\, dV$; (d) $\oint dq_{rev}/T$; (e) $\oint H\, dT$; (f) $\oint dU$; (g) $\oint dq_{rev}$; (h) $\oint dq_P$; (i) $\oint dw_{rev}$; (j) $\oint dw_{rev}/P$.

3.35 Consider the following quantities: C_P, $C_{P,m}$, R (the gas constant), k (Boltzmann's constant), q, U/T. (a) Which have the same dimensions as S? (b) Which have the same dimensions as S_m?

3.36 What is the relevance to thermodynamics of the following refrain from the Gilbert and Sullivan operetta *H.M.S. Pinafore*? "What, never? No, never! What, *never*? Well, hardly ever!"

3.37 In the tropics, water at the surface of the ocean is warmer than water well below the surface. Someone proposes to draw heat from the warm surface water, convert part of it to work, and discard the remainder to cooler water below the surface. Does this proposal violate the second law?

3.38 Use (3.15) to show that it is impossible to attain the absolute zero of temperature.

3.39 Suppose that an infinitesimal crystal of ice is added to 10.0 g of supercooled liquid water at -10.0°C in an adiabatic container and the system reaches equilibrium at a fixed pressure of 1 atm. (a) What is ΔH for the process? (b) The equilibrium state will contain some ice and will therefore consist either of ice plus liquid at 0°C or of ice at or below 0°C. Use the answer to (a) to deduce exactly what is present at equilibrium. (c) Calculate ΔS for the process. (See Prob. 2.48 for data.)

3.40 Give the SI units of (a) S; (b) S_m; (c) q; (d) P; (e) M_r (molecular weight); (f) M (molar mass).

3.41 Which of the following statements can be proved from the second law of thermodynamics? (a) For any closed system, equilibrium corresponds to the position of maximum entropy of the system. (b) The entropy of an isolated system must remain constant. (c) For a system enclosed in impermeable adiabatic walls, the system's entropy is maximized at equilibrium. (d) The entropy of a closed system can never decrease. (e) The entropy of an isolated system can never decrease.

3.42 True or false? (a) For every process in an isolated system, $\Delta T = 0$. (b) For every process in an isolated system that has no macroscopic kinetic or potential energy, $\Delta U = 0$. (c) For every process in an isolated system, $\Delta S = 0$. (d) If a closed system undergoes a reversible process for which $\Delta V = 0$, then the P-V work done on the system in this process must be zero. (e) ΔS when 1 mol of $N_2(g)$ goes irreversibly from 25°C and 10 L to 25°C and 20 L must be the same as ΔS when 1 mol of $N_2(g)$ goes reversibly from 25°C and 10 L to 25°C and 20 L. (f) $\Delta S = 0$ for every adiabatic process in a closed system. (g) For every reversible process in a closed system, $\Delta S = \Delta H/T$. (h) A closed-system process that has $\Delta T = 0$, must have $\Delta U = 0$. (i) For every isothermal process in a closed system, $\Delta S = \Delta H/T$. (j) $q = 0$ for every isothermal process in a closed system. (k) In every cyclic process, the final and initial states of the system are the same and the final and initial states of the surroundings are the same.

CHAPTER

4

Material Equilibrium

The zeroth, first, and second laws of thermodynamics give us the state functions T, U, and S. The second law enables us to determine whether a given process is possible. A process that decreases S_{univ} is impossible; one that increases S_{univ} is possible and irreversible. Reversible processes have $\Delta S_{univ} = 0$. Such processes are possible in principle but difficult to achieve in practice. Our aim in this chapter is to use this entropy criterion to derive specific conditions for material equilibrium in a nonisolated system. These conditions will be formulated in terms of state functions of the system.

4.1   MATERIAL EQUILIBRIUM

Material equilibrium (Sec. 1.2) means that in each phase of the closed system, the number of moles of each substance present remains constant in time. Material equilibrium is subdivided into (*a*) **reaction equilibrium,** which is equilibrium with respect to conversion of one set of chemical species to another set, and (*b*) **phase equilibrium,** which is equilibrium with respect to transport of matter between phases of the system without conversion of one species to another. (Recall from Sec. 1.2 that a phase is a homogeneous portion of a system.) The condition for material equilibrium will be derived in Sec. 4.7 and will be applied to phase equilibrium in Sec. 4.8 and to reaction equilibrium in Sec. 4.9.

To aid in discussing material equilibrium, we shall introduce two new state functions in Sec. 4.4, the Helmholtz function $A \equiv U - TS$ and the Gibbs function $G \equiv H - TS$. It turns out that the conditions for reaction equilibrium and phase equilibrium are most conveniently formulated in terms of state functions called the chemical potentials (Sec. 4.7), which are closely related to G.

A second theme of this chapter is the use of the combined first and second laws to derive expressions for thermodynamic quantities in terms of readily measured properties (Secs. 4.5 and 4.6).

     The initial application of the laws of thermodynamics to material equilibrium is largely the work of Josiah Willard Gibbs (1839–1903). Gibbs received his doctorate in engineering from Yale in 1863 with a thesis on gear design. From 1866 to 1869 Gibbs studied mathematics and physics in Europe. In 1871 he was appointed Professor of Mathematical Physics, without salary, at Yale. At that time his only published work was a railway brake patent. In 1876–1878 he published in the *Transactions of the Connecticut Academy of Arts and Sciences* a 300-page monograph titled "On the Equilibrium of Heterogeneous Substances." This work used the first and second laws of thermodynamics to deduce the conditions of material equilibrium. Gibbs' second major contribution was his book *Elementary Principles in Statistical Mechanics* (1902), which laid much of the foundation of statistical mechanics. Gibbs also developed vector analysis. Gibbs' life was rather uneventful; he never married and lived in his family's house until his death. Ostwald wrote of Gibbs: "To physical chemistry he gave form and content for a hundred years." Planck wrote that Gibbs "will ever be reckoned among the most renowned theoretical physicists of all times. . . ."

4.2 THERMODYNAMIC PROPERTIES OF NONEQUILIBRIUM SYSTEMS

This chapter deals with systems in which a chemical reaction or transport of matter from one phase to another is occurring. Since such systems are not in thermodynamic equilibrium, we first examine to what extent we can assign definite values of thermodynamic properties to nonequilibrium systems.

Consider a system that is not in material equilibrium but is in mechanical and thermal equilibrium, with P and T uniform throughout the system. We shall assume that within each phase of the system the composition is uniform. We thus assume that (*a*) the rate of diffusion within a phase is rapid compared with the rate of transport of components from one phase to another, and (*b*) any chemical reactions do not occur at an explosive rate, which would destroy thermal and mechanical equilibrium. Our aim is to find the equilibrium *position*; thermodynamics can give no information on the *rate* of a process. Since the final equilibrium position is independent of the rate at which equilibrium is reached (provided this rate is nonzero), we are free to make convenient assumptions about rates.

We first consider systems not in phase equilibrium. For example, the system of Fig. 4.1 initially consists of a large crystal of NaCl separated by a partition from an unsaturated solution of NaCl in water, with P and T held fixed. Since U and S are extensive, we have

$$U = U_{soln} + U_{NaCl}, \qquad S = S_{soln} + S_{NaCl} \qquad (4.1)$$

Now the frictionless partition is removed. It requires only an infinitesimal force to do this, and the removal is done reversibly and adiabatically. q and w for the removal of the partition are zero. Therefore ΔU and ΔS are zero for its removal. Thus, immediately after the removal of the partition, Eq. (4.1) still holds. The instant after the partition is removed, we no longer have phase equilibrium, since the solid NaCl starts to dissolve in the unsaturated solution. Despite this lack of phase equilibrium, we have shown that it is still meaningful to ascribe values to U and S for the system, namely, the values (4.1). Of course, as NaCl dissolves in the solution, the values of U and S change, but at any concentration of dissolved NaCl, we can imagine replacing the partition without changing U and S, and then we can apply (4.1). Therefore (4.1) is valid at any stage of the solution process. Even though the system is not in phase equilibrium when the partition is absent, we can still ascribe values of U and S to it, namely, the values we would assign if the partition were present.

Now consider systems not in reaction equilibrium. Suppose H_2, O_2, and H_2O gases are mixed. Provided no catalyst is present and the temperature is moderate, the gases will not react when mixed. We can use the first law to measure the ΔU of the mixing. Also, using semipermeable membranes (Sec. 3.4), we can do the mixing reversibly and hence measure ΔS for the mixing process. It therefore makes sense to ascribe definite values of U and S to the mixture for any composition whatever. However, the mixture is not necessarily at reaction equilibrium. If we add the appropriate catalyst, the gases will react, changing the mixture's composition. At any point during the reaction, we can withdraw the catalyst, stopping the reaction. At this new

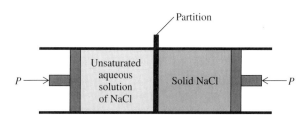

Figure 4.1

When the partition is removed, the system is not in phase equilibrium.

composition, we can ascribe new values to U and S of the mixture. These values can be determined by measurements done in a reversible separation process.

We conclude that *values of U and S can be assigned to a system that is in mechanical and thermal equilibrium and has a uniform composition in each phase, even though the system is not in material equilibrium.* Of course, such systems also have well-defined values of P, V, and T.

We can go even further. Thus suppose a system lacks thermal equilibrium and has T varying from one end to the other. We can imagine the system cut into "infinitesimal" slices such that the temperature within each slice is essentially constant. We can then assign values of thermodynamic variables (T, P, V, U, S, composition) to each slice. The total S and U of the system is the sum of the values for the slices. Since thermodynamics is a macroscopic science, each thin slice must contain enough molecules to make it meaningful to assign it a macroscopic property like temperature. The number of molecules in each slice should be much, much greater than 1 but much, much less than 10^{23}. For a system with a concentration gradient, a similar imaginary division into tiny parts can be done.

4.3 ENTROPY AND EQUILIBRIUM

Consider an *isolated* system that is not at material equilibrium. The spontaneous chemical reactions or transport of matter between phases that are occurring in this system are irreversible processes that increase the entropy. These processes continue until the system's entropy is maximized. Once S is maximized, any further processes can only decrease S, which would violate the second law. The criterion for equilibrium in an *isolated* system is maximization of the system's entropy S.

When we deal with material equilibrium in a closed system, the system is ordinarily not isolated. Instead, it can exchange heat and work with its surroundings. Under these conditions, we can take the system itself *plus* the surroundings with which it interacts to constitute an isolated system, and *the condition for material equilibrium in the system is then maximization of the total entropy of the system plus its surroundings:*

$$S_{\text{syst}} + S_{\text{surr}} \text{ a maximum at equilib.} \qquad (4.2)^*$$

Chemical reactions and transport of matter between phases continue in a system until $S_{\text{syst}} + S_{\text{surr}}$ has been maximized.

It is usually most convenient to deal with properties of the system and not have to worry about changes in the thermodynamic properties of the surroundings as well. Thus, although the criterion (4.2) for material equilibrium is perfectly valid and general, it will be more useful to have a criterion for material equilibrium that refers only to thermodynamic properties of the system itself. Since S_{syst} is a maximum at equilibrium only for an isolated system, consideration of the entropy of the system does not furnish us with an equilibrium criterion. We must look for another system state function to find the equilibrium criterion.

Reaction equilibrium is ordinarily studied under one of two conditions. For reactions that involve gases, the chemicals are usually put in a container of fixed volume, and the system is allowed to reach equilibrium at constant T and V in a constant-temperature bath. For reactions in liquid solutions, the system is usually held at atmospheric pressure and allowed to reach equilibrium at constant T and P.

To find equilibrium criteria for these conditions, consider Fig. 4.2. The closed system at temperature T is placed in a bath also at T. The system and surroundings are isolated from the rest of the world. The system is not in material equilibrium but is in mechanical and thermal equilibrium. The surroundings are in material, mechanical, and thermal equilibrium. System and surroundings can exchange energy (as heat and

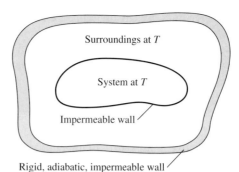

Figure 4.2

A closed system that is in mechanical and thermal equilibrium but not in material equilibrium.

work) but not matter. Let chemical reaction or transport of matter between phases or both be occurring in the system at rates small enough to maintain thermal and mechanical equilibrium. Let heat dq_{syst} flow into the system as a result of the changes that occur in the system during an infinitesimal time period. For example, if an endothermic chemical reaction is occurring, dq_{syst} is positive. Since system and surroundings are isolated from the rest of the world, we have

$$dq_{surr} = -dq_{syst} \qquad (4.3)$$

Since the chemical reaction or matter transport within the nonequilibrium system is irreversible, dS_{univ} must be positive [Eq. (3.39)]:

$$dS_{univ} = dS_{syst} + dS_{surr} > 0 \qquad (4.4)$$

for the process. (Recall from Sec. 4.2 that one can meaningfully assign an entropy to a system that is not in material equilibrium.) The surroundings are in thermodynamic equilibrium throughout the process. Therefore, as far as the surroundings are concerned, the heat transfer is reversible, and [Eq. (3.20)]

$$dS_{surr} = dq_{surr}/T \qquad (4.5)$$

However, the system is not in thermodynamic equilibrium, and the process involves an irreversible change in the system. Therefore $dS_{syst} \neq dq_{syst}/T$. Equations (4.3) to (4.5) give $dS_{syst} > -dS_{surr} = -dq_{surr}/T = dq_{syst}/T$. Therefore

$$dS_{syst} > dq_{syst}/T$$
$$dS > dq_{irrev}/T \qquad \text{closed syst. in therm. and mech. equilib.} \qquad (4.6)$$

where we dropped the subscript syst from S and q since, by convention, unsubscripted symbols refer to the system. [Note that the thermal- and mechanical-equilibrium condition in (4.6) does not necessarily mean that T and P are held constant. For example, an exothermic reaction can raise the temperature of the system and the surroundings, but thermal equilibrium can be maintained provided the reaction is extremely slow.]

When the system has reached material equilibrium, any infinitesimal process is a change from a system at equilibrium to one infinitesimally close to equilibrium and hence is a reversible process. Thus, at material equilibrium we have

$$dS = dq_{rev}/T \qquad (4.7)$$

Combining (4.7) and (4.6), we have

$$dS \geq \frac{dq}{T} \qquad \text{material change, closed syst. in mech. and therm. equilib.} \qquad (4.8)$$

where the equality sign holds only when the system is in material equilibrium. For a reversible process, dS equals dq/T. For an irreversible chemical reaction or phase change, dS is greater than dq/T because of the extra disorder created in the system by the irreversible material change.

The first law for a closed system is $dq = dU - dw$. Multiplication of (4.8) by T (which is positive) gives $dq \leq T\, dS$. Hence for a closed system in mechanical and thermal equilibrium, we have $dU - dw \leq T\, dS$, or

$$dU \leq T\, dS + dw \qquad \begin{array}{l}\text{material change, closed syst. in}\\ \text{mech. and therm. equilib.}\end{array} \qquad (4.9)$$

where the equality sign applies only at material equilibrium.

4.4 THE GIBBS AND HELMHOLTZ ENERGIES

We now use (4.9) to deduce conditions for material equilibrium in terms of state functions of the system. We first examine material equilibrium in a system held at constant T and V. Here $dV = 0$ and $dT = 0$ throughout the irreversible approach to equilibrium. The inequality (4.9) involves dS and dV, since $dw = -P\, dV$ for P-V work only. To introduce dT into (4.9), we add and subtract $S\, dT$ on the right. Note that $S\, dT$ has the dimensions of entropy times temperature, the same dimensions as the term $T\, dS$ that appears in (4.9), so we are allowed to add and subtract $S\, dT$. We have

$$dU \leq T\, dS + S\, dT - S\, dT + dw \qquad (4.10)$$

The differential relation $d(uv) = u\, dv + v\, du$ [Eq. (1.28)] gives $d(TS) = T\, dS + S\, dT$, and Eq. (4.10) becomes

$$dU \leq d(TS) - S\, dT + dw \qquad (4.11)$$

The relation $d(u + v) = du + dv$ [Eq. (1.28)] gives $dU - d(TS) = d(U - TS)$, and (4.11) becomes

$$d(U - TS) \leq -S\, dT + dw \qquad (4.12)$$

If the system can do only P-V work, then $dw = -P\, dV$ (we use dw_{rev} since we are assuming mechanical equilibrium). We have

$$d(U - TS) \leq -S\, dT - P\, dV \qquad (4.13)$$

At constant T and V, we have $dT = 0 = dV$ and (4.13) becomes

$$d(U - TS) \leq 0 \qquad \begin{array}{l}\text{const. } T \text{ and } V, \text{ closed syst. in}\\ \text{therm. and mech. equilib., } P\text{-}V \text{ work only}\end{array} \qquad (4.14)$$

where the equality sign holds at material equilibrium.

Therefore, for a closed system held at constant T and V, the state function $U - TS$ continually decreases during the spontaneous, irreversible processes of chemical reaction and matter transport between phases until material equilibrium is reached. At material equilibrium, $d(U - TS)$ equals 0, and $U - TS$ has reached a minimum. Any spontaneous change at constant T and V away from equilibrium (in either direction) would mean an increase in $U - TS$, which, working back through the preceding equations from (4.14) to (4.4), would mean a decrease in $S_{univ} = S_{syst} + S_{surr}$. This decrease would violate the second law. The approach to and achievement of material equilibrium is a consequence of the second law.

The condition for material equilibrium in a closed system capable of doing only P-V work and held at constant T and V is minimization of the system's state function $U - TS$. This state function is called the **Helmholtz free energy,** the **Helmholtz energy,** the **Helmholtz function,** or the **work function** and is symbolized by A:

$$A \equiv U - TS \qquad (4.15)^*$$

Now consider material equilibrium for constant T and P conditions, $dP = 0$, $dT = 0$. To introduce dP and dT into (4.9) with $dw = -P\,dV$, we add and subtract $S\,dT$ and $V\,dP$:

$$dU \leq T\,dS + S\,dT - S\,dT - P\,dV + V\,dP - V\,dP$$

$$dU \leq d(TS) - S\,dT - d(PV) + V\,dP$$

$$d(U + PV - TS) \leq -S\,dT + V\,dP$$

$$d(H - TS) \leq -S\,dT + V\,dP$$

Therefore, for a material change at constant T and P in a closed system in mechanical and thermal equilibrium and capable of doing only P-V work, we have

$$d(H - TS) \leq 0 \qquad \text{const. } T, P \tag{4.16}$$

where the equality sign holds at material equilibrium.

Thus, the state function $H - TS$ continually decreases during material changes at constant T and P until equilibrium is reached. The condition for material equilibrium at constant T and P in a closed system doing P-V work only is minimization of the system's state function $H - TS$. This state function is called the **Gibbs function,** the **Gibbs energy,** or the **Gibbs free energy** and is symbolized by G:

$$G \equiv H - TS \equiv U + PV - TS \tag{4.17}*$$

G decreases during the approach to equilibrium at constant T and P, reaching a minimum at equilibrium (Fig. 4.3). As G of the system decreases at constant T and P, S_{univ} increases [see Eq. (4.21)]. Since U, V, and S are extensive, G is extensive.

Both A and G have units of energy (J or cal). However, they are not energies in the sense of being conserved. $G_{\text{syst}} + G_{\text{surr}}$ need not be constant in a process, nor need $A_{\text{syst}} + A_{\text{surr}}$ remain constant. Note that A and G are defined for any system to which meaningful values of U, T, S, P, V can be assigned, not just for systems held at constant T and V or constant T and P.

Summarizing, we have shown that:

In a closed system capable of doing only P-V work, the constant-T-and-V material-equilibrium condition is the minimization of the Helmholtz function A, and the constant-T-and-P material-equilibrium condition is the minimization of the Gibbs function G:

$$dA = 0 \qquad \text{at equilib., const. } T, V \tag{4.18}*$$

$$dG = 0 \qquad \text{at equilib., const. } T, P \tag{4.19}*$$

where dG is the infinitesimal change in G due to an infinitesimal amount of chemical reaction or phase change at constant T and P.

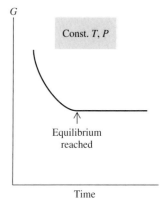

Figure 4.3

For a closed system with P-V work only, the Gibbs energy is minimized if equilibrium is reached under conditions of constant T and P.

EXAMPLE 4.1 ΔG and ΔA for a phase change

Calculate ΔG and ΔA for the vaporization of 1.00 mol of H_2O at 1.00 atm and 100°C. Use data from Prob. 2.48.

We have $G \equiv H - TS$. For this process, T is constant and $\Delta G = G_2 - G_1 = H_2 - TS_2 - (H_1 - TS_1) = \Delta H - T\,\Delta S$:

$$\Delta G = \Delta H - T\,\Delta S \qquad \text{const. } T \tag{4.20}$$

The process is reversible and isothermal, so $dS = dq/T$ and $\Delta S = q/T$. Since P is constant and only P-V work is done, we have $\Delta H = q_P = q$. Therefore (4.20) gives $\Delta G = q - T(q/T) = 0$. The result $\Delta G = 0$ makes sense because a reversible (equilibrium) process in a system at constant T and P has $dG = 0$ [Eq. (4.19)].

From $A \equiv U - TS$, we get $\Delta A = \Delta U - T \Delta S$ at constant T. Use of $\Delta U = q + w$ and $\Delta S = q/T$ gives $\Delta A = q + w - q = w$. The work is reversible P-V work at constant pressure, so $w = -\int_1^2 P\, dV = -P\,\Delta V$. From the 100°C density in Prob. 2.48, the molar volume of $H_2O(l)$ at 100°C is 18.8 cm³/mol. We can accurately estimate V_m of the gas from the ideal-gas law: $V_m = RT/P = 30.6 \times 10^3$ cm³/mol. Therefore $\Delta V = 30.6 \times 10^3$ cm³ and

$$w = (-30.6 \times 10^3 \text{ cm}^3 \text{ atm})(8.314 \text{ J})/(82.06 \text{ cm}^3 \text{ atm}) = -3.10 \text{ kJ} = \Delta A$$

Exercise

Find ΔG and ΔA for the freezing of 1.00 mol of H_2O at 0°C and 1 atm. Use data from Prob. 2.48. (*Answer:* 0, -0.16_5 J.)

What is the relation between the minimization-of-G equilibrium condition at constant T and P and the maximization-of-S_{univ} equilibrium condition? Consider a system in mechanical and thermal equilibrium undergoing an irreversible chemical reaction or phase change at constant T and P. Since the surroundings undergo a reversible isothermal process, $\Delta S_{surr} = q_{surr}/T = -q_{syst}/T$. Since P is constant, $q_{syst} = \Delta H_{syst}$ and $\Delta S_{surr} = -\Delta H_{syst}/T$. We have $\Delta S_{univ} = \Delta S_{surr} + \Delta S_{syst}$ and

$$\Delta S_{univ} = -\Delta H_{syst}/T + \Delta S_{syst} = -(\Delta H_{syst} - T \Delta S_{syst})/T = -\Delta G_{syst}/T$$

$$\Delta S_{univ} = -\Delta G_{syst}/T \qquad \text{closed syst., const. } T \text{ and } P, P\text{-}V \text{ work only} \qquad (4.21)$$

where (4.20) was used. *The decrease in G_{syst} as the system proceeds to equilibrium at constant T and P corresponds to a proportional increase in S_{univ}.* The occurrence of a reaction is favored by having ΔS_{syst} positive and by having ΔS_{surr} positive. Having ΔH_{syst} negative (an exothermic reaction) favors the reaction's occurrence because the heat transferred to the surroundings increases the entropy of the surroundings ($\Delta S_{surr} = -\Delta H_{syst}/T$).

The names "work function" and "Gibbs free energy" arise as follows. Let us drop the restriction that only P-V work be performed. From (4.12) we have for a closed system in thermal and mechanical equilibrium that $dA \leq -S\,dT + dw$. For a constant-temperature process in such a system, $dA \leq dw$. For a finite isothermal process, $\Delta A \leq w$. Our convention is that w is the work done *on* the system. The work w_{by} done *by* the system on its surroundings is $w_{by} = -w$, and $\Delta A \leq -w_{by}$ for an isothermal process. Multiplication of an inequality by -1 reverses the direction of the inequality; therefore

$$w_{by} \leq -\Delta A \qquad \text{const. } T, \text{ closed syst.} \qquad (4.22)$$

The term "work function" (*Arbeitsfunktion*) for A arises from (4.22). The work done by the system in an isothermal process is less than or equal to the negative of the change in the state function A. The equality sign in (4.22) holds for a reversible process. Moreover, $-\Delta A$ is a fixed quantity for a given change of state. Hence the maximum work output by a closed system for an *isothermal* process between two given states is obtained when the process is carried out reversibly.

Note that the work w_{by} done by a system can be greater than or less than $-\Delta U$, the internal energy decrease of the system. For any process in a closed system, $w_{by} = -\Delta U + q$. The heat q that flows into the system is the source of energy that allows w_{by} to differ from $-\Delta U$. Recall the Carnot cycle, where $\Delta U = 0$ and $w_{by} > 0$.

Now consider G. From $G = A + PV$, we have $dG = dA + P\,dV + V\,dP$, and use of (4.12) for dA gives $dG \le -S\,dT + dw + P\,dV + V\,dP$ for a closed system in thermal and mechanical equilibrium. For a process at constant T and P in such a system

$$dG \le dw + P\,dV \qquad \text{const. } T \text{ and } P \text{, closed syst.} \tag{4.23}$$

Let us divide the work into P-V work and non-P-V work $w_{\text{non-}P\text{-}V}$. (The most common kind of $w_{\text{non-}P\text{-}V}$ is electrical work.) If the P-V work is done in a mechanically reversible manner, then $dw = -P\,dV + dw_{\text{non-}P\text{-}V}$; Eq. (4.23) becomes $dG \le dw_{\text{non-}P\text{-}V}$ or $\Delta G \le w_{\text{non-}P\text{-}V} = -w_{\text{by,non-}P\text{-}V}$. Therefore

$$\Delta G \le w_{\text{non-}P\text{-}V} \quad \text{and} \quad w_{\text{by, non-}P\text{-}V} \le -\Delta G \qquad \text{const. } T \text{ and } P \text{, closed syst.} \tag{4.24}$$

For a reversible change, the equality sign holds and $w_{\text{by,non-}P\text{-}V} = -\Delta G$. In many cases (for example, a battery, a living organism), the P-V expansion work is not useful work, but $w_{\text{by,non-}P\text{-}V}$ is the useful work output. The quantity $-\Delta G$ *equals the maximum possible nonexpansion work output* $w_{\text{by,non-}P\text{-}V}$ *done by a system in a constant-T-and-P process.* Hence the term "free energy." (Of course, for a system with P-V work only, $dw_{\text{by,non-}P\text{-}V} = 0$ and $dG = 0$ for a reversible, isothermal, isobaric process.) Examples of nonexpansion work in biological systems are the work of contracting muscles and of transmitting nerve impulses (Sec. 14.16).

Summary

The maximization of S_{univ} leads to the following equilibrium conditions. When a closed system capable of only P-V work is held at constant T and V, the condition for material equilibrium (meaning phase equilibrium and reaction equilibrium) is that the Helmholtz function A (defined by $A \equiv U - TS$) is minimized. When such a system is held at constant T and P, the material-equilibrium condition is the minimization of the Gibbs function $G \equiv H - TS$.

4.5 THERMODYNAMIC RELATIONS FOR A SYSTEM IN EQUILIBRIUM

In the last section we introduced two new thermodynamic state functions, A and G. We shall apply the conditions (4.18) and (4.19) for material equilibrium in Sec. 4.7. Before doing so, we investigate the properties of A and G. In fact, in this section we shall consider the broader question of the thermodynamic relations between all state functions in systems in equilibrium. Since a system undergoing a reversible process is passing through only equilibrium states, we shall be considering reversible processes in this section.

Basic Equations

All thermodynamic state-function relations can be derived from six basic equations. The first law for a closed system is $dU = dq + dw$. If only P-V work is possible, and if the work is done reversibly, then $dw = dw_{\text{rev}} = -P\,dV$. For a reversible process, the relation $dS = dq_{\text{rev}}/T$ [Eq. (3.20)] gives $dq = dq_{\text{rev}} = T\,dS$. Hence, under these conditions, $dU = T\,dS - P\,dV$. This is the first basic equation; it combines the first and second laws. The next three basic equations are the definitions of H, A, and G [Eqs. (2.45), (4.15), and (4.17)]. Finally, we have the C_P and C_V equations $C_V = dq_V/dT = (\partial U/\partial T)_V$ and $C_P = dq_P/dT = (\partial H/\partial T)_P$ [Eqs. (2.51) to (2.53)]. The six basic equations are

$$dU = T\,dS - P\,dV \qquad \text{closed syst., rev. proc., } P\text{-}V \text{ work only} \tag{4.25}*$$

$$H \equiv U + PV \tag{4.26}*$$

$$A \equiv U - TS \tag{4.27}*$$

$$G \equiv H - TS \tag{4.28}*$$

$$C_V = \left(\frac{\partial U}{\partial T} \right)_V \qquad \text{closed syst. in equilib., } P\text{-}V \text{ work only} \tag{4.29}*$$

$$C_P = \left(\frac{\partial H}{\partial T} \right)_P \qquad \text{closed syst. in equilib., } P\text{-}V \text{ work only} \tag{4.30}*$$

The heat capacities C_V and C_P have alternative expressions that are also basic equations. Consider a reversible flow of heat accompanied by a temperature change dT. By definition, $C_X = dq_X/dT$, where X is the variable (P or V) held constant. But $dq_{\text{rev}} = T \, dS$, and we have $C_X = T \, dS/dT$, where dS/dT is for constant X. Putting X equal to V and P, we have

$$C_V = T \left(\frac{\partial S}{\partial T} \right)_V, \qquad C_P = T \left(\frac{\partial S}{\partial T} \right)_P \qquad \text{closed syst. in equilib.} \tag{4.31}*$$

The heat capacities C_P and C_V are key properties since they allow us to find the rates of change of U, H, and S with respect to temperature [Eqs. (4.29) to (4.31)].

The relation $dU = T \, dS - P \, dV$ in (4.25) applies to a reversible process in a closed system. Let us consider processes that change the system's composition. There are two ways the composition can change. First, one can add or remove one or more substances. However, the requirement of a closed system ($dU \neq dq + dw$ for an open system) rules out addition or removal of matter. Second, the composition can change by chemical reactions or by transport of matter from one phase to another in the system. The usual way of carrying out a chemical reaction is to mix the chemicals and allow them to reach equilibrium. This spontaneous chemical reaction is irreversible, since the system passes through nonequilibrium states. The requirement of reversibility ($dq \neq T \, dS$ for an irreversible chemical change) rules out a chemical reaction as ordinarily conducted. Likewise, if we put several phases together and allow them to reach equilibrium, we have an irreversible composition change. For example, if we throw a handful of salt into water, the solution process goes through nonequilibrium states and is irreversible. The equation $dU = T \, dS - P \, dV$ does not apply to such irreversible composition changes in a closed system.

We can, if we like, carry out a composition change reversibly in a closed system. If we start with a system that is initially in material equilibrium and reversibly vary the temperature or pressure, we generally get a shift in the equilibrium position, and this shift is reversible. For example, if we have an equilibrium mixture of N_2, H_2, and NH_3 (together with a catalyst) and we slowly and reversibly vary T or P, the position of chemical-reaction equilibrium shifts. This composition change is reversible, since the closed system passes through equilibrium states only. For such a reversible composition change, $dU = T \, dS - P \, dV$ does apply.

This section deals only with reversible processes in closed systems. Most commonly, the system's composition is fixed, but the equations of this section also apply to processes where the composition of the closed system changes reversibly, with the system passing through equilibrium states only.

The Gibbs Equations
We now derive expressions for dH, dA, and dG that correspond to $dU = T \, dS - P \, dV$ [Eq. (4.25)] for dU. From $H \equiv U + PV$ and $dU = T \, dS - P \, dV$, we have

$$dH = d(U + PV) = dU + d(PV) = dU + P \, dV + V \, dP$$

$$= (T \, dS - P \, dV) + P \, dV + V \, dP$$

$$dH = T \, dS + V \, dP \tag{4.32}$$

Similarly,

$$dA = d(U - TS) = dU - T\,dS - S\,dT = T\,dS - P\,dV - T\,dS - S\,dT$$

$$= -S\,dT - P\,dV$$

$$dG = d(H - TS) = dH - T\,dS - S\,dT = T\,dS + V\,dP - T\,dS - S\,dT$$

$$= -S\,dT + V\,dP$$

where (4.32) was used.

Collecting the expressions for dU, dH, dA, and dG, we have

$$dU = T\,dS - P\,dV \qquad \qquad \text{(4.33)}*$$

$$dH = T\,dS + V\,dP \qquad \qquad \text{(4.34)}$$

$$\left. \begin{array}{l} \\ \\ \\ \end{array} \right\} \begin{array}{l} \text{closed syst., rev. proc.,} \\ \text{P-V work only} \end{array}$$

$$dA = -S\,dT - P\,dV \qquad \qquad \text{(4.35)}$$

$$dG = -S\,dT + V\,dP \qquad \qquad \text{(4.36)}*$$

These are the **Gibbs equations.** The first can be written down from the first law $dU = dq + dw$ and knowledge of the expressions for dw_{rev} and dq_{rev}. The other three can be quickly derived from the first by use of the definitions of H, A, and G. Thus they need not be memorized. The expression for dG is used so often, however, that it saves time to memorize it.

The Gibbs equation $dU = T\,dS - P\,dV$ implies that U is being considered a function of the variables S and V. From $U = U(S, V)$, we have [Eq. (1.30)]

$$dU = \left(\frac{\partial U}{\partial S} \right)_V dS + \left(\frac{\partial U}{\partial V} \right)_S dV$$

Since dS and dV are arbitrary and independent of each other, comparison of this equation with $dU = T\,dS - P\,dV$ gives

$$\left(\frac{\partial U}{\partial S} \right)_V = T, \qquad \left(\frac{\partial U}{\partial V} \right)_S = -P \qquad \qquad \text{(4.37)}$$

A quick way to get these two equations is to first put $dV = 0$ in $dU = T\,dS - P\,dV$ to give $(\partial U/\partial S)_V = T$ and then put $dS = 0$ in $dU = T\,dS - P\,dV$ to give $(\partial U/\partial V)_S = -P$. [Note from the first equation in (4.37) that an increase in internal energy at constant volume will always increase the entropy.] The other three Gibbs equations (4.34) to (4.36) give in a similar manner $(\partial H/\partial S)_P = T$, $(\partial H/\partial P)_S = V$, $(\partial A/\partial T)_V = -S$, $(\partial A/\partial V)_T = -P$, and

$$(\partial G/\partial T)_P = -S, \qquad (\partial G/\partial P)_T = V \qquad \qquad \text{(4.38)}$$

Our aim is to be able to express any thermodynamic property of an equilibrium system in terms of easily measured quantities. *The power of thermodynamics is that it enables properties that are difficult to measure to be expressed in terms of easily measured properties.* The easily measured properties most commonly used for this purpose are [Eqs. (1.43) and (1.44)]

$$C_P(T, P), \qquad \alpha(T, P) \equiv \frac{1}{V}\left(\frac{\partial V}{\partial T} \right)_P, \qquad \kappa(T, P) \equiv -\frac{1}{V}\left(\frac{\partial V}{\partial P} \right)_T \qquad \text{(4.39)}*$$

Since these are state functions, they are functions of T, P, and composition. We are considering mainly constant-composition systems, so we omit the composition dependence. Note that α and κ can be found from the equation of state $V = V(T, P)$ if this is known.

The Euler Reciprocity Relation

To relate a desired property to C_P, α, and κ, we use the basic equations (4.25) to (4.31) and mathematical partial-derivative identities. Before proceeding, there is another partial-derivative identity we shall need. If z is a function of x and y, then [Eq. (1.30)]

$$dz = \left(\frac{\partial z}{\partial x}\right)_y dx + \left(\frac{\partial z}{\partial y}\right)_x dy \equiv M\, dx + N\, dy \qquad (4.40)$$

where we defined the functions M and N as

$$M \equiv (\partial z/\partial x)_y, \qquad N \equiv (\partial z/\partial y)_x \qquad (4.41)$$

From Eq. (1.36), the order of partial differentiation does not matter:

$$\frac{\partial}{\partial y}\left(\frac{\partial z}{\partial x}\right) = \frac{\partial}{\partial x}\left(\frac{\partial z}{\partial y}\right) \qquad (4.42)$$

Hence if $dz = M\, dx + N\, dy$, Eqs. (4.40) to (4.42) give

$$\left(\frac{\partial M}{\partial y}\right)_x = \left(\frac{\partial N}{\partial x}\right)_y \qquad \textbf{(4.43)*}$$

Equation (4.43) is the **Euler reciprocity relation.**

The Maxwell Relations

The Gibbs equation (4.33) for dU is

$$dU = T\, dS - P\, dV = M\, dx + N\, dy \quad \text{where } M \equiv T, N \equiv -P, x \equiv S, y \equiv V$$

The Euler relation $(\partial M/\partial y)_x = (\partial N/\partial x)_y$ gives

$$(\partial T/\partial V)_S = [\partial(-P)/\partial S]_V = -(\partial P/\partial S)_V$$

Application of the Euler relation to the other three Gibbs equations gives three more thermodynamic relations. We find (Prob. 4.5)

$$\left(\frac{\partial T}{\partial V}\right)_S = -\left(\frac{\partial P}{\partial S}\right)_V, \qquad \left(\frac{\partial T}{\partial P}\right)_S = \left(\frac{\partial V}{\partial S}\right)_P \qquad (4.44)$$

$$\left(\frac{\partial S}{\partial V}\right)_T = \left(\frac{\partial P}{\partial T}\right)_V, \qquad \left(\frac{\partial S}{\partial P}\right)_T = -\left(\frac{\partial V}{\partial T}\right)_P \qquad (4.45)$$

These are the **Maxwell relations** (after James Clerk Maxwell, one of the greatest of nineteenth-century physicists). The first two Maxwell relations are little used. The last two are extremely valuable, since they relate the isothermal pressure and volume variations of entropy to measurable properties.

The equations in (4.45) are examples of the powerful and remarkable relationships that thermodynamics gives us. Suppose we want to know the effect of an isothermal pressure change on the entropy of a system. We cannot check out an entropy meter from the stockroom to monitor S as P changes. However, the relation $(\partial S/\partial P)_T = -(\partial V/\partial T)_P$ in (4.45) tells us that all we have to do is measure the rate of change of the system's volume with temperature at constant P, and this simple measurement enables us to calculate the rate of change of the system's entropy with respect to pressure at constant T.

Dependence of State Functions on T, P, and V

We now find the dependence of U, H, S, and G on the variables of the system. The most common independent variables are T and P. We shall relate the temperature and pressure variations of H, S, and G to the directly measurable properties C_P, α, and κ.

For U, the quantity $(\partial U/\partial V)_T$ occurs more often than $(\partial U/\partial P)_T$, so we shall find the temperature and volume variations of U.

Volume Dependence of U

We want $(\partial U/\partial V)_T$. The Gibbs equation (4.33) gives $dU = T\,dS - P\,dV$. The partial derivative $(\partial U/\partial V)_T$ corresponds to an isothermal process. For an isothermal process, the equation $dU = T\,dS - P\,dV$ becomes

$$dU_T = T\,dS_T - P\,dV_T \qquad (4.46)$$

where the T subscripts indicate that the infinitesimal changes dU, dS, and dV are for a constant-T process. Since $(\partial U/\partial V)_T$ is wanted, we divide (4.46) by dV_T, the infinitesimal volume change at constant T, to give

$$\frac{dU_T}{dV_T} = T\,\frac{dS_T}{dV_T} - P$$

From the definition of a partial derivative, the quantity dU_T/dV_T is the partial derivative $(\partial U/\partial V)_T$, and we have

$$\left(\frac{\partial U}{\partial V}\right)_T = T\left(\frac{\partial S}{\partial V}\right)_T - P$$

Application of the Euler reciprocity relation (4.43) to the Gibbs equation $dA = -S\,dT - P\,dV$ [Eq. (4.35)] gives the Maxwell relation $(\partial S/\partial V)_T = (\partial P/\partial T)_V$ [Eq. (4.45)]. Therefore

$$\left(\frac{\partial U}{\partial V}\right)_T = T\left(\frac{\partial P}{\partial T}\right)_V - P = \frac{\alpha T}{\kappa} - P \qquad (4.47)$$

where $(\partial P/\partial T)_V = \alpha/\kappa$ [Eq. (1.45)] was used. Equation (4.47) is the desired expression for $(\partial U/\partial V)_T$ in terms of easily measured properties.

Temperature Dependence of U

The basic equation (4.29) is the desired relation: $(\partial U/\partial T)_V = C_V$.

Temperature Dependence of H

The basic equation (4.30) is the desired relation: $(\partial H/\partial T)_P = C_P$.

Pressure Dependence of H

We want $(\partial H/\partial P)_T$. Starting with the Gibbs equation $dH = T\,dS + V\,dP$ [Eq. (4.34)], imposing the condition of constant T, and dividing by dP_T, we get $dH_T/dP_T = T\,dS_T/dP_T + V$ or

$$\left(\frac{\partial H}{\partial P}\right)_T = T\left(\frac{\partial S}{\partial P}\right)_T + V$$

Application of the Euler reciprocity relation to $dG = -S\,dT + V\,dP$ gives $(\partial S/\partial P)_T = -(\partial V/\partial T)_P$ [Eq. (4.45)]. Therefore

$$\left(\frac{\partial H}{\partial P}\right)_T = -T\left(\frac{\partial V}{\partial T}\right)_P + V = -TV\alpha + V \qquad (4.48)$$

Temperature Dependence of S

The basic equation (4.31) for C_P is the desired relation:

$$\left(\frac{\partial S}{\partial T}\right)_P = \frac{C_P}{T} \qquad (4.49)$$

Pressure Dependence of S

The Euler reciprocity relation applied to the Gibbs equation $dG = -S\,dT + V\,dP$ gives

$$\left(\frac{\partial S}{\partial P}\right)_T = -\left(\frac{\partial V}{\partial T}\right)_P = -\alpha V \qquad (4.50)$$

as already noted in Eq. (4.45).

Temperature and Pressure Dependences of G

In $dG = -S\,dT + V\,dP$, we set $dP = 0$ to get $(\partial G/\partial T)_P = -S$. In $dG = -S\,dT + V\,dP$, we set $dT = 0$ to get $(\partial G/\partial P)_T = V$. Thus [Eq. (4.38)]

$$\left(\frac{\partial G}{\partial T}\right)_P = -S, \qquad \left(\frac{\partial G}{\partial P}\right)_T = V \qquad (4.51)$$

Summary on Finding T, P, and V Dependences of State Functions

To find $(\partial/\partial P)_T$, $(\partial/\partial V)_T$, $(\partial/\partial T)_V$, or $(\partial/\partial T)_P$ of U, H, A, or G, one starts with the Gibbs equation for dU, dH, dA, or dG [Eqs. (4.33) to (4.36)], imposes the condition of constant T, V, or P, divides by dP_T, dV_T, dT_V, or dT_P, and, if necessary, uses one of the Maxwell relations (4.45) or the heat-capacity relations (4.31) to eliminate $(\partial S/\partial V)_T$, $(\partial S/\partial P)_T$, $(\partial S/\partial T)_V$, or $(\partial S/\partial T)_P$. To find $(\partial U/\partial T)_V$ and $(\partial H/\partial T)_P$, it is faster to simply write down the C_V and C_P equations (4.29) and (4.30).

In deriving thermodynamic identities, it is helpful to remember that the temperature dependences of S [the derivatives $(\partial S/\partial T)_P$ and $(\partial S/\partial T)_V$] are related to C_P and C_V [Eq. (4.31)] and the volume and pressure dependences of S [the derivatives $(\partial S/\partial P)_T$ and $(\partial S/\partial V)_T$] are given by the Maxwell relations (4.45). Equation (4.45) need not be memorized, since it can quickly be found from the Gibbs equations for dA and dG by using the Euler reciprocity relation.

As a reminder, the equations of this section apply to a closed system of fixed composition and also to closed systems where the composition changes reversibly.

Magnitudes of T, P, and V Dependences of U, H, S, and G

We have $(\partial U_m/\partial T)_V = C_{V,m}$ and $(\partial H_m/\partial T)_P = C_{P,m}$. The heat capacities $C_{P,m}$ and $C_{V,m}$ are always positive and usually are not small. Therefore U_m and H_m increase rapidly with increasing T (see Fig. 5.11). An exception is at very low T, since $C_{P,m}$ and $C_{V,m}$ go to zero as T goes to absolute zero (Secs. 2.11 and 5.7).

Using (4.47) and experimental data, one finds (as discussed later in this section) that $(\partial U/\partial V)_T$ (which is a measure of the strength of intermolecular forces) is zero for ideal gases, is small for real gases at low and moderate pressures, is substantial for gases at high pressures, and is very large for liquids and solids.

Using (4.48) and typical experimental data (Prob. 4.8), one finds that $(\partial H_m/\partial P)_T$ is rather small for solids and liquids. It takes very high pressures to produce substantial changes in the internal energy and enthalpy of a solid or liquid. For ideal gases $(\partial H_m/\partial P)_T = 0$ (Prob. 4.21), and for real gases $(\partial H_m/\partial P)_T$ is generally small.

From $(\partial S/\partial T)_P = C_P/T$, it follows that the entropy S increases rapidly as T increases (see Fig. 5.11).

We have $(\partial S_m/\partial P)_T = -\alpha V_m$. As noted in Sec. 1.7, α is somewhat larger for gases than for condensed phases. Moreover, V_m at usual temperatures and pressures is about 10^3 times as great for gases as for liquids and solids. Thus, the variation in entropy with pressure is small for liquids and solids but is substantial for gases. Since α is positive for gases, the entropy of a gas decreases rapidly as the pressure increases (and the volume decreases); recall Eq. (3.29) for ideal gases.

For G, we have $(\partial G_m/\partial P)_T = V_m$. For solids and liquids, the molar volume is relatively small, so G_m for condensed phases is rather insensitive to moderate changes in pressure, a fact we shall use frequently. For gases, V_m is large and G_m increases rapidly as P increases (due mainly to the decrease in S as P increases).

We also have $(\partial G/\partial T)_P = -S$. However, thermodynamics does not define absolute entropies, only entropy differences. The entropy S has an arbitrary additive constant. Thus $(\partial G/\partial T)_P$ has no physical meaning in thermodynamics, and it is impossible to measure $(\partial G/\partial T)_P$ of a system. However, from $(\partial G/\partial T)_P = -S$, we can derive $(\partial \Delta G/\partial T)_P = -\Delta S$. This equation has physical meaning.

In summary: For *solids* and *liquids*, temperature changes usually have significant effects on thermodynamic properties, but pressure effects are small unless very large pressure changes are involved. For *gases* not at high pressure, temperature changes usually have significant effects on thermodynamic properties and pressure changes have significant effects on properties that involve the entropy (for example, S, A, G) but usually have only slight effects on properties not involving S (for example, U, H, C_P).

Joule–Thomson Coefficient

We now express some more thermodynamic properties in terms of easily measured quantities. We begin with the Joule–Thomson coefficient $\mu_{JT} \equiv (\partial T/\partial P)_H$. Equation (2.65) gives $\mu_{JT} = -(\partial H/\partial P)_T/C_P$. Substitution of (4.48) for $(\partial H/\partial P)_T$ gives

$$\mu_{JT} = (1/C_P)[T(\partial V/\partial T)_P - V] = (V/C_P)(\alpha T - 1) \qquad (4.52)$$

which relates μ_{JT} to α and C_P.

Heat-Capacity Difference

Equation (2.60) gives $C_P - C_V = [(\partial U/\partial V)_T + P](\partial V/\partial T)_P$. Substitution of $(\partial U/\partial V)_T = \alpha T/\kappa - P$ [Eq. (4.47)] gives $C_P - C_V = (\alpha T/\kappa)(\partial V/\partial T)_P$. Use of $\alpha \equiv V^{-1}(\partial V/\partial T)_P$ gives

$$C_P - C_V = TV\alpha^2/\kappa \qquad (4.53)$$

For a condensed phase (liquid or solid), C_P is readily measured, but C_V is hard to measure. Equation (4.53) gives a way to calculate C_V from the measured C_P.

Note the following: (1) As $T \to 0$, $C_P \to C_V$. (2) The compressibility κ can be proved to be always positive (*Zemansky and Dittman*, sec. 14-9). Hence $C_P \geq C_V$. (3) If $\alpha = 0$, then $C_P = C_V$. For liquid water at 1 atm, the molar volume reaches a minimum at 3.98°C (Fig. 1.5). Hence $(\partial V/\partial T)_P = 0$ and $\alpha = 0$ for water at this temperature. Thus $C_P = C_V$ for water at 1 atm and 3.98°C.

EXAMPLE 4.2 $C_P - C_V$

For water at 30°C and 1 atm: $\alpha = 3.04 \times 10^{-4}\ K^{-1}$, $\kappa = 4.52 \times 10^{-5}\ atm^{-1} = 4.46 \times 10^{-10}\ m^2/N$, $C_{P,m} = 17.99\ cal/(mol\ K)$, $V_m = 18.1\ cm^3/mol$. Find $C_{V,m}$ of water at 30°C and 1 atm.

Division of (4.53) by the number of moles of water gives $C_{P,m} - C_{V,m} = TV_m\alpha^2/\kappa$. We find

$$\frac{TV_m\alpha^2}{\kappa} = \frac{(303\ K)(18.1 \times 10^{-6}\ m^3\ mol^{-1})(3.04 \times 10^{-4}\ K^{-1})^2}{4.46 \times 10^{-10}\ m^2/N}$$

$$TV_m\alpha^2/\kappa = 1.14\ J\ mol^{-1}\ K^{-1} = 0.27\ cal\ mol^{-1}\ K^{-1}$$

$$C_{V,m} = 17.72\ cal/(mol\ K) \qquad (4.54)$$

For liquid water at 1 atm and 30°C, there is little difference between $C_{P,m}$ and $C_{V,m}$. This is due to the rather small α value of 30°C water; α is zero at 4°C and is still small at 30°C.

Exercise

For water at 95.0°C and 1 atm: $\alpha = 7.232 \times 10^{-4}$ K^{-1}, $\kappa = 4.81 \times 10^{-5}$ bar^{-1}, $c_P = 4.210$ J/(g K), and $\rho = 0.96189$ g/cm^3. Find c_V for water at 95.0°C and 1 atm. [*Answer:* 3.794 J/(g K).]

The use of (4.53) and experimental $C_{P,m}$ values to find $C_{V,m}$ for solids and liquids gives the following results at 25°C and 1 atm:

Substance	Cu(s)	NaCl(s)	I$_2$(s)	C$_6$H$_6$(l)	CS$_2$(l)	CCl$_4$(l)
$C_{V,m}$/[J/(mol K)]	23.8	47.7	48	95	47	91
$C_{P,m}$/[J/(mol K)]	24.4	50.5	54	136	76	132

$C_{P,m}$ and $C_{V,m}$ usually do not differ by much for solids *but differ greatly for liquids.*

Ideal-Gas $(\partial U/\partial V)_T$

An ideal gas obeys the equation of state $PV = nRT$, whereas a perfect gas obeys both $PV = nRT$ *and* $(\partial U/\partial V)_T = 0$. For an ideal gas, $(\partial P/\partial T)_V = nR/V$, and Eq. (4.47) gives $(\partial U/\partial V)_T = nRT/V - P = P - P = 0$.

$$(\partial U/\partial V)_T = 0 \qquad \text{ideal gas} \tag{4.55}$$

We have proved that *all ideal gases are perfect,* so there is no distinction between an ideal gas and a perfect gas. From now on, we shall drop the term "perfect gas."

$(\partial U/\partial V)_T$ of Solids, Liquids, and Nonideal Gases

The internal pressure $(\partial U/\partial V)_T$ is, as noted in Sec. 2.6, a measure of intermolecular interactions in a substance. The relation $(\partial U/\partial V)_T = \alpha T/\kappa - P$ [Eq. (4.47)] enables one to find $(\partial U/\partial V)_T$ from experimental data. For solids, the typical values $\alpha = 10^{-4.5}$ K^{-1} and $\kappa = 10^{-5.5}$ atm^{-1} (Sec. 1.7) give at 25°C and 1 atm

$$(\partial U/\partial V)_T \approx (10^{-4.5}\,\text{K}^{-1})(300\,\text{K})(10^{5.5}\,\text{atm}) - 1\,\text{atm} \approx 3000\,\text{atm} \approx 300\,\text{J/cm}^3$$

For liquids, the typical α and κ values give at 25°C and 1 atm

$$(\partial U/\partial V)_T \approx (10^{-3}\,\text{K}^{-1})(300\,\text{K})(10^4\,\text{atm}) \approx 3000\,\text{atm} \approx 300\,\text{J/cm}^3$$

The large $(\partial U/\partial V)_T$ values indicate strong intermolecular forces in solids and liquids.

EXAMPLE 4.3 $(\partial U/\partial V)_T$

Estimate $(\partial U/\partial V)_T$ for N$_2$ gas at 25°C and 1 atm using the van der Waals equation and the van der Waals constants of Sec. 8.4.

The van der Waals equation (1.39) is

$$(P + an^2/V^2)(V - nb) = nRT \tag{4.56}$$

We have $(\partial U/\partial V)_T = T(\partial P/\partial T)_V - P$ [Eq. (4.47)]. Solving the van der Waals equation for P and taking $(\partial/\partial T)_V$, we have

$$P = \frac{nRT}{V - nb} - \frac{an^2}{V^2} \quad \text{and} \quad \left(\frac{\partial P}{\partial T}\right)_V = \frac{nR}{V - nb}$$

$$\left(\frac{\partial U}{\partial V}\right)_T = T\left(\frac{\partial P}{\partial T}\right)_V - P = \frac{nRT}{V - nb} - \left(\frac{nRT}{V - nb} - \frac{an^2}{V^2}\right) = \frac{an^2}{V^2} \tag{4.57}$$

From Sec. 8.4, $a = 1.35 \times 10^6 \ cm^6 \ atm \ mol^{-2}$ for N_2. At 25°C and 1 atm, the gas is nearly ideal and V/n can be found from $PV = nRT$ with little error. We get $V/n = 24.5 \times 10^3 \ cm^3/mol$. Thus

$$(\partial U/\partial V)_T = (1.35 \times 10^6 \ cm^6 \ atm/mol^2)/(24.5 \times 10^3 \ cm^3/mol)^2$$

$$= (0.0022 \ atm)(8.314 \ J)/(82.06 \ cm^3 \ atm) = 0.00023 \ J/cm^3$$

$$= 0.23 \ J/L$$

The smallness of $(\partial U/\partial V)_T$ indicates the smallness of intermolecular forces in N_2 gas at 25°C and 1 atm.

Exercise

Use the van der Waals equation and data in Sec. 8.4 to estimate $(\partial U/\partial V)_T$ for HCl(g) at 25°C and 1 atm. Why is $(\partial U/\partial V)_T$ larger for HCl(g) than for $N_2(g)$? [*Answer:* 0.0061 atm = 0.62 J/L.]

We can get an approximate expression for $U_{intermol}$, the contribution of intermolecular interactions to U, as follows. As V changes at constant T, the average distance between molecules changes and hence the intermolecular-interaction energy $U_{intermol}$ changes. The translational, rotational, vibrational, and electronic contributions to U depend on T but not on V (Sec. 2.11). Infinite volume corresponds to infinite average distance between molecules and hence to $U_{intermol} = 0$. Therefore, $U(T, V) - U(T, \infty) = U_{intermol}(T, V)$. Integration of $dU_T = (\partial U/\partial V)_T \ dV_T$ at constant T gives

$$\int_{\infty}^{V'} (\partial U/\partial V)_T \ dV = \int_{\infty}^{V'} dU = U(T, V') - U(T, \infty) = U_{intermol}(T, V')$$

where V' is some particular volume. For a van der Waals gas, Eq. (4.57) gives $(\partial U/\partial V)_T = an^2/V^2$, so $\int_{\infty}^{V'} (\partial U/\partial V)_T \ dV = \int_{\infty}^{V'} (an^2/V^2) \ dV = -an^2/V'$. Dropping the prime from V, we have for a van der Waals gas: $U_{intermol} = -an^2/V$. In terms of molar quantities,

$$U_{intermol,m} = -a/V_m \qquad \text{for a van der Waals gas} \qquad (4.58)$$

Equation (4.58) is only a rough approximation to $U_{intermol,m}$ of real gases. First of all, (4.58) includes only the effect of intermolecular attractions, which lower U; at high densities, intermolecular repulsions, which tend to increase U, become significant. Also there is a dependence of $U_{intermol,m}$ on temperature at constant V_m, since at very high temperatures the molecules collide with high energy, which increases the effects of intermolecular repulsions and raises U.

By fitting $(\partial U/\partial V)_T$ data to an algebraic expression and integrating this expression, one can obtain $U_{intermol,m}$ for a liquid or gas [for details, see A. F. M. Barton, *J. Chem. Educ.*, **48**, 156 (1971)]. Figure 4.4 plots the result for $(C_2H_5)_2O$. This figure is

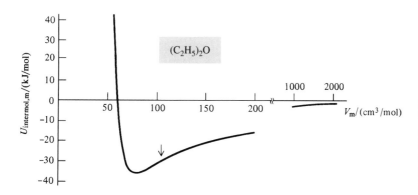

Figure 4.4

$U_{intermol,m}$ of $(C_2H_5)_2O$ versus V_m. Note the break and change of scale on the horizontal axis. The arrow indicates V_m of the liquid at 25°C and 1 atm.

approximate and assumes that $U_{intermol,m}$ depends only on V_m. As V_m decreases from infinity, $U_{intermol,m}$ and the molar internal energy U_m decrease at first, because of intermolecular attractions, but ultimately increase when the molecules are squeezed close together and the strong intermolecular repulsions dominate.

The arrow at 104 cm³/mol corresponds to V_m of the liquid at 25°C and 1 atm. The observed liquid molar volume of 104 cm³/mol is substantially larger than the volume 79 cm³/mol that corresponds to the minimum in $U_{intermol,m}$, so the liquid exists in a state where $U_{intermol,m}$ decreases with decreasing volume. At 104 cm³/mol, the attractive intermolecular forces are much larger than the repulsive intermolecular forces, and bringing the molecules closer together increases the net attraction and lowers the internal energy. This makes $(\partial U/\partial V)_T$ positive. We noted earlier in this subsection that $(\partial U/\partial V)_T$ is large and positive for typical liquids.

One's first impulse might be to expect a liquid's observed V to correspond to the minimum in $U_{intermol}$, but this is not so. The constant-T-and-P equilibrium condition is minimization of G, not minimization of U. We have $G = U + PV - TS$. The larger V is, the greater will be the entropy S. Roughly speaking, the molecules are more disordered at higher V. More precisely, (4.45) and (1.45) give $(\partial S/\partial V)_T = (\partial P/\partial T)_V = \alpha/\kappa$. With a few rare exceptions α and κ are both positive, so S nearly always increases as V increases. The $-TS$ term in G causes the minimum in G to occur for a V larger than that corresponding to the minimum in $U_{intermol}$. The observed V of the liquid is a compromise between minimizing the internal energy (which favors smaller V) and maximizing the entropy (which favors larger V).

$U_{intermol}$ of a liquid can be estimated as $-\Delta U$ of vaporization.

4.6 CALCULATION OF CHANGES IN STATE FUNCTIONS

Section 2.9 discussed calculation of ΔU and ΔH in a process, and Sec. 3.4 discussed calculation of ΔS. These discussions were incomplete, since we did not have expressions for $(\partial U/\partial V)_T$, for $(\partial H/\partial P)_T$, and for $(\partial S/\partial P)_T$ in paragraph 8 of Sec. 3.4. We now have expressions for these quantities. Knowing how U, H, and S vary with T, P, and V, we can find ΔU, ΔH, and ΔS for an arbitrary process in a closed system of constant composition. We shall also consider calculation of ΔA and ΔG.

Calculation of ΔS
Suppose a closed system of constant composition goes from state (P_1, T_1) to state (P_2, T_2) by any path, including, possibly, an irreversible path. The system's entropy is a function of T and P; $S = S(T, P)$, and

$$dS = \left(\frac{\partial S}{\partial T}\right)_P dT + \left(\frac{\partial S}{\partial P}\right)_T dP = \frac{C_P}{T} dT - \alpha V \, dP \qquad (4.59)$$

where (4.49) and (4.50) were used. Integration gives

$$\Delta S = S_2 - S_1 = \int_1^2 \frac{C_P}{T} dT - \int_1^2 \alpha V \, dP \qquad (4.60)$$

Since C_P, α, and V depend on both T and P, these are line integrals [unlike the integral in the perfect-gas ΔS equation (3.29)].

Since S is a state function, ΔS is independent of the path used to connect states 1 and 2. A convenient path (Fig. 4.5) is first to hold P constant at P_1 and change T from T_1 to T_2. Then T is held constant at T_2, and P is changed from P_1 to P_2. For step (a), $dP = 0$, and (4.60) gives

$$\Delta S_a = \int_{T_1}^{T_2} \frac{C_P}{T} dT \qquad \text{const. } P = P_1 \qquad (4.61)$$

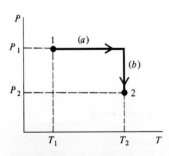

Figure 4.5

Path for calculating ΔS or ΔH.

With P held constant, C_P in (4.61) depends only on T, and we have an ordinary integral, which is easily evaluated if we know how C_P varies with T. For step (b), $dT = 0$, and (4.60) gives

$$\Delta S_b = -\int_{P_1}^{P_2} \alpha V \, dP \qquad \text{const. } T = T_2 \qquad (4.62)$$

With T held constant, α and V in (4.62) are functions of P only, and the integral is an ordinary integral. ΔS for the process $(P_1, T_1) \rightarrow (P_2, T_2)$ equals $\Delta S_a + \Delta S_b$.

If the system undergoes a phase transition in a process, we must make separate allowance for this change. For example, to calculate ΔS for heating ice at $-5°C$ and 1 atm to liquid water at $5°C$ and 1 atm, we use (4.61) to calculate the entropy change for warming the ice to $0°C$ and for warming the water from $0°C$ to $5°C$, but we must also add in the entropy change [Eq. (3.25)] for the melting process. During melting, $C_P \equiv dq_P/dT$ is infinite, and Eq. (4.61) doesn't apply.

EXAMPLE 4.4 ΔS when both T and P change

Calculate ΔS when 2.00 mol of water goes from $27°C$ and 1 atm to $37°C$ and 40 atm. Use data in Example 4.2 and neglect the pressure and temperature variations of $C_{P,m}$, α, and V_m.

Equation (4.61) gives $\Delta S_a = \int_{300\,K}^{310\,K} (nC_{P,m}/T) \, dT$, where the integration is at $P = P_1 = 1$ atm. Neglecting the slight temperature dependence of $C_{P,m}$, we have

$$\Delta S_a = (2.00 \text{ mol})[17.99 \text{ cal/(mol K)}] \ln(310/300) = 1.18 \text{ cal/K} = 4.94 \text{ J/K}$$

Equation (4.62) gives $\Delta S_b = -\int_{1\,atm}^{40\,atm} \alpha n V_m \, dP$, where the integration is at $T = T_2 = 310$ K. Neglecting the pressure variation in α and V_m and assuming their $30°C$ values are close to their $37°C$ values, we have

$$\Delta S_b = -(0.000304 \text{ K}^{-1})(2.00 \text{ mol})(18.1 \text{ cm}^3/\text{mol})(39 \text{ atm})$$

$$= -0.43 \text{ cm}^3 \text{ atm/K} = -(0.43 \text{ cm}^3 \text{ atm/K})(8.314 \text{ J})/(82.06 \text{ cm}^3 \text{ atm})$$

$$= -0.04 \text{ J/K}$$

$$\Delta S = \Delta S_a + \Delta S_b = 4.94 \text{ J/K} - 0.04 \text{ J/K} = 4.90 \text{ J/K}$$

Note the smallness of the pressure effect.

Exercise

Suppose that $H_2O(l)$ goes from $29.0°C$ and 1 atm to $31.0°C$ and pressure P_2. What value of P_2 would make $\Delta S = 0$ for this process? State any approximations made. (*Answer:* 8.9×10^2 atm.)

Calculation of ΔH and ΔU

The use of Eqs. (4.30) and (4.48) in $dH = (\partial H/\partial T)_P \, dT + (\partial H/\partial P)_T \, dP$ followed by integration gives

$$\Delta H = \int_1^2 C_P \, dT + \int_1^2 (V - TV\alpha) \, dP \qquad (4.63)$$

The line integrals in (4.63) are readily evaluated by using the path of Fig. 4.5. As usual, separate allowance must be made for phase changes. ΔH for a constant-pressure phase change equals the heat of the transition.

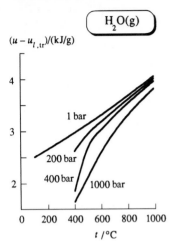

$(u - u_{l,tr})/(kJ/g)$

$H_2O(g)$

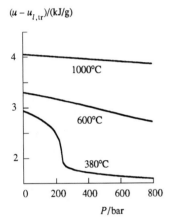

$(u - u_{l,tr})/(kJ/g)$

Figure 4.6

Specific internal energy of $H_2O(g)$ versus T and versus P.

ΔU can be easily found from ΔH using $\Delta U = \Delta H - \Delta(PV)$. Alternatively, we can write down an equation for ΔU similar to (4.63) using either T and V or T and P as variables.

Figures 4.6 and 4.7 plot $u - u_{tr,l}$ and $s - s_{tr,l}$ for $H_2O(g)$ versus T and P, where $u_{tr,l}$ and $s_{tr,l}$ are the specific internal energy and specific entropy of liquid water at the triple point (Sec. 1.5), and $u \equiv U/m$, $s \equiv S/m$, where m is the mass. The points on these curves can be calculated using Eqs. (4.60) and (4.63), and Δu and Δs of vaporization of water. These curves are discussed in Sec. 8.6.

Calculation of ΔG and ΔA

From $G \equiv H - TS$ and the equation after (2.48), we have $G_2 - G_1 = \Delta G = \Delta H - \Delta(TS) = \Delta H - T_1 \Delta S - S_1 \Delta T - \Delta S \Delta T$. However, thermodynamics does not define entropies but only gives entropy *changes*. Thus S_1 is undefined in the expression for ΔG. Therefore ΔG is undefined unless $\Delta T = 0$. For an *isothermal* process, $G = H - TS$ gives [Eq. (4.20)]

$$\Delta G = \Delta H - T \Delta S \qquad \text{const. } T \qquad (4.64)$$

Thus ΔG is defined for an isothermal process. To calculate ΔG for an isothermal process, we first calculate ΔH and ΔS (Secs. 2.9, 3.4, and 4.6) and then use (4.64). Alternatively, ΔG for an isothermal process that does not involve an irreversible composition change can be found from $(\partial G/\partial P)_T = V$ [Eq. (4.51)] as

$$\Delta G = \int_{P_1}^{P_2} V \, dP \qquad \text{const. } T \qquad (4.65)$$

A special case is ΔG for a reversible process at constant T and P in a system with P-V work only. Here, $\Delta H = q$ and $\Delta S = q/T$. Equation (4.64) gives

$$\Delta G = 0 \qquad \text{rev. proc. at const. } T \text{ and } P; P\text{-}V \text{ work only} \qquad (4.66)$$

An important example is a reversible phase change. For example, $\Delta G = 0$ for melting ice or freezing water at 0°C and 1 atm (but $\Delta G \neq 0$ for the freezing of supercooled water at -10°C and 1 atm). Equation (4.66) is no surprise, since the equilibrium condition for a closed system (P-V work only) held at constant T and P is the minimization of G ($dG = 0$).

As with ΔG, we are interested in ΔA only for processes with $\Delta T = 0$, since ΔA is undefined if T changes. We use $\Delta A = \Delta U - T \Delta S$ or $\Delta A = -\int_1^2 P \, dV$ to find ΔA for an isothermal process.

4.7   CHEMICAL POTENTIALS AND MATERIAL EQUILIBRIUM

The basic equation $dU = T \, dS - P \, dV$ and the related equations (4.34) to (4.36) for dH, dA, and dG do not apply when the composition is changing due to interchange of matter with the surroundings or to irreversible chemical reaction or irreversible interphase transport of matter within the system. We now develop equations that hold during such processes.

The Gibbs Equations for Nonequilibrium Systems

Consider a one-phase system that is in thermal and mechanical equilibrium but not necessarily in material equilibrium. Since thermal and mechanical equilibrium exist, T and P have well-defined values and the system's thermodynamic state is defined by the values of $T, P, n_1, n_2, \ldots, n_k$, where the n_i's ($i = 1, 2, \ldots, k$) are the mole numbers of the k components of the one-phase system. As shown in Sec. 4.2, even though the system is not in material equilibrium, we can still assign meaningful values to U and S of the system (relative to their values in some chosen reference state). Since T, P, V, U, and S have values, H, A, and G also have values. The state functions U, H, A, and G can each be expressed as functions of T, P, and the n_i's.

At any instant during a chemical process in the system, the Gibbs energy is

$$G = G(T, P, n_1, \ldots, n_k) \tag{4.67}$$

Let T, P, and the n_i's change by the infinitesimal amounts $dT, dP, dn_1, \ldots, dn_k$ as the result of an irreversible chemical reaction or irreversible transport of matter into the system. We want dG for this infinitesimal process. Since G is a state function, we shall replace the actual irreversible change by a reversible change and calculate dG for the reversible change. We imagine using an anticatalyst to "freeze out" any chemical reactions in the system. We then reversibly add dn_1 moles of substance 1, dn_2 moles of 2, etc., and reversibly change T and P by dT and dP.

To add substance 1 to a system reversibly, we use a rigid membrane permeable to substance 1 only. If pure substance 1 is on one side of the membrane and the system is on the other side, we can adjust the pressure of pure 1 so that there is no tendency for component 1 to flow between system and surroundings. An infinitesimal change in the pressure of pure 1 then reversibly changes n_1 in the system.

The total differential of (4.67) is

$$dG = \left(\frac{\partial G}{\partial T}\right)_{P,n_i} dT + \left(\frac{\partial G}{\partial P}\right)_{T,n_i} dP + \left(\frac{\partial G}{\partial n_1}\right)_{T,P,n_{j\neq1}} dn_1 + \cdots + \left(\frac{\partial G}{\partial n_k}\right)_{T,P,n_{j\neq k}} dn_k \tag{4.68}$$

where the following conventions are used: the subscript n_i on a partial derivative means that all mole numbers are held constant; the subscript $n_{j\neq i}$ on a partial derivative means that all mole numbers except n_i are held fixed. For a reversible process where no change in composition occurs, Eq. (4.36) reads

$$dG = -S\,dT + V\,dP \qquad \text{rev. proc., } n_i \text{ fixed, } P\text{-}V \text{ work only} \tag{4.69}$$

It follows from (4.69) that

$$\left(\frac{\partial G}{\partial T}\right)_{P,n_i} = -S, \qquad \left(\frac{\partial G}{\partial P}\right)_{T,n_i} = V \tag{4.70}$$

where we added the subscripts n_i to emphasize the constant composition. Substitution of (4.70) in (4.68) gives for dG in a reversible process in a one-phase system with only P-V work:

$$dG = -S\,dT + V\,dP + \sum_{i=1}^{k} \left(\frac{\partial G}{\partial n_i}\right)_{T,P,n_{j\neq i}} dn_i \tag{4.71}$$

Now suppose the state variables change because of an irreversible material change. Since G is a state function, dG is independent of the process that connects states $(T, P, n_1, n_2, \ldots)$ and $(T + dT, P + dP, n_1 + dn_1, n_2 + dn_2, \ldots)$. Therefore dG for the irreversible change is the same as dG for a reversible change that connects these two states. Hence Eq. (4.71) gives dG for the irreversible material change. Note also that all the state functions in (4.71) are defined for the system during the irreversible composition change (Sec. 4.2). Thus (4.71) is the desired relation for dG.

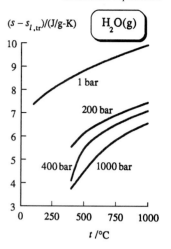

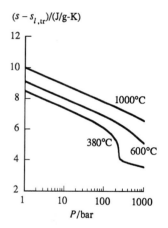

Figure 4.7

Specific entropy of $H_2O(g)$ versus T and versus P.

To save time in writing, we define the **chemical potential** μ_i (mu eye) of substance i in the one-phase system as

$$\mu_i \equiv \left(\frac{\partial G}{\partial n_i} \right)_{T,P,n_{j \neq i}} \qquad \text{one-phase syst.} \qquad \textbf{(4.72)}*$$

where G is the Gibbs energy of the one-phase system. Equation (4.71) then becomes

$$dG = -S\,dT + V\,dP + \sum_i \mu_i\,dn_i \qquad \begin{array}{l} \text{one-phase syst. in therm.} \\ \text{and mech. equilib., } P\text{-}V \text{ work only} \end{array} \qquad \textbf{(4.73)}*$$

Equation (4.73) is the key equation of chemical thermodynamics. It applies to a process in which the single-phase system is in thermal and mechanical equilibrium but is not necessarily in material equilibrium. Thus (4.73) holds during an irreversible chemical reaction and during transport of matter into or out of the system. Our previous equations were for closed systems, but we now have an equation applicable to open systems.

Let us obtain the equation for dU that corresponds to (4.73). From $G \equiv U + PV - TS$, we have $dU = dG - P\,dV - V\,dP + T\,dS + S\,dT$. The use of (4.73) gives

$$dU = T\,dS - P\,dV + \sum_i \mu_i\,dn_i \qquad (4.74)$$

This equation may be compared with $dU = T\,dS - P\,dV$ for a reversible process in a closed system.

From $H = U + PV$ and $A = U - TS$, together with (4.74), we can obtain expressions for dH and dA for irreversible chemical changes. Collecting together the expressions for dU, dH, dA, and dG, we have

$$dU = T\,dS - P\,dV + \sum_i \mu_i\,dn_i \qquad \qquad \textbf{(4.75)}*$$

$$dH = T\,dS + V\,dP + \sum_i \mu_i\,dn_i \qquad \text{one-phase syst.} \qquad (4.76)$$

$$dA = -S\,dT - P\,dV + \sum_i \mu_i\,dn_i \qquad \begin{array}{l}\text{in mech. and therm.}\\ \text{equilib., } P\text{-}V \text{ work only}\end{array} \qquad (4.77)$$

$$dG = -S\,dT + V\,dP + \sum_i \mu_i\,dn_i \qquad \qquad \textbf{(4.78)}*$$

These equations are the extensions of the Gibbs equations (4.33) to (4.36) to processes involving exchange of matter with the surroundings or irreversible composition changes. The extra terms $\sum_i \mu_i\,dn_i$ in (4.75) to (4.78) allow for the effect of the composition changes on the state functions U, H, A, and G. Equations (4.75) to (4.78) are also called the **Gibbs equations.**

Equations (4.75) to (4.78) are for a one-phase system. Suppose the system has several phases. Just as the letter i in (4.78) is a general index denoting any one of the chemical species present in the system, let α (alpha) be a general index denoting any one of the phases of the system. Let G^α be the Gibbs energy of phase α, and let G be the Gibbs energy of the entire system. The state function $G \equiv U + PV - TS$ is extensive. Therefore we add the Gibbs energy of each phase to get G of the multiphase system: $G = \sum_\alpha G^\alpha$. If the system has three phases, then $\sum_\alpha G^\alpha$ has three terms. The relation $d(u + v) = du + dv$ shows that the differential of a sum is the sum of the differentials. Therefore, $dG = d(\sum_\alpha G^\alpha) = \sum_\alpha dG^\alpha$. The one-phase Gibbs equation (4.78) written for phase α reads

$$dG^\alpha = -S^\alpha\,dT + V^\alpha\,dP + \sum_i \mu_i^\alpha\,dn_i^\alpha$$

Substitution of this equation into $dG = \sum_{\alpha} dG^{\alpha}$ gives

$$dG = -\sum_{\alpha} S^{\alpha} \, dT + \sum_{\alpha} V^{\alpha} \, dP + \sum_{\alpha} \sum_{i} \mu_i^{\alpha} \, dn_i^{\alpha} \qquad (4.79)$$

where S^{α} and V^{α} are the entropy and volume of phase α, μ_i^{α} is the chemical potential of chemical species i in phase α, and n_i^{α} is the number of moles of i in phase α. Equation (4.72) written for phase α reads

$$\mu_i^{\alpha} \equiv \left(\frac{\partial G^{\alpha}}{\partial n_i^{\alpha}} \right)_{T,P,n_{j \neq i}^{\alpha}} \qquad \textbf{(4.80)*}$$

(We have taken T of each phase to be the same and P of each phase to be the same. This will be true for a system in mechanical and thermal equilibrium provided no rigid or adiabatic walls separate the phases.) Since S and V are extensive, the sums over the entropies and volumes of the phases equal the total entropy S of the system and the total volume V of the system, and (4.79) becomes

$$dG = -S \, dT + V \, dP + \sum_{\alpha} \sum_{i} \mu_i^{\alpha} \, dn_i^{\alpha} \qquad \begin{array}{l} \text{syst. in mech. and therm.} \\ \text{equilib., } P\text{-}V \text{ work only} \end{array} \qquad \textbf{(4.81)*}$$

Equation (4.81) is the extension of (4.78) to a several-phase system. Don't be intimidated by the double sum in (4.81). It simply tells us to add up $\mu \, dn$ for each species in each phase of the system. For example, for a system consisting of a liquid phase l and a vapor phase v, each of which contains only water (w) and acetone (ac), we have $\sum_{\alpha} \sum_{i} \mu_i^{\alpha} \, dn_i^{\alpha} = \mu_w^l \, dn_w^l + \mu_{ac}^l \, dn_{ac}^l + \mu_w^v \, dn_w^v + \mu_{ac}^v \, dn_{ac}^v$, where μ_w^l is the chemical potential of water in the liquid phase.

Material Equilibrium

We now derive the condition for material equilibrium, including both phase equilibrium and reaction equilibrium. Consider a closed system in mechanical and thermal equilibrium and held at constant T and P as it proceeds to material equilibrium. We showed in Sec. 4.4 that, during an irreversible chemical reaction or interphase transport of matter in a closed system at constant T and P, the Gibbs function G is decreasing ($dG < 0$). At equilibrium, G has reached a minimum, and $dG = 0$ for any infinitesimal change at constant T and P [Eq. (4.19)]. At constant T and P, $dT = 0 = dP$, and from (4.81) the equilibrium condition $dG = 0$ becomes

$$\sum_{\alpha} \sum_{i} \mu_i^{\alpha} \, dn_i^{\alpha} = 0 \qquad \begin{array}{l} \text{material equilib., closed syst.,} \\ P\text{-}V \text{ work only, const. } T, P \end{array} \qquad (4.82)$$

This is the desired relation.

Now consider material equilibrium in a closed system held at constant T and V. Generalizing (4.77) to a several-phase system, we have (using $dV = \sum_{\alpha} dV^{\alpha}$)

$$dA = -S \, dT - P \, dV + \sum_{\alpha} \sum_{i} \mu_i^{\alpha} \, dn_i^{\alpha} \qquad \begin{array}{l} \text{syst. in mech. and therm.} \\ \text{equilib., } P\text{-}V \text{ work only} \end{array} \qquad (4.83)$$

The Helmholtz energy A is a minimum for chemical equilibrium at constant T and V. Hence $dA = 0$ for constant-T-and-V equilibrium, and (4.83) gives

$$\sum_{\alpha} \sum_{i} \mu_i^{\alpha} \, dn_i^{\alpha} = 0 \qquad \begin{array}{l} \text{material equilib., closed syst.,} \\ P\text{-}V \text{ work only, const. } T, V \end{array} \qquad (4.84)$$

which is the same as (4.82) for material equilibrium at constant T and P.

Not only is the material-equilibrium condition (4.82) valid for equilibrium reached under conditions of constant T and P or constant T and V, but it holds no matter how the closed system reaches equilibrium. To show this, consider an infinitesimal reversible process in a closed system with P-V work only. Equation (4.81) applies. Also, Eq. (4.36), which reads $dG = -S\,dT + V\,dP$, applies. Subtraction of $dG = -S\,dT + V\,dP$ from (4.81) gives

$$\sum_\alpha \sum_i \mu_i^\alpha \, dn_i^\alpha = 0 \qquad \text{rev. proc., closed syst., } P\text{-}V \text{ work only} \qquad (4.85)$$

Equation (4.85) must hold for any reversible process in a closed system with P-V work only. An infinitesimal process in a system that is in equilibrium is a reversible process (since it connects an equilibrium state with one infinitesimally close to equilibrium). Hence (4.85) must hold for any infinitesimal change in a system that has reached material equilibrium. Therefore (4.85) holds for any closed system in material equilibrium. If the system reaches material equilibrium under conditions of constant T and P, then G is minimized at equilibrium. If equilibrium is reached under conditions of constant T and V, then A is minimized at equilibrium. If equilibrium is reached under other conditions, then neither A nor G is necessarily minimized at equilibrium, but in all cases, Eq. (4.85) holds at equilibrium. Equation (4.85) is the desired general condition for material equilibrium. This equation will take on simpler forms when we apply it to phase and reaction equilibrium in the following sections.

Chemical Potentials

The chemical potential μ_i of substance i in a one-phase system is $\mu_i \equiv (\partial G/\partial n_i)_{T,P,n_{j\neq i}}$ [Eq. (4.72)]. Since G is a function of T, P, n_1, n_2, ..., its partial derivative $\partial G/\partial n_i \equiv \mu_i$ is also a function of these variables:

$$\mu_i = \mu_i(T, P, n_1, n_2, \ldots) \qquad \text{one-phase syst.}$$

The chemical potential of substance i in the phase is a state function that depends on the temperature, pressure, and composition of the phase. Since μ_i is the ratio of infinitesimal changes in two extensive properties, it is an intensive property. From $\mu_i \equiv (\partial G/\partial n_i)_{T,P,n_{j\neq i}}$, the chemical potential of substance i gives the rate of change of the Gibbs energy G of the phase with respect to the moles of i added at constant T, P, and other mole numbers. The state function μ_i was introduced into thermodynamics by Gibbs.

Because chemical potentials are intensive properties, we can use mole fractions instead of moles to express the composition dependence of μ. For a several-phase system, the chemical potential of substance i in phase α is

$$\mu_i^\alpha = \mu_i^\alpha(T^\alpha, P^\alpha, x_1^\alpha, x_2^\alpha, \ldots)$$

Note that, even if substance i is absent from phase α ($n_i^\alpha = 0$), its chemical potential μ_i^α in phase α is still defined. There is always the possibility of introducing substance i into the phase. When dn_i^α moles of i is introduced at constant T, P, and $n_{j\neq i}$, the Gibbs energy of the phase changes by dG^α and μ_i^α is given by dG^α/dn_i^α.

The simplest possible system is a single phase of pure substance i, for example, solid copper or liquid water. Let $G_{m,i}(T, P)$ be the molar Gibbs energy of pure i at the temperature and pressure of the system. By definition, $G_{m,i} \equiv G/n_i$, so the Gibbs energy of the pure, one-phase system is $G = n_i G_{m,i}(T, P)$. Partial differentiation of this equation gives

$$\mu_i \equiv (\partial G/\partial n_i)_{T,P} = G_{m,i} \qquad \text{one-phase pure substance} \qquad (4.86)^*$$

For a pure substance, μ_i is the molar Gibbs free energy. However, μ_i in a one-phase mixture need not equal G_m of pure i.

4.8   PHASE EQUILIBRIUM

The two kinds of material equilibrium are phase equilibrium and reaction equilibrium (Sec. 4.1). A phase equilibrium involves the same chemical species present in different phases [for example, $C_6H_{12}O_6(s) \rightleftharpoons C_6H_{12}O_6(aq)$]. A reaction equilibrium involves different chemical species, which may or may not be present in the same phase [for example, $CaCO_3(s) \rightleftharpoons CaO(s) + CO_2(g)$ and $N_2(g) + 3H_2(g) \rightleftharpoons 2NH_3(g)$]. Phase equilibrium is considered in this section, reaction equilibrium in the next.

The condition for material equilibrium in a closed system with P-V work only is given by Eq. (4.85) as $\Sigma_\alpha \Sigma_i \mu_i^\alpha \, dn_i^\alpha = 0$, which holds for any possible infinitesimal change in the mole numbers n_i^α. Consider a several-phase system that is in equilibrium, and suppose that dn_j moles of substance j were to flow from phase β (beta) to phase δ (delta) (Fig. 4.8). For this process, Eq. (4.85) becomes

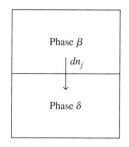

Figure 4.8

dn_j moles of substance j flows from phase β to phase δ.

$$\mu_j^\beta \, dn_j^\beta + \mu_j^\delta \, dn_j^\delta = 0 \qquad (4.87)$$

From Fig. 4.8, we have $dn_j^\beta = -dn_j$ and $dn_j^\delta = dn_j$. Therefore $-\mu_j^\beta \, dn_j + \mu_j^\delta \, dn_j = 0$, and

$$(\mu_j^\delta - \mu_j^\beta) \, dn_j = 0$$

Since $dn_j \neq 0$, we must have $\mu_j^\delta - \mu_j^\beta = 0$, or

$$\mu_j^\beta = \mu_j^\delta \qquad \text{phase equilib. in closed syst., } P\text{-}V \text{ work only} \qquad \textbf{(4.88)}^*$$

For a closed system with P-V work only in thermal and mechanical equilibrium, the phase equilibrium condition is that the chemical potential of a given substance is the same in every phase of the system.

Now suppose the closed system (which is in thermal and mechanical equilibrium and is capable of P-V work only) has not yet reached phase equilibrium. Let dn_j moles of substance j flow spontaneously from phase β to phase δ. For this irreversible process, the equation preceding (4.16) gives $dG < -S \, dT + V \, dP$. But dG for this process is given by (4.81) as $dG = -S \, dT + V \, dP + \Sigma_\alpha \Sigma_i \mu_i^\alpha \, dn_i^\alpha$. Hence the inequality $dG < -S \, dT + V \, dP$ becomes

$$-S \, dT + V \, dP + \sum_\alpha \sum_i \mu_i^\alpha \, dn_i^\alpha < -S \, dT + V \, dP$$

$$\sum_\alpha \sum_i \mu_i^\alpha \, dn_i^\alpha < 0$$

For the spontaneous flow of dn_j moles of j from phase β to phase δ, we have $\Sigma_\alpha \Sigma_i \mu_i^\alpha \, dn_i^\alpha = \mu_j^\beta \, dn_j^\beta + \mu_j^\delta \, dn_j^\delta = -\mu_j^\beta \, dn_j + \mu_j^\delta \, dn_j < 0$, and

$$(\mu_j^\delta - \mu_j^\beta) \, dn_j < 0 \qquad (4.89)$$

Since dn_j is positive, (4.89) requires that $\mu_j^\delta - \mu_j^\beta$ be negative: $\mu_j^\delta < \mu_j^\beta$. The spontaneous flow was assumed to be from phase β to phase δ. We have thus shown that for a system in thermal and mechanical equilibrium:

Substance j flows spontaneously from a phase with higher chemical potential μ_j to a phase with lower chemical potential μ_j.

This flow will continue until the chemical potential of substance j has been equalized in all the phases of the system. Similarly for the other substances. (As a substance flows from one phase to another, the compositions of the phases are changed and hence the chemical potentials in the phases are changed.) Just as a difference in temperature is the driving force for the flow of heat from one phase to another, a difference in chemical potential μ_i is the driving force for the flow of chemical species i from one phase to another.

If $T^\beta > T^\delta$, heat flows spontaneously from phase β to phase δ until $T^\beta = T^\delta$. If $P^\beta > P^\delta$, work "flows" from phase β to phase δ until $P^\beta = P^\delta$. If $\mu_j^\beta > \mu_j^\delta$, substance j flows spontaneously from phase β to phase δ until $\mu_j^\beta = \mu_j^\delta$. The state function T determines whether there is thermal equilibrium between phases. The state function P determines whether there is mechanical equilibrium between phases. The state functions μ_i determine whether there is material equilibrium between phases.

One can prove from the laws of thermodynamics that the chemical potential μ_j^δ of substance j in phase δ must increase when the mole fraction x_j^δ of j in phase δ is increased by the addition of j at constant T and P (see *Kirkwood and Oppenheim*, sec. 6-4):

$$(\partial \mu_j^\delta / \partial x_j^\delta)_{T,P,n_{i\neq j}^\delta} > 0 \tag{4.90}$$

EXAMPLE 4.5 Change in μ_i when a solid dissolves

A crystal of ICN is added to pure liquid water and the system is held at 25°C and 1 atm. Eventually a saturated solution is formed, and some solid ICN remains undissolved. At the start of the process, is μ_{ICN} greater in the solid phase or in the pure water? What happens to μ_{ICN} in each phase as the crystal dissolves? (See if you can answer these questions before reading further.)

At the start of the process, some ICN "flows" from the pure solid phase into the water. Since substance j flows from a phase with higher μ_j to one with lower μ_j, the chemical potential μ_{ICN} in the solid must be greater than μ_{ICN} in the pure water. (Recall from Sec. 4.7 that μ_{ICN} is defined for the pure-water phase even though there is no ICN in the water.) Since μ is an intensive quantity and the temperature, pressure, and mole fraction in the pure-solid phase do not change as the solid dissolves, $\mu_{ICN(s)}$ remains constant during the process. As the crystal dissolves, x_{ICN} in the aqueous phase increases and (4.90) shows that $\mu_{ICN(aq)}$ increases. This increase continues until $\mu_{ICN(aq)}$ becomes equal to $\mu_{ICN(s)}$. The system is then in phase equilibrium, no more ICN dissolves, and the solution is saturated.

Exercise

The equilibrium vapor pressure of water at 25°C is 24 torr. Is the chemical potential of $H_2O(l)$ at 25°C and 20 torr less than, equal to, or greater than μ of $H_2O(g)$ at this T and P? (*Answer:* greater than.)

Just as temperature is an intensive property that governs the flow of heat, chemical potentials are intensive properties that govern the flow of matter from one phase to another. Temperature is less abstract than chemical potential because we have experience using a thermometer to measure temperature and can visualize temperature as a measure of average molecular energy. One can get some feeling for chemical potential by viewing it as a measure of escaping tendency. The greater the value of μ_j^δ, the greater the tendency of substance j to leave phase δ and flow into an adjoining phase where its chemical potential is lower.

There is one exception to the phase-equilibrium condition $\mu_j^\beta = \mu_j^\delta$, which we now examine. We found that a substance flows from a phase where its chemical potential is higher to a phase where its chemical potential is lower. Suppose that substance j is initially absent from phase δ. Although there is no j in phase δ, the chemical potential μ_j^δ is a defined quantity, since we could, in principle, introduce dn_j moles of j into δ and measure $(\partial G^\delta / \partial n_j^\delta)_{T,P,n_{i\neq j}} = \mu_j^\delta$ (or use statistical mechanics to calculate μ_j^δ). If initially $\mu_j^\beta > \mu_j^\delta$, then j flows from phase β to phase δ until phase equilibrium is reached.

However, if initially $\mu_j^\delta > \mu_j^\beta$, then j cannot flow out of δ (since it is absent from δ). The system will therefore remain unchanged with time and hence is in equilibrium. Therefore when a substance is absent from a phase, the equilibrium condition becomes

$$\mu_j^\delta \geq \mu_j^\beta \qquad \text{phase equilib., } j \text{ absent from } \delta \tag{4.91}$$

for all phases β in equilibrium with δ. In the preceding example of ICN(s) in equilibrium with a saturated aqueous solution of ICN, the species H_2O is absent from the pure solid phase, so all we can say is that μ_{H_2O} in the solid phase is greater than or equal to μ_{H_2O} in the solution.

The principal conclusion of this section is:

In a closed system in thermodynamic equilibrium, the chemical potential of any given substance is the same in every phase in which that substance is present.

EXAMPLE 4.6 Conditions for phase equilibrium

Write the phase-equilibrium conditions for a liquid solution of acetone and water in equilibrium with its vapor.

Acetone (ac) and water (w) are each present in both phases, so the equilibrium conditions are $\mu_{ac}^l = \mu_{ac}^v$ and $\mu_w^l = \mu_w^v$, where μ_{ac}^l and μ_{ac}^v are the chemical potentials of acetone in the liquid phase and in the vapor phase, respectively.

Exercise

Write the phase-equilibrium conditions for a crystal of NaCl in equilibrium with an aqueous solution of NaCl. (*Answer:* $\mu_{NaCl}^s = \mu_{NaCl}^{aq}$.)

4.9   REACTION EQUILIBRIUM

We now apply the material-equilibrium condition to reaction equilibrium. Let the reaction be

$$aA_1 + bA_2 + \cdots \rightarrow eA_m + fA_{m+1} + \cdots \tag{4.92}$$

where $A_1, A_2, \ldots$ are the reactants, $A_m, A_{m+1}, \ldots$ are the products, and $a, b, \ldots, e, f, \ldots$ are the coefficients. For example, for the reaction

$$2C_6H_6 + 15O_2 \rightarrow 12CO_2 + 6H_2O$$

$A_1 = C_6H_6, A_2 = O_2, A_3 = CO_2, A_4 = H_2O$ and $a = 2, b = 15, e = 12, f = 6$. The substances in the reaction (4.92) need not all occur in the same phase, since the material-equilibrium condition applies to several-phase systems.

We adopt the convention of transposing the reactants in (4.92) to the right side of the equation to get

$$0 \rightarrow -aA_1 - bA_2 - \cdots + eA_m + fA_{m+1} + \cdots \tag{4.93}$$

We now let

$$\nu_1 \equiv -a, \quad \nu_2 \equiv -b, \quad \ldots, \quad \nu_m \equiv e, \quad \nu_{m+1} \equiv f, \quad \ldots$$

and write (4.93) as

$$0 \rightarrow \nu_1 A_1 + \nu_2 A_2 + \cdots + \nu_m A_m + \nu_{m+1} A_{m+1} + \cdots$$

$$0 \rightarrow \sum_i \nu_i A_i \tag{4.94}$$

where the **stoichiometric coefficients** ν_i (nu i) are negative for reactants and positive for products. For example, the reaction $2C_6H_6 + 15O_2 \rightarrow 12CO_2 + 6H_2O$ becomes $0 \rightarrow -2C_6H_6 - 15O_2 + 12CO_2 + 6H_2O$, and the stoichiometric coefficients are $\nu_{C_6H_6} = -2$, $\nu_{O_2} = -15$, $\nu_{CO_2} = 12$, and $\nu_{H_2O} = 6$. The stoichiometric coefficients are pure numbers with no units.

During a chemical reaction, the change Δn in the number of moles of each substance is proportional to its stoichiometric coefficient ν, where the proportionality constant is the same for all species. This proportionality constant is called the **extent of reaction** ξ (xi). For example, in the reaction $N_2 + 3H_2 \rightarrow 2NH_3$, suppose that 20 mol of N_2 reacts. Then 60 mol of H_2 will have reacted, and 40 mol of NH_3 will have been formed. We have $\Delta n_{N_2} = -20$ mol $= -1(20$ mol$)$, $\Delta n_{H_2} = -60$ mol $= -3(20$ mol$)$, $\Delta n_{NH_3} = 40$ mol $= 2(20$ mol$)$, where the numbers -1, -3, and 2 are the stoichiometric coefficients. The extent of reaction here is $\xi = 20$ mol. If x moles of N_2 reacts, then $3x$ moles of H_2 will react and $2x$ moles of NH_3 will be formed; here $\xi = x$ mol and $\Delta n_{N_2} = -x$ mol, $\Delta n_{H_2} = -3x$ mol, $\Delta n_{NH_3} = 2x$ mol.

For the general chemical reaction $0 \rightarrow \Sigma_i \nu_i A_i$ [Eq. (4.94)] undergoing a definite amount of reaction, the change in moles of species i, Δn_i, equals ν_i multiplied by the proportionality constant ξ:

$$\Delta n_i \equiv n_i - n_{i,0} = \nu_i \xi \qquad (4.95)^*$$

where $n_{i,0}$ is the number of moles of substance i present at the start of the reaction. ξ measures how much reaction has occurred. Since ν_i is dimensionless and Δn_i has units of moles, ξ has units of moles. ξ is positive if the reaction has proceeded left to right, and negative if it has proceeded right to left.

EXAMPLE 4.7 Extent of reaction

Suppose 0.6 mol of O_2 reacts according to $3O_2 \rightarrow 2O_3$. Find ξ.

The change in number of moles of species i during a reaction is proportional to its stoichiometric coefficient ν_i, where the proportionality constant is the extent of reaction ξ; $\Delta n_i = \nu_i \xi$. Since $\nu_{O_2} = -3$ and $\Delta n_{O_2} = -0.6$ mol, we have -0.6 mol $= -3\xi$ and $\xi = 0.2$ mol.

Exercise

In the reaction $2NH_3 \rightarrow N_2 + 3H_2$, suppose that initially 0.80 mol of NH_3, 0.70 mol of H_2, and 0.40 mol of N_2 are present. At a later time t, 0.55 mol of H_2 is present. Find ξ and find the moles of NH_3 and N_2 present at t. (*Answer:* -0.05 mol, 0.90 mol, 0.35 mol.)

The material equilibrium condition is $\Sigma_i \Sigma_\alpha \mu_i^\alpha \, dn_i^\alpha = 0$ [Eq. (4.85)]. Section 4.8 showed that at equilibrium the chemical potential of species i is the same in every phase that contains i, so we can drop the phase superscript α from μ_i^α and write the material equilibrium condition as

$$\sum_i \sum_\alpha \mu_i^\alpha \, dn_i^\alpha = \sum_i \mu_i \left(\sum_\alpha dn_i^\alpha \right) = \sum_i \mu_i \, dn_i = 0 \qquad (4.96)$$

where dn_i is the change in the total number of moles of i in the closed system and μ_i is the chemical potential of i in any phase that contains i.

For a finite extent of reaction ξ, we have $\Delta n_i = \nu_i \xi$ [Eq. (4.95)]. For an infinitesimal extent of reaction $d\xi$, we have

$$dn_i = \nu_i \, d\xi \qquad (4.97)$$

Substitution of $dn_i = \nu_i \, d\xi$ into the equilibrium condition $\Sigma_i \, \mu_i \, dn_i = 0$ [Eq. (4.96)] gives $(\Sigma_i \, \nu_i \mu_i) \, d\xi = 0$. This equation must hold for arbitrary infinitesimal values of $d\xi$. Hence

The condition for chemical-reaction equilibrium in a closed system is that $\Sigma_i \, \nu_i\mu_i = 0$.

When the reaction $0 \rightarrow \Sigma_i \, \nu_i A_i$ has reached equilibrium, then

$$\sum_i \nu_i\mu_i = 0 \qquad \text{reaction equilib. in closed syst., } P\text{-}V \text{ work only} \qquad \textbf{(4.98)}*$$

where ν_i and μ_i are the stoichiometric coefficient and chemical potential of species A_i. The relation of (4.98) to the more familiar concept of the equilibrium constant will become clear in later chapters. *Note that (4.98) is valid no matter how the closed system reaches equilibrium.* For example, it holds for equilibrium reached in a system held at constant T and P, or at constant T and V, or in an isolated system.

The equilibrium condition (4.98) is easily remembered by noting that it is obtained by simply replacing each substance in the reaction equation (4.92) by its chemical potential.

EXAMPLE 4.8 Condition for reaction equilibrium

Write out the equilibrium condition (4.98) for

(a) $2C_6H_6 + 15O_2 \rightarrow 12CO_2 + 6H_2O$

(b) $aA + bB \rightarrow cC + dD$

Since reactants have negative stoichiometric coefficients, the equilibrium condition for (a) is $\Sigma_i \, \nu_i\mu_i = -2\mu_{C_6H_6} - 15\mu_{O_2} + 12\mu_{CO_2} + 6\mu_{H_2O} = 0$ or

$$2\mu_{C_6H_6} + 15\mu_{O_2} = 12\mu_{CO_2} + 6\mu_{H_2O}$$

which has the same form as the chemical reaction. For the general reaction (b), the equilibrium condition (4.98) is

$$a\mu_A + b\mu_B = c\mu_C + d\mu_D$$

Exercise

Write the equilibrium condition for $2H_2 + O_2 \rightarrow 2H_2O$. (*Answer:* $2\mu_{H_2} + \mu_{O_2} = 2\mu_{H_2O}$.)

The equilibrium condition $\Sigma_i \, \nu_i\mu_i = 0$ looks abstract, but it simply says that at reaction equilibrium, the chemical potentials of the products "balance" those of the reactants.

If the reacting system is held at constant T and P, the Gibbs energy G is minimized at equilibrium. Note from the Gibbs equation (4.78) for dG that the sum $\Sigma_i \, \mu_i \, dn_i$ in (4.96) equals dG at constant T and P; $dG_{T,P} = \Sigma_i \, \mu_i \, dn_i$. Use of $dn_i = \nu_i \, d\xi$ [Eq. (4.97)] gives $dG_{T,P} = \Sigma_i \, \nu_i\mu_i \, d\xi$:

$$\frac{dG}{d\xi} = \sum_i \nu_i\mu_i \qquad \text{const. } T, P \qquad (4.99)$$

At equilibrium, $dG/d\xi = 0$, and G is minimized. The μ_i's in (4.99) are the chemical potentials of the substances in the reaction mixture, and they depend on the composition of the mixture (the n_i's). Hence the chemical potentials vary during the reaction.

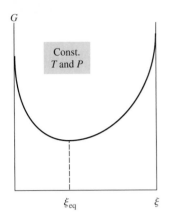

Figure 4.9

Gibbs energy versus extent of reaction in a system held at constant T and P.

This variation continues until G (which depends on the μ_i's and the n_i's at constant T and P) is minimized (Fig. 4.3) and (4.98) is satisfied. Figure 4.9 sketches G versus ξ for a reaction run at constant T and P. For constant T and V, G is replaced by A in the preceding discussion.

The quantity $\sum_i \nu_i \mu_i$ in the equilibrium condition (4.98) is often written as $\Delta_r G$ (where r stands for reaction) or as ΔG, so with this notation (4.98) becomes $\Delta_r G = 0$, where $\Delta_r G \equiv \sum_i \nu_i \mu_i$. However, $\sum_i \nu_i \mu_i$ is not the actual change in G in the reacting system and the Δ_r in $\Delta_r G$ really means $(\partial/\partial \xi)_{T,P}$. (See Sec. 11.9 for further discussion.)

Note the resemblance of the reaction-equilibrium condition (4.98) to the phase-equilibrium condition (4.88). If we regard the movement of substance A_i from phase β to phase δ to be the chemical reaction $A_i^\beta \rightarrow A_i^\delta$, then $\nu = -1$ for A_i^β and $\nu = 1$ for A_i^δ. Equation (4.98) gives $-\mu_i^\beta + \mu_i^\delta = 0$, which is the same as (4.88).

4.10 ENTROPY AND LIFE

The second law of thermodynamics is the law of increase in entropy. Increasing entropy means increasing disorder. Living organisms maintain a high degree of internal order. Hence one might ask whether life processes violate the second law.

We first ask whether we can meaningfully define the entropy of a living organism. Processes are continually occurring in a living organism, and the organism is not in an equilibrium state. However, the discussion of Sec. 4.2 shows that the entropy of nonequilibrium systems can be defined. We therefore can consider whether living systems obey the second law.

The first thing to note is that the statement $\Delta S \geq 0$ applies only to systems that are both closed and thermally isolated from their surroundings; see Eq. (3.37). Living organisms are open systems, since they take in and expel matter; further, they exchange heat with their surroundings. According to the second law, we must have $\Delta S_{\text{syst}} + \Delta S_{\text{surr}} \geq 0$ for an organism, but ΔS_{syst} (ΔS of the organism) can be positive, negative, or zero. Any decrease in S_{syst} must, according to the second law, be compensated for by an increase in S_{surr} that is at least as great as the magnitude of the decrease in S_{syst}. For example, during the freezing of water to the more ordered state of ice, S_{syst} decreases, but the heat flow from system to surroundings increases S_{surr}.

Entropy changes in open systems can be analyzed as follows. Let dS_{syst} be the entropy change of any system (open or closed) during an infinitesimal time interval dt. Let dS_i be the system's entropy change due to processes that occur entirely within the system during dt. Let dS_e be the system's entropy change due to exchanges of energy and matter between system and surroundings during dt. Any heat flow dq into or out of the system that occurs as a result of chemical reactions in the system is considered to contribute to dS_e. We have $dS_{\text{syst}} = dS_i + dS_e$. As far as internal changes in the system are concerned, we can consider the system to be isolated from its surroundings; hence Eqs. (3.38) and (3.35) give $dS_i \geq 0$, where the inequality sign holds for irreversible internal processes. However, dS_e can be positive, negative, or zero, and dS_{syst} can be positive, negative, or zero.

The state of a fully grown living organism remains about the same from day to day. The organism is not in an equilibrium state, but it is approximately in a steady state. Thus over a 24-hr period, ΔS_{syst} of a fully grown organism is about zero: $\Delta S_{\text{syst}} \approx 0$. The internal processes of chemical reaction, diffusion, blood flow, etc., are irreversible; hence $\Delta S_i > 0$ for the organism. Thus ΔS_e must be negative to compensate for the positive ΔS_i. We can break ΔS_e into a term due to heat exchange with the surroundings and a term due to matter exchange with the surroundings. The sign of q, and hence the sign of that part of ΔS_e due to heat exchange, can be positive or negative, depending on whether the surroundings are hotter or colder than the organism. We shall concentrate on that part of ΔS_e that is due to matter exchange. The organism takes in highly ordered large molecules such as proteins, starch, and sugars, whose entropy per unit mass is low. The organism excretes waste products that contain smaller, less ordered molecules, whose entropy per unit mass is high. Thus the entropy of the food intake is less than the entropy of the excretion products returned to the surroundings; this keeps ΔS_e negative. The organism discards matter with a greater entropy content than the matter it takes in, thereby losing entropy to the environment to compensate for the entropy produced in internal irreversible processes.

The preceding analysis shows there is no reason to believe that living organisms violate the second law. A quantitative demonstration of this by measurement of $\Delta S_{\text{syst}} + \Delta S_{\text{surr}}$ for an organism and its surroundings would be a difficult task. [By the use of metabolic data and equations of irreversible thermodynamics, dS_i/dt has been estimated for chick embryos as a function of time and found to be positive; see D. Briedis and R. C. Seagrave, *J. Theor. Biol.,* **110,** 173 (1984).]

4.11　SUMMARY

The Helmholtz energy A and the Gibbs energy G are state functions defined by $A \equiv U - TS$ and $G \equiv H - TS$. The condition that the total entropy of system plus surroundings be maximized at equilibrium leads to the condition that A or G of a closed system with only P-V work be minimized if equilibrium is reached in a system held at fixed T and V or fixed T and P, respectively.

The first law $dU = dq + dw$ combined with the second-law expression $dq_{\text{rev}} = T\,dS$ gives $dU = T\,dS - P\,dV$ (the Gibbs equation for dU) for a reversible change in a closed system with P-V work only. This equation, the definitions $H \equiv U + PV$, $A \equiv U - TS$, $G \equiv H - TS$, and the heat-capacity equations $C_P = (\partial H/\partial T)_P = T\,(\partial S/\partial T)_P$ and $C_V = (\partial U/\partial T)_V = T\,(\partial S/\partial T)_V$ are the basic equations for a closed system in equilibrium.

From the Gibbs equations for dU, dH, and dG, expressions for the variations in U, H, and G with respect to T, P, and V were found in terms of the readily measured properties C_P, α, and κ. Application of the Euler reciprocity relation to $dG = -S\,dT + V\,dP$ gives $(\partial S/\partial P)_T = -(\partial V/\partial T)_P$; $(\partial S/\partial V)_T$ is found similarly from the Gibbs equation for dA. These relations allow calculation of ΔU, ΔH, and ΔS for arbitrary changes of state.

For a system (open or closed) in mechanical and thermal equilibrium with P-V work only, one has $dG = -S\,dT + V\,dP + \sum_\alpha \sum_i \mu_i^\alpha\,dn_i^\alpha$, where the chemical potential of substance i in phase α is defined as $\mu_i^\alpha \equiv (\partial G^\alpha/\partial n_i^\alpha)_{T,P,n_{j\neq i}^\alpha}$. This expression for dG applies during an irreversible chemical reaction or transport of matter between phases.

The condition for equilibrium between phases is that, for each substance i, the chemical potential μ_i must be the same in every phase in which i is present: $\mu_i^\alpha = \mu_i^\beta$. The condition for reaction equilibrium is that $\sum_i \nu_i\mu_i = 0$, where the ν_i's are the reaction's stoichiometric coefficients, negative for reactants and positive for products. The chemical potentials are the key properties in chemical thermodynamics, since they determine phase and reaction equilibrium.

Important kinds of calculations dealt with in this chapter include:

- Calculation of ΔU, ΔH, and ΔS for changes in system temperature and pressure and calculation of ΔG and ΔA for isothermal processes (Sec. 4.6).
- Calculation of $C_P - C_V$, $(\partial U/\partial V)_T$, $(\partial H/\partial P)_T$, $(\partial S/\partial T)_P$, $(\partial S/\partial P)_T$, etc., from readily measured properties (C_P, α, κ) (Sec. 4.5).

Although this has been a long, mathematical chapter, it has presented concepts and results that lie at the heart of chemical thermodynamics and that will serve as a foundation for the remaining thermodynamics chapters.

FURTHER READING

Zemansky and Dittman, chaps. 9, 14; *Denbigh,* chap. 2; *Andrews* (1971), chaps. 13, 15, 20, 21; *Van Wylen and Sonntag,* chap. 10; *Lewis and Randall,* App. 6; *McGlashan,* chaps. 6, 8.

PROBLEMS

Section 4.4

4.1 True or false? (a) The quantities U, H, A, and G all have the same dimensions. (b) The relation $\Delta G = \Delta H - T \Delta S$ is valid for all processes. (c) $G = A + PV$. (d) For every closed system in thermal and mechanical equilibrium and capable of only P-V work, the state function G is minimized when material equilibrium is reached. (e) The Gibbs energy of 12 g of ice at 0°C and 1 atm is less than the Gibbs energy of 12 g of liquid water at 0°C and 1 atm. (f) The quantities $S\,dT$, $T\,dS$, $V\,dP$, and $\int_1^2 V\,dP$ all have dimensions of energy.

4.2 Calculate ΔG, ΔA, and ΔS_{univ} for each of the following processes and state any approximations made: (a) reversible melting of 36.0 g of ice at 0°C and 1 atm (use data from Prob. 2.48); (b) reversible vaporization of 39 g of C_6H_6 at its normal boiling point of 80.1°C and 1 atm; (c) adiabatic expansion of 0.100 mol of a perfect gas into vacuum (Joule experiment) with initial temperature of 300 K, initial volume of 2.00 L, and final volume of 6.00 L.

Section 4.5

4.3 Express each of the following rates of change in terms of state functions. (a) The rate of change of U with respect to temperature in a system held at constant volume. (b) The rate of change of H with respect to temperature in a system held at constant pressure. (c) The rate of change of S with respect to temperature in a system held at constant pressure.

4.4 The relation $(\partial U/\partial S)_V = T$ [Eq. (4.37)] is notable because is relates the three fundamental thermodynamic state functions U, S, and T. The reciprocal of this relation, $(\partial S/\partial U)_V = 1/T$, shows that entropy always increases when internal energy increases at constant volume. Use the Gibbs equation for dU to show that $(\partial S/\partial V)_U = P/T$.

4.5 Verify the Maxwell relations (4.44) and (4.45).

4.6 For water at 30°C and 1 atm, use data preceding Eq. (4.54) to find (a) $(\partial U/\partial V)_T$; (b) μ_{JT}.

4.7 Given that, for $CHCl_3$ at 25°C and 1 atm, $\rho = 1.49$ g/cm^3, $C_{P,\text{m}} = 116$ J/(mol K), $\alpha = 1.33 \times 10^{-3}$ K^{-1}, and $\kappa = 9.8 \times 10^{-5}$ atm^{-1}, find $C_{V,\text{m}}$ for $CHCl_3$ at 25°C and 1 atm.

4.8 For a liquid with the typical values $\alpha = 10^{-3}$ K^{-1}, $\kappa = 10^{-4}$ atm^{-1}, $V_{\text{m}} = 50$ cm^3/mol, $C_{P,\text{m}} = 40$ cal/mol-K, calculate at 25°C and 1 atm (a) $(\partial H_{\text{m}}/\partial T)_P$; (b) $(\partial H_{\text{m}}/\partial P)_T$; (c) $(\partial U/\partial V)_T$; (d) $(\partial S_{\text{m}}/\partial T)_P$; (e) $(\partial S_{\text{m}}/\partial P)_T$; (f) $C_{V,\text{m}}$; (g) $(\partial A/\partial V)_T$.

4.9 Show that $(\partial U/\partial P)_T = -TV\alpha + PV\kappa$ (a) by starting from the Gibbs equation for dU; (b) by starting from (4.47) for $(\partial U/\partial V)_T$.

4.10 Show that $(\partial U/\partial T)_P = C_P - PV\alpha$ (a) by starting from $dU = T\,dS - P\,dV$; (b) by substituting (4.26) into (4.30).

4.11 Starting from $dH = T\,dS + V\,dP$, show that $(\partial H/\partial V)_T = \alpha T/\kappa - 1/\kappa$.

4.12 Consider solids, liquids, and gases not at high pressure. For which of these is $C_{P,\text{m}} - C_{V,\text{m}}$ usually largest? Smallest?

4.13 Verify that $[\partial(G/T)/\partial T]_P = -H/T^2$. This is the *Gibbs–Helmholtz equation.*

4.14 Derive the equations in (4.31) for $(\partial S/\partial T)_P$ and $(\partial S/\partial T)_V$ from the Gibbs equations (4.33) and (4.34) for dU and dH.

4.15 Show that $\mu_J = (P - \alpha T\kappa^{-1})/C_V$, where μ_J is the Joule coefficient.

4.16 A certain gas obeys the equation of state $PV_{\text{m}} = RT(1 + bP)$, where b is a constant. Prove that for this gas (a) $(\partial U/\partial V)_T = bP^2$; (b) $C_{P,\text{m}} - C_{V,\text{m}} = R(1 + bP)^2$; (c) $\mu_{JT} = 0$.

4.17 Use Eqs. (4.30), (4.42), and (4.48) to show that $(\partial C_P/\partial P)_T = -T(\partial^2 V/\partial T^2)_P$. The volumes of substances increase approximately linearly with T, so $\partial^2 V/\partial T^2$ is usually quite small. Consequently, the pressure dependence of C_P can usually be neglected unless one is dealing with high pressures.

4.18 The volume of Hg in the temperature range 0 to 100°C at 1 atm is given by $V = V_0(1 + at + bt^2)$, where $a = 0.18182 \times 10^{-3}$ °C^{-1}, $b = 0.78 \times 10^{-8}$ °C^{-2}, and where V_0 is the volume at 0°C and t is the Celsius temperature. The density of mercury at 1 atm and 0°C is 13.595 g/cm^3. (a) Use the result of Prob. 4.17 to calculate $(\partial C_{P,\text{m}}/\partial P)_T$ for Hg at 25°C and 1 atm. (b) Given that $C_{P,\text{m}} = 6.66$ cal mol^{-1} K^{-1} for Hg at 1 atm and 25°C, estimate $C_{P,\text{m}}$ of Hg at 25°C and 10^4 atm.

4.19 For a liquid obeying the equation of state $V_{\text{m}} = c_1 + c_2T + c_3T^2 - c_4P - c_5PT$ [Eq. (1.40)], find expressions for each of the following properties in terms of the c's, C_P, P, T, and V: (a) $C_P - C_V$; (b) $(\partial U/\partial V)_T$; (c) $(\partial S/\partial P)_T$; (d) μ_{JT}; (e) $(\partial S/\partial T)_P$; (f) $(\partial G/\partial P)_T$.

4.20 A reversible adiabatic process is an isentropic (constant-entropy) process. (a) Let $\alpha_S \equiv V^{-1}(\partial V/\partial T)_S$. Use the first Maxwell equation in (4.44) and Eqs. (1.32), (1.35), and (4.31) to show that $\alpha_S = -C_V\kappa/TV\alpha$. (b) Evaluate α_S for a perfect gas. Integrate the result, assuming that C_V is constant, and verify that you obtain Eq. (2.76) for a reversible adiabatic process in a perfect gas. (c) The adiabatic compressibility is $\kappa_S \equiv -V^{-1}(\partial V/\partial P)_S$. Starting from $(\partial V/\partial P)_S = (\partial V/\partial T)_S (\partial T/\partial P)_S$, prove that $\kappa_S = C_V\kappa/C_P$.

4.21 Since all ideal gases are perfect (Sec. 4.5) and since for a perfect gas $(\partial H/\partial P)_T = 0$ [Eq. (2.70)], it follows that $(\partial H/\partial P)_T = 0$ for an ideal gas. Verify this directly from (4.48).

4.22 From (4.58), $U_{intermol,m} = -a/V_m$ for a gas obeying the van der Waals equation of state. For common small and medium-size molecules, a is typically 10^6 to 10^7 cm^6 atm mol^{-2} (Sec. 8.4). Calculate the typical range of $U_{intermol,m}$ in a gas at 25°C and 1 atm. Repeat for 25°C and 40 atm.

4.23 (a) For $(C_2H_5)_2O$, use Fig. 4.4 to estimate ΔU_m for vaporization of the liquid to a gas at 1 atm and 35°C, given that $V_m = 107$ cm^3/mol for the liquid at 35°C and 1 atm. (b) The experimental ΔH_m of vaporization of $(C_2H_5)_2O$ is 6.4 kcal/mol at 35°C. Calculate the experimental ΔU_m of vaporization.

4.24 (a) For liquids at 1 atm, the attractive intermolecular forces make the main contribution to $U_{intermol}$. Use the van der Waals expression (4.58) and the van der Waals a value of 1.34×10^6 cm^6 atm mol^{-2} for Ar to show that for liquid or gaseous Ar,

$$U_m \approx -(1.36 \times 10^5 \text{ J cm}^3/\text{mol}^2)/V_m + (12.5 \text{ J/mol-K})T + \text{const.}$$

(b) Calculate the translational and intermolecular energies in liquid and in gaseous Ar at 1 atm and 87.3 K (the normal boiling point). The liquid's density is 1.38 g/cm^3 at 87 K. (c) Estimate ΔU_m for the vaporization of Ar at its normal boiling point and compare the result with the experimental value 5.8 kJ/mol.

Section 4.6

4.25 True or false? (a) ΔG is undefined for a process in which T changes. (b) $\Delta G = 0$ for a reversible phase change at constant T and P.

4.26 Calculate ΔG and ΔA when 2.50 mol of a perfect gas with $C_{V,m} = 1.5R$ goes from 28.5 L and 400 K to 42.0 L and 400 K.

4.27 For the processes of Probs. 2.44a, b, d, e, and f, state whether each of ΔA and ΔG is positive, zero, or negative.

4.28 Calculate ΔA and ΔG when a mole of water vapor initially at 200°C and 1 bar undergoes a cyclic process for which $w = 145$ J.

4.29 (a) Find ΔG for the fusion of 50.0 g of ice at 0°C and 1 atm. (b) Find ΔG for the supercooled-water freezing process of Prob. 3.15.

4.30 Find ΔA and ΔG when 0.200 mol of He(g) is mixed at constant T and P with 0.300 mol of $O_2(g)$ at 27°C. Assume ideal gases.

4.31 Suppose 1.00 mol of water initially at 27°C and 1 atm undergoes a process whose final state is 100°C and 50 atm. Use data given preceding Eq. (4.54) and the approximation that the temperature and pressure variations of α, κ, and C_P can be neglected to calculate: (a) ΔH; (b) ΔU; (c) ΔS.

4.32 Calculate ΔG for the isothermal compression of 30.0 g of water from 1.0 atm to 100.0 atm at 25°C; neglect the variation of V with P.

4.33 A certain gas obeys the equation of state $PV_m = RT(1 + bP + cP^2)$, where b and c are constants. Find expressions for ΔH_m and ΔS_m for a change of state of this gas from (P_1, T_1) to (P_2, T_2); neglect the temperature and pressure dependence of $C_{P,m}$.

4.34 If 1.00 mol of water at 30.00°C is reversibly and adiabatically compressed from 1.00 to 10.00 atm, calculate the final volume by using expressions from Prob. 4.20 and neglecting the temperature and pressure variation in κ_S. Next calculate the final temperature. Then use the first law and the $(\partial V/\partial P)_S$ expression in Prob. 4.20 to calculate ΔU; compare the result with the approximate answer of Prob. 2.46. See Eq. (4.54) and data preceding it.

4.35 Use a result of the example after Eq. (4.55) to derive an expression for ΔU for a gas obeying the van der Waals equation and undergoing a change of state.

Section 4.7

4.36 True or false? (a) The chemical potential μ_i is a state function. (b) μ_i is an intensive property. (c) μ_i in a phase must remain constant if T, P, and x_i remain constant in the phase. (d) The SI units of μ_i are J/mol. (e) The definition of μ_i for a single-phase system is $\mu_i = (\partial G_i/\partial n_i)_{T,P,n_{j\neq i}}$. (f) The chemical potential of pure liquid acetone at 300 K and 1 bar equals G_m of liquid acetone at 300 K and 1 bar. (g) The chemical potential of benzene in a solution of benzene and toluene at 300 K and 1 bar must be equal to G_m of pure benzene at 300 K and 1 bar.

4.37 Show that $\mu_i = (\partial U/\partial n_i)_{S,V,n_{j\neq i}} = (\partial H/\partial n_i)_{S,P,n_{j\neq i}} = (\partial A/\partial n_i)_{T,V,n_{j\neq i}}$.

4.38 Use Eq. (4.75) to show that $dq = T\,dS + \Sigma_i \mu_i\,dn_i$ for a one-phase closed system with P-V work only in mechanical and thermal equilibrium. This expression gives dq during a chemical reaction. Since the reaction is irreversible, $dq \neq T\,dS$.

Section 4.8

4.39 True or false? (a) The chemical potential of benzene in a solution of benzene and toluene must equal the chemical potential of toluene in that solution. (b) The chemical potential of sucrose in a solution of sucrose in water at 300 K and 1 bar must equal the molar Gibbs energy of solid sucrose at 300 K and 1 bar. (c) The chemical potential of sucrose in a saturated solution of sucrose in water at 300 K and 1 bar must equal the molar Gibbs energy of solid sucrose at 300 K and 1 bar. (d) If phases α and β are in equilibrium with each other, the chemical potential of phase α must equal the chemical potential of phase β.

4.40 For each of the following closed systems, write the condition(s) for material equilibrium between phases: (a) ice in equilibrium with liquid water; (b) solid sucrose in equilibrium with a saturated aqueous solution of sucrose; (c) a two-phase system consisting of a saturated solution of ether in water and a saturated solution of water in ether; (d) ice in equilibrium with an aqueous solution of sucrose.

4.41 For each of the following pairs of substances, state which substance, if either, has the higher chemical potential: (a) $H_2O(l)$ at 25°C and 1 atm vs. $H_2O(g)$ at 25°C and 1 atm; (b) $H_2O(s)$ at 0°C and 1 atm vs. $H_2O(l)$ at 0°C and 1 atm; (c) $H_2O(s)$ at −5°C and 1 atm vs. supercooled $H_2O(l)$ at −5°C and 1 atm; (d) $C_6H_{12}O_6(s)$ at 25°C and 1 atm vs. $C_6H_{12}O_6(aq)$ in an unsaturated aqueous solution at 25°C and 1 atm; (e) $C_6H_{12}O_6(s)$ at 25°C and 1 atm vs. $C_6H_{12}O_6(aq)$ in a saturated solution at 25°C and 1 atm; (f) $C_6H_{12}O_6(s)$ at 25°C and 1 atm vs. $C_6H_{12}O_6(aq)$ in a supersaturated solution at 25°C and 1 atm. (g) Which substance in (a) has the higher G_m?

4.42 Show that for ice in equilibrium with liquid water at 0°C and 1 atm the condition of equality of chemical potentials is equivalent to $\Delta G = 0$ for $H_2O(s) \rightarrow H_2O(l)$.

Section 4.9

4.43 Give the value of the stoichiometric coefficient ν for each species in the reaction $C_3H_8(g) + 5O_2(g) \rightarrow 3CO_2(g) + 4H_2O(l)$.

4.44 Write the reaction equilibrium condition for $N_2 + 3H_2 \rightleftharpoons 2NH_3$ in a closed system.

4.45 Suppose that in the reaction $2O_3 \rightarrow 3O_2$, a closed system initially contains 5.80 mol O_2 and 6.20 mol O_3. At some later time, 7.10 mol of O_3 is present. What is ξ at this time?

General

4.46 For $H_2O(s)$ at 0°C and 1 atm and $H_2O(l)$ at 0°C and 1 atm, which of the following quantities must be equal for the two phases? (a) S_m; (b) U_m; (c) H_m; (d) G_m; (e) μ; (f) V_m.

4.47 Consider a two-phase system that consists of liquid water in equilibrium with water vapor; the system is kept in a constant-temperature bath. (a) Suppose we reversibly increase the system's volume, holding T and P constant, causing some of the liquid to vaporize. State whether each of ΔH, ΔS, ΔS_{univ}, and ΔG is positive, zero, or negative. (b) Suppose we suddenly remove some of the water vapor, holding T and V constant. This reduces the pressure below the equilibrium vapor pressure of water, and liquid water will evaporate at constant T and V until the equilibrium vapor pressure is restored. For this evaporation process state whether each of ΔU, ΔS, ΔS_{univ}, and ΔA is positive, zero, or negative.

4.48 For each of the following processes, state which of ΔU, ΔH, ΔS, ΔS_{univ}, ΔA, and ΔG must be zero. (a) A nonideal gas undergoes a Carnot cycle. (b) Hydrogen is burned in an adiabatic calorimeter of fixed volume. (c) A nonideal gas undergoes a Joule–Thomson expansion. (d) Ice is melted at 0°C and 1 atm.

4.49 Give an example of a liquid with a negative $(\partial U/\partial V)_T$.

4.50 Give the name of each of these Greek letters and state the thermodynamic quantity that each stands for: (a) ν; (b) μ; (c) ξ; (d) α; (e) κ; (f) ρ.

4.51 Give the conditions of applicability of each of these equations: (a) $dU = dq + dw$; (b) $dU = T\,dS − P\,dV$; (c) $dU = T\,dS − P\,dV + \Sigma_i \Sigma_\alpha \mu_i^\alpha \, dn_i^\alpha$.

4.52 Give the SI units of (a) ΔG; (b) ΔS_m; (c) C_P; (d) μ_i.

4.53 For a closed system with P-V work only, (a) write the equation that gives the condition of phase equilibrium; (b) write the equation that gives the condition of reaction equilibrium. (c) Explain why $dG = 0$ is not the answer to (a) and (b).

4.54 For a closed system with P-V work only and held at constant T and P, show that $dS = dq/T − dG/T$ for an irreversible material change. (*Hint:* Start with $G \equiv H − TS$.)

4.55 An equation for G_m of a pure substance as a function of T and P (or of A_m as a function of T and V) is called a **fundamental equation of state.** From a fundamental equation of state, one can calculate all thermodynamic properties of a substance. Express each of the following properties in terms of G_m, T, P, $(\partial G_m/\partial T)_P$, $(\partial G_m/\partial P)_T$, $(\partial^2 G_m/\partial T^2)_P$, $(\partial^2 G_m/\partial P^2)_T$, and $\partial^2 G_m/\partial P\partial T$. (a) S_m; (b) V_m; (c) H_m; (d) U_m; (e) $C_{P,m}$; (f) $C_{V,m}$; (g) α; (h) κ. [Using Eqs. like (4.60) and (4.63) for ΔH and ΔS and experimental C_P, α, and κ data, one can construct a fundamental equation of state of the form $G_m = f(T, P)$, where U and S have each been arbitrarily assigned a value of zero in some reference state, which is usually taken as the liquid at the triple point. Accurate fundamental equations of state have been constructed for several fluids. For fluid H_2O, fundamental equations of state contain about 50 parameters whose values are adjusted to give good fits of experimental data; see A. Saul and W. Wagner, *J. Phys. Chem. Ref. Data*, **18**, 1537 (1989); P. G. Hill, *ibid.*, **19**, 1233 (1990).]

4.56 When 3.00 mol of a certain gas is heated reversibly from 275 K and 1 bar to 375 K and 1 bar, ΔS is 20.0 J/K. If 3.00 mol of this gas is heated irreversibly from 275 K and 1 bar to 375 K and 1 bar, will ΔS be less than, the same as, or greater than 20.0 J/K?

4.57 For each of the following sets of quantities, all the quantities except one have something in common. State what they have in common and state which quantity does not belong with the others. (In some cases, more than one answer for the property in common might be possible.) (a) C_V, C_P, U, T, S, G, A, V; (b) H, U, G, S, A.

4.58 For each of the following statements, tell which state function(s) is (are) being described. (a) It enables one to find the rates of change of enthalpy and of entropy with respect to temperature at constant pressure. (b) They determine whether substance i in phase α is in phase equilibrium with i in phase β. (c) It enables one to find the rates of change of U and of S with respect to T at constant V. (d) It is maximized when an isolated system reaches equilibrium. (e) It is maximized when a system reaches equilibrium. (f) It is minimized when a closed system capable of P-V work only and held at constant T and P reaches equilibrium.

4.59 True or false? (a) $C_{P,m} − C_{V,m} = R$ for all gases. (b) $C_P − C_V = TV\alpha^2/\kappa$ for any substance. (c) ΔG is always zero for a reversible process in a closed system capable of *P-V* work only. (*d*) The Gibbs energy of a closed system with *P-V* work only is always minimized at equilibrium. (*e*) The work done by a closed system can exceed the decrease in the system's internal energy. (*f*) For an irreversible, isothermal, isobaric process in a closed system with *P-V* work only, ΔG must be negative. (*g*) G_{syst} + G_{surr} is constant for any process. (*h*) ΔS is positive for every irreversible process. (*i*) $\Delta S_{\text{syst}} + \Delta S_{\text{surr}}$ is positive for every irreversible process. (*j*) $\Delta(TS) = S\,\Delta T + T\,\Delta S$. (*k*) $\Delta(U - TS) = \Delta U - \Delta(TS)$. (*l*) $(\partial V/\partial T)_P = \Delta V/\Delta T$ for a constant-pressure process. (*m*) If a system remains in thermal and mechanical equilibrium during a process, then its *T* and *P* are constant during the process. (*n*) The entropy *S* of a closed system with *P-V* work only is always maximized at equilibrium. (*o*) If $a > b$, then we must have $ka > kb$, where *k* is a nonzero constant.

CHAPTER 5

Standard Thermodynamic Functions of Reaction

For the chemical reaction $aA + bB \rightleftharpoons cC + dD$, we found the condition for reaction equilibrium to be $a\mu_A + b\mu_B = c\mu_C + d\mu_D$ [Eq. (4.98)]. To effectively apply this condition to reactions, we will need tables of thermodynamic properties (such as G, H, and S) for individual substances. The main topic of this chapter is how one uses experimental data to construct such tables. In these tables, the properties are for substances in a certain state called the standard state, so this chapter begins by defining the standard state (Sec. 5.1). From tables of standard-state thermodynamic properties, one can calculate the changes in standard-state enthalpy, entropy, and Gibbs energy for chemical reactions. Chapters 6 and 11 show how equilibrium constants for reactions can be calculated from such standard-state property changes.

5.1 STANDARD STATES OF PURE SUBSTANCES

The **standard state** of a pure substance is defined as follows. For a pure solid or a pure liquid, the standard state is defined as the state with pressure $P = 1$ bar [Eq. (1.11)] and temperature T, where T is some temperature of interest. Thus for each value of T there is a single standard state for a pure substance. The symbol for a standard state is a degree superscript (read as "naught," "zero," or "standard"), with the temperature written as a subscript. For example, the molar volume of a pure solid or liquid at 1 bar and 200 K is symbolized by $V_{m,200}^\circ$, where the degree superscript indicates the standard pressure of 1 bar and 200 stands for 200 K. For a pure gas, the standard state at temperature T is chosen as the state where $P = 1$ bar and the gas behaves as an ideal gas. Since real gases do not behave ideally at 1 bar, the standard state of a pure gas is a fictitious state. Calculation of properties of the gas in the fictitious standard state from properties of the real gas is discussed in Sec. 5.4. Summarizing, the standard states for pure substances are:

Solid or liquid: $P = 1$ bar, T

Gas: $P = 1$ bar, T, gas ideal

(5.1)*

The standard-state pressure is denoted by P°:

$$P^\circ \equiv 1 \text{ bar}$$

(5.2)*

Standard states for components of solutions are discussed in Chapters 9 and 10.

5.2   STANDARD ENTHALPY OF REACTION

For a chemical reaction, we define the **standard enthalpy (change) of reaction** ΔH_T° as the enthalpy change for the process of transforming stoichiometric numbers of moles of the pure, separated reactants, each in its standard state at temperature T, to stoichiometric numbers of moles of the pure, separated products, each in its standard state at the same temperature T. Often ΔH_T° is called the *heat of reaction.* (Sometimes the symbol $\Delta_r H_T^\circ$ is used for ΔH_T°, where the r subscript stands for "reaction.") The quantity ΔU_T° is defined in a similar manner.

For the reaction

$$a\text{A} + b\text{B} \rightarrow c\text{C} + d\text{D}$$

the standard enthalpy change ΔH_T° is

$$\Delta H_T^\circ \equiv c H_{m,T}^\circ(\text{C}) + d H_{m,T}^\circ(\text{D}) - a H_{m,T}^\circ(\text{A}) - b H_{m,T}^\circ(\text{B})$$

where $H_{m,T}^\circ(\text{C})$ is the molar enthalpy of substance C in its standard state at temperature T. For the general reaction [Eq. (4.94)]

$$0 \rightarrow \sum_i \nu_i \text{A}_i$$

we have

$$\Delta H_T^\circ \equiv \sum_i \nu_i H_{m,T,i}^\circ \qquad (5.3)*$$

where the ν_i's are the stoichiometric coefficients (positive for products and negative for reactants) and $H_{m,T,i}^\circ$ is the molar enthalpy of A_i in its standard state at T. For example, ΔH_T° for $2\text{C}_6\text{H}_6(l) + 15\text{O}_2(g) \rightarrow 12\text{CO}_2(g) + 6\text{H}_2\text{O}(l)$ is

$$\Delta H_T^\circ = 12 H_{m,T}^\circ(\text{CO}_2, g) + 6 H_{m,T}^\circ(\text{H}_2\text{O}, l) - 2 H_{m,T}^\circ(\text{C}_6\text{H}_6, l) - 15 H_{m,T}^\circ(\text{O}_2, g)$$

The letters l and g denote the liquid and gaseous states.

Since the stoichiometric coefficients ν_i in (5.3) are dimensionless, the units of ΔH_T° are the same as those of $H_{m,T,i}^\circ$, namely, J/mol or cal/mol. The subscript T in ΔH_T° is often omitted. Since ΔH_T° is a molar quantity, it is best written as $\Delta H_{m,T}^\circ$. However, the m subscript is usually omitted, and we won't bother to include it.

Note that ΔH° depends on how the reaction is written. For

$$2\text{H}_2(g) + \text{O}_2(g) \rightarrow 2\text{H}_2\text{O}(l) \qquad (5.4)$$

the standard enthalpy of reaction ΔH_T° [Eq. (5.3)] is twice that for

$$\text{H}_2(g) + \tfrac{1}{2}\text{O}_2(g) \rightarrow \text{H}_2\text{O}(l) \qquad (5.5)$$

since each stoichiometric coefficient ν_i in (5.4) is twice the corresponding ν_i in (5.5). Although we can't have half a molecule, we can have half a mole of O_2, so (5.5) is a valid way of writing a reaction in chemical thermodynamics. For (5.4), one finds $\Delta H_{298}^\circ = -572$ kJ/mol, whereas for (5.5) $\Delta H_{298}^\circ = -286$ kJ/mol, where 298 stands for 298.15 K. The factor mol^{-1} in ΔH° indicates that we are giving the standard enthalpy change per mole of reaction as written, where the amount of reaction that has occurred is measured by ξ, the extent of reaction (Sec. 4.9). A ΔH° value is for $\xi = 1$ mol. Since $\Delta n_i = \nu_i \xi$ [Eq. (4.95)], when $\xi = 1$ mol for (5.4), 2 mol of H_2O is produced; whereas when $\xi = 1$ mol for (5.5), 1 mol of H_2O is produced.

We want to be able to calculate ΔH° of a reaction from tabulated thermodynamic data for the reactants and products. The definition (5.3) of ΔH_T° contains the standard-state molar enthalpy $H_{m,T}^\circ$ of each species at T. However, the laws of thermodynamics allow us to measure only *changes* in enthalpies, internal energies, and entropies (ΔH, ΔU, and ΔS). Therefore, thermodynamics does not provide absolute values of U, H, and S, but only relative values, and we cannot tabulate absolute enthalpies of substances. Instead, we tabulate standard enthalpies of formation. The next section defines the standard enthalpy of formation $\Delta_f H^\circ_{T,i}$ of substance i and shows that ΔH°_T of Eq. (5.3) is given by $\Delta H^\circ_T = \Sigma_i \nu_i \Delta_f H^\circ_{T,i}$.

> It is possible in principle to assign an absolute value for U (and hence for H) of a substance by using Einstein's formula $E = mc^2$, where m is the relativistic mass of a system with energy E, and c is the speed of light in a vacuum. For a system at rest in the absence of external fields, E equals the internal energy U. Thus to find the absolute internal energy of a mole of water at a given T and P, we simply multiply the molar mass by c^2. However, such a calculation is useless for chemical thermodynamics because the mass change that accompanies a change of state in chemical thermodynamics is far too small to be detected. For example, the increase in mass on melting one mole of ice is only 7×10^{-11} g. [From Prob. 2.48a, $\Delta U = 6$ kJ for the melting, and $\Delta m = \Delta U/c^2 = 6000 \text{ J}/(3 \times 10^8 \text{ m/s})^2 = 7 \times 10^{-14}$ kg.] Hence we cannot find absolute values of U and H that are accurate enough to be useful in chemical thermodynamics.

Phase Abbreviations

The letters s, l, and g stand for solid, liquid, and gas. Solids that have an ordered structure at the molecular level are called crystalline (abbreviated cr), whereas solids with a disordered structure are called amorphous (abbreviated am); see Sec. 24.1. The term **condensed phase** (abbreviated cd) means either a solid or a liquid; **fluid phase** (abbreviated fl) means either a liquid or a gas.

5.3   STANDARD ENTHALPY OF FORMATION

The **standard enthalpy of formation** (or **standard heat of formation**) $\Delta_f H^\circ_T$ of a pure substance at temperature T is ΔH° for the process in which one mole of the substance in its standard state at T is formed from the corresponding separated elements at T, each element being in its reference form. The **reference form** (or **reference phase**) of an element at temperature T is usually taken as the form of the element that is most stable at T and 1-bar pressure.

For example, the standard enthalpy of formation of gaseous formaldehyde $H_2CO(g)$ at 307 K, symbolized by $\Delta_f H^\circ_{307,H_2CO(g)}$, is the standard enthalpy change ΔH°_{307} for the process

$$C(\text{graphite}, 307 \text{ K}, P^\circ) + H_2(\text{ideal gas}, 307 \text{ K}, P^\circ) + \tfrac{1}{2}O_2(\text{ideal gas}, 307 \text{ K}, P^\circ) \rightarrow$$

$$H_2CO(\text{ideal gas}, 307 \text{ K}, P^\circ)$$

The gases on the left are in their standard states, which means they are unmixed, each in its pure state at standard pressure $P^\circ = 1$ bar and 307 K. At 307 K and 1 bar, the stable forms of hydrogen and oxygen are $H_2(g)$ and $O_2(g)$, so $H_2(g)$ and $O_2(g)$ are taken as the reference forms of hydrogen and oxygen. At 307 K and 1 bar, the most stable form of carbon is graphite, not diamond, so graphite appears in the formation reaction.

Consider $\Delta_f H^\circ$ of HBr(g). At 1 bar, Br_2 boils at 331.5 K. Therefore, $\Delta_f H^\circ_{330}$ of HBr(g) involves *liquid* Br_2 at 330 K and 1 bar reacting with standard-state $H_2(g)$, whereas $\Delta_f H^\circ_{335}$ of HBr(g) involves *gaseous* standard-state Br_2 reacting.

Since $\Delta_f H^\circ$ values are *changes* in enthalpies, they can be found from experimental data and thermodynamics equations; for details, see Sec. 5.4.

For an element in its reference form, $\Delta_f H^\circ_T$ is zero. For example, $\Delta_f H^\circ_{307}$ of graphite is, by definition, ΔH° of the reaction C(graphite, 307 K, P°) $\rightarrow$ C(graphite,

307 K, $P°$). Nothing happens in this "process," so its $\Delta H°$ is zero. For diamond, $\Delta_f H°_{307}$ is not zero, but is $\Delta H°$ of C(graphite, 307 K, $P°$) → C(diamond, 307 K, $P°$), which experiment gives as 1.9 kJ/mol.

Even though a particular form of a substance may not be stable at temperature T and 1 bar, one can still use experimental data and thermodynamics equations to find $\Delta_f H°_T$ of that form. For example, $H_2O(g)$ is not stable at 25°C and 1 bar, but we can use the measured heat of vaporization of liquid water at 25°C to find $\Delta_f H°_{298}$ of $H_2O(g)$; see Sec. 5.10 for details.

We now prove that the standard enthalpy change $\Delta H°_T$ for a chemical reaction is given by

$$\Delta H°_T = \sum_i \nu_i \, \Delta_f H°_{T,i} \qquad (5.6)*$$

where ν_i is the stoichiometric coefficient of substance i in the reaction and $\Delta_f H°_{T,i}$ is the standard enthalpy of formation of substance i at temperature T.

To prove (5.6), consider the reaction $aA + bB \rightarrow cC + dD$, where a, b, c, and d are the unsigned stoichiometric coefficients and A, B, C, and D are substances. Figure 5.1 shows two different isothermal paths from reactants to products in their standard states. Step 1 is a direct conversion of reactants to products. Step 2 is a conversion of reactants to standard-state elements in their reference forms. Step 3 is a conversion of elements to products. (Of course, the same elements produced by the decomposition of the reactants will form the products.) Since H is a state function, ΔH is independent of path and $\Delta H_1 = \Delta H_2 + \Delta H_3$. We have $\Delta H_1 = \Delta H°_T$ for the reaction. The reverse of process 2 would form $aA + bB$ from their elements; hence,

$$-\Delta H_2 = a \, \Delta_f H°_T(A) + b \, \Delta_f H°_T(B)$$

where $\Delta_f H°_T(A)$ is the standard enthalpy of formation of substance A at temperature T. Step 3 is the formation of $cC + dD$ from their elements, so

$$\Delta H_3 = c \, \Delta_f H°_T(C) + d \, \Delta_f H°_T(D)$$

The relation $\Delta H_1 = \Delta H_2 + \Delta H_3$ becomes

$$\Delta H°_T = -a \, \Delta_f H°_T(A) - b \, \Delta_f H°_T(B) + c \, \Delta_f H°_T(C) + d \, \Delta_f H°_T(D)$$

which is Eq. (5.6) for the reaction $aA + bB \rightarrow cC + dD$, since the stoichiometric coefficients ν_i are negative for reactants.

There are many more chemical reactions than there are chemical substances. Rather than having to measure and tabulate $\Delta H°$ for every possible chemical reaction, we can use (5.6) to calculate $\Delta H°$ from tabulated $\Delta_f H°$ values of the substances involved, provided we have determined $\Delta_f H°$ of each substance. The next section tells how $\Delta_f H°$ is measured.

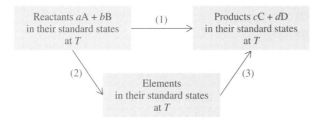

Figure 5.1

Steps used to relate $\Delta H°$ of a reaction to $\Delta_f H°$ of reactants and products.

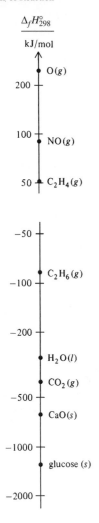

$\Delta_f H^\circ_{298}$

kJ/mol

- 200 — O(g)

- 100 — NO(g)

- 50 — C$_2$H$_4$(g)

- −50

- −100 — C$_2$H$_6$(g)

- −200

- H$_2$O(l)

- CO$_2$(g)

- −500 — CaO(s)

- −1000

- glucose (s)

- −2000

Figure 5.2

$\Delta_f H^\circ_{298}$ values. The scales are logarithmic.

5.4 DETERMINATION OF STANDARD ENTHALPIES OF FORMATION AND REACTION

Measurement of $\Delta_f H^\circ$

The quantity $\Delta_f H^\circ_{T,i}$ is ΔH° for isothermally converting pure standard-state elements in their reference forms to standard-state substance i. To find $\Delta_f H^\circ_{T,i}$, we must carry out the following steps:

1. If any of the elements involved are gases at T and 1 bar, we calculate ΔH for the hypothetical transformation of each gaseous element from an ideal gas at T and 1 bar to a real gas at T and 1 bar. This step is necessary because the standard state of a gas is the hypothetical ideal gas at 1 bar, whereas only real gases exist at 1 bar. The procedure for this calculation is given at the end of this section.
2. We measure ΔH for mixing the pure elements at T and 1 bar.
3. We use $\Delta H = \int_1^2 C_P \, dT + \int_1^2 (V - TV\alpha) \, dP$ [Eq. (4.63)] to find ΔH for bringing the mixture from T and 1 bar to the conditions under which we plan to carry out the reaction to form substance i. (For example, in the combustion of an element with oxygen, we might want the initial pressure to be 30 atm.)
4. We use a calorimeter (see after Example 5.1) to measure ΔH for the reaction in which the compound is formed from the mixed elements.
5. We use (4.63) to find ΔH for bringing the compound from the state in which it is formed in step 4 to T and 1 bar.
6. If compound i is a gas, we calculate ΔH for the hypothetical transformation of i from a real gas to an ideal gas at T and 1 bar.

The net result of these six steps is the conversion of standard-state elements at T to standard-state i at T. The standard enthalpy of formation $\Delta_f H^\circ_{T,i}$ is the sum of these six ΔH's. The main contribution by far comes from step 4, but in precise work one includes all the steps.

Once $\Delta_f H^\circ_i$ has been found at one temperature, its value at any other temperature can be calculated using C_P data for i and its elements (see Sec. 5.5). Nearly all thermodynamics tables list $\Delta_f H^\circ$ at 298.15 K (25°C). Some tables list $\Delta_f H^\circ$ at other temperatures. Some values of $\Delta_f H^\circ_{298}$ are plotted in Fig. 5.2. A table of $\Delta_f H^\circ_{298}$ is given in the Appendix. Once we have built up such a table, we can use Eq. (5.6) to find ΔH°_{298} for any reaction whose species are listed.

EXAMPLE 5.1 Calculation of ΔH° from $\Delta_f H^\circ$ data

Find ΔH°_{298} for the combustion of one mole of the simplest amino acid, glycine, NH$_2$CH$_2$COOH, according to

$$\text{NH}_2\text{CH}_2\text{COOH}(s) + \tfrac{9}{4}\text{O}_2(g) \rightarrow 2\text{CO}_2(g) + \tfrac{5}{2}\text{H}_2\text{O}(l) + \tfrac{1}{2}\text{N}_2(g) \quad (5.7)$$

Substitution of Appendix $\Delta_f H^\circ_{298}$ values into $\Delta H^\circ_{298} = \Sigma_i \nu_i \Delta_f H^\circ_{298,i}$ [Eq. (5.6)] gives ΔH°_{298} as

$$[\tfrac{1}{2}(0) + \tfrac{5}{2}(-285.830) + 2(-393.509) - (-528.10) - \tfrac{9}{4}(0)] \text{ kJ/mol}$$
$$= -973.49 \text{ kJ/mol}$$

Exercise

Use Appendix data to find ΔH°_{298} for the combustion of one mole of sucrose, C$_{12}$H$_{22}$O$_{11}$(s), to CO$_2$(g) and H$_2$O(l). (*Answer:* −5644.5 kJ/mol.)

Calorimetry

To carry out step 4 of the preceding procedure to find $\Delta_f H°$ of a compound, we must measure ΔH for the chemical reaction that forms the compound from its elements. For certain compounds, this can be done in a calorimeter. We shall consider measurement of ΔH for chemical reactions in general, not just for formation reactions.

The most common type of reaction studied calorimetrically is combustion. One also measures heats of hydrogenation, halogenation, neutralization, solution, dilution, mixing, phase transitions, etc. Heat capacities are also determined in a calorimeter. Reactions where some of the species are gases (for example, combustion reactions) are studied in a constant-volume calorimeter. Reactions not involving gases are studied in a constant-pressure calorimeter.

The **standard enthalpy of combustion** $\Delta_c H_T°$ of a substance is $\Delta H_T°$ for the reaction in which one mole of the substance is burned in O_2. For example, $\Delta_c H°$ for solid glycine is $\Delta H°$ for reaction (5.7). Some $\Delta_c H_{298}°$ values are plotted in Fig. 5.3.

An **adiabatic bomb calorimeter** (Fig. 5.4) is used to measure heats of combustion. Let R stand for the mixture of reactants, P for the product mixture, and K for the bomb walls plus the surrounding water bath. Suppose we start with the reactants at 25°C. Let the measured temperature rise due to the reaction be ΔT. Let the system be the bomb, its contents, and the surrounding water bath. This system is thermally insulated and does no work on its surroundings (except for a completely negligible amount of work done by the expanding water bath when its temperature rises). Therefore $q = 0$ and $w = 0$. Hence $\Delta U = 0$ for the reaction, as noted in step (*a*) of Fig. 5.4.

After the temperature rise ΔT due to the reaction is accurately measured, one cools the system back to 25°C. Then one measures the amount of electrical energy U_{el} that must be supplied to raise the system's temperature from 25°C to 25°C + ΔT; this is step (*b*) in Fig. 5.4. We have $\Delta U_b = U_{el} = VIt$, where V, I, and t are the voltage, current, and time.

The desired quantity $\Delta_r U_{298}$ (where r stands for reaction) is shown as step (*c*). The change in the state function U must be the same for path (*a*) as for path (*c*) + (*b*), since these paths connect the same two states. Thus $\Delta U_a = \Delta U_c + \Delta U_b$ and $0 = \Delta_r U_{298} + U_{el}$. Hence $\Delta_r U_{298} = -U_{el}$, and the measured U_{el} enables $\Delta_r U_{298}$ to be found.

Instead of using U_{el}, we could use an alternative procedure. We have seen that $\Delta_r U_{298} = -\Delta U_b$ (Fig. 5.4*b*). If we imagine carrying out step (*b*) by supplying heat q_b

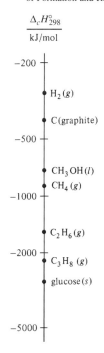

$$\frac{\Delta_c H_{298}°}{kJ/mol}$$

Figure 5.3

Standard enthalpies of combustion at 25°C. The scale is logarithmic. The products are $CO_2(g)$ and $H_2O(l)$.

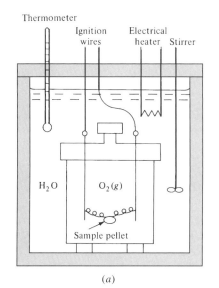

(*a*)

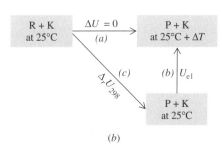

(*b*)

Figure 5.4

(*a*) An adiabatic bomb calorimeter. The shaded walls are adiabatic. (*b*) Energy relations for this calorimeter.

to the system K + P (instead of using electrical energy), then we would have $\Delta U_b = q_b = C_{K+P} \Delta T$, where C_{K+P} is the average heat capacity of the system K + P over the temperature range. Thus

$$\Delta_r U_{298} = -C_{K+P} \Delta T \qquad (5.8)$$

To find C_{K+P}, we repeat the combustion experiment in the same calorimeter using benzoic acid, whose ΔU of combustion is accurately known. For burning the benzoic acid, let $\Delta_r U'_{298}$, P', and $\Delta T'$ denote ΔU_{298} of reaction, the reaction products, and the temperature rise. Similar to Eq. (5.8), we have $\Delta_r U'_{298} = -C_{K+P'} \Delta T'$. Measurement of $\Delta T'$ and calculation of $\Delta_r U'_{298}$ from the known ΔU of combustion of benzoic acid then gives us the heat capacity $C_{K+P'}$. The temperature ranges over which the two combustions are carried out are very similar. Also, the main contribution to $C_{K+P'}$ and C_{K+P} comes from the bomb walls and the water bath. For these reasons, it is an excellent approximation to take $C_{K+P'} = C_{K+P}$. (In precise work, the difference between the two is calculated using the known heat capacities of the combustion products.) Knowing C_{K+P}, we find $\Delta_r U_{298}$ from Eq. (5.8).

To find the standard internal energy change ΔU_{298}° for the reaction, we must allow for the changes in U_R and U_P that occur when the reactants and products are brought from the states that occur in the calorimeter to their standard states. This correction is typically about 0.1 percent for combustion reactions.

(An analysis similar to Fig. 5.4b enables one to estimate the temperature of a flame. See Prob. 5.62.)

For reactions that do not involve gases, one can use an adiabatic constant-pressure calorimeter. The discussion is similar to that for the adiabatic bomb calorimeter, except that P is held fixed instead of V, and ΔH of reaction is measured instead of ΔU.

EXAMPLE 5.2 Calculation of $\Delta_c U^\circ$ from calorimetric data

Combustion of 2.016 g of solid glucose ($C_6H_{12}O_6$) at 25°C in an adiabatic bomb calorimeter with heat capacity 9550 J/K gives a temperature rise of 3.282°C. Find $\Delta_c U_{298}^\circ$ of solid glucose.

With the heat capacity of the products neglected, Eq. (5.8) gives $\Delta U = -(9550 \text{ J/K})(3.282 \text{ K}) = -31.34 \text{ kJ}$ for combustion of 2.016 g of glucose. The experimenter burned (2.016 g)/(180.16 g/mol) = 0.01119 mol. Hence ΔU per mole of glucose burned is $(-31.34 \text{ kJ})/(0.01119 \text{ mol}) = -2801 \text{ kJ/mol}$, and this is $\Delta_c U_{298}^\circ$ if the difference between conditions in the calorimeter and standard-state conditions is neglected.

Exercise

If 1.247 g of glucose is burned in an adiabatic bomb calorimeter whose heat capacity is 11.45 kJ/K, what will be the temperature rise? (*Answer:* 1.693 K.)

Relation between ΔH° and ΔU°

Calorimetric study of a reaction gives either ΔU° or ΔH°. Use of $H \equiv U + PV$ allows interconversion between ΔH° and ΔU°. For a process at constant pressure, $\Delta H = \Delta U + P \Delta V$. Since the standard pressure P° [Eq. (5.2)] is the same for all substances, conversion of pure standard-state reactants to products is a constant-pressure process, and for a reaction we have

$$\Delta H^\circ = \Delta U^\circ + P^\circ \Delta V^\circ \qquad (5.9)$$

Similar to $\Delta H° = \Sigma_i \nu_i H°_{m,i}$ [Eq. (5.3)], the changes in standard-state volume and internal energy for a reaction are given by $\Delta V° = \Sigma_i \nu_i V°_{m,i}$ and $\Delta U° = \Sigma_i \nu_i U°_{m,i}$. A sum like $\Sigma_i \nu_i U°_{m,i}$ looks abstract, but when we see $\Sigma_i \nu_i \cdots$, we can translate this into "products minus reactants," since the stoichiometric coefficient ν_i is positive for products and negative for reactants.

The molar volumes of gases at 1 bar are much greater than those of liquids or solids, so it is an excellent approximation to consider only the gaseous reactants and products in applying (5.9). For example, consider the reaction

$$aA(s) + bB(g) \rightarrow cC(g) + dD(g) + eE(l)$$

Neglecting the volumes of the solid and liquid substances A and E, we have $\Delta V° = cV°_{m,C} + dV°_{m,D} - bV°_{m,B}$. The standard state of a gas is an ideal gas, so $V°_m = RT/P°$ for each of the gases C, D, and B. Hence $\Delta V° = (c + d - b)RT/P°$. The quantity $c + d - b$ is the total number of moles of product gases minus the total number of moles of reactant gases. Thus, $c + d - b$ is the change in the number of moles of gas for the reaction. We write $c + d - b = \Delta n_g/\text{mol}$, where n_g stands for moles of gas. Since $c + d - b$ is a dimensionless number, we divided Δn_g by the unit "mole" to make it dimensionless. We thus have $\Delta V° = (\Delta n_g/\text{mol})RT/P°$, and (5.9) becomes

$$\Delta H°_T = \Delta U°_T + \Delta n_g RT/\text{mol} \qquad (5.10)$$

For example, the reaction $C_3H_8(g) + 5O_2(g) \rightarrow 3CO_2(g) + 4H_2O(l)$ has $\Delta n_g/\text{mol} = 3 - 1 - 5 = -3$ and (5.10) gives $\Delta H°_T = \Delta U°_T - 3RT$. At 300 K, $\Delta H° - \Delta U° = -7.48$ kJ/mol for this reaction, which is small but not negligible.

EXAMPLE 5.3 Calculation of $\Delta_f U°$ from $\Delta_f H°$

For $CO(NH_2)_2(s)$, $\Delta_f H°_{298} = -333.51$ kJ/mol. Find $\Delta_f U°_{298}$ of $CO(NH_2)_2(s)$.
The formation reaction is

$$C(\text{graphite}) + \tfrac{1}{2}O_2(g) + N_2(g) + 2H_2(g) \rightarrow CO(NH_2)_2(s)$$

and has $\Delta n_g/\text{mol} = 0 - 2 - 1 - \tfrac{1}{2} = -\tfrac{7}{2}$. Equation (5.10) gives

$$\Delta_f U°_{298} = -333.51 \text{ kJ/mol} - (-\tfrac{7}{2})(8.314 \times 10^{-3} \text{ kJ/mol-K})(298.15 \text{ K})$$
$$= -324.83 \text{ kJ/mol}$$

Exercise

For $CF_2ClCF_2Cl(g)$, $\Delta_f H°_{298} = -890.4$ kJ/mol. Find $\Delta_f U°_{298}$ of $CF_2ClCF_2Cl(g)$. (*Answer:* -885.4 kJ/mol.)

Exercise

In Example 5.2, $\Delta_c U°_{298}$ of glucose was found to be -2801 kJ/mol. Find $\Delta_c H°_{298}$ of glucose. (*Answer:* -2801 kJ/mol.)

For reactions not involving gases, Δn_g is zero, and $\Delta H°$ is essentially the same as $\Delta U°$ to within experimental error. For reactions involving gases, the difference between $\Delta H°$ and $\Delta U°$, though certainly not negligible, is usually not great. The quantity RT in (5.10) equals 2.5 kJ/mol at 300 K and 8.3 kJ/mol at 1000 K, and $\Delta n_g/\text{mol}$ is usually a small integer. These RT values are small compared with typical $\Delta H°$ values, which are hundreds of kJ/mol (see the $\Delta_f H°$ values in the Appendix). In qualitative reasoning, chemists often don't bother to distinguish between $\Delta H°$ and $\Delta U°$.

Hess's Law

Suppose we want the standard enthalpy of formation $\Delta_f H^\circ_{298}$ of ethane gas at 25°C. This is ΔH°_{298} for $2C(\text{graphite}) + 3H_2(g) \rightarrow C_2H_6(g)$. Unfortunately, we cannot react graphite with hydrogen and expect to get ethane, so the heat of formation of ethane cannot be measured directly. This is true for most compounds. Instead, we determine the heats of combustion of ethane, hydrogen, and graphite, these heats being readily measured. The following values are found at 25°C:

$$C_2H_6(g) + \tfrac{7}{2}O_2(g) \rightarrow 2CO_2(g) + 3H_2O(l) \qquad \Delta H^\circ_{298} = -1560 \text{ kJ/mol} \quad (1)$$

$$C(\text{graphite}) + O_2(g) \rightarrow CO_2(g) \qquad \Delta H^\circ_{298} = -393\tfrac{1}{2} \text{ kJ/mol} \quad (2)$$

$$H_2(g) + \tfrac{1}{2}O_2(g) \rightarrow H_2O(l) \qquad \Delta H^\circ_{298} = -286 \text{ kJ/mol} \quad (3)$$

Multiplying the definition $\Delta H^\circ = \Sigma_i \, \nu_i H^\circ_{m,i}$ [Eq. (5.3)] by -1, 2, and 3 for reactions (1), (2), and (3), respectively, we get

$$-(-1560 \text{ kJ/mol}) = -2H^\circ_m(CO_2) - 3H^\circ_m(H_2O) + H^\circ_m(C_2H_6) + 3.5H^\circ_m(O_2)$$

$$2(-393\tfrac{1}{2} \text{ kJ/mol}) = 2H^\circ_m(CO_2) - 2H^\circ_m(O_2) - 2H^\circ_m(C)$$

$$3(-286 \text{ kJ/mol}) = 3H^\circ_m(H_2O) - 3H^\circ_m(H_2) - 1.5H^\circ_m(O_2)$$

where the subscript 298 on the H°_m's is understood. Addition of these equations gives

$$-85 \text{ kJ/mol} = H^\circ_m(C_2H_6) - 2H^\circ_m(C) - 3H^\circ_m(H_2) \qquad (5.11)$$

But the quantity on the right side of (5.11) is ΔH° for the desired formation reaction

$$2C(\text{graphite}) + 3H_2(g) \rightarrow C_2H_6(g) \qquad (5.12)$$

Therefore $\Delta_f H^\circ_{298} = -85$ kJ/mol for ethane.

We can save time in writing if we just look at chemical reactions (1) to (3), figure out what factors are needed to multiply each reaction so that they add up to the desired reaction (5.12), and apply these factors to the ΔH° values. Thus, the desired reaction (5.12) has 2 moles of C on the left, and multiplication of reaction (2) by 2 will give 2 moles of C on the left. Similarly, we multiply reaction (1) by -1 to give 1 mole of C_2H_6 on the right and multiply reaction (3) by 3 to give 3 moles of H_2 on the left. Multiplication of reactions (1), (2), and (3) by -1, 2, and 3, followed by addition, gives reaction (5.12). Hence ΔH°_{298} for (5.12) is $[-(-1560) + 2(-393\tfrac{1}{2}) + 3(-286)]$ kJ/mol. The procedure of combining heats of several reactions to obtain the heat of a desired reaction is *Hess's law*. Its validity rests on the fact that H is a state function, so ΔH° is independent of the path used to go from reactants to products. ΔH° for the path elements $\rightarrow$ ethane is the same as ΔH° for the path

$$\text{elements} + \text{oxygen} \rightarrow \text{combustion products} \rightarrow \text{ethane} + \text{oxygen}$$

Since the reactants and products are not ordinarily in their standard states when we carry out a reaction, the actual enthalpy change ΔH_T for a reaction differs somewhat from ΔH°_T. However, this difference is small, and ΔH_T and ΔH°_T are unlikely to have different signs. For the discussion of this paragraph, we shall assume that ΔH_T and ΔH°_T have the same sign. If this sign is positive, the reaction is said to be **endothermic;** if this sign is negative, the reaction is **exothermic.** For a reaction run at constant pressure in a system with P-V work only, ΔH equals q_P, the heat flowing into the system.

The quantities ΔH_T and ΔH°_T correspond to enthalpy differences between products and reactants *at the same temperature T*; $\Delta H_T = H_{\text{products},T} - H_{\text{reactants},T}$. Therefore, when a reaction is run under constant-T-and-P conditions (in a constant-temperature bath), the heat q absorbed by the system equals ΔH_T. For an exothermic reaction ($\Delta H_T < 0$) run at constant T and P, q is negative and the system gives off heat to its surroundings. When an endothermic reaction is run at constant T and P, heat flows into the system. If an exothermic reaction is run under adiabatic and constant-P conditions, then $q = 0$ (since the process is adiabatic) and $\Delta H \equiv H_{\text{products}} - H_{\text{reactants}} = 0$ (since $\Delta H = q_P$); here, the products will be at a higher temperature than the reactants (Fig. 5.5). For an exothermic reaction run under conditions that are neither adiabatic nor isothermal, some heat flows to the surroundings and the temperature of the system rises by an amount that is less than ΔT under adiabatic conditions.

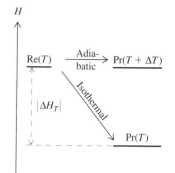

EXAMPLE 5.4 Calculation of $\Delta_f H°$ from $\Delta_c H°$

The standard enthalpy of combustion $\Delta_c H°_{298}$ of $C_2H_6(g)$ to $CO_2(g)$ and $H_2O(l)$ is -1559.8 kJ/mol. Use this $\Delta_c H°$ and Appendix data on $CO_2(g)$ and $H_2O(l)$ to find $\Delta_f H°_{298}$ of $C_2H_6(g)$.

Combustion means burning in oxygen. The combustion reaction for one mole of ethane is

$$C_2H_6(g) + \tfrac{7}{2}O_2(g) \rightarrow 2CO_2(g) + 3H_2O(l)$$

The relation $\Delta H° = \Sigma_i \nu_i \Delta_f H°_i$ [Eq. (5.6)] gives for this combustion

$$\Delta_c H° = 2\,\Delta_f H°(CO_2, g) + 3\,\Delta_f H°(H_2O, l) - \Delta_f H°(C_2H_6, g) - \tfrac{7}{2}\,\Delta_f H°(O_2, g)$$

Substitution of the values of $\Delta_f H°$ of $CO_2(g)$ and $H_2O(l)$ and $\Delta_c H°$ gives at 298 K

$$-1559.8 \text{ kJ/mol} = 2(-393.51 \text{ kJ/mol}) + 3(-285.83 \text{ kJ/mol})$$
$$- \Delta_f H°(C_2H_6, g) - 0$$
$$\Delta_f H°(C_2H_6, g) = -84.7 \text{ kJ/mol}$$

Note that this example essentially repeats the preceding Hess's law calculation. Reactions (2) and (3) above are the formation reactions of $CO_2(g)$ and $H_2O(l)$.

Exercise

$\Delta_c H°_{298}$ of crystalline buckminsterfullerene, $C_{60}(cr)$, is -2.589×10^4 kJ/mol [H. P. Diogo et al., *J. Chem. Soc. Faraday Trans.*, **89**, 3541 (1993)]. With the aid of Appendix data, find $\Delta_f H°_{298}$ of $C_{60}(cr)$. (*Answer*: 2.28×10^3 kJ/mol.)

Figure 5.5

Enthalpy changes for adiabatic versus isothermal transformations of reactants (Re) to products (Pr) at constant pressure. The reaction is exothermic.

Calculation of $H_{\text{id}} - H_{\text{re}}$

The standard state of a gas is the hypothetical ideal gas at 1 bar. To find $\Delta_f H°$ of a gaseous compound or a compound formed from gaseous elements, we must be able to calculate the difference between the standard-state ideal-gas enthalpy and the enthalpy of the real gas (steps 1 and 6 in the first part of Sec. 5.4). Let $H_{\text{re}}(T, P°)$ be the enthalpy of a (real) gaseous substance at T and $P°$, and let $H_{\text{id}}(T, P°)$ be the enthalpy of the corresponding fictitious ideal gas at T and $P°$, where $P° \equiv 1$ bar. $H_{\text{id}}(T, P°)$ is the enthalpy of a hypothetical gas in which each molecule has the same structure (bond distances and angles and conformation) as in the real gas but in which there are no forces between the molecules. To find $H_{\text{id}} - H_{\text{re}}$, we use the following hypothetical isothermal process at T:

$$\text{Real gas at } P° \xrightarrow{(a)} \text{real gas at 0 bar} \xrightarrow{(b)} \text{ideal gas at 0 bar} \xrightarrow{(c)} \text{ideal gas at } P° \quad (5.13)$$

In step (*a*), we isothermally reduce the pressure of the real gas from 1 bar to zero. In step (*b*), we wave a magic wand that eliminates intermolecular interactions, thereby changing the real gas into an ideal gas at zero pressure. In step (c), we isothermally increase the pressure of the ideal gas from 0 to 1 bar. The overall process converts the real gas at 1 bar and T into an ideal gas at 1 bar and T. For this process,

$$\Delta H = H_{id}(T, P°) - H_{re}(T, P°) = \Delta H_a + \Delta H_b + \Delta H_c \qquad (5.14)$$

The enthalpy change ΔH_a for step (a) is calculated from the integrated form of Eq. (4.48), $(\partial H/\partial P)_T = V - TV\alpha$ [Eq. (4.63) with $dT = 0$]:

$$\Delta H_a = H_{re}(T, 0 \text{ bar}) - H_{re}(T, P°) = \int_{P°}^{0} (V - TV\alpha)\, dP$$

For step (b), $\Delta H_b = H_{id}(T, 0 \text{ bar}) - H_{re}(T, 0 \text{ bar})$. The quantity $U_{re} - U_{id}$ (both at the same T) is $U_{intermol}$ (Sec. 2.11), the contribution of intermolecular interactions to the internal energy. Since intermolecular interactions go to zero as P goes to zero in the real gas, we have $U_{re} = U_{id}$ in the zero-pressure limit. Also, as P goes to zero, the equation of state for the real gas approaches that for the ideal gas. Therefore $(PV)_{re}$ equals $(PV)_{id}$ in the zero-pressure limit. Hence $H_{re} \equiv U_{re} + (PV)_{re}$ equals H_{id} in the zero-pressure limit:

$$H_{re}(T, 0 \text{ bar}) = H_{id}(T, 0 \text{ bar}) \quad \text{and} \quad \Delta H_b = 0 \qquad (5.15)$$

For step (c), ΔH_c is zero, since H of an ideal gas is independent of pressure. Equation (5.14) becomes

$$H_{id}(T, P°) - H_{re}(T, P°) = \int_{0}^{P°} \left[T\left(\frac{\partial V}{\partial T}\right)_P - V \right] dP \qquad \text{const. } T \qquad (5.16)$$

where $\alpha \equiv V^{-1}(\partial V/\partial T)_P$ was used. The integral in (5.16) is evaluated using P-V-T data or an equation of state (Sec. 8.8) for the real gas. The difference $H_{m,re} - H_{m,id}$ is quite small at 1 bar (since intermolecular interactions are quite small in a 1-bar gas) but is included in precise work. Some values of $H_{m,re} - H_{m,id}$ at 298 K and 1 bar are -7 J/mol for Ar, -17 J/mol for Kr, and -61 J/mol for C_2H_6. Figure 5.6 plots $H_{m,re}$ and $H_{m,id}$ versus P for $N_2(g)$ at 25°C, with $H_{m,id}$ arbitrarily set equal to zero. Steps (a) and (c) of the process (5.13) are indicated in the figure. Intermolecular attractions make U_{re} and H_{re} slightly less than U_{id} and H_{id}, respectively, at 1 bar.

Conventional Enthalpies. Instead of tabulating $\Delta_f H°$ values and using these to find $\Delta H°$ of reactions, one can construct a table of *conventional* (or *relative*) standard-state enthalpies of substances and use these to calculate $\Delta H°$ of reactions from $\Delta H° = \Sigma_i \nu_i H°_{m,i}$, where $H°_{m,i}$ is the conventional standard-state molar enthalpy of substance i. To construct such a table, we (1) pick sufficient reference substances and arbitrarily assign an enthalpy value to each reference substance in a specified reference state; (2) use thermodynamic equations and experimental data to calculate ΔH for the change from reference substances in their reference states to the desired substance in its standard state.

The most convenient choice of reference substances is the pure elements. The enthalpy reference state usually chosen is the standard state of the most stable form at 25°C and 1 bar. For each element in its most stable form at 25°C, the conventional standard-state enthalpy is arbitrarily set equal to zero:

$$H°_{m,298} = 0 \qquad \text{for each element in its stable form} \qquad (5.17)$$

Although the actual absolute enthalpies of different elements do differ, the convention (5.17) cannot lead to error in chemical reactions because elements are never interconverted in chemical reactions. Knowing the conventional $H°_{m,298}$ of an element, we can use $\Delta H = \int_1^2 C_P\, dT$ at constant P to find the conventional $H°_m$ of an element at any temperature T. If any phase changes occur between 298.15 K and T, we make separate allowance for them.

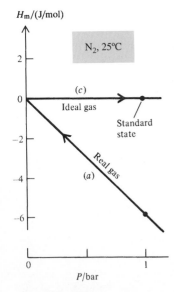

$H_m/(\text{J/mol})$

N_2, 25°C

(c)
Ideal gas

Standard state

Real gas

(a)

P/bar

Figure 5.6

Change in H_m with P for the isothermal conversion of real-gas N_2 to ideal-gas N_2 at 25°C.

So far, only elements have been considered. Suppose we want the conventional enthalpy of liquid water at T. The formation reaction is $H_2 + \frac{1}{2}O_2 \rightarrow H_2O$. Therefore $\Delta_f H_T^\circ(H_2O, l) = H_{m,T}^\circ(H_2O, l) - H_{m,T}^\circ(H_2, g) - \frac{1}{2}H_{m,T}^\circ(O_2, g)$. Knowing the conventional enthalpies $H_{m,T}^\circ$ for the elements H_2 and O_2, we use the experimental $\Delta_f H_T^\circ$ of $H_2O(l)$ (determined as discussed earlier) to find the conventional $H_{m,T}^\circ$ of $H_2O(l)$. Similarly, we can find conventional enthalpies of other compounds.

5.5 TEMPERATURE DEPENDENCE OF REACTION HEATS

Suppose we have determined ΔH° for a reaction at temperature T_1 and we want ΔH° at T_2. Differentiation of $\Delta H^\circ = \sum_i \nu_i H_{m,i}^\circ$ [Eq. (5.3)] with respect to T gives $d\,\Delta H^\circ/dT = \sum_i \nu_i dH_{m,i}^\circ/dT$, since the derivative of a sum equals the sum of the derivatives. (The derivatives are not partial derivatives. Since P is fixed at the standard-state value 1 bar, $H_{m,i}^\circ$ and ΔH° depend only on T.) The use of $(\partial H_{m,i}/\partial T)_P = C_{P,m,i}$ [Eq. (4.30)] gives

$$\frac{d\,\Delta H^\circ}{dT} = \sum_i \nu_i C_{P,m,i}^\circ \equiv \Delta C_P^\circ \tag{5.18}$$

where $C_{P,m,i}^\circ$ is the molar heat capacity of substance i in its standard state at the temperature of interest, and where we defined the **standard heat-capacity change** ΔC_P° for the reaction as equal to the sum in (5.18). More informally, if pr and re stand for stoichiometric numbers of moles of products and reactants, respectively, then

$$\frac{d\,\Delta H^\circ}{dT} = \frac{d(H_{pr}^\circ - H_{re}^\circ)}{dT} = \frac{dH_{pr}^\circ}{dT} - \frac{dH_{re}^\circ}{dT} = C_{P,pr}^\circ - C_{P,re}^\circ = \Delta C_P^\circ$$

Equation (5.18) is easy to remember since it resembles $(\partial H/\partial T)_P = C_P$.

Integration of (5.18) between the limits T_1 and T_2 gives

$$\Delta H_{T_2}^\circ - \Delta H_{T_1}^\circ = \int_{T_1}^{T_2} \Delta C_P^\circ \, dT \tag{5.19}$$

which is the desired relation (*Kirchhoff's law*).

An easy way to see the validity of (5.19) is from the following diagram:

$$\text{Standard-state reactants at } T_2 \overset{(a)}{\rightarrow} \text{standard-state products at } T_2$$

$$\downarrow (b) \qquad\qquad\qquad\qquad\qquad \uparrow (d)$$

$$\text{Standard-state reactants at } T_1 \overset{(c)}{\rightarrow} \text{standard-state products at } T_1$$

We can go from reactants to products at T_2 by a path consisting of step (a) or by a path consisting of steps $(b) + (c) + (d)$. Since enthalpy is a state function, ΔH is independent of path and $\Delta H_a = \Delta H_b + \Delta H_c + \Delta H_d$. The use of $\Delta H = \int_{T_1}^{T_2} C_P \, dT$ [Eq. (2.79)] to find ΔH_d and ΔH_b then gives Eq. (5.19).

Over a short temperature range, the temperature dependence of ΔC_P° in (5.19) can often be neglected to give $\Delta H_{T_2}^\circ \approx \Delta H_{T_1}^\circ + \Delta C_{P,T_1}^\circ (T_2 - T_1)$. This equation is useful if we have $C_{P,m}^\circ$ data at T_1 only, but can be seriously in error if $T_2 - T_1$ is not small.

The standard-state molar heat capacity $C_{P,m}^\circ$ of a substance depends on T only and is commonly expressed by a power series of the form

$$C_{P,m}^\circ = a + bT + cT^2 + dT^3 \tag{5.20}$$

where the coefficients a, b, c, and d are found by a least-squares fit of the experimental $C_{P,m}^\circ$ data. Such power series are valid only in the temperature range of the data used to find the coefficients. The temperature dependence of C_P was discussed in Sec. 2.11 (see Fig. 2.15).

EXAMPLE 5.5 Change in $\Delta H°$ with temperature

Use Appendix data and the approximation that $\Delta C_P°$ is independent of T to estimate $\Delta H°_{1200}$ for the reaction

$$2CO(g) + O_2(g) \rightarrow 2CO_2(g)$$

Equation (5.19) gives

$$\Delta H°_{1200} - \Delta H°_{298} = \int_{298\ K}^{1200\ K} \Delta C_P° \, dT \qquad (5.21)$$

Appendix $\Delta_f H°_{298}$ and $\Delta C_P°$ data give

$$\Delta H°_{298}/(\text{kJ/mol}) = 2(-393.509) - 2(-110.525) - 0 = -565.968$$

$$\Delta C_{P,298}°/(\text{J/mol-K}) = 2(37.11) - 2(29.116) - 29.355 = -13.37$$

With the approximation $\int_{T_1}^{T_2} \Delta C_P° \, dT \approx \Delta C_{P,T_1}° \int_{T_1}^{T_2} dT$, Eq. (5.21) becomes

$$\Delta H°_{1200} = -565968\ \text{J/mol} + (-13.37\ \text{J/mol-K})(1200\ \text{K} - 298.15\ \text{K})$$

$$= -578.03\ \text{kJ/mol}$$

Exercise

Use Appendix data and neglect the temperature dependence of $\Delta C_P°$ to estimate $\Delta H°_{1000}$ for $O_2(g) \rightarrow 2O(g)$. (*Answer:* 508.50 kJ/mol.)

EXAMPLE 5.6 Change in $\Delta H°$ with T

The $C_{P,m}°$'s of the gases O_2, CO, and CO_2 in the range 298 to 1500 K can each be represented by Eq. (5.20) with these coefficients:

	a/(J/mol-K)	b/(J/mol-K^2)	c/(J/mol-K^3)	d/(J/mol-K^4)
$O_2(g)$	25.67	0.01330	-3.764×10^{-6}	-7.310×10^{-11}
$CO(g)$	28.74	-0.00179	1.046×10^{-5}	-4.288×10^{-9}
$CO_2(g)$	21.64	0.06358	-4.057×10^{-5}	9.700×10^{-9}

Use these data and Appendix data to find for the reaction $2CO(g) + O_2(g) \rightarrow 2CO_2(g)$ an expression for $\Delta H_T°$ in the range 298 to 1500 K and calculate $\Delta H°_{1200}$. Was the approximation made in Example 5.5 justified?

We use Eq. (5.21). We have

$$\Delta C_P° = 2C_{P,m,CO_2}° - 2C_{P,m,CO}° - C_{P,m,O_2}°$$

Substitution of the series (5.20) for each $C_{P,m}°$ gives

$$\Delta C_P° = \Delta a + T \, \Delta b + T^2 \, \Delta c + T^3 \, \Delta d$$

where $\Delta a \equiv 2a_{CO_2} - 2a_{CO} - a_{O_2}$ with similar equations for Δb, Δc, and Δd. Substitution of $\Delta C_P°$ into (5.19) and integration gives

$$\Delta H_{T_2}° - \Delta H_{T_1}° = \Delta a(T_2 - T_1) + \tfrac{1}{2}\,\Delta b(T_2^2 - T_1^2) + \tfrac{1}{3}\,\Delta c(T_2^3 - T_1^3) + \tfrac{1}{4}\,\Delta d(T_2^4 - T_1^4)$$

Substitution of values in the table gives

$$\Delta a/(\text{J/mol-K}) = 2(21.64) - 2(28.74) - 25.67 = -39.87$$

$$\Delta b/(\text{J/mol-K}^2) = 0.11744, \quad \Delta c/(\text{J/mol-K}^3) = -9.8296 \times 10^{-5},$$

$$\Delta d/(\text{J/mol-K}^4) = 2.8049 \times 10^{-8}$$

With $T_1 = 298.15$ K, Example 5.5 gives $\Delta H^\circ_{T_1} = -565.968$ kJ/mol. We can thus use the $\Delta H^\circ_{T_2} - \Delta H^\circ_{T_1}$ equation to find $\Delta H^\circ_{T_2}$. Substitution of numerical values gives at $T_2 = 1200$ K,

$$\Delta H^\circ_{1200}/(\text{J/mol}) = -565968 - 39.87(901.85) + \tfrac{1}{2}(0.11744)(1.3511 \times 10^6)$$

$$+ \tfrac{1}{3}(-9.8296 \times 10^{-5})(1.7015 \times 10^9)$$

$$+ \tfrac{1}{4}(2.8049 \times 10^{-8})(2.0657 \times 10^{12})$$

$$\Delta H^\circ_{1200} = -563.85 \text{ kJ/mol}$$

The value -578.03 kJ/mol found in Example 5.5 with the approximation of taking ΔC°_P as constant is greatly in error, as might be expected since the temperature interval from 298 to 1200 K is large. The ΔC°_P-versus-T polynomial equation shows that $\Delta C^\circ_P/(\text{J/mol-K})$ is -13 at 298 K, -7 at 400 K, and 8 at 1200 K and is far from being constant.

Exercise

For O(g) in the range 298 to 1500 K, $C^\circ_{P,m}$ is given by the polynomial equation (5.20) with $a = 23.34$ J/(mol K), $b = -0.006584$ J/(mol K^2), $c = 5.902 \times 10^{-6}$ J/(mol K^3), and $d = -1.757 \times 10^{-9}$ J/(mol K^4). Find ΔH°_{1000} for O$_2(g) \rightarrow 2\text{O}(g)$. What is unusual about $C^\circ_{P,m}$ for O(g)? (*Answer:* 505.23 kJ/mol. It decreases with increasing T in this range.)

Note that ΔH°_{1200} in this example is not greatly changed from ΔH°_{298}. Usually, ΔH° and ΔS° for reactions not in solution change slowly with T (provided no species undergo phase changes in the temperature interval). The enthalpies and entropies of all reactants and products increase with T (Sec. 4.5), but the increases of products tend to cancel those of reactants, making ΔH° and ΔS° vary slowly with T.

5.6 USE OF A SPREADSHEET TO OBTAIN A POLYNOMIAL FIT

One often wants to fit a given set of data to a polynomial. This is easily done with a spreadsheet such as Excel, Lotus 1-2-3, or Quattro Pro. For example, $C^\circ_{P,m}/(\text{J/mol-K})$ values for CO(g) at 298.15, 400, 500, . . . , 1500 K are 29.142, 29.342, 29.794, 30.443, 31.171, 31.899, 32.577, 33.183, 33.710, 34.175, 34.572, 34.920, and 35.217. Suppose we want to find the coefficients in the cubic polynomial (5.20) that best fits these values. The following directions are for the Excel spreadsheet, which is part of the Microsoft Office suite of programs, and is widely available in student computer labs of colleges. The directions are for Excel 2000 and may need to be modified slightly for other versions.

Enter a title in cell A1. (To enter something in a cell, **select** the cell by clicking on it with the mouse, type the entry, and press Enter.) Enter the label T/K in cell A2 and the label Cp in cell B2. The temperatures are entered in cells A3 to A15 and the $C^\circ_{P,m}$ values in cells B3 to B15 (Fig. 5.7). Select all the data by dragging the mouse over cells A3 to B15. Click on the chart icon on the toolbar or chose Chart from the Insert menu. Go through the Chart Wizard dialog boxes, choosing XY (Scatter) as the type of plot and a plot showing data points only as the subtype. Choose Series in columns, omit titles and a legend, and place the chart as an object in the sheet with the data.

Figure 5.7

Cubic polynomial fit to $C_{P,m}^\circ$ of $CO(g)$.

	A	B	C	D	E	F	G
1	CO Cp polynomial fit			a	b	c	d
2	T/K	Cp	Cpfit	28.74	-0.00179	1.05E-05	-4.29E-09
3	298.15	29.143	29.022				
4	400	29.342	29.422				
5	500	29.794	29.923				
6	600	30.443	30.504				
7	700	31.171	31.14				
8	800	31.899	31.805				
9	900	32.577	32.474				
10	1000	33.183	33.12				
11	1100	33.71	33.718				
12	1200	34.175	34.242				
13	1300	34.572	34.667				
14	1400	34.92	34.967				
15	1500	35.217	35.115				

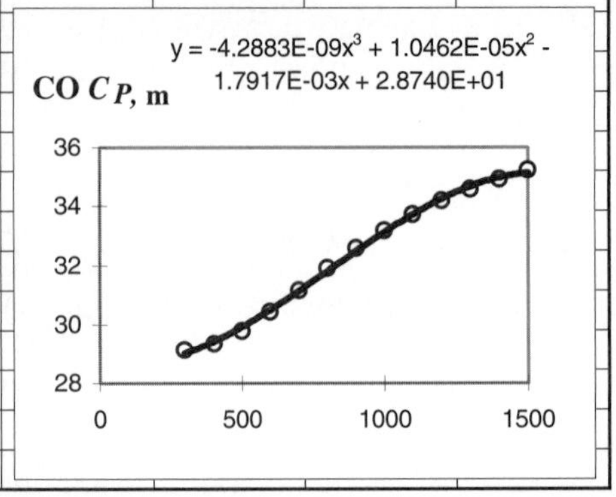

Chart: CO $C_{P,m}$, with displayed equation $y = -4.2883\text{E-}09x^3 + 1.0462\text{E-}05x^2 - 1.7917\text{E-}03x + 2.8740\text{E+}01$

After the plot appears, click on a data point on the chart, thereby highlighting all the data points. From the Chart menu, choose Add Trendline. In the Trendline dialog box, click on the polynomial picture and change the Order of the polynomial to 3. Click the Options tab of the Trendline dialog box and click in the box Display equation on chart. Then click OK. You will see the cubic-fit equation displayed on the chart, with coefficients equal to the values given in Example 5.6. (The coefficients are chosen so as to minimize the sum of the squares of the deviations of the experimental $C_{P,m}^\circ$ values from the values calculated with the polynomial.) You can adjust the number of significant figures visible in the coefficients by double-clicking on the trendline equation on the chart and clicking on the Number tab of the Format Data Labels box; then choose the Scientific category and change the number of decimal places.

To see how well values calculated from the polynomial equation fit the data, enter the labels a, b, c, d in cells D1, E1, F1, G1; enter the values of the coefficients in cells D2, E2, F2, G2; enter the label Cpfit in cell C2; and enter the formula

$$=\$D\$2+\$E\$2*A3+\$F\$2*A3^2+\$G\$2*A3^3$$

into cell C3. The equals sign tells the Excel spreadsheet that this is a **formula,** meaning that what is displayed in cell C3 will be the result of a calculation rather than the text of what is typed in C3. The * denotes multiplication and the ^ denotes exponentiation. The $ signs are explained below. When this formula is entered into C3, the number 29.022 appears in C3. This is the value of the polynomial (5.20) at 298.15 K, the value in A3. (To see the formula that lies behind a number in a cell, we can select the cell and look in the formula bar that lies above the spreadsheet.) Then click on cell C3 to highlight it and choose Copy from the Edit menu. Then click on cell C4 and drag the mouse from C4 to C16 to highlight these cells. Then choose Paste from the Edit menu. This will paste the polynomial formula into these cells but with the temperature cell A3 in the formula changed to A4 in cell C4, to A5 in cell C5, etc. Cells C3 to C15 will then contain the polynomial-fit values. (For an alternative procedure, see Probs. 5.28 and 5.29.)

The $ signs in the formula prevent D2, E2, F2, and G2 from being changed when the formula is copied from C3 into C4 to C16. A cell address with $ signs is called an *absolute* reference, whereas one without $ signs is a *relative* reference. When a formula is copied from one row to the row below, the row numbers of all relative references are increased by 1, while absolute references do not change. When the formula is copied to the second row below the original row, the row numbers of relative references are increased by 2; and so on. Figure 5.8 shows some of the formulas in column

	C
1	
2	Cpfit
3	=$D$2+$E$2*A3+$F$2*A3^2+$G$2*A3^3
4	=$D$2+$E$2*A4+$F$2*A4^2+$G$2*A4^3
5	=$D$2+$E$2*A5+$F$2*A5^2+$G$2*A5^3
6	=$D$2+$E$2*A6+$F$2*A6^2+$G$2*A6^3
7	=$D$2+$E$2*A7+$F$2*A7^2+$G$2*A7^3

Figure 5.8

Some of the formulas in column C of the Fig. 5.7 spreadsheet.

C. [To display all the formulas in their cells, click in the blank gray rectangle in the upper left corner of the spreadsheet to select all the cells and then hold down the Control key (the Command key on the Macintosh) while pressing the backquote (grave accent) key `. Pressing these keys again will restore the usual display.]

Excel is easy to use and very useful for solving many scientific problems. However, tests of the "reliability of Excel [97] in three areas: estimation, random number generation, and statistical distributions" concluded: "Excel [97] has been found inadequate in all three areas" [B. D. McCullough and B. Wilson, *Comput. Statist. Data Anal.,* **31,** 27 (1999)]. An assessment of the situation by the statistician Neil Cox (www1.agresearch.cri.nz:8000/Science/Statistics/exceluse.htm) concluded that "Excel [97] uses algorithms that are not robust and can lead to errors in extreme cases" and "The errors are very unlikely to arise in typical scientific data analysis." Links to resources on the use of Microsoft Excel in statistics can be found at www.mailbase.ac.uk/lists/assume/files/.

5.7   CONVENTIONAL ENTROPIES AND THE THIRD LAW

Conventional Entropies

The second law of thermodynamics tells us how to measure changes in entropy but does not provide absolute entropies. We could tabulate entropies of formation $\Delta_f S^\circ$, but this is not generally done. Instead, one tabulates conventional (or relative) entropies of substances. To set up a table of conventional standard-state entropies, we (1) assign an arbitrary entropy value to each element in a chosen reference state, and (2) find ΔS for the change from elements in their reference states to the desired substance in its standard state.

The choice of the entropy reference state is the pure element in its stable condensed form (solid or liquid) at 1 bar in the limit $T \to 0$ K. We arbitrarily set the molar entropy S_m for each *element* in this state equal to zero:

$$S^\circ_{m,0} = \lim_{T \to 0} S^\circ_{m,T} = 0 \qquad \text{element in stable condensed form} \qquad \textbf{(5.22)*}$$

The degree superscript in (5.22) indicates the standard pressure of 1 bar. The subscript zero indicates a temperature of absolute zero. As we shall see, absolute zero is unattainable, so we use the limit in (5.22). Helium remains a liquid as T goes to zero at 1 bar. All other elements are solids in this limit. Since elements are never interconverted in chemical reactions, we are free to make the arbitrary assignment (5.22) for each element.

To find the conventional $S^\circ_{m,T}$ for an element at any T, we use (5.22) and the constant-P equation $\Delta S = \int_{T_1}^{T_2} (C_P/T)\, dT$ [Eq. (3.30)], including also the ΔS of any phase changes that occur between absolute zero and T.

How do we find the conventional entropy of a *compound*? We saw that ΔU or ΔH values for reactions are readily measured as q_V or q_P for the reactions, and these ΔH values then allow us to set up a table of conventional enthalpies (or enthalpies of formation) for compounds. However, ΔS for a chemical reaction is not so easily measured. We have

$\Delta S = q_{rev}/T$ for constant temperature. However, a chemical reaction is an irreversible process, and measurement of the isothermal irreversible heat of a reaction does not give ΔS for the reaction. As we shall see in Chapter 14, one can carry out a chemical reaction reversibly in an electrochemical cell and use measurements on such cells to find ΔS values for reactions. Unfortunately, the number of reactions that can be carried out in an electrochemical cell is too limited to enable us to set up a complete table of conventional entropies of compounds, so we have a problem.

The Third Law of Thermodynamics

The solution to our problem is provided by the third law of thermodynamics. About 1900, T. W. Richards measured $\Delta G°$ as a function of temperature for several chemical reactions carried out reversibly in electrochemical cells. Walther Nernst pointed out that Richards's data indicated that the slope of the $\Delta G°$-versus-T curve for a reaction goes to zero as T goes to absolute zero. Therefore in 1907 Nernst postulated that for any change

$$\lim_{T \to 0} (\partial \Delta G/\partial T)_P = 0 \tag{5.23}$$

From (4.51), we have $(\partial G/\partial T)_P = -S$; hence $(\partial \Delta G/\partial T)_P = \partial(G_2 - G_1)/\partial T = \partial G_2/\partial T - \partial G_1/\partial T = -S_2 + S_1 = -\Delta S$. Thus (5.23) implies that

$$\lim_{T \to 0} \Delta S = 0 \tag{5.24}$$

Nernst believed (5.24) to be valid for any process. However, later experimental work by Simon and others showed (5.24) to hold only for changes involving substances in internal equilibrium. Thus (5.24) does not hold for a transition involving a supercooled liquid, which is not in internal equilibrium. (See also Sec. 22.9.)

We therefore adopt as the **Nernst–Simon statement of the third law of thermodynamics:**

For any isothermal process that involves only substances in internal equilibrium, the entropy change goes to zero as T goes to zero:

$$\lim_{T \to 0} \Delta S = 0 \tag{5.25}*$$

The Nernst–Simon statement is often restricted to pure substances but in fact it is valid for mixtures. (See J. A. Beattie and I. Oppenheim, *Principles of Thermodynamics,* Elsevier, 1979, secs. 11.18, 11.19, and 11.24.) The ideal-gas entropy-of-mixing formula (3.33) gives a nonzero isothermal ΔS of mixing that is independent of T and so seems to contradict the third law. The term *ideal gas* as used so far in this book means a *classical* ideal gas, which is a gas with no intermolecular interactions and with the molecules obeying classical mechanics. For such gases, $PV = nRT$ and the mixing formula (3.33) are obeyed. In reality, molecules obey quantum mechanics, not classical mechanics. Provided T is not close to absolute zero, it is an adequate approximation to use classical mechanics to treat the molecular motions. When T is close to absolute zero, one must use quantum mechanics. It is found that quantum ideal gases do obey the third law.

Although the third law does apply to mixtures, it is hard to achieve the required condition of internal equilibrium in solid mixtures at very low T, so to avoid error, it is safest to apply the third law only to pure substances.

Determination of Conventional Entropies

To see how (5.25) is used to find conventional entropies of compounds, consider the process

$$H_2(s) + \tfrac{1}{2}O_2(s) \to H_2O(s) \tag{5.26}$$

where the pure, separated elements at 1 bar and T are converted to the compound H_2O at 1 bar and T. For this process,

$$\Delta S = S_m^\circ(H_2O) - S_m^\circ(H_2) - \tfrac{1}{2}S_m^\circ(O_2) \tag{5.27}$$

Our arbitrary choice of the entropy of each element as zero at 0 K and 1 bar [Eq. (5.22)] gives $\lim_{T\to 0} S_m^\circ(H_2) = 0$ and $\lim_{T\to 0} S_m^\circ(O_2) = 0$. The third law, Eq. (5.25), gives for the process (5.26): $\lim_{T\to 0} \Delta S = 0$. In the limit $T \to 0$, Eq. (5.27) thus becomes $\lim_{T\to 0} S_m^\circ(H_2O) = 0$, which we write more concisely as $S_{m,0}^\circ(H_2O) = 0$.

Exactly the same argument applies for any compound. Hence $S_{m,0}^\circ = 0$ for any element or compound in internal equilibrium. The third law (5.25) shows that an isothermal pressure change of a substance in internal equilibrium in the limit of absolute zero has $\Delta S = 0$. Hence we can drop the superscript degree (which indicates $P = 1$ bar). Also, if $S_{m,0} = 0$, then $S_0 = 0$ for any amount of the substance. Our conclusion is that the conventional entropy of any element or compound in internal equilibrium is zero in the limit $T \to 0$:

$$S_0 = 0 \qquad \text{element or compound in int. equilib.} \tag{5.28}*$$

Now that we have the conventional standard-state entropies of substances at $T = 0$, their conventional standard-state entropies at any other T are readily found by using the constant-P equation $S_{T_2} - S_0 = S_{T_2} = \int_0^{T_2} (C_P/T)\, dT$ [Eq. (3.30)], with inclusion also of the ΔS of any phase changes between absolute zero and T_2. For example, for a substance that is a liquid at T_2 and 1 bar, to get S_{m,T_2}° we add the entropy changes for (a) warming the solid from 0 K to the melting point T_{fus}, (b) melting the solid at T_{fus} [Eq. (3.25)], and (c) warming the liquid from T_{fus} to T_2:

$$S_{m,T_2}^\circ = \int_0^{T_{fus}} \frac{C_{P,m}^\circ(s)}{T}\, dT + \frac{\Delta_{fus}H_m^\circ}{T_{fus}} + \int_{T_{fus}}^{T_2} \frac{C_{P,m}^\circ(l)}{T}\, dT \tag{5.29}$$

where $\Delta_{fus}H_m$ is the molar enthalpy change on melting (fusion) and $C_{P,m}(s)$ and $C_{P,m}(l)$ are the molar heat capacities of the solid and liquid forms of the substance. Since the standard pressure is 1 bar, each term in (5.29) is for a pressure of 1 bar. Thermodynamic properties of solids and liquids change very slowly with pressure (Sec. 4.5), and the difference between 1-bar and 1-atm properties of solids and liquids is experimentally undetectable, so it doesn't matter whether P is 1 bar or 1 atm in (5.29). At 1 atm, T_{fus} is the normal melting point of the solid (Sec. 7.2).

Frequently a solid undergoes one or more phase transitions from one crystalline form to another before the melting point is reached. For example, the stable low-T form of sulfur is orthorhombic sulfur; at 95°C, solid orthorhombic sulfur is transformed to solid monoclinic sulfur (whose melting point is 119°C). The entropy contribution of each such solid–solid phase transition must be included in (5.29) as an additional term $\Delta_{trs}H_m/T_{trs}$, where $\Delta_{trs}H_m$ is the molar enthalpy change of the phase transition at temperature T_{trs}.

For a substance that is a gas at 1 bar and T_2, we include the ΔS_m of vaporization at the boiling point T_b and the ΔS_m of heating the gas from T_b to T_2.

In addition, since the standard state is the ideal gas at 1 bar $\equiv P^\circ$, we include the small correction for the difference between ideal-gas and real-gas entropies. The quantity $S_{id}(T, P^\circ) - S_{re}(T, P^\circ)$ is calculated from the hypothetical isothermal three-step process (5.13). For step (a) of (5.13), we use $(\partial S/\partial P)_T = -(\partial V/\partial T)_P$ [Eq. (4.50)] to write $\Delta S_a = -\int_{P^\circ}^0 (\partial V/\partial T)_P\, dP = \int_0^{P^\circ} (\partial V/\partial T)_P\, dP$. For step (b) of (5.13), we use a result of statistical mechanics that shows that the entropy of a real gas and the entropy of the corresponding ideal gas (no intermolecular interactions) become equal in the limit of zero density (see Prob. 22.92). Therefore $\Delta S_b = 0$. For step (c), the use of

$(\partial S/\partial P)_T = -(\partial V/\partial T)_P$ [Eq. (4.50)] and $PV = nRT$ gives $\Delta S_c = -\int_0^{P^\circ} (nR/P)\, dP$. The desired ΔS is the sum $\Delta S_a + \Delta S_b + \Delta S_c$; per mole of gas, we have

$$S_{m,id}(T, P^\circ) - S_{m,re}(T, P^\circ) = \int_0^{P^\circ} \left[\left(\frac{\partial V_m}{\partial T} \right)_P - \frac{R}{P} \right] dP \qquad (5.30)$$

where the integral is evaluated at constant T. Knowledge of the P-V-T behavior of the real gas allows calculation of the contribution (5.30) to S_m°, the conventional standard-state molar entropy of the gas. (See Sec. 8.8.) Some values of $S_{m,id} - S_{m,re}$ in J/(mol K) at 25°C and 1 bar are 0.15 for $C_2H_6(g)$ and 0.67 for $n\text{-}C_4H_{10}(g)$.

The first integral in (5.29) presents a problem in that $T = 0$ is unattainable (Sec. 5.11). Also, it is impractical to measure $C_{P,m}^\circ(s)$ below a few degrees Kelvin. Debye's statistical-mechanical theory of solids (Sec. 24.12) and experimental data show that specific heats of nonmetallic solids at very low temperatures obey

$$C_{P,m}^\circ \approx C_{V,m}^\circ = aT^3 \qquad \text{very low } T \qquad (5.31)$$

where a is a constant characteristic of the substance. At the very low temperatures to which (5.31) applies, the difference $TV\alpha^2/\kappa$ between C_P and C_V [Eq. (4.53)] is negligible, because both T and α vanish (see Prob. 5.60) in the limit of absolute zero. For metals, a statistical-mechanical treatment (*Kestin and Dorfman,* sec. 9.5.2) and experimental data show that at very low temperatures

$$C_{P,m}^\circ \approx C_{V,m}^\circ = aT^3 + bT \qquad \text{metal at very low } T \qquad (5.32)$$

where a and b are constants. (The term bT arises from the conduction electrons.) One uses measured values of $C_{P,m}^\circ$ at very low temperatures to determine the constant(s) in (5.31) or (5.32). Then one uses (5.31) or (5.32) to extrapolate $C_{P,m}^\circ$ to $T = 0$ K. Note that C_P vanishes as T goes to zero.

For example, let $C_{P,m}^\circ(T_{low})$ be the observed value of $C_{P,m}^\circ$ of a nonconductor at the lowest temperature for which $C_{P,m}^\circ$ is conveniently measurable (typically about 10 K). Provided T_{low} is low enough for (5.31) to apply, we have

$$aT_{low}^3 = C_{P,m}^\circ(T_{low}) \qquad (5.33)$$

We write the first integral in (5.29) as

$$\int_0^{T_{fus}} \frac{C_{P,m}^\circ}{T}\, dT = \int_0^{T_{low}} \frac{C_{P,m}^\circ}{T}\, dT + \int_{T_{low}}^{T_{fus}} \frac{C_{P,m}^\circ}{T}\, dT \qquad (5.34)$$

The first integral on the right of (5.34) is evaluated by use of (5.31) and (5.33):

$$\int_0^{T_{low}} \frac{C_{P,m}^\circ}{T}\, dT = \int_0^{T_{low}} \frac{aT^3}{T}\, dT = \frac{aT^3}{3}\bigg|_0^{T_{low}} = \frac{aT_{low}^3}{3} = \frac{C_{P,m}^\circ(T_{low})}{3} \qquad (5.35)$$

To evaluate the second integral on the right side of (5.34) and the integral from T_{fus} to T_2 in (5.29), we can fit polynomials like (5.20) to the measured $C_{P,m}^\circ(T)$ data and then integrate the resulting expressions for $C_{P,m}^\circ/T$. Alternatively, we can use graphical integration: We plot the measured values of $C_{P,m}^\circ(T)/T$ against T between the relevant temperature limits, draw a smooth curve joining the points, and measure the area under the curve to evaluate the integral. Equivalently, since $(C_P/T)\, dT = C_P\, d\ln T$, we can plot C_P versus $\ln T$ and measure the area under the curve. Still another approach is numerical integration (Probs. 5.33 and 5.34).

EXAMPLE 5.7 Calculation of $S^\circ_{m,298}$

For SO_2, the normal melting and boiling points are 197.6 and 263.1 K. The heats of fusion and vaporization are 1769 and 5960 cal/mol, respectively, at the normal melting and boiling points. $C_{P,m}$ at 1 atm is graphed versus ln T in Fig. 5.9 from 15 K to 298 K; at 15.0 K, $C_{P,m} = 0.83$ cal/(mol K). [Data are mainly from W. F. Giauque and C. C. Stephenson, *J. Am. Chem. Soc.,* **60,** 1389 (1938).] The use of Eq. (5.30) gives $S_{m,id} - S_{m,re} = 0.07$ cal/(mol K) at 298 K and 1 atm (Prob. 8.25). Estimate $S^\circ_{m,298}$ of $SO_2(g)$.

Since the data are for 1 atm, we shall carry out the integrations at 1 atm pressure and include at the end the ΔS for changing the gas from 1 atm to 1 bar pressure.

From Eq. (5.35), integration of $(C_P/T) \, dT = C_P \, d \ln T$ from 0 to 15 K contributes [0.83 cal/(mol K)]/3 = 0.28 cal/(mol K).

The integral of $C_P \, d \ln T$ from 15 K to the melting point 197.6 K equals the area under the line labeled "Solid" in Fig. 5.9. This area is approximately a right triangle whose height is 16 cal/(mol K) and whose base is ln 197.6 − ln 15.0 = 5.286 − 2.708 = 2.58. The area of this triangle is $\frac{1}{2}$(2.58) [16 cal/(mol K)] = $20._6$ cal/(mol K). [An accurate evaluation using numerical integration gives 20.12 cal/(mol K); see Prob. 5.34.]

$\Delta_{fus}S_m$ equals $\Delta_{fus}H_m/T_{fus}$ = (1769 cal/mol)/(197.6 K) = 8.95 cal/(mol K).

The integral of $C_P \, d \ln T$ of the liquid from the melting point 197.6 K to the boiling point 263.1 K equals the area under the "Liquid" line. This area is approximately a rectangle of height 21 cal/(mol K) and base ln 263.1 − ln 197.6 = 0.286. The rectangle's area is [21 cal/(mol K)] (0.286) = 6.0 cal/(mol K). [Accurate evaluation gives 5.96 cal/(mol K); Prob. 5.34.]

ΔS_m of vaporization is (5960 cal/mol)/(263.1 K) = 22.65 cal/(mol K).

From Fig. 5.9, the integral of $C_P \, d \ln T$ for the gas from 263.1 to 298.15 K is the product of 10 cal/(mol K) and ln 298.15 − ln 263.1 = 0.125. This integral equals 1.2_5 cal/(mol K). [Accurate evaluation gives 1.22 cal/(mol K).]

So far, we have gone from the solid at 0 K and 1 atm to the real gas at 298.15 K and 1 atm. We next add in the given value $S_{m,id} - S_{m,re} = 0.07$ cal/(mol K) to reach the ideal gas at 298.15 K and 1 atm. The final step is to include ΔS_m for changing the ideal gas from 1 atm to 1 bar at 298.15 K. For an isothermal ideal-gas process, Eq. (3.29) and Boyle's law give $\Delta S_m = R \ln (V_2/V_1) = R \ln (P_1/P_2)$. The ΔS_m for going from 1 atm to 1 bar ($\approx$ 750 torr) is thus $R \ln (760/750) = 0.03$ cal/(mol K).

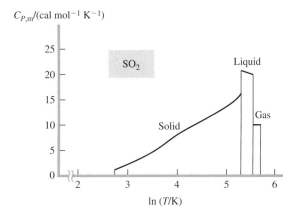

Figure 5.9

Integration of $C_{P,m} \, d \ln T$ for SO_2 at 1 atm.

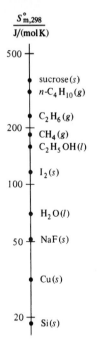

$$\frac{S^{\circ}_{m,298}}{J/(mol\,K)}$$

Figure 5.10

$S^{\circ}_{m,298}$ values. The scale is
logarithmic.

Adding everything, we get

$S^{\circ}_{m,298}$
$$\approx (0.28 + 20._6 + 8.95 + 6.0 + 22.65 + 1.2_5 + 0.07 + 0.03)\ \text{cal/(mol K)}$$

$$S^{\circ}_{m,298} \approx 59._8\ \text{cal/(mol K)}$$

[The accurate values give $S^{\circ}_{m,298} = 59.28$ cal/(mol K) $= 248.0$ J/(mol K).]

Exercise

Use Fig. 5.9 to estimate $S^{\circ}_{m,148} - S^{\circ}_{m,55}$ for $SO_2(s)$. (*Answer:* 11 cal mol^{-1} K^{-1}.)

Figure 5.10 plots some conventional $S^{\circ}_{m,298}$ values. The Appendix tabulates $S^{\circ}_{m,298}$ for various substances. Diamond has the lowest $S^{\circ}_{m,298}$ of any substance. The Appendix $S^{\circ}_{m,298}$ values show that (*a*) molar entropies of gases tend to be higher than those of liquids; (*b*) molar entropies of liquids tend to be higher than those of solids; (*c*) molar entropies tend to increase with increasing number of atoms in a molecule.

Conventional entropies are often called absolute entropies. However, this name is inappropriate in that these entropies are not absolute entropies but relative (conventional) entropies. Since full consideration of this question requires statistical mechanics, we postpone its discussion until Sec. 22.9.

Since $C_{P,m} = (\partial H_m/\partial T)_P$, integration of $C^{\circ}_{P,m}$ from 0 K to T with the addition of ΔH°_m for all phase transitions that occur between 0 and T gives $H^{\circ}_{m,T} - H^{\circ}_{m,0}$, where $H^{\circ}_{m,T}$ and $H^{\circ}_{m,0}$ are the standard-state molar enthalpies of the substance at T and of the corresponding solid at 0 K. For solids and liquids, $H^{\circ}_{m,T} - H^{\circ}_{m,0}$ is essentially the same as $U^{\circ}_{m,T} - U^{\circ}_{m,0}$. Figure 5.11 plots $H^{\circ}_{m,T} - H^{\circ}_{m,0}$ versus T and plots $S^{\circ}_{m,T}$ versus T for SO_2. Both H_m and S_m increase as T increases. Note the large increases in S and H that occur on melting and vaporization.

Standard Entropy of Reaction

For a reaction with stoichiometric coefficients ν_i, the **standard entropy change** is

$$\Delta S^{\circ}_T = \sum_i \nu_i S^{\circ}_{m,T,i} \tag{5.36}$$

which is similar to $\Delta H^{\circ} = \sum_i \nu_i H^{\circ}_{m,i}$ [Eq. (5.3)]. Using (5.36), we can calculate ΔS°_{298} from tabulated conventional entropies $S^{\circ}_{m,298}$.

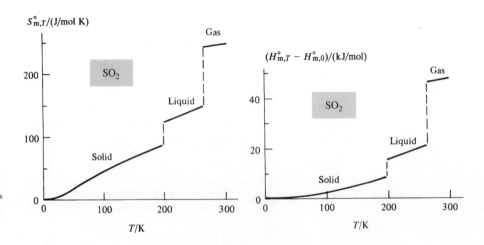

Figure 5.11

$S^{\circ}_{m,T}$ and $H^{\circ}_{m,T} - H^{\circ}_{m,0}$ versus T for SO_2, where $H^{\circ}_{m,0}$ is for solid SO_2.

Differentiation of (5.36) with respect to T and use of $(\partial S_i/\partial T)_P = C_{P,i}/T$ [Eq. (4.49)], followed by integration, give (Prob. 5.40)

$$\Delta S_{T_2}^{\circ} - \Delta S_{T_1}^{\circ} = \int_{T_1}^{T_2} \frac{\Delta C_P^{\circ}}{T} \, dT \tag{5.37}$$

which enables ΔS° at any T to be calculated from ΔS_{298}°. Note that (5.37) and (5.19) apply only if no species undergoes a phase change in the temperature interval.

EXAMPLE 5.8 ΔS° for a reaction

Use data in the Appendix to find ΔS_{298}° for the reaction $4NH_3(g) + 3O_2(g) \rightarrow 2N_2(g) + 6H_2O(l)$.

Substitution of Appendix $S_{m,298}^{\circ}$ values into (5.36) gives

$$\Delta S_{298}^{\circ}/[J/(mol\ K)] = 2(191.61) + 6(69.91) - 4(192.45) - 3(205.138)$$

$$= -582.53$$

Gases tend to have higher entropies than liquids, and the very negative ΔS° for this reaction results from the decrease of 5 moles of gases in the reaction.

Exercise

Find ΔS_{298}° for $2CO(g) + O_2(g) \rightarrow 2CO_2(g)$. (*Answer:* -173.01 J mol^{-1} K^{-1}.)

5.8   STANDARD GIBBS ENERGY OF REACTION

The **standard Gibbs energy (change)** ΔG_T° for a chemical reaction is the change in G for converting stoichiometric numbers of moles of the separated pure reactants, each in its standard state at T, into the separated pure products in their standard states at T. Similar to $\Delta H_T^{\circ} = \sum_i \nu_i H_{m,T,i}^{\circ}$ [Eq. (5.3)], we have

$$\Delta G_T^{\circ} = \sum_i \nu_i G_{m,T,i}^{\circ} \tag{5.38}$$

If the reaction is one of formation of a substance from its elements in their reference forms, then ΔG_T° is the **standard Gibbs energy of formation** $\Delta_f G_T^{\circ}$ of the substance. For an element in its reference form at T, $\Delta_f G_T^{\circ}$ is zero, since formation of an element from itself is no change at all. Recall from Sec. 4.6 that ΔG is physically meaningful only for processes with $\Delta T = 0$. The same reasoning that gave $\Delta H_T^{\circ} = \sum_i \nu_i \Delta_f H_{T,i}^{\circ}$ [Eq. (5.6)] shows that

$$\Delta G_T^{\circ} = \sum_i \nu_i \Delta_f G_{T,i}^{\circ} \tag{5.39}*$$

How do we get $\Delta_f G^{\circ}$ values? From $G \equiv H - TS$, we have $\Delta G = \Delta H - T\, \Delta S$ for an isothermal process. If the process is the formation reaction for substance i, then

$$\Delta_f G_{T,i}^{\circ} = \Delta_f H_{T,i}^{\circ} - T\, \Delta_f S_{T,i}^{\circ} \tag{5.40}$$

The standard entropy of formation $\Delta_f S_{T,i}^{\circ}$ is calculated from tabulated entropy values $S_{m,T}^{\circ}$ for substance i and its elements. Knowing $\Delta_f H_{T,i}^{\circ}$ and $\Delta_f S_{T,i}^{\circ}$, we can calculate and then tabulate $\Delta_f G_{T,i}^{\circ}$.

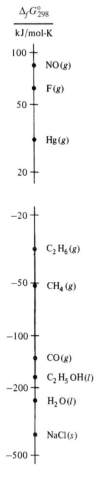

Figure 5.12

$\Delta_f G^\circ_{298}$ values. The scale is logarithmic.

EXAMPLE 5.9 Calculation of $\Delta_f G^\circ_{298}$

Use Appendix $\Delta_f H^\circ_{298}$ and $S^\circ_{m,298}$ data to calculate $\Delta_f G^\circ_{298}$ for $H_2O(l)$ and compare with the listed value.

The formation reaction is $H_2(g) + \frac{1}{2}O_2(g) \rightarrow H_2O(l)$, so

$$\Delta_f S^\circ_{298,H_2O(l)} = S^\circ_{m,298,H_2O(l)} - S^\circ_{m,298,H_2(g)} - \frac{1}{2}S^\circ_{m,298,O_2(g)}$$

$$\Delta_f S^\circ_{298} = [69.91 - 130.684 - \tfrac{1}{2}(205.138)] \text{ J/(mol K)} = -163.34_3 \text{ J/(mol K)}$$

$\Delta_f H^\circ_{298}$ is -285.830 kJ/mol, and (5.40) gives

$$\Delta_f G^\circ_{298} = -285.830 \text{ kJ/mol} - (298.15 \text{ K})(-0.16334_3 \text{ kJ/mol-K})$$

$$= -237.129 \text{ kJ/mol}$$

which agrees with the value listed in the Appendix.

Exercise

Use Appendix $\Delta_f H^\circ$ and S°_m data to calculate $\Delta_f G^\circ_{298}$ for $MgO(c)$ and compare with the listed value. (*Answer:* -569.41 kJ/mol.)

Figure 5.12 plots some $\Delta_f G^\circ_{298}$ values, and the Appendix lists $\Delta_f G^\circ_{298}$ for many substances. From tabulated $\Delta_f G^\circ_T$ values, we can find ΔG°_T for a reaction using (5.39).

EXAMPLE 5.10 ΔG° for a reaction

Find ΔG°_{298} for $4NH_3(g) + 3O_2(g) \rightarrow 2N_2(g) + 6H_2O(l)$ from Appendix data.

Substitution of Appendix $\Delta_f G^\circ_{298}$ values into (5.39) gives ΔG°_{298} as

$$[2(0) + 6(-237.129) - 3(0) - 4(-16.45)] \text{ kJ/mol} = -1356.97 \text{ kJ/mol}$$

Exercise

Use Appendix data to find ΔG°_{298} for $C_3H_8(g) + 5O_2(g) \rightarrow 3CO_2(g) + 4H_2O(l)$. (*Answer:* -2108.22 kJ/mol.)

Suppose we want ΔG° for a reaction at a temperature other than 298.15 K. We previously showed how to find ΔS° and ΔH° at temperatures other than 298.15 K. The use of $\Delta G^\circ_T = \Delta H^\circ_T - T \Delta S^\circ_T$ then gives ΔG° at any temperature T.

We have discussed calculation of thermodynamic properties from calorimetric data. We shall see in Chapter 22 that statistical mechanics allows thermodynamic properties of an ideal gas to be calculated from molecular data (bond distances and angles, vibrational frequencies, etc.).

An alternative to tabulating $\Delta_f G^\circ$ values is to tabulate conventional standard-state Gibbs energies $G^\circ_{m,T}$, defined by $G^\circ_{m,T} \equiv H^\circ_{m,T} - TS^\circ_{m,T}$, where $H^\circ_{m,T}$ and $S^\circ_{m,T}$ are conventional enthalpy and entropy values (Secs. 5.4 and 5.7). For an element in its reference form, the conventional $H^\circ_{m,298}$ is zero [Eq. (5.17)], but $S^\circ_{m,298}$ is not zero. ($S^\circ_{m,0}$ is zero.) Therefore the conventional $G^\circ_{m,298}$ of an element is not zero.

5.9 THERMODYNAMICS TABLES

Tabulations of thermodynamic data most commonly list $\Delta_f H^\circ_{298}$, $S^\circ_{m,298}$, $\Delta_f G^\circ_{298}$, and $C^\circ_{P,m,298}$. Older tables usually use the thermochemical calorie (= 4.184 J) as the energy unit. (Some physicists and engineers use the international-steam-table calorie, defined as 4.1868 J.) Newer tables use the joule.

Prior to 1982, the recommended standard-state pressure $P°$ was 1 atm, and values in older tables are for $P° = 1$ atm. In 1982, the International Union of Pure and Applied Chemistry (IUPAC) changed the recommended standard-state pressure to 1 bar, since 1 bar (= 10^5 Pa) is more compatible with SI units than 1 atm. Most newer tables use $P° = 1$ bar. Thermodynamic properties of solids and liquids vary very slowly with pressure (Sec. 4.5), and the change from 1 atm (760 torr) to 1 bar (750.062 torr) has a negligible effect on tabulated thermodynamic properties of solids and liquids. For a gas, the standard state is an ideal gas. For an ideal gas, H_m and $C_{P,m}$ depend on T only and are independent of pressure. Therefore $\Delta_f H°$ and $C°_{P,m}$ of gases are unaffected by the change to 1 bar. The effect of an isothermal pressure change on an ideal-gas entropy is given by (3.29) and Boyle's law as $S_2 - S_1 = nR \ln(P_1/P_2)$, so

$$S_{m,T,1\,bar} - S_{m,T,1\,atm} = (8.314 \text{ J/mol-K}) \ln(760/750.062) = 0.1094 \text{ J/(mol K)} \quad (5.41)$$

The change from 1 atm to 1 bar adds 0.109 J/(mol K) to $S°_m$ of a gas. This change is small but not negligible. Since $S°_m$ is changed, so is $\Delta_f G°$ if any species in the formation reaction is a gas (see Prob. 5.51). For a full discussion of the effects of the 1-atm to 1-bar change, see R. D. Freeman, *J. Chem. Educ.,* **62,** 681 (1985).

The tabulated values of $\Delta_f G°_T$ and $\Delta_f H°_T$ depend on the reference forms chosen for the elements at temperature T. There is a major exception to the rule that the reference form is the most stable form at T and 1 bar. For elements that are gases at 25°C and 1 bar, most thermodynamics tables choose the reference form as a gas for all temperatures below 25°C, even though the stable form might be the liquid or solid element. In mixing $\Delta_f G°$ and $\Delta_f H°$ data from two tables, one must be sure the same reference forms are used in both tables. Otherwise, error can result.

$\Delta H°$, $\Delta S°$, and $\Delta G°$ at temperatures other than 25°C can be calculated from tables of $\Delta_f H°$, $S°_m$, and $\Delta_f G°$ at various temperatures. Instead of tabulating $\Delta_f H°$ and $\Delta_f G°$ versus T, some tables list $H°_{m,T} - H°_{m,298}$ (or $H°_{m,T} - H°_{m,0}$) versus T and $(G°_{m,T} - H°_{m,298})/T$ [or $(G°_{m,T} - H°_{m,0})/T$] versus T. To find $\Delta H°_T$ and $\Delta G°_T$ from such tables, we use

$$\Delta H°_T = \Delta H°_{298} + \sum_i \nu_i (H°_{m,T} - H°_{m,298})_i \quad (5.42)$$

$$\Delta G°_T = \Delta H°_{298} + T \sum_i \nu_i [(G°_{m,T} - H°_{m,298})/T]_i \quad (5.43)$$

Equation (5.42) follows from $\sum_i \nu_i (H°_{m,T} - H°_{m,298})_i = \sum_i \nu_i H°_{m,T,i} - \sum_i \nu_i H°_{m,298,i} = \Delta H°_T - \Delta H°_{298}$. Equation (5.43) is proved similarly.

EXAMPLE 5.11 $\Delta G°_T$

At $T = 1000$ K, some values of $-(G°_{m,T} - H°_{m,298})/T$ (note the minus sign) in J/(mol K) are 220.877 for $O_2(g)$, 212.844 for $CO(g)$, and 235.919 for $CO_2(g)$. Find $\Delta G°_{1000}$ for $2CO(g) + O_2(g) \rightarrow 2CO_2(g)$.

Using Appendix $\Delta_f H°_{298}$ data, we find $\Delta H°_{298} = -565.968$ kJ/mol (as in Example 5.5 in Sec. 5.5). Substitution in (5.43) gives

$$\Delta G°_{1000} = -565.968 \text{ kJ/mol} + (1000 \text{ K}) [2(-235.919) - 2(-212.844)$$

$$- (-220.877)]10^{-3} \text{ kJ/(mol K)}$$

$$= -391.241 \text{ kJ/mol}$$

Exercise

For C(graphite), $H^\circ_{m,1000} - H^\circ_{m,298} = 11.795$ kJ/mol and $S^\circ_{m,1000} = 24.457$ J/(mol K). Use these data and data in the example to find ΔG°_{1000} for C(graphite) $+ O_2(g) \rightarrow CO_2(g)$. (*Answer:* −395.89 kJ/mol.)

The quantity $H^\circ_{m,T} - H^\circ_{m,298}$ is found by integrating $C^\circ_{P,m}$ data from 25°C to T, since $(\partial H/\partial T)_P = C_P$. We have

$$(G^\circ_{m,T} - H^\circ_{m,298})/T = (H^\circ_{m,T} - TS^\circ_{m,T})/T - H^\circ_{m,298}/T$$

$$= (H^\circ_{m,T} - H^\circ_{m,298})/T - S^\circ_{m,T}$$

so $(G^\circ_{m,T} - H^\circ_{m,298})/T$ is found from $H^\circ_{m,T} - H^\circ_{m,298}$ and $S^\circ_{m,T}$ data.

The reason for dividing $G^\circ_{m,T} - H^\circ_{m,298}$ by T is to make the function vary slowly with T, which enables accurate interpolation in the table. Tabulating $H^\circ_{m,T} - H^\circ_{m,298}$ and $(G^\circ_{m,T} - H^\circ_{m,298})/T$ is convenient because these quantities can be found from properties of one substance only (in contrast to $\Delta_f H^\circ$ and $\Delta_f G^\circ$, which also depend on properties of the elements), and these quantities are more accurately known than $\Delta_f H^\circ$ and $\Delta_f G^\circ$; moreover, these quantities for ideal gases can be accurately calculated using statistical mechanics (Chapter 22) if the molecular structure and vibration frequencies are known.

If we have thermodynamic data at only 25°C for the reaction species, we need expressions for $C^\circ_{P,m}$ of the species to find ΔH°, ΔS°, and ΔG° at other temperatures. $C^\circ_{P,m}$ polynomials [Eq. (5.20)] are given in O. Knacke et al., *Thermochemical Properties of Inorganic Substances,* 2d ed., Springer-Verlag, 1991, for 900 inorganic substances; in *Lide and Kehiaian* for 216 substances; and in *Reid, Prausnitz, and Poling* for 618 gases. Such polynomials are easily generated from $C^\circ_{P,m}$-versus-T data using a spreadsheet (Sec. 5.6).

A widely used tabulation of thermodynamic data for inorganic compounds, one- and two-carbon organic compounds, and species (including ions) in aqueous solution is D. D. Wagman et al., *The NBS Tables of Chemical Thermodynamic Properties,* 1982, published by the American Chemical Society and the American Institute of Physics for the National Bureau of Standards (vol. 11, supp. 2, of *J. Phys. Chem. Ref. Data*). These tables list $\Delta_f H^\circ_{298}$, $\Delta_f G^\circ_{298}$, $S^\circ_{m,298}$, and $C^\circ_{P,m,298}$ for about 10000 substances. Thermodynamic data for inorganic and organic compounds at 25°C and at other temperatures are given in *Landolt-Börnstein,* 6th ed., vol. II, pt. 4, pp. 179–474. $\Delta_f H^\circ_{298}$ data for many organic compounds are tabulated in J. B. Pedley et al., *Thermochemical Data of Organic Compounds,* 2d ed., Chapman and Hall, 1986; J. B. Pedley, *Thermochemical Data and Structures of Organic Compounds* (TRC Data Series), Springer-Verlag, 1994. $S^\circ_{m,298}$ and $C^\circ_{P,m,298}$ data for 2500 condensed-phase organic compounds are given in E. S. Domalski and E. D. Hearing, *J. Phys. Chem. Ref. Data,* **25,** 1 (1996).

The National Institute of Standards and Technology (NIST) Chemistry Webbook (webbook.nist.gov/) gives 25°C thermodynamic data for 5000 organic and inorganic compounds and gives C_P polynomial expressions for some substances.

Thermodynamic data over wide temperature ranges are tabulated for mainly inorganic compounds in (*a*) M. W. Chase et al., *NIST-JANAF Thermochemical Tables,* 4th ed., 1998, published by the American Chemical Society and the American Institute of Physics for the National Institute of Standards and Technology; (*b*) I. Barin, *Thermochemical Data of Pure Substances,* 3d ed., VCH, 1995; (*c*) O. Knacke et al., *Thermochemical Properties of Inorganic Substances,* 2d ed., Springer-Verlag, 1991; (*d*) O. Kubaschewski and C. B. Alcock, *Metallurgical Thermochemistry,* 5th ed., Pergamon, 1979.

Thermodynamic data over a range of T are given for organic compounds in (a) D. R. Stull et al., *The Chemical Thermodynamics of Organic Compounds,* Wiley, 1969 (gas-phase data); (b) *Selected Values of Properties of Hydrocarbons and Related Compounds,* 1966–1985, *Selected Values of Properties of Chemical Compounds,* 1966–1985, *TRC Thermodynamic Tables—Hydrocarbons,* 1985–, *TRC Thermodynamic Tables—Non-Hydrocarbons,* 1985–, all published in loose-leaf form by the Thermodynamics Research Center, Texas A&M University; (c) M. Frenkel et al., *Thermodynamics of Organic Compounds in the Gas State,* Vols. I and II (TRC Data Series), Springer-Verlag, 1994.

Thermodynamic data for biochemical compounds are tabulated by R. C. Wilhoit in chap. 2 of H. D. Brown (ed.), *Biochemical Microcalorimetry,* Academic Press, 1969; see also H.-J. Hinz (ed.), *Thermodynamic Data for Biochemistry and Biotechnology,* Springer-Verlag, 1986.

5.10 ESTIMATION OF THERMODYNAMIC PROPERTIES

About 10^7 chemical compounds are known, and it is likely that $\Delta_f H°$, $S°_m$, $C°_{P,m}$, and $\Delta_f G°$ for most known compounds will never be measured. Several methods have been proposed for estimating thermodynamic properties of a compound for which data do not exist. Chemical engineers often use estimation methods. It's a lot cheaper and faster to estimate needed unknown thermodynamic quantities than to measure them, and quantities obtained by estimation methods are sufficiently reliable to be useful for many purposes. An outstanding compilation of reliable estimation methods for thermodynamic and transport properties (Chapter 16) of liquids and gases is *Reid, Prausnitz, and Poling.*

Bond Additivity

Many properties can be estimated as the sum of contributions from the chemical bonds. One uses experimental data on compounds for which data exist to arrive at typical values for the bond contributions to the property in question. These bond contributions are then used to estimate the property in compounds for which data are unavailable. It should be emphasized that this approach is only an approximation.

Bond additivity methods work best for *ideal-gas* thermodynamic properties and usually cannot be applied to liquids or solids because of the unpredictable effects of intermolecular forces. For a compound that is a liquid or solid at 25°C and 1 bar, the ideal-gas state (like a supercooled liquid state) is not stable. Let P_{vp} be the liquid's vapor pressure at 25°C. To relate observable thermodynamic properties of the liquid at 25°C and 1 bar to ideal-gas properties at 25°C and 1 bar, we use the following isothermal process at 25°C (Fig. 5.13): (a) change the liquid's pressure from 1 bar to P_{vp}; (b) reversibly vaporize the liquid at 25°C and P_{vp}; (c) reduce the gas pressure to zero; (d) wave a magic wand that transforms the real gas to an ideal gas; (e) compress the ideal gas to $P = 1$ bar. Since the differences between real-gas and ideal-gas properties at 1 bar are quite small, one usually replaces steps (c), (d), and (e) with a compression of the gas (assumed to behave ideally) from pressure P_{vp} to 1 bar. Also, step (a) usually has a negligible effect on the liquid's properties. Thus, knowledge of ΔH_m of vaporization enables estimates of enthalpies and entropies of the liquid to be found from estimated ideal-gas enthalpies and entropies. Methods for estimation of $\Delta_{vap} H_m$ are discussed in *Reid, Prausnitz, and Poling,* chap. 7.

Benson and Buss constructed a table of bond contributions to $C°_{P,m,298}$, $S°_{m,298}$, and $\Delta_f H°_{298}$ for compounds *in the ideal-gas state* [S. W. Benson and J. H. Buss, *J. Chem. Phys.,* **29,** 546 (1958)]. Addition of these contributions enables one to estimate ideal-gas $S°_{m,298}$ and $C°_{P,m,298}$ values with typical errors of 1 to 2 cal/(mol K) and $\Delta_f H°_{298}$ values with typical errors of 3 to 6 kcal/mol. It should be noted that a contribution to

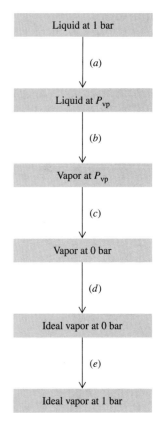

Figure 5.13

Conversion of a liquid at 25°C and 1 bar to an ideal gas at 25°C and 1 bar.

$S^\circ_{m,298}$ that arises from the symmetry of the molecule must be included to obtain valid results (see the discussion of the symmetry number in Chapter 22).

For example, some bond additivity contributions to $\Delta_f H^\circ_{298}$/(kcal/mol) are

C—C	C—H	C—O	O—H
2.73	−3.83	−12.0	−27.0

$\Delta_f H^\circ_{298}$/(kcal/mol) of $C_2H_6(g)$ and $C_4H_{10}(g)$ are then predicted to be $2.73 + 6(-3.83)$ $= -20.2$ and $3(2.73) + 10(-3.83) = -30.1$, as compared with the experimental values -20.0 for ethane, -30.4 for butane, and -32.1 for isobutane. Since the $\Delta_f H^\circ_{298}$ values are for formation from graphite and H_2, the bond-contribution values have built-in allowances for the enthalpy changes of the processes C(graphite) $\rightarrow$ C(g) and $H_2(g) \rightarrow 2H(g)$.

Bond Energies

Closely related to the concept of bond contributions to $\Delta_f H^\circ$ is the concept of **average bond energy.** Suppose we want to estimate ΔH°_{298} of a *gas*-phase reaction using molecular properties. We have $\Delta H^\circ_{298} = \Delta U^\circ_{298} + \Delta(PV)^\circ_{298}$. As noted in Sec. 5.4, the $\Delta(PV)^\circ$ term is generally substantially smaller than the ΔU° term, and ΔH° generally varies slowly with T. Therefore, ΔH°_{298} will usually be pretty close to ΔU°_0, the reaction's change in ideal-gas internal energy in the limit of absolute zero. Intermolecular forces don't contribute to ideal-gas internal energies, and at absolute zero, molecular translational and rotational energies are zero. Therefore ΔU°_0 is due to changes in molecular electronic energy and in molecular zero-point vibrational energy (Sec. 2.11). We shall see in Chapter 21 that electronic energies are much larger than vibrational energies, so it is a good approximation to neglect the change in zero-point vibrational energy. Therefore ΔU°_0 and ΔH°_{298} are largely due to changes in molecular electronic energy. To estimate this change, we imagine the reaction occurring by the following path:

$$\text{Gaseous reactants} \xrightarrow{(a)} \text{gaseous atoms} \xrightarrow{(b)} \text{gaseous products} \qquad (5.44)$$

In step (a), we break all bonds in the molecule and form separated atoms. It seems plausible that the change in electronic energy for step (a) can be *estimated* as the sum of the energies associated with each bond in the reacting molecules. In step (b), we form products from the atoms and we estimate the energy change as minus the sum of the bond energies in the products.

To show how bond energies are found from experimental data, consider the gas-phase atomization process

$$CH_4(g) \rightarrow C(g) + 4H(g) \qquad (5.45)$$

(**Atomization** is the dissociation of a substance into gas-phase atoms.) We define the average C—H bond energy in methane as one-fourth of ΔH°_{298} for the reaction (5.45). From the Appendix, $\Delta_f H^\circ_{298}$ of CH_4 is -74.8 kJ/mol. ΔH°_{298} for sublimation of graphite to C(g) is 716.7 kJ/mol. Hence $\Delta_f H^\circ_{298}$ of C(g) is 716.7 kJ/mol, as listed in the Appendix. (Recall that $\Delta_f H^\circ$ is zero for the *stable* form of an element. At 25°C, the stable form of carbon is graphite and not gaseous carbon atoms.) $\Delta_f H^\circ_{298}$ of H(g) is listed as 218.0 kJ/mol. [This is ΔH°_{298} for $\frac{1}{2}H_2(g) \rightarrow H(g)$.] For (5.45) we thus have

$$\Delta H^\circ_{298} = [716.7 + 4(218.0) - (-74.8)]\,\text{kJ/mol} = 1663.5\,\text{kJ/mol}$$

Hence the average C—H bond energy in CH_4 is 416 kJ/mol.

To arrive at a carbon–carbon single-bond energy, consider the process $C_2H_6(g) \rightarrow$ $2C(g) + 6H(g)$. Appendix $\Delta_f H^\circ_{298}$ values give $\Delta H^\circ_{298} = 2826$ kJ/mol for this reaction.

This ΔH°_{298} is taken as the sum of contributions from six C—H bonds and one C—C bond. Use of the CH_4 value 416 kJ mol^{-1} for the C—H bond, gives the C—C bond energy as $[2826 - 6(416)]$ kJ/mol = 330 kJ/mol.

The average-bond-energy method would then estimate the heat of atomization of propane $CH_3CH_2CH_3(g)$ at 25°C as $[8(416) + 2(330)]$ kJ/mol = 3988 kJ/mol. We break the formation of propane into two steps:

$$3C(\text{graphite}) + 4H_2(g) \rightarrow 3C(g) + 8H(g) \rightarrow C_3H_8(g)$$

The Appendix $\Delta_f H^{\circ}$ data give ΔH°_{298} for the first step as 3894 kJ/mol. We have estimated ΔH°_{298} for the second step as -3988 kJ/mol. Hence the average-bond-energy estimate of $\Delta_f H^{\circ}_{298}$ of propane is -94 kJ/mol. The experimental value is -104 kJ/mol, so we are off by 10 kJ/mol.

Some values for average bond energies are listed in Table 20.1 in Sec. 20.1. The C—H and C—C values listed differ somewhat from the ones calculated above, so as to give better overall agreement with experiment.

The bond-additivity-contribution method and the average-bond-energy method of finding $\Delta_f H^{\circ}_{298}$ are equivalent to each other. Each bond contribution to $\Delta_f H^{\circ}_{298}$ of a hydrocarbon is a combination of bond energies and the enthalpy changes of the processes $C(\text{graphite}) \rightarrow C(g)$ and $H_2(g) \rightarrow 2H(g)$ (see Prob. 5.57).

To estimate ΔH°_{298} for a gas-phase reaction, one uses (5.44) to write $\Delta H^{\circ}_{298} = \Delta_{at}H^{\circ}_{298,re} - \Delta_{at}H^{\circ}_{298,pr}$, where $\Delta_{at}H^{\circ}_{re}$ and $\Delta_{at}H^{\circ}_{pr}$, the heats of atomization of the reactants and products, can be found by adding up the bond energies. Corrections for strain energies in small-ring compounds, resonance energies in conjugated compounds, and steric energies in bulky compounds are often included.

Thus, the main contribution to ΔH° of a gas-phase reaction comes from the change in electronic energy that occurs when bonds are broken and new bonds formed. Changes in translational, rotational, and vibrational energies make much smaller contributions.

Group Additivity

Bond additivity and bond-energy calculations usually give reasonable estimates of gas-phase enthalpy changes, but can be significantly in error. An improvement on bond additivity is the method of group contributions. Here, one estimates thermodynamic quantities as the sum of contributions from groups in the molecule. Corrections for ring strain and for certain nonbonded interactions (such as the repulsion between two methyl groups that are bonded to adjacent carbons and that are in a *gauche* conformation) are included. A **group** consists of an atom in the molecule together with the atoms bonded to it. However, an atom bonded to only one atom is not considered to produce a group. The molecule $(CH_3)_3CCH_2CH_2Cl$ contains three $C–(H)_3(C)$ groups, one $C–(C)_4$ group, one $C–(C)_2(H)_2$ group, and one $C–(C)(H)_2(Cl)$ group, where the central atom of each group is listed first.

The group-contribution method requires tables with many more entries than the bond-contribution method. Tables of gas-phase group contributions to $\Delta_f H^{\circ}$, $C^{\circ}_{P,m}$, and S°_m for 300 to 1500 K are given in S. W. Benson et al., *Chem. Rev.*, **69**, 279 (1969), and S. W. Benson, *Thermochemical Kinetics*, 2d ed., Wiley-Interscience, 1976. See also N. Cohen and S. W. Benson, *Chem. Rev.*, **93**, 2419 (1993). These tables give $C^{\circ}_{P,m}$ and S°_m ideal-gas values with typical errors of 1 cal/(mol K) and $\Delta_f H^{\circ}$ ideal-gas values with typical errors of 1 or 2 kcal/mol. Some gas-phase group additivity values for $\Delta_f H^{\circ}_{298}$/(kJ/mol) are

$C–(C)(H)_3$	$C–(C)_2(H)_2$	$C–(C)_3H$	$C–(C)_4$	$O–(C)(H)$	$O–(C)_2$	$C–(C)(H)_2O$	$C–(H)_3(O)$
-41.8	-20.9	-10.0	-0.4	-158.6	-99.6	-33.9	-41.8

Group-additivity values for $\Delta_f H^\circ_{298}$ have been tabulated for solid, for liquid, and for gaseous C-H-O compounds in N. Cohen, *J. Phys. Chem. Ref. Data,* **25,** 1411 (1996). The average absolute errors are 1.3 kcal/mol for gases, 1.3 kcal/mol for liquids, and 2.2 kcal/mol for solids. (A few compounds with large errors were omitted in calculating these errors.)

The computer programs CHETAH (www.normas.com/ASTM/BOOKS/DS51C .html), NIST Therm/Est (www.esm-software.com/nist-thermest), and NIST Organic Structures and Properties (www.esm-software.com/nist-struct-prop) use Benson's group-additivity method to estimate thermodynamic properties of organic compounds.

Sign of ΔS°

Now consider ΔS°. Entropies of gases are substantially higher than those of liquids or solids, and substances with molecules of similar size have similar entropies. Therefore, for reactions involving only gases, pure liquids, and pure solids, the sign of ΔS° will usually be determined by the change in total number of moles of gases. If the change in moles of gases is positive, ΔS° will be positive; if this change is negative, ΔS° will be negative; if this change is zero, ΔS° will be small. For example, for $2H_2(g) + O_2(g) \rightarrow 2H_2O(l)$, the change in moles of gases is -3, and this reaction has $\Delta S^\circ_{298} = -327$ J/(mol K).

Other Estimation Methods

Thermodynamic properties of gas-phase compounds can often be rather accurately calculated by combining statistical-mechanics formulas with quantum-mechanical calculations (Secs. 22.6, 22.7, 22.8) or molecular-mechanics calculations (Sec. 20.13).

5.11 THE UNATTAINABILITY OF ABSOLUTE ZERO

Besides the Nernst–Simon formulation of the third law, another formulation of this law is often given, the *unattainability formulation.* In 1912, Nernst gave a "derivation" of the unattainability of absolute zero from the second law of thermodynamics (see Prob. 3.38). However, Einstein showed that Nernst's argument was fallacious, so the unattainability statement cannot be derived from the second law. [For details, see P. S. Epstein, *Textbook of Thermodynamics,* Wiley, 1937, pp. 244–245; F. E. Simon, *Z. Naturforsch.,* **6a,** 397 (1951); P. T. Landsberg, *Rev. Mod. Phys.,* **28,** 363 (1956); M. L. Boas, *Am. J. Phys.,* **28,** 675 (1960).]

The unattainability of absolute zero is usually regarded as a formulation of the third law of thermodynamics, equivalent to the entropy formulation (5.25). Supposed proofs of this equivalence are given in several texts. However, careful studies of the question show that the unattainability and entropy formulations of the third law are not equivalent [P. T. Landsberg, *Rev. Mod. Phys.,* **28,** 363 (1956); R. Haase, pp. 86–96, in *Eyring, Henderson, and Jost,* vol. I]. Haase concluded that the unattainability of absolute zero follows as a consequence of the first and second laws plus the Nernst–Simon statement of the third law. However, Landsberg disagreed with this conclusion and work by Wheeler also indicates that the unattainability formulation does not follow from the first and second laws plus the Nernst–Simon statement [J. C. Wheeler, *Phys. Rev. A,* **43,** 5289 (1991); **45,** 2637 (1992)]. Landsberg states that the third law of thermodynamics should be regarded as consisting of two nonequivalent statements: the Nernst–Simon entropy statement and the unattainability statement [P. T. Landsberg, *Am. J. Phys.,* **65,** 269 (1997)].

Although absolute zero is unattainable, temperatures as low as 2×10^{-8} K have been reached. One can use the Joule–Thomson effect to liquefy helium gas. By pumping away the helium vapor above the liquid, thereby causing the liquid helium to evaporate rapidly, one can attain temperatures of about 1 K. To reach lower temperatures, adiabatic demagnetization can be used. For details, see *Zemansky and Dittman,* chaps. 18 and 19, and P. V. E. McClintock et al., *Matter at Low Temperatures,* Wiley, 1984.

The lowest temperature reached in bulk matter by using adiabatic demagnetization is 1.2×10^{-5} K [K. Gloos et al., *J. Low. Temp. Phys.*, **73**, 101 (1988); *Discover*, June 1989, p. 16]. Using a combination of laser light, an applied inhomogeneous magnetic field, and applied radiofrequency radiation, physicists cooled a sample of 2000 low-pressure ^{87}Rb gas-phase atoms to 2×10^{-8} K [M. H. Anderson et al., *Science*, **269**, 198 (1995); jilawww.colorado.edu/bec/). Silver nuclei have been cooled to a nuclear-spin temperature of 2×10^{-9} K by adiabatic demagnetization (O. V. Lounasmaa, *Physics Today*, October 1989, p. 26).

5.12 SUMMARY

The standard state (symbolized by the ° superscript) of a pure liquid or solid at temperature T is defined as the state with $P = 1$ bar; for a pure gas, the standard state has $P = 1$ bar and the gas behaving ideally.

The standard changes in enthalpy, entropy, and Gibbs energy for the chemical reaction $0 \to \Sigma_i \nu_i A_i$ are defined as $\Delta H_T^\circ \equiv \Sigma_i \nu_i H_{m,T,i}^\circ$, $\Delta S_T^\circ \equiv \Sigma_i \nu_i S_{m,T,i}^\circ$, and $\Delta G_T^\circ \equiv \Sigma_i \nu_i G_{m,T,i}^\circ$ and are related by $\Delta G_T^\circ = \Delta H_T^\circ - T \Delta S_T^\circ$. ΔH° and ΔG° of a reaction can be calculated from tabulated $\Delta_f H^\circ$ and $\Delta_f G^\circ$ values of the species involved by using $\Delta H_T^\circ = \Sigma_i \nu_i \Delta_f H_{T,i}^\circ$ and $\Delta G_T^\circ = \Sigma_i \nu_i \Delta_f G_{T,i}^\circ$, where the standard enthalpy and Gibbs energy of formation $\Delta_f H_i^\circ$ and $\Delta_f G_i^\circ$ correspond to formation of one mole of substance i from its elements in their reference forms.

The convention that $S_0^\circ = 0$ for all elements and the third law of thermodynamics ($\Delta S_0 = 0$ for changes involving only substances in internal equilibrium) lead to a conventional S_0° value of zero for every substance. The conventional $S_{m,T}^\circ$ value of a substance can then be found by integration of $C_{P,m}^\circ / T$ from absolute zero with inclusion of ΔS of any phase transitions.

Using ΔH° (or ΔS°) at one temperature and C_P° data, one can calculate ΔH° (or ΔS°) at another temperature.

To avoid confusion, it is essential to pay close attention to thermodynamic symbols, including the subscripts and superscripts. The quantities H, ΔH, ΔH°, and $\Delta_f H^\circ$ generally have different meanings.

Important kinds of calculations discussed in this chapter include:

- Determination of ΔH° of a reaction by combining ΔH° values of other reactions (Hess's law).
- Calculation of $\Delta_r U$ from adiabatic bomb calorimetry data.
- Calculation of ΔH° from ΔU°, and vice versa.
- Calculation of S_m° of a pure substance from $C_{P,m}^\circ$ data, enthalpies of phase changes, and the Debye T^3 law.
- Calculation of ΔH°, ΔS°, and ΔG° of chemical reactions from tabulated $\Delta_f H^\circ$, S_m°, and $\Delta_f G^\circ$ data.
- Determination of ΔH° (or ΔS°) at one temperature from ΔH° (or ΔS°) at another temperature and $C_{P,m}^\circ(T)$ data.
- Estimation of ΔH° using bond energies.
- Use of a spreadsheet to fit equations to data.

FURTHER READING

Heats of reaction and calorimetry: *McGlashan*, pp. 17–25, 48–71; *Rossiter, Hamilton, and Baetzold*, vol. VI, chap. 7; S. Sunner and M. Mansson (eds.), *Combustion Calorimetry*, Pergamon, 1979. The third law: *Eyring, Henderson, and Jost*, vol. I, pp. 86–96, 436–486.

For data sources, see Sec. 5.9.

PROBLEMS

Section 5.1

5.1 True or false? (*a*) The term *standard state* implies that the temperature is 0°C. (*b*) The term *standard state* implies that the temperature is 25°C. (*c*) The standard state of a pure gas is the pure gas at a pressure of 1 bar and temperature *T*.

Section 5.2

5.2 True or false? (*a*) The SI units of $\Delta H°$ for a reaction are J. (*b*) Doubling the coefficients of a reaction doubles its $\Delta H°$. (*c*) $\Delta H°$ depends on temperature. (*d*) The reaction $N_2 + 3H_2 \rightarrow 2NH_3$ has $\Sigma_i \nu_i = -2$.

5.3 For $2H_2S(g) + 3O_2(g) \rightarrow 2H_2O(l) + 2SO_2(g)$, express $\Delta H°_T$ in terms of standard-state molar enthalpies $H°_{m,i}$ of the species involved.

5.4 For $Na(s) + HCl(g) \rightarrow NaCl(s) + \frac{1}{2}H_2(g)$, $\Delta H°_{298}$ is -319 kJ mol^{-1}. Find $\Delta H°_{298}$ for:
(*a*) $2Na(s) + 2HCl(g) \rightarrow 2NaCl(s) + H_2(g)$
(*b*) $4Na(s) + 4HCl(g) \rightarrow 4NaCl(s) + 2H_2(g)$
(*c*) $NaCl(s) + \frac{1}{2}H_2(g) \rightarrow Na(s) + HCl(g)$

Section 5.3

5.5 True or false? (*a*) $\Delta_f H°_{298}$ is zero for $O(g)$. (*b*) $\Delta_f H°_{298}$ is zero for $O_2(g)$. (*c*) $\Delta_f H°_{400}$ is zero for $O_2(g)$.

5.6 For each of the following, write the reaction of formation from reference-form elements at room temperature: (*a*) $CCl_4(l)$; (*b*) $NH_2CH_2COOH(s)$; (*c*) $H(g)$; (*d*) $N_2(g)$.

5.7 For which elements is the reference form at 25°C (*a*) a liquid; (*b*) a gas?

Section 5.4

5.8 Write balanced reactions for the combustion of one mole of each of the following to $CO_2(g)$ and $H_2O(l)$. (*a*) $C_4H_{10}(g)$; (*b*) $C_2H_5OH(l)$.

5.9 True or false? (*a*) When sucrose is burned in an adiabatic constant-volume calorimeter, $\Delta U = 0$ for the combustion process, where the system is the calorimeter contents. (*b*) The reaction $N_2(g) + 3H_2(g) \rightarrow 2NH_3(g)$ has $\Delta H°_T < \Delta U°_T$. (*c*) The reaction $N_2(g) \rightarrow 2N(g)$ is endothermic. (*d*) When an exothermic reaction is carried out in an adiabatic container, the products are at a higher temperature than the reactants. (*e*) For $CH_3OH(l)$, $\Delta_f H°_{298} - \Delta_f U°_{298}$ equals $\Delta_c H°_{298} - \Delta_c U°_{298}$ [where $H_2O(l)$ is formed in the combustion reaction].

5.10 Use data in the Appendix to find $\Delta H°_{298}$ for:
(*a*) $2H_2S(g) + 3O_2(g) \rightarrow 2H_2O(l) + 2SO_2(g)$
(*b*) $2H_2S(g) + 3O_2(g) \rightarrow 2H_2O(g) + 2SO_2(g)$
(*c*) $2HN_3(g) + 2NO(g) \rightarrow H_2O_2(l) + 4N_2(g)$

5.11 (*a*) Use Appendix data to find $\Delta_c H°_{298}$ and $\Delta_c U°_{298}$ of α-D-glucose(*c*), $C_6H_{12}O_6$, to $CO_2(g)$ and $H_2O(l)$. (*b*) 0.7805 g of α-D-glucose is burned in the adiabatic bomb calorimeter of Fig. 5.4. The bomb is surrounded by 2.500 L of H_2O at 24.030°C. The bomb is made of steel and weighs 14.05 kg. Specific heats at constant pressure of water and steel at 24°C are 4.180 and

0.450 J/(g °C), respectively. The density of water at 24°C is 0.9973 g/cm^3. Assuming the heat capacity of the chemicals in the bomb is negligible compared with the heat capacity of the bomb and surrounding water, find the final temperature of the system. Neglect the temperature dependence of c_P. Neglect the changes in thermodynamic functions that occur when the reactants and products are brought from their standard states to those that occur in the calorimeter.

5.12 Repeat Prob. 5.11*b*, taking account of the heat capacity of the bomb contents. The bomb has an interior volume of 380 cm^3 and is initially filled with $O_2(g)$ at 30 atm pressure.

5.13 When 0.6018 g of naphthalene, $C_{10}H_8(s)$, was burned in an adiabatic bomb calorimeter, a temperature rise of 2.035 K was observed and 0.0142 g of fuse wire (used to ignite the sample) was burned. In the same calorimeter, combustion of 0.5742 g of benzoic acid produced a temperature rise of 1.270 K, and 0.0121 g of fuse wire was burned. The ΔU for combustion of benzoic acid under typical bomb conditions is known to be -26.434 kJ/g, and the ΔU for combustion of the wire is -6.28 kJ/g. (*a*) Find the average heat capacity of the calorimeter and its contents. Neglect the difference in heat capacity between the chemicals in the two experiments. (*b*) Neglecting the changes in thermodynamic functions that occur when species are brought from their standard states to those that occur in the calorimeter, find $\Delta_c U°$ and $\Delta_c H°$ of naphthalene.

5.14 The reaction $2A(g) + 3B(l) \rightarrow 5C(g) + D(g)$ is carried out in an adiabatic bomb calorimeter. An excess of A is added to 1.450 g of B. The molecular weight of B is 168.1. The reaction goes essentially to completion. The initial temperature is 25.000°C. After the reaction, the temperature is 27.913°C. A direct current of 12.62 mA (milliamperes) flowing through the calorimeter heater for 812 s is needed to bring the product mixture from 25.000°C to 27.913°C, the potential drop across the heater being 8.412 V. Neglecting the changes in thermodynamic functions that occur when the reactants and products are brought from their standard states to the states that occur in the calorimeter, estimate $\Delta U°_{298}$ and $\Delta H°_{298}$ for this reaction. (One watt = one volt × one ampere = one joule per second.)

5.15 For $H_2(g) + \frac{1}{2}O_2(g) \rightarrow H_2O(l)$, find $\Delta H°_{298} - \Delta U°_{298}$ (*a*) neglecting $V°_{m,H_2O(l)}$; (*b*) not neglecting $V°_{m,H_2O(l)}$.

5.16 The standard enthalpy of combustion at 25°C of liquid acetone $(CH_3)_2CO$ to $CO_2(g)$ and $H_2O(l)$ is -1790 kJ/mol. Find $\Delta_f H°_{298}$ and $\Delta_f U°_{298}$ of $(CH_3)_2CO(l)$.

5.17 The standard enthalpy of combustion of the solid amino acid alanine, $NH_2CH(CH_3)COOH$, to $CO_2(g)$, $H_2O(l)$, and $N_2(g)$ at 25°C is -1623 kJ/mol. Find $\Delta_f H°_{298}$ and $\Delta_f U°_{298}$ of solid alanine. Use data in the Appendix.

5.18 Given the following $\Delta H°_{298}$ values in kcal/mol, where *gr* stands for graphite,

$$Fe_2O_3(s) + 3C(gr) \rightarrow 2Fe(s) + 3CO(g) \qquad 117$$
$$FeO(s) + C(gr) \rightarrow Fe(s) + CO(g) \qquad 37$$

$$2CO(g) + O_2(g) \rightarrow 2CO_2(g) \qquad -135$$

$$C(gr) + O_2(g) \rightarrow CO_2(g) \qquad -94$$

find $\Delta_f H^\circ_{298}$ of FeO(s) and of $Fe_2O_3(s)$.

5.19 Given the following ΔH°_{298}/(kJ/mol) values,

$$4NH_3(g) + 5O_2(g) \rightarrow 4NO(g) + 6H_2O(l) \quad -1170$$

$$2NO(g) + O_2(g) \rightarrow 2NO_2(g) \qquad -114$$

$$3NO_2(g) + H_2O(l) \rightarrow 2HNO_3(l) + NO(g) \qquad -72$$

find ΔH°_{298} for $NH_3(g) + 2O_2(g) \rightarrow HNO_3(l) + H_2O(l)$ without using Appendix data.

5.20 Apply $\Delta H^\circ = \Sigma_i \nu_i \Delta_f H^\circ_i$ to Eq. (1) preceding Eq. (5.11) and use data in Eqs. (1), (2), and (3) to find $\Delta_f H^\circ_{298}$ of $C_2H_6(g)$.

5.21 (a) A gas obeys the equation of state $P(V_m - b) = RT$, where b is a constant. Show that, for this gas, $H_{m,id}(T, P) - H_{m,re}(T, P) = -bP$. (b) If $b = 45$ cm³/mol, calculate $H_{m,id} - H_{m,re}$ at 25°C and 1 bar.

5.22 Use Appendix data to find the conventional H°_m of (a) $H_2(g)$ at 25°C; (b) $H_2(g)$ at 35°C; (c) $H_2O(l)$ at 25°C; (d) $H_2O(l)$ at 35°C. Neglect the temperature dependence of C_P.

Section 5.5

5.23 True or false? (a) The rate of change of ΔH° with respect to temperature is equal to ΔC°_P. (b) The rate of change of ΔH° with respect to pressure is zero. (c) For a reaction involving only ideal gases, ΔC°_P is independent of temperature. (d) $\int_{T_1}^{T_2} T \, dT = \frac{1}{2}(T_2 - T_1)^2$.

5.24 Use data in the Appendix and the approximation of neglecting the temperature dependence of $C^\circ_{P,m}$ to estimate ΔH°_{370} for the reactions of Prob. 5.10.

5.25 Compute $\Delta_f H^\circ_{1000}$ of HCl(g) from Appendix data and these $C^\circ_{P,m}$/[J/(mol K)] expressions, which hold from 298 to 1500 K.

$$27.14 + 0.009274(T/K) - 1.381(10^{-5} \, T^2/K^2)$$
$$+ 7.645(10^{-9} \, T^3/K^3)$$

$$26.93 + 0.03384(T/K) - 3.896(10^{-5} \, T^2/K^2)$$
$$+ 15.47(10^{-9} \, T^3/K^3)$$

$$30.67 - 0.007201(T/K) - 1.246(10^{-5} \, T^2/K^2)$$
$$- 3.898(10^{-9} \, T^3/K^3)$$

for $H_2(g)$, $Cl_2(g)$, and HCl(g), in that order.

Section 5.6

5.26 Set up a spreadsheet and verify the CO C_P fit given in Sec. 5.6.

5.27 Values of $C^\circ_{P,m}$/(J/mol-K) for $O_2(g)$ at T/K values of 298.15, 400, 500, . . . , 1500 are 29.376, 30.106, 31.091, 32.090, 32.981, 33.733, 34.355, 34.870, 35.300, 35.667, 35.988, 36.277, and 36.544. Use a spreadsheet to fit a cubic polynomial [Eq. (5.20)] to these data.

5.28 Instead of inserting a trendline, another Excel procedure to fit a cubic function to C_P data is as follows. Enter the C_P data in cells A3 to A15; enter the T values in B3 to B15; enter the T^2 values in C3 to C15 by entering the formula =B3^2 in C3 and copying and pasting this formula to C4 to C15; enter the T^3 values in D3 to D15. From the Tools menu, choose Data Analysis. (If Data Analysis is not visible on the Tools menu, choose Add-Ins on the Tools menu, check Analysis ToolPak and click OK.) In the Data Analysis box choose Regression and click OK. In Input Y Range enter A3:A15 (the colon indicates a range); in Input X Range enter B3:D15; click in the Residuals Box, the Line Fit Plots box, and the Residual Plots box; then click OK. On a new sheet in the workbook, you will get output that includes the desired coefficients in a column labeled Coefficients. The predicted C_P values and their errors (the **residuals**) will also be listed. (You can go from one sheet of a workbook to another by clicking on the tab for the desired sheet at the bottom of the screen.) Carry out this procedure for the CO C_P data and verify that the same results are found as in Sec. 5.5. The Regression procedure allows one to find the coefficients $A, B, C, D, \ldots$ in the fit $g(x) = A + Bf_1(x) + Cf_2(x) + Df_3(x) + \cdots$, where $f_1, f_2, f_3, \ldots$ are functions that do not contain unknown constants. In this example, the f's are T, T^2, and T^3.

5.29 Another form besides (5.20) used to fit C_P data is $A + BT + CT^2 + D/T^2$. Use the Regression procedure of Prob. 5.28 to find the coefficients $A, B, C,$ and D that fit the CO data. You will need a column containing $1/T^2$ values. Use the spreadsheet to calculate the sum of the squares of the residuals for this fit and compare with the fit given by (5.20). Entering the Excel formula =SUM(K3:K15) into a cell will put the sum of the numbers in cells K3 to K15 into that cell.

Section 5.7

5.30 True or false? For the combustion of glucose, ΔS°_T equals $\Delta H^\circ_T/T$.

5.31 For solid 1,2,3-trimethylbenzene, $C^\circ_{P,m} = 0.62$ J mol⁻¹ K⁻¹ at 10.0 K. Find S°_m at 10.0 K for this substance. Find $C^\circ_{P,m}$ and S°_m at 6.0 K for this substance.

5.32 Substance Y melts at 200 K and 1 atm with $\Delta_{fus}H_m = 1450$ J/mol. For solid Y, $C^\circ_{P,m} = cT^3 + dT^4$ for 10 K $\leq T \leq$ 20 K and $C^\circ_{P,m} = e + fT + gT^2 + hT^3$ for 20 K $\leq T \leq$ 200 K. For liquid Y, $C^\circ_{P,m} = i + jT + kT^2 + lT^3$ for 200 K $\leq T \leq$ 300 K. (a) Express $S^\circ_{m,300}$ of liquid Y in terms of the constants $c, d, e, \ldots, l$. (b) Express $H^\circ_{m,300} - H^\circ_{m,0}$ of liquid Y in terms of these constants. Neglect the difference between 1-atm and 1-bar properties of the solid and liquid.

5.33 The definite integral $\int_a^b f(x) \, dx$ can be estimated as follows. The interval from a to b is divided into n subintervals each of width w. Let $f(a) = f_0, f(a + w) = f_1, f(a + 2w) = f_2, \ldots, f(a + nw) = f(b) = f_n$. The *trapezoidal rule* is

$$\int_a^b f(x) \, dx \approx w\left(\tfrac{1}{2}f_0 + f_1 + f_2 + \cdots + f_{n-1} + \tfrac{1}{2}f_n\right)$$

The more accurate *Simpson's rule* requires that n be even and is

$$\int_a^b f(x)\, dx \approx \tfrac{1}{3} w(f_0 + 4f_1 + 2f_2 + 4f_3$$

$$+ \cdots + 2f_{n-2} + 4f_{n-1} + f_n)$$

(*a*) Derive the trapezoidal rule as follows. Join the points $f_0, f_1,$ $f_2, \ldots , f_n$ on the curve $f(x)$ by straight-line segments; then estimate the area under the curve as the sum of the areas under the line segments. A *trapezoid* is a quadrilateral with two parallel sides; its area equals $\tfrac{1}{2}(c + d)s$, where c and d are the lengths of the parallel sides and s is the perpendicular distance between them. (*b*) Estimate $\int_1^2 x^{-1}\, dx$ using the trapezoidal rule first with $n = 10$ and then with $n = 20$; then use Simpson's rule with $n = 10$. Compare with the exact value.

5.34 $C_{P,m}$ values at 1 atm for SO_2 [mainly from Giauque and Stephenson, *J. Am. Chem. Soc.*, **60**, 1389 (1938)] are as follows, where the first number in each pair is T/K and the second number (in boldface type) is $C_{P,m}$ in cal/(mol K). *Solid:* 15, **0.83**; 20, **1.66**; 25, **2.74**; 30, **3.79**; 35, **4.85**; 40, **5.78**; 45, **6.61**; 50, **7.36**; 55, **8.02**; 60, **8.62**; 70, **9.57**; 80, **10.32**; 90, **10.93**; 100, **11.49**; 110, **11.97**; 120, **12.40**; 130, **12.83**; 140, **13.31**; 150, **13.82**; 160, **14.33**; 170, **14.85**; 180, **15.42**; 190, **16.02**; 197.64, **16.50**. *Liquid:* 197.64, **20.98**; 200, **20.97**; 220, **20.86**; 240, **20.76**; 260, **20.66**; 263.1, **20.64**. *Gas:* 263.1, **9.65**; 280, **9.71**; 298.15, **9.80**. (*a*) Use the trapezoidal rule (Prob. 5.33) to evaluate the contribution to S_m made by $\int (C_P/T)\, dT$ from 15 K to the melting point 197.6 K; apply the rule separately to the intervals from 15 to 60 K, from 60 to 190 K, and from 190 to 197.6 K. (*b*) Repeat (*a*) using Simpson's rule from 15 to 55 K and from 60 to 180 K and the trapezoidal rule for the rest. (*c*) Use the trapezoidal rule to evaluate $\int (C_P/T)\, dT$ for the liquid between its freezing point and its boiling point 263.1 K; repeat for the gas between 263.1 and 298.15 K.

5.35 (This problem is only for masochists.) Use data in Prob. 5.34 and graphical integration to evaluate $\int (C_P/T)\, dT$ for solid SO_2 from 15 K to the melting point. Graphical integration can be done by counting the number of squares under the curve (estimating the fractions of squares partly under the curve) or by cutting out the area under the curve, weighing it, and weighing a known number of squares.

5.36 Suppose that instead of the convention (5.22), we had taken $S_{m,0}^\circ$ of graphite, $H_2(s)$, and $O_2(s)$ to be a, b, and c, respectively, where a, b, and c are certain constants. (*a*) How would $S_{m,298}^\circ$ for graphite, $H_2(g)$, $O_2(g)$, $CH_4(g)$, $H_2O(l)$, and $CO_2(g)$ be changed from their values listed in the Appendix? (*b*) How would ΔS_{298}° for $CH_4(g) + 2O_2(g) \rightarrow CO_2(g) + 2H_2O(l)$ be changed from its value calculated from Appendix data?

5.37 Use data in the Appendix and data preceding Eq. (4.54) and make certain approximations to calculate the conventional S_m of $H_2O(l)$ at (*a*) 298.15 K and 1 bar; (*b*) 348.15 K and 1 bar; (*c*) 298.15 K and 100 bar; (*d*) 348.15 K and 100 bar.

5.38 For the reactions of Prob. 5.10, find ΔS_{298}° from data in the Appendix.

5.39 For the reactions in Prob. 5.10, find ΔS_{370}°; neglect the temperature variation in ΔC_P°.

5.40 Derive Eq. (5.37) for $\Delta S_{T_2}^\circ - \Delta S_{T_1}^\circ$.

5.41 (*a*) Use $S_{m,298}^\circ$ Appendix data and the expression for $\Delta C_P^\circ(T)$ in Example 5.6 in Sec. 5.5 to find ΔS_{1000}° for $2CO(g) + O_2(g) \rightarrow 2CO_2(g)$. (*b*) Repeat the calculation using $C_{P,m,298}$ data and assuming ΔC_P° is independent of T.

5.42 For reasonably low pressures, a good equation of state for gases is the truncated virial equation (Sec. 8.2) $PV_m/RT = 1 + f(T)P$, where $f(T)$ is a function of T (different for different gases). Show that for this equation of state

$$S_{m,\text{id}}(T, P) - S_{m,\text{re}}(T, P) = RP[f(T) + Tf'(T)]$$

Section 5.8

5.43 For urea, $CO(NH_2)_2(c)$, $\Delta_f H_{298}^\circ = -333.51$ kJ/mol and $S_{298}^\circ = 104.60$ J/(mol K). With the aid of Appendix data, find $\Delta_f G_{298}^\circ$ of urea.

5.44 For the reactions in Prob. 5.10, find ΔG_{298}° using (*a*) the results of Probs. 5.10 and 5.38; (*b*) $\Delta_f G_{298}^\circ$ values in the Appendix.

5.45 For the reactions of Prob. 5.10, use the results of Probs. 5.24 and 5.39 to find ΔG_{370}°.

5.46 Use Appendix data to find the conventional $G_{m,298}^\circ$ for (*a*) $O_2(g)$; (*b*) $H_2O(l)$.

Section 5.9

5.47 Look up in one of the references cited near the end of Sec. 5.9 $\Delta_f G^\circ$ data at 1000 K to find ΔG_{1000}° for $2CH_4(g) \rightarrow C_2H_6(g) + H_2(g)$.

5.48 Some values of $(H_{m,2000}^\circ - H_{m,298}^\circ)/(kJ/mol)$ are 52.93 for $H_2(g)$, 56.14 for $N_2(g)$, and 98.18 for $NH_3(g)$. Use these data and Appendix data to find ΔH_{2000}° for $N_2(g) + 3H_2(g) \rightarrow 2NH_3(g)$.

5.49 For $T = 2000$ K, some values of $-(G_{m,T}^\circ - H_{m,298}^\circ)/T$ in J/(mol K) are 161.94 for $H_2(g)$, 223.74 for $N_2(g)$, and 242.08 for $NH_3(g)$. Use these and Appendix data to find $\Delta_f G_{2000}^\circ$ of $NH_3(g)$.

5.50 Verify Eq. (5.43) for ΔG_T°.

5.51 (*a*) If ΔG_T^{bar} and ΔG_T^{atm} are ΔG_T° values based on 1-bar and 1-atm standard-state pressures, respectively, use Eq. (5.41) to show that

$$\Delta G_T^{\text{bar}} - \Delta G_T^{\text{atm}} = -T[0.1094 \text{ J/(mol K)}]\, \Delta n_g/\text{mol}$$

where $\Delta n_g/\text{mol}$ is the change in number of moles of gases for the reaction. (*b*) Calculate this difference for $\Delta_f G_{298}^\circ$ of $H_2O(l)$.

Section 5.10

5.52 (*a*) Use bond energies listed in Sec. 20.1 to estimate ΔH_{298}° for $CH_3CH_2OH(g) \rightarrow CH_3OCH_3(g)$. Compare with the true value 51 kJ/mol. (*b*) Repeat (*a*) using bond-additivity values. (*c*) Repeat (*a*) using group-additivity values.

5.53 (*a*) Use Appendix data and bond energies in Sec. 20.1 to estimate $\Delta_f H^\circ_{298}$ of $CH_3OCH_2CH_3(g)$. (*b*) Repeat (*a*) using bond-additivity values. (*c*) Repeat (*a*) using group-additivity values.

5.54 Look up the Benson–Buss bond contribution method (Sec. 5.10) and use it to estimate $S^\circ_{m,298}$ of $COF_2(g)$; be sure to include the symmetry correction. Compare with the correct value in the Appendix.

5.55 The vapor pressure of liquid water at 25°C is 23.8 torr, and its molar enthalpy of vaporization at 25°C and 23.8 torr is 10.5 kcal/mol. Assume the vapor behaves ideally, neglect the effect of a pressure change on H and S of the liquid, and calculate ΔH°_{298}, ΔS°_{298}, and ΔG°_{298} for the vaporization of water; use only data in this problem. Compare your results with values found from data in the Appendix.

5.56 For $CH_3OH(l)$ at 25°C, the vapor pressure is 125 torr, ΔH_m of vaporization is 37.9 kJ/mol, $\Delta_f H^\circ$ is −238.7 kJ/mol, and S°_m is 126.8 J/(mol K). Making reasonable approximations, find $\Delta_f H^\circ_{298}$ and $S^\circ_{m,298}$ of $CH_3OH(g)$.

5.57 Let D_{CC} and D_{CH} be the C—C and C—H bond energies and b_{CC} and b_{CH} be the $\Delta_f H^\circ_{298}$ bond-additivity values for these bonds. (*a*) Express $\Delta_f H^\circ_{298}$ of $C_nH_{2n+2}(g)$ in terms of b_{CC} and b_{CH}. (*b*) Express $\Delta_f H^\circ_{298}$ of $C_nH_{2n+2}(g)$ in terms of D_{CC}, D_{CH}, $\Delta_f H^\circ_{298}[H(g)]$ and $\Delta_f H^\circ_{298}[C(g)]$. (*c*) Equate the expressions in (*a*) and (*b*) to each other and then set $n = 1$ and $n = 2$ to show that $b_{CC} = -D_{CC} + 0.5\,\Delta_f H^\circ_{298}[C(g)]$ and $b_{CH} = -D_{CH} + \Delta_f H^\circ_{298}[H(g)] + 0.25\,\Delta_f H^\circ_{298}[C(g)]$. Substitute these two equations for b_{CC} and b_{CH} into the equation found by equating the expressions in (*a*) and (*b*) and verify that this equation is satisfied.

General

5.58 Give the SI units of (*a*) pressure; (*b*) enthalpy; (*c*) molar entropy; (*d*) Gibbs energy; (*e*) molar volume; (*f*) temperature.

5.59 If ΔH° for a reaction is independent of T, prove that the reaction's ΔS° is independent of T. [*Hint:* Use Eq. (5.18).]

5.60 (*a*) Show that for any substance $\lim_{T\to 0} \alpha = 0$. (*Hint:* Use one of the Maxwell relations.) (*b*) Verify that α for an ideal gas does not obey the result in (*a*). Hence, (classical) ideal gases do not obey the third law (as noted in Sec. 5.7).

5.61 Without consulting tables, state whether or not each of the following must be equal to zero. (*Note: S* is the conventional entropy.) (*a*) $\Delta_f H^\circ_{298}(N_2O_5, g)$; (*b*) $\Delta_f H^\circ_{298}(Cl, g)$; (*c*) $\Delta_f H^\circ_{298}(Cl_2, g)$; (*d*) $S^\circ_{m,298}(Cl_2, g)$; (*e*) $S^\circ_{m,0}(N_2O_5, c)$; (*f*) $\Delta_f S^\circ_{350}(N_2, g)$; (*g*) $\Delta_f G^\circ_{400}(N_2, g)$; (*h*) $C^\circ_{P,m,0}(NaCl, c)$; (*i*) $C^\circ_{P,m,298}(O_2, g)$.

5.62 The *adiabatic flame temperature* is the temperature that would be reached in a flame if no heating of the surroundings occurred during the combustion, so that ΔU of the reaction is used entirely to raise the temperature of the reaction products and to do expansion work. To estimate this temperature, use the scheme of Fig. 5.4*b* with the following changes. Since the combustion is at constant P and is assumed adiabatic, we have $\Delta H = q_P = 0$, so step (*a*) has $\Delta H = 0$ instead of $\Delta U = 0$. Likewise, ΔH replaces ΔU in steps (*b*) and (*c*). For combustion in air, the calorimeter K is replaced by 3.76 moles of $N_2(g)$ for each mole of $O_2(g)$. Estimate the adiabatic flame temperature for combustion of methane, $CH_4(g)$, in air initially at 25°C, assuming that O_2 and CH_4 are present in stoichiometric amounts. Use Appendix data. Note that step (*b*) involves vaporization of water. Proper calculation of ΔH_b requires integrating C_P of the products. Instead, assume that an average C_P of the products can be used over the temperature range involved and that this average is found by combining the following 1000-K $C^\circ_{P,m}$ values, given in J/(mol K): 32.7 for $N_2(g)$, 41.2 for $H_2O(g)$, 54.3 for $CO_2(g)$.

5.63 Without using tables, state which of each of the following pairs has the greater $S^\circ_{m,298}$: (*a*) $C_2H_6(g)$ or n-$C_4H_{10}(g)$; (*b*) $H_2O(l)$ or $H_2O(g)$; (*c*) $H(g)$ or $H_2(g)$; (*d*) $C_{10}H_8(s)$ or $C_{10}H_8(g)$.

5.64 Without using thermodynamics tables, predict the sign of ΔS°_{298} and ΔH°_{298} for each of the following. You can use Table 20.1. (*a*) $(C_2H_5)_2O(l) \to (C_2H_5)_2O(g)$; (*b*) $Cl_2(g) \to 2Cl(g)$; (*c*) $C_{10}H_8(g) \to C_{10}H_8(s)$; (*d*) combustion of $(COOH)_2(s)$ to $CO_2(g)$ and $H_2O(l)$; (*e*) $C_2H_4(g) + H_2(g) \to C_2H_6(g)$.

5.65 The Tennessee Valley Authority's coal-burning Paradise power plant (by no means the world's largest) produces 1000 MW of power and has an overall thermal efficiency of 39%. The overall thermal efficiency is defined as the work output divided by the absolute value of the heat of combustion of the fuel. [Since only 85 to 90% of the heat of combustion is transferred to the steam, the overall thermal efficiency is not the same as the efficiency defined by Eq. (3.1).] The typical enthalpy of combustion of coal is −10000 British thermal units (Btu) per pound, where 1 Btu equals 1055 J. How many pounds of coal does the Paradise plant burn in (*a*) one minute; (*b*) one day; (*c*) one year?

5.66 True or false? (*a*) When an exothermic reaction in a closed system with *P-V* work only is run under isobaric and adiabatic conditions, $\Delta H = 0$. (*b*) When a substance is in its thermodynamic standard state, the substance must be at 25°C. (*c*) *G* of an element in its stable form and in its standard state at 25°C is taken to be zero. (*d*) If an exothermic reaction occurs in an isolated system, ΔT of the system must be positive. (*e*) If an exothermic reaction is run under isothermal conditions, *q* of the system must be negative.

5.67 What experimental measurements are needed to determine $\Delta_f H^\circ_{298}$, $S^\circ_{m,298}$, and $\Delta_f G^\circ_{298}$ of a newly synthesized liquid hydrocarbon?

5.68 Use Appendix $\Delta_f H^\circ_{298}$ and $S^\circ_{m,298}$ data to calculate $\Delta_f G^\circ_{298}$ of $C_2H_5OH(l)$. Compare with the value listed in the Appendix.

Reaction Equilibrium in Ideal Gas Mixtures

The second law of thermodynamics led us to conclude that the entropy of system plus surroundings is maximized at equilibrium. From this entropy maximization condition we found that the condition for reaction equilibrium in a closed system is $\Sigma_i \nu_i \mu_i = 0$ [Eq. (4.98)], where the ν_i's are the stoichiometric coefficients in the reaction and the μ_i's are the chemical potentials of the species in the reaction. Section 6.2 applies this equilibrium condition to a reaction in an ideal gas mixture and shows that for the ideal-gas reaction $a A + b B \rightleftharpoons c C + d D$, the partial pressures of the gases at equilibrium must be such that the quantity $(P_C/P°)^c(P_D/P°)^d/(P_A/P°)^a(P_B/P°)^b$ (where $P° \equiv 1$ bar) is equal to the equilibrium constant for the reaction, where the equilibrium constant can be calculated from $\Delta G°$ of the reaction. (We learned in Chapter 5 how to use thermodynamics tables to find $\Delta G°$ from $\Delta_f G°$ data.) Section 6.3 shows how the ideal-gas equilibrium constant changes with temperature. Sections 6.4 and 6.5 show how to calculate the equilibrium composition of an ideal-gas reaction mixture from the equilibrium constant and the initial composition. Section 6.6 discusses shifts in ideal-gas equilibria.

Chapter 6 gives us the power to calculate the equilibrium composition for an ideal-gas reaction from the initial composition, the temperature and pressure, and $\Delta_f G°$ data.

To apply the equilibrium condition $\Sigma_i \nu_i \mu_i = 0$ to an ideal-gas reaction, we need to relate the chemical potential μ_i of a component of an ideal gas mixture to observable properties. This is done in Sec. 6.1.

Chapter 6 is concerned only with ideal-gas equilibria. Reaction equilibrium in nonideal gases and in liquid solutions is treated in Chapter 11.

In a particular system with chemical reactions, reaction equilibrium might or might not hold. When the reaction system is not in equilibrium, we need to use chemical kinetics (Chapter 17) to find the composition (which changes with time). In gas-phase reactions, equilibrium is often reached if the temperature is high (so reaction rates are high) or if the reaction is catalyzed. High-temperature reactions occur in rockets and reaction equilibrium is often assumed in rocketry calculations. (Recall the NIST-JANAF tables of thermodynamic data mentioned in Sec. 5.9. JANAF stands for Joint Army–Navy–Air Force; these tables originated to provide thermodynamic data for rocketry calculations.) Industrial gas-phase reactions that are run at elevated temperatures in the presence of solid-phase catalysts include the synthesis of NH_3 from N_2 and H_2, the conversion of SO_2 to SO_3 for use in preparation of H_2SO_4, and the synthesis of CH_3OH from CO and H_2. Equilibria involving such species as H, H^+, e^-, H^- H_2, He, He^+, and He^{2+} determine the composition at the sun's surface (the photosphere), which is at 5800 K and about 1 atm.

Even if equilibrium is not reached, knowing the equilibrium constant is important since this enables us to find the maximum possible yield of a desired product under given conditions.

In aqueous solutions, reactions that involve ions are generally fast and equilibrium is usually assumed; recall acid–base and complex-ion equilibrium calculations done in general and analytical chemistry. Equilibrium analysis is important in environmental-chemistry studies of the composition of water systems such as lakes and in the study of air pollution. Figure 6.5 shows the formation of significant amounts of NO in heated air at equilibrium, and the formation of NO in automobile engines and in industrial burning of coal and oil in power plants pollutes the atmosphere. (Figure 6.5 is not quantitatively applicable to automobile engines because the combustion of the fuel depletes the air of oxygen and because there is not enough time for equilibrium to be reached, so NO formation must be analyzed kinetically. The equilibrium constant determines the maximum amount of NO that can be formed.)

6.1   CHEMICAL POTENTIALS IN AN IDEAL GAS MIXTURE

Before dealing with μ_i of a component of an ideal gas mixture, we find an expression for μ of a pure ideal gas.

Chemical Potential of a Pure Ideal Gas

The chemical potential is an intensive property, so μ for a pure gas depends on T and P only. Since reaction equilibrium is usually studied in systems held at constant temperature while the amounts and partial pressures of the reacting gases vary, we are most interested in the variation of μ with pressure. The Gibbs equation for dG for a fixed amount of substance is $dG = -S\,dT + V\,dP$ [Eq. (4.36)], and division by the number of moles of the pure ideal gas gives $dG_m = d\mu = -S_m\,dT + V_m\,dP$, since the chemical potential μ of a pure substance equals G_m [Eq. (4.86)]. For constant T, this equation becomes

$$d\mu = V_m\,dP = (RT/P)\,dP \qquad \text{const. } T, \text{ pure ideal gas}$$

If the gas undergoes an isothermal change of state from pressure P_1 to P_2, integration of this equation gives

$$\int_1^2 d\mu = RT \int_{P_1}^{P_2} \frac{1}{P}\,dP$$

$$\mu(T, P_2) - \mu(T, P_1) = RT \ln (P_2/P_1) \qquad \text{pure ideal gas} \qquad (6.1)$$

Let P_1 be the standard pressure $P^\circ \equiv 1$ bar. Then $\mu(T, P_1)$ equals $\mu^\circ(T)$, the gas's standard-state chemical potential at temperature T, and (6.1) becomes $\mu(T, P_2) = \mu^\circ(T) + RT \ln (P_2/P^\circ)$. The subscript 2 is not needed, so the chemical potential $\mu(T, P)$ of a pure ideal gas at T and P is

$$\mu = \mu^\circ(T) + RT \ln (P/P^\circ) \qquad \text{pure ideal gas, } P^\circ \equiv 1 \text{ bar} \qquad (6.2)$$

Figure 6.1 plots $\mu - \mu^\circ$ versus P at fixed T for a pure ideal gas. For a pure ideal gas, $\mu = G_m = H_m - TS_m$, and H_m is independent of pressure [Eq. (2.70)], so the pressure dependence of μ in Fig. 6.1 is due to the change of S_m with P. In the zero-pressure, infinite-volume limit, the entropy of an ideal gas becomes infinite, and μ goes to $-\infty$.

Chemical Potentials in an Ideal Gas Mixture

To find the chemical potentials in an ideal gas mixture, we give a fuller definition of an ideal gas mixture than we previously gave. An **ideal gas mixture** is a gas mixture

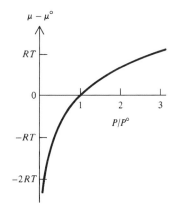

Figure 6.1

Variation of the chemical potential μ of a pure ideal gas with pressure at constant temperature. μ° is the standard-state chemical potential, corresponding to $P = P^\circ = 1$ bar.

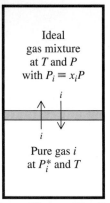

At equilibrium, $P_i^* = P_i$.

Figure 6.2

An ideal gas mixture separated from pure gas i by a membrane permeable to i only.

having the following properties: (1) The equation of state $PV = n_{tot}RT$ [Eq. (1.22)] is obeyed for all temperatures, pressures, and compositions, where n_{tot} is the total number of moles of gas. (2) If the mixture is separated from pure gas i (where i is any one of the mixture's components) by a thermally conducting rigid membrane permeable to gas i only (Fig. 6.2), then at equilibrium the partial pressure $P_i \equiv x_i P$ [Eq. (1.23)] of gas i in the mixture is equal to the pressure of the pure-gas-i system.

This definition makes sense from a molecular viewpoint. Since there are no intermolecular interactions either in the pure ideal gases or in the ideal gas mixture, we expect the mixture to obey the same equation of state obeyed by each pure gas, and condition (1) holds. If two samples of pure ideal gas i at the same T were separated by a membrane permeable to i, equilibrium (equal rates of passage of i through the membrane from each side) would be reached with equal pressures of i on each side. Because there are no intermolecular interactions, the presence of other gases on one side of the membrane has no effect on the net rate of passage of i through the membrane, and condition (2) holds.

The **standard state** of component i of an ideal gas mixture at temperature T is defined to be pure ideal gas i at T and pressure $P^\circ \equiv 1$ bar.

In Fig. 6.2, let μ_i be the chemical potential of gas i in the mixture, and let μ_i^* be the chemical potential of the pure gas in equilibrium with the mixture through the membrane. (We shall often use an asterisk to distinguish thermodynamic properties of a pure substance from those of a component of a mixture.) The condition for phase equilibrium between the mixture and pure i is $\mu_i = \mu_i^*$ (Sec. 4.8). The mixture is at temperature T and pressure P, and has mole fractions $x_1, x_2, \ldots, x_i, \ldots$. The pure gas i is at temperature T and pressure P_i^*. But from condition (2) of the definition of an ideal gas mixture, P_i^* at equilibrium equals the partial pressure $P_i \equiv x_i P$ of i in the mixture. Therefore the phase-equilibrium condition $\mu_i = \mu_i^*$ becomes

$$\mu_i(T, P, x_1, x_2, \ldots) = \mu_i^*(T, x_i P) = \mu_i^*(T, P_i) \qquad \text{ideal gas mixture} \qquad (6.3)$$

Equation (6.3) states that the chemical potential μ_i of component i of an ideal gas mixture at T and P equals the chemical potential μ_i^* of pure gas i at T and P_i (its partial pressure in the mixture). This result makes sense; since intermolecular interactions are absent, the presence of other gases in the mixture has no effect on μ_i.

From Eq. (6.2), the chemical potential of pure gas i at pressure P_i is $\mu_i^*(T, P_i) = \mu_i^\circ(T) + RT \ln (P_i/P^\circ)$, and Eq. (6.3) becomes

$$\mu_i = \mu_i^\circ(T) + RT \ln (P_i/P^\circ) \qquad \text{ideal gas mixture, } P^\circ \equiv 1 \text{ bar} \qquad \textbf{(6.4)*}$$

Equation (6.4) is the fundamental thermodynamic equation for an ideal gas mixture. In (6.4), μ_i is the chemical potential of component i in an ideal gas mixture, P_i is the partial pressure of gas i in the mixture, and $\mu_i^\circ(T) [= G_{m,i}^\circ(T)]$ is the chemical potential of pure ideal gas i at the standard pressure of 1 bar and at the same temperature T as the mixture. Since the standard state of a component of an ideal gas mixture was defined to be pure ideal gas i at 1 bar and T, μ_i° is the standard-state chemical potential of i in the mixture. μ_i° depends only on T because the pressure is fixed at 1 bar for the standard state.

Equation (6.4) shows that the graph in Fig. 6.1 applies to a component of an ideal gas mixture if μ and μ° are replaced by μ_i and μ_i°, and P is replaced by P_i.

Equation (6.4) can be used to derive the thermodynamic properties of an ideal gas mixture. The result (Prob. 9.20) is that *each of U, H, S, G, and C_P for an ideal gas mixture is the sum of the corresponding thermodynamic functions for the pure gases calculated for each pure gas occupying a volume equal to the mixture's volume at a pressure equal to its partial pressure in the mixture and at a temperature equal to its temperature in the mixture.* These results make sense from the molecular picture in which each gas has no interaction with the other gases in the mixture.

6.2   IDEAL-GAS REACTION EQUILIBRIUM

The equilibrium condition for the reaction $0 \rightleftharpoons \Sigma_i \nu_i A_i$ (where ν_i is the stoichiometric coefficient of species A_i) is $\Sigma_i \nu_i \mu_i = 0$ [Eq. (4.98)]. We now specialize to the case where all reactants and products are ideal gases.

For the ideal-gas reaction

$$aA + bB \rightleftharpoons cC + dD$$

the equilibrium condition $\Sigma_i \nu_i \mu_i = 0$ is

$$a\mu_A + b\mu_B = c\mu_C + d\mu_D$$

$$c\mu_C + d\mu_D - a\mu_A - b\mu_B = 0$$

Each chemical potential in an ideal gas mixture is given by Eq. (6.4) as $\mu_i = \mu_i^\circ + RT \ln (P_i/P^\circ)$, and substitution in the equilibrium condition gives

$$c\mu_C^\circ + cRT \ln (P_C/P^\circ) + d\mu_D^\circ + dRT \ln (P_D/P^\circ)$$
$$- a\mu_A^\circ - aRT \ln (P_A/P^\circ) - b\mu_B^\circ - bRT \ln (P_B/P^\circ) = 0$$

$$c\mu_C^\circ + d\mu_D^\circ - a\mu_A^\circ - b\mu_B^\circ =$$
$$-RT[c \ln (P_C/P^\circ) + d \ln (P_D/P^\circ) - a \ln (P_A/P^\circ) - b \ln (P_B/P^\circ)] \quad (6.5)$$

Since $\mu = G_m$ for a pure substance, the quantity on the left side of (6.5) is the standard Gibbs energy change ΔG_T° for the reaction [Eq. (5.38)]

$$\Delta G_T^\circ \equiv \sum_i \nu_i G_{m,T,i}^\circ = \sum_i \nu_i \mu_i^\circ(T) = c\mu_C^\circ + d\mu_D^\circ - a\mu_A^\circ - b\mu_B^\circ$$

The equilibrium condition (6.5) becomes

$$\Delta G^\circ = -RT[\ln (P_C/P^\circ)^c + \ln (P_D/P^\circ)^d - \ln (P_A/P^\circ)^a - \ln (P_B/P^\circ)^b]$$

$$\Delta G^\circ = -RT \ln \frac{(P_{C,eq}/P^\circ)^c (P_{D,eq}/P^\circ)^d}{(P_{A,eq}/P^\circ)^a (P_{B,eq}/P^\circ)^b} \quad (6.6)$$

where the identities $a \ln x = \ln x^a$, $\ln x + \ln y = \ln xy$, and $\ln x - \ln y = \ln (x/y)$ were used, and where the eq subscripts emphasize that these are partial pressures at equilibrium. Defining the standard equilibrium constant K_P° for the ideal-gas reaction $aA + bB \rightarrow cC + dD$ as

$$K_P^\circ \equiv \frac{(P_{C,eq}/P^\circ)^c (P_{D,eq}/P^\circ)^d}{(P_{A,eq}/P^\circ)^a (P_{B,eq}/P^\circ)^b}, \qquad P^\circ \equiv 1 \text{ bar} \quad (6.7)$$

we have for Eq. (6.6)

$$\Delta G^\circ = -RT \ln K_P^\circ$$

We now repeat the derivation for the general ideal-gas reaction $0 \rightarrow \Sigma_i \nu_i A_i$. Substitution of the expression $\mu_i = \mu_i^\circ + RT \ln (P_i/P^\circ)$ for a component of an ideal gas mixture into the equilibrium condition $\Sigma_i \nu_i \mu_i = 0$ gives

$$\sum_i \nu_i \mu_i = \sum_i \nu_i[\mu_i^\circ + RT \ln (P_{i,eq}/P^\circ)] = 0$$

$$\sum_i \nu_i \mu_i^\circ(T) + RT \sum_i \nu_i \ln (P_{i,eq}/P^\circ) = 0 \quad (6.8)$$

where the sum identities $\Sigma_i (a_i + b_i) = \Sigma_i a_i + \Sigma_i b_i$ and $\Sigma_i ca_i = c \Sigma_i a_i$ [Eq. (1.50)] were used. We have $\mu_i^\circ(T) = G_{m,T,i}^\circ$. Therefore

$$\Delta G_T^\circ = \sum_i \nu_i G_{m,T,i}^\circ = \sum_i \nu_i \mu_i^\circ(T) \quad (6.9)$$

and (6.8) becomes

$$\Delta G_T^\circ = -RT \sum_i \nu_i \ln (P_{i,\text{eq}}/P^\circ) = -RT \sum_i \ln (P_{i,\text{eq}}/P^\circ)^{\nu_i} \tag{6.10}$$

where $k \ln x = \ln x^k$ was used. The sum of logarithms equals the log of the product:

$$\sum_{i=1}^{n} \ln a_i = \ln a_1 + \ln a_2 + \cdots + \ln a_n = \ln (a_1 a_2 \cdots a_n) = \ln \prod_{i=1}^{n} a_i$$

where the large capital pi denotes a product:

$$\prod_{i=1}^{n} a_i \equiv a_1 a_2 \cdots a_n \tag{6.11}*$$

As with sums, the limits are often omitted when they are clear from the context. Use of $\Sigma_i \ln a_i = \ln \Pi_i a_i$ in (6.10) gives

$$\Delta G_T^\circ = -RT \ln \left[\prod_i (P_{i,\text{eq}}/P^\circ)^{\nu_i} \right] \tag{6.12}$$

We define K_P° as the product that occurs in (6.12):

$$K_P^\circ \equiv \prod_i (P_{i,\text{eq}}/P^\circ)^{\nu_i} \quad \text{ideal-gas reaction equilib.} \tag{6.13}*$$

Equation (6.12) becomes

$$\Delta G^\circ = -RT \ln K_P^\circ \quad \text{ideal-gas reaction equilib.} \tag{6.14}*$$

Recall that if $y = \ln_e x$, then $x = e^y$ [Eq. (1.65)]. Thus (6.14) can be written as

$$K_P^\circ = e^{-\Delta G^\circ/RT} \tag{6.15}$$

Equation (6.9) shows that ΔG° depends only on T. It therefore follows from (6.15) that K_P° *for a given ideal-gas reaction is a function of T only* and is independent of the pressure, the volume, and the amounts of the reaction species present in the mixture: $K_P^\circ = K_P^\circ(T)$. At a given temperature, K_P° is a constant for a given reaction. K_P° is the **standard equilibrium constant** (or the **standard pressure equilibrium constant**) for the ideal-gas reaction.

Summarizing, for the ideal-gas reaction $0 \rightleftharpoons \Sigma_i \nu_i A_i$, we started with the general condition for reaction equilibrium $\Sigma_i \nu_i \mu_i = 0$ (where the ν_i's are the stoichiometric coefficients); we replaced each μ_i with the ideal-gas-mixture expression $\mu_i = \mu_i^\circ + RT \ln (P_i/P^\circ)$ for the chemical potential μ_i of component i and found that $\Delta G^\circ = -RT \ln K_P^\circ$. This equation relates the standard Gibbs energy change ΔG° [defined by (6.9)] to the equilibrium constant K_P° [defined by (6.13)] for the ideal-gas reaction.

Because the stoichiometric coefficients ν_i are negative for reactants and positive for products, K_P° has the products in the numerator and the reactants in the denominator. Thus, for the ideal-gas reaction

$$N_2(g) + 3H_2(g) \rightarrow 2NH_3(g) \tag{6.16}$$

we have $\nu_{N_2} = -1$, $\nu_{H_2} = -3$, and $\nu_{NH_3} = 2$, so

$$K_P^\circ = [P(NH_3)_{\text{eq}}/P^\circ]^2 [P(N_2)_{\text{eq}}/P^\circ]^{-1} [P(H_2)_{\text{eq}}/P^\circ]^{-3} \tag{6.17}$$

$$K_P^\circ = \frac{[P(NH_3)_{\text{eq}}/P^\circ]^2}{[P(N_2)_{\text{eq}}/P^\circ][P(H_2)_{\text{eq}}/P^\circ]^3} \tag{6.18}$$

where the pressures are the equilibrium partial pressures of the gases in the reaction mixture. At any given temperature, the equilibrium partial pressures must be such as to satisfy (6.18). If the partial pressures do not satisfy (6.18), the system is not in reaction equilibrium and its composition will change until (6.18) is satisfied.

For the ideal-gas reaction $aA + bB \rightleftharpoons cC + dD$, the standard (pressure) equilibrium constant is given by (6.7).

Since $P_i/P°$ in (6.13) is dimensionless, the standard equilibrium constant $K_P°$ is dimensionless. In (6.14), the log of $K_P°$ is taken; one can take the log of a dimensionless number only. It is sometimes convenient to work with an equilibrium constant that omits the $P°$ in (6.13). We define the *equilibrium constant* (or *pressure equilibrium constant*) K_P as

$$K_P \equiv \prod_i (P_{i,\text{eq}})^{\nu_i} \qquad (6.19)$$

K_P has dimensions of pressure raised to the change in mole numbers for the reaction as written. For example, for (6.16), K_P has dimensions of pressure^{-2}.

The existence of a standard equilibrium constant $K_P°$ that depends only on T is a rigorous deduction from the laws of thermodynamics. The only assumption is that we have an ideal gas mixture. Our results are a good approximation for real gas mixtures at low densities.

EXAMPLE 6.1 Finding $K_P°$ and $\Delta G°$ from the equilibrium composition

A mixture of 11.02 mmol (millimoles) of H_2S and 5.48 mmol of CH_4 was placed in an empty container along with a Pt catalyst, and the equilibrium

$$2H_2S(g) + CH_4(g) \rightleftharpoons 4H_2(g) + CS_2(g) \qquad (6.20)$$

was established at 700°C and 762 torr. The reaction mixture was removed from the catalyst and rapidly cooled to room temperature, where the rates of the forward and reverse reactions are negligible. Analysis of the equilibrium mixture found 0.711 mmol of CS_2. Find $K_P°$ and $\Delta G°$ for the reaction at 700°C.

Since 0.711 mmol of CS_2 was formed, 4(0.711 mmol) = 2.84 mmol of H_2 was formed. For CH_4, 0.711 mmol reacted, and 5.48 mmol − 0.71 mmol = 4.77 mmol was present at equilibrium. For H_2S, 2(0.711 mmol) reacted, and 11.02 mmol − 1.42 mmol = 9.60 mmol was present at equilibrium. To find $K_P°$, we need the partial pressures P_i. We have $P_i \equiv x_i P$, where $P = 762$ torr and the x_i's are the mole fractions. Omitting the eq subscript to save writing, we have at equilibrium

$$n_{H_2S} = 9.60 \text{ mmol}, n_{CH_4} = 4.77 \text{ mmol}, n_{H_2} = 2.84 \text{ mmol}, n_{CS_2} = 0.711 \text{ mmol}$$

$$x_{H_2S} = 9.60/17.92 = 0.536, x_{CH_4} = 0.266, x_{H_2} = 0.158, x_{CS_2} = 0.0397$$

$$P_{H_2S} = 0.536(762 \text{ torr}) = 408 \text{ torr}, P_{CH_4} = 203 \text{ torr},$$

$$P_{H_2} = 120 \text{ torr}, P_{CS_2} = 30.3 \text{ torr}$$

The standard pressure $P°$ in $K_P°$ is 1 bar ≈ 750 torr [Eq. (1.12)], and (6.13) gives

$$K_P° = \frac{(P_{H_2}/P°)^4(P_{CS_2}/P°)}{(P_{H_2S}/P°)^2(P_{CH_4}/P°)} = \frac{(120 \text{ torr}/750 \text{ torr})^4(30.3 \text{ torr}/750 \text{ torr})}{(408 \text{ torr}/750 \text{ torr})^2(203 \text{ torr}/750 \text{ torr})}$$

$$= 0.000331$$

The use of $\Delta G° = -RT \ln K_P°$ [Eq. (6.14)] at 700°C = 973 K gives

$$\Delta G_{973}° = -[8.314 \text{ J/(mol K)}](973 \text{ K}) \ln 0.000331 = 64.8 \text{ kJ/mol}$$

In working this problem, we assumed an ideal gas mixture, which is a good assumption at the T and P of the experiment.

Exercise

If 0.1500 mol of $O_2(g)$ is placed in an empty container and equilibrium is reached at 3700 K and 895 torr, one finds that 0.1027 mol of $O(g)$ is present. Find $K_P°$ and $\Delta G°$ for $O_2(g) \rightleftharpoons 2O(g)$ at 3700 K. Assume ideal gases. (*Answers:* 0.634, 14.0 kJ/mol.)

Exercise

If 0.1500 mol of $O_2(g)$ is placed in an empty 32.80-L container and equilibrium is established at 4000 K, one finds the pressure is 2.175 atm. Find $K_P°$ and $\Delta G°$ for $O_2(g) \rightleftharpoons 2O(g)$ at 4000 K. Assume ideal gases. (*Answers:* 2.22, −26.6 kJ/mol.)

Concentration and Mole-Fraction Equilibrium Constants

Gas-phase equilibrium constants are sometimes expressed using concentrations instead of partial pressures. For n_i moles of ideal gas i in a mixture of volume V, the partial pressure is $P_i = n_i RT/V$ [Eq. (1.24)]. Defining the (**molar**) **concentration** c_i of species i in the mixture as

$$c_i \equiv n_i/V \tag{6.21}*$$

we have

$$P_i = n_i RT/V = c_i RT \qquad \text{ideal gas mixture} \tag{6.22}$$

Use of (6.22) in (6.7) gives for the ideal-gas reaction $aA + bB \rightleftharpoons fF + dD$

$$K_P° = \frac{(c_{F,eq}RT/P°)^f(c_{D,eq}RT/P°)^d}{(c_{A,eq}RT/P°)^a(c_{B,eq}RT/P°)^b} = \frac{(c_{F,eq}/c°)^f(c_{D,eq}/c°)^d}{(c_{A,eq}/c°)^a(c_{B,eq}/c°)^b}\left(\frac{c°RT}{P°}\right)^{f+d-a-b} \tag{6.23}$$

where $c°$, defined as $c° \equiv 1$ mol/liter = 1 mol/dm³, was introduced to make all fractions on the right side of (6.23) dimensionless. Note that $c°RT$ has the same dimensions as $P°$. The quantity $f + d - a - b$ is the change in number of moles for the reaction as written, which we symbolize by $\Delta n/\text{mol} \equiv f + d - a - b$. Since $f + d - a - b$ is dimensionless and Δn has units of moles, Δn was divided by the unit "mole" in the definition. For $N_2(g) + 3H_2(g) \rightleftharpoons 2NH_3(g)$, $\Delta n/\text{mol} = 2 - 1 - 3 = -2$. Defining the *standard concentration equilibrium constant* $K_c°$ as

$$K_c° \equiv \prod_i (c_{i,eq}/c°)^{\nu_i} \qquad \text{where } c° \equiv 1 \text{ mol/liter} \equiv 1 \text{ mol/dm}^3 \tag{6.24}$$

we have for (6.23)

$$K_P° = K_c°(RTc°/P°)^{\Delta n/\text{mol}} \tag{6.25}$$

Knowing $K_P°$, we can find $K_c°$ from (6.25). $K_c°$ is, like $K_P°$, dimensionless. Since $K_P°$ depends only on T, and $c°$ and $P°$ are constants, Eq. (6.25) shows that $K_c°$ is a function of T only.

One can also define a *mole-fraction equilibrium constant* K_x:

$$K_x \equiv \prod_i (x_{i,eq})^{\nu_i} \tag{6.26}$$

The relation between K_x and K_P° is (Prob. 6.7)

$$K_P^\circ = K_x(P/P^\circ)^{\Delta n/\text{mol}} \tag{6.27}$$

Except for reactions with $\Delta n = 0$, the equilibrium constant K_x depends on P as well as on T and so is not as useful as K_P°.

Introduction of K_c° and K_x is simply a convenience, and any ideal-gas equilibrium problem can be solved using only K_P°. Since the standard state is defined as having 1 bar pressure, ΔG° is directly related to K_P° by $\Delta G^\circ = -RT \ln K_P^\circ$ [Eq. (6.14)] but is only indirectly related to K_c° and K_x through (6.25) and (6.27).

Qualitative Discussion of Chemical Equilibrium

The following discussion applies in a general way to all kinds of reaction equilibria, not just ideal-gas reactions.

The standard equilibrium constant K_P° is the product and quotient of positive numbers and must therefore be positive: $0 < K_P^\circ < \infty$. If K_P° is very large ($K_P^\circ \gg 1$), its numerator must be much greater than its denominator, and this means that the equilibrium pressures of the products are usually greater than those of the reactants. Conversely, if K_P° is very small ($K_P^\circ \ll 1$), its denominator is large compared with its numerator and the reactant equilibrium pressures are usually larger than the product equilibrium pressures. A moderate value of K_P° usually means substantial equilibrium pressures of both products and reactants. (The word "usually" has been used because it is not the pressures that appear in the equilibrium constant but the pressures raised to the stoichiometric coefficients.) A large value of the equilibrium constant favors products; a small value favors reactants.

We have $K_P^\circ = 1/e^{\Delta G^\circ/RT}$ [Eq. (6.15)]. If $\Delta G^\circ \gg 0$, then $e^{\Delta G^\circ/RT}$ is very large and K_P° is very small. If $\Delta G^\circ \ll 0$, then $K_P^\circ = e^{-\Delta G^\circ/RT}$ is very large. If $\Delta G^\circ \approx 0$, then $K_P^\circ \approx 1$. A large positive value of ΔG° favors reactants; a large negative ΔG° favors products. More precisely, it is $\Delta G^\circ/RT$, and not ΔG°, that determines K_P°. If $\Delta G^\circ = 12RT$, then $K_P^\circ = e^{-12} = 6 \times 10^{-6}$. If $\Delta G^\circ = -12RT$, then $K_P^\circ = e^{12} = 2 \times 10^5$. If $\Delta G^\circ = 50RT$, then $K_P^\circ = 2 \times 10^{-22}$. Because of the exponential relation between K_P° and ΔG°, unless ΔG° is in the approximate range $-12RT < \Delta G^\circ < 12RT$, the equilibrium constant will be very large or very small. At 300 K, $RT = 2.5$ kJ/mol and $12RT = 30$ kJ/mol, so unless $|\Delta G_{300}^\circ| < 30$ kJ/mol, the equilibrium amounts of products or of reactants will be very small. The Appendix data show $\Delta_f G_{298}^\circ$ values are typically a couple of hundred kJ/mol, so for the majority of reactions, ΔG° will not lie in the range $-12RT$ to $+12RT$ and K_P° will be very large or very small. Figure 6.3 plots K_P° versus ΔG° at two temperatures using a logarithmic scale for K_P°. A small change in ΔG° produces a large change in $K_P^\circ = e^{-\Delta G^\circ/RT}$. For example, at 300 K, a decrease of only 10 kJ/mol in ΔG° increases K_P° by a factor of 55.

Since $G^\circ = H^\circ - TS^\circ$, we have for an isothermal process

$$\Delta G^\circ = \Delta H^\circ - T\,\Delta S^\circ \qquad \text{const. } T \tag{6.28}$$

so ΔG° is determined by ΔH°, ΔS°, and T. If T is low, the factor T in (6.28) is small and the first term on the right side of (6.28) is dominant. The fact that ΔS° goes to zero as T goes to zero (the third law) adds to the dominance of ΔH° over $T\,\Delta S^\circ$ at low temperatures. Thus in the limit $T \to 0$, ΔG° approaches ΔH°. For low temperatures, we have the following rough relation:

$$\Delta G^\circ \approx \Delta H^\circ \qquad \text{low } T \tag{6.29}$$

For an exothermic reaction, ΔH° is negative, and hence from (6.29) ΔG° is negative at low temperatures. Thus at low T, products of an exothermic reaction are favored over reactants. (Recall from Sec. 4.4 that a negative ΔH increases the entropy of the surroundings.) For the majority of reactions, the values of ΔH° and $T\,\Delta S^\circ$ are such that

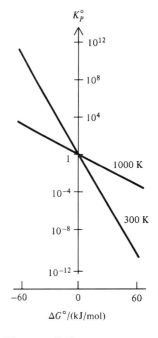

Figure 6.3

Variation of K_P° with ΔG° for two temperatures. The vertical scale is logarithmic.

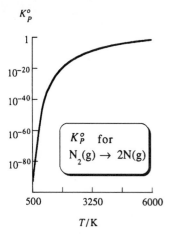

Figure 6.4

K_P° versus T for $N_2(g) \rightleftharpoons 2N(g)$.
The vertical scale is logarithmic.

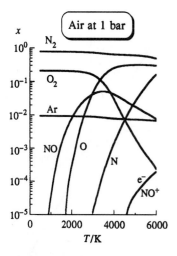

Figure 6.5

Mole-fraction equilibrium
composition of dry air versus T at
1 bar pressure. CO_2 and other
minor components are omitted.
The vertical scale is logarithmic.
The mole fraction of Ar ($x_{Ar} = n_{Ar}/n_{tot}$) decreases above 3000 K
because the dissociation of O_2
increases the total number of
moles present. Above 6000 K,
formation of O^+ and N^+ becomes
significant. At 15000 K, only
charged species are present in
significant amounts.

at room temperature (and below) the first term on the right side of (6.28) dominates. Thus, for most exothermic reactions, products are favored at room temperature. However, ΔH° alone does not determine the equilibrium constant, and there are many *endo*thermic reactions with ΔG° negative and products favored at room temperature, because of the $-T \Delta S^\circ$ term.

For very high temperatures, the factor T makes the second term on the right side of (6.28) the dominant one, and we have the following rough relation:

$$\Delta G^\circ \approx -T \Delta S^\circ \qquad \text{high } T \tag{6.30}$$

At high temperatures, a reaction with a positive ΔS° has a negative ΔG°, and products are favored.

Consider the breaking of a chemical bond, for example, $N_2(g) \rightleftharpoons 2N(g)$. Since a bond is broken, the reaction is highly endothermic ($\Delta H^\circ \gg 0$). Therefore at reasonably low temperatures, ΔG° is highly positive, and N_2 is not significantly dissociated at low temperatures (including room temperature). For $N_2(g) \rightleftharpoons 2N(g)$, the number of moles of gases increases, so we expect this reaction to have a positive ΔS° (Sec. 5.10). (Appendix data give $\Delta S_{298}^\circ = 115$ J mol^{-1} K^{-1} for this reaction.) Thus for high temperatures, we expect from (6.30) that ΔG° for $N_2 \rightleftharpoons 2N$ will be negative, favoring dissociation to atoms.

Figure 6.4 plots K_P° versus T for $N_2(g) \rightleftharpoons 2N(g)$. At 1 bar, significant dissociation occurs only above 3500 K. Calculation of the composition of nitrogen gas above 6000 K must also take into account the ionization of N_2 and N to $N_2^+ + e^-$ and $N^+ + e^-$, respectively. Calculation of the high-T composition of air must take into account the dissociation of O_2 and N_2, the formation of NO, and the ionization of the molecules and atoms present. Figure 6.5 plots the composition of dry air at 1 bar versus T. Thermodynamic data for gaseous ions and for $e^-(g)$ can be found in the NIST-JANAF tables (Sec. 5.9).

6.3 TEMPERATURE DEPENDENCE OF THE EQUILIBRIUM CONSTANT

The ideal-gas equilibrium constant K_P° is a function of temperature only. Let us derive its temperature dependence. Equation (6.14) gives $\ln K_P^\circ = -\Delta G^\circ/RT$. Differentiation with respect to T gives

$$\frac{d \ln K_P^\circ}{dT} = \frac{\Delta G^\circ}{RT^2} - \frac{1}{RT}\frac{d(\Delta G^\circ)}{dT} \tag{6.31}$$

Use of $\Delta G^\circ \equiv \Sigma_i \nu_i G_{m,i}^\circ$ [Eq. (6.9)] gives

$$\frac{d}{dT}\Delta G^\circ = \frac{d}{dT}\sum_i \nu_i G_{m,i}^\circ = \sum_i \nu_i \frac{dG_{m,i}^\circ}{dT} \tag{6.32}$$

From $dG_m = -S_m \, dT + V_m \, dP$, we have $(\partial G_m/\partial T)_P = -S_m$ for a pure substance. Hence

$$dG_{m,i}^\circ/dT = -S_{m,i}^\circ \tag{6.33}$$

The degree superscript indicates the pressure of pure ideal gas i is fixed at the standard value 1 bar. Hence $G_{m,i}^\circ$ depends only on T, and the partial derivative becomes an ordinary derivative. Using (6.33) in (6.32), we have

$$\frac{d \Delta G^\circ}{dT} = -\sum_i \nu_i S_{m,i}^\circ = -\Delta S^\circ \tag{6.34}$$

where $\Delta S°$ is the reaction's standard entropy change, Eq. (5.36). Hence (6.31) becomes

$$\frac{d \ln K_P°}{dT} = \frac{\Delta G°}{RT^2} + \frac{\Delta S°}{RT} = \frac{\Delta G° + T\,\Delta S°}{RT^2} \qquad (6.35)$$

Since $\Delta G° = \Delta H° - T\,\Delta S°$, we end up with

$$\frac{d \ln K_P°}{dT} = \frac{\Delta H°}{RT^2} \qquad \textbf{(6.36)*}$$

This is the **van't Hoff equation**. [Since $\ln K_P° = -\Delta G°/RT$, Eq. (6.36) follows from the Gibbs–Helmholtz equation (Prob. 4.13) $(\partial(G/T)/\partial T)_P = -H/T^2$.] In (6.36), $\Delta H° = \Delta H_T°$ is the standard enthalpy change for the ideal-gas reaction at temperature T [Eq. (5.3)]. The greater the value of $|\Delta H°|$, the faster the equilibrium constant $K_P°$ changes with temperature.

The degree superscript in (6.36) is actually unnecessary, since H of an ideal gas is independent of pressure and the presence of other ideal gases. Therefore, ΔH per mole of reaction in the ideal gas mixture is the same as $\Delta H°$. However, S of an ideal gas depends strongly on pressure, so ΔS and ΔG *per mole of reaction in the mixture differ quite substantially from* $\Delta S°$ *and* $\Delta G°$.

Multiplication of (6.36) by dT and integration from T_1 to T_2 gives

$$\ln \frac{K_P°(T_2)}{K_P°(T_1)} = \int_{T_1}^{T_2} \frac{\Delta H°(T)}{RT^2}\,dT \qquad (6.37)$$

To evaluate the integral in (6.37), we need $\Delta H°$ as a function of T. $\Delta H°(T)$ can be found by integration of $\Delta C_P°$ (Sec. 5.5). Evaluation of the integral in Eq. (5.19) leads to an equation with the typical form (see Example 5.6 in Sec. 5.5)

$$\Delta H_T° = A + BT + CT^2 + DT^3 + ET^4 \qquad (6.38)$$

where A, B, C, D, and E are constants. Substitution of (6.38) into (6.37) allows $K_P°$ at any temperature T_2 to be found from its known value at T_1.

$\Delta H°$ for gas-phase reactions usually varies slowly with T, so if $T_2 - T_1$ is reasonably small, it is generally a good approximation to neglect the temperature dependence of $\Delta H°$. Moving $\Delta H°$ outside the integral sign in (6.37) and integrating, we get

$$\ln \frac{K_P°(T_2)}{K_P°(T_1)} \approx \frac{\Delta H°}{R}\left(\frac{1}{T_1} - \frac{1}{T_2}\right) \qquad (6.39)$$

EXAMPLE 6.2 Change of $K_P°$ with T

Find $K_P°$ at 600 K for $N_2O_4(g) \rightleftharpoons 2NO_2(g)$ (a) using the approximation that $\Delta H°$ is independent of T; (b) using the approximation that $\Delta C_P°$ is independent of T; (c) using the NIST-JANAF tables (Sec. 5.9).

(a) If $\Delta H°$ is assumed independent of T, then integration of the van't Hoff equation gives (6.39). Appendix data for $NO_2(g)$ and $N_2O_4(g)$ give $\Delta H_{298}° = 57.20$ kJ/mol and $\Delta G_{298}° = 4730$ J/mol. From $\Delta G° = -RT \ln K_P°$, we find $K_{P,298}° = 0.148$. Substitution in (6.39) gives

$$\ln \frac{K_{P,600}°}{0.148} \approx \frac{57200\text{ J/mol}}{8.314\text{ J/mol-K}}\left(\frac{1}{298.15\text{ K}} - \frac{1}{600\text{ K}}\right) = 11.609$$

$$K_{P,600}° \approx 1.63 \times 10^4$$

(b) If ΔC_P° is assumed independent of T, then Eq. (5.19) gives $\Delta H^\circ(T) \approx \Delta H^\circ(T_1) + \Delta C_P^\circ(T_1)(T - T_1)$. Substitution of this equation into (6.37) gives an equation for $\ln[K_P^\circ(T_2)/K_P^\circ(T_1)]$ that involves $\Delta C_P^\circ(T_1)$ as well as $\Delta H^\circ(T_1)$; see Prob. 6.15. Appendix data give $\Delta C_{P,298}^\circ = -2.88$ J/mol-K, and substitution in the equation of Prob. 6.15 gives (Prob. 6.15) $K_{P,600}^\circ \approx 1.52 \times 10^4$.

(c) From the NIST-JANAF tables, one finds $\Delta G_{600}^\circ = -47.451$ kJ/mol, from which one finds $K_{P,600}^\circ = 1.35 \times 10^4$.

Exercise

Find K_P° for $O_2(g) \rightleftharpoons 2O(g)$ at 25°C, at 1000 K, and at 3000 K using Appendix data and the approximation that ΔH° is independent of T. Compare with the NIST-JANAF-tables values 2.47×10^{-20} at 1000 K and 0.0128 at 3000 K. (Answers: 6.4×10^{-82}, 1.2×10^{-20}, 0.0027.)

Since $d(T^{-1}) = -T^{-2}\,dT$, the van't Hoff equation (6.36) can be written as

$$\frac{d\ln K_P^\circ}{d(1/T)} = -\frac{\Delta H^\circ}{R} \qquad (6.40)$$

The derivative dy/dx at a point x_0 on a graph of y versus x is equal to the slope of the y-versus-x curve at x_0 (Sec. 1.6). Therefore, Eq. (6.40) tells us that the slope of a graph of $\ln K_P^\circ$ versus $1/T$ at a particular temperature equals $-\Delta H^\circ/R$ at that temperature. If ΔH° is essentially constant over the temperature range of the plot, the graph of $\ln K_P^\circ$ versus $1/T$ is a straight line.

If K_P° is known at several temperatures, use of (6.40) allows ΔH° to be found. This gives another method for finding ΔH°, useful if $\Delta_f H^\circ$ of all the species are not known. ΔG_T° can be found from K_P° using $\Delta G_T^\circ = -RT\ln K_P^\circ(T)$. Knowing ΔG° and ΔH°, we can calculate ΔS° from $\Delta G^\circ = \Delta H^\circ - T\Delta S^\circ$. Therefore, *measurement of K_P° over a temperature range allows calculation of ΔG°, ΔH°, and ΔS° of the reaction for temperatures in that range.*

If ΔH° is essentially constant over the temperature range, one can use (6.39) to find ΔH° from only two values of K_P° at different temperatures. Students therefore sometimes wonder why it is necessary to go to the trouble of plotting $\ln K_P^\circ$ versus $1/T$ for several K_P° values and taking the slope. There are several reasons for making a graph. First, ΔH° might change significantly over the temperature interval, and this will be revealed by nonlinearity of the graph. Even if ΔH° is essentially constant, there is always some experimental error in the K_P° values, and the graphed points will show some scatter about a straight line. Using all the data to make a graph and taking the line that gives the best fit to the points results in a ΔH° value more accurate than one calculated from only two data points.

The slope and intercept of the best straight line through the points can be found using the method of least squares (Prob. 6.57), which is readily done on many calculators. Even if a least-squares calculation is done, it is still useful to make a graph, since the graph will show if there is any systematic deviation from linearity due to temperature variation of ΔH° and will show if any point lies way off the best straight line because of a blunder in measurements or calculations.

Figure 6.6a plots ΔH°, ΔS°, ΔG°, and $R\ln K_P^\circ$ versus T for $N_2(g) + 3H_2(g) \rightleftharpoons 2NH_3(g)$. Note that (Sec. 5.5) ΔH° and ΔS° vary only slowly with T, except for low T, where ΔS° goes to zero in accord with the third law. ΔG° increases rapidly and almost linearly with increasing T; this is due to the factor T that multiplies ΔS° in $\Delta G^\circ = \Delta H^\circ - T\Delta S^\circ$. Since ΔH° is negative, $\ln K_P^\circ$ decreases as T increases [Eq. (6.36)]. The rate of decrease of $\ln K_P^\circ$ with respect to T decreases as T increases, because of the $1/T^2$ factor in $d\ln K_P^\circ/dT = \Delta H^\circ/RT^2$.

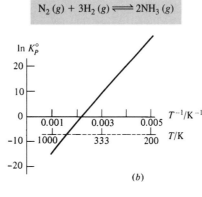

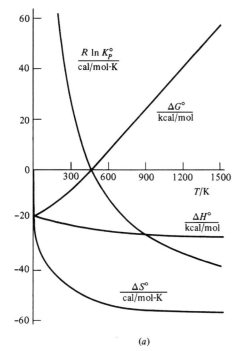

Figure 6.6

Thermodynamic quantities for $N_2(g) + 3H_2(g) \rightleftharpoons 2NH_3(g)$. In the $T \to 0$ limit, $\Delta S° \to 0$.

At high temperatures, $-RT \ln K_P° = \Delta G° \approx -T\,\Delta S°$, so $R \ln K_P° \approx \Delta S°$ in the high-T limit—note the approach of the $R \ln K_P°$ curve to the $\Delta S°$ curve at high T. At low T, $-RT \ln K_P° = \Delta G° \approx \Delta H°$, so $\ln K_P° \approx -\Delta H°/RT$. Therefore $\ln K_P°$ and $K_P°$ go to infinity as $T \to 0$. The products $2NH_3(g)$ have a lower enthalpy and lower internal energy (since $\Delta U° = \Delta H°$ in the $T = 0$ limit) than the reactants $N_2(g) + 3H_2(g)$, and in the $T = 0$ limit, the equilibrium position corresponds to complete conversion to the low-energy species, the products. *The low-T equilibrium position is determined by the internal-energy change $\Delta U°$. The high-T equilibrium position is determined by the entropy change $\Delta S°$.*

Figure 6.6b plots $\ln K_P°$ versus $1/T$ for $N_2(g) + 3H_2(g) \rightleftharpoons 2NH_3(g)$ for the range 200 to 1000 K. The plot shows a very slight curvature, as a result of the small temperature variation of $\Delta H°$.

EXAMPLE 6.3 $\Delta H°$ from $K_P°$ versus T data

Use Fig. 6.6b to estimate $\Delta H°$ for $N_2(g) + 3H_2(g) \rightleftharpoons 2NH_3(g)$ for temperatures in the range 300 to 500 K.

Since only an estimate is required, we shall ignore the slight curvature of the plot and treat it as a straight line. The line goes through the two points

$$T^{-1} = 0.0040 \text{ K}^{-1}, \ln K_P° = 20.0 \quad \text{and} \quad T^{-1} = 0.0022 \text{ K}^{-1}, \ln K_P° = 0$$

Hence the slope (Sec. 1.6) is $(20.0 - 0)/(0.0040 \text{ K}^{-1} - 0.0022 \text{ K}^{-1}) = 1.11 \times 10^4$ K. Note that the slope has units. From Eq. (6.40), the slope of a $\ln K_P°$-versus-$1/T$ plot equals $-\Delta H°/R$, so

$$\Delta H° = -R \times \text{slope} = -(1.987 \text{ cal mol}^{-1} \text{ K}^{-1})(1.11 \times 10^4 \text{ K})$$

$$= -22 \text{ kcal/mol}$$

in agreement with Fig. 6.6a.

Exercise

In Fig. 6.6a, draw the tangent line to the $R \ln K_P^\circ$ curve at 1200 K and from the slope of this line calculate ΔH_{1200}°. (*Answer:* -27 kcal/mol.)

6.4   IDEAL-GAS EQUILIBRIUM CALCULATIONS

For an ideal-gas reaction, once we know the value of K_P° at a given temperature, we can find the equilibrium composition of any given reaction mixture at that temperature and a specified pressure or volume. K_P° can be determined by chemical analysis of a single mixture that has reached equilibrium at the temperature of interest. However, it is generally simpler to determine K_P° from ΔG°, using $\Delta G^\circ = -RT \ln K_P^\circ$. In Chapter 5, we showed how calorimetric measurements (heat capacities and heats of phase transitions of pure substances, and heats of reaction) allow one to find $\Delta_f G_T^\circ$ values for a great many compounds. Once these values are known, we can calculate ΔG_T° for any chemical reaction between these compounds, and from ΔG° we get K_P°.

Thus thermodynamics enables us to find K_P° for a reaction without making any measurements on an equilibrium mixture. This knowledge is of obvious value in finding the maximum possible yield of product in a chemical reaction. If ΔG_T° is found to be highly positive for a reaction, this reaction will not be useful for producing the desired product. If ΔG_T° is negative or only slightly positive, the reaction *may* be useful. Even though the equilibrium position yields substantial amounts of products, we must still consider the *rate* of the reaction (a subject outside the scope of thermodynamics). Often, a reaction with a negative ΔG° is found to proceed extremely slowly. Hence we may have to search for a catalyst to speed up attainment of equilibrium. Often, several different reactions can occur for a given set of reactants, and we must then consider the rates and the equilibrium constants of several simultaneous reactions.

We now examine equilibrium calculations for ideal-gas reactions. We shall use K_P° in all our calculations. K_c° could also have been used, but consistent use of K_P° avoids having to learn any formulas with K_c°. We shall assume the density is low enough to allow the gas mixture to be treated as ideal.

The equilibrium composition of an ideal-gas reaction mixture is a function of T and P (or T and V) and the initial composition (mole numbers) $n_{1,0}, n_{2,0}, \ldots$ of the mixture. The equilibrium composition is related to the initial composition by a single variable, the equilibrium extent of reaction ξ_{eq}. We have [Eq. (4.95)] $\Delta n_i \equiv n_{i,eq} - n_{i,0} = \nu_i \xi_{eq}$. Thus our aim in an ideal-gas equilibrium calculation is to find ξ_{eq}. We do this by expressing the equilibrium partial pressures in K_P° in terms of the equilibrium mole numbers $n_i = n_{i,0} + \nu_i \xi$, where, for simplicity, the eq subscripts have been omitted.

The specific steps to find the equilibrium composition of an ideal-gas reaction mixture are as follows.

1. Calculate ΔG_T° of the reaction using $\Delta G_T^\circ = \Sigma_i \nu_i \Delta_f G_{T,i}^\circ$ and a table of $\Delta_f G_T^\circ$ values.
2. Calculate K_P° using $\Delta G^\circ = -RT \ln K_P^\circ$. [If $\Delta_f G^\circ$ data at the temperature T of the reaction are unavailable, K_P° at T can be estimated using the form (6.39) of the van't Hoff equation, which assumes ΔH° is constant.]
3. Use the stoichiometry of the reaction to express the equilibrium mole numbers n_i in terms of the initial mole numbers $n_{i,0}$ and the equilibrium extent of reaction ξ_{eq}, according to $n_i = n_{i,0} + \nu_i \xi_{eq}$.
4. (*a*) If the reaction is run at fixed T and P, use $P_i = x_i P = (n_i / \Sigma_i n_i) P$ and the expression for n_i from step 3 to express each equilibrium partial pressure P_i in terms of ξ_{eq}.

(b) If the reaction is run at fixed T and V, use $P_i = n_i RT/V$ to express each P_i in terms of ξ_{eq}. Thus:

$$P_i = x_i P = \frac{n_i}{\sum\limits_i n_i} P \quad \text{if } P \text{ is known}; \qquad P_i = \frac{n_i RT}{V} \quad \text{if } V \text{ is known}$$

5. Substitute the P_i's (expressed as functions of ξ_{eq}) into the equilibrium-constant expression $K_P^\circ = \Pi_i (P_i/P^\circ)^{\nu_i}$ and solve for ξ_{eq}.
6. Calculate the equilibrium mole numbers from ξ_{eq} and the expressions for n_i in step 3.

As an example, consider the reaction

$$N_2(g) + 3H_2(g) \rightleftharpoons 2NH_3(g)$$

with the initial composition of 1.0 mol of N_2, 2.0 mol of H_2, and 0.50 mol of NH_3. To do step 3, let $z \equiv \xi_{eq}$ be the equilibrium extent of reaction. Constructing a table like that used in general-chemistry equilibrium calculations, we have

	N_2	H_2	NH_3
Initial moles	1.0	2.0	0.50
Change	$-z$	$-3z$	$2z$
Equilibrium moles	$1.0 - z$	$2.0 - 3z$	$0.50 + 2z$

where $\Delta n_i = \nu_i \xi$ [Eq. (4.95)] was used to calculate the changes. The total number of moles at equilibrium is $n_{tot} = 3.5 - 2z$. If P is held fixed, we express the equilibrium partial pressures as $P_{N_2} = x_{N_2} P = [(1.0 - z)/(3.5 - 2z)]P$, etc., where P is known. If V is held fixed, we use $P_{N_2} = n_{N_2} RT/V = (1.0 - z)RT/V$, etc., where T and V are known. One then substitutes the expressions for the partial pressures into the K_P° expression (6.18) to get an equation with one unknown, the equilibrium extent of reaction z. One then solves for z and uses the result to calculate the equilibrium mole numbers.

The equilibrium extent of reaction might be positive or negative. We define the **reaction quotient** Q_P for the ammonia synthesis reaction as

$$Q_P \equiv \frac{P_{NH_3}^2}{P_{N_2} P_{H_2}^3} \tag{6.41}$$

where the partial pressures are those present in the mixture at some particular time, not necessarily at equilibrium. If the initial value of Q_P is less than K_P [Eq. (6.19)], then the reaction must proceed to the right to produce more products and increase Q_P until it becomes equal to K_P at equilibrium. Hence if $Q_P < K_P$ then $\xi_{eq} > 0$. If $Q_P > K_P$, then $\xi_{eq} < 0$.

To find the maximum and minimum possible values of the equilibrium extent of reaction z in the preceding NH_3 example, we use the condition that the equilibrium mole numbers can never be negative. The relation $1.0 - z > 0$ gives $z < 1.0$. The relation $2.0 - 3z > 0$ gives $z < \frac{2}{3}$. The relation $0.50 + 2z > 0$ gives $z > -0.25$. Hence $-0.25 < z < 0.667$. The K_P° equation has z^4 as the highest power of z (this comes from $P_{N_2} P_{H_2}^3$ in the denominator) and so has four roots. Only one of these will lie in the range -0.25 to 0.667.

EXAMPLE 6.4 Equilibrium composition at fixed T and P

Suppose that a system initially contains 0.300 mol of $N_2O_4(g)$ and 0.500 mol of $NO_2(g)$, and the equilibrium

$$N_2O_4(g) \rightleftharpoons 2NO_2(g)$$

is attained at 25°C and 2.00 atm. Find the equilibrium composition.

Carrying out step 1 of the preceding scheme, we use Appendix data to get

$$\Delta G^\circ_{298}/(\text{kJ/mol}) = 2(51.31) - 97.89 = 4.73$$

For step 2, we have $\Delta G^\circ = -RT \ln K^\circ_P$ and

$$4730 \text{ J/mol} = -(8.314 \text{ J/mol-K})(298.1 \text{ K}) \ln K^\circ_P$$

$$\ln K^\circ_P = -1.908 \quad \text{and} \quad K^\circ_P = 0.148$$

For step 3, let x moles of N_2O_4 react to reach equilibrium. By the stoichiometry, $2x$ moles of NO_2 will be formed and the equilibrium mole numbers will be

$$n_{N_2O_4} = (0.300 - x) \text{ mol} \quad \text{and} \quad n_{NO_2} = (0.500 + 2x) \text{ mol} \quad (6.42)$$

[Note that the equilibrium extent of reaction is $\xi = x$ mol and the equations in (6.42) satisfy $n_i = n_{i,0} + \nu_i \xi$.]

Since T and P are fixed, we use Eq. (6.41) in step 4(a) to write

$$P_{NO_2} = x_{NO_2}P = \frac{0.500 + 2x}{0.800 + x}P, \quad P_{N_2O_4} = x_{N_2O_4}P = \frac{0.300 - x}{0.800 + x}P$$

since $\Sigma_i n_i = (0.300 - x) \text{ mol} + (0.500 + 2x) \text{ mol} = (0.800 + x) \text{ mol}$.

Performing step 5, we have

$$K^\circ_P = \frac{[P_{NO_2}/P^\circ]^2}{P_{N_2O_4}/P^\circ}$$

$$0.148 = \frac{(0.500 + 2x)^2(P/P^\circ)^2}{(0.800 + x)^2} \frac{0.800 + x}{(0.300 - x)(P/P^\circ)} = \frac{0.250 + 2x + 4x^2}{0.240 - 0.500x - x^2} \frac{P}{P^\circ}$$

The reaction occurs at $P = 2.00$ atm $= 1520$ torr, and $P^\circ = 1$ bar $= 750$ torr. Thus $0.148(P^\circ/P) = 0.0730$. Clearing of fractions, we find

$$4.0730x^2 + 2.0365x + 0.2325 = 0$$

The quadratic formula $x = [-b \pm (b^2 - 4ac)^{1/2}]/2a$ for the solutions of $ax^2 + bx + c = 0$ gives

$$x = -0.324 \quad \text{and} \quad x = -0.176$$

The number of moles of each substance present at equilibrium must be positive. Thus, $n(N_2O_4) = (0.300 - x)$ mol > 0, and x must be less than 0.300. Also, $n(NO_2) = (0.500 + 2x)$ mol > 0, and x must be greater than -0.250. We have $-0.250 < x < 0.300$. The root $x = -0.324$ must therefore be discarded. Thus $x = -0.176$, and step 6 gives

$$n(N_2O_4) = (0.300 - x) \text{ mol} = 0.476 \text{ mol}$$

$$n(NO_2) = (0.500 + 2x) \text{ mol} = 0.148 \text{ mol}$$

Exercise

For O(g) at 4200 K, $\Delta_f G^\circ = -26.81$ kJ/mol. For a system whose initial composition is 1.000 mol of $O_2(g)$, find the equilibrium composition at 4200 K and 3.00 bar. (*Answer:* 0.472 mol of O_2, 1.056 mol of O.)

EXAMPLE 6.5 Equilibrium composition at fixed T and V

$K_P^\circ = 6.51$ at 800 K for the ideal-gas reaction $2A + B \rightleftharpoons C + D$. If 3.000 mol of A, 1.000 mol of B, and 4.000 mol of C are placed in an 8000-cm^3 vessel at 800 K, find the equilibrium amounts of all species.

Proceeding to step 3 of the preceding scheme, we suppose that x moles of B react to reach equilibrium. Then at equilibrium,

$$n_B = (1 - x) \text{ mol}, \quad n_A = (3 - 2x) \text{ mol}, \quad n_C = (4 + x) \text{ mol}, \quad n_D = x \text{ mol}$$

The reaction is run at constant T and V. Using $P_i = n_i RT/V$ according to step 4(b) and substituting into K_P°, we get

$$K_P^\circ \equiv \frac{(P_C/P^\circ)(P_D/P^\circ)}{(P_A/P^\circ)^2(P_B/P^\circ)} = \frac{(n_C RT/V)(n_D RT/V)P^\circ}{(n_A RT/V)^2(n_B RT/V)} = \frac{n_C n_D}{n_A^2 n_B} \frac{VP^\circ}{RT}$$

where $P^\circ \equiv 1$ bar. Use of 1 atm = 760 torr, 1 bar = 750.06 torr, and $R = 82.06$ cm^3 atm mol^{-1} K^{-1} gives $R = 83.14$ cm^3 bar mol^{-1} K^{-1}. Substitution for the n_i's gives

$$6.51 = \frac{(4 + x)x \text{ mol}^2}{(3 - 2x)^2(1 - x) \text{ mol}^3} \frac{8000 \text{ cm}^3 \text{ bar}}{(83.14 \text{ cm}^3 \text{ bar mol}^{-1} \text{ K}^{-1})(800 \text{ K})}$$

$$x^3 - 3.995x^2 + 5.269x - 2.250 = 0 \qquad (6.43)$$

where we divided by the coefficient of x^3. We have a cubic equation to solve. The formula for the roots of a cubic equation is quite complicated. Moreover, equations of degree higher than quartic often arise in equilibrium calculations, and there is no formula for the roots of such equations. Hence we shall solve (6.43) by trial and error. The requirements $n_B > 0$ and $n_D > 0$ show that $0 < x < 1$. For $x = 0$, the left side of (6.43) equals -2.250; for $x = 1$, the left side equals 0.024. Hence x is much closer to 1 than to 0. Guessing $x = 0.9$, we get -0.015 for the left side. Hence the root is between 0.9 and 1.0. Interpolation gives an estimate of $x = 0.94$. For $x = 0.94$, the left side equals 0.003, so we are still a bit high. Trying $x = 0.93$, we get -0.001 for the left side. Hence the root is 0.93 (to two places). The equilibrium amounts are then $n_A = 1.14$ mol, $n_B = 0.07$ mol, $n_C = 4.93$ mol, and $n_D = 0.93$ mol.

Exercise

$K_P^\circ = 3.33$ at 400 K for the ideal-gas reaction $2R + 2S \rightleftharpoons V + W$. If 0.400 mol of R and 0.400 mol of S are placed in an empty 5.000-L vessel at 400 K, find the equilibrium amounts of all species. (*Hint:* To avoid solving a quartic equation, take the square root of both sides of the equation.) (*Answer:* 0.109 mol of R, 0.109 mol of S, 0.145 mol of V, 0.145 mol of W.)

Some electronic calculators can automatically find the roots of an equation. Use of such a calculator gives the roots of Eq. (6.43) as $x = 0.9317...$ and two imaginary numbers.

A spreadsheet can also be used to solve (6.43). The directions will be given for Excel. Begin by entering a guess for x in cell A1. Since we know x is between 0 and 1, we guess x as 0.5. Then enter the formula =A1^3-3.995*A1^2+5.269*A1-2.250 into cell B1.

Excel contains a program called the Solver that will adjust the values in user-specified cells so as to make the values in other cells satisfy conditions set by the user. (The corresponding programs in Lotus 1-2-3 and Quattro Pro are called the Solver and the Optimizer, respectively.) To invoke the Solver, choose Solver on the Tools menu. (If you don't see Solver on the Tools menu, choose Add-Ins on the Tools menu, click the box for Solver Add-In, and click OK. If you don't see Solver Add-In in the Add Ins box, you need to click Browse and find the Solver.xla file.) In the Solver Parameters dialog box that opens, enter B1 in the Set Target Cell box, click Value of after Equal To and enter 0 after Value of. In the By Changing Cells box, enter A1 to tell Excel that the number in A1 (the value of x) is to be varied. To have Excel solve the problem, just click on Solve in the Solver Parameters box. After a moment, Excel displays the Solver Results box telling you that it has found a solution. Click OK. Cell B1 now has a value very close to zero and cell A1 has the desired solution 0.9317....

If you again choose Solver from the Tools menu and click Options in the Solver Parameters box, you will see the default value 0.000001 in the Precision box. Excel stops and declares that it has found a solution when all the required conditions are satisfied within the specified precision. With the default precision, the value in B1 will be something like 3×10^{-7}. To verify the accuracy of the solution, it's a good idea to change the Precision from 10^{-6} to 10^{-10}, rerun the Solver, and verify that this does not significantly change the answer in A1. Then restore the default precision.

Examples 6.4 and 6.5 used general procedures applicable to all ideal-gas equilibrium calculations. Example 6.6 considers a special kind of ideal-gas reaction: isomerization.

EXAMPLE 6.6 Equilibrium composition in isomerization

Suppose the gas-phase isomerization reactions $A \rightleftharpoons B$, $A \rightleftharpoons C$, and $B \rightleftharpoons C$ reach equilibrium at a fixed T. Express the equilibrium mole fractions of A, B, and C in terms of equilibrium constants.

Let $K_{B/A} \equiv K_P^\circ$ for $A \rightleftharpoons B$, and let $K_{C/A} \equiv K_P^\circ$ for $A \rightleftharpoons C$. We have

$$K_{B/A} = \frac{P_B/P^\circ}{P_A/P^\circ} = \frac{x_B P/P^\circ}{x_A P/P^\circ} = \frac{x_B}{x_A} \quad \text{and} \quad K_{C/A} = \frac{x_C}{x_A} \qquad (6.44)$$

The sum of the mole fractions is 1, and use of (6.44) gives

$$x_A + x_B + x_C = 1$$
$$x_A + x_A K_{B/A} + x_A K_{C/A} = 1$$
$$x_A = \frac{1}{1 + K_{B/A} + K_{C/A}} \qquad (6.45)$$

From $x_B = K_{B/A}x_A$ and $x_C = K_{C/A}x_A$, we get

$$x_B = \frac{K_{B/A}}{1 + K_{B/A} + K_{C/A}} \quad \text{and} \quad x_C = \frac{K_{C/A}}{1 + K_{B/A} + K_{C/A}} \qquad (6.46)$$

Using these equations, one finds (Prob. 6.33) the equilibrium mole fractions in a gas-phase mixture (assumed ideal) of pentane, isopentane, and neopentane at various temperatures to be as shown in Fig. 6.7.

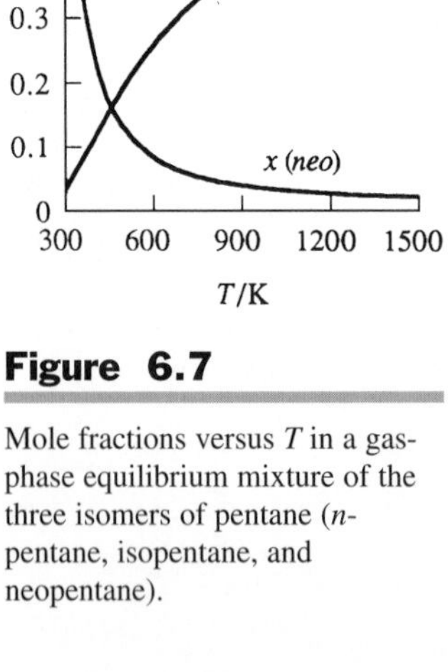

Exercise

A 300-K gas-phase equilibrium mixture of the isomers A, B, and C contains 0.16 mol of A, 0.24 mol of B, and 0.72 mol of C. Find $K_{B/A}$ and $K_{C/A}$ at 300 K. (*Answers:* 1.5, 4.5.)

Figure 6.7

Mole fractions versus T in a gas-phase equilibrium mixture of the three isomers of pentane (n-pentane, isopentane, and neopentane).

Since the standard pressure P° appears in the definition of K_P°, the 1982 change of P° from 1 atm to 1 bar affects K_P° values slightly. See Prob. 6.37.

When $|\Delta G°|$ is large, $K_P°$ is very large or very small. For example, if $\Delta G°_{298} = 137$ kJ/mol, then $K_{P,298}° = 10^{-24}$. From this value of $K_P°$, we might well calculate that at equilibrium only a few molecules or even only a fraction of one molecule of a product is present. When the number of molecules of a species is small, thermodynamics is not rigorously applicable and the system shows continual fluctuations about the thermodynamically predicted number of molecules (Sec. 3.7).

Tables of data often list $\Delta_f H°$ and $\Delta_f G°$ values to 0.01 kJ/mol. However, experimental errors in measured $\Delta_f H°$ values typically run $\frac{1}{2}$ to 2 kJ/mol, although they may be substantially smaller or larger. An error in $\Delta G°_{298}$ of 2 kJ/mol corresponds to a factor of 2 in $K_P°$. Thus, the reader should take equilibrium constants calculated from thermodynamic data with a grain of NaCl(s).

In the preceding examples, the equilibrium composition for a given set of conditions (constant T and V or constant T and P) was calculated from $K_P°$ and the initial composition. For a system reaching equilibrium while T and P are constant, the equilibrium position corresponds to the minimum in the Gibbs energy G. Figure 6.8 plots the conventional values (Chapter 5) of G, H, and TS (where $G = H - TS$) versus extent of reaction for the ideal-gas reaction $N_2 + 3H_2 \rightleftharpoons 2NH_3$ run at the fixed T and P of 500 K and 4 bar with initial composition 1 mol N_2 and 3 mol H_2. At equilibrium, $\xi_{eq} = 0.38$. The plot is made using the fact that each of G, H, and S of the ideal gas mixture is the sum of contributions from each pure gas (Sec. 6.1). See Prob. 6.58 for details.

6.5 SIMULTANEOUS EQUILIBRIA

This section shows how to solve a system with several simultaneous ideal-gas reactions that have species in common. Suppose the following two ideal-gas reactions occur:

$$(1) \quad CH_4 + H_2O \rightleftharpoons CO + 3H_2$$
$$(2) \quad CH_4 + 2H_2O \rightleftharpoons CO_2 + 4H_2$$
(6.47)

Let the initial (0 subscript) numbers of moles be

$$n_{0,CH_4} = 1 \text{ mol}, \quad n_{0,H_2O} = 1 \text{ mol}, \quad n_{0,CO_2} = 1 \text{ mol}, \quad n_{0,H_2} = 1 \text{ mol}, \quad n_{0,CO} = 2 \text{ mol}$$

[The reactions (6.47) are industrially important in the production of hydrogen from natural gas. The reverse of reaction (1) is one of the reactions in the Fischer–Tropsch process that converts CO and H_2 (formed from the reaction of coal with air and water vapor) to hydrocarbons and water. During World War II, Germany was cut off from oil supplies and used the Fischer–Tropsch process to produce gasoline.]

Using $P_i/P° = x_i P/P° = n_i P/n_{tot} P°$, we get as the equilibrium conditions for these reactions:

$$K_{P,1}° = \frac{n_{CO}(n_{H_2})^3}{n_{CH_4} n_{H_2O}} \left(\frac{P}{P° n_{tot}}\right)^2, \qquad K_{P,2}° = \frac{n_{CO_2}(n_{H_2})^4}{n_{CH_4}(n_{H_2O})^2} \left(\frac{P}{P° n_{tot}}\right)^2$$
(6.48)

With simultaneous equilibria, it is often simpler to use conservation-of-matter conditions for each element instead of using the extents of reaction. The number of moles of each element are:

$$n_C = n_{CH_4} + n_{CO_2} + n_{CO}, \quad n_H = 4n_{CH_4} + 2n_{H_2O} + 2n_{H_2}, \quad n_O = n_{H_2O} + 2n_{CO_2} + n_{CO}$$

Using the initial composition, we have $n_C = 1 + 1 + 2 = 4$, $n_H = 8$, and $n_O = 5$ throughout the reaction. At 900 K, the NIST-JANAF tables (Sec. 5.9) $\Delta_f G°$ data give the equilibrium constants for the reactions as $K_{P,1}° = 1.30$ and $K_{P,2}° = 2.99$. We shall find the equilibrium composition at a fixed pressure of 0.01 bar. We have five unknowns, the numbers of moles of each of the five species in the reactions. The five equations

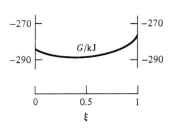

320 —

300 —

TS/kJ

40 —

20 —

—| 220

—| 200

H/kJ

—| -60

—| -80

-270 —

G/kJ

—| -270

-290 —

—| -290

0   0.5   1

ξ

Figure 6.8

Variation of G, H, and TS with extent of reaction ξ in the synthesis of $NH_3(g)$ at 500 K and 4 bar for an initial composition of 1 mol of N_2 and 3 mol of H_2. The H-versus-ξ curve is linear. Since Δn is negative for the reaction, S decreases as ξ increases. (Of course, S_{univ} reaches a maximum when G reaches a minimum.)

to be satisfied are the two equilibrium conditions and the three conservation-of-moles conditions for each element.

Computer-algebra programs such as Mathcad, Scientific Notebook, Maple, and Mathematica can be used to solve the equations. We shall use the Excel spreadsheet to do this. (Lotus 1-2-3 or Quattro Pro could also be used.) The initial setup of the spreadsheet is shown in Fig. 6.9a. The reactions are entered into cells B1 and B2. The initial numbers of moles are entered into cells B5 to F5. An initial guess is needed for the equilibrium composition. We use the initial composition as our guess and enter this in cells B6 to F6. Cell G6 will contain n_{tot} at equilibrium. To produce this number, we enter the formula =SUM(B6:F6) into cell G6. The SUM(B6:F6) function adds the numbers in cells B6 through F6. (The colon denotes a range of cells.)

To get the initial setup of the spreadsheet, enter data, labels, and formulas as shown in Fig. 6.9b. The spreadsheet will appear as shown in Fig. 6.9a. Cells B10, C10, and D10 contain the initial moles of the elements. Cells F11 and G11 contain the equilibrium constants calculated from the current values of the equilibrium mole numbers and $P/n_{tot}P°$ using the equations in (6.48). Cells B12, C12, and D12 contain the fractional errors in the mole numbers of each element and cells F12 and G12 contain the fractional errors in the calculated equilibrium constants. Make sure you understand all

	A	B	C	D	E	F	G
1		CH4+H2O=CO+3H2	K1=		1.3	T/K=	900
2		CH4+2H2O=CO2+4H2	K2=		2.99	P/bar=	0.01
3							
4		CH4	H2O	CO2	H2	CO	
5	initial mol	1	1	1	1	2	ntot
6	eq mol	1	1	1	1	2	6
7							(P/bar)/ntot
8							0.0016667
9		carbon	hydrogen	oxygen			
10	initial	4	8	5		K1calc	K2calc
11	equilib	4	8	5		5.556E-06	2.778E-06
12	fractnl error	0	0	0	fractnl erro	-0.9999957	-0.9999991

(a)

	B	C	D	E	F	G
4	CH4	H2O	CO2	H2	CO	
5	1	1	1	1	2	ntot
6	1	1	1	1	2	=SUM(B6:F6)
7						(P/bar)/ntot
8						=G2/G6
9	carbon	hydrogen	oxygen			
10	=B5+D5+F5	=4*B5+2*C5+2*E5	=C5+2*D5+F5		K1calc	K2calc
11	=B6+D6+F6	=4*B6+2*C6+2*E6	=C6+2*D6+F6		=G8^2*F6*E6^3/(C6*B6)	=G8^2*D6*E6^4/(B6*C6^2)
12	=(B11-B10)/B10	=(C11-C10)/C10	=(D11-D10)/D10	fractnl erro	=(F11-E1)/E1	=(G11-E2)/E2

(b)

Figure 6.9

Initial setup of the Excel spreadsheet for finding the equilibrium composition of the reaction system (6.47) at 900 K and a specified pressure. (a) Values view (the default view). (b) Formula view of part of the sheet.

the formulas in Fig. 6.9*b*. To solve the problem, we need to make cells B12, C12, D12, F12, and G12 differ negligibly from zero.

We shall use the Solver (Sec. 6.4) in Excel to do this. (Mathcad has what is called a solve block to solve a system of simultaneous equations subject to specified constraints. Maple V has the function fsolve that will solve simultaneous equations for roots that lie in specified ranges; Mathematica has the function FindRoot.)

After setting up the Excel spreadsheet as in Figs. 6.9*a* and 6.9*b*, choose Solver on the Tools menu. In the Solver Parameters dialog box that opens, enter F12 in the Set Target Cell box, click Value of after Equal To and enter 0 after Equal To. In the By Changing Cells box enter B6:F6 to tell Excel that the numbers in these five cells (the equilibrium mole numbers) are to be varied. To enter the remaining conditions to be satisfied, click Add below Subject to the Constraints. In the Add Constraint box that opens, enter G12 under Cell Reference, choose = in the drop-down list in the middle, and enter 0 at the right. Click Add. Then enter B12:D12 under Cell Reference, choose =, and enter 0 at the right. We have now specified the five conditions to be satisfied, but it is also desirable to give Excel some guidance on the unknown mole numbers. These numbers cannot be negative or zero, so we shall require them to each be larger than some very small number, say, 10^{-14}. Therefore enter B6:F6 under Cell Reference in the Add Constraint box, choose >=, and enter 1E-14 at the right. Then click OK to close the Add Constraint box. In the Solver Parameters box, you will see the constraints listed. (The $ signs can be ignored.)

Now click on Solve in the Solver Parameters box. When Excel displays the Solver Results box telling you that it has found a solution, click OK. The spreadsheet now looks like Fig. 6.10. The desired solution is shown in cells B6 to F6. The fractional errors in F12 and G12 are less than 10^{-6}, which is the default value of the Precision parameter in Excel.

To save the results, select cells B6 to G6 by dragging over them with the mouse, choose Copy from the Edit menu, click on cell B15, and choose Paste from the Edit menu to paste the results into cells B15 to G15. Also, enter the pressure value 0.01 in A15 and in row 14 put labels for the data.

We will now change the pressure to 0.1 bar and redo the calculation. Enter 0.1 in cell G2. Note that this changes the value in cell G8, whose formula (Fig. 6.9*b*) depends on G2, and changes the values in F11, F12, G11, and G12. Now choose Solver from the Tools menu and click on Solve. The Solver then gives the solution at the new pressure. We can copy and paste this solution into B16 to G16. After a few more runs, we have a table of composition data versus *P* that can be graphed.

To make the graph, first select the block of data to be graphed by dragging with the mouse. Then choose Chart from the Insert menu or click on the Chart button on the toolbar. In the series of boxes that follow, choose XY (Scatter) as the chart type,

	A	B	C	D	E	F	G
1		CH4+H2O=CO+3H2		K1=	1.3	T/K=	900
2		CH4+2H2O=CO2+4H2		K2=	2.99	P/bar=	0.01
3							
4		CH4	H2O	CO2	H2	CO	
5	initial mol	1	1	1	1	2	ntot
6	eq mol	0.0006101	0.3248051	0.675805	3.6739747	3.3235849	7.9987798
7							(P/bar)/ntot
8							0.0012502
9		carbon	hydrogen	oxygen			
10	initial	4	8	5		K1calc	K2calc
11	equilib	4	8	5		1.2999999	2.9899998
12	fractnl erro	0	0	0	fractnl error	-1.06E-07	-7.62E-08

Figure 6.10

The spreadsheet of Fig. 6.9 after running the Solver to make the fractional errors close to zero.

n/mol

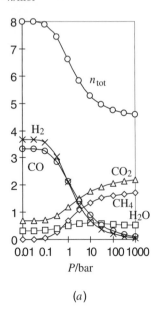

(a)

$n(H_2O)$/mol

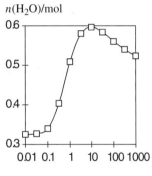

P/bar

(b)

Figure 6.11

Excel graphs of the equilibrium
composition of the reaction system
of Fig. 6.9 versus pressure. (The
gas mixture is assumed ideal,
which is a poor approximation at
high pressure.)

data points connected by smoothed lines as the subtype, and Series in Columns. Figure 6.11a shows the composition versus pressure. (The horizontal axis has been made logarithmic by first selecting the chart by clicking on it with the mouse, then double clicking on the x axis to open the Format Axis box, selecting the Scale tab, and clicking the Logarithmic scale box.) Figure 6.11b shows n_{H_2O} versus P. The surprising appearance of Fig. 6.11b is discussed in Sec. 6.6.

> Excel contains a program called Visual Basic for Applications (VBA), which enables you to automate varying the pressure over a range of values, running the Solver at each value, and copying and pasting the results.

The Solver is not guaranteed to find a solution. If our initial guesses for the composition are very far from the equilibrium values, the Solver might wander off in the wrong direction and be unable to find the correct solution. If we want to calculate the composition at pressures ranging from 0.01 to 1000 bar, the Solver has the best chance of succeeding if we do all the calculations in order of increasing pressure, using the previous pressure's results as the initial guess for the new equilibrium composition, rather than jumping around. If the Solver does fail to find a solution, try a different initial guess for the equilibrium composition. For a description of the programs in the Solver, see D. Fylstra et al., *Interfaces,* **28,** 29 (1998).

In a system with two or more reactions, the set of reactions one can deal with is not unique. For example, in this system, instead of using reactions (1) and (2), we could use the reactions

$$(3) \quad CO_2 + H_2 \rightleftharpoons CO + H_2O$$

$$(4) \quad 4CO + 2H_2O \rightleftharpoons 3CO_2 + CH_4$$

where if R_1, R_2, R_3, and R_4 denote the reactions, we have $R_3 = R_1 - R_2$ and $R_4 = 3R_2 - 4R_1$. Use of (3) and (4) instead of (1) and (2) will give the same equilibrium composition. In finding the equilibrium composition of a system with multiple equilibria, one deals only with **independent reactions,** where the word independent means that no reaction of the set of reactions can be written as a combination of the other reactions of the set. A method to find the number of independent reactions from the chemical species present in the system is given in sec. 4.16 of *Denbigh*.

SHIFTS IN IDEAL-GAS REACTION EQUILIBRIA

If T, P, or the composition of an ideal gas mixture in equilibrium is changed, the equilibrium position may shift. We now examine the direction of such shifts.

To help figure out the direction of the equilibrium shift, we define the **reaction quotient** Q_P for the reaction $0 \to \Sigma_i \nu_i \mu_i$ at some instant of time as

$$Q_P \equiv \prod_i P_i^{\nu_i} \tag{6.49}$$

where P_i is the partial pressure of gas i in the system at a particular time, and the system is not necessarily in equilibrium. We imagine that the change to the equilibrium system is made instantaneously, so the system has no time to react while the change is being made. We then compare the value of Q_P the instant after the change is made with the value of K_P. If $Q_P < K_P$, the equilibrium position will shift to the right so as to produce more products (which appear in the numerator of Q_P) and increase Q_P until it equals K_P at the new equilibrium position. If we find $Q_P = K_P$, then the system is in equilibrium after the change and the change produces no shift in the equilibrium position. If $Q_P > K_P$, the equilibrium shifts to the left. Alternatively, we can compare K_P° with $Q_P^\circ \equiv \Pi_i \, (P_i/P^\circ)^{\nu_i}$.

Isobaric Temperature Change

Suppose we change T, keeping P constant. Since $d \ln y = (1/y) \, dy$, Eq. (6.36) gives $dK_P^\circ/dT = K_P^\circ \, \Delta H^\circ/RT^2$. Since K_P° and RT^2 are positive, the sign of dK_P°/dT is the same as the sign of ΔH°.

If ΔH° is positive, then dK_P°/dT is positive; for a temperature increase ($dT > 0$), dK_P° is then positive, and K_P° increases. Since $P_i = x_i P$, and P is held fixed during the change in T, the instant after T increases but before any shift in composition occurs, all partial pressures are unchanged and Q_P° is unchanged. Therefore the instant after the temperature increase, we have $K_P^\circ > Q_P^\circ$ and the equilibrium must shift to the right to increase Q_P°. Thus for an endothermic reaction ($\Delta H^\circ > 0$), an increase in temperature at constant pressure will shift the equilibrium to the right.

If ΔH° is negative (an exothermic reaction), then dK_P°/dT is negative and a positive dT gives a negative dK_P°. An isobaric temperature increase shifts the equilibrium to the left for an exothermic reaction.

These results can be summarized in the rule that *an increase in T at constant P in a closed system shifts the equilibrium in the direction in which the system absorbs heat from the surroundings.* Thus, for an endothermic reaction, the equilibrium shifts to the right as T increases.

Isothermal Pressure Change

Consider the ideal-gas reaction A $\rightleftharpoons$ 2B. Let equilibrium be established, and suppose we then double the pressure at constant T by isothermally compressing the mixture to half its original volume; thereafter, P is held constant at its new value. The equilibrium constant K_P is unchanged since T is unchanged. Since $P_i = x_i P$, this doubling of P doubles P_A and doubles P_B (before any shift in equilibrium occurs). This quadruples the numerator of $Q_P \equiv P_B^2/P_A$ and doubles its denominator; thus Q_P is doubled. Before the pressure increase, Q_P was equal to K_P, but after the pressure increase, Q_P has been increased and is greater than K_P. The system is no longer in equilibrium, and Q_P will have to decrease to restore equilibrium. $Q_P \equiv P_B^2/P_A$ decreases when the equilibrium shifts to the left, thereby decreasing P_B and increasing P_A. Thus a pressure increase shifts the gas-phase equilibrium A $\rightleftharpoons$ 2B to the left, the side with fewer moles in the balanced reaction.

Generalizing to the ideal-gas reaction aA $+ b$B $+ \cdots \rightleftharpoons e$E $+ f$F $+ \cdots$, we see that if the total moles $e + f + \cdots$ on the right is larger than the total moles $a + b + \cdots$ on the left, an isothermal increase in pressure will increase the numerator of the reaction quotient

$$Q_P \equiv \frac{P_E^e P_F^f \cdots}{P_A^a P_B^b \cdots} = \frac{x_E^e x_F^f \cdots}{x_A^a x_B^b \cdots} \frac{P^{e+f+\cdots}}{P^{a+b+\cdots}}$$

more than the denominator and will therefore shift the equilibrium to the left (the side with fewer moles) to reduce Q_P to K_P. If $e + f + \cdots$ is less than $a + b + \cdots$, a pressure increase shifts the equilibrium to the right. If $e + f + \cdots$ equals $a + b + \cdots$, a pressure increase has no effect on Q_P and does not shift the equilibrium. Since the system's volume is proportional to the total number of moles of gas present, we have the rule that *an increase in P at constant T in a closed system shifts the equilibrium in the direction in which the system's volume decreases.*

Although K_P depends on T only, the equilibrium composition of an ideal-gas reaction mixture depends on *both* T and P, except for reactions with $\Delta n = 0$. Figure 6.12

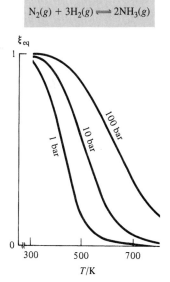

$N_2(g) + 3H_2(g) \rightleftharpoons 2NH_3(g)$

Figure 6.12

Equilibrium extent of reaction versus T at several pressures for the ammonia-synthesis reaction with an initial composition of 1 mol N_2 and 3 mol H_2. A pressure increase at fixed T increases the yield of NH_3.

plots the equilibrium extent of reaction versus T at three pressures for $N_2(g) + 3H_2(g)$ $\rightleftharpoons 2NH_3(g)$, assuming ideal-gas behavior.

The italicized rules for shifts produced by an isothermal pressure change and by an isobaric temperature change constitute *Le Châtelier's principle*. These two rules can be proved valid for any reaction, not just ideal-gas reactions (see *Kirkwood and Oppenheim*, pp. 108–109).

Isochoric Addition of Inert Gas

Suppose we add some inert gas to an equilibrium mixture, holding V and T constant. Since $P_i = n_i RT/V$, the partial pressure of each gas taking part in the reaction is unaffected by such an addition of an inert gas. Hence the reaction quotient $Q_P \equiv \prod_i P_i^{\nu_i}$ is unaffected and remains equal to K_P. Thus, there is no shift in ideal-gas equilibrium for isochoric, isothermal addition of an inert gas. This makes sense because in the absence of intermolecular interactions, the reacting ideal gases have no way of knowing whether there is any inert gas present.

Addition of a Reactant Gas

Suppose that for the reaction $A + B \rightleftharpoons 2C + D$ we add some A to an equilibrium mixture of A, B, C, and D while holding T and V constant. Since $P_i = n_i RT/V$, this addition increases P_A and does not change the other partial pressures. Since P_A appears in the denominator of the reaction quotient (6.49) (ν_A is negative), addition of A at constant T and V makes Q_P less than K_P. The equilibrium must then shift to the right in order to increase the numerator of Q_P and make Q_P equal to K_P again. Thus, addition of A at constant T and V shifts the equilibrium to the right, thereby consuming some of the added A. Similarly, addition of a reaction product at constant T and V shifts the equilibrium to the left, thereby consuming some of the added substance. Removal of some of a reaction product from a mixture held at constant T and V shifts the equilibrium to the right, producing more product.

It might be thought that the same conclusions apply to addition of a reactant while holding T and P constant. Surprisingly, however, there are circumstances where constant-T-and-P addition of a reactant will shift the equilibrium so as to produce more of the added species. For example, consider the ideal-gas equilibrium $N_2 + 3H_2$ $\rightleftharpoons 2NH_3$. Suppose equilibrium is established at a T and P for which K_x [Eq. (6.26)] is 8.33; $K_x = 8.33 = [x(NH_3)]^2/x(N_2)[x(H_2)]^3$. Let the amounts $n(N_2) = 3.00$ mol, $n(H_2) = 1.00$ mol, and $n(NH_3) = 1.00$ mol be present at this T and P. Defining Q_x as $Q_x \equiv \prod_i (x_i)^{\nu_i}$, we find that, for these amounts, $Q_x = (0.2)^2/0.6(0.2)^3 = 8.33$. Since $Q_x = K_x$, the system is in equilibrium. Now, holding T and P constant, we add 0.1 mol of N_2. Because T and P are constant, K_x is still 8.33. After the N_2 is added, but before any shift in equilibrium occurs, we have

$$Q_x = \frac{(1/5.1)^2}{(3.1/5.1)(1/5.1)^3} = 8.39$$

Q_x now exceeds K_x, and the equilibrium must therefore shift to the left in order to reduce Q_x to 8.33; this shift produces more N_2. Addition of N_2 under these conditions shifts the equilibrium to produce more N_2. Although the addition of N_2 increases $x(N_2)$, it decreases $x(H_2)$ [and $x(NH_3)$], and the fact that $x(H_2)$ is cubed in the denominator outweighs the increase in $x(N_2)$ and the decrease in $x(NH_3)$. Therefore, in this case, Q_x increases on addition of N_2. For the general conditions under which addition of a reagent at constant T and P shifts the equilibrium to produce more of the added species, see Prob. 6.48. In this discussion, we assumed that Q_x always decreases when the reaction shifts to the left and increases when the reaction shifts to the right. For a proof of this, see L. Katz, *J. Chem. Educ.,* **38,** 375 (1961).

Le Châtelier's principle is often stated as follows: In a system at equilibrium, a change in one of the variables that determines the equilibrium will shift the equilibrium in the direction counteracting the change in that variable. The example just given shows this statement is *false*. A change in a variable may or may not shift the equilibrium in a direction that counteracts the change.

Some advocates of the "counteracting change" formulation of Le Châtelier's principle claim that if the principle is restricted to intensive variables (such as temperature, pressure, and mole fraction), it becomes valid. In the NH_3 example just given, although the equilibrium shifts to produce more N_2 when N_2 is added at constant T and P, this shift does decrease the N_2 mole fraction. [After the N_2 is added but before the shift occurs, we have $n(N_2) = 3.1$, $n(H_2) = 1$, $n(NH_3) = 1$, $n_{tot} = 5.1$, and $x(N_2) = 0.607843$. When the equilibrium shifts, one finds (Prob. 6.49) $\xi = -0.0005438$, $n(N_2) = 3.1 - \xi = 3.1005438$, $n_{tot} = 5.1 - 2\xi = 5.1010876$, and $x(N_2) = 0.607820 < 0.607843$.] However, consider a system at equilibrium with $n(N_2) = 2$ mol, $n(H_2) = 4$ mol, and $n(NH_3) = 4$ mol. Here $K_x = (0.4)^2/0.2(0.4)^3 = 12.5$. Now suppose we add 10 mol of N_2 at constant T and P to give a system with $n(N_2) = 12$ mol, $x(N_2) = 0.6$, $x(H_2) = 0.2 = x(NH_3)$, and $Q_x = 8.33....$ When the shift to the new equilibrium occurs, one finds (Prob. 6.49) $\xi = 0.12608$, $n(N_2) = 11.8739$, $n(H_2) = 3.62175$, $n(NH_3) = 4.2522$, $n_{tot} = 19.7478$, and $x(N_2) = 0.6013 > 0.6$. Thus, addition of N_2 to this system at constant T and P produces a shift that further increases the intensive variable $x(N_2)$. Hence, Le Châtelier's principle can fail even when restricted to intensive variables. [These failures were pointed out in K. Posthumus, *Rec. Trav. Chim.,* **52,** 25 (1933); **53,** 308 (1933).]

If the Le Châtelier "counteracting change" statement is carefully formulated and restricted to changes in intensive variables brought about by infinitesimal changes in the system and subsequent shifts in equilibrium, then it is valid [see J. de Heer, *J. Chem. Educ.,* **34,** 375 (1957); M. Hillert, *J. Phase Equilib.,* **16,** 403 (1995); Z.-K. Liu et al., *Fluid Phase Equilib.,* **121,** 167 (1996)], but changes in the real world are always finite rather than infinitesimal.

Shifts in Systems with More Than One Reaction

Predicting the effect of a change such as an isothermal pressure increase in a system with more than one reaction is tricky. Each of the reactions in (6.47) has more moles of products than reactants, and we might therefore expect that an isothermal pressure increase on a system with these two reactions in equilibrium would always shift both reactions (1) and (2) in (6.47) to the left, the side with fewer moles. Thus, one might expect an isothermal pressure increase to always increase the number of moles of water vapor present at equilibrium. However, Fig. 6.11*b* shows that above 10 bar, an increase in P at 900 K will *decrease* the equilibrium amount of water vapor [as was pointed out in I. Fishtik et al., *J. Chem. Soc. Farad. Trans.,* **91,** 259 (1995); I. Nagypál et al., *Pure Appl. Chem.,* **70,** 583 (1998)]. (Note, however, from Fig. 6.11*a* that an isothermal increase in P always decreases n_{tot}.)

The change in the extent of reaction (1) equals the change in moles of CO, since this species appears only in reaction (1); thus, $\Delta\xi_1 = \Delta n_{CO}$. Similarly, $\Delta\xi_2 = \Delta n_{CO_2}$. When P is increased from 10 bar to 30 bar at 900 K, we find (Prob. 6.41) $\Delta n_{CO} = \Delta\xi_1 = 0.553$ mol $- 0.886$ mol $= -0.333$ mol and $\Delta n_{CO_2} = \Delta\xi_2 = 1.931$ mol $- 1.758$ mol $= 0.173$ mol. The changes in amounts of H_2O due to reactions 1 and 2 are $\Delta n_{H_2O,1} = -\Delta\xi_1 = 0.333$ mol and $\Delta n_{H_2O,2} = -2\Delta\xi_2 = -0.346$ mol. Thus $\Delta n_{H_2O} = -0.013$ mol. Since the stoichiometry of each of the reactions (1) and (2) has an increase of 2 moles, we have $\Delta n_{tot} = 2(-0.333$ mol$) + 2(0.173$ mol$) = -0.320$ mol.

Thus, above 10 bar, an isothermal pressure increase shifts reaction (2) toward the side with the greater number of moles of gas. What about the reasoning in the Isothermal Pressure Change subsection, which predicted a shift to the side with the smaller number of moles? The answer is that that reasoning assumed the system had only one reaction. When two reactions with species in common are present, the two reactions influence each other and the situation is complex and not easily analyzed by simply looking at the stoichiometry of the reactions. (See also Prob. 6.44, where it is shown that the extent of one reaction depends on the choice of the second reaction.) However, it can be proved that no matter how many reactions occur in the ideal-gas system, an isothermal pressure increase will always produce a shift toward smaller n_{tot} and smaller V.

Other bizarre shifts in systems with several reactions (for example, dilution leading to precipitation) are discussed in the references given in I. Nagypál et al., *Pure Appl. Chem.,* **70,** 583 (1998).

6.7 SUMMARY

The chemical potential of gas i at partial pressure P_i in an ideal gas mixture is $\mu_i = \mu_i^\circ(T) + RT \ln (P_i/P^\circ)$, where $\mu_i^\circ(T)$, the standard-state chemical potential of i, equals $G_{m,i}^\circ(T)$, the molar Gibbs energy of pure gas i at $P^\circ \equiv 1$ bar and T.

For the ideal-gas reaction $0 \rightleftharpoons \Sigma_i \nu_i A_i$, use of this expression for μ_i in the equilibrium condition $\Sigma_i \nu_i \mu_i = 0$ leads to $\Delta G^\circ = -RT \ln K_P^\circ$, where $\Delta G^\circ \equiv \Sigma_i \nu_i \mu_i^\circ$ and the standard equilibrium constant $K_P^\circ \equiv \Pi_i \ (P_{i,eq}/P^\circ)^{\nu_i}$ is a function of T only. The temperature dependence of the standard equilibrium constant is given by $d \ln K_P^\circ/dT = \Delta H^\circ/RT^2$.

Important kinds of ideal-gas equilibrium calculations dealt with in this chapter include:

- Calculation of K_P° and ΔG° from the observed equilibrium composition.
- Calculation of K_P° from ΔG° using $\Delta G^\circ = -RT \ln K_P^\circ$.
- Calculation of the equilibrium composition from K_P° and the initial composition for constant-T-and-P or constant-T-and-V conditions.
- Calculation of K_P° at T_2 from K_P° at T_1 using $d \ln K_P^\circ/dT = \Delta H^\circ/RT^2$.
- Calculation of ΔH°, ΔG°, and ΔS° from K_P° versus T data using $\Delta G^\circ = -RT \ln K_P^\circ$ to get ΔG°, $d \ln K_P^\circ/dT = \Delta H^\circ/RT^2$ to get ΔH°, and $\Delta G^\circ = \Delta H^\circ - T \Delta S^\circ$ to get ΔS°.

FURTHER READING

Denbigh, chap. 4; *Zemansky and Dittman,* chap. 15; *de Heer,* chaps. 19 and 20.

PROBLEMS

Section 6.1

6.1 Use $\mu_i = \mu_i^\circ + RT \ln (P_i/P^\circ)$ to calculate ΔG when the pressure of 3.00 mol of a pure ideal gas is isothermally decreased from 2.00 bar to 1.00 bar at 400 K.

6.2 True or false? (*a*) The chemical potential of ideal gas i in an ideal gas mixture at temperature T and partial pressure P_i equals the chemical potential of pure gas i at temperature T and pressure P_i. (*b*) μ of a pure ideal gas goes to $-\infty$ as $P \to 0$ and goes to $+\infty$ as $P \to \infty$. (*c*) The entropy of a mixture of N_2 and O_2 gases (assumed ideal) is equal to the sum of the entropies of the pure gases, each at the same temperature and volume as the mixture.

Section 6.2

6.3 For the gas-phase reaction $2SO_2 + O_2 \rightleftharpoons 2SO_3$, observed mole fractions for a certain equilibrium mixture at 1000 K and 1767 torr are $x_{SO_2} = 0.310$, $x_{O_2} = 0.250$, and $x_{SO_3} = 0.440$. (*a*) Find K_P° and ΔG° at 1000 K, assuming ideal gases. (*b*) Find K_P at 1000 K. (*c*) Find K_c° at 1000 K.

6.4 An experimenter places 15.0 mmol of A and 18.0 mmol of B in a container. The container is heated to 600 K, and the gas-phase equilibrium A + B $\rightleftharpoons$ 2C + 3D is established. The equilibrium mixture is found to have pressure 1085 torr and to contain 10.0 mmol of C. Find K_P° and ΔG° at 600 K, assuming ideal gases.

6.5 A 1055-cm^3 container was evacuated, and 0.01031 mol of NO and 0.00440 mol of Br$_2$ were placed in the container; the equilibrium $2NO(g) + Br_2(g) \rightleftharpoons 2NOBr(g)$ was established at 323.7 K, and the final pressure was measured as 231.2 torr. Find K_P° and ΔG° at 323.7 K, assuming ideal gases. (*Hint:* Calculate n_{tot}.)

6.6 The reaction $N_2(g) \rightleftharpoons 2N(g)$ has $K_P^\circ = 3 \times 10^{-6}$ at 4000 K. A certain gas mixture at 4000 K has partial pressures $P_{N_2} = 720$ torr, $P_N = 0.12$ torr, and $P_{He} = 320$ torr. Is the mixture in reaction equilibrium? If not, will the amount of $N(g)$ increase or decrease as the system proceeds to equilibrium at 4000 K in a fixed volume?

6.7 Derive Eq. (6.27) relating K_x and K_P°.

6.8 Evaluate $\Pi_{j=1}^4 j(j+1)$.

6.9 True or false for ideal-gas reactions? (*a*) K_P° is always dimensionless. (*b*) K_P is always dimensionless. (*c*) K_P is never dimensionless. (*d*) K_P° for the reverse reaction is the negative of K_P° for the forward reaction. (*e*) K_P° for the reverse reaction is the reciprocal of K_P° for the forward reaction. (*f*) Doubling the coefficients doubles K_P°. (*g*) Doubling the coefficients squares K_P°. (*h*) K_P° for a particular reaction is a function of temperature but is independent of pressure and of the initial composition of the reaction mixture.

Section 6.3
6.10 For the reaction $N_2O_4(g) \rightleftharpoons 2NO_2(g)$, measurements of the composition of equilibrium mixtures gave $K_P^\circ = 0.144$ at 25.0°C and $K_P^\circ = 0.321$ at 35.0°C. Find ΔH°, ΔS°, and ΔG° at 25°C for this reaction. State any assumptions made. Do not use Appendix data.

6.11 For $PCl_5(g) \rightleftharpoons PCl_3(g) + Cl_2(g)$, observed equilibrium constants (from measurements on equilibrium mixtures at low pressure) vs. T are

K_P°	0.245	1.99	4.96	9.35
T/K	485	534	556	574

(*a*) Using *only* these data, find ΔH°, ΔG°, and ΔS° at 534 K for this reaction. (*b*) Repeat for 574 K.

6.12 For the ideal-gas reaction $PCl_5(g) \rightleftharpoons PCl_3(g) + Cl_2(g)$, use Appendix data to estimate K_P° at 400 K; assume that ΔH° is independent of T.

6.13 The ideal-gas reaction $CH_4(g) + H_2O(g) \rightleftharpoons CO(g) + 3H_2(g)$ at 600 K has $\Delta H^\circ = 217.9$ kJ/mol, $\Delta S^\circ = 242.5$ J/(mol K), and $\Delta G^\circ = 72.4$ kJ/mol. Estimate the temperature at which $K_P^\circ = 26$ for this reaction. State approximations made.

6.14 For the reaction $N_2O_4(g) \rightleftharpoons 2NO_2(g)$ in the range 298 to 900 K,

$$K_P^\circ = a(T/K)^b e^{-c/(T/K)}$$

where $a = 1.09 \times 10^{13}$, $b = -1.304$, and $c = 7307$. (*a*) Find expressions for ΔG°, ΔH°, ΔS°, and ΔC_P° as functions of T for this reaction. (*b*) Calculate ΔH° at 300 and at 600 K.

6.15 Complete the work of part (*b*) of Example 6.2 in Sec. 6.3 as follows. Show that if ΔC_P° is assumed independent of T, then

$$\ln \frac{K_P^\circ(T_2)}{K_P^\circ(T_1)} \approx \frac{\Delta H^\circ(T_1)}{R}\left(\frac{1}{T_1} - \frac{1}{T_2}\right)$$
$$+ \frac{\Delta C_P^\circ(T_1)}{R}\left(\ln\frac{T_2}{T_1} + \frac{T_1}{T_2} - 1\right)$$

Use this equation and Appendix data to estimate $K_{P,600}^\circ$ for $N_2O_4(g) \rightleftharpoons 2NO_2(g)$.

6.16 (*a*) Replacing T_2 by T and considering T_1 as a fixed temperature, we can write the approximate equation (6.39) in the form $\ln K_P^\circ(T) \approx -\Delta H^\circ/RT + C$, where the constant C equals $\ln K_P^\circ(T_1) + \Delta H^\circ/RT_1$. Derive the following exact equation:

$$\ln K_P^\circ(T) = -\Delta H_T^\circ/RT + \Delta S_T^\circ/R$$

The derivation is very short. (*b*) Use the equation derived in (*a*) and the approximation that ΔH° and ΔS° are independent of T to derive Eq. (6.39).

6.17 (*a*) For $2CO(g) + O_2(g) \rightleftharpoons 2CO_2(g)$, assume ideal-gas behavior and use data in the Appendix and the expression for ΔH° found in Example 5.6 in Sec. 5.5 to find an expression for $\ln K_P^\circ(T)$ valid from 300 to 1500 K. (*b*) Calculate K_P° at 1000 K for this reaction.

6.18 Consider the ideal-gas dissociation reaction $A \rightleftharpoons 2B$. For A and B, we have $C_{P,m,A}^\circ = a + bT + cT^2$ and $C_{P,m,B}^\circ = e + fT + gT^2$, where a, b, c, e, f, g are known constants and these equations are valid over the temperature range from T_1 to T_2. Further, suppose that $\Delta H_{T_1}^\circ$ and $K_P^\circ(T_1)$ are known. Find an expression for $\ln K_P^\circ(T)$ valid between T_1 and T_2.

6.19 Prove that for an ideal-gas reaction

$$\frac{d \ln K_c^\circ}{dT} = \frac{\Delta U^\circ}{RT^2}$$

6.20 Prove that for an ideal-gas reaction

$$\left(\frac{\partial \ln K_x}{\partial T}\right)_P = \frac{\Delta H^\circ}{RT^2}, \quad \left(\frac{\partial \ln K_x}{\partial P}\right)_T = -\frac{\Delta n/mol}{P}$$

6.21 True or false? (*a*) If ΔH° is positive, then K_P° must increase as T increases. (*b*) For an ideal-gas reaction, ΔH° must be independent of T.

Section 6.4
6.22 A certain gas mixture held at 395°C has the following initial partial pressures: $P(Cl_2) = 351.4$ torr; $P(CO) = 342.0$ torr; $P(COCl_2) = 0$. At equilibrium, the total pressure is 439.5 torr. V is held constant. Find K_P° at 395°C for $CO + Cl_2 \rightleftharpoons COCl_2$. [$COCl_2$ (phosgene) was used as a poison gas in World War I.]

6.23 Suppose 1.00 mol of CO_2 and 1.00 mol of COF_2 are placed in a very large vessel at 25°C, and a catalyst for the gas-phase reaction $2COF_2 \rightleftharpoons CO_2 + CF_4$ is added. Use data in the Appendix to find the equilibrium amounts.

6.24 For the ideal-gas reaction $A + B \rightleftharpoons 2C + 2D$, it is given that $\Delta G^\circ_{500} = 1250$ cal mol^{-1}. (a) If 1.000 mol of A and 1.000 mol of B are placed in a vessel at 500 K and P is held fixed at 1200 torr, find the equilibrium amounts. (b) If 1.000 mol of A and 2.000 mol of B are placed in a vessel at 500 K and P is held fixed at 1200 torr, find the equilibrium amounts.

6.25 Suppose 0.300 mol of H_2 and 0.100 mol of D_2 are placed in a 2.00-L vessel at 25°C together with a catalyst for the isotope-exchange reaction $H_2(g) + D_2(g) \rightleftharpoons 2HD(g)$, where $D \equiv {}^2H$ is deuterium. Use Appendix data to find the equilibrium composition.

6.26 At 400 K, $K^\circ_P = 36$ for $N_2(g) + 3H_2(g) \rightleftharpoons 2NH_3(g)$. Find the equilibrium amounts of all species if the following amounts are placed in a 2.00-L vessel at 400 K, together with a catalyst. (a) 0.100 mol of N_2 and 0.300 mol of H_2. (*Hint:* Solving a quartic equation can be avoided in this part of the problem.) (b) 0.200 mol of N_2, 0.300 mol of H_2, and 0.100 mol of NH_3.

6.27 For the gas-phase reaction $N_2 + 3H_2 \rightleftharpoons 2NH_3$, a closed system initially contains 4.50 mol of N_2, 4.20 mol of H_2, and 1.00 mol of NH_3. Give the maximum and minimum possible values at equilibrium of each of the following quantities: ξ; n_{N_2}; n_{H_2}; n_{NH_3}.

6.28 (a) For the ideal-gas reaction $A \rightleftharpoons 2B$ reaching equilibrium at constant T and P, show that $K^\circ_P = [x_B^2/(1 - x_B)](P/P^\circ)$, where x_B is the equilibrium mole fraction. (b) Use the result of (a) to show that $x_B = \frac{1}{2}[(z^2 + 4z)^{1/2} - z]$, where $z \equiv K^\circ_P P^\circ/P$. (c) A system that is initially composed of 0.200 mol of O_2 reaches equilibrium at 5000 K and 1.50 bar. Find the equilibrium mole fraction and moles of O_2 and O, given that $K^\circ_P = 49.3$ for $O_2(g) \rightleftharpoons 2O(g)$ at 5000 K. (d) Find the equilibrium mole fractions in an equilibrium mixture of NO_2 and N_2O_4 gases at 25°C and 2.00 atm. Use Appendix data.

6.29 At 727°C, $K^\circ_P = 3.42$ for $2SO_2(g) + O_2(g) \rightleftharpoons 2SO_3(g)$. If 2.65 mmol of SO_2, 3.10 mmol of O_2, and 1.44 mmol of SO_3 are placed in an empty 185-cm^3 vessel held at 727°C, find the equilibrium amounts of all species and find the equilibrium pressure.

6.30 For the ideal-gas reaction $A + B \rightleftharpoons C$, a mixture with $n_A = 1.000$ mol, $n_B = 3.000$ mol, and $n_C = 2.000$ mol is at equilibrium at 300 K and 1.000 bar. Suppose the pressure is isothermally increased to 2.000 bar; find the new equilibrium amounts.

6.31 For the reaction $PCl_5(g) \rightleftharpoons PCl_3(g) + Cl_2(g)$, use data in the Appendix to find K°_P at 25°C and at 500 K. Assume ideal-gas behavior and neglect the temperature variation in ΔH°. If we start with pure PCl_5, calculate the equilibrium mole fractions of all species at 500 K and 1.00 bar.

6.32 At 400 K, $K^\circ_P = 36$ for $N_2(g) + 3H_2(g) \rightleftharpoons 2NH_3(g)$. Find K°_P at 400 K for (a) $\frac{1}{2}N_2(g) + \frac{3}{2}H_2(g) \rightleftharpoons NH_3(g)$; (b) $2NH_3(g) \rightleftharpoons N_2(g) + 3H_2(g)$.

6.33 Given the $\Delta_f G^\circ_{1000}$ gas-phase values 84.31 kcal/mol for n-pentane, 83.64 kcal/mol for isopentane, and 89.21 kcal/mol

for neopentane, find the mole fractions present in an equilibrium mixture of these gases at 1000 K and 0.50 bar.

6.34 Use $\Delta_f G^\circ$ data in the NIST-JANAF tables (Sec. 5.9) to find K°_P at 6000 K for $N(g) \rightleftharpoons N^+(g) + e^-(g)$.

6.35 Suppose that for a certain ideal-gas reaction, the error in ΔG°_{300} is 2.5 kJ/mol. What error in K°_P does this cause?

6.36 At high temperatures, I_2 vapor is partially dissociated to I atoms. Let P^* be the expected pressure of I_2 calculated ignoring dissociation, and let P be the observed pressure. Some values for I_2 samples are:

T/K	973	1073	1173	1274
P^*/atm	0.0576	0.0631	0.0684	0.0736
P/atm	0.0624	0.0750	0.0918	0.1122

(a) Show that the equilibrium mole fractions are $x_I = 2(P - P^*)/P$ and $x_{I_2} = (2P^* - P)/P$. (b) Show that $K^\circ_P = 4(P - P^*)^2/(2P^* - P)P^\circ$, where $P^\circ \equiv 1$ bar. (c) Find ΔH° for $I_2(g) \rightleftharpoons 2I(g)$ at 1100 K.

6.37 If K^{bar}_P and K^{atm}_P are the K°_P values with $P^\circ \equiv 1$ bar and with $P^\circ \equiv 1$ atm, respectively, show that $K^{bar}_P = K^{atm}_P \times (1.01325)^{\Delta n/mol}$.

6.38 For the ideal-gas reaction $N_2 + 3H_2 \rightleftharpoons 2NH_3$, suppose 1 mol of N_2 and 3 mol of H_2 are present initially in a system held at constant T and P; no other gases are present initially. Let x be the number of moles of N_2 that have reacted when equilibrium is reached. ($x = \xi_{eq}$.) Show that

$$x = 1 - [1 - s/(s + 4)]^{1/2} \quad \text{where } s \equiv (27K^\circ_P)^{1/2}P/P^\circ$$

6.39 When the ideal-gas reaction $A + B \rightleftharpoons C + D$ has reached equilibrium, state whether or not each of the following relations must be true (all quantities are the values at equilibrium). (a) $n_C + n_D = n_A + n_B$; (b) $P_C + P_D = P_A + P_B$; (c) $n_A = n_B$; (d) $n_C = n_A$; (e) $n_C = n_D$; (f) if only A and B are present initially, then $n_C = n_A$; (g) if only A and B are present initially, then $n_C = n_D$; (h) if only A and B are present initially, then $n_C + n_D = n_A + n_B$; (i) $\mu_A + \mu_B = \mu_C + \mu_D$ no matter what the initial composition.

6.40 If in a gas-phase closed system, all the N_2 and H_2 come from the dissociation of NH_3 according to $2NH_3 \rightleftharpoons N_2 + 3H_2$, which one of the following statements is true at any time during the reaction? (a) $x_{N_2} = 3x_{H_2}$; (b) $3x_{N_2} = x_{H_2}$; (c) neither (a) nor (b) is necessarily true.

Section 6.5

6.41 (a) Set up the spreadsheet of Fig. 6.9 and compute the 900 K equilibrium composition of this system at 0.01, 0.1, 1.0, 10, 30, 100, and 1000 bar. Use the spreadsheet to graph the results. (b) Revise the spreadsheet of Fig. 6.9 to calculate the equilibrium composition at 1200 K and 0.20 bar.

6.42 In the Fig. 6.9 spreadsheet, the Solver was used to make the fractional errors in the calculated equilibrium constants very small. Explain why the alternative procedure of having the Solver make the absolute errors very small might produce very inaccurate results in certain circumstances.

6.43 For reactions (1) and (2) of (6.47) and the initial composition of Fig. 6.9, give the minimum and maximum possible values of ξ_1 and of ξ_2 and give the minimum and maximum possible numbers of moles of each species. (These conditions could be added as constraints.)

6.44 (a) For the 0.01 bar calculation shown in Fig. 6.10, find ξ_{eq} for reactions (1) and (2) in (6.47). (b) Suppose that instead of reactions (1) and (2), we describe the system by reaction (1) and the reaction $CO_2 + H_2 \rightleftharpoons CO + H_2O$, which is reaction (1) minus (2). What is ξ_{eq} for reaction (1) with this choice?

6.45 (a) For air up to 4000 K, one must consider the reactions $N_2 \rightleftharpoons 2N$, $O_2 \rightleftharpoons 2O$, and $N_2 + O_2 \rightleftharpoons 2NO$. Suppose someone suggests that the reactions $N + O \rightleftharpoons NO$ and $N + O_2 \rightleftharpoons NO + O$ should also be included. Show that each of these reactions can be written as a combination of the first three reactions and so these two reactions need not be included. (b) NIST-JANAF-table $\Delta_f G^\circ_{4000}$ values for $N(g)$, $O(g)$, and $NO(g)$ are 210.695, -13.270, and 40.132 kJ/mol, respectively. Use a spreadsheet (or Mathcad, Maple V, or Mathematica) to calculate the composition of dry air at 4000 K and 1 bar. Take the initial composition as 0.78 mol N_2, 0.21 mol O_2, and 0.01 mol Ar. Neglect the ionization of NO. (c) Vary the pressure over the range 0.001 bar to 1000 bar and plot the results.

Section 6.6

6.46 For the ideal-gas reaction $PCl_5(g) \rightleftharpoons PCl_3(g) + Cl_2(g)$, state whether the equilibrium shifts to the right, left, or neither when each of the following changes is made in an equilibrium mixture at 25°C. You may use Appendix data. (a) T is decreased at constant P. (b) V is decreased at constant T. (c) Some PCl_5 is removed at constant T and V. (d) $He(g)$ is added at constant T and V. (e) $He(g)$ is added at constant T and P.

6.47 Suppose the temperature of an equilibrium ideal-gas reaction mixture is increased at constant volume. Under what condition does the equilibrium shift to the right? (*Hint:* Use the result of an earlier problem in this chapter.)

6.48 (a) Show that

$$\left(\frac{\partial \ln Q_x}{\partial n_j} \right)_{n_{i \neq j}} = \frac{1}{Q_x} \left(\frac{\partial Q_x}{\partial n_j} \right)_{n_{i \neq j}} = \frac{\nu_j - x_j \, \Delta n/\text{mol}}{n_j}$$

where $Q_x \equiv \prod_i (x_i)^{\nu_i}$. (b) Use the result of part (a) to show that addition at constant T and P of a small amount of reacting species j to an ideal-gas equilibrium mixture will shift the equilibrium to produce more j when the following two conditions are both satisfied: (1) The species j appears on the side of the reaction equation that has the greater sum of the coefficients; (2) the equilibrium mole fraction x_j is greater than $\nu_j/(\Delta n/\text{mol})$. (c) For the reaction $N_2 + 3H_2 \rightleftharpoons 2NH_3$, when will addition of N_2 to an equilibrium mixture held at constant T and P shift the equilibrium to produce more N_2? Answer the same question for H_2 and for NH_3. Assume ideal behavior.

6.49 For the gas-phase ammonia-synthesis reaction: (a) Suppose a system is in equilibrium with 3 mol of N_2, 1 mol of H_2, and 1 mol of NH_3. If 0.1 mol of N_2 is added at constant T and P, find $n(N_2)$ and $x(N_2)$ at the new equilibrium position. (You can use the Solver in a spreadsheet.) (b) Suppose the system is in equilibrium with 2 mol of N_2, 4 mol of H_2, and 4 mol of NH_3. If 10 mol of N_2 is added at constant T and P, find $n(N_2)$ and $x(N_2)$ at the new equilibrium position.

General

6.50 The synthesis of ammonia from N_2 and H_2 is an exothermic reaction. Hence the equilibrium yield of ammonia decreases as T increases. Explain why the synthesis of ammonia from its elements (Haber process) is typically run at the high temperature of 800 K rather than at a lower temperature. (Haber developed the use of Cl_2 as a poison gas in World War I. His wife, also a chemist, tried to dissuade him from this work, but failed. She then committed suicide.)

6.51 For the gas-phase reaction

$$I_2 + \text{cyclopentene} \rightleftharpoons \text{cyclopentadiene} + 2HI$$

measured K_P° values in the range 450 to 700 K are fitted by $\log K_P^\circ = 7.55 - (4.83 \times 10^3)(\text{K}/T)$. Calculate ΔG°, ΔH°, ΔS°, and ΔC_P° for this reaction at 500 K. Assume ideal gases.

6.52 A certain ideal-gas dissociation reaction $A \rightleftharpoons 2B$ has $\Delta G^\circ_{1000} = 4000$ J mol^{-1}, which gives $K_P^\circ = 0.6$ at 1000 K. If pure A is put in a vessel at 1000 K and 1 bar and held at constant T and P, then A will partially dissociate to give some B. Someone presents the following chain of reasoning. "The second law of thermodynamics tells us that a process in a closed system at constant T and P that corresponds to $\Delta G > 0$ is forbidden [Eq. (4.16)]. The standard Gibbs free-energy change for the reaction $A \rightleftharpoons 2B$ is positive. Therefore, any amount of dissociation of A to B at constant T and P corresponds to an increase in G and is forbidden. Hence gas A held at 1000 K and 1 bar will not give any B at all." Point out the fallacy in this argument.

6.53 An ideal-gas reaction mixture is in a constant-temperature bath. State whether each of the following will change the value of K_P°. (a) Addition of a reactant. (b) Addition of an inert gas. (c) Change in pressure for a reaction with $\Delta n \neq 0$. (d) Change in temperature of the bath.

6.54 Suppose we have a mixture of ideal gases reacting according to $A + B \rightleftharpoons C + 2D$. The mixture is held at constant T and at a constant (total) pressure of 1 bar. Let one mole of A react. (a) Is the observed ΔH per mole of reaction in the mixture equal to ΔH° for the reaction? (b) Is the observed ΔS per mole of reaction equal to ΔS°? (c) Is the observed ΔG per mole of reaction equal to ΔG°?

6.55 Suppose that the standard pressure had been chosen as 1000 torr instead of 1 bar. With this definition, what would be the values of K_P and K_P° at 25°C for $N_2O_4(g) \rightleftharpoons 2NO_2(g)$? Use Appendix data.

6.56 For cis-EtHC=CHPr$(g) \rightleftharpoons trans$-EtHC=CHPr$(g)$ (where Et is C_2H_5 and Pr is $CH_3CH_2CH_2$), $\Delta H^\circ_{300} = -0.9$ kcal/mol, $\Delta S^\circ_{300} = 0.6$ cal/(mol K), and $\Delta C^\circ_{P,300} = 0$. [K. W. Egger, *J. Am. Chem. Soc.*, **89**, 504 (1967).] Assume that $\Delta C_P^\circ = 0$ for all temperatures above 300 K so that ΔH° and ΔS° remain constant as T increases. (a) The equilibrium amount of which isomer increases as T increases? (b) In the limit of very high T, which isomer is present in the greater amount? (c) Explain any apparent contradiction between the answers to (a) and (b). (d) For this reaction, state whether each of these quantities increases or decreases as T increases: $\Delta G°$, $K_P°$, and $\Delta G°/T$. (e) Is it possible for $\Delta G°$ of a reaction to increase with T while at the same time $K_P°$ also increases with T?

6.57 Given the data points (x_i, y_i), where $i = 1, \ldots, n$, we want to find the slope m and intercept b of the straight line $y = mx + b$ that gives the best fit to the data. We assume that (1) there is no significant error in the x_i values; (2) the y_i measurements each have essentially the same relative precision; (3) the errors in the y_i values are randomly distributed according to the normal distribution law. With these assumptions, it can be shown that the best values of m and b are found by minimizing the sum of the squares of the deviations of the experimental y_i values from the calculated y values. Show that minimization of $\Sigma_i (y_i - mx_i - b)^2$ (by setting $\partial/\partial m$ and $\partial/\partial b$ of the sum equal to zero) leads to $mD = n \Sigma_i x_iy_i - \Sigma_i x_i \Sigma_i y_i$ and $bD = \Sigma_i x_i^2 \times \Sigma_i y_i - \Sigma_i x_i \Sigma_i x_iy_i$, where $D \equiv n \Sigma_i x_i^2 - (\Sigma_i x_i)^2$. Condition (1) is usually met in physical chemistry because the x_i's are things like reciprocals of temperature or time, and these quantities are easily measured accurately. However, condition (2) is often not met because the y_i values are things like ln $K_P°$, whereas it is the $K_P°$ values that have been measured and that have the same precision. Therefore, don't put too much faith in least-squares-calculated quantities.

6.58 Consider the ideal-gas reaction $N_2 + 3H_2 \rightleftharpoons 2NH_3$ run at constant T and P with $T = 500$ K and $P = 4$ bar, and with the initial composition $n_{N_2} = 1$ mol, $n_{H_2} = 3$ mol, $n_{NH_3} = 0$. (a) Express the mole fractions in terms of the extent of reaction ξ. (b) Use the italicized statement at the end of Sec. 6.1 and Eq. (6.4) for μ_i to express G and H of the reaction mixture in terms of the $\mu_i°$'s, ξ, P, T, and the $H_{m,i}°$'s. (c) Conventional values of $G_{m,i}° = \mu_i°$ (Sec. 5.8) at 500 K are -97.46 kJ/mol for N_2, -66.99 kJ/mol for H_2, and -144.37 kJ/mol for NH_3. Conventional values of $H_{m,i}°$ (Sec. 5.4) at 500 K are 5.91 kJ/mol for N_2, 5.88 kJ/mol for H_2, and -38.09 kJ/mol for NH_3. Calculate G and H of the reaction mixture for ξ values of 0, 0.2, 0.3, 0.4, 0.6, 0.8, 1.0. Then use $G = H - TS$ to calculate TS. Check your results against Fig. 6.8. Part (c) is a lot more fun if done on a computer or programmable calculator.

6.59 Give a specific example of an ideal-gas reaction for which (a) the equilibrium position is independent of pressure; (b) the equilibrium position is independent of temperature.

6.60 (a) Give a specific example of a gas-phase reaction mixture for which the mole fraction of one of the reactants increases when the reaction proceeds a small extent to the right. If you can't think of an example, see part (b) of this problem. (b) For a reaction mixture containing only gases that participate in the reaction, use $x_i = n_i/n_{tot}$ and $dn_i = \nu_i \, d\xi$ [Eq. (4.97)] to show that the infinitesimal change dx_i in the mole fraction of gas i due to a change $d\xi$ in the extent of reaction is $dx_i = n_{tot}^{-1}[\nu_i - x_i(\Delta n/\text{mol})] \, d\xi$.

6.61 Rodolpho states that the equation $d \ln K_P°/dT = \Delta H°/RT^2$ shows that the sign of $\Delta H°$ determines whether $K_P°$ increases or decreases as T increases. Mimi states that the equations $\Delta G° = -RT \ln K_P°$ and $d \Delta G°/dT = -\Delta S°$ show that the sign of $\Delta S°$ determines whether $K_P°$ increases or decreases as T increases. Who is right? What error did the other person make?

6.62 Which of the following quantities can never be negative? (a) $\Delta_r G°$; (b) $K_P°$; (c) $\Delta_f G°$; (d) ξ_{eq}.

6.63 True or false? (a) If $\Delta G° > 0$, then no amount of products can be formed when the reaction is run at constant T and P in a closed system capable of P-V work only. (b) In any closed system with P-V work only, G is always minimized at equilibrium. (c) If the partial pressure P_i increases in an ideal gas mixture held at constant T, then μ_i increases in the mixture. (d) Addition of a reactant gas to an ideal-gas reaction mixture always shifts the equilibrium to use up some of the added gas. (e) S of a closed system is always maximized at equilibrium. (f) It is possible for the entropy of a closed system to decrease substantially in an irreversible process. (g) $\Pi_{i=1}^n ca_i = c^n \Pi_{i=1}^n a_i$. (h) The equilibrium position of an ideal-gas reaction is always independent of pressure. (i) $\Delta G°$ for an ideal-gas reaction is a function of pressure. (j) $\Delta G°$ for an ideal-gas reaction is a function of temperature. (k) For an ideal-gas reaction with $\Delta n \neq 0$, the change in standard-state pressure from 1 atm to 1 bar changed the value of $K_P°$ but did not change the value of K_P. (l) For an ideal-gas reaction at temperature T, $\Delta_r S° = \Delta_r H°/T$. (m) The chemical potential μ_i of substance i in a phase is a function of T, P, and x_i, but is always independent of the mole fractions $x_{j \neq i}$. (n) The chemical potential μ_i of component i of an ideal gas mixture is a function of T, P, and x_i, but is always independent of the mole fractions $x_{j \neq i}$.

One-Component Phase Equilibrium

The two kinds of material equilibrium are reaction equilibrium and phase equilibrium (Sec. 4.1). We studied reaction equilibrium in ideal gases in Chapter 6. We now begin the study of phase equilibrium. The phase-equilibrium condition (4.88) and (4.91) is that for each species, the chemical potential of that species must be the same in every phase in which the species is present.

The main topics of Chapter 7 are the phase rule and one-component phase equilibrium. Section 7.1 derives the phase rule, which tells us how many intensive variables are needed to specify the thermodynamic state of a system apart from specification of the sizes of the phases. The rest of Chapter 7 is restricted to systems with one component and discusses phase diagrams for such systems. A one-component phase diagram shows the region of temperature and pressure in which each of the various phases of a substance is stable. Since the equilibrium condition at fixed T and P is the minimization of the Gibb's energy G, the most stable phase of a pure substance at a given T and P is the phase with the lowest value of $G_m = \mu$. (Recall that for a pure substance, $G_m = \mu$.) Section 7.2 discusses the typical features of one-component phase diagrams and Sec. 7.3 derives the Clapeyron equation, which gives the slopes of the phase-equilibrium lines on a P-versus-T one-component phase diagram. Sections 7.4 and 7.5 discuss special kinds of phase transitions (solid–solid and higher-order).

Phase equilibrium and phase transitions occur widely in the world around us, from the boiling of water in a teakettle to the melting of Antarctic glaciers. The water cycle of evaporation, condensation to form clouds, and rainfall plays a key role in the ecology of the planet. Laboratory and industrial applications of phase transitions abound, and include such processes as distillation, precipitation, crystallization, and adsorption of gases on the surfaces of solid catalysts. The universe is believed to have undergone phase transitions in its early history as it expanded and cooled after the big bang (M. J. Rees, *Before the Beginning,* Perseus, 1998, p. 205), and some physicists have speculated that the big bang that gave birth to the universe was a phase transition produced by random fluctuations in a preexisting quantum vacuum (A. H. Guth, *The Inflationary Universe,* Perseus, 1997, pp. 12–14 and chap. 17).

7.1   THE PHASE RULE

Recall from Sec. 1.2 that a **phase** is a homogeneous portion of a system. A system may have several solid phases and several liquid phases but usually has at most one gas phase. (For systems with more than one gas phase, see Sec. 12.7.) In Secs. 7.2 to 7.5 we shall consider phase equilibrium in systems that have only one component. Before specializing to one-component systems, we want to answer the general question of how many independent variables are needed to define the equilibrium state of a multiphase, multicomponent system.

To describe the equilibrium state of a system with several phases and several chemical species, we can specify the mole numbers of each species in each phase and the temperature and pressure, T and P. Provided no rigid or adiabatic walls separate phases, T and P are the same in all phases at equilibrium. Specifying mole numbers is not what we shall do, however, since the mass of each phase of the system is of no real interest. The mass or size of each phase does not affect the phase-equilibrium position, since the equilibrium position is determined by equality of chemical potentials, which are intensive variables. (For example, in a two-phase system consisting of an aqueous solution of NaCl and solid NaCl at fixed T and P, the equilibrium concentration of dissolved NaCl in the saturated solution is independent of the mass of each phase.) We shall therefore deal with the mole fractions of each species in each phase, rather than with the mole numbers. The mole fraction of species j in phase α is $x_j^\alpha \equiv n_j^\alpha/n_{\text{tot}}^\alpha$, where n_j^α is the number of moles of substance j in phase α and n_{tot}^α is the total number of moles of all substances (including j) in phase α.

The number of **degrees of freedom** (or the *variance*) f of an equilibrium system is defined as the number of independent intensive variables needed to specify its intensive state. Specification of the **intensive state** of a system means specification of its thermodynamic state except for the sizes of the phases. The equilibrium intensive state is described by specifying the intensive variables P, T, and the mole fractions in each of the phases. As we shall see, these variables are not all independent.

We initially make two assumptions, which will later be eliminated: (1) No chemical reactions occur. (2) Every chemical species is present in every phase.

Let the number of different chemical species in the system be denoted by c, and let p be the number of phases present. From assumption 2, there are c chemical species in each phase and hence a total of pc mole fractions. Adding in T and P, we have

$$pc + 2 \tag{7.1}$$

intensive variables to describe the intensive state of the equilibrium system. However, these $pc + 2$ variables are not all independent; there are relations between them. First of all, the sum of the mole fractions in each phase must be 1:

$$x_1^\alpha + x_2^\alpha + \cdots + x_c^\alpha = 1 \tag{7.2}$$

where x_1^α is the mole fraction of species 1 in phase α, etc. There is an equation like (7.2) for each phase, and hence there are p such equations. We can solve these equations for $x_1^\alpha, x_1^\beta, \ldots$, thereby eliminating p of the intensive variables.

In addition to the relations (7.2), there are the conditions for equilibrium. We have already used the conditions for thermal and mechanical equilibrium by taking the same T and the same P for each phase. For material equilibrium, the following phase-equilibrium conditions [Eq. (4.88)] hold for the chemical potentials:

$$\mu_1^\alpha = \mu_1^\beta = \mu_1^\gamma = \cdots \tag{7.3}$$

$$\mu_2^\alpha = \mu_2^\beta = \mu_2^\gamma = \cdots \tag{7.4}$$

$$\cdots\cdots\cdots\cdots\cdots\cdots\cdots \tag{7.5}$$

$$\mu_c^\alpha = \mu_c^\beta = \mu_c^\gamma = \cdots \tag{7.6}$$

Since there are p phases, (7.3) contains $p - 1$ equality signs and $p - 1$ independent equations. Since there are c different chemical species, there are a total of $c(p - 1)$ equality signs in the set of equations (7.3) to (7.6). We thus have $c(p - 1)$ independent relations between chemical potentials. Each chemical potential is a function of T, P, and the composition of the phase (Sec. 4.7); for example, $\mu_1^\alpha = \mu_1^\alpha(T, P, x_1^\alpha, \ldots, x_c^\alpha)$. Hence the $c(p - 1)$ equations (7.3) to (7.6) provide $c(p - 1)$ simultaneous relations between T, P, and the mole fractions, which we can solve for $c(p - 1)$ of these variables, thereby eliminating $c(p - 1)$ intensive variables.

We started out with $pc + 2$ intensive variables in (7.1). We eliminated p of them using (7.2) and $c(p - 1)$ of them using (7.3) to (7.6). Therefore the number of independent intensive variables (which, by definition, is the number of degrees of freedom f) is

$$f = pc + 2 - p - c(p - 1)$$

$$f = c - p + 2 \qquad \text{no reactions} \tag{7.7}$$

Equation (7.7) is the **phase rule,** first derived by Gibbs.

Now let us drop assumption 2 and allow for the possibility that one or more chemical species might be absent from one or more phases. An example is a saturated aqueous salt solution in contact with pure solid salt. If species i is absent from phase δ, the number of intensive variables is reduced by 1, since x_i^δ is identically zero and is not a variable. However, the number of relations between the intensive variables is also reduced by 1, since we drop μ_i^δ from the set of equations (7.3) to (7.6). Recall that when substance i is absent from phase δ, μ_i^δ need not equal the chemical potential of i in the other phases [Eq. (4.91)]. Therefore, the phase rule (7.7) still holds when some species do not appear in every phase.

EXAMPLE 7.1 The phase rule

Find f for a system consisting of solid sucrose in equilibrium with an aqueous solution of sucrose.

The system has two chemical species (water and sucrose), so $c = 2$. The system has two phases (the saturated solution and the solid sucrose), so $p = 2$. Hence

$$f = c - p + 2 = 2 - 2 + 2 = 2$$

Two degrees of freedom make sense, since once T and P are specified, the equilibrium mole fraction (or concentration) of sucrose in the saturated solution is fixed.

Exercise

Find f for a system consisting of a liquid solution of methanol and ethanol in equilibrium with a vapor mixture of methanol and ethanol. Give a reasonable choice for the independent intensive variables. (*Answer:* 2; T and the liquid-phase ethanol mole fraction.)

Once the f degrees of freedom have been specified, then any scientist can prepare the system and get the same value for a measured intensive property of each phase of the system as any other scientist would get. Thus, once the temperature and pressure of an aqueous saturated sucrose solution have been specified, then the solution's density, refractive index, thermal expansivity, molarity, and specific heat capacity are all fixed, but the volume of the solution is not fixed.

An error students sometimes make is to consider a chemical species present in two phases as contributing 2 to c. For example, they will consider sucrose(s) and sucrose(aq) as two chemical species. From the derivation of the phase rule, it is clear that a chemical species present in several phases contributes only 1 to c, the number of chemical species present.

The Phase Rule in Systems with Reactions

We now drop assumption 1 and suppose that chemical reactions can occur. For each independent chemical reaction, there is an equilibrium condition $\Sigma_i \nu_i \mu_i = 0$ [Eq. (4.98)], where the μ_i's and ν_i's are the chemical potentials and stoichiometric coefficients of the reacting species. Each independent chemical reaction provides one relation between the chemical potentials, and, like relations (7.3) to (7.6), each such relation can be used to eliminate one variable from T, P, and the mole fractions. If the number of independent chemical reactions is r, then the number of independent intensive variables is reduced by r and the phase rule (7.7) becomes

$$f = c - p + 2 - r \tag{7.8}$$

By independent chemical reactions, we mean that no reaction can be written as a combination of the others (Sec. 6.5).

In addition to reaction-equilibrium relations, there may be other restrictions on the intensive variables of the system. For example, suppose we have a gas-phase system containing only NH_3; we then add a catalyst to establish the equilibrium $2NH_3 \rightleftharpoons N_2 + 3H_2$; further, we refrain from introducing any N_2 or H_2 from outside. Since all the N_2 and H_2 comes from the dissociation of NH_3, we must have $n_{H_2} = 3n_{N_2}$ and $x_{H_2} = 3x_{N_2}$. This stoichiometry condition is an additional relation between the intensive variables besides the equilibrium relation $2\mu_{NH_3} = \mu_{N_2} + 3\mu_{H_2}$. In ionic solutions, the condition of electrical neutrality provides such an additional relation.

If, besides the r reaction-equilibrium conditions of the form $\sum_i \nu_i \mu_i = 0$, there are a additional restrictions on the mole fractions arising from stoichiometric and electroneutrality conditions, then the number of degrees of freedom f is reduced by a and the phase rule (7.8) becomes

$$f = c - p + 2 - r - a \tag{7.9}*$$

where c is the number of chemical species, p is the number of phases, r is the number of independent chemical reactions, and a is the number of additional restrictions.

We can preserve the simple form (7.7) for the phase rule by defining the number of **independent components** c_{ind} as

$$c_{ind} \equiv c - r - a \tag{7.10}$$

Equation (7.9) then reads

$$f = c_{ind} - p + 2 \tag{7.11}*$$

Many books call c_{ind} simply the number of components.

EXAMPLE 7.2   The phase rule

For an aqueous solution of the weak acid HCN, write the reaction equilibrium conditions and find f and c_{ind}.

The system has the five chemical species H_2O, HCN, H^+, OH^-, and CN^-, so $c = 5$. The two independent reactions $H_2O \rightleftharpoons H^+ + OH^-$ and $HCN \rightleftharpoons H^+ + CN^-$ give two equilibrium conditions: $\mu_{H_2O} = \mu_{H^+} + \mu_{OH^-}$ and $\mu_{HCN} = \mu_{H^+} + \mu_{CN^-}$. The system has $r = 2$. In addition, there is the electroneutrality condition $n_{H^+} = n_{CN^-} + n_{OH^-}$; division by n_{tot} gives the mole-fraction relation $x_{H^+} = x_{CN^-} + x_{OH^-}$. (See also Prob. 7.6.) Thus, $a = 1$. The phase rule (7.9) gives

$$f = c - p + 2 - r - a = 5 - 1 + 2 - 2 - 1 = 3$$

$$c_{ind} = c - r - a = 5 - 2 - 1 = 2$$

The result $f = 3$ makes sense, since once the three intensive variables T, P, and the HCN mole fraction are specified, all the remaining mole fractions can be calculated using the H_2O and HCN dissociation equilibrium constants. The two independent components are most conveniently considered to be H_2O and HCN.

Exercise

Find f and c_{ind} for (a) an aqueous solution of HCN and KCN; (b) an aqueous solution of HCN and KCl; (c) an aqueous solution of the weak diprotic acid H_2SO_3. [*Answers:* (a) 4, 3; (b) 4, 3; (c) 3, 2.]

EXAMPLE 7.3 The phase rule

Find f in a system consisting of $CaCO_3(s)$, $CaO(s)$, and $CO_2(g)$, where all the CaO and CO_2 come from the reaction $CaCO_3(s) \rightleftharpoons CaO(s) + CO_2(g)$.

A phase is a homogeneous portion of a system, and this system has three phases: $CaCO_3(s)$, $CaO(s)$, and $CO_2(g)$. The system has three chemical species. There is one reaction-equilibrium condition, $\mu_{CaCO_3(s)} = \mu_{CaO(s)} + \mu_{CO_2(g)}$, so $r = 1$. Are there any additional restrictions on the mole fractions? It is true that the number of moles of CaO(s) must equal the number of CO_2 moles: $n_{CaO(s)} = n_{CO_2(g)}$. However, this equation cannot be converted into a relation between the mole fractions in each phase, and it does not provide an additional relation between intensive variables. Hence

$$c_{ind} = c - r - a = 3 - 1 - 0 = 2$$

$$f = c_{ind} - p + 2 = 2 - 3 + 2 = 1$$

The value $f = 1$ makes sense, since once T is fixed the pressure of CO_2 gas in equilibrium with the $CaCO_3$ is fixed by the reaction-equilibrium condition, and so P of the system is fixed.

Exercise

Find c_{ind} and f for a gas-phase mixture of O_2, O, O^+, and e^-, in which all the O comes from the dissociation of O_2 and all the O^+ and e^- come from the ionization of O. Give the most reasonable choice of independent intensive variables. (*Answer:* 1, 2; T and P.)

In doubtful cases, rather than applying (7.9) or (7.11), it is often best to first list the intensive variables and then list all the independent restrictive relations between them. Subtraction gives f. For example, for the $CaCO_3$–CaO–CO_2 example just given, the intensive variables are T, P, and the mole fractions in each phase. Since each phase is pure, we know that in each phase, the mole fraction of each of $CaCO_3$, CaO, and CO_2 is either 0 or 1; hence the mole fractions are fixed and are not variables. There is one independent relation between intensive variables, namely, the already stated reaction-equilibrium condition. Therefore $f = 2 - 1 = 1$. Knowing f, we can then calculate c_{ind} from (7.11) if c_{ind} is wanted.

In $dG = -S\,dT + V\,dP + \Sigma_i \mu_i\,dn_i$, the Gibbs equation (4.78) for a phase, the sum is over all the actual chemical species in the phase. Provided the phase is in reaction equilibrium, it is possible to show that this equation remains valid if the sum is taken over only the independent components of the phase; see Prob. 7.53. This is a useful result, because one often does not know the nature or the amounts of some of the chemical species actually present in the phase. For example, in a solution, the solute might be solvated by an unknown number of solvent molecules, and the solvent might be dissociated or associated to an unknown extent. Despite these reactions that produce new species, we need only extend the sum $\Sigma_i \mu_i\,dn_i$ over the two independent components, the solute and the solvent, and evaluate dn of the solute and the solvent ignoring solvation, association, or dissociation.

> Note the following restrictions on the applicability of the phase rule (7.9). There must be no walls between phases. We equated the temperatures of the phases, the pressures of the phases, and the chemical potentials of a given component in the phases. These equalities need not hold if adiabatic, rigid, or impermeable walls separate phases. The system must be capable of P-V work only. If, for example, we can do electrical work on the system by applying an electric field, then the electric field strength is an additional intensive variable that must be specified to define the system's state. For systems containing a pair of optical isomers, exceptions to the phase rule can occur [J. C. Wheeler, *J. Chem. Phys.*, **73**, 5771 (1980)].

7.2   ONE-COMPONENT PHASE EQUILIBRIUM

In the rest of this chapter, we specialize to phase equilibrium in systems with one independent component. (Chapter 12 deals with multicomponent phase equilibrium.) We shall be concerned in this chapter with pure substances.

An example is a one-phase system of pure liquid water. If we ignore the dissociation of H_2O, we would say that only one species is present ($c = 1$), and there are no reactions or additional restrictions ($r = 0$, $a = 0$); hence $c_{ind} = 1$ and $f = 2$. If we take account of the dissociation $H_2O \rightleftharpoons H^+ + OH^-$, the system has three chemical species ($c = 3$), one reaction-equilibrium condition [$\mu(H_2O) = \mu(H^+) + \mu(OH^-)$], and one electroneutrality or stoichiometry condition [$x(H^+) = x(OH^-)$]. Therefore $c_{ind} = 3 - 1 - 1 = 1$, and $f = 2$. Thus, whether or not we take dissociation into account, the system has one independent component and 2 degrees of freedom (the temperature and pressure).

With $c_{ind} = 1$, the phase rule (7.11) becomes

$$f = 3 - p \qquad \text{for } c_{ind} = 1$$

If $p = 1$, then $f = 2$; if $p = 2$, then $f = 1$; if $p = 3$, then $f = 0$. The maximum f is 2. For a one-component system, specification of at most two intensive variables describes the intensive state. We can represent any intensive state of a one-component system by a point on a two-dimensional P-versus-T diagram, where each point corresponds to a definite T and P. Such a diagram is a **phase diagram.**

A P-T phase diagram for pure water is shown in Fig. 7.1. The one-phase regions are the open areas. Here $p = 1$ and there are 2 degrees of freedom, in that both P and T must be specified to describe the intensive state.

Along the lines (except at point A), two phases are present in equilibrium. Hence $f = 1$ along a line. Thus, with liquid and vapor in equilibrium, we can vary T anywhere along the line AC, but once T is fixed, then P, the **(equilibrium) vapor pressure** of liquid water at temperature T, is fixed. The **boiling point** of a liquid at a given pressure P is the temperature at which its equilibrium vapor pressure equals P. The **normal boiling point** is the temperature at which the liquid's vapor pressure is 1 atm. *Line* AC *gives the boiling point of water as a function of pressure.* The H_2O normal boiling point is not precisely 100°C; see Sec. 1.5. If T is considered to be the independent variable, *line* AC *gives the vapor pressure of liquid water as a function of temperature.* Figure 7.1 shows that the boiling point at a given pressure is the maximum temperature at which a stable liquid can exist at that pressure.

The change in 1982 of the thermodynamic standard-state pressure from 1 atm to 1 bar did not affect the definition of the normal-boiling-point pressure, which remains at 1 atm.

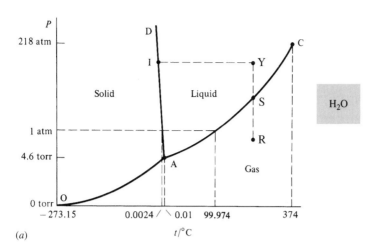

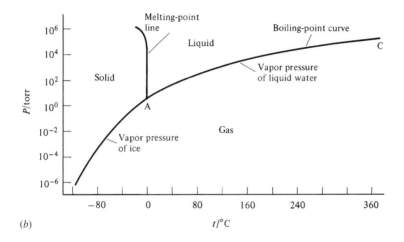

Figure 7.1

The H_2O phase diagram at low
and moderate pressures. (*a*)
Caricature of the diagram. (*b*) The
diagram drawn accurately. The
vertical scale is logarithmic. (For
the H_2O phase diagram at high
pressures, see Fig. 7.9*b*.)

Point A is the **triple point.** Here solid, liquid, and vapor are in mutual equilibrium, and $f = 0$. Since there are no degrees of freedom, the triple point occurs at a definite T and P. Recall that the water triple point is used as the reference temperature for the thermodynamic temperature scale. By definition, the water triple-point temperature is exactly 273.16 K. The water triple-point pressure is found to be 4.585 torr. The present definition of the Celsius scale t is $t \equiv T - 273.15°$ [Eq. (1.16)]. Hence the water triple-point temperature is exactly 0.01°C.

The **melting point** of a solid at a given pressure P is the temperature at which solid and liquid are in equilibrium for pressure P. Line AD in Fig. 7.1 is the solid–liquid equilibrium line for H_2O and gives the melting point of ice as a function of pressure. Note that the melting point of ice decreases slowly with increasing pressure. The **normal melting point** of a solid is the melting point at $P = 1$ atm. For water, the normal melting point is 0.0024°C. The ice point (Secs. 1.3 and 1.5), which occurs at 0.0000°C, is the equilibrium temperature of ice and *air-saturated* liquid water at 1 atm pressure. The equilibrium temperature of ice and *pure* liquid water at 1 atm pressure is 0.0024°C. (The dissolved N_2 and O_2 lower the freezing point compared with that of pure water; see Sec. 12.3.) For a pure substance, the **freezing point** of the liquid at a given pressure equals the melting point of the solid.

Along line OA, there is equilibrium between solid and vapor. Ice heated at a pressure below 4.58 torr will sublime to vapor rather than melt to liquid. *Line* OA *is the vapor-pressure curve of the solid.* Statistical mechanics shows that the vapor pressure

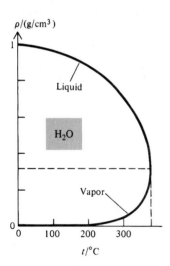

Figure 7.2

Densities of liquid water and water vapor in equilibrium with each other plotted versus temperature. At the critical temperature 374°C, these densities become equal.

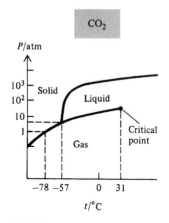

Figure 7.3

The CO_2 phase diagram. The CO_2 triple-point pressure of 5.1 atm is one of the highest known. For most substances, the triple-point pressure is below 1 atm. The vertical scale is logarithmic.

of a solid goes to zero as $T \to 0$ (Prob. 24.42), so the solid–vapor line on a P-T phase diagram intersects the origin (the point $P = 0$, $T = 0$).

Suppose liquid water is placed in a closed container fitted with a piston, the system is heated to 300°C, and the system's pressure is set at 0.5 atm. These T and P values correspond to point R in Fig. 7.1. The equilibrium phase at R is gaseous H_2O, so the system consists entirely of $H_2O(g)$ at 300°C and 0.5 atm. If the piston pressure is now slowly increased while T is held constant, the system remains gaseous until the pressure of point S is reached. At S, the vapor starts to condense to liquid, and this condensation continues at constant T and P until all the vapor has condensed. During condensation, the system's volume V decreases (Fig. 8.4), but its intensive variables remain fixed. The amounts of liquid and vapor present at S can be varied by varying V. After all the vapor has condensed at S, let the pressure of the liquid be increased isothermally to reach point Y. If the system is now cooled at constant pressure, its temperature will eventually fall to the temperature at point I, where the liquid begins to freeze. The temperature will remain fixed until all the liquid has frozen. Further cooling simply lowers the temperature of the ice.

Suppose we now start at point S with liquid and vapor in equilibrium and slowly heat the closed system, adjusting the volume (if necessary) to maintain the presence of liquid and vapor phases in equilibrium. The system moves from point S along the liquid–vapor line toward point C, with both T and P increasing. During this process, the liquid-phase density decreases because of the thermal expansion of the liquid, and the vapor-phase density increases because of the rapid increase in liquid vapor pressure with T. Eventually, point C is reached, at which the liquid and vapor densities (and all other intensive properties) become equal to each other. See Fig. 7.2. At point C, the two-phase system becomes a one-phase system, and the liquid–vapor line ends.

Point C is the **critical point.** The temperature and pressure at this point are the **critical temperature** and the **critical pressure,** T_c and P_c. For water, $T_c = 647$ K = 374°C and $P_c = 218$ atm. At any temperature above T_c, liquid and vapor phases cannot coexist in equilibrium, and isothermal compression of the vapor will not cause condensation, in contrast to compression below T_c. Note that it is possible to go from point R (vapor) to point Y (liquid) without condensation occurring by varying T and P so as to go around the critical point C without crossing the liquid–vapor line AC. In such a process, the density changes continuously, and there is a continuous transition from vapor to liquid, rather than a sudden transition as in condensation.

The phase diagram for CO_2 is shown in Fig. 7.3. For CO_2, an increase in pressure increases the melting point. The triple-point pressure of CO_2 is 5.1 atm. Therefore at 1 atm, solid CO_2 will sublime to vapor when warmed rather than melt to liquid; hence the name "dry ice."

The liquid–vapor line on a P-T phase diagram ends in a critical point. Above T_c, there is no distinction between liquid and vapor. One might ask whether the solid–liquid line ends in a critical point at high pressure. No solid–liquid critical point has ever been found, and such a critical point is believed to be impossible.

Since the equilibrium condition at constant T and P is the minimization of G, *the stable phase at any point on a one-component P-T phase diagram is the one with the lowest G_m (the lowest μ).*

For example, at point S in Fig. 7.1a, liquid and vapor coexist and have equal chemical potentials. Since $(\partial G_m/\partial P)_T = V_m$ [Eq. (4.51)] and $V_{m,gas} \gg V_{m,liq}$, an isothermal decrease in P lowers substantially the chemical potential of the vapor but has only a small effect on μ of the liquid. Therefore, lowering P makes the vapor have the lower chemical potential, and vapor is the stable phase at point R.

We can also look at phase equilibrium in terms of enthalpy (or energy) and entropy effects. We have $\mu_{gas} - \mu_{liq} = H_{m,gas} - H_{m,liq} - T(S_{m,gas} - S_{m,liq})$. The ΔH_m term favors the liquid, which has a lower H_m than the gas (because of intermolecular attractions in the liquid). The $-T \Delta S_m$ term favors the gas, which has a higher entropy S_m. At low T, the ΔH_m term dominates and the liquid is more stable than the gas. At high T, the $-T \Delta S_m$ term dominates and the gas is more stable. At low pressures, the increase of $S_{m,gas}$ with decreasing P (and increasing V_m) makes the gas more stable than the liquid.

Enthalpies and Entropies of Phase Changes

A phase change at constant T and P is generally accompanied by an enthalpy change, often called the **(latent) heat** of the transition. (Certain special phase changes have $\Delta H = 0$; see Sec. 7.5.) One has **enthalpies** or **heats of fusion** (solid → liquid), **sublimation** (solid → gas), **vaporization** (liquid → gas), and **transition** (solid → solid—see Sec. 7.4), symbolized by $\Delta_{fus}H$, $\Delta_{sub}H$, $\Delta_{vap}H$, and $\Delta_{trs}H$.

Figure 7.1 shows that fusion, sublimation, and vaporization equilibria each exist over a range of T (and P). ΔH values for these processes change as the temperature of the phase equilibrium changes. For example, ΔH_m of vaporization of water for points along the liquid–vapor equilibrium line AC in Fig. 7.1 is plotted in Fig. 7.4 as a function of the liquid–vapor equilibrium temperature. Note the rapid drop in $\Delta_{vap}H_m$ as the critical temperature 374°C is approached.

We have $\Delta_{vap}H = \Delta_{vap}U + P \Delta_{vap}V$, where usually $P \Delta_{vap}V \ll \Delta_{vap}U$. The $\Delta_{vap}U$ term is the difference between intermolecular interaction energies of the gas and liquid: $\Delta_{vap}U = U_{intermol,gas} - U_{intermol,liq}$. If P is low or moderate (well below the critical-point pressure), then $U_{m,intermol,gas} \approx 0$ and $\Delta_{vap}H_m \approx \Delta_{vap}U_m \approx -U_{m,intermol,liq}$. Therefore $\Delta_{vap}H_m$ is a measure of the strength of intermolecular interactions in the liquid. For substances that are liquids at room temperature, $\Delta_{vap}H_m$ values at the normal boiling point run 20 to 50 kJ/mol. Each molecule in a liquid interacts with several other molecules, so the molar energy for interaction between two molecules is substantially less than $\Delta_{vap}H_m$. For example, the main interaction between H_2O molecules is hydrogen bonding. If we assume that at 0°C each H atom in $H_2O(l)$ participates in a hydrogen bond, then there are twice as many H bonds as H_2O molecules, and the 45-kJ/mol $\Delta_{vap}H_m$ value indicates a 22-kJ/mol energy for each H bond. $\Delta_{vap}H_m$ values run substantially less than chemical-bond energies, which are 150 to 800 kJ/mol (Table 20.1).

An approximate rule for relating enthalpies and entropies of liquids to those of gases is **Trouton's rule,** which states that $\Delta_{vap}S_{m,nbp}$ for vaporization of a liquid at its normal boiling point (nbp) is roughly $10\frac{1}{2}R$:

$$\Delta_{vap}S_{m,nbp} = \Delta_{vap}H_{m,nbp}/T_{nbp} \approx 10\tfrac{1}{2}R = 21 \text{ cal/(mol K)} = 87 \text{ J/(mol K)}$$

Trouton's rule fails for highly polar liquids (especially hydrogen-bonded liquids) and for liquids boiling below 150 K or above 1000 K (see Table 7.1). The accuracy of Trouton's rule can be improved substantially by taking

$$\Delta_{vap}S_{m,nbp} \approx 4.5R + R \ln (T_{nbp}/K) \tag{7.12}$$

For $T_{nbp} \approx 400$ K, Eq. (7.12) gives $\Delta_{vap}S_{m,nbp} \approx 4.5R + R \ln 400 = 10.5R$, which is Trouton's rule. Equation (7.12) has been discovered several times by various workers and is the *Trouton–Hildebrand–Everett rule*. See L. K. Nash, *J. Chem. Educ.,* **61,** 981 (1984) for its history. The physical content of Eq. (7.12) is that $\Delta_{vap}S_m$ is approximately the same for nonassociated liquids when they are evaporated to the same molar volume in the gas phase. The $R \ln (T_{nbp}/K)$ term corrects for different molar volumes of the gases at the different boiling points; see Prob. 7.20.

Table 7.1 gives ΔS_m and ΔH_m data for fusion (fus) at the normal melting point (nmp) and vaporization at the normal boiling point (nbp). The listed values of $\Delta_{vap}S_{m,nbp}$ predicted by the Trouton–Hildebrand–Everett (THE) rule indicate that this rule works

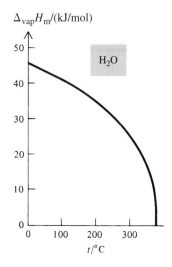

$\Delta_{vap}H_m/(kJ/mol)$

Figure 7.4

Molar enthalpy of vaporization of liquid water versus temperature. At the critical temperature 374°C, $\Delta_{vap}H$ becomes zero.

TABLE 7.1

Enthalpies and Entropies of Fusion and Vaporization[a]

Substance	T_{nmp}	$\Delta_{fus}H_m$	$\Delta_{fus}S_m$	T_{nbp}	$\Delta_{vap}H_m$	$\Delta_{vap}S_m$	$\Delta_{vap}S_m^{THE}$
	K	kJ/mol	J/(mol K)	K	kJ/mol	J/(mol K)	J/(mol K)
Ne	24.5	0.335	13.6	27.1	1.76	65.0	64.8
N_2	63.3	0.72	11.4	77.4	5.58	72.1	73.6
Ar	83.8	1.21	14.4	87.3	6.53	74.8	74.6
C_2H_6	89.9	2.86	31.8	184.5	14.71	79.7	80.8
$(C_2H_5)_2O$	156.9	7.27	46.4	307.7	26.7	86.8	85.1
NH_3	195.4	5.65	28.9	239.7	23.3	97.4	83.0
CCl_4	250.	2.47	9.9	349.7	30.0	85.8	86.1
H_2O	273.2	6.01	22.0	373.1	40.66	109.0	86.7
I_2	386.8	15.5	40.1	457.5	41.8	91.4	88.3
Zn	693.	7.38	10.7	1184.	115.6	97.6	96.3
NaCl	1074.	28.2	26.2	1738.	171.	98.4	99.4

[a] $\Delta_{fus}H_m$ and $\Delta_{fus}S_m$ are at the normal melting point (nmp). $\Delta_{vap}H_m$ and $\Delta_{vap}S_m$ are at the normal boiling point (nbp). $\Delta_{vap}S_m^{THE}$ is the normal-boiling-point $\Delta_{vap}S_m$ value predicted by the Trouton–Hildebrand–Everett rule.

well for liquids boiling at low, moderate, and high temperatures but fails for hydrogen-bonded liquids. As intermolecular attractions increase, both $\Delta_{vap}H_m$ and T_{nbp} increase.

$\Delta_{vap}H_{m,nbp}$ is usually substantially larger than $\Delta_{fus}H_{m,nmp}$. $\Delta_{fus}S_{m,nmp}$ varies greatly from compound to compound, in contrast to $\Delta_{vap}S_{m,nbp}$. Amazingly, $\Delta_{fus}H$ for ^3He between 0 and 0.3 K is slightly negative; to freeze liquid ^3He at constant T and P below 0.3 K, one must heat it.

Although $H_2O(g)$ is not thermodynamically stable at 25°C and 1 bar, one can use the experimental vapor pressure of $H_2O(l)$ at 25°C to calculate $\Delta_f G^\circ_{298}$ of $H_2O(g)$. See Probs. 7.50 and 8.36.

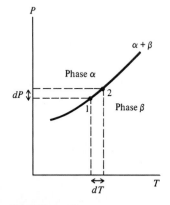

Figure 7.5

Two neighboring points on a two-phase line of a one-component system.

7.3 THE CLAPEYRON EQUATION

The Clapeyron equation gives the slope dP/dT of a two-phase equilibrium line on a P-T phase diagram of a one-component system. To derive it, we consider two infinitesimally close points 1 and 2 on such a line (Fig. 7.5). The line in Fig. 7.5 might involve solid–liquid, solid–vapor, or liquid–vapor equilibrium or even solid–solid equilibrium (Sec. 7.4). We shall call the two phases involved α and β. The condition for phase equilibrium is $\mu^\alpha = \mu^\beta$. No subscript is needed because we have only one component. For a pure substance, μ equals G_m [Eq. (4.86)]. Therefore $G_m^\alpha = G_m^\beta$ for any point on the α-β equilibrium line. *The molar Gibbs energies of one-component phases in equilibrium are equal.* At point 1 in Fig. 7.5, we thus have $G_{m,1}^\alpha = G_{m,1}^\beta$. Likewise, at point 2, $G_{m,2}^\alpha = G_{m,2}^\beta$, or $G_{m,1}^\alpha + dG_m^\alpha = G_{m,1}^\beta + dG_m^\beta$, where dG_m^α and dG_m^β are the infinitesimal changes in molar Gibbs energies of phases α and β as we go from point 1 to point 2. Use of $G_{m,1}^\alpha = G_{m,1}^\beta$ in the last equation gives

$$dG_m^\alpha = dG_m^\beta \tag{7.13}$$

For a single phase, we have $dG = -S\,dT + V\,dP + \Sigma_i \mu_i\,dn_i$ [Eq. (4.78)], and for a pure (one-component) phase, the sum has only one term:

$$dG = -S\,dT + V\,dP + \mu\,dn \qquad \text{pure phase} \tag{7.14}$$

We have $G_m \equiv G/n$ and $G = nG_m$. Therefore $dG = n\,dG_m + G_m\,dn$, and (7.14) becomes $n\,dG_m + G_m\,dn = -S\,dT + V\,dP + \mu\,dn$. Since $\mu = G_m$ for a pure phase, we get $n\,dG_m = -S\,dT + V\,dP$. Division by n gives

$$dG_m = -S_m\,dT + V_m\,dP \qquad \text{one-phase, one-comp. syst.} \qquad (7.15)$$

Equation (7.15) applies to both open and closed systems. A quick way to obtain (7.15) is to divide $dG = -S\,dT + V\,dP$ by n. Although $dG = -S\,dT + V\,dP$ applies to a closed system, G_m is an intensive property and is unaffected by a change in system size.

Use of (7.15) in (7.13) gives

$$-S_m^\alpha\,dT + V_m^\alpha\,dP = -S_m^\beta\,dT + V_m^\beta\,dP \qquad (7.16)$$

where dT and dP are the infinitesimal changes in T and P on going from point 1 to point 2 along the α-β equilibrium line. Rewriting (7.16), we have

$$(V_m^\alpha - V_m^\beta)\,dP = (S_m^\alpha - S_m^\beta)\,dT$$

$$\frac{dP}{dT} = \frac{S_m^\alpha - S_m^\beta}{V_m^\alpha - V_m^\beta} = \frac{\Delta S_m}{\Delta V_m} = \frac{\Delta S}{\Delta V} \qquad \textbf{(7.17)*}$$

where ΔS and ΔV are the entropy and volume changes for the phase transition $\beta \to \alpha$. For the transition $\alpha \to \beta$, ΔS and ΔV are each reversed in sign, and their quotient is unchanged, so it doesn't matter which phase we call α.

For a reversible (equilibrium) phase change, we have $\Delta S = \Delta H/T$, Eq. (3.25). Equation (7.17) becomes

$$\frac{dP}{dT} = \frac{\Delta H_m}{T\,\Delta V_m} = \frac{\Delta H}{T\,\Delta V} \qquad \text{one component two-phase equilib.} \qquad \textbf{(7.18)*}$$

Equation (7.18) is the **Clapeyron equation,** also called the *Clausius–Clapeyron equation.* Its derivation involved no approximations, and (7.18) is an exact result for a one-component system.

For a liquid-to-vapor transition, both ΔH and ΔV are positive; hence dP/dT is positive. The liquid–vapor line on a one-component P-T phase diagram has positive slope. The same is true of the solid–vapor line. For a solid-to-liquid transition, ΔH is virtually always positive; ΔV is usually positive but is negative in a few cases, for example, H_2O, Ga, and Bi. Because of the volume decrease for the melting of ice, the solid–liquid equilibrium line slopes to the left in the water P-T diagram (Fig. 7.1). For nearly all other substances, the solid–liquid line has positive slope (as in Fig. 7.3). The fact that the melting point of ice is lowered by a pressure increase is in accord with Le Châtelier's principle (Sec. 6.6), which predicts that a pressure increase will shift the equilibrium to the side with the smaller volume. Liquid water has a smaller volume than the same mass of ice.

For melting, ΔV_m is much smaller than for sublimation or vaporization. Hence the Clapeyron equation (7.18) shows that the solid–liquid equilibrium line on a P-versus-T phase diagram will have a much steeper slope than the solid–vapor or liquid–vapor lines (Fig. 7.1).

Liquid–Vapor and Solid–Vapor Equilibrium

For phase equilibrium between a gas and a liquid or solid, $V_{m,\text{gas}}$ is much greater than $V_{m,\text{liq}}$ or $V_{m,\text{solid}}$ unless T is near the critical temperature, in which case the vapor and liquid densities are close (Fig. 7.2). Thus, when one of the phases is a gas, $\Delta V_m = V_{m,\text{gas}} - V_{m,\text{liq or solid}} \approx V_{m,\text{gas}}$. If the vapor is assumed to behave approximately ideally, then $V_{m,gas} \approx RT/P$. These two approximations give $\Delta V_m \approx RT/P$, and the Clapeyron equation (7.18) becomes

$$dP/dT \approx P\,\Delta H_m/RT^2$$

$$\frac{d \ln P}{dT} \approx \frac{\Delta H_m}{RT^2} \qquad \text{solid–gas or liq.–gas equilib. not near } T_c \qquad \textbf{(7.19)*}$$

since $dP/P = d \ln P$. Note the resemblance to the van't Hoff equation (6.36). Equation (7.19) does not hold at temperatures near the critical temperature T_c, where the gas density is high, the vapor is far from ideal, and the liquid's volume is not negligible compared with the gas's volume. Equation (7.19) is called the **Clausius–Clapeyron equation** in most physical chemistry texts. However, most physics and engineering thermodynamics texts use the name Clausius–Clapeyron equation to refer to Eq. (7.18).

Since $d(1/T) = -(1/T^2)\,dT$, Eq. (7.19) can be written as

$$d \ln P/d(1/T) \approx -\Delta H_m/R \qquad \text{solid–gas or liq.–gas equilib. not near } T_c \qquad (7.20)$$

The quantity $\Delta H_m = H_{m,gas} - H_{m,liq}$ (or $H_{m,gas} - H_{m,solid}$) depends on the temperature of the phase transition. Once T of the transition is specified, the transition pressure is fixed, so P is not an independent variable along the equilibrium line. From (7.20), a plot of $\ln P$ versus $1/T$ has slope $-\Delta H_{m,T}/R$ at temperature T, and measurement of this slope at various temperatures allows ΔH_m of vaporization or sublimation to be found at each temperature. If the temperature interval is not large and if we are not near T_c, ΔH_m will vary only slightly and the plot will be nearly linear (Fig. 7.6). Strictly speaking, we cannot take the log of a quantity with units. To get around this, note that $d \ln P = d \ln (P/P^\dagger)$, where $P^\dagger$ is any convenient fixed pressure such as 1 torr, 1 bar, or 1 atm; we thus plot $\ln (P/P^\dagger)$ versus $1/T$.

If we make a third approximation and take ΔH_m to be constant along the equilibrium line, integration of (7.19) gives

$$\int_1^2 d \ln P \approx \Delta H_m \int_1^2 \frac{1}{RT^2}\,dT$$

$$\ln \frac{P_2}{P_1} \approx -\frac{\Delta H_m}{R}\left(\frac{1}{T_2} - \frac{1}{T_1}\right) \qquad \text{solid–gas or liq.–gas equilib. not near } T_c \qquad (7.21)$$

If P_1 is 1 atm, then T_1 is the normal boiling point T_{nbp}. Dropping the unnecessary subscript 2 from (7.21), we have

$$\ln (P/\text{atm}) \approx -\Delta H_m/RT + \Delta H_m/RT_{nbp} \qquad \text{liq.–gas equilib. not near } T_c \qquad (7.22)$$

Actually, $\Delta_{vap}H_m$ is reasonably constant over only a short temperature range (Fig. 7.4), and (7.21) and (7.22) must not be applied over a large range of T. The integration of (7.18) taking into account the temperature variation of ΔH_m, gas nonideality, and the liquid's volume is discussed in *Reid, Prausnitz, and Poling,* chap. 7; see also *Denbigh,* secs. 6.3 and 6.4.

Equation (7.22) gives $P/\text{atm} \approx Be^{-\Delta H_m/RT}$, where $B \equiv e^{\Delta H_m/RT_{nbp}}$ for liquids. The exponential function in this equation gives a rapid increase in vapor pressure with temperature for solids and liquids. Vapor-pressure data for ice and liquid water are plotted in Fig. 7.1*b*. As T goes from $-111°C$ to $-17°C$, the vapor pressure of ice increases by a factor of 10^6, going from 10^{-6} to 1 torr. The vapor pressure of liquid water goes from 4.6 torr at the triple-point temperature $0.01°C$ to 760 torr at the normal boiling point $99.97°C$ to 165000 torr at the critical temperature $374°C$. As T increases, the fraction of molecules in the liquid or solid with enough kinetic energy

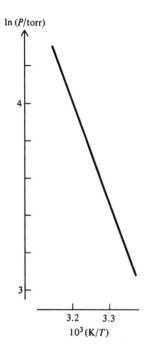

ln (P/torr)

4

3

3.2   3.3

$10^3\,(K/T)$

Figure 7.6

Plot of $\ln P$ (where P is the vapor pressure) versus $1/T$ for water for temperatures from 45°C to 25°C. If $10^3(K/T) = 3.20$, then $1/T = 0.00320$ K^{-1} and $T = 312$ K.

to escape from the attractions of surrounding molecules increases rapidly, giving a rapid increase in vapor pressure.

Vapor pressures of liquids are measured with a manometer. The low vapor pressures of solids can be found by measuring the rate of mass decrease due to vapor escaping through a tiny hole of known area—see Sec. 15.6.

Vapor pressures are affected slightly by an applied external pressure such as that of the air in a room; see Prob. 7.49.

EXAMPLE 7.4 Change of vapor pressure with temperature

The normal boiling point of ethanol is 78.3°C, and at this temperature $\Delta_{vap}H_m = 38.9$ kJ/mol. To what value must P be reduced if we want to boil ethanol at 25.0°C in a vacuum distillation?

The boiling point is the temperature at which the liquid's vapor pressure equals the applied pressure P on the liquid. The desired value of the applied pressure P is thus the vapor pressure of ethanol at 25°C. To solve the problem, we must find the vapor pressure of ethanol at 25°C. We know that the vapor pressure at the normal boiling point is 760 torr. The variation of vapor pressure with temperature is given by the approximate form (7.19) of the Clapeyron equation: $d \ln P/dT \approx \Delta H_m/RT^2$. If the temperature variation of $\Delta_{vap}H_m$ is neglected, integration gives [Eq. (7.21)]

$$\ln \frac{P_2}{P_1} \approx -\frac{\Delta H_m}{R}\left(\frac{1}{T_2} - \frac{1}{T_1}\right)$$

Let state 2 be the normal-boiling-point state with $T_2 = (78.3 + 273.2)$ K $= 351.5$ K and $P_2 = 760$ torr. We have $T_1 = (25.0 + 273.2)$ K $= 298.2$ K and

$$\ln \frac{760 \text{ torr}}{P_1} \approx -\frac{38.9 \times 10^3 \text{ J/mol}}{8.314 \text{ J mol}^{-1} \text{ K}^{-1}}\left(\frac{1}{351.5 \text{ K}} - \frac{1}{298.2 \text{ K}}\right) = 2.38$$

$$760 \text{ torr}/P_1 \approx 10.8, \qquad P_1 \approx 70 \text{ torr}$$

The experimental vapor pressure of ethanol at 25°C is 59 torr. The substantial error in our result is due to nonideality of the vapor (which results mainly from hydrogen-bonding forces between vapor molecules) and to the temperature variation of $\Delta_{vap}H_m$; at 25°C, $\Delta_{vap}H_m$ of ethanol is 42.5 kJ/mol, substantially higher than its 78.3°C value. For an improved calculation, see Prob. 7.25.

Exercise

The normal boiling point of Br_2 is 58.8°C, and its vapor pressure at 25°C is 0.2870 bar. Estimate the average $\Delta_{vap}H_m$ of Br_2 in this temperature range. (*Answer:* 30.7 kJ/mol.)

Exercise

Use data in Table 7.1 to estimate the boiling point of Ar at 1.50 atm. (*Answer:* 91.4 K.)

Exercise

Use Fig. 7.6 to find the slope of a $\ln P$-versus-$1/T$ plot for the vaporization of H_2O near 35°C. Then use this slope to find $\Delta_{vap}H_m$ of H_2O at 35°C. (*Answers:* -5400 K, $10._7$ kcal/mol $= 45$ kJ/mol.)

Solid–Liquid Equilibrium

For a solid–liquid transition, Eq. (7.19) does not apply. For fusion (melting), the Clapeyron equation (7.18) and (7.17) reads $dP/dT = \Delta_{fus}S/\Delta_{fus}V = \Delta_{fus}H/(T\,\Delta_{fus}V)$. Multiplication by T and integration gives

$$\int_1^2 dP = \int_1^2 \frac{\Delta_{fus}S}{\Delta_{fus}V}\,dT = \int_1^2 \frac{\Delta_{fus}H}{T\,\Delta_{fus}V}\,dT \qquad (7.23)$$

The quantities $\Delta_{fus}S$ ($\equiv S_{liq} - S_{solid}$), $\Delta_{fus}H$, and $\Delta_{fus}V$ change along the solid–liquid equilibrium line due to changes in both T_{fus} and P_{fus} along this line. However, the steepness of the slope of the P-versus-T fusion line (Fig. 7.1b) means that unless $P_2 - P_1$ is very large, the change in melting-point temperature T_{fus} will be quite small. Moreover, the properties of solids and liquids change only slowly with pressure (Sec. 4.5). Therefore, unless $P_2 - P_1$ is quite large, we can approximate $\Delta_{fus}S$, $\Delta_{fus}H$, and $\Delta_{fus}V$ as constant. To integrate (7.23), we can assume either that $\Delta_{fus}S/\Delta_{fus}V$ is constant or that $\Delta_{fus}H/\Delta_{fus}V$ is constant. For small changes in freezing point, these two approximations give similar results and either can be used. For substantial changes in freezing point, neither approximation is accurate for solid–liquid transitions. However, the equilibrium lines for many solid–solid transitions (Sec. 7.4) on P-versus-T phase diagrams are observed to be nearly straight over wide temperature ranges. The constant slope $dP/dT = \Delta_{trs}S/\Delta_{trs}V$ for such solid–solid transitions means that taking $\Delta_{trs}S/\Delta_{trs}V$ is often a good approximation here.

If we approximate $\Delta_{fus}S/\Delta_{fus}V$ as constant, then (7.23) becomes

$$P_2 - P_1 \approx \frac{\Delta_{fus}S}{\Delta_{fus}V}(T_2 - T_1) = \frac{\Delta_{fus}H}{T_1\,\Delta_{fus}V}(T_2 - T_1) \qquad \text{solid–liq. eq., } T_2 - T_1 \text{ small}$$

$$(7.24)$$

EXAMPLE 7.5 Effect of pressure on melting point

Find the melting point of ice at 100 atm. Use data from Prob. 2.48.

Since this is a solid–liquid equilibrium, Eq. (7.24) applies. [A frequent student error is to apply Eq. (7.19) to solid–liquid equilibria.] For 1 g of ice, $\Delta_{fus}H = 79.7$ cal and the densities give $\Delta_{fus}V \equiv V_{liq} - V_{solid} = 1.000$ cm^3 $- 1.091$ cm^3 $= -0.091$ cm^3. Let state 1 be the normal melting point. Then $P_2 - P_1 = 100$ atm $- 1$ atm $= 99$ atm, and (7.24) becomes

$$99 \text{ atm} \approx \frac{79.7 \text{ cal}}{(273.15 \text{ K})(-0.091 \text{ cm}^3)}\Delta T$$

$$\Delta T = -30.9 \text{ K cm}^3 \text{ atm/cal}$$

We now use two values of R to convert cm^3 atm to cal, so as to eliminate cm^3 atm/cal.

$$\Delta T = -30.9 \text{ K} \frac{\text{cm}^3 \text{ atm}}{\text{cal}} \frac{1.987 \text{ cal mol}^{-1} \text{ K}^{-1}}{82.06 \text{ cm}^3 \text{ atm mol}^{-1} \text{ K}^{-1}} = -0.75 \text{ K}$$

Hence, $T_2 = 273.15$ K $- 0.75$ K $= 272.40$ K. The pressure increase of 99 atm has lowered the melting point by only 0.75 K to $-0.75°$C.

Exercise

Repeat this problem assuming that $\Delta_{fus}H/\Delta_{fus}V$ is constant. (*Answer:* 272.40 K.)

Exercise

At the normal melting point of NaCl, 801°C, its enthalpy of fusion is 28.8 kJ/mol, the density of the solid is 2.16_5 g/cm³, and the density of the liquid is 1.73_3 g/cm³. What pressure increase is needed to raise the melting point by 1.00°C? (*Answer:* 39 atm.)

Equation (7.24) is generally accurate up to a few hundred atmospheres but fails for larger pressure differences, due to the changes in $\Delta_{fus}S$, $\Delta_{fus}V$, and $\Delta_{fus}H$ with changes in T and P along the equilibrium line. For example, on the H_2O solid–liquid equilibrium line, the following data are observed:

t	0°C	−5°C	−20°C
P/atm	1	590	1910
$(\Delta_{fus}H/\Delta_{fus}V)$/(kJ/cm³)	−3.71	−3.04	−1.84
$(\Delta_{fus}S/\Delta_{fus}V)$/(J/cm³-K)	−13.6	−11.3	−7.26

For most substances, $\Delta_{fus}V \equiv V_{liq} - V_{solid}$ is positive. Liquids are more compressible than solids, so as P_{fus} increases, V_{liq} decreases faster than V_{solid} and $\Delta_{fus}V$ decreases. For a few substances, $\Delta_{fus}V$ is positive at low P_{fus} values and becomes negative at high P_{fus}. Here, the slope of the solid–liquid line changes sign at high pressures, producing a maximum in melting point at the pressure where $\Delta_{fus}V = 0$. Figure 7.7 shows the melting-point line on the P-versus-T phase diagram of europium.

In applying the Clapeyron equation $dP/dT = \Delta H/(T\,\Delta V)$ to phase transitions involving only condensed phases (solid–liquid or solid–solid), we approximated ΔV as constant and calculated ΔV from the experimental densities of the two phases. In applying the Clapeyron equation to transitions involving a gas phase (solid–gas or liquid–gas), we neglected V of the condensed phase and approximated ΔV as V_{gas}, where we used the ideal-gas approximation for the gas volume; these approximations are valid well below the critical point.

$\Delta_{vap}H_m$ from Linear and Nonlinear Least-Squares Fits

Example 7.6 shows how a spreadsheet is used to find $\Delta_{vap}H_m$ from vapor-pressure data.

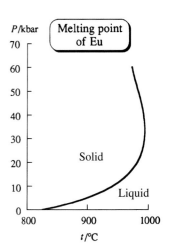

Figure 7.7

Melting point of europium versus pressure. (The melting-point line of graphite also shows a temperature maximum.)

EXAMPLE 7.6 $\Delta_{vap}H_m$ from linear and nonlinear least-squares fits

Accurate vapor-pressure data for water are [H. F. Stimson, *J. Res. Natl. Bur. Stand.*, **73A,** 493 (1969)]

t_{68}/°C	40	50	60	70	80
P/torr	55.364	92.592	149.510	233.847	355.343

where the temperatures and pressures are estimated to be accurate to within about $10^{-4}\%$ and $10^{-3}\%$, respectively, and t_{68} denotes the now obsolete International Temperature Scale of 1968. Find $\Delta_{vap}H_m$ of water at 60°C.

If the three approximations that give Eq. (7.19) are made, then Eqs. (7.20) and (7.22) show that a plot of ln (P/torr) versus $1/T$ will be a straight line with slope $-\Delta H_m/R$. We use appendix II of *Quinn* to convert the temperatures to ITS-90 (Sec. 1.5). The data are entered into a spreadsheet (Fig. 7.8) and ln (P/torr) and $1/T$ are calculated in columns D and E. For columns B, D, and E, only the formulas in row 3 need be typed; the others are produced using Copy and Paste (or in Excel by dragging the tiny rectangle at the lower right of a selected cell). To

Figure 7.8

Spreadsheet for finding $\Delta_{vap}H_m$. The lower view shows the formulas.

	A	B	C	D	E	F	G
1	vapor pressure H2O CC eqn lst sqs						
2	t_{90}/C	T/K	P/torr	ln P/torr	1/T	P(exp fit)	res-sqs
3	39.99	313.14	55.364	4.01393	0.003193	55.58574	0.04917
4	49.987	323.137	92.592	4.5282	0.003095	92.37595	0.046677
5	59.984	333.134	149.51	5.00736	0.003002	148.9069	0.363762
6	69.982	343.132	233.847	5.45467	0.002914	233.4571	0.152036
7	79.979	353.129	355.343	5.87308	0.002832	356.7985	2.118484
8						sumsqres	2.73013
9						b	m
10						20.43627	-5141.24

	A	B	C	D	E	F	G
1	vapor press						
2	t_{90}/C	T/K	P/torr	ln P/torr	1/T	P(exp fit)	res-sqs
3	=40-0.01	=A3+273.15	55.364	=LN(C3)	=1/B3	=EXP($F$10+$G$10/B3)	=(F3-C3)^2
4	=50-0.013	=A4+273.15	92.592	=LN(C4)	=1/B4	=EXP($F$10+$G$10/B4)	=(F4-C4)^2
5	=60-0.016	=A5+273.15	149.51	=LN(C5)	=1/B5	=EXP($F$10+$G$10/B5)	=(F5-C5)^2
6	=70-0.018	=A6+273.15	233.847	=LN(C6)	=1/B6	=EXP($F$10+$G$10/B6)	=(F6-C6)^2
7	=80-0.021	=A7+273.15	355.343	=LN(C7)	=1/B7	=EXP($F$10+$G$10/B7)	=(F7-C7)^2
8						sumsqres	=SUM(G3:G7)
9						b	m
10						20.43627	-5141.24

use Excel to get the coefficients m and b that give the best least-squares fit to the straight line $y = mx + b$, choose Data Analysis from the Tools menu and then choose Regression in the scroll-down list. (If Data Analysis is not on the Tools menu, choose Add-Ins on the Tools menu, check Analysis ToolPak and click OK.) In the dialog box that opens, enter D3 : D7 as the Input Y range and E3 : E7 as the Input X range. Click the Output range button and enter a cell such as A14 as the upper left cell of the least-squares-fit output data. Click the Residuals box and the Residual Plots box, and click on OK. You get a host of statistical data as output. The desired constants b and m are the two numbers listed under the heading Coefficients. One finds 20.4363 as the intercept b and -5141.24 as the slope m (the coefficient of the X Variable). (You are also told that there is a 95% probability that the slope lies in the range -5093 to -5190.) Although the **residuals** [the deviations of the experimental ln (P/torr) values from the values calculated using the straight-line fit] are small, note that the graph of the residuals is roughly parabolic with positive residuals for the middle three points and negative residuals for the first and last points. This indicates that the data fit a curved line a bit better than a straight line. (We know that ΔH_m is not really constant, and two other approximations have been made.) For data that have random errors on top of a straight-line fit, the residuals are randomly positive and negative. [Another way in Excel to get the coefficients of a straight-line fit is to enter the formulas =SLOPE(D3:D7,E3:E7) and =INTERCEPT(D3:D7,E3:E7) into two empty cells. A third way is to graph the data using an XY (Scatter) plot that plots only the data points. Then click on a graph data point and choose Add Trendline from the Chart menu; see Sec. 5.6.]

Since we plotted ln (P/torr) versus $1/T$, the slope has units K^{-1}. Thus $-5141.2\ K^{-1} = -\Delta H_m/R$, and

$$\Delta H_m = (5141.2\ K^{-1})(8.3145\ J\ mol^{-1}\ K^{-1}) = 42.75\ kJ/mol$$

As noted in Prob. 6.57, a least-squares fit assumes that the y data points have about the same relative precision. Since it is the values of P and not the values of ln (P/torr) that were measured and that have about the same relative precision, the procedure of linearizing the data by taking logarithms is not the best way to get an accurate ΔH_m. The best way to treat the data is to minimize the sums of the squares of the deviations of the P values. This is easily done in Excel using the Solver (Sec. 6.4) with the parameters from the straight-line fit as the initial guesses. We found from the linear fit that ln $(P/\text{torr}) \approx m(1/T) + b$, so $P/\text{torr} \approx e^{b+m/T}$. Enter the straight-line-fit values $b = 20.43627$ and $m = -5141.24$ in cells F10 and G10. Enter the formula =EXP($F$10+$G$10/B3) in F3 and copy it to F4 through F7. The fitted and experimental values of P are in columns F and C, respectively, so to get the squares of the residuals, we enter =(F3-C3)^2 into G3 and copy it to G4 through G7. We put the sum of the squares of the residuals into G8. Then we use the Solver to minimize G8 by changing F10 and G10. The result is something like $m = -5111.26$ and $b = 20.34821$. The sum of the squares of the residuals has been reduced from 2.73 for the linear-fit parameters to 0.98, so a much better fit has been obtained. With the revised slope, we get $\Delta H_m = 42.50\ kJ/mol$, which is in better agreement with the best literature value 42.47 kJ/mol.

[An alternative to using the Solver is to use a weighted linear least-squares fit; see R. de Levie, *J. Chem. Educ.,* **63,** 10 (1986).]

Exercise

Set up the spreadsheet of Fig. 7.8 and use the Solver to verify the nonlinear-fit m and b values given in this example.

7.4 SOLID–SOLID PHASE TRANSITIONS

Many substances have more than one solid form. Each such form has a different crystal structure and is thermodynamically stable over certain ranges of T and P. This phenomenon is called *polymorphism.* Polymorphism in elements is called *allotropy.* Recall from Sec. 5.7 that in finding the conventional entropy of a substance, we must take any solid–solid phase transitions into account.

Part of the phase diagram for sulfur is shown in Fig. 7.9a. At 1 atm, slow heating of (solid) orthorhombic sulfur transforms it at 95°C to (solid) monoclinic sulfur. The normal melting point of monoclinic sulfur is 119°C. The stability of monoclinic sulfur is confined to a closed region of the P-T diagram. Note the existence of three *triple points* (three-phase points) in Fig. 7.9a: orthorhombic–monoclinic–vapor equilibrium at 95°C, monoclinic–liquid–vapor equilibrium at 119°C, and orthorhombic–monoclinic–liquid equilibrium at 151°C. At pressures above those in Fig. 7.9a, 10 more solid phases of sulfur have been observed (*Young,* sec. 10.3).

If orthorhombic sulfur is heated rapidly at 1 atm, it melts at 114°C to liquid sulfur, without first being transformed to monoclinic sulfur. Although orthorhombic sulfur is thermodynamically unstable between 95°C and 114°C at 1 atm, it can exist for short periods under these conditions, where its G_m is greater than that of monoclinic sulfur.

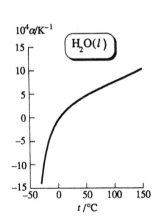

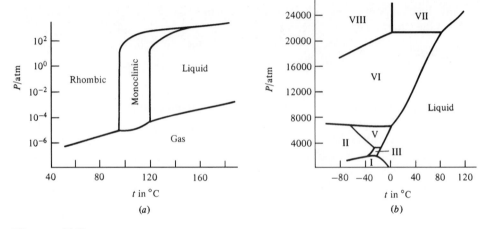

Figure 7.9

(a) Part of the sulfur phase diagram. The vertical scale is logarithmic. (Orthorhombic sulfur is commonly, but inaccurately, called rhombic sulfur.) (b) A portion of the H_2O phase diagram at high pressure.

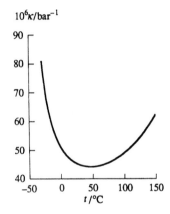

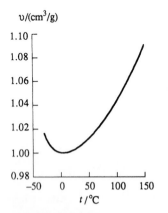

Figure 7.10

Thermal expansivity α, isothermal compressibility κ, and specific volume v of liquid water at 1 atm plotted versus temperature. Below 0°C, the water is supercooled.

Phase α is said to be **metastable** with respect to phase β at a given T and P if $G_m^\alpha > G_m^\beta$ at that T and P and if the rate of conversion of α to β is slow enough to allow α to exist for a significant period of time. Another example of metastability besides orthorhombic sulfur is diamond. Appendix data show that G_m of diamond is greater than G_m of graphite at 25°C and 1 atm. Other examples are liquids cooled below their freezing points (**supercooled** liquids) or heated above their boiling points (**super-heated** liquids) and gases cooled below their condensation temperatures (*supersaturated* vapors). One can supercool water to −40°C and superheat it to 280°C at 1 atm. Figure 7.10 plots some properties of liquid water at 1 atm for the temperature range −30°C to 150°C. In the absence of dust particles, one can compress water vapor at 0°C to five times the liquid's vapor pressure before condensation occurs.

The temperature of the lowest region of the atmosphere (the *troposphere*) decreases with increasing altitude (Fig. 15.17). The droplets of liquid water in most tropospheric clouds above 2 or 3 km altitude are supercooled. Only when the temperature reaches −15°C do cloud droplets begin to freeze in significant amounts.

A solid usually cannot be superheated above its melting point. Computer simulations of solids (Sec. 24.14) indicate that the surface of a solid begins to melt below the melting point to give a thin liquidlike surface film whose properties are intermediate between those of the solid and those of the liquid, and whose thickness increases as the melting point is approached. This *surface melting* has been observed in Pb, Ar, O_2, CH_4, H_2O, and biphenyl. For the (110) surface of Pb (see Sec. 24.7 for an explanation of the notation), the liquid surface film is 10 Å thick at 10 K below the Pb melting point and is 25 Å thick at 1 K below the melting point. See J. W. M. Frenken et al., *Phys. Rev. B.*, **34**, 7506 (1986); *Phys. Rev. Lett.*, **60**, 1727 (1988); B. Pluis et al., *Phys. Rev. Lett.*, **59**, 2678 (1987); R. Lipowsky et al., *Phys. Rev. Lett.*, **62**, 913 (1989).

Superheating of solids that do not have a free external surface has been observed. Examples are gold-plated silver crystals and ice grains in the presence of $CH_4(g)$ at 250 bar [L. A. Stern et al., *J. Phys. Chem. B*, **102**, 2627 (1998)]. The methane forms a hydrate compound at the surface of the ice, thereby allowing superheating.

The bubbling that occurs during boiling is a nonequilibrium phenomenon. Bubbles appear at places where the liquid's temperature exceeds the boiling point (that is, the liquid is superheated) and the liquid's vapor pressure exceeds the pressure in the liquid. (Surface tension makes the pressure inside a bubble exceed the pressure of its surroundings; see Sec. 13.2.) Liquid helium II has a very high thermal conductivity, which prevents local hot spots from developing. When helium II is vaporized at its boiling point, it does not bubble.

Some physicists have speculated that the universe might currently exist in a metastable high-energy false-vacuum state that is separated by an energy barrier from the lower-energy true vacuum state. If this is so, there is a slight probability for the universe to spontaneously undergo a phase transition to the true vacuum state. The transition would start at a particular location and would propagate throughout the universe at nearly the speed of light. In the true vacuum state, the laws of physics would differ from those in the false-vacuum state (P. Davies, *The Last Three Minutes,* BasicBooks, 1994, chap. 10). "Vacuum decay is the ultimate ecological catastrophe . . . after vacuum decay . . . life as we know it [is] impossible [S. Coleman and F. DeLuccia, *Phys. Rev. D,* **21,** 3305 (1980)]. It has even been suggested that just as a tiny crystal of ice dropped into supercooled water nucleates the formation of ice, a high concentration of energy produced in a collision experiment by particle physicists might nucleate a phase transition to the true vacuum state, thereby destroying the universe as we know it. Since the energies produced by particle physicists are less than the highest energies that occur naturally in cosmic rays, such a catastrophic laboratory accident is extremely unlikely [P. Hut and M. Rees, *Nature,* **302,** 508 (1983)].

The phase diagram of water is actually far more complex than the one shown in Fig. 7.1. At high pressures, the familiar form of ice is not stable and other forms exist (Fig. 7.9*b*). Note the existence of several triple points at high pressure and the high freezing point of water at high P.

Ordinary ice is ice Ih (where the h is for hexagonal and describes the crystal structure). Not shown in Fig. 7.9*b* are the metastable forms ice Ic (cubic) (obtained by condensation of water vapor below $-80°C$) and ices IV and XII [C. Lobban et al., *Nature,* **391,** 268 (1998)], which exist in the same region as ice V. Also not shown are the low-temperature forms ice IX and ice XI, and the very-high-pressure form ice X, all of whose phase boundaries are not well established. In addition to the crystalline forms ices I to XII, at least two amorphous forms of ice exist. Structures of the various forms of ice are discussed in *Franks,* vol. 1, pp. 116–129; V. F. Petrenko and R. W. Whitworth, *Physics of Ice,* Oxford University Press, 1999, chap. 11.

The plot of Kurt Vonnegut's novel *Cat's Cradle* (Dell, 1963), written when only ices I to VIII were known, involves the discovery of ice IX, a form supposed to exist at 1 atm with a melting point of 114°F, relative to which liquid water is unstable. Ice IX brings about the destruction of life on earth. (Kurt Vonnegut's brother Bernard helped develop the method of seeding supercooled clouds with AgI crystals to induce ice formation and increase the probability of snow or rain.)

The properties of matter at high pressures are of obvious interest to geologists. Some pressures in the earth are 10^3 bar at the deepest part of the ocean, 10^4 bar at the boundary between the crust and mantle, 1.4×10^6 bar at the boundary between the mantle and core, and 3.6×10^6 bar at the center of the earth. The pressure at the center of the sun is 10^{11} bar.

Matter has been studied in the laboratory at pressures exceeding 10^6 bar. Such pressures are produced in a *diamond-anvil cell,* in which the sample is mechanically compressed between the polished faces of two diamonds. Because diamonds are transparent, the optical properties of the compressed sample can be studied. The diamond-anvil cell is small enough to be held in one's hand and the diamond faces that compress the sample are less than 1 mm in diameter. To detect a phase transition and find the structure of the new phase formed in a diamond-anvil cell, one commonly uses x-ray diffraction (Sec. 24.9). The pressure can be found from the pressure-induced shift in spectral lines of a tiny chip of ruby that is included in the sample cell. Pressures of 5 megabars have been obtained with a diamond-anvil cell [A. L. Ruoff et al., *Rev. Sci. Instrum.*, **63**, 4342 (1992)]. Theoretical calculations indicate that, at sufficiently high pressures, every solid is converted to a metallic form. This has been verified for I_2, CsI, Xe, S, and oxygen.

Metallic solid hydrogen has been called "the holy grail of high-pressure physics." Theoretical estimates of the required pressure vary widely. Solid hydrogen has been compressed to 3.4 megabars without being metallized [C. Narayana et al., *Nature*, **393**, 46 (1998)]. It has been speculated that once metallic solid hydrogen is formed, it might remain in the metastable metallic form when the pressure is released and might be usable as a lightweight structural material to make such things as automobiles. Solid metallic hydrogen might be a superconductor at low temperatures. Although metallic solid hydrogen has not been achieved, metallic liquid hydrogen has been formed very briefly at 1.4 Mbar and 2600 K by shock-wave compression [W. J. Nellis et al., *Phys. Rev. B*, **59**, 3434 (1999)]. The planet Jupiter is 90% hydrogen and at the very high pressures and temperatures of its interior, much of this hydrogen likely exists in a metallic liquid state, giving rise to the magnetic field of Jupiter [www-phys.llnl.gov/H_Div/GG/Nellis.html].

EXAMPLE 7.7 Phase stability

At 25°C and 1 bar, the densities of diamond and graphite are $\rho_{di} = 3.52$ g/cm^3 and $\rho_{gr} = 2.25$ g/cm^3. Use Appendix data to find the minimum pressure needed to convert graphite to diamond at 25°C. State any approximations made.

As noted in Sec. 7.2, the stable phase is the one with the lowest G_m. Appendix $\Delta_f G°$ values show that for the transformation diamond → graphite at 25°C and 1 bar,

$$\Delta G° = -2.90 \text{ kJ/mol} = G_{m,gr}(1 \text{ bar}) - G_{m,di}(1 \text{ bar})$$

Thus at room T and P, graphite is the stable phase and diamond is metastable.

How does changing the pressure affect G_m and affect the relative stability of the two forms? From $dG_m = -S_m \, dT + V_m \, dP$, we have $(\partial G_m/\partial P)_T = V_m = M/\rho$, where M is the molar mass. The smaller density of graphite makes V_m of graphite greater than V_m of diamond, so G_m of graphite increases faster than G_m of diamond as P is increased, and eventually diamond becomes the more-stable phase. At the pressure P_2 at which the graphite-to-diamond phase transition occurs, we have $G_{m,gr}(P_2) = G_{m,di}(P_2)$.

Integrating $dG_m = (M/\rho) \, dP$ (T const.) at constant T and neglecting the change of ρ with pressure, we have

$$G_m(P_2) = G_m(P_1) + (M/\rho)(P_2 - P_1)$$

Substitution of this equation into $G_{m,gr}(P_2) = G_{m,di}(P_2)$ gives

$$G_{m,gr}(P_1) + (M/\rho_{gr})(P_2 - P_1) = G_{m,di}(P_1) + (M/\rho_{di})(P_2 - P_1)$$

$$P_2 - P_1 = \frac{G_{m,gr}(P_1) - G_{m,di}(P_1)}{M(1/\rho_{di} - 1/\rho_{gr})} = \frac{(-2900 \text{ J/mol})}{(12.01 \text{ g/mol})(3.52^{-1} - 2.25^{-1})(\text{cm}^3/\text{g})}$$

$$= 1506 \text{ J/cm}^3$$

$$P_2 - 1 \text{ bar} = 1506 \frac{\text{J}}{\text{cm}^3} \frac{82.06 \text{ cm}^3 \text{ atm mol}^{-1} \text{ K}^{-1}}{8.314 \text{ J mol}^{-1} \text{ K}^{-1}} = 14900 \text{ atm}$$

$$P_2 = 14900 \text{ atm} = 15100 \text{ bar}$$

Thus, above 15.1 kbar, diamond is predicted to be the more-stable phase. In actuality, just as diamond will persist indefinitely at room T and P even though it is metastable with respect to graphite, graphite will persist indefinitely at room temperature and pressures above 15.1 kbar. Conversion of graphite to diamond is done in the laboratory by increasing both P and T in the presence of a catalyst. There is a large activation-energy barrier (Sec. 17.8) involved in converting the "infinite" two-dimensional covalent-bond structure of graphite to the "infinite" three-dimensional covalent structure of diamond (Fig. 24.19). Thermodynamics cannot tell us about rates of processes.

Exercise

The solid–liquid–gas triple point of carbon is at 5000 K and 100 bar and the diamond–graphite–liquid triple point is at 4900 K and 10^5 bar. The graphite melting line shows a maximum temperature and the diamond melting line has a positive dP/dT. The critical point is at roughly 6800 K and 2×10^4 bar. Sketch the phase diagram of carbon using a logarithmic scale for pressure. [*Answer:* See F. P. Bundy et al., *Carbon,* **34,** 141 (1996).]

One-component solid–solid transitions between different structural forms are common. One-component liquid–liquid phase transitions are rare but occur in ^{3}He and ^{4}He (Sec. 7.5) and there is some evidence for such phase transitions in liquid carbon, sulfur, selenium, and iodine [M. Togaya, *Phys. Rev. Lett.,* **79,** 2474 (1997); V. V. Brazhkin et al., *Physica B,* **265,** 64 (1999)].

7.5 HIGHER-ORDER PHASE TRANSITIONS

For the equilibrium phase transitions at constant T and P discussed in Secs. 7.2 to 7.4, the transition is accompanied by a transfer of heat $q_P \neq 0$ between system and surroundings; also, the system generally undergoes a volume change. Such transitions with $\Delta H \neq 0$ are called **first order** or **discontinuous.**

For a first-order transition, $C_P = (\partial H/\partial T)_P$ of the two phases is observed to differ. C_P may either increase (as in the transition of ice to water) or decrease (as in water $\rightarrow$ steam) on going from the low-T to high-T phase (see Fig. 2.15). Right at the transition temperature, $C_P = dq_P/dT$ is infinite, since the nonzero latent heat is absorbed by the system with no temperature change (Fig. 7.11a).

Certain special phase transitions occur with $q_P = \Delta H = T \Delta S = 0$ and with $\Delta V = 0$. These are called **higher-order** or **continuous** transitions. For such a transition, the Clapeyron equation $dP/dT = \Delta H/(T \Delta V)$ is meaningless. For a higher-order transition, $\Delta U = \Delta(H - PV) = \Delta H - P \Delta V = 0$. The known higher-order transitions are either second-order transitions or lambda transitions.

A *second-order* transition is defined as one where $\Delta H = T \Delta S = 0$, $\Delta V = 0$, and C_P does not become infinite at the transition temperature but does change by a finite amount (Fig. 7.11b). The only known second-order transitions are those between liquid ^{3}He B and liquid ^{3}He N, between liquid ^{3}He A and liquid ^{3}He N (J. Wilks and D. S. Betts, *An Introduction to Liquid Helium,* 2d ed., Oxford, 1987, chap. 9), and between normal conductivity and superconductivity in certain metals. Some metals, for example, Hg, Sn, Pb, Al, on being cooled to characteristic temperatures (4.2 K for Hg at 1 atm) become superconductors with zero electrical resistance.

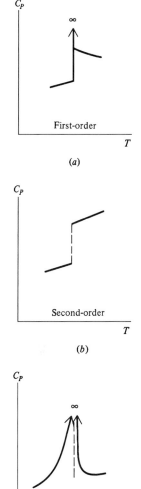

Figure 7.11

C_P versus T in the region of (*a*) a first-order transition; (*b*) a second-order transition; (*c*) a lambda transition. For some lambda transitions, C_P goes to a very large finite value (rather than ∞) at the transition temperature.

A *lambda transition* is one where $\Delta H = T \Delta S = 0 = \Delta V$ at the lambda-point temperature T_λ and C_P shows one of the following two behaviors: either (a) C_P goes to infinity as T_λ is approached from above and from below (Fig. 7.11c), or (b) C_P goes to a very large finite value as T_λ is approached from above and below and the slope $\partial C_P / \partial T$ is infinite at T_λ (see Prob. 7.42). The shape of the C_P-versus-T curve resembles the Greek letter λ (lambda). Examples of lambda transitions include the transition between liquid helium I and liquid helium II in ^4He; the transition between ferromagnetism and paramagnetism in metals like Fe or Ni; and order–disorder transitions in certain alloys, for example, β-brass, and in certain compounds, for example, NH_4Cl, HF, and CH_4. (Some people use the term second-order transition as meaning the same thing as higher-order transition.)

When liquid ^4He is cooled, as the temperature falls below the lambda temperature T_λ (whose value depends somewhat on pressure and is about 2 K), a substantial fraction of the atoms enter a superfluid state in which they flow without internal friction. The lower the temperature below T_λ, the greater the fraction of atoms in the superfluid state. (This is a quantum-mechanical effect.) Helium below T_λ is called the helium II phase. C_P of liquid ^4He was measured to within 2×10^{-9} K of T_λ in a 1992 experiment on the U.S. space shuttle *Columbia,* thus eliminating the perturbing effects of gravity. For the results, see Prob. 7.42.

β-brass is a nearly equimolar mixture of Zn and Cu; for simplicity, let us assume an exactly equimolar mixture. The crystal structure has each atom surrounded by eight nearest neighbors that lie at the corners of a cube. Interatomic forces are such that the lowest-energy arrangement of atoms in the crystal is a completely ordered structure with each Zn atom surrounded by eight Cu atoms and each Cu atom surrounded by eight Zn atoms. (Imagine two interpenetrating cubic arrays, one of Cu atoms and one of Zn atoms.) In the limit of absolute zero, this is the crystal structure. As the alloy is warmed from T near zero, part of the added energy is used to interchange Cu and Zn atoms randomly. The degree of disorder increases as T increases. This increase is a *cooperative* phenomenon, in that the greater the disorder, the energetically easier it is to produce further disorder. The rate of change in the degree of disorder with respect to T increases as the lambda-point temperature $T_\lambda = 739$ K is approached, and this rate becomes infinite at T_λ, thereby making C_P infinite at T_λ.

At T_λ, all the long-range order in the solid has disappeared, meaning that an atom located at a site that was originally occupied by a Cu atom at 0 K is now as likely to be a Zn atom as a Cu atom. However, at T_λ there still remains some short-range order, meaning that it is still somewhat more than 50% probable that a given neighbor of a Cu atom will be a Zn atom (see Fig. 7.12 and Prob. 7.40). The short-range order finally disappears at a temperature somewhat above T_λ; when this happens, the eight atoms that surround a Cu atom will have an average of four Zn atoms and four Cu atoms. At T_λ, the rate of change of both the short-range order and the long-range order with respect to T is infinite.

In solid NH_4Cl, each NH_4^+ ion is surrounded by eight Cl^- ions at the corners of a cube. The four protons of an NH_4^+ ion lie on lines going from N to four of the eight Cl^- ions. There are two equivalent orientations of an NH_4^+ ion with respect to the surrounding Cl^- ions. At very low T, all the NH_4^+ ions have the same orientation. As T is increased, the NH_4^+ orientations become more and more random. This order–disorder transition is a lambda transition. [The situation in NH_4Cl is complicated by the fact that a very small first-order transition is superimposed on the lambda transition; see B. W. Weiner and C. W. Garland, *J. Chem. Phys.,* **56**, 155 (1972).]

Note that the effects of a lambda transition occur over a range of temperatures. (In β-brass, significant excess heat capacity is observed from about 450 K to about 850 K. Many other lambda transitions show effects over much shorter temperature ranges.) At any time during the course of the lambda transition, *only one phase is present.* The

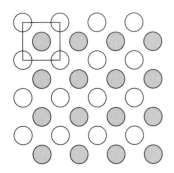

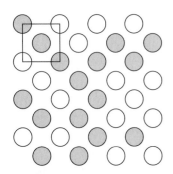

Figure 7.12

Two-dimensional analog of β-brass. The upper figure is at absolute zero, where the four nearest neighbors of each Cu atom (shaded) are Zn atoms (unshaded). The lower figure is at T_λ, where half the sites that were occupied by Cu atoms at $T = 0$ are now occupied by Zn atoms (and vice versa). However, some short-range order remains at T_λ, in that more than half the nearest-neighbor pairs contain one Cu atom and one Zn atom.

nature of this phase changes in a continuous manner as T is increased. In contrast, the effects of a first-order phase transition occur at a single temperature, and during the transition, *two phases* (with different structures, different V_m values, and different H_m values) *are present.*

7.6   SUMMARY

The phase rule $f = c - p + 2 - r - a$ gives the number of degrees of freedom f for an equilibrium system containing c chemical species and p phases and having r independent chemical reactions and a additional restrictions on the mole fractions. f is the number of intensive variables needed to specify the intensive state of the system.

The stable phase of a one-component system at a given T and P is the phase with the lowest $G_m = \mu$ at that T and P.

The Clapeyron equation $dP/dT = \Delta H/(T \Delta V)$ gives the slopes of the lines on a one-component P-T phase diagram. The Clapeyron equation tells (a) how the vapor pressure of a solid varies with T (line OA in Fig. 7.1a); (b) how the vapor pressure of a liquid varies with T or, equivalently, how the boiling point of a liquid varies with P (line AC in Fig. 7.1a); (c) how the melting point of a solid varies with P (line AD in Fig. 7.1a).

For phase equilibrium between a one-component gas and a solid or liquid, neglect of the volume of the condensed phase and approximation of the gas as ideal converts the Clapeyron equation into $d \ln P/dT \approx \Delta H_m/RT^2$.

Important kinds of calculations dealt with in this chapter include:

- Use of the phase rule to find the number of degrees of freedom f.
- Use of $d \ln P/dT \approx \Delta H_m/RT^2$ and vapor-pressure data to find $\Delta_{vap}H_m$ or $\Delta_{sub}H_m$ of a pure substance.
- Use of $d \ln P/dT \approx \Delta H_m/RT^2$ and the vapor pressure at one temperature to find the vapor pressure at another temperature.
- Use of $d \ln P/dT \approx \Delta H_m/RT^2$ to find the boiling point at a given pressure from the normal boiling point.
- Use of the Clapeyron equation to find the change in melting point with pressure.
- Use of $dG_m = -S_m\, dT + V_m\, dP$ and $\Delta_f G°$ data to find the transition P or T for converting one form of a solid to another.

FURTHER READING AND DATA SOURCES

Denbigh, chap. 5; *de Heer,* chaps. 18, 21; *Zemansky and Dittman,* chap. 16; *Andrews* (1971), chap. 25.

Vapor pressures; enthalpies and entropies of phase transitions. *Landolt-Börnstein,* 6th ed., vol. II, part 2a, pp. 1–184; vol. II, part 4, pp. 179–430. D. E. Gray (ed.), *American Institute of Physics Handbook,* 3d ed., McGraw-Hill, 1972, pp. 4-261 to 4-315; pp. 4-222 to 4-261. *Reid, Prausnitz, and Poling,* pp. 656–732. I. Barin and O. Knacke, *Thermochemical Properties of Inorganic Substances,* Springer-Verlag, 1973. TRC Thermodynamic Tables (see Sec. 5.9 for the full references). O. Kubaschewski and C. B. Alcock, *Metallurgical Thermochemistry,* 5th ed., Pergamon, 1979. J. Timmermans, *Physico-Chemical Constants of Pure Organic Compounds,* vols. I and II, Elsevier, 1950, 1965. *Lide and Kehiaian,* sec. 2.1. NIST Chemistry Webbook at webbook.nist.gov/.

PROBLEMS

Section 7.1

7.1 True or false? (*a*) Since the three possible phases are solid, liquid, and gas, the maximum possible value of the number of phases p in the phase rule is 3. (*b*) The number of degrees of freedom f is the number of variables needed to specify the thermodynamic state of a system.

7.2 For each of the following equilibrium systems, find the number of degrees of freedom f and give a reasonable choice of the independent intensive variables. (Omit consideration of water ionization.) (*a*) An aqueous solution of sucrose. (*b*) An aqueous solution of sucrose and ribose. (*c*) Solid sucrose and an aqueous solution of sucrose and ribose. (*d*) Solid sucrose, solid ribose, and an aqueous solution of sucrose and ribose. (*e*) Liquid water and water vapor. (*f*) An aqueous sucrose solution and water vapor. (*g*) Solid sucrose, an aqueous sucrose solution, and water vapor. (*h*) Liquid water, liquid benzene (these two liquids are essentially immiscible), and a mixture of the vapors of these two liquids.

7.3 (*a*) If a system has c_{ind} independent components, what is the maximum number of phases that can exist in equilibrium? (*b*) In the book *Regular Solutions* by J. H. Hildebrand and R. L. Scott (Prentice-Hall, 1962), there is a photograph of a system with 10 liquid phases in equilibrium. What must be true about the number of independent components in this system?

7.4 (*a*) For an aqueous solution of H_3PO_4, write down the reaction-equilibrium conditions and the electroneutrality condition. What is f? (*b*) For an aqueous solution of KBr and NaCl, write down the stoichiometric relations between ion mole fractions. Does the electroneutrality condition give an independent relation? What is f?

7.5 Find f for the following systems and give a reasonable choice for the independent intensive variables: (*a*) a gaseous mixture of N_2, H_2, and NH_3 with no catalyst present (so that the rate of reaction is zero); (*b*) a gaseous mixture of N_2, H_2, and NH_3 with a catalyst present to establish reaction equilibrium; (*c*) the system of (*b*) with the added condition that all the N_2 and H_2 must come from the dissociation of the NH_3; (*d*) A gas-phase mixture of N_2 and N in reaction equilibrium with the condition that all the N comes from the dissociation of N_2; (*e*) a system formed by heating pure $CaCO_3(s)$ to partially decompose it into $CaO(s)$ and $CO_2(g)$, where, in addition, some of each of the solids $CaCO_3$ and CaO has sublimed to vapor. (No CaO or CO_2 is added from the outside.)

7.6 In the HCN(aq) example in Sec. 7.1, the relation $n_{H^+} = n_{OH^-} + n_{CN^-}$ was considered to be an electroneutrality relation. Show that this equation can be considered to be a stoichiometry relation.

7.7 For a system of NaCl(s) and NaCl(aq) in equilibrium, find f if the solid is considered to consist of the single species NaCl and the solute species present in solution (*a*) is considered to be NaCl(aq); (*b*) are taken to be Na$^+$(aq) and Cl$^-$(aq).

7.8 (*a*) For pure liquid water, calculate c_{ind} and f if we consider that the chemical species present are H_2O, H^+, OH^-, and hydrogen-bonded dimers $(H_2O)_2$ formed by the association reaction $2H_2O \rightleftharpoons (H_2O)_2$. (*b*) What happens to c_{ind} and f if we add hydrogen-bonded trimers $(H_2O)_3$ to the list of species?

7.9 Find the relation between f, c_{ind}, and p if (*a*) rigid, permeable, thermally conducting walls separate all the phases of a system; (*b*) movable, impermeable, thermally conducting walls separate all the phases of a system.

Section 7.2

7.10 True or false? (*a*) The normal boiling point is the temperature at which the vapor pressure of a liquid equals 1 atm. (*b*) At the critical point of a pure substance, the densities of the liquid and the vapor are equal. (*c*) The minimum possible value of f in the phase rule is 1. (*d*) The normal boiling point of pure water is precisely 100°C. (*e*) The enthalpy of vaporization of a liquid becomes zero at the critical point. (*f*) Along a line in a one-component phase diagram, $f = 1$. (*g*) At the solid–liquid–gas triple point of a one-component system, $f = 0$. (*h*) CO_2 has no normal boiling point. (*i*) Ice melts above 0.00°C if the pressure is 100 torr.

7.11 For each of the following conditions, state which phase (solid, liquid, or gas) of H_2O has the lowest chemical potential. (*a*) 25°C and 1 atm; (*b*) 25°C and 0.1 torr; (*c*) 0°C and 500 atm; (*d*) 100°C and 10 atm; (*e*) 100°C and 0.1 atm.

7.12 For the H_2O phase diagram of Fig. 7.1*a*, state the number of degrees of freedom (*a*) along the line AC; (*b*) in the liquid area; (*c*) at the triple point A.

7.13 The vapor pressure of water at 25°C is 23.76 torr. (*a*) If 0.360 g of H_2O is placed in an empty rigid container at 25°C with $V = 10.0$ L, state what phase(s) are present at equilibrium and the mass of H_2O in each phase. (*b*) The same as (*a*), except that $V = 20.0$ L. State any approximations you make.

7.14 Ar has normal melting and boiling points of 83.8 and 87.3 K; its triple point is at 83.8 K and 0.7 atm, and its critical temperature and pressure are 151 K and 48 atm. State whether Ar is a solid, liquid, or gas under each of the following conditions: (*a*) 0.9 atm and 90 K; (*b*) 0.7 atm and 80 K; (*c*) 0.8 atm and 88 K; (*d*) 0.8 atm and 84 K; (*e*) 1.2 atm and 83.5 K; (*f*) 1.2 atm and 86 K; (*g*) 0.5 atm and 84 K.

7.15 Figure 3.7 shows a reversible isobaric path from liquid water at −10°C and 1 atm to ice at −10°C and 1 atm. Use Fig. 7.1 to help devise a reversible isothermal path between these two states.

7.16 For each pair, state which substance has the greater $\Delta_{vap}H_m$ at its normal boiling point: (*a*) Ne or Ar; (*b*) H_2O or H_2S; (*c*) C_2H_6 or C_3H_8.

7.17 The normal boiling point of CS_2 is 319.4 K. Estimate $\Delta_{vap}H_m$ and $\Delta_{vap}S_m$ of CS_2 at the normal boiling point using (*a*) Trouton's rule; (*b*) the Trouton–Hildebrand–Everett rule.

7.18 Use the relation between entropy and disorder to explain why the normal-boiling-point $\Delta_{vap}S_m$ of a hydrogen-bonded liquid exceeds the Trouton–Hildebrand–Everett-rule value.

7.19 Given the normal boiling points 81.7 K for CO, 614 K for anthracene, 1691 K for $MgCl_2$, and 2846 K for Cu, (a) estimate $\Delta_{vap}H_{m,nbp}$ of each of these substances as accurately as you can; (b) use the end-of-chapter data sources to find the experimental $\Delta_{vap}H_{m,nbp}$ values and calculate the percent errors in your estimates.

7.20 Consider the following reversible isothermal two-step process: vaporization of one mole of liquid i at $T_{nbp,i}$ and 1 atm to gaseous i with molar volume $V_{m,i}$; volume change of gas i from $V_{m,i}$ to a certain fixed molar volume $V_m^\dagger$. Show that if ΔS_m for the two-step process is assumed to be the same for any liquid, one obtains the Trouton–Hildebrand–Everett rule $\Delta_{vap}S_{m,nbp} = a + R \ln (T_{nbp}/K)$, where a is a constant.

Section 7.3

7.21 True or false? (a) For a reversible phase change at constant T and P, $\Delta S = \Delta H/T$. (b) The relation $d \ln P/dT \approx \Delta H_m/RT^2$ should not be applied to solid–liquid transitions. (c) The relation $d \ln P/dT \approx \Delta H_m/RT^2$ should not be applied to solid–vapor transitions. (d) The relation $d \ln P/dT \approx \Delta H_m/RT^2$ should not be applied near the critical point. (e) $\int_{T_1}^{T_2} (1/T) \, dT = \ln (T_2 - T_1)$. (f) $\int_{T_1}^{T_2} (1/T) \, dT = (\ln T_2)/(\ln T_1)$.

7.22 The normal boiling point of diethyl ether ("ether") is 34.5°C, and its $\Delta_{vap}H_{m,nbp}$ is 6.38 kcal/mol. Find the vapor pressure of ether at 25.0°C. State any approximations made.

7.23 Use the Clapeyron equation and data from Prob. 2.48 to find the pressure at which water freezes at (a) −1.00°C; (b) −10.00°C. (c) The experimental values of these pressures are 131 atm and 1090 atm. Explain why the value you found in (b) is greatly in error.

7.24 The heat of fusion of Hg at its normal melting point, −38.9°C, is 2.82 cal/g. The densities of Hg(s) and Hg(l) at −38.9°C and 1 atm are 14.193 and 13.690 g/cm³, respectively. Find the melting point of Hg at (a) 100 atm; (b) 500 atm.

7.25 (a) Repeat the ethanol example of Sec. 7.3 using the average of the 25°C and 78.3°C $\Delta_{vap}H_m$ values instead of the 78.3°C value. Compare the result with the experimental 25°C vapor pressure. (b) The actual molar volumes of ethanol vapor in the temperature and pressure ranges of this example are less than those predicted by $PV_m = RT$. Will inclusion of nonideality of the vapor improve or worsen the agreement of the result of (a) with the experimental 25°C vapor pressure?

7.26 The average enthalpy of sublimation of $C_{60}(s)$ (buckminsterfullerene) over the range 600 to 800 K was determined by allowing the vapor in equilibrium with the solid at a fixed temperature to leak into a mass spectrometer and measuring the integrated intensity I of the C_{60} peaks. The graph of $\ln (IT/K)$ versus T^{-1} was found to have an average slope of -2.18×10^4 K [C. K. Mathews et al., *J. Phys. Chem.*, **96**, 3566 (1992)]. The solid's vapor pressure can be shown to be proportional to IT (see Prob. 15.34). Find $\Delta_{sub}H_m$ of $C_{60}(s)$ in this temperature range.

7.27 The vapor pressure of water at 25°C is 23.76 torr. Calculate the average value of ΔH_m of vaporization of water over the temperature range 25°C to 100°C.

7.28 ΔH of vaporization of water is 539.4 cal/g at the normal boiling point. (a) Many bacteria can survive at 100°C by forming spores. Most bacterial spores die at 120°C. Hence, autoclaves used to sterilize medical and laboratory instruments are pressurized to raise the boiling point of water to 120°C. At what pressure does water boil at 120°C? (b) What is the boiling point of water at the top of Pike's Peak (altitude 14100 ft), where the atmospheric pressure is typically 446 torr?

7.29 Some vapor pressures of liquid Hg are:

t	80.0°C	100.0°C	120.0°C	140.0°C
P/torr	0.08880	0.2729	0.7457	1.845

(a) Find the average ΔH_m of vaporization over this temperature range from a plot of $\ln P$ versus $1/T$. (b) Find the vapor pressure at 160°C. (c) Estimate the normal boiling point of Hg. (d) Repeat (b) using a spreadsheet Solver to minimize the sums of the squares of the deviations of the calculated P values from the observed values.

7.30 Some vapor pressures of solid CO_2 are:

t	−120.0°C	−110.0°C	−100.0°C	−90.0°C
P/torr	9.81	34.63	104.81	279.5

(a) Find the average ΔH_m of sublimation over this temperature range. (b) Find the vapor pressure at −75°C.

7.31 Use Trouton's rule to show that the change ΔT in normal boiling point T_{nbp} due to a small change ΔP in pressure is roughly $\Delta T \approx T_{nbp} \Delta P/(10\frac{1}{2} \text{ atm})$.

7.32 (a) At 0.01°C, $\Delta_{vap}H_m$ of H_2O is 45.06 kJ/mol, and $\Delta_{fus}H_m$ of ice is 6.01 kJ/mol. Find ΔH_m for sublimation of ice at 0.01°C. (b) Compute the slope dP/dT of each of the three lines at the H_2O triple point. See Prob. 2.48 for further data. State any approximations made.

7.33 Vapor-pressure data vs. temperature are often represented by the *Antoine equation*

$$\ln (P/\text{torr}) = A - B/(T/K + C)$$

where A, B, and C are constants chosen to fit the data and K = 1 kelvin. The Antoine equation is very accurate over a limited vapor-pressure range, typically 10 torr to 1500 torr. For H_2O in the temperature range 11°C to 168°C, Antoine constants are $A = 18.3036$, $B = 3816.44$, $C = -46.13$. (a) Use the Antoine equation to find vapor pressures of H_2O at 25°C and 150°C and compare with the experimental values 23.77 torr and 3569 torr. (b) Use the Antoine equation to calculate $\Delta_{vap}H_m$ of H_2O at 100°C. State any approximations made. (For more accurate results, see Prob. 8.43.)

7.34 The vapor pressure of $SO_2(s)$ is 1.00 torr at 177.0 K and 10.0 torr at 195.8 K. The vapor pressure of $SO_2(l)$ is 33.4 torr at 209.6 K and 100.0 torr at 225.3 K. (a) Find the temperature and pressure of the SO_2 triple point. State any approximations made. (b) Find ΔH_m of fusion of SO_2 at the triple point.

7.35 The normal melting point of Ni is 1452°C. The vapor pressure of liquid Ni is 0.100 torr at 1606°C and 1.00 torr at 1805°C. The molar heat of fusion of Ni is 4.2_5 kcal/mol. Making reasonable approximations, estimate the vapor pressure of solid Ni at 1200°C.

Section 7.4

7.36 At 1000 K and 1 bar, V_m of graphite is 1.97 cm³/mol greater than that of diamond, and $\Delta_f G^\circ_{di} = 6.07$ kJ/mol. Find the pressure of the 1000 K point on the diamond–graphite phase-transition line.

7.37 The stable form of tin at room temperature and pressure is white tin, which has a metallic crystal structure. When tin is used for construction in cold climates, it may be gradually converted to the allotropic gray form, whose structure is nonmetallic. Use Appendix data to find the temperature below which gray tin is the stable form at 1 bar. State any approximations made.

7.38 In Example 7.7 in Sec. 7.4, we found 15100 bar as the 25°C graphite–diamond transition pressure. At 25°C, graphite is more compressible than diamond. If this were taken into account, would we get a transition pressure greater or less than 15100 bar?

Section 7.5

7.39 Sketch H versus T for (a) a first-order transition; (b) a second-order transition; (c) a lambda transition where C_P is infinite at T_λ. [Hint: Use Eq. (4.30).] Repeat the problem for S versus T.

7.40 For β-brass, let the sites occupied by Cu atoms at $T = 0$ be called the C_0 sites and let the $T = 0$ Zn sites be called Z_0. At any temperature T, let r be the number of atoms in right positions (a Cu atom on a C_0 site or a Zn on a Z_0) and let w be the number of atoms on wrong positions (Cu on a Z_0 site or Zn on C_0). The *long-range-order* parameter σ_l is defined as $\sigma_l \equiv (r - w)/(r + w)$. (a) What is σ_l at $T = 0$? (b) What is σ_l if all atoms are on wrong sites? Would this situation be highly ordered or highly disordered? (c) What is σ_l at T_λ, where the number of rightly and the number of wrongly located atoms are equal? (d) What is σ_l in each of the drawings in Fig. 7.12? Let n_p be the total number of nearest-neighbor pairs of atoms in β-brass and let n_{rp} be the number of right nearest-neighbor pairs (Cu–Zn or Zn–Cu). The *short-range-order* parameter σ_s is defined as $\sigma_s \equiv 2n_{rp}/n_p - 1$. (e) What is σ_s at $T = 0$? (f) What is σ_s in the $T \to \infty$ limit of complete disorder? (g) What is σ_s in each of the drawings in Fig. 7.12? Note: Count only nearest-neighbor pairs; these are pairs with one atom at the center of a square and one at the corner of that square. (h) Sketch σ_l and σ_s versus T/T_λ using the results of this problem and the information in Sec. 7.5.

7.41 Explain what is happening in the order–disorder lambda transition in HI(s). (Hint: This molecule is polar.)

7.42 For $T_\lambda - 10^{-2}$ K $< T < T_\lambda - 10^{-9}$ K, measured C_P values of liquid He obey the relation [J. A. Lipa et al., *Phys. Rev. Lett.*, **76**, 944 (1996)]

$$C_P/(J/mol\text{-}K) = (A'/\alpha)t^{-\alpha}(1 + Dt^{1/2} + Et) + B$$

where $A' = 5.7015$, $\alpha = -0.01285$, $D = -0.0228$, $E = 0.323$, $B = 456.28$, and $t \equiv 1 - T/T_\lambda$. For $T_\lambda + 10^{-6}$ K $> T > T_\lambda + 10^{-8}$ K, the same equation holds except that A' is replaced by $A = 6.094$ and t is replaced by $s \equiv T/T_\lambda - 1$. (a) Assume that these expressions hold up to T_λ and find C_P at T_λ. (b) Find $\partial C_P/\partial T$ at T_λ. (c) Use a spreadsheet or other program to graph C_P versus $(T - T_\lambda)/T_\lambda$ in the region close to T_λ. (α is called a *critical exponent*. There is a substantial theory devoted to prediction of α and other critical exponents for continuous phase transitions. See J. J. Binney et al., *The Theory of Critical Phenomena*, Oxford, 1992.)

General

7.43 Which has the higher vapor pressure at −20°C, ice or supercooled liquid water? Explain.

7.44 At the solid–liquid–vapor triple point of a pure substance, which has the greater slope, the solid–vapor or the liquid–vapor line? Explain.

7.45 A beaker at sea level contains pure water. Calculate the difference between the freezing point of water at the surface and water 10 cm below the surface.

7.46 On the sea bottom at the Galápagos Rift, water heated to 350°C gushes out of hydrothermal vents at a depth of 3000 m. Will this water boil or remain liquid at this depth? The vapor pressure of water is 163 atm at 350°C. (The heat of this water is used as an energy source by sulfide-oxidizing bacteria contained in the tissues of tube worms living on the ocean floor.)

7.47 In Prob. 4.29b, we found that ΔG is −2.76 cal/g for the conversion of supercooled water to ice, both at −10°C and 1 atm. The vapor pressure of ice is 1.950 torr at −10°C. Find vapor pressure of supercooled water at −10°C. Neglect the effect of pressure changes on G_m of condensed phases.

7.48 The vapor pressure of liquid water at 0.01°C is 4.585 torr. Find the vapor pressure of ice at 0.01°C.

7.49 (a) Consider a two-phase system, where one phase is pure liquid A and the second phase is an ideal gas mixture of A vapor with inert gas B (assumed insoluble in liquid A). The presence of gas B changes μ_A^l, the chemical potential of liquid A, because B increases the total pressure on the liquid phase. However, since the vapor is assumed ideal, the presence of B does not affect μ_A^g, the chemical potential of A in the vapor phase [see Eq. (6.4)]. Because of its effect on μ_A^l, gas B affects the liquid–vapor equilibrium position, and its presence changes the equilibrium vapor pressure of A. Imagine an isothermal infinitesimal change dP in the total pressure P of the system. Show that this causes a change dP_A in the vapor pressure of A given by

$$\frac{dP_A}{dP} = \frac{V^l_{m,A}}{V^g_{m,A}} = \frac{V^l_{m,A}P_A}{RT} \qquad \text{const. } T \qquad (7.25)$$

Equation (7.25) is often called the Gibbs equation. Because $V^l_{m,A}$ is much less than $V^g_{m,A}$, the presence of gas B at low or moderate pressures has only a small effect on the vapor pressure of A.

(b) The vapor pressure of water at 25°C is 23.76 torr. Calculate the vapor pressure of water at 25°C in the presence of 1 atm of inert ideal gas insoluble in water.

7.50 The vapor pressure of water at 25°C is 23.766 torr. Calculate ΔG_{298}° for the process $H_2O(l) \rightarrow H_2O(g)$. Assume the vapor is ideal. Compare with the value found from data in the Appendix.

7.51 Benzene obeys Trouton's rule, and its normal boiling point is 80.1°C. (a) Use (7.22) to derive an equation for the vapor pressure of benzene as a function of T. (b) Find the vapor pressure of C_6H_6 at 25°C. (c) Find the boiling point of C_6H_6 at 620 torr.

7.52 Some vapor pressures for $H_2O(l)$ are 4.258 torr at −1.00°C, 4.926 torr at 1.00°C, 733.24 torr at 99.00°C, and 787.57 torr at 101.00°C. (a) Calculate ΔH_m, ΔS_m, and ΔG_m for the equilibrium vaporization of $H_2O(l)$ at 0°C and at 100°C. Explain why the calculated 100°C $\Delta_{vap} H_m$ value differs slightly from the true value. (b) Calculate ΔH° and ΔS° for vaporization of $H_2O(l)$ at 0°C; make reasonable approximations. The vapor pressure of water at 0°C is 4.58 torr.

7.53 A solution is prepared by mixing n_A^s moles of the solvent A with n_B^s moles of the solute B. The s stands for stoichiometric and indicates that these mole numbers need not be the number of moles of A and B actually present in the solution, since A and B may undergo reactions in the solution. In the solution, A and B react to form E and F according to

$$aA + bB \rightleftharpoons eE + fF$$

If $b = 0$, this reaction might be dissociation or association of the solvent. If $b \neq 0$, the reaction might be solvation of the solute B. No E or F is added from the outside. Therefore, $c_{ind} = c - r - a = 4 - 1 - 1 = 2$.

Let dn_A^s and dn_B^s moles of A and B be added to the solution from outside. This addition shifts the reaction equilibrium to the right, and the extent of reaction changes by $d\xi$. Since $dn_i = \nu_i \, d\xi$ for a reaction, the actual change in number of moles of A is $dn_A = dn_A^s + \nu_A \, d\xi = dn_A^s - a \, d\xi$. The quantity $\Sigma_i \, \mu_i \, dn_i$ that occurs in the Gibbs equation for dG is $\Sigma_i \, \mu_i \, dn_i = \mu_A \, dn_A + \mu_B \, dn_B + \mu_E \, dn_E + \mu_F \, dn_F$. (a) Show that

$$\sum_i \mu_i \, dn_i = \mu_A \, dn_A^s + \mu_B \, dn_B^s$$

provided reaction equilibrium is maintained. Thus we can take the sum over only the independent components A and B and ignore solvation, association, or dissociation. (b) Use the result of part (a) to show that

$$\mu_A = (\partial G / \partial n_A^s)_{T,P,n_B^s}$$

7.54 For pressures under 10^4 atm, which of the following phases can exist at temperatures arbitrarily close to absolute zero: (a) $H_2O(s)$; (b) $H_2O(l)$; (c) $H_2O(g)$?

7.55 True or false? (a) For a one-component system, the maximum number of phases that can coexist in equilibrium is three. (b) The equation $dP/dT = \Delta H/(T \, \Delta V)$ is exact. (c) The equation $d \ln P/dT = \Delta H_m/RT^2$ is exact. (d) When three phases coexist in equilibrium in a one-component system, one of the phases must be a gas, one must be a liquid, and one must be a solid. (e) For a one-component system, the most stable phase at a given T and P is the phase with the lowest G_m. (f) Solid H_2O cannot exist at 100°C as a stable phase. (g) For a pure substance, the vapor pressure of the solid is equal to the vapor pressure of the liquid at the triple-point temperature. (h) Liquid water cannot exist at 1 atm and 150°C. (i) If phases α and β of a closed system are in equilibrium, then μ^{α} must equal μ^{β}.

CHAPTER

8

Real Gases

8.1 COMPRESSION FACTORS

An ideal gas obeys the equation of state $PV_m = RT$. This chapter examines the P-V-T behavior of real gases.

As a measure of the deviation from ideality of the behavior of a real gas, we define the **compressibility factor** or **compression factor** Z of a gas as

$$Z(P, T) \equiv PV_m/RT \qquad (8.1)$$

Do not confuse the compressibility factor Z with the isothermal compressibility κ. Since V_m in (8.1) is a function of T and P (Sec. 1.7), Z is a function of T and P. For an ideal gas, $Z = 1$ for all temperatures and pressures. Figure 8.1a shows the variation of Z with P at 0°C for several gases. Figure 8.1b shows the variation of Z with P for CH_4 at several temperatures. Note that $Z = V_m/V_m^{id}$ and that $Z = P/P_{id}$, where V_m^{id} is the molar volume of an ideal gas at the same T and P as the real gas, and P_{id} is the pressure of an ideal gas at the same T and V_m as the real gas. When $Z < 1$, the gas exerts a lower pressure than an ideal gas would. Figure 8.1b shows that, at high pressures, P of a gas can easily be 2 or 3 times larger or smaller than P_{id}.

The curves of Fig. 8.1 show that ideal behavior ($Z = 1$) is approached in the limit $P \to 0$ and also in the limit $T \to \infty$. For each of these limits, the gas volume goes to infinity for a fixed quantity of gas, and the density goes to zero. Deviations from ideality are due to intermolecular forces and to the nonzero volume of the molecules themselves. At zero density, the molecules are infinitely far apart, and intermolecular forces are zero. At infinite volume, the volume of the molecules themselves is negligible compared with the infinite volume the gas occupies. Hence the ideal-gas equation of state is obeyed in the limit of zero gas density.

Figure 8.1

(a) Compression factors of some gases at 0°C. (b) Methane compression factors at several temperatures.

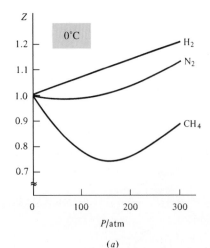

(a)

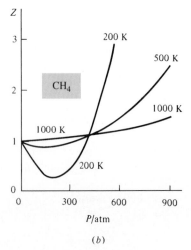

(b)

A real gas then obeys $PV = ZnRT$. Numerical tables of $Z(P, T)$ are available for many gases.

REAL-GAS EQUATIONS OF STATE

An algebraic formula for the equation of state of a real gas is more convenient to use than numerical tables of Z. The best-known such equation is the **van der Waals equation**

$$\left(P + \frac{a}{V_{\mathrm{m}}^2} \right)(V_{\mathrm{m}} - b) = RT \qquad \text{or} \qquad P = \frac{RT}{V_{\mathrm{m}} - b} - \frac{a}{V_{\mathrm{m}}^2} \qquad (8.2)$$

where the first equation was divided by $V_{\mathrm{m}} - b$ to solve for P. In addition to the gas constant R, the van der Waals equation contains two other constants, a and b, whose values differ for different gases. A method for determining a and b values is given in Sec. 8.4. The term a/V_{m}^2 in (8.2) is meant to correct for the effect of intermolecular attractive forces on the gas pressure. This term decreases as V_{m} and the average intermolecular distance increase. The nonzero volume of the molecules themselves makes the volume available for the molecules to move in less than V, so some volume b is subtracted from V_{m}. The volume b is roughly the same as the molar volume of the solid or liquid, where the molecules are close together; b is roughly the volume excluded by intermolecular repulsive forces. The van der Waals equation is a major improvement on the ideal-gas equation but is unsatisfactory at very high pressures and its overall accuracy is mediocre.

A quite accurate two-parameter equation of state for gases is the **Redlich–Kwong equation** [O. Redlich and J. N. S. Kwong, *Chem. Rev.*, **44**, 233 (1949)]:

$$P = \frac{RT}{V_{\mathrm{m}} - b} - \frac{a}{V_{\mathrm{m}}(V_{\mathrm{m}} + b)T^{1/2}} \qquad (8.3)$$

which is useful over very wide ranges of T and P. The Redlich–Kwong parameters a and b differ in value for any given gas from the van der Waals a and b.

Statistical mechanics shows (see Sec. 22.11) that the equation of state of a real gas not at very high pressure can be expressed as the following power series in $1/V_{\mathrm{m}}$:

$$PV_{\mathrm{m}} = RT \left[1 + \frac{B(T)}{V_{\mathrm{m}}} + \frac{C(T)}{V_{\mathrm{m}}^2} + \frac{D(T)}{V_{\mathrm{m}}^3} + \cdots \right] \qquad (8.4)$$

This is the **virial equation of state.** The coefficients $B, C, \ldots$, which are functions of T only, are the **second, third, . . . virial coefficients.** They are found from experimental P-V-T data of gases (Probs. 8.38 and 10.65). Usually, the limited accuracy of the data allows evaluation of only $B(T)$ and sometimes $C(T)$. Figure 8.2 plots the typical behavior of B and C versus T. Some values of $B(T)$ for Ar are

$B/(\mathrm{cm}^3/\mathrm{mol})$	-251	-184	-86	-47	-28	-16	-1	7	12	22
T/K	85	100	150	200	250	300	400	500	600	1000

Statistical mechanics gives equations relating the virial coefficients to the potential energy of intermolecular forces.

A form of the virial equation equivalent to (8.4) uses a power series in P:

$$PV_{\mathrm{m}} = RT[1 + B^{\dagger}(T)P + C^{\dagger}(T)P^2 + D^{\dagger}(T)P^3 + \cdots] \qquad (8.5)$$

The relations between the coefficients $B^{\dagger}, C^{\dagger}, \ldots$ and $B, C, \ldots$ in (8.4) are worked out in Prob. 8.4. One finds

$$B = B^{\dagger}RT, \qquad C = (B^{\dagger 2} + C^{\dagger})R^2T^2 \qquad (8.6)$$

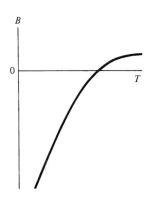

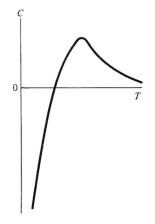

Figure 8.2

Typical temperature variation of the second and third virial coefficients $B(T)$ and $C(T)$.

If P is not high, terms beyond C/V_m^2 or $C^\dagger P^2$ in (8.4) and (8.5) are usually negligible and can be omitted. At high pressures, the higher terms become important. At very high pressures, the virial equation fails. For gases at pressures up to a few atmospheres, one can drop terms after the second term in (8.4) and (8.5), provided T is not very low; Eq. (8.5) becomes

$$V_m = RT/P + B \qquad \text{low } P \tag{8.7}$$

where (8.6) was used. Equation (8.7) gives a convenient and accurate way to correct for gas nonideality at low P. Equation (8.7) shows that at low P, the second virial coefficient $B(T)$ is the correction to the ideal-gas molar volume RT/P. For example, for $Ar(g)$ at 250.00 K and 1.0000 atm, the truncated virial equation (8.7) and the preceding table of Ar B values gives $V_m = RT/P + B = 20515 \text{ cm}^3/\text{mol} - 28 \text{ cm}^3/\text{mol} = 20487 \text{ cm}^3/\text{mol}$.

Multiplication of the van der Waals equation (8.2) by V_m/RT gives the compression factor $Z \equiv PV_m/RT$ of a van der Waals (vdW) gas as

$$\frac{PV_m}{RT} = Z = \frac{V_m}{V_m - b} - \frac{a}{RTV_m} = \frac{1}{1 - b/V_m} - \frac{a}{RTV_m} \qquad \text{vdW gas}$$

where the numerator and denominator of the first fraction were divided by V_m. Since $1/(1 - b/V_m)$ is greater than 1, intermolecular repulsions (represented by b) tend to make Z greater than 1 and P greater than P_{id}. Since $-a/RTV_m$ is negative, intermolecular attractions (represented by a) tend to decrease Z and make P less than P_{id}.

b is approximately the liquid's molar volume, so we will have $b < V_m$ for the gas and $b/V_m < 1$. We can therefore use the following expansion for $1/(1 - b/V_m)$:

$$\frac{1}{1 - x} = 1 + x + x^2 + x^3 + \cdots \qquad \text{for } |x| < 1 \tag{8.8}$$

You may recall (8.8) from your study of geometric series. Equation (8.8) can also be derived as a Taylor series (Sec. 8.9). The use of (8.8) with $x = b/V_m$ gives

$$\frac{PV_m}{RT} = Z = 1 + \left(b - \frac{a}{RT}\right)\frac{1}{V_m} + \frac{b^2}{V_m^2} + \frac{b^3}{V_m^3} + \cdots \qquad \text{vdW gas} \tag{8.9}$$

The van der Waals equation now has the same form as the virial equation (8.4). The van der Waals prediction for the second virial coefficient is $B(T) = b - a/RT$.

At low pressures, V_m is much larger than b and the b^2/V_m^2, b^3/V_m^3, . . . terms can be neglected to give $Z \approx 1 + (b - a/RT)/V_m$. At low T (and low P), we have $a/RT > b$, so $b - a/RT$ is negative, Z is less than 1, and P is less than P_{id} (as in the low-P parts of the 200-K and 500-K CH_4 curves in Fig. 8.1b). At low T, intermolecular attractions (van der Waals a) are more important than intermolecular repulsions (van der Waals b) in determining P. At high T (and low P), we have $b - a/RT > 0$, $Z > 1$, and $P > P_{id}$ (as in the 1000-K curve in Fig. 8.1b). At high T, the molecules smash into each other harder than at low T, which increases the influence of repulsions on P.

Similar to the effect of T on Z at low P, the Joule–Thomson coefficient μ_{JT} changes from positive to negative as T increases (Fig. 2.9), because of the increased influence of repulsions; see Prob. 8.39.

A comparison of equations of state for gases [K. K. Shah and G. Thodos, *Ind. Eng. Chem.*, **57**(3), 30 (1965)] concluded that the Redlich–Kwong equation is the best two-parameter equation of state. Because of its simplicity and accuracy, the Redlich–Kwong equation has been widely used.

Gas Mixtures

So far we have considered pure real gases. For a real gas mixture, V depends on the mole fractions, as well as on T and P. One approach to the P-V-T behavior of real gas mixtures is to use a two-parameter equation of state like the van der Waals or Redlich–Kwong with the parameters a and b taken as functions of the mixture's composition. For a mixture of two gases, 1 and 2, one often takes

$$a = x_1^2 a_1 + 2x_1 x_2 (a_1 a_2)^{1/2} + x_2^2 a_2 \quad \text{and} \quad b = x_1 b_1 + x_2 b_2 \qquad (8.10)$$

where x_1 and x_2 are the mole fractions of the components. b is related to the molecular size, so b is taken as a weighted average of b_1 and b_2. The parameter a is related to intermolecular attractions. The quantity $(a_1 a_2)^{1/2}$ is an estimate of what the intermolecular interaction between gas 1 and gas 2 molecules might be. In applying an equation of state to a mixture, V_m is interpreted as the **mean molar volume** of the system, defined by

$$V_m \equiv V/n_{tot} \qquad (8.11)$$

For the virial equation of state, the second virial coefficient for a mixture of two gases is $B = x_1^2 B_1 + 2x_1 x_2 B_{12} + x_2^2 B_2$, where B_{12} is best determined from experimental data on the mixture, but can be crudely estimated as $B_{12} \approx \frac{1}{2}(B_1 + B_2)$.

The **mixing rule** (8.10) works well only if the molecules of gases 1 and 2 are similar (for example, two hydrocarbons). To improve performance, a in (8.10) is often modified to $a = x_1^2 a_1 + 2x_1 x_2 (1 - k_{12})(a_1 a_2)^{1/2} + x^2 a_2$, where k_{12} is a constant whose value is found by fitting experimental data for gases 1 and 2 and differs for different pairs of gases. Many other mixing rules have been proposed [see P. Ghosh, *Chem. Eng. Technol.*, **22**, 379 (1999)].

8.3 CONDENSATION

Provided T is below the critical temperature, any real gas condenses to a liquid when the pressure is increased sufficiently. Figure 8.3 plots several isotherms for H_2O on a P-V diagram. (These isotherms correspond to vertical lines on the P-T phase diagram of Fig. 7.1.) For temperatures below 374°C, the gas condenses to a liquid when P is increased. Consider the 300°C isotherm. To go from R to S, we slowly push in the piston, decreasing V and V_m and increasing P, while keeping the gas in a constant-temperature bath. Having reached S, we now observe that pushing the piston further in causes some of the gas to liquefy. As the volume is further decreased, more of the

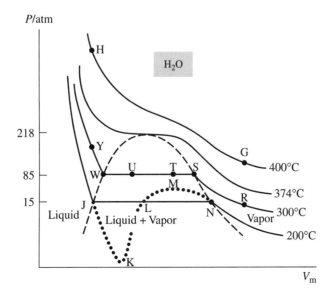

Figure 8.3

Isotherms of H_2O (solid lines). Not drawn to scale. The dashed line separates the two-phase region from one-phase regions. The critical point is at the top of the dashed line and has $V_m = 56$ cm³/mol. For the two-phase region, $V_m = V/n_{tot}$. (The dotted curve shows the behavior of a van der Waals or Redlich–Kwong isotherm in the two-phase region; see Sec. 8.4.)

Figure 8.4

Condensation of a gas. The system is surrounded by a constant-T bath (not shown).

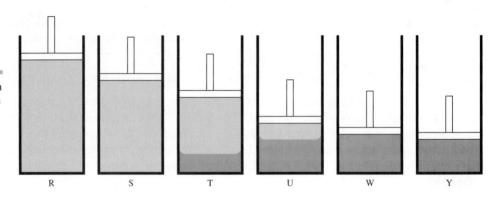

R   S   T   U   W   Y

gas liquefies until at point W we have all liquid. See Fig. 8.4. For all points between S and W on the isotherm, two phases are present. Moreover, the gas pressure above the liquid (its vapor pressure) remains constant for all points between S and W. (The terms **saturated vapor** and **saturated liquid** refer to a gas and liquid in equilibrium with each other; from S to W, the vapor and liquid phases are saturated.) Going from W to Y by pushing in the piston still further, we observe a steep increase in pressure with a small decrease in volume; liquids are relatively incompressible. The isotherm RSTUWY in Fig. 8.3 corresponds to the vertical line RSY in Fig. 7.1.

Above the critical temperature (374°C for water) no amount of compression will cause the separation out of a liquid phase in equilibrium with the gas. As we approach the critical isotherm from below, the length of the horizontal portion of an isotherm where liquid and gas coexist decreases until it reaches zero at the critical point. The molar volumes of saturated liquid and gas at 300°C are given by the points W and S. As T is increased, the difference between molar volumes of saturated liquid and gas decreases, becoming zero at the critical point (Fig. 7.2).

The pressure, temperature, and molar volume at the critical point are the **critical pressure** P_c, the **critical temperature** T_c, and the **critical (molar) volume** $V_{m,c}$. Table 8.1 lists some data.

For most substances, T_c is roughly 1.6 times the absolute temperature T_{nbp} of the normal boiling point: $T_c \approx 1.6T_{nbp}$. Also, $V_{m,c}$ is usually about 2.7 times the normal-boiling-point molar volume $V_{m,nbp}$. P_c is typically 10 to 100 atm. Above T_c, the molecular kinetic energy (whose average value is $\frac{3}{2}kT$ per molecule) is large enough to overcome the forces of intermolecular attraction, and no amount of pressure will liquefy the gas. At T_{nbp}, the fraction of molecules having sufficient kinetic energy to escape from intermolecular attractions is large enough to make the vapor pressure equal to 1 atm. Both T_c and T_{nbp} are determined by intermolecular forces, so T_c and T_{nbp} are correlated.

Usually one thinks of converting a gas to a liquid by a process that involves a sudden change in density between gas and liquid, so that we go through a two-phase region in the liquefaction process. For example, for the isotherm RSTUWY in Fig. 8.3, two phases are present for points between S and W: a gas phase of molar volume V_{mS} and a liquid phase of molar volume V_{mW}. (Because T and P are constant along SW, the gas and liquid molar volumes each remain constant along SW. The actual amounts of gas and liquid change in going from S to W, so the actual volumes of gas and liquid vary along SW.) Since $V_{mS} > V_{mW}$, the gas density is less than the liquid density. However, as noted in Sec. 7.2, one can change a gas into a liquid by a process in which there is always present only a single phase whose density shows no discontinuous changes. For example, in Fig. 8.3, we could go vertically from R to G, then isothermally to H, and finally vertically to Y. We end up with a liquid at Y but, during the process RGHY, the system's properties vary continuously and there is no point at which we could say that the system changes from gas to liquid.

TABLE 8.1

Critical Constants

Species	T_c/K	P_c/atm	$V_{m,c}$/(cm³/mol)	Species	T_c/K	P_c/atm	$V_{m,c}$/(cm³/mol)
Ne	44.4	27.2	41.7	CO_2	304.2	72.88	94.0
Ar	150.9	48.3	74.6	HCl	324.6	82.0	81.
N_2	126.2	33.5	89.5	CH_3OH	512.5	80.8	117.
H_2O	647.1	217.8	56.0	$n\text{-}C_8H_{18}$	568.8	24.5	492.
D_2O	643.9	213.9	56.2	C_3H_8	369.8	41.9	203.
H_2S	373.2	88.2	98.5	I_2	819.	115.	155.
				Ag	7480.	5000.	58.

Thus there is a continuity between the gaseous and the liquid states. In recognition of this continuity, the term **fluid** is used to mean either a liquid or a gas. What is ordinarily called a liquid can be viewed as a very dense gas. Only when both phases are present in the system is there a clear-cut distinction between liquid and gaseous states. However, for a single-phase fluid system it is customary to define as a **liquid** a fluid whose temperature is below the critical temperature T_c and whose molar volume is less than $V_{m,c}$ (so that its density is greater than the critical density). If these two conditions are not met, the fluid is called a **gas.** Some people make a further distinction between *gas* and *vapor*, but we shall use these words interchangeably.

A **supercritical fluid** is one whose temperature T and pressure P satisfy $T > T_c$ and $P > P_c$. A supercritical fluid usually has a liquidlike density but its viscosity (Sec. 16.3) is much lower than typical for a liquid and diffusion coefficients (Sec. 16.4) in it are much higher than in liquids. Supercritical CO_2 is used commercially as a solvent to decaffeinate coffee. Supercritical and near-critical water are good solvents for organic compounds and are being studied as environmentally friendly solvents for organic reactions (*Chem. Eng. News,* Jan. 3, 2000, p. 26).

8.4 CRITICAL DATA AND EQUATIONS OF STATE

Critical-point data can be used to find values for parameters in equations of state such as the van der Waals equation. Along a horizontal two-phase line such as WS in Fig. 8.3, the isotherm has zero slope; $(\partial P/\partial V_m)_T = 0$ along WS. The critical point is the limiting point of a series of such horizontal two-phase lines. Therefore $(\partial P/\partial V_m)_T = 0$ holds at the critical point. Figure 8.3 shows that along the critical isotherm (374°C) the slope $(\partial P/\partial V_m)_T$ is zero at the critical point and is negative on both sides of it. Hence the function $(\partial P/\partial V_m)_T$ is a maximum at the critical point. When a function of V_m is a maximum at a point, its derivative with respect to V_m is zero at that point. Therefore, $(\partial/\partial V_m)_T(\partial P/\partial V_m)_T \equiv (\partial^2 P/\partial V_m^2)_T = 0$ at the critical point. Thus

$$(\partial P/\partial V_m)_T = 0 \quad \text{and} \quad (\partial^2 P/\partial V_m^2)_T = 0 \qquad \text{at the critical point} \qquad (8.12)$$

These conditions enable us to determine parameters in equations of state.

For example, differentiating the van der Waals equation (8.2), we get

$$\left(\frac{\partial P}{\partial V_m}\right)_T = -\frac{RT}{(V_m - b)^2} + \frac{2a}{V_m^3} \quad \text{and} \quad \left(\frac{\partial^2 P}{\partial V_m^2}\right)_T = \frac{2RT}{(V_m - b)^3} - \frac{6a}{V_m^4}$$

Application of the conditions (8.12) then gives

$$\frac{RT_c}{(V_{m,c} - b)^2} = \frac{2a}{V_{m,c}^3} \quad \text{and} \quad \frac{RT_c}{(V_{m,c} - b)^3} = \frac{3a}{V_{m,c}^4} \qquad (8.13)$$

Moreover, the van der Waals equation itself gives at the critical point

$$P_c = \frac{RT_c}{V_{m,c} - b} - \frac{a}{V_{m,c}^2} \tag{8.14}$$

Division of the first equation in (8.13) by the second yields $V_{m,c} - b = 2V_{m,c}/3$, or

$$V_{m,c} = 3b \tag{8.15}$$

Use of $V_{m,c} = 3b$ in the first equation in (8.13) gives $RT_c/4b^2 = 2a/27b^3$, or

$$T_c = 8a/27Rb \tag{8.16}$$

Substitution of (8.15) and (8.16) into (8.14) gives $P_c = (8a/27b)/2b - a/9b^2$, or

$$P_c = a/27b^2 \tag{8.17}$$

We thus have *three* equations [(8.15) to (8.17)] relating the three critical constants P_c, $V_{m,c}$, T_c to the *two* parameters to be determined, a and b. If the van der Waals equation were accurately obeyed in the critical region, it would not matter which two of the three equations were used to solve for a and b. However, this is not the case, and the values of a and b obtained depend on which two of the three critical constants are used. It is customary to choose P_c and T_c, which are more accurately known than $V_{m,c}$. Solving (8.16) and (8.17) for a and b, we get

$$b = RT_c/8P_c, \qquad a = 27R^2T_c^2/64P_c \qquad \text{vdW gas} \tag{8.18}$$

Some van der Waals a and b values calculated from Eq. (8.18) and P_c and T_c data of Table 8.1 are:

Gas	Ne	N_2	H_2O	HCl	CH_3OH	n-C_8H_{18}
$10^{-6}a/(\text{cm}^6 \text{ atm mol}^{-2})$	0.21	1.35	5.46	3.65	9.23	37.5
$b/(\text{cm}^3 \text{ mol}^{-1})$	16.7	38.6	30.5	40.6	65.1	238

From (8.15), $V_{m,c} = 3b$. Also, $V_{m,c} \approx 2.7V_{m,nbp}$ (Sec 8.3), where $V_{m,nbp}$ is the liquid's molar volume at its normal boiling point. Therefore b is roughly the same as $V_{m,nbp}$ (as noted in Sec. 8.2). $V_{m,nbp}$ is a bit more than the volume of the molecules themselves. Note from the tabulated b values that the larger the molecule, the greater the b value. Recall that the van der Waals a is related to intermolecular attractions. The greater the intermolecular attraction, the greater the a value.

Combination of (8.15) to (8.17) shows that the van der Waals equation predicts for the compressibility factor at the critical point

$$Z_c \equiv P_cV_{m,c}/RT_c = \tfrac{3}{8} = 0.375 \tag{8.19}$$

This may be compared with the ideal-gas prediction $P_cV_{m,c}/RT_c = 1$. Of the known Z_c values, 80% lie between 0.25 and 0.30, significantly less than predicted by the van der Waals equation. The smallest known Z_c is 0.12 for HF; the largest is 0.46 for CH_3NHNH_2.

For the Redlich–Kwong equation, a similar treatment gives (the algebra is complicated, so the derivation is omitted)

$$a = R^2T_c^{5/2}/9(2^{1/3} - 1)P_c = 0.42748R^2T_c^{5/2}/P_c \tag{8.20}$$

$$b = (2^{1/3} - 1)RT_c/3P_c = 0.08664RT_c/P_c \tag{8.21}$$

$$P_cV_{m,c}/RT_c = \tfrac{1}{3} = 0.333 \tag{8.22}$$

To use a two-parameter equation of state, we need to know the substance's critical pressure and temperature, so as to evaluate the parameters. If P_c and T_c are unknown, they can be estimated to within a few percent by group-contribution methods (Sec. 5.10); see *Reid, Prausnitz, and Poling,* sec. 2-2.

Since there is a continuity between the liquid and gaseous states, it should be possible to develop an equation of state that would apply to liquids as well as gases. The van der Waals equation fails to reproduce the isotherms in the liquid region of Fig. 8.3. The Redlich–Kwong equation does work fairly well in the liquid region for some liquids. Of course, this equation does not reproduce the horizontal portion of isotherms in the two-phase region of Fig. 8.3. The slope $(\partial P/\partial V_m)_T$ is discontinuous at points S and W in the figure. A simple algebraic expression like the Redlich–Kwong equation will not have such discontinuities in $(\partial P/\partial V_m)_T$. What happens is that a Redlich–Kwong isotherm oscillates in the two-phase region (Fig. 8.3). The *Peng–Robinson* equation of state (Prob. 8.16) is an improvement on the Redlich–Kwong equation and works well for liquids as well as gases; see D.-Y. Peng and D. B. Robinson, *Ind. Eng. Chem. Fundam.,* **15,** 59 (1976).

Hundreds of equations of state have been proposed in recent years, especially by chemical engineers. Many of these are modifications of the Redlich–Kwong equation. An equation that is superior for predicting *P-V-T* behavior of gases may be inferior for predicting vapor–liquid equilibrium behavior, so it is difficult to identify one equation of state as the best overall. For reviews of equations of state, see C. Tsonopoulos and J. L. Heidman, *Fluid Phase Equil.,* **24,** 1 (1985); A. Anderko, *Fluid Phase Equil.,* **61,** 145 (1990).

The van der Waals and Redlich–Kwong equations are **cubic equations of state,** meaning that when they are cleared of fractions, V_m appears in terms proportional to V_m^3, V_m^2, and V_m only. A cubic algebraic equation always has three roots. Hence when a cubic equation of state (eos) is solved for V_m at a fixed T and P, three values of V_m will satisfy the equation. At a temperature above the critical temperature T_c, two of the roots will be complex numbers and one will be a real number so there is a single real V_m that satisfies the eos. At T_c, the eos has three equal real roots. Below T_c, there will be three unequal real roots. A cubic eos isotherm in the two-phase region below T_c will resemble the dotted line in Fig. 8.3, which has three values of V_m that satisfy the eos at the fixed condensation pressure, namely, the V_m values at points J, L, and N. The V_m values at J and N correspond to V_m of the liquid and V_m of the gas, respectively, that are in equilibrium with each other. The value of V_m at L has no physical significance.

The portion of the eos dotted-line isotherm from J to the minimum at K corresponds to liquid that is at 200°C but is at a pressure lower than the 200°C vapor pressure of 15 atm. Such a point lies below the liquid–vapor equilibrium line in Fig. 7.1 and hence the liquid is in a metastable superheated state (Sec. 7.4) at points between J and K. Likewise, the dotted-line isotherm portion NM corresponds to supercooled vapor. The isotherm portion KLM has $(\partial P/\partial V_m)_T > 0$. As noted after Eq. (1.44), $(\partial V_m/\partial P)_T = 1/(\partial P/\partial V_m)_T$ must be negative, so the portion KLM has no physical significance.

At some temperatures, part of the JK Redlich–Kwong or van der Waals isotherm goes below $P = 0$, indicating negative pressures for the superheated liquid. This is nothing to be alarmed about. In fact, liquids can exist in a metastable state under tension, which corresponds to a negative pressure. For water, negative pressures of hundreds of atmospheres have been observed. Sap in plants is at a negative pressure (P. G. Debenedetti, *Metastable Liquids,* Princeton, 1996, sec. 1.2.3). According to the *cohesion–tension theory* of sap ascent in plants, water in plants is pulled upward by negative pressures created by evaporation of water from the leaves; the term *cohesion* refers to intermolecular hydrogen bonding that holds the water molecules together in the liquid, allowing for large tensions. Direct measurements of negative pressures in plants support the cohesion–tension theory [C. F. Wei et al., *Plant Physiol.,* **121,** 1191 (1999); *Trends in Plant Sci.,* **4,** 372 (1999)].

8.5   CALCULATION OF LIQUID–VAPOR EQUILIBRIA

At any given temperature T, an equation of state can be used to predict the vapor pressure P, the molar volumes V_m^l and V_m^v of the liquid and vapor in equilibrium, and the enthalpy of vaporization of a substance.

For the 200°C isotherm in Fig. 8.3, the points J and N correspond to liquid and vapor in equilibrium. The phase-equilibrium condition is the equality of chemical potentials of the substance in the two phases: $\mu_J^l = \mu_N^v$ or $G_{m,J}^l = G_{m,N}^v$, since $\mu = G_m$ for a pure substance. Dropping the J and N subscripts, we have $G_m^l = G_m^v$ or in terms of the Helmholtz function A:

$$A_m^l + PV_m^l = A_m^v + PV_m^v$$

$$P(V_m^v - V_m^l) = -(A_m^v - A_m^l) \tag{8.23}$$

The Gibbs equation $dA_m = -S_m\, dT - P\, dV_m$ at constant T gives $dA_m = -P\, dV_m$ and integration from point J to N along the path JKLMN gives

$$A_m^v - A_m^l = -\int_{V_m^l}^{V_m^v} P_{eos}\, dV_m \qquad \text{const. } T$$

where eos indicates that the integral is evaluated along the equation-of-state isotherm JKLMN. Equation (8.23) becomes

$$P(V_m^v - V_m^l) = \int_{V_m^l}^{V_m^v} P_{eos}\, dV_m \qquad \text{const. } T \tag{8.24}$$

The left side of (8.24) is the area of a rectangle whose top edge is the horizontal line JLN of length $(V_m^v - V_m^l)$ in Fig. 8.3 and whose bottom edge lies on the $P = 0$ (horizontal) axis. The right side of (8.24) is the area under the dotted line JKLMN. This area will equal the rectangular area only if the areas of the regions labeled I and II in Fig. 8.5 are equal (*Maxwell's equal-area rule*).

For the Redlich–Kwong equation (8.3), Eq. (8.24) becomes

$$P(V_m^v - V_m^l) = \int_{V_m^l}^{V_m^v} \left[\frac{RT}{V_m - b} - \frac{a}{V_m(V_m + b)T^{1/2}} \right] dV_m \qquad \text{const. } T$$

$$P = \frac{1}{V_m^v - V_m^l} \left[RT \ln \frac{V_m^v - b}{V_m^l - b} - \frac{a}{bT^{1/2}} \ln \frac{V_m^v(V_m^l + b)}{(V_m^v + b)V_m^l} \right] \tag{8.25}$$

where the identity $\int [v(v + b)]^{-1}\, dv = b^{-1} \ln [v/(v + b)]$ was used. In addition to satisfying (8.25), the Redlich–Kwong equation (8.3) must be satisfied at point J for the liquid and at point N for the vapor, giving the equations

$$P = \frac{RT}{V_m^l - b} - \frac{a}{V_m^l(V_m^l + b)T^{1/2}} \quad \text{and} \quad P = \frac{RT}{V_m^v - b} - \frac{a}{V_m^v(V_m^v + b)T^{1/2}} \tag{8.26}$$

We have to solve the three simultaneous equations (8.25) and (8.26) for the three unknowns: the vapor pressure P and the liquid and vapor molar volumes V_m^l and V_m^v. Example 8.1 shows how this is done using the Excel spreadsheet.

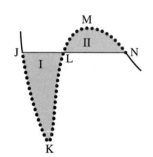

Figure 8.5

JKLMN is a cubic-equation-of-state isotherm in the liquid–vapor region on a P-versus-V_m plot (Fig. 8.3). Areas I and II must be equal.

EXAMPLE 8.1   Prediction of vapor pressure from an equation of state

Use the Redlich–Kwong equation to estimate the vapor pressure and the saturated liquid and vapor molar volumes of C_3H_8 at 25°C.

Equations (8.20) and (8.21) and the critical constants in Table 8.1 give the propane Redlich–Kwong constants as $a = 1.80_7 \times 10^8$ cm^6 atm K$^{1/2}$ mol^{-2} and $b = 62.7$ cm^3/mol. To get initial estimates of the unknowns (which are needed to use the Solver), we graph the propane 25°C Redlich–Kwong isotherm. The values of a, b, R, and T are entered on the spreadsheet (Fig. 8.6) using a consistent set of units (in this case, atm, cm^3, mol, and K). The volumes are entered in column A and the Redlich–Kwong formula (8.3) for the pressure is entered into cell B9 and copied to the cells below B9. The Redlich–Kwong pressure in (8.3) becomes infinite at $V_m = b$ and the liquid's volume must be somewhat greater than the b value of 62.7 cm^3/mol. If we start the V_m column with 65 cm^3/mol in A9, we get a pressure of 9377 atm in B9. The propane critical constants in Table 8.1 show that 25°C is below T_c and the 25°C vapor pressure must be below $P_c = 42$ atm. We therefore increase V_m in A9 until a more reasonable pressure is found. At 95 cm^3/mol, we get a 59 atm pressure, which is a reasonable starting point. When the graph is made, one finds that as V_m is increased above 95 cm^3/mol, P initially changes rapidly and then more slowly. Hence to get a good graph, we initially use a smaller interval ΔV_m. Cell A10 contains the formula =A9+5, which is copied to A11 through A18. A19 contains the formula =A18+15, which is copied to cells below. At higher V_m, the interval can be further increased.

The table and graph for the isotherm show a local maximum pressure (corresponding to point M in Fig. 8.5) of 19 atm and a minimum pressure of -55 atm. The vapor pressure must be above zero, and Fig. 8.5 shows that it must be below the maximum at M of 19 atm. We shall arbitrarily average these limits and take an initial estimate of the vapor pressure as 9.5 atm. (One could also try and draw the horizontal line JN to satisfy the equal-area rule, but this is not so easy

	A	B	C	D	E	F	G
1	Propane	a=	1.807E+08	b =	62.7	Rr =	82.06
2	Redlich-		T =	298.15			
3	Kwong	P/atm =	9.5	Vv/cm3/mol =	2150	VL/cm3/mol	100
4	isotherm	P1/atm =	10.744183	P2/atm =	12.7194	P3/atm =	9.5216705
5		P1err=	0.1309666	P2err =	0.338884	P3err=	0.0022811
6							
7	Vm/	P/					
8	cm3/mol	atm					
9	95	58.9365					
10	100	12.7194					
11	105	-15.9204					
12	110	-33.6228					
13	115	-44.2961					
14	120	-50.3483					
15	125	-53.3167					
16	130	-54.21					
17	135	-53.7049					
18	140	-52.2633					
19	155	-45.0624					
20	170	-36.5258					

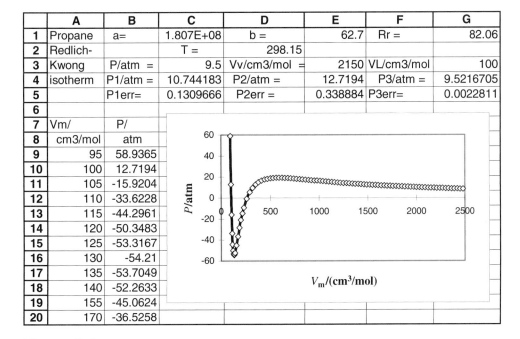

Figure 8.6

Spreadsheet for finding vapor pressure from an equation of state.

to do.) To get the initial estimates of the liquid and vapor molar volumes, we need the isotherm's minimum and maximum volumes that correspond to $P = 9.5$ atm (points J and N). The spreadsheet table shows that at 100 cm³/mol, P is 12.7 atm, which is fairly close to 9.5 atm, so we take 100 cm³/mol as the initial guess for V_m^l. The table also shows that at 2150 cm³/mol, P is close to 9.5 atm, which gives the initial guess for V_m^v. [Of course, the 320 cm³/mol value where $P = 9.5$ atm is ignored (point L).]

These initial guesses for the three unknowns are entered into cells C3, E3, and G3 (Fig. 8.6). The right side of Eq. (8.25) is entered as a formula in C4, and the right sides of the equations in (8.26) are entered as formulas in E4 and G4. For example, G4 contains the formula `=G1*D2/(E3-E1)-C1/(E3*(E3+E1)*D2^0.5)`. The formulas `=(C4-C3)/C3`, `=(E4-C3)/C3`, `=(G4-C3)/C3` for the errors are entered in C5, E5, and G5. The Solver is set up to make C5 equal to zero by changing C3, E3, and G3, subject to the constraints that E5 and G5 equal zero, that C3 be positive and less than 19 atm, that G3 be greater than 95 and less than 105 (the value in the table where the pressure first becomes negative), and that E3 be greater than 600 (the volume at the maximum point M).

With these settings, the Solver quickly converges to the solution $P = 10.85$ atm, $V_m^l = 100.3$ cm³/mol, $V_m^v = 1823$ cm³/mol. The experimental values are 9.39 atm, 89.5 cm³/mol, and 2136 cm³/mol. The Redlich–Kwong results are neither very bad nor very good. If the van der Waals equation is used, the results (16.6 atm, 141.6 cm³/mol, 1093 cm³/mol) are very poor (Prob. 8.14). If the Peng–Robinson equation is used, the results (9.39 atm, 86.1 cm³/mol, 2140 cm³/mol) are quite good (Prob. 8.16). The Peng–Robinson and the Soave–Redlich–Kwong (Prob. 8.15) equations are widely used to predict liquid–vapor equilibrium properties of mixtures.

Exercise

Set up the spreadsheet and verify the results in this example. Then repeat the calculation starting with an initial guess of 16 atm for the vapor pressure and corresponding guesses for the molar volumes and see if the Solver finds the answer. Then repeat with an initial guess of 4 atm and corresponding volumes. Then repeat the calculation for propane at 0°C. The experimental values are 4.68 atm, 83.4 cm³/mol, 4257 cm³/mol. (*Answer:* 5.65 atm, 91.3 cm³/mol, 3492 cm³/mol.)

The procedure for finding $\Delta_{vap}H$ from an eos is outlined in Prob. 8.17. The Redlich–Kwong equation predicts 13.4 kJ/mol for propane at 25°C compared with the experimental value 14.8 kJ/mol.

The minimum of a Redlich–Kwong isotherm can be used to predict the maximum tension that a liquid can be subjected to (Prob. 8.18).

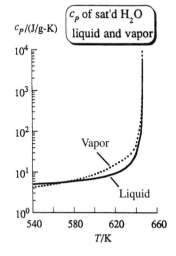

c_p of sat'd H$_2$O liquid and vapor

c_P/(J/g-K)

Vapor

Liquid

540   580   620   660

T/K

Figure 8.7

Specific heat capacity of saturated liquid water and water vapor versus T. The vertical scale is logarithmic. As the critical temperature 647 K is approached, these specific heats go to infinity.

8.6 THE CRITICAL STATE

A fluid at its critical point is said to be in the **critical state.** As noted at the beginning of Sec. 8.4, $(\partial P/\partial V_m)_T = 0$ at the critical point, and $(\partial P/\partial V_m)_T$ is negative on either side of the critical point. Hence, $(\partial V_m/\partial P)_T = -\infty$ at the critical point [Eq. (1.32)]. The isothermal compressibility is $\kappa \equiv -(\partial V_m/\partial P)_T/V_m$, so $\kappa = \infty$ at the critical point. We have $(\partial P/\partial T)_{V_m} = \alpha/\kappa$ [Eq. (1.45)]. Experiment shows $(\partial P/\partial T)_{V_m}$ is finite and positive at the critical point. Therefore, $\alpha = \infty$ at the critical point. We have $C_{P,m} = C_{V,m} + TV_m\alpha^2/\kappa = C_{V,m} + TV_m\alpha(\partial P/\partial T)_{V_m}$ [Eq. (4.53)]. Since $\alpha = \infty$ at the critical point, it follows that $C_{P,m} = \infty$ at the critical point. Figure 8.7 plots c_P for saturated liquid water and for saturated water vapor versus T. (Recall Fig. 7.2, which plots ρ of each of the saturated phases.) As the critical point (374°C, 218 atm) is approached, $C_{P,m}$ of each phase goes to infinity. For points close to the critical point, $C_{P,m}$ is quite large. This explains the large maxima in c_P of $H_2O(g)$ on the 400°C isotherm and the 300-bar isobar in Fig. 2.5.

Figure 8.8 plots the specific volume v versus P for H_2O for isotherms in the region of T_c. (These curves are similar to those in Fig. 8.3, except that the axes are interchanged and the isotherms in Fig. 8.8 are accurately drawn.) On an isotherm below $T_c = 374$°C, we see condensation and a sudden change in v at a fixed pressure. On the 380°C isotherm above T_c, although there is not a sudden change in v, we do see a rather rapid change in v over a small range of P. For the 380°C isotherm, this is the part of the curve from a to b.

The solid line in Fig. 8.9 shows the liquid–vapor equilibrium line for H_2O, which ends at the critical point, point C. The nonvertical dashed line in Fig. 8.9 is an isochore (line of constant V_m and constant density) corresponding to the critical molar volume $V_{m,c}$. The vertical dashed line in Fig. 8.9 corresponds to the 380°C isotherm in Fig. 8.8. Points a and b correspond to points a and b in Fig. 8.8. Thus, when the isochore corresponding to $V_{m,c}$ is approached and crossed close to the critical point, the fluid shows a rather rapid change from a gaslike to a liquidlike density and compressibility. Moreover, one will see similar rapid changes from gaslike to liquidlike entropy and internal energy, as shown by the 380°C isotherms and 400-bar isobars in Figs. 4.6 and 4.7. As the temperature is increased well above T_c, these regions of rapid change from gaslike to liquidlike properties gradually disappear.

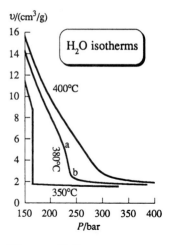

Figure 8.8

Accurately plotted isotherms of H_2O in the critical region.

8.7   THE LAW OF CORRESPONDING STATES

The (dimensionless) **reduced pressure** P_r, **reduced temperature** T_r, and **reduced volume** V_r of a gas in the state (P, V_m, T) are defined as

$$P_r \equiv P/P_c, \qquad V_r \equiv V_m/V_{m,c}, \qquad T_r \equiv T/T_c \qquad (8.27)$$

where P_c, $V_{m,c}$, T_c are the critical constants of the gas. Van der Waals pointed out that, if one uses reduced variables to express the states of gases, then, to a pretty good approximation, all gases show the same P-V_m-T behavior. In other words, if two different gases are each at the same P_r and T_r, they have nearly the same V_r values. This observation is called the **law of corresponding states.** Mathematically,

$$V_r = f(P_r, T_r) \qquad (8.28)$$

where approximately the same function f applies to any gas.

A two-parameter equation of state like the van der Waals or Redlich–Kwong can be expressed as an equation of the form (8.28) with the constants a and b eliminated. For example, for the van der Waals equation (8.2), use of (8.18) to eliminate a and b and (8.19) to eliminate R gives (Prob. 8.19)

$$(P_r + 3/V_r^2)(V_r - \tfrac{1}{3}) = \tfrac{8}{3}T_r \qquad (8.29)$$

If we multiply the law of corresponding states (8.28) by P_r/T_r, we get $P_rV_r/T_r = P_rf(P_r, T_r)/T_r$. The right side of this equation is some function of P_r and T_r, which we shall call $g(P_r, T_r)$. Thus

$$P_rV_r/T_r = g(P_r, T_r) \qquad (8.30)$$

where the function g is approximately the same for all gases.

Since every gas obeys $PV_m = RT$ in the limit of zero density, then for any gas $\lim_{V\to\infty} (PV_m/RT) = 1$. If this equation is multiplied by $RT_c/P_cV_{m,c}$ and (8.27) and

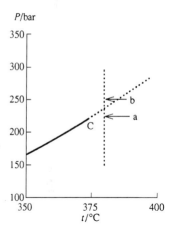

Figure 8.9

The solid curve is the P-versus-T liquid–vapor equilibrium line of H_2O, which ends at the critical point C at 374°C. The dashed curve from 374 to 400°C is an isochore with molar volume equal to the critical molar volume.

(8.30) are used, we get $\lim (P_r V_r / T_r) = RT_c / P_c V_{m,c}$ and $\lim g = 1/Z_c$. Since g is the same function for every gas, its limiting value as V goes to infinity must be the same constant for every gas. Calling this constant K, we have the prediction that $Z_c = 1/K$ for every gas. The law of corresponding states predicts that the critical compression factor is the same for every gas. Actually, Z_c varies from 0.12 to 0.46 (Sec. 8.4), so this prediction is false.

Multiplication of (8.30) by $P_c V_{m,c}/RT_c$ gives $PV_m/RT = Z_c g(P_r, T_r) \equiv G(P_r, T_r)$ or

$$Z = G(P_r, T_r) \tag{8.31}$$

Since the law of corresponding states predicts Z_c to be the same constant for all gases and g to be the same function for all gases, the function G, defined as $Z_c g$, is the same for all gases. Thus the law of corresponding states predicts that the compression factor Z is a universal function of P_r and T_r. To apply (8.31), a graphical approach is often used. One takes data for a representative sample of gases and calculates average Z values at various values of P_r and T_r. These average values are then plotted, with the result shown in Fig. 8.10. Such graphs (see *Reid, Prausnitz, and Poling*, chap. 3) can predict *P-V-T* data for gases to within a few percent, except for compounds with large dipole moments.

The law of corresponding states can be explained by the fact that the interaction between two gas molecules can be roughly approximated by a potential-energy function containing only two parameters, one measuring the strength of intermolecular attraction and one related to intermolecular repulsion (molecular size). An example is the Lennard-Jones potential (22.136). Starting from a two-parameter intermolecular potential-energy function, one can use statistical mechanics to deduce an equation of state that contains only two parameters. This equation can then be put in reduced form to give V_r as some universal function of P_r and T_r.

The assumption of an intermolecular potential-energy function with only two parameters is most valid when the interaction energy between two molecules is a function of only the distance between the centers of the molecules, being independent of the relative orientation of the two molecules. This holds best for approximately spherical molecules with low dipole moments [for example (Z_c values in parentheses), Ar (0.29), CH_4 (0.29), CF_4 (0.28), N_2 (0.29), O_2 (0.29), CO (0.30)] and does not hold well for molecules with high dipole moments, including those showing hydrogen bonding [for example, H_2O (0.23), HCN (0.20), HF (0.12), CH_3CN (0.18)], and for highly nonspherical molecules [for example, n-$C_{16}H_{34}$ (0.22)]. Z_c is close to 0.29 for gases obeying the law of corresponding states.

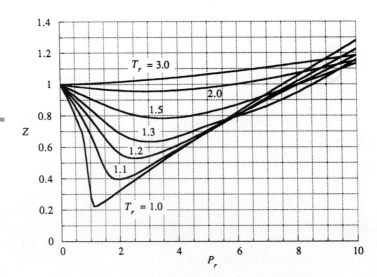

Figure 8.10

Average compression factor as a function of reduced variables.

8.8 DIFFERENCES BETWEEN REAL-GAS AND IDEAL-GAS THERMODYNAMIC PROPERTIES

Sections 8.1 to 8.4 consider the difference between real-gas and ideal-gas P-V-T behavior. Besides P-V-T behavior, one is often interested in the difference between real-gas and ideal-gas thermodynamic properties such as U, H, A, S, and G at a given T and P. For example, since the standard state of a gas at a given T is the hypothetical ideal gas at T and 1 bar (Sec. 5.1), one needs these differences to find the standard-state thermodynamic properties of gases from experimental data for real gases. Recall the calculation of S°_m for SO_2 in Sec. 5.7. Another use for such differences is as follows. Reliable methods exist for estimating thermodynamic properties in the ideal-gas state (Sec. 5.10). After using such an estimation method, one would want to correct the results to correspond to the real-gas state. This is especially important at high pressures. Industrial processes often involve gases at pressures of hundreds of atmospheres, so chemical engineers are keenly interested in differences between real-gas and ideal-gas properties. For a full discussion of such differences (which are called *residual functions* or *departure functions*), see *Reid, Prausnitz, and Poling*, chap. 5.

Let $H^{id}_m(T, P) - H_m(T, P)$ be the difference between ideal- and real-gas molar enthalpies at T and P. Unsuperscripted thermodynamic properties refer to the real gas. Equations (5.16) and (5.30) give $H^{id}_m(T, P) - H_m(T, P) = \int_0^P [T(\partial V_m/\partial T)_P - V_m] \, dP'$ and $S^{id}_m(T, P) - S_m(T, P) = \int_0^P [(\partial V_m/\partial T)_P - R/P'] \, dP'$, where the integrals are at constant T and the prime was added to the integration variable to avoid using the symbol P with two meanings. Figures 8.11 and 8.12 plot enthalpy and entropy departure functions of $CH_4(g)$ versus T and P.

If we have a reliable equation of state for the gas, we can use it to find $(\partial V_m/\partial T)_P$ and V_m and thus evaluate $H^{id}_m - H_m$ and $S^{id}_m - S_m$. The virial equation of state in the form (8.5) is especially convenient for this purpose, since it gives V_m and $(\partial V_m/\partial T)_P$ as functions of P, allowing the integrals to be easily evaluated. For the results, see Prob. 8.23.

Unfortunately, the Redlich–Kwong and van der Waals equations are cubics in V_m and cannot be readily used in these formulas. One way around this difficulty is to expand these equations of state into a virial form involving powers of $1/V_m$ [for example, Eq. (8.9) for the van der Waals equation] and then use (8.6) to put the equation into the virial form (8.5) involving powers of P. This approach is useful at low pressures. See Probs. 8.24 and 8.25. A more general approach is to use T and V as the variables, instead of T and P. This allows expressions valid at all pressures to be found from the equation of state. For details, see Prob. 8.26.

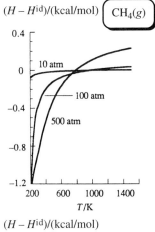

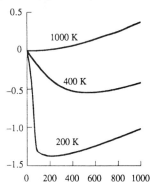

Figure 8.11

Difference between real- and ideal-gas molar enthalpy of CH_4 plotted versus T and versus P.

8.9 TAYLOR SERIES

In Sec. 8.2, the Taylor series expansion (8.8) of $1/(1 - x)$ was used. We now discuss Taylor series.

Let $f(x)$ be a function of the real variable x, and let f and all its derivatives exist at the point $x = a$ and in some neighborhood of a. It may then be possible to express $f(x)$ as the following **Taylor series** in powers of $(x - a)$:

$$f(x) = f(a) + \frac{f'(a)(x - a)}{1!} + \frac{f''(a)(x - a)^2}{2!} + \frac{f'''(a)(x - a)^3}{3!} + \cdots$$

$$f(x) = \sum_{n=0}^{\infty} \frac{f^{(n)}(a)}{n!}(x - a)^n \qquad \textbf{(8.32)}*$$

$(S - S^{id})/(\text{cal/mol-K})$

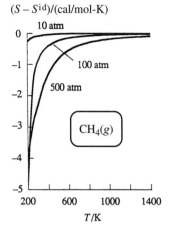

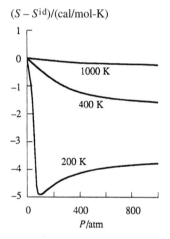

Figure 8.12

Difference between real- and ideal-gas molar entropy for CH_4 plotted versus T and versus P.

In (8.32), $f^{(n)}(a)$ is the nth derivative $d^n f(x)/dx^n$ evaluated at $x = a$. The zeroth derivative of f is defined to be f itself. The factorial function is defined by

$$n! \equiv n(n-1)(n-2)\cdots 2 \cdot 1 \quad \text{and} \quad 0! \equiv 1 \qquad \textbf{(8.33)*}$$

where n is a positive integer. The derivation of (8.32) is given in most calculus texts.

To use (8.32), we must know for what range of values of x the infinite series represents $f(x)$. The infinite series in (8.32) will converge to $f(x)$ for all values of x within some interval centered at $x = a$:

$$a - c < x < a + c \qquad (8.34)$$

where c is some positive number. The value of c can often be found by taking the distance between the point a and the real singularity of $f(x)$ nearest to a. A *singularity* of f is a point where f or one of its derivatives doesn't exist. For example, the function $1/(1 - x)$ expanded about $a = 0$ gives the Taylor series (8.8). The nearest real singularity to $x = 0$ is at $x = 1$, since $1/(1 - x)$ becomes infinite at $x = 1$. For this function, $c = 1$, and the Taylor series (8.8) converges to $1/(1 - x)$ for all x in the range $-1 < x < 1$. In some cases, c is less than the distance to the nearest real singularity. The general method of finding c is given in Prob. 8.33.

EXAMPLE 8.2 Taylor series

Find the Taylor series for $\sin x$ with $a = 0$.

To find $f^{(n)}(a)$ in (8.32), we differentiate $f(x)$ n times and then set $x = a$. For $f(x) = \sin x$ and $a = 0$, we get

$$f(x) = \sin x \qquad f(a) = \sin 0 = 0$$

$$f'(x) = \cos x \qquad f'(a) = \cos 0 = 1$$

$$f''(x) = -\sin x \qquad f''(a) = -\sin 0 = 0$$

$$f'''(x) = -\cos x \qquad f'''(a) = -\cos 0 = -1$$

$$f^{(iv)}(x) = \sin x \qquad f^{(iv)}(a) = \sin 0 = 0$$

$\cdots\cdots\cdots\cdots\cdots\cdots\cdots\cdots\cdots\cdots\cdots\cdots\cdots\cdots\cdots\cdots$

The values of $f^{(n)}(a)$ are the set of numbers $0, 1, 0, -1$ repeated again and again. The Taylor series (8.32) is

$$\sin x = 0 + \frac{1(x-0)}{1!} + \frac{0(x-0)^2}{2!} + \frac{(-1)(x-0)^3}{3!} + \frac{0(x-0)^4}{4!} + \cdots$$

$$\sin x = x - x^3/3! + x^5/5! - x^7/7! + \cdots \qquad \text{for all } x \qquad (8.35)$$

The function $\sin x$ has no singularities for real values of x. A full mathematical investigation shows that (8.35) is valid for all values of x.

Exercise

Use (8.32) to find the first four nonzero terms of the Taylor series for $\cos x$ with $a = 0$. (*Answer:* $1 - x^2/2! + x^4/4! - x^6/6! + \cdots$.)

Another example is $\ln x$. Since $\ln 0$ doesn't exist, we cannot take $a = 0$ in (8.32). A convenient choice is $a = 1$. We find (Prob. 8.29)

$$\ln x = (x - 1) - (x - 1)^2/2 + (x - 1)^3/3 - \cdots \qquad \text{for } 0 < x < 2 \qquad (8.36)$$

The nearest singularity to $a = 1$ is at $x = 0$ (where f doesn't exist), and the series (8.36) converges to $\ln x$ for $0 < x < 2$. Two other important Taylor series are

$$e^x = 1 + x + \frac{x^2}{2!} + \frac{x^3}{3!} + \cdots = \sum_{n=0}^{\infty} \frac{x^n}{n!} \qquad \text{for all } x \qquad (8.37)$$

$$\cos x = 1 - x^2/2! + x^4/4! - x^6/6! + \cdots \qquad \text{for all } x \qquad (8.38)$$

Taylor series are useful in physical chemistry when x in (8.32) is close to a, so that only the first few terms in the series need be included. For example, at low pressures, V_m of a gas is large and $b/V_m \, (= x)$ in (8.9) is close to zero. In general, Taylor series are useful under limiting conditions such as low P in a gas or low concentration in a solution.

8.10 SUMMARY

The compression factor of a gas is defined by $Z \equiv PV_m/RT$ and measures the deviation from ideal-gas P-V-T behavior. In the van der Waals equation of state for gases, $(P + a/V_m^2)(V_m - b) = RT$, the a/V_m^2 term represents intermolecular attractions and b represents the volume excluded by intermolecular repulsions. The Redlich–Kwong equation is an accurate two-parameter equation of state for gases. The parameters in these equations of state are evaluated from critical-point data. The virial equation, derived from statistical mechanics, expresses Z as a power series in $1/V_m$, where the expansion coefficients are related to intermolecular forces.

Important kinds of calculations dealt with in this chapter include:

- Use of nonideal equations of state such as the van der Waals, the Redlich–Kwong, and the virial equations to calculate P or V of a pure gas or a gas mixture.
- Calculation of constants in the van der Waals equation from critical-point data.
- Calculation of differences between real-gas and ideal-gas thermodynamic properties using an equation of state.
- Use of an equation of state to calculate vapor pressures and saturated liquid and vapor molar volumes.

FURTHER READING AND DATA SOURCES

Reid, Prausnitz, and Poling, chaps. 3 and 4; *Van Ness and Abbott,* chap. 4; *McGlashan,* chap. 12.

Compression factors: *Landolt-Börnstein,* 6th ed., vol. II, pt. 1, pp. 72–270.

Critical constants: A. P. Kudchadker et al., *Chem. Rev.,* **68,** 659 (1968) (organic compounds); J. F. Mathews, *Chem. Rev.,* **72,** 71 (1972) (inorganic compounds); *Reid, Prausnitz, and Poling,* pp. 656–732; *Landolt-Börnstein,* 6th ed., vol. II, pt. 1, pp. 331–356; K. H. Simmrock, R. Janowsky, and A. Ohnsorge, *Critical Data of Pure Substances,* DECHEMA, 1986; *Lide and Kehiaian,* Table 2.1.1; NIST Chemistry Webbook at webbook.nist.gov/.

Virial coefficients: J. H. Dymond and E. B. Smith, *The Virial Coefficients of Pure Gases and Mixtures,* Oxford University Press, 1980.

PROBLEMS

Section 8.2

8.1 Give the SI units of (a) a and b in the van der Waals equation; (b) a and b in the Redlich–Kwong equation; (c) $B(T)$ in the virial equation.

8.2 Verify that the van der Waals, the virial, and the Redlich–Kwong equations all reduce to $PV = nRT$ in the limit of zero density.

8.3 For C_2H_6 at 25°C, $B = -186$ cm^3/mol and $C = 1.06 \times 10^4$ cm^6/mol^2. (a) Use the virial equation (8.4) to calculate the pressure of 28.8 g of $C_2H_6(g)$ in a 999-cm^3 container at 25°C. Compare with the ideal-gas result. (b) Use the virial equation (8.5) to calculate the volume of 28.8 g of C_2H_6 at 16.0 atm and 25°C. Compare with the ideal-gas result.

8.4 Use the following method to verify Eq. (8.6) for the virial coefficients. Solve equation (8.4) for P, substitute the result into the right side of (8.5), and compare the coefficient of each power of $1/V_m$ with that in (8.4).

8.5 Use Eq. (8.7) and data in Sec. 8.2 to find V_m of Ar(g) at 200 K and 1 atm.

8.6 At 25°C, $B = -42$ cm^3/mol for CH_4 and $B = -732$ cm^3/mol for n-C_4H_{10}. For a mixture of 0.0300 mol of CH_4 and 0.0700 mol of n-C_4H_{10} at 25°C in a 1.000-L vessel, calculate the pressure using the virial equation and (a) the approximation $B_{12} \approx \frac{1}{2}(B_1 + B_2)$; (b) the fact that for this mixture, $B_{12} = -180$ cm^3/mol. Compare the results with the ideal-gas-equation result.

Section 8.4

8.7 For ethane, $P_c = 48.2$ atm and $T_c = 305.4$ K. Calculate the pressure exerted by 74.8 g of C_2H_6 in a 200-cm^3 vessel at 37.5°C using (a) the ideal-gas law; (b) the van der Waals equation; (c) the Redlich–Kwong equation; (d) the virial equation, given that for ethane $B = -179$ cm^3/mol and $C = 10400$ cm^6/mol^2 at 30°C, and $B = -157$ cm^3/mol and $C = 9650$ cm^6/mol^2 at 50°C.

8.8 For a mixture of 0.0786 mol of C_2H_4 and 0.1214 mol of CO_2 in a 700.0-cm^3 container at 40°C, calculate the pressure using (a) the ideal-gas equation; (b) the van der Waals equation, data in Table 8.1, and the C_2H_4 critical data $T_c = 282.4$ K, $P_c = 49.7$ atm; (c) the experimental compression factor $Z = 0.9689$.

8.9 Show that if all terms after C/V_m^2 are omitted from the virial equation (8.4), this equation predicts $Z_c = \frac{1}{3}$.

8.10 (a) Calculate the van der Waals a and b of Ar from data in Table 8.1. (b) Use Eq. (8.9) to calculate the van der Waals second virial coefficient B for Ar at 100, 200, 300, 500, and 1000 K and compare with the experimental values in Sec. 8.2.

8.11 Equation (4.58) gives $U_{m,intermol} = -a/V_m$ for a fluid that obeys the van der Waals equation. Taking $U_{m,intermol} \approx 0$ for the gas phase, we can use $a/V_{m,nbp,liq}$ to estimate ΔU_m of vaporization at the normal boiling point (nbp). The temperature and density at the normal boiling point are 77.4 K and 0.805 g/cm^3 for

N_2 and 188.1 K and 1.193 g/cm^3 for HCl. Use the van der Waals constants listed in Sec. 8.4 to estimate $\Delta_{vap}H_{m,nbp}$ of N_2, HCl, and H_2O. Compare with the experimental values 1.33 kcal/mol for N_2, 3.86 kcal/mol for HCl, and 9.7 kcal/mol for H_2O.

Section 8.5

8.12 Use the spreadsheet of Fig. 8.6 to find the Redlich–Kwong estimates of the vapor pressure and saturated liquid and vapor molar volumes of propane at −20°C.

8.13 Use a spreadsheet and Table 8.1 data to find the Redlich–Kwong estimates of the vapor pressure and saturated liquid and vapor molar volumes of CO_2 at 0°C. The experimental values are 34.4 atm, 47.4 cm^3/mol, and 452 cm^3/mol.

8.14 Use the van der Waals equation to estimate the vapor pressure and saturated liquid and vapor molar volumes of propane at 25°C.

8.15 The **Soave–Redlich–Kwong equation** is

$$P = \frac{RT}{V_m - b} - \frac{a(T)}{V_m(V_m + b)}$$

where $b = 0.08664RT_c/P_c$ (as in the Redlich–Kwong equation) and $a(T)$ is the following function of temperature:

$$a(T) = 0.42748(R^2T_c^2/P_c)\{1 + m[1 - (T/T_c)^{0.5}]\}^2$$

$$m \equiv 0.480 + 1.574\omega - 0.176\omega^2$$

The quantity ω is the **acentric factor** of the gas, defined as

$$\omega \equiv -1 - \log_{10}(P_{vp}/P_c)|_{T/T_c=0.7}$$

where P_{vp} is the vapor pressure of the liquid at $T = 0.7T_c$. The acentric factor is close to zero for gases with approximately spherical molecules of low polarity. A tabulation of ω values is given in Appendix A of *Reid, Prausnitz, and Poling*. The Soave–Redlich–Kwong equation has two parameters a and b, but evaluation of these parameters requires knowing three properties of the gas: T_c, P_c, and ω. For propane $\omega = 0.153$. (a) Show that $a(T) = 1.08_2 \times 10^7$ atm cm^6 mol^{-2} for propane at 25°C. (b) Use the Soave–Redlich–Kwong equation to find the vapor pressure and saturated liquid and vapor molar volumes of propane at 25°C. The Redlich–Kwong spreadsheet of Fig. 8.6 can be used if the $T^{1/2}$ factors in the denominators of all formulas are deleted.

8.16 The **Peng–Robinson equation** is

$$P = \frac{RT}{V_m - b} - \frac{a(T)}{V_m(V_m + b) + b(V_m - b)}$$

where

$$b = 0.07780RT_c/P_c$$

$$a(T) = 0.45724(R^2T_c^2/P_c)\{1 + k[1 - (T/T_c)^{1/2}]\}^2$$

$$k \equiv 0.37464 + 1.54226\omega - 0.26992\omega^2$$

where ω is defined in Prob. 8.15. (a) Use data in Prob. 8.15 to show that for propane at 25°C, $a(T) = 1.13_3 \times 10^7$ atm cm^6

mol^{-2}. (*b*) Use the Peng–Robinson equation to predict the vapor pressure and saturated liquid and vapor molar volumes of propane at 25°C. You will need the integral

$$\int \frac{1}{x^2 + sx + c} \, dx = \frac{1}{(s^2 - 4c)^{1/2}} \ln \frac{2x + s - (s^2 - 4c)^{1/2}}{2x + s + (s^2 - 4c)^{1/2}}$$

8.17 To calculate $\Delta_{vap}H_m$ from a cubic equation of state, we integrate $(\partial U_m/\partial V_m)_T = T(\partial P/\partial T)_{V_m} - P$ [Eq. (4.47)] along JKLMN in Fig. 8.5 to get

$$\Delta_{vap}U_m \equiv U_m^v - U_m^l$$

$$= \int_{V_m^l}^{V_m^v} \left[T\left(\frac{\partial P_{eos}}{\partial T}\right)_{V_m} - P_{eos} \right] dV_m \qquad \text{const. } T$$

where P_{eos} and $(\partial P_{eos}/\partial T)_{V_m}$ are found from the equation of state. Then we use

$$\Delta_{vap}H_m = \Delta_{vap}U_m + P(V_m^v - V_m^l)$$

where the vapor pressure P and the saturated molar volumes are found from the equation of state, as in Sec. 8.5. (*a*) Show that the Redlich–Kwong equation gives

$$\Delta_{vap}H_m = \frac{3a}{2bT^{1/2}} \ln \frac{V_m^v(V_m^l + b)}{V_m^l(V_m^v + b)} + P(V_m^v - V_m^l)$$

(*b*) Calculate $\Delta_{vap}H_m$ for propane at 25°C using the Redlich–Kwong equation and the results of Example 8.1.

8.18 For diethyl ether, $P_c = 35.9$ atm and $T_c = 466.7$ K. The lowest observed negative pressure that liquid diethyl ether can be subjected to at 403 K is -14 atm. Use a spreadsheet to plot the 403 K Redlich–Kwong isotherm; find the pressure minimum (point K in Fig. 8.5) for the superheated liquid and compare with -14 atm.

Section 8.7

8.19 Verify the reduced van der Waals equation (8.29) by substituting (8.18) for a and b and (8.19) for R in (8.2).

8.20 The Berthelot equation of state for gases is

$$(P + a/TV_m^2)(V_m - b) = RT$$

(*a*) Show that the Berthelot parameters are $a = 27R^2T_c^3/64P_c$ and $b = RT_c/8P_c$. (*b*) What value of Z_c is predicted? (*c*) Write the Berthelot equation in reduced form.

8.21 For C_2H_6, $V_{m,c} = 148$ cm^3/mol. Use the reduced van der Waals equation (8.29) to answer Prob. 8.7. Note that the result is very different from that of Prob. 8.7*b*.

8.22 For gases obeying the law of corresponding states, the second virial coefficient B is accurately given by the equation (*McGlashan*, p. 203)

$$BP_c/RT_c = 0.597 - 0.462e^{0.7002T_c/T}$$

Use this equation and Table 8.1 data to calculate B of Ar at 100, 200, 300, 500, and 1000 K and compare with the experimental values in Sec. 8.2.

Section 8.8

8.23 Use the virial equation in the form (8.5) to show that at T and P

$$H_m^{id} - H_m = RT^2\left[\frac{dB^\dagger}{dT}P + \frac{1}{2}\frac{dC^\dagger}{dT}P^2 + \cdots\right]$$

$$S_m^{id} - S_m = R\left[\left(B^\dagger + T\frac{dB^\dagger}{dT}\right)P\right.$$

$$\left. + \frac{1}{2}\left(C^\dagger + T\frac{dC^\dagger}{dT}\right)P^2 + \cdots\right]$$

$$G_m^{id} - G_m = -RT\left[B^\dagger P + \tfrac{1}{2}C^\dagger P^2 + \cdots\right]$$

8.24 (*a*) Use the results of Prob. 8.23 and Eqs. (8.9) and (8.6) to show that, for a van der Waals gas at T and P, $H_m^{id} - H_m = (2a/RT - b)P + \cdots$ and $S_m^{id} - S_m = (a/RT^2)P + \cdots$. (*b*) For C_2H_6, $T_c = 305.4$ K and $P_c = 48.2$ atm. Calculate the values of $H_m^{id} - H_m$ and $S_m^{id} - S_m$ predicted by the van der Waals equation for C_2H_6 at 298 K and 1 bar. (At 1 bar, powers of P higher than the first can be neglected with negligible error.) Compare with the experimental values 15 cal/mol and 0.035 cal/(mol K).

8.25 Although the overall performance of the Berthelot equation (Prob. 8.20) is quite poor, it does give pretty accurate estimates of $H_m^{id} - H_m$ and $S_m^{id} - S_m$ for many gases at low pressures. Expand the Berthelot equation into virial form and use the approach of Prob. 8.24*a* to show that the Berthelot equation gives at T and P: $H_m^{id} - H_m = (3a/RT^2 - b)P + \cdots$ and $S_m^{id} - S_m = (2a/RT^3)P + \cdots$. (*b*) Neglect terms after P and use the results for Prob. 8.20*a* to show that the Berthelot equation predicts $H_m^{id} - H_m \approx 81RT_c^3P/64T^2P_c - RT_cP/8P_c$ and $S_m^{id} - S_m \approx 27RT_c^3P/32T^3P_c$. (*c*) Use the Berthelot equation to calculate $H_m^{id} - H_m$ and $S_m^{id} - S_m$ for C_2H_6 at 298 K and 1 bar and compare with the experimental values. See Prob. 8.24*b* for data. (*d*) Use the Berthelot equation to calculate $S_m^{id} - S_m$ for SO_2 ($T_c = 430.8$ K, $P_c = 77.8$ atm) at 298 K and 1 atm.

8.26 (*a*) Let V_m be the molar volume of a real gas at T and P and let V_m^{id} be the ideal-gas molar volume at T and P. In the process (5.13), note that $V_m \to \infty$ as $P \to 0$. Use a modification of the process (5.13) in which step (*c*) is replaced by two steps, a contraction from infinite molar volume to molar volume V_m followed by a volume change from V_m to V_m^{id}, to show that

$$A_m^{id}(T, P) - A_m(T, P) = \int_\infty^{V_m}\left(P' - \frac{RT}{V_m'}\right)dV_m' - RT \ln\frac{V_m^{id}}{V_m}$$

where the integral is at constant T and $V_m^{id} \equiv RT/P$. This formula is convenient for use with equations like the Redlich–Kwong and van der Waals, which give P as a function of V_m. Formulas for $S_m^{id} - S_m$ and $H_m^{id} - H_m$ are not needed, since these differences are easily derived from $A_m^{id} - A_m$ using $(\partial A_m/\partial T)_V = -S_m$ and $A_m = U_m - TS_m = H_m - PV_m - TS_m$. (*b*) For the Redlich–Kwong equation, show that, at T and P, $A_m^{id} - A_m = RT \ln(1 - b/V_m) + (a/bT^{1/2}) \ln(1 + b/V_m) - RT \ln(V_m^{id}/V_m)$. (*c*) From (*b*), derive expressions for $S_m^{id} - S_m$ and $U_m^{id} - U_m$ for a Redlich–Kwong gas.

8.27 Use the corresponding-states equation for B in Prob. 8.22, data in Prob. 8.24, and the results of Prob. 8.23 to estimate $H_m^{id} - H_m$ and $S_m^{id} - S_m$ for C_2H_6 at 298 K and 1 bar and compare with the experimental values.

Section 8.9

8.28 Use (8.32) to verify the Taylor series (8.8) for $1/(1 - x)$.

8.29 Verify the Taylor series (8.36) for $\ln x$.

8.30 Verify the Taylor series (8.37) for e^x.

8.31 Derive the Taylor series (8.38) for $\cos x$ by differentiating (8.35).

8.32 Use (8.35) to calculate the sine of 35° to four significant figures. Before beginning, decide whether x in (8.35) is in degrees or in radians.

8.33 This problem is only for those familiar with the notion of the complex plane (in which the real and imaginary parts of a number are plotted on the horizontal and vertical axes). The *radius of convergence c* in (8.34) for the Taylor series (8.32) can be shown to equal the distance between point a and the singularity in the complex plane that is nearest to a (see *Sokolnikoff and Redheffer,* sec. 8.10). Find the radius of convergence for the Taylor-series expansion $1/(x^2 + 4)$ about $a = 0$.

8.34 Use a programmable calculator or computer to calculate the truncated e^x Taylor series $\sum_{n=0}^{m} x^n/n!$ for $m = 5, 10,$ and 20 and (*a*) $x = 1$; (*b*) $x = 10$. Compare the results in each case with e^x.

General

8.35 The normal boiling point of benzene is 80°C. The density of liquid benzene at 80°C is 0.81 g/cm³. Estimate P_c, T_c, and $V_{m,c}$ for benzene.

8.36 The vapor pressure of water at 25°C is 23.766 torr. Calculate ΔG_{298}° for the process $H_2O(l) \rightarrow H_2O(g)$; do not assume ideal vapor; instead use the results of Prob. 8.24*a* and data in Sec. 8.4 to correct for nonideality. Compare your answer with that to Prob. 7.50 and with the value found from $\Delta_f G_{298}^\circ$ values in the Appendix.

8.37 (*a*) Use the virial equation (8.5) to show that

$$\mu_{JT} = \frac{RT^2}{C_{P,m}}\left(\frac{dB^\dagger}{dT} + \frac{dC^\dagger}{dT}P + \frac{dD^\dagger}{dT}P^2 + \cdots\right)$$

$$\lim_{P\to 0} \mu_{JT} = (RT^2/C_{P,m})(dB^\dagger/dT) \neq 0$$

Thus, even though the Joule–Thomson coefficient of an ideal gas is zero, the Joule–Thomson coefficient of a real gas does not become zero in the limit of zero pressure. (*b*) Use (8.4) to show that, for a real gas, $(\partial U/\partial V)_T \rightarrow 0$ as $P \rightarrow 0$.

8.38 Use the virial equation (8.4) to show that for a real gas

$$\lim_{P\to 0} (V_m - V_m^{id}) = B(T)$$

8.39 At low P, all terms but the first in the μ_{JT} series in Prob. 8.37 can be omitted. (*a*) Show that the van der Waals equation

(8.9) predicts $\mu_{JT} = (2a/RT - b)/C_{P,m}$ at low P. (*b*) At low temperatures, the attractive term $2a/RT$ is greater than the repulsive term b and the low-P μ_{JT} is positive. At high temperature, $b > 2a/RT$ and $\mu_{JT} < 0$. The temperature at which μ_{JT} is zero in the $P \rightarrow 0$ limit is the low-pressure inversion temperature $T_{i,P\to 0}$. For N_2, use data in Sec. 8.4 and the Appendix to calculate the van der Waals predictions for $T_{i,P\to 0}$ and for μ_{JT} at 298 K and low P. Compare with the experimental values 638 K and 0.222 K/atm. (Better results can be obtained with a more accurate equation of state—for example, the Redlich–Kwong.)

8.40 For each of the following pairs, state which species has the greater van der Waals a, which has the greater van der Waals b, which has the greater T_c, and which has the greater $\Delta_{vap}H_m$ at the normal boiling point. (*a*) He or Ne; (*b*) C_2H_6 or C_3H_8; (*c*) H_2O or H_2S.

8.41 The van der Waals equation is a cubic in V_m, which makes it tedious to solve for V_m at a given T and P. One way to find V_m is by successive approximations. We write $V_m = b + RT/(P + a/V_m^2)$. To obtain an initial estimate V_{m0} of V_m, we neglect a/V_m^2 to get $V_{m0} = b + RT/P$. An improved estimate is $V_{m1} = b + RT/(P + a/V_{m0}^2)$. From V_{m1}, we get V_{m2}, etc. Use successive approximations to find the van der Waals V_m for CH_4 at 273 K and 100 atm, given that $T_c = 190.6$ K and $P_c = 45.4$ atm for CH_4. (The calculation is more fun if done on a programmable calculator.) Compare with the experimental V_m found from Fig. 8.1.

8.42 Use Fig. 8.10 to find V_m for CH_4 at 286 K and 91 atm. See Prob. 8.41 for data.

8.43 In Prob. 7.33, the Antoine equation was used to find $\Delta_{vap}H_m$ of H_2O at 100°C. The result was inaccurate due to neglect of gas nonideality. We now obtain an accurate result. For H_2O at 100°C, the second virial coefficient is -452 cm³/mol. (*a*) Use the Antoine equation and Prob. 7.33 data to find dP/dT for H_2O at 100°C, where P is the vapor pressure. (*b*) Use the Clapeyron equation $dP/dT = \Delta H_m/(T \Delta V_m)$ to find $\Delta_{vap}H_m$ of H_2O at 100°C; calculate ΔV_m using the truncated virial equation (8.7) and the saturated liquid's 100°C molar volume, which is 19 cm³/mol. Compare your result with the accepted value 40.66 kJ/mol.

8.44 Some V_m versus P data for $CH_4(g)$ at -50°C are

P/atm	5	10	20	40	60
V_m/(cm³/mol)	3577	1745	828	365	206

For the virial equation (8.4) with terms after C omitted, use a spreadsheet to find the B and C values that minimize the sums of the squares of the deviations of the calculated pressures from the observed pressures.

8.45 True or false? (*a*) The parameter a in the van der Waals equation has the same value for all gases. (*b*) The parameter a in the van der Waals equation for N_2 has the same value as a in the Redlich–Kwong equation for N_2.

Solutions

Much of chemistry and biochemistry takes place in solution. A **solution** is a homogeneous mixture; that is, a solution is a one-phase system with more than one component. The phase may be solid, liquid, or gas. Much of this chapter deals with liquid solutions, but most of the equations of Secs. 9.1 to 9.4 apply to all solutions.

Section 9.1 defines the ways of specifying solution composition. The thermodynamics of solutions is formulated in terms of partial molar properties. Their definitions, interrelations, and experimental determination are discussed in Secs. 9.2 and 9.4. Just as the behavior of gases is discussed in terms of departures from the behavior of a simple model (the ideal gas) that holds under a limiting condition (that of low density and therefore negligible intermolecular interactions), the behavior of liquid solutions is discussed in terms of departures from one of two models: (*a*) the ideal solution, which holds in the limit of almost negligible differences in properties between the solution components (Secs. 9.5 and 9.6); (*b*) the ideally dilute solution, which holds in the limit of a very dilute solution (Secs. 9.7 and 9.8). Nonideal solutions are discussed in Chapters 10 and 11.

9.1   SOLUTION COMPOSITION

The composition of a solution can be specified in several ways. The **mole fraction** x_i of species i is defined by Eq. (1.6) as $x_i \equiv n_i/n_{\text{tot}}$, where n_i is the number of moles of i and n_{tot} is the total number of moles of all species in the solution. The **(molar) concentration** (or *volume concentration*) c_i of species i is defined by (6.21) as

$$c_i \equiv n_i/V \qquad \textbf{(9.1)*}$$

where V is the solution's volume. For liquid solutions, the molar concentration of a species in moles per liter (dm^3) is called the **molarity.** The **mass concentration** ρ_i of species i in a solution of volume V is

$$\rho_i \equiv m_i/V \qquad \textbf{(9.2)*}$$

where m_i is the mass of i present.

For liquid and solid solutions, it is often convenient to treat one substance (called the **solvent**) differently from the others (called the **solutes**). Usually, the solvent mole fraction is greater than the mole fraction of each solute. We adopt the convention that *the solvent is denoted by the letter* A.

The **molality** m_i of species i in a solution is defined as the number of moles of i divided by the mass of the solvent. Let a solution contain n_B moles of solute B (plus certain amounts of other solutes) and n_A moles of solvent A. Let M_A be the solvent molar mass. From Eq. (1.4), the solvent mass w_A equals $n_A M_A$. We use w for mass, to avoid confusion with molality. The solute molality m_B is

$$m_B \equiv \frac{n_B}{w_A} = \frac{n_B}{n_A M_A} \qquad \textbf{(9.3)*}$$

M_A in (9.3) is the solvent molar mass (*not* molecular weight) and must have the proper dimensions. The molecular weight is dimensionless, whereas M_A has units of mass per mole (Sec. 1.4). The units of M_A are commonly either grams per mole or kilograms per mole. Chemists almost always use moles per kilogram as the unit of molality. Therefore it is desirable that M_A in (9.3) be in kg/mol. Note that it is the mass of the *solvent* (and not the solution's mass) that appears in the molality definition (9.3).

The **weight percent** of species B in a solution is $(w_B/w) \times 100\%$, where w_B is the mass of B and w is the mass of the solution. The *weight fraction* of B is w_B/w.

Since V of a solution depends on T and P, the concentrations c_i change with changing T and P. Mole fractions and molalities are independent of T and P.

EXAMPLE 9.1 Solution composition

An aqueous $AgNO_3$ solution that is 12.000% $AgNO_3$ by weight has a density 1.1080 g/cm³ at 20°C and 1 atm. Find the mole fraction, the molar concentration at 20°C and 1 atm, and the molality of the solute $AgNO_3$.

The unknowns are intensive properties and do not depend on the size of the solution. We are therefore free to choose a convenient fixed quantity of solution to work with. We take 100.00 g of solution. In 100.00 g of solution there are 12.00 g of $AgNO_3$ and 88.00 g of H_2O. Converting to moles, we find $n(AgNO_3) = 0.07064$ mol and $n(H_2O) = 4.885$ mol. Therefore $x(AgNO_3) = 0.07064/4.955_6 = 0.01425$. The volume of 100.00 g of solution is $V = m/\rho = (100.00 \text{ g})/(1.1080 \text{ g/cm}^3) = 90.25$ cm³. The definitions $c_i = n_i/V$ and $m_i = n_i/w_A$ [Eqs. (9.1) and (9.3)] give

$$c(AgNO_3) = (0.07064 \text{ mol})/(90.25 \text{ cm}^3) = 7.827 \times 10^{-4} \text{ mol/cm}^3$$

$$= (7.827 \times 10^{-4} \text{ mol/cm}^3)(10^3 \text{ cm}^3/1 \text{ L}) = 0.7827 \text{ mol/L}$$

$$m(AgNO_3) = (0.07064 \text{ mol})/(88.0 \text{ g}) = 0.8027 \times 10^{-3} \text{ mol/g}$$

$$= (0.8027 \times 10^{-3} \text{ mol/g})(10^3 \text{ g/kg}) = 0.8027 \text{ mol/kg}$$

In this example, the weight percent was known, and it was convenient to work with 100 g of solution. If the molarity is known, a convenient amount of solution to take is 1 L. If the molality is known, it is convenient to work with an amount of solution that contains 1 kg of solvent.

Exercise

A solution is prepared by dissolving 555.5 g of sucrose, $C_{12}H_{22}O_{11}$, in 600 mL of water and diluting with water to a final volume of 1.0000 L. The density of the final solution is found to be 1.2079 g/cm³. Find the sucrose mole fraction, molality, and weight percent in this solution. (*Answers:* 0.04289, 2.488 mol/kg, 45.99%.)

9.2 PARTIAL MOLAR QUANTITIES

Partial Molar Volumes

Suppose we form a solution by mixing at constant temperature and pressure n_1, n_2, ..., n_r moles of substances 1, 2, ..., r. Let $V^*_{m,1}, ..., V^*_{m,r}$ be the molar volumes of pure substances 1, 2, ..., r at T and P, and let V^* be the total volume of the unmixed (pure) components at T and P. *The star indicates a property of a pure substance or a collection of pure substances.* We have

$$V^* = n_1 V_{m,1}^* + n_2 V_{m,2}^* + \cdots + n_r V_{m,r}^* = \sum_i n_i V_{m,i}^* \qquad (9.4)$$

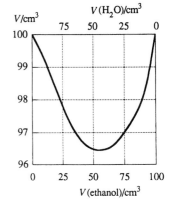

After mixing, one finds that the volume V of the solution is *not* in general equal to the unmixed volume; $V \neq V^*$. For example, addition of 50.0 cm^3 of water to 50.0 cm^3 of ethanol at 20°C and 1 atm gives a solution whose volume is only 96.5 cm^3 at 20°C and 1 atm (Fig. 9.1). The difference between V of the solution and V^* results from (a) differences between intermolecular forces in the solution and those in the pure components; (b) differences between the packing of molecules in the solution and the packing in the pure components, due to differences in sizes and shapes of the molecules being mixed.

Figure 9.1

Volume V of a solution formed by mixing a volume V_{ethanol} of pure ethanol with a volume $(100 \text{ cm}^3 - V_{\text{ethanol}})$ of pure water at 20°C and 1 atm.

We can write an equation like (9.4) for any extensive property, for example, U, H, S, G, and C_P. One finds that each of these properties generally changes on mixing the components at constant T and P.

We want expressions for the volume V of the solution and for its other extensive properties. Each such property is a function of the solution's state, which can be specified by the variables T, P, n_1, n_2, ..., n_r. Therefore

$$V = V(T, P, n_1, \ldots, n_r), \qquad U = U(T, P, n_1, \ldots, n_r) \qquad (9.5)$$

with similar equations for H, S, etc. The total differential of V in (9.5) is

$$dV = \left(\frac{\partial V}{\partial T}\right)_{P,n_i} dT + \left(\frac{\partial V}{\partial P}\right)_{T,n_i} dP + \left(\frac{\partial V}{\partial n_1}\right)_{T,P,n_{i\neq1}} dn_1 + \cdots + \left(\frac{\partial V}{\partial n_r}\right)_{T,P,n_{i\neq r}} dn_r$$
$$(9.6)$$

The subscript n_i in the first two partial derivatives indicates that all mole numbers are held constant; the subscript $n_{i\neq1}$ indicates that all mole numbers except n_1 are held constant. We define the **partial molar volume** $\bar{V}_j$ of substance j in the solution as

$$\bar{V}_j \equiv \left(\frac{\partial V}{\partial n_j}\right)_{T,P,n_{i\neq j}} \qquad \text{one-phase syst.} \qquad \textbf{(9.7)}^*$$

where V is the solution's volume and where the partial derivative is taken with T, P, and all mole numbers except n_j held constant. (The bar in $\bar{V}_j$ does not signify an average value.) Equation (9.6) becomes

$$dV = \left(\frac{\partial V}{\partial T}\right)_{P,n_i} dT + \left(\frac{\partial V}{\partial P}\right)_{T,n_i} dP + \sum_i \bar{V}_i \, dn_i \qquad (9.8)$$

Equation (9.8) gives the infinitesimal volume change dV that occurs when the temperature, pressure, and mole numbers of the solution are changed by dT, dP, dn_1, dn_2, ...

From (9.7), a partial molar volume is the ratio of infinitesimal changes in two extensive properties and so is an intensive property. Like any intensive property, $\bar{V}_i$ depends on T, P, and the mole fractions in the solution:

$$\bar{V}_i = \bar{V}_i(T, P, x_1, x_2, \ldots) \qquad (9.9)$$

From (9.7), if dV is the infinitesimal change in solution volume that occurs when dn_j moles of substance j is added to the solution with T, P, and all mole numbers except n_j held constant, then $\bar{V}_j$ equals dV/dn_j. See Fig. 9.2. $\bar{V}_j$ is the rate of change of solution volume with respect to n_j at constant T and P. The partial molar volume $\bar{V}_j$ of substance j in the solution tells how the solution's volume V responds to the constant-T-and-P addition of j to the solution; dV equals $\bar{V}_j \, dn_j$ when j is added at constant T and P.

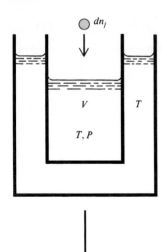

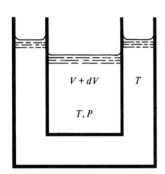

Figure 9.2

Addition of dn_j moles of substance j to a solution held at constant T and P produces a change dV in the solution's volume. The partial molar volume $\bar{V}_j$ of j in the solution equals dV/dn_j.

The volume of pure substance j is $V_j^* = n_j V_{m,j}^*(T, P)$, where $V_{m,j}^*$ is the molar volume of pure j. If we view a pure substance as a special case of a solution, then the definition (9.7) of $\bar{V}_j$ gives $\bar{V}_j^* \equiv (\partial V/\partial n_j)_{T,P,n_{i\neq j}} = (\partial V_j^*/\partial n_j)_{T,P} = V_{m,j}^*$. Thus

$$\bar{V}_j^* = V_{m,j}^* \tag{9.10}*$$

The partial molar volume of a pure substance is equal to its molar volume. However, the partial molar volume of component j of a solution is *not* necessarily equal to the molar volume of pure j.

EXAMPLE 9.2 Partial molar volumes in an ideal gas mixture

Find the partial molar volume of a component of an ideal gas mixture.
  We have

$$V = (n_1 + n_2 + \cdots + n_i + \cdots + n_r)RT/P$$

$$\bar{V}_i = (\partial V/\partial n_i)_{T,P,n_{j\neq i}} = RT/P \qquad \text{ideal gas mixture} \tag{9.11}$$

Of course, RT/P is the molar volume of pure gas i at the T and P of the mixture, so $\bar{V}_i = V_{m,i}^*$ for an ideal gas mixture, a result not true for nonideal gas mixtures.

Exercise

A certain two-component gas mixture obeys the equation of state $P(V - n_1 b_1 - n_2 b_2) = (n_1 + n_2)RT$, where b_1 and b_2 are constants. Find $\bar{V}_1$ and $\bar{V}_2$ for this mixture. (*Answer:* $RT/P + b_1$, $RT/P + b_2$.)

Relation between Solution Volume and Partial Molar Volumes

We now find an expression for the volume V of a solution. V depends on temperature, pressure, and the mole numbers. For fixed values of T, P, and solution mole fractions x_i, the volume, which is an extensive property, is directly proportional to the total number of moles n in the solution. (If we double all the mole numbers at constant T and P, then V is doubled; if we triple the mole numbers, then V is tripled; etc.) Since V is proportional to n for fixed T, P, $x_1, x_2, \ldots, x_r$, the equation for V must have the form

$$V = nf(T, P, x_1, x_2, \ldots) \tag{9.12}$$

where $n \equiv \Sigma_i n_i$ and where f is some function of T, P, and the mole fractions. Differentiation of (9.12) at constant T, P, $x_1, \ldots, x_r$ gives

$$dV = f(T, P, x_1, x_2, \ldots)\, dn \qquad \text{const. } T, P, x_i \tag{9.13}$$

Equation (9.8) becomes for constant T and P

$$dV = \sum_i \bar{V}_i\, dn_i \qquad \text{const. } T, P \tag{9.14}$$

We have $x_i = n_i/n$ or $n_i = x_i n$. Therefore $dn_i = x_i\, dn + n\, dx_i$. At fixed x_i, we have $dx_i = 0$, and $dn_i = x_i\, dn$. Substitution into (9.14) gives

$$dV = \sum_i x_i \bar{V}_i\, dn \qquad \text{const. } T, P, x_i \tag{9.15}$$

Comparison of the expressions (9.13) and (9.15) for dV gives (after division by dn): $f = \Sigma_i x_i \bar{V}_i$. Equation (9.12) becomes $V = nf = n \Sigma_i x_i \bar{V}_i$ or (since $x_i = n_i/n$)

$$V = \sum_i n_i \bar{V}_i \qquad \text{one-phase syst.} \tag{9.16}*$$

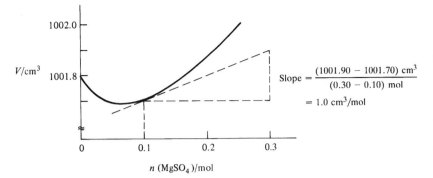

Figure 9.3

Volumes at 20°C and 1 atm of solutions containing 1000 g of water and n moles of $MgSO_4$. The dashed lines are used to find that $\bar{V}_{MgSO_4} = 1.0$ cm³/mol at molality 0.1 mol/kg.

This key result expresses the solution's volume V in terms of the partial molar volumes $\bar{V}_i$ of the components of the solution, where each $\bar{V}_i$ [Eq. (9.9)] is evaluated at the temperature, pressure, and mole fractions of the solution.

The change in volume on mixing the solution from its pure components at constant T and P is given by the difference of (9.16) and (9.4):

$$\Delta_{mix}V \equiv V - V^* = \sum_i n_i(\bar{V}_i - V^*_{m,i}) \qquad \text{const. } T, P \qquad (9.17)$$

where mix stands for mixing (and not for mixture).

Equation (9.16) is sometimes written as $V_m = \sum_i x_i \bar{V}_i$, where the **mean molar volume** V_m of the solution is [Eq. (8.11)] $V_m \equiv V/n$, with $n \equiv \sum_i n_i$.

Measurement of Partial Molar Volumes

Consider a solution composed of substances A and B. To measure $\bar{V}_B \equiv (\partial V/\partial n_B)_{T,P,n_A}$, we prepare solutions at the desired T and P all of which contain a fixed number of moles of component A but varying values of n_B. We then plot the measured solution volumes V versus n_B. The slope of the V-versus-n_B curve at any composition is then $\bar{V}_B$ for that composition. The slope at any point on a curve is found by drawing the tangent line at that point and measuring its slope.

Once $\bar{V}_B$ has been found by the *slope method,* $\bar{V}_A$ can be calculated using $V = n_A\bar{V}_A + n_B\bar{V}_B$ [Eq. (9.16)]; we have $\bar{V}_A = (V - n_B\bar{V}_B)/n_A$.

Figure 9.3 plots V versus $n(MgSO_4)$ for $MgSO_4(aq)$ solutions that contain a fixed amount (1000 g or 55.5 mol) of the solvent (H_2O) at 20°C and 1 atm. For 1000 g of solvent, n_B is numerically equal to the solute molality in mol/kg.

EXAMPLE 9.3 The slope method for partial molar volume

Use Fig. 9.3 to find $\bar{V}_{MgSO_4}$ and $\bar{V}_{H_2O}$ in $MgSO_4(aq)$ at 20°C and 1 atm with molality 0.1 mol/kg.

Drawing the tangent line at 0.1 mol of $MgSO_4$ per kg of H_2O, one finds its slope to be 1.0 cm³/mol, as shown in Fig. 9.3. Hence $\bar{V}_{MgSO_4} = 1.0$ cm³/mol at $m_{MgSO_4} = 0.1$ mol/kg. At $m_{MgSO_4} = 0.1$ mol/kg, the solution's volume is $V = 1001.70$ cm³. This solution has 0.10 mol of $MgSO_4$ and 1000 g of H_2O, which is 55.51 mol of H_2O. Use of $V = n_A\bar{V}_A + n_B\bar{V}_B$ gives

$$V = 1001.70 \text{ cm}^3 = (55.51 \text{ mol})\bar{V}_{H_2O} + (0.10 \text{ mol})(1.0 \text{ cm}^3/\text{mol})$$

$$\bar{V}_{H_2O} = 18.04 \text{ cm}^3/\text{mol}$$

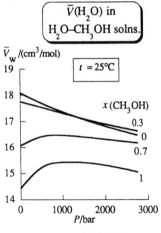

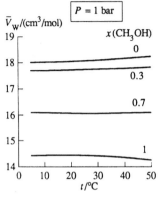

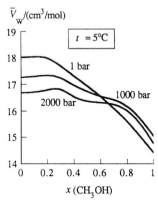

Figure 9.4

Partial molar volumes $\bar{V}_W$ of water in solutions of water (W) and methanol (M). The $x_M = 1$ curves are infinite-dilution values. [Data from A. J. Easteal and L. A. Woolf, *J. Chem. Thermodyn.*, **17**, 49 (1985).]

Exercise

Find $\bar{V}_{MgSO_4}$ and $\bar{V}_{H_2O}$ in 0.20 mol/kg $MgSO_4(aq)$ at 20°C and 1 atm. (*Answers:* 2.2 cm³/mol, 18.04 cm³/mol.)

Because of strong attractions between the solute ions and the water molecules, the solution's volume V in Fig. 9.3 initially decreases with increasing n_{MgSO_4} at fixed n_{H_2O}. The negative slope means that the partial molar volume $\bar{V}_{MgSO_4}$ is negative for molalities less than 0.07 mol/kg. The tight packing of water molecules in the solvation shells around the ions makes the volume of a dilute $MgSO_4$ solution less than the volume of the pure water used to prepare the solution, and $\bar{V}_{MgSO_4}$ is negative.

The value of $\bar{V}_i$ in the limit as the concentration of solute i goes to zero is the **infinite-dilution** partial molar volume of i and is symbolized by $\bar{V}_i^\infty$. To find $\bar{V}_i^\infty$ of $MgSO_4$ in water at 20°C, one draws in Fig. 9.3 the line tangent to the curve at $n_{MgSO_4} = 0$ and takes its slope. Some $\bar{V}_i^\infty$ values for solutes in aqueous solution at 25°C and 1 atm compared with the molar volumes $V_{m,i}^*$ of the pure solutes are:

Solute	NaCl	Na₂SO₄	MgSO₄	H₂SO₄	CH₃OH	n-C₃H₇OH
$\bar{V}_i^\infty/(cm^3/mol)$	16.6	11.6	−7.0	14.1	38.7	70.7
$V_{m,i}^*/(cm^3/mol)$	27.0	53.0	45.3	53.5	40.7	75.1

For a two-component solution, there is only one independent mole fraction, so $\bar{V}_A = \bar{V}_A(T, P, x_A)$ and $\bar{V}_B = \bar{V}_B(T, P, x_A)$ [Eq. (9.9)], in agreement with the fact that a two-component, one-phase system has 3 degrees of freedom. For solutions of water (W) and methanol (M), Fig. 9.4 shows temperature, pressure, and composition dependences of the partial molar volume $\bar{V}_W$. At $x_M = 0$, the solution is pure water, and the value $\bar{V}_W = 18.07$ cm³/mol at 25°C and 1 bar in Fig. 9.4 is V_m^* of pure H_2O at 25°C and 1 bar.

Other Partial Molar Quantities

The ideas just developed for the volume V apply to any extensive property of the solution. For example, the solution's internal energy U is a function of $T, P, n_1, \ldots, n_r$ [Eq. (9.5)], and by analogy with $\bar{V}_i \equiv (\partial V/\partial n_i)_{T,P,n_{j\neq i}}$ [Eq. (9.7)], the **partial molar internal energy** $\bar{U}_i$ of component i in the solution is defined by

$$\bar{U}_i \equiv (\partial U/\partial n_i)_{T,P,n_{j\neq i}} \qquad \text{one-phase syst.} \qquad (9.18)$$

The same arguments that gave $V = \Sigma_i\, n_i \bar{V}_i$ [Eq. (9.16)] give (simply replace the symbol V by U in all the equations of the derivation)

$$U = \sum_i n_i \bar{U}_i \qquad \text{one-phase syst.} \qquad (9.19)$$

where U is the internal energy of the solution.

We also have partial molar enthalpies $\bar{H}_i$, partial molar entropies $\bar{S}_i$, partial molar Helmholtz energies $\bar{A}_i$, partial molar Gibbs energies $\bar{G}_i$, and partial molar heat capacities $\bar{C}_{P,i}$:

$$\bar{H}_i \equiv (\partial H/\partial n_i)_{T,P,n_{j\neq i}}, \qquad \bar{S}_i \equiv (\partial S/\partial n_i)_{T,P,n_{j\neq i}} \qquad (9.20)$$

$$\bar{G}_i \equiv (\partial G/\partial n_i)_{T,P,n_{j\neq i}}, \qquad \bar{C}_{P,i} \equiv (\partial C_P/\partial n_i)_{T,P,n_{j\neq i}} \qquad (9.21)$$

where H, S, G, and C_P are the solution's enthalpy, entropy, Gibbs energy, and heat capacity. *All partial molar quantities are defined with T, P, and $n_{j\neq i}$ held constant.*

The partial molar Gibbs energy is especially important since it is identical to the chemical potential [Eq. (4.72)]:

$$\bar{G}_i \equiv \left(\frac{\partial G}{\partial n_i}\right)_{T,P,n_{j\neq i}} \equiv \mu_i \qquad \text{one-phase syst.} \qquad \textbf{(9.22)}^*$$

Analogous to (9.16) and (9.19), the Gibbs energy G of a solution is

$$G = \sum_i n_i \bar{G}_i \equiv \sum_i n_i \mu_i \qquad \text{one-phase syst.} \qquad (9.23)$$

Equations like (9.23) and (9.19) show the key role of partial molar properties in solution thermodynamics. Each extensive property of a solution is expressible in terms of partial molar quantities.

If Y is any extensive property of a solution, the corresponding partial molar property of component i of the solution is defined by

$$\bar{Y}_i \equiv (\partial Y/\partial n_i)_{T,P,n_{j \neq i}} \qquad \qquad \textbf{(9.24)*}$$

Partial molar quantities are the ratio of two infinitesimal extensive quantities and so are intensive properties. Analogous to (9.8), dY is

$$dY = \left(\frac{\partial Y}{\partial T}\right)_{P,n_i} dT + \left(\frac{\partial Y}{\partial P}\right)_{T,n_i} dP + \sum_i \bar{Y}_i \, dn_i \qquad (9.25)$$

The same reasoning that led to (9.16) gives for the Y value of the solution

$$Y = \sum_i n_i \bar{Y}_i \qquad \text{one-phase syst.} \qquad \textbf{(9.26)*}$$

Equation (9.26) suggests that we view $n_i \bar{Y}_i$ as the contribution of solution component i to the extensive property Y of the phase. However, such a view is oversimplified. The partial molar quantity $\bar{Y}_i$ is a function of T, P, and the solution mole fractions. Because of intermolecular interactions, $\bar{Y}_i$ is a property of the solution as a whole, and not a property of component i alone.

As noted at the end of Sec. 7.1, for a system in equilibrium, the equation $dG = -S \, dT + V \, dP + \sum_i \mu_i \, dn_i$ is valid whether the sum is taken over all species actually present or over only the independent components. Similarly, the relations $G = \sum_i n_i \bar{G}_i$ and $Y = \sum_i n_i \bar{Y}_i$ [Eqs. (9.23) and (9.26)] are valid if the sum is taken over all chemical species, using the actual number of moles of each species present, or over only the independent components, using the apparent numbers of moles present and ignoring chemical reactions. The proof is essentially the same as given in Prob. 7.53.

Relations between Partial Molar Quantities

For most of the thermodynamic relations between extensive properties of a homogeneous system, there are corresponding relations with the extensive variables replaced by partial molar quantities. For example, G, H, and S of a solution satisfy

$$G = H - TS \qquad (9.27)$$

If we differentiate (9.27) partially with respect to n_i at constant T, P, and $n_{j \neq i}$ and use the definitions (9.20) to (9.22) of $\bar{H}_i$, $\bar{G}_i$, and $\bar{S}_i$, we get

$$(\partial G/\partial n_i)_{T,P,n_{j \neq i}} = (\partial H/\partial n_i)_{T,P,n_{j \neq i}} - T(\partial S/\partial n_i)_{T,P,n_{j \neq i}}$$

$$\mu_i \equiv \bar{G}_i = \bar{H}_i - T\bar{S}_i \qquad (9.28)$$

which corresponds to (9.27).

Another example is the first equation of (4.70):

$$\left(\frac{\partial G}{\partial T}\right)_{P,n_j} = -S \qquad (9.29)$$

Partial differentiation of (9.29) with respect to n_i gives

$$-\left(\frac{\partial S}{\partial n_i}\right)_{T,P,n_{j \neq i}} = \left(\frac{\partial}{\partial n_i}\left(\frac{\partial G}{\partial T}\right)_{P,n_j}\right)_{T,P,n_{j \neq i}} = \left(\frac{\partial}{\partial T}\left(\frac{\partial G}{\partial n_i}\right)_{T,P,n_{j \neq i}}\right)_{P,n_j}$$

where $\partial^2 z/(\partial x\, \partial y) = \partial^2 z/(\partial y\, \partial x)$ [Eq. (1.36)] was used. Use of (9.20) and (9.22) gives

$$\left(\frac{\partial \mu_i}{\partial T}\right)_{P,n_j} \equiv \left(\frac{\partial \bar{G}_i}{\partial T}\right)_{P,n_j} = -\bar{S}_i \qquad (9.30)$$

which corresponds to (9.29) with extensive variables replaced by partial molar quantities. Similarly, partial differentiation with respect to n_i of $(\partial G/\partial P)_{T,n_j} = V$ leads to

$$\left(\frac{\partial \mu_i}{\partial P}\right)_{T,n_j} \equiv \left(\frac{\partial \bar{G}_i}{\partial P}\right)_{T,n_j} = \bar{V}_i \qquad (9.31)$$

The subscript n_j in (9.31) indicates that all mole numbers are held constant.

Importance of the Chemical Potentials

The chemical potentials are the key properties in chemical thermodynamics. The μ_i's determine reaction equilibrium and phase equilibrium [Eqs. (4.88) and (4.98)]. Moreover, *all other partial molar properties and all thermodynamic properties of the solution can be found from the μ_i's* if we know the chemical potentials as functions of T, P, and composition. The partial derivatives of μ_i with respect to T and P give $-\bar{S}_i$ and $\bar{V}_i$ [Eqs. (9.30) and (9.31)]. The use of $\mu_i = \bar{H}_i - T\bar{S}_i$ [Eq. (9.28)] then gives $\bar{H}_i$. The use of $\bar{U}_i = \bar{H}_i - P\bar{V}_i$ (Prob. 9.19) and $\bar{C}_{P,i} = (\partial \bar{H}_i/\partial T)_{P,n_j}$ gives $\bar{U}_i$ and $\bar{C}_{P,i}$. Once we know the partial molar quantities μ_i, $\bar{S}_i$, $\bar{V}_i$, etc., we get the solution properties as $G = \sum_i n_i \bar{G}_i$, $S = \sum_i n_i \bar{S}_i$, $V = \sum_i n_i \bar{V}_i$, etc. [Eq. (9.26)]. Note that knowing V as a function of T, P, and composition means we know the equation of state of the solution.

EXAMPLE 9.4 Use of μ_i to get $\bar{V}_i$

Find $\bar{V}_i$ for a component of an ideal gas mixture starting from μ_i.

The chemical potential of a component of an ideal gas mixture is [Eq. (6.4)]

$$\mu_i = \mu_i^\circ(T) + RT \ln (P_i/P^\circ) = \mu_i^\circ(T) + RT \ln (x_i P/P^\circ)$$

Use of $\bar{V}_i = (\partial \mu_i/\partial P)_{T,n_j}$ [Eq. (9.31)] gives, in agreement with (9.11),

$$\bar{V}_i = RT(\partial[\ln (x_i P/P^\circ)]/\partial P)_{T,n_j} = RT/P$$

Exercise

Use the result $\bar{V}_i = RT/P$ to verify the relation $V = \sum_i n_i \bar{V}_i$ [Eq. (9.16)] for an ideal gas mixture.

Summary

The partial molar volume $\bar{V}_i$ of component i in a solution of volume V is defined as $\bar{V}_i \equiv (\partial V/\partial n_i)_{T,P,n_{j\neq i}}$. The solution's volume is given by $V = \sum_i n_i \bar{V}_i$. Similar equations hold for other extensive properties (U, H, S, G, etc.). Relations between $\bar{G}_i$, $\bar{H}_i$, $\bar{S}_i$, and $\bar{V}_i$ were found; these resemble corresponding relations between G, H, S, and V. All thermodynamic properties of a solution can be obtained if the chemical potentials $\mu_i \equiv \bar{G}_i$ are known as functions of T, P, and composition.

9.3 MIXING QUANTITIES

Similar to defining $\Delta_{\mathrm{mix}} V \equiv V - V^*$ at constant T and P [Eq. (9.17)], one defines other mixing quantities for a solution. For example,

$$\Delta_{\mathrm{mix}} H \equiv H - H^*, \qquad \Delta_{\mathrm{mix}} S \equiv S - S^*, \qquad \Delta_{\mathrm{mix}} G \equiv G - G^*$$

where H, S, and G are properties of the solution and H^*, S^*, and G^* are properties of the pure unmixed components at the same T and P as the solution.

The key mixing quantity is $\Delta_{\text{mix}}G = G - G^*$. The Gibbs energy G of the solution is given by Eq. (9.23) as $G = \sum_i n_i\bar{G}_i$ (where $\bar{G}_i$ is a partial molar quantity). The Gibbs energy G^* of the unmixed components is $G^* = \sum_i n_i G^*_{\text{m},i}$ (where $G^*_{\text{m},i}$ is the molar Gibbs energy of pure substance i). Therefore

$$\Delta_{\text{mix}}G \equiv G - G^* = \sum_i n_i(\bar{G}_i - G^*_{\text{m},i}) \qquad \text{const. } T, P \qquad (9.32)$$

which is similar to (9.17) for $\Delta_{\text{mix}}V$. We have

$$\Delta_{\text{mix}}G = \Delta_{\text{mix}}H - T\,\Delta_{\text{mix}}S \qquad \text{const. } T, P \qquad (9.33)$$

which is a special case of $\Delta G = \Delta H - T\,\Delta S$ at constant T.

Just as $\bar{S}_i$ and $\bar{V}_i$ can be found as partial derivatives of $\bar{G}_i$ [Eqs. (9.30) and (9.31)], $\Delta_{\text{mix}}S$ and $\Delta_{\text{mix}}V$ can be found as partial derivatives of $\Delta_{\text{mix}}G$. Taking $(\partial/\partial P)_{T,n_j}$ of (9.32), we have

$$\left(\frac{\partial \Delta_{\text{mix}}G}{\partial P}\right)_{T,n_j} = \frac{\partial}{\partial P}\sum_i n_i(\bar{G}_i - G^*_{\text{m},i}) = \sum_i n_i\left[\left(\frac{\partial \bar{G}_i}{\partial P}\right)_{T,n_j} - \left(\frac{\partial G^*_{\text{m},i}}{\partial P}\right)_T\right]$$

$$= \sum_i n_i(\bar{V}_i - V^*_{\text{m},i})$$

$$\left(\frac{\partial \Delta_{\text{mix}}G}{\partial P}\right)_{T,n_j} = \Delta_{\text{mix}}V \qquad (9.34)$$

where (9.31), (4.51), and (9.17) were used.

Similarly, taking $(\partial/\partial T)_{P,n_j}$ of (9.32), one finds (Prob. 9.21)

$$\left(\frac{\partial \Delta_{\text{mix}}G}{\partial T}\right)_{P,n_j} = -\Delta_{\text{mix}}S \qquad (9.35)$$

The partial molar relations and mixing relations of the last section and this one are easily written down, since they resemble equations involving G. Thus, (9.28) and (9.33) resemble $G = H - TS$, (9.30) and (9.35) resemble $(\partial G/\partial T)_P = -S$ [Eq. (4.51)], and (9.31) and (9.34) resemble $(\partial G/\partial P)_T = V$ [Eq. (4.51)].

The changes $\Delta_{\text{mix}}V$, $\Delta_{\text{mix}}U$, $\Delta_{\text{mix}}H$, and $\Delta_{\text{mix}}C_P$ that accompany solution formation are due entirely to changes in intermolecular interactions (both energetic and structural). However, changes in S, A, and G result not only from changes in intermolecular interactions but also from the unavoidable increase in entropy that accompanies the constant-T-and-P mixing of substances and the simultaneous increase in volume each component occupies. Even if the intermolecular interactions in the solution are the same as in the pure substances, $\Delta_{\text{mix}}S$ and $\Delta_{\text{mix}}G$ will still be nonzero.

It might be thought that $\Delta_{\text{mix}}S$ at constant T and P will always be positive, since a solution seems intuitively to be more disordered than the separated pure components. It is true that the contribution of the volume increase of each component to $\Delta_{\text{mix}}S$ is always positive. However, the contribution of changing intermolecular interactions can be either positive or negative and sometimes is sufficiently negative to outweigh the contribution of the volume increases. For example, for mixing 0.5 mol H_2O and 0.5 mol $(C_2H_5)_2NH$ at 49°C and 1 atm, experiment gives $\Delta_{\text{mix}}S = -8.8$ J/K. This can be ascribed to stronger hydrogen bonding between the amine and water than the average of the hydrogen-bond strengths in the pure components. The mixing here is highly exothermic, so that ΔS_{surr} is larger than $|\Delta S_{\text{syst}}|$, ΔS_{univ} is positive, and $\Delta_{\text{mix}}G = \Delta_{\text{mix}}H - T\,\Delta_{\text{mix}}S$ is negative (Fig. 9.5).

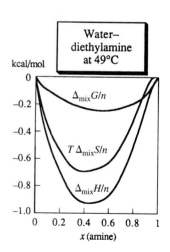

Figure 9.5

Thermodynamic mixing quantities for solutions of water + diethylamine at 49°C and 1 atm. Note that $\Delta_{\text{mix}}S$ is negative. n is the total number of moles.

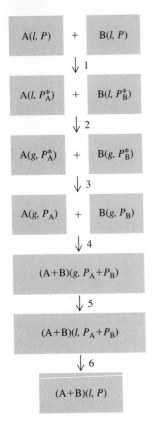

Figure 9.6

Six-step isothermal process to
convert pure liquids A and B at
pressure P to a solution of A + B
at P. P_A^* and P_B^* are the vapor
pressures of pure A and pure B,
and P_A and P_B are the partial
vapor pressures of the solution of
A + B.

Mixing quantities such as $\Delta_{\text{mix}}V$, $\Delta_{\text{mix}}H$, and $\Delta_{\text{mix}}S$ tell us about intermolecular interactions in the solution as compared with those in the pure components. Unfortunately, interpretation of mixing quantities of liquids in terms of molecular interactions is difficult; see *Rowlinson and Swinton*, chap. 5, for examples.

Experimental Determination of Mixing Quantities

$\Delta_{\text{mix}}V$ is easily found from density measurements on the solution and the pure components or from direct measurement of the volume change on isothermal mixing of the components. $\Delta_{\text{mix}}H$ at constant T and P is easily measured in a constant-pressure calorimeter.

How do we get $\Delta_{\text{mix}}G$? $\Delta_{\text{mix}}G$ is calculated from vapor-pressure measurements. One measures the partial pressures P_A and P_B of A and B in the vapor in equilibrium with the solution and measures the vapor pressures P_A^* and P_B^* of pure A and pure B at the temperature of the solution. The hypothetical isothermal path of Fig. 9.6 starts with the pure liquids A and B at T and P and ends with the liquid solution at T and P. Therefore ΔG for this six-step process equals $\Delta_{\text{mix}}G$. One uses thermodynamic relations to express ΔG of each step in terms of P_A, P_B, P_A^*, and P_B^*, thereby obtaining $\Delta_{\text{mix}}G$ in terms of these vapor pressures. If the gases A and B are assumed ideal and the slight changes in G in steps 1 and 6 are neglected, the result is (Prob. 9.62)

$$\Delta_{\text{mix}}G = n_A RT \ln (P_A/P_A^*) + n_B RT \ln (P_B/P_B^*)$$

$\Delta_{\text{mix}}S$ is found from $\Delta_{\text{mix}}G$ and $\Delta_{\text{mix}}H$ using $\Delta_{\text{mix}}G = \Delta_{\text{mix}}H - T\,\Delta_{\text{mix}}S$.

9.4   DETERMINATION OF PARTIAL MOLAR QUANTITIES

Partial Molar Volumes

A method for finding partial molar volumes in a two-component solution that is more accurate than the slope method of Fig. 9.3 in Sec. 9.2 is the following. Let $n \equiv n_A + n_B$ be the total number of moles in the solution. One plots $\Delta_{\text{mix}}V/n$ [where $\Delta_{\text{mix}}V$ is defined by (9.17)] against the B mole fraction x_B. One draws the tangent line to the curve at some particular composition x_B' (see Fig. 9.7). The intercept of this tangent line with the $\Delta_{\text{mix}}V/n$ axis (at $x_B = 0$ and $x_A = 1$) gives $\overline{V}_A - V_{m,A}^*$ at the composition x_B'; the intersection of this tangent line with the vertical line $x_B = 1$ gives $\overline{V}_B - V_{m,B}^*$ at x_B'. (Proof of these statements is outlined in Prob. 9.26.) Since the pure-component molar volumes $V_{m,A}^*$ and $V_{m,B}^*$ are known, we can then find the partial molar volumes $\overline{V}_A$ and $\overline{V}_B$ at x_B'.

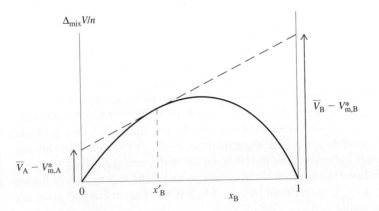

Figure 9.7

Accurate method to determine
partial molar volumes in a two-
component solution.

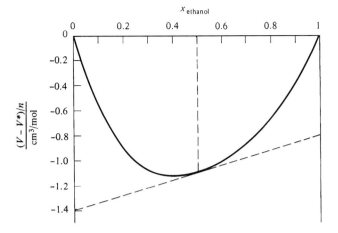

Figure 9.8

$\Delta_{mix}V/n$ for water–ethanol solutions at 20°C and 1 atm. The tangent line is used to find the partial molar volumes at ethanol mole fraction equal to 0.5.

EXAMPLE 9.5 Intercept method for $\bar{V}_i$

Figure 9.8 plots $\Delta_{mix}V/n$ against $x_{C_2H_5OH}$ for water–ethanol solutions at 20°C and 1 atm. Use this plot to find the partial molar volumes of water (W) and ethanol (E) in a solution with $x_E = 0.5$, given that at 20°C and 1 atm, V_m is 18.05 cm³/mol for water and 58.4 cm³/mol for ethanol.

The tangent line to the curve is drawn at $x_E = 0.5$. Its intercept at $x_E = 0$ is at −1.35 cm³/mol, so $\bar{V}_W - V^*_{m,W} = -1.35$ cm³/mol and $\bar{V}_W = 18.05$ cm³/mol − 1.35 cm³/mol = 16.7 cm³/mol at $x_E = 0.5$. The tangent line intersects $x_E = 1$ at −0.8 cm³/mol, so $\bar{V}_E - V^*_{m,E} = -0.8$ cm³/mol and $\bar{V}_E = 57.6$ cm³/mol at $x_E = 0.5$.

Exercise

Use these results for $\bar{V}_E$ and $\bar{V}_W$ and the equation $V = \Sigma_i\, n_i\bar{V}_i$ to calculate the volume of a mixture of 0.50 mol of water and 0.50 mol of ethanol at 20°C and 1 atm. Use $V = (V - V^*) + V^*$ and Fig. 9.8 to calculate this volume and compare the results. (*Answers:* 37.1_5 cm³, 37.1_4 cm³.)

Exercise

Use Fig. 9.8 to find $\bar{V}_E$ and $\bar{V}_W$ in a solution composed of 3.50 mol of ethanol and 1.50 mol of water at 20°C and 1 atm. (*Answers:* 58.0 cm³/mol, 16.0_5 cm³/mol.)

Drawing tangents at several solution compositions in Fig. 9.8 and using the intercepts to find the partial molar volumes at these compositions, one obtains the results shown in Fig. 9.9. This figure plots $\bar{V}_E$ and $\bar{V}_W$ against solution composition. Note that when $\bar{V}_E$ is decreasing, $\bar{V}_W$ is increasing, and vice versa. We will see in Sec. 10.3 that $d\bar{V}_A$ and $d\bar{V}_B$ must have opposite signs at constant T and P in a two-component solution. The limiting value of $\bar{V}_E$ at $x_E = 1$ is the molar volume of pure ethanol.

A third way to find partial molar volumes is to fit solution volume data for fixed n_A to a polynomial in n_B. Differentiation then gives $\bar{V}_B$. See Prob. 9.25.

Partial Molar Enthalpies, Entropies, and Gibbs Energies

Similar to $V = \Sigma_i\, n_i\bar{V}_i$, the enthalpy H of a solution is given by $H = \Sigma_i\, n_i\bar{H}_i$ [Eq. (9.26)], where the partial molar enthalpy $\bar{H}_i$ of substance i is $\bar{H}_i \equiv (\partial H/\partial n_i)_{T,P,n_{j\neq i}}$ [Eq. (9.20)]. The enthalpy of mixing to form the solution from its pure components at con-

Figure 9.9

Partial molar volumes in water–ethanol solutions at 20°C and 1 atm.

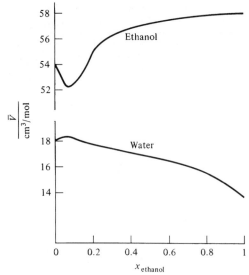

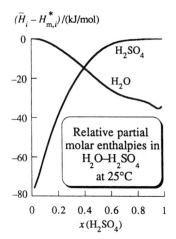

Figure 9.10

Relative partial molar enthalpies in H_2O–H_2SO_4 solutions at 25°C and 1 atm. [Data from F. J. Zeleznik, *J. Phys. Chem. Ref. Data*, **20**, 1157 (1991).]

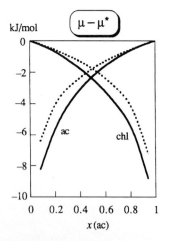

Figure 9.11

Relative partial molar Gibbs energies (chemical potentials) in acetone–chloroform solutions at 35°C and 1 atm.

stant T and P is $\Delta_{\text{mix}}H = H - H^* = \sum_i n_i(\bar{H}_i - H^*_{\text{m},i})$, which is similar to (9.17) for $\Delta_{\text{mix}}V$. For a two-component solution

$$\Delta_{\text{mix}}H = n_A(\bar{H}_A - H^*_{\text{m,A}}) + n_B(\bar{H}_B - H^*_{\text{m,B}}) \qquad (9.36)$$

Although we can measure the volume V of a solution, we cannot measure its enthalpy H, since only enthalpy differences can be measured. We therefore deal with the enthalpy of the solution relative to the enthalpy of some reference system, which we may take to be the unmixed components.

Similar to the procedure of Fig. 9.8, we plot $\Delta_{\text{mix}}H/n$ against x_B and draw the tangent line at some composition x'_B. The intercepts of the tangent line at $x_B = 0$ and $x_B = 1$ give $\bar{H}_A - H^*_{\text{m,A}}$ and $\bar{H}_B - H^*_{\text{m,B}}$, respectively, at x'_B. We thus determine partial molar enthalpies relative to the molar enthalpies of the pure components. Figure 9.10 shows relative partial molar enthalpies in H_2O–H_2SO_4 solutions at 25°C and 1 atm.

From experimental $\Delta_{\text{mix}}S$ and $\Delta_{\text{mix}}G$ data, one obtains relative partial molar entropies and Gibbs energies $\bar{S}_i - S^*_{\text{m},i}$ and $\mu_i - \mu_i^*$ by the same procedure as for $\bar{H}_i - H^*_{\text{m},i}$. The solid lines in Fig. 9.11 show $\mu_i - \mu_i^*$ for the components in acetone–chloroform solutions at 35°C and 1 atm. Note that μ_i goes to $-\infty$ as x_i goes to 0. One finds that $\bar{S}_i$ goes to $+\infty$ as x_i goes to 0 [see Eq. (9.28)]. This behavior is explained in Sec. 9.8, where the meaning of the dashed lines in Fig. 9.11 is given.

Integral and Differential Heats of Solution

For a two-component solution, the quantity $\Delta_{\text{mix}}H/n_B$ is called the *integral heat of solution per mole of* B in the solvent A and is symbolized by $\Delta H_{\text{int,B}}$:

$$\Delta H_{\text{int,B}} \equiv \Delta_{\text{mix}}H/n_B \qquad (9.37)$$

where $\Delta_{\text{mix}}H$ is given by (9.36). $\Delta H_{\text{int,B}}$ is an intensive property that depends on T, P, and x_B. Physically, $\Delta H_{\text{int,B}}$ is numerically equal to the heat absorbed by the system when 1 mole of pure B is added at constant T and P to enough pure A to produce a solution of the desired mole fraction x_B. The limit of $\Delta H_{\text{int,B}}$ as the solvent mole fraction x_A goes to 1 is the *integral heat of solution at infinite dilution* $\Delta H^\infty_{\text{int,B}}$ *per mole of* B in A. The quantity $\Delta H^\infty_{\text{int,B}}$ equals the heat absorbed by the system when 1 mole of solute B is dissolved in an infinite amount of solvent A at constant T and P. Figure 9.12 plots $\Delta H_{\text{int,H}_2\text{SO}_4}$ versus $x_{\text{H}_2\text{SO}_4}$ for H_2SO_4 in water at 25°C and 1 atm. At $x_B = 1$, $\Delta H_{\text{int,B}} = 0$, since $\Delta_{\text{mix}}H = 0$ and $n_B \neq 0$ at $x_B = 1$.

The integral heat of solution per mole of B involves the addition of 1 mole of B to pure A to produce the solution, a process in which the B mole fraction changes from zero to its final value x_B. Suppose, instead, that we add (at constant T and P) 1 mole of B to an infinite volume of solution whose B mole fraction is x_B. The solution composition will remain fixed during this process. The enthalpy change per mole of added B when B is added at constant T and P to a solution of fixed composition is called the *differential heat of solution* of B in A and is symbolized by $\Delta H_{diff,B}$. The quantity $\Delta H_{diff,B}$ is an intensive function of T, P, and solution composition. From the preceding definitions, it follows that at infinite dilution the differential and integral heats of solution become equal: $\Delta H_{int,B}^{\infty} = \Delta H_{diff,B}^{\infty}$ [see Figs. 9.10 and 9.12 and Eq. (9.38)].

Rather than imagine a solution of infinite volume, we can imagine adding at constant T and P an infinitesimal amount dn_B of B to a solution of finite volume and with composition x_B. If dH is the enthalpy change for this infinitesimal process, then $\Delta H_{diff,B} = dH/dn_B$ at the composition x_B. When dn_B moles of pure B is added to the solution at constant T and P, the solution's enthalpy changes by $dH_{soln} = \bar{H}_B \, dn_B$ [this follows from the definition $\bar{H}_B \equiv (\partial H_{soln}/\partial n_B)_{T,P,n_A}$] and the enthalpy of pure B changes by $dH_B^* = -H_{m,B}^* \, dn_B$ (since $H_B^* = n_B H_{m,B}^*$). The overall enthalpy change for this addition is then $dH = \bar{H}_B \, dn_B - H_{m,B}^* \, dn_B$, and $\Delta H_{diff} \equiv dH/dn_B = \bar{H}_B - H_{m,B}^*$. Thus

$$\Delta H_{diff,B} = \bar{H}_B - H_{m,B}^* \qquad (9.38)$$

The differential heat of solution of B equals the partial molar enthalpy of B in the solution minus the molar enthalpy of pure B. Determination of the relative partial molar enthalpy $\bar{H}_B - H_{m,B}^*$ was discussed earlier in this section. Figure 9.10 plots differential heats of solution in H_2O–H_2SO_4 solutions at 25°C; here, either H_2O or H_2SO_4 can be regarded as the solvent.

Some values of differential heats of solution (relative partial molar enthalpies) of solutes in aqueous solution at infinite dilution at 25°C and 1 bar are:

Solute	NaCl	K_2SO_4	LiOH	CH_3COOH	CH_3OH	$CO(NH_2)_2$
$(\bar{H}_B^{\infty} - H_{m,B}^*)/(kJ/mol)$	3.9	23.8	−23.6	−1.5	−7.3	15.1

If B is a solid at 25°C, $H_{m,B}^*$ in this table refers to solid B. Dissolving a tiny amount of NaCl in water at 25°C is an endothermic process, whereas dissolving a tiny amount of LiOH in water is exothermic.

The NBS thermodynamics tables cited in Sec. 5.9 list the *apparent* enthalpy of formation for several solutes in aqueous solutions of various concentrations. To see what these apparent enthalpies mean, consider H_2SO_4. The NBS tables give $\Delta_f H_{298}^{\circ} = -813.99$ kJ/mol for pure $H_2SO_4(l)$ and $\Delta_f H_{298}^{\circ} = -855.44$ kJ/mol for H_2SO_4 dissolved in 2 moles of water. This means that, for mixing 1 mol of $H_2SO_4(l)$ and 2 mol of $H_2O(l)$ to form a solution, the enthalpy change is $[-855.44 - (-813.99)]$ kJ/mol $= -41.45$ kJ/mol at 25°C and 1 bar. In other words, the integral heat of solution per mole of H_2SO_4 for formation of an H_2O–H_2SO_4 solution with $x_{H_2SO_4} = \frac{1}{3}$ is $\Delta H_{int,H_2SO_4} = -41.45$ kJ/mol at 25°C and 1 bar. The exothermicity of mixing water and sulfuric acid is well known.

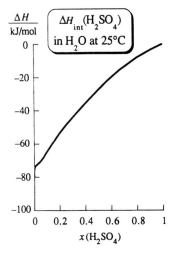

Figure 9.12

Integral heat of solution of H_2SO_4 in water at 25°C and 1 atm versus H_2SO_4 mole fraction.

EXAMPLE 9.6 Differential heats of solution

For aqueous H_2SO_4 solutions at 25°C and 1 bar, integral heats of solution per mole of H_2SO_4 versus H_2SO_4 mole fraction are (B $\equiv H_2SO_4$ and A $\equiv H_2O$):

$\Delta H_{int,B}/(kJ/mol)$	−15.55	−27.80	−35.90	−41.45	−48.92
x_B	0.6667	0.500	0.400	0.3333	0.250

Find the differential heats of solution (relative partial molar enthalpies) of H_2O and H_2SO_4 at $x_{H_2SO_4} = 0.400$.

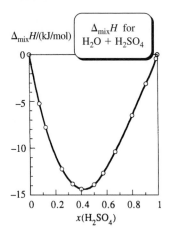

Figure 9.13

$\Delta_{mix}H/n$ for $H_2O + H_2SO_4$ solutions at 25°C and 1 atm.

The differential heats $\bar{H}_A - H^*_{m,A}$ and $\bar{H}_B - H^*_{m,B}$ [Eq. (9.38)] can be found from a plot of $\Delta_{mix}H/n$ versus x_B (similar to Fig. 9.8). The $\Delta H_{int,B}$ values given are $\Delta_{mix}H/n_B$ [Eq. (9.37)]. Therefore, we multiply $\Delta H_{int,B} = \Delta_{mix}H/n_B$ by $n_B/n = x_B$ to get $\Delta_{mix}H/n$. We get (of course, $\Delta_{mix}H$ is zero at $x_B = 0$ and at $x_B = 1$):

$(\Delta_{mix}H/n)/(kJ/mol)$	0	−10.37	−13.90	−14.36	−13.82	−12.23	0
x_B	1	0.6667	0.500	0.400	0.3333	0.250	0

Plotting the data (Fig. 9.13) and drawing the tangent line at $x_B = 0.400$, one finds (Prob. 9.27) the intercept at $x_B = 0$ to be −13.2 kJ/mol $= \Delta H_{diff,H_2O}$ and the intercept at $x_B = 1$ to be −16.2 kJ/mol $= \Delta H_{diff,H_2SO_4}$.

Exercise

Show that for a two-component solution, $\Delta_{mix}H = n_A \Delta H_{diff,A} + n_B \Delta H_{diff,B}$. Then verify that the ΔH_{diff} values calculated in this example satisfy this equation for a solution of 0.400 mol of H_2SO_4 and 0.600 mol of H_2O.

9.5   IDEAL SOLUTIONS

The discussion in Secs. 9.1 to 9.4 applies to all solutions. The rest of this chapter deals with special kinds of solutions. This section and the next consider ideal solutions.

The molecular picture of an ideal gas mixture is one with no intermolecular interactions. For a condensed phase (solid or liquid), the molecules are close together, and we could never legitimately assume no intermolecular interactions. Our molecular picture of a liquid or solid **ideal solution** (also called an **ideal mixture**) will be *a solution where the molecules of the various species are so similar to one another that replacing molecules of one species with molecules of another species will not change the spatial structure or the intermolecular interaction energy in the solution.*

Consider a solution of two species B and C. To prevent change in spatial structure of the liquids (or solids) on mixing B and C, the B molecules must be essentially the same size and shape as the C molecules. To prevent change in the intermolecular interaction energy on mixing, the intermolecular interaction energies should be essentially the same for B-B, B-C, and C-C pairs of molecules.

The closest resemblance occurs for isotopic species; for example, a mixture of $^{12}CH_3I$ and $^{13}CH_3I$. [Strictly speaking, even here there would be very slight departures from ideal behavior. The difference in isotopic masses leads to a difference in the magnitudes of molecular zero-point vibrations (Chapter 21), which causes the bond lengths and the dipole moments of the two isotopic species to differ very slightly. Hence the molecular sizes and intermolecular forces will differ *very* slightly for the isotopic species.] Apart from isotopic species, there are some pairs of liquids for which we would expect quite similar B-B, B-C, and C-C intermolecular interactions and quite similar B and C molecular volumes and hence would expect nearly ideal-solution behavior. Examples include benzene–toluene, n-C_7H_{16}–n-C_8H_{18}, C_2H_5Cl–C_2H_5Br, and $C(CH_3)_4$–$Si(CH_3)_4$.

The ideal-solution model serves as a reference point for discussing the behavior of real solutions. Deviations from ideal-solution behavior are due to differing B-B, B-C, and C-C intermolecular forces and to differing sizes and shapes of the B and C molecules, and these deviations can tell us something about the intermolecular interactions in the solution.

The preceding molecular definition of an ideal solution is not acceptable in thermodynamics, which is a macroscopic science. To arrive at a thermodynamic definition of an ideal solution, we examine $\Delta_{mix}G$ data. One finds that, when two liquids B and

C whose molecules resemble each other closely are mixed at constant T and P, the experimental $\Delta_{\text{mix}}G$ data (as found from vapor-pressure measurements—Sec. 9.3) satisfy the following equation for all solution compositions:

$$\Delta_{\text{mix}}G = RT(n_B \ln x_B + n_C \ln x_C) \qquad \text{ideal soln., const. } T, P \qquad (9.39)$$

where n_B, n_C, x_B, and x_C are the mole numbers and mole fractions of B and C in the solution and R is the gas constant. For example, $\Delta_{\text{mix}}G$ data for solutions of cyclopentane (C_5H_{10}) plus cyclohexane (C_6H_{12}) at 25°C and 1 atm as compared with the ideal-solution values $\Delta_{\text{mix}}G^{\text{id}}$ calculated from (9.39) are [M. B. Ewing and K. N. Marsh, *J. Chem. Thermodyn.*, **6**, 395 (1974)]:

$x_{C_6H_{12}}$	0.1	0.2	0.3	0.4	0.5	0.6	0.8
$(\Delta_{\text{mix}}G/n)/(\text{J/mol})$	−807	−1242	−1517	−1672	−1722	−1672	−1242
$(\Delta_{\text{mix}}G^{\text{id}}/n)/(\text{J/mol})$	−806	−1240	−1514	−1668	−1718	−1668	−1240

where $n \equiv n_C + n_B$. For a solution of C_6H_6 plus C_6D_6 (where $D \equiv {}^2H$) with $x_{C_6H_6} = 0.5$, experimental versus ideal-solution $\Delta_{\text{mix}}G$ values at various temperatures are [G. Jakli et al., *J. Chem. Phys.*, **68**, 3177 (1978)]:

t	10°C	25°C	50°C	80°C
$(\Delta_{\text{mix}}G/n)/(\text{J/mol})$	−1631.2	−1717.7	−1861.8	−2034.7
$(\Delta_{\text{mix}}G^{\text{id}}/n)/(\text{J/mol})$	−1631.8	−1718.3	−1862.3	−2035.2

We can show why (9.39) would very likely hold for ideal solutions. From the molecular definition, it is clear that formation of an ideal solution from the pure components at constant T and P is accompanied by no change in energy or volume: $\Delta_{\text{mix}}U = 0$ and $\Delta_{\text{mix}}V = 0$. Therefore $\Delta_{\text{mix}}H = \Delta_{\text{mix}}U + P\Delta_{\text{mix}}V = 0$.

What about $\Delta_{\text{mix}}S$? $\Delta_{\text{mix}}S$ is ΔS for the process of Fig. 9.14. We found in Sec. 3.7 that for a process in a closed system, $\Delta S \equiv S_2 - S_1 = k \ln (p_2/p_1)$ [Eq. (3.53)], where p_1 and p_2 are the probabilities of the initial and final states and k is Boltzmann's constant. The initial state has all the B molecules in the left portion of the container and all the C molecules in the right portion. The final state has the B and C molecules uniformly distributed throughout the container, with no change in T or P. The only difference between the initial and final states is in the spatial distribution of the molecules. Because the B and C molecules have no differences in intermolecular interactions or sizes and shapes, the B and C molecules have no preference as to their locations and will be distributed at random in the container. We want the probability ratio p_2/p_1 for the condition of random spatial distribution in state 2, with each molecule having no preference as to which molecules are its neighbors.

We could use probability theory to calculate p_1 and p_2, but this is unnecessary because we previously dealt with the same situation of the random distribution of two species in a container. When two ideal gases mix at constant T and P, there is random distribution of the B and C molecules. For both an ideal gas mixture and an ideal solution, the probability that any given molecule is in the left part of the mixture equals $V_B^*/(V_B^* + V_C^*) = V_B^*/V$, where V_B^* and V_C^* are the unmixed volumes of B and C and V is the mixture's volume. Therefore p_1 and p_2 will be the same for an ideal solution as for an ideal gas mixture, and $\Delta_{\text{mix}}S$, which equals $k \ln (p_2/p_1)$, will be the same for ideal solutions and ideal gas mixtures.

For ideal gases, Eq. (3.32) gives

$$\Delta_{\text{mix}}S = -n_B R \ln (V_B^*/V) - n_C R \ln (V_C^*/V)$$

and this equation gives $\Delta_{\text{mix}}S$ for ideal solutions. Since B and C molecules have the same size and the same intermolecular forces, B and C have equal molar volumes: $V_{m,B}^* = V_{m,C}^*$. Substitution of $V_B^* = n_B V_{m,B}^*$, $V_C^* = n_C V_{m,C}^* = n_C V_{m,B}^*$, and $V = V_B^* + V_C^*$

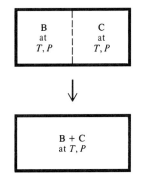

Figure 9.14

Mixing of two liquids at constant T and P.

$= (n_B + n_C)V^*_{m,B}$ into the above $\Delta_{mix}S$ equation gives $\Delta_{mix}S = -n_B R \ln x_B - n_C R \ln x_C$ for an ideal solution. (For a more rigorous statistical-mechanical derivation of this result, see *Rowlinson and Swinton,* p. 280.)

Substitution of $\Delta_{mix}H = 0$ and $\Delta_{mix}S = -n_B R \ln x_B - n_C R \ln x_C$ into $\Delta_{mix}G = \Delta_{mix}H - T\Delta_{mix}S$ then gives the experimentally observed $\Delta_{mix}G$ equation (9.39) for ideal solutions.

It might seem puzzling that an equation like (9.39) that applies to ideal *liquid* mixtures and *solid* mixtures would contain the *gas* constant R. However, R is a far more fundamental constant than simply the zero-pressure limit of PV/nT of a gas. R (in the form $R/N_A = k$) occurs in the fundamental equation (3.52) for entropy and occurs in other fundamental equations of statistical mechanics (Chapter 22).

As noted in Sec. 9.2, the chemical potentials μ_i in the solution are the key thermodynamic properties, so we now derive them from $\Delta_{mix}G$ of (9.39). We have $\Delta_{mix}G = G - G^* = \sum_i n_i\mu_i - \sum_i n_i\mu_i^*$ [Eq. (9.32)]. For an ideal solution, $\Delta_{mix}G = RT \sum_i n_i \ln x_i$ [Eq. (9.39)]. Equating these $\Delta_{mix}G$ expressions, we get

$$\sum_i n_i\mu_i = \sum_i n_i(\mu_i^* + RT \ln x_i) \tag{9.40}$$

where the sum identities (1.50) were used. The only way this last equation can hold for all n_i values is if (see Prob. 9.61 for a rigorous derivation)

$$\mu_i = \mu_i^*(T, P) + RT \ln x_i \qquad \text{ideal soln.} \tag{9.41}$$

where (since $\Delta_{mix}G$ is at constant T and P), $\mu_i^*(T, P)$ is the chemical potential of pure substance i at the temperature T and pressure P of the solution.

We shall adopt (9.41) as the thermodynamic definition of an **ideal solution.** *A solution is ideal if the chemical potential of every component in the solution obeys (9.41) for all solution compositions and for a range of T and P.*

Just as the ideal-gas law $PV = nRT$ is approached in the limit as the gas density goes to zero, the ideal-solution law (9.41) is approached in the limit as the solution components resemble one another more and more closely, without, however, becoming identical.

Figure 9.15 plots μ_i versus x_i at fixed T and P for an ideal solution, where $\mu_i = \mu_i^* + RT \ln x_i$. As $x_i \to 0$, $\mu_i \to -\infty$. As x_i increases, μ_i increases, reaching the chemical potential μ_i^* of pure i in the limit $x_i = 1$. Recall the general result that μ_i of a substance in a phase must increase as the i mole fraction x_i increases at constant T and P [Eq. (4.90)].

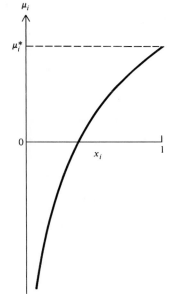

Figure 9.15

The chemical potential μ_i of a component of an ideal solution plotted versus x_i at fixed T and P.

Summary

$\Delta_{mix}G$ data (as found from vapor-pressure measurements) and statistical-mechanical arguments show that in solutions where the molecules of different species resemble one another extremely closely in size, shape, and intermolecular interactions, the chemical potential of each species is given by $\mu_i = \mu_i^*(T, P) + RT \ln x_i$; such a solution is called an ideal solution.

9.6   THERMODYNAMIC PROPERTIES OF IDEAL SOLUTIONS

In the last section we started with the molecular definition of an ideal solution and arrived at the thermodynamic definition (9.41). This section uses the chemical potentials (9.41) to derive thermodynamic properties of ideal solutions. Before doing so, we first define standard states for ideal-solution components.

Standard States

Standard states were defined for pure substances in Sec. 5.1 and for components of an ideal gas mixture in Sec. 6.1. The **standard state** of each component i of an ideal liquid solution is defined to be pure liquid i at the temperature T and pressure P of the solution. For solid solutions, we use the pure solids. We have $\mu_i^\circ = \mu_i^*(T, P)$, where, as always, *the degree superscript denotes the standard state and the star superscript indicates a pure substance.* The ideal-solution definition (9.41) then is

$$\mu_i = \mu_i^* + RT \ln x_i \qquad \text{ideal soln.} \qquad \textbf{(9.42)*}$$

$$\mu_i^\circ \equiv \mu_i^*(T, P) \qquad \text{ideal soln.} \qquad \textbf{(9.43)*}$$

where μ_i is the chemical potential of component i present with mole fraction x_i in an ideal solution at temperature T and pressure P, and μ_i^* is the chemical potential of pure i at the temperature and pressure of the solution.

Mixing Quantities

If we know mixing quantities such as $\Delta_{\text{mix}}G$, $\Delta_{\text{mix}}V$, and $\Delta_{\text{mix}}H$, then we know the values of G, V, H, etc., for the solution relative to values for the pure components. All the mixing quantities are readily obtained from the chemical potentials (9.42).

We have $\Delta_{\text{mix}}G = G - G^* = \Sigma_i n_i(\mu_i - \mu_i^*)$ [Eqs. (9.32) and (9.22)]. Equation (9.42) gives $\mu_i - \mu_i^* = RT \ln x_i$. Therefore

$$\Delta_{\text{mix}}G = RT \sum_i n_i \ln x_i \qquad \text{ideal soln., const. } T, P \qquad (9.44)$$

which is the same as (9.39). Since $0 < x_i < 1$, we have $\ln x_i < 0$ and $\Delta_{\text{mix}}G < 0$, as must be true for an irreversible (spontaneous) process at constant T and P.

From (9.34), $\Delta_{\text{mix}}V = (\partial \Delta_{\text{mix}}G/\partial P)_{T,n_i}$. But the ideal-solution $\Delta_{\text{mix}}G$ in (9.44) does not depend on P. Therefore

$$\Delta_{\text{mix}}V = 0 \qquad \text{ideal soln., const. } T, P \qquad (9.45)$$

There is no volume change on forming an ideal solution from its components at constant T and P, as expected from the molecular definition (Sec. 9.5).

From (9.35), $\Delta_{\text{mix}}S = -(\partial \Delta_{\text{mix}}G/\partial T)_{P,n_i}$. Taking $\partial/\partial T$ of (9.44), we get

$$\Delta_{\text{mix}}S = -R \sum_i n_i \ln x_i \qquad \text{ideal soln., const. } T, P \qquad (9.46)$$

which is positive. $\Delta_{\text{mix}}S$ is the same for ideal solutions as for ideal gases [Eq. (3.33)].

From $\Delta_{\text{mix}}G = \Delta_{\text{mix}}H - T \Delta_{\text{mix}}S$ and (9.44) and (9.46), we find

$$\Delta_{\text{mix}}H = 0 \qquad \text{ideal soln., const. } T, P \qquad (9.47)$$

There is no heat of mixing on formation of an ideal solution at constant T and P.

From $\Delta_{\text{mix}}H = \Delta_{\text{mix}}U + P \Delta_{\text{mix}}V$ at constant P and T and Eqs. (9.45) and (9.47), we have $\Delta_{\text{mix}}U = 0$ for forming an ideal solution at constant T and P, as expected from the molecular picture.

Figure 9.16 plots $\Delta_{\text{mix}}G/n$, $\Delta_{\text{mix}}H/n$, and $T \Delta_{\text{mix}}S/n$ for an ideal two-component solution against the B mole fraction x_B at 25°C, where $n \equiv n_B + n_C$.

Vapor Pressure

If the applied pressure on an ideal liquid solution is reduced until the solution begins to vaporize, we obtain a two-phase system of solution in equilibrium with its vapor. As we shall see, the mole fractions in the vapor phase will generally differ from those in the liquid phase. Let $x_1^v, x_2^v, \ldots, x_i^v, \ldots$ be the mole fractions in the vapor phase in equilibrium at temperature T with an ideal liquid solution whose mole fractions are x_1^l, $x_2^l, \ldots, x_i^l, \ldots$ (Fig. 9.17). The vapor pressure is P and equals the sum of the partial pressures of the gases: $P = P_1 + P_2 + \cdots + P_i + \cdots$, where $P_i \equiv x_i^v P$ [Eq. (1.23)]. The system's pressure equals the vapor pressure P. We now derive the vapor-pressure equation for an ideal solution.

Chapter 9

Figure 9.16

Mixing quantities for a two-component ideal solution as a function of composition at 25°C.

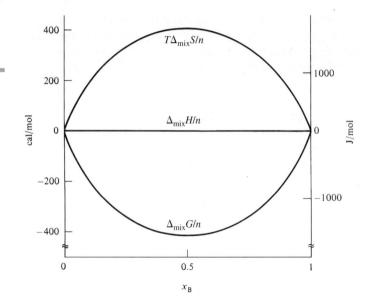

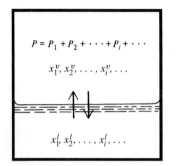

Figure 9.17

An ideal solution in equilibrium with its vapor (v).

The condition for phase equilibrium between the ideal solution and its vapor is $\mu_i^l = \mu_i^v$ [Eq. (4.88)] for each substance i, where μ_i^l and μ_i^v are the chemical potentials of i in the liquid solution and in the vapor, respectively. We shall assume that the vapor is an ideal gas mixture, which is a pretty good assumption at the low or moderate pressures at which solutions are usually studied. In an ideal gas mixture, $\mu_i^v = \mu_i^{\circ v} + RT \ln (P_i/P^\circ)$ [Eq. (6.4)], where $\mu_i^{\circ v}$ is the chemical potential of pure ideal gas i at T and $P^\circ \equiv 1$ bar, and P_i is the partial pressure of i in the vapor in equilibrium with the solution. Substitution of this expression for μ_i^v and of $\mu_i^l = \mu_i^{*l} + RT \ln x_i^l$ [Eq. (9.42)] for the ideal solution μ_i^l into the equilibrium condition $\mu_i^l = \mu_i^v$ gives

$$\mu_i^l = \mu_i^v$$

$$\mu_i^{*l}(T, P) + RT \ln x_i^l = \mu_i^{\circ v}(T) + RT \ln (P_i/P^\circ) \tag{9.48}$$

Let P_i^* be the vapor pressure of pure liquid i at temperature T. For equilibrium between pure liquid i and its vapor, we have $\mu_i^{*l}(T, P_i^*) = \mu_i^{*v}(T, P_i^*)$ or [Eq. (6.4)]

$$\mu_i^{*l}(T, P_i^*) = \mu_i^{\circ v}(T) + RT \ln (P_i^*/P^\circ) \tag{9.49}$$

Subtraction of (9.49) from (9.48) gives

$$\mu_i^{*l}(T, P) - \mu_i^{*l}(T, P_i^*) + RT \ln x_i^l = RT \ln (P_i/P_i^*) \tag{9.50}$$

For liquids, μ_i^* (which equals $G_{m,i}^*$) varies very slowly with pressure (Sec. 4.5), so it is an excellent approximation to take $\mu_i^{*l}(T, P) = \mu_i^{*l}(T, P_i^*)$ (unless the pressure is very high). Equation (9.50) then simplifies to $RT \ln x_i^l = RT \ln (P_i/P_i^*)$. If $\ln a = \ln b$, then $a = b$. Therefore $x_i^l = P_i/P_i^*$ and

$$P_i = x_i^l P_i^* \qquad \text{ideal soln., ideal vapor, } P \text{ not very high} \qquad (9.51)*$$

In **Raoult's law** (9.51), P_i is the partial pressure of substance i in the vapor in equilibrium with an ideal liquid solution at temperature T, x_i^l is the mole fraction of i in the ideal solution, and P_i^* is the vapor pressure of pure liquid i at the same temperature T as the solution. Note that as x_i^l in (9.51) goes to 1, P_i goes to P_i^*, as it should. As x_i^l increases, both the chemical potential μ_i^l (Fig. 9.15) and the partial vapor pressure P_i increase. Recall that μ_i is a measure of the escaping tendency of i from a phase. Raoult's law can be written as

$$x_i^v P = x_i^l P_i^* \tag{9.52}$$

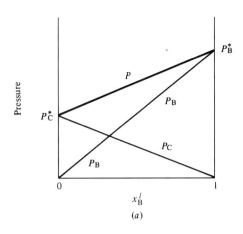

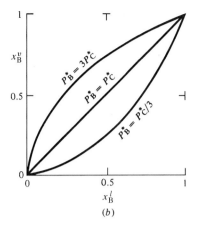

Figure 9.18

(a) Partial pressures P_B and P_C and (total) vapor pressure $P = P_B + P_C$ above an ideal solution as a function of composition at fixed T. (b) Vapor-phase mole fraction of B versus x_B^l for an ideal solution of B + C plotted for three different ratios P_B^*/P_C^* of the pure-component vapor pressures.

where P is the (total) vapor pressure of the ideal solution.

The vapor pressure P in equilibrium with an ideal solution is the sum of the partial pressures. For a two-component solution, Raoult's law gives

$$P = P_B + P_C = x_B^l P_B^* + x_C^l P_C^* = x_B^l P_B^* + (1 - x_B^l) P_C^* \qquad (9.53)$$

$$P = (P_B^* - P_C^*) x_B^l + P_C^* \qquad (9.54)$$

At fixed temperature, P_B^* and P_C^* are constants, and the two-component ideal-solution vapor pressure P varies linearly with x_B^l. For $x_B^l = 0$, we have pure C, and $P = P_C^*$. For $x_B^l = 1$, the solution is pure B, and $P = P_B^*$. Figure 9.18a shows the Raoult's law partial pressures P_B and P_C [Eq. (9.51)] and the total vapor pressure P of an ideal solution as a function of composition at fixed T. A nearly ideal solution such as benzene–toluene shows a vapor-pressure curve that conforms closely to Fig. 9.18a. Figure 9.18b plots x_B^v versus x_B^l in an ideal two-component solution for the three cases $P_B^* = 3P_C^*$, $P_B^* = P_C^*$, and $P_B^* = P_C^*/3$. Note that the vapor is richer than the liquid in the more volatile component. For example, if $P_B^* > P_C^*$, then $x_B^v > x_B^l$. The curves are calculated from Eqs. (9.52) and (9.54); see Prob. 9.39.

> In deriving Raoult's law, we neglected the effect of a pressure change on G_m^* of the liquid components and we assumed ideal gases. At the pressures of 0 to 1 atm at which solutions are usually studied, the effect of pressure changes on G_m^* of liquids is negligible; the effect of nonideality of the vapor, although small, is usually *not* negligible and should be included in precise work. See Sec. 10.11, V. Fried, *J. Chem. Educ.*, **45**, 720 (1968), and *McGlashan*, sec. 16.7.

EXAMPLE 9.7 Raoult's law

The vapor pressure of benzene is 74.7 torr at 20°C, and the vapor pressure of toluene is 22.3 torr at 20°C. A certain solution of benzene and toluene at 20°C has a vapor pressure of 46.0 torr. Find the benzene mole fraction in this solution and in the vapor above this solution.

Benzene (b) and toluene (t) molecules resemble each other closely, so it is a good approximation to assume an ideal solution and use Raoult's law (9.51). The vapor pressure of the solution is

$$46.0 \text{ torr} = P_b + P_t = x_b^l P_b^* + x_t^l P_t^* = x_b^l (74.7 \text{ torr}) + (1 - x_b^l)(22.3 \text{ torr})$$

Solving, we find $x_b^l = 0.452$. The benzene partial vapor pressure is $P_b = x_b^l P_B^* = 0.452(74.7 \text{ torr}) = 33.8$ torr. The benzene vapor-phase mole fraction is $x_b^v = P_b/P = 33.8/46.0 = 0.735$ [Eq. (1.23)].

Exercise

A solution at 20°C is composed of 1.50 mol of benzene and 3.50 mol of toluene. Find the pressure and the benzene mole fraction for the vapor in equilibrium with this solution. In this exercise and the next, use data in the above example. (*Answer:* 38.0 torr, 0.589.)

Exercise

The vapor in equilibrium with a certain solution of benzene and toluene at 20°C has a benzene mole fraction of 0.300. Find the benzene mole fraction in the liquid solution and find the vapor pressure of the solution. (*Answer:* 0.113, 28.2 torr.)

For a two-component solution, vapor-pressure problems involve four mole fractions and five pressures. Two of the four mole fractions x_B^l, x_C^l, x_B^v, and x_C^v can be eliminated using $x_B^l + x_C^l = 1$ and $x_B^v + x_C^v = 1$. The five vapor pressures are the vapor pressures P_B^* and P_C^* of the pure liquids, the vapor pressure P of the solution, and the partial pressures P_B and P_C in the vapor in equilibrium with the solution. The pressures satisfy the relations $P_B \equiv x_B^v P$ and $P_C \equiv x_C^v P$ (from which it follows that $P_B + P_C = P$) and if the solution is ideal, the Raoult's law equations $P_B = x_B^l P_B^*$ and $P_C = x_C^l P_C^*$. We have seven unknowns (the five unknown pressures and two unknown independent mole fractions) and four independent equations. To solve the problem, we need three pieces of information; for example, the values of P_B^*, P_C^*, and x_B^l (or P or x_B^v).

Partial Molar Properties

Expressions for partial molar properties of an ideal solution are easily derived from the chemical potentials $\mu_i = \mu_i^*(T, P) + RT \ln x_i$ by using $\bar{S}_i = -(\partial\mu_i/\partial T)_{P,n_j}$, $\bar{V}_i = (\partial\mu_i/\partial P)_{T,n_j}$, and $\bar{H}_i = \mu_i + T\bar{S}_i$ [Eqs. (9.30), (9.31), and (9.28)]. One finds (Prob. 9.43)

$$\bar{S}_i = S_{m,i}^* - R \ln x_i, \qquad \bar{V}_i = V_{m,i}^*, \qquad \bar{H}_i = H_{m,i}^* \qquad \text{ideal soln.} \qquad (9.55)$$

These results are consistent with $\Delta_{\text{mix}}V = 0$, $\Delta_{\text{mix}}H = 0$, and $\Delta_{\text{mix}}S \neq 0$.

Ideal Gas Mixtures

We have thought in terms of liquid and solid ideal solutions in this section. However, it is clear that an ideal gas mixture meets the molecular definition of an ideal solution, since mixing ideal gases produces no energetic or structural changes. Moreover, one can show (Prob. 9.44) that the chemical potentials in an ideal gas mixture can be put in the form (9.41) defining an ideal solution. An ideal gas mixture is an ideal solution.

9.7 IDEALLY DILUTE SOLUTIONS

An ideal solution occurs in the limit where the molecules of the different species resemble one another very closely. A different kind of limit is where the solvent mole fraction approaches 1, so that all solutes are present in very low concentrations. Such a solution is called an **ideally dilute** (or **ideal-dilute**) **solution.** *In an ideally dilute solution, solute molecules interact essentially only with solvent molecules,* because of the high dilution of the solutes.

Consider such a very dilute solution of nonelectrolytes. (In electrolyte solutions, the strong interionic forces give substantial solute–solute interactions even at very high dilutions; hence the ideally dilute solution model is not useful for electrolyte solutions. Also, each electrolyte gives two or more ions in solution, and so the chemical potential μ_i of an electrolyte solute differs in form from μ_i of a nonelectrolyte, even in the limit of infinite dilution. Electrolyte solutions are treated in Chapter 10.) We shall use A to denote the solvent and i to signify any one of the solutes. The condition of high dilution is that the solvent mole fraction x_A is very close to 1. For such a very dilute solution, solute molecules are generally surrounded by only solvent molecules, so that all solute molecules are in an essentially uniform environment; see Fig. 9.19.

To arrive at a thermodynamic definition of an ideally dilute solution, one uses vapor-pressure data for highly dilute nonelectrolyte solutions to arrive at an equation for $\Delta_{diln}G$, the Gibbs energy change that occurs when an ideally dilute solution is diluted by the addition of a certain amount of the solvent A. One then derives the chemical potentials μ_i and μ_A in the ideally dilute solution from the $\Delta_{diln}G$ equation in the same way that the chemical potentials (9.41) in an ideal solution were derived from the $\Delta_{mix}G$ equation (9.39). The details of the derivation of μ_i and μ_A from $\Delta_{diln}G$ are given in Prob. 9.46. One finds

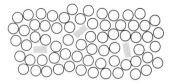

Figure 9.19

In an ideally dilute solution, solute molecules (shaded) interact only with solvent molecules.

$$\mu_i = RT \ln x_i + f_i(T, P) \qquad \text{solute in ideally dil. soln.} \qquad (9.56)$$

$$\mu_A = \mu_A^*(T, P) + RT \ln x_A \qquad \text{solvent in ideally dil. soln.} \qquad (9.57)$$

where R is the gas constant, $f_i(T, P)$ is some function of T and P, $\mu_A^*(T, P) \equiv G_{m,A}^*(T, P)$ is the chemical potential of pure liquid solvent A at the T and P of the solution, and x_i and x_A are the mole fractions of solute i and solvent A in the solution. Statistical-mechanical derivations of (9.56) and (9.57) are given in E. A. Guggenheim, *Mixtures,* Oxford, 1952, sec. 5.04; A. J. Staverman, *Rec. Trav. Chim.,* **60**, 76 (1941). The laws of thermodynamics are general and cannot supply us with the explicit forms of equations of state or chemical potentials for specific systems. Such information must be obtained by appeal to molecular (statistical-mechanical) arguments or to experimental data (as in the use of $PV = nRT$ for low-density gases).

We adopt as the thermodynamic definition: *An **ideally dilute** solution is one in which the solute and solvent chemical potentials are given by (9.56) and (9.57) for a range of composition with x_A close to 1 and for a range of T and P.*

As a real solution becomes more dilute, the chemical potentials approach (9.56) and (9.57) more closely. Just how dilute a solution must be in order to be considered ideally dilute depends on how accurately one wants to represent the solution's thermodynamic properties. A rough rule for nonelectrolyte solutions is that $z_i x_i$ should be less than 0.1, where z_i is the average number of nearest neighbors for solute i. For approximately spherical solute and solvent molecules of similar size, z_i is roughly 10. For solutes with large molecules (for example, polymers), z_i can be much larger. A polymer solution becomes ideally dilute at a much lower solute mole fraction than a nonpolymer solution, since much higher dilutions are required to ensure that a polymer solute molecule is, to a high probability, surrounded only by solvent molecules.

Ideal solutions and ideally dilute solutions are different and must not be confused with each other. Unfortunately, people sometimes use the term "ideal solution" when what is meant is an ideally dilute solution.

At the high dilutions for which (9.56) applies, the mole fraction x_i is proportional to the molar concentration c_i and to the molality m_i to a high degree of approximation (Prob. 9.8). Hence (9.56) can be written as $\mu_i = RT \ln c_i + h_i(T, P)$ or as $\mu_i = RT \ln m_i + k_i(T, P)$, where h_i and k_i are functions related to f_i. Therefore, molalities or molar concentrations can be used instead of mole fractions in dealing with solutes in ideally dilute solutions.

Summary

$\Delta_{diln}G$ data (as found from vapor-pressure measurements) and statistical-mechanical arguments show that in the limit of high dilution of a solution (x_A close to 1), the solute chemical potentials are given by $\mu_i = f_i(T, P) + RT \ln x_i$ and the solvent chemical potential is $\mu_A = \mu_A^*(T, P) + RT \ln x_A$. This is an ideally dilute solution.

THERMODYNAMIC PROPERTIES OF IDEALLY DILUTE SOLUTIONS

Before deriving thermodynamic properties of ideally dilute solutions from the chemical potentials (9.56) and (9.57), we define the standard states for components of ideally dilute solutions.

Standard States

The **standard state of the solvent** A in an ideally dilute solution is defined to be pure A at the temperature T and pressure P of the solution. Therefore the solvent standard-state chemical potential is $\mu_A^\circ \equiv \mu_A^*(T, P)$, and (9.57) can be written as $\mu_A = \mu_A^\circ + RT \ln x_A$ for the solvent.

Now consider the solutes. From (9.56), we have $\mu_i = f_i(T, P) + RT \ln x_i$. The standard state of solute i is defined so as to make its standard-state chemical potential μ_i° equal to $f_i(T, P)$; $\mu_i^\circ \equiv f_i(T, P)$. This definition of μ_i° gives

$$\mu_i = \mu_i^\circ + RT \ln x_i \qquad \text{solute in ideally dil. soln.} \qquad (9.58)$$

What choice of solute standard state is implied by taking μ_i° equal to $f_i(T, P)$? When x_i becomes 1 in (9.58), the log term vanishes and the equation gives μ_i (at $x_i = 1$) as equal to μ_i°. It might therefore be thought that the standard state of solute i is pure i at the temperature and pressure of the solution. This supposition is *wrong*. The ideally dilute solution relation (9.58) is valid *only* for high dilution (where x_i is much less than 1), and we cannot legitimately take the limit of this relation as x_i goes to 1.

However, one could imagine a *hypothetical* case in which $\mu_i = \mu_i^\circ + RT \ln x_i$ holds for all values of x_i. In this hypothetical case, μ_i would become equal to μ_i° in the limit $x_i \to 1$. The choice of solute standard state uses this hypothetical situation. The **standard state for solute** i in an ideally dilute solution is defined to be the fictitious state at the temperature and pressure of the solution that arises by supposing that $\mu_i = \mu_i^\circ + RT \ln x_i$ holds for all values of x_i and setting $x_i = 1$. This hypothetical state is an extrapolation of the properties of solute i in the very dilute solution to the limit $x_i \to 1$.

The solid line in Fig. 9.20 shows μ_i versus $\ln x_i$ at fixed T and P for a typical nonelectrolyte solution. At high dilutions ($x_i < 0.01$ and $\ln x_i < -4$), the solution is essentially ideally dilute and μ_i varies essentially linearly with $\ln x_i$ according to $\mu_i = \mu_i^\circ + RT \ln x_i$. As $\ln x_i$ increases above -4, the solution deviates more and more from ideally dilute behavior. The dashed line shows the hypothetical case where ideally dilute behavior holds as $x_i \to 1$ and $\ln x_i \to 0$. The equation of the dashed line is $\mu_i = \mu_i^\circ + RT \ln x_i$. For the dashed line, μ_i becomes equal to μ_i° when x_i reaches 1. Thus μ_i° can be found by extrapolating the high-dilution behavior of the solution to $x_i = 1$. For an actual example of this, see the discussion of Fig. 9.21.

Since the properties of i in the dilute solution depend very strongly on the solvent (which provides the environment for the i molecules), the fictitious standard state of solute i depends on what the solvent is. The properties of the standard state also depend on T and P, and μ_i° is a function of T and P but not of the mole fractions: $\mu_i^\circ = \mu_i^\circ(T, P)$. We might write $\mu_i^\circ = \mu_i^{\circ,A}(T, P)$ to indicate that the solute standard state depends on the solvent, but we won't do so unless we are dealing with solutions of i in two different solvents.

The fictitious standard state of solute i is a state in which i is pure, but in which, by some magical means, each i molecule experiences the same intermolecular forces it experiences in the ideally dilute solution, where it is surrounded by solvent molecules.

In summary, the solute chemical potentials μ_i and the solvent chemical potential μ_A in an ideally dilute solution are

$$\mu_i = \mu_i^\circ(T, P) + RT \ln x_i \qquad \text{for } i \neq A, \qquad \text{ideally dil. soln.} \qquad \textbf{(9.59)*}$$

$$\mu_A = \mu_A^\circ + RT \ln x_A, \qquad \mu_A^\circ \equiv \mu_A^*(T, P) \qquad \text{ideally dil. soln.} \qquad \textbf{(9.60)*}$$

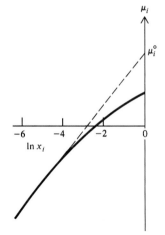

μ_i

μ_i°

$-6 \qquad -4 \qquad -2 \qquad 0$
$\ln x_i$

Figure 9.20

Chemical potential μ_i plotted versus $\ln x_i$ for a typical nonelectrolyte solute. The dashed line extrapolates the ideally dilute behavior to the limit $x_i \to 1$.

provided x_A is close to 1. The solvent standard state is pure liquid A at the temperature and pressure T and P of the solution. The standard state of solute i is the fictitious state at T and P obtained by taking the limit $x_i \to 1$ while pretending that (9.59) holds for all concentrations.

Although (9.59) and (9.60) look like (9.42) and (9.43) for an ideal solution, ideally dilute solutions and ideal solutions are not the same. Equations (9.59) and (9.60) hold only for high dilution, whereas (9.42) holds for all solution compositions. Moreover, the standard state for every component of an ideal solution is the actual state of the pure component at T and P of the solution, whereas the standard state of each solute in an ideally dilute solution is fictitious.

Some workers choose the standard state of solution components to have a pressure of 1 bar, rather than the pressure of the solution as we have done. Since μ of solids and liquids is insensitive to pressure changes, this difference in choice of standard states is of little significance unless high pressures are involved.

Vapor Pressure

Let P_i be the partial pressure of solute i in the vapor in equilibrium with an ideally dilute solution at temperature T and pressure P, where P equals the (total) vapor pressure above the solution. The chemical potential μ_i^l of i in the solution is given by (9.59). We shall assume the vapor to be an ideal gas mixture, so the chemical potential of i in the vapor (v) is $\mu_i^v = \mu_i^{\circ v}(T) + RT \ln (P_i/P^\circ)$ [Eq. (6.4)]. Equating μ_i in the solution to μ_i^v, we have

$$\mu_i^l = \mu_i^v$$

$$\mu_i^{\circ l} + RT \ln x_i^l = \mu_i^{\circ v} + RT \ln (P_i/P^\circ)$$

$$(\mu_i^{\circ l} - \mu_i^{\circ v})/RT = \ln (P_i/x_i^l P^\circ)$$

$$P_i/x_i^l P^\circ = \exp[(\mu_i^{\circ l} - \mu_i^{\circ v})/RT] \qquad (9.61)$$

where $\exp z \equiv e^z$. Since $\mu_i^{\circ l}$ depends on T and P, and $\mu_i^{\circ v}$ depends on T, the right side of (9.61) is a function of T and P. Defining K_i as

$$K_i(T, P) \equiv P^\circ \exp[(\mu_i^{\circ l} - \mu_i^{\circ v})/RT] \qquad \text{where } P^\circ \equiv 1 \text{ bar} \qquad (9.62)$$

we have for (9.61)

$$P_i = K_i x_i^l \qquad \text{solute in ideally dil. soln., ideal vapor} \qquad \textbf{(9.63)}*$$

Henry's law (9.63) states that the vapor partial pressure of solute i above an ideally dilute solution is proportional to the mole fraction of i in the solution.

The **Henry's law constant** K_i is constant with respect to variations in solution composition over the range for which the solution is ideally dilute. K_i has the dimensions of pressure. Since the standard-state chemical potential $\mu_i^{\circ l}$ of solute i in the solution depends on the nature of the solvent (as well as the solute), K_i differs for the same solute in different solvents.

The pressure dependence of K_i arises from the dependence of $\mu_i^{\circ l}$ on pressure. As noted previously, the chemical potentials in condensed phases vary only slowly with pressure. Hence, K_i depends only weakly on pressure, and its pressure dependence can be neglected, except at quite high pressures. We thus take K_i to depend only on T. This approximation corresponds to a similar approximation made in deriving Raoult's law (9.51).

Henry's law (9.63) resembles Raoult's law (9.51). In both laws, the vapor-phase partial pressure of the species is proportional to its mole fraction in the solution. However, the proportionality constant in Henry's law is not equal to the vapor pressure of the pure component, as it is for Raoult's law. This is because molecules of solute i in the ideally dilute solution are in an environment different from their environment in

Figure 9.21

Partial and total vapor pressures for (a) acetone–chloroform solutions at 35°C; (b) acetone–CS_2 solutions at 29°C.

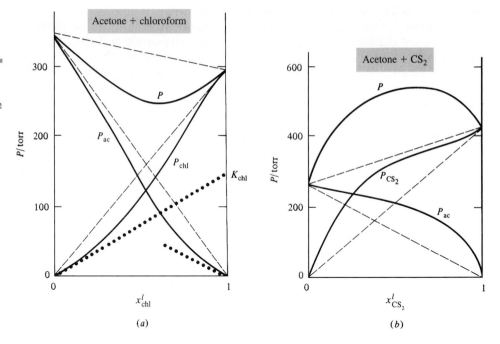

pure i. In contrast, in an ideal solution, the environment surrounding a molecule is similar to that in the pure substance.

What about the solvent vapor pressure? Equation (9.60) for the solvent chemical potential μ_A in an ideally dilute solution is the same as Eqs. (9.42) and (9.43) for the chemical potential of a component of an ideal solution. Therefore, the same derivation that gave Raoult's law (9.51) for the vapor partial pressure of an ideal-solution component gives as the vapor partial pressure of the solvent in an ideally dilute solution

$$P_A = x_A^l P_A^* \qquad \text{solvent in ideally dil. soln., ideal vapor} \qquad \textbf{(9.64)*}$$

Of course, (9.64) and (9.63) hold only for the concentration range of high dilution.

In an ideally dilute solution, the solvent obeys Raoult's law and the solutes obey Henry's law.

At sufficiently high dilutions, all nonelectrolyte solutions become ideally dilute. For less dilute solutions, the solution is no longer ideally dilute and shows deviations from Raoult's and Henry's laws. Two systems that show large deviations are graphed in Fig. 9.21.

The solid lines in Fig. 9.21a show the observed partial and total vapor pressures above solutions of acetone (ac) plus chloroform (chl) at 35°C. The three upper dashed lines show the partial and total vapor pressures that would occur for an ideal solution, where Raoult's law is obeyed by both species (Fig. 9.18a). In the limit $x_{chl}^l \rightarrow 1$, the solution becomes ideally dilute with chloroform as the solvent and acetone as the solute. For $x_{chl}^l \rightarrow 0$, the solution becomes ideally dilute with acetone as the solvent and chloroform as the solute. Hence, near $x_{chl}^l = 1$, the observed chloroform partial pressure approaches the Raoult's law line very closely, whereas near $x_{chl}^l = 0$, the observed acetone partial pressure approaches the Raoult's law line very closely. Near $x_{chl}^l = 1$, the partial pressure of the solute acetone varies nearly linearly with mole fraction (Henry's law). Near $x_{chl}^l = 0$, the partial pressure of the solute chloroform varies nearly linearly with mole fraction.

The two lower dotted lines show the Henry's law lines extrapolated from the observed limiting slopes of P_{chl} near $x_{chl}^l = 0$ and P_{ac} near $x_{chl}^l = 1$. The dotted line that starts from the origin is the Henry's law line for chloroform as solute and is drawn tangent to the P_{chl} curve at $x_{chl}^l = 0$. This dotted line plots $P_{chl}^{id\text{-}dil}$ versus x_{chl}^l, where $P_{chl}^{id\text{-}dil}$ is the chloroform partial vapor pressure the solution would have if it were ideally dilute. The equation of this dotted line is given by (9.63) as $P_{chl}^{id\text{-}dil} = K_{chl}x_{chl}^l$, so at $x_{chl}^l = 1$ we have $P_{chl}^{id\text{-}dil} = K_{chl}$. Therefore the intersection of the chloroform Henry's law line with the right-hand vertical line $x_{chl}^l = 1$ equals K_{chl}, the Henry's law constant for the solute chloroform in the solvent acetone. From the figure, $K_{chl} = 145$ torr. The Henry's law constant K_{chl} for chloroform in the solvent acetone is what the vapor pressure of pure chloroform would be if ideally dilute behavior held as $x_{chl}^l \to 1$. The actual vapor pressure of pure chloroform at 35°C is 293 torr (the intersection of the P and P_{chl} curves with $x_{chl}^l = 1$ in Fig. 9.21a). Similarly, the intersection of the acetone Henry's law line with $x_{chl}^l = 0$ gives K_{ac}.

Once we have found K_{chl}, we can use $K_i \equiv P° \exp[(\mu_i^{°l} - \mu_i^{°v})/RT]$ [Eq. (9.62)] to find $\mu_i^{°l}$ of the solute chloroform relative to $\mu_i^{°v}$ of chloroform vapor. From $K_{chl} = 145$ torr and $P° \equiv 1$ bar ≈ 750 torr, one finds (Prob. 9.51) $\mu_{chl}^{°l} - \mu_{chl}^{°v} = -4.21$ kJ/mol for chloroform in acetone at 35°C. If the conventional value (Sec. 5.8) of $\mu_{chl}^{°v}$ is known, then the conventional value of $\mu_{chl}^°$ in the solution is known.

For all compositions, the partial and total vapor pressures in Fig. 9.21a are below those predicted by Raoult's law. The solution is said to show *negative deviations* from Raoult's law. The acetone–CS$_2$ system in Fig. 9.21b shows *positive deviations* from Raoult's law at all compositions. For certain systems, one component shows a positive deviation, while the second component shows a negative deviation at the same composition [M. L. McGlashan, *J. Chem. Educ.,* **40,** 516 (1963)].

Solubility of Gases in Liquids

For gases that are sparingly soluble in a given liquid, the concentration of the dissolved gas is usually low enough for the solution to be approximately ideally dilute, and Henry's law (9.63) holds well. Therefore

$$x_i^l = K_i^{-1}P_i \qquad P \text{ not very high} \qquad (9.65)$$

where x_i^l is the mole fraction of dissolved gas in the solution at a given temperature and P_i is the partial pressure of gas i above the solution. The gas solubility (as measured by x_i^l) is proportional to P_i above the solution, provided the solution is ideally dilute. Figure 9.22 plots the mole fraction x_i^l of dissolved N$_2$ (and H$_2$) in water at 50°C versus N$_2$ (or H$_2$) partial pressure above the solution. Up to 100 atm, the N$_2$ plot obeys Henry's law $x_i^l = K_i^{-1}P_i$ and is essentially linear. Above 100 atm, the N$_2$ plot shows increasing deviations from the Henry's law line (the dotted line) because of the dependence of K_i on pressure and deviations of the gas from ideal-gas behavior. H$_2$ obeys Henry's law up to 200 atm.

At the low solute concentrations for which Henry's law applies, the solute's molality m_i and molar concentration c_i are each essentially proportional to its mole fraction x_i (Prob. 9.8). Therefore molalities or concentrations can be used instead of mole fractions in Henry's law: $P_i = K_{i,m}m_i$ or $P_i = K_{i,c}c_i$, where $K_{i,m}$ and $K_{i,c}$ are constants related to K_i in (9.65).

Some values of K_i for gases in water and in benzene at 25°C are

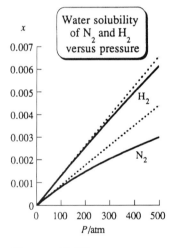

Figure 9.22

Mole-fraction solubilities of H$_2$ and N$_2$ in water at 50°C versus gas partial pressure. The dotted lines are the Henry's law lines.

i	H$_2$	N$_2$	O$_2$	CO	Ar	CH$_4$	C$_2$H$_6$
K_{i,H_2O}/kbar	71.7	86.4	44.1	58.8	40.3	40.4	30.3
K_{i,C_6H_6}/kbar	3.93	2.27	1.24	1.52	1.15	0.49	0.068

From (9.65), the larger the K_i value, the smaller the solubility of the gas. Note the much greater solubility of these gases in benzene as compared with water.

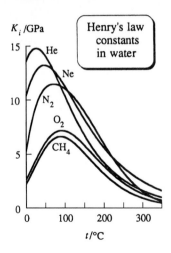

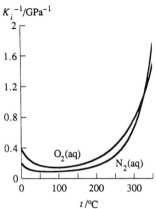

Figure 9.23

Henry's law constant K_i (at 1 bar) for several gases in water plotted versus T (upper figure) and $1/K_i$ for O_2 and N_2 in water versus T.

The solubility of most nonpolar gases (and liquids) in water goes through a minimum as T increases. Figure 9.23 plots K_i at 1 bar for several gases in water versus T. The maxima in K_i correspond to minima in solubility since the solubility is proportional to K_i^{-1}. Also plotted are K_i^{-1} for O_2 and N_2 in water versus T. The solubilities increase strongly as the critical temperature 374°C of water is approached.

Henry's law does not apply to a dilute aqueous HCl solution. Even in the limit of infinite dilution, μ_i of a strong electrolyte such as HCl does not have the form $\mu_i = \mu_i^\circ + RT \ln x_i$ used in deriving Henry's law. See Prob. 10.72 for this case.

Partial Molar Quantities

The partial molar properties of the solution's components are derived from their chemical potentials. For the solvent in an ideally dilute solution, μ_A in (9.60) has the same form as the chemical potential (9.42) and (9.43) for a component of an ideal solution. Therefore the solvent's partial molar properties are the same as for an ideal-solution component, and (9.55) gives

$$\bar{V}_A = V^*_{m,A}, \quad \bar{H}_A = H^*_{m,A}, \quad \bar{S}_A = S^*_{m,A} - R \ln x_A \quad \text{solvent in ideally dil. soln.}$$
(9.66)

For the solute partial molar properties, one finds (Probs. 9.54 and 9.55)

$$\bar{V}_i = \bar{V}_i^\circ = \bar{V}_i^\infty \quad \text{solute in ideally dil. soln.} \quad (9.67)$$

$$\bar{H}_i = \bar{H}_i^\circ = \bar{H}_i^\infty, \quad \bar{S}_i = \bar{S}_i^\circ - R \ln x_i \quad \text{solute in ideally dil. soln.} \quad (9.68)$$

where ° denotes the standard state and ∞ denotes infinite dilution. For example, $\bar{V}_i^\infty \equiv \lim_{x_A \to 1} \bar{V}_i$.

Although *some* of the solute standard-state partial molar properties are the same as the corresponding infinite-dilution values, the solute standard state is *not* the same as the state of infinite dilution. From $\mu_i = \mu_i^\circ + RT \ln x_i$, we have $\mu_i^\infty = \mu_i^\circ - \infty = -\infty$ (since $\ln 0 = -\infty$). Therefore $\mu_i^\infty \neq \mu_i^\circ$ and the standard state differs from the infinite-dilution state.

$\Delta_{\text{mix}}V$ and $\Delta_{\text{mix}}H$ are not zero for an ideally dilute solution (Prob. 9.56).

Since a real solution becomes ideally dilute in the limit $x_A \to 1$, nonelectrolyte solute chemical potentials obey $\mu_i = \mu_i^\circ + RT \ln x_i$ at high dilutions; therefore, $\mu_i \to -\infty$ as $x_i \to 0$ (Fig. 9.15). From (9.68), $\bar{S}_i \to \infty$ as $x_i \to 0$.

The dashed lines in Fig. 9.11 show $\mu_i - \mu_i^* = RT \ln x_i$ for an ideal solution. The negative deviations from ideal-solution behavior shown by μ_{ac} and μ_{chl} in Fig. 9.11 correspond to the negative deviations shown by P_{ac} and P_{chl} in Fig. 9.21a.

Reaction Equilibrium

For a chemical reaction in an ideally dilute solution, we can substitute $\mu_i = \mu_i^\circ + RT \ln x_i$ into the equilibrium condition $\sum_i \nu_i \mu_i = 0$ to derive a mole-fraction equilibrium constant $K_x \equiv \prod_i (x_{i,\text{eq}})^{\nu_i}$, where $x_{i,\text{eq}}$ is the equilibrium mole fraction of species i; see Prob. 9.57 for details.

For most equilibria in aqueous solutions, some of the reacting species are ions, which makes the ideally dilute solution approximation poor. Ionic equilibria are considered in Chapter 11.

9.9   SUMMARY

The volume of a solution is given by $V = \sum_i n_i \bar{V}_i$, where the partial molar volume of component i in the solution is defined by $\bar{V}_i \equiv (\partial V / \partial n_i)_{T,P,n_{j \neq i}}$. Similar equations hold for other extensive properties of the solution (for example, U, H, S, G, C_P). The partial molar properties $\bar{G}_i (\equiv \mu_i)$, $\bar{H}_i$, $\bar{S}_i$, and $\bar{V}_i$ obey relations analogous to the relations between the corresponding molar properties G, H, S, and V of pure substances. The chemical potentials μ_i are the key thermodynamic properties of a solution.

The volume change $\Delta_{\text{mix}}V$ for forming a solution of volume V from its pure components at constant T and P is $\Delta_{\text{mix}}V \equiv V - V^* = \Sigma_i\, n_i(\bar{V}_i - V^*_{\text{m},i})$. The mixing quantities $\Delta_{\text{mix}}G$, $\Delta_{\text{mix}}H$, $\Delta_{\text{mix}}S$, and $\Delta_{\text{mix}}V$ obey relations analogous to the relations between the corresponding properties of pure substances [Eqs. (9.33) to (9.35)].

An ideal solution is one in which the molecules of each species are so similar to one another that molecules of one species can replace molecules of another species without changing the solution's spatial structure or intermolecular interaction energy. The thermodynamic definition of an ideal solution is a solution in which the chemical potential of each species is given by $\mu_i = \mu_i^*(T, P) + RT \ln x_i$ for all compositions and a range of T and P. The standard state of an ideal-solution component is the pure substance at T and P of the solution. For an ideal solution, $\Delta_{\text{mix}}H = 0$, $\Delta_{\text{mix}}V = 0$, and $\Delta_{\text{mix}}S$ is the same as for an ideal gas mixture. By equating the chemical potentials of i in the solution and in the vapor (assumed ideal), one finds the partial pressures in the vapor in equilibrium with an ideal solution to be $P_i = x_i^l P_i^*$ (Raoult's law).

An ideally dilute (or ideal-dilute) solution is one so dilute that solute molecules interact essentially only with solvent molecules (molecular definition). In an ideally dilute solution, the solute chemical potentials are $\mu_i = \mu_i^\circ(T, P) + RT \ln x_i$ and the solvent chemical potential is $\mu_A = \mu_A^*(T, P) + RT \ln x_A$ for a small range of compositions with x_A close to 1 (thermodynamic definition). For an ideally dilute solution, the solute standard state is the fictitious state at T and P of the solution in which the solute is pure but its molecules experience the same intermolecular forces they experience when surrounded by solvent molecules in the ideally dilute solution. The solvent standard state is pure A at the T and P of the solution. The solute and solvent partial pressures in the vapor in equilibrium with an ideally dilute solution are given by Henry's law $P_i = K_i x_i^l$ and by Raoult's law $P_A = x_A^l P_A^*$, respectively.

The following superscripts are used in this chapter: $^\circ \equiv$ standard state, $^* \equiv$ pure substance, $^\infty \equiv$ infinite dilution.

Important kinds of calculations discussed in this chapter include:

- Calculation of solution mole fractions, molalities, and molar concentrations.
- Calculation of a solution's volume from its partial molar volumes using $V = \Sigma_i\, n_i \bar{V}_i$ and similar calculations for other extensive properties.
- Determination of partial molar volumes relative to the molar volumes of the pure components ($\bar{V}_i - V^*_{\text{m},i}$) using intercepts of a tangent line to the $\Delta_{\text{mix}}V/n$ curve, and similar determination of other partial molar properties.
- Calculation of mixing quantities for ideal solutions.
- Calculation of vapor partial pressures of ideal solutions using Raoult's law $P_i = x_i^l P_i^*$.
- Calculation of vapor partial pressures of ideally dilute solutions using Raoult's and Henry's laws $P_A = x_A^l P_A^*$ and $P_i = K_i x_i^l$.
- Use of dilute-solution vapor pressures to find the Henry's law constant K_i.
- Use of Henry's law to find gas solubilities in liquids.

FURTHER READING AND DATA SOURCES

McGlashan, secs. 2.7 to 2.11, chaps. 16, 18; *de Heer,* chaps. 25 and 26; *Denbigh,* secs. 2.13 and 2.14, chap. 8; *Prigogine and Defay,* chaps. 20 and 21.

Mixing quantities: C. P. Hicks in *Specialist Periodical Reports, Chemical Thermodynamics,* vol. 2, Chemical Society, London, 1978, chap. 9; *Landolt-Börnstein,* New Series, Group IV, vol. 2.

Vapor pressures and vapor compositions of solutions: *Landolt-Börnstein,* 6th ed., vol. II, part 2a, pp. 336–711 and vol. IV, part 4b, pp. 1–120; *Landolt-Börnstein,* New Series, Group IV, vol. 3; M. Hirata et al., *Computer Aided Data Book of Vapor–Liquid Equilibria,* Elsevier, 1975.

Solubility of gases in liquids: *Landolt-Börnstein,* vol. II, part 2b and vol. IV, parts 4c1 and 4c2.

PROBLEMS

Section 9.1

9.1 Give the SI units of each of these solution-composition quantities: (a) c_i; (b) m_i (molality); (c) x_i.

9.2 Which of the three quantities in Prob. 9.1 change if T changes? If P changes?

9.3 Calculate the number of moles of the solute HCl in each of the following aqueous solutions. (a) 145 mL of a solution with HCl molarity 0.800 mol/dm³; (b) 145 g of a 10.0 weight percent HCl solution; (c) 145 g of a solution whose HCl molality is 4.85 mol/kg.

9.4 In an aqueous solution of CH_3OH that is 30.00% CH_3OH by weight, the CH_3OH molarity at 20°C and 1 atm is 8.911 mol/dm³. (a) Find the solution's density at 20°C and 1 atm. (b) Find the CH_3OH molality. (c) Find the CH_3OH mass concentration.

9.5 Find the NH_3 molality and mole fraction in an aqueous solution of NH_3 that is 0.800% NH_3 by weight.

9.6 When 2.296 mol of CsCl is dissolved in 450 mL of water and the resulting solution is diluted to a volume of 1.0000 L at 20°C and 1 atm, the final solution has a density of 1.2885 g/cm³. Find the CsCl molality in the final solution.

9.7 The density of a KI(aq) solution with molality 1.506 mol/kg is 1.1659 g/cm³ at 20°C and 1 atm. Find the KI molarity.

9.8 Show that in a very dilute solution of density ρ with solvent mole fraction close to 1, the solutes' molar concentrations and molalities are $c_i \approx \rho x_i/M_A$ and $m_i \approx x_i/M_A$ and that $c_i \approx \rho m_i$.

9.9 Show that $m_B = (1000 n_B/n_A M_{r,A})$ mol/kg, where m_B is the molality of solute B and $M_{r,A}$ is the molecular weight (relative molecular mass) of the solvent.

Section 9.2

9.10 True or false? (a) $\bar{V}_i \equiv (\partial V_i/\partial n_i)_{T,P,n_{j\neq i}}$. (b) The volume of a solution at T and P equals the sum of the volumes of its pure components at T and P. (c) $\bar{V}_i$ in a solution must equal $V_{m,i}^*$. (d) The SI units of $\bar{V}_i$ are m³/mol. (e) If half of a solution is poured down the sink, the partial molar volumes in the remaining solution are equal to those in the original solution. (f) The volume of a solution cannot be less than the volume of the pure solvent used to prepare the solution. (g) $\bar{H}_i \equiv (\partial H_i/\partial n_i)_{T,P,n_{j\neq i}}$. (h) μ_i is a partial molar quantity. (i) In a solution of water plus ethanol, each of the quantities $\bar{V}_i$, $\bar{S}_i$, and $\bar{G}_i$ is a function of T, P, and x_{H_2O} and no other variables.

9.11 At 25°C and 1 atm, a 0.5000-mol/kg solution of NaCl in water has $\bar{V}_{NaCl} = 18.63$ cm³/mol and $\bar{V}_{H_2O} = 18.062$ cm³/mol. Find the volume at 25°C and 1 atm of a solution prepared by dissolving 0.5000 mol of NaCl in 1000.0 g of water.

9.12 In an aqueous 0.1000-mol/kg NaCl solution at 25°C and 1 atm, $\bar{C}_{P,H_2O} = 17.992$ cal/(mol K) and $\bar{C}_{P,NaCl} = -17.00$ cal/(mol K). Find C_P of 1000.0 g of such a solution. Note that this amount of solution does *not* contain 0.1000 mol of NaCl.

9.13 At 25°C and 1 atm, a solution of 72.061 g of H_2O and 192.252 g of CH_3OH has a volume of 307.09 cm³. In this solution, $\bar{V}_{H_2O} = 16.488$ cm³/mol. Find $\bar{V}_{CH_3OH}$ in this solution.

9.14 The density of a methanol–water solution that is 12.000 weight percent methanol is 0.97942 g/cm³ at 15°C and 1 atm. For a solution that is 13.000 weight percent methanol, the density is 0.97799 g/cm³ at this T and P. Since the change in solution composition is small, we can estimate $\bar{V}_A$ by

$$\bar{V}_A \equiv (\partial V/\partial n_A)_{T,P,n_B} \approx (\Delta V/\Delta n_A)_{T,P,n_B}$$

Calculate $\bar{V}(CH_3OH)$ for a methanol–water solution at 15°C and 1 atm that is $12\frac{1}{2}\%$ CH_3OH by weight. Then calculate $\bar{V}(H_2O)$ for this solution.

9.15 Use Fig. 9.3 to find (a) the molality at which $\bar{V}_{MgSO_4} = 0$; (b) the partial molar volume of $MgSO_4(aq)$ in the limit of an infinitely dilute solution; (c) $\bar{V}$ of $MgSO_4$ and $\bar{V}$ of H_2O in a 0.05 mol/kg $MgSO_4(aq)$ solution.

9.16 At infinite dilution, the ions of an electrolyte are infinitely far apart and do not interact with one another. Therefore, $\bar{V}_i^\infty$ of a strong electrolyte in solution is the sum of $\bar{V}^\infty$ values for the ions. Some $\bar{V}_i^\infty$ values for aqueous solutions at 25°C and 1 atm are 16.6 cm³/mol for NaCl, 38.0 cm³/mol for KNO_3, and 27.8 cm³/mol for $NaNO_3$. (a) Find $\bar{V}_i^\infty$ for KCl in water at 25°C and 1 atm. (b) Find $(\partial\mu_i/\partial P)_{T,n_j}^\infty$ for KCl in water at 25°C.

9.17 Prove that the internal energy of a phase satisfies $U = -PV + TS + \Sigma_i n_i\mu_i$. The proof is very short.

9.18 Write the defining equation for the partial molar Helmholtz energy of substance i in a solution and state fully what every symbol in your definition stands for.

9.19 Show that $\bar{H}_i = \bar{U}_i + P\bar{V}_i$.

9.20 (a) Use $G = \Sigma_i n_i\mu_i$, $\mu_i = \mu_i^\circ + RT \ln (P_i/P^\circ)$ [Eqs. (9.23) and (6.4)] and Eq. (4.65) applied to pure gas i to show that G of an ideal gas mixture at T is given by $G = \Sigma_i G_i^*(T, P_i, n_i)$, where P_i and n_i are the partial pressure and number of moles of gas i in the mixture, and G_i^* is the Gibbs energy of n_i moles of pure gas i at temperature T and pressure P_i. (This result was mentioned in Sec. 6.1.) (b) Use $(\partial G/\partial T)_{P,n_i} = -S$ and the result of (a) to show that for an ideal gas mixture $S = \Sigma_i \, S_i^*(T, P_i, n_i)$. (c) Use $G = H - TS$ to show that for an ideal gas mixture $H = \Sigma_i \, H_i^*(T, n_i)$. (d) Show that for an ideal gas mixture $C_P = \Sigma_i \, C_{P,i}^*(T, n_i)$ and $U = \Sigma_i \, U_i^*(T, n_i)$. (e) Find C_P at 25°C and 500 torr of a mixture of 0.100 mol of $O_2(g)$ and 0.300 mol of $CO_2(g)$, using Appendix data. State any assumptions made.

Section 9.3

9.21 Verify (9.35) for $(\partial\Delta_{mix}G/\partial T)_{P,n_j}$.

Section 9.4

9.22 Use Fig. 9.9 to calculate the volume at 20°C and 1 atm of a solution formed from 20.0 g of H_2O and 45.0 g of C_2H_5OH.

9.23 Use Fig. 9.8 to find the partial molar volumes at ethanol mole fraction 0.400.

9.24 The densities of H_2O and CH_3OH at 25°C and 1 atm are 0.99705 and 0.78706 g/cm³, respectively. For solutions of these two compounds at 25°C and 1 atm, $\Delta_{mix}V/n$-vs.-x_{H_2O} data are:

$(\Delta_{mix}V/n)/(cm^3/mol)$	−0.34	−0.60	−0.80
x_{H_2O}	0.1	0.2	0.3
$(\Delta_{mix}V/n)/(cm^3/mol)$	−0.94₅	−1.01	−0.98
x_{H_2O}	0.4	0.5	0.6
$(\Delta_{mix}V/n)/(cm^3/mol)$	−0.85	−0.61₅	−0.31
x_{H_2O}	0.7	0.8	0.9

Use the intercept method (Fig. 9.8) to find the partial molar volumes at x_{H_2O} values of (a) 0; (b) 0.4; (c) 0.6.

9.25 Let V be the volume of an aqueous solution of NaCl at 25°C and 1 atm that contains 1000 g of water and n_B moles of NaCl. One finds that the following empirical formula accurately reproduces the experimental data:

$$V = a + bn_B + cn_B^{3/2} + kn_B^2 \qquad \text{for } n_A M_A = 1 \text{ kg}$$

$$a = 1002.96 \text{ cm}^3, \qquad b = 16.6253 \text{ cm}^3/\text{mol}$$

$$c = 1.7738 \text{ cm}^3/\text{mol}^{3/2}, \qquad k = 0.1194 \text{ cm}^3/\text{mol}^2$$

(a) Show that the NaCl partial molar volume $\bar{V}_B$ is

$$\bar{V}_B = b + (3c/2)n_B^{1/2} + 2kn_B \qquad \text{for } n_A M_A = 1 \text{ kg}$$

(b) Find $\bar{V}_{NaCl}$ for a solution with NaCl molality $m_B = 1.0000$ mol/kg. (c) Use (9.16) to show that the partial molar volume of the water in the solution is

$$\bar{V}_A = (M_A/1000 \text{ g})(a - \tfrac{1}{2}cn_B^{3/2} - kn_B^2) \qquad \text{for } n_A M_A = 1 \text{ kg}$$

(d) Show that the results for (a) and (c) can be written as

$$\bar{V}_B = b + (3c/2)(m_B \text{ kg})^{1/2} + 2km_B \text{ kg}$$

$$\bar{V}_A = (M_A/1000 \text{ g})(a - \tfrac{1}{2}cm_B^{3/2} \text{ kg}^{3/2} - km_B^2 \text{ kg}^2)$$

Since $\bar{V}_A$, $\bar{V}_B$, and m_B are all intensive quantities, we need not specify n_A in these equations. (e) Find $\bar{V}_{H_2O}$ for a solution with $m_B = 1.0000$ mol/kg. (f) Find $\bar{V}_{NaCl}^\infty$.

9.26 Prove the validity of the intercept method (Fig. 9.7) of determining partial molar volumes in a two-component solution as follows. (All the equations of this problem are for fixed T and P.) (a) Let $z \equiv \Delta_{mix}V/n$, where $n = n_A + n_B$. Verify that $V = (n_A + n_B)z + n_A V_{m,A}^* + n_B V_{m,B}^*$. (b) Take $(\partial/\partial n_A)_{n_B}$ of the equation in (a) to show that $\bar{V}_A = n(\partial z/\partial n_A)_{n_B} + z + V_{m,A}^*$. (c) Use $(\partial z/\partial n_A)_{n_B} = (dz/dx_B)(\partial x_B/\partial n_A)_{n_B}$ [Eq. (1.35)] and the result for (b) to show that $dz/dx_B = (V_{m,A}^* - \bar{V}_A + z)/x_B$. Also, explain why the n_B subscript can be omitted from $(\partial z/\partial x_B)_{n_B}$. Let $y = mx_B + b$ be the equation of the tangent line to the z-versus-x_B curve at the point with $x_B = x_B'$ and $z = z'$, and let $\bar{V}_A'$ and $\bar{V}_B'$ be the partial molar volumes at x_B'. Recall that, for the straight line $y = mx_B + b$, the slope is m and the intercept at $x_B = 0$ is b. The slope m is given by the result for (c) as $m = (V_{m,A}^* - \bar{V}_A' + z')/x_B'$. Also, since the tangent line passes through the point (x_B', z'), we have $z' = mx_B' + b = (V_{m,A}^* - \bar{V}_A' + z') + b$. Therefore, $b = \bar{V}_A' - V_{m,A}^*$, which is what we wanted to prove. (d) Verify that the tangent line's intercept at $x_B = 1$ gives $\bar{V}_B' - V_{m,B}^*$.

9.27 For the H_2SO_4 solutions given in Example 9.6, use Fig. 9.13 to find the differential heats of solution at $x_{H_2SO_4} = 0.4$ and at $x_{H_2SO_4} = 0.333$.

9.28 Prove that in a two-component solution, $\Delta H_{diff,B} = (\partial\Delta_{mix}H/\partial n_B)_{T,P,n_A}$.

9.29 The NBS tables (Sec. 5.9) give at 25°C $\Delta_f H_{NaCl(s)}^\circ = -411.153$ kJ/mol and give the following apparent $\Delta_f H^\circ$ data in kJ/mol for aqueous NaCl solutions at 25°C:

n_{H_2O}/n_{NaCl}	9	15	25	50
$\Delta_f H_{NaCl(aq)}^\circ$	−409.279	−408.806	−408.137	−407.442

Calculate and plot $\Delta_{mix}H/n$ versus x_{NaCl} and use the intercept method to find $\bar{H}_{NaCl} - H_{m,NaCl}^*$ and $\bar{H}_{H_2O} - H_{m,H_2O}^*$ at $x_{NaCl} = 0.05$, where $H_{m,NaCl}^*$ is for solid NaCl.

9.30 Look up apparent $\Delta_f H^\circ$ data for HCl solutions in the NBS thermodynamics tables and find ΔH_{diff} for HCl and for H_2O at $x_{HCl} = 0.30$, $T = 298$ K, and $P = 1$ bar.

Section 9.5

9.31 True or false? (a) Intermolecular interactions are negligible in an ideal solution. (b) If B is a component of an ideal solution, μ_B cannot be greater than μ_B^*. (c) If B is a component of a solution, μ_B cannot be greater than μ_B^*. (d) A solution of water plus ethanol is nearly ideal.

9.32 Would a liquid mixture of the two optical isomers of CHFClBr be an ideal solution? Explain.

Section 9.6

9.33 True or false? (a) At constant T and P, $\Delta_{mix}G$ must be negative for an ideal solution. (b) At constant T and P, $\Delta_{mix}G$ must be negative for every solution. (c) At constant T and P,

$\Delta_{mix}S = \Delta_{mix}H/T$ for an ideal solution. (*d*) For equilibrium between a solution and its vapor, μ of the solution must equal μ of the vapor. (*e*) For equilibrium between an ideal solution and an ideal vapor, x_B^l must equal x_B^v. (*f*) In an ideal solution, the partial molar volume of a component equals the molar volume of the pure substance.

9.34 State the two approximations that are made when Raoult's law is derived from the ideal-solution chemical potentials.

9.35 Find $\Delta_{mix}G$, $\Delta_{mix}V$, $\Delta_{mix}S$, and $\Delta_{mix}H$ for mixing 100.0 g of benzene with 100.0 g of toluene at 20°C and 1 atm. Assume an ideal solution.

9.36 Benzene (C_6H_6) and toluene ($C_6H_5CH_3$) form nearly ideal solutions. At 20°C the vapor pressure of benzene is 74.7 torr, and that of toluene is 22.3 torr. (*a*) Find the equilibrium partial vapor pressures above a 20°C solution of 100.0 g of benzene plus 100.0 g of toluene. (*b*) Find the mole fractions in the vapor phase that is in equilibrium with the solution of part (*a*).

9.37 At 100°C the vapor pressures of hexane and octane are 1836 and 354 torr, respectively. A certain liquid mixture of these two compounds has a vapor pressure of 666 torr at 100°C. Find the mole fractions in the liquid mixture and in the vapor phase. Assume an ideal solution.

9.38 A solution of hexane and heptane at 30°C with hexane mole fraction 0.305 has a vapor pressure of 95.0 torr and a vapor-phase hexane mole fraction of 0.555. Find the vapor pressures of pure hexane and heptane at 30°C. State any approximations made.

9.39 (*a*) Use Raoult's law to show that for an ideal solution of B and C, the B mole fraction in the vapor phase in equilibrium with the solution is

$$x_B^v = \frac{x_B^l P_B^*/P_C^*}{1 + x_B^l(P_B^*/P_C^* - 1)}$$

(*b*) At 20°C the vapor pressure of benzene (C_6H_6) is 74.7 torr and that of toluene ($C_6H_5CH_3$) is 22.3 torr. For solutions of benzene plus toluene (assumed ideal) in equilibrium with vapor at 20°C, plot x_B^v versus x_B^l for benzene. Repeat for toluene.

9.40 At 20°C and 1 atm, the density of benzene is 0.8790 g/cm^3 and that of toluene is 0.8668 g/cm^3. Find the density of a solution of 33.33 g of benzene and 33.33 g of toluene at 20°C and 1 atm. Assume an ideal solution.

9.41 (*a*) Show that $\Delta_{mix}C_P = 0$ for an ideal solution. (*b*) At 25°C and 1 atm, $C_{P,m} = 136$ J/(mol K) for benzene (C_6H_6) and $C_{P,m} = 156$ J/(mol K) for toluene ($C_6H_5CH_3$). Find C_P of a solution of 100.0 g of benzene and 100.0 g of toluene at 25°C and 1 atm. Assume an ideal solution.

9.42 Draw tangents to the $\Delta_{mix}G/n$ curve of Fig. 9.16 to find $\mu_A - \mu_A^*$ and $\mu_B - \mu_B^*$ at $x_B = 0.50$ and at $x_B = 0.25$. Compare your results with those calculated from $\mu_i = \mu_i^* + RT \ln x_i$.

9.43 Derive the ideal-solution partial-molar properties (9.55) from the chemical potentials (9.42).

9.44 Consider an ideal gas mixture at T and P; show that for component i, $\mu_i = \mu_i^*(T, P) + RT \ln x_i$. Therefore an ideal gas mixture is an ideal solution. Of course, an ideal solution is not necessarily an ideal gas mixture. Note also the different choice of standard state for an ideal-solution component and an ideal-gas-mixture component.

9.45 Let phases α and β, each composed of liquids 1 and 2, be in equilibrium with each other. Show that if substances 1 and 2 form ideal solutions, then $x_1^\alpha = x_1^\beta$ and $x_2^\alpha = x_2^\beta$. Therefore, the two phases have the same composition and are actually one phase. Hence liquids that form ideal solutions are miscible in all proportions.

Section 9.7

9.46 Consider the constant-T-and-P dilution process of adding $n_{A,2} - n_{A,1}$ moles of solvent A to an ideally dilute solution (solution 1) that contains n_i moles of solute i and $n_{A,1}$ moles of A to give an ideally dilute solution of n_i moles of i and $n_{A,2}$ moles of A. Experimental vapor-pressure data for highly dilute solutions show that ΔG for this process is given by

$$\Delta G = n_i RT(\ln x_{i,2} - \ln x_{i,1})$$
$$+ RT(n_{A,2} \ln x_{A,2} - n_{A,1} \ln x_{A,1}) \qquad (9.69)$$

where $x_{i,2}$, $x_{i,1}$, $x_{A,2}$, and $x_{A,1}$ are the final and initial mole fractions of the solute and the solvent in the solution. (*a*) Use Eq. (9.23) to show that for this process

$$\Delta G = n_i(\mu_{i,2} - \mu_{i,1}) + n_{A,2}\mu_{A,2} - n_{A,1}\mu_{A,1}$$
$$- (n_{A,2} - n_{A,1})\mu_A^* \qquad (9.70)$$

where $\mu_{i,2}$, $\mu_{i,1}$, $\mu_{A,2}$, and $\mu_{A,1}$ are the final and initial chemical potentials of the solute and the solvent in the solution. Comparison of the coefficient of n_i in (9.70) with that in (9.69) gives

$$\mu_{i,2} - \mu_{i,1} = RT(\ln x_{i,2} - \ln x_{i,1}) \qquad \text{const. } T, P \quad (9.71)$$

The only way (9.71) can hold is if

$$\mu_i = RT \ln x_i + f_i(T, P) \qquad (9.72)$$

where $f_i(T, P)$ is some function of T and P, which cancels in $\mu_{i,2} - \mu_{i,1}$ at constant T and P. (*b*) Use Eqs. (9.69) to (9.71) to show that

$$n_{A,2}\mu_{A,2} - n_{A,1}\mu_{A,1} = n_{A,2}(\mu_A^* + RT \ln x_{A,2})$$
$$- n_{A,1}(\mu_A^* + RT \ln x_{A,1})$$
$$\text{const. } T, P \qquad (9.73)$$

The only way (9.73) can hold is if

$$\mu_A = \mu_A^* + RT \ln x_A \qquad (9.74)$$

Section 9.8

9.47 A solution of ethanol (eth) and chloroform (chl) at 45°C with $x_{eth} = 0.9900$ has a vapor pressure of 177.95 torr. At this high dilution of chloroform, the solution can be assumed to be essentially ideally dilute. The vapor pressure of pure ethanol at 45°C is 172.76 torr. (*a*) Find the partial pressures of the gases in equilibrium with the solution. (b) Find the mole fractions in the vapor phase. (c) Find the Henry's law constant for chloroform in ethanol at 45°C. (d) Predict the vapor pressure and vapor-phase mole fractions at 45°C for a chloroform–ethanol solution with $x_{eth} = 0.9800$. Compare with the experimental values $P = 183.38$ torr and $x_{eth}^v = 0.9242$.

9.48 The vapor in equilibrium with a solution of ethanol (eth) and chloroform (chl) at 45°C with $x_{chl}^l = 0.9900$ has a pressure of 438.59 torr and has $x_{chl}^v = 0.9794$. The solution can be assumed to be essentially ideally dilute. (a) Find the vapor-phase partial pressures. (b) Calculate the vapor pressure of pure chloroform at 45°C. (c) Find the Henry's law constant for ethanol in chloroform at 45°C.

9.49 Use Fig. 9.21 to find (a) the vapor pressure of CS_2 at 29°C; (b) x_{chl}^v in the vapor in equilibrium with a 35°C acetone–chloroform solution with $x_{chl}^l = 0.40$. (The horizontal scale is linear.)

9.50 From Fig. 9.21b, estimate K_i for acetone in CS_2 and for CS_2 in acetone at 29°C.

9.51 Use the definition (9.62) of K_i and $K_{chl} = 145$ torr (Fig. 9.21a) to find $\mu_{chl}^{\circ l} - \mu_{chl}^{\circ v}$ for chloroform in acetone at 35°C.

9.52 At 20°C, 0.164 mg of H_2 dissolves in 100.0 g of water when the H_2 pressure above the water is 1.000 atm. (a) Find the Henry's law constant for H_2 in water at 20°C. (b) Find the mass of H_2 that will dissolve in 100.0 g of water at 20°C when the H_2 pressure is 10.00 atm. Neglect the pressure variation in K_i.

9.53 Air is 21% O_2 and 78% N_2 by mole fraction. Find the masses of O_2 and N_2 dissolved in 100.0 g of water at 20°C that is in equilibrium with air at 760 torr. For aqueous solutions at 20°C, $K_{O_2} = 2.95 \times 10^7$ torr and $K_{N_2} = 5.75 \times 10^7$ torr.

9.54 (a) Use Eq. (9.31) to show that $\bar{V}_i = V_i^\circ$ for a solute in an ideally dilute solution. Explain why V_i° is independent of concentration in the ideally dilute range and why $V_i^\circ = V_i^\infty$. (b) Use Fig. 9.8 to find $\bar{V}_{H_2O}^\infty$ in a water–ethanol solution at 20°C and 1 atm.

9.55 Derive (9.68) for $\bar{H}_i$ and $\bar{S}_i$ in ideally dilute solutions.

9.56 Show that for an ideally dilute solution

$$\Delta_{mix}V = \sum_{i \neq A} n_i (\bar{V}_i^\circ - V_{m,i}^*), \qquad \Delta_{mix}H = \sum_{i \neq A} n_i (\bar{H}_i^\circ - H_{m,i}^*)$$

9.57 Substitute $\mu_i = \mu_i^\circ + RT \ln x_i$ into the equilibrium condition $\sum_i \nu_i \mu_i = 0$ to derive $\Delta G^\circ = -RT \ln K_x$ for an ideally dilute solution, where $\Delta G^\circ \equiv \sum_i \nu_i \mu_i^\circ$ and $K_x \equiv \prod_i (x_{i,eq})^{\nu_i}$.

9.58 The definition (9.62) of the Henry's law constant K_i shows that if we know K_i in a solvent A, we can find $\mu_i^{\circ l} - \mu_i^{\circ v} = \bar{G}_i^{\circ l} - \bar{G}_i^{\circ v}$, the change in standard-state partial molar Gibbs energy of gas i when it dissolves in liquid A. If we know K_i as a function of T, we can find $\bar{H}_i^{\circ l} - \bar{H}_i^{\circ v}$ using Eq. (9.75) of Prob. 9.59. Knowing $\bar{G}_i^{\circ l} - \bar{G}_i^{\circ v}$ and $\bar{H}_i^{\circ l} - \bar{H}_i^{\circ v}$, we can find $\bar{S}_i^{\circ l} - \bar{S}_i^{\circ v}$. (a) For O_2 in water, $K_i = 2.95 \times 10^7$ torr at 20°C and $K_i = 3.52 \times 10^7$ torr at 30°C. Does the solubility of O_2 in water increase or decrease from 20 to 30°C? (b) Use (9.75) to estimate $\bar{H}_i^{\circ l} -$ $\bar{H}_i^{\circ v}$ for O_2 in water in the range 20 to 30°C. (c) Use data in Sec. 9.8 to find $\bar{G}_i^{\circ l} - \bar{G}_i^{\circ v}$ for O_2 in water at 25°C. (d) Estimate $\bar{S}_i^{\circ l} - \bar{S}_i^{\circ v}$ for O_2 in water at 25°C.

9.59 Show that the temperature and pressure variations of the Henry's law constant are

$$\left(\frac{\partial \ln K_i}{\partial T} \right)_P = \frac{\bar{H}_i^{\circ v} - \bar{H}_i^{\circ l}}{RT^2} = \frac{\bar{H}_i^{\circ v} - \bar{H}_i^{\infty l}}{RT^2} \qquad (9.75)$$

$$\left(\frac{\partial \ln K_i}{\partial P} \right)_T = \frac{\bar{V}_i^{\circ l}}{RT} = \frac{\bar{V}_i^{\infty l}}{RT} \qquad (9.76)$$

General

9.60 The normal boiling points of benzene and toluene are 80.1 and 110.6°C, respectively. Both liquids obey Trouton's rule well. For a benzene–toluene liquid solution at 120°C with $x_{C_6H_6}^l = 0.68$, estimate the vapor pressure and $x_{C_6H_6}^v$. State any approximations made. (The experimental values are 2.38 atm and 0.79.)

9.61 Derive (9.41) for the chemical potentials in an ideal solution by taking $(\partial / \partial n_C)_{T,P,n_B}$ of the $\Delta_{mix}G$ equation (9.39), noting that $\Delta_{mix}G = G - G^* = G - n_B \mu_B^* - n_C \mu_C^*$.

9.62 The process of Fig. 9.6 enables calculation of $\Delta_{mix}G$. (a) Find expressions for ΔG of each step in Fig. 9.6, assuming all gases are ideal. To find ΔG_4, use a result stated at the end of Sec. 6.1 and derived in Prob. 9.20. (b) Explain why ΔG_1 and ΔG_6 are quite small unless P is very high. (c) Show that if ΔG_1 and ΔG_6 are assumed negligible, then

$$\Delta_{mix}G = n_A RT \ln (P_A/P_A^*) + n_B RT \ln (P_B/P_B^*) \quad (9.77)$$

(d) Verify that if Raoult's law is obeyed, (9.77) reduces to the ideal-solution $\Delta_{mix}G$ equation.

9.63 For ethanol(eth)–chloroform(chl) solutions at 45°C, vapor pressures and vapor-phase ethanol mole fractions as a function of solution composition are [G. Scatchard and C. L. Raymond, *J. Am. Chem. Soc.*, **60**, 1278 (1938)]:

x_{eth}	0.2000	0.4000	0.6000	0.8000
x_{eth}^v	0.1552	0.2126	0.2862	0.4640
P/torr	454.53	435.19	391.04	298.18

At 45°C, $P_{eth}^* = 172.76$ torr and $P_{chl}^* = 433.54$ torr. Use Eq. (9.77) of Prob. 9.62 to calculate and plot $\Delta_{mix}G/(n_A + n_B)$.

9.64 A *simple* two-component solution is one for which

$$\Delta_{mix}G = n_A RT \ln x_A + n_B RT \ln x_B + (n_A + n_B) x_A x_B W(T, P)$$

at constant T and P, where $W(T, P)$ is a function of T and P. Statistical mechanics indicates that when the A and B molecules are approximately spherical and have similar sizes, the solution will be approximately simple. For a simple solution, (a) find expressions for $\Delta_{mix}H$, $\Delta_{mix}S$, and $\Delta_{mix}V$; (b) show that $\mu_A = \mu_A^* + RT \ln x_A + W x_B^2$, with a similar equation for μ_B; (c) find expressions for the vapor partial pressures P_A and P_B, assuming ideal vapor.

9.65 Show that well below the critical point, the vapor pressure P_i of pure liquid i is given by $P_i = P° \exp(-\Delta_{vap}G°/RT)$, where $P° \equiv 1$ bar and $\Delta_{vap}G°$ is the standard Gibbs energy of vaporization. State any approximations made.

9.66 Which of the partial molar properties $\bar{V}_i$, $\bar{U}_i$, $\bar{H}_i$, $\bar{S}_i$, and $\bar{G}_i$ are equal to their corresponding pure-component molar properties for (a) a component of an ideal solution; (b) the solvent in an ideally dilute solution; (c) a solute in an ideally dilute solution?

9.67 State whether each of the following equations applies to no solutions, to all solutions, to ideal solutions, to ideally dilute solutions, or to both ideal and ideally dilute solutions. (a) $G = \sum_i n_i\mu_i$; (b) $\mu_i = \mu_i^* + RT \ln x_i$ for all components; (c) $\mu_i = \mu_i° + RT \ln x_i$ for all components; (d) $V = \sum_i n_i\bar{V}_i$; (e) $V = \sum_i n_i V_{m,i}^*$; (f) $P_i = x_i^l P_i^*$ for all components; (g) $\Delta_{mix}H = 0$; (h) $\Delta_{mix}G = 0$.

9.68 For each component of each of the following liquid solutions, state whether it will approximately obey Raoult's law, Henry's law, or neither: (a) $x_{CCl_4} = 0.5$, $x_{CH_3OH} = 0.5$; (b) $x_{CCl_4} = 0.99$, $x_{CH_3OH} = 0.01$; (c) $x_{CCl_4} = 0.01$, $x_{CH_3OH} = 0.99$; (d) $x_{CCl_4} = 0.4$, $x_{SiCl_4} = 0.6$.

9.69 True or false? (a) $\Delta_{mix}G$ at constant T and P must be negative. (b) $\Delta_{mix}S$ at constant T and P must be positive. (c) Intermolecular interactions are negligible in an ideal solution. (d) Solute–solute interactions are negligible in an ideally dilute solution. (e) The standard state of the solute in an ideally dilute solution is the state of infinite dilution at the T and P of the solution. (f) When 30.0 mL of a 15.0 weight percent $HCl(aq)$ solution is added to 50.0 mL of a 15.0 weight percent $HCl(aq)$ solution at constant T and P, the final volume must be 80.0 mL.

Nonideal Solutions

Using molecular arguments and experimental data, we obtained expressions for the chemical potentials μ_i in ideal gas mixtures (Chapter 6) and in ideal and ideally dilute solutions (Chapter 9). All thermodynamic properties follow from these chemical potentials. For example, we derived the reaction-equilibrium conditions for ideal gases and ideally dilute solutions (the K_P° and K_x equilibrium constants), the conditions for phase equilibrium between an ideal or ideally dilute solution and its vapor (Raoult's law, Henry's law), and the differences between the thermodynamic properties of an ideal solution and the properties of the pure components ($\Delta_{mix}V$, $\Delta_{mix}H$, $\Delta_{mix}S$, $\Delta_{mix}G$).

We therefore know how to deal with ideal solutions. However, all solutions in the real world are nonideal. What happens when the system is not ideal? This chapter deals with (a) nonideal liquid and solid solutions of nonelectrolytes (Secs. 10.1 to 10.5), (b) solutions of electrolytes (Secs. 10.6 to 10.10), and (c) nonideal gas mixtures (Sec. 10.11). Chapter 11 considers reaction equilibrium in nonideal systems. Deviations from ideality are often quite large and must be included for accurate results in biochemical, environmental, and industrial applications of thermodynamics.

The chemical potentials in nonideal systems are usually expressed in terms of activities and activity coefficients, so our first task is to define these quantities and tell how they are measured.

10.1 ACTIVITIES AND ACTIVITY COEFFICIENTS

The chemical potentials are the key thermodynamic properties, since all other thermodynamic properties can be derived from the μ_i's. For an ideal (id) or ideally dilute liquid or solid solution of nonelectrolytes, the chemical potential of each component is [Eqs. (9.42), (9.43), (9.59), and (9.60)]

$$\mu_i^{id} = \mu_i^\circ + RT \ln x_i \qquad \text{ideal or ideally dil. soln.} \qquad (10.1)^*$$

where μ_i° is the chemical potential in the appropriately defined standard state. Equation (10.1) gives $\ln x_i = (\mu_i^{id} - \mu_i^\circ)/RT$, or

$$x_i = \exp[(\mu_i^{id} - \mu_i^\circ)/RT] \qquad \text{ideal or ideally dil. soln.} \qquad (10.2)$$

A **nonideal solution** is defined as one that is neither ideal nor ideally dilute. We shall discuss the behavior of nonideal-solution components in terms of departures from ideal or ideally dilute behavior. To make it easy to compare nonideal and ideal behavior, we choose to express the nonideal chemical potentials μ_i in a form that closely resembles the ideal chemical potentials in (10.1). For each component i of a nonideal solution, we choose a standard state and symbolize the **standard-state chemical potential** of i by μ_i°. (The standard state will be chosen to correspond to the standard state used in either an ideal or ideally dilute solution; see below.) We then define the **activity** a_i of substance i in any solution (nonideal or ideal) by

$$a_i \equiv \exp[(\mu_i - \mu_i^\circ)/RT] \qquad \text{every soln.} \qquad (10.3)$$

The defining equation (10.3) for a_i is chosen to resemble (10.2) for ideal and ideally dilute solutions, so as to lead to a nonideal μ_i expression that can be readily compared with (10.1). Taking logs of (10.3), we get $\ln a_i = (\mu_i - \mu_i^\circ)/RT$, or

$$\mu_i = \mu_i^\circ + RT \ln a_i \qquad \text{every soln.} \qquad (10.4)^*$$

Thus, the activity a_i replaces the mole fraction x_i in the expression for μ_i in a non-ideal solution. From (10.1) and (10.4) we see that $a_i = x_i$ in an ideal or ideally dilute solution. When solution component i is in its standard state, μ_i equals μ_i° and, from (10.3), its activity a_i equals 1 ($a_i^\circ = 1$).

The difference between the real-solution chemical potential μ_i in (10.4) and the corresponding ideal-solution μ_i^{id} in (10.1) is

$$\mu_i - \mu_i^{\text{id}} = RT \ln a_i - RT \ln x_i = RT \ln (a_i/x_i)$$

The ratio a_i/x_i is thus a measure of the departure from ideal behavior. We therefore define the **activity coefficient** γ_i (gamma i) of component i as $\gamma_i \equiv a_i/x_i$, so that

$$a_i = \gamma_i x_i \qquad \text{every soln.} \qquad (10.5)^*$$

The activity coefficient γ_i measures the degree of departure of substance i's behavior from ideal or ideally dilute behavior. The activity a_i can be viewed as being obtained from the mole fraction x_i by correcting for nonideality. In an ideal or ideally dilute solution, the activity coefficients γ_i are 1. From (10.4) and (10.5), the chemical potentials in a nonideal solution of nonelectrolytes are

$$\mu_i = \mu_i^\circ + RT \ln \gamma_i x_i \qquad (10.6)^*$$

Since μ_i depends on T, P, and the mole fractions, the activity a_i in (10.3) and the activity coefficient $\gamma_i \equiv a_i/x_i$ depend on these variables:

$$a_i = a_i(T, P, x_1, x_2, \ldots), \qquad \gamma_i = \gamma_i(T, P, x_1, x_2, \ldots)$$

Note from (10.3) and (10.5) that a_i and γ_i are dimensionless and nonnegative.

The task of thermodynamics is to show how a_i and γ_i can be found from experimental data; see Sec. 10.3. The task of statistical mechanics is to find a_i and γ_i from the intermolecular interactions in the solution.

The activity a_i of species i is $a_i \equiv e^{\mu_i/RT} e^{-\mu_i^\circ/RT}$ [Eq. (10.3)]. If the composition of the solution is varied at fixed T and P, the factor $e^{-\mu_i^\circ/RT}$ remains constant and a_i varies in proportion to $e^{\mu_i/RT}$. The activity a_i is a measure of the chemical potential μ_i in the solution. As μ_i increases, a_i increases. If we add some of substance i to a solution at fixed T and P, the chemical potential μ_i must increase [Eq. (4.90)]. Therefore, constant-T-and-P addition of i to a solution must increase the activity a_i. Like the chemical potential, a_i is a measure of the escaping tendency of i from the solution.

The activity a_i is more convenient to use in numerical calculations than μ_i because (a) we cannot determine absolute values of μ_i (only relative values); (b) $\mu_i \to -\infty$ as $x_i \to 0$; (c) a_i can be compared with x_i (and γ_i with 1) to judge the degree of nonideality.

Standard States for Nonideal-Solution Components

To complete the definitions (10.3) and (10.5) of a_i and γ_i, we must specify the standard state of each solution component. Two different standard-state conventions are used with Eq. (10.6).

Convention I For a solution where the mole fractions of all components can be varied over a considerable range, one usually uses Convention I. The most common case is a solution of two or more liquids (for example, ethanol plus water). The

Convention I standard state of each solution component i is taken as pure liquid i at the temperature and pressure of the solution:

$$\mu_{I,i}^\circ \equiv \mu_i^*(T, P) \qquad \text{for all components} \qquad (10.7)^*$$

where the subscript I indicates the Convention I choice of standard states, the degree indicates the standard state, and the star indicates a pure substance. *Convention I is the same convention as that used for ideal solutions* (Sec. 9.6).

The value of the chemical potential $\mu_i \equiv (\partial G/\partial n_i)_{T,P,n_{j\neq i}}$ is clearly independent of the choice of standard state. However, the value of μ_i° depends on the choice of standard state. Therefore, $a_i \equiv \exp[(\mu_i - \mu_i^\circ)/RT]$ [Eq. (10.3)] depends on this choice; hence $\gamma_i \equiv a_i/x_i$ also depends on the choice of standard state. We use the subscript I to denote Convention I activities, activity coefficients, and standard-state chemical potentials, writing them as $a_{I,i}$, $\gamma_{I,i}$, and $\mu_{I,i}^\circ$. An alternative notation for $\gamma_{I,i}$ is f_i (which can be confused with the quantity fugacity, defined in Sec. 10.11).

Since the Convention I standard state is the same as the ideal-solution standard state, μ_i° in the ideal-solution equation $\mu_i^{id} = \mu_i^\circ + RT \ln x_i$ is the same as $\mu_{I,i}^\circ$ in the Convention I nonideal equation $\mu_i = \mu_{I,i}^\circ + RT \ln \gamma_{I,i} x_i$. It follows that for an ideal solution $\gamma_{I,i} = 1$. For a nonideal solution, *the deviations of the $\gamma_{I,i}$'s from* 1 *measure the deviation of the solution's behavior from ideal-solution behavior.*

Equations (10.6) and (10.7) give $\mu_i = \mu_i^* + RT \ln \gamma_{I,i} x_i$. As x_i goes to 1 at constant T and P, the chemical potential μ_i goes to μ_i^*, since the solution becomes pure i. Hence the $x_i \to 1$ limit of this last equation is $\mu_i^* = \mu_i^* + RT \ln \gamma_{I,i}$ or $\ln \gamma_{I,i} = 0$ and $\gamma_{I,i} = 1$:

$$\gamma_{I,i} \to 1 \text{ as } x_i \to 1 \qquad \text{for each } i \qquad (10.8)^*$$

The Convention I activity coefficient of species i goes to 1 as the solution composition approaches pure i (see Fig. 10.3a).

Since the Convention I standard state of each solution component is the pure substance, the Convention I standard-state thermodynamic properties of i equal the corresponding properties of pure i. Convention I puts all the components on the same footing and does not single out one component as the solvent. Therefore, Convention I is often called the **symmetrical convention.**

Convention II Convention II (also called the **unsymmetrical convention**) is used when one wants to treat one solution component (the *solvent* A) differently from the other components (the *solutes i*). Common cases are solutions of solids or gases in a liquid solvent.

The Convention II standard state of the solvent A is pure liquid A at the T and P of the solution. With $\mu_{II,A}^\circ = \mu_A^*(T, P)$, Eq. (10.6) becomes $\mu_A = \mu_A^* + RT \ln \gamma_{II,A} x_A$. Taking the limit of this equation as $x_A \to 1$, we find [as in (10.8)] that $\gamma_{II,A} \to 1$ as $x_A \to 1$. Thus

$$\mu_{II,A}^\circ = \mu_A^*(T, P), \qquad \gamma_{II,A} \to 1 \text{ as } x_A \to 1 \qquad (10.9)^*$$

For each solute $i \neq A$, Convention II chooses the standard state so that $\gamma_{II,i}$ goes to 1 in the limit of infinite dilution:

$$\gamma_{II,i} \to 1 \text{ as } x_A \to 1 \qquad \text{for each } i \neq A \qquad (10.10)^*$$

Note that the limit in (10.10) is taken as the *solvent* mole fraction x_A goes to 1 (and hence $x_i \to 0$), which is quite different from (10.8), where the limit is taken as $x_i \to 1$. We choose a Convention II standard state that is consistent with (10.10) as follows. Setting μ_i in (10.6) equal to μ_i°, we get $0 = RT \ln \gamma_i x_i$, so $\gamma_{II,i} x_i$ must equal 1 in the standard state. When x_A is near 1 and the solute mole fractions are small, then by (10.10) the activity coefficient $\gamma_{II,i}$ is close to 1. We choose the standard state of each solute i as the *fictitious* state obtained as follows. We pretend that the behavior of μ_i that holds in the limit of infinite dilution (namely, $\mu_i = \mu_i^\circ + RT \ln x_i$) holds for all values of x_i, and we take the limit as $x_i \rightarrow 1$ (Fig. 9.20). This gives a fictitious standard state with $\gamma_{\text{II},i} = 1$, $x_i = 1$, and $\mu_i = \mu_i^\circ$. This fictitious state corresponds to pure solute i in which each i molecule experiences the same intermolecular forces it experiences in an ideally dilute solution in the solvent A.

The Convention II solute standard state is the same as that used for solutes in an ideally dilute solution (Sec. 9.8 and Fig. 9.20), so the Convention II standard-state thermodynamic properties are the same as for solutes in an ideally dilute solution. Thus, for each solute i, we have [Eqs. (9.67) and (9.68)]

$$\bar{V}_{\text{II},i}^\circ = \bar{V}_i^\infty, \qquad \bar{H}_{\text{II},i}^\circ = \bar{H}_i^\infty, \qquad \bar{S}_{\text{II},i}^\circ = (\bar{S}_i + R \ln x_i)^\infty$$

where the infinity superscript denotes the infinite-dilution limit. The equation for $\bar{S}_{\text{II},i}^\circ$ is valid because in the infinite-dilution limit the solution becomes ideally dilute and $\bar{S}_i = \bar{S}_i^\circ - R \ln x_i$ [Eq. (9.68)] then holds. Convention II standard-state solute properties that don't involve the entropy are equal to their infinite-dilution values.

The Convention II solute and solvent standard states are the same as those used for ideally dilute solutions. Therefore (by the same reasoning used earlier for Convention I and ideal solutions), in an ideally dilute solution, $\gamma_{\text{II},A} = 1$ and $\gamma_{\text{II},i} = 1$. *The deviations of $\gamma_{\text{II},A}$ and $\gamma_{\text{II},i}$ from 1 measure the deviations of the solution's behavior from ideally dilute behavior.*

> The concepts of activity and activity coefficient were introduced by the American chemist G. N. Lewis. (Recall Lewis dot structures, the Lewis octet rule, Lewis acids and bases.) Lewis spent the early part of his career at Harvard and M.I.T. In 1912, he became head of the chemistry department at the University of California at Berkeley. In 1916 he proposed that a chemical bond consists of a shared pair of electrons, a novel idea at the time. He measured $\Delta_f G^\circ$ for many compounds and cataloged the available free-energy data, drawing the attention of chemists to the usefulness of such data. The concept of partial molar quantities is due to Lewis. His 1923 book *Thermodynamics* (written with Merle Randall) made thermodynamics accessible to chemists. Lewis was resentful that his early speculative ideas on chemical bonding and the nature of light were not appreciated by Harvard chemists, and in 1929 he refused an honorary degree from Harvard. His later years were spent working on relativity and photochemistry.

Summary

The chemical potential of solution component i is expressed in terms of the activity a_i and the activity coefficient γ_i, where a_i and γ_i are defined so that $\mu_i = \mu_i^\circ + RT \ln a_i$, where $a_i = \gamma_i x_i$. Convention I chooses the standard state of each solution component as the pure substance at the T and P of the solution; Convention I activity coefficients measure deviations from ideal-solution behavior. Convention II uses the same standard states as for an ideally dilute solution, and deviations of Convention II activity coefficients from 1 measure deviations from ideally dilute behavior. Whereas each $\gamma_{\text{I},i}$ goes to 1 as x_i goes to 1, Convention II activity coefficients all go to 1 as the solvent mole fraction $x_A \rightarrow 1$.

10.2   EXCESS FUNCTIONS

The thermodynamic properties of a solution of two liquids are often expressed in terms of excess functions. The *excess Gibbs energy* G^E of a mixture of liquids is defined as the difference between the actual Gibbs energy G of the solution and the Gibbs energy G^{id} of a hypothetical ideal solution with the same T, P, and composition as the actual solution: $G^E \equiv G - G^{\text{id}}$. Similar definitions hold for other excess properties:

$$G^E \equiv G - G^{\text{id}}, \qquad H^E \equiv H - H^{\text{id}}, \qquad S^E \equiv S - S^{\text{id}}, \qquad V^E \equiv V - V^{\text{id}} \qquad (10.11)$$

Subtraction of $G^{id} = H^{id} - TS^{id}$ from $G = H - TS$ gives $G^E = H^E - TS^E$.

We have $G = \Sigma_i n_i \mu_i = \Sigma_i n_i(\mu_i^* + RT \ln \gamma_{I,i} x_i)$ [Eqs. (9.23), (10.6), and (10.7)] and (since $\gamma_{I,i} = 1$ in an ideal solution) $G^{id} = \Sigma_i n_i(\mu_i^* + RT \ln x_i)$. Subtraction gives

$$G - G^{id} = G^E = RT \sum_i n_i \ln \gamma_{I,i} \qquad (10.12)$$

so G^E can be found from the activity coefficients. Conversely, if G^E is known as a function of solution composition, the activity coefficients can be calculated from G^E (see Prob. 10.4).

Excess functions are found from mixing quantities. We have

$$G^E \equiv G - G^{id} = G - G^{id} + G^* - G^* = G - G^* - (G^{id} - G^*)$$

$$G^E = \Delta_{mix}G - \Delta_{mix}G^{id}$$

The same argument holds for other excess properties, and (since $\Delta_{mix}H^{id} = 0$ and $\Delta_{mix}V^{id} = 0$)

$$G^E = \Delta_{mix}G - \Delta_{mix}G^{id}, \qquad S^E = \Delta_{mix}S - \Delta_{mix}S^{id},$$

$$H^E = \Delta_{mix}H, \qquad V^E = \Delta_{mix}V$$

where $\Delta_{mix}G^{id}$ and $\Delta_{mix}S^{id}$ are given by (9.44) and (9.46).

Figure 10.1 shows typical curves of G, G^{id}, $\Delta_{mix}G$, $\Delta_{mix}G^{id}$, and G^E versus composition at constant T and P for solutions of two liquids B and C that show positive deviations from ideality. In drawing the curves, it was arbitrarily assumed that $G_{m,C} = 0$ and $G_{m,B} = 10$ kJ/mol.

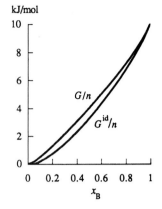

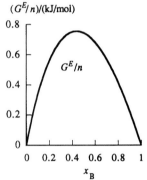

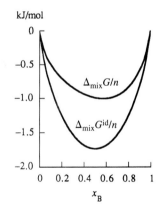

10.3 DETERMINATION OF ACTIVITIES AND ACTIVITY COEFFICIENTS

The formalism of Sec. 10.1 leads nowhere unless we can determine activity coefficients. Once these are known, the chemical potentials μ_i are known, since $\mu_i = \mu_i^\circ + RT \ln \gamma_i x_i$ [Eq. (10.6)]. From the chemical potentials, the other thermodynamic properties can be found.

Activity coefficients are usually found from data on phase equilibria, most commonly from vapor-pressure measurements. The condition for phase equilibrium between the solution and its vapor is that for each species i the chemical potential μ_i in the solution must equal the chemical potential μ_i^v of i in the vapor phase. We shall assume the vapor in equilibrium with the solution to be an ideal gas mixture. Departures from ideality in gases are ordinarily much smaller than are departures from ideal-solution behavior in liquids. (See Sec. 10.11 for allowance for gas nonideality.) Since μ_i^v depends on the vapor partial pressure P_i and since μ_i in solution depends on γ_i, measurement of P_i allows the activity coefficient γ_i to be found. The vapor partial pressure P_i allows us to probe the escaping tendency of i from the solution.

Figure 10.1

Typical curves of G, G^E, and $\Delta_{mix}G$ at 25°C for two liquids that show positive deviations from ideality. G^{id} and $\Delta_{mix}G^{id}$ are the corresponding quantities for ideal solutions. Of course, $G_{id}^E = 0$. n is the total number of moles.

Convention I

Suppose we want a solution's activities $a_{I,i}$ and activity coefficients $\gamma_{I,i}$ for the Convention I choice of standard states. Recall that for an ideal solution we started from $\mu_i = \mu_i^\circ + RT \ln x_i^l$ and derived Raoult's law $P_i = x_i^l P_i^*$ (Sec. 9.6). For a real solution, the activity replaces the mole fraction in μ_i, and we have $\mu_i = \mu_{I,i}^\circ + RT \ln a_{I,i}$. Also, the Convention I standard states are the same as the ideal-solution standard states, so μ_i° has the same meaning in these two expressions for μ_i. Therefore, exactly the same steps that gave Raoult's law $P_i = x_i^l P_i^*$ in Sec. 9.6 will give for a nonideal solution

$$P_i = a_{I,i}P_i^* \qquad \text{ideal vapor, } P \text{ not very high} \qquad (10.13)^*$$

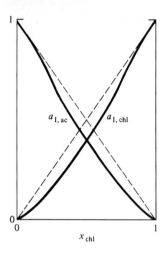

Figure 10.2

Convention I activities versus composition for acetone–chloroform solutions at 35°C. The dashed lines are for an ideal solution.

Thus $a_{I,i} = P_i/P_i^*$, where P_i is the partial vapor pressure of i above the solution and P_i^* is the vapor pressure of pure i at the temperature of the solution.

At a given temperature, P_i^* is a constant, so (10.13) shows that the activity $a_{I,i}$ of a substance in a solution is proportional to the vapor partial pressure P_i of the solution. Therefore a plot of P_i versus x_i^l is, except for a change in scale, the same as a plot of $a_{I,i}$ versus x_i^l. Figure 9.21a plots P_{ac} and P_{chl} versus x_{chl}^l for acetone–chloroform solutions at 35°C. To change these plots to activity plots, we divide P_{ac} by P_{ac}^* (which is a constant) and divide P_{chl} by P_{chl}^*, since $a_{I,ac} = P_{ac}/P_{ac}^*$ and $a_{I,chl} = P_{chl}/P_{chl}^*$. Figure 10.2 shows the resulting activity curves, which have the same shapes as the vapor-pressure curves in Fig. 9.21a. The plots of Fig. 10.2 agree with the result that a_i must increase as x_i increases (Sec. 10.1). The dashed lines in Fig. 10.2 show the hypothetical ideal-solution activities $a_i^{id} = x_i$.

Since $a_{I,i} = \gamma_{I,i} x_i$, Eq. (10.13) becomes

$$P_i = \gamma_{I,i} x_i^l P_i^* \qquad \text{or} \qquad x_i^v P = \gamma_{I,i} x_i^l P_i^* \qquad (10.14)$$

where x_i^l is the mole fraction of i in the liquid (or solid) solution, x_i^v is its mole fraction in the vapor above the solution, and P is the vapor pressure of the solution. To find $a_{I,i}$ and $\gamma_{I,i}$ we measure the solution vapor pressure and analyze the vapor and liquid to find x_i^v and x_i^l. For a two-component solution, the vapor composition can be found by condensing a portion of it, measuring the density or refractive index of the condensate, and comparing with values for solutions of known composition. Equation (10.14) is Raoult's law modified to allow for solution nonideality.

Since the partial pressure P_i^{id} above an ideal solution is given by Raoult's law as $P_i^{id} = x_i^l P_i^*$, Eq. (10.14) can be written as $\gamma_{I,i} = P_i/P_i^{id}$. The Convention I activity coefficient is the ratio of the actual partial vapor pressure to what the partial vapor pressure would be if the solution were ideal. If component i shows a positive deviation (Sec. 9.8) from ideality ($P_i > P_i^{id}$), then its activity coefficient $\gamma_{I,i}$ is greater than 1. A negative deviation from ideality ($P_i < P_i^{id}$) means $\gamma_{I,i} < 1$. In Fig. 9.21a, γ_I for acetone and γ_I for chloroform are less than 1 for all solution compositions. In Fig. 9.21b, the γ_I's are greater than 1.

Note from (10.6) and (10.1) that having the γ_I's less than 1 means the chemical potentials are less than the corresponding ideal-solution chemical potentials μ^{id}. Therefore G (which equals $\sum_i n_i \mu_i$) is less than G^{id}, and the solution is more stable than the corresponding ideal solution. Negative deviations mean that the components of the solution feel friendly toward each other and have a smaller tendency to escape each other's close company by vaporizing, where the comparison is with an ideal solution, in which the components have the same feelings for each other as for molecules of their own kind. Solutions with positive deviations are less stable than the corresponding ideal solutions. If the positive deviations become large enough, the solution will separate into two liquid phases whose compositions differ from each other and whose total G is less than that of the solution (partial miscibility—Sec. 12.7).

EXAMPLE 10.1 Convention I activity coefficients

For solutions of acetone (ac) plus chloroform (chl) at 35.2°C, vapor pressures P and acetone vapor-phase mole fractions x_{ac}^v are given in Table 10.1 as functions of the liquid-phase acetone mole fraction x_{ac}^l. (These data are graphed in Fig. 9.21.) (a) Find the Convention I activity coefficients in these solutions. (b) Find $\Delta_{mix}G$ when 0.200 mol of acetone and 0.800 mol of chloroform are mixed at 35.2°C and 1 bar.

TABLE 10.1

Vapor Pressures and Vapor Compositions for Acetone–Chloroform Solutions at 35.2°C

x^l_{ac}	x^v_{ac}	P/torr	x^l_{ac}	x^v_{ac}	P/torr
0.0000	0.0000	293	0.6034	0.6868	267
0.0821	0.0500	$279._5$	0.7090	0.8062	286
0.2003	0.1434	262	0.8147	0.8961	307
0.3365	0.3171	249	0.9397	0.9715	332
0.4188	0.4368	248	1.0000	1.0000	$344._5$
0.5061	0.5625	255			

(a) For $x^l_{ac} = 0.0821$, Eq. (10.14) gives

$$\gamma_{I,ac} = \frac{x^v_{ac}P}{x^l_{ac}P^*_{ac}} = \frac{0.0500(279.5 \text{ torr})}{0.0821(344.5 \text{ torr})} = 0.494$$

$$\gamma_{I,chl} = \frac{x^v_{chl}P}{x^l_{chl}P^*_{chl}} = \frac{0.9500(279.5 \text{ torr})}{0.9179(293 \text{ torr})} = 0.987$$

Similar treatment of the other data and use of (10.8) give:

x_{ac}	0	0.082	0.200	0.336	0.506	0.709	0.815	0.940	1
$\gamma_{I,ac}$		0.494	0.544	0.682	0.824	0.943	0.981	0.997	1
$\gamma_{I,chl}$	1	0.987	0.957	0.875	0.772	0.649	0.588	0.536	
x_{chl}	1	0.918	0.800	0.664	0.494	0.291	0.185	0.060	0

Figure 10.3a plots the activity coefficients γ_I versus solution composition.

(b) The mixing is at 1 bar, whereas at $x_{ac} = 0.200$, the solution is under a pressure of 262 torr (its vapor pressure), and the γ_I's are for this pressure.

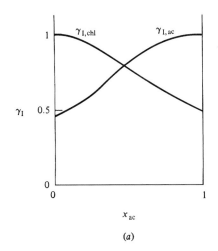

(a)

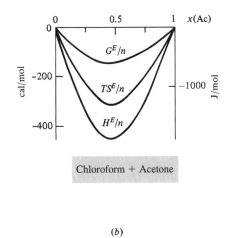

(b)

Figure 10.3

Properties of acetone–chloroform solutions at 35°C. (a) Convention I activity coefficients. (b) Excess functions (n is the total number of moles).

However, for liquid solutions, the activity coefficients (like the chemical potentials) change very slowly with pressure, and the effect of this pressure change on the γ_I's can be ignored. We have

$$\Delta_{mix}G = G - G^* = \sum_i n_i(\mu_i - \mu_i^*) = \sum_i n_i(\mu_{I,i}^\circ + RT \ln \gamma_{I,i}x_i - \mu_i^*)$$

$$\Delta_{mix}G = \sum_i n_i RT \ln \gamma_{I,i}x_i$$

since $\mu_{I,i}^\circ = \mu_i^*$ [Eq. (10.7)]. Then

$$\Delta_{mix}G = [8.314 \text{ J/(mol K)}](308.4 \text{ K})$$
$$\times \{(0.200 \text{ mol}) \ln [(0.544)(0.200)] + (0.800 \text{ mol}) \ln [(0.957)(0.800)]\}$$
$$= -1685 \text{ J}$$

Exercise

For an acetone–chloroform solution at 35.2°C with $x_{ac}^l = 0.4188$, use Table 10.1 to find $\gamma_{I,ac}$ and $\gamma_{I,chl}$. Find $\Delta_{mix}G$ when 0.4188 mol of acetone and 0.5812 mol of chloroform are mixed at 35.2°C and 1 bar. (*Answers:* 0.751, 0.820, −2347 J.)

From the activity coefficients calculated in this example, we can find relative partial molar Gibbs energies (Fig. 9.11) in the solutions using $\mu_i - \mu_i^* = RT \ln \gamma_{I,i}x_i$.

Use of $P_i = \gamma_{I,i}x_iP_i^*$ in the $\Delta_{mix}G$ equation in this example gives $\Delta_{mix}G = \sum_i n_iRT \ln (P_i/P_i^*)$. This equation (previously given in Prob. 9.62) allows $\Delta_{mix}G$ to be calculated directly from vapor-pressure data (Sec. 9.3).

The large negative deviations of γ_I from 1 in this example (and the corresponding large negative deviations in Fig. 9.21*a* from Raoult's law) indicate large deviations from ideal-solution behavior. Nuclear magnetic resonance spectra indicate that these deviations are due to hydrogen bonding between acetone and chloroform according to $Cl_3C—H \cdots O=C(CH_3)_2$ [C. M. Huggins et al., *J. Chem. Phys.*, **23**, 1244 (1955); A. Apelblat et al., *Fluid Phase Equilibria*, **4**, 229 (1980).] The hydrogen bonding makes the acetone–chloroform intermolecular attraction stronger than the average of the acetone–acetone and chloroform–chloroform attractions. Therefore, $\Delta_{mix}H$ is negative, compared with zero for formation of an ideal solution. The hydrogen bonding gives a significant degree of order in the mixture, making $\Delta_{mix}S$ less than for formation of an ideal solution. The enthalpy effect turns out to outweigh the entropy effect, and $\Delta_{mix}G = \Delta_{mix}H - T \Delta_{mix}S$ is less than $\Delta_{mix}G$ for formation of an ideal solution. Figure 10.3*b* shows H^E/n, TS^E/n, and G^E/n for acetone–chloroform solutions at 35°C, where the excess quantities are (Sec. 10.2) $H^E = \Delta_{mix}H$, $S^E = \Delta_{mix}S - \Delta_{mix}S^{id}$, and $G^E = \Delta_{mix}G - \Delta_{mix}G^{id}$.

Both energy effects ($\Delta_{mix}H$) and entropy effects ($\Delta_{mix}S$) contribute to deviations from ideal-solution behavior. Sometimes the entropy effect is more important than the energy effect. For example, for a solution of 0.5 mol of ethanol plus 0.5 mol of water at 25°C, $H^E = \Delta_{mix}H$ is negative ($\Delta_{mix}H = -400$ J/mol), but TS^E is much more negative than H^E ($TS^E = -1200$ J/mol), so $G^E = H^E - TS^E$ is highly positive. The solution is less stable than the corresponding ideal solution. This solution shows positive deviations from ideality, even though the mixing is exothermic.

Convention II

Now suppose we want the Convention II activity coefficients. The standard states in Convention II are the same as for an ideally dilute solution. Whereas $\mu_i = \mu_i^\circ +$

$RT \ln x_i$ for an ideally dilute solution, we have $\mu_i = \mu^\circ_{\text{II},i} + RT \ln a_{\text{II},i}$ in a nonideal solution. Hence, exactly the same steps that led to Henry's law $P_i = K_i x^l_i$ [Eq. (9.63)] for the solutes and Raoult's law $P_A = x^l_A P^*_A$ [Eq. (9.64)] for the solvent lead to modified forms of these laws with mole fractions replaced by Convention II activities. Therefore, the partial vapor pressures of any solution are

$$P_i = K_i a_{\text{II},i} = K_i \gamma_{\text{II},i} x^l_i \qquad \text{for } i \neq A, \text{ ideal vapor} \qquad (10.15)$$

$$P_A = a_{\text{II},A} P^*_A = \gamma_{\text{II},A} x^l_A P^*_A \qquad \text{ideal vapor, } P \text{ not very high} \qquad (10.16)$$

where A is the solvent. To apply (10.15), we need the Henry's law constant K_i. This can be found by measurements on very dilute solutions where $\gamma_{\text{II},i} = 1$. Thus, vapor-pressure measurements give the Convention II activities and activity coefficients. Equations (10.15) and (10.16) give $\gamma_{\text{II},i} = P_i/P^{\text{id-dil}}_i$ and $\gamma_{\text{II},A} = P_A/P^{\text{id-dil}}_A$, where id-dil stands for ideally dilute.

EXAMPLE 10.2 Convention II activity coefficients

Find the Convention II activity coefficients at 35.2°C for acetone–chloroform solutions, taking acetone as the solvent. Use Table 10.1.

Ordinarily, one would use Convention I for acetone–chloroform solutions, but for illustrative purposes we use Convention II. Equation (10.16) for the solvent Convention II activity coefficient $\gamma_{\text{II},A}$ is the same as the Convention I equation (10.14), so $\gamma_{\text{II},A} = \gamma_{\text{I},A}$. Since acetone has been designated as the solvent, we have $\gamma_{\text{II},ac} = \gamma_{\text{I},ac}$. The $\gamma_{\text{I},ac}$ values were found in Example 10.1.

For the solute chloroform, Eq. (10.15) gives $\gamma_{\text{II},chl} = P_{chl}/K_{chl}x^l_{chl}$. We need the Henry's law constant K_{chl}. In Fig. 9.21a, the Henry's law dotted line for chloroform intersects the right-hand axis at 145 torr, and this is K_{chl} in acetone. (A more accurate value of K_{chl} can be found by plotting P_{chl}/x^l_{chl} versus x^l_{chl} and extrapolating to $x^l_{chl} = 0$. See also Prob. 10.10.) The Table 10.1 data and $K_{chl} = 145$ torr then allow calculation of $\gamma_{\text{II},chl}$. Time can be saved by noting that $\gamma_{\text{I},i} = P_i/x^l_i P^*_i$, so $\gamma_{\text{II},i}/\gamma_{\text{I},i} = (P_i/K_i x^l_i) \div (P_i/x^l_i P^*_i) = P^*_i/K_i = (293 \text{ torr})/(145 \text{ torr}) = 2.02$. Thus $\gamma_{\text{II},chl} = 2.02\gamma_{\text{I},chl}$. Using the $\gamma_{\text{I},chl}$ values from Example 10.1 and (10.10), we find:

x_{ac}	0	0.082	0.200	0.336	0.506	0.709	0.815	0.940	1
$\gamma_{\text{II},chl}$	2.02	1.99	1.93	1.77	1.56	1.31	1.19	1.08	1
$\gamma_{\text{II},ac}$		0.494	0.544	0.682	0.824	0.943	0.981	0.997	1

The γ_{II}'s are plotted in Fig. 10.4. Both γ_{II}'s go to 1 as the solvent mole fraction $x_{ac} \to 1$, whereas $\gamma_{\text{I},chl} \to 1$ as $x_{chl} \to 1$ and $\gamma_{\text{I},ac} \to 1$ as $x_{ac} \to 1$ (Fig. 10.3a).

Exercise

Use Table 10.1 to find $\gamma_{\text{II},ac}$ and $\gamma_{\text{II},chl}$ in a 35.2°C acetone–chloroform solution with $x^l_{ac} = 0.4188$ if acetone is considered to be the solvent. (*Answer:* 0.751, 1.65_6.)

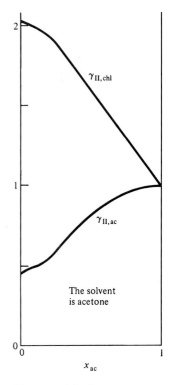

Figure 10.4

Convention II activity coefficients versus composition for acetone–chloroform solutions at 35°C with acetone taken as the solvent.

Note that $\gamma_{\text{II},chl} > 1$ with acetone as the solvent, whereas $\gamma_{\text{I},chl} < 1$ (Fig. 10.3a). This corresponds to the fact that P_{chl} in Fig. 9.21a is less than the corresponding Raoult's law (ideal-solution) dashed-line partial pressure, and P_{chl} is greater than the corresponding Henry's law (ideally dilute solution) partial pressure. γ_{I} measures deviations from ideal-solution behavior; γ_{II} measures deviations from ideally dilute solution behavior.

Since $\gamma_{\mathrm{II,chl}} > 1$ and $\gamma_{\mathrm{II,ac}} < 1$ for acetone as solvent, μ_{chl} in Eq. (10.6) is greater than $\mu_{\mathrm{chl}}^{\mathrm{id\text{-}dil}}$, the chloroform chemical potential in a hypothetical ideally dilute solution of the same composition, and $\mu_{\mathrm{ac}} < \mu_{\mathrm{ac}}^{\mathrm{id\text{-}dil}}$. In a hypothetical ideally dilute solution, the chloroform molecules interact only with the solvent acetone, and this is a favorable interaction due to the hydrogen bonding discussed earlier. In the real solution, $CHCl_3$ molecules also interact with other $CHCl_3$ molecules, which is a less favorable interaction than with acetone molecules; this increases μ_{chl} above $\mu_{\mathrm{chl}}^{\mathrm{id\text{-}dil}}$. In an ideally dilute solution, the interaction of the solvent acetone with the solute chloroform has an insignificant effect on $\mu_{\mathrm{ac}}^{\mathrm{id\text{-}dil}}$. In the real solution, the acetone–chloroform interaction is significant, and since this interaction is favorable, μ_{ac} is less than $\mu_{\mathrm{ac}}^{\mathrm{id\text{-}dil}}$.

The Gibbs–Duhem Equation

Activity coefficients of nonvolatile solutes can be found from vapor-pressure data using the Gibbs–Duhem equation, which we now derive. Taking the total differential of $G = \Sigma_i\, n_i\mu_i$ [Eq. (9.23)], we find as the change in G of the solution in any infinitesimal process (including processes that change the amounts of the components of the solution)

$$dG = d\sum_i n_i\mu_i = \sum_i d(n_i\mu_i) = \sum_i (n_i\, d\mu_i + \mu_i\, dn_i) = \sum_i n_i\, d\mu_i + \sum_i \mu_i\, dn_i$$

The use of $dG = -S\, dT + V\, dP + \Sigma_i\, \mu_i\, dn_i$ [Eq. (4.73)] gives

$$-S\, dT + V\, dP + \sum_i \mu_i\, dn_i = \sum_i n_i\, d\mu_i + \sum_i \mu_i\, dn_i$$

$$\sum_i n_i\, d\mu_i + S\, dT - V\, dP = 0 \qquad (10.17)$$

This is the **Gibbs–Duhem equation.** Its most common application is to a constant-T-and-P process ($dT = 0 = dP$), where it becomes

$$\sum_i n_i\, d\mu_i \equiv \sum_i n_i\, d\bar{G}_i = 0 \qquad \text{const. } T, P \qquad (10.18)$$

Equation (10.18) can be generalized to any partial molar quantity as follows. If Y is any extensive property of a solution, then $Y = \Sigma_i\, n_i\bar{Y}_i$ [Eq. (9.26)] and $dY = \Sigma_i\, n_i\, d\bar{Y}_i + \Sigma_i\, \bar{Y}_i\, dn_i$. Equation (9.25) with $dT = 0 = dP$ reads $dY = \Sigma_i\, \bar{Y}_i\, dn_i$. Equating these two expressions for dY, we get

$$\sum_i n_i\, d\bar{Y}_i = 0 \quad \text{or} \quad \sum_i x_i\, d\bar{Y}_i = 0 \qquad \text{const. } T, P \qquad (10.19)$$

where the form involving the mole fractions x_i was found by dividing by the total number of moles. The Gibbs–Duhem equation (10.19) shows that the $\bar{Y}_i$'s are not all independent. Knowing the values of $r - 1$ of the $\bar{Y}_i$'s as functions of composition for a solution of r components, we can integrate (10.19) to find $\bar{Y}_r$.

For a two-component solution, Eq. (10.19) with $Y = V$ (the volume) reads $x_A\, d\bar{V}_A + x_B\, d\bar{V}_B = 0$ or $d\bar{V}_A = -(x_B/x_A)\, d\bar{V}_B$ at constant T and P. Thus, $d\bar{V}_A$ and $d\bar{V}_B$ must have opposite signs, as in Fig. 9.9. Similarly, $d\mu_A$ and $d\mu_B$ must have opposite signs when the solution's composition changes at constant T and P.

Activity Coefficients of Nonvolatile Solutes

For a solution of a solid in a liquid solvent, the vapor partial pressure of the solute over the solution is usually immeasurably small and cannot be used to find the solute's activity coefficient. Measurement of the vapor pressure as a function of solution composition gives P_A, the solvent partial pressure, and hence allows calculation of the solvent activity coefficient γ_A as a function of composition. We then use the integrated Gibbs–Duhem equation to find the solute activity coefficient γ_B.

After division by $n_A + n_B$ the Gibbs–Duhem equation (10.18) gives

$$x_A\, d\mu_A + x_B\, d\mu_B = 0 \qquad \text{const. } T, P \qquad (10.20)$$

From (10.6) we have $\mu_A = \mu_A^\circ(T, P) + RT \ln \gamma_A + RT \ln x_A$ and

$$d\mu_A = RT\, d \ln \gamma_A + (RT/x_A)\, dx_A \qquad \text{const. } T, P$$

Similarly, $d\mu_B = RT\, d \ln \gamma_B + (RT/x_B)\, dx_B$ at constant T and P. Substitution for $d\mu_A$ and $d\mu_B$ in (10.20) gives after division by RT:

$$x_A\, d \ln \gamma_A + dx_A + x_B\, d \ln \gamma_B + dx_B = 0 \qquad \text{const. } T, P$$

Since $x_A + x_B = 1$, we have $dx_A + dx_B = 0$, and the last equation becomes

$$d \ln \gamma_B = -(x_A/x_B)\, d \ln \gamma_A \qquad \text{const. } T, P \qquad (10.21)$$

Integrating between states 1 and 2, and choosing Convention II, we get

$$\ln \gamma_{II,B,2} - \ln \gamma_{II,B,1} = -\int_1^2 \frac{x_A}{1 - x_A}\, d \ln \gamma_{II,A} \qquad \text{const. } T, P \qquad (10.22)$$

Let state 1 be pure solvent A. Then $\gamma_{II,B,1} = 1$ [Eq. (10.10)] and $\ln \gamma_{II,B,1} = 0$. We plot $x_A/(1 - x_A)$ versus $\ln \gamma_{II,A}$. The area under the curve from $x_A = 1$ to $x_A = x_{A,2}$ gives $-\ln \gamma_{II,B,2}$. Even though the integrand $x_A/(1 - x_A) \to \infty$ as $x_A \to 1$, the area under the curve is finite; but the infinity makes it hard to accurately evaluate the integral in (10.22) graphically. A convenient way to avoid this infinity is discussed after Eq. (10.59). [Equation (10.22) is for a two-component solution. Surprisingly, if activity-coefficient data for one component of a multicomponent solution are available over the full range of compositions, one can find the activity coefficients of all the other components; see *Pitzer* (1995) pp. 220, 250, 300.]

Some activity coefficients for aqueous solutions of sucrose at 25°C calculated from vapor-pressure measurements and the Gibbs–Duhem equation are:

$x(H_2O)$	0.999	0.995	0.980	0.960	0.930	0.900
$\gamma_{II}(H_2O)$	1.0000	0.9999	0.998	0.990	0.968	0.939
$\gamma_{II}(C_{12}H_{22}O_{11})$	1.009	1.047	1.231	1.58	2.31	3.23

Note from (10.21) that $\gamma_{II,sucrose}$ must increase when γ_{II,H_2O} decreases at constant T and P. Because of the large size of a sucrose molecule (molecular weight 342) compared with a water molecule, the mole-fraction values can mislead one into thinking that a solution is more dilute than it actually is. For example, in an aqueous sucrose solution with $x(\text{sucrose}) = 0.10$, 62% of the atoms are in sucrose molecules, and the solution is extremely concentrated. Even though only 1 molecule in 10 is sucrose, the large size of sucrose molecules makes it highly likely for a given sucrose molecule to be close to several other sucrose molecules, and $\gamma_{II,sucrose}$ deviates greatly from 1.

Aqueous sucrose solutions have $\gamma_{II,i} > 1$ and $\gamma_{II,A} < 1$. The same reasoning used for acetone–chloroform solutions shows that sucrose–H_2O interactions are more favorable than sucrose–sucrose interactions.

Other Methods for Finding Activity Coefficients

Some other phase-equilibrium properties that can be used to find activity coefficients are freezing points of solutions (Sec. 12.3) and osmotic pressures of solutions (Sec. 12.4). Activity coefficients of electrolytes in solution can be found from galvanic-cell data (Sec. 14.10).

In industrial processes, liquid mixtures are often separated into their pure components by distillation. The efficient design of distillation apparatus requires knowledge of the partial vapor pressures of the mixture's components, which in turn requires knowledge of the activity coefficients in the mixture. Therefore, chemical engineers have devised several methods for estimating activity coefficients from limited data; see Sec. 10.5 and *Reid, Prausnitz, and Poling,* chap. 8.

10.4 ACTIVITY COEFFICIENTS ON THE MOLALITY AND MOLAR CONCENTRATION SCALES

So far in this chapter, we have expressed solution compositions using mole fractions and have written the chemical potential of each solute i as

$$\mu_i = \mu_{\text{II},i}^\circ + RT \ln \gamma_{\text{II},i} x_i \qquad \text{where } \gamma_{\text{II},i} \to 1 \text{ as } x_A \to 1 \qquad (10.23)$$

where A is the solvent. However, for solutions of solids or gases in a liquid, the solute chemical potentials are usually expressed in terms of molalities. The molality of solute i is $m_i = n_i/n_A M_A$ [Eq. (9.3)]. Division of numerator and denominator by n_{tot} gives $m_i = x_i/x_A M_A$ and $x_i = m_i x_A M_A$. The μ_i expression becomes

$$\mu_i = \mu_{\text{II},i}^\circ + RT \ln \left(\gamma_{\text{II},i} m_i x_A M_A m^\circ/m^\circ\right)$$

$$\mu_i = \mu_{\text{II},i}^\circ + RT \ln \left(M_A m^\circ\right) + RT \ln \left(x_A \gamma_{\text{II},i} m_i/m^\circ\right)$$

where, to keep later equations dimensionally correct, the argument of the logarithm was multiplied and divided by m°, where m° is defined by $m^\circ \equiv 1$ mol/kg. We can take the log of a dimensionless number only. The quantity $M_A m^\circ$ is dimensionless. For example, for H_2O, $M_A m^\circ = (18 \text{ g/mol}) \times (1 \text{ mol/kg}) = 0.018$.

We now define $\mu_{m,i}^\circ$ and $\gamma_{m,i}$ as

$$\mu_{m,i}^\circ \equiv \mu_{\text{II},i}^\circ + RT \ln \left(M_A m^\circ\right), \qquad \gamma_{m,i} \equiv x_A \gamma_{\text{II},i} \qquad (10.24)$$

With these definitions, μ_i becomes

$$\mu_i = \mu_{m,i}^\circ + RT \ln \left(\gamma_{m,i} m_i/m^\circ\right), \qquad m^\circ \equiv 1 \text{ mol/kg}, \qquad i \neq A \quad \textbf{(10.25)}\text{*}$$

$$\gamma_{m,i} \to 1 \text{ as } x_A \to 1 \qquad (10.26)$$

where the limiting behavior of $\gamma_{m,i}$ follows from (10.24) and (10.10). The motive for the definitions in (10.24) is to produce an expression for μ_i in terms of m_i that has the same form as the expression for μ_i in terms of x_i. Note the similarity between (10.25) and (10.23). We call $\gamma_{m,i}$ the **molality-scale activity coefficient** of solute i and $\mu_{m,i}^\circ$ the molality-scale standard-state chemical potential of i. Since $\mu_{\text{II},i}^\circ$ in (10.24) is a function of T and P only, $\mu_{m,i}^\circ$ is a function of T and P only.

What is the molality-scale standard state? Setting μ_i in (10.25) equal to $\mu_{m,i}^\circ$, we see that this standard state has $\gamma_{m,i} m_i/m^\circ = 1$. We shall take the standard-state molality as $m_i = m^\circ = 1$ mol/kg (as is implied in the notation m° for 1 mol/kg), and we must then have $\gamma_{m,i} = 1$ in the standard state. The molality-scale solute standard state is thus the fictitious state (at the T and P of the solution) with $m_i = 1$ mol/kg and $\gamma_{m,i} = 1$. This state involves an extrapolation of the behavior of the ideally dilute solution (where $\gamma_{m,i} = 1$) to a molality of 1 mol/kg (see Fig. 10.5).

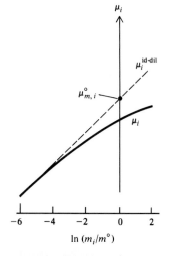

Figure 10.5

Chemical potential μ_i of a nonelectrolyte solution plotted versus $\ln (m_i/m^\circ)$. The dashed line extrapolates the ideally dilute solution behavior to higher molalities. The solute's standard state corresponds to the point on the dashed line where $m_i = m^\circ = 1$ mol/kg and $\ln (m_i/m^\circ) = 0$.

Using (10.24) one finds (Prob. 10.14) the following solute molality-scale standard-state properties:

$$\bar{V}_{m,i}^\circ = \bar{V}_i^\infty, \qquad \bar{H}_{m,i}^\circ = \bar{H}_i^\infty, \qquad \bar{S}_{m,i}^\circ = (\bar{S}_i + R \ln m_i/m^\circ)^\infty, \qquad i \neq A \quad (10.27)$$

Although (10.25) is used for each solute, the mole-fraction scale is used for the solvent:

$$\mu_A = \mu_A^\circ + RT \ln \gamma_A x_A, \qquad \mu_A^\circ = \mu_A^*(T, P), \qquad \gamma_A \to 1 \text{ as } x_A \to 1 \qquad (10.28)$$

Solute chemical potentials are sometimes expressed in terms of molar concentrations c_i instead of molalities, as follows:

$$\mu_i = \mu_{c,i}^\circ + RT \ln \left(\gamma_{c,i} c_i / c^\circ \right) \qquad \text{for } i \neq A \qquad (10.29)$$

$$\gamma_{c,i} \to 1 \text{ as } x_A \to 1 \qquad c^\circ \equiv 1 \text{ mol/dm}^3$$

which have the same forms as (10.25) and (10.26). The relations between $\mu_{c,i}^\circ$ and $\mu_{II,i}^\circ$ and between $\gamma_{c,i}$ and $\gamma_{II,i}$ are worked out in Prob. 10.15. As always, the mole-fraction scale is used for the solvent.

Equations (10.4), (10.25), and (10.29) give as the activities on the molality and molar-concentration scales

$$a_{m,i} = \gamma_{m,i} m_i / m^\circ, \qquad a_{c,i} = \gamma_{c,i} c_i / c^\circ \qquad \mathbf{(10.30)^*}$$

which may be compared with $a_i = \gamma_i x_i$ [Eq. (10.5)].

Some values of γ_{II}, γ_m, and γ_c for sucrose in water at 25°C and 1 atm are plotted in Fig. 10.6.

For solute i, one has the choice of expressing μ_i using the mole-fraction scale (Convention II), the molality scale, or the molar concentration scale. None of these scales is more fundamental than the others (see *Franks*, vol. 4, pp. 4, 7–8), and which scale is used is simply a matter of convenience. In dilute solutions, γ_{II}, γ_m, and γ_c are nearly equal to one another, and each measures the deviation from ideally dilute behavior (no solute–solute interactions). In concentrated solutions, these activity coefficients differ from one another, and it is not meaningful to say which one is the best measure of such deviations.

The solute activity coefficients $\gamma_{II,i}$ (often denoted by $\gamma_{x,i}$), $\gamma_{m,i}$, and $\gamma_{c,i}$) are sometimes called *Henry's-law activity coefficients* since they measure deviations from Henry's law. The activity coefficient $\gamma_{I,i}$ is called a *Raoult's-law activity coefficient*.

Table 11.1 in Sec. 11.8 summarizes the standard states used for solutions and pure substances.

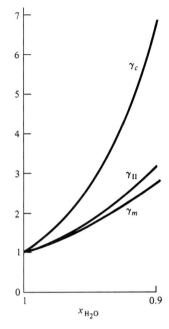

Figure 10.6

γ_c, γ_{II}, and γ_m of the solute sucrose in water at 25°C and 1 atm plotted versus solution composition.

10.5 MODELS FOR NONELECTROLYTE ACTIVITY COEFFICIENTS

This section discusses methods to find activity coefficients in solutions of two liquids when experimental data are limited or missing.

For two liquids whose molecules have similar sizes and shapes, an approximation that sometimes works reasonably well is to assume the liquids form a simple solution. A **simple solution** of liquids B and C is one where the Convention I activity coefficients depend on composition according to

$$RT \ln \gamma_{I,B} = W x_C^2, \qquad RT \ln \gamma_{I,C} = W x_B^2 \qquad (10.31)$$

where W depends on T and P but not on the mole fractions. [Expressions for the activity coefficients in a solution must satisfy the Gibbs–Duhem equation (10.21); see Prob. 10.16.]

Equation (10.31) can be derived from a *lattice model* of the solution. Here, one assumes that B and C molecules have similar sizes and shapes and that each molecule occupies a single site in a three-dimensional array. Let z be the number of nearest neighbors of each molecule in the array. The intermolecular-interaction energy is approximated as the sum of nearest-neighbor interactions. If ε_{BB}, ε_{CC}, and ε_{BC} are the interaction energies for adjacent B-B, C-C, and B-C pairs, respectively, then one finds the energy of mixing to be (*Denbigh*, sec.14.2)

$$\Delta_{mix} U = z N_A [\varepsilon_{BC} - \tfrac{1}{2}(\varepsilon_{BB} + \varepsilon_{CC})] n_B n_C / (n_B + n_C) \equiv W n_B n_C / (n_B + n_C)$$

where $W \equiv z N_A [\varepsilon_{BC} - \frac{1}{2}(\varepsilon_{BB} + \varepsilon_{CC})]$. Since the molecules are of similar size and shape, we can approximate $\Delta_{mix}V$ as zero, so $\Delta_{mix}H = \Delta_{mix}U$. If we make the further approximation that $\Delta_{mix}S = \Delta_{mix}S^{id}$, then the excess entropy S^E (Sec. 10.2) is zero, and $G^E = H^E - TS^E \approx H^E = \Delta_{mix}H = \Delta_{mix}U$, so

$$G^E = \frac{n_B n_C}{n_B + n_C} W(T, P)$$

Using $RT \ln \gamma_{I,i} = (\partial G^E / \partial n_i)_{T,P,n_{j \neq i}}$ (Prob. 10.4), we get Eq. (10.31) (Prob. 10.17).

The simple-solution model, which has only one parameter, W, occasionally works well, but accurate modeling of activity coefficients usually requires an equation with more than one parameter. A widely used representation of G^E of a solution of liquids B and C is the **Redlich–Kister equation** (also called the Margules equation):

$$
\begin{aligned}
G_m^E &\equiv G^E/(n_B + n_C) \\
&= x_B x_C RT [A_1 + A_2(x_B - x_C) + A_3(x_B - x_C)^2 + \cdots + A_n(x_B - x_C)^{n-1}]
\end{aligned}
$$

(10.32)

where the A coefficients (which depend on T and P) are found by fitting G^E data or activity-coefficient data (Prob. 10.19). Usually, three terms in brackets in (10.32) are enough to give an accurate fit to the data. Differentiation of G_m^E gives the activity coefficients in terms of the A's (Prob. 10.23). Since activity coefficients vary slowly with T, the temperature dependence of the A's is sometimes neglected over a short temperature range. If only one term is taken in (10.32), we get the simple-solution model. Note that the Redlich–Kister equation is not symmetric in B and C, so the values of the A coefficients depend on which liquid we call B.

Special techniques that use gas–liquid chromatography or differential ebulliometry (in which the boiling-point difference between a pure solvent and a dilute solution is measured to 0.001 K at a variable applied pressure) enable quick and accurate determination of the infinite-dilution activity coefficients $\gamma_{I,B}^\infty$ and $\gamma_{I,C}^\infty$, where $\gamma_{I,B}^\infty = \lim_{x_B \to 0} \gamma_{I,B}$. These two activity coefficients can then be used to fix the values A_1 and A_2 in the two-parameter version of the Redlich–Kister equation, thereby enabling activity coefficients to be estimated over the full composition range (Prob. 10.24).

Chemical engineers have devised group-contribution methods to estimate activity coefficients in the complete absence of data. Here, the activity coefficients are expressed as functions of the mole fractions and of parameters for interactions between various chemical groups in the solution components' molecules. The parameter values were chosen to give good fits to known activity-coefficients. (See *Reid, Prausnitz, and Poling*, pp. 311–332.) Such group-contribution methods (with names like ASOG and UNIFAC) often work fairly well, but sometimes yield very inaccurate results.

10.6   SOLUTIONS OF ELECTROLYTES

Electrolyte Solutions

An **electrolyte** is a substance that yields ions in solution, as evidenced by the solution's showing electrical conductivity. A *polyelectrolyte* is an electrolyte that is a polymer. Ionization of acidic groups in DNA and acidic and basic groups in proteins makes these molecules polyelectrolytes (see Sec. 16.6). For a given solvent, an electrolyte is classified as **weak** or **strong**, according to whether its solution is a poor or good conductor of electricity at moderate concentrations. For water as the solvent, some weak electrolytes are NH_3, CO_2, and CH_3COOH, and some strong electrolytes are NaCl, HCl, and $MgSO_4$.

An alternative classification, based on structure, is into true electrolytes and potential electrolytes. A *true electrolyte* consists of ions in the pure state. Most salts are true electrolytes. A crystal of NaCl, $CuSO_4$, or MgS consists of positive and negative ions. When an ionic crystal dissolves in a solvent, the ions break off from the crystal and go into solution as solvated ions. The term **solvated** indicates that each ion in solution is surrounded by a sheath of a few solvent molecules bound to the ion by electrostatic forces and traveling through the solution with the ion. When the solvent is water, solvation is called **hydration** (Fig. 10.7).

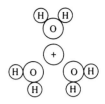

Some salts of certain transition metals and of Al, Sn, and Pb have largely covalent bonding and are not true electrolytes. Thus, for $HgCl_2$, the interatomic distances in the crystal show the presence of $HgCl_2$ molecules (rather than Hg^{2+} and Cl^- ions), and $HgCl_2$ is largely molecular in aqueous solution, as evidenced by the low electrical conductivity. In contrast, $HgSO_4$, $Hg(NO_3)_2$, and HgF_2 are ionic.

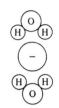

Figure 10.7

Hydration of ions in solution.

A *potential* electrolyte consists of uncharged molecules in the pure state, but when dissolved in a solvent, it reacts with the solvent to some extent to yield ions. Thus, acetic acid reacts with water according to $HC_2H_3O_2 + H_2O \rightleftharpoons H_3O^+ + C_2H_3O_2^-$, yielding hydronium and acetate ions. Hydrogen chloride reacts with water according to $HCl + H_2O \rightleftharpoons H_3O^+ + Cl^-$. For the strong electrolyte HCl, the equilibrium lies far to the right. For the weak electrolyte acetic acid, the equilibrium lies far to the left, except in very dilute solutions.

In the pure liquid state, a true electrolyte is a good conductor of electricity. In contrast, a potential electrolyte is a poor conductor in the pure liquid state.

Because of the strong long-range forces between ions in solution, the use of activity coefficients in dealing with electrolyte solutions is essential, even for quite dilute solutions. Positive and negative ions occur together in solutions, and we cannot readily make observations on the positive ions alone to determine their activity. Hence, a special development of electrolyte activity coefficients is necessary. *Our aim is to derive an expression for the chemical potential of an electrolyte in solution in terms of experimentally measurable quantities.*

For simplicity, we consider a solution composed of a nonelectrolyte solvent A, for example, H_2O or CH_3OH, and a single electrolyte that yields only two kinds of ions in solution, for example, Na_2SO_4, $MgCl_2$, or HNO_3 but not $KAl(SO_4)_2$. Let the electrolyte i have the formula $M_{\nu_+}X_{\nu_-}$, where ν_+ and ν_- are integers, and let i yield the ions M^{z+} and X^{z-} in solution:

$$M_{\nu_+}X_{\nu_-}(s) \rightarrow \nu_+ M^{z+}(sln) + \nu_- X^{z-}(sln) \qquad (10.33)$$

where *sln* indicates species in solution. For example, for $Ba(NO_3)_2$ and $BaSO_4$:

$Ba(NO_3)_2$: $M = Ba$, $X = NO_3$; $\nu_+ = 1$, $\nu_- = 2$; $z_+ = 2$, $z_- = -1$

$BaSO_4$: $M = Ba$, $X = SO_4$; $\nu_+ = 1$, $\nu_- = 1$; $z_+ = 2$, $z_- = -2$

When $z_+ = 1$ and $|z_-| = 1$, we have a 1:1 electrolyte. $Ba(NO_3)_2$ is a 2:1 electrolyte; Na_2SO_4 is a 1:2 electrolyte; $MgSO_4$ is a 2:2 electrolyte.

Don't be intimidated by the notation. In the following discussion, the z's are charges, the ν's (nu's) are numbers of ions in the chemical formula, the μ's (mu's) are chemical potentials, and the γ's (gamma's) are activity coefficients.

Chemical Potentials in Electrolyte Solutions

We shall restrict the treatment in this section to strong electrolytes. Let the solution be prepared by dissolving n_i moles of electrolyte i with formula $M_{\nu_+}X_{\nu_-}$ in n_A moles of solvent A. The species present in solution are A molecules, M^{z+} ions, and X^{z-} ions. Let n_A, n_+, and n_- and μ_A, μ_+, and μ_- be the numbers of moles and the chemical potentials of A, M^{z+}, and X^{z-}, respectively.

The quantity μ_+ is by definition [Eq. (4.72)]

$$\mu_+ \equiv (\partial G/\partial n_+)_{T,P,n_{j\neq +}} \tag{10.34}$$

where G is the Gibbs energy of the solution. In (10.34), we must vary n_+ while holding fixed the amounts of all other species, including n_-. However, the requirement of electrical neutrality of the solution prevents varying n_+ while n_- is held fixed. We can't readily vary $n(Na^+)$ in an NaCl solution while holding $n(Cl^-)$ fixed. The same situation holds for μ_-. There is thus no simple way to determine μ_+ and μ_- experimentally. (Chemical potentials of single ions in solution can, however, be estimated theoretically using statistical mechanics; see Sec. 10.8.)

Since μ_+ and μ_- are not measurable, we define μ_i, the **chemical potential of the electrolyte as a whole** (in solution), by

$$\mu_i \equiv (\partial G/\partial n_i)_{T,P,n_A} \tag{10.35}$$

where G is G of the solution. The number of moles n_i of dissolved electrolyte can readily be varied at constant n_A, so μ_i can be experimentally measured (relative to its value in some chosen standard state). Definitions similar to (10.35) hold for other partial molar properties of the electrolyte as a whole. For example, $\bar{V}_i \equiv (\partial V/\partial n_i)_{T,P,n_A}$, where V is the solution's volume. $\bar{V}_i$ for $MgSO_4$ in water was mentioned in Sec. 9.2.

To relate μ_+ and μ_- to μ_i, we use the Gibbs dG equation (4.73), which for the electrolyte solution is

$$dG = -S\,dT + V\,dP + \mu_A\,dn_A + \mu_+\,dn_+ + \mu_-\,dn_- \tag{10.36}$$

The numbers of moles of cations and anions from $M_{\nu_+}X_{\nu_-}$ is given by Eq. (10.33) as $n_+ = \nu_+ n_i$ and $n_- = \nu_- n_i$. Therefore

$$dG = -S\,dT + V\,dP + \mu_A\,dn_A + (\nu_+\mu_+ + \nu_-\mu_-)\,dn_i \tag{10.37}$$

Setting $dT = 0$, $dP = 0$, and $dn_A = 0$ in (10.37) and using $\mu_i \equiv (\partial G/\partial n_i)_{T,P,n_A}$ [Eq. (10.35)], we get

$$\mu_i = \nu_+\mu_+ + \nu_-\mu_- \tag{10.38*}$$

which relates the chemical potential μ_i of the electrolyte as a whole to the chemical potentials μ_+ and μ_- of the cation and anion. For example, the chemical potential of $CaCl_2$ in aqueous solution is $\mu(CaCl_2, aq) = \mu(Ca^{2+}, aq) + 2\mu(Cl^-, aq)$.

We now consider the explicit expressions for μ_A and μ_i. The chemical potential μ_A of the solvent can be expressed on the mole-fraction scale [Eq. (10.28)]:

$$\mu_A = \mu_A^*(T, P) + RT \ln \gamma_{x,A} x_A, \qquad (\gamma_{x,A})^\infty = 1 \tag{10.39}$$

where $\gamma_{x,A}$ is the mole-fraction activity coefficient and the ∞ superscript denotes infinite dilution.

The electrolyte chemical potentials μ_i, μ_+, and μ_- are usually expressed on the molality scale. Let m_+ and m_- be the molalities of the ions M^{z+} and X^{z-}, and let γ_+ and γ_- be the molality-scale activity coefficients of these ions. The m subscript is omitted from the γ's since *in this section we shall use only the molality scale for solute species*. Equations (10.25) and (10.26) give as the chemical potentials of the ions:

$$\mu_+ = \mu_+^\circ + RT \ln (\gamma_+ m_+/m^\circ), \qquad \mu_- = \mu_-^\circ + RT \ln (\gamma_- m_-/m^\circ) \tag{10.40}$$

$$m^\circ \equiv 1 \text{ mol/kg}, \qquad \gamma_+^\infty = \gamma_-^\infty = 1 \tag{10.41}$$

where μ_+° and μ_-° are the ions' molality-scale standard-state chemical potentials.

Substitution of (10.40) for μ_+ and μ_- into (10.38) gives μ_i as

$$\mu_i = \nu_+\mu_+^\circ + \nu_-\mu_-^\circ + \nu_+ RT \ln (\gamma_+ m_+/m^\circ) + \nu_- RT \ln (\gamma_- m_-/m^\circ)$$

$$\mu_i = \nu_+\mu_+^\circ + \nu_-\mu_-^\circ + RT \ln [(\gamma_+)^{\nu_+}(\gamma_-)^{\nu_-}(m_+/m^\circ)^{\nu_+}(m_-/m^\circ)^{\nu_-}] \tag{10.42}$$

Because μ_+ and μ_- can't be determined experimentally, the single-ion activity coefficients γ_+ and γ_- in (10.40) can't be measured. The combination $(\gamma_+)^{\nu_+}(\gamma_-)^{\nu_-}$ that occurs in the experimentally measurable quantity μ_i in (10.42) is measurable. Therefore, to get a measurable activity coefficient, we define the **molality-scale mean ionic activity coefficient** $\gamma_\pm$ of the electrolyte $M_{\nu_+}X_{\nu_-}$ as

$$(\gamma_\pm)^{\nu_+ + \nu_-} \equiv (\gamma_+)^{\nu_+}(\gamma_-)^{\nu_-} \qquad \textbf{(10.43)*}$$

For example, for $BaCl_2$, $(\gamma_\pm)^3 = (\gamma_+)(\gamma_-)^2$ and $\gamma_\pm = (\gamma_+)^{1/3}(\gamma_-)^{2/3}$. The definition (10.43) applies also to solutions of several electrolytes. For a solution of NaCl and KCl, there is a $\gamma_\pm$ for the ions K^+ and Cl^- and a different $\gamma_\pm$ for Na^+ and Cl^-.

To simplify the appearance of (10.42) for μ_i, we define $\mu_i^\circ(T, P)$ (the electrolyte's standard-state chemical potential) and ν (the total number of ions in the electrolyte's formula) as

$$\mu_i^\circ \equiv \nu_+ \mu_+^\circ + \nu_- \mu_-^\circ \qquad (10.44)$$

$$\nu \equiv \nu_+ + \nu_- \qquad \textbf{(10.45)*}$$

With the definitions (10.43) to (10.45) of $\gamma_\pm$, μ_i°, and ν, Eq. (10.42) for the electrolyte's chemical potential μ_i becomes

$$\mu_i = \mu_i^\circ + RT \ln \left[(\gamma_\pm)^\nu (m_+/m^\circ)^{\nu_+}(m_-/m^\circ)^{\nu_-} \right] \qquad (10.46)$$

$$\gamma_\pm^\infty = 1 \qquad (10.47)$$

where the infinite-dilution behavior of $\gamma_\pm$ follows from (10.43) and (10.41).

What is the relation between the ionic molalities m_+ and m_- in (10.46) and the electrolyte's molality? The **stoichiometric molality** m_i of electrolyte i is defined as

$$m_i \equiv n_i/w_A \qquad (10.48)$$

where the solution is prepared by dissolving n_i moles of electrolyte in a mass w_A of solvent. To express μ_i in (10.46) as a function of m_i, we shall relate m_+ and m_- to m_i. The formula of the strong electrolyte $M_{\nu_+}X_{\nu_-}$ contains ν_+ cations and ν_- anions, so the ionic molalities are $m_+ = \nu_+ m_i$ and $m_- = \nu_- m_i$, where m_i is the electrolyte's stoichiometric molality (10.48). The molality factor in (10.46) is then

$$(m_+)^{\nu_+}(m_-)^{\nu_-} = (\nu_+ m_i)^{\nu_+}(\nu_- m_i)^{\nu_-} = (\nu_+)^{\nu_+}(\nu_-)^{\nu_-}m_i^\nu \qquad (10.49)$$

where $\nu \equiv \nu_+ + \nu_-$ [Eq. (10.45)]. We define $\nu_\pm$ [analogous to $\gamma_\pm$ in (10.43)] as

$$(\nu_\pm)^\nu \equiv (\nu_+)^{\nu_+}(\nu_-)^{\nu_-} \qquad (10.50)$$

For example, for $Mg_3(PO_4)_2$, $\nu_\pm = (3^3 \times 2^2)^{1/5} = 108^{1/5} = 2.551$. If $\nu_+ = \nu_-$, then $\nu_\pm = \nu_+ = \nu_-$ (Prob. 10.29). With the definition (10.50), Eq. (10.49) becomes $(m_+)^{\nu_+}(m_-)^{\nu_-} = (\nu_\pm m_i)^\nu$. The quantity in brackets in (10.46) and the expression (10.46) for μ_i become

$$\left[(\gamma_\pm)^\nu (m_+/m^\circ)^{\nu_+}(m_-/m^\circ)^{\nu_-} \right] = (\nu_\pm \gamma_\pm m_i/m^\circ)^\nu$$

$$\mu_i = \mu_i^\circ + \nu RT \ln (\nu_\pm \gamma_\pm m_i/m^\circ) \qquad \text{strong electrolyte} \qquad (10.51)$$

where $\ln x^y = y \ln x$ was used. Equation (10.51) expresses the electrolyte's chemical potential μ_i in terms of its stoichiometric molality m_i.

Setting μ_i in (10.51) equal to μ_i°, we see that the standard state of the electrolyte i as a whole has $\nu_\pm \gamma_i m_i/m^\circ = 1$. The **standard state** of i as a whole is taken as the fictitious state with $\gamma_i = 1$ and $\nu_\pm m_i/m^\circ = 1$. This standard state has $m_i = (1/\nu_\pm)$ mol/kg.

The activity a_i of electrolyte i as a whole is defined so that $\mu_i = \mu_i^\circ + RT \ln a_i$ [Eq. (10.4)] holds. Therefore (10.51) gives for an electrolyte

$$a_i = (\nu_\pm \gamma_\pm m_i/m^\circ)^\nu \qquad (10.52)$$

Equation (10.51) is the desired expression for the electrolyte's chemical potential in terms of experimentally measurable quantities. The expression (10.51) for μ_i of an electrolyte differs from the expression $\mu_i = \mu_i^{\circ} + RT \ln(\gamma_i m_i / m^{\circ})$ [Eq. (10.25)] for a nonelectrolyte by the presence of ν, $\nu_{\pm}$, and the expression for $\gamma_{\pm}$. Even in the infinite-dilution limit where $\gamma_{\pm} = 1$, the electrolyte and nonelectrolyte forms of μ_i differ.

Gibbs Energy of an Electrolyte Solution

Equations (10.37) and (10.38) give

$$dG = -S\,dT + V\,dP + \mu_A\,dn_A + \mu_i\,dn_i \tag{10.53}$$

which has the same form as (4.73). Hence, the same reasoning that gave (9.23) and (10.18) gives for an electrolyte solution

$$G = n_A\mu_A + n_i\mu_i \tag{10.54}$$

$$n_A\,d\mu_A + n_i\,d\mu_i = 0 \qquad \text{const. } T, P \tag{10.55}$$

Equation (10.55) is the Gibbs–Duhem equation for an electrolyte solution.

Summary

For a solution of n_i moles of the strong electrolyte $M_{\nu_+}X_{\nu_-}$ in the solvent A, we defined the chemical potential μ_i of the electrolyte as a whole by $\mu_i \equiv (\partial G / \partial n_i)_{T,P,n_A}$ and found that $\mu_i = \nu_+\mu_+ + \nu_-\mu_-$, where μ_+ and μ_- are the chemical potentials of the cation and anion. The electrolyte's chemical potential in solution was found to be $\mu_i = \mu_i^{\circ} + \nu RT \ln(\nu_{\pm}\gamma_{\pm}m_i / m^{\circ})$, where $\nu \equiv \nu_+ + \nu_-$, $\nu_{\pm}$ is defined by $(\nu_{\pm})^{\nu} \equiv (\nu_+)^{\nu_+}(\nu_-)^{\nu_-}$, and the mean molal ionic activity coefficient $\gamma_{\pm}$ is defined by $(\gamma_{\pm})^{\nu} \equiv (\gamma_+)^{\nu_+}(\gamma_-)^{\nu_-}$.

10.7 DETERMINATION OF ELECTROLYTE ACTIVITY COEFFICIENTS

The Gibbs–Duhem equation was used in Sec. 10.3 to find the activity coefficient of a nonvolatile nonelectrolyte solute from known values of the solvent activity coefficient; see Eq. (10.22). A similar procedure applies to a solution of a nonvolatile electrolyte. We restrict the discussion to a solution of a single strong nonvolatile electrolyte i with the formula $M_{\nu_+}X_{\nu_-}$.

The solvent's chemical potential can be written as $\mu_A = \mu_A^{*} + RT \ln a_A$, where the mole-fraction scale is used [Eq. (10.39)]. This expression for μ_A is the same as (10.4) and (10.7). Therefore, the vapor-pressure equation (10.13), which follows from (10.4) and (10.7), holds for the solvent in an electrolyte solution:

$$P_A = a_A P_A^{*} \qquad \text{ideal vapor, } P \text{ not very high} \tag{10.56}$$

Since the electrolyte solute is assumed nonvolatile, P_A equals the vapor pressure of the solution, and (10.56) allows the solvent activity and activity coefficient to be found from vapor-pressure measurements. Substitution of the constant-T-and-P differentials $d\mu_A$ [found from (10.39)] and $d\mu_i$ [found from (10.51)] into the Gibbs–Duhem equation (10.55) followed by integration then allows the electrolyte's mean activity coefficient $\gamma_{\pm}$ in (10.51) to be found as a function of composition from the known solvent activity coefficient as a function of composition. Electrolyte activity coefficients can also be found from galvanic-cell data; Sec. 14.10.

Some experimental values of $\gamma_{\pm}$ for aqueous electrolyte solutions at 25°C and 1 atm ($m^{\circ} \equiv 1$ mol/kg) are given in Table 10.2 and plotted in Fig. 10.8. Even at $m_i = 0.001$ mol/kg, the electrolyte activity coefficients in Table 10.2 deviate substantially from 1 because of the long-range interionic forces. For comparison, for the nonelectrolyte $CH_3(CH_2)_2OH$ in water at 25°C and 1 atm, $\gamma_{m,i} = 0.9999$ at $m_i = 0.001$ mol/kg,

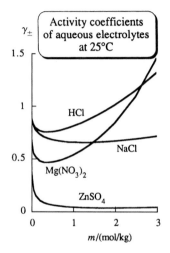

Activity coefficients of aqueous electrolytes at 25°C

$\gamma_{\pm}$

1.5

1

HCl

NaCl

0.5

Mg(NO$_3$)$_2$

ZnSO$_4$

0

0   1   2   3

$m/$(mol/kg)

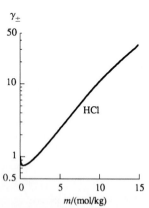

$\gamma_{\pm}$

50

10

HCl

1

0.5

0   5   10   15

$m/$(mol/kg)

Figure 10.8

Activity coefficients of some electrolytes in aqueous solutions at 25°C and 1 atm.

TABLE 10.2

Activity Coefficients $\gamma_\pm$ of Electrolytes in Water At 25°C and 1 atm

$m_i/m°$	LiBr	HCl	$CaCl_2$	$Mg(NO_3)_2$	Na_2SO_4	$CuSO_4$
0.001	0.965	0.965	0.888	0.882	0.886	0.74
0.01	0.905	0.905	0.729	0.712	0.712	0.44
0.1	0.797	0.797	0.517	0.523	0.446	0.154
0.5	0.754	0.759	0.444	0.470	0.268	0.062
1	0.803	0.810	0.496	0.537	0.204	0.043
5	2.70	2.38	5.91		0.148	
10	20.0	10.4	43.1			
20	486.					

$\gamma_{m,i} = 0.9988$ at $m_i = 0.01$ mol/kg, and $\gamma_{m,i} = 0.988$ at $m_i = 0.1$ mol/kg. In concentrated electrolyte solutions, both very large and very small values of $\gamma_\pm$ can occur. For example, in aqueous solution at 25°C and 1 atm, $\gamma_\pm[UO_2(ClO_4)_2] = 1510$ at $m_i = 5.5$ mol/kg and $\gamma_\pm(CdI_2) = 0.017$ at $m_i = 2.5$ mol/kg.

The Practical Osmotic Coefficient. Although the solvent's chemical potential μ_A can be expressed using the mole-fraction scale [Eq. (10.39)], it is customary in work with electrolyte solutions to express μ_A in terms of the *(solvent) practical osmotic coefficient* ϕ (phi). The definition of ϕ for a solution of a strong electrolyte is

$$\phi \equiv -\frac{\ln a_A}{M_A \nu m_i} \equiv \frac{\mu_A^* - \mu_A}{RTM_A \nu m_i} \tag{10.57}$$

where $\nu = \nu_+ + \nu_-$, m_i is the electrolyte's stoichiometric molality (10.48), and M_A is the solvent's molar mass. In (10.57), a_A is the solvent's activity on the mole-fraction scale. The equivalence of the two forms of ϕ in (10.57) follows from $\mu_A = \mu_A^* + RT \ln a_A$. Since $a_A = P_A/P_A^*$ [Eq. (10.56)], we have

$$\phi = \frac{1}{M_A \nu m_i} \ln \frac{P_A^*}{P_A} \qquad \text{ideal vapor, } P \text{ not very high} \tag{10.58}$$

and ϕ can be found from vapor-pressure measurements. ϕ goes to 1 as $x_A \to 1$ (Prob. 10.33).

Rewriting (10.57), we have as the osmotic-coefficient expression for the solvent chemical potential: $\mu_A = \mu_A^* - \phi RTM_A \nu m_i$. The use of this equation and of (10.51) for μ_i in the Gibbs–Duhem equation (10.55) allows the electrolyte activity coefficient $\gamma_\pm$ to be related to the solvent osmotic coefficient ϕ. The derivation is similar to that of (10.22) and is left as an exercise (Prob. 10.33). One finds

$$\ln \gamma_\pm(m) = \phi(m) - 1 + \int_0^m \frac{\phi(m_i) - 1}{m_i} \, dm_i \qquad \text{const. } T, P \tag{10.59}$$

Values of ϕ are available from (10.58). By plotting $(\phi - 1)/m_i$ versus m_i and taking the area under the curve, we can evaluate the integral in (10.59) and obtain $\gamma_\pm$ at the electrolyte molality m (see Prob. 10.34).

The reason for using ϕ instead of the solvent activity coefficient is that, in dilute electrolyte solutions, the solvent activity coefficient may be very close to 1 even though the solute activity coefficient deviates substantially from 1 and the solution is far from ideally dilute. It is inconvenient to work with activity coefficients with values like 1.0001.

For a solution of a single nonelectrolyte i, the practical osmotic coefficient ϕ is defined by an equation like (10.57), except that ν is omitted. A nonelectrolyte solute can be viewed as a special case of an electrolyte solute for which $\nu = 1$ and $\nu_+ = 1$; note that (10.51) with $\nu = 1 = \nu_\pm$ reduces to (10.25). Therefore, Eq. (10.58) with ν taken as 1 is valid for a nonelectrolyte solution; also, Eq. (10.59) with $\gamma_\pm$ replaced by γ_i is valid for a nonelectrolyte solution.

For a nonelectrolyte solution, the McMillan–Mayer statistical-mechanical theory of solutions shows that ϕ can be written in terms of the solute molality according to $\phi = 1 + c_1 m + c_2 m^2 + \cdots$ [*Pitzer* (1995), p. 250], where the c's are functions of T and P. Therefore, at high dilution, $\phi - 1 \propto m$ and the integrand in (10.59) remains finite as $m_i \to 0$ for a nonelectrolyte [unlike the integrand in (10.22)]. Once the molality-scale activity coefficient γ_m has been found from (10.59), γ_{II} can be found from (10.24). The most convenient way to evaluate the integral in (10.59) is to fit a power series to ϕ (Prob. 10.32).

10.8 THE DEBYE–HÜCKEL THEORY OF ELECTROLYTE SOLUTIONS

In 1923, Debye and Hückel used a highly simplified model of an electrolyte solution and statistical mechanics to derive theoretical expressions for the ionic activity coefficients γ_+ and γ_-. In their model, the ions are taken to be uniformly charged hard spheres of diameter a. The difference in size between the positive and negative ions is ignored, and a is interpreted as the mean ionic diameter. The solvent A is treated as a structureless medium with dielectric constant $\varepsilon_{r,A}$ (epsilon r, A). [If $\mathbf{F}$ is the force between two charges in vacuum and $\mathbf{F}_A$ is the net force between the same charges immersed in the dielectric medium A, then $\mathbf{F}_A/\mathbf{F} = 1/\varepsilon_{r,A}$; see (14.89).]

The Debye–Hückel treatment assumes that the solution is very dilute. This restriction allows several simplifying mathematical and physical approximations to be made. At high dilution, the main deviation from ideally dilute behavior comes from the long-range Coulomb's law attractions and repulsions between the ions. Debye and Hückel assumed that all the deviation from ideally dilute behavior is due to interionic Coulombic forces.

An ion in solution is surrounded by an atmosphere of solvent molecules and other ions. On the average, each positive ion will have more negative ions than positive ions in its immediate vicinity. Debye and Hückel used the Boltzmann distribution law of statistical mechanics (Sec. 22.5) to find the average distribution of charges in the neighborhood of an ion.

They then calculated the activity coefficients as follows. Let the electrolyte solution be held at constant T and P. Imagine that we have the magical ability to vary the charges on the ions in the solution. We first reduce the charges on all the ions to zero; the Coulombic interactions between the ions disappear, and the solution becomes ideally dilute. We now reversibly increase all the ionic charges from zero to their values in the actual electrolyte solution. Let w_{el} be the electrical work done on the system in this constant-T-and-P charging process. Equation (4.24) shows that for a reversible constant-T-and-P process, $\Delta G = w_{\mathrm{non}\text{-}P\text{-}V}$; in this case, $w_{\mathrm{non}\text{-}P\text{-}V} = w_{\mathrm{el}}$. Debye and Hückel calculated w_{el} from the electrostatic potential energy of interaction between each ion and the average distribution of charges in its neighborhood during the charging process. Since the charging process starts with an ideally dilute solution and ends with the actual electrolyte solution, ΔG is $G - G^{\mathrm{id}\text{-}\mathrm{dil}}$, where G is the actual Gibbs energy of the solution and $G^{\mathrm{id}\text{-}\mathrm{dil}}$ is the Gibbs energy the solution would have if it were ideally dilute. Therefore $G - G^{\mathrm{id}\text{-}\mathrm{dil}} = w_{\mathrm{el}}$.

$G^{\mathrm{id}\text{-}\mathrm{dil}}$ is known from $G^{\mathrm{id}\text{-}\mathrm{dil}} = \sum_j n_j \mu_j^{\mathrm{id}\text{-}\mathrm{dil}}$, and $G - G^{\mathrm{id}\text{-}\mathrm{dil}}$ is known from calculation of w_{el}. Therefore G of the solution is known. Taking $\partial G/\partial n_+$ and $\partial G/\partial n_-$, one gets the ionic chemical potentials μ_+ and μ_-, so the activity coefficients γ_+ and γ_- in (10.40) are known. (For a full derivation, see *Bockris and Reddy*, sec. 3.3.)

Debye and Hückel's final result is

$$\ln \gamma_+ = -\frac{z_+^2 A I_m^{1/2}}{1 + BaI_m^{1/2}}, \qquad \ln \gamma_- = -\frac{z_-^2 A I_m^{1/2}}{1 + BaI_m^{1/2}} \qquad (10.60)$$

where A, B, and I_m are defined as

$$A \equiv (2\pi N_A \rho_A)^{1/2} \left(\frac{e^2}{4\pi\varepsilon_0 \varepsilon_{r,A} kT}\right)^{3/2}, \qquad B \equiv e\left(\frac{2N_A \rho_A}{\varepsilon_0 \varepsilon_{r,A} kT}\right)^{1/2} \qquad (10.61)$$

$$I_m \equiv \frac{1}{2}\sum_j z_j^2 m_j \qquad \textbf{(10.62)}*$$

In these equations (which are in SI units), a is the mean ionic diameter, γ_+ and γ_- are the molality-scale activity coefficients for the ions M^{z+} and X^{z-}, respectively, N_A is the Avogadro constant, k is Boltzmann's constant [Eq. (3.57)], e is the proton charge, ε_0 is the permittivity of vacuum (ε_0 occurs as a proportionality constant in Coulomb's law; see Sec. 14.1), ρ_A is the solvent density, $\varepsilon_{r,A}$ is the solvent dielectric constant, and T is the absolute temperature. I_m is called the (**molality-scale**) **ionic strength;** the sum in (10.62) goes over all ions in solution, m_j being the molality of ion j with charge z_j.

Although the Debye–Hückel theory gives γ of each ion, we cannot measure γ_+ or γ_- individually. Hence, we express the Debye–Hückel result in terms of the mean ionic activity coefficient $\gamma_\pm$. Taking the log of $(\gamma_\pm)^{\nu_+ + \nu_-} \equiv (\gamma_+)^{\nu_+}(\gamma_-)^{\nu_-}$ [Eq. (10.43)], we get

$$\ln \gamma_\pm = \frac{\nu_+ \ln \gamma_+ + \nu_- \ln \gamma_-}{\nu_+ + \nu_-} \qquad (10.63)$$

Since the electrolyte $M_{\nu_+} X_{\nu_-}$ is electrically neutral, we have

$$\nu_+ z_+ + \nu_- z_- = 0 \qquad (10.64)$$

Multiplication of (10.64) by z_+ yields $\nu_+ z_+^2 = -\nu_- z_+ z_-$; multiplication of (10.64) by z_- yields $\nu_- z_-^2 = -\nu_+ z_+ z_-$. Addition of these two equations gives

$$\nu_+ z_+^2 + \nu_- z_-^2 = -z_+ z_-(\nu_+ + \nu_-) = z_+ |z_-|(\nu_+ + \nu_-) \qquad (10.65)$$

since z_- is negative. Substitution of the Debye–Hückel equations (10.60) into (10.63) followed by use of (10.65) gives

$$\ln \gamma_\pm = -z_+ |z_-| \frac{A I_m^{1/2}}{1 + BaI_m^{1/2}} \qquad (10.66)$$

Using the SI values for N_A, k, e, and ε_0, and $\varepsilon_r = 78.38$, $\rho = 997.05$ kg/m³ for H_2O at 25°C and 1 atm, we have for (10.61)

$$A = 1.1744 \ (\text{kg/mol})^{1/2}, \qquad B = 3.285 \times 10^9 \ (\text{kg/mol})^{1/2} \ \text{m}^{-1}$$

Substituting the numerical values for B and A into (10.66) and dividing A by 2.3026 to convert to base 10 logs, we get

$$\log_{10} \gamma_\pm = -0.510 z_+ |z_-| \frac{(I_m/m°)^{1/2}}{1 + 0.328(a/\text{Å})(I_m/m°)^{1/2}} \qquad \text{dil. 25°C aq. soln.} \quad (10.67)$$

where $1 \ \text{Å} \equiv 10^{-10}$ m and $m° \equiv 1$ mol/kg. I_m in (10.62) has units of mol/kg, and the ionic diameter a has units of length, so $\log \gamma_\pm$ is dimensionless, as it must be.

For very dilute solutions, I_m is very small, and the second term in the denominator in (10.67) can be neglected compared with 1. Therefore,

$$\ln \gamma_\pm = -z_+ |z_-| A I_m^{1/2} \qquad \text{very dil. soln.} \qquad (10.68)$$

$$\log_{10} \gamma_\pm = -0.510 z_+ |z_-| (I_m/m°)^{1/2} \qquad \text{very dil. aq. soln., 25°C} \quad (10.69)$$

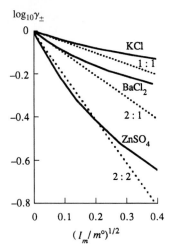

Figure 10.9

Plots of $\log_{10} \gamma_{\pm}$ versus square root of ionic strength for some aqueous electrolytes at 25°C and 1 atm. The dotted lines show the predictions of the Debye–Hückel limiting law (10.68).

Equation (10.68) is called the **Debye–Hückel limiting law**, since it is valid only in the limit of infinite dilution. (Actually, most of the laws of science are limiting laws.)

How well does the Debye–Hückel theory work? Experimental data show that Eq. (10.69) does give the correct limiting behavior for electrolyte solutions as $I_m \to 0$ (Fig. 10.9). Equation (10.69) is found to be accurate when $I_m \leq 0.01$ mol/kg. For a 2:2 electrolyte this corresponds to a molality of $0.01/4 \approx 0.002$. (It is sometimes unkindly said that the Debye–Hückel theory applies to slightly contaminated distilled water.) The more complete equation (10.67) is reasonably accurate for aqueous solutions with $I_m \leq 0.1$ mol/kg if we choose the ionic diameter a to give a good fit to the data. Values of a so found typically range from 3 to 9 Å for common inorganic salts, which are reasonable values for hydrated ions. At a given ionic strength, the theory works better as $z_+|z_-|$ decreases; for example, at $I_m = 0.1$ mol/kg, the Debye–Hückel theory is more reliable for 1:1 electrolytes than for 2:2 electrolytes. Part of the reason for this lies in ionic association (Sec. 10.9).

To eliminate the empirically determined ionic diameter a from (10.67), we note that, for $a \approx 3$ Å, we have $0.328(a/\text{Å}) \approx 1$. Hence one often simplifies (10.67) to

$$\log_{10} \gamma_{\pm} = -0.510 z_+ |z_-| \frac{(I_m/m°)^{1/2}}{1 + (I_m/m°)^{1/2}} \qquad \text{dil. aq. soln., 25°C} \quad (10.70)$$

The properties of very dilute electrolyte solutions frequently cannot be measured with the required accuracy. Hence, even though the range of validity of the Debye–Hückel theory is limited to quite dilute solutions, the theory is of great practical importance since it allows measured properties of electrolyte solutions to be reliably extrapolated into the region of very low concentrations.

Knowing the dilute-solution activity coefficients γ_+, γ_-, and $\gamma_{\pm}$ from the Debye–Hückel equation, we know the chemical potentials μ_+, μ_-, and μ_i [Eqs. (10.40) and (10.51)]. From these chemical potentials, we can derive limiting laws for all other ionic and electrolyte thermodynamic properties, for example, $\bar{V}_i$, $\bar{H}_i$, and $\bar{S}_i$.

Figure 10.8 shows that, as the electrolyte's molality m_i increases from zero, its activity coefficient $\gamma_{\pm}$ first decreases from the ideally dilute value 1 and then increases. The fact that $\gamma_{\pm}$ is less than 1 in dilute solutions of the electrolyte means that the electrolyte's chemical potential μ_i is less than it would be in a hypothetical ideally dilute solution (no solute–solute interactions) with the same composition, and this means a lower solute contribution to G than for an ideally dilute solution [see Eqs. (10.51) and (10.54)]. Each ion in the solution tends to surround itself with ions of opposite charge, and the electrostatic attractions between the oppositely charged ions stabilize the solution and lower its G.

The increase in the electrolyte's $\gamma_{\pm}$ at higher molalities may be due to hydration of ions. Hydration reduces the amount of free water molecules, thereby reducing the effective concentration of water in the solution and increasing the effective molality of the electrolyte, an increase that is reflected in the increase in $\gamma_{\pm}$. For example, for NaCl, experimental evidence (*Bockris and Reddy,* sec. 2.8) indicates the Na^+ ion carries four H_2O molecules along with itself as it moves through the solution and the Cl^- ion carries two H_2O molecules as it moves. Thus, each mole of NaCl in solution ties up 6 moles of H_2O. One kilogram of water contains 55.5 moles. In a 0.1 mol/kg aqueous NaCl solution, there are $55.5 - 6(0.1) = 54.9$ moles of free water per kilogram of solvent, so here the effect of hydration is slight. However, in a 3 mol/kg aqueous NaCl solution, there are only $55 - 18 = 37$ moles of free water per kilogram of solvent, which is a very substantial reduction.

Electrolyte Activity Coefficients at Higher Concentrations

Several methods have been proposed to calculate electrolyte activity coefficients at higher concentrations than the very dilute solutions to which the Debye–Hückel equation applies.

It has been found empirically that addition of a term linear in I_m to the Debye–Hückel equation (10.70) improves agreement with experiment in less dilute solutions. Davies proposed the following expression containing no adjustable parameters (*Davies*, pp. 39–43):

$$\log_{10} \gamma_{\pm} = -0.51 z_+ |z_-| \left[\frac{(I_m/m^{\circ})^{1/2}}{1 + (I_m/m^{\circ})^{1/2}} - 0.30(I_m/m^{\circ}) \right] \quad \text{in } H_2O \text{ at } 25^{\circ}C$$

(10.71)

The Davies equation for $\log_{10} \gamma_+$ (or $\log_{10} \gamma_-$) is obtained by replacement of $z_+|z_-|$ in (10.71) with z_+^2 (or z_-^2). The Davies modification of the Debye–Hückel equation is typically in error by $1\frac{1}{2}\%$ at $I_m/m^{\circ} = 0.1$. The linear term in (10.71) causes $\gamma_{\pm}$ to go through a minimum and then increase as I_m increases, in agreement with the behavior in Fig. 10.8. As I_m/m° increases above 0.1, agreement of the Davies equation with experiment decreases; at $I_m/m^{\circ} = 0.5$, the error is typically 5 to 10%. It is best to use experimental values of $\gamma_{\pm}$, especially for ionic strengths above 0.1 mol/kg, but in the absence of experimental data, the Davies equation can serve to estimate $\gamma_{\pm}$. The Davies equation predicts that $\gamma_{\pm}$ will have the same value at a given I_m for any 1:1 electrolyte. In reality, $\gamma_{\pm}$ values for 1:1 electrolytes are equal only in the limit of high dilution.

In using the Debye–Hückel or the Davies equation in a solution containing several electrolytes, note that *all* the ions in the solution contribute to I_m in (10.62), but that z_+ and $|z_-|$ in (10.70) and (10.71) refer to the ionic charges of the particular electrolyte for which $\gamma_{\pm}$ is being calculated.

Certain information about the Davies equation has been suppressed in this section. For the full story, see Sec. 10.9.

EXAMPLE 10.3 The Davies equation

Use the Davies equation to estimate $\gamma_{\pm}$ for aqueous $CaCl_2$ solutions at $25^{\circ}C$ with molalities 0.001, 0.01, and 0.1 mol/kg.

We have $I_m \equiv \frac{1}{2} \sum_j z_j^2 m_j = \frac{1}{2}(z_+^2 m_+ + z_-^2 m_-) = \frac{1}{2}(4m_+ + m_-)$ [Eq. (10.62)]. We have $m_+ = m_i$ and $m_- = 2m_i$, where m_i is the $CaCl_2$ stoichiometric molality. The ionic strength is $I_m = \frac{1}{2}(4m_i + 2m_i) = 3m_i$. The Davies equation (10.71) with $z_+ = 2$, $|z_-| = 1$, and $I_m = 3m_i$ becomes

$$\log_{10} \gamma_{\pm} = -1.02(3m_i/m^{\circ})^{1/2}/[1 + (3m_i/m^{\circ})^{1/2}] + 0.92m_i/m^{\circ}$$

Substitution of $m_i/m^{\circ} = 0.001$, 0.01, and 0.1, gives $\gamma_{\pm} = 0.887$, 0.722, and 0.538, respectively. These calculated values can be compared with the experimental values 0.888, 0.729, 0.517 listed in Sec. 10.7. [For comparison, the Debye–Hückel equation (10.70) gives 0.885, 0.707, and 0.435.]

Exercise

Use the Davies equation to estimate $\gamma_{\pm}$ at $25^{\circ}C$ for (*a*) 0.001 mol/kg $AlCl_3(aq)$; (*b*) 0.001 mol/kg $CuSO_4(aq)$. [*Answers*: (*a*) 0.781; (*b*) 0.761.]

In the 1970s, Meissner and coworkers found that $\gamma_{\pm}$ of a strong electrolyte in water at $25^{\circ}C$ could be rather accurately represented up to ionic strengths of 10 or 20 mol/kg by an empirical equation that contains only one parameter (symbolized by q) whose value is specific to the electrolyte (*Tester and Modell*, sec. 12.6). Thus if a single $\gamma_{\pm}$ value is known at a nondilute molality, activity coefficients can be calculated over a wide molality range. The Meissner equation (also called the Kusik–Meissner equation) is given in Prob. 10.40. Meissner and coworkers also developed procedures to allow for the temperature dependence of $\gamma_\pm$ and to calculate $\gamma_\pm$ values in solutions of several electrolytes.

In the 1970s, Pitzer and coworkers developed equations for calculating $\gamma_\pm$ values in concentrated aqueous electrolyte solutions [*Pitzer* (1995), chaps. 17 and 18; *Pitzer* (1991), chap. 3]. Although Pitzer's approach is based on a statistical-mechanical theory of interactions between ions in the solution, his equations have a substantial dose of empiricism, in that the mathematical forms of some of the terms in the equations were chosen by seeing what forms give the best fit to data. Moreover, the equations contain parameters whose values are not calculated theoretically but are chosen to fit activity-coefficient or osmotic-coefficient data for the electrolyte(s) in question.

For an aqueous solution of a single strong electrolyte, the Pitzer equation gives $\ln \gamma_\pm$ as a function of I_m and contains three parameters (called $\beta^{(0)}$, $\beta^{(1)}$, C^ϕ) for 1:1, 2:1, 3:1, 1:2, and 1:3 electrolytes and four parameters ($\beta^{(0)}$, $\beta^{(1)}$, $\beta^{(2)}$, C^ϕ) for a 2:2 electrolyte. Parameter values for about 300 electrolytes at 25°C have been found. To calculate $\gamma_\pm$ in a solution containing several electrolytes, the following additional parameters are needed: $\theta_{cc'}$ for each pair of unlike cations, $\theta_{aa'}$ for each pair of unlike anions, $\psi_{cc'a}$ for each triplet of two unlike cations and one anion, and $\psi_{aa'c}$ for each triplet of two unlike anions and one cation.

For example, if one wanted the activity coefficients in an aqueous solution containing the ions Na^+, K^+, Mg^{2+}, Cl^-, and SO_4^{2-}, one uses data on single-electrolyte solutions of $NaCl$, Na_2SO_4, KCl, K_2SO_4, $MgCl_2$, and $MgSO_4$ to find $\beta^{(0)}$, $\beta^{(1)}$, $\beta^{(2)}$, and C^ϕ for each cation–anion pair and one uses data on the two-electrolyte solutions $NaCl$–Na_2SO_4, KCl–K_2SO_4, $MgCl_2$–$MgSO_4$, $NaCl$–KCl, Na_2SO_4–K_2SO_4, $NaCl$–$MgCl_2$, Na_2SO_4–$MgSO_4$, KCl–$MgCl_2$, and K_2SO_4–$MgSO_4$ (all possible pairs of electrolytes with one ion in common) to evaluate such parameters as $\theta_{Na,K}$ and $\psi_{Na,K,Cl}$ (found from $NaCl$–KCl data). Data for the θ and ψ parameters are much more limited than for the single-electrolyte parameters. In the absence of θ and ψ data, one can take these parameters as zero. The Pitzer equations usually give fairly accurate results with θ and ψ omitted, but for good accuracy they are required.

The Pitzer equations are widely used and have been applied to study reaction and solubility equilibria in such systems as seawater, the Dead Sea, lakes, oil-field brines, and acidic mine-drainage waters with excellent results. The value of the Pitzer equations is for dealing with solutions of several electrolytes, where their performance is usually better than the Meissner model (J. F. Zemaitis et al., *Handbook of Aqueous Electrolytes,* Design Institute for Physical Property Data, 1986).

For multicomponent electrolyte solutions with ionic strengths above 10 or 15 mol/kg, the Pitzer equations usually do not apply. Such high ionic strengths occur in atmospheric aerosols, such as sea-spray aerosols, where evaporation of water produces solutions supersaturated in $NaCl$. (The flux of sea salt between the oceans and the atmosphere is estimated at 10^{15} g per year.) Pitzer and coworkers developed a version of the Pitzer equations that is based on mole fractions instead of molalities and that applies at extremely high concentrations [see *Pitzer* (1995), pp. 308–316].

10.9   IONIC ASSOCIATION

In Section 10.6, a strong electrolyte in aqueous solution was assumed to exist entirely in the form of ions. Actually, this picture is incorrect, and (except for 1:1 electrolytes) there is a significant amount of association between oppositely charged ions in solution to yield ion pairs. For a true electrolyte, we start out with ions in the crystal, get solvated ions in solution as the crystal dissolves, and then get some degree of association of solvated ions to form ion pairs in solution. The equilibrium for ion-pair formation is

$$M^{z+}(sln) + X^{z-}(sln) \rightleftharpoons MX^{z_++z_-}(sln) \tag{10.72}$$

For example, in an aqueous (aq) $Ca(NO_3)_2$ solution, (10.72) reads $Ca^{2+}(aq) + NO_3^-(aq)$ $\rightleftharpoons Ca(NO_3)^+(aq)$.

The concept of ion pairs was introduced in 1926 by Bjerrum (*Bockris and Reddy*, vol. I, sec. 3.8; *Davies,* chap. 15). Bjerrum proposed (rather arbitrarily) that two oppositely charged ions close enough to make the potential energy of attraction between them larger in magnitude than $2kT$ [where k is Boltzmann's constant (3.57)] be considered an ion pair. Bjerrum used a model similar to that of Debye and Hückel to find a theoretical expression for the degree of association to ion pairs as a function of the electrolyte concentration, z_+, z_-, T, ε_r, and the mean ionic diameter a. His theory indicated that in water, association to ion pairs is usually negligible for 1:1 electrolytes but can be quite substantial for electrolytes with higher $z_+|z_-|$ values, even at low concentrations. As z_+ and $|z_-|$ increase, the magnitude of the interionic electrostatic attraction increases and ion pairing increases.

The solvent H_2O has a high dielectric constant (because of the polarity of the water molecule). In solvents with lower values of ε_r, the magnitude of the electrostatic attraction energy is greater than in aqueous solutions. Hence, ion-pair formation in these solvents is greater than in water. Even for 1:1 electrolytes, ion-pair formation is important in solvents with low dielectric constants.

Ionic association reduces the number of ions in solution and hence reduces the electrical conductivity of the solution. For example, in $CaSO_4$ solutions, Ca^{2+} and SO_4^{2-} associate to form neutral $CaSO_4$ ion pairs; in MgF_2 solutions, Mg^{2+} and F^- form MgF^+ ion pairs. The degree of association in an electrolyte solution can best be determined from conductivity measurements on the solution (Sec. 16.6). For example, one finds that $MgSO_4$ in water at 25°C is 10% associated at 0.001 mol/kg; $CuSO_4$ in water at 25°C is 35% associated at 0.01 mol/kg and is 57% associated at 0.1 mol/kg. From conductivity data, one can calculate the equilibrium constant for the ionic association reaction $M^{z+} + X^{z-} \rightleftharpoons MX^{z_++z_-}$. Conductivity data indicate that for 1:1 electrolytes ionic association is unimportant in dilute aqueous solutions but is sometimes significant in concentrated aqueous solutions; for most electrolytes with higher $z_+|z_-|$ values, ionic association is important in both dilute and concentrated aqueous solutions. In the limit of infinite dilution, the degree of association goes to zero. Figure 10.10 plots the percentage of cations that exist in ion pairs versus molality in aqueous solutions for typical 1:1, 2:1, 2:2, and 3:1 electrolytes.

Conductivity measurements have shown that the qualitative conclusions of the Bjerrum theory are generally correct, but quantitative agreement with experiment is sometimes lacking.

Ion pairs should be distinguished from complex ions. Complex-ion formation is common in aqueous solutions of transition-metal halides. In the complex ions $AgCl(aq)$ and $AgCl_2^-(aq)$, the Cl^- ions are in direct contact with the central Ag^+ ion and each Ag—Cl bond has a substantial amount of covalent character. In contrast, the positive and negative ions of an ion pair usually retain at least part of their solvent sheaths and are held together by ionic (electrostatic) forces. The absorption spectrum of the solution can frequently be used to distinguish between ion-pair and complex-ion formation. In some solutions, both ion pairs and complex ions are present.

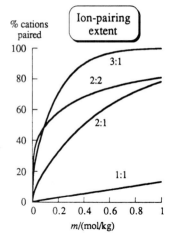

Figure 10.10

Typical ion-pairing extents in water at 25°C versus molality. (See Probs. 11.17 and 11.22 for the method used to calculate these curves.)

Thermodynamics of Ion Pairing

With allowance for formation of $MX^{z_++z_-}$ ion pairs [Eq. (10.72)], the Sec. 10.6 thermodynamic treatment of a solution containing n_i moles of the strong electrolyte $M_{\nu_+}X_{\nu_-}$ in solvent A is modified as follows.

Let n_A, n_+, n_-, and n_{IP} and μ_A, μ_+, μ_-, and μ_{IP} be the numbers of moles and the chemical potentials of A, M^{z+}, X^{z-}, and $MX^{z_++z_-}$, respectively. With ion pairing included, the Gibbs equation (10.36) becomes

$$dG = -S\,dT + V\,dP + \mu_A\,dn_A + \mu_+\,dn_+ + \mu_-\,dn_- + \mu_{IP}\,dn_{IP} \quad (10.73)$$

If no ion pairs were formed, the numbers of moles of cations and anions from $M_{\nu_+}X_{\nu_-}$ would be $n_+ = \nu_+ n_i$ and $n_- = \nu_- n_i$ [Eq. (10.33)]. With ion-pair formation, the number of moles of cations and anions are each reduced by n_{IP} [Eq. (10.72)]:

$$n_+ = \nu_+ n_i - n_{IP}, \qquad n_- = \nu_- n_i - n_{IP} \qquad (10.74)$$

$$dG = -S\,dT + V\,dP + \mu_A\,dn_A + \mu_+(\nu_+\,dn_i - dn_{IP}) + \mu_-(\nu_-\,dn_i - dn_{IP})$$
$$+ \mu_{IP}\,dn_{IP}$$

Use of the equilibrium condition $\mu_{IP} = \mu_+ + \mu_-$ for the ion-pair formation reaction (10.72) simplifies dG to

$$dG = -S\,dT + V\,dP + \mu_A\,dn_A + (\nu_+\mu_+ + \nu_-\mu_-)\,dn_i \qquad (10.75)$$

which is the same as (10.37) in the absence of ion pairs.

The chemical potential $\mu_i \equiv (\partial G/\partial n_i)_{T,P,n_{j\neq i}}$ of the electrolyte as a whole is thus the same as (10.38). Therefore, Eq. (10.46) for μ_i is still valid. However, the molalities m_+ and m_- in (10.46) are changed. Let α be the fraction of the ions M^{z+} that do *not* associate with X^{z-} ions to form ion pairs. If no ion pairing occurred, the number of moles of M^{z+} in solution would be $\nu_+ n_i$, where n_i is the number of moles of $M_{\nu_+}X_{\nu_-}$ used to prepare the solution. With ion pairing, the number of moles of M^{z+} in solution is $n_+ = \alpha\nu_+ n_i$. There are a total of $\nu_+ n_i$ moles of M in the solution, present partly as M^{z+} ions and partly in $MX^{z_+ + z_-}$ ion pairs [Eq. (10.74)]. Hence the number of moles of the ion pair is

$$n_{IP} = \nu_+ n_i - n_+ = \nu_+ n_i - \alpha\nu_+ n_i = (1 - \alpha)\nu_+ n_i$$

A total of $\nu_- n_i$ moles of X is present, partly as X^{z-} and partly in the ion pairs. Hence the number of moles of X^{z-} is

$$n_- = \nu_- n_i - n_{IP} = [\nu_- - (1 - \alpha)\nu_+]n_i$$

Dividing the equations for n_+ and n_- by the solvent mass, we get as the molalities

$$m_+ = \alpha\nu_+ m_i, \qquad m_- = [\nu_- - (1 - \alpha)\nu_+]m_i$$

Substitution of these molalities into Eq. (10.46) for μ_i and the use of (10.45) and (10.50) for ν and $\nu_\pm$ gives (after a bit of algebra—Prob. 10.44)

$$\mu_i = \mu_i^\circ + \nu RT \ln \left(\nu_\pm \gamma_\pm^\dagger m_i/m^\circ\right) \qquad \text{strong electrolyte} \qquad (10.76)$$

$$\gamma_\pm^\dagger \equiv \alpha^{\nu_+/\nu}[1 - (1 - \alpha)(\nu_+/\nu_-)]^{\nu_-/\nu}\gamma_\pm, \qquad (\gamma_\pm^\dagger)^\infty = 1 \qquad (10.77)$$

where the dagger indicates that $\gamma_\pm^\dagger$ allows for ion pairing. The infinite-dilution value of $\gamma_\pm^\dagger$ follows from its definition in (10.77) and from $\gamma_\pm^\infty = 1$ [Eq. (10.47)] and the fact that the extent of ion pairing goes to zero at infinite dilution ($\alpha^\infty = 1$). If $\nu_+ = \nu_-$, then $\gamma_\pm^\dagger = \alpha\gamma_\pm$ (Prob. 10.43).

Equation (10.76) differs from (10.51) for an electrolyte with no ion pairing by the replacement of $\gamma_\pm$ with $\gamma_\pm^\dagger$. Comparison of (10.51) and (10.76) shows that

$$\gamma_\pm^\dagger = \gamma_\pm \qquad \text{if no ion pairing}$$

This result also follows by putting $\alpha = 1$ in (10.77).

The degree of ion pairing is not always known, so α in (10.77) may not be known. One therefore measures (see Sec. 10.7) and tabulates $\gamma_\pm^\dagger$, rather than $\gamma_\pm$, for strong electrolytes. The activity coefficient $\gamma_\pm^\dagger$ deviates from 1 because of (a) deviations of the solution from ideally dilute behavior and (b) ion-pair formation, which makes α in (10.77) less than 1. Although it is actually $\gamma_\pm^\dagger$ that is tabulated for strong electrolytes, tables (such as Table 10.2) use the symbol $\gamma_\pm$ for $\gamma_\pm^\dagger$. Strictly speaking $\gamma_\pm = \gamma_\pm^\dagger$ only for no ion-pair formation.

Taking ionic association into account improves the accuracy of the Debye–Hückel equation. Formation of ion pairs reduces the number of ions in the solution. In calculating the ionic strength, one does not include ions that are associated to form neutral ion pairs. Also, one uses (10.77) to relate the activity coefficient $\gamma_\pm$ calculated by the Debye–Hückel theory to the experimentally observed activity coefficient $\gamma_\pm^\dagger$. These procedures should also be followed when using the Davies equation. (See Prob. 10.47.) The Davies-equation example (Example 10.3) ignored ion pairing. This is a reasonably good approximation for the dilute solutions in that example.

Taking ion pairing into account considerably complicates calculations in a solution of several electrolytes since many ion-pairing equilibria have to be taken into account and solved for. The equilibrium constants for ion-pairing reactions are often not accurately known. Moreover, ion pairing cannot completely account for all effects of ionic association because at high concentrations of ions with $|z| > 1$, three ions may associate with one another to form triple ions. To avoid the complications introduced by ion pairing, some workers prefer to ignore ion pairing. Thus, the Pitzer model of Sec. 10.8 assumes there are no ion pairs. The Pitzer parameters are chosen to fit experimental activity coefficient data, so the values of these parameters implicitly incorporate the effects of ion pairing. Some workers have used the Pitzer equations together with explicit allowance for ion pairing to obtain better results than given by the Pitzer equations with ion pairing ignored [see *Pitzer* (1991), pp. 294, 306, 307]. When this is done, the Pitzer parameters are modified from their usual values.

The Pitzer and Meissner equations do not explicitly consider ion pairing and so one calculates I_m in these equations by assuming each strong electrolyte is present solely as ions, and these equations are designed to yield the experimentally observed activity coefficient $\gamma_\pm^\dagger$.

10.10 STANDARD-STATE THERMODYNAMIC PROPERTIES OF SOLUTION COMPONENTS

To deal with chemical equilibrium in solution, we want to tabulate standard-state thermodynamic properties of substances in solution. How are such properties determined?

Nonelectrolyte Solutions

For nonelectrolyte solutions where Convention I is used, the standard states are the pure substances, and we know how to determine standard-state properties of pure substances (see Chapter 5).

For solid and gaseous solutes, the molality scale is most commonly used. The standard Gibbs energy of formation and standard enthalpy of formation of substance i in solution at temperature T are defined by

$$\Delta_f G_T^\circ(i, sln) \equiv \mu_{m,i}^\circ(T, P^\circ) - G_{\text{elem}}^\circ(T) \tag{10.78}$$

$$\Delta_f H_T^\circ(i, sln) \equiv \bar{H}_{m,i}^\circ(T, P^\circ) - H_{\text{elem}}^\circ(T) \tag{10.79}$$

where i, sln indicates substance i in solution in some particular solvent, $\mu_{m,i}^\circ$ and $\bar{H}_{m,i}^\circ$ are the molality-scale standard-state partial molar Gibbs energy and enthalpy of i in solution, and G_{elem}° and H_{elem}° are the standard-state Gibbs energy and enthalpy of the pure, separated elements needed to form 1 mole of i. One way to determine standard-state molality-scale thermodynamic properties is from solubility data—the fact that μ_i in a saturated solution equals μ_i^* enables us to relate properties in solution to pure-substance properties. The following example shows how this is done.

EXAMPLE 10.4 Standard-state properties of a solute

The molality of a saturated solution of sucrose in water at 25°C and 1 bar is 6.05 mol/kg. Vapor-pressure measurements and the Gibbs–Duhem equation give $\gamma_m(C_{12}H_{22}O_{11}) = 2.87$ in the saturated solution. For pure sucrose at 25°C, $\Delta_f G° = -1544$ kJ/mol, $\Delta_f H° = -2221$ kJ/mol, and $S_m° = 360$ J/(mol K). At 25°C and 1 bar, the differential heat of solution of sucrose in water at infinite dilution is 5.9 kJ/mol. Find $\Delta_f G_{298}°$, $\Delta_f H_{298}°$, and $\bar{S}_{298}°$ for $C_{12}H_{22}O_{11}(aq)$.

Using the phase-equilibrium condition of equality of chemical potentials, we equate μ of pure sucrose to μ of sucrose in the saturated solution. μ of sucrose in solution is given by $\mu_i = \mu_{m,i}° + RT \ln (\gamma_{m,i} m_i / m°)$ [Eq. (10.25)], and we have

$$G_{m,i}^*(T, P°) = \mu_{m,i}°(T, P°) + RT \ln (\gamma_{m,i,sat} m_{i,sat} / m°) \qquad (10.80)$$

where $\gamma_{m,i,sat}$ and $m_{i,sat}$ are the sucrose activity coefficient and molality in the saturated solution, $m° \equiv 1$ mol/kg, and $G_{m,i}^*$ is the molar G of pure sucrose. Subtraction of $G_{elem}(T, P°)$ from each side of (10.80) gives

$$G_{m,i}^* - G_{elem} = \mu_{m,i}° - G_{elem} + RT \ln (\gamma_{m,i,sat} m_{i,sat} / m°)$$

The left side of this equation is by definition $\Delta_f G°$ of pure sucrose; $\mu_{m,i}° - G_{elem}$ on the right side is $\Delta_f G°$ of sucrose(aq) [Eq. (10.78)]. Therefore

$$\Delta_f G°(i*) = \Delta_f G°(i, sln) + RT \ln (\gamma_{m,i,sat} m_{i,sat} / m°) \qquad (10.81)$$

$$-1544 \text{ kJ/mol} = \Delta_f G°(\text{sucrose}, aq) + [8.314 \text{ J/(mol K)}](298 \text{ K}) \ln (2.87 \times 6.05)$$

$$\Delta_f G_{298}°(\text{sucrose}, aq) = -1551 \text{ kJ/mol}$$

To find $\Delta_f H°$ of sucrose(aq), we start with ΔH_{diff}^∞ of sucrose, which is the difference at 1 bar between the partial molar enthalpy of sucrose in water at infinite dilution and the molar enthalpy of pure sucrose: $\Delta H_{diff,i}^\infty = \bar{H}_i^\infty - H_{m,i}^*$ [Eq. (9.38)]. But $\bar{H}_i^\infty = \bar{H}_{m,i}°$ [Eq. (10.27)], so

$$\Delta H_{diff,i}^\infty = \bar{H}_i^\infty - H_{m,i}^* = \bar{H}_{m,i}° - H_{m,i}^* = (\bar{H}_{m,i}° - H_{elem}°) - (H_{m,i}^* - H_{elem}°)$$

$$\Delta H_{diff,i}^\infty = \Delta_f H°(i, sln) - \Delta_f H°(i*) \qquad (10.82)$$

$$5.9 \text{ kJ/mol} = \Delta_f H°(\text{sucrose}, aq) - (-2221 \text{ kJ/mol})$$

$$\Delta_f H°(\text{sucrose}, aq) = -2215 \text{ kJ/mol}$$

Using

$$\Delta_f G°(i, sln) = \Delta_f H°(i, sln) - T \Delta_f S°(i, sln) \qquad (10.83)$$

one finds $\bar{S}_{298}°(\text{sucrose}, aq) = 408$ J/(mol K) (Prob. 10.49).

Exercise

Subtract standard-state thermodynamic properties of the elements from $\mu_i° = \bar{H}_i° - T\bar{S}_i°$ [Eq. (9.28)] to derive Eq. (10.83).

Electrolyte Solutions

Standard-state thermodynamic properties for electrolyte solutes can be found by the same method as used for sucrose(aq) in the preceding example. For an electrolyte in solution, $\mu_i = \mu_{m,i}° + \nu RT \ln (\nu_\pm \gamma_\pm m_i / m°)$ [Eq. (10.51)]. Equating μ_i in a saturated solution to μ of the pure solid electrolyte leads to the following equation [analogous to (10.81)] for electrolyte i:

$$\Delta_f G^\circ(i^*) = \Delta_f G^\circ(i, sln) + \nu RT \ln\left(\nu_\pm \gamma_{\pm,\text{sat}} m_{i,\text{sat}}/m^\circ\right) \qquad (10.84)$$

From (10.82) and (10.83), we can find $\Delta_f H^\circ$ and $\bar{S}^\circ$ of electrolyte i in solution.

For electrolyte solutions, we can work with thermodynamic properties (μ_i, $\bar{H}_i$, $\bar{S}_i$, etc.) of the electrolyte as a whole, and these properties are experimentally determinable. Suppose we have 30 common cations and 30 common anions. This means we must measure thermodynamic properties for 900 electrolytes in water. If we could determine single-ion chemical potentials μ_+ and μ_-, we would then need to measure values for only 60 ions, since μ_i for an electrolyte can be determined from $\mu_i = \nu_+\mu_+ + \nu_-\mu_-$ [Eq. (10.38)]. Unfortunately, single-ion chemical potentials cannot readily be measured. What is done is to assign arbitrary values for thermodynamic properties to the aqueous H^+ ion. Thermodynamic properties of other aqueous ions are then tabulated relative to $H^+(aq)$.

The convention adopted is that $\Delta_f G^\circ$ of the aqueous H^+ ion is zero at every temperature:

$$\Delta_f G_T^\circ[H^+(aq)] = 0 \qquad \text{by convention} \qquad (10.85)$$

The reaction of formation of $H^+(aq)$ in its standard state at temperature T and pressure 1 bar $\equiv P^\circ$ from H_2 gas in its standard state is

$$\tfrac{1}{2}H_2(\text{ideal gas}, P^\circ) \to H^+(aq, m = m^\circ, \gamma_m = 1) + e^-(ss) \qquad (10.86)$$

where $e^-(ss)$ indicates 1 mole of electrons in some particular standard state, which we shall leave unspecified. Whatever the value of ΔG° for (10.86) actually is, this value will cancel in calculating thermodynamic-property changes for ionic reactions in aqueous solutions. The value of ΔG° for (10.86) will not cancel in calculations on reactions that involve transport of ions from one phase to another, for example, the reaction $H^+(g) \to H^+(aq)$, or on half-reactions, for example, (10.86). Hence, the convention (10.85) cannot be used to calculate thermodynamic quantities for ion-transport reactions or half-reactions. Such reactions are not readily studied experimentally but can be discussed theoretically using statistical mechanics.

We have $d\,\Delta G^\circ/dT = -\Delta S^\circ$. Since ΔG° for (10.86) is taken as zero at every temperature, $d\,\Delta G^\circ/dT$ for (10.86) equals zero and ΔS° for the $H^+(aq)$ formation reaction (10.86) is zero at every temperature:

$$\Delta_f S_T^\circ[H^+(aq)] = 0 \qquad \text{by convention} \qquad (10.87)$$

We also have for the reaction (10.86), $\Delta H^\circ = \Delta G^\circ + T\,\Delta S^\circ = 0 + 0 = 0$. Hence

$$\Delta_f H_T^\circ[H^+(aq)] = 0 \qquad \text{by convention} \qquad (10.88)$$

In tables of thermodynamic properties of ions, the standard-state entropy and heat capacity of $H^+(aq)$ at every temperature are taken as zero by convention:

$$\bar{S}_T^\circ[H^+(aq)] = 0 \qquad \text{by convention} \qquad (10.89)$$

$$\bar{C}_{P,T}^\circ[H^+(aq)] = 0 \qquad \text{by convention} \qquad (10.90)$$

Having adopted conventions for $H^+(aq)$, we can find thermodynamic properties of aqueous ions relative to those of $H^+(aq)$, as follows. Equation (10.44) gives for the electrolyte i with formula $M_{\nu_+}X_{\nu_-}$ in solution: $\mu_i^\circ = \nu_+\mu_+^\circ + \nu_-\mu_-^\circ$, where the molality scale is used. Subtraction of G_{elem}° from each side of this equation and the use of (10.78) and corresponding equations for the ions gives

$$\Delta_f G^\circ[i(aq)] = \nu_+\,\Delta_f G_+^\circ + \nu_-\,\Delta_f G_-^\circ \qquad (10.91)$$

where $\Delta_f G_+^\circ$ and $\Delta_f G_-^\circ$ are $\Delta_f G^\circ$ of the cation and anion in solution. For example, $\Delta_f G_T^\circ[\mathrm{BaCl}_2(aq)] = \Delta_f G_T^\circ[\mathrm{Ba}^{2+}(aq)] + 2\,\Delta_f G_T^\circ[\mathrm{Cl}^-(aq)]$.

Similar relations can be derived from (10.44) for $\bar{S}^\circ$ and $\Delta_f H^\circ$ (Prob. 10.55):

$$\bar{S}^\circ[i(aq)] = \nu_+ \bar{S}_+^\circ + \nu_- \bar{S}_-^\circ \tag{10.92}$$

$$\Delta_f H^\circ[i(aq)] = \nu_+ \Delta_f H_+^\circ + \nu_- \Delta_f H_-^\circ \tag{10.93}$$

The relation $\bar{V}_i^\infty = \nu_+ \bar{V}_+^\infty + \nu_- \bar{V}_-^\infty$ (equivalent to $\bar{V}_i^\circ = \nu_+ \bar{V}_+^\circ + \nu_- \bar{V}_-^\circ$) was used in Prob. 9.16. In these equations, the standard state for an ion is the fictitious state with m/m° and γ_m of that ion equal to 1.

The properties of the electrolyte i on the left sides of Eqs. (10.91) to (10.93) can be found experimentally. Observations on the electrolyte $\mathrm{H}_{\nu_-}\mathrm{X}_{\nu_-}(aq)$ together with the $\mathrm{H}^+(aq)$ conventions give the thermodynamic properties of $\mathrm{X}^{z-}(aq)$ relative to those of $\mathrm{H}^+(aq)$. Observations on $\mathrm{M}_{\nu_+}\mathrm{X}_{\nu_-}(aq)$ then give the properties of $\mathrm{M}^{z+}(aq)$. For examples, see Probs. 10.58 and 10.59.

Values of standard-state conventional thermodynamic properties of some ions in water at 25°C are listed in the Appendix.

The thermodynamic properties (10.91) to (10.93) for an electrolyte as a whole in aqueous solution are indicated by the state designation ai (standing for aqueous, ionized) in the NBS tables (Sec. 5.9). Thermodynamic properties for ion pairs, complex ions, and simple ions in aqueous solution are indicated by ao (aqueous, undissociated). Thus, the NBS tables give at 298 K $\Delta_f G^\circ[\mathrm{ZnSO}_4(ai)] = -891.6$ kJ/mol for the ionized electrolyte ZnSO_4 [Eq. (10.91)] and $\Delta_f G^\circ[\mathrm{ZnSO}_4(ao)] = -904.9$ kJ/mol for the $\mathrm{ZnSO}_4(aq)$ ion pair (Sec. 10.9).

Because of the strong ordering produced by hydration of ions, ΔS° for dissolving a salt in water is sometimes negative, even though a highly ordered low-entropy crystal is destroyed and a mixture produced from two pure substances.

10.11 NONIDEAL GAS MIXTURES

Nonideal Gas Mixtures

The **standard state** of component i of a nonideal gas mixture is taken as pure gas i at the temperature T of the mixture, at 1 bar pressure, and such that i exhibits ideal-gas behavior. This is the same choice of standard state as that made in Sec. 5.1 for a pure nonideal gas and in Sec. 6.1 for a component of an ideal gas mixture. Thermodynamic properties of this fictitious standard state can be calculated once the behavior of the real gas is known.

The *activity* a_i of a component of a nonideal gas mixture is defined as in (10.3):

$$a_i \equiv \exp[(\mu_i - \mu_i^\circ)/RT] \tag{10.94}$$

where μ_i is the chemical potential of gas i in the mixture and μ_i° is the chemical potential of i in its standard state. Taking logs, we have, similar to (10.4),

$$\mu_i = \mu_i^\circ(T) + RT \ln a_i \tag{10.95}$$

The choice of standard state (with $P = 1$ bar) makes μ_i° depend only on T for a component of a nonideal gas mixture.

The **fugacity** f_i of a component of any gas mixture is defined as $f_i \equiv a_i \times 1$ bar:

$$f_i/P^\circ = a_i \qquad \text{where } P^\circ \equiv 1 \text{ bar} \tag{10.96}$$

Since a_i is dimensionless, f_i has units of pressure. Since μ_i in (10.94) is an intensive property that depends on T, P, and the mixture's mole fractions, f_i is a function of these variables: $f_i = f_i(T, P, x_1, x_2, \dots)$. Equation (10.95) becomes

$$\mu_i = \mu_i^\circ(T) + RT \ln (f_i/P^\circ) \tag{10.97}^*$$

For an ideal gas mixture, (6.4) reads

$$\mu_i^{id} = \mu_i^\circ + RT \ln (P_i/P^\circ)$$

Comparison with (10.97) shows that *the fugacity f_i plays the same role in a nonideal gas mixture as the partial pressure P_i in an ideal gas mixture.* Statistical mechanics shows that, in the limit of zero pressure, μ_i approaches μ_i^{id}. Moreover, μ_i° in (10.97) is the same as μ_i° in an ideal gas mixture. Therefore f_i in (10.97) must approach P_i in the limit as the mixture's pressure P goes to zero and the gas becomes ideal:

$$f_i \rightarrow P_i \text{ as } P \rightarrow 0 \qquad \text{or} \qquad \lim_{P \rightarrow 0} (f_i/P_i) = 1 \qquad (10.98)$$

The partial pressure P_i of gas i in a nonideal (or ideal) gas mixture is defined as $P_i \equiv x_i P$ [Eq. (1.23)]. The deviation of the fugacity f_i of i from the partial pressure P_i in a gas mixture is measured by the **fugacity coefficient** ϕ_i (phi i) of gas i. The definition of ϕ_i is $\phi_i \equiv f_i/P_i \equiv f_i/x_i P$, so

$$f_i = \phi_i P_i = \phi_i x_i P \qquad \textbf{(10.99)*}$$

Like f_i, ϕ_i is a function of T, P, and the mole fractions. (There is no connection between the fugacity coefficient ϕ_i and the osmotic coefficient ϕ of Sec. 10.7.)

For an ideal gas mixture, $f_i = P_i$ and $\phi_i = 1$ for each component.

Subtraction of $\mu_i^{id} = \mu_i^\circ + RT \ln (P_i/P^\circ)$ from (10.97) gives $\ln (f_i/P_i) = (\mu_i - \mu_i^{id})/RT$ or $f_i = P_i \exp[(\mu_i - \mu_i^{id})/RT]$. The standard pressure P° does not appear in this equation, so f_i and ϕ_i ($\equiv f_i/P_i$) are independent of the choice of standard-state pressure P°.

The formalism of fugacities and fugacity coefficients is of no value unless the f_i's and ϕ_i's can be found from experimental data. We now show how to do this. Equation (9.31) reads $(\partial \mu_i/\partial P)_{T,n_j} = \bar{V}_i$. Hence $d\mu_i = \bar{V}_i \, dP$ at constant T and n_j, where constant n_j means that all mole numbers including n_i are held fixed. Equation (10.97) gives $d\mu_i = RT \, d \ln f_i$ at constant T and n_j. Equating these two expressions for $d\mu_i$, we have $RT \, d \ln f_i = \bar{V}_i \, dP$ and

$$d \ln f_i = (\bar{V}_i/RT) \, dP \qquad \text{const. } T, n_j \qquad (10.100)$$

Since $f_i = \phi_i x_i P$ [Eq. (10.99)], we have $\ln f_i = \ln \phi_i + \ln x_i + \ln P$ and $d \ln f_i = d \ln \phi_i + d \ln P = d \ln \phi_i + (1/P) \, dP$, since x_i is constant at constant composition. Thus (10.100) becomes

$$d \ln \phi_i = (\bar{V}_i/RT) \, dP - (1/P) \, dP$$

$$\ln \frac{\phi_{i,2}}{\phi_{i,1}} = \int_{P_1}^{P_2} \left(\frac{\bar{V}_i}{RT} - \frac{1}{P} \right) dP \qquad \text{const. } T, n_j$$

where we integrated from state 1 to 2. In the limit $P_1 \rightarrow 0$, we have $\phi_{i,1} \rightarrow 1$ [Eq. (10.98)], and our final result is

$$\ln \phi_{i,2} = \int_0^{P_2} \left(\frac{\bar{V}_i}{RT} - \frac{1}{P} \right) dP \qquad \text{const. } T, n_j \qquad (10.101)$$

To determine the fugacity coefficient of gas i in a mixture at temperature T, (total) pressure P_2, and a certain composition, we measure the partial molar volume $\bar{V}_i$ in the mixture as a function of pressure. We then plot $\bar{V}_i/RT - 1/P$ versus P and measure the area under the curve from $P = 0$ to $P = P_2$. Once $\phi_{i,2}$ is known, $f_{i,2}$ is known from (10.99). Once the fugacities f_i have been found for a range of T, P, and composition, the chemical potentials μ_i in Eq. (10.97) are known for this range. From the μ_i's, we can calculate all the thermodynamic properties of the mixture.

Pure Nonideal Gas

For the special case of a one-component nonideal gas mixture, that is, a pure nonideal gas, the partial molar volume $\bar{V}_i$ becomes the molar volume V_m of the gas, and Eqs. (10.97), (10.98), (10.99), and (10.101) become

$$\mu = \mu^\circ(T) + RT \ln (f/P^\circ) \tag{10.102}$$

$$f = \phi P, \qquad f \rightarrow P \text{ as } P \rightarrow 0 \tag{10.103}$$

$$\ln \phi_2 = \int_0^{P_2} \left(\frac{V_m}{RT} - \frac{1}{P} \right) dP \qquad \text{const. } T \tag{10.104}$$

where f is a function of T and P; $f = f(T, P)$.

The integral in (10.104) can be evaluated from measured values of V_m versus P (Prob. 10.65) or from an equation of state. Use of the virial equation (8.5) in (10.104) gives the simple result (Prob. 10.61)

$$\ln \phi = B^\dagger(T)P + \tfrac{1}{2}C^\dagger(T)P^2 + \tfrac{1}{3}D^\dagger(T)P^3 + \cdots \tag{10.105}$$

In accord with the law of corresponding states (Sec. 8.7), different gases at the same reduced temperature and reduced pressure have approximately the same fugacity coefficient. Figure 10.11 shows some plots of ϕ (averaged over several nonpolar gases) versus P_r at several T_r values. [For more plots and for tables, see R. H. Newton, *Ind. Eng. Chem.*, **27**, 302 (1935); R. H. Perry and C. H. Chilton, *Chemical Engineers' Handbook*, 5th ed., McGraw-Hill, 1973, p. **4**-52; *Reid, Prausnitz, and Poling*, sec. 5-4.]

Figure 10.12 shows plots of ϕ and f versus P for CH_4 at 50°C. When ϕ is less than 1, the gas's $G_m = \mu = \mu^\circ + RT \ln (\phi P/P^\circ)$ is less than the corresponding ideal-gas $G_m^{id} = \mu^{id} = \mu^\circ + RT \ln (P/P^\circ)$; the gas is more stable than the corresponding ideal gas, due to intermolecular attractions.

The concept of fugacity is sometimes extended to liquid and solid phases, but this is an unnecessary duplication of the activity concept.

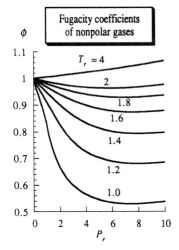

Figure 10.11

Typical fugacity coefficients ϕ of nonpolar gases as a function of reduced variables.

Determination of Mixture Fugacities

Equation (10.97) gives μ_i for each component of a nonideal gas mixture in terms of the fugacity f_i of i. Since all thermodynamic properties follow from μ_i, we have in principle solved the problem of the thermodynamics of a nonideal gas mixture. However, experimental evaluation of the fugacity coefficients from (10.101) requires a tremendous amount of work, since the partial molar volume $\bar{V}_i$ of each component must be determined as a function of P. Moreover, the fugacities so obtained apply to only one particular mixture composition. Usually, we want the f_i's for various mixture compositions, and for each such composition, the $\bar{V}_i$'s must be measured as a function of P and the integrations performed.

The fugacity coefficients in a gas mixture can be roughly estimated from the fugacity coefficients of the pure gases (which are comparatively easy to measure) using the *Lewis–Randall rule:* $\phi_i \approx \phi_i^*(T, P)$, where ϕ_i is the fugacity coefficient of gas i in the mixture and $\phi_i^*(T, P)$ is the fugacity coefficient of pure gas i at the temperature T and (total) pressure P of the mixture. For example, for air at 1 bar and 0°C, the fugacity coefficient of N_2 would be estimated by the fugacity coefficient of pure N_2 at 0°C and 1 bar.

Taking ϕ_i in the mixture at T and P equal to $\phi_i^*(T, P)$ amounts to assuming that the intermolecular interactions in the gas mixture are the same as those in the pure gas, so the i molecules aren't aware of any difference in environment between the mixture and the pure gas. With the same intermolecular interactions between all species, we have an ideal solution (Sec. 9.6). In an ideal solution, $\bar{V}_i(T, P) = V_{m,i}^*(T, P)$ [Eq. (9.55)] and comparison of (10.101) and (10.104) shows that ϕ_i in the mixture at T and P

equals $\phi_i^*(T, P)$. The Lewis–Randall rule clearly works best in mixtures where the molecules have similar size and similar intermolecular forces. When the intermolecular forces for different pairs of molecules differ substantially (which happens quite often), the rule can be greatly in error. Despite its inaccuracy, the Lewis–Randall rule is often used because it is easy to apply.

A much better approach than the Lewis–Randall rule is to find an expression for $\bar{V}_i$ from a reliable equation of state for the mixture (for example, the Redlich–Kwong equation of Sec. 8.2) and use this expression in (10.101) to find ϕ_i. To apply an equation of state to a mixture, one uses rules to express the parameters in the mixture's equation of state in terms of the parameters of the pure gases; see Sec. 8.2. Explicit equations for $\ln \phi_i$ for several accurate equations of state are given in *Reid, Prausnitz, and Poling,* sec. 5-8.

Liquid–Vapor Equilibrium

The expression for μ_i in a nonideal gas mixture is obtained by replacing P_i in the ideal-gas-mixture μ_i with f_i. Therefore, to take gas nonideality into account, all the pressures and partial pressures in the liquid–vapor equilibrium equations of Sec. 10.3 are replaced by fugacities. For example, the partial vapor pressures of a solution are given by $P_i = \gamma_{\mathrm{I},i} x_i^l P_i^*$ [Eq. (10.14)] if the vapor is ideal. For nonideal vapor, this equation becomes

$$f_i = \gamma_{\mathrm{I},i} x_i^l f_i^*$$

where f_i^* is the fugacity of the vapor in equilibrium with pure liquid i at the temperature of the solution, x_i^l and $\gamma_{\mathrm{I},i}$ are the mole-fraction and Convention I activity coefficient of i in the solution, and f_i is the fugacity of i in the nonideal gas mixture in equilibrium with the solution. For the pressure range of zero to a few atmospheres, the truncated virial equation $V_\mathrm{m} = RT/P + B$ [Eq. (8.7) and the paragraph after (8.11)] is widely used to correct for gas nonideality in liquid–vapor studies.

10.12   SUMMARY

The all-important chemical potentials μ_i of components of a nonideal solid or liquid solution are expressed in terms of activities and activity coefficients. One defines a standard state for each component i and then defines its activity a_i so that $\mu_i = \mu_i^\circ + RT \ln a_i$, where μ_i° is the standard-state chemical potential of i.

All standard states are at the T and P of the solution. If the mole-fraction scale is used, the activity a_i is expressed as $a_i = \gamma_i x_i$, where γ_i is the mole-fraction activity coefficient. Two different choices of mole-fraction standard states give Convention I and Convention II. Convention I (used for mixtures of two liquids) takes the standard state of each component as the pure component. Convention II (used for solutions of a solid or gas solute in a liquid solvent) takes the solvent standard state as the pure solvent and takes each solute standard state as the fictitious state of pure solute with solute molecules experiencing the same intermolecular forces they experience in the infinitely dilute solution. If the molality scale is used, each solute activity is $a_{m,i} = \gamma_{m,i} m_i/m^\circ$, where $\gamma_{m,i}$ is the molality-scale activity coefficient; the solvent activity is $a_\mathrm{A} = \gamma_\mathrm{A} x_\mathrm{A}$; the solute standard state is the fictitious state with $\gamma_{m,i} = 1$ and $m_i = 1$ mol/kg; the solvent standard state is the pure solvent. The molar concentration scale is similar to the molality scale, but with c_i used instead of m_i.

For Convention I, the activity coefficient $\gamma_{\mathrm{I},i}$ of substance i goes to 1 as the solution becomes pure i. For Convention II, the molality scale, and the molar concentration scale, all activity coefficients go to 1 in the infinite-dilution limit of pure solvent. Deviations of the $\gamma_{\mathrm{I},i}$'s from 1 measure deviations from ideal-solution behavior, whereas $\gamma_{\mathrm{II},i}$, $\gamma_{m,i}$, and $\gamma_{c,i}$ measure deviations from ideally dilute solution behavior.

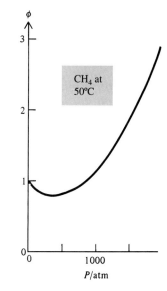

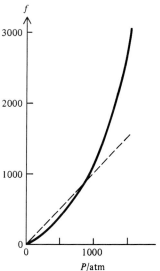

Figure 10.12

Fugacity coefficient ϕ and fugacity f plotted versus P for CH_4 at 50°C. The dashed line corresponds to ideal-gas behavior with $f = P$ and $\phi = 1$.

Electrolyte solutions require special treatment. For the strong electrolyte $M_{\nu_+}X_{\nu_-}$, the electrolyte chemical potential in solution is given by $\mu_i = \mu_i^\circ + \nu RT \ln (\nu_\pm \gamma_\pm m_i/m^\circ)$, where the meanings of the quantities in this equation are summarized at the end of Sec. 10.6. Electrolyte activity coefficients in very dilute solutions can be found from the Debye–Hückel theory.

From measurements of activity coefficients, solubilities, and heats of solution, one can find standard-state partial molar properties of solution components. Thermodynamics tables take the standard state of solute i in aqueous solution to be the fictitious molality-scale standard state with $m_i = 1$ mol/kg and $\gamma_{m,i} = 1$ for nonelectrolytes and for single ions. Single-ion properties in thermodynamics tables are based on the conventions of taking $\bar{S}^\circ$, $\Delta_f G^\circ$, $\Delta_f H^\circ$, and $\bar{C}_P^\circ$ of $H^+(aq)$ as zero.

In a nonideal gas mixture, the fugacities f_i and fugacity coefficients ϕ_i are defined so that the chemical potentials have the form $\mu_i = \mu_i^\circ + RT \ln (f_i/P^\circ) = \mu_i^\circ + RT \ln (\phi_i P_i/P^\circ)$, where $P_i \equiv x_i P$ and the standard state of each component is the hypothetical pure ideal gas at 1 bar and the temperature of the mixture. Fugacity coefficients in a mixture can be found from P-V-T data for the mixture or can be estimated from an equation of state for the mixture.

Important kinds of calculations discussed in this chapter include:

- Calculation of activity coefficients from vapor-pressure data using $P_i = \gamma_{I,i} x_i^l P_i^*$ or $P_i = K_i \gamma_{II,i} x_i^l$ and $P_A = \gamma_{II,A} x_A^l P_A^*$.
- Calculation of activity coefficients of a nonvolatile solute from solvent vapor-pressure data and the Gibbs–Duhem equation.
- Calculation of electrolyte activity coefficients from the Debye–Hückel equation or the Davies equation.
- Calculation of fugacity coefficients from P-V-T data or an equation of state.

FURTHER READING AND DATA SOURCES

Denbigh, chaps. 7, 9, 10; *McGlashan*, chaps. 16, 18, 20; *Prigogine and Defay*, chaps. 20, 21, 27; *Eyring, Henderson, and Jost*, vol. I, pp. 320–352; *Lewis and Randall*, chaps. 20, 21, 22; *Robinson and Stokes*, chaps. 1, 2, 12 to 15; *Bockris and Reddy*, chap. 3; *Davies*.

Excess quantities: M. L. McGlashan (ed.), *Specialist Periodical Reports, Chemical Thermodynamics*, vol. 2, Chemical Society, 1978, pp. 247–538.

Electrolyte-solution activity coefficients and osmotic coefficients: *Robinson and Stokes*, appendices; R. Parsons, *Handbook of Electrochemical Constants*, Academic Press, 1959.

Fugacities of gases: *Landolt-Börnstein*, vol. II, part 1, pp. 310–327; *TRC Thermodynamic Tables* (see Sec. 5.9 for the full reference).

PROBLEMS

Section 10.1

10.1 True or false? (*a*) When a solution component is in its standard state, its activity is 1. (*b*) If μ_i increases in an isothermal, isobaric process, then a_i must increase. (*c*) a_i and γ_i are intensive properties. (*d*) The Convention I standard states are the same as those for an ideal solution and the Convention II standard states are the same as for an ideally dilute solution. (*e*) All activity coefficients γ_i go to 1 in the limit $x_i \to 1$.

10.2 For each of the following quantities, state whether its value depends on the choice of standard state for i: (*a*) μ_i; (*b*) μ_i°; (*c*) γ_i; (*d*) a_i.

10.3 Prove that $0 \leq a_{I,i} \leq 1$. [*Hint:* Use (4.90).]

Section 10.2

10.4 (a) Take $\partial/\partial n_i$ of $G^E \equiv G - G^{id}$ to show that $(\partial G^E/\partial n_i)_{T,P,n_{j\neq i}} = \mu_i - \mu_i^{id}$. (b) Use the result of (a) to show that

$$RT \ln \gamma_i = (\partial G^E/\partial n_i)_{T,P,n_{j\neq i}} \qquad (10.106)$$

Hence γ_i can be found from G^E data. (c) For a solution of the liquids B and C, the *mean molar* G^E is $G_m^E \equiv G^E/(n_B + n_C)$. Differentiate $G^E \equiv (n_B + n_C)G_m^E$ with respect to n_B at constant n_C and use $\partial G_m^E/\partial n_B = (\partial G_m^E/\partial x_B)(\partial x_B/\partial n_B)_{n_C}$ to show that

$$RT \ln \gamma_{I,B} = G_m^E + x_C(\partial G_m^E/\partial x_B)_{T,P}$$

Section 10.3

10.5 True or false? (a) For the solvent in a solution, $\gamma_{II,A} = \gamma_{I,A}$. (b) For the solvent in a solution, $a_{II,A} = a_{I,A}$.

10.6 At 35°C, the vapor pressure of chloroform is 295.1 torr, and that of ethanol (eth) is 102.8 torr. A chloroform–ethanol solution at 35°C with $x_{eth}^l = 0.200$ has a vapor pressure of 304.2 torr and a vapor composition of $x_{eth}^v = 0.138$. (a) Calculate γ_I and a_I for chloroform and for ethanol in this solution. (b) Calculate $\mu_i - \mu_i^*$ for each component of this solution. (c) Calculate ΔG for the mixing of 0.200 mol of liquid ethanol and 0.800 mol of liquid chloroform at 35°C. (d) Calculate $\Delta_{mix}G$ for the corresponding ideal solution.

10.7 For solutions of water (w) and H_2O_2 (hp) at 333.15 K, some liquid and vapor compositions and (total) vapor pressures are

x_w^l	0.100	0.300	0.500	0.700	0.900
x_w^v	0.301	0.696	0.888	0.967	0.995
P/kPa	3.00	5.03	8.33	12.86	17.77

The pure-liquid 333.15 K vapor pressures are $P_w^* = 19.92$ kPa and $P_{hp}^* = 2.35$ kPa. For a solution at 333.15 K with $x_w^l = 0.300$, calculate (a) γ_I and a_I of water and of H_2O_2; (b) a_{II} and γ_{II} of water and of H_2O_2 if the solvent is taken as water; (c) $\Delta_{mix}G$ to form 125 g of this solution from the pure components at constant T and P.

10.8 Activity coefficients of Zn in solutions of Zn in Hg at 25°C were determined by measurements on electrochemical cells. Hg is taken as the solvent. The data are fitted by $\gamma_{II,Zn} = 1 - 3.92x_{Zn}$ for solutions up to saturation. (a) Show that

$$\ln \gamma_{Hg} = (2.92)^{-1}[3.92 \ln (1 - x_{Zn}) - \ln (1 - 3.92x_{Zn})]$$

for this composition range. Use a table of integrals. (b) Calculate $\gamma_{II,Zn}$, $a_{II,Zn}$, γ_{Hg}, and a_{Hg} for $x_{Zn} = 0.0400$.

10.9 Use activity-coefficient data in Sec. 10.3 to calculate G^E/n versus composition for acetone–chloroform solutions at 35.2°C, where n is the total number of moles. Compare the results with Fig. 10.3b.

10.10 (a) Show that $\gamma_{II,i} = \gamma_{I,i}/\gamma_{I,i}^\infty$. (b) Use the result for (a) and Fig. 10.3a to find the relation between γ_{II} and γ_I for chloroform in the solvent acetone at 35.2°C.

10.11 (a) A certain aqueous solution of sucrose at 25°C has a vapor pressure of 23.34 torr. The vapor pressure of water at 25°C is 23.76 torr. Find the activity of the solvent water in this sucrose solution. (b) A 2.00 mol/kg aqueous sucrose solution has a vapor pressure of 22.75 torr at 25°C. Find the activity and activity coefficient of the solvent water in this solution.

Section 10.4

10.12 Which of these activity coefficients must go to 1 as the solvent mole fraction x_A goes to 1: $\gamma_{m,i}$, $\gamma_{c,i}$, $\gamma_{II,i}$, $\gamma_{II,A}$?

10.13 For a 1.50 mol/kg 25°C sucrose solution in water, $\gamma_m = 1.292$ for the solute sucrose. For this solution, find γ_{II}, a_{II}, and a_m for sucrose.

10.14 Use Eq. (10.24) and the expressions in Sec. 10.1 for Convention II standard-state solute properties to derive the solute molality-scale standard-state properties (10.27).

10.15 Equate the concentration-scale expression (10.29) for μ_i to (10.23). Then take the limit as $x_A \to 1$ to show that $\mu_{c,i}^\circ = \mu_{II,i}^\circ + RT \ln V_{m,A}^* c^\circ$, where $V_{m,A}^*$ is the solvent's molar volume. Use this result to show that $\gamma_{c,i} = (x_i/V_{m,A}^* c_i)\gamma_{II,i} = (\rho_A m_i/c_i)\gamma_{m,i}$, where ρ_A is the density of the pure solvent. Then show that $a_{c,i} = (\rho_A m^\circ/c^\circ)a_{m,i}$. For H_2O at 25°C and 1 bar, $\rho_A = 0.997$ kg/dm³, so here $a_{c,i} = 0.997a_{m,i}$.

Section 10.5

10.16 Use the Gibbs–Duhem equation to show that if a solution of B + C has $RT \ln \gamma_{I,B} = W(T, P)x_C^2$, then $RT \ln \gamma_{I,C} = Wx_B^2$.

10.17 Verify that application of Eq. (10.106) of Prob. 10.4b to $G^E = Wn_B n_C/(n_B + n_C)$ gives (10.31) as the activity coefficients.

10.18 Assume that acetone and chloroform form simple solutions at 35.2°C. Use the value of $\gamma_I(CHCl_3)$ at $x(CHCl_3) = 0$ in Fig. 10.3 to evaluate W. Then calculate γ_I for each component at $x(CHCl_3) = 0.494$ and compare with the experimental values.

10.19 (a) Some activity coefficients in solutions of 1-chlorobutane (chl) and heptane (hep) at 323.20 K are

x_{chl}	0.100	0.300	0.500	0.700	0.900
$\gamma_{I,chl}$	1.340	1.169	1.081	1.032	1.004
$\gamma_{I,hep}$	1.005	1.039	1.093	1.173	1.311

Calculate G_m^E for these five solutions. (b) Use a spreadsheet Solver to fit the parameters in the three-parameter Redlich–Kister equation to this G_m^E data. Take B as chl. How good is the fit? (c) Use the results of Prob. 10.23 to predict $\gamma_{I,chl}$ and $\gamma_{I,hep}$ at $x_{chl} = 0$ and at $x_{chl} = 0.4$.

10.20 Repeat Prob. 10.19b using first the two-parameter Redlich–Kister equation and then the four-parameter Redlich–Kister equation. Compare these fits with the three-parameter fit.

10.21 Use the acetone–chloroform γ_I data of Example 10.1 of Sec. 10.3 to calculate G_m^E values. Then use a spreadsheet Solver to do two-, three-, and four-parameter Redlich–Kister fits to G_m^E and comment on the quality of the fits. Use the best fit to predict γ_I^∞ values.

10.22 If the designations of B and C in the Redlich–Kister equation (10.32) are interchanged, how are the values of the A coefficients changed?

10.23 Use the equation of Prob. 10.4c to show that if G_m^E is fitted with the three-parameter Redlich–Kister equation, the activity coefficients are given by

$$\ln \gamma_{I,B} = (A_1 + 3A_2 + 5A_3)x_C^2 - (4A_2 + 16A_3)x_C^3 + 12A_3x_C^4$$
$$\ln \gamma_{I,C} = (A_1 - 3A_2 + 5A_3)x_B^2 + (4A_2 - 16A_3)x_B^3 + 12A_3x_B^4$$

10.24 (a) Suppose the infinite-dilution activity coefficients $\gamma_{I,B}^\infty$ and $\gamma_{I,C}^\infty$ have been measured and we want to fit the two-parameter Redlich–Kister equation. Use the equations of Prob. 10.23 with $A_3 = 0$ to show that

$$A_1 = \tfrac{1}{2}(\ln \gamma_{I,B}^\infty + \ln \gamma_{I,C}^\infty), \qquad A_2 = \tfrac{1}{2}(\ln \gamma_{I,C}^\infty - \ln \gamma_{I,B}^\infty)$$

(b) For solutions of CCl_4 and $SiCl_4$ at 50°C, $\gamma_{I,CCl_4}^\infty = 1.12_9$, $\gamma_{I,SiCl_4}^\infty = 1.14_0$, and the two-parameter Redlich–Kwong equation fits well. Calculate A_1 and A_2 for this mixture and predict the γ_I values at $x_{SiCl_4} = 0.4$.

Section 10.6

10.25 For each of these electrolytes, give the values of ν_+, ν_-, z_+, and z_-: (a) KCl; (b) $MgCl_2$; (c) $MgSO_4$; (d) $Ca_3(PO_4)_2$. (e) Which of the electrolytes (a) to (d) are 1:1 electrolytes?

10.26 Write the expression for $\gamma_\pm$ in terms of γ_+ and γ_- for each of the electrolytes in Prob. 10.25.

10.27 Express μ_i for $ZnCl_2$ in terms of μ_+ and μ_-.

10.28 Calculate $\nu_\pm$ for each electrolyte in Prob. 10.25.

10.29 Show that if $\nu_+ = \nu_-$, then $\nu_\pm = \nu_+ = \nu_-$.

10.30 Express a_i for $MgCl_2(aq)$ in terms of m_i.

10.31 At 25°C, the vapor pressure of a 4.800 mol/kg solution of KCl in water is 20.02 torr. The vapor pressure of pure water at 25°C is 23.76 torr. For the solvent in this KCl solution, find (a) ϕ; (b) γ_A and a_A if x_A is calculated using H_2O, $K^+(aq)$, and $Cl^-(aq)$ as the solution constituents; (c) γ_A and a_A if x_A is calculated using H_2O and KCl(aq) as the solution constituents.

10.32 Robinson and Stokes used vapor-pressure data to find practical osmotic coefficients in aqueous solutions of sucrose (supplied by the Colonial Sugar Refining Company) at 25°C. The following expression reproduces their results:

$$\phi = 1 + am/m° + b(m/m°)^2 + c(m/m°)^3 + d(m/m°)^4$$

where $a = 0.07028$, $b = 0.01847$, $c = -0.004045$, $d = 0.000228$, m is the sucrose molality, and $m° \equiv 1$ mol/kg. (a) Use (10.59) with $\gamma_\pm$ replaced by γ_i to show that for sucrose

$$\ln \gamma_m = 2am/m° + \tfrac{3}{2}b(m/m°)^2 + \tfrac{4}{3}c(m/m°)^3 + \tfrac{5}{4}d(m/m°)^4$$

(b) Calculate γ_m and γ_{II} for sucrose in a 6.00 mol/kg aqueous solution at 25°C.

10.33 (a) Use (10.57) and (10.51) in the Gibbs–Duhem equation (10.55) to show that

$$d \ln \gamma_\pm = d\phi + [(\phi - 1)/m_i] dm_i \qquad \text{const. } T, P$$

(b) Use (10.57), (10.39), (8.36), and (10.48) to show that $\phi \to 1$ as $x_A \to 1$. (c) Show that integration of the result from (a) gives Eq. (10.59).

10.34 The Debye–Hückel theory shows that at very high dilutions of electrolyte i, $\phi - 1 \propto m_i^{1/2}$ (*McGlashan*, sec. 20.4), so the integrand in (10.59) becomes infinite as $m_i \to 0$ and the integral is hard to evaluate graphically. Show that if we define $w_i \equiv m_i^{1/2}$, then this integral becomes $2\int_0^{w_i^2} [(\phi - 1)/w_i] dw_i$. Since $\phi - 1 \propto w_i$, there is no infinity in this integrand.

Section 10.8

10.35 Calculate the ionic strength I_m in a solution that contains 0.0100 mol KCl, 0.0050 mol $MgCl_2$, 0.0020 mol $MgSO_4$, and 100 g H_2O.

10.36 For a solution of a single strong electrolyte, show that $I_m = \tfrac{1}{2}z_+|z_-|\nu m_i$.

10.37 Use the Davies equation to estimate (a) $\gamma_\pm$ for 0.02 mol/kg $CaCl_2(aq)$ at 25°C; (b) $\gamma_\pm$ for $CaCl_2$ in an aqueous 25°C solution that has $CaCl_2$ molality 0.02 mol/kg, $CuSO_4$ molality 0.01 mol/kg, and $Al(NO_3)_3$ molality 0.005 mol/kg; (c) γ_+ and γ_- for the solution of part (a).

10.38 (a) For a 0.001 mol/kg 25°C $CaCl_2(aq)$ solution, what value of a in the Debye–Hückel equation (10.67) is required to give agreement with the experimental $\gamma_\pm$ of Table 10.2? (b) Use the value of a from part (a) and Eq. (10.67) to estimate $\gamma_\pm$ of 0.01 mol/kg 25°C $CaCl_2(aq)$.

10.39 Calculate $\gamma_\pm$ in a 0.0200 mol/kg HCl solution in CH_3OH at 25°C and 1 atm. For CH_3OH at 25°C and 1 atm, the dielectric constant is 32.6 and the density is 0.787 g/cm³. Assume $a = 3$ Å.

10.40 For a 25°C aqueous solution of a single strong electrolyte, the Meissner equation is

$$\log_{10} \gamma_\pm = -0.5107z_+|z_-| \frac{I^{1/2}}{1 + cI^{1/2}}$$
$$+ z_+|z_-| \log_{10}[1 - b + b(1 + 0.1I)^q]$$
$$b \equiv 0.75 - 0.065q, \qquad c \equiv 1 + 0.055qe^{-0.023I^3}$$

where $I \equiv I_m/m°$. For $Na_2SO_4(aq)$, $q = -0.19$. Calculate the Meissner-predicted $\gamma_\pm$ of $Na_2SO_4(aq)$ at 0.1 mol/kg and 1 mol/kg and compare with the values in Table 10.2.

10.41 For the $CaCl_2(aq)$ Table 10.2 data: (a) use the 0.1 mol/kg $\gamma_\pm$ and a spreadsheet Solver to find the Meissner q. (q can be positive or negative and a reasonable first guess for q is 0.) Then predict $\gamma_\pm$ at 5 mol/kg and at 10 mol/kg and compare with the experimental values; (b) find q by fitting the five lowest molalities using the Solver and then predict the 5 mol/kg and 10 mol/kg values.

10.42 The main ions in seawater are Na^+, K^+, Mg^{2+}, Ca^{2+}, Cl^-, and SO_4^{2-}. Suppose we wanted to use the Pitzer equations to calculate activity coefficients in seawater. (a) For how many single-electrolyte solutions and how many two-electrolyte solutions do we need $\gamma_\pm$ data so as to find the Pitzer parameters?

(b) How many θ and how many ψ parameters are there for seawater? (c) For each two-electrolyte solution of (a), state which θ and ψ parameters are found from data on that solution.

Section 10.9

10.43 Show from (10.77) that $\gamma_\pm^\dagger = \alpha\gamma_\pm$ if $\nu_+ = \nu_-$.

10.44 Verify Eqs. (10.76) and (10.77) for μ_i and $\gamma_\pm^\dagger$.

10.45 Verify that $\gamma_\pm^\dagger = (m_+/m_+^\infty)^{\nu_+/\nu}(m_-/m_-^\infty)^{\nu_-/\nu}\gamma_\pm$, where $m_+^\infty \equiv \nu_+ m_i$ and $m_-^\infty \equiv \nu_- m_i$ are the maximum possible molalities of the cation and anion and occur in the limit of infinite dilution, where there is no ion pairing and $\alpha = 1$. (From this result, it follows that $\gamma_\pm^\dagger \le \gamma_\pm$.)

10.46 Starting from $G = n_A\mu_A + n_+\mu_+ + n_-\mu_- + n_{IP}\mu_{IP}$, derive Eq. (10.54) for an electrolyte solution.

10.47 For $Pb(NO_3)_2$, the fraction of Pb^{2+} ions that associate with NO_3^- ions to form ion pairs is known to be $1 - \alpha = 0.43$ in a 0.100-mol/kg aqueous solution at 25°C. (a) Calculate I_m of this solution. Note that the ion pair is charged. (b) Use the Davies equation to calculate $\gamma_\pm$ for this solution. Then calculate $\gamma_\pm^\dagger$. The experimental $\gamma_\pm^\dagger$ is 0.395.

10.48 Use Fig. 14.24 to decide whether ion pairing will increase or decrease in water as T increases.

Section 10.10

10.49 Use (10.83) and the $\Delta_f G^\circ$ and $\Delta_f H^\circ$ results of the Sec. 10.10 example to find $\bar{S}_{298}^\circ$(sucrose, aq).

10.50 Use data in the Appendix to find ΔG_{298}°, ΔH_{298}°, and ΔS_{298}° for (a) $H^+(aq) + OH^-(aq) \to H_2O(l)$; (b) $CO_3^{2-}(aq) + 2H^+(aq) \to H_2O(l) + CO_2(g)$.

10.51 (a) Use Appendix data to find $\Delta_f G^\circ$, $\Delta_f H^\circ$, and $\bar{S}^\circ$ at 25°C for $Cu(NO_3)_2(aq)$. (b) Use Appendix data to find ΔH_{298}° for $NaCl(s) \to NaCl(aq)$.

10.52 The NBS tables give $\Delta_f G_{298}^\circ = -1010.61$ kJ/mol for $NaSO_4^-(aq)$. With the aid of Appendix data, find ΔG_{298}° for the ion-pair formation reaction $Na^+(aq) + SO_4^{2-}(aq) \to NaSO_4^-(aq)$.

10.53 The NBS tables (Sec. 5.9) list the following $\Delta_f G_{298}^\circ$ values: -108.74 kJ/mol for $NO_3^-(ao)$ and -111.25 kJ/mol for $HNO_3(ai)$. Without looking up any data, explain why at least one of these numbers must be in error.

10.54 Find the conventional value of $\bar{S}_i^\circ$ of $H_3O^+(aq)$ at 25°C. (Hint: Consider the two equivalent ways of writing the ionization of water: $H_2O \rightleftharpoons H^+ + OH^-$ and $2H_2O \rightleftharpoons H_3O^+ + OH^-$.)

10.55 Derive Eqs. (10.92) and (10.93).

10.56 (a) The solubility of $O_2(g)$ in water at 25°C and 1 bar pressure of O_2 above the solution is 1.26 mmol per kilogram of water. Find $\Delta_f G_{298}^\circ$ for O_2 in water. The molality-scale standard state is used for the solute O_2. (b) Use a Henry's law constant of C_2H_6 in Sec. 9.8 to find $\Delta_f G_{298}^\circ$ for $C_2H_6(aq)$.

10.57 Derive the following equations for partial molar properties of a solute in a nonelectrolyte solution:

$$\bar{S}_i = \bar{S}_{m,i}^\circ - R\ln(\gamma_{m,i}m_i/m^\circ) - RT(\partial\ln\gamma_{m,i}/\partial T)_{P,n_j}$$

$$\bar{V}_i = \bar{V}_{m,i}^\circ + RT(\partial\ln\gamma_{m,i}/\partial P)_{T,n_j}$$

$$\bar{H}_i = \bar{H}_{m,i}^\circ - RT^2(\partial\ln\gamma_{m,i}/\partial T)_{P,n_j}$$

10.58 Measurements on electrochemical cells (Sec. 14.10) give for $HCl(aq)$ that $\Delta_f G_{298}^\circ = -131.23$ kJ/mol and $\Delta_f H_{298}^\circ = -167.16$ kJ/mol. Use these data, Appendix entropy data for $H_2(g)$ and $Cl_2(g)$, and the $H^+(aq)$ conventions to find $\Delta_f G_{298}^\circ$, $\Delta_f H_{298}^\circ$, and $\bar{S}_{298}^\circ$ of $Cl^-(aq)$. Start with Eq. (10.91).

10.59 At 25°C and 1 bar, the differential heat of solution of KCl in water at infinite dilution is 17.22 kJ/mol. A saturated 25°C aqueous KCl solution has KCl molality 4.82 mol/kg and activity coefficient $\gamma_\pm = 0.588$. For pure $KCl(s)$ at 25°C, $\Delta_f G^\circ = -409.14$ kJ/mol, $\Delta_f H^\circ = -436.75$ kJ/mol, and $\bar{S}^\circ = 82.59$ J/(mol K). Find $\Delta_f G_{298}^\circ$, $\Delta_f H_{298}^\circ$, and $\bar{S}_{298}^\circ$ for $K^+(aq)$ using these data and the results found for $Cl^-(aq)$ in Prob. 10.58.

10.60 For ions in aqueous solution, would you expect $\bar{S}_i^\circ$ to increase or decrease as the absolute value $|z_i|$ of the ionic charge increases? Explain your answer. Check your answer by consulting Appendix data.

Section 10.11

10.61 (a) For a pure gas that obeys the virial equation (8.5), derive (10.105) for $\ln\phi$. (b) Use (8.6) and (8.9) to show that for a van der Waals gas

$$\ln\phi = \frac{bRT - a}{R^2T^2}P + \frac{2abRT - a^2}{2R^4T^4}P^2 + \cdots$$

10.62 (a) For CO_2, the critical temperature and pressure are 304.2 K and 72.8 atm. Assume CO_2 obeys the van der Waals equation and use the result from Prob. 10.61b to estimate ϕ for CO_2 at 1.00 atm and 75°C and at 25.0 atm and 75°C. Compare with the experimental values 0.9969 at 1 atm and 0.92 at 25 atm. (b) Use the Lewis–Randall rule to estimate the fugacity and fugacity coefficient of CO_2 in a mixture of 1.00 mol CO_2 and 9.00 mol O_2 at 75°C and 25.0 atm.

10.63 For a pure gas, show that $\ln\phi = (G_m - G_m^{id})/RT$, where G_m^{id} is the molar Gibbs energy of the corresponding ideal gas at the same T and P.

10.64 (a) Calculate ΔG when 1.000 mol of an ideal gas at 0°C is isothermally compressed from 1.000 to 1000 atm. (b) For N_2 at 0°C, $\phi = 1.84$ at 1000 atm and $\phi = 0.9996$ at 1 atm. Calculate ΔG when 1.000 mol of N_2 is isothermally compressed from 1.000 to 1000 atm at 0°C.

10.65 For CH_4 at -50°C, measured V_m values as a function of P are

V_m/(cm³/mol)	3577	1745	828	365
P/atm	5	10	20	40

V_m/(cm³/mol)	206	127.0	90.1	75.4
P/atm	60	80	100	120

(*a*) Use a graph to find the fugacity and fugacity coefficient of CH_4 at $-50°C$ and 120 atm. See Probs. 5.33 and 5.35. Note from Prob. 8.38 that $V_m - RT/P$ does *not* go to zero as the gas pressure goes to zero. (*b*) Give the value of the second virial coefficient B for CH_4 at $-50°C$. (*c*) Instead of using a graph, answer (*a*) by using the Solver to fit the data with Eq. (8.5) and find $B^†$, $C^†$, and $D^†$; then use (10.105).

10.66 (*a*) Use the law-of-corresponding-states equation in Prob. 8.22 and Table 8.1 to estimate the second virial coefficient B for N_2 at $0°C$. (*b*) Use Eq. (10.105) with terms after $B^†P$ omitted to estimate ϕ at $0°C$ of N_2 for $P = 1.00$ atm and for $P = 25$ atm. Compare with the experimental values 0.99955 at 1 atm and 0.9895 at 25 atm.

10.67 A liquid mixture of carbon tetrachloride (car) and chloroform (chl) at $40.0°C$ with $x_{chl} = 0.5242$ has a vapor pressure of 301.84 torr and has vapor-phase composition $x_{chl}^v = 0.6456$. The pure-liquid $40°C$ vapor pressures are $P_{chl} = 360.51$ torr and $P_{car} = 213.34$ torr. The $40°C$ second virial coefficients of the pure gases are $B_{chl} = -1040$ cm³/mol and $B_{car} = -1464$ cm³/mol. (*a*) Use the Lewis–Randall rule and Eq. (10.105) with terms after $B^†P$ omitted to estimate the fugacity coefficients ϕ_{chl} and ϕ_{car} in the saturated vapor mixture and in the pure saturated vapors. (*b*) Calculate the activity coefficients $\gamma_{I,chl}$ and $\gamma_{I,car}$ in the liquid mixture using the fugacity coefficients found in (*a*). (*c*) Calculate the activity coefficients $\gamma_{I,chl}$ and $\gamma_{I,car}$ assuming the vapor mixture and the pure vapors are ideal.

General

10.68 Verify that the expressions for the ideal-solution chemical potentials in (9.42) obey the Gibbs–Duhem equation (10.18).

10.69 For a 1.0 mol/dm³ $NaCl(aq)$ solution, pretend that the ions are uniformly distributed in space and calculate the average distance between centers of nearest-neighbor ions. (See Prob. 2.54.)

10.70 Suppose that A and B molecules have similar sizes and shapes and that A-A and B-B intermolecular attractions are stronger than A-B attractions. State whether each of the following quantities for a solution of A + B is likely to be larger or smaller than the corresponding quantity for an ideal solution. (*a*) $\Delta_{mix}H$; (*b*) $\Delta_{mix}S$; (*c*) $\Delta_{mix}G$.

10.71 Answer the following without looking up any formulas. For a dilute electrolyte solution with $\gamma_+ < 1$, would you expect γ_+ to increase or to decrease (*a*) if the ionic charge z_+ increases; (*b*) if the ionic diameter a increases; (*c*) if the ionic strength I_m increases; (*d*) if the solvent's dielectric constant increases; (*e*) if the temperature increases. Explain each of your answers.

10.72 (*a*) Use (10.51) for μ_i to show that for an electrolyte solute i in a solution in equilibrium with vapor (assumed ideal), the equation corresponding to Henry's law (9.63) is

$$P_i = K_i(\nu_\pm \gamma_\pm m_i/m°)^\nu$$

where K_i is defined as in (9.62) except that $\mu_i°$ is replaced by $\mu_{m,i}°$. Show that for $HCl(aq)$, this equation becomes $P_i = K_i(\gamma_\pm m_i/m°)^2$. (*b*) Use data in Table 10.2 and in the Appendix to find the HCl partial pressure in equilibrium with a 0.10 mol/kg $25°C$ $HCl(aq)$ solution.

10.73 For a solution of ethanol (E) in water (W), state whether each of the following activity coefficients is equal to 1 if water is considered the solvent whenever a solvent is to be specified. (*a*) $\gamma_{I,W}$, $\gamma_{I,E}$, $\gamma_{II,W}$, $\gamma_{II,E}$, and $\gamma_{m,E}$, each evaluated in the limit $x_W \to 1$; (*b*) the activity coefficients of (*a*) evaluated in the limit $x_E \to 1$.

10.74 True or false? (*a*) When a solution component is in its standard state, its activity is 1. (*b*) If a solution component has its activity equal to 1, the component must be in its standard state. (*c*) The activity a_i is never negative. (*d*) Activity coefficients are never negative. (*e*) $\gamma_\pm = (\gamma_+)^{\nu_+}(\gamma_-)^{\nu_-}$. (*f*) $\Delta S_{298}°$ for dissolving a salt in water is always positive.

Reaction Equilibrium in Nonideal Systems

As noted at the start of Chapter 6, reaction equilibrium calculations have important industrial, environmental, biochemical, and geochemical applications. Chapter 6 dealt with equilibrium in ideal-gas reactions and Sec. 9.8 mentioned equilibrium in ideally dilute solutions. Equilibria in aqueous solutions commonly involve ionic species, for which the ideally dilute solution approximation is poor. Some key industrial gas-phase reactions are run at high pressures, where the gases are far from ideal. It is therefore essential to know how to compute equilibrium compositions in nonideal systems, which is what Chapter 11 is about.

11.1 THE EQUILIBRIUM CONSTANT

For the chemical reaction $0 \rightleftharpoons \Sigma_i \nu_i A_i$ with stoichiometric coefficients ν_i, the reaction equilibrium condition is $\Sigma_i \nu_i \mu_{i,\text{eq}} = 0$ [Eq. (4.98)], where $\mu_{i,\text{eq}}$ is the equilibrium value of the chemical potential (partial molar Gibbs energy) of the ith species.

To obtain a convenient expression for μ_i, we choose a standard state for each species i and define the **activity** a_i of i in the reaction mixture by

$$a_i \equiv e^{(\mu_i - \mu_i^\circ)/RT} \tag{11.1}$$

where μ_i is the chemical potential of i in the reaction mixture and μ_i° is its standard-state chemical potential. The activity a_i depends on the choice of standard state and is meaningless unless the standard state has been specified. From (11.1), a_i depends on the same variables as μ_i. The activity a_i is a dimensionless intensive property. Comparison of (11.1) with (10.3) and (10.94) shows that a_i in (11.1) is what we previously defined to be the activity of a species in a solid, liquid, or gaseous mixture. Table 11.1 in Sec. 11.8 summarizes the choices of standard states. Taking logs of (11.1), we get

$$\mu_i = \mu_i^\circ + RT \ln a_i \tag{11.2}*$$

Substitution of (11.2) into the equilibrium condition $\Sigma_i \nu_i \mu_{i,\text{eq}} = 0$ gives

$$\sum_i \nu_i \mu_i^\circ + RT \sum_i \nu_i \ln a_{i,\text{eq}} = 0 \tag{11.3}$$

where $a_{i,\text{eq}}$ is the equilibrium value of the activity a_i. The first sum in this equation is defined to be ΔG°, the **standard Gibbs energy change** for the reaction (reactants and products each in standard states). We have $\Sigma_i \nu_i \ln a_{i,\text{eq}} = \Sigma_i \ln (a_{i,\text{eq}})^{\nu_i} = \ln \Pi_i (a_{i,\text{eq}})^{\nu_i}$ [Eqs. (1.68) and (1.67)], so (11.3) becomes

$$\Delta G^\circ + RT \ln \prod_i (a_{i,\text{eq}})^{\nu_i} = 0$$

Defining $K°$ to be the product in this last equation, we have

$$\Delta G° = -RT \ln K° \tag{11.4}*$$

$$\Delta G° \equiv \sum_i \nu_i \mu_i° \tag{11.5}*$$

$$K° \equiv \prod_i (a_{i,\text{eq}})^{\nu_i} \tag{11.6}*$$

$K°$ is called the **standard equilibrium constant,** the **activity equilibrium constant,** or simply the **equilibrium constant.** We always choose the standard states so that $\mu_i°$ depends at most on T and P. (For gases, $\mu_i°$ depends on T only.) Hence $\Delta G°$ depends at most on T and P, and $K°$, which equals $\exp(-\Delta G°/RT)$, depends at most on T and P, and not on the mole fractions. We have thus "solved" the problem of chemical equilibrium in an arbitrary system. The equilibrium position occurs when the activities are such that $\prod_i (a_i)^{\nu_i}$ equals the equilibrium constant $K°$, where $K°$ is found from (11.4) as $\exp(-\Delta G°/RT)$. To solve the problem in a practical sense, we must express the activities in terms of experimentally observable quantities.

11.2 REACTION EQUILIBRIUM IN NONELECTROLYTE SOLUTIONS

To apply the results of the last section to solutions of nonelectrolytes, we choose one of the conventions of Chapter 10 and introduce the appropriate expressions for the activities a_i into the equilibrium constant $K°$ of (11.6).

Most commonly, one component of the solution is designated the solvent. For the solvent, we use the mole-fraction scale [Eq. (10.28)]. For the solutes, one can use the mole-fraction scale, the molality scale, or the molar concentration scale.

If the mole-fraction scale is used for the solutes, the activity $a_{x,i}$ of species i is $a_{x,i} = \gamma_{\text{II},i} x_i$ [Eq. (10.5)], where γ_{II} denotes the Convention II activity coefficient (Sec. 10.1), which goes to 1 at infinite dilution. The subscript x on a reminds us that the activity depends on which scale is used. The equilibrium constant $K°$ in Eq. (11.6) then becomes $K_x = \prod_i (\gamma_{\text{II},i} x_i)^{\nu_i}$, where the subscript on K denotes use of the mole-fraction scale and where the eq subscript has been omitted for simplicity. Equations (11.4) and (11.5) become $\Delta G_x° \equiv \sum_i \nu_i \mu_{\text{II},i}° = -RT \ln K_x$.

Thermodynamic data for species in aqueous solutions are usually tabulated for the molality-scale standard state. Therefore, one most commonly uses the molality scale for solutes. From (10.30), the activity $a_{m,i}$ of solute i on the molality scale is $a_{m,i} = \gamma_{m,i} m_i/m°$ ($i \neq A$, where A is the solvent), where the standard molality $m°$ equals 1 mol/kg. The equilibrium constant (11.6) becomes

$$K_m° = (\gamma_{x,A} x_A)^{\nu_A} \prod_{i \neq A} (\gamma_{m,i} m_i/m°)^{\nu_i} \tag{11.7}$$

The degree superscript on K indicates a dimensionless equilibrium constant. The mole-fraction scale is retained for the solvent A, so the solvent activity in (11.7) is different in form from the solute activities. The solvent stoichiometric coefficient ν_A is zero if the solvent does not appear in the chemical reaction. If the solution is dilute, both x_A and $\gamma_{x,A}$ are close to 1 and it is a good approximation to omit the factor $(\gamma_{x,A} x_A)^{\nu_A}$ from $K_m°$. For the molality scale, (11.4) and (11.5) become

$$\Delta G_m° = -RT \ln K_m° \tag{11.8}$$

$$\Delta G_m° \equiv \nu_A \mu_{x,A}° + \sum_{i \neq A} \nu_i \mu_{m,i}° \tag{11.9}$$

Note that $\mu°$ of the solvent cannot be omitted from $\Delta G_m°$ (unless $\nu_A = 0$) even if the solution is very dilute.

The molar concentration scale is occasionally used for solute activities. Here, $a_{c,i}$ = $\gamma_{c,i} c_i / c°$ [Eq. (10.30)]. The equations for $K_c°$ and $\Delta G_c°$ are the same as (11.7) to (11.9) except that the letter m is replaced by c everywhere.

K_x, $K_c°$, and $K_m°$ have different values for the same reaction. Likewise, $\Delta G_x°$, $\Delta G_c°$, and $\Delta G_m°$ differ for the same reaction, since the value of the standard-state quantity $\mu_i°$ depends on the choice of standard state for species i. Thus, in using Gibbs free-energy data to calculate equilibrium compositions, one must be clear what the choice of standard state is for the tabulated data.

To apply these expressions to calculate equilibrium compositions, we use the procedures discussed in Chapter 10 to determine activity coefficients. If the nonelectrolyte solution is dilute, we can approximate each activity coefficient as 1. The equilibrium constant K_x then reduces to the expression $K_x = \Pi_i (x_{i,eq})^{\nu_i}$ for ideally dilute solutions (Sec. 9.8).

Since $\mu_{m,i}°$ for solute i depends on what the solvent is, the equilibrium constant $K_m°$ for a given reaction is different in different solvents. Also, the activity coefficients are different in different solvents because of different intermolecular interactions. Thus the equilibrium amounts differ in different solvents.

11.3   REACTION EQUILIBRIUM IN ELECTROLYTE SOLUTIONS

The most commonly studied solution equilibria are ionic equilibria in aqueous solutions. As well as being important in inorganic chemistry, ionic equilibria are significant in biochemistry. For the majority of biologically important reactions, at least some of the species involved are ions. Examples include the organic phosphates (such as adenosine triphosphate, ATP) and the anions of certain acids (such as citric acid) involved in metabolic energy transformations; inorganic ions such as H_3O^+ and Mg^{2+} participate in many biochemical reactions.

Since thermodynamic data for ionic species are usually tabulated for the molality-scale standard state, we shall use the molality-scale equilibrium constant $K_m°$ of Eq. (11.7) for electrolytes.

Many ionic reactions in solution are acid–base reactions. We adopt the Brønsted definition of an **acid** as a proton donor and a **base** as a proton acceptor.

The water molecule is *amphoteric*, meaning that water can act as either an acid or a base. In pure liquid water and in aqueous solutions, the following ionization reaction occurs to a slight extent:

$$H_2O + H_2O \rightleftharpoons H_3O^+ + OH^- \tag{11.10}$$

For (11.10), the activity equilibrium constant (11.6) is

$$K_w° = \frac{a(H_3O^+)a(OH^-)}{[a(H_2O)]^2} \tag{11.11}$$

where the subscript w (for water) is traditional. The standard state of the solvent H_2O is pure H_2O, so $a(H_2O) = 1$ for pure H_2O [Eq. (11.1)]. In aqueous solutions, $a(H_2O)$ = $\gamma_x(H_2O)x(H_2O)$. For *dilute* aqueous solutions, the mole fraction $x(H_2O)$ is close to 1, and (since H_2O is uncharged) $\gamma_x(H_2O)$ is close to 1. Therefore, we usually approximate $a(H_2O)$ as 1 in dilute aqueous solutions. Use of $a(H_2O) \approx 1$, and $a_i = \gamma_i m_i / m°$ [Eq. (10.30)] for each ion gives

$$K_w° = a(H_3O^+)a(OH^-) = [\gamma(H_3O^+)m(H_3O^+)/m°][\gamma(OH^-)m(OH^-)/m°]$$

where $m° \equiv 1$ mol/kg and where the subscript m was omitted from γ. *All unsubscripted activity coefficients in this section are on the molality scale.* This expression differs from (11.7) by the omission of the solvent's activity. The mean molal ionic activity coefficient $\gamma_{\pm}$ is defined by $(\gamma_{\pm})^{\nu_+ + \nu_-} = (\gamma_+)^{\nu_+}(\gamma_-)^{\nu_-}$ [Eq. (10.43)]. For the H_2O ionization, $\nu_+ = 1 = \nu_-$, $\gamma_{\pm}^2 = \gamma_+ \gamma_-$, and K_w° becomes

$$K_w^\circ = \gamma_{\pm}^2 m(H_3O^+) m(OH^-)/(m^\circ)^2 \qquad \text{dil. aq. soln.} \qquad (11.12)*$$

Experiment (Prob. 14.50) gives 1.00×10^{-14} for K_w° at 25°C and 1 atm. Approximating $\gamma_{\pm}$ as 1 in pure water, we get $m(H_3O^+) = m(OH^-) = 1.00 \times 10^{-7}$ mol/kg in pure water at 25°C. This gives an ionic strength $I_m = 1.00 \times 10^{-7}$ mol/kg. The Davies equation (10.71) then gives $\gamma_{\pm} = 0.9996$ in pure water, which is essentially equal to 1. Hence the H_3O^+ and OH^- molalities are accurately equal to 1.00×10^{-7} mol/kg in pure water at 25°C. In an aqueous solution that is not extremely dilute, $\gamma_{\pm}$ in (11.12) will probably not be close to 1.

Since μ° for each species in solution depends on pressure, ΔG° for the reaction depends on pressure and the equilibrium constant for a reaction in solution depends on pressure. However, this dependence is weak. Ordinarily, equilibrium constants in solution are determined for P near 1 bar, and this value of P is assumed throughout this section.

Next, consider the ionization of the weak acid HX in aqueous solution. The ionization reaction and the molality-scale equilibrium constant (11.7) are

$$HX + H_2O \rightleftharpoons H_3O^+ + X^- \qquad (11.13)$$

$$K_a^\circ = \frac{[\gamma(H_3O^+)m(H_3O^+)/m^\circ][\gamma(X^-)m(X^-)/m^\circ]}{\gamma(HX)m(HX)/m^\circ} \qquad (11.14)$$

where the subscript a (for acid) is traditional and where the activity of the solvent H_2O is approximated as 1 in dilute solutions. Figure 11.1 plots K_a° at 25°C and 1 bar for some acids in water. In most applications, the HX molality is rather low, and it is a good approximation to take $\gamma = 1$ for the uncharged species HX. However, even though the X^- and H_3O^+ molalities are usually much less than the HX molality, we cannot set $\gamma = 1$ for these ions. γ for an ion deviates significantly from 1 even in quite dilute solutions. Using (10.43) to introduce $\gamma_{\pm}$, we have

$$K_a = \frac{\gamma_{\pm}^2 m(H_3O^+) m(X^-)}{m(HX)} \qquad \text{dil. soln.} \qquad (11.15)$$

where $\gamma_{\pm}$ is for the pair of ions H_3O^+ and X^- and differs from $\gamma_{\pm}$ in (11.12). In (11.15) we have omitted dividing each molality by the standard molality m° (= 1 mol/kg), so K_a has the dimensions of molality (mol/kg). Correspondingly, the degree superscript on K_a is omitted.

EXAMPLE 11.1 Weak-acid ionization

$K_a = 1.75 \times 10^{-5}$ mol/kg for acetic acid ($HC_2H_3O_2$) in water at 25°C. Find the H_3O^+ and OH^- molalities in a 0.200-mol/kg 25°C aqueous solution of acetic acid.

To solve (11.15) for $m(H_3O^+)$, we need $\gamma_{\pm}$. To use the Davies equation (10.71) to estimate $\gamma_{\pm}$, we need the ionic strength I_m, which can't be calculated until $m(H_3O^+)$ is known. The solution to this dilemma is to first estimate $m(H_3O^+)$ and $m(X^-)$ by setting $\gamma_{\pm} = 1$ in (11.15) and solving for the ionic molalities. With these approximate molalities, we calculate an approximate I_m and then use the Davies equation to find an approximate $\gamma_{\pm}$, which we use in (11.15) to find a more accurate value for the molalities. If necessary, we can then use these more accurate molalities to find a more accurate I_m, and so on.

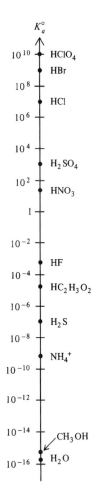

K_a°

10^{10}	HClO$_4$
	HBr
10^8	
	HCl
10^6	
10^4	
	H$_2$SO$_4$
10^2	
	HNO$_3$
1	
10^{-2}	
	HF
10^{-4}	
	HC$_2$H$_3$O$_2$
10^{-6}	
	H$_2$S
10^{-8}	
	NH$_4^+$
10^{-10}	
10^{-12}	
10^{-14}	CH$_3$OH
10^{-16}	H$_2$O

Figure 11.1

Ionization constants of acids in water at 25°C and 1 atm. The values for strong acids are approximate. For consistency with Eq. (11.15), K_a for H_2O is $\gamma_{\pm}^2 m(H_3O^+)m(OH^-)/m(H_2O)$, which differs from K_w. The scale is logarithmic. (Data from J. March, *Advanced Organic Chemistry*, 3d ed., Wiley, 1985, pp. 220–222.)

Let $m(X^-) = x$. Equation (11.13) gives $m(HX) = 0.200$ mol/kg $- x$ and $m(H_3O^+) = x$, since the H_3O^+ formed by the water ionization (11.10) is negligible compared with that from the acetic acid. Setting $\gamma_\pm = 1$ in (11.15), we have

$$1.75 \times 10^{-5} \text{ mol/kg} \approx \frac{x^2}{0.200 \text{ mol/kg} - x} \qquad (11.16)$$

We can solve (11.16) using the quadratic formula, but a faster method is an iterative solution, as follows. Since K_a is much less than the acid's stoichiometric molality (0.200 mol/kg), the degree of ionization will be slight and 0.200 mol/kg $- x$ can be well approximated by 0.200 mol/kg. (In extremely dilute solutions the degree of ionization is substantial, and this approximation cannot be made.) Therefore $x^2/(0.200 \text{ mol/kg}) \approx 1.75 \times 10^{-5}$ mol/kg, and $x \approx 1.87 \times 10^{-3}$ mol/kg. With this value of x, the denominator in (11.16) becomes 0.200 mol/kg $- 0.002$ mol/kg $= 0.198$ mol/kg. Hence $x^2/(0.198 \text{ mol/kg}) \approx 1.75 \times 10^{-5}$ mol/kg, and we get $x \approx 1.86 \times 10^{-3}$ mol/kg.

Thus, with $\gamma_\pm$ taken as 1, we find $m(H_3O^+) = m(X^-) \approx 1.86 \times 10^{-3}$ mol/kg and $I_m \equiv \frac{1}{2}\Sigma_j z_j^2 m_j \approx 1.86 \times 10^{-3}$ mol/kg [Eq. (10.62)]. The Davies equation (10.71) then gives $\gamma_\pm = 0.953$. Equation (11.15) becomes

$$1.75 \times 10^{-5} \text{ mol/kg} = \frac{(0.953)^2 x^2}{0.200 \text{ mol/kg} - x}$$

Solving iteratively, as above, we get $x = 1.95 \times 10^{-3}$ mol/kg $= m(H_3O^+) = I_m$. With this I_m, the Davies equation gives $\gamma_\pm = 0.952$. With this $\gamma_\pm$, the equilibrium-constant expression gives $x = 1.96 \times 10^{-3}$ mol/kg $= m(H_3O^+)$. This I_m gives $\gamma_\pm = 0.952$ again, so the calculation is finished.

We have $K_w = \gamma_\pm^2 m(H_3O^+)m(OH^-)$. The ionic strength is set by the acetic acid ionization and is 1.96×10^{-3} mol/kg. The Davies equation gives the same value for $\gamma_\pm$ of the pair H_3O^+ and OH^- as for the pair H_3O^+ and acetate, namely, 0.952. Hence, $m(OH^-) = (1.00 \times 10^{-14})/[(0.952)^2(1.96 \times 10^{-3})]$ mol/kg $= 5.63 \times 10^{-12}$ mol/kg.

An alternative solution method is to use a spreadsheet (Fig. 11.2). One sets up the Solver to set the error `Kerr` to zero by varying x subject to the constraint that x is between 0 and m. The initial guess for x is the value found by hand calculation with $\gamma_\pm$ omitted.

	A	B	C	D	E	F
1	weak acid	Ka =	1.75E-05	m =	0.2	
2	x =	1.87E-03	I =	1.87E-03		
3	g =	0.953117	Kcalc =	1.603E-05	Kerr =	-0.0838

	A	B	C	D	E	F
1	weak acid	Ka =	0.0000175	m =	0.2	
2	x =	0.00187	I =	=B2		
3	g =	=10^(-0.51*(I^0.5/(1+I^0.5)-0.3*I))	Kcalc =	=g^2*x^2/(m-x)	Kerr =	=(D3-C1)/C1

Figure 11.2

Spreadsheet for weak-acid ionization.

Exercise

Set up the spreadsheet of Fig. 11.2 and solve for x.

Exercise

Find $m(H_3O^+)$ in 1.00 mol/kg $HC_2H_3O_2(aq)$ at 25°C. (*Answer:* 4.49 × 10^{-3} mol/kg.)

In Example 11.1, I_m is quite low so it doesn't make much difference whether we include $\gamma_\pm$. This is not true in Example 11.2.

EXAMPLE 11.2 Buffer solution

Find $m(H_3O^+)$ in a 25°C aqueous solution with the stoichiometric molalities [Eq. (10.48)] $m(HC_2H_3O_2) = 0.100$ mol/kg and $m(NaC_2H_3O_2) = 0.200$ mol/kg (a buffer solution).

The salt $NaC_2H_3O_2$ (which is a 1:1 electrolyte) exists virtually entirely in the form of positive and negative ions in solution. Also, the ionization of $HC_2H_3O_2$ will contribute little to I_m in comparison with the contribution from the $NaC_2H_3O_2$. Therefore, $I_m = \frac{1}{2}(0.200 + 0.200)$ mol/kg $= 0.200$ mol/kg. The Davies equation (10.71) then gives $\gamma_\pm \approx 0.746$. Substitution into (11.15) gives

$$1.75 \times 10^{-5} \text{ mol/kg} = \frac{\gamma_\pm^2 m(H_3O^+)m(C_2H_3O_2^-)}{m(HC_2H_3O_2)}$$

$$= \frac{(0.746)^2 m(H_3O^+)(0.200 \text{ mol/kg})}{0.100 \text{ mol/kg}}$$

where $m(C_2H_3O_2^-)$ is well approximated by considering only the acetate ion from the $NaC_2H_3O_2$ and where we set $m(HC_2H_3O_2)$ equal to 0.100 mol/kg since the degree of ionization of acetic acid is far less than in the previous example, because of the added sodium acetate (*common-ion effect*). Solving, we find $m(H_3O^+) = 1.5_7 \times 10^{-5}$ mol/kg. Note that, if $\gamma_\pm$ were omitted in this example, we would get $8.7_5 \times 10^{-6}$ mol/kg for $m(H_3O^+)$, which is in error by a whopping 44%. Except for solutions of quite low ionic strength, ionic equilibrium calculations that omit activity coefficients are likely to give only qualitatively correct answers. Many of the ionic equilibrium calculations you did in first-year chemistry are correct only in the exponent of 10, because of neglect of activity coefficients (and neglect of ion-pair formation in salt solutions).

Exercise

Find $m(H_3O^+)$ in a 25°C aqueous solution with the stoichiometric molalities $m(HC_2H_3O_2) = 1.00$ mol/kg and $m(NaCl) = 0.200$ mol/kg. (*Answer:* 5.60 × 10^{-3} mol/kg.)

EXAMPLE 11.3 Very dilute weak-acid ionization

For HOI, $K_a = 2.3 \times 10^{-11}$ mol/kg in water at 25°C. Find $m(H_3O^+)$ in a 1.0×10^{-4} mol/kg 25°C aqueous solution of HOI.

We have

$$HOI + H_2O \rightleftharpoons H_3O^+ + OI^- \tag{11.17}$$

$$K_a = \gamma_\pm^2 m(H_3O^+)m(OI^-)/m(HOI) \tag{11.18}$$

where $\gamma(HOI)$ and $a(H_2O)$ have each been taken as 1. Because of the extremely low value of K_a, the ionic strength is extremely low, and we can take $\gamma_\pm = 1$. If we proceed as we did with $HC_2H_3O_2$, we have $m(H_3O^+) = m(OI^-) = x$ and $m(HOI) = 0.00010$ mol/kg $- x$. Since K_a is far less than the stoichiometric molality, we can set 0.00010 mol/kg $- x$ equal to 0.00010 mol/kg. We then have 2.3×10^{-11} mol/kg $= x^2/(0.00010$ mol/kg$)$, which gives $x = 4.8 \times 10^{-8}$ mol/kg $= m(H_3O^+)$. However, this answer cannot be correct. We know that $m(H_3O^+)$ equals 1.0×10^{-7} mol/kg in pure water. A solution with $m(H_3O^+) = 4.8 \times 10^{-8}$ mol/kg would have a lower H_3O^+ molality than pure water and would be basic. However, HOI is an acid. The error here is failure to consider the contribution to $m(H_3O^+)$ from the ionization (11.10) of water. In Examples 11.1 and 11.2, the H_3O^+ from the weak-acid ionization far exceeded that from the ionization of water, but this is not true here. We must consider the two simultaneous equilibria (11.17) and (11.10).

The ionization of water can be allowed for as follows. Let m be the stoichiometric HOI molality, which is 1.0×10^{-4} mol/kg in this problem. As above, we can approximate $m(HOI)$ as m and $\gamma_\pm$ as 1. Equation (11.18) becomes $K_a = m(H_3O^+)m(OI^-)/m$. The electroneutrality condition is

$$m(H_3O^+) = m(OH^-) + m(OI^-) = K_w/m(H_3O^+) + m(OI^-)$$

so $m(OI^-) = m(H_3O^+) - K_w/m(H_3O^+)$. Substitution of this $m(OI^-)$ expression into $K_a = m(H_3O^+)m(OI^-)/m$ gives $K_a = m(H_3O^+)^2/m - K_w/m$, so $m(H_3O^+) = (K_w + mK_a)^{1/2}$. Substitution of numerical values gives $m(H_3O^+) = 1.1 \times 10^{-7}$ mol/kg. The solution is slightly acidic, as expected.

A systematic procedure for dealing with simultaneous equilibria is outlined in Prob. 11.19.

Exercise

At what stoichiometric molality will a 25°C aqueous HOI solution have $m(H_3O^+) = 2.0 \times 10^{-7}$ mol/kg? (*Answer:* 0.0013 mol/kg.)

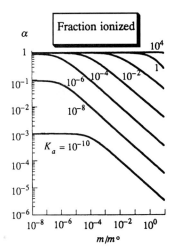

For an aqueous solution of the weak acid HX with stoichiometric molality m, the *degree of dissociation α* is defined as

$$\alpha \equiv \frac{m(X^-)}{m} = \frac{m(X^-)}{m(X^-) + m(HX)} = \frac{1}{1 + m(HX)/m(X^-)} = \frac{1}{1 + \gamma_\pm^2 m(H_3O^+)/K_a}$$

where (11.15) was used. As m goes to zero, $\gamma_\pm$ goes to 1. Also, as m goes to zero, the contribution of the ionization of HX to $m(H_3O^+)$ becomes negligible and all the H_3O^+ comes from the ionization of water. Hence in the limit of infinite dilution, $m(H_3O^+)$ becomes $K_w^{1/2}$ and the degree of dissociation of HX approaches

$$\alpha^\infty = \frac{1}{1 + K_w^{1/2}/K_a} = \frac{1}{1 + (10^{-7} \text{ mol kg}^{-1})/K_a} \qquad \text{at 25°C in } H_2O \qquad (11.19)$$

At 25°C, an acid with $K_a = 10^{-5}$ mol/kg is 99% dissociated at infinite dilution. However, an acid with $K_a = 10^{-7}$ mol/kg is only 50% dissociated at infinite dilution. The H_3O^+ from water partially suppresses the ionization of the weak acid at infinite dilution.

Figure 11.3 plots α and $m(H_3O^+)$ versus stoichiometric molality m of the acid HX in water at 25°C for several K_a values.

Other types of aqueous ionic equilibria include reactions of cationic and anionic acids and bases (for example, NH_4^+, $C_2H_3O_2^-$, CO_3^{2-}) with water (Prob. 11.13); solubility

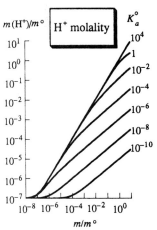

Figure 11.3

Degree of dissociation and H^+ molality versus acid HX stoichiometric molality in aqueous solutions at 25°C for several K_a° values. The Davies equation was used to estimate $\gamma_\pm$.

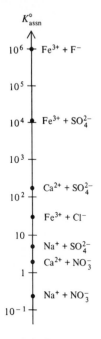

Figure 11.4

Equilibrium constants for
association of ions to form ion
pairs in water at 25°C and 1 atm.
The scale is logarithmic.

equilibria (Sec. 11.4); association equilibria involving complex ions [the equilibrium constants for the reactions $Ag^+ + NH_3 \rightleftharpoons Ag(NH_3)^+$ and $Ag(NH_3)^+ + NH_3 \rightleftharpoons Ag(NH_3)_2^+$ are called *association* or *stability* constants, and those for the reverse reactions are called *dissociation* constants]; association equilibria to form ion pairs (Sec. 10.9). See Fig. 11.4. Instead of listing the equilibrium constant K, tables often give pK, where p$K \equiv -\log_{10} K$.

In first-year-chemistry equilibrium calculations, molar concentrations (and not molalities) are used. It turns out (Prob. 11.21) that, in dilute aqueous solutions, the molality-scale and the concentration-scale equilibrium constants are nearly equal numerically, and the equilibrium molar concentrations and molalities are nearly equal numerically. It therefore makes little difference whether the molar-concentration scale or the molality scale is used for dilute aqueous solutions.

11.4 REACTION EQUILIBRIA INVOLVING PURE SOLIDS OR PURE LIQUIDS

So far in this chapter, we have considered only reactions that occur in a single phase. However, many reactions involve one or more pure solids or pure liquids. An example is $CaCO_3(s) \rightleftharpoons CaO(s) + CO_2(g)$. The equilibrium condition $\Sigma_i \nu_i \mu_{i,\text{eq}} = 0$ applies whether or not all species are in the same phase. To apply the equilibrium relation $K° = \Pi_i (a_{i,\text{eq}})^{\nu_i}$, we want an expression for the activity of a pure solid or liquid. The activity a_i satisfies $\mu_i = \mu_i° + RT \ln a_i$ [Eq. (11.2)], so

$$RT \ln a_i = \mu_i - \mu_i° \qquad \text{pure sol. or liq.} \qquad (11.20)$$

As in Sec. 5.1, we choose the standard state of a pure solid or liquid to be the state with $P = 1$ bar $\equiv P°$ and T equal to the temperature of the reaction mixture. Therefore $\mu_i°$ is a function of T only. To find $\ln a_i$ in (11.20), we need $\mu_i - \mu_i°$. For a pure substance, $\mu_i - \mu_i° = \mu_i^*(T, P) - \mu_i^*(T, P°)$, since the standard state is at the same temperature as the system. The pressure dependence of μ for a pure substance is found from $d\mu = dG_m = -S_m dT + V_m dP$ as $d\mu_i = V_{m,i} dP$ at constant T. Integration from the standard pressure $P°$ to an arbitrary pressure P gives

$$\mu_i(T, P) - \mu_i°(T) = \int_{P°}^{P} V_{m,i} dP' \qquad \text{const. } T, \text{ pure sol. or liq.} \qquad (11.21)$$

where the prime was added to the dummy integration variable to avoid the use of P with two different meanings in the same equation. Substitution of (11.21) into (11.20) gives

$$\ln a_i = \frac{1}{RT} \int_{P°}^{P} V_{m,i} dP' \qquad \text{pure sol. or liq., } T \text{ const.} \qquad (11.22)$$

where $V_{m,i}$ is the molar volume of pure i. Since solids and liquids are rather incompressible, it is a good approximation to take $V_{m,i}$ as independent of P and remove it from the integral to give

$$\ln a_i \approx (P - P°)V_{m,i}/RT \qquad \text{pure sol. or liq.} \qquad (11.23)$$

At the standard pressure of 1 bar, the activity of a pure solid or liquid is 1 (since the substance is in its standard state). G is relatively insensitive to pressure for condensed phases (Sec. 4.5). Hence we expect a_i to be rather insensitive to pressure for solids and liquids. For example, a solid with molecular weight 200 and density 2.00 g/cm³ has $V_{m,i} = 100$ cm³/mol. From (11.23), we find at $P = 20$ bar and $T = 300$ K that $a_i = 1.08$, which is pretty close to 1. Provided P remains below, say, 20 bar, we can approximate the activity of most pure solids and liquids as 1. This approximation is not valid for a substance with a large V_m, such as a polymer.

As an example, consider the equilibrium

$$CaCO_3(s) \rightleftharpoons CaO(s) + CO_2(g) \tag{11.24}$$

Equation (11.6) gives $K° = a[CaO(s)]a[CO_2(g)]/a[CaCO_3(s)]$. If P is not high, we can take the activity of each solid as 1 and the activity of the gas (assumed ideal) as $P(CO_2)/P°$ [Eq. (10.96) with fugacity replaced by pressure]. Therefore

$$K° \approx a[CO_2(g)] \approx P(CO_2)/P° \qquad \text{where } P° \equiv 1 \text{ bar} \tag{11.25}$$

Thus, at a given T, the CO_2 pressure above $CaCO_3(s)$ is constant. Note, however, that in the calculation of $\Delta G°$ using $\Delta G° = \Sigma_i \nu_i \mu_i°$, it would be wrong to omit G_m for the solids $CaCO_3$ and CaO. The fact that a for each solid is nearly 1 means that $\mu - \mu°$ ($= RT \ln a$) of each solid is nearly 0. However, $\mu°$ for each solid is nowhere near zero and must be included in calculating $\Delta G°$.

Now consider the equilibrium between a solid salt $M_{\nu_+} X_{\nu_-}$ and a saturated aqueous solution of the salt. The reaction is

$$M_{\nu_+} X_{\nu_-}(s) \rightleftharpoons \nu_+ M^{z+}(aq) + \nu_- X^{z-}(aq) \tag{11.26}$$

where z_+ and z_- are the charges on the ions and ν_+ and ν_- are the numbers of positive and negative ions. Choosing the molality scale for the solute species, we have as the equilibrium constant for (11.26)

$$K° = \frac{(a_+)^{\nu_+}(a_-)^{\nu_-}}{a[M_{\nu_+} X_{\nu_-}(s)]} = \frac{(\gamma_+ m_+/m°)^{\nu_+}(\gamma_- m_-/m°)^{\nu_-}}{a[M_{\nu_+} X_{\nu_-}(s)]}$$

where a_+, γ_+, and m_+ are the activity, molality-scale activity coefficient, and molality of the ion $M^{z+}(aq)$. Provided the system is not at high pressure, we can take $a = 1$ for the pure solid salt. Dropping $m°$ from $K°$ and using $(\gamma_\pm)^{\nu_+ + \nu_-} \equiv (\gamma_+)^{\nu_+}(\gamma_-)^{\nu_-}$ [Eq. (10.43)], we have as the **solubility product** (sp) equilibrium constant

$$K_{sp} = (\gamma_\pm)^{\nu_+ + \nu_-}(m_+)^{\nu_+}(m_-)^{\nu_-} \tag{11.27}*$$

Equation (11.27) is valid for any salt, but its main application is to salts only slightly soluble in water. For a highly soluble salt, the ionic strength of a saturated solution is high, the mean ionic activity coefficient $\gamma_\pm$ differs substantially from 1, and its value may not be accurately known. Moreover, ion-pair formation may be substantial in concentrated solutions of a salt.

EXAMPLE 11.4 Solubility-product equilibria

K_{sp} for AgCl in water is 1.78×10^{-10} mol^2/kg^2 at 25°C. Find the solubility of AgCl at 25°C in (a) pure water; (b) a 0.100 mol/kg $KNO_3(aq)$ solution; (c) a 0.100 mol/kg KCl(aq) solution.

(a) For AgCl(s) $\rightleftharpoons$ Ag$^+$(aq) + Cl$^-$(aq), we have

$$K_{sp} = \gamma_\pm^2 m(Ag^+)m(Cl^-)$$

Because of the very small value of K_{sp}, the ionic strength of a saturated AgCl solution is extremely low and $\gamma_\pm$ can be taken as 1. Since $m(Ag^+) = m(Cl^-)$ in a solution containing only dissolved AgCl, we have 1.78×10^{-10} mol^2/kg^2 = $[m(Ag^+)]^2$. Hence $m(Ag^+) = 1.33 \times 10^{-5}$ mol/kg. The solubility of AgCl in pure water at 25°C is 1.33×10^{-5} mole per kilogram of solvent.

(b) The ionic strength of a 0.100 mol/kg KNO_3 solution is 0.100 mol/kg. The Davies equation (10.71) gives $\gamma_\pm = 0.78$. Setting $m(Ag^+) = m(Cl^-)$, we have 1.78×10^{-10} mol^2/kg^2 = $(0.78)^2[m(Ag^+)]^2$. Hence $m(Ag^+) = 1.71 \times 10^{-5}$

mol/kg. Note the 29% increase in solubility compared with that in pure water. The added KNO_3 reduces $\gamma_\pm$ and increases the solubility, a phenomenon called the *salt effect*.

(c) In 0.100 mol/kg KCl, the ionic strength is 0.100 mol/kg and the Davies equation gives $\gamma_\pm = 0.78$. The Cl^- from AgCl is negligible compared with that from the KCl. Setting $m(Cl^-) = 0.100$ mol/kg, we have

$$1.78 \times 10^{-10}\ mol^2/kg^2 = (0.78)^2 m(Ag^+)(0.100\ mol/kg)$$

Therefore $m(Ag^+) = 2.9 \times 10^{-9}$ mol/kg. Note the sharp decrease in solubility compared with either pure water or the KNO_3 solution (common-ion effect).

Exercise

Find the solubility of AgCl in 0.0200 mol/kg $Ag_2SO_4(aq)$ at 25°C. Neglect ion pairing. (*Answer:* 6.77×10^{-9} mol/kg.)

In Example 11.4, we ignored the possibility of ion-pair formation and assumed that all the silver chloride in solution existed as Ag^+ and Cl^- ions. This is a good assumption for dilute solutions of a 1:1 electrolyte. However, in working with K_{sp} for other than 1:1 electrolytes, substantial error can frequently result if ion-pair formation is not taken into account; see Prob. 11.28 and L. Meites, J. S. F. Pode, and H. C. Thomas, *J. Chem. Educ.*, **43**, 667 (1966).

Although ion-pair formation can be neglected in Example 11.4, complex-ion formation often cannot be neglected in AgCl solutions. The ions Ag^+ and Cl^- react in aqueous solution to form a series of four complex ions: $Ag^+ + Cl^- \rightleftharpoons AgCl(aq)$, $AgCl(aq) + Cl^- \rightleftharpoons AgCl_2^-$, $AgCl_2^- + Cl^- \rightleftharpoons AgCl_3^{2-}$, $AgCl_3^{2-} + Cl^- \rightleftharpoons AgCl_4^{3-}$. Inclusion of complex-ion formation shows that, although the results for (a) and (b) in the above example are correct, the result for (c) is in error.

For a homogeneous reaction such as $N_2(g) + 3H_2(g) \rightleftharpoons 2NH_3(g)$ or $HCN(aq) + H_2O \rightleftharpoons H_3O^+(aq) + CN^-(aq)$, there will always be some of each species present at equilibrium. In contrast, reactions involving pure solids have the possibility of going to completion. For example, for $CaCO_3(s) \rightleftharpoons CaO(s) + CO_2(g)$, $K° = P(CO_2)/P°$ [Eq. (11.25)]. At 800°C, $K° = 0.24$ for this reaction. If we place $CaCO_3(s)$ into an evacuated container at 800°C, the $CaCO_3$ will decompose until $P(CO_2)$ reaches 0.24 bar. If the container volume is large enough, all the $CaCO_3$ may decompose before this equilibrium pressure can be attained. Similarly, if a crystal of AgCl is added to a large enough volume of water, all the AgCl can dissolve without having $\gamma_\pm^2 m(Ag^+)m(Cl^-)$ reach K_{sp}.

EXAMPLE 11.5 Calculation of K_{sp}

The $\Delta_f G°_{298}$ values for $Ag_2SO_4(s)$, $Ag^+(aq)$, and $SO_4^{2-}(aq)$ are −618.41, 77.11, and −744.53 kJ/mol, respectively. Find K_{sp} for Ag_2SO_4 in water at 25°C.

The reaction is $Ag_2SO_4(s) \rightleftharpoons 2Ag^+(aq) + SO_4^{2-}(aq)$. We calculate $\Delta G°_{298} = 28.10$ kJ/mol. Use of $\Delta G° = -RT \ln K°$ gives

$$K_{sp}° = 1.2 \times 10^{-5} \quad and \quad K_{sp} = 1.2 \times 10^{-5}\ mol^3/kg^3$$

Exercise

$\Delta_f G°_{298}$ values for $K^+(aq)$, $Cl^-(aq)$, and KCl(s) are −283.27, −131.228, and −409.14 kJ/mol, respectively. Find K_{sp} for KCl in water at 25°C. (*Answer:* 8.68 mol^2/kg^2.)

11.5　REACTION EQUILIBRIUM IN NONIDEAL GAS MIXTURES

The activity a_i of component i of a nonideal gas mixture is [Eqs. (10.96) and (10.99)]

$$a_i = f_i/P° = \phi_i P_i/P° = \phi_i x_i P/P° \qquad \text{where } P° \equiv 1 \text{ bar} \qquad (11.28)$$

where f_i, ϕ_i, P_i, and x_i are the fugacity, fugacity coefficient, partial pressure, and mole fraction of gas i, and P is the pressure of the mixture. Substitution into $K° = \Pi_i (a_i)^{\nu_i}$ [Eq. (11.6)] gives at equilibrium in a gas-phase reaction with stoichiometric coefficients ν_i

$$K° = \prod_i \left(\frac{f_i}{P°} \right)^{\nu_i} = \prod_i \left(\frac{\phi_i x_i P}{P°} \right)^{\nu_i} \qquad (11.29)$$

The standard state for each gas has the pressure fixed at 1 bar, so $\Delta G°$ depends only on T. Hence the equilibrium constant $K°$, which equals $\exp(-\Delta G°/RT)$ [Eq. (11.4)], depends only on T. Using the identity $\Pi_i (a_i b_i) = \Pi_i a_i \Pi_i b_i$, we rewrite (11.29) as

$$\frac{K°}{\prod_i (\phi_i)^{\nu_i}} = \prod_i \left(\frac{x_i P}{P°} \right)^{\nu_i} \qquad (11.30)$$

To calculate the equilibrium composition at a given T and P of a reacting nonideal gas mixture, the following approximate procedure is often used. Tables of $\Delta_f G_T°$ for the reacting gases are used to calculate $\Delta G_T°$ for the reaction. The equilibrium constant $K°$ is then calculated from $\Delta G_T°$. The fugacity coefficients $\phi_i^*(T, P)$ of the pure gases are found using either law-of-corresponding-states charts of ϕ_i^* as a function of reduced temperature and pressure (Sec. 10.11) or tabulations of $\phi_i^*(T, P)$ for the individual gases. The Lewis–Randall rule $\phi_i \approx \phi_i^*(T, P)$ (Sec. 10.11) is then used to estimate ϕ_i for each gas in the mixture. The quantity on the left side of (11.30) is calculated, and (11.30) is then used to find the equilibrium composition by the procedures of Sec. 6.4.

A better, but more complicated, procedure is to use an equation of state for the mixture. One initially sets all the ϕ_i's equal to 1 and solves (11.30) for the initial estimate of the equilibrium composition. One uses the mixture's equation of state to calculate each ϕ_i from Eq. (10.101) at this composition. These ϕ_i's are used in (11.30) to solve for an improved estimate of the equilibrium composition, which is then used with the equation of state to find improved ϕ_i's; and so on. One continues until no further change in composition is found. For an example, see H. F. Gibbard and M. R. Emptage, *J. Chem. Educ.,* **53,** 218 (1976).

11.6　COMPUTER PROGRAMS FOR EQUILIBRIUM CALCULATIONS

A natural-water system such as a lake or stream might contain one or two dozen dissolved chemicals, which can react with one another to form hundreds of possible dissolved species or solid precipitates. To deal with such a complex system, a computer program is essential. There are two common methods of computer solution of multiple-equilibria problems at constant T and P. One approach uses equilibrium constants and finds the species amounts that satisfy the equilibrium-constant expressions and the stoichiometry (conservation of matter) requirements. An alternative approach writes G of each phase as $G = \Sigma_i n_i \mu_i$, where each μ_i is expressed as a function of composition. One then minimizes G of the system by varying the composition subject to the stoichiometry requirements (see W. R. Smith and R. W. Missen, *Chemical Reaction Equilibrium Analysis,* Krieger, 1991, for details). Most of the following programs have a built-in database of free-energy data and parameters for estimating ionic activity coefficients.

The programs MINTEQA2 (www.scisoftware.com and www.cosmiclink.co.za/iscp/models/geochem/minteqa2/minteqa2.htm), MINEQL+ (www.mineql.com/), and PHREEQC (water.usgs.gov/software/phreeqc.html) are computationally similar programs for calculations on natural and laboratory aqueous systems of ionic strengths less than 0.5 mol/kg. Ion activity coefficients are estimated from the Debye–Hückel formula (10.60) with a term proportional to I_m added, using parameters fitted for each species. If these parameters are not available, the Davies equation is used. Activity coefficients for uncharged solute species are estimated from the formula

$$\log_{10} \gamma_i = 0.1 \, (I_m/m^\circ)$$

The experimentally determined activity coefficients of several uncharged solutes in aqueous solution have been found to fit the equation $\log_{10} \gamma_i = b_i(I_m/m^\circ)$, where b_i differs for different species. For several species in water at 25°C, b_i is roughly 0.1. MINTEQA2 and PHREEQC are DOS programs and can be downloaded at no cost (follow the links at www.cosmiclink.co.za/iscp/models/geochem/geochem.htm). MINEQL+ has DOS and Windows versions.

The program EQS4WIN (www.mathtrek.com) solves multiple-equilibria problems that can involve several phases but no activity or fugacity coefficients are used, the phases being assumed to be ideal. A free demonstration version is available.

ChemSage [www.esm-software.com/chemsage/; G. Eriksson and K. Hack, *Metallurg. Trans. B,* **21B,** 1013 (1990)] is a general-purpose DOS program that does nonideal chemical equilibrium calculations in multiphase systems at temperatures up to 6000 K and pressures to 1 Mbar. For nonideal gases, fugacity coefficients are found using the virial equation. For aqueous electrolyte solutions, the Pitzer equations are used. For nonelectrolyte solutions, the user can choose from several models for estimating activity coefficients. For this powerful program, you pay a powerful price ($2495 or $1295 academic price in 2001). FACT-Win (www.crct.polymtl.ca/FACT/FACTWin/FactWin.htm) is a Windows program that uses ChemSage to calculate equilibria.

HSC Chemistry (www.esm-software.com/hsc/) is a Windows program for multiphase chemical-equilibrium calculations. Gases are assumed ideal and the user must specify activity coefficients for species in solution.

You can use the Internet to do chemical equilibrium calculations at no charge. The program TCC (blue.caltech.edu/tcc) does ideal-gas equilibrium calculations. The user selects the desired compounds to include from a database of several hundred gases and selects T and P and the initial composition. The program then considers all possible reactions involving the selected compounds and finds the equilibrium composition. EQUILIB-Web and AQUALIB-Web (www.crct.polymtl.ca/FACT/web/factweb.htm) do equilibrium calculations involving ideal gases or dilute aqueous solutions and are limited to no more than three reactants. INFRA (www.com2com.ru/~maikov/) does ideal-gas and real-gas calculations on mixtures containing up to 20 species. Fugacity coefficients are found from the Peng–Robinson equation (Prob. 8.16).

11.7 TEMPERATURE AND PRESSURE DEPENDENCES OF THE EQUILIBRIUM CONSTANT

From $\Delta G^\circ = -RT \ln K^\circ$ [Eq. (11.4)], we have

$$\ln K^\circ = -\Delta G^\circ/RT \tag{11.31}$$

where $\Delta G^\circ \equiv \Sigma_i \, \nu_i \mu_i^\circ$ is the standard change in Gibbs energy (all species in their standard states). For gases and for pure liquids and solids, we chose a fixed-pressure standard state ($P^\circ = 1$ bar), so that here ΔG° and hence K° are independent of pressure and depend only on T. For liquid and solid solutions, we chose a variable-pressure standard state, with standard-state pressure equal to the actual pressure of the solution, so that here $\Delta G°$ and $K°$ are functions of both T and P.

Differentiation of (11.31) with respect to T gives

$$\left(\frac{\partial \ln K°}{\partial T}\right)_P = \frac{\Delta G°}{RT^2} - \frac{(\partial \Delta G°/\partial T)_P}{RT} = \frac{\Delta G°}{RT^2} + \frac{\Delta S°}{RT} = \frac{\Delta G° + T\,\Delta S°}{RT^2}$$

$$\left(\frac{\partial \ln K°}{\partial T}\right)_P = \frac{\Delta H°}{RT^2} \qquad\qquad \textbf{(11.32)}*$$

In deriving (11.32), we used $(\partial \Delta G°/\partial T)_P = (\partial/\partial T)_P \sum_i \nu_i \mu_i° = \sum_i \nu_i (\partial \mu_i°/\partial T)_P = -\sum_i \nu_i \bar{S}_i° = -\Delta S°$, since $(\partial \mu_i/\partial T)_P = -\bar{S}_i$ [Eq. (9.30)]. When liquid or solid solutions are not involved, the partial derivative in (11.32) becomes an ordinary derivative. $\Delta H°$ is equal to $\sum_i \nu_i \bar{H}_i°$, where the ν_i's are the stoichiometric coefficients and the $\bar{H}_i°$'s are the standard-state molar or partial molar enthalpies. For the application of (11.32) to reactions in solution, see Prob. 11.41.

Figure 11.5 plots the molality-scale ionization constant $K_w°$ of saturated liquid water (water in equilibrium with water vapor) versus temperature. The pressure is not constant for this plot, but below 250°C the effect of the pressure variation on $K_w°$ is slight. At the 220°C maximum in $K_w°$, $\partial \ln K_w°/\partial T$ is zero and $\Delta H°$ of ionization is zero [Eq. (11.32)]. The strong temperature dependence of $\Delta H°$ of water ionization (which goes from $+60$ kJ/mol at 0°C to -100 kJ/mol at 300°C) is an example of the fact that for many ionic reactions in aqueous solution, $\Delta H°$ depends strongly on T (in contrast to gas-phase reactions, where $\Delta H°$ usually varies quite slowly with T).

Consider a reaction in which all the reactants and products are in a liquid or solid solution. Differentiation of (11.31) with respect to P gives

$$\left(\frac{\partial \ln K°}{\partial P}\right)_T = -\frac{1}{RT}\left(\frac{\partial \Delta G°}{\partial P}\right)_T = -\frac{1}{RT}\left(\frac{\partial}{\partial P}\right)_T \sum_i \nu_i \mu_i° = -\frac{\Delta V°}{RT}$$

where $(\partial \mu_i/\partial P)_T = \bar{V}_i$ [Eq. (9.31)] was used. If a reaction involves species in a liquid or solid solution and species not in a liquid or solid solution (for example, a solubility product), then in calculating $\Delta V°$ we consider only species in the solution. Species not in solution have pressure-independent standard states and make no contribution to $\partial \Delta G°/\partial P$. (However, we must allow for the effect of pressure on the activity of such species in $K°$.) Therefore for any reaction

$$\left(\frac{\partial \ln K°}{\partial P}\right)_T = -\frac{\Delta V°_{\text{soln}}}{RT} \qquad\qquad (11.33)$$

where the subscript is a reminder to include only species in solution in calculating $\Delta V°_{\text{soln}}$. Usually $\Delta V°_{\text{soln}}$ is small, and the pressure dependence of $K°$ is slight unless high pressures are involved.

Figure 11.6 plots the ionization constant $K_w°$ [Eq. (11.12)] for water at 25°C as a function of pressure. An increase in P from 1 to 200 bar increases $K_w°$ by 18%, and an increase from 1 to 1000 bar roughly doubles $K_w°$.

Although $K°$ depends only weakly on pressure, it usually depends strongly on temperature, since $\Delta H°$ in (11.32) is usually large. For example, the reaction $N_2(g) + 3H_2(g) \rightleftharpoons 2NH_3(g)$ has $\Delta H° \approx -25$ kcal/mol, and its equilibrium constant $K°$ decreases from 3×10^{13} at 200 K to 3×10^{-7} at 1000 K (Fig. 6.6).

Another example is the denaturation (unfolding) of a protein. A protein molecule is a long-chain polymer of amino acids. Enzymes are globular proteins. In a globular protein, certain portions of the chain are coiled into helical segments that are stabilized by hydrogen bonds between one turn of a helix and the next. The partly coiled protein folds on itself to give a roughly ellipsoidal overall shape. The folding is not random,

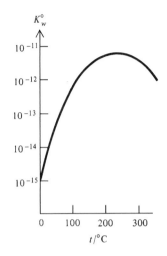

Figure 11.5

Ionization constant $K_w° = \gamma_\pm^2 m_+ m_- /(m°)^2$ for saturated liquid water versus temperature. The vertical scale is logarithmic. [Data from H. L. Clever, *J. Chem. Educ.*, **45**, 231 (1968).]

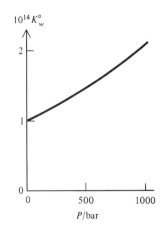

Figure 11.6

Water ionization constant $K_w°$ versus pressure at 25°C. [Data from D. A. Lown et al., *Trans. Faraday Soc.*, **64**, 2073 (1968).]

Figure 11.7

Typical temperature dependences
of thermodynamic quantities for
protein denaturation in water.

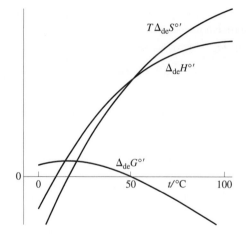

but is determined in part by hydrogen bonds, van der Waals forces (Sec. 22.10), and S—S covalent bonds between sulfur-containing amino acids. In the denaturation reaction, the protein unfolds into a random conformation, called a random coil.

The breaking of hydrogen bonds during denaturation requires energy and produces a more disordered protein structure, and therefore makes positive contributions to $\Delta_{de}H°$ and $\Delta_{de}S°$ (where de stands for denaturation). In addition, the interactions between the two forms of the protein and the solvent water lead to negative contributions to $\Delta_{de}H°$ and $\Delta_{de}S°$, which rapidly become less important as T increases. Therefore, $\Delta_{de}H°$ and $\Delta_{de}S°$ increase rapidly with increasing T (Fig. 11.7). The net result is the $\Delta_{de}G°$-versus-T curve in Fig. 11.7, which shows that denaturation occurs when T is raised. In the temperature range in which denaturation occurs, $\Delta_{de}H°$ is large (typically 200 to 600 kJ/mol), so denaturation occurs over a small range of T. For example, the digestive enzyme chymotrypsin in aqueous solution at pH 2 is 97% in its native (globular) form at 37°C and is 96% denatured at 50°C. Note from Fig. 11.7 that $\Delta_{de}H°$ and $\Delta_{de}S°$ can each be positive or negative, depending on T. The parabolic shape of the $\Delta_{de}G°$ curve indicates that $\Delta_{de}G°$ becomes negative at some temperature below 0°C; in fact, denaturation of proteins in supercooled water has been observed (P. G. Debenedetti, *Metastable Liquids,* Princeton, 1996, sec. 1.2.2).

11.8   SUMMARY OF STANDARD STATES

The equilibrium constant for a reaction is $K° = \Pi_i\,(a_{i,eq})^{\nu_i}$ [Eq. (11.6)]. The activity of species i is $a_i = \exp[(\mu_i - \mu_i°)/RT]$ [Eq. (11.1)], where $\mu_i°$ is the standard-state chemical potential of i. The choice of standard state therefore determines a_i and determines the form of the equilibrium constant.

Table 11.1 summarizes the choices of standard states made in earlier sections and lists the forms of the chemical potentials.

11.9   GIBBS ENERGY CHANGE FOR A REACTION

The term *Gibbs energy change for a reaction* has at least three different meanings, which we now discuss.

1. $\Delta G°$.   The standard molar Gibbs energy change $\Delta G°$ for a reaction is defined by (11.5) as $\Delta G° \equiv \Sigma_i\, \nu_i\mu_i°$, where $\mu_i°$ is the value of the chemical potential of substance i in its standard state. Since $\mu_i°$ is an intensive quantity and ν_i is a dimensionless number, $\Delta G°$ is an intensive quantity with units J/mol or cal/mol. For a

TABLE 11.1

Summary of Standard States and Chemical Potentials[a]

Substance	Standard state	$\mu_i = \mu_i^\circ + RT \ln a_i$
Gas (pure or in gas mixture)	Pure ideal gas at 1 bar and T	$\mu_i = \mu_i^\circ(T) + RT \ln (f_i/P^\circ)$
Pure liquid or pure solid	Pure substance at 1 bar and T	$\mu_i = \mu_i^\circ(T) + \int_{P^\circ}^{P} V_{m,i} \, dP'$
Solution component, Convention I	Pure i at T and P of solution	$\mu_i = \mu_i^*(T, P) + RT \ln (\gamma_{I,i} x_i)$
Solvent A	Pure A at T and P of solution	$\mu_A = \mu_A^*(T, P) + RT \ln (\gamma_A x_A)$
Nonelectrolyte solute:		
Convention II	Fictitious state with $x_i = 1 = \gamma_{II,i}$	$\mu_i = \mu_{II,i}^\circ(T, P) + RT \ln (\gamma_{II,i} x_i)$
molality scale	Fictitious state with $m_i/m^\circ = 1 = \gamma_{m,i}$	$\mu_i = \mu_{m,i}^\circ(T, P) + RT \ln (\gamma_{m,i} m_i/m^\circ)$
concentration scale	Fictitious state with $c_i/c^\circ = 1 = \gamma_{c,i}$	$\mu_i = \mu_{c,i}^\circ(T, P) + RT \ln (\gamma_{c,i} c_i/c^\circ)$
Electrolyte solute:		
molality scale	Fictitious state with $\gamma_\pm = 1 = \nu_+ m_i/m^\circ$	$\mu_i = \mu_i^\circ(T, P) + RT \ln (\nu_+ \gamma_\pm m_i/m^\circ)^\nu$

[a]Limiting behaviors: $\phi_i \to 1$ as $P_i \to 0$, where $f_i = \phi_i x_i P$; $\gamma_{I,i} \to 1$ as $x_i \to 1$; $\gamma_A^\infty = 1$; $\gamma_{II,i}^\infty = 1$; $\gamma_{m,i}^\infty = 1$; $\gamma_{c,i}^\infty = 1$; $\gamma_\pm^\infty = 1$.

gas-phase reaction, the standard state of each gas is the hypothetical pure ideal gas at 1 bar. For a reaction in liquid solution with use of the molality scale, the standard state of each nonelectrolyte solute is the hypothetical state with $m_i = 1$ mol/kg and $\gamma_{m,i} = 1$. These standard states do not correspond to the states of the reactants in the reaction mixture. Therefore, ΔG° (and ΔH°, ΔS°, etc.) refer not to the actual change in the reaction mixture but to a hypothetical change from standard states of the separated reactants to standard states of the separated products.

2. $(\partial G/\partial \xi)_{T,P}$. Equation (4.99) reads $dG/d\xi = \Sigma_i \nu_i \mu_i$ at constant T and P, where ξ is the extent of reaction, the μ_i's are the actual chemical potentials in the reaction mixture at some particular value of ξ, and dG is the infinitesimal change in Gibbs energy of the reaction mixture due to a change in the extent of reaction from ξ to $\xi + d\xi$:

$$\left(\frac{\partial G}{\partial \xi}\right)_{T,P} = \sum_i \nu_i \mu_i \qquad (11.34)$$

The sum on the right side of (11.34) is frequently denoted by $\Delta_r G$ or ΔG, but this notation is misleading in that $\Sigma_i \nu_i \mu_i$ is not the change in G of the system as the reaction occurs but is the instantaneous rate of change in G with respect to ξ. If the reaction mixture were of infinite mass, so that a finite change in ξ would not change the μ_i's in the mixture, then $\Sigma_i \nu_i \mu_i \times 1$ mol would be ΔG for a change $\Delta \xi = 1$ mol. Note that $(\partial G/\partial \xi)_{T,P}$ is the slope of the G-versus-ξ curve (Fig. 4.9).

3. ΔG. From Eq. (9.23), the Gibbs energy G of a homogeneous reaction mixture at a given instant is equal to $\Sigma_i n_i \mu_i$, where n_i (not to be confused with the stoichiometric coefficient ν_i) is the number of moles of i in the mixture and μ_i is its chemical potential in the mixture. If at times t_1 and t_2 these quantities are $n_{i,1}$, $\mu_{i,1}$ and $n_{i,2}$, $\mu_{i,2}$, respectively, then the actual change ΔG in the Gibbs energy of the reacting system from time t_1 to time t_2 is $\Delta G = \Sigma_i n_{i,2} \mu_{i,2} - \Sigma_i n_{i,1} \mu_{i,1}$.

The quantities $\Delta G°$ and $(\partial G/\partial \xi)_{T,P}$ are related. Substitution of $\mu_i = \mu_i° + RT \ln a_i$ [Eq. (11.2)] into (11.34) gives

$$\left(\frac{\partial G}{\partial \xi}\right)_{T,P} = \sum_i \nu_i \mu_i° + RT \sum_i \nu_i \ln a_i$$

Using $\Delta G° \equiv \sum_i \nu_i \mu_i°$ [Eq. (11.5)] and $\sum_i \nu_i \ln a_i = \sum_i (\ln a_i)^{\nu_i} = \ln \prod_i (a_i)^{\nu_i}$ [Eqs. (1.68) and (1.67)], we get

$$\left(\frac{\partial G}{\partial \xi}\right)_{T,P} = \Delta G° + RT \ln Q, \qquad Q \equiv \prod_i (a_i)^{\nu_i} \qquad (11.35)$$

where Q is the **reaction quotient** (first used in Sec. 6.4). Since $\Delta G° = -RT \ln K°$ [Eq. (11.4)], Eq. (11.35) can be written as

$$(\partial G/\partial \xi)_{T,P} = RT \ln (Q/K°) \qquad (11.36)$$

The activities of the products appear in the numerator of Q. At the start of the reaction when no products are present, $Q = 0$ and (11.36) gives $(\partial G/\partial \xi)_{T,P} = -\infty$. (Note the negatively infinite slope of the G-versus-ξ curve in Fig. 4.9 at $\xi = 0$.) Before equilibrium is reached, we have $Q < K°$ and $(\partial G/\partial \xi)_{T,P} < 0$. At equilibrium, $Q = K°$ and $(\partial G/\partial \xi)_{T,P} = 0$. For a system with $Q > K°$, (11.36) gives $(\partial G/\partial \xi)_{T,P} > 0$ and the reaction proceeds in reverse; Q decreases until at equilibrium $Q = K°$, $(\partial G/\partial \xi)_{T,P} = 0$, and G is minimized. The direction of spontaneous reaction at constant T and P is determined by the sign of $(\partial G/\partial \xi)_{T,P}$.

A given reaction at fixed T and P has a single value of $\Delta G°$, but $(\partial G/\partial \xi)_{T,P}$ in the system can have any value from $-\infty$ to $+\infty$.

Instead of $\Delta G°$, biochemists often use the quantity $\Delta G°'$, defined as

$$\Delta G°' \equiv \sum_{i \neq H^+} \nu_i \mu_i° + \nu(H^+) \mu°'(H^+) \qquad (11.37)$$

where $\mu°'(H^+)$ is the chemical potential of H^+ at an H^+ activity of 10^{-7}. Because biological fluids have H^+ molalities close to 10^{-7} mol/kg, $\Delta G°'$ values are more relevant to reactions in living organisms than are $\Delta G°$ values, which have a 1-mol/kg standard-state molality.

11.10 COUPLED REACTIONS

Suppose two chemical reactions are occurring in a system, and there is a chemical species that takes part in both reactions. The reactions are then *coupled*, since one reaction will influence the equilibrium position of the second reaction. Thus, suppose species M is a product of reaction 1 and is a reactant in reaction 2:

Reaction 1: $\qquad\qquad\qquad\qquad$ A + B $\rightleftharpoons$ M + D

Reaction 2: $\qquad\qquad\qquad\qquad$ M + R $\rightleftharpoons$ S + T

If reaction 1 has $\Delta G° \gg 0$ and equilibrium constant $K \ll 1$, then at ordinary concentrations very little A and B will react to produce M and D in the absence of reaction 2. If reaction 2 has $\Delta G° \ll 0$ and $K \gg 1$, then if both reactions occur, reaction 2 will use up large amounts of M, thereby allowing reaction 1 to proceed to a substantial degree.

An example of coupled reactions is the observation that most phosphate salts are virtually insoluble in water but quite soluble in aqueous solutions of strong acids. In acidic solutions, the rather strong Brønsted base PO_4^{3-} reacts to a very great extent with H_3O^+ to yield the extremely weak acid HPO_4^{2-} (which reacts further with H_3O^+

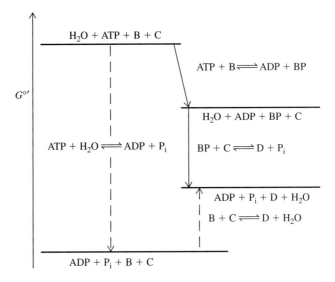

Figure 11.8

Biochemical coupling.

Figure 11.9

$G^{\circ\prime}$ for species involved in the coupled reactions (11.38).

to yield $H_2PO_4^-$), thereby greatly reducing the PO_4^{3-} concentration in solution and shifting the solubility-product equilibrium (11.26) of the phosphate salt to the right.

A different kind of reaction coupling is important in biology. The hydrolysis of adenosine triphosphate (ATP) to adenosine diphosphate (ADP) plus inorganic phosphate (P_i) has $\Delta G^\circ < 0$ and is thermodynamically favored. In living organisms, this hydrolysis is coupled to such thermodynamically unfavored processes as the synthesis of large biochemicals (for example, amino acids, proteins, RNA, and DNA) from small molecules, the transport of chemical species from regions of low to regions of high chemical potential (active transport—Sec. 12.4), and muscle contraction to perform mechanical work. The thermodynamically unfavored resynthesis of ATP from ADP is made to occur by being coupled with the oxidation of glucose, for which $\Delta G^\circ < 0$. See Fig. 11.8.

As an example of a biochemical coupling scheme, the thermodynamically unfavored synthesis $B + C \rightleftharpoons D + H_2O$ can be brought about by coupling it with ATP hydrolysis according to the enzyme-catalyzed reactions

$$B + ATP \rightleftharpoons BP + ADP$$
$$BP + C \rightleftharpoons D + P_i \tag{11.38}$$

where the intermediate species BP (phosphorylated B) is common to the two reactions. The net reaction for the sequence (11.38) is

$$B + C + ATP \rightleftharpoons D + ADP + P_i$$

which is the sum of $B + C \rightleftharpoons D + H_2O$ and $ATP + H_2O \rightleftharpoons ADP + P_i$. In this scheme, $\bar{G}^\circ$ of $BP + H_2O$ is substantially greater than $\bar{G}^\circ$ of $B + P_i$, so the equilibrium constant of $BP + C \rightleftharpoons D + P_i$ is much greater than that of the uncoupled reaction $B + C \rightleftharpoons D + H_2O$. See Fig. 11.9. (All species in these reactions are in aqueous solution.)

Reaction equations like ATP + $H_2O \rightleftharpoons$ ADP + P_i or those in (11.38) are called *biochemical reaction equations.* In a biochemical reaction equation, the hydrogen atoms and the charges are not balanced, and symbols like ATP and P_i stand for the sum of all forms in which ATP and inorganic phosphate exist. Thus, P_i includes such species as PO_4^{3-}, HPO_4^{2-}, $H_2PO_4^-$, and $MgHPO_4$, which exist in equilibrium with one another; ATP includes such species as ATP^{4-}, $HATP^{3-}$, H_2ATP^{2-}, $MgATP^{2-}$, etc. To discuss the details of biochemical processes, one uses *chemical reaction equations,* such as $ATP^{4-} + H_2O \rightleftharpoons ADP^{3-} + HPO_4^{2-} + H^+$, in which atoms and charges are balanced. The relation between equilibrium constants for biochemical reactions and those for chemical reactions is discussed in R. A. Alberty and A. Cornish-Bowden, *Trends Biochem. Sci.,* **18,** 288 (1993); R. N. Goldberg et al., *J. Phys. Chem. Ref. Data,* **22,** 515 (1993); R. A. Alberty, *Pure Appl. Chem.,* **66,** 1641 (1994).

> One must use caution in applying thermodynamics to living organisms. Organisms and the cells that compose them are open (rather than closed) systems and are not at equilibrium. The rates of the chemical reactions may thus be more relevant than the values of the equilibrium constants. For discussion on these and related points, see B. E. C. Banks, *Chem. Brit.,* **5,** 514 (1969); L. Pauling, ibid., **6,** 468 (1970); D. Wilkie, ibid., **6,** 472; A. F. Huxley, ibid., **6,** 477; R. A. Ross and C. A. Vernon, ibid., **6,** 541; B. E. C. Banks and C. A. Vernon, *J. Theor. Biol.,* **29,** 301 (1970).

11.11   SUMMARY

The activity a_i of species i in a system is defined to satisfy $\mu_i = \mu_i^\circ + RT \ln a_i$. Substitution into the reaction-equilibrium condition $\Sigma_i \, \nu_i \mu_i = 0$ leads to $\Delta G^\circ = -RT \ln K^\circ$, which relates the standard Gibbs energy change ΔG° to the equilibrium constant $K^\circ \equiv \Pi_i \, (a_{i,\text{eq}})^{\nu_i}$. For reactions in solution, the mole-fraction scale is used for the solvent (A) and the molality scale is most commonly used for each solute (i); $a_A = \gamma_{x,A} x_A$ and $a_i = \gamma_{m,i} m_i / m^\circ$, where the γ's are activity coefficients. In dilute solutions, it is reasonable to approximate a_A as 1 and $\gamma_{m,i}$ as 1 for nonelectrolytes. For ions, the Davies equation can be used to estimate γ_m. The activity of a pure solid or pure liquid not at high pressure can be approximated as 1. Calculations for ionization of weak acids and for solubility-product equilibria of salts were discussed. For gases, $a_i = f_i / P^\circ = \phi_i x_i P / P^\circ$. The Lewis–Randall rule $\phi_i \approx \phi_i^*(T, P)$ or an equation of state can be used to estimate fugacity coefficients in gas mixtures. The temperature and pressure dependences of K° are given by (11.32) and (11.33).

Important kinds of calculations discussed in this chapter include:

- Calculation of equilibrium constants for nonideal systems from $\Delta_f G^\circ$ data using $\Delta G^\circ = -RT \ln K^\circ$.
- Calculation of equilibrium molalities in electrolyte equilibria (e.g., weak-acid ionization, solubility product) with use of the Davies equation to estimate activity coefficients.
- Calculation of nonideal-gas equilibria.
- Calculation of changes in K° with temperature and pressure changes.

FURTHER READING AND DATA SOURCES

Denbigh, secs. 4.5, 4.9, chap. 10; *McGlashan,* secs. 12.13, 12.14, 12.15, 18.10, 20.9–20.11; J. E. Ricci, *Hydrogen Ion Concentration,* Princeton, 1952; J. N. Butler and D. R. Cogley, *Ionic Equilibrium,* Wiley, 1998.

Equilibrium constants in water (including ionization constants of acids and bases, solubility products, stability constants of complex ions, and constants for ion-pair formation): A. E. Martell and R. M. Smith, *Critical Stability Constants,* vols. 1–6, Plenum, 1974–1989.

PROBLEMS

Where appropriate, use the Davies equation to estimate activity coefficients.

Section 11.1

11.1 True or false? (*a*) Activities are dimensionless. (*b*) The standard-state activity a_i° equals 1.

Section 11.3

11.2 True or false? (*a*) The product of H_3O^+ and OH^- molalities in water at 25°C and 1 bar is always 1.0×10^{-14} mol^2/kg^2. (*b*) For the weak acid HX(*aq*), $\gamma_+\gamma_- = \gamma_\pm$. (*c*) The degree of ionization of $HC_2H_3O_2(aq)$ is unchanged if some NaCl is dissolved in the solution. (*d*) In a solution prepared by adding *m* moles of $HC_2H_3O_2$ to 1 kg of water, the H^+ molality can never exceed *m* mol/kg.

11.3 In the expression (11.15) for the ionization constant K_a, $m(H_3O^+)$ includes (*a*) only the hydronium ions that come from the ionization of HX; (*b*) all the hydronium ions in the solution, no matter what their source.

11.4 For formic acid, HCOOH, $K_a = 1.80 \times 10^{-4}$ mol/kg in water at 25°C and 1 bar. (*a*) For a solution of 4.603 g of HCOOH in 500.0 g of H_2O at 25°C and 1 bar, find the H^+ molality as accurately as possible. (*b*) Repeat the calculation if 0.1000 mol of KCl is added to the solution of (*a*). (*c*) Find the H^+ molality in a solution at 25°C and 1 bar prepared by adding 0.1000 mol of formic acid and 0.2000 mol of potassium formate to 500.0 g of water.

11.5 (*a*) Write a computer program to calculate $m(H^+)$ in a solution of the acid HX for a range of stoichiometric molalities from m_1 to m_2 in steps of Δm. The input is K_a°, m_1, m_2, and Δm. Use the Davies equation to estimate $\gamma_\pm$. Assume that the H^+ from H_2O ionization is negligible. Test your program by running it for $K_a^\circ = 0.01$ and compare the results with the Fig. 11.3 value $m(H^+) = 0.00173$ mol/kg at $m = 0.00200$ mol/kg. (*b*) Explain why your program will not give accurate results at extremely low molalities and at high molalities.

11.6 Given the following $\Delta_f G_{298}^\circ$/(kJ/mol) values from the NBS tables (Sec. 5.9): -27.83 for un-ionized $H_2S(aq)$, 12.08 for $HS^-(aq)$, and 85.8 for $S^{2-}(aq)$, calculate the ionization constants K_a° for the acids H_2S and HS^- in water at 25°C and 1 bar. Compare with the experimental values (Prob. 11.14). If the discrepancy surprises you, reread the next-to-last paragraph of Sec. 6.4.

11.7 Find the H^+ molality in a 1.00×10^{-5} mol/kg aqueous HCN solution at 25°C and 1 bar, given that $K_a = 6.2 \times 10^{-10}$ mol/kg for HCN at 25°C.

11.8 Calculate $m(H_3O^+)$ in a 0.20 mol/kg aqueous solution of NaCl at 25°C.

11.9 The human body is typically at 98.6°F = 37.0°C. (*a*) Use the expression given in Prob. 11.38 to calculate $m(H_3O^+)$ in pure water at 37°C. (*b*) Using only Appendix data, estimate K_w° at 37°C and compare with the value found in part (*a*). State any approximation made.

11.10 Find $m(H_3O^+)$ in a 1.00×10^{-8} mol/kg aqueous HCl solution at 25°C.

11.11 A 0.200 mol/kg solution of the acid HX is found to have $m(H_3O^+) = 1.00 \times 10^{-2}$ mol/kg. Find K_a for this acid.

11.12 Estimate $a(H_2O)$ in a 0.50 mol/kg aqueous NaCl solution; take $\gamma(H_2O) = 1$.

11.13 Find $m(H_3O^+)$ in a 0.10 mol/kg solution of $NaC_2H_3O_2$ in water at 25°C, given that $K_a = 1.75 \times 10^{-5}$ mol/kg for $HC_2H_3O_2$ at 25°C. (*Hint:* The acetate ion is a base and reacts with water as follows: $C_2H_3O_2^- + H_2O \rightleftharpoons HC_2H_3O_2 + OH^-$.) Show that the equilibrium constant for this reaction is $K_b = K_w/K_a$. Neglect the OH^- coming from the ionization of water.

11.14 For H_2S, the ionization constant is 1.0×10^{-7} mol/kg in water at 25°C. For HS^- in water at 25°C, the ionization constant is 1×10^{-17} mol/kg. [Reported values for this constant range from 10^{-12} to 10^{-19} mol/kg. The value given here is from B. Meyer et al., *Inorg. Chem.,* **22,** 2345 (1983).] (*a*) Ignoring activity coefficients, calculate the H_3O^+, HS^-, and S^{2-} molalities in a 0.100 mol/kg aqueous H_2S solution at 25°C, making reasonable approximations to simplify the calculation. (*b*) The same as (*a*), except that activity coefficients are to be included in the calculations. For the ionization of HS^-, use the form of the Davies equation that corresponds to (10.60).

11.15 Given these $\Delta_f G_{298}^\circ$/(kJ/mol) values: -454.8 for $Mg^{2+}(aq)$, -128.0 for $IO_3^-(aq)$, and -587.0 for the ion pair $MgIO_3^+(aq)$, find K° at 25°C for $Mg^{2+}(aq) + IO_3^-(aq) \rightleftharpoons MgIO_3^+(aq)$.

11.16 For $CuSO_4$, the equilibrium constant for association to form $CuSO_4$ ion pairs has been found from conductivity measurements to be 230 kg/mol in aqueous solution at 25°C. Use the Davies equation (10.71) to calculate the Cu^{2+} molality, $\gamma_\pm$, and $\gamma_\pm^\dagger$ [Eq. (10.77)] in a 0.0500 mol/kg aqueous $CuSO_4$ solution at 25°C. (*Hint:* First estimate $\gamma_\pm$ by neglecting ion association; then use this estimated $\gamma_\pm$ to calculate an approximate Cu^{2+} molality; then calculate improved I_m and $\gamma_\pm$ values; and then recalculate the Cu^{2+} molality. Repeat as many times as necessary to obtain convergence.)

11.17 Write a computer program that will calculate the molality of $MX^{z_++z_-}(aq)$ ion pairs in a solution of the strong electrolyte $M_{\nu_+}X_{\nu_-}$ at a given electrolyte stoichiometric molality using an inputted value of the equilibrium constant for ion-pair formation. Use the Davies equation to find γ_+ and γ_- [see the sentence after (10.71)]. See Prob. 11.16 for help. Check your program against a few values in Fig. 10.10.

11.18 Set up a spreadsheet and use the Solver to solve the HOI example (Example 11.3 in Sec. 11.3). Assume activity coefficients equal 1 but make no other approximations. Avoid algebraic manipulations. Have the equilibrium-constant expressions, the electroneutrality condition, and conservation of matter for the IO group satisfied.

11.19 When two or more simultaneous ionic equilibria occur, the following systematic procedure can be used. **1.** Write down the equilibrium-constant expression for each reaction. **2.** Write down the condition for electrical neutrality of the solution. **3.** Write down relations that express the conservation of matter for substances added to the solution. **4.** Solve the resulting set of simultaneous equations, making judicious approximations where possible. For a dilute aqueous solution of the weak acid HX with stoichiometric molality m: (*a*) perform steps 1 and 2, assuming that $a(H_2O) = 1$ and $\gamma = 1$ for each ion (do not neglect the ionization of water); (*b*) perform step 3 for the X group of atoms (which occurs in HX and in X^-); (*c*) manipulate the resulting set of four simultaneous equations in four unknowns to eliminate all molalities except $m(H_3O^+)$ to show that

$$y^3 + K_a y^2 - (K_w + mK_a)y - K_a K_w = 0$$

where $y \equiv m(H_3O^+)$. This is a cubic equation that can be solved to give $m(H_3O^+)$.

11.20 Solve the HOI example (Example 11.3 in Sec. 11.3) by using the Solver in a spreadsheet to solve the cubic equation in Prob. 11.19.

11.21 Let $K_{c,a}$ and $K_{m,a}$ be the concentration-scale and the molality-scale equilibrium constants for ionization of the acid HX. (*a*) Use the relation $\gamma_{c,i}c_i = \rho_A \gamma_{m,i}m_i$ (proved in Prob. 10.15) to show that $K_{c,a}/K_{m,a} = \rho_A$. Since $\rho_A = 0.997$ kg/dm³ for water at 25°C, $K_{c,a}^\circ$ and $K_{m,a}^\circ$ have essentially the same numerical values for aqueous solutions. (*b*) Show that in a dilute solution, $c_i/m_i \approx \rho_A$. Therefore, the molality in mol/kg and the concentration in mol/dm³ are nearly equal numerically for each solute in dilute aqueous solutions. (*c*) Show that $\gamma_{c,i} \approx \gamma_{m,i}$ in dilute aqueous solutions.

11.22 Fuoss's theory of ion-pair formation gives the following expression (*in SI units*) for the concentration-scale equilibrium constant for the ion-association reaction $M^{z+} + X^{z-} \rightleftharpoons MX^{z_++z_-}$ in solution:

$$K_c = \tfrac{4}{3}\pi a^3 N_A \exp b \tag{11.39}$$

where N_A is the Avogadro constant, a is the mean ionic diameter (as in the Debye–Hückel theory), and

$$b \equiv z_+|z_-|e^2/4\pi\varepsilon_0\varepsilon_{r,A}akT \tag{11.40}$$

where the symbols in (11.40) are defined following (10.62). [For the derivation of (11.39), see R. M. Fuoss, *J. Am. Chem. Soc.*, **80**, 5059 (1958).] For the value $a = 4.5$ Å, use the Fuoss equation to calculate the ion-association equilibrium constant K_c in aqueous solution at 25°C for (*a*) 1:1 electrolytes; (*b*) 2:1 electrolytes; (*c*) 2:2 electrolytes; (*d*) 3:2 electrolytes. (*Hint:* Be careful with the units of a. Note that the traditional units of K_c are dm³/mol.) Conductivity measurements show that in aqueous solutions at 25°C, the ion-association equilibrium constant is typically of the order of magnitude 0.3 dm³/mol for 1:1 electrolytes, 5 dm³/mol for 2:1 electrolytes, 200 dm³/mol for 2:2 electrolytes, and 4000 dm³/mol for 3:2 electrolytes. How well does the Fuoss equation agree with these experimental values?

Section 11.4

11.23 Calculate the activity at 25°C of NaCl(*s*) at 1, 10, 100, and 1000 bar. The density of NaCl at 25°C and 1 bar is 2.16 g/cm³.

11.24 For $AgBrO_3$ in water at 25°C and 1 bar, $K_{sp} = 5.38 \times 10^{-5}$ mol²/kg². Calculate the solubility of $AgBrO_3$ in water at 25°C. Neglect ion pairing.

11.25 For CaF_2 in water at 25°C and 1 bar, $K_{sp} = 3.2 \times 10^{-11}$. Calculate the solubility of CaF_2 in water at 25°C and 1 bar. In this dilute solution, ion pairing can be neglected.

11.26 Find K_{sp} for BaF_2 in water at 25°C and 1 bar, given these $\Delta_f G^\circ_{298}$/(kJ/mol) values: -560.77 for $Ba^{2+}(aq)$, -278.79 for $F^-(aq)$, -1156.8 for $BaF_2(s)$.

11.27 (*a*) Use $\Delta_f G^\circ$ data in the Appendix to calculate K_{sp} for KCl in water at 25°C. (*b*) A saturated solution of KCl in water at 25°C has a molality 4.82 mol/kg. Calculate $\gamma_\pm$ of KCl in a saturated aqueous solution at 25°C.

11.28 For $CaSO_4$ in water at 25°C, the equilibrium constant for the formation of ion pairs is 190 kg/mol. The solubility of $CaSO_4$ in water at 25°C is 2.08 g per kilogram of water. Calculate K_{sp} for $CaSO_4$ in water at 25°C. (*Hint:* Get an initial estimate of the ion-pair molality and the ion molalities by ignoring activity coefficients. Get an initial estimate of I_m and use this to get an initial estimate of $\gamma_\pm$. Then recalculate the ionic molalities. Then calculate an improved $\gamma_\pm$ value and recalculate the ionic molalities. Keep repeating the calculations until convergence is obtained. Then calculate K_{sp}.)

11.29 Use data in the Appendix to calculate the equilibrium pressure of CO_2 above $CaCO_3$(calcite) at 25°C.

11.30 The equilibrium constant for the reaction $Fe_3O_4(s) + CO(g) \rightleftharpoons 3FeO(s) + CO_2(g)$ is 1.15 at 600°C. If a mixture of 2.00 mol Fe_3O_4, 3.00 mol CO, 4.00 mol FeO, and 5.00 mol CO_2 is brought to equilibrium at 600°C, find the equilibrium composition. Assume the pressure is low enough for the gases to behave ideally.

11.31 (*a*) If 5.0 g of $CaCO_3(s)$ is placed in a 4000-cm³ container at 1073 K, give the final amounts of $CaCO_3(s)$, CaO(*s*), and $CO_2(g)$ present. See Prob. 11.37 for K°. (*b*) The same as (*a*), except that 0.50 g of $CaCO_3$ is placed in the container.

11.32 The reaction $CaCO_3(s) \rightleftharpoons CaO(s) + CO_2(g)$ has $K° = 0.244$ at 800°C. A 4.00-L vessel at 800°C initially contains only $CO_2(g)$ at pressure P. If 0.500 g of $CaO(s)$ is added to the container, find the equilibrium amounts of $CaCO_3(s)$, $CaO(s)$, and $CO_2(g)$ if the initial CO_2 pressure P is (a) 125 torr; (b) 235 torr; (c) 825 torr.

Section 11.5

11.33 Use Appendix data to find $K°_{298}$ for the nonideal-gas reaction $2HCl(g) \rightleftharpoons H_2(g) + Cl_2(g)$.

11.34 At 450°C and 300 bar, fugacity coefficients estimated from law-of-corresponding-states graphs are $\phi_{N_2} = 1.14$, $\phi_{H_2} = 1.09$, and $\phi_{NH_3} = 0.91$. The equilibrium constant for $N_2(g) + 3H_2(g) \rightleftharpoons 2NH_3(g)$ at 450°C is $K° = 4.6 \times 10^{-5}$. Using the Lewis–Randall rule to estimate mixture fugacity coefficients, calculate the equilibrium composition of a system that initially consists of 1.00 mol of N_2 and 3.00 mol of H_2 and that is held at 450°C and 300 bar. (*Hint:* The quartic equation that results can be reduced to a quadratic equation by taking the square root of both sides.)

11.35 For NH_3, N_2, and H_2, the critical temperatures are 405.6, 126.2, and 33.3 K, respectively, and the critical pressures are 111.3, 33.5, and 12.8 atm, respectively. $\Delta_f G°_{700}$ for NH_3 is 6.49 kcal/mol. Use the Lewis–Randall rule and law-of-corresponding-states graphs of fugacity coefficients (Sec. 10.11) to calculate the equilibrium composition at 700 K of a system that initially consists of 1.00 mol of NH_3 if P is held fixed at 500 atm. *Note:* For H_2, to improve the fit of the observed fugacity coefficients to the law-of-corresponding-states graphs, one uses $T/(T_c + 8$ K$)$ and $P/(P_c + 8$ atm$)$ in place of the usual expressions for reduced temperature and pressure. (*Hint:* The quartic equation obtained in solving this problem can be reduced to a quadratic equation by taking the square root of both sides.)

Section 11.6

11.36 Using the approximation $\log_{10} \gamma_i = 0.1 I_m/m°$ for uncharged solutes in aqueous 25°C solutions, redo Examples 11.1 and 11.2 in Sec. 11.3, starting from the solutions already found.

Section 11.7

11.37 Measured CO_2 equilibrium pressures above mixtures of $CaCO_3(s)$ and $CaO(s)$ at various temperatures are

P/torr	23.0	70	183	381	716
T/K	974	1021	1073	1125	1167

(a) At 800°C (1073 K), find $\Delta G°$, $\Delta H°$, and $\Delta S°$ for $CaCO_3(s) \rightleftharpoons CaO(s) + CO_2(g)$. Do not use Appendix data. (b) Estimate the CO_2 pressure above a $CaCO_3$–CaO mixture at 1000°C.

11.38 The molality-scale ionization constant of water can be represented as the following function of temperature:

$$\log K°_w = 948.8760 - 24746.26(\text{K}/T) - 405.8639 \log (T/\text{K})$$
$$+ 0.48796(T/\text{K}) - 0.0002371(T/\text{K})^2$$

[See H. L. Clever, *J. Chem. Educ.*, **45**, 231 (1968) for a review of experimental work on K_w.] Calculate $\Delta G°$, $\Delta S°$, and $\Delta H°$ for the ionization of water at 25°C. Do not use Appendix data.

11.39 Use the 25°C estimated $\bar{V}_i^\infty$ values -5.4 and 1.4 cm³/mol for $H^+(aq)$ and $OH^-(aq)$, respectively, and the H_2O density 0.997 g/cm³ to estimate $K°_w$ for $H_2O \rightleftharpoons H^+(aq) + OH^-(aq)$ at 25°C and 200 bar. Take H_2O as a species in solution, so that H_2O contributes to $\Delta V°_{\text{soln}}$. State any approximations made. Compare with the experimental value 1.18×10^{-14}.

11.40 For acetic acid in water at 25°C, the ionization-constant ratio $K°_a(400$ bar$)/K°_a(1$ bar$)$ is 1.191 [D. A. Lown et al., *Trans. Faraday Soc.*, **64**, 2073 (1968)]. (a) Find $\Delta V°$ for the CH_3COOH ionization. State any approximations made. (b) Estimate the pressure needed to double $K°_a$.

11.41 (a) Use (11.32) to show that

$$\left(\frac{\partial \ln K°_m}{\partial T} \right)_P = \frac{\nu_A H^*_{m,A} + \Sigma_{i \neq A} \nu_i \bar{H}_i^\infty}{RT^2} = \frac{\Delta H^\infty}{RT^2}$$

where A is the solvent. (b) Show that $K_c/K_m = \rho_A^b$, where $b \equiv \Sigma_{i \neq A} \nu_i$. (Use a result from Prob. 10.15.) (c) Use the results for (a) and (b) to show that

$$\left(\frac{\partial \ln K°_c}{\partial T} \right)_P = \frac{\Delta H^\infty}{RT^2} - \alpha_A \sum_{i \neq A} \nu_i$$

(d) Use (11.32) and the result for (c) to show that $\bar{H}°_{c,i} = \bar{H}_i^\infty - RT^2 \alpha_A$ for $i \neq A$.

11.42 For which of the following reactions is the equilibrium constant a function of pressure? (a) $N_2(g) + 3H_2(g) \rightleftharpoons 2NH_3(g)$; (b) $CaCO_3(s) \rightleftharpoons CaO(s) + CO_2(g)$; (c) $NH_3(aq) + H_2O \rightleftharpoons NH_4^+(aq) + OH^-(aq)$.

Section 11.9

11.43 Show that $\Delta G°' = \Delta G° - 16.118\nu(H^+)RT$, where $\Delta G°'$ is defined by (11.37).

11.44 For NH_3, $\Delta_f G°_{500}$ is 4.83 kJ/mol. For a mixture of 4.00 mol H_2, 2.00 mol N_2, and 1.00 mol NH_3 held at 500 K and 3.00 bar, find $(\partial G/\partial \xi)_{T,P}$ for the reaction $N_2(g) + 3H_2(g) \rightleftharpoons 2NH_3(g)$. Assume ideal gases. For this mixture, will the reaction proceed spontaneously to the right or to the left?

General

11.45 We saw in Sec. 6.6 that the constant-T-and-P addition of a reactant to a gas-phase equilibrium might shift the equilibrium so as to produce more of the added species. For a single reaction in a *dilute* liquid solution, could the addition of a *solute* shift the equilibrium to produce more of that solute? (Assume that activity coefficients can be approximated as 1 and use the result for Prob. 6.48.)

11.46 (a) Verify that $\Delta G°$ for the process $i(sln) \rightarrow i(g)$ satisfies $\Delta G° = -RT \ln (K_i/P°)$, where K_i is the Henry's law constant for substance i in the solvent. Thus $K_i/P°$, which equals $(P_i/P°)/x_i^l$ for an ideally dilute solution [Eq. (9.63)], can be viewed as the equilibrium constant for $i(sln) \rightarrow i(g)$. (b) Given the following $\Delta_f G^\circ_{298}/(\text{kJ/mol})$ data from the NBS tables, calculate $K_{i,m}$ for O_2 in water and for CH_4 in water at 25°C and 1 bar: 16.4 for $O_2(aq)$, -50.72 for $CH_4(g)$, -34.33 for $CH_4(aq)$. Because the molality-scale standard state is used for solutes in aqueous solution, the calculated Henry's law constant is $K_{i,m}$, where $P_i = K_{i,m}m_i$ (Sec. 9.8).

11.47 True or false? (a) The addition at constant T and V of a chemically inert gas (for example, He) to a gas-phase mixture in reaction equilibrium will never shift the equilibrium. (b) For a closed-system reaction mixture held at constant T and P, the sign of $(\partial G/\partial \xi)_{T,P}$ determines in which direction the reaction proceeds; if $(\partial G/\partial \xi)_{T,P} < 0$, the reaction proceeds in the forward direction, whereas if $(\partial G/\partial \xi)_{T,P} > 0$, the reaction proceeds in the reverse direction. (c) A weak acid is completely dissociated in the limit of infinite dilution in aqueous solution. (d) If ΔG° for a reaction is positive, no reaction whatever will occur when reactants are mixed and held at constant T and P. (e) The standard state of a species is always chosen as a pure substance. (f) ΔG° always refers to a transition from *pure* standard-state reactants to *pure* standard-state products. (g) $\Delta_r S^\circ = \Delta_r H^\circ / T$.

Multicomponent Phase Equilibrium

One-component phase equilibrium was discussed in Chapter 7. We now consider multicomponent phase equilibria, which have important applications in chemistry, chemical engineering, materials science, and geology.

12.1 COLLIGATIVE PROPERTIES

We begin with a group of interrelated properties of solutions that are called **colligative properties** (from the Latin *colligatus*, meaning "bound together"). When a solute is added to a pure solvent A, the A mole fraction decreases. The relation $(\partial \mu_A/\partial x_A)_{T,P,n_{i \neq A}} > 0$ [Eq. (4.90)] shows that a decrease in x_A ($dx_A < 0$) must decrease the chemical potential of A ($d\mu_A < 0$). Therefore, addition of a solute at constant T and P lowers the solvent chemical potential μ_A below μ_A^*. This change in solvent chemical potential changes the vapor pressure, the normal boiling point, and the normal freezing point and causes the phenomenon of osmotic pressure. These four properties are the colligative properties. Each involves an equilibrium between phases.

The chemical potential μ_A is a measure of the escaping tendency of A from the solution, so the decrease in μ_A means the vapor partial pressure P_A of the solution is less than the vapor pressure P_A^* of pure A. The next section discusses this vapor-pressure lowering.

12.2 VAPOR-PRESSURE LOWERING

Consider a solution of a nonvolatile solute in a solvent. A **nonvolatile** solute is one whose contribution to the vapor pressure of the solution is negligible. This condition will hold for most solid solutes but not for liquid or gaseous solutes. The solution's vapor pressure P is then due to the solvent A alone. For simplicity, we shall assume pressures are low enough to treat all gases as ideal. If this is not so, pressures are to be replaced by fugacities.

From Eq. (10.16) for nonelectrolyte solutions and Eq. (10.56) for electrolyte solutions, the solution's vapor pressure is

$$P = P_A = \gamma_A x_A P_A^* \qquad \text{nonvolatile solute} \qquad (12.1)$$

where the mole-fraction scale is used for the solvent activity coefficient γ_A. The change in vapor pressure ΔP compared with pure A is $\Delta P = P - P_A^*$. Use of (12.1) gives

$$\Delta P = (\gamma_A x_A - 1)P_A^* \qquad \text{nonvol. solute} \qquad (12.2)$$

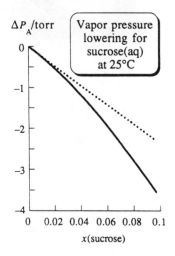

Figure 12.1

Vapor-pressure lowering ΔP versus sucrose mole fraction for aqueous sucrose solutions at 25°C (solid line). The dotted line is for an ideally dilute solution.

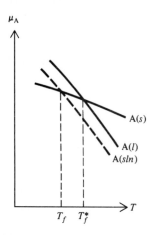

Figure 12.2

Chemical potential of A as a function of T (at fixed P) for pure solid A, pure liquid A, and A in solution (the dashed line). The lowering of μ_A by addition of solute to A(l) lowers the freezing point from T_f^* to T_f.

As noted in Sec. 10.3, measurement of solution vapor pressures enables determination of γ_A. Use of the Gibbs–Duhem equation then gives γ of the solute.

If the solution is very dilute, then $\gamma_A \approx 1$ and

$$\Delta P = (x_A - 1)P_A^* \qquad \text{ideally dil. soln., nonvol. solute} \qquad (12.3)$$

For a single nondissociating solute, $1 - x_A$ equals the solute mole fraction x_B and $\Delta P = -x_B P_A^*$. Under these conditions, ΔP is independent of the nature of B and depends only on its mole fraction in solution. Figure 12.1 plots ΔP versus x_B for sucrose(aq) at 25°C. The dotted line is for an ideally dilute solution.

12.3 FREEZING-POINT DEPRESSION AND BOILING-POINT ELEVATION

The normal boiling point (Chapter 7) of a pure liquid or solution is the temperature at which its vapor pressure equals 1 atm. A nonvolatile solute lowers the vapor pressure (Sec. 12.2). Hence it requires a higher temperature for the solution's vapor pressure to reach 1 atm, and the normal boiling point of the solution is elevated above that of the pure solvent.

Addition of a solute to A usually lowers the freezing point. Figure 12.2 plots μ_A for pure solid A, pure liquid A, and A in solution (sln) versus temperature at a fixed pressure of 1 atm. At the normal freezing point T_f^* of pure A, the phases A(s) and A(l) are in equilibrium and their chemical potentials are equal: $\mu_{A(s)}^* = \mu_{A(l)}^*$. Below T_f^*, pure solid A is more stable than pure liquid A, and $\mu_{A(s)}^* < \mu_{A(l)}^*$, since the most stable pure phase is the one with the lowest μ (Sec. 7.2). Above T_f^*, A(l) is more stable than A(s), and $\mu_{A(l)}^* < \mu_{A(s)}^*$. Addition of solute to A(l) at constant T and P always lowers μ_A (Sec. 12.1), so $\mu_{A(sln)} < \mu_{A(l)}^*$ at any given T, as shown in the figure. This makes the intersection of the A(sln) and A(s) curves occur at a lower T than the intersection of the A(l) and A(s) curves. The solution's freezing point T_f (which occurs when $\mu_{A(sln)} = \mu_{A(s)}^*$, provided pure A freezes out of the solution) is thus less than the freezing point T_f^* of pure A(l). The lowering of μ_A stabilizes the solution and decreases the tendency of A to escape from the solution by freezing out.

We now calculate the freezing-point depression due to solute B in solvent A. We shall assume that only pure solid A freezes out of the solution when it is cooled to its freezing point (Fig. 12.3). This is the most common situation. For other cases, see Sec. 12.8. The equilibrium condition at the normal (that is, 1-atm) freezing point is that the chemical potentials of pure solid A and of A in the solution must be equal. μ_A in the solution is $\mu_{A(sln)} = \mu_{A(l)}^\circ + RT \ln a_A = \mu_{A(l)}^* + RT \ln a_A$ [Eqs. (10.4) and (10.9)], where $\mu_{A(l)}^*$ is the chemical potential of pure liquid A and a_A is the activity of A in the solution. Equating $\mu_{A(s)}^*$ and $\mu_{A(sln)}$ at the solution's normal freezing point T_f, we have

$$\mu_{A(s)}^*(T_f, P) = \mu_{A(sln)}(T_f, P)$$

$$\mu_{A(s)}^*(T_f, P) = \mu_{A(l)}^*(T_f, P) + RT_f \ln a_A$$

where P is 1 atm. The chemical potential μ^* of a pure substance equals its molar Gibbs energy G_m^* [Eq. (4.86)], so

$$\ln a_A = \frac{G_{m,A(s)}^*(T_f) - G_{m,A(l)}^*(T_f)}{RT_f} = -\frac{\Delta_{fus}G_{m,A}(T_f)}{RT_f} \qquad (12.4)$$

where $\Delta_{fus}G_{m,A} \equiv G_{m,A(l)}^* - G_{m,A(s)}^*$ is ΔG_m for fusion of A. Since P is fixed at 1 atm, the pressure dependence of G_m^* is omitted.

The solution's freezing point T_f is a function of the activity a_A of A in solution. Alternatively, we can consider T_f to be the independent variable and view a_A as a function of T_f. We now differentiate (12.4) with respect to T_f at constant P. In Chapter 6, we differentiated $\ln K_P^\circ = -\Delta G^\circ/RT$ [Eq. (6.14)] with respect to T to get $(d/dT)(\ln K_P^\circ) = (d/dT)(-\Delta G^\circ/RT) = \Delta H^\circ/RT^2$ [Eq. (6.36)] for a chemical reaction. We can consider the fusion process A(s) $\rightarrow$ A(l) at pressure P° and temperature T_f as a reaction with A(s) as the reactant and A(l) as the product. Therefore the same derivation that gave $d(-\Delta G^\circ/RT)/dT = \Delta H^\circ/RT^2$ can be applied to the fusion process to give

$$\frac{d}{dT_f}\left(\frac{-\Delta_{\text{fus}}G_{\text{m,A}}(T_f)}{RT_f}\right) = \frac{\Delta_{\text{fus}}H_{\text{m,A}}(T_f)}{RT_f^2}$$

Taking $(\partial/\partial T_f)$ of (12.4), we thus get

$$\left(\frac{\partial \ln a_A}{\partial T_f}\right)_P = \frac{\Delta_{\text{fus}}H_{\text{m,A}}(T_f)}{RT_f^2} \tag{12.5}$$

$$d \ln a_A = (\Delta_{\text{fus}}H_{\text{m,A}}/RT_f^2)\, dT_f \qquad P \text{ const.} \tag{12.6}$$

where $\Delta_{\text{fus}}H_{\text{m,A}}(T_f)$ is the molar enthalpy of fusion of pure A at T_f and 1 atm. [Since the activity of pure solid A at 1 atm is 1 (Sec. 11.4), a_A can be viewed as the equilibrium constant K° of Eq. (11.6) for A(s) $\rightleftharpoons$ A(sln) and (12.5) is the van't Hoff equation (11.32) for A(s) $\rightleftharpoons$ A(sln); the standard state of A(sln) is pure liquid A, so $\Delta_{\text{fus}}H_{\text{m,A}}$ is ΔH° for A(s) $\rightleftharpoons$ A(sln).]

Integration of (12.6) from state 1 to state 2 gives

$$\ln \frac{a_{A,2}}{a_{A,1}} = \int_1^2 \frac{\Delta_{\text{fus}}H_{\text{m,A}}(T_f)}{RT_f^2}\, dT_f$$

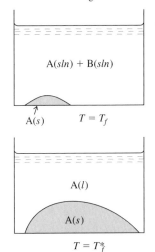

Figure 12.3

The upper figure shows solid A in equilibrium with a solution of A + B at the solution's freezing temperature T_f. The lower figure shows solid A in equilibrium with pure liquid A at the freezing point T_f^* of pure A.

Let state 1 be pure A. Then $T_{f,1} = T_f^*$, the freezing point of pure A, and $a_{A,1} = 1$, since μ_A (which equals $\mu_A^* + RT \ln a_A$) becomes equal to μ_A^* when $a_A = 1$. Let state 2 be a general state with activity $a_{A,2} = a_A$ and $T_{f,2} = T_f$. Using $a_A = \gamma_A x_A$ [Eq. (10.5)], where x_A and γ_A are the solvent mole fraction and mole-fraction-scale activity coefficient in the solution whose freezing point is T_f, we have

$$\ln \gamma_A x_A = \int_{T_f^*}^{T_f} \frac{\Delta_{\text{fus}}H_{\text{m,A}}(T)}{RT^2}\, dT \qquad P \text{ const.} \tag{12.7}$$

where the dummy integration variable (Sec. 1.8) was changed from T_f to T.

If there is only one solute B in the solution, and if B is neither associated nor dissociated, then $x_A = 1 - x_B$ and

$$\ln \gamma_A x_A = \ln \gamma_A + \ln x_A = \ln \gamma_A + \ln (1 - x_B) \tag{12.8}$$

The Taylor series for $\ln x$ is [Eq. (8.36)]: $\ln x = (x - 1) - (x - 1)^2/2 + \cdots$. With $x = 1 - x_B$, this series becomes

$$\ln (1 - x_B) = -x_B - x_B^2/2 - \cdots$$

Statistical-mechanical theories of solutions and experimental data show that $\ln \gamma_A$ can be expanded as (*Kirkwood and Oppenheim*, pp. 176–177):

$$\ln \gamma_A = B_2 x_B^2 + B_3 x_B^3 + \cdots \qquad \text{nonelectrolyte solution} \tag{12.9}$$

where $B_2, B_3, \ldots$ are functions of T and P. Substitution of these two series into (12.8) gives

$$\ln \gamma_A x_A = -x_B + (B_2 - \tfrac{1}{2})x_B^2 + \cdots \tag{12.10}$$

We now specialize to ideally dilute solutions. Here x_B is very small, and terms in x_B^2 and higher powers in (12.10) are negligible compared with the $-x_B$ term. (If $x_B = 10^{-2}$, then $x_B^2 = 10^{-4}$.) Thus

$$\ln \gamma_A x_A = -x_B \qquad \text{ideally dil. soln.} \qquad (12.11)$$

For a very dilute solution, the freezing-point change $T_f - T_f^*$ will be very small and T will vary only slightly in the integral in (12.7). The quantity $\Delta_{fus}H_{m,A}(T)$ will thus vary only slightly, and we can approximate it as constant and equal to $\Delta_{fus}H_{m,A}$ at T_f^*. Substituting (12.11) into (12.7), taking $\Delta_{fus}H_{m,A}/R$ outside the integral, and using $\int (1/T^2)\, dT = -1/T$, we get for (12.7)

$$-x_B = \frac{\Delta_{fus}H_{m,A}(T_f^*)}{R}\left(\frac{1}{T_f^*} - \frac{1}{T_f}\right) = \frac{\Delta_{fus}H_{m,A}}{R}\left(\frac{T_f - T_f^*}{T_f^* T_f}\right) \qquad (12.12)$$

The quantity $T_f - T_f^*$ is the **freezing-point depression** ΔT_f:

$$\Delta T_f \equiv T_f - T_f^* \qquad (12.13)$$

Since T_f is close to T_f^*, the quantity $T_f^* T_f$ in (12.12) can be replaced with $(T_f^*)^2$ with negligible error for ideally dilute solutions (Prob. 12.8); Eq. (12.12) becomes

$$\Delta T_f = -x_B R(T_f^*)^2 / \Delta_{fus}H_{m,A} \qquad (12.14)$$

We have $x_B = n_B/(n_A + n_B) \approx n_B/n_A$, since $n_B \ll n_A$. The solute molality is $m_B = n_B/n_A M_A$, where M_A is the solvent molar mass. Hence for this very dilute solution, we have $x_B = M_A m_B$, and (12.14) becomes

$$\Delta T_f = -\frac{M_A R(T_f^*)^2}{\Delta_{fus}H_{m,A}} m_B$$

$$\Delta T_f = -k_f m_B \qquad \text{ideally dil. soln., pure A freezes out} \qquad \textbf{(12.15)*}$$

where the solvent's **molal freezing-point-depression constant** k_f is defined by

$$k_f \equiv M_A R(T_f^*)^2 / \Delta_{fus}H_{m,A} \qquad (12.16)$$

Note from the derivation of (12.15) that its validity does not require the solute to be nonvolatile.

For water, $\Delta_{fus}H_m$ at $0°C$ is 6007 J/mol and

$$k_f = \frac{(18.015 \times 10^{-3}\ \text{kg/mol})(8.3145\ \text{J mol}^{-1}\ \text{K}^{-1})(273.15\ \text{K})^2}{6007\ \text{J mol}^{-1}} = 1.860\ \text{K kg/mol}$$

Some other k_f values in K kg/mol are benzene, 5.1; acetic acid, 3.8; camphor, 40.

Equation (12.7) enables the solvent activity coefficient γ_A to be found from the freezing-point depression. To obtain accurate results, one must allow for the temperature dependence of $\Delta_{fus}H_{m,A}$ (*Lewis and Randall,* chap. 26) unless ΔT_f is very small.

Another application of freezing-point-depression data is to find molecular weights of nonelectrolytes. To find the molecular weight of B, one measures ΔT_f for a dilute solution of B in solvent A and calculates the B molality m_B from (12.15). Use of $m_B = n_B/w_A$ [Eq. (9.3)], where w_A is the solvent mass, then gives n_B, the number of moles of B in solution. The molar mass M_B is then found from $M_B = w_B/n_B$ [Eq. (1.4)], where w_B is the known mass of B in solution. Since (12.15) applies only in an ideally dilute solution, an accurate determination of molecular weight requires that ΔT_f be found for a few molalities. One then plots the calculated M_B values versus m_B and extrapolates to $m_B = 0$. Practical applications of freezing-point depression include the use of salt to melt ice and snow and the addition of antifreeze (ethylene glycol, $HOCH_2CH_2OH$) to the water in automobile radiators.

Some organisms that live in below-0°C environments use freezing-point depression to prevent their body fluids from freezing. Freezing-point-depressing solutes synthesized by organisms in response to cold include glycerol [$HOCH_2CH(OH)CH_2OH$], ethylene glycol, and various sugars. For example, the glycerol concentration in larvae of the goldenrod gall moth is near zero in summer months but increases to 19% by weight in winter months. An alternative strategy used by many insects, fish, and mammals is to have antifreeze proteins keep their fluids in a metastable supercooled liquid state (Sec. 7.4) from 1 to 10 K below the freezing point. The mechanism of action of antifreeze proteins is under debate. (See *Debenedetti*, sec. 1.2.1.)

EXAMPLE 12.1 Molecular weight from freezing-point depression

The molal freezing-point-depression constant of benzene is 5.07 K kg/mol. A 0.450% solution of monoclinic sulfur in benzene freezes 0.088 K below the freezing point of pure benzene. Find the molecular formula of the sulfur in benzene.

The solution is very dilute, and we shall assume it to be ideally dilute. 100.000 g of solution contains 0.450 g of sulfur and 99.550 g of benzene. From $\Delta T_f = -k_f m_B$, the sulfur molality is

$$m_B = -\frac{\Delta T_f}{k_f} = -\frac{-0.088 \text{ K}}{5.07 \text{ K kg/mol}} = 0.0174 \text{ mol/kg}$$

But $m_B = n_B/w_A$ [Eq. (9.3)], so the number of moles of sulfur is

$$n_B = m_B w_A = (0.0174 \text{ mol/kg})(0.09955 \text{ kg}) = 0.00173 \text{ mol}$$

The sulfur molar mass is

$$M_B = w_B/n_B = (0.450 \text{ g})/(0.00173 \text{ mol}) = 260 \text{ g/mol}$$

The atomic weight of S is 32.06. Since $260/32.06 = 8.1 \approx 8$, the molecular formula is S_8.

Exercise

For D_2O (where D $\equiv$ ^{2}H), the normal freezing point is 3.82°C and $\Delta_{fus}H_m(T_f^*) =$ 1507 cal/mol. (a) Find k_f for D_2O. (b) Find the freezing point of a solution of 0.954 g of CH_3COCH_3 in 68.40 g of D_2O. Explain why your answer is approximate. [*Answers:* (a) 2.02_6 K kg/mol; (b) 3.33°C.]

As a pure substance freezes at fixed pressure, the temperature of the system remains constant until all the liquid has frozen. As a dilute solution of B in solvent A freezes at fixed pressure, the freezing point keeps dropping, since as pure A freezes out, the molality of B in the solution keeps increasing. To determine the freezing point of a solution, one can use the method of cooling curves (Sec. 12.8).

Freezing points are usually measured with the system open to the air. The dissolved air slightly lowers the freezing points of both pure A and the solution, but the depression due to the dissolved air will be virtually the same for pure A and for the solution and will cancel in the calculation of ΔT_f.

If there are several species in solution, then x_A in (12.8) equals $1 - \Sigma_{i \neq A} x_i$, where the sum goes over all solute species. Equation (12.11) becomes $\ln \gamma_A x_A \approx -\Sigma_{i \neq A} x_i$. For a dilute solution, we have $x_i \approx M_A m_i$, and Eq. (12.15) becomes for several solute species

$$\Delta T_f = -k_f m_{tot} \qquad \text{ideally dil. soln., pure A freezes out} \qquad (12.17)$$

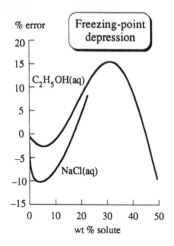

Figure 12.4

Percent errors of freezing points calculated from the ideally-dilute-solution equation (12.17) for NaCl(*aq*) and C₂H₅OH(*aq*) solutions at 1 atm.

where the total solute molality is $m_{\text{tot}} \equiv \Sigma_{i \neq A}\, m_i$. Note that ΔT_f is independent of the nature of the species in solution and depends only on the total molality, provided the solution is dilute enough to be considered ideally dilute.

For electrolyte solutions, one cannot use (12.17), since an electrolyte solution only becomes ideally dilute at molalities too low to produce a measurable ΔT_f. One must retain γ_A in (12.7) for electrolyte solutions. (See Prob. 12.13.) In the crudest approximation with $\gamma_A = 1$, we would expect from (12.17) that an electrolyte like NaCl that yields two ions in solution would give roughly twice the freezing-point depression as a nonelectrolyte at the same molality.

Figure 12.4 plots the percent deviations $100(\Delta T_f^{\text{id-dil}} - \Delta T_f)/\Delta T_f$ of ideally dilute solution freezing-point depressions $\Delta T_f^{\text{id-dil}}$ $(= -k_f m_{\text{tot}})$ from observed values ΔT_f for aqueous solutions of C₂H₅OH and NaCl. The deviations result from the approximation $\ln \gamma_A x_A \approx -x_B \approx n_B/n_A$ [Eq. (12.11)], from neglect of the temperature dependence of $\Delta_{\text{fus}} H_{\text{m,A}}$, and from the replacement of $T_f T_f^*$ by $(T_f^*)^2$.

The boiling-point-elevation formula is found the same way as for freezing-point depression. We start with an equation like that preceding (12.4) except that $\mu_{A(s)}^*$ is replaced by $\mu_{A(v)}^*$ (where v is for vapor) and T_f is replaced by T_b, the solution's boiling point. Equation (12.4) for freezing-point depression is $RT_f \ln a_A = -\Delta_{\text{fus}} G_{\text{m,A}}(T_f)$, whereas the analog of (12.4) for boiling-point elevation is $RT_b \ln a_A = \Delta_{\text{vap}} G_{\text{m,A}}(T_b)$ with no minus sign. Going through the same steps as for freezing-point depression, one derives equations that correspond to (12.15) and (12.16):

$$\Delta T_b = k_b m_B \qquad \text{ideally dil. soln., nonvol. solute} \qquad \textbf{(12.18)}*$$

$$k_b \equiv M_A R (T_b^*)^2 / \Delta_{\text{vap}} H_{\text{m,A}} \qquad (12.19)$$

where $\Delta T_b \equiv T_b - T_b^*$ is the boiling-point elevation for the ideally dilute solution and T_b^* is the boiling point of pure solvent A. The assumption in (12.15) that only pure A freezes out of the solution corresponds to the assumption in (12.18) that only pure A vaporizes out of the solution, which means that the solute is nonvolatile. For water, $k_b = 0.513$ °C kg/mol. Boiling-point elevation can be used to find molecular weights but is less accurate than freezing-point depression.

Currently, molecular weights of nonpolymers are most often determined using mass spectrometry (MS). The molecular weight is the mass number of the parent peak. Special mass spectrometric techniques can accurately measure protein molecular weights. In matrix-assisted laser desorption ionization (MALDI), the protein in a low concentration in a solid matrix of a compound such as 2,5-dihydroxybenzoic acid is exposed to a pulse of laser radiation. Part of the matrix is vaporized, thereby bringing protein molecules into the gas phase and ionizing them. MALDI MS can accurately determine molecular weights up to 500000. In electrospray ionization (ESI), a solution of the protein is sprayed into the mass spectrometer and heated flowing gas evaporates the solvent from the spray droplets. ESI MS can determine molecular weights up to 200000.

12.4 OSMOTIC PRESSURE

Osmotic Pressure

Semipermeable membranes exist that allow only certain chemical species to pass through them. Imagine a box divided into two chambers by a rigid, thermally conducting, semipermeable membrane that allows solvent A to pass through it but does not allow the passage of solute B. In the left chamber, we put pure A, and in the right, a solution of B in A (Fig. 12.5). We restrict A to be a nonelectrolyte.

Suppose the initial heights of the liquids in the two capillary tubes are equal. The chambers are thus initially at equal pressures: $P_L = P_R$, where the subscripts stand for

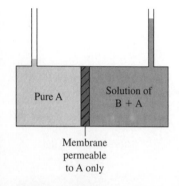

Figure 12.5

Setup for measurement of osmotic pressure.

left and right. Since the membrane is thermally conducting, thermal equilibrium is maintained: $T_L = T_R = T$. The chemical potential of A on the left is μ_A^*. With equal T and P in the two liquids, the presence of solute B in the solution on the right makes μ_A on the right less than μ_A^* (Sec. 12.1). Substances flow from high to low chemical potential (Sec. 4.8), and we have $\mu_A^* = \mu_{A,L} > \mu_{A,R}$. Therefore substance A will flow through the membrane from left (the pure solvent) to right (the solution). The liquid in the right tube rises, thereby increasing the pressure in the right chamber. We have $(\partial \mu_A/\partial P)_T = \bar{V}_A$ [Eq. (9.31)]. Since $\bar{V}_A$ is positive in a dilute solution [Eq. (9.66)], the increase in pressure increases $\mu_{A,R}$ until eventually equilibrium is reached with $\mu_{A,R} = \mu_{A,L}$. Since the membrane is impermeable to B, there is no equilibrium relation for μ_B. If the membrane were permeable to both A and B, the equilibrium condition would have equal concentrations of B and equal pressures in the two chambers.

Let the equilibrium pressures in the left and right chambers be P and $P + \Pi$, respectively. We call Π the **osmotic pressure.** It is the extra pressure that must be applied to the solution to make μ_A in the solution equal to μ_A^* so as to achieve membrane equilibrium for species A between the solution and pure A. In the solution, we have $\mu_A = \mu_A^* + RT \ln \gamma_A x_A$ [Eqs. (10.6) and (10.9)], and at equilibrium

$$\mu_{A,L} = \mu_{A,R} \tag{12.20}$$

$$\mu_A^*(P, T) = \mu_A^*(P + \Pi, T) + RT \ln \gamma_A x_A \tag{12.21}$$

where we do not assume an ideally dilute solution. Note that γ_A in (12.21) is the value at $P + \Pi$ of the solution. From $d\mu_A^* = dG_{m,A}^* = -S_{m,A}^* \, dT + V_{m,A}^* \, dP$, we have $d\mu_A^* = V_{m,A}^* \, dP$ at constant T. Integration from P to $P + \Pi$ gives

$$\mu_A^*(P + \Pi, T) - \mu_A^*(P, T) = \int_P^{P+\Pi} V_{m,A}^* \, dP' \qquad \text{const. } T \tag{12.22}$$

where a prime was added to the dummy integration variable to avoid the use of the symbol P with two different meanings. Substitution of (12.22) into (12.21) gives

$$RT \ln \gamma_A x_A = -\int_P^{P+\Pi} V_{m,A}^* \, dP' \qquad \text{const. } T \tag{12.23}$$

$V_{m,A}^*$ of a liquid varies very slowly with pressure and can be taken as constant unless very high osmotic pressures are involved. The right side of (12.23) then becomes $-V_{m,A}^*(P + \Pi - P) = -V_{m,A}^*\Pi$, and (12.23) becomes $RT \ln \gamma_A x_A = -V_{m,A}^*\Pi$, or

$$\Pi = -(RT/V_{m,A}^*) \ln \gamma_A x_A \tag{12.24}$$

For an ideally dilute solution of a solute B that is neither associated nor dissociated, γ_A is 1 and $\ln \gamma_A x_A \approx -x_B$ [Eq. (12.11)]. Hence

$$\Pi = (RT/V_{m,A}^*)x_B \qquad \text{ideally dil. soln.} \tag{12.25}$$

Since the solution is quite dilute, we have $x_B = n_B/(n_A + n_B) \approx n_B/n_A$ and

$$\Pi = \frac{RT}{V_{m,A}^*} \frac{n_B}{n_A} \qquad \text{ideally dil. soln.} \tag{12.26}$$

where n_A and n_B are the moles of solvent and solute in the solution that is in membrane equilibrium with pure solvent A. Since the solution is very dilute, its volume V is very nearly equal to $n_A V_{m,A}^*$, and (12.26) becomes $\Pi = RTn_B/V$, or

$$\Pi = c_B RT \qquad \text{ideally dil. soln.} \tag{12.27}$$

where the molar concentration c_B equals n_B/V. Note the formal resemblance to the equation of state for an ideal gas, $P = cRT$, where $c = n/V$. Equation (12.27), which is called *van't Hoff's law,* is valid in the limit of infinite dilution.

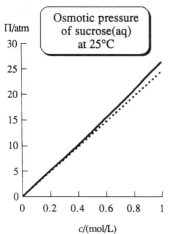

Figure 12.6

Osmotic pressure Π of aqueous sucrose solutions at 25°C plotted versus sucrose concentration. The dotted line is for an ideally dilute solution.

Since solute molality, molar concentration, and mole fraction are proportional to one another in an ideally dilute solution (Prob. 9.8), dilute-solution colligative properties can be expressed using any of these composition measures. Equation (12.25) uses mole fraction, (12.26) uses molality (since $n_B/n_A = M_A m_B$), and (12.27) uses molar concentration.

Figure 12.6 plots Π versus solute concentration for aqueous sucrose solutions at 25°C. The dotted line is for an ideally dilute solution.

For solutions that are not ideally dilute, Eq. (12.24) holds. However, a different (but equivalent) expression for Π in nonideally dilute solutions is often more convenient than (12.24). In 1945, McMillan and Mayer developed a statistical-mechanical theory for nonelectrolyte solutions (see *Hill*, chap. 19). They proved that the osmotic pressure in a nonideally dilute nonelectrolyte two-component solution is given by

$$\Pi = RT(M_B^{-1}\rho_B + A_2\rho_B^2 + A_3\rho_B^3 + \cdots) \tag{12.28}$$

where M_B is the solute molar mass and ρ_B is the solute mass concentration: $\rho_B \equiv w_B/V$ [Eq. (9.2)], where w_B is the mass of solute B. The quantities $A_2, A_3, \ldots$ are related to the solute–solute intermolecular forces in solvent A and are functions of T (and weakly of P). Note the formal resemblance of (12.28) to the virial equation (8.4) for gases. In the limit of infinite dilution, $\rho_B \to 0$ and (12.28) becomes $\Pi = RT\rho_B/M_B = RTw_B/M_B V = RTn_B/V = c_B RT$, which is the van't Hoff law.

Osmotic pressure is sometimes misunderstood. Consider a 0.01 mol/kg solution of glucose in water at 25°C and 1 atm. When we say the freezing point of this solution is $-0.02°C$, we do not imply that the solution's temperature is actually $-0.02°C$. The freezing point is that temperature at which the solution would be in equilibrium with pure solid water at 1 atm. Likewise, when we say that the osmotic pressure of this solution is 0.24 atm (see Example 12.2), we do not imply that the pressure in the solution is 0.24 atm (or 1.24 atm). Instead, the osmotic pressure is the extra pressure that would have to be applied to the solution so that, if it were placed in contact with a membrane permeable to water but not glucose, it would be in membrane equilibrium with pure water, as in Fig. 12.7.

EXAMPLE 12.2 Osmotic pressure

Find the osmotic pressure at 25°C and 1 atm of a 0.0100 mol/kg solution of glucose ($C_6H_{12}O_6$) in water.

It is a good approximation to consider this dilute nonelectrolyte solution as ideally dilute. Almost all the contribution to the solution's mass and volume comes from the water, and the density of water is nearly 1.00 g/cm³. Therefore an amount of this solution that contains 1 kg of water will have a volume very close to 1000 cm³ = 1 L, and the glucose molar concentration is well approximated as 0.0100 mol/dm³. (See also Prob. 11.21b.) Substitution in (12.27) gives

$$\Pi = c_B RT = (0.0100 \text{ mol/dm}^3)(82.06 \times 10^{-3} \text{ dm}^3 \text{ atm mol}^{-1} \text{ K}^{-1})(298.1 \text{ K})$$

$$\Pi = 0.245 \text{ atm} = 186 \text{ torr}$$

Figure 12.7

Pure water in equilibrium with water in a glucose solution.

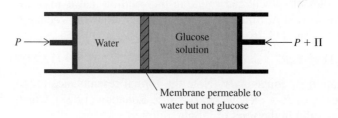

where the volume unit in R was converted to dm^3 to match that in c_B. Alternatively, Eq. (12.25) or (12.26) can be used to find Π. These equations give answers very close to that found from (12.27).

Exercise

At 25°C, a solution prepared by dissolving 82.7 mg of a nonelectrolyte in water and diluting to 100.0 mL has an osmotic pressure of 83.2 torr. Find the molecular weight of the nonelectrolyte. (*Answer:* 185.)

Note the substantial value of Π for the very dilute 0.01 mol/kg glucose solution in this example. Since the density of water is 1/13.6 times that of mercury, an osmotic pressure of 186 torr (186 mmHg) corresponds to a height of 18.6 cm $\times$ 13.6 = 250 cm = 2.5 m = 8.2 ft of liquid in the right-hand tube in Fig. 12.5. In contrast, an aqueous 0.01 mol/kg solution will show a freezing-point depression of only 0.02 K. The large value of Π results from the fact (noted many times previously) that the chemical potential of a component of a condensed phase is rather insensitive to pressure. Hence it takes a large value of Π to change the chemical potential of A in the solution so that it equals the chemical potential of pure A at pressure P.

The mechanism of osmotic flow is not the business of thermodynamics, but we shall mention three commonly cited mechanisms: (1) The size of the membrane's pores may allow small solvent molecules to pass through but not allow large solute molecules to pass. (2) The volatile solvent may vaporize into the pores of the membrane and condense out on the other side, but the nonvolatile solute does not do so. (3) The solvent may dissolve in the membrane.

Polymer Molecular Weights

The substantial value of Π given by dilute solutions makes osmotic-pressure measurements valuable in finding molecular weights of high-molecular-weight substances such as polymers. For such substances, the freezing-point depression is too small to be useful. For example, if $M_B = 10^4$ g/mol, a solution of 1.0 g of B in 100 g of water has $\Delta T_f = -0.002$°C and has $\Pi = 19$ torr at 25°C.

Polymer solutions show large deviations from ideally dilute behavior even at very low molalities. The large size of the molecules causes substantial solute–solute interactions in dilute polymer solutions, so it is essential to measure Π at several dilute concentrations and extrapolate to infinite dilution to find the polymer's true molecular weight. Π is given by the McMillan–Mayer expression (12.28). In dilute solutions it is often adequate to terminate the series after the A_2 term. Thus, $\Pi/RT = \rho_B/M_B + A_2\rho_B^2$, or

$$\Pi/\rho_B = RT/M_B + RTA_2\rho_B \qquad \text{dil. soln.} \qquad (12.29)$$

A plot of Π/ρ_B versus ρ_B gives a straight line with intercept RT/M_B at $\rho_B = 0$. In some cases, the A_3 term is not negligible in dilute solutions (see Prob. 12.22).

A synthetic polymer usually consists of molecules of varying chain length, since chain termination in a polymerization reaction is a random process. We now find the expression for the apparent molecular weight of such a solute as determined by osmotic-pressure measurements. If there are several solute species in the solution, the solvent mole fraction x_A equals $1 - \Sigma_{i\neq A} x_i$, where the sum goes over the various solute species. Use of the Taylor series (8.36) gives

$$\ln x_A = \ln\left(1 - \sum_{i\neq A} x_i\right) \approx -\sum_{i\neq A} x_i \approx -\frac{1}{n_A}\sum_{i\neq A} n_i \qquad (12.30)$$

Hence in place of $\Pi = c_B RT$ [Eq. (12.27)], we get

$$\Pi = RT \sum_{i \neq A} c_i = \frac{RT}{V} \sum_{i \neq A} n_i \qquad \text{ideally dil. soln.} \qquad (12.31)$$

If we pretended that there was only one solute species B, with molar mass M_B, we would use data extrapolated to infinite dilution to calculate M_B from $\Pi = c_B RT = w_B RT / M_B V$ [Eq. (12.27)] as $M_B = w_B RT / \Pi V$, where w_B is the solute mass. Substitution of (12.31) for Π gives

$$M_B = \frac{w_B}{\displaystyle\sum_{i \neq A} n_i} = \frac{\displaystyle\sum_{i \neq A} w_i}{\displaystyle\sum_{i \neq A} n_i} = \frac{\displaystyle\sum_{i \neq A} n_i M_i}{\displaystyle\sum_{i \neq A} n_i} \qquad (12.32)$$

where w_i, n_i, and M_i are the mass, the number of moles, and the molar mass of solute i. The quantity on the right side of (12.32) is the **number average molar mass.** The number of moles n_i is proportional to the number of molecules of species i. Hence each value of M_i in (12.32) is weighted according to the number of molecules having that molecular weight. The same result is found for the molecular weight calculated from the other colligative properties.

If we consider the collection of solute molecules only, the denominator on the right side of (12.32) is n_{tot}, the total number of moles of solute, and n_i/n_{tot} is the mole fraction x_i of solute species i in the collection of solute molecules. (Of course, x_i is not the mole fraction of species i in the solution. We are now considering the solute species apart from the solvent.) Introducing the symbol M_n for the number average molar mass, we rewrite (12.32) as

$$M_n = \sum_i x_i M_i = \frac{w_{tot}}{n_{tot}} \qquad (12.33)$$

where the sum goes over all solute species and where w_{tot} and n_{tot} are the total mass and total number of moles of solute species.

Osmosis

In Fig. 12.7, the additional externally applied pressure Π produces membrane equilibrium between the solution and the pure solvent. If the pressure on the solution were less than $P + \Pi$, then μ_A would be less in the solution than in the pure solvent and there would be a net flow of solvent from the pure solvent on the left to the solution on the right, a process called **osmosis.** If, however, the pressure on the solution is increased above $P + \Pi$, then μ_A in the solution becomes greater than μ_A in the pure solvent and there is a net flow of solvent from the solution to the pure solvent, a phenomenon called **reverse osmosis.** Reverse osmosis is used to desalinate seawater. Here, one requires a membrane that is nearly impermeable to salt ions, strong enough to withstand the pressure difference, and permeable to water. Membranes of cellulose acetate or hollow nylon fibers are used in desalination plants.

Osmosis is of fundamental importance in biology. Cell membranes are permeable to H_2O, CO_2, O_2, and N_2 and to certain organic molecules (for example, amino acids, glucose) and are impermeable to proteins and polysaccharides. Inorganic ions and disaccharides (for example, sucrose) generally pass quite slowly through cell membranes. The cells of an organism are bathed by body fluids (for example, blood, lymph, sap) containing various solutes.

The situation is more complex than in Fig. 12.5 since solutes are present on both sides of the membrane, which is permeable to water and some solutes (which we symbolize by B, C, . . .) but impermeable to others (which we symbolize by L, M, . . .). In the absence of active transport (discussed shortly), water and solutes B, C, . . . will move through the cell membrane until the chemical potentials of H_2O, of B, of C, . . . are equalized on each side of the membrane. If the fluid surrounding a cell is more concentrated in solutes L, M, . . . than the cell fluid is, the cell will lose water by osmosis; the surrounding fluid is said to be *hypertonic* with respect to the cell. If the surrounding fluid is less concentrated in L, M, . . . than the cell, the cell gains water from the surrounding *hypotonic* fluid. When there is no net transfer of water between cell and surroundings, the two are *isotonic*.

Blood and lymph are approximately isotonic to the cells of an organism. Intravenous feeding and injections use a salt solution that is isotonic with blood. If water were injected, the red blood cells would gain water by osmosis and might burst. Plant roots absorb water from the surrounding hypotonic soil fluids by osmosis.

Living cells are able to transport a chemical species through a cell membrane from a region of low chemical potential of that solute to a region of high chemical potential, a direction opposite that of spontaneous flow. Such transport (called **active transport**) is accomplished by coupling the transport with a process for which ΔG is negative (Sec. 11.10). For example, a certain protein in cell membranes simultaneously (*a*) actively transports K^+ ions into cells from surrounding fluids having lower K^+ concentrations, (*b*) actively transports Na^+ ions out of cells, and (*c*) hydrolyzes ATP to ADP (a reaction for which G decreases—Fig. 11.8). About one-third of the ATP consumed by a resting animal is used for active transport ("pumping") of Na^+ and K^+ across membranes. A resting human consumes about 40 kg of ATP in 24 hr, and this ATP must be continually resynthesized from ADP. The active transport of Na^+ out of cells makes possible the spontaneous, passive flow of Na^+ into cells, and this spontaneous inward flow of Na^+ is coupled to and drives the active transport of glucose and amino acids into cells.

> There is a logical gap in our derivation of the osmotic pressure. In Sec. 4.8, we derived μ_A^α $= \mu_A^\beta$ (the equality of chemical potentials in different phases in equilibrium) under the suppositions that $T^\alpha = T^\beta$ and $P^\alpha = P^\beta$. However, in osmotic equilibrium, $P^\alpha \neq P^\beta$ (where α and β are the phases separated by the semipermeable membrane). Hence we must show that $\mu_A^\alpha = \mu_A^\beta$ at equilibrium even when the phases are at different pressures. The proof is outlined in Prob. 12.27. [In a steady-state system with a temperature gradient (for example, Fig. 16.1), the chemical potential of a species differs in regions at different T. Thus μ_A^α $= \mu_A^\beta$ only if T is uniform.]

12.5   TWO-COMPONENT PHASE DIAGRAMS

Phase diagrams for one-component systems were discussed in Chapter 7. Phase diagrams for two-component systems are discussed in Secs. 12.5 to 12.10 and for three-component systems in Sec. 12.12.

With $c_{ind} = 2$, the phase rule $f = c_{ind} - p + 2$ becomes $f = 4 - p$. For a one-phase, two-component system, $f = 3$. The three independent intensive variables are P, T, and one mole fraction. For convenience, we usually keep P or T constant and plot a two-dimensional phase diagram, which is a cross section of a three-dimensional plot. The restriction of constant T or P in the two-dimensional plot reduces f by 1 in this plot. A two-component system is called a **binary** system.

Multicomponent phase equilibria have important applications in chemistry, geology, and materials science. *Materials science* studies the structure, properties, and applications of scientific and industrial materials. The main classes of materials are *metals, semiconductors, polymers, ceramics*, and *composites*. Traditionally, the term "ceramic" referred to materials produced by baking moist clay to form hard solids. Nowadays, the term is broadened to include all inorganic, nonmetallic materials processed or used at high temperature. Most ceramics are compounds of one or more metals with a nonmetal (commonly oxygen) and are mechanically strong and resistant to heat and chemicals. Some examples of ceramics are sand, porcelain, cement, glass, bricks, diamond, SiC, Si_3N_4, Al_2O_3, MgO, and $MgSiO_4$; many ceramics are silicates. A composite material is made of two or more materials and may possess properties not present in any one component. Bone is a composite of the soft, strong, polymeric protein collagen and the hard, brittle mineral hydroxyapatite [approximate formula $3Ca_3(PO_4)_2 \cdot Ca(OH)_2$]. Fiberglass is a composite containing a plastic strengthened by the addition of glass fibers.

12.6 TWO-COMPONENT LIQUID–VAPOR EQUILIBRIUM

Instead of plotting complete phase diagrams, we shall usually consider only one portion of the phase diagram at a time. This section deals with the liquid–vapor part of the phase diagram of a two-component system, which is important in laboratory and industrial separations of liquids by distillation.

Ideal Solution at Fixed Temperature

Consider two liquids B and C that form an ideal solution. We hold the temperature fixed at some value T that is above the freezing points of B and C. We shall plot the system's pressure P against x_B, the **overall mole fraction** of B in the system:

$$x_B \equiv \frac{n_{B,\text{total}}}{n_{\text{total}}} = \frac{n_B^l + n_B^v}{n_B^l + n_B^v + n_C^l + n_C^v} \qquad (12.34)$$

where n_B^l and n_B^v are the number of moles of B in the liquid and vapor phases, respectively. For a closed system x_B is fixed, although n_B^l and n_B^v may vary.

Let the system be enclosed in a cylinder fitted with a piston and immersed in a constant-temperature bath (Fig. 12.8a). To see what the P-versus-x_B phase diagram looks like, let us initially set the external pressure on the piston high enough for the system to be entirely liquid (point A in Fig. 12.8b). As the pressure is lowered below that at A, the system eventually reaches a pressure where the liquid just begins to vaporize (point D). At point D, the liquid has composition x_B^l, where x_B^l at D is equal to the overall mole fraction x_B since only an infinitesimal amount of liquid has vaporized. What is the composition of the first vapor that comes off? Raoult's law $P_B \equiv x_B^v P = x_B^l P_B^*$ [Eq. (9.52)] relates the vapor-phase mole fractions to the liquid composition as follows:

$$x_B^v = x_B^l P_B^* / P \quad \text{and} \quad x_C^v = x_C^l P_C^* / P \qquad (12.35)$$

where P_B^* and P_C^* are the vapor pressures of pure B and pure C at T, where the system's pressure P equals the sum $P_B + P_C$ of the partial pressures, where $x_B^l \equiv n_B^l/(n_B^l + n_C^l)$, and the vapor is assumed ideal.

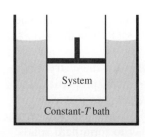

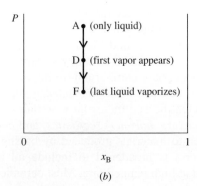

Figure 12.8

(a) A system held at constant T.
(b) Points on the P-versus-x_B phase diagram of the system in (a).

From (12.35) we have

$$\frac{x_B^v}{x_C^v} = \frac{x_B^l}{x_C^l} \frac{P_B^*}{P_C^*} \qquad \text{ideal soln.} \tag{12.36}$$

Let B be the more volatile component, meaning that $P_B^* > P_C^*$. Equation (12.36) then shows that $x_B^v/x_C^v > x_B^l/x_C^l$. The vapor above an ideal solution is richer than the liquid in the more volatile component (Fig. 9.18*b*). Equations (12.35) and (12.36) apply at any pressure where liquid–vapor equilibrium exists, not just at point D.

Now let us isothermally lower the pressure below point D, causing more liquid to vaporize. Eventually, we reach point F in Fig. 12.8*b*, where the last drop of liquid vaporizes. Below F, we have only vapor. For any point on the line between D and F, liquid and vapor phases coexist in equilibrium.

We can repeat this experiment many times, each time starting with a different composition for the closed system. For composition x_B', we get points D′ and F′; for composition x_B'', we get points D″ and F″; and so on. We then plot the points D, D′, D″, . . . and join them, and do the same for F, F′, F″, . . . (Fig. 12.9).

What is the equation of the curve DD′D″? For each of these points, liquid of composition x_B^l [or $(x_B^l)'$, etc.] is just beginning to vaporize. The vapor pressure of this liquid is $P = P_B + P_C = x_B^l P_B^* + x_C^l P_C^* = x_B^l P_B^* + (1 - x_B^l) P_C^*$, and

$$P = P_C^* + (P_B^* - P_C^*) x_B^l \qquad \text{ideal soln.} \tag{12.37}$$

This is the same as Eq. (9.54) and is the equation of a straight line that starts at P_C^* for $x_B^l = 0$ and ends at P_B^* for $x_B^l = 1$. All along the line DD′D″, the liquid is just beginning to vaporize, so the overall mole fraction x_B is equal to the mole fraction of B in the liquid x_B^l. DD′D″ is thus a plot of the total vapor pressure P versus x_B^l [Eq. (12.37)].

What is the equation of the curve FF′F″? Along this curve, the last drop of liquid is vaporizing, so the overall x_B (which is what is plotted on the abscissa) will now equal x_B^v, the mole fraction of B in the vapor. FF′F″ is then a plot of the total vapor pressure P versus x_B^v. To obtain P as a function of x_B^v, we must express x_B^l in (12.37) as a function of x_B^v. To do this, we use Raoult's law $P_B \equiv x_B^v P = x_B^l P_B^*$ to write $x_B^l = x_B^v P/P_B^*$. Substitution of this x_B^l expression in (12.37) gives $P = P_C^* + (P_B^* - P_C^*) x_B^v P/P_B^*$. Solving this equation for P, we get

$$P = \frac{P_B^* P_C^*}{x_B^v (P_C^* - P_B^*) + P_B^*} \qquad \text{ideal soln.} \tag{12.38}$$

This is the desired equation for P versus x_B^v and is the FF′F″ curve.

We now redraw the phase diagram in Fig. 12.10. From the preceding discussion, *the upper line is the P-versus-x_B^l curve and the lower line is the P-versus-x_B^v curve.*

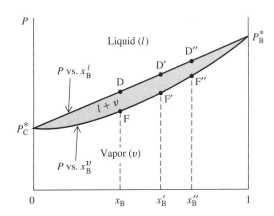

Figure 12.9

Pressure-versus-composition liquid–vapor phase diagram for an ideal solution at fixed *T*.

Figure 12.10

Pressure-versus-composition liquid–vapor phase diagram for an ideal solution at fixed T. The lower line is the P-versus-x_B^v curve, and the upper line is the P-versus-x_B^l curve.

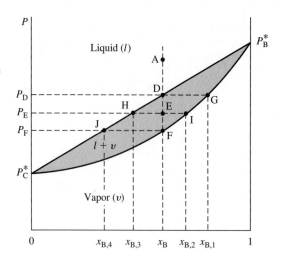

Consider again the process of starting at point A (where P is high enough for only liquid to be present) and isothermally lowering the pressure. The system is closed, so (even though the composition of the liquid and vapor phases may vary) the overall mole fraction of B remains fixed at x_B throughout the process. Hence the process is represented by a vertical line on the P-versus-x_B diagram. At point D with system pressure P_D, the liquid just begins to vaporize. What is the composition of the first vapor that comes off? What we want is the value of x_B^v when liquid–vapor equilibrium exists and when the system's pressure P (which is also the total vapor pressure) equals P_D. The lower curve on the phase diagram is a plot of Eq. (12.38) and gives P as a function of x_B^v. Alternatively, we can view the lower curve as giving x_B^v as a function of P. Hence, to find x_B^v when P equals P_D, we find the point on the lower curve that corresponds to pressure P_D. This is point G and gives the composition (labeled $x_{B,1}$) of the first vapor that comes off.

As P is lowered further, it reaches P_E. For point E on the phase diagram (which lies between points D and F), the system consists of two phases, a liquid and a vapor in equilibrium. What are the compositions of these phases? The upper curve in Fig. 12.10 relates P to x_B^l and the lower curve relates P to x_B^v. Hence at point E with pressure P_E, we have $x_B^v = x_{B,2}$ (point I) and $x_B^l = x_{B,3}$ (point H). Finally, at point F with pressure P_F, the last liquid vaporizes. Here, $x_B^v = x_B$ and $x_B^l = x_{B,4}$ (point J). Below F, we have vapor of composition x_B. As the pressure is lowered and the liquid vaporizes in the closed system, x_B^l falls from D to J, that is, from x_B to $x_{B,4}$. This is because substance B is more volatile than C. Also, as the liquid vaporizes, x_B^v falls from G to F, that is, from $x_{B,1}$ to x_B. This is because liquid vaporized later is richer in substance C. For states where both liquid and vapor phases are present, the system's pressure P equals the vapor pressure of the liquid.

A line of constant overall composition, for example, ADEF, is an *isopleth*.

The P-versus-x_B liquid–vapor phase diagram at constant T of two liquids that form an ideal solution thus has three regions. At any point above both curves in Fig. 12.10, only liquid is present. At any point below both curves, only vapor is present. At a typical point E between the two curves, two phases are present: a liquid whose composition is given by point H ($x_B^l = x_{B,3}$) and a vapor whose composition is given by point I ($x_B^v = x_{B,2}$). The *overall* composition of the two-phase system is given by the x_B value at point E. The overall x_B value at E is given by Eq. (12.34) and differs from the mole fractions of B in the two phases in equilibrium. Confusion about this point is a frequent source of student error.

The horizontal line HEI is called a tie line. A **tie line** on a phase diagram is a line whose endpoints correspond to the compositions of two phases in equilibrium with each other. The endpoints of a tie line lie at the boundaries of the two-phase region.

The two-phase region between the liquid and vapor curves is a gap in the phase diagram in which a single homogeneous phase cannot exist. Two-phase regions are shown shaded in this chapter.

A point in a two-phase region of a two-component phase diagram gives the system's overall composition, and the compositions of the two phases in equilibrium are given by the points at the ends of the tie line through that point.

EXAMPLE 12.3 Phase compositions at a point on a tie line

For a liquid–vapor system whose state corresponds to point E in the Fig. 12.10 phase diagram, find the overall mole fraction of B in the system and find the mole fraction of B in each phase present in the system. Assume the scales in the phase diagram are linear.

The overall x_B [Eq. (12.34)] corresponds to the x_B value at point E. The length from $x_B = 0$ to $x_B = 1$ in Fig. 12.10 is 5.98 cm. The distance from $x_B = 0$ to the intersection of the vertical line from E with the x_B axis is 3.59 cm. Therefore the overall x_B for the system at E is $x_B = 3.59/5.98 = 0.60_0$. The system at E consists of liquid and vapor phases in equilibrium. The liquid-phase composition is given by point H at the left end of the tie line HEI as $x_{B,3}$. The distance from $x_B = 0$ to $x_B = x_{B,3}$ is 2.79 cm, so $x_{B,3} = 2.79/5.98 = 0.46_7 = x_B^l$. The vapor-phase composition is given by point I on the HEI tie line as $x_{B,2} = 4.22/5.98 = 0.70_6 = x_B^v$.

Exercise

For a liquid–vapor system whose state corresponds to the point that is the intersection of the vertical line from H with line JF, find the overall x_B, x_B^l, and x_B^v. (*Answer:* $x_B = 0.46_7$, $x_B^l = 0.33_2$, $x_B^v = 0.60_0$.)

For a two-phase, two-component system, the number of degrees of freedom is $f = c_{ind} - p + 2 = 2 - 2 + 2 = 2$. In the Fig. 12.10 phase diagram, T is held fixed and this reduces f to 1 in the two-phase region of Fig. 12.10. Hence, once P is fixed, f is 0 in this two-phase region. For a fixed P, both x_B^v and x_B^l are thus fixed. For example, at pressure P_E in Fig. 12.10, x_B^v is fixed as $x_{B,2}$ and x_B^l is fixed as $x_{B,3}$. The overall x_B depends on the relative amounts of the liquid and vapor phases that are present in equilibrium. Recall that the masses of the phases, which are extensive variables, are not considered in calculating f.

Different relative amounts of liquid and vapor phases at the pressure P_E in Fig. 12.10 correspond to different points along the tie line HEI with different values of the overall mole fraction x_B but the same value of x_B^l and the same value of x_B^v. We now relate the location of point E on the tie line to the relative amounts of the two phases present. For the two-phase, two-component system, let n_B, n^l, and n^v be the total number of moles of B, the total number of moles in the liquid phase, and the total number of moles in the vapor phase, respectively. The overall mole fraction of B is $x_B = n_B/(n^l + n^v)$, so $n_B = x_B n^l + x_B n^v$. Also, $n_B = n_B^l + n_B^v = x_B^l n^l + x_B^v n^v$. Equating these two expressions for n_B, we get

$$x_B n^l + x_B n^v = x_B^l n^l + x_B^v n^v$$
$$n^l(x_B - x_B^l) = n^v(x_B^v - x_B) \tag{12.39}$$
$$n^l \overline{EH} = n^v \overline{EI} \tag{12.40}$$

where $\overline{EH}$ and $\overline{EI}$ are the lengths of the lines from E to the liquid and vapor curves in Fig. 12.10 and n^l and n^v are the total numbers of moles in the liquid and vapor phases, respectively. Equation (12.40) is the **lever rule.** Note its resemblance to the lever law of physics: $m_1 l_1 = m_2 l_2$, where m_1 and m_2 are masses that balance each other on a

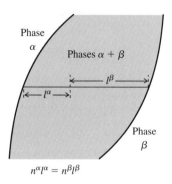

$$n^\alpha l^\alpha = n^\beta l^\beta$$

Figure 12.11

The lever rule gives the ratio of moles present in each phase of a two-phase binary system as n^α/n^β $= l^\beta/l^\alpha$, where l^β and l^α are the distances from the point corresponding to the system's overall mole fraction to the endpoints of the tie line.

seesaw with fulcrum a distance l_1 from mass m_1 and l_2 from m_2. When point E in Fig. 12.10 is close to point H on the liquid line, $\overline{EH}$ is less than $\overline{EI}$ and (12.40) tells us that n^l is greater than n^v. When E coincides with H, then $\overline{EH}$ is zero and n^v must be zero; only liquid is present.

The above derivation of the lever rule clearly applies to any two-phase, two-component system, not just to liquid–vapor equilibrium. Therefore, if α and β are the two phases present, n^α and n^β are the *total* numbers of moles in phase α and in phase β, respectively, and l^α and l^β are the lengths of the lines from a point in a two-phase region of the phase diagram to the phase-α and phase-β lines, then by analogy to (12.40) we have (Fig. 12.11)

$$n^\alpha l^\alpha = n^\beta l^\beta \qquad (12.41)^*$$

A common student error is to write $n_B^\alpha l^\alpha = n_B^\beta l^\beta$ instead of (12.41).

If the overall weight fraction of B (instead of x_B) is used as the abscissa of the phase diagram, the masses replace the numbers of moles in the above derivation, and the lever rule becomes

$$m^\alpha l^\alpha = m^\beta l^\beta \qquad (12.42)$$

where m^α and m^β are the masses of phases α and β.

EXAMPLE 12.4 Phase compositions in a two-phase region

Let the two-component system of Fig. 12.10 contain 10.00 mol of B and 6.66 mol of C and have pressure P_E. How many phases are present in the system? Find the number of moles of B present in each phase.

The overall B mole fraction is $x_B = 10.00/(10.00 + 6.66) = 0.600$. The length from $x_B = 0$ to $x_B = 1$ in Fig. 12.10 is 5.98 cm. We have 0.600(5.98 cm) = 3.59 cm, so the overall x_B lies 3.59 cm to the right of $x_B = 0$. The system's pressure is P_E, so the system's state must lie on the horizontal line at P_E. Locating the point on this line that lies 3.59 cm to the right of $x_B = 0$, we arrive at point E. Since point E is in the two-phase region, the system has two phases.

At pressure P_E, the endpoints H and I of the tie line HEI give the mole-fraction compositions of the liquid and vapor phases in equilibrium. In Example 12.3, we found that at H, $x_B^l = x_{B,3} = 0.46_7$ and at I, $x_B^v = x_{B,2} = 0.70_6$. To solve this problem, we can use the conservation-of-matter equation $n_B = n_B^\alpha + n_B^\beta$ or we can use the lever rule. We have

$$n_B = n_B^l + n_B^v = x_B^l n^l + x_B^v n^v$$
$$10.0 \text{ mol} = 0.46_7 n^l + 0.70_6(16.66 \text{ mol} - n^l)$$
$$n^l = 7.3_7 \text{ mol}$$
$$n_B^l = x_B^l n^l = 0.46_7(7.3_7 \text{ mol}) = 3.4_4 \text{ mol},$$
$$n_B^v = 10.00 \text{ mol} - 3.4_4 \text{ mol} = 6.5_6 \text{ mol}$$

In working this problem, it is essential to avoid confusing the quantities n_B^l (the number of moles of B in the liquid phase), n^l (the total number of moles in the liquid phase), and n_B (the total number of moles of B in the system).

An alternative solution uses the lever rule (12.41): $n^l \overline{EH} = n^v \overline{EI}$. We have $\overline{EH}$ $= 0.60_0 - 0.46_7 = 0.13_3$ and $\overline{EI} = 0.70_6 - 0.60_0 = 0.10_6$, so the lever rule gives $n^l(0.133) = (16.66 \text{ mol} - n^l)0.106$ and $n^l = 7.3_9$ mol. Then $n_B^l = x_B^l n^l =$ etc.

Exercise

If the system of Fig. 12.10 contains 0.400 mol of B and 0.600 mol of C and is at pressure P_F, find the number of moles of B present in each phase. (*Answer:* n_B^l $= 0.24_8$, $n_B^v = 0.15_2$.)

Ideal Solution at Fixed Pressure

Now consider the fixed-pressure liquid–vapor phase diagram of two liquids that form an ideal solution. The explanation is quite similar to the fixed-temperature case just discussed in great and somewhat repetitious detail, so we can be brief here. We plot T versus x_B, the overall mole fraction of one component. The phase diagram is Fig. 12.12.

T_C^* and T_B^* are the normal boiling points of pure C and pure B if the fixed pressure is 1 atm. The lower curve gives T as a function of x_B^l (or vice versa) for a system with liquid and vapor phases in equilibrium and is the boiling-point curve of the ideal solution. The upper curve gives T as a function of x_B^v (or vice versa) for a system with liquid–vapor equilibrium. The vapor curve lies above the liquid curve on a T-versus-x_B diagram but lies below the liquid curve on a P-versus-x_B diagram (Fig. 12.10). This is obvious since the vapor phase is favored by high T and by low P.

If we draw a horizontal tie line across the width of the $l + v$ region, the endpoints of the tie line give the compositions of the liquid and vapor phases in equilibrium with each other at the tie-line temperature and at the fixed pressure of Fig. 12.12. For example, line LQ is the tie line at temperature T_1, and the liquid and vapor compositions in equilibrium at T_1 and the pressure of the diagram are x_B^l and $x_{B,1}$.

If we isobarically heat a closed system of composition x_B^l, vapor will first appear at point L. As we raise the temperature and vaporize more of the liquid, the liquid will become richer in the less volatile, higher-boiling component C. Eventually, we reach point N, where the last drop of liquid vaporizes.

The first vapor that comes off when a solution of composition x_B^l is boiled has a value of x_B^v given by point Q. If we remove this vapor from the system and condense it, we get liquid of composition $x_{B,1}$. Vaporization of this liquid gives vapor of initial composition $x_{B,2}$ (point R). Thus by successively condensing and revaporizing the mixture, we can ultimately separate C from B. This procedure is called *fractional distillation*. We get the maximum enrichment in B by taking just the first bit of vapor that comes off. This maximum degree of enrichment for any one distillation step is said to represent one *theoretical plate*. By packing the distillation column, we in effect get many successive condensations and revaporizations, giving a column with several theoretical plates. Industrial distillation columns are up to 75 m (250 ft) high and may have hundreds of theoretical plates.

How do we plot the two curves in Fig. 12.12? We start with $P_B^*(T)$ and $P_C^*(T)$, the known vapor pressures of pure B and pure C as functions of T. Let the fixed pressure be $P^\#$. We have $P^\# = P_B + P_C$, where P_B and P_C are the partial pressures of B and C in the vapor. Raoult's law gives $P^\# = x_B^l P_B^*(T) + (1 - x_B^l) P_C^*(T)$, and

$$x_B^l = \frac{P^\# - P_C^*(T)}{P_B^*(T) - P_C^*(T)} \qquad \text{ideal soln.} \qquad (12.43)$$

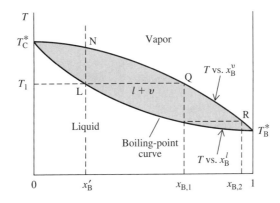

Figure 12.12

Temperature-versus-composition liquid–vapor phase diagram for an ideal solution at fixed pressure. T_B^* and T_C^* are the boiling points of pure B and pure C at the pressure of the diagram.

Since P_B^* and P_C^* are known functions of T, we can use (12.43) to find x_B^l at any given T and thereby plot the lower (liquid) curve. To plot the vapor curve, we use $x_B^v = P_B/P^\#$ $= x_B^l P_B^*/P^\#$; substitution of (12.43) gives

$$x_B^v = \frac{P_B^*(T)}{P^\#} \frac{P^\# - P_C^*(T)}{P_B^*(T) - P_C^*(T)} \qquad \text{ideal soln.} \qquad (12.44)$$

which is the desired equation for x_B^v as a function of T. Equations (12.43) and (12.44) are the same as (12.37) and (12.38), except that P is now fixed at $P^\#$ and T is regarded as a variable.

Nonideal Solutions

Having examined liquid–vapor equilibrium for ideal solutions, we now consider non-ideal solutions. Liquid–vapor phase diagrams for nonideal systems are obtained by measurement of the pressure and composition of the vapor in equilibrium with liquid of known composition. If the solution is only slightly nonideal, the curves resemble those for ideal solutions. If, however, the solution has a great enough deviation from ideality to give a maximum or minimum in the P-versus-x_B^l curve (as in Fig. 9.21), a new phenomenon appears.

Suppose the system shows a maximum in the P-versus-x_B^l curve, which is the upper curve in the P-versus-x_B phase diagram. What does the lower curve (the vapor curve) look like? Suppose we imagine the phase diagram to look like Fig. 12.13. Let point D in Fig. 12.13 be the maximum on the liquid curve. If we start at point A in a closed system and isothermally reduce the pressure, we shall reach point D, where the liquid just begins to vaporize. What is the composition of the first vapor that comes off? To answer this, we want the value of x_B^v that corresponds to the pressure (desig-nated P_{max} in Fig. 12.13) at point D. However, there is no point on the vapor curve (the lower curve) in Fig. 12.13 with pressure P_{max}. Hence, the phase diagram cannot look like Fig. 12.13. The only way (consistent with the requirement that the vapor phase be favored by low pressure and therefore always lie below the liquid phase) we can draw the phase diagram so that there is a point on the vapor curve with pressure P_{max} is to have the vapor curve touch the liquid curve at P_{max}, as in Fig. 12.14a.

What does the fixed-pressure T-versus-x_B phase diagram that corresponds to Fig. 12.14a look like? Let T' be the temperature for which Fig. 12.14a is drawn, and let $x_{B,1}$ be the value of x_B that corresponds to P_{max}. If P is fixed at P_{max}, liquid with x_B^l equal to $x_{B,1}$ will boil at temperature T'. However, liquid with x_B^l less than or greater than $x_{B,1}$ will not have sufficient vapor pressure to boil at T' and will boil at higher temperatures. Therefore, a maximum on the P-x_B phase diagram will correspond to a minimum on the T-x_B diagram. The T-versus-x_B phase diagram will look like Fig. 12.14b.

Let the minimum in Fig. 12.14b occur at composition x_B'. (If the fixed value of P for Fig. 12.14b equals P_{max} in Fig. 12.14a, then x_B' in Fig. 12.14b equals $x_{B,1}$ in Fig. 12.14a. Usually, P is fixed at 1 atm, so x_B' and $x_{B,1}$ usually differ.) A liquid of

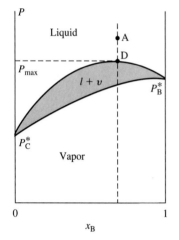

Figure 12.13

Erroneous pressure-versus-composition liquid–vapor phase diagram with a maximum.

Figure 12.14

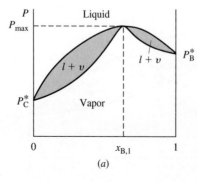
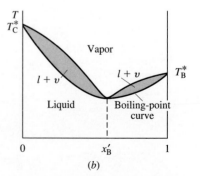

(a) Pressure-versus-composition liquid–vapor phase diagram with a maximum. (b) The corresponding temperature-versus-composition diagram.

composition x'_B when boiled will yield vapor with the same composition as the liquid. Since vaporization does not change the liquid's composition, the entire sample of liquid will boil at a constant temperature. Such a constant-boiling solution is called an **azeotrope.** The boiling behavior of an azeotropic solution resembles that of a pure compound and contrasts with that of most solutions of two liquids, which boil over a temperature range. However, since the azeotrope's composition depends on the pressure, a mixture that shows azeotropic behavior at one pressure will boil over a temperature range at a different P. Thus, an azeotrope can be distinguished from a compound.

Drawing lines in Fig. 12.14*b* similar to those in Fig. 12.12, we see that fractional distillation of a solution of two substances that form an azeotrope leads to separation into either pure B and azeotrope (if $x^l_B > x'_B$) or pure C and azeotrope (if $x^l_B < x'_B$). A liquid–vapor phase diagram with an azeotrope resembles two nonazeotropic liquid–vapor diagrams placed side by side.

The most famous azeotrope is that formed by water and ethanol. At 1 atm, the azeotropic composition is 96 percent C_2H_5OH by weight (192 proof) and the boiling point is 78.2°C, which is below the normal boiling points of water and ethanol. Absolute (100%) ethanol cannot be prepared by distillation at 1 atm of a dilute aqueous solution of ethanol.

A tabulation of known azeotropes is L. H. Horsley, Azeotropic Data III, *Adv. Chem. Ser.* 116, American Chemical Society, 1973. About half of the binary systems examined show azeotropes.

Figure 12.12 shows that, when no azeotrope is formed, the vapor in equilibrium with a liquid is always richer in the lower-boiling (more volatile) component than the liquid. When, however, a minimum-boiling azeotrope is formed, Fig. 12.14*b* shows that for some liquid compositions the vapor is richer in the higher-boiling component.

A negative deviation from Raoult's law large enough to give a minimum in the P-versus-x^l_B curve gives a maximum on the T-x_B phase diagram and a maximum-boiling azeotrope.

If the positive deviation from ideality is large enough, the two liquids may become only partially miscible with each other. Liquid–liquid equilibrium for partially miscible liquids is discussed in Sec. 12.7; liquid–vapor equilibrium for this case is considered in Prob. 12.67.

Figure 12.15 summarizes key points about tie lines in a two-phase region. The principles illustrated in this figure also apply to the binary phase diagrams in later sections

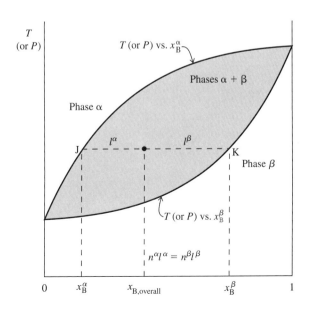

Figure 12.15

For a two-phase two-component system, the overall mole fraction $x_{B,overall}$ of B in the system is given by the location on the tie line of the point that describes the system's state. The mole fractions x^α_B and x^β_B of B in each phase are given by the endpoints of the tie line. The lever rule $n^\alpha l^\alpha = n^\beta l^\beta$ gives the ratio n^α/n^β of the total moles in each phase.

of this chapter. All points between J and K on the tie line correspond to states with the same value of x_B^α and the same x_B^β.

12.7 TWO-COMPONENT LIQUID–LIQUID EQUILIBRIUM

When any amounts of ethanol and water are shaken together in a separatory funnel at room temperature, one always obtains a single-phase liquid system. Ethanol and water are soluble in each other to unlimited extents and are said to be **completely miscible.** When roughly equal amounts of 1-butanol and water are shaken together at room temperature, one obtains a system consisting of two liquid phases: one phase is water containing a small amount of dissolved 1-butanol, and the other is 1-butanol containing a small amount of dissolved water. These two liquids are **partially miscible,** meaning that each is soluble in the other to a limited extent.

With P held fixed (typically at 1 atm), the most common form of the T-versus-x_B liquid–liquid phase diagram for two partially miscible liquids B and C looks like Fig. 12.16. To understand this diagram, imagine we start with pure C and gradually add B while keeping the temperature fixed at T_1. The system's state starts at point F (pure C) and moves horizontally to the right. Along FG, one phase is present, a dilute solution of solute B in solvent C. At point G, we have reached the maximum solubility of liquid B in liquid C at T_1. Addition of more B then produces a two-phase system for all points between G and E: phase 1 is a dilute saturated solution of B in C and has composition $x_{B,1}$; phase 2 is a dilute saturated solution of C in B and has composition $x_{B,2}$. The overall composition of the two-phase system at a typical point D is $x_{B,3}$. The relative amounts of the two phases present in equilibrium are given by the lever rule. At D, there is more of phase 1 than phase 2. As we continue to add more B, the overall composition eventually reaches point E. At E, there is just enough B present to allow all the C to dissolve in B to form a saturated solution of C in B. The system therefore again becomes a single phase at E. From E to H we are just diluting the solution of C in B. To actually reach H requires the addition of an infinite amount of B.

With two components and two phases present in equilibrium, the number of degrees of freedom is 2. However, since both P and T are fixed along line GE, f is 0 on GE. Two points on GE have the same value for each of the intensive variables P, T, $x_{C,1}$, $x_{B,1}$, $x_{C,2}$, $x_{B,2}$.

As the temperature is raised, the region of liquid–liquid immiscibility decreases, until at T_c (the **critical solution temperature**) it shrinks to zero. Above T_c, the liquids are completely miscible. The critical point at the top of the two-phase region in Fig.

Figure 12.16

Temperature-versus-composition liquid–liquid phase diagram for two partially miscible liquids. P is held fixed.

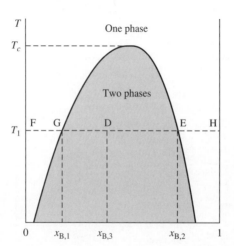

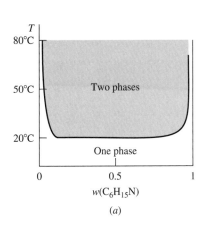

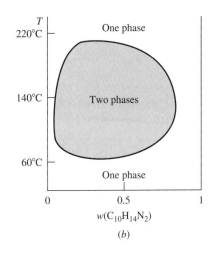

Figure 12.17

Temperature-versus-composition
liquid–liquid phase diagrams
for (a) water–triethylamine;
(b) water–nicotine. The horizontal
axis is the weight fraction of the
organic liquid. In (b), the pressure
of the system equals the vapor
pressure of the solution(s) and so
is not fixed.

12.16 is similar to the liquid–vapor critical point of a pure substance, discussed in Sec. 8.3. In both cases, as the critical point is approached, the properties of two phases in equilibrium become more and more alike, until at the critical point the two phases become identical, yielding a one-phase system.

For certain pairs of liquids, decreasing temperature leads to greater miscibility, and the liquid–liquid diagram resembles Fig. 12.17a. An example is water–triethylamine. Occasionally, a system shows a combination of the behaviors in Figs. 12.16 and 12.17a, and the phase diagram resembles Fig. 12.17b. Such systems have lower and upper critical solution temperatures. Examples are nicotine–water and m-toluidine–glycerol. The lower critical solution temperatures in Fig. 12.17 are due to an increase in the hydrogen bonding between water and the amine as T decreases; see J. S. Walker and C. A. Vause, *Scientific American,* May 1987, p. 98.

The two-phase regions in Figs. 12.16 and 12.17 are called **miscibility gaps.**

Although it is often stated that gases are miscible in all proportions, in fact several cases of gas–gas miscibility gaps are known. Examples include CO_2–H_2O, NH_3–CH_4, and He–Xe. These gaps occur at temperatures above the critical temperatures of both components and hence by the conventional terminology of Sec. 8.3 involve two gases. Most such gaps occur at rather high pressures and liquidlike densities; however, n-butane–helium shows a miscibility gap at pressures as low as 40 atm. See R. P. Gordon, *J. Chem. Educ.,* **49,** 249 (1972).

EXAMPLE 12.5 Phase compositions in a two-phase region

Figure 12.18 shows the liquid–liquid phase diagram of water (W) plus 1-butanol (B) at the vapor pressure of the system. Find the number of moles of each substance in each phase if 4.0 mol of W and 1.0 mol of B are shaken together at 30°C.

The overall x_B is (1.0 mol)/(5.0 mol) = 0.20. At 30°C, the point $x_B = 0.20$ lies in the two-phase region. Drawing a tie line at 30°C across the width of the two-phase region, we get line RS. Let α and β denote the phases present. Point R lies at $x_B^\alpha = 0.02$. Point S lies at $x_B^\beta = 0.48$. We have

$$n_B = n_B^\alpha + n_B^\beta = x_B^\alpha n^\alpha + x_B^\beta n^\beta$$

$$1.0 \text{ mol} = 0.02 n^\alpha + 0.48(5.0 - n^\alpha)$$

$$n^\alpha = 3.04 \text{ mol}, \qquad n^\beta = 5.00 \text{ mol} - 3.04 \text{ mol} = 1.9_6 \text{ mol}$$

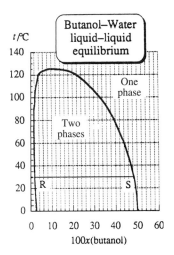

Figure 12.18

Butanol–water liquid–liquid phase diagram at 1 atm.

$$n_B^\alpha = x_B^\alpha n^\alpha = 0.02(3.0_4 \text{ mol}) = 0.06 \text{ mol}, \quad n_B^\beta = 0.48(1.9_6 \text{ mol}) = 0.94 \text{ mol}$$

$$n_W^\alpha = n^\alpha - n_B^\alpha = 3.0_4 \text{ mol} - 0.06 \text{ mol} = 2.9_8 \text{ mol}$$

$$n_W^\beta = n_W - n_W^\alpha = (4.0 - 2.9_8) \text{ mol} = 1.0_2 \text{ mol}$$

Alternatively, the lever rule can be used.

Exercise

Solve this problem using the lever rule.

Exercise

Repeat this example for 3.0 mol of W and 1.0 mol of B shaken together at 90°C.
(*Answer:* $n_W^\alpha = 1.3$ mol, $n_B^\alpha = 0.02_6$ mol, $n_W^\beta = 1.7$ mol, $n_B^\beta = 0.98$ mol.)

Partition Coefficients

Suppose solvents A and B are partly miscible at temperature T and form the phases α (a dilute solution of B in solvent A) and β (a dilute solution of A in B) when shaken at T. If we add solute i to the system, it will distribute itself between the phases α and β so as to satisfy $\mu_i^\alpha = \mu_i^\beta$. Using the concentration scale, we have [Eq. (10.29)]

$$\mu_{c,i}^{\circ,\alpha} + RT \ln \left(\gamma_{c,i}^\alpha c_i^\alpha / c^\circ \right) = \mu_{c,i}^{\circ,\beta} + RT \ln \left(\gamma_{c,i}^\beta c_i^\beta / c^\circ \right)$$

$$\ln \left(\gamma_{c,i}^\alpha c_i^\alpha / \gamma_{c,i}^\beta c_i^\beta \right) = -(\mu_{c,i}^{\circ,\alpha} - \mu_{c,i}^{\circ,\beta})/RT$$

$$K_{AB,i} \equiv \frac{c_i^\alpha}{c_i^\beta} = \frac{\gamma_{c,i}^\beta}{\gamma_{c,i}^\alpha} \exp\left[-(\mu_{c,i}^{\circ,\alpha} - \mu_{c,i}^{\circ,\beta})/RT \right] \tag{12.45}$$

The quantity $K_{AB,i} \equiv c_i^\alpha / c_i^\beta$ is the **partition coefficient** (or **distribution coefficient**) for solute i in solvents A and B. (Recall separatory-funnel extractions done in organic chem lab). $K_{AB,i}$ is not accurately equal to the ratio of solubilities of i in A and B because phases α and β are not pure A and pure B. The exponential in (12.45) is a function of T and weakly a function of P. The equation preceding (12.45) is the relation $\Delta G^\circ = -RT \ln K^\circ$ for the "reaction" $i(\beta) \rightarrow i(\alpha)$.

As the amounts of i in phases α and β change, the activity-coefficient ratio in (12.45) changes, and the concentrations of B in phase α and A in phase β also change (see Sec. 12.12). Therefore $K_{AB,i}$ depends on how much i was added to the system and is not a true constant at fixed T and P, unless α and β are ideally dilute solutions. The $K_{AB,i}$ value tabulated in the literature is the value corresponding to very dilute solutions of i in α and β, where the activity coefficients are very close to 1 and the compositions of phases α and β are very close to what they would be in the absence of solute i.

The **octanol/water partition coefficient** K_{ow} of a solute between the phases formed by 1-octanol and water is c^{oct}/c^{wat}, where oct denotes the octanol-rich phase. K_{ow} is widely used in drug and environmental studies as a measure of how a solute distributes itself between an organic phase and an aqueous phase. The octanol–water liquid–liquid phase diagram resembles Fig. 12.18; at 25°C, the phases in equilibrium have $x_{octanol}^\alpha = 0.793$ and $x_{H_2O}^\beta = 0.993$.

A drug with too high a K_{ow} will tend to accumulate in fatty tissue of the body and might not reach its intended target. A drug with too low a K_{ow} will not readily go through cell membranes (which are lipidlike).

Fish swimming in polluted water may have concentrations of a pollutant such as DDT that are thousands of times the concentration in the water, due to the high solubility of the pollutant in the fish's fatty tissue. The *bioconcentration factor* BCF is defined by the equilibrium concentration ratio: BCF $\equiv c_{organism}/c_{water}$. Measurement of a BCF (which varies with species of fish) is time-consuming and costly ($30000), and one can roughly estimate BCF of an organic compound with low polarity from K_{ow} using $\log_{10} BCF \approx \log_{10} K_{ow} - 1.32$ if $1.5 < \log K_{ow} < 6.5$ [D. Mackay, *Environ. Sci. Technol.*, **16**, 274 (1982); for better equations see W. M. Meylan et al., *Environ. Toxicol. Chem.*, **18**, 664 (1999)]. High K_{ow} values are also correlated with high values of preferential absorption of organic pollutants in soil. The main substances in soil that absorb organic pollutants are mixtures of organic compounds.

TWO-COMPONENT SOLID–LIQUID EQUILIBRIUM

We now discuss binary solid–liquid diagrams. The effect of pressure on solids and liquids is slight, and unless one is interested in high-pressure phenomena, one holds P fixed at 1 atm and examines the T-x_B solid–liquid phase diagram.

Liquid-Phase Miscibility and Solid-Phase Immiscibility

Let substances B and C be miscible in all proportions in the liquid phase and completely immiscible in the solid phase. Mixing any amounts of liquids B and C will produce a single-phase system that is a solution of B plus C. Since solids B and C are completely insoluble in each other, cooling a liquid solution of B and C will cause either pure B or pure C to freeze out of the solution.

The typical appearance of the solid–liquid phase diagram for this case is shown in Fig. 12.19. T_B^* and T_C^* are the freezing points of pure B and pure C.

The origin of the regions on this diagram is as follows. In the low-temperature limit, we have a two-phase mixture of pure solid B plus pure solid C, since the solids are immiscible. In the high-T limit, we have a one-phase liquid solution of B plus C, since the liquids are miscible. Now consider cooling a liquid solution of B and C that has x_B^l close to 1 (the right side of the diagram). Eventually, we reach a temperature where the solvent B begins to freeze out, giving a two-phase region with solid B in equilibrium with a liquid solution of B and C. The curve DE thus gives the depression of the freezing point of B due to solute C. Likewise, if we cool a liquid solution of B plus C that has x_C^l close to 1 (the left side of the diagram), we eventually get pure C freezing out and AFGE is the freezing-point-depression curve of C due to solute B. If we cool a two-phase mixture of solution plus either solid, the solution will eventually all freeze, giving a mixture of solid B and solid C.

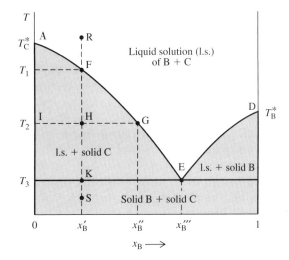

Figure 12.19

Solid–liquid phase diagram for complete liquid miscibility and solid immiscibility. P is held fixed.

The two freezing-point curves intersect at point E. For a solution with x_B^l to the left of E, solid C will freeze out as T is lowered. For x_B^l to the right of E, solid B will freeze out. At the values of T and x_B^l corresponding to point E, the chemical potentials of B and C in the solution equal the chemical potentials of pure solid B and C, respectively, and both B and C freeze out when a solution with the eutectic composition x_B''' is cooled. Point E is the **eutectic point** (Greek *eutektos,* "easily melted").

The portion of line DE with x_B very close to 1 can be calculated from the ideally dilute–solution equation (12.14) with A and B replaced by B and C, respectively. Likewise, the portion of line AFGE with x_B very close to 0 (and hence x_C very close to 1) can be calculated from (12.14) with A and B replaced by C and B, respectively. Away from the ends of these lines, Eq. (12.7) with A replaced by B or C applies. This exact equation is hard to use to find the freezing point T_f as a function of x_B or x_C. To get a rough idea of the shape of the curves DE and AE, we neglect the temperature dependences of $\Delta_{fus}H_{m,B}$ and $\Delta_{fus}H_{m,C}$; and we approximate γ_B and γ_C as 1 over the full range of solution composition (this ideal-solution approximation is usually quite poor). Thus the very approximate equation for DE is

$$R \ln x_B \approx \Delta_{fus}H_{m,B}\left(\frac{1}{T_B^*} - \frac{1}{T}\right) \qquad \text{for DE} \qquad (12.46)$$

The approximate equation for AE is obtained by replacing B with C in (12.46).

Suppose we start at point R in Fig. 12.19 and isobarically cool a liquid solution of B and C with composition x_B'. The overall composition of the closed system remains constant at x_B', and we proceed vertically down from R. When T reaches T_1, solid C starts to freeze out. As C freezes out, x_B^l increases and (since B is the solute here) the freezing point is lowered further. To freeze out more of the solvent (C), we therefore must lower the temperature further. At a typical temperature T_2, there is an equilibrium between a solution whose composition is given by point G as x_B'' and solid C, whose composition is given by point I as $x_B = 0$. As usual, the points at the ends of the tie line (GHI) give the compositions of the two phases in equilibrium. The lever rule gives $n_C^s \overline{HI} = (n_B^l + n_C^l)\overline{HG}$, where n_C^s is the number of moles of solid C in equilibrium with a solution of n_B^l moles of B plus n_C^l moles of C. At point F, the lever rule gives $n_C^s = 0$. As T drops along the line FHK, the horizontal distance to the line AFGE increases, indicating an increase in n_C^s.

As T is lowered further, we finally reach the **eutectic temperature** T_3 at point K. Here, the solution has composition x_B''' (point E), and now both solid C and solid B freeze out, since both solids freeze out when a solution with the eutectic composition is cooled. The relative amounts of B and C that freeze out at E correspond to the eutectic composition x_B''', and the entire remaining solution freezes at T_3 with no further change in composition. At K, three phases are in equilibrium (solution, solid B, and solid C), so the lever rule (12.41) does not apply. With three phases, we have $f = 2 - 3 + 2 = 1$ degree of freedom; this degree of freedom has been eliminated by the specification that P is fixed at 1 atm. Hence there are no degrees of freedom for the three-phase system, and the temperature must remain constant at T_3 until all the solution has frozen and the number of phases has dropped to 2. Below T_3, we are simply cooling a mixture of solid B plus solid C. The endpoints of a horizontal tie line drawn through point S lie at $x_B = 0$ and $x_B = 1$, and these are the compositions of the phases [pure C(s) and pure B(s)] present at S.

If we reverse the process and start at point S with solid B plus solid C, the first liquid formed will have the eutectic composition x_B'''. The system will remain at point K until all the B has melted, along with enough C to give a solution with eutectic composition. Then the remaining solid C will melt over the temperature range T_3 to T_1. (Sharpness of melting point is one test organic chemists use for the purity of a compound.) A solid mixture that has the eutectic composition will melt entirely at one temperature (T_3). A solution of B and C that has the eutectic composition will freeze entirely at temperature T_3 to produce a eutectic mixture of solids B and C. However, a eutectic mixture is not a compound. Microscopic examination will show the eutectic solid to be an intimate mixture of crystals of B and crystals of C.

Systems with the solid–liquid phase diagram of Fig. 12.19 are called *simple eutectic systems*. Examples include Pb–Sb, benzene–naphthalene, Si–Al, KCl–AgCl, Bi–Cd, C_6H_6–CH_3Cl, and chloroform–aniline.

Solid Solutions

Certain pairs of substances form solid solutions. In a solid solution of B and C, there are no individual crystals of B or C. Instead, the molecules or atoms or ions are mixed together at the molecular level, and the composition of the solution can be varied continuously over a certain range. Solid solutions can be prepared by condensing a vapor of B plus C or by cooling a liquid solution of B and C. Two solids might be completely miscible, partly miscible, or completely immiscible.

In an *interstitial* solid solution, the B molecules or atoms (which must be small) occupy interstices (holes) in the crystal structure of substance C. For example, steel is a solution in which carbon atoms occupy interstices in the Fe crystal structure. In a *substitutional* solid solution, molecules or atoms or ions of B substitute for those of C at random locations in the crystal structure. Examples include Cu–Ni, Na_2CO_3–K_2CO_3, and *p*-dichlorobenzene–*p*-dibromobenzene. Substitutional solids are formed by substances with atoms, molecules, or ions of similar size and structure.

Analysis of a transition-metal oxide or sulfide frequently shows an apparent violation of the law of definite proportions. For example, ZnO usually has a Zn/O mole ratio slightly greater than 1. The explanation is that the "zinc oxide" is actually an interstitial solid solution of Zn in ZnO.

Liquid-Phase Miscibility and Solid-Phase Miscibility

Some pairs of substances are completely miscible in the solid state. Examples include Cu–Ni, Sb–Bi, Pd–Ni, KNO_3–$NaNO_3$, and *d*-carvoxime–*l*-carvoxime. With complete miscibility in both the liquid and the solid phases, the T-x_B binary phase diagram may look like Fig. 12.20, which is for Cu–Ni.

If a melt of Cu and Ni with any composition is cooled, a solid solution begins to freeze out. This solid solution is richer in Ni than the liquid solution. As the two-phase system of solid plus melt is cooled further, the mole fraction of Ni decreases in both the solid solution and the liquid melt. Eventually, a solid solution is formed that has the same composition as the liquid melt we started with.

Note that the freezing point of Cu is *raised* by the presence of a small amount of Ni. In discussing freezing-point depression in Sec. 12.3, we assumed solid-phase

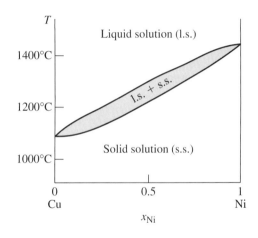

Figure 12.20

The Cu–Ni solid–liquid phase diagram at 1 atm.

immiscibility, so that only pure solid solvent froze out. When the solids are miscible, the freezing point of the lower-melting component may be raised by the presence of the second component. An analogous situation is boiling-point elevation. When the solute is involatile, the solvent boiling point is elevated. However, if the solute is more volatile than the solvent, the solvent boiling point may be depressed. Note the resemblance between the solid–liquid T-x_B diagrams in Figs. 12.20 and 12.21a and the liquid–vapor T-x_B diagrams in Figs. 12.12 and 12.14b, respectively.

When the two miscible solids form an approximately ideal solid solution, the solid–liquid phase diagram resembles Fig. 12.20. However, when there are large deviations from ideality, the solid–liquid phase diagram may show a minimum or a maximum. Figure 12.21a for Cu–Au shows a minimum. Figure 12.21b for the optical isomers d-carvoxime–l-carvoxime ($C_{10}H_{14}NOH$) shows a maximum. Here, the freezing point of each compound is elevated by the presence of the other. The strong negative deviation from ideality indicates that in the solid state, d-carvoxime molecules prefer to associate with l-carvoxime molecules rather than with their own kind.

Liquid-Phase Miscibility and Solid-Phase Partial Miscibility

When B and C are completely miscible in the liquid phase and partly miscible in the solid phase, the T-x_B diagram looks like Fig. 12.22, which is for Cu–Ag.

If a liquid melt (solution) of Cu and Ag with $x_{Cu} = 0.2$ is cooled, at point S a solid phase (called the α phase) that is a saturated solution of Cu in Ag begins to separate out. The initial composition of this solid solution is given by point Y at the end of the SY tie line. As the two-phase mixture of liquid solution plus solid solution is cooled further, the percentage of Cu in the solid solution that is in equilibrium with the melt

Figure 12.21

Solid–liquid T-x_B phase diagrams at 1 atm for (a) Cu–Au; (b) d–carvoxime–l-carvoxime.

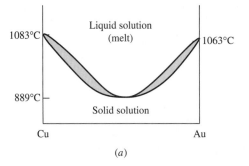
(a)

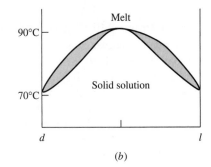
(b)

Figure 12.22

Solid–liquid phase diagram for Cu–Ag at 1 atm.

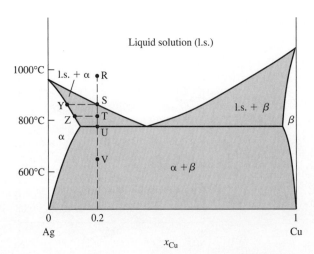

increases. At point U, the melt has the eutectic composition, and two solid phases now freeze out—the α phase (solid Ag saturated with Cu) and the β phase (solid Cu saturated with Ag). Examination of the solid at V will show large crystals of phase α (which formed before point U was reached) and tiny crystals of phases α and β (which formed at U).

One complication is that diffusion of molecules, atoms, and ions through solids is quite slow, and it takes a long time for equilibrium to be reached in a solid phase. At point T, the solid in equilibrium with the melt has a composition given by point Z, whereas the first solid frozen out had a composition given by Y. It may be necessary to hold the system at point T for a long time before the solid phase becomes homogeneous with composition Z throughout.

The rate of diffusion in solids depends on the temperature. At elevated temperatures not greatly below the melting points of the solids, solid-state diffusion is generally rapid enough to allow equilibrium to be attained in a few days. At room temperature, diffusion is so slow that many years may be required to reach equilibrium in a solid. In Wright Park in Manhattan, there stands the sculpture *3000* A.D. by Terry Fugate-Wilcox. This artwork is a 36-foot-high tower consisting of alternating slabs of aluminum and magnesium bolted together. Supposedly by the year 3000, solid-state diffusion will have transformed the sculpture into a homogeneous alloy of the two metals.

The two-phase region labeled $\alpha + \beta$ in Fig. 12.22 is a miscibility gap (Sec. 12.7). The two-phase regions α + liquid solution and β + liquid solution make up a **phase-transition loop.** The two-phase regions in Figs. 12.10 and 12.20 illustrate the simplest kind of phase-transition loop. Figures 12.22 and 12.14*b* each show a phase-transition loop with a minimum. Figure 12.22 shows the intersection of a miscibility gap with a phase-transition loop that has a minimum. Figure 12.23 shows how we can imagine a phase diagram like Fig. 12.22 to arise by having a solid-phase miscibility gap approach and ultimately intersect a solid–liquid phase-transition loop that has a minimum. The condensed-phase diagram of Ni–Au resembles Fig. 12.23*b*.

Certain solid–liquid phase diagrams result from the intersection of a solid-phase miscibility gap with a simple solid–liquid phase-transition loop like the one in Fig. 12.20. This gives a phase diagram like Fig. 12.24. The α phase is a solid solution of B in the C crystal structure; the β phase is a solid solution of C in B. If solid α of composition F is heated, it starts to melt at point G, forming a two-phase mixture of α and liquid solution of initial composition N. However, when point H is reached, the remaining portion of phase α "melts" to form liquid of composition M plus solid phase β of composition R; $\alpha(s) \rightarrow \beta(s)$ + liquid solution. During this transition, the three phases α, β, and liquid are present, and the number of degrees of freedom is $f = 2 - 3 + 2 = 1$; however, since P is held fixed at 1 atm, the system has 0 degrees of freedom, and the

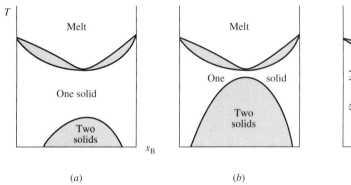

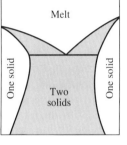

Figure 12.23

A solid-phase miscibility gap approaches and in (*c*) intersects a solid–liquid phase-transition loop.

Figure 12.24

A solid–liquid phase diagram with
a peritectic temperature.

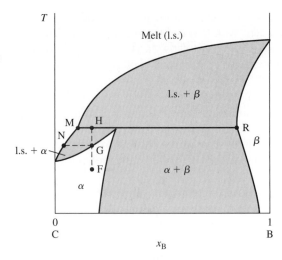

transition from α to β + liquid must occur at a fixed temperature (called the *peritectic temperature*). Further heating after the transition at H brings us first into a two-phase region of β plus liquid solution and finally into a one-phase region of liquid solution. A *peritectic phase transition* (for example, the transition at H) is one where heating transforms a solid phase to a liquid phase plus a second solid phase: $\text{solid}_1 \rightarrow \text{liquid} + \text{solid}_2$. In contrast, a eutectic phase transition has the pattern on heating: $\text{solid}_1 + \text{solid}_2 \rightarrow \text{liquid}$.

Compound Formation—Liquid-Phase Miscibility and Solid-Phase Immiscibility

Fairly commonly, substances B and C form a solid compound that can exist in equilibrium with the liquid. Figure 12.25 shows the solid–liquid phase diagram for phenol (P) plus aniline (A), which form the compound $C_6H_5OH \cdot C_6H_5NH_2$ (PA). The aniline mole fraction x_A on the abscissa is calculated pretending that only aniline and phenol (and no addition compound) are present. Although the system has $c = 3$ (instead of 2), the number of degrees of freedom is unchanged by compound formation, since we now have the equilibrium restriction $\mu_P + \mu_A = \mu_{PA}$. Thus, $c - r - a = c_{ind}$ in Eq. (7.10) is still 2, and the system is binary.

Figure 12.25 can be understood qualitatively by imagining it to consist of a simple eutectic diagram for phenol–PA adjacent to a simple eutectic diagram for PA–aniline. The liquid solution at the top of the diagram is an equilibrium mixture of P, A, and PA. Depending on the solution's composition, solid phenol, solid PA, or solid aniline will separate out on cooling, until one of the two eutectic temperatures is reached, at which time a second solid also freezes out. If a solution with $x_A = 0.5$ is cooled, only pure solid PA separates out and the solution freezes entirely at one temperature (31°C), the melting point of PA. Although the freezing-point-depression curves for P and for A each start off with nonzero slope, the freezing-point-depression curve of PA has zero slope at the PA melting point (proof of this is given in *Haase and Schönert*, p. 101).

As usual, the composition of each phase present in a two-phase region is given by the endpoints of a tie line drawn across the width of that region. For example, a (horizontal) tie line drawn in one of the l.s. + PA(*s*) regions of Fig. 12.25 extends from the vertical line at $x(A) = 0.5$ [corresponding to the phase PA(*s*)] to the curved boundary line between the l.s. + PA(*s*) region and the l.s. region.

Some systems exhibit formation of several compounds. If n compounds are formed, the solid–liquid phase diagram can be viewed as consisting of $n + 1$ adjacent

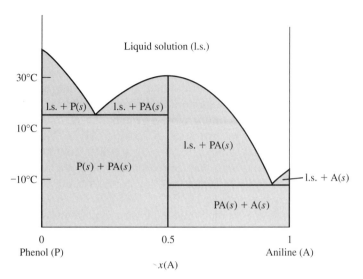

Figure 12.25

Phenol–aniline solid–liquid phase diagram at 1 atm. The symbols P(s), PA(s), and A(s) denote solid phenol, solid addition compound, and solid aniline, respectively.

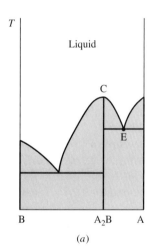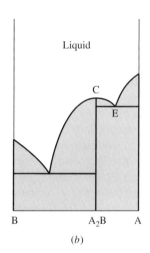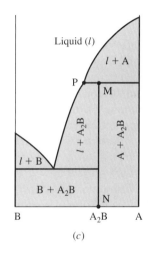

Figure 12.26

Origin of a peritectic point.

simple-eutectic phase diagrams (provided there are no peritectic points—see the next two paragraphs). For an example, see Prob. 12.51.

Compound Formation with Incongruent Melting—Liquid-Phase Miscibility and Solid-Phase Immiscibility. Figure 12.26a shows a phase diagram with formation of the solid compound A_2B. Now let the melting point of A be increased to give Fig. 12.26b. A further increase in the A melting point will yield Fig. 12.26c. In Fig. 12.26c, the freezing-point-depression curve of A no longer intersects the right-hand freezing-point-depression curve of A_2B (curve CE in Fig. 12.26a and b), so the eutectic point between the compound A_2B and A is eliminated. Instead, the intersection at point P produces the phase diagram of Fig. 12.26c. (The system K–Na has the phase diagram of Fig. 12.26c; the compound formed is Na_2K.)

Line MN is pure solid A_2B. If solid A_2B is heated, it melts sharply at temperature T_P to give a liquid solution (whose composition is given by point P) in equilibrium with pure solid A; $A_2B(s) \rightarrow A(s) +$ solution. Thus, at least some decomposition of the compound occurs on melting. Since the liquid solution formed has a different x_A value than the compound, the compound is said to melt *incongruently*. (The compound in Fig. 12.25 melts *congruently* to give a liquid with the same x_A as the solid compound.) Point P is called a *peritectic point*. When several compounds are formed, there is the possibility of more than one peritectic point. In the system Cu–La, the compounds LaCu and $LaCu_4$ melt incongruently, and the compounds $LaCu_2$ and $LaCu_6$ melt congruently.

Experimental Methods

One way to determine a solid–liquid phase diagram experimentally is by *thermal analysis.* Here, one allows a liquid solution (melt) of the two components to cool and measures the system's temperature as a function of time; this is repeated for several liquid compositions to give a set of cooling curves. The time variable t is approximately proportional to the amount of heat q lost from the system, so the slope dT/dt of a cooling curve is approximately proportional to the reciprocal of the system's heat capacity $C_P = dq_P/dT$. Typical cooling curves for the simple eutectic system of Fig. 12.19 are shown in Fig. 12.27.

When pure C is cooled (curve 1), the temperature remains constant at the freezing point T_C^* while the entire sample freezes. The heat capacity of the system $C(s) + C(l)$ at T_C^* is infinite (Sec. 7.5). The slight dip below the freezing point is due to supercooling. After the sample is frozen, the temperature drops as solid C is cooled. Curve 2 is for a liquid mixture with the composition of point R in Fig. 12.19. Here, when solid C begins to freeze out at T_1, the slope of the cooling curve changes. This slope change is called a **break.** The break occurs because the heat capacity of the system $C(s) + $ liq. soln. is greater than that of the system consisting of liquid solution only, since much of the heat removed from the former system serves to convert liquid C to solid C rather than to decrease the system's temperature. When the system reaches the eutectic temperature T_3, the entire remaining liquid freezes at a constant temperature and the cooling curve becomes horizontal, exhibiting what is called a eutectic **halt.** By plotting the temperatures of the observed cooling-curve breaks against x_B, we generate the freezing-point-depression curves AE and DE of Fig. 12.19.

Another way to determine phase diagrams is to hold a system of known overall composition at a fixed temperature long enough for equilibrium to be attained. The phases present are then separated and analyzed chemically. This is repeated for many different compositions and temperatures to generate the phase diagram.

Solid–liquid equilibria are commonly studied with the system open to the atmosphere. The solubility of air in the solid and liquid phases is generally slight enough to be ignored, and the atmosphere simply acts as a piston that provides a constant external pressure of 1 atm. The atmosphere is not part of the system, and in applying the phase rule to the system, one therefore does not count the atmosphere as one of the phases or include O_2 and N_2 as components of the system. Also, since the pressure is fixed, the number of degrees of freedom is reduced by 1.

Figure 12.27

Two cooling curves for Fig. 12.19. Curve 1 is for pure C. Curve 2 is for a solution of B in the solvent C.

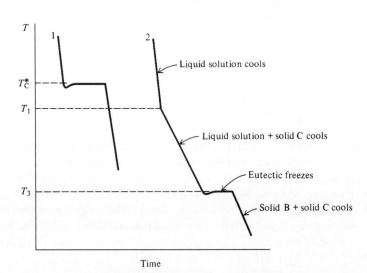

12.9   STRUCTURE OF PHASE DIAGRAMS

The one-component (unary) *P-T* phase diagrams of Chapter 7 contain *one-phase areas* separated by *two-phase lines;* the two-phase lines intersect in *three-phase points* (triple points). A phase transition (for example, solid to liquid) corresponds to moving from one one-phase area to another. During the course of the transition, the system contains two phases in equilibrium, and its state is represented by a point on a two-phase line. At a triple point, three phases coexist in equilibrium.

The binary *T-x_B* phase diagrams of Secs. 12.6 to 12.8 consist of *one-phase areas, two-phase areas, one-phase vertical lines,* and *three-phase horizontal lines.*

The two-phase areas are either miscibility gaps or phase-transition loops. Examples of miscibility gaps are the two-phase areas in Figs. 12.16 and 12.17, the $\alpha + \beta$ regions in Figs. 12.22 and 12.24, the solid B + solid C region of Fig. 12.19, and the regions at the bottom left and bottom right of Fig. 12.25. Examples of phase-transition loops are the $l + v$ regions in Figs. 12.12 and 12.14*b*, the l.s. + solid C and l.s. + solid B regions of Fig. 12.19, and the l.s. + α and l.s. + β regions of Fig. 12.22.

When the system consists of one phase, this phase is either a pure substance or a solution. If the phase is a pure substance, it has a fixed value of x_B and so corresponds to a vertical line on the *T-x_B* diagram. Examples are the lines at $x_B = 0$ and $x_B = 1$ in Fig. 12.19 and the lines at $x_A = 0$, 0.5, and 1 in Fig. 12.25. If the phase is a solution (solid, liquid, or gaseous), x_B can be varied continuously over a certain range, as can the temperature, so we get a one-phase area. Examples are the liquid regions at the top of Figs. 12.19 and 12.20, the vapor region in Fig. 12.12, the solid regions at the bottoms of Figs. 12.20 and 12.21, and the α phase and the β phase in Fig. 12.22. To repeat, a vertical one-phase line corresponds to a pure substance; a one-phase area corresponds to a solution.

When a binary system has three phases in equilibrium, $f = c_{ind} - p + 2 = 2 - 3 + 2 = 1$, but since *P* is fixed, *f* is reduced to 0. Therefore *T* is fixed, and the existence of three phases in equilibrium in a binary system must correspond to a horizontal (isothermal) line on the *T-x_B* diagram. Examples are the horizontal lines in Figs. 12.19, 12.22, 12.24, and 12.25.

The compositions of the phases in equilibrium on a three-phase line are given by the two points at the ends of the line and a third point that occurs at the intersection of a vertical line or phase boundary lines with the three-phase line.

When the system crosses the boundary between a one-phase area and a two-phase area, the cooling curve shows a break. The cooling curve shows a halt at a three-phase line and at the transition from a one-phase region (line or area) to another one-phase region (for example, at the freezing points of pure substances such as P, A, and PA in Fig. 12.25).

Each point on a phase diagram represents an equilibrium state. The location of the point on the diagram tells us the temperature, pressure, and mole fractions in each phase. For a two-phase system, the location of the point along the tie line tells us the ratio of the total number of moles in one phase to that in the other.

12.10   SOLUBILITY

In a solubility equilibrium, we have pure solid C in equilibrium with a saturated liquid solution of C plus D, where C is labeled the solute and D the solvent (Fig. 12.28). In a freezing-point-depression equilibrium, we have pure solid C in equilibrium with a liquid solution of C plus D, where C is labeled the solvent and D the solute (Figs. 12.28 and 12.3). Since the labeling of solute and solvent is arbitrary, the freezing-point-depression and solubility situations are fundamentally the same, and we can apply the ideas and equations of freezing-point depression to solubility simply by interchanging the designations of solute and solvent.

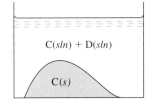

Figure 12.28

A system with pure solid C in equilibrium with a solution of C + D. We can interpret the system as giving the solubility of solid C in liquid D or as giving the freezing-point depression of liquid C due to added D.

For example, Fig. 12.29 shows the two-component solid–liquid T-versus-x phase diagram of the system benzene (ben) plus naphthalene (nap) at 1 atm. Curve CE is the freezing-point-depression curve of the solvent naphthalene (whose freezing point is 80°C) due to added solute benzene. Alternatively, CE can be interpreted as the solubility curve of the solute naphthalene in the solvent benzene. A point on the dashed 25°C tie line to the left of point S corresponds to solid naphthalene in equilibrium with a liquid solution having $x_{ben} = 0.7$ and $x_{nap} = 0.3$ (point S). The diagram shows that the freezing point of naphthalene is lowered from 80°C to 25°C when enough benzene has been added to liquid naphthalene to reduce x_{nap} to 0.3. The diagram also shows that the solubility of solid naphthalene in liquid benzene at 25°C and 1 atm is 0.3 mol of naphthalene per 0.7 mol of benzene. Note from curve CE that the solubility of naphthalene in benzene goes to infinity as the solution's temperature approaches the 80°C melting point of naphthalene. (The liquids are completely miscible.)

Since the solubility curve CE is the freezing-point depression curve of liquid naphthalene, in a solution of naphthalene in benzene at 25°C, the solute naphthalene is best considered to be in the liquid state. Note also that in the expression $\mu_i = \mu_i^\circ + RT \ln \gamma_i x_i$ for naphthalene in the solution (where Convention I is used), $\mu_i^\circ = \mu_i^*$ is the chemical potential of pure (supercooled) *liquid* naphthalene at the T and P of the solution.

What about the solubility of a salt in water? Figure 12.30 shows the T-x_A diagram for $NaNO_3$–H_2O. Figure 12.30 cheats a bit, since above 100°C the system's pressure is not held constant but is set at the vapor pressure of the solution by keeping the system in a sealed container. If the system were kept open to the atmosphere, it would boil away above 100°C. Since the effect of pressure on the chemical potentials of species in condensed phases is slight, the fact that P varies in part of the phase diagram is of little consequence.

Curve CE in Fig. 12.30 gives the freezing-point depression of water due to added $NaNO_3$. Alternatively, we can interpret CE as giving the solubility of solid water in supercooled liquid $NaNO_3$. This second interpretation sounds bizarre but is quite valid. Curve DE gives the freezing-point depression of liquid $NaNO_3$ due to added water. Alternatively, we can interpret DE as giving the solubility of solid $NaNO_3$ in water at various temperatures. The second interpretation sounds more natural, since we ordinarily prepare a liquid solution by adding solid $NaNO_3$ to liquid water (instead of cooling molten $NaNO_3 + H_2O$); also, the curve DE is ordinarily not investigated above 100°C.

Suppose we start with an equilibrium mixture of liquid water and a large amount of ice at 0°C (point C in Fig. 12.30) in a thermally insulated container. Let some solid

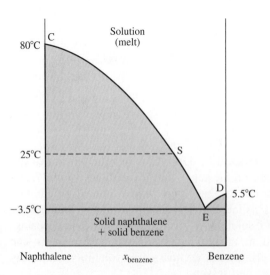

Figure 12.29

Solid–liquid phase diagram for benzene–naphthalene at 1 atm.

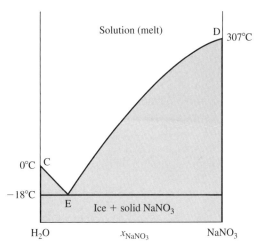

Figure 12.30

Solid–liquid phase diagram for $NaNO_3$–H_2O. (Not drawn to scale.)

NaNO$_3$ be added. The $NaNO_3$ dissolves in the liquid water and lowers the freezing point of the water to some temperature T_f that is below 0°C. The solution and the ice at 0°C are no longer in equilibrium, and the ice starts to melt. Since the system is adiabatically enclosed, its internal energy U must remain constant and the energy to melt the ice must come at the expense of the kinetic energy of the molecules of the system. The decrease in molecular kinetic energy means that the system's temperature falls. The temperature will fall until it reaches the freezing point T_f of the solution and equilibrium is attained.

The lowest temperature attainable by addition of $NaNO_3$ to ice plus water is the eutectic temperature −18°C (point E), where ice, solid $NaNO_3$, and a saturated aqueous $NaNO_3$ solution exist in equilibrium. Such a system will remain at −18°C until heat leaking into the system melts all the ice and the system moves into the region between the eutectic line and DE. Eutectic mixtures are used to provide a cold constant-temperature bath. For $NaCl$–H_2O, the eutectic temperature is −21°C; for $CaCl_2 \cdot 6H_2O$–H_2O it is −50°C.

Line ED in Fig. 12.30 shows the solubility of $NaNO_3$ in water to increase continually with increasing temperature. This behavior is commonly observed for solids in water, but there are exceptions. For example, the solubility of Li_2SO_4 in water decreases with increasing temperature for temperatures up to 160°C. The appearance of the phase diagram is considered in Prob. 12.54.

12.11   COMPUTER CALCULATION OF PHASE DIAGRAMS

Experimental determination of phase diagrams is time-consuming, and computer calculations can often help a lot. To calculate a phase diagram, one expresses G of the system as the sum of G of each phase: $G = \sum_\alpha G^\alpha = \sum_\alpha \sum_i n_i^\alpha \mu_i^\alpha$ [Eq. (9.23)]. Each chemical potential is expressed as $\mu_i^\alpha = \mu_i^{\circ,\alpha} + RT \ln \gamma_i^\alpha x_i^\alpha$ [Eq. (10.6)]. One uses a combination of theoretical models, experimental data, and estimation methods to provide an expression for each activity coefficient as a function of the phase composition, temperature, and pressure. One chooses a temperature, pressure, and an initial set of mole numbers and varies the amounts n_i^α in each phase to minimize G at fixed T and P. (Recall that minimization of G is one method used in computer solution of multiple-equilibrium problems; Sec 11.6.) One then changes some or all of the quantities T, P, and mole numbers and again minimizes G. After enough calculations have been done, one can plot the phase diagram. Example 12.6 shows how this works in a simple case.

EXAMPLE 12.6 Calculation of a liquid–liquid phase diagram

For a solution of the liquids D and E at 350 K and 1 bar, the infinite-dilution Convention I activity coefficient of D is measured to be $\gamma_{I,D}^{\infty} = 5.55_3$. Assuming that D and E form a simple solution (Sec. 10.5) and that W in (10.31) is independent of T, calculate the liquid–liquid phase diagram of D and E at 1 bar.

From Eq. (10.31), $RT \ln \gamma_{I,D} = W x_E^2$. At infinite dilution of D, $x_E = 1$ and $W = R(350 \text{ K}) \ln 5.55_3 = R(600 \text{ K})$. The two liquids might be completely miscible or partly miscible, depending on the temperature. Let α and β denote the two phases present. We have

$$G = G^{\alpha} + G^{\beta} = n_D^{\alpha} \mu_D^{\alpha} + n_E^{\alpha} \mu_E^{\alpha} + n_D^{\beta} \mu_D^{\beta} + n_E^{\beta} \mu_E^{\beta} \qquad (12.47)$$

$$\mu_D^{\alpha} = \mu_D^* + RT \ln \gamma_{I,D}^{\alpha} x_D^{\alpha} = \mu_D^* + RT \ln x_D^{\alpha} + W(x_E^{\alpha})^2 \qquad (12.48)$$

where (10.31) was used and where similar equations hold for the other three chemical potentials.

We use the Solver in Excel to vary the mole numbers in (12.47) so as to minimize G. Figure 12.31 shows the completed spreadsheet. The sizes of the phases are irrelevant and for simplicity it was assumed that the system contains one mole of D and one mole of E. Therefore cells F2 and F3 contain the formulas =1–$D$2 and =1–$D$3 for n_D^{β} and n_E^{β}. Cell B7 contains the formula in Eq. (12.48) for μ_D^{α}, and D7, F7, and H7 contain similar formulas. Cell B8 contains the formula (12.47). The mole fractions are calculated in B6 to E6. The values of μ_D^* and μ_E^* are irrelevant to minimizing G (Prob. 12.59), and any values can be placed in cells H3 and H4. One starts by assigning a temperature in D1; 220 K was the first value used. The Solver is set up to minimize G in B8 by varying D2 and D3 (n_D^{α} and n_E^{α}) subject to the constraints that n_D^{α} and n_E^{α} each must lie between 10^{-10} and 0.999999999 [this avoids the problem of taking the log of zero in (12.48)] and that $x_D^{\alpha} \le x_D^{\beta}$. One enters values between 0 and 1 into D2 and D3 as initial guesses and runs the Solver. Always take the initial guesses to have $n_D^{\alpha} < n_E^{\alpha}$. After the Solver finds the equilibrium mole fractions at 220 K,

	A	B	C	D	E	F	G	H
1	phase		T/K =	300	Rr =	8.3145	W =	4988.7
2	diagram		nDa =	0.129302	nDb =	0.8707	W/R =	600
3	simple		nEa =	0.129302	nEb =	0.8707	muD*=	500
4	mixture		na =	0.258604	nb =	1.7414	muE*=	0
5	T	xDa	xDb	xEa	xEb			
6	300	0.5	0.5	0.5	0.5			
7	muDa=	18.223	muEa=	-481.777	muDb=	18.223	muEb=	-481.8
8	G =	-463.6						
9								
10	T	xDa	xDb					
11	220	0.1028	0.89721					
12	240	0.1448	0.85521					
13	260	0.201	0.79896					
14	280	0.2824	0.71757					
15	290	0.344	0.65599					
16	295	0.3889	0.61106					
17	298	0.4295	0.57052					
18	300	0.5	0.5					

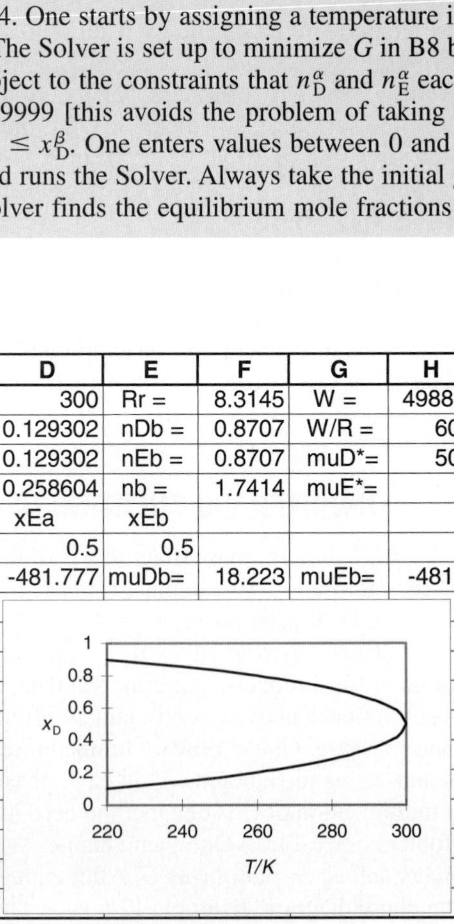

Figure 12.31

Spreadsheet for liquid–liquid phase diagram for liquids forming a simple solution.

one uses Paste Special on the Edit menu to paste only the values of cells A6, B6, and C6 to A11, B11, and C11. The temperature in D1 is then increased and the Solver used again. When the Solver is run at 300 K or at any temperature above 300 K, it will give the result that x_D^{α} and x_D^{β} are essentially equal to each other. When the two phases have the same composition, they are in fact one phase, so 300 K is the critical solution temperature and the liquids are completely miscible above 300 K. The phase diagram is most conveniently plotted with T on the horizontal axis, and so is rotated by $90°$ compared with those in Sec. 12.7. Select cells A11 through C18 as the data for the plot.

Exercise

(*a*) Set up the spreadsheet and verify Fig. 12.31. Verify that at each temperature, the spreadsheet shows $\mu_D^{\alpha} = \mu_D^{\beta}$ at equilibrium. (*b*) Change W to $(800\text{ K})R$ and find the new phase diagram.

The possibility for a lower critical solution temperature and the relation between W and T_c are explored in Probs. 12.55 to 12.58.

Some programs for calculating phase diagrams are ChemSage (which also calculates reaction equilibria—Sec. 11.6), Thermo-Calc (www.thermocalc.se; a free Windows version with limited capabilities is available for academic users), and MTDATA (www.npl.co.uk/npl/cmmt/mtdata/mtdata.html). The journal *CALPHAD* (standing for calculation of phase diagrams) reports research in this area.

One often does a computer calculation of a phase diagram as a preliminary to experimental work. The calculated diagram will suggest temperature and composition regions worthy of detailed investigation. After some data have been gathered, one can enter the data into the program to produce an improved calculated diagram.

12.12 THREE-COMPONENT SYSTEMS

A three-component (or *ternary*) system has $f = 3 - p + 2 = 5 - p$. For $p = 1$, there are 4 degrees of freedom. To make a two-dimensional plot, we must hold two variables fixed (instead of one, as with binary systems). We shall hold both T and P fixed. For a one-phase system, the two variables will be taken as x_A and x_B, the mole fractions of components A and B. For multiphase systems, x_A and x_B will be taken as the *overall* mole fractions of A and B in the system. Once x_A and x_B are fixed, x_C is fixed. We could use a rectangular plot with x_A and x_B as the variables on the two axes (Prob. 12.61). However, Gibbs suggested the use of an equilateral-triangle plot, and this has become standard for ternary systems.

The triangular coordinate system is based on the following theorem. Let D be an arbitrary point inside an equilateral triangle. If perpendiculars are drawn from D to the sides of the triangle (Fig. 12.32*a*), the sum of the lengths of these three lines is a constant equal to the triangle's height h; $\overline{DE} + \overline{DF} + \overline{DG} = h$. Plane-geometry fans will find suggestions for a proof of this in Prob. 12.62. We take the height h to be 1 and take the lengths $\overline{DE}$, $\overline{DF}$, and $\overline{DG}$ equal to the mole fractions of components A, B, and C, respectively. If more convenient, we can use weight fractions instead.

Thus, the perpendicular distance from point D to the side of the triangle opposite the A vertex is the mole fraction x_A of component A at point D. Similarly for components B and C. Any overall composition of the system can be represented by a point in or on the triangle, and we get Fig. 12.32*b*. In this figure, equally spaced lines have been drawn parallel to each side. On a line parallel to side BC (which is opposite to vertex A), the overall mole fraction of A is constant. The point marked with a dot represents

Figure 12.32

(a) $\overline{DE}$, $\overline{DF}$, and $\overline{DG}$ are perpendicular to the sides of the equilateral triangle and give the mole fractions of the three components. (b) Triangular coordinate system used in ternary phase diagrams.

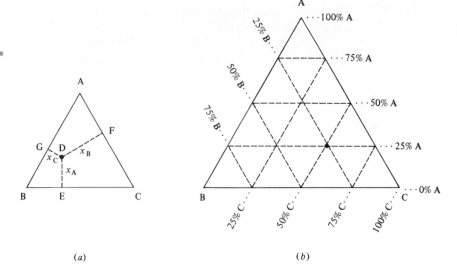

(a)

(b)

Figure 12.33

Liquid–liquid phase diagram for water–acetone–ether at 30°C and 1 atm. The coordinates are the mole fractions.

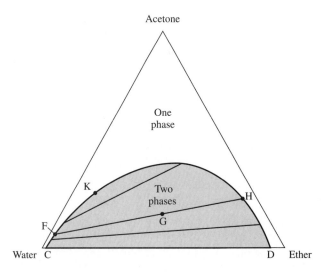

50 mole percent C, 25 mole percent A, 25 mole percent B. Along edge AC, the percentage of B present is zero; points on AC correspond to the binary system A + C. At vertex A, we have 100% A. At this point, the distance to the side opposite vertex A is a maximum. Note that, once x_A and x_B are fixed, the location of the point in the triangle is fixed as the intersection of the two lines corresponding to the given values of x_A and x_B.

We shall consider only ternary liquid–liquid equilibrium. Consider the system acetone–water–diethyl ether ("ether") at 1 atm and 30°C. Under these conditions, water and acetone are completely miscible with each other, ether and acetone are completely miscible with each other, and water and ether are partly miscible. Addition of sufficient acetone to a two-phase mixture of water and ether will produce a one-phase solution. The ternary phase diagram is Fig. 12.33.

The region above curve CFKHD is a one-phase area. For a point in the region below this curve, the system consists of two liquid phases in equilibrium. The lines in this region are tie lines whose endpoints give the compositions of the two phases in equilibrium. For a binary system, the tie lines on a T-x_A or P-x_A diagram are horizontal, since the two phases in equilibrium have the same T and P. On a ternary x_A-x_B-x_C triangular phase diagram, there is clearly no need for tie lines to be horizontal, and the phase diagram is incomplete unless tie lines are drawn in the two-phase regions. The locations of the tie lines are determined by chemical analysis of pairs of phases in equilibrium. In Fig. 12.33, a system of overall composition G consists of a water-rich, ether-poor phase α of composition F and an ether-rich, water-poor phase β of composition H. The slope of the tie line FGH shows that phase α has a smaller acetone mole fraction than phase β. Point K, the limiting point approached by the tie lines as the two phases in equilibrium become more and more alike, is called the *plait point* or the *isothermal critical point*. From the locations of F and H and the densities of the phases at F and H, we can calculate the partition coefficient (12.45) for acetone between the two phases at F and H.

What about the lever rule in ternary systems? The derivation of Eq. (12.39) is readily seen to be valid for *any* two-phase system; the number of components and the nature of the phases (solid, liquid, gas) are irrelevant. Rewriting (12.39) in a general form, we have

$$n^{\alpha}(x_B - x_B^{\alpha}) = n^{\beta}(x_B^{\beta} - x_B) \qquad \text{two-phase syst.} \qquad (12.49)$$

where n^{α} and n^{β} are the total number of moles of all species in phases α and β and where x_B, x_B^{α}, and x_B^{β} are the overall mole fraction of B, the mole fraction of B in phase α, and the mole fraction of B in phase β. Starting from (12.49) and using some trigonometry, one can show (Prob. 12.65) that $l^{\alpha}n^{\alpha} = l^{\beta}n^{\beta}$, where l^{α} and l^{β} are the tie-line lengths to the points that give the compositions of phases α and β. For example, in Fig. 12.33, this equation becomes $\overline{FG}n_F = \overline{GH}n_H$, where n_F and n_H are the total numbers of moles in the phases whose compositions are given by point F and H, respectively. The lever rule thus applies in two-phase regions of ternary (as well as binary) systems.

12.13 SUMMARY

The four colligative properties, vapor-pressure lowering, freezing-point depression, boiling-point elevation, and osmotic pressure, all result from the decrease in solvent chemical potential produced by the addition of a solute. Colligative properties are used to determine molecular weights and activity coefficients. In an ideally dilute solution, the freezing-point depression is proportional to the total molality of the solutes: $\Delta T_f = -k_f \sum_{i \neq A} m_i$ if only pure solvent freezes out. Osmotic pressure is especially useful in finding polymer molecular weights.

Two-component phase diagrams are plotted as T (or P) versus overall mole fraction x_B of one component, where P (or T) is held constant. In a two-component phase diagram, areas are either one-phase or two-phase regions, and horizontal lines contain three phases. A two-phase area is either a miscibility gap or a phase-transition loop. In a two-phase area, the compositions of the two phases in equilibrium are given by the points at the ends of a horizontal tie line that extends across the width of the two-phase region, and the overall composition is given by the location of the point on the tie line. The following kinds of phase diagrams were discussed: liquid–vapor P-x_B and T-x_B diagrams for completely miscible liquids; liquid–liquid T-x_B diagrams for partially miscible liquids; solid–liquid T-x_B diagrams with liquid-phase miscibility and solid-phase immiscibility, partial miscibility, and miscibility, including the complications introduced by compound formation.

Three-component phase diagrams are plotted on a triangular-coordinate system with T and P held constant.

Important kinds of calculations discussed in this chapter include:

- Calculation of freezing-point depression and boiling-point elevation from $\Delta T_f = -k_f m_B$ and $\Delta T_b = k_b m_B$.
- Calculation of molecular weights from freezing-point-depression data.

- Calculation of osmotic pressure from the van't Hoff equation $\Pi = c_B RT$ in ideally dilute solutions.
- Calculation of molecular weights from osmotic-pressure data using the van't Hoff equation or the McMillan–Mayer equation.
- Use of the phase diagram (and perhaps the lever rule $n^\alpha l^\alpha = n^\beta l^\beta$) to calculate the amounts of components present in each of two phases in equilibrium, given the overall amounts of each component present and the temperature (or pressure).
- Spreadsheet calculation of a binary liquid–liquid phase diagram from an assumed model for the activity coefficients.

FURTHER READING AND DATA SOURCES

Ricci; Dickerson, chap. 6; *de Heer,* chap. 22; *Denbigh,* secs. 8.8–8.13; *Kirkwood and Oppenheim,* sec. 11-7; *Haase and Schönert;* A. M. Alper (ed.), *Phase Diagrams,* vols. I–V, Academic Press, 1970–1978.

Freezing-point-depression, boiling-point-elevation, and osmotic-pressure data: *Landolt-Börnstein,* 6th ed., vol. II, pt. 2a, pp. 844–974.

Phase diagrams: *Landolt–Börnstein,* 6th ed., vol. II, pts. 2a, 2b, 2c, 3; M. Hansen, *Constitution of Binary Alloys,* 2d ed., McGraw-Hill, 1958; R. P. Elliot, *Constitution of Binary Alloys,* first supplement, McGraw-Hill, 1965; E. M. Levin et al. (eds.), *Phase Diagrams for Ceramists,* vols. 1–9, American Ceramic Society, 1964–1992; J. M. Sørensen and W. Arlt, *Liquid–Liquid Equilibrium Data Collection,* pts. 1–3 [vol. V of D. Behrens and R. Eckermann (eds.), *DECHEMA Chemistry Data Series*], DECHEMA, 1979–1980.

PROBLEMS

Section 12.1

12.1 True or false? (*a*) Addition of a solute at constant T and P to a pure solvent A always decreases μ_A. (*b*) Addition of a solute at constant T and P to a solution containing solvent A always decreases μ_A.

Section 12.2

12.2 True or false? (*a*) Addition of a nonvolatile solute to a pure solvent at constant T always lowers the vapor pressure. (*b*) The vapor pressure of a solution of A and B at temperature T is always less than the vapor pressure of pure A at T.

12.3 The vapor pressure of water at 110°C is 1074.6 torr. Find the vapor pressure at 110°C of a 2.00 wt % sucrose $(C_{12}H_{22}O_{11})$ solution in water. State any approximations made.

Section 12.3

12.4 True or false? (*a*) The system's temperature remains constant as pure water freezes at constant pressure. (*b*) The system's temperature remains constant as an aqueous solution of sucrose freezes at constant pressure. (*c*) The equation $\Delta T_f = -k_f m_B$ assumes that the solute B is nonvolatile. (*d*) In a solution with solvent A, μ_A must be lower than μ_A of pure A at the same T and P provided the solution is not supercooled or super-

saturated. (*e*) μ_A of solvent A in a supercooled solution is higher than μ_A of pure A at the same T and P. (*f*) The higher the molecular weight of a nonelectrolyte solute, the smaller the freezing-point depression produced by one gram of that solute in 1000 g of solvent. (*g*) If a solute partly dimerizes in a solvent, the freezing-point depression is less than it would be if the solute did not dimerize. (*h*) If ΔT_f is -1.45°C, then $\Delta T_f = -1.45$ K.

12.5 For cyclohexane, C_6H_{12}, the normal melting point is 6.47°C and the heat of fusion at this temperature is 31.3 J/g. Find the freezing point of a solution of 226 mg of pentane, C_5H_{12}, in 16.45 g of cyclohexane. State any assumptions or approximations made.

12.6 The freezing point of a solution of 2.00 g of maltose in 98.0 g of water is -0.112°C. Estimate the molecular weight of maltose. See Prob. 12.7 for a more accurate result.

12.7 Let W be the weight percent of maltose in an aqueous solution. The following freezing-point depressions are observed for maltose(*aq*) solutions:

W	3.00	6.00	9.00	12.00
ΔT_f/°C	-0.169	-0.352	-0.550	-0.765

(a) Show that the equation $\Delta T_f = -k_f m_B$ gives $M_B = -k_f w_B/(\Delta T_f w_A)$, where w_B and w_A are the masses of B and A in the solution. (b) Plot the calculated molecular weights vs. W and extrapolate to zero concentration to find the true molecular weight.

12.8 In deriving $\Delta T_f = -k_f m_B$, we replaced $1/T_f^* T_f$ by $1/(T_f^*)^2$. Use the $1/(1-x)$ Taylor series (8.8) to show that

$$\frac{1}{T_f^* T_f} = \frac{1}{(T_f^*)^2}\left[1 - \frac{\Delta T_f}{T_f^*} + \left(\frac{\Delta T_f}{T_f^*}\right)^2 - \cdots\right]$$

where $\Delta T_f \equiv T_f - T_f^*$. If $\Delta T_f \ll T_f^*$, terms after the 1 in the series can be neglected.

12.9 When 1.00 g of urea $[CO(NH_2)_2]$ is dissolved in 200 g of solvent A, the A freezing point is lowered by 0.250°C. When 1.50 g of the nonelectrolyte Y is dissolved in 125 g of the same solvent A, the A freezing point is lowered by 0.200°C. (a) Find the molecular weight of Y. (b) The freezing point of A is 12°C, and its molecular weight is 200. Find $\Delta_{fus}H_m$ of A.

12.10 When 542 mg of the nonelectrolyte compound Z is dissolved in a certain mass of the solvent A, the A freezing point is depressed by 1.65 times the depression observed when 679 mg of $CO(NH_2)_2$ is dissolved in the same mass of A. Find the molecular weight of Z.

12.11 Given that $\Delta_{vap}H_m = 40.66$ kJ/mol for water at 100°C, calculate k_b of H_2O.

12.12 The boiling point of $CHCl_3$ is 61.7°C. For a solution of 0.402 g of naphthalene $(C_{10}H_8)$ in 26.6 g of $CHCl_3$, the boiling point is elevated by 0.455 K. Find $\Delta_{vap}H_m$ of $CHCl_3$.

12.13 (a) Use (10.57) and (12.16) to verify that, for a solution of an electrolyte, Eq. (12.7) with $\Delta_{fus}H_m$ assumed constant becomes

$$\Delta T_f \approx -k_f \phi \nu m_i$$

where m_i is the electrolyte's stoichiometric molality. Note the resemblance to (12.17), except for the presence of ϕ to correct for nonideality. For historical reasons, the quantity $\phi \nu$ is called the *van't Hoff i factor*: $i \equiv \phi \nu$. (b) The freezing-point depression of an aqueous 4.00 wt % K_2SO_4 solution is -0.950°C. Calculate the osmotic coefficient ϕ in this solution at -1°C.

12.14 Phenol (C_6H_5OH) is partially dimerized in the solvent bromoform. When 2.58 g of phenol is dissolved in 100 g of bromoform, the bromoform freezing point is lowered by 2.37°C. Pure bromoform freezes at 8.3°C and has $k_f = 14.1$°C kg mol^{-1}. Calculate the equilibrium constant K_m for the dimerization reaction of phenol in bromoform at 6°C, assuming an ideally dilute solution.

12.15 Suppose that 6.0 g of a mixture of naphthalene $(C_{10}H_8)$ and anthracene $(C_{14}H_{10})$ is dissolved in 300 g of benzene. When the solution is cooled, it begins to freeze at a temperature 0.70°C below the freezing point (5.5°C) of pure benzene. Find the composition of the mixture, given that k_f for benzene is 5.1 °C kg mol^{-1}.

12.16 (a) For an 8.000 wt % aqueous sucrose solution, calculate the freezing point using $\Delta T_f = -k_f m_B$. (b) The observed freezing point of this solution is 0.485°C below that of water. Use Eq. (12.7) with $\Delta_{fus}H_{m,A}$ assumed constant to calculate a_{H_2O} and γ_{H_2O} in this solution. (c) Use (10.57) with $\nu = 1$ to find the osmotic coefficient ϕ in this solution.

Section 12.4

12.17 True or false? The osmotic pressure is the pressure exerted on the semipermeable membrane by the solute molecules.

12.18 In a 0.300 mol/kg aqueous solution of sucrose, the $C_{12}H_{22}O_{11}$ molarity is 0.282 mol/dm^3 at 20°C and 1 atm. The density of water is 0.998 g/cm^3 at 20°C and 1 atm. (a) Estimate the osmotic pressure of this solution using the van't Hoff equation. (b) The observed osmotic pressure of this solution is 7.61 atm. Use Eq. (12.24) to find a_{H_2O} and γ_{H_2O} in this solution. *Note:* γ_A in (12.24) is at the pressure $P + \Pi$, but the pressure dependence of γ_A is slight and can be neglected here.

12.19 The osmotic pressure of an aqueous solution of bovine serum albumin with $\rho_B = 0.0200$ g/cm^3 is 6.1 torr at 0°C. Estimate the molecular weight of this protein. Explain why your answer is only an estimate.

12.20 Let the left chamber in Fig. 12.5 contain pure water and the right chamber 1.0 g of $C_{12}H_{22}O_{11}$ plus 100 g of water at 25°C. Estimate the height of the liquid in the right capillary tube at equilibrium. Assume that the volume of liquid in the capillary tube is negligible compared with that in the right chamber.

12.21 For a certain sample of a synthetic poly(amino acid) in water at 30°C (density 0.996 g/cm^3), osmotic-pressure determinations gave the following values for the difference in height Δh between the liquids in the capillary tubes in Fig. 12.5:

Δh/cm	2.18	3.58	6.13	9.22
$\rho_B/$(g/dm^3)	3.71	5.56	8.34	11.12

Convert the height readings to pressures and find the number average molecular weight of the polymer.

12.22 Sometimes a polymer-solution Π/ρ_B-vs.-ρ_B plot shows significant nonlinearity due to a nonnegligible contribution of the A_3 term in (12.28). In this case, polymer-solution theories and experimental data indicate that A_3 can usually be approximated by $A_3 \approx \frac{1}{4}M_B A_2^2$. With this approximation, show that (12.28) with terms after A_3 neglected gives $(\Pi/\rho_B)^{1/2} = (RT/M_B)^{1/2} + \frac{1}{2}A_2(RTM_B)^{1/2}\rho_B$. Therefore a plot of $(\Pi/\rho_B)^{1/2}$ vs. ρ_B is linear with intercept $(RT/M_B)^{1/2}$ at $\rho_B = 0$. When the Π/ρ_B-vs.-ρ_B curve is nonlinear, a more accurate molecular weight can be obtained from a $(\Pi/\rho_B)^{1/2}$-vs.-ρ_B curve.

12.23 (a) The osmotic pressure of human blood is 7 atm at 37°C. Pretend that NaCl forms ideally dilute solutions in water and use (12.31) to estimate the molarity of a saline (NaCl) solution that is isotonic to blood at 37°C. Compare with the value 0.15 mol/dm^3 actually used for intravenous injections. (b) The principal solute molalities (in mol/kg) in seawater are NaCl,

0.460; $MgCl_2$, 0.034; $MgSO_4$, 0.019; $CaSO_4$, 0.009. Pretend that seawater is an ideally dilute solution, ignore ion pairing, and estimate the osmotic pressure of seawater at 20°C.

12.24 Dry air at sea level has the following composition by mole fraction: 78% N_2, 21% O_2, 1% Ar. Calculate the number average molecular weight of air.

12.25 Use (12.24) and (10.57) to show that for an electrolyte solution the osmotic pressure is given by

$$\Pi = \phi RT\nu n_i / n_A V_{m,A}^*$$

where ϕ is for the solution at $P + \Pi$. Comparison with Eq. (12.26) explains the name (practical) osmotic coefficient for ϕ.

12.26 (a) Solution 1 contains solvent A at mole fraction $x_{A,1}$ plus solute B. Solution 2 contains A at mole fraction $x_{A,2}$ (where $x_{A,2} < x_{A,1}$) plus B. A membrane permeable only to species A separates the two solutions. The solutions are assumed ideally dilute. Show that to achieve equilibrium an osmotic pressure

$$\Pi = RT(x_{B,2} - x_{B,1})/V_{m,A}^* \qquad \text{ideally dil. solns.}$$

must be applied to solution 2. Verify that this equation reduces to (12.25) when solution 1 is pure solvent. (b) Suppose a 0.100 mol/kg aqueous sucrose solution is separated from a 0.0200 mol/kg aqueous sucrose solution by a membrane permeable only to water. Find the value of Π needed to achieve equilibrium at 25°C.

12.27 Use (4.12) with $dT = 0$, $dw = -P^\alpha dV^\alpha - P^\beta dV^\beta$, and $dA = dA^\alpha + dA^\beta$ and substitute (4.77) at constant T for dA^α and for dA^β to show that the phase equilibrium condition $\mu_i^\alpha = \mu_i^\beta$ holds even when $P^\alpha \neq P^\beta$.

Section 12.5
12.28 What is the maximum number of phases that can coexist in a binary system?

Section 12.6
12.29 For the liquid–vapor phase diagram of Fig. 12.12, suppose we know the system is at temperature T_1 and is at the pressure of the diagram, and that both liquid and vapor phases are present. (a) Can we find the mole fractions in the liquid phase and in the vapor phase? If your answer is yes, find these quantities. (b) Can we find the overall mole fraction? If your answer is yes, find it.

12.30 From the data in Table 10.1, plot the acetone–chloroform liquid–vapor P-x_B phase diagram at 35°C.

12.31 For the system of Fig. 12.12, suppose that a liquid solution with B mole fraction 0.30 is distilled at the pressure of the diagram using a column with an efficiency of two theoretical plates. Give the composition of the first drop of distillate.

12.32 For the binary system whose phase diagram at a certain fixed T is Fig. 12.10, (a) find the composition of the vapor in equilibrium with liquid whose composition is $x_B = 0.720$; (b) find the mole fractions in each phase if the system's pressure is P_D and the system's overall B mole fraction is $x_{B,2}$. Assume the scales in Fig. 12.10 are linear.

12.33 If the system of Fig. 12.12 is at temperature T_1 and contains 2.00 mol of B and consists entirely of vapor, give an inequality that the number of moles of C present must satisfy.

12.34 If the system of Fig. 12.12 is at temperature T_1 and contains 4.00 mol of B and 3.00 mol of C, find the number of moles of B and C present in each phase.

12.35 Benzene (ben) and toluene (tol) form nearly ideal solutions. The 20°C vapor pressures are $P_{ben}^* = 74.7$ torr and $P_{tol}^* = 22.3$ torr. Plot the P-vs.-x_{ben} liquid–vapor phase diagram for benzene–toluene solutions at 20°C.

12.36 For the system of Fig. 12.12, suppose that a liquid solution with B mole fraction 0.30 is heated in a closed system held at the constant pressure of the diagram. (a) Give the composition of the first vapor formed. (b) Give the composition of the last drop of liquid vaporized. (c) Give the composition of each phase present when half the moles of liquid have been vaporized.

Section 12.7
12.37 Professor Blitzstein asked students to figure out how many degrees of freedom there are in the two-phase region of Fig. 12.16. Regina answered: "With two phases, two components, and no reactions or stoichiometric restrictions, f would be $2 - 2 + 2 = 2$, but Fig. 12.16 is only a cross section of the complete three-dimensional diagram and P is restricted to be constant in this cross section, so f is 1." Horace answered: "This phase-rule stuff makes my head spin. All I know is that we can go up and down in the two-phase region of Fig. 12.16 and this makes T vary, and we can go left and right and this varies x_B. Since the two intensive variables T and x_B can vary in this region, f must be 2." Who is right? What error did the other student make?

12.38 Use Fig. 12.17b to find the masses of water and nicotine present in each phase if 10 g of nicotine and 10 g of water are mixed at 80°C and 1 atm.

12.39 For a certain liquid mixture of equal weights of water and nicotine at 80°C, the mass of the water-rich phase is 20 g. Use Fig. 12.17b to find the mass of water and the mass of nicotine in the water-poor phase.

12.40 Water and phenol are partially miscible at 50°C. When these two liquids are mixed at 50°C and 1 atm, at equilibrium one phase is 89% water by weight and the other is $37\frac{1}{2}$% water by weight. If 6.00 g of phenol and 4.00 g of water are mixed at 50°C and 1 atm, find the mass of water and the mass of phenol in each phase at equilibrium by using (a) the lever rule; (b) conservation of matter (without using the lever rule).

12.41 For DDT ($C_{14}H_9Cl_5$) at 25°C, $\log K_{ow} = 6.91$. If 100 mg of DDT, 80 g of water, and 10 g of 1-octanol are equilibrated at 25°C, find the mass of each substance present in each phase. Use data in Sec. 12.7. The density of 1-octanol at 25°C is 0.83 g/cm³. To find the density of each phase, assume $\Delta_{mix}V = 0$.

12.42 For naphthalene at 25°C, $\log K_{ow} = 3.30$. Find ΔG_c° for the transfer of naphthalene from the water-rich phase to the octanol-rich phase.

Section 12.8

12.43 Without referring to the text, sketch a constant-P solid–liquid T-versus-x_B phase diagram if B and C form completely miscible liquids and completely immiscible solids. For each area and for the horizontal line, state what phase or phases are present and which substances are in each phase and state the number of degrees of freedom.

12.44 Given the following melting points and heats of fusion: benzene, 5.5°C, 30.4 cal/g; cyclohexane (C_6H_{12}), 6.6°C, 7.47 cal/g. Plot the T-x_B solid–liquid phase diagram for these two compounds and find the eutectic temperature and the eutectic composition. Assume that the liquid solutions are ideal and that no solid solutions are formed; neglect the temperature dependence of $\Delta_{fus}H$. Compare your values with the experimental eutectic values $-42\frac{1}{2}$°C and $73\frac{1}{2}$ mole percent cyclohexane.

12.45 When melts of Zn + Mg are cooled, breaks and halts are observed at the following temperatures in °C, where W_{Zn} is the wt % of zinc:

W_{Zn}	0	10	20	30	40
Break	$\cdots$	623°	566°	530°	443°
Halt	651°	344°	343°	347°	344°

W_{Zn}	50	60	70	80	84.3
Break	356°	437°	517°	577°	$\cdots$
Halt	346°	346°	347°	343°	595°

W_{Zn}	90	95	97	97.5	100
Break	557°	456°	$\cdots$	379°	$\cdots$
Halt	368°	367°	368°	368°	419°

Because of experimental errors, the temperature of the eutectic halt varies slightly from run to run. Plot the phase diagram of T versus weight percent Zn and label all areas.

12.46 (*a*) Use the following facts to sketch the H_2O–NaCl T-vs.-weight percent NaCl solid–liquid phase diagram up to 100°C; label all areas as to the phases present. The components form the compound NaCl $\cdot$ 2H_2O, which melts incongruently at the peritectic temperature of 0.1°C. The melting point of ice is (surprise!) 0.0°C. The eutectic temperature for liquid + $H_2O(s)$ + NaCl $\cdot$ 2$H_2O(s)$ is −21°C, and the eutectic point occurs at 23 wt % NaCl. The freezing point of a 13 wt % NaCl aqueous solution is −9°C. The solubility of NaCl in water is 26 g NaCl per 74 g H_2O at 0.1°C and increases to 28 g NaCl per 72 g H_2O at 100°C. (*b*) If an aqueous solution of NaCl is evaporated to dryness at 20°C, what solid(s) is (are) obtained? (*c*) Describe what happens when a system at 20°C with overall composition of 80 wt % NaCl is slowly cooled to −10°C. Is any ice present at −10°C?

12.47 Sketch several cooling curves for the system Cu–Ag of Fig. 12.22 showing the different types of behavior observed.

12.48 Sketch several cooling curves for the system of Fig. 12.26*c* showing the different types of behavior observed.

12.49 Bi and Te form the solid compound Bi_2Te_3, which melts congruently at about 600°C. Bi and Te melt at about 300°C and 450°C, respectively. Solid Bi_2Te_3 is partially misci-ble at all temperatures with solid Bi and is partially miscible at all temperatures with solid Te. Sketch the appearance of the Bi–Te T-x_B solid–liquid phase diagram; label all regions.

12.50 The Fe–Au solid–liquid T-x_B phase diagram can be viewed as the intersection of a solid-phase miscibility gap with a solid–liquid phase-transition loop having a minimum at $x_{Au} = 0.8$. The miscibility gap intersects the phase-transition loop at $x_{Au} = 0.1$ and at $x_{Au} = 0.3$. Fe has a higher melting point than Au. Sketch the phase diagram and label all areas.

12.51 The solid–liquid phase diagram of water–nitric acid shows formation of the congruently melting compounds (melting points in parentheses) $HNO_3 \cdot 3H_2O$ (−18°C) and $HNO_3 \cdot H_2O$ (−38°C). The melting point of HNO_3 is −41°C. As one goes across the phase diagram from H_2O to HNO_3, the eutectic temperatures are −43°C, −42°C, and −66°C. All the solids are completely immiscible. Draw the solid–liquid phase diagram.

Section 12.10

12.52 (*a*) The heat of fusion of naphthalene is 147 J/g, and its melting point is 80°C. Estimate the mole-fraction solubility of naphthalene in benzene at 25°C and compare with the experimental value 0.296. Use a version of Eq. (12.46). (*b*) Estimate the mole-fraction solubility of naphthalene in toluene at 25°C and compare with the experimental value $x(C_{10}H_8) = 0.286$. (*c*) For anthracene ($C_{14}H_{10}$), $\Delta_{fus}H = 162$ J/g and the melting point is 216°C. Estimate the solubility of anthracene in benzene at 60°C.

12.53 Use Fig. 12.29 to find how many moles of naphthalene will dissolve in 1.00 kg of benzene at $52\frac{1}{2}$°C.

12.54 The eutectic temperature for water plus Li_2SO_4 is −23°C. The solubility of Li_2SO_4 in water decreases as T increases from −23°C to 160°C and increases above 160°C. Sketch the portion of the solid–liquid phase diagram below 200°C.

Section 12.11

12.55 Change W to $-(600 \text{ K})R$ in the Fig. 12.31 spreadsheet and find the liquid–liquid phase diagram. Does the result make sense in terms of the physical meaning of W? Explain.

12.56 Let W in the Fig. 12.31 spreadsheet vary with temperature according to $W = R(600 \text{ K} - 0.36T)$. Find the liquid–liquid phase diagram.

12.57 (*a*) Let W in the Fig. 12.31 spreadsheet vary with temperature according to $W = R(3T - 200 \text{ K})$. Find the liquid–liquid phase diagram. (*b*) Repeat for $W = R(3T - 265 \text{ K})$.

12.58 (*a*) Examine the results of Example 12.6 and of Probs. 12.55, 12.56, and 12.57, and formulate a hypothesis relating the critical solution temperature and the value of W/R at the critical solution temperature for a simple solution. (*b*) Change W in Fig. 12.31 to $W = R[3.5T - 200 \text{ K} - (0.0020 \text{ K})(T/\text{K})^2]$ and find the phase diagram over the range 200 to 600 K. Do your results satisfy the hypothesis you formulated in part (*a*)?

12.59 Show that in the final expression for G of the system of the Fig. 12.31 spreadsheet, the quantities μ_D^* and μ_E^* occur in terms that remain constant as n_D^α and n_E^α vary. Hence the μ_D^* and μ_E^* values cannot affect the minimization of G.

Section 12.12

12.60 Let $K_{ew,ac}$ be the partition coefficient for acetone between ether-rich and water-rich phases at 30°C. Use Fig. 12.33 to answer the following questions. (*a*) Does $K_{ew,ac}$ increase, decrease, or remain constant as the acetone concentration in the two phases increases? (*b*) Are the $K_{ew,ac}$ values less than, equal to, or greater than 1?

12.61 For the ternary system of Fig. 12.33, set up rectangular coordinates with x(ether) on the y axis and $x(H_2O)$ on the x axis and sketch the general appearance of the phase diagram in these coordinates. Chemical engineers often use rectangular rather than triangular coordinates for ternary systems.

12.62 Prove that $\overline{DE} + \overline{DF} + \overline{DG} = h$ in Fig. 12.32*a*. (*Hint:* Draw lines DA, DB, and DC and recall that the area of a triangle equals half the product of the base and altitude.)

12.63 (*a*) From Fig. 12.33, use a ruler to estimate the mole fractions in the phases present at point G. (*b*) Suppose the ternary system at point G has a total of 40 moles present. Find the number of moles of each component in each phase.

12.64 For the system water (1) plus ethyl acetate (2) plus acetone (3) at 30°C and 1 atm, mole-fraction compositions of pairs of liquid phases α and β in equilibrium are:

x_2^α	x_3^α	x_2^β	x_3^β
0.016	0.000	0.849	0.000
0.018	0.011	0.766	0.061
0.020	0.034	0.618	0.157
0.026	0.068	0.496	0.241
0.044	0.117	0.320	0.292
0.103	0.206	0.103	0.206

where the last set of data gives the isothermal critical solution point. (*a*) Plot the ternary phase diagram including tie lines; use commercially available triangular coordinate paper. (*b*) Suppose 0.10 mole of acetone, 0.20 mole of ethyl acetate, and 0.20 mole of water are mixed at 30°C and 1 atm. Find the mass of each component present in each phase at equilibrium.

12.65 In Fig. 12.33, draw vertical lines from F, G, and H, and horizontal lines from F and G; then use (12.49) to show that $\overline{FG}n_F = \overline{GH}n_H$.

General

12.66 (*a*) Beaker A contains 20 cm^3 of pure H_2O; beaker B contains 20 cm^3 of a 5 wt % NaCl solution. The beakers each have volumes of 400 cm^3 and are in a sealed thermally conducting box. Describe the equilibrium state of this system. (*b*) Beaker A contains 0.0100 mol of sucrose dissolved in 100 g of water. Beaker B contains 0.0300 mol of sucrose in 100 g of water. Give the contents of each beaker at equilibrium if the beakers each have volumes of 400 cm^3 and are in a sealed thermally conducting box.

12.67 Binary liquid-state partial miscibility corresponds to very large positive deviations from ideality, so the liquid–vapor phase-transition loop has a maximum on the P-x_A diagram and a minimum on the T-x_A diagram. The T-x_A liquid–vapor phase diagram when the liquids are partially miscible therefore shows the intersection of a miscibility gap with a phase-transition loop that has a minimum. This liquid–vapor phase diagram resembles Fig. 12.22, which also shows the intersection of a miscibility gap and a minimum-containing phase-transition loop. Sketch the appearance of a binary liquid–vapor T-x_A diagram for liquid-phase partial miscibility; label all areas and three-phase lines.

12.68 Systems A and B are each at the same temperature T_1. The two systems are mixed in an adiabatically enclosed container. Is it possible for the final temperature to be less than T_1? If so, give one or more examples.

12.69 Give the number of degrees of freedom (*a*) in the liquid region of Fig. 12.12; (*b*) in the two-phase region of Fig. 12.16; (*c*) along the horizontal line in Fig. 12.22. In each case, state which intensive variables constitute the degrees of freedom.

12.70 The liquid–vapor T-versus-x_B phase diagram of ethanol plus ethyl acetate at $P = 1.00$ atm shows an azeotrope that boils at 71.8°C. Find the Convention I activity coefficients of ethanol and ethyl acetate in this azeotropic liquid mixture, given that the pure-component vapor pressures at 71.8°C are 581 torr for ethanol and 631 torr for ethyl acetate.

12.71 For the liquid–liquid system of water (w) plus benzene (ben) at 25°C and 1 atm, the benzene mole fractions in the two liquid phases α and β in equilibrium with each other are $x_{ben}^\alpha = 0.000405$ and $x_{ben}^\beta = 0.99700$. (*a*) Using reasonable approximations, estimate the Convention I activity coefficients of water and benzene in each liquid phase in equilibrium at 25°C and 1 atm. (*Hint:* Certain activity coefficients can be approximated as 1.) (*b*) Find the vapor pressure of a saturated solution of water in benzene at 25°C, given that the pure-component vapor pressures at 25°C are $P_w^* = 23.8$ torr and $P_{ben}^* = 95.2$ torr. (*c*) Find the vapor pressure of a saturated solution of benzene in water at 25°C.

12.72 When water is shaken with benzene, one obtains two liquid phases in equilibrium: a saturated solution of a small amount of benzene in water and a saturated solution of a small amount of water in benzene. Show that the partial vapor pressure of benzene in equilibrium with a saturated solution of benzene in water is equal to the partial vapor pressure of benzene in equilibrium with a saturated solution of water in benzene at the same temperature, provided the vapors are assumed ideal. (*Hint:* Think in terms of chemical potentials.)

12.73 A certain aqueous solution of a low-molecular-weight solid nonelectrolyte freezes at −0.64°C. For this solution, estimate (*a*) the normal boiling point; (*b*) the vapor pressure at 25°C; (*c*) the osmotic pressure at 20°C. The vapor pressure of pure water is 23.76 torr at 25°C.

12.74 For a T-versus-x_B phase diagram at constant P for the binary system B + C, consider two different points R and S on a tie line in a two-phase ($\alpha + \beta$) region. State whether each of the following quantities has the same or different values in the two states corresponding to points R and S: (a) T; (b) P; (c) x_B^α; (d) x_B^β; (e) the overall x_B; (f) n^α/n^β.

12.75 True or false? (a) Addition of a tiny amount of a soluble impurity to a pure liquid always lowers the freezing point. (b) Addition of a tiny amount of a soluble impurity to a pure liquid must lower the freezing point if only pure solvent freezes out. (c) A liquid solution of two substances will always freeze entirely at one temperature. (d) A liquid solution of two substances will never freeze entirely at one temperature. (e) The partial pressure P_B of B vapor (assumed ideal) in equilibrium with a nonideal solution of B plus C must always increase when the B mole fraction in the solution is increased at constant temperature. (f) For a binary two-phase system, the closer the point on a tie line is to a phase, the more of that phase is present. (g) For a binary (B + C) two-phase ($\alpha + \beta$) system, $n_B^\alpha l^\alpha = n_B^\beta l^\beta$, where l^α and l^β are the distances of a point on a tie line to the ends of the tie line. (h) For a constant-T (or constant-P) phase diagram of the binary system B + C, two different points P and Q on the same tie line in a two-phase region containing phases α and β correspond to states with the same value of x_B^β but different values of overall x_B. (i) A system at equilibrium with substances i and k present in phase β must have $\mu_i^\beta = \mu_k^\beta$. (j) If a two-phase system in equilibrium contains a liquid phase and a vapor phase that are not separated by anything, and if the effects of the earth's gravitational field are neglected, then the pressure of the vapor equals the pressure in the interior of the liquid. (k) In a two-phase, two-component liquid–vapor system, $x_{B,overall} = x_B^l + x_B^v$.

CHAPTER
13
Surface Chemistry

Molecules at the surface of a phase are in a different environment than those in the interior of the phase. Surface chemistry deals with systems where surface effects are important. Surface effects are of tremendous industrial and biological significance. Many reactions occur most readily on the surfaces of catalysts, and heterogeneous catalysis is important in the synthesis of industrial chemicals. Such subjects as lubrication, corrosion, adhesion, detergency, and reactions in electrochemical cells involve surface effects. Many industrial products are colloids (Sec. 13.6) with large surface areas. The problem of how biological cell membranes function belongs to surface science.

13.1   THE INTERPHASE REGION

So far in this book, each phase of a thermodynamic system has been considered to be strictly homogeneous, with its intensive properties constant throughout the phase. However, when surface effects are considered, it is clear that a phase is not strictly homogeneous throughout. For example, in a system composed of the phases α and β (Fig. 13.1a), molecules at or very near the region of contact of phases α and β have a different molecular environment than molecules in the interior of either phase. The three-dimensional region of contact between phases α and β in which molecules interact with molecules of both phases is called the **interfacial layer, surface layer,** or **interphase region.** This region is a few molecules thick if ions are not present. (Intermolecular forces between neutral molecules are negligible beyond about 3 molecular diameters; see Sec. 2.11.) The term **interface** refers to the apparent two-dimensional geometrical boundary surface separating the two phases.

Figure 13.1b is a schematic drawing of a cross section of a two-phase system with a planar interface. All molecules between the planes VW and AB have the same environment and are part of the **bulk phase** α. All molecules between planes CD and RS have the same environment and are part of the bulk phase β. The interfacial layer (whose thickness is grossly exaggerated in the figure) consists of the molecules between planes AB and CD.

Since the interfacial layer is only a few molecular diameters thick, usually only an extremely small fraction of the system's molecules are in this layer and the influence of surface effects on the system's properties is essentially negligible. This chapter considers systems where surface effects are significant; for example, colloidal systems,

Figure 13.1

(a) A two-phase system. (b) The interfacial layer between two bulk phases.

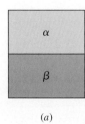

(a)

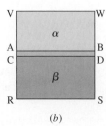

(b)

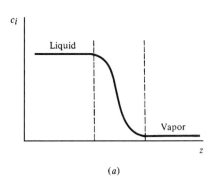

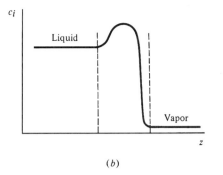

(a)   (b)

Figure 13.2

Change in concentration of a component in going from the bulk liquid phase to the bulk vapor phase.

where the surface-to-volume ratio is high, and gas–solid systems, where substantial amounts of gas can be adsorbed at the solid's surface.

The interfacial layer is a transition region between the bulk phases α and β and is not homogeneous. Instead, its properties vary from those characteristic of the bulk phase α to those characteristic of the bulk phase β. For example, if β is a liquid solution and α is the vapor in equilibrium with the solution, approximate statistical-mechanical calculations and physical arguments indicate that the concentration c_i of component i may vary with z (the vertical coordinate in Fig. 13.1b) in one of the ways shown in Fig. 13.2. The dashed lines mark the boundaries of the interfacial layer and correspond to planes AB and CD in Fig. 13.1b. Statistical-mechanical calculations and study of light reflected from interfaces indicate that the interfacial layer between a pure liquid and its vapor is typically about three molecular diameters thick. For solid–solid, solid–liquid, and solid–gas interfaces, the transition between the bulk phases is usually more abrupt than for the liquid–vapor interface of Fig. 13.2.

Because of differences in intermolecular interactions, molecules in the interphase region have a different average intermolecular interaction energy than molecules in either bulk phase. An adiabatic change in the area of the interface between α and β will therefore change the system's internal energy U.

For example, consider a liquid in equilibrium with its vapor (Fig. 13.3). Recall from the arrow in Fig. 4.4 that intermolecular interactions in a liquid lower the internal energy. Molecules at the surface of the liquid experience fewer attractions from other liquid-phase molecules compared with molecules in the bulk liquid phase and so have a higher average energy than molecules in the bulk liquid phase. The concentration of molecules in the vapor phase is so low that we can ignore interactions between vapor-phase molecules and molecules at the surface of the liquid. It requires work to increase the area of the liquid–vapor interface in Fig. 13.3, since such an increase means fewer molecules in the bulk liquid phase and more in the surface layer. It is generally true that positive work is required to increase the area of an interface between two phases. For this reason, systems tend to assume a configuration of minimum surface area. Thus an isolated drop of liquid is spherical, since a sphere is the shape with a minimum ratio of surface area to volume.

Let $\mathcal{A}$ be the area of the interface between phases α and β. The number of molecules in the interphase region is proportional to $\mathcal{A}$. Suppose we reversibly increase the area of the interface by $d\mathcal{A}$. The increase in the number of molecules in the interphase region is proportional to $d\mathcal{A}$, and so the work needed to increase the interfacial area is proportional to $d\mathcal{A}$. Let the proportionality constant be symbolized by $\gamma^{\alpha\beta}$, where the superscripts indicate that the value of this constant depends on the nature of the phases in contact. The reversible work needed to increase the interfacial area is then $\gamma^{\alpha\beta} d\mathcal{A}$. The quantity $\gamma^{\alpha\beta}$ is called the **interfacial tension** or the **surface tension.** When one phase is a gas, the term "surface tension" is more commonly used. Since it requires

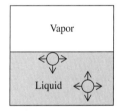

Figure 13.3

Attractive forces on molecules in a liquid.

positive work to increase $\mathcal{A}$, the quantity $\gamma^{\alpha\beta}$ is positive. The stronger the intermolecular attractions in a liquid, the greater the work needed to bring molecules from the bulk liquid to the surface and therefore the greater the value of $\gamma^{\alpha\beta}$.

In addition to the work $\gamma^{\alpha\beta}\, d\mathcal{A}$ required to change the interfacial area, there is the work $-P\, dV$ associated with a reversible volume change, where P is the pressure in each bulk phase and V is the system's total volume. Thus the work done on the closed system of phases α and β is

$$dw_{\text{rev}} = -P\, dV + \gamma^{\alpha\beta}\, d\mathcal{A} \qquad \text{plane interface} \qquad (13.1)^*$$

We shall take (13.1) as the definition of $\gamma^{\alpha\beta}$ for a closed two-phase system with a planar interface. The reason for the restriction to a planar interface will become clear in the next section. From (13.1), if the piston in Fig. 13.5 is slowly moved an infinitesimal distance, work $-P\, dV + \gamma^{\alpha\beta}\, d\mathcal{A}$ is done on the system.

The term **surface tension of liquid** α refers to the interfacial tension $\gamma^{\alpha\beta}$ for the system of liquid α in equilibrium with its vapor β. Surface tensions of liquids are often measured against air. When phase β is an inert gas at low or moderate pressure, the value of $\gamma^{\alpha\beta}$ is nearly independent of the composition of β.

Since we shall be considering systems with only one interface, from here on, $\gamma^{\alpha\beta}$ will be symbolized simply by γ.

The surface tension γ has units of work (or energy) divided by area. The cgs unit of γ is ergs/cm^2, which equals dyn/cm, since 1 erg = 1 dyn cm. The SI unit of γ is J/m^2 = N/m. The reader can verify that

$$1\ \text{erg/cm}^2 = 1\ \text{dyn/cm} = 10^{-3}\ \text{J/m}^2 = 10^{-3}\ \text{N/m} = 1\ \text{mN/m} = 1\ \text{mJ/m}^2 \qquad (13.2)$$

For most organic and inorganic liquids, γ at room temperature ranges from 15 to 50 dyn/cm. For water, γ has the high value of 73 dyn/cm at 20°C, because of the strong intermolecular forces associated with hydrogen bonding. Liquid metals have very high surface tensions; that of Hg at 20°C is 490 dyn/cm. For a liquid–liquid interface with each liquid saturated with the other, γ is generally less than γ of the pure liquid with the higher γ. Measurement of γ is discussed in Sec. 13.2.

As the temperature of a liquid in equilibrium with its vapor is raised, the two phases become more and more alike until at the critical temperature T_c the liquid–vapor interface disappears and only one phase is present. At T_c, the value of γ must therefore become 0, and we expect that γ of a liquid will continually decrease as T is raised to the critical temperature. The following empirical equation (due to Katayama and Guggenheim) reproduces the $\gamma(T)$ behavior of many liquids:

$$\gamma = \gamma_0 (1 - T/T_c)^{11/9} \qquad (13.3)$$

where γ_0 is an empirical parameter characteristic of the liquid. Since 11/9 is close to 1, we have $\gamma \approx \gamma_0 - \gamma_0 T/T_c$, and γ decreases approximately linearly as T increases. Figure 13.4 plots γ versus T for some liquids.

The quantity P in (13.1) is the pressure in each of the bulk phases α and β of the system. However, because of the surface tension, P is not equal to the pressure exerted by the piston in Fig. 13.5 when the system and piston are in equilibrium. Let the system be contained in a rectangular box of dimensions l_x, l_y, and l_z, where the x, y, and z axes are shown in Fig. 13.5. Let the piston move a distance dl_y in the process of doing work dw_{rev} on the system, and let the piston exert a force F_{pist} on the system. The work done by the piston is $dw_{\text{rev}} = F_{\text{pist}}\, dl_y$ [Eq. (2.10)]. Use of (13.1) gives $F_{\text{pist}}\, dl_y = -P\, dV + \gamma\, d\mathcal{A}$. The system's volume is $V = l_x l_y l_z$, and $dV = l_x l_z\, dl_y$. The area of the interface between phases α and β is $\mathcal{A} = l_x l_y$, and $d\mathcal{A} = l_x\, dl_y$. Therefore $F_{\text{pist}}\, dl_y = -P l_x l_z\, dl_y + \gamma l_x\, dl_y$ and

$$F_{\text{pist}} = -P l_x l_z + \gamma l_x \qquad (13.4)$$

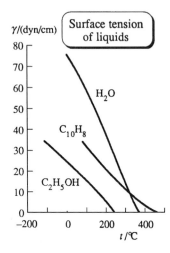

Figure 13.4

Temperature dependence of the surface tension of some liquids. γ becomes zero at the critical point. $C_{10}H_8$ is naphthalene.

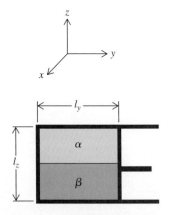

Figure 13.5

A two-phase system confined by a piston.

The pressure P_{pist} exerted by the piston is $-F_{pist}/\mathscr{A}_{pist} = -F_{pist}/l_x l_z$, where $\mathscr{A}_{pist}$ is the piston's area. F_{pist} is in the negative y direction and so is negative; pressure is a positive quantity, so the minus sign has been added. Division of (13.4) by $\mathscr{A}_{pist} = l_x l_z$ gives

$$P_{pist} = P - \gamma/l_z \qquad (13.5)$$

The term γ/l_z is ordinarily very small compared with P. For the typical values $l_z = 10$ cm and $\gamma = 50$ dyn/cm, one finds $\gamma/l_z = 5 \times 10^{-6}$ atm (Prob. 13.7).

Since the force exerted by body A on body B is the negative of the force of B on A (Newton's third law), Eq. (13.4) shows that the system exerts a force $Pl_x l_z - \gamma l_x$ on the piston. The presence of the interface causes a force γl_x to be exerted by the system on the piston, and this force is in a direction opposite that associated with the system's pressure P. The quantity l_x is the length of the line of contact of the interface and the piston, so γ *is the force per unit length* exerted on the piston as a result of the existence of the interphase region. Mechanically, the system acts as if the two bulk phases were separated by a thin membrane under tension. This is the origin of the name "surface tension" for γ. Insects that skim over a water surface take advantage of surface tension.

In the bulk phases α and β in Figs. 13.1 and 13.5, the pressure is uniform and equal to P in all directions. In the interphase region, the pressure in the z direction equals P, but the pressure in the x and y directions is not equal to P. Instead, the fact that the pressure (13.5) on the piston is less than the pressure P in the bulk phases tells us that P_y (the system's pressure in the y direction) in the interphase region is less than P. By symmetry, $P_x = P_y$ in the interphase region. The interphase region is not homogeneous, and the pressures P_x and P_y in this region are functions of the z coordinate. Because the interphase region is extremely thin, it is an approximation to talk of a macroscopic property like pressure for this region.

The relations to be developed in this chapter apply in principle to any kind of interface in equilibrium. However, measurement of the surface tension of a solid is hard.

13.2   CURVED INTERFACES

When the interface between phases α and β is curved, the surface tension causes the equilibrium pressures in the bulk phases α and β to differ. This can be seen from Fig. 13.6a. If the lower piston is reversibly pushed in to force more of phase α into the conical region (while some of phase β is pushed out of the conical region through the top channel), the curved interface moves upward, thereby increasing the area $\mathscr{A}$ of the interface between α and β. Since it requires work to increase $\mathscr{A}$, it requires a greater force to push in the lower piston than to push in the upper piston (which would decrease $\mathscr{A}$). We have shown that $P^\alpha > P^\beta$, where α is the phase on the concave side of the curved interface. (Alternatively, if we imagine phases α and β to be separated

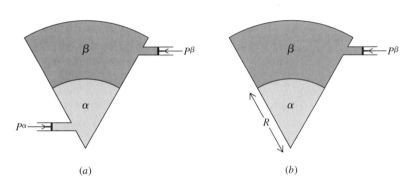

Figure 13.6

Two-phase systems with a curved interface.

by a thin membrane under tension, this hypothetical membrane would exert a net downward force on phase α, making P^α exceed P^β.)

To allow for this pressure difference, we rewrite the definition (13.1) of γ as

$$dw_{\text{rev}} = -P^\alpha \, dV^\alpha - P^\beta \, dV^\beta + \gamma \, d\mathcal{A} \qquad \textbf{(13.6)*}$$

$$V = V^\alpha + V^\beta$$

where $-P^\alpha \, dV^\alpha$ is the P-V work done on the bulk phase α, V^α and V^β are the volumes of phases α and β, and V is the total volume of the system. Since the volume of the interphase region is negligible compared with that of a bulk phase, we have taken $V^\alpha + V^\beta = V$.

To derive the relation between P^α and P^β, consider the modified setup of Fig. 13.6b. We shall assume the interface to be a segment of a sphere. Let the piston be reversibly pushed in slightly, changing the system's total volume by dV. From the definition of work as the product of force and displacement, which equals (force/area) $\times$ (displacement $\times$ area) = pressure $\times$ volume change, the work done on the system by the piston is $-P^\dagger \, dV$, where $P^\dagger$ is the pressure at the interface between system and surroundings, where the force is being exerted. Since $P^\dagger = P^\beta$, we have

$$dw_{\text{rev}} = -P^\beta \, dV = -P^\beta \, d(V^\alpha + V^\beta) = -P^\beta \, dV^\alpha - P^\beta \, dV^\beta \qquad (13.7)$$

Equating (13.7) and (13.6), we get

$$-P^\beta \, dV^\alpha - P^\beta \, dV^\beta = -P^\alpha \, dV^\alpha - P^\beta \, dV^\beta + \gamma \, d\mathcal{A}$$

$$P^\alpha - P^\beta = \gamma (d\mathcal{A}/dV^\alpha) \qquad (13.8)$$

Let R be the distance from the apex of the cone to the interface between α and β in Fig. 13.6b, and let the solid angle at the cone's apex be Ω. The total solid angle around a point in space is 4π steradians. Hence, V^α equals $\Omega/4\pi$ times the volume $\frac{4}{3}\pi R^3$ of a sphere of radius R, and $\mathcal{A}$ equals $\Omega/4\pi$ times the area $4\pi R^2$ of a sphere. (In Fig. 13.6b, all of phase α is within the cone.) We have

$$V^\alpha = \Omega R^3/3, \qquad \mathcal{A} = \Omega R^2$$

$$dV^\alpha = \Omega R^2 \, dR, \qquad d\mathcal{A} = 2\Omega R \, dR$$

Hence $d\mathcal{A}/dV^\alpha = 2/R$ and (13.8) for the pressure difference between two bulk phases separated by a spherical interface becomes

$$P^\alpha - P^\beta = \frac{2\gamma}{R} \qquad \text{spherical interface} \qquad (13.9)$$

Equation (13.9) was derived independently by Young and by Laplace about 1805. As $R \rightarrow \infty$ in (13.9), the pressure difference goes to zero, as it should for a planar interface. The pressure difference (13.9) is substantial only when R is small. For example, for a water–air interface at 20°C, $P^\alpha - P^\beta$ is 0.1 torr for $R = 1$ cm and is 10 torr for $R = 0.01$ cm. The pressure-difference equation for a nonspherical curved interface is more complicated than (13.9) and is omitted.

One consequence of (13.9) is that the pressure inside a bubble of gas in a liquid is greater than the pressure of the liquid. Another consequence is that the vapor pressure of a tiny drop of liquid is slightly higher than the vapor pressure of the bulk liquid; see Prob. 13.35.

Equation (13.9) is the basis for the **capillary-rise** method of measuring the surface tension of liquid–vapor and liquid–liquid interfaces. Here, a capillary tube is inserted in the liquid, and measurement of the height to which the liquid rises in the tube allows calculation of γ. You have probably observed that the water–air interface of an aqueous solution in a glass tube is curved rather than flat. The shape of the interface depends on the relative magnitudes of the adhesive forces between the liquid and the glass and the internal cohesive forces in the liquid. Let the liquid make a **contact angle** θ with the glass (Fig. 13.7). When the adhesive forces exceed the cohesive forces, θ lies in the range $0° \leq \theta < 90°$ (Fig. 13.7a). When the cohesive forces exceed the adhesive forces, then $90° < \theta \leq 180°$.

Suppose that $0° \leq \theta < 90°$. Figure 13.8a shows the situation immediately after a capillary tube has been inserted into a wide dish of liquid β. Points 1 and 6 are at the same height in phase α (which is commonly either air or vapor of liquid β), so $P_1 = P_6$. Points 2 and 5 are located an equal distance below points 1 and 6 in phase α, so $P_2 = P_5$. Points 2 and 3 are just above and just below the planar interface outside the capillary tube, so $P_2 = P_3$. Hence, $P_5 = P_3$. Because the interface in the capillary tube is curved, we know from (13.9) that $P_4 < P_5 = P_3$. Since $P_4 < P_3$, phase β is not in equilibrium, and fluid will flow from the high-pressure region around point 3 into the low-pressure region around point 4, causing fluid β to rise into the capillary tube.

The equilibrium condition is shown in Fig. 13.8b. Here, $P_1 = P_6$, and since points 8 and 5 are an equal distance below points 1 and 6, respectively, $P_8 = P_5$. Also, $P_3 = P_4$, since phase β is now in equilibrium. Subtraction gives $P_8 - P_3 = P_5 - P_4$. The pressures P_2 and P_3 are equal, so

$$P_8 - P_2 = P_5 - P_4 = (P_5 - P_7) + (P_7 - P_4) \qquad (13.10)$$

where P_7 was added and subtracted. Equation (1.9) gives $P_2 - P_8 = \rho_\alpha g h$ and $P_4 - P_7 = \rho_\beta g h$, where ρ_α and ρ_β are the densities of phases α and β and h is the capillary rise. Provided the capillary tube is narrow, the interface can be considered to be a segment of a sphere, and (13.9) gives $P_5 - P_7 = 2\gamma/R$, where R is the sphere's radius. Substitution in (13.10) gives $-\rho_\alpha g h = 2\gamma/R - \rho_\beta g h$ and

$$\gamma = \tfrac{1}{2}(\rho_\beta - \rho_\alpha)ghR \qquad (13.11)$$

When phases β and α are a liquid and a gas, the contact angle on clean glass is usually 0 (liquid Hg is an exception). For $\theta = 0$, the liquid is said to *wet the glass completely*. With a zero contact angle and with a spherically shaped interface, the interface is a hemisphere, and the radius R becomes equal to the radius r of the capillary tube (Fig. 13.9b). Here,

$$\gamma = \tfrac{1}{2}(\rho_\beta - \rho_\alpha)ghr \qquad \text{for } \theta = 0 \qquad (13.12)$$

(For a slightly more accurate equation, see Prob. 13.13.) For $\theta \neq 0$, we see from Fig. 13.9a that $r = R \cos \theta$, so $\gamma = \tfrac{1}{2}(\rho_\beta - \rho_\alpha)ghr/\cos \theta$. Since contact angles are hard to measure accurately, the capillary-rise method is only accurate when $\theta = 0$.

For liquid mercury on glass, the liquid–vapor interface looks like Fig. 13.7b with $\theta \approx 140°$. Here, we get a capillary depression instead of a capillary rise.

In the determination of osmotic pressure in Fig. 12.5, a correction for capillary rise must be made.

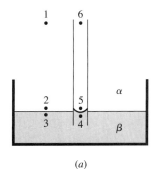

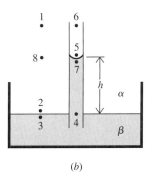

(a) (b)

Figure 13.7

Contact angles between a liquid and a glass capillary tube.

(a) (b)

Figure 13.8

Capillary rise.

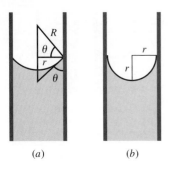

Figure 13.9

Contact angles: (a) $\theta \neq 0$;
(b) $\theta = 0$.

Capillary action is familiar from such things as the spreading of a liquid dropped onto cloth. The spaces between the fibers of the cloth act as capillary tubes into which the liquid is drawn. When fabrics are made water-repellent, a chemical (for example, a silicone polymer) is applied that makes the contact angle θ exceed 90°, so that water is not drawn into the fabric.

EXAMPLE 13.1 Capillary rise

For a water–air interface at 25°C and 1 atm, calculate the capillary rise in a glass capillary tube with inside diameter 0.200 mm. The surface tension of water at 25°C is 72.0 dyn/cm. The densities of air and water at 25°C and 1 atm are 0.001 g/cm³ and 0.997 g/cm³.

Substitution in (13.12) gives

$$72.0 \text{ dyn/cm} = \tfrac{1}{2}(0.997 - 0.001)(\text{g/cm}^3)(981 \text{ cm/s}^2)h(0.0100 \text{ cm})$$

$$h = 14.7 \text{ cm}$$

since 1 dyn = 1 g cm/s². The large value of h is due to the tiny diameter of the capillary tube.

Exercise

Find the inside diameter of a glass capillary in which water shows a capillary rise of 88 mm at 25°C. (*Answer:* 0.33 mm.)

13.3 THERMODYNAMICS OF SURFACES

There are two different approaches to the thermodynamics of systems in which surface effects are significant. Guggenheim in 1940 treated the interfacial layer as a three-dimensional thermodynamic phase having a certain volume, internal energy, entropy, etc. Gibbs in 1878 replaced the actual system by a hypothetical one in which the presence of the interphase region is allowed for by a two-dimensional surface phase that has zero volume but nonzero values of other thermodynamic properties. The Guggenheim method is easier to visualize than the Gibbs model and more closely corresponds to the actual physical situation. However, the Gibbs method is more widely used and is the one we shall adopt.

In the Gibbs approach, the actual system of Fig. 13.10a (which consists of the bulk phases α and β plus the interphase region) is replaced by the hypothetical model system of Fig. 13.10b. In the model system, phases α and β are separated by a surface of zero thickness, the **Gibbs dividing surface.** Phases α and β on either side of this

Figure 13.10

(a) A two-phase system. (b) The corresponding Gibbs model system.

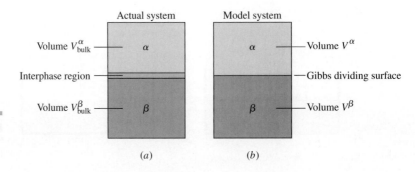

dividing surface are defined to have the same intensive properties as the bulk phases α and β, respectively, in the actual system. The location of the dividing surface in the model system is somewhat arbitrary but generally corresponds to a location within or very close to the interphase region of the actual system. Experimentally measurable quantities must be independent of the choice of location of the dividing surface, which is just a mental construct. We restrict the treatment to a planar interface. The thermodynamics of curved interfaces is the subject of controversy; see L. Boruvka et al., *J. Phys. Chem.,* **89,** 2714 (1985); **90,** 125 (1986); S. M. Oversteegen et al., *Phys. Chem. Chem. Phys.,* **1,** 4987 (1999).

The Gibbs model ascribes to the dividing surface whatever values of thermodynamic properties are necessary to make the model system have the same total volume, internal energy, entropy, and amounts of components that the actual system has. We shall use a superscript σ (sigma) to denote a thermodynamic property of the dividing surface. The dividing surface has zero thickness and zero volume: $V^\sigma = 0$. If V is the volume of the actual system, and V^α and V^β are the volumes of phases α and β in the model system, we require that $V = V^\alpha + V^\beta + V^\sigma$, so

$$V = V^\alpha + V^\beta \tag{13.13}$$

Let U^α_{bulk} and V^α_{bulk} be the energy and volume of the bulk phase α in the actual system. The ratio $U^\alpha_{\text{bulk}}/V^\alpha_{\text{bulk}}$ is the energy per unit volume (the *energy density*) in the bulk phase α. By definition, the energy density in phase α of the model system equals the energy density $U^\alpha_{\text{bulk}}/V^\alpha_{\text{bulk}}$ in the bulk phase α of the actual system. Since phase α of the model system has volume V^α, the energy U^α of the model phase α is

$$U^\alpha = \left(\frac{U^\alpha_{\text{bulk}}}{V^\alpha_{\text{bulk}}} \right) V^\alpha \tag{13.14}$$

with a similar equation for the energy U^β of the model phase β. The total internal energy of the model system is $U^\alpha + U^\beta + U^\sigma$, where U^σ (the **surface excess internal energy**) is the internal energy ascribed to the dividing surface. By definition, this total energy must equal the total internal energy U of the actual system:

$$U = U^\alpha + U^\beta + U^\sigma \quad \text{or} \quad U^\sigma = U - U^\alpha - U^\beta \tag{13.15}$$

Exactly the same arguments hold for entropy, so

$$S^\alpha = (S^\alpha_{\text{bulk}}/V^\alpha_{\text{bulk}})V^\alpha, \qquad S^\beta = (S^\beta_{\text{bulk}}/V^\beta_{\text{bulk}})V^\beta, \qquad S^\sigma = S - S^\alpha - S^\beta \tag{13.16}$$

where S is the total entropy of the actual system and S^α, S^β, and S^σ are the entropies of the model phases α and β and the dividing surface.

The same arguments hold for the amount of component i, so

$$n^\alpha_i = c^\alpha_i V^\alpha, \qquad n^\beta_i = c^\beta_i V^\beta \tag{13.17}$$

$$n_i = n^\alpha_i + n^\beta_i + n^\sigma_i \quad \text{or} \quad n^\sigma_i = n_i - n^\alpha_i - n^\beta_i \tag{13.18}$$

where c^α_i is the molar concentration of component i in the bulk phase α of the actual system (and, by definition, in phase α of the model system), n^α_i and n^β_i are the numbers of moles of i in phases α and β of the model system, n^σ_i is the number of moles of i in the dividing surface, and n_i is the total number of moles of i in the actual system (and in the model system). n^σ_i, called the **surface excess amount** of component i, can be positive, negative, or zero. The definition

$$n^\sigma_i \equiv n_i - (n^\alpha_i + n^\beta_i) = n_i - (c^\alpha_i V^\alpha + c^\beta_i V^\beta) \tag{13.19}$$

states that the surface excess amount n^σ_i is the difference between the amount of i in the actual system and the amount of i that would be in the system if the homogeneity of the bulk phases α and β persisted right up to the dividing surface.

Figure 13.11

Variation of the concentration of
component i with the z coordinate.

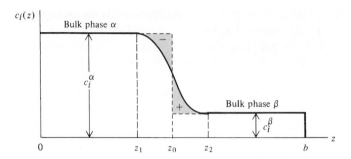

The value of n_i^σ depends on the location of the dividing surface, as we now show. Let the concentration c_i of component i in the actual system vary with the z coordinate as shown by the $c_i(z)$ curve of Fig. 13.11. The interphase region is between z_1 and z_2, and the dividing surface has been placed at z_0.

Imagine the system (which extends from $z = 0$ to $z = b$) to be cut into infinitesimally thin slices taken parallel to the planar interface. Let a given slice contain dn_i moles of component i and have thickness dz, cross-sectional area $\mathcal{A}$, and volume $dV = \mathcal{A}\, dz$. Then $c_i = dn_i/dV = dn_i/(\mathcal{A}\, dz)$, and $dn_i = c_i \mathcal{A}\, dz$. The total number of moles n_i of i in the system is obtained by summing up the infinitesimal amounts dn_i for the infinite number of slices into which the system is cut. This sum is by definition the definite integral from 0 to b of $dn_i = c_i \mathcal{A}\, dz$, and $n_i = \mathcal{A} \int_0^b c_i\, dz$. The integral $\int_0^b c_i\, dz$ is the area under the solid-line curve in Fig. 13.11.

If the homogeneity of the bulk phases α and β persisted up to the dividing surface at z_0, the concentration of i would be given by the upper horizontal line to the left of z_0 and by the lower horizontal line to the right of z_0. The same argument used to show that $n_i = \mathcal{A} \int_0^b c_i\, dz$ shows that the model-system amounts n_i^α and n_i^β in (13.19) are equal to $\mathcal{A}$ times the areas under the upper and lower horizontal lines, respectively.

Therefore, the surface excess amount n_i^σ in (13.19) is equal to $\mathcal{A}$ times the difference between the area under the $c_i(z)$ curve and the areas under the horizontal lines c_i^α and c_i^β. This area difference equals the shaded area to the right of z_0 in Fig. 13.11 minus the shaded area to the left of z_0. In Fig. 13.11, the positive and negative areas are roughly equal, so n_i^σ is approximately zero for this choice of dividing surface. If the dividing surface in Fig. 13.11 is moved to the right, then the negative area will exceed the positive area and n_i^σ becomes negative; if the dividing surface is moved to the left, then n_i^σ becomes positive.

Similar arguments show that U^σ and S^σ also depend on the location of the dividing surface. Since n_i^σ, U^σ, and S^σ depend on the location of the dividing surface, these quantities are not in general physically measurable. The dividing surface is a hypothetical entity and is not intended to represent the actual interphase region.

The first law of thermodynamics is $dU = dq + dw$ for a closed system. For a reversible process, $dq = T\, dS$. In a two-phase system, Eq. (13.1) gives $dw_{\text{rev}} = -P\, dV + \gamma\, d\mathcal{A}$. Therefore

$$dU = T\, dS - P\, dV + \gamma\, d\mathcal{A} \qquad \text{rev. proc., closed syst., planar interface} \qquad (13.20)$$

For an open system, the arguments of Sec. 4.7 require that the terms

$$\sum_i \mu_i^\alpha\, dn_i^\alpha + \sum_i \mu_i^\beta\, dn_i^\beta + \sum_i \mu_i^\sigma\, dn_i^\sigma \qquad (13.21)$$

be added to (13.20), where μ_i^α, μ_i^β, and μ_i^σ are the chemical potentials of i in phase α, phase β, and the surface phase of the model system. At equilibrium, $\mu_i^\alpha = \mu_i^\beta = \mu_i^\sigma$.

Let μ_i denote the chemical potential of i anywhere in the system. Expression (13.21) becomes at equilibrium

$$\sum_i \mu_i \, dn_i^\alpha + \sum_i \mu_i \, dn_i^\beta + \sum_i \mu_i \, dn_i^\sigma = \sum_i \mu_i \, d(n_i^\alpha + n_i^\beta + n_i^\sigma) = \sum_i \mu_i \, dn_i$$

where (13.18) was used. Hence for an open two-phase system in equilibrium

$$dU = T \, dS - P \, dV + \gamma \, d\mathcal{A} + \sum_i \mu_i \, dn_i \qquad \text{rev. proc., planar interface} \qquad (13.22)$$

The presence of the interface leads to the additional term $\gamma \, d\mathcal{A}$ in dU.

For phases α and β in the Gibbs model system, Eq. (4.75) gives

$$dU^\alpha = T \, dS^\alpha - P \, dV^\alpha + \sum_i \mu_i \, dn_i^\alpha, \qquad dU^\beta = T \, dS^\beta - P \, dV^\beta + \sum_i \mu_i \, dn_i^\beta$$

$$(13.23)$$

Equation (13.15) gives $dU^\sigma = dU - dU^\alpha - dU^\beta$. The use of (13.22) and (13.23) together with $dS^\sigma = dS - dS^\alpha - dS^\beta$, $dV = dV^\alpha + dV^\beta$, and $dn_i^\sigma = dn_i - dn_i^\alpha - dn_i^\beta$ gives

$$dU^\sigma = T \, dS^\sigma + \gamma \, d\mathcal{A} + \sum_i \mu_i \, dn_i^\sigma \qquad \text{rev. proc.} \qquad (13.24)$$

Equation (13.24) is now integrated for a process in which the size of the model system is increased at constant T, P, and concentrations in the phases, starting from state 1 and ending at state 2. Under these conditions, the intensive variables T, γ, and the μ_i's are constant and can be taken outside the integral sign. Therefore

$$\int_1^2 dU^\sigma = T \int_1^2 dS^\sigma + \gamma \int_1^2 d\mathcal{A} + \sum_i \mu_i \int_1^2 dn_i^\sigma \qquad \text{const. } T, P, \text{ conc.}$$

$$U_2^\sigma - U_1^\sigma = T(S_2^\sigma - S_1^\sigma) + \gamma(\mathcal{A}_2 - \mathcal{A}_1) + \sum_i \mu_i (n_{i,2}^\sigma - n_{i,1}^\sigma)$$

Let state 1 be the limiting state obtained as the size of the model system goes to zero. All extensive properties are zero in this state, so the terms with the subscript 1 drop out. State 2 is a general state, and omitting the subscript 2, we have

$$U^\sigma = TS^\sigma + \gamma \mathcal{A} + \sum_i \mu_i n_i^\sigma \qquad (13.25)$$

Equation (13.25) is valid for any state of the system.

The total differential of (13.25) is

$$dU^\sigma = T \, dS^\sigma + S^\sigma \, dT + \gamma \, d\mathcal{A} + \mathcal{A} \, d\gamma + \sum_i \mu_i \, dn_i^\sigma + \sum_i n_i^\sigma \, d\mu_i \quad (13.26)$$

Equating the right sides of (13.24) and (13.26), we get

$$S^\sigma \, dT + \mathcal{A} \, d\gamma + \sum_i n_i^\sigma \, d\mu_i = 0 \qquad (13.27)$$

Equation (13.27) is the analog of the Gibbs–Duhem equation (10.17) for the hypothetical surface phase in the Gibbs model system.

At constant temperature, (13.27) becomes $\mathcal{A} \, d\gamma = -\sum_i n_i^\sigma \, d\mu_i$, which is called the **Gibbs adsorption isotherm.** The **surface (excess) concentration** Γ_i^σ (capital gamma) of component i is defined as

$$\Gamma_i^\sigma \equiv n_i^\sigma / \mathcal{A} \qquad (13.28)$$

where $\mathscr{A}$ is the area of the interface. The Gibbs adsorption isotherm becomes

$$d\gamma = -\sum_i \Gamma_i^\sigma \, d\mu_i \qquad \text{const. } T \qquad (13.29)$$

As noted earlier, the n_i^σ's (and hence the Γ_i^σ's) depend on the choice of dividing surface and are not experimentally observable quantities. To obtain physically meaningful quantities, we choose one particular dividing surface and refer the Γ_i^σ's to that surface. The dividing surface chosen is the one that makes n_1^σ (and hence Γ_1^σ) zero, where component 1 is a particular component of the system, usually the solvent. Let $\Gamma_{i(1)}$ (called the **relative adsorption** of component i with respect to component 1) denote the value of $\Gamma_i^\sigma = n_i^\sigma/\mathscr{A}$ for the dividing surface that makes $n_1^\sigma = 0$. One finds (Prob. 13.20) that $\Gamma_{i(1)}$ is a function of c_i^α, c_i^β, c_1^α, c_1^β, n_i, n_1, $\mathscr{A}$, and V. All these quantities are experimentally measurable properties of the actual system and are independent of the location of the hypothetical dividing surface. Therefore $\Gamma_{i(1)}$ is experimentally measurable.

For the dividing surface that makes n_1^σ and Γ_1^σ zero, the Gibbs adsorption isotherm (13.29) becomes

$$d\gamma = -\sum_{i \neq 1} \Gamma_{i(1)} \, d\mu_i \qquad \text{const. } T \qquad (13.30)$$

All quantities in this equation are experimentally measurable.

The most common applications of the Gibbs adsorption isotherm are to two-phase systems in which the concentrations of components 1 and i in phase β are much smaller than those in phase α; $c_1^\beta \ll c_1^\alpha$ and $c_i^\beta \ll c_i^\alpha$. Examples include (a) a liquid–vapor system in which the vapor pressure is low or moderate, so that the vapor-phase concentrations are much less than the corresponding liquid-phase concentrations; (b) a liquid–liquid system in which solvent 1 and solute i of phase α are virtually insoluble in phase β; (c) a solid–liquid system in which solvent 1 and solute i of the liquid are insoluble in the solid (this case is important in electrochemistry). For such systems, one finds that (Prob. 13.21)

$$\Gamma_{i(1)} = \frac{n_1^s}{\mathscr{A}}\left(\frac{n_i^s}{n_1^s} - \frac{n_{i,\text{bulk}}^\alpha}{n_{1,\text{bulk}}^\alpha}\right) \qquad \text{when } c_i^\alpha \gg c_i^\beta, c_1^\alpha \gg c_1^\beta \qquad (13.31)$$

where n_i^s and n_1^s are the numbers of moles of substances i and 1 in the interphase region of the *actual* system (not the model system) and $n_{i,\text{bulk}}^\alpha$ and $n_{1,\text{bulk}}^\alpha$ are the numbers of moles of i and 1 in the bulk phase α of the actual system (Fig. 13.10a). When the relative adsorption $\Gamma_{i(1)}$ of solute i is positive, the ratio n_i^s/n_1^s of the amount of solute to the amount of solvent in the interphase region of the system is greater than the corresponding ratio $n_{i,\text{bulk}}^\alpha/n_{1,\text{bulk}}^\alpha$ in bulk phase α and component i is said to be (*positively*) adsorbed at the interface (see Fig. 13.2b). When $\Gamma_{i(1)}$ is negative, i is *negatively adsorbed* at the interface. **Adsorption** is the enrichment of a component in the interphase region compared with a bulk region.

Having considered the meaning of $\Gamma_{i(1)}$, we now return to the Gibbs adsorption isotherm. For a two-component system, (13.30) reads

$$d\gamma = -\Gamma_{2(1)} \, d\mu_2 \qquad \text{const. } T, \text{ binary syst.} \qquad (13.32)$$

At least one of the two bulk phases must be a solid or liquid; let us call this phase α. For this phase, $\mu_2 = \mu_2^{\circ,\alpha}(T, P) + RT \ln a_2^\alpha$ [Eq. (10.4)]. The pressure dependence of $\mu_2^{\circ,\alpha}$ is slight for a condensed phase; moreover, the surface tension is often measured in the presence of air at the constant pressure of 1 atm. Hence, at constant T, we can take $d\mu_2 = RT \, d \ln a_2^\alpha$, and (13.32) becomes

$$\Gamma_{2(1)} = -\frac{1}{RT}\left(\frac{\partial \gamma}{\partial \ln a_2^\alpha}\right)_T \qquad \text{binary syst.} \qquad (13.33)$$

If the molar concentration scale is used for solute 2, its activity in phase α is $a_2^\alpha = \gamma_{c,2}^\alpha c_2^\alpha/c^\circ$ [Eq. (10.30)]. If phase α is dilute enough to be considered ideally dilute, then $a_2^\alpha = c_2^\alpha/c^\circ$ (where $c^\circ \equiv 1$ mol/dm^3), and (13.33) becomes

$$\Gamma_{2(1)} = -\frac{1}{RT}\left(\frac{\partial \gamma}{\partial \ln (c_2^\alpha/c^\circ)}\right)_T \qquad \text{binary syst., ideally dil. soln.} \qquad (13.34)$$

The slope of a plot of the solution's surface tension γ versus $\ln (c_2^\alpha/c^\circ)$ at a given temperature equals $-RT\Gamma_{2(1)}$ and allows calculation of $\Gamma_{2(1)}$. If the solution is not ideally dilute, activity-coefficient data are needed to find $\Gamma_{2(1)}$.

Equation (13.34) states that $\Gamma_{2(1)}$ is positive if the surface tension decreases with increasing solute concentration and is negative if γ increases with increasing c_2^α. The observed behavior of solutes in dilute aqueous solutions can be classified into three types (Fig. 13.12). Type I solutes produce a small rate of increase in γ with increasing concentration; examples include most inorganic salts and sucrose. The increase in γ for salt solutions can be explained by saying that the increased opportunity for attractions between oppositely charged ions in the bulk phase compared with the surface layer (Fig. 13.3) reduces the number of ions in the surface layer and the negative adsorption increases γ.

Type II solutes give a substantial and steady rate of decrease in γ with increasing concentration; examples include the majority of organic compounds that have some water solubility. Water-soluble organic compounds usually contain a polar part (for example, an OH or COOH group) and a nonpolar hydrocarbon part. Such molecules tend to accumulate in the surface layer, where they are oriented with their polar parts pointing toward, and interacting with, the polar water molecules of the bulk solution and their nonpolar parts pointing away from the bulk solution (see Sec. 13.4 and Fig. 13.14). The resulting positive adsorption decreases γ.

For Type III solutes, γ shows a very rapid drop followed by a sudden leveling off as the concentration is increased. Examples are salts of medium-chain-length organic acids (soaps, RCOO$^-$Na$^+$), alkyl sulfate salts (ROSO$_2$O$^-$Na$^+$), quaternary amine salts [(CH$_3$)$_3$RN$^+$Cl$^-$], and polyoxyethylene compounds [R(OCH$_2$CH$_2$)$_n$OH, where n is 5 to 15]. Type III solutes are strongly adsorbed at the interface. (The leveling off of γ occurs at the critical micelle concentration; see Sec. 13.6.)

A solute that significantly lowers the surface tension is said to be a *surface-active agent* or *surfactant*. Type III solutes act as detergents and are outstanding surfactants. The lowering of γ helps remove oily dirt particles from solid surfaces.

The surface layer of natural bodies of water (oceans, lakes, bays) is greatly enriched in organic compounds (long-chain organic acids, esters, proteins, etc.) originating mainly from decomposed plants and animals. This surface region (called the **surface microlayer** or *biofilm*) contains much greater concentrations of organic pollutants (pesticides, polyaromatic hydrocarbons, PCBs, etc.) than the underlying bulk liquid. The thickness of the surface microlayer is not readily measurable and varies according to circumstances and as to how one defines this region. Thicknesses ranging from 1 to 1000 μm have been quoted, with 50 μm being a typical value.

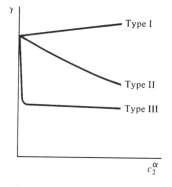

Figure 13.12

Typical surface-tension-versus-concentration curves for aqueous solutions.

13.4 SURFACE FILMS ON LIQUIDS

In 1774, Benjamin Franklin read a paper to the Royal Society describing the result of pouring olive oil on the surface of a London pond: "The oil, though not more than a teaspoonful, produced an instant calm over a space several yards square, which spread amazingly, and extended itself gradually till it reached the leeside, making all that quarter of the pond, perhaps half an acre, as smooth as a looking-glass." A calculation from Franklin's data gives the thickness of the oil film as 24 Å (Prob. 13.23), which is the order of magnitude of the length of an olive-oil molecule and indicates a one-molecule-thick (monomolecular) surface film.

Many water-insoluble organic compounds that contain a medium-length hydrocarbon chain with a polar group at one end will spread spontaneously over a water surface to give a surface film. Examples are $CH_3(CH_2)_{16}COOH$ (stearic acid), $CH_3(CH_2)_{11}OH$ (lauryl alcohol), and $CH_3(CH_2)_{14}COOC_2H_5$ (ethyl palmitate). Evidence (cited later in this section) shows these films to be generally one molecule thick. Such monomolecular surface films are called *Langmuir films* or *spread monolayers*.

The rather long hydrocarbon chain makes the solubility of these compounds in water extremely low. At room temperature, these compounds are solids or high-boiling liquids with very low vapor pressures. Therefore the amounts of solute i present in the bulk phases (water and air) are negligible compared with the amount of i present in the interphase region as a monolayer. With $n_i^\alpha = 0 = n_i^\beta$, Eqs. (13.18) and (13.28) become $n_i^\sigma = n_i$ and $\Gamma_i^\sigma = n_i/\mathscr{A}$, where n_i is (to a good approximation) the total number of moles of i in the system. For such systems, Γ_i^σ is independent of the location of the Gibbs dividing surface and has a direct physical significance. It follows that the relative adsorption $\Gamma_{i(1)}$ equals Γ_i^σ and is positive. Therefore (Sec. 13.3) the surface tension is lowered by the presence of the film.

Surface films are studied with a *surface balance* (Fig. 13.13a); a floating barrier (*float*) separates a clean water surface from a water surface containing the monolayer. The force on the float is measured by a torsion wire attached to it.

Let γ^* and γ be the surface tensions of pure water and of water covered with the monolayer, respectively. The discussion after Eq. (13.5) shows that the surface tension of the pure water produces a force γ^* per unit length that pulls the float toward the right in Fig. 13.13a, and the surface tension of the monolayer-covered water produces a force γ per unit length that pulls the float toward the left. Since γ is less than γ^*, there is a net force per unit length on the float of $\gamma^* - \gamma$ toward the right. This force per unit length is called the *surface pressure* π; thus $\pi \equiv \gamma^* - \gamma$.

Suppose the adjustable barrier in Fig. 13.13a is moved to the right. This decreases the area $\mathscr{A}$ available to the monolayer molecules and increases their adsorption $\Gamma_i^\sigma = n_i/\mathscr{A}$. The increased adsorption of these surface-active molecules further lowers γ and increases the surface pressure π. A typical π-versus-$\mathscr{A}$ curve at a fixed temperature is shown in Fig. 13.13b.

In Fig. 13.13b, π increases gradually with decreasing area until point C is reached, after which further decrease in $\mathscr{A}$ sharply increases π. At point C, the film has been compressed sufficiently to bring its molecules almost in contact with one another, and the film now strongly resists further compression. If the area $\mathscr{A}_0$ at point C (called the *Pockels point*) is divided by the number of molecules N_i of i present in the monolayer, we get an estimate of the cross-sectional area per molecule. [Agnes Pockels was an amateur scientist who did much of her work in her kitchen. See M. E. Derrick, *J. Chem. Educ.*, **59**, 1030 (1982).]

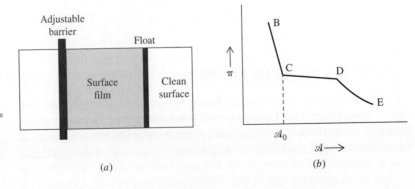

Figure 13.13

(a) Schematic diagram of a surface balance. (b) Typical curve of surface pressure versus area for a surface film.

For each of the acids $CH_3(CH_2)_nCOOH$ with $n = 14$, 16, and 24, Langmuir found that $\mathscr{A}_0/N_i$ is 21 Å^2. This value agrees with that obtained by x-ray diffraction of crystals for the cross-sectional area per molecule. The independence of $\mathscr{A}_0/N_i$ of the chain length shows that at the Pockels point the monolayer molecules are oriented vertically; the polar COOH end points toward the water phase (and thus can interact with the polar water molecules), and the nonpolar hydrocarbon part points toward the vapor phase (Fig. 13.14). The polar part of the molecule is said to be **hydrophilic** ("water-loving") and the hydrocarbon part **hydrophobic** ("water-hating"). Molecules with both hydrophilic and hydrophobic groups are called **amphipathic** or **amphiphilic** (from the Greek *amphi*, "dual," *pathos*, "feeling," *philos*, "loving"). The polar part likes a polar solvent, and the nonpolar part likes a nonpolar solvent. The surfactant molecules mentioned in Sec. 13.3 are amphiphilic.

For the portion DE of the isotherm (Fig. 13.13b) at low surface pressures, the acid molecules in the monolayer are reasonably far apart from one another, and there is little interaction between them. The state of the monolayer acid molecules along DE is analogous to that of a two-dimensional gas. At high surface pressures, along CB, the acid molecules are quite close together, and their state is analogous to that of a two-dimensional liquid. The approximately horizontal portion CD of the isotherm can be interpreted as states where some of the acid molecules are in a two-dimensional liquid state and the remainder in a two-dimensional gaseous state. Note the resemblance of the curve BCDE to the isotherms of a three-dimensional fluid below its critical point (Fig. 8.3).

A practical application of monolayers is to reduce the rate of evaporation of water from reservoirs. Cetyl alcohol $[CH_3(CH_2)_{15}OH]$ is commonly used.

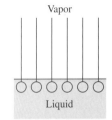

Figure 13.14

Amphiphilic molecules in a surface film at the Pockels point. The hydrophilic polar group in each molecule points toward the liquid-water phase.

13.5 ADSORPTION OF GASES ON SOLIDS

In this section, the interphase region of an essentially nonvolatile solid (A) in contact with a gas (B) is considered. The industrially important catalytic activity of such solids as finely divided Pt, Pd, and Ni results from adsorption of gases. Commonly used gases in adsorption studies include He, H_2, N_2, O_2, CO, CO_2, CH_4, C_2H_6, C_2H_4, NH_3, and SO_2. Commonly used solids include metals, metal oxides, silica gel (SiO_2), and carbon in the form of charcoal. The solid on whose surface adsorption occurs is called the **adsorbent** or *substrate*. The adsorbed gas is the **adsorbate**. Adsorption occurs at the solid–gas interface and is to be distinguished from absorption, in which the gas penetrates throughout the bulk solid phase. An example of absorption is the reaction of water vapor with anhydrous $CaCl_2$ to form a hydrate compound.

A complication in gas–solid studies is that the surfaces of solids are rough, and it is hard to determine the surface area of a solid reliably.

Chemisorption and Physical Adsorption

Adsorption on solids is classified into **physical adsorption** (or *physisorption*) and *chemical adsorption* (or **chemisorption**). The dividing line between the two is not always sharp. In physical adsorption, the gas molecules are held to the solid's surface by relatively weak intermolecular van der Waals forces. In chemisorption, a chemical reaction occurs at the solid's surface, and the gas is held to the surface by relatively strong chemical bonds.

Physical adsorption is nonspecific. For example, N_2 will be physically adsorbed on any solid provided the temperature is low enough. Chemisorption is similar to ordinary chemical reactions in that it is highly specific. For example, N_2 is chemisorbed at room temperature on Fe, W, Ca, and Ti but not on Ni, Ag, Cu, or Pb. Solid Au chemisorbs O_2, C_2H_2, and CO but not H_2, CO_2, or N_2.

The enthalpy changes for chemisorption are usually substantially greater in magnitude than those for physical adsorption. Typically $\Delta \bar{H}$ for chemisorption lies in the range -40 to -800 kJ/mol (-10 to -200 kcal/mol), whereas $\Delta \bar{H}$ for physical adsorption is usually from -4 to -40 kJ/mol (-1 to -10 kcal/mol), similar to enthalpies of gas condensation. Chemical bonds may be broken as well as formed in chemisorption (for example, H_2 is chemisorbed on metals as H atoms). One therefore might expect ΔH to show both positive and negative values for chemisorption, similar to ΔH for ordinary chemical reactions. However, we expect that ΔS for chemisorption of a gas onto a solid will be quite negative, so ΔH of chemisorption must be significantly negative to make ΔG negative and produce a significant amount of chemisorption. An exception is the chemisorption of $H_2(g)$ onto glass. Here, the two moles of adsorbed H atoms formed from one mole of $H_2(g)$ have substantial mobility on the solid's surface and have a greater entropy than the $H_2(g)$, and ΔH of chemisorption is slightly positive; see J. H. de Boer, *Adv. Catal.,* **9,** 472 (1957).

For chemisorption, once a monolayer of adsorbed gas covers the solid's surface, no further chemical reaction between the gas (species B) and the solid (species A) can occur. For physical adsorption, once a monolayer has formed, intermolecular interactions between adsorbed B molecules in the monolayer and gas-phase B molecules can lead to formation of a second layer of adsorbed gas. The enthalpy change for formation of the first layer of physically adsorbed molecules is determined by solid–gas (A-B) intermolecular forces, whereas the enthalpy change for formation of the second, third, . . . physically adsorbed layers is determined by B-B intermolecular forces and is about the same as the ΔH of condensation of gas B to a liquid. Although only one layer can be chemically adsorbed, physical adsorption of further layers on top of a chemisorbed monolayer sometimes occurs.

Substantial physical adsorption usually occurs only at temperatures near or below the boiling point of the gas.

The chemical reactions that occur with chemisorption have been determined for several systems. When H_2 is chemisorbed on metals, H atoms are formed on the surface and bond to metal atoms, as evidenced by the fact that metals that chemisorb H_2 will catalyze the exchange reaction $H_2 + D_2 \rightarrow 2HD$. Chemisorption of C_2H_6 on metals occurs mainly by breakage of a C—H bond, and to a lesser extent by breakage of the C—C bond; evidence for this is from a comparison of the rates of the metal-catalyzed exchange and cracking reactions $C_2H_6 + D_2 \rightarrow C_2H_5D + HD$ and $C_2H_6 + H_2 \rightarrow 2CH_4$. The chemisorbed structures are

$$
\begin{array}{ccc}
\text{H} \quad \text{CH}_2\text{CH}_3 & & \text{H}_3\text{C} \quad \text{CH}_3 \\
| \quad\quad | & & | \quad\quad | \\
-\text{M}-\text{M}- & \text{and} & -\text{M}-\text{M}- \\
| \quad\quad | & & | \quad\quad |
\end{array}
$$

where M is a surface metal atom. Chemisorption of CO_2 on metal oxides probably occurs with formation of carbonate ions: $CO_2 + O^{2-} \rightarrow CO_3^{2-}$. Comparison of the infrared spectra of CO chemisorbed on metals with those of metal carbonyl compounds suggests that one or both of the kinds of bonding shown in Fig. 13.15 occur, depending on which metal is used. In some cases, the carbon of a chemisorbed CO molecule bonds to three M atoms simultaneously.

Species like CO, NH_3, and C_2H_4 that have unshared electron pairs or multiple bonds can be chemisorbed without dissociating (**nondissociative** or **molecular adsorption**). In contrast, species like H_2, CH_4, and C_2H_6 usually dissociate when chemisorbed (**dissociative adsorption**); for exceptions, see Sec. 20.6. Some gases (for example, CO, N_2) undergo both dissociative and nondissociative adsorption, depending on the adsorbent used.

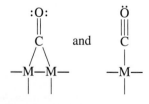

Figure 13.15

CO chemisorbed on a metal surface.

Figure 13.16a shows H_2 physically adsorbed on the surface of metal M. Figure 13.16b shows H_2 dissociatively chemisorbed on M.

Adsorption Isotherms

Under ordinary conditions, the surface of a solid is covered with adsorbed species (such as carbon, hydrocarbons, oxygen, sulfur, and water) derived mainly from atmospheric gases. A solid–gas adsorption study requires an initially clean solid surface. To produce a clean surface, one may heat the solid strongly in high vacuum, a procedure called *outgassing*. However, such heating may not desorb all the surface contamination. A better procedure is to vaporize the solid in vacuum and condense it as a thin film on a solid surface. Another cleaning method is to bombard the surface with Ar^+ ions. Alternatively, a crystal may be cleaved in vacuum to produce a clean surface.

In adsorption studies, the amount of gas adsorbed at a given temperature is measured as a function of the gas pressure P in equilibrium with the solid. The container holding the adsorbent is in a constant-temperature bath and is separated by a stopcock from the adsorbate gas. The number of moles n of gas adsorbed by a solid sample can be calculated (using the ideal-gas law) from the observed gas-pressure change when the sample is exposed to the gas or can be found by observing the stretching of a spring from which the adsorbent container is suspended.

By repeating the experiment with different initial pressures, one generates a series of values of moles adsorbed n versus equilibrium gas pressure P at fixed adsorbent temperature. If m is the adsorbent mass, a plot of n/m (the moles of gas adsorbed per gram of adsorbent) versus P at fixed T is an **adsorption isotherm.** For no good reason, it has been traditional to plot adsorption isotherms with the amount adsorbed expressed as the gas volume v (corrected to 0°C and 1 atm) adsorbed per gram of adsorbent. v is directly proportional to n/m. Figure 13.17 shows two typical isotherms.

For O_2 on charcoal at 90 K (Fig. 13.17a), the amount adsorbed increases with P until a limiting value is reached. This isotherm is said to be of Type I, and its interpretation is that at the adsorption limit the solid's surface is covered by a monolayer of O_2 molecules, after which no further O_2 is adsorbed. Type I isotherms are typical for chemisorption. The isotherm in Fig. 13.17b is of Type II. Here, after formation of a monolayer of adsorbed gas is substantially complete, further increases in gas pressure cause formation of a second layer of adsorbed molecules, then a third layer, etc. (multilayer adsorption). Type II isotherms are typical for physical adsorption.

In 1918 Langmuir used a simple model of the solid surface to derive an equation for an isotherm. He assumed that the solid has a uniform surface, that adsorbed molecules don't interact with one another, that the adsorbed molecules are localized at specific sites, and that only a monolayer can be adsorbed.

At equilibrium, the rates of adsorption and desorption of molecules from the surface are equal. Let N be the number of adsorption sites on the bare solid surface. (For

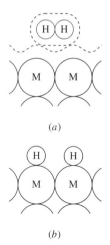

Figure 13.16

(a) H_2 physically adsorbed on a metal surface. The dashed lines are drawn using the van der Waals radii (Sec. 24.6) of the atoms. (b) H_2 dissociatively chemisorbed on a metal surface.

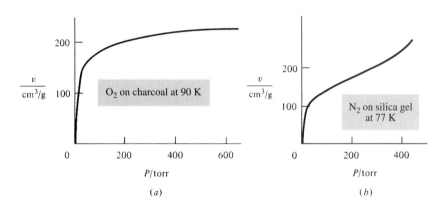

Figure 13.17

(a) Adsorption isotherm of O_2 on charcoal at 90 K. (b) Adsorption isotherm of N_2 on silica gel at 77 K.

example, for chemisorption of CO_2 on a metal oxide to produce CO_3^{2-}, N is the number of oxide ions at the surface.) Let θ be the fraction of adsorption sites occupied by adsorbate at equilibrium. The rate of desorption is proportional to the number θN of adsorbed molecules and equals $k_d \theta N$, where k_d is a constant at fixed temperature. The adsorption rate is proportional to the rate of collisions of gas-phase molecules with unoccupied adsorption sites, since only a monolayer can be formed. The rate of collisions of gas molecules with the surface is proportional to the gas pressure P [Eq. (15.56)], and the number of unoccupied sites is $(1 - \theta)N$. The adsorption rate is therefore $k_a P(1 - \theta)N$, where k_a is a constant at fixed T. Equating the adsorption and desorption rates and solving for θ, we get

$$k_a P(1 - \theta)N = k_d \theta N$$

$$\theta = \frac{k_a P}{k_d + k_a P} = \frac{(k_a/k_d)P}{1 + (k_a/k_d)P} = \frac{bP}{1 + bP} \qquad \text{where } b(T) \equiv k_a/k_d \qquad (13.35)$$

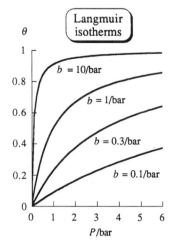

Since the rate constants k_a and k_d depend on temperature, b depends on T. The fraction θ of sites occupied at pressure P equals v/v_{mon}, where v is the volume adsorbed at P (as previously defined) and v_{mon} is the volume adsorbed in the high-pressure limit when a monolayer covers the entire surface. Equation (13.35) becomes

$$v = \frac{v_{\text{mon}} bP}{1 + bP} \qquad (13.36)$$

The shape of the **Langmuir isotherm** (13.36) resembles Fig. 13.17a. In the low-P limit, bP in the denominator in (13.35) can be neglected and θ increases linearly with P, according to $\theta \approx bP$. In the high-P limit, $\theta \rightarrow 1$. Figure 13.18 plots θ versus P according to the Langmuir isotherm (13.35) for several values of b. (Compare this figure with Fig. 13.19.)

To test whether (13.36) fits a given set of data, we take the reciprocal of each side to give $1/v = 1/v_{\text{mon}} bP + 1/v_{\text{mon}}$. A plot of $1/v$ versus $1/P$ gives a straight line if the Langmuir isotherm is obeyed. The Langmuir isotherm is found to work reasonably well for many (but far from all) cases of chemisorption.

In deriving the Langmuir isotherm, we assumed that only one gas is chemisorbed and that this adsorption is nondissociative. If two gases A and B undergo nondissociative adsorption on the same surface, the Langmuir assumptions give (Prob. 13.29)

$$\theta_A = \frac{b_A P_A}{1 + b_A P_A + b_B P_B} \quad \text{and} \quad \frac{v}{v_{\text{mon}}} = \frac{b_A P_A + b_B P_B}{1 + b_A P_A + b_B P_B} \qquad (13.37)$$

where θ_A is the fraction of adsorption sites occupied by A molecules and b_A and b_B are constants.

If a single gas is dissociatively adsorbed according to $A_2(g) \rightleftharpoons 2A(ads)$ (where *ads* stands for adsorbed), the Langmuir isotherm becomes (Prob. 17.96)

$$\theta = \frac{b^{1/2} P^{1/2}}{1 + b^{1/2} P^{1/2}} \qquad (13.38)$$

Most of Langmuir's assumptions are false. The surfaces of most solids are not uniform, and the desorption rate depends on the location of the adsorbed molecule. The force between adjacent adsorbed molecules is often substantial, as shown by changes in the heat of adsorption with increasing θ. There is much evidence that adsorbed molecules can move about on the surface. This mobility is much greater for physically adsorbed molecules than for chemisorbed ones and increases as T increases. Multilayer adsorption is common in physical adsorption. Thus Langmuir's derivation of (13.36) cannot be taken too seriously. Statistical-mechanical derivations of the Langmuir isotherm require fewer assumptions than Langmuir's derivation.

Figure 13.18

Langmuir isotherms of fractional surface coverage versus gas pressure for several values of the rate-constants ratio $b \equiv k_a/k_d$. At $bP = 1$, the surface is 50% covered.

The *Freundlich isotherm*

$$v = kP^a \tag{13.39}$$

where k and a are constants (with $0 < a < 1$), was suggested on empirical grounds in the nineteenth century. Equation (13.39) gives $\log v = \log k + a \log P$. The intercept and slope of a plot of $\log v$ versus $\log P$ give $\log k$ and a. The Freundlich isotherm can be derived by modifying the Langmuir assumptions to allow for several kinds of adsorption sites on the solid, each kind having a different heat of adsorption. The Freundlich isotherm is not valid at very high pressures but is frequently more accurate than the Langmuir isotherm for intermediate pressures.

The Freundlich equation is often applied to adsorption of solutes from liquid solutions onto solids. Here, the solute's concentration c replaces P, and the mass adsorbed per unit mass of adsorbent replaces v.

The Langmuir and Freundlich isotherms apply to Type I isotherms only. In 1938, Brunauer, Emmett, and Teller modified Langmuir's assumptions to give an isotherm for multilayer physical adsorption (Type II). Their result is

$$\frac{P}{v(P^* - P)} = \frac{1}{v_{mon}c} + \frac{c-1}{v_{mon}c}\frac{P}{P^*} \tag{13.40}$$

where v is defined above, v_{mon} is the v corresponding to a monolayer, c is a constant at fixed T, and P^* is the vapor pressure of the adsorbate at the temperature of the experiment. (For $P \geq P^*$, the gas condenses to a liquid.) The constants c and v_{mon} can be obtained from the slope and intercept of a plot of $P/v(P^* - P)$ versus P/P^*. The *Brunauer–Emmett–Teller* (BET) isotherm fits many Type II isotherms well, especially for intermediate pressures. Once v_{mon} has been obtained from the BET isotherm, the number of molecules needed to form a monolayer is known and the surface area of the solid adsorbent can be estimated by using an estimated value for the surface area occupied by one adsorbed molecule.

Adsorption is nearly always exothermic, and as the temperature increases, the amount adsorbed at a given P nearly always decreases, in accord with Le Châtelier's principle. See Fig. 13.19. From a set of isotherms, one can read off the pressure on each isotherm that corresponds to a fixed value of v and hence corresponds to a fixed surface coverage θ. This tells us how the pressure P of gas in equilibrium with the solid surface varies with T at fixed θ. Provided the gas is ideal, a thermodynamic analysis gives the following relation between these variables (for a derivation, see *Defay, Prigogine, Bellemans, and Everett,* pp. 48–50):

$$\left(\frac{\partial \ln P}{\partial T}\right)_\theta = -\frac{\Delta \bar{H}_a}{RT^2} \tag{13.41}$$

where $\Delta \bar{H}_a$, the *differential molar enthalpy of adsorption,* equals dH/dn, where dH is the infinitesimal enthalpy change when dn moles are adsorbed at coverage θ. Just as the differential heat of solution (Sec. 9.4) depends on the concentration of the solution, $\Delta \bar{H}_a$ depends on the fraction θ of surface coverage. Note the resemblance of (13.41) to the Clausius–Clapeyron equation $d \ln P/dT = \Delta_{vap}H_m/RT^2$ for the vapor pressure of a solid or liquid in equilibrium with an ideal gas.

The quantity $-\Delta \bar{H}_a$ is the *isosteric heat of adsorption,* where "isosteric" refers to the constancy of θ. A plot of $\ln P$ versus $1/T$ at fixed θ has a slope of $\Delta \bar{H}_a/R$, since $d(1/T) = -(1/T^2)\,dT$. One finds that $|\Delta \bar{H}_a|$ usually decreases significantly as θ increases (Fig. 13.20). This occurs because the strongest binding sites tend to be filled first and because repulsions between adsorbed species increase as θ increases.

Besides studying adsorption isotherms, one can obtain information on adsorption by heating an adsorbate-covered solid and monitoring the pressure of the desorbed gas as T rises, a process called *thermal desorption*; see Sec. 17.18.

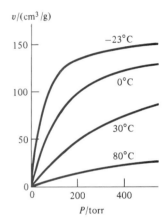

$v/(\text{cm}^3/\text{g})$

Figure 13.19

Adsorption isotherms of NH_3 on charcoal at several temperatures.

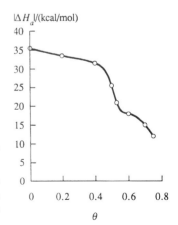

$|\Delta H_a|/(\text{kcal/mol})$

Figure 13.20

Isosteric heat of adsorption of CO on the (111) surface of Pd. (The notation is explained in Sec. 24.7.)

There is currently an explosive growth in research on the surface structure of solids and substances adsorbed on solids, topics important to heterogeneous catalysis and microelectronics. For discussion see Secs. 17.18, 21.9, and 24.10.

13.6   COLLOIDS

When an aqueous solution containing Cl^- ion is added to one containing Ag^+ ion, under certain conditions the solid AgCl precipitate may form as extremely tiny crystals that remain suspended in the liquid instead of settling out as a filterable precipitate. This is an example of a colloidal system.

Colloidal Systems

A **colloidal system** consists of particles that have in at least one direction a dimension lying in the approximate range 20 to 10^4 Å (2 to 1000 nm) and a medium in which the particles are dispersed. The particles are called **colloidal particles** or the *dispersed phase*. The medium is called the **dispersion medium** or the *continuous phase*. The colloidal particles may be in the solid, liquid, or gaseous state, or they may be individual molecules. The dispersion medium may be solid, liquid, or gas. The term **colloid** can mean either the colloidal system of particles plus dispersion medium or just the colloidal particles.

A **sol** is a colloidal system whose dispersion medium is a liquid or gas. When the dispersion medium is a gas, the sol is called an **aerosol.** Fog is an aerosol with liquid particles. Smoke is an aerosol with liquid or solid particles. Tobacco smoke has liquid particles. The earth's atmosphere contains an aerosol of aqueous H_2SO_4 and $(NH_4)_2SO_4$ droplets resulting from the burning of sulfur-containing fuels and volcanic eruptions. This sulfate aerosol produces acid rain and reflects some of the incident sunlight, thereby cooling the earth [*Science,* **255,** 682 (1992); R. J. Charlson and T. M. L. Wigley, *Scientific American,* Feb. 1994, p. 48]. A sol that consists of a liquid dispersed in a liquid is an **emulsion.** A sol that consists of solid particles suspended in a liquid is a **colloidal suspension.** An example is the aqueous AgCl system previously mentioned.

A **foam** is a colloidal system in which gas bubbles are dispersed in a liquid or solid. Although the diameters of the bubbles usually exceed 10^4 Å, the distance between bubbles is usually less than 10^4 Å, so foams are classified as colloidal systems; in foams, the dispersion medium is in the colloidal state. Foams are familiar to anyone who uses soap, drinks beer, or goes to the beach. Pumice stone is a foam with air bubbles dispersed in rock of volcanic origin.

Colloidal systems can be classified into those in which the dispersed particles are single molecules (monomolecular particles) and those in which the particles are aggregates of many molecules (polymolecular particles). Colloidal dispersions of AgCl, As_2S_3, and Au in water contain polymolecular particles, and the system has two phases: water and the dispersed particles. The tiny size of the particles results in a very large interfacial area, and surface effects (for example, adsorption on the colloidal particles) are of major importance in determining the system's properties. On the other hand, in a polymer solution (for example, a solution of a protein in water) the colloidal particle is a single molecule, and the system has one phase. Here, there are no interfaces, but solvation of the polymer molecules is significant. The large size of the solute molecules causes a polymer solution to resemble a colloidal dispersion of polymolecular particles in such properties as scattering of light and sedimentation in a centrifuge, so polymer solutions are classified as colloidal systems.

Lyophilic Colloids

When a protein crystal is dropped into water, the polymer molecules spontaneously dissolve to produce a colloidal dispersion. Colloidal dispersions that can be formed by spontaneous dispersion of the dry bulk material of the colloidal particles in the dispersion medium are called **lyophilic** ("solvent-loving"). A lyophilic sol is thermodynamically more stable than the two-phase system of dispersion medium and bulk colloid material.

Certain compounds in solution yield lyophilic colloidal systems as a result of spontaneous association of their molecules to form colloidal particles. If one plots the osmotic pressure of an aqueous solution of a soap (a compound with the formula $RCOO^-M^+$, where R is a straight chain with 10 to 20 carbons, and M is Na or K) versus the solute's stoichiometric concentration, one finds that at a certain concentration (called the **critical micelle concentration,** cmc) the solution shows a sharp drop in the slope of the Π-versus-c curve. Starting at the cmc, the solution's light-scattering ability (turbidity) rises sharply. These facts indicate that above the cmc a substantial portion of the solute ions are aggregated to form units of colloidal size. Such aggregates are called **micelles.** Dilution of the solution below the cmc eliminates the micelles, so micelle formation is reversible. Light-scattering data show that a micelle is approximately spherical and contains from 20 to a few hundred monomer units, depending on the compound. Figure 13.21*a* shows the structure of a soap micelle in aqueous solution. The hydrocarbon part of each monomer anion is directed toward the center, and the polar COO^- group is on the outside. Many of the micelle's COO^- groups have solvated Na^+ ions bound to them (ion pairing, Sec. 10.9). At high concentrations of dissolved surfactant, micelles with nonspherical shapes are formed. Such shapes include cylinders (with their ends capped by hemispheres) and disks.

Intestinal absorption of fats is aided by solubilization of the fat molecules in micelles formed by anions of bile acids. Solubilization of cholesterol in these bile-salt micelles aids in excretion of cholesterol from the body.

Although a micelle-containing system is sometimes treated as having two phases, it is best considered as a one-phase solution in which the reversible equilibrium $nL \rightleftharpoons L_n$ exists, where L is the monomer and L_n the micelle. That micelle formation does not correspond to separation of a second phase is shown by the fact that the cmc does not have a precisely defined value but corresponds to a narrow range of concentrations. Figure 13.21*b* shows the variation of monomer and micelle concentrations with the solute stoichiometric concentration. The rather sudden rise in micelle concentration at the cmc results from the large value of n; see Prob. 13.33. The limit $n \to \infty$ would correspond to a phase change occurring at a precisely defined concentration to give a two-phase system.

Lyophobic Colloids
When solid AgCl is brought in contact with water, it does not spontaneously disperse to form a colloidal system. Sols that cannot be formed by spontaneous dispersion are

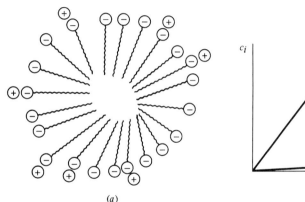

(a)

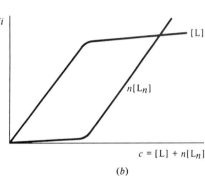

(b)

Figure 13.21

(*a*) A soap micelle in aqueous solution. (*b*) Monomer (L) and micelle (L_n) concentrations versus stoichiometric concentration c.

called **lyophobic** ("solvent-hating"). Lyophobic sols are thermodynamically unstable with respect to separation into two unmixed bulk phases (recall that the stable state of a system is one of minimum interfacial area), but the rate of separation may be extremely small. Gold sols prepared by Faraday are on exhibit in the British Museum.

The long life of lyophobic sols is commonly due to adsorbed ions on the colloidal particles; repulsion between like charges keeps the particles from aggregating. The presence of adsorbed ions can be shown by the migration of the colloidal particles in an applied electric field (a phenomenon called *electrophoresis*). A lyophobic sol can also be stabilized by the presence of a polymer (for example, the protein gelatin) in the solution. The polymer molecules become adsorbed on and surround each colloidal particle, thereby preventing coagulation of the particles.

Many lyophobic colloids can be prepared by precipitation reactions. Precipitation in either very dilute or very concentrated solutions tends to produce colloids. Lyophobic sols can also be produced by mechanically breaking down a bulk substance into tiny particles and dispersing them in a medium. For example, emulsions can be prepared by vigorous shaking of two essentially immiscible liquids in the presence of an emulsifying agent (defined shortly).

Sedimentation

The particles in a noncolloidal suspension of a solid in a liquid will eventually settle out under the influence of gravity, a process called **sedimentation.** For colloidal particles whose size is well below 10^3 Å, accidental thermal convection currents and the random collisions between the colloidal particles and molecules of the dispersion medium prevent sedimentation. A sol with larger colloidal particles will show sedimentation with time.

Emulsions

The liquids in most emulsions are water and an oil, where "oil" denotes an organic liquid essentially immiscible with water. Such emulsions are classified as either oil-in-water (O/W) emulsions, in which water is the continuous phase and the oil is present as tiny droplets, or water-in-oil (W/O) emulsions, in which the oil is the continuous phase. Emulsions are lyophobic colloids. They are stabilized by the presence of an *emulsifying agent,* which is commonly an amphiphilic species that forms a surface film at the interface between each colloidal droplet and the dispersion medium, thereby lowering the interfacial tension and preventing coagulation. The cleansing action of soaps and other detergents results in part from their acting as emulsifying agents to keep tiny droplets of grease suspended in water. Milk is an O/W emulsion of butterfat droplets in water; the emulsifying agent is the protein casein. Many pharmaceutical preparations and cosmetics (salves, ointments, cold cream) are emulsions.

Gels

A **gel** is a semirigid colloidal system of at least two components in which both components extend continuously throughout the system. An inorganic gel typically consists of water trapped within a three-dimensional network of tiny crystals of an inorganic solid. The crystals are held together by van der Waals forces, and the water is both adsorbed on the crystals and mechanically enclosed by them. Recall the white gelatinous precipitate of $Al(OH)_3$ obtained in the qualitative-analysis scheme. In contrast to a gel, the solid particles in a colloidal suspension are well separated from one another and move about freely in the liquid.

When an aqueous solution of the protein gelatin is cooled, a polymer gel is formed. Here, water is trapped within a network formed by the long-chain polymer molecules. In this network, polymer chains are entangled with one another and are held together by van der Waals forces, by hydrogen bonds, and perhaps by some covalent bonds. (Include lots of sugar and some artificial flavor and color with the gelatin, and you've got Jell-O.) The polysaccharide agar forms a polymer gel with water, which is used as a culture medium for bacteria.

If the liquid phase of a gel is removed by heating and pressurizing the gel above the critical temperature and pressure of the liquid (supercritical conditions; Sec 8.3) and allowing the fluid to vent, one obtains an **aerogel.** An aerogel is a strong, low-density solid whose volume is only a bit less than that of the original gel. The space formerly filled by the liquid in a gel contains air in the aerogel, so the aerogel is permeated by tiny pores. The most-studied aerogel is silica aerogel, where the solid is SiO_2 (silica), which is a covalent-network solid (Sec. 24.3) with a three-dimensional array of bonded Si and O atoms. (Silica occurs in nature as sand and is the main ingredient in glass.) The original gel can be made by the reaction $Si(OC_2H_5)_4 + 2H_2O \rightarrow SiO_2(s) + 4C_2H_5OH$ carried out in the solvent ethanol and yielding a gel with ethanol as the liquid. Some silica aerogel properties are: density typically 0.1 g/cm^3 but can be as low as 0.003 g/cm^3; internal surface area (determined by N_2 adsorption) typically 800 m^2/g; internal free volume typically 95% but can be as high as 99.9%; typical thermal conductivity 0.00015 J s^{-1} cm^{-1} K^{-1} (which is extremely low for a solid; see Fig. 16.2); mean pore diameter 20 nm. Aerogels may find uses in catalysis and in thermal insulation.

The spacecraft *Stardust,* launched in 1999 and returning to earth in January 2006, will reach the comet Wild 2 in 2004. Dust from the comet will be collected by impact with blocks of low-density silica aerogel, which will also collect (on their opposite sides) interstellar dust (stardust.jpl.nasa.gov/).

13.7   SUMMARY

Molecules in the interphase region experience different forces and have different average energies than molecules in either bulk phase. It therefore requires work $\gamma \, d\mathcal{A}$ to reversibly change the area of the interface between two phases by $d\mathcal{A}$, where γ is the surface tension.

For a spherically shaped interface, the existence of surface tension leads to a pressure difference between the two bulk phases given by $\Delta P = 2\gamma/R$, where R is the radius of the spherical interface. The phase on the concave side of the interface is at the higher pressure. Since the liquid–vapor interface in a capillary tube is curved, this pressure difference will produce a capillary rise of the liquid, given by Eq. (13.12) for zero contact angle.

The thermodynamics of systems where surface effects are significant is based on replacement of the actual system by a model system consisting of two homogeneous phases α and β separated by a zero-thickness surface σ (the Gibbs dividing surface). Extensive thermodynamic properties are assigned to the dividing surface according to $n_i^\sigma = n_i - n_i^\alpha - n_i^\beta$, $U^\sigma = U - U^\alpha - U^\beta$, etc., where n_i and U are the moles of i and the internal energy of the entire system. The thermodynamic properties of the hypothetical dividing surface σ depend on the location of σ and therefore are not in general experimentally observable. The relative adsorption $\Gamma_{i(1)}$ of component i with respect to component 1 is defined as $\Gamma_{i(1)} = n_i^\sigma/\mathcal{A}$, where n_i^σ is for the particular dividing surface that makes $n_1^\sigma = 0$ and $\mathcal{A}$ is the area of the planar interface. The relative adsorption $\Gamma_{i(1)}$ is experimentally observable and is related to the rate of change in the surface tension γ with respect to the log of the activity of i, according to the Gibbs adsorption isotherm (13.33) and (13.34).

Certain amphiphilic molecules (containing both hydrophilic and hydrophobic parts) can form one-molecule-thick spread monolayers on water.

Gases can be chemisorbed or physically adsorbed on solids. A gas–solid adsorption isotherm plots v (the gas volume, corrected to 0°C and 1 atm, adsorbed per gram of solid) versus gas pressure P. The Langmuir isotherm (13.36) works fairly well for chemisorption.

A colloidal system contains particles whose dimension in at least one direction is in the range 20 Å to 10^4 Å.

Important kinds of calculations discussed in this chapter include:

- Calculation of the pressure difference across a spherical interface from $\Delta P = 2\gamma/R$.
- Calculation of the surface tension from the capillary rise using Eq. (13.12).
- Calculation of the relative adsorption $\Gamma_{2(1)}$ from the Gibbs adsorption isotherm (13.34).
- Calculation of constants in the Langmuir isotherm (13.36) from a plot of $1/v$ versus $1/P$.
- Calculation of surface areas of solids using the BET isotherm (13.40) and adsorption data.
- Calculation of isosteric heats of adsorption from Eq. (13.41).

FURTHER READING AND DATA SOURCES

Adamson; Aveyard and Haydon; Defay, Prigogine, Bellemans, and Everett; Gasser.
 Surface and interfacial tensions: J. J. Jasper, *J. Phys. Chem. Ref. Data,* **1,** 841 (1972); *Landolt-Börnstein,* vol. II, pt. 3, pp. 420–468.

PROBLEMS

Section 13.1

13.1 True or false? (*a*) Increasing the area of a liquid–vapor interface increases U of the system. (*b*) The surface tension of a liquid goes to zero as the critical temperature is approached.

13.2 Give the SI units of surface tension.

13.3 (*a*) Calculate the surface area of a 1.0-cm^3 sphere of gold. (*b*) Calculate the surface area of a colloidal dispersion of 1.0 cm^3 of gold in which each gold particle is a sphere of radius 300 Å.

13.4 Calculate the minimum work needed to increase the area of the surface of water from 2.0 cm^2 to 5.0 cm^2 at 20°C. The surface tension of water is 73 dyn/cm at 20°C.

13.5 The surface tension of ethyl acetate at 0°C is 26.5 mN/m, and its critical temperature is 523.2 K. Estimate its surface tension at 50°C. The experimental value is 20.2 mN/m.

13.6 J. R. Brock and R. B. Bird [*Am. Inst. Chem. Eng. J.,* **1,** 174 (1955)] found that for liquids that are not highly polar or hydrogen-bonded the constant γ_0 in (13.3) is usually well approximated by

$$\gamma_0 = (P_c/\text{atm})^{2/3}(T_c/\text{K})^{1/3}(0.432/Z_c - 0.951)\,\text{dyn/cm}$$

where P_c, T_c, and Z_c are the critical pressure, temperature, and compressibility factor. For ethyl acetate, $P_c = 37.8$ atm, $T_c = 523.2$ K, and $Z_c = 0.252$. Calculate the percent error of the Brock–Bird predicted value of γ for ethyl acetate at 0°C. The experimental value is 26.5 dyn/cm.

13.7 Calculate γ/l_z in Eq. (13.5) for the typical values $l_z = 10$ cm and $\gamma = 50$ dyn/cm; express your answer in atmospheres.

Section 13.2

13.8 True or false? (*a*) At equilibrium in a closed system with no walls between phases, all phases must be at the same temperature and at the same pressure. (*b*) For a two-phase system with a curved interface, the phase on the concave side is at higher pressure than the other phase.

13.9 Calculate the pressure inside a bubble of gas in water at 20°C if the pressure of the water is 760 torr and the bubble radius is 0.040 cm. See Prob. 13.4 for γ.

13.10 At 20°C, the capillary rise at sea level for methanol in contact with air in a tube with inside diameter 0.350 mm is 3.33 cm. The contact angle is zero. The densities of methanol and air at 20°C are 0.7914 and 0.0012 g/cm^3. Find γ for CH$_3$OH at 20°C.

13.11 For the Hg–air interface on glass, $\theta = 140°$. Find the capillary depression of Hg in contact with air at 20°C in a glass tube with inside diameter 0.350 mm. For Hg at 20°C, $\rho = 13.59$ g/cm^3 and $\gamma = 490$ ergs/cm^2. (See Prob. 13.10.)

13.12 At 20°C, the interfacial tension for the liquids *n*-hexane and water is 52.2 ergs/cm^2. The densities of *n*-hexane and water at 20°C are 0.6599 and 0.9982 g/cm^3. Assuming a zero contact angle, calculate the capillary rise at 20°C in a 0.350-mm inside diameter tube inserted into a two-phase *n*-hexane–water system.

13.13 (*a*) In Eq. (13.12), *h* is the height of the bottom of the meniscus. Hence, Eq. (13.12) neglects the pressure due to the small amount of liquid β that is above the bottom of the meniscus. Show that, if this liquid is taken into account, then $\gamma = \frac{1}{2}(\rho_\beta - \rho_\alpha)gr(h + \frac{1}{3}r)$ for $\theta = 0$. (*b*) Rework Prob. 13.10 using this more accurate equation.

13.14 Two capillary tubes with inside radii 0.600 and 0.400 mm are inserted into a liquid with density 0.901 g/cm³ in contact with air of density 0.001 g/cm³. The difference between the capillary rises in the tubes is 1.00 cm. Find γ. Assume a zero contact angle.

Section 13.3

13.15 For dilute solutions, the surface tension often varies linearly with solute molar concentration: $\gamma = \gamma^* - bc$, where *b* is a constant. Show that here $\Gamma_{2(1)} = (\gamma^* - \gamma)/RT$.

13.16 At 21°C, surface tensions of aqueous solutions of $C_6H_5CH_2CH_2COOH$ vs. solute molality are

m/(mmol/kg)	11.66	15.66	19.99	27.40	40.8
γ/(dyn/cm)	61.3	59.2	56.1	52.5	47.2

Find $\Gamma_{2(1)}$ for a solution with 20 mmol of solute per kilogram of water. Use the equation that corresponds to (13.34) for the molality scale.

13.17 Show that, for a binary solution of an electrolyte, the Gibbs adsorption isotherm is

$$\Gamma_{2(1)} = -\frac{1}{\nu RT}\left(\frac{\partial \gamma}{\partial \ln (\gamma_2 m_2/m^\circ)}\right)_T$$

where γ_2 and m_2 are the electrolyte's mean ionic activity coefficient and molality and ν and m° are defined by (10.45) and (10.41). The differences between this equation and (13.33) are the presence of ν and the fact that activity coefficients cannot be neglected in dilute electrolyte solutions.

13.18 A certain solution of solute *i* in water has mole fraction $x_i = 0.10$. The interphase region contains 2.0×10^{-8} mol of *i* and 45×10^{-8} mol of H_2O. The solution's surface area is 100 cm². Find $\Gamma_{i(1)}$.

13.19 Show that $\gamma = (\partial A/\partial \mathcal{A})_{T,V,n_i}$, where the Helmholtz energy of the system is $A \equiv U - TS$. [*Hint:* Use (13.22).]

13.20 (*a*) Let z_0 be the position of the dividing surface that makes $n_1^\sigma = 0$. Use (13.19) with *i* replaced by 1 to show that $z_0 = (c_1^\beta V - n_1)/\mathcal{A}(c_1^\beta - c_1^\alpha)$. (*b*) Substitute the result of (*a*) into (13.19) for n_i^σ to show that

$$\Gamma_{i(1)} = \frac{1}{\mathcal{A}}\left[(n_i - c_i^\beta V) - (n_1 - c_1^\beta V)\frac{c_i^\beta - c_i^\alpha}{c_1^\beta - c_1^\alpha}\right]$$

13.21 (*a*) Show that, when $c_1^\beta \ll c_1^\alpha$ and $c_i^\beta \ll c_i^\alpha$, then $n_i - c_i^\beta V = n_{i,\text{bulk}}^\alpha + n_i^s$ and $n_1 - c_1^\beta V = n_{1,\text{bulk}}^\alpha + n_1^s$, where the quantities on the right are defined following Eq. (13.31). Start by writing n_i as the sum of the moles of *i* in the two bulk phases and the interphase region. (*b*) Substitute the result for (*a*) into the $\Gamma_{i(1)}$ equation of Prob. 13.20*b* to derive (13.31).

Section 13.4

13.22 For stearic acid $[CH_3(CH_2)_{16}COOH]$ the area per molecule at the Pockels point is 20 Å², and the density is 0.94 g/cm³ at 20°C. Estimate the length of a stearic acid molecule.

13.23 One acre = 4057 m². One teaspoonful $\approx$ 4.8 cm³. (*a*) From Franklin's data in Sec. 13.4, calculate the thickness of the olive-oil film. Assume he used exactly 1 teaspoonful of oil. (*b*) Olive oil is mainly glycerol trioleate $[(C_{17}H_{33}COO)_3C_3H_5]$ with a density at room temperature of 0.90 g/cm³. Calculate the area occupied by each olive-oil molecule in Franklin's film. (*c*) Calculate $\Gamma_{2(1)}$ for Franklin's monolayer.

Section 13.5

13.24 For N_2 adsorbed on a certain sample of charcoal at −77°C, adsorbed volumes (recalculated to 0°C and 1 atm) per gram of charcoal vs. N_2 pressure are

P/atm	3.5	10.0	16.7	25.7	33.5	39.2
v/(cm³/g)	101	136	153	162	165	166

(*a*) Fit the data with the Langmuir isotherm (13.36) and give the values of v_{mon} and *b*. (*b*) Fit the data with the Freundlich isotherm and give the values of *k* and *a*. (*c*) Calculate *v* at 7.0 atm using both the Langmuir isotherm and the Freundlich isotherm.

13.25 The *Temkin isotherm* for gas adsorption on solids is $v = r \ln sP$, where *r* and *s* are constants. (*a*) What should be plotted against what to give a straight line if the Temkin isotherm is obeyed? (*b*) Fit the data of Prob. 13.24 to the Temkin isotherm and evaluate *r* and *s*.

13.26 Besides plotting $1/v$ vs. $1/P$, there is another way to plot the Langmuir isotherm (13.36) to yield a straight line. What is this way?

13.27 For N_2 adsorbed on a certain sample of ZnO powder at 77 K, adsorbed volumes (recalculated to 0°C and 1 atm) per gram of ZnO vs. N_2 pressure are

P/torr	v/(cm³/g)	P/torr	v/(cm³/g)	P/torr	v/(cm³/g)
56	0.798	183	1.06	442	1.71
95	0.871	223	1.16	533	2.08
145	0.978	287	1.33	609	2.48

The normal boiling point of N_2 is 77 K. (*a*) Plot *v* versus *P* and decide whether the Langmuir or the BET equation is more appropriate. (*b*) Use the equation you decided on in (*a*) to find the volume v_{mon} needed to form a monolayer; also find the other constant in the isotherm equation. (*c*) Assume that an adsorbed N_2 molecule occupies an area of 16 Å² and calculate the surface area of 1.00 g of the ZnO powder.

13.28 Show that for $\theta \ll 1$, the Langmuir isotherm (13.36) reduces to the Freundlich isotherm with $a = 1$.

13.29 Show that the Langmuir assumptions lead to (13.37) for a mixture of gases A and B in nondissociative-adsorption equilibrium with a solid.

13.30 For H_2 adsorbed on W powder, the following data were found:

θ	0.005	0.005	0.10	0.10	0.10
P/torr	0.0007	0.03	8	23	50
t/°C	500	600	500	600	700

where t is the Celsius temperature and P is the H_2 pressure in equilibrium with the tungsten at fractional surface coverage θ. (a) For $\theta = 0.005$, find the average $\Delta \bar{H}_a$ over the range 500°C to 600°C. (b) For $\theta = 0.10$, find the average $\Delta \bar{H}_a$ over each of the ranges 500°C to 600°C and 600°C to 700°C.

13.31 (a) Write the BET isotherm in the form $v/v_{mon} = f(P)$. (b) Show that if $P \ll P^*$, the BET isotherm reduces to the Langmuir isotherm.

13.32 When CO chemisorbed on W is heated gradually, a substantial amount of gas is evolved in the temperature range 400 to 600 K and a substantial amount is evolved in the range 1400 to 1800 K, with not much evolved at other temperatures. What does this suggest about CO chemisorbed on tungsten?

Section 13.6

13.33 Let K_c° be the concentration-scale standard equilibrium constant for the equilibrium $nL \rightleftharpoons L_n$ between monomers and micelles in solution, where L is an uncharged species (for example, a polyoxyethylene). (a) Let c be the stoichiometric concentration of the solute (that is, the number of moles of monomer used to prepare a liter of solution) and let x be the concentration of micelles at equilibrium: $x = [L_n]$, where the brackets denote concentration. Show that $c = nx + (x/K_c)^{1/n}$. Assume all activity coefficients are 1. (b) Let f be the fraction of L present as monomer. Show that $f = 1 - nx/c$. (c) For $n = 50$ and $K_c^\circ = 10^{200}$, calculate and graph [L], $n[L_n]$, and f as functions of c. (Hint: Calculate c for various assumed values of x, rather than the reverse.) (d) If the cmc is taken as the value of c for which $f = 0.5$, give the value of the cmc.

General

13.34 For aqueous solutions of saturated aliphatic acids at 18°C, the surface tensions are fitted by

$$\gamma = [73.0 - 29.9 \log_{10}(ac + 1)]\,\text{dyn/cm}$$

where a is a constant and c is the acid's concentration. (a) Find $\Gamma_{2(1)}$ as a function of concentration. (b) Which gas–solid adsorption isotherm does the expression in (a) resemble? (c) For butanoic acid, $a = 19.64$ dm³/mol. Plot $\Gamma_{2(1)}$ vs. c for aqueous solutions of this acid for the concentration range 0 to 1 mol/dm³.

13.35 From Eq. (13.9), a drop of liquid of radius r is at a higher pressure than the vapor it is in equilibrium with. This increased pressure affects the chemical potential of the liquid and raises its vapor pressure slightly. (a) Use the integrated form of Eq. (7.25) in Prob. 7.49 to show that the vapor pressure P_r of such a drop is

$$P_r = P \exp(2\gamma V_{m,l}/rRT)$$

where $V_{m,l}$ is the molar volume of the liquid and P its bulk vapor pressure. This is the *Kelvin equation*. (b) The vapor pressure and surface tension of water at 20°C are 17.535 torr and 73 dyn/cm. Calculate the 20°C vapor pressure of a drop of water of radius 1.00×10^{-5} cm.

Electrochemical Systems

This chapter deals with the thermodynamics of electrochemical systems, which are systems with a difference in electric potential between two or more phases. (A familiar example is a battery.) Electric potential is defined in Sec. 14.1, which reviews electrostatics. Electrical forces, fields, potentials, and potential energy are important not only in the thermodynamics of electrochemical systems but throughout chemistry. The properties of an atom or molecule are the result of electrical interactions among the electrons and nuclei. To write the fundamental equation for dealing with molecules (the Schrödinger equation), we need to know the equation for the potential energy of interaction between two charges.

Forces between molecules are also electrical in nature. Two molecular properties that mainly determine intermolecular forces are the molecular dipole moment and the polarizability. These properties are discussed in Sec. 14.15.

The major part of Chapter 14 (Secs. 14.4 to 14.12) deals with galvanic cells. As well as their practical use to supply electric power, galvanic cells enable us to find ΔH°, ΔG°, ΔS°, and K° for reactions and to find activity coefficients of electrolytes.

14.1 ELECTROSTATICS

Before developing the thermodynamics of electrochemical systems, we review *electrostatics*, which is the physics of electric charges at rest. In this chapter, all equations are written in a form valid for SI units.

Coulomb's Law

The SI unit of electric charge Q is the **coulomb** (C), defined in Sec. 16.5. Two kinds of charges exist, positive and negative. Like kinds of charges repel each other, and unlike kinds attract. The magnitude F of the force that a point electric charge Q_1 exerts on a second charge Q_2 is given by Coulomb's law as $F = K|Q_1 Q_2|/r^2$, where r is the distance between the charges and K is a proportionality constant. The absolute-value signs are present because the magnitude of a vector cannot be negative. The direction of $\mathbf{F}$ is along the line joining the charges. In the SI system, the proportionality constant K is written as $1/4\pi\varepsilon_0$:

$$F = \frac{1}{4\pi\varepsilon_0} \frac{|Q_1 Q_2|}{r^2} \qquad (14.1)*$$

Experiment gives the constant ε_0 (the **permittivity of vacuum**) as

$$\varepsilon_0 = 8.854 \times 10^{-12} \text{ C}^2 \text{ N}^{-1} \text{ m}^{-2} = 8.854 \times 10^{-12} \text{ C}^2 \text{ kg}^{-1} \text{ m}^{-3} \text{ s}^2$$

$$1/4\pi\varepsilon_0 = 8.988 \times 10^9 \text{ N m}^2 \text{ C}^{-2} \qquad (14.2)$$

The Electric Field

To avoid the notion of action at a distance, the concept of electric field is introduced. An electric charge Q_1 is said to produce an **electric field** in the space around itself, and this field exerts a force on any charge Q_2 that is present in the space around Q_1. The **electric field (strength) E** at a point P in space is defined as the electrical force per unit charge experienced by a test charge Q_t at rest at point P:

$$\mathbf{E} \equiv \mathbf{F}/Q_t \qquad \text{where } Q_t \text{ is part of syst.} \qquad (14.3)*$$

Equation (14.3) associates a vector **E** with each point in space.

Equation (14.3) defines the electric field that exists at P when the charge Q_t is present in the system. However, the presence of Q_t may affect the surrounding charges, thereby making **E** depend on the nature of the test charge. For example, if Q_t is placed in or near a material body, it may change the distribution of charges in the body. Therefore, if Q_t in (14.3) is not part of the system under discussion and we want to know what **E** is at a given point in the system in the absence of Q_t, we rewrite (14.3) as

$$\mathbf{E} \equiv \lim_{Q_t \to 0} \mathbf{F}/Q_t \qquad \text{where } Q_t \text{ is not part of syst.} \qquad (14.4)$$

An infinitesimally small test charge does not disturb the charge distribution in the system, so (14.4) gives the value of **E** in the absence of the test charge. Since **F** in (14.4) is proportional to Q_t, the field **E** in (14.4) is independent of Q_t.

EXAMPLE 14.1 Electric field of a point charge

Find **E** in the space around a point charge Q if no other charges are present in the system.

Let a tiny test charge dQ_t be put at a distance r from Q. Then the magnitude of **E** is given by (14.4) as $E = dF/|dQ_t|$, where the magnitude of the force on dQ_t is given by (14.1) as $dF = |Q\,dQ_t|/4\pi\varepsilon_0 r^2$. Therefore the magnitude of **E** is

$$E = \frac{1}{4\pi\varepsilon_0} \frac{|Q|}{r^2} \qquad (14.5)$$

From (14.4), the direction of **E** is the same as the direction of **F** on a positive test charge, so **E** at point P is in the direction of the line from the charge Q to point P. The vector **E** points outward if Q is positive; inward if Q is negative. Figure 14.1 shows **E** at several points around a positive charge. The arrows farther away from Q are shorter, since **E** falls off as $1/r^2$.

Exercise

If E is 80 N/C at a distance of 5.00 cm from a certain charge, at what distance would E be 800 N/C? (*Answer:* 1.58 cm.)

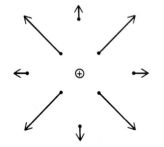

Figure 14.1

The electric field vector at several points in the space around a positive charge. **E** falls off as $1/r^2$.

Electric Potential

Instead of describing things in terms of the electric field, it is often more convenient to use the electric potential ϕ (phi). The **electric potential difference** $\phi_b - \phi_a$ between points b and a in an electric field is defined as the work per unit charge to move a test charge reversibly from a to b:

$$\phi_b - \phi_a \equiv \lim_{Q_t \to 0} w_{a \to b}/Q_t \equiv dw_{a \to b}/dQ_t \qquad (14.6)$$

where $dw_{a \to b}$ is the reversible electrical work done by an external agent that moves an infinitesimal test charge dQ_t from a to b. The word "reversible" indicates that the force exerted by the agent differs only infinitesimally from the force exerted by the system's electric field on dQ_t. By assigning a value to the electric potential ϕ_a at point a, we have then defined the **electric potential** ϕ_b at any point b. The usual convention is to choose point a at infinity (where the test charge interacts with no other charges) and to define ϕ at infinity as zero. Equation (14.6) then becomes

$$\phi_b \equiv \lim w_{\infty \to b}/Q_t \qquad (14.7)$$

The SI unit of electric potential is the **volt** (V), defined as one joule per coulomb:

$$1 \text{ V} \equiv 1 \text{ J/C} = 1 \text{ N m C}^{-1} = 1 \text{ kg m}^2 \text{ s}^{-2} \text{ C}^{-1} \qquad (14.8)$$

since $1 \text{ J} = 1 \text{ N m} = 1 \text{ kg m}^2 \text{ s}^{-2}$ [Eq. (2.14)].

The SI unit of E in (14.4) is the newton per coulomb. Use of (14.8) gives $1 \text{ N/C} = 1 \text{ N V J}^{-1} = 1 \text{ N V N}^{-1} \text{ m}^{-1} = 1 \text{ V/m}$, and E is usually expressed in volts per meter or volts per centimeter:

$$1 \text{ N/C} = 1 \text{ V/m} = 10^{-2} \text{ V/cm} \qquad (14.9)$$

When we do reversible work $w_{\infty \to b}$ in moving a charge from infinity to b in an electric field, we change the charge's potential energy V by $w_{\infty \to b}$ (just as we change the potential energy of a mass by changing its height in the earth's gravitational field). Thus, $\Delta V = V_b - V_\infty = V_b = w_{\infty \to b}$, where V_∞ has been taken as zero. The use of (14.7) gives the electric potential energy of charge Q_t at point b as

$$V_b = \phi_b Q_t \qquad (14.10)$$

The electric field $\mathbf{E}$ *is the force per unit charge. The electric potential* ϕ *is the potential energy per unit charge* [Eqs. (14.3) and (14.10)].

Equation (2.21) gives $F_x = -\partial V/\partial x$. Division by Q_t gives $F_x/Q_t = -\partial(V/Q_t)/\partial x$. Use of (14.10) and the x component of (14.3) transforms this equation into $E_x = -\partial \phi/\partial x$. The same arguments hold for the y and z coordinates, so

$$E_x = -\partial \phi/\partial x, \qquad E_y = -\partial \phi/\partial y, \qquad E_z = -\partial \phi/\partial z \qquad (14.11)$$

Equation (14.11) shows that the electric field $\mathbf{E}$ at a point in space can be found if the electric potential ϕ is known as a function of x, y, and z. Conversely, $\phi(x, y, z)$ can be found from $\mathbf{E}$ by integration of (14.11). The integration constant is determined by setting $\phi = 0$ at a convenient location (usually infinity). From (14.11), the electric field is related to the spatial rate of change of ϕ. Hence the units V/m for E [Eq. (14.9)].

EXAMPLE 14.2 Electric potential due to a point charge

(a) Find the expression for the electric potential ϕ at an arbitrary point P in the space around a point charge Q. Take $\phi = 0$ at infinity. (b) Calculate ϕ and E 1.00 Å away from a proton. The proton charge is 1.6×10^{-19} C.

(a) Let the coordinate origin be at Q and let the x axis run along the line from Q to P. The electric field at P is then in the x direction: $E = E_x = Q/4\pi\varepsilon_0 x^2$, $E_y = 0$, $E_z = 0$, where (14.5) was used. Substitution in the first equation of (14.11) and integration give $\phi = -\int E_x \, dx = -\int (Q/4\pi\varepsilon_0 x^2) \, dx = Q/4\pi\varepsilon_0 x + c = Q/4\pi\varepsilon_0 r + c$, where r is the distance between charge Q and point P and c is an integration constant. In general, c could be a function of y and z, but the fact that $E_y = 0 = E_z$ together with Eq. (14.11) requires that c be independent of y

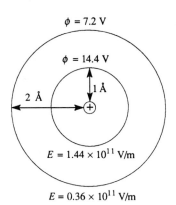

$\phi = 7.2$ V

$\phi = 14.4$ V

1 Å

2 Å

(+)

$E = 1.44 \times 10^{11}$ V/m

$E = 0.36 \times 10^{11}$ V/m

Figure 14.2

E and ϕ at distances of 1 and 2 Å from a proton. ϕ falls off as $1/r$.

and z. Defining ϕ as zero at $r = \infty$, we get $c = 0$. Therefore the potential due to a point charge Q is

$$\phi = \frac{1}{4\pi\varepsilon_0} \frac{Q}{r} \tag{14.12}$$

(b) Substitution in (14.12) and use of (14.2) for $1/4\pi\varepsilon_0$ give

$$\phi = (8.99 \times 10^9 \text{ N m}^2/\text{C}^2)(1.6 \times 10^{-19} \text{ C})/(1.0 \times 10^{-10} \text{ m}) = 14 \text{ V}$$

Substitution in (14.5) gives $E = 1.4 \times 10^{11}$ V/m.

Figure 14.2 shows ϕ and E on spherical surfaces centered on a proton and having radii of 1 and 2 Å.

Exercise

Find ϕ and E 10.0 cm from a charge of 1.00 C. Take $\phi = 0$ at infinity. (*Answer:* 9.0×10^{10} V, 9.0×10^{11} V/m.)

Equations (14.12) and (14.5) also hold for the field and electric potential outside a spherically symmetric charge distribution whose total charge is Q; here, r is the distance to the center of the charge distribution.

From (14.12), ϕ increases as one moves closer to a positive charge. A negative charge moves spontaneously toward a positive charge, so *electrons move spontaneously from regions of low to regions of high electric potential within a phase.*

Experiment shows that the electric field of a system of charges equals the vector sum of the electric fields due to the individual charges. The electric potential equals the sum of the electric potentials due to the individual charges.

In discussing electric fields and electric potentials at a "point" in matter, we generally mean the average field and the average potential in a volume containing far, far fewer than 10^{23} molecules but far, far more than one molecule. The electric field within a single molecule shows very sharp variations.

Consider a single phase that is an electrical conductor (for example, a metal, an electrolyte solution) and is in thermodynamic equilibrium. Since the phase is in equilibrium, no currents are flowing. (Flow of a noninfinitesimal current is an irreversible process, because of the heat generated by the current.) It follows that the electric field at all points in the interior of the phase must be zero. Otherwise, the charges of the phase would experience electrical forces, and a net current would flow. Since **E** is zero, Eq. (14.11) shows that ϕ *is constant in the bulk phase* (Sec. 13.1) *of a conductor where no currents are flowing.* If this phase has a net electrical charge, at equilibrium this charge will be distributed over the surface of the phase. This is so because the repulsion of like charges will cause them to move to the surface, where they are as far apart as possible.

Summary

The magnitude of the force between two electric charges is $F = |Q_1 Q_2|/4\pi\varepsilon_0 r^2$. The electric field strength **E** at a point in space is defined as the force per unit charge: $\mathbf{E} \equiv \mathbf{F}/Q$. The electric potential ϕ at a point in space is the potential energy per unit charge: $\phi = V/Q$. From ϕ, one can find **E** using (14.11).

14.2   ELECTROCHEMICAL SYSTEMS

Previous chapters considered systems with electrically neutral phases and no differences in electric potential between phases. However, when a system contains charged species and at least one charged species cannot penetrate all the phases of the system,

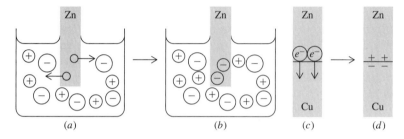

Figure 14.3

Development of electric potential differences between Zn(s) and ZnSO$_4$(aq) and between Cu(s) and Zn(s).

some of the phases can become electrically charged. For example, suppose a membrane permeable to K$^+$ ions but not to Cl$^-$ ions separates an aqueous KCl solution from pure water. Diffusion of K$^+$ ions through the membrane will produce net charges on each phase and a potential difference between the phases.

Another example is a piece of Zn dipping into an aqueous ZnSO$_4$ solution (Fig. 14.3a) held at constant T and P. The Zn metal can be viewed as composed of Zn^{2+} ions and mobile valence electrons. Zn^{2+} ions can be transferred between the metal and the solution, but the metal's electrons cannot enter the solution. Suppose the ZnSO$_4$ solution is extremely dilute. Then the initial rate at which Zn^{2+} ions leave the metal and enter the solution is greater than the rate at which Zn^{2+} ions enter the metal from the solution. This net loss of Zn^{2+} from the metal produces a negative charge (excess of electrons) on the Zn. The negative charge slows the rate of the process Zn^{2+}(*metal*) $\rightarrow$ Zn^{2+}(*aq*) and increases the rate of Zn^{2+}(*aq*) $\rightarrow$ Zn^{2+}(*metal*). Eventually an equilibrium is reached in which the rates of these opposing processes are equal and the Gibbs energy G of the system is a minimum. At equilibrium, the Zn has a net negative charge, and a potential difference $\Delta\phi$ exists between Zn and the solution (Fig. 14.3b).

Techniques of electrode kinetics (see *Bockris and Reddy,* p. 892) show that at equilibrium between Zn and 1 mol/dm^3 ZnSO$_4$(*aq*) at 20°C and 1 atm, the Zn^{2+} ions that cross 1 cm^2 of the metal–solution interface in each direction in 1 s carry a charge of 2 × 10^{-5} C. This equilibrium current flow in each direction is called the *exchange current.* How many moles of Zn^{2+} carry this charge of 2 × 10^{-5} C?

The charge on one proton is $e = 1.60218 \times 10^{-19}$ C. The charge per mole of protons is $N_A e$ (where N_A is the Avogadro constant) and is called the **Faraday constant** F. Use of $N_A = 6.02214 \times 10^{23}$ mol^{-1} gives

$$F \equiv N_A e = 96485 \text{ C/mol} \qquad \textbf{(14.13)}^*$$

The charge on one particle (ion, molecule, or electron) of species i is $z_i e$, where e is the proton charge and the **charge number** z_i of species i is an integer. For example, $z_i = 2$ for Zn^{2+}, $z_i = -1$ for an electron (e$^-$), and $z_i = 0$ for H$_2$O. Since F is the charge per mole of protons and a particle of species i has a charge that is z_i times the proton charge, *the charge per mole of species i is $z_i F$.* The charge on n_i moles of i is thus

$$Q_i = z_i F n_i \qquad (14.14)$$

The 2 × 10^{-5} C of Zn^{2+} ions thus corresponds to

$$n_i = Q_i/z_i F = (2 \times 10^{-5} \text{ C})/2(96485 \text{ C/mol}) = 1 \times 10^{-10} \text{ mol}$$

of Zn^{2+} entering and leaving 1 cm^2 of the metal each second.

The Zn–ZnSO$_4$(*aq*) example indicates that at any metal–solution interface at equilibrium, a potential difference $\Delta\phi$ exists. The magnitude and sign of $\Delta\phi$ depend on T, P, the nature of the metal, the nature of the solvent, and the concentrations of metal ions in solution.

Another example of an interphase potential difference is a piece of Cu in contact with a piece of Zn (Fig. 14.3c). Diffusion in solids is extremely slow at room

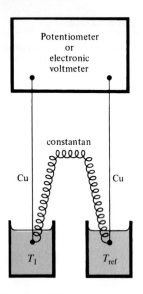

Figure 14.4

A thermocouple. The electric potential difference between Cu and constantan (an alloy of Cu and Ni) depends on temperature, so if T_1 differs from T_{ref}, there is a nonzero potential difference between the two Cu wires, whose value depends on T_1 and enables T_1 to be found. T_{ref} is commonly taken as the ice point.

temperature, so the Cu^{2+} and Zn^{2+} ions do not move between the phases to any significant extent. However, electrons are free to move from one metal to the other, and they do so, resulting at equilibrium (minimum G) in a net negative charge on the Cu and a net positive charge on the Zn (Fig. 14.3d). This charge can be detected by separating the metals and touching one of them to the terminal of an electroscope. (In an electroscope, two pieces of metal foil attached to the same terminal repel each other when they become charged.) The development of charge by two different metals in contact was discovered by Galvani and Volta in the 1790s. In one experiment, Galvani discharged this charge through the nerve of a dead frog's leg muscles, causing the muscles to contract. The magnitude of the interphase potential difference between two metals depends on temperature. A *thermocouple* uses this temperature dependence to measure temperature (Fig. 14.4).

The transfer of charge between two phases α and β produces a difference in electric potential between the phases at equilibrium: $\phi^\alpha \neq \phi^\beta$, where ϕ^α and ϕ^β are the potentials in the bulk phases (Fig. 13.1b) α and β. (The electric potential in the bulk of a phase is sometimes called the *inner potential* or the *Galvani potential*.) We define an **electrochemical system** as a heterogeneous system in which there is a difference of electric potential between two or more phases.

Besides interphase charge transfer, other effects contribute to $\phi^\alpha - \phi^\beta$. For example, in Fig. 14.3b, water molecules in the immediate vicinity of the Zn metal will tend to be oriented with their positively charged hydrogen atoms toward the negative Zn. Moreover, the negative charge on the Zn metal will distort (or polarize) the distribution of electrons within each adjacent water molecule. Also, Zn^{2+} ions will tend to predominate over SO_4^{2-} ions in the immediate vicinity of the negative Zn metal. The orientation of water molecules, the polarization of electronic charge in the water molecules, and the nonuniform distribution of ions all affect $\phi^\alpha - \phi^\beta$.

A difference in potential between phases can even occur without transfer of charge between the phases. An example is a two-phase system of liquid water plus liquid benzene, which are nearly immiscible. There will be a preferred orientation of the water molecules at the interface between phases, because of different interactions between C_6H_6 molecules and the negative and positive sides of the water molecules. This makes $\phi^\alpha - \phi^\beta$ nonzero. Interphase potential differences arising without charge transfer between phases are relatively small (typical estimates are tens of millivolts) compared with those arising with charge transfer (typically, a volt or two).

Our main interest will lie in electrochemical systems all of whose phases are electrical conductors. Such phases include metals, semiconductors, molten salts, and liquid solutions containing ions.

A significant point is that the potential difference $\Delta\phi$ between two phases in contact cannot be readily measured. Suppose we wanted to measure $\Delta\phi$ between a piece of Zn and an aqueous $ZnCl_2$ solution. If we make electrical contact with these phases with two wires of a voltmeter or potentiometer (Sec. 14.4), we create at least one new interface in the system, that between the voltmeter wire and the $ZnCl_2$ solution. The potential difference measured by the meter includes the potential difference between the meter wire and the solution, and we have not measured what we set out to measure. The kind of potential difference that is readily measured is the potential difference between two phases having the same chemical composition. Attachment of voltmeter wires to these phases creates potential differences between the wires and the phases, but these potential differences are equal in magnitude and cancel each other if the two wires are made of the same metal.

Although $\Delta\phi$ between phases in contact cannot readily be measured, it can be calculated from a statistical-mechanical model of the system. $\Delta\phi$ can be calculated if the distribution of charges and dipoles in the interphase region is known.

Summary

When two different electrically conducting phases come in contact, a difference in electric potential ϕ is usually established between them as a result of transfer of charges between phases and of nonuniform distribution of ions, orientation of molecules with dipole moments, and distortion of charge distributions in molecules near the interface. Potential differences can be measured only between two phases that have the same chemical composition.

14.3   THERMODYNAMICS OF ELECTROCHEMICAL SYSTEMS

We now develop the thermodynamics of electrochemical systems composed of phases that are electrical conductors. The treatment applies only to systems in which there is at most an infinitesimal flow of current, since equilibrium thermodynamics does not apply to irreversible processes.

In an electrochemical system, the phases generally have nonzero net charges, and electric potential differences exist between phases. These electric potential differences are typically a few volts or less (see Sec. 14.7). How much transfer of charged matter between phases occurs when a potential difference of, say, 10 V exists between phases? To get an order-of-magnitude answer, we consider an isolated spherical phase of radius 10 cm that is at an electric potential of $\phi = 10$ V with respect to infinity. Let Q be the net charge on the phase. The electric potential at the edge of the phase of radius r is given by (14.12) as $\phi = Q/4\pi\varepsilon_0 r$, and

$$Q = 4\pi\varepsilon_0 r\phi = 4\pi(8.8 \times 10^{-12}\,\text{C}^2\,\text{N}^{-1}\,\text{m}^{-2})(0.1\,\text{m})(10\,\text{V}) = 1 \times 10^{-10}\,\text{C}$$

Suppose this charge is due to an excess of Cu^{2+} ions. We have $Q_i = z_i F n_i$ [Eq. (14.14)], and the amount of excess Cu^{2+} is

$$n_i = Q_i/z_i F = (1 \times 10^{-10}\,\text{C})/2(96485\,\text{C/mol}) = 5 \times 10^{-16}\,\text{mol}$$

which is a mere 3×10^{-14} g of Cu^{2+}. We conclude that *the net charges of phases of electrochemical systems are due to transfers of amounts of matter far too small to be detected chemically.*

The presence of electric potential differences between phases affects the thermodynamic equations because the internal energy of a charged species depends on the electric potential of the phase it is in. When the phases of an electrochemical system are brought together to form the system, tiny amounts of charge transfer between phases produce potential differences between phases. Let us imagine a hypothetical system in which these charge transfers have not occurred, so that all the phases have an electric potential of zero: $\phi^\alpha = \phi^\beta = \cdots = 0$. If we add dn_j moles of j to phase α of this hypothetical system, the Gibbs equation (4.75) gives the change in internal energy of phase α as

$$dU^\alpha = T\,dS^\alpha - P\,dV^\alpha + \mu_j^\alpha\,dn_j^\alpha \qquad \text{for } \phi^\alpha = 0 \qquad (14.15)$$

where the chemical potential μ_j^α is a function of T, P, and the composition of the phase: $\mu_j^\alpha = \mu_j^\alpha(T, P, x_1^\alpha, x_2^\alpha, \ldots)$.

Now consider the actual system in which charge transfers between phases do occur to produce phases with electric potentials $\phi^\alpha, \phi^\beta, \ldots$. As noted earlier in this section, these charge transfers correspond to negligible amounts of chemical species transferred, so we can consider each phase of the actual electrochemical system to have the same composition as the corresponding phase of the hypothetical system with electric potentials equal to zero.

Suppose we add dn_j^α moles of substance j to phase α of the electrochemical system. How does dU^α for this process compare with dU^α in (14.15) for addition of dn_j^α

to the system with $\phi^\alpha = 0$? The chemical composition of phase α is the same for both processes. The only difference is that the hypothetical system has $\phi^\alpha = 0$, whereas the actual system has $\phi^\alpha \neq 0$. The electric potential energy of a charge Q at a location where the electric potential is ϕ is equal to ϕQ [Eq. (14.10)]. If dQ_j^α is the charge on the added dn_j^α moles, then this charge will have zero electric potential energy in the hypothetical system where $\phi^\alpha = 0$ and will have an electric potential energy of $\phi^\alpha \, dQ_j^\alpha$ in the actual system. This electric potential energy contributes to the change dU^α for the addition process, so dU^α for the actual system will equal dU^α of (14.15) plus $\phi^\alpha \, dQ_j^\alpha$:

$$dU^\alpha = T \, dS^\alpha - P \, dV^\alpha + \mu_j^\alpha \, dn_j^\alpha + \phi^\alpha \, dQ_j^\alpha \tag{14.16}$$

The charge dQ_j^α is $dQ_j^\alpha = z_j F \, dn_j^\alpha$ [Eq. (14.14)], and (14.16) becomes

$$dU^\alpha = T \, dS^\alpha - P \, dV^\alpha + (\mu_j^\alpha + z_j F \phi^\alpha) \, dn_j^\alpha \tag{14.17}$$

Note that μ_j^α is the same in (14.15) and (14.17), since μ_j^α is a function of T, P, and composition, all of which are the same in the two systems. Thus the expressions for μ_j^α derived in previous chapters hold for μ_j^α in (14.17).

If we consider addition of infinitesimal amounts of other species to phase α, the same reasoning gives

$$dU^\alpha = T \, dS^\alpha - P \, dV^\alpha + \sum_i (\mu_i^\alpha + z_i F \phi^\alpha) \, dn_i^\alpha \tag{14.18}$$

Equation (14.18) shows that the presence of a nonzero electric potential ϕ^α in phase α causes the chemical potential μ_i^α to be replaced by $\mu_i^\alpha + z_i F \phi^\alpha$ in the Gibbs equation for dU^α. The quantity $\mu_i^\alpha + z_i F \phi^\alpha$ is called the **electrochemical potential** $\tilde{\mu}_i^\alpha$:

$$\tilde{\mu}_i^\alpha \equiv \mu_i^\alpha + z_i F \phi^\alpha \tag{14.19}*$$

(The symbol over the μ is called a tilde.) Since $z_i F$ is the molar charge of species i, Eq. (14.10) shows that *the electrochemical potential $\tilde{\mu}_i^\alpha$ is the sum of the chemical potential μ_i^α and the molar electrostatic potential energy $z_i F \phi^\alpha$ of species i in phase α.*

Using the definitions $H \equiv U + PV$, $A \equiv U - TS$, and $G \equiv U + PV - TS$ and (14.18), we see that $\tilde{\mu}_i^\alpha$ replaces μ_i^α in the Gibbs equations for dH, dA, and dG. Thus, to find the correct thermodynamic equations for an electrochemical system, we take the thermodynamic equations for the corresponding nonelectrochemical system (all ϕ's equal to 0) and replace the chemical potentials μ_i^α with the electrochemical potentials $\tilde{\mu}_i^\alpha = \mu_i^\alpha + z_i F \phi^\alpha$. When $\phi^\alpha = 0$ in (14.19), $\tilde{\mu}_i^\alpha$ reduces to the ordinary chemical potential μ_i^α.

For nonelectrochemical systems, the phase- and reaction-equilibrium conditions are $\mu_i^\alpha = \mu_i^\beta$ and $\Sigma_i \nu_i \mu_i = 0$ [Eqs. (4.88) and (4.98)]. Since μ_i^α is to be replaced by $\tilde{\mu}_i^\alpha$ in all thermodynamic equations for electrochemical systems, we conclude that:

In a closed electrochemical system, the phase-equilibrium condition for two phases α and β in contact is that

$$\tilde{\mu}_i^\alpha = \tilde{\mu}_i^\beta \tag{14.20}*$$

for each substance i present in both phases. In a closed electrochemical system, the reaction-equilibrium condition is

$$\sum_i \nu_i \tilde{\mu}_i = 0 \tag{14.21}$$

where the ν_i's are the stoichiometric coefficients in the reaction.

If substance i is absent from phase β but present in phase α, then $\tilde{\mu}_i^\beta$ need not equal $\tilde{\mu}_i^\alpha$ at phase equilibrium [Eq. (4.91)]. If i is present in phases α and δ, but these phases are separated by a phase in which i is absent, then $\tilde{\mu}_i^\alpha$ need not equal $\tilde{\mu}_i^\delta$ at phase equilibrium. An example is two pieces of metal dipping in the same solution but not in direct contact; $\tilde{\mu}$ of the electrons in one metal need not equal $\tilde{\mu}$ of the electrons in the second metal.

Since $\tilde{\mu}_i^\alpha = \mu_i^\alpha + z_i F \phi^\alpha$ [Eq. (14.19)], the phase-equilibrium condition (14.20) gives

$$\mu_i^\alpha + z_i F \phi^\alpha = \mu_i^\beta + z_i F \phi^\beta \tag{14.22}$$

$$\mu_i^\alpha - \mu_i^\beta = z_i F(\phi^\beta - \phi^\alpha) \tag{14.23}$$

This important equation relates $\mu_i^\alpha - \mu_i^\beta$ (the difference in equilibrium chemical potentials of species i in phases α and β) to $\phi^\beta - \phi^\alpha$ (the electric potential difference between the phases). If the interphase potential difference is zero, then $\mu_i^\alpha = \mu_i^\beta$ at equilibrium, as in earlier chapters. For uncharged species, z_i is zero and $\mu_i^\alpha = \mu_i^\beta$. For charged species, the greater the value of $|\phi^\beta - \phi^\alpha|$, the greater the difference $|\mu_i^\alpha - \mu_i^\beta|$. In Fig. 14.3$a$ and b, a piece of Zn dips into a very dilute $ZnSO_4$ solution. Zinc ions flow from the metal to the solution, producing a potential difference between the phases. The flow continues until $\phi^\beta - \phi^\alpha$ is large enough to satisfy (14.23), making the electrochemical potentials of Zn^{2+} equal in the two phases.

The reaction equilibrium condition (14.21) is $\Sigma_i \nu_i \tilde{\mu}_i = 0$. Consider the special case where all the charged species that participate in the reaction occur in the same phase, phase α. Substitution of $\tilde{\mu}_i^\alpha = \mu_i^\alpha + z_i F \phi^\alpha$ [Eq. (14.19)] into $\Sigma_i \nu_i \tilde{\mu}_i = 0$ gives $\Sigma_i \nu_i \mu_i + F\phi^\alpha \Sigma_i \nu_i z_i = 0$. The total charge is unchanged in a chemical reaction, so $\Sigma_i \nu_i z_i = 0$. [For example, for $2Fe^{3+}(aq) + Zn(s) \rightleftharpoons Zn^{2+}(aq) + 2Fe^{2+}(aq)$, we have $\Sigma_i \nu_i z_i = -2(3) - 1(0) + 1(2) + 2(2) = 0$.] Therefore the reaction equilibrium condition is

$$\sum_i \nu_i \mu_i = 0 \qquad \text{all charged species in same phase} \tag{14.24}$$

The value of ϕ^α is thus irrelevant when all charged species occur in the same phase. This makes sense, since the reference level of electric potential is arbitrary and we can take $\phi^\alpha = 0$ if we like.

In Chapters 10 and 11, we considered chemical potentials and reaction equilibrium for ions in electrolyte solutions. All charged species were present in the same phase, so there was no need to consider the electrochemical potentials.

Summary

In an electrochemical system (one with electric potential differences between phases), the electrochemical potentials $\tilde{\mu}_i^\alpha$ replace the chemical potentials in all thermodynamic equations. For example, the phase-equilibrium condition is the equality of electrochemical potentials: $\tilde{\mu}_i^\alpha = \tilde{\mu}_i^\beta$. The electrochemical potential of substance i in phase α is given by $\tilde{\mu}_i^\alpha = \mu_i^\alpha + z_i F \phi^\alpha$, where $z_i F$ is the molar charge of species i, ϕ^α is the electric potential of phase α, and μ_i^α is the chemical potential of i in α; z_i (an integer) is the charge number of species i, and F is the Faraday constant (the charge per mole of protons). Since the changes in chemical composition that accompany the development of interphase potential differences are extremely small, the chemical potential μ_i^α in an electrochemical system is the same as the chemical potential μ_i^α in the corresponding chemical system with no potential differences between phases. For example, μ_i^α of an ion in solution in an electrochemical system is given by Eq. (10.40).

14.4   GALVANIC CELLS

Galvanic Cells

If we attach a piece of wire to a device that produces an electric current in the wire, we can use the current to do useful work. For example, we might put the current-carrying wire in a magnetic field; this produces a force on the wire, giving us a motor. For a wire of resistance R carrying a current I, there is an electric potential difference $\Delta\phi$ between its ends, where $\Delta\phi$ is given by "Ohm's law" [Eq. (16.54)] as $|\Delta\phi| = IR$. This difference of potential corresponds to an electric field in the wire, which causes electrons to flow. To generate a current in the wire, we require a device that will maintain an electric potential difference between its output **terminals.** Any such device is called a **seat of electromotive force (emf).** Attaching a wire to the terminals of a seat of emf produces a current I in the wire (Fig. 14.5).

The **electromotive force (emf)** $\mathscr{E}$ of a seat of emf is defined as the potential difference between its terminals when the resistance R of the load attached to the terminals goes to infinity and hence the current goes to zero. *The emf is thus the open-circuit potential difference between the terminals.* (The difference of potential $\Delta\phi$ between the terminals in Fig. 14.5 depends on the value of the current I that flows through the circuit, because the seat of emf has an internal resistance R_{int} and the potential drop IR_{int} reduces $\Delta\phi$ between the terminals below the open-circuit $\Delta\phi$.)

One kind of seat of emf is an electric generator. Here, a mechanical force moves a metal wire through a magnetic field. This field exerts a force on the electrons in the metal, producing an electric current and a potential difference between the ends of the wire. An electric generator converts mechanical energy into electrical energy.

Another kind of seat of emf is a **galvanic** (or **voltaic**) **cell.** This is a multiphase electrochemical system in which the interphase potential differences result in a net potential difference between the terminals. The potential differences between phases result from the transfer of chemical species between phases, and a galvanic cell converts chemical energy into electrical energy. The phases of a galvanic cell must be electrical conductors; otherwise, a continuous current could not flow in Fig. 14.5.

Since only potential differences between chemically identical pieces of matter are readily measurable (Sec. 14.2), we specify that the two terminals of a galvanic cell are to be made of the same metal. Otherwise, we could not measure the cell emf, which is the open-circuit potential difference between the terminals. Suppose the terminals α and δ of a cell are made of copper and the potential difference between the terminals (the "voltage") is 2 V. Strictly speaking, the chemical compositions of the terminals differ, since the charges on the terminals differ. However, as shown in Sec. 14.3, the difference in chemical composition is so slight that we can ignore it and take the compositions of the terminals to be the same. Since μ_i^α is a function of T, P, and composition (but not of ϕ^α), we conclude that in a galvanic cell whose terminals α and δ are made of the same metal and are at the same T and P, the chemical potential of a species is the same in each terminal: $\mu_i^\alpha = \mu_i^\delta$.

The metal terminals of a galvanic cell are *electronic conductors,* meaning that the current is carried by electrons. Suppose all phases of the galvanic cell were electronic conductors. For example, the cell might be Cu′|Zn|Ag|Cu″, which is shorthand for a copper terminal Cu′ attached to a piece of Zn attached to a piece of Ag attached to a second copper terminal Cu″. Since electrons are free to move between all phases, the phase-equilibrium condition $\tilde{\mu}_i^\alpha = \tilde{\mu}_i^\beta$ [Eq. (14.20)] shows that $\tilde{\mu}(e^-)$ (the electrochemical potential of electrons) is the same in all phases of the open-circuit cell. In particular, $\tilde{\mu}(e^-$ in Cu′$) = \tilde{\mu}(e^-$ in Cu″$)$. The use of $\tilde{\mu}_i^\alpha = \mu_i^\alpha + z_i F\phi^\alpha$ [Eq. (14.19)] gives

$$\mu(e^-\text{ in Cu}') - F\phi(\text{Cu}') = \mu(e^-\text{ in Cu}'') - F\phi(\text{Cu}'')$$

Figure 14.5

A seat of emf attached to a load.

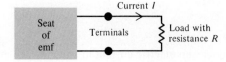

Since the terminals Cu' and Cu'' have the same chemical composition, it follows that $\mu(e^-$ in Cu'$) = \mu(e^-$ in Cu''$)$. Therefore, $\phi(\text{Cu}') = \phi(\text{Cu}'')$. The terminals have the same open-circuit electric potential, and the cell emf is zero. We conclude that a galvanic cell must have at least one phase that is impermeable to electrons. This allows $\tilde{\mu}(e^-)$ to differ in the two terminals.

The current in the phase that is impermeable to electrons must be carried by ions. Most commonly, the ionic conductor in a galvanic cell is an electrolyte solution. Other possibilities include a molten salt and a solid salt at a temperature high enough to allow ions to move through the solid at a useful rate. Batteries for heart pacemakers commonly use solid LiI as the ionic conductor.

Summarizing, a galvanic cell has terminals made of the same metal, has all phases being electrical conductors, has at least one phase that is an ionic conductor (but not an electronic conductor), and allows electric charge to be readily transferred between phases. We can symbolize the galvanic cell by T-E-I-E'-T' (Fig. 14.6), where T and T' are the terminals, I is the ionic conductor, and E and E' are two pieces of metal (called **electrodes**) that make contact with the ionic conductor. The current is carried by electrons in T, T', E, and E' and by ions in I.

The Daniell Cell

An example of a galvanic cell is the Daniell cell (Fig. 14.7), used in the early days of telegraphy. In this cell, a porous ceramic barrier separates a compartment containing a Zn rod in a $ZnSO_4$ solution from a compartment containing a Cu rod in a $CuSO_4$ solution. The Cu and Zn *electrodes* are attached to the wires Cu'' and Cu', which are the *terminals*. The porous barrier prevents extensive mixing of the solutions by convection currents but allows ions to pass from one solution to the other.

Consider first the open-circuit state, with the terminals not connected to a load (Fig. 14.7a). At the Zn electrode, an equilibrium is set up between aqueous Zn^{2+} ions and Zn^{2+} ions in the metal (as discussed in Sec. 14.2): $Zn^{2+}(Zn) \rightleftharpoons Zn^{2+}(aq)$. Adding this equation to the equation for the equilibrium between zinc ions and zinc atoms in the zinc metal, $Zn \rightleftharpoons Zn^{2+}(Zn) + 2e^-(Zn)$, we can write the equilibrium at the Zn–$ZnSO_4(aq)$ interface as $Zn \rightleftharpoons Zn^{2+}(aq) + 2e^-(Zn)$. Since the potential difference between the Zn electrode and the $ZnSO_4$ solution is not measurable, we do not know whether the equilibrium position for a given $ZnSO_4$ concentration leaves the Zn at a higher or lower potential than the solution. Let us assume that there is a net loss of Zn^{2+} to the solution, leaving a negative charge on the Zn and leaving the Zn at a lower

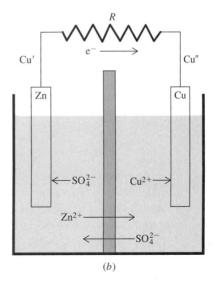

Figure 14.6

A galvanic cell consists of terminals T and T', electrodes E and E', and an ionic conductor I.

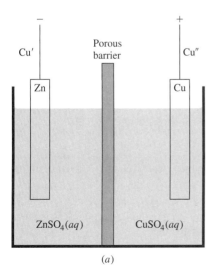

Figure 14.7

The Daniell cell. (*a*) Open-circuit state. (*b*) Closed-circuit state.

potential than the solution: $\phi(\text{Zn}) < \phi(\text{aq. ZnSO}_4)$. Although the potential difference $\phi(\text{aq. ZnSO}_4) - \phi(\text{Zn})$ is not known, the emfs in Table 14.1 in Sec. 14.7 indicate that this potential difference is typically on the order of a volt or two. As noted in Sec. 14.3, the amount of Zn^{2+} transferred between the metal and the solution is far too small to be detected by chemical analysis.

A similar equilibrium occurs at the $\text{Cu–CuSO}_4(aq)$ interface. However, Cu is a less active metal than Zn and has much less tendency to go into solution. [If a Zn rod is dipped into a CuSO_4 solution, metallic Cu immediately plates out on the Zn and Zn goes into solution: $\text{Cu}^{2+}(aq) + \text{Zn} \rightarrow \text{Cu} + \text{Zn}^{2+}(aq)$. If a Cu rod is dipped into a ZnSO_4 solution, no detectable amount of Zn plates out on the Cu. The equilibrium constant for the reaction $\text{Cu}^{2+}(aq) + \text{Zn} \rightleftharpoons \text{Cu} + \text{Zn}^{2+}(aq)$ is extremely large.] We thus expect that for comparable concentrations of CuSO_4 and ZnSO_4, the Cu electrode at equilibrium will have a smaller negative charge than the Zn electrode and might even have a positive charge (corresponding to a net gain of Cu^{2+} ions from the solution). Let us therefore assume that the equilibrium electric potential of Cu is greater than that of the aqueous CuSO_4 solution: $\phi(\text{Cu}) > \phi(\text{aq. CuSO}_4)$.

At the junction between the Cu′ terminal and the Zn electrode in Fig. 14.7a, there is an equilibrium exchange of electrons producing a potential difference between these phases. Since the potential difference between two phases of different composition is not measurable, the value of this potential difference is not known, but it is likely that $\phi(\text{Cu}') < \phi(\text{Zn})$, as in Fig. 14.3$d$.

There is no potential difference between the Cu electrode and the Cu″ terminal, since they are in contact and have the same chemical composition. More formally, $\mu(\text{e}^- \text{ in Cu}) = \mu(\text{e}^- \text{ in Cu}'')$; Eq. (14.23) then gives $\phi(\text{Cu}) = \phi(\text{Cu}'')$.

There is a potential difference at the junction of the ZnSO_4 and CuSO_4 solutions. This potential difference is small compared with the other interphase potential differences in the cell, and we shall neglect it for now, taking $\phi(\text{aq. ZnSO}_4) = \phi(\text{aq. CuSO}_4)$. See Sec. 14.9 for discussion of this liquid-junction potential difference.

The cell emf is defined as the open-circuit potential difference between the cell's terminals: $\mathcal{E} \equiv \phi(\text{Cu}'') - \phi(\text{Cu}') = \phi(\text{Cu}) - \phi(\text{Cu}')$. Adding and subtracting $\phi(\text{aq. CuSO}_4)$, $\phi(\text{aq. ZnSO}_4)$, and $\phi(\text{Zn})$ on the right side of this equation, we get

$$\mathcal{E} = [\phi(\text{Cu}) - \phi(\text{aq. CuSO}_4)] + [\phi(\text{aq. CuSO}_4) - \phi(\text{aq. ZnSO}_4)]$$
$$+ [\phi(\text{aq. ZnSO}_4) - \phi(\text{Zn})] + [\phi(\text{Zn}) - \phi(\text{Cu}')] \qquad (14.25)$$

The cell emf is the sum of the potential differences at the following interfaces between phases: $\text{Cu–CuSO}_4(aq)$, $\text{CuSO}_4(aq)\text{–ZnSO}_4(aq)$, $\text{ZnSO}_4(aq)\text{–Zn}$, $\text{Zn–Cu}'$. From the preceding discussion, the first term in brackets on the right side of (14.25) is positive, the second term is negligible, the third term is positive, and the fourth term is positive. Therefore $\mathcal{E} \equiv \phi(\text{Cu}'') - \phi(\text{Cu}')$ is positive, and the terminal attached to the Cu electrode is at a higher potential than the terminal attached to Zn. This is indicated by the $+$ and $-$ signs in Fig. 14.7.

We can write down an expression for the open-circuit potential difference at each interface. Equation (14.23) applied to Cu^{2+} ions at the $\text{Cu–CuSO}_4(aq)$ interface gives

$$\phi(\text{Cu}) - \phi(\text{aq. CuSO}_4) = [\mu^{\text{aq}}(\text{Cu}^{2+}) - \mu^{\text{Cu}}(\text{Cu}^{2+})]/2F$$

where the superscripts aq and Cu indicate the aqueous CuSO_4 and the Cu phases. Note that $\Delta\phi$ at the $\text{Cu–CuSO}_4(aq)$ phase boundary is determined by the chemical potentials of Cu^{2+} in Cu and in aqueous CuSO_4 and these chemical potentials are independent of the electrical state of the phases. Therefore the equilibrium $\Delta\phi$ for these two phases is independent of the presence or absence of contacts with other phases.

Now consider what happens when the circuit for the Daniell cell is completed by attaching a metal resistor R between the terminals (Fig. 14.7b). The Cu′ terminal (attached to Zn) is at a lower potential than the Cu″ terminal (attached to Cu), so electrons are forced to flow through R from Cu′ to Cu″. [It was noted after Eq. (14.12) that electrons move spontaneously from regions of low to regions of high electric potential, provided the regions have the same chemical composition, so that only the difference in electric potential influences the flow.] When electrons leave the Cu′ terminal, the equilibrium at the Cu′–Zn interface is disturbed, causing electrons to flow out of Zn into Cu′. This disturbs the equilibrium Zn $\rightleftharpoons$ Zn$^{2+}(aq)$ + 2e$^-$(Zn) at the Zn–ZnSO$_4(aq)$ interface and causes more Zn to go into solution, leaving electrons behind on the Zn to make up for the electrons that are leaving the Zn. The flow of electrons into the Cu electrode from the external circuit causes Cu^{2+} ions from the CuSO$_4$ solution to combine with electrons in the Cu metal and deposit as Cu atoms on the Cu electrode: Cu$^{2+}(aq)$ + 2e$^-$(Cu) → Cu.

In the region around the Cu electrode, the CuSO$_4$ solution is being depleted of positive ions (Cu^{2+}) while the region around the Zn electrode is being enriched in positive ions (Zn^{2+}). This causes a flow of positive ions through the solutions from the Zn electrode to the Cu electrode; simultaneously, negative ions move toward the Zn electrode (Fig. 14.7b). The current is carried through the solution by the Zn^{2+}, Cu^{2+}, and SO$_4^{2-}$ ions.

During the operation of the cell, the electrochemical reactions Zn → Zn$^{2+}(aq)$ + 2e$^-$(Zn) and Cu$^{2+}(aq)$ + 2e$^-$(Cu) → Cu occur. We call these the **half-reactions** of the cell. There is also the electron-flow process 2e$^-$(Zn) → 2e$^-$(Cu). Addition of this flow process and the two half-reactions gives the overall galvanic-cell reaction: Zn + Cu$^{2+}(aq)$ → Zn$^{2+}(aq)$ + Cu. The Zn electrode plus its associated ZnSO$_4$ solution form a **half-cell**; likewise, Cu and aqueous CuSO$_4$ form a second half-cell. So far, we have used the word "electrode" to mean the piece of metal that dips into a solution in a half-cell. Often, however, the term **electrode** is used to refer to a half-cell consisting of metal plus solution.

Oxidation is a loss of electrons. **Reduction** is a gain of electrons. The half-reaction Zn → Zn$^{2+}(aq)$ + 2e$^-$(Zn) is an oxidation. The half-reaction Cu$^{2+}(aq)$ + 2e$^-$(Cu) → Cu is a reduction. If we were to bring the species Cu, Zn, Cu$^{2+}(aq)$, and Zn$^{2+}(aq)$ in contact with one another, the oxidation–reduction (redox) reaction Zn + Cu$^{2+}(aq)$ → Cu + Zn$^{2+}(aq)$ would occur. In the Daniell cell, the oxidation and reduction parts of this reaction occur at different locations connected by a wire through which electrons are forced to flow. Separation of the oxidation and reduction half-reactions allows the chemical energy of the reaction to be converted into electrical energy.

We define the **anode** as the electrode at which oxidation occurs and the **cathode** as the electrode at which reduction occurs. In the Daniell cell, Zn is the anode.

> The open-circuit condition (Fig. 14.7a) of the Daniell cell is not a stable situation. The slow diffusion of Cu^{2+} into the ZnSO$_4$ solution will eventually allow the Cu^{2+} ions to come in contact with the Zn electrode, causing the spontaneous redox reaction Cu$^{2+}(aq)$ + Zn → Cu + Zn$^{2+}(aq)$ to occur directly, without flow of electrons through a wire. This would destroy the cell. For this reason, the Daniell cell cannot be left on open circuit. Instead, a resistor is kept connected between the terminals. Note from Fig. 14.7b that, as the cell operates, the electric field in the solution forces Cu^{2+} ions away from the ZnSO$_4$ solution, preventing them from getting at the Zn electrode. Many modern galvanic cells (batteries) have half-reactions that involve insoluble salts (see Sec. 14.11). This allows the cell to be kept on the shelf on open circuit.

Cell Diagrams and IUPAC Conventions

A galvanic cell is represented by a **diagram** in which the following conventions are used. A vertical line indicates a phase boundary. The phase boundary between two miscible liquids is indicated by a dashed or dotted vertical line. Two species present in the same phase are separated by a comma.

The diagram of the Daniell cell (Fig. 14.7) is

$$Cu' | Zn | ZnSO_4(aq) \vdots CuSO_4(aq) | Cu \qquad (14.26)$$

(The Cu″ terminal and the Cu electrode form a single phase.) The Cu′ terminal is often omitted from the cell diagram. For completeness, the $ZnSO_4$ and $CuSO_4$ molalities can be given in the diagram.

The following IUPAC conventions define the cell emf and cell reaction for a given cell diagram:

(A) The cell emf $\mathcal{E}$ is defined as

$$\mathcal{E} \equiv \phi_R - \phi_L \qquad \textbf{(14.27)*}$$

where ϕ_R and ϕ_L are the open-circuit electric potentials of the terminals on the right side and left side of the cell diagram. "Right" and "left" have nothing to do with the physical arrangement of the cell on the laboratory bench.

(B) The cell reaction is defined to involve oxidation at the electrode on the left side of the cell diagram and reduction at the electrode on the right.

For the cell diagram (14.26), Convention A gives $\mathcal{E} = \phi(Cu) - \phi(Cu')$. We saw earlier that $\phi(Cu)$ is greater than $\phi(Cu')$, so $\mathcal{E}$ for (14.26) is positive. For $CuSO_4$ and $ZnSO_4$ molalities close to 1 mol/kg, experiment gives $\mathcal{E}_{(14.26)} = 1.1$ V. For (14.26), Convention B gives the half-reaction at the left electrode as $Zn \rightarrow Zn^{2+} + 2e^-$ and that at the right electrode as $Cu^{2+} + 2e^- \rightarrow Cu$. The overall reaction for (14.26) is $Zn + Cu^{2+} \rightarrow Zn^{2+} + Cu$ (which is the spontaneous cell reaction when the Daniell cell is connected to a load—Fig. 14.7b).

Suppose we had written the cell diagram as

$$Cu | CuSO_4(aq) \vdots ZnSO_4(aq) | Zn | Cu' \qquad (14.28)$$

Then Convention A gives $\mathcal{E}_{(14.28)} = \phi(Cu') - \phi(Cu)$. Since $\phi(Cu) > \phi(Cu')$, the emf for this diagram is negative: $\mathcal{E}_{(14.28)} = -1.1$ V. Convention B gives the half-reactions for (14.28) as $Cu \rightarrow Cu^{2+} + 2e^-$ and $Zn^{2+} + 2e^- \rightarrow Zn$. The overall reaction for (14.28) is $Zn^{2+} + Cu \rightarrow Zn + Cu^{2+}$, which is the reverse of the spontaneous Daniell-cell reaction.

A positive emf for a cell diagram means that the cell reaction corresponding to this diagram will occur spontaneously when the cell is connected to a load. This is because oxidation (loss of electrons) at the left electrode sends electrons flowing out of this electrode to the right electrode, and electrons flow spontaneously from low to high ϕ; therefore $\phi_R > \phi_L$ and $\mathcal{E} > 0$.

Measurement of Cell Emfs

The emf of a galvanic cell can be accurately measured using a *potentiometer* (Fig. 14.8). Here, the emf $\mathcal{E}_X$ of cell X is balanced by an opposing potential difference $\Delta\phi_{opp}$, so as to make the current through the cell zero. Measurement of $\Delta\phi_{opp}$ gives $\mathcal{E}_X$.

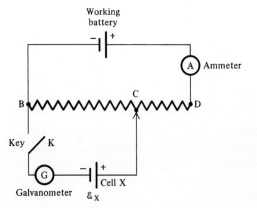

Figure 14.8

A potentiometer.

The resistor between B and D is a uniform slide-wire of total resistance R. The contact point C is adjusted until the galvanometer G shows no deflection when the tap key K is closed, indicating zero current passing through cell X. When no current flows through the cell when the key is closed, the negative terminal of the cell is at the same potential as point B and the positive terminal of the cell is at the same potential as point C. Hence, when balance is achieved, the potential drop across the resistor CB equals the zero-current potential drop across the cell's terminals, which is the cell emf $\mathscr{E}_X$. Ohm's law (16.54) gives $\mathscr{E}_X = |\Delta\phi_{opp}| = IR_X$, where I is the current in the upper part of the circuit and R_X is the resistance of the wire between B and C; we have $R_X = (\overline{BC}/\overline{BD})R$. Measurement of I and R_X enables $\mathscr{E}_X$ to be found.

In practice, one balances the circuit twice, once with cell X and once with a standard cell S of accurately known emf $\mathscr{E}_S$ in place of X. Let R_S and R_X be the resistances needed to balance $\mathscr{E}_S$ and $\mathscr{E}_X$. Then $\mathscr{E}_S = IR_S$ and $\mathscr{E}_X = IR_X$. (Since no current flows through S or X, the current I is unchanged when the cell is changed.) We have $\mathscr{E}_X/\mathscr{E}_S = R_X/R_S$, which enables $\mathscr{E}_X$ to be found.

When the potentiometer in Fig. 14.8 is only infinitesimally out of balance, an infinitesimal current flows through the cell X. Equilibrium is maintained at each phase boundary in the cell, and the cell reaction occurs reversibly. The rate of the reversible cell reaction is infinitesimal, and it takes an infinite time to carry out a noninfinitesimal amount of reaction. When a noninfinitesimal current is drawn from the cell, as in Fig. 14.7b, the cell reaction occurs irreversibly.

Potentiometers have been made obsolete by electronic digital voltmeters, which can measure cell emfs while drawing a negligible current.

Electrolytic Cells

In a galvanic cell, a chemical reaction produces a flow of electric current; chemical energy is converted into electrical energy. In an **electrolytic cell,** a flow of current produces a chemical reaction; electrical energy from an external source is converted into chemical energy.

Figure 14.9 shows an electrolytic cell. Two Pt electrodes are attached to the terminals of a seat of emf (for example, a galvanic cell or a dc generator). The Pt electrodes dip into an aqueous NaOH solution. Electrons flow into the negative Pt electrode from the seat of emf, and H_2 is liberated at this electrode: $2H_2O + 2e^- \rightarrow H_2 + 2OH^-$. At the positive electrode, O_2 is liberated: $4OH^- \rightarrow 2H_2O + O_2 + 4e^-$. Doubling the first half-reaction and adding it to the second, we get the overall electrolysis reaction $2H_2O \rightarrow 2H_2 + O_2$.

The same definitions of anode and cathode are used for electrolytic cells as for galvanic cells. Therefore, the cathode in Fig. 14.9 is the negative electrode. In a galvanic cell, the cathode is the positive electrode.

The elements Al, Na, and F_2 are commercially prepared by the electrolysis of molten Al_2O_3, molten NaCl, and liquid HF. Electrolysis is also used to plate one metal on another.

The term **electrochemical cell** indicates either a galvanic or an electrolytic cell. Galvanic and electrolytic cells are quite different from each other, and this chapter deals mainly with galvanic cells.

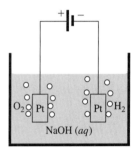

Figure 14.9

An electrolytic cell. Electrons flow from the seat of emf into the right-hand electrode where a reduction produces H_2.

14.5   TYPES OF REVERSIBLE ELECTRODES

Equilibrium thermodynamics applies only to reversible processes. To apply thermodynamics to galvanic cells (Sec. 14.6), we require that the cell be reversible. Consider a cell with its emf balanced in a potentiometer (Fig. 14.8). If the cell is reversible, the processes that occur in the cell when the contact point C is moved slightly to the right must be the reverse of the processes that occur when C is moved slightly to the left.

For the Daniell cell, when C is moved slightly to the left, the potential drop across BC becomes slightly less than the cell's emf and the cell functions as a galvanic cell with Zn going into solution as Zn^{2+} at the zinc electrode and Cu^{2+} plating out as Cu at the copper electrode. When C is moved slightly to the right, the externally applied emf is slightly greater than the emf of the Daniell cell, so the direction of current flow through the cell is reversed. The cell then functions as an electrolytic cell, with Zn being plated out at the zinc electrode and Cu going into solution at the copper electrode. Thus, the electrode reactions are reversed.

Despite this, the Daniell cell is not reversible. The irreversibility arises at the liquid junction. (A **liquid junction** is the interface between two miscible electrolyte solutions.) When the Daniell cell functions as a galvanic cell, Zn^{2+} ions move into the $CuSO_4$ solution (Fig. 14.7b). However, when the cell's emf is overridden by an external emf that reverses the current direction, the reversal of current in the solution means that Cu^{2+} ions will move into the $ZnSO_4$ solution. Since these processes at the liquid junction are not the reverse of each other, the cell is irreversible.

To have reversibility at an electrode, all reactants and products of the electrode half-reaction must be present at the electrode. For example, if we had a cell one of whose electrodes was Zn dipping into an aqueous solution of NaCl, then when electrons are moving out of this electrode, the half-reaction is $Zn \rightarrow Zn^{2+}(aq) + 2e^-$, whereas when the potentiometer slide-wire is moved in the opposite direction and electrons are moving into the Zn electrode, the half-reaction is $2H_2O + 2e^- \rightarrow H_2 + 2OH^-(aq)$, since there is no Zn^{2+} to plate out of the solution. Reversibility requires the presence of Zn^{2+} in the solution around the Zn electrode.

The main types of reversible electrodes (half-cells) are

1. **Metal–metal-ion electrodes.** Here, a metal M is in electrochemical equilibrium with a solution containing M^{z+} ions. The half-reaction is $M^{z+} + z_+e^- \rightleftharpoons M$. Examples include $Cu^{2+}|Cu$, $Hg_2^{2+}|Hg$, $Ag^+|Ag$, $Pb^{2+}|Pb$, and $Zn^{2+}|Zn$. Metals that react with the solvent cannot be used. Group IA and most group IIA metals (Na, Ca, . . .) react with water; zinc reacts with aqueous acidic solutions. For certain metals one must use N_2 to remove air from the cell to prevent oxidation of the metal by dissolved O_2.

2. **Amalgam electrodes.** An *amalgam* is a solution of a metal in liquid Hg. In an amalgam electrode, an amalgam of metal M is in equilibrium with a solution containing M^{z+} ions. The mercury does not participate in the electrode reaction, which is $M^{z+}(sln) + z_+e^- \rightleftharpoons M(Hg)$, where M(Hg) indicates M dissolved in Hg. Active metals like Na or Ca can be used in an amalgam electrode.

3. **Redox electrodes.** Every electrode involves an oxidation–reduction half-reaction. However, custom dictates that the term "redox electrode" refer only to an electrode whose redox half-reaction is between two species present in the same solution. The metal that dips into this solution serves only to supply or accept electrons. For example, a Pt wire dipping into a solution containing Fe^{2+} and Fe^{3+} is a redox electrode whose half-reaction is $Fe^{3+} + e^- \rightleftharpoons Fe^{2+}$. The half-cell diagram is $Pt|Fe^{3+}, Fe^{2+}$. Another example is $Pt|MnO_4^-, Mn^{2+}$.

4. **Metal–insoluble-salt electrodes.** Here, a metal M is in contact with one of its very slightly soluble salts $M_{\nu_+}X_{\nu_-}$ and with a solution that is saturated with $M_{\nu_+}X_{\nu_-}$ and that contains a soluble salt or acid with the anion X^{z-}.

   For example, the **silver–silver chloride electrode** (Fig. 14.10a) consists of Ag metal, solid AgCl, and a solution that contains Cl^- ions (from, say, KCl or HCl) and is saturated with AgCl. There are three phases present, and the electrode is usually symbolized by $Ag|AgCl(s)|Cl^-(aq)$. One way to prepare this electrode is by electrodeposition of a layer of Ag on a piece of Pt, followed by electrolytic conversion of part of the Ag to AgCl. The Ag is in electrochemical equilibrium with the Ag^+ in the solution: $Ag \rightleftharpoons Ag^+(aq) + e^-$. Since the solution is saturated with

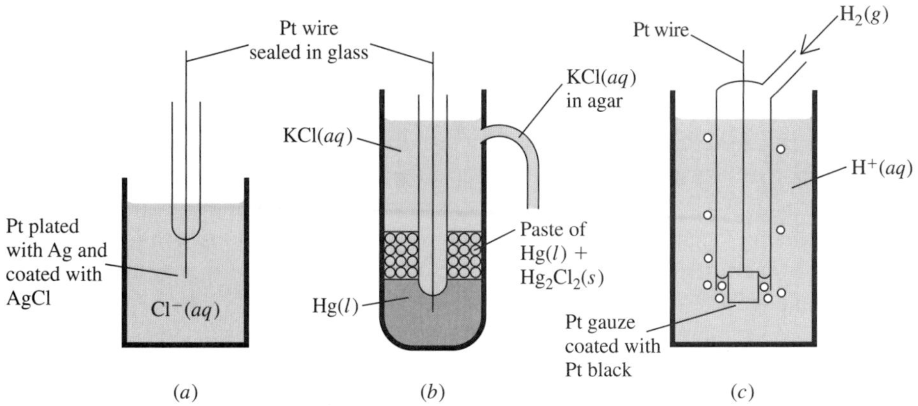

Figure 14.10

(*a*) The Ag–AgCl electrode.
(*b*) The calomel electrode. (*c*) The hydrogen electrode.

AgCl, any Ag^+ added to the solution reacts as follows: $Ag^+(aq) + Cl^-(aq) \rightleftharpoons$ $AgCl(s)$. The net electrode half-reaction is the sum of these two reactions:

$$Ag(s) + Cl^-(aq) \rightarrow AgCl(s) + e^- \qquad (14.29)$$

The *calomel electrode* (Fig. 14.10*b*) is $Hg|Hg_2Cl_2(s)|KCl(aq)$. The half-reaction is $2Hg + 2Cl^- \rightleftharpoons Hg_2Cl_2(s) + 2e^-$, which is the sum of $2Hg \rightleftharpoons$ $Hg_2^{2+}(aq) + 2e^-$ and $Hg_2^{2+}(aq) + 2Cl^-(aq) \rightleftharpoons Hg_2Cl_2(s)$. (Calomel is Hg_2Cl_2.) When the solution is saturated with KCl, we have the *saturated calomel electrode*.

The diagrams for metal–insoluble-salt half-cells can be misleading. Thus, the diagram $Hg|Hg_2Cl_2(s)|KCl(aq)$ might seem to suggest that the Hg is not in contact with the aqueous solution, when in fact all three phases are in contact with one another.

5. **Gas electrodes.** Here, a gas is in equilibrium with ions in solution. For example, the **hydrogen electrode** is $Pt|H_2(g)|H^+(aq)$, and its half-reaction is

$$H_2(g) \rightleftharpoons 2H^+(aq) + 2e^- \qquad (14.30)$$

H_2 is bubbled over the Pt, which dips into an acidic solution (Fig. 14.10*c*). The Pt contains a coat of electrolytically deposited colloidal Pt particles (*platinum black*), which catalyze the forward and reverse reactions in (14.30), enabling the equilibrium to be rapidly established. The H_2 gas is chemisorbed as H atoms on the platinum: $H_2(g) \rightleftharpoons 2H(Pt) \rightleftharpoons 2H^+(aq) + 2e^-(Pt)$.

The *chlorine electrode* is $Pt|Cl_2(g)|Cl^-(aq)$ with half-reaction $Cl_2 + 2e^- \rightleftharpoons$ $2Cl^-(aq)$. A reversible oxygen electrode is extremely difficult to prepare, because of formation of an oxide layer on the metal and other problems.

6. **Nonmetal nongas electrodes.** The most important examples are the bromine and iodine electrodes: $Pt|Br_2(l)|Br^-(aq)$ and $Pt|I_2(s)|I^-(aq)$. In these electrodes, the solution is saturated with dissolved Br_2 or I_2.

7. **Membrane electrodes.** These are defined and discussed in Sec. 14.12.

A galvanic cell formed from half-cells that have different electrolyte solutions contains a liquid junction where these solutions meet and is therefore irreversible. An example is the Daniell cell (14.26). If we tried to get around this irreversibility by having the Cu and Zn rods dip into a common solution that has both $CuSO_4$ and $ZnSO_4$, the Cu^{2+} ions would react with the Zn rod and the attempt would fail.

A reversible galvanic cell requires two half-cells that use the same electrolyte solution. An example is the cell

$$Pt|H_2(g)|HCl(aq)|AgCl(s)|Ag|Pt' \qquad (14.31)$$

composed of a hydrogen electrode and an Ag–AgCl electrode, each dipping in the same HCl solution. An advantage of metal–insoluble-salt electrodes is that they can be used to make cells without liquid junctions.

Printed Electrodes

Cheap, disposable electrodes can be produced by printing an "ink" made of suitable material onto a supporting strip of polymeric or ceramic material. A graphite-containing ink yields a graphite electrode. An ink containing Ag and AgCl yields a silver–silver chloride electrode. People with diabetes test their blood glucose level by putting a drop of blood on a test strip, which is inserted into a handheld meter. One type of glucose meter uses a test strip containing two printed electrodes and the enzyme glucose oxidase, which is specific for the oxidation of glucose. Oxidation of the glucose and subsequent reactions lead to a current flow whose magnitude is proportional to the glucose concentration.

The glucose monitor is an example of a **biosensor,** which is a device that contains a biological material (for example, an enzyme, an antibody, a cell, or a tissue) that interacts with the substance being tested for (called the *substrate* or *analyte*). The result of this interaction is then converted ("transduced") into a measurable physical signal (for example, a current flow or an electric potential difference), whose magnitude is proportional to the substrate's concentration. The bananatrode, an electrode containing a slice of banana or banana mixed with graphite, is a biosensor that detects the neurotransmitter dopamine. See B. Eggins, *Biosensors,* Wiley–Teubner, 1996.

14.6   THERMODYNAMICS OF GALVANIC CELLS

In this section, we use thermodynamics to relate the emf (the open-circuit potential difference between the terminals) of a reversible galvanic cell to the chemical potentials of the species in the cell reaction. Consider such a cell with its terminals on open circuit. For example, the cell might be

$$\text{Pt}_L|\text{H}_2(g)|\text{HCl}(aq)|\text{AgCl}(s)|\text{Ag}|\text{Pt}_R \tag{14.32}$$

where the subscripts L and R indicate the left and right terminals. The IUPAC conventions (Sec. 14.4) give the half-reactions and overall reaction as

$$\text{H}_2(g) \rightleftharpoons 2\text{H}^+ + 2\text{e}^-(\text{Pt}_L)$$

$$\frac{[\text{AgCl}(s) + \text{e}^-(\text{Pt}_R) \rightleftharpoons \text{Ag} + \text{Cl}^-] \times 2}{2\text{AgCl}(s) + \text{H}_2(g) + 2\text{e}^-(\text{Pt}_R) \rightleftharpoons 2\text{Ag} + 2\text{H}^+ + 2\text{Cl}^- + 2\text{e}^-(\text{Pt}_L)} \tag{14.33}$$

Since the terminals are on open circuit, flow of electrons from Pt_L to Pt_R cannot occur. Therefore the electrons have been included in the overall reaction. We call (14.33) the cell's **electrochemical reaction** to distinguish it from the cell's **chemical reaction,** which is

$$2\text{AgCl}(s) + \text{H}_2(g) \rightleftharpoons 2\text{Ag} + 2\text{H}^+ + 2\text{Cl}^- \tag{14.34}$$

Electrochemical Equilibrium in a Galvanic Cell

When an open-circuit reversible cell is assembled from its component phases, tiny amounts of charge are transferred between phases until electrochemical equilibrium is reached. In the open-circuit Daniell cell of Fig. 14.7a, electrochemical equilibrium exists between the Zn electrode and the ZnSO_4 solution, between the Cu electrode and the CuSO_4 solution, and between the Cu′ terminal and the Zn electrode. However, the liquid junction introduces irreversibility (as noted in Sec. 14.5), and there is no electrochemical equilibrium between the two solutions. In the reversible cell (14.32), there is no liquid junction and all adjacent phases are in electrochemical equilibrium.

We first consider reversible galvanic cells. When the phases of the open-circuit reversible cell (14.32) are brought together, the two half-reactions occur until electrochemical equilibrium is reached and the overall electrochemical reaction (14.33) is in equilibrium. The equilibrium condition in a closed electrochemical system is $\Sigma_i \nu_i \tilde{\mu}_i = 0$ [Eq. (14.21)], where the $\tilde{\mu}_i$'s are the electrochemical potentials, the ν_i's are the stoichiometric coefficients, and the sum goes over all species in the electrochemical reaction; this is the equilibrium condition for any reversible open-circuit galvanic cell. We write the sum $\Sigma_i \nu_i \tilde{\mu}_i$ as a sum over electrons plus a sum over all other species:

$$0 = \sum_i \nu_i \tilde{\mu}_i = \sum_{e^-} \nu(e^-)\tilde{\mu}(e^-) + \sum_i' \nu_i \tilde{\mu}_i \qquad (14.35)$$

where the prime on the second sum indicates that it does not include electrons. For example, for the cell reaction (14.33)

$$\sum_{e^-} \nu(e^-)\tilde{\mu}(e^-) = -2\tilde{\mu}[e^-(Pt_R)] + 2\tilde{\mu}[e^-(Pt_L)] \qquad (14.36)$$

$$\sum_i' \nu_i \tilde{\mu}_i = -2\tilde{\mu}(AgCl) - \tilde{\mu}(H_2) + 2\tilde{\mu}(Ag) + 2\tilde{\mu}(H^+) + 2\tilde{\mu}(Cl^-)$$

where $\tilde{\mu}[e^-(Pt_R)]$ is the electrochemical potential of electrons in the Pt_R terminal.

Let T_R and T_L denote the right- and left-hand terminals of the cell. Let n, the **charge number of the cell reaction,** be defined as the number of electrons transferred for the cell electrochemical reaction as written. For example, n is 2 for the cell reaction (14.33). The charge number n is a positive, dimensionless number. The sum over electrons in (14.35) can be written as [see (14.36)]

$$\sum_{e^-} \nu(e^-)\tilde{\mu}(e^-) = -n\tilde{\mu}[e^-(T_R)] + n\tilde{\mu}[e^-(T_L)] \qquad (14.37)$$

By the IUPAC conventions, oxidation (loss of electrons) occurs at T_L, so $e^-(T_L)$ appears on the right side of the cell's electrochemical reaction, as in (14.33), and $e^-(T_L)$ has a positive stoichiometric coefficient: $\nu[e^-(T_L)] = +n$, as in (14.37).

Use of $\tilde{\mu}_i^\alpha = \mu_i^\alpha + z_i F \phi^\alpha$ [Eq. (14.19)] with $i = e^-$ and $z_i = -1$ gives for (14.37)

$$\sum_{e^-} \nu(e^-)\tilde{\mu}(e^-) = n\mu[e^-(T_L)] - n\mu[e^-(T_R)] + nF(\phi_R - \phi_L)$$

where ϕ_R and ϕ_L are the potentials of the right and left terminals. The chemical potentials μ_i depend on T, P, and composition, and the terminals have the same T, P, and composition. Therefore the chemical potentials of the electrons in the terminals are equal: $\mu[e^-(T_L)] = \mu[e^-(T_R)]$. We now have

$$\sum_{e^-} \nu(e^-)\tilde{\mu}(e^-) = nF(\phi_R - \phi_L) = nF\mathscr{E} \qquad (14.38)$$

where the definition $\mathscr{E} \equiv \phi_R - \phi_L$ of the cell emf was used.

Since the cell is reversible, there are no liquid junctions and all ions in the cell reaction occur in the same phase, namely, the phase that is an ionic conductor (an electrolyte solution, a fused salt, etc.). In deriving Eq. (14.24), we proved that $\Sigma_j \nu_j \tilde{\mu}_j = \Sigma_j \nu_j \mu_j$ when all charged species in the sum occur in the same phase. This condition holds for the sum $\Sigma_i' \nu_i \tilde{\mu}_i$ on the right side of (14.35), so $\Sigma_i' \nu_i \tilde{\mu}_i = \Sigma_i' \nu_i \mu_i$. Substitution of this relation and of (14.38) into (14.35) gives

$$\sum_i' \nu_i \mu_i = -nF\mathscr{E} \qquad \text{rev. cell} \qquad (14.39)$$

Equation (14.39) relates the cell emf to the chemical potentials of the species in the cell's chemical reaction. For example, for the cell (14.32) with chemical reaction (14.34), we have $-2\mu(AgCl) - \mu(H_2) + 2\mu(Ag) + 2\mu(H^+) + 2\mu(Cl^-) = -2F\mathscr{E}$.

The quantity $\Sigma_i' \nu_i \mu_i$ is called ΔG in many texts, and (14.39) is written as $\Delta G = -nF\mathscr{E}$. However, as noted in Sec. 11.9, the symbol ΔG has several meanings. A better designation for this sum is $(\partial G/\partial \xi)_{T,P}$ [Eq. (11.34)], where G is the Gibbs energy of the species in the cell's chemical reaction.

The Nernst Equation

We now express the chemical potentials in (14.39) in terms of activities. The definition of the activity a_i of species i gives $\mu_i = \mu_i^\circ + RT \ln a_i$ [Eq. (11.2)], where μ_i° is the chemical potential of i in its chosen standard state. Carrying out the same manipulations used to derive (6.14) and (11.4), we have

$$\sum_i {}' \nu_i \mu_i = \sum_i {}' \nu_i \mu_i^\circ + RT \sum_i {}' \nu_i \ln a_i = \Delta G^\circ + RT \ln \left[\prod_i {}' (a_i)^{\nu_i} \right] \quad (14.40)$$

where $\Delta G^\circ \equiv \Sigma_i' \nu_i \mu_i^\circ$ is the standard molar Gibbs energy change [Eq. (11.5)] for the cell's chemical reaction. Substitution of (14.40) in (14.39) gives

$$\mathscr{E} = -\frac{\Delta G^\circ}{nF} - \frac{RT}{nF} \ln \left[\prod_i {}' (a_i)^{\nu_i} \right] \quad (14.41)$$

If all the chemical species were in their standard states, it follows from $\mu_i = \mu_i^\circ + RT \ln a_i$ that all activities would equal 1. From (14.41), the cell emf would then equal $-\Delta G^\circ/nF$. Hence, $-\Delta G^\circ/nF$ is called **standard emf** $\mathscr{E}^\circ$ of the cell (or the *standard potential* of the cell's chemical reaction). We have $\mathscr{E}^\circ \equiv -\Delta G^\circ/nF$ and

$$\Delta G^\circ = -nF\mathscr{E}^\circ \quad \textbf{(14.42)*}$$

$$\mathscr{E} = \mathscr{E}^\circ - \frac{RT}{nF} \ln \left[\prod_i {}' (a_i)^{\nu_i} \right] = \mathscr{E}^\circ - \frac{RT}{nF} \ln Q \qquad \text{rev. cell} \quad \textbf{(14.43)*}$$

In the **reaction quotient** (or **activity quotient**) $Q \equiv \Pi_i' (a_i)^{\nu_i}$, the product goes over all species in the cell's chemical reaction but does not include electrons. The **Nernst equation** (14.43) relates the cell's emf $\mathscr{E}$ to the activities a_i of the substances in the cell's chemical reaction and to the standard potential $\mathscr{E}^\circ$ of the reaction; $\mathscr{E}^\circ$ is related to ΔG° of the cell's chemical reaction [Eq. (14.42)].

Equations (14.40) to (14.43) are ambiguous in that the scale for solute activities has not been specified. Activities a_i and standard-state chemical potentials μ_i° differ on the molality and molar concentration scales (Sec. 10.4). For electrochemical cells, the molality scale is most commonly used for solutes in aqueous solution. *All aqueous-solution activities and activity coefficients in this chapter are on the molality scale.* (In aqueous solutions, the molality and molar-concentration activities $a_{m,i}$ and $a_{c,i}$ are nearly equal, as shown in Prob. 10.15.)

The product $\Pi_i' (a_i)^{\nu_i}$ in the Nernst equation (14.43) is not in general equal to the ordinary chemical equilibrium constant K° that appears in (11.6). Although it is true that a reversible cell sitting on open circuit is in equilibrium, this equilibrium is an *electrochemical* equilibrium, and the cell's electrochemical reaction [for example, (14.33)] involves transfer of electrons between phases that differ in electric potential. The activities in (14.43) are equal to whatever values are used when the cell is set up, since the attainment of electrochemical equilibrium between phases involves negligible changes in concentrations.

Let us rewrite the Nernst equation in terms of the chemical equilibrium constant K°. Equations (14.42) and (11.4) give $\mathscr{E}^\circ = -\Delta G^\circ/nF = (RT \ln K^\circ)/nF$. Equation (14.43) becomes $\mathscr{E} = (RT \ln K^\circ)/nF - (RT \ln Q)/nF$, or

$$\mathscr{E} = \frac{RT}{nF} \ln \frac{K^\circ}{Q} \quad (14.44)$$

When Q equals K°, the cell emf is zero. The greater the departure of Q from K°, the greater the magnitude of the emf.

Let the cell diagram be written with the terminal that is at the higher potential put on the right, so that $\mathscr{E} \equiv \phi_R - \phi_L$ is positive. From (14.44), positive $\mathscr{E}$ means that $Q < K°$, where Q and $K°$ are for the reaction corresponding to the cell diagram (oxidation at the left electrode). Suppose the cell is connected to a load (as in Fig. 14.7b). Since $\mathscr{E}$ is positive, the spontaneous reaction that occurs is the same as the reaction corresponding to the cell diagram, as noted after Eq. (14.28). This means that the amounts of products in the cell-diagram reaction increase as the cell operates, which increases Q. As Q increases toward $K°$, the cell emf decreases, reaching zero when Q equals $K°$.

EXAMPLE 14.3 Cell emf in terms of activities

Use the Nernst equation to express the emf of the cell (14.32) in terms of activities. Assume the pressure does not greatly differ from 1 bar.

The cell's overall chemical reaction is given by Eq. (14.34) as

$$2AgCl(s) + H_2(g) \rightleftharpoons 2Ag(s) + 2H^+(aq) + 2Cl^-(aq)$$

The Nernst equation (14.43) gives

$$\mathscr{E} = \mathscr{E}° - \frac{RT}{2F} \ln \frac{[a(H^+)]^2[a(Cl^-)]^2[a(Ag)]^2}{[a(AgCl)]^2 a(H_2)}$$

At or near 1 bar, the activities of the pure solids Ag and AgCl are 1 (Sec. 11.4). Equation (10.96) gives $a(H_2) = f(H_2)/P°$, where f is the H_2 fugacity and $P° \equiv 1$ bar. For pressures close to 1 bar, we can replace $f(H_2)$ with $P(H_2)$ with negligible error (Prob. 14.18). Therefore

$$\mathscr{E} = \mathscr{E}° - \frac{RT}{2F} \ln \frac{[a(H^+)]^2[a(Cl^-)]^2}{P(H_2)/P°} \qquad (14.45)$$

Exercise

For $Pt|Cl_2(g)|HCl(aq)|AgCl(s)|Ag(s)|Pt'$ at pressures that are not high, express $\mathscr{E}$ in terms of activities. [*Answer:* $\mathscr{E} = \mathscr{E}° - (RT/2F) \ln [P(Cl_2)/P°]$.]

We want to express the HCl activity product $a(H^+)a(Cl^-)$ in (14.45) in terms of measurable quantities. Rather than dealing directly with this particular activity product, we shall consider the general strong electrolyte $M_{\nu_+} X_{\nu_-}$, for which the activity product $(a_+)^{\nu_+}(a_-)^{\nu_-}$ occurs, where a_+ and a_- are the cation and anion activities. Using $a_+ = \gamma_+ m_+/m°$ and $a_- = \gamma_- m_-/m°$ [Eq. (10.30)], we have

$$(a_+)^{\nu_+}(a_-)^{\nu_-} = (\gamma_+)^{\nu_+}(\gamma_-)^{\nu_-}(m_+/m°)^{\nu_+}(m_-/m°)^{\nu_-}$$

The ionic molalities from $M_{\nu_+} X_{\nu_-}$ are $m_+ = \nu_+ m_i$ and $m_- = \nu_- m_i$, where m_i is the electrolyte's stoichiometric molality [Eq. (10.48)], and we have

$$(a_+)^{\nu_+}(a_-)^{\nu_-} = (\gamma_+)^{\nu_+}(\gamma_-)^{\nu_-}(\nu_+)^{\nu_+}(\nu_-)^{\nu_-}(m_i/m°)^{\nu_+ + \nu_-} = (\nu_\pm \gamma_\pm m_i/m°)^{\nu}$$

$$(a_+)^{\nu_+}(a_-)^{\nu_-} = (\nu_\pm \gamma_\pm m_i/m°)^{\nu} = (\nu_+)^{\nu_+}(\nu_-)^{\nu_-}(\gamma_\pm m_i/m°)^{\nu_+ + \nu_-} \qquad (14.46)$$

where $\nu \equiv \nu_+ + \nu_-$, $(\gamma_\pm)^{\nu} = (\gamma_+)^{\nu_+}(\gamma_-)^{\nu_-}$, and $(\nu_\pm)^{\nu} \equiv (\nu_+)^{\nu_+}(\nu_-)^{\nu_-}$ [Eqs. (10.45), (10.43), and (10.50)]. Equation (14.46) relates the activity product of the electrolyte $M_{\nu_+} X_{\nu_-}$ to its stoichiometric molality m_i and activity coefficient $\gamma_\pm$. [Another derivation of (14.46) is given in Prob. 14.22.]

If ion pairing (Sec. 10.9) is explicitly allowed for, then $\gamma_\pm$ in (14.46) is replaced by $\gamma_\pm^\dagger$, as in Eq. (10.76). Whatever symbol is used, the electrolyte activity coefficient in (14.46) is the experimentally observed value for $M_{\nu_+}X_{\nu_-}$.

The electrolyte HCl has $\nu_+ = \nu_- = 1$, and Eq. (14.46) gives

$$a(H^+)a(Cl^-) = \gamma_\pm^2 (m/m^\circ)^2 \qquad (14.47)$$

where m is the HCl stoichiometric molality and $m^\circ = 1$ mol/kg. Substituting for the activities in (14.45), we get for the cell (14.32)

$$\mathscr{E} = \mathscr{E}^\circ - \frac{RT}{2F} \ln \frac{(\gamma_\pm m/m^\circ)^4}{P(H_2)/P^\circ} \qquad (14.48)$$

Determination of $\mathscr{E}^\circ$

How is $\mathscr{E}^\circ$ in the Nernst equation (14.43) found? If all chemical species in the cell are in their standard states, then all the a_i's are 1 and the logarithm term in (14.43) vanishes, making $\mathscr{E}$ equal to $\mathscr{E}^\circ$. However, the molality-scale standard states of solutes are fictitious states, unattainable in reality. Hence, it is generally impossible to prepare a cell with all species in their standard states. Even though the species are not in their standard states, it still might happen that the activities of all species in the cell are 1, in which case $\mathscr{E} = \mathscr{E}^\circ$. This, however, is not a practical way to find $\mathscr{E}^\circ$, since activity coefficients are not generally known to high accuracy; one can't be sure that $\gamma_\pm m_i = 1$ mol/kg for each species in solution.

$\mathscr{E}^\circ$ can be found from $\Delta G^\circ = -nF\mathscr{E}^\circ$ [Eq. (14.42)] if the standard-state partial molar Gibbs energies μ_i° of the species in the cell reaction are known.

$\mathscr{E}^\circ$ can be determined from emf measurements on the cell by an extrapolation procedure. For example, consider the cell (14.32), whose emf is given by (14.48). Rewriting (14.48), we have

$$\mathscr{E} + \frac{2RT}{F} \ln (m/m^\circ) - \frac{RT}{2F} \ln [P(H_2)/P^\circ] = \mathscr{E}^\circ - \frac{2RT}{F} \ln \gamma_\pm \qquad (14.49)$$

All quantities on the left side of (14.49) are known. In the limit $m \to 0$, the activity coefficient $\gamma_\pm$ goes to 1 and $\ln \gamma_\pm$ goes to 0. Therefore, extrapolation of the left side to $m = 0$ gives $\mathscr{E}^\circ$. The Debye–Hückel equation (10.69) shows that $\ln \gamma_\pm$ is proportional to $m^{1/2}$ in very dilute solutions. Hence, for very low molalities a plot of the left side of (14.49) versus $m^{1/2}$ gives a straight line whose intercept at $m = 0$ is $\mathscr{E}^\circ$ (Fig. 14.11).

Irreversible Galvanic Cells

The derivation of the Nernst equation assumed thermodynamic equilibrium, which means that the cell must be reversible. The emf given by the Nernst equation (14.43) is the sum of the potential differences at the phase boundaries of a cell without a liquid junction. When the cell has a liquid junction, the observed cell emf includes the additional potential difference between the two electrolyte solutions. [For example, see Eq. (14.25).] We call this additional potential difference the **liquid-junction potential** $\mathscr{E}_J$:

$$\mathscr{E}_J \equiv \phi_{\text{soln},R} - \phi_{\text{soln},L} \qquad (14.50)$$

where $\phi_{\text{soln},R}$ is the potential of the electrolyte solution of the half-cell on the right of the cell diagram. For example, for the Daniell cell (14.26), $\mathscr{E}_J = \phi(\text{aq. CuSO}_4) - \phi(\text{aq. ZnSO}_4)$. The Nernst equation gives the sums of the potential differences at all interfaces except at the liquid junction. Hence, the emf $\mathscr{E}$ of a cell with a junction equals $\mathscr{E}_J + \mathscr{E}_{\text{Nernst}}$, where $\mathscr{E}_{\text{Nernst}}$ is given by (14.43). Therefore,

$$\mathscr{E} = \mathscr{E}_J + \mathscr{E}^\circ - \frac{RT}{nF} \ln \left[\prod_i' (a_i)^{\nu_i} \right] \qquad \text{cell with liq. junct.} \qquad (14.51)$$

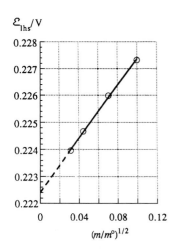

$\mathscr{E}_{\text{lhs}}/V$

Figure 14.11

Extrapolation to get $\mathscr{E}^\circ$ of the cell (14.32) at 25°C. $\mathscr{E}_{\text{lhs}}$ is the value of the left side of Eq. (14.49).

[For a proof of (14.51) for the Daniell cell, see Prob. 14.21.]

Junction potentials are small, but they cannot be neglected in accurate work. By connecting the two electrolyte solutions with a salt bridge, the junction potential can be minimized (but not completely eliminated). A **salt bridge** consists of a gel (Sec. 13.6) made by adding agar to a concentrated aqueous KCl solution. The gel permits diffusion of ions but eliminates convection currents. A cell with a salt bridge has two liquid junctions, the sum of whose potentials turns out to be quite small (see Sec. 14.9). A salt bridge is symbolized by two vertical lines (solid, dotted, or dashed, according to the whim of the writer). Thus, the diagram

$$Au_L|Zn|ZnSO_4(aq)\!\vdots\!CuSO_4(aq)|Cu|Au_R \qquad (14.52)$$

symbolizes a Daniell cell with gold terminals and with a salt bridge separating the two solutions (Fig. 14.12).

The Nernst equation (14.43) and its modification (14.51) give the open-circuit potential difference (emf) of the cell terminals and also give the potential difference of the terminals when the cell is balanced in a potentiometer. However, these equations do not give the potential difference between the terminals in the highly irreversible situation where the cell is sending a noninfinitesimal current through a load. The potential difference between the terminals when current flows belongs to the subject of electrode kinetics. Unfortunately, the Nernst equation has often been used in irreversible situations, where it does not apply.

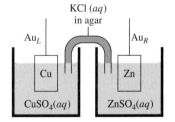

Figure 14.12

The Daniell cell (14.52). A salt bridge joins the solutions.

Summary

A galvanic cell's *electrochemical* reaction includes the transfer of electrons between the cell terminals, as in (14.33); its *chemical* reaction omits electrons, as in (14.34). Application of the equilibrium condition $\Sigma_i \nu_i \tilde{\mu}_i = 0$ to the electrochemical reaction of an open-circuit cell at electrochemical equilibrium gives the emf of a reversible galvanic cell as $\mathscr{E} = \mathscr{E}° - (RT/nF) \ln Q$. In this equation (the Nernst equation), $\mathscr{E}° \equiv -\Delta G°/nF$, n is the number of electrons transferred in the cell's electrochemical reaction, and $Q \equiv \Pi_i' (a_i)^{\nu_i}$, where a_i and ν_i are the activity and stoichiometric coefficient of species i, and the product goes over all species in the cell's chemical reaction. $\Delta G°$ is the change in standard molar Gibbs energy for the cell's chemical reaction. The activity product for the ions of a strong electrolyte (which occurs in the Nernst equation for certain cells) is given by (14.46). The standard potential $\mathscr{E}°$ can be found by extrapolation to infinite dilution of an expression involving measured $\mathscr{E}$ values or can be calculated if $\Delta G°$ is known.

14.7 STANDARD ELECTRODE POTENTIALS

In this section, the term "electrode" is used synonymously with "half-cell."

If we have 100 different electrodes, they can be combined to give $100(99)/2 = 4950$ different galvanic cells. However, to determine $\mathscr{E}°$ for these cells requires far fewer than 4950 measurements. All we have to do is pick one reversible electrode as a reference and measure $\mathscr{E}°$ for the 99 cells composed of the reference electrode and each of the remaining electrodes. We shall see that these 99 $\mathscr{E}°$ values enable calculation of all 4950 $\mathscr{E}°$ values.

The reference electrode chosen for work in aqueous solutions is the hydrogen electrode $Pt|H_2(g)|H^+(aq)$. The *standard potential of an electrode reaction* (abbreviated to **standard electrode potential**) at temperature T and pressure P is defined to be the standard potential $\mathscr{E}°$ for the cell at T and P that has the hydrogen electrode on the left of its diagram and the electrode in question on the right. For example, the standard electrode potential for the $Cu^{2+}|Cu$ electrode is $\mathscr{E}°$ for the cell

$$Cu'|Pt|H_2(g)|H^+(aq)\!\vdots\!Cu^{2+}(aq)|Cu \qquad (14.53)$$

which from (14.42) equals $-\Delta G^\circ/2F$ for the chemical reaction $H_2(g) + Cu^{2+}(aq) \rightarrow 2H^+(aq) + Cu$. Experiment gives $\mathscr{E}^\circ = 0.34$ V for this cell at 25°C and 1 bar. Recall that the standard states of species in solution involve a variable pressure. Unless otherwise specified, a pressure of 1 bar will be understood.

The standard electrode potential of electrode i is defined using a cell with i on the right side of the cell diagram, and by the IUPAC cell-diagram convention (Sec. 14.4), reduction occurs at the right electrode. Therefore the standard electrode potential for electrode i corresponds to a chemical reaction in which reduction occurs at electrode i. *All standard electrode potentials are reduction potentials.*

Suppose we have measured the standard electrode potentials of all electrodes of interest. We now ask for $\mathscr{E}^\circ$ for a cell composed of any two electrodes. For example, we might ask for $\mathscr{E}^\circ$ of the cell

$$Cu|Cu^{2+} \vdots Ga^{3+}|Ga|Cu' \tag{14.54}$$

To simplify the derivation, we shall write all cell reactions and half-reactions so as to make the cell-reaction charge number n equal to 1. This is all right, since the potential difference between the two terminals is independent of the choice of stoichiometric coefficients in the cell reaction. For the cell (14.54), the half-reactions are $\frac{1}{2}Cu \rightleftharpoons \frac{1}{2}Cu^{2+} + e^-$ and $\frac{1}{3}Ga^{3+} + e^- \rightleftharpoons \frac{1}{3}Ga$, and the cell reaction is

$$\tfrac{1}{2}Cu + \tfrac{1}{3}Ga^{3+} \rightleftharpoons \tfrac{1}{2}Cu^{2+} + \tfrac{1}{3}Ga \tag{14.55}$$

Let $\mathscr{E}^\circ_R$ and $\mathscr{E}^\circ_L$ denote the standard electrode potentials of the electrodes on the right and left of (14.54); that is, $\mathscr{E}^\circ_R$ is $\mathscr{E}^\circ$ for the cell

$$Ga|Pt|H_2(g)|H^+|Ga^{3+}|Ga' \tag{14.56}$$

and $\mathscr{E}^\circ_L$ is $\mathscr{E}^\circ$ for (14.53). The chemical reaction of the cell (14.54) is the difference between the reactions of the cells (14.56) and (14.53):

Cell (14.56):	$\tfrac{1}{3}Ga^{3+} + \tfrac{1}{2}H_2 \rightleftharpoons \tfrac{1}{3}Ga + H^+$
$-$Cell (14.53):	$-(\tfrac{1}{2}Cu^{2+} + \tfrac{1}{2}H_2 \rightleftharpoons \tfrac{1}{2}Cu + H^+)$
Cell (14.54):	$\tfrac{1}{2}Cu + \tfrac{1}{3}Ga^{3+} \rightleftharpoons \tfrac{1}{2}Cu^{2+} + \tfrac{1}{3}Ga$

Therefore, ΔG° for the reaction of cell (14.54) is the difference between the ΔG°'s of the cells (14.56) and (14.53): $\Delta G^\circ_{(14.54)} = \Delta G^\circ_{(14.56)} - \Delta G^\circ_{(14.53)}$. Use of $\Delta G^\circ = -nF\mathscr{E}^\circ$ [Eq. (14.42)] with $n = 1$ gives $-F\mathscr{E}^\circ_{(14.54)} = -F\mathscr{E}^\circ_{(14.56)} + F\mathscr{E}^\circ_{(14.53)}$. Division by $-F$ gives $\mathscr{E}^\circ_{(14.54)} = \mathscr{E}^\circ_{(14.56)} - \mathscr{E}^\circ_{(14.53)}$, or

$$\mathscr{E}^\circ = \mathscr{E}^\circ_R - \mathscr{E}^\circ_L \tag{14.57}*$$

where $\mathscr{E}^\circ_R$ and $\mathscr{E}^\circ_L$ are the standard electrode potentials of the right and left half-cells of a cell whose standard emf is $\mathscr{E}^\circ$. *Both $\mathscr{E}^\circ_R$ and $\mathscr{E}^\circ_L$ are reduction potentials.* Since the cell reaction involves oxidation at the left-hand electrode, $\mathscr{E}^\circ_L$ appears with a minus sign in the expression (14.57) for $\mathscr{E}^\circ$ of the cell's reaction.

Although a particular cell was used to derive (14.57), the same reasoning shows it to be valid for any cell. Equation (14.57) applies at any fixed temperature and allows $\mathscr{E}^\circ$ for any cell to be found from a tabulation of standard electrode potentials at the temperature of interest. Table 14.1 lists some standard electrode potentials in aqueous solutions at 25°C and 1 bar.

The standard electrode potential for the hydrogen electrode is zero, since $\mathscr{E}^\circ$ is zero for the cell $Pt|H_2|H^+|H_2|Pt$. The reaction for this cell is $H_2 + 2H^+ \rightleftharpoons 2H^+ + H_2$; this reaction clearly has $\Delta G^\circ = 0 = -nF\mathscr{E}^\circ$, and $\mathscr{E}^\circ = 0$.

Because standard electrode potentials are relative to the hydrogen electrode, which involves $H_2(g)$, the 1982 change in standard pressure from 1 atm to 1 bar affects most standard electrode potentials. For an electrode that does not involve gases, one finds $\mathscr{E}^{\circ,bar}_{298} = \mathscr{E}^{\circ,atm}_{298} - 0.00017$ V.

EXAMPLE 14.4 $\mathscr{E}°$ of a cell

For the cell $Cu|Cu^{2+}(aq)\vdots Ag^{+}(aq)|Ag|Cu'$, write the cell reaction and use Table 14.1 to find $\mathscr{E}°$ at 25°C and 1 bar. When the activities are close to 1: which terminal is at the higher potential? what is the spontaneous cell reaction? which electrode is the anode for the spontaneous reaction? into which electrode do electrons flow? Is it possible to change conditions so that the spontaneous reaction is the reverse of the reaction when the activities are close to 1?

According to the IUPAC conventions (Sec. 14.4), we write the cell reaction with oxidation occurring at the left electrode of the cell diagram. Hence the half-reactions are $Cu(s) \rightarrow Cu^{2+}(aq) + 2e^-$ and $Ag^+(aq) + e^- \rightarrow Ag(s)$. Multiplying the second half-reaction by 2 to balance the electrons, and adding it to the first, we get the cell reaction as

$$Cu(s) + 2Ag^+(aq) \rightarrow Cu^{2+}(aq) + 2Ag(s)$$

From (14.57) and Table 14.1, $\mathscr{E}° = \mathscr{E}°_R - \mathscr{E}°_L = 0.799 \text{ V} - 0.339 \text{ V} = 0.460 \text{ V}$. *Note that even though one half-reaction was multiplied by 2, we do not multiply its reduction potential by 2.* When activities are close to 1, the Nernst equation tells us that the emf $\mathscr{E}$ is close to $\mathscr{E}°$, which is positive. Therefore, $\mathscr{E} \equiv \phi_R - \phi_L > 0$ and the right terminal Cu' is at the higher potential. Since $\mathscr{E}$ is positive, the preceding cell-diagram reaction is the same as the spontaneous reaction, as noted in the paragraph after Eq. (14.28). The anode is where oxidation occurs and is the copper electrode, Ag being the cathode. Electrons flow into the Ag electrode where they reduce Ag^+ ions. To make the spontaneous reaction be the reverse of the preceding cell reaction, we need to make the cell emf negative. The Nernst equation shows that $\mathscr{E}$ contains the term $-(RT/2F) \ln \{a(Cu^{2+})/[a(Ag^+)]^2\}$. If we make $[a(Ag^+)]^2$ much, much, much less than $a(Cu^{2+})$, this term will become sufficiently negative to make $\mathscr{E}$ negative.

Exercise

Use Table 14.1 to find $\mathscr{E}°_{298}$ for $3Cu(s) + 2Fe^{3+}(aq) \rightarrow 3Cu^{2+}(aq) + 2Fe(s)$. (*Answer:* −0.38 V.)

TABLE 14.1

Standard Electrode Potentials in H_2O at 25°C and 1 bar

Half-cell reaction	$\mathscr{E}°/\text{V}$	Half-cell reaction	$\mathscr{E}°/\text{V}$
$K^+ + e^- \rightarrow K$	−2.936	$2D^+ + 2e^- \rightarrow D_2$	−0.01
$Ca^{2+} + 2e^- \rightarrow Ca$	−2.868	$2H^+ + 2e^- \rightarrow H_2$	0
$Na^+ + e^- \rightarrow Na$	−2.714	$AgBr(c) + e^- \rightarrow Ag + Br^-$	0.073
$Mg^{2+} + 2e^- \rightarrow Mg$	−2.360	$AgCl(c) + e^- \rightarrow Ag + Cl^-$	0.2222
$Al^{3+} + 3e^- \rightarrow Al$	−1.677	$Hg_2Cl_2(c) + 2e^- \rightarrow 2Hg(l) + 2Cl^-$	0.2680
$2H_2O + 2e^- \rightarrow H_2(g) + 2OH^-$	−0.828	$Cu^{2+} + 2e^- \rightarrow Cu$	0.339
$Zn^{2+} + 2e^- \rightarrow Zn$	−0.762	$Cu^+ + e^- \rightarrow Cu$	0.518
$Ga^{3+} + 3e^- \rightarrow Ga$	−0.549	$I_2(c) + 2e^- \rightarrow 2I^-$	0.535
$Fe^{2+} + 2e^- \rightarrow Fe$	−0.44	$Hg_2SO_4(c) + 2e^- \rightarrow 2Hg(l) + SO_4^{2-}$	0.615
$Cd^{2+} + 2e^- \rightarrow Cd$	−0.402	$Fe^{3+} + e^- \rightarrow Fe^{2+}$	0.771
$PbI_2(c) + 2e^- \rightarrow Pb + 2I^-$	−0.365	$Ag^+ + e^- \rightarrow Ag$	0.7992
$PbSO_4(c) + 2e^- \rightarrow Pb + SO_4^{2-}$	−0.356	$Br_2(l) + 2e^- \rightarrow 2Br^-$	1.078
$Sn^{2+} + 2e^- \rightarrow Sn(white)$	−0.141	$O_2(g) + 4H^+ + 4e^- \rightarrow 2H_2O$	1.229
$Pb^{2+} + 2e^- \rightarrow Pb$	−0.126	$Cl_2(g) + 2e^- \rightarrow 2Cl^-$	1.360
$Fe^{3+} + 3e^- \rightarrow Fe$	−0.04	$Au^+ + e^- \rightarrow Au$	1.69

The process of combining two half-cell $\mathscr{E}°$ values to find a third half-cell $\mathscr{E}°$ is considered in Prob. 14.31.

Standard electrode potentials are sometimes called single electrode potentials, but this name is highly misleading. Every number in Table 14.1 is an $\mathscr{E}°$ value for a complete cell. For example, the value of 0.339 V listed for the half-reaction $Cu^{2+} + 2e^- \rightarrow Cu$ is $\mathscr{E}°$ for the cell (14.53). Even if the potential difference across the interface (H_2 adsorbed on Pt)–$H^+(aq)$ with H_2 and H^+ at unit activity happened to be zero at 25°C, the listed value 0.339 V would not give the potential difference across the $Cu^{2+}(aq)$–Cu interface, since the cell (14.53) also contains a potential difference across the Cu'–Pt interface.

EXAMPLE 14.5 Cell emf

Vapor-pressure measurements give the mean ionic activity coefficient of $CdCl_2$ in a 0.100 mol/kg aqueous $CdCl_2$ solution at 25°C and 1 bar as $\gamma_\pm = 0.228$. Find $\mathscr{E}°$ and $\mathscr{E}$ at 25°C and 1 bar for the cell

$$Cu_L | Cd(s) | CdCl_2(aq, 0.100 \text{ mol/kg}) | AgCl(s) | Ag(s) | Cu_R$$

By convention, the left-hand electrode involves oxidation, so the half-reactions and overall chemical reaction are

$$Cd \rightleftharpoons Cd^{2+} + 2e^-$$
$$\underline{(AgCl + e^- \rightleftharpoons Ag + Cl^-) \times 2}$$
$$Cd(s) + 2AgCl(s) \rightleftharpoons 2Ag(s) + Cd^{2+}(aq) + 2Cl^-(aq)$$

Equation (14.57) and Table 14.1 give at 25°C

$$\mathscr{E}° = \mathscr{E}°_R - \mathscr{E}°_L = 0.2222 \text{ V} - (-0.402 \text{ V}) = 0.624 \text{ V}$$

The Nernst equation (14.43) gives

$$\mathscr{E} = \mathscr{E}° - \frac{RT}{2F} \ln \frac{[a(Ag)]^2 a(Cd^{2+})[a(Cl^-)]^2}{a(Cd)[a(AgCl)]^2}$$

$$\mathscr{E} = \mathscr{E}° - (RT/2F) \ln \{a(Cd^{2+})[a(Cl^-)]^2\} \qquad (14.58)$$

since the activities of the pure solids are 1 at 1 bar. The ion activity product in (14.58) is evaluated by use of Eq. (14.46) with $\nu_+ = 1$ and $\nu_- = 2$:

$$a(Cd^{2+})[a(Cl^-)]^2 = 1^1 \cdot 2^2 \cdot [(0.228)(0.100)]^3 = 4.74 \times 10^{-5}$$

Substitution in (14.58) gives

$$\mathscr{E} = 0.624 \text{ V} - \frac{(8.314 \text{ J mol}^{-1} \text{ K}^{-1})(298.15 \text{ K})}{2(96485 \text{ C mol}^{-1})} \ln (4.74 \times 10^{-5})$$

$$\mathscr{E} = 0.624 \text{ V} - (-0.128 \text{ V}) = 0.752 \text{ V}$$

since 1 J/C = 1 V [Eq. (14.8)]. Note that, since the volt is an SI unit, R must be expressed using joules, the SI energy unit. (If $\gamma_\pm$ were taken as 1, the result would be an emf of 0.695 V, so $\gamma_\pm$ affects $\mathscr{E}$ substantially.)

Exercise

For $Cu_L | Zn(s) | ZnBr_2(aq, 0.20 \text{ mol/kg}) | AgBr(s) | Ag(s) | Cu_R$, find $\mathscr{E}_{298}$ at 1 bar, given that $\gamma_\pm = 0.462$ in the $ZnBr_2$ solution. (*Answer:* 0.909 V.)

For the cell $Cu_L|Ag|AgCl(s)|CdCl_2(0.100 \text{ mol/kg})|Cd|Cu_R$, which interchanges the electrodes compared with the diagram in the example, $\mathscr{E}°$ would be -0.624 V and $\mathscr{E}$ would be -0.752 V.

Suppose we want to calculate the emf of the Daniell cell (14.52) with the assumption that the salt bridge makes the liquid-junction potentials negligible. The Nernst equation would contain the log of

$$a(Zn^{2+})/a(Cu^{2+}) = \gamma(Zn^{2+})m(Zn^{2+})/\gamma(Cu^{2+})m(Cu^{2+})$$

If both solutions were dilute, we could use the Davies equation to calculate the ionic activity coefficients. Also, we would have to know the equilibrium constants for ion-pair formation in $CuSO_4$ and $ZnSO_4$ solutions, so as to calculate the ionic molalities from the stoichiometric molalities of the salts. If the solutions are not dilute, we can't find the single-ion activity coefficients and hence can't calculate $\mathscr{E}$.

The Nernst equation contains the term $-(RT/nF)2.3026 \log_{10} Q$. At 25°C, one finds $2.3026RT/F = 0.05916$ V.

CONCENTRATION CELLS

To form a galvanic cell, we bring two half-cells together. If the electrochemical reactions in the half-cells differ, the overall cell reaction is a chemical reaction and the cell is a *chemical cell.* Examples are the cells (14.52) and (14.32). If the electrochemical reactions in the two half-cells are the same but one species B is at a different concentration in each half-cell, the cell will have a nonzero emf and its overall reaction will be a physical reaction that amounts to the transfer of B from one concentration to the other. This is a **concentration cell.** An example is a cell composed of two chlorine electrodes with different pressures of Cl_2:

$$Pt_L|Cl_2(P_L)|HCl(aq)|Cl_2(P_R)|Pt_R \qquad (14.59)$$

where P_L and P_R are the Cl_2 pressures at the left and right electrodes. Adding the two half-reactions $2Cl^- \rightarrow Cl_2(P_L) + 2e^-$ and $Cl_2(P_R) + 2e^- \rightarrow 2Cl^-$, we obtain as the overall cell reaction $Cl_2(P_R) \rightarrow Cl_2(P_L)$. Equation (14.57) gives $\mathscr{E}° = \mathscr{E}°_R - \mathscr{E}°_L = 1.36$ V $- 1.36$ V $= 0$. For any concentration cell, $\mathscr{E}°$ is zero, since $\mathscr{E}°_R$ equals $\mathscr{E}°_L$. The Nernst equation (14.43) with fugacities approximated by pressures gives for (14.59)

$$\mathscr{E} = -(RT/2F) \ln (P_L/P_R) \qquad (14.60)$$

Another example of a concentration cell is

$$Cu_L|CuSO_4(m_L)|CuSO_4(m_R)|Cu_R \qquad (14.61)$$

LIQUID-JUNCTION POTENTIALS

To see how a liquid-junction potential arises, consider the Daniell cell (Fig. 14.7) with its emf balanced in a potentiometer, so no current flows. For simplicity, let the $CuSO_4$ and $ZnSO_4$ molalities be the same, giving equal SO_4^{2-} concentrations in the two solutions. At the junction between the solutions, ions from each solution diffuse into the other solution. It happens that Cu^{2+} ions in water are slightly more mobile than Zn^{2+} ions, so the Cu^{2+} ions diffuse into the $ZnSO_4$ solution faster than the Zn^{2+} ions diffuse into the $CuSO_4$ solution. This produces a small excess of positive charge on the $ZnSO_4$ side of the boundary and a small excess of negative charge on the $CuSO_4$ side. The negative charge on the $CuSO_4$ side speeds up the diffusion of the Zn^{2+} ions. The negative charge builds up until a steady state is reached with the Zn^{2+} and Cu^{2+} ions migrating at equal rates across the boundary. The steady-state charges on each side of the boundary produce a potential difference $\phi(\text{aq. ZnSO}_4) - \phi(\text{aq. CuSO}_4) \equiv \mathscr{E}_J$, which contributes to the measured cell emf.

In some cases, one can estimate liquid-junction potentials from emf measurements. An example is the cell

$$\text{Ag}|\text{AgCl}(s)|\text{LiCl}(m)\vdots\text{NaCl}(m)|\text{AgCl}(s)|\text{Ag} \qquad (14.62)$$

where $m(\text{LiCl}) = m(\text{NaCl})$. The half-reactions $\text{Ag} + \text{Cl}^-(\text{in aq. LiCl}) \to \text{AgCl} + e^-$ and $\text{AgCl} + e^- \to \text{Ag} + \text{Cl}^-(\text{in aq. NaCl})$ give the overall cell reaction as $\text{Cl}^-(\text{in aq. LiCl}) \to \text{Cl}^-(\text{in aq. NaCl})$. For this cell, $\mathscr{E}°$ is zero, and Eq. (14.51) gives

$$\mathscr{E} = \mathscr{E}_J - \frac{RT}{F} \ln \frac{\gamma(\text{Cl}^- \text{ in aq. NaCl})}{\gamma(\text{Cl}^- \text{ in aq. LiCl})}$$

At low molalities, $\gamma(\text{Cl}^-)$ will be very nearly the same in NaCl and LiCl solutions of equal molality (see the Debye–Hückel equation). Therefore, to a good approximation, $\mathscr{E} = \mathscr{E}_J$, and the measured emf is that due to the liquid junction.

Some observed approximate liquid-junction potentials at 25°C for cells like (14.62) with various electrolyte pairs at $m = 0.01$ mol/kg are -2.6 mV for LiCl–NaCl, -7.8 mV for LiCl–CsCl, 27.0 mV for HCl–NH_4Cl, and 33.8 mV for HCl–LiCl. The larger values for junctions involving H^+ are due to the very high mobility of $\text{H}^+(aq)$ relative to that of other cations; see Sec. 16.6. We see that liquid-junction potentials are of the order of magnitude 10 or 20 mV. This is small but far from negligible, since cell emfs are routinely measured to 0.1 mV = 0.0001 V or better.

To see how effective a salt bridge is in reducing $\mathscr{E}_J$, consider the cell

$$\text{Hg}|\text{Hg}_2\text{Cl}_2(s)|\text{HCl}(0.1 \text{ mol/kg})\vdots\text{KCl}(m)\vdots\text{KCl}(0.1 \text{ mol/kg})|\text{Hg}_2\text{Cl}_2(s)|\text{Hg}$$

where the KCl(m) solution is a salt bridge with molality m. When $m = 0.1$ mol/kg, the cell resembles the cell (14.62) and its emf (which is observed to be 27 mV) is a good approximation to $\mathscr{E}_J$ between 0.1 mol/kg HCl and 0.1 mol/kg KCl. Salt bridges use concentrated KCl solutions. When the KCl molality m is increased to 3.5 mol/kg, the cell emf drops to 1 mV, which is a good approximation to the sum of the junction potentials at the interfaces HCl(0.1 mol/kg)–KCl(3.5 mol/kg) and KCl(3.5 mol/kg)–KCl(0.1 mol/kg). We can expect that a cell with a concentrated KCl salt bridge will typically have a net junction potential of 1 or 2 mV.

The liquid-junction potential between a concentrated aqueous KCl solution and any dilute aqueous solution is quite small, for the following reasons. Because the KCl solution is concentrated, the junction potential is determined mainly by the ions of this solution. The mobilities of the isoelectronic ions $_{19}\text{K}^+$ and $_{17}\text{Cl}^-$ in water are nearly equal, so these ions diffuse out of the salt bridge into the dilute solution at nearly equal rates and the junction potential is therefore small.

Most cells with salt bridges contain two junctions between concentrated KCl and dilute solutions, and here $\mathscr{E}_J$ is further reduced by a near cancellation of oppositely directed junction potentials.

14.10 APPLICATIONS OF EMF MEASUREMENTS

Determination of $\Delta G°$ and $K°$

Once $\mathscr{E}°$ for a cell has been found [either by extrapolation of emf data as in Fig. 14.11 or by combining $\mathscr{E}°$ values for half-cell reactions (Table 14.1)], $\Delta G°$ and the equilibrium constant $K°$ of the cell's chemical reaction can be found from $\Delta G° = -nF\mathscr{E}°$ [Eq. (14.42)] followed by $\Delta G° = -RT \ln K°$ [Eq. (11.4)].

EXAMPLE 14.6 Calculation of ΔG° and K° from $\mathscr{E}^\circ$

Use standard electrode potentials (Table 14.1) to find ΔG_{298}° and K_{298}° for $Cu^{2+}(aq) + Zn(s) \rightarrow Cu(s) + Zn^{2+}(aq)$.

The oxidation and reduction half-reactions are $Zn(s) \rightarrow Zn^{2+}(aq) + 2e^-$ and $Cu^{2+}(aq) + 2e^- \rightarrow Cu(s)$. By convention, a cell's reaction involves reduction at the half-cell on the right side of the cell diagram. Hence, the desired half-reactions correspond to a cell with a $Zn|Zn^{2+}$ electrode at the left side of the cell diagram and a $Cu^{2+}|Cu$ electrode on the right, as in the cell (14.52). The relation $\mathscr{E}^\circ = \mathscr{E}_R^\circ - \mathscr{E}_L^\circ$ [Eq. (14.57)] gives $\mathscr{E}^\circ$ for the desired redox reaction. From Table 14.1, $\mathscr{E}_L^\circ = -0.762$ V and $\mathscr{E}_R^\circ = 0.339$ V, so

$$\mathscr{E}_{298}^\circ = \mathscr{E}_R^\circ - \mathscr{E}_L^\circ = 0.339 \text{ V} - (-0.762 \text{ V}) = 1.101 \text{ V}$$

$$\Delta G_{298}^\circ = -nF\mathscr{E}^\circ = -2(96485 \text{ C/mol})(1.101 \text{ V}) = -212._5 \text{ kJ/mol}$$

since 1 V = 1 J/C [Eq. (14.8)]. Use of $\Delta G^\circ = -RT \ln K^\circ$ gives

$$\ln K^\circ = -\frac{\Delta G_{298}^\circ}{RT} = \frac{212500 \text{ J/mol}}{(8.314 \text{ J/mol-K})(298.15 \text{ K})} = 85.7_3, \qquad K^\circ = 2 \times 10^{37}$$

At equilibrium, virtually no Cu^{2+} remains in solution.

Exercise

Use Table 14.1 to find ΔG_{298}° and K_{298}° for $I_2(c) + 2Br^-(aq) \rightarrow 2I^-(aq) + Br_2(l)$ and for the reverse reaction. (*Answer:* 104.8 kJ/mol, 4×10^{-19}, -104.8 kJ/mol, 2×10^{18}.)

Substitution of $\Delta G^\circ = -nF\mathscr{E}^\circ$ into $\Delta G^\circ = -RT \ln K^\circ$ gives

$$\ln K^\circ = nF\mathscr{E}^\circ/RT \qquad (14.63)$$

Equation (14.63) gives $K^\circ = \exp(nF\mathscr{E}^\circ/RT)$. For $n = 1$, we find that each difference of 0.1 V between the half-reaction standard potentials contributes a factor of 49 to K° at 25°C. The more positive $\mathscr{E}^\circ$ is, the larger K° is. A large negative $\mathscr{E}^\circ$ indicates a very small K°. See Fig. 14.13.

The more negative the reduction potential $\mathscr{E}^\circ$ for the half-reaction $M^{z+} + z_+e^- \rightarrow M$, the greater the tendency for metal M to be oxidized. Thus, a metal will tend to replace in solution those metals that lie below it in Table 14.1. For example, Zn replaces Cu^{2+} from aqueous solutions ($Zn + Cu^{2+} \rightarrow Zn^{2+} + Cu$). Metals lying above the hydrogen electrode in Table 14.1 replace H^+ from acidic solutions and dissolve readily in aqueous acids, generating H_2. Metals near the top of the table, for example, Na, K, Ca, replace H^+ from water.

Although the anode reaction in a cell is an oxidation and the cathode reaction is a reduction, the overall cell reaction is not necessarily an oxidation–reduction reaction (as can be seen from the AgCl example, Example 14.7), so (14.63) is not limited to redox reactions. Equilibrium constants that have been determined from cell emf measurements include redox K° values, solubility products, dissociation constants of complex ions, the ionization constant of water, ionization constants of weak acids, and ion-pair-formation equilibrium constants; see Probs. 14.50 and 14.51.

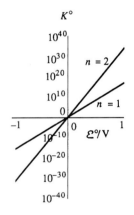

Figure 14.13

Plot of equilibrium constant K° versus standard emf $\mathscr{E}^\circ$ at 25°C. The vertical scale is logarithmic. A small change in $\mathscr{E}^\circ$ corresponds to a large change in the equilibrium constant.

EXAMPLE 14.7 Calculation of K_{sp} from $\mathscr{E}^\circ$

Devise a cell whose overall reaction is $AgCl(s) \rightarrow Ag^+(aq) + Cl^-(aq)$ and use its $\mathscr{E}_{298}^\circ$ value to find $K_{sp,298}^\circ$ for AgCl.

Such a cell is

$$Ag|Ag^+ \vdots Cl^-|AgCl(s)|Ag \qquad (14.64)$$

The half-reactions are $Ag \to Ag^+ + e^-$ and $AgCl(s) + e^- \to Ag + Cl^-$, and the overall reaction is $AgCl(s) \to Ag^+ + Cl^-$. At the anode, Ag is oxidized, and at the cathode Ag (in AgCl) is reduced, so the overall cell reaction is not a redox reaction. Table 14.1 and Eq. (14.63) give $\mathscr{E}° = 0.2222$ V $- 0.7992$ V $= -0.5770$ V, and $K_{sp}° = 1.76 \times 10^{-10}$ at 25°C and 1 bar. Note that there is no need to set up and measure $\mathscr{E}°$ for the cell (14.64), since its $\mathscr{E}°$ can be found by combining measured standard electrode potentials of the $Ag^+|Ag$ and Ag–AgCl electrodes.

Exercise

Use data in Table 14.1 to find $K_{sp}°$ for $PbSO_4(aq)$ at 25°C. (*Answer:* 1.7×10^{-8}.)

Determination of $\Delta S°$, $\Delta H°$, and $\Delta C_P°$

Using $(\partial \mu_i°/\partial T)_P = -\bar{S}_i°$ [Eq. (9.30)], we have $[\partial(\Delta G°)/\partial T]_P = (\partial/\partial T)_P \sum_i \nu_i \mu_i° = -\sum_i \nu_i \bar{S}_i° = -\Delta S°$. Substitution of $-nF\mathscr{E}°$ for $\Delta G°$ gives

$$\Delta S° = nF\left(\frac{\partial \mathscr{E}°}{\partial T}\right)_P \qquad (14.65)$$

Evaluation of the temperature derivative of $\mathscr{E}°$ enables the standard-state molar entropy change $\Delta S°$ of the cell's reaction to be found. Recall the role of galvanic-cell measurements in establishing the third law of thermodynamics.

$\Delta H°$ can then be found from $\Delta G° = \Delta H° - T \Delta S°$.

Since $\bar{C}_{P,i}° = T(\partial \bar{S}_i°/\partial T)_P$, we have $\Delta C_P° = T[\partial(\Delta S°)/\partial T]_P$ and (14.65) gives

$$\Delta C_P° = nFT(\partial^2 \mathscr{E}°/\partial T^2)_P \qquad (14.66)$$

The derivatives in (14.65) and (14.66) are found by measuring $\mathscr{E}°$ at several temperatures and fitting the observed values to the truncated Taylor series

$$\mathscr{E}° = a + b(T - T_0) + c(T - T_0)^2 + d(T - T_0)^3 \qquad (14.67)$$

where a, b, c, and d are constants and T_0 is some fixed temperature in the range of the measurements. Differentiation of (14.67) then allows $\Delta S°$, $\Delta H°$, and $\Delta C_P°$ to be calculated. Since each differentiation decreases the accuracy of the data, highly accurate $\mathscr{E}°$ values are required for an accurate $\Delta C_P°$ to be found. Figure 14.14 plots $\mathscr{E}°$ versus T for the cell (14.31).

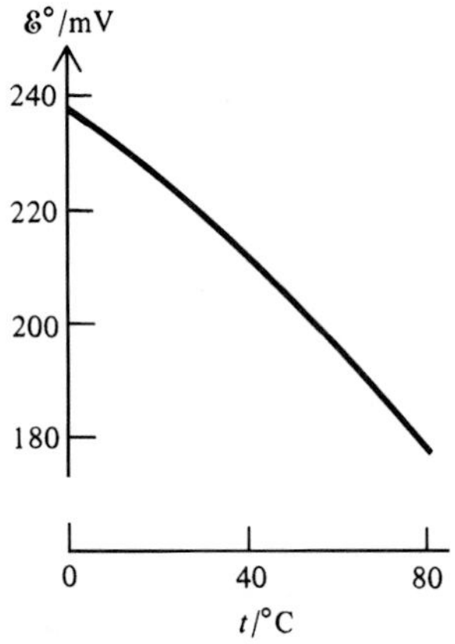

Figure 14.14

$\mathscr{E}°$ versus temperature at 1 bar for the cell (14.31), consisting of a hydrogen electrode and an Ag–AgCl electrode.

EXAMPLE 14.8 Calculation of $\Delta S°$ from $\mathscr{E}°(T)$

The chemical reaction $H_2(g) + 2AgCl(s) \to 2Ag(s) + 2HCl(aq)$ [Eq. (14.34)] occurs in the cell (14.32). Measured $\mathscr{E}°$ values (Fig. 14.14) for this cell in the temperature range 0°C to 90°C at 1 bar are well fitted by Eq. (14.67) with (see Prob. 14.43)

$$T_0 = 273.15 \text{ K}, \qquad a = 0.23643 \text{ V}, \qquad 10^4 b = -4.8621 \text{ V/K}$$

$$10^6 c = -3.4205 \text{ V/K}^2, \qquad 10^9 d = 5.869 \text{ V/K}^3 \qquad (14.68)$$

Find $\Delta S_{273}°$ for this reaction.

Substitution of (14.67) into (14.65) gives

$$\Delta S° = nF[b + 2c(T - T_0) + 3d(T - T_0)^2] \qquad (14.69)$$

Substitution of numerical values gives at 0°C

$$\Delta S_{273}^\circ = nFb = 2(96485 \text{ C/mol})(-4.8621 \times 10^{-4} \text{ V/K})$$

$$= -93.82 \text{ J/mol-K}$$

Exercise

Find ΔS° and ΔC_P° for the reaction (14.34) at 15°C. (*Answer:* -112.9 J/mol-K and -351 J/mol-K.)

Determination of Activity Coefficients

Since the emf of a cell depends on the activities of the ions in solution, it is easy to use measured emf values to calculate activity coefficients. For example, for the cell (14.32) with HCl as its electrolyte, the cell reaction is (14.34) and the emf is given by Eq. (14.48). Once $\mathscr{E}^\circ$ has been found by extrapolation of (14.49) to $m = 0$, the activity coefficient $\gamma_\pm$ of HCl(aq) at any molality m can be calculated from the measured emf $\mathscr{E}$ at that molality by using (14.48). Some results for $\gamma_\pm$ of HCl(aq) at 25°C and 1 bar are 0.905 at 0.01 mol/kg, 0.796 at 0.1 mol/kg, and 0.809 at 1 mol/kg (Fig. 10.8).

Determination of pH

The symbol pb means $-\log_{10} b$, where b is some physical quantity: p$b \equiv -\log_{10} b$. For example,

$$pc(\text{H}^+) \equiv -\log_{10}\left[c(\text{H}^+)/c^\circ\right], \qquad pm(\text{H}^+) \equiv -\log_{10}\left[m(\text{H}^+)/m^\circ\right]$$

$$pa_c(\text{H}^+) \equiv -\log_{10} a_c(\text{H}^+), \qquad pa_m(\text{H}^+) \equiv -\log_{10} a_m(\text{H}^+) \tag{14.70}$$

where $c^\circ \equiv 1 \text{ mol/dm}^3$ and $m^\circ \equiv 1 \text{ mol/kg}$ have been inserted to make the arguments of the logarithms dimensionless (as they must be). In (14.70), $a_c(\text{H}^+)$ and $a_m(\text{H}^+)$ are the concentration- and molality-scale activities of H^+ [Eq. (10.30)].

Over the years, each of the quantities in (14.70) has been called "the pH" of a solution. The present definition of pH is none of these. Instead pH is defined operationally to yield a quantity that is easily and reproducibly measured and as closely equal to $-\log_{10} a_m(\text{H}^+)$ as present theory allows.

To understand the current definition of pH, consider the cell

$$\text{Pt}|\text{H}_2(g)|\text{soln. X}\vdots\text{KCl(sat.)}|\text{Hg}_2\text{Cl}_2(s)|\text{Hg}|\text{Pt}' \tag{14.71}$$

which consists of a saturated calomel electrode and a hydrogen electrode dipping into an aqueous solution X whose molality-scale activity of H^+ is $a_\text{X}(\text{H}^+)$. The cell reaction and emf $\mathscr{E}_\text{X}$ [Eq. (14.51)] are

$$\tfrac{1}{2}\text{H}_2(g) + \tfrac{1}{2}\text{Hg}_2\text{Cl}_2(s) \rightleftharpoons \text{Hg}(l) + \text{H}^+(aq, \text{X}) + \text{Cl}^-(aq)$$

$$\mathscr{E}_\text{X} = \mathscr{E}_{J,\text{X}} + \mathscr{E}^\circ - RTF^{-1}\left[\ln a_\text{X}(\text{H}^+) + \ln a(\text{Cl}^-) - \tfrac{1}{2}\ln f(\text{H}_2)/P^\circ\right]$$

where $\mathscr{E}_{J,\text{X}}$ is the junction potential between solution X and the saturated KCl solution. Provided the ionic strength of solution X is reasonably low, $\mathscr{E}_{J,\text{X}}$ should be small, because of the concentrated KCl solution (see Sec. 14.9).

If a second cell is set up identical to (14.71) except that solution X is replaced by solution S, then the emf $\mathscr{E}_\text{S}$ of this cell will be

$$\mathscr{E}_\text{S} = \mathscr{E}_{J,\text{S}} + \mathscr{E}^\circ - RTF^{-1}\left[\ln a_\text{S}(\text{H}^+) + \ln a(\text{Cl}^-) - \tfrac{1}{2}\ln f(\text{H}_2)/P^\circ\right]$$

where $a_S(H^+)$ is the activity of H^+ in solution S. Subtraction gives

$$\mathscr{E}_X - \mathscr{E}_S = \mathscr{E}_{J,X} - \mathscr{E}_{J,S} - RTF^{-1}[\ln a_X(H^+) - \ln a_S(H^+)]$$

$$RTF^{-1}(\ln 10)[-\log_{10} a_X(H^+) + \log_{10} a_S(H^+)] = \mathscr{E}_X - \mathscr{E}_S + \mathscr{E}_{J,S} - \mathscr{E}_{J,X}$$

$$pa_X(H^+) = pa_S(H^+) + \frac{\mathscr{E}_X - \mathscr{E}_S}{RTF^{-1}\ln 10} + \frac{\mathscr{E}_{J,S} - \mathscr{E}_{J,X}}{RTF^{-1}\ln 10} \qquad (14.72)$$

where we used $pa_X(H^+) \equiv -\log_{10} a_X(H^+)$ and $\ln x = (\ln 10)(\log_{10} x)$ [Eq. (1.69)].

If solutions X and S are reasonably similar, the junction potentials $\mathscr{E}_{J,X}$ and $\mathscr{E}_{J,S}$ will be approximately equal and the last term in (14.72) can be neglected. By analogy to (14.72) with the last term omitted, the pH of solution X is defined as

$$pH(X) \equiv pH(S) + \frac{\mathscr{E}_X - \mathscr{E}_S}{RTF^{-1}\ln 10} \qquad (14.73)$$

In this equation, pH(S) is the assigned pH value for a standard solution. The pH(S) values are numbers chosen to be as closely equal to $-\log_{10} a_S(H^+)$ as present knowledge allows. We have $a_S(H^+) = \gamma_S(H^+)m(H^+)/m°$. For a solution with low ionic strength, a reasonably accurate value of $\gamma_S(H^+)$ can be calculated from an extended form of the Debye–Hückel equation, allowing $-\log_{10} a_S(H^+)$ to be accurately estimated for the solution S of known composition. Assigned pH(S) values are listed in *Pure Appl. Chem.,* **57,** 531 (1985). As an example, the assigned pH(S) of 0.0500 mol/kg aqueous potassium hydrogen phthalate at 25°C and 1 bar is 4.005.

The defined pH in Eq. (14.73) differs from $pa_{m,X}(H^+)$ because $\mathscr{E}_{J,S} - \mathscr{E}_{J,X}$ is not precisely zero and because the assigned standard pH(S) values are not precisely equal to $pa_S(H^+)$. The current definition of pH gives a quantity that is easily measured rather than a quantity that has a precise thermodynamic significance. For solutions of low ionic strength and pH in the range 2 to 12, the operationally defined pH in (14.73) is probably within 0.02 of $-\log_{10} a_m(H^+)$. In aqueous solution, $a_m(H^+)$ differs only slightly from $a_c(H^+)$; see Prob. 10.15.

In commercially available pH meters, the hydrogen electrode in (14.71) is replaced by a glass electrode (Sec. 14.12).

Potentiometric Titrations

In an acid–base titration, the pH changes rapidly as the neutralization point is reached. The slope of a plot of pH versus volume of added reagent is a maximum at the endpoint (Fig. 14.15). Monitoring the pH with a pH meter enables the endpoint to be determined. The solution being titrated is solution X in cell (14.71). Redox titrations can be done potentiometrically by making the solution being titrated part of a galvanic cell whose emf is monitored.

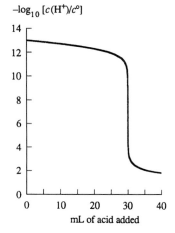

$-\log_{10}[c(H^+)/c°]$

Figure 14.15

Titration of a strong base with a strong acid. The pH changes rapidly at the endpoint.

14.11 BATTERIES

The term **battery** means either a single galvanic cell or several galvanic cells connected in series, in which case the emfs are additive.

The lead storage battery used in cars consists of three or six galvanic cells in series and has an emf of 6 or 12 V. Each cell is

$$Pb|PbSO_4(s)|H_2SO_4(aq)|PbSO_4(s)|PbO_2(s)|Pb' \qquad (14.74)$$

The reactions are

$$Pb + HSO_4^- \rightarrow PbSO_4(s) + H^+ + 2e^-$$

$$\underline{PbO_2(s) + 3H^+ + HSO_4^- + 2e^- \rightarrow PbSO_4(s) + 2H_2O}$$

$$Pb + PbO_2(s) + 2H^+ + 2HSO_4^- \rightarrow 2PbSO_4(s) + 2H_2O$$

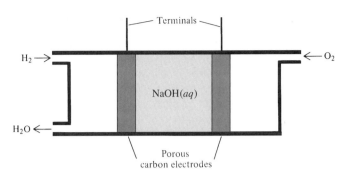

Figure 14.16

A hydrogen–oxygen fuel cell.

The cell is reversible and is readily recharged.

The United States space program and the desire for electrically powered automobiles have spurred the development of many new batteries.

A *fuel cell* is a galvanic cell in which the reactants are continuously fed to each electrode from outside the cell. Figure 14.16 shows a hydrogen–oxygen fuel cell whose diagram is

$$C|H_2(g)|NaOH(aq)|O_2(g)|C'$$

The electrodes are made of porous graphite (which is a good conductor). The H_2 and O_2 gases are continually fed in and diffuse into the electrode pores. The electrolyte solution also diffuses part way into the pores. Each electrode is impregnated with a catalyst to speed the oxidation or reduction half-reaction. In the anode pores, H_2 is oxidized to H^+, according to $H_2 \rightarrow 2H^+ + 2e^-$. The H^+ is neutralized by the OH^- of the electrolyte ($2H^+ + 2OH^- \rightarrow 2H_2O$), so the net anode reaction is $H_2 + 2OH^- \rightarrow 2H_2O + 2e^-$. At the cathode, oxygen is reduced: $O_2 + 2H_2O + 4e^- \rightarrow 4OH^-$. The net reaction is $2H_2 + O_2 \rightarrow 2H_2O$. Hydrogen–oxygen fuel cells are used in United States spacecraft to supply power for heat, light, and radio communication.

14.12 ION-SELECTIVE MEMBRANE ELECTRODES

An **ion-selective membrane electrode** contains a glass, crystalline, or liquid membrane for which the potential difference between the membrane and an electrolyte solution it is in contact with is determined by the activity of one particular ion.

The most widely used membrane electrode is the **glass electrode** (Fig. 14.17a), which contains a very thin glass membrane of special composition. Glass contains a three-dimensional network of covalently bound Si and O atoms with a net negative charge, plus positive ions, for example, Na^+, Li^+, Ca^{2+}, in the spaces in the Si–O network. The positive ions of the alkali metals can move through the glass, giving it a very weak electrical conductivity. The thinness (0.005 cm) of the membrane reduces

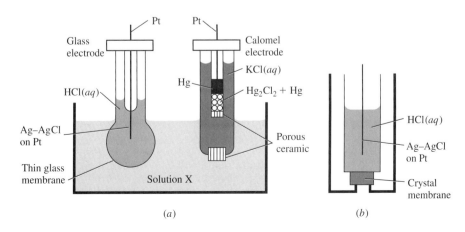

Figure 14.17

(a) Measurement of pH using a glass electrode. (b) A crystal-membrane electrode.

its resistance. Even so, the resistances of glass electrodes run 10^7 to 10^9 ohms. The high resistance makes emf measurements with a potentiometer (Fig. 14.8) inaccurate, because the galvanometer can't readily detect the extremely small currents involved, so an electronic voltmeter is used. An Ag–AgCl electrode and an internal filling solution of aqueous HCl are sealed in as part of the glass electrode.

The main application of glass electrodes is to pH measurement. To measure the pH of solution X, we set up the cell (Fig. 14.17a)

$$\text{Pt}|\text{Ag}|\text{AgCl}(s)|\text{HCl}(aq)|\text{glass}|\text{soln. X} \vdots \text{KCl(sat)}|\text{Hg}_2\text{Cl}_2(s)|\text{Hg}|\text{Pt}'$$

Let this be cell X with emf $\mathcal{E}_X$.

Before a freshly made glass electrode is used, it is immersed in water for a few hours. Monovalent cations, for example, Na^+, at and near the surface of the glass are replaced by H^+ ions from the water. When the glass electrode is immersed in solution X, an equilibrium between H^+ ions in solution and H^+ ions in the glass surface is set up. This charge transfer between glass and solution produces a potential difference between the glass and the solution. Equation (14.23) gives

$$\phi(\text{X}) - \phi(\text{glass}) = [\mu^{\text{gl}}(H^+) - \mu^{\text{X}}(H^+)]/F \qquad (14.75)$$

The emf $\mathcal{E}_X$ of cell X equals (14.75) plus the $\Delta\phi$'s at all the other interfaces.

Let solution X be replaced by a standard solution S, to give cell S with emf $\mathcal{E}_S$. Similar to (14.75), we have

$$\phi(\text{S}) - \phi(\text{glass}) = [\mu^{\text{gl}}(H^+) - \mu^{\text{S}}(H^+)]/F \qquad (14.76)$$

If we assume that the liquid-junction potential $\mathcal{E}_{J,X}$ between solution X and the calomel electrode equals the junction potential $\mathcal{E}_{J,S}$ between solution S and the calomel electrode, then the $\Delta\phi$'s for the cells X and S are the same at all interfaces except at the glass–solution X or S interface. Therefore, $\mathcal{E}_X - \mathcal{E}_S$ equals (14.75) minus (14.76), and $\mathcal{E}_X - \mathcal{E}_S = \mu^{\text{S}}(H^+)/F - \mu^{\text{X}}(H^+)/F$. Substitution of $\mu(H^+) = \mu^{\circ,\text{aq}}(H^+) + RT \ln a(H^+)$ for $\mu^{\text{S}}(H^+)$ and $\mu^{\text{X}}(H^+)$ and the use of (1.69) give

$$\mathcal{E}_X - \mathcal{E}_S = RTF^{-1}(\ln 10)\left[\log_{10} a^{\text{S}}(H^+) - \log_{10} a^{\text{X}}(H^+)\right]$$

where $a^{\text{S}}(H^+)$ and $a^{\text{X}}(H^+)$ are the activities of H^+ in solutions S and X. We now replace $\log_{10} a^{\text{S}}(H^+)$ by $-\text{pH(S)}$, where the defined pH(S) of the standard solution approximates closely $-\log_{10} a^{\text{S}}(H^+)$. Since we have used the same approximations as in Sec. 14.10 [namely, assuming equal liquid-junction potentials and using a defined pH(S) value], we replace $\log_{10} a^{\text{X}}(H^+)$ by $-\text{pH(X)}$, to get

$$\text{pH(X)} = \text{pH(S)} + (\mathcal{E}_X - \mathcal{E}_S)F(RT \ln 10)^{-1}$$

which is the same as Eq. (14.73). Therefore a glass electrode can replace the hydrogen electrode in pH measurements.

Glass is made by cooling a molten mixture of SiO_2 and metal oxides. By varying the composition of the glass, one can produce a glass electrode that is sensitive to an ion other than H^+. Examples of such ions are Na^+, K^+, Li^+, NH_4^+, and Tl^+.

The glass membrane can be replaced by a crystal of a salt that is "insoluble" in water and that has significant ionic conductivity at room temperature. Figure 14.17b shows a crystal-membrane electrode. As an example, a crystal of LaF_3 gives a membrane that is sensitive to F^-; there is an equilibrium between F^- adsorbed at the crystal surface and F^- in the solution the electrode dips into, and an equation like (14.75) holds with F^- replacing H^+ and $-F$ replacing F.

Ion-selective membrane electrodes enable measurement of the activities of certain ions that are hard to determine by traditional analytical methods. Examples are Na^+, K^+, Ca^{2+}, NH_4^+, Mg^{2+}, F^-, NO_3^-, and ClO_4^-.

Glass microelectrodes are used to measure activities of H^+, Na^+, and K^+ in biological tissues.

14.13   MEMBRANE EQUILIBRIUM

Consider two KCl solutions (α and β) separated by a membrane permeable to K^+ but impermeable to Cl^- and to the solvent(s) (Fig. 14.18). Let solution α be more concentrated than β. The K^+ ions will tend to diffuse through the membrane from α to β. This produces a net positive charge on the β side of the membrane and a net negative charge on the α side. The negative charge on solution α slows the diffusion of K^+ from α to β and speeds up diffusion of K^+ from β to α. Eventually an equilibrium is reached in which the K^+ diffusion rates are equal. At equilibrium, solution β is at a higher electric potential than α, as a result of transfer of a chemically undetectable amount of K^+.

To derive an expression for the potential difference across the membrane, consider two electrolyte solutions α and β that are separated by a membrane permeable to ion k and possibly to some (but not all) of the other ions present; the membrane is impermeable to the solvent(s). At equilibrium, $\tilde{\mu}_k^\alpha = \tilde{\mu}_k^\beta$ [Eq. (14.20)]. Use of $\tilde{\mu}_k^\alpha = \mu_k^\alpha + z_k F \phi^\alpha = \mu_k^{\circ,\alpha} + RT \ln a_k^\alpha + z_k F \phi^\alpha$ [Eqs. (14.19) and (11.2)] gives

$$\mu_k^{\circ,\alpha} + RT \ln a_k^\alpha + z_k F \phi^\alpha = \mu_k^{\circ,\beta} + RT \ln a_k^\beta + z_k F \phi^\beta$$

$$\Delta\phi \equiv \phi^\beta - \phi^\alpha = -\frac{\mu_k^{\circ,\beta} - \mu_k^{\circ,\alpha}}{z_k F} - \frac{RT}{z_k F} \ln \frac{a_k^\beta}{a_k^\alpha} \qquad (14.77)$$

$\Delta\phi$ is the **membrane** (or **transmembrane**) **potential.** Note the resemblance of (14.77) to the Nernst equation (14.41).

If the solvents in solutions α and β are the same, then $\mu_k^{\circ,\alpha} = \mu_k^{\circ,\beta}$ and (14.77) becomes

$$\phi^\beta - \phi^\alpha = -\frac{RT}{z_k F} \ln \frac{a_k^\beta}{a_k^\alpha} = \frac{RT}{z_k F} \ln \frac{\gamma_k^\alpha m_k^\alpha}{\gamma_k^\beta m_k^\beta} \qquad (14.78)$$

If the membrane is permeable to several ions, the equilibrium activities and the equilibrium value of $\Delta\phi$ must be such that (14.78) is satisfied for each ion that can pass through the membrane.

The preceding situation where the membrane is impermeable to the solvent is called *nonosmotic membrane equilibrium.* More commonly, the membrane is permeable to the solvent, as well as to one or more of the ions. The requirements of equal electrochemical potentials in the two phases for the solvent and for the permeating ions lead to a pressure difference between the two solutions at equilibrium. The equation for $\Delta\phi$ in *osmotic membrane equilibrium* (also called *Donnan membrane equilibrium*) is more complicated than (14.78). (See *Guggenheim*, sec. 8.08, for the treatment.) However, for dilute solutions, Eq. (14.78) turns out to be a good approximation for osmotic membrane equilibrium.

The potential difference between solutions α and β can be reasonably accurately measured by setting up the cell

$$Ag_L|AgCl(s)|KCl(aq)\,\vdots\,\alpha|\text{membrane}|\beta\,\vdots\,KCl(aq)|AgCl(s)|Ag_R \qquad (14.79)$$

in which concentrated-KCl salt bridges connect each solution to an Ag–AgCl electrode. Provided α and β are reasonably dilute, the sum of the liquid-junction potentials will be small (1 or 2 mV). Hence, to a good approximation, the cell emf is $\mathscr{E} \equiv \phi(Ag_R) - \phi(Ag_L) = \phi(\text{soln. } \beta) - \phi(\text{soln. } \alpha)$, since the potential differences at the interfaces to the left of α in the diagram cancel with those to the right of β.

The Nernst equation (14.78) only applies when membrane equilibrium exists. It does not apply to cell membrane potentials in living organisms, because the solutions in living organisms are not in equilibrium. (At equilibrium, nothing is happening. If you're in equilibrium, you're dead.) For discussion of nerve-cell membrane potentials, see Sec. 14.16.

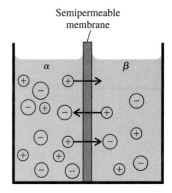

Semipermeable
membrane

Figure 14.18

Two KCl(aq) solutions separated by a membrane permeable to only K^+. Diffusion of K^+ through the membrane produces a transmembrane potential.

Figure 14.19

(a) Electrical double layer in the Stern model. (b) Electric potential versus distance in the Stern model.

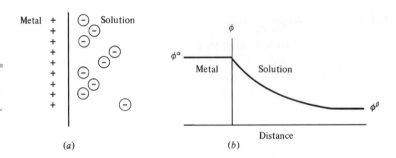

(a)   (b)

14.14   THE ELECTRICAL DOUBLE LAYER

We saw in Sec. 14.2 that the interphase region between two bulk phases usually contains a complex distribution of electric charge resulting from (a) charge transfer between phases, (b) unequal adsorption of positive and negative ions, (c) orientation of molecules with permanent dipole moments, and (d) distortion (polarization) of electronic charge in molecules. For historical reasons, the charge distribution in the interphase region is called the **electrical double layer.**

We shall consider the double layer at the interface between a metal electrode and an aqueous electrolyte solution, for example, between Cu and $CuSO_4(aq)$. Suppose the electrode is positively charged because of a net gain of Cu^{2+} ions from the solution. In 1924, Stern proposed that some of the excess negative ions in the solution are adsorbed on the electrode and held at a fixed distance determined by the ionic radius while thermal motion distributes the remainder of the excess negative ions diffusely in the interphase region (Fig. 14.19a).

Figure 14.19b shows the variation in the electric potential ϕ with distance from the electrode, as calculated from the Stern model. As we go from phase α to phase β, the electric potential in the interphase region (Sec. 13.1) gradually changes from ϕ^α to ϕ^β. If z is the direction perpendicular to the α-β interface, the derivative $\partial\phi/\partial z$ is nonzero in the interphase region, so the electric field $E_z = -\partial\phi/\partial z$ [Eq. (14.11)] is nonzero in the interphase region. An ion or electron with charge Q that moves from the bulk phase α to the bulk phase β experiences an electric force in the interphase region and has its electrical energy changed by $(\phi^\beta - \phi^\alpha)Q$ [Eq. (14.10)]. This is reflected in the $z_i F\phi^\alpha$ term in the electrochemical potential (14.19).

The Stern model is probably essentially correct. However, Stern did not explicitly consider the orientation of water dipoles at the electrode. Most of the electrode surface is covered with a layer of adsorbed water molecules. If the electrode is positively charged, most of the water molecules in the adsorbed layer will have their negative (oxygen) sides in contact with the electrode. This orientation of the water dipole moments affects ϕ in the interphase region.

The electric field in the electrode–solution interphase region is extremely high. Table 14.1 indicates that electrode–solution potential differences are typically about 1 V. The electrode–solution interphase region is of the order of 50 Å thick. If z is the direction perpendicular to the interface, then $|E_z| = d\phi/dz \approx \Delta\phi/\Delta z \approx (1 \text{ V})/(50 \times 10^{-8} \text{ cm})$ $= 2 \times 10^6$ V/cm.

14.15   DIPOLE MOMENTS AND POLARIZATION

As noted in Secs. 14.2 and 14.14, orientation of molecules with dipole moments and distortion (polarization) of molecular charge distributions contribute to interphase potential differences in electrochemical systems such as galvanic cells. This section discusses molecular dipole moments and polarization.

Dipole Moments

Two charges Q and $-Q$ equal in magnitude and opposite in sign and separated by a distance d that is small compared with the distances from the charges to an observer constitute an **electric dipole** (Fig. 14.20a). The **electric dipole moment** μ of an electric dipole is defined as a vector of magnitude

$$\mu \equiv Qd \qquad (14.80)*$$

(a)

and direction from the negative to the positive charge. The SI unit of μ is the coulomb-meter (C m).

Let ϕ be the electric potential produced by a dipole. From Eq. (14.12) and Fig. 14.20b, we have $\phi = Q/4\pi\varepsilon_0 r_2 - Q/4\pi\varepsilon_0 r_1$. In accord with the above definition of an electric dipole, we assume that $r \gg d$, where r is the distance to the point P at which ϕ is being calculated. For $r \gg d$, the use of some geometry shows (Prob. 14.58) that the electric potential ϕ of a dipole is well approximated as

$$\phi = \frac{1}{4\pi\varepsilon_0} \frac{\mu \cos \theta}{r^2} \qquad \text{for } r \gg d \qquad (14.81)$$

where the angle θ is defined in Fig. 14.20b.

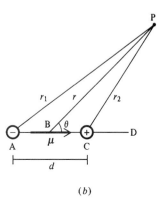

(b)

Figure 14.20

(a) An electric dipole.
(b) Calculation of the electric potential of the dipole.

Note that ϕ for a dipole falls off as $1/r^2$, in contrast to ϕ for a single charge (a monopole), which falls off as $1/r$ [Eq. (14.12)]. Since the electric field is found by differentiation of ϕ [Eq. (14.11)], the electric field of a dipole falls off as $1/r^3$, compared with $1/r^2$ for a single charge. *The force field of a dipole is of relatively short range compared with the long-range force field of a single charge.*

Consider a distribution of several electric charges Q_i, with the total charge being zero: $\Sigma_i Q_i = 0$. If one calculates the distribution's electric potential at any point whose distance from the distribution is much larger than the distance between any two charges of the distribution, one finds that the potential is given by (14.81) provided the **electric dipole moment** of the charge distribution is defined as

$$\mu \equiv \sum_i Q_i \mathbf{r}_i \qquad (14.82)$$

where $\mathbf{r}_i$ is the vector from the origin (chosen arbitrarily) to charge Q_i. (Proof of this statement is given in E. R. Peck, *Electricity and Magnetism*, McGraw-Hill, 1953, p. 29.) Thus, the electric potential produced by a neutral molecule at a point well outside the molecule is given by (14.81) and (14.82). Since the charges of the molecule are moving, $\mathbf{r}_i$ must be interpreted as an average location for charge Q_i. For an H_2O molecule, which has 10 electrons and 3 nuclei, the sum in (14.82) has 13 terms.

For a distribution that consists of a charge $-Q$ with (x, y, z) coordinates $(-\frac{1}{2}d, 0, 0)$ and a charge Q with coordinates $(\frac{1}{2}d, 0, 0)$, Eq. (14.82) gives $\mu_x = \Sigma_i Q_i x_i = -Q(-\frac{1}{2}d) + Q(\frac{1}{2}d) = Qd$ and $\mu_y = 0 = \mu_z$, in agreement with (14.80).

The sum in (14.82) can be written as a sum over the negative charges plus a sum over the positive charges. A molecule has a nonzero dipole moment if the effective centers of negative charge and positive charge do not coincide. Some molecules with $\mu \neq 0$ are HCl, H_2O (which is nonlinear), and CH_3Cl. Some molecules with $\mu = 0$ are H_2, CO_2 (which is linear), CH_4, and C_6H_6. A molecule is said to be **polar** or **nonpolar** according to whether $\mu \neq 0$ or $\mu = 0$, respectively. Dipole moments of some molecules are listed in Sec. 21.7.

Polarization

When a molecule with a zero dipole moment is placed in an external electric field $\mathbf{E}$, the field shifts the centers of positive and negative charge, **polarizing** the molecule and giving it an **induced dipole moment** μ_{ind}. For example, if a positive charge is placed above the plane of a benzene molecule, the average positions of the electrons will shift upward, giving the molecule a dipole moment whose direction is perpendicular to the molecular plane. If the molecule has a nonzero dipole moment $\boldsymbol{\mu}$ (in the absence of any external field), the induced moment $\boldsymbol{\mu}_{ind}$ produced by the field will add to the permanent dipole moment $\boldsymbol{\mu}$. The induced dipole moment is proportional to the electric field $\mathbf{E}$ experienced by the molecule:

$$\boldsymbol{\mu}_{ind} = \alpha\mathbf{E} \qquad (14.83)$$

where the proportionality constant α is called the molecule's (**electric**) **polarizability.** Actually, α is a function of direction in the molecule. For example, the polarizability of HCl along the bond differs from its polarizability in a direction perpendicular to the bond axis. For liquids and gases, where the molecules are rotating rapidly, we use an α averaged over direction.

The electric field at points well outside a neutral isolated molecule is determined by the magnitude and orientation of the molecule's electric dipole moment $\boldsymbol{\mu}$. When the molecule is interacting with other molecules, it has an induced dipole moment $\boldsymbol{\mu}_{ind} = \alpha\mathbf{E}$ in addition to its permanent moment $\boldsymbol{\mu}$. The molecular dipole moment $\boldsymbol{\mu}$ and molecular polarizability α largely determine intermolecular interactions in non-hydrogen-bonded substances (see Sec. 22.10).

The molecular polarizability increases with increasing number of electrons and increases as the electrons become less tightly held by the nuclei. From (14.82), (14.83), and (14.3), the SI units of α are $(C\ m)/(N\ C^{-1}) = C^2\ m\ N^{-1}$. From (14.2), the units of $\alpha/4\pi\varepsilon_0$ are m^3, which is a unit of volume. Polarizabilities are often tabulated as $\alpha/4\pi\varepsilon_0$ values. Some values are shown in Fig. 14.21.

A small volume in a piece of matter is said to be electrically **unpolarized** or **polarized** according to whether the net dipole moment of the volume is zero or nonzero, respectively. A small volume within the bulk phase of pure liquid water is unpolarized; the molecular dipoles are oriented randomly, so their vector sum is zero. In an electrolyte solution, the water in the vicinity of each ion is polarized, because of the orientation of the H_2O dipoles and because of the induced dipole moments.

Consider two flat, parallel metal plates with opposite charges that are equal in magnitude. This is a *capacitor* or *condenser* (Fig. 14.22). The electric field E in the region between the plates is constant. (See *Halliday and Resnick,* sec. 30-2, for the proof.) Let x be the direction perpendicular to the plates. Integration of $E_x = -\partial\phi/\partial x$ with $E_x = E = $ const. gives for the potential difference $\Delta\phi$ between the plates

$$|\Delta\phi| = Ed \qquad (14.84)$$

where d is the distance between the plates.

When a nonconducting substance (a **dielectric**) is placed between the plates, the nonconductor becomes polarized, as a result of two effects: (1) The electric field of the plates tends to orient the permanent dipoles of the dielectric so that the negative ends of the moments lie toward the positive plate. The degree of orientation shown in

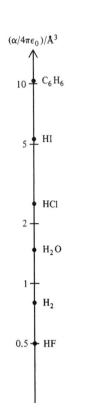

$(\alpha/4\pi\varepsilon_0)/\text{Å}^3$

Figure 14.21

Polarizabilities (divided by $4\pi\varepsilon_0$) for some molecules. The scale is logarithmic.

Figure 14.22

(a) A capacitor. (b) Orientation of molecular dipoles of a dielectric in a capacitor. (c) Surface charges resulting from dielectric polarization.

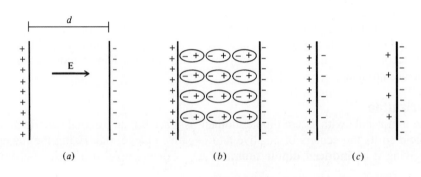

Fig. 14.22b is greatly exaggerated. The orientation is far from complete, since it is opposed by the random thermal motion of the molecules. (2) The electric field of the plates produces induced dipole moments $\boldsymbol{\mu}_{ind}$ that are oriented with their negative ends toward the positive plate. For a dielectric whose molecules have zero permanent dipole moment, the **orientation polarization** (effect 1) is absent. The **induced** (or **distortion**) **polarization** (effect 2) is always present.

For any small volume in the bulk phase of the polarized dielectric, the net charge is zero. However, because of the polarization, there is a negative charge at the dielectric surface in contact with the positive plate and a positive charge on the opposite surface of the dielectric (Fig. 14.22c). These surface charges partly cancel the effect of the charges on the metal plates, thereby reducing the electric field in the region between the plates and reducing the potential difference between the plates.

Dielectric Constants

The **dielectric constant** (or **relative permittivity**) ε_r of a dielectric is defined by $\varepsilon_r \equiv E_0/E$, where E_0 and E are the electric fields in the space between the plates of a condenser when the plates are separated by a vacuum and by the dielectric, respectively. Equation (14.84) gives $E_0/E = \Delta\phi_0/\Delta\phi$, where $\Delta\phi_0$ and $\Delta\phi$ are the potential differences between the plates in the absence and in the presence of the dielectric. Therefore

$$\varepsilon_r \equiv E_0/E = \Delta\phi_0/\Delta\phi \qquad (14.85)$$

Let Q be the absolute value of the charge on one of the plates, and let $\mathscr{A}$ be its area. In the absence of a dielectric, the electric field between the plates is $E_0 = Q/\varepsilon_0\mathscr{A}$ [eq. (30-2) in *Halliday and Resnick*]. With a dielectric between the plates, let Q_P be the absolute value of the charge on one surface of the polarized dielectric. Q_P neutralizes part of the charge on each plate, so the field is now $E = (Q - Q_P)/\varepsilon_0\mathscr{A}$. Hence, $E_0/E = Q/(Q - Q_P) = \varepsilon_r$, where (14.85) was used. Therefore

$$Q - Q_P = Q/\varepsilon_r \qquad (14.86)$$

The deviation of ε_r from 1 is due to two effects: the induced polarization and the orientation of the permanent dipole moments. Therefore, ε_r increases as the molecular polarizability α increases and ε_r increases as the molecular electric dipole moment μ increases. Using the Boltzmann distribution law (Chapter 22) to describe the orientations of the dipoles in the applied electric field, one can show that for pure gases (polar or nonpolar) at low or moderate pressure and for nonpolar liquids or solids

$$\frac{\varepsilon_r - 1}{\varepsilon_r + 2} \frac{M}{\rho} = \frac{N_A}{3\varepsilon_0}\left(\alpha + \frac{\mu^2}{3kT}\right) \qquad (14.87)$$

[For the derivation, see *McQuarrie* (1973), chap. 13.] In the **Debye–Langevin equation** (14.87), M is the molar mass (not the molecular weight), ρ is the density, k is Boltzmann's constant (3.57), N_A is the Avogadro constant, and T is the temperature. One can show that $(\varepsilon_r - 1)/(\varepsilon_r + 2)$ increases as ε_r increases (Prob. 14.65).

The dielectric constant ε_r can be measured using Eq. (14.85). From (14.87), a plot of $M\rho^{-1}(\varepsilon_r - 1)/(\varepsilon_r + 2)$ versus $1/T$ is a straight line with slope $N_A\mu^2/9\varepsilon_0 k$ and intercept $N_A\alpha/3\varepsilon_0$. Such a plot allows α and μ of the gas molecules to be found.

Because of solute–solvent interactions, the Debye–Langevin equation is not strictly applicable to liquid solutions of polar molecules in a nonpolar solvent, but is often applied in modified form to such solutions to obtain approximate μ values. Many different such modifications exist [see H. B. Thompson, *J. Chem. Educ.,* **43,** 66 (1966)], and they typically yield dipole moments with errors of about 10 percent. For a more accurate procedure to calculate dipole moments from dielectric constants of solutions, see M. Barón, *J. Phys. Chem,.* **89,** 4873 (1985).

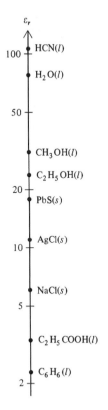

Figure 14.23

Dielectric constants for some liquids and solids at 25°C and 1 atm. The scale is logarithmic.

Some dielectric constants for liquids and solids at 25°C and 1 bar are plotted in Fig. 14.23. The high dielectric constants of H_2O and HCN are due to their high dipole moments. The dielectric constant of a single crystal depends on its orientation in the capacitor. Values of ε_r for solids are usually given for a mixture of small crystals with random orientations.

For gases, ε_r is quite close to 1, and $\varepsilon_r + 2$ can be taken as 3 in (14.87). Since ρ is proportional to P, Eq. (14.87) shows that $\varepsilon_r - 1$ for a gas increases essentially linearly with P at constant T. Some ε_r values at 20°C and 1 atm are 1.00054 for air, 1.00092 for CO_2, 1.0031 for HBr, and 1.0025 for n-pentane.

For polar liquids, ε_r decreases as T increases (Fig. 14.24). At higher temperatures the random thermal motion decreases the orientation polarization. For nonpolar liquids, the orientation polarization is absent, so ε_r varies only slightly with T. This variation is due to the change in ρ with T in Eq. (14.87).

Consider two charges Q_1 and Q_2 immersed in a dielectric fluid with dielectric constant ε_r, and let the charges be separated by at least several molecules of dielectric. The charge Q_1 polarizes the dielectric in its immediate neighborhood. Let Q_1 be positive. The negative charges of the oriented dipoles immediately adjacent to Q_1 partly "neutralize" Q_1, giving it an effective charge $Q_{\text{eff}} = Q_1 - Q_P$, where $-Q_P$ is the charge on the spherical "surface" of the dielectric immediately surrounding Q_1. It turns out that Eq. (14.86) gives the correct result for the effective charge: $Q_{\text{eff}} = Q_1 - Q_P = Q_1/\varepsilon_r$. (See F. W. Sears, *Electricity and Magnetism*, Addison-Wesley, 1951, p. 190.) At points not too close to Q_1, the electric field E due to Q_1 and the induced charges surrounding it equals $Q_{\text{eff}}/4\pi\varepsilon_0 r^2$. Therefore

$$E = \frac{1}{4\pi\varepsilon_0\varepsilon_r}\frac{Q_1}{r^2} = \frac{1}{4\pi\varepsilon}\frac{Q_1}{r^2} \qquad (14.88)$$

where the *permittivity* ε of the medium is defined as $\varepsilon \equiv \varepsilon_r\varepsilon_0$.

Now consider the force on Q_2. This force is due to (a) the charge Q_1, (b) the induced charge $-Q_P$ around Q_1, and (c) the induced charge around Q_2. The induced charge around Q_2 is spherically distributed around Q_2 and produces no net force on Q_2. Therefore the force F on Q_2 is found from the field (14.88), which results from charges (a) and (b). Equations (14.3) and (14.88) give

$$F = \frac{1}{4\pi\varepsilon_0\varepsilon_r}\frac{|Q_1Q_2|}{r^2} \qquad (14.89)$$

In a fluid with dielectric constant ε_r, the force on Q_2 is reduced by the factor $1/\varepsilon_r$ compared with the force in vacuum.

Since intermolecular forces are electrical, the dielectric constant ε_r of a solvent affects equilibrium constants and reaction rate constants. Recall the higher degree of ion pairing of electrolytes in solvents with low ε_r values (Sec. 10.9).

14.16 BIOELECTROCHEMISTRY

The fluids in and surrounding the cells of living organisms contain significant amounts of dissolved electrolytes. The total electrolyte molarity is typically 0.3 mol/dm³ for mammalian fluids.

Figure 14.25 shows a piece of animal tissue pinned to the bottom of a chamber filled with a solution having the same composition as the extracellular fluid of the organism. By penetrating the membrane of a single cell with a micro salt bridge, one sets up the electrochemical cell (14.79), where phase β is the interior of the biological cell, the membrane is the cell's membrane, and phase α is the bathing solution. The potential difference measured is that between the interior and exterior of the biological cell and is the transmembrane potential. The membrane potential is displayed on an oscilloscope. The micro salt bridge consists of glass drawn to a very fine tip and filled with concentrated aqueous KCl.

Cells show a potential difference $\phi^{int} - \phi^{ext}$ of -30 to -100 mV across their membranes, the cell's interior being at a lower potential than the exterior. Typical values are -90 mV for resting muscle cells, -70 mV for resting nerve cells, and -40 mV for liver cells. Since interphase potential differences exist in living organisms, living organisms are electrochemical systems (Sec. 14.2).

When an impulse propagates along a nerve cell or when a muscle cell contracts, the transmembrane potential $\phi^{int} - \phi^{ext}$ changes, becoming momentarily positive. Nerve impulses are transmitted by changes in nerve-cell membrane potentials. Muscles are caused to contract by changes in muscle-cell membrane potentials. Our perception of the external world through the senses of sight, hearing, touch, etc., our thought processes, and our voluntary and involuntary muscular contractions are all intimately connected to interphase potential differences. An understanding of life requires an understanding of how these potential differences are maintained and how they are changed.

The existence of transmembrane potential differences means that there is an electrical double layer at the membrane of each cell. The double layer is approximately equivalent to a distribution of electric dipoles at the cell's surface. Consider the heart muscles. As these muscles contract and relax, the potential differences across their cell membranes continually change, and hence the total dipole moment of the heart changes, and so do the electric field and electric potential produced by the heart. An electrocardiogram (ECG) measures the difference in electric potential between points on the surface of the body as a function of time. Changes in these potential differences arise from the changes in the heart dipole moment. An electroencephalogram (EEG) records the time-varying potential difference between two points on the scalp and reflects the electrical activity of nerve cells in the brain.

In 1943, Goldman used a nonequilibrium approach and the assumption of a linear variation of ϕ within the membrane to derive the following expression for the transmembrane potential of a biological cell:

$$\phi^{int} - \phi^{ext} \approx \frac{RT}{F} \ln \frac{P(K^+)[K^+]^{ext} + P(Na^+)[Na^+]^{ext} + P(Cl^-)[Cl^-]^{int}}{P(K^+)[K^+]^{int} + P(Na^+)[Na^+]^{int} + P(Cl^-)[Cl^-]^{ext}} \quad (14.90)$$

where $[K^+]^{ext}$ and $[K^+]^{int}$ are the K^+ molar concentrations outside and inside the cell and $P(K^+)$ is the permeability of the membrane to K^+ ions. $P(K^+)$ is defined as $D(K^+)/\tau$, where $D(K^+)$ is the diffusion coefficient of K^+ through the membrane of thickness τ. Diffusion coefficients are defined in Sec. 16.4. For the derivation of (14.90), see *Eyring, Henderson, and Jost,* vol. IXB, chap. 11.

A nerve-cell membrane is permeable to all three ions K^+, Na^+, and Cl^-. For a resting nerve cell of a squid, one finds that $P(K^+)/P(Cl^-) \approx 2$ and $P(K^+)/P(Na^+) \approx 25$. For squid nerve cells, the observed concentrations in mmol/dm^3 are $[K^+]^{int} = 410$, $[Na^+]^{int} = 49$, $[Cl^-]^{int} = 40$, $[K^+]^{ext} = 10$, $[Na^+]^{ext} = 460$, $[Cl^-]^{ext} = 540$. These three ions are the main inorganic ions. Inside the cell, there is also a substantial concentration of

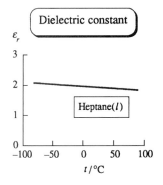

Dielectric constant

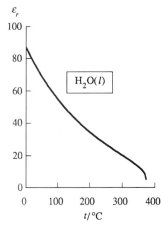

Figure 14.24

Dielectric constant versus temperature for liquid heptane at 1 atm and for saturated liquid water (liquid water under its own vapor pressure; Sec. 8.3). Water near its critical point is a good solvent for organic compounds.

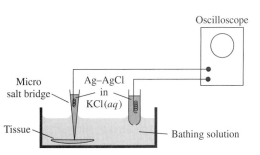

Figure 14.25

Measurement of the transmembrane potential.

organic anions (charged proteins, organic phosphates, and anions of organic acids); these have quite low permeabilities.

We can use the Nernst equation (14.78) with activity coefficients omitted to see which ions are in electrochemical equilibrium across the membrane. The above concentrations give the following *equilibrium* $\Delta\phi$ values at 25°C:

$$\Delta\phi_{eq}(K^+) = -95 \text{ mV}, \qquad \Delta\phi_{eq}(Na^+) = +57 \text{ mV}, \qquad \Delta\phi_{eq}(Cl^-) = -67 \text{ mV}$$

The observed transmembrane potential for a resting squid nerve cell is -70 mV at 25°C. Hence, Cl^- is in electrochemical equilibrium, but K^+ and Na^+ are not.

For the -70-mV membrane potential, Eq. (14.78) gives the following equilibrium concentration ratios at 25°C: $c^{ext}/c^{int} = 1:15$ for $z = +1$ ions; $c^{ext}/c^{int} = 15:1$ for $z = -1$ ions. The actual concentration ratios are 1:41 for K^+, 9:1 for Na^+, and 14:1 for Cl^-. Hence Na^+ continuously flows spontaneously into the cell and K^+ flows spontaneously out. [We have $\tilde{\mu}^{ext}(Na^+) > \tilde{\mu}^{int}(Na^+)$.] The observed steady-state concentrations of Na^+ and K^+ are maintained by an *active-transport* process that uses some of the cell's metabolic energy to continually "pump" Na^+ out of the cell and K^+ into it (Sec. 12.4).

A nerve impulse is a brief (1 ms) change in the transmembrane potential. This change travels along the nerve fiber at 10^3 to 10^4 cm/s, depending on the species and the kind of nerve. The change in $\Delta\phi$ is initiated by a local increase in the membrane's permeability to Na^+, with $P(Na^+)/P(K^+)$ reaching about 20. With $P(Na^+)$ much greater than both $P(K^+)$ and $P(Cl^-)$ in (14.90), the membrane potential moves toward the $\Delta\phi_{eq}(Na^+)$ value of $+60$ mV. The observed peak value of $\Delta\phi$ is $+40$ or 50 mV during the passage of a nerve impulse. After this peak is reached, $P(Na^+)$ decreases and the potential returns toward its resting value of -70 mV.

14.17   SUMMARY

The electric potential difference $\phi_b - \phi_a$ between two points is the reversible work per unit charge to move a charge from a to b.

An electrochemical system is one with a difference in electric potential between two or more of its phases. Such potential differences are due to charge transfer between phases, to orientation and polarization of molecules in the interphase region, and to unequal adsorption of positive and negative ions in the interphase region. The potential difference between phases is measurable only if the phases have the same chemical composition.

The existence of potential differences between phases of an electrochemical system requires that the electrochemical potential $\tilde{\mu}_i$ replace the chemical potential μ_i in all thermodynamic equations. We have $\tilde{\mu}_i^\alpha = \mu_i^\alpha + z_i F\phi^\alpha$, where $z_i F$ is the molar charge of species i and ϕ^α is the electric potential of phase α. The phase-equilibrium condition is $\tilde{\mu}_i^\alpha = \tilde{\mu}_i^\beta$.

The phases of a galvanic cell can be symbolized by T-E-I-E'-T', where I is an ionic conductor (for example, an electrolyte solution or two electrolyte solutions connected by a salt bridge), E and E' are the electrodes, and T and T' are the terminals, which are made of the same metal. The potential difference between T and T' is the sum of the potential differences between the adjacent phases of the cell. The emf $\mathcal{E}$ of a galvanic cell is defined as $\mathcal{E} \equiv \phi_R - \phi_L$, where ϕ_R and ϕ_L are the open-circuit electric potentials of the terminals on the right and left of the cell diagram. The emf of a reversible galvanic cell is given by the Nernst equation (14.43). If the cell contains a liquid junction, the liquid-junction potential $\mathcal{E}_J$ is added to the right side of the Nernst equation. The cell's standard potential $\mathcal{E}°$ satisfies $\Delta G° = -nF\mathcal{E}°$, where $\Delta G°$ is for the cell's chemical reaction and n is the number of electrons in the cell's electrochemical reaction.

The standard electrode potential for an electrode half-reaction is defined to be the standard potential $\mathscr{E}°$ of a cell with the hydrogen electrode on the left of its diagram and the electrode in question on the right. The standard electrode potential of the hydrogen electrode is 0. The standard emf of any cell is given by $\mathscr{E}° = \mathscr{E}°_R - \mathscr{E}°_L$, where $\mathscr{E}°_R$ and $\mathscr{E}°_L$ are the standard electrode potentials (reduction potentials) for the right and left half-cells in the cell diagram.

Cell emfs and their temperature derivatives can be used to determine activity coefficients of electrolytes, pH, and $\Delta G°$, $\Delta H°$, $\Delta S°$, and $K°$ of reactions.

The electric dipole moment $\boldsymbol{\mu}$ of a neutral molecule is $\boldsymbol{\mu} \equiv \Sigma_i Q_i \mathbf{r}_i$, where $\mathbf{r}_i$ is the vector from the origin to charge Q_i. When a molecule is placed in an electric field **E,** its charge distribution is polarized and it acquires an induced dipole moment $\boldsymbol{\mu}_{\text{ind}}$ given by $\boldsymbol{\mu}_{\text{ind}} = \alpha\mathbf{E}$, where α is the molecule's electric polarizability.

When two charges are immersed in a nonconducting (dielectric) fluid having dielectric constant ε_r, polarization of the fluid reduces the force on each charge by the factor $1/\varepsilon_r$ as compared with when the charges are in vacuum. A molecule's dipole moment and polarizability can be found from measurements of the dielectric constant as a function of temperature by using the Debye–Langevin equation.

Important kinds of calculations discussed in this chapter include:

- Calculation of $\mathscr{E}°$ of a cell's reaction using $\Delta G° = -nF\mathscr{E}°$.
- Calculation of $\mathscr{E}°$ from a table of standard electrode potentials using $\mathscr{E}° = \mathscr{E}°_R - \mathscr{E}°_L$.
- Calculation of the emf $\mathscr{E}$ of a reversible galvanic cell using the Nernst equation $\mathscr{E} = \mathscr{E}° - (RT/nF) \ln Q$, where $Q \equiv \Pi'_i (a_i)^{\nu_i}$. The electrolyte activity product that occurs in the Nernst equation is evaluated using Eq. (14.46).
- Calculation of $\Delta G°$, $\Delta S°$, and $\Delta H°$ of a cell's reaction from $\mathscr{E}°$ versus T data.
- Calculation of equilibrium constants from $\mathscr{E}°$ data using $\Delta G° = -nF\mathscr{E}°$ and $\Delta G° = -RT \ln K°$.
- Calculation of electrolyte activity coefficients from cell emf data using the Nernst equation.
- Calculation of μ and α from ε_r versus T data using the Debye–Langevin equation.

FURTHER READING AND DATA SOURCES

Kirkwood and Oppenheim, chap. 13; *Guggenheim,* chap. 8; *Denbigh,* secs. 4.14, 4.15, and 10.15; *McGlashan,* chap. 19; *Bockris and Reddy,* chaps. 7 and 8; *Ives and Janz; Bates; Robinson and Stokes,* chap. 8.

Standard electrode potentials: S. G. Bratsch, *J. Phys. Chem. Ref. Data,* **18,** 1 (1989).

Dielectric constants: *Landolt-Börnstein,* 6th ed., vol. II, pt. 6, pp. 449–908.

Polarizabilities: *Landolt-Börnstein,* 6th ed., vol. I, pt. 3, pp. 509–512.

PROBLEMS

Section 14.1

14.1 Which of these are vectors? (*a*) The electric field. (*b*) The electric potential.

14.2 True or false? (*a*) The electric field due to a positive charge is directed away from the charge and the electric field due to a negative charge is directed toward the charge. (*b*) The electric potential increases as one comes closer to a positive charge. (*c*) The electric potential at a point midway between a proton and an electron is zero.

14.3 Calculate the force that a He nucleus exerts on an electron 1.0 Å away.

14.4 Calculate the magnitude of the electric field of a proton at a distance of (*a*) 2.0 Å; (*b*) 4.0 Å.

14.5 Calculate the electric potential difference between two points that are 4.0 and 2.0 Å away from a proton.

Section 14.2
14.6 True or false? (a) The Faraday constant equals the charge per mole of electrons. (b) In this chapter, e stands for the charge on an electron.

14.7 Calculate the charge on (a) 3.00 mol of Hg_2^{2+} ions; (b) 0.600 mol of electrons.

Section 14.3
14.8 True or false? In an electrochemical system, $(\partial G^\alpha/\partial n_i^\alpha)_{T,P,n_{j\neq i}^\alpha} = \tilde{\mu}_i^\alpha$.

14.9 Theoretical calculations indicate that, for Li and Rb in contact at 25°C, the potential difference is $\phi(\text{Li}) - \phi(\text{Rb}) \approx 0.1$ V. Estimate the difference in chemical potential between electrons in Li and electrons in Rb.

Section 14.4
14.10 True or false? (a) The emf of the Daniell cell equals the open-circuit potential difference between the piece of copper that dips into the $CuSO_4$ solution and the Zn that dips into the $ZnSO_4$ solution. (b) The emf of a galvanic cell is the open circuit potential difference between two phases whose chemical compositions differ negligibly from each other. (c) In the spontaneous chemical reaction of a galvanic cell, electrons flow from the cathode to the anode.

14.11 For a certain open-circuit Daniell cell with the diagram (14.26), suppose the following is true: For the Cu electrode, ϕ is 0.3 V higher than ϕ for the $CuSO_4(aq)$ solution; for the $ZnSO_4(aq)$ solution, ϕ is 0.1 V higher than ϕ for the $CuSO_4(aq)$ solution; the Zn electrode and the $ZnSO_4(aq)$ solution are at the same electric potential; for the Cu' terminal, ϕ is 0.2 V lower than ϕ for the Zn electrode. Find the emf of this cell.

Section 14.6
14.12 True or false? (a) Increasing the activity of a product in the cell's chemical reaction must decrease the emf of the cell. (b) The cell-reaction's charge number n is a positive number with no units. (c) If we double all the coefficients in a cell's reaction, the charge number n is doubled and the emf is unchanged. (d) The standard emf $\mathscr{E}°$ of a galvanic cell is the limiting value of $\mathscr{E}$ taken as all molalities go to zero.

14.13 Give the charge number n for each of these reactions: (a) $H_2 + Br_2 \rightarrow 2HBr$; (b) $\frac{1}{2}H_2 + \frac{1}{2}Br_2 \rightarrow HBr$; (c) $2HBr \rightarrow H_2 + Br_2$; (d) $3Zn + 2Al^{3+} \rightarrow 3Zn^{2+} + 2Al$; (e) $Hg_2Cl_2 + H_2 \rightarrow 2Hg + 2Cl^- + 2H^+$.

14.14 Use data in the Appendix to find $\mathscr{E}_{298}°$ for $N_2O_4(g) + Cu^{2+}(aq) + 2H_2O(l) \rightarrow Cu + 4H^+(aq) + 2NO_3^-(aq)$.

14.15 Suppose we add a pinch of salt (NaCl) to the $CuSO_4$ solution of cell (14.52) thermostated at 25°C. (a) Is $\mathscr{E}$ changed? Explain. (b) Is $\mathscr{E}°$ changed? Explain.

14.16 Express the emf of the cell

$$\text{Pt}|\text{In}(s)|\text{In}_2(\text{SO}_4)_3(aq, m)|\text{Hg}_2\text{SO}_4(s)|\text{Hg}(l)|\text{Pt}'$$

in terms of $\mathscr{E}°$, T, and $\gamma_\pm$ and m of $In_2(SO_4)_3(aq)$.

14.17 For the cell (14.32), emfs at 60°C and 1 bar H_2 pressure as a function of HCl molality m are:

$m/(\text{mol kg}^{-1})$	0.001	0.002	0.005	0.1
$\mathscr{E}/\text{V}$	0.5951	0.5561	0.5050	0.3426

(a) Use a graphical method to find $\mathscr{E}°$ at 60°C. (b) Calculate the 60°C HCl(aq) mean ionic activity coefficients at $m = 0.005$ mol/kg and 0.1 mol/kg.

14.18 The second virial coefficient of $H_2(g)$ at 25°C is $B = 14.0$ cm³/mol. (a) Use the $\ln \phi$ expression (10.105) to calculate the fugacity of $H_2(g)$ at 25°C and 1 bar; neglect terms after $B^\dagger P$. (b) Calculate the error in $\mathscr{E}_{298}$ of a cell that uses the hydrogen electrode at 25°C and 1 bar if f_{H_2} is replaced by P_{H_2} in the Nernst equation.

14.19 It was specified in Sec. 14.4 that the terminals of a galvanic cell must be made of the same metal. One might wonder whether the cell emf depends on the identity of this metal. Explain how Eq. (14.39) shows that the emf of a cell is independent of what metal is used for the terminals.

14.20 Suppose that a cell's electrochemical reaction is multiplied by 2. What effect does this have on each of the following quantities in the Nernst equation: (a) n; (b) Q; (c) $\ln Q$; (d) $\mathscr{E}$?

14.21 Consider the Daniell cell (14.26). (a) Apply Eq. (14.21) to the electrochemical equilibrium $Cu \rightleftharpoons Cu^{2+}(aq) + 2e^-(Cu)$ at the $Cu-CuSO_4(aq)$ interface to show that

$$\phi(\text{Cu}) - \phi(\text{aq. CuSO}_4) =$$
$$[\mu^{aq}(\text{Cu}^{2+}) - \mu(\text{Cu}) + 2\mu^{Cu}(e^-)]/2F$$

(b) Find a similar equation for $\phi(\text{Zn}) - \phi(\text{aq. ZnSO}_4)$. (c) Find a similar equation for $\phi(\text{Cu}') - \phi(\text{Zn})$. (d) Substitute the results of (a), (b), and (c) into Eq. (14.25) and use Eq. (11.2) to show that the result for the cell emf is Eq. (14.51).

14.22 For the electrolyte $M_{\nu_+}X_{\nu_-}$, use Eqs. (10.4), (10.38), and (10.44) to show that $a_i = (a_+)^{\nu_+}(a_-)^{\nu_-}$. Combining this equation with the equation $a_i = (\nu_\pm \gamma_\pm m_i/m°)^\nu$ [Eq. (10.52)], we get Eq. (14.46).

Section 14.7
14.23 True or false? (a) When a half-reaction is multiplied by 2, its standard reduction potential $\mathscr{E}°$ is multiplied by 2. (b) In the equation $\mathscr{E}° = \mathscr{E}_R° - \mathscr{E}_L°$, both $\mathscr{E}_R°$ and $\mathscr{E}_L°$ are reduction potentials.

14.24 (a) Use data in the Appendix to find $\mathscr{E}_{298}°$ for the reaction $3Cu^{2+}(aq) + 2Fe(s) \rightarrow 2Fe^{3+}(aq) + 3Cu(s)$. (b) Use data in Table 14.1 to answer the question in (a).

14.25 (a) $\mathscr{E}°$ of the cell calomel|nonesuch composed of a calomel electrode and a nonesuch electrode is -1978 mV at 25°C. Find the standard electrode potential of the nonesuch electrode at 25°C. (b) At 43°C, the cell calomel|nonpareil has $\mathscr{E}° = -0.80$ V and the cell nonesuch|calomel has $\mathscr{E}° = 1.70$ V. Find $\mathscr{E}°$ for the cell nonpareil|nonesuch at 43°C.

14.26 What values of the activity quotient Q are required for the cell (14.32) to have the following emfs at 25°C: (a) -1.00 V; (b) 1.00 V?

14.27 If the cell (14.32) has $a(HCl) = 1.00$, what value of $P(H_2)$ is needed to make the cell emf at 25°C equal to (*a*) -0.300 V; (*b*) 0.300 V?

14.28 For the cell

$$Pt_L|Fe^{2+}(a = 2.00), Fe^{3+}(a = 1.20)\ \vdots\ I^-(a = 0.100)|I_2(s)|Pt_R$$

(*a*) write the cell reaction; (*b*) calculate $\mathscr{E}_{298}$ assuming the net liquid-junction potential is negligible. (*c*) Which terminal is at the higher potential? (*d*) When the cell is connected to a load, into which terminal do electrons flow from the load?

14.29 For the cell

$$Cu|CuSO_4(1.00\ mol/kg)|Hg_2SO_4(s)|Hg|Cu'$$

(*a*) write the cell reaction; (*b*) calculate $\mathscr{E}$ at 25°C and 1 bar given that $\gamma_\pm$ of $CuSO_4$ is 0.043 for these conditions; (*c*) calculate the erroneous value of $\mathscr{E}$ that would be obtained if the $CuSO_4$ activity coefficient were taken as 1.

14.30 Calculate $\mathscr{E}_{298}$ for the cell

$$Cu_L|Zn|ZnCl_2(0.0100\ mol/kg)|AgCl(s)|Ag|Pt|Cu_R$$

given that $\gamma_\pm$ of $ZnCl_2$ is 0.708 at this molality and temperature.

14.31 Calculation of $\mathscr{E}°$ of a *half*-reaction from $\mathscr{E}°$ values of two related half-reactions is a bit tricky. Given that at 25°C, $\mathscr{E}° = -0.424$ V for $Cr^{3+}(aq) + e^- \rightarrow Cr^{2+}(aq)$ and $\mathscr{E}° = -0.90$ V for $Cr^{2+}(aq) + 2e^- \rightarrow Cr$, find $\mathscr{E}°$ at 25°C for $Cr^{3+}(aq) + 3e^- \rightarrow Cr$. (*Hint:* Combine the two half-reactions to get the third, and combine the $\Delta G°$ values; then find $\mathscr{E}°$.)

14.32 Consider the Daniell cell

$$Cu'|Zn|ZnSO_4(m_1)\ \vdots\ CuSO_4(m_2)|Cu$$

with $m_1 = 0.00200$ mol/kg and $m_2 = 0.00100$ mol/kg. The cell's chemical reaction is $Zn + Cu^{2+}(aq) \rightarrow Zn^{2+}(aq) + Cu$. Estimate $\mathscr{E}$ at 25°C of this cell using the Davies equation to estimate activity coefficients and assuming that the salt bridge makes $\mathscr{E}_J$ negligible; neglect ion pairing.

Section 14.8

14.33 Using half-cells listed in Table 14.1, write the diagram of a chemical cell without a liquid junction whose electrolyte is (*a*) $KCl(aq)$; (*b*) $H_2SO_4(aq)$.

14.34 Using half-cells listed in Table 14.1, write the diagrams of three different chemical cells without liquid junctions that have $HCl(aq)$ as the electrolyte.

14.35 For the cell

$$Ag_L|AgNO_3(0.0100\ mol/kg)\ \vdots\ AgNO_3(0.0500\ mol/kg)|Ag_R$$

(*a*) use the Davies equation to find $\mathscr{E}_{298}$; neglect ion pairing and assume the salt bridge makes the net liquid-junction potential negligible. (*b*) Which terminal is at the higher potential? (*c*) When the cell is connected to a load, into which terminal do electrons flow from the load?

14.36 Calculate the emf of the cell (14.59) at 85°C if $P_L = 2521$ torr, $P_R = 666$ torr, and $m(HCl) = 0.200$ mol/kg.

Section 14.10

14.37 True or false? (*a*) Doubling the coefficients in a chemical reaction will square the value of the equilibrium constant, will double $\Delta G°$, and will not change $\mathscr{E}°$. (*b*) The chemical reaction of a galvanic cell must be an oxidation–reduction reaction.

14.38 For the cell at 25°C and 1 bar

$$Pt|Ag|AgCl(s)|HCl(aq)|Hg_2Cl_2(s)|Hg|Pt'$$

(*a*) write the cell reaction; (*b*) use Table 14.1 to find the emf if the HCl molality is 0.100 mol/kg; (*c*) find the emf if the HCl molality is 1.00 mol/kg. (*d*) For this cell, $(\partial\mathscr{E}/\partial T)_P = 0.338$ mV/K at 25°C and 1 bar. Find $\Delta G°$, $\Delta H°$, and $\Delta S°$ for the cell reaction at 25°C.

14.39 Find $K°$ at 25°C for $2H^+(aq) + D_2 \rightleftharpoons H_2 + 2D^+(aq)$ using data in Table 14.1.

14.40 Use data in Table 14.1 and the convention (10.85) to determine $\Delta_f G°_{298}$ for (*a*) $Na^+(aq)$; (*b*) $Cl^-(aq)$; (*c*) $Cu^{2+}(aq)$.

14.41 Use data in Table 14.1 to calculate $K°_{sp}$ of PbI_2 at 25°C.

14.42 Use Table 14.1 to calculate $\Delta G°$ and $K°$ at 298 K for (*a*) $Cl_2(g) + 2Br^-(aq) \rightleftharpoons 2Cl^-(aq) + Br_2(l)$; (*b*) $\frac{1}{2}Cl_2(g) + Br^-(aq) \rightleftharpoons Cl^-(aq) + \frac{1}{2}Br_2(l)$; (*c*) $2Ag + Cl_2(g) \rightleftharpoons 2AgCl(s)$; (*d*) $2AgCl(s) \rightleftharpoons 2Ag + Cl_2(g)$; (*e*) $3Fe^{2+}(aq) \rightleftharpoons Fe + 2Fe^{3+}(aq)$.

14.43 Measured values of $\mathscr{E}°/V$ for the hydrogen, Ag–AgCl cell (14.32) at 0°C, 10°C, 20°C, . . ., 70°C are 0.23638, 0.23126, 0.22540, 0.21887, 0.21190, 0.20431, 0.19630, 0.18762. Use a spreadsheet to do a least-squares fit of this data to Eq. (14.67). The values of the coefficients you find will differ from those in (14.68) because the fit (14.68) uses additional data.

14.44 Using only electrodes listed in Table 14.1, devise three different cells that each have the cell reaction $3Fe^{2+}(aq) \rightarrow 2Fe^{3+}(aq) + Fe(s)$. Calculate $\mathscr{E}°$, $n\mathscr{E}°$, and $\Delta G°$ for each of these cells at 25°C and 1 bar.

14.45 For the cell $Pt|Fe|Fe^{2+}\ \vdots\ Fe^{2+}, Fe^{3+}|Pt'$, one finds $(\partial\mathscr{E}°/\partial T)_P = 1.14$ mV/K at 25°C. (*a*) Write the cell reaction using the smallest possible whole numbers as the stoichiometric coefficients. (*b*) With the aid of data in Table 14.1, calculate $\Delta S°$, $\Delta G°$, and $\Delta H°$ for the cell reaction at 25°C.

14.46 Use data in Eq. (14.68) to find $\Delta G°$, $\Delta H°$, $\Delta S°$, and $\Delta C_P°$ at 10°C for the reaction $H_2(g) + 2AgCl(s) \rightarrow 2Ag(s) + 2HCl(aq)$.

14.47 The solubility product for AgI in water at 25°C is 8.2×10^{-17}. Use data in Table 14.1 to find $\mathscr{E}°$ for the Ag–AgI electrode at 25°C.

14.48 The cell

$$Pt|H_2(1\ bar)|HBr(aq)|AgBr(s)|Ag|Pt'$$

at 25°C with HBr molality 0.100 mol/kg has $\mathscr{E} = 0.200$ V. Find the activity coefficient $\gamma_\pm$ of $HBr(aq)$ at this molality.

14.49 Use data in Table 14.1 to calculate $\Delta_f G°_{298}$ of $HCl(aq)$ and of $Cl^-(aq)$.

14.50 Consider the cell at 1 bar H_2 pressure

$$Pt|H_2(g)|NaOH(m_1), NaCl(m_2)|AgCl(s)|Ag|Pt'$$

(a) Show that $\mathscr{E} = \mathscr{E}° - RTF^{-1} \ln a(H^+)a(Cl^-)$ and that

$$\mathscr{E} = \mathscr{E}° - \frac{RT}{F} \ln \frac{K_w° a(H_2O)\gamma(Cl^-)m(Cl^-)}{\gamma(OH^-)m(OH^-)}$$

where $K_w°$ is the ionization constant of water. (b) For this cell at 25°C, it is found that

$$\mathscr{E} - \mathscr{E}° + RTF^{-1} \ln\left[m(Cl^-)/m(OH^-)\right]$$

approaches the limit 0.8279 V as the ionic strength goes to zero. Calculate $K_w°$ at 25°C.

14.51 Consider the cell at 1 bar H_2 pressure

$$Pt|H_2(g)|HX(m_1), NaX(m_2), NaCl(m_3)|AgCl(s)|Ag|Pt'$$

where the anion X^- is acetate, $C_2H_3O_2^-$. (a) Show that

$$\mathscr{E} = \mathscr{E}° - \frac{RT}{F} \ln \frac{\gamma(Cl^-)m(Cl^-)\gamma(HX)m(HX)K_a°}{\gamma(X^-)m(X^-)m°}$$

where $K_a°$ is the ionization constant of the weak acid HX and $m° \equiv 1$ mol/kg. (b) The zero-ionic-strength limit of

$$\mathscr{E} - \mathscr{E}° + RTF^{-1} \ln\left[m(HX)m(Cl^-)/m(X^-)m°\right]$$

at 25°C is 0.2814 V. Calculate $K_a°$ for acetic acid at 25°C.

14.52 An excess of Sn powder is added to a 0.100 mol/kg aqueous $Pb(NO_3)_2$ solution at 25°C. Neglecting ion pairing and omitting activity coefficients, estimate the equilibrium molalities of Pb^{2+} and Sn^{2+}. Explain why omission of the activity coefficients is a reasonably good approximation here.

14.53 For the cell (14.71), the observed emf at 25°C was 612 mV. When solution X was replaced by a standard phosphate buffer solution whose assigned pH is 6.86, the emf was 741 mV. Find the pH of solution X.

Section 14.13
14.54 A solution containing 0.100 mol/kg NaCl and 0.200 mol/kg KBr is separated by a membrane permeable only to Na^+ from a solution that is 0.150 mol/kg in $NaNO_3$ and 0.150 mol/kg in KNO_3. Calculate the transmembrane potential at 25°C; state and justify any approximations you make.

Section 14.15
14.55 The electric dipole moment of HCl is 3.57×10^{-30} C m, and its bond length is 1.30 Å. If we pretend that the molecule consists of charges $+\delta$ and $-\delta$ separated by 1.30 Å, find δ. Also, calculate δ/e, where e is the proton charge.

14.56 Calculate the magnitude and the direction of the electric dipole moment of each of these systems: (a) a charge $2e$ at the origin, a charge $-0.5e$ at $(-1.5$ Å$, 0, 0)$, and a charge $-1.5e$ at $(1.0$ Å$, 0, 0)$; (b) a charge $2e$ at the origin, a charge $-e$ at $(1.0$ Å$, 0, 0)$, and a charge $-e$ at $(0, 1.0$ Å$, 0)$. (c) Repeat the

calculation of system (a) but put the origin at the $-1.5e$ charge and verify that the result is unchanged.

14.57 Prove that for a neutral system, the dipole moment is unchanged by a change in origin.

14.58 Derive Eq. (14.81) for the electric potential of a dipole as follows. (a) Show that

$$1/r_2 - 1/r_1 = (r_1^2 - r_2^2)/r_1r_2(r_1 + r_2)$$

(b) It is clear that, for $r \gg d$ in Fig. 14.20b, we have $r_1 \approx r_2 \approx r$, so the denominator in (a) is approximately $2r^3$. Also angles PAD, PBD, and PCD are approximately equal. Use the law of cosines (see any trigonometry text) to show that $r_1^2 - r_2^2 \approx 2rd\cos\theta$. (c) Use the results of (a) and (b) to verify (14.81).

14.59 Calculate the work needed to increase the distance between a K^+ ion and a Cl^- ion from 10 to 100 Å in (a) a vacuum; (b) water at 25°C (see Sec. 10.8 for data).

14.60 For $CCl_4(l)$ at 20°C and 1 atm, $\varepsilon_r = 2.24$ and $\rho = 1.59$ g/cm³. Calculate α and $\alpha/4\pi\varepsilon_0$ for CCl_4.

14.61 (a) For $CH_4(g)$ at 0°C and 1.000 atm, $\varepsilon_r = 1.00094$. Calculate α and $\alpha/4\pi\varepsilon_0$ for CH_4. (b) Calculate ε_r for CH_4 at 100°C and 10.0 atm.

14.62 Values of $10^5(\varepsilon_r - 1)$ for $H_2O(g)$ at 1.000 atm as a function of T are:

T/K	384.3	420.1	444.7	484.1	522.0
$10^5(\varepsilon_r - 1)$	546	466	412	353	302

Use a graphical method to find the dipole moment and polarizability of H_2O.

14.63 State whether each of the following is a molecular property or a macroscopic property: (a) μ; (b) α; (c) ε_r. Give the SI units of each of these properties.

14.64 For each of the following pairs of liquids, state which one has the larger dielectric constant at a given temperature. (a) CS_2 or CSe_2; (b) n-C_6H_{14} or n-$C_{10}H_{22}$; (c) o-dichlorobenzene or p-dichlorobenzene.

14.65 (a) Use calculus to show that $(\varepsilon_r - 1)/(\varepsilon_r + 2)$ increases as ε_r increases. (b) What are the minimum and maximum possible values of $(\varepsilon_r - 1)/(\varepsilon_r + 2)$?

Section 14.16
14.66 Use Eq. (14.90) and data following this equation to calculate the membrane potential of a resting squid nerve cell at 25°C. Compare with the experimental value −70 mV.

General
14.67 Consider the cell

$$Ag_L|AgCl(s)|HCl(m_1)\vdots HCl(m_2)|AgCl(s)|Ag_R$$

with $m_1 = 0.0100$ mol/kg and $m_2 = 0.100$ mol/kg. A theoretical equation gives the following estimate of the liquid-junction potential: $\mathscr{E}_J = -38$ mV. Use the single-ion form of the Davies equation to calculate the emf of this cell at 25°C.

14.68 Use the Davies equation to estimate $-\log_{10} a(H^+)$ in a 0.100-mol/kg aqueous HCl solution at 25°C. Compare with the assigned pH of 1.09 for this solution.

14.69 It was shown in Prob. 10.15 that $a_{c,i} = 0.997 a_{m,i}$ in water at 25°C and 1 bar. Use this result to calculate the difference $\mathscr{E}_m^\circ - \mathscr{E}_c^\circ$ at 25°C for the reaction $2Ag + Cu^{2+}(aq) \rightarrow 2Ag^+(aq) + Cu$, where $\mathscr{E}_m^\circ$ uses molality-scale standard states and $\mathscr{E}_c^\circ$ uses molar-concentration-scale standard states. (*Hint:*

Use K°.) Given that errors in $\mathscr{E}^\circ$ are typically a couple of millivolts, is the difference $\mathscr{E}_m^\circ - \mathscr{E}_c^\circ$ significant?

14.70 Give the SI unit of (*a*) charge; (*b*) length; (*c*) electric field; (*d*) emf; (*e*) electric potential difference; (*f*) dipole moment; (*g*) dielectric constant; (*h*) electrochemical potential.

14.71 The concentration cell (14.61) has the same kinds of half-cells, namely, $Cu|CuSO_4(aq)$, but has a nonzero emf due to differences in $CuSO_4$ molalities. Explain how it is possible for the galvanic cell $Cu_L|CuSO_4(aq)|Cu_R$, which contains only one $CuSO_4$ solution, to have a nonzero emf.

CHAPTER 15

Kinetic Theory of Gases

This chapter derives properties of an ideal gas based on a model of the gas as consisting of spherical molecules obeying classical mechanics. Properties derived include the equation of state (Secs. 15.2 and 15.3), the distribution of molecular speeds (Sec. 15.4), the average molecular speed (Sec. 15.5), molecular collision rates, and the average distance traveled between collisions (Sec. 15.7). These properties are important in discussing the rates of gas-phase reactions (Chapters 17 and 23) and in dealing with transport properties (for example, heat flow) in gases (Chapter 16).

15.1 KINETIC–MOLECULAR THEORY OF GASES

Chapters 1 to 12 use mainly a macroscopic approach. Chapters 13 and 14 use both macroscopic and molecular approaches. The remaining chapters use mainly a molecular approach to physical chemistry.

This chapter and several sections of the next discuss the **kinetic–molecular theory of gases** (**kinetic theory**, for short). The kinetic theory of gases pictures a gas as composed of a huge number of molecules whose size is small compared with the average distance between molecules. The molecules move freely and rapidly through space. Although this picture seems obvious nowadays, it wasn't until about 1850 that the kinetic theory began to win acceptance.

Kinetic theory began with Daniel Bernoulli's 1738 derivation of Boyle's law using Newton's laws of motion applied to molecules. Bernoulli's work was ignored for over 100 years. In 1845 John Waterston submitted a paper to the Royal Society of England that correctly developed many concepts of kinetic theory. Waterston's paper was rejected as "nonsense." Joule's experiments demonstrating that heat is a form of energy transfer made the ideas of kinetic theory seem plausible, and from 1848 to 1898, Joule, Clausius, Maxwell, and Boltzmann developed the kinetic theory of gases.

From 1870 to 1910 a controversy raged between the schools of Energetics and Atomism. The Atomists (led by Boltzmann) held that atoms and molecules were real entities, whereas the Energetists (Mach, Ostwald, Duhem) denied the existence of atoms and molecules and argued that the kinetic theory of gases was a mechanical model that imitated the properties of gases but did not correspond to the true structure of matter. [The Freudian-oriented sociologist Lewis Feuer speculates that Mach's opposition to atomism was an unconscious expression of revolt against his father; the shape of atoms unconsciously reminded Mach of testicles, and "a reality deatomized was a projection, we might infer, in which the father himself was unmanned." (L. Feuer, *Einstein and the Generations of Science,* Basic Books, 1974, p. 39.)]

The attacks on the kinetic theory of gases led Boltzmann to write in 1898: "I am conscious of being only an individual struggling weakly against the stream of time. But it still remains in my power to contribute in such a way that when the theory of gases is again revived, not too much will have to be rediscovered." (*Lectures on Gas Theory,* trans. S. G. Brush, University of California Press, 1964.) Some people have attributed Boltzmann's suicide in 1906 to depression resulting from attacks on the kinetic theory.

In 1905, Einstein applied kinetic theory to the Brownian motion of a tiny particle suspended in a fluid (Sec. 3.7). Einstein's theoretical equations were confirmed by Perrin's experiments in 1908, thereby convincing the Energetists of the reality of atoms and molecules.

The kinetic theory of gases uses a molecular picture to derive macroscopic properties of matter and is therefore a branch of statistical mechanics.

This chapter considers gases at low pressures (ideal gases). Since the molecules are far apart at low pressures, we ignore intermolecular forces (except at the moment of collision between two molecules; see Sec. 15.7). The kinetic theory of gases assumes the molecules to obey Newton's laws of motion. Actually, molecules obey quantum mechanics (Chapter 18). The use of classical mechanics leads to incorrect results for the heat capacities of gases (Sec. 15.10) but is an excellent approximation when dealing with properties such as pressure and diffusion.

15.2 PRESSURE OF AN IDEAL GAS

The pressure exerted by a gas on the walls of its container is due to bombardment of the walls by the gas molecules. The number of molecules in a gas is huge (2×10^{19} in 1 cm^3 at 1 atm and 25°C), and the number of molecules that hit a container wall in a tiny time interval is huge [3×10^{17} impacts with a 1-cm^2 wall in 1 microsecond for O_2 at 1 atm and 25°C; see Eq. (15.57)], so the individual impacts of molecules result in an apparently steady pressure on the wall.

Let the container be a rectangular box with sides of length l_x, l_y, and l_z. Let **v** be the *velocity* [Eq. (2.2)] of a given molecule. The components of **v** in the x, y, and z directions are v_x, v_y, and v_z. To find these components, we slide the vector **v** so that its tail lies at the coordinate origin and take the projections of **v** on the x, y, and z axes. The particle's **speed** v is the magnitude (length) of the vector **v**. Application of the Pythagorean theorem twice in Fig. 15.1 gives $v^2 = \overline{OC}^2 = \overline{OB}^2 + v_z^2 = v_x^2 + v_y^2 + v_z^2$; thus

$$v^2 = v_x^2 + v_y^2 + v_z^2 \tag{15.1}*$$

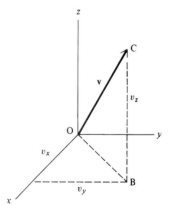

Figure 15.1

Velocity components of a molecule.

The velocity **v** is a vector. The speed v and the velocity components v_x, v_y, v_z are scalars. A velocity component like v_x can be positive, negative, or zero (corresponding to motion in the positive x direction, motion in the negative x direction, or no motion in the x direction), but v must by definition be positive or zero.

The kinetic energy ε_{tr} (epsilon$_{tr}$) of motion of a molecule of mass m through space is

$$\varepsilon_{tr} \equiv \tfrac{1}{2}mv^2 = \tfrac{1}{2}mv_x^2 + \tfrac{1}{2}mv_y^2 + \tfrac{1}{2}mv_z^2 \tag{15.2}*$$

We call ε_{tr} the **translational energy** of the molecule (Fig. 2.14).

Let the gas be in thermodynamic equilibrium. Because the gas and its surroundings are in thermal equilibrium, there is no net transfer of energy between them. We therefore assume that in a collision with the wall, a gas molecule does not change its translational energy.

In reality, a molecule colliding with the wall may undergo a change in ε_{tr}. However, for each gas molecule that loses translational energy to the wall molecules in a wall collision, another gas molecule will gain translational energy in a wall collision. Besides translational energy, the gas molecules also have rotational and vibrational energies (Sec. 2.11). In a wall collision, some of the molecule's translational energy may be transformed into rotational energy and vibrational energy, or vice versa. On the average, such transformations balance out, and wall collisions cause no net transfer of energy between translation and vibration–rotation in a gas in equilibrium. Since pressure is a property averaged over many wall collisions, we assume that in any one wall collision, there is no change in the molecule's translational kinetic energy. Although this assumption is false, it is "true" averaged over all the molecules, and hence gives the correct result for the pressure.

Let $\langle F \rangle$ denote the average value of some time-dependent property $F(t)$. To aid in finding the expression for the gas pressure, we shall find an equation for the average value of $F(t)$ over the time interval from t_1 to t_2. The average value of a quantity is the sum of its observed values divided by the number of observations:

$$\langle F \rangle = \frac{1}{n} \sum_{i=1}^{n} F_i \tag{15.3}$$

where the F_i's are the observed values. For the function $F(t)$, there are an infinite number of values, since there are an infinite number of times in the interval from t_1 to t_2. We therefore divide this interval into a large number n of subintervals, each of duration Δt, and take the limit as $n \to \infty$ and $\Delta t \to 0$. Multiplying and dividing each term in (15.3) by Δt, we have

$$\langle F \rangle = \lim_{n \to \infty} \frac{1}{n\,\Delta t} \left[F(t_1)\,\Delta t + F(t_1 + \Delta t)\Delta t + F(t_1 + 2\,\Delta t)\Delta t + \cdots + F(t_2)\Delta t \right]$$

The limit of the quantity in brackets is by Eq. (1.59) the definite integral of F from t_1 to t_2. Also, $n\,\Delta t = t_2 - t_1$. Therefore, the time average of $F(t)$ is

$$\langle F \rangle = \frac{1}{t_2 - t_1} \int_{t_1}^{t_2} F(t)\,dt \tag{15.4}$$

Throughout this chapter, angle brackets will denote an average, whether it is a time average, as in (15.4), or an average over the molecules, as in (15.8) and (15.10).

Figure 15.2 shows a molecule i colliding with wall W where W is parallel to the xz plane. Let i have the velocity components $v_{x,i}$, $v_{y,i}$, $v_{z,i}$ before the collision. For simplicity, we assume the molecule is reflected off the wall at the same angle it hit the wall. (Since the walls are not actually smooth but are made of molecules, this assumption does not reflect reality.) The collision thus changes $v_{y,i}$ to $-v_{y,i}$ and leaves $v_{x,i}$ and $v_{z,i}$ unchanged. This leaves the molecule's speed $v_i^2 = v_{x,i}^2 + v_{y,i}^2 + v_{z,i}^2$ unchanged and its translational energy $\frac{1}{2} m v_i^2$ unchanged.

To get the pressure on wall W, we need the average perpendicular force exerted by the molecules on this wall. Consider the motion of molecule i. It collides with W and then moves to the right, eventually colliding with wall W′, then moves to the left to collide again with W, etc. Collisions with the top, bottom, and side walls may occur between collisions with W and W′, but these collisions do not change $v_{y,i}$. For

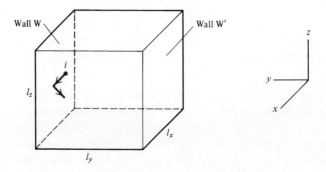

Figure 15.2

Molecule i colliding with container wall W.

our purposes, one "cycle" of motion of molecule i will extend from a time t_1 that just precedes a collision with W to a time t_2 that just precedes the next collision with W. During the very short time that i is colliding with W, Newton's second law $F_y = ma_y$ gives the y component of the force on i as

$$F_{y,i} = ma_{y,i} = m \frac{dv_{y,i}}{dt} = \frac{d}{dt}(mv_{y,i}) = \frac{dp_{y,i}}{dt} \tag{15.5}$$

where the y component of the (linear) momentum is defined by $p_y \equiv mv_y$. [The (**linear**) **momentum p** is a vector defined by $\mathbf{p} \equiv m\mathbf{v}$.] Let the collision of i with W extend from time t' to t''. Equation (15.5) gives $dp_{y,i} = F_{y,i}\, dt$. Integrating from t' to t'', we get $p_{y,i}(t'') - p_{y,i}(t') = \int_{t'}^{t''} F_{y,i}\, dt$. The y momentum of i before the wall collision is $p_{y,i}(t') = mv_{y,i}$, and the y momentum after the collision is $p_{y,i}(t'') = -mv_{y,i}$. Hence, $-2mv_{y,i} = \int_{t'}^{t''} F_{y,i}\, dt$.

Let $F_{W,i}$ be the perpendicular force on wall W due to collision with molecule i. Newton's third law (action = reaction) gives $F_{W,i} = -F_{y,i}$, so $2mv_{y,i} = \int_{t'}^{t''} F_{W,i}\, dt$. For times between t_1 and t_2 but outside the collision interval from t' to t'', the force $F_{W,i}$ is zero, since molecule i is not colliding with W during such times. Therefore the integration can be extended over the whole time interval t_1 to t_2 to get $2mv_{y,i} = \int_{t_1}^{t_2} F_{W,i}\, dt$. Use of (15.4) gives

$$2mv_{y,i} = \langle F_{W,i}\rangle(t_2 - t_1) \tag{15.6}$$

where $\langle F_{W,i}\rangle$ is the average perpendicular force exerted on wall W by molecule i.

The time $t_2 - t_1$ is the time needed for i to travel a distance $2l_y$ in the y direction, so as to bring it back to W. Since $\Delta y = v_y\, \Delta t$, we have $t_2 - t_1 = 2l_y/v_{y,i}$ and (15.6) becomes

$$\langle F_{W,i}\rangle = mv_{y,i}^2/l_y$$

The time average of the total force on wall W is found by summing the average forces of the individual molecules. If the number of gas molecules present is N, then

$$\langle F_W\rangle = \sum_{i=1}^{N} \langle F_{W,i}\rangle = \sum_{i=1}^{N} \frac{mv_{y,i}^2}{l_y} = \frac{m}{l_y}\sum_{i=1}^{N} v_{y,i}^2$$

We shall see in Sec. 15.4 that the molecules do not all move at the same speed. The average value of v_y^2 for all the molecules is by definition [Eq. (15.3)] given by $\langle v_y^2\rangle = N^{-1}\sum_i v_{y,i}^2$. Therefore $\langle F_W\rangle = mN\langle v_y^2\rangle/l_y$.

The pressure P on W equals the average perpendicular force $\langle F_W\rangle$ divided by the area $l_x l_z$ of W. We have $P = \langle F_W\rangle/l_x l_z$ and

$$P = mN\langle v_y^2\rangle/V \qquad \text{ideal gas} \tag{15.7}$$

where $V = l_x l_y l_z$ is the container volume.

There is nothing special about the y direction, so the properties of the gas must be the same in any direction. Therefore

$$\langle v_x^2\rangle = \langle v_y^2\rangle = \langle v_z^2\rangle \tag{15.8}$$

Furthermore, $\langle v^2\rangle$, the average of the square of the molecular speed, is [see Eqs. (15.1) and (15.3)]

$$\langle v^2\rangle = \langle v_x^2 + v_y^2 + v_z^2\rangle \equiv \frac{1}{N}\sum_{i=1}^{N}(v_{x,i}^2 + v_{y,i}^2 + v_{z,i}^2)$$

$$= \frac{1}{N}\sum_{i=1}^{N} v_{x,i}^2 + \frac{1}{N}\sum_{i=1}^{N} v_{y,i}^2 + \frac{1}{N}\sum_{i=1}^{N} v_{z,i}^2 \tag{15.9}$$

$$\langle v^2\rangle = \langle v_x^2\rangle + \langle v_y^2\rangle + \langle v_z^2\rangle = 3\langle v_y^2\rangle \tag{15.10}$$

where (15.8) was used. Therefore (15.7) becomes

$$P = \frac{mN\langle v^2 \rangle}{3V} \qquad \text{ideal gas} \tag{15.11}$$

Equation (15.11) expresses the macroscopic property of pressure in terms of the molecular properties m, N (the number of gas molecules), and $\langle v^2 \rangle$.

The translational kinetic energy ε_{tr} of molecule i is $\frac{1}{2}mv_i^2$. The average translational energy per molecule is

$$\langle \varepsilon_{tr} \rangle = \tfrac{1}{2}m\langle v^2 \rangle \tag{15.12}$$

This equation gives $\langle v^2 \rangle = 2\langle \varepsilon_{tr}\rangle/m$, so Eq. (15.11) can be written as $PV = \frac{2}{3}N\langle \varepsilon_{tr}\rangle$. The quantity $N\langle \varepsilon_{tr}\rangle$ is the *total translational kinetic energy* E_{tr} of the gas molecules. Therefore

$$PV = \tfrac{2}{3}E_{tr} \qquad \text{ideal gas} \tag{15.13}$$

The treatment just given assumed a pure gas with all molecules having the same mass m. If instead we have a mixture of gases b, c, and d, then at low pressure the gas molecules act independently of one another and the pressure P is the sum of pressures due to each kind of molecule: $P = P_b + P_c + P_d$ (Dalton's law). From (15.11), $P_b = \frac{1}{3}N_b m_b\langle v_b^2\rangle/V$, with similar equations for P_c and P_d.

15.3 TEMPERATURE

Consider two fluid (liquid or gaseous) thermodynamic systems 1 and 2 in contact. If the molecules of system 1 have an average translational kinetic energy $\langle \varepsilon_{tr}\rangle_1$ that is greater than the average translational energy $\langle \varepsilon_{tr}\rangle_2$ of system 2 molecules, the more energetic molecules of system 1 will tend to lose translational energy to the molecules of 2 in collisions with them. This transfer of energy at the molecular level will correspond to a flow of heat from 1 to 2 at the macroscopic level. Only if $\langle \varepsilon_{tr}\rangle_1$ equals $\langle \varepsilon_{tr}\rangle_2$ will there be no tendency for a net transfer of energy to occur in 1-2 collisions. But if there is no heat flow between 1 and 2, these systems are in thermal equilibrium and by the thermodynamic definition of temperature (Sec. 1.3) systems 1 and 2 have equal temperatures. Thus, when $\langle \varepsilon_{tr}\rangle_1 = \langle \varepsilon_{tr}\rangle_2$, we have $T_1 = T_2$; when $\langle \varepsilon_{tr}\rangle_1 > \langle \varepsilon_{tr}\rangle_2$, we have $T_1 > T_2$. This argument indicates that there is a correspondence between $\langle \varepsilon_{tr}\rangle$ and the macroscopic property T. The system's temperature is thus some function of the average translational energy per molecule: $T = T(\langle \varepsilon_{tr}\rangle)$. The ideal-gas kinetic–molecular equation (15.13) reads $PV = \frac{2}{3}E_{tr} = \frac{2}{3}N\langle \varepsilon_{tr}\rangle$. Since T is some function of $\langle \varepsilon_{tr}\rangle$, at constant temperature $\langle \varepsilon_{tr}\rangle$ is constant. Hence (15.13) says that PV of an ideal gas is constant at constant temperature. Thus Boyle's law has been derived from the kinetic–molecular theory.

The equation relating T and $\langle \varepsilon_{tr}\rangle$ cannot be found solely from the kinetic–molecular theory because the temperature scale is arbitrary and can be chosen in many ways (Sec. 1.3). The choice of the temperature scale will determine the relation between $\langle \varepsilon_{tr}\rangle$ and T. We *defined* the absolute temperature T in Sec. 1.5 in terms of properties of ideal gases. The ideal-gas equation $PV = nRT$ incorporates the definition of T. Comparison of $PV = nRT$ with $PV = \frac{2}{3}E_{tr}$ [Eq. (15.13)] gives

$$E_{tr} = \tfrac{3}{2}nRT \tag{15.14}$$

Had some other definition of temperature been chosen, a different relation between E_{tr} and temperature would have been obtained.

We have $E_{tr} = N\langle \varepsilon_{tr}\rangle$. Also, the number of moles is $n = N/N_A$, where N_A is the Avogadro constant and N is the number of gas molecules. Equation (15.14) becomes

$N\langle\varepsilon_{tr}\rangle = \frac{3}{2}NRT/N_A$, and $\langle\varepsilon_{tr}\rangle = \frac{3}{2}RT/N_A = \frac{3}{2}kT$, where $k \equiv R/N_A$ is **Boltzmann's constant** [Eq. (3.57)]. Thus

$$\langle\varepsilon_{tr}\rangle = \frac{3}{2}kT \tag{15.15}*$$

$$k \equiv R/N_A \tag{15.16}*$$

Equation (15.15) is the explicit relation between absolute temperature and average molecular translational energy. Although we derived (15.15) by considering an ideal gas, the discussion at the beginning of this section indicates it to be valid for any fluid system. [If system 1 is an ideal gas and system 2 is a general fluid system, the relation $\langle\varepsilon_{tr}\rangle_1 = \langle\varepsilon_{tr}\rangle_2$ when $T_1 = T_2$ shows that (15.15) holds for system 2.] The absolute temperature of a fluid (as defined by the ideal-gas scale and the thermodynamic scale) turns out to be directly proportional to the average translational kinetic energy per molecule: $T = \frac{2}{3}k^{-1}\langle\varepsilon_{tr}\rangle$. (See also Sec. 22.11.)

A fuller version of the argument given in the first paragraph of this section (see *Tabor,* sec. 3.4.1) shows that the argument depends on the validity of applying classical mechanics to molecular motions. A classical-mechanical description of the translational motion in a liquid or gas is usually accurate, but the motion of a molecule about its equilibrium position in a solid is not well described by classical mechanics. For solids, it turns out that the average kinetic energy of vibration of a molecule about its equilibrium position is equal to $\frac{3}{2}kT$ only in the high-temperature limit and differs from $\frac{3}{2}kT$ at lower temperatures, because of quantum-mechanical effects (Sec. 24.12). For liquids at very low temperatures (for example, He at 4 K or H_2 and Ne at 20 K), quantum-mechanical effects produce deviations from $\langle\varepsilon_{tr}\rangle = \frac{3}{2}kT$.

Besides translational energy, a molecule has rotational, vibrational, and electronic energies (Sec. 2.11). Monatomic molecules (for example, He or Ar) have no rotational or vibrational energy and ideal gases have no intermolecular energy. Therefore the thermodynamic internal energy U of an ideal gas of monatomic molecules is the sum of the total molecular translational energy E_{tr} and the total molecular electronic energy E_{el}:

$$U = E_{tr} + E_{el} = \frac{3}{2}nRT + E_{el} \qquad \text{ideal monatomic gas} \tag{15.17}$$

The heat capacity at constant volume is $C_V = (\partial U/\partial T)_V$ [Eq. (2.53)]. Provided the temperature is not extremely high, the molecular electrons will not be excited to higher energy levels and the electronic energy will remain constant as T varies. Therefore $C_V = \partial U/\partial T = \partial E_{tr}/\partial T = \frac{3}{2}nR$, and the molar C_V is

$$C_{V,m} = \frac{3}{2}R \qquad \text{ideal monatomic gas, } T \text{ not extremely high} \tag{15.18}$$

The use of $C_{P,m} - C_{V,m} = R$ [Eq. (2.72)] gives

$$C_{P,m} = \frac{5}{2}R \qquad \text{ideal monatomic gas, } T \text{ not extremely high} \tag{15.19}$$

These equations are well obeyed by monatomic gases at low densities. For example, for Ar at 1 atm, $C_{P,m}/R$ values are 2.515 at 200 K, 2.506 at 300 K, 2.501 at 600 K, and 2.500 at 2000 K. The small deviations from (15.19) are due to nonideality (intermolecular forces) and disappear in the limit of zero density.

Equation (15.15) enables us to estimate how fast molecules move. We have $\frac{3}{2}kT = \langle\varepsilon_{tr}\rangle = \frac{1}{2}m\langle v^2\rangle$, so

$$\langle v^2\rangle = 3kT/m \tag{15.20}$$

The square root of $\langle v^2\rangle$ is called the **root-mean-square speed** v_{rms}:

$$v_{rms} \equiv \langle v^2\rangle^{1/2} \tag{15.21}*$$

We shall see in Sec. 15.5 that v_{rms} differs slightly from the average speed $\langle v\rangle$. The quantity k/m in (15.20) equals $k/m = R/N_A m = R/M$, since the molar mass M (Sec. 1.4) equals the mass of one molecule times the number of molecules per mole. Recall that M is not the molecular weight. The molecular weight is dimensionless, whereas M has units of mass per mole. The square root of (15.20) is

$$v_{\text{rms}} = \left(\frac{3RT}{M} \right)^{1/2} \tag{15.22}$$

Equation (15.22) need not be memorized, since it can be quickly derived from $\langle \varepsilon_{\text{tr}} \rangle = \frac{3}{2}kT$ [Eq. (15.15)]. Equations (15.11) and (15.13) are also easily derived from (15.15).

It will be helpful to keep in mind the following notation:

m = mass of one gas molecule, $\qquad M$ = molar mass of the gas

N = number of gas molecules, $\qquad N_A$ = the Avogadro constant

15.4 DISTRIBUTION OF MOLECULAR SPEEDS IN AN IDEAL GAS

There is no reason to assume that all the molecules in a gas move at the same speed, and we now derive the distribution law for molecular speeds in an ideal gas in equilibrium.

What is meant by the *distribution* of molecular speeds? One might answer that we want to know how many molecules have any given speed v. But this approach makes no sense. Thus, suppose we ask how many molecules have speed 585 m/s. The answer is zero, since the chance that any molecule has a speed of exactly 585.000. . . m/s is vanishingly small. The only sensible approach is to ask how many molecules have a speed lying in some tiny range of speeds, for example, from 585.000 to 585.001 m/s.

We take an infinitesimal interval dv of speed, and we ask: How many molecules have a speed in the range from v to $v + dv$? Let this number be dN_v. The number dN_v is infinitesimal compared with 10^{23} but large compared with 1. The fraction of molecules having speeds in the range v to $v + dv$ is dN_v/N, where N is the total number of gas molecules. This fraction will obviously be proportional to the width of the infinitesimal interval of speeds: $dN_v/N \propto dv$. It will also depend on the location of the interval, that is, on the value of v. (For example, the number of molecules with speeds in the range of 627.400 to 627.401 m/s differs from the number with speeds in the range 585.000 to 585.001 m/s.) Therefore

the fraction of molecules with speeds between v and $v + dv = dN_v/N = G(v)\,dv$

$$\tag{15.23}$$

where $G(v)$ is some to-be-determined function of v.

The function $G(v)$ is the **distribution function** for molecular speeds. $G(v)\,dv$ gives the fraction of molecules with speed in the range v to $v + dv$. The fraction dN_v/N is the *probability* that a molecule will have its speed between v and $v + dv$. Therefore $G(v)\,dv$ is a probability. The distribution function $G(v)$ is also called a **probability density**, since it is a probability per unit interval of speed.

Let $\text{Pr}(v_1 \le v \le v_2)$ be the probability that a molecule's speed lies between v_1 and v_2. To find this probability (which equals the fraction of molecules with speeds in the range v_1 to v_2), we divide the interval from v_1 to v_2 into infinitesimal intervals each of width dv and sum the probabilities of being in each tiny interval:

$$\text{Pr}(v_1 \le v \le v_2) = G(v_1)\,dv + G(v_1 + dv)\,dv + G(v_1 + 2\,dv)\,dv$$
$$+ \cdots + G(v_2)\,dv$$

But the infinite sum of infinitesimals is the definite integral of $G(v)$ from v_1 to v_2 [see Eq. (1.59)], so

$$\text{Pr}(v_1 \le v \le v_2) = \int_{v_1}^{v_2} G(v)\,dv \tag{15.24}$$

A molecule must have its speed in the interval $0 \leq v \leq \infty$, so the probability (15.24) becomes 1 when $v_1 = 0$ and $v_2 = \infty$. Therefore $G(v)$ must satisfy

$$\int_0^\infty G(v)\, dv = 1 \qquad (15.25)$$

We now derive $G(v)$. This was first done by Maxwell in 1860. Amazingly enough, the only assumptions needed are: (**1**) the velocity distribution is independent of direction; (**2**) the value of v_y or v_z that a molecule has does not affect the probabilities of its having various values of v_x. Assumption 1 must be true because all directions of space are equivalent when no external electric or gravitational fields are present. Assumption 2 is discussed at the end of this section.

Distribution Function for v_x

To aid in finding $G(v)$, we first derive the distribution function for v_x, the x component of the velocity. Let g denote this function, so that $dN_{v_x}/N = g\, dv_x$, where dN_{v_x} is the number of molecules in the gas that have their x component of velocity lying between v_x and $v_x + dv_x$, with no specification being made about the v_y or v_z values of these molecules. The functions g and G are different functions, so we use different symbols for them. Since the v_y and v_z values are not specified for these dN_{v_x} molecules, the function g depends on v_x only, and

the fraction of molecules with x component of velocity between v_x and $v_x + dv_x =$

$$dN_{v_x}/N = g(v_x)\, dv_x \qquad (15.26)$$

The range of v_x is $-\infty$ to ∞, and, similar to (15.25), g must satisfy

$$\int_{-\infty}^\infty g(v_x)\, dv_x = 1 \qquad (15.27)$$

There are also distribution functions for v_y and v_z. Because the velocity distribution is independent of direction (assumption 1), the functional form of the v_y and v_z distribution functions is the same as that for the v_x distribution. Therefore

$$dN_{v_y}/N = g(v_y)\, dv_y \quad \text{and} \quad dN_{v_z}/N = g(v_z)\, dv_z \qquad (15.28)$$

where g is the same function in all three equations of (15.26) and (15.28).

We now ask: What is the probability that a molecule will simultaneously have its x component of velocity in the range v_x to $v_x + dv_x$, its y component of velocity in the range v_y to $v_y + dv_y$, and its z component of velocity in the range v_z to $v_z + dv_z$? By assumption 2, the probabilities for the various v_x values are independent of v_y and v_z. Therefore, we are dealing with probabilities of independent events. The probability that three independent events will all happen is equal to the product of the probabilities of the three events. [This theorem was previously used after Eq. (3.48).] Therefore the desired probability equals $g(v_x)\, dv_x \times g(v_y)\, dv_y \times g(v_z)\, dv_z$. Let $dN_{v_x v_y v_z}$ denote the number of molecules whose x, y, and z components of velocity all lie in the above ranges. Then

$$dN_{v_x v_y v_z}/N = g(v_x)g(v_y)g(v_z)\, dv_x\, dv_y\, dv_z \qquad (15.29)$$

The function $G(v)$ in (15.23) is the distribution function for *speeds*. The function $g(v_x)g(v_y)g(v_z)$ in (15.29) is the distribution function for *velocities*. The vector **v** is specified by giving its three components v_x, v_y, v_z, and the distribution function in (15.29) specifies these three components.

Let us set up a coordinate system whose axes give the values of v_x, v_y, and v_z (Fig. 15.3). The "space" defined by this coordinate system is called *velocity space* and is an abstract mathematical space rather than a physical space.

Figure 15.3

An infinitesimal box in velocity space.

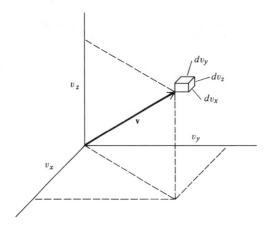

The probability $dN_{v_x v_y v_z}/N$ in (15.29) is the probability that a molecule has the tip of its velocity vector lying in a rectangular box located at (v_x, v_y, v_z) in velocity space and having edges dv_x, dv_y, and dv_z (Fig. 15.3). By assumption 1, the velocity distribution is independent of direction. Hence the probability $dN_{v_x v_y v_z}/N$ cannot depend on the direction of the velocity vector but only on its magnitude, which is the speed v. In other words, the probability that the tip of the velocity vector **v** lies in a tiny box with edges dv_x, dv_y, dv_z is the same for all boxes that lie at the same distance from the origin in Fig. 15.3. This makes sense, because all directions in space in the gas are equivalent, and the probability $dN_{v_x v_y v_z}/N$ cannot depend on the direction of motion of a molecule. Therefore the probability density $g(v_x)g(v_y)g(v_z)$ in (15.29) must be a function of v only. Calling this function $\phi(v)$, we have

$$g(v_x)g(v_y)g(v_z) = \phi(v) \tag{15.30}$$

[$\phi(v)$ is not the same function as $G(v)$ in (15.23). See the following discussion for the relation between ϕ and G.] We also have $v^2 = v_x^2 + v_y^2 + v_z^2$, Eq. (15.1). Equations (15.30) and (15.1) are sufficient to determine g. Before reading on, you might try to think of a function g that would have the property (15.30).

To find g, we take $(\partial/\partial v_x)_{v_y, v_z}$ of (15.30), obtaining

$$g'(v_x)g(v_y)g(v_z) = \frac{d\phi(v)}{dv}\frac{\partial v}{\partial v_x}$$

where the chain rule was used to find $\partial\phi/\partial v_x$. From $v^2 = v_x^2 + v_y^2 + v_z^2$ [Eq. (15.1)], we get $2v\,dv = 2v_x\,dv_x + 2v_y\,dv_y + 2v_z\,dv_z$, so $\partial v/\partial v_x = v_x/v$. This also follows by direct differentiation of $v = (v_x^2 + v_y^2 + v_z^2)^{1/2}$. We have

$$g'(v_x)g(v_y)g(v_z) = \phi'(v) \cdot v_x/v$$

Dividing this equation by $v_x g(v_x)g(v_y)g(v_z) = v_x \phi(v)$, we get

$$\frac{g'(v_x)}{v_x g(v_x)} = \frac{1}{v}\frac{\phi'(v)}{\phi(v)} \tag{15.31}$$

Since v_x, v_y, and v_z occur symmetrically in (15.30) and (15.1), taking $\partial/\partial v_y$ and $\partial/\partial v_z$ of (15.30) will give equations similar to (15.31):

$$\frac{g'(v_y)}{v_y g(v_y)} = \frac{1}{v}\frac{\phi'(v)}{\phi(v)} \quad \text{and} \quad \frac{g'(v_z)}{v_z g(v_z)} = \frac{1}{v}\frac{\phi'(v)}{\phi(v)} \tag{15.32}$$

Equations (15.31) and (15.32) give

$$\frac{g'(v_x)}{v_x g(v_x)} = \frac{g'(v_y)}{v_y g(v_y)} \equiv b \tag{15.33}$$

where we defined the quantity b. Since b equals $g'(v_y)/v_y g(v_y)$, b must be independent of v_x and v_z. But since b equals $g'(v_x)/v_x g(v_x)$, b must be independent of v_y and v_z. Therefore b is independent of v_x, v_y, and v_z and is a constant.

Equation (15.33) reads $bv_x = (dg/dv_x)/g$. Separating the variables g and v_x, we have $dg/g = bv_x \, dv_x$. Integration gives $\ln g = \frac{1}{2}bv_x^2 + c$, where c is an integration constant. Hence $g = \exp(\frac{1}{2}bv_x^2) \exp c$, where $\exp c \equiv e^c$. We have

$$g = A \exp\left(\tfrac{1}{2}bv_x^2\right) \tag{15.34}$$

where $A \equiv \exp c$ is a constant. We have found the distribution function g for v_x, and as a check, we note that it satisfies (15.30), since

$$g(v_x)g(v_y)g(v_z) = A^3 \exp\left(\tfrac{1}{2}bv_x^2\right) \exp\left(\tfrac{1}{2}bv_y^2\right) \exp\left(\tfrac{1}{2}bv_z^2\right)$$

$$= A^3 \exp\left[\tfrac{1}{2}b(v_x^2 + v_y^2 + v_z^2)\right] = A^3 e^{bv^2/2}$$

We still must evaluate the constants A and b in (15.34). To evaluate A, we substitute (15.34) into $\int_{-\infty}^{\infty} g(v_x)\, dv_x = 1$ [Eq. (15.27)] to get

$$A \int_{-\infty}^{\infty} e^{bv_x^2/2} \, dv_x = 1 \tag{15.35}$$

(b must be negative; otherwise the integral doesn't exist.)

Table 15.1 lists some definite integrals useful in the kinetic theory of gases. (The derivation of these integrals is outlined in Probs. 15.18 and 15.19.) Recall that

$$n! \equiv n(n-1)(n-2)\cdots 1 \quad \text{and} \quad 0! \equiv 1$$

where n is a positive integer. Integrals 2 and 5 in the table are special cases ($n = 0$) of integrals 3 and 6, respectively.

We must evaluate the integral in (15.35). Since the integration variable in a definite integral is a dummy variable (Sec. 1.8), we can change v_x in (15.35) to x. We must evaluate $\int_{-\infty}^{\infty} e^{bx^2/2} \, dx$. Using first integral 1 in Table 15.1 with $n = 0$ and $a = -b/2$ and then integral 2 with $a = -b/2$, we have

$$\int_{-\infty}^{\infty} e^{bx^2/2} \, dx = 2 \int_{0}^{\infty} e^{bx^2/2} \, dx = 2 \frac{\pi^{1/2}}{2(-b/2)^{1/2}} = \left(-\frac{2\pi}{b}\right)^{1/2}$$

Equation (15.35) becomes $A(-2\pi/b)^{1/2} = 1$ and $A = (-b/2\pi)^{1/2}$. The $g(v_x)$ distribution function in (15.34) becomes

$$g(v_x) = (-b/2\pi)^{1/2} e^{bv_x^2/2} \tag{15.36}$$

TABLE 15.1

Integrals Occurring in the Kinetic Theory of Gases

Even powers of x	Odd powers of x
1. $\displaystyle\int_{-\infty}^{\infty} x^{2n}e^{-ax^2}\, dx = 2\int_{0}^{\infty} x^{2n}e^{-ax^2}\, dx$	**4.** $\displaystyle\int_{-\infty}^{\infty} x^{2n+1}e^{-ax^2}\, dx = 0$
2. $\displaystyle\int_{0}^{\infty} e^{-ax^2}\, dx = \frac{\pi^{1/2}}{2a^{1/2}}$	**5.** $\displaystyle\int_{0}^{\infty} xe^{-ax^2}\, dx = \frac{1}{2a}$
3. $\displaystyle\int_{0}^{\infty} x^{2n}e^{-ax^2}\, dx = \frac{(2n)!\,\pi^{1/2}}{2^{2n+1}n!a^{n+1/2}}$	**6.** $\displaystyle\int_{0}^{\infty} x^{2n+1}e^{-ax^2}\, dx = \frac{n!}{2a^{n+1}}$

where $a > 0$ and $n = 0, 1, 2, \ldots$

To evaluate b, we use the relation $\langle \varepsilon_{tr} \rangle = \frac{3}{2}kT$, Eq. (15.15). The average translational kinetic energy of a molecule is $\langle \varepsilon_{tr} \rangle = \frac{1}{2}m\langle v^2 \rangle = \frac{3}{2}m\langle v_x^2 \rangle$, where we used (15.10) with v_y replaced by v_x. Therefore $\frac{3}{2}m\langle v_x^2 \rangle = \frac{3}{2}kT$, and

$$\langle v_x^2 \rangle = kT/m \qquad (15.37)$$

We now calculate $\langle v_x^2 \rangle$ from the distribution function (15.36) and compare the result with (15.37) to find b.

To evaluate $\langle v_x^2 \rangle$, the following theorem is used. Let $g(w)$ be the distribution function for the continuous variable w; that is, the probability that this variable lies between w and $w + dw$ is $g(w)\,dw$. Then the average value of any function $f(w)$ is

$$\langle f(w) \rangle = \int_{w_{min}}^{w_{max}} f(w)g(w)\,dw \qquad (15.38)*$$

where w_{min} and w_{max} are the minimum and maximum values of w. In using (15.38), keep in mind these ranges for v and v_x:

$$0 \le v < \infty \quad \text{and} \quad -\infty < v_x < \infty \qquad (15.39)*$$

The proof of (15.38) follows.

We first consider a variable that takes on only discrete values (rather than a continuous range of values as is true for w). Suppose a class of seven students takes a five-question quiz, and the scores are 20, 40, 40, 80, 80, 80, and 100. The average of the square of the scores, $\langle s^2 \rangle$, is

$$\langle s^2 \rangle = (20^2 + 40^2 + 40^2 + 80^2 + 80^2 + 80^2 + 100^2)/7$$

$$= [0(0)^2 + 1(20)^2 + 2(40)^2 + 0(60)^2 + 3(80)^2 + 1(100)^2]/7$$

$$= \frac{1}{N}\sum_s n_s s^2 = \sum_s \frac{n_s}{N} s^2$$

where the s's are the possible scores (0, 20, 40, 60, 80, 100), n_s is the number of people who got score s, N is the total number of people, and the sum goes over all the possible scores. If N is very large (as it is for molecules), then n_s/N is the probability $p(s)$ of getting score s. Therefore $\langle s^2 \rangle = \sum_s p(s)s^2$.

The same argument holds for the average value of any function of s. For example, $\langle s \rangle = \sum_s p(s)s$ and $\langle 2s^3 \rangle = \sum_s p(s)2s^3$. If $f(s)$ is any function of s, then

$$\langle f(s) \rangle = \sum_s p(s)f(s) \qquad (15.40)*$$

where $p(s)$ is the probability of observing the value s for a variable that takes on discrete values.

For a variable w that takes on a continuous range of values, Eq. (15.40) must be modified. The probability $p(s)$ of having the value s is replaced by the probability that w lies in the infinitesimal range from w to $w + dw$. This probability is $g(w)\,dw$, where $g(w)$ is the distribution function (probability density) for w. For a continuous variable, (15.40) becomes $\langle f(w) \rangle = \sum_w f(w)g(w)\,dw$. But the infinite sum over infinitesimal quantities is the definite integral over the full range of w. Therefore we have proved (15.38).

It readily follows from (15.38) that the average of a sum equals the sum of the averages. If f_1 and f_2 are any two functions of w, then (Prob. 15.16)

$$\langle f_1(w) + f_2(w) \rangle = \langle f_1(w) \rangle + \langle f_2(w) \rangle \qquad (15.41)*$$

However, the average of a product is not necessarily equal to the product of the averages (see Prob. 15.24). If c is a constant, then (Prob. 15.16)

$$\langle cf(w)\rangle = c\langle f(w)\rangle$$

Returning to the evaluation of $\langle v_x^2\rangle$ in (15.37), we use (15.38) with $w = v_x$, $f(w) = v_x^2$, $w_{min} = -\infty$, $w_{max} = \infty$, and $g(v_x)$ given by Eq. (15.36) to get

$$\langle v_x^2\rangle = \int_{-\infty}^{\infty} v_x^2 g(v_x)\, dv_x = \int_{-\infty}^{\infty} v_x^2 \left(\frac{-b}{2\pi}\right)^{1/2} e^{bv_x^2/2}\, dv_x$$

Changing the dummy variable v_x in the integral to x and using the Table 15.1 integrals 1 and 3 with $n = 1$ and $a = -b/2$, we get

$$\langle v_x^2\rangle = 2\left(\frac{-b}{2\pi}\right)^{1/2} \int_0^{\infty} x^2 e^{bx^2/2}\, dx = 2\left(\frac{-b}{2\pi}\right)^{1/2} \frac{2!\,\pi^{1/2}}{2^3 1!(-b/2)^{3/2}} = \frac{1}{-b}$$

Comparison with (15.37) gives $-1/b = kT/m$, and $b = -m/kT$.

The distribution function (15.36) for v_x is therefore [see also (15.26)]

$$\frac{1}{N}\frac{dN_{v_x}}{dv_x} = g(v_x) = \left(\frac{m}{2\pi kT}\right)^{1/2} e^{-mv_x^2/2kT} \tag{15.42}*$$

where dN_{v_x} is the number of molecules with x component of velocity between v_x and $v_x + dv_x$.

Equation (15.42) looks complicated but is reasonably easy to remember since it has the form $g = \text{const.} \times e^{-\varepsilon_{tr,x}/kT}$, where $\varepsilon_{tr,x} = \frac{1}{2}mv_x^2$ is the kinetic energy of motion in the x direction and m is the mass of one molecule. The constant that multiplies the exponential is determined by the requirement that $\int_{-\infty}^{\infty} g\, dv_x = 1$. Note the presence of kT as a characteristic energy in (15.42) and (15.15); this is a general occurrence in statistical mechanics.

The equations for $g(v_y)$ and $g(v_z)$ are obtained from (15.42) by replacing v_x by v_y and v_z.

Distribution Function for v

Now that $g(v_x)$ has been found, we can find the distribution function $G(v)$ for the speed. $G(v)\, dv$ is the probability that a molecule's speed lies between v and $v + dv$; that is, $G(v)\, dv$ is the probability that in the v_x, v_y, v_z coordinate system (Fig. 15.3) the tip of the molecule's velocity vector $\mathbf{v}$ lies within a thin spherical shell of inner radius v and outer radius $v + dv$. Consider a tiny rectangular box lying within this shell and having edges dv_x, dv_y, and dv_z (Fig. 15.4). The probability that the tip of $\mathbf{v}$ lies in this tiny box is given by (15.29) as

$$g(v_x)g(v_y)g(v_z)\, dv_x\, dv_y\, dv_z = (m/2\pi kT)^{3/2} e^{-m(v_x^2+v_y^2+v_z^2)/2kT}\, dv_x\, dv_y\, dv_z$$

$$g(v_x)g(v_y)g(v_z)\, dv_x\, dv_y\, dv_z = \left(\frac{m}{2\pi kT}\right)^{3/2} e^{-mv^2/2kT}\, dv_x\, dv_y\, dv_z \tag{15.43}$$

where (15.42) and the analogous equations for $g(v_y)$ and $g(v_z)$ were used.

The probability that the tip of $\mathbf{v}$ lies in the thin spherical shell is the sum of the probabilities (15.43) for all the tiny rectangular boxes that make up the thin shell:

$$G(v)\, dv = \sum_{shell} \left(\frac{m}{2\pi kT}\right)^{3/2} e^{-mv^2/2kT}\, dv_x\, dv_y\, dv_z$$

$$= \left(\frac{m}{2\pi kT}\right)^{3/2} e^{-mv^2/2kT} \sum_{shell} dv_x\, dv_y\, dv_z$$

Figure 15.4

A thin spherical shell in velocity space and an infinitesimal box within this shell.

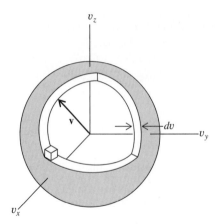

since the function $e^{-mv^2/2kT}$ is constant within the shell. (v varies only infinitesimally in the shell.) The quantity $dv_x\, dv_y\, dv_z$ is the volume of one of the tiny rectangular boxes, and the sum of these volumes over the shell is the volume of the shell. The shell has outer and inner radii $v + dv$ and v, so the shell volume is

$$\tfrac{4}{3}\pi(v + dv)^3 - \tfrac{4}{3}\pi v^3 = \tfrac{4}{3}\pi[v^3 + 3v^2\, dv + 3v\,(dv)^2 + (dv)^3] - \tfrac{4}{3}\pi v^3 = 4\pi v^2\, dv$$

since $(dv)^2$ and $(dv)^3$ are negligible compared with dv. [The quantity $4\pi v^2\, dv$ is the differential of the spherical volume $\tfrac{4}{3}\pi v^3 \equiv V$. This is true because $dV = (dV/dv)\, dv$, where dV is the sphere's infinitesimal volume change produced by an increase in radius in velocity space from v to $v + dv$.] Use of $4\pi v^2\, dv$ for the shell volume gives the final result for the distribution function $G(v)$ in (15.23) as

$$\frac{dN_v}{N} = G(v)\, dv = \left(\frac{m}{2\pi kT}\right)^{3/2} e^{-mv^2/2kT} 4\pi v^2\, dv \qquad (15.44)$$

Since $G(v)\, dv$ is a probability and probabilities are dimensionless, $G(v)$ has dimensions of v^{-1} and SI units of s/m.

Equation (15.44) is for a pure gas. In a mixture of gases b and c, each gas has its own distribution of speeds with N and m in (15.44) replaced by N_b and m_b or by N_c and m_c.

Summarizing, we have shown that the fraction of ideal-gas molecules with x component of velocity in the range v_x to $v_x + dv_x$ is $dN_{v_x}/N = g(v_x)\, dv_x$, where $g(v_x)$ is given by (15.42). The fraction of molecules with speed in the range v to $v + dv$ is $dN_v/N = G(v)\, dv$, where $G(v)\, dv$ is given by (15.44). The relation between g and G is

$$G(v) = g(v_x)g(v_y)g(v_z) \cdot 4\pi v^2 \qquad \textbf{(15.45)*}$$

where the factor $4\pi v^2$ comes from the volume $4\pi v^2\, dv$ of the thin spherical shell.

Equations (15.42) and (15.44) are the **Maxwell distribution laws** for v_x and v in a gas and are the key results of this section. (We shall see at the end of Sec. 22.11 that the Maxwell distribution laws hold in liquids as well as in gases.)

In using the Maxwell distribution laws, it is helpful to note that $m/k = N_A m/N_A k = M/R$, where M is the molar mass and R is the gas constant:

$$m/k = M/R \qquad \textbf{(15.46)*}$$

EXAMPLE 15.1 Number of molecules with speeds in a tiny interval

For 1.00 mol of $CH_4(g)$ at 0°C and 1 atm, find the number of molecules whose speed lies in the range 90.000 m/s to 90.002 m/s.

The most accurate way to answer this question is to use Eq. (15.24), which reads $\Pr(v_1 \leq v \leq v_2) = \int_{v_1}^{v_2} G(v)\, dv$. However, because the interval from v_1 to v_2 in this problem is very small, a simpler way to proceed is to consider the interval as infinitesimal and use Eq. (15.44). (See also Prob. 15.14.) To evaluate $m/2kT$ in (15.44), we use (15.46):

$$\frac{m}{2kT} = \frac{M}{2RT} = \frac{16.0 \text{ g mol}^{-1}}{2(8.314 \text{ J mol}^{-1} \text{ K}^{-1})(273 \text{ K})} = \frac{0.0160 \text{ kg}}{4540 \text{ J}} = 3.52 \times 10^{-6} \text{ s}^2/\text{m}^2$$

since 1 J = 1 kg m^2/s^2. Note especially that the units of M were changed from g/mol to kg/mol, to match the units of R. A mole has 6.02×10^{23} molecules, $dv = 0.002$ m/s, and (15.44) gives

$$dN_v = (6.02 \times 10^{23})[(3.52 \times 10^{-6} \text{ s}^2/\text{m}^2)/\pi]^{3/2}$$

$$\times e^{-(3.52 \times 10^{-6} \text{ s}^2/\text{m}^2)(90.0 \text{ m/s})^2} 4\pi (90.0 \text{ m/s})^2 (0.002 \text{ m/s})$$

$$dN_v = 1.4 \times 10^{17}$$

Exercise

How many molecules in a sample of 10.0 g of He(g) at 300 K and 1 bar have speeds in the range 3000.000 to 3000.001 m/s? (*Answer:* 1.6×10^{16}.)

The function $g(v_x)$ has the form $A \exp(-av_x^2)$, and has its maximum at $v_x = 0$. Figure 15.5 plots $g(v_x)$ for N$_2$ at two temperatures. As T increases, $g(v_x)$ broadens. This distribution law, called the **normal** or **gaussian distribution,** occurs in many contexts besides kinetic theory (for example, the distribution of heights in a population, the distribution of random errors of measurement).

The distribution function $G(v)$ for speeds is plotted for N$_2$ in Fig. 15.6 at two temperatures. For very small v, the exponential $e^{-mv^2/2kT}$ is close to 1, and $G(v)$ increases as v^2. For very large v, the exponential factor dominates the v^2 factor, and $G(v)$ decreases rapidly with increasing v. As T increases, the distribution curve broadens and shifts to higher speeds. Figure 15.7a shows $G(v)$ for several gases at a fixed T. The curves differ because of the quantity $m/k = M/R$ that occurs in the Maxwell distribution (15.44). Recall from Eq. (15.15) that $\langle \varepsilon_{tr} \rangle \equiv \frac{1}{2}m\langle v^2 \rangle$ is the same for any gas at a given T. Therefore, lighter molecules tend to move faster at a given T.

To get a physical feeling for the numerical values of $G(v)$ in these plots, note from (15.44) that, for a mole of gas, $(6 \times 10^{20})G(v)/(\text{s/m})$ is the number of gas molecules with speed in the range v to $v + 0.001$ m/s.

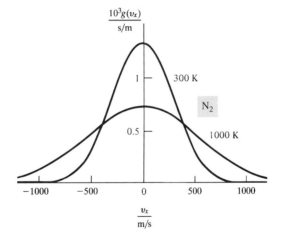

Figure 15.5

Distribution functions for v_x in N$_2$ gas at 300 K and at 1000 K. (N$_2$ has molecular weight 28. For H$_2$ with molecular weight 2, these curves apply at 21 K and 71 K.)

Figure 15.6

Distribution functions for speeds in N_2 at 300 K and at 1000 K.

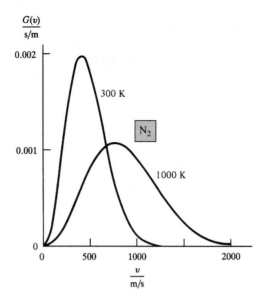

Figure 15.7

(a) Distribution functions for v in several gases at 300 K. Molecular weights are given in parentheses. (b) Probability-density plot for velocities.

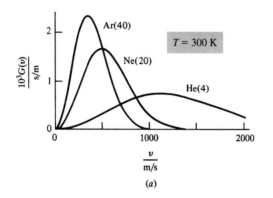

(a)

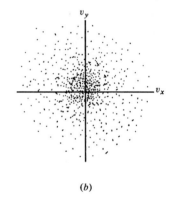

(b)

The most-probable value of v_x is zero (Fig. 15.5). Likewise, the most-probable values of v_y and v_z are zero. Hence if we make a three-dimensional plot of the probabilities of the various values of v_x, v_y, and v_z by putting dots at points in velocity space so that the density of dots in each region is proportional to the probability that $\mathbf{v}$ lies in that region, the maximum density of dots will occur at the origin ($v_x = v_y = v_z = 0$). Figure 15.7b shows a two-dimensional cross section of such a plot.

Although the most-probable value of each component of the velocity is zero, Fig. 15.6 shows that the most-probable value of the speed is not zero. This apparent contradiction is reconciled by realizing that, although the probability density $g(v_x)g(v_y)g(v_z)$, which is proportional to $e^{-mv^2/2kT}$ [Eq. (15.43)], is a maximum at the origin (where $v = 0$ and $v_x = v_y = v_z = 0$) and decreases with increasing v, the volume $4\pi v^2\,dv$ in (15.44) of the thin spherical shell increases with increasing v. These two opposing factors then give a nonzero most-probable value for v. Although the number of dots in any small region of fixed size will decrease as we move away from the origin in Fig. 15.7b, the number of dots in a thin shell of fixed thickness in Fig. 15.4 will initially increase, then reach a maximum, and finally decrease as v (the distance to the origin) increases.

Maxwell derived Eq. (15.44) in 1860. It took until 1955 for the first accurate direct experimental verification of this distribution law to be made [R. C. Miller and P. Kusch, *Phys. Rev.,* **99,** 1314 (1955)]. Miller and Kusch measured the distribution of

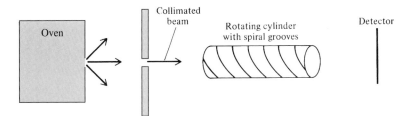

Figure 15.8

Apparatus for testing the Maxwell distribution law.

speeds in a beam of gas molecules emerging from a small hole in an oven (Fig. 15.8). The rotating cylinder with spiral grooves acts as a speed selector, since only molecules with a certain value of v will pass through the grooves without striking the walls of the grooves. Changing the rate of cylinder rotation changes the speed selected. These workers found excellent agreement with the predictions of the Maxwell law. (The velocity distribution in the beam is not Maxwellian: There are no negative values of v_x; moreover, there is a higher fraction of fast molecules in the beam than in the oven, since fast molecules hit the walls more often than slow ones and so are more likely to escape from the oven. Measurement of the distribution of speeds in the beam enables calculation of the distribution in the oven.)

The derivation of the Maxwell distribution given in this section is Maxwell's original 1860 derivation. This derivation does not bring out the physical process by which the distribution of speeds is attained. If two gas samples at different temperatures are mixed, eventually equilibrium will be attained at some intermediate temperature T', and a Maxwellian distribution characteristic of T' is established. The physical mechanism that brings about the Maxwellian distribution is the collisions between molecules. In 1872, Boltzmann derived the Maxwell distribution by a method based on the dynamics of molecular collisions.

A significant defect of Maxwell's original derivation is the use of assumption 2, which asserts that the variables v_x, v_y, and v_z are statistically independent of one another. This assumption is not really obvious, so the derivation cannot be considered truly satisfactory. Later derivations by Maxwell and by Boltzmann used more plausible assumptions. Perhaps the best theoretical justification of the Maxwell speed distribution law is to consider it as a special case of the more general Boltzmann distribution law, which will be derived in Sec. 22.5 from general statistical-mechanical principles. Under conditions of extremely low temperature or extremely high density, the Maxwell distribution law does not hold because of quantum-mechanical effects, and the velocity components v_x, v_y, and v_z are no longer statistically independent; see Secs. 22.3, 22.5, and Prob. 22.23.

15.5   APPLICATIONS OF THE MAXWELL DISTRIBUTION

The Maxwell distribution function $G(v)$ can be used to calculate the average value of any function of v, since Eqs. (15.38) and (15.39) give

$$\langle f(v) \rangle = \int_0^\infty f(v)G(v)\,dv$$

For example, the **average speed** $\langle v \rangle$ is

$$\langle v \rangle = \int_0^\infty vG(v)\,dv = 4\pi\left(\frac{m}{2\pi kT}\right)^{3/2}\int_0^\infty e^{-mv^2/2kT}v^3\,dv$$

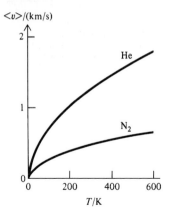

Figure 15.9

Average speed versus temperature in He and in N_2.

where $G(v)$ was taken from Eq. (15.44). Use of integral 6 in Table 15.1 with $n = 1$ and $a = m/2kT$ gives

$$\langle v \rangle = 4\pi \left(\frac{m}{2\pi kT} \right)^{3/2} \frac{1}{2(m/2kT)^2} = \left(\frac{8kT}{\pi m} \right)^{1/2} = \left(\frac{8N_A kT}{\pi N_A m} \right)^{1/2}$$

$$\langle v \rangle = \left(\frac{8RT}{\pi M} \right)^{1/2} \qquad \textbf{(15.47)*}$$

The average molecular speed is proportional to $T^{1/2}$ and inversely proportional to $M^{1/2}$ (Fig. 15.9).

We already know [Eqs. (15.22) and (15.21)] that $v_{rms} \equiv \langle v^2 \rangle^{1/2} = (3RT/M)^{1/2}$, but we could check this by calculating $\int_0^\infty v^2 G(v)\, dv$ (Prob. 15.22).

The **most probable speed** v_{mp} is the speed for which $G(v)$ is a maximum (see Fig. 15.6). Setting $dG(v)/dv = 0$, one finds (Prob. 15.21)

$$v_{mp} = (2RT/M)^{1/2}$$

The speeds v_{mp}, $\langle v \rangle$, v_{rms} stand in the ratio $2^{1/2}:(8/\pi)^{1/2}:3^{1/2} = 1.414:1.596:1.732$. Thus $v_{mp}:\langle v \rangle:v_{rms} = 1:1.128:1.225$.

EXAMPLE 15.2 Calculation of $\langle v \rangle$

Find $\langle v \rangle$ for (a) O_2 at 25°C and 1 bar; (b) H_2 at 25°C and 1 bar.

Substitution in (15.47) gives for O_2

$$\langle v \rangle = \left[\frac{8(8.314 \text{ J mol}^{-1} \text{ K}^{-1})(298 \text{ K})}{\pi(0.0320 \text{ kg mol}^{-1})} \right]^{1/2} = \left[\frac{8(8.314 \text{ kg m}^2 \text{ s}^{-2})298}{3.14(0.0320 \text{ kg})} \right]^{1/2}$$

$$\langle v \rangle = 444 \text{ m/s} \qquad \text{for } O_2 \text{ at } 25°\text{C} \qquad (15.48)$$

since $1 \text{ J} = 1 \text{ kg m}^2 \text{ s}^{-2}$ [Eq. (2.14)]. Because the SI unit of the joule was used in R, the molar mass M was expressed in kg mol^{-1}. For the reader who prefers cgs units, the calculation would be

$$\langle v \rangle = \left[\frac{8(8.314 \times 10^7 \text{ ergs mol}^{-1} \text{ K}^{-1})(298 \text{ K})}{3.14(32.0 \text{ g mol}^{-1})} \right]^{1/2} = 44400 \text{ cm/s}$$

since $1 \text{ J} = 10^7$ ergs and $1 \text{ erg} = 1 \text{ g cm}^2 \text{ s}^{-2}$. The speed 444 m/s is 993 mi/hr, so at room temperature the gas molecules are not sluggards.

For H_2, one obtains $\langle v \rangle = 1770$ m/s at 25°C. At the same temperature, the H_2 and O_2 molecules have the same average kinetic energy $\langle \frac{1}{2}mv^2 \rangle = \frac{3}{2}kT$. Hence the H_2 molecules must move faster on the average to compensate for their lighter mass. The H_2 molecule is one-sixteenth as heavy as an O_2 molecule and moves four times as fast, on the average.

Exercise

Find v_{rms} in He(g) at 0°C. (*Answer:* 1300 m/s.)

At 25°C and 1 bar, the speed of sound in O_2 is 330 m/s and in H_2 is 1330 m/s, so $\langle v \rangle$ is the same order of magnitude as the speed of sound in the gas. This is reasonable, since the propagation of sound is closely connected with the motions of the gas molecules. It can be shown that the speed of sound in an ideal gas equals $(\gamma RT/M)^{1/2}$, where $\gamma \equiv C_P/C_V$; see *Zemansky and Dittman*, sec. 5-7.

The fraction of molecules whose x component of velocity lies between 0 and v_x' is given by (15.24) with v replaced with v_x and (15.42) as

$$\frac{N(0 \leq v_x \leq v_x')}{N} = \int_0^{v_x'} g(v_x) \, dv_x = \left(\frac{m}{2\pi kT}\right)^{1/2} \int_0^{v_x'} e^{-mv_x^2/2kT} \, dv_x \quad (15.49)$$

where $N(0 \leq v_x \leq v_x')$ is the number of molecules with v_x in the range 0 to v_x'. Let $s \equiv (m/kT)^{1/2} v_x$. Then $ds = (m/kT)^{1/2} \, dv_x$, and

$$\frac{N(0 \leq v_x \leq v_x')}{N} = \frac{1}{(2\pi)^{1/2}} \int_0^u e^{-s^2/2} \, ds \quad \text{where } u \equiv \left(\frac{m}{kT}\right)^{1/2} v_x' \quad (15.50)$$

The indefinite integral $\int e^{-s^2/2} \, ds$ cannot be expressed in terms of elementary functions. However, by expanding the integrand $e^{-s^2/2}$ in a Taylor series and integrating term by term, one can express the integral as an infinite series, thereby allowing the definite integral in (15.50) to be evaluated for any desired value of u (Prob. 15.27). The *Gauss error integral* $I(u)$ is defined as

$$I(u) \equiv \frac{1}{(2\pi)^{1/2}} \int_0^u e^{-s^2/2} \, ds \quad (15.51)$$

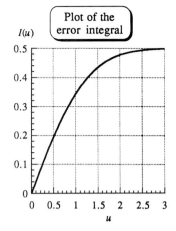

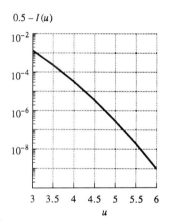

Equation (15.50) becomes $N(0 \leq v_x \leq v_x')/N = I(v_x'\sqrt{m/kT})$. The function $I(u)$ is graphed in Fig. 15.10. Tables of $I(u)$ are given in statistics handbooks.

The fraction of molecules with speed v lying in the range 0 to v' can be expressed in terms of the function I (see Prob. 15.28). Because of the exponential falloff of $G(v)$ with increasing v, only a small fraction of molecules have speeds that greatly exceed v_{mp}. For example, the fraction of molecules with speeds exceeding $3v_{mp}$ is 0.0004, only 1 molecule in 10^{15} has a speed exceeding $6v_{mp}$, and it is virtually certain that in a mole of gas not a single molecule has a speed exceeding $9v_{mp}$.

The Maxwell distribution can be rewritten in terms of the translational kinetic energy $\varepsilon_{tr} = \frac{1}{2}mv^2$. We have $v = (2\varepsilon_{tr}/m)^{1/2}$ and $dv = (1/2m)^{1/2}\varepsilon_{tr}^{-1/2} \, d\varepsilon_{tr}$. If a molecule has a speed between v and $v + dv$, its translational energy is between $\frac{1}{2}mv^2$ and $\frac{1}{2}m(v + dv)^2 = \frac{1}{2}m[v^2 + 2v \, dv + (dv)^2] = \frac{1}{2}mv^2 + mv \, dv$; that is, its translational energy is between ε_{tr} and $\varepsilon_{tr} + d\varepsilon_{tr}$, where $\varepsilon_{tr} = \frac{1}{2}mv^2$ and $d\varepsilon_{tr} = mv \, dv$. Replacing v and dv in (15.44) by their equivalents, we get as the distribution function for ε_{tr}

$$\frac{dN_{\varepsilon_{tr}}}{N} = 2\pi \left(\frac{1}{\pi kT}\right)^{3/2} \varepsilon_{tr}^{1/2} e^{-\varepsilon_{tr}/kT} \, d\varepsilon_{tr} \quad (15.52)$$

where $dN_{\varepsilon_{tr}}$ is the number of molecules with translational energy between ε_{tr} and $\varepsilon_{tr} + d\varepsilon_{tr}$. Figure 15.11 plots the distribution function (15.52). Note the difference in shape between this figure and Fig. 15.6.

Figure 15.10

Plots of the Gauss error integral $I(u)$ and of $0.5 - I(u)$.

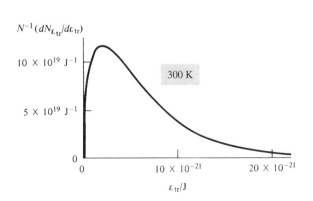

Figure 15.11

Distribution function for kinetic energy at 300 K. This curve applies to any gas.

We previously found that the *average* translational kinetic energy is the same for all gases at the same T. Equation (15.52), in which m is absent, shows that the ε_{tr} distribution function is the same for all gases at the same T.

15.6 COLLISIONS WITH A WALL AND EFFUSION

We now calculate the rate of collisions of gas molecules with a container wall. Let dN_W be the number of molecules that hit wall W in Fig. 15.2 during the infinitesimal time interval dt. The collision rate dN_W/dt will clearly be proportional to the area $\mathcal{A}$ of the wall. What molecular properties of the gas will influence the collision rate with the wall? Clearly, this rate will be proportional to the average speed of the molecules and will be proportional to the number of molecules per unit volume, N/V. Thus we expect $dN_W/dt = c\mathcal{A}\langle v\rangle N/V$, where c is a constant. As a check, note that dN_W/dt has dimensions of time^{-1} and $\mathcal{A}\langle v\rangle N/V$ also has dimensions of time^{-1}. We now find the explicit formula for dN_W/dt.

Consider a molecule with y component of speed in the range v_y to $v_y + dv_y$. For this molecule to hit wall W in the interval dt, it (a) must be moving to the left (that is, it must have $v_y > 0$) and (b) must be close enough to W to reach it in time dt or less—the molecule travels a distance $v_y\,dt$ in the y direction (which is perpendicular to the wall) in time dt and must be within this distance from W.

The number of molecules with y component of velocity in the range v_y to $v_y + dv_y$ is [Eq. (15.42)] $dN_{v_y} = Ng(v_y)\,dv_y$. The molecules are evenly distributed throughout the container, so the fraction of molecules within distance $v_y\,dt$ of W is $v_y\,dt/l_y$. The product $dN_{v_y}(v_y\,dt/l_y)$ gives the number of molecules that have y component of velocity in the range v_y to $v_y + dv_y$ and are within distance $v_y\,dt$ of W. This number is $[Ng(v_y)v_y/l_y]\,dv_y\,dt$. Dividing by the area $l_x l_z$ of the wall, we get as the number of collisions per unit area of W in time dt due to molecules with y velocity between v_y and $v_y + dv_y$

$$(N/V)g(v_y)v_y\,dv_y\,dt \tag{15.53}$$

where $V = l_x l_y l_z$ is the container volume. To get the total number of collisions with W in time dt, we sum (15.53) over all positive values of v_y and multiply by the area $\mathcal{A}$ of W. [A molecule with $v_y < 0$ does not satisfy requirement (a) and cannot hit the wall in time dt. Therefore only positive v_y's are summed over.] The infinite sum of infinitesimal quantities is the definite integral over v_y from 0 to ∞, and the number of molecules that hit W in time dt is

$$dN_W = \mathcal{A}\frac{N}{V}\left[\int_0^\infty g(v_y)v_y\,dv_y\right]dt \tag{15.54}$$

The use of (15.42) for $g(v_y)$ and integral 5 in Table 15.1 gives

$$\int_0^\infty g(v_y)v_y\,dv_y = \left(\frac{m}{2\pi kT}\right)^{1/2}\int_0^\infty v_y e^{-mv_y^2/2kT}\,dv_y = \left(\frac{RT}{2\pi M}\right)^{1/2} = \tfrac{1}{4}\langle v\rangle$$

where (15.47) for $\langle v\rangle$ was used. Also, since $PV = nRT = (N/N_A)RT$, the number of molecules per unit volume is

$$N/V = PN_A/RT \tag{15.55}$$

Equation (15.54) becomes

$$\frac{1}{\mathcal{A}}\frac{dN_W}{dt} = \frac{1}{4}\frac{N}{V}\langle v\rangle = \frac{1}{4}\frac{PN_A}{RT}\left(\frac{8RT}{\pi M}\right)^{1/2} \tag{15.56}$$

where dN_W/dt is the rate of molecular collisions with a wall of area $\mathcal{A}$.

EXAMPLE 15.3 Wall collision rate

Calculate the number of wall collisions per second per square centimeter in O_2 at 25°C and 1.00 atm.

Using (15.48) for $\langle v \rangle$, we get

$$\frac{1}{\mathscr{A}} \frac{dN_W}{dt} = \frac{1}{4} \frac{(1.00 \text{ atm})(6.02 \times 10^{23} \text{ mol}^{-1})}{(82.06 \text{ cm}^3 \text{ atm mol}^{-1} \text{ K}^{-1})(298 \text{ K})} \ 4.44 \times 10^4 \text{ cm s}^{-1}$$

$$= 2.7 \times 10^{23} \text{ cm}^{-2} \text{ s}^{-1} \tag{15.57}$$

Exercise

For N_2 at 350 K and 2.00 atm, find the number of molecular collisions with a container wall of area 1.00 cm² that occur in 1.00 s. (*Answer:* 5.4×10^{23}.)

Suppose there is a *tiny* hole of area $\mathscr{A}_{\text{hole}}$ in the wall and that outside the container is a vacuum. [If the hole is not tiny, the gas will escape rapidly, thereby destroying the Maxwellian distribution of velocities used to derive (15.56).] Molecules hitting the hole escape, and the rate of escape is given by (15.56) as

$$\frac{dN}{dt} = -\frac{PN_A\mathscr{A}_{\text{hole}}}{(2\pi MRT)^{1/2}} \tag{15.58}$$

The minus sign is present because dN, the infinitesimal change in the number of molecules in the container, is negative. Escape of a gas through a tiny hole is called **effusion.** The rate of effusion is proportional to $M^{-1/2}$ (*Graham's law of effusion*). Differences in effusion rates can be used to separate isotopic species. During World War II, repeated effusion of $UF_6(g)$ was used to separate $^{235}UF_6$ from $^{238}UF_6$ to obtain the fissionable ^{235}U for use in an atomic bomb.

Let λ be the average distance a gas molecule travels between collisions with other gas molecules. For (15.58) to apply, the hole's diameter d_{hole} must be substantially less than λ. Otherwise molecules will collide with one another near the hole, thereby developing a collective flow through the hole, which is a departure from the Maxwell distribution of Fig. 15.5. This bulk flow arises because the escape of molecules through the hole depletes the region near the hole of gas molecules, lowering the pressure near the hole. Therefore, molecules in the region of the hole experience fewer collisions on the side near the hole than on the side away from the hole and experience a net force toward the hole. (In effusion, there is no net force toward the hole.) Flow of gas or liquid due to a pressure difference is called *viscous flow, convective flow,* or *bulk flow* (see Sec. 16.3). Effusion is an example of *free-molecule* (or *Knudsen) flow;* here λ is so large that intermolecular collisions can be ignored.

Another requirement for the applicability of (15.58) is that the wall be thin. Otherwise, escaping gas molecules colliding with the sides of the hole can be reflected back into the container.

In the Knudsen method for finding the vapor pressure of a solid, the weight loss due to effusion of vapor in equilibrium with the solid in a container with a tiny hole is measured. Since P is constant, we have $dN/dt = \Delta N/\Delta t$. Measurement of $\Delta N/\Delta t$ allows the vapor pressure P to be calculated from (15.58); see Prob. 15.31.

For a liquid in equilibrium with its vapor, the rate of evaporation of liquid molecules equals the rate of condensation of gas molecules. A reasonable assumption is that virtually every vapor molecule that strikes the liquid's surface condenses to liquid. Equation (15.58) with P equal to the vapor pressure allows calculation of the rate at which vapor molecules hit the surface and hence allows calculation of the rate at which liquid molecules evaporate (see Prob. 15.33).

Figure 15.12

Colliding molecules.

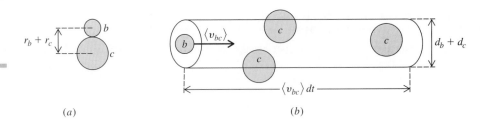

(a)

(b)

MOLECULAR COLLISIONS AND MEAN FREE PATH

Kinetic theory allows the rate of intermolecular collisions to be calculated. We adopt the crude model of a molecule as a hard sphere of diameter d. We assume that no intermolecular forces exist except at the moment of collision, when the molecules bounce off each other like two colliding billiard balls. At high gas pressures, intermolecular forces are substantial, and the equations derived in this section do not apply. Intermolecular collisions are important in reaction kinetics, since molecules must collide with each other in order to react. Intermolecular collisions also serve to maintain the Maxwell distribution of speeds.

The rigorous derivation of the collision rate is complicated, so we shall give only a nonrigorous treatment. (The full derivation is given in *Present*, sec. 5-2.)

We shall consider collisions in a pure gas and also in a mixture of two gases b and c. For either pure gas b or a mixture of b and c, let $z_b(b)$ be the number of collisions per unit time that one particular b molecule makes with other b molecules, and let Z_{bb} be the total number of all b-b collisions per unit time and per unit volume of gas. For a mixture of b and c, let $z_b(c)$ be the number of collisions per unit time that one particular b molecule makes with c molecules, and let Z_{bc} be the total number of b-c collisions per unit time per unit volume. Thus

$$z_b(c) \equiv \text{collision rate for one particular } b \text{ molecule with } c \text{ molecules}$$

$$Z_{bc} \equiv \text{total } b\text{-}c \text{ collision rate per unit volume}$$

Let N_b and N_c be the numbers of b and c molecules present.

To calculate $z_b(c)$, we shall pretend that all molecules are at rest except one particular b molecule, which moves at the constant speed $\langle v_{bc} \rangle$, where $\langle v_{bc} \rangle$ is the average speed of b molecules relative to c molecules in the actual gas with all molecules in motion. Let d_b and d_c be the diameters of the b and c molecules, and let r_b and r_c be their radii. The moving b molecule will collide with a c molecule whenever the distance between the centers of the pair is within $\frac{1}{2}(d_b + d_c) = r_b + r_c$ (Fig. 15.12a). Imagine a cylinder of radius $r_b + r_c$ centered on the moving b molecule (Fig. 15.12b). In time dt, the moving molecule will travel a distance $\langle v_{bc} \rangle \, dt$ and will sweep out a cylinder of volume $\pi(r_b + r_c)^2 \cdot \langle v_{bc} \rangle \, dt \equiv V_{\text{cyl}}$. The moving b molecule will collide with all c molecules whose centers lie within this cylinder. Since the stationary c molecules are uniformly distributed throughout the container volume V, the number of c molecules with centers in the cylinder is $(V_{\text{cyl}}/V)N_c$, and this is the number of collisions between a particular b molecule and c molecules in time dt. The number of such collisions per unit time is thus $z_b(c) = (V_{\text{cyl}}/V)N_c/dt$, and

$$z_b(c) = (N_c/V)\pi(r_b + r_c)^2 \langle v_{bc} \rangle \tag{15.59}$$

To complete the derivation, we need $\langle v_{bc} \rangle$, the average speed of b molecules relative to c molecules. Figure 15.13a shows the displacement vectors $\mathbf{r}_b$ and $\mathbf{r}_c$ of molecules b and c from the coordinate origin. The position of b relative to c is specified by the vector $\mathbf{r}_{bc}$. These three vectors form a triangle, and the usual rule of vector addition gives $\mathbf{r}_c + \mathbf{r}_{bc} = \mathbf{r}_b$, or $\mathbf{r}_{bc} = \mathbf{r}_b - \mathbf{r}_c$. Differentiating this equation with respect to t and using $\mathbf{v} \equiv d\mathbf{r}/dt$ [Eq. (2.2)], we get $\mathbf{v}_{bc} = \mathbf{v}_b - \mathbf{v}_c$. This equation shows that the vectors $\mathbf{v}_b$, $\mathbf{v}_c$, and $\mathbf{v}_{bc}$ form a triangle (Fig. 15.13b).

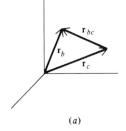

(a)

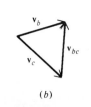

(b)

Figure 15.13

Molecular displacement vectors and velocity vectors.

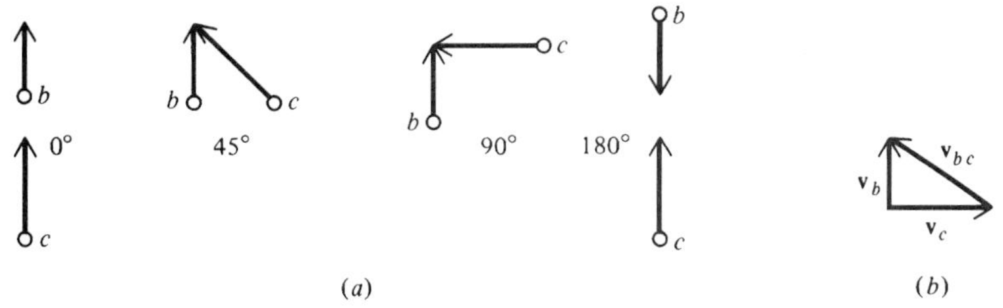

Figure 15.14

Colliding molecules.

In a b-c collision, the molecules can approach each other at any angle from 0 to 180° (Fig. 15.14a). The average approach angle is 90°, so to calculate $\langle v_{bc} \rangle$ we imagine a 90° collision between a b molecule with speed $\langle v_b \rangle$ and a c molecule with speed $\langle v_c \rangle$, where these average speeds are given by (15.47). The triangle formed by the vectors in Fig. 15.13b is a right triangle for a 90° collision (Fig. 15.14b), and the Pythagorean theorem gives

$$\langle v_{bc} \rangle^2 = \langle v_b \rangle^2 + \langle v_c \rangle^2 = 8RT/\pi M_b + 8RT/\pi M_c$$

where M_b and M_c are the molar masses of gases b and c. Substitution in (15.59) gives

$$z_b(c) = \pi(r_b + r_c)^2 [\langle v_b \rangle^2 + \langle v_c \rangle^2]^{1/2}(N_c/V)$$

$$z_b(c) = \pi(r_b + r_c)^2 \left[\frac{8RT}{\pi} \left(\frac{1}{M_b} + \frac{1}{M_c} \right) \right]^{1/2} \frac{N_c}{V} \qquad (15.60)$$

To find $z_b(b)$, we put $c = b$ in (15.60) to get

$$z_b(b) = 2^{1/2}\pi d_b^2 \langle v_b \rangle \frac{N_b}{V} = 2^{1/2}\pi d_b^2 \left(\frac{8RT}{\pi M_b} \right)^{1/2} \frac{P_b N_A}{RT} \qquad (15.61)$$

where $N_b/V = P_b N_A/RT$ [Eq. (15.55)] was used. Equation (15.61) is valid whether b is a pure gas or a component of an ideal gas mixture.

The total b-c collision rate equals the collision rate $z_b(c)$ of a particular b molecule with c molecules multiplied by the number of b molecules. Hence the total b-c collision rate per unit volume is $Z_{bc} = N_b z_b(c)/V$, and

$$Z_{bc} = \pi(r_b + r_c)^2 \left[\frac{8RT}{\pi} \left(\frac{1}{M_b} + \frac{1}{M_c} \right) \right]^{1/2} \left(\frac{N_b}{V} \right) \left(\frac{N_c}{V} \right) \qquad (15.62)$$

If we were to calculate the total b-b collision rate by multiplying the b-b collision rate $z_b(b)$ of one b molecule by the number of b molecules, we would be counting each b-b collision twice. For example, the collision between molecules b_1 and b_2 would be counted once as one of the collisions made by b_1 and once as one of the collisions made by b_2. (See also Prob. 15.36.) Therefore a factor of $\frac{1}{2}$ must be included, and the total b-b collision rate per unit volume is $Z_{bb} = \frac{1}{2} N_b z_b(b)/V$, or

$$Z_{bb} = \frac{1}{2^{1/2}} \pi d_b^2 \langle v_b \rangle \left(\frac{N_b}{V} \right)^2 = \frac{1}{2^{1/2}} \pi d_b^2 \left(\frac{8RT}{\pi M_b} \right)^{1/2} \left(\frac{P_b N_A}{RT} \right)^2 \qquad (15.63)$$

EXAMPLE 15.4 Intermolecular collision rates

For O_2 at 25°C and 1.00 atm, estimate $z_b(b)$ and Z_{bb}. The bond distance in O_2 is 1.2 Å.

The O_2 molecule is neither hard nor spherical, but a reasonable estimate of the diameter d in the hard-sphere model might be twice the bond length: $d \approx 2.4$ Å.

Equations (15.61) and (15.48) give the collision rate of one particular O_2 molecule as

$$z_b(b) \approx 2^{1/2}\pi(2.4 \times 10^{-8}\ \text{cm})^2(44400\ \text{cm/s})\frac{(1.00\ \text{atm})(6.02 \times 10^{23}\ \text{mol}^{-1})}{(82.06\ \text{cm}^3\text{-atm/mol-K})(298\ \text{K})}$$

$$z_b(b) \approx 2.8 \times 10^9\ \text{collns./s} \qquad \text{for } O_2 \text{ at } 25°C \text{ and } 1\ \text{atm} \qquad (15.64)$$

Even though the gas molecules are far apart from each other compared with the molecular diameter, the very high average molecular speed causes a molecule to make very many collisions per second. Substitution in (15.63) gives the total number of collisions per second per cubic centimeter of gas as

$$Z_{bb} \approx 3.4 \times 10^{28}\ \text{cm}^{-3}\ \text{s}^{-1} \qquad \text{for } O_2 \text{ at } 25°C \text{ and } 1\ \text{atm} \qquad (15.65)$$

Exercise

Verify (15.65) for Z_{bb}.

The **mean free path** λ (lambda) is the average distance a molecule travels between two successive intermolecular collisions. In a mixture of gases b and c, λ_b differs from λ_c. The gas molecules have a distribution of speeds. (More precisely, each species has its own distribution.) The speed of a given b molecule changes many times each second as a result of intermolecular collisions. Over a long time t, the average speed of a given b molecule is $\langle v_b \rangle$, the distance it travels is $\langle v_b \rangle t$, and the number of collisions it makes is $[z_b(b) + z_b(c)]t$. Therefore the average distance traveled by a b molecule between collisions is $\lambda_b = \langle v_b \rangle t/[z_b(b) + z_b(c)]t$, and

$$\lambda_b = \langle v_b \rangle/[z_b(b) + z_b(c)] \qquad (15.66)$$

where $\langle v_b \rangle$, $z_b(b)$, and $z_b(c)$ are given by (15.47), (15.61), and (15.60). (We made the plausible assumption that the time-average speed of a single b molecule equals the average speed of all b molecules in the gas at a particular moment.)

In pure gas b, there are no b-c collisions, $z_b(c) = 0$, and

$$\lambda = \frac{\langle v_b \rangle}{z_b(b)} = \frac{1}{2^{1/2}\pi d^2(N/V)} = \frac{1}{2^{1/2}\pi d^2}\frac{RT}{PN_A} \qquad \text{pure gas} \qquad (15.67)$$

As P increases, the mean free path λ decreases (Fig. 15.15), because of the increase in molecular collision rate $z_b(b)$ [Eq. (15.61)].

For O_2 at $25°C$ and 1 atm, the use of (15.67), (15.64), and (15.48) gives

$$\lambda = \frac{4.44 \times 10^4\ \text{cm s}^{-1}}{2.8 \times 10^9\ \text{s}^{-1}} = 1.6 \times 10^{-5}\ \text{cm} = 1600\ \text{Å} \qquad \text{for } O_2 \text{ at } 25°C \text{ and } 1\ \text{atm}$$

The average time between collisions is $\lambda/\langle v \rangle = 1/z_b(b)$, which equals 4×10^{-10} s for O_2 at $25°C$ and 1 atm.

Note that at 1 atm and $25°C$: (a) λ is small compared with macroscopic dimensions (1 cm), so the molecules collide with each other far more often than with the container walls; (b) λ is large compared with molecular dimensions (10^{-8} cm), so a molecule moves a distance of many molecular diameters before colliding with another molecule; (c) λ is large compared with the average distance between gas molecules (about 35 Å—Sec. 2.11 and Prob. 2.54).

A good vacuum is 10^{-6} torr $\approx 10^{-9}$ atm. Since λ is inversely proportional to P, the mean free path in O_2 at $25°C$ and 10^{-9} atm is 1.6×10^{-5} cm $\times 10^9 = 160$ m $= 0.1$ mi, which is large compared with the usual container dimensions. In a good vacuum, the gas molecules collide far more often with the container walls than with one another. At 10^{-9} atm and $25°C$, a given O_2 molecule makes only an average of 2.8 collisions per second with other gas molecules.

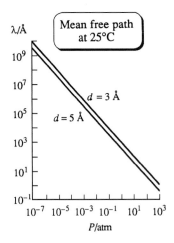

λ/Å

Mean free path at 25°C

10^9

10^7

$d = 3$ Å

10^5

$d = 5$ Å

10^3

10^1

10^{-1}

10^{-7} 10^{-5} 10^{-3} 10^{-1} 10^1 10^3

P/atm

Figure 15.15

Mean free path versus pressure in a gas at 25°C for two different molecular diameters. Both scales are logarithmic.

The mean free path plays a key role in transport properties of gases (Chapter 16), which depend on molecular collisions.

The very high rate of intermolecular collisions in a gas at 1 atm raises the following point. Chemical reactions between gases often proceed quite slowly. Equation (15.64) shows that if every b-c collision in a gas mixture caused a chemical reaction, gas-phase reactions at 1 atm would be over in a fraction of a second, which is contrary to experience. Usually, only a very small fraction of collisions result in reaction, since the colliding molecules must have a certain minimum energy of relative motion in order to react and must be properly oriented. Because of the exponential falloff of the distribution function $G(v)$ at high v, only a small fraction of molecules may have enough energy to react.

The direction of motion of a molecule changes at each collision, and the short mean free path ($\approx 10^{-5}$ cm) at 1 atm makes the molecular path resemble Fig. 3.14 for Brownian motion.

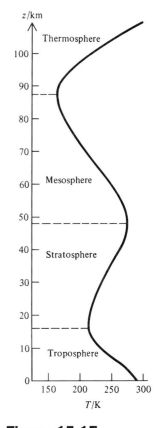

In deriving (15.7) for the gas pressure, we ignored intermolecular collisions and assumed that a given molecule changes its v_y value only when it collides with wall W or wall W′ in Fig. 15.2. Actually (unless P is extremely low), a gas molecule makes many, many collisions with other gas molecules between two successive wall collisions. These intermolecular collisions produce random changes in v_y of molecule i in Fig. 15.2. For a rigorous derivation of (15.7) that allows for collisions, see *Present*, pp. 18–19.

Figure 15.16

A thin layer of gas in a gravitational field.

15.8   THE BAROMETRIC FORMULA

For an ideal gas in the earth's gravitational field, the gas pressure decreases with increasing altitude. Consider a thin layer of gas at altitude z above the earth's surface (Fig. 15.16). Let the layer have thickness dz, mass dm, and cross-sectional area $\mathcal{A}$. The upward force F_{up} on this layer results from the pressure P of the gas just below the layer, and $F_{up} = P\mathcal{A}$. The downward force on the layer results from the gravitational force $dm\, g$ [Eq. (2.6)] and the pressure $P + dP$ of the gas just above the layer. (dP is negative.) Thus, $F_{down} = g\, dm + (P + dP)\mathcal{A}$. Since the layer is in mechanical equilibrium, these forces balance: $P\mathcal{A} = (P + dP)\mathcal{A} + g\, dm$, and

$$dP = -(g/\mathcal{A})\, dm \tag{15.68}$$

The ideal-gas law gives $PV = P(\mathcal{A}\, dz) = (dm/M)RT$, so $dm = (PM\mathcal{A}/RT)\, dz$, and (15.68) becomes after separating the variables P and z

$$dP/P = -(Mg/RT)\, dz \tag{15.69}$$

Let P_0 be the pressure at zero altitude (the earth's surface) and P' be the pressure at altitude z'. Integration of (15.69) gives

$$\ln \frac{P'}{P_0} = -\int_0^{z'} \frac{Mg}{RT}\, dz \tag{15.70}$$

Since the thickness of the earth's atmosphere is much less than the earth's radius, we can neglect the variation of the gravitational acceleration g with altitude z. The temperature of the atmosphere changes substantially as z changes (Fig. 15.17), but we shall make the rough approximation of assuming an isothermal atmosphere. With the z dependence of g and T neglected, Mg/RT can be taken outside the integral in (15.70) to give $\ln(P'/P_0) = -Mgz'/RT$; dropping the primes, we have

$$P = P_0 e^{-Mgz/RT} \qquad \text{const. } T, g \tag{15.71}$$

The gravitational force on the molecules increases the concentration of molecules at lower levels, and the gas pressure and density (which is proportional to P) decrease exponentially with increasing altitude.

Figure 15.17

Average earth atmospheric temperature versus altitude during June at 40° N latitude. (This figure explains why mountaintops are often snow-covered year-round.)

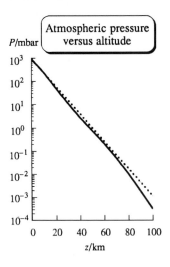

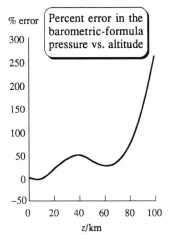

Figure 15.18

The solid line in the upper graph shows pressure versus altitude in the earth's atmosphere. The dotted line plots the barometric formula (15.71). The lower graph plots the percent error of the barometric formula versus altitude.

A slightly more general form of (15.71) is $P_2/P_1 = e^{-Mg(z_2-z_1)/RT}$, where P_2 and P_1 are the pressures at altitudes z_2 and z_1. When $Mg(z_2 - z_1)/RT$ equals 1, P_2 is $1/e \approx 1/2.7$ of P_1. For air with $M = 29$ g/mol and $T = 250$ K, the equation $Mg \, \Delta z/RT = 1$ gives $\Delta z = 7.3$ km. For each altitude increase of 7.3 km (4.5 mi), the atmospheric pressure falls to about 1/2.7 of the preceding value. Because of the exponential decrease of density with altitude, over 99 percent of the mass of the earth's atmosphere lies below 35 km altitude.

In a mixture of ideal gases, each gas i has its own molar mass M_i. One can show (see Prob. 15.56 and *Guggenheim*, sec. 9.08) that an equation like (15.71) applies to each gas independently, and $P_i = P_{i,0}e^{-M_igz/RT}$, where P_i and $P_{i,0}$ are partial pressures of gas i at altitudes z and 0. The larger M_i is, the faster P_i should decrease with altitude. However, convection currents, turbulence, and other phenomena keep the gases rather well mixed in the earth's lower and middle atmosphere, and the number average molar mass of air remains essentially constant at 29 g/mol (Prob. 12.24) up to 90 km. Above 90 km, the mole fractions of the lighter gases increase, and above 800 km, the predominant species are H, H_2, and He.

Because of the dependence of T on z, Eq. (15.71) is only a rough approximation to actual atmospheric pressures. Figure 15.18 plots observed atmospheric pressures P_{obs} and pressures P_{calc} calculated from (15.71) using an average temperature $T = 250$ K (Fig. 15.17), $M = 29$ g/mol, $g = 9.81$ m/s², and $P_0 = 1013$ mbar [Eqs. (1.10) and (1.11)].

EXAMPLE 15.5 Change of gas pressure with altitude

A container of O_2 at 1.00 bar and 25°C is 100 cm high and is at sea level. Calculate the difference in pressure between the gas at the bottom of the container and the gas at the top.

From (15.71), $P_{top} = P_{bot}e^{-Mgz/RT}$, so $P_{bot} - P_{top} = P_{bot}(1 - e^{-Mgz/RT})$. We have

$$\frac{Mgz}{RT} = \frac{(0.0320 \text{ kg mol}^{-1})(9.81 \text{ m s}^{-2})(1.00 \text{ m})}{(8.314 \text{ J mol}^{-1} \text{ K}^{-1})(298 \text{ K})} = 0.000127$$

$$P_{bot} - P_{top} = (1.00 \text{ bar})(1 - e^{-0.000127}) = 0.000127 \text{ bar} = 0.095 \text{ torr}$$

Exercise

If the air pressure on the top floor of a building where each floor is 10 feet higher than the preceding one is 4.0 torr less than the air pressure on the bottom floor, how many floors does the building have? Make reasonable assumptions. (*Answer:* 16.)

15.9 THE BOLTZMANN DISTRIBUTION LAW

Since $P = NRT/N_AV$, the ratio P/P_0 in the barometric equation (15.71) is equal to $N(z)/N(0)$, where $N(z)$ and $N(0)$ are the numbers of molecules in thin layers of equal volumes at heights z and 0. Also, $Mgz/RT = N_Amgz/N_AkT = mgz/kT$, where m is the mass of a molecule. The quantity mgz equals $\varepsilon_p(z) - \varepsilon_p(0) \equiv \Delta\varepsilon_p$, where $\varepsilon_p(z)$ and $\varepsilon_p(0)$ are the potential energies of molecules at heights z and 0 [see Eq. (2.26)]. Therefore (15.71) can be written as

$$N(z)/N(0) = e^{-[\varepsilon_p(z)-\varepsilon_p(0)]/kT} = e^{-\Delta\varepsilon_p/kT} \qquad (15.72)$$

The v_x distribution law (15.42) reads $dN_{v_x}/N = Ae^{-\varepsilon_{tr}/kT}\,dv_x$, where $\varepsilon_{tr} = \frac{1}{2}mv_x^2$. Let dN_1 and dN_2 be the numbers of molecules whose v_x values lie in the ranges $v_{x,1}$ to $v_{x,1} + dv_x$ and $v_{x,2}$ to $v_{x,2} + dv_x$, respectively. Then

$$dN_2/dN_1 = e^{-(\varepsilon_{tr,2}-\varepsilon_{tr,1})/kT} = e^{-\Delta\varepsilon_{tr}/kT} \qquad (15.73)$$

Equations (15.72) and (15.73) are each special cases of the more general **Boltzmann distribution law:**

$$\frac{N_2}{N_1} = e^{-\Delta\varepsilon/kT} \qquad \text{where } \Delta\varepsilon \equiv \varepsilon_2 - \varepsilon_1 \qquad \textbf{(15.74)}*$$

In this equation, N_1 is the number of molecules in state 1 and ε_1 is the energy of state 1; N_2 is the number of molecules in state 2. The **state** of a particle is defined in classical mechanics by specifying its position and velocity to within infinitesimal ranges. Equation (15.74) is a result of statistical mechanics for a system in thermal equilibrium and will be derived in Sec. 22.5. If ε_2 is greater than ε_1, then $\Delta\varepsilon$ is positive and (15.74) says that N_2 is less than N_1. *The number of molecules in a state decreases with increasing energy of the state.*

The factor v^2 in the Maxwell distribution of speeds, Eq. (15.44), might seem to contradict the Boltzmann distribution (15.74), but this is not so. N_1 in (15.74) is the number of molecules in a given state, and many different states may have the same energy. Thus, molecules that have the same speed but different velocities are in different states but have the same translational energy. The velocity vectors for such molecules point in different directions but have the same lengths, and their tips all lie in the thin spherical shell in Fig. 15.4. The factor $4\pi v^2$ in (15.44) arises from molecules having the same speed v but different velocities $\mathbf{v}$. Use of the velocity distribution function of Eqs. (15.29) and (15.43) for $dN_{v_xv_yv_z}/N$ gives a result like (15.73), in agreement with the Boltzmann distribution law.

15.10 HEAT CAPACITIES OF IDEAL POLYATOMIC GASES

In Sec. 15.3, we used the kinetic theory to show that $C_{V,m} = \frac{3}{2}R$ for an ideal monatomic gas. What about polyatomic gases? Thermal energy added to a gas of polyatomic molecules can appear as rotational and vibrational (as well as translational) energies of the gas molecules (Sec. 2.11), so the molar heat capacity $C_{V,m} = (\partial U_m/\partial T)_V$ of a polyatomic gas exceeds that of a monatomic gas. There is also molecular electronic energy, but this is usually not excited except at very high temperatures, typically 10^4 K. This section outlines the classical-mechanical kinetic-theory treatment of the contributions of molecular vibration and rotation to U_m and $C_{V,m}$.

The energy of a monatomic molecule in an ideal gas is

$$\varepsilon = \tfrac{1}{2}mv_x^2 + \tfrac{1}{2}mv_y^2 + \tfrac{1}{2}mv_z^2 + \varepsilon_{el} = p_x^2/2m + p_y^2/2m + p_z^2/2m + \varepsilon_{el} \qquad (15.75)$$

where the momentum components are $p_x \equiv mv_x$, $p_y \equiv mv_y$, $p_z \equiv mv_z$, and where ε_{el} is the electronic energy.

The molar internal energy U_m of a gas is the sum of the energies of the individual molecules and so is equal to the Avogadro constant N_A times the average molecular energy: $U_m = N_A\langle\varepsilon\rangle$. From Eq. (15.37), we have $\langle v_x^2\rangle = kT/m$ and $\langle\frac{1}{2}mv_x^2\rangle = \frac{1}{2}kT$. Per mole of gas, we have $N_A\langle\frac{1}{2}mv_x^2\rangle = \frac{1}{2}N_AkT = \frac{1}{2}RT$ as the contribution of translational motion in the x direction to the molar internal energy U_m. Since $C_{V,m} = \partial U_m/\partial T$, we get a contribution of $\frac{1}{2}R$ to $C_{V,m}$ due to molecular translational energy in the x direction and a total contribution of $\frac{3}{2}R$ to $C_{V,m}$ due to the x, y, and z translational energies, as noted in Eq. (15.18).

For a polyatomic molecule, the translational energy is the same as in (15.75) and so contributes $\frac{3}{2}R$ to $C_{V,m}$ of a gas of polyatomic molecules, the same as for a gas of monatomic molecules. However, polyatomic molecules also have rotational and vibrational energies. The classical-mechanical expressions for the molecular rotational energy and vibrational energy turn out to be sums of terms that are each proportional to the square of a momentum or the square of a coordinate; these terms are said to be *quadratic* in the momentum or coordinate. One can use the Boltzmann distribution law (15.74) to show that $\langle cw^2 \rangle = \frac{1}{2}kT$, where c is a constant and w is either a momentum or a coordinate (Prob. 15.47). Thus, classical statistical mechanics predicts that: Each term in the expression for the molecular energy that is quadratic in a momentum or a coordinate will contribute $\frac{1}{2}RT$ to U_m and $\frac{1}{2}R$ to $C_{V,m}$. This is the *equipartition-of-energy principle*.

The molecular energy is the sum of translational, rotational, vibrational, and electronic energies: $\varepsilon = \varepsilon_{tr} + \varepsilon_{rot} + \varepsilon_{vib} + \varepsilon_{el}$. The translational energy is given by (15.75) as $\varepsilon_{tr} = p_x^2/2m + p_y^2/2m + p_z^2/2m$. Since ε_{tr} has three terms, each quadratic in a momentum, the equipartition principle says that ε_{tr} contributes $3 \times \frac{1}{2}R = \frac{3}{2}R$ to $C_{V,m}$. This result agrees with data on monatomic gases.

Rotational and vibrational motions will be considered in Chapter 21. At this point, we simply quote some results from that chapter without justifying them. For a linear molecule (for example, H_2, CO_2, C_2H_2), the classical-mechanical expression for ε_{rot} has two quadratic terms and that for ε_{vib} has $2(3\mathcal{N} - 5)$ quadratic terms, where $\mathcal{N}$ is the number of atoms in the molecule. For a nonlinear molecule (for example, H_2O, CH_4), ε_{rot} has three quadratic terms and ε_{vib} has $2(3\mathcal{N} - 6)$ quadratic terms. A linear molecule thus has $3 + 2 + 2(3\mathcal{N} - 5) = 6\mathcal{N} - 5$ quadratic terms in ε, and $C_{V,m}$ is predicted to be $(3\mathcal{N} - 2.5)R$ for an ideal gas of linear molecules, assuming T is not high enough to excite electronic energy. A nonlinear molecule has $3 + 3 + 2(3\mathcal{N} - 6) = 6\mathcal{N} - 6$ quadratic terms in ε, and $C_{V,m}$ is predicted to be $(3\mathcal{N} - 3)R$ for an ideal gas of nonlinear molecules, except at extremely high T. These values should apply at all temperatures below, say, 10^4 K.

For CO_2 considered as an ideal gas, the equipartition theorem predicts $C_{V,m} = [3(3) - 2.5]R = 6.5R$, independent of T. Experimental $C_{V,m}$ values for CO_2 at low pressure are plotted in Fig. 15.19, and these differ from the equipartition prediction. Similar disagreements between experimental $C_{V,m}$ values and those predicted by the equipartition theorem occur for other polyatomic gases. In general, $C_{V,m}$ increases with T and attains the equipartition value only at high T. The data show that the equipartition theorem is *false*. The equipartition principle fails because it uses the classical-mechanical expression for the molecular energy, whereas molecular vibrations and rotations obey quantum mechanics, not classical mechanics. The quantum theory of heat capacities is discussed in Chapters 22 and 24. Except at extraordinarily low temperatures, the classical-mechanical treatment of molecular translational energy is a highly accurate approximation and translational motion contributes $\frac{3}{2}R$ to $C_{V,m}$ of a gas.

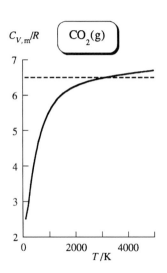

$C_{V,m}/R$ | $CO_2(g)$

Figure 15.19

Molar C_V of $CO_2(g)$ at 1 atm plotted versus T. The equipartition-theorem value is $C_{V,m} = 6.5R$.

SUMMARY

We found an expression for the pressure exerted by an ideal gas by considering the impacts of molecules on the container walls. Comparison of this expression with $PV = nRT$ showed that the total molecular translational energy in the gas is $E_{tr} = \frac{3}{2}nRT$. It follows that the average translational energy per gas molecule is $\langle \varepsilon_{tr} \rangle = \frac{3}{2}kT$. Use of $\varepsilon_{tr} = \frac{1}{2}mv^2$ gives the rms molecular speed as $v_{rms} = (3RT/M)^{1/2}$.

Starting from the assumption that the velocity distribution is independent of direction and that the v_x, v_y, and v_z values of a molecule are statistically independent of one another, we derived the Maxwell distribution laws for the speed v and velocity components v_x, v_y, and v_z of gas molecules (Figs. 15.5 and 15.6). The fraction of molecules with x component of velocity between v_x and $v_x + dv_x$ is $dN_{v_x}/N = g(v_x)\,dv_x$, where $g(v_x) = (m/2\pi kT)^{1/2}\exp(-\frac{1}{2}mv_x^2/kT)$. The fraction of molecules with speed between v and $v + dv$ is $dN_v/N = G(v)\,dv$, where $G(v) = 4\pi v^2 g(v_x)g(v_y)g(v_z)$ and is given by (15.44). The use of $G(v)$ allows calculation of the average value of any function of v as $\langle f(v)\rangle = \int_0^\infty f(v)G(v)\,dv$. For example, we found $\langle v\rangle = (8RT/\pi M)^{1/2}$.

The rate of collisions of gas molecules with a wall and the rate of effusion through a tiny hole are given by Eqs. (15.56) and (15.58). Expressions were found for the collision rate $z_b(c)$ of a single b molecule with the c molecules and for the total b-c collision rate per unit volume, Z_{bc}. The mean free path λ is the average distance a molecule travels between successive collisions and is given by (15.67).

The distribution of molecules among their possible states is given by the Boltzmann distribution law $N_2/N_1 = e^{-(\varepsilon_2 - \varepsilon_1)/kT}$ [Eq. (15.74)].

The classical equipartition theorem for the heat capacities of polyatomic gases predicts incorrect results.

Important kinds of calculations discussed in this chapter include:

- Calculation of the average molecular translational energy $\langle \varepsilon_{tr}\rangle = \frac{3}{2}kT$.
- Calculation of the heat capacity of an ideal monatomic gas, $C_{V,m} = \frac{3}{2}R$.
- Calculation of the root-mean-square speed of gas molecules, $v_{rms} = (3RT/M)^{1/2}$.
- Use of the distribution function $G(v)$ to calculate the probability $G(v)\,dv$ that a gas molecule has its speed between v and $v + dv$.
- Use of the distribution functions $G(v)$ and $g(v_x)$ to calculate average values of functions of v or v_x from $\langle f(v)\rangle = \int_0^\infty f(v)G(v)\,dv$ or $\langle f(v_x)\rangle = \int_{-\infty}^\infty f(v_x)g(v_x)\,dv_x$.
- Evaluation of kinetic-theory integrals of the form $\int_{-\infty}^\infty x^m e^{-ax^2}\,dx$ using the integrals in Table 15.1.
- Calculation of the collision rate with a wall using (15.56).
- Calculation of the rate of escape from a hole and of the vapor pressure of a solid using (15.58).
- Calculation of the individual collision rate $z_b(c)$ and the total collision rate per unit volume, Z_{bc}, using (15.60) and (15.62).
- Calculation of the mean free path (15.67).
- Calculation of pressures in an isothermal atmosphere using $P = P_0 e^{-Mgz/RT}$.
- Calculation of high-temperature $C_{V,m}$ values of gases using the equipartition theorem.

FURTHER READING

Kauzmann; Kennard; Present; Tabor.

PROBLEMS

Section 15.2

15.1 True or false? (*a*) The speed v of a molecule is never negative. (*b*) The velocity component v_x of a molecule is never negative.

Section 15.3

15.2 True or false? (*a*) $C_{P,m,500}^\circ$ is the same for He(*g*) and Ne(*g*). (*b*) v_{rms} at 400 K and 1 bar is the same for He(*g*) and Ne(*g*).

(*c*) $\langle \varepsilon_{tr}\rangle$ is the same for $N_2(g)$ and He(*g*) at 300 K and 1 bar. (*d*) At 300 K and 1 bar, v_{rms} is greater in He(*g*) than in Ne(*g*).

15.3 Calculate the total molecular translational energy at 25°C and 1.0 atm for (*a*) 1.00 mol of O_2; (*b*) 1.00 mol of CO_2; (*c*) 470 mg of CH_4.

15.4 Calculate the average translational energy of one molecule at 298°C and 1 bar for (*a*) O_2; (*b*) CO_2.

15.5 Calculate $\langle \varepsilon_{tr}(100°C) \rangle / \langle \varepsilon_{tr}(0°C) \rangle$ for an ideal gas.

15.6 Calculate $v_{rms}(Ne)/v_{rms}(He)$ at 0°C.

15.7 At what temperature will H_2 molecules have the same rms speed as O_2 molecules have at 20°C? Solve this problem without calculating v_{rms} for O_2.

15.8 Calculate the total translational kinetic energy of the air molecules in a 5.0 m × 6.0 m × 3.0 m room at 20°C and 1.00 atm. Neglect the volume occupied by furniture and people. Repeat the calculation for 40°C and 1.00 atm.

Section 15.4

15.9 (a) Give the range of v. (b) Give the range of v_x. (c) Without doing any calculations, give the value of $\langle v_y \rangle$ in $O_2(g)$ at 298 K and 1 bar. (d) For the formula $dN_v/N = G(v) \, dv$, state in words what dN_v is.

15.10 True or false? (a) At 300 K and 1 bar, $v_{x,mp}$ is the same in He(g) and $N_2(g)$. (b) At 300 K and 1 bar, v_{mp} is the same in He(g) and $N_2(g)$. (c) At 300 K and 1 bar, $\langle \varepsilon_{tr} \rangle$ is the same in He(g) and $N_2(g)$. (d) The distribution function $G(v)$ is dimensionless.

15.11 For 1.00 mol of O_2 at 300 K and 1.00 atm, calculate (a) the number of molecules whose speed lies in the range 500.000 to 500.001 m/s (because this speed interval is very small, the distribution function changes only very slightly over this interval, and the interval can be considered infinitesimal); (b) the number of molecules with v_z in the range 150.000 to 150.001 m/s; (c) the number of molecules that simultaneously have v_z in the range 150.000 to 150.001 m/s and have v_x in the range 150.000 to 150.001 m/s.

15.12 For $CH_4(g)$ at 300 K and 1 bar, calculate the probability that a molecule picked at random has its speed in the range 400.000 to 400.001 m/s. This interval is small enough to be considered infinitesimal.

15.13 Use Fig. 15.6 to estimate the number of molecules whose speed lies in the range 500.0000 to 500.0002 m/s for 1.00 mol of $N_2(g)$ at (a) 300 K; (b) 1000 K.

15.14 (a) Show that if v_2 is close enough to v_1 for $G(v)$ to be considered as essentially constant in the interval from v_1 to v_2, then Eq. (15.24) gives $\text{Pr}(v_1 \leq v \leq v_2) \approx G(v)(v_2 - v_1)$. (b) Calculate the percent change in $G(v)$ over the speed interval in the Sec. 15.4 example.

15.15 (a) For O_2 at 25°C and 1 atm, calculate the ratio of the probability that a molecule has its speed in an infinitesimal interval dv located at 500 m/s to the probability that its speed lies in an infinitesimal interval dv at 1500 m/s. (b) At what temperature does $O_2(g)$ have $G(v)$ at $v = 500$ m/s equal to $G(v)$ at $v = 1500$ m/s?

15.16 (a) Verify Eq. (15.41) for the average of a sum. (b) Show that $\langle cf(w) \rangle = c\langle f(w) \rangle$, where c is a constant.

15.17 In the mythical world of Flatland, everything is two-dimensional. Find the expression for the probability that a molecule in a two-dimensional ideal gas has its speed in the range v to $v + dv$.

15.18 (a) Use Fig. 15.5 to explain why $\int_{-\infty}^{\infty} e^{-ax^2} \, dx = 2 \int_0^{\infty} e^{-ax^2} \, dx$. This integral is integral 1 in Table 15.1 with $n = 0$, and a similar argument shows the validity of integral 1. (b) Sketch the function xe^{-ax^2} and explain why $\int_{-\infty}^{\infty} xe^{-ax^2} \, dx = 0$. A similar argument shows the validity of integral 4.

15.19 (a) Derive integral 5 in Table 15.1 by changing to a new variable. (b) The definite integral 5 in Table 15.1 is a function of the parameter a, which occurs in the integrand. For such an integral, one can show (see *Sokolnikoff and Redheffer*, p. 348) that

$$\frac{\partial}{\partial a} \int_c^d f(x, a) \, dx = \int_c^d \frac{\partial f(x, a)}{\partial a} \, dx$$

Differentiate integral 5 with respect to a to show that $\int_0^{\infty} x^3 e^{-ax^2} \, dx = 1/2a^2$. (c) Find $\int_0^{\infty} x^5 e^{-ax^2} \, dx$. Repeated differentiation gives integral 6.

Section 15.5

15.20 For CO_2 at 500 K, calculate (a) v_{rms}; (b) $\langle v \rangle$; (c) v_{mp}.

15.21 Show that $v_{mp} = (2RT/M)^{1/2}$.

15.22 Use the Maxwell distribution to verify that $\langle v^2 \rangle = 3RT/M$.

15.23 Use the distribution function for v to find $\langle v^3 \rangle$ for ideal-gas molecules. Does $\langle v^3 \rangle$ equal $\langle v \rangle \langle v^2 \rangle$?

15.24 (a) Find $\langle v_x \rangle$ for ideal-gas molecules. Give a physical explanation of the result. (b) $\langle v_x^2 \rangle$ is given by (15.37). Explain why $\langle v_x \rangle^2 \neq \langle v_x^2 \rangle$. (c) Find $v_{x,rms}$. (d) What is $v_{x,mp}$?

15.25 Find $\langle v_x^4 \rangle$ for ideal-gas molecules.

15.26 Find the most probable molecular translational energy $\varepsilon_{tr,mp}$ for an ideal gas. Compare with $\langle \varepsilon_{tr} \rangle$.

15.27 (a) Show $(2\pi)^{1/2}I(u) = \sum_{n=0}^{\infty}(-1)^n u^{2n+1}/[(2n + 1)2^n n!] = u - u^3/6 + u^5/40 - u^7/336 + \cdots$, where $I(u)$ is defined by (15.51). [*Hint:* Use (8.37) with $x = -s^2/2$.] (b) Use the series to verify that $I(0.30) = 0.118$. (c) If you have a calculator that does numerical integration, use it to find $I(0.30)$.

15.28 (a) Use integration by parts to show that the fraction of molecules whose speed is in the range 0 to v' is

$$2I(2^{1/2}v'/v_{mp}) - 2(v'/v_{mp})\pi^{-1/2}e^{-(v'/v_{mp})^2}$$

where the function I is defined by (15.51). (b) Use Fig. 15.10 to help find the fraction of molecules whose speed exceeds $4.243v_{mp}$.

15.29 We took the range of v as 0 to ∞. (a) Explain why this is incorrect and give the correct range of v. (b) Explain why the error in taking the range as 0 to ∞ is utterly negligible.

Section 15.6

15.30 Find the molecular formula of a hydrocarbon gas that effuses 0.872 times as fast as O_2 through a small hole, the temperatures and pressures being equal.

15.31 A container holding solid scandium in equilibrium with its vapor at 1690 K shows a weight loss of 10.5 mg in 49.5 min through a circular hole of diameter 0.1763 cm. (a) Find the vapor pressure of Sc at 1690 K in torr. (b) Is $\lambda \gg d_{\text{hole}}$?

15.32 Dry air contains 0.033 percent CO_2 by volume. Calculate the total mass of CO_2 that strikes 1 cm² of one side of a green leaf in 1 s in dry air at 25°C and 1.00 atm.

15.33 For Octoil, $C_6H_4(COOC_8H_{17})_2$, the vapor pressure is 0.010 torr at 393 K. Calculate the number of Octoil molecules that evaporate into vacuum from a 1.0 cm² surface of the liquid at 393 K in 1.0 s; also calculate the mass that evaporates.

15.34 This problem justifies a statement made in Prob. 7.26 about the relation between C_{60} vapor pressure P and mass-spectrometer integrated peak intensity I. The integrated intensity I of the C_{60} peaks is proportional to the rate at which molecules effuse into the mass spectrometer and is also proportional to the probability a given molecule gets ionized by the electron beam in the mass spectrometer. This probability is proportional to the average time it takes a molecule to pass through the electron beam and thus is inversely proportional to the average molecular speed in the direction of motion of the C_{60} molecular beam. This average speed in the direction of beam motion in an effusive molecular beam can be shown to be proportional to $T^{1/2}$ (*Kauzmann*, p. 162). Use these facts to show that $I \propto P/T$ so $P \propto IT$.

Section 15.7

15.35 True or false? (a) $Z_{bc} = Z_{cb}$. (b) $z_b(c) = z_c(b)$.

15.36 A container of volume 1×10^{-5} cm³ holds three molecules of gas b, which we label b_1, b_2, and b_3. In 1 s, there are two b_1-b_2 collisions, two b_1-b_3 collisions, and two b_2-b_3 collisions. (a) Find $z_b(b)$. (b) Find Z_{bb} without using the result for $z_b(b)$. Is Z_{bb} equal to $N_b z_b(b)/V$? Is Z_{bb} equal to $\frac{1}{2}N_b z_b(b)/V$?

15.37 (a) A certain sample of a pure gas has $\langle v \rangle = 450$ m/s and the average time between two successive collisions of a given molecule with other molecules is 4.0×10^{-10} s. Find the mean free path in this gas. (b) For a certain mixture of gases B and C, the following are true: $\langle v_B \rangle = 385$ m/s; a given B molecule has an average time of 5.0×10^{-10} s between two successive collisions with B molecules and has an average time of 8.0×10^{-10} s between successive collisions with C molecules. Which molecules have the larger diameter, B or C? Find λ_B.

15.38 For N_2 (collision diameter = 3.7 Å; see Sec. 16.3) at 25°C and 1.00 atm, calculate (a) the number of collisions per second made by one molecule; (b) the number of collisions per second per cubic centimeter. (c) Repeat (a) and (b) for N_2 at 25°C and 1.0×10^{-6} torr (a typical "vacuum" pressure); save time by using the results of (a) and (b).

15.39 The average surface temperature of Mars is 220 K, and the surface pressure is 4.7 torr. The Martian atmosphere is mainly CO_2 and N_2, with smaller amounts of Ar, O_2, CO, H_2O, and Ne. Considering only the two main components, we approximate the Martian atmospheric composition as $x_{CO_2} \approx 0.97$ and $x_{N_2} \approx 0.03$. The collision diameters are (Sec. 16.3) $d_{CO_2} = 4.6$ Å and $d_{N_2} = 3.7$ Å. For gas at 220 K at the Martian surface, calculate (a) the collision rate for one particular CO_2 molecule with other CO_2 molecules; (b) the collision rate for one partic-

ular N_2 molecule with CO_2 molecules; (c) the number of collisions per second made by one particular N_2 molecule; (d) the number of CO_2-N_2 collisions per second in 1.0 cm³.

15.40 For $N_2(g)$ with collision diameter 3.7 Å, calculate the mean free path at 300 K and (a) 1.00 bar; (b) 1.00 torr; (c) 1.0 $\times 10^{-6}$ torr.

Section 15.8

15.41 The top of Pike's Peak is 14100 ft above sea level. Neglecting the variation of T with altitude and using the average surface temperature of 290 K and the number average molecular weight of 29 for air, calculate the atmospheric pressure at the top of this mountain. Compare with the observed average value 446 torr.

15.42 Calculate the difference in barometer readings between the first and fourth floors of a building at sea level if each floor is 10 ft high.

15.43 For what increase in altitude is the earth's atmospheric pressure cut in half? Take $T = 250$ K and $M = 29$ g/mol.

15.44 Using data in Prob. 15.39 and an average temperature of 180 K for the lower Martian atmosphere, estimate the atmospheric pressure at an altitude of 40 km on Mars. On Mars, $g = 3.7$ m/s².

15.45 Figure 15.17 shows that the temperature in the troposphere up to 12 km decreases approximately linearly with altitude. (a) Use the assumption $T = T_0 - az$, where a is a constant, to show that $P = P_0(1 - az/T_0)^{Mg/aR}$, where P and z are the air pressure and the altitude up to 12 km. (b) Estimate T_0 and a from Fig. 15.17. (c) Calculate P at 10 km using the formula of (a) and compare with the value given by (15.71) with $T = 250$ K.

Section 15.10

15.46 (a) For CH_4 at 400 K, what value of $C_{P,m}$ is predicted by the equipartition principle? (b) Would CH_4 actually have this value of $C_{P,m}$ at 400 K? (c) Under what condition(s) would CH_4 have the equipartition-principle value of $C_{P,m}$?

15.47 Let w stand for any one of the coordinates or momenta in the expressions for the molecular translational, rotational, or vibrational energy; let c denote the coefficient of w^2 in the energy expression; and let $\varepsilon_w \equiv cw^2$. To evaluate $\langle \varepsilon_w \rangle = \langle cw^2 \rangle$, we need the distribution function $g(w)$ for w. Let dN_1 molecules have a value of w in the range w_1 to $w_1 + dw$, with a similar definition for dN_2. The definition of g gives $dN_1/N = g(w_1)\,dw$ and $dN_2/N = g(w_2)\,dw$, where N is the total number of gas molecules. Taking the ratio of these equations, we have $dN_2/dN_1 = g(w_2)/g(w_1)$. The Boltzmann distribution law (15.74) gives

$$\frac{dN_2}{dN_1} = \frac{g(w_2)}{g(w_1)} = e^{-[\varepsilon_w(w_2) - \varepsilon_w(w_1)]/kT} = \frac{e^{-\varepsilon_w(w_2)/kT}}{e^{-\varepsilon_w(w_1)/kT}}$$

The only way this equation can hold for all w_1 and w_2 values is for $g(w)$ to have the form $g(w) = Ae^{-\varepsilon_w/kT}$, where the constant A is determined by the requirement that the total probability is 1; $\int_{w_{\text{min}}}^{w_{\text{max}}} g(w)\,dw = 1$. The minimum and maximum values of w are

$-\infty$ and $+\infty$ whether w is a cartesian coordinate or a momentum. (a) Show that $A = (c/\pi kT)^{1/2}$. (b) Show that $\langle \varepsilon_w \rangle = \frac{1}{2}kT$.

General

15.48 Calculate v_{rms} at 25°C for a dust particle of mass 1.0×10^{-10} g suspended in air, assuming that the particle can be treated as a giant molecule.

15.49 Is $\langle v^2 \rangle$ equal to $\langle v \rangle^2$ for ideal-gas molecules?

15.50 Let s be the number obtained when a single cubic die is thrown. Assuming the die is not loaded, use (15.40) to calculate $\langle s \rangle$ and $\langle s^2 \rangle$. Is $\langle s \rangle^2$ equal to $\langle s^2 \rangle$?

15.51 The *standard deviation* σ_x of a variable x can be defined by $\sigma_x^2 \equiv \langle x^2 \rangle - \langle x \rangle^2$. (a) Show that $\sigma_{v_x} = (kT/m)^{1/2}$ for the Maxwell distribution of v_x. (b) What fraction of ideal-gas molecules have v_x within ± 1 standard deviation from the mean (average) value $\langle v_x \rangle$?

15.52 Apply the method used to derive (15.34) from (15.30) to show that (3.51) is the only function that satisfies (3.50).

15.53 What region in the velocity space of Fig. 15.3 corresponds to molecules that have v_x values between b and c, where b and c are two constants?

15.54 (a) What is the area under the 300-K curve in Fig. 15.6? Answer without doing any calculations or integrations. (b) Use the figure to give a *rough* calculation of this area.

15.55 Consider samples of pure $H_2(g)$ and pure $O_2(g)$, each at 300 K and 1 atm. For each of the following properties, state which gas (if any) has the greater value. As much as possible, answer without looking up formulas. (a) v_{rms}; (b) average ε_{tr}; (c) density; (d) mean free path; (e) collision rate with a wall of unit area.

15.56 Just as the phase equilibrium condition in an electrochemical system is $\mu_i^\alpha + z_i F \phi^\alpha = \mu_i^\beta + z_i F \phi^\beta$, where $z_i F \phi^\alpha$ is the molar electrostatic potential energy of species i in phase α, the phase equilibrium condition between isothermal layers at altitudes z^α and z^β in a gravitational field is (see *Denbigh*, p. 87, for a proof) $\mu_i^\alpha + M_i g z^\alpha = \mu_i^\beta + M_i g z^\beta$, where $M_i g z^\alpha$ is the molar gravitational potential energy of species i at z^α. Use this equation to show that for an isothermal ideal gas mixture in a gravitational field, $P_i^\beta = P_i^\alpha \exp[-M_i g(z^\beta - z^\alpha)/RT]$.

15.57 (a) Without looking at the figures in this chapter, sketch $G(v)$ vs. v on the same plot for gases B and C at the same T, where $M_B > M_C$. (b) Repeat (a) for a single gas at temperatures T_1 and T_2, where $T_2 > T_1$.

15.58 Which of these quantities depend on the molecular diameter d (answer without looking up any formulas)? (a) λ; (b) $\langle v \rangle$; (c) $z_b(b)$; (d) Z_{bb}.

15.59 (a) Which of the four branches of physical chemistry (Fig. 1.1) does the material of Chapter 15 belong to? (b) Does Chapter 15 deal with systems in equilibrium?

15.60 For ideal gases, classify each of the following statements as true or false. (a) The most probable speed is zero. (b) The most probable v_x value is zero. (c) The most probable v_z value is zero. (d) In a pure gas that is at a constant temperature, all the molecules travel at the same speed. (e) If two different pure gases are at the same temperature, the average speed of the molecules in one gas must be the same as the average speed of the molecules in the second gas. (f) If two different pure gases are at the same temperature, the average kinetic energy of the molecules in one gas must be the same as the average kinetic energy of the molecules in the second gas. (g) Doubling the absolute temperature will double the average speed of the molecules. (h) The total molecular translational kinetic energy E_{tr} in 1 dm^3 of ideal gas at 1 bar and 200 K equals E_{tr} in 1 dm^3 of ideal gas at 1 bar and 400 K. (i) If we know the temperature of an ideal gas, we can predict the translational energy of an individual molecule picked at random.

Transport Processes

16.1 KINETICS

So far, we have discussed only equilibrium properties of systems. Processes in systems in equilibrium are reversible and are comparatively easy to treat. This chapter and the next deal with nonequilibrium processes, which are irreversible and hard to treat. The rate of a reversible process is infinitesimal. Irreversible processes occur at nonzero rates.

The study of rate processes is called **kinetics** or **dynamics.** Kinetics is one of the four branches of physical chemistry (Fig. 1.1). A system may be out of equilibrium because matter or energy or both are being transported between the system and its surroundings or between one part of the system and another. Such processes are **transport processes,** and the branch of kinetics that studies the rates and mechanisms of transport processes is **physical kinetics.** Even though neither matter nor energy is being transported through space, a system may be out of equilibrium because certain chemical species in the system are reacting to produce other species. The branch of kinetics that studies the rates and mechanisms of chemical reactions is **chemical kinetics** or **reaction kinetics.** Physical kinetics is discussed in Chapter 16 and chemical kinetics in Chapter 17.

There are several kinds of transport processes. If temperature differences exist between the system and surroundings or within the system, it is not in thermal equilibrium and heat energy flows. *Thermal conduction* is studied in Sec. 16.2. If unbalanced forces exist in the system, it is not in mechanical equilibrium and parts of the system move. The flow of fluids is the subject of *fluid dynamics* (or *fluid mechanics*). Some aspects of fluid dynamics are treated in Sec. 16.3 on viscosity. If differences in concentrations of substances exist between different regions of a solution, the system is not in material equilibrium and matter flows until the concentrations and the chemical potentials have been equalized. This flow differs from the bulk flow that arises from pressure differences and is called *diffusion* (Sec. 16.4). When an electric field is applied to a system, electrically charged particles (electrons and ions) experience a force and may move through the system, producing an electric current. *Electrical conduction* is studied in Secs. 16.5 and 16.6.

We shall see that the laws describing thermal conduction, fluid flow, diffusion, and electrical conduction all have the same form, namely, that the rate of transport is proportional to the spatial derivative (gradient) of some property.

Transport properties are important in determining how fast pollutants spread in the environment (see chap. 4 of D. G. Crosby, *Environmental Toxicology and Chemistry,* Oxford, 1998). Biological examples of transport phenomena include the flow of blood, the diffusion of solute molecules in cells and through cell membranes, and the diffusion of neurotransmitters between nerve cells. Transport phenomena such as migration of charged species in an electric field are used to separate biomolecules and played a key role in sequencing the human genome (Sec. 16.6).

Figure 16.1

Conduction of heat through a substance.

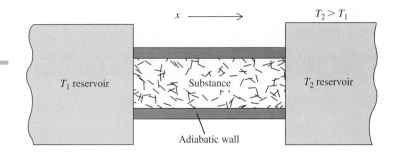

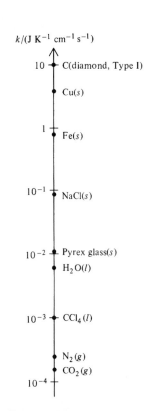

$k/(\text{J K}^{-1}\,\text{cm}^{-1}\,\text{s}^{-1})$

Figure 16.2

Thermal conductivities of substances at 25°C and 1 atm. The scale is logarithmic.

16.2   THERMAL CONDUCTIVITY

Figure 16.1 shows a substance in contact with two heat reservoirs at different temperatures. A steady state will eventually be reached in which there is a uniform temperature gradient dT/dx in the substance, and the temperature between the reservoirs varies linearly with x from T_1 at the left end to T_2 at the right end. (The **gradient** of a quantity is its rate of change with respect to a spatial coordinate.) The rate of heat flow dq/dt across any plane perpendicular to the x axis and lying between the reservoirs will also be uniform and will clearly be proportional to $\mathscr{A}$, the substance's cross-sectional area in a plane perpendicular to the x axis. Experiment shows that dq/dt is also proportional to the temperature gradient dT/dx. Thus

$$\frac{dq}{dt} = -k\mathscr{A}\,\frac{dT}{dx} \qquad (16.1)$$

where the proportionality constant k is the substance's **thermal conductivity** and dq is the heat energy that in time dt crosses a plane with area $\mathscr{A}$ and perpendicular to the x axis. The minus sign occurs because dT/dx is positive but dq/dt is negative (the heat flows to the left in the figure). Equation (16.1) is **Fourier's law** of heat conduction. This law also holds when the temperature gradient in the substance is nonuniform; in this case, dT/dx has different values at different places on the x axis, and dq/dt varies from place to place. (Fourier, discoverer of the laws of heat conduction, apparently suffered from a thyroid disorder and wore an overcoat in summer.)

k is an intensive property whose value depends on T, P, and composition. Values of k for some substances at 25°C and 1 atm are shown in Fig. 16.2. Metals are good conductors of heat because of the electrical-conduction electrons, which move relatively freely through the metal. Most nonmetals are poor conductors of heat. Gases are very poor conductors because of the low density of molecules. Diamond has the highest room-temperature thermal conductivity of any substance at 300 K.

Although the system in Fig. 16.1 is not in thermodynamic equilibrium, we assume that any tiny portion of the system can be assigned values of thermodynamic variables, such as T, U, S, and P, and that all the usual thermodynamic relations between such variables hold in each tiny subsystem. (Recall the discussion in Sec. 4.2.) This assumption, called the **principle of local state** or the **hypothesis of local equilibrium,** holds well in most (but not all) systems of interest.

The thermal conductivity k is a function of the local thermodynamic state of the system and therefore depends on T and P for a pure substance (Fig. 16.3). For solids and liquids, k may either decrease or increase with increasing T. For gases, k increases with increasing T (Fig. 16.7). The pressure dependence of k for gases is discussed later in this section.

Thermal conduction is due to molecular collisions. Molecules in a higher-temperature region have a higher average energy than molecules in an adjacent lower-temperature region. In intermolecular collisions, it is very probable for molecules with higher energy to lose energy to lower-energy molecules. This results in a flow of molecular energy from high-T to low-T regions. In gases, the molecules move relatively freely, and the flow of molecular energy in thermal conduction occurs by an actual transfer of molecules from one region of space to an adjacent region, where they undergo collisions. In liquids and solids, the molecules do not move freely, and the molecular energy is transferred by successive collisions between molecules in adjacent layers, without substantial transfer of molecules between regions.

Besides conduction, heat can be transferred by convection and by radiation. In **convection,** heat is transferred by a current of fluid moving between regions that differ in temperature. This bulk convective flow arises from differences in pressure or in density in the fluid and should be distinguished from the random molecular motion involved in thermal conduction in gases. In **radiative transfer** of heat, a warm body emits electromagnetic waves (Sec. 21.1), some of which are absorbed by a cooler body (for example, the sun and the earth). Equation (16.1) assumes the absence of convection and radiation. In measuring k for fluids, great care must be taken to avoid convection currents.

Kinetic Theory of Thermal Conductivity of Gases

The kinetic theory of gases yields theoretical expressions for the thermal conductivity and other transport properties of gases, and the results agree reasonably well with experiment. The rigorous equations underlying transport processes in gases were worked out in the 1860s and 1870s by Maxwell and by Boltzmann, but it wasn't until 1917 that Sydney Chapman and David Enskog, working independently, solved these extremely complicated equations. (The Chapman–Enskog theory is so severely mathematical that Chapman remarked that reading an exposition of the theory is "like chewing glass.") Instead of presenting rigorous analyses, this chapter gives very crude treatments based on the assumption of hard-sphere molecules with a mean free path given by Eq. (15.67). The mean-free-path method (given by Maxwell in 1860) gives results that are qualitatively correct but quantitatively wrong.

We shall assume that the gas pressure is neither very high nor very low. Our treatment is based on collisions between two molecules and assumes no intermolecular forces except at the moment of collision. At high pressures, intermolecular forces in the intervals between collisions become important, and the mean-free-path formula (15.67) does not apply. At very low pressures, the mean free path λ becomes comparable to, or larger than, the dimensions of the container, and wall collisions become important. Thus our treatment applies only for pressures such that $d \ll \lambda \ll L$, where d is the molecular diameter and L is the smallest dimension of the container. In Sec. 15.7, we found λ to be about 10^{-5} cm at 1 atm and room temperature. Since λ is inversely proportional to pressure, λ is 10^{-7} cm at 10^2 atm and is 10^{-2} cm at 10^{-3} atm. Thus our treatment applies to the pressure range from 10^{-2} or 10^{-3} atm to 10^1 or 10^2 atm.

We make the following assumptions: (1) The molecules are rigid, nonattracting spheres of diameter d. (2) All molecules in a given region move at the same speed $\langle v \rangle$ characteristic of the temperature of that region and travel the same distance λ between successive collisions. (3) The direction of molecular motion after a collision is completely random. (4) Complete adjustment of the molecular energy ε occurs at each collision; this means that a gas molecule moving in the x direction and colliding with a molecule in a plane located at $x = x'$ will take on the average energy ε' characteristic of molecules in the plane at x'.

Assumptions 1 and 2 are false. Assumption 3 is also inaccurate, in that, after a collision, a molecule is somewhat more likely to be moving in or close to its original direction of motion than in other directions. Assumption 4 is not bad for translational energy but is very inaccurate for rotational and vibrational energies.

Let a steady state be established in Fig. 16.1 and consider a plane perpendicular to the x axis and located at $x = x_0$ (Fig. 16.4). To calculate k, we must find the rate of

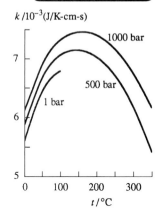

Thermal conductivity of liquid H_2O

$k/10^{-3}$(J/K-cm-s)

Figure 16.3

Thermal conductivity k of liquid water versus temperature at several pressures.

Figure 16.4

Three planes separated by $\frac{2}{3}\lambda$, where λ is the mean free path in the gas.

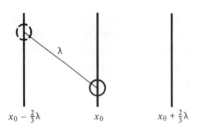

$$x_0 - \tfrac{2}{3}\lambda \qquad\qquad x_0 \qquad\qquad x_0 + \tfrac{2}{3}\lambda$$

flow of heat energy through this plane. The net heat flow dq through the x_0 plane in time dt is

$$dq = \varepsilon_L\, dN_L - \varepsilon_R\, dN_R \tag{16.2}$$

where dN_L is the number of molecules coming from the left that cross the x_0 plane in time dt and ε_L is the average energy (translational, rotational, and vibrational) of each of these molecules; dN_R and ε_R are the corresponding quantities for molecules crossing the x_0 plane from the right.

Since we are assuming no convection, there is no net flow of gas, and $dN_L = dN_R$. To find dN_L, we think of the plane at x_0 as an invisible "wall," and use Eq. (15.56), which gives the number of molecules hitting a wall in time dt as $dN = \frac{1}{4}(N/V)\langle v\rangle \mathscr{A}\, dt$. Therefore

$$dN_L = dN_R = \tfrac{1}{4}(N/V)\langle v\rangle \mathscr{A}\, dt \tag{16.3}$$

where N/V is the number of molecules per unit volume at the x_0 plane, whose cross-sectional area is $\mathscr{A}$.

The molecules coming from the left have traveled an average distance λ since their last collision. The molecules move into the x_0 plane at various angles. By averaging over the angles, one finds that the average perpendicular distance from the x_0 plane to the point of last collision is $\frac{2}{3}\lambda$ (see *Kennard*, pp. 139–140, for the proof). Figure 16.4 shows an "average" molecule moving into the x_0 plane from the left. Molecules moving into the x_0 plane from the left will, by assumption 4, have an average energy that is characteristic of molecules in the plane at $x_0 - \frac{2}{3}\lambda$. Thus, $\varepsilon_L = \varepsilon_-$, where ε_- is the average molecular energy in the plane at $x_0 - \frac{2}{3}\lambda$. Similarly, $\varepsilon_R = \varepsilon_+$, where ε_+ is the average molecular energy at the $x_0 + \frac{2}{3}\lambda$ plane. Equation (16.2) becomes $dq = \varepsilon_-\, dN_L - \varepsilon_+\, dN_R$, and substitution of (16.3) for dN_L and dN_R gives

$$dq = \tfrac{1}{4}(N/V)\langle v\rangle \mathscr{A}(\varepsilon_- - \varepsilon_+)\, dt \tag{16.4}$$

The energy difference $\varepsilon_- - \varepsilon_+$ is directly related to the temperature difference $T_- - T_+$ between the $x_0 - \frac{2}{3}\lambda$ and $x_0 + \frac{2}{3}\lambda$ planes. Letting $d\varepsilon$ denote this energy difference, we have

$$\varepsilon_- - \varepsilon_+ \equiv d\varepsilon = \frac{d\varepsilon}{dT}\, dT = \frac{d\varepsilon}{dT}\frac{dT}{dx}\, dx \tag{16.5}$$

where $dT \equiv T_- - T_+$ and

$$dx = \left(x_0 - \tfrac{2}{3}\lambda\right) - \left(x_0 + \tfrac{2}{3}\lambda\right) = -\tfrac{4}{3}\lambda \tag{16.6}$$

Since we assumed no intermolecular forces except at the instant of collision, the total energy is the sum of the energies of the individual gas molecules, and the local molar thermodynamic internal energy is $U_{\mathrm{m}} = N_A\varepsilon$, where N_A is the Avogadro constant. Therefore

$$\frac{d\varepsilon}{dT} = \frac{d(U_{\mathrm{m}}/N_A)}{dT} = \frac{1}{N_A}\frac{dU_{\mathrm{m}}}{dT} = \frac{C_{V,\mathrm{m}}}{N_A} \tag{16.7}$$

since $C_{V,m} = dU_m/dT$ for an ideal gas [Eq. (2.68)]. Use of (16.7) and (16.6) in (16.5) gives

$$\varepsilon_- - \varepsilon_+ = -\frac{4C_{V,m}\lambda}{3N_A}\frac{dT}{dx} \tag{16.8}$$

and (16.4) becomes

$$dq = -\frac{N}{3N_A V}\langle v\rangle \mathscr{A} C_{V,m}\lambda \frac{dT}{dx}dt$$

We have $N/N_A V = n/V = m/MV = \rho/M$, where n, m, ρ, and M are the number of moles of gas, the mass of the gas, the gas density, and the gas molar mass. Therefore

$$\frac{dq}{dt} = -\frac{\rho\langle v\rangle C_{V,m}\lambda}{3M}\mathscr{A}\frac{dT}{dx}$$

Comparison with Fourier's law $dq/dt = -k\mathscr{A}\,dT/dx$ [Eq. (16.1)] gives

$$k \approx \tfrac{1}{3}C_{V,m}\lambda\langle v\rangle\rho/M \qquad \text{hard spheres} \tag{16.9}$$

Because of the crudity of assumptions 2 to 4, the numerical coefficient in this equation is wrong. A rigorous theoretical treatment (*Kennard*, pp. 165–180) for hard-sphere monatomic molecules gives

$$k = \frac{25\pi}{64}\frac{C_{V,m}\lambda\langle v\rangle\rho}{M} \qquad \text{hard spheres, monatomic} \tag{16.10}$$

The rigorous extension of (16.10) to polyatomic gases is a very difficult problem that has not yet been fully solved. Experiments on intermolecular energy transfer show that rotational and vibrational energy is not as easily transferred in collisions as translational energy. The heat capacity $C_{V,m}$ is the sum of a translational part and a vibrational and rotational part [see Sec. 15.10 and Eq. (15.18)]:

$$C_{V,m} = C_{V,m,tr} + C_{V,m,vib+rot} = \tfrac{3}{2}R + C_{V,m,vib+rot} \tag{16.11}$$

Because vibrational and rotational energy is less easily transferred than translational, it contributes less to k. Therefore, in the expression for k, the coefficient of $C_{V,m,vib+rot}$ should be less than the value $25\pi/64$, which is correct for $C_{V,m,tr}$ [Eq. (16.10)]. Eucken gave nonrigorous arguments for taking the coefficient of $C_{V,m,vib+rot}$ as two-fifths that of $C_{V,m,tr}$, and doing so leads to fairly good agreement with experiment. Thus, for polyatomic molecules, $25\pi C_{V,m}/64$ in (16.10) is replaced by

$$\frac{25\pi}{64}C_{V,m,tr} + \frac{2}{5}\frac{25\pi}{64}C_{V,m,vib+rot} = \frac{25\pi}{64}\frac{3R}{2} + \frac{5\pi}{32}\left(C_{V,m} - \frac{3R}{2}\right)$$
$$= \frac{5\pi}{32}\left(C_{V,m} + \frac{9}{4}R\right)$$

The thermal conductivity of a gas of polyatomic (or monatomic) hard-sphere molecules is then predicted to be

$$k = \frac{5\pi}{32}\left(C_{V,m} + \frac{9}{4}R\right)\frac{\lambda\langle v\rangle\rho}{M} = \frac{5}{16}\left(C_{V,m} + \frac{9}{4}R\right)\left(\frac{RT}{\pi M}\right)^{1/2}\frac{1}{N_A d^2} \qquad \text{hard spheres} \tag{16.12}$$

where (15.67) and (15.47) for λ and $\langle v\rangle$ and the ideal-gas law $\rho = PM/RT$ were used. (Other approaches to the calculation of thermal conductivities are considered in *Reid, Prausnitz, and Poling*, chap. 10.)

Use of (16.12) to calculate k requires knowledge of the molecular diameter d. Even a truly spherical molecule like He does not have a well-defined size, so it is hard to say what value of d should be used in (16.12). In the next section, we shall use experimental gas viscosities to get d values appropriate to the hard-sphere model [see (16.25) and (16.26)]. Using d values calculated from 0°C viscosities, one finds the following ratios of theoretical gas thermal conductivities predicted by (16.12) to experimental values at 0°C: 1.05 for He, 0.99 for Ar, 0.96 for O_2, and 0.97 for C_2H_6.

How does k in (16.12) depend on T and P? The heat capacity $C_{V,m}$ varies slowly with T and very slowly with P. Hence (16.12) predicts $k \propto T^{1/2}P^0$. Surprisingly, k is predicted to be independent of pressure. As P increases, the number of heat carriers (molecules) per unit volume increases, thereby tending to increase k. However, this increase is nullified by the decrease in λ in (16.10) with increasing P. As λ decreases, each molecule goes a shorter average distance between collisions and is therefore less effective in transporting heat.

Data show that k for gases does increase with increasing T but faster than the $T^{1/2}$ behavior predicted by the rather crude hard-sphere model. Molecules are actually "soft" rather than hard. Moreover, they attract one another over significant distances. Use of improved expressions for intermolecular forces gives better agreement with the observed T dependence of k (*Kauzmann*, pp. 218–231).

The prediction that k is independent of P holds well, provided P is not too high or too low. (Recall the restriction $d \ll \lambda \ll L$.) Values of k versus P are plotted for some gases at 50°C in Fig. 16.5. k is nearly constant for pressures up to about 50 atm.

At very low pressures (below, say, 0.01 torr), the gas molecules in Fig. 16.1 travel back and forth between the reservoirs, making very few collisions with one another. At pressures low enough to make λ substantially larger than the separation between the heat reservoirs, heat is transferred by molecules moving directly from one reservoir to the other, and the rate of heat flow is proportional to the rate of molecular collisions with the reservoir walls. Since the rate of wall collisions is proportional to the pressure, dq/dt becomes proportional to P at very low pressures and goes to zero as P goes to zero. One finds that Fourier's law (16.1) does not hold in this very-low-pressure range (see *Kauzmann*, p. 206), and so k is not defined here. Between the pressure range where dq/dt is independent of P and the range where it is proportional to P, there is a transition range in which k falls off from its moderate-pressure value. The falloff of k begins at 10 to 50 torr, depending on the gas.

The pressure dependence of dq/dt at very low pressures is the basis of the Pirani gauge and the thermocouple gauge used to measure pressure in vacuum systems. These gauges have a heated wire sealed into the vacuum system. The temperature T and hence the resistance R of this wire vary with P of the surrounding gas, and monitoring T or R of a properly calibrated gauge gives us P.

A theoretical equation for k of liquids is given in Prob. 16.5.

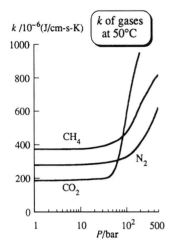

Figure 16.5

Thermal conductivities of some gases versus P at 50°C.

16.3 VISCOSITY

Viscosity

This section deals with the bulk flow of fluids (liquids and gases) under a pressure gradient. Some fluids flow more easily than others. The property that characterizes a fluid's resistance to flow is its *viscosity* η (eta). We shall see that the speed of flow through a tube is inversely proportional to the viscosity.

To get a precise definition of η, consider a fluid flowing steadily between two large plane parallel plates (Fig. 16.6). Experiment shows that the speed v_y of the fluid flow is a maximum midway between the plates and decreases to zero at each plate. The arrows in the figure indicate the magnitude of v_y as a function of the vertical coordinate x. The condition of zero flow speed at the boundary between a solid and a fluid, called the **no-slip condition,** is an experimental fact. (Evidence for the no-slip condi-

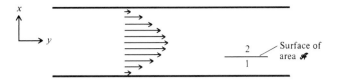

Figure 16.6

A fluid flowing between two
planar plates.

tion is the layer of dust and dirt that accumulates on the blades of a fan.) Adjacent horizontal layers of fluid flow at different speeds and "slide over" one another. As two adjacent layers slip past each other, each exerts a frictional resistive force on the other, and this internal friction gives rise to viscosity.

Consider an imaginary surface of area $\mathcal{A}$ drawn between and parallel to the plates (Fig. 16.6). Whether the fluid is at rest or in motion, the fluid on one side of this surface exerts a force of magnitude $P\mathcal{A}$ in the x direction on the fluid on the other side, where P is the local pressure in the fluid. Moreover, because of the change in flow speed as x changes, the fluid on one side of the surface exerts a frictional force in the y direction on the fluid on the other side. Let F_y be the frictional force exerted by the slower-moving fluid on one side of the surface (side 1 in the figure) on the faster-moving fluid (side 2). Experiments on fluid flow show that F_y is proportional to the surface area of contact and to the gradient dv_y/dx of flow speed. The proportionality constant is the fluid's **viscosity** η (sometimes called the *dynamic viscosity*):

$$F_y = -\eta \mathcal{A}\, \frac{dv_y}{dx} \qquad (16.13)$$

The minus sign shows that the viscous force on the faster-moving fluid is in the direction opposite its motion. By Newton's third law of motion (action = reaction), the faster-moving fluid exerts a force $\eta \mathcal{A}(dv_y/dx)$ in the positive y direction on the slower-moving fluid. The viscous force tends to slow down the faster-moving fluid and speed up the slower-moving fluid.

Equation (16.13) is **Newton's law of viscosity.** Experiments show it to be well obeyed by gases and by most liquids, provided the flow rate is not too high. When Eq. (16.13) applies, we have **laminar** (or *streamline*) flow. At high rates of flow, (16.13) does not hold, and the flow is called **turbulent.** Both laminar flow and turbulent flow are types of bulk (or viscous) flow. In contrast, for flow of a gas at very low pressures, the mean free path is long, and the molecules flow independently of one another; this is *molecular flow,* and it is not a type of bulk flow.

> When flow is turbulent, addition of extremely small amounts of a long-chain polymer solute to the liquid reduces substantially the resistance to flow through pipes (*New York Times,* Jan. 12, 1988, p. C1; *Physics Today,* March 1978, p. 17). Firefighters use this method to increase the flow rate of water through firehoses. Dolphins apparently secrete a polymeric fluid that reduces the resistance to their motion through water. Blood flow is mainly laminar. Turbulent flow is noisy and can be detected with a stethoscope. Turbulence-produced noises (murmurs and bruits) heard with a stethoscope indicate abnormalities. The onset and cessation of noise is used to measure systolic and diastolic blood pressure with a stethoscope and blood-pressure cuff. Atherosclerotic plaques "usually . . . form where the arteries branch off—presumably because the constant turbulence at these areas injures the arterial wall, making it more susceptible" to plaque formation (*Merck Manual of Medical Information: Home Edition,* Merck, 1997, chap. 26).

A *Newtonian fluid* is one for which η is independent of dv_y/dx. For a *non-Newtonian fluid,* η in (16.13) changes as dv_y/dx changes. Gases and most pure non-polymeric liquids are Newtonian. Polymer solutions, liquid polymers, and colloidal suspensions are often non-Newtonian. An increase in flow rate and in dv_y/dx may change the shape of flexible polymer molecules, facilitating flow and reducing η.

Figure 16.7

Viscosity η, thermal conductivity k, and self-diffusion coefficient D versus T at 1 atm for (a) $H_2O(l)$; (b) $Ar(g)$.

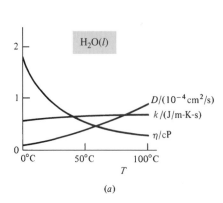

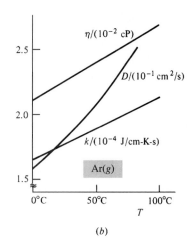

(a)

(b)

From (16.13), the SI units of η are $N\ m^{-2}\ s = Pa\ s = kg\ m^{-1}\ s^{-1}$, since $1\ N = 1\ kg\ m\ s^{-2}$. The cgs units of η are $dyn\ cm^{-2}\ s = g\ cm^{-1}\ s^{-1}$, and $1\ dyn\ cm^{-2}\ s$ is called one **poise** (P). Since $1\ dyn = 10^{-5}\ N$, we have

$$1\ P \equiv 1\ dyn\ cm^{-2}\ s = 0.1\ N\ m^{-2}\ s \qquad (16.14)$$

Some values of η in centipoises for liquids and gases at 25°C and 1 atm are

Substance	C_6H_6	H_2O	H_2SO_4	olive oil	glycerol	O_2	CH_4
η/cP	0.60	0.89	19	80	954	0.021	0.011

Gases are much less viscous than liquids. The viscosity of liquids generally decreases rapidly with increasing temperature. (Molasses flows faster at higher temperatures.) The viscosity of liquids increases with increasing pressure. The earth has a solid inner core surrounded by a liquid outer core. The outer core is at very high pressure (1 to 3 Mbar) and is barely a liquid; its viscosity is 10^9 P [D. E. Smylie, *Science,* **255,** 1678 (1992)].

Figure 16.7a plots η versus T for $H_2O(l)$ at 1 atm. Also plotted are water's thermal conductivity k (Sec. 16.2) and self-diffusion coefficient D (Sec. 16.4). Figure 16.7b plots these quantities for $Ar(g)$ at 1 atm.

Strong intermolecular attractions in a liquid hinder flow and make η large. Therefore, liquids of high viscosity have high boiling points and high heats of vaporization. Viscosities of liquids decrease as T increases, because the higher translational kinetic energy allows intermolecular attractions to be overcome more easily. In gases, intermolecular attractions are much less significant in determining η than in liquids.

The viscosities of liquids are also influenced by the molecular shape. Long-chain liquid polymers are highly viscous, because the chains become tangled with one another, hindering flow. The viscosity of liquid sulfur shows an extraordinary ten-thousandfold *increase* with temperature in the range 155°C to 185°C (Fig. 16.8). Below 150°C, liquid sulfur consists of S_8 rings. Near 155°C, the rings begin to break, producing S_8 radicals, which polymerize to long-chain molecules containing an average of 10^5 S_8 units.

Since $F_y = ma_y = m(dv_y/dt) = d(mv_y)/dt = dp_y/dt$, Newton's law of viscosity (16.13) can be written as

$$\frac{dp_y}{dt} = -\eta \mathcal{A} \frac{dv_y}{dx} \qquad (16.15)$$

where dp_y/dt is the time rate of change in the y component of momentum of a layer on one side of a surface in the fluid due to its interaction with fluid on the other side. The

η/cP Viscosity of liquid sulfur

Figure 16.8

Viscosity of liquid sulfur versus temperature at 1 atm. The vertical scale is logarithmic.

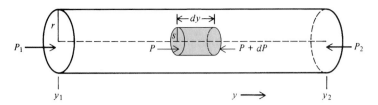

Figure 16.9

A fluid flowing in a cylindrical tube. The shaded portion of fluid is used in the derivation of Poiseuille's law (Prob. 16.12).

molecular explanation of viscosity is that it is due to a *transport of momentum* across planes perpendicular to the x axis in Fig. 16.6. Molecules in adjacent layers of the fluid have different average values of p_y, since adjacent layers are moving at different speeds. In gases, the random molecular motion brings some molecules from the faster-moving layer into the slower-moving layer, where they collide with slower-moving molecules and impart extra momentum to them, thereby tending to speed up the slower layer. Similarly, slower-moving molecules moving into the faster layer tend to slow down this layer. In liquids, the momentum transfer between layers occurs mainly by collisions between molecules in adjacent layers, without actual transfer of molecules between layers.

Flow Rate of Fluids

Newton's viscosity law (16.13) allows the rate of flow of a fluid through a tube to be determined. Figure 16.9 shows a fluid flowing in a cylindrical tube. The pressure P_1 at the left end of the tube is greater than the pressure P_2 at the right end, and the pressure drops continually along the tube. The flow speed v_y is zero at the walls (the no-slip condition) and increases toward the center of the pipe. By the symmetry of the tube, v_y can depend only on the distance s from the tube's center (and not on the angle of rotation about the tube's axis); thus v_y is a function of s only; $v_y = v_y(s)$ (see also Prob. 16.13). The liquid flows in infinitesimally thin cylindrical layers, a layer with radius s flowing with speed $v_y(s)$.

Using Newton's viscosity law, one finds (see Prob. 16.12a for the derivation) that $v_y(s)$ for laminar flow of a fluid in a cylindrical tube of radius r is

$$v_y = \frac{1}{4\eta}\left(r^2 - s^2\right)\left(-\frac{dP}{dy}\right) \qquad \text{laminar flow} \qquad (16.16)$$

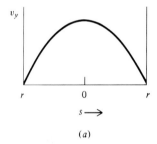

where dP/dy (which is negative) is the pressure gradient. Equation (16.16) shows that $v_y(s)$ is a parabolic function for laminar flow in a pipe; see Fig. 16.10a. (For turbulent flow, there are random fluctuations of velocity with time, and portions of the fluid move perpendicularly to the pipe axis as well as in the axial direction. The time-average velocity profile for turbulent flow looks like Fig. 16.10b.)

Application of (16.16) to a liquid shows that (see Prob. 16.12b) for laminar (non-turbulent) flow of a liquid in a tube of radius r, the flow rate is

$$\frac{V}{t} = \frac{\pi r^4}{8\eta}\frac{P_1 - P_2}{y_2 - y_1} \qquad \text{laminar flow of liquid} \qquad (16.17)$$

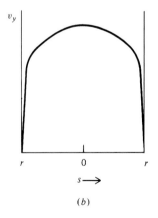

where V is the volume of liquid that passes a cross section of the tube in time t and $(P_2 - P_1)/(y_2 - y_1)$ is the pressure gradient along the tube (Fig. 16.9). Equation (16.17) is **Poiseuille's law.** [The French physician Poiseuille (1799–1869) was interested in blood flow in capillaries and measured flow rates of liquids in narrow glass tubes. Blood flow is a complex process that is not fully described by Poiseuille's law.

Figure 16.10

Velocity profiles for fluid flow in a cylindrical pipe: (a) laminar flow; (b) turbulent flow. $s = 0$ corresponds to the center of the pipe.

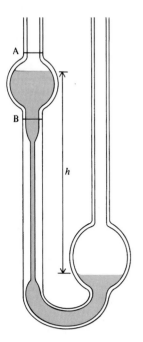

Figure 16.11

Ostwald viscometer. One measures the time for the liquid to fall from level A to level B. The pressure difference driving the liquid through the tube is $\rho g h$, where h is shown and ρ and g are the liquid's density and the gravitational acceleration.

For the biophysics of blood flow, see G. J. Hademenos, *American Scientist,* **85,** 226 (1997).] Note the very strong dependence of flow rate on tube radius and the inverse dependence on fluid viscosity η. (A vasodilator drug such as nitroglycerin increases the radius of blood vessels, thereby reducing the resistance to flow and the load on the heart. This relieves the pain of angina pectoris.) For a gas (assumed ideal), Poiseuille's law is modified to (see Prob. 16.12c)

$$\frac{dn}{dt} \approx \frac{\pi r^4}{16\eta RT} \frac{P_1^2 - P_2^2}{y_2 - y_1} \qquad \text{laminar isothermal flow of ideal gas} \qquad (16.18)$$

where dn/dt is the flow rate in moles per unit time and P_1 and P_2 are the inlet and outlet pressures at y_1 and y_2. Equation (16.18) is accurate only if P_1 and P_2 don't differ greatly from each other (see Prob. 16.13).

Measurement of Viscosity

Measurement of the flow rate through a capillary tube of known radius allows η of a liquid or gas to be found from (16.17) or (16.18).

A convenient way to determine the viscosity of a liquid is to use an **Ostwald viscometer** (Fig. 16.11). Here, one measures the time t it takes for the liquid level to fall from the mark at A to the mark at B as the liquid flows through the capillary tube. One then refills the viscometer with a liquid of known viscosity using the same liquid volume as before, and again measures t. The pressure driving the liquid through the tube is $\rho g h$ (where ρ is the liquid density, g the gravitational acceleration, and h the difference in liquid levels between the two arms of the viscometer), and $\rho g h$ replaces $P_1 - P_2$ in Poiseuille's law (16.17). Since h varies during the experiment, the flow rate varies. From (16.17), the time t needed for a given volume to flow is directly proportional to η and inversely proportional to ΔP. Since $\Delta P \propto \rho$, we have $t \propto \eta/\rho$, where the proportionality constant depends on the geometry of the viscometer. Hence $\rho t/\eta$ is a constant for all runs. For two different liquids a and b, we thus have $\rho_a t_a/\eta_a = \rho_b t_b/\eta_b$ and

$$\frac{\eta_b}{\eta_a} = \frac{\rho_b t_b}{\rho_a t_a} \qquad (16.19)$$

where η_a, ρ_a, and t_a and η_b, ρ_b, and t_b are the viscosities, densities, and flow times for liquids a and b. If η_a, ρ_a, and ρ_b are known, one can find η_b.

Another way to find η of a liquid is to measure the rate of fall of a spherical solid through the liquid. The layer of fluid in contact with the ball moves along with it (no-slip condition), and a gradient of speed develops in the fluid surrounding the sphere. This gradient generates a viscous force F_{fr} resisting the sphere's motion. This viscous force F_{fr} is found to be proportional to the moving body's speed v (provided v is not too high)

$$F_{\text{fr}} = fv \qquad (16.20)$$

where f is a constant called the **friction coefficient.** Stokes proved that, for a solid sphere of radius r moving at speed v through a Newtonian fluid of viscosity η,

$$F_{\text{fr}} = 6\pi\eta rv \qquad (16.21)$$

provided v is not too high. This equation applies to motion through a gas, provided r is much greater than the mean free path λ and there is no slip. For a derivation of **Stokes' law** (16.21), see *Bird, Stewart, and Lightfoot,* pp. 132–133. (The force F_{fr} on a solid moving through a fluid is called the **drag,** and is of obvious interest to fish and birds. See S. Vogel, *Life in Moving Fluids,* 2nd ed., Princeton U. Press, 1994.)

A spherical body falling through a fluid experiences a downward gravitational force mg, an upward frictional force given by (16.21), and an upward buoyant force

F_{buoy} that results from the greater fluid pressure below the body than above it [Eq. (1.9)]. To find F_{buoy}, imagine that the immersed object with volume V is replaced by fluid of equal volume. The buoyant force doesn't depend on the object being buoyed up, so the buoyant force on the fluid of volume V equals that on the original immersed object. However, the fluid is at rest, so the upward buoyant force on it equals the downward gravitational force, which is its weight. Therefore, an object of volume V immersed in a fluid is buoyed up by a force equal to the weight of fluid of volume V. This is *Archimedes' principle* (allegedly discovered while he was bathing).

Let m_{fl} be the mass of fluid of volume V. The falling sphere will reach a terminal speed at which the downward and upward forces on it balance. Equating the downward and upward forces on the sphere, we have $mg = 6\pi\eta rv + m_{\text{fl}}g$ and

$$6\pi\eta rv = (m - m_{\text{fl}})g = (\rho - \rho_{\text{fl}})gV = (\rho - \rho_{\text{fl}})g\tfrac{4}{3}\pi r^3$$

$$v = 2(\rho - \rho_{\text{fl}})gr^2/9\eta \qquad (16.22)$$

where ρ and ρ_{fl} are the densities of the sphere and the fluid, respectively. Measurement of the terminal speed of fall allows η to be found.

Kinetic Theory of Gas Viscosity

The kinetic-theory derivation of η for gases is very similar to the derivation of the thermal conductivity, except that momentum [Eq. (16.15)] rather than heat energy is transported. Replacing dq by dp_y and ε by mv_y in (16.4), we get

$$dp_y = \tfrac{1}{4}(N/V)\langle v\rangle \mathcal{A}(mv_{y,-} - mv_{y,+})\,dt$$

where $mv_{y,-}$ is the y momentum of a molecule in the plane at $x_0 - \tfrac{2}{3}\lambda$ (Fig. 16.4) and $mv_{y,+}$ is the corresponding quantity for the $x_0 + \tfrac{2}{3}\lambda$ plane; dp_y is the net momentum flow across a surface of area $\mathcal{A}$ in time dt. We have $dv_y = (dv_y/dx)\,dx = -(dv_y/dx)\cdot\tfrac{4}{3}\lambda$ [Eq. (16.6)], where $dv_y = v_{y,-} - v_{y,+}$. Also, $Nm/V = \rho$, where m is the mass of one molecule. Hence

$$dp_y/dt = -\tfrac{1}{3}\rho\langle v\rangle\lambda\mathcal{A}(dv_y/dx) \qquad (16.23)$$

Comparison with Newton's viscosity law $dp_y/dt = -\eta\mathcal{A}(dv_y/dx)$ [Eq. (16.15)] gives

$$\eta \approx \tfrac{1}{3}\rho\langle v\rangle\lambda \qquad \text{hard spheres} \qquad (16.24)$$

Because of the crudity of assumptions 2 to 4 of Sec. 16.2, the coefficient in (16.24) is wrong. The rigorous result for hard-sphere molecules is (*Present*, sec. 11-2)

$$\eta = \frac{5\pi}{32}\rho\langle v\rangle\lambda = \frac{5}{16\pi^{1/2}}\frac{(MRT)^{1/2}}{N_A d^2} \qquad \text{hard spheres} \qquad (16.25)$$

where (15.47) and (15.67) for $\langle v\rangle$ and λ, and $PM = \rho RT$, were used.

EXAMPLE 16.1 Viscosity and molecular diameter

The viscosity of $HCl(g)$ at $0°C$ and 1 atm is 0.0131 cP. Calculate the hard-sphere diameter of an HCl molecule.

Use of 1 P = 0.1 N s m^{-2} [Eq. (16.14)] gives $\eta = 1.31 \times 10^{-5}$ N s m^{-2}. Substitution in (16.25) gives

$$d^2 = \frac{5}{16\pi^{1/2}}\frac{[(36.5 \times 10^{-3}\,\text{kg mol}^{-1})(8.314\,\text{J mol}^{-1}\,\text{K}^{-1})(273\,\text{K})]^{1/2}}{(6.02 \times 10^{23}\,\text{mol}^{-1})(1.31 \times 10^{-5}\,\text{N s m}^{-2})}$$

$$d^2 = 2.03 \times 10^{-19}\,\text{m}^2 \quad \text{and} \quad d = 4.5 \times 10^{-10}\,\text{m} = 4.5\,\text{Å}$$

Exercise

The viscosity of water vapor at 100°C and 1 bar is 123 μP. Calculate the hard-sphere diameter of an H_2O molecule. (*Answer:* 4.22 Å.)

Exercise

Show that (16.25) and (16.12) predict that $k = (C_{V,m} + \frac{9}{4}R)\eta/M$ for a gas of hard-sphere molecules.

Some hard-sphere molecular diameters calculated from (16.25) using η at 0°C and 1 atm are:

Molecule	He	H_2	N_2	O_2	CH_4	C_2H_4	H_2O	CO_2	
$d/Å$	2.2	2.7	3.7	3.6	4.1	4.9	3.2	4.6	(16.26)

Because the hard-sphere model is a poor representation of intermolecular forces, d values calculated from (16.25) vary with temperature (Prob. 16.15).

Equation (16.25) predicts the viscosity of a gas to increase with increasing temperature and to be independent of pressure. Both these predictions are surprising, in that (by analogy with liquids) one might expect the gas to flow more easily at higher T and less easily at higher P.

When Maxwell derived (16.24) in 1860, there were virtually no data on the temperature and pressure dependences of gas viscosities, so Maxwell and his wife Katherine (née Dewar) measured η as a function of T and P for gases. (In a postcard to a scientific colleague, Maxwell wrote: "My better $\frac{1}{2}$, who did all the real work of the kinetic theory is at present engaged in other researches. When she is done, I will let you know her answer to your enquiry [about experimental data]." The experimental results were that indeed η of a gas did increase with increasing T and was essentially independent of P. This provided strong early confirmation of the kinetic theory.

As with the thermal conductivity, η increases with T substantially faster than the $T^{1/2}$ prediction of (16.25), because of the crudity of the hard-sphere model. For example, Fig. 16.7*b* shows a near linear increase with T for Ar(*g*). Use of a more realistic model of intermolecular forces than the hard-sphere model gives much better agreement with experiment (*Reid, Prausnitz, and Poling,* chap. 9).

Data for η (in micropoises) are plotted versus P in Fig. 16.12 for some gases at 50°C. As with k, the viscosity is nearly independent of P up to 50 or 100 atm. At very low pressures, where the mean free path is comparable to, or larger than, the dimensions of the container, Newton's viscosity law (16.13) does not hold. (See *Kauzmann,* p. 207.)

For liquids (unlike gases), there is no satisfactory theory that allows prediction of viscosities. Empirical estimation methods give rather poor predictions of liquid viscosities (see *Reid, Prausnitz, and Poling,* chap. 9).

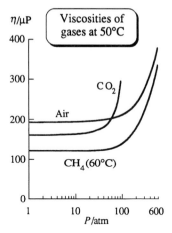

Figure 16.12

Viscosities versus P for some gases at 50°C.

Viscosity of Polymer Solutions

A molecule of a long-chain synthetic polymer usually exists in solution as a *random coil.* There is nearly free rotation about the single bonds of the chain, so we can crudely picture the polymer as composed of a large number of links with random orientations between adjacent links. This picture is essentially the same as the random motion of a particle undergoing Brownian motion, each "step" of Brownian motion corresponding to a chain link. A polymer random coil therefore resembles the path of a particle undergoing Brownian motion (Fig. 3.14). The degree of compactness of the coil depends on the relative strengths of the intermolecular forces between the polymer and solvent molecules as compared with the forces between two parts of the polymer chain. The compactness therefore varies from solvent to solvent for a given polymer.

We can expect the viscosity of a polymer solution to depend on the size and shape (and hence on the molecular weight and the degree of compactness) of the polymer molecules in the solution. If we restrict ourselves to a given kind of synthetic polymer in a given solvent, then the degree of compactness remains the same, and the polymer molecular weight can be determined by viscosity measurements. Solutions of polyethylene $(CH_2CH_2)_n$ will show different viscosity properties in a given solvent, depending on the degree of polymerization n.

The *relative viscosity* (or *viscosity ratio*) η_r of a polymer solution is defined as $\eta_r \equiv \eta/\eta_A$, where η and η_A are the viscosities of the solution and the pure solvent A. Note that η_r is a dimensionless number. Of course, η_r depends on concentration, approaching 1 in the limit of infinite dilution. Addition of a polymer to a solvent increases the viscosity, so η_r is greater than 1. Because polymer solutions are often non-Newtonian, one measures their viscosities at low flow rates, so that the flow rate has little effect on the molecular shape and on the viscosity.

The *intrinsic viscosity* (or *limiting viscosity number*) $[\eta]$ of a polymer solution is

$$[\eta] \equiv \lim_{\rho_B \to 0} \frac{\eta_r - 1}{\rho_B} \qquad \text{where } \eta_r \equiv \eta/\eta_A \qquad (16.27)$$

where $\rho_B \equiv m_B/V$ is the mass concentration [Eq. (9.2)] of the polymer, m_B and V being the mass of polymer in the solution and the solution volume. One finds that $[\eta]$ depends on the solvent as well as on the polymer. In 1942, Huggins showed that $(\eta_r - 1)/\rho_B$ is a linear function of ρ_B in dilute solutions, so a plot of $(\eta_r - 1)/\rho_B$ versus ρ_B allows one to obtain $[\eta]$ by extrapolation to $\rho_B = 0$.

Experimental data show that for a given kind of synthetic polymer in a given solvent, the following relation is well obeyed at fixed temperature:

$$[\eta] = K(M_B/M^\circ)^a \qquad (16.28)$$

where M_B is the molar mass of the polymer, K and a are empirical constants, and $M^\circ \equiv 1$ g/mol. For example, for polyisobutylene in benzene at 24°C, one finds $a = 0.50$ and $K = 0.083$ cm³/g. Typically, a lies between 0.5 and 1.1. (Data on synthetic polymers are tabulated in J. Brandrup et al., *Polymer Handbook,* 4th ed., Wiley, 1999.) To apply (16.28), one must first determine K and a for the polymer and the solvent using polymer samples whose molecular weights have been found by some other method (such as osmotic-pressure measurements). Once K and a are known, the molar mass of a given sample of the polymer can be found by viscosity measurements.

A particular protein has a definite molecular weight. In contrast, preparation of a synthetic polymer produces molecules with a distribution of molecular weights, since chain termination can occur with any length of chain.

Let n_i and x_i be the number of moles and the mole fraction of polymer species i with molar mass M_i present in a polymer sample. The **number average molar mass** M_n of the sample is defined by Eqs. (12.32) and (12.33) as $M_n \equiv m/n = \Sigma_i x_i M_i$, where the sum goes over all the polymer species and m and n are the total mass and the total number of moles of polymeric material.

In M_n, the molar mass of each species has a weighting factor given by x_i, its mole fraction; x_i is proportional to the relative number of i molecules present. In the **weight** (or **mass**) **average molar mass** M_w, the molar mass of each species has a weighting factor given by w_i, its mass (or weight) fraction in the polymer mixture, where $w_i \equiv m_i/m$ (m_i is the mass of species i present in the mixture). Thus

$$M_w \equiv \sum_i w_i M_i = \frac{\sum_i m_i M_i}{\sum_i m_i} = \frac{\sum_i n_i M_i^2}{\sum_i n_i M_i} = \frac{\sum_i x_i M_i^2}{\sum_i x_i M_i} \qquad (16.29)$$

For a polymer with a distribution of molecular weights, Eq. (16.28) yields a *viscosity average molar mass* M_v, where $M_v = [\Sigma_i w_i M_i^a]^{1/a}$.

16.4 DIFFUSION AND SEDIMENTATION

Diffusion

Figure 16.13 shows two fluid phases 1 and 2 separated by a removable impermeable partition. The system is held at constant T and P. Each phase contains only substances j and k but with different initial molar concentrations: $c_{j,1} \neq c_{j,2}$ and $c_{k,1} \neq c_{k,2}$, where $c_{j,1}$ is the concentration of j in phase 1. One or both phases might be pure. When the partition is removed, the two phases are in contact, and the random molecular motion of j and k molecules will reduce and ultimately eliminate the concentration differences between the two solutions. This spontaneous decrease in concentration differences is **diffusion.**

Diffusion is a macroscopic motion of components of a system that arises from concentration differences. If $c_{j,1} < c_{j,2}$, there is a net flow of j from phase 2 to phase 1 and a net flow of k from phase 1 to 2. This flow continues until the concentrations and chemical potentials of j and k are constant throughout the cell. Diffusion differs from the macroscopic bulk flow that arises from pressure differences (Sec. 16.3). In bulk flow in the y direction (Fig. 16.9), the flowing molecules have an additional component of velocity v_y that is superimposed on the random distribution of velocities. In diffusion, all the molecules have only random velocities. However, because the concentration c_j on the right of a plane perpendicular to the diffusion direction is greater than the concentration to the left of this plane, more j molecules cross this plane from the right than from the left, giving a net flow of j from right to left. Figure 16.14 shows how the j concentration profile along the diffusion cell varies with time during a diffusion experiment.

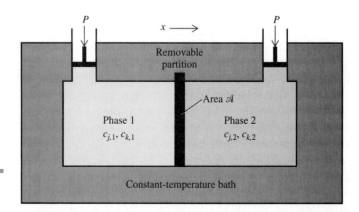

Figure 16.13

When the partition is removed, diffusion occurs.

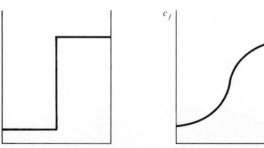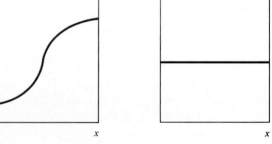

Figure 16.14

Concentration profiles during a diffusion experiment.

(*a*) $t = 0$ (*b*) Intermediate time   (*c*) $t = \infty$

Experiment shows that the following equations are obeyed in diffusion:

$$\frac{dn_j}{dt} = -D_{jk}\mathcal{A}\frac{dc_j}{dx} \quad \text{and} \quad \frac{dn_k}{dt} = -D_{kj}\mathcal{A}\frac{dc_k}{dx} \qquad (16.30)$$

In (16.30), which is **Fick's first law of diffusion,** dn_j/dt is the net rate of flow of j (in moles per unit time) across a plane P of area $\mathcal{A}$ perpendicular to the x axis; dc_j/dx is the value at plane P of the rate of change of the molar concentration of j with respect to the x coordinate; and D_{jk} is called the (**mutual**) **diffusion coefficient.** The diffusion rate is proportional to $\mathcal{A}$ and to the concentration gradient. As time goes on, dc_j/dx at a given plane changes, eventually becoming zero. Diffusion then stops.

The diffusion coefficient D_{jk} is a function of the local state of the system and therefore depends on T, P, and the local composition of the solution. In a diffusion experiment, one measures the concentrations as functions of distance x at various times t. If the two solutions differ substantially in initial concentrations, then, since the diffusion coefficients are functions of concentration, D_{jk} varies substantially with distance x along the diffusion cell and with time as the concentrations change, so the experiment yields some sort of complicated average D_{jk} for the concentrations involved. If the initial concentrations in phase 1 are made close to those in phase 2, the variation of D_{jk} with concentration can be neglected and one obtains a D_{jk} value corresponding to the average composition of 1 and 2.

If solutions 1 and 2 mix with no volume change, then one can show (Prob. 16.31) that D_{jk} and D_{kj} in (16.30) are equal: $D_{jk} = D_{kj}$. For gases, volume changes are negligible for constant-T-and-P mixing. For liquids, volume changes on mixing are not always negligible, but by having solutions 1 and 2 differ only slightly in composition, we can satisfy the condition of negligible volume change.

For a given pair of gases, one finds that D_{jk} varies only slightly with composition, increases as T increases, and decreases as P increases. Values for several gas pairs at 0°C and 1 atm are:

Gas pair	H_2–O_2	He–Ar	O_2–N_2	O_2–CO_2	CO_2–CH_4	CO–C_2H_4
$D_{jk}/(\text{cm}^2\text{ s}^{-1})$	0.70	0.64	0.18	0.14	0.15	0.12

In liquid solutions, D_{jk} varies strongly with composition and increases as T increases. Figure 16.15 plots D_{jk} versus ethanol mole fraction for H_2O–ethanol solutions at 25°C and 1 atm. The values at $x(\text{ethanol}) = 0$ and 1 are extrapolations.

Let D_{iB}^{∞} denote the value of D_{iB} for a very dilute solution of solute i in solvent B. For example, Fig. 16.15 gives $D_{H_2O,C_2H_5OH}^{\infty} = 2.4 \times 10^{-5}$ cm^2 s^{-1} at 25°C and 1 atm. Some D^{∞} values at 25°C and 1 atm for the solvent H_2O are:

i	N_2	LiBr	NaCl	n-C_4H_9OH	sucrose	hemoglobin
$10^5 D_{i,H_2O}^{\infty}/(\text{cm}^2\text{ s}^{-1})$	1.6	1.4	2.2	0.56	0.52	0.07

Mutual diffusion coefficients for solids depend on concentration and increase rapidly as T increases. Some solid-phase diffusion coefficients at 1 atm are:

i–B	Bi–Pb	Sb–Ag	Al–Cu	Ni–Cu	Ni–Cu	Cu–Ni
Temperature	20°C	20°C	20°C	630°C	1025°C	1025°C
$D_{iB}^{\infty}/(\text{cm}^2\text{ s}^{-1})$	10^{-16}	10^{-21}	10^{-30}	10^{-13}	10^{-9}	10^{-11}

Suppose solutions 1 and 2 in Fig. 16.13 have the same composition ($c_{j,1} = c_{j,2}$ and $c_{k,1} = c_{k,2}$), and we add a tiny amount of radioactively labeled species j to solution 2. The diffusion coefficient of the labeled j in the otherwise homogeneous mixture of j and k is called the **tracer diffusion coefficient** $D_{T,j}$ of j in the mixture. If $c_{k,1} = 0 = c_{k,2}$, then we are measuring the diffusion coefficient of a tiny amount of radioactively labeled j in pure j; this is the **self-diffusion coefficient** D_{jj}.

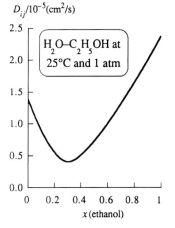

$D_{ij}/10^{-5}(\text{cm}^2/\text{s})$

H$_2$O–C$_2$H$_5$OH at 25°C and 1 atm

x(ethanol)

Figure 16.15

Mutual diffusion coefficient versus composition for water–ethanol solutions at 25°C and 1 atm.

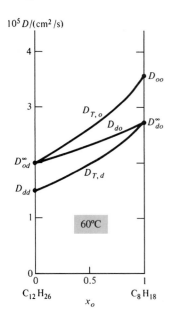

Figure 16.16

Tracer diffusion coefficients $D_{T,d}$ and $D_{T,o}$ and mutual diffusion coefficient D_{do} versus composition for liquid solutions of octane (o) plus dodecane (d) at 60°C and 1 atm. [Data from A. L. Van Geet and A. W. Adamson, *J. Phys. Chem.*, **68**, 238 (1964).]

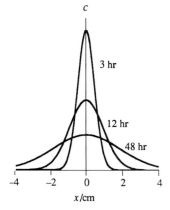

Figure 16.17

Diffusion of a solute with $D = 10^{-5}$ cm²/s, the typical value in a liquid. The solute is located in the $x = 0$ plane initially, and its distribution in the x direction is shown after 3, 12, and 48 hr.

For liquid mixtures of octane (o) and dodecane (d) at 60°C and 1 atm, Fig. 16.16 plots the mutual diffusion coefficient $D_{od} = D_{do}$ and the tracer diffusion coefficients $D_{T,o}$ and $D_{T,d}$ versus octane mole fraction x_o. Note that the tracer diffusion coefficient $D_{T,o}$ of octane in the mixture goes to the self-diffusion coefficient D_{oo} in the limit as $x_o \to 1$ and goes to the infinite-dilution mutual diffusion coefficient D_{od}^∞ as $x_o \to 0$.

Some self-diffusion coefficients at 1 atm are:

Gas (0°C)	H_2	O_2	N_2	HCl	CO_2	C_2H_6	Xe
$D_{jj}/(\text{cm}^2\ \text{s}^{-1})$	1.5	0.19	0.15	0.12	0.10	0.09	0.05

Liquid (25°C)	H_2O	C_6H_6	Hg	CH_3OH	C_2H_5OH	$n\text{-}C_3H_7OH$
$10^5 D_{jj}/(\text{cm}^2\ \text{s}^{-1})$	2.4	2.2	1.7	2.3	1.0	0.6

Diffusion coefficients at 1 atm and 25°C are typically 10^{-1} cm² s⁻¹ for gases and 10^{-5} cm² s⁻¹ for liquids; they are extremely small for solids.

Net Displacement of Diffusing Molecules

An early objection to the kinetic theory of gases was that if gases really consisted of molecules moving about freely at supersonic speeds, mixing of gases should take place almost instantaneously. This does not occur. If a chemistry lecturer generates Cl_2, it may take a couple of minutes for those in the back of the room to smell the gas. The reason mixing of gases is slow relative to the speeds of gas molecules is that at ordinary pressures a gas molecule goes only a very short distance (about 10^{-5} cm at 1 atm and 25°C; see Sec. 15.7) before colliding with another molecule; at each collision, the direction of motion changes, and each molecule has a zigzag path (Fig. 3.14). The net motion in any given direction is quite small because of these continual changes in direction.

How far on the average does a molecule undergoing random diffusional motion travel in a given direction in time t? For a diffusing molecule, let Δx be the net displacement in the x direction that occurs in time t. Since the motion is random, Δx is as likely to be positive as negative, so the average value $\langle \Delta x \rangle$ is zero (provided no boundary wall prevents diffusion in a particular direction). We therefore consider $\langle (\Delta x)^2 \rangle$, the average of the square of the x displacement. In 1905, Einstein proved that

$$\langle (\Delta x)^2 \rangle = 2Dt \tag{16.31}$$

where D is the diffusion coefficient. A derivation of the **Einstein–Smoluchowski equation** (16.31) is given in *Kennard*, pp. 286–287. See also Prob. 16.32.

The quantity

$$(\Delta x)_{\text{rms}} \equiv \langle (\Delta x)^2 \rangle^{1/2} = (2Dt)^{1/2} \tag{16.32}$$

is the root-mean-square net displacement of a diffusing molecule in the x direction in time t. Taking t to be 60 s and D to be 10^{-1}, 10^{-5}, and 10^{-20} cm² s⁻¹, we find the typical rms x displacements in 1 min of molecules at room temperature and 1 atm to be only 3 cm in gases, 0.03 cm in liquids, and less than 1 Å in solids. In 1 min, a typical gas molecule of molecular weight 30 travels a total distance of 3×10^6 cm at room temperature and pressure [Eq. (15.48)], but its rms net displacement in any given direction is only 3 cm, because of collisions. Of course, there is a distribution of Δx values, and many molecules go shorter or longer distances than $(\Delta x)_{\text{rms}}$. This distribution turns out to be gaussian (Fig. 16.17), so a substantial fraction of molecules go 2 or 3 times $(\Delta x)_{\text{rms}}$, but a negligible fraction go 7 or 8 times $(\Delta x)_{\text{rms}}$. See Prob. 16.34 for a spreadsheet simulation of diffusion.

If $(\Delta x)_{\text{rms}}$ is only 3 cm in 1 min in a gas at room T and P, why does a student in the back of the room smell the Cl_2 generated at the front of the room in only a couple of minutes? The answer is that under uncontrolled conditions, convection currents due to pressure and density differences are much more effective in mixing gases than is diffusion.

Although diffusion in liquids is slow on a macroscopic scale, it is fairly rapid on the scale of biological-cell distances. A typical diffusion coefficient for a protein in water at body temperature is 10^{-6} cm^2/s, and a typical diameter of a eukaryotic cell (one with a nucleus) is 10^{-3} cm = 10^5 Å. The typical time required for a protein molecule to diffuse this distance is given by Eq. (16.31) as $t = (10^{-3}$ cm$)^2/2(10^{-6}$ cm^2/s$)$ = 0.5 s. Nerve cells are up to 100 cm long, and diffusion of a chemical would clearly not be an effective way to transmit a signal along a nerve cell. However, diffusion of certain chemicals (neurotransmitters) is used to transmit signals from one nerve cell to another over the very short (typically, 500 Å) gap (synapse) between them.

Brownian Motion

Diffusion results from the random thermal motion of molecules. This random motion can be observed indirectly by its effect on colloidal particles suspended in a fluid. These particles undergo a random Brownian motion (Sec. 3.7) as a result of microscopic fluctuations in pressure in the fluid. Brownian motion is the perpetual dance of the molecules made visible. The colloidal particle can be considered to be a giant "molecule," and its Brownian motion is really a diffusion process.

A colloidal particle of mass m in a fluid of viscosity η experiences a time-varying force $\mathbf{F}(t)$ due to random collisions with molecules of the fluid. Let $F_x(t)$ be the x component of this random force. In addition, the particle experiences a frictional force $\mathbf{F}_{\text{fr}}$ that results from the liquid's viscosity and opposes the motion of the particle. The x component of $\mathbf{F}_{\text{fr}}$ is given by Eq. (16.20) as $F_{\text{fr},x} = -fv_x = -f(dx/dt)$, where f is the friction coefficient. The minus sign is present because when v_x (the particle's x component of velocity) is positive, $F_{\text{fr},x}$ is in the negative x direction. Newton's second law $F_x = ma_x = m(d^2x/dt^2)$ when multiplied by x gives

$$xF_x(t) - fx(dx/dt) = mx(d^2x/dt^2) \qquad (16.33)$$

Einstein averaged (16.33) over many colloidal particles. Assuming that the colloidal particles have an average kinetic energy equal to the average translational energy $\frac{3}{2}kT$ of the molecules of the surrounding fluid [Eq. (15.15)], he found the particles' average square displacement in the x direction to increase with time according to

$$\langle(\Delta x)^2\rangle = 2kTf^{-1}t \qquad (16.34)$$

The derivation of (16.34) from (16.33) is outlined in Prob. 16.30.

If the colloidal particles are spheres each with radius r, then Stokes' law (16.21) gives $|F_{\text{fr},x}| = 6\pi\eta rv_x$ and the friction coefficient is $f = 6\pi\eta r$. Equation (16.34) becomes

$$\langle(\Delta x)^2\rangle = \frac{kT}{3\pi\eta r}t \qquad \text{spherical particles} \qquad (16.35)$$

Equation (16.35) was derived by Einstein in 1905 and verified experimentally by Perrin. Measurement of $\langle(\Delta x)^2\rangle$ for colloidal particles of known size enables $k = R/N_A$ to be calculated and hence enables Avogadro's number to be found.

Theory of Diffusion in Liquids

Consider a very dilute solution of solute i in solvent B. The Einstein–Smoluchowski equation (16.31) gives the mean square x displacement of an i molecule in time t as $\langle(\Delta x)^2\rangle = 2D_{i\text{B}}^{\infty}t$, where $D_{i\text{B}}^{\infty}$ is the diffusion coefficient for a very dilute solution of i in B. Equation (16.34) gives $\langle(\Delta x)^2\rangle = (2kT/f)t$. Therefore $(2kT/f)t = 2D_{i\text{B}}^{\infty}t$, or

$$D_{i\text{B}}^{\infty} = kT/f \qquad (16.36)$$

where f is the friction coefficient [Eq. (16.20)] for motion of i molecules in the solvent B. Equation (16.36) is the *Nernst–Einstein equation.*

Application of the macroscopic concept of a viscous resisting force to the motion of a particle of colloidal size through a fluid is valid, but its application to the motion of individual molecules through a fluid is open to doubt, unless the solute molecules are much larger than the solvent molecules, for example, a solution of a polymer in water. Therefore, (16.36) is nonrigorous.

If we assume that the i molecules are spherical with radius r_i and assume that Stokes' law (16.21) can be applied to the motion of i molecules through the solvent B, then $f = 6\pi\eta_B r_i$ and (16.36) becomes

$$D_{iB}^\infty \approx \frac{kT}{6\pi\eta_B r_i} \qquad \text{for } i \text{ spherical, } r_i > r_B, \text{ liquid soln.} \qquad (16.37)$$

Equation (16.37) is the **Stokes–Einstein equation.** As noted after (16.21), Stokes' law is not valid for motion in gases when r is very small, so (16.37) applies only to liquids.

We can expect (16.37) to work best when r_i is substantially larger than r_B. The use of Stokes' law assumes that there is no slip at the surface of the diffusing particle. Fluid dynamics shows that when there is no tendency for the fluid to stick at the surface of the diffusing particle, Stokes' law is replaced by $F_{fr} = 4\pi\eta_B r_i v_i$. Data on diffusion coefficients in solution indicate that for solute molecules of size similar to that of the solvent molecules, the 6 in Eq. (16.37) should be replaced by a 4:

$$D_{iB}^\infty \approx \frac{kT}{4\pi\eta_B r_i} \qquad \text{for } i \text{ spherical, } r_i \approx r_B, \text{ liquid soln.} \qquad (16.38)$$

For $r_i < r_B$, the 4 should be replaced by a smaller number.

A study of diffusion coefficients in water [J. T. Edward, *J. Chem. Educ.,* **47,** 261 (1970)] showed that (16.37) and (16.38) work surprisingly well. The molecular radii were calculated from the van der Waals radii of the atoms (Sec. 24.6).

For a theoretical equation for the self-diffusion coefficient D_{jj} in a pure liquid, see Prob. 16.26.

Kinetic Theory of Diffusion in Gases

The mean-free-path kinetic theory of diffusion in gases is similar to that of thermal conductivity and viscosity, except that matter, rather than energy or momentum, is transported. Consider first a mixture of species j with an isotopic tracer species $j^\#$, which has the same diameter and nearly the same mass as j. Let there be a concentration gradient $dc^\#/dx$ of $j^\#$. Molecules of $j^\#$ cross a plane at x_0 coming from the left and from the right. We take the concentration of $j^\#$ molecules crossing from either side as the concentration in the plane where (on the average) they made their last collision. These planes are at a distance $\frac{2}{3}\lambda$ from x_0 (Sec. 16.2). The number of molecules moving into the x_0 plane in time dt from one side is $\frac{1}{4}(N/V)\langle v\rangle \mathcal{A}\, dt$ [Eq. (15.56)]. Since $N/V = N_A n/V = N_A c$, the net number of $j^\#$ molecules that cross the x_0 plane in time dt is

$$dN^\# = \tfrac{1}{4}\langle v\rangle \mathcal{A} N_A(c_-^\# - c_+^\#)\, dt \qquad (16.39)$$

where $c_-^\#$ and $c_+^\#$ are the concentrations of $j^\#$ at the $x_0 - \frac{2}{3}\lambda$ and $x_0 + \frac{2}{3}\lambda$ planes. We have $c_-^\# - c_+^\# = dc^\# = (dc^\#/dx)\, dx = -(dc^\#/dx)\frac{4}{3}\lambda$, and (16.39) becomes

$$\frac{dn^\#}{dt} = -\tfrac{1}{3}\lambda\langle v\rangle \mathcal{A}\, \frac{dc^\#}{dx} \qquad (16.40)$$

Comparison with Fick's law (16.30) gives for the self-diffusion coefficient

$$D_{jj} \approx \tfrac{1}{3}\lambda\langle v\rangle \qquad \text{hard spheres} \qquad (16.41)$$

As usual, the numerical coefficient is wrong, and a rigorous treatment gives for hard spheres (*Present*, sec. 8-3)

$$D_{jj} = \frac{3\pi}{16}\lambda\langle v\rangle = \frac{3}{8\pi^{1/2}}\left(\frac{RT}{M}\right)^{1/2}\frac{1}{d^2(N/V)} \qquad \text{hard spheres} \qquad (16.42)$$

where (15.67) and (15.47) for λ and $\langle v\rangle$ were used. Use of $PV = (N_{total}/N_A)RT$ in (16.42) gives $D_{jj} \propto T^{3/2}/P$. The inverse dependence of D on pressure is due to the higher collision rate at higher pressure, and this inverse proportionality is usually well obeyed (Fig. 16.18).

The simple mean-free-path treatment applied to a mixture of gases j and k predicts the mutual diffusion coefficient D_{jk} to be a strong function of the j mole fraction, whereas experiment shows D_{jk} for gases to be almost independent of x_j. (The reasons for this failure are discussed in *Present*, pp. 50–51.) A rigorous treatment for hard spheres (*Present*, sec. 8-3) predicts D_{jk} to be independent of the relative proportions of j and k present.

Sedimentation of Polymer Molecules in Solution

Recall from Sec. 15.8 that the molecules of a gas in the earth's gravitational field show an equilibrium distribution in accord with the Boltzmann distribution law, the concentration of molecules decreasing exponentially with increasing altitude. A similar distribution holds for solute molecules in a solution in the earth's gravitational field. For a solution in which the distribution of solute molecules is initially uniform, there will be a net downward drift of solute molecules, until the equilibrium distribution is attained.

Consider a polymer molecule of mass M_i/N_A (where M_i is the molar mass and N_A the Avogadro constant) in a solvent of density less than that of the polymer. The polymer molecule will tend to drift downward (sediment). The polymer molecule is acted on by the following forces: (*a*) a downward force equal to the molecule's weight $M_iN_A^{-1}g$, where g is the gravitational constant; (*b*) an upward viscous force fv_{sed}, where f is the friction coefficient and v_{sed} is the downward drift speed; (*c*) an upward buoyant force that equals the weight of the displaced fluid (Sec. 16.3). The effective volume of the polymer molecule in solution depends on the solvent (Sec. 16.3), and we may take $\bar{V}_i/N_A$ as the effective volume of a molecule, where $\bar{V}_i$ is the partial molar volume of i in the solution. The buoyant force is therefore $(\rho\bar{V}_i/N_A)g$, where ρ is the density of the solvent.

The polymer's molecular weight may not be known, so $\bar{V}_i$ may not be known. We therefore define the *partial specific volume* $\bar{v}_i$ as $\bar{v}_i \equiv (\partial V/\partial m_i)_{T,P,m_B}$, where V is the solution's volume, m_i is the mass of polymer in solution, and B is the solvent. Since $m_i = M_in_i$, we have $\partial V/\partial m_i = (\partial V/\partial n_i)(\partial n_i/\partial m_i) = (\partial V/\partial n_i)/M_i$, or $\bar{v}_i = \bar{V}_i/M_i$. The buoyant force is then $\rho\bar{v}_iM_iN_A^{-1}g$.

The molecule will reach a terminal sedimentation speed v_{sed} at which the downward and upward forces balance:

$$M_iN_A^{-1}g = fv_{sed} + \rho\bar{v}_iM_iN_A^{-1}g \qquad (16.43)$$

Although sedimentation of relatively large colloidal particles in the earth's gravitational field is readily observed (Sec. 13.6), the gravitational field is actually too weak to produce observable sedimentation of polymer molecules in solution. Instead, one uses an ultracentrifuge, a device that spins the polymer solution at very high speed.

A particle revolving at constant speed v in a circle of radius r is undergoing an acceleration v^2/r directed toward the center of the circle, a *centripetal acceleration* (*Halliday and Resnick*, eq. 11-10). The speed is given by $v = r\omega$, where the *angular speed* ω (omega) is defined as $d\theta/dt$, where θ is the rotational angle in radians. The centripetal acceleration is therefore $r\omega^2$, where ω is 2π times the number of revolutions per unit time. The centripetal force is, by Newton's second law, $mr\omega^2$, where m is the particle's mass.

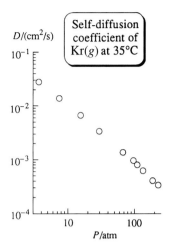

Figure 16.18

Self-diffusion coefficient of Kr(*g*) at 35°C versus *P*. Both scales are logarithmic.

Just as a marble on a merry-go-round tends to move outward, so the protein molecules tend to sediment outward in the revolving tube in the ultracentrifuge. If we use a coordinate system that revolves along with the solution, then in this coordinate system, the centripetal acceleration $r\omega^2$ disappears, and in its place one must introduce a fictitious centrifugal force $mr\omega^2$ acting outward on the particle (*Halliday and Resnick,* sec. 6-4 and supplementary topic I). In the revolving coordinate system, Newton's second law is not obeyed unless this fictitious force is introduced. $F = ma$ holds only in a nonaccelerating coordinate system.

Comparison of the fictitious centrifugal force $mr\omega^2$ in a centrifuge with the gravitational force mg in a gravitational field shows that $r\omega^2$ corresponds to g. Therefore, replacing g in Eq. (16.43) by $r\omega^2$, we get

$$M_i N_A^{-1} r\omega^2 = f v_{\text{sed}} + \rho \bar{v}_i M_i N_A^{-1} r\omega^2 \qquad (16.44)$$

The buoyant force arises, as in a gravitational field, from a pressure gradient in the fluid. The friction coefficient f can be found from diffusion data. The Nernst–Einstein equation (16.36) gives for a very dilute solution: $f = kT/D_{iB}^\infty$, where D_{iB}^∞ is the infinite-dilution diffusion coefficient of the polymer in the solvent. Using $f = kT/D_{iB}^\infty$ and $R = N_A k$, we find from (16.44) that

$$M_i = \frac{RT v_{\text{sed}}^\infty}{D_{iB}^\infty r\omega^2 (1 - \rho \bar{v}_i)} \qquad (16.45)$$

Measurement of v_{sed} extrapolated to infinite dilution and of D_{iB}^∞ enables the polymer molar mass to be found. Special optical techniques are used to measure v_{sed} in the revolving solution. The quantity $v_{\text{sed}}/r\omega^2$ is the **sedimentation coefficient** s of the polymer in the solvent. The SI unit of s is seconds (s), but sedimentation coefficients are often expressed using the *svedberg* (symbol Sv or S), defined as 10^{-13} s.

16.5   ELECTRICAL CONDUCTIVITY

Electrical conduction is a transport phenomenon in which electrical charge (carried by electrons or ions) moves through the system. The **electric current** I is defined as the rate of flow of charge through the conducting material:

$$I \equiv dQ/dt \qquad \textbf{(16.46)*}$$

where dQ is the charge that passes through a cross section of the conductor in time dt. The **electric current density** j is the electric current per unit cross-sectional area:

$$j \equiv I/\mathscr{A} \qquad \textbf{(16.47)*}$$

where $\mathscr{A}$ is the conductor's cross-sectional area. The SI unit of current is the **ampere** (A) and equals one coulomb per second:

$$1 \text{ A} = 1 \text{ C/s} \qquad \textbf{(16.48)*}$$

Although the charge Q is more fundamental than the current I, it is easier to measure current than charge. The SI system therefore takes the ampere as one of its fundamental units. The ampere is defined as the current that when flowing through two long, straight parallel wires exactly one meter apart will produce a force per unit length between the wires of exactly 2×10^{-7} N/m. (One current produces a magnetic field that exerts a force on the moving charges in the other wire.) The force between two current-carrying wires can be measured accurately using a current balance (see *Halliday and Resnick,* sec. 34-4).

The coulomb is defined as the charge transported in one second by a one-ampere current: $1 \text{ C} \equiv 1 \text{ A s}$. Equation (16.48) then follows from this definition.

To avoid confusion, we shall use only SI units for electrical quantities in this chapter, and all electrical equations will be written in a form valid for SI units.

Charge flows because it experiences an electric force, so there must be an electric field **E** in a current-carrying conductor. The **conductivity** (formerly called the *specific conductance*) κ (kappa) of a substance is defined by

$$\kappa \equiv j/E \quad \text{or} \quad j = \kappa E \qquad \textbf{(16.49)*}$$

where E is the magnitude of the electric field. The higher the conductivity κ, the greater the current density j that flows for a given applied electric field. The reciprocal of the conductivity is the **resistivity** ρ:

$$\rho = 1/\kappa \qquad \textbf{(16.50)*}$$

Let the x direction be the direction of the electric field in the conductor. Equation (14.11) gives $E_x = -d\phi/dx$, where ϕ is the electric potential at a point in the conductor. Hence (16.49) can be written as $I/\mathscr{A} = -\kappa(d\phi/dx)$. The use of (16.46) gives

$$\frac{dQ}{dt} = -\kappa\mathscr{A}\frac{d\phi}{dx} \qquad (16.51)$$

Current flows in a conductor only when there is a gradient of electric potential in the conductor. Such a gradient can be produced by attaching the ends of the conductor to the terminals of a battery.

Note the resemblance of (16.51) to the transport equations (16.1), (16.15), and (16.30) (Fourier's, Newton's, and Fick's laws) for thermal conduction, viscous flow, and diffusion. Each of these equations has the form

$$\frac{1}{\mathscr{A}}\frac{dW}{dt} = -L\frac{dB}{dx} \qquad (16.52)$$

where $\mathscr{A}$ is the cross-sectional area, W is the physical quantity being transported (q in thermal conduction, p_y in viscous flow, n_j in diffusion, Q in electrical conduction), L is a constant (k, η, D_{jk}, or κ), and dB/dx is the gradient of a physical quantity (T, v_y, c_j, or ϕ) along the direction x in which W flows. The quantity $(1/\mathscr{A})(dW/dt)$ is called the **flux** of W and is the rate of transport of W through unit area perpendicular to the flow direction. In all four transport equations, *the flux is proportional to a gradient*.

Consider a current-carrying conductor that has a homogeneous composition and a constant cross-sectional area $\mathscr{A}$. Then the current density j will be constant at every point in the conductor. From $j = \kappa E$ [Eq. (16.49)], the field strength E is constant at every point, and the equation $E = -d\phi/dx$ integrates to $\phi_2 - \phi_1 = -E(x_2 - x_1)$. Hence $E = -d\phi/dx = -\Delta\phi/\Delta x$. Equation (16.49) becomes $I/\mathscr{A} = \kappa(-\Delta\phi)/\Delta x$. Let $\Delta x = l$, where l is the length of the conductor. Then $|\Delta\phi|$ is the magnitude of the **electric potential difference** between the ends of the conductor, and we have $|\Delta\phi| = Il/\kappa\mathscr{A}$ or

$$|\Delta\phi| = (\rho l/\mathscr{A})I \qquad (16.53)$$

The quantity $|\Delta\phi|$ is often called the "voltage." The **resistance** R of the conductor is defined by

$$R \equiv |\Delta\phi|/I \quad \text{or} \quad |\Delta\phi| = IR \qquad \textbf{(16.54)*}$$

Equations (16.53) and (16.54) give

$$R = \rho l/\mathscr{A} \qquad (16.55)$$

From (16.54), R has units of volts per ampere. The SI unit of resistance is the **ohm** (symbol Ω, capital omega):

$$1\ \Omega \equiv 1\ \text{V/A} = 1\ \text{kg m}^2\ \text{s}^{-1}\ \text{C}^{-2} \qquad (16.56)$$

where (14.8) and (16.48) were used. From (16.55), the resistivity ρ has units of ohms times a length and is usually given in Ω cm or Ω m. The conductivity $\kappa = 1/\rho$ has units

Ω^{-1} cm^{-1} or Ω^{-1} m^{-1}. The unit Ω^{-1} is sometimes written as mho, which is ohm spelled backward; however, the correct SI name for the reciprocal ohm is the *siemens* (S): 1 S $\equiv$ 1 Ω^{-1}.

The conductivity κ and its reciprocal ρ depend on the composition of the conductor but not on its dimensions. From (16.55), the resistance R depends on the dimensions of the conductor as well as the material that composes it.

For many substances, κ in (16.49) is independent of the magnitude of the applied electric field E and hence is independent of the magnitude of the current density. Such substances are said to obey Ohm's law. **Ohm's law** is the statement that κ remains constant as E changes. For a substance that obeys Ohm's law, a plot of j versus E is a straight line with slope κ. Metals obey Ohm's law. Solutions of electrolytes obey Ohm's law, provided E is not extremely high and provided steady-state conditions are maintained (see Section 16.6). Many books state that Ohm's law is Eq. (16.54). This is inaccurate. Equation (16.54) is simply the *definition* of R, and this definition applies to all substances. Ohm's law is the statement that R is independent of $|\Delta\phi|$ (and of I) and does not apply to all substances.

Some resistivity and conductivity values for substances at 20°C and 1 atm are:

Substance	Cu	KCl(aq, 1 mol/dm^3)	CuO	glass
$\rho/(\Omega\ \text{cm})$	2×10^{-6}	9	10^5	10^{14}
$\kappa/(\Omega^{-1}\ \text{cm}^{-1})$	6×10^5	0.1	10^{-5}	10^{-14}

Metals have very low ρ values and very high κ values. Concentrated aqueous solutions of strong electrolytes have rather low ρ values. An **electrical insulator** (for example, glass) is a substance with a very low κ. A **semiconductor** (for example, CuO) is a substance with κ intermediate between κ of metals and insulators. Semiconductors and insulators generally do not obey Ohm's law; their conductivity increases with increasing applied potential difference $|\Delta\phi|$.

16.6 ELECTRICAL CONDUCTIVITY OF ELECTROLYTE SOLUTIONS

Electrolysis

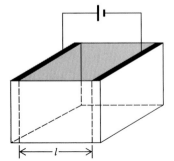

Figure 16.19

An electrolysis cell.

Figure 16.19 shows two metal electrodes at each end of a cell filled with a solution of electrolyte. A potential difference is applied to the electrodes by connecting them to a battery. Electrons carry the current through the metal wires and the metal electrodes. Ions carry the current through the solution. At each electrode–solution interface, an electrochemical reaction occurs that transfers electrons either to or from the electrode, allowing charge to flow completely around the circuit. For example, if both electrodes are Cu and the electrolyte solute is $CuSO_4$, the electrode reactions are $Cu^{2+}(aq) + 2e^- \rightarrow Cu$ and $Cu \rightarrow Cu^{2+}(aq) + 2e^-$.

For 1 mole of Cu to be deposited from solution, 2 moles of electrons must flow through the circuit. (A mole of electrons is Avogadro's number of electrons.) If the current I is kept constant, the charge that flows is $Q = It$ [Eq. (16.46)]. Experiment shows that to deposit 1 mole of Cu requires the flow of 192970 C, so the absolute value of the total charge on 1 mole of electrons is 96485 C. The absolute value of the charge per mole of electrons is the **Faraday constant** $F = 96485$ C/mol. We have [Eq. (14.13)] $F = N_A e$, where e is the proton charge and N_A is the Avogadro constant. To deposit 1 mole of the metal M from a solution containing the ion M^{z+} requires the flow of z_+ moles of electrons. The number of moles of M deposited by a flow of charge Q is therefore $Q/z_+ F$, and the mass m of metal M deposited is

$$m = QM/z_+ F \tag{16.57}$$

where M is the molar mass of the metal M.

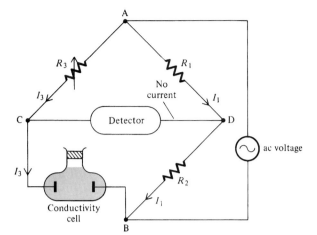

Figure 16.20

Measurement of the conductivity
of an electrolyte solution using a
Wheatstone bridge.

The total charge that flows through a circuit during time t' is given by integration of (16.46) as $Q = \int_0^{t'} I\, dt$, which equals It' if I is constant. It isn't easy to keep I constant, and a good way to measure Q is to put an electrolysis cell in series in the circuit, weigh the metal deposited, and calculate Q from (16.57). Such a device is called a *coulometer*. Silver is the metal most often used.

Measurement of Conductivity

The resistance R of an electrolyte solution cannot be reliably measured using direct current, because changes in concentration of the electrolyte and buildup of electrolysis products at the electrodes change the resistance of the solution. To eliminate these effects, one uses an alternating current and uses platinum electrodes coated with colloidal platinum black. The colloidal Pt adsorbs any gases produced during each half cycle of the alternating current.

The conductivity cell (surrounded by a constant-T bath) is placed in one arm of a Wheatstone bridge (Fig. 16.20). The resistance R_3 is adjusted until no current flows through the detector between points C and D. These points are then at equal potential. From "Ohm's law" (16.54), we have $|\Delta\phi|_{AD} = I_1 R_1$, $|\Delta\phi|_{AC} = I_3 R_3$, $|\Delta\phi|_{DB} = I_1 R_2$, and $|\Delta\phi|_{CB} = I_3 R$. Since $\phi_D = \phi_C$, we have $|\Delta\phi|_{AC} = |\Delta\phi|_{AD}$ and $|\Delta\phi|_{CB} = |\Delta\phi|_{DB}$. Therefore $I_3 R_3 = I_1 R_1$, and $I_3 R = I_1 R_2$. Dividing the second equation by the first, we get $R/R_3 = R_2/R_1$, from which R can be found. [This discussion is oversimplified, since it ignores the capacitance of the conductivity cell; see J. Braunstein and G. D. Robbins, *J. Chem. Educ.*, **48**, 52 (1971).] R is found to be independent of the magnitude of the applied ac potential difference, so Ohm's law is obeyed.

Once R is known, the conductivity can be calculated from (16.55) and (16.50) as $\kappa = 1/\rho = l/\mathcal{A}R$, where $\mathcal{A}$ and l are the area of and the separation between the electrodes. The *cell constant* K_{cell} is defined as $l/\mathcal{A}$, and $\kappa = K_{cell}/R$. Instead of measuring l and $\mathcal{A}$, it is more accurate to determine K_{cell} for the apparatus by measuring R for a KCl solution of known conductivity. Accurate κ values for KCl at various concentrations have been determined by measurements in cells of accurately known dimensions. Extremely pure solvent is used in conductivity work, since traces of impurities can significantly affect κ. The conductivity of the pure solvent is subtracted from that of the solution to get κ of the electrolyte.

Molar Conductivity

Since the number of charge carriers per unit volume usually increases with increasing electrolyte concentration, the solution's conductivity κ usually increases as the electrolyte's concentration increases. To get a measure of the current-carrying ability of a given amount of electrolyte, one defines the **molar conductivity** Λ_m (capital lambda em) of an electrolyte in solution as

$$\Lambda_m \equiv \kappa/c \qquad (16.58)*$$

where c is the electrolyte's stoichiometric molar concentration.

EXAMPLE 16.2 Molar conductivity

The conductivity κ of a 1.00 mol/dm^3 aqueous KCl solution at 25°C and 1 atm is 0.112 Ω^{-1} cm^{-1}. Find the KCl molar conductivity in this solution.

Substitution in (16.58) gives

$$\Lambda_{m,KCl} = \frac{\kappa}{c} = \frac{0.112\ \Omega^{-1}\ cm^{-1}}{1.00\ mol\ dm^{-3}} \frac{10^3\ cm^3}{1\ dm^3} = 112\ \Omega^{-1}\ cm^2\ mol^{-1}$$

which also equals 0.0112 Ω^{-1} m^2 mol^{-1}.

Exercise

For 0.10 mol/L CuSO$_4$(aq) at 25°C and 1 atm, calculate Λ_m from the κ value in Fig. 16.21a. Check your answer by using Fig. 16.21b. (*Answer:* 90 Ω^{-1} cm^2 mol^{-1}.)

For a strong electrolyte with no ion pairing, the concentration of ions is directly proportional to the stoichiometric concentration of the electrolyte, so it might be thought that dividing κ by c would give a quantity that is independent of concentration. However, the Λ_m's of NaCl(aq), KBr(aq), etc., do vary with concentration. This is because interactions between ions affect the conductivity κ, and these interactions change as c changes.

Λ_m depends on the solvent as well as on the electrolyte. We shall consider mainly aqueous solutions.

Values of κ and Λ_m for KCl in aqueous solution at various concentrations at 25°C and 1 atm are:

$c/(\text{mol dm}^{-3})$	0	0.001	0.01	0.1	1
$\kappa/(\Omega^{-1}\ \text{cm}^{-1})$	0	0.000147	0.00141	0.0129	0.112
$\Lambda_m/(\Omega^{-1}\ \text{cm}^2\ \text{mol}^{-1})$	(150)	147	141	129	112

The Λ_m value at zero concentration is obtained by extrapolation. Let Λ_m^∞ denote the infinite-dilution value: $\Lambda_m^\infty = \lim_{c \to 0} \Lambda_m$.

Figure 16.21 plots κ versus c and Λ_m versus $c^{1/2}$ for some electrolytes in aqueous solution. The rapid increase in Λ_m for CH$_3$COOH as $c \to 0$ is due to an increase in the degree of dissociation of this weak acid as c decreases. The slow decrease in Λ_m of HCl and KCl as c increases is due to attractions between oppositely charged ions, which reduce the conductivity. Λ_m for CuSO$_4$ decreases more rapidly than for HCl or KCl partly because of the increased degree of ion pairing (Sec. 10.9) as c of this 2:2 electrolyte increases. The higher κ and Λ_m of HCl as compared with KCl result from a special mechanism of transport of H$_3$O$^+$ ions, discussed later in this section. At very high concentrations, the conductivity κ of solutions of most strong electrolytes actually decreases with increasing concentration (Fig. 16.22).

For the electrolyte M$_{\nu_+}$X$_{\nu_-}$ yielding the ions M^{z+} and X^{z-} in solution, the *equivalent conductivity* Λ_{eq} is defined as

$$\Lambda_{eq} \equiv \kappa/\nu_+ z_+ c \equiv \Lambda_m/\nu_+ z_+ \qquad (16.59)$$

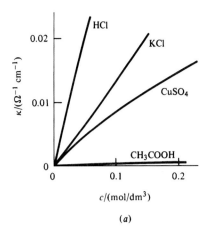

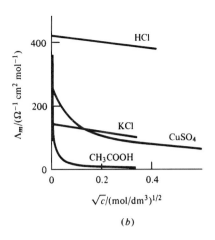

(a)

(b)

Figure 16.21

(a) Conductivity κ versus
concentration c for some aqueous
electrolytes at 25°C and 1 atm.
(b) Molar conductivity Λ_m versus
$c^{1/2}$ for these solutions.

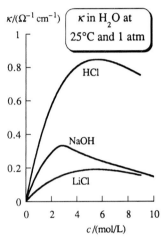

Figure 16.22

Conductivity versus concentration
for some strong electrolytes in
water at 25°C and 1 atm.

(A solution containing 1 mole of completely dissociated electrolyte would contain $\nu_+ z_+$ moles of positive charge.) For example, for $Cu_3(PO_4)_2(aq)$ we have $\nu_+ = 3$, $z_+ = 2$, and $\Lambda_{eq} = \Lambda_m/6$. Most tables in the literature list Λ_{eq}. The IUPAC has recommended discontinuing the use of equivalent conductivity. The concept of equivalents serves no purpose except to confuse chemistry students.

The subscripts m and eq can be omitted if the species to which Λ refers is specified. Thus, for $CuSO_4(aq)$, experiment gives $\Lambda_m^\infty = 266.8 \ \Omega^{-1} \ cm^2 \ mol^{-1}$ at 25°C and 1 atm. Since $\nu_+ z_+ = 2$, we have $\Lambda_{eq}^\infty = 133.4 \ \Omega^{-1} \ cm^2 \ equiv^{-1}$. We therefore write $\Lambda^\infty(CuSO_4) = 266.8 \ \Omega^{-1} \ cm^2 \ mol^{-1}$ and $\Lambda^\infty(\frac{1}{2}CuSO_4) = 133.4 \ \Omega^{-1} \ cm^2 \ mol^{-1}$.

Contributions of Individual Ions to the Current

The current in an electrolyte solution is the sum of the currents carried by the individual ions. Consider a solution with only two kinds of ions, positive ions with charge $z_+ e$ and negative ions with charge $z_- e$, where e is the proton charge. When a potential difference is applied to the electrodes, the cations feel an electric field E, which accelerates them. The viscous frictional force exerted by the solvent on the ions is proportional to the speed of the ions and opposes their motion. This force increases as the ions are accelerated. When the viscous force balances the electric-field force, the cations are no longer accelerated and travel at a constant terminal speed v_+, called the **drift speed.** We shall later see that the terminal speed is reached in about 10^{-13} s, which is virtually instantaneously.

Let there be N_+ cations in the solution. In time dt, the cations move a distance $v_+ \, dt$, and all cations within this distance from the negative electrode will reach the electrode in time dt. The number of cations within this distance of the electrode is $(v_+ \, dt/l)N_+$, where l is the separation between the electrodes (Fig. 16.19). Each cation has charge $z_+ e$, so the positive charge dQ_+ crossing a plane parallel to the electrodes in time dt is $dQ_+ = (z_+ e v_+ N_+/l) \, dt$. The current density j_+ due to the cations is $j_+ \equiv I_+/\mathscr{A} = \mathscr{A}^{-1} \, dQ_+/dt$, so

$$j_+ = z_+ e v_+ N_+/V$$

where $V = \mathscr{A}l$ is the solution's volume. Similarly, the anions contribute a current density $j_- = |z_-| e v_- N_-/V$, where we adopt the convention that both v_+ and v_- are considered positive. We have $eN_+/V = eN_A n_+/V = Fc_+$, where n_+ is the number of moles of the cation M^{z+} in the solution, F is the Faraday constant, and $c_+ = n_+/V$ is the molar concentration of M^{z+}. Hence $j_+ = z_+ F v_+ c_+$. Similarly, $j_- = |z_-| F v_- c_-$. The observed current density j is

$$j = j_+ + j_- = z_+ F v_+ c_+ + |z_-| F v_- c_- \qquad (16.60)$$

If several kinds of ions are present in the solution, the current density j_B due to ion B and the total current density j are

$$j_B = |z_B| F v_B c_B \qquad \text{and} \qquad j = \sum_B j_B = \sum_B |z_B| F v_B c_B \qquad (16.61)$$

The B current density j_B is proportional to the molar charge $z_B F$, the drift speed v_B, and the concentration c_B.

The drift speed v_B of an ion depends on the electric-field strength, the ion, the solvent, T, P, and the concentrations of all the ions in the solution.

Electric Mobilities of Ions

Since $j = \kappa E$, the conductivity of an electrolyte solution is [Eq. (16.61)] $\kappa = \sum_B |z_B| F(v_B/E) c_B$. For a given solution with fixed values of the concentrations c_B, experiment shows Ohm's law to be obeyed, meaning that κ is independent of E. This implies that, for fixed concentrations in the solution, each ratio v_B/E is equal to a constant that is characteristic of the ion B but independent of the electric-field strength E. We call this constant the **electric mobility** u_B of ion B:

$$u_B \equiv v_B/E \qquad \text{or} \qquad v_B = u_B E \qquad \textbf{(16.62)*}$$

The drift speed v_B of an ion is proportional to the applied field E, and the proportionality constant is the ion's mobility u_B.

The preceding expression for κ becomes

$$\kappa = \sum_B |z_B| F u_B c_B = \sum_B \kappa_B \qquad (16.63)$$

where κ_B is the contribution of ion B to the conductivity. For a solution with only two kinds of ions,

$$\kappa = z_+ F u_+ c_+ + |z_-| F u_- c_- \qquad (16.64)$$

Each small portion of the conducting solution must remain electrically neutral, since even a tiny departure from electrical neutrality would generate a huge electric field (Sec. 14.3). Electroneutrality requires that $z_+ e c_+ + z_- e c_- = 0$, or $z_+ c_+ = |z_-| c_-$. Hence (16.64) and (16.60) become for a solution with two kinds of ions

$$\kappa = z_+ F c_+ (u_+ + u_-) \quad \text{and} \quad j = z_+ F c_+ (v_+ + v_-) \qquad (16.65)$$

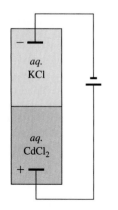

Figure 16.23

Moving-boundary apparatus for determining ionic mobility.

Ionic mobilities can be measured by the *moving-boundary method*. Figure 16.23 shows a solution of KCl placed over a solution of $CdCl_2$ in an electrolysis tube of cross-sectional area $\mathscr{A}$. The solutions used must have an ion in common. When the current flows, the K^+ ions migrate upward to the negative electrode, as do the Cd^{2+} ions. For the experiment to work, the cations of the lower solution must have a lower mobility than the cations of the upper solution: $u(Cd^{2+}) < u(K^+)$.

The speed $v(K^+)$ of migration of the K^+ ions is found by measuring the distance x that the boundary moves in time t. The boundary between the solutions is visible because of a difference in refractive index of the two solutions. We have $v(K^+) = x/t$. The electric mobility $u(K^+)$ is given by (16.62) as $u(K^+) = v(K^+)/E$. From $\kappa \equiv j/E \equiv I/\mathscr{A}E$ [Eqs. (16.47) and (16.49)], we have

$$E = I/\kappa \mathscr{A} \qquad (16.66)$$

Therefore

$$u(K^+) = x \kappa \mathscr{A}/It \qquad (16.67)$$

where κ is the conductivity of the KCl solution (assumed known). The product It equals the charge Q that flows and is measured by a coulometer. For the reasons why the boundary remains sharp and why the experiment measures $u(K^+)$ but not $u(Cl^-)$, see M. Spiro in *Rossiter, Hamilton, and Baetzold,* vol. II, sec. 5.3.

To measure $u(Cl^-)$, we could use solutions of KCl and KNO_3.

Some observed mobilities as a function of electrolyte concentration for Na^+ and Cl^- ions in NaCl(*aq*) at 25°C and 1 atm are plotted in Fig. 16.24. The decreases in u as c increases are due to interionic attractions.

For a 0.20 mol/dm^3 aqueous NaCl solution at 25°C and 1 atm, one finds $u(Cl^-)$ $= 65.1 \times 10^{-5} \ cm^2 \ V^{-1} \ s^{-1}$. This value differs slightly from the $u(Cl^-)$ value $65.6 \times 10^{-5} \ cm^2 \ V^{-1} \ s^{-1}$ in a 0.20 mol/dm^3 KCl solution, because of slight differences in Na^+–Cl^- interactions compared with K^+–Cl^- interactions.

Experimental electric mobilities extrapolated to infinite dilution for ions in water at 25°C and 1 atm are:

Ion	H_3O^+	Li^+	Na^+	Mg^{2+}	OH^-	Cl^-	Br^-	NO_3^-
$10^5 u^\infty/(cm^2 \ V^{-1} \ s^{-1})$	363	40.2	51.9	55.0	206	79.1	81.0	74.0

Since interionic forces vanish at infinite dilution, $u^\infty(Na^+)$ is the same for solutions of NaCl, Na_2SO_4, etc.

For small inorganic ions, u^∞ in aqueous solutions at 25°C and 1 atm usually lies in the range 40 to $80 \times 10^{-5} \ cm^2 \ V^{-1} \ s^{-1}$. However, $H_3O^+(aq)$ and $OH^-(aq)$ show abnormally high mobilities. These high mobilities are due to a special *jumping* mechanism that operates in addition to the usual motion through the solvent. A proton from an H_3O^+ ion can jump to a neighboring H_2O molecule, a process that has the same effect as the motion of H_3O^+ through the solution:

$$H-\underset{\overset{|}{\underset{(+)}{O}}}{\overset{H}{|}}-H \ + \ \overset{H}{\underset{|}{O}}-H \longrightarrow H-\overset{H}{\underset{|}{O}} \ + \ H-\underset{\overset{|}{\underset{(+)}{O}}}{\overset{H}{|}}-H$$

The high mobility of OH^- is due to a transfer of a proton from an H_2O molecule to an OH^- ion, which is equivalent to the motion of OH^- in the opposite direction:

$$\underset{\overset{|}{\underset{(-)}{O}}}{\overset{H}{|}} \ + \ H-\overset{H}{\underset{|}{O}} \longrightarrow \overset{H}{\underset{|}{O}}-H \ + \ \underset{\overset{|}{\underset{(-)}{O}}}{\overset{H}{|}} \tag{16.68}$$

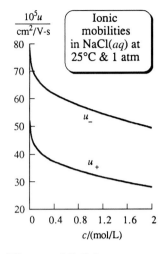

Figure 16.24

Anion and cation ionic mobilities versus concentration for aqueous NaCl at 25°C and 1 atm.

EXAMPLE 16.3 Drift speed

A typical electric-field strength for an electrolysis experiment is 10 V/cm. (*a*) Calculate the drift speed for Mg^{2+} ions in this field in dilute aqueous solution at 25°C and 1 atm. (*b*) Compare the result of (*a*) with the rms speed of random thermal motion of these ions. (*c*) Compare the distance traveled by Mg^{2+} ions in one second due to the electric field with the diameter of a solvent molecule.

(*a*) Equation (16.62) and the preceding table of u^∞ values give

$$v = uE = (55 \times 10^{-5} \ cm^2 \ V^{-1} \ s^{-1})(10 \ V/cm) = 0.0055 \ cm/s$$

(*b*) The average translational kinetic energy of random thermal motion of the Mg^{2+} ions is $\frac{3}{2}kT = \frac{1}{2}m\langle v^2 \rangle$, so the rms speed of random thermal motion is $v_{rms} = (3RT/M)^{1/2}$ and

$$v_{rms} = [3(8.3 \ J/mol\text{-}K)(298 \ K)/(0.024 \ kg/mol)]^{1/2} = 560 \ m/s = 56000 \ cm/s$$

The speed of migration toward the electrode is far, far smaller than the average speed of random motion.

(c) With a drift speed of 0.0055 cm/s, the electric field produces a displacement of 0.0055 cm in one second. The diameter of a water molecule is listed in (16.26) as 3.2 Å. The one-second displacement is 1.7×10^5 times the solvent diameter.

Exercise

Consider a 0.100 M NaCl(aq) solution at 25°C and 1 atm undergoing electrolysis with an electric-field strength of 15 V/cm. (a) Find the drift speed of the Cl^- ions. (b) How many current-carrying Cl^- ions cross a 1.00-cm²-area plane parallel to the electrodes in 1.00 s? (*Answers:* (a) 0.010 cm/s; (b) 6.0×10^{17}.)

Ionic mobilities at infinite dilution can be estimated theoretically as follows. At extremely high dilution, interionic forces are negligible, so the only electric force an ion experiences is due to the applied electric field E. From (14.3), the electric force on an ion with charge $z_B e$ has the magnitude $|z_B|eE$. This force is opposed by the frictional force $f v_B^\infty$, where f is the friction coefficient [Eq. (16.20)]. When the terminal speed has been reached, the electric and frictional forces balance: $|z_B|eE = f v_B^\infty$, and the terminal speed is $v_B^\infty = |z_B|eE/f$. The infinite-dilution mobility $u_B^\infty = v_B^\infty/E$ is then

$$u_B^\infty = |z_B|e/f \tag{16.69}$$

A rough estimate of the friction coefficient f can be obtained by assuming that the solvated ions are spherical and that Stokes' law (16.21) applies to their motion through the solvent. (Because the ions are solvated, they are substantially larger than the solvent molecules.) Stokes' law gives $f = 6\pi\eta r_B$, and

$$u_B^\infty \approx \frac{|z_B|e}{6\pi\eta r_B} \tag{16.70}$$

Equation (16.70) attributes the differences in infinite-dilution mobilities of ions entirely to the differences in their charges and radii. Of course, this equation can't be used for H_3O^+ or OH^-.

The smaller values of u^∞ for cations than for anions (H_3O^+ and OH^- excepted) indicate that cations are more hydrated than anions. The smaller size of cations produces a more intense electric field surrounding them, and they therefore hold on to more H_2O molecules than anions. The average number of water molecules that move with an ion in solution is called the *hydration number* n_h of the ion. Some values of n_h estimated using electric mobilities and other methods are [J. O'M. Bockris and P. P. S. Saluja, *J. Phys. Chem.*, **76**, 2140 (1972); ibid., **77**, 1598 (1973); ibid., **79**, 1230 (1975); R. W. Impey et al., *J. Phys. Chem.*, **87**, 5071 (1983)]:

Ion	Li^+	Na^+	K^+	Mg^{2+}	F^-	Cl^-	Br^-	I^-
n_h	$4\frac{1}{2}$	4	3	12	4	2	1	1

The methods used to find n_h involve assumptions of uncertain accuracy, so these values are approximate. The hydration number n_h should be distinguished from the (average) coordination number of an ion in solution. The *coordination number* is the average number of water molecules that are nearest neighbors of the ion (whether or not they move with the ion) and may be estimated from x-ray diffraction data of the solution. Some values are 6 for each of Na^+, K^+, Cl^-, and Mg^{2+}.

Aside from the approximation of using Stokes' law, it is difficult to use (16.70) to predict u values because the radius r_B of the solvated ion is not accurately known. What is often done is to use (16.70) to calculate r_B from u_B^∞.

EXAMPLE 16.4 Ionic radii in solution

Estimate the radii of $Li^+(aq)$ and $Na^+(aq)$, given that the viscosity of water at 25°C is 0.89 cP.

Equation (16.70), the u^∞ value of $Li^+(aq)$ in the table earlier in this section, and the relation $1\ P = 0.1\ N\ s\ m^{-2}$ [Eq. (16.14)] give

$$r(Li^+) \approx \frac{1(1.6 \times 10^{-19}\ C)}{6\pi(0.89 \times 10^{-3}\ N\ s\ m^{-2})[40 \times 10^{-5}(10^{-2}\ m)^2\ V^{-1}\ s^{-1}]}$$

$$\approx 2.4 \times 10^{-10}\ m = 2.4\ Å$$

Na^+ and Li^+ have the same charges, and (16.70) gives the radii as inversely proportional to the mobilities. Therefore, $r(Na^+) \approx (40/52)(2.4\ Å) = 1.8\ Å$. The larger size of $Li^+(aq)$ (despite the smaller atomic number of Li) is due to the larger n_h value of Li.

Exercise

In CH_3OH at 25°C and 1 atm, $u^\infty(Li^+) = 4.13 \times 10^{-4}\ cm^2/V\text{-}s$, $u^\infty(Na^+) = 4.69 \times 10^{-4}\ cm^2/V\text{-}s$, and $\eta = 0.55$ cP. Estimate the radii of Li^+ and Na^+ ions in methanol and compare with the values in water. (*Answers:* 3.7 Å and 3.3 Å.)

Stokes' law can be used to estimate how long it takes an ion to reach its terminal speed after the electric field is applied. From (16.70), the terminal speed equals $|z|eE/6\pi\eta r$. The force due to the electric field is $|z|eE$. If we neglect the frictional resistance, Newton's second law $F = ma = m\ dv/dt$ gives $|z|eE \approx m\ dv/dt$, which integrates to $v \approx |z|eEt/m$. Setting v equal to the terminal speed, we get $|z|eEt/m \approx |z|eE/6\pi\eta r$. The time needed to reach the terminal speed is then $t \approx m/6\pi\eta r$. For $m = 10^{-22}$ g, $\eta = 10^{-2}$ g s^{-1} cm^{-1}, and $r = 10^{-8}$ cm, we get $t \approx 10^{-13}$ s. Since we neglected F_{fr}, the actual time required is somewhat longer.

Electrophoresis

The migration of charged polymeric molecules (*polyelectrolytes*) and charged colloidal particles in an electric field is called **electrophoresis.** Electrophoresis can separate different proteins and different nucleic acids and is commonly done with a polymer gel (Sec. 13.6) as the medium. Electrophoresis "is the most important physical technique available" in biochemistry and molecular biology (K. E. van Holde et al., *Principles of Physical Biochemistry,* Prentice-Hall, 1998, sec. 5.3).

When electrophoresis is done in a free solvent, heating of the solvent by the electric current will produce convectional flow, which destroys the desired separation. Use of a gel eliminates the undesirable effects of convection. One commonly used gel is an agarose gel, which contains an aqueous medium dispersed in the pores of a three-dimensional network formed by a polysaccharide obtained from agar.

In a DNA (deoxyribonucleic acid) molecule, each phosphate group that links two deoxyriboses has one acidic hydrogen (hence the A in DNA). Ionization of these hydrogens gives DNA a negative charge in aqueous solution. The R side chains of 3 of the 20 amino acids $NH_2CHRCOOH$ that occur in proteins contain an amine group and the R chains of two contain a COOH group. In a buffered highly basic (high pH) solution of a protein, neutralization of the COOH groups produces COO^- groups that give the protein a negative charge. In buffered low-pH solutions, protonation of the amine groups gives the protein a positive charge. At a certain intermediate pH (the *isoelectric point*), the protein is uncharged.

In gel electrophoresis, upper and lower buffer solutions are connected by a slab of the gel (which contains the buffer in its pores). Each buffer solution contains an electrode. A solution of the macromolecules to be separated is layered into a notch in the upper edge of the gel. The gel edge contains several notches, so several samples can be run simultaneously in parallel lanes.

From Eq. (16.69), the mobility in a free solvent is directly proportional to the charge of the migrating molecule and is inversely proportional to the friction coefficient. The charge on a DNA fragment is proportional to its length, and the friction coefficient of the DNA is also proportional to its length. Therefore, the mobility of a DNA fragment in free solvent is essentially independent of the length of the DNA fragment and electrophoresis in free solvent does not separate DNA fragments of different lengths. In a polymer gel, shorter DNA fragments are able to move faster through the pores than longer DNA fragments, so separation according to size is readily achieved.

The electrophoretic mobility u of a biomolecule depends on its charge, its size and shape, the nature and concentrations of other charged species in the solution, the solvent viscosity, and the nature of the gel (which acts as a molecular sieve). Theoretical prediction of the ratio u/u_0, where u is the mobility in the gel and u_0 is the mobility in the free solvent, is a challenging problem. Various models (none fully successful) for this mobility ratio are discussed in J.-L. Viovy, *Rev. Mod. Phys.,* **72,** 813 (2000).

In DNA fingerprinting used in criminal investigations and genetics studies, a DNA sample is treated with an enzyme that cuts DNA at specific sites, producing fragments whose lengths vary from person to person. The fragments are separated by gel electrophoresis, and the resulting pattern is made visible by (*a*) blotting the separated fragments onto a membrane, (*b*) treating the membrane with a radioactive probe that binds to the fragments, and (*c*) exposing radiation-sensitive photographic film to the membrane.

Separation of DNA molecules with more than 10^5 base pairs by conventional gel electrophoresis fails, because such large molecules migrate at a rate that is essentially independent of size. In *pulsed-field electrophoresis,* the direction of the electric field is periodically reversed for a brief time, and this repeated reversal causes very large DNA fragments to migrate at a rate that depends on size. The theory of pulsed-field electrophoresis is the subject of debate.

In *isoelectric focusing,* one establishes a pH gradient within the gel. Each protein migrates until it reaches its isoelectric point, thereby allowing separation of proteins with different isoelectric points.

In **capillary electrophoresis,** instead of moving through a gel slab, the polyelectrolyte molecules move through narrow (0.01-cm inside diameter) quartz capillaries. The capillaries can be filled with a gel. More commonly, the capillaries are filled with a solution of a polymer (such as polyacrylamide) at high concentration. Interactions between the migrating polyelectrolyte molecules and the polymer molecules lead to separation according to size. Detection of the migrating molecules is by ultraviolet absorption or by fluorescence (Sec. 21.11). The high resistance of the medium in a capillary reduces the magnitude of the current that flows and reduces the heating that occurs. This allows a higher voltage to be applied than when using a gel slab, thereby speeding up the separation. Capillary electrophoresis is particularly suitable for automated procedures. The ABI Prism 3700 Automated DNA Analyzer used in sequencing the human genome is a capillary electrophoresis machine.

Transport Numbers

The **transport number** (or **transference number**) t_B of ion B in an electrolyte solution is defined as the fraction of the current that it carries:

$$t_B \equiv j_B/j \qquad\qquad (16.71)^*$$

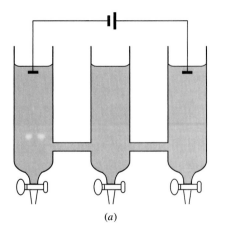

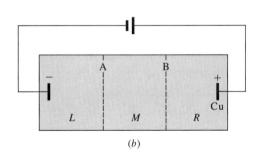

Figure 16.25

(*a*) Hittorf apparatus for
measuring transport numbers of
ions in solution. (*b*) Electrolysis of
$Cu(NO_3)_2(aq)$ with a Cu anode
and an inert cathode.

where j_B is the current density of ion B and j is the total current density. Using (16.61)
and (16.49) for j_B and j, we have $t_B = j_B/j = |z_B|Fv_Bc_B/\kappa E$. Since $v_B/E = u_B$ [Eq.
(16.62)], we get

$$t_B = |z_B|Fc_Bu_B/\kappa \qquad (16.72)$$

The transport number of an ion can be calculated from its mobility and κ. The sum of
the transport numbers of all the ionic species in solution must be 1.

For a solution containing only two kinds of ions, Eqs. (16.71) and (16.60) give t_+
$= j_+/j = j_+/(j_+ + j_-) = z_+v_+c_+/(z_+v_+c_+ + |z_-|v_-c_-)$. The use of the electroneutral-
ity condition $z_+c_+ = |z_-|c_-$ gives $t_+ = v_+/(v_+ + v_-)$. The use of $v_+ = u_+E$ and $v_- =$
u_-E [Eq. (16.62)] then gives $t_+ = u_+/(u_+ + u_-)$. Thus

$$t_+ = \frac{j_+}{j_+ + j_-} = \frac{v_+}{v_+ + v_-} = \frac{u_+}{u_+ + u_-}, \qquad t_- = \frac{j_-}{j_+ + j_-} = \text{etc.} \quad (16.73)$$

Transport numbers can be measured by the *Hittorf method* (Fig. 16.25*a*). After
electrolysis has proceeded for a while, one drains the solutions in each of the com-
partments and analyzes them. The results allow t_+ and t_- to be found.

Figure 16.25 shows what happens in the electrolysis of $Cu(NO_3)_2$ with a Cu anode
and an inert cathode. Let a total charge Q flow during the experiment. Then Q/F moles
of electrons flow. The anode reaction is $Cu \rightarrow Cu^{2+}(aq) + 2e^-$, so $Q/2F$ moles of
Cu^{2+} enter the right compartment R from the anode. The total number of moles of
charge on the ions that pass plane B during the experiment is Q/F. The Cu^{2+} ions carry
a fraction t_+ of the current, and the charge on the Cu^{2+} ions moving from R to M during the experiment is t_+Q. Therefore $t_+Q/2F$ moles of Cu^{2+} pass out of R into M during the experiment. The net change in the number of moles of Cu^{2+} in compartment
R is $Q/2F - t_+Q/2F$:

$$\Delta n_R(Cu^{2+}) = (1 - t_+)Q/2F = t_-Q/2F \qquad (16.74)$$

Since NO_3^- carries a fraction t_- of the current, the magnitude of the charge on the
nitrate ions moving from M to R during the experiment is t_-Q, and t_-Q/F moles of
NO_3^- move into R:

$$\Delta n_R(NO_3^-) = t_-Q/F \qquad (16.75)$$

Equations (16.74) and (16.75) are consistent with the requirement that R remain electrically neutral.

The charge Q is measured with a coulometer, and chemical analysis gives the
Δn's. Therefore t_+ and t_- can be found from (16.74).

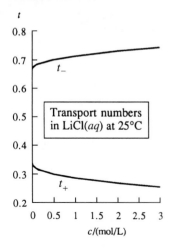

Figure 16.26

Cation and anion transport numbers versus concentration for LiCl(aq) at 25°C and 1 atm.

Since the mobilities u_+ and u_- depend on concentration and do not necessarily change at the same rate as c changes, the transport numbers t_+ and t_- depend on concentration. Transport numbers in LiCl(aq) at 25°C and 1 atm are plotted versus c in Fig. 16.26.

Observed t^∞ values lie between 0.3 and 0.7 for most ions. H_3O^+ and OH^- have unusually high t^∞ values in aqueous solution, because of their high mobilities. Some values for aqueous solutions at 25°C and 1 atm are $t^\infty(H^+) = 0.82$ and $t^\infty(Cl^-) = 0.18$ for HCl; $t^\infty(K^+) = 0.49$ and $t^\infty(Cl^-) = 0.51$ for KCl; $t^\infty(Ca^{2+}) = 0.44$ and $t^\infty(Cl^-) = 0.56$ for $CaCl_2$.

In the preceding discussion, we considered only the ions Cu^{2+} and NO_3^-. However, a $Cu(NO_3)_2$ solution has a significant concentration of $Cu(NO_3)^+$ ion pairs, and these carry part of the current. When a compartment is chemically analyzed for Cu^{2+}, it is the total amount of Cu present in solution that is determined, and individual amounts present as Cu^{2+} and $Cu(NO_3)^+$ are not found. Therefore the values t_+ and t_- obtained in the Hittorf method are not, strictly speaking, the transport numbers of the actual ions. Instead they are what are called *ion-constituent transport numbers*. The ion-constituent Cu(II) exists in a $Cu(NO_3)_2$ solution as Cu^{2+} ions and as $Cu(NO_3)^+$ ions. Similarly, in the moving-boundary method, one obtains mobilities and transport numbers of ion-constituents. At infinite dilution, there is no ion pairing, so t^∞ and u^∞ values do apply to the ions Cu^{2+} and NO_3^-. See M. Spiro, *J. Chem. Educ.*, **33**, 464 (1956); M. Spiro in *Rossiter, Hamilton and Baetzold*, vol. II, chap. 8.

Molar Conductivities of Ions

The molar conductivity of an electrolyte in solution is $\Lambda_m \equiv \kappa/c$ [Eq. (16.58)]. By analogy, we define the **molar conductivity** $\lambda_{m,B}$ of ion B as

$$\lambda_{m,B} \equiv \kappa_B/c_B \qquad (16.76)*$$

where κ_B is the contribution of ion B to the solution's conductivity and c_B is its molar concentration. Note that c_B is the actual concentration of ion B in the solution, whereas c is the electrolyte's stoichiometric concentration. Equation (16.63) gives

$$\kappa_B = |z_B|Fu_B c_B \qquad (16.77)$$

$$\lambda_{m,B} = |z_B|Fu_B \qquad (16.78)*$$

since $\lambda_{m,B} = \kappa_B/c_B$ [Eq. (16.76)]. The molar conductivity of an ion can therefore be found from its mobility. (The *equivalent conductivity* of ion B is $\lambda_{eq,B} \equiv \lambda_{m,B}/|z_B| = Fu_B$.)

Substitution of $\Lambda_m = \kappa/c$ and (16.76) in $\kappa = \Sigma_B \kappa_B$ gives

$$\Lambda_m = \frac{1}{c} \sum_B c_B \lambda_{m,B} \qquad (16.79)$$

which relates Λ_m of the electrolyte to the λ_m's of the ions. For a strong electrolyte $M_{\nu_+} X_{\nu_-}$ that is completely dissociated, (16.79) becomes

$$\Lambda_m = c^{-1}(c_+\lambda_{m,+} + c_-\lambda_{m,-}) = c^{-1}(\nu_+c\lambda_{m,+} + \nu_-c\lambda_{m,-}) \qquad (16.80)$$

$$\Lambda_m = \nu_+\lambda_{m,+} + \nu_-\lambda_{m,-} \quad \text{strong electrolyte, no ion pairs} \qquad (16.81)$$

For example, $\Lambda_m(MgCl_2) = \lambda_m(Mg^{2+}) + 2\lambda_m(Cl^-)$, provided there are no ion pairs. For a weak acid HX whose degree of dissociation is α, Eq. (16.79) gives

$$\Lambda_m = c^{-1}(c_+\lambda_{m,+} + c_-\lambda_{m,-}) = c^{-1}(\alpha c\lambda_{m,+} + \alpha c\lambda_{m,-}) \qquad (16.82)$$

$$\Lambda_m = \alpha(\lambda_{m,+} + \lambda_{m,-}) \quad \text{for 1:1 weak electrolyte} \qquad (16.83)$$

For a weak acid in water, $\alpha^\infty \neq 1$ (Sec. 11.3); therefore, $\Lambda_m^\infty \neq \lambda_{m,+}^\infty + \lambda_{m,-}^\infty$ for the weak acid HX in water.

Since the mobility $u(Cl^-)$ in an NaCl solution differs slightly from $u(Cl^-)$ in a KCl solution at nonzero concentrations, $\lambda_m(Cl^-)$ in NaCl and KCl solutions differ. However, in the limit of infinite dilution, interionic forces go to zero and the ions move independently. Therefore $\lambda_m^\infty(Cl^-)$ is the same for all chloride salts. Figure 16.27 plots $\lambda_m(Cl^-)$ versus $c^{1/2}$ for NaCl(aq) and KCl(aq) at 25°C and 1 atm.

Some λ_m^∞ values in water at 25°C and 1 atm are (M. Spiro in *Rossiter, Hamilton, and Baetzold,* vol. II, p. 784):

Cation	H_3O^+	NH_4^+	K^+	Na^+	Ag^+	Ca^{2+}	Mg^{2+}
$\lambda_m^\infty/(\Omega^{-1}\,cm^2\,mol^{-1})$	350.0	73.5	73.5	50.1	62.1	118.0	106.1

Anion	OH^-	Br^-	Cl^-	NO_3^-	CH_3COO^-	SO_4^{2-}
$\lambda_m^\infty/(\Omega^{-1}\,cm^2\,mol^{-1})$	199.2	78.1	76.3	71.4	40.8	159.6

The $|z_B|$ factor in (16.78) tends to make λ_m for +2 and −2 ions larger than for +1 and −1 ions.

From tabulated λ_m^∞ values, we can calculate Λ_m^∞ for a strong electrolyte as [Eq. (16.81)]:

$$\Lambda_m^\infty = \nu_+\lambda_{m,+}^\infty + \nu_-\lambda_{m,-}^\infty \qquad \text{strong electrolyte} \qquad \textbf{(16.84)}^*$$

since there is no ion pairing at infinite dilution. Mobilities u^∞ can be calculated from λ_m^∞ values using $\lambda_{m,B}^\infty = |z_B|Fu_B^\infty$ [Eq. (16.78)]. From the u^∞ values, t^∞ can be found using $t_+ = u_+/(u_+ + u_-)$ [Eq. (16.73)].

Infinite-dilution transport numbers and molar conductivities are related. For the strong electrolyte $M_{\nu_+}X_{\nu_-}$, we have $t_+ = j_+/j = \kappa_+E/\kappa E = \kappa_+/\kappa = \lambda_{m,+}c_+/\Lambda_m c$. In the infinite-dilution limit, there are no ion pairs and $c_+ = \nu_+c$. Therefore

$$t_+^\infty = \frac{\nu_+\lambda_{m,+}^\infty}{\Lambda_m^\infty} = \frac{\nu_+\lambda_{m,+}^\infty}{\nu_+\lambda_{m,+}^\infty + \nu_-\lambda_{m,-}^\infty} \qquad \text{strong electrolyte} \qquad (16.85)$$

with a similar equation for t_-^∞.

Since the solvent's viscosity decreases as the temperature increases, the ionic mobility $u_B^\infty \approx |z_B|e/6\pi\eta r_B$ [Eq. (16.70)] increases as T increases. Therefore $\lambda_{m,B}^\infty = |z_B|Fu_B^\infty \approx z_B^2 Fe/6\pi\eta r_B$ increases as T increases (Fig. 16.28).

The fundamental molecular quantity that governs an ion's motion in an applied electric field is its mobility $u_B \equiv v_B/E$. The molar conductivity of an ion is the product of its mobility and the magnitude of its molar charge: $\lambda_{m,B} = |z_B|Fu_B$. The molar conductivity Λ_m of the electrolyte $M_{\nu_+}X_{\nu_-}$ is the sum of contributions from the cation and anion molar conductivities: $\Lambda_m = \nu_+\lambda_{m,+} + \nu_-\lambda_{m,-}$ for a strong electrolyte with no ion pairing. The solution's conductivity κ is related to Λ_m by $\Lambda_m \equiv \kappa/c$. The fraction of the current carried by the cations is the cation transport number $t_+ = u_+/(u_+ + u_-)$.

Concentration Dependence of Molar Conductivities

Some Λ_m data for NaCl and $HC_2H_3O_2$ in water at 25°C and 1 atm are:

$c/(mol\,dm^{-3})$	0	10^{-4}	10^{-3}	10^{-2}	10^{-1}
$\Lambda_m(NaCl)/(\Omega^{-1}\,cm^2\,mol^{-1})$	(126.4)	125.5	123.7	118.4	106.7
$\Lambda_m(CH_3COOH)/(\Omega^{-1}\,cm^2\,mol^{-1})$		134.6	49.2	16.2	5.2

The relation $\Lambda_m = \Sigma_B(c_B/c)\lambda_{m,B}$ [Eq. (16.79)] shows that Λ_m of an electrolyte changes with electrolyte concentration for two reasons: (*a*) The ionic concentrations c_B may not be proportional to the electrolyte stoichiometric concentration c, and (*b*) the ionic molar conductivities $\lambda_{m,B}$ change with concentration.

The sharp increase in Λ_m of a weak acid like acetic acid as c goes to zero (Fig. 16.21*b*) is due mainly to the rapid increase in the degree of dissociation as c goes

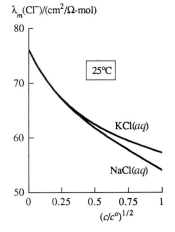

$\lambda_m(Cl^-)/(cm^2/\Omega\text{-mol})$

Figure 16.27

λ_m of Cl^- versus $c^{1/2}$ for KCl(aq) and NaCl(aq) at 25°C.

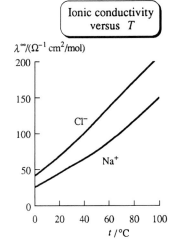

$\lambda^\infty/(\Omega^{-1}\,cm^2/mol)$

Figure 16.28

λ_m^∞ versus temperature for $Na^+(aq)$ and $Cl^-(aq)$.

to zero; see Eq. (16.83). This rapid increase in Λ_m makes extrapolation to $c = 0$ very difficult for weak electrolytes. For strong electrolytes other than 1:1 electrolytes, part of the decrease in Λ_m with increasing c is due to formation of ion pairs, which reduces the ionic concentrations. However, even for 1:1 electrolytes, which do not show significant ion pairing in water, Λ_m decreases as c increases. This decrease arises from interionic forces. One finds that for a strong electrolyte, a plot of Λ_m versus $c^{1/2}$ is linear at very high dilutions, and this allows reliable extrapolation to $c = 0$.

From (16.78), the ionic molar conductivity $\lambda_{m,B}$ equals $|z_B|Fu_B$. If the mobility $u_B = v_B/E$ were independent of concentration, λ_m would be independent of c. However, the ion drift speed v_B depends on c because of interionic interactions.

Debye and Hückel applied their theory of ionic interactions to calculate the electric mobility of ions in very dilute solution. Their treatment was improved by Onsager in 1927 to give the (*Debye–Hückel–*) *Onsager limiting law*. For the special case of an electrolyte yielding two kinds of ions with $z_+ = |z_-|$, the Onsager equation for Λ_m in water at 25°C and 1 atm is

$$\Lambda_m = (c_+/c)\{\lambda_{m,+}^\infty + \lambda_{m,-}^\infty - [az_+^3 + bz_+^3(\lambda_{m,+}^\infty + \lambda_{m,-}^\infty)](c_+/c^\circ)^{1/2}\} \quad (16.86)$$

$$a \equiv 60.6 \; \Omega^{-1} \, cm^2 \, mol^{-1}, \quad b \equiv 0.230, \quad z_+ = |z_-|, \quad \text{in } H_2O \text{ at } 25°C$$

where c_+ is the actual concentration of the cation, c is the stoichiometric concentration of the electrolyte, and $c^\circ \equiv 1$ mol/dm^3. (For the derivation and the formula when $z_+ \neq |z_-|$, see *Eyring, Henderson, and Jost,* vol. IXA, chap. 1.)

For a strong electrolyte with no ion pairing and with $z_+ = |z_-|$, we have $c_+ = c$, and (16.86) becomes

$$\Lambda_m = \Lambda_m^\infty - (az_+^3 + bz_+^3\Lambda_m^\infty)(c/c^\circ)^{1/2} \quad \text{for } z_+ = |z_-|, \text{ strong electrolyte} \quad (16.87)$$

where (16.84) was used. Note the $c^{1/2}$ dependence, in agreement with experimental data for strong electrolytes. The a and b terms correct for interionic interactions. With these terms omitted, (16.87) gives the no-interaction result $\Lambda_m = \Lambda_m^\infty$.

The Onsager equation is well obeyed by solutions of 1:1 electrolytes with c_+ less than 0.002 mol/dm^3 and by very dilute solutions of higher-valency electrolytes if ion pairing is taken into account in calculating c_+ and c_-. (For conductivity equations applicable at higher concentrations than the Onsager equation, see M. Spiro in *Rossiter, Hamilton, and Baetzold,* vol. II, pp. 673–679.)

Applications of Conductivity

Substitution of $\lambda_{m,B} \equiv \kappa_B/c_B$ into $\kappa = \Sigma_B \, \kappa_B$ gives

$$\kappa = \sum_B \lambda_{m,B} c_B$$

Measurement of κ enables the endpoint of a titration to be found, since the plot of κ versus volume of added reagent changes slope at the endpoint. For example, if an aqueous HCl solution is titrated with NaOH, κ decreases before the endpoint because H_3O^+ ions are being replaced by Na^+ ions, and κ increases after the endpoint as a result of the increase in Na^+ and OH^- concentrations.

Conductivity measurements can give the concentration changes during a chemical reaction between ions in solution, enabling the reaction rate to be followed.

Conductivity measurements can be used to determine ionic equilibrium constants such as dissociation constants of weak acids, solubility-product constants, the ionization constant of water, and association constants for ion-pair formation. Consider a very dilute solution of the electrolyte MX in which there is an ionic equilibrium. Using the measured value of $\Lambda_m \equiv \kappa/c$, we can solve the Onsager equation (16.86) for the ionic concentration c_+ (see the next paragraph). From c_+, the electrolyte stoichiometric concentration c, and activity coefficients calculated using the Debye–Hückel equation (10.70), we can find the ionic equilibrium constant K_c (see Chapter 11).

It is convenient to use $\Lambda_m = \kappa/c$ [Eq. (16.58)] to rewrite (16.86) as

$$\kappa = c_+[\lambda_{m,+}^\infty + \lambda_{m,-}^\infty - S(c_+/c^\circ)^{1/2}] \qquad \text{for } z_+ = |z_-| \qquad (16.88)$$

$$S \equiv az_+^3 + bz_+^3(\lambda_{m,+}^\infty + \lambda_{m,-}^\infty), \qquad c^\circ \equiv 1 \text{ mol/dm}^3 \qquad (16.89)$$

where S incorporates the Onsager corrections to the conductivity. Equation (16.88) is a cubic equation in $c_+^{1/2}$. For hand calculations, it is fastest to solve it by successive approximations. We rewrite the equation as

$$c_+ = \frac{\kappa}{\lambda_{m,+}^\infty + \lambda_{m,-}^\infty - S(c_+/c^\circ)^{1/2}} \qquad \text{for } z_+ = |z_-| \qquad (16.90)$$

At the high dilutions to which the Onsager equation applies, the interionic-forces correction term $S(c_+/c^\circ)^{1/2}$ is much less than $\lambda_{m,+}^\infty + \lambda_{m,-}^\infty$, so as an initial approximation we can set $c_+ = 0$ in the denominator on the right of (16.90). Equation (16.90) is then used to calculate an improved value of the cation concentration c_+, which is then substituted in the right side of (16.90) to find a further improved c_+ value. The calculation is repeated until the answer converges.

Problems 16.58 to 16.61 outline applications of (16.90) to ionic equilibria.

To use (16.90), $\lambda_{m,+}^\infty$ and $\lambda_{m,-}^\infty$ must be known. They are found by extrapolation of mobility or transport-number measurements on strong electrolytes, as discussed earlier. See also Probs. 16.46 and 16.47.

16.7   SUMMARY

The rate of flow per unit cross-sectional area (the flux) of heat, momentum, matter, and charge in thermal conduction, viscous flow, diffusion, and electrical conduction is proportional to the gradient of temperature, speed, concentration, and electric potential, respectively, in the flow direction. The proportionality constants are the thermal conductivity k, the viscosity η, the diffusion coefficient D, and the electrical conductivity κ.

The kinetic theory together with a hard-sphere model for intermolecular interactions gives expressions for k, η, and D in gases at pressures neither very high nor very low. These expressions work pretty well, deviations from experiment being due mainly to the inadequacy of the hard-sphere model in representing intermolecular forces.

Integration of Newton's viscosity law yields expressions for the flow rates of liquids and gases under pressure gradients. Measurement of such flow rates gives η.

Polymer molecular weights can be determined from measured polymer-solution viscosities and sedimentation rates.

The rms displacement in a given direction for a diffusing molecule is given by the Einstein–Smoluchowski equation as $(\Delta x)_{\text{rms}} = (2Dt)^{1/2}$, where D and t are the diffusion coefficient and the time. Diffusion coefficients in liquids can be estimated by the Stokes–Einstein equation $D_{iB}^\infty \approx kT/6\pi\eta_B r_i$ [Eq. (16.37)].

The electric current I is defined as the rate of charge flow dQ/dt. The electric current density is $j \equiv I/\mathcal{A}$, where $\mathcal{A}$ is the conductor's cross-sectional area. The conductivity κ of a substance is an intensive property defined by $\kappa \equiv j/E$, where E is the magnitude of the electric field producing the current flow.

The molar conductivity of an electrolyte solution with stoichiometric concentration c is $\Lambda_m \equiv \kappa/c$. Ions move through a current-carrying electrolyte solution with a drift speed v_B that is proportional to the electric field strength: $v_B = u_B E$, where u_B is the electric mobility of ion B. The conductivity of an electrolyte solution is given by (16.63) as $\kappa = \Sigma_B \kappa_B = \Sigma_B |z_B| F u_B c_B$. The contribution of ion B to κ is proportional to its molar charge $|z_B| F$, its mobility u_B, and its concentration c_B. The transport number of an ion is the fraction of current that it carries: $t_B = j_B/j = \kappa_B/\kappa$. The molar conductivity $\lambda_{m,B}$ of ion B is $\lambda_{m,B} \equiv \kappa_B/c_B = |z_B| F u_B$. The electrolyte's molar conductivity Λ_m is related to the molar conductivities and concentrations of its ions; see Eqs. (16.79) to (16.84).

Molar conductivities decrease with increasing electrolyte concentration as a result of interionic forces. In dilute solutions, this decrease can be calculated from the Onsager equation (16.86). The Onsager equation allows ionic concentrations to be found from measured conductivities and therefore yields ionic equilibrium constants.

Important kinds of calculations discussed in this chapter include:

- Calculation of the thermal conductivity k, the viscosity η, and the self-diffusion coefficient D_{jj} of a gas from the hard-sphere kinetic-theory equations (16.12), (16.25), and (16.42).
- Calculation of the hard-sphere diameter of a gas from its viscosity using (16.25).
- Use of Poiseuille's law (16.17) or (16.18) to calculate the flow rate of a liquid or gas in a pipe from the viscosity, or to calculate the viscosity from the flow rate.
- Calculation of viscosity-average molecular weights of polymers from viscosity data using (16.28).
- Calculation of $(\Delta x)_{rms}$ in diffusion using the Einstein–Smoluchowski equation $(\Delta x)_{rms} = (2Dt)^{1/2}$.
- Estimation of diffusion coefficients in liquids using the Stokes–Einstein equations (16.37) and (16.38).
- Calculation of polymer molecular weights from sedimentation coefficients using (16.45).
- Calculation of the conductivity κ of an electrolyte solution from the resistance of the solution and the resistance in the same cell of a KCl solution of known κ.
- Calculation of the molar conductivity $\Lambda_m \equiv \kappa/c$.
- Estimation of ionic radii in solution from mobilities using (16.70).
- Calculation of ionic λ_m values from mobilities using $\lambda_{m,B} = |z_B| F u_B$.
- Calculation of Λ_m^∞ from ionic λ_m^∞ values using (16.84).
- Calculation of ionic concentrations and equilibrium constants from conductivities in very dilute solutions using the Onsager equation.

FURTHER READING AND DATA SOURCES

Present, chaps. 3 and 4; *Kauzmann,* chap. 5; *Kennard,* chap. 4; *Reid, Prausnitz, and Poling,* chaps. 9–11; *Bird, Stewart, and Lightfoot; Robinson and Stokes,* chaps. 5, 6, 7, 10; *Bockris and Reddy,* chap. 4.

Thermal conductivity: *Landolt-Börnstein,* 6th ed., vol. II, pt. 5*b*, pp. 39–203; Y. S. Touloukian and C. Y. Ho (eds.), *Thermophysical Properties of Matter,* vols. 1–3, Plenum, 1970–1976.

Viscosity: *Landolt-Börnstein,* 6th ed., vol. II, pt. 5*a*, pp. 1–512; J. Timmermans, *Physico-Chemical Constants of Pure Organic Compounds,* vols. 1 and 2, Elsevier, 1950, 1965.

Diffusion coefficients: *Landolt-Börnstein,* 6th ed., vol. II, pt. 5*a*, pp. 513–725 and pt. 5*b*, pp. 1–39; T. R. Marrero and E. A. Mason, *J. Phys. Chem. Ref. Data,* **1,** 3 (1972).

Molar conductivities and transport numbers: *Landolt-Börnstein,* 6th ed., vol. II, pt. 7, pp. 27–726; M. Spiro in *Rossiter, Hamilton, and Baetzold,* vol. II, pp. 782–785.

PROBLEMS

Section 16.2

16.1 True or false? (a) The thermal conductivity k of a phase is an intensive property that depends on T, P, and the composition of the phase. (b) The heat flow rate dq/dt across the yz plane is proportional to the temperature gradient dT/dx at that plane. (c) Fourier's law of heat conduction holds at very low gas pressures.

16.2 If the distance between the reservoirs in Fig. 16.1 is 200 cm, the reservoir temperatures are 325 and 275 K, the substance is an iron rod with cross-sectional area 24 cm^2, $k = 0.80$ J K^{-1} cm^{-1} s^{-1}, and a steady state is present, calculate (a) the heat that flows in 60 s; (b) ΔS_{univ} in 60 s.

16.3 Use the d value in (16.26) to calculate the thermal conductivity of He at 1 atm and 0°C and at 10 atm and 100°C. The experimental value at 0°C and 1 atm is 1.4×10^{-3} J cm^{-1} K^{-1} s^{-1}.

16.4 Use data in (16.26) and the Appendix to calculate the thermal conductivity of CH$_4$ at 25°C and 1 atm. The experimental value is 34×10^{-5} J cm^{-1} K^{-1} s^{-1}.

16.5 Bridgman derived the following kinetic-theory equation for the thermal conductivity of a liquid (see *Bird, Stewart, and Lightfoot,* p. 260, for the derivation):

$$k = \frac{3R}{N_A^{1/3} V_m^{2/3}} \left(\frac{C_{P,m}}{C_{V,m} \rho \kappa} \right)^{1/2}$$

where R is the gas constant and ρ, κ, and V_m are the density, isothermal compressibility, and molar volume of the liquid. This equation works surprisingly well, especially if the factor 3 is replaced by 2.8. This is the well-known scientific principle of the fudge factor. Use this equation with the 3 changed to 2.8 to estimate the thermal conductivity of water at 30°C and 1 atm; use data in and preceding Eq. (4.54). The experimental value is 6.13 mJ cm^{-1} K^{-1} s^{-1}.

Section 16.3

16.6 True or false? (a) For laminar fluid flow in a cylindrical pipe, the flow speed is the same at all points in a plane perpendicular to the axis of the pipe. (b) For laminar fluid flow in a cylindrical pipe, the maximum flow speed is at the center of the pipe. (c) As temperature increases, the viscosity of liquids usually decreases and the viscosity of gases usually increases. (d) Newton's viscosity law fails at extremely high flow rates.

16.7 The *Reynolds number Re* is defined by $Re \equiv \rho \langle v_y \rangle d / \eta$, where ρ and η are the density and viscosity of a fluid flowing with average speed $\langle v_y \rangle$ in a tube of diameter d. (The two-letter symbol Re stands for a single physical quantity.) Experience indicates that, when $Re < 2000$, the flow is laminar. For water flowing in a pipe of diameter 1.00 cm, calculate the maximum value of $\langle v_y \rangle$ for laminar flow at 25°C ($\eta = 0.89$ cP).

16.8 (a) For a certain liquid flowing through a cylindrical pipe of inside diameter 0.200 cm and length 24.0 cm, a volume of 148 cm^3 is discharged in 120 s when the pressure drop between the pipe ends is 32.0 torr. The liquid's density is 1.35

g/cm^3. Find the liquid's viscosity. (b) Calculate the Reynolds number (Prob. 16.7) and check that the flow is laminar. (*Hint:* Show that $\langle v_y \rangle = V/\mathcal{A}t$, where V/t is the flow rate and $\mathcal{A}$ is the cross-sectional area.)

16.9 The body-temperature viscosity and density of human blood are 4 cP and 1.0 g/cm^3. The flow rate of blood from the heart through the aorta is 5 L/min in a resting human. The aorta's diameter is typically 2.5 cm. For this flow rate, (a) find the pressure gradient along the aorta; (b) find the average speed of flow (see Prob. 16.8b); (c) find the Reynolds number (Prob. 16.7) and decide if the flow is laminar or turbulent. Repeat (c) for a flow rate of 30 L/min, the maximum flow rate during physical activity.

16.10 The viscosity of O$_2$ at 0°C and pressures within an order of magnitude of 1 atm is 1.92×10^{-4} P. Calculate the flow rate (in g/s) of O$_2$ at 0°C through a tube of inside diameter 0.420 mm and length 220 cm when the inlet and outlet pressures are 1.20 and 1.00 atm.

16.11 When 10.0 mL of water at 20°C is placed in an Ostwald viscometer, it takes 136.5 s for the liquid level to drop from the first mark to the second. For 10.0 mL of hexane at 20°C in the same viscometer, the corresponding time is 67.3 s. Find the viscosity of hexane at 20°C and 1 atm. Data at 20°C and 1 atm: $\eta_{H_2O} = 1.002$ cP, $\rho_{H_2O} = 0.998$ g/cm^3, $\rho_{C_6H_{14}} = 0.659$ g/cm^3.

16.12 Derive Poiseuille's law as follows. (a) Consider a solid-cylindrical portion C of fluid of length dy, radius s, and axis coinciding with the pipe's axis (Fig. 16.9). Let P and $P + dP$ be the pressures at the left and right ends of C, respectively. (dP is negative.) As noted in Sec. 16.3, each infinitesimally thin cylindrical layer of fluid within C flows at constant speed and so is not being accelerated. Therefore the total force on C is zero. The force on C is the sum of the fluid-pressure forces at each end of C and the viscous force on the outer curved surface of C, due to slower-moving fluid just outside C. The area of the curved surface of C equals its circumference $2\pi s$ times its length dy. Equate the total force on C to zero to show that

$$dv_y/ds = (s/2\eta)(dP/dy)$$

Integrate this equation and use the no-slip condition $v_y = 0$ at $s = r$ to show that

$$v_y = (1/4\eta)(r^2 - s^2)(-dP/dy)$$

(b) Consider a thin shell of fluid between cylinders of radii s and $s + ds$. All fluid in this shell moves at speed v_y. Show that the volume of fluid in the shell that passes a given location in time dt is $2\pi s v_y \, ds \, dt$. To get the total volume dV that flows through a given cross section of the pipe in time dt, integrate this expression over all shells from $s = 0$ to $s = r$, using the result of (a) for v_y and $dm = \rho \, dV$ to show that the rate of mass flow through the pipe is

$$\frac{dm}{dt} = \frac{\pi r^4 \rho}{8\eta} \left(-\frac{dP}{dy} \right) \qquad (16.91)$$

By conservation of mass, dm/dt is constant along the pipe. For a liquid, ρ can be considered essentially constant along the pipe. Separate the variables P and y in (16.91), integrate from one end of the pipe to the other, and use $dm = \rho\, dV$ to obtain Poiseuille's law (16.17). (c) For a gas, ρ is not constant along the pipe, since P varies along the pipe. Substitute $\rho = PM/RT$ into (16.91), separate P and y, and integrate to obtain (16.18). (See also Prob. 16.13.)

16.13 In deriving Poiseuille's law (16.17), we asserted that v_y depends on s only. One might wonder whether v_y depends also on the distance y along the cylindrical tube. Consider a thin shell of fluid between cylinders of radii s and $s + ds$. Show that the mass dm of fluid in this thin shell that passes a fixed location in time dt is $dm = \rho v_y \, d\mathscr{A}\, dt$, where $d\mathscr{A}$ is the shell's cross-sectional area. By conservation of mass, dm/dt must be constant along the tube. For a liquid, ρ is nearly independent of P, so v_y is essentially constant along the tube. For a gas, ρ varies strongly with P. Hence, for flow of a gas, v_y does depend on y. In deriving (16.18) in Prob. 16.12, we assumed the fluid was traveling at constant v_y along the tube (no acceleration). This assumption is false for flow of a gas, so Eq. (16.18) is an approximation, valid if P_1 and P_2 are reasonably close to each other.

16.14 Calculate the terminal speed of fall in water at 25°C of a spherical steel ball of diameter 1.00 mm and density 7.8 g/cm³. Repeat for glycerol (density 1.25 g/cm³). Use data following (16.14).

16.15 Some viscosities of $CO_2(g)$ at 1 atm are 139, 330, and 436 μP (micropoise) at 0°C, 490°C, and 850°C, respectively. Calculate the apparent hard-sphere diameter of CO_2 at each of these temperatures.

16.16 The viscosity of H_2 at 0°C and 1 atm is 8.53×10^{-5} P. Find the viscosity of D_2 at 0°C and 1 atm.

16.17 (a) Find M_n and M_w for a polymer sample that is an equimolar mixture of species with molecular weights 2.0×10^5 and 6.0×10^5. (b) Find M_n and M_w for a polymer sample that is a mixture of equal weights of species with molecular weights 2.0×10^5 and 6.0×10^5.

16.18 For solutions of polystyrene in benzene at 25°C, the following relative viscosities were measured as a function of polystyrene mass concentration ρ_B:

$\rho_B/(\mathrm{g/dm^3})$	1.000	3.000	4.500	6.00
η_r	1.157	1.536	1.873	2.26

For polystyrene in benzene at 25°C, the constants in (16.28) are $K = 0.034$ cm³/g and $a = 0.65$. Find the viscosity average molecular weight of the polystyrene sample.

Section 16.4

16.19 True or false? (a) Diffusion arises from pressure differences. (b) Diffusion does not occur in solids. (c) At 1 atm, diffusion coefficients are much greater in gases than in liquids. (d) The rms net displacement of diffusing molecules is proportional to the diffusion time. (e) For a collection of diffusing molecules with no boundary walls, $\langle \Delta x \rangle$ is zero. (f) For a collection of diffusing molecules with no boundary walls, $\langle (\Delta x)^2 \rangle$ is zero.

16.20 (a) For Sb diffusing into Ag at 20°C, how many years will it take for $(\Delta x)_{rms}$ to reach 1 cm? See Sec. 16.4 for D. (b) Repeat (a) for Al diffusing into Cu at 20°C.

16.21 Calculate $(\Delta x)_{rms}$ for a sucrose molecule in a dilute aqueous solution at 25°C for times of (a) 1 min; (b) 1 hr; (c) 1 day (see Sec. 16.4 for D).

16.22 If r is the net displacement of a diffusing molecule in time t, show that $r_{rms} = (6Dt)^{1/2}$.

16.23 Observations by Perrin on spherical particles of gamboge (a gum resin obtained from trees native to Cambodia) with average radius 2.1×10^{-5} cm suspended in water at 17°C (for which $\eta = 0.011$ P) gave $10^4(\Delta x)_{rms}$ as 7.1, 10.6, and 11.3 cm for time intervals of 30, 60, and 90 s, respectively. Calculate values of Avogadro's number from these data.

16.24 Suppose the following Δx values (in μm) are found in observations over equal time intervals on particles undergoing Brownian motion: $-5.3, +3.4, -1.9, -0.4, +0.5, +3.1, -0.2, -3.5, +1.4, +0.3, -1.0, +2.6$. Calculate $\langle \Delta x \rangle$ and $(\Delta x)_{rms}$.

16.25 Use (16.26) to calculate D_{jj} for O_2 at 0°C and (a) 1.00 atm; (b) 10.0 atm. The experimental value at 0°C and 1 atm is 0.19 cm²/s.

16.26 (a) To get a theoretical equation for the self-diffusion coefficient D_{jj} of a pure liquid, suppose that the liquid volume can be divided into cubical cells, each cell having an edge length $2r_j$ and containing one spherical j molecule of radius r_j. If $V_{m,j}$ is the molar volume of liquid j, show that $r_j = \frac{1}{2}(V_{m,j}/N_A)^{1/3}$, so that (16.38) becomes $D_{jj} \approx (kT/2\pi\eta_j)(N_A/V_{m,j})^{1/3}$, an equation due to Li and Chang. This equation gives D_{jj} with errors of typically 10%. (b) Estimate D_{jj} for water at 25°C and 1 atm ($\eta = 0.89$ cP) and compare with the experimental value 2.4×10^{-5} cm²/s.

16.27 Calculate D_{iB}^{∞} for N_2 in water at 25°C and 1 atm; use data from Sec. 16.3. The experimental value is 1.6×10^{-5} cm²/s.

16.28 (a) Verify that the rigorous theory predicts for hard-sphere gas molecules $D_{jj} = 6\eta/5\rho$. (b) For Ne at 0°C and 1 atm, $\eta = 2.97 \times 10^{-4}$ P. Predict D_{jj} at 0°C and 1.00 atm. The experimental value is 0.44 cm²/s.

16.29 Calculate D_{iB} for hemoglobin in water at 25°C ($\eta = 0.89$ cP), given that $V_m = 48000$ cm³/mol for hemoglobin. Assume the molecules are spherical and estimate the volume of a molecule as V_m/N_A. The experimental value is 7×10^{-7} cm²/s.

16.30 (a) Verify that Eq. (16.33) can be written as $xF_x(t) - \frac{1}{2}f\, d(x^2)/dt = \frac{1}{2}m\, d^2(x^2)/dt^2 - m(dx/dt)^2$. (b) Take the average of the equation in (a) over many colloidal particles, noting that $\langle xF_x \rangle = 0$, because F_x and x vary independently of each other and each is as likely to be positive as negative. Show that $\langle m(dx/dt)^2 \rangle = 2\langle \varepsilon_x \rangle = kT$, where $\langle \varepsilon_x \rangle$ is the particles' average kinetic energy in the x direction, and where it is assumed that $\langle \varepsilon \rangle = \frac{3}{2}kT$. (c) Show that $m\, ds/dt + fs = 2kT$, where $s \equiv$

$d\langle x^2 \rangle/dt$. (d) Show that the equation in (c) integrates to $2kT - fs = e^{-fc/m}e^{-ft/m}$, where c is an integration constant. The exponential $e^{-ft/m}$ is essentially zero for a noninfinitesimal value of t (see part e), so $s \equiv d\langle x^2\rangle/dt = 2kT/f$. Show that this equation integrates to (16.34) if we let $x = 0$ at $t = 0$. (e) In (d), we set $e^{-ft/m}$ equal to 0. For a spherical colloidal particle of radius 10^{-5} cm and density 3 g/cm^3 undergoing Brownian motion in water at room temperature ($\eta = 0.01$ P), calculate $e^{-ft/m}$ for $t = 1$ s.

16.31 Consider constant-T-and-P diffusion in solutions of substances A and B, under the assumption that $\Delta V = 0$ for mixing of all solutions of A and B. Since $\Delta_{\mathrm{mix}}V = 0$, the partial molar volumes $\bar{V}_A$ and $\bar{V}_B$ are constants. From (9.14), the volume change in time dt in the solution on one side of a plane through which diffusion is occurring is $dV = \bar{V}_A\,dn_A + \bar{V}_B\,dn_B$, where dn_A and dn_B are the numbers of moles passing through this plane in time dt. (a) Solve the equations of (16.30) for dn_A and dn_B and substitute the results in $dV = 0$ to show that $D_{AB}\bar{V}_A(dc_A/dx) + D_{BA}\bar{V}_B(dc_B/dx) = 0$. ($b$) Use (9.16) to show that $c_A\bar{V}_A + c_B\bar{V}_B = 1$. Differentiate this equation with respect to x and combine the result with that of (a) to show that $D_{AB} = D_{BA}$.

16.32 "Derive" the Einstein–Smoluchowski equation $(\Delta x)_{\mathrm{rms}} = (2Dt)^{1/2}$ as follows. Consider three planes L, M, and R (for left, middle, and right) perpendicular to the x axis and separated by a distance $(\Delta x)_{\mathrm{rms}}$ between L and M and between M and R, where $(\Delta x)_{\mathrm{rms}}$ is for time t. Let the concentration gradient dc/dx be constant, and let c_L and c_R be the average concentrations of the diffusing species in the regions between L and M and between M and R, respectively. All molecules of the diffusing species between L and M are within a distance $(\Delta x)_{\mathrm{rms}}$ of M and are therefore capable of crossing it in time t. However, half these molecules have a positive Δx and half have a negative Δx, so only half these molecules cross M in time t. Similarly for molecules between M and R. Show that the net rate of flow of the diffusing species to the right through plane M is

$$dn/dt = \Delta n/\Delta t = \tfrac{1}{2}(c_L - c_R)\mathscr{A}(\Delta x)_{\mathrm{rms}}/t$$

where $\mathscr{A}$ is the area of M. Show that the concentration gradient at M is $dc/dx = (c_R - c_L)/(\Delta x)_{\mathrm{rms}}$. Substitute these two expressions into Fick's law (16.30) and obtain $(\Delta x)_{\mathrm{rms}} = (2Dt)^{1/2}$.

16.33 For human hemoglobin in water at 20°C, one finds $\bar{v}^{\infty} = 0.749$ cm^3/g, $D^{\infty} = 6.9 \times 10^{-7}$ cm^2/s, and $s^{\infty} = 4.47 \times 10^{-13}$ s. The density of water at 20°C is 0.998 g/cm^3. Calculate the molecular weight of human hemoglobin.

16.34 (a) Diffusion of a solute initially located in the $x = 0$ plane (Fig. 16.17) can be simulated on the Excel spreadsheet as follows. Fill cells C3, D3, ..., GT3, GU3 with the integers $-100, -99, \ldots, 99, 100$ by an efficient method. Put zero in the row-5 cells C5, D5, ..., GU5 and then change the CY5 entry to 10000. Imagine the x axis divided into tiny intervals, each of the same width. The row 3 entries give the distance (in arbitrary units) of each interval from the origin at CY3. The row 5 numbers give the number of molecules initially present in each interval. Suppose that random diffusion during each small time interval δt causes one-third of the molecules in the ith x-axis

interval to move to interval $i - 1$, one-third to remain in interval i, and one-third to move to interval $i + 1$. Enter the formula =(B5+C5+D5)/3 into cell C6. Select C6 and click on Copy on the Edit menu. From the Edit menu, choose Go To; then in the Reference box enter C6:GU1005 and click OK. This selects the rectangular block of cells from C6 to GU1005. From the Edit menu, choose Paste and click OK. Examine some of the formulas in the C6 to GU1005 block and convince yourself that each cell entry is one-third the sum of the entries of the three cells in the previous row that are nearest the cell in question. Thus each successive row gives the distribution of molecules after one additional time interval. (You can eliminate the annoying decimal places by choosing Cells on the Format menu and then clicking the Number tab and choosing the integer format.) After preparing the spreadsheet, do a Column chart of the data in row 1005. (b) Use the spreadsheet to calculate $\langle (\Delta x)^2 \rangle$ after 10, 100, and 1000 time intervals using $\langle s^2 \rangle = \Sigma_s (n_s s^2/N)$, the equation after (15.39). How well is the relation $\langle (\Delta x)^2 \rangle \propto t$ obeyed? (d) If each spreadsheet time interval is 1 s and each x-axis interval is 10^{-6} cm, calculate the diffusion coefficient D, using $\langle (\Delta x)^2 \rangle$ after 1000 time intervals. (e) Suppose we assume that after each time interval, 40% of the molecules remain in the same x interval, 25% move one interval to the left, 5% move two intervals to the left, 25% move one interval to the right, and 5% move two intervals to the right. Revise the spreadsheet in accord with this rule and repeat all the preceding calculations.

Section 16.5

16.35 For a current of 1.0 A in a metal wire with cross-sectional area 0.02 cm^2, how many electrons pass through a cross section in 1.0 s?

16.36 Calculate the resistance at 20°C of a copper wire with length 250 cm and cross-sectional area 0.0400 cm^2, given that the resistivity of Cu at 20°C is 1.67×10^{-6} Ω cm.

16.37 Calculate the current in a 100-Ω resistor when the potential difference between its ends is 25 V.

16.38 In an electrolysis experiment, a current of 0.10 A flows through a solution of conductivity $\kappa = 0.010$ Ω^{-1} cm^{-1} and cross-sectional area 10 cm^2. Find the electric-field strength in the solution.

Section 16.6

16.39 True or false? (a) Λ_m for a strong electrolyte decreases as the electrolyte concentration increases. (b) At low concentrations, κ for a strong electrolyte increases as the electrolyte concentration increases. (c) κ for a strong electrolyte always increases as the electrolyte concentration increases. (d) For the weak acid HX, $\Lambda_m^{\infty} = \lambda_{m,+}^{\infty} + \lambda_{m,-}^{\infty}$. ($e$) For NaCl($aq$), $\Lambda_m^{\infty} = \lambda_{m,+}^{\infty} + \lambda_{m,-}^{\infty}$.

16.40 Calculate the mass of Cu deposited in 30.0 min from a CuSO$_4$ solution by a 2.00-A current.

16.41 The following resistances are observed at 25°C in a conductivity cell filled with various solutions: 411.82 Ω for a 0.741913 wt % KCl solution; 10.875 kΩ for a 0.001000 mol/dm^3 solution of MCl$_2$; 368.0 kΩ for the deionized water used to prepare the solutions. The conductivity of a 0.741913% KCl solution is known to be 0.012856 (U.S. int. Ω)$^{-1}$ cm^{-1} at 25°C. The U.S. international ohm is an obsolete unit equal to 1.000495 Ω. Calculate (a) the cell constant; (b) κ of MCl$_2$ in a 25°C aqueous 10^{-3} mol/dm^3 solution; (c) Λ_m of MCl$_2$ in this solution; (d) Λ_{eq} of MCl$_2$ in this solution.

16.42 For a 5.000 mmol/dm^3 aqueous solution of SrCl$_2$ at 25°C, the conductivity is 1.242×10^{-3} Ω^{-1} cm^{-1}. For SrCl$_2$ in this solution, calculate (a) Λ_m; (b) Λ_{eq}.

16.43 The moving-boundary method was applied to a 0.02000 mol/dm^3 aqueous NaCl solution at 25°C using CdCl$_2$ as the following solution. For a current held constant at 1.600 mA, Longsworth found that the boundary moved 10.00 cm in 3453 s in a tube of average cross-sectional area 0.1115 cm^2. The conductivity of this NaCl solution at 25°C is 2.313×10^{-3} Ω^{-1} cm^{-1}. Calculate $u(Na^+)$ and $t(Na^+)$ in this solution.

16.44 (a) Show that the transport number t_B can be calculated from moving-boundary data using

$$t_B = |z_B|Fc_B\mathscr{A}x/Q$$

where Q is the charge that flows when the boundary moves a distance x. (b) The moving-boundary method was applied to a 33.27 mmol/dm^3 aqueous solution of GdCl$_3$ at 25°C using LiCl as the following solution. For a constant current of 5.594 mA, it took 4406 s for the boundary to travel between two marks on the tube; the volume between these marks was known to be 1.111 cm^3. Find the cation and anion transport numbers in this GdCl$_3$ solution.

16.45 A 0.14941 wt % aqueous KCl solution at 25°C was electrolyzed in a Hittorf apparatus using two Ag–AgCl electrodes. The cathode reaction was $AgCl(s) + e^- \rightarrow Ag(s) + Cl^-(aq)$; the anode reaction was the reverse of this. After the experiment, it was found that 160.24 mg of Ag had been deposited in a coulometer connected in series with the apparatus and that the cathode compartment contained 120.99 g of solution that was 0.19404% KCl by weight. Calculate t_+ and t_- in the KCl solution used in the experiment. Neglect the transport of water by ions. (*Hint:* Use the fact that the mass of water in the cathode compartment remains constant.)

16.46 (a) The following Λ_m^∞ values in Ω^{-1} cm^2 mol^{-1} are found for 25°C solutions in the solvent methanol: KNO$_3$, 114.5; KCl, 105.0; LiCl, 90.9. Using only these data, calculate Λ_m^∞ for LiNO$_3$ in CH$_3$OH at 25°C. (b) The following Λ_m^∞ values in Ω^{-1} cm^2 mol^{-1} are found for 25°C aqueous solutions: HCl, 426; NaCl, 126; NaC$_2$H$_3$O$_2$, 91. Using only these data, calculate $\lambda_{m,+}^\infty + \lambda_{m,-}^\infty$ for HC$_2$H$_3$O$_2$ in water at 25°C.

16.47 To find λ_m^∞ values in a given solvent, we need only one accurate transport number in the solvent. For the solvent CH$_3$OH at 25°C, the cation transport number t_+^∞ in NaCl has been found to be 0.463. Observed Λ_m^∞ values in Ω^{-1} cm^2 mol^{-1} in CH$_3$OH at 25°C are 96.9 for NaCl; 106.4 for NaNO$_3$; 100.2 for LiNO$_3$; 107.0 for NaCNS; 192 for HCl; 244 for Ca(CNS)$_2$. Find λ_m^∞ in CH$_3$OH at 25°C for Na$^+$, Cl$^-$, NO$_3^-$, Li$^+$, CNS$^-$, H$^+$, and Ca^{2+}.

16.48 For ClO$_4^-$ in water at 25°C, $\lambda_m^\infty = 67.2$ Ω^{-1} cm^2 mol^{-1}. (a) Calculate $u^\infty(ClO_4^-)$ in water at 25°C. (b) Calculate the drift speed $v^\infty(ClO_4^-)$ in water at 25°C in a field of 24 V/cm. (c) Estimate the radius of the hydrated perchlorate ion.

16.49 From λ_m^∞ data tabulated in Sec. 16.6, calculate Λ_m^∞ for the following electrolytes in water at 25°C: (a) NH$_4$NO$_3$; (b) (NH$_4$)$_2$SO$_4$; (c) MgSO$_4$; (d) Ca(OH)$_2$.

16.50 Use λ_m^∞ data in Sec. 16.6 to calculate $t^\infty(Mg^{2+})$ and $t^\infty(NO_3^-)$ for Mg(NO$_3$)$_2$(aq) at 25°C.

16.51 For Na$_2$SO$_4$(aq) at 25°C and 1 atm, $\Lambda_m^\infty = 259.8$ Ω^{-1} cm^2 mol^{-1} and $t_+^\infty = 0.386$. Using only these numbers, find λ_m^∞ of Na$^+$(aq) and of SO$_4^{2-}$(aq) at 25°C and 1 atm.

16.52 A very dilute solution of AgNO$_3$(aq) is electrolyzed using inert electrodes, and 1.00 mmol of Ag is deposited. Use infinite-dilution data in Sec. 16.6 to estimate the number of moles of NO$_3^-$(aq) that crossed the plane midway between the electrodes during the electrolysis.

16.53 The charge of Mg^{2+} is twice that of Na$^+$, and from Eq. (16.70) one might therefore expect Mg^{2+}(aq) to have a much greater u^∞ than Na$^+$(aq). Actually, these ions have very similar mobilities. Explain why.

16.54 (a) Which of the following quantities must be the same for CaCl$_2$(aq) as for NaCl(aq) at the same temperature and pressure: $\lambda_m^\infty(Cl^-)$, $t^\infty(Cl^-)$, $u^\infty(Cl^-)$? (b) Must $u(Cl^-)$ be the same in 1.00 mol/dm^3 NaCl and KCl solutions at the same T and P?

16.55 For an NaCl(aq) solution, which of the following quantities go to zero as the NaCl concentration goes to zero? Assume the solvent's contribution to the conductivity has been subtracted off. (a) Λ_m; (b) κ; (c) $\lambda_m(Na^+)$; (d) $t(Na^+)$.

16.56 (a) Use (16.70) to show that

$$\frac{d \ln \lambda_m^\infty}{dT} \approx -\frac{1}{\eta}\frac{d\eta}{dT}$$

(b) Viscosities of water at 1 atm and 24°C, 25°C, and 26°C are 0.9111, 0.8904, and 0.8705 cP, respectively. Approximate $d\eta/dT$ by $\Delta\eta/\Delta T$ and show that the equation in (a) predicts $d \ln \lambda_m^\infty/dT \approx 0.023$ K^{-1} for all ions in water at 25°C. Experimental values for this quantity are typically 0.018 to 0.022 K^{-1}. (c) Estimate λ_m^∞ for NO$_3^-$(aq) at 35°C and 1 atm from the tabulated 25°C value.

16.57 (a) Use the Onsager equation to calculate Λ_m and κ for a 0.00200 mol/dm^3 solution of KNO$_3$ in water at 25°C and 1 atm. (b) Find the resistance of this solution in a conductivity cell with electrodes of area 1.00 cm^2 and separation 10.0 cm.

16.58 The conductivity of pure water at 25°C and 1 atm is $5.4_7 \times 10^{-8}$ Ω^{-1} cm^{-1}. [H. C. Duecker and W. Haller, *J. Phys. Chem.*, **66**, 225 (1962).] Use (16.90) to find K_c for the ionization of water at 25°C.

16.59 The conductivity of a saturated aqueous CaSO$_4$ solution at 25°C is 2.21×10^{-3} Ω^{-1} cm^{-1}. (a) Use (16.90) to find the concentration-scale K_{sp} for CaSO$_4$ in water at 25°C.

(b) Does the existence of $CaSO_4$ ion pairs in the solution cause error in the result for (a)?

16.60 The conductivity of a 0.001028 mol/dm³ aqueous solution of $HC_2H_3O_2$ at 25°C is $4.95 \times 10^{-5}\ \Omega^{-1}\ cm^{-1}$. Use (16.90) to find K_c for the ionization of acetic acid in water at 25°C.

16.61 The conductivity of a 2.500×10^{-4} mol/dm³ aqueous solution of $MgSO_4$ at 25°C is $6.156 \times 10^{-5}\ \Omega^{-1}\ cm^{-1}$. Use (16.90) to calculate K_c for the ion-pair-formation reaction $Mg^{2+}(aq) + SO_4^{2-}(aq) \rightleftharpoons MgSO_4(aq)$ at 25°C.

16.62 Verify that, if the correction term $S(c_+/c°)^{1/2}$ is omitted from (16.90), the degree of dissociation of the weak acid HX is given by $\alpha \approx \Lambda_m/(\lambda_{m,+}^\infty + \lambda_{m,-}^\infty)$, an equation due to Arrhenius.

16.63 Find Λ_m^∞ for $SrCl_2$ in water at 25°C from the following data on aqueous 25°C $SrCl_2$ solutions.

$c/(\text{mmol/dm}^3)$	0.25	0.50	2.50
$\Lambda_m/(\Omega^{-1}\ cm^2/mol)$	263.8	260.7	248.5

16.64 Write a computer program that will use Eq. (16.90) to calculate c_+ from κ and $\lambda_{m,+}^\infty + \lambda_{m,-}^\infty$ using successive approximations.

General
16.65 For the ion $Mg^{2+}(aq)$ at 25°C, estimate the rms distances traveled in the x direction in 1 s and in 10 s as a result of random thermal motion.

16.66 State whether each of the following properties increases or decreases as intermolecular attractions increase: (a) viscosity of a liquid; (b) surface tension of a liquid; (c) normal boiling point; (d) molar heat of vaporization; (e) critical temperature; (f) van der Waals a.

16.67 As noted, transport properties obey the equation $(1/\mathcal{A})(dW/dt) = -L(dB/dt)$. For each of the four transport properties studied in this chapter, give the symbols and SI units for W, L, and B.

CHAPTER

17 Reaction Kinetics

17.1 REACTION KINETICS

We began our study of nonequilibrium processes in Chapter 16, which deals with physical kinetics (the rates and mechanisms of transport processes). We now move on to chemical kinetics.

Chemical kinetics, also called **reaction kinetics,** is the study of the rates and mechanisms of chemical reactions. A reacting system is not in equilibrium, so reaction kinetics is not part of thermodynamics but is a branch of kinetics (Sec. 16.1). This chapter deals mainly with experimental aspects of reaction kinetics. The theoretical calculation of reaction rates is discussed in Chapter 23.

Applications of reaction kinetics abound. In the industrial synthesis of compounds, reaction rates are as important as equilibrium constants. The thermodynamic equilibrium constant tells us the maximum possible yield of NH_3 obtainable at any given T and P from N_2 and H_2, but if the reaction rate between N_2 and H_2 is too low, the reaction will not be economical to carry out. Frequently, in organic preparative reactions, several possible competing reactions can occur, and the relative rates of these reactions usually influence the yield of each product. What happens to pollutants released to the atmosphere can be understood only by a kinetic analysis of atmospheric reactions. An automobile works because the rate of oxidation of hydrocarbons is negligible at room temperature but rapid at the high temperature of the engine. Many of the metals and plastics of modern technology are thermodynamically unstable with respect to oxidation, but the rate of this oxidation is slow at room temperature. Reaction rates are fundamental to the functioning of living organisms. Biological catalysts (enzymes) control the functioning of an organism by selectively speeding up certain reactions. In summary, to understand and predict the behavior of a chemical system, one must consider both thermodynamics and kinetics.

We begin with some definitions. A **homogeneous reaction** is one that occurs entirely in one phase. A **heterogeneous reaction** involves species present in two or more phases. Sections 17.1 to 17.17 deal with homogeneous reactions. Section 17.18 deals with heterogeneous reactions. Homogeneous reactions are divided into gas-phase reactions and reactions in (liquid) solutions. Sections 17.1 to 17.14 apply to both gas-phase and solution kinetics. Section 17.15 deals with aspects of kinetics unique to reactions in solution.

Rate of Reaction

Consider the homogeneous reaction

$$aA + bB + \cdots \rightarrow eE + fF + \cdots \tag{17.1}$$

where $a, b, \ldots, e, f, \ldots$ are the coefficients in the balanced chemical equation and A, B, $\ldots$, E, F, $\ldots$ are the chemical species. In this chapter, it is assumed that the reaction occurs in a closed system. The rate at which any reactant is consumed is proportional to its stoichiometric coefficient. Therefore

$$\frac{dn_A/dt}{dn_B/dt} = \frac{a}{b} \quad \text{and} \quad \frac{1}{a}\frac{dn_A}{dt} = \frac{1}{b}\frac{dn_B}{dt}$$

where t is the time and n_A is the number of moles of A present at time t. The **rate of conversion** J for the homogeneous reaction (17.1) is defined as

$$J \equiv -\frac{1}{a}\frac{dn_A}{dt} = -\frac{1}{b}\frac{dn_B}{dt} = \cdots = \frac{1}{e}\frac{dn_E}{dt} = \frac{1}{f}\frac{dn_F}{dt} = \cdots \qquad (17.2)$$

Since A is disappearing, dn_A/dt is negative and J is positive. At equilibrium, $J = 0$.

Actually, the relation $-a^{-1}\,dn_A/dt = e^{-1}\,dn_E/dt$ need not hold if the reaction has more than one step. For a multistep reaction, the reactant A may first be converted to some reaction intermediate rather than directly to a product. Thus the instantaneous relation between dn_A/dt and dn_E/dt may be complicated. If, as is commonly true, the concentrations of all reaction intermediates are very small throughout the reaction, their effect on the stoichiometry can be neglected.

The conversion rate J is an extensive quantity and depends on the system's size. The conversion rate per unit volume, J/V, is called the **rate of reaction** r:

$$r \equiv \frac{J}{V} = \frac{1}{V}\left(-\frac{1}{a}\frac{dn_A}{dt}\right) \qquad (17.3)$$

r is an intensive quantity and depends on T, P, and the concentrations in the homogeneous system. In most (but not all) systems studied, the volume either is constant or changes by a negligible amount. When V is essentially constant, we have $(1/V)(dn_A/dt) = d(n_A/V)/dt = dc_A/dt = d[A]/dt$, where $c_A \equiv [A]$ is the **molar concentration** [Eq. (9.1)] of A. Thus, for the reaction (17.1)

$$r = -\frac{1}{a}\frac{d[A]}{dt} = -\frac{1}{b}\frac{d[B]}{dt} = \cdots = \frac{1}{e}\frac{d[E]}{dt} = \frac{1}{f}\frac{d[F]}{dt} = \cdots \quad \text{const. } V \quad \textbf{(17.4)}*$$

We shall assume constant volume in this chapter. Common units for r are mol dm^{-3} s^{-1} (where 1 dm^3 = 1 L) and mol cm^{-3} s^{-1}.

EXAMPLE 17.1 Rate expressions

For the homogeneous reactions (a) $N_2 + 3H_2 \rightarrow 2NH_3$ and (b) $0 \rightarrow \Sigma_i \nu_i A_i$ [Eq. (4.94)], express r in terms of rates of change of concentrations. Under what conditions do your answers apply?

(a) From (17.4), $r = -d[N_2]/dt = -\frac{1}{3}d[H_2]/dt = \frac{1}{2}d[NH_3]/dt$.

(b) The quantities $-a, -b, \ldots, e, f, \ldots$ in (17.4) are the stoichiometric coefficients (Sec. 4.9) for the reaction (17.1). Therefore (17.4) gives $r = (1/\nu_i)\,d[A_i]/dt$ as the rate of the general reaction $0 \rightarrow \Sigma_i \nu_i A_i$.

For these expressions to be valid, V must be constant and any reaction intermediates must have negligible concentrations.

Rate Laws

For many (but not all) reactions, the rate r at time t is experimentally found to be related to the concentrations of species present at that time t by an expression of the form

$$r = k[A]^\alpha[B]^\beta \cdots [L]^\lambda \qquad \textbf{(17.5)}*$$

where the exponents $\alpha, \beta, \ldots, \lambda$ are usually integers or half-integers $(\frac{1}{2}, \frac{3}{2}, \ldots)$. The proportionality constant k, called the **rate constant** or **rate coefficient,** is a function of temperature and pressure. The pressure dependence of k is small and is usually ignored. The reaction is said to have **order** α with respect to A, order β with respect to B, etc. The exponents $\alpha, \beta, \ldots$ are also called **partial orders.** The sum $\alpha + \beta + \cdots + \lambda \equiv n$ is the **overall order** (or simply the **order**) of the reaction. Since r has units of concentration over time, k in (17.5) has units concentration^{1-n} time^{-1}. Most commonly, k is given in $(dm^3/mol)^{n-1}\ s^{-1}$. A first-order $(n = 1)$ rate constant has units s^{-1} and is independent of the units used for concentration.

In gas-phase kinetics, reaction rates and rate constants are sometimes defined in terms of molecular concentrations rather than molar concentrations; see Prob. 17.7.

The expression for r as a function of concentrations at a fixed temperature is called the **rate law.** A rate law has the form $r = f([A], [B], \ldots)$ at fixed T, where f is some function of the concentrations. Some observed rate laws for homogeneous reactions are

$$
\begin{array}{lll}
(1) & H_2 + Br_2 \rightarrow 2HBr & r = \dfrac{k[H_2][Br_2]^{1/2}}{1 + j[HBr]/[Br_2]} \\[2ex]
(2) & 2N_2O_5 \rightarrow 4NO_2 + O_2 & r = k[N_2O_5] \\[1ex]
(3) & H_2 + I_2 \rightarrow 2HI & r = k[H_2][I_2] \\[1ex]
(4) & 2NO + O_2 \rightarrow 2NO_2 & r = k[NO]^2[O_2] \\[1ex]
(5) & CH_3CHO \rightarrow CH_4 + CO & r = k[CH_3CHO]^{3/2} \qquad\qquad (17.6) \\[1ex]
(6) & 2SO_2 + O_2 \xrightarrow{NO} 2SO_3 & r = k[O_2][NO]^2 \\[1ex]
(7) & H_2O_2 + 2I^- + 2H^+ \rightarrow 2H_2O + I_2 & r = k_1[H_2O_2][I^-] \\[0.5ex]
 & & \quad + k_2[H_2O_2][I^-][H^+] \\[2ex]
(8) & Hg_2^{2+} + Tl^{3+} \rightarrow 2Hg^{2+} + Tl^+ & r = k\dfrac{[Hg_2^{2+}][Tl^{3+}]}{[Hg^{2+}]}
\end{array}
$$

where the values of k depend strongly on temperature and differ from one reaction to another. In reaction (1), j is a constant. Reactions (1) to (6) are gas phase; reactions (7) and (8) are in aqueous solution. For reaction (1), the concept of order does not apply. Each of the two terms in the rate law of reaction (7) has an order, but the reaction rate itself does not have an order. Reaction (5) has order $\frac{3}{2}$. In reaction (6), the species NO speeds up the reaction but does not appear in the overall chemical equation and is therefore a **catalyst.** In reaction (8), the order with respect to Hg^{2+} is -1. Note from reactions (1), (2), (5), (6), (7), and (8) that the exponents in the rate law can differ from the coefficients in the balanced chemical equation. *Rate laws must be determined from measurements of reaction rates and cannot be deduced from the reaction stoichiometry.*

Actually, the use of concentrations in the rate law is strictly correct only for ideal systems. For nonideal systems, see Sec. 17.10.

Suppose all the concentrations in (17.5) have the order of magnitude 1 mol/dm^3. With the order-of-magnitude approximation $d[E]/dt \approx \Delta[E]/\Delta t$, Eqs. (17.4) and (17.5) give for 1-mol/dm^3 concentrations: $\Delta[E]/\Delta t \approx k(1\ \text{mol/dm}^3)^n$, where n is the reaction order and e has been omitted since it doesn't affect the order of magnitude of things. For a substantial amount of reaction, $\Delta[E]$ will have the same order of magnitude as [A], namely, 1 mol/dm^3. Therefore, $1/k$ [multiplied by $(dm^3/mol)^{n-1}$ to make things dimensionally correct] gives the order of magnitude of the time needed for a substantial amount of reaction to occur when the concentrations have the order of magnitude 1 mol/dm^3. For example, reaction (3) in (17.6) has $k = 0.0025 \text{ dm}^3$ $\text{mol}^{-1} \text{ s}^{-1}$ at 629 K; it will take on the order of 400 s for a substantial amount of this reaction to occur when the concentrations are 1 mol/dm^3, which corresponds to partial pressures of 50 atm.

Reaction Mechanisms

Equation (17.1) gives the overall stoichiometry of the reaction but does not tell us the process, or **mechanism,** by which the reaction actually occurs. For example, the gas-phase NO-catalyzed oxidation of SO_2 [reaction (6) in (17.6)] has been postulated to occur by the following two-step process:

$$O_2 + 2NO \rightarrow 2NO_2$$
$$NO_2 + SO_2 \rightarrow NO + SO_3 \tag{17.7}$$

Since two NO_2 molecules are produced in the first step, the second step must occur twice each time the first step occurs once. Adding twice the second step to the first, one gets the overall stoichiometry as $2SO_2 + O_2 \rightarrow 2SO_3$. The overall reaction does not contain the intermediate species NO_2 or the catalyst NO (which is consumed in the first step and regenerated in the second). A species like NO_2 in (17.7) that is formed in one step of a mechanism and consumed in a subsequent step so that it does not appear in the overall reaction is a **reaction intermediate.**

There is good evidence that the gas-phase decomposition of N_2O_5 with overall reaction $2N_2O_5 \rightarrow 4NO_2 + O_2$ occurs by the following multistep mechanism:

Step (a): $\qquad\qquad\qquad N_2O_5 \rightleftharpoons NO_2 + NO_3$

Step (b): $\qquad\qquad\quad NO_2 + NO_3 \rightarrow NO + O_2 + NO_2 \tag{17.8}$

Step (c): $\qquad\qquad\qquad NO + NO_3 \rightarrow 2NO_2$

Here, there are two reaction intermediates, NO_3 and NO. Any proposed mechanism must add up to give the observed overall reaction stoichiometry. Step (c) consumes the NO molecule produced in step (b), so step (c) must occur once for each occurrence of (b). Steps (b) and (c) together consume two NO_3's. Since step (a) produces only one NO_3, the forward reaction of step (a) must twice for each occurrence of steps (b) and (c). Taking 2 times step (a) plus 1 times step (b) plus 1 times step (c), we get $2N_2O_5 \rightarrow 4NO_2 + O_2$, as we should.

The number of times a given step in the mechanism occurs for each occurrence of the overall reaction as written is the **stoichiometric number** s of the step. For the overall reaction $2N_2O_5 \rightarrow 4NO_2 + O_2$, the stoichiometric numbers of steps (a), (b), and (c) in the mechanism (17.8) are 2, 1, and 1, respectively. Don't confuse the stoichiometric number s of a step with the stoichiometric coefficient ν of a chemical species.

Each step in the mechanism of a reaction is called an **elementary reaction.** A **simple reaction** consists of a single elementary step. A **complex** (or **composite**) **reaction** consists of two or more elementary steps. The N_2O_5 decomposition reaction is complex. The Diels–Alder addition of ethylene to butadiene to give cyclohexene is believed to be simple, occurring as the single step $CH_2{=}CH_2 + CH_2{=}CHCH{=}CH_2 \rightarrow C_6H_{10}$. *Most chemical reactions are composite.*

The form of the rate law is a consequence of the mechanism of the reaction; see Sec. 17.6. For some reactions, the form of the rate law changes with temperature, indicating a change in mechanism. Sometimes one finds that data for a given homogeneous reaction are fitted by an expression like $r = k[A]^{1.38}$. This indicates that the reaction probably proceeds by two different simultaneously occurring mechanisms that produce different orders. Thus, it is likely that one could get as good a fit or better with an expression like $r = k'[A] + k''[A]^2$.

Pseudo Order

For the hydrolysis of sucrose,

$$C_{12}H_{22}O_{11} + H_2O \rightarrow \underset{\text{Glucose}}{C_6H_{12}O_6} + \underset{\text{Fructose}}{C_6H_{12}O_6} \qquad (17.9)$$

one finds the rate to be given by $r = k[C_{12}H_{22}O_{11}]$. However, since the solvent H_2O participates in the reaction, one would expect the rate law to have the form $r = k'[C_{12}H_{22}O_{11}]^w[H_2O]^v$. Because H_2O is always present in great excess, its concentration remains nearly constant during a given run and from one run to another. Therefore $[H_2O]^v$ is essentially constant, and the rate law is apparently $r = k[C_{12}H_{22}O_{11}]$, where $k = k'[H_2O]^v$. This reaction is said to be *pseudo* first order. It is hard to determine v, but kinetic data indicate $v \approx 6$. (This can be explained by a reaction mechanism that involves a hexahydrate of sucrose.)

Pseudo order is involved in catalyzed reactions. A catalyst affects the rate without being consumed during the reaction. The hydrolysis of sucrose is acid-catalyzed. During a given run, the H_3O^+ concentration remains fixed. However, when $[H_3O^+]$ is varied from one run to another, it is found that the hydrolysis rate is in fact first order with respect to H_3O^+. Thus, the correct rate law for the hydrolysis of sucrose is $r = k''[C_{12}H_{22}O_{11}][H_2O]^6[H_3O^+]$, and the reaction has order 8. During a given run, however, its apparent (or pseudo) order is 1.

17.2   MEASUREMENT OF REACTION RATES

To measure the reaction rate r [Eq. (17.4)], one must follow the concentration of a reactant or product as a function of time. In the *chemical method,* one places several reaction vessels with identical initial compositions in a constant-temperature bath. At intervals, one withdraws samples from the bath, slows down or stops the reaction, and rapidly analyzes the mixture chemically. Methods for slowing the reaction include cooling the sample, removing a catalyst, greatly diluting the reaction mixture, and adding a species that quickly combines with one of the reactants. Gas samples are frequently analyzed with a mass spectrometer or a gas chromatograph.

Physical methods are usually more accurate and less tedious than chemical methods. Here, one measures a physical property of the reacting system as a function of time. This allows the reaction to be followed continuously as it proceeds. For a gas-phase reaction with a change in total number of moles, the gas pressure P can be followed. The danger in this procedure is that, if side reactions occur, the total pressure will not correctly indicate the progress of the reaction being studied. For a liquid-phase reaction that occurs with a measurable volume change, V can be followed by running the reaction in a dilatometer, a vessel capped with a graduated capillary tube. Liquid-phase addition polymerization reactions show substantial volume decreases, and dilatometry is the most common way of measuring polymerization rates. If one of the species has a characteristic spectroscopic absorption band, the intensity of this band can be followed. If at least one of the species is optically active [as is true for the sucrose hydrolysis (17.9)], the optical rotation can be followed. Ionic reactions in solution can be followed by measuring the electrical conductivity. The refractive index of a liquid solution can be measured as a function of time.

Most commonly, the reactants are mixed and kept in a closed vessel; this is the *static method.* In the *flow method,* the reactants continuously flow into the reaction vessel (which is maintained at constant temperature), and products continuously flow out. After the reaction has run for a while, a steady state (Sec. 1.2) is reached in the reaction vessel and the concentrations at the outlet remain constant with time. The rate law and rate constant can be found by measuring the outlet concentrations for several different inlet concentrations and flow rates. Flow systems are widely used in industrial chemical production.

The already mentioned "classical" methods of kinetics are limited to reactions with half-lives of at least a few seconds. (The half-life is the time it takes for the concentration of a reactant to be cut in half.) Many important chemical reactions have half-lives (for typical reactant concentrations) in the range 10^0 to 10^{-11} s, and are called *fast reactions*. Examples include gas-phase reactions where one reactant is a free radical and many aqueous-solution reactions that involve ions. (A **free radical** is a species with one or more unpaired electrons; examples are CH_3 and Br.) Many reactions in biological systems are fast. For example, the rate constants for formation of complexes between an enzyme and a small molecule (*substrate*) are typically 10^6 to 10^9 dm^3 mol^{-1} s^{-1}. Methods used to study the rates of fast reactions are discussed in Sec. 17.14.

There are many pitfalls in kinetics work. A review of solution kinetics stated: "It is a sad fact that there are many kinetic data in the literature that are worthless and many more that are wrong in some important respect. These faulty data can be found in papers old and new, authored by chemists of small reputation and by some of the best known kineticists . . ." (J. F. Bunnett in *Bernasconi*, pt. I, p. 230).

One must be sure the reactants and products are known. One must check for side reactions. The reagents and solvent must be carefully purified. Certain reactions are sensitive to trace impurities. For example, dissolved O_2 from the air may strongly affect the rate and products of a free-radical reaction. Traces of metal ions catalyze certain reactions. Traces of water strongly affect certain reactions in nonaqueous solvents. To be sure that the rate law has been correctly determined, it is best to make a series of runs varying all concentrations and following the reaction as far to completion as possible. Data that appear to lie on a straight line for the first 50% of a reaction may deviate greatly from this line when the reaction is followed to 70 or 80% of completion.

Because of the many possible sources of error, reported rate constants are often inaccurate.

Benson has developed methods to estimate the order of magnitude of gas-phase rate constants. See S. W. Benson and D. M. Golden in *Eyring, Henderson, and Jost*, vol. VII, pp. 57–124; S. W. Benson, *Thermochemical Kinetics*, 2d ed., Wiley-Interscience, 1976.

17.3 INTEGRATION OF RATE LAWS

Kinetics experiments (Sec. 17.2) yield the concentrations [A], [B], . . . of reacting species as functions of time at a fixed temperature. The rate law (Sec. 17.1) that governs a reaction is a differential equation that gives the rates of change $d[A]/dt$, etc., in the concentrations of the reacting species. Methods for deducing the rate law from kinetics data are discussed in the next section. Most of these methods compare concentrations of reacting species predicted by the possible rate laws with the experimental data. To obtain the concentrations versus time predicted by a rate law, one must integrate the rate law. Therefore, this section integrates the commonly occurring rate laws. In this section, we start with an assumed rate law $r \equiv -(1/a) \, d[A]/dt = f([A], [B], . . .)$, where f is a known function, and we integrate it to find [A] as a function of time: $[A] = g(t)$, where g is some function.

In the following discussion, it is assumed unless stated otherwise that: (*a*) The reaction is carried out at constant temperature. With T constant, the rate constant k is constant. (*b*) The volume is constant. With V constant, the reaction rate r is given by (17.4). (*c*) The reaction is "irreversible," meaning that no significant amount of reverse reaction occurs. This will be true if the equilibrium constant is very large or if one studies only the initial rate.

First-Order Reactions

Suppose the reaction $a\text{A} \rightarrow$ products is first-order with $r = k[\text{A}]$. From (17.4) and (17.5), the rate law is

$$r = -\frac{1}{a} \frac{d[\text{A}]}{dt} = k[\text{A}] \tag{17.10}$$

Defining k_A as $k_\text{A} \equiv ak$, we have

$$d[\text{A}]/dt = -k_\text{A}[\text{A}] \qquad \text{where } k_\text{A} \equiv ak \tag{17.11}$$

The subscript in k_A reminds us that this rate constant refers to the rate of change in the concentration of A. Chemists are inconsistent in their definitions of rate constants, so in using measured k values one must be sure of the definition.

The variables in (17.11) are [A] and t. To solve this differential equation, we rearrange it to separate [A] and t on opposite sides. We have $[\text{A}]^{-1} d[\text{A}] = -k_\text{A} dt$. Integration gives $\int_1^2 [\text{A}]^{-1} d[\text{A}] = -\int_1^2 k_\text{A} dt$, and

$$\ln ([\text{A}]_2/[\text{A}]_1) = -k_\text{A}(t_2 - t_1) \tag{17.12}$$

Equation (17.12) holds for any two times during the reaction. If state 1 is the state at the start of the reaction when $[\text{A}] = [\text{A}]_0$ and $t = 0$, then (17.12) becomes

$$\ln \frac{[\text{A}]}{[\text{A}]_0} = -k_\text{A}t \tag{17.13}$$

where [A] is the concentration at time t. Use of (1.65) gives $[\text{A}]/[\text{A}]_0 = e^{-k_\text{A}t}$, and

$$[\text{A}] = [\text{A}]_0 e^{-k_\text{A}t} \tag{17.14}*$$

For a first-order reaction, [A] *decreases exponentially with time* (Fig. 17.1a). Equations (17.10) and (17.14) give $r = k[\text{A}] = k[\text{A}]_0 e^{-k_\text{A}t}$, so the rate r decreases exponentially with time for a first-order reaction.

If the reaction is first-order, Eq. (17.13) (multiplied by -1) shows that a plot of $\ln ([\text{A}]_0/[\text{A}])$ versus t gives a straight line of slope k_A.

The time needed for [A] to drop to half its value is called the reaction's **half-life** $t_{1/2}$. Setting $[\text{A}] = \frac{1}{2}[\text{A}]_0$ and $t = t_{1/2}$ in (17.13) or $[\text{A}]_2/[\text{A}]_1 = \frac{1}{2}$ and $t_2 - t_1 = t_{1/2}$ in (17.12), we get $-k_\text{A}t_{1/2} = \ln \frac{1}{2} = -0.693$. For a first-order reaction,

$$k_\text{A}t_{1/2} = 0.693 \qquad \text{first-order reaction} \tag{17.15}*$$

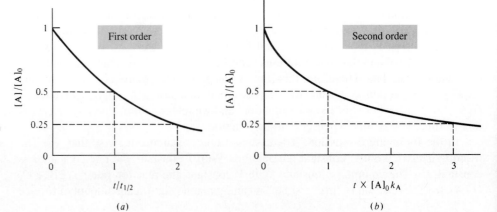

Figure 17.1

Reactant concentration versus time in (*a*) a first-order reaction; (*b*) a second-order reaction. The half-life is independent of initial concentration for a first-order reaction.

Second-Order Reactions

The most common forms of second-order rate laws are $r = k[A]^2$ and $r = k[A][B]$, where A and B are two different reactants.

Suppose the reaction $aA \rightarrow$ products is second-order with $r = k[A]^2$. Then $r = -a^{-1} d[A]/dt = k[A]^2$. Defining $k_A \equiv ak$ as in (17.11) and separating variables, we have

$$\frac{d[A]}{dt} = -k_A[A]^2 \qquad \text{and} \qquad \int_1^2 \frac{1}{[A]^2}\, d[A] = -k_A \int_1^2 dt$$

$$\frac{1}{[A]_1} - \frac{1}{[A]_2} = -k_A(t_2 - t_1) \qquad \text{or} \qquad \frac{1}{[A]} - \frac{1}{[A]_0} = k_A t \qquad (17.16)$$

$$[A] = \frac{[A]_0}{1 + k_A t[A]_0}, \qquad k_A \equiv ak \qquad (17.17)$$

From (17.16), a plot of $1/[A]$ versus t gives a straight line of slope k_A if $r = k[A]^2$.

The half-life is found by setting $[A] = \frac{1}{2}[A]_0$ and $t = t_{1/2}$ in (17.16), to give

$$t_{1/2} = 1/[A]_0 k_A \qquad \text{second-order reaction with } r = k[A]^2$$

For a second-order reaction, $t_{1/2}$ depends on the initial A concentration, which is in contrast to a first-order reaction; $t_{1/2}$ doubles when the A concentration is cut in half. Thus it takes twice as long for the reaction to go from 50 to 75% completion as from 0 to 50% completion (Fig. 17.1*b*).

Now suppose the reaction is $aA + bB \rightarrow$ products with rate law $r = k[A][B]$. Then (17.4) gives

$$\frac{1}{a} \frac{d[A]}{dt} = -k[A][B] \qquad (17.18)$$

Equation (17.18) has three variables: $[A]$, $[B]$, and t. To integrate (17.18), we must eliminate $[B]$ by relating it to $[A]$. The amounts of B and A that react are proportional to their coefficients b and a in the reaction, so $\Delta n_B / \Delta n_A = b/a$. Division by the volume gives $b/a = \Delta[B]/\Delta[A] = ([B] - [B]_0)/([A] - [A]_0)$, where $[B]_0$ and $[A]_0$ are the initial concentrations of B and A. Solving for $[B]$, we find

$$[B] = [B]_0 - ba^{-1}[A]_0 + ba^{-1}[A] \qquad (17.19)$$

Substituting (17.19) into (17.18), separating $[A]$ and t, and integrating, we get

$$\frac{1}{a} \int_1^2 \frac{1}{[A]([B]_0 - ba^{-1}[A]_0 + ba^{-1}[A])}\, d[A] = -\int_1^2 k\, dt \qquad (17.20)$$

A table of integrals gives

$$\int \frac{1}{x(p + sx)}\, dx = -\frac{1}{p} \ln \frac{p + sx}{x} \qquad \text{for } p \neq 0 \qquad (17.21)$$

To verify this relation, differentiate the right side of (17.21). Using (17.21) with $p = [B]_0 - ba^{-1}[A]_0$, $s = ba^{-1}$, and $x = [A]$, we get for (17.20)

$$\frac{1}{a} \frac{1}{[B]_0 - ba^{-1}[A]_0} \ln \frac{[B]_0 - ba^{-1}[A]_0 + ba^{-1}[A]}{[A]} \bigg|_1^2 = k(t_2 - t_1)$$

Use of (17.19) gives

$$\frac{1}{a[B]_0 - b[A]_0} \ln \frac{[B]}{[A]} \bigg|_1^2 = k(t_2 - t_1)$$

$$\frac{1}{a[B]_0 - b[A]_0} \ln \frac{[B]/[B]_0}{[A]/[A]_0} = kt \qquad (17.22)$$

In Eq. (17.22), [A] and [B] are the concentrations at time t, and $[A]_0$ and $[B]_0$ are the concentrations at time 0. A plot of the left side of (17.22) versus t gives a straight line of slope k. The concept of reaction half-life does not apply to (17.22), since when $[B] = \frac{1}{2}[B]_0$, [A] will not equal $\frac{1}{2}[A]_0$, unless the reactants are mixed in stoichiometric proportions.

A special case of (17.18) is where A and B are initially present in stoichiometric proportions, so that $[B]_0/[A]_0 = b/a$. Equation (17.22) does not apply here, since $a[B]_0 - b[A]_0$ in (17.22) becomes zero. To deal with this case, we recognize that B and A will remain in stoichiometric proportions throughout the reaction: $[B]/[A] = b/a$ at any time. This follows from (17.19) with $[B]_0 = (b/a)[A]_0$. Equation (17.18) becomes $(1/b[A]^2)\, d[A] = -k\, dt$. Integration gives [similar to (17.16)]

$$\frac{1}{[A]} - \frac{1}{[A]_0} = bkt \tag{17.23}$$

Third-Order Reactions

The most common third-order rate laws are $r = k[A]^3$, $r = k[A]^2[B]$, and $r = k[A][B][C]$. The tedious details of the integrations are left as exercises for the reader (Probs. 17.17 and 17.24).

The rate law $d[A]/dt = -k_A[A]^3$ integrates to

$$\frac{1}{[A]^2} - \frac{1}{[A]_0^2} = 2k_A t \qquad \text{or} \qquad [A] = \frac{[A]_0}{(1 + 2k_A t[A]_0^2)^{1/2}} \tag{17.24}$$

The rate laws $a^{-1}\, d[A]/dt = -k[A]^2[B]$ and $a^{-1}\, d[A]/dt = -k[A][B][C]$ yield complicated expressions that are given in Prob. 17.24.

*n*th-Order Reaction

Of the many nth-order rate laws, we consider only

$$d[A]/dt = -k_A[A]^n \tag{17.25}$$

Integration gives

$$\int_1^2 [A]^{-n}\, d[A] = -k_A \int_1^2 dt \tag{17.26}$$

$$\frac{[A]^{-n+1} - [A]_0^{-n+1}}{-n + 1} = -k_A t \qquad \text{for } n \neq 1 \tag{17.27}$$

Multiplication of both sides by $(1 - n)[A]_0^{n-1}$ gives

$$\left(\frac{[A]}{[A]_0} \right)^{1-n} = 1 + [A]_0^{n-1}(n - 1)k_A t \qquad \text{for } n \neq 1 \tag{17.28}$$

Setting $[A] = \frac{1}{2}[A]_0$ and $t = t_{1/2}$, we get as the half-life

$$t_{1/2} = \frac{2^{n-1} - 1}{(n - 1)[A]_0^{n-1}k_A} \qquad \text{for } n \neq 1 \tag{17.29}$$

Note that (17.28) and (17.29) apply to all values of n except 1. In particular, these equations hold for $n = 0$, $n = \frac{1}{2}$, and $n = \frac{3}{2}$. For $n = 1$, integration of (17.25) gives a logarithm. Equations (17.14) and (17.15) give for this case

$$[A] = [A]_0 e^{-k_A t}, \qquad t_{1/2} = 0.693/k_A \qquad \text{for } n = 1 \tag{17.30}$$

Reversible First-Order Reactions

So far, we have neglected the **reverse** (or **back**) reaction, an assumption that is strictly valid only if the equilibrium constant is infinite but that holds well during the early stages of a reaction. We now allow for the reverse reaction.

Let the reversible reaction $A \rightleftharpoons C$ (with stoichiometric coefficients of 1) be first order in both the forward (f) and back (b) directions, so that $r_f = k_f[A]$ and $r_b = k_b[C]$. The stoichiometric coefficient of A is -1 for the forward reaction and 1 for the reverse reaction. If $(d[A]/dt)_f$ denotes the rate of change of [A] due to the forward reaction, then $-(d[A]/dt)_f = r_f = k_f[A]$. The rate of change of [A] due to the reverse reaction is $(d[A]/dt)_b = r_b = k_b[C]$ (assuming negligible concentrations of any intermediates). Then

$$d[A]/dt = (d[A]/dt)_f + (d[A]/dt)_b = -k_f[A] + k_b[C] \qquad (17.31)$$

We have $\Delta[C] = -\Delta[A]$, so $[C] - [C]_0 = -([A] - [A]_0)$. Substitution of $[C] = [C]_0 + [A]_0 - [A]$ into (17.31) gives

$$d[A]/dt = k_b[C]_0 + k_b[A]_0 - (k_f + k_b)[A] \qquad (17.32)$$

Before integrating this equation, we simplify its appearance. In the limit as $t \to \infty$, the system reaches equilibrium, the rates of the forward and reverse reactions having become equal. At equilibrium the concentration of each species is constant, and $d[A]/dt$ is 0. Let $[A]_{eq}$ be the equilibrium concentration of A. Setting $d[A]/dt = 0$ and $[A] = [A]_{eq}$ in (17.32), we get

$$k_b[C]_0 + k_b[A]_0 = (k_f + k_b)[A]_{eq} \qquad (17.33)$$

The use of (17.33) in (17.32) gives $d[A]/dt = (k_f + k_b)([A]_{eq} - [A])$. Using the identity $\int (x + s)^{-1} dx = \ln (x + s)$ to integrate this equation, we get

$$\ln \frac{[A] - [A]_{eq}}{[A]_0 - [A]_{eq}} = -(k_f + k_b)t$$

$$[A] - [A]_{eq} = ([A]_0 - [A]_{eq})e^{-jt} \qquad \text{where } j \equiv k_f + k_b \qquad (17.34)$$

where $[A]_{eq}$ can be found from (17.33). Note the close resemblance to the first-order rate law (17.14). Equation (17.14) is a special case of (17.34) with $[A]_{eq} = 0$ and $k_b = 0$. A plot of [A] versus t resembles Fig. 17.1a, except that [A] approaches $[A]_{eq}$ rather than 0 as $t \to \infty$ (Fig. 17.2).

Discussion of opposing reactions with orders greater than 1 is omitted.

Consecutive First-Order Reactions

Frequently a product of one reaction becomes a reactant in a subsequent reaction. This is true in multistep reaction mechanisms. We shall consider only the simple case of two consecutive irreversible first-order reactions: $A \to B$ with rate constant k_1, and $B \to C$ with rate constant k_2:

$$A \xrightarrow{k_1} B \xrightarrow{k_2} C \qquad (17.35)$$

where for simplicity we have assumed stoichiometric coefficients of unity. Since the reactions were assumed to be first-order, the rates of the first and second reactions are $r_1 = k_1[A]$ and $r_2 = k_2[B]$. The rates of change of [B] due to the first reaction and to the second reaction are $(d[B]/dt)_1 = k_1[A]$ and $(d[B]/dt)_2 = -k_2[B]$, respectively. Thus

$$d[B]/dt = (d[B]/dt)_1 + (d[B]/dt)_2 = k_1[A] - k_2[B]$$

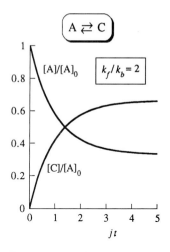

Figure 17.2

Concentrations versus time for the reversible first-order reaction $A \rightleftharpoons C$ with forward and reverse rate constants k_f and k_b plotted for the case $k_f/k_b = 2$. As $t \to \infty$, $[C]/[A] \to 2$, which is the equilibrium constant for the reaction.

We have

$$d[A]/dt = -k_1[A], \qquad d[B]/dt = k_1[A] - k_2[B], \qquad d[C]/dt = k_2[B] \qquad (17.36)$$

Let only A be present in the system at $t = 0$:

$$[A]_0 \neq 0, \qquad [B]_0 = 0, \qquad [C]_0 = 0 \qquad (17.37)$$

We have three coupled differential equations. The first equation in (17.36) is the same as (17.11), and use of (17.14) gives

$$[A] = [A]_0 e^{-k_1 t} \qquad (17.38)$$

Substitution of (17.38) into the second equation of (17.36) gives

$$d[B]/dt = k_1[A]_0 e^{-k_1 t} - k_2[B] \qquad (17.39)$$

The integration of (17.39) is outlined in Prob. 17.19. The result is

$$[B] = \frac{k_1[A]_0}{k_2 - k_1}\left(e^{-k_1 t} - e^{-k_2 t}\right) \qquad (17.40)$$

To find [C], we use conservation of matter. The total number of moles present is constant with time, so $[A] + [B] + [C] = [A]_0$. The use of (17.38) and (17.40) gives

$$[C] = [A]_0\left(1 - \frac{k_2}{k_2 - k_1}e^{-k_1 t} + \frac{k_1}{k_2 - k_1}e^{-k_2 t}\right) \qquad (17.41)$$

Figure 17.3 plots [A], [B], and [C] for two values of k_2/k_1. Note the maximum in the intermediate species [B].

Competing First-Order Reactions

Frequently a species can react in different ways to give a variety of products. For example, toluene can be nitrated at the ortho, meta, or para positions. We shall consider the simplest case, that of two competing irreversible first-order reactions:

$$A \xrightarrow{k_1} C \quad \text{and} \quad A \xrightarrow{k_2} D \qquad (17.42)$$

where the stoichiometric coefficients are taken as 1 for simplicity. The rate law is

$$d[A]/dt = -k_1[A] - k_2[A] = -(k_1 + k_2)[A] \qquad (17.43)$$

This equation is the same as (17.11) with k_A replaced by $k_1 + k_2$. Hence (17.14) gives $[A] = [A]_0 e^{-(k_1 + k_2)t}$.

For C, we have $d[C]/dt = k_1[A] = k_1[A]_0 e^{-(k_1 + k_2)t}$. Multiplication by dt and integration from time 0 (where $[C]_0 = 0$) to an arbitrary time t gives

$$[C] = \frac{k_1[A]_0}{k_1 + k_2}\left(1 - e^{-(k_1 + k_2)t}\right) \qquad (17.44)$$

Figure 17.3

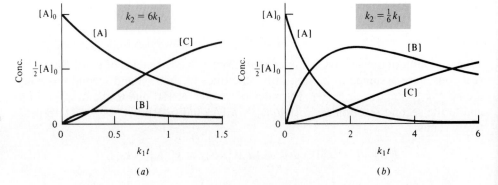

Concentrations versus time for the consecutive first-order reactions $A \to B \to C$ with rate constants k_1 and k_2. (a) $k_2 = 6k_1$; (b) $k_2 = \frac{1}{6}k_1$.

Similarly, integration of $d[D]/dt = k_2[A]$ gives

$$[D] = \frac{k_2[A]_0}{k_1 + k_2}\left(1 - e^{-(k_1+k_2)t}\right) \tag{17.45}$$

The sum of the rate constants $k_1 + k_2$ appears in the exponentials for both [C] and [D]. Figure 17.4 plots [A], [C], and [D] versus t for $k_1 = 2k_2$.

Division of (17.44) by (17.45) gives at any time during the reaction

$$[C]/[D] = k_1/k_2 \tag{17.46}$$

The amounts of C and D obtained depend on the relative rates of the two competing reactions. Measurement of [C]/[D] allows k_1/k_2 to be found.

In this example we assumed the competing reactions to be irreversible. In general, this will not be true, and we must also consider the reverse reactions

$$C \xrightarrow{k_{-1}} A \quad \text{and} \quad D \xrightarrow{k_{-2}} A \tag{17.47}$$

Moreover, the product C may well react to give the product D, and vice versa: $C \rightleftharpoons D$. If we wait an infinite amount of time, the system will reach equilibrium and the ratio [C]/[D] will be determined by the ratio K_1/K_2 of concentration-scale equilibrium constants for the reactions in (17.42): $K_1/K_2 = ([C]/[A]) \div ([D]/[A]) = [C]/[D]$ at $t = \infty$, where an ideal system was assumed. This situation is called **thermodynamic control** of products. Here, the product with the most negative $G°$ is favored. On the other hand, during the early stages of the reaction when any reverse reaction or interconversion of C and D can be neglected, Eq. (17.46) will apply and we have **kinetic control** of products. If the rate constants k_{-1} and k_{-2} for the reverse reactions (17.47) and the rate constants for interconversion of the products C and D are all much, much less than k_1 and k_2 for the forward reactions (17.42), the products will be kinetically controlled even when A has been nearly all consumed. It often happens that $k_1/k_2 \gg 1$ and $K_1/K_2 \ll 1$, so C is favored kinetically and D is favored thermodynamically. The relative yield of products then depends on whether there is kinetic or thermodynamic control.

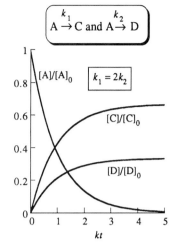

Figure 17.4

Concentrations versus time for the competing first-order reactions A → C and A → D with rate constants k_1 and k_1 plotted for the case $k_1/k_2 = 2$. For all times during the reaction, [C]/[D] = $k_1/k_2 = 2$. The constant k is $k \equiv k_1 + k_2$.

Numerical Integration of Rate Equations

The rate law gives $d[A]/dt$ and when integrated enables us to find [A] at some arbitrary time t from its known value $[A]_0$ at time zero. The integration can be done numerically rather than analytically. See Sec. 17.7.

17.4 DETERMINATION OF THE RATE LAW

Experimental data give species concentrations at various times during the reaction. This section discusses how the rate law $r = f([A], [B], \dots)$ (where f is some function) is found from experimental concentration-versus-time data. The discussion is restricted to cases where the rate law has the form

$$r = k[A]^\alpha[B]^\beta \cdots [L]^\lambda \tag{17.48}$$

as in Eq. (17.5). It is usually best to find the orders $\alpha, \beta, \dots, \lambda$ first and then find the rate constant k. Four methods for finding the orders follow.

1. **Half-life method.** This method applies when the rate law has the form $r = k[A]^n$. Then Eqs. (17.29) and (17.30) apply. If $n = 1$, then $t_{1/2}$ is independent of $[A]_0$. If $n \neq 1$, then (17.29) gives

$$\log_{10} t_{1/2} = \log_{10}\frac{2^{n-1} - 1}{(n-1)k_A} - (n-1)\log_{10}[A]_0 \tag{17.49}$$

A plot of $\log_{10} t_{1/2}$ versus $\log_{10} [A]_0$ gives a straight line of slope $1 - n$. This statement is also valid for $n = 1$. To use the method, one plots [A] versus t for a run. One picks any [A] value, say [A]′, and finds the point where [A] has fallen to $\frac{1}{2}[A]′$. The time interval between these two points is $t_{1/2}$ for the initial concentration [A]′. One then picks another point [A]″ and determines $t_{1/2}$ for this A concentration. After repeating this process several times, one plots $\log_{10} t_{1/2}$ versus the log of the corresponding initial A concentration and measures the slope.

The half-life method has the disadvantage that, if data from a single run are used, the reaction must be followed to a high percentage of completion. An improvement is the use of the **fractional life** t_α, defined as the time required for $[A]_0$ to fall to $\alpha[A]_0$. (For the half-life, $\alpha = \frac{1}{2}$.) If $r = k[A]^n$, the result of Prob. 17.26 shows that a plot of $\log_{10} t_\alpha$ versus $\log_{10} [A]_0$ is a straight line with slope $1 - n$. A convenient value of α is 0.75.

EXAMPLE 17.2 Half-life method

Data for the dimerization $2A \rightarrow A_2$ of a certain nitrile oxide (compound A) in ethanol solution at 40°C follow:

[A]/(mmol/dm³)	68.0	50.2	40.3	33.1	28.4	22.3	18.7	14.5
t/min	0	40	80	120	160	240	300	420

Find the reaction order using the half-life method.

Figure 17.5a plots [A] versus t. We pick the initial [A] values 68, 60, 50, 40, and 30 mmol/dm³, read off the times corresponding to these points, and read off the times corresponding to half of each of these [A] values. The results are

[A]/(mmol/dm³)	$68 \rightarrow 34$	$60 \rightarrow 30$	$50 \rightarrow 25$	$40 \rightarrow 20$	$30 \rightarrow 15$
t/min	$0 \rightarrow 114$	$14 \rightarrow 146$	$42 \rightarrow 205$	$82 \rightarrow 280$	$146 \rightarrow 412$

(The fact that the half-life is not constant tells us that the reaction is not first order.) The $[A]_0$ and corresponding $t_{1/2}$ values and their logs are:

$[A]_0$/(mmol/dm³)	68	60	50	40	30
$t_{1/2}$/min	114	132	163	198	266
$\log_{10}(t_{1/2}/\text{min})$	2.057	2.121	2.212	2.297	2.425
$\log_{10}\{[A]_0/(\text{mmol/dm}^3)\}$	1.833	1.778	1.699	1.602	1.477

Figure 17.5b plots $\log_{10} t_{1/2}$ versus $\log_{10} [A]_0$. The slope of this plot is $(2.415 - 2.150)/(1.500 - 1.750) = -1.06 = 1 - n$, and $n = 2.06$. The reaction is second-order.

Exercise

For the reaction $3ArSO_2H \rightarrow ArSO_2SAr + ArSO_3H + H_2O$ in acetic acid solution at 70°C (where Ar stands for the p-tolyl group), the following data were obtained:

[ArSO₂H]/(mmol/L)	100	84.3	72.2	64.0	56.8	38.7	29.7	19.6
t/min	0	15	30	45	60	120	180	300

Use the half-life method to find the reaction order. (*Answer:* $n = 2$.)

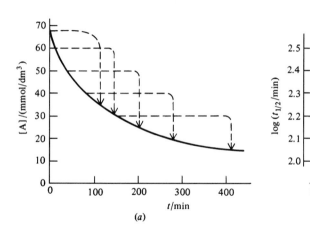

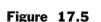

Figure 17.5

Determination of reaction order by the half-life method.

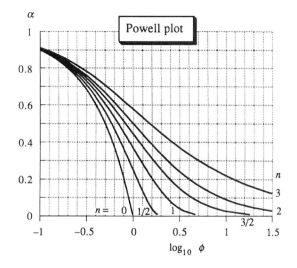

2. **Powell-plot method.** This method (*Moore and Pearson,* pp. 20–21) applies when the rate law has the form $r = k[A]^n$. Let the dimensionless parameters α and ϕ be defined as

$$\alpha \equiv [A]/[A]_0, \qquad \phi \equiv k_A[A]_0^{n-1}t \qquad (17.50)$$

α is the fraction of A unreacted. Equations (17.28) and (17.13) become

$$\alpha^{1-n} - 1 = (n - 1)\phi \qquad \text{for } n \neq 1$$

$$\ln \alpha = -\phi \qquad \text{for } n = 1 \qquad (17.51)$$

For a given n, there is a fixed relation between α and ϕ for every reaction of order n. These equations are used to plot α versus $\log_{10} \phi$ for commonly occurring values of n to give a series of master curves (Fig. 17.6). From the data for a run in a kinetics experiment, one plots α versus $\log_{10} t$ on translucent graph paper *using the same scales as in the master plots.* (Ordinary graph paper can be made translucent by applying a little oil. Alternatively, one can make a transparency of the master curves using a copying machine that gives copies with exactly the same size as the original.) Since $\log_{10} \phi$ differs from $\log_{10} t$ by $\log_{10} (k_A[A]_0^{n-1})$, which is a constant for a given run, the experimental curve will be shifted along the horizontal axis by a constant from the applicable master curve. One slides the experimental curve back and forth (while keeping the horizontal $\log_{10} \phi$ and $\log_{10} t$ axes of the master and experimental plots superimposed) until it coincides with one of the master curves. This gives n. The Powell-plot method requires the initial

Figure 17.6

Powell-plot master curves.

investment of time needed to make the master plots, but once these are prepared, the method is quick, easy, and fun to use. Table 17.1 gives the data needed to make the master plots.

It might seem that methods 1 and 2 are of little value, since they apply only when $r = k[A]^n$. However, by taking the initial concentrations of reactants in stoichiometric proportions, the rate law (17.48) is reduced to this special form, provided the products do not appear in the rate law. Thus, if the reaction is $aA + bB \rightarrow$ products, and if $[A]_0 = as$, $[B]_0 = bs$, where s is a constant, then the relation $\Delta n_i = \nu_i \xi$ [Eq. (4.95)] gives $[A] = [A]_0 - a\xi/V = a(s - \xi/V)$ and $[B] = b(s - \xi/V)$. The rate law (17.48) becomes $r = ka^\alpha b^\beta (s - \xi/V)^{\alpha+\beta} = \text{const.} \times [A]^{\alpha+\beta}$. Methods 1 and 2 will give the overall order when the reactants are mixed in stoichiometric proportions.

3. **Initial-rate method.** Here, one measures the initial rate r_0 for several runs, varying the initial concentration of one reactant at a time. Suppose we measure r_0 for the two different initial A concentrations $[A]_{0,1}$ and $[A]_{0,2}$ while keeping $[B]_0$, $[C]_0$, ... fixed. With only $[A]_0$ changed and with the rate law assumed to have the form $r = k[A]^\alpha [B]^\beta \cdots [L]^\lambda$, the ratio of initial rates for runs 1 and 2 is $r_{0,2}/r_{0,1} = ([A]_{0,2}/[A]_{0,1})^\alpha$, from which α is readily found. For example, if tripling $[A]_0$ is found to multiply the initial rate by 9, then $9 = 3^\alpha$, and $\alpha = 2$. A more reliable result can be found by making several runs in which only $[A]$ is varied over a wide range. Since $\log r_0 = \log k + \alpha \log [A]_0 + \beta \log [B]_0 + \cdots$, a plot of $\log r_0$ versus $\log [A]_0$ at constant $[B]_0$, ... has slope α. [See J. P. Birk, *J. Chem. Educ.*, **53**, 704 (1976) for further discussion.] The orders $\beta, \gamma, \ldots$ are found similarly.

The initial rate $r_0 \equiv -a^{-1} d[A]/dt|_{t=0}$ [Eq. (17.4)] can be found by plotting $[A]$ versus t, drawing the tangent line at $t = 0$, and finding its slope or by numerical differentiation of $[A]$ versus t data (*Chapra and Canale*, Chap. 23). Finding accurate initial rates is not easy.

4. **Isolation method.** Here, one makes the initial concentration of reactant A much less than the concentrations of all other species: $[B]_0 \gg [A]_0$, $[C]_0 \gg [A]_0$, etc. Thus, we can make $[A]_0 = 10^{-3}$ mol/dm^3 and all other concentrations at least 0.1 mol/dm^3. Then the concentrations of all reactants except A will be essentially constant with time. The rate law (17.48) becomes

$$r = k[A]^\alpha [B]_0^\beta \cdots [L]_0^\lambda = j[A]^\alpha \qquad \text{where } j \equiv k[B]_0^\beta \cdots [L]_0^\lambda \qquad (17.52)$$

where j is essentially constant. The reaction has the pseudo order α under these conditions. One then analyzes the data from the run using method 1 or 2 to find α. To find β, we can make $[B]_0 \ll [A]_0$, $[B]_0 \ll [C]_0$, ... and proceed as we did in finding α. Alternatively, we can keep $[A]_0$ fixed at a value much less than all other concentrations but change $[B]_0$ to a new value $[B]_0'$. With only $[B]_0$ changed,

TABLE 17.1

Values of $\log_{10} \phi$ for Powell-Plot Master Curves

n	α	0.9	0.8	0.7	0.6	0.5	0.4	0.3	0.2	0.1
0		−1.000	−0.699	−0.523	−0.398	−0.301	−0.222	−0.155	−0.097	−0.046
$\frac{1}{2}$		−0.989	−0.675	−0.486	−0.346	−0.232	−0.134	−0.044	0.044	0.136
1		−0.977	−0.651	−0.448	−0.292	−0.159	−0.038	0.081	0.207	0.362
$\frac{3}{2}$		−0.966	−0.627	−0.408	−0.235	−0.082	0.065	0.218	0.393	0.636
2		−0.954	−0.602	−0.368	−0.176	0.000	0.176	0.368	0.602	0.954
3		−0.931	−0.551	−0.284	−0.051	0.176	0.419	0.704	1.079	

we evaluate the apparent rate constant j' that corresponds to $[B]_0'$; Eq. (17.52) gives $j/j' = ([B]_0/[B]_0')^\beta$, so β can be found from j and j'.

Many books suggest that the overall order be obtained by trial and error, as follows. If the rate law is $r = k[A]^n$, one plots $\ln [A]$ versus t, $[A]^{-1}$ versus t, and $[A]^{-2}$ versus t and obtains a straight line for one of these plots according to whether $n = 1$, 2, or 3, respectively [see Eqs. (17.13), (17.16), and (17.24)]. This method is dangerous to use, since it is often hard to decide which of these plots is most nearly linear. (See Prob. 17.29.) Thus, the wrong order can be obtained, especially if the reaction has not been followed very far; see Fig. 17.7. Moreover, if the order is $\frac{3}{2}$, it is not hard to conclude erroneously that $n = 1$ or $n = 2$.

Once the order with respect to each species has been obtained by one of the above methods, the rate constant k is found from the slope of an appropriate graph. For example, if the rate law has been found to be $r = k[A]$, Eq. (17.13) shows that a plot of $\ln [A]$ versus t gives a straight line of slope $-k_A$. If $r = k[A][B]$, Eq. (17.22) shows that a plot of $\ln ([B]/[A])$ versus t gives a straight line of slope $k(a[B]_0 - b[A]_0)$. The best straight-line fit can be found by a least-squares treatment (Prob. 6.57). A proper least-squares treatment when the variables have been changed to yield a linear equation requires the use of statistical weights (which chemists often fail to do); see sec. 3.6 of R. J. Cvetanovic et al., *J. Phys. Chem.*, **83,** 50 (1979), which is an excellent reference on treatment of kinetic data.

Rather than using the slope of an unweighted linear graph, a more accurate value of k can be found by using the Excel spreadsheet Solver to vary k so as to minimize the deviations between calculated and experimental values of $[A]$.

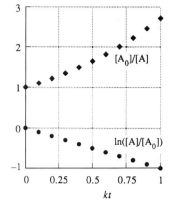

Figure 17.7

Plots of $\ln [A]$ and $1/[A]$ versus time for a first-order reaction with rate constant k. Note that the $1/[A]$-versus-t plot is nearly linear for small t, which can lead one to erroneously conclude that the reaction is second-order. From (17.14) we have $[A]_0/[A] = e^{kt} \approx 1 + kt$ for small t [Eq. (8.37)].

EXAMPLE 17.3 Finding k

Find the rate constant for the dimerization reaction $2A \rightarrow A_2$ of Example 17.2.

This reaction was found to be second-order. From (17.16), a plot of $1/[A]$ versus t will be linear with slope k_A, where $k_A \equiv ak = 2k$. Figure 17.8 plots $1/[A]$ versus t (where $[A]$ was converted to mol/L). The least-squares-fit slope is 0.12_9 L mol^{-1} min^{-1} = 0.0021_5 L mol^{-1} s^{-1} = $2k$ and $k = 0.0010_7$ L mol^{-1} s^{-1}.

To find k more accurately, we set up a spreadsheet with the experimental $[A]$ values in column A and the t values in column B. We designate a cell for k_A and enter an initial guess for it. (A good initial guess would be the value found from the slope of the straight-line graph, but the initial guess of 0 also works in this case.) The formula $[A]_0/(1 + k_A t[A]_0)$ [Eq. (17.17)] for $[A]$ is entered in column C. The squares of the deviations between corresponding column A and C values are calculated in column D. The Solver is used to minimize the sum of the column D values by varying k_A (subject to the constraint $k_A \geq 0$). The result is 0.128_1 L mol^{-1} min^{-1} (Prob. 17.36b) with a sum of squares of deviations equal to 0.238 mol^2/L^2 compared with 0.295 mol^2/L^2 for the 0.128_9 linear-plot value.

Exercise

Use the $ArSO_2H$ reaction data in the exercise in Example 17.2 to find k for this reaction. (*Answer:* 0.00076 L/mol-s from a straight-line graph; 0.00072 L/mol-s from minimization of the squares of concentration deviations.)

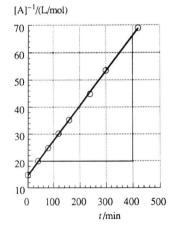

Figure 17.8

Plot of $1/[A]$ versus t used to find a second-order rate constant. The slope is $(66 - 20)$ (L/mol) $\div$ $(400 - 43)$ min = 0.12_9 L mol^{-1} min^{-1}.

The procedure of using the integrated rate law to calculate values of k from pairs of successive observations (or from the initial conditions and each observation) and then averaging these k's yields quite inaccurate results and is best avoided; see *Moore and Pearson*, p. 69; *Bamford and Tipper*, vol. 1, pp. 362–365.

RATE LAWS AND EQUILIBRIUM CONSTANTS FOR ELEMENTARY REACTIONS

The examples in Eq. (17.6) show that the orders in the rate law of an overall reaction often differ from the stoichiometric coefficients. An overall reaction occurs as a series of elementary steps, these steps constituting the **mechanism** of the reaction. We now consider the rate law for an elementary reaction. Section 17.6 shows how the rate law for an overall reaction follows from its mechanism.

The number of molecules that react in an elementary step is the **molecularity** of the elementary reaction. Molecularity is defined only for elementary reactions and should not be used to describe overall reactions that consist of more than one elementary step. The elementary reaction A → products is **unimolecular.** The elementary reactions A + B → products and 2A → products are **bimolecular.** The elementary reactions A + B + C → products, 2A + B → products, and 3A → products are **trimolecular** (or **termolecular**). No elementary reactions involving more than three molecules are known, because of the very low probability of near-simultaneous collision of more than three molecules. Most elementary reactions are unimolecular or bimolecular. Trimolecular reactions are rare because of the low probability of three-body collisions.

Consider a bimolecular elementary reaction A + B → products, where A and B may be the same or different molecules. Although not every collision between A and B will produce products, the rate of reaction $r = J/V$ in (17.3) will be proportional to Z_{AB}, the rate of A-B collisions per unit volume. Equations (15.62) and (15.63) show that in an ideal gas, Z_{AB} is proportional to $(n_A/V)(n_B/V)$, where $n_A/V \equiv [A]$ is the molar concentration of A. Therefore r for an elementary bimolecular ideal-gas reaction will be proportional to $[A][B]$; $r = k[A][B]$, where k is the proportionality constant. Similarly, for a trimolecular elementary reaction in an ideal gas, the reaction rate will be proportional to the rate of three-body collisions per unit volume and therefore will be proportional to $[A][B][C]$; see Sec. 23.1.

For the unimolecular ideal-gas reaction B → products, there is a fixed probability that any particular B molecule will decompose or isomerize to products in unit time. Therefore, the number of molecules reacting in unit time is proportional to the number N_B present, and the reaction rate $r = J/V$ is proportional to N_B/V and hence to $[B]$; we have $r = k[B]$. A fuller treatment of unimolecular reactions is given in Sec. 17.11.

Similar considerations apply to reactions in an ideal or ideally dilute solution.

In summary, in an ideal system, the rate law for the elementary reaction aA + bB → products is $r = k[A]^a[B]^b$, where $a + b$ is 1, 2, or 3. *For an elementary reaction, the orders in the rate law equal the coefficients of the reactants.* Don't overlook the word *elementary* in this sentence.

The rate law for elementary reactions in nonideal systems is discussed in Sec. 17.10. Kinetics data are usually not accurate enough to cause worry about deviations from ideality, except for ionic reactions.

We now examine the relation between the equilibrium constant for a reversible elementary reaction and the rate constants for the forward and reverse reactions. Consider the reversible *elementary* reaction

$$a\text{A} + b\text{B} \underset{k_b}{\overset{k_f}{\rightleftharpoons}} c\text{C} + d\text{D}$$

in an ideal system. The rate laws for the forward (f) and back (b) elementary reactions are $r_f = k_f[A]^a[B]^b$ and $r_b = k_b[C]^c[D]^d$. At equilibrium these opposing rates are equal: $r_{f,\text{eq}} = r_{b,\text{eq}}$, or

$$k_f([A]_{\text{eq}})^a([B]_{\text{eq}})^b = k_b([C]_{\text{eq}})^c([D]_{\text{eq}})^d \quad \text{and} \quad \frac{k_f}{k_b} = \frac{([C]_{\text{eq}})^c([D]_{\text{eq}})^d}{([A]_{\text{eq}})^a([B]_{\text{eq}})^b}$$

But the quantity on the right side of this last equation is the concentration-scale equilibrium constant K_c for the reaction. Therefore

$$K_c = k_f/k_b \qquad \text{elementary reaction} \qquad (17.53)^*$$

If $k_f \gg k_b$, then $K_c \gg 1$ and the equilibrium position favors products. The relation between k_f, k_b, and K_c for a composite reaction is discussed in Sec. 17.9.

> For a reaction in solution where the solvent is a reactant or product, the equilibrium constant K_c in (17.53) differs from the K_c used in thermodynamics. In thermodynamics, the mole-fraction scale (rather than the concentration scale) is used for the solvent; for reactions in an ideally dilute solution, the solvent is usually omitted from the thermodynamic K_c, since its mole fraction is approximately 1. Therefore, if ν_A is the stoichiometric coefficient of the solvent A in the elementary reaction, then $K_{c,\text{kin}} = K_{c,\text{td}}[A]^{\nu_A}$, where $K_{c,\text{td}}$ is the thermodynamic K_c and $K_{c,\text{kin}}$ is the kinetic K_c, as in Eq. (17.53). If the reaction orders with respect to solvent are unknown, the solvent is omitted from the forward and reverse rate laws, and $K_{c,\text{kin}} = K_{c,\text{td}}$.

<h2>17.6   REACTION MECHANISMS</h2>

The observed rate law provides information on the mechanism of a reaction, in that any proposed mechanism must yield the observed rate law. Usually an exact deduction of the rate law from the differential rate equations of a multistep mechanism is not possible, because of mathematical difficulties in dealing with a system of several interrelated differential equations. Therefore, one of two approximation methods is generally used, the rate-determining-step approximation or the steady-state approximation.

The Rate-Determining-Step Approximation

In the **rate-determining-step approximation** (also called the **rate-limiting-step approximation** or the **equilibrium approximation**), the reaction mechanism is assumed to consist of one or more reversible reactions that stay close to equilibrium during most of the reaction, followed by a relatively slow rate-determining step, which in turn is followed by one or more rapid reactions. In special cases, there may be no equilibrium steps before the rate-determining step or no rapid reactions after the rate-determining step.

As an example, consider the following mechanism composed of unimolecular (elementary) reactions

$$A \underset{k_{-1}}{\overset{k_1}{\rightleftharpoons}} B \underset{k_{-2}}{\overset{k_2}{\rightleftharpoons}} C \underset{k_{-3}}{\overset{k_3}{\rightleftharpoons}} D \qquad (17.54)$$

where step 2 ($B \rightleftharpoons C$) is assumed to be the rate-determining step. For this assumption to be valid, we must have $k_{-1} \gg k_2$. The slow rate of $B \rightarrow C$ compared with $B \rightarrow A$ ensures that most B molecules go back to A rather than going to C, thereby ensuring that step 1 ($A \rightleftharpoons B$) remains close to equilibrium. Furthermore, we must have $k_3 \gg k_2$ and $k_3 \gg k_{-2}$ to ensure that step 2 acts as a "bottleneck" and that product D is rapidly formed from C. The overall rate is then controlled by the rate-determining step $B \rightarrow C$. (Note that since $k_3 \gg k_{-2}$, the rate-limiting step is not in equilibrium.) Since we are examining the rate of the forward reaction $A \rightarrow D$, we further assume that $k_2[B] \gg k_{-2}[C]$. During the early stages of the reaction, the concentration of C will be low compared with B, and this condition will hold. Thus we neglect the reverse reaction for step 2. Since the rate-controlling step is taken to be essentially irreversible, it is irrelevant whether the rapid steps after the rate-limiting step are reversible or not. The observed rate law will depend only on the nature of the equilibria that precede the rate-determining step and on this step itself. See Example 17.4.

The relative magnitude of k_1 compared with k_2 is irrelevant to the validity of the rate-limiting-step approximation. Hence, the rate constant k_2 of the rate-determining step might be larger than k_1. However, the rate $r_2 = k_2[B]$ of the rate-determining step must be much smaller than the rate $r_1 = k_1[A]$ of the first step. This follows from $k_2 \ll k_{-1}$ and $k_1/k_{-1} \approx [B]/[A]$ (the conditions for step 1 to be nearly in equilibrium).

For the reverse overall reaction, the rate-determining step is the reverse of that for the forward reaction. For example, for the reverse of (17.54), the rate-determining step is C → B. This follows from the above inequalities $k_{-2} \ll k_3$ (which ensures that step D ⇌ C is in equilibrium) and $k_{-1} \gg k_2$ (which ensures that B → A is rapid).

EXAMPLE 17.4 The rate-determining-step approximation

The rate law for the Br^--catalyzed aqueous reaction

$$H^+ + HNO_2 + C_6H_5NH_2 \xrightarrow{Br^-} C_6H_5N_2^+ + 2H_2O$$

is observed to be

$$r = k[H^+][HNO_2][Br^-] \tag{17.55}$$

A proposed mechanism is

$$H^+ + HNO_2 \underset{k_{-1}}{\overset{k_1}{\rightleftharpoons}} H_2NO_2^+ \qquad \text{rapid equilib.}$$

$$H_2NO_2^+ + Br^- \xrightarrow{k_2} ONBr + H_2O \qquad \text{slow} \tag{17.56}$$

$$ONBr + C_6H_5NH_2 \xrightarrow{k_3} C_6H_5N_2^+ + H_2O + Br^- \qquad \text{fast}$$

Deduce the rate law for this mechanism and relate the observed rate constant k in (17.55) to the rate constants in the assumed mechanism (17.56).

The second step in (17.56) is rate-limiting. Since step 3 is much faster than step 2, we can take $d[C_6H_5N_2^+]/dt$ as equal to the rate of formation of ONBr in step 2. Therefore the reaction rate is

$$r = k_2[H_2NO_2^+][Br^-] \tag{17.57}$$

(Since step 2 is an *elementary* reaction, its rate law is determined by its stoichiometry, as noted in Sec. 17.5.) The species $H_2NO_2^+$ in (17.57) is a reaction intermediate, and we want to express r in terms of reactants and products. Since step 1 is in near equilibrium, Eq. (17.53) gives

$$K_{c,1} = \frac{k_1}{k_{-1}} = \frac{[H_2NO_2^+]}{[H^+][HNO_2]} \quad \text{and} \quad [H_2NO_2^+] = (k_1/k_{-1})[H^+][HNO_2]$$

Substitution in (17.57) gives

$$r = (k_1k_2/k_{-1})[H^+][HNO_2][Br^-]$$

in agreement with (17.55). We have $k = k_1k_2/k_{-1} = K_{c,1}k_2$. The observed rate constant contains the equilibrium constant for step 1 and the rate constant for the rate-determining step 2. The rate law does not contain the reactant $C_6H_5NH_2$, which occurs in the rapid step after the rate-determining step.

Note from Eq. (17.57) that the rate of the overall reaction equals the rate of the rate-determining step. This is true in general, provided the stoichiometric number of the rate-determining step equals 1, so that the overall reaction occurs once for each occurrence of the rate-determining step.

Exercise

For the reaction $H_2O_2 + 2H^+ + 2I^- \rightarrow I_2 + 2H_2O$ in acidic aqueous solution, the rate law (7) in (17.6) indicates the reaction proceeds by two simultaneous mechanisms. Suppose one mechanism is

$$H^+ + I^- \rightleftharpoons HI \qquad \text{rapid equilib.}$$

$$HI + H_2O_2 \rightarrow H_2O + HOI \qquad \text{slow}$$

$$HOI + I^- \rightarrow I_2 + OH^- \qquad \text{fast}$$

$$OH^- + H^+ \rightarrow H_2O \qquad \text{fast}$$

Verify that this mechanism adds to the correct overall reaction. Find the rate law predicted by this mechanism. (*Answer*: $r = k[H_2O_2][H^+][I^-]$.)

The Steady-State Approximation

Multistep reaction mechanisms usually involve one or more intermediate species that do not appear in the overall equation. For example, the postulated species $H_2NO_2^+$ in the mechanism (17.56) is such a reaction intermediate. Frequently, these intermediates are very reactive and therefore do not accumulate to any significant extent during the reaction; that is, $[I] \ll [R]$ and $[I] \ll [P]$ during most of the reaction, where I is an intermediate and R and P are reactants and products. Oscillations in the concentration of a species during a reaction are rare, so we can assume that $[I]$ will start at 0, rise to a maximum, $[I]_{max}$, and then fall back to 0. If $[I]$ remains small during the reaction, $[I]_{max}$ will be small compared with $[R]_{max}$ and $[P]_{max}$ and the curves of $[R]$, $[I]$, and $[P]$ versus t will resemble those of Fig. 17.3a, where the reactant R is A, the intermediate I is B, and the product P is C. Note that, except for the initial period of time (called the *induction period*) when B is rising rapidly, the slope of the B curve is much less than the slopes of the A and C curves. In the R, I, P notation, we have $d[I]/dt \ll d[R]/dt$ and $d[I]/dt \ll d[P]/dt$.

It is therefore frequently a good approximation to take $d[I]/dt = 0$ for each reactive intermediate. This is the **steady-state** (or **stationary-state**) **approximation.** The steady-state approximation assumes that (after the induction period) the rate of formation of a reaction intermediate essentially equals its rate of destruction, so as to keep it at a near-constant steady-state concentration.

EXAMPLE 17.5 The steady-state approximation

Apply the steady-state approximation to the mechanism (17.56), dropping the assumptions that steps 1 and -1 are in near equilibrium and that step 2 is slow.

We have

$$r = d[C_6H_5N_2^+]/dt = k_3[ONBr][C_6H_5NH_2]$$

The intermediates are ONBr and $H_2NO_2^+$. To eliminate the intermediate ONBr from the rate expression, we apply the steady-state approximation $d[ONBr]/dt = 0$. The species ONBr is formed by the elementary step 2 at the rate

$$(d[ONBr]/dt)_2 = k_2[H_2NO_2^+][Br^-]$$

and is consumed by step 3 with

$$(d[ONBr]/dt)_3 = -k_3[ONBr][C_6H_5NH_2]$$

The net rate of change of [ONBr] equals $(d[ONBr]/dt)_2 + (d[ONBr]/dt)_3$, and we have

$$d[ONBr]/dt = 0 = k_2[H_2NO_2^+][Br^-] - k_3[ONBr][C_6H_5NH_2]$$

$$[ONBr] = k_2[H_2NO_2^+][Br^-]/k_3[C_6H_5NH_2]$$

Substitution of this expression for [ONBr] in the above equation for r gives

$$r = k_2[H_2NO_2^+][Br^-] \tag{17.58}$$

To eliminate the intermediate $H_2NO_2^+$ from r, we use the steady-state approximation $d[H_2NO_2^+]/dt = 0$. Since $H_2NO_2^+$ is formed by step 1 in (17.56) and consumed by steps -1 and 2, we have

$$d[H_2NO_2^+]/dt = 0 = k_1[H^+][HNO_2] - k_{-1}[H_2NO_2^+] - k_2[H_2NO_2^+][Br^-]$$

$$[H_2NO_2^+] = \frac{k_1[H^+][HNO_2]}{k_{-1} + k_2[Br^-]}$$

Substitution in (17.58) gives

$$r = \frac{k_1 k_2[H^+][HNO_2][Br^-]}{k_{-1} + k_2[Br^-]} \tag{17.59}$$

which is the rate law predicted by the steady-state approximation. To obtain agreement with the observed rate law (17.55), we must further assume that $k_{-1} \gg k_2[Br^-]$, in which case (17.59) reduces to (17.55). The assumption $k_{-1} \gg k_2[Br^-]$ means that the rate $k_{-1}[H_2NO_2^+]$ of reversion of $H_2NO_2^+$ back to H^+ and HNO_2 is much greater than the rate $k_2[H_2NO_2^+][Br^-]$ of reaction of $H_2NO_2^+$ with Br^-. This is the condition for steps 1 and -1 of (17.56) to be in near equilibrium, as in the rate-limiting-step approximation.

Exercise

Apply the steady-state approximation to the mechanism for $H_2O_2 + 2H^+ + 2I^- \rightarrow I_2 + 2H_2O$ given in the preceding exercise to find the predicted rate law. (*Answer:* $r = k_1 k_2[H^+][I^-][H_2O_2]/(k_{-1} + k_2[H_2O_2])$.)

In summary, to apply the rate-determining-step approximation: (*a*) take the reaction rate r as equal to the rate of the rate-determining step (divided by the stoichiometric number s_{rds} of the rate-determining step, if $s_{rds} \neq 1$); (*b*) eliminate the concentrations of any reaction intermediates that occur in the rate expression obtained in (*a*) by using equilibrium-constant expressions for the equilibria that precede the rate-determining step.

To apply the steady-state approximation: (*a*) take the reaction rate r as equal to the rate of formation of a product; (*b*) eliminate the concentrations of any reaction intermediates that occur in the rate expression obtained in (*a*) by using $d[I]/dt = 0$ to find the concentration of each such intermediate I; (*c*) if step (*b*) introduces concentrations of other intermediates, apply $d[I]/dt = 0$ to these intermediates to eliminate their concentrations.

The steady-state approximation usually gives more complicated rate laws than the rate-limiting-step approximation. In a given reaction, one or the other or both or neither of these approximations may be valid. Noyes has analyzed the conditions of validity of each approximation (R. M. Noyes, in *Bernasconi,* pt. I, chap. V).

Computer programs that accurately numerically integrate the differential equations of a multistep mechanism have shown that the widely used steady-state approximation can sometimes lead to substantial error. [See L. A. Farrow and D. Edelson, *Int. J. Chem. Kinet.*, **6**, 787 (1974); T. Turányi et al., *J. Phys. Chem.*, **97**, 163 (1993).]

From Rate Law to Mechanism

So far in this section, we have been considering how, starting from an assumed mechanism for a reaction, one deduces the rate law implied by this mechanism. We now examine the reverse process, namely, how, starting with the experimentally observed rate law, one devises possible mechanisms that are consistent with this rate law.

The following rules help in finding mechanisms that fit an observed rate law. [See also J. O. Edwards, E. F. Greene, and J. Ross, *J. Chem. Educ.*, **45**, 381 (1968); H. Taube, ibid., **36**, 451 (1959); J. P. Birk, ibid., **47**, 805 (1970); J. F. Bunnett, in *Bernasconi*, pt. I, sec. 3.5.] Rules 1 to 3 apply only when the rate-limiting-step approximation is valid.

1a. If the rate law is $r = k[A]^\alpha[B]^\beta \cdots [L]^\lambda$, where $\alpha, \beta, \ldots, \lambda$ are positive integers, the total composition of the reactants in the rate-limiting step is $\alpha A + \beta B + \cdots + \lambda L$. Specification of the "total composition" of the reactants in the rate-determining step means specification of the total number of reactant atoms of each type and of the total charge on the reactants. However, the actual species that react in the rate-limiting step cannot be deduced from the rate law.

EXAMPLE 17.6 Devising a mechanism

The gas-phase reaction $2NO + O_2 \rightarrow 2NO_2$ has the observed rate law $r = k[NO]^2[O_2]$. Devise some mechanisms for this reaction that have a rate-determining step and that lead to this rate law.

For the observed rate law $r = k[NO]^2[O_2]$, rule *1a* with A = NO, $\alpha = 2$, B = O_2, $\beta = 1$ gives the total reactant rate-determining-step composition as $2NO + O_2$, which equals N_2O_4. Any mechanism whose rate-determining step has its total reactant composition equal to N_2O_4 will yield the correct rate law. Some possible rate-determining steps with total reactant composition N_2O_4 are (a) $N_2O_2 + O_2 \rightarrow$ products; (b) $NO_3 + NO \rightarrow$ products; (c) $2NO + O_2 \rightarrow$ products; and (d) $N_2 + 2O_2 \rightarrow$ products.

Reaction (a) contains the intermediate N_2O_2, so in a mechanism with (a) as its rate-determining step, the rate-determining step must be preceded by a step that forms N_2O_2. A plausible mechanism with the rate-determining step (a) is

$$2NO \rightleftharpoons N_2O_2 \qquad \text{equilib.}$$
$$N_2O_2 + O_2 \rightarrow 2NO_2 \qquad \text{slow} \tag{17.60}$$

A mechanism with (b) as its rate-determining step is

$$NO + O_2 \rightleftharpoons NO_3 \qquad \text{equilib.}$$
$$NO_3 + NO \rightarrow 2NO_2 \qquad \text{slow} \tag{17.61}$$

A third possible mechanism is the one-step trimolecular reaction

$$2NO + O_2 \rightarrow 2NO_2 \tag{17.62}$$

which has (c) as its rate-determining step.

Each of the mechanisms (17.60), (17.61), and (17.62) adds up to the correct overall stoichiometry $2NO + O_2 \rightarrow 2NO_2$, and each has N_2O_4 as the total reactant composition in the rate-limiting step. The reader can verify (Prob. 17.54) that each mechanism leads to the rate law $r = k[NO]^2[O_2]$. Which of these mechanisms is the correct one is not known.

1b. If the rate law is $r = k[\text{A}]^\alpha[\text{B}]^\beta \cdots [\text{L}]^\lambda/[\text{M}]^\mu[\text{N}]^\nu \cdots [\text{R}]^\rho$, where $\alpha, \beta, \ldots, \lambda, \mu, \nu, \ldots, \rho$ are positive integers, the total composition of the reactants in the rate-limiting step is $\alpha\text{A} + \beta\text{B} + \cdots + \lambda\text{L} - \mu\text{M} - \nu\text{N} - \cdots - \rho\text{R}$. Moreover, the species $\mu\text{M}, \nu\text{N}, \ldots, \rho\text{R}$ appear as products in equilibria that precede the rate-limiting step, and these species do not enter into the rate-limiting step.

EXAMPLE 17.7 Devising a mechanism

The reaction $\text{Hg}_2^{2+} + \text{Tl}^{3+} \rightarrow 2\text{Hg}^{2+} + \text{Tl}^+$ in aqueous solution has the rate law

$$r = k\frac{[\text{Hg}_2^{2+}][\text{Tl}^{3+}]}{[\text{Hg}^{2+}]} \qquad (17.63)$$

(a) Devise a mechanism consistent with this rate law. (b) Is the rate of this reaction infinite at the start of the reaction when $[\text{Hg}^{2+}] = 0$?

(a) According to rule 1b, the rate-determining-step reactants have the total composition $\text{Hg}_2^{2+} + \text{Tl}^{3+} - \text{Hg}^{2+}$, which is HgTl^{3+}; also, the species Hg^{2+} is not a reactant in the rate-limiting step but is a product in an equilibrium that precedes the rate-limiting step. One possible mechanism (see also Prob. 17.50) is

$$\text{Hg}_2^{2+} \underset{k_{-1}}{\overset{k_1}{\rightleftharpoons}} \text{Hg}^{2+} + \text{Hg} \qquad \text{equilib.}$$

$$\text{Hg} + \text{Tl}^{3+} \overset{k_2}{\rightarrow} \text{Hg}^{2+} + \text{Tl}^+ \qquad \text{slow}$$

The second step is rate-limiting, so $r = k_2[\text{Hg}][\text{Tl}^{3+}]$. To eliminate the reaction intermediate Hg, we use the equilibrium condition for the first step:

$$K_{c,1} = \frac{k_1}{k_{-1}} = \frac{[\text{Hg}^{2+}][\text{Hg}]}{[\text{Hg}_2^{2+}]} \qquad \text{and} \qquad [\text{Hg}] = \frac{k_1[\text{Hg}_2^{2+}]}{k_{-1}[\text{Hg}^{2+}]}$$

Therefore $r = k_1k_2[\text{Tl}^{3+}][\text{Hg}_2^{2+}]/k_{-1}[\text{Hg}^{2+}]$, in agreement with (17.63).

(b) An apparently puzzling thing about the rate law (17.63) is that it seems to predict $r = \infty$ at the start of the reaction when the concentration of the product Hg^{2+} is zero. Actually, Eq. (17.63) is not valid at the start of the reaction. In deriving (17.63) from the mechanism, we used the equilibrium expression for step 1. Hence (17.63) is valid only for times after the equilibrium $\text{Hg}_2^{2+} \rightleftharpoons \text{Hg}^{2+} + \text{Hg}$ has been established. Since this equilibrium is rapidly established compared with the rate-limiting second step, any deviation of the rate from (17.63) during the first few instants of the reaction will have no significant influence on the observed kinetics.

2. If, as is usually true, the order with respect to the solvent (S) is unknown, the total reactant composition of the rate-limiting step is $\alpha\text{A} + \beta\text{B} + \cdots + \lambda\text{L} - \mu\text{M} - \nu\text{N} - \cdots - \rho\text{R} + x\text{S}$, where the rate law is as in rule 1b, and x can be 0, ±1, $\pm2, \ldots$.

For example, the reaction $\text{H}_3\text{AsO}_4 + 3\text{I}^- + 2\text{H}^+ \rightarrow \text{H}_3\text{AsO}_3 + \text{I}_3^- + \text{H}_2\text{O}$ in aqueous solution has the rate law $r = k[\text{H}_3\text{AsO}_4][\text{I}^-][\text{H}^+]$, where the order with respect to H_2O is unknown. The total composition of the rate-limiting-step reactants is then $\text{H}_3\text{AsO}_4 + \text{I}^- + \text{H}^+ + x\text{H}_2\text{O} = \text{AsIH}_{4+2x}\text{O}_{4+x}$. Any value of x less than -2 would give a negative number of H atoms, so $x \geq -2$ and the rate-limiting-step reactants have one As atom, one I atom, at least two O atoms, and possibly some H atoms.

3. If the rate law has the factor $[\text{B}]^{1/2}$, the mechanism probably involves splitting a B molecule into two species before the rate-limiting step.

An example is a reaction that is catalyzed by H^+, where the source of the H^+ is the ionization of a weak acid: $CH_3COOH \rightleftharpoons H^+ + CH_3COO^-$. Suppose the rate-limiting step is $H^+ + A \rightarrow E + C$, where the catalyst H^+ is regenerated in a subsequent rapid step. Then $r = k[H^+][A]$. Since H^+ is regenerated, its concentration remains essentially constant during the reaction and we have $[H^+] = [CH_3COO^-]$. Letting K_c be the acetic acid ionization constant, we have $K_c = [H^+]^2/[CH_3COOH]$, and $[H^+] = K_c^{1/2}[CH_3COOH]^{1/2}$. Therefore $r = kK_c^{1/2}[A][CH_3COOH]^{1/2}$.

Half-integral orders often occur in chain reactions (Sec. 17.13). The rate-limiting-step approximation is usually not applicable to chain reactions, but the half-integral orders still result from splitting of a molecule, usually as the first step in the chain reaction.

4. A rate law with a sum of terms in the denominator indicates a mechanism with one or more reactive intermediates to which the steady-state approximation is applicable (rather than a rate-limiting-step mechanism). An example is (17.59).
5. Elementary reactions are usually unimolecular or bimolecular, rarely trimolecular, and never of molecularity greater than 3.

There are usually a few plausible mechanisms that are compatible with a given observed rate law. Confirming evidence for a proposed mechanism can be obtained by detecting the postulated reaction intermediate(s). If the intermediates are relatively stable, the reaction can be slowed drastically by cooling or dilution, and the reaction mixture analyzed chemically for the suspected intermediates. For example, rapid chilling of a bunsen-burner flame followed by chemical analysis shows the presence of CH_2O and peroxides, indicating that these species are intermediates in hydrocarbon combustion.

Usually, the intermediates are too unstable to be isolated. Reactive intermediates can often be detected spectroscopically. Thus, certain bands in the emission spectrum of an H_2–O_2 flame have been shown to be due to OH radicals. The absorption spectrum of the blue intermediate species NO_3 in the N_2O_5 decomposition [Eq. (17.8)] has been observed in shock-tube studies.

Many gas-phase intermediates (for example, OH, CH_2, CH_3, H, O, C_6H_5) have been detected by allowing some of the reaction mixture to leak into a mass spectrometer. Species with unpaired electrons (free radicals) can be detected by electron-spin-resonance spectroscopy (Sec. 21.13). Frequently, radicals are "trapped" by condensing a mixture of the radical and an inert gas on a cold surface. *Most gas-phase reactions have been found to be composite and to involve radicals as intermediates.*

Evidence for a postulated reaction intermediate can often be obtained by seeing how the addition of a suitably chosen species affects the reaction rate and the products. For example, the hydrolysis of certain alkyl halides in the mixed solvent acetone–water according to $RCl + H_2O \rightarrow ROH + H^+ + Cl^-$ is found to have $r = k[RCl]$. The proposed mechanism is the slow, rate-determining step $RCl \rightarrow R^+ + Cl^-$, followed by the rapid step $R^+ + H_2O \rightarrow ROH + H^+$. Evidence for the existence of the carbonium ion R^+ is the fact that in the presence of N_3^-, the rate constant and rate law are unchanged, but substantial amounts of RN_3 are formed. The N_3^- combines with R^+ after the rate-determining dissociation of RCl and does not affect the rate.

Isotopically substituted species can help clarify a mechanism. For example, isotopic tracers show that in a reaction between a primary or secondary alcohol and an organic acid to give an ester and water, the oxygen in the water usually comes from the acid: $R'C(O)^{16}OH + R^{18}OH \rightarrow R'C(O)^{18}OR + H^{16}OH$. The mechanism must involve breaking the C—OH bond of the acid.

The stereochemistry of reactants and products is an important clue to the reaction mechanism. For example, addition of Br_2 to a cycloalkene gives a product in which the two Br atoms are trans to each other. This indicates that Br_2 does not add to the double bond in a single elementary reaction.

If the mechanism of a reaction has been determined, the mechanism of the reverse reaction is known, since the reverse reaction must proceed by a series of steps that is the exact reverse of the forward mechanism, provided the conditions of temperature, solvent, etc., are unchanged. This must be so, because the laws of particle motion are symmetric with respect to time reversal (Sec. 3.8). Thus, if the mechanism of the reaction $H_2 + I_2 \rightarrow 2HI$ happened to be $I_2 \rightleftharpoons 2I$, followed by $2I + H_2 \rightarrow 2HI$, the decomposition of HI to H_2 and I_2 at the same temperature would proceed by the mechanism $2HI \rightarrow 2I + H_2$, followed by $2I \rightleftharpoons I_2$.

Once the mechanism of a reaction has been determined, it may be possible to use kinetic data to deduce rate constants for some of the elementary steps in the mechanism. In a reaction with a rate-limiting step, the observed rate constant k is usually the product of the rate constant of the rate-limiting step and the equilibrium constants of the steps preceding this step. Equilibrium constants for gas-phase reactions can often be found from thermodynamic data or statistical mechanics (Sec. 22.8), thereby allowing the rate constant of the rate-limiting step to be calculated. Knowledge of rate constants for elementary reactions allows the various theories of reaction rates to be tested.

It is wise to retain some skepticism toward proposed reaction mechanisms. The reaction $H_2 + I_2 \rightarrow 2HI$ was found to have the rate law $r = k[H_2][I_2]$ by Bodenstein in the 1890s. Until 1967, most kineticists believed the mechanism to be the single bimolecular step $H_2 + I_2 \rightarrow 2HI$. In 1967 Sullivan presented strong experimental evidence that the mechanism involves I atoms and might well consist of the rapid equilibrium $I_2 \rightleftharpoons 2I$ followed by the rate-limiting trimolecular reaction $2I + H_2 \rightarrow 2HI$. (The reader can verify that this mechanism leads to the observed rate law.) Sullivan's work convinced most people that the one-step bimolecular mechanism was wrong. However, further theoretical discussion of Sullivan's data has led some workers to argue that these data are not inconsistent with the bimolecular mechanism. Thus, the mechanism of the H_2–I_2 reaction is not fully settled. [See G. G. Hammes and B. Widom, *J. Am. Chem. Soc.,* **96,** 7621 (1974); R. M. Noyes, ibid., **96,** 7623.]

Some compilations of proposed reaction mechanisms are *Bamford and Tipper,* vols. 4–25; M. V. Twigg (ed.), *Mechanisms of Inorganic and Organometallic Reactions,* vols. 1–, Plenum, 1983–; A. G. Sykes (ed.), *Advances in Inorganic and Bioinorganic Reaction Mechanisms,* vols. 1–, Academic, 1983–; A. C. Knipe and W. E. Watts (eds.), *Organic Reaction Mechanisms, 1981–*; Wiley, 1983–.

Let us summarize the steps used to investigate the kinetics of a reaction. (*a*) The reactants and products are established, so that the reaction's overall stoichiometry is known. (*b*) One observes concentrations as functions of time for several kinetics runs with different initial concentrations. (*c*) The data of (*b*) are analyzed to find the rate law. (*d*) Plausible mechanisms that are consistent with the rate law are thought up, and these mechanisms are tested by, for example, attempting to detect reaction intermediates.

17.7   COMPUTER INTEGRATION OF RATE EQUATIONS

Reaction mechanisms may have many steps and may involve many species, each of which may appear in several steps. It is therefore often impossible to analytically integrate the simultaneous rate equations for a mechanism. Hence, numerical integration of rate equations is an important tool in investigating complicated reaction mechanisms. Numerical integration is widely used in the kinetics of atmospheric reactions, combustion reactions, and biochemical metabolic pathways.

Although the real use of numerical integration is for several simultaneous differential equations, we start by considering a system with a single equation. Suppose the reaction $aA + bB \rightarrow$ products has the rate law $d[A]/dt = f([A], [B])$, where f is some function of the concentrations. Using the stoichiometry relation (17.19), we can express [B] in terms of [A], thereby expressing $d[A]/dt$ as some function of [A] only:

$$d[A]/dt = g([A]) \qquad (17.64)$$

where g is a known function. Over a very short time interval Δt, the derivative $d[A]/dt$ can be replaced by $\Delta[A]/\Delta t$ to give the approximation

$$\Delta[A] = g([A]) \, \Delta t \qquad (17.65)$$

Starting from the known values $[A]_0$ and $[B]_0$ at $t = 0$, we repeatedly use (17.65) to successively calculate the concentrations $[A]_1$, $[A]_2$, $[A]_3$, ... at times $t_1 = \Delta t$, $t_2 = 2\Delta t$, $t_3 = 3\Delta t$, ... according to

$$[A]_1 = [A]_0 + g([A]_0) \, \Delta t, \quad [A]_2 = [A]_1 + g([A]_1) \, \Delta t,$$

$$[A]_3 = [A]_2 + g([A]_2) \, \Delta t, \quad \ldots$$

This approximation, called the **Euler method,** becomes more accurate the smaller the value of Δt, but the smaller Δt is, the more calculations are required to integrate over a fixed time interval, so the Euler method is very inefficient. To test the accuracy of the calculation, one repeats it with Δt cut in half and sees if this changes the [A]-versus-t curve significantly. The Euler method is easily implemented on a spreadsheet (Prob. 17.55). It might be thought that by making Δt small enough, unlimited accuracy can be attained in the Euler method, but this is not so. The smaller the step size, the greater the number of computations that must be done. Since computer calculations are done with a fixed maximum number of significant figures, each time a calculation is done there is a small roundoff error. These roundoff errors tend to accumulate, producing a substantial cumulative roundoff error in later points. Because of inefficiency and roundoff errors, the Euler method is not of practical use in kinetics.

A substantial improvement on the Euler method is the **modified Euler** (or **midpoint**) method. The Euler method uses the approximation $[A]_{n+1} = [A]_n + d[A]/dt|_n \, \Delta t$, where $d[A]/dt|_n = g([A]_n)$ is the derivative at the beginning point of the time interval. If we replace the derivative at the beginning point with the derivative at the midpoint of each time interval, we will get an improved estimate for $\Delta[A]$ over the interval. Let $n + 1/2$ denote the midpoint of the nth interval. Since [A] is unknown at the interval midpoint, we can't evaluate $g([A]_{n+1/2})$ to find $d[A]/dt$ at the midpoint. We therefore use the unmodified Euler method to estimate [A] at the midpoint and use this estimated [A] and the rate law (17.64) to estimate $d[A]/dt$ at the midpoint. The Euler estimate of the midpoint concentration is $[A]_{n+1/2} = [A]_n + g([A]_n) \, \Delta t/2$. The estimated derivative at the midpoint is given by (17.64) as $g([A]_{n+1/2})$, where g is a known function. The modified-Euler formulas are therefore

$$[A]_{n+1} = [A]_n + g([A]_{n+1/2}) \, \Delta t, \qquad \text{where} \quad [A]_{n+1/2} = [A]_n + g([A]_n) \, \Delta t/2$$

where $n = 0, 1, 2, \ldots$. Spreadsheet implementation of the modified Euler method is considered in Prob. 17.57.

The modified Euler method is an example of a second-order Runge–Kutta method. **Runge–Kutta** methods use the formula $[A]_{n+1} = [A]_n + \phi([A]_n) \, \Delta t$, where the function ϕ is chosen to give an estimate of the typical slope over the interval. For a kth-order Runge–Kutta method, ϕ is chosen so that if one expands the function [A] in a Taylor series about the value $[A]_n$, the Runge–Kutta estimate of $[A]_{n+1}$ agrees with the Taylor-series estimate up through terms of order $(\Delta t)^k$. Fourth-order Runge–Kutta methods are widely used to solve differential equations since they are usually accurate and computationally efficient. To further improve the efficiency of Runge–Kutta methods, one can use a formula to estimate the error at each point and adjust the step size at that point so that Δt is as large as possible while keeping the error within a preset maximum allowed value. The term *adaptive* denotes use of a variable step size.

Now suppose we have several simultaneous rate equations to integrate:

$$d[A]/dt = f([A], [B], \ldots), \qquad d[B]/dt = g([A], [B], \ldots), \quad \ldots$$

where $f, g, \ldots$ are known functions of concentrations at time t in the system. To use the modified Euler (midpoint) method to find concentrations at point $n + 1$ from concentrations at point n, we first use the Euler formula to estimate the midpoint concentrations as

$$[A]_{n+1/2} = [A]_n + f([A]_n, [B]_n, \ldots)\,\Delta t/2,$$

$$[B]_{n+1/2} = [B]_n + g([A]_n, [B]_n, \ldots)\,\Delta t/2, \quad \ldots$$

and then use the estimated midpoint concentrations to estimate the concentrations at point $n + 1$ as

$$[A]_{n+1} = [A]_n + f([A]_{n+1/2}, [B]_{n+1/2}, \ldots)\,\Delta t,$$

$$[B]_{n+1} = [B]_n + g([A]_{n+1/2}, [B]_{n+1/2}, \ldots)\,\Delta t, \quad \ldots$$

Simultaneous rate equations often have rate constants that differ by several orders of magnitude from one another, so the solution contains at least one component that varies much more rapidly than other components. Mathematicians call such a system of differential equations a *stiff* system. A stiff system requires an extremely small time step Δt in order to properly integrate it numerically. Several methods have been devised to deal with stiff systems. The *Rosenbrock* method is a modified Runge–Kutta method that is widely used to deal with stiff systems.

Some Texas Instruments graphing calculators can solve simultaneous differential equations using an adaptive third-order Runge–Kutta method. Some versions of the program Mathcad have Runge–Kutta and Rosenbrock differential-equation solvers.

For further details on computer solution of differential equations see *Steinfeld, Francisco, and Hase,* sec. 2.6; W. H. Press et al., *Numerical Recipes in Fortran,* 2nd ed., Cambridge, 1992 (www.nr.com); *Chapra and Canale,* chaps. 25 and 26.

Computer Programs for Solution of Rate Equations

Several free programs for solving simultaneous rate equations on personal computers are available. These include the Windows program KinTekSim (www.kintek-corp .com/kinteksim.htm), the Windows and Macintosh program CKS (Chemical Kinetics Simulator) (www.almaden.ibm.com/st/msim/), and the Windows program Gepasi (www.gepasi.org).

17.8   TEMPERATURE DEPENDENCE OF RATE CONSTANTS

Rate constants depend strongly on temperature, typically increasing rapidly with increasing T (Fig. 17.9a). A rough rule, valid for many reactions in solution, is that, near room temperature, k doubles or triples for each 10-°C increase in T.

In 1889 Arrhenius noted that the $k(T)$ data for many reactions fit the equation

$$k = Ae^{-E_a/RT} \tag{17.66}*$$

where A and E_a are constants characteristic of the reaction and R is the gas constant. E_a is the **Arrhenius activation energy** and A is the **pre-exponential factor** or the

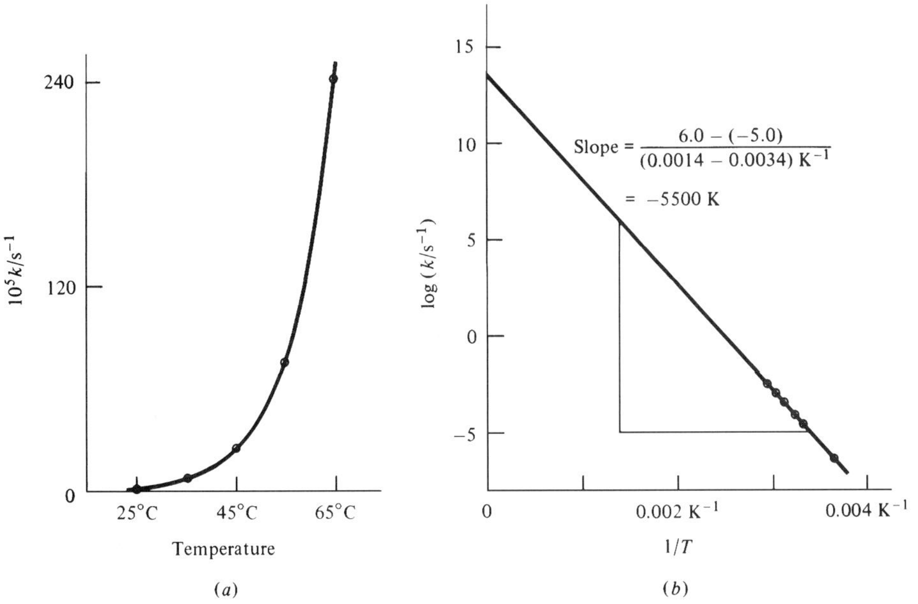

Figure 17.9

(a) Rate constant versus temperature for the gas-phase first-order decomposition reaction $2N_2O_5 \rightarrow 4NO_2 + O_2$.
(b) Arrhenius plot of $\log_{10} k$ versus $1/T$ for this reaction. Note the long extrapolation needed to find A.

Arrhenius A factor. The units of A are the same as those of k. The units of E_a are the same as those of RT, namely, energy per mole; E_a is usually expressed in kJ/mol or kcal/mol. Arrhenius arrived at (17.66) by arguing that the temperature dependence of rate constants would probably resemble the temperature dependence of equilibrium constants. By analogy to (6.36), (11.32), and the equation in Prob. 6.19, Arrhenius wrote $d \ln k/dT = E_a/RT^2$, which integrates to (17.66) if E_a is assumed independent of T.

Taking logs of (17.66), we get

$$\ln k = \ln A - \frac{E_a}{RT} \qquad \text{or} \qquad \log_{10} k = \log_{10} A - \frac{E_a}{2.303RT} \qquad (17.67)$$

If the Arrhenius equation is obeyed, a plot of $\log_{10} k$ versus $1/T$ is a straight line with slope $-E_a/2.303R$ and intercept $\log_{10} A$. This enables E_a and A to be found. [For more accurate procedures, see sec. 7 of R. J. Cvetanovic et al., *J. Phys. Chem.*, **83**, 50 (1979).] A typical experimental error in E_a is 1 kcal/mol and in A is a factor of 3.

EXAMPLE 17.8 E_a and A from $k(T)$ data

Use Fig. 17.9 to find A and E_a for $2N_2O_5 \rightarrow 4NO_2 + O_2$.

Equation (17.67) is slightly defective in that one can take the log of only a dimensionless number. From the label on the vertical axis of Fig. 17.9a, k for this (first-order) reaction has units of s^{-1}. Therefore A in (17.66) has units of s^{-1}. Rewriting (17.67) in a dimensionally correct form for a first-order reaction, we have $\log_{10} (k/s^{-1}) = \log_{10} (A/s^{-1}) - E_a/2.303RT$. The intercept of the $\log_{10} (k/s^{-1})$-versus-$1/T$ plot is $\log_{10} (A/s^{-1})$. Figure 17.9b gives this intercept as 13.5. Therefore $\log_{10} (A/s^{-1}) = 13.5$, $A/s^{-1} = 3 \times 10^{13}$, and $A = 3 \times 10^{13} \text{ s}^{-1}$. The slope in Fig. 17.9b is -5500 K, so $-5500 \text{ K} = -E_a/2.303R$, which gives $E_a = 25$ kcal/mol $= 105$ kJ/mol.

For further points about finding E_a and A from $k(T)$ data, see Prob. 17.71.

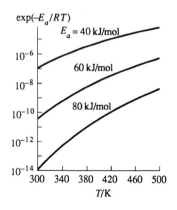

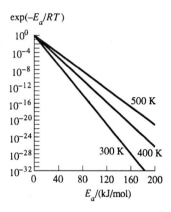

Figure 17.10

The fraction of collisions with relative kinetic energy exceeding E_a plotted versus E_a and versus T. For a typical activation energy of 20 kcal/mol $\approx$ 80 kJ/mol, only a tiny fraction of collisions has kinetic energy exceeding E_a. The vertical scales are logarithmic.

Exercise

For the reaction $C_2H_5I + OH^- \rightarrow C_2H_5OH + I^-$ in ethanol, rate-constant data versus T are

$10^4 k/(\text{L mol}^{-1}\text{ s}^{-1})$	0.503	3.68	67.1	1190
T/K	289.0	305.2	332.9	363.8

Use a graph to find E_a and A for this reaction. (*Answer:* $21_{.6}$ kcal/mol, 1×10^{12} L mol^{-1} s^{-1}.)

The Arrhenius equation (17.66) holds rather well for nearly all elementary homogeneous reactions and for most composite reactions. A simple interpretation of (17.66) is that two colliding molecules require a certain minimum kinetic energy of relative motion to initiate the breaking of the appropriate bonds and allow new compounds to be formed. (For a unimolecular reaction, a certain minimum energy is needed to isomerize or decompose the molecule; the source of this energy is collisions; see Sec. 17.11.) The Maxwell distribution law (15.52) contains a factor $e^{-\varepsilon/kT}$, and one finds (Sec. 23.1) that the fraction of collisions in which the relative kinetic energy of the molecules along the line of the collision exceeds the value ε_a is equal to $e^{-\varepsilon_a/kT} = e^{-E_a/RT}$ where $E_a = N_A\varepsilon_a$ is the molecular kinetic energy expressed on a per-mole basis. Figure 17.10 plots the fraction $e^{-E_a/RT}$ versus T and versus E_a.

Note from (17.66) that *a low activation energy means a fast reaction and a high activation energy means a slow reaction.* The rapid increase in k as T increases is due mainly to the increase in the number of collisions whose energy exceeds the activation energy.

In the Arrhenius equation (17.66), both A and E_a are constants. Sophisticated theories of reaction rates (Sec. 23.4) yield an equation similar to (17.66), except that A and E_a both depend on temperature. When $E_a \gg RT$ (which is true for most chemical reactions), the temperature dependences of E_a and A are usually too small to be detected by the rather inaccurate kinetic data available, unless a wide temperature range is studied.

The general definition of the **activation energy** E_a of any rate process, applicable whether or not E_a varies with T, is

$$E_a \equiv RT^2 \frac{d\ln k}{dT} \qquad (17.68)$$

(which resembles the equation in Prob. 6.19). If E_a is independent of T, integration of (17.68) yields (17.66), where A is also independent of T. Whether or not E_a depends on T, the **pre-exponential factor** A for any rate process is defined, in analogy with (17.66), as

$$A \equiv k e^{E_a/RT} \qquad (17.69)$$

From (17.69), we get $k = A e^{-E_a/RT}$, a generalized version of (17.66), in which both A and E_a may depend on T. A simple physical interpretation of E_a in (17.68) is provided by the following theorem. In a gas-phase elementary bimolecular reaction, $\varepsilon_a \equiv E_a/N_A$ is equal to the average total energy (relative translational plus internal) of those pairs of reactant molecules that are undergoing reaction minus the average total energy of all pairs of reactant molecules. For proofs, see the references in sec. 3.1.2 of *Laidler* (1987).

Observed activation energies lie in the range 0 to 80 kcal/mol (330 kJ/mol) for most elementary chemical reactions and tend to be lower for bimolecular than unimolecular reactions. Unimolecular decompositions of compounds with strong bonds have very high E_a values. For example, E_a is 100 kcal/mol for the gas-phase decomposition

$CO_2 \rightarrow CO + O$. An upper limit on observed E_a values is set by the fact that reactions with extremely high activation energies are too slow to observe.

For unimolecular reactions, A is typically 10^{12} to 10^{15} s^{-1}. For bimolecular reactions, A is typically 10^8 to 10^{12} dm^3 mol^{-1} s^{-1}.

Recombination of two radicals to form a stable polyatomic molecule requires no bonds to be broken, and most such gas-phase reactions have zero activation energies. Examples are $2CH_3 \rightarrow C_2H_6$ and $CH_3 + Cl \rightarrow CH_3Cl$. (For the recombination of atoms, see Sec. 17.12.) With zero activation energy, the rate constant is essentially independent of T. If R denotes a free radical and M a closed-shell molecule, then for exothermic bimolecular gas-phase reactions, $E_a \approx 0$ for $R_1 + R_2$ reactions, E_a is usually in the range 0 to 15 kcal/mol for $R + M$ reactions, and E_a is usually in the range 20 to 50 kcal/mol for $M + M$ reactions. [The E_a ranges for endothermic bimolecular reactions can be found from Eq. (17.71).]

EXAMPLE 17.9 E_a from $k(T)$

Calculate E_a for a reaction whose rate constant at room temperature is doubled by a 10-°C increase in T. Then repeat the calculation for a reaction whose rate constant is tripled.

The Arrhenius equation (17.66) gives

$$\frac{k(T_2)}{k(T_1)} = \frac{Ae^{-E_a/RT_2}}{Ae^{-E_a/RT_1}} = \exp\left(\frac{E_a}{R}\frac{T_2 - T_1}{T_1T_2}\right)$$

Taking logs, we get

$$E_a = RT_1T_2(\Delta T)^{-1} \ln\left[k(T_2)/k(T_1)\right]$$

$$= (1.987 \text{ cal mol}^{-1}\text{ K}^{-1})(298 \text{ K})(308 \text{ K})(10 \text{ K})^{-1} \ln (2 \text{ or } 3)$$

$$E_a = \begin{cases} 13 \text{ kcal/mol} = 53 \text{ kJ/mol} & \text{for doubling} \\ 20 \text{ kcal/mol} = 84 \text{ kJ/mol} & \text{for tripling} \end{cases}$$

Exercise

If $E_a = 30$ kcal/mol, find the effect of a 10-°C temperature increase on the room-temperature rate constant. (*Answer:* Rate constant is quintupled.)

EXAMPLE 17.10 Effect of E_a on k

Calculate the room-temperature ratio of the rate constants for two reactions that have the same A value but have E_a values that differ by (*a*) 1 kcal/mol; (*b*) 10 kcal/mol.

The Arrhenius equation (17.66) gives

$$\frac{k_1}{k_2} = \frac{Ae^{-E_{a,1}/RT}}{Ae^{-E_{a,2}/RT}} = \exp\frac{E_{a,2} - E_{a,1}}{RT}$$

$$= \exp\frac{1 \text{ kcal mol}^{-1} \quad \text{or} \quad 10 \text{ kcal mol}^{-1}}{(1.987 \times 10^{-3} \text{ kcal mol}^{-1}\text{ K}^{-1})(298 \text{ K})}$$

$$= \begin{cases} 5.4 & \text{for part } (a) \\ 2 \times 10^7 & \text{for part } (b) \end{cases}$$

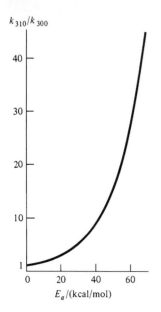

Figure 17.11

Ratio of rate constants at 310 K and 300 K versus activation energy. The larger the E_a value, the faster the rate constant increases with T.

Each 1 kcal/mol decrease in E_a multiplies the room-temperature rate by 5.4.

Exercise

If reactions 1 and 2 have $A_1 = 5A_2$ and $k_1 = 100k_2$ at room temperature, find $E_{a,1} - E_{a,2}$. (*Answer:* −1.8 kcal/mol.)

Figure 17.11 plots k_{310}/k_{300}, the ratio of rate constants at 310 and 300 K, versus the activation energy E_a, using the equation from Example 17.9. The greater the E_a value, the faster k increases with T, as is obvious from $d \ln k/dT = E_a/RT^2$ [Eq. (17.68)]; see also Fig. 17.10.

The rates of many physiological processes vary with T according to the Arrhenius equation. Examples include the rate of chirping of tree crickets ($E_a = 12$ kcal/mol), the rate of firefly light flashes ($E_a = 12$ kcal/mol), and the frequency of human alpha brain waves ($E_a = 7$ kcal/mol). [See K. J. Laidler, *J. Chem. Educ.*, **49**, 343 (1972).]

Let k_f and k_b be the forward and reverse rate constants of an *elementary* reaction, and let $E_{a,f}$ and $E_{a,b}$ be the corresponding activation energies. Equation (17.53) gives $k_f/k_b = K_c$, where K_c is the concentration-scale equilibrium constant of the reaction. Hence, $\ln k_f - \ln k_b = \ln K_c$. Differentiation with respect to T gives

$$d \ln k_f/dT - d \ln k_b/dT = d \ln K_c/dT \qquad (17.70)$$

Equation (17.68) gives $d \ln k_f/dT = E_{a,f}/RT^2$ and $d \ln k_b/dT = E_{a,b}/RT^2$. The result for Prob. 6.19 gives $d \ln K_c/dT = \Delta U°/RT^2$ for an ideal-gas reaction, where $\Delta U°$ is the change in standard-state molar internal energy for the reaction and is related to $\Delta H°$ by (5.10). Hence, for an ideal-gas elementary reaction, Eq. (17.70) becomes

$$E_{a,f} - E_{a,b} = \Delta U° \qquad \text{elementary reaction} \qquad (17.71)$$

Figure 17.12 illustrates Eq. (17.71) for positive and negative $\Delta U°$ values.

For reactions in solution, things are a bit more complicated. The K_c in (17.70) is $K_{c,\text{kin}}$ and includes the concentration of the solvent A raised to its stoichiometric coefficient ν_A (recall the discussion at the end of Sec. 17.5). From the relation $K_{c,\text{kin}} = K_{c,\text{td}}[A]^{\nu_A}$ and the equation in Prob. 11.41c, we get for a reaction in solution that $\partial \ln K_{c,\text{kin}}/\partial T = \Delta H^\infty/RT^2 - \alpha_A \Sigma_i \nu_i$, where α_A is the thermal expansivity of the solvent and the sum goes over all species. The term $\alpha_A \Sigma_i \nu_i$ is generally quite small compared with $\Delta H^\infty/RT^2$. Moreover, ΔH^∞ (which is the change in standard-state enthalpy values using the molality scale for solutes) differs negligibly from $\Delta U°$ for a condensed-phase reaction. Hence we can take $d \ln K_c/dT = \Delta U°/RT^2$ for reactions in solution, with no significant error. (In the spirit of neglecting the pressure dependence of rate constants, we take K_c to depend on T only.) Thus Eq. (17.71) holds well for elementary reactions in solution.

Figure 17.12

Relation between forward and back activation energies and $\Delta U°$ for an elementary reaction.

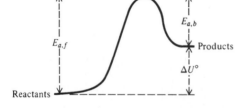

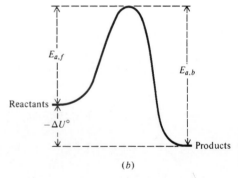

(a)   (b)

Now consider the temperature dependence of the rate constant k of an overall reaction composed of several elementary steps. If the rate-limiting-step approximation is valid, k will typically have the form k_1k_2/k_{-1}, where k_1 and k_{-1} are the forward and reverse rate constants of the equilibrium step preceding the rate-determining step 2. (See Example 17.4 in Sec. 17.6.) Using the Arrhenius equation, we have

$$k = \frac{k_1k_2}{k_{-1}} = \frac{A_1e^{-E_{a,1}/RT}A_2e^{-E_{a,2}/RT}}{A_{-1}e^{-E_{a,-1}/RT}} = \frac{A_1A_2}{A_{-1}} e^{-(E_{a,1}-E_{a,-1}+E_{a,2})/RT}$$

Therefore, writing k in the form $k = Ae^{-E_a/RT}$, we have for the overall activation energy $E_a = E_{a,1} - E_{a,-1} + E_{a,2}$.

If a reaction proceeds by two competing mechanisms, the overall rate constant need not obey the Arrhenius equation. For example, suppose the reaction $A \rightarrow C$ proceeds by the mechanisms

$$A \xrightarrow{k_1} C \qquad \text{and} \qquad A \xrightarrow{k_2} D \xrightarrow{k_3} C \qquad (17.72)$$

where the first step ($A \rightarrow D$) of the second mechanism is rate-determining for that mechanism. Then $r = -d[A]/dt = k_1[A] + k_2[A] = (k_1 + k_2)[A]$, and the overall rate constant is $k = k_1 + k_2 = A_1e^{-E_{a,1}/RT} + A_2e^{-E_{a,2}/RT}$, which does not have the Arrhenius form (17.66).

In Sec. 23.4, we shall see that a theoretical expression for the temperature dependence of the rate constant is [Eq. (23.21)] $k = CT^m e^{-E'/RT}$, where C, m, and E' are constants. When accurate rate-constant data are available over a wide range of T, this three-parameter equation generally gives a better fit than the two-parameter Arrhenius equation.

17.9 RELATION BETWEEN RATE CONSTANTS AND EQUILIBRIUM CONSTANTS FOR COMPOSITE REACTIONS

For an elementary reaction, K_c equals k_f/k_b [Eq. (17.53)]. For a composite reaction consisting of several elementary steps, this simple relation need not hold. A thorough discussion of the relation between k_f, k_b, and K_c is complicated [see R. K. Boyd, *Chem. Rev.*, **77**, 93 (1977)], so the treatment of this section is restricted to reactions where the rate-limiting-step approximation is valid and where the system is ideal.

Let the overall reaction be $aA + bB \rightleftharpoons cC + dD$. When the rate-limiting-step approximation holds, the forward and reverse reaction rates will have the forms $r_f = k_f[A]^{\alpha_f}[B]^{\beta_f}[C]^{\gamma_f}[D]^{\delta_f}$ and $r_b = k_b[A]^{\alpha_b}[B]^{\beta_b}[C]^{\gamma_b}[D]^{\delta_b}$. Both reactants and products can occur in the forward rate law, as shown by the example (17.63). r_f is the observed rate when the concentrations of products are much less than those of reactants; r_b is the rate when the concentrations of reactants are much less than those of products.

Before presenting the general relation between k_f, k_b, and K_c, we look at a specific example. The reaction

$$2Fe^{2+} + 2Hg^{2+} \rightleftharpoons 2Fe^{3+} + Hg_2^{2+} \qquad (17.73)$$

in $HClO_4(aq)$ has the forward rate law $r_f = k_f[Fe^{2+}][Hg^{2+}]$. A plausible mechanism compatible with this rate law is

$$Fe^{2+} + Hg^{2+} \underset{k_{-1}}{\overset{k_1}{\rightleftharpoons}} Fe^{3+} + Hg^+ \qquad \text{slow}$$

$$\qquad (17.74)$$

$$2Hg^+ \underset{k_{-2}}{\overset{k_2}{\rightleftharpoons}} Hg_2^{2+} \qquad \text{rapid}$$

where step 1 is rate-limiting. Then

$$r_f \equiv \tfrac{1}{2}\, d[Fe^{3+}]/dt = \tfrac{1}{2}k_1[Fe^{2+}][Hg^{2+}] = k_f[Fe^{2+}][Hg^{2+}] \qquad (17.75)$$

The factor $\tfrac{1}{2}$ occurs because the stoichiometric coefficient of Fe^{3+} is 2 in the overall reaction (17.73); see the definition (17.4) of r. Since we are examining the forward rate with product Fe^{3+} present in only small amounts, we don't include the reverse of the rate-limiting step 1; see the discussion in Sec. 17.6. Likewise in considering the reverse reaction below, we consider only the reverse of step 1 with rate constant k_{-1}.

For the reverse of reaction (17.73), the mechanism is the reverse of (17.74), namely, the rapid equilibrium $Hg_2^{2+} \rightleftharpoons 2Hg^+$ followed by the rate-limiting step $Fe^{3+} + Hg^+ \rightarrow Fe^{2+} + Hg^{2+}$ with rate constant k_{-1}. The rate of the reverse reaction is $r_b = \tfrac{1}{2}\, d[Fe^{2+}]/dt = \tfrac{1}{2}k_{-1}[Fe^{3+}][Hg^+]$. From the equilibrium step, we get the relation $[Hg^+] = (k_{-2}/k_2)^{1/2}[Hg_2^{2+}]^{1/2}$. Therefore

$$r_b = \tfrac{1}{2}k_{-1}(k_{-2}/k_2)^{1/2}[Fe^{3+}][Hg_2^{2+}]^{1/2} = k_b[Fe^{3+}][Hg_2^{2+}]^{1/2} \qquad (17.76)$$

At equilibrium, $r_f = r_b$, and (17.75) and (17.76) give

$$\frac{k_f}{k_b} = \frac{[Fe^{3+}]_{eq}([Hg_2^{2+}]_{eq})^{1/2}}{[Fe^{2+}]_{eq}[Hg^{2+}]_{eq}} = K_c^{1/2}$$

where K_c is the equilibrium constant for (17.73).

In 1957 Horiuti proved that for a reaction with a rate-limiting step,

$$k_f/k_b = K_c^{1/s_{rds}} \qquad (17.77)$$

where s_{rds} is the stoichiometric number (Sec. 17.1) of the rate-determining step. For the reaction (17.73), the rate-limiting step is step 1 of (17.74). Step 1 must occur twice for each occurrence of (17.73) since step 1 consumes one Fe^{2+} ion and the overall reaction (17.73) consumes two Fe^{2+} ions. Thus $s_{rds} = 2$ for the mechanism (17.74) and the overall reaction (17.73); Eq. (17.77) becomes $k_f/k_b = K_c^{1/2}$, as found above. Only when $s_{rds} = 1$ is $K_c = k_f/k_b$. For a proof of (17.77), see J. Horiuti and T. Nakamura, *Adv. Catal.,* **17,** 1 (1967).

The relation between the equilibrium constant for the overall reaction and the rate constants of the elementary steps of the mechanism (valid whether or not the rate-determining-step approximation applies) is given in Prob. 17.76.

17.10　THE RATE LAW IN NONIDEAL SYSTEMS

For an elementary reaction in an *ideal* system, the rates of the forward and reverse reactions contain the concentrations of the reacting species, as does the equilibrium constant K_c. In a *nonideal* system, the equilibrium constant for the elementary reaction $aA + bB \rightleftharpoons cC + dD$ is $K^\circ = (a_{C,eq})^c(a_{D,eq})^d/(a_{A,eq})^a(a_{B,eq})^b$ [Eq. (11.6)], where $a_{A,eq}$ is the activity of A at equilibrium. It seems reasonable therefore that the rate laws for an elementary reaction in a nonideal system would be

$$r_f \overset{?}{=} k_f a_A^a a_B^b \quad \text{and} \quad r_b \overset{?}{=} k_b a_C^c a_D^d \qquad (17.78)$$

Setting $r_f = r_b$ at equilibrium and using (17.78), we get $K^\circ = k_f/k_b$.

In the 1920s it was generally believed that (17.78) was correct. However, suppose that instead of (17.78) we write

$$r_f = k_f Y a_A^a a_B^b \quad \text{and} \quad r_b = k_b Y a_C^c a_D^d \qquad \text{elem. react.} \qquad (17.79)$$

where Y is some unspecified function of T, P, and the concentrations. Then at equilibrium, $r_f = r_b$, and (17.79) also leads to $K^\circ = k_f/k_b$, since Y cancels. In fact, kinetic data on ionic reactions in aqueous solution clearly show that (17.78) is *wrong* and that the correct form of the rate law is (17.79).

Thus, in a nonideal solution, the rate law for the elementary reaction $aA + bB \rightarrow$ products is

$$r = k^{\infty}Y(\gamma_A[A])^a(\gamma_B[B])^b \equiv k_{app}[A]^a[B]^b \qquad \text{elem. react.} \qquad (17.80)$$

where the γ's are the concentration-scale activity coefficients, Y is a parameter that depends on T, P, and the concentrations, and the apparent rate constant is defined by $k_{app} \equiv k^{\infty}Y(\gamma_A)^a(\gamma_B)^b$. The reason for the ∞ superscript on k will become clear shortly. In the limit of an infinitely dilute solution, ideal behavior is reached, and r must equal $k^{\infty}[A]^a[B]^b$ as discussed in Sec. 17.5. Since the γ's become 1 at infinite dilution, Y in (17.80) must go to 1 in the infinite-dilution limit. Therefore, the true rate constant k^{∞} can be determined by measuring k_{app} as a function of concentration and extrapolating to infinite dilution. Once k^{∞} is known, Y can be calculated for any solution composition from $Y = k_{app}/k^{\infty}(\gamma_A)^a(\gamma_B)^b$.

Section 23.8 gives a physical interpretation of Y. The accuracy of rate data is usually too low to detect deviations from ideality except for ionic reactions.

17.11 UNIMOLECULAR REACTIONS

Most elementary reactions are either bimolecular (A + B $\rightarrow$ products) or unimolecular (A $\rightarrow$ products). Unimolecular reactions are either *isomerizations*, for example, *cis*-CHCl=CHCl $\rightarrow$ *trans*-CHCl=CHCl, or *decompositions,* for example, CH_3CH_2I $\rightarrow CH_2=CH_2 + HI$. It is easy to understand how a bimolecular elementary reaction occurs: The molecules A and B collide, and if their relative kinetic energy exceeds the activation energy, the collision can lead to the breaking of bonds and the formation of new bonds. But what about a unimolecular reaction? Why should a molecule spontaneously break apart or isomerize? It seems reasonable that an A molecule would acquire the necessary activation energy by collision with another molecule. However, a collisional activation seems to imply second-order kinetics, in contrast to the observed first-order kinetics of unimolecular reactions. The answer to this problem was given by Lindemann in 1922.

Lindemann proposed the following detailed mechanism to explain the unimolecular reaction A $\rightarrow$ B (+ C).

$$A + M \underset{k_{-1}}{\overset{k_1}{\rightleftharpoons}} A^* + M$$

$$A^* \overset{k_2}{\rightarrow} B \; (+ \; C) \qquad (17.81)$$

In this scheme, A* is an A molecule that has enough vibrational energy to decompose or isomerize (its vibrational energy exceeds the activation energy for the reaction A $\rightarrow$ products). A* is called an *energized* molecule. [The species A* is not an activated complex (this term will be defined in Chapter 23) but is simply an A molecule in a high vibrational energy level.] The energized species A* is produced by collision of A with an M molecule (step 1). In this collision, kinetic energy of M is transferred into vibrational energy of A. Any molecule M can excite A to a higher vibrational level. Thus M might be another A molecule, or a product molecule, or a molecule of a species present in the gas or solution but not appearing in the overall unimolecular reaction A $\rightarrow$ products. Once A* has been produced, it can either (*a*) be de-energized back to A by a collision in which the vibrational energy of A* is transferred to kinetic energy of an M molecule (step -1), or (*b*) be transformed to products B + C by having the extra vibrational energy break the appropriate chemical bond(s) to cause decomposition or isomerization (step 2).

The reaction rate is $r = d[B]/dt = k_2[A^*]$. Applying the steady-state approximation to the reactive species A^*, we have

$$d[A^*]/dt = 0 = k_1[A][M] - k_{-1}[A^*][M] - k_2[A^*]$$

$$[A^*] = \frac{k_1[A][M]}{k_{-1}[M] + k_2}$$

Substituting in $r = k_2[A^*]$, we get

$$r = \frac{k_1 k_2[A][M]}{k_{-1}[M] + k_2} \tag{17.82}$$

The rate law (17.82) has no definite order.

There are two limiting cases for (17.82). If $k_{-1}[M] \gg k_2$, the k_2 term in the denominator can be dropped to give

$$r = (k_1 k_2/k_{-1})[A] \qquad \text{for } k_{-1}[M] \gg k_2 \tag{17.83}$$

If $k_2 \gg k_{-1}[M]$, the $k_{-1}[M]$ term is omitted and

$$r = k_1[A][M] \qquad \text{for } k_2 \gg k_{-1}[M] \tag{17.84}$$

In gas-phase reactions, Eq. (17.83) is called the high-pressure limit, since at high pressures, the concentration $[M]$ is large and $k_{-1}[M]$ is much larger than k_2. Equation (17.84) is the low-pressure limit.

The high-pressure rate law (17.83) is first-order. The low-pressure rate law (17.84) is second-order but is more subtle than it looks. The concentration $[M]$ is the total concentration of all species present. If the overall reaction is an isomerization, $A \to B$, the total concentration $[M]$ remains constant as the reaction progresses and one observes pseudo-first-order kinetics. If the overall reaction is a decomposition, $A \to B + C$, then $[M]$ increases as the reaction progresses. However, the decomposition products B and C are generally less efficient (that is, have smaller k_1 values) than A in energizing A (see Prob. 17.79), and this approximately compensates for the increase in $[M]$. Thus, $k_1[M]$ remains approximately constant during the reaction, and again we have pseudo-first-order kinetics.

In the high-pressure limit where $k_{-1}[M] \gg k_2$, the rate $k_{-1}[A^*][M]$ of the de-energization reaction $A^* + M \to A + M$ is much greater than the rate $k_2[A^*]$ of $A^* \to B + C$, and steps 1 and -1 are essentially in equilibrium. The unimolecular step 2 is then rate-controlling, and we get first-order kinetics [Eq. (17.83)]. In the low-pressure limit $k_{-1}[M] \ll k_2$, the reaction $A^* \to B + C$ is much faster than the de-energization reaction, the rate-limiting step is the bimolecular energization reaction $A + M \to A^* + M$ (which is relatively slow because of the low values of $[M]$ and $[A]$), and we get second-order kinetics [Eq. (17.84)].

A key idea in the Lindemann mechanism is the time lag that exists between the energization of A to A^* and the decomposition of A^* to products. This time lag allows A^* to be de-energized back to A, and the near equilibrium of steps 1 and -1 produces first-order kinetics. In the limit of zero lifetime of A^*, the reaction would become $A + M \to B (+ C)$ and would be second-order. Note also that in the limit $k_2 \to \infty$, Eq. (17.82) gives second-order kinetics. The vibrationally excited species A^* has a nonzero lifetime because the molecule has several bonds, and it takes time for the vibrational energy to concentrate in the particular bond that breaks in the reaction $A \to$ products. It follows that a molecule with only one bond (for example, I_2) cannot decompose by a unimolecular reaction (see also Sec. 17.12).

The experimental unimolecular rate constant k_{uni} is defined by $r = k_{uni}[A]$, where r is the observed rate. Equation (17.82) gives

$$k_{uni} = \frac{k_1 k_2[M]}{k_{-1}[M] + k_2} = \frac{k_1 k_2}{k_{-1} + k_2/[M]} \tag{17.85}$$

The high-pressure limit of k_{uni} is $\lim_{P \to \infty} k_{uni} \equiv k_{uni,P=\infty} = k_1 k_2 / k_{-1}$. As the initial pressure P_0 for a run is decreased, k_{uni} decreases, since [M] decreases. At very low initial pressures, k_{uni} equals $k_1[M]$, and k_{uni} decreases linearly with decreasing P_0. This predicted falloff of k_{uni} with decreasing P_0 has been experimentally confirmed for unimolecular gas-phase reactions by measuring the initial rate r_0 as a function of the initial gas pressure P_0. Figure 17.13 shows a typical result. Significant falloff of k_{uni} from its high-pressure value usually begins in the range 10 to 200 torr.

The Lindemann expression (17.85) gives $1/k_{uni} = k_{-1}/k_1 k_2 + 1/k_1[M]$, which predicts that a plot of $1/k_{uni}$ versus $1/P_0$ will be linear. Figure 17.14 plots $1/k_{uni}$ versus $1/P_0$ for $CH_3NC \to CH_3CN$ at 230°C. The plot shows substantial nonlinearity. This is because the Lindemann scheme oversimplifies things, taking k_2 to be a constant for all A* molecules, whereas in reality, the greater the vibrational energy of A*, the greater the probability that it will isomerize or decompose.

The Lindemann mechanism also applies to reactions in liquid solutions. However, in solutions, it is not possible to observe a falloff of k_{uni} since the presence of the solvent keeps [M] high. Thus the rate law is (17.83) in solution.

Steps 1 and −1 of the Lindemann mechanism (17.81) are not elementary *chemical* reactions (since no new compounds are formed) but are elementary *physical* reactions in which energy is transferred. Such energy-transfer processes occur continually in any system. The reasons for considering steps 1 and −1 in addition to the unimolecular elementary chemical reaction of step 2 are (*a*) to explain how collisional activation can produce first-order kinetics and (*b*) to deal with the low-*P* falloff of gas-phase unimolecular rate constants. Unless one is dealing with a gas-phase system in the low-*P* range, it is not necessary to consider steps 1 and −1 explicitly, and a unimolecular reaction can be written simply as A → products.

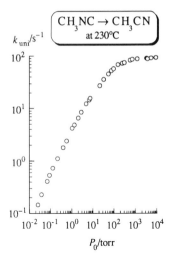

Figure 17.13

Observed rate constant for the gas-phase unimolecular reaction $CH_3NC \to CH_3CN$ at 230°C as a function of the initial pressure P_0. The scales are logarithmic.

17.12 TRIMOLECULAR REACTIONS

Trimolecular (elementary) reactions are rare. The best examples of gas-phase trimolecular reactions are the recombinations of two atoms to form a diatomic molecule. The energy released on formation of the chemical bond becomes vibrational energy of the diatomic molecule, and unless a third body is present to carry away this energy, the molecule will dissociate back to atoms during its first vibration. Thus, the recombination of two I atoms occurs as the single elementary step

$$I + I + M \to I_2 + M \qquad (17.86)$$

where M can be any atom or molecule. The observed rate law is $r = k[I]^2[M]$. A reaction like $CH_3 + CH_3 \to C_2H_6$ does not require a third body, since the extra vibrational energy attained on formation of the C_2H_6 molecule can be distributed among vibrations of several bonds and no bond vibration need be energetic enough to break that bond. A few special cases are known in which the recombination of atoms can occur in the absence of a third body, the excess energy being removed by emission of light from an excited state of the molecule.

The rate constant for (17.86) has been measured as a function of T in flash-photolysis experiments (Sec. 17.14). Since no bonds are broken in (17.86), one would expect it to have zero activation energy. Actually, the rate constant *decreases* with increasing T, indicating a negative E_a [see Eq. (17.68)]. Increasing the temperature increases the trimolecular collision rate. However, as the energy of a trimolecular collision increases, the probability decreases that a given I + I + M collision will result in transfer of energy to M accompanied by formation of I_2. The activation energy for the recombination of atoms (A + B + M → AB + M) is typically 0 to −4 kcal/mol (0 to −17 kJ/mol).

The decomposition of I_2 (or any diatomic molecule) must occur by the reverse of (17.86) (see Sec. 17.6). Thus, the diatomic molecule AB decomposes by the

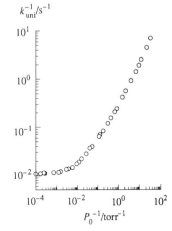

Figure 17.14

Plot of $1/k$ versus $1/P_0$ for $CH_3NC \to CH_3CN$ at 230°C.

bimolecular reaction $AB + M \rightarrow A + B + M$, where the atoms A and B may be the same (as in I_2) or different (as in HCl). Since E_a for (17.86) is slightly negative, Eq. (17.71) shows that E_a for the decomposition of a diatomic molecule is slightly less than $\Delta U°$ for the decomposition. For decomposition of a polyatomic molecule to two radicals, E_a equals $\Delta U°$ since E_a for the recombination is zero (as noted in Sec. 17.8).

A recombination reaction to give a triatomic molecule often requires a third body M to carry away energy. A triatomic molecule has only two bonds, and the extra vibrational energy produced by recombination might rapidly concentrate in one bond and dissociate the molecule unless a third body is present. For example, the recombination of O_2 and O is $O + O_2 + M \rightarrow O_3 + M$. Reversal of this elementary reaction shows that O_3 decomposes by a bimolecular step (see Prob. 17.52). Molecules with several bonds can decompose by a unimolecular reaction and do not require a third body when they are formed in a recombination reaction.

The gas-phase reactions of NO with Cl_2, Br_2, and O_2 are kinetically third-order. Some people believe the mechanism to be a single elementary trimolecular step (for example, $2NO + Cl_2 \rightarrow 2NOCl$), but others consider the mechanism to be two bimolecular steps, as in (17.60) or (17.61).

In solutions, trimolecular (elementary) reactions are also uncommon.

17.13   CHAIN REACTIONS AND FREE-RADICAL POLYMERIZATIONS

A **chain reaction** contains a series of steps in which a reactive intermediate is consumed, reactants are converted to products, and the intermediate is regenerated. Regeneration of the intermediate allows this cycle to be repeated over and over again. Thus a small amount of intermediate produces a large amount of product. *Most combustions, explosions, and addition polymerizations are chain reactions and usually involve free radicals as intermediates.*

One of the best-understood chain reactions is that between H_2 and Br_2. The overall stoichiometry is $H_2 + Br_2 \rightarrow 2HBr$. The observed rate law for this gas-phase reaction in the temperature range 500 to 1500 K is

$$r = \frac{1}{2} \frac{d[HBr]}{dt} = \frac{k[H_2][Br_2]^{1/2}}{1 + j[HBr]/[Br_2]} \tag{17.87}$$

where k and j are constants. Since the stoichiometric coefficient is 2 for HBr, a factor $\frac{1}{2}$ is included in (17.87) [see Eq. (17.4)]. The constant j has only a very slight temperature dependence and equals 0.12. Since an increase in [HBr] decreases r in (17.87), the product HBr is said to *inhibit* the reaction.

The $\frac{1}{2}$-power dependence of r on $[Br_2]$ suggests that the mechanism involves splitting a Br_2 molecule (recall rule 3 in Sec. 17.6). Br_2 can split only into two Br atoms. A Br atom can then react with H_2 to give HBr and H. Each H atom produced can then react with Br_2 to give HBr and Br, thereby regenerating the reactive intermediate Br. The mechanism of the reaction is thus believed to be

$$Br_2 + M \underset{k_{-1}}{\overset{k_1}{\rightleftharpoons}} 2Br + M$$

$$Br + H_2 \underset{k_{-2}}{\overset{k_2}{\rightleftharpoons}} HBr + H \tag{17.88}$$

$$H + Br_2 \overset{k_3}{\rightarrow} HBr + Br$$

In step 1, a Br_2 molecule collides with any species M, thereby gaining the energy to dissociate into two Br atoms. Step -1 is the reverse process, in which two Br atoms

Figure 17.15

Scheme for the gas-phase H_2 +
$Br_2 \rightarrow 2HBr$ chain reaction.

recombine to form Br_2, the third body M being needed to carry away part of the bond energy released (see Sec. 17.12). Step 1 is the **initiation step,** since it generates the chain-carrying reactive radical Br. Step -1 is the **termination** (or *chain-breaking*) **step,** since it removes Br.

Steps 2 and 3 form a **chain** that consumes Br, converts H_2 and Br_2 into HBr, and regenerates Br (Fig. 17.15). Steps 2 and 3 are **propagation steps.** Step -2 (HBr + H $\rightarrow$ Br + H_2) is an **inhibition step,** since it destroys the product HBr and therefore decreases r. Note from steps -2 and 3 that HBr and Br_2 compete for H atoms. This competition leads to the $j[HBr]/[Br_2]$ term in the denominator of r. For each Br atom produced by step 1, we get many repetitions of steps 2 and 3 (we shall see below that $k_1 \ll k_2$ and $k_1 \ll k_3$). The reactive intermediates H and Br that occur in the chain-propagating steps are called **chain carriers.** Adding steps 2 and 3, we get Br + H_2 + $Br_2 \rightarrow 2HBr$ + Br, which agrees with the overall stoichiometry $H_2 + Br_2 \rightarrow 2HBr$. Other possible steps (for example, the reverse of reaction 3) are too slow to contribute to the mechanism; see Prob. 17.82.

The mechanism (17.88) gives the rate of product formation as

$$d[HBr]/dt = r_2 - r_{-2} + r_3 = k_2[Br][H_2] - k_{-2}[HBr][H] + k_3[H][Br_2] \quad (17.89)$$

where r_2, r_{-2}, and r_3 are the rates of steps 2, -2, and 3. Equation (17.89) contains the concentrations of the free-radical intermediates H and Br. Applying the steady-state approximation to these reactive intermediates, we get

$$d[H]/dt = 0 = r_2 - r_{-2} - r_3 \quad (17.90)$$

$$d[Br]/dt = 0 = 2r_1 - 2r_{-1} - r_2 + r_{-2} + r_3 \quad (17.91)$$

The factors of 2 are present because Eq. (17.4) gives $r_1 = \frac{1}{2}(d[Br]/dt)_1$ for step 1 and, similarly, for step -1. Addition of (17.90) and (17.91) gives $0 = 2r_1 - 2r_{-1}$. Thus

$$r_1 = r_{-1}$$

$$k_1[Br_2][M] = k_{-1}[Br]^2[M]$$

$$[Br] = (k_1/k_{-1})^{1/2}[Br_2]^{1/2} \quad (17.92)$$

The equation $r_1 = r_{-1}$ states that the initiation rate equals the termination rate. This is a consequence of the steady-state approximation. To find [H], we use (17.90), which gives

$$0 = k_2[Br][H_2] - k_{-2}[HBr][H] - k_3[H][Br_2]$$

Substituting (17.92) for [Br] into this equation and solving for [H], we get

$$[H] = \frac{k_2(k_1/k_{-1})^{1/2}[Br_2]^{1/2}[H_2]}{k_3[Br_2] + k_{-2}[HBr]} = \frac{k_2(k_1/k_{-1})^{1/2}[H_2][Br_2]^{-1/2}}{k_3 + k_{-2}[HBr]/[Br_2]} \quad (17.93)$$

By substituting (17.93) and (17.92) into (17.89), we can find $d[HBr]/dt$ as a function of $[H_2]$, $[Br_2]$, and [HBr]. To avoid the algebra involved, we note from (17.90) that $r_2 = r_{-2} + r_3$. Substitution of this expression into (17.89) gives $d[HBr]/dt = 2r_3 = 2k_3[H][Br_2]$. Substitution of (17.93) for [H] gives the desired result:

$$r = \frac{1}{2}\frac{d[HBr]}{dt} = \frac{k_2(k_1/k_{-1})^{1/2}[H_2][Br_2]^{1/2}}{1 + (k_{-2}/k_3)[HBr]/[Br_2]} \quad (17.94)$$

which agrees with the observed form of the rate law, Eq. (17.87). We have

$$k = k_2(k_1/k_{-1})^{1/2} \qquad \text{and} \qquad j = k_{-2}/k_3 \qquad (17.95)$$

The activation energies for the steps in this mechanism can be estimated. The trimolecular recombination reaction, step -1, must have an E_a that is essentially zero or even slightly negative (see Sec. 17.12); we thus take $E_{a,-1} \approx 0$. For step 1, thermodynamic data give $\Delta U° = 45$ kcal/mol, and Eq. (17.71) therefore gives $E_{a,1} \approx 45$ kcal/mol. The ratio k_1/k_{-1} in (17.95) is the equilibrium constant $K_{c,1}$ for the elementary reaction $Br_2 \rightleftharpoons 2Br$, and $K_{c,1}(T)$ can be found from thermodynamic data. The rate constant k in (17.87) is known as a function of T from measurement of r at different temperatures. The first equation in (17.95) then enables us to find the elementary rate constant k_2 as a function of T. The use of (17.68) then gives the activation energy for step 2. The result is $E_{a,2} = 18$ kcal/mol. Thermodynamic data give $\Delta U°$ for reaction 2 as 17 kcal/mol; Eq. (17.71) then gives $E_{a,-2} = 1$ kcal/mol. For the constant j, we have from (17.95) and (17.66):

$$j = k_{-2}/k_3 = (A_{-2}/A_3)e^{(E_{a,3} - E_{a,-2})/RT}$$

Since j is observed to be essentially independent of T, we must have $E_{a,3} \approx E_{a,-2} = 1$ kcal/mol. Note that $E_{a,1}$ (45 kcal/mol) is much greater than $E_{a,2}$ (18 kcal/mol) and $E_{a,3}$ (1 kcal/mol).

In the mechanism (17.88), the chain reaction is *thermally* initiated by heating the reaction mixture to a temperature at which some Br_2–M collisions have enough relative kinetic energy to dissociate Br_2 to the chain carrier Br. The H_2–Br_2 chain reaction can also be initiated *photochemically* (at lower temperatures than required for the thermal reaction) by absorption of light, which dissociates Br_2 to 2Br. Still another way to initiate a chain reaction is by addition of a substance (called an **initiator**) that reacts to produce chain carriers. For example, Na vapor added to an H_2–Br_2 mixture will react with Br_2 to give the chain carrier Br: $Na + Br_2 \rightarrow NaBr + Br$.

Since each atom or molecule of chain carrier produces many product molecules, a small amount of any substance that destroys chain carriers will greatly slow down (or *inhibit*) a chain reaction. For example, NO can combine with the chain-carrying radical CH_3 to give CH_3NO. O_2 is an inhibitor in the H_2–Cl_2 chain reaction, since it combines with Cl atoms to give ClO_2. Vitamin E inhibits the chain-reaction peroxidation of lipids (fats) in organisms by reacting with the radical ROO.

Each of the steps 2 and 3 in the chain of the H_2–Br_2 reaction consumes one chain carrier and produces one chain carrier. In certain chain reactions, the chain produces more chain carriers than it consumes. This is a **branching chain reaction.** For a branching chain reaction, the reaction rate may increase rapidly as the reaction proceeds, and this increase may lead to an explosion. Obviously, the steady-state approximation doesn't apply in such a situation. One of the most studied branching chain reactions is the combustion of hydrogen: $2H_2 + O_2 \rightarrow 2H_2O$. The chain-branching steps include $H + O_2 \rightarrow OH + O$ and $O + H_2 \rightarrow OH + H$. Each of these reactions produces two chain carriers and consumes only one.

A highly exothermic reaction that is not a chain reaction may lead to an explosion if the heat of the reaction is not transferred rapidly enough to the surroundings. The increase in system temperature increases the reaction rate until the system explodes.

Gaseous hydrocarbon combustions are very complicated branching chain reactions. The combustion of CH_4 involves at least 22 elementary reactions and 12 species, including CH_3, CH_3O, CH_2O, HCO, H, O, OH, and OOH (see *Steinfeld, Francisco, and Hase*, sec. 14.3).

Formation of atmospheric smog involves chain reactions that oxidize hydrocarbons. Initiation occurs photochemically when the pollutant gas NO_2 absorbs light and dissociates to give NO and O. The O atoms react with O_2 to give O_3. Dissociation of O_3 by high-frequency ultraviolet light then gives O_2 and O*, where O* denotes O atoms in a certain excited electronic state. The reaction $O^* + H_2O \rightarrow 2OH$ produces chain-carrying OH radicals that attack pollutant gaseous hydrocarbons. ("The key to understanding tropospheric chemistry [lies] in the reactions of the hydroxyl radical"; J. H. Seinfeld and S. Pandis, *Atmospheric Chemistry and Physics,* Wiley, 1998, p. 240.)

Addition polymers are formed by chain reactions. Thus, polymerization of ethylene can be initiated by an organic radical $R\cdot$ (the radical being formed by thermal decomposition of, for example, an organic peroxide): $R\cdot + CH_2{=}CH_2 \rightarrow RCH_2CH_2\cdot$. The reactive radical product then attacks another monomer, generating another radical: $RCH_2CH_2\cdot + CH_2{=}CH_2 \rightarrow RCH_2CH_2CH_2CH_2\cdot$. Addition of monomer continues until terminated by, for example, the combination of two polymeric radicals.

Free-Radical Polymerization

We now develop the kinetics of liquid-phase free-radical addition polymerizations. These are carried out either in a solvent or in the pure monomer, with some initiator added. Let I and M stand for the initiator and the monomer. The reaction mechanism is

$$I \xrightarrow{\;k_i\;} 2R\cdot$$

$$R\cdot + M \xrightarrow{\;k_a\;} RM\cdot$$

$$RM\cdot + M \xrightarrow{\;k_{p1}\;} RM_2\cdot, \quad RM_2\cdot + M \xrightarrow{\;k_{p2}\;} RM_3\cdot, \ldots$$

$$RM_m\cdot + RM_n\cdot \xrightarrow{\;k_{t,mn}\;} RM_{m+n}R \quad \text{for } m = 0, 1, 2, \ldots, \quad n = 0, 1, 2, \ldots$$

In the initiation step with rate constant k_i, the initiator thermally decomposes to a small extent to yield $R\cdot$ radicals. An example is the decomposition of benzoyl peroxide: $(C_6H_5COO)_2 \rightarrow 2C_6H_5COO\cdot$. In the addition step $R\cdot + M \rightarrow RM\cdot$ with rate constant k_a, $R\cdot$ adds to the monomer. In the propagation steps with rate constants $k_{p1}, k_{p2}, \ldots$, monomers add to the growing chain. In the termination steps, chains combine to yield polymer molecules. In some cases, termination occurs mainly by transfer of an H atom between $RM_n\cdot$ and $RM_m\cdot$ (*disproportionation*) to yield two polymer molecules, one of which has a terminal double bond.

To keep life simple, one assumes that the radical reactivities are independent of size so that all propagation steps have the same rate constant, which we call k_p:

$$k_{p1} = k_{p2} = \cdots \equiv k_p$$

(For discussion of this approximation, see *Allcock and Lampe,* pp. 283–284.) Likewise, we assume that radical size does not affect the termination rate constants. However, $k_{t,mn}$ does depend on whether or not m equals n. The rates $d[RM_{m+n}R]/dt$ and $d[RM_{2n}R]/dt$ for the elementary termination reactions $RM_m\cdot + RM_n\cdot \rightarrow RM_{m+n}R$ and $2RM_n\cdot \rightarrow RM_{2n}R$ are proportional to the rate at which the reactant radicals encounter each other in unit volume of solution. Just as the collision rate per unit volume Z_{bb} of Eq. (15.63) is obtained by putting $c = b$ in Z_{bc} and multiplying by $\frac{1}{2}$, the encounter rate per unit volume for like radicals contains an extra factor of $\frac{1}{2}$ as compared with the encounter rate per unit volume for unlike radicals. Hence the rate constant for termination between like radicals is $\frac{1}{2}$ that for termination between unlike radicals: $k_{t,nn} = \frac{1}{2}k_{t,mn}$ for $m \neq n$. Letting k_t denote the termination rate constant for like radicals, we have

$$k_t = k_{t,nn} \quad \text{for all } n \qquad \text{and} \qquad k_{t,mn} = 2k_t \quad \text{for } m \neq n \qquad (17.96)$$

The rate r_M of consumption of the monomer is

$$r_M = -d[M]/dt = k_a[R\cdot][M] + k_p[RM\cdot][M] + k_p[RM_2\cdot][M] + \cdots$$

$$-\frac{d[M]}{dt} \approx k_p[M]\sum_{n=0}^{\infty}[RM_n\cdot] \equiv k_p[M][R_{tot}\cdot] \qquad (17.97)$$

where $[R_{tot}\cdot]$ is the total concentration of all radicals. Since there are hundreds or thousands of terms of significant magnitude in (17.97), the approximation of changing k_a to k_p in the first term is of no consequence.

To find $[R_{tot}\cdot]$ and hence $-d[M]/dt$, we apply the steady-state approximation to each radical: $d[R\cdot]/dt = 0$, $d[RM\cdot]/dt = 0$, $d[RM_2\cdot]/dt = 0$, ... Addition of these equations gives

$$d[R_{tot}\cdot]/dt = 0 \qquad (17.98)$$

where $[R_{tot}\cdot] \equiv \sum_{n=0}^{\infty} [RM_n\cdot]$.

In the addition step $R\cdot + M \to RM\cdot$ and in each propagation step, one radical is consumed and one radical is produced. Hence these steps do not affect $[R_{tot}\cdot]$ and $d[R_{tot}\cdot]/dt$. We therefore need consider only the initiation and termination steps in applying the steady-state condition $d[R_{tot}\cdot]/dt = 0$.

The contribution of the initiation step to $d[R_{tot}\cdot]/dt$ equals $(d[R\cdot]/dt)_i$, the rate at which $R\cdot$ is formed in the initiation step. Not all radicals $R\cdot$ initiate polymer chains. Some recombine to I in the solvent "cage" (Sec. 17.15) that surrounds them, and others are lost by reacting with the solvent. Therefore one writes

$$(d[R_{tot}\cdot]/dt)_i = (d[R\cdot]/dt)_i = 2fk_i[I] \qquad (17.99)$$

where f is the fraction of radicals $R\cdot$ that react with M. Typically, $0.3 < f < 0.8$.

Termination of $RM_n\cdot$ occurs by $2RM_n\cdot \to RM_{2n}R$ or by $RM_n\cdot + RM_m\cdot \to RM_{n+m}R$ with $m = 0, 1, 2, \ldots$, but $m \neq n$. The contribution of the termination steps to the rate of disappearance of those radicals that contain n monomers is

$$\left(\frac{d[RM_n\cdot]}{dt}\right)_t = -2k_{t,nn}[RM_n\cdot]^2 - k_{t,mn}[RM_n\cdot]\sum_{m \neq n}[RM_m\cdot]$$

$$= -2k_t[RM_n\cdot]\sum_{m=0}^{\infty}[RM_m\cdot] = -2k_t[RM_n\cdot][R_{tot}\cdot] \quad (17.100)$$

where (17.96) was used. The total rate of consumption of radicals in the termination steps is found by summing over (17.100):

$$\left(\frac{d[R_{tot}\cdot]}{dt}\right)_t = \sum_{n=0}^{\infty}\left(\frac{d[RM_n\cdot]}{dt}\right)_t = -2k_t[R_{tot}\cdot]\sum_{n=0}^{\infty}[RM_n\cdot] = -2k_t[R_{tot}\cdot]^2 \quad (17.101)$$

Adding (17.99) and (17.101) and applying the steady-state approximation, we have

$$d[R_{tot}\cdot]/dt = (d[R_{tot}\cdot]/dt)_i + (d[R_{tot}\cdot]/dt)_t = 2fk_i[I] - 2k_t[R_{tot}\cdot]^2 = 0$$

$$[R_{tot}\cdot] = (fk_i/k_t)^{1/2}[I]^{1/2} \qquad (17.102)$$

Substitution in (17.97) gives

$$-d[M]/dt = k_p(fk_i/k_t)^{1/2}[M][I]^{1/2} \qquad (17.103)$$

The reaction is first-order in monomer and $\frac{1}{2}$-order in initiator.

The **degree of polymerization** DP of a polymer molecule is the number of monomers in the polymer. In a tiny time dt during the polymerization reaction, suppose that 10^4 monomer molecules M are consumed and 10 polymer molecules with various chain lengths are produced. By the steady-state approximation, the concentrations of the intermediates $RM\cdot$, $RM_2\cdot$, etc., are not changing significantly. Therefore, by conservation of matter, the 10 polymer molecules must contain a total of 10^4 monomer units, and the average degree of polymerization during this time interval is $\langle DP \rangle = 10^4/10 = 10^3$. We see that $\langle DP \rangle = -d[M]/d[P_{tot}]$, where $[P_{tot}]$ is the total concentration of polymer molecules:

$$\langle DP \rangle = \frac{-d[M]}{d[P_{tot}]} = \frac{-d[M]/dt}{d[P_{tot}]/dt} \qquad (17.104)$$

Since one polymer molecule is formed when two radicals combine, the rate of polymer formation is half the rate of consumption of radicals in the termination steps:

$$d[P_{tot}]/dt = -\tfrac{1}{2}(d[R_{tot}\cdot]/dt)_t = k_t[R_{tot}\cdot]^2 = fk_i[I] \qquad (17.105)$$

where (17.101) and (17.102) were used.

Substitution of (17.103) and (17.105) in (17.104) gives

$$\langle DP \rangle = \frac{k_p[M]}{(fk_ik_t)^{1/2}[I]^{1/2}} \qquad \text{for termination by combination} \qquad (17.106)$$

If the dominant mode of termination is disproportionation, $\langle DP \rangle$ is half (17.106). A low initiator concentration compared with monomer favors a high $\langle DP \rangle$.

Liquid-phase polymerization reactions are usually run at temperatures for which k_i is in the range 10^{-5} to 10^{-6} s^{-1}. At 50°C, k_t is typically 10^6 to 10^9 dm^3 mol^{-1} s^{-1} (the high values are due to the high reactivity of radicals with one another), and k_p is typically 10^2 to 10^4 dm^3 mol^{-1} s^{-1}. For [M] = 5 mol/dm^3 and [I] = 0.01 mol/dm^3, we find that typically $[R_{tot}\cdot] = 10^{-8}$ mol/dm^3 and $\langle DP \rangle = 7000$. Despite the fact that $k_t \gg k_p$, the very low concentration of radicals compared with monomers makes it much more likely for a radical to react with a monomer than with another radical. Therefore the chain grows quite long before termination.

17.14 FAST REACTIONS

Rate constants span an enormous range of values. In aqueous solution, the fastest known second-order elementary reaction is $H_3O^+(aq) + OH^-(aq) \rightarrow 2H_2O$, for which $k = 1.4 \times 10^{11}$ dm^3 mol^{-1} s^{-1} at 25°C. For gas-phase reactions, an upper limit to rate constants is set by the collision rate. Using (15.62) for Z_{BC}, one finds (Prob. 17.110) that if reaction occurs at every collision (this is not true for most reactions), k for the elementary reaction $B(g) + C(g) \rightarrow$ products will be about 10^{11} dm^3 mol^{-1} s^{-1} at 300 K. Gas-phase radical combination reactions (for example, $2Cl_3C\cdot \rightarrow C_2Cl_6$) often have k values of this magnitude. There is no lower limit on rate constants. The rates of extremely slow reactions can be measured by radioactively labeling a reactant, allowing the reaction to run for a few weeks, separating a radioactive product from the reaction mixture, and measuring its radioactivity. With this method, second-order rate constants as small as 10^{-12} dm^3 mol^{-1} s^{-1} have been measured, and a first-order reaction with a half-life of 10^5 years was followed; see the references in E. S. Lewis et al., *J. Org. Chem.*, **34**, 255 (1969).

Experimental Methods for Fast Reactions

Many reactions are too fast to follow by the classical methods discussed in Sec. 17.2. One way to study fast reactions is with *rapid-flow* methods. A schematic diagram of a liquid-phase *continuous-flow* system is given in Fig. 17.16. Reactants A and B are rapidly driven into the mixing chamber M by pushing in the plungers of the syringes. Mixing occurs in $\tfrac{1}{2}$ to 1 ms. The reaction mixture then flows through the narrow observation tube. At point P along the tube one measures the light absorption at a wavelength

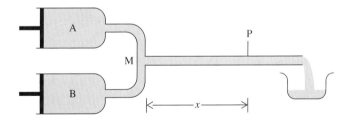

Figure 17.16

A continuous-flow system with rapid mixing of reactants.

Figure 17.17

A stopped-flow system.

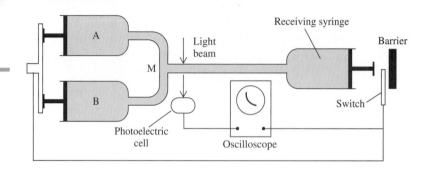

at which one species absorbs to determine the concentration of that species. For gas-phase reactions, the syringes are replaced by bulbs of gases A and B, and flow is caused by pumping at the exit of the observation tube.

Let the speed at which the mixture flows through the observation tube be v and the distance between the mixing chamber M and the observation point P be x. Then the rule "distance equals rate times time" gives the time t after the reaction started as $t = x/v$. For the typical values $v = 1000$ cm/s and $x = 10$ cm, observation at P gives the concentration of a species 10 ms after the reaction has begun. Because the mixture at point P is continuously replenished with newly mixed reactants, the concentrations of species remain constant at P. By varying the observation distance x and the flow speed v, one obtains reactant concentrations at various times.

A modification of the continuous-flow method is the *stopped-flow* method (Fig. 17.17). Here, the reactants are mixed at M and rapidly flow through the observation tube into the receiving syringe, driving its plunger against a barrier and thereby stopping the flow. This plunger also hits a switch which stops the motor-driven plungers and triggers the oscilloscope sweep. One observes the light absorption at P as a function of time, using a photoelectric cell, which converts the light signal to an electrical signal that is displayed on the oscilloscope screen. Because of the rapid mixing and flow and the short distance between M and P, the reaction is observed essentially from its start. The stopped-flow method is actually a static method with rapid mixing, rather than a flow method.

The continuous-flow and stopped-flow methods are applicable to reactions with half-lives in the range 10^1 to 10^{-3} s.

The main limitation on the rapid-flow techniques is set by the time required to mix the reactants. The mixing problem is eliminated in *relaxation* methods. Here, one takes a system in reaction equilibrium and suddenly changes one of the variables that determine the equilibrium position. By following the approach of the system to its new equilibrium position, one can determine rate constants. Details of the calculation are given later in this section. Relaxation methods are useful mainly for liquid-phase reactions. The scientific meaning of **relaxation** is the approach of a system to a new equilibrium position after it has been perturbed.

The most common relaxation method is the *temperature-jump (T-jump) method*. Here, one abruptly discharges a high-voltage capacitor through the solution, raising its temperature from T_1 to T_2 in about 1 μs (1 microsecond = 10^{-6} s). Typically, $T_2 - T_1$ is 3°C to 10°C. Discharge of the capacitor triggers the oscilloscope sweep, which displays the light absorption of the solution as a function of time, as in the stopped-flow method. The oscilloscope trace is photographed. Another way to heat the solution in 1 μs is by a pulse of microwave radiation (recall microwave ovens). This method has the advantage of being applicable to nonconducting solutions and the disadvantage of producing only a small temperature rise ($\approx$ 1°C). Provided $\Delta H°$ is not zero, Eq. (11.32) shows that the equilibrium constant $K(T_2)$ differs from $K(T_1)$.

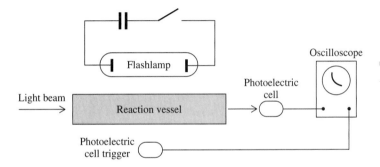

Figure 17.18

A flash-photolysis experiment.

In the *pressure-jump method,* a sudden change in P shifts the equilibrium [see Eq. (11.33)]. In the *electric-field-jump method,* a suddenly applied electric field shifts the equilibrium of a reaction that involves a change in total dipole moment.

A limitation on relaxation methods is that the reaction must be reversible, with detectable amounts of all species present at equilibrium.

Rapid-flow and relaxation techniques have been used to measure the rates of proton-transfer (acid–base) reactions, electron-transfer (redox) reactions, complex-ion-formation reactions, ion-pair-formation reactions, and enzyme–substrate-complex formation reactions.

Relaxation methods apply rather small perturbations to a system and do not generate new chemical species. The flash-photolysis and shock-tube methods apply a large perturbation to a system, thereby generating one or more reactive species whose reactions are then followed.

In *flash photolysis* (Fig. 17.18), one exposes a gaseous or liquid system to a high-intensity, short-duration flash of visible and ultraviolet light. Molecules that absorb the light either dissociate to radicals or are excited to high-energy states. The reactions of these species are followed by measurement of light absorption. The oscilloscope sweep in Fig. 17.18 is triggered by a photoelectric cell that detects light from the flash. With a flashlamp, the light flash lasts about 10 μs. Use of a laser instead of a flashlamp typically gives a pulse of 10-ns duration (1 nanosecond $= 10^{-9}$ s). With special techniques, one can generate a laser light pulse with the incredibly short duration of 0.1 ps (1 picosecond $= 10^{-12}$ s), allowing rate processes in the picosecond range to be studied. Picosecond spectroscopy is used to study the mechanisms of photosynthesis and vision.

In a *shock tube,* the reactant gas mixture at low pressure is separated by a thin diaphragm from an inert gas at high pressure. The diaphragm is punctured, causing a shock wave to travel down the tube. The sudden great increase in pressure and temperature produces excited states and free radicals. The reactions of these species are then followed by observation of their absorption spectra.

Nuclear-magnetic-resonance spectroscopy can be used to measure the rates of certain rapid isomerization and exchange reactions (see Sec. 21.12).

Kinetics of Relaxation

In relaxation methods, a system in equilibrium suffers a small perturbation that changes the equilibrium constant. The system then relaxes to its new equilibrium position. We now integrate the rate law for a typical case.

Consider the reversible *elementary* reaction

$$A + B \underset{k_b}{\overset{k_f}{\rightleftharpoons}} C$$

where the forward and reverse rate laws are $r_f = k_f[A][B]$ and $r_b = k_b[C]$. Suppose the temperature is suddenly changed. For all times after the T jump, we have

$$d[A]/dt = -k_f[A][B] + k_b[C] \qquad (17.107)$$

Let $[A]_{eq}$, $[B]_{eq}$, and $[C]_{eq}$ be the equilibrium concentrations at the new temperature T_2, and let $x \equiv [A]_{eq} - [A]$. For every mole of A that reacts, 1 mole of B reacts and 1 mole of C is formed. Hence $[B]_{eq} - [B] = x$ and $[C]_{eq} - [C] = -x$. Also, $d[A]/dt = -dx/dt$. Equation (17.107) becomes

$$-dx/dt = -k_f([A]_{eq} - x)([B]_{eq} - x) + k_b([C]_{eq} + x)$$

$$dx/dt = k_f[A]_{eq}[B]_{eq} - k_b[C]_{eq} - xk_f([A]_{eq} + [B]_{eq} + k_b k_f^{-1} - x) \qquad (17.108)$$

When equilibrium is reached, $d[A]/dt = 0$, and (17.107) gives

$$k_f[A]_{eq}[B]_{eq} - k_b[C]_{eq} = 0 \qquad (17.109)$$

Since the perturbation is small, the deviation x of [A] from its equilibrium value is small and $x \ll [A]_{eq} + [B]_{eq}$. Using (17.109) and neglecting the x in parentheses in (17.108), we get

$$dx/dt = -\tau^{-1}x \qquad \text{where } \tau \equiv \{k_f([A]_{eq} + [B]_{eq}) + k_b\}^{-1} \qquad (17.110)$$

This integrates to $x = x_0 e^{-t/\tau}$, where x_0 is the value of x the instant after the T jump is applied at $t = 0$. Since $x = [A]_{eq} - [A] = [C] - [C]_{eq}$, we have

$$[A] - [A]_{eq} = ([A]_0 - [A]_{eq})e^{-t/\tau}$$

with similar equations holding for [B] and [C]. Thus the approach of each species to its new equilibrium value is first-order with rate constant $1/\tau$ [see Eq. (17.14)]. This result holds when any elementary reaction is subjected to a *small* perturbation from equilibrium. However, the definition of τ depends on the stoichiometry of the elementary reaction. See Prob. 17.87 for another example. The constant τ is called the *relaxation time*; τ is the time it takes the deviation $[A] - [A]_{eq}$ to drop to $1/e$ of its initial value.

EXAMPLE 17.11 Relaxation

For the elementary reaction $H^+ + OH^- \rightleftharpoons H_2O$, the relaxation time has been measured as 36 μs at 25°C. Find k_f.

The reaction has the form $A + B \rightleftharpoons C$, so the preceding treatment applies. Equations (17.110) and (17.109) give

$$\tau^{-1} = k_f([H^+]_{eq} + [OH^-]_{eq}) + k_b \quad \text{and} \quad k_b[H_2O]_{eq} = k_f[H^+]_{eq}[OH^-]_{eq}$$

Eliminating k_b and using $[OH^-]_{eq} = [H^+]_{eq}$, we get

$$\tau^{-1} = k_f(2[H^+]_{eq} + [H^+]_{eq}^2/[H_2O]_{eq})$$

Using $[H^+]_{eq} = 1.0 \times 10^{-7}$ mol/dm³ and $[H_2O]_{eq} = 55.5$ mol/dm³, we find $k_f = 1.4 \times 10^{11}$ dm³ mol⁻¹ s⁻¹.

Exercise

If the temperature of pure water is suddenly raised from 20°C to 25°C at 1 atm, how long will it take for $[H^+]$ to reach 0.99×10^{-7} mol/L? $10^{14} K^{\circ}_{c,w}$ is 1.00 at 25°C and is 0.67 at 20°C. (*Answer:* 1.0×10^{-4} s.)

17.15   REACTIONS IN LIQUID SOLUTIONS

Most of the ideas of the previous sections of this chapter apply to both gas-phase and liquid-phase kinetics. We now examine aspects of reaction kinetics unique to reactions in liquid solutions.

Solvent Effects on Rate Constants

The difference between a gas-phase and a liquid-phase reaction is the presence of the solvent. The reaction rate can depend strongly on the solvent used. For example, rate constants at 25°C for the second-order substitution reaction $CH_3I + Cl^- \rightarrow CH_3Cl + I^-$ in three different amide solvents are 0.00005, 0.00014, and 0.4 dm^3 mol^{-1} s^{-1} in $HC(O)NH_2$, $HC(O)N(H)CH_3$, and $HC(O)N(CH_3)_2$, respectively. Thus the rate constant k for a given reaction is a function of the solvent as well as the temperature.

Solvent effects on reaction rates have many sources. The reacting species are usually **solvated** (that is, are bound to one or more solvent molecules), and the degree of solvation changes with change in solvent, thus affecting k. Certain solvents may catalyze the reaction. Most reactions in solution involve ions or polar molecules as reactants or reaction intermediates, and here the electrostatic forces between the reacting species depend on the solvent's dielectric constant. The rate of very fast reactions in solution may be limited by the rate at which two reactant molecules can diffuse through the solvent to encounter each other, and here the solvent's viscosity influences k. Hydrogen bonding between solvent and a reactant can affect k.

For a reaction that can occur by two competing mechanisms, the rates of these mechanisms may be affected differently by a change in solvent, so the mechanism can differ from one solvent to another.

For certain unimolecular reactions and certain bimolecular reactions between species of low polarity, k is essentially unchanged on going from one solvent to another. For example, rate constants at 50°C for the bimolecular Diels–Alder dimerization of cyclopentadiene ($2C_5H_6 \rightarrow C_{10}H_{12}$) are 6×10^{-6} dm^3 mol^{-1} s^{-1} in the gas phase; 6×10^{-6}, 10×10^{-6}, and 20×10^{-6} dm^3 mol^{-1} s^{-1} in the solvents CS_2, C_6H_6, and C_2H_5OH, respectively.

When the solvent is a reactant, it is usually not possible to determine the order with respect to solvent.

Ionic Reactions

In gas-phase kinetics, reactions involving ions are rare. In solution, ionic reactions are abundant. The difference is due to solvation of ions in solution, which sharply reduces $\Delta H°$ (and hence $\Delta G°$) for ionization.

Ionic gas-phase reactions occur when the energy needed to ionize molecules is supplied by outside sources. In a mass spectrometer, bombardment by an electron beam knocks electrons out of gas-phase molecules, and the kinetics of ionic reactions in gases can be studied in a mass spectrometer. In the earth's upper atmosphere, absorption of light produces O_2^+, N_2^+, O^+, and He^+ ions, which then react. In gases at 10^4 K and above, collisions produce very substantial ionization. Such a gas is called a **plasma** (examples include sparks and atmospheres of stars).

Information about the effects of a solvent on an ionic reaction can be obtained by studying the reaction in the gas phase using a technique called *ion cyclotron resonance* (ICR) and comparing the results with those in solution. For example, to study the gas-phase reaction $Cl^- + CH_3Br \rightarrow CH_3Cl + Br^-$, one passes a pulse of electrons through CCl_4 vapor to produce Cl^- ions. CH_3Br vapor is then introduced. After a timed interval, the amount of Cl^- present is determined by measuring the amount of energy these ions absorb from an oscillating applied electric field while the ions move on a circle in an applied magnetic field.

One finds that the gas-phase rate constant for $Cl^- + CH_3Br \rightarrow CH_3Cl + Br^-$ is 3×10^9 times that in acetone and 10^{15} times that in water. In water, the solvation shells around the Cl^- and CH_3Br species must be partly disrupted before these species can come in contact, and this requires a substantial amount of activation energy, making the reaction far slower than in the gas phase. The reaction is faster in acetone than in water, because the degree of solvation is less in acetone. For more on ICR, see R. T. McIver, *Scientific American,* Nov. 1980, p. 186; *Bernasconi,* pt. I, chap. XIV.

For ionic reactions in solution, activity coefficients must be used in analyzing kinetic data (Sec. 17.10). Since the activity coefficients are frequently unknown, one often adds a substantial amount of an inert salt to keep the ionic strength (and hence the activity coefficients) essentially constant during the reaction. The apparent rate constant obtained then depends on the ionic strength.

Many ionic reactions in solution are extremely fast; see the discussion below on diffusion-controlled reactions.

Encounters, Collisions, and the Cage Effect

In a gas at low or moderate pressure, the molecules are far apart and move freely between collisions. In a liquid, there is little empty space between molecules, and they cannot move freely. Instead, a given molecule can be viewed as being surrounded by a **cage** formed by other molecules. A given molecule vibrates against the "walls" of this cage many times before it "squeezes" through the closely packed surrounding molecules and diffuses out of the cage. A liquid's structure thus resembles somewhat the structure of a solid.

This reduced mobility in liquids hinders two reacting solute molecules B and C from getting to each other in solution. However, once B and C do meet, they will be surrounded by a cage of solvent molecules that keeps them close together for a relatively long time, during which they collide repeatedly with each other and with the cage walls of solvent molecules. A process in which B and C diffuse together to become neighbors is called an **encounter.** Each encounter in solution involves many *collisions* between B and C while they remain trapped in the solvent cage (Fig. 17.19). In a gas, there is no distinction between a collision and an encounter.

Theoretical estimates indicate that in water at room temperature, two molecules in a solvent cage will collide 20 to 200 times with each other before they diffuse out of the cage. [See Table I in A. J. Benesi, *J. Phys. Chem.,* **86,** 4926 (1982).] The number of collisions per encounter will be greater the greater the viscosity of the solvent. Although the rate of encounters per unit volume between pairs of solute molecules in a liquid solution is much less than the corresponding rate of collisions in a gas, the compensating effect of a large number of collisions per encounter in solution makes the collision rate roughly the same in solution as in a gas at comparable concentrations of reactants. Direct evidence for this is the near constancy of rate constants for certain reactions on going from the gas phase to a solution (see the above data on the C_5H_6 dimerization). Although the collision rate is about the same in a gas and in solution, the pattern of collisions is quite different, with collisions in solution grouped into sets, with short time intervals between successive collisions of any one set and long intervals between successive sets of collisions (Fig. 17.20).

What experimental evidence exists for the cage effect? In 1961, Lyon and Levy photochemically decomposed mixtures of the isotopic species CH_3NNCH_3 and CD_3NNCD_3. Absorption of light dissociates the molecules to N_2 and $2CH_3$ or $2CD_3$.

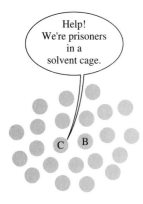

Figure 17.19

Molecules B and C in a solvent cage.

Figure 17.20

Collision pattern in a liquid.

Time ⟶

The methyl radicals then combine to give ethane. When the reaction was carried out in the gas phase, the ethane formed consisted of CH_3CH_3, CH_3CD_3, and CD_3CD_3 in proportions indicating random mixing of CH_3 and CD_3 before recombination. When the reaction was carried out in the inert solvent isooctane, only CH_3CH_3 and CD_3CD_3 were obtained. The absence of CH_3CD_3 showed that the two methyl radicals formed from a given parent molecule were kept together by the solvent cage until they recombined.

Diffusion-Controlled Reactions

Suppose the activation energy for the bimolecular elementary reaction $B + C \rightarrow$ products in solution is very low, so that there is a substantial probability for reaction to occur at each collision. Since each encounter in solution consists of many collisions, B and C might well react every time they encounter each other. The reaction rate will then be given by the rate of B-C encounters, and the reaction rate will be determined solely by how fast B and C can diffuse toward each other through the solvent. A reaction that occurs whenever B and C encounter each other in solution is called a **diffusion-controlled reaction.**

In 1917, Smoluchowski derived the following theoretical expression for the rate constant k_D of the elementary diffusion-controlled reaction $B + C \rightarrow$ products:

$$k_D = 4\pi N_A (r_B + r_C)(D_B + D_C) \qquad \text{where } B \neq C, \text{nonionic} \quad (17.111)$$

Here, N_A is the Avogadro constant, r_B and r_C are the radii of B and C molecules (assumed to be spherical, for simplicity), and D_B and D_C are the diffusion coefficients (Sec. 16.4) of B and C in the solvent. For a derivation of (17.111), see Sec. 23.8.

The rates $d[E]/dt$ of the diffusion-controlled elementary reactions $B + C \rightarrow E + F$ and $B + B \rightarrow E + G$ are each proportional to the encounter rate per unit volume. (The proportionality constant is $1/N_A$, which converts from molecules to moles.) Just as the collision rate per unit volume Z_{bb} of Eq. (15.63) is obtained by putting $c = b$ in Z_{bc} and multiplying by $\frac{1}{2}$, the encounter rate per unit volume for like molecules contains an extra factor of $\frac{1}{2}$, as compared with the encounter rate per unit volume for unlike molecules. Hence the diffusion-controlled rate constant for like molecules is found by multiplying (17.111) by $\frac{1}{2}$:

$$k_D = 2\pi N_A (r_B + r_C)(D_B + D_C) \qquad \text{where } B = C, \text{nonionic} \quad (17.112)$$

where $r_B = r_C$ and $D_B = D_C$.

Equations (17.111) and (17.112) apply when B and C are uncharged. However, if B and C are ions, the strong Coulombic attraction or repulsion will clearly affect the encounter rate. Debye in 1942 showed that for ionic diffusion-controlled reactions in very dilute solutions

$$k_D = 4\pi N_A (r_B + r_C)(D_B + D_C)\frac{W}{e^W - 1} \qquad \text{where } B \neq C, \text{ionic} \quad (17.113)$$

$$W \equiv \frac{z_B z_C e^2}{4\pi \varepsilon_0 \varepsilon_r kT(r_B + r_C)}$$

In this definition of W, SI units are used, ε_r is the solvent dielectric constant, z_B and z_C are the charge numbers of B and C, k is Boltzmann's constant, e is the proton charge, and ε_0 is the permittivity of vacuum. $r_B + r_C$ is the same as a in the Debye–Hückel equation (10.67). Since a typically ranges from 3 to 8 Å, a reasonable value for $r_B + r_C$ is 5 Å $= 5 \times 10^{-10}$ m. Using the SI values of k, e, and ε_0 and the value $\varepsilon_r = 78.4$ for water at 25°C, one finds for H_2O at 25°C and $r_B + r_C = 5$ Å:

z_B, z_C	1, 1	2, 1	2, 2	1, −1	2, −1	2, −2	3, −1
$W/(e^W - 1)$	0.45	0.17	0.019	1.9	3.0	5.7	4.3

Equations (17.111) to (17.113) set upper limits on the rate constants for reactions in solution. To test them, we need reactions with essentially zero activation energy. Recombination of two radicals to form a stable molecule has $E_a \approx 0$. Using flash photolysis, one can produce such species in solution and measure their recombination rate. A precise test of the equations for k_D is usually not possible, since the diffusion coefficients of such radicals in solution are not known. However, D_B and D_C can be estimated by analogy with stable species having similar structures. Recombination reactions studied include $I + I \rightarrow I_2$ in CCl_4, $OH + OH \rightarrow H_2O_2$ in H_2O, and $2CCl_3 \rightarrow C_2Cl_6$ in cyclohexene. Radical recombination rates in solution are generally in good agreement with the theoretically calculated values for diffusion-controlled reactions.

To decide whether a reaction is diffusion controlled, one compares the observed k with k_D calculated from one of the above equations. Rate constants for many very fast reactions in solution have been measured using relaxation techniques. Reactions of H_3O^+ with bases (for example, OH^-, $C_2H_3O_2^-$) are found to be diffusion controlled. Most termination reactions in liquid-phase free-radical polymerizations (Sec. 17.13) are diffusion controlled. The majority of the reactions of the hydrated electron $e^-(aq)$, produced by irradiating an aqueous solution with a brief pulse of high-energy electrons or x-rays, are diffusion controlled.

Equations (17.111) to (17.113) for k_D can be simplified by using the Stokes–Einstein equation (16.37) relating the diffusion coefficient of a molecule to the viscosity of the medium it moves through: $D_B \approx kT/6\pi\eta r_B$ and $D_C \approx kT/6\pi\eta r_C$, where η is the solvent's viscosity and k is Boltzmann's constant. Equation (17.111) becomes

$$k_D \approx \frac{2RT}{3\eta}\frac{(r_B + r_C)^2}{r_B r_C} = \frac{2RT}{3\eta}\left(2 + \frac{r_B}{r_C} + \frac{r_C}{r_B}\right) \qquad \text{where } B \neq C, \text{ nonionic}$$

The value of $2 + r_B/r_C + r_C/r_B$ is rather insensitive to the ratio r_B/r_C. Since the treatment is approximate, we might as well set $r_B = r_C$ to get

$$k_D \approx \begin{cases} 8RT/3\eta & \text{where } B \neq C, \text{ nonionic} \qquad (17.114) \\ 4RT/3\eta & \text{where } B = C, \text{ nonionic} \qquad (17.115) \end{cases}$$

For water at 25°C, $\eta = 8.90 \times 10^{-4}\ kg\ m^{-1}\ s^{-1}$, and substitution in (17.114) gives $k_D \approx 0.7 \times 10^{10}\ dm^3\ mol^{-1}\ s^{-1}$ for a nonionic diffusion-controlled reaction with $B \neq C$. The $W/(e^W - 1)$ factor multiplies k_D by 2 to 10 for oppositely charged ions and by 0.5 to 0.01 for like charged ions. Thus, k_D is 10^8 to $10^{11}\ dm^3\ mol^{-1}\ s^{-1}$ in water at 25°C, depending on the charges and sizes of the reacting species.

The majority of reactions in liquid solution are not diffusion controlled. Instead, only a small fraction of encounters lead to reaction. Such reactions are called **chemically controlled,** since their rate depends on the probability that an encounter will lead to chemical reaction.

Activation Energies

Gas-phase reactions are commonly studied at temperatures up to 1500 K, whereas reactions in solution are studied up to 400 or 500 K. Hence reactions with high activation energies will proceed at negligible rates in solution. Therefore, most reactions observed in solution have activation energies in the range 2 to 35 kcal/mol (8 to 150 kJ/mol), compared with −3 to 100 kcal/mol (−15 to 400 kJ/mol) for gas-phase reactions. The 10-°C doubling or tripling rule (Sec. 17.8) indicates that many reactions in solution have E_a in the range 13 to 20 kcal/mol.

For a nonionic diffusion-controlled reaction, Eqs. (17.114), (17.115), and (17.68) show that E_a involves $\eta^{-1}\ d\eta/dT$. For water at 25°C and 1 atm, one finds (Prob. 17.90) a theoretical prediction of $E_a \approx 4\frac{1}{2}$ kcal/mol = 19 kJ/mol for such reactions.

A **catalyst** is a substance that increases the rate of a reaction and can be recovered chemically unchanged at the end of the reaction. The rate of a reaction depends on the rate constants in the elementary steps of the reaction mechanism. A catalyst provides an alternate mechanism that is faster than the mechanism in the absence of the catalyst. Moreover, although the catalyst participates in the mechanism, it must be regenerated. A simple scheme for a catalyzed reaction is

$$R_1 + C \rightarrow I + P_1$$
$$I + R_2 \rightarrow P_2 + C \qquad (17.116)$$

where C is the catalyst, R_1 and R_2 are reactants, P_1 and P_2 are products, and I is an intermediate. The catalyst is consumed to form an intermediate, which then reacts to regenerate the catalyst and give products. The mechanism (17.116) is faster than the mechanism in the absence of C. In most cases, the catalyzed mechanism has a lower activation energy than that of the uncatalyzed mechanism. In a few cases, the catalyzed mechanism has a higher E_a (and a higher A factor); see J. A. Campbell, *J. Chem. Educ.,* **61,** 40 (1984).

An example of (17.116) is the mechanism (17.7). In (17.7), R_1 is O_2, R_2 is SO_2, the catalyst C is NO, the intermediate I is NO_2, there is no P_1, and P_2 is SO_3. Another example is (17.117), below, in which the catalyst is Cl and the intermediate is ClO. In many cases, the catalyzed mechanism has several steps and more than one intermediate.

In **homogeneous catalysis,** the catalyzed reaction occurs in one phase. In **heterogeneous catalysis** (Sec. 17.18), it occurs at the interface between two phases.

The equilibrium constant for the overall reaction $R_1 + R_2 \rightleftharpoons P_1 + P_2$ is determined by $\Delta G°$ (according to $\Delta G° = -RT \ln K°$) and is therefore independent of the reaction mechanism. Hence a *catalyst cannot alter the equilibrium constant of a reaction.* This being so, a catalyst for a forward reaction must be a catalyst for the reverse reaction also. Note that reversing the mechanism (17.116) gives a mechanism whereby catalyst is consumed to produce an intermediate which then reacts to regenerate the catalyst. Since the hydrolysis of esters is catalyzed by H_3O^+, the esterification of alcohols must also be catalyzed by H_3O^+.

Although a catalyst cannot change the equilibrium constant, a homogeneous catalyst can change the equilibrium composition of a system. The mole-fraction equilibrium constant is $K° = \Pi_i (a_{i,\text{eq}})^{\nu_i}$, where $a_i = \gamma_i x_i$. A catalyst present in the same phase as the reactants and products will change the activity coefficients γ_i. Unless the changes in reactant and product γ_i values happen to offset each other, the presence of the homogeneous catalyst will change the equilibrium mole fractions of reactants and products. Since a catalyst is usually present in small amounts, its effect on the equilibrium composition is usually small.

The rate law for a homogeneously catalyzed reaction often has the form

$$r = k_0[A]^\alpha \cdots [L]^\lambda + k_{\text{cat}}[A]^{\alpha'} \cdots [L]^{\lambda'}[\text{cat.}]^\sigma$$

where k_0 is the rate constant in the absence of catalyst ([cat.] = 0) and k_{cat} is the rate constant for the catalyzed mechanism. The order with respect to catalyst is commonly 1. If the lowering of E_a is substantial, the first term in r is negligible compared with the second, unless [cat.] is extremely small. The activation energies for the uncatalyzed and catalyzed reactions can be found from the temperature dependences of k_0 and k_{cat}. The reaction $2H_2O_2(aq) \rightarrow 2H_2O + O_2$ has the following activation energies: 17 kcal/mol when uncatalyzed; 10 kcal/mol when catalyzed by Fe^{2+}; 12 kcal/mol when catalyzed by colloidal Pt particles; 2 kcal/mol when catalyzed by the enzyme liver catalase. From the last example in Sec. 17.8, a decrease from 17 to 2 kcal/mol in

E_a increases the room-temperature rate constant by a factor of $(5.4)^{15} = 10^{11}$, assuming the A factor is not significantly changed.

Many reactions in solution are catalyzed by acids or bases or both. The hydrolysis of esters is catalyzed by H_3O^+ and by OH^- (but not by other Brønsted acids or bases). The rate law for ester hydrolysis generally has the form

$$r = k_0[RCOOR'] + k_{H^+}[H_3O^+][RCOOR'] + k_{OH^-}[OH^-][RCOOR']$$

where the rate constants include the concentration of water raised to unknown powers. Strictly speaking, OH^- is not a catalyst in ester hydrolysis, but is a reactant since it reacts with the product RCOOH.

An **autocatalytic reaction** is one where a product speeds up the reaction. An example is the H_3O^+-catalyzed hydrolysis of esters, $RCOOR' + H_2O \rightarrow RCOOH + R'OH$; here, H_3O^+ from the ionization of the product RCOOH increases the H_3O^+ concentration as the reaction proceeds, and this tends to speed up the reaction. Another kind of autocatalysis occurs in the elementary reaction $A + B \rightarrow C + 2A$. The rate law is $r = k[A][B]$. During the reaction, the A concentration increases, and this increase offsets the decrease in r produced by the decrease in [B]. A spectacular example is an atomic bomb. Here A is a neutron. The reaction sequence $A + B \rightarrow C + D$ followed by $C + E \rightarrow 2A + F$ is autocatalytic.

For certain complex reactions that contain autocatalytic reaction steps, one observes repeated oscillations in the concentrations of one or more species (intermediates or catalysts) as a function of time. In 1921, Bray reported that the homogeneous closed-system decomposition of aqueous H_2O_2 in the presence of IO_3^- and I_2 shows repeated oscillations in the I_2 concentration. Bray's work was dismissed by many chemists who erroneously believed that oscillations in the concentration of a species would violate the second-law requirement that G must continually decrease as a reaction goes to equilibrium at constant T and P in a closed system. Eventually, it was realized that oscillations in concentration during a reaction need not violate the second law, and the experimental work of Belousov and Zhabotinskii and the theoretical work of Prigogine and others in the years 1950–1970 firmly established the existence of oscillating reactions.

In a closed system, the oscillations will eventually die out as equilibrium is approached. Oscillations can be maintained indefinitely in an open system. (Of course, oscillations about the equilibrium position in a closed system are forbidden by the second law. The observed oscillations occur as G decreases and the system approaches equilibrium.) If an oscillating reaction mixture is not stirred, the interaction between autocatalysis and diffusion may produce patterns in the solution due to spatial variations in the concentrations of intermediates or catalysts. Oscillating reactions may have considerable significance in biological systems (they may be involved in such phenomena as the heart beating). For more on oscillating reactions, see I. R. Epstein, *Chem. Eng. News,* March 30, 1987, pp. 24–36; R. J. Field and F. W. Schneider, *J. Chem. Educ.,* **66,** 195 (1989).

An **inhibitor** (or *negative catalyst*) is a substance that decreases the rate of a reaction when added in small quantities. Inhibitors may destroy a catalyst present in the system or may react with reaction intermediates in a chain reaction.

The catalytic destruction of ozone in the earth's stratosphere (the portion of the atmosphere from 10 or 15 km to 50 km—Fig. 15.17) is of major current concern. Stratospheric ozone is formed when O_2 absorbs ultraviolet radiation and dissociates to O atoms ($O_2 + h\nu \rightarrow 2O$, where $h\nu$ denotes a photon of ultraviolet radiation—see Sec. 18.2); the O atoms combine with O_2 to form O_3 ($O + O_2 + M \rightarrow O_3 + M$—see Sec. 17.12). The O_3 can break down to O_2 by absorption of ultraviolet radiation ($O_3 + h\nu \rightarrow O_2 + O$) and by reaction with O ($O_3 + O \rightarrow 2O_2$). The net result of these reactions is an approximately steady-state O_3 stratospheric concentration of a few parts per million. The stratosphere contains only 10% of the atmosphere's mass but has 90% of the atmosphere's ozone.

In 1974, F. Sherwood Rowland and Mario Molina proposed that Cl atoms catalyze the decomposition of stratospheric O_3 by the mechanism

$$Cl + O_3 \rightarrow ClO + O_2$$
$$ClO + O \rightarrow Cl + O_2$$
(17.117)

The net reaction is $O_3 + O \rightarrow 2O_2$. Depletion of ozone is undesirable, since it would increase the amount of ultraviolet radiation reaching us, thereby increasing the incidence of skin cancer and cataracts, reducing crop yields, damaging some marine life, and altering the climate.

The chlorofluorocarbons $CFCl_3$ and CF_2Cl_2 have been used as working fluids in refrigerators and air conditioners, as solvents for cleaning electronic circuit boards, and as blowing agents for producing insulating foams. When released to the atmosphere, these gases slowly diffuse to the stratosphere, where they produce Cl atoms by absorbing ultraviolet radiation ($CFCl_3 + h\nu \rightarrow CFCl_2 + Cl$). Some of the Cl atoms produced react with CH_4 to produce HCl. Also, ClO radicals produced by the first reaction in (17.117) react with NO_2 to give $ClONO_2$. The reactions $Cl + CH_4 \rightarrow CH_3 + HCl$ and $ClO + NO_2 \rightarrow ClONO_2$ tie up most of the stratospheric chlorine in the "reservoir" species HCl and $ClONO_2$, which do not significantly destroy O_3. Thus, if only homogeneous gas-phase reactions occurred, stratospheric O_3 depletion would be minimal. Unfortunately, this is not the case.

During the Antarctic winter (June to August), much of Antarctica is in darkness and the cold temperatures lead to the formation of stratospheric clouds. The nature of the particles in polar stratospheric clouds (PSCs) is not fully established. A mass spectrometer aboard a balloon gondola found that a large fraction of arctic PSCs consist of solid particles of $HNO_3 \cdot 3H_2O$ (Prob. 12.51) [C. Voigt et al., *Science*, **290**, 1756 (2000)]. Particles of $H_2O(s)$ and supercooled liquid particles composed of a ternary solution of H_2O–HNO_3–H_2SO_4 are also present in PSCs. HCl condenses on the surfaces of solid PSC particles and dissolves in liquid PSC particles, and the following reaction occurs on and in the particles:

$$ClONO_2 + HCl \rightarrow Cl_2 + HNO_3$$
(17.118)

This reaction converts chlorine from the reservoir species HCl and $ClONO_2$ to Cl_2. Moreover, it converts nitrogen into HNO_3, which does not readily react with Cl or ClO. Also, the reaction $N_2O_5 + H_2O \rightarrow 2HNO_3$ occurs on stratospheric cloud surfaces. Since N_2O_5 is a source of NO_2 [recall (17.8)], conversion of N_2O_5 to HNO_3 depletes stratospheric NO_2. Sedimentation of some stratospheric clouds to lower atmospheric levels removes nitrogen from the stratosphere.

When spring comes to Antarctica (September and October), the sun reappears and the Cl_2 produced by the reaction (17.118) is readily decomposed to Cl atoms by ultraviolet radiation. The Cl atoms react with O_3 to give ClO. Since NO_2 has been depleted by the wintertime stratospheric clouds, the ClO will not be tied up as $ClONO_2$. The concentration of O atoms in the Antarctic stratosphere is quite low, so the mechanism (17.117) contributes only about 5% to the observed O_3 destruction. About 75% of the Antarctic O_3 depletion is due to the catalytic mechanism

$$Cl + O_3 \rightarrow ClO + O_2$$
$$2ClO + M \rightarrow (ClO)_2 + M$$
$$(ClO)_2 + h\nu \rightarrow Cl + ClOO$$
$$ClOO + M \rightarrow Cl + O_2 + M$$
(17.119)

where M is a third body and the stoichiometric number of the first step is 2. The net reaction for (17.119) is $2O_3 \rightarrow 3O_2$. About 20% of the Antarctic ozone loss is due to a mechanism involving BrO and ClO. Some ozone depletion occurs in the Arctic stratosphere, but much less than over Antarctica. The Arctic winter stratosphere has fewer and shorter-lived stratospheric clouds than the Antarctic stratosphere.

During September and October, about 70% of the Antarctic stratospheric ozone is destroyed. The depletion is largely confined to Antarctica because of the presence of the Antarctic vortex, a ring of rapidly circulating air that tends to isolate the Antarctic atmosphere. In November, the vortex breaks down and ozone-rich air from outside Antarctica replenishes the Antarctic ozone.

Significant stratospheric ozone depletion due to chlorofluorocarbons is now being observed outside the Antarctic and Arctic regions. Smoothed spectroscopic measurements of total atmospheric ozone above the Swiss ski resort at Arosa for the period 1926–1997 show an average O_3 level that is constant from 1926 to 1973 and that declines by 2.9% per decade from 1973 to 1997 (www.epa.gov/ozone/science/arosa.html). Satellite observations show similar declines at other midlatitude locations.

These midlatitude depletions are attributed to (a) ozone-poor air from the polar regions spreading to midlatitudes; (b) stratospheric aerosols composed of droplets of H_2O–H_2SO_4 solutions, which convert N_2O_5 into HNO_3 and convert $ClONO_2$ and HCl to Cl_2 and HOCl, which are broken down to Cl and ClO by ultraviolet sunlight; (c) the possible processing of midlatitude air through polar stratospheric clouds.

A 1992 international treaty eliminated nearly all chlorofluorocarbon production. However, the long life of chlorofluorocarbons in the atmosphere means that the Antarctic ozone hole and midlatitude ozone depletion will last until about 2050.

For more on ozone depletion, see J. W. Anderson et al., *Science,* **251,** 39 (1991); O. B. Toon and R. P. Turco, *Scientific American,* June 1991, pp. 68–74; P. Hamill and O. B. Toon, *Physics Today,* Dec. 1991, pp. 34–42; S. Solomon, *Rev. Geophys.,* **37,** 275 (1999).

17.17 ENZYME CATALYSIS

Most of the reactions that occur in living organisms are catalyzed by molecules called **enzymes.** Most enzymes are proteins. (Certain RNA molecules also act as enzymes.) An enzyme is specific in its action. Many enzymes catalyze only the conversion of a particular reactant to a particular product (and the reverse reaction); other enzymes catalyze only a certain class of reactions (for example, ester hydrolysis). Enzymes speed up reaction rates very substantially, and in their absence most biochemical reactions occur at negligible rates. The molecule an enzyme acts on is called the **substrate.** The substrate binds to a specific *active site* on the enzyme to form an **enzyme–substrate complex.** While bound to the enzyme, the substrate is converted to product, which is then released from the enzyme. Some physiological poisons act by binding to the active site of an enzyme, thereby blocking (or *inhibiting*) the action of the enzyme. The structure of an inhibitor may resemble the structure of the enzyme's substrate. Cyanide acts by blocking the enzyme cytochrome oxidase.

The single-celled *Escherichia coli,* a bacterium that flourishes in human colons, contains about 2500 different enzymes and a total of 10^6 enzyme molecules.

There are many possible schemes for enzyme catalysis, but we shall consider only the simplest mechanism, which is

$$\text{E} + \text{S} \underset{k_{-1}}{\overset{k_1}{\rightleftharpoons}} \text{ES} \underset{k_{-2}}{\overset{k_2}{\rightleftharpoons}} \text{E} + \text{P} \tag{17.120}$$

where E is the free enzyme, S is the substrate, ES is the enzyme–substrate complex, and P is the product. The overall reaction is $\text{S} \rightarrow \text{P}$. The enzyme is consumed in step 1 and regenerated in step 2.

In most experimental studies on enzyme kinetics, the enzyme concentration is much less than the substrate concentration: $[E] \ll [S]$. Hence the concentration of the intermediate ES is much less than that of S, and the steady-state approximation can be used for ES:

$$d[ES]/dt = 0 = k_1[E][S] - k_{-1}[ES] - k_2[ES] + k_{-2}[E][P] \quad (17.121)$$

If $[E]_0$ is the initial enzyme concentration, then $[E]_0 = [E] + [ES]$. Since the enzyme concentration $[E]$ during the reaction is generally not known while $[E]_0$ is known, we replace $[E]$ by $[E]_0 - [ES]$:

$$0 = ([E]_0 - [ES])(k_1[S] + k_{-2}[P]) - (k_{-1} + k_2)[ES]$$

$$[ES] = \frac{k_1[S] + k_{-2}[P]}{k_{-1} + k_2 + k_1[S] + k_{-2}[P]} [E]_0 \quad (17.122)$$

The reaction rate is $r = -d[S]/dt$, and (17.120) gives

$$r = k_1[E][S] - k_{-1}[ES] = k_1([E]_0 - [ES])[S] - k_{-1}[ES]$$

$$r = k_1[E]_0[S] - (k_1[S] + k_{-1})[ES] \quad (17.123)$$

Since the concentration of the intermediate ES is very small, we have $-d[S]/dt = d[P]/dt$. Substitution of (17.122) into (17.123) gives

$$r = \frac{k_1 k_2[S] - k_{-1}k_{-2}[P]}{k_1[S] + k_{-2}[P] + k_{-1} + k_2} [E_0] \quad (17.124)$$

Usually, the reaction is followed only to a few percent completion and the initial rate determined. Setting the product concentration $[P]$ equal to 0 and $[S]$ equal to $[S]_0$, we get as the initial rate r_0

$$r_0 = \frac{k_1 k_2[S]_0[E]_0}{k_1[S]_0 + k_{-1} + k_2} = \frac{k_2[E]_0[S]_0}{K_M + [S]_0} \quad (17.125)$$

where the *Michaelis constant* K_M is defined by $K_M \equiv (k_{-1} + k_2)/k_1$. The reciprocal of (17.125) is

$$\frac{1}{r_0} = \frac{K_M}{k_2[E]_0} \frac{1}{[S]_0} + \frac{1}{k_2[E]_0} \quad (17.126)$$

Equation (17.125) is the *Michaelis–Menten equation*, and (17.126) is the *Lineweaver–Burk equation*. One measures r_0 for several $[S]_0$ values with $[E]_0$ held fixed. The constants k_2 and K_M are found from the intercept and slope of a plot of $1/r_0$ versus $1/[S]_0$, since $[E]_0$ is known. Strictly speaking, r_0 is not the rate at $t = 0$, since there is a short induction period before steady-state conditions are established. However, the induction period is generally too short to detect.

Figure 17.21 plots r_0 in (17.125) against $[S]_0$ for fixed $[E]_0$. In the limit of high concentration of substrate, virtually all the enzyme is in the form of the ES complex,

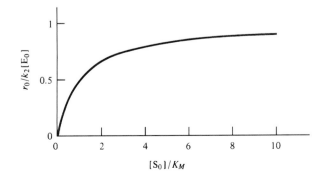

Figure 17.21

Initial rate versus initial substrate concentration for the Michaelis–Menten mechanism.

and the rate becomes a maximum that is independent of substrate concentration. Equation (17.125) gives $r_{0,max} = k_2[E]_0$ for $[S]_0 \gg K_M$. At low substrate concentrations, (17.125) gives $r_0 = (k_2/K_M)[E]_0[S]_0$, and the reaction is second-order. Equation (17.124) predicts that r is always proportional to $[E]_0$ provided that $[E]_0 \ll [S]_0$, so steady-state conditions hold.

The quantity $r_{0,max}/[E]_0$ is the *turnover number* of the enzyme. The turnover number is the maximum number of moles of product produced in unit time by 1 mole of enzyme and is also the maximum number of molecules of product produced in unit time by one enzyme molecule. From the preceding paragraph, $r_{0,max} = k_2[E]_0$, so the turnover number for the simple model (17.120) is k_2. Turnover numbers for enzymes range from 10^{-2} to 10^6 molecules per second, with 10^3 s^{-1} being typical. One molecule of the enzyme carbonic anhydrase will dehydrate 6×10^5 H$_2$CO$_3$ molecules per second; the reaction $H_2CO_3(aq) \rightleftharpoons H_2O + CO_2(aq)$ is important in the excretion of CO_2 from the capillaries of the lungs. For comparison, a typical turnover rate in heterogeneous catalysis (Sec. 17.18) is 1 s^{-1}.

Although many experimental studies on enzyme kinetics give a rate law in agreement with the Michaelis–Menten equation, the mechanism (17.120) is grossly oversimplified. For one thing, there is much evidence that, while the substrate is bound to the enzyme, it generally undergoes a chemical change before being released as product. Hence a better model is

$$E + S \rightleftharpoons ES \rightleftharpoons EP \rightleftharpoons E + P \qquad (17.127)$$

The model (17.127) gives a rate law that has the same form as the Michaelis–Menten equation, but the constants k_2 and K_M are replaced by constants with different significances. Another defect of (17.120) is that it takes the catalytic reaction as $S \rightleftharpoons P$, whereas most enzyme-catalyzed reactions involve two substrates and two products: $A + B \rightleftharpoons P + Q$. The enzyme then has two active sites, one for each substrate. With two substrates, there are many possible mechanisms. For details, see A. R. Schulz, *Enzyme Kinetics,* Cambridge Univ. Press, 1994.

Enzyme reactions are quite fast but can be studied using "classical" methods by keeping [E] and [S] very low. Typical values are [E] = 10^{-9} mol dm^{-3} and [S] = 10^{-5} mol dm^{-3}. The ratio [S]/[E] must be large to ensure steady-state conditions. Modern methods of studying fast reactions (for example, rapid flow, relaxation) provide more information than the classical methods, in that rate constants for individual steps in a multistep mechanism can be determined.

17.18   HETEROGENEOUS CATALYSIS

Many industrial chemical reactions are run in the presence of solid catalysts. Examples are the Fe-catalyzed synthesis of NH$_3$ from N$_2$ and H$_2$; the SiO$_2$/Al$_2$O$_3$-catalyzed cracking of relatively high-molecular-weight hydrocarbons to gasoline; the Pt-catalyzed (or V$_2$O$_5$-catalyzed) oxidation of SO$_2$ to SO$_3$, which is then reacted with water to produce H$_2$SO$_4$, the leading industrial chemical (annual U.S. production, 10^{11} lb). The liquid catalyst H$_3$PO$_4$, distributed on diatomaceous earth, is used in the polymerization of alkenes.

Solid-state catalysis can lower activation energies substantially. For 2HI → H$_2$ + I$_2$, the activation energy is 44 kcal/mol for the uncatalyzed homogeneous reaction, is 25 kcal/mol when catalyzed by Au, and is 14 kcal/mol when catalyzed by Pt. The A factor is also changed.

For a solid catalyst to be effective, one or more of the reactants must be chemisorbed on the solid (Sec. 13.5). Physical adsorption is only significant in heterogeneous catalysis in a few special cases such as recombination of radicals.

The mechanisms of only a few heterogeneously catalyzed reactions are known. In writing such mechanisms, the adsorption site is often indicated by a star. Perhaps the best understood heterogeneously catalyzed reaction is the oxidation of CO on a Pt or Pd catalyst (a reaction important in automobile catalytic converters); the mechanism is

$$CO(g) + * \rightleftharpoons \underset{*}{CO} \qquad \text{and} \qquad O_2(g) + 2* \rightarrow \underset{*}{2O}$$

$$\underset{*}{O} + \underset{*}{CO} \rightarrow CO_2(g) + 2*$$

where possible structures for adsorbed CO are shown in Fig. 13.15. Each star is a metal atom on the surface of the solid. [Under certain conditions, this reaction shows oscillations in rate due to a reversible change in the metal's surface structure that is induced by adsorbed CO; see R. Imbihl and G. Ertl, *Chem. Rev.,* **95,** 697 (1995).]

Most heterogeneous catalysts are metals, metal oxides, or acids. Common metal catalysts include Fe, Co, Ni, Pd, Pt, Cr, Mn, W, Ag, and Cu. Many metallic catalysts are transition metals with partly vacant d orbitals that can be used in bonding to the chemisorbed species. Common metal oxide catalysts are Al_2O_3, Cr_2O_3, V_2O_5, ZnO, NiO, and Fe_2O_3. Common acid catalysts are H_3PO_4 and H_2SO_4.

A good catalyst should have moderate values for the enthalpies of adsorption of the reactants. If $|\Delta \bar{H}_{ads}|$ is very small, there will be little adsorption and hence a slow reaction. If $|\Delta \bar{H}_{ads}|$ is very large, the reactants will be held very tightly at their adsorption sites and will have little tendency to react with each other.

To increase the exposed surface area, the catalyst is often distributed on the surface of a porous *support* (or *carrier*). Common supports are silica gel (SiO_2), alumina (Al_2O_3), carbon (in the form of charcoal), and diatomaceous earth. The support may be inert or may contribute catalytic activity.

The activity of a catalyst may be increased and its lifetime extended by addition of substances called *promoters*. The iron catalyst used in NH_3 synthesis contains small amounts of the oxides of K, Ca, Al, Si, Mg, Ti, Zr, and V. The Al_2O_3 acts as a barrier that prevents the tiny crystals of Fe from joining together (sintering); formation of larger crystals decreases the surface area and the catalytic activity.

Small amounts of certain substances that bond strongly to the catalyst can inactivate (or *poison*) it. These poisons may be present as impurities in the reactants or may be formed as reaction by-products. Catalytic poisons include compounds of S, N, and P having lone pairs of electrons (for example, H_2S, CS_2, HCN, PH_3, CO) and certain metals (for example, Hg, Pb, As). Because lead is a catalytic poison, lead-free gasoline must be used in cars equipped with catalytic converters used to remove pollutants from the exhaust.

The amount of poison needed to eliminate the activity of a catalyst is usually much less than needed to cover the catalyst's surface completely. This indicates that the catalyst's activity is largely confined to a fraction of surface sites, called *active sites* (or *active centers*). The surface of a solid is not smooth and uniform but is rough on an atomic scale. The surface of a metal catalyst contains steplike jumps that join relatively smooth planes (Fig. 24.26); hydrocarbon bonds break mainly at these steps and not on the smooth planes (*Chem. Eng. News*, Dec. 8, 1975, p. 23).

The following steps occur in fluid-phase reactions catalyzed by solids: (1) diffusion of reactant molecules to the solid's surface; (2) chemisorption of at least one reactant species on the surface; (3) chemical reaction between molecules adsorbed on adjacent sites or between an adsorbed molecule and fluid-phase molecules colliding with the surface; (4) desorption of products from the surface; (5) diffusion of products into the bulk fluid. For reactions that occur between two adsorbed molecules, migration of adsorbed molecules on the surface may occur between steps 2 and 3.

A general treatment involves the rates for all five steps and is complicated. In many cases, one of these steps is much slower than all the others, and only the rate of the slow step need be considered. We shall consider mainly solid-catalyzed reactions of gases where step 3 is much slower than all other steps.

If step 3 is between species chemisorbed on the surface, the reaction is said to occur by a **Langmuir–Hinshelwood** mechanism. If step 3 involves a chemisorbed species reacting with a fluid-phase species, the mechanism is called **Rideal–Eley.** Langmuir–Hinshelwood mechanisms are believed to be more common than Rideal–Eley ones.

Step 3 may consist of more than one elementary chemical reaction. Since the detailed mechanism of the surface reaction is usually unknown, we adopt the simplifying assumption of taking step 3 to consist of a single unimolecular or bimolecular elementary reaction or a slow (rate-determining) elementary reaction followed by one or more rapid steps. This assumption may be compared with the assumption of the grossly oversimplified mechanism (17.120) for enzyme catalysis.

Since we are assuming the adsorption and desorption rates to be much greater than the chemical-reaction rate for each species, adsorption–desorption equilibrium is maintained for each species during the reaction. We can therefore use the Langmuir isotherm, which is derived by equating the adsorption and desorption rates for a given species (see Sec. 13.5). The Langmuir isotherm assumes a uniform surface, which is far from true in heterogeneous catalysis, so use of the Langmuir isotherm is one more oversimplification in the treatment.

The *conversion rate* J of a heterogeneously catalyzed reaction is defined by (17.2) as $\nu_B^{-1} \, dn_B/dt$, where ν_B is the stoichiometric coefficient (Sec. 4.9) of any species B in the overall reaction. Since the chemical reaction occurs on the catalyst's surface, J will clearly be proportional to the catalyst's surface area $\mathcal{A}$. Let r_s be the **conversion rate per unit surface area** of the catalyst. Then

$$r_s \equiv \frac{J}{\mathcal{A}} \equiv \frac{1}{\mathcal{A}} \frac{1}{\nu_B} \frac{dn_B}{dt} \tag{17.128}$$

If $\mathcal{A}$ is unknown, one uses the rate per unit mass of catalyst.

Suppose the elementary reaction on the surface is the unimolecular step A → C + D. Then r_s, the conversion rate per unit surface area, will be proportional to the number of adsorbed A molecules per unit surface area $(n_A/\mathcal{A})$, and this in turn will be proportional to θ_A, the fraction of adsorption sites occupied by A molecules. Therefore, $r_s = k\theta_A$, where k is a rate constant with units mol cm^{-2} s^{-1}. Since the products C and D might compete with A for adsorption sites, we use the form of the Langmuir isotherm that applies when more than one species is adsorbed. Equation (13.37) generalized to several nondissociatively adsorbed species is

$$\theta_A = \frac{b_A P_A}{1 + \Sigma_i b_i P_i} \tag{17.129}$$

where the sum goes over all species. The rate law $r_s = k\theta_A$ becomes

$$r_s = k \frac{b_A P_A}{1 + b_A P_A + b_C P_C + b_D P_D} \tag{17.130}$$

If the products are very weakly adsorbed ($b_C P_C$ and $b_D P_D \ll 1 + b_A P_A$), then

$$r_s = k \frac{b_A P_A}{1 + b_A P_A} \tag{17.131}$$

The low-pressure and high-pressure limits of (17.131) are

$$r_s = \begin{cases} kb_A P_A & \text{at low } P \\ k & \text{at high } P \end{cases}$$

At low P, the reaction is first-order; at high P, zero-order. At high pressure, the surface is fully covered with A, so an increase in P_A has no effect on the rate. Note the resemblance to the low-substrate and high-substrate limits of the Michaelis–Menten equation (17.125). In fact, Eqs. (17.125) and (17.131) have essentially the same form. Compare also Figs. 17.21 and 13.17a. In both enzyme catalysis and heterogeneous catalysis, there is binding to a limited number of active sites.

The W-catalyzed decomposition of PH_3 at 700°C follows the rate law (17.131), being first-order below 10^{-2} torr and zero-order above 1 torr. The decomposition of N_2O on Mn_3O_4 has $r_s = kP_{N_2O}/(1 + bP_{N_2O} + cP_{O_2}^{1/2})$; this rate law is similar to (17.130) except that P_{O_2} appears to the $\frac{1}{2}$ power, indicating dissociative adsorption of O_2—compare Eq. (13.38). The product O_2 is adsorbed as O atoms and competes with N_2O for the active sites, thereby inhibiting the reaction.

Suppose the elementary surface reaction is bimolecular, $A + B \rightarrow C + D$, and both reactants are adsorbed on the surface. In a liquid or gas, molecules diffuse through the fluid until they collide and possibly react, and the elementary reaction rate is proportional to the product of the volume concentrations $(n_A/V)(n_B/V)$. Similarly, reactant molecules adsorbed on a solid surface can migrate or diffuse from one adsorption site to the next, until they meet and possibly react, and the reaction rate is proportional to the product of the surface concentrations $(n_A/\mathcal{A})(n_B/\mathcal{A})$, which in turn is proportional to $\theta_A \theta_B$. Thus $r_s = k\theta_A \theta_B$. Use of the Langmuir isotherm (17.129) gives for nondissociatively adsorbed species

$$r_s = k \frac{b_A b_B P_A P_B}{(1 + b_A P_A + b_B P_B + b_C P_C + b_D P_D)^2} \tag{17.132}$$

Suppose the reactant B is adsorbed much more strongly than all other species: $b_B P_B \gg 1 + b_A P_A + b_C P_C + b_D P_D$. Then, $r_s = kb_A P_A/b_B P_B$, and the *reactant* B inhibits the reaction. This seemingly paradoxical situation can be understood by realizing that when reactant B is much more strongly adsorbed than reactant A, the fraction of surface occupied by A goes to zero; hence $r_s = k\theta_A \theta_B$ goes to zero. The maximum rate occurs when the two reactants are equally adsorbed. An example of inhibition by a reactant is the Pt-catalyzed reaction $2CO + O_2 \rightarrow 2CO_2$ (discussed earlier in this section), whose rate is inversely proportional to the CO pressure. CO also binds more strongly than O_2 to the Fe atom in hemoglobin and so is a physiological poison.

For the Rideal–Eley bimolecular mechanism $A(ads) + B(g) \rightarrow$ products, the rate is proportional to $\theta_A P_B$, since the rate of collisions of B with the surface is proportional to P_B. Use of the Langmuir isotherm (17.129) for θ_A gives a rate law that differs from (17.132).

The rate law for NH_3 synthesis on promoted iron catalysts cannot be fitted to a Langmuir-type equation. Here, the rate-determining step is the adsorption of N_2 (step 2 in the above scheme). See *Wilkinson,* pp. 246–247. In some cases, desorption of the product is the rate-determining step.

Kinetics of Adsorption, Desorption, and Surface Migration of Gases on Solids

A full understanding of heterogeneous catalysis requires knowledge of the kinetics of adsorption, desorption, and surface migration.

The rate $r_{ads} = -(1/\mathcal{A})(dn_{B(g)}/dt)$ of the adsorption reaction $B(g) \rightarrow B(ads)$ (nondissociative adsorption) or $B(g) \rightarrow C(ads) + D(ads)$ (dissociative adsorption) is $r_{ads} = k_{ads} f(\theta)[B(g)]$, where $f(\theta)$ is a function of the fraction θ of occupied adsorption sites. In the Langmuir treatment (Sec. 13.5), $f(\theta) = 1 - \theta$ for nondissociative adsorption and $f(\theta) = (1 - \theta)^2$ for dissociative adsorption, where two adjacent vacant adsorption sites are needed. The adsorption rate constant is $k_{ads} = A_{ads} e^{-E_{a,ads}/RT}$, where

$E_{a,\text{ads}}$ is the activation energy for adsorption—the minimum energy a B molecule needs to be adsorbed. For most common gases on *clean* metal surfaces, chemisorption is found to be *nonactivated,* meaning that $E_{a,\text{ads}} \approx 0$.

The very high rate of collisions of gas molecules with a surface at ordinary pressures and the fact that $E_{a,\text{ads}}$ is often zero means that chemisorption is very rapid at ordinary pressures. To study its kinetics, contaminating background gases must be at extremely low pressures (10^{-10} torr or less), and the initial pressure of the gas being studied must be very low (typically, 10^{-7} torr). By monitoring the pressure versus time of contact between gas and solid held at a fixed T, one can measure the rate of adsorption.

Adsorption rates are usually expressed in terms of the *sticking coefficient* (or *sticking probability*) s, defined as

$$s \equiv \frac{\text{rate of adsorption per unit area}}{\text{rate of gas–solid collisions per unit area}} = \frac{r_{\text{ads}}}{P(2\pi MRT)^{-1/2}}$$

where Eq. (15.56) (divided by N_A to convert it from a molecular rate to a molar rate) was used. Measurement of $r_{\text{ads}} \equiv -(1/\mathscr{A})\, dn_{\text{B}(g)}/dt$ gives s. The sticking coefficient s depends on the gas, the solid, which crystal face of the solid is exposed, the temperature, and the fractional coverage θ of the surface. s becomes zero when $\theta = 1$, because only a monolayer can be chemisorbed. In the Langmuir treatment (Sec. 13.5), s and r_{ads} are taken as proportional to $1 - \theta$ for nondissociative adsorption, but experimental data usually show a more complex dependence of s on θ (Fig. 17.22). $s(\theta)$ is usually greater than the Langmuir expression. We can explain this by assuming that a molecule hitting an occupied adsorption site can be physisorbed there and then migrate to a nearby vacant site where it is chemisorbed. For H_2, O_2, CO, and N_2 on clean metals at temperatures where chemisorption occurs, s_0, the sticking coefficient at zero surface coverage ($\theta = 0$), is typically 0.1 to 1, but occasionally is much smaller. In most cases, s_0 either decreases or remains about the same with increasing temperature. In a few cases, s_0 increases with T; here, chemisorption is activated: $E_{a,\text{ads}} > 0$.

The rate of the desorption reaction B(ads) $\rightarrow$ B(g) is $r_{\text{des}} = -(1/\mathscr{A})\, dn_{\text{B}(ads)}/dt = -d[\text{B}]_s/dt = k_{\text{des}}[\text{B}]_s$, where $[\text{B}]_s \equiv n_{\text{B}(ads)}/\mathscr{A}$ is the surface concentration of B(ads). The desorption rate constant is $k_{\text{des}} = A_{\text{des}}e^{-E_{a,\text{des}}/RT}$, where $E_{a,\text{des}}$ is the activation energy for desorption. The typical lifetime of B(ads) on the surface can be taken as the half-life $t_{1/2} = 0.693/k_{\text{des}}$ for this first-order reaction. For the bimolecular desorption C(ads) + D(ads) $\rightarrow$ B(g), we have $r_{\text{des}} = -d[\text{C}]_s/dt = k_{\text{des}}[\text{C}]_s^2$, since $[\text{D}]_s = [\text{C}]_s$.

The kinetics of desorption can be studied by *thermal-desorption* experiments. Here, a solid with adsorbed gas is heated at a known rate in a vacuum system with a known pumping rate, and the system pressure is monitored versus time. A mass spectrometer is used to identify the desorbed gas(es). If the temperature increase is rapid (dT/dt in the range 10 to 1000 K/s), we have *flash desorption,* if slow (10 K/min to 10 K/s), *temperature-programmed desorption.* Analysis of P-versus-T desorption curves (see *Gasser,* pp. 67–71) gives $E_{a,\text{des}}$. Since A_{des} is often hard to determine from the data, a value $A_{\text{des}} = 10^{13}\ \text{s}^{-1}$ is often assumed for unimolecular desorptions. P-versus-T desorption curves often show more than one peak, indicating more than one kind of surface bonding, each with its own $E_{a,\text{des}}$. Recall the discussion in Sec. 13.5 on bonding of CO to metals.

For the common case where $E_{a,\text{ads}}$ is zero, the relation (17.71) shows that $E_{a,\text{des}}$ equals $-\Delta U^\circ_{\text{ads}}$, which is approximately equal to the absolute value of the enthalpy (heat) of adsorption. $E_{a,\text{des}}$ may depend on the surface coverage θ.

Surface migration of an adsorbed species can be studied by field-emission microscopy (*Gasser,* pp. 153–157), which allows one to follow the motions of individual atoms adsorbed on a metal surface. The results are expressed in terms of a diffusion coefficient (Sec. 16.4) D for surface migration, where $D = D_0 e^{-E_{a,\text{mig}}/RT}$, with

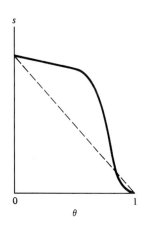

Figure 17.22

Typical curve of sticking probability versus fractional surface coverage. The dashed line is the Langmuir assumption that s is proportional to $1 - \theta$.

$E_{a,\text{mig}}$ being the activation energy for surface diffusion; this is the minimum energy an adsorbed species needs to move to an adjacent adsorption site. For chemisorbed species, $E_{a,\text{mig}}$ is typically 10 to 20% of $E_{a,\text{des}}$, so it is a lot easier for adsorbed species to move from one adsorption site to another than it is to be desorbed. The rms displacement d in a given direction of an adsorbed species in time t is $d \approx (2Dt)^{1/2}$ [Eq. (16.32)].

17.19   NUCLEAR DECAY

The decay of a radioactive isotope follows first-order kinetics. Each nucleus of a particular radioactive isotope has a certain probability of breaking down in unit time. The number $-dN$ of nuclei that decay in a small time dt is therefore proportional to the number N of radioactive nuclei currently present in the system and also to the length of the small time interval: $-dN \propto N\, dt$, or

$$dN/dt = -\lambda N \tag{17.133}$$

where the proportionality constant λ is called the **decay constant.** The number of radioactive nuclei is decreasing with time, so dN is negative. For radioactive decay, the decay rate is expressed in terms of the number of nuclei, rather than in terms of molar concentrations. Equation (17.133) has the same form as (17.11). By analogy to (17.14), Eq. (17.133) integrates to

$$N = N_0 e^{-\lambda t} \tag{17.134}$$

where N_0 is the number of radioactive nuclei present at $t = 0$.

Since the kinetics is first-order, the isotope's half-life, given by Eq. (17.15), is

$$t_{1/2} = 0.693/\lambda \tag{17.135*}$$

Some half-lives and decay modes are:

	^1_0n	^{12}N	^{14}C	^{238}U
Half-life	15 min	0.01 s	5730 yr	4.5×10^9 yr
Decay mode	β^-	β^+	β^-	α

An α particle consists of two protons and two neutrons and is a ^{4}He nucleus. The decay of ^{238}U is $^{238}_{92}\text{U} \to {}^4_2\text{He} + {}^{234}_{90}\text{Th}$. The ^{234}Th nucleus is itself unstable and decays by beta emission to ^{234}Pa, which is unstable; the ultimate decay product of ^{238}U is ^{206}Pb. The **mass number** A of a nucleus equals the number of protons plus neutrons and is written as a left superscript. The **atomic number** Z equals the number of protons and is written as a left subscript.

A β^- particle is an electron. The free neutron is unstable, decaying into a proton, an electron, and an antineutrino (symbol $\bar{\nu}$): ${}^1_0\text{n} \to {}^1_1\text{p} + {}^0_{-1}\text{e} + {}^0_0\bar{\nu}$. The antineutrino has no charge and was formerly believed to have zero rest mass and to travel at the speed of light. Evidence now suggests that neutrinos do have a very small nonzero mass. The emission of an electron from a nucleus can be described as the breakdown of a neutron in the nucleus into a proton, which remains in the nucleus, and an electron and antineutrino, which are ejected from the nucleus. The decay equation for ^{14}C is ${}^{14}_6\text{C} \to {}^{14}_7\text{N} + {}^0_{-1}\text{e} + {}^0_0\bar{\nu}$. The symbol ^{14}N in this equation stands for a *nucleus* of ^{14}N; emission of a beta particle by ^{14}C produces a ^{14}N$^+$ ion, which eventually picks up an electron from the surroundings to become a neutral ^{14}N atom.

A β^+ particle is a positron. A positron has the same mass as an electron and a charge equal in magnitude but opposite in sign to the electron charge. The positron is the **antiparticle** of the electron. When a positron collides with an electron, they annihilate each other, producing two high-energy photons. (Photons are discussed in Sec. 18.2.) Positrons are not constituents of ordinary matter. Several synthetic nuclei decay by positron emission. An example is $^{12}_{7}N \rightarrow ^{12}_{6}C + ^{0}_{1}e + ^{0}_{0}\nu$, where ν is a neutrino. The particles ν and $\bar{\nu}$ are antiparticles of each other.

Just as the electrons in atoms can exist in excited states (Chapter 19), the nucleus possesses excited states. An excited nucleus can lose energy by emission of a high-energy photon (a gamma ray). In gamma emission, the charge and mass number of the nucleus remain unchanged. The nucleus simply goes from a high energy level to a lower one.

The **activity** A of a radioactive sample is defined as the number of disintegrations per second: $A \equiv -dN/dt$. Equation (17.133) gives

$$A = \lambda N \tag{17.136}*$$

where N is the number of radioactive nuclei present. Substitution in (17.134) gives

$$A = A_0 e^{-\lambda t} \tag{17.137}$$

Radioactive decay is a probabilistic process, and the discussion of Sec. 3.7 shows that fluctuations on the order of $A^{1/2}$ are to be expected in the observed activity. The activity of a 10-mg sample of ^{238}U is only 100 dis/s, and easily observed fluctuations of 10 dis/s, or 10%, occur in the decay rate. In a chemical reaction, the number of particles reacting per unit time is extremely large, and fluctuations in rate are unobservable.

EXAMPLE 17.12 Half-life and decay calculations

A 1.00-g sample of ^{226}Ra emits 3.7×10^{10} alpha particles per second. Find λ and $t_{1/2}$. Find A after 999 years.

We have

$$N_0 = 1.00 \text{ g} \; \frac{1 \text{ mol Ra}}{226 \text{ g}} \; \frac{6.02 \times 10^{23} \text{ atoms}}{1 \text{ mol}} = 2.66 \times 10^{21} \text{ atoms}$$

$$\lambda = A_0/N_0 = (3.7 \times 10^{10} \text{ s}^{-1})/(2.66 \times 10^{21}) = 1.39 \times 10^{-11} \text{ s}^{-1}$$

$$t_{1/2} = 0.693/\lambda = 5.0 \times 10^{10} \text{ s} = 1600 \text{ years}$$

$$A = A_0 e^{-\lambda t} = (3.7 \times 10^{10} \text{ s}^{-1}) \exp\left[(-1.39 \times 10^{-11} \text{ s}^{-1})(3.15 \times 10^{10} \text{ s})\right]$$

$$= 2.4 \times 10^{10} \text{ dis/s}$$

Exercise

The only radioactive naturally occurring isotope of K is ^{40}K, with $t_{1/2} = 1.28 \times 10^9$ yr and a natural isotopic abundance of 0.00117 percent. Find the activity of a sample of 10.0 g of KCl and find its activity after 2.00×10^8 yr. (*Answers:* 16.2 dis/s and 14.5 dis/s.)

17.20   SUMMARY

The rate r of the homogeneous reaction $0 \rightarrow \Sigma_i \nu_i A_i$ in a system with negligible volume change and negligible concentrations of intermediates is $r = (1/\nu_i) \, d[A_i]/dt$, where ν_i is the stoichiometric coefficient of species A_i (negative for reactants and positive for products) and $d[A_i]/dt$ is the rate of change in the concentration of A_i. The expression for r as a function of concentrations at fixed temperature is called the rate law. Most commonly, the rate law has the form $r = k[A]^\alpha[B]^\beta \cdots [L]^\lambda$, where the rate constant k depends strongly on temperature (and very weakly on pressure) and α, β, $\ldots$, λ (the partial orders) are usually integers or half-integers. The overall order is $\alpha + \beta + \cdots + \lambda$. The partial orders in the rate law may differ from the coefficients in the chemical reaction and must be found by experiment.

Reaction rates are measured by following the concentrations of species versus time using physical or chemical methods. Special techniques (for example, flow systems, relaxation, flash photolysis) are used to follow very fast reactions.

Various forms of the rate law were integrated in Sec. 17.3. For a first-order reaction, the half-life is independent of initial reactant concentration.

If the rate law has the form $r = k[A]^n$, the order n can be determined by the fractional-life method or the Powell-plot method. By making the concentration of reactant A much less than that of the other reactants (the isolation method), one reduces the rate law to the form $r = j[A]^\alpha$, and the partial order α can be found using the fractional-life or Powell-plot method. The partial orders can also be found from the changes in initial rates produced by changes in the initial reactant concentrations. Once the partial orders have been found, the rate constant is evaluated from the slope of the appropriate straight-line plot.

The majority of chemical reactions are composite, meaning that they consist of a series of steps, each step being called an elementary reaction. The series of steps is called the reaction mechanism. The mechanism generally involves one or more reaction intermediates that are produced in one step and consumed in a later step.

The rate law for the *elementary* reaction $aA + bB \rightarrow$ products is $r = k[A]^a[B]^b$ in an ideal system. The quantity $a + b$ (which may be 1, 2, or, rarely, 3) is the molecularity of the elementary reaction. The equilibrium constant for an *elementary* reaction equals the ratio of the forward rate constant to the reverse rate constant: $K_c = k_f/k_b$. None of these statements is necessarily true for a composite reaction.

To arrive at the rate law predicted by a mechanism, one usually uses either the rate-determining-step or the steady-state approximation. The rate-determining-step approximation assumes the mechanism contains a relatively slow rate-determining step; this step may be preceded by steps that are in near equilibrium and may be followed by rapid steps. One sets the overall rate equal to the rate of the rate-determining step (provided this step has a stoichiometric number equal to 1) and eliminates any reaction intermediates from the rate law by solving for their concentrations using the equilibria that precede the rate-determining step. In the steady-state approximation, one assumes that after a brief induction period the concentration of each reaction intermediate I is essentially constant; one sets $d[I]/dt = 0$, solves for $[I]$, and uses the result to find the rate law.

Rules for devising mechanisms consistent with an observed rate law were given. If the reaction has a rate-determining step, then the total reactant composition of this step is found by summing the species in the rate law (rule 1 in Sec. 17.6).

The temperature dependence of the rate constants of elementary reactions and most complex reactions can be represented by the Arrhenius equation $k = Ae^{-E_a/RT}$, where A and E_a are the pre-exponential factor and the Arrhenius activation energy.

In unimolecular reactions, molecules receive the energy needed to decompose or isomerize by collisional activation into states of high vibrational energy. An activated molecule either loses its extra vibrational energy in a collision or reacts to form products.

A chain reaction contains an initiation step that produces a reactive intermediate (often a free radical), one or more propagation steps that consume the reactive intermediate, produce products, and regenerate the intermediate, and a termination step that consumes the intermediate. The kinetics of the $H_2 + Br_2$ chain reaction and of free-radical addition polymerization were discussed.

For reactions in liquid solutions, the solvent may strongly affect the rate constant. Each encounter between two reacting species A and B in solution involves many collisions between A and B while they are trapped in a surrounding cage of solvent molecules. If A and B molecules react whenever they encounter each other, the reaction rate is controlled by the rate at which A and B diffuse toward each other. Theoretical equations for the rates of such diffusion-controlled reactions were given.

A catalyst speeds up the rate of a reaction but does not affect the equilibrium constant. The catalyst participates in the reaction mechanism but is regenerated unchanged at the end of the reaction. The functioning of biological organisms depends on catalysis by enzymes. A simple model (the Michaelis–Menten model) of enzyme kinetics was given. Many industrial reactions are run in the presence of solid catalysts. The Langmuir isotherm was used to rationalize observed kinetic behavior in heterogeneous catalysis. The kinetics of gas–solid adsorption, surface migration, and desorption was examined.

Radioactive decay follows first-order kinetics.

Important kinds of calculations discussed in this chapter include:

- Calculation of the amounts of reactants and products present at a given time from the integrated rate law and the initial composition.
- Determination of reaction orders from kinetics data.
- Determination of the rate constant from kinetics data.
- Use of the Arrhenius equation $k = Ae^{-E_a/RT}$ to calculate A and E_a from k-versus-T data or to calculate $k(T_2)$ from $k(T_1)$ and A and E_a.
- Calculation of $\langle DP \rangle$ in radical addition polymerizations.
- Calculation of k for a diffusion-controlled reaction.
- Calculation of the parameters in the Michaelis–Menten equation (17.125) from rate-versus-concentration data.
- Calculation of the activity of a radioactive sample at time t from its half-life and initial activity using $\lambda t_{1/2} = 0.693$ and $A = A_0 e^{-\lambda t}$.

FURTHER READING AND DATA SOURCES

Moore and Pearson; Gardiner; Wilkinson; Nicholas; Laidler (1987); *Espenson; Steinfeld, Francisco, and Hase; Bamford and Tipper; Hague; Robinson and Holbrook; Bernasconi; Gasser; Allcock and Lampe,* chaps. 3 and 12; *Billmeyer,* chap. 3.

Rate constants, pre-exponential factors, and activation energies: *Landolt–Börnstein,* vol. II, pt. 5b, pp. 247–336; S. W. Benson and H. E. O'Neal, Kinetic Data on Gas Phase Unimolecular Reactions, *Nat. Bur. Stand. U.S. Publ. NSRDS-NBS* 21, 1970; A. F. Trotman-Dickenson and G. S. Milne, Tables of Bimolecular Gas Phase Reactions, *Nat. Bur. Stand. U.S. Publ. NSRDS-NBS* 9, 1967; E. T. Denisov, *Liquid-Phase Reaction Rate Constants,* Plenum, 1974; J. Brandrup and E. H. Immergut (eds.), *Polymer Handbook,* 3d ed., Wiley, 1989, sec. II; L. H. Gevantman and D. Garvin, *Int. J. Chem. Kinet,* **5,** 213 (1973); R. F. Hampson and D. Garvin, *J. Phys. Chem.,* **81,** 2317 (1977); N. Cohen and K. R. Westberg, *J. Phys. Chem. Ref. Data,* **12,** 531 (1983); **20,** 1211 (1991); R. Atkinson et al., *J. Phys. Chem Ref. Data,* **26,** 521 (1997); **28,** 191 (1999). The National Institute of Standards and Technology (www.nist.gov/srd/kinet.htm) publishes (a) the *NIST Chemical Kinetics Database,* which contains data on over 15000 gas-phase reactions; (b) the *NDRL/NIST Solution Kinetics Database,* which contains data on 10800 reactions involving radicals in solution.

Sticking coefficients, surface diffusion coefficients, and desorption activation energies: *Bamford and Tipper,* vol. 19, pp. 42–49, 126–140, 158–162.

PROBLEMS

Section 17.1

17.1 True or false? (*a*) Every reaction has an order. (*b*) All rate constants have the same dimensions. (*c*) Homogeneous reaction rates have the dimensions of concentration divided by time. (*d*) Partial orders are always integers. (*e*) Rate constants depend on temperature. (*f*) Partial orders are never negative. (*g*) Rate constants are never negative. (*h*) Every species that appears in the rate law of a reaction must be a reactant or product in that reaction.

17.2 Give the commonly used units of the rate constant for (*a*) a first-order reaction; (*b*) a second-order reaction; (*c*) a third-order reaction.

17.3 For the reaction $2A + B \rightarrow$ products, which statement is true? (*a*) $dn_A/dt = 2 dn_B/dt$; (*b*) $2 dn_A/dt = dn_B/dt$.

17.4 If the reaction $N_2 + 3H_2 \rightarrow 2NH_3$ has $d[H_2]/dt = -0.006$ mol/L-s at a certain instant, what is $d[N_2]/dt$ at that instant?

17.5 For the gas-phase reaction $2N_2O_5 \rightarrow 4NO_2 + O_2$, the rate constant k is 1.73×10^{-5} s^{-1} at 25°C. The observed rate law is $r = k[N_2O_5]$. (*a*) Calculate r and J for this reaction in a 12.0-dm^3 container with $P(N_2O_5) = 0.10$ atm at 25°C. (*b*) Calculate $d[N_2O_5]/dt$ for the conditions of part (*a*). (*c*) Calculate the number of N_2O_5 molecules that decompose in 1 s for the conditions of (*a*). (*d*) What are k, r, and J for the conditions of (*a*) if the reaction is written $N_2O_5 \rightarrow 2NO_2 + \frac{1}{2}O_2$?

17.6 Verify that $J = d\xi/dt$, where J is the rate of conversion and ξ is the extent of reaction.

17.7 The *number concentration* C_B of species B in a phase of volume V is defined as $C_B = N_B/V$, where N_B is the number of B molecules in the phase. For the reaction (17.1) in a constant-volume system, the reaction rate r_C based on number concentrations is defined as $r_C \equiv -(1/b) dC_B/dt$ and the number-concentration rate constant k_C satisfies

$$r_C = k_C C_A^\alpha C_B^\beta \cdots C_L^\lambda$$

Show that $k_C = k/N_A^{n-1}$ for an nth-order reaction, where N_A is the Avogadro constant. [The commonly used units of k_C are (cm^3)$^{n-1}$ s^{-1}, which is usually written as (cm^3/molecule)$^{n-1}$ s^{-1} for clarity; of course, "molecule" is not a unit.]

17.8 In gas-phase kinetics, pressures instead of concentrations are sometimes used in rate laws. Suppose that for $aA \rightarrow$ products, one finds that $-a^{-1} dP_A/dt = k_P P_A^n$, where k_P is a constant and P_A is the partial pressure of A. (*a*) Show that $k_P = k(RT)^{1-n}$. (*b*) Is this relation valid for any nth-order reaction?

17.9 Reactions 1 and 2 are each first-order, and $k_1 > k_2$ at a certain temperature T. Must r_1 be greater than r_2 at T?

17.10 For the mechanism

$$A + B \rightarrow C + D$$
$$2C \rightarrow F$$
$$F + B \rightarrow 2A + G$$

(*a*) give the stoichiometric number of each step and give the overall reaction; (*b*) classify each species as reactant, product, intermediate, or catalyst.

17.11 The gas-phase reaction $2NO_2 + O_3 \rightarrow N_2O_5 + O_2$ has the rate constant $k = 2.0 \times 10^4$ dm^3 mol^{-1} s^{-1} at 300 K. What is the order of this reaction?

Section 17.3

17.12 For the reaction scheme (17.35) with only A present initially, sketch the rates r_1 and r_2 of reactions 1 and 2 versus time.

17.13 For each of the following sets of first-order reactions, write expressions for $d[B]/dt$, $d[E]/dt$, and $d[F]/dt$ in terms of rate constants and concentrations. (*a*) $B \xrightarrow{k_1} E$; (*b*) $B \xrightarrow{k_1} E \xrightarrow{k_2} F$; (*c*) $B \underset{k_{-1}}{\overset{k_1}{\rightleftharpoons}} E \xrightarrow{k_2} F$; (*d*) $B \underset{k_{-1}}{\overset{k_1}{\rightleftharpoons}} E$ and $B \underset{k_{-3}}{\overset{k_3}{\rightleftharpoons}} F$.

17.14 For the reaction scheme $A \rightarrow B \rightarrow C$ where the concentrations of any intermediates are negligible, which of the following statements hold during the reaction? (*a*) $[A] = -[B]$; (*b*) $\Delta[A] = -\Delta[B]$; (*c*) $\Delta[A] + \Delta[B] + \Delta[C] = 0$.

17.15 The first-order reaction $2A \rightarrow 2B + C$ is 35% complete after 325 s. (*a*) Find k and k_A, where k_A is defined in (17.11). (*b*) How long will it take for the reaction to be 70% complete? 90% complete?

17.16 (*a*) Use information in Prob. 17.5 to calculate the half-life for the N_2O_5 decomposition at 25°C. (*b*) Calculate $[N_2O_5]$ after 24.0 hr if $[N_2O_5]_0 = 0.010$ mol dm^{-3} and the system is at 25°C.

17.17 Derive the integrated rate law (17.24).

17.18 For the gas-phase reaction $2NO_2 + F_2 \rightarrow 2NO_2F$, the rate constant k is 38 dm^3 mol^{-1} s^{-1} at 27°C. The reaction is first-order in NO_2 and first-order in F_2. (*a*) Calculate the number of moles of NO_2, F_2, and NO_2F present after 10.0 s if 2.00 mol of NO_2 is mixed with 3.00 mol of F_2 in a 400-dm^3 vessel at 27°C. (*b*) For the system of (*a*), calculate the initial reaction rate and the rate after 10.0 s.

17.19 (*a*) The differential equation $dy/dx = f(x) + g(x)y$, where f and g are functions of x, has as its solution

$$y = e^{w(x)}\left[\int e^{-w(x)} f(x)\, dx + c\right], \qquad w(x) \equiv \int g(x)\, dx$$

where c is an arbitrary constant. Prove this result by substituting the proposed solution into the differential equation. (*b*) Use the result for (*a*) to solve the differential equation (17.39); use (17.37) to evaluate c.

17.20 Does the term "reversible" have the same meaning in kinetics as in thermodynamics?

17.21 Let the reaction $aA \rightarrow$ products have rate law $r = k[A]^2$. Write down the equation that gives r of this reaction as a function of time.

17.22 If the reaction A → products is zero-order, sketch [A] versus t.

17.23 For the rate law $r = k[A]^n$, for what values of n does the reaction go to completion in a finite time?

17.24 Let $\delta_{ab} \equiv a[B]_0 - b[A]_0$, $\delta_{ac} \equiv a[C]_0 - c[A]_0$, and $\delta_{bc} \equiv b[C]_0 - c[B]_0$. For the reaction $aA + bB \to$ products, show that the rate law $a^{-1} d[A]/dt = -k[A]^2[B]$ integrates to (use a table of integrals)

$$\frac{1}{\delta_{ab}}\left(\frac{1}{[A]_0} - \frac{1}{[A]}\right) + \frac{b}{\delta_{ab}} \ln \frac{[B]/[B]_0}{[A]/[A]_0} = -kt$$

where [B] is given by (17.19). (For $aA + bB + cC \to$ products, the rate law $a^{-1} d[A]/dt = -k[A][B][C]$ integrates to

$$\frac{a}{\delta_{ab}\delta_{ac}} \ln \frac{[A]}{[A]_0} - \frac{b}{\delta_{ab}\delta_{bc}} \ln \frac{[B]}{[B]_0} + \frac{c}{\delta_{ac}\delta_{bc}} \ln \frac{[C]}{[C]_0} = -kt$$

but it's not worthwhile to spend the time to derive this.)

Section 17.4

17.25 (a) If $r = k[A]^2[B]$ for a reaction, by what factor is the initial rate multiplied if the initial A concentration is multiplied by 1.5 and the initial B concentration is tripled? (b) If tripling the initial A concentration multiplies the initial rate by 27, what is the order with respect to A?

17.26 Show that, if $r = k_A[A]^n$, then

$$\log_{10} t_\alpha = \log_{10} \frac{\alpha^{1-n} - 1}{(n-1)k_A} - (n-1)\log_{10}[A]_0 \quad \text{for } n \neq 1$$

$$t_\alpha = -(\ln \alpha)/k_A \quad \text{for } n = 1$$

where t_α is the fractional life.

17.27 For $\alpha = 0.05$ in (17.50), calculate the Powell-plot parameter $\log_{10} \phi$ for $n = 0, \frac{1}{2}, 1, \frac{3}{2}, 2,$ and 3.

17.28 For the decomposition of $(CH_3)_2O$ (species A) at 777 K, the time required for $[A]_0$ to fall to $0.69[A]_0$ as a function of $[A]_0$ is:

$10^3[A]_0/(mol/dm^3)$	8.13	6.44	3.10	1.88
$t_{0.69}/s$	590	665	900	1140

(a) Find the order of the reaction. (b) Find k_A in $d[A]/dt = -k_A[A]^n$.

17.29 It was noted in Sec. 17.4 that the trial-and-error method of determining reaction orders is poor. Data for the decomposition of $(CH_3)_3COOC(CH_3)_3(g)$, species A, at 155°C are (where $c° \equiv 1$ mol/dm³):

t/min	0	3	6	9
$10^3[A]/c°$	6.35	5.97	5.64	5.31

t/min	12	15	18	21
$10^3[A]/c°$	5.02	4.74	4.46	4.22

(a) Plot $\log_{10} 10^3[A]$ versus t and $(10^3[A])^{-1}$ versus t and see if you can decide which plot is more nearly linear. (b) Make a Powell plot and see if this allows the order to be determined.

17.30 The reaction $n\text{-}C_3H_7Br + S_2O_3^{2-} \to C_3H_7S_2O_3^- + Br^-$ in aqueous solution is first order in C_3H_7Br and first-order in $S_2O_3^{2-}$. At 37.5°C, the following data were obtained (where $c° \equiv 1$ mol/dm³ and 1 ks = 10^3 s):

$10^3[S_2O_3^{2-}]/c°$	96.6	90.4	86.3	76.6	66.8
t/ks	0	1.110	2.010	5.052	11.232

The initial C_3H_7Br concentration was 39.5 mmol/dm³. Find the rate constant using a graphical method.

17.31 At $t = 0$, butadiene was introduced into an empty vessel at 326°C and the dimerization reaction $2C_4H_6 \to C_8H_{12}$ followed by monitoring the pressure P. The following data were obtained (1 ks = 10^3 s):

t/ks	$P/torr$	t/ks	$P/torr$	t/ks	$P/torr$
0	632.0	1.751	535.4	5.403	453.3
0.367	606.6	2.550	509.3	7.140	432.8
0.731	584.2	3.652	482.8	10.600	405.3
1.038	567.3				

(a) Find the reaction order using a Powell plot or the fractional-life method. (b) Evaluate the rate constant.

17.32 Initial rates r_0 for the reaction $2A + C \to$ products at 300 K at various sets of initial concentrations are as follows (where $c° \equiv 1$ mol/dm³):

$[A]_0/c°$	0.20	0.60	0.20	0.60
$[B]_0/c°$	0.30	0.30	0.90	0.30
$[C]_0/c°$	0.15	0.15	0.15	0.45
$100r_0/(c°/s)$	0.60	1.81	5.38	1.81

(a) Assume that the rate law has the form (17.5) and determine the partial orders. (b) Evaluate the rate constant. (c) Explain why determining a rate law and rate constant using only initial-rate data can sometimes give erroneous results. (*Hint:* See Sec. 17.1.)

17.33 For the reaction $A + B \to C + D$, a run with $[A]_0 = 400$ mmol dm⁻³ and $[B]_0 = 0.400$ mmol dm⁻³ gave the following data (where $c° \equiv 1$ mol/dm³):

t/s	0	120	240	360	∞
$10^4[C]/c°$	0	2.00	3.00	3.50	4.00

and a run with $[A]_0 = 0.400$ mmol dm⁻³ and $[B]_0 = 1000$ mmol dm⁻³ gave

$10^{-3}t/s$	0	69	208	485	∞
$10^4[C]/c°$	0	2.00	3.00	3.50	4.00

Find the rate law and the rate constant. The numbers have been chosen to make determination of the orders simple.

17.34 For the reaction A → products, data for a run with $[A]_0 = 0.600$ mol dm⁻³ are:

t/s	$[A]/[A]_0$	t/s	$[A]/[A]_0$
0	1	400	0.511
100	0.829	600	0.385
200	0.688	1000	0.248
300	0.597		

(a) Find the order of the reaction. (b) Find the rate constant.

17.35 For the reaction $2A + B \rightarrow C + D + 2E$, data for a run with $[A]_0 = 800$ mmol/L and $[B]_0 = 2.00$ mmol/L are

t/ks	8	14	20	30	50	90
$[B]/[B]_0$	0.836	0.745	0.680	0.582	0.452	0.318

and data for a run with $[A]_0 = 600$ mmol/L and $[B]_0 = 2.00$ mmol/L are

t/ks	8	20	50	90
$[B]/[B]_0$	0.901	0.787	0.593	0.453

Find the rate law and rate constant.

17.36 For Example 17.3, use a spreadsheet to find k_A (a) from the regression line of a $1/[A]$ versus t plot; (b) by a least-squares fit on the $[A]$ versus t data, as described in Example 17.3. (c) Repeat (a) and (b) for the Exercise in Example 17.3.

Section 17.5

17.37 True or false? (a) The rate law for the elementary reaction $A + B \rightarrow$ products in an ideal system must be $r = k[A][B]$. (b) The rate law for the composite reaction $C + D \rightarrow$ products in an ideal system might not be $r = k[C][D]$.

17.38 The rate constant for the elementary gas-phase reaction $N_2O_4 \rightarrow 2NO_2$ is 4.8×10^4 s^{-1} at 25°C. Use data in the Appendix to calculate the rate constant at 25°C for $2NO_2 \rightarrow N_2O_4$.

17.39 For the elementary reaction $A + B \rightarrow 2C$ with rate constant k, express $d[A]/dt$ and $d[C]/dt$ in terms of the reaction rate r; then express $d[A]/dt$ and $d[C]/dt$ in terms of k and molar concentrations.

Section 17.6

17.40 True or false? (a) If we know the mechanism of a reaction including the values of the elementary rate constants, we can find the rate law (assuming the differential equations can be solved). (b) If we know the rate law of a reaction, we can deduce what its mechanism must be.

17.41 Explain why the step $Hg_2^{2+} \rightarrow 2Hg^{2+}$ cannot occur in a reaction mechanism.

17.42 For the mechanism given in Example 17.7, explain why the statements $k_1 > k_2$ or $k_1 < k_2$ are meaningless.

17.43 Devise another mechanism besides (17.56) that gives the rate law (17.55) for the reaction in Example 17.4 and that has a rate-determining step.

17.44 For the reaction $OCl^- + I^- \rightarrow OI^- + Cl^-$ in aqueous solution at 25°C, initial rates r_0 as a function of initial concentrations (where $c^\circ \equiv 1$ mol/dm^3) are:

$10^3[ClO^-]/c^\circ$	4.00	2.00	2.00	2.00
$10^3[I^-]/c^\circ$	2.00	4.00	2.00	2.00
$10^3[OH^-]/c^\circ$	1000	1000	1000	250
$10^3 r_0/(c^\circ$ s$^{-1})$	0.48	0.50	0.24	0.94

(a) Find the rate law and the rate constant. (b) Devise a mechanism consistent with the observed rate law.

17.45 The gas-phase reaction $2NO_2Cl \rightarrow 2NO_2 + Cl_2$ has $r = k[NO_2Cl]$. Devise two mechanisms consistent with this rate law.

17.46 The reaction $2Cr^{2+} + Tl^{3+} \rightarrow 2Cr^{3+} + Tl^+$ in aqueous solution has $r = k[Cr^{2+}][Tl^{3+}]$. Devise two mechanisms consistent with this rate law.

17.47 The gas-phase reaction $2NO_2 + F_2 \rightarrow 2NO_2F$ has $r = k[NO_2][F_2]$. Devise a mechanism consistent with this rate law.

17.48 The gas-phase reaction $XeF_4 + NO \rightarrow XeF_3 + NOF$ has $r = k[XeF_4][NO]$. Devise a mechanism consistent with this rate law.

17.49 The gas-phase reaction $2Cl_2O + 2N_2O_5 \rightarrow 2NO_3Cl + 2NO_2Cl + O_2$ has the rate law $r = k[N_2O_5]$. Devise a mechanism consistent with this rate law.

17.50 For the reaction $Hg_2^{2+} + Tl^{3+} \rightarrow 2Hg^{2+} + Tl^+$, devise another mechanism besides the one in Example 17.7 that gives the observed rate law (17.63).

17.51 Explain why it is virtually certain that the homogeneous gas-phase reaction $2NH_3 \rightarrow N_2 + 3H_2$ does not occur by a one-step mechanism.

17.52 The gas-phase decomposition of ozone, $2O_3 \rightarrow 3O_2$, is believed to have the mechanism

$$O_3 + M \underset{k_{-1}}{\overset{k_1}{\rightleftharpoons}} O_2 + O + M$$

$$O + O_3 \overset{k_2}{\rightarrow} 2O_2$$

where M is any molecule. (a) Verify that $d[O_2]/dt = 2k_2[O][O_3] + k_1[O_3][M] - k_{-1}[O_2][O][M]$. Write down a similar expression for $d[O_3]/dt$. (b) Use the steady-state approximation for [O] to simplify the expressions in (a) to $d[O_2]/dt = 3k_2[O_3][O]$ and $d[O_3]/dt = -2k_2[O_3][O]$. (c) Show that, when the steady-state approximation for [O] is substituted into either $d[O_2]/dt$ or $d[O_3]/dt$, one obtains

$$r = \frac{k_1 k_2 [O_3]^2}{k_{-1}[O_2] + k_2[O_3]/[M]}$$

(d) Assume step 1 is in near equilibrium so that step 2 is rate-determining, and derive an expression for r. *Hint:* Because O_2 appears as a product in both the rate-determining step 2 and the preceding step 1, this problem is tricky. From the overall stoichiometry, we have $r = \frac{1}{3}d[O_2]/dt$. The O_2 production rate in the rate-determining step 2 is $(d[O_2]/dt)_2 = 2k_2[O][O_3]$. However, for each time step 2 occurs, step 1 occurs once and produces one O_2 molecule. Hence three O_2 molecules are produced each time the rate-determining step occurs, and the total O_2 production rate is $d[O_2]/dt = 3k_2[O][O_3]$. (e) Under what condition does the steady-state approximation reduce to the equilibrium approximation?

17.53 (a) Apply the steady-state approximation to the N_2O_5 decomposition mechanism (17.8) and show that $r = k[N_2O_5]$, where $k = k_a k_b/(k_{-a} + 2k_b)$. (*Hint:* Use the steady-state approximation for both intermediates.) (b) Apply the rate-determining-step approximation to the N_2O_5 mechanism, assuming that step b is slow compared with steps $-a$ and c. (c) Under what condition does the rate law in (a) reduce to that in (b)? (d) The rate constant for the reaction in Prob. 17.49 is numerically equal to the rate constant for the N_2O_5 decomposition. Devise a mechanism for the reaction in Prob. 17.49 that will explain this fact.

17.54 Verify that each of the mechanisms (17.60), (17.61), and (17.62) gives $r = k[NO]^2[O_2]$.

Section 17.7

17.55 Consider a reaction with $d[A]/dt = -k[A]^n$ with $n = 1$, $k = 0.15$ s^{-1}, and $[A]_0 = 1.0000$ mol/L. Use a spreadsheet to apply the Euler method to find [A] at 1.00 s and at 3.00 s, taking Δt as 0.2 s. Then repeat with $\Delta t = 0.1$ s. Compare your results with the exact values. (*Hints:* Designate cells for n, Δt, and k. Put the t values in column A and the Euler-calculated [A] values in column B.)

17.56 Repeat Prob. 17.55 with $n = 2$, $[A]_0 = 1.0000$ mol/L, and $k = 0.15$ L/mol-s.

17.57 Repeat Prob. 17.55 using the modified Euler method. (*Hints:* Put the t values in column A, the $[A]_{n+1/2}$ values in column B, and the $[A]_n$ values in column C. Note that the column B formulas will refer to column C cells and so column B numbers will not show up until column C is completed. The top row of the array will contain t_0, $[A]_{1/2}$, and $[A]_0$.)

17.58 Use one of the programs mentioned near the end of Sec. 17.7 to solve for concentrations versus time in the system of reactions (17.35) with $[A]_0 = 1.00$ mol/L, $[B]_0 = [C]_0 = 0$, $k_1 = 0.02$ s^{-1}, and $k_2 = k_1/6$. Compare a few of the [B] values with those found from the exact solution (17.40).

Section 17.8

17.59 True or false? (*a*) Because the Arrhenius equation contains the gas constant R, the Arrhenius equation applies only to gas-phase reactions. (*b*) The Arrhenius equation holds exactly. (*c*) The pre-exponential factor A has the same units for all reactions.

17.60 The reaction $2DI \rightarrow D_2 + I_2$ has $k = 1.2 \times 10^{-3}$ dm^3 mol^{-1} s^{-1} at 660 K and $E_a = 177$ kJ/mol. Calculate k at 720 K for this reaction.

17.61 Rate constants for the gas-phase reaction $H_2 + I_2 \rightarrow 2HI$ at various temperatures are ($c^\circ \equiv 1$ mol/dm^3):

$10^3k/(c^{\circ-1}\ s^{-1})$	0.54	2.5	14	25	64
T/K	599	629	666	683	700

Find E_a and A from a graph.

17.62 For the reaction $2HI \rightarrow H_2 + I_2$, values of k are 1.2×10^{-3} and 3.0×10^{-5} dm^3 mol^{-1} s^{-1} at 700 and 629 K, respectively. Estimate E_a and A.

17.63 What value of k is predicted by the Arrhenius equation for $T \rightarrow \infty$? Is this result physically reasonable?

17.64 The number of chirps per minute of a snowy tree cricket (*Oecanthus fultoni*) at several temperatures is 178 at 25.0°C, 126 at 20.3°C, and 100 at 17.3°C. (*a*) Find the activation energy for the chirping process. (*b*) Find the chirping rate to be expected at 14.0°C. Compare the result with the rule that

the Fahrenheit temperature equals 40 plus the number of cricket chirps in 15 seconds.

17.65 The gas-phase reaction $2N_2O_5 \rightarrow 4NO_2 + O_2$ has

$$k = 2.05 \times 10^{13} \exp\left(-24.65 \text{ kcal mol}^{-1}/RT\right) \text{ s}^{-1}$$

(*a*) Give the values of A and E_a. (*b*) Find $k(0°C)$. (*c*) Find $t_{1/2}$ at $-50°C$, 0°C, and 50°C.

17.66 For $T = 300$ K, 310 K, and 320 K, calculate the fraction of collisions in which the relative kinetic energy along the line of the collision exceeds (80 kJ/mol)/N_A.

17.67 For a system with the two competing mechanisms in (17.72) with the first step of the second mechanism being rate-determining, show that the observed activation energy is

$$E_a = \frac{k_1(T)E_{a,1} + k_2(T)E_{a,2}}{k_1(T) + k_2(T)}$$

17.68 For the elementary gas-phase reaction $CO + NO_2 \rightarrow CO_2 + NO$, one finds that $E_a = 116$ kJ/mol. Use data in the Appendix to find E_a for the reverse reaction.

17.69 For the mechanism (1) $A + B \rightleftharpoons C + D$; (2) $2C \rightarrow G + H$, step 2 is rate-determining. Given the activation energies $E_{a,1} = 120$ kJ/mol, $E_{a,-1} = 96$ kJ/mol, and $E_{a,2} = 196$ kJ/mol, find E_a for the overall reaction.

17.70 (*a*) Find the activation energy of a reaction whose rate constant is multiplied by 6.50 when T is increased from 300.0 K to 310.0 K. (*b*) For a reaction with $E_a = 19$ kJ/mol (4.5 kcal/mol), by what factor is k multiplied when T increases from 300.0 K to 310.0 K?

17.71 For the gas-phase reaction $2N_2O_5 \rightarrow 4NO_2 + O_2$ some rate constants are

$10^5k/s^{-1}$	1.69	6.73	24.9	75.0	243
t/°C	25	35	45	55	65

(*a*) Use a spreadsheet to plot ln k versus $1/T$ and from the regression line find E_a and A. Use these values in the Arrhenius equation to find $k_{i,calc}$ at each temperature. Then calculate the percent error for each k_i and calculate $\Sigma_i (k_{i,calc} - k_i)^2$. (*b*) Use the Solver to find E_a and A by minimizing $\Sigma_i (k_{i,calc} - k_i)^2$. To help the Solver, do the following: Take the initial E_a and A guesses as the values found in (*a*). In the Solver Options box, (1) check Use Automatic Scaling (this choice is appropriate when quantities involved in the optimization differ by several orders of magnitude); (2) check Central Derivatives (this gives a more accurate estimation of the derivatives used in the minimization); (3) change the Solver default Precision and Convergence values to values that are at least 10^5 times as small. Repeat the minimization several times, each time starting from different E_a and A guesses, and select the E_a and A pair that gives the lowest $\Sigma_i (k_{i,calc} - k_i)^2$. Compare $\Sigma_i (k_{i,calc} - k_i)^2$ with the value found in (*a*) and compare the percent errors in $k_{i,calc}$ with those in (*a*). Minimization of $\Sigma_i (k_{i,calc} - k_i)^2$ gives greater weight to the larger (higher-T) values of k_i and gives a good fit to these values at the expense of fitting the smaller k_i values.

The proper procedure to find E_a and A is to do several kinetics runs at each T so that a standard deviation σ_i can be calculated for k at each T. Statistical weights ω_i are calculated as $\omega_i = 1/\sigma_i^2$ and the quantitiy $\Sigma_i\, \omega_i(k_{i,\text{calc}} - k_i)^2$ is minimized by using a program such as the Excel Solver. Alternatively, one can transform the Arrhenius equation to linear form by taking the log of both sides. When this linearization is done, the statistical weights ω_i must be adjusted to new values ω_i'. For the transformation from k to $\ln k$, it turns out that $\omega_i' = \omega_i k_i^2 = k_i^2/\sigma_i^2$. If only a single measured rate constant is available at each T, a not unreasonable assumption is that each k_i value has about the same percent error. This means that the standard deviation σ_i (which is a measure of the typical error) is approximately proportional to k_i; $\sigma_i = ck_i$, where c is a constant. In this case, the weights for the linear plot become $\omega_i' = k_i^2/\sigma_i^2 = 1/c^2$, so all points of the linear plot have the same weight. The procedure of part (a) is then appropriate.

Section 17.9

17.72 $\Delta_f G_{700}^\circ$ of HI is -11.8 kJ/mol. For $H_2 + I_2 \rightarrow 2HI$, use data in Probs. 17.61 and 17.62 to find (a) the stoichiometric number of the rate-determining step; (b) K_c at 629 K.

17.73 For the aqueous-solution reaction

$$BrO_3^- + 3SO_3^{2-} \rightarrow Br^- + 3SO_4^{2-}$$

one finds $r = k[BrO_3^-][SO_3^{2-}][H^+]$. Give the expression for the rate law of the reverse reaction if the rate-determining step has stoichiometric number (a) 1; (b) 2.

17.74 For a gas-phase reaction whose rate-limiting step has stoichiometric number s, show that $E_{a,f} - E_{a,b} = \Delta U^\circ/s$.

17.75 Show that (17.77) is valid when the designations of forward and reverse reactions are interchanged.

17.76 When an overall reaction is at equilibrium, the forward rate of a given elementary step must equal the reverse rate of that step. Also, the elementary steps multiplied by their stoichiometric numbers add to the overall reaction. Use these facts to show that the equilibrium constant K_c for an overall reaction whose mechanism has m elementary steps is related to the elementary rate constants by $K_c = \Pi_{i=1}^m (k_i/k_{-i})^{s_i}$, where k_i, k_{-i}, and s_i are the forward and reverse rate constants and the stoichiometric number for the ith elementary step.

17.77 For the N_2O_5 mechanism (17.8), what is the rate law for the reverse reaction $4NO_2 + O_2 \rightarrow 2N_2O_5$ if step (b) is the rate-determining step?

Section 17.11

17.78 For the unimolecular isomerization of cyclopropane to propylene, values of k_{uni} versus initial pressure P_0 at 470°C are

P_0/torr	110	211	388	760
$10^5 k_{\text{uni}}/\text{s}^{-1}$	9.58	10.4	10.8	11.1

Take the reciprocal of Eq. (17.85) and plot these data in a way that gives a straight line. From the slope and intercept, evaluate $k_{\text{uni},P=\infty}$ and the Lindemann parameters k_1 and k_{-1}/k_2.

17.79 Explain why the products B and C in the unimolecular decomposition A $\rightarrow$ B + C are each less effective than A in energizing A.

Section 17.13

17.80 For the $H_2 + Br_2$ mechanism (17.88), write expressions for $d[Br_2]/dt$ and $d[Br]/dt$ in terms of concentrations and rate constants (do not eliminate intermediates).

17.81 An oversimplified version of the CH_3CHO decomposition mechanism is

$$CH_3CHO \xrightarrow{(1)} CH_3 + CHO$$

$$CH_3 + CH_3CHO \xrightarrow{(2)} CH_4 + CH_3CO$$

$$CH_3CO \xrightarrow{(3)} CO + CH_3$$

$$2CH_3 \xrightarrow{(4)} C_2H_6$$

(The CHO reacts to form minor amounts of various species.) (a) Identify the initiation, propagation, and termination steps. (b) What is the overall reaction, neglecting minor products formed in initiation and termination steps? (c) Show that $r = k[CH_3CHO]^{3/2}$, where $k = k_2(k_1/2k_4)^{1/2}$.

17.82 In the treatment of the $H_2 + Br_2$ chain reaction, the following elementary reactions were not considered: (I) $H_2 + M \rightarrow 2H + M$; (II) $Br + HBr \rightarrow H + Br_2$; (III) $H + Br + M \rightarrow HBr + M$. Use qualitative reasoning involving activation energies and concentrations to explain why the rate of each of these reactions is negligible compared with the rates of those in (17.88).

17.83 For the reversible reaction $CO + Cl_2 \rightleftharpoons COCl_2$, the mechanism is believed to be

Step 1:	$Cl_2 + M \rightleftharpoons 2Cl + M$
Step 2:	$Cl + CO + M \rightleftharpoons COCl + M$
Step 3:	$COCl + Cl_2 \rightleftharpoons COCl_2 + Cl$

(a) Identify the initiation, propagation, and termination steps. (b) Assume steps 1 and 2 each to be in equilibrium, and find the rate law for the *forward* reaction. (c) What is the rate law for the reverse reaction?

17.84 Let E_a be the activation energy for the rate constant k in (17.95). (a) Relate E_a to $E_{a,1}$, $E_{a,-1}$, and $E_{a,2}$. (b) Measurement of $k(T)$ gives $E_a = 40.6$ kcal/mol and $A = 1.6 \times 10^{11}$ dm$^{3/2}$ mol$^{-1/2}$ s^{-1}. Use data in the Appendix to evaluate $E_{a,2}$ and find an expression for the elementary rate constant k_2 as a function of T.

17.85 (a) For a free-radical addition polymerization with $k_i = 5 \times 10^{-5}$ s^{-1}, $f = 0.5$, $k_t = 2 \times 10^7$ dm^3 mol^{-1} s^{-1}, and $k_p = 3 \times 10^3$ dm^3 mol^{-1} s^{-1}, and with initial concentrations [M] = 2 mol/dm^3 and [I] = 0.008 mol/dm^3, calculate the following quantities for the early stages of the reaction when [M] and [I] are close to their initial values: $[R_{\text{tot}} \cdot]$, $\langle DP \rangle$, $-d[M]/dt$, and $d[P_{\text{tot}}]/dt$; assume that termination is by combination. (b) Repeat the calculations of (a) when termination is by disproportionation.

17.86 For some free-radical addition polymerizations, one need not include an initiator substance I. Rather, heating the monomer produces free radicals that initiate the polymerization. Suppose that I is absent and that the initiation reaction is $2M \rightarrow 2R\cdot$ with rate constant k_i. Find expressions for $-d[M]/dt$, $[R_{tot}\cdot]$, and $\langle DP \rangle$ by modifying the treatment in the text. Assume termination is by combination.

Section 17.14

17.87 For the elementary reaction $A \rightleftharpoons 2C$, show that if a system in equilibrium is subjected to a small perturbation, then $[A] - [A]_{eq}$ is given by the equation following (17.110) if τ is defined as $\tau^{-1} \equiv k_f + 4k_b[C]_{eq}$.

Section 17.15

17.88 For the photolysis of $CH_3NNC_2H_5$, what products will be obtained if the reaction is carried out (a) in the gas phase; (b) in solution in an inert solvent?

17.89 For I in CCl_4 at 25°C, the diffusion coefficient is estimated to be 4.2×10^{-5} cm^2 s^{-1}, and the radius of I is about 2 Å. Calculate k_D for $I + I \rightarrow I_2$ in CCl_4 at 25°C and compare with the observed value 0.8×10^{10} dm^3 mol^{-1} s^{-1}.

17.90 (a) Show that for a nonionic diffusion-controlled reaction, $E_a \approx RT - RT^2\eta^{-1} d\eta/dT$. (b) Use data in Prob. 16.56 to calculate E_a for such a reaction in water at 25°C.

Section 17.16

17.91 True or false? (a) In homogeneous catalysis, the catalyst does not appear in the rate law. (b) A catalyst does not appear in the overall reaction. (c) In homogeneous catalysis, doubling the catalyst concentration will not change the rate. (d) In homogeneous catalysis, the catalyst does not appear in any of the steps of the mechanism.

Section 17.17

17.92 The reaction $CO_2(aq) + H_2O \rightarrow H^+ + HCO_3^-$ catalyzed by the enzyme bovine carbonic anhydrase was studied in a stopped-flow apparatus at pH 7.1 and temperature 0.5°C. For an initial enzyme concentration of 2.8×10^{-9} mol dm^{-3}, initial rates as a function of $[CO_2]_0$ are (where $c° \equiv 1$ mol/dm^3):

$10^3[CO_2]_0/c°$	1.25	2.50	5.00	20.0
$10^4r_0/(c°$ s$^{-1})$	0.28	0.48	0.80	1.5$_5$

Find k_2 and K_M from a Lineweaver–Burk plot.

17.93 For the Michaelis–Menten mechanism, show that, when $[S]_0$ is equal to K_M and $[P]$ is negligible, then $[ES]/[E]_0 = 0.5$ (that is, half the enzyme active sites are filled) and $r_0 = \frac{1}{2}r_{0,max}$.

Section 17.18

17.94 Observed half-lives for the W-catalyzed decomposition of NH_3 at 1100°C as a function of initial NH_3 pressure P_0 for a fixed mass of catalyst and a fixed container volume are 7.6, 3.7, and 1.7 min for P_0 values of 265, 130, and 58 torr, respectively. Find the reaction order.

17.95 It is believed that N_2 and H_2 are chemisorbed on Fe as N and H atoms, which then react stepwise to give NH_3. What would be the stoichiometric number of the rate-determining step in the Fe-catalyzed synthesis of NH_3 if the rate-determining step were: (a) $N_2 + 2* \rightarrow 2N*$; (b) $H_2 + 2* \rightarrow 2H*$; (c) $N* + H* \rightarrow *NH + *$; (d) $*NH + H* \rightarrow *NH_2 + *$; (e) $*NH_2 + H* \rightarrow *NH_3 + *$; (f) $*NH_3 \rightarrow NH_3 + *$? Rate measurements using isotopic tracers indicate that the stoichiometric number of the rate-determining step is probably 1 for the NH_3 synthesis on iron. What does this indicate about the rate-determining step? Write the overall reaction with the smallest possible integers.

17.96 Derive the Langmuir isotherm (13.38) for dissociative adsorption $A_2(g) \rightarrow 2A(ads)$, using a procedure similar to that used to derive the nondissociative isotherm (13.35).

17.97 When $CO(g)$ is nondissociatively chemisorbed on the (111) plane of a Pt crystal at 300 K, the maximum amount of CO adsorbed is 2.3×10^{-9} mol per cm^2 of surface. (a) How many adsorption sites does this surface have per cm^2? (b) The product Pt of the gas pressure and the time the solid surface is exposed to this pressure is often measured in units of langmuirs, where one *langmuir* (L) equals 10^{-6} torr $\cdot$ s. When a clean Pt(111) surface of area 5.00 cm^2 is exposed to 0.43 langmuir of $CO(g)$ at 300 K, 9.2×10^{-10} mol of CO is chemisorbed. Find the fraction θ of occupied sites. Estimate s_0, the sticking coefficient at $\theta = 0$.

17.98 For CO nondissociatively adsorbed on the (111) plane of Ir, $A_{des} = 2.4 \times 10^{14}$ s^{-1} and $E_{a,des} = 151$ kJ/mol. Find the half-life of CO chemisorbed on Ir(111) at (a) 300 K; (b) 700 K.

17.99 For nitrogen atoms chemisorbed on the (110) plane of W, $D_0 = 0.014$ cm^2/s and $E_{a,mig} = 88$ kJ/mol. Find the rms displacement in a given direction of such a chemisorbed N atom in 1 s and in 100 s at 300 K.

17.100 For a chemisorbed molecule, a typical A_{des} value might be 10^{15} s^{-1}. For a molecule whose chemisorption is non-activated, estimate the half-life on the adsorbent surface at 300 K if $|\Delta H_{ads}|$ is (a) 50 kJ/mol; (b) 100 kJ/mol; (c) 200 kJ/mol.

17.101 Show that for a half-reaction at an electrode of a galvanic or electrolytic cell, the conversion rate per unit surface area is $r_s = j/nF$, where n is the number of electrons in the half-reaction and $j \equiv I/\mathcal{A}$ is the current density.

Section 17.19

17.102 A sample of 0.420 mg of $^{233}UF_6$ shows an activity of 9.88×10^4 counts per second. Find $t_{1/2}$ of ^{233}U.

17.103 (a) The half-life of ^3H is 12.4 years. Calculate the activity of 20.0 g of HNO_3 containing 0.200 mole percent $^3HNO_3$. (b) Find the activity after 6.20 years.

17.104 The nuclide X has two decay modes: $X \rightarrow B$ with decay constant λ_1 and $X \rightarrow C$ with decay constant λ_2. Express $t_{1/2}$ of X in terms of λ_1 and λ_2.

17.105 The half-life of ^{14}C is 5730 years. The activity of carbon in living beings is 12.5 counts per minute per gram of carbon. (a) Calculate the percent of carbon in living beings that is ^{14}C. (b) Calculate the activity of carbon from the remains of an organism that died 50000 years ago. (c) Find the age of wood from an Egyptian tomb that shows an activity of 7.0 counts per minute per gram of carbon.

17.106 The uranium present in earth today is 99.28% ^{238}U and 0.72% ^{235}U. The half-lives are 4.51×10^9 years for ^{238}U and 7.0×10^8 years for ^{235}U. How long ago was this uranium 50% ^{238}U and 50% ^{235}U? Isotopic abundances are given on an atom-percent basis.

17.107 Show that the activity of a radioactive sample is given by $A = A_0(\frac{1}{2})^{t/t_{1/2}}$.

General

17.108 The dominant mechanism of nuclear fusion of hydrogen to helium in the sun is believed to be

$$\ce{^1_1H + ^1_1H -> ^2_1H + ^0_1e + \nu}$$

$$\ce{^2_1H + ^1_1H -> ^3_2He + \gamma}$$

$$\ce{^3_2He + ^3_2He -> ^4_2He + 2^1_1H}$$

$$\ce{^0_1e + ^0_{-1}e -> 2\gamma}$$

where the last reaction is electron–positron annihilation. (a) What is the overall reaction? What is the stoichiometric number of each step in the mechanism? (b) The isothermal ΔU_m for this fusion reaction is -2.6×10^9 kJ/mol. The sun radiates 3.9×10^{26} J/s. How many moles of ^4He are produced each second in the sun? (c) The earth is on the average 1.5×10^8 km from the sun. Find the number of neutrinos that hit a square centimeter of the earth in 1 s. Consider the square centimeter to be perpendicular to the earth–sun line. (Experiments indicate the actual neutrino flux from the sun to be far less than the theoretically calculated value, so a revision in current views of solar structure or neutrino properties may be required.)

17.109 One molecule of $^{12}C^1H_4$ has how many (a) electrons; (b) nuclei; (c) protons; (d) neutrons? Answer the same questions for one molecule of $^{13}C^1H_4$.

17.110 (a) For the elementary reaction $B(g) + C(g) \rightarrow$ products, show that if reaction occurs at every collision, then $k_{max} = Z_{BC}/N_A[B][C]$. (b) Calculate k_{max} at 300 K for the typical values $M_B = 30$ g/mol, $M_C = 50$ g/mol, $r_B + r_C = 4$ Å.

17.111 True or false? (a) The half-life is independent of initial concentration only for first-order reactions. (b) The units of a first-order rate constant are s^{-1}. (c) Changing the temperature changes the rate constant. (d) Elementary reactions with molecularity greater than 3 generally don't occur. (e) For a homogeneous reaction, $J = d\xi/dt$, where J is the conversion rate and ξ is the extent of reaction. (f) $K_c = k_f/k_b$ for every reaction in an ideal system. (g) If the partial orders differ from the coefficients in the balanced reaction, the reaction must be composite. (h) If the partial orders are equal to the corresponding coefficients in the balanced reaction, the reaction must be simple. (i) For an elementary reaction, the partial orders are determined by the reaction stoichiometry. (j) For a reaction with $E_a > 0$, the greater the activation energy, the more rapidly the rate constant increases with temperature. (k) The presence of a homogeneous catalyst cannot change the equilibrium composition of a system. (l) Since the concentrations of reactants decrease with time, the rate r of a reaction always decreases as time increases. (m) Knowledge of the rate law of a reaction allows us to decide unambiguously what the mechanism is. (n) Activation energies are never negative.

We now begin the study of **quantum chemistry,** which applies quantum mechanics to chemistry. Chapter 18 deals with **quantum mechanics,** the laws governing the behavior of microscopic particles such as electrons and nuclei. Chapters 19 and 20 apply quantum mechanics to atoms and molecules. Chapter 21 applies quantum mechanics to spectroscopy, the study of the absorption and emission of electromagnetic radiation. Quantum mechanics is used in statistical mechanics (Chapter 22) and in theoretical chemical kinetics (Chapter 23).

Unlike thermodynamics, quantum mechanics deals with systems that are not part of everyday macroscopic experience, and the formulation of quantum mechanics is quite mathematical and abstract. This abstractness takes a while to get used to, and it is natural to feel somewhat uneasy when first reading Chapter 18.

In an undergraduate physical chemistry course, it is not possible to give a full presentation of quantum mechanics. Derivations of results that are given without proof may be found in quantum chemistry texts listed in the Bibliography.

Sections 18.1 to 18.4 give the historical background of quantum mechanics. Section 18.5 discusses the uncertainty principle, a key concept that underlies the differences between quantum mechanics and classical (Newtonian) mechanics. Quantum mechanics describes the state of a system using a state function (or wave function) Ψ. Sections 18.6 and 18.7 describe the meaning of Ψ and the time-dependent and time-independent Schrödinger equations used to find Ψ. Sections 18.8, 18.9, 18.10, 18.12, 18.13, and 18.14 consider the Schrödinger equation, the wave functions, and the allowed quantum-mechanical energy levels for several systems. Sections 18.11 and 18.16 discuss operators, which are used extensively in quantum mechanics. Section 18.15 introduces some of the approximation methods used to apply quantum mechanics to chemistry.

Essentially all of chemistry is a consequence of the laws of quantum mechanics. If we want to understand chemistry at the fundamental level of electrons, atoms, and molecules, we must understand quantum mechanics. Quantities such as the heat of combustion of octane, the 25°C entropy of liquid water, the reaction rate of N_2 and H_2 gases at specified conditions, the equilibrium constants of chemical reactions, the absorption spectra of coordination compounds, the NMR spectra of organic compounds, the nature of the products formed when organic compounds react, the shape a protein molecule folds into when it is formed in a cell, the structure and function of DNA are all a consequence of quantum mechanics.

In 1929, Dirac, one of the founders of quantum mechanics, wrote that "The general theory of quantum mechanics is now almost complete . . . . The underlying physical laws necessary for the mathematical theory of . . . the whole of chemistry are thus completely known, and the difficulty is only that the exact application of these laws leads to equations much too complicated to be soluble." After its discovery, quantum mechanics was used to develop many concepts that helped explain chemical properties. However, because of the very difficult calculations needed to apply quantum mechanics to chemical systems, quantum mechanics was of little practical value in accurately calculating the properties of chemical systems for many years after its discovery. Nowadays, however, the extraordinary computational power of modern computers allows quantum-mechanical calculations to give accurate chemical predictions in many systems of real chemical interest. As computers become even more powerful and applications of quantum mechanics in chemistry increase, the need for all chemists to be familiar with quantum mechanics will increase.

18.1 BLACKBODY RADIATION AND ENERGY QUANTIZATION

Classical physics is the physics developed before 1900. It consists of classical mechanics (Sec. 2.1), Maxwell's theory of electricity, magnetism, and electromagnetic radiation (Sec. 21.1), thermodynamics, and the kinetic theory of gases (Chapters 15 and 16). In the late nineteenth century, some physicists believed that the theoretical structure of physics was complete, but in the last quarter of the nineteenth century, various experimental results were obtained that could not be explained by classical physics. These results led to the development of quantum theory and the theory of relativity. An understanding of atomic structure, chemical bonding, and molecular spectroscopy must be based on quantum theory, which is the subject of this chapter.

One failure of classical physics was the incorrect $C_{V,m}$ values of polyatomic molecules predicted by the kinetic theory of gases (Sec. 15.10). A second failure was the inability of classical physics to explain the observed frequency distribution of radiant energy emitted by a hot solid.

When a solid is heated, it emits light. Classical physics pictures light as a wave consisting of oscillating electric and magnetic fields, an **electromagnetic wave.** (See Sec. 21.1 for a fuller discussion.) The frequency ν (nu) and wavelength λ (lambda) of an electromagnetic wave traveling through vacuum are related by

$$\lambda\nu = c \qquad (18.1)*$$

where $c = 3.0 \times 10^8$ m/s is the speed of light in vacuum. The human eye is sensitive to electromagnetic waves whose frequencies lie in the range 4×10^{14} to 7×10^{14} cycles/s. However, electromagnetic radiation can have any frequency (see Fig. 21.2). We shall use the term "light" as synonymous with electromagnetic radiation, not restricting it to visible light.

Different solids emit radiation at different rates at the same temperature. To simplify things, one deals with the radiation emitted by a blackbody. A **blackbody** is a body that absorbs all the electromagnetic radiation that falls on it. A good approximation to a blackbody is a cavity with a tiny hole. Radiation that enters the hole is repeatedly reflected within the cavity (Fig. 18.1a). At each reflection, a certain fraction of the radiation is absorbed by the cavity walls, and the large number of reflections causes virtually all the incident radiation to be absorbed. When the cavity is heated, its walls emit light, a tiny portion of which escapes through the hole. It can be shown that the rate of radiation emitted per unit surface area of a blackbody is a function of only

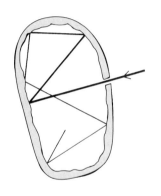

(a)

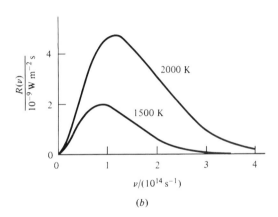

(b)

Figure 18.1

(a) A cavity acting as a blackbody. (b) Frequency distribution of blackbody radiation at two temperatures. (The visible region is from 4×10^{14} to 7×10^{14} s^{-1}.)

its temperature and is independent of the material of which the blackbody is made. (See *Zemansky and Dittman,* sec. 4-14, for a proof.)

By using a prism to separate the various frequencies emitted by the cavity, one can measure the amount of blackbody radiant energy emitted in a given narrow frequency range. Let the *frequency distribution* of the emitted blackbody radiation be described by the function $R(\nu)$, where $R(\nu)\,d\nu$ is the energy with frequency in the range ν to $\nu + d\nu$ that is radiated per unit time and per unit surface area. (Recall the discussion of distribution functions in Sec. 15.4.) Figure 18.1*b* shows some experimentally observed $R(\nu)$ curves. As T increases, the maximum in $R(\nu)$ shifts to higher frequencies. When a metal rod is heated, it first glows red, then orange-yellow, then white, then blue-white. (White light is a mixture of all colors.) Our bodies are not hot enough to emit visible light, but we do emit infrared radiation.

In June 1900, Lord Rayleigh attempted to derive the theoretical expression for the function $R(\nu)$. Using the equipartition-of-energy theorem (Sec. 15.10), he found that classical physics predicted $R(\nu) = (2\pi kT/c^2)\nu^2$, where k and c are Boltzmann's constant and the speed of light. But this result is absurd, since it predicts that the amount of energy radiated would increase without limit as ν increases. In actuality, $R(\nu)$ reaches a maximum and then falls off to zero as ν increases (Fig. 18.1*b*). Thus, classical physics fails to predict the spectrum of blackbody radiation.

On October 19, 1900, the physicist Max Planck announced to the German Physical Society his discovery of a formula that gave a highly accurate fit to the observed curves of blackbody radiation. Planck's formula was $R(\nu) = a\nu^3/(e^{b\nu/T} - 1)$, where a and b are constants with certain numerical values. Planck had obtained this formula by trial and error and at that time had no theory to explain it. On December 14, 1900, Planck presented to the German Physical Society a theory that yielded the blackbody-radiation formula he had found empirically a few weeks earlier. Planck's theory gave the constants a and b as $a = 2\pi h/c^2$ and $b = h/k$, where h was a new constant in physics and k is Boltzmann's constant [Eq. (3.57)]. Planck's theoretical expression for the frequency distribution of blackbody radiation is then

$$R(\nu) = \frac{2\pi h}{c^2} \frac{\nu^3}{e^{h\nu/kT} - 1} \tag{18.2}$$

Planck considered the walls of the blackbody to contain electric charges that oscillated (vibrated) at various frequencies [Maxwell's electromagnetic theory of light (Sec. 21.1) shows that electromagnetic waves are produced by accelerated electric charges. A charge oscillating at frequency ν will emit radiation at that frequency.] In order to derive (18.2), Planck found that he had to assume that the energy of each oscillating charge could take on only the possible values 0, $h\nu$, $2h\nu$, $3h\nu$, ..., where ν is the frequency of the oscillator and h is a constant (later called **Planck's constant**) with the dimensions of energy $\times$ time. This assumption then leads to Eq. (18.2). (For Planck's derivation, see M. Jammer, *The Conceptual Development of Quantum Mechanics,* McGraw-Hill, 1966, sec. 1.2.) Planck obtained a numerical value of h by fitting the formula (18.2) to the observed blackbody curves. The modern value is

$$h = 6.626 \times 10^{-34}\ \text{J} \cdot \text{s} = 6.626 \times 10^{-27}\ \text{erg} \cdot \text{s} \tag{18.3}*$$

In classical physics, energy takes on a continuous range of values, and a system can lose or gain any amount of energy. In direct contradiction to classical physics, Planck restricted the energy of each oscillating charge to a whole-number multiple of $h\nu$ and hence restricted the amount of energy each oscillator could gain or lose to an integral multiple of $h\nu$. Planck called the quantity $h\nu$ a **quantum** of energy (the Latin word *quantum* means "how much"). *In classical physics, energy is a continuous variable. In quantum physics, the energy of a system is* **quantized,** meaning that the energy can take on only certain values. Planck introduced the idea of energy quantization in one case, the emission of blackbody radiation. In the years 1900–1926, the concept of energy quantization was gradually extended to all microscopic systems.

Planck's assumption of energy quantization was originally intended only as a calculational device, and he planned to take the limit $h \to 0$ at the end of the derivation. He found, however, that taking this limit gave the wrong result, but with $h \neq 0$ he obtained the correct formula (18.2). Expanding the exponential in (18.2) in a Taylor series, we get $-1 + e^{h\nu/kT} = -1 + 1 + h\nu/kT + h^2\nu^2/2k^2T^2 + \cdots$, which goes to $h\nu/kT$ as $h \to 0$. Hence, (18.2) approaches $2\pi\nu^2 kT/c^2$ as $h \to 0$, which is the (erroneous) classical result of Rayleigh.

The concept of quantization of energy was a revolutionary departure from classical physics, and most physicists were very reluctant to accept this idea. One of the most reluctant was Planck himself, whose conservative temperament was offended by energy quantization. In the years following 1900, Planck tried repeatedly to derive (18.2) without using energy quantization, but he failed.

██ 18.2 ██   THE PHOTOELECTRIC EFFECT AND PHOTONS

The person who recognized the value of Planck's idea was Einstein, who applied the concept of energy quantization to electromagnetic radiation and showed that this explained the experimental observations in the photoelectric effect.

In the **photoelectric effect,** a beam of electromagnetic radiation (light) shining on a metal surface causes the metal to emit electrons; electrons absorb energy from the light beam, thereby acquiring enough energy to escape from the metal. A practical application is the photoelectric cell, used to measure light intensities, to prevent elevator doors from crushing people, and in smoke detectors (light scattered by smoke particles causes electron emission, which sets off an alarm).

Experimental work around 1900 had shown that (*a*) Electrons are emitted only when the frequency of the light exceeds a certain minimum frequency ν_0 (the *threshold frequency*). The value of ν_0 differs for different metals and lies in the ultraviolet for most metals. (*b*) Increasing the intensity of the light increases the number of electrons emitted but does not affect the kinetic energy of the emitted electrons. (*c*) Increasing the frequency of the radiation increases the kinetic energy of the emitted electrons.

These observations on the photoelectric effect cannot be understood using the classical picture of light as a wave. The energy in a wave is proportional to its intensity but is independent of its frequency, so one would expect the kinetic energy of the emitted electrons to increase with an increase in light intensity and to be independent of the light's frequency. Moreover, the wave picture of light would predict the photoelectric effect to occur at any frequency, provided the light is sufficiently intense.

In 1905 Einstein explained the photoelectric effect by extending Planck's concept of energy quantization to electromagnetic radiation. (Planck had applied energy quantization to the oscillators in the blackbody but had considered the electromagnetic radiation to be a wave.) Einstein proposed that in addition to having wavelike properties, light could also be considered to consist of particlelike entities (quanta), each quantum of light having an energy $h\nu$, where h is Planck's constant and ν is the frequency of the light. These entities were later named **photons,** and the energy of a photon is

$$E_{\text{photon}} = h\nu \qquad \qquad \textbf{(18.4)}*$$

The energy in a light beam is the sum of the energies of the individual photons and is therefore quantized.

Let electromagnetic radiation of frequency ν fall on a metal. The photoelectric effect occurs when an electron in the metal is hit by a photon. The photon disappears, and its energy $h\nu$ is transferred to the electron. Part of the energy absorbed by the electron is used to overcome the forces holding the electron in the metal, and the remainder appears as kinetic energy of the emitted electron. Conservation of energy therefore gives

$$h\nu = \Phi + \tfrac{1}{2}mv^2 \qquad (18.5)$$

where the *work function* Φ is the minimum energy needed by an electron to escape the metal and $\tfrac{1}{2}mv^2$ is the kinetic energy of the free electron. The valence electrons in metals have a distribution of energies (Sec. 24.11), so some electrons need more energy than others to leave the metal. The emitted electrons therefore show a distribution of kinetic energies, and $\tfrac{1}{2}mv^2$ in (18.5) is the maximum kinetic energy of emitted electrons.

Einstein's equation (18.5) explains all the observations in the photoelectric effect. If the light frequency is such that $h\nu < \Phi$, a photon does not have enough energy to allow an electron to escape the metal and no photoelectric effect occurs. The minimum frequency ν_0 at which the effect occurs is given by $h\nu_0 = \Phi$. (The work function Φ differs for different metals, being lowest for the alkali metals.) Equation (18.5) shows the kinetic energy of the emitted electrons to increase with ν and to be independent of the light intensity. An increase in intensity with no change in frequency increases the energy of the light beam and hence increases the number of photons per unit volume in the light beam, thereby increasing the rate of emission of electrons.

Einstein's theory of the photoelectric effect agreed with the qualitative observations, but it wasn't until 1916 that R. A. Millikan made an accurate quantitative test of Eq. (18.5). The difficulty in testing (18.5) is the need to maintain a very clean surface of the metal. Millikan found accurate agreement between (18.5) and experiment.

At first, physicists were very reluctant to accept Einstein's hypothesis of photons. Light shows the phenomena of diffraction and interference (*Halliday and Resnick,* chaps. 45 and 46), and these effects are shown only by waves, not by particles. Eventually, physicists became convinced that the photoelectric effect could be understood only by viewing light as being composed of photons. However, diffraction and interference can be understood only by viewing light as a wave and not as a collection of particles.

Thus, light seems to exhibit a dual nature, behaving like waves in some situations and like particles in other situations. This apparent duality is logically contradictory, since the wave and particle models are mutually exclusive. Particles are localized in space, but waves are not. The photon picture gives a quantization of the light energy, but the wave picture does not. In Einstein's equation $E_{\text{photon}} = h\nu$, the quantity E_{photon} is a particle concept, but the frequency ν is a wave concept, so this equation is, in a sense, self-contradictory. An explanation of these apparent contradictions is given in Sec. 18.4.

In 1907 Einstein applied the concept of energy quantization to the vibrations of the atoms in a solid, thereby showing that the heat capacity of a solid goes to zero as T goes to zero, a result in agreement with experiment but in disagreement with the classical equipartition theorem. See Sec. 24.12 for details.

18.3 THE BOHR THEORY OF THE HYDROGEN ATOM

The next major application of energy quantization was the Danish physicist Niels Bohr's 1913 theory of the hydrogen atom. A heated gas of hydrogen atoms emits electromagnetic radiation containing only certain distinct frequencies (Fig. 21.36). During 1885 to 1910, Balmer, Rydberg, and others found that the following empirical formula correctly reproduces the observed H-atom spectral frequencies:

$$\frac{\nu}{c} = \frac{1}{\lambda} = R\left(\frac{1}{n_b^2} - \frac{1}{n_a^2}\right) \qquad n_b = 1, 2, 3, \dots; \quad n_a = 2, 3, \dots; \quad n_a > n_b \qquad (18.6)$$

where the *Rydberg constant R* equals 1.096776×10^5 cm^{-1}. There was no explanation for this formula until Bohr's work.

If one accepts Einstein's equation $E_{photon} = h\nu$, the fact that only certain frequencies of light are emitted by H atoms indicates that contrary to classical ideas, a hydrogen atom can exist only in certain energy states. Bohr therefore postulated that the energy of a hydrogen atom is quantized: (1) An atom can take on only certain distinct energies $E_1, E_2, E_3, \ldots$. Bohr called these allowed states of constant energy the *stationary states* of the atom. This term is not meant to imply that the electron is at rest in a stationary state. Bohr further assumed that (2) An atom in a stationary state does not emit electromagnetic radiation. To explain the line spectrum of hydrogen, Bohr assumed that (3) When an atom makes a transition from a stationary state with energy E_{upper} to a lower-energy stationary state with energy E_{lower}, it emits a photon of light. Since $E_{photon} = h\nu$, conservation of energy gives

$$E_{upper} - E_{lower} = h\nu \qquad (18.7)^*$$

where $E_{upper} - E_{lower}$ is the energy difference between the atomic states involved in the transition and ν is the frequency of the light emitted. Similarly, an atom can make a transition from a lower-energy to a higher-energy state by absorbing a photon of frequency given by (18.7). The Bohr theory provided no description of the transition process between two stationary states. Of course, transitions between stationary states can occur by means other than absorption or emission of electromagnetic radiation. For example, an atom can gain or lose electronic energy in a collision with another atom.

Equations (18.6) and (18.7) with upper and lower replaced by a and b give $E_a - E_b = Rhc(1/n_b^2 - 1/n_a^2)$, which strongly indicates that the energies of the H-atom stationary states are given by $E = -Rhc/n^2$, with $n = 1, 2, 3, \ldots$. Bohr then introduced further postulates to derive a theoretical expression for the Rydberg constant. He assumed that (4) The electron in an H-atom stationary state moves in a circle around the nucleus and obeys the laws of classical mechanics. The energy of the electron is the sum of its kinetic energy and the potential energy of the electron–nucleus electrostatic attraction. Classical mechanics shows that the energy depends on the radius of the orbit. Since the energy is quantized, only certain orbits are allowed. Bohr used one final postulate to select the allowed orbits. Most books give this postulate as (5) The allowed orbits are those for which the electron's angular momentum $m_e v r$ equals $nh/2\pi$, where m_e and v are the electron's mass and speed, r is the radius of the orbit, and $n = 1, 2, 3, \ldots$. Actually, Bohr used a different postulate which is less arbitrary than 5 but less simple to state. The postulate Bohr used is equivalent to 5 and is omitted here. (If you're curious, see *Karplus and Porter*, sec. 1.4.)

With his postulates, Bohr derived the following expression for the H-atom energy levels: $E = -m_e e^4/8\varepsilon_0^2 h^2 n^2$, where e is the proton charge and ε_0 occurs in Coulomb's law (14.1). Therefore, Bohr predicted that $Rhc = m_e e^4/8\varepsilon_0^2 h^2$ and $R = m_e e^4/8\varepsilon_0^2 h^3 c$. Substitution of the values of m_e, e, h, ε_0, and c gave a result in good agreement with the experimental value of the Rydberg constant, indicating that the Bohr model gave the correct energy levels of H.

Although the Bohr theory is historically important for the development of quantum theory, postulates 4 and 5 are in fact *false*, and the Bohr theory was superseded in 1926 by the Schrödinger equation, which provides a correct picture of electronic behavior in atoms and molecules. Although postulates 4 and 5 are false, postulates 1, 2, and 3 are consistent with quantum mechanics.

18.4   THE DE BROGLIE HYPOTHESIS

In the years 1913 to 1925, attempts were made to apply the Bohr theory to atoms with more than one electron and to molecules. However, all attempts to derive the spectra

Second overtone

First overtone

Fundamental

Figure 18.2

Fundamental and overtone vibrations of a string.

of such systems using extensions of the Bohr theory failed. It gradually became clear that there was a fundamental error in the Bohr theory. The fact that the Bohr theory works for H is something of an accident.

A key idea toward resolving these difficulties was advanced by the French physicist Louis de Broglie (1892–1987) in 1923. The fact that a heated gas of atoms or molecules emits radiation of only certain frequencies shows that the energies of atoms and molecules are quantized, only certain energy values being allowed. Quantization of energy does not occur in classical mechanics; a particle can have any energy in classical mechanics. Quantization does occur in wave motion. For example, a string held fixed at each end has quantized modes of vibration (Fig. 18.2). The string can vibrate at its fundamental frequency ν, at its first overtone frequency 2ν, at its second overtone frequency 3ν, etc. Frequencies lying between these integral multiples of ν are not allowed.

De Broglie therefore proposed that just as light shows both wave and particle aspects, matter also has a "dual" nature. As well as showing particlelike behavior, an electron could also show wavelike behavior, the wavelike behavior manifesting itself in the quantized energy levels of electrons in atoms and molecules. Holding the ends of a string fixed quantizes its vibrational frequencies. Similarly, confining an electron in an atom quantizes its energies.

De Broglie obtained an equation for the wavelength λ to be associated with a material particle by reasoning in analogy with photons. We have $E_{\text{photon}} = h\nu$. Einstein's special theory of relativity gives the photon energy as $E_{\text{photon}} = mc^2$, where c is the speed of light. In this equation, m is the relativistic mass of the photon. A photon has zero rest mass, but photons always move at speed c in vacuum and are never at rest. At speed c, the photon has a nonzero mass m. Equating the two expressions for E_{photon}, we get $h\nu = mc^2$. But $\nu = c/\lambda$, where λ is the wavelength of the light. Hence, $hc/\lambda = mc^2$ and $\lambda = h/mc$ for a photon. By analogy, de Broglie proposed that a material particle with mass m and speed v would have a wavelength λ given by

$$\lambda = h/mv \tag{18.8}$$

Note that $mv = p$, where p is the particle's momentum.

The de Broglie wavelength of an electron moving at 1.0×10^6 m/s is

$$\lambda = \frac{6.6 \times 10^{-34} \text{ J s}}{(9.1 \times 10^{-31} \text{ kg})(1.0 \times 10^6 \text{ m/s})} = 7 \times 10^{-10} \text{ m} = 7 \text{ Å}$$

This wavelength is on the order of magnitude of molecular dimensions and indicates that wave effects are important in electronic motions in atoms and molecules. For a macroscopic particle of mass 1.0 g moving at 1.0 cm/s, a similar calculation gives $\lambda = 7 \times 10^{-27}$ cm. The extremely small size of λ (which results from the smallness of Planck's constant h in comparison with mv) indicates that quantum effects are unobservable for the motion of macroscopic objects.

De Broglie's bold hypothesis was experimentally confirmed in 1927 by Davisson and Germer, who observed diffraction effects when an electron beam was reflected from a crystal of Ni; G. P. Thomson observed diffraction effects when electrons were passed through a thin sheet of metal. See Fig. 18.3. Similar diffraction effects have been observed with neutrons, protons, helium atoms, and hydrogen molecules, indicating that the de Broglie hypothesis applies to all material particles, not just electrons. An application of the wavelike behavior of microscopic particles is the use of electron diffraction and neutron diffraction to obtain molecular structures (Secs. 24.9 and 24.10).

Electrons show particlelike behavior in some experiments (for example, the cathode-ray experiments of J. J. Thomson, Sec. 19.2) and wavelike behavior in other experiments. As noted in Sec. 18.2, the wave and particle models are incompatible

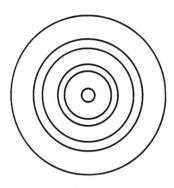

Figure 18.3

Diffraction rings observed when electrons are passed through a thin polycrystalline metal sheet.

with each other. An entity cannot be both a wave and a particle. How can we explain the apparently contradictory behavior of electrons? The source of the difficulty is the attempt to describe microscopic entities like electrons by using concepts developed from our experience in the macroscopic world. The particle and wave concepts were developed from observations on large-scale objects, and there is no guarantee that they will be fully applicable on the microscopic scale. Under certain experimental conditions, an electron behaves like a particle. Under other conditions, it behaves like a wave. However, an electron is neither a particle nor a wave. It is something that cannot be adequately described in terms of a model we can visualize.

A similar situation holds for light, which shows wave properties in some situations and particle properties in others. Light originates in the microscopic world of atoms and molecules and cannot be fully understood in terms of models visualizable by the human mind.

Although both electrons and light exhibit an apparent "wave–particle duality," there are significant differences between these entities. Light travels at speed c in vacuum, and photons have zero rest mass. Electrons always travel at speeds less than c and have a nonzero rest mass.

18.5 THE UNCERTAINTY PRINCIPLE

The apparent wave–particle duality of matter and of radiation imposes certain limitations on the information we can obtain about a microscopic system. Consider a microscopic particle traveling in the y direction. Suppose we measure the x coordinate of the particle by having it pass through a narrow slit of width w and fall on a fluorescent screen (Fig. 18.4). If we see a spot on the screen, we can be sure the particle passed through the slit. Therefore, we have measured the x coordinate at the time of passing the slit to an accuracy w. Before the measurement, the particle had zero velocity v_x and zero momentum $p_x = mv_x$ in the x direction. Because the microscopic particle has wavelike properties, it will be diffracted at the slit. Photographs of electron-diffraction patterns at a single slit and at multiple slits are given in C. Jönsson, *Am. J. Phys.*, **42**, 4 (1974).

Diffraction is the bending of a wave around an obstacle. A classical particle would go straight through the slit, and a beam of such particles would show a spread of length w in where they hit the screen. A wave passing through the slit will spread out to give a diffraction pattern. The curve in Fig. 18.4 shows the intensity of the wave at various points on the screen. The maxima and minima result from constructive and destructive interference between waves originating from various parts of the slit. **Interference** results from the superposition of two waves traveling through the same region of space. When the waves are in phase (crests occurring together), constructive interference occurs, with the amplitudes adding to give a stronger wave. When the waves are out of phase (crests of one wave coinciding with troughs of the second wave), destructive interference occurs and the intensity is diminished.

The first minima (points P and Q) in the single-slit diffraction pattern occur at places on the screen where waves originating from the top of the slit travel one-half wavelength less or more than waves originating from the middle of the slit. These waves are then exactly out of phase and cancel each other. Similarly, waves originating from a distance d below the top of the slit cancel waves originating a distance d below the center of the slit. The condition for the first diffraction minimum is then $\overline{DP} - \overline{AP} = \frac{1}{2}\lambda = \overline{CD}$ in Fig. 18.4, where C is located so that $\overline{CP} = \overline{AP}$. Because the distance from the slit to the screen is much greater than the slit width, angle APC is nearly zero and angles PAC and ACP are each nearly 90°. Hence, angle ACD is

Figure 18.4

Diffraction at a slit.

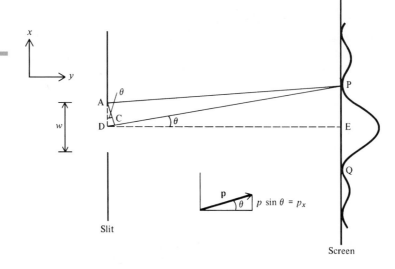

essentially 90°. Angles PDE and DAC each equal 90° minus angle ADC. These two angles are therefore equal and have been marked θ. We have $\sin \theta = \overline{DC}/\overline{AD} = \frac{1}{2}\lambda/\frac{1}{2}w = \lambda/w$. The angle θ at which the first diffraction minimum occurs is given by $\sin \theta = \lambda/w$.

For a microscopic particle passing through the slit, diffraction at the slit will change the particle's direction of motion. A particle diffracted by angle θ and hitting the screen at P or Q will have an x component of momentum $p_x = p \sin \theta$ at the slit (Fig. 18.4), where p is the particle's momentum. The intensity curve in Fig. 18.4 shows that the particle is most likely to be diffracted by an angle lying in the range $-\theta$ to $+\theta$, where θ is the angle to the first diffraction minimum. Hence, the position measurement produces an uncertainty in the p_x value given by $p \sin \theta - (-p \sin \theta) = 2p \sin \theta$. We write $\Delta p_x = 2p \sin \theta$, where Δp_x gives the uncertainty in our knowledge of p_x at the slit. We saw in the previous paragraph that $\sin \theta = \lambda/w$, so $\Delta p_x = 2p\lambda/w$. The de Broglie relation (18.8) gives $\lambda = h/p$, so $\Delta p_x = 2h/w$. The uncertainty in our knowledge of the x coordinate is given by the slit width, so $\Delta x = w$. Therefore, $\Delta x \, \Delta p_x = 2h$.

Before the measurement, we had no knowledge of the particle's x coordinate, but we knew that it was traveling in the y direction and so had $p_x = 0$. Thus, before the measurement, $\Delta x = \infty$ and $\Delta p_x = 0$. The slit of width w gave the x coordinate to an uncertainty w ($\Delta x = w$) but introduced an uncertainty $\Delta p_x = 2h/w$ in p_x. By reducing the slit width w, we can measure the x coordinate as accurately as we please, but as $\Delta x = w$ becomes smaller, $\Delta p_x = 2h/w$ becomes larger. The more we know about x, the less we know about p_x. The measurement introduces an uncontrollable and unpredictable disturbance in the system, changing p_x by an unknown amount.

Although we have analyzed only one experiment, analysis of many other experiments leads to the same conclusion: the product of the uncertainties in x and p_x of a particle is on the order of magnitude of Planck's constant or greater:

$$\Delta x \, \Delta p_x \gtrsim h \qquad \text{(18.9)*}$$

This is the **uncertainty principle,** discovered by Heisenberg in 1927. A general quantum-mechanical proof of (18.9) was given by Robertson in 1929. Similarly we have $\Delta y \, \Delta p_y \gtrsim h$ and $\Delta z \, \Delta p_z \gtrsim h$.

The small size of h makes the uncertainty principle of no consequence for macroscopic particles.

18.6   QUANTUM MECHANICS

The fact that electrons and other microscopic "particles" show wavelike as well as particlelike behavior indicates that electrons do not obey classical mechanics. Classical mechanics was formulated from the observed behavior of macroscopic objects and does not apply to microscopic particles. The form of mechanics obeyed by microscopic systems is called **quantum mechanics,** since a key feature of this mechanics is the quantization of energy. The laws of quantum mechanics were discovered by Heisenberg, Born, and Jordan in 1925 and by Schrödinger in 1926. Before discussing these laws, we consider some aspects of *classical* mechanics.

Classical Mechanics

The motion of a one-particle, one-dimensional classical-mechanical system is governed by Newton's second law $F = ma = m\, d^2x/dt^2$. To obtain the particle's position x as a function of time, this differential equation must be integrated twice with respect to time. The first integration gives dx/dt, and the second integration gives x. Each integration introduces an arbitrary integration constant. Therefore, integration of $F = ma$ gives an equation for x that contains two unknown constants c_1 and c_2; we have $x = f(t, c_1, c_2)$, where f is some function. To evaluate c_1 and c_2, we need two pieces of information about the system. If we know that at a certain time t_0, the particle was at the position x_0 and had speed v_0, then c_1 and c_2 can be evaluated from the equations $x_0 = f(t_0, c_1, c_2)$ and $v_0 = f'(t_0, c_1, c_2)$, where f' is the derivative of f with respect to t. Thus, provided we know the force F and the particle's initial position and velocity (or momentum), we can use Newton's second law to predict the position of the particle at any future time. A similar conclusion holds for a three-dimensional many-particle classical system.

The **state** of a system in classical mechanics is defined by specifying all the forces acting and all the positions and velocities (or momenta) of the particles. We saw in the preceding paragraph that knowledge of the present state of a classical-mechanical system enables its future state to be predicted with certainty.

The Heisenberg uncertainty principle, Eq. (18.9), shows that simultaneous specification of position and momentum is impossible for a microscopic particle. Hence, the very knowledge needed to specify the classical-mechanical state of a system is unobtainable in quantum theory. The state of a quantum-mechanical system must therefore involve less knowledge about the system than in classical mechanics.

Quantum Mechanics

In quantum mechanics, the **state** of a system is defined by a mathematical function Ψ (capital psi) called the **state function** or the **time-dependent wave function.** (As part of the definition of the state, the potential-energy function V must also be specified.) Ψ is a function of the coordinates of the particles of the system and (since the state may change with time) is also a function of time. For example, for a two-particle system, $\Psi = \Psi(x_1, y_1, z_1, x_2, y_2, z_2, t)$, where x_1, y_1, z_1 and x_2, y_2, z_2 are the coordinates of particles 1 and 2, respectively. The state function is in general a complex quantity; that is, $\Psi = f + ig$, where f and g are real functions of the coordinates and time and $i \equiv \sqrt{-1}$. The state function is an abstract entity, but we shall later see how Ψ is related to physically measurable quantities.

The state function changes with time. For an n-particle system, quantum mechanics postulates that the equation governing how Ψ changes with t is

$$-\frac{\hbar}{i}\frac{\partial \Psi}{\partial t} = -\frac{\hbar^2}{2m_1}\left(\frac{\partial^2 \Psi}{\partial x_1^2} + \frac{\partial^2 \Psi}{\partial y_1^2} + \frac{\partial^2 \Psi}{\partial z_1^2}\right) - \cdots$$

$$-\frac{\hbar^2}{2m_n}\left(\frac{\partial^2 \Psi}{\partial x_n^2} + \frac{\partial^2 \Psi}{\partial y_n^2} + \frac{\partial^2 \Psi}{\partial z_n^2}\right) + V\Psi \qquad (18.10)$$

In this equation, $\hbar$ (h bar) is Planck's constant divided by 2π,

$$\hbar \equiv h/2\pi \qquad \textbf{(18.11)*}$$

i is $\sqrt{-1}$; $m_1, \ldots, m_n$ are the masses of particles $1, \ldots, n$; x_1, y_1, z_1 are the spatial coordinates of particle 1; and V is the potential energy of the system. Since the potential energy is energy due to the particles' positions, V is a function of the particles' coordinates. Also, V can vary with time if an externally applied field varies with time. Hence, V is in general a function of the particles' coordinates and the time. V is derived from the forces acting in the system; see Eq. (2.21). The dots in Eq. (18.10) stand for terms involving the spatial derivatives of particles $2, 3, \ldots, n - 1$.

Equation (18.10) is a complicated partial differential equation. For most of the problems dealt with in this book, it will not be necessary to use (18.10), so don't panic.

The concept of the state function Ψ and Eq. (18.10) were introduced by the Austrian physicist Erwin Schrödinger (1887–1961) in 1926. Equation (18.10) is the **time-dependent Schrödinger equation.** Schrödinger was inspired by the de Broglie hypothesis to search for a mathematical equation that would resemble the differential equations that govern wave motion and that would have solutions giving the allowed energy levels of a quantum system. Using the de Broglie relation $\lambda = h/p$ and certain plausibility arguments, Schrödinger proposed Eq. (18.10) and the related time-independent equation (18.24) below. These plausibility arguments have been omitted in this book. It should be emphasized that these arguments can at best make the Schrödinger equation seem plausible. They can in no sense be used to derive or prove the Schrödinger equation. The Schrödinger equation is a fundamental postulate of quantum mechanics and cannot be derived. The reason we believe it to be true is that its predictions give excellent agreement with experimental results. "One could argue that the Schrödinger equation has had more to do with the evolution of twentieth-century science and technology than any other discovery in physics." (Jeremy Bernstein, *Cranks, Quarks, and the Cosmos,* Basic Books, 1993, p. 54.)

In 1925, several months before Schrödinger's work, Werner Heisenberg (1901–1976), Max Born (1882–1970), and Pascual Jordan (1902–1980) developed a form of quantum mechanics based on mathematical entities called matrices. A matrix is a rectangular array of numbers; matrices are added and multiplied according to certain rules. The *matrix mechanics* of these workers turns out to be fully equivalent to the Schrödinger form of quantum mechanics (which is often called *wave mechanics*). We shall not discuss matrix mechanics.

Schrödinger also contributed to statistical mechanics, relativity, and the theory of color vision and was deeply interested in philosophy. In an epilogue to his 1944 book, *What Is Life?*, Schrödinger wrote: "So let us see whether we cannot draw the correct, non-contradictory conclusion from the following two premises: (i) My body functions as a pure mechanism according to the Laws of Nature. (ii) Yet I know, by incontrovertible direct experience, that I am directing its motions . . . . The only possible inference from these two facts is, I think, that I—I in the widest meaning of the word, that is to say, every conscious mind that has ever said or felt 'I'—am the person, if any, who controls the 'motion of the atoms' according to the Laws of Nature." Schrödinger's life and loves are chronicled in W. Moore, *Schrödinger, Life and Thought,* Cambridge University Press, 1989.

The time-dependent Schrödinger equation (18.10) contains the first derivative of Ψ with respect to t, and a single integration with respect to time gives us Ψ. Integration of (18.10) therefore introduces only one integration constant, which can be evaluated if Ψ is known at some initial time t_0. Therefore, knowing the initial quantum-mechanical state $\Psi(x_1, \ldots, z_n, t_0)$ and the potential energy V, we can use (18.10) to predict the future quantum-mechanical state. The time-dependent Schrödinger equation is the quantum-mechanical analog of Newton's second law, which allows the future state of a classical-mechanical system to be predicted from its present state. We shall soon see, however, that knowledge of the state in quantum mechanics usually involves a knowledge of only probabilities, rather than certainties, as in classical mechanics.

What is the relation between quantum mechanics and classical mechanics? Experiment shows that macroscopic bodies obey classical mechanics (provided their speed is much less than the speed of light). We therefore expect that in the classical-mechanical limit of taking $h \to 0$, the time-dependent Schrödinger equation ought to reduce to Newton's second law. This was shown by Ehrenfest in 1927; for Ehrenfest's proof, see *Park*, sec. 3.3.

Physical Meaning of the State Function Ψ

Schrödinger originally conceived of Ψ as the amplitude of some sort of wave that was associated with the system. It soon became clear that this interpretation was wrong. For example, for a two-particle system, Ψ is a function of the six spatial coordinates x_1, y_1, z_1, x_2, y_2, and z_2, whereas a wave moving through space is a function of only three spatial coordinates. The correct physical interpretation of Ψ was given by Max Born in 1926. Born postulated that $|\Psi|^2$ gives the **probability density** for finding the particles at given locations in space. (Probability densities for molecular speeds were discussed in Sec. 15.4.) To be more precise, suppose a one-particle system has the state function $\Psi(x, y, z, t')$ at time t'. Consider the probability that a measurement of the particle's position at time t' will find the particle with its x, y, and z coordinates in the infinitesimal ranges x_a to $x_a + dx$, y_a to $y_a + dy$, and z_a to $z_a + dz$, respectively. This is the probability of finding the particle in a tiny rectangular-box-shaped region located at point (x_a, y_a, z_a) in space and having edges dx, dy, and dz (Fig. 18.5). Born's postulate is that the probability is given by

$$\Pr(x_a \leq x \leq x_a + dx, y_a \leq y \leq y_a + dy, z_a \leq z \leq z_a + dz)$$
$$= |\Psi(x_a, y_a, z_a, t')|^2 \, dx \, dy \, dz \qquad \textbf{(18.12)*}$$

where the left side of (18.12) denotes the probability the particle is found in the box of Fig. 18.5.

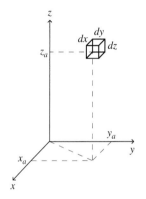

Figure 18.5

An infinitesimal box in space.

EXAMPLE 18.1 Probability for finding a particle

Suppose that at time t' the state function of a one-particle system is

$$\Psi = (2/\pi c^2)^{3/4} e^{-(x^2+y^2+z^2)/c^2} \qquad \text{where } c = 2 \text{ nm}$$

[One nanometer (nm) $\equiv 10^{-9}$ m.] Find the probability that a measurement of the particle's position at time t' will find the particle in the tiny cubic region with its center at $x = 1.2$ nm, $y = -1.0$ nm, and $z = 0$ and with edges each of length 0.004 nm.

The distance 0.004 nm is much less than the value of c and a change of 0.004 nm in one or more of the coordinates will not change the probability density $|\Psi|^2$ significantly. It is therefore a good approximation to consider the interval 0.004 nm as infinitesimal and to use (18.12) to write the desired probability as

$$|\Psi|^2 \, dx \, dy \, dz = (2/\pi c^2)^{3/2} e^{-2(x^2+y^2+z^2)/c^2} \, dx \, dy \, dz$$
$$= [2/(4\pi \text{ nm}^2)]^{3/2} e^{-2[(1.2)^2+(-1)^2+0^2]/4}(0.004 \text{ nm})^3$$
$$= 1.200 \times 10^{-9}$$

Exercise

(a) At what point is the probability density a maximum for the Ψ of this example? Answer by simply looking at $|\Psi|^2$. (b) Redo the calculation with x changed to its minimum value in the tiny cubic region and then with x changed to its maximum value in the region. Compare the results with that found when the central value of x is used. [*Answers:* (a) At the origin. (b) 1.203×10^{-9}, 1.197×10^{-9}.]

The state function Ψ is a complex quantity, and $|\Psi|$ is the absolute value of Ψ. Let $\Psi = f + ig$, where f and g are real functions and $i \equiv \sqrt{-1}$. The **absolute value** of Ψ is defined by $|\Psi| \equiv (f^2 + g^2)^{1/2}$. For a real quantity, g is zero, and the absolute value becomes $(f^2)^{1/2}$, which is the usual meaning of absolute value for a real quantity. The **complex conjugate** Ψ^* of Ψ is defined by

$$\Psi^* \equiv f - ig, \qquad \text{where } \Psi = f + ig \qquad \textbf{(18.13)*}$$

To get Ψ^*, we replace i by $-i$ wherever it occurs in Ψ. Note that

$$\Psi^*\Psi = (f - ig)(f + ig) = f^2 - i^2 g^2 = f^2 + g^2 = |\Psi|^2 \qquad \text{(18.14)}$$

since $i^2 = -1$. Therefore, instead of $|\Psi|^2$, we can write $\Psi^*\Psi$. The quantity $|\Psi|^2 = \Psi^*\Psi = f^2 + g^2$ is real and nonnegative, as a probability density must be.

In a two-particle system, $|\Psi(x_1, y_1, z_1, x_2, y_2, z_2, t')|^2 \, dx_1 \, dy_1 \, dz_1 \, dx_2 \, dy_2 \, dz_2$ is the probability that, at time t', particle 1 is in a tiny rectangular-box-shaped region located at point (x_1, y_1, z_1) and having dimensions dx_1, dy_1, and dz_1, and particle 2 is simultaneously in a box-shaped region at (x_2, y_2, z_2) with dimensions dx_2, dy_2, and dz_2. Born's interpretation of Ψ gives results fully consistent with experiment.

For a one-particle, one-dimensional system, $|\Psi(x, t)|^2 \, dx$ is the probability that the particle is between x and $x + dx$ at time t. The probability that it is in the region between a and b is found by summing the infinitesimal probabilities over the interval from a to b to give the definite integral $\int_a^b |\Psi|^2 \, dx$. Thus

$$\Pr(a \leq x \leq b) = \int_a^b |\Psi|^2 \, dx \qquad \text{one-particle, one-dim. syst.} \qquad \textbf{(18.15)*}$$

The probability of finding the particle somewhere on the x axis must be 1. Hence, $\int_{-\infty}^{\infty} |\Psi|^2 \, dx = 1$. When Ψ satisfies this equation, it is said to be **normalized.** The normalization condition for a one-particle, three-dimensional system is

$$\int_{-\infty}^{\infty} \int_{-\infty}^{\infty} \int_{-\infty}^{\infty} |\Psi(x, y, z, t)|^2 \, dx \, dy \, dz = 1 \qquad \text{(18.16)}$$

For an n-particle, three-dimensional system, the integral of $|\Psi|^2$ over all $3n$ coordinates $x_1, \ldots, z_n$, each integrated from $-\infty$ to ∞, equals 1.

The integral in (18.16) is a multiple integral. In a double integral like $\int_a^b \int_c^d f(x, y) \, dx \, dy$, one first integrates $f(x, y)$ with respect to x (while treating y as a constant) between the limits c and d, and then integrates the result with respect to y. For example, $\int_0^1 \int_0^4 (2xy + y^2) \, dx \, dy = \int_0^1 (x^2 y + xy^2) \, |_0^4 \, dy = \int_0^1 (16y + 4y^2) \, dy = 28/3$. To evaluate a triple integral like (18.16), we first integrate with respect to x while treating y and z as constants, then integrate with respect to y while treating z as constant, and finally integrate with respect to z.

The normalization requirement is often written

$$\int |\Psi|^2 \, d\tau = 1 \qquad \textbf{(18.17)*}$$

where $\int d\tau$ is a shorthand notation that stands for the *definite* integral over the full ranges of all the spatial coordinates of the system. For a one-particle, three-dimensional system, $\int d\tau$ implies a triple integral over x, y, and z from $-\infty$ to ∞ for each coordinate [Eq. (18.16)].

By substitution, it is easy to see that, if Ψ is a solution of (18.10), then so is $c\Psi$, where c is an arbitrary constant. Thus, there is always an arbitrary multiplicative constant in each solution to (18.10). The value of this constant is chosen so as to satisfy the normalization requirement (18.17).

From the state function Ψ, we can calculate the probabilities of the various possible outcomes when a measurement of position is made on the system. In fact, Born's work is more general than this. It turns out that Ψ gives information on the outcome of a measurement of *any* property of the system, not just position. For example, if Ψ is known, we can calculate the probability of each possible outcome when a measurement of p_x, the x component of momentum, is made. The same is true for a measurement of energy, or angular momentum, etc. [The procedure for calculating these probabilities from Ψ is discussed in *Levine* (2000), sec. 7.6.]

The state function Ψ is not to be thought of as a physical wave. Instead Ψ is an abstract mathematical entity that gives information about the state of the system. Everything that can be known about the system in a given state is contained in the state function Ψ. Instead of saying "the state described by the function Ψ," we can just as well say "the state Ψ." The information given by Ψ is the probabilities for the possible outcomes of measurements of the system's physical properties.

The state function Ψ describes a physical system. In Chapters 18 to 21, the system will usually be a particle, atom, or molecule. One can also consider the state function of a system that contains a large number of molecules, for example, a mole of some compound; this will be done in Chapter 22 on statistical mechanics.

Classical mechanics is a deterministic theory in that it allows us to predict the exact paths taken by the particles of the system and tells us where they will be at any future time. In contrast, quantum mechanics gives only the probabilities of finding the particles at various locations in space. The concept of a path for a particle becomes rather fuzzy in a time-dependent quantum-mechanical system and disappears in a time-independent quantum-mechanical system.

> Some philosophers have used the Heisenberg uncertainty principle and the nondeterministic nature of quantum mechanics as arguments in favor of human free will.
>
> The probabilistic nature of quantum mechanics disturbed many physicists, including Einstein, Schrödinger, and de Broglie. (Einstein wrote in 1926: "Quantum mechanics ... says a lot, but does not really bring us any closer to the secret of the Old One. I, at any rate, am convinced that He does not throw dice." When someone pointed out to Einstein that Einstein himself had introduced probability into quantum theory when he interpreted a light wave's intensity in each small region of space as being proportional to the probability of finding a photon in that region, Einstein replied, "A good joke should not be repeated too often.") These scientists believed that quantum mechanics does not furnish a complete description of physical reality. However, attempts to replace quantum mechanics by an underlying deterministic theory have failed. There appears to be a fundamental randomness in nature at the microscopic level.

Summary

The state of a quantum-mechanical system is described by its state function Ψ, which is a function of time and the spatial coordinates of the particles of the system. The state function provides information on the probabilities of the outcomes of measurements on the system. For example, when a position measurement is made on a one-particle system at time t', the probability that the particle's coordinates are found to be in the ranges x to $x + dx$, y to $y + dy$, z to $z + dz$ is given by $|\Psi(x, y, z, t')|^2 \, dx \, dy \, dz$. The function $|\Psi|^2$ is the probability density for position. Because the total probability of finding the particles somewhere is 1, the state function is normalized, meaning that the definite integral of $|\Psi|^2$ over the full range of all the spatial coordinates is equal to 1. The state function Ψ changes with time according to the time-dependent Schrödinger equation (18.10), which allows the future state (function) to be calculated from the present state (function).

18.7 THE TIME-INDEPENDENT SCHRÖDINGER EQUATION

For an isolated atom or molecule, the forces acting depend only on the coordinates of the charged particles of the system and are independent of time. Therefore, the potential energy V is independent of t for an isolated system. For systems where V is independent of time, the time-dependent Schrödinger equation (18.10) has solutions of the form $\Psi(x_1, \ldots, z_n, t) = f(t)\psi(x_1, \ldots, z_n)$, where ψ (lowercase psi) is a function of the $3n$ coordinates of the n particles and f is a certain function of time. We shall demonstrate this for a one-particle, one-dimensional system.

For a one-particle, one-dimensional system with V independent of t, Eq. (18.10) becomes

$$-\frac{\hbar^2}{2m}\frac{\partial^2 \Psi}{\partial x^2} + V(x)\Psi = -\frac{\hbar}{i}\frac{\partial \Psi}{\partial t} \tag{18.18}$$

Let us look for those solutions of (18.18) that have the form

$$\Psi(x, t) = f(t)\psi(x) \tag{18.19}$$

We have $\partial^2\Psi/\partial x^2 = f(t) \, d^2\psi/dx^2$ and $\partial\Psi/\partial t = \psi(x) \, df/dt$. Substitution into (18.18) followed by division by $f\psi = \Psi$ gives

$$-\frac{\hbar^2}{2m}\frac{1}{\psi(x)}\frac{d^2\psi}{dx^2} + V(x) = -\frac{\hbar}{i}\frac{1}{f(t)}\frac{df(t)}{dt} \equiv E \tag{18.20}$$

where the parameter E was defined as $E \equiv -(\hbar/i)f'(t)/f(t)$.

From the definition of E, it is equal to a function of t only and hence is independent of x. However, (18.20) shows that $E = -(\hbar^2/2m)\psi''(x)/\psi(x) + V(x)$, which is a function of x only and is independent of t. Hence, E is independent of t as well as independent of x and must therefore be a constant. Since the constant E has the same dimensions as V, it has the dimensions of energy. Quantum mechanics postulates that E is in fact the energy of the system.

Equation (18.20) gives $df/f = -(iE/\hbar) \, dt$, which integrates to $\ln f = -iEt/\hbar + C$. Therefore $f = e^C e^{-iEt/\hbar} = Ae^{-iEt/\hbar}$, where $A \equiv e^C$ is an arbitrary constant. The constant A can be included as part of the $\psi(x)$ factor in (18.19), so we omit it from f. Thus

$$f(t) = e^{-iEt/\hbar} \tag{18.21}$$

Equation (18.20) also gives

$$-\frac{\hbar^2}{2m}\frac{d^2\psi(x)}{dx^2} + V(x)\psi(x) = E\psi(x) \tag{18.22}$$

which is the **(time-independent) Schrödinger equation** for a one-particle, one-dimensional system. Equation (18.22) can be solved for ψ when the potential-energy function $V(x)$ has been specified.

For an n-particle, three-dimensional system, the same procedure that led to Eqs. (18.19), (18.21), and (18.22) gives

$$\Psi = e^{-iEt/\hbar}\psi(x_1, y_1, z_1, \ldots, x_n, y_n, z_n) \tag{18.23}$$

where the function ψ is found by solving

$$-\frac{\hbar^2}{2m_1}\left(\frac{\partial^2\psi}{\partial x_1^2} + \frac{\partial^2\psi}{\partial y_1^2} + \frac{\partial^2\psi}{\partial z_1^2}\right) - \cdots - \frac{\hbar^2}{2m_n}\left(\frac{\partial^2\psi}{\partial x_n^2} + \frac{\partial^2\psi}{\partial y_n^2} + \frac{\partial^2\psi}{\partial z_n^2}\right) + V\psi = E\psi$$

$$(18.24)*$$

The solutions ψ to the time-independent Schrödinger equation (18.24) are the (time-independent) wave functions. States for which Ψ is given by (18.23) are called stationary states. We shall see that for a given system there are many different solutions to (18.24), different solutions corresponding to different values of the energy E. In general, quantum mechanics gives only probabilities and not certainties for the outcome of a measurement. However, when a system is in a stationary state, a measurement of its energy is certain to give the particular energy value that corresponds to the wave function ψ of the system. Different systems have different forms for the potential-energy function $V(x_1, \ldots, z_n)$, and this leads to different sets of allowed wave functions and energies when (18.24) is solved for different systems. All this will be made clearer by the examples in the next few sections.

For a stationary state, the probability density $|\Psi|^2$ becomes

$$|\Psi|^2 = |f\psi|^2 = (f\psi)^*f\psi = f^*\psi^*f\psi = e^{iEt/\hbar}\psi^*e^{-iEt/\hbar}\psi = e^0\psi^*\psi = |\psi|^2 \quad (18.25)$$

where we used (18.19), (18.21) and the identity

$$(f\psi)^* = f^*\psi^*$$

(Prob. 18.19). Hence, for a stationary state, $|\Psi|^2 = |\psi|^2$, which is independent of time. For a stationary state, the probability density and the energy are constant with time. There is no implication, however, that the particles of the system are at rest in a stationary state.

It turns out that the probabilities for the outcomes of measurements of any physical property involve $|\Psi|$, and since $|\Psi| = |\psi|$, these probabilities are independent of time for a stationary state. Thus, the $e^{-iEt/\hbar}$ factor in (18.23) is of little consequence, and *the essential part of the state function for a stationary state is the time-independent wave function $\psi(x_1, \ldots, z_n)$*. For a stationary state, the normalization condition (18.17) becomes $\int |\psi|^2 \, d\tau = 1$, where $\int d\tau$ denotes the definite integral over all space.

The wave function ψ of a stationary state of energy E must satisfy the time-independent Schrödinger equation (18.24). However, quantum mechanics postulates that not all functions that satisfy (18.24) are allowed as wave functions for the system. In addition to being a solution of (18.24), a wave function must meet the following three conditions: (a) The wave function must be **single-valued**. (b) The wave function must be **continuous**. (c) The wave function must be **quadratically integrable**. Condition (a) means that ψ has one and only one value at each point in space. The function of Fig. 18.6a, which is multiple-valued at some points, is not a possible wave function for a one-particle, one-dimensional system. Condition (b) means that ψ makes no sudden jumps in value. A function like that in Fig. 18.6b is ruled out. Condition (c) means that the integral over all space $\int |\psi|^2 \, d\tau$ is a finite number. The function x^2 (Fig. 18.6c) is not quadratically integrable, since $\int_{-\infty}^{\infty} x^4 \, dx = (x^5/5)\big|_{-\infty}^{\infty} = \infty - (-\infty) = \infty$. Condition (c) allows the wave function to be multiplied by a constant that normalizes it, that is, that makes $\int |\psi|^2 \, d\tau = 1$. [If ψ is a solution of the Schrödinger equation (18.24), then so is $k\psi$, where k is any constant; see Prob. 18.20.] A function obeying conditions (a), (b), and (c) is said to be **well-behaved**.

Since E occurs as an undetermined parameter in the Schrödinger equation (18.24), the solutions ψ that are found by solving (18.24) will depend on E as a parameter: $\psi = \psi(x_1, \ldots, z_n; E)$. It turns out that ψ is well-behaved only for certain particular values of E, and it is these values that are the allowed energy levels. An example is given in the next section.

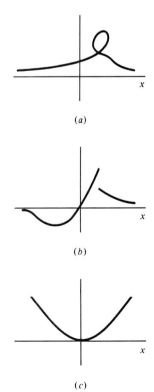

(a)

(b)

(c)

Figure 18.6

(a) A multivalued function. (b) A discontinuous function. (c) A function that is not quadratically integrable.

We shall mainly be interested in the stationary states of atoms and molecules, since these give the allowed energy levels. For a collision between two molecules or for a molecule exposed to the time-varying electric and magnetic fields of electromagnetic radiation, the potential energy V depends on time, and one must deal with the time-dependent Schrödinger equation and with nonstationary states.

Summary

In an isolated atom or molecule, the potential energy V is independent of time and the system can exist in a stationary state, which is a state of constant energy and time-independent probability density. For a stationary state, the probability density for the particles' locations is given by $|\psi|^2$, where the time-independent wave function ψ is a function of the coordinates of the particles of the system. The possible stationary-state wave functions and energies for a system are found by solving the time-independent Schrödinger equation (18.24) and picking out only those solutions that are single-valued, continuous, and quadratically integrable.

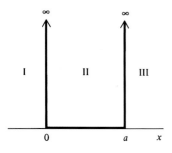

Figure 18.7

Potential-energy function for a particle in a one-dimensional box.

18.8   THE PARTICLE IN A ONE-DIMENSIONAL BOX

The introduction to quantum mechanics in the last two sections is quite abstract. To help make the ideas of quantum mechanics more understandable, this section examines the stationary states of a simple system, **a particle in a one-dimensional box.** By this is meant a single microscopic particle of mass m moving in one dimension x and subject to the potential-energy function of Fig. 18.7. The potential energy is zero for x between 0 and a (region II) and is infinite elsewhere (regions I and III):

$$V = \begin{cases} 0 & \text{for } 0 \le x \le a \\ \infty & \text{for } x < 0 \text{ and for } x > a \end{cases}$$

This potential energy confines the particle to move in the region between 0 and a on the x axis. No real system has a V as simple as Fig. 18.7, but the particle in a box can be used as a crude model for dealing with pi electrons in conjugated molecules (Sec. 20.11).

We restrict ourselves to considering the states of constant energy, the stationary states. For these states, the (time-independent) wave functions ψ are found by solving the Schrödinger equation (18.24), which for a one-particle, one-dimensional system is

$$-\frac{\hbar^2}{2m}\frac{d^2\psi}{dx^2} + V\psi = E\psi \tag{18.26}$$

Since a particle cannot have infinite energy, there must be zero probability of finding the particle in regions I and III, where V is infinite. Therefore, the probability density $|\psi|^2$ and hence ψ must be zero in these regions: $\psi_{\text{I}} = 0$ and $\psi_{\text{III}} = 0$, or

$$\psi = 0 \quad \text{for } x < 0 \text{ and for } x > a \tag{18.27}$$

Inside the box (region II), V is zero and (18.26) becomes

$$\frac{d^2\psi}{dx^2} = -\frac{2mE}{\hbar^2}\psi \quad \text{for } 0 \le x \le a \tag{18.28}$$

To solve this equation, we need a function whose second derivative gives us the same function back again, but multiplied by a constant. Two functions that behave this way are the sine function and the cosine function, so let us try as a solution

$$\psi = A \sin rx + B \cos sx$$

where A, B, r, and s are constants. Differentiation of ψ gives $d^2\psi/dx^2 = -Ar^2 \sin rx - Bs^2 \cos sx$. Substitution of the trial solution in (18.28) gives

$$-Ar^2 \sin rx - Bs^2 \cos sx = -2mE\hbar^{-2}A \sin rx - 2mE\hbar^{-2}B \cos sx \qquad (18.29)$$

If we take $r = s = (2mE)^{1/2} \hbar^{-1}$, Eq. (18.29) is satisfied. The solution of (18.28) is therefore

$$\psi = A \sin\left[(2mE)^{1/2}\hbar^{-1}x\right] + B \cos\left[(2mE)^{1/2}\hbar^{-1}x\right] \qquad \text{for } 0 \le x \le a \qquad (18.30)$$

A more formal derivation than we have given shows that (18.30) is indeed the general solution of the differential equation (18.28).

As noted in Sec. 18.7, not all solutions of the Schrödinger equation are acceptable wave functions. Only well-behaved functions are allowed. The solution of the particle-in-a-box Schrödinger equation is the function defined by (18.27) and (18.30), where A and B are arbitrary constants of integration. For this function to be continuous, the wave function inside the box must go to zero at the two ends of the box, since ψ equals zero outside the box. We must require that ψ in (18.30) go to zero as $x \to 0$ and as $x \to a$. Setting $x = 0$ and $\psi = 0$ in (18.30), we get $0 = A \sin 0 + B \cos 0 = A \cdot 0 + B \cdot 1$, so $B = 0$. Therefore

$$\psi = A \sin\left[(2mE)^{1/2}\hbar^{-1}x\right] \qquad \text{for } 0 \le x \le a \qquad (18.31)$$

Setting $x = a$ and $\psi = 0$ in (18.31), we get $0 = \sin[(2mE)^{1/2}\hbar^{-1}a]$. The function $\sin w$ equals zero when w is 0, $\pm\pi$, $\pm 2\pi$, $\ldots$, $\pm n\pi$, so we must have

$$(2mE)^{1/2}\hbar^{-1}a = \pm n\pi \qquad (18.32)$$

Substitution of (18.32) in (18.31) gives $\psi = A \sin(\pm n\pi x/a) = \pm A \sin(n\pi x/a)$, since $\sin(-z) = -\sin z$. The use of $-n$ instead of n multiplies ψ by -1. Since A is arbitrary, this doesn't give a solution different from the $+n$ solution, so there is no need to consider the $-n$ values. Also, the value $n = 0$ must be ruled out, since it would make $\psi = 0$ everywhere (Prob. 18.26), meaning there is no probability of finding the particle in the box. The allowed wave functions are therefore

$$\psi = A \sin(n\pi x/a) \qquad \text{for } 0 \le x \le a, \qquad \text{where } n = 1, 2, 3, \ldots \qquad (18.33)$$

The allowed energies are found by solving (18.32) for E to get

$$E = \frac{n^2 h^2}{8ma^2}, \qquad n = 1, 2, 3, \ldots \qquad \textbf{(18.34)}*$$

where $\hbar \equiv h/2\pi$ was used. Only these values of E make ψ a well-behaved (continuous) function. For example, Fig. 18.8 plots ψ of (18.27) and (18.31) for $E = (1.1)^2 h^2/8ma^2$. Because of the discontinuity at $x = a$, this is not an acceptable wave function.

Confining the particle to be between 0 and a requires that ψ be zero at $x = 0$ and $x = a$, and this quantizes the energy. An analogy is the quantization of the vibrational modes of a string that occurs when the string is held fixed at both ends. The energy levels (18.34) are proportional to n^2, and the separation between adjacent levels increases as n increases (Fig. 18.9).

The magnitude of the constant A in ψ in (18.33) is found from the normalization condition (18.17) and (18.25): $\int |\psi|^2 \, d\tau = 1$. Since $\psi = 0$ outside the box, we need only integrate from 0 to a, and

$$1 = \int_{-\infty}^{\infty} |\psi|^2 \, dx = \int_{0}^{a} |\psi|^2 \, dx = |A|^2 \int_{0}^{a} \sin^2\left(\frac{n\pi x}{a}\right) dx$$

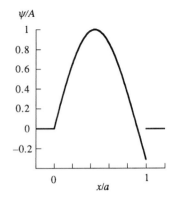

Figure 18.8

Plot of the solution to the particle-in-a-box Schrödinger equation for $E = (1.1)^2 h^2/8ma^2$. This solution is discontinuous at $x = a$.

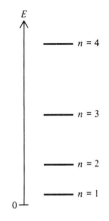

Figure 18.9

Lowest four energy levels of a particle in a one-dimensional box.

A table of integrals gives $\int \sin^2 cx \, dx = x/2 - (1/4c) \sin 2cx$, and we find $|A| = (2/a)^{1/2}$. The **normalization constant** A can be taken as any number having absolute value $(2/a)^{1/2}$. We could take $A = (2/a)^{1/2}$, or $A = -(2/a)^{1/2}$, or $A = i(2/a)^{1/2}$ (where $i = \sqrt{-1}$), etc. Choosing $A = (2/a)^{1/2}$, we get

$$\psi = \left(\frac{2}{a}\right)^{1/2} \sin \frac{n\pi x}{a} \quad \text{for } 0 \leq x \leq a, \quad \text{where } n = 1, 2, 3, \dots \quad (18.35)$$

For a one-particle, one-dimensional system, $|\psi(x)|^2 \, dx$ is a probability. Since probabilities have no units, $\psi(x)$ must have dimensions of length$^{-1/2}$, as is true for ψ in (18.35).

The state functions for the stationary states of the particle in a box are given by (18.19), (18.21), and (18.35) as $\Psi = e^{-iEt/\hbar}(2/a)^{1/2} \sin(n\pi x/a)$, for $0 \leq x \leq a$, where $E = n^2h^2/8ma^2$ and $n = 1, 2, 3, \dots$.

EXAMPLE 18.2 Calculation of a transition wavelength

Find the wavelength of the light emitted when a 1×10^{-27} g particle in a 3-Å one-dimensional box goes from the $n = 2$ to the $n = 1$ level.

The wavelength λ can be found from the frequency ν. The quantity $h\nu$ is the energy of the emitted photon and equals the energy *difference* between the two levels involved in the transition [Eq. (18.7)]:

$$h\nu = E_{\text{upper}} - E_{\text{lower}} = 2^2 h^2/8ma^2 - 1^2 h^2/8ma^2 \quad \text{and} \quad \nu = 3h/8ma^2$$

where (18.34) was used. Use of $\lambda = c/\nu$ and $1 \text{ Å} \equiv 10^{-10}$ m [Eq. (2.87)] gives

$$\lambda = \frac{8ma^2c}{3h} = \frac{8(1 \times 10^{-30} \text{ kg})(3 \times 10^{-10} \text{ m})^2(3 \times 10^8 \text{ m/s})}{3(6.6 \times 10^{-34} \text{ J s})} = 1 \times 10^{-7} \text{ m}$$

(The mass m is that of an electron, and the wavelength lies in the ultraviolet.)

Exercise

(*a*) For a particle of mass 9.1×10^{-31} kg in a certain one-dimensional box, the $n = 3$ to $n = 2$ transition occurs at $\nu = 4.0 \times 10^{14}$ s^{-1}. Find the length of the box. (*Answer:* 1.07 nm.) (*b*) Show that the frequency of the $n = 3$ to 2 particle-in-a-one-dimensional-box transition is 5/3 times the frequency of the 2 to 1 transition.

Let us contrast the quantum-mechanical and classical pictures. Classically, the particle can rattle around in the box with any nonnegative energy; $E_{\text{classical}}$ can be any number from zero on up. (The potential energy is zero in the box, so the particle's energy is entirely kinetic. Its speed v can have any nonnegative value, so $\frac{1}{2}mv^2$ can have any nonnegative value.) Quantum-mechanically, the energy can take on only the values (18.34). The energy is quantized in quantum mechanics, whereas it is continuous in classical mechanics.

Classically, the minimum energy is zero. Quantum-mechanically, the particle in a box has a minimum energy that is greater than zero. This energy, $h^2/8ma^2$, is the **zero-point energy.** Its existence is a consequence of the uncertainty principle. Suppose the particle could have zero energy. Since its energy is entirely kinetic, its speed v_x and momentum $mv_x = p_x$ would then be zero. With p_x known to be zero, the uncertainty Δp_x is zero, and the uncertainty principle $\Delta x \, \Delta p_x \gtrsim h$ gives $\Delta x = \infty$. However, we know the particle to be somewhere between $x = 0$ and $x = a$, so Δx cannot exceed a. Hence, a zero energy is impossible for a particle in a box.

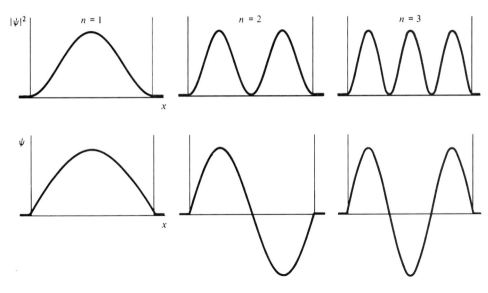

Figure 18.10

Wave functions and probability densities for the lowest three particle-in-a-box stationary states.

The stationary states of a particle in a box are specified by giving the value of the integer n in (18.35). n is called a **quantum number.** The lowest-energy state ($n = 1$) is the **ground state.** States higher in energy than the ground state are **excited states.**

Figure 18.10 plots the wave functions ψ and the probability densities $|\psi|^2$ for the first three particle-in-a-box stationary states. For $n = 1$, $n\pi x/a$ in the wave function (18.35) goes from 0 to π as x goes from 0 to a, so ψ is half of one cycle of a sine function.

Classically, all locations for the particle in the box are equally likely. Quantum-mechanically, the probability density is not uniform but shows oscillations. In the limit of a very high quantum number n, the oscillations in $|\psi|^2$ come closer and closer together and ultimately become undetectable; this corresponds to the classical result of uniform probability density. The relation $8ma^2E/h^2 = n^2$ shows that for a macroscopic system (E, m, and a having macroscopic magnitudes), n is very large, so the limit of large n is the classical limit.

A point at which $\psi = 0$ is called a **node.** The number of nodes increases by 1 for each increase in n. The existence of nodes is surprising from a classical viewpoint. For example, for the $n = 2$ state, it is hard to understand how the particle can be found in the left half of the box or in the right half but never at the center. The behavior of microscopic particles (which have a wave aspect) cannot be rationalized in terms of a visualizable model.

The wave functions ψ and probability densities $|\psi|^2$ are spread out over the length of the box, much like a wave (compare Figs. 18.10 and 18.2). However, quantum mechanics does not assert that the particle itself is spread out like a wave; a measurement of position will give a definite location for the particle. It is the wave function ψ (which gives the probability density $|\psi|^2$) that is spread out in space and obeys a wave equation.

EXAMPLE 18.3 Probability calculations

(a) For the ground state of a particle in a one-dimensional box of length a, find the probability that the particle is within $\pm 0.001a$ of the point $x = a/2$. (b) For the particle-in-a-box stationary state with quantum number n, write down (but do not evaluate) an expression for the probability that the particle will be found between $a/4$ and $a/2$. (c) For a particle-in-a-box stationary state, what is the probability that the particle will be found in the left half of the box?

(a) The probability density (the probability per unit length) equals $|\psi|^2$. Figure 18.10 shows that $|\psi|^2$ for $n = 1$ is essentially constant over the very small interval $0.002a$, so we can consider this interval to be infinitesimal and take $|\psi|^2 \, dx$ as the desired probability. For $n = 1$, Eq. (18.35) gives $|\psi|^2 = (2/a) \sin^2 (\pi x/a)$. With $x = a/2$ and $dx = 0.002a$, the probability is $|\psi|^2 \, dx = (2/a) \sin^2 (\pi/2) \times 0.002a = 0.004$.

(b) From Eq. (18.15), the probability that the particle is between points c and d is $\int_c^d |\Psi|^2 \, dx$. But $|\Psi|^2 = |\psi|^2$ for a stationary state [Eq. (18.25)], so the probability is $\int_c^d |\psi|^2 \, dx$. The desired probability is $\int_{a/4}^{a/2} (2/a) \sin^2 (n\pi x/a) \, dx$, where (18.35) was used for ψ.

(c) For each particle-in-a-box stationary state, the graph of $|\psi|^2$ is symmetric about the midpoint of the box, so the probabilities of being in the left and right halves are equal and are each equal to 0.5.

Exercise

For the $n = 2$ state of a particle in a box of length a, (a) find the probability the particle is within $\pm 0.0015a$ of $x = a/8$; (b) find the probability the particle is between $x = 0$ and $x = a/8$. (*Answers:* (a) 0.0030; (b) $1/8 - 1/4\pi = 0.0454$.)

If ψ_i and ψ_j are particle-in-a-box wave functions with quantum numbers n_i and n_j, one finds (Prob. 18.29) that

$$\int_0^a \psi_i^* \psi_j \, d\tau = 0 \qquad \text{for } n_i \neq n_j \qquad (18.36)$$

where $\psi_i = (2/a)^{1/2} \sin (n_i \pi x/a)$ and $\psi_j = (2/a)^{1/2} \sin (n_j \pi x/a)$. The functions f and g are said to be **orthogonal** when $\int f^* g \, d\tau = 0$, where the integral is a definite integral over the full range of the spatial coordinates. One can show that two wave functions that correspond to different energy levels of a quantum-mechanical system are orthogonal (Sec. 18.16).

18.9 THE PARTICLE IN A THREE-DIMENSIONAL BOX

The particle in a three-dimensional box is a single particle of mass m confined to remain within the volume of a box by an infinite potential energy outside the box. The simplest box shape to deal with is a rectangular parallelepiped. The potential energy is therefore $V = 0$ for points such that $0 \leq x \leq a$, $0 \leq y \leq b$, and $0 \leq z \leq c$ and $V = \infty$ elsewhere. The dimensions of the box are a, b, and c. In Secs. 21.3 and 22.6, this system will be used to give the energy levels for translational motion of ideal-gas molecules in a container.

Let us solve the time-independent Schrödinger equation for the stationary-state wave functions and energies. Since $V = \infty$ outside the box, ψ is zero outside the box, just as for the corresponding one-dimensional problem. Inside the box, $V = 0$, and the Schrödinger equation (18.24) becomes

$$-\frac{\hbar^2}{2m} \left(\frac{\partial^2 \psi}{\partial x^2} + \frac{\partial^2 \psi}{\partial y^2} + \frac{\partial^2 \psi}{\partial z^2} \right) = E\psi \qquad (18.37)$$

Let us assume that solutions of (18.37) exist that have the form $X(x)Y(y)Z(z)$, where $X(x)$ is a function of x only and Y and Z are functions of y and z. For an arbitrary partial differential equation, it is not in general possible to find solutions in which the variables are present in separate factors. However, it can be proved mathematically that, if we succeed in finding well-behaved solutions to (18.37) that have the form $X(x)Y(y)Z(z)$, then there are no other well-behaved solutions, so we shall have found the general solution of (18.37). Our assumption is then

$$\psi = X(x)Y(y)Z(z) \qquad (18.38)$$

Partial differentiation of (18.38) gives

$$\partial^2\psi/\partial x^2 = X''(x)Y(y)Z(z), \quad \partial^2\psi/\partial y^2 = X(x)Y''(y)Z(z), \quad \partial^2\psi/\partial z^2 = X(x)Y(y)Z''(z)$$

Substitution in (18.37) followed by division by $X(x)Y(y)Z(z) = \psi$ gives

$$-\frac{\hbar^2}{2m}\frac{X''(x)}{X(x)} - \frac{\hbar^2}{2m}\frac{Y''(y)}{Y(y)} - \frac{\hbar^2}{2m}\frac{Z''(z)}{Z(z)} = E \qquad (18.39)$$

Let $E_x \equiv -(\hbar^2/2m)X''(x)/X(x)$. Then (18.39) gives

$$E_x \equiv -\frac{\hbar^2}{2m}\frac{X''(x)}{X(x)} = E + \frac{\hbar^2}{2m}\frac{Y''(y)}{Y(y)} + \frac{\hbar^2}{2m}\frac{Z''(z)}{Z(z)} \qquad (18.40)$$

From its definition, E_x is a function of x only. However, the relation $E_x = E + \hbar^2 Y''/2mY + \hbar^2 Z''/2mZ$ in (18.40) shows E_x to be independent of x. Therefore E_x is a constant, and we have from (18.40)

$$-(\hbar^2/2m)X''(x) = E_x X(x) \qquad \text{for } 0 \le x \le a \qquad (18.41)$$

Equation (18.41) is the same as the Schrödinger equation (18.28) for a particle in a one-dimensional box if X and E_x in (18.41) are identified with ψ and E, respectively, in (18.28). Moreover, the condition that $X(x)$ be continuous requires that $X(x) = 0$ at $x = 0$ and at $x = a$, since the three-dimensional wave function is zero outside the box. These are the same requirements that ψ in (18.28) must satisfy. Therefore, the well-behaved solutions of (18.41) and (18.28) are the same. Replacing ψ and E in (18.34) and (18.35) by X and E_x, we get

$$X(x) = \left(\frac{2}{a}\right)^{1/2}\sin\frac{n_x\pi x}{a}, \qquad E_x = \frac{n_x^2 h^2}{8ma^2}, \qquad n_x = 1, 2, 3, \ldots \qquad (18.42)$$

where the quantum number is called n_x.

Equation (18.39) is symmetric with respect to x, y, and z, so the same reasoning that gave (18.42) gives

$$Y(y) = \left(\frac{2}{b}\right)^{1/2}\sin\frac{n_y\pi y}{b}, \qquad E_y = \frac{n_y^2 h^2}{8mb^2}, \qquad n_y = 1, 2, 3, \ldots \qquad (18.43)$$

$$Z(z) = \left(\frac{2}{c}\right)^{1/2}\sin\frac{n_z\pi z}{c}, \qquad E_z = \frac{n_z^2 h^2}{8mc^2}, \qquad n_z = 1, 2, 3, \ldots \qquad (18.44)$$

where, by analogy to (18.40),

$$E_y \equiv -\frac{\hbar^2}{2m}\frac{Y''(y)}{Y(y)}, \qquad E_z \equiv -\frac{\hbar^2}{2m}\frac{Z''(z)}{Z(z)} \qquad (18.45)$$

We assumed in Eq. (18.38) that the wave function ψ is the product of separate factors $X(x)$, $Y(y)$, and $Z(z)$ for each coordinate. Having found X, Y, and Z [Eqs. (18.42), (18.43), and (18.44)], we have as the stationary-state wave functions for a particle in a three-dimensional rectangular box

$$\psi = \left(\frac{8}{abc}\right)^{1/2}\sin\frac{n_x\pi x}{a}\sin\frac{n_y\pi y}{b}\sin\frac{n_z\pi z}{c} \qquad \text{inside the box} \qquad (18.46)$$

Outside the box, $\psi = 0$.

Figure 18.11

Probability densities for three states of a particle in a two-dimensional box whose dimensions have a 2:1 ratio. The states are the ψ_{11}, ψ_{12}, and ψ_{21} states, where the subscripts give the n_x and n_y values.

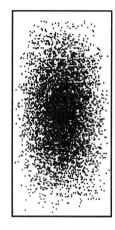
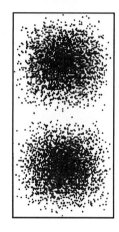
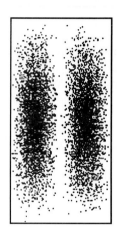

$n_x = 1$, $n_y = 1$ $n_x = 1$, $n_y = 2$ $n_x = 2$, $n_y = 1$

Equations (18.39), (18.40), and (18.45) give $E = E_x + E_y + E_z$, and use of (18.42) to (18.44) for E_x, E_y, and E_z gives the allowed energy levels as

$$E = \frac{h^2}{8m}\left(\frac{n_x^2}{a^2} + \frac{n_y^2}{b^2} + \frac{n_z^2}{c^2}\right) \qquad (18.47)$$

The quantities E_x, E_y, and E_z are the kinetic energies associated with motion in the x, y, and z directions.

The procedure used to solve (18.37) is called **separation of variables.** The conditions under which it works are discussed in Sec. 18.11.

The wave function has three quantum numbers because this is a three-dimensional problem. The quantum numbers n_x, n_y, and n_z vary independently of one another. The state of the particle in the box is specified by giving the values of n_x, n_y, and n_z. The ground state is $n_x = 1$, $n_y = 1$, and $n_z = 1$.

The Particle in a Two-Dimensional Box

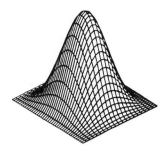

For a particle in a two-dimensional rectangular box with sides a and b, the same procedure that gave (18.46) and (18.47) gives

$$\psi = (4/ab)^{1/2} \sin(n_x\pi x/a) \sin(n_y\pi y/b) \qquad \text{for } 0 \le x \le a, 0 \le y \le b \quad (18.48)$$

and $E = (h^2/8m)(n_x^2/a^2 + n_y^2/b^2)$. For a two-dimensional box with $b = 2a$, Fig. 18.11 shows the variation of the probability density $|\psi|^2$ in the box for three states. The greater the density of dots in a region, the greater the value of $|\psi|^2$. Figure 18.12 shows three-dimensional graphs of $|\psi|^2$ for the lowest two states. The height of the surface above the xy plane gives the value of $|\psi|^2$ at point (x, y). Figure 18.13 is a three-dimensional graph of ψ for the $n_x = 1$, $n_y = 2$ state; ψ is positive in half the box, negative in the other half, and zero on the line that separates these two halves. Figure 18.14 shows contour plots of constant $|\psi|$ for the $n_x = 1$, $n_y = 2$ state; the contours shown are those for which $|\psi|/|\psi|_{\text{max}} = 0.9$ (the two innermost loops), 0.7, 0.5, 0.3, and 0.1, where $|\psi|_{\text{max}}$ is the maximum value of $|\psi|$. These contours correspond to $|\psi|^2/|\psi|^2_{\text{max}} = 0.81$, 0.49, 0.25, 0.09, and 0.01.

Figure 18.12

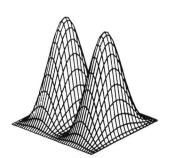

Three-dimensional plot of $|\psi|^2$ for the ψ_{11} and ψ_{12} states of a two-dimensional box with $b = 2a$.

18.10 DEGENERACY

Suppose the sides of the three-dimensional box of the last section have equal lengths: $a = b = c$. Then (18.46) and (18.47) become

$$\psi = (2/a)^{3/2} \sin(n_x\pi x/a) \sin(n_y\pi y/a) \sin(n_z\pi z/a) \qquad (18.49)$$

Figure 18.13

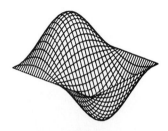

Three-dimensional plot of ψ_{12} for a particle in a two-dimensional box with $b = 2a$.

$$E = (n_x^2 + n_y^2 + n_z^2)h^2/8ma^2 \qquad (18.50)$$

Let us use numerical subscripts on ψ to specify the n_x, n_y, and n_z values. The lowest-energy state is ψ_{111} with $E = 3h^2/8ma^2$. The states ψ_{211}, ψ_{121}, and ψ_{112} each have energy $6h^2/8ma^2$. Even though they have the same energy, these are different states. With $n_x = 2$, $n_y = 1$, and $n_z = 1$ in (18.49), we get a different wave function than with $n_x = 1$, $n_y = 2$, and $n_z = 1$. The ψ_{211} state has zero probability density of finding the particle at $x = a/2$ (see Fig. 18.10), but the ψ_{121} state has a maximum probability density at $x = a/2$.

The terms "state" and "energy level" have different meanings in quantum mechanics. A **stationary state** is specified by giving the wave function ψ. Each different ψ is a different state. An **energy level** is specified by giving the value of the energy. Each different value of E is a different energy level. The three different particle-in-a-box states ψ_{211}, ψ_{121}, and ψ_{112} belong to the same energy level, $6h^2/8ma^2$. Figure 18.15 shows the lowest few stationary states and energy levels of a particle in a cubic box.

An energy level that corresponds to more than one state is said to be **degenerate.** The number of different states belonging to the level is the **degree of degeneracy** of the level. The particle-in-a-cubic-box level $6h^2/8ma^2$ is threefold degenerate. The particle-in-a-box degeneracy arises when the dimensions of the box are made equal. Degeneracy usually arises from the symmetry of the system.

Figure 18.14

Contour plot of constant $|\psi|$ for the state of Fig. 18.13.

18.11 OPERATORS

Operators

Quantum mechanics is most conveniently formulated in terms of operators. An **operator** is a rule for transforming a given function into another function. For example, the operator d/dx transforms a function into its first derivative: $(d/dx)f(x) = f'(x)$. Let $\hat{A}$ symbolize an arbitrary operator. (We shall use a circumflex to denote an operator.) If $\hat{A}$ transforms the function $f(x)$ into the function $g(x)$, we write $\hat{A}f(x) = g(x)$. If $\hat{A}$ is the operator d/dx, then $g(x) = f'(x)$. If $\hat{A}$ is the operator "multiplication by $3x^2$," then $g(x) = 3x^2f(x)$. If $\hat{A} = \log$, then $g(x) = \log f(x)$.

The **sum** of two operators $\hat{A}$ and $\hat{B}$ is defined by

$$(\hat{A} + \hat{B})f(x) \equiv \hat{A}f(x) + \hat{B}f(x) \qquad \textbf{(18.51)}\textbf{*}$$

For example, $(\ln + d/dx)f(x) = \ln f(x) + (d/dx) f(x) = \ln f(x) + f'(x)$. Similarly, $(\hat{A} - \hat{B})f(x) \equiv \hat{A}f(x) - \hat{B}f(x)$.

The **square** of an operator is defined by $\hat{A}^2f(x) \equiv \hat{A}[\hat{A}f(x)]$. For example,

$$(d/dx)^2 f(x) = (d/dx) [(d/dx) f(x)] = (d/dx) [f'(x)] = f''(x) = (d^2/dx^2) f(x)$$

Therefore, $(d/dx)^2 = d^2/dx^2$.

The **product** of two operators is defined by

$$(\hat{A}\hat{B})f(x) \equiv \hat{A}[\hat{B}f(x)] \qquad \textbf{(18.52)}\textbf{*}$$

The notation $\hat{A}[\hat{B} f(x)]$ means that we first apply the operator $\hat{B}$ to the function $f(x)$ to get a new function, and then we apply the operator $\hat{A}$ to this new function.

Two operators are **equal** if they produce the same result when operating on an arbitrary function: $\hat{B} = \hat{C}$ if and only if $\hat{B}f = \hat{C}f$ for every function f.

EXAMPLE 18.4 Operator algebra

Let the operators $\hat{A}$ and $\hat{B}$ be defined as $\hat{A} \equiv x \cdot$ and $\hat{B} \equiv d/dx$. (a) Find $(\hat{A} + \hat{B})(x^3 + \cos x)$. (b) Find $\hat{A}\hat{B}f(x)$ and $\hat{B}\hat{A}f(x)$. Are the operators $\hat{A}\hat{B}$ and $\hat{B}\hat{A}$ equal? (c) Find $\hat{A}\hat{B} - \hat{B}\hat{A}$.

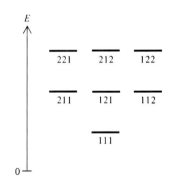

Figure 18.15

Lowest seven stationary states (and lowest three energy levels) for a particle in a cubic box. The numbers are the values of the quantum numbers n_x, n_y, and n_z.

(a) Using the definition (18.51) of the sum of operators, we have

$$(\hat{A} + \hat{B})(x^3 + \cos x) = (x + d/dx)(x^3 + \cos x)$$
$$= x(x^3 + \cos x) + (d/dx)(x^3 + \cos x)$$
$$= x^4 + x\cos x + 3x^2 - \sin x$$

(b) The definition (18.52) of the operator product gives

$$\hat{A}\hat{B}f(x) \equiv \hat{A}[\hat{B}f(x)] = x[(d/dx)f(x)] = x[f'(x)] = xf'(x)$$
$$\hat{B}\hat{A}f(x) \equiv \hat{B}[\hat{A}f(x)] = (d/dx)[xf(x)] = xf'(x) + f(x)$$

In this example, $\hat{A}\hat{B}$ and $\hat{B}\hat{A}$ produce different results when they operate on $f(x)$, so $\hat{A}\hat{B}$ and $\hat{B}\hat{A}$ are not equal in this case. In multiplication of numbers, the order doesn't matter. In multiplication of operators, the order may matter.

(c) To find the operator $\hat{A}\hat{B} - \hat{B}\hat{A}$, we examine the result of applying it to an arbitrary function $f(x)$. We have $(\hat{A}\hat{B} - \hat{B}\hat{A})f(x) = \hat{A}\hat{B}f - \hat{B}\hat{A}f = xf' - (xf' + f) = -f$, where the definition of the difference of operators and the results of (b) were used. Since $(\hat{A}\hat{B} - \hat{B}\hat{A})f(x) = -1 \cdot f(x)$ for all functions $f(x)$, the definition of equality of operators gives

$$\hat{A}\hat{B} - \hat{B}\hat{A} = -1$$

where the multiplication sign after the -1 is omitted, as is customary.

The operator $\hat{A}\hat{B} - \hat{B}\hat{A}$ is called the **commutator** of $\hat{A}$ and $\hat{B}$ and is symbolized by $[\hat{A}, \hat{B}]$;

$$[\hat{A}, \hat{B}] \equiv \hat{A}\hat{B} - \hat{B}\hat{A}$$

Exercise

Let $\hat{R} \equiv x^2$ and $\hat{S} \equiv d^2/dx^2$. (a) Find $(\hat{R} + \hat{S})(x^4 + 1/x)$. (b) Find $\hat{R}\hat{S}f(x)$ and $\hat{S}\hat{R}f(x)$. (c) Find $[\hat{R}, \hat{S}]$. [Answers: (a) $x^6 + 12x^2 + x + 2/x^3$; (b) $x^2f''(x)$, $2f(x) + 4xf'(x) + x^2f''(x)$; (c) $-2 - 4x(d/dx)$.]

Operators in Quantum Mechanics

In quantum mechanics, each physical property of a system has a corresponding operator. The operator that corresponds to p_x, the x component of momentum of a particle, is postulated to be $(\hbar/i)(\partial/\partial x)$, with similar operators for p_y and p_z:

$$\hat{p}_x = \frac{\hbar}{i}\frac{\partial}{\partial x}, \qquad \hat{p}_y = \frac{\hbar}{i}\frac{\partial}{\partial y}, \qquad \hat{p}_z = \frac{\hbar}{i}\frac{\partial}{\partial z} \qquad (18.53)*$$

where $\hat{p}_x$ is the quantum-mechanical operator for the property p_x and $i \equiv \sqrt{-1}$. The operator that corresponds to the x coordinate of a particle is multiplication by x, and the operator that corresponds to $f(x, y, z)$, where f is any function, is multiplication by that function. Thus,

$$\hat{x} = x\times, \qquad \hat{y} = y\times, \qquad \hat{z} = z\times, \qquad \hat{f}(x, y, z) = f(x, y, z)\times \qquad (18.54)*$$

To find the operator that corresponds to any other physical property, we write down the classical-mechanical expression for that property as a function of cartesian coordinates and corresponding momenta and then replace the coordinates and momenta by their corresponding operators (18.53) and (18.54). For example, the energy of a one-particle system is the sum of its kinetic and potential energies:

$$E = K + V = \tfrac{1}{2}m(v_x^2 + v_y^2 + v_z^2) + V(x, y, z, t)$$

To express E as a function of the momenta and coordinates, we note that $p_x = mv_x$, $p_y = mv_y$, $p_z = mv_z$. Therefore,

$$E = \frac{1}{2m}(p_x^2 + p_y^2 + p_z^2) + V(x, y, z, t) \equiv H \qquad (18.55)*$$

The expression for the energy as a function of coordinates and momenta is called the system's **Hamiltonian** H [after W. R. Hamilton (1805–1865), who reformulated Newton's second law in terms of H]. The use of Eq. (18.53) and $i^2 = -1$ gives

$$\hat{p}_x^2 f(x, y, z) = (\hbar/i)(\partial/\partial x)[(\hbar/i)(\partial/\partial x)f] = (\hbar^2/i^2)\,\partial^2 f/\partial x^2 = -\hbar^2\,\partial^2 f/\partial x^2$$

so $\hat{p}_x^2 = -\hbar^2\,\partial^2/\partial x^2$ and $\hat{p}_x^2/2m = -(\hbar^2/2m)\,\partial^2/\partial x^2$. From (18.54), the potential-energy operator is simply multiplication by $V(x, y, z, t)$. (Time is a parameter in quantum mechanics, and there is no time operator.) Replacing p_x^2, p_y^2, p_z^2, and V in (18.55) by their operators, we get as the energy operator, or **Hamiltonian operator,** for a one-particle system

$$\hat{E} = \hat{H} = -\frac{\hbar^2}{2m}\left(\frac{\partial^2}{\partial x^2} + \frac{\partial^2}{\partial y^2} + \frac{\partial^2}{\partial z^2}\right) + V(x, y, z, t) \times \qquad (18.56)$$

To save time in writing, we define the **Laplacian operator** ∇^2 (read as "del squared") by $\nabla^2 \equiv \partial^2/\partial x^2 + \partial^2/\partial y^2 + \partial^2/\partial z^2$ and write the one-particle Hamiltonian operator as

$$\hat{H} = -(\hbar^2/2m)\nabla^2 + V \qquad (18.57)$$

where the multiplication sign after V is understood.

For a many-particle system, we have $\hat{p}_{x,1} = (\hbar/i)\,\partial/\partial x_1$ for particle 1, and the Hamiltonian operator is readily found to be

$$\hat{H} = -\frac{\hbar^2}{2m_1}\nabla_1^2 - \frac{\hbar^2}{2m_2}\nabla_2^2 - \cdots - \frac{\hbar^2}{2m_n}\nabla_n^2 + V(x_1, \ldots, z_n, t) \qquad (18.58)*$$

$$\nabla_1^2 \equiv \frac{\partial^2}{\partial x_1^2} + \frac{\partial^2}{\partial y_1^2} + \frac{\partial^2}{\partial z_1^2} \qquad (18.59)*$$

with similar definitions for $\nabla_2^2, \ldots, \nabla_n^2$. The terms in (18.58) are the operators for the kinetic energies of particles $1, 2, \ldots, n$ and the potential energy of the system.

From (18.58), we see that the time-dependent Schrödinger equation (18.10) can be written as

$$-\frac{\hbar}{i}\frac{\partial\Psi}{\partial t} = \hat{H}\Psi \qquad (18.60)$$

and the time-independent Schrödinger equation (18.24) can be written as

$$\hat{H}\psi = E\psi \qquad (18.61)*$$

where V in (18.61) is independent of time. Since there is a whole set of allowed stationary-state wave functions and energies, (18.61) is often written as $\hat{H}\psi_j = E_j\psi_j$, where the subscript j labels the various wave functions (states) and their energies.

When an operator $\hat{B}$ applied to the function f gives the function back again but multiplied by the constant c, that is, when

$$\hat{B}f = cf$$

one says that f is an **eigenfunction** of $\hat{B}$ with **eigenvalue** c. (However, the function $f = 0$ everywhere is not allowed as an eigenfunction.) The wave functions ψ in (18.61) are eigenfunctions of the Hamiltonian operator $\hat{H}$, the eigenvalues being the allowed energies E.

Operator algebra differs from ordinary algebra. From $\hat{H}\psi = E\psi$ [Eq. (18.61)], one cannot conclude that $\hat{H} = E$. $\hat{H}$ is an operator and E is a number, and the two are not equal. Note, for example, that $(d/dx)e^{2x} = 2e^{2x}$, but $d/dx \neq 2$. In Example 18.4, we found that $(\hat{A}\hat{B} - \hat{B}\hat{A})f(x) = -1 \cdot f(x)$ (for $\hat{A} = x\cdot$ and $\hat{B} = d/dx$) and concluded that $\hat{A}\hat{B} - \hat{B}\hat{A} = -1\cdot$. Because this equation applies to all functions $f(x)$, it is valid to delete the $f(x)$ here. However, the relation $(d/dx)e^{2x} = 2e^{2x}$ applies only to the function e^{2x}, and this function cannot be deleted.

EXAMPLE 18.5 Eigenfunctions

Verify directly that $\hat{H}\psi = E\psi$ for the particle in a one-dimensional box.

Inside the box (Fig. 18.7), $V = 0$ and Eq. (18.56) gives $\hat{H} = -(\hbar^2/2m)\, d^2/dx^2$ for this one-dimensional problem. The wave functions are given by (18.35) as $\psi = (2/a)^{1/2} \sin(n\pi x/a)$. We have, using (1.27) and (18.11):

$$\hat{H}\psi = -\frac{\hbar^2}{2m}\frac{d^2}{dx^2}\left[\left(\frac{2}{a}\right)^{1/2}\sin\frac{n\pi x}{a}\right] = -\frac{h^2}{4\pi^2(2m)}\left(\frac{2}{a}\right)^{1/2}\left(-\frac{n^2\pi^2}{a^2}\right)\sin\frac{n\pi x}{a}$$

$$= \frac{n^2h^2}{8ma^2}\left(\frac{2}{a}\right)^{1/2}\sin\frac{n\pi x}{a} = E\psi$$

since $E = n^2h^2/8ma^2$ [Eq. (18.34)].

Exercise

Verify that the function Ae^{ikx}, where A and k are constants, is an eigenfunction of the operator $\hat{p}_x$. What is the eigenvalue? (*Answer: $k\hbar$.*)

The operators that correspond to physical quantities in quantum mechanics are linear. A **linear operator** $\hat{L}$ is one that satisfies the following two equations for all functions f and g and all constants c:

$$\hat{L}(f + g) = \hat{L}f + \hat{L}g \quad \text{and} \quad \hat{L}(cf) = c\hat{L}f$$

The operator $\partial/\partial x$ is linear, since $(\partial/\partial x)(f + g) = \partial f/\partial x + \partial g/\partial x$ and $(\partial/\partial x)(cf) = c\, \partial f/\partial x$. The operator $\sqrt{\ }$ is nonlinear, since $\sqrt{f + g} \neq \sqrt{f} + \sqrt{g}$.

If the function ψ satisfies the time-independent Schrödinger equation $\hat{H}\psi = E\psi$, then so does the function $c\psi$, where c is any constant. Proof of this follows from the fact that the Hamiltonian operator $\hat{H}$ is a linear operator. We have $\hat{H}(c\psi) = c\hat{H}\psi = cE\psi = E(c\psi)$. The freedom to multiply ψ by a constant enables us to normalize ψ.

Measurement

Multiplication of $\hat{H}\psi = E\psi$ [Eq. (18.61)] by $e^{-iEt/\hbar}$ gives $e^{-iEt/\hbar}\hat{H}\psi = Ee^{-iEt/\hbar}\psi$. For a stationary state, $\hat{H}$ does not involve time and $e^{-iEt/\hbar}\hat{H}\psi = \hat{H}(e^{-iEt/\hbar}\psi)$. Using $\Psi = e^{-iEt/\hbar}\psi$ [Eq. (18.23)], we have

$$\hat{H}\Psi = E\Psi$$

so Ψ is an eigenfunction of $\hat{H}$ with eigenvalue E for a stationary state. A stationary state has a definite energy, and measurement of the system's energy will always give a single predictable value when the system is in a stationary state. For example, for the $n = 2$ particle-in-a-box stationary state, measurement of the energy will always give the result $2^2h^2/8ma^2$ [Eq. (18.34)].

What about properties other than the energy? Let the operator $\hat{M}$ correspond to the property M. Quantum mechanics postulates that *if the system's state function Ψ happens to be an eigenfunction of $\hat{M}$ with eigenvalue c (that is, if $\hat{M}\Psi = c\Psi$), then a* *measurement of M is certain to give the value c as the result.* (Examples will be given when we consider angular momentum in Sec. 19.4). If Ψ is not an eigenfunction of $\hat{M}$, then the result of measuring M cannot be predicted. (However, the probabilities of the various possible outcomes of a measurement of M can be calculated from Ψ, but discussion of how this is done is omitted.) For stationary states, the essential part of Ψ is the time-independent wave function ψ, and ψ replaces Ψ in the italicized statement in this paragraph.

Average Values

From (15.38), the average value of x for a one-particle, one-dimensional quantum-mechanical system equals $\int_{-\infty}^{\infty} xg(x)\,dx$, where $g(x)$ is the probability density for finding the particle between x and $x + dx$. But the Born postulate (Sec. 18.6) gives $g(x) = |\Psi(x)|^2$. Hence, $\langle x \rangle = \int_{-\infty}^{\infty} x|\Psi(x)|^2\,dx$. Since $|\Psi|^2 = \Psi^*\Psi$, we have $\langle x \rangle = \int_{-\infty}^{\infty} \Psi^*x\Psi\,dx = \int_{-\infty}^{\infty} \Psi^*\hat{x}\Psi\,dx$, where (18.54) was used.

What about the average value of an arbitrary physical property M for a general quantum-mechanical system? Quantum mechanics *postulates* that the average value of any physical property M in a system whose state function is Ψ is given by

$$\langle M \rangle = \int \Psi^*\hat{M}\Psi\,d\tau \tag{18.62}$$

where $\hat{M}$ is the operator for the property M and the integral is a definite integral over all space. In (18.62), $\hat{M}$ operates on Ψ to produce the result $\hat{M}\Psi$, which is a function. The function $\hat{M}\Psi$ is then multiplied by Ψ^*, and the resulting function $\Psi^*\hat{M}\Psi$ is integrated over the full range of the spatial coordinates of the system. For example, Eq. (18.53) gives the p_x operator as $\hat{p}_x = (\hbar/i)\,\partial/\partial x$, and the average value of p_x for a one-particle, three-dimensional system whose state function is Ψ is $\langle p_x \rangle = (\hbar/i)\int_{-\infty}^{\infty}\int_{-\infty}^{\infty}\int_{-\infty}^{\infty} \Psi^*(\partial\Psi/\partial x)\,dx\,dy\,dz$.

The average value of M is the average of the results of a very large number of measurements of M made on identical systems, each of which is in the same state Ψ just before the measurement.

If Ψ happens to be an eigenfunction of $\hat{M}$ with eigenvalue c, then $\hat{M}\Psi = c\Psi$ and (18.62) becomes $\langle M \rangle = \int \Psi^*\hat{M}\Psi\,d\tau = \int \Psi^*c\Psi\,d\tau = c\int \Psi^*\Psi\,d\tau = c$, since Ψ is normalized. This result makes sense since, as noted in the last subsection, c is the only possible result of a measurement of M if $\hat{M}\Psi = c\Psi$.

For a stationary state, Ψ equals $e^{-iEt/\hbar}\psi$ [Eq. (18.23)]. Since $\hat{M}$ doesn't affect the $e^{-iEt/\hbar}$ factor, we have

$$\Psi^*\hat{M}\Psi = e^{iEt/\hbar}\psi^*\hat{M}e^{-iEt/\hbar}\psi = e^{iEt/\hbar}e^{-iEt/\hbar}\psi^*\hat{M}\psi = \psi^*\hat{M}\psi$$

Therefore, for a stationary state,

$$\langle M \rangle = \int \psi^*\hat{M}\psi\,d\tau \tag{18.63}*$$

EXAMPLE 18.6 Average value

For a particle in a one-dimensional-box stationary state, give the expression for $\langle x^2 \rangle$.

For a one-particle, one-dimensional problem, $d\tau = dx$. Since $\hat{x}^2 = x^2 \cdot$, we have

$$\langle x^2 \rangle = \int_{-\infty}^{\infty} \psi^*x^2\psi\,dx = \int_{-\infty}^{0} x^2|\psi|^2\,dx + \int_{0}^{a} x^2|\psi|^2\,dx + \int_{a}^{\infty} x^2|\psi|^2\,dx$$

since $\psi^*\psi = |\psi|^2$ [Eq. (18.14)]. For $x < 0$ and $x > a$, we have $\psi = 0$ [Eq. (18.27)] and inside the box $\psi = (2/a)^{1/2} \sin(n\pi x/a)$ [Eq. (18.35)]. Therefore

$$\langle x^2 \rangle = \frac{2}{a} \int_0^a x^2 \sin^2 \frac{n\pi x}{a} \, dx$$

Evaluation of the integral is left as a homework problem (Prob. 18.41).

Exercise

Evaluate $\langle p_x \rangle$ for a particle in a one-dimensional-box stationary state. [*Answer:* $(2n\pi\hbar/ia^2) \int_0^a \sin(n\pi x/a) \cos(n\pi x/a) \, dx = 0$.]

Separation of Variables

Let $q_1, q_2, \ldots, q_r$ be the coordinates of a system. For example, for a two-particle system, $q_1 = x_1, q_2 = y_1, \ldots, q_6 = z_2$. Suppose the Hamiltonian operator has the form

$$\hat{H} = \hat{H}_1 + \hat{H}_2 + \cdots + \hat{H}_r \qquad (18.64)$$

where the operator $\hat{H}_1$ involves only q_1, the operator $\hat{H}_2$ involves only q_2, etc. An example is the particle in a three-dimensional box, where one has $\hat{H} = \hat{H}_x + \hat{H}_y + \hat{H}_z$, with $\hat{H}_x \equiv -(\hbar^2/2m) \, \partial^2/\partial x^2$, etc. We saw in Sec. 18.9 that, for this case, $\psi = X(x)Y(y)Z(z)$ and $E = E_x + E_y + E_z$, where $\hat{H}_x X(x) = E_x X(x)$, $\hat{H}_y Y(y) = E_y Y(y)$, $\hat{H}_z Z(z) = E_z Z(z)$ [Eqs. (18.41) and (18.45)].

The same type of argument used in Sec. 18.9 shows (Prob. 18.42) that when $\hat{H}$ is the sum of separate terms for each coordinate, as in (18.64), then each stationary-state wave function is the product of separate factors for each coordinate and each stationary-state energy is the sum of energies for each coordinate:

$$\psi = f_1(q_1)f_2(q_2) \cdots f_r(q_r) \qquad (18.65)^*$$

$$E = E_1 + E_2 + \cdots + E_r \qquad (18.66)^*$$

where $E_1, E_2, \ldots$ and the functions $f_1, f_2, \ldots$ are found by solving

$$\hat{H}_1 f_1 = E_1 f_1, \quad \hat{H}_2 f_2 = E_2 f_2, \quad \ldots, \quad \hat{H}_r f_r = E_r f_r \qquad (18.67)$$

The equations in (18.67) are, in effect, separate Schrödinger equations, one for each coordinate.

Noninteracting Particles

An important case where separation of variables applies is a system of n noninteracting particles, meaning that the particles exert no forces on one another. For such a system, the classical-mechanical energy is the sum of the energies of the individual particles, so the classical Hamiltonian H and the quantum-mechanical Hamiltonian operator $\hat{H}$ have the forms $H = H_1 + H_2 + \cdots + H_n$ and $\hat{H} = \hat{H}_1 + \hat{H}_2 + \cdots + \hat{H}_n$, where $\hat{H}_1$ involves only the coordinates of particle 1, $\hat{H}_2$ involves only particle 2, etc. Here, by analogy to (18.65) to (18.67), we have

$$\psi = f_1(x_1, y_1, z_1)f_2(x_2, y_2, z_2) \cdots f_n(x_n, y_n, z_n) \qquad (18.68)^*$$

$$E = E_1 + E_2 + \cdots + E_n \qquad (18.69)^*$$

$$\hat{H}_1 f_1 = E_1 f_1, \quad \hat{H}_2 f_2 = E_2 f_2, \quad \ldots, \quad \hat{H}_n f_n = E_n f_n \qquad (18.70)^*$$

For a system of noninteracting particles, there is a separate Schrödinger equation for each particle, the wave function is the product of wave functions of the individual particles, and the energy is the sum of the energies of the individual particles. (For noninteracting particles, the probability density $|\psi|^2$ is the product of probability densities for each particle: $|\psi|^2 = |f_1|^2|f_2|^2 \cdots |f_n|^2$. This is in accord with the theorem that the probability that several independent events will all occur is the product of the probabilities of the separate events.)

18.12   THE ONE-DIMENSIONAL HARMONIC OSCILLATOR

The one-dimensional **harmonic oscillator** is a useful model for treating the vibration of a diatomic molecule (Sec. 21.3) and is also relevant to vibrations of polyatomic molecules (Sec. 21.8) and crystals (Sec. 24.12).

Classical Treatment

Before examining the quantum mechanics of a harmonic oscillator, we review the classical treatment. Consider a particle of mass m that moves in one dimension and is attracted to the coordinate origin by a force proportional to its displacement from the origin: $F = -kx$, where k is called the **force constant.** When x is positive, the force is in the $-x$ direction, and when x is negative, F is in the $+x$ direction. A physical example is a mass attached to a frictionless spring, x being the displacement from the equilibrium position. From (2.21), $F = -dV/dx$, where V is the potential energy. Hence $-dV/dx = -kx$, and $V = \frac{1}{2}kx^2 + c$. The choice of zero of potential energy is arbitrary. Choosing the integration constant c as zero, we have (Fig. 18.16)

$$V = \tfrac{1}{2}kx^2 \qquad (18.71)$$

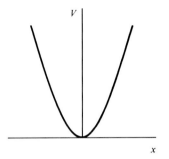

Figure 18.16

The potential-energy function for a one-dimensional harmonic oscillator.

Newton's second law $F = ma$ gives $m\,d^2x/dt^2 = -kx$. The solution to this differential equation is

$$x = A \sin\left[(k/m)^{1/2}t + b\right] \qquad (18.72)$$

as can be verified by substitution in the differential equation (Prob. 18.51). In (18.72), A and b are integration constants. The maximum and minimum values of the sine function are $+1$ and -1, so the particle's x coordinate oscillates back and forth between $+A$ and $-A$. A is the *amplitude* of the motion.

The *period* τ (tau) of the oscillator is the time required for one complete cycle of oscillation. For one cycle of oscillation, the argument of the sine function in (18.72) must increase by 2π, since 2π is the period of a sine function. Hence the period satisfies $(k/m)^{1/2}\tau = 2\pi$, and $\tau = 2\pi(m/k)^{1/2}$. The **frequency** ν is the reciprocal of the period and equals the number of vibrations per second ($\nu = 1/\tau$); thus

$$\nu = \frac{1}{2\pi}\left(\frac{k}{m}\right)^{1/2} \qquad \textbf{(18.73)*}$$

The energy of the harmonic oscillator is $E = K + V = \frac{1}{2}mv_x^2 + \frac{1}{2}kx^2$. The use of (18.72) for x and of $v_x = dx/dt = (k/m)^{1/2}A\cos[(k/m)^{1/2}t + b]$ leads to (Prob. 18.51)

$$E = \tfrac{1}{2}kA^2 \qquad (18.74)$$

Equation (18.74) shows that the classical energy can have any nonnegative value. As the particle oscillates, its kinetic energy and potential energy continually change, but the total energy remains constant at $\frac{1}{2}kA^2$.

Classically, the particle is limited to the region $-A \le x \le A$. When the particle reaches $x = A$ or $x = -A$, its speed is zero (since it reverses its direction of motion at $+A$ and $-A$) and its potential energy is a maximum, being equal to $\frac{1}{2}kA^2$. If the particle were to move beyond $x = \pm A$, its potential energy would increase above $\frac{1}{2}kA^2$. This is impossible for a classical particle. The total energy is $\frac{1}{2}kA^2$ and the kinetic energy is nonnegative, so the potential energy $(V = E - K)$ cannot exceed the total energy.

Quantum-Mechanical Treatment

Now for the quantum-mechanical treatment. Substitution of $V = \frac{1}{2}kx^2$ in (18.26) gives the time-independent Schrödinger equation as

$$-\frac{\hbar^2}{2m}\frac{d^2\psi}{dx^2} + \frac{1}{2}kx^2\psi = E\psi \qquad (18.75)$$

Solution of the harmonic-oscillator Schrödinger equation (18.75) is complicated and is omitted in this book (see any quantum chemistry text). Here, we examine the results. One finds that quadratically integrable (Sec. 18.7) solutions to (18.75) exist only for the following values of E:

$$E = (v + \tfrac{1}{2})h\nu \qquad \text{where } v = 0, 1, 2, \ldots \qquad \textbf{(18.76)}^*$$

where the vibrational frequency ν is given by (18.73) and the quantum number v takes on nonnegative integral values. [Don't confuse the typographically similar symbols ν (nu) and v (vee).] The energy is quantized. The allowed energy levels (Fig. 18.17) are equally spaced (unlike the particle in a box). The zero-point energy is $\frac{1}{2}h\nu$. (For a collection of harmonic oscillators in thermal equilibrium, all the oscillators will fall to the ground state as the temperature goes to absolute zero; hence the name *zero-point energy*.) For all values of E other than (18.76), one finds that the solutions to (18.75) go to infinity as x goes to $\pm\infty$, so these solutions are not quadratically integrable and are not allowed as wave functions.

The well-behaved solutions to (18.75) turn out to have the form

$$\psi_v = \begin{cases} e^{-\alpha x^2/2}\,(c_0 + c_2 x^2 + \cdots + c_v x^v) & \text{for } v \text{ even} \\ e^{-\alpha x^2/2}\,(c_1 x + c_3 x^3 + \cdots + c_v x^v) & \text{for } v \text{ odd} \end{cases}$$

where $\alpha \equiv 2\pi\nu m/\hbar$. The polynomial that multiplies $e^{-\alpha x^2/2}$ contains only even powers of x or only odd powers, depending on whether the quantum number v is even or odd. The explicit forms of the lowest few wave functions $\psi_0, \psi_1, \psi_2,$ and ψ_3 (where the subscript on ψ gives the value of the quantum number v) are given in Fig. 18.18, which plots these ψ's. As with the particle in a one-dimensional box, the number of nodes increases by one for each increase in the quantum number. Note the qualitative resemblance of the wave functions in Figs. 18.18 and 18.10.

The harmonic-oscillator wave functions fall off exponentially to zero as $x \to \pm\infty$. Note, however, that even for very large values of x, the wave function ψ and the probability density $|\psi|^2$ are not zero. There is some probability of finding the particle at an indefinitely large value of x. For a classical-mechanical harmonic oscillator with energy $(v + \frac{1}{2})h\nu$, Eq. (18.74) gives $(v + \frac{1}{2})h\nu = \frac{1}{2}kA^2$, and $A = [(2v + 1)h\nu/k]^{1/2}$. A classical oscillator is confined to the region $-A \leq x \leq A$. However, a quantum-mechanical oscillator has some probability of being found in the *classically forbidden* regions $x > A$ and $x < -A$, where the potential energy is greater than the particle's total energy. This penetration into classically forbidden regions is called **tunneling.** Tunneling occurs more readily the smaller the particle's mass and is most important in chemistry for electrons, protons, and H atoms. Tunneling influences the rates of reactions involving these species (see Secs. 23.3 and 23.4). Electron tunneling is the basis for the scanning tunneling microscope, a remarkable device that gives pictures of the atoms on the surface of a solid (Sec. 24.10). Tunneling makes possible the fusion of hydrogen nuclei to helium nuclei in the sun, despite the electrical repulsion between two hydrogen nuclei.

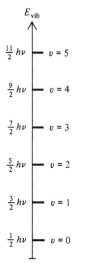

Figure 18.17

Energy levels of a one-dimensional harmonic oscillator.

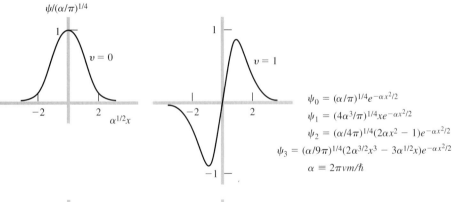

$$\psi_0 = (\alpha/\pi)^{1/4}e^{-\alpha x^2/2}$$
$$\psi_1 = (4\alpha^3/\pi)^{1/4}xe^{-\alpha x^2/2}$$
$$\psi_2 = (\alpha/4\pi)^{1/4}(2\alpha x^2 - 1)e^{-\alpha x^2/2}$$
$$\psi_3 = (\alpha/9\pi)^{1/4}(2\alpha^{3/2}x^3 - 3\alpha^{1/2}x)e^{-\alpha x^2/2}$$

$$\alpha \equiv 2\pi vm/\hbar$$

Figure 18.18

Wave functions for the lowest four harmonic-oscillator stationary states.

18.13   TWO-PARTICLE PROBLEMS

Consider a two-particle system where the coordinates of the particles are x_1, y_1, z_1 and x_2, y_2, z_2. The **relative** (or **internal**) **coordinates** x, y, z are defined by

$$x \equiv x_2 - x_1, \qquad y \equiv y_2 - y_1, \qquad z \equiv z_2 - z_1 \qquad (18.77)$$

These are the coordinates of particle 2 in a coordinate system whose origin is attached to particle 1 and moves with it.

In most cases, the potential energy V of the two-particle system depends only on the relative coordinates x, y, and z. For example, if the particles are electrically charged, the Coulomb's law potential energy of interaction between the particles depends only on the distance r between them, and $r = (x^2 + y^2 + z^2)^{1/2}$. Let us assume that $V = V(x, y, z)$. Let X, Y, and Z be the coordinates of the center of mass of the system; X is given by $(m_1 x_1 + m_2 x_2)/(m_1 + m_2)$, where m_1 and m_2 are the masses of the particles (*Halliday and Resnick*, sec. 9-1). If one expresses the classical energy (that is, the classical Hamiltonian) of the system in terms of the internal coordinates x, y, and z and the center-of-mass coordinates X, Y, and Z, instead of x_1, y_1, z_1, x_2, y_2, and z_2, it turns out (see Prob. 18.54) that

$$H = \left[\frac{1}{2\mu}(p_x^2 + p_y^2 + p_z^2) + V(x, y, z) \right] + \left[\frac{1}{2M}(p_X^2 + p_Y^2 + p_Z^2) \right] \quad (18.78)$$

where M is the system's total mass ($M = m_1 + m_2$), the **reduced mass** μ is defined by

$$\mu \equiv \frac{m_1 m_2}{m_1 + m_2} \qquad \textbf{(18.79)}*$$

and the momenta in (18.78) are defined by

$$p_x \equiv \mu v_x, \qquad p_y \equiv \mu v_y, \qquad p_z \equiv \mu v_z$$

$$p_X \equiv M v_X, \qquad p_Y \equiv M v_Y, \qquad p_Z \equiv M v_Z \qquad (18.80)$$

where $v_x = dx/dt$, etc., and $v_X = dX/dt$, etc.

Equation (18.55) shows that the Hamiltonian (18.78) is the sum of a Hamiltonian for a fictitious particle of mass μ and coordinates x, y, and z that has the potential energy $V(x, y, z)$ and a Hamiltonian for a second fictitious particle of mass $M = m_1 + m_2$ and coordinates X, Y, and Z that has $V = 0$. Moreover, there is no term for any interaction between these two fictitious particles. Hence, Eqs. (18.69) and (18.70) show that the quantum-mechanical energy E of the two-particle system is given by $E = E_\mu + E_M$, where E_μ and E_M are found by solving

$$\hat{H}_\mu \psi_\mu(x, y, z) = E_\mu \psi_\mu(x, y, z) \quad \text{and} \quad \hat{H}_M \psi_M(X, Y, Z) = E_M \psi_M(X, Y, Z)$$

The Hamiltonian operator $\hat{H}_\mu$ is formed from the terms in the first pair of brackets in (18.78), and $\hat{H}_M$ is formed from the terms in the second pair of brackets.

Introduction of the relative coordinates x, y, and z and the center-of-mass coordinates X, Y, and Z reduces the two-particle problem to two separate one-particle problems. We solve a Schrödinger equation for a fictitious particle of mass μ moving subject to the potential energy $V(x, y, z)$, and we solve a separate Schrödinger equation for a fictitious particle whose mass is M $(= m_1 + m_2)$ and whose coordinates are the system's center-of-mass coordinates X, Y, and Z. The Hamiltonian $\hat{H}_M$ involves only kinetic energy. If the two particles are confined to a box, we can use the particle-in-a-box energies (18.47) for E_M. The energy E_M is translational energy of the two-particle system as a whole. The Hamiltonian $\hat{H}_\mu$ involves the kinetic energy and potential energy of motion of the particles relative to each other, so E_μ is the energy associated with this relative or "internal" motion.

The system's total energy E is the sum of its translational energy E_M and its internal energy E_μ. For example, the energy of a hydrogen atom in a box is the sum of the atom's translational energy through space and the atom's internal energy, which is composed of potential energy of interaction between the electron and the proton and kinetic energy of motion of the electron relative to the proton.

18.14   THE TWO-PARTICLE RIGID ROTOR

The **two-particle rigid rotor** consists of particles of masses m_1 and m_2 constrained to remain a fixed distance d from each other. This is a useful model for treating the rotation of a diatomic molecule; see Sec. 21.3. The system's energy is wholly kinetic, and $V = 0$. Since $V = 0$ is a special case of V being a function of only the relative coordinates of the particles, the results of the last section apply. The quantum-mechanical energy is the sum of the translational energy of the system as a whole and the energy of internal motion of one particle relative to the other. The interparticle distance is constant, so the internal motion consists entirely of changes in the spatial orientation of the interparticle axis. The internal motion is a rotation of the two-particle system.

Solution of the Schrödinger equation for internal motion is complicated, so we shall just quote the results without proof. [For a derivation, see, for example, *Levine* (2000), sec. 6.4.] The allowed rotational energies turn out to be

$$E_{\text{rot}} = J(J + 1)\frac{\hbar^2}{2I} \qquad \text{where} \quad J = 0, 1, 2, \ldots \qquad \textbf{(18.81)}^*$$

where the rotor's **moment of inertia** I is

$$I = \mu d^2 \qquad \textbf{(18.82)}^*$$

with $\mu = m_1 m_2/(m_1 + m_2)$. The spacing between adjacent rotational energy levels increases with increasing quantum number J (Fig. 18.19). There is no zero-point rotational energy.

The rotational wave functions are most conveniently expressed in terms of the angles θ and ϕ that give the spatial orientation of the rotor (Fig. 18.20). One finds $\psi_{rot} = \Theta_{JM_J}(\theta)\Phi_{M_J}(\phi)$, where Θ_{JM_J} is a function of θ whose form depends on the two quantum numbers J and M_J, and Φ_{M_J} is a function of ϕ whose form depends on M_J. These functions won't be given here but will be discussed in Sec. 19.3.

Ordinarily, the wave function for internal motion of a two-particle system is a function of three coordinates. However, since the interparticle distance is held fixed in this problem, ψ_{rot} is a function of only two coordinates, θ and ϕ. Since there are two coordinates, there are two quantum numbers, J and M_J. The possible values of M_J turn out to range from $-J$ to J in steps of 1:

$$M_J = -J, -J + 1, \ldots, J - 1, J \qquad (18.83)*$$

For example, if $J = 2$, then $M_J = -2, -1, 0, 1, 2$. For a given J, there are $2J + 1$ values of M_J. The quantum numbers J and M_J determine the rotational wave function, but E_{rot} depends only on J. Hence, each rotational level is $(2J + 1)$-fold degenerate. For example, the value $J = 1$ corresponds to one energy level ($E_{rot} = \hbar^2/I$) and corresponds to the three M_J values $-1, 0, 1$. Therefore for $J = 1$ there are three different ψ_{rot} functions, that is, three different rotational states.

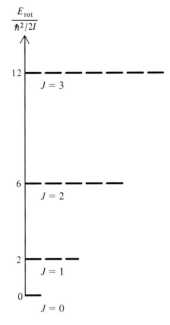

Figure 18.19

Lowest four energy levels of a two-particle rigid rotor. Each energy level consists of $2J + 1$ states.

EXAMPLE 18.7 Rotational energy levels

Find the two lowest rotational energy levels of the $^1H^{35}Cl$ molecule, treating it as a rigid rotor. The bond distance is 1.28 Å in HCl. Atomic masses are listed in a table inside the back cover.

The rotational energy [Eqs. (18.81) and (18.82)] depends on the reduced mass μ of Eq. (18.79). The atomic mass m_1 in μ equals the molar mass M_1 divided by the Avogadro constant N_A. Using the table of atomic masses, we have

$$\mu = \frac{m_1 m_2}{m_1 + m_2} = \frac{[(1.01 \text{ g/mol})/N_A][(35.0 \text{ g/mol})/N_A]}{[(1.01 \text{ g/mol}) + (35.0 \text{ g/mol})]/N_A} = \frac{0.982 \text{ g/mol}}{6.02 \times 10^{23}/\text{mol}}$$

$$= 1.63 \times 10^{-24} \text{ g}$$

$$I = \mu d^2 = (1.63 \times 10^{-27} \text{ kg})(1.28 \times 10^{-10} \text{ m})^2 = 2.67 \times 10^{-47} \text{ kg m}^2$$

The two lowest rotational levels have $J = 0$ and $J = 1$, and (18.81) gives $E_{J=0} = 0$ and

$$E_{J=1} = \frac{J(J + 1)\hbar^2}{2I} = \frac{1(2)(6.63 \times 10^{-34} \text{ J s})^2}{2(2\pi)^2(2.67 \times 10^{-47} \text{ kg m}^2)} = 4.17 \times 10^{-22} \text{ J}$$

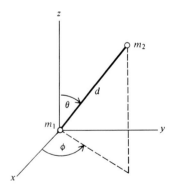

Figure 18.20

A two-particle rigid rotor.

Exercise

The separation between the two lowest rotational levels of $^{12}C^{32}S$ is 3.246×10^{-23} J. Calculate the bond distance in $^{12}C^{32}S$. (*Answer:* 1.538 Å.)

18.15 APPROXIMATION METHODS

For a many-electron atom or molecule, the interelectronic repulsion terms in the potential energy V make it impossible to solve the Schrödinger equation (18.24) exactly. One must resort to approximation methods.

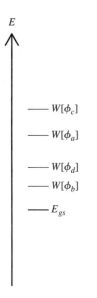

Figure 18.21

The variational integral cannot be less than the true ground-state energy E_{gs}. The quantities $W[\phi_a]$, $W[\phi_b]$, $W[\phi_c]$, and $W[\phi_d]$ are the values of the variational integral in (18.84) for the normalized functions ϕ_a, ϕ_b, ϕ_c, and ϕ_d. Of these functions, ϕ_b gives the lowest W and so its W is closest to E_{gs}.

The Variation Method

The most widely used approximation method is the **variation method.** From the postulates of quantum mechanics, one can deduce the following theorem (for the proof, see any quantum chemistry text). Let $\hat{H}$ be the time-independent Hamiltonian operator of a quantum-mechanical system. If ϕ is any *normalized, well-behaved* function of the coordinates of the particles of the system, then

$$\int \phi^* \hat{H} \phi \, d\tau \geq E_{gs} \qquad \text{for } \phi \text{ normalized} \tag{18.84}$$

where E_{gs} is the system's true ground-state energy and the definite integral goes over all space. (Do not confuse the variation *function* ϕ with the *angle* ϕ in Fig. 18.20.)

To apply the variation method, one takes many different normalized, well-behaved functions ϕ_1, ϕ_2, ..., and for each of them one computes the **variational integral** $\int \phi^* \hat{H} \phi \, d\tau$. The variation theorem (18.84) shows that the function giving the lowest value of $\int \phi^* \hat{H} \phi \, d\tau$ provides the closest approximation to the ground-state energy (Fig. 18.21). This function can serve as an approximation to the true ground-state wave function and can be used to compute approximations to ground-state molecular properties in addition to the energy (for example, the dipole moment).

Suppose we were lucky enough to guess the true ground-state wave function ψ_{gs}. Substitution of $\phi = \psi_{gs}$ in (18.84) and the use of (18.61) and (18.17) give the variational integral as $\int \psi_{gs}^* \hat{H} \psi_{gs} \, d\tau = \int \psi_{gs}^* E_{gs} \psi_{gs} \, d\tau = E_{gs} \int \psi_{gs}^* \psi_{gs} \, d\tau = E_{gs}$. We would then get the true ground-state energy.

If the variation function ϕ is not normalized, it must be multiplied by a normalization constant N before being used in (18.84). The normalization condition is $1 = \int |N\phi|^2 \, d\tau = |N|^2 \int |\phi|^2 \, d\tau$. Hence,

$$|N|^2 = \frac{1}{\int |\phi|^2 \, d\tau} \tag{18.85}$$

Use of the normalized function $N\phi$ in place of ϕ in (18.84) gives $\int N^* \phi^* \hat{H}(N\phi) \, d\tau = |N|^2 \int \phi^* \hat{H} \phi \, d\tau \geq E_{gs}$, where we used the linearity of $\hat{H}$ (Sec. 18.11) to write $\hat{H}(N\phi) = N\hat{H}\phi$. Substitution of (18.85) into the last inequality gives

$$\frac{\int \phi^* \hat{H} \phi \, d\tau}{\int \phi^* \phi \, d\tau} \geq E_{gs} \tag{18.86}*$$

where ϕ need not be normalized but must be well behaved.

EXAMPLE 18.8 Trial variation function

Devise a trial variation function for the particle in a one-dimensional box and use it to estimate E_{gs}.

The particle in a box is exactly solvable, and there is no need to resort to an approximate method. For instructional purposes, let's pretend we don't know how to solve the particle-in-a-box Schrödinger equation. We know that the true ground-state wave function is zero outside the box, so we take the variation function ϕ to be zero outside the box. Equations (18.84) and (18.86) are valid only if ϕ is a well-behaved function, and this requires that ϕ be continuous. For ϕ to be continuous at the ends of the box, it must be zero at $x = 0$ and at $x = a$, where a is the box length. Perhaps the simplest way to get a function that vanishes at 0 and a is to take $\phi = x(a - x)$ for the region inside the box. As noted above, $\phi =$

0 outside the box. Since we did not normalize ϕ, Eq. (18.86) must be used. For the particle in a box, $V = 0$ and $\hat{H} = -(\hbar^2/2m)\,d^2/dx^2$ inside the box. We have

$$\int \phi^*\hat{H}\phi\, d\tau = \int_0^a x(a-x)\left(\frac{-\hbar^2}{2m}\right)\frac{d^2}{dx^2}\left[x(a-x)\right] dx$$

$$= \frac{-\hbar^2}{2m}\int_0^a x(a-x)(-2)\,dx = \frac{\hbar^2 a^3}{6m}$$

Also, $\int \phi^*\phi\, d\tau = \int_0^a x^2(a-x)^2\, dx = a^5/30$. The variation theorem (18.86) becomes $(\hbar^2 a^3/6m) \div (a^5/30) \geq E_{gs}$, or

$$E_{gs} \leq 5h^2/4\pi^2 ma^2 = 0.12665 h^2/ma^2$$

From (18.34), the true ground-state energy is $E_{gs} = h^2/8ma^2 = 0.125h^2/ma^2$. The variation function $x(a-x)$ gives a 1.3% error in E_{gs}.

Figure 18.22 plots the normalized variation function $(30/a^5)^{1/2}x(a-x)$ and the true ground-state wave function $(2/a)^{1/2}\sin(\pi x/a)$. Figure 18.22 also plots the percent deviation of the variation function from the true wave function versus x.

Exercise

Which of the following functions could be used as trial variation functions for the particle in a box? All functions are zero outside the box and the expression given applies only inside the box. (a) $-x^2(a-x)^2$; (b) x^2; (c) x^3; (d) $\sin(\pi x/a)$; (e) $\cos(\pi x/a)$; (f) $x(a-x)\sin(\pi x/a)$. [Answer: (a), (d), (f).]

If the normalized variation function ϕ contains the parameter c, then the variational integral $W \equiv \int \phi^*\hat{H}\phi\, d\tau$ will be a function of c, and one minimizes W by setting $\partial W/\partial c = 0$.

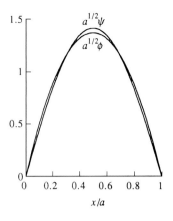

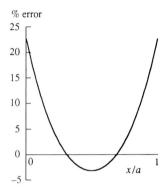

Figure 18.22

The upper figure plots the variation function $\phi = (30/a^5)^{1/2}x(a-x)$ and the true ground-state wave function ψ_{gs} for the particle in a one-dimensional box. The lower figure plots the percent deviation of this ϕ from the true ψ_{gs}.

EXAMPLE 18.9 Variation function with a parameter

Apply the variation function e^{-cx^2} to the harmonic oscillator, where c is a parameter whose value is chosen to minimize the variational integral.

The harmonic-oscillator potential energy (18.71) is $\frac{1}{2}kx^2$ and the Hamiltonian operator (18.56) is $\hat{H} = -(\hbar^2/2m)(d^2/dx^2) + \frac{1}{2}kx^2$. We have

$$\hat{H}\phi = -\frac{\hbar^2}{2m}\frac{d^2(e^{-cx^2})}{dx^2} + \frac{1}{2}kx^2 e^{-cx^2} = -\frac{\hbar^2}{2m}(4c^2x^2 - 2c)e^{-cx^2} + \frac{1}{2}kx^2 e^{-cx^2}$$

$$\int \phi^*\hat{H}\phi\, d\tau = \int_{-\infty}^{\infty}\left[-\frac{\hbar^2}{2m}(4c^2x^2 - 2c)e^{-2cx^2} + \frac{1}{2}kx^2 e^{-2cx^2}\right] dx$$

$$= \frac{\hbar^2}{m}\left(\frac{\pi c}{8}\right)^{1/2} + \frac{k}{4}\left(\frac{\pi}{8c^3}\right)^{1/2}$$

where Table 15.1 of Sec. 15.4 was used to evaluate the integrals. Also,

$$\int \phi^*\phi\, d\tau = \int_{-\infty}^{\infty} e^{-2cx^2}\, dx = \left(\frac{\pi}{2c}\right)^{1/2}$$

$$W \equiv \frac{\int \phi^*\hat{H}\phi\, d\tau}{\int \phi^*\phi\, d\tau} = \frac{\hbar^2 c}{2m} + \frac{k}{8c}$$

We now find the value of c that minimizes W:

$$0 = \frac{\partial W}{\partial c} = \frac{\hbar^2}{2m} - \frac{k}{8c^2}$$

We have $c^2 = mk/4\hbar^2$ and $c = \pm(mk)^{1/2}/2\hbar$. The negative value for c would give a positive exponent in the variation function $\phi = e^{-cx^2}$; ϕ would go to infinity as x goes to $\pm\infty$ and ϕ would not be quadratically integrable. We therefore reject the negative value of c. With $c = (mk)^{1/2}/2\hbar$, the variational integral W becomes

$$W = \frac{\hbar^2 c}{2m} + \frac{k}{8c} = \frac{\hbar k^{1/2}}{4m^{1/2}} + \frac{\hbar k^{1/2}}{4m^{1/2}} = \frac{hk^{1/2}}{4\pi m^{1/2}} = \frac{h\nu}{2}$$

where $\nu = (1/2\pi)(k/m)^{1/2}$ [Eq. (18.73)] was used. The value $h\nu/2$ is the true ground-state energy of the harmonic oscillator [Eq. (18.76)], and with $c = (mk)^{1/2}/2\hbar = \pi\nu m/\hbar$, the trial function e^{-cx^2} is the same as the unnormalized ground-state wave function of the harmonic oscillator (Fig. 18.18).

Exercise

Verify the integration results in this example.

A common form for variational functions in quantum mechanics is the **linear variation function**

$$\phi = c_1 f_1 + c_2 f_2 + \cdots + c_n f_n$$

where $f_1, \ldots, f_n$ are functions and $c_1, \ldots, c_n$ are variational parameters whose values are determined by minimizing the variational integral. Let W be the left side of (18.86). Then the conditions for a minimum in W are $\partial W/\partial c_1 = 0$, $\partial W/\partial c_2 = 0$, $\ldots$, $\partial W/\partial c_n = 0$. These conditions lead to a set of equations that allows the c's to be found. It turns out that there are n different sets of coefficients $c_1, \ldots, c_n$ that satisfy $\partial W/\partial c_1 = \cdots = \partial W/\partial c_n = 0$, so we end up with n different variational functions $\phi_1, \ldots, \phi_n$ and n different values for the variational integral $W_1, \ldots, W_n$, where $W_1 = \int \phi_1^* \hat{H} \phi_1 \, d\tau / \int \phi_1^* \phi_1 \, d\tau$, etc. If these W's are numbered in order of increasing energy, it can be shown that $W_1 \geq E_{gs}$, $W_2 \geq E_{gs+1}$, etc., where E_{gs}, E_{gs+1}, $\ldots$ are the true energies of the ground state, the next-lowest state, etc. Thus, use of the linear variation function $c_1 f_1 + \cdots + c_n f_n$ gives us approximations to the energies and wave functions of the lowest n states in the system. (In using this method, one deals separately with wave functions of different symmetry.)

Perturbation Theory

In recent years, the perturbation-theory approximation method has become important in molecular electronic structure calculations. Let $\hat{H}$ be the time-independent Hamiltonian operator of a system whose Schrödinger equation $\hat{H}\psi_n = E_n\psi_n$ we seek to solve. In the perturbation-theory approximation, one divides $\hat{H}$ into two parts:

$$\hat{H} = \hat{H}^0 + \hat{H}' \tag{18.87}$$

where $\hat{H}^0$ is the Hamiltonian operator of a system whose Schrödinger equation can be solved exactly and $\hat{H}'$ is a term whose effects one hopes are small. The system with Hamiltonian $\hat{H}^0$ is called the *unperturbed system*, $\hat{H}'$ is called the *perturbation*, and the system with Hamiltonian $\hat{H} = \hat{H}^0 + \hat{H}'$ is called the *perturbed system*. One finds that the energy E_n of state n of the perturbed system can be written as

$$E_n = E_n^{(0)} + E_n^{(1)} + E_n^{(2)} + \cdots \tag{18.88}$$

where $E_n^{(0)}$ is the energy of state n of the unperturbed system, and $E_n^{(1)}, E_n^{(2)}, \ldots$ are called the first-order, second-order, $\ldots$ corrections to the energy. (For derivations of this and other perturbation-theory equations, see a quantum chemistry text.) If the problem is suitable for perturbation theory, the quantities $E_n^{(1)}, E_n^{(2)}, E_n^{(3)}, \ldots$ decrease as the order of the perturbation correction increases.

To find $E_n^{(0)}$ we solve the Schrödinger equation $\hat{H}^0 \psi_n^{(0)} = E_n^{(0)} \psi_n^{(0)}$ of the unperturbed system.

Perturbation theory shows that the first-order energy correction $E_n^{(1)}$ is given by

$$E_n^{(1)} = \int \psi_n^{(0)*} \hat{H}' \psi_n^{(0)} \, d\tau \qquad (18.89)$$

Since $\psi_n^{(0)}$ is known, $E_n^{(1)}$ is easily calculated. The formulas for $E_n^{(2)}, E_n^{(3)}, \ldots$ are complicated and are omitted.

EXAMPLE 18.10 Perturbation theory

Suppose a one-particle, one-dimensional system has

$$\hat{H} = -(\hbar^2/2m) \, d^2/dx^2 + \tfrac{1}{2}kx^2 + bx^4$$

where b is small. Apply perturbation theory to obtain an approximation to the stationary-state energies of this system.

If we take $\hat{H}^0 = -(\hbar^2/2m) \, d^2/dx^2 + \tfrac{1}{2}kx^2$ and $\hat{H}' = bx^4$, then the unperturbed system is a harmonic oscillator, whose energies and wave functions are known (Sec. 18.12). From (18.76), we have $E_n^{(0)} = (n + \tfrac{1}{2})h\nu$, with $n = 0, 1, 2, \ldots$ and $\nu = (1/2\pi)(k/m)^{1/2}$, where the quantum-number symbol was changed from v to n to conform with the notation of this section. Including only the first-order correction to E_n, we have from (18.76) and (18.89)

$$E_n \approx E_n^{(0)} + E_n^{(1)} = (n + \tfrac{1}{2})h\nu + b \int_{-\infty}^{\infty} \psi_{n,\mathrm{ho}}^* x^4 \psi_{n,\mathrm{ho}} \, dx \qquad (18.90)$$

where $\psi_{n,\mathrm{ho}}$ is the harmonic-oscillator wave function with quantum number n. Substitution of the known $\psi_{n,\mathrm{ho}}$ functions (Fig. 18.18) enables $E_n^{(1)}$ to be found.

Exercise

Evaluate $E_n^{(1)}$ in (18.90) for the ground state. Use Table 15.1. (*Answer:* $3bh^2/64\pi^4\nu^2m^2$.)

18.16 HERMITIAN OPERATORS

Section 18.11 noted that operators in quantum mechanics are linear. Quantum-mechanical operators that correspond to a physical property must have another property besides linearity, namely, they must be Hermitian. This section discusses Hermitian operators and their properties. The material of this section is important for a thorough understanding of quantum mechanics, but is not essential to understanding the material in the remaining chapters of this book and so may be omitted if time does not allow its inclusion. The abstract material of this section can induce dizziness in susceptible individuals and is best studied in small doses.

Hermitian Operators

The quantum-mechanical average value $\langle M \rangle$ of the physical quantity M must be a real number. To take the complex conjugate of a number, we replace i by $-i$ wherever it occurs. A real number does not contain i, so a real number equals its complex conjugate:

$z = z^*$ if z is real. Hence $\langle M \rangle = \langle M \rangle^*$. We have $\langle M \rangle = \int \Psi^* \hat{M} \Psi \, d\tau$ [Eq. (18.62)] and $\langle M \rangle^* = \int (\Psi^* \hat{M} \Psi)^* \, d\tau = \int (\Psi^*)^* (\hat{M} \Psi)^* \, d\tau = \int \Psi (\hat{M} \Psi)^* \, d\tau$. Therefore

$$\int \Psi^* \hat{M} \Psi \, d\tau = \int \Psi (\hat{M} \Psi)^* \, d\tau \qquad (18.91)$$

Equation (18.91) must hold for all possible state functions Ψ, that is, for all functions that are continuous, single-valued, and quadratically integrable. A linear operator that obeys (18.91) for all well-behaved functions is called a **Hermitian operator.** If $\hat{M}$ is a Hermitian operator, it follows from (18.91) that (Prob. 18.62)

$$\int f^* \hat{M} g \, d\tau = \int g (\hat{M} f)^* \, d\tau \qquad \textbf{(18.92)*}$$

where f and g are arbitrary well-behaved functions (not necessarily eigenfunctions of any operator) and the integrals are definite integrals over all space. Although (18.92) looks like a more stringent requirement than (18.91), it is actually a consequence of (18.91). Thus a Hermitian operator obeys (18.92). The Hermitian property (18.92) is readily verified for the quantum-mechanical operators $x \cdot$ and $(\hbar/i)(\partial/\partial x)$ (Prob. 18.63).

Eigenvalues of Hermitian Operators

Section 18.11 noted that when Ψ is an eigenfunction of $\hat{M}$ with eigenvalue c, a measurement of M will give the value c. Since measured values are real, we expect c to be a real number. We now prove that *the eigenvalues of a Hermitian operator are real numbers.* To prove the theorem, we take the special case of (18.92) where f and g are the same function and this function is an eigenfunction of $\hat{M}$ with eigenvalue b. With $f = g$ and $\hat{M} f = bf$, (18.92) becomes

$$\int f^* bf \, d\tau = \int f (bf)^* \, d\tau$$

Using $(bf)^* = b^* f^*$ and taking the constants outside the integrals, we get $b \int f^* f \, d\tau = b^* \int ff^* \, d\tau$ or

$$(b - b^*) \int |f|^2 \, d\tau = 0 \qquad (18.93)$$

The quantity $|f|^2$ is never negative. The only way the definite integral $\int |f|^2 \, d\tau$ (which is the infinite sum of the nonnegative infinitesimal quantities $|f|^2 \, d\tau$) could be zero would be if the function f were zero everywhere. However, the function $f = 0$ is not allowed as an eigenfunction (Sec. 18.11). Therefore (18.93) requires that $b - b^* = 0$ and $b = b^*$. Only a real number is equal to its complex conjugate, so the eigenvalue b must be real.

Orthogonality of Eigenfunctions

We noted in Eq. (18.36) that the particle-in-a-one-dimensional-box stationary-state wave functions, which are eigenfunctions of $\hat{H}$, are orthogonal, meaning that $\int \psi_i^* \psi_j \, d\tau = 0$ when $i \neq j$. This is an example of the theorem that *two eigenfunctions of a Hermitian operator that correspond to different eigenvalues are orthogonal.* The proof is as follows.

The Hermitian property (18.92) holds for all well-behaved functions. In particular, it holds if we take f and g as two of the eigenfunctions of the Hermitian operator $\hat{M}$. With $\hat{M} f = bf$ and $\hat{M} g = cg$, the Hermitian property $\int f^* \hat{M} g \, d\tau = \int g (\hat{M} f)^* \, d\tau$ becomes

$$c \int f^* g \, d\tau = \int g (bf)^* \, d\tau = \int g b^* f^* \, d\tau = b \int g f^* \, d\tau$$

since a Hermitian operator has real eigenvalues. We have

$$(c - b) \int f^*g \, d\tau = 0$$

If the eigenvalues c and b are different ($c \neq b$), then $\int f^*g \, d\tau = 0$, and the theorem is proved.

If the eigenvalues b and c happen to be equal, then orthogonality need not necessarily hold. Recall that we saw examples of different eigenfunctions of $\hat{H}$ having the same eigenvalue when we discussed the degenerate energy levels of the particle in a three-dimensional cubic box and the rigid two-particle rotor (Secs. 18.10 and 18.14). Because the quantum-mechanical operator $\hat{M}$ is linear, one can show (Prob. 18.64) that if the functions f_1 and f_2 are eigenfunctions of $\hat{M}$ with the same eigenvalue, that is, if $\hat{M}f_1 = bf_1$ and $\hat{M}f_2 = bf_2$, then any linear combination $c_1 f_1 + c_2 f_2$ (where c_1 and c_2 are constants) is an eigenfunction of $\hat{M}$ with eigenvalue b. This freedom to take linear combinations of eigenfunctions with the same eigenvalue enables us to choose the constants c_1 and c_2 so as to give orthogonal eigenfunctions (Prob. 18.65). From here on, we shall assume that this has been done, so that all eigenfunctions of a Hermitian operator that we deal with will be orthogonal.

Let the set of functions $g_1, g_2, g_3, \ldots$ be the eigenfunctions of a Hermitian operator. Since these functions are (or can be chosen to be) orthogonal, we have $\int g_j^* g_k \, d\tau = 0$ when $j \neq k$ (that is, when g_j and g_k are different eigenfunctions). We shall always normalize eigenfunctions of operators, so $\int g_j^* g_j \, d\tau = 1$. These two equations expressing orthogonality and normalization can be written as the single equation

$$\int g_j^* g_k \, d\tau = \delta_{jk} \tag{18.94}$$

where the **Kronecker delta** δ_{jk} is a special symbol defined to equal 1 when $j = k$ and to equal 0 when j and k differ:

$$\delta_{jk} \equiv 1 \quad \text{when } j = k, \qquad \delta_{jk} \equiv 0 \quad \text{when } j \neq k \tag{18.95}$$

A set of functions that are orthogonal and normalized is an **orthonormal set.**

Complete Sets of Eigenfunctions

A set of functions $g_1, g_2, g_3, \ldots$ is said to be a **complete set** if every well-behaved function that depends on the same variables as the g's and obeys the same boundary conditions as the g's can be expressed as the sum $\sum_i c_i g_i$, where the c's are constants whose values depend on the function being expressed. The sets of eigenfunctions of many of the Hermitian operators that occur in quantum mechanics have been proved to be complete, and quantum mechanics assumes that *the set of eigenfunctions of a Hermitian operator that represents a physical quantity is a complete set.* If F is a well-behaved function and the set $g_1, g_2, g_3, \ldots$ is the set of eigenfunctions of the Hermitian operator $\hat{R}$ that corresponds to the physical property R, then

$$F = \sum_k c_k g_k \tag{18.96}$$

and one says that F has been *expanded* in terms of the set of g's.

How do we find the coefficients c_k in the expansion (18.96)? Multiplication of (18.96) by g_j^* gives $g_j^* F = \sum_k c_k g_j^* g_k$. Integration of this equation over the full range of all the coordinates gives

$$\int g_j^* F \, d\tau = \int \sum_k c_k g_j^* g_k \, d\tau = \sum_k \int c_k g_j^* g_k \, d\tau = \sum_k c_k \int g_j^* g_k \, d\tau = \sum_k c_k \delta_{jk}$$

where the orthonormality of the eigenfunctions of a Hermitian operator [Eq. (18.94)] and the fact that the integral of a sum equals the sum of the integrals were used. The

Kronecker delta δ_{jk} is always zero except when k equals j. Therefore every term in the sum $\sum_k c_k \delta_{jk}$ is zero except for the one term where k becomes equal to j: thus, $\sum_k c_k \delta_{jk} = c_j \delta_{jj} = c_j$ [Eq. (18.95)]. Therefore

$$c_j = \int g_j^* F \, d\tau$$

Changing j to k in this equation and substituting in the expansion (18.96), we have

$$F = \sum_k \left(\int g_k^* F \, d\tau \right) g_k \qquad (18.97)$$

where the definite integrals $\int g_j^* F \, d\tau$ are constants. Equation (18.97) shows how to expand any function F in terms of a known complete set of functions $g_1, g_2, g_3, \ldots$.

Suppose we are unable to solve the Schrödinger equation for a system we are interested in. We can express the unknown ground-state wave function as $\psi_{gs} = \sum_k c_k g_k$, where the g's are a known complete set of functions. We then use the linear variation method (Sec. 18.15) to solve for the coefficients c_k, thereby obtaining ψ_{gs}. The difficulty with this approach is that a complete set of functions usually contains an infinite number of functions. We are therefore forced to limit ourselves to a finite number of functions in the expansion sum, thereby introducing error into our determination of ψ_{gs}. Most methods of calculating wave functions for molecules use expansions, as we shall see in Chapter 20.

Consider an example. Let the function F be defined as $F = x^2(a - x)$ for x between 0 and a and $F = 0$ elsewhere. Could we use the particle-in-a-box stationary-state wave functions $\psi_n = (2/a)^{1/2} \sin(n\pi x/a)$ [Eq. (18.35)] to expand F? The function F is well-behaved and satisfies the same boundary conditions as ψ_n, namely, F is zero at the ends of the box. The functions ψ_n are the eigenfunctions of a Hermitian operator (the particle-in-a-box Hamiltonian $\hat{H}$) and so are a complete set. Therefore we can express F as $F = \sum_{n=1}^{\infty} c_n \psi_n$, where the coefficients c_n are given in (18.97) as

$$c_n = \int \psi_n^* F \, d\tau = \int_0^a \left(\frac{2}{a} \right)^{1/2} \sin \frac{n\pi x}{a} x^2(a - x) \, dx \qquad (18.98)$$

Problem 18.66 evaluates c_n and shows how the sum $\sum_n c_n \psi_n$ becomes a more and more accurate representation of F as more terms are included in the sum.

Summary

Quantum-mechanical operators that correspond to physical properties are Hermitian, meaning that they satisfy (18.92) for all well-behaved functions f and g. The eigenvalues of a Hermitian operator are real. The eigenfunctions of a Hermitian operator are (or can be chosen to be) orthogonal. The eigenfunctions of a Hermitian operator form a complete set, meaning that any well-behaved function can be expanded in terms of them.

18.17　SUMMARY

Electromagnetic waves of frequency ν and wavelength λ travel at speed $c = \lambda\nu$ in vacuum. Processes involving absorption or emission of electromagnetic radiation (for example, blackbody radiation, the photoelectric effect, spectra of atoms and molecules) can be understood by viewing the electromagnetic radiation to be composed of photons, each photon having an energy $h\nu$, where h is Planck's constant. When an atom or molecule absorbs or emits a photon, it makes a transition between two energy levels E_a and E_b whose energy difference is $h\nu$; $E_a - E_b = h\nu$.

De Broglie proposed that microscopic particles such as electrons have wavelike properties, and this was confirmed by observation of electron diffraction. Because of this wave–particle duality, simultaneous measurement of the precise position and momentum of a microscopic particle is impossible (the Heisenberg uncertainty principle).

The state of a quantum-mechanical system is described by the state function Ψ, which is a function of the particles' coordinates and the time. The change in Ψ with time is governed by the time-dependent Schrödinger equation (18.10) [or (18.60)], which is the quantum-mechanical analog of Newton's second law in classical mechanics. The probability density for finding the system's particles is $|\Psi|^2$. For example, for a two-particle, one-dimensional system, $|\Psi(x_1, x_2, t)|^2 \, dx_1 \, dx_2$ is the probability of simultaneously finding particle 1 between x_1 and $x_1 + dx_1$ and particle 2 between x_2 and $x_2 + dx_2$ at time t.

When the system's potential energy V is independent of time, the system can exist in one of many possible stationary states. For a stationary state, the state function is $\Psi = e^{-iEt/\hbar}\psi$. The (time-independent) wave function ψ is a function of the particles' coordinates and is one of the well-behaved solutions of the (time-independent) Schrödinger equation $\hat{H}\psi = E\psi$, where E is the energy and the Hamiltonian operator $\hat{H}$ is the quantum-mechanical operator that corresponds to the classical quantity E. To find the operator corresponding to a classical quantity M, one writes down the classical-mechanical expression for M in terms of cartesian coordinates and momenta and then replaces the coordinates and momenta by their corresponding quantum-mechanical operators: $\hat{x}_1 = x_1 \times$, $\hat{p}_{x,1} = (\hbar/i) \, \partial/\partial x_1$, etc. For a stationary state, $|\Psi|^2 = |\psi|^2$ and the probability density and energy are independent of time.

In accord with the probability interpretation, the state function is normalized to satisfy $\int |\Psi|^2 \, d\tau = 1$, where $\int d\tau$ denotes the definite integral over the full range of the particles' coordinates. For a stationary state, the normalization condition becomes $\int |\psi|^2 \, d\tau = 1$.

The average value of property M for a system in stationary state ψ is $\langle M \rangle = \int \psi^* \hat{M}\psi \, d\tau$, where $\hat{M}$ is the quantum-mechanical operator for property M.

The stationary-state wave functions and energies were found for the following systems. (a) Particle in a one-dimensional box ($V = 0$ for x between 0 and a; $V = \infty$ elsewhere): $E = n^2h^2/8ma^2$, $\psi = (2/a)^{1/2} \sin (n\pi x/a)$, $n = 1, 2, 3, \ldots$. (b) Particle in a three-dimensional rectangular box with dimensions a, b, c: $E = (h^2/8m) \cdot (n_x^2/a^2 + n_y^2/b^2 + n_z^2/c^2)$. (c) One-dimensional harmonic oscillator ($V = \frac{1}{2}kx^2$): $E = (v + \frac{1}{2})h\nu$, $\nu = (1/2\pi)(k/m)^{1/2}$, $v = 0, 1, 2, \ldots$. (d) Two-particle rigid rotor (particles at fixed distance d and energy entirely kinetic): $E = J(J + 1)\hbar^2/2I$, $I = \mu d^2$, $J = 0, 1, 2, \ldots$; $\mu \equiv m_1 m_2/(m_1 + m_2)$ is the reduced mass.

When more than one state function corresponds to the same energy level, that energy level is said to be degenerate. There is degeneracy for the particle in a cubic box and for the two-particle rigid rotor.

For a system of noninteracting particles, the stationary-state wave functions are products of wave functions for each particle and the energy is the sum of the energies of the individual particles.

The variation theorem states that for any well-behaved trial variation function ϕ, one has $\int \phi^* \hat{H}\phi \, d\tau / \int \phi^* \phi \, d\tau \geq E_{gs}$, where $\hat{H}$ is the system's Hamiltonian operator and E_{gs} is its true ground-state energy.

Important kinds of calculations discussed in this chapter include:

- Use of $\lambda\nu = c$ to calculate the wavelength of light from the frequency, and vice versa.
- Use of $E_{\text{upper}} - E_{\text{lower}} = h\nu$ to calculate the frequency of the photon emitted or absorbed when a quantum-mechanical system makes a transition between two states.
- Use of energy-level formulas such as $E = n^2h^2/8ma^2$ for the particle in a box or $E = (v + \frac{1}{2})h\nu$ for the harmonic oscillator to calculate energy levels of quantum-mechanical systems.

- For a one-particle, one-dimensional, stationary-state system, use of $|\psi|^2 \, dx$ to calculate the probability of finding the particle between x and $x + dx$ and of $\int_a^b |\psi|^2 \, dx$ to calculate the probability of finding the particle between a and b.
- Use of $\langle M \rangle = \int \psi^* \hat{M} \psi \, d\tau$ to calculate average values.
- Use of the variation theorem to estimate the ground-state energy of a quantum-mechanical system.

FURTHER READING

Hanna, chap. 3; *Karplus and Porter,* chap. 2; *Levine* (2000), chaps. 1–4, 8, 9; *Lowe*, chaps. 1–3, 7; *McQuarrie* (1983), chaps. 1–5; *Atkins and Friedman*, chaps. 1–5, 8.

PROBLEMS

Section 18.1

18.1 (*a*) Let ν_{max} be the frequency at which the blackbody-radiation function (18.2) is a maximum. Show that $\nu_{max} = kTx/h$, where x is the nonzero solution of $x + 3e^{-x} = 3$. Since x is a constant, ν_{max} increases linearly with T. (*b*) Use a calculator with an e^x key to solve the equation in (*a*) by trial and error. To save time, use interpolation after you have found the successive integers that x lies between. Alternatively, use the Solver in Excel. (*c*) Calculate ν_{max} for a blackbody at 300 K and at 3000 K. Refer to Fig. 21.2 to state in which portions of the electromagnetic spectrum these frequencies lie. (*d*) The light emitted by the sun conforms closely to the blackbody radiation law and has $\nu_{max} = 3.5 \times 10^{14} \text{ s}^{-1}$. Estimate the sun's surface temperature. (*e*) The skin temperature of humans is 33°C, and the emission spectrum of human skin at this temperature conforms closely to blackbody radiation. Find ν_{max} for human skin at 33°C. What region of the electromagnetic spectrum is this in?

18.2 (*a*) Use the fact that $\int_0^\infty [z^3/(e^z - 1)] \, dz = \pi^4/15$ to show that the total radiant energy emitted per second by unit area of a blackbody is $2\pi^5 k^4 T^4/15c^2h^3$. Note that this quantity is proportional to T^4 (*Stefan's law*). (*b*) The sun's diameter is 1.4×10^9 m and its effective surface temperature is 5800 K. Assume the sun is a blackbody and estimate the rate of energy loss by radiation from the sun. (*c*) Use $E = mc^2$ to calculate the relativistic mass of the photons lost by radiation from the sun in 1 year.

Section 18.2

18.3 The work function of K is 2.2 eV and that of Ni is 5.0 eV, where 1 eV = 1.60×10^{-19} J. (*a*) Calculate the threshold frequencies and wavelengths for these two metals. (*b*) Will violet light of wavelength 4000 Å cause the photoelectric effect in K? In Ni? (*c*) Calculate the maximum kinetic energy of the electrons emitted in (*b*).

18.4 Calculate the energy of a photon of red light of wavelength 700 nm. (1 nm = 10^{-9} m.)

18.5 A 100-W sodium-vapor lamp emits yellow light of wavelength 590 nm. Calculate the number of photons emitted per second.

18.6 Millikan found the following data for the photoelectric effect in Na:

$10^{12} K_{max}$/ergs	3.41	2.56	1.95	0.75
λ/Å	3125	3650	4047	5461

where K_{max} is the maximum kinetic energy of emitted electrons and λ is the wavelength of the incident radiation. Plot K_{max} versus ν. From the slope and intercept, calculate h and the work function for Na.

Section 18.4

18.7 Calculate the de Broglie wavelength of (*a*) a neutron moving at 6.0×10^6 cm/s; (*b*) a 50-g particle moving at 120 cm/s.

Section 18.5

18.8 A beam of electrons traveling at 6.0×10^8 cm/s falls on a slit of width 2400 Å. The diffraction pattern is observed on a screen 40 cm from the slit. The x and y axes are defined as in Fig. 18.4. Find (*a*) the angle θ to the first diffraction minimum; (*b*) the width of the central maximum of the diffraction pattern on the screen; (*c*) the uncertainty Δp_x at the slit.

18.9 Estimate the minimum uncertainty in the x component of velocity of an electron whose position is measured to an uncertainty of 1×10^{-10} m.

Section 18.6

18.10 For a system containing three particles, what are the variables on which the state function Ψ depends?

18.11 True or false? (*a*) In the equation $\int |\Psi|^2 \, d\tau = 1$, the integral is an indefinite integral. (*b*) The state function Ψ takes on only real values. (*c*) If z is a complex number, then $zz^* = |z|^2$. (*d*) If z is a complex number, then $z + z^*$ is always a real number. (*e*) If $z = a + bi$, where a and b are real numbers, and we plot a on the x axis and b on the y axis, the distance of the point (a, b) from the origin in the xy plane is equal to $|z|$. (*f*) The absolute value $|z|$ of a complex number must be a real nonnegative number.

18.12 Find the absolute value of (*a*) -2; (*b*) $3 - 2i$; (*c*) $\cos \theta + i \sin \theta$; (*d*) $-3e^{-i\pi/5}$.

18.13 Verify that, if Ψ is a solution of the time-dependent Schrödinger equation (18.10), then $c\Psi$ is also a solution, where *c* is any constant.

18.14 Show that

$$\int_a^b \int_c^d \int_s^t f(r)g(\theta)h(\phi) \, dr \, d\theta \, d\phi$$

$$= \int_s^t f(r) \, dr \int_c^d g(\theta) \, d\theta \int_a^b h(\phi) \, d\phi$$

where the limits are constants.

18.15 Verify that Ψ in Example 18.1 is normalized.

Section 18.7
18.16 For a system consisting of three particles, on what variables does the time-independent wave function ψ depend?

18.17 True or false? For a stationary state, (*a*) $|\psi| = |\Psi|$; (*b*) $\psi = \Psi$; (*c*) the probability density is independent of time; (*d*) the energy is a constant.

18.18 Which is more general, the time-dependent Schrödinger equation or the time-independent Schrödinger equation?

18.19 Prove that $(fg)^* = f^*g^*$, where *f* and *g* are complex quantities.

18.20 Verify that if ψ is a solution of the time-independent Schrödinger equation (18.24), then $k\psi$ is also a solution, where *k* is any constant.

18.21 State whether each of the functions (*a*) to (*d*) is quadratically integrable. (*a* and *b* are positive constants.) (*a*) e^{-ax^2} (*Hint:* See Table 15.1.) (*b*) e^{-bx}. (*c*) $1/x$. (*Hint:* In (*c*) and (*d*), write the integral as the sum of two integrals.) (*d*) $1/|x|^{1/4}$. (*e*) True or false? A function that becomes infinite at a point must not be quadratically integrable.

Section 18.8
18.22 Calculate the wavelength of the photon emitted when a 1.0×10^{-27} g particle in a box of length 6.0 Å goes from the *n* = 5 to the *n* = 4 level.

18.23 (*a*) For a particle in the stationary state *n* of a one-dimensional box of length *a*, find the probability that the particle is in the region $0 \leq x \leq a/4$. (*b*) Calculate this probability for *n* = 1, 2, and 3.

18.24 For a 1.0×10^{-26} g particle in a box whose ends are at $x = 0$ and $x = 2.000$ Å, calculate the probability that the particle's *x* coordinate is between 1.6000 and 1.6001 Å if (*a*) *n* = 1; (*b*) *n* = 2.

18.25 Sketch ψ and $|\psi|^2$ for the *n* = 4 and *n* = 5 states of a particle in a one-dimensional box.

18.26 Solve Eq. (18.28) for the special case $E = 0$. Then apply the continuity requirement at each end of the box to eval-

uate the two integration constants and thus show that $\psi = 0$ for $E = 0$. From (18.32), $E = 0$ corresponds to $n = 0$, so $n = 0$ is not allowed.

18.27 For an electron in a certain one-dimensional box, the lowest observed transition frequency is 2.0×10^{14} s^{-1}. Find the length of the box.

18.28 If the *n* = 3 to 4 transition for a certain particle-in-a-box system occurs at 4.00×10^{13} s^{-1}, find the frequency of the *n* = 6 to 9 transition in this system.

18.29 Verify the orthogonality equation (18.36) for particle-in-a-box wave functions.

18.30 For the particle in a box, check that the wave functions (18.35) satisfy the Schrödinger equation (18.28) by substituting (18.35) into (18.28).

Section 18.9
18.31 For a particle in a two-dimensional box with sides of equal lengths, draw rough sketches of contours of constant $|\psi|$ for the states (*a*) $n_x = 2$, $n_y = 1$; (*b*) $n_x = 2$, $n_y = 2$. At what points in the box is $|\psi|$ a maximum for each state? (*Hint:* The maximum value of $|\sin \theta|$ is 1.)

Section 18.10
18.32 For a particle in a cubic box of edge *a*: (*a*) How many states have energies in the range 0 to $16h^2/8ma^2$? (*b*) How many energy levels lie in this range?

Section 18.11
18.33 True or false? (*a*) $(\hat{A} + \hat{B})f(x)$ is always equal to $\hat{A}f(x) + \hat{B}f(x)$. (*b*) $\hat{A}[f(x) + g(x)]$ is always equal to $\hat{A}f(x) + \hat{A}g(x)$. (*c*) $\hat{B}\hat{C}f(x)$ is always equal to $\hat{C}\hat{B}f(x)$. (*d*) $[\hat{A}f(x)]/f(x)$ is always equal to $\hat{A}$, provided $f(x) \neq 0$. (*e*) $3x$ is an eigenvalue of $\hat{x}$. (*f*) $3x$ is an eigenfunction of $\hat{x}$. (*g*) $e^{ikx/\hbar}$, where *k* is a constant, is an eigenfunction of $\hat{p}_x$ with eigenvalue *k*.

18.34 If *f* is a function, state whether or not each of the following expressions is equal to $f^*\hat{B}f$. (*a*) $f^*(\hat{B}f)$. (*b*) $\hat{B}(f^*f)$. (*c*) $(\hat{B}f)f^*$. (*d*) $f^*f\hat{B}$.

18.35 If the energy of a particle in the *n* = 5 stationary state of a particle in a one-dimensional box of length *a* is measured, state the possible result(s).

18.36 Let $\hat{A} = d^2/dx^2$ and $\hat{B} = x \times$. (*a*) Find $\hat{A}\hat{B}f(x) - \hat{B}\hat{A}f(x)$. (*b*) Find $(\hat{A} + \hat{B})(e^{x^2} + \cos 2x)$.

18.37 (*a*) Classify each of these operators as linear or nonlinear: $\partial^2/\partial x^2$, $2 \, \partial/\partial z$, $3z^2 \times$, $(\ )^2$, $(\ )^*$. (*b*) Verify that $\hat{H}$ in (18.58) is linear.

18.38 State whether each of the following entities is an operator or a function: (*a*) $\hat{A}\hat{B}g(x)$; (*b*) $\hat{A}\hat{B} + \hat{B}\hat{A}$; (*c*) $\hat{B}^2f(x)$; (*d*) $g(x)\hat{A}$; (*e*) $g(x)\hat{A}f(x)$.

18.39 Find the quantum-mechanical operator for (*a*) p_x^3; (*b*) p_z^4.

18.40 (*a*) Which of the functions $\sin 3x$, $6 \cos 4x$, $5x^3$, $1/x$, $3e^{-5x}$, $\ln 2x$ are eigenfunctions of d^2/dx^2? (*b*) For each eigenfunction, state the eigenvalue.

18.41 For a particle in a one-dimensional-box stationary state, show that (a) $\langle p_x \rangle = 0$; (b) $\langle x \rangle = a/2$; (c) $\langle x^2 \rangle = a^2(1/3 - 1/2n^2\pi^2)$.

18.42 Verify the separation-of-variables equations of Sec. 18.11 as follows. For the Hamiltonian (18.64), write the time-independent Schrödinger equation. Assume solutions of the form (18.65) and substitute into the Schrödinger equation to derive (18.66) and (18.67).

18.43 For a system of two noninteracting particles of masses m_1 and m_2 in a one-dimensional box of length a, give the formulas for the stationary-state wave functions and energies.

Section 18.12

18.44 For each of the following, state whether it is a nu or a vee and whether it is a quantum number or a frequency. (a) ν; (b) v.

18.45 Calculate the frequency of radiation emitted when a harmonic oscillator of frequency 6.0×10^{13} s^{-1} goes from the $v = 8$ to the $v = 7$ level.

18.46 Draw rough sketches of ψ^2 for the $v = 0$, 1, 2, and 3 harmonic-oscillator states.

18.47 Find the most probable value(s) of x for a harmonic oscillator in the state (a) $v = 0$; (b) $v = 1$.

18.48 Verify that ψ_0 in Fig. 18.18 is a solution of the Schrödinger equation (18.75).

18.49 Verify that the harmonic-oscillator ψ_1 in Fig. 18.18 is normalized. (See Table 15.1.)

18.50 For the ground state of a harmonic oscillator, calculate (a) $\langle x \rangle$; (b) $\langle x^2 \rangle$; (c) $\langle p_x \rangle$. See Table 15.1.

18.51 (a) Verify Eq. (18.74). (b) Verify by substitution that (18.72) satisfies the differential equation $m \, d^2x/dt^2 = -kx$.

18.52 A mass of 45 g on a spring oscillates at the frequency of 2.4 vibrations per second with an amplitude 4.0 cm. (a) Calculate the force constant of the spring. (b) What would be the quantum number v if the system were treated quantum-mechanically?

18.53 For a three-dimensional harmonic oscillator, $V = \frac{1}{2}k_x x^2 + \frac{1}{2}k_y y^2 + \frac{1}{2}k_z z^2$, where the three force constants k_x, k_y, k_z are not necessarily equal. (a) Write down the expression for the energy levels of this system. Define all symbols. (b) What is the zero-point energy?

Section 18.13

18.54 Substitute (18.79), (18.80), and $M = m_1 + m_2$ into (18.78) and verify that H reduces to $p_1^2/2m_1 + p_2^2/2m_2 + V$, where p_1 is the momentum of particle 1.

Section 18.14

18.55 Consider the $^{12}C^{16}O$ molecule to be a two-particle rigid rotor with m_1 and m_2 equal to the atomic masses and the interparticle distance fixed at the CO bond length 1.13 Å. (a) Find the reduced mass. (b) Find the moment of inertia. (c) Find the energies of the four lowest rotational levels and give the degeneracy of each of these levels. (d) Calculate the frequency of the radiation absorbed when a $^{12}C^{16}O$ molecule goes from the $J = 0$ level to the $J = 1$ level. Repeat for $J = 1$ to $J = 2$.

Section 18.15

18.56 The unnormalized particle-in-a-box variation function $x(a - x)$ for x between 0 and a was used in Example 18.8. (a) Use work in that example to show that $(30/a^5)^{1/2}x(a - x)$ is the normalized form of this function. (b) Calculate $\langle x^2 \rangle$ using the normalized function in (a) and compare the result with the exact ground-state $\langle x^2 \rangle$ (Prob. 18.41).

18.57 For the particle in a one-dimensional box, one finds that use of the normalized variation function $\phi = Nx^k(a - x)^k$, where k is a parameter, gives $\int \phi^* \hat{H} \phi \, d\tau = (\hbar^2/ma^2) \cdot (4k^2 + k)/(2k - 1)$. Find the value of k that minimizes the variational integral W and find the value of W for this k value. Compare the percent error in the ground-state energy with that for the function $Nx(a - x)$ used in Example 18.8.

18.58 (a) Apply the variation function $x^2(a - x)^2$ for x between 0 and a to the particle in a box and estimate the ground-state energy. Calculate the percent error in E_{gs}. (b) Explain why the function x^2 (for x between 0 and a) cannot be used as a variation function for the particle in a box.

18.59 Consider a one-particle, one-dimensional system with $V = \infty$ for $x < 0$, $V = \infty$ for $x > a$, and $V = kx$ for $0 \le x \le a$, where k is small. Treat the system as a perturbed particle in a box and find $E_n^{(0)} + E_n^{(1)}$ for the state with quantum number n. Use a table of integrals.

Section 18.16

18.60 True or false? (a) All the eigenvalues of a Hermitian operator are real numbers. (b) Two eigenfunctions of the same Hermitian operator are always orthogonal. (c) $\delta_{jk} = \delta_{kj}$. (d) A Hermitian operator cannot contain the imaginary number i. (e) $\sum_n b_m c_m \delta_{mn} = b_n c_n$.

18.61 Prove that the sum of two Hermitian operators is a Hermitian operator.

18.62 Use the following procedure to show that for the Hermitian operator $\hat{M}$, Eq. (18.92) is a consequence of (18.91). (a) Set $\Psi = f + cg$ in (18.91), where c is an arbitrary constant. Use (18.91) to cancel some terms in the resulting equation, thereby getting

$$c^* \int g^* \hat{M} f \, d\tau + c \int f^* \hat{M} g \, d\tau$$
$$= c \int g(\hat{M}f)^* \, d\tau + c^* \int f(\hat{M}g)^* \, d\tau$$

(18.99)

(b) First set $c = 1$ in (18.99). Then set $c = i$ in (18.99) and divide the resulting equation by i. Add these two equations, thereby proving (18.92).

18.63 Verify that if f and g are functions of x and $d\tau = dx$, then (a) the Hermitian property (18.92) holds for $\hat{x}$; (b) (18.92) holds for $\hat{p}_x$. [*Hint:* For part (b), use integration by parts and the fact that a quadratically integrable function must go to zero as x goes to $\pm\infty$.]

18.64 Given that $\hat{M}$ is a linear operator and that $\hat{M}f_1 = bf_1$ and $\hat{M}f_2 = bf_2$, prove that $c_1f_1 + c_2f_2$, where c_1 and c_2 are constants, is an eigenfunction of $\hat{M}$ with eigenvalue b.

18.65 If $\hat{M}$ is a linear operator with $\hat{M}f_1 = bf_1$ and $\hat{M}f_2 = bf_2$, and we define g_1 and g_2 as $g_1 \equiv f_1$ and $g_2 \equiv f_2 + kf_1$, where $k \equiv -\int f_1^* f_2 \, d\tau / \int f_1^* f_1 \, d\tau$, verify that g_1 and g_2 are orthogonal.

18.66 Let $F \equiv x^2(a - x)$ for x between 0 and a. Let $G \equiv \sum_{n=1}^{m} c_n \psi_n$, where ψ_n is a particle-in-a-one-dimensional box wave function with quantum number n and c_n is given by (18.98). To make things a bit simpler, take the length a as equal to 1. (a) Use a table of integrals to find c_n. (b) Use a spreadsheet to calculate F and G for $m = 3$ and plot them on the same graph. (c) Repeat (b) for $m = 5$ and comment on the results.

General

18.67 What are the SI units of a stationary-state wave function ψ for (a) a one-particle, one-dimensional system; (b) a one-particle, three-dimensional system; (c) a two-particle, three-dimensional system?

18.68 By fitting experimental blackbody-radiation curves using Eq. (18.2), Planck not only obtained a value for h, but also obtained the first reasonably accurate values of k, N_A, and the proton charge e. Explain how Planck obtained values for these constants.

18.69 State quantitatively the effect on the system's energy levels of each of the following: (a) doubling the box length for a particle in a one-dimensional box; (b) doubling the interparticle distance of a two-particle rigid rotor; (c) doubling the mass of a harmonic oscillator.

18.70 True or false? (a) In classical mechanics, knowledge of the present state of an isolated system allows the future state to be predicted with certainty. (b) In quantum mechanics, knowledge of the present state of an isolated system allows the future state to be predicted with certainty. (c) For a stationary state, Ψ is the product of a function of time and a function of the coordinates. (d) An increase in the particle mass would decrease the ground-state energy of both the particle in a box and the harmonic oscillator. (e) For a system of noninteracting particles, each stationary-state wave function is equal to the sum of wave functions for each particle. (f) The one-dimensional harmonic-oscillator energy levels are nondegenerate. (g) Ψ must be real. (h) The energies of any two photons must be equal. (i) In the variation method, the variational function ϕ must be an eigenfunction of $\hat{H}$.

18.71 *Physicist trivia question.* Name the physicist (or physicists) referred to in each of the following descriptions. The same name can be used more than once and all names appear in Chapter 18. (a) On Christmas vacation in Arosa, Switzerland,

this 38-year-old professor of theoretical physics at the University of Zurich began work on a series of papers titled "Quantisierung als Eigenwertproblem" ("Quantization as an Eigenvalue Problem"), papers described by Born as "of a grandeur unsurpassed in theoretical physics." (b) On the tiny North Sea island of Helgoland, where he had gone to recover from a severe attack of hay fever, this 23-year-old lecturer at the University of Göttingen conceived the ideas embodied in his paper "Über Quantentheoretische Umdeutung Kinematischer und Mechanischer Beziehungen" ("On a Quantum-Theoretical Interpretation of Kinematical and Mechanical Relations") that marks the birth of modern quantum mechanics. He later wrote: "it was almost three o'clock in the morning before the final result of my computations lay before me . . . I could no longer doubt the mathematical consistency and coherence of the kind of quantum mechanics to which my calculations pointed. At first, I was deeply alarmed. I had the feeling that, through the surface of atomic phenomena, I was looking at a strangely beautiful interior . . . I was far too excited to sleep, and so, as a new day dawned, I made for the southern tip of the island, where I had been longing to climb a rock jutting out into the sea. I now did so without too much trouble, and waited for the sun to rise." (c) He had the middle names Karl Ernst Ludwig. (d) He headed Germany's atom-bomb project during World War II. (e) In September 1943, under cover of darkness, he crossed the Öresund strait by boat from German-occupied Denmark to Sweden. A few days later, a British bomber flew him to Scotland. The plane had no passenger seat and he flew in the bomb bay. He failed to use the oxygen mask supplied him and lost consciousness during the flight, but recovered when the plane landed. (f) In a discussion with Bohr about quantum mechanics, he said "If we are still going to have to put up with these damn quantum jumps, I am sorry I ever had anything to do with quantum theory." (g) The play *Copenhagen* (winner of the Tony award for best new drama on Broadway in 2000) deals with a 1941 meeting between these two physicists in which they discussed the possibility of developing an atomic bomb. (h) He spent 1940–1956 working mainly on a unified field theory at the Dublin Institute for Advanced Studies; during the early years of this period he lived in a two-story house with his wife Anny, his mistress Hilde March (wife of the physicist Arthur March), and his daughter Ruth (who did not find out she was his daughter until she was 17). (i) This nineteenth-century mathematician and physicist reformulated classical mechanics in a form especially suitable for formulating quantum mechanics. He was appointed Professor of Astronomy at Trinity College while still an undergraduate student. (j) He used the pseudonym Nicholas **B**aker when he worked on the atomic bomb in Los Alamos, New Mexico. (k) In a 1999 poll of physicists he was chosen as the greatest physicist of all time and was chosen by *Time* magazine as "Person of the Twentieth Century." (l) His son Erwin was found guilty of complicity in the 1944 plot to kill Hitler and was executed in February 1945. Shortly thereafter, when the town he was living in became a battlefield, this 87-year-old physicist had to hide in the woods and sleep in haystacks. (m) He believed in the Hindu philosophy of Vedanta, which he summarized as "we living beings all belong to one another . . . we are all actually members or aspects of a single being, which we may in western terminology call God, while in the Upanishads it is called Brahman." (*n*) His 1919 divorce settlement with his first wife provided that in addition to providing child support, if he were to win a Nobel Prize, the prize money would be given to his ex-wife. She received the prize money in 1923. In a letter he praised a friend's ability to have a happy, long-lasting marriage, "an undertaking in which I failed twice rather disgracefully." (*o*) He helped develop quantum mechanics and from 1957 to 1961 was a member of West Germany's parliament. (*p*) In a letter to a father whose daughter had died, he wrote: "A human being is a part of the whole, called by us "Universe," a part limited in time and space. He experiences himself, his thoughts and feelings as something separated from the rest—a kind of optical delusion of his consciousness. This delusion is a prison for us, restricting us to our personal desires and to affection for a few persons nearest to us. Our task must be to free ourselves from this prison by widening our circle of compassion to embrace all living creatures and the whole of nature in its beauty."

Atomic Structure

Chapter 18 introduced some of the main ideas of quantum chemistry and looked at the solutions of the time-independent Schrödinger equation for the particle in a box, the rigid two-particle rotor, and the harmonic oscillator. We now use quantum mechanics to discuss the electronic structure of atoms (Chapter 19) and molecules (Chapter 20).

19.1 UNITS

The forces acting between the particles in atoms and molecules are electrical (Prob. 19.70). To formulate the Hamiltonian operator of an atom or molecule, we need the expression for the potential energy of interaction between two charged particles. Equation (14.12) gives the electric potential ϕ at a distance r from charge Q_1 as $\phi = Q_1/4\pi\varepsilon_0 r$. Equation (14.10) gives the potential energy V of interaction of a second charge Q_2 with this electric potential as $V = \phi Q_2 = (Q_1/4\pi\varepsilon_0 r)Q_2$. The potential energy of interaction between two charges separated by a distance r is therefore

$$V = \frac{1}{4\pi\varepsilon_0} \frac{Q_1 Q_2}{r} \qquad (19.1)^*$$

Equation (19.1) is in SI units, with Q_1 and Q_2 in coulombs (C), r in meters, and V in joules. The values of the constants ε_0 and $1/4\pi\varepsilon_0$ are listed in Eq. (14.2) and in the fundamental-constants table inside the back cover. The $1/4\pi\varepsilon_0$ factor in (19.1) is a bit of a pain, but if you learn the value of the speed of light c, you can avoid looking up $1/4\pi\varepsilon_0$ by using the relation (Prob. 19.3)

$$\frac{1}{4\pi\varepsilon_0} = 10^{-7}c^2 \text{ N s}^2/\text{C}^2 \qquad (19.2)$$

SI units are used in this chapter. Older books on atomic and molecular structure often use gaussian units, in which Coulomb's law is written as $F = |Q_1'Q_2'|/r^2$, where F is in dynes, r in centimeters, and Q_1' and Q_2' in statcoulombs (statC) and the primes denote use of gaussian units. One coulomb corresponds to 2.9979×10^9 statC.

Atomic and molecular energies are very small. A convenient unit for these energies is the **electronvolt** (eV), defined as the kinetic energy acquired by an electron accelerated through a potential difference of one volt. From (14.10), the change in potential energy for this process is $-e(1 \text{ V})$, where $-e$ is the electron's charge. This loss in potential energy is matched by a gain of kinetic energy equal to $e(1 \text{ V})$. Substitution of (19.4) for e gives $1 \text{ eV} = (1.6022 \times 10^{-19} \text{ C}) (1 \text{ V})$; the use of $1 \text{ V} = 1 \text{ J/C}$ [Eq. (14.8)] gives

$$1 \text{ eV} = 1.6022 \times 10^{-19} \text{ J} \qquad (19.3)$$

| 19.2 | **HISTORICAL BACKGROUND** |

In a low-pressure gas-discharge tube, bombardment of the negative electrode (the cathode) by positive ions causes the cathode to emit what nineteenth-century physicists called *cathode rays*. In 1897, J. J. Thomson measured the deflection of cathode rays in simultaneously applied electric and magnetic fields of known strengths. His experiment allowed calculation of the charge-to-mass ratio Q/m of the cathode-ray particles. Thomson found that Q/m was independent of the metal used for the cathode, and his experiments mark the discovery of the electron. [G. P. Thomson, who was one of the first people to observe diffraction effects with electrons (Sec. 18.4), was J. J.'s son. It has been said that J. J. Thomson got the Nobel Prize for proving the electron to be a particle, and G. P. Thomson got the Nobel Prize for proving the electron to be a wave.]

Let the symbol e denote the charge on the proton. The electron charge is then $-e$. Thomson found $e/m_e = 1.7 \times 10^8$ C/g, where m_e is the electron's mass.

The first accurate measurement of the electron's charge was made by R. A. Millikan and Harvey Fletcher in the period 1909–1913 (see H. Fletcher, *Physics Today*, June 1982, p. 43). They observed the motion of charged oil drops in oppositely directed electric and gravitational fields and found that all observed values of the charge Q of a drop satisfied $|Q| = ne$, where n was a small integer and was clearly the number of extra or missing electrons on the charged oil drop. The smallest observed difference between values of $|Q|$ could then be taken as the magnitude of the charge of the electron. The currently accepted value of the proton charge is

$$e = 1.6022 \times 10^{-19} \text{ C} \tag{19.4}$$

From the values of e and the Faraday constant F, an accurate value of the Avogadro constant N_A can be obtained. Equation (14.13) gives $N_A = F/e = $ (96485 C mol^{-1})/(1.6022 $\times$ 10^{-19} C) = 6.022 $\times$ 10^{23} mol^{-1}.

From the values of e and e/m_e, the electron mass m_e can be found. The modern value is 9.1094×10^{-28} g. The ^{1}H atom mass is 1.0078 g/6.022 $\times$ 10^{23} = 1.6735 $\times$ 10^{-24} g. This is 1837 times m_e, and so a proton is 1836 times as heavy as an electron. Nearly all the mass of an atom is in its nucleus.

The existence of the atomic nucleus was demonstrated by the 1909–1911 experiments of Rutherford, Geiger, and Marsden, who allowed a beam of alpha particles (He^{2+} nuclei) to fall on a very thin gold foil. Although most of the alpha particles passed nearly straight through the foil, a few were deflected through large angles. Since the very light electrons of the gold atoms cannot significantly deflect the alpha particles (in a collision between a truck and a bicycle, it is the bicycle that is deflected), one need consider only the force between the alpha particle and the positive charge of a gold atom. This force is given by Coulomb's law (14.1). To get a force large enough to produce the observed large deflections, Rutherford found that r in (14.1) had to be in the range 10^{-12} to 10^{-13} cm, which is much less than the known radius of an atom (10^{-8} cm). Rutherford therefore concluded in 1911 that the positive charge of an atom was not distributed throughout the atom but was concentrated in a tiny central region, the nucleus.

In 1913, Bohr proposed his theory of the hydrogen atom (Sec. 18.3). By the early 1920s, physicists realized that the Bohr theory was not correct.

In January 1926, Erwin Schrödinger formulated the Schrödinger equation. He solved the time-independent Schrödinger equation for the hydrogen atom in his first paper on quantum mechanics, obtaining energy levels in agreement with the observed spectrum. In 1929, Hylleraas used the quantum-mechanical variational method (Sec. 18.15) to obtain a ground-state energy for helium in accurate agreement with experiment.

19.3 THE HYDROGEN ATOM

The simplest atom is hydrogen. The Schrödinger equation can be solved exactly for the H atom but not for atoms with more than one electron. Ideas developed in treating the H atom provide a basis for dealing with many-electron atoms.

The hydrogen atom is a two-particle system in which a nucleus and an electron interact according to Coulomb's law. Instead of dealing only with the H atom, we shall consider the slightly more general problem of the **hydrogenlike atom**; this is an atom with one electron and Z protons in the nucleus. The values $Z = 1, 2, 3, \ldots$ give the species H, He$^+$, Li^{2+}, ... With the nuclear charge Q_1 set equal to Ze and the electron charge Q_2 set equal to $-e$, Eq. (19.1) gives the potential energy as $V = -Ze^2/4\pi\varepsilon_0 r$, where r is the distance between the electron and the nucleus.

The potential-energy function depends only on the relative coordinates of the two particles, and so the conclusions of Sec. 18.13 apply. The total energy E_{tot} of the atom is the sum of its translational energy and the energy of internal motion of the electron relative to the proton. The translational energy levels can be taken as the particle-in-a-box levels (18.47). The box is the container holding the gas of H atoms. We now focus on the energy E of internal motion. The Hamiltonian H for the internal motion is given by the terms in the first pair of brackets in (18.78), and the corresponding Hamiltonian operator $\hat{H}$ for the internal motion is

$$\hat{H} = -\frac{\hbar^2}{2\mu}\left(\frac{\partial^2}{\partial x^2} + \frac{\partial^2}{\partial y^2} + \frac{\partial^2}{\partial z^2}\right) - \frac{Ze^2}{4\pi\varepsilon_0 r} \tag{19.5}$$

where x, y, and z are the coordinates of the electron relative to the nucleus and $r = (x^2 + y^2 + z^2)^{1/2}$. The reduced mass μ is $\mu = m_1 m_2/(m_1 + m_2)$ [Eq. (18.79)], where m_1 and m_2 are the nuclear and electron masses. For an H atom, $m_{\text{nucleus}} = m_{\text{proton}} = 1836.15 m_e$ and

$$\mu_{\text{H}} = \frac{1836.15 m_e^2}{1837.15 m_e} = 0.999456 m_e \tag{19.6}$$

which differs only slightly from m_e.

The H-atom Schrödinger equation $\hat{H}\psi = E\psi$ is difficult to solve in cartesian coordinates but is relatively easy to solve in spherical coordinates. The spherical coordinates r, θ, and ϕ of the electron relative to the nucleus are defined in Fig. 19.1. (Math books usually interchange θ and ϕ.) The projection of r on the z axis is $r \cos\theta$, and its projection on the xy plane is $r \sin\theta$. The relation between cartesian and spherical coordinates is therefore

$$x = r\sin\theta\cos\phi, \qquad y = r\sin\theta\sin\phi, \qquad z = r\cos\theta \tag{19.7}$$

Note that $x^2 + y^2 + z^2 = r^2$. The ranges of the coordinates are

$$0 \le r \le \infty, \quad 0 \le \theta \le \pi, \quad 0 \le \phi \le 2\pi \tag{19.8}*$$

To solve the H-atom Schrödinger equation, one transforms the partial derivatives in the Hamiltonian operator (19.5) to derivatives with respect to r, θ, and ϕ and then uses the separation-of-variables procedure (Sec. 18.11). The details (which can be found in quantum chemistry texts) are omitted, and only an outline of the solution process will be given. The H-atom Schrödinger equation in spherical coordinates is found to be separable when the substitution

$$\psi = R(r)\,\Theta(\theta)\,\Phi(\phi)$$

is made, where R, Θ, and Φ are functions of r, θ, and ϕ, respectively. One obtains three separate differential equations, one for each coordinate.

The differential equation for $\Phi(\phi)$ is found to have solutions of the form $\Phi(\phi) = A e^{im\phi}$, where $i = \sqrt{-1}$, A is an integration constant whose value is chosen to

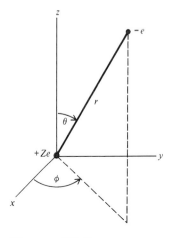

Figure 19.1

Coordinates of the electron relative to the nucleus in a hydrogenlike atom.

normalize Φ, and m (not to be confused with a mass) is a constant introduced in the process of separating the ϕ differential equation (recall the introduction of the separation constants E_x, E_y, and E_z in solving the problem of the particle in a three-dimensional box in Sec. 18.9). Since addition of 2π to the coordinate ϕ brings us back to the same point in space, the requirement that the wave function be single-valued (Sec. 18.7) means that we must have $\Phi(\phi) = \Phi(\phi + 2\pi)$. One finds (Prob. 19.20) that this equation is satisfied only if m is an integer (positive, negative, or zero).

The solution to the differential equation for $\Theta(\theta)$ is a complicated function of θ that involves a separation constant l and also the integer m that occurs in the ϕ equation. The $\Theta(\theta)$ solutions are not quadratically integrable except for values of l that satisfy $l = |m|$, $|m| + 1$, $|m| + 2$, . . . , where $|m|$ is the absolute value of the integer m. Thus l is an integer with minimum possible value 0, since this is the minimum possible value of $|m|$. The condition $l \geq |m|$ means that m ranges from $-l$ to $+l$ in steps of 1.

The differential equation for $R(r)$ for the H atom contains the energy E of internal motion as a parameter and also contains the quantum number l. The choice of zero level of energy is arbitrary. The potential energy $V = -Ze^2/4\pi\varepsilon_0 r$ in (19.5) takes the zero level to correspond to infinite separation of the electron and nucleus, which is an ionized atom. If the internal energy E is less than zero, the electron is bound to the nucleus. If the internal energy is positive, the electron has enough energy to escape the attraction of the nucleus and is free. One finds that for negative E, the function $R(r)$ is not quadratically integrable except for values of E that satisfy $E = -Z^2 e^4 \mu/(4\pi\varepsilon_0)^2 2n^2\hbar^2$, where n is an integer such that $n \geq l + 1$. Since the minimum l is zero, the minimum n is 1. Also, l cannot exceed $n - 1$. Further, one finds that all positive values of E are allowed. When the electron is free, its energy is continuous rather than quantized.

In summary, the hydrogenlike-atom wave functions have the form

$$\psi = R_{nl}(r)\Theta_{lm}(\theta)\Phi_m(\phi) \tag{19.9}$$

where the radial function $R_{nl}(r)$ is a function of r whose form depends on the quantum numbers n and l, the theta factor depends on l and m, and the phi factor is

$$\Phi_m(\phi) = (2\pi)^{-1/2} e^{im\phi}, \qquad i \equiv \sqrt{-1} \tag{19.10}$$

Since there are three variables, the solutions involve three quantum numbers: the **principal quantum number** n, the **angular-momentum quantum number** l, and the **magnetic quantum number** m (often symbolized by m_l). For ψ to be well behaved, the quantum numbers are restricted to the values

$$n = 1, 2, 3, \ldots \tag{19.11}*$$

$$l = 0, 1, 2, \ldots, n - 1 \tag{19.12}*$$

$$m = -l, -l + 1, \ldots, l - 1, l \tag{19.13}*$$

For example, for $n = 2$, l can be 0 or 1. For $l = 0$, m is 0. For $l = 1$, m can be $-1, 0$, or 1. The allowed bound-state energy levels are

$$E = -\frac{Z^2}{n^2} \frac{e^2}{(4\pi\varepsilon_0)2a} \qquad \text{where } a \equiv \frac{\hbar^2(4\pi\varepsilon_0)}{\mu e^2} \tag{19.14}$$

where $n = 1, 2, 3, \ldots$. Also, all values $E \geq 0$ are allowed, corresponding to an ionized atom. Figure 19.2 shows some of the allowed energy levels and the potential-energy function.

The following letter code is often used to specify the l value of an electron:

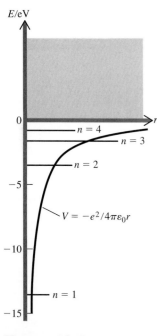

Figure 19.2

Energy levels and potential-energy function of the hydrogen atom. The shading indicates that all positive energies are allowed.

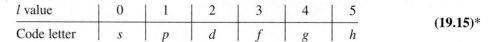

l value	0	1	2	3	4	5	
Code letter	s	p	d	f	g	h	**(19.15)***

The value of n is given as a prefix to the l code letter, and the m value is added as a subscript. Thus, $2s$ denotes the $n = 2$, $l = 0$ state; $2p_{-1}$ denotes the $n = 2$, $l = 1$, $m = -1$ state.

The hydrogenlike-atom energy levels (19.14) depend only on n, but the wave functions (19.9) depend on all three quantum numbers n, l, and m. Therefore, there is degeneracy. For example, the $n = 2$ H-atom level is fourfold degenerate (spin considerations omitted), the wave functions (states) $2s$, $2p_1$, $2p_0$, and $2p_{-1}$ all having the same energy.

The defined quantity a in (19.14) has the dimensions of length. For a hydrogen atom, substitution of numerical values gives (Prob. 19.15) $a = 0.5295$ Å.

If the reduced mass μ in the definition of a is replaced by the electron mass m_e, we get the **Bohr radius** a_0:

$$a_0 = \hbar^2(4\pi\varepsilon_0)/m_e e^2 = 0.5292 \text{ Å} \qquad (19.16)$$

a_0 was the radius of the $n = 1$ circle in the Bohr theory.

EXAMPLE 19.1 Ground-state energy of H

Calculate the ground-state hydrogen-atom energy E_{gs}. Also, express E_{gs} in electronvolts.

Setting $n = 1$ and $Z = 1$ in (19.14), we get

$$E_{gs} = -\frac{e^2}{(4\pi\varepsilon_0)2a} = -\frac{(1.6022 \times 10^{-19} \text{ C})^2}{4\pi(8.854 \times 10^{-12} \text{ C}^2 \text{ J}^{-1} \text{ m}^{-1})2(0.5295 \times 10^{-10} \text{ m})}$$

$$= -2.179 \times 10^{-18} \text{ J}$$

The use of the conversion factor (19.3) gives for an H atom

$$E_{gs} = -e^2/(4\pi\varepsilon_0)(2a) = -13.60 \text{ eV} \qquad (19.17)$$

Exercise

Find the wavelength of the longest-wavelength absorption line for a gas of ground-state hydrogen atoms. (*Answer*: 121.6 nm.)

$|E_{gs}|$ is the minimum energy needed to remove the electron from an H atom and is the **ionization energy** of H. The **ionization potential** of H is 13.60 V.

From (19.17), the energy levels (19.14) can be written as

$$E = -(Z^2/n^2)(13.60 \text{ eV}) \qquad \text{H-like atom} \qquad \textbf{(19.18)}\text{*}$$

Although the reduced mass μ in (19.14) differs for different hydrogenlike species (H, He$^+$, Li^{2+}, . . .), the differences are very slight and have been ignored in (19.18).

Quantum chemists often use a system called *atomic units*, in which energies are reported in *hartrees* and distances in *bohrs*. These quantities are defined as

$$1 \text{ bohr} \equiv a_0 = 0.5292 \text{ Å}, \qquad 1 \text{ hartree} \equiv e^2/(4\pi\varepsilon_0)a_0 = 27.211 \text{ eV}$$

The ground-state energy (19.17) of H would be $-\frac{1}{2}$ hartree if a were approximated by a_0.

The first few $R_{nl}(r)$ and $\Theta_{lm}(\theta)$ factors in the wave functions (19.9) are

$$R_{1s} = 2(Z/a)^{3/2}e^{-Zr/a}$$

$$R_{2s} = 2^{-1/2}(Z/a)^{3/2}(1 - Zr/2a)e^{-Zr/2a} \qquad (19.19a)$$

$$R_{2p} = (24)^{-1/2}(Z/a)^{5/2} r e^{-Zr/2a}$$

$$\Theta_{s_0} = 1/\sqrt{2}, \qquad \Theta_{p_0} = \tfrac{1}{2}\sqrt{6}\,\cos\theta, \qquad \Theta_{p_1} = \Theta_{p_{-1}} = \tfrac{1}{2}\sqrt{3}\,\sin\theta \qquad (19.19b)$$

where the code (19.15) was used for l. The general form of R_{nl} is $r^l e^{-Zr/na}$ times a polynomial of degree $n - l - 1$ in r:

$$R_{nl}(r) = r^l e^{-Zr/na}(b_0 + b_1 r + b_2 r^2 + \cdots + b_{n-l-1} r^{n-l-1})$$

where $b_0, b_1, \ldots$ are certain constants whose values depend on n and l. As n increases, $e^{-Zr/na}$ dies off more slowly as r increases, so the average radius $\langle r \rangle$ of the atom increases as n increases. For the ground state, one finds (Prob. 19.17) $\langle r \rangle = 3a/2Z$, which is 0.79 Å for H. In (19.19) and (19.10), e is the base of natural logarithms, and not the proton charge.

Figure 19.3 shows some plots of $R_{nl}(r)$. The radial factor in ψ has $n - l - 1$ nodes (not counting the node at the origin for $l \neq 0$).

For s states ($l = 0$), Eqs. (19.10) and (19.19) give the angular factor in ψ as $1/\sqrt{4\pi}$, which is independent of θ and ϕ. For s states, ψ depends only on r and is therefore said to be **spherically symmetric.** For $l \neq 0$, the angular factor is not constant, and ψ is not spherically symmetric. Note from Fig. 19.3 that R_{nl}, and hence ψ, is nonzero at the nucleus ($r = 0$) for s states.

The ground-state wave function is found by multiplying R_{1s} in (19.19) by the s-state angular factor $(4\pi)^{-1/2}$ to give

$$\psi_{1s} = \pi^{-1/2}(Z/a)^{3/2} e^{-Zr/a} \qquad (19.20)$$

Wave Functions of a Degenerate Energy Level

To deal with the $2p$ wave functions of the H atom, we need to use a quantum-mechanical theorem about wave functions of a degenerate level. By a **linear combination** of the functions $g_1, g_2, \ldots, g_k$, one means a function of the form $c_1 g_1 + c_2 g_2 + \cdots + c_k g_k$, where the c's are constants. Any linear combination of two or more stationary-state wave functions that belong to the same degenerate energy level is an

Figure 19.3

Radial factors in some hydrogen-atom wave functions.

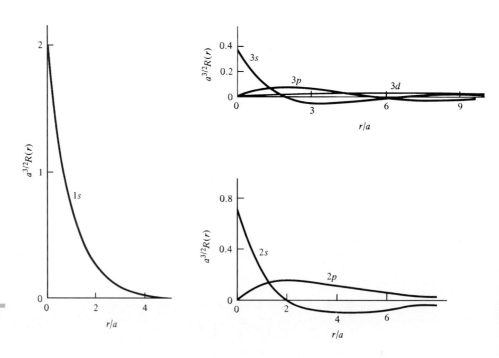

eigenfunction of the Hamiltonian operator with the same energy value as that of the degenerate level. In other words, if $\hat{H}\psi_1 = E_1\psi_1$ and $\hat{H}\psi_2 = E_1\psi_2$, then $\hat{H}(c_1\psi_1 + c_2\psi_2) = E_1(c_1\psi_1 + c_2\psi_2)$. The linear combination $c_1\psi_1 + c_2\psi_2$ (when multiplied by a normalization constant) is therefore also a valid wave function, meaning that it is an eigenfunction of $\hat{H}$ and therefore a solution of the Schrödinger equation. The proof follows from the fact that $\hat{H}$ is a linear operator (Sec. 18.11).

$$\hat{H}(c_1\psi_1 + c_2\psi_2) = \hat{H}(c_1\psi_1) + \hat{H}(c_2\psi_2) = c_1\hat{H}\psi_1 + c_2\hat{H}\psi_2$$
$$= c_1E_1\psi_1 + c_2E_1\psi_2 = E_1(c_1\psi_1 + c_2\psi_2)$$

Note that this theorem does not apply to wave functions belonging to two different energy levels. If $\hat{H}\psi_5 = E_5\psi_5$ and $\hat{H}\psi_6 = E_6\psi_6$ with $E_5 \neq E_6$, then $c_1\psi_5 + c_2\psi_6$ is not an eigenfunction of $\hat{H}$.

Real Wave Functions

The Φ factor (19.10) in the H-atom wave function (19.9) contains i and so is complex. Chemists often find it convenient to work with real wave functions instead. To get real functions, we use the theorem just stated.

From (19.9), (19.10), and (19.19), the complex $2p$ functions are

$$2p_{+1} = be^{-Zr/2a}r \sin \theta \, e^{i\phi}, \qquad 2p_{-1} = be^{-Zr/2a}r \sin \theta \, e^{-i\phi}$$

$$2p_0 = ce^{-Zr/2a}r \cos \theta$$

where $b \equiv (1/8\pi^{1/2})(Z/a)^{5/2}$ and $c \equiv \pi^{-1/2}(Z/2a)^{5/2}$. The $2p_0$ function is real as it stands. Equation (19.7) gives $r \cos \theta = z$, so the $2p_0$ function is also written as

$$2p_z \equiv 2p_0 = cze^{-Zr/2a}$$

(Don't confuse the nuclear charge Z with the z spatial coordinate.) The $2p_{+1}$ and $2p_{-1}$ functions are each eigenfunctions of $\hat{H}$ with the same energy eigenvalue, so we can take any linear combination of them and have a valid wave function.

As a preliminary, we note that

$$e^{i\phi} = \cos \phi + i \sin \phi \qquad \textbf{(19.21)}*$$

and $e^{-i\phi} = (e^{i\phi})^* = \cos \phi - i \sin \phi$. For a proof of (19.21), see Prob. 19.18.

We define the linear combinations $2p_x$ and $2p_y$ as

$$2p_x \equiv (2p_1 + 2p_{-1})/\sqrt{2}, \qquad 2p_y \equiv (2p_1 - 2p_{-1})/i\sqrt{2} \qquad (19.22)$$

The $1/\sqrt{2}$ factors normalize these functions. Using (19.21) and its complex conjugate, we find (Prob. 19.19) that

$$2p_x = cxe^{-Zr/2a}, \qquad 2p_y = cye^{-Zr/2a} \qquad (19.23)$$

where $c \equiv \pi^{-1/2}(Z/2a)^{5/2}$. The $2p_x$ and $2p_y$ functions have the same n and l values as the $2p_1$ and $2p_{-1}$ functions (namely, $n = 2$ and $l = 1$) but do not have a definite value of m. Similar linear combinations give real wave functions for higher H-atom states. The real functions (which have directional properties) are more suitable than the complex functions for use in treating the bonding of atoms to form molecules. Table 19.1 lists the $n = 1$ and $n = 2$ real hydrogenlike functions.

Orbitals

An **orbital** is a one-electron spatial wave function. Since a hydrogenlike atom has one electron, all the hydrogenlike wave functions are orbitals. The use of (one-electron) orbitals in many-electron atoms is considered later in this chapter.

TABLE 19.1

Real Hydrogenlike Wave Functions for $n = 1$ and $n = 2$

$1s = \pi^{-1/2}(Z/a)^{3/2}e^{-Zr/a}$

$2s = \frac{1}{4}(2\pi)^{-1/2}(Z/a)^{3/2}(2 - Zr/a)e^{-Zr/2a}$

$2p_x = \frac{1}{4}(2\pi)^{-1/2}(Z/a)^{5/2}re^{-Zr/2a}\sin\theta\,\cos\phi$

$2p_y = \frac{1}{4}(2\pi)^{-1/2}(Z/a)^{5/2}re^{-Zr/2a}\sin\theta\,\sin\phi$

$2p_z = \frac{1}{4}(2\pi)^{-1/2}(Z/a)^{5/2}re^{-Zr/2a}\cos\theta$

The **shape** of an orbital is defined as a surface of constant probability density that encloses some large fraction (say 90%) of the probability of finding the electron. The probability density is $|\psi|^2$. When $|\psi|^2$ is constant, so is $|\psi|$. Hence $|\psi|$ *is constant on the surface of an orbital.*

For an *s* orbital, ψ depends only on r, and $|\psi|$ is constant on the surface of a sphere with center at the nucleus. An *s* orbital has a spherical shape.

The volume element in spherical coordinates (see any calculus text) is

$$d\tau = r^2 \sin\theta\,dr\,d\theta\,d\phi \qquad (19.24)*$$

This is the volume of an infinitesimal solid for which the spherical coordinates lie in the ranges r to $r + dr$, θ to $\theta + d\theta$, and ϕ to $\phi + d\phi$.

EXAMPLE 19.2 $1s$ orbital radius

Find the radius of the $1s$ orbital in H using the 90 percent probability definition.

The probability that a particle will be in a given region is found by integrating the probability density $|\psi|^2$ over the volume of the region. The region being considered here is a sphere of radius r_{1s}. For this region, θ and ϕ go over their full ranges 0 to π and 0 to 2π, respectively, and r goes from 0 to r_{1s}. Also, $\psi_{1s} = \pi^{-1/2}(Z/a)^{3/2}e^{-Zr/a}$ (Table 19.1). Using (19.24) for $d\tau$, we have as the probability that the electron is within distance r_{1s} from the nucleus:

$$0.90 = \int_0^{2\pi}\int_0^{\pi}\int_0^{r_{1s}} \pi^{-1}\left(\frac{Z}{a}\right)^3 e^{-2Zr/a}r^2\sin\theta\,dr\,d\theta\,d\phi$$

$$0.90 = \frac{Z^3}{\pi a^3}\int_0^{2\pi}d\phi\int_0^{\pi}\sin\theta\,d\theta\int_0^{r_{1s}}e^{-2Zr/a}r^2\,dr$$

where the integral identity of Prob. 18.14 was used. One next evaluates the integrals and uses trial and error or the Excel Solver to find the value of r_{1s} that satisfies this equation with $Z = 1$. The remaining work is left as an exercise. One finds $r_{1s} = 1.4$ Å for H.

Exercise

Evaluate the integrals in this example (use a table of integrals for the r integral) and show that the result for $Z = 1$ is $e^{-2w}(2w^2 + 2w + 1) - 0.1 = 0$, where $w \equiv r_{1s}/a$. Solve this equation to show that $r_{1s} = 1.41$ Å.

Consider the shapes of the real $2p$ orbitals. The $2p_z$ orbital is $2p_z = cze^{-Zr/2a}$, where c is a constant. The $2p_z$ function is zero in the xy plane (where $z = 0$), is positive above this nodal plane (where z is positive), and is negative below this plane. A detailed investigation (Prob. 19.71) gives the curves shown in Fig. 19.4 as the contours of constant $|\psi_{2p_z}|$ in the yz plane. The curves shown are for $|\psi/\psi_{max}| = 0.9$ (the two innermost ovals), 0.7, 0.5, 0.3, and 0.1, where ψ_{max} is the maximum value of ψ_{2p_z}. The three-dimensional shape of the $2p_z$ orbital is obtained by rotating a cross section around the z axis. This gives two distorted ellipsoids, one above and one below the xy plane. The ellipsoids do not touch each other. This is obvious from the fact that ψ has opposite signs on each ellipsoid. The absolute value $|\psi|$ is the same on each ellipsoid of the $2p_z$ orbital. The $2p_x$, $2p_y$, and $2p_z$ orbitals have the same shape but different orientations in space. The two distorted ellipsoids are located on the x axis for the $2p_x$ orbital, on the y axis for the $2p_y$ orbital, and on the z axis for the $2p_z$ orbital.

The $2p_z$ wave function is a function of the three spatial coordinates: $\psi_{2p_z} = \psi_{2p_z}(x, y, z)$. Just as two dimensions are needed to graph a function of one variable, it would require four dimensions to graph $\psi_{2p_z}(x, y, z)$. Figure 19.5 shows a three-dimensional graph of $\psi_{2p_z}(0, y, z)$. In this graph, the value of ψ_{2p_z} at each point in the yz plane is given by the height of the graph above this plane. Note the resemblance to Fig. 18.13 for the $n_x = 1$, $n_y = 2$ particle-in-a-two-dimensional-box state.

Figure 19.6 shows some hydrogen-atom orbital shapes. The plus and minus signs in Fig. 19.6 give the algebraic signs of ψ and have nothing to do with electric charge. The $3p_z$ orbital has a spherical node (shown by the dashed line in Fig. 19.6). The $3d_{z^2}$ orbital has two nodal cones (dashed lines). The other four $3d$ orbitals have the same shape as one another but different orientations; each of these orbitals has two nodal planes separating the four lobes.

Probability Density

The electron probability density for the hydrogenlike-atom ground state (Table 19.1) is $|\psi_{1s}|^2 = (Z^3/\pi a^3)e^{-2Zr/a}$. The $1s$ probability density is a maximum at the nucleus ($r = 0$). Figure 19.7 is a schematic indication of this, the density of the dots indicating

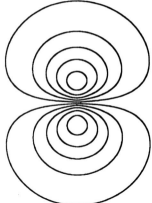

Figure 19.4

Contours of the $2p_z$ orbital in the yz plane. The z axis is vertical.

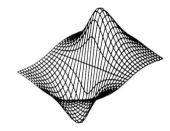

Figure 19.5

Three-dimensional graph of values of ψ_{2p_z} in the yz plane.

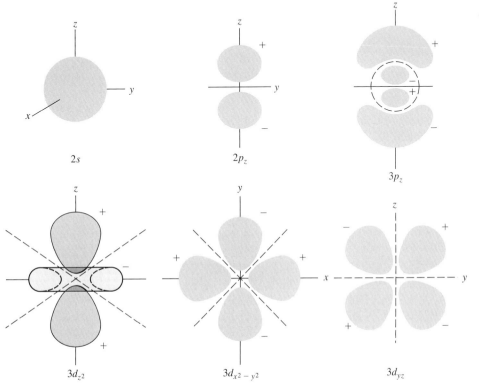

2s

$2p_z$

$3p_z$

$3d_{z^2}$

$3d_{x^2-y^2}$

$3d_{yz}$

Figure 19.6

Shapes of some hydrogen-atom orbitals. (Not drawn to scale.) Note the different orientation of the axes in the $3d_{x^2-y^2}$ figure compared with the others. Not shown are the $3d_{xy}$ and $3d_{xz}$ orbitals; these have their lobes between the x and y axes and between the x and z axes, respectively.

Figure 19.7

Probability densities in three
hydrogen-atom states. (Not drawn
to scale.)

 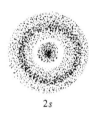 

$1s$ $2s$ $2p_z$

the relative probability densities in various regions. Since $|\psi_{1s}|^2$ is nonzero every-where, the electron can be found at any location in the atom (in contrast to the Bohr theory, where it had to be at a fixed r). Figure 19.7 also indicates the variation in probability density for the $2s$ and $2p_z$ states. Note the nodal sphere in the $2s$ function.

Radial Distribution Function

Suppose we want the probability $\Pr(r \rightarrow r + dr)$ that the electron–nucleus distance is between r and $r + dr$. This is the probability of finding the electron in a thin spherical shell whose center is at the nucleus and whose inner and outer radii are r and $r + dr$. For an s orbital, ψ is independent of θ and ϕ and so is essentially constant in the thin shell. Hence, the desired probability is found by multiplying $|\psi_s|^2$ (the probability per unit volume) by the volume of the thin shell. This volume is $\frac{4}{3}\pi(r + dr)^3 - \frac{4}{3}\pi r^3 = 4\pi r^2\,dr$, where the terms in $(dr)^2$ and $(dr)^3$ are negligible. Therefore, for an s state $\Pr(r \rightarrow r + dr) = 4\pi r^2|\psi_s|^2\,dr$.

For a non-s state, ψ depends on the angles, so $|\psi|^2$ is not constant in the thin shell. Let us divide the shell into tiny volume elements such that the spherical coordinates range from r to $r + dr$, from θ to $\theta + d\theta$, and from ϕ to $\phi + d\phi$ in each tiny element. The volume $d\tau$ of each such element is given by (19.24), and the probability that the electron is in an element is $|\psi|^2\,d\tau = |\psi|^2 r^2 \sin\theta\,dr\,d\theta\,d\phi$. To find $\Pr(r \rightarrow r + dr)$, we must sum these infinitesimal probabilities over the thin shell. Since the shell goes over the full range of θ and ϕ, the desired sum is the definite integral over the angles. Hence, $\Pr(r \rightarrow r + dr) = \int_0^{2\pi} \int_0^{\pi} |\psi|^2 r^2 \sin\theta\,dr\,d\theta\,d\phi$. The use of $\psi = R\Theta\Phi$ [Eq. (19.9)] and a result similar to that in Prob. 18.14 gives

$$\Pr(r \rightarrow r + dr) = |R|^2 r^2\,dr \int_0^{\pi} |\Theta|^2 \sin\theta\,d\theta \int_0^{2\pi} |\Phi|^2\,d\phi$$

$$\Pr(r \rightarrow r + dr) = [R_{nl}(r)]^2 r^2\,dr \qquad (19.25)$$

since the multiplicative constants in the Θ and Φ functions have been chosen to nor-malize Θ and Φ; $\int_0^{\pi} |\Theta|^2 \sin\theta\,d\theta = 1$ and $\int_0^{2\pi} |\Phi|^2\,d\phi = 1$ (Prob. 19.28). Equation (19.25) holds for both s and non-s states. The function $[R(r)]^2 r^2$ in (19.25) is the **radial distribution function** and is plotted in Fig. 19.8 for several states. For the ground state, the radial distribution function is a maximum at $r = a/Z$ (Prob. 19.26), which is 0.53 Å for H.

For the hydrogen-atom ground state, the probability density $|\psi|^2$ is a maximum at the origin (nucleus), but the radial distribution function $R^2 r^2$ is zero at the nucleus because of the r^2 factor; the most probable value of r is 0.53 Å. A little thought shows that these facts are not contradictory. In finding $\Pr(r \rightarrow r + dr)$, we find the probability that the electron is in a thin shell. This thin shell ranges over all values of θ and ϕ and so is composed of many volume elements. As r increases, the thin-shell volume $4\pi r^2\,dr$ increases. This increase, combined with the decrease in the probability density $|\psi|^2$ as r increases, gives a maximum in $\Pr(r \rightarrow r + dr)$ for a value of r between 0 and ∞. The radial distribution function is zero at the nucleus because the thin-shell volume $4\pi r^2\,dr$ is zero here. (Note the resemblance to the discussion of the distribution function for speeds in a gas; Sec. 15.4.)

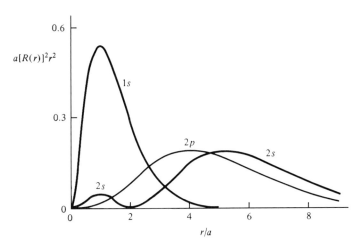

Figure 19.8

Radial distribution functions for
some hydrogen-atom states.

Average Values

To find the average value of any property M of a stationary-state hydrogen atom, one uses $\langle M \rangle = \int \psi^* \hat{M} \psi \, d\tau$, Eq. (18.63).

EXAMPLE 19.3 Finding $\langle r \rangle$

Find the average value of the electron–nucleus separation in a hydrogenlike atom in the $2p_z$ state.

We have $\langle r \rangle = \int \psi^* \hat{r} \psi \, d\tau$. The $2p_z$ wave function is given in Table 19.1 and is real, so $\psi^* = \psi$. The operator $\hat{r}$ is multiplication by r. Thus, $\psi^* \hat{r} \psi = \psi^2 r$. The volume element is $d\tau = r^2 \sin \theta \, dr \, d\theta \, d\phi$ [Eq. (19.24)], and (19.8) gives the coordinate limits. Therefore

$$\langle r \rangle = \int \psi^* \hat{r} \psi \, d\tau$$

$$\langle r \rangle = \frac{1}{16(2\pi)} \left(\frac{Z}{a} \right)^5 \int_0^{2\pi} \int_0^{\pi} \int_0^{\infty} r^2 e^{-Zr/a} \cos^2 \theta \, (r) r^2 \sin \theta \, dr \, d\theta \, d\phi$$

$$= \frac{1}{32\pi} \left(\frac{Z}{a} \right)^5 \int_0^{2\pi} d\phi \int_0^{\pi} \cos^2 \theta \sin \theta \, d\theta \int_0^{\infty} r^5 e^{-Zr/a} \, dr$$

where the integral identity of Prob. 18.14 was used. A table of definite integrals gives $\int_0^{\infty} x^n e^{-bx} \, dx = n!/b^{n+1}$ for $b > 0$ and n a positive integer. Evaluation of the integrals (Prob. 19.25) gives $\langle r \rangle = 5a/Z$, where a is defined by (19.14) and equals 0.53 Å. As a partial check, note that $5a/Z$ has units of length.

Exercise

Find $\langle z^2 \rangle$ for the $2p_z$ H-atom state. (*Answer:* $18a^2 = 5.05$ Å^2.)

19.4   ANGULAR MOMENTUM

The H-atom quantum numbers l and m are related to the angular momentum of the electron. The (linear) momentum **p** of a particle of mass m and velocity **v** is defined classically by $\mathbf{p} \equiv m\mathbf{v}$. Don't confuse the mass m with the m quantum number. Let **r**

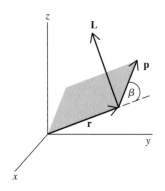

Figure 19.9

The angular-momentum vector **L** of a particle is perpendicular to the vectors **r** and **p** and has magnitude $rp \sin \beta$.

be the vector from the origin of a coordinate system to the particle. The particle's **angular momentum L** with respect to the coordinate origin is defined classically as a vector of length $rp \sin \beta$ (where β is the angle between **r** and **p**) and direction perpendicular to both **r** and **p**; see Fig. 19.9. More concisely, $\mathbf{L} \equiv \mathbf{r} \times \mathbf{p}$, where $\times$ indicates the vector cross-product.

To deal with angular momentum in quantum mechanics, one uses the quantum-mechanical operators for the components of the **L** vector. We shall omit the quantum-mechanical treatment [see *Levine* (2000), chaps. 5 and 6] and simply state the results. There are two kinds of angular momentum in quantum mechanics. **Orbital angular momentum** is the quantum-mechanical analog of the classical quantity **L** and is due to the motion of a particle through space. In addition to orbital angular momentum, many particles have an intrinsic angular momentum called *spin* angular momentum; this will be discussed in the next section.

The H-atom stationary-state wave functions $\psi_{nlm} = R_{nl}(r)\Theta_{lm}(\theta)\Phi_m(\phi)$ [Eq. (19.9)] are eigenfunctions of the energy operator $\hat{H}$ with eigenvalues given by Eq. (19.14); $\hat{H}\psi_{nlm} = E_n\psi_{nlm}$ [Eq. (18.61)], where $E_n = -(Z^2/n^2)(e^2/8\pi\varepsilon_0 a)$. This means that a measurement of the energy of an H atom in the state ψ_{nlm} must give the result E_n. One can show that the H-atom functions ψ_{nlm} are also eigenfunctions of the angular-momentum operators $\hat{L}^2$ and $\hat{L}_z$, where $\hat{L}^2$ is the operator for the square of the magnitude of the electron's orbital angular momentum **L** with respect to the nucleus, and $\hat{L}_z$ is the operator for the z component of **L**. The eigenvalues are $l(l+1)\hbar^2$ for $\hat{L}^2$ and $m\hbar$ for $\hat{L}_z$:

$$\hat{L}^2\psi_{nlm} = l(l+1)\hbar^2\psi_{nlm}, \qquad \hat{L}_z\psi_{nlm} = m\hbar\psi_{nlm} \qquad (19.26)$$

where the quantum numbers l and m are given by (19.12) and (19.13). These eigenvalue equations mean that the magnitude $|\mathbf{L}|$ and the z component L_z of the electron's orbital angular momentum in the H-atom state ψ_{nlm} are

$$|\mathbf{L}| = \sqrt{l(l+1)}\,\hbar, \qquad L_z = m\hbar \qquad (19.27)^*$$

For s states ($l = 0$), the electronic orbital angular momentum is zero (a result quite difficult to understand classically). For p states ($l = 1$), the magnitude of **L** is $\sqrt{2}\hbar$, and L_z can be $\hbar$, 0, or $-\hbar$. The possible orientations of **L** for $l = 1$ and $m = 1$, 0, and -1 are shown in Fig. 19.10. The quantum number l specifies the magnitude $|\mathbf{L}|$ of **L,** and m specifies the z component L_z of **L.** When $|\mathbf{L}|$ and L_z are specified in a quantum-mechanical system, it turns out that L_x and L_y cannot be specified, so **L** can lie anywhere on the surface of a cone about the z axis. For $m = 0$, the cone becomes a circle in the xy plane.

Figure 19.10

Allowed spatial orientations of the electronic orbital-angular-momentum vector **L** for $l = 1$ and $m = -1$, 0, and 1.

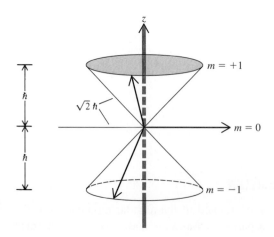

When an external magnetic field is applied to a hydrogen atom, the energies of the states depend on the m quantum number as well as on n.

The H-atom quantum numbers l and m are analogous to the two-particle-rigid-rotor quantum numbers J and M_J (Sec. 18.14). The functions Θ and Φ in the two-particle-rigid-rotor wave functions ψ_{rot} are the same functions as Θ and Φ in the H-atom wave functions (19.9).

The H-atom wave functions ψ_{nlm} are simultaneously eigenfunctions of the H-atom Hamiltonian operator $\hat{H}$ and of the angular-momentum operators $\hat{L}^2$ and $\hat{L}_z$ [Eq. (19.26)]. A theorem of quantum mechanics shows this is possible because the operators $\hat{H}$, $\hat{L}^2$, and $\hat{L}_z$ all **commute** with one another, meaning that the commutators (Example 18.4) $[\hat{H}, \hat{L}^2]$, $[\hat{H}, \hat{L}_z]$, and $[\hat{L}^2, \hat{L}_z]$ are all equal to zero. However, one finds that $[\hat{L}_z, \hat{L}_x] \neq 0$ and $[\hat{L}_z, \hat{L}_y] \neq 0$. Hence the ψ_{nlm} functions are not eigenfunctions of $\hat{L}_x$ and $\hat{L}_y$, and the quantities L_x and L_y cannot be specified for the states ψ_{nlm}. The $l = 0$ states are an exception. When $l = 0$, the orbital-angular-momentum magnitude $|\mathbf{L}|$ in (19.27) is zero and every component L_x, L_y, and L_z has the definite value of zero.

19.5 ELECTRON SPIN

The Schrödinger equation is a nonrelativistic equation and fails to account for certain relativistic phenomena. In 1928, the British physicist P. A. M. Dirac discovered the correct relativistic quantum-mechanical equation for a one-electron system. Dirac's relativistic equation predicts the existence of electron spin. Electron spin was first proposed by Uhlenbeck and Goudsmit in 1925 to explain certain observations in atomic spectra. In the nonrelativistic Schrödinger version of quantum mechanics that we are using, the existence of electron spin must be added to the theory as an additional postulate.

What is spin? **Spin** is an intrinsic (built-in) angular momentum possessed by elementary particles. This intrinsic angular momentum is in addition to the orbital angular momentum (Sec. 19.4) the particle has as a result of its motion through space. In a crude way, one can think of this intrinsic (or spin) angular momentum as being due to the particle's spinning about its own axis, but this picture should not be considered to represent reality. Spin is a nonclassical effect.

Quantum mechanics shows that the magnitude of the orbital angular momentum $\mathbf{L}$ of any particle can take on only the values $[l(l + 1)]^{1/2}\hbar$, where $l = 0, 1, 2, \ldots$; the z component L_z can take on only the values $m\hbar$, where $m = -l, \ldots, +l$. We mentioned this for the electron in the H atom, Eq. (19.27).

Let $\mathbf{S}$ be the spin-angular-momentum vector of an elementary particle. By analogy to orbital angular momentum [Eqs. (19.13) and (19.27)], we postulate that the magnitude of $\mathbf{S}$ is

$$|\mathbf{S}| = [s(s + 1)]^{1/2}\hbar \qquad \textbf{(19.28)}*$$

and that S_z, the component of the spin angular momentum along the z axis, can take on only the values

$$S_z = m_s\hbar \qquad \text{where } m_s = -s, -s + 1, \ldots, s - 1, s \qquad \textbf{(19.29)}*$$

The spin-angular-momentum quantum numbers s and m_s are analogous to the orbital-angular-momentum quantum numbers l and m, respectively. The analogy is not complete, since one finds that a given species of elementary particle can have only one value for s and this value may be half-integral $(\frac{1}{2}, \frac{3}{2}, \ldots)$ as well as integral $(0, 1, \ldots)$. Experiment shows that electrons, protons, and neutrons all have $s = \frac{1}{2}$. Therefore, $m_s = -\frac{1}{2}$ or $+\frac{1}{2}$ for these particles.

$$s = \tfrac{1}{2}, \qquad m_s = +\tfrac{1}{2}, -\tfrac{1}{2} \qquad \text{for an electron} \qquad \textbf{(19.30)}*$$

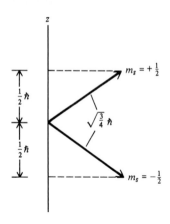

Figure 19.11

Orientations of the electron spin vector **S** with respect to the z axis. For $m_s = +\frac{1}{2}$, the vector **S** must lie on the surface of a cone about the z axis; similarly for $m_s = -\frac{1}{2}$.

With $s = \frac{1}{2}$, the magnitude of the electron spin-angular-momentum vector is $|\mathbf{S}| = [s(s + 1)]^{1/2}\hbar = (3/4)^{1/2}\hbar$ and the possible values of S_z are $\frac{1}{2}\hbar$ and $-\frac{1}{2}\hbar$. Figure 19.11 shows the orientations of **S** for these two spin states. Chemists often use the symbols $\uparrow$ and $\downarrow$ to indicate the $m_s = +\frac{1}{2}$ and $m_s = -\frac{1}{2}$ states, respectively.

> Photons have $s = 1$. However, because photons are relativistic entities traveling at speed c, it turns out that they don't obey (19.29). Instead, photons can have only $m_s = +1$ or $m_s = -1$. These two m_s values correspond to left- and right-circularly polarized light.

The wave function is supposed to describe the state of the system as fully as possible. An electron has two possible spin states, namely, $m_s = +\frac{1}{2}$ and $m_s = -\frac{1}{2}$, and the wave function should indicate which spin state the electron is in. We therefore postulate the existence of two spin functions α and β that indicate the electron's spin state: α means that m_s is $+\frac{1}{2}$; β means that m_s is $-\frac{1}{2}$. The spin functions α and β can be considered to be functions of some hypothetical internal coordinate ω (omega) of the electron: $\alpha = \alpha(\omega)$ and $\beta = \beta(\omega)$. Since nothing is known of the internal structure of an electron (or even whether it has an internal structure), ω is purely hypothetical.

Since the spin function α has $m_s = \frac{1}{2}$ and β has $m_s = -\frac{1}{2}$ and both have $s = \frac{1}{2}$, by analogy to (19.26), we write

$$\hat{S}^2\alpha = \tfrac{3}{4}\hbar^2\alpha, \qquad \hat{S}^2\beta = \tfrac{3}{4}\hbar^2\beta, \qquad \hat{S}_z\alpha = \tfrac{1}{2}\hbar\alpha, \qquad \hat{S}_z\beta = -\tfrac{1}{2}\hbar\beta \quad (19.31)$$

where these equations are purely symbolic, in that we have not specified forms for the spin functions α and β or for the operators $\hat{S}^2$ and $\hat{S}_z$.

For a one-electron system, the spatial wave function $\psi(x, y, z)$ is multiplied by either α or β to form the complete wave function including spin. To a very good approximation, the spin has no effect on the energy of a one-electron system. For the hydrogen atom, the electron spin simply doubles the degeneracy of each level. For the H-atom ground level, there are two possible wave functions, $1s\alpha$ and $1s\beta$, where $1s = \pi^{-1/2}(Z/a)^{3/2}e^{-Zr/a}$. A one-electron wave function like $1s\alpha$ or $1s\beta$ that includes both spatial and spin functions is called a **spin-orbital.**

With inclusion of electron spin in the wave function, ψ of an n-electron system becomes a function of $4n$ variables: $3n$ spatial coordinates and n **spin variables** or **spin coordinates.** The normalization condition (18.17) must be modified to include an integration or summation over the spin variables as well as an integration over the spatial coordinates. If one uses the hypothetical spin coordinate ω, one integrates over ω. A common alternative is to take the spin quantum number m_s of each electron as being the spin variable of that electron. In this case, one sums over the two possible m_s values of each electron in the normalization equation. Such sums or integrals over the spin variables are, like (19.31), purely symbolic.

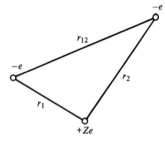

Figure 19.12

Interparticle distances in the heliumlike atom.

19.6   THE HELIUM ATOM AND THE PAULI PRINCIPLE

The Helium Atom

The helium atom consists of two electrons and a nucleus (Fig. 19.12). Separation of the translational energy of the atom as a whole from the internal motion is more complicated than for a two-particle problem and won't be gone into here. We shall just assume that it is possible to separate the translational motion from the internal motions.

The Hamiltonian operator for the internal motions in a heliumlike atom is

$$\hat{H} = -\frac{\hbar^2}{2m_e}\nabla_1^2 - \frac{\hbar^2}{2m_e}\nabla_2^2 - \frac{Ze^2}{4\pi\varepsilon_0 r_1} - \frac{Ze^2}{4\pi\varepsilon_0 r_2} + \frac{e^2}{4\pi\varepsilon_0 r_{12}} \qquad (19.32)$$

The first term is the operator for the kinetic energy of electron 1. In this term, $\nabla_1^2 \equiv \partial^2/\partial x_1^2 + \partial^2/\partial y_1^2 + \partial^2/\partial z_1^2$ (where x_1, y_1, and z_1 are the coordinates of electron 1, the origin being taken at the nucleus) and m_e is the electron mass. It would be more accurate to replace m_e by the reduced mass μ, but μ differs almost negligibly from m_e for He and heavier atoms. The second term is the operator for the kinetic energy of electron 2, and $\nabla_2^2 \equiv \partial^2/\partial x_2^2 + \partial^2/\partial y_2^2 + \partial^2/\partial z_2^2$. The third term is the potential energy of interaction between electron 1 and the nucleus and is obtained by putting $Q_1 = -e$ and $Q_2 = Ze$ in $V = Q_1Q_2/4\pi\varepsilon_0 r$ [Eq. (19.1)]. For helium, the atomic number Z is 2. In the third term, r_1 is the distance between electron 1 and the nucleus: $r_1^2 = x_1^2 + y_1^2 + z_1^2$. The fourth term is the potential energy of interaction between electron 2 and the nucleus. The last term is the potential energy of interaction between electrons 1 and 2 separated by distance r_{12} and is found by putting $Q_1 = Q_2 = -e$ in $V = Q_1Q_2/4\pi\varepsilon_0 r$. There is no term for kinetic energy of the nucleus, because we are considering only the internal motion of the electrons relative to the nucleus.

The Schrödinger equation is $\hat{H}\psi = E\psi$, where ψ is a function of the spatial coordinates of the electrons relative to the nucleus: $\psi = \psi(x_1, y_1, z_1, x_2, y_2, z_2)$, or $\psi = \psi(r_1, \theta_1, \phi_1, r_2, \theta_2, \phi_2)$ if spherical coordinates are used. Electron spin is being ignored for now and will be taken care of later.

Because of the interelectronic repulsion term $e^2/4\pi\varepsilon_0 r_{12}$, the helium-atom Schrödinger equation can't be solved exactly. As a crude approximation, we can ignore the $e^2/4\pi\varepsilon_0 r_{12}$ term. The Hamiltonian (19.32) then has the approximate form $\hat{H}_{approx} = \hat{H}_1 + \hat{H}_2$, where $\hat{H}_1 \equiv -(\hbar^2/2m_e)\nabla_1^2 - Ze^2/4\pi\varepsilon_0 r_1$ is a hydrogenlike Hamiltonian for electron 1 and $\hat{H}_2 \equiv -(\hbar^2/2m_e)\nabla_2^2 - Ze^2/4\pi\varepsilon_0 r_2$ is a hydrogenlike Hamiltonian for electron 2. Since $\hat{H}_{approx}$ is the sum of Hamiltonians for two noninteracting particles, the approximate energy is the sum of energies of each particle and the approximate wave function is the product of wave functions for each particle [Eqs. (18.68) to (18.70)]:

$$E \approx E_1 + E_2 \qquad \text{and} \qquad \psi \approx \psi_1(r_1, \theta_1, \phi_1)\psi_2(r_2, \theta_2, \phi_2) \qquad (19.33)$$

where $\hat{H}_1\psi_1 = E_1\psi_1$ and $\hat{H}_2\psi_2 = E_2\psi_2$. Since $\hat{H}_1$ and $\hat{H}_2$ are hydrogenlike Hamiltonians, E_1 and E_2 are hydrogenlike energies and ψ_1 and ψ_2 are hydrogenlike wave functions (orbitals).

Let us check the accuracy of this approximation. Equations (19.14) and (19.18) give $E_1 = -(Z^2/n_1^2)(e^2/8\pi\varepsilon_0 a) = -(Z^2/n_1^2)(13.6 \text{ eV})$, where n_1 is the principal quantum number of electron 1 and a has been replaced by the Bohr radius a_0, since the reduced mass μ was replaced by the electron mass in (19.32). A similar equation holds for E_2. For the helium-atom ground state, the principal quantum numbers of the electrons are $n_1 = 1$ and $n_2 = 1$; also, $Z = 2$. Hence,

$$E \approx E_1 + E_2 = -4(13.6 \text{ eV}) - 4(13.6 \text{ eV}) = -108.8 \text{ eV}$$

The experimental first and second ionization energies of He are 24.6 and 54.4 eV, so the true ground-state energy is -79.0 eV. (The first and second ionization energies are the energy changes for the processes He $\rightarrow$ He$^+$ + e$^-$ and He$^+$ $\rightarrow$ He^{2+} + e$^-$, respectively.) The approximate result -108.8 eV is grossly in error, as might be expected from the fact that the $e^2/4\pi\varepsilon_0 r_{12}$ term we ignored is not small.

The approximate ground-state wave function is given by Eq. (18.68) and Table 19.1 as

$$\psi \approx (Z/a_0)^{3/2}\pi^{-1/2}e^{-Zr_1/a_0} \cdot (Z/a_0)^{3/2}\pi^{-1/2}e^{-Zr_2/a_0} \qquad (19.34)$$

with $Z = 2$. We shall abbreviate (19.34) as

$$\psi \approx 1s(1)1s(2) \qquad (19.35)$$

where $1s(1)$ indicates that electron 1 is in a $1s$ hydrogenlike orbital (one-electron spatial wave function). We have the familiar He ground-state configuration $1s^2$.

Two-Electron Spin Functions

To be fully correct, electron spin must be included in the wave function. One's first impulse might be to write down the following four spin functions for two-electron systems:

$$\alpha(1)\alpha(2), \qquad \beta(1)\beta(2), \qquad \alpha(1)\beta(2), \qquad \beta(1)\alpha(2) \qquad (19.36)$$

where the notation $\beta(1)\alpha(2)$ means electron 1 has its spin quantum number m_{s1} equal to $-\frac{1}{2}$ and electron 2 has $m_{s2} = +\frac{1}{2}$. However, the last two functions in (19.36) are not valid spin functions because they distinguish between the electrons. Electrons are identical to one another, and there is no way of experimentally determining which electron has $m_s = +\frac{1}{2}$ and which has $m_s = -\frac{1}{2}$. In classical mechanics, we can distinguish two identical particles from each other by following their paths. However, the Heisenberg uncertainty principle makes it impossible to follow the path of a particle in quantum mechanics. Therefore, the wave function must not distinguish between the electrons. Thus, the fourth spin function in (19.36), which says that electron 1 has spin β and electron 2 has spin α, cannot be used. Instead of the third and fourth spin functions in (19.36), it turns out (see below for the justification) that one must use the functions $2^{-1/2}[\alpha(1)\beta(2) - \beta(1)\alpha(2)]$ and $2^{-1/2}[\alpha(1)\beta(2) + \beta(1)\alpha(2)]$. For each of these functions, electron 1 has both spin α and spin β, and so does electron 2. The $2^{-1/2}$ in these functions is a normalization constant.

The proper two-electron spin functions are therefore

$$\alpha(1)\alpha(2), \qquad \beta(1)\beta(2), \qquad 2^{-1/2}[\alpha(1)\beta(2) + \beta(1)\alpha(2)] \qquad \textbf{(19.37)*}$$

$$2^{-1/2}[\alpha(1)\beta(2) - \beta(1)\alpha(2)] \qquad \textbf{(19.38)*}$$

The three spin functions in (19.37) are unchanged when electrons 1 and 2 are interchanged. For example, interchanging the electrons in the third function gives $2^{-1/2}[\alpha(2)\beta(1) + \beta(2)\alpha(1)]$, which equals the original function. These three spin functions are said to be **symmetric** with respect to electron interchange. The spin function (19.38) is multiplied by -1 when the electrons are interchanged, since interchange gives

$$2^{-1/2}[\alpha(2)\beta(1) - \beta(2)\alpha(1)] = -2^{-1/2}[\alpha(1)\beta(2) - \beta(1)\alpha(2)]$$

The function (19.38) is **antisymmetric,** meaning that interchange of the coordinates of two particles multiplies the function by -1.

The Pauli Principle

A particle whose spin quantum number s is half-integral ($\frac{1}{2}$ or $\frac{3}{2}$ or $\frac{5}{2}$ or . . .) is called a **fermion** (after the Italian-American physicist Enrico Fermi). A particle whose s is integral (0 or 1 or 2 or . . .) is called a **boson** (after the Indian physicist S. N. Bose). Electrons have $s = \frac{1}{2}$ and are fermions.

Since two identical particles cannot be distinguished from each other in quantum mechanics, interchange of two identical particles in the wave function must leave all physically observable properties unchanged. In particular, the probability density $|\psi|^2$ must be unchanged. We therefore expect that ψ itself would be multiplied by either $+1$ or -1 by such an interchange or relabeling. It turns out that only one of these possibilities occurs, depending on the nature of the particles. Experimental evidence shows the validity of the following statement:

The complete wave function (including both spatial and spin coordinates) of a system of identical fermions must be antisymmetric with respect to interchange of all the coordinates (spatial and spin) of any two particles. For a system of identical bosons, the complete wave function must be symmetric with respect to such interchange.

This fact, discovered by Dirac and by Heisenberg in 1926, is called the **Pauli principle.** In 1940, Pauli deduced the Pauli principle from relativistic quantum field theory. In the nonrelativistic version of quantum mechanics that we are using, the Pauli principle must be regarded as an additional postulate.

We are now ready to include spin in the ground-state He wave function. The approximate ground-state spatial function $1s(1)1s(2)$ of (19.35) is symmetric with respect to electron interchange, since $1s(2)1s(1) = 1s(1)1s(2)$. Since electrons are fermions, the Pauli principle demands that the complete wave function be antisymmetric. To get an antisymmetric ψ, we must multiply $1s(1)1s(2)$ by the antisymmetric function (19.38). Use of the symmetric spin functions in (19.37) would give a symmetric wave function, which is forbidden for fermions. With inclusion of spin, the approximate ground-state He wave function becomes

$$\psi \approx 1s(1)1s(2) \cdot 2^{-1/2}[\alpha(1)\beta(2) - \beta(1)\alpha(2)] \qquad (19.39)$$

Interchange of the electrons multiplies ψ by -1, so (19.39) is antisymmetric. Note that the two electrons in the $1s$ orbital have opposite spins.

The wave function (19.39) can be written as the determinant

$$\psi \approx \frac{1}{\sqrt{2}} \begin{vmatrix} 1s(1)\alpha(1) & 1s(1)\beta(1) \\ 1s(2)\alpha(2) & 1s(2)\beta(2) \end{vmatrix} \qquad (19.40)$$

A second-order determinant is defined by

$$\begin{vmatrix} a & b \\ c & d \end{vmatrix} \equiv ad - bc \qquad \textbf{(19.41)*}$$

The use of (19.41) in (19.40) gives (19.39).

The justification for replacing the third and fourth spin functions in (19.36) by the linear combinations in (19.37) and (19.38) is that the latter two functions are the only normalized linear combinations of $\alpha(1)\beta(2)$ and $\beta(1)\alpha(2)$ that are either symmetric or antisymmetric with respect to electron interchange and that therefore do not distinguish between the electrons.

Improved Ground-State Wave Functions for Helium

For one- and two-electron systems, the wave function is a product of a spatial factor and a spin factor. The atomic Hamiltonian (to a very good approximation) contains no terms involving spin. Because of these facts, the spin part of the wave function need not be explicitly included in calculating the energy of one- and two-electron systems and will be omitted in the calculations in this section.

We saw above that ignoring the $e^2/4\pi\varepsilon_0 r_{12}$ term in $\hat{H}$ and taking E as the sum of two hydrogenlike energies gave a 38% error in the ground-state He energy. To improve on this dismal result, we can use the variation method. The most obvious choice of variational function is the $1s(1)1s(2)$ function of (19.34) and (19.35), which is a normalized product of hydrogenlike $1s$ orbitals. The variational integral in (18.84) is then $W = \int 1s(1)1s(2)\hat{H}1s(1)1s(2)\,d\tau$, where $\hat{H}$ is the true Hamiltonian (19.32) and $d\tau = d\tau_1\,d\tau_2$, with $d\tau_1 = r_1^2 \sin\theta_1\,dr_1\,d\theta_1\,d\phi_1$. Since $e^2/4\pi\varepsilon_0 r_{12}$ is part of $\hat{H}$, the effect of the interelectronic repulsion will be included in an average way, rather than being ignored as it was in (19.33). Evaluation of the variational integral is complicated and is omitted here. The result is $W = -74.8$ eV, which is reasonably close to the true ground-state energy -79.0 eV.

A further improvement is to use a variational function having the same form as (19.34) and (19.35) but with the nuclear charge Z replaced by a variational parameter ζ (zeta). We then vary ζ to minimize the variational integral $W = \int \phi^* \hat{H} \phi\,d\tau$, where the normalized variation function ϕ is $(\zeta/a_0)^3\pi^{-1}e^{-\zeta r_1/a_0}e^{-\zeta r_2/a_0}$. Substitution of the He Hamiltonian (19.32) with $Z = 2$ and evaluation of the integrals leads to

$$W = (\zeta^2 - 27\zeta/8)e^2/4\pi\varepsilon_0 a$$

[See *Levine* (2000), sec. 9.4, for the details.] The minimization condition $\partial W/\partial\zeta = 0$ then gives $0 = (2\zeta - 27/8)e^2/4\pi\varepsilon_0 a$ and the optimum value of ζ is 27/16 = 1.6875. This value of ζ gives $W = -2.848(e^2/4\pi\varepsilon_0 a) = -2.848(2 \times 13.6 \text{ eV}) = -77.5 \text{ eV}$, where (19.17) was used. This result is only 2% above the true ground-state energy -79.0 eV.

The parameter ζ is called an *orbital exponent*. The fact that ζ is less than the atomic number $Z = 2$ can be attributed to the **shielding** or **screening** of one electron from the nucleus by the other electron. When electron 1 is between electron 2 and the nucleus, the repulsion between electrons 1 and 2 subtracts from the attraction between electron 2 and the nucleus. Thus, ζ can be viewed as the "effective" nuclear charge for the $1s$ electrons. Since both electrons are in the same orbital, the screening effect is not great and ζ is only 0.31 less than Z.

By using complicated variational functions, workers have obtained agreement to 1 part in 2 million between the theoretical and the experimental ionization energies of ground-state He. [C. L. Pekeris, *Phys. Rev.,* **115,** 1216 (1959); C. Schwartz, *Phys. Rev.,* **128,** 1146 (1962).]

Excited-State Wave Functions for Helium

We saw that the approximation of ignoring the interelectronic repulsion in the Hamiltonian gives the helium wave functions as products of two hydrogenlike functions [Eq. (19.33)]. The hydrogenlike $2s$ and $2p$ orbitals have the same energy, and we might expect the approximate spatial wave functions for the lowest excited energy level of He to be $1s(1)2s(2)$, $1s(2)2s(1)$, $1s(1)2p_x(2)$, $1s(2)2p_x(1)$, $1s(1)2p_y(2)$, $1s(2)2p_y(1)$, $1s(1)2p_z(2)$, and $1s(2)2p_z(1)$, where $1s(2)2p_z(1)$ is a function with electron 2 in the $1s$ orbital and electron 1 in the $2p_z$ orbital. Actually, these functions are incorrect, in that they distinguish between the electrons. As we did above with the spin functions, we must take linear combinations to give functions that don't distinguish between the electrons. Analogous to the linear combinations in (19.37) and (19.38), the correct normalized approximate spatial functions are

$$2^{-1/2}[1s(1)2s(2) + 1s(2)2s(1)] \tag{19.42}$$

$$2^{-1/2}[1s(1)2s(2) - 1s(2)2s(1)] \tag{19.43}$$

$$2^{-1/2}[1s(1)2p_x(2) + 1s(2)2p_x(1)] \quad \text{etc.} \tag{19.44}$$

$$2^{-1/2}[1s(1)2p_x(2) - 1s(2)2p_x(1)] \quad \text{etc.} \tag{19.45}$$

where each "etc." indicates two similar functions with $2p_x$ replaced by $2p_y$ or $2p_z$.

If ψ_k is the true wave function of state k of a system, then $\hat{H}\psi_k = E_k\psi_k$, where E_k is the energy of state k. We therefore have $\int\psi_k^*\hat{H}\psi_k \, d\tau = \int\psi_k^* E_k\psi_k \, d\tau = E_k \int\psi_k^*\psi_k \, d\tau = E_k$, since ψ_k is normalized. This result suggests that if we have an approximate wave function $\psi_{k,\text{approx}}$ for state k, an approximate energy can be obtained by replacing ψ_k with $\psi_{k,\text{approx}}$ in the integral:

$$E_k \approx \int\psi_{k,\text{approx}}^*\hat{H}\psi_{k,\text{approx}}^* \, d\tau \tag{19.46}$$

where $\hat{H}$ is the true Hamiltonian, including the interelectronic repulsion term(s).

Use of the eight approximate functions (19.42) to (19.45) in Eq. (19.46) then gives approximate energies for these states. Because of the difference in signs, the two states (19.42) and (19.43) that arise from the $1s2s$ configuration will clearly have different

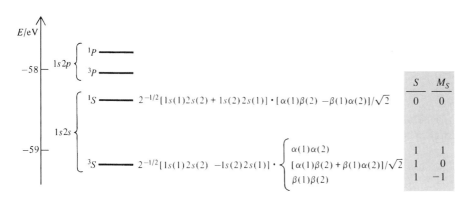

Figure 19.13

Energies of the terms arising from the helium-atom $1s2s$ and $1s2p$ electron configurations.

energies. The state (19.43) turns out to be lower in energy. The approximate wave functions are real, and the contribution of the $e^2/4\pi\varepsilon_0 r_{12}$ interelectronic repulsion term in $\hat{H}$ to the integral in Eq. (19.46) is $\int \psi_{k,\text{approx}}^2 e^2/4\pi\varepsilon_0 r_{12} \, d\tau$. One finds that the integral $\int (19.42)^2 e^2/4\pi\varepsilon_0 r_{12} \, d\tau$ differs in value from $\int (19.44)^2 e^2/4\pi\varepsilon_0 r_{12} \, d\tau$, where "(19.42)" and "(19.44)" stand for the functions in Eqs. (19.42) and (19.44). Hence, the $1s2p$ state in (19.44) differs in energy from the corresponding $1s2s$ state in (19.42). Likewise, the states (19.43) and (19.45) differ in energy from each other. Since the $2p_x$, $2p_y$, and $2p_z$ orbitals have the same shape, changing $2p_x$ to $2p_y$ or $2p_z$ in (19.44) or (19.45) doesn't affect the energy.

Thus, the states of the $1s2s$ configuration give two different energies and the states of the $1s2p$ configuration give two different energies, for a total of four different energies (Fig. 19.13). The $1s2s$ states turn out to lie lower in energy than the $1s2p$ states. Although the $2s$ and $2p$ orbitals have the same energy in one-electron (hydrogenlike) atoms, the interelectronic repulsions in atoms with two or more electrons remove the $2s$-$2p$ degeneracy. The reason the $2s$ orbital lies below the $2p$ orbital can be seen from Figs. 19.8 and 19.3. The $2s$ orbital has more probability density near the nucleus than the $2p$ orbital. Thus, a $2s$ electron is more likely than a $2p$ electron to penetrate within the probability density of the $1s$ electron. When it penetrates, it is no longer shielded from the nucleus and feels the full nuclear charge and its energy is thereby lowered. Similar penetration effects remove the l degeneracy in higher orbitals. For example, $3s$ lies lower than $3p$, which lies lower than $3d$ in atoms with more than one electron.

What about electron spin? The function (19.42) is symmetric with respect to electron interchange and so must be combined with the antisymmetric two-electron spin function (19.38) to give an overall ψ that is antisymmetric: $\psi \approx (19.42) \times (19.38)$. The function (19.43) is antisymmetric and so must be combined with one of the symmetric spin functions in (19.37). Because there are three symmetric spin functions, inclusion of spin in (19.43) gives three different wave functions, each having the same spatial factor. Since the spin factor doesn't affect the energy, there is a threefold spin degeneracy associated with the function (19.43). Similar considerations hold for the $1s2p$ states.

Figure 19.13 shows the energies and some of the approximate wave functions for the states arising from the $1s2s$ and $1s2p$ configurations. The labels ${}^3S$, ${}^1S$, ${}^3P$, and ${}^1P$ and the S and M_S values are explained shortly. The atomic energies shown in Fig. 19.13 are called **terms**, rather than energy levels, for a reason to be explained later. The ${}^3S$ term of the $1s2s$ configuration is threefold degenerate, because of the three symmetric spin functions. The ${}^1S$ term is onefold degenerate (that is, nondegenerate), since there is only one wave function for this term. The approximate wave functions for the ${}^3P$ term are obtained from those of the ${}^3S$ term by replacing $2s$ by $2p_x$, by $2p_y$, and by $2p_z$. Each of these replacements gives three wave functions (due to the three symmetric spin functions), so the ${}^3P$ term is ninefold degenerate. The ${}^1P$ term is threefold degenerate, since three functions are obtained by replacement of $2s$ in the ^1S function with $2p_x$, with $2p_y$, and with $2p_z$.

Note that the helium wave functions (19.39) and (19.42) to (19.45) are only approximations, since at best they take account of the interelectronic repulsion in only an average way. Thus, to say that the helium ground state has the electron configuration $1s^2$ is only approximately true. The use of orbitals (one-electron wave functions) in many-electron atoms is only an approximation.

19.7   TOTAL ORBITAL AND SPIN ANGULAR MOMENTA

The magnitude of the orbital angular momentum of the electron in a one-electron atom is given by (19.27) as $[l(l + 1)]^{1/2}\hbar$, where $l = 0, 1, 2, \ldots$. For an atom with more than one electron, the orbital angular momentum vectors $\mathbf{L}_i$ of the individual electrons add to give a **total electronic orbital angular momentum L,** given by $\mathbf{L} = \Sigma_i \mathbf{L}_i$. Quantum mechanics shows that the magnitude of $\mathbf{L}$ is given by

$$|\mathbf{L}| = [L(L + 1)]^{1/2}\hbar, \qquad \text{where } L = 0, 1, 2, \ldots \qquad (19.47)$$

The value of the total electronic orbital-angular-momentum quantum number L is indicated by a code letter similar to (19.15), except that capital letters are used:

L value	0	1	2	3	4	5
Code letter	S	P	D	F	G	H

For the $1s2s$ configuration of the He atom, both electrons have $l = 0$. Hence, the total orbital angular momentum is zero, and the code letter S is used for each of the two terms that arise from the $1s2s$ configuration (Fig. 19.13). For the $1s2p$ configuration, one electron has $l = 0$ and one has $l = 1$. Hence, the total-orbital-angular-momentum quantum number L equals 1, and the code letter P is used.

The **total electronic spin angular momentum S** of an atom (or molecule) is the vector sum of the spin angular momenta of the individual electrons: $\mathbf{S} = \Sigma_i \mathbf{S}_i$. The magnitude of $\mathbf{S}$ has the possible values $[S(S + 1)]^{1/2}\hbar$, where the total electronic spin quantum number S can be $0, \frac{1}{2}, 1, \frac{3}{2}, \ldots$. (Don't confuse the spin quantum number S with the orbital-angular-momentum code letter S.) The component of $\mathbf{S}$ along the z axis has the possible values $M_S\hbar$, where $M_S = -S, -S + 1, \ldots, S - 1, S$.

For a two-electron system such as He, each electron has spin quantum number $s = \frac{1}{2}$, and the total spin quantum number S can be 0 or 1, depending on whether the two electron spin vectors point in opposite directions or in approximately the same direction. (See Prob. 19.47.) For $S = 1$, the total spin angular momentum is $[S(S + 1)]^{1/2}\hbar = (1 \cdot 2)^{1/2}\hbar = 1.414\hbar$. The spin angular momentum of each electron is $(\frac{1}{2} \cdot \frac{3}{2})^{1/2}\hbar = 0.866\hbar$. The algebraic sum of the spin angular momenta of the individual electrons is $0.866\hbar + 0.866\hbar = 1.732\hbar$, which is greater than the magnitude of the total spin angular momentum. Hence, the two spin-angular-momentum vectors of the electrons cannot be exactly parallel; see, for example, Fig. 19.14a.

For spin quantum number $S = 0$, M_S must be zero, and there is only one possible spin state. This spin state corresponds to the antisymmetric spin function (19.38).

For $S = 1$, M_S can be -1, 0, or $+1$. The $M_S = -1$ spin state arises when each electron has $m_s = -\frac{1}{2}$ and so corresponds to the symmetric spin function $\beta(1)\beta(2)$ in (19.37). The $M_S = +1$ spin state corresponds to the function $\alpha(1)\alpha(2)$. For the $M_S = 0$ state, one electron must have $m_s = +\frac{1}{2}$ and the other $m_s = -\frac{1}{2}$. This is the function $2^{-1/2}[\alpha(1)\beta(2) + \beta(1)\alpha(2)]$ in (19.37). Although the z components of the two electron spins are in opposite directions, the two spin vectors can still add to give a total electronic spin with $S = 1$, as shown in Fig. 19.14a. The three symmetric spin functions in (19.37) thus correspond to $S = 1$.

$S = 1, M_S = 0$

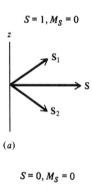

(a)

$S = 0, M_S = 0$

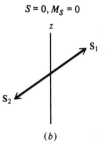

(b)

Figure 19.14

Spin orientations corresponding to the spin functions
(a) $2^{-1/2}[\alpha(1)\beta(2) + \beta(1)\alpha(2)]$ and
(b) $2^{-1/2}[\alpha(1)\beta(2) - \beta(1)\alpha(2)]$.
$\mathbf{S}$ is the total electronic spin angular momentum.

The quantity $2S + 1$ (where S is the total spin quantum number) is called the **spin multiplicity** of an atomic term and is written as a left superscript to the code letter for L. The lowest term in Fig. 19.13 has spin wave functions that correspond to total spin quantum number $S = 1$. Hence, $2S + 1$ equals 3 for this term, and the term is designated ^3S (read as "triplet ess"). The second-lowest term in Fig. 19.13 has the $S = 0$ spin function and so has $2S + 1 = 1$. This is a ^1S ("singlet ess") term. The S and M_S total spin quantum numbers are listed in Fig. 19.13 for each state (wave function) of the ^1S and ^3S terms.

Note that the triplet term of the $1s2s$ configuration lies lower than the singlet term. The same is true for the terms of the $1s2p$ configuration. This illustrates **Hund's rule:** *For a set of terms arising from a given electron configuration, the lowest-lying term is generally the one with the maximum spin multiplicity.* There are several exceptions to Hund's rule. The theoretical basis for Hund's rule is discussed in R. L. Snow and J. L. Bills, *J. Chem. Educ.*, **51**, 585 (1974); I. Shim and J. P. Dahl, *Theor. Chim. Acta*, **48**, 165 (1978); J. W. Warner and R. S. Berry, *Nature*, **313**, 160 (1985).

The maximum spin multiplicity is produced by having the maximum number of electrons with parallel spins. Two electrons are said to have **parallel** spins when their spin-angular-momentum vectors point in approximately the same direction, as, for example, in Fig. 19.14*a*. Two electrons have **antiparallel** spins when their spin vectors point in opposite directions to give a net spin angular momentum of zero, as in Fig. 19.14*b*. The ^3S and ^1S terms of the $1s2s$ configuration can be represented by the diagrams

$$^3S: \quad \frac{\uparrow \quad \uparrow}{1s \quad 2s} \qquad ^1S: \quad \frac{\uparrow \quad \downarrow}{1s \quad 2s}$$

where the spins are parallel in ^3S and antiparallel in ^1S.

Electrons in a filled subshell (for example, the electrons in $2p^6$) have all their spins paired and contribute zero to the total electronic spin angular momentum. For each electron in a closed subshell with a positive value for the m quantum number, there is an electron with the corresponding negative value of m. (For example, in $2p^6$, there are two electrons with $m = +1$ and two with $m = -1$.) Therefore, electrons in a filled subshell contribute zero to the total electronic orbital angular momentum. Hence, electrons in closed subshells can be ignored when finding the possible values of the quantum numbers L and S for the total orbital and spin angular momenta. For example, the $1s^2 2s^2 2p^6 3s3p$ electron configuration of Mg gives rise to the same terms as the $1s2p$ configuration of He, namely, ^3P and ^1P.

The He-atom Hamiltonian (19.32) is not quite complete in that it omits a term called the *spin–orbit interaction* arising from the interaction between the spin and orbital motions of the electrons. The spin–orbit interaction is very small (except in heavy atoms), but it partly removes the degeneracy of a term, splitting an atomic term into a number of closely spaced *energy levels*. (See Prob. 19.48.) For example, the ^3P term in Fig. 19.13 is split into three closely spaced levels; the other three terms are each slightly shifted in energy by the spin–orbit interaction but are not split. Because of this spin–orbit splitting, the energies shown in Fig. 19.13 do not quite correspond to the actual pattern of atomic energy levels, and the energies in this figure are therefore called terms rather than energy levels.

An atomic term corresponds to definite values of the total orbital angular-momentum quantum number L and the total spin angular-momentum quantum number S. The L value is indicated by a code letter (S, P, D, . . .), and the S value is indicated by writing the value of $2S + 1$ as a left superscript to the L code letter.

For further discussion on the addition of angular momenta and on the way the terms arising from a given atomic electron configuration are found, see Probs. 19.47 and 19.48.

Lithium and the Pauli Exclusion Principle

As we did with helium, we can omit the interelectronic repulsion terms $(e^2/4\pi\varepsilon_0) \cdot (1/r_{12} + 1/r_{13} + 1/r_{23})$ from the Li-atom Hamiltonian to give an approximate Hamiltonian that is the sum of three hydrogenlike Hamiltonians. The approximate wave function is then the product of hydrogenlike (one-electron) wave functions. For the ground state, we might expect the approximate wave function $1s(1)1s(2)1s(3)$. However, we have not taken account of electron spin or the Pauli principle. The symmetric spatial function $1s(1)1s(2)1s(3)$ must be multiplied by an antisymmetric three-electron spin function. One finds, however, that it is impossible to write an antisymmetric spin function for three electrons. With three or more electrons, the antisymmetry requirement of the Pauli principle cannot be satisfied by writing a wave function that is the product of separate spatial and spin factors.

The clue to constructing an antisymmetric wave function for three or more electrons lies in Eq. (19.40), which shows that the ground-state wave function of helium can be written as a determinant. The reason a determinant gives an antisymmetric wave function follows from the theorem: *Interchange of two rows of a determinant changes the sign of the determinant.* (For a proof, see *Sokolnikoff and Redheffer,* app. A.) Interchange of rows 1 and 2 of the determinant in (19.40) amounts to interchanging the electrons. Thus a determinantal ψ is multiplied by -1 by such an interchange and therefore satisfies the Pauli antisymmetry requirement.

Let f, g, and h be three spin-orbitals. (Recall that a spin-orbital is the product of a spatial orbital and a spin factor; Sec. 19.5.) We can get an antisymmetric three-electron wave function by writing the following determinant (called a **Slater determinant**)

$$\frac{1}{\sqrt{6}} \begin{vmatrix} f(1) & g(1) & h(1) \\ f(2) & g(2) & h(2) \\ f(3) & g(3) & h(3) \end{vmatrix} \tag{19.48}$$

The $1/\sqrt{6}$ is a normalization constant; there are six terms in the expansion of this determinant.

A third-order determinant is defined by

$$\begin{vmatrix} a & b & c \\ d & e & f \\ g & h & i \end{vmatrix} \equiv a \begin{vmatrix} e & f \\ h & i \end{vmatrix} - b \begin{vmatrix} d & f \\ g & i \end{vmatrix} + c \begin{vmatrix} d & e \\ g & h \end{vmatrix}$$

$$= aei - ahf - bdi + bgf + cdh - cge \tag{19.49}$$

where (19.41) was used. The second-order determinant that multiplies a in the expansion is found by striking out the row and the column that contains a in the third-order determinant; similarly for the multipliers of $-b$ and c. The reader can verify that interchange of two rows multiplies the determinant's value by -1.

To get an antisymmetric approximate wave function for Li, we use (19.48). Let us try to put all three electrons into the $1s$ orbital by taking the spin-orbitals to be $f = 1s\alpha$, $g = 1s\beta$, and $h = 1s\alpha$. The determinant (19.48) becomes

$$\frac{1}{\sqrt{6}} \begin{vmatrix} 1s(1)\alpha(1) & 1s(1)\beta(1) & 1s(1)\alpha(1) \\ 1s(2)\alpha(2) & 1s(2)\beta(2) & 1s(2)\alpha(2) \\ 1s(3)\alpha(3) & 1s(3)\beta(3) & 1s(3)\alpha(3) \end{vmatrix} \tag{19.50}$$

Expansion of this determinant using (19.49) shows it to equal zero. This can be seen without multiplying out the determinant by using the following theorem (*Sokolnikoff and Redheffer,* app. A): *If two columns of a determinant are identical, the determinant equals zero.* The first and third columns of (19.50) are identical, and so (19.50) vanishes.

If any two of the spin-orbitals f, g, and h in (19.48) are the same, two columns of the determinant are the same and the determinant vanishes. Of course, zero is ruled out as a possible wave function, since there would then be no probability of finding the electrons.

The Pauli-principle requirement that the electronic wave function be antisymmetric thus leads to the conclusion:

No more than one electron can occupy a given spin-orbital.

This is the **Pauli exclusion principle,** first stated by Pauli in 1925. An orbital (or one-electron spatial wave function) is defined by giving its three quantum numbers (n, l, m in an atom). A spin-orbital is defined by giving the three quantum numbers of the orbital and the m_s quantum number ($+\frac{1}{2}$ for spin function α, $-\frac{1}{2}$ for β). Thus, in an atom, the exclusion principle requires that no two electrons have the same values for all four quantum numbers n, l, m, and m_s.

> Some physicists have speculated that small violations of the Pauli principle might occur. To test this, Ramberg and Snow passed a large current through a copper strip and searched for x-rays that would occur if an electron in the current dropped into the $1s$ orbital of a Cu atom to give a $1s^3$ atom. No such x-rays were found, and the experiment showed that the probability that a new electron introduced into copper would violate the Pauli principle is less than 2×10^{-26} [E. Ramberg and G. A. Snow, *Phys. Lett. B*, **238**, 438 (1990)].

The antisymmetry requirement holds for any system of identical fermions (Sec. 19.6), so *in a system of identical fermions each spin-orbital can hold no more than one fermion.* In contrast, ψ is symmetric for bosons, so *there is no limit to the number of bosons that can occupy a given spin-orbital.*

Coming back to the Li ground state, we can put two electrons with opposite spins in the $1s$ orbital ($f = 1s\alpha$, $g = 1s\beta$), but to avoid violating the exclusion principle, the third electron must go in the $2s$ orbital ($h = 2s\alpha$ or $2s\beta$). The approximate Li ground-state wave function is therefore

$$\psi \approx \frac{1}{\sqrt{6}} \begin{vmatrix} 1s(1)\alpha(1) & 1s(1)\beta(1) & 2s(1)\alpha(1) \\ 1s(2)\alpha(2) & 1s(2)\beta(2) & 2s(2)\alpha(2) \\ 1s(3)\alpha(3) & 1s(3)\beta(3) & 2s(3)\alpha(3) \end{vmatrix} \qquad (19.51)$$

When (19.51) is multiplied out, it becomes a sum of six terms, each containing a spatial and a spin factor, so ψ cannot be written as a single spatial factor times a single spin factor. Because the $2s$ electron could have been given spin β, the ground state is doubly degenerate. The elements in each row of the Slater determinant (19.51) involve the same electron. The elements in each column involve the same spin-orbital.

Since all the electrons are s electrons with $l = 0$, the total-orbital-angular-momentum quantum number L is 0. The $1s$ electrons have antiparallel spins, so the total electronic spin of the atom is due to the $2s$ electron and the total-electronic-spin quantum number S is $\frac{1}{2}$. The spin multiplicity $2S + 1$ is 2, and the ground term of Li is designated ^2S.

A variational treatment using (19.51) would replace Z in the $1s$ function in Table 19.1 by a parameter ζ_1 and Z in the $2s$ function by a parameter ζ_2. These parameters are "effective" atomic numbers that allow for electron screening. One finds the optimum values to be $\zeta_1 = 2.69$ and $\zeta_2 = 1.78$. As expected, the $2s$ electron is much better screened from the $Z = 3$ nucleus than the $1s$ electrons. The calculated variational energy turns out to be -201.2 eV, compared with the true ground-state energy -203.5 eV.

The Periodic Table

A qualitative and semiquantitative understanding of atomic structure can be obtained from the orbital approximation. As we did with He and Li, we write an approximate wave function that assigns the electrons to hydrogenlike spin-orbitals. In each orbital, the nuclear charge is replaced by a variational parameter that represents an effective nuclear charge Z_{eff} and allows for electron screening. To satisfy the Pauli principle, the wave function is written as a Slater determinant. For some atomic states, the wave function must be written as a linear combination of a few Slater determinants, but we won't worry about this complication.

Since an electron has two possible spin states (α or β), the exclusion principle requires that no more than two electrons occupy the same orbital in an atom or molecule. Two electrons in the same orbital must have antiparallel spins, and such electrons are said to be **paired.** A set of orbitals with the same n value and the same l value constitutes a **subshell.** The lowest few subshells are $1s$, $2s$, $2p$, $3s$, ... An s subshell has $l = 0$ and $m = 0$ and hence can hold at most two electrons without violating the exclusion principle. A p subshell has $l = 1$ and the three possible m values -1, 0, $+1$; hence, a p subshell has a capacity of 6 electrons; d and f subshells hold a maximum of 10 and 14 electrons, respectively.

The hydrogenlike energy formula (19.18) can be modified to approximate crudely the energy ε of a given atomic orbital as

$$\varepsilon \approx -(Z_{eff}^2/n^2)(13.6 \text{ eV}) \tag{19.52}$$

where n is the principal quantum number and the effective nuclear charge Z_{eff} differs for different subshells in the same atom. We write $Z_{eff} = Z - s$, where Z is the atomic number and the **screening constant** s for a given subshell is the sum of contributions from other electrons in the atom.

Figure 19.15 shows orbital energies for neutral atoms calculated using an approximate method. The scales in this figure are logarithmic. Note that the energy of an orbital depends strongly on the atomic number, decreasing with increasing Z, as would be expected from Eq. (19.52). Note the square-root sign on the vertical axis. Because Z_{eff} for the $1s$ electrons in Ne ($Z = 10$) is almost 10 times as large as Z_{eff} for the $1s$ electron in H, the $1s$ orbital energy ε_{1s} for Ne is roughly 100 times ε_{1s} for H. (Of course, ε_{1s} is negative.) As mentioned earlier, the l degeneracy that exists for $Z = 1$ is removed in many-electron atoms. For most values of Z, the $3s$ and $3p$ orbitals are much closer together than the $3p$ and $3d$ orbitals, and we get the familiar stable octet of outer electrons (ns^2np^6). For Z between 7 and 21, the $4s$ orbital lies below the $3d$ (s orbitals are more penetrating than d orbitals), but for $Z > 21$, the $3d$ lies lower.

We saw that the ground-state configuration of Li is $1s^2 2s$, where the superscripts give the numbers of electrons in each subshell and a superscript of 1 is understood for the $2s$ subshell. We expect Li to readily lose one electron (the $2s$ electron) to form the Li^+ ion, and this is the observed chemical behavior.

The ground-state configurations of Be and B are $1s^2 2s^2$ and $1s^2 2s^2 2p$, respectively. For C, the ground-state configuration is $1s^2 2s^2 2p^2$. A given electron configuration can give rise to more than one atomic term. For example, the He $1s2s$ configuration produces the two terms ^3S and ^1S (Fig. 19.13). Figuring out the terms arising from the $1s^2 2s^2 2p^2$ configuration is complicated and is omitted. Hund's rule tells us that the lowest-lying term will have the two $2p$ spins parallel:

$$\underset{1s}{\uparrow\downarrow} \quad \underset{2s}{\uparrow\downarrow} \quad \underset{2p}{\uparrow \quad \uparrow \quad \underline{}}$$

Putting the two $2p$ electrons in different orbitals minimizes the electrostatic repulsion between them. The $2p$ subshell is filled at $_{10}$Ne, whose electron configuration is $1s^2 2s^2 2p^6$. Like helium, neon does not form chemical compounds.

Sodium has the ground-state configuration $1s^2 2s^2 2p^6 3s$, and its chemical and physical properties resemble those of Li (ground-state configuration $1s^2 2s$), its predecessor

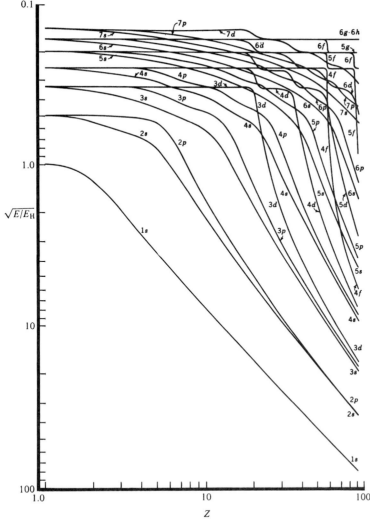

Figure 19.15

Approximate orbital energies
versus atomic number Z in neutral
atoms. $E_H = -13.6$ eV, the
ground-state H-atom energy.
[Redrawn by M. Kasha from R.
Latter, *Phys. Rev.,* **99,** 510 (1955).]

in group 1 of the periodic table. The periodic table is a consequence of the hydrogen-like energy-level pattern, the allowed electronic quantum numbers, and the exclusion principle. The third period ends with Ar, whose ground-state configuration is $1s^2 2s^2 2p^6 3s^2 3p^6$.

For $Z = 19$ and 20, the $4s$ subshell lies below the $3d$ (Fig. 19.15) and K and Ca have the outer electron configurations $4s$ and $4s^2$, respectively. With $Z = 21$, the $3d$ subshell begins to fill, giving the first series of *transition elements*. The $3d$ subshell is filled at $_{30}$Zn, outer electron configuration $3d^{10}4s^2$ ($3d$ now lies lower than $4s$). Filling the $4p$ subshell then completes the fourth period. For discussion of the electron configurations of transition-metal atoms and ions, see L. G. Vanquickenborne et al., *J. Chem. Educ.,* **71,** 469 (1994); *Levine* (2000), sec. 11.2.

The rare earths (lanthanides) and actinides in the sixth and seventh periods correspond to filling the $4f$ and $5f$ subshells.

The order of filling of subshells in the periodic table is given by the $n + l$ rule: Subshells fill in order of increasing $n + l$ values; for subshells with equal $n + l$ values, the one with the lower n fills first.

Niels Bohr rationalized the periodic table in terms of filling the atomic energy levels, and the familiar long form of the periodic table is due to him.

Atomic Properties

The **first, second, third, . . . ionization energies** of atom A are the energies required for the processes $A \rightarrow A^+ + e^-$, $A^+ \rightarrow A^{2+} + e^-$, $A^{2+} \rightarrow A^{3+} + e^-$, . . . , where A, A^+, etc., are isolated atoms or ions in their ground states. Ionization energies are traditionally expressed in eV. The corresponding numbers in volts are called the **ionization potentials.** Some first, second, and third ionization energies in eV are (C. E. Moore, Ionization Potentials and Ionization Limits, *Nat. Bur. Stand. U.S. Publ. NSRDS-NBS 34,* 1970):

H	He	Li	Be	B	C	N	O	F	Ne	Na
13.6	24.6	5.4	9.3	8.3	11.3	14.5	13.6	17.4	21.6	5.1
	54.4	75.6	18.2	25.2	24.4	29.6	35.1	35.0	41.0	47.3
		122.5	153.9	37.9	47.9	47.4	54.9	62.7	63.4	71.6

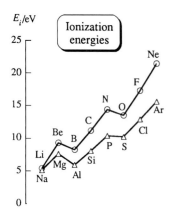

Ionization energies

Figure 19.16

Ionization energies of second- and third-period elements.

Note the low value for removal of the $2s$ electron from Li and the high value for removal of a $1s$ electron from Li^+. Ionization energies clearly show the "shell" structure of atoms.

The first ionization energy decreases going down a group in the periodic table because the increase in quantum number n of the valence electron increases the average distance of the electron from the nucleus, making it easier to remove. The first ionization energy generally increases going across a period (Fig. 19.16). As we go across a period, the nuclear charge increases, but the electrons being added have the same or a similar value of n and so don't screen one another very effectively; the effective nuclear charge $Z_{\text{eff}} = Z - s$ in (19.52) increases across a period, since Z is increasing faster than s, and the valence electrons become more tightly bound. Metals have lower ionization energies than nonmetals.

The **electron affinity** of atom A is the energy released in the process $A + e^- \rightarrow A^-$. Some values in eV are [T. Andersen et al., *J. Phys. Chem. Ref. Data,* **28,** 1511 (1999)]:

H	He	Li	Be	B	C	N	O	F	Ne	Na
0.8	<0	0.6	<0	0.3	1.3	−0.1	1.5	3.4	<0	0.5

Note the opposite convention in the definitions of ionization energy and electron affinity. The ionization energy is ΔE accompanying loss of an electron. The electron affinity is $-\Delta E$ accompanying gain of an electron.

The motion of an electric charge produces a magnetic field. The orbital angular momentum of atomic electrons with $l \neq 0$ therefore produces a magnetic field. The "spinning" of an electron about its own axis is a motion of electric charge and also produces a magnetic field. Because of the existence of electron spin, an electron acts like a tiny magnet. (Magnetic interactions between electrons are much smaller than the electrical forces and can be neglected in the Hamiltonian except in very precise calculations.) The magnetic fields of electrons with opposite spins cancel each other. It follows that an atom in a state with $L \neq 0$ and/or $S \neq 0$ produces a magnetic field and is said to be **paramagnetic.** In a magnetized piece of iron, the majority of electron spins are aligned in the same direction to produce the observed magnetic field.

The **radius** of an atom is not a well-defined quantity, as is obvious from Figs. 19.7 and 19.8. From observed bond lengths in molecules and interatomic distances in crystals, various kinds of atomic radii can be deduced. (See Secs. 20.1 and 24.6.) Atomic radii decrease going across a given period because of the increase in Z_{eff} and increase going down a given group because of the increase in n.

The energies of excited states of most atoms of the periodic table have been determined from atomic spectral data and are tabulated in C. E. Moore, Atomic Energy Levels, *Nat. Bur. Stand. U.S. Circ.* 467, vols. I, II, and III, 1949, 1952, and 1958; C. E. Moore, Atomic Energy Levels, *Nat. Bur. Stand. Publ. NSRDS-NBS 35,* vols. I, II, and III, 1971.

Figure 19.17 shows some of the term energies (as determined from emission and absorption spectra) of the Na atom, whose ground-state electron configuration is

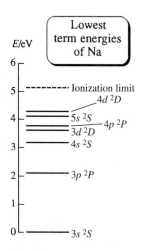

E/eV

Lowest term energies of Na

Figure 19.17

Some term energies of Na. The orbital of the excited electron is given preceding each term symbol.

$1s^2 2s^2 2p^6 3s$. The zero level of energy has been taken at the ground state. This choice differs from the convention used in Fig. 19.2 and Eq. (19.5), where the zero level of energy corresponds to all charges being infinitely far from one another. (Each ^2P, ^2D, ^2F, . . . term in Na is split by spin–orbit interaction into two closely spaced energy levels, which are not shown.)

Each excited term in Fig. 19.17 arises from an electron configuration with the $3s$ valence electron excited to a higher orbital, which is written preceding the term symbol. Terms that correspond to electron configurations with an inner-shell Na electron excited have energies higher than the first ionization energy of Na (the dashed line), and such terms are not readily observed spectroscopically. Because Na has only one valence electron, each electron configuration with filled inner shells gives rise to only one term, and the Na term-energy diagram is simple. Note from Fig. 19.17 that the $4s$ ^2S term of Na lies below the $3d$ ^2D term, as might be expected from Fig. 19.15.

19.9 HARTREE–FOCK AND CONFIGURATION-INTERACTION WAVE FUNCTIONS

Hartree–Fock Wave Functions

Wave functions like (19.39) for He and (19.51) for Li are approximations. How can these approximate wave functions be improved? One way is by not restricting the one-electron spatial functions to hydrogenlike functions. Instead, for the He ground state, we take as a trial variation function

$$\phi(1)\phi(2)2^{-1/2}[\alpha(1)\beta(2) - \beta(1)\alpha(2)] \qquad (19.53)$$

and we look for the function ϕ that minimizes the variational integral, where ϕ need not be a hydrogenlike $1s$ orbital but can have any form. For the Li ground state, we use a function like (19.51) but with the $1s$ and $2s$ functions replaced by unknown functions f and g, and we look for those functions f and g that minimize the variational integral. These variation functions are still antisymmetrized products of one-electron spin-orbitals, so the functions ϕ, f, and g are still atomic orbitals.

In the period 1927–1930, the English physicist Hartree and the Russian physicist Fock developed a systematic procedure for finding the best possible forms for the orbitals. A variational wave function that is an antisymmetrized product of the best possible orbitals is called a **Hartree–Fock wave function.** For each state of a given system, there is a single Hartree–Fock wave function. Hartree and Fock showed that the Hartree–Fock orbitals ϕ_i satisfy the equation

$$\hat{F}\phi_i = \varepsilon_i \phi_i \qquad (19.54)$$

where the Hartree–Fock operator $\hat{F}$ is a complicated operator whose form is discussed later in this section. Each of the spatial orbitals ϕ_i is a function of the three spatial coordinates; ε_i is the energy of orbital i.

The Hartree–Fock operator $\hat{F}$ in (19.54) is peculiar in that its form depends on what the eigenfunctions ϕ_i are. Since the orbitals ϕ_i are not known before the Hartree–Fock equations (19.54) are solved, the operator $\hat{F}$ is initially unknown. To solve (19.54), one starts with an initial guess for the orbitals ϕ_i, which allows one to calculate an initial guess for $\hat{F}$. One uses this initial estimate of $\hat{F}$ to solve (19.54) for an improved set of orbitals, and then uses these orbitals to calculate an improved $\hat{F}$, which is then used to solve for further improved orbitals, etc. The process is continued until no further significant change in the orbitals occurs from one iteration to the next.

Each spatial orbital in (19.54) is a function of the three spatial coordinates of a single electron. Likewise, the Hartree–Fock operator $\hat{F}$ contains the coordinates of a single electron and derivatives with respect to those coordinates. If we use the number

1 to label the electron in (19.54), we can write (19.54) as $\hat{F}(1)\phi_i(1) = \varepsilon_i\phi_i(1)$. The operator $\hat{F}(1)$ for an atom is the sum of the following terms: (a) The kinetic-energy operator $-(\hbar^2/2m)\nabla_1^2$ for electron 1. (b) The potential energy $-Ze^2/4\pi\varepsilon_0 r_1$ of attraction between electron 1 and the nucleus. (c) The potential energy of repulsion between electron 1 and a hypothetical continuous spatial distribution of negative charge whose charge density (charge per unit volume) is calculated by imagining that each of the other electrons in the atom is smeared out into a charge cloud whose charge density at each point is $-e|\phi_j|^2$, where ϕ_j is the orbital occupied by the hypothetical smeared-out electron. One adds the charge densities of the smeared-out electrons to get a charge cloud whose interaction with electron 1 is calculated. Part c of $\hat{F}(1)$ is called the *Coulomb operator*. (d) An *exchange operator* that involves the occupied orbitals and is present so as to make the overall wave function antisymmetric with respect to exchange of electrons. Parts c and d of $\hat{F}$ depend on the occupied orbitals, which are unknown at the start of the calculation.

Originally, Hartree–Fock orbitals were calculated numerically, and the results expressed as a table of values of ϕ_i at various points in space. In 1951, Roothaan showed that the most convenient way to express Hartree–Fock orbitals is as linear combinations of a set of functions called **basis functions.** A set of functions is said to be a **complete set** if *every* well-behaved function can be written as a linear combination of the members of the complete set. If the functions $g_1, g_2, g_3, \ldots$ form a complete set, then every well-behaved function f can be expressed as

$$f = \sum_k c_k g_k \tag{19.55}$$

where the coefficients c_k are constants that depend on what the function f is. It generally requires an infinite number of functions $g_1, g_2, \ldots$ to have a complete set (but not every infinite set of functions is complete). The basis functions g_k used to express the Hartree–Fock orbitals ϕ_i must be a complete set. We have

$$\phi_i = \sum_k b_k g_k \tag{19.56}$$

An orbital ϕ_i is specified by stating what the set of basis functions g_k is and giving the coefficients b_k. Roothaan showed how to calculate the b_k's that give the best possible orbitals. A different set of coefficients in (19.56) is used to express each of the orbitals.

A complete set of basis functions commonly used in atomic Hartree–Fock calculations is the set of **Slater-type orbitals** (STOs). An STO has the form $G_n(r)\Theta_{lm}(\theta)\Phi_m(\phi)$, where $\Theta_{lm}(\theta)$ and $\Phi_m(\phi)$ are the same functions as in the hydrogenlike orbitals (19.9). The radial factor has the form $G_n(r) = Nr^{n-1}e^{-\zeta r/a_0}$, where N is a normalization constant, n is the principal quantum number, and ζ is a variational parameter (the *orbital exponent*). The function $G_n(r)$ differs from a hydrogenlike radial factor in containing r^{n-1} in place of r^l times a polynomial in r. It has been shown that the set of STOs with n, l, and m given by (19.11) to (19.13) and with all possible positive values of ζ forms a complete set. Although, in principle, one needs an infinite number of basis functions to express a Hartree–Fock orbital, in practice, each atomic Hartree–Fock orbital can be very accurately approximated using only a few well-chosen STOs.

For example, for the helium ground state, Clementi expressed the Hartree–Fock orbital ϕ of the electrons as a linear combination of five $1s$ STOs that differ in their values of ζ. Thus $\phi = \sum_{k=1}^5 b_k g_k$, where the coefficients b_k are found by solving the Hartree–Fock equation (19.54) and each g_k function is a $1s$ STO with the form $g_k = N_k \exp(-\zeta_k r/a_0)$. Each N_k is a normalization constant, and each g_k function has a fixed value of ζ_k. (The ζ_k values used were 1.417, 2.377, 4.396, 6.527, and 7.943, and the values found for b_k were 0.768, 0.223, 0.041, −0.010, and 0.002, respectively.) The Hartree–Fock helium ground-state wave function is then (19.53) with the function ϕ given by the five-term sum just discussed.

To solve (19.54) for the Hartree–Fock orbitals of an atom or molecule with many electrons requires a huge amount of computation, and it wasn't until the advent of high-speed computers in the 1960s that such calculations became practicable. Hartree–Fock wave functions have been computed for the ground states and certain excited states of the first 54 atoms of the periodic table. Hartree–Fock wave functions play a key role in the quantum chemistry of molecules (Chapter 20).

Although a Hartree–Fock wave function is an improvement on one that uses hydrogenlike orbitals, it is still only an approximation to the true wave function. The Hartree–Fock wave function assigns each electron pair to its own orbital. The forms of these orbitals are computed to take interelectronic repulsions into account in an average way. However, electrons are not actually smeared out into a static distribution of charge but interact with one another instantaneously. An orbital wave function cannot account for these instantaneous interactions, so the true wave function cannot be expressed as an antisymmetrized product of orbitals.

For helium, the use of a hydrogenlike $1s$ orbital with a variable orbital exponent gives a ground-state energy of −77.5 eV (Sec. 19.6) compared with the true value −79.0 eV. The Hartree–Fock wave function for the helium ground state gives an energy of −77.9 eV, which is still in error by 1.1 eV. The energy error of the Hartree–Fock wave function is called the **correlation energy,** since it results from the fact that the Hartree–Fock wave function neglects the instantaneous correlations in the motions of the electrons. Electrons repel one another and correlate their motions to avoid being close together; this phenomenon is called **electron correlation.**

Configuration Interaction

A method used to improve a Hartree–Fock wave function is configuration interaction. When a Hartree–Fock ground-state wave function of an atom or molecule is calculated, one also obtains expressions for unoccupied excited-state orbitals. It is possible to show that the set of functions obtained by making all possible assignments of electrons to the available orbitals is a complete set. Hence, the true wave function ψ of the ground state can be expressed as

$$\psi = \sum_j a_j \psi_{\text{orb},j} \tag{19.57}$$

where the $\psi_{\text{orb},j}$'s are approximate orbital wave functions that differ in the assignment of electrons to orbitals. Each $\psi_{\text{orb},j}$ is a Slater determinant of spin-orbitals. The functions $\psi_{\text{orb},j}$ are called *configuration functions* (or *configurations*). One uses a variational procedure to find the values of the coefficients a_j that minimize the variational integral. This type of calculation is called **configuration interaction** (CI).

For the helium ground state, the term with the largest coefficient in the CI wave function will be a Slater determinant with both electrons in orbitals resembling $1s$ orbitals, but Slater determinants with electrons in $2s$-like and higher orbitals will also contribute. A CI wave function for the He ground state has the form $\psi = a_1\psi(1s^2) + a_2\psi(1s2s) + a_3\psi(1s3s) + a_4\psi(2s^2) + a_5\psi(2p^2) + a_6\psi(2s3s) + a_7\psi(3s^2) + a_8\psi(2p3p) + a_9\psi(3d^2) + \cdots$, where the a's are numerical coefficients and $\psi(1s2s)$ indicates a Slater determinant with one electron in a $1s$-like orbital and one in a $2s$-like orbital. (This last statement is inaccurate; see Prob. 19.63.)

CI computer calculations are extremely time-consuming, since it often requires a linear combination of thousands or even millions of configuration functions to give an accurate representation of ψ.

19.10   SUMMARY

The potential energy of interaction between two charges separated by distance r is $V = Q_1Q_2/4\pi\varepsilon_0 r$.

The Schrödinger equation for the H-atom internal motion is separable in spherical coordinates r, θ, and ϕ. The stationary-state hydrogenlike-atom wave functions and bound-state energies are $\psi = R_{nl}(r)\Theta_{lm}(\theta)\Phi_m(\phi)$ and $E = -(Z^2/n^2)(e^2/8\pi\varepsilon_0 a)$, where $a \equiv \hbar^2 4\pi\varepsilon_0/\mu e^2$ (μ is the reduced mass) and the quantum numbers take the values $n = 1, 2, 3, \ldots$; $l = 0, 1, 2, \ldots, n - 1$; $m = -l, \ldots, +l$. The letters s, p, d, f, $\ldots$ indicate $l = 0, 1, 2, 3, \ldots$, respectively. The magnitude of the electron's orbital angular momentum with respect to the nucleus is $|\mathbf{L}| = [l(l + 1)]^{1/2}\hbar$, and L_z equals $m\hbar$. An orbital is a one-electron spatial wave function. The shape of an orbital is defined as a surface of constant $|\psi|$ that encloses some large fraction of the probability density. Figure 19.6 shows some H-atom orbital shapes.

The average value of any function f of r for a hydrogen-atom stationary state is given by $\langle f(r) \rangle = \int |\psi|^2 f(r) \, d\tau$. Use of $\psi = R\Theta\Phi$, $d\tau = r^2 \sin\theta \, dr \, d\theta \, d\phi$, and the limits $0 \leq r \leq \infty$, $0 \leq \theta \leq \pi$, $0 \leq \phi \leq 2\pi$ gives

$$\langle f(r) \rangle = \int_0^\infty f(r)|R(r)|^2 r^2 \, dr \int_0^\pi |\Theta(\theta)|^2 \sin\theta \, d\theta \int_0^{2\pi} |\Phi(\phi)|^2 \, d\phi$$

Electrons and other elementary particles have a built-in angular momentum (spin angular momentum) $\mathbf{S}$ of magnitude $[s(s + 1)]^{1/2}\hbar$, where $s = \frac{1}{2}$ for an electron. The z component of $\mathbf{S}$ is $m_s\hbar$, where $m_s = \pm\frac{1}{2}$ for an electron. The symbols α and β denote spin functions with $m_s = +\frac{1}{2}$ and $m_s = -\frac{1}{2}$, respectively. A product of one-electron spatial and spin functions is called a spin-orbital. Particles are classified as fermions or bosons, according to whether s is half-integral or integral, respectively. The complete wave function of a system of identical fermions must be antisymmetric with respect to interchange of all coordinates (spatial and spin) of any two particles, meaning that such an interchange multiplies the wave function by -1. The wave function of a system of identical bosons must be symmetric.

There are three symmetric two-electron spin functions [Eq. (19.37)] and one antisymmetric one [Eq. (19.38)]. An approximate ground-state wave function for He is $1s(1)1s(2)$ times (19.38).

In a many-electron atom, the total electronic orbital and spin angular momenta are $[L(L + 1)]^{1/2}\hbar$ and $[S(S + 1)]^{1/2}\hbar$, respectively, where the quantum number L can be $0, 1, 2, \ldots$ and the quantum number S can be $0, \frac{1}{2}, 1, \frac{3}{2}, \ldots$. The L value is indicated using the letter code $S, P, D, F, \ldots$, and the value of $2S + 1$ (the spin multiplicity) is written as a left superscript to the L code letter. Atomic states that correspond to the same electron configuration and have the same L and S values belong to the same atomic term. Usually, the lowest-energy term of a given electron configuration is the term with the largest S value (Hund's rule).

An approximate (antisymmetric) wave function for a many-electron atom can be written as a Slater determinant of spin-orbitals. In such an approximate wave function, no more than one electron can occupy a given spin-orbital (the Pauli exclusion principle). The variation of atomic-orbital energy with atomic number is given in Fig. 19.15. The periodic table, ionization energies, and electron affinities were discussed.

The best possible (that is, lowest-energy) wave function that assigns each electron to a single spin-orbital is called the Hartree–Fock wave function. Hartree–Fock orbitals are expressed as linear combinations of basis functions. The Hartree–Fock wave function is still an approximation to the true wave function. In a configuration-interaction (CI) calculation, the wave function is written as a linear combination of the Hartree–Fock wave function and functions in which some of the electrons occupy excited orbitals. A CI wave function can approach the true wave function if enough configuration functions are included.

FURTHER READING

Hanna, chap. 6; *Karplus and Porter,* chaps. 3, 4; *Levine* (2000), chaps. 10, 11; *Lowe,* chaps. 4, 5; *McQuarrie* (1983), chap. 8; *Atkins and Friedman,* chap. 7.

PROBLEMS

Section 19.1

19.1 True or false? (*a*) Doubling the distance between two charges multiplies the force between them by one-half. (*b*) Doubling the distance between two charges multiplies the potential energy of their interaction by one-half. (*c*) One joule is many orders of magnitude larger than one electronvolt.

19.2 Use the relation $V = Q_1 Q_2 / 4\pi\varepsilon_0 r$ to deduce the SI units of $4\pi\varepsilon_0$.

19.3 Maxwell's electromagnetic theory of light (Sec. 21.1) shows that the speed of light in vacuum is $c = (\mu_0\varepsilon_0)^{-1/2}$, where ε_0 occurs in the proportionality constant in Coulomb's law and μ_0 occurs in the proportionality constant in Ampère's law for the magnetic field produced by an electric current. μ_0 is arbitrarily assigned the value $\mu_0 \equiv 4\pi \times 10^{-7}$ N s^2/C^2 in the SI system. Use $c = (\mu_0\varepsilon_0)^{-1/2}$ to verify Eq. (19.2) for $1/4\pi\varepsilon_0$.

19.4 (*a*) Calculate the electrostatic potential energy of two electrons separated by 3.0 Å in vacuum. Express your answer in joules and in electronvolts. (*b*) Calculate the electrostatic potential energy in eV of a system of two electrons and a proton in vacuum if the electrons are separated by 3.0 Å and the electron–proton distances are 4.0 and 5.0 Å.

Section 19.2

19.5 Explain why the observed charge-to-mass ratio of electrons decreases when the electrons are accelerated to very high speeds.

19.6 What fraction of the volume of an atom of radius 10^{-8} cm is occupied by its nucleus if the nuclear radius is 10^{-12} cm?

19.7 The density of gold is 19.3 g/cm^3. If gold atoms were cubes, what would the length of each side of a cubic atom be?

Section 19.3

19.8 True or false? (*a*) The photon emitted in an $n = 3$ to $n = 2$ transition in the H atom has a lower frequency than the photon for an $n = 2$ to $n = 1$ H-atom transition. (*b*) The ground-state energy of He$^+$ is about 4 times the ground-state energy of H. (*c*) ψ is zero at the nucleus for all H-atom stationary states. (*d*) $|\psi|^2$ is a maximum at the nucleus in the H-atom ground state. (*e*) The most probable value of the electron–nucleus distance in a ground-state H atom is zero. (*f*) The smallest allowed value of the atomic quantum number n is 0. (*g*) For H-atom stationary states with $l = 0$, ψ is independent of θ and ϕ. (*h*) For the H-atom ground state, the electron is confined to move on the surface of a sphere centered around the nucleus. (*i*) For the H-atom ground state, the electron is confined to move within a sphere of fixed radius.

19.9 Match each of the spherical coordinates r, θ, and ϕ with each of the following descriptions and give the range of each coordinate. (*a*) Angle between the positive z axis and the radius vector. (*b*) Distance to the origin. (*c*) Angle between the positive x axis and the projection of the radius vector in the xy plane.

19.10 True or false? For the hydrogen atom, (*a*) the allowed energy levels are $E = -(13.60 \text{ eV})/n^2$ and $E \geq 0$; (*b*) any photon with energy $E_{\text{photon}} \geq 13.60$ eV can ionize a hydrogen atom in the $n = 1$ state; (*c*) Any photon with $E_{\text{photon}} \geq 0.75(13.60 \text{ eV})$ can cause a hydrogen atom to go from the $n = 1$ state to the $n = 2$ state.

19.11 Give the allowed values of (*a*) l for $n = 5$ and (*b*) m if $l = 5$.

19.12 Omitting spin considerations, give the degeneracy of the hydrogenlike energy level with (*a*) $n = 1$; (*b*) $n = 2$; (*c*) $n = 3$.

19.13 Calculate the ionization potential in V of (*a*) He$^+$; (*b*) Li^{2+}.

19.14 Calculate the wavelength of the photon emitted when an electron goes from the $n = 3$ to $n = 2$ level of a hydrogen atom.

19.15 Calculate a in Eq. (19.14).

19.16 Positronium is a species consisting of an electron bound to a positron (Sec. 17.19). Calculate its ionization potential.

19.17 Show that $\langle r \rangle = 3a/2Z$ for a ground-state hydrogenlike atom. Use a table of integrals.

19.18 Use the Taylor-series expansions about $\phi = 0$ for $e^{i\phi}$, $\sin\phi$, and $\cos\phi$ to verify that $e^{i\phi} = \cos\phi + i\sin\phi$.

19.19 Verify Eq. (19.23) for $2p_x$ and $2p_y$.

19.20 (*a*) Let $z_1 = a_1 + ib_1$ and $z_2 = a_2 + ib_2$, where $i = \sqrt{-1}$ and the a's and b's are real. If $z_1 = z_2$, what must be true about the a's and b's? (*b*) Verify that the requirement that $\Phi(\phi) = \Phi(\phi + 2\pi)$ leads to the requirement that m in (19.10) be an integer.

19.21 Draw a rough graph (not to scale) of the value of ψ_{2p_z} along the z axis versus z. Then do the same for $|\psi_{2p_z}|^2$.

19.22 Find r_{2s} for H using the 90% probability definition.

19.23 Verify that the $1s$ wave function in Table 19.1 is an eigenfunction of the hydrogenlike Hamiltonian operator. (Use the chain rule to find the partial derivatives.)

19.24 Show that the average potential energy $\langle V \rangle$ for a ground-state hydrogenlike atom is $-Z^2 e^2 / 4\pi\varepsilon_0 a$.

19.25 (*a*) Complete Example 19.3 in Sec. 19.3 and find $\langle r \rangle$ for a hydrogen atom in the $2p_z$ state. (*b*) Without doing any calculations, give the value of $\langle r \rangle$ for a $2p_x$ H atom. (*c*) Verify your answer to (*b*) by evaluating the appropriate triple integral.

19.26 Show that the maximum in the radial distribution function of a ground-state hydrogenlike atom is at a/Z.

19.27 For a hydrogen atom in a 1s state, calculate the probability that the electron is between 0 and 2.00 Å from the nucleus.

19.28 Verify that $\int_0^{2\pi} |\Phi|^2 \, d\phi = 1$, where Φ is given by (19.10).

Section 19.4

19.29 True or false for the classical-mechanical angular momentum $\mathbf{L}$? (*a*) $\mathbf{L}$ of a particle depends on which point is chosen as the origin. (*b*) For a particle vibrating back and forth on a straight line through the origin, $\mathbf{L}$ is zero. (*c*) For a particle revolving around the origin on a circle, $\mathbf{L}$ is nonzero.

19.30 Calculate the angles the three angular-momentum vectors make with the z axis in Fig. 19.10.

19.31 (*a*) From the definition of angular momentum in Sec. 19.4, show that for a classical particle of mass m moving on a circle of radius r, the magnitude of the angular momentum with respect to the circle's center is mvr. (*b*) What is the direction of the $\mathbf{L}$ vector for this system?

19.32 Give the magnitude of the ground-state orbital angular momentum of the electron in a hydrogen atom according to (*a*) quantum mechanics; (*b*) the Bohr theory.

19.33 Calculate the magnitude of the orbital angular momentum of a 3p electron in a hydrogenlike atom.

Section 19.5

19.34 Calculate in SI units the magnitude of the spin angular momentum of an electron.

19.35 Calculate the angles between the spin vectors and the z axis in Fig. 19.11.

19.36 State what physical property is associated with each of the following quantum numbers in a one-electron atom and give the value of this physical property in terms of the quantum number. (*a*) l; (*b*) m; (*c*) s; (*d*) m_s.

19.37 For a particle with $s = 3/2$: (*a*) sketch the possible orientations of the $\mathbf{S}$ vector with the z axis; (*b*) calculate the smallest possible angle between $\mathbf{S}$ and the z axis.

Section 19.6

19.38 True or false? (*a*) The spatial factor in the ground-state wave function of He is antisymmetric. (*b*) All two-electron spin functions are antisymmetric. (*c*) The wave function of every system of identical particles must be antisymmetric with respect to exchange of all coordinates of any two particles. (*d*) Interchange of electrons 1 and 2 in the He-atom Hamiltonian (19.32) does not change this Hamiltonian.

19.39 State whether each of these functions is symmetric, antisymmetric, or neither: (*a*) $f(1)g(2)$; (*b*) $g(1)g(2)$; (*c*) $f(1)g(2) - g(1)f(2)$; (*d*) $r_1^2 - 2r_1 r_2 + r_2^2$; (*e*) $(r_1 - r_2)e^{-br_{12}}$, where r_{12} is the distance between particles 1 and 2.

19.40 A professor does a variational calculation on the ground state of He and finds that the variational integral equals -86.7 eV. Explain why it is certain that the professor made an error.

Section 19.7

19.41 Give the term symbol for the term arising from each of the following H-atom electron configurations: (*a*) 1s; (*b*) 3p; (*c*) 3d.

19.42 Give the values of L and S for a ^4F term.

19.43 State what physical property is associated with each of the following quantum numbers in a many-electron atom and give the value of this property in terms of the quantum number: (*a*) L; (*b*) S; (*c*) M_S.

19.44 For a ^3D term, give the value of (*a*) the total electronic orbital angular momentum; (*b*) the total electronic spin angular momentum.

19.45 Give the terms arising from each of the following electron configurations of K: (*a*) $1s^2 2s^2 2p^6 3s^2 3p^6 3d$; (*b*) $1s^2 2s^2 2p^6 3s^2 3p^6 4p$.

19.46 Draw a sketch like Fig. 19.14 that shows the orientations of $\mathbf{S}_1$, $\mathbf{S}_2$, and $\mathbf{S}$ for the spin function $\alpha(1)\alpha(2)$. (*Hint:* Begin by finding the angles between the z axis and each of $\mathbf{S}_1$, $\mathbf{S}_2$, and $\mathbf{S}$.)

19.47 Consider two angular momenta $\mathbf{M}_1$ and $\mathbf{M}_2$ (these can be orbital or spin angular momenta) whose magnitudes are $[j_1(j_1 + 1)]^{1/2}\hbar$ and $[j_2(j_2 + 1)]^{1/2}\hbar$, respectively. Let $\mathbf{M}_1$ and $\mathbf{M}_2$ combine with each other to give a total angular momentum $\mathbf{M}$, which is the vector sum of $\mathbf{M}_1$ and $\mathbf{M}_2$; $\mathbf{M} = \mathbf{M}_1 + \mathbf{M}_2$. The magnitude of $\mathbf{M}$ can be shown to be $[J(J + 1)]^{1/2}\hbar$, where the quantum number J has the possible values [*Levine* (2000), sec. 11.4]

$$j_1 + j_2, \; j_1 + j_2 - 1, \; j_1 + j_2 - 2, \ldots, \; |j_1 - j_2|$$

For example, when the spins of two electrons with spin quantum numbers $s_1 = \frac{1}{2}$ and $s_2 = \frac{1}{2}$ combine to give a total electronic spin, the possible values of the total spin quantum number are $\frac{1}{2} + \frac{1}{2} = 1$ and $|\frac{1}{2} - \frac{1}{2}| = 0$. (*a*) For terms arising from the electron configuration $1s^2 2s^2 2p^6 3s^2 3p3d$, give the possible values of the total electronic orbital-angular-momentum quantum number L (electrons in filled subshells contribute zero to the total orbital angular momentum and to the total spin angular momentum and so can be ignored) and give the possible values of the total electronic spin quantum number S. (*b*) Pair each possible value of L with each possible value of S to give the terms that arise from the . . . $3p3d$ electron configuration. [*Note*: For an electron configuration like $1s^2 2s^2 2p^2$ that has two or more electrons in a partly filled subshell, the Pauli exclusion principle restricts the possible terms, and special techniques must be used to find the terms in this case. See *Levine* (2000), sec. 11.5.]

19.48 When spin–orbit interaction splits an atomic term into energy levels, each energy level can be characterized by a total electronic angular momentum **J** that is the vector sum of the total electronic orbital and spin angular momenta: **J** = **L** + **S**. The magnitude of **J** is $[J(J + 1)]^{1/2}\hbar$, where the possible values of the quantum number J are given by the angular-momentum addition rule in Prob. 19.47 as

$$J = L + S, L + S - 1, L + S - 2, \ldots |L - S|$$

Each level is indicated by writing its J value as a subscript on the term symbol. For example, the Na electron configuration $1s^2 2s^2 2p^6 3p$ gives rise to the term ^2P with $L = 1$ and $S = \frac{1}{2}$. With $L = 1$ and $S = \frac{1}{2}$, the possible J values are $1 + \frac{1}{2} = \frac{3}{2}$ and $|1 - \frac{1}{2}| = \frac{1}{2}$. Therefore, a ^2P term has two energy levels, $^2P_{3/2}$ and $^2P_{1/2}$. Give the levels that arise from each of the following terms: (a) ^2S; (b) ^4P; (c) ^5F; (d) ^3D.

Section 19.8

19.49 Write down the Hamiltonian operator for the internal motion in Li.

19.50 For a system of two electrons in a one-dimensional box, write down the approximate wave functions (interelectronic repulsion ignored) including spin for states that have one electron with $n = 1$ and one electron with $n = 2$. Which of these states has (have) the lowest energy?

19.51 Write down an approximate wave function for the Be ground state.

19.52 Which of the first 10 elements in the periodic table have paramagnetic ground states?

19.53 Calculate the eighteenth ionization potential of Ar.

19.54 Use the ionization-potential data in Sec. 19.8 to calculate Z_{eff} for the $2s$ electrons in (a) Li; (b) Be.

19.55 (a) Suppose the electron had spin quantum number $s = \frac{3}{2}$. What would be the ground-state configurations of atoms with 3, 9, and 17 electrons? (b) Suppose the electron had $s = 1$. What would be the ground-state configurations of atoms with 3, 9, and 17 electrons?

19.56 For each pair, state which would have the higher first ionization potential (refer to a periodic table): (a) Na, K; (b) K, Ca; (c) Cl, Br; (d) Br, Kr.

19.57 Use Fig. 19.15 to calculate Z_{eff} for the $1s$, $2s$, and $2p$ electrons in Ne.

19.58 True or false? (a) The $2s$ orbital energy in K is lower than the $1s$ orbital energy in H. (b) Interchange of two rows of a determinant multiplies the determinant's value by -1.

19.59 Of the elements with $Z \leq 10$, which one has the largest number of unpaired electrons in its ground state?

19.60 Consider the systems (a) $Na^+ + 2e^-$, (b) $Na + e^-$, (c) Na^-, where in each system the Na atom or ion and the electron(s) are at infinite separation from one another. Use data in Sec. 19.8 to decide which system has the lowest energy and which has the highest energy.

19.61 Which species in each of the following pairs has the larger atomic radius: (a) Ca, Sr; (b) F, Ne; (c) Ar, K; (d) C, O; (e) Cl^-, Ar?

General

19.62 Write out the explicit form for the five-term ground-state helium-atom Hartree–Fock orbital given in Sec. 19.9. The only nonnumerical constants in your expression should be a_0 and π.

19.63 Let D_1, D_2, D_3, and D_4 be two-row Slater determinants that contain the following spin-orbitals: $1s\alpha$ and $2s\alpha$ in D_1; $1s\alpha$ and $2s\beta$ in D_2; $1s\beta$ and $2s\alpha$ in D_3; $1s\beta$ and $2s\beta$ in D_4. Consider the four helium-atom approximate wave functions given in Fig. 19.13 for states of the $1s2s$ configuration. Show that two of these wave functions are each equal to one of the determinants D_1, D_2, D_3, D_4 but that the other two wave functions must each be expressed as a linear combination of two of these determinants. Thus, an orbital wave function for a state with partly filled orbitals must sometimes be expressed as a linear combination of more than one Slater determinant. [In the CI wave function (19.57), each $\psi_{orb,j}$ should have the same S and M_S spin quantum numbers as the wave function ψ. Thus, for a CI wave function for the He ground state, the $\psi_{orb,j}$ that corresponds to the $1s2s$ configuration must have the spin function (19.38) (Fig. 19.13). As shown in this problem, this $\psi_{orb,j}$ is a linear combination of two Slater determinants.]

19.64 Derive the formula for the volume of a sphere by integrating the spherical coordinate volume element (19.24) over the sphere's volume.

19.65 For each of the following pairs, state which quantity (if any) is larger: (a) the ground-state energy of H or He^+; (b) the ionization energy of K or K^+; (c) the wavelength of the longest-wavelength electronic absorption of ground-state H or He^+; (d) the ionization energy of Cl^- or the electron affinity of Cl?

19.66 Give an example of a quantum-mechanical system for which the spacing between energy levels: (a) increases as E increases; (b) remains the same as E increases; (c) decreases as E increases.

19.67 For the ground state of the hydrogen atom, find the probability the electron is in a tiny spherical region of radius 1.0×10^{-3} Å if this sphere is centered at a point that is (a) at the origin (nucleus); (b) a distance 0.50 Å from the nucleus; (c) a distance 5.0 Å from the nucleus. Consider the tiny sphere to be infinitesimal.

19.68 For the ground state of the hydrogen atom, find the probability that the distance between the electron and the proton lies in each of the following ranges (treat each range as infinitesimal): (a) 0.100 and 0.101 Å; (b) 0.500 and 0.501 Å; (c) 1.000 and 1.001 Å; (d) 5.000 and 5.001 Å.

19.69 For each of the following systems, give the expression for $d\tau$ in the equation $\int |\psi|^2 d\tau = 1$ and give the limits on each coordinate: (a) one-dimensional harmonic oscillator; (b) particle in a three-dimensional rectangular box with edges a, b, and c; (c) the hydrogen atom internal motion using spherical coordinates.

19.70 Is there a gravitational attraction between the electron and the proton in the H atom? If there is, why is this not taken into account in the Hamiltonian? Do a calculation to support your answer.

19.71 (a) Show that the maximum value of ψ_{2p_z} for $Z = 1$ is $\psi_{\text{max}} = 1/(2a)^{3/2}\pi^{1/2}e$. (b) Write a computer program that will vary z/a from 0.01 to 10 in steps of 0.01 and for each value of z/a will calculate values of y/a for which $|\psi_{2p_z}/\psi_{\text{max}}|$ is equal to a certain constant k, where the value of k is input at the start of the program. Note that for some values of z/a, there are no values of y/a that satisfy the condition. Be careful that spurious values of y/a are eliminated. (The output of this program can be used as input to a graphing program to graph contours of the $2p_z$ orbital.)

19.72 *Physicist trivia question.* Name the physicist referred to in each of the following descriptions. All names appear in Chapter 19. Two of these physicists have elements named after them. (a) This experimental physicist (rated the 10th greatest physicist of all times in a 1999 poll) was weak in mathematics. Norman Ramsey took a course given by him in the 1930s and found that when this physicist tried to derive in class the formula for Rutherford scattering of alpha particles, "he got completely fouled up in the math, and he finally ended up telling us to go home and work it out for ourselves." Later, Ramsey came to recognize the great physical insight this physicist had and Ramsey concluded that "an ability to make a formal mathematical derivation was not the criterion of being a good physicist." (b) He was friends with the Swiss psychoanalyst Carl Jung and contributed a chapter to a book written by Jung. Jung published analyses of many of the dreams of this physicist; the number 4 often occurred in these dreams. (c) He was one of the few twentieth-century physicists who did outstanding work in both experiments and theory. In the mid-1930s, he and coworkers bombarded many elements with neutrons and produced radioactive products. He found that uranium irradiated with neutrons gave products whose atomic numbers did not lie in the range 86 to 92 and concluded that he had produced new elements with atomic numbers of 93 and 94, which he called ausenium and hasperium, respectively. He received a Nobel Prize in physics "for his demonstration of the existence of new radioactive elements produced by neutron irradiation, and for his related discovery of nuclear reactions brought about by slow neutrons." In fact, he had not prepared elements with $Z > 92$. One month after he received his Nobel Prize, Hahn and Strassmann published work showing that neutron irradiation of uranium gave barium as one product. Meitner and Frisch used Bohr's liquid-drop model of the nucleus to interpret the Hahn–Strassmann results as the fission of a uranium nucleus to produce two lighter nuclei. On December 2, 1942, the first human-produced self-sustaining nuclear-fission chain reaction was achieved on a squash court at the University of Chicago in a uranium pile constructed under the direction of the subject of this question. The success of the experiment was reported in a coded telephone conversation with the words "The Italian navigator has just landed in the New World."

19.73 True or false? (a) In this chapter, e stands for the charge on an electron. (b) The exact helium-atom ground-state wave function is a product of wave functions for each electron. (c) The wave function of every system of fermions must be antisymmetric with respect to interchange of all coordinates of any two particles. (d) The spin quantum number s of an electron has the possible values $\pm\frac{1}{2}$. (e) The shape of a $2p_z$ orbital is two tangent spheres. (f) All states belonging to the same electron configuration of a given atom must have the same energy. (g) Every solution of the time-independent Schrödinger equation is a possible stationary state. (h) The ground-state wave function of a lithium atom cannot be expressed as a spatial factor times a spin factor. (i) Every linear combination of two solutions of the time-independent Schrödinger equation is a solution of this equation.

Molecular Electronic Structure

A full and correct treatment of molecules must be based on quantum mechanics. Indeed, the stability of a covalent bond cannot be understood without quantum mechanics. Because of the mathematical difficulties involved in the application of quantum mechanics to molecules, chemists have developed a variety of empirical concepts to describe bonding. Section 20.1 discusses some of these concepts. Section 20.2 describes how the molecular Schrödinger equation is separated into Schrödinger equations for electronic motion and for nuclear motion. The one-electron molecule H_2^+ is discussed in Sec. 20.3 to develop some ideas about electron orbitals in molecules. A major approximation method used in describing molecular electronic structure is the molecular-orbital method, developed in Secs. 20.4 to 20.6. Section 20.8 shows how molecular properties are calculated from electronic wave functions. Section 20.9 discusses some of the remarkable advances in calculation of molecular electronic structure made in recent years. The currently most widely used method for calculating molecular properties, density-functional theory, is presented in Sec. 20.10. Section 20.11 discusses semiempirical methods of calculation, which can treat large molecules. Section 20.12 gives details on how electronic-structure calculations are done. Section 20.13 presents the molecular-mechanics method, a nonquantum-mechanical method that can be applied to very large molecules.

20.1 CHEMICAL BONDS

Bond Radii

The length of a bond in a molecule is the distance between the nuclei of the two atoms forming the bond. Spectroscopic and diffraction methods (Chapters 21 and 24) enable bond lengths to be measured accurately. Bond lengths range from 0.74 Å in H_2 to 4.65 Å in Cs_2 and are usually in the range 1–2 Å for bonds between elements in the first, second, and third periods. The length of a given kind of bond is found to be approximately constant from molecule to molecule. For example, the carbon–carbon single-bond length in most nonconjugated molecules lies in the range 1.53 to 1.54 Å. Moreover, one finds that the bond length d_{AB} between atoms A and B is *approximately* equal to $\frac{1}{2}(d_{AA} + d_{BB})$, where d_{AA} and d_{BB} are the typical A—A and B—B bond lengths. For example, let A and B be Cl and C. The bond length in Cl_2 is 1.99 Å, and $\frac{1}{2}(d_{AA} + d_{BB}) = \frac{1}{2}(1.99 + 1.54)$ Å $= 1.76$ Å, in good agreement with the observed bond length 1.76_6 Å in CCl_4.

One can therefore take $\frac{1}{2}d_{AA}$ as the **bond radius** (or **covalent radius**) r_A for atom A and use a table of bond radii to estimate the bond length d_{AB} as $r_A + r_B$. Double and triple bonds are shorter than the corresponding single bonds, and so different bond radii are used for single, double, and triple bonds. Some bond radii (due mainly to Pauling) in angstroms (Å) are (1 Å $\equiv 10^{-8}$ cm $\equiv 10^{-10}$ m):

	H	C	N	O	F	P	S	Cl	Br	I
Single	0.30	0.77	0.70	0.66	0.64	1.10	1.04	0.99	1.14	1.33
Double		0.67	0.60	0.56		1.00	0.94			
Triple		0.60	0.55							

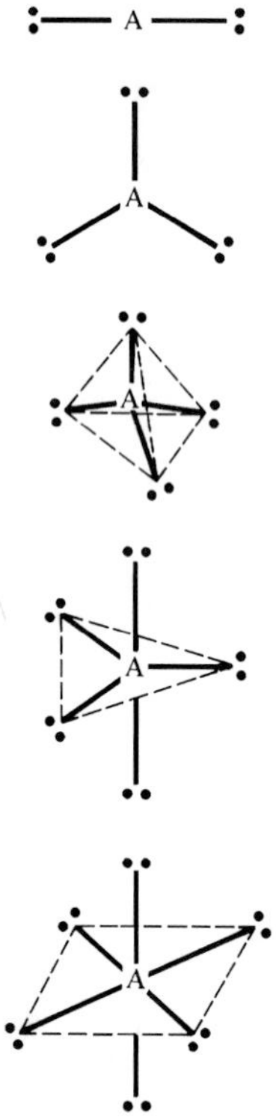

The bond length 0.74 Å in H_2 indicates $r_A = 0.37$ Å for H, but the listed value 0.30 Å works better in predicting bond lengths between H and other elements.

When atoms A and B differ substantially in electronegativity, the observed bond length is often shorter than $r_A + r_B$.

The carbon–carbon bond length in benzene is 1.40 Å. This lies between the carbon–carbon single-bond length 1.54 Å and double-bond length 1.34 Å, which indicates that the benzene bonds are intermediate between single and double bonds.

Bond Angles

The VSEPR (valence-shell electron-pair repulsion) method estimates bond angles at an atom A by counting the number of valence electron pairs that surround atom A in the molecule's Lewis electron-dot formula. The valence pairs around A are arranged in space to minimize electrostatic repulsions between pairs. The VSEPR arrangements for various numbers of pairs are (Fig. 20.1):

Number of pairs	2	3	4	5	6
Arrangement	linear	trigonal planar	tetrahedral	trigonal bipyramidal	octahedral
Angles	180°	120°	109.5°	90°, 120°	90°

For five valence electron pairs, lone pairs are placed in the equatorial position(s) (Fig. 20.2). A double bond or a triple bond is counted as one pair for the purposes of the VSEPR method. Lone pairs are more spread out in space than bonding pairs, so the lone pairs push the bonding pairs on an atom together a bit, making the bond angle(s) at that atom a bit less than the values listed in the table just given. For example, the bond angle in H_2O (which has two bonding pairs and two lone pairs on O) is 104.5° instead of 109.5°, and the bond angles in ClF_3 (Fig. 20.2) are a bit less than 90°.

Bond Energies

Section 5.10 explained how experimental ΔH°_{298} values for gas-phase atomization processes can be used to give average bond energies (which can be used to estimate ΔH°_{298} for gas-phase reactions). Table 20.1 lists some average bond energies. The values listed for H—H, F—F, Cl—Cl, O=O, and N≡N are ΔH°_{298} for dissociation of the

Figure 20.1

Arrangements of valence electron pairs around a central atom.

TABLE 20.1

Average Bond Energies in kJ/mol[a]

C—H	C—C	C—O	C—N	C—S	C—F	C—Cl	C—Br	C—I	F—F
415	344	350	292	259	441	328	276	240	158

N—H	O—H	S—H	S—S	N—O	O—O	N—N	N—Cl	H—H	Cl—Cl
391	463	368	266	175	143	159	200	436	243

C=C	C=O	C=N	N=N	O=O	C≡C	C≡N	N≡N		
615	725	615	418	498	812	890	946		

[a]Data from L. Pauling, *General Chemistry*, 3d ed., Freeman, 1970, p. 913.

appropriate gas-phase diatomic molecule. Double and triple bonds are stronger than single bonds. The N—N, O—O, and N—O single bonds are quite weak.

The fact that the bond energy of a carbon–carbon double bond is less than twice the energy of a carbon–carbon single bond makes vinyl addition polymerizations possible. The reaction $RCH_2CH_2 \cdot + CH_2{=}CH_2 \rightarrow RCH_2CH_2CH_2CH_2 \cdot$ has $\Delta S°$ negative, since two molecules are replaced by one, and has $\Delta H°$ negative, since one C=C bond is replaced by two C—C bonds.

Tabulated bond energies are on a per-mole basis. To convert to a per-molecule basis, we divide by the Avogadro constant. One kJ/mol corresponds to $(1 \text{ kJ/mol})/N_A$ = 1.66054×10^{-21} J per molecule. Since 1 eV = 1.60218×10^{-19} J [Eq. (19.3)], 1 kJ/mol corresponds to 0.010364 eV per molecule. Thus

$$1 \text{ eV/molecule corresponds to } 96.485 \text{ kJ/mol } (23.061 \text{ kcal/mol}) \qquad (20.1)$$

Bond Moments

The **electric dipole moment** $\boldsymbol{\mu}$ of a charge distribution is defined by Eq. (14.82). Molecular dipole moments can be found by microwave spectroscopy (Sec. 21.7) or by dielectric-constant measurements (Sec. 14.15). From (14.82), the SI unit of μ is the coulomb-meter (C m). Molecular dipole moments are usually quoted in units of **debyes** (D), where

$$1 \text{ D} \equiv 3.335641 \times 10^{-30} \text{ C m} \qquad (20.2)$$

[The debye was originally defined as 10^{-18} statC cm, where 1 statC (Sec. 19.1) corresponds to 3.335641×10^{-10} C.] For example μ of HCl is 1.07 D = 3.57×10^{-30} C m.

The dipole moment of a molecule can be roughly estimated by taking the *vector* sum of assigned **bond dipole moments** for the bonds. Some bond moments in debyes are:

H—O	H—N	H—C	C—Cl	C—Br	C—O	C=O	C—N	C≡N
1.5	1.3	0.4	1.5	1.4	0.8	2.5	0.5	3.5

where the first-listed atom is the positive end of the bond moment. The value for H—C is an assumed one, and the other moments involving C depend on the magnitude and sign of this assumed value. The above table uses the traditionally assumed polarity H^+—C^-, but this may be incorrect; see A. E. Reed and F. Weinhold, *J. Chem. Phys.,* **84,** 2428 (1986); K. B. Wiberg et al., *Science,* **252,** 1266 (1991).

The H—O and H—N bond moments are calculated from the observed dipole moments of H_2O and NH_3 without explicitly considering the contributions of the lone pairs to the dipole moment. Their contributions are absorbed into the values calculated for the OH and NH moments. For example, the observed μ for H_2O is 1.85 D, and the bond angle is 104.5°; Fig. 20.3 gives $2\mu_{OH} \cos 52.2° = 1.85$ D, and the O—H bond moment is $\mu_{OH} = 1.5$ D. The other moments listed are calculated from the experimental dipole moments and geometries of CH_3Cl, CH_3Br, CH_3OH, $(CH_3)_2CO$,

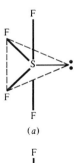

(a)

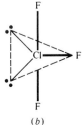

(b)

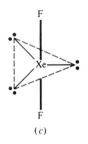

(c)

Figure 20.2

Some molecules with five valence electron pairs around the central atom.

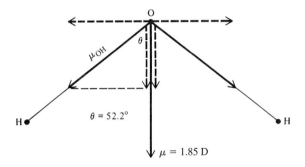

Figure 20.3

Calculation of the OH bond moment in H_2O. The dashed vectors are the bond-moment components along $\boldsymbol{\mu}$ and perpendicular to $\boldsymbol{\mu}$.

$(CH_3)_3N$, and CH_3CN using the assumed CH bond moment and the OH and NH moments.

A shortcut in bond-moment calculations is to note that the vector sum of the three CH bond moments of a tetrahedral CH_3 group equals the moment of one CH bond. This follows from the zero dipole moment of methane (HCH_3).

Electronegativity

The **electronegativity** x of an element is a measure of the ability of an atom of that element to attract the electrons in a bond. The degree of polarity of an A—B bond is related to the difference in the electronegativities of the atoms forming the bond.

Many electronegativity scales have been proposed [J. Mullay, *Structure and Bonding,* **66**, 1 (1987); L. C. Allen, *Acc. Chem. Res.,* **23**, 175 (1990)]. The best-known is the Pauling scale, based on bond energies. Pauling observed that the A—B average bond energy generally exceeds the mean of the A—A and B—B average bond energies by an amount that increases with increasing polarity of the A—B bond. The Pauling electronegativity scale defines the electronegativity difference between elements A and B as

$$|x_A - x_B| \equiv 0.102(\Delta_{AB}/\text{kJ mol}^{-1})^{1/2} \tag{20.3}$$

where $\Delta_{AB} \equiv E(A—B) - \frac{1}{2}[E(A—A) + E(B—B)]$ and where the E's are average single-bond energies. The electronegativity of H is arbitrarily set at 2.2.

The exothermicity of the combustion of hydrocarbons can be explained in terms of electronegativities. The large electronegativity differences between C and O and between O and H lead to highly polar bonds in the products CO_2 and H_2O, whereas the C—H, C—C, and O=O bonds in the reactants have low or no polarity. Therefore the total bond energy of the products is substantially greater than that of the reactants and the reaction is very exothermic.

The Allred–Rochow scale [A. L. Allred and E. G. Rochow, *J. Inorg. Nucl. Chem.,* **5**, 264, 269 (1958)] defines the electronegativity x_A of element A as

$$x_A \equiv 0.359 Z_{eff}/(r_A/\text{Å})^2 + 0.744 \tag{20.4}$$

where r_A is the bond radius of A and Z_{eff} is the effective nuclear charge [Eq. (19.52)] that would act on an electron added to the valence shell of a neutral A atom. The quantity $Z_{eff}e^2/4\pi\varepsilon_0 r_A^2$ is the average force exerted by atom A on an added electron. The constants 0.359 and 0.744 were chosen to make the scale as consistent as possible with the Pauling scale.

The Allen scale [L. C. Allen, *J. Am. Chem. Soc.,* **111**, 9003 (1989); J. B. Mann et al., *J. Am. Chem. Soc.,* **122**, 2780, 5132 (2000)] takes the electronegativity x of an atom as proportional to the average ionization energy $\langle E_{i,\text{val}} \rangle$ of the valence-shell electrons of the ground-state free atom: $x = 0.169\langle E_{i,\text{val}} \rangle/\text{eV}$.

The Nagle scale [J. K. Nagle, *J. Am. Chem. Soc.,* **112**, 4741 (1990)] defines the electronegativity x in terms of the polarizability α (Sec. 14.15) of the atom: $x = 1.66[n(4\pi\varepsilon_0 \text{ Å}^3/\alpha)]^{1/3} + 0.37$, where n is the number of valence electrons of the atom and $1 \text{ Å} = 10^{-10}$ m. Nagle assumed $n = 2$ (the valence s electrons) for each transition element.

Some electronegativities on the Pauling scale are given in Table 20.2. Electronegativities tend to decrease going down a group in the periodic table (because of the increasing distance of the valence electrons from the nucleus) and to increase going across a period (mainly because of the increasing Z_{eff} resulting from the lesser screening by electrons added to the same shell). Although electronegativity is an imprecise concept, electronegativities on various scales generally agree well. Defects of the Pauling scale are discussed in L. R. Murphy et al., *J. Phys. Chem. A,* **104**, 5867 (2000).

TABLE 20.2

Some Pauling Electronegativities[a]

H	Li	Be	B	C	N	O	F
2.2	1.0	1.6	2.0	2.5	3.0	3.4	4.0
	Na	Mg	Al	Si	P	S	Cl
	0.9	1.3	1.6	1.9	2.2	2.6	3.2
	K	Ca	Ga	Ge	As	Se	Br
	0.8	1.0	1.8	2.0	2.2	2.6	3.0
	Rb	Sr	In	Sn	Sb	Te	I
	0.8	0.9	1.8	2.0	2.1		2.7

[a]Data from A. L. Allred, *J. Inorg. Nucl. Chem.*, **17**, 215 (1961).

20.2   THE BORN–OPPENHEIMER APPROXIMATION

All molecular properties are, in principle, calculable by solving the Schrödinger equation for the molecule. Because of the great mathematical difficulties involved in solving the molecular Schrödinger equation, one must make approximations. Until about 1960, the level of approximations was such that the calculations gave only qualitative and not quantitative information. Since then, the use of computers has made molecular wave-function calculations accurate enough to give reliable quantitative information in many cases.

The Hamiltonian operator for a molecule is

$$\hat{H} = \hat{K}_N + \hat{K}_e + \hat{V}_{NN} + \hat{V}_{Ne} + \hat{V}_{ee} \tag{20.5}$$

where $\hat{K}_N$ and $\hat{K}_e$ are the kinetic-energy operators for the nuclei and the electrons, respectively, $\hat{V}_{NN}$ is the potential energy of repulsions between the nuclei, $\hat{V}_{Ne}$ is the potential energy of attractions between the electrons and the nuclei, and $\hat{V}_{ee}$ is the potential energy of repulsions between the electrons.

The Born–Oppenheimer Approximation

The molecular Schrödinger equation $\hat{H}\psi = E\psi$ is extremely complicated, and it would be almost hopeless to attempt an exact solution, even for small molecules. Fortunately, the fact that nuclei are much heavier than electrons allows the use of a very accurate approximation that greatly simplifies things. In 1927, Max Born and J. Robert Oppenheimer showed that it is an excellent approximation to treat the electronic and nuclear motions separately. The mathematics of the Born–Oppenheimer approximation is complicated, and so we shall give only a qualitative physical discussion.

Because of their much greater masses, the nuclei move far more slowly than the electrons, and the electrons carry out many "cycles" of motion in the time it takes the nuclei to move a short distance. The electrons see the heavy, slow-moving nuclei as almost stationary point charges, whereas the nuclei see the fast-moving electrons as essentially a three-dimensional distribution of charge.

One therefore assumes a fixed configuration of the nuclei, and for this configuration one solves an electronic Schrödinger equation to find the molecular electronic energy and wave function. This process is repeated for many different fixed nuclear configurations to give the electronic energy as a function of the positions of the nuclei. The nuclear configuration that corresponds to the minimum value of the electronic energy is the equilibrium geometry of the molecule. Having found how the electronic energy varies as a function of the nuclear configuration, one then uses this electronic energy function as the potential-energy function in a Schrödinger equation for the nuclear motion, thereby obtaining the molecular vibrational and rotational energy levels for a given electronic state.

The electronic Schrödinger equation is formulated for a fixed set of locations for the nuclei. Therefore, the nuclear kinetic-energy operator $\hat{K}_N$ in (20.5) is omitted from the Hamiltonian, and the **electronic Hamiltonian** $\hat{H}_e$ and **electronic Schrödinger equation** are

$$\hat{H}_e = \hat{K}_e + \hat{V}_{Ne} + \hat{V}_{ee} + \hat{V}_{NN} \tag{20.6}$$

$$\hat{H}_e\psi_e = E_e\psi_e \tag{20.7}$$

E_e is the **electronic energy,** including the energy V_{NN} of nuclear repulsion. Note that V_{NN} in (20.6) is a constant, since the nuclei are held fixed. The electronic wave function ψ_e is a function of the $3n$ spatial and n spin coordinates (Sec. 19.5) of the n electrons of the molecule. The electronic energy E_e contains potential and kinetic energy of the electrons and potential energy of the nuclei.

Consider a diatomic (two-atom) molecule with nuclei A and B with atomic numbers Z_A and Z_B. The spatial configuration of the nuclei is specified by the distance R between the two nuclei. The potential-energy operator $\hat{V}_{Ne}$ depends on R as a parameter, as does the internuclear repulsion $\hat{V}_{NN}$, which equals $Z_A Z_B e^2/4\pi\varepsilon_0 R$ [Eq. (19.1)]. (A **parameter** is a quantity that is constant for one set of circumstances but may vary for other circumstances.) Hence, at each value of R, we get a different electronic wave function and energy. These quantities depend on R as a parameter and vary continuously as R varies. We therefore have $\psi_e = \psi_e(q_1, \ldots, q_n; R)$ and $E_e = E_e(R)$, where q_n stands for the spatial coordinates and spin coordinate of electron n. For a polyatomic molecule, ψ_e and E_e will depend parametrically on the locations of all the nuclei:

$$\psi_e = \psi_e(q_1, \ldots, q_n; Q_1, \ldots, Q_N), \qquad E_e = E_e(Q_1, \ldots, Q_N) \tag{20.8}$$

where the Q's are the coordinates of the N nuclei.

Of course, a molecule has many different possible electronic states. For each such state, there is a different electronic wave function and energy, which vary as the nuclear configuration varies. Figure 20.4 shows $E_e(R)$ curves for the ground electronic state and some excited states of H_2. Since $E_e(R)$ is the potential-energy function for motion of the nuclei, a state with a minimum in the $E_e(R)$ curve is a bound state, with the atoms bonded to each other. For an electronic state with no minimum, $E_e(R)$ increases continually as R decreases. This means that the atoms repel each other as they come together, and this is not a bound state. The colliding atoms simply bounce off each other. The two lowest electronic states in Fig. 20.4 each dissociate to two ground-state ($1s$) hydrogen atoms. [Note from (19.18) that -27.2 eV is the energy of two $1s$ hydrogen atoms.] The ground electronic state of H_2 dissociates to (and arises from) $1s$ H atoms with opposite electronic spins, whereas the repulsive first excited electronic state arises from $1s$ H atoms with parallel electron spins.

The internuclear distance R_e at the minimum in the E_e curve for a bound electronic state is the **equilibrium bond length** for that state. (Because of molecular zero-point vibrations, R_e is not quite the same as the observed bond length.) As R goes to zero, E_e goes to infinity, because of the internuclear repulsion V_{NN}. As R goes to infinity, E_e goes to the sum of the energies of the separated atoms into which the molecule decomposes. The difference $E_e(\infty) - E_e(R_e)$ is the **equilibrium dissociation energy** D_e of the molecule (Fig. 20.4). Some D_e and R_e values (found by spectroscopy) for the ground electronic states of diatomic molecules are given in Table 20.3. Note the high D_e values of CO and N_2, which have triple bonds.

We now resume consideration of the Born–Oppenheimer approximation. Having solved the electronic Schrödinger equation (20.7) to obtain the electronic energy

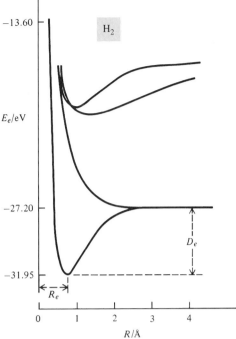

Figure 20.4

Potential-energy curves for the
lowest few electronic states of H_2.
R_e and D_e of the ground electronic
state are shown.

TABLE 20.3

Diatomic-Molecule Ground-State D_e and R_e Values

	H_2^+	H_2	He_2^+	Li_2	C_2	N_2	O_2	F_2
D_e/eV	2.8	4.75	2.5	1.1	6.3	9.9	5.2	1.7
R_e/Å	1.06	0.74	1.1	2.7	1.24	1.10	1.21	1.41

	CH	CO	NaCl	OH	HCl	CaO	NaH	NaK
D_e/eV	3.6	11.2	4.3	4.6	4.6	4.8	2.0	0.6
R_e/Å	1.12	1.13	2.36	0.97	1.27	1.82	1.89	3.59

$E_e(Q_1, \ldots, Q_N)$ as a function of the nuclear coordinates, we use this as the potential-energy function in the Schrödinger equation for nuclear motion:

$$(\hat{K}_N + E_e)\psi_N \equiv \hat{H}_N\psi_N = E\psi_N \qquad (20.9)$$

The Hamiltonian $\hat{H}_N$ for nuclear motion equals the nuclear kinetic-energy operator $\hat{K}_N$ plus the electronic energy function E_e, so E in (20.9) includes both electronic and nuclear energies and is the total energy of the molecule. The nuclear wave function ψ_N is a function of the $3N$ spatial and N spin coordinates of the N nuclei.

E_e is the potential energy for nuclear vibration. As the relatively sluggish nuclei vibrate, the rapidly moving electrons almost instantaneously adjust their wave function ψ_e and energy E_e to follow the nuclear motion. The electrons act somewhat like springs connecting the nuclei. As the internuclear distances change, the energy stored in the "springs" (that is, in the electronic motions) changes.

The nuclear kinetic-energy operator $\hat{K}_N$ involves vibrational, rotational, and translational kinetic energies. (Rotational and translational motions do not change the electronic energy E_e.) We shall deal with nuclear vibrations and rotations in Chapter 21. The remainder of this chapter deals with the electronic wave function and energy.

The Born–Oppenheimer treatment shows that the complete molecular wave function ψ is to a very good approximation equal to the product of electronic and nuclear wave functions: $\psi = \psi_e \psi_N$.

In addition to making the Born–Oppenheimer approximation, one usually neglects relativistic effects in treating molecules. This is a very good approximation for molecules composed of light atoms, but it is not good for molecules containing heavy atoms. Inner-shell electrons in atoms of high atomic number move at very high speeds and are significantly affected by the relativistic increase of mass with speed. The valence electrons can undergo relativistic effects due to interactions with the inner-shell electrons and to the portion of the valence-electrons' probability density that deeply penetrates the inner-shell electrons. Relativistic effects have substantial influence on bond lengths and binding energies of molecules containing atoms of high atomic number (for example, Au). See P. Pyykkö, *Chem. Rev.*, **88**, 563 (1988).

Ionic and Covalent Bonding

A bound electronic state of a diatomic molecule has a minimum in its curve of electronic energy E_e versus internuclear distance R (Fig. 20.4). Why is E_e lower in the molecule than in the separated atoms?

An ionic molecule like NaCl is held together by the Coulombic attraction between the ions. Solid NaCl consists of an array of alternating Na^+ and Cl^- ions, and one cannot pick out individual NaCl molecules. However, gas-phase NaCl consists of individual ionic NaCl molecules. (In aqueous solution, hydration of the ions makes the separated hydrated ions more stable than Na^+Cl^- molecules.) Ionic molecules dissociate to neutral atoms in the gas phase. Consider, for example, NaCl. The ionization energy of Na is 5.14 eV, whereas the electron affinity of Cl is only 3.61 eV. Hence, isolated Na and Cl atoms are more stable than isolated Na^+ and Cl^- ions. Thus as the internuclear distance R increases, the bonding in NaCl shifts from ionic to covalent at very large R values.

EXAMPLE 20.1 D_e and μ of NaCl

Use the model of an NaCl molecule as consisting of nonoverlapping spherical Na^+ and Cl^- ions separated by the experimentally observed distance $R_e = 2.36$ Å (Table 20.3) to estimate the equilibrium dissociation energy D_e and the dipole moment of NaCl.

Equation (19.1) gives the potential energy of interaction between two charges as $V = Q_1 Q_2 / 4\pi\varepsilon_0 r$. Therefore the energy needed to take the Na^+ and Cl^- ions from a 2.36-Å separation to an infinite separation (where $V = 0$) is $e^2 / 4\pi\varepsilon_0 R_e$. The use of (19.4) for e and of (19.3) gives

$$\frac{e^2}{4\pi\varepsilon_0 R_e} = \frac{(1.602 \times 10^{-19}\,C)^2}{4\pi(8.854 \times 10^{-12}\,C^2/\text{N-m}^2)(2.36 \times 10^{-10}\,m)}$$

$$= 9.77 \times 10^{-19}\,J = 6.10\,eV$$

However, this is not the estimate of D_e, since (as already noted) NaCl dissociates to neutral atoms. Breaking the dissociation into two hypothetical steps, we have

$$NaCl \xrightarrow{(a)} Na^+ + Cl^- \xrightarrow{(b)} Na + Cl$$

where the two ions (and the two atoms) are at infinite separation from each other. We estimated the energy change for step (*a*) as 6.10 eV. Addition of an electron to Na^+ lowers the energy by the Na ionization energy 5.14 eV, and removal of an electron from Cl^- raises the energy by the Cl electron affinity 3.61 eV. Hence, the nonoverlapping-spherical-ion model gives the energy needed to dissociate NaCl into Na + Cl as

$$6.10 \text{ eV} - 5.14 \text{ eV} + 3.61 \text{ eV} = 4.57 \text{ eV}$$

which is only 7 percent away from the experimental value $D_e = 4.25$ eV. The error results from neglect of the repulsion between the slightly overlapping electron probability densities of the Na^+ and Cl^- ions, which makes the molecule less stable than calculated.

The dipole moment of a charge distribution is given by Eq. (14.82) as $\mu = \Sigma_i Q_i \mathbf{r}_i$. The charge on Na^+ equals the proton charge e. Taking the coordinate origin at the center of the Cl^- ion, we estimate μ as

$$\mu = eR_e = (1.602 \times 10^{-19} \text{ C})(2.36 \times 10^{-10} \text{ m}) = 3.78 \times 10^{-29} \text{ C m} = 11.3 \text{ D}$$

where (20.2) was used. This value is not far from the experimental value 9.0 D. The error can be attributed to polarization of one ion by the other.

Exercise

The LiF molecule has $R_e = 1.56$ Å. Estimate D_e and μ of LiF. Use data in Chapter 19. (*Answers*: 7.2 eV and 7.5 D.)

The ionic bonding in NaCl can be contrasted with the nonpolar covalent bonding in H_2 and other homonuclear diatomic molecules. Here, the bonding electrons are shared equally. For a diatomic molecule formed from different nonmetals (for example, HCl, BrCl) or from different metals (for example, NaK), the bonding is polar covalent, the more electronegative atom having a greater share of the electrons and a partial negative charge. Bonds between metals with relatively high electronegativities and nonmetals are sometimes polar covalent, rather than ionic, as noted in Sec. 10.6.

The physical reason for the stability of a covalent bond is not a fully settled question. A somewhat oversimplified statement is that the stability is due to the decrease in the average potential energy of the electrons forming the bond. This decrease results from the greater electron–nuclear attractions in the molecule compared with those in the separated atoms. The electrons in the bond can feel the simultaneous attractions of two nuclei. This decrease in electronic potential energy outweighs the increases in interelectronic repulsions and internuclear repulsions that occur as the atoms come together.

20.3 THE HYDROGEN MOLECULE ION

The simplest molecule is H_2^+, which consists of two protons and one electron.

Adopting the Born–Oppenheimer approximation, we hold the nuclei at a fixed distance R and deal with the electronic Schrödinger equation $\hat{H}_e \psi_e = E_e \psi_e$ [Eq. (20.7)]. The electronic Hamiltonian including nuclear repulsion for H_2^+ is given by Eqs. (20.6), (19.1), and (18.58) as

$$\hat{H}_e = -\frac{\hbar^2}{2m_e}\nabla^2 - \frac{e^2}{4\pi\varepsilon_0 r_A} - \frac{e^2}{4\pi\varepsilon_0 r_B} + \frac{e^2}{4\pi\varepsilon_0 R} \qquad (20.10)$$

where r_A and r_B are the distances from the electron to nuclei A and B and R is the internuclear distance (Fig. 20.5). The first term on the right side of (20.10) is the operator for the kinetic energy of the electron. The second and third terms are the potential energies of attraction between the electron and the nuclei. The last term is the repulsion

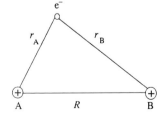

Figure 20.5

Interparticle distances in the H_2^+ molecule.

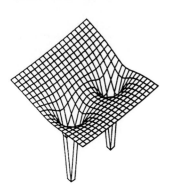

Figure 20.6

Three-dimensional plot of the potential energy of attraction between the electron and the nuclei of H_2^+ in a plane containing the nuclei.

between the nuclei. Since H_2^+ has only one electron, there is no interelectronic repulsion. Figure 20.6 is a three-dimensional plot of values of $-e^2/4\pi\varepsilon_0 r_A - e^2/4\pi\varepsilon_0 r_B$ in a plane containing the nuclei.

The electronic Schrödinger equation $\hat{H}_e\psi_e = E_e\psi_e$ can be solved exactly for H_2^+, but the solutions are complicated. For our purposes, an approximate treatment will suffice. The lowest electronic state of H_2^+ will dissociate to a ground-state ($1s$) H atom and a proton as R goes to infinity. Suppose the electron in H_2^+ is close to nucleus A and rather far from nucleus B. The H_2^+ electronic wave function should then resemble a ground-state H-atom wave function for atom A; that is, ψ_e will be approximately given by the function (Table 19.1 in Sec. 19.3)

$$1s_A \equiv (1/a_0)^{3/2}\pi^{-1/2}e^{-r_A/a_0} \qquad (20.11)$$

where the Bohr radius a_0 is used since the nuclei are fixed. Similarly, when the electron is close to nucleus B, ψ_e can be roughly approximated by

$$1s_B \equiv (1/a_0)^{3/2}\pi^{-1/2}e^{-r_B/a_0}$$

This suggests as an approximate wave function for the H_2^+ ground electronic state:

$$\phi = c_A 1s_A + c_B 1s_B = a_0^{-3/2}\pi^{-1/2}(c_A e^{-r_A/a_0} + c_B e^{-r_B/a_0}) \qquad (20.12)$$

which is a linear combination of the $1s_A$ and $1s_B$ atomic orbitals. When the electron is very close to nucleus A, then r_A is much less than r_B and e^{-r_A/a_0} is much greater than e^{-r_B/a_0}. Hence, the $1s_A$ term in (20.12) dominates, and the wave function resembles that of an H atom at nucleus A, as it should. Similarly for the electron close to nucleus B. One multiplies the spatial function of (20.12) by a spin function (either α or β) to get the complete approximate wave function.

The function (20.12) can be regarded as a variation function and the constants c_A and c_B chosen to minimize the variational integral $W = \int \phi^* \hat{H}\phi\, d\tau / \int \phi^*\phi\, d\tau$. The function (20.12) is a linear combination of two functions, and, as noted in Sec. 18.15, the conditions $\partial W/\partial c_A = 0 = \partial W/\partial c_B$ will be satisfied by two sets of values of c_A and c_B. These sets will yield approximate wave functions and energies for the lowest two electronic states of H_2^+. We need not go through the details of evaluating W and setting $\partial W/\partial c_A = 0 = \partial W/\partial c_B$, since the fact that the nuclei are identical requires that the electron probability density be the same on each side of the molecule. Restricting ourselves to a real variation function, the electron probability density is $\phi^2 = c_A^2(1s_A)^2 + c_B^2(1s_B)^2 + 2c_A c_B 1s_A 1s_B$. To have ϕ^2 be the same at corresponding points on each side of the molecule, we must have either $c_B = c_A$ or $c_B = -c_A$. For $c_B = c_A$, we have

$$\phi = c_A(1s_A + 1s_B), \qquad \phi^2 = c_A^2(1s_A^2 + 1s_B^2 + 2\cdot 1s_A 1s_B) \qquad (20.13)$$

For $c_B = -c_A$,

$$\phi' = c_A'(1s_A - 1s_B), \qquad \phi'^2 = c_A'^2(1s_A^2 + 1s_B^2 - 2\cdot 1s_A 1s_B) \qquad (20.14)$$

The constants c_A and c_A' are found by requiring that ϕ and ϕ' be normalized.

The normalization condition for the function in (20.13) is

$$1 = \int \phi^2\, d\tau = c_A^2\left(\int 1s_A^2\, d\tau + \int 1s_B^2\, d\tau + 2\int 1s_A 1s_B\, d\tau\right)$$

The H-atom wave functions are normalized, so $\int 1s_A^2\, d\tau = \int 1s_B^2\, d\tau = 1$. Defining the **overlap integral** S as

$$S \equiv \int 1s_A 1s_B\, d\tau$$

we get $1 = c_A^2(2 + 2S)$ and $c_A = (2 + 2S)^{-1/2}$. Similarly, one finds $c_A' = (2 - 2S)^{-1/2}$. Hence the normalized approximate wave functions for the lowest two H_2^+ electronic states are

$$\phi = (2 + 2S)^{-1/2}(1s_A + 1s_B), \qquad \phi' = (2 - 2S)^{-1/2}(1s_A - 1s_B) \qquad (20.15)$$

For completeness, each spatial function should be multiplied by a one-electron spin function, either α or β.

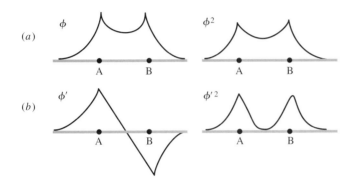

Figure 20.7

Graphs of (a) ground state and (b) first excited state H_2^+ approximate wave functions for points on the internuclear axis. (Not drawn to scale.)

The value of the overlap integral $\int 1s_A 1s_B \, d\tau$ depends on how much the functions $1s_A$ and $1s_B$ overlap each other. Only regions of space where both $1s_A$ and $1s_B$ are of significant magnitude will contribute substantially to S. The main contribution to S therefore comes from the region between the nuclei. The value of S clearly depends on the internuclear distance R. For $R = 0$, we have $1s_A = 1s_B$ and $S = 1$. For $R = \infty$, the $1s_A$ and $1s_B$ atomic orbitals don't overlap, and $S = 0$. For R between 0 and ∞, S is between 0 and 1 and is easily evaluated from the expressions for $1s_A$ and $1s_B$.

The probability density ϕ^2 in (20.13) can be written as $c_A^2(1s_A^2 + 1s_B^2)$ plus $2c_A^2 1s_A 1s_B$. The $1s_A^2 + 1s_B^2$ part of ϕ^2 is proportional to the probability density due to two separate noninteracting $1s$ H atoms. The term $2c_A^2 1s_A 1s_B$ is large only in regions where both $1s_A$ and $1s_B$ are reasonably large. This term therefore increases the electron probability density in the region between the nuclei. This buildup of probability density between the nuclei (at the expense of regions outside the internuclear region) causes the electron to feel the attractions of both nuclei at once, thereby lowering its average potential energy and providing a stable covalent bond. The bonding is due to the overlap of the atomic orbitals $1s_A$ and $1s_B$.

Figure 20.7a graphs ϕ and ϕ^2 of (20.13) for points along the line joining the nuclei. The probability-density buildup between the nuclei is evident.

For the function ϕ' of (20.14), the term $-2c_A'^2 1s_A 1s_B$ decreases the electron probability density between the nuclei. At any point on a plane midway between the nuclei and perpendicular to the internuclear axis we have $r_A = r_B$ and $1s_A = 1s_B$. Hence $\phi' = 0 = \phi'^2$ on this plane, which is a *nodal plane* for the function ϕ'. Figure 20.7b shows ϕ' and ϕ'^2 for points along the internuclear axis.

The functions ϕ and ϕ' depend on the internuclear distance R, since r_A and r_B in $1s_A$ and $1s_B$ depend on R (see Fig. 20.5). The variational integral W is therefore a function of R. When W is evaluated for ϕ and ϕ', one finds that ϕ gives an electronic energy curve $W(R) \approx E_e(R)$ with a minimum; see the lower curve in Fig. 20.8. In contrast, the $W(R)$-versus-R curve for the H_2^+ function ϕ' in 20.15 has no minimum (Fig. 20.8), indicating an unbound electronic state. These facts are understandable from the preceding electron-probability-density discussion.

The true values of R_e and D_e for H_2^+ are 1.06 Å and 2.8 eV. The function (20.13) gives $R_e = 1.32$ Å and $D_e = 1.8$ eV, which is rather poor. Substantial improvement can be obtained if a variational parameter ζ is included in the exponentials, so that $1s_A$ and $1s_B$ become proportional to $e^{-\zeta r_A/a_0}$ and $e^{-\zeta r_B/a_0}$. One then finds $R_e = 1.07$ Å and $D_e = 2.35$ eV (Fig. 20.9). The parameter ζ depends on R and is found to equal 1.24 at R_e.

An orbital is a one-electron spatial wave function. H_2^+ has but one electron, and the approximate wave functions ϕ and ϕ' in (20.13) and (20.14) are approximations to the orbitals of the two lowest electronic states of H_2^+. An orbital for an atom is called an **atomic orbital** (AO). An orbital for a molecule is a **molecular orbital** (MO). Just as the wave function of a many-electron atom can be approximated by use of AOs, the

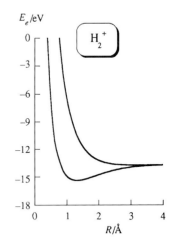

Figure 20.8

Electronic energy (including internuclear repulsion) versus R for the ground state and first excited state of H_2^+ as calculated from the approximate wave functions $N(1s_A + 1s_B)$ and $N'(1s_A - 1s_B)$ [Eq. (20.15)].

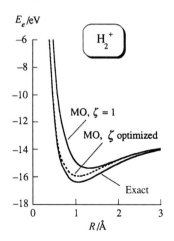

Figure 20.9

Electronic energy including
internuclear repulsion for the H_2^+
ground electronic state. The curves
are calculated from the exact wave
function, from the LCAO MO
function with optimized orbital
exponent ζ, and from the LCAO
MO function with $\zeta = 1$ (as in the
lower curve in Fig. 20.8).

wave function of a many-electron molecule can be approximated by use of MOs. Each
MO can hold two electrons of opposite spin.

The situation is more complicated for molecules than for atoms, in that the number of nuclei varies from molecule to molecule. Whereas hydrogenlike orbitals with
effective nuclear charges are useful for all many-electron atoms, the H_2^+-like orbitals
with effective nuclear charges are directly applicable only to molecules with two identical nuclei, that is, **homonuclear** diatomic molecules. We shall later see, however,
that since a molecule is held together by bonds and since (with some exceptions) each
bond is between two atoms, we can construct an approximate molecular wave function using bond orbitals (and lone-pair and inner-shell orbitals), where the bond
orbitals resemble diatomic-molecule MOs.

Let us consider further excited electronic states of H_2^+. We expect such states to
dissociate to a proton and a $2s$ or $2p$ or $3s$ or . . . H atom. Therefore, analogous to the
functions (20.13) and (20.14), we write as approximate wave functions (molecular
orbitals) for excited H_2^+ states

$$N(2s_A + 2s_B), \quad N(2s_A - 2s_B), \quad N(2p_{xA} + 2p_{xB}), \quad N(2p_{xA} - 2p_{xB}), \quad \text{etc.} \quad (20.16)$$

where the normalization constant N differs for different states. Actually, because of the
degeneracy of the $2s$ and $2p$ states in the H atom, we should expect extensive mixing
together of $2s$ and $2p$ AOs in the H_2^+ MOs. Since we are mainly interested in H_2^+ MOs
for use in many-electron molecules, and since the $2s$ and $2p$ levels are not degenerate
in many-electron atoms, we shall ignore such mixing for now. A wave function like
$2s_A + 2s_B$ expresses the fact that there is a 50-50 probability as to which nucleus the
electron will go with when the molecule dissociates ($R \rightarrow \infty$).

The MOs in (20.15) and (20.16) are *l*inear *c*ombinations of *a*tomic *o*rbitals and so
are called LCAO MOs. There is no necessity for MOs to be expressed as linear combinations of AOs, but this approximate form is a very convenient one. Let us see what
these MOs look like.

Since the functions $1s_A$ and $1s_B$ are both positive in the internuclear region, the
function $1s_A + 1s_B$ shows a buildup of probability density between the nuclei, whereas
the linear combination $1s_A - 1s_B$ has a nodal plane between the nuclei. Figure 20.10
shows contours of constant probability density for the two MOs (20.13) and (20.14)
formed from $1s$ AOs. The three-dimensional shape of these orbitals is obtained by
rotating the contours about the line joining the nuclei. See also Fig. 20.11.

A word about terminology. The component of electronic orbital angular momentum along the internuclear (z) axis of H_2^+ can be shown to have the possible values L_z
$= m\hbar$, where $m = 0, \pm 1, \pm 2, \ldots$. (Unlike the H atom, there is no l quantum number in H_2^+, since the magnitude of the total electronic orbital angular momentum is not

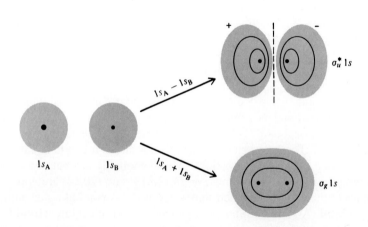

Figure 20.10

Formation of homonuclear
diatomic MOs from $1s$ AOs. The
dashed line indicates a nodal
plane.

fixed in H_2^+. This is because there is spherical symmetry in H but only axial symmetry in H_2^+.) The following code letters are used to indicate the $|m|$ value:

$|m|$	0	1	2	3	$\cdots$
Letter	σ	π	δ	ϕ	$\cdots$

(20.17)

These are the Greek equivalents of s, p, d, f.

The AOs $1s_A$ and $1s_B$ have zero electronic orbital angular momentum along the molecular axis, and so the two MOs formed from these AOs have $m = 0$ and from (20.17) are σ (sigma) MOs. We call these the $\sigma_g 1s$ MO and the $\sigma_u^* 1s$ MO. The $1s$ indicates that they originate from separated-atom $1s$ AOs. The star indicates the **antibonding** character of the $1s_A - 1s_B$ MO, associated with the nodal plane and the charge depletion between the nuclei.

The g subscript (from the German *gerade*, "even") means that the orbital has the same value at two points that are on diagonally opposite sides of the center of the molecule and equidistant from the center. The u subscript (*ungerade*, "odd") means that the values of the orbital differ by a factor -1 at two such points. [The point diagonally opposite (x, y, z) is at $(-x, -y, -z)$. An **even function** of x, y, and z is one for which $f(-x, -y, -z) = f(x, y, z)$. An **odd function** is one that satisfies $f(-x, -y, -z) = -f(x, y, z)$.]

The linear combinations $2s_A + 2s_B$ and $2s_A - 2s_B$ give the $\sigma_g 2s$ and $\sigma_u^* 2s$ MOs, whose shapes resemble those of the $\sigma_g 1s$ and $\sigma_u^* 1s$ MOs.

Let the molecular axis be the z axis. Because of the opposite signs of the right lobe of $2p_{zA}$ and the left lobe of $2p_{zB}$ (Fig. 20.12), the linear combination $2p_{zA} + 2p_{zB}$ has a nodal plane midway between the nuclei, as indicated by the dashed line. The charge depletion between the nuclei makes this an antibonding MO. The linear combination $2p_{zA} - 2p_{zB}$ gives charge buildup between the nuclei and is a bonding MO. The $2p_z$ AO has atomic quantum number $m = 0$ (Sec. 19.3) and so has $L_z = 0$. The MOs formed from $2p_z$ AOs are therefore σ MOs, the $\sigma_g 2p$ and $\sigma_u^* 2p$ MOs. Their three-dimensional shapes are obtained by rotating the contours in Fig. 20.12 about the internuclear (z) axis.

Formation of homonuclear diatomic MOs from the $2p_x$ AOs is shown in Fig. 20.13. The p_x AO is a linear combination of $m = 1$ and $m = -1$ AOs [see Eq. (19.22)] and has $|m| = 1$. Therefore the MOs made from the $2p_x$ AOs have $|m| = 1$ and are π MOs [Eq. (20.17)]. The linear combination $N(2p_{xA} + 2p_{xB})$ has charge buildup in the internuclear regions above and below the z axis and is therefore bonding. This MO has opposite signs at the diagonally opposite points c and d in Fig. 20.13 and so is a u MO, the $\pi_u 2p_x$ MO. The linear combination $N(2p_{xA} - 2p_{xB})$ gives the antibonding $\pi_g^* 2p_x$ MO.

The σ MOs in Figs. 20.10 and 20.12 are symmetric about the internuclear axis; the orbital shapes are figures of rotation about the z axis. In contrast, the $\pi_u 2p_x$ and $\pi_g^* 2p_x$ MOs consist of blobs of probability density above and below the yz plane, which is a nodal plane for these MOs.

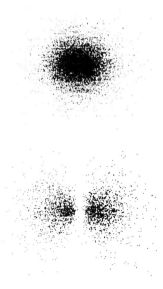

Figure 20.11

Electron probability density in a plane containing the nuclei for the ground and first excited states of H_2^+.

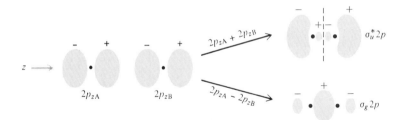

Figure 20.12

Formation of homonuclear diatomic MOs from $2p_z$ AOs.

Figure 20.13

Formation of homonuclear
diatomic MOs from $2p_x$ AOs.

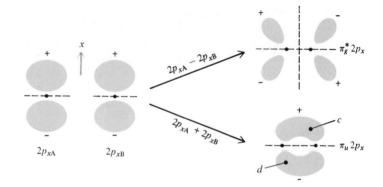

The linear combinations $2p_{yA} + 2p_{yB}$ and $2p_{yA} - 2p_{yB}$ give the $\pi_u 2p_y$ and $\pi_g^* 2p_y$ MOs. These MOs have the same shapes as the $\pi_u 2p_x$ and $\pi_g^* 2p_x$ MOs but are rotated by 90° about the internuclear axis compared with the $\pi 2p_x$ MOs. Since they have the same shapes, the $\pi_u 2p_x$ and $\pi_u 2p_y$ MOs have the same energy. Likewise, the $\pi_g^* 2p_x$ and $\pi_g^* 2p_y$ MOs have the same energy (see Fig. 20.14).

The σ MOs have no nodal planes containing the internuclear axis. (Some σ MOs have a nodal plane or planes perpendicular to the internuclear axis.) Each π MO has one nodal plane containing the internuclear axis. This is true provided one uses the real $2p$ AOs to form the MOs, as we have done. It turns out that δ MOs have two nodal planes containing the internuclear axis (see Fig. 20.28c). We shall later use the number of nodal planes to classify bond orbitals in polyatomic molecules.

20.4   THE SIMPLE MO METHOD FOR DIATOMIC MOLECULES

MOs for Homonuclear Diatomic Molecules

Just as we constructed approximate wave functions for many-electron atoms by feeding electrons two at a time into hydrogenlike AOs, we shall construct approximate wave functions for many-electron homonuclear diatomic molecules by feeding electrons two at a time into H_2^+-like MOs. Figure 20.14 shows the lowest-lying H_2^+-like MOs (Sec. 20.3). Similar to AO energies (Fig. 19.15), the energies of these MOs vary from molecule to molecule. They also vary with varying internuclear distance in the same molecule. The energy order shown in the figure is the order in which the MOs are filled in going through the periodic table, as shown by spectroscopic observations.

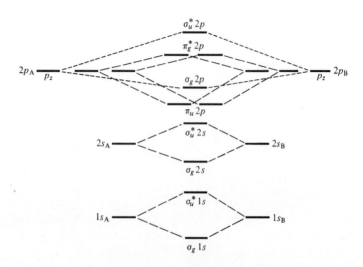

Figure 20.14

Lowest-lying homonuclear
diatomic MOs. The dashed lines
show which AOs contribute to
each MO.

The AOs at the sides are connected by dashed lines to the MOs to which they contribute. Note that each pair of AOs leads to the formation of two MOs, a bonding MO with energy lower than that of the AOs and an antibonding MO with energy higher than that of the AOs.

The Hydrogen Molecule

H_2 consists of two protons (A and B) and two electrons (1 and 2); see Fig. 20.15. The electronic Hamiltonian (including nuclear repulsion) is [Eqs. (20.6), (19.1), and (18.58)]

$$\hat{H}_e = -\frac{\hbar^2}{2m_e}\nabla_1^2 - \frac{\hbar^2}{2m_e}\nabla_2^2$$

$$-\frac{e^2}{4\pi\varepsilon_0 r_{1A}} - \frac{e^2}{4\pi\varepsilon_0 r_{1B}} - \frac{e^2}{4\pi\varepsilon_0 r_{2A}} - \frac{e^2}{4\pi\varepsilon_0 r_{2B}} + \frac{e^2}{4\pi\varepsilon_0 r_{12}} + \frac{e^2}{4\pi\varepsilon_0 R}$$

$$(20.18)$$

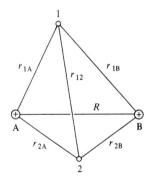

Figure 20.15

Interparticle distances in the H_2 molecule.

where r_{1A} is the distance between electron 1 and nucleus A, r_{12} is the distance between the electrons, and R is the distance between the nuclei. The first two terms are the kinetic-energy operators for electrons 1 and 2, the next four terms are the potential energy of attractions between the electrons and the nuclei, $e^2/4\pi\varepsilon_0 r_{12}$ is the potential energy of repulsion between the electrons, and $e^2/4\pi\varepsilon_0 R$ is the potential energy of internuclear repulsion. R is held fixed.

Because of the interelectronic repulsion term $e^2/4\pi\varepsilon_0 r_{12}$, the electronic Schrödinger equation $\hat{H}_e\psi_e = E_e\psi_e$ cannot be solved exactly for H_2. If this term is ignored, we get an approximate electronic Hamiltonian that is the sum of two H_2^+-like electronic Hamiltonians, one for electron 1 and one for electron 2. [This isn't quite true, because the internuclear repulsion $e^2/4\pi\varepsilon_0 R$ is the same in (20.10) and (20.18). However, $e^2/4\pi\varepsilon_0 R$ is a constant and therefore only shifts the energy by $e^2/4\pi\varepsilon_0 R$ but does not affect the wave functions; see Prob. 20.32.] The approximate electronic wave function for H_2 is then the product of two H_2^+-like electronic wave functions, one for each electron [Eq. (18.68)]. This is exactly analogous to approximating the He wave function by the product of two H-like wave functions in Sec. 19.6.

The function $(2 + 2S)^{-1/2}(1s_A + 1s_B)$ in Eq. (20.15) is an approximate wave function for the H_2^+ ground electronic state, and so the MO approximation to the H_2 ground-electronic-state spatial wave function is

$$\sigma_g 1s(1) \cdot \sigma_g 1s(2) = N[1s_A(1) + 1s_B(1)] \cdot [1s_A(2) + 1s_B(2)] \qquad (20.19)$$

where the normalization constant N is $(2 + 2S)^{-1}$. The numbers in parentheses refer to the electrons. For example, $1s_A(2)$ is proportional to e^{-r_{2A}/a_0}. The MO wave function (20.19) is analogous to the He ground-state AO wave function $1s(1)1s(2)$ in Eq. (19.35); the MO $\sigma_g 1s$ replaces the AO $1s$.

To take care of spin and the Pauli principle, the symmetric two-electron spatial function (20.19) must be multiplied by the antisymmetric spin function (19.38). The approximate MO ground-state wave function for H_2 is then

$$\sigma_g 1s(1)\sigma_g 1s(2)2^{-1/2}[\alpha(1)\beta(2) - \beta(1)\alpha(2)] = \frac{1}{\sqrt{2}}\begin{vmatrix} \sigma_g 1s(1)\alpha(1) & \sigma_g 1s(1)\beta(1) \\ \sigma_g 1s(2)\alpha(2) & \sigma_g 1s(2)\beta(2) \end{vmatrix}$$

$$(20.20)$$

where we introduced the Slater determinant (Sec. 19.8). Just as the ground-state electron configuration of He is $1s^2$, the ground-state electron configuration of H_2 is $(\sigma_g 1s)^2$; compare (20.20) with (19.40).

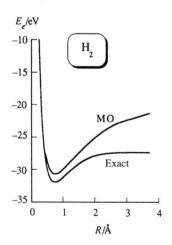

E_e/eV

Figure 20.16

Ground-state electronic energy
including internuclear repulsion
for H_2 as calculated from the
LCAO-MO wave function (20.19)
with an optimized orbital exponent
compared with the exact ground-
state electronic energy curve. Note
the incorrect behavior of the MO
function as $R \rightarrow \infty$.

We have put each electron in H_2 into an MO. This allows for interelectronic repulsion only in an average way, so the treatment is an approximate one. We are using the crudest possible version of the MO approximation.

Using the approximate wave function (20.20), one evaluates the variational integral W to get W as a function of R. Since evaluation of molecular quantum-mechanical integrals is complicated, we shall just quote the results. With inclusion of a variable orbital exponent, the function (20.20) and (20.19) gives a $W(R)$ curve (Fig. 20.16) with a minimum at $R = 0.73$ Å, which is close to the observed value $R_e = 0.74$ Å in H_2. The calculated D_e is 3.49 eV, which is far from the experimental value 4.75 eV. This is the main failing of the MO method; molecular dissociation energies are not accurately calculated.

We approximated the $\sigma_g 1s$ MO in the H_2 ground-state approximate wave function (20.19) by the linear combination $N(1s_A + 1s_B)$. To improve the MO wave function, we can look for the best possible form for the $\sigma_g 1s$ MO, still writing the spatial wave function as the product of an orbital for each electron. The best possible MO wave function is the Hartree–Fock wave function (Secs. 19.9 and 20.5). Finding the Hartree–Fock wave function for H_2 is not too difficult. The H_2 Hartree–Fock wave function predicts $R_e = 0.73$ Å and $D_e = 3.64$ eV; D_e is still substantially in error. As noted in Sec. 19.9, the Hartree–Fock wave function is not the true wave function, because of neglect of electron correlation.

In the 1960s, Kolos and Wolniewicz used very complicated variational functions that go beyond the Hartree–Fock approximation. With the inclusion of relativistic corrections and corrections for deviations from the Born-Oppenheimer approximation, they calculated $D_0/hc = 36117.9$ cm^{-1} for H_2. (D_0 differs from D_e by the zero-point vibrational energy; see Chapter 21.) At the time the calculation was completed, the experimental D_0/hc was 36114 ± 1 cm^{-1}, and the 4 cm^{-1} discrepancy was a source of embarrassment to the theoreticians. Finally, reinvestigations of the spectrum of H_2 showed that the experimental result was in error and gave the new experimental value 36118.1 ± 0.2 cm^{-1}, in excellent agreement with the value calculated from quantum mechanics.

What about excited electronic states for H_2? The lowest-lying excited H_2 MO is the $\sigma_u^* 1s$ MO. Just as the lowest excited electron configuration of He is $1s2s$, the lowest excited electron configuration of H_2 is $(\sigma_g 1s)(\sigma_u^* 1s)$, with one electron in each of the MOs $\sigma_g 1s$ and $\sigma_u^* 1s$. Like the He $1s2s$ configuration, the $(\sigma_g 1s)(\sigma_u^* 1s)$ H_2 configuration gives rise to two terms, a singlet with total spin quantum number $S = 0$ and a triplet with total spin quantum number $S = 1$. In accord with Hund's rule, the triplet lies lower and is therefore the lowest excited electronic level of H_2. In analogy with (19.43), the triplet has the MO wave functions

$$2^{-1/2}[\sigma_g 1s(1)\sigma_u^* 1s(2) - \sigma_g 1s(2)\sigma_u^* 1s(1)] \times \text{spin function} \qquad (20.21)$$

where the spin function is one of the three symmetric spin functions (19.37). With one electron in a bonding orbital and one in an antibonding orbital, we expect no net bonding. This is borne out by experiment and by accurate theoretical calculations, which show the $E_e(R)$ curve to have no minimum (Fig. 20.4).

The H_2 levels (20.20) and (20.21) both dissociate into two H atoms in $1s$ states. The bonding level (20.20) has the electrons paired with opposite spins and a net spin of zero. The repulsive level (20.21) has the electrons unpaired with approximately parallel spins. Whether two approaching $1s$ H atoms attract or repel each other depends on whether their spins are antiparallel or parallel.

Other Homonuclear Diatomic Molecules

The simple MO treatment of He_2 places the four electrons into the two lowest available MOs to give the ground-state configuration $(\sigma_g 1s)^2(\sigma_u^* 1s)^2$. The MO wave function is a Slater determinant with four rows and four columns. With two bonding and two antibonding electrons, we expect no net bonding and no stability for the ground electronic state. This is in agreement with experiment. When two ground-state He atoms approach each other, the electronic energy curve $E_e(R)$ resembles the second lowest curve of Fig. 20.4. Since $E_e(R)$ is the potential energy for nuclear motion, two $1s^2$ He atoms strongly repel each other. Actually, in addition to the strong, relatively short-range repulsion, there is a very weak attraction at relatively large values of R that produces a very slight minimum in the He-He potential-energy curve. This attraction is responsible for the liquefaction of He at very low temperature and produces an extremely weakly bound ground-state He_2 molecule ($D_0 = 10^{-7}$ eV) that has been detected at $T = 10^{-3}$ K (see Sec. 22.10). At ordinary temperatures, the He_2 concentration is negligible.

Similar to the repulsion between two $1s^2$ He atoms is the observed repulsion whenever two closed-shell atoms or molecules approach each other closely. This repulsion is important in chemical kinetics, since it is related to the activation energy of chemical reactions (see Sec. 23.2). Part of this repulsion is attributable to the Coulombic repulsion between electrons, but a major part of the repulsion is a consequence of the Pauli principle, as we now show. Let $\psi(q_1, q_2, q_3, \ldots)$ be the wave function for a system of electrons, where q_1 stands for the four coordinates (three spatial and one spin) of electron 1. The Pauli antisymmetry principle (Sec. 19.6) requires that interchange of the coordinates of electrons 1 and 2 multiply ψ by -1. Therefore, $\psi(q_2, q_1, q_3, \ldots) = -\psi(q_1, q_2, q_3, \ldots)$. Now suppose that electrons 1 and 2 have the same spin coordinate (both α or both β) and the same spatial coordinates. Then $q_1 = q_2$, and $\psi(q_1, q_1, q_3, \ldots) = -\psi(q_1, q_1, q_3, \ldots)$. Hence, $2\psi(q_1, q_1, q_3, \ldots) = 0$, and $\psi(q_1, q_1, q_3, \ldots) = 0$.

The vanishing of $\psi(q_1, q_1, q_3, \ldots)$ shows that there is zero probability for two electrons to have the same spatial and spin coordinates. Two electrons that have the same spin (both with $m_s = \frac{1}{2}$ or both with $m_s = -\frac{1}{2}$) have zero probability of being at the same point in space. Moreover, because ψ is a continuous function, the probability that two electrons with the same spin will approach each other closely must be very small. Electrons with the same spin tend to avoid each other and act as if they repelled each other over and above the Coulombic repulsion. This apparent extra repulsion of electrons with like spins is called the **Pauli repulsion.** (It is sometimes mistakenly said that the Pauli repulsion is due to magnetic forces between spins. This is not so. Magnetic forces are weak and generally can be neglected in atoms and molecules.) The Pauli repulsion is not a real physical force. It is an apparent force that is a consequence of the antisymmetry requirement of the wave function.

When two $1s^2$ He atoms approach, the antisymmetry requirement causes an apparent Pauli repulsion between the spin-α electron on one atom and the spin-α electron on the other atom; likewise for the spin-β electrons. As the He atoms approach each other, there is a depletion of electron probability density in the region between the nuclei (and a corresponding buildup of probability density in regions outside the nuclei) and the atoms repel each other.

The ground-state electron configurations of Li_2, Be_2, etc., are formed by filling in the homonuclear diatomic MOs in Fig. 20.14 (see Prob. 20.29). For example, O_2 has 16 electrons, and Fig. 20.14 gives the ground-state configuration

$$(\sigma_g 1s)^2 (\sigma_u^* 1s)^2 (\sigma_g 2s)^2 (\sigma_u^* 2s)^2 (\pi_u 2p)^4 (\sigma_g 2p)^2 (\pi_g^* 2p)^2$$

Actually, spectroscopic evidence shows that in O_2 the $\sigma_g 2p$ MO lies slightly lower than the $\pi_u 2p$ MOs, so $\sigma_g 2p$ precedes $\pi_u 2p$ in the electron configuration. Figure 20.17 shows the distribution of the valence electrons in MOs in the O_2 ground state. In accord with Hund's rule of maximum multiplicity for the ground state, the two antibonding π electrons are placed in separate orbitals to allow a triplet ground state. This

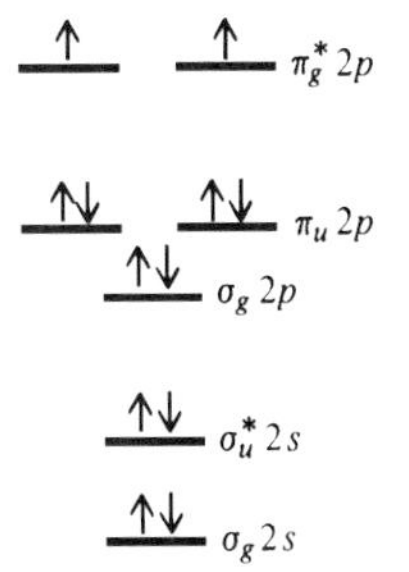

Figure 20.17

Occupied valence MOs in the O_2 ground electronic state. (Not to scale.)

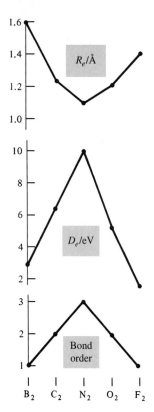

Figure 20.18

Correlation between bond orders and bond lengths and dissociation energies for some homonuclear diatomic molecules.

agrees with the observed paramagnetism of ground-state O_2. In O_2, there are four more bonding electrons than antibonding electrons, and so the MO theory predicts a double bond (composed of one σ bond and one π bond) for O_2. Note the higher D_e for O_2 compared with the single-bonded species F_2 and Li_2 (Table 20.3). The double bond makes R_e of O_2 less than R_e of Li_2. R_e of O_2 is greater than R_e of H_2 because of the presence of the inner-shell $1s$ electrons on the O atoms.

In O_2, the high nuclear charge draws the $1s$ orbitals on each atom in close to the nuclei, and there is virtually no overlap between these AOs. Therefore, the $\sigma_g 1s$ and $\sigma_u^* 1s$ MO energies in O_2 are each nearly the same as the $1s$ AO energy in an O atom. Inner-shell electrons play no real part in chemical bonding, other than to screen the valence electrons from the nuclei.

Figure 20.18 plots R_e, D_e, and the MO *bond order* (defined as half the difference between the number of bonding and antibonding electrons) for some second-row homonuclear diatomic molecules. The higher the bond order, the greater is D_e and the smaller is R_e.

Instead of the separated-atoms notation for homonuclear diatomic MOs, quantum chemists prefer a notation in which the lowest σ_g MO is called $1\sigma_g$, the next lowest σ_g MO is called $2\sigma_g$, etc. In this notation, the MOs in Fig. 20.14 are called (in order of increasing energy) $1\sigma_g$, $1\sigma_u$, $2\sigma_g$, $2\sigma_u$, $1\pi_u$, $3\sigma_g$, $1\pi_g$, $3\sigma_u$.

Heteronuclear Diatomic Molecules

The MO method feeds the electrons of a heteronuclear diatomic molecule into molecular orbitals. In the crudest approximation, each bonding MO is taken as a linear combination of two AOs, one from each atom. In constructing MOs, one uses the principle that *only AOs of reasonably similar energies contribute substantially to a given MO*.

As an example, consider HF. Figure 19.15 shows that the energy of a $2p$ AO in $_9F$ is reasonably close to the $1s$ AO energy in H, but the $2s$ AO in F is substantially lower in energy than the $1s$ H AO. (The logarithmic scale makes the fluorine $2s$ level appear closer to the $2p$ level than it actually is.) The $2p$ AO in F lies somewhat lower than the $1s$ AO in H because the five $2p$ electrons in F screen one another rather poorly, giving a large Z_{eff} for the $2p$ electrons [Eq. (19.52)]; this large Z_{eff} makes F more electronegative than H [Eq. (20.4)].

Let the HF molecular axis be the z axis, and let F2p and H1s denote a $2p$ AO on F and a $1s$ AO on H. The $F2p_z$ AO has quantum number $m = 0$ and has no nodal plane containing the internuclear axis. The overlap of this AO with the H1s AO, which also has $m = 0$ and no nodal plane containing the z axis, therefore gives rise to a σ MO (Fig. 20.19). We therefore form the linear combination $c_1 H1s + c_2 F2p_z$. Minimization of the variational integral will lead to two sets of values for c_1 and c_2, one set giving a bonding MO and the other an antibonding MO:

$$\sigma = c_1 H1s + c_2 F2p_z \quad \text{and} \quad \sigma^* = c_1' H1s - c_2' F2p_z \qquad (20.22)$$

The σ MO in (20.22) has c_1 and c_2 both positive and is bonding because of the charge buildup between the nuclei. The antibonding σ^* MO in (20.22) has opposite signs for the coefficients of the AOs and so has charge depletion between the nuclei. This MO is unoccupied in the HF ground state. The g, u designation does not apply to heteronuclear diatomics.

Figure 20.19

Formation of the bonding MO in HF.

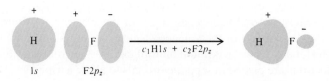

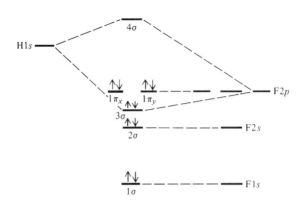

Figure 20.20

MO energies in HF. (Not to scale.)

In contrast to the $F2p_z$ AO, the $F2p_x$ and $F2p_y$ AOs have $|m| = 1$ and have one nodal plane containing the internuclear (z) axis. These AOs will therefore be used to form π MOs in HF. Since H has no valence-shell AOs with $|m| = 1$, the π MOs in HF will consist entirely of F AOs, and these MOs are $\pi_x = F2p_x$ and $\pi_y = F2p_y$.

The $1s$ and $2s$ AOs in F are too low in energy to take a substantial part in the bonding and therefore form nonbonding σ MOs in HF. Don't confuse a nonbonding MO with an antibonding MO. A nonbonding MO shows neither charge depletion nor charge buildup between the nuclei.

In the standard notation for heteronuclear diatomic molecules, the lowest σ MO is called the 1σ MO, the next lowest σ MO is the 2σ MO, etc. The lowest π energy level is called the 1π level, etc. In our crude approximation, the occupied MOs in hydrogen fluoride are

$$1\sigma = F1s, \qquad 2\sigma = F2s, \qquad 3\sigma = c_1 H1s + c_2 F2p_z$$
$$1\pi_x = F2p_x, \qquad 1\pi_y = F2p_y \tag{20.23}$$

where $1\pi_x$ and $1\pi_y$ have the same energy. Since F is more electronegative than H, we expect $|c_2| > |c_1|$ in the 3σ MO; the electrons of the bond are more likely to be found close to F than to H.

Figure 20.20 shows the energy-level scheme for HF in the simple approximation (20.23). The 1π MOs are lone-pair AOs on F and have nearly the same energy as $F2p$ AOs. The 2σ MO is also a lone-pair orbital.

An H atom is special, since it has no p valence orbitals. Consider a polar-covalent heteronuclear diatomic molecule AB, where both A and B are from the second or a higher period and hence have s and p valence levels. Let B be somewhat more electronegative than A. We draw Fig. 20.21 similar to Figs. 20.14 and 20.20 to show the

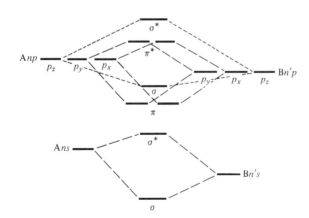

Figure 20.21

MOs formed from valence s and p AOs of atoms A and B with rather similar electronegativities.

formation of valence MOs from the ns and np valence AOs of A and the $n's$ and $n'p$ valence AOs of B; n and n' are the principal quantum numbers of the valence electrons and equal the periods of A and B in the periodic table. [It is assumed that B and A don't differ greatly in electronegativity. If B were much more electronegative than A (as, for example, in BF), then the valence p level of B might lie close to the valence s level of A, and the p_z AO of B would combine mainly with the s valence AO of A.] The MO shapes are similar to those in Figs. 20.10 to 20.13 for homonuclear diatomics, except that in each bonding MO the probability density is greater around the more electronegative element B than around A, and each bonding MO contour is therefore larger around B than around A. In each antibonding MO, the probability density is larger around A, since more of the atom-B AO has been "used up" in forming the corresponding bonding MO.

To get the valence MO configuration of molecules like CN, NO, CO, or ClF, we feed the valence electrons into the MOs of Fig. 20.21. For example, CO has 10 valence electrons and has the configuration $(\sigma_s)^2(\sigma_s^*)^2(\pi)^4(\sigma_p)^2$. With six more bonding than antibonding electrons, the molecule has a triple bond (composed of one σ and two π bonds), in accord with the dot structure :C≡O:. The lowest two MOs in CO are the 1σ and 2σ MOs, formed from linear combinations of C1s and O1s AOs, and the complete MO configuration of CO in the standard notation is $1\sigma^2 2\sigma^2 3\sigma^2 4\sigma^2 1\pi^4 5\sigma^2$.

20.5   SCF AND HARTREE–FOCK WAVE FUNCTIONS

The best possible wave function with electrons assigned to orbitals is the Hartree–Fock wave function. Starting in the 1960s, the use of electronic computers allowed Hartree–Fock wave functions for many molecules to be calculated. The Hartree–Fock orbitals ϕ_i of a molecule must be found by solving the Hartree–Fock equations (19.54): $\hat{F}\phi_i = \varepsilon_i\phi_i$. The terms in the Fock operator $\hat{F}$ for an atom were discussed after (19.54). $\hat{F}$ for a molecule is the same as $\hat{F}$ for an atom except that the electron–nucleus attraction $-Ze^2/4\pi\varepsilon_0 r_1$ in an atom (term b) is replaced by $-\sum_\alpha Z_\alpha e^2/4\pi\varepsilon_0 r_{1\alpha}$, which gives the potential energy of the attractions between electron 1 and all the nuclei; $r_{1\alpha}$ is the distance between electron 1 and nucleus α.

As is done for atoms, each Hartree–Fock MO is expressed as a linear combination of a set of functions called basis functions. If enough basis functions are included, one can get MOs that differ negligibly from the true Hartree–Fock MOs. Any functions can be used as basis functions, so long as they form a complete set (as defined in Sec. 19.9). Since molecules are made of bonded atoms, it is most convenient to use atomic orbitals as the basis functions. Each MO is then written as a linear combination of the basis-set AOs, and the coefficients of the AOs are found by solving the Hartree–Fock equations.

To have an accurate representation of an MO requires that the MO be expressed as a linear combination of a complete set of functions. This means that all the AOs of a given atom, whether occupied or unoccupied in the free atom, contribute to the MOs. To simplify the calculation, one frequently solves the Hartree–Fock equations using in the basis set only those AOs from each atom whose principal quantum number does not exceed the principal quantum number of the atom's valence electrons. Such a basis set limited to inner-shell and valence-shell AOs is called a **minimal basis set.** Use of a minimal basis set gives only an approximation to the Hartree–Fock MOs. Any wave function found by solving the Hartree–Fock equations is called a **self-consistent-field (SCF) wave function.** Only if the basis set is very large is an SCF wave function accurately equal to the Hartree–Fock wave function.

Which AOs contribute to a given MO is determined by the symmetry properties of the MO. For example, we saw at the end of Sec. 20.3 that MOs of a diatomic molecule can be classified as $\sigma, \pi, \delta, \ldots$ according to whether they have 0, 1, 2, $\ldots$ nodal planes containing the internuclear axis. Only AOs that have 0 such nodal planes contribute to a σ MO; only AOs that have 1 such nodal plane contribute to a π MO; etc. In Sec. 20.4, we took each diatomic MO as a linear combination of only two AOs. This is the crudest approximation and does not give an accurate representation of MOs. In actuality, all σ AOs of the two atoms contribute to each σ MO; similarly for π MOs. (By a σ AO is meant one with no nodal plane containing the internuclear axis.)

Consider, for example, a minimal-basis-set calculation of HF. The valence electron in H has $n = 1$, so we use only the H1s AO. The valence electrons in F have $n = 2$, so we use the $1s$, $2s$, $2p_x$, $2p_y$, and $2p_z$ AOs of F. This gives a total of six basis functions. The H1s, F1s, F2s, and F2p_z AOs each have 0 nodal planes containing the internuclear (z) axis, so each σ MO of the HF molecule is a linear combination of these four AOs. Solution of the Hartree–Fock equations using this minimal basis set gives the occupied σ MOs as [B. J. Ransil, *Rev. Mod. Phys.*, **32**, 245 (1960)]

$$1\sigma = 1.000(\text{F}1s) + 0.012(\text{F}2s) + 0.002(\text{F}2p_z) - 0.003(\text{H}1s)$$

$$2\sigma = -0.018(\text{F}1s) + 0.914(\text{F}2s) + 0.090(\text{F}2p_z) + 0.154(\text{H}1s) \quad (20.24)$$

$$3\sigma = -0.023(\text{F}1s) - 0.411(\text{F}2s) + 0.711(\text{F}2p_z) + 0.516(\text{H}1s)$$

The 1σ MO has a significant contribution only from the F1s AO. The 2σ MO has a significant contribution only from the F2s AO. This is in accord with the simple arguments that led to the very approximate MOs in (20.23). The 3σ MO has its largest contributions from H1s and F2p_z but [unlike the crude approximation of (20.23)] also has a significant contribution from the F2s AO. The mixing together of two or more AOs on the same atom is called **hybridization.** [Each AO in (20.24) and in other equations in this section and the next is actually an approximate AO whose form is given by a single Slater-type orbital (Sec. 19.9).]

The F2p_x and F2p_y AOs each have one nodal plane containing the internuclear axis, and these AOs form the occupied π MOs of HF:

$$1\pi_x = \text{F}2p_x, \qquad 1\pi_y = \text{F}2p_y \quad (20.25)$$

The 1π level is doubly degenerate, so any two linear combinations of the orbitals in (20.25) could be used. (Recall the theorem about degenerate levels in Sec. 19.3.)

For the molecule F_2, a minimal-basis-set SCF calculation (Ransil, op. cit.) gives the MO we previously called the $\sigma_g 2p$ MO as $-0.005(1s_A + 1s_B) - 0.179(2s_A + 2s_B) + 0.648(2p_{zA} - 2p_{zB})$. This can be compared with the simple expression $N(2p_{zA} - 2p_{zB})$ used earlier. When a larger basis set is used, this F_2 MO is found to have small contributions also from $3s$, $3d\sigma$, and $4f\sigma$ AOs, where $3d\sigma$ and $4f\sigma$ signify AOs with no nodal planes containing the molecular axis.

To reach the true molecular wave function, one must go beyond the Hartree–Fock approximation. Methods that do this are discussed in Secs. 20.8 and 20.9.

20.6 THE MO TREATMENT OF POLYATOMIC MOLECULES

As with diatomic molecules, one expresses the MOs of a polyatomic molecule as linear combinations of basis functions. Most commonly, AOs of the atoms forming the molecule are used as the basis functions. To find the coefficients in the linear combinations, one solves the Hartree–Fock equations (19.54). Which AOs contribute to a given MO is determined by the symmetry of the molecule.

The BeH$_2$ Molecule
We shall apply the MO method to BeH_2. Since the valence shell of Be has $n = 2$, a minimal-basis-set calculation uses the Be1s, Be2s, Be2p_x, Be2p_y, Be2p_z AOs and the H$_A$1s and H$_B$1s AOs, where H$_A$ and H$_B$ are the two H atoms. The molecule has six electrons, and these will fill the lowest three MOs in the ground state.

Figure 20.22

Linear combinations of H-atom $1s$ AOs in BeH_2 that have suitable symmetry.

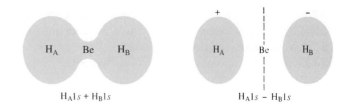

$H_A 1s + H_B 1s$ $H_A 1s - H_B 1s$

Accurate theoretical calculations show that the equilibrium geometry is linear and symmetric (HBeH), and we shall assume this structure. Each MO of this linear molecule can be classified as σ, π, δ, . . . according to whether it has 0, 1, 2, . . . nodal planes containing the internuclear axis. Further, since the molecule has a center of symmetry at the Be nucleus, we can classify each MO as g or u (as we did with homonuclear diatomics), according to whether it has the same or opposite signs on diagonally opposite sides of the Be atom.

The Be$1s$ AO has a much lower energy than all the other AOs in the basis set (Fig. 19.15), so the lowest MO will be nearly identical to the Be$1s$ AO. The function Be$1s$ has no nodal planes containing the internuclear axis and is a σ function; it also has g symmetry. We therefore write

$$1\sigma_g = \text{Be}1s \qquad (20.26)$$

where the 1 indicates that this is the lowest σ_g MO.

The $2s$ and $2p$ valence AOs of Be and the $1s$ valence AOs of H_A and H_B have similar energies and will be combined to form the remaining occupied MOs. In forming these MOs one must take the symmetry of the molecule into account. An MO without either g or u symmetry could not be a solution of the BeH_2 Hartree–Fock equations. The proof of this requires group theory and is omitted.

For a BeH_2 MO to have g or u symmetry (that is, for the square of the MO to have the same value at corresponding diagonally opposite points on each side of the central Be atom) the squares of the coefficients of the $H_A 1s$ and $H_B 1s$ AOs must be equal in each BeH_2 MO. Just as the $1s_A$ and $1s_B$ AOs in a homonuclear diatomic molecule's MOs occur as the linear combinations $1s_A + 1s_B$ and $1s_A - 1s_B$, the $H_A 1s$ and $H_B 1s$ AOs in BeH_2 can occur only as the linear combinations $H_A 1s + H_B 1s$ and $H_A 1s - H_B 1s$ in the BeH_2 MOs that satisfy the Hartree–Fock equations. Both these linear combinations have no nodal plane containing the internuclear axis. Hence these linear combinations will contribute to σ MOs. The linear combination $H_A 1s + H_B 1s$ has equal values at points diagonally opposite the center of the molecule (Fig. 20.22) and so has σ_g symmetry. The linear combination $H_A 1s - H_B 1s$ has opposite signs at points diagonally opposite the molecular center and thus has σ_u symmetry.

What about the Be AOs? The Be$2s$ AO has σ_g symmetry. Calling the internuclear axis the z axis, we see from Fig. 20.23 that the Be$2p_z$ AO has σ_u symmetry. The Be$2p_x$ and Be$2p_y$ AOs each have π_u symmetry.

The basis-set functions and their symmetries are thus

Be$1s$	Be$2s$	Be$2p_z$	Be$2p_x$	Be$2p_y$	$H_A 1s + H_B 1s$	$H_A 1s - H_B 1s$
σ_g	σ_g	σ_u	π_u	π_u	σ_g	σ_u

Combining functions that have σ_g symmetry and comparable energies, we form a σ_g MO as follows:

$$2\sigma_g = c_1 \text{Be}2s + c_2(H_A 1s + H_B 1s) \qquad (20.27)$$

The 2 in $2\sigma_g$ indicates that this is the second lowest σ_g MO, the lowest being (20.26). The very-low-energy Be$1s$ AO will make a very slight contribution to $2\sigma_g$, which we shall neglect. With c_1 and c_2 both positive, this MO has probability-density buildup between Be and H_A and between Be and H_B and is therefore bonding (Fig. 20.24).

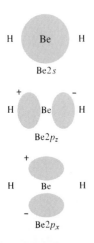

Be$2s$

Be$2p_z$

Be$2p_x$

Figure 20.23

Be AOs in BeH_2.

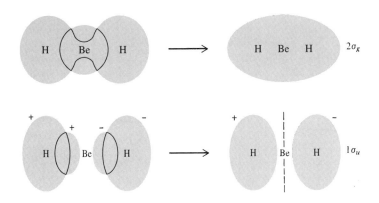

Figure 20.24

Formation of bonding MOs in
BeH_2.

Similarly, we form a bonding σ_u MO as (Fig. 20.24)

$$1\sigma_u = c_3 Be2p_z + c_4(H_A 1s - H_B 1s) \tag{20.28}$$

with c_3 and c_4 both positive. This bonding MO has its energy below the energies of the
$Be2p_z$ and $H1s$ AOs from which it is formed.

The coefficients c_1, c_2, c_3, c_4 are found by solving the Hartree–Fock equations.
The $Be2p_x$ and $Be2p_y$ AOs form two π_u MOs:

$$1\pi_{u,x} = Be2p_x, \quad 1\pi_{u,y} = Be2p_y \tag{20.29}$$

These two MOs have the same energy and constitute the doubly degenerate $1\pi_u$
energy level. The nonbonding π MOs of (20.29) have nearly the same energy as the
$Be2p_x$ and $Be2p_y$ AOs and so lie above the bonding $2\sigma_g$ and $1\sigma_u$ MOs. The π MOs
are therefore unoccupied in the ground state of this six-electron molecule.

A minimal-basis-set SCF calculation on BeH_2 [R. G. A. R. Maclagan and G. W.
Schnuelle, *J. Chem. Phys.*, **55,** 5431 (1971)] gave the occupied MOs as

$$1\sigma_g = 1.00(Be1s) + 0.016(Be2s) - 0.002(H_A 1s + H_B 1s)$$

$$2\sigma_g = -0.09(Be1s) + 0.40(Be2s) + 0.45(H_A 1s + H_B 1s) \tag{20.30}$$

$$1\sigma_u = 0.44(Be2p_z) + 0.44(H_A 1s - H_B 1s)$$

The $1\sigma_g$ MO is essentially a $Be1s$ AO, as anticipated in (20.26). The $2\sigma_g$ and $1\sigma_u$ MOs
have essentially the forms of (20.27) and (20.28).

There are also two antibonding MOs $3\sigma_g^*$ and $2\sigma_u^*$ formed from the same AOs as
the two bonding MOs (20.27) and (20.28):

$$3\sigma_g^* = c_1' Be2s - c_2'(H_A 1s + H_B 1s), \quad 2\sigma_u^* = c_3' Be2p_z - c_4'(H_A 1s - H_B 1s)$$

$$\tag{20.31}$$

Figure 20.25 sketches the AO and MO energies for BeH_2. Of course, this molecule has many higher unoccupied MOs that are not shown in the figure. These MOs
are formed from higher AOs of Be and the H's. The BeH_2 ground-state configuration
is $(1\sigma_g)^2(2\sigma_g)^2(1\sigma_u)^2$. There are four bonding electrons and hence two bonds. The $1\sigma_g$
electrons are nonbonding inner-shell electrons.

Note that a bonding MO has a lower energy than the AOs from which it is formed,
an antibonding MO has a higher energy than the AOs from which it is formed, and a
nonbonding MO has approximately the same energy as the AO or AOs from which it
is formed.

Localized MOs

The BeH_2 bonding MOs $2\sigma_g$ and $1\sigma_u$ in Fig. 20.24 are each delocalized over the entire
molecule. The two electrons in the $2\sigma_g$ MO move over the entire molecule, as do the

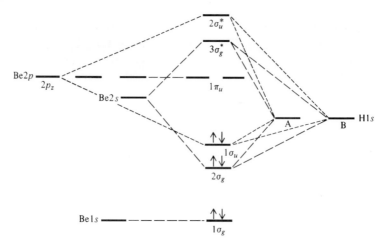

Figure 20.25

Energy-level scheme for BeH_2 MOs. (Not to scale.)

two in the $1\sigma_u$ MO. This is puzzling to a chemist, who likes to think in terms of individual bonds: H—Be—H or H:Be:H. The existence of bond energies, bond moments, and bond vibrational frequencies (Sec. 21.9) that are roughly the same for a given kind of bond in different molecules shows that there is much validity in the picture of individual bonds. How can we reconcile the existence of individual bonds with the delocalized MOs found by solving the Hartree–Fock equations?

Actually, we *can* use the MO method to arrive at a picture in accord with chemical experience, as we now show. The MO ground-state wave function for BeH_2 is a 6 × 6 Slater determinant (Sec. 19.8). The first two rows of this Slater determinant are

$$1\sigma_g(1)\alpha(1) \quad 1\sigma_g(1)\beta(1) \quad 2\sigma_g(1)\alpha(1) \quad 2\sigma_g(1)\beta(1) \quad 1\sigma_u(1)\alpha(1) \quad 1\sigma_u(1)\beta(1)$$

$$1\sigma_g(2)\alpha(2) \quad 1\sigma_g(2)\beta(2) \quad 2\sigma_g(2)\alpha(2) \quad 2\sigma_g(2)\beta(2) \quad 1\sigma_u(2)\alpha(2) \quad 1\sigma_u(2)\beta(2)$$

The third row involves electron 3, etc. Each column has the same spin-orbital. Now it is a well-known theorem (*Sokolnikoff and Redheffer*, app. A) that addition of a constant times one column of a determinant to another column leaves the determinant unchanged in value. For example, if we add three times column 1 of the determinant in (19.41) to column 2, we get

$$\begin{vmatrix} a & b + 3a \\ c & d + 3c \end{vmatrix} = a(d + 3c) - c(b + 3a) = ad - bc = \begin{vmatrix} a & b \\ c & d \end{vmatrix}$$

Thus, if we like, we can add a multiple of one column of the Slater-determinant MO wave function to another column without changing the wave function. This addition will "mix" together different MOs, since each column is a different spin-orbital. We can therefore take linear combinations of MOs to form new MOs without changing the overall wave function. Of course, the new MOs should each be normalized and, for computational convenience, should also be orthogonal to one another.

The BeH_2 MOs $1\sigma_g$, $2\sigma_g$, and $1\sigma_u$ satisfy the Hartree–Fock equations (19.54) and have the symmetry of the molecule. Because they have the molecular symmetry, they are delocalized over the whole molecule. (More accurately, the $2\sigma_g$ and $1\sigma_u$ MOs are delocalized, but the inner-shell $1\sigma_g$ is localized on the central Be atom.) These delocalized MOs satisfying the Hartree–Fock equations and having the symmetry of the molecule are called the **canonical MOs.** The canonical MOs are unique (except for the possibility of taking linear combinations of degenerate MOs).

As just shown, we can take linear combinations of the canonical MOs to form a new set of MOs that will give the same overall wave function. The new MOs will not individually be solutions of the Hartree–Fock equations $\hat{F}\phi_i = \varepsilon_i\phi_i$, but the wave function formed from these MOs will have the same energy and the same total probability density as the wave function formed from the canonical MOs.

Of the many possible sets of MOs that can be formed, we want to find a set that will have each MO classifiable as one of the following: a **bonding** (*b*) orbital localized between two atoms and having charge buildup between the atoms, an **inner-shell** (*i*) orbital, or a **lone-pair** (*l*) orbital. We call such a set of MOs **localized MOs.** Each localized MO will not have the symmetry of the molecule, but the localized MOs will correspond closely to a chemist's picture of bonding. Since localized MOs are not eigenfunctions of the Hartree–Fock operator $\hat{F}$, in a certain sense each such MO does not correspond to a definite orbital energy. However, one can calculate an average energy of a localized MO by averaging over the orbital energies of the canonical MOs that form the localized MO.

Consider BeH_2. The $1\sigma_g$ canonical MO is an inner-shell (*i*) AO on Be and can therefore be taken as one of the localized MOs: $i(Be) = 1\sigma_g = Be1s$. The $2\sigma_g$ and $1\sigma_u$ canonical MOs are delocalized. Figure 20.24 shows that the $1\sigma_u$ MO has opposite signs in the two halves of the molecule, whereas $2\sigma_g$ is essentially positive throughout the molecule. Hence, by taking linear combinations that are the sum and difference of these two canonical MOs, we get MOs that are each largely localized between only two atoms (Fig. 20.26). Thus, we take the localized bonding MOs b_1 and b_2 as

$$b_1 = 2^{-1/2}(2\sigma_g + 1\sigma_u), \qquad b_2 = 2^{-1/2}(2\sigma_g - 1\sigma_u) \qquad (20.32)$$

where the $2^{-1/2}$ is a normalization constant. The b_1 localized MO corresponds to a bond between Be and H_A. The b_2 MO gives the Be—H_B bond.

Using these localized MOs, we write the BeH_2 MO wave function as a 6×6 Slater determinant whose first row is

$$i(1)\alpha(1) \quad i(1)\beta(1) \quad b_1(1)\alpha(1) \quad b_1(1)\beta(1) \quad b_2(1)\alpha(1) \quad b_2(1)\beta(1)$$

This localized-MO wave function is equal to the wave function that uses delocalized (canonical) MOs.

Equation (20.32) expresses the localized bonding MOs b_1 and b_2 in terms of the canonical MOs. Substitution of (20.27) and (20.28) into (20.32) gives

$$b_1 = 2^{-1/2}[c_1Be2s + c_3Be2p_z + (c_2 + c_4)H_A1s + (c_2 - c_4)H_B1s]$$
$$b_2 = 2^{-1/2}[c_1Be2s - c_3Be2p_z + (c_2 - c_4)H_A1s + (c_2 + c_4)H_B1s] \qquad (20.33)$$

As a rough approximation, we see from (20.27), (20.28), and (20.30) that $c_2 \approx c_4$ and $c_1 \approx c_3$. These approximations give

$$b_1 \approx 2^{-1/2}[c_1(Be2s + Be2p_z) + 2c_2H_A1s]$$
$$b_2 \approx 2^{-1/2}[c_1(Be2s - Be2p_z) + 2c_2H_B1s] \qquad (20.34)$$

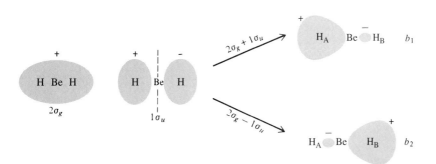

Figure 20.26

Formation of localized bonding MOs in BeH_2 from linear combinations of delocalized (canonical) MOs.

The approximate MOs (20.34) are each fully localized between Be and one H atom, but the more accurate expressions (20.33) show that the Be—H_A bonding MO b_1 has a small contribution from the H_B1s AO and so is not fully localized between the two atoms forming the bond.

Note that [unlike the canonical MOs (20.27) and (20.28)] the b_1 and b_2 localized MOs each have the Be2s and Be2p_z AOs mixed together, or *hybridized*. **Hybridization** is the mixing of different AOs of the same atom. The precise degree of hybridization depends on the values of c_1 and c_3 in (20.33). In the approximation of (20.34), the MOs b_1 and b_2 would each contain equal amounts of the Be2s and Be2p_z AOs. The two normalized linear combinations

$$2^{-1/2}(2s + 2p_z) \quad \text{and} \quad 2^{-1/2}(2s - 2p_z) \qquad (20.35)$$

are called *sp* **hybrid AOs.** Comparison of (20.30) with (20.27) and (20.28) gives $c_1 = 0.40$ and $c_3 = 0.44$, so c_1 and c_3 are not precisely equal, but are nearly equal. Thus, the Be AOs in the BeH_2 bonding MOs are not precisely *sp* hybrids but are nearly *sp* hybrids.

Note from Eq. (20.33) and Fig. 20.26 that the localized bonding MOs b_1 and b_2 in BeH_2 are **equivalent** to each other. By this we mean that b_1 and b_2 have the same shapes and are interchanged by a rotation that interchanges the two equivalent chemical bonds in BeH_2. If we rotate b_1 and b_2 180° about an axis through Be and perpendicular to the molecular axis (thereby interchanging H_A1s and H_B1s and changing Be2p_z to $-Be2p_z$), then b_1 is changed to b_2, and vice versa. Because of the symmetry of the molecule, we expect b_1 and b_2 to be equivalent orbitals. It is possible to show that the linear combinations in (20.32) are the only linear combinations of the $2\sigma_g$ and $1\sigma_u$ canonical MOs that meet the requirements of being normalized, equivalent, and orthogonal.

We arrived at the approximately *sp* hybrid Be AOs in the localized BeH_2 bonding MOs by first finding the delocalized canonical MOs and then transforming to localized MOs. A simpler (and more approximate) approach is often preferred by chemists for qualitative discussions of bonding. In this procedure, one omits consideration of the canonical MOs. Instead, one forms the required hybrid AOs on the free Be atom and then uses these hybrids to form localized bonding MOs with the H1s AOs. For BeH_2 with its 180° bond angle, we need two equivalent hybrid AOs on Be that point in opposite directions. The valence AOs of Be are 2s and 2p. Figure 20.27 shows that the linear combinations (20.35) give two equivalent hybridized AOs oriented 180° apart. It is possible to show that the *sp* hybrids (20.35) are the only linear combinations that give AOs at 180° that are equivalent, normalized, and orthogonal in the free atom. We then overlap each of these *sp* hybrid AOs with an H1s AO to form the two bonds. This gives the approximate localized bonding MOs of Eq. (20.34).

Although the *sp* hybrids (20.35) are the only linear combinations of 2s and 2p_z that give equivalent orbitals in the free Be atom, we must expect that in the BeH_2 molecule the interaction between the Be hybrids and the H atoms will alter the nature of these hybrids somewhat. What is really wanted is equivalent, normalized, orthogonal MOs in the BeH_2 molecule and not equivalent, normalized, orthogonal AOs in the Be

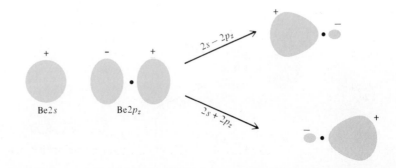

Figure 20.27

Formation of *sp* hybrid AOs in the Be atom.

atom. The equivalent MOs in BeH_2 are (20.33), and as noted above, these contain not precisely sp hybrids but only approximately sp hybrids. Another approximation involved in the use of sp Be hybrids to form the MOs (20.34) is neglect of the small contribution of the $H_B 1s$ AO to the bonding MO between Be and H_A. This is the term $(c_2 - c_4)H_B 1s = 0.01(H_B 1s)$ in (20.33).

Energy-Localized MOs

For BeH_2, the symmetry of the molecule enables one to determine what linear combination of canonical MOs to use to get localized bonding MOs. For less symmetric molecules, one cannot use symmetry, since the localized MOs need not be equivalent to one another. Several methods have been suggested for finding localized MOs from the canonical MOs. A widely accepted approach is that of Edmiston and Ruedenberg, who defined the **energy-localized MOs** as those orthogonal MOs that minimize the total of the Coulombic repulsions between the various pairs of localized MOs considered as charge distributions in space. This gives localized MOs that are separated as far as possible from one another.

In most cases, the energy-localized MOs agree with what one would expect from the Lewis dot formula. For example, for H_2O, the energy-localized MOs turn out to be one inner-shell MO, two bonding MOs, and two lone-pair MOs, in agreement with the dot formula H:Ö:H. The inner-shell MO is nearly identical to the $O1s$ AO. One bonding energy-localized MO is largely localized in the $O—H_A$ region, and the other is largely localized in the $O—H_B$ region. The angle between the bonding localized MOs is $103°$, which is nearly the same as the $104.5°$ experimental bond angle in water. The angle between the lone-pair localized orbitals is $114°$. Each bonding localized MO is mainly a linear combination of $2s$ and $2p$ oxygen AOs and a $1s$ hydrogen AO. Each lone-pair MO is mainly a hybrid of $2s$ and $2p$ AOs on oxygen. [See *Levine* (2000), sec. 15.9 and W. von Niessen, *Theor. Chim. Acta*, **29**, 29 (1973).]

Sigma, Pi, and Delta Bonds

In most cases, each localized bonding MO of a molecule contains substantial contributions from AOs of only two atoms, the atoms forming the bond. In analogy with the classification used for diatomic molecules, each localized bonding MO of a polyatomic molecule is classified as σ, π, δ, . . . according to whether the MO has 0, 1, 2, . . . nodal planes containing the axis between the two bonded atoms. The BeH_2 MOs b_1 and b_2 in Fig. 20.26 are clearly σ MOs. One finds that a single bond between two atoms nearly always corresponds to a σ localized MO. Nearly always, a double bond between two atoms is composed of one σ localized MO and one π localized MO. Nearly always, a triple bond is composed of one σ-bond orbital and two π-bond orbitals. A quadruple bond is composed of one σ bond, two π bonds, and one δ bond.

A σ bond is formed by overlap of two AOs that have no nodal planes containing the bond axis. Figure 20.28*a* shows some kinds of AO overlap that produce σ localized bond MOs. Figure 20.28*b* shows some overlaps that lead to π bonds. Figure 20.28*c* shows formation of a δ bond.

Chemists have known about σ and π bonds since the 1930s. In 1964, Cotton pointed out that the $Re_2Cl_8^{2-}$ ion has a quadruple bond between the two Re atoms, as shown by an abnormally short Re–Re bond distance. This bond is composed of one σ bond, two π bonds, and one δ bond, the δ bond being formed by overlap of two $d_{x^2-y^2}$ AOs, one on each Re atom. Several other transition-metal species have quadruple bonds. [F. A. Cotton, *Chem. Soc. Rev.*, **4**, 27 (1975); F. A. Cotton and R. A. Walton, *Multiple Bonds between Metal Atoms*, Wiley, 1982.]

Methane, Ethylene, and Acetylene

For CH_4, the canonical occupied MOs are found to consist of an MO that is a nearly pure $C1s$ AO and four delocalized bonding MOs that each extend over much of the

Figure 20.28

Overlap of AOs to form (a) σ bonds; (b) π bonds; (c) a δ bond. The lobes in (c) are in front of and behind the paper.

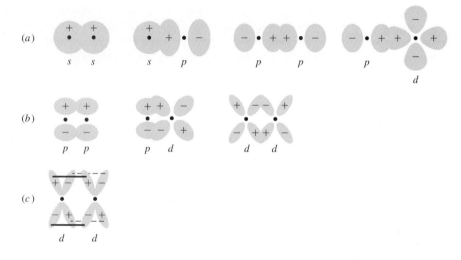

molecule. When the canonical MOs are transformed to energy-localized MOs, the localized MOs are found to consist of an inner-shell MO that is essentially a pure $C1s$ AO and four localized bonding MOs, each bonding MO pointing toward one of the H atoms of the tetrahedral molecule. The localized bonding MO between C and atom H_A is [R. M. Pitzer, *J. Chem. Phys.*, **46,** 4871 (1967)]

$$0.02(C1s) + 0.292(C2s) + 0.277(C2p_x + C2p_y + C2p_z)$$

$$+ 0.57(H_A1s) - 0.07(H_B1s + H_C1s + H_D1s)$$

The carbon $2s$ and $2p$ AOs make nearly equal contributions, and the hybridization on carbon is approximately sp^3. It would be exactly sp^3 if the coefficient of $C2s$ equaled that of the $C2p$ AOs. Atom H_A is in the positive octant of space, and the combination $C2p_x + C2p_y + C2p_z$ (which is proportional to $x + y + z$) has its maximum probability density along the line running through C and H_A and on both sides of the C nucleus. Addition of $C2s$ to $C2p_x + C2p_y + C2p_z$ cancels most of the probability density on the side of C that is away from H_A and reinforces the probability density in the region between C and H_A. (This is the same thing that occurs in Fig. 20.27 for the BeH_2 sp hybrids.) Overlap of the C hybrid AO with the H_A1s AO then forms the bond. Each bonding localized MO has no nodal planes containing the axis between the bonded atoms and is a σ MO.

Consider ethylene ($H_2C{=}CH_2$). The molecule is planar, with the bond angles at each carbon close to 120°. A minimal basis set consists of four $H1s$ AOs and two each of $C1s$, $C2s$, $C2p_x$, $C2p_y$, and $C2p_z$. Let the molecular plane be the yz plane. One way to form localized MOs for C_2H_4 is to use linear combinations of the $C2s$, $C2p_y$, and $C2p_z$ AOs at each carbon to form three sp^2 hybrid AOs at each carbon. These hybrids make 120° angles with one another. Overlap of two of the three sp^2 hybrids at each carbon with $H1s$ AOs forms the C—H single bonds, and these are σ bonds. Overlap of the third sp^2 hybrid of one carbon with the third sp^2 hybrid of the second carbon gives a σ bonding MO between the two carbons (Fig. 20.29a). Overlap of the $2p_x$ AOs of each carbon gives a localized π bonding MO between the carbons (Fig. 20.29b). This π MO has a nodal plane coinciding with the molecular plane and containing the C—C axis.

Ethylene has 16 electrons. Four of them fill the two localized inner-shell MOs, each of which is a $1s$ AO on one of the carbons; eight electrons fill the four C—H bond MOs; two fill the C—C σ-bond MO, and two fill the C—C π-bond MO. (Unlike that in diatomic molecules, the ethylene π-bond MO is nondegenerate.) In this picture, the carbon–carbon double bond consists of one σ bond and one π bond.

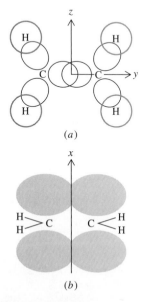

Figure 20.29

Bonding in ethylene: (a) σ bonds; (b) π bond.

The above description of localized MOs for C_2H_4 is the traditional one. However, calculation of the energy-localized MOs in ethylene shows that two bent, equivalent "banana" bonds (Fig. 20.30) between the two carbons are more localized than the traditional σ-π description [U. Kaldor, *J. Chem. Phys.*, **46**, 1981 (1967)].

The traditional description of HC≡CH uses two sp hybrids on each carbon to overlap the H1s AOs and to form a σ bond between the carbons. The linear combinations $C_A 2p_x + C_B 2p_x$ and $C_A 2p_y + C_B 2p_y$ (where the z axis is the molecular axis) give two π bonds between C_A and C_B. In this picture, the triple bond consists of one σ bond and two π bonds. Again, actual calculation shows the energy-localized MOs to consist of three equivalent bent banana bonds.

Benzene

The carbons in benzene (C_6H_6) form a regular hexagon with $120°$ bond angles. We can use three sp^2 hybrid AOs at each carbon to form localized σ bonds with one hydrogen and with two adjacent carbons. This leaves a $2p_z$ AO at each carbon (where the z axis is perpendicular to the molecular plane). For benzene, two equivalent Lewis dot formulas can be written; the carbon–carbon single bonds and double bonds are interchanged in the two formulas. Moreover, $\Delta_f H°$ for benzene (Prob. 20.39) and the chemical behavior of benzene differ from what is expected of a species with localized double bonds. Hence, it would not be suitable to form three localized π MOs by pairwise interactions of the six $2p_z$ AOs. Instead, all six $2p_z$ AOs must be considered to interact with one another, and one uses delocalized (canonical) MOs to form the π bonds. The six $2p_z$ AOs give six delocalized π MOs, three bonding and three antibonding. These six MOs are linear combinations of the six $2p_z$ AOs; their forms are fully determined by the symmetry of benzene. [See Prob. 20.40 and *Levine* (2000), sec. 16.3.]

Three of the four valence electrons of each carbon go into the three bonding MOs formed from the sp^2 hybrids, leaving one electron at each carbon to go into the π MOs. These remaining six electrons fill the three bonding π MOs. Each π MO has a nodal plane that coincides with the molecular plane (since each $2p_z$ AO has such a plane). This nodal plane is analogous to the nodal plane of a localized π bond joining two doubly bonded atoms (for example, as in ethylene), and so these benzene MOs are called π MOs.

A similar situation holds for other planar conjugated organic compounds. (A **conjugated** molecule has a framework consisting of alternating carbon–carbon single and double bonds.) One can form in-plane localized σ-bond MOs using sp^2 hybrids on each carbon, but one uses delocalized (canonical) MOs for the π MOs.

Three-Center Bonds

B_2H_6 has 12 valence electrons, which is not enough to allow one to write a Lewis dot structure with two electrons shared between each pair of bonded atoms. Calculation of the energy-localized MOs of B_2H_6 shows that two of the localized MOs each extend over two B atoms and one H atom to give two three-center bonds. [E. Switkes et al., *J. Chem. Phys.*, **51**, 2085 (1969).] The H atoms of the three-center bonds lie above and below the plane of the remaining six atoms and midway between the borons. Three-center bonds also occur in higher boron hydrides.

Multicenter bonding occurs in chemisorption. For example, experimental evidence shows that an ethylene molecule can bond to a metal's surface by forming two σ bonds with metal atoms (the stars in Fig. 20.31a) or by forming a π complex (Fig. 20.31b) in which electron probability density of the carbon–carbon π bond bonds the molecule to a metal atom as well as bonding the carbons to each other (*Bamford and Tipper*, vol. 20, pp. 22–23). Such a π complex is stabilized by donation of electron density from the C_2H_4 π electrons into vacant metal-atom orbitals and by donation of

Figure 20.30

Equivalent "banana" bonds in ethylene. The view is the same as shown in Fig. 20.29b.

(a)

(b)

(c)

(d)

Figure 20.31

Species chemisorbed on the surface of a solid.

electron density from filled metal-atom orbitals into the vacant antibonding π MO in C_2H_4 (back bonding). π complexes occur in such organometallic compounds as dibenzenechromium, $C_6H_6CrC_6H_6$, ferrocene, $C_5H_5FeC_5H_5$, and the complex ion $[Pt(C_2H_4)Cl_3]^-$; see *DeKock and Gray*, sec. 6-4.

Cyclopropane can be weakly chemisorbed by forming a complex in which the electron probability density of a carbon–carbon bond bonds the molecule to a surface metal atom (*Bamford and Tipper*, vol. 20, p. 102), as shown in Fig. 20.31*c*. H_2 can be nondissociatively weakly chemisorbed to form the complex in Fig. 20.31*d*. In $[Cr(CO)_5(H_2)]$ and a few other coordination compounds, H_2 forms a three-center bond with the central metal atom; see G. Wilkinson (ed.), *Comprehensive Coordination Chemistry*, vol. 2, Pergamon, 1987, pp. 690–691.

Ligand-Field Theory

The application of MO theory to transition-metal complexes gives what is called *ligand-field theory*. See *DeKock and Gray*, sec. 6-7.

Canonical versus Localized MOs

For accurate quantitative calculations of molecular properties, one solves the Hartree–Fock equations (19.54) and obtains the canonical (delocalized) MOs. Since each canonical MO corresponds to a definite orbital energy, these MOs are also useful in discussing transitions to excited electronic states and ionization.

For qualitative discussion of bonding in the ground electronic state of a molecule, it is usually simplest to describe things in terms of localized bonding MOs (constructed from suitably hybridized AOs of pairs of bonded atoms), lone-pair MOs, and inner-shell MOs. One can usually get a reasonably good idea of the localized MOs without going through the difficult computations involved in first finding the canonical MOs and then using the Edmiston–Ruedenberg criterion to transform the canonical MOs to localized MOs. Localized MOs are approximately transferable from molecule to molecule; for example, the C—H localized MOs in CH_4 and C_2H_6 are very similar to each other. Canonical MOs are not transferable.

20.7　THE VALENCE-BOND METHOD

So far our discussion of molecular electronic structure has been based on the MO approximation. Historically, the first quantum-mechanical treatment of molecular bonding was the 1927 Heitler–London treatment of H_2. Their approach was extended by Slater and by Pauling to give the **valence-bond (VB) method.**

Heitler and London started with the idea that a ground-state H_2 molecule is formed from two $1s$ H atoms. If all interactions between the H atoms were ignored, the wave function for the system of two atoms would be the product of the separate wave functions of each atom. Hence, the first approximation to the H_2 spatial wave function is $1s_A(1)1s_B(2)$, where $1s_A(1) = \pi^{-1/2}a_0^{-3/2}e^{-r_{1A}/a_0}$. This product wave function is unsatisfactory, since it distinguishes between the identical electrons, saying that electron 1 is on nucleus A and electron 2 is on nucleus B. To take care of electron indistinguishability, we must write the approximation to the ground-state H_2 spatial wave function as the linear combination $N'[1s_A(1)1s_B(2) + 1s_A(2)1s_B(1)]$. This function is symmetric with respect to electron interchange and therefore requires the antisymmetric two-electron spin function (19.38). The ground-state H_2 Heitler–London VB wave function is then

$$N'[1s_A(1)1s_B(2) + 1s_A(2)1s_B(1)] \cdot 2^{-1/2}[\alpha(1)\beta(2) - \beta(1)\alpha(2)] \quad (20.36)$$

Introducing a variable orbital exponent and using (20.36) in the variational integral, one finds a predicted D_e of 3.78 eV compared with the experimental value 4.75 eV and the Hartree–Fock value 3.64 eV.

The Heitler–London function (20.36) is a linear combination of two determinants:

$$N\begin{vmatrix} 1s_A(1)\alpha(1) & 1s_B(1)\beta(1) \\ 1s_A(2)\alpha(2) & 1s_B(2)\beta(2) \end{vmatrix} - N\begin{vmatrix} 1s_A(1)\beta(1) & 1s_B(1)\alpha(1) \\ 1s_A(2)\beta(2) & 1s_B(2)\alpha(2) \end{vmatrix} \quad (20.37)$$

The two determinants differ by giving different spins to the AOs $1s_A$ and $1s_B$ involved in the bonding.

When multiplied out, the MO spatial function (20.19) for H_2 equals

$$N[1s_A(1)1s_A(2) + 1s_B(1)1s_B(2) + 1s_A(1)1s_B(2) + 1s_B(1)1s_A(2)]$$

Because of the terms $1s_A(1)1s_A(2)$ and $1s_B(1)1s_B(2)$, the MO function gives a 50% probability that an H_2 molecule will dissociate into $H^- + H^+$, and a 50% probability for dissociation into $H + H$. In actuality, a ground-state H_2 molecule always dissociates to two neutral H atoms. This incorrect dissociation prediction is related to the poor dissociation energies predicted by the MO method. In contrast, the VB function (20.36) correctly predicts dissociation into $H + H$.

If, instead of the symmetric spatial function in (20.36), one uses the antisymmetric spatial function $N[1s_A(1)1s_B(2) - 1s_A(2)1s_B(1)]$ multiplied by one of the three symmetric spin functions in (19.37), one gets the VB functions for the first excited electronic level (a triplet level) of H_2. The minus sign produces charge depletion between the nuclei, and the atoms repel each other as they come together.

To apply the VB method to polyatomic molecules, one writes down all possible ways of pairing up the unpaired electrons of the atoms forming the molecule. Each way of pairing gives one of the **resonance structures** of the molecule. For each resonance structure, one writes down a function (called a *bond eigenfunction*) resembling (20.37), and the molecular wave function is taken as a linear combination of the bond eigenfunctions. The coefficients in the linear combination are found by minimizing the variational integral. Besides covalent pairing structures, one also includes ionic structures. For example, for H_2, the only covalent pairing structure is H—H, but one also has the ionic resonance structures $H^+ H^-$ and $H^- H^+$. These ionic structures correspond to the bond eigenfunctions $1s_A(1)1s_A(2)$ and $1s_B(1)1s_B(2)$. By symmetry, the two ionic structures contribute equally, so with inclusion of ionic structures, the VB spatial wave function for H_2 becomes

$$c_1[1s_A(1)1s_B(2) + 1s_B(1)1s_A(2)] + c_2[1s_A(1)1s_A(2) + 1s_B(1)1s_B(2)]$$

One says there is *ionic–covalent resonance*. One expects $c_2 \ll c_1$ for this nonpolar molecule.

In many cases, one uses hybrid atomic orbitals to form the bond eigenfunctions. For example, for the tetrahedral molecule CH_4, one combines four sp^3 hybrid AOs on carbon with the $1s$ AOs of the hydrogens.

For polyatomic molecules, the VB wave function is cumbersome. For example, CH_4 has four bonds, and the bond eigenfunction corresponding to the single most important resonance structure (the one with each $H1s$ AO paired with one of the carbon sp^3 hybrids) turns out to be a linear combination of $2^4 = 16$ determinants. Inclusion of other resonance structures further complicates the wave function.

The calculations of the VB method turn out to be more difficult than those of the MO method. The various MO approaches have overshadowed the VB method when it comes to actual computation of molecular wave functions and properties. However, the language of VB theory provides organic chemists with a simple qualitative tool for rationalizing many observed trends.

20.8 CALCULATION OF MOLECULAR PROPERTIES

This section considers the calculation of molecular properties from approximate molecular wave functions.

Molecular Geometry

The equilibrium geometry of a molecule is the spatial configuration of the nuclei for which the electronic energy (including nuclear repulsion) E_e in the electronic Schrödinger equation (20.7) is a minimum. To determine the equilibrium geometry theoretically, one calculates the molecular wave function and electronic energy for many different configurations of the nuclei, varying the bond distances, bond angles, and dihedral angles to find the minimum-energy configuration. A very efficient way to find the equilibrium geometry involves calculating the derivatives of the electronic energy with respect to each of the nuclear coordinates (this set of derivatives is called the energy **gradient**) for an initially guessed geometry. One then uses the values of these derivatives to change the nuclear coordinates to new values that are likely to be closer to the equilibrium geometry, and one then calculates the wave function, energy, and energy gradient at the new geometry. This process is repeated until the components of the energy gradient are all very close to zero, indicating that the energy minimum has been found.

One finds that even though the Hartree–Fock MO wave function differs significantly from the true wave function, it gives generally accurate bond distances and bond angles. Some examples of calculated Hartree–Fock (or approximate Hartree–Fock) geometries are (experimental values in parentheses):

H_2O: $r(OH) = 0.94$ Å $(0.96$ Å$)$,   HOH angle $= 106.1°$ $(104.5°)$

H_2CO: $r(CH) = 1.10$ Å $(1.12$ Å$)$, $r(CO) = 1.22$ Å $(1.21$ Å$)$

         HCH angle $= 114.8°$ $(116.5°)$

C_6H_6: $r(CC) = 1.39$ Å $(1.40$ Å$)$, $r(CH) = 1.08$ Å $(1.08$ Å$)$

It has been found that to obtain an accurate geometry, one needs only an approximation to the Hartree–Fock wave function. Minimal-basis SCF wave functions (Sec. 20.5) usually give accurate geometries, but occasionally show large errors. A somewhat larger than minimal-basis-set SCF calculation is needed to obtain a reliable geometry.

Dipole Moments

The classical expression for the dipole moment of a charge distribution is $\boldsymbol{\mu} = \Sigma_i\, Q_i \mathbf{r}_i$ [Eq. (14.82)]. To calculate the dipole moment of a molecule from its equilibrium-geometry electronic wave function ψ, we use the right side of (14.82) as an operator in the average-value expression (18.63) to get

$$\boldsymbol{\mu} = \int \psi^* \sum_i Q_i \mathbf{r}_i \psi \, d\tau = \sum_i Q_i \int |\psi|^2 \mathbf{r}_i \, d\tau \qquad (20.38)$$

where the sum goes over all the electrons and nuclei, the integral is over the electronic coordinates, and Q_i is the charge on particle i. Evaluation of (20.38) once ψ is known is easy. The hard thing is to get a reasonably accurate approximation to ψ.

One finds that Hartree–Fock MO wave functions give generally accurate molecular dipole moments. Some values of the Hartree–Fock and the experimental dipole moments are:

	HCN	H_2O	LiH	NaCl	CO	NH_3
μ_{HF}/D	3.29	1.98	6.00	9.18	-0.11	1.66
μ_{exp}/D	2.98	1.85	5.88	9.00	$+0.27$	1.48

The experimental CO dipole moment has the carbon negative (Prob. 20.19); the calculated Hartree–Fock CO dipole moment is in the wrong direction. A calculation using a CI wave function gives the proper polarity for CO.

Minimal-basis-set SCF wave functions give only fairly accurate dipole moments, and larger than minimal basis sets are needed to get good accuracy.

Ionization Energies

The molecular ionization energy I is the energy needed to remove the most loosely held electron from the molecule. T. C. Koopmans (cowinner of the 1975 Nobel Prize in economics) proved in 1933 that the energy needed to remove an electron from an orbital of a closed-shell atom or molecule is well approximated by minus the Hartree–Fock orbital energy ε_i in (19.54). The molecular ionization energy can therefore be estimated by taking $-\varepsilon_i$ of the highest occupied MO. One finds pretty good agreement between Koopmans'-theorem Hartree–Fock ionization energies and experimental ionization energies. Some results are (experimental values in parentheses): 17.4 eV (15.6 eV) for N_2, 13.8 eV (12.6 eV) for H_2O, 9.1 eV (9.3 eV) for C_6H_6.

Dissociation Energies

To calculate D_e theoretically, one subtracts the calculated Hartree–Fock molecular energy at the equilibrium geometry from the Hartree–Fock energies of the separated atoms that form the molecule. Hartree–Fock wave functions give poor D_e values. Some results for binding energies are:

	H_2	BeO	N_2	CO	F_2	CO_2	H_2O	N_2O
$D_{e,HF}$/eV	3.64	2.0	5.3	7.9	−1.4	11.3	6.9	4.0
$D_{e,exp}$/eV	4.75	4.7	9.9	11.2	1.6	16.8	10.1	11.7

The Hartree–Fock wave functions predict the separated atoms F + F to be more stable than the F_2 molecule at R_e.

The **equilibrium dissociation energy** D_e is not directly observable because the nuclei vibrate about the equilibrium geometry. Recall that a one-dimensional harmonic oscillator has a ground-state zero-point vibrational energy of $\frac{1}{2}h\nu$ (Sec. 18.12). The molecule's energy in its ground vibrational level is higher than the equilibrium-geometry energy by the zero-point energy (ZPE) and so the observed **ground-state dissociation energy** D_0 is less than D_e, the difference being the ZPE. Figure 21.9 shows this for H_2.

The **atomization energy** of a species is the thermodynamic internal energy change for the atomization process of gas-phase molecules dissociating to gas-phase atoms. For example (5.45) is the atomization process for $CH_4(g)$. Most commonly, one deals with $\Delta_{at}U^\circ$ at 0 K or at 298 K. Since $\Delta_{at}U^\circ$ is a per-mole quantity, we have $\Delta_{at}U_0^\circ = N_A D_0$, where N_A is the Avogadro constant. One can calculate $\Delta_{at}U_{298}^\circ$ from $\Delta_{at}U_0^\circ$ using statistical mechanics (Sec. 22.8).

Rotational Barriers

The equilibrium conformation of ethane, H_3C—CH_3, has the hydrogens of one CH_3 group staggered with respect to the hydrogens of the other CH_3 group. The barrier to internal rotation about the single bond in ethane is quite low, being 0.13 eV, which corresponds to 3 kcal/mol. The barrier can be determined experimentally from thermodynamic data or the infrared spectrum. To calculate the rotational barrier B theoretically, one calculates wave functions and energies for the staggered and eclipsed geometries and takes the energy difference. SCF wave functions give pretty accurate rotational barriers, provided one uses a substantially larger than minimal basis set. Some results are:

	C_2H_6	CH_3CHO	CH_3OH	CH_3NH_2	CH_3SiH_3
B_{exp}/(kcal/mol)	2.9	1.2	1.1	2.0	1.7
B_{calc}/(kcal/mol)	3.2	1.1	1.4	2.4	1.4

Quantum-mechanical analyses of the C_2H_6 rotational barrier have given two different explanations of its origin [P. v. R. Schleyer et al., *J. Am. Chem. Soc.,* **114,** 6791 (1992)].

The Hartree–Fock method does well on barrier calculations because no bonds are broken or formed in going from the staggered to the eclipsed conformation and the correlation energy (which is the energy error in the Hartree–Fock method) is nearly the same for the two conformations. In contrast, when a molecule dissociates, bonds are broken and the correlation energy changes substantially. The Hartree–Fock wave function therefore cannot deal with dissociation.

Relative Energies of Isomers

Although energies of dissociation of molecules to atoms calculated by the Hartree–Fock method are very inaccurate, *relative* energies of isomeric molecules are generally predicted accurately by Hartree–Fock wave functions. Relative energies of isomers are calculated the same way rotational barriers are calculated. Minimal-basis-set SCF wave functions don't give accurate relative energies of isomers, and one must use a larger than minimal basis set to get good results. As an example, SCF calculations with a basis set substantially larger than minimal (but not large enough to give a near-Hartree–Fock wave function) gave the following ground-state electronic energies in kcal/mol of C_3H_4 isomers relative to that of propyne (experimental values in parentheses): allene, 1.7 (1.6); cyclopropene, 25.4 (21.9); they gave the following electronic energies of C_4H_6 isomers relative to 1,3-butadiene: 2-butyne, 6.7 (8.6); cyclobutene, 12.4 (11.2); methylenecyclopropane, 20.2 (21.7); bicyclobutane, 30.4 (25.6); 1-methylcyclopropene, 31.4 (32.1) [see M. C. Flanigan et al. in G. A. Segal (ed.), *Semiempirical Methods of Electronic Structure Calculation,* pt. B, Plenum, 1977, p. 1].

A major application of SCF energy and geometry calculations is to reaction intermediates, which often are too short-lived to have their structures determined by spectroscopy. For example, the relative energies and the geometries of carbonium ions containing up to eight carbon atoms have been calculated (*Hehre* et al., sec. 7.3).

Relative Energies of Conformers

A **conformation** of a molecule is defined by specifying the dihedral angles of rotation about the single bonds. A conformation that corresponds to an energy minimum is called a **conformer.** Figure 20.32 shows the *gauche* and *trans* conformers of butane, $CH_3CH_2CH_2CH_3$, which occur at CCCC dihedral angles of about 65° (as shown by quantum-mechanical calculations) and 180°, respectively. Energy differences between conformers of a molecule are usually small (typically 0 to 2 kcal/mol for conformers that differ by rotation about one bond), and internal rotation barriers to interconversion of conformers are usually small. Therefore, different conformers of a molecule are usually not isolable, and a molecule with internal rotation about single bonds consists of a mixture of conformers whose relative amounts are determined by the Boltzmann distribution law (Secs. 15.9, 22.5, and 22.8). Minimal-basis-set SCF calculations are unreliable for predicting energy differences between conformers, and large-basis-set SCF calculations are needed to obtain reasonably reliable results here. Figure 20.33 plots the electronic energy of butane versus CCCC dihedral angle.

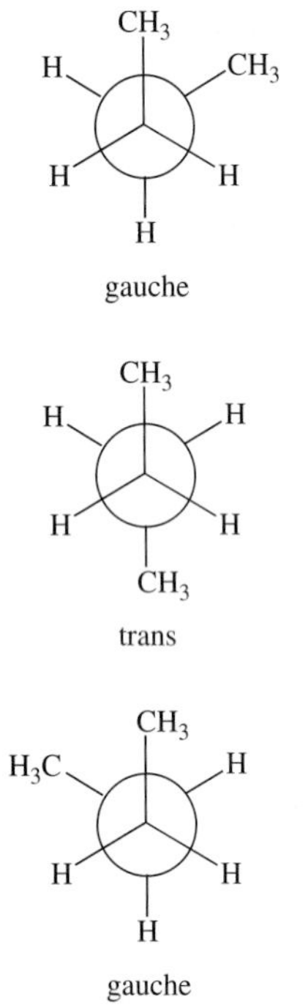

Electron Probability Density

Let $\rho(x, y, z)\, dx\, dy\, dz$ be the probability of finding an electron of a many-electron molecule in the box-shaped region located at x, y, z and having edges dx, dy, dz; by "an electron," we mean any electron, not a particular one. The electron probability density $\rho(x, y, z)$ can be calculated theoretically from the molecular electronic wave function ψ_e by integrating $|\psi_e|^2$ over the spin coordinates of all electrons and over the spatial coordinates of all but one electron and multiplying the result by the number of electrons in the molecule. One can find ρ experimentally by analyzing x-ray diffraction

gauche

trans

gauche

Figure 20.32

Conformers of butane, $CH_3CH_2CH_2CH_3$.

data of crystals (Chapter 24). Electron densities calculated from Hartree–Fock wave functions for small molecules agree well with experimentally determined densities; see P. Coppens and M. B. Hall (eds.), *Electron Distributions and the Chemical Bond*, Plenum, 1982, pp. 265, 331. ρ plays a key role in density-functional theory (Sec. 20.10), one of the most important methods for calculating molecular properties.

EXAMPLE 20.2 Particle Probability Density

Find the particle probability density function $\rho(x)$ for a system of two identical noninteracting particles each of which has spin $s = 0$ in a stationary state of a one-dimensional box of length a if the quantum numbers for the two particles are equal.

The particles do not have spin so the wave function is a function of only the spatial coordinates. Since the particles are noninteracting, the wave function is the product of a particle-in-a-box wave function for each particle. Particles with $s = 0$ are bosons and require a symmetric wave function. The normalized wave function is thus [similar to the He ground-state approximate wave function (19.35)]

$$\psi = f(1)f(2)$$

where f is a normalized particle-in-a-box wave function with quantum number n and the numbers in parentheses are the labels of the particles. Thus, $f(1) = (2/a)^{1/2} \sin(n\pi x_1/a)$ inside the box [Eq. (18.35)]. Integrating $|\psi|^2$ over the spatial coordinates of all particles except particle 1, and multiplying it by the number of particles, we have

$$\rho = 2[f(1)]^2 \int_0^a [f(2)]^2 \, dx_2 = 2[f(1)]^2$$

since $f(2)$ is normalized. Dropping the unnecessary subscript 1, we have

$$\rho(x) = 2(2/a) \sin^2 (n\pi x/a)$$

inside the box and $\rho = 0$ outside the box. The case where the two particles have different quantum numbers is considered in Prob. 20.44.

Exercise

Find $\int_{-\infty}^{\infty} \rho(x) \, dx$ for this problem. (*Answer*: 2.)

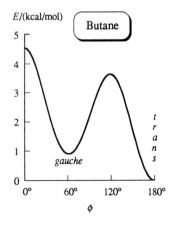

Figure 20.33

Electronic energy including nuclear repulsion of butane (Fig. 20.32) versus angle ϕ of internal rotation. This is the potential-energy function for torsional (twisting) vibration about the central C—C bond; it was found from vibrational transition frequencies in the Raman spectrum (Sec. 21.10) [D. A. C. Compton, S. Montero, and W. F. Murphy, *J. Phys. Chem.*, **84**, 3587 (1980)].

20.9 ACCURATE CALCULATION OF MOLECULAR ELECTRONIC WAVE FUNCTIONS AND PROPERTIES

The major sources of error in quantum-mechanical calculations of ground-state molecular properties are (*a*) inadequacy of the basis set; (*b*) neglect or incomplete treatment of electron correlation; (*c*) neglect of relativistic effects.

Relativistic effects strongly influence the properties of molecules containing very heavy atoms such as Au, Hg, and Pb, but for molecules composed of atoms with atomic number less than 54, relativistic effects can usually be neglected.

Basis Sets

Most quantum-mechanical calculations use a basis set to express the molecular orbitals [Eq. (19.56)]. As noted in Sec. 20.8, SCF calculations that use a minimal basis set (containing only inner- and valence-shell AOs) cannot be relied on to give accurate molecular properties. Calculations using the very large basis sets needed to come close to the Hartree–Fock wave function are feasible only for rather small molecules. For medium-size molecules, one must limit the basis-set size, and this is a major source of error in calculated properties.

Although the Slater-type orbitals (STOs) of Sec. 19.9 are often used as basis sets in atomic calculations, use of STOs as basis functions in polyatomic-molecule calculations produces integrals that are very time-consuming to evaluate on a computer. Therefore, most molecular quantum-mechanical calculations use Gaussian functions instead of STOs as the basis functions. A **Gaussian function** contains the factor $e^{-\zeta r^2}$ instead of the $e^{-\zeta r}$ factor in an STO, and molecular integrals with Gaussian basis functions are evaluated very rapidly on a computer. However, the factor $e^{-\zeta r^2}$ is not as accurate a representation of the actual behavior of an AO as the factor $e^{-\zeta r}$, so one must use a linear combination of a few Gaussian functions to represent an AO.

Many Gaussian basis sets have been devised for use in molecular calculations. Some of the most widely used are the basis sets contained in the molecular-electronic structure program *Gaussian* (which exists in several versions, each labeled by the year in which the version was released). These basis sets (listed in order of increasing size) include the STO-3G, 3-21G, 3-21G$^{(*)}$, 6-31G*, and 6-31G** sets, where the numbers and symbols are related to the number of basis functions on each atom. STO-3G is a minimal basis set and often gives unreliable results, so this set is essentially obsolete. The 6-31G** basis set, also called 6-31G(d, p), is not a particularly large set and many larger sets have been formed by adding functions to 6-31G**. For details of these and other basis sets, see *Levine* (2000), sec. 15.4.

Configuration Interaction

In order to calculate molecular properties to high accuracy, one must usually go beyond the Hartree–Fock (SCF) method and include electron correlation. One method to do this is configuration interaction (CI)—Sec. 19.9. The steps in a CI calculation are: (1) One chooses a set of basis functions $g_1, g_2, \ldots, g_i, \ldots$ (2) Each molecular orbital ϕ_j is written as a linear combination of the basis functions: $\phi_j = \Sigma_i c_i g_i$, where the expansion coefficients c_i are to be determined. This expression for the MOs is substituted into the Hartree–Fock equations $\hat{F}\phi_j = \varepsilon_j \phi_j$ [Eq. (19.54)], and the SCF iterative procedure [described in the paragraph after Eq. (19.54)] is carried out to solve for the coefficients c_i, which determine the MOs ϕ_j. Just as the simple linear variation function $c_A 1s_A + c_B 1s_B$ for H_2^+ gave two MOs, one occupied and one unoccupied in the ground state (Sec. 20.3), the SCF procedure will give expressions for many MOs, some occupied and some unoccupied. The unoccupied orbitals are called *virtual orbitals*. The ground-state SCF wave function is a Slater determinant with the lowest MOs occupied. (3) The molecular wave function ψ is then expressed as a linear combination of the SCF wave function found in step 2 and functions with one or more electrons placed in virtual orbitals: $\psi = \Sigma_j a_j \psi_{\text{orb},j}$ [Eq. (19.57)], and the variation method is used to find the coefficients a_j that minimize the variational integral.

If the sum $\Sigma_j a_j \psi_{\text{orb},j}$ includes functions with all possible assignments of electrons to virtual orbitals (this is called *full* CI), and if the basis set $g_1, g_2, \ldots$ used to expand the MOs is a complete set, then one can prove that the exact wave function will be obtained. In practice, the basis set must be limited in size (and is therefore incomplete), and full CI takes far too much computer time to be feasible (except for small molecules and small basis sets).

Experience shows that CI calculations with small basis sets do not give highly accurate results for molecular properties, and basis sets at least as large as 6-31G* should be used in CI calculations.

Each of the functions $\psi_{\text{orb},j}$ in the CI wave function is classified as *singly excited, doubly excited, triply excited,* . . . according to whether one, two, three, . . . electrons are in virtual orbitals. Since full CI is impractical, one must limit the number of terms in the CI wave function. The most common kind of CI calculation is CISD, where the letters SD indicate that all possible singly and doubly excited configuration functions $\psi_{orb,j}$ are included, but configuration functions with 3 or more electrons in virtual orbitals are omitted. For molecules with up to 10 or 20 electrons, CISD with an adequate-size basis set yields very accurate results. However, as the number of electrons increases, CISD becomes more and more inaccurate. A CISDTQ calculation (which includes all singly, doubly, triply, and quadruply excited configuration functions) with an adequate-size basis set will yield quite accurate results for molecules with up to 50 electrons. However, CISDTQ calculations with large basis sets require monumental amounts of computer time and are impractical to perform except for small molecules.

Møller–Plesset Perturbation Theory

Because of the inefficiency of CI calculations, quantum chemists have developed other methods to include electron correlation. One such method is Møller–Plesset (MP) perturbation theory. The Hartree–Fock wave function is an antisymmetrized product of spin-orbitals, where each MO ϕ_i is found by solving the Hartree–Fock equations $\hat{F}\phi_i = \varepsilon_i \phi_i$ [Eq. (19.54)]. MP perturbation theory writes the molecular electronic Hamiltonian $\hat{H}$ as the sum of an unperturbed Hamiltonian $\hat{H}^0$ and a perturbation $\hat{H}'$, where $\hat{H}^0$ is taken as the sum of the Hartree–Fock operators $\hat{F}$ for the electrons in the molecule. (See Sec. 18.15 for a general discussion of perturbation theory.) With this choice of $\hat{H}^0$, one finds that the unperturbed wave function $\psi^{(0)}$ is the Hartree–Fock wave function and the energy $E^{(0)} + E^{(1)}$ is the Hartree–Fock energy [see *Levine* (2000), sec. 15.18 for details]. To improve on the Hartree–Fock energy, one then calculates the higher-order energy corrections $E^{(2)}$, $E^{(3)}$, etc. MP calculations are designated MP2, MP3, MP4, according to whether the highest-order energy correction included is $E^{(2)}$, $E^{(3)}$, or $E^{(4)}$. MP2 calculations can be done much faster than CI calculations and are a common way to include correlation in calculations on ground-state molecules.

An MP calculation begins by choosing a basis set to express the MOs and then finds the SCF wave function corresponding to this basis set—steps 1 and 2 in the above CI discussion. The expressions for the MP energy corrections $E^{(2)}$, $E^{(3)}$, . . . involve integrals over the MOs (including the virtual orbitals), and these integrals can be related to integrals over the basis functions. Evaluation of these integrals and substitution in the relevant formulas gives $E^{(2)}$, $E^{(3)}$, etc. Most MP calculations are MP2 calculations. Almost never are calculations done beyond MP4. The energy gradient can be readily found in the MP method, and this allows the MP equilibrium geometry to be readily found.

The Coupled-Cluster Method

The coupled-cluster (CC) method is a third way to allow for electron correlation. Like the CI and MP methods, there are various levels of CC calculations: CCD, CCSD, CCSDT, . . . , where CCSDT stands for coupled-cluster singles, doubles, and triples. Because CCSDT calculations usually require prohibitive amounts of computer time, an approximation to CCSDT called CCSD(T) is more commonly used than CCSDT. CCSD(T) gives very accurate results but is limited to quite small molecules. For details, see *Levine* (2000), sec. 15.19.

Density-Functional Theory

This highly popular method deserves its own section and is discussed in Sec. 20.10.

Applications

Quantum-chemistry calculations have advanced to the point where they can help answer many chemically significant questions.

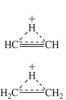

Figure 20.34

Equilibrium structures of the vinyl $(C_2H_3^+)$ and ethyl $(C_2H_5^+)$ cations.

For example, calculations using large basis sets and allowing for electron correlation by the MP, CISD, and CC methods have shown that the equilibrium structures of the vinyl cation and the ethyl cation are not the classical structures $CH_2{=}CH^+$ and $CH_3CH_2^+$, but are the bridged structures shown in Fig. 20.34, which have three-center bonds [B. Ruscic et al., *J. Chem. Phys.,* **91**, 114 (1989); C. Liang, T. P. Hamilton, and H. F. Schaefer, III, *J. Chem. Phys.,* **92**, 3653 (1990)]. The calculated energy differences between the classical and bridged structures are about 4 kcal/mol for $C_2H_3^+$ and 6 kcal/mol for $C_2H_5^+$. Spectroscopic observations confirm that the equilibrium structures are the bridged structures.

Spectroscopic observations of 1,3-butadiene show the presence of two conformers. Two possible conformations are the *s-trans* and *s-cis* configurations:

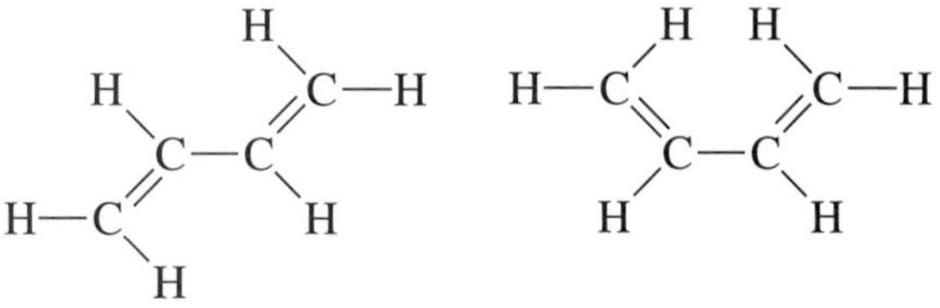

s-trans   *s-cis*

Spectroscopic evidence shows that the *s-trans* form (CCCC dihedral angle of 180°) is the predominant conformation. The second conformer is either the planar *s-cis* form (with CCCC dihedral angle of 0°), which is favored by conjugation between the double bonds and disfavored by steric repulsion between two H atoms, or the nonplanar *gauche* form with a CCCC dihedral angle somewhere between 0° and 90°. One gas-phase spectroscopic study concluded that the second conformer is the *gauche* form [K. B. Wiberg and R. E. Rosenberg, *J. Am. Chem. Soc.,* **112**, 1509 (1990)], whereas a spectroscopic study of butadiene in a frozen Ar matrix concluded that in this environment the second conformer is the *s-cis* form [B. R. Arnold et al., *J. Am. Chem. Soc.,* **112**, 1808 (1990)]. Virtually all SCF, MP2, MP4, CISD, and CC calculations with adequate-size basis sets predict the second conformer to be the *gauche* form with a CCCC dihedral angle of 38° ± 10° and an energy that is 0.5 to 1 kcal/mol below that of the *s-cis* form, which the calculations show to be at an energy maximum (Fig. 20.35) [M. A. Murcko et al., *J. Phys. Chem.,* **100**, 16162 (1996)].

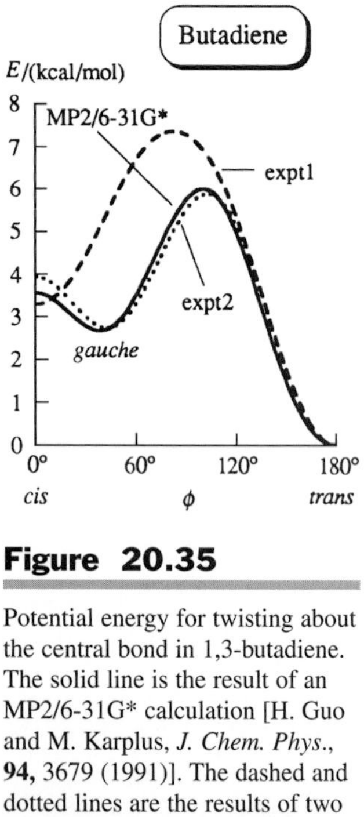

An important question is the energy of the hydrogen bond between two H_2O molecules. Study of the temperature and pressure dependences of the thermal conductivity k of water vapor showed that the reaction $2H_2O(g) \rightarrow (H_2O)_2(g)$ has $\Delta H_{373}^\circ = -3.6 \pm 0.5$ kcal/mol [L. A. Curtiss et al., *J. Chem. Phys.,* **71**, 2703 (1979)], which corresponds to an electronic energy change of $\Delta E_e = -5.4 \pm 0.7$ kcal/mol for nonvibrating, nonrotating molecules. Three sets of MP, CI, and CC calculations with very large basis sets yielded estimates of ΔE_e of dimerization of -4.7 ± 0.35 kcal/mol [K. Szalewicz et al., *J. Chem. Phys.,* **89**, 3662 (1988)], -5.1 kcal/mol [D. Feller, *J. Chem. Phys.,* **96**, 6104 (1992)], and -5.02 ± 0.05 kcal/mol [W. Klopper et al., *Phys. Chem. Chem. Phys.,* **2**, 2227 (2000)]. Theoretical calculations have determined the dimerization energy with a much smaller uncertainty than the experimental result.

Figure 20.35

Potential energy for twisting about the central bond in 1,3-butadiene. The solid line is the result of an MP2/6-31G* calculation [H. Guo and M. Karplus, *J. Chem. Phys.,* **94**, 3679 (1991)]. The dashed and dotted lines are the results of two different possible interpretations of butadiene vibrational-spectroscopy data [J. R. Durig et al., *Can. J. Phys.,* **53**, 1832 (1975)].

20.10   DENSITY-FUNCTIONAL THEORY (DFT)

In the period 1995–2000, density-functional theory (DFT) showed a meteoric rise to popularity in quantum-chemistry calculations: "a substantial majority of the [quantum chemistry] papers published today are based on applications of density functional theory" [K. Raghavachari, *Theor. Chem. Acc.,* **103**, 361 (2000)].

The Hohenberg–Kohn Theorem

In DFT, one does not attempt to calculate the molecular wave function. Instead, one works with the electron probability density $\rho(x, y, z)$ (Sec. 20.8). DFT is based on a theorem proved in 1964 by Pierre Hohenberg and Walter Kohn that states that *the energy and all other properties of a ground-state molecule are uniquely determined by the ground-state electron probability density $\rho(x, y, z)$*. One therefore says that the ground-state electronic energy E_{gs} is a functional of ρ and one writes $E_{gs} = E_{gs}[\rho(x, y, z)]$ or simply

$$E_{gs} = E_{gs}[\rho] \tag{20.39}$$

where the square brackets denote a functional relation.

What is a functional? Recall that a *function* $y = f(x)$ is a rule that associates a number (the value of y) with each value of the independent variable x. For example, the function $y = 3x^2 - 2$ associates the value $y = 73$ with $x = 5$. A **functional** $z = G[f]$ is a rule that associates a number with each function f. For example, if $z = \int_0^1 f(x)\, dx$, then this functional associates the value $z = 3$ with the function $f(x) = 6x^2 + 1$. The variational integral $W = \int \phi^* \hat{H} \phi \, d\tau / \int \phi^* \phi \, d\tau$ (Sec. 18.15) associates a number with each well-behaved function ϕ and we can write $W = W[\phi]$.

The Kohn–Sham Method

Unfortunately, the functional $E_{gs}[\rho]$ in (20.39) is unknown, so the Hohenberg–Kohn theorem does not tell us *how* to calculate E_{gs} from ρ or how to find ρ without first finding the ground-state molecular electronic wave function. In 1965, Kohn and Sham devised a practical method to find ρ and to calculate E_{gs} from ρ. The Kohn–Sham equations do contain an unknown functional, but using a combination of physical insight and guesswork, physicists and chemists have developed approximations to this functional that allow accurate calculations of molecular properties, thereby turning DFT into a key method in quantum chemistry.

The Kohn–Sham (KS) method uses a fictitious **reference system** (usually denoted by the subscript s) that contains the same number of electrons (n) as the molecule we are dealing with, but that differs from the molecule in that (a) the electrons in the reference system do not exert forces on one another; (b) each electron i ($i = 1, 2, \ldots, n$) in the reference system experiences a potential energy $v_s(x_i, y_i, z_i)$, where v_s is the same function for each electron and is such as to make the electron probability density ρ_s in the reference system exactly equal to the ground-state electron probability density ρ in the real molecule: $\rho = \rho_s$. The actual form of v_s is not known. (In the real molecule, the electrons experience attractions to the nuclei, but these are not present in the reference system.)

Because the electrons in the reference system do not interact with one another, the Hamiltonian $\hat{H}_s$ of the reference system is the sum of Hamiltonians of the individual electrons:

$$\hat{H}_s = -\frac{\hbar^2}{2m_e} \sum_{i=1}^n \nabla_i^2 + \sum_{i=1}^n v_s(x_i, y_i, z_i) \equiv \sum_{i=1}^n \hat{h}_i^{KS}$$

$$\hat{h}_i^{KS} \equiv -\frac{\hbar^2}{2m_e} \nabla_i^2 + v_s(x_i, y_i, z_i) \tag{20.40}$$

$\hat{h}_i^{KS}$ is the one-electron Kohn–Sham Hamiltonian. Since the reference system s consists of noninteracting particles, its wave function, if spin and the Pauli principle are neglected, is the product of one-electron spatial wave functions, each of which is an eigenfunction of $\hat{h}_i^{KS}$ [Eqs. (18.68) and (18.70)]. To allow for spin and the Pauli-principle antisymmetry requirement, the ground-state wave function of the reference system is a Slater determinant (Sec. 19.8) of spin-orbitals, one for each electron. Each spin-orbital is the product of a spatial orbital θ_i^{KS} and a spin function (either α or β). The Kohn–Sham orbitals θ_i^{KS} are eigenfunctions of $\hat{h}_i^{KS}$:

$$\hat{h}_i^{KS}\theta_i^{KS} = \varepsilon_i^{KS}\theta_i^{KS} \tag{20.41}$$

where ε_i^{KS} is the Kohn–Sham orbital energy of θ_i^{KS}. Each Kohn–Sham orbital holds two electrons of opposite spin.

One can prove that the probability density ρ for a wave function that is a Slater determinant (such as the wave function of the KS reference system) is the sum of the probability densities $|\theta_i^{KS}|^2$ of the individual orbitals. Also, by definition, the reference system has the same ρ as the molecule:

$$\rho = \rho_s = \sum_{i=1}^{n} |\theta_i^{KS}|^2 \tag{20.42}$$

The Kohn–Sham Energy Expression

We now need the expression for the molecule's ground-state electronic energy E_e. (The reference system and the molecule have the same probability density function but they do not have the same ground-state energy.) Kohn and Sham derived the following exact equation for E_e:

$$E_e = \langle K_{e,s}\rangle + \langle V_{Ne}\rangle + J + V_{NN} + E_{xc}[\rho] \tag{20.43}$$

We now discuss the meanings of the terms in (20.43). Complicated equations are given in small print at the end of this subsection, so they can be skimmed over, if desired.

$\langle K_{e,s}\rangle$ is the average electronic kinetic energy in the reference system. Its value can be calculated from the Kohn–Sham orbitals θ_i^{KS} of the reference system [see Eq. (20.46)].

$\langle V_{Ne}\rangle$ is the average potential energy of attractions between the electrons and the nuclei in the molecule. Its value can be calculated from the electron probability density $\rho(x, y, z)$ [see Eq. (20.47)]. $\rho(x, y, z)$ is the same for the molecule as for the reference system and is calculated from the Kohn–Sham orbitals θ_i^{KS} using Eq. (20.42).

J is the classical energy of electrical repulsion that arises between the infinitesimal charge elements of a hypothetical smeared-out electron charge cloud whose probability density is $\rho(x, y, z)$. J can be calculated from $\rho(x, y, z)$ [see Eq. (20.48)].

The internuclear repulsion energy V_{NN} is a constant that depends on the nuclear charges and the internuclear distances and is calculated from the molecular geometry at which the calculation is being done; one sums the potential energies (19.1) for each internuclear repulsion.

$E_{xc}[\rho]$ in (20.43), called the **exchange–correlation energy functional,** is a functional of ρ that is defined as

$$E_{xc}[\rho] \equiv \langle K_e\rangle - \langle K_{e,s}\rangle + \langle V_{ee}\rangle - J \tag{20.44}$$

$E_{xc}[\rho]$ is the sum of two differences: (a) the difference $\langle K_e\rangle - \langle K_{e,s}\rangle$ between the average electronic kinetic energy in the molecule and in the reference system; and (b) the difference $\langle V_{ee}\rangle - J$ between the average potential energy of interelectronic repulsion $\langle V_{ee}\rangle$ in the molecule and the classical charge-cloud self-repulsion energy J. The values of $\langle K_e\rangle$ and $\langle K_{e,s}\rangle$ are expected to be similar to each other; also, the values of $\langle V_{ee}\rangle$ and J are similar. Hence, the two differences in (20.44) are relatively small quantities. These two differences are not zero because the Pauli-principle requirement that the wave function be antisymmetric with respect to exchange produces *exchange* effects on the energy, and because the instantaneous correlations between the motions of the electrons produces *correlation* effects on the energy.

Equation (20.43) is an exact expression for the molecular electronic energy. However, no one knows what the true mathematical expression for the functional E_{xc}

is. Hence, approximations to E_{xc} must be used. The term $E_{xc}[\rho]$ can be found if we know ρ and if we know what the functional E_{xc} is. Approximations to E_{xc} are discussed later in this section, and for now we shall assume that a good approximation to the functional E_{xc} is known.

Note that if (20.44) is substituted into (20.43), we get

$$E_e = \langle K_e \rangle + \langle V_{Ne} \rangle + \langle V_{ee} \rangle + V_{NN} \qquad (20.45)$$

an equation that follows from $\hat{H}_e = \hat{K}_e + \hat{V}_{Ne} + \hat{V}_{ee} + \hat{V}_{NN}$ [Eq. (20.6)] if we take the average value of each side of (20.6) (see Prob. 20.47).

We now give the explicit formulas for some of the terms in (20.43). $\langle K_{e,s} \rangle$ and $\langle V_{Ne} \rangle$ can be shown to be [see *Levine* (2000), pp. 579, 575]

$$\langle K_{e,s} \rangle = -\frac{\hbar^2}{2m_e} \sum_{i=1}^{n} \int_{-\infty}^{\infty} \int_{-\infty}^{\infty} \int_{-\infty}^{\infty} \theta_i^{KS}(1)^* \nabla_1^2 \theta_i^{KS}(1) \, dx_1 \, dy_1 \, dz_1 \qquad (20.46)$$

$$\langle V_{Ne} \rangle = -\sum_{\alpha} \frac{Z_\alpha e^2}{4\pi\varepsilon_0} \int_{-\infty}^{\infty} \int_{-\infty}^{\infty} \int_{-\infty}^{\infty} \frac{\rho(x, y, z)}{r_\alpha} \, dx \, dy \, dz \qquad (20.47)$$

where $r_\alpha = [(x - x_\alpha)^2 + (y - y_\alpha)^2 + (z - z_\alpha)^2]^{1/2}$ is the distance from point (x, y, z) to nucleus α, located at $(x_\alpha, y_\alpha, z_\alpha)$. If we pretended that the electrons were smeared out into a static continuous distribution of charge whose electron density [defined as dn/dV, where dn is the infinitesimal fractional number of electrons in the infinitesimal volume dV located at point (x, y, z) in space] is ρ, then a term in the sum in (20.47) is the potential energy of attraction between nucleus α and the smeared-out electron charge cloud. J is given by (Prob. 20.49)

$$J \equiv \frac{1}{2} e^2 \int_{-\infty}^{\infty} \int_{-\infty}^{\infty} \int_{-\infty}^{\infty} \int_{-\infty}^{\infty} \int_{-\infty}^{\infty} \int_{-\infty}^{\infty} \frac{\rho(x_1, y_1, z_1)\rho(x_2, y_2, z_2)}{4\pi\varepsilon_0 r_{12}} \, dx_1 \, dy_1 \, dz_1 \, dx_2 \, dy_2 \, dz_2$$

$$(20.48)$$

Finding the Kohn–Sham Orbitals

As noted in the discussion after (20.43), the quantities $\langle K_{e,s} \rangle$, $\langle V_{Ne} \rangle$, J, and E_{xc} can all be calculated if we know the orbitals θ_i^{KS}, and V_{NN} is easily calculated from the locations of the nuclei. Thus, we can find the molecular electronic ground-state energy E_e if the Kohn–Sham orbitals are known. How are the Kohn–Sham orbitals θ_i^{KS} found? Hohenberg and Kohn proved that the true ground-state electron probability density ρ minimizes the functional $E_e[\rho]$ for the energy. Since ρ is determined by the KS orbitals θ_i^{KS} [Eq. (20.42)], one can vary the orbitals θ_i^{KS} so as to minimize E_e in (20.43).

It turns out (see *Parr and Yang*, Sec. 7.2 for the proof) that the orthogonal and normalized KS orbitals that minimize E_e satisfy

$$\hat{h}_i^{KS} \theta_i^{KS} = \varepsilon_i^{KS} \theta_i^{KS}$$

[Eq. (20.41)] with the one-electron KS Hamiltonian $\hat{h}_i^{KS}$ being the sum of the following four one-electron terms: (a) the one-electron kinetic-energy operator $-(\hbar^2/2m_e)\nabla_1^2$; (b) the potential energy $-\sum_\alpha Z_\alpha e^2/4\pi\varepsilon_0 r_{1\alpha}$ of attractions between electron 1 and the nuclei; (c) the potential energy of repulsion between electron 1 and a hypothetical charge cloud of electron density ρ due to smeared out electrons; (d) an *exchange–correlation potential* $v_{xc}(x_1, y_1, z_1)$ whose form is discussed below. [The sum of terms b, c, and d equals what is called v_s in (20.40).]

The first three terms in $\hat{h}_i^{KS}$ are in fact identical to the first three terms a, b, and c of Secs. 19.9 and 20.5 in the Fock operator $\hat{F}$ of the Hartree–Fock equations (19.54). The only difference between the Hartree–Fock equations (19.54) for the Hartree–Fock orbitals and the Kohn–Sham equations (20.41) for the Kohn–Sham orbitals is that the exchange operator (term d) of the Hartree–Fock equations is replaced by the **exchange–correlation potential** v_{xc}. The expression for v_{xc} is found to be

$$v_{xc}(x, y, z) = \frac{\delta E_{xc}[\rho]}{\delta \rho} \qquad (20.49)$$

where the notation $\delta E_{xc}/\delta \rho$ indicates the *functional derivative* of the functional E_{xc} that occurs in the energy equation (20.43). The functional derivative $\delta E_{xc}/\delta \rho$ at (x, y, z) depends on how much the functional $E_{xc}[\rho]$ changes when the function ρ changes by a tiny amount in a tiny region centered at (x, y, z). We shall not worry about the precise definition of the functional derivative but simply note that if $E_{xc}[\rho]$ is known, its functional derivative v_{xc} can be easily found (see Prob. 20.50). The fourth term (term d) in the Fock operator $\hat{F}$ of the Hartree–Fock equations allows for the effects of electron exchange (the Pauli principle) but does not allow for electron correlation. The fourth term v_{xc} in the KS operator $\hat{h}_i^{KS}$ allows for both exchange and correlation.

Carrying Out a Density-Functional Calculation

Assuming that we have a reasonable approximation for the functional $E_{xc}[\rho]$, how do we do a density-functional calculation? One starts with an initial guess for the molecule's electron density $\rho(x, y, z)$ found by superimposing calculated electron densities of the individual atoms at the nuclear geometry chosen for the calculation. From the initial ρ, one finds $E_{xc}[\rho]$ and then finds its functional derivative to get an initial estimate of v_{xc} [Eq. (20.49)]. This v_{xc} is used in the Kohn–Sham equations $\hat{h}_i^{KS}\theta_i^{KS} = \varepsilon_i^{KS}\theta_i^{KS}$ to solve for initial estimates of the orbitals θ_i^{KS}. (As is done in solving the Hartree–Fock equations, one usually expands the unknown orbitals using a basis set. Many of the same basis sets used for Hartree–Fock calculations are also used for KS calculations.) The initial orbitals θ_i^{KS} are used to calculate an improved probability density ρ from (20.42). This improved ρ is used to find an improved $E_{xc}[\rho]$, from which an improved v_{xc} is found. The improved v_{xc} is used in the KS equations (20.41) to find improved orbitals; etc. One continues the iterations until no further significant change is found from one cycle to the next. The molecular energy is then found from (20.43) using the final orbitals and ρ.

The Exchange–Correlation Energy Functional E_{xc}

Kohn and Sham suggested the use of a certain form for $E_{xc}[\rho]$ called the *local (spin) density approximation* (LDA or LSDA) that theory shows to be accurate when the electron density ρ varies very slowly with position. In a molecule, ρ does not vary very slowly with position. One finds that LSDA KS DFT calculations give good results for molecular geometries, dipole moments, and vibrational frequencies, but rather poor results for atomization energies. The LSDA E_{xc} is a definite integral of a certain function of ρ.

For convenience in devising approximations to E_{xc}, E_{xc} is usually split into an exchange part and a correlation part:

$$E_{xc}[\rho] = E_x[\rho] + E_c[\rho] \qquad (20.50)$$

and people devise separate approximations for E_x and E_c.

In the late 1980s, Becke showed that by taking E_{xc} as an integral of a certain function of ρ *and* the derivatives $\partial\rho/\partial x$, $\partial\rho/\partial y$, $\partial\rho/\partial z$ (these derivatives constitute the *gradient* of ρ), one gets greatly improved results for molecular atomization energies. Such a functional is called a *gradient-corrected functional* and use of a gradient-corrected functional gives the **generalized-gradient approximation** (GGA). In 1993, Becke proposed a further improvement in E_{xc}^{GGA} by adding to it a term aE_x^{HF}, where E_x^{HF} has the form of the expression used for the exchange energy in Hartree–Fock calculations but is evaluated using KS rather than Hartree–Fock orbitals, and a is an empirical parameter whose value was chosen to optimize the performance of E_{xc} in calculations on a test series of molecules. A GGA E_{xc} that includes a contribution from E_x^{HF} is called a **hybrid functional.** Hybrid functionals seem to give the best overall performance, but this conclusion is subject to change, since people are continually proposing new functionals.

The most widely used functional in DFT calculations done in the period 1995–2000 has been the hybrid functional called B3LYP, where B indicates that it includes a term for E_x^{GGA} devised by Becke, LYP indicates a term for E_c^{GGA} devised by Lee, Yang, and Parr, and the 3 indicates that it contains three empirical parameters whose values were chosen to optimize its performance. The hybrid functional B3PW91 is similar to B3LYP except that it uses the Perdew–Wang 1991 expression for E_c instead of the LYP formula. In 1997, Becke devised a hybrid E_{xc} called B97, which contains 10 empirical parameters whose values were chosen to optimize its performance.

The performance of several E_{xc} functionals was assessed using a rather large basis set [A. J. Cohen and N. C. Handy, *Chem. Phys. Lett.,* **316,** 160 (2000)]. For a set of 97 energies (mostly atomization energies) mean absolute errors were 3.4 kcal/mol for B3LYP and 2.5 kcal/mol for B97. B3LYP and B97 gave similar good performances for molecular geometries. The performances of B3LYP and B3PW91 were similar.

Performance of DFT

The time needed to do a DFT calculation is roughly the same as the time needed to do a Hartree–Fock calculation on the same molecule with the same basis set. A Hartree–Fock calculation (which restricts the molecular wave function to be a Slater determinant of spin-orbitals) yields only an approximate wave function and energy, no matter how large the basis set is. In contrast, a DFT calculation is, in principle, capable of yielding the exact values of the energy and other molecular properties. (No approximations were made in the equations of KS DFT.) In practice, because the true functional $E_{xc}[\rho]$ is unknown, DFT calculations yield approximate results. The quality of the results depends on how good the E_{xc} used in the calculation is. With present-day E_{xc} functionals, DFT calculations yield substantially more-accurate results than Hartree–Fock calculations. DFT calculations should not be done with basis sets smaller than 6-31G*.

One finds that Hartree–Fock calculations are often unreliable for transition-metal compounds. In contrast, DFT structures and relative energies of transition-metal compounds are usually reliable and "Computational transition metal chemistry today is almost synonymous with DFT for medium-sized molecules" [E. R. Davidson, *Chem. Rev.,* **100,** 351 (2000)].

For a sample of 108 molecules, average absolute errors in bond lengths, bond angles, dipole moments, and atomization energies were [A. C. Scheiner et al., *J. Comput. Chem.,* **18,** 775 (1997)]:

	HF/6-31G**	MP2/6-31G**	B3PW91/6-31G**
Bond lengths	0.021 Å	0.015 Å	0.011 Å
Bond angles	1.3°	1.1°	1.0°
Dipole moments	0.23 D	0.20 D	0.16 D
$\Delta_{at}U/(kcal/mol)$	119.2	22.0	6.8

The hybrid DFT method outperforms the Hartree–Fock (HF) and the MP2 methods.

Although the Kohn–Sham orbitals are calculated for the fictitious reference system (and not for the real molecule), one finds that Kohn–Sham orbitals closely resemble

Hartree–Fock SCF MOs calculated for the real molecule [R. Stowasser and R. Hoffmann, *J. Am. Chem. Soc.,* **121,** 3414 (1999)]. Hartree–Fock MOs have been used to provide qualitative explanations and predictions of many chemical phenomena, and Kohn–Sham orbitals can be used for the same purpose.

Despite its many successes, DFT does have some drawbacks. DFT is basically a ground-state theory. People are currently working to extend it to excited states. Because an approximate E_{xc} is used, it is possible for a DFT calculation to give an energy that is less than the ground-state energy. In ab initio calculations (this term is defined in Sec. 20.11), one knows in principle how to improve the accuracy of the calculation, namely, one uses larger basis sets and goes to higher levels of theory (for example, HF, MP2, MP4, . . . or HF, CISD, CISDT, CISDTQ, . . . or HF, CCSD, CCSDT, . . .). (In practice, high-level ab initio calculations are limited to rather small molecules.) In DFT, the performance of a given functional cannot be predicted and one must try it out on a variety of molecules and properties to assess its performance. There is no systematic way in DFT to devise better functionals. The currently available functionals do not give good accuracy when dealing with van der Waals intermolecular interactions; nor can they match the performance of very high-level ab initio methods, such as CCSD(T), when dealing with small molecules. DFT predictions for activation energies of chemical reactions are fairly often inaccurate.

For more on DFT, see *Parr and Yang; Levine* (2000), sec. 15.20; W. Koch and M. C. Holthausen, *A Chemist's Guide to Density Functional Theory,* Wiley-VCH, 2000.

20.11 SEMIEMPIRICAL METHODS

Classification of Methods

Quantum-mechanical methods of treating molecules are classified as ab initio, density-functional, or semiempirical. An **ab initio** calculation uses the true molecular Hamiltonian and does not use empirical data in the calculation. (Ab initio means "from the beginning" in Latin.) The Hartree–Fock method calculates the antisymmetrized product Φ of spin-orbitals that minimizes the variational integral $\int \Phi^* \hat{H} \Phi \, d\tau$, where $\hat{H}$ is the true molecular Hamiltonian. Therefore, a Hartree–Fock calculation is an ab initio one (as is an SCF calculation that gives only an approximation to the Hartree–Fock wave function because of the limited size of the basis set). Of course, because of the restricted form of Φ, the Hartree–Fock method does not give the true wave function. CI, MP, and CC calculations (See. 20.9) are also ab initio methods and are capable in principle of converging to the exact wave function and energy provided the basis set is large enough and the level of the calculation is high enough. Note that the term "ab initio" does not guarantee high accuracy. For example, ab initio Hartree–Fock calculations give wildly inaccurate molecular dissociation energies.

A **semiempirical** method uses a simpler Hamiltonian than the true one, uses empirical data to assign values to some of the integrals that occur in the calculation, and neglects some of the integrals. The reason for resorting to semiempirical methods is that accurate ab initio calculations on large molecules cannot be done at present. Semiempirical methods were originally developed for conjugated organic molecules and later were extended to encompass all molecules.

Density-functional calculations are hard to classify as either ab initio or semiempirical and are usually considered as a third category of quantum-chemistry methods.

Semiempirical Methods for Conjugated Molecules

For a planar or near-planar conjugated organic compound, each MO can be classified as σ or π. Each σ MO is unchanged on reflection in the molecular plane (which is not a nodal plane for a σ MO), whereas each π MO changes sign on reflection in the molecular plane (which is a nodal plane for each π MO). (Recall the discussion of benzene in Sec. 20.6.) The σ MOs have electron probability density strongly concentrated in the region of the molecular plane. The π MOs have blobs of probability density above and below the molecular plane. The σ MOs can be taken as either delocalized or localized. However, the π MOs in a conjugated compound are best taken as delocalized. In conjugated molecules, the highest-energy occupied MOs are usually π MOs.

Because of the different symmetries of σ and π MOs, one can make the *approximation* of treating the π electrons separately from the σ electrons. One imagines the σ electrons to produce some sort of effective potential in which the π electrons move.

The simplest semiempirical treatment of conjugated molecules is the *free-electron molecular-orbital (FE MO) method*. The FE MO method deals only with the π electrons. It assumes that each π electron is free to move along the length of the molecule (potential energy $V = 0$) but cannot move beyond the ends of the molecule (potential energy $V = \infty$). This is the particle-in-a-box potential energy, and the FE MO method feeds the π electrons into particle-in-a-one-dimensional-box MOs, each such occupied MO holding two electrons of opposite spin. The FE MO method is extremely crude and is of only historical interest. Problem 20.51 outlines an application of the FE MO method.

A somewhat more sophisticated method than the FE MO method is the **Hückel method** for conjugated hydrocarbons (developed in the 1930s). The Hückel MO method deals only with the π electrons. It takes each π MO as a linear combination of the $2p_z$ AOs of the conjugated carbon atoms (where the z axis is perpendicular to the molecular plane). These linear combinations are used in the variational integral, which is expressed as a sum of integrals involving the various $2p_z$ AOs. The Hückel method approximates many of these integrals as zero and leaves others as parameters whose values are picked to give the best fit to experimental data. Details may be found in most quantum-chemistry texts. The Hückel method was a mainstay of theoretically inclined organic chemists for many years but, because of the development of improved semiempirical methods (discussed below), is now only rarely used.

For certain purposes, all one is interested in is the relative signs of the AOs that contribute to the π MOs. (An example is the Woodward–Hoffmann rules for deducing the steric course of certain organic reactions; see any modern organic chemistry text.) To deduce these signs, we use the idea that the π MOs will show a pattern of nodes perpendicular to the molecular plane that will resemble the nodal pattern for the particle in a one-dimensional box and the harmonic oscillator.

Consider butadiene, $CH_2{=}CH{-}CH{=}CH_2$, for example. We take the π MOs as linear combinations of the four $2p_z$ carbon AOs, where the z axis is perpendicular to the molecular plane. (This is a minimal-basis-set treatment; Sec. 20.5.) Let p_1, p_2, p_3, p_4 denote these AOs. Figure 18.10 shows that the lowest π MO will have no nodes perpendicular to the molecular plane, the next lowest π MO will have one such node (located at the midpoint of the molecule), etc. To form the lowest π MO, we must therefore combine the four $2p_z$ AOs all with the same signs: $c_1 p_1 + c_2 p_2 + c_3 p_3 + c_4 p_4$, where the c's are all positive. For the purpose of determining the relative signs we won't worry about the fact that c_1 and c_2 differ in value (since the end and interior carbons are not equivalent); we shall simply write $p_1 + p_2 + p_3 + p_4$ for the lowest π MO. To have a single node in the center of the molecule, we must take $p_1 + p_2 - p_3 - p_4$ as the second lowest π MO; this is the highest occupied π MO in the ground state, since there are four π electrons in butadiene. To get two symmetrically placed nodes, we take $p_1 - p_2 - p_3 + p_4$ as the third lowest π MO. To get three nodes, we take $p_1 - p_2 + p_3 - p_4$ as the fourth lowest π MO. See Fig. 20.36. Butadiene has four π electrons, two in each double bond. We put two electrons of opposite spin into each π MO. Hence the ground electronic state of butadiene will have the lowest two MOs of Fig. 20.36 occupied and the other two vacant.

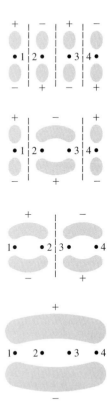

Figure 20.36

Rough sketches of the four lowest π MOs in butadiene.

General Semiempirical Methods

An improved version of the Hückel method, applicable to both conjugated and non-conjugated molecules, is the *extended Hückel (EH) method*, developed in the 1950s and 1960s by Wolfsberg and Helmholz and by Hoffmann. The EH method treats all the valence electrons of a molecule and neglects fewer integrals than the Hückel method. The calculations of the EH method are relatively easy to perform (thanks to the many simplifying approximations made). The quantitative predictions of the EH method are generally rather poor, and the main value of the method is the qualitative insights it provides into chemical bonding.

The Hückel and extended Hückel methods are quite crude, in that they use a very simplified Hamiltonian that contains no repulsion terms between electrons. Several improved semiempirical theories have been developed that include some of the electron repulsions in the Hamiltonian operator and that apply to both conjugated and nonconjugated molecules. The most widely used semiempirical methods are AM1 and PM3.

The AM1 (Austin Model 1) method (named after the University of Texas at Austin, where the method was developed) was devised by Dewar and co-workers [M. J. S. Dewar et al., *J. Am. Chem. Soc.*, **107**, 3902 (1985)]. AM1 treats only the valence electrons. It solves equations resembling the Hartree–Fock equations (19.54) to find self-consistent MOs, but since an approximate Hamiltonian is used and rather drastic approximations are made for many of the integrals that occur, the MOs found are only rough approximations to Hartree–Fock MOs. Dewar's aim was not to have a method giving approximations to Hartree–Fock results but one giving accurate molecular geometries and molecular dissociation energies. It might seem unreasonable to expect a theory that involves more approximations than the Hartree–Fock method to succeed in an area (calculation of dissociation energies) where the Hartree–Fock method fails. However, by choosing the values of the parameters in the AM1 method so as to reproduce known heats of atomization of many compounds, Dewar was able to build in compensation for the neglect of electron correlation that occurs in the Hartree–Fock theory.

AM1 is parametrized for compounds containing H, B, Al, C, Si, Ge, Sn, N, P, O, S, F, Cl, Br, I, Zn, and Hg. AM1 has a different set of parameters for each element. The number of parameters varies somewhat from element to element but is on the order of 15 per element. The parameter values were fixed by varying the parameters so that the theory gives a good least-squares fit to known gas-phase $\Delta_f H^\circ_{298}$ values (found from AM1-calculated atomization energies and experimental heats of formation of gaseous atoms) and values of molecular geometries and dipole moments. PM3 (parametric method 3) is similar to AM1 except that it uses a different set of parameters [J. J. P. Stewart, *J. Comput. Chem.*, **10**, 209, 221 (1989); **11**, 543 (1990); **12**, 329 (1991)].

The AM1 and PM3 methods usually give satisfactory bond lengths and bond angles, but their predictions of dihedral angles are not so good. Calculated dipole moments are fairly reliable. AM1 and PM3 predictions of gas-phase $\Delta_f H^\circ_{298}$ values are of only fair accuracy. For samples of 460 bond lengths (R_{AB}), 196 bond angles ($\angle ABC$), 16 dihedral angles ($\angle ABCD$), 125 dipole moments (μ), and 886 gas-phase $\Delta_f H^\circ_{298}$ values, average absolute errors are [J. J. P. Stewart, *op. cit.*]:

	R_{AB}	$\angle ABC$	$\angle ABCD$	μ	$\Delta_f H^\circ_{298}$
AM1	0.051 Å	3.8°	12.5°	0.35 D	14.2 kcal/mol
PM3	0.037 Å	4.3°	14.9°	0.38 D	9.6 kcal/mol

Geometries and dipole moments are less accurate than for ab initio and DFT methods. AM1 and PM3 show large percentage errors in predicting barriers to internal rotation and do not give accurate predictions of conformational-energy differences.

When AM1 or PM3 $\Delta_f H^\circ_{298}$ values are combined to predict ΔH° of a reaction, errors for each compound are multiplied by the stoichiometric coefficients in the reaction and errors for different compounds can add in a random manner, so the prediction of ΔH° is unreliable. Thus, the quantitative accuracy of AM1 and PM3 is mediocre. Because of the many approximations made, AM1 and PM3 are much faster than ab initio and DFT methods and can deal with larger molecules (up to 1000 atoms) than can ab initio and DFT methods (up to 100 atoms). The molecular-mechanics method (Sec. 20.13) can deal with molecules with 10000 atoms.

20.12 PERFORMING QUANTUM CHEMISTRY CALCULATIONS

The ab initio HF (Hartree–Fock) and MP methods, the DFT method, and semiempirical methods are widely used in chemistry, not just by theoretical chemists but by all kinds of chemists as an aid to predict and interpret experimental results.

In describing a molecular electronic-structure calculation, one specifies the method used followed by the basis set. For example, HF/6-31G* denotes a Hartree–Fock calculation done with the 6-31G* basis set. (Any calculation that solves the Hartree–Fock equations (19.54) is called a Hartree–Fock calculation even though the basis set might be substantially smaller than that needed to give a truly accurate approximation to the Hartree–Fock wave function.) B3LYP/6-31G** denotes a DFT calculation done with the B3LYP exchange–correlation functional and the 6-31G** basis set. The AM1 and PM3 methods use a fixed minimal-basis set of Slater-type orbitals. Since one cannot vary the basis set, no basis set is specified when describing an AM1 or PM3 calculation.

A **single-point calculation** is one done only at a single fixed molecular geometry specified by the user. In a **geometry-optimization,** the quantum-mechanical program will vary the locations of the nuclei so as to locate a minimum in the electronic energy E_e of (20.7). A geometry-optimization calculation consists of many single-point calculations, with each single-point energy calculation followed by an energy-gradient calculation (Sec. 20.8) to help the program decide on the next geometry to try. The geometry-optimization calculation continues until the magnitude of the gradient is very close to zero, indicating that an energy minimum has been found. In a **vibrational-frequency calculation,** the program calculates the molecular vibration frequencies (Section 21.8). A vibrational-frequency calculation must be preceded by a geometry-optimization, since vibrational frequencies calculated for a geometry that is not at an energy minimum are meaningless. A **transition-state optimization** attempts to find the geometry and electronic energy of the transition state in a chemical reaction (Sec. 23.4).

Geometry-optimization calculations for large molecules are too time consuming to be done with high-level methods. Since an accurate geometry can usually be found with a low-level method, a common procedure is to do a low-level calculation to find the geometry and then use this geometry in a single-point high-level calculation of the energy.

The geometry-optimization process locates the energy minimum that is closest to the starting geometry. For example, if one does a geometry-optimization calculation of butane (Figs. 20.32 and 20.33) and inputs an initial geometry that has a CCCC dihedral angle close to 60°, the program will converge to the geometry of the gauche conformer, whereas if one starts with the CCCC dihedral angle close to 180°, the program converges to the trans conformer. The trans conformer is the **global minimum** for butane, meaning that it has the lowest energy of all the conformers, whereas the gauche conformer is only a **local minimum,** meaning that its energy is the minimum energy for all geometries in its immediate vicinity. Every conformer (Sec. 20.8) lies at a local minimum.

Large molecules may have huge numbers of conformers, making it extraordinarily difficult to find the global minimum and those local minima whose energy is low enough to have them significantly populated at room temperature. For example, for the cycloalkane $C_{17}H_{34}$, a study found 262 conformers with energy in the range 0–3 kcal/mole above the global minimum, 1368 conformers in the range 3–5 kcal/mol, 8165 conformers in the range 5–10 kcal/mol, and 2718 conformers in the range 10–20 kcal/mol [I. Kolossváry and W. C. Guida, *J. Am. Chem. Soc.,* **118,** 5011 (1996)]. The backbone of each amino acid residue in a polypeptide chain has two dihedral angles, and each such dihedral angle has three likely potential-energy minima. Hence a polypeptide with 40 amino acid residues has at least $3^{2(40)} = 3^{80} \approx 10^{38}$ possible conformers. Many special methods of **conformational searching** exist, whose aim is to find low-energy conformers [see *Leach,* chap. 8; *Levine* (2000), sec. 15.12]. Because of the huge numbers of conformers involved, the energy calculations in conformational searching on large molecules are usually done with the nonquantum-mechanical molecular-mechanics method (Sec. 20.13).

Programs

Many programs for molecular electronic-structure calculations exist.

Gaussian (www.gaussian.com) is the most widely used program for ab initio and density functional calculations, and can also do semiempirical calculations. *Gaussian* exists in versions for UNIX workstations and Windows personal computers. The first version of *Gaussian* was released in 1970 and *Gaussian 98* was released in 1998. *Gaussian* was developed by John Pople and co-workers and has been a key force in the growing use of quantum-chemistry calculations by chemists, since it is an easy-to-use program that allows a very wide variety of calculations to be done by virtually every available quantum-mechanical method.

Some other quantum-chemistry programs are the free program GAMESS (www .msg.ameslab.gov/GAMESS), Q-Chem (www.q-chem.com), Jaguar (www.schrodinger .com), SPARTAN (www.wavefun.com), and HyperChem (www.hyper.com), and the semiempirical programs MOPAC 2000 (www.schrodinger.com) and AMPAC (www .semichem.com). The free Windows program ArgusLab (www.planaria-software.com) does Hartree–Fock, AM1, and PM3 calculations.

Input

The input to a quantum-chemistry program specifies the kind of calculation to be done, the method and basis set to be used, and the initial set of locations for the nuclei. Figure 20.37 shows the input to the Windows version of *Gaussian* for a calculation on the methanol molecule, CH_3OH. In the Route Section, HF/3-21G tells the program to use the Hartree–Fock (SCF) method and to use the 3-21G basis set. The keyword Opt specifies a geometry optimization calculation by minimizing the energy. (If Opt is omitted, a single-point calculation is done.) The information in the Title Section has no effect on the program. The 0 and 1 tell the program that the molecule is uncharged and has spin multiplicity $2S + 1$ equal to 1; that is, $S = 0$.

The Molecule Specification section gives the initial geometry specified by the user. In Fig. 20.37, the geometry is specified by an array called a **Z-matrix.** To write down the Z-matrix, we first sketch the molecule (Fig. 20.38). The first column of the Z-matrix specifies the atoms present in the molecule. The numbers 1 to 6 after the element symbols are for the convenience of the user and can be omitted. The first row of the Z-matrix says that the first atom is a carbon atom. The second row tells the program to place an oxygen atom at a distance 1.43 Å from atom 1 (meaning the atom on row 1 of the Z-matrix). The value 1.43 Å was chosen to be the sum of the C and O single-bond radii in Sec. 20.1; other bond distances in the Z-matrix were chosen in the same way. The third row of the Z-matrix places an H atom at a distance 0.96 Å from

```
Existing File Job Edit
File  Edit  Check-Route  Set-Start
[                                                    ] [ ]  [ Additional Steps ]  [ 0 ]

        % Section  [                                    ]
                   [                                    ]

     Route Section  [ # HF/3-21G Opt                     ]
                    [                                    ]

     Title Section  [ staggered methanol Hf/3-21G  optimization ]
                    [                                    ]

   Charge , Multipl. [ 0 1 ]
                        Molecule Specification
   C1
   O2 1 1.43
   H3 2 0.96 1 107.0
   H4 1 1.07 2 109.5 3 -60.0
   H5 1 1.07 2 109.5 3 180.0
   H6 1 1.07 2 109.5 3 60.0
```

Figure 20.37

Input to Windows version of
Gaussian for HF/3-21G geometry
optimization of CH_3OH.

the row-2 atom (oxygen) with the angle H3–O2–C1 equal to 107.0° (note the 3, 2, and 1 in row 3).

The guess 107.0° for the COH bond angle was found using the VSEPR method (Sec. 20.1). The CH_3OH Lewis structure has four valence electron pairs on O. The VSEPR method arranges these pairs tetrahedrally. Since two of the four pairs on oxygen are lone pairs, the COH angle is predicted to be a bit less than the tetrahedral angle of 109.5°. Hence the 107° guess.

Row 4 of the Z-matrix places atom H4 1.07 Å from C1 with an angle of 109.5° (the tetrahedral angle predicted by VSEPR) for H4–C1–O2 (note the 4, 1, 2 on row 4). The -60.0 on row 4 means that the dihedral angle defined by atoms 4–1–2–3 is −60.0°. It takes one-third of a complete rotation of the CH_3 group to bring H4 to where H6 was and one-third of a rotation is 360°/3 = 120°. To bring H4 into alignment with H3, we need one-half of the rotation just described, namely, 60°. The sign of the dihedral angle 4–1–2–3 is defined as negative if a counterclockwise rotation of the first-listed atom (atom 4) is needed to make it align with atom 3. Rows 5 and 6 of the Z-matrix are constructed similar to row 4. Of course, the Z-matrix for a molecule is not unique and many different valid Z-matrices can be written for methanol. In Fig. 20.38, the OH hydrogen was staggered with respect to the methyl hydrogens in accord with the rule that the conformation about a bond that joins two atoms each of which has tetrahedral bond angles is usually staggered.

Some rules in constructing a Z-matrix are: Bond angles must be greater than 0° and less than 180° (see Prob. 20.54). All angles and lengths must contain a decimal point. The atom numbers in columns 2, 4, and 6 must be atoms whose locations were specified in previous rows.

Note that although the bond distances were used as a convenience in specifying the Z-matrix, the Z-matrix does not actually specify which atoms are bonded to each other. All it does is tell *Gaussian* the atomic number and location of each nucleus. In fact, instead of using a Z-matrix, one can specify the initial geometry by giving the symbol for each nucleus followed by the three cartesian coordinates of that nucleus on the same line as the symbol.

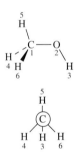

Figure 20.38

The methanol molecule. The lower figure is a Newman projection with the oxygen behind the carbon.

In some programs (for example, SPARTAN, HyperChem, and ArgusLab) the user inputs the starting structure using a molecule builder that builds the model on the computer screen. The user selects the desired atoms and bonds and the program uses built-in typical bond lengths and angles and rules for conformations to produce the initial structure. The user can modify the conformation if desired. Although *Gaussian* itself does not have this feature, there are programs that can serve as graphical interfaces to *Gaussian*. Two such programs are GaussView (www.gaussian.com) and Chem3D (www.camsoft.com).

You can use the program CORINA to generate a molecular structure at no charge. Go to www2.ccc.uni-erlangen.de/services/3d.html and enter the structure as a SMILES string. A SMILES string omits hydrogens and uses an equals sign to denote a double bond. For example, CC=O denotes CH_3CHO and C1CC1 denotes cyclopropane, where the numeral 1's indicate atoms bonded together. The program will generate a rotatable model of the molecule. If you save the structure to your computer and open the saved file in a word processor, you will see the nuclear cartesian coordinates. These can then be used as input to a quantum-chemistry program.

Output

Coming back to the methanol calculation, *Gaussian* begins the SCF calculation by using a semiempirical method to generate an initial guess for the MOs. It then solves the Hartree–Fock equations (19.54) using the specified basis set. The words SCF Done in the output are followed by E(RHF) = −114.3962071 A.U., which is the Hartree–Fock energy in atomic units (hartrees—Sec. 19.3). The orbital energies are given. The dipole moment is also calculated for this initial structure. The program calculates derivatives of the energy with respect to the nuclear coordinates. If these derivatives are all very small, the initial structure is close enough to a local minimum in the energy and the calculation is done. If not, the nuclei are moved to new locations (whose values are chosen based on the values of the energy derivatives and estimates of the energy second derivatives) and the SCF calculation is repeated at the new geometry. For this calculation, the energy found at the new geometry is −114.3978271 A.U., which is lower than the initial value, indicating we are closer to a minimum. The next two geometries yield energies of −114.3980171 and −114.3980192 A.U., respectively. For this last energy the energy derivatives are all small enough to declare that the optimization is completed and *Gaussian* reports the final optimized bond distances, bond angles, and dihedral angles and the final dipole moment. The optimized HF/3-21G geometry compared with the initially guessed geometry is:

	R(CO)	R(OH)	R(CH6)	R(CH5)	∠COH	∠OCH6	∠OCH5	∠H6COH3	∠H5COH3
Init.	1.43	0.96	1.07	1.07	107°	109.5°	109.5°	60°	180°
Opt.	1.441	0.966	1.085	1.079	110.3°	112.2°	106.3°	61.4°	180°

20.13   THE MOLECULAR-MECHANICS (MM) METHOD

The **molecular-mechanics** (or *empirical-force-field*) **method** can handle very large organic and organometallic ground-state molecules (up to 10^4 atoms). Molecular mechanics (MM) is an empirical nonquantum-mechanical method and does not use a Hamiltonian or wave function. Instead, the molecule is viewed as atoms held together by bonds, and the molecular electronic energy is expressed as the sum of bond-stretching, bond-bending, and other kinds of energies.

The Steric Energy

Rather than dealing with the molecular electronic energy E_e of Eqs. (20.7) and (20.8), the MM method uses a quantity called the **steric energy** V_{steric} that (as will be explained below) differs from E_e by an unknown constant C, whose value may differ for different molecules: $V_{\text{steric}} = E_e + C$. Recall that in the Schrödinger equation (20.9) for nuclear motion, the electronic energy E_e serves as the potential energy. As noted after Eq. (2.21), the potential energy always has an arbitrary additive constant. Therefore, V_{steric}, which differs from E_e by only a constant, is also a valid potential energy for nuclear motion.

The MM expression for V_{steric} is

$$V_{\text{steric}} = V_{\text{str}} + V_{\text{bend}} + V_{\text{tors}} + V_{\text{cross}} + V_{\text{vdW}} + V_{\text{es}} \qquad (20.51)$$

The specific forms for each term in (20.51) and the parameter values used in these terms define a molecular-mechanics **force field,** since the forces on the nuclei can be found from the derivatives of the potential energy V_{steric} [Eq. (2.21)].

In (20.51), V_{str} is the change in electronic energy due to bond stretching. In the simplest force fields, V_{str} is taken as the sum of harmonic-oscillator terms [Eqs. (18.71) and (21.18)] for each pair of bonded atoms:

$$V_{\text{str}} = \sum_{1,2} V_{\text{str},ij} \qquad \text{where} \qquad V_{\text{str},ij} = \tfrac{1}{2} k_{IJ}(l_{ij} - l_{IJ}^0)^2 \qquad (20.52)$$

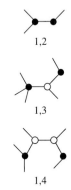

1,2

1,3

1,4

Figure 20.39

1,2, 1,3, and 1,4 atoms.

Atoms bonded to each other are called 1,2 atoms (Fig. 20.39). Atoms separated by two bonds are 1,3 atoms, and atoms separated by three bonds are 1,4 atoms. The sum in (20.52) is over all pairs of 1,2 atoms. The parameter k_{IJ} is the force constant for stretching the bond between atoms i and j, the parameter l_{IJ}^0 is a fixed *reference length,* and l_{ij} is the distance between the bonded atoms i and j for a particular molecular geometry. The harmonic-oscillator expression in (20.52) is not highly accurate for molecular vibration and some force fields use a more complicated and more accurate expression for $V_{\text{str},ij}$

An MM force field classifies each atom in the molecule into an **atom type,** depending on its atomic number and on how it is bonded in the molecule. Some typical atom types are sp^3 carbon (carbon bonded to four atoms) nonaromatic sp^2 carbon, sp carbon, aromatic carbon, hydrogen bonded to carbon, hydrogen bonded to oxygen, hydrogen bonded to nitrogen, etc. A typical force field for organic compounds has 60 to 70 atom types. In the expression $V_{\text{str},ij} = \tfrac{1}{2} k_{IJ}(l_{ij} - l_{IJ}^0)^2$, the capital letters I and J stand for the atom types of atoms i and j in the molecule. Thus in $CH_3CH_2CH_2OCH{=}CH_2$, the $CH_3{-}CH_2$ bond and the $CH_2{-}CH_2$ bond have the same k_{IJ} and the same l_{IJ}^0, whereas the $CH_3{-}CH_2$ bond and the $CH{=}CH_2$ bond have different k_{IJ} values and different l_{IJ}^0 values. The reference length l_{IJ}^0 is close to the typical bond length between atoms of types I and J.

The parameters k_{IJ} and l_{IJ}^0 in V_{str} (and the parameters in the other terms) are found as follows. One picks initial values of the parameters guided by experimental data (such as bond-length data and force constants found from analysis of vibrational spectra) and by the results of ab initio calculations. One then varies the parameter values in the force field so as to minimize the errors in molecular geometries and other properties predicted by the force field for a set of test molecules (the training set). Finding the optimum parameters in a force field is similar to finding the best values for the parameters in a semiempirical theory (Sec. 20.11), to finding the nuclear coordinates that minimize the electronic energy in a molecular-geometry optimization calculation (Sec. 20.12), and to finding parameters in a function so as to minimize the sum of squares of the deviations from experimental data points (Example 7.6 in Sec. 7.3). Similar mathematical procedures are used for all these *optimization* processes.

The term V_{bend} in the steric energy (20.51) is the energy change due to bending of bonds, and in the simplest force fields has the form

$$V_{\text{bend}} = \sum V_{\text{bend},ijk} \qquad \text{where} \qquad V_{\text{bend},ijk} = \tfrac{1}{2} k_{IJK}(\theta_{ijk} - \theta_{IJK}^0)^2$$

The sum is over all bond angles; θ_{ijk} is the ijk bond angle, and k_{IJK} and θ_{IJK}^0 are parameters.

The term V_{tors} accounts for the change in electronic energy with rotation (torsion) about a bond. (Figure 20.33 shows V_{tors} for rotation about the central CC bond in butane.) The form of V_{tors} is $V_{tors} = \Sigma_{1,4} V_{tors,ijkl}$, where the sum is over all 1,4 atom pairs and $V_{tors,ijkl}$ is a trigonometric function of the torsion dihedral angle defined by atoms $ijkl$. $V_{tors,ijkl}$ contains one or more parameters. In CH_3CH_3, each H atom on the left carbon has a 1,4 relation with each of the three H atoms on the right carbon, so there are nine terms in V_{tors} for ethane.

V_{cross} contains cross terms that allow for interactions between stretching, bending, and torsion.

V_{es} allows for electrostatic attractions and repulsions between nonbonded atoms and usually has the form

$$V_{es} = \sum_{1,\geq 4} V_{es,ij} \quad \text{where} \quad V_{es,ij} = \frac{Q_i Q_j}{4\pi\varepsilon_0 R_{ij}}$$

where the sum goes over all 1,4, 1,5, 1,6, ... atom pairs. R_{ij} is the distance between atoms i and j, and Q_i and Q_j are the (partial) electrical charges on atoms i and j in the molecule. The partial atomic charge Q_i on an atom in a molecule is an ill-defined quantity that can neither be measured experimentally nor calculated theoretically in a unique manner. Rather, many different theoretical methods have been proposed for arriving at Q_i values. A force field might use Q_i values based on the atom type and what atoms are bonded to atom i, or it might take Q_i values found on similar atoms in ab initio calculations on small molecules using some scheme for partitioning the electronic charge among the atoms.

The term V_{vdW} is the contribution of van der Waals nonbonded interactions (Sec. 22.10) between atoms in the molecule and is taken as $V_{vdW} = \Sigma_{1,\geq 4} V_{vdW,ij}$, where $V_{vdW,ij}$ usually has the form of the Lennard-Jones 6–12 potential [Eq. (22.136)].

A typical force field for organic molecules might specify values for 5000 parameters. The bond-bending parameters k_{IJK} and θ_{IJK}^0 involve three atom types and the torsion parameters involve four atom types, so with perhaps 60 different atom types in the force field, a huge number of parameters (far greater than 10^4) are needed to cover all possible situations. If one applies molecular mechanics to somewhat unusual molecules, one often runs into the problem of needing parameters whose values are missing from the force field. When a parameter is missing, the MM program will estimate its value, based on a parameter whose value is likely to be similar to the missing value. Such estimates limit the accuracy of MM calculations.

Performing an MM Calculation

Unlike ab initio, semiempirical, and density-functional calculations, the input to a molecular-mechanics calculation must specify not only the initial locations of the nuclei but also which atoms are bonded to each other and how they are bonded (the atom types). This is most conveniently done by building a molecular model on a computer screen.

After the molecule and its initial geometry are specified, the molecular-mechanics program varies the molecular geometry to minimize the steric energy V_{steric} (just as quantum-mechanical methods minimize E_e in a geometry-optimization calculation). Thus one obtains the geometry and V_{steric} for the conformer closest to the starting geometry. Many MM programs can do automatic conformational searches to locate many different conformers and find their V_{steric} values.

The most time-consuming step in calculating V_{steric} of a large molecule is the evaluation of V_{es} and V_{vdW}. For a molecule with 10^4 atoms, V_{es} and V_{vdW} are each the sum of about $10^8/2$ $V_{es,ij}$ and $V_{vdW,ij}$ terms, and these terms must be recalculated for each geometry change in the geometry-optimization process. To reduce the calculation time, people often use cutoffs, meaning that the $V_{es,ij}$ and $V_{vdW,ij}$ terms are omitted for atoms separated by more than a specified cutoff distance. For $V_{vdW,ij}$ a cutoff of 8 Å has often been used in the past. Although van der Waals interactions are weak and short range, the number of such interactions increases rapidly with increasing distance from a given atom, and it turns out that to obtain valid results, the cutoff for $V_{vdW,ij}$ should be at least 18 Å. Although people have used cutoffs of 10 or 15 Å for $V_{es,ij}$, electrostatic interactions are long-range and no cutoff should be used for them. Various special techniques have been devised to speed up evaluation of V_{es}.

In a quantum-mechanical calculation, the molecular electronic energy E_e has a well-defined meaning, namely, it is the energy for a given fixed configuration of the nuclei where the zero level of energy corresponds to a situation with all the electrons and nuclei at infinite separations from one another and at rest. In contrast, V_{steric} in a molecular-mechanics calculation does not have a well-defined meaning, since it refers to a hypothetical molecule in which all the bond distances and angles have their reference values and all torsional, electrostatic, and van der Waals interactions are absent. Thus V_{steric} differs from E_e by an unknown constant C. Assuming accurate modeling of V_{steric} and accurate quantum-mechanical calculation of E_e, minimization of V_{steric} will give the same geometry as minimization of E_e.

The constant C will be the same for different conformers of the same molecule since the zero level of V_{steric} is the same for such species. Therefore, the differences between V_{steric} for different conformers of the same molecule are valid estimates of the differences in E_e for these conformers, provided one uses the same force field to calculate all the V_{steric} values. (V_{steric} values of different force fields cannot be meaningfully compared.)

The difference in V_{steric} for the isomers CH_3CH_2OH and CH_3OCH_3 cannot be used to estimate ΔE_e for these molecules, since the nature of the bonds differs. However, we could use ΔV_{steric} to estimate ΔE_e for $CH_3CH_2CH_2OH$ and $CH_3CH(OH)CH_3$, since here the bonding is the same.

After finding the equilibrium geometry by an MM calculation, one can use the partial atomic charges Q_i to estimate the dipole moment $\boldsymbol{\mu} = \Sigma_i \, Q_i \mathbf{r}_i$. From the second derivatives of V_{steric}, one can find molecular vibration frequencies. By combining V_{steric} with empirical bond-energy parameters, one can estimate the gas-phase $\Delta_f H^\circ_{298}$, but most MM programs don't do this.

Force Fields and Programs

Some widely used MM force fields are MM2 for small and moderate-size organic compounds; MM3 for organic compounds, polypeptides, and proteins [N. L. Allinger and L. Yan, *J. Am. Chem. Soc.,* **115,** 11918 (1993) and references cited therein]; MMFF94 for small organic compounds, proteins, and nucleic acids [T. A. Halgren, *J. Comput. Chem.,* **20,** 730 (1999) and references cited therein]; and AMBER [W. D. Cornell et al., *J. Am. Chem. Soc.,* **117,** 5179 (1995); www.amber.ucsf.edu] and CHARMM [A. D. Mackerell et al., *J. Am. Chem. Soc.,* **117,** 11946 (1995); *J. Phys. Chem.,* **102,** 3586 (1998); yuri.harvard.edu] for polypeptides, proteins, and nucleic acids.

Many MM programs are available for personal computers. The program Hyper-Chem has the force fields MM+ (a version of MM2), AMBER, BIO+ (a version of CHARMM), and OPLS. PC Spartan Pro and MacSpartan Pro have the MMFF94 force field. CHEM3D (www.camsoft.com) contains a version of the MM2 force field; you can download a free demonstration version Chem3D Net that can do MM calculations on molecules with no more than six nonhydrogen atoms. The free MM program B has the AMBER force field and can handle organic molecules and biopolymers; it can be run over the Internet (www.scripps.edu/~nwhite/Biomer/).

Performance and Applications

A properly parametrized MM force field applied to compounds similar to those used in the parametrization will give very good results. For example, for a sample of 30 organic compounds, MMFF94 gave root-mean-square errors of 0.014 Å in bond lengths and 1.2° in bond angles [T. A. Halgren, *J. Comput. Chem.,* **10,** 982 (1989)]. MM3 gave an average absolute error of 0.6 kcal/mol in gas-phase $\Delta_f H^\circ_{298}$ values for a sample of 45 alcohols and ethers [N. L. Allinger et al., *J. Am. Chem. Soc.,* **112,** 8293 (1990)]. However, when, as often happens, missing parameters must be estimated, the accuracy of the results is impaired.

Because of its ability to handle large molecules, molecular mechanics is widely used to deal with biological molecules. For example, folding of a small protein in solution has been modeled using the AMBER force field (see Sec. 24.14).

Molecular mechanics is not suitable for dealing with chemical reactions in which bonds are broken. Several versions of combined quantum-mechanical and molecular mechanics methods (QM/MM) have been developed. To treat an enzyme-catalyzed reaction by a QM/MM method, one uses a quantum-mechanical method to treat the active site of the enzyme and uses MM for the rest of the enzyme.

20.14 FUTURE PROSPECTS

The use of electronic computers has brought remarkable advances in the ability of quantum chemists to deal with problems of real chemical interest. For example, quantum-chemistry calculations are now being used to study chemisorption on metal catalysts and hydration of ions in solution. Whereas quantum-mechanical calculations used to be confined to small molecules and were published in journals read mainly by physical chemists and chemical physicists, such calculations now deal with medium-sized and even fairly large molecules and appear regularly in the *Journal of the American Chemical Society*, read by all kinds of chemists.

The accuracy of quantum-mechanical predictions of molecular properties and the size of molecules treatable will increase as faster computers are developed and as new calculational methods are devised. Quantum-mechanical calculations will clearly play an expanding role in physical chemistry.

The 1998 Nobel Prize for chemistry was awarded to Walter Kohn, one of the developers of density-functional theory, and John A. Pople, one of the developers of the *Gaussian* series of programs. The Nobel committee stated that computational quantum chemistry "is revolutionising the whole of chemistry."

20.15 SUMMARY

Covalent bond distances can be estimated as the sum of atomic covalent radii. ΔH° of gas-phase reactions can be estimated from bond energies. Molecular dipole moments can be estimated as vector sums of bond dipole moments. The electronegativity of an element is a measure of the ability of an atom of that element to attract the electrons in a chemical bond.

Since nuclei are much heavier than electrons, one can deal separately with electronic and nuclear motions in a molecule (the Born–Oppenheimer approximation). One first solves an electronic Schrödinger equation with the nuclei in fixed positions. This gives an electronic energy and wave function that depend on the nuclear positions as parameters. One then uses this electronic energy as the potential energy in the Schrödinger equation for the nuclear motion (rotation and vibration).

Approximate electronic wave functions for H_2^+ can be written as linear combinations of atomic orbitals of each H atom. These one-electron wave functions can be used as molecular orbitals for homonuclear diatomic molecules. Each MO of a diatomic molecule is classified as $\sigma, \pi, \delta, \phi, \ldots$ according to whether m, the quantum number for the electronic orbital-angular-momentum component along the molecular axis has absolute value 0, 1, 2, 3, ... Homonuclear diatomic MOs are further classified as g or u according to whether the orbital has the same or the opposite sign on diagonally opposite sides of the molecule's center. In the LCAO MO approximation, the H_2 electronic ground-state wave function is given by $N[1s_A(1) + 1s_B(1)][1s_A(2) + 1s_B(2)] \cdot [\alpha(1)\beta(2) - \alpha(2)\beta(1)]/2^{1/2}$.

The Pauli antisymmetry principle leads to an apparent extra repulsion between electrons of like spin (Pauli repulsion).

An SCF wave function is one in which each electron is assigned to a single spin-orbital, the spatial part of each spin-orbital is expressed as a linear combination of basis functions, and the coefficients in these linear combinations are found by solving the Hartree–Fock equations; the SCF wave function is an antisymmetrized product (Slater determinant) of spin-orbitals. If the basis set is large enough so that the MOs found are the best possible orbitals, the SCF wave function is the Hartree–Fock wave function. To reach the true wave function, one must go beyond the Hartree–Fock wave function, for example, by using configuration interaction or Møller–Plesset perturbation theory.

MOs that satisfy the Hartree–Fock equations are called canonical MOs. Canonical MOs have the symmetry of the molecule and are spread out (delocalized) over the molecule. For example, the bonding MOs of BeH_2 in Fig. 20.24 extend over all three atoms. By taking linear combinations of the canonical MOs, one can form localized MOs. Each localized MO is classifiable as an inner-shell, lone-pair, or bonding MO. The MO wave function that uses localized MOs is equal to the wave function that uses canonical MOs. In an unconjugated molecule, each localized bonding MO is largely (but not completely) confined to two bonded atoms (molecules with three-center bonds are an exception). The AOs in a bonding MO are often hybridized extensively. For example, the hybridizations in BeH_2, C_2H_4, and CH_4 are approximately sp, sp^2, and sp^3, respectively.

Ab initio SCF wave functions give generally accurate values for molecular geometries, dipole moments, ionization energies, rotational barriers, relative energies of isomers, and electron probability densities (provided an adequate-size basis set is used), but give poor values of molecular dissociation energies.

Gaussian basis sets are usually used for molecular electronic-structure calculations.

The main methods for improving ab initio SCF wave functions are Møller–Plesset (MP) perturbation theory, which takes the unperturbed wave function as the SCF wave function; configuration interaction (CI), which is usually limited to inclusion of functions with one or two electrons in virtual orbitals; and the coupled-cluster (CC) method.

Density-functional theory (DFT) is based on the Hohenberg–Kohn theorem that the ground-state electron probability density ρ determines the ground-state energy and all other molecular properties. The Kohn–Sham (KS) version of DFT uses a fictitious reference system of noninteracting electrons whose probability density ρ is the same as that of the ground-state molecule. The KS theory is exact in principle, but approximate in practice, since the true form of the key quantity the exchange–correlation functional E_{xc} is unknown. Gradient-corrected approximations to E_{xc} (especially hybrid functionals) provide generally accurate results for small and medium-sized molecules, but are not as accurate as high-level ab initio methods such as CC. DFT is currently the most widely used calculation method in quantum chemistry.

The AM1 and PM3 semiempirical methods have a fair degree of accuracy for many (but not all) molecular properties and can be applied to molecules too large to treat by density-functional or ab initio methods.

The input to a quantum-mechanical calculation includes the identity and location of each nucleus. A geometry-optimization calculation finds the local energy minimum that is nearest to the starting geometry. One way to specify the starting geometry is with a Z-matrix.

The molecular-mechanics method is an empirical nonquantum-mechanical method that treats the molecule as atoms held together by bonds and writes the molecular steric energy as the sum of contributions from bond bending, bond stretching, bond torsion, and van der Waals and electrostatic interactions of nonbonded atoms. It can be applied to very large molecules and yields good results when applied to molecules similar to those for which the force field was parametrized.

FURTHER READING

Karplus and Porter, chaps. 5, 6; *Atkins and Friedman,* chap. 10; *Levine* (2000), chaps. 13–17; *Lowe,* chaps. 8, 10, 11, 14; *McQuarrie* (1983), chap. 9; *DeKock and Gray; Leach; Jensen.*

PROBLEMS

Section 20.1

20.1 Estimate the bond lengths in (a) CH_3OH; (b) HCN.

20.2 Explain why the observed boron–fluorine bond length in BF_3 is substantially less than the sum of the B and F single-bond radii.

20.3 Predict the shape and bond angles of (a) $TeBr_2$: (b) $HgCl_2$; (c) $SnCl_2$; (d) XeF_2; (e) ClO_2^-.

20.4 Predict the shape of (a) BrF_3; (b) GaI_3; (c) H_3O^+; (d) PCl_3.

20.5 Predict the shape of (a) SnH_4; (b) SeF_4; (c) XeF_4; (d) BH_4^-; (e) BrF_4^-.

20.6 Predict the shape of (a) $AsCl_5$; (b) BrF_5; (c) $SnCl_6^{2-}$.

20.7 Predict the shape of (a) O_3; (b) NO_3^-; (c) SO_3; (d) SO_2; (e) SO_2Cl_2; (f) $SOCl_2$; (g) IO_3^-; (h) SOF_4; (i) XeO_3; (j) $XeOF_4$.

20.8 Estimate the bond angles in (a) CH_3CN; (b) CH_2=$CHCH_3$; (c) CH_3NH_2; (d) CH_3OH; (e) FOOF.

20.9 Does O_3 have a dipole moment? Explain your answer.

20.10 Use average bond energies to estimate ΔH_{298}° for the following gas-phase reactions: (a) $C_2H_2 + 2H_2 \rightarrow C_2H_6$; (b) $N_2 + 3H_2 \rightarrow 2NH_3$. Compare with the true values found from data in the Appendix.

20.11 The dipole moments of CH_3F and CH_3I are 1.85 and 1.62 D, respectively. Use the H—C bond moment listed in Sec. 20.1 to estimate the C—F and C—I bond moments.

20.12 Use bond moments to estimate the dipole moments of (a) CH_3Cl; (b) CH_3CCl_3; (c) $CHCl_3$; (d) Cl_2C=CH_2. Assume tetrahedral angles at singly bonded carbons and 120° angles at doubly bonded carbons. Compare with the experimental values, which are (a) 1.87 D; (b) 1.78 D; (c) 1.01 D; (d) 1.34 D.

20.13 What would the bond moment of C≡N be if one assumed the H—C moment was 0.4 but had the polarity H^-—C^+?

20.14 Use Table 20.1 to compute the Pauling electronegativity differences for the following pairs of elements: (a) C, H; (b) C, O; (c) C, Cl. Compare with the electronegativity differences found from the Pauling values in Table 20.2. The discrepancies are due to use of average bond energies different from those listed in Table 20.1.

20.15 Calculate the Allen-scale electronegativities of (a) H; (b) Li; (c) Be; (d) Na.

20.16 Some values of $\alpha/(4\pi\varepsilon_0 \text{ Å}^3)$ are 0.667 for H, $24._3$ for Li, 5.6_0 for Be, 3.0_3 for B, 1.7_6 for C, 1.1_0 for N, 0.80_2 for O, and 0.55_7 for F. Calculate the Nagle-scale electronegativities of these elements.

20.17 (a) If A, B, and C are elements whose electronegativities satisfy $x_A > x_B > x_C$, show that if the Pauling electronegativity scale is valid, then $\Delta_{AC}^{1/2} = \Delta_{AB}^{1/2} + \Delta_{BC}^{1/2}$. (b) Test this relation for C, N, and O.

20.18 (a) Write a Lewis dot formula for H_2SO_4 that has eight electrons around S. (b) What formal charge does this dot formula give the S atom? (*The formal charge* is found by dividing the electrons of each bond equally between the two bonded atoms.) How reasonable is this formal charge? (c) Write the Lewis dot formula for SF_6. (d) Write a dot formula for H_2SO_4 that gives S a zero formal charge. (e) Explain why the observed sulfur–oxygen bond lengths in SO_4^{2-} in metal sulfates are 1.5 to 1.6 Å, whereas the sum of the single-bond radii of S and O is 1.70 Å.

20.19 Draw the Lewis dot structure for CO. What is the formal charge (Prob. 20.18b) on carbon?

20.20 Predict the sign of $\Delta H°$ of each of the following reactions without using any thermodynamic data or bond-energy data: (a) $H_2(g) + Cl_2(g) \rightarrow 2HCl(g)$; (b) $CH_4(g) + Cl_2(g) \rightarrow CH_3Cl(g) + HCl(g)$.

Section 20.2

20.21 True or false? (a) In the electronic Schrödinger equation, V_{NN} is a constant. (b) V_{ee} is absent from the electronic Schrödinger equation. (c) $\hat{K}_e$ is absent from the electronic Schrödinger equation. (d) $\hat{K}_N$ is absent from the electronic Schrödinger equation. (e) $\hat{K}_N$ is present in the Schrödinger equation for nuclear motion.

20.22 For the H_2 molecule, give the explicit forms for each of the operators $\hat{K}_e$, $\hat{K}_N$, $\hat{V}_{NN}$, $\hat{V}_{Ne}$, and $\hat{V}_{ee}$. Use capital letters for the nuclei and numbers for the electrons. Use r_{1A} to denote the distance between electron 1 and nucleus A.

20.23 (a) The KF molecule has $R_e = 2.17$ Å. The ionization potential of K is 4.34 V, and the electron affinity of F is 3.40 eV. Use the model of nonoverlapping spherical ions to estimate D_e for KF. (The experimental value is 5.18 eV.) (b) Estimate the dipole moment of KF. (The experimental value is 8.60 D.) (c) Explain why KCl has a larger dipole moment than KF.

20.24 For an ionic molecule like NaCl, the electronic energy $E_e(R)$ equals the Coulomb's law potential energy $-e^2/4\pi\varepsilon_0 R$ plus a term that allows for the Pauli-principle repulsion due to the overlap of the ions' probability densities. This repulsion term can be very crudely estimated by the function B/R^{12}, where B is a positive constant. (See the discussion of the Lennard-Jones potential in Sec. 22.10.) Thus, $E_e \approx B/R^{12} - e^2/4\pi\varepsilon_0 R$ for an ionic molecule. (a) Use the fact that E_e is a minimum at $R = R_e$ to show that $B = R_e^{11}e^2/48\pi\varepsilon_0$. (b) Use the above expression for E_e and the Na and Cl ionization potential and electron affinity to estimate D_e for NaCl ($R_e = 2.36$ Å). (c) D_e for NaCl is 4.25 eV. Does the function B/R^{12} overestimate or underestimate the Pauli repulsion? What value of m gives agreement with the observed D_e if the Pauli repulsion is taken as A/R^m, where A and m are constants?

Section 20.3

20.25 True or false? (a) There is no interelectronic repulsion in the H_2^+ molecule. (b) The wave function $N(1s_A + 1s_B)$ is the exact electronic wave function for the ground electronic state of H_2^+. (c) The ground electronic state of H_2^+ is a bound state and the first excited electronic state of H_2^+ is an unbound state. (d) The plane perpendicular to the molecular axis at the midpoint between the nuclei in H_2^+ is a node for the first excited state.

20.26 In one dimension, an even function satisfies $f(-x) = f(x)$ and an odd function satisfies $f(-x) = -f(x)$. State whether each of the following functions is even, odd, or neither. (a) $3x^2 + 4$; (b) $2x^2 + 2x$; (c) x; (d) the $v = 0$ harmonic-oscillator wave function; (e) the $v = 1$ harmonic-oscillator wave function.

20.27 For the ground electronic state of H_2^+ with the nuclei at their equilibrium separation, use the approximate wave function in (20.15) to calculate the probability of finding the electron in a tiny box of volume 10^{-6} Å^3 if the box is located (a) at

one of the nuclei; (b) at the midpoint of the internuclear axis; (c) on the internuclear axis and one-third of the way from nucleus A to nucleus B. Use Table 20.3 and the equation $S = e^{-R/a_0}(1 + R/a_0 + R^2/3a_0^2)$ [*Levine* (2000), sec. 13.5].

Section 20.4

20.28 Write down the MO wave function for the (repulsive) ground electronic state of He$_2$.

20.29 Give the ground-state MO electronic configuration for (a) He$_2^+$; (b) Li$_2$; (c) Be$_2$; (d) C$_2$; (e) N$_2$; (f) F$_2$. Which of these species are paramagnetic?

20.30 Give the bond order of each molecule in Prob. 20.29.

20.31 Use the MO electron configurations to predict which of each of the following sets has the highest D_e; (a) N$_2$ or N$_2^+$; (b) O$_2$, O$_2^+$, or O$_2^-$.

20.32 Let ψ be an eigenfunction of $\hat{H}$; that is, let $\hat{H}\psi = E\psi$. Show that $(\hat{H} + c)\psi = (E + c)\psi$, where c is any constant. Hence, ψ is an eigenfunction of $\hat{H} + c$ with eigenvalue $E + c$.

20.33 For each of the species NCl, NCl$^+$, and NCl$^-$, use the MO method to (a) write the valence-electron configuration; (b) find the bond order; (c) decide whether the species is paramagnetic.

Section 20.5

20.34 (a) List the minimal-basis-set AOs for a calculation on CO. (b) If the internuclear axis is the z axis, which of the AOs in (a) can contribute to σ MOs? To π MOs?

Section 20.6

20.35 Sketch the two antibonding MOs (20.31) of BeH$_2$.

20.36 For the linear BeH$_2$ molecule with the z axis taken as the molecular axis, classify each of the following AOs as g or u and as σ, π, or δ: (a) Be$3d_{z^2}$; (b) Be$3d_{x^2-y^2}$; (c) Be$3d_{xy}$; (d) Be$3d_{xz}$.

20.37 Let the line between atom A and atom B of a polyatomic molecule be the z axis. For each of the following atom A atomic orbitals, state whether it will contribute to a σ, π, or δ localized MO in the molecule: (a) s; (b) p_x; (c) p_y; (d) p_z; (e) d_{z^2}; (f) $d_{x^2-y^2}$; (g) d_{xy}; (h) d_{xz}; (i) d_{yz}.

20.38 (a) For H$_2$CO, list all the AOs that go into a minimal-basis-set MO calculation. (b) Use these AOs to form localized MOs for H$_2$CO. For each localized MO, state which AOs make the main contributions to it and state whether it is inner-shell, lone-pair, or bonding. State whether each localized bonding MO is σ or π. Take the z axis along the CO bond and the x axis perpendicular to the molecule. Use the σ-π description of the double bond.

20.39 (a) Use average-bond-energy data and data on C(g) and H(g) in the Appendix to estimate $\Delta_f H°_{298}$ of C$_6$H$_6$(g) on the assumption that benzene contains three carbon–carbon single bonds and three carbon–carbon double bonds. Compare the result with the experimental value. (b) Repeat (a) for cyclohexene(g) (one double bond).

20.40 Let $p_1, \ldots, p_6$ be the $2p_z$ AOs of the carbons in benzene. The unnormalized forms of the occupied π MOs in benzene are $p_1 + p_2 + p_3 + p_4 + p_5 + p_6, p_2 + p_3 - p_5 - p_6$, and $2p_1 + p_2 - p_3 - 2p_4 - p_5 + p_6$. (a) Sketch these MOs. (b) Which of the three is lowest in energy?

20.41 Plot the value along the z axis of the $2s + 2p_z$ hybrid AO in (20.35) versus z/a. Take the nuclear charge as 1. Note that the outer portion of the $2s$ AO in (20.35) is assumed positive (as in Fig. 20.27). This convention is opposite that used in Table 19.1, so multiply the $2s$ AO in Table 19.1 by -1 before adding it to $2p_z$.

Section 20.8

20.42 True or false? (a) D_e is always greater than D_0. (b) The molecular electron probability density is experimentally observable. (c) The molecular wave function is experimentally observable.

20.43 For each of the following, state whether specification of all bond distances and all bond angles fully specifies the molecular geometry. (a) H_2O; (b) H_2O_2; (c) NH_3; (d) $ClCH_2CH_2Cl$; (e) $CH_2{=}CH_2$.

20.44 Find the particle probability density function $\rho(x)$ for a system of two identical noninteracting particles each of which has spin $s = 0$ in a stationary state of a one-dimensional box of length a if the quantum numbers for the two particles are unequal ($n_1 \neq n_2$). Also find $\int_{-\infty}^{\infty} \rho(x)\, dx$. (*Hint*: Make sure you use a symmetric wave function.)

Section 20.10

20.45 Which of these are functionals? (a) $\int_3^8 [g(x)]^2\, dx$; (b) $df(x)/dx$; (c) $df(x)/dx|_{x=0}$; (d) $\int_0^2 [f(x) + d^2f/dx^2]\, dx$.

20.46 Which of the following are numbers and which are functions of x, y, and z? (a) E_e; (b) ρ; (c) E_{xc}; (d) v_{xc}.

20.47 Show that multiplication of the electronic Schrödinger equation $\hat{H}_e \psi_e = E_e \psi_e$ by ψ_e^* followed by integration over all space gives Eq. (20.45) for E_e.

20.48 True or false? (a) A DFT calculation does not find the wave function of the molecule. (b) The KS orbitals are for the reference system of noninteracting electrons. (c) The KS reference system has the same ground-state energy as the molecule. (d) The KS reference system has the same ground-state probability density ρ as the molecule. (e) The exact form of E_{xc} is unknown.

20.49 Show that J in (20.48) is the classical expression for the electrostatic self-repulsion energy of a continuous distribution of electrical charge whose charge density (charge per unit volume) is $-e\rho$. Start by writing the repulsion between two infinitesimal elements of charge dQ_1 and dQ_2 of the continuous distribution.

20.50 For the functional

$$F[\rho] \equiv \int_e^t \int_c^d \int_a^b g(x, y, z, \rho, \rho_x, \rho_y, \rho_z)\, dx\, dy\, dz$$

where $\rho_x \equiv \partial\rho/\partial x$, etc., one can show that the functional derivative is

$$\frac{\delta F}{\delta \rho} = \frac{\partial g}{\partial \rho} - \frac{\partial}{\partial x}\frac{\partial g}{\partial \rho_x} - \frac{\partial}{\partial y}\frac{\partial g}{\partial \rho_y} - \frac{\partial}{\partial z}\frac{\partial g}{\partial \rho_z}$$

In the LDA approximation, the exchange part of E_{xc} is given by

$$E_x^{LDA} = -\frac{3}{4}\left(\frac{3}{\pi}\right)^{1/3} \int_{-\infty}^{\infty}\int_{-\infty}^{\infty}\int_{-\infty}^{\infty} [\rho(x, y, z)]^{4/3}\, dx\, dy\, dz$$

Find $v_x^{LDA} = \delta E_x^{LDA}/\delta\rho$.

Section 20.11

20.51 To illustrate the FE MO method, consider the ions

$$(CH_3)_2\overset{+}{N}{=}CH({-}CH{=}CH)_k{-}\ddot{N}(CH_3)_2 \qquad (20.53)$$

where k, the number of $-CH{=}CH$ groups in the ion, can be 0, 1, 2, ... Each ion has an equivalent Lewis structure with the charge on the right-hand nitrogen and all carbon–carbon single and double bonds interchanged. All the carbon–carbon bond lengths are equal, and the π electrons, which form the second bond of each double bond, are reasonably free to move along the molecule. (a) Use the FE MO method to calculate the wavelength of the longest-wavelength electronic absorption band of the ion (20.53) with $k = 1$. Assume the carbon–carbon and conjugated carbon–nitrogen bond distances are 1.40 Å (as in benzene) and add in one extra bond length at each end of the ion to get the "box" length. Begin by deciding how many π electrons there are. (Note that the lone pair on nitrogen takes part in the π bonding.) Assign two π electrons to each π MO and use the quantum numbers of the highest-occupied and lowest-vacant π MOs. Compare your result with the experimental value 312 nm. (b) For the ion (20.53), show that the FE MO method predicts the longest-wavelength absorption to be at $\lambda = (2k + 4)^2(64.6\text{ nm})/(2k + 5)$.

20.52 Although ab initio SCF calculations do not give accurate molecular dissociation energies, whereas some semiempirical methods do give pretty good estimates of gas-phase $\Delta_f H°$ values, one can combine the results of an ab initio SCF calculation with empirical parameters to obtain a gas-phase $\Delta_f H°_{298}$ value that is usually more accurate than those obtained from semiempirical calculations. The equation used is

$$\Delta_f H°_{298} = N_A\left(E_e - \sum_a n_a \alpha_a\right)$$

where the Avogadro constant N_A converts from a per molecule to a per mole basis, E_e is the ab initio SCF electronic energy of the molecule, the sum goes over the different kinds of atoms in the molecule, n_a is the number of a atoms in the molecule, and the α_a's are empirical parameters for the various kinds of atoms. The α_a values are found by fitting known $\Delta_f H°$ values and depend somewhat on the basis set used. For HF/6-31G* calculations, some α_a values in hartrees (1 hartree = 27.2114 eV = 4.35974×10^{-18} J) are

$$\alpha_H = -0.57077, \quad \alpha_C = -37.88449,$$
$$\alpha_N = -54.46414, \quad \alpha_O = -74.78852$$

Some HF/6-31G* energies in hartrees are -40.19517 for CH_4 and -150.76479 for H_2O_2. Use these results to estimate $\Delta_f H^\circ_{298}$ for $CH_4(g)$ and $H_2O_2(g)$. The experimental values are -74.8 kJ/mol for $CH_4(g)$ and -136.3 kJ/mol for $H_2O_2(g)$.

Section 20.12

20.53 Draw the conformation of the molecule described by the following Z-matrix.

C1

C2	1	1.54				
H3	2	1.09	1	120.0		
H4	1	1.09	2	109.5	3	180.0
H5	1	1.09	2	109.5	3	-60.0
H6	1	1.09	2	109.5	3	60.0
O7	2	1.23	1	120.0	4	0.0

20.54 (a) Devise a Z-matrix for each of the two planar conformers of formic acid, $HC(=O)OH$. (b) To avoid $180°$ angles (which are forbidden in a Z-matrix), one includes a dummy atom, symbolized by X, in the Z-matrix. *Gaussian* uses the dummy atom to define the locations of the nondummy atoms but ignores X in the quantum-mechanical calculations. Devise a Z-matrix for CO_2 that has a dummy atom placed 1.0 Å from C with the line XC making a $90°$ angle with the molecular axis. (A dihedral angle 4–2–1–3 with atoms 4 and 3 both bonded to atom 2 is allowed.)

20.55 (a) Using suitable software, do HF/6-31G* and HF/6-31G** geometry optimizations for methanol and compare with the HF/3-21G geometry. Also compare the dipole moments found with these three basis sets. (b) Do a B3LYP/6-31G* geometry optimization for CH_3OH.

20.56 (a) Using suitable software, do HF/6-31G* geometry optimizations for each of the two planar conformers of HCOOH and find the HF/6-31G* energy difference in kcal/mol (omitting zero-point energy). (b) Repeat (a) using B3LYP/6-31G* calculations.

Section 20.13

20.57 Using suitable software and as many different force fields as you have access to, do MM geometry optimizations for CH_3OH.

20.58 (a) Do MM geometry optimizations to find the predicted energy difference between the gauche and trans conformers of butane.

20.59 (a) Do MM geometry optimizations to find the predicted energy difference between the cis and trans isomers of 1,2-difluoroethylene.

20.60 For C_2Cl_6, how many terms are there in each of the following: (a) V_{str}; (b) V_{bend}; (c) V_{tors}; (d) V_{vdW}; (e) V_{es}?

General

20.61 Arrange the following molecular energies in order of increasing magnitude: (a) the typical energy of a covalent single bond; (b) the average molecular kinetic energy in a fluid at room temperature; (c) the rotational barrier in C_2H_6; (d) the typical energy of a double bond; (e) the ionization energy of H.

20.62 True or false? (a) The maximum electron probability density in the ground electronic state of H_2^+ occurs at each nucleus. (b) If sufficient basis functions are used, the Hartree–Fock wave function of a many-electron molecule will reach the true wave function. (c) In a homonuclear diatomic molecule, combining two $2p$ AOs always yields a π MO. (d) An ab initio calculation will give an accurate prediction of every molecular property. (e) In homonuclear diatomic molecules, all u MOs are antibonding. (f) The H_2^+ ground electronic state has spin quantum number $S = 0$. (g) The H_2 ground electronic state has spin quantum number $S = 0$. (h) Two electrons with the same spin have zero probability of being at the same point in space. (i) All triatomic molecules are planar.

20.63 Which scientist mentioned in Chapter 20 is described by each of the following? (a) He headed the U.S. atom-bomb project in World War II. (b) He won the Nobel Prize for chemistry and the Nobel Prize for peace. In 1952, he was unable to attend a scientific meeting in Britain when the U.S. State Department refused to issue him a passport. In 1953, he published an erroneous model for DNA. Some people have speculated that had he gone to the 1952 meeting, he might have seen Rosalind Franklin's x-ray diffraction photos of DNA crystals, and, being an expert on crystallography, might have been able to beat out Watson and Crick in arriving at the correct DNA structure.

Spectroscopy and Photochemistry

Most of our experimental information on the energy levels of atoms and molecules comes from spectroscopy, the study of the absorption and the emission of electromagnetic radiation (light) by matter. Section 21.1 examines the nature of light. Section 21.2 is a general discussion of spectroscopy. This is followed by Secs. 21.3 to 21.10 on the rotational and vibrational spectra of diatomic and polyatomic molecules. Electronic spectra are considered in Sec. 21.11, and magnetic resonance spectra in Secs. 21.12 and 21.13. Sections 21.14 and 21.15 discuss other branches of spectroscopy. Closely related to spectroscopy is photochemistry (Sec. 21.16) the study of chemical reactions caused or catalyzed by light. The full application of molecular symmetry to spectroscopy and other areas of quantum chemistry is based on the branch of mathematics called group theory, which is discussed in Sec. 21.17.

Spectroscopy is the major experimental tool for investigations at the molecular level. Spectroscopy is used to find molecular structures (conformations, bond lengths, and angles) and molecular vibration frequencies. Organic chemists use nuclear-magnetic-resonance spectroscopy in structural investigations. Analytical chemists use spectroscopy to find the composition of a sample. In kinetics, spectroscopy is used to follow the concentrations of reacting species as functions of time, and to detect reaction intermediates. Biochemists use spectroscopy extensively to study the structure and dynamics of biological molecules. Astronomers use spectroscopy to investigate the composition of stars, planets, and interstellar dust. Spectroscopy is used to determine the levels of pollutants in air.

21.1   ELECTROMAGNETIC RADIATION

In 1801, Thomas Young observed interference of light when a light beam was diffracted at two pinholes, thereby showing the wave nature of light. A wave involves a vibration in space and in time, and so the question arises: What physical quantity is vibrating in a light wave? The answer was provided by Maxwell.

In the 1860s, Maxwell showed that the known laws of electricity and magnetism can all be derived from a set of four differential equations. These equations (called *Maxwell's equations*) interrelate the electric and magnetic field vectors **E** and **B**, the electric charge, and the electric current. Maxwell's equations are the fundamental equations of electricity and magnetism, just as Newton's laws are the fundamental equations of classical mechanics.

In addition to containing all the laws of electricity and magnetism known in the 1860s, Maxwell's equations predicted something that was unknown at that time, namely, that an accelerated electric charge will emit energy in the form of electromagnetic waves traveling at a speed v_{em} in vacuum, where

$$v_{em} = (4\pi\varepsilon_0 \times 10^{-7}\,\text{N s}^2\,\text{C}^{-2})^{-1/2} \tag{21.1}$$

Recall that ε_0 (the permittivity of vacuum) occurs in Coulomb's law (14.1). Substitution of the experimental value (14.2) of ε_0 gives $v_{em} = 2.998 \times 10^8$ m/s, which equals the experimentally observed speed of light in vacuum. Maxwell therefore proposed that light consists of electromagnetic waves.

Maxwell's prediction of the existence of electromagnetic waves was confirmed by Hertz in 1887. Hertz produced electromagnetic waves by the oscillations of electrons in the metal wires of a tuned ac circuit; he detected these waves using a loop of wire (just as the antenna in your TV set detects electromagnetic waves emitted by the transmitters of TV stations). The oscillating electric field of the electromagnetic wave exerts a time-varying force on the electrons in the wires of the detector circuit, thereby producing an alternating current in these wires.

Maxwell's equations show that **electromagnetic waves** consist of oscillating electric and magnetic fields. The electric and magnetic field vectors **E** and **B** are perpendicular to each other and are perpendicular to the direction of travel of the wave. Figure 21.1 shows an electromagnetic wave traveling in the y direction. The vectors shown give the values of **E** and **B** at points on the y axis at one instant of time. As time passes, the crests (peaks) and troughs (valleys) move to the right. This figure is not a complete description of the wave, since it gives the values of **E** and **B** only at points lying on the y axis. To describe an electromagnetic wave fully we must give the values of six numbers (the three components of **E** and the three components of **B**) at every point in the region of space through which the wave is moving.

The wave shown in Fig. 21.1 is **plane-polarized,** meaning that the **E** vectors all lie in the same plane. Such a plane-polarized wave would be produced by the back-and-forth oscillation of electrons in a straight wire. The light emitted by a collection of heated atoms or molecules (for example, sunlight) is *unpolarized*, with the electric-field vectors pointing in different directions at different points in space. This is because the molecules act independently of one another and the radiation produced has random orientations of the **E** vector at various points in space. For an unpolarized wave, **E** is still perpendicular to the direction of travel.

The **wavelength** λ of a wave is the distance between successive crests. One **cycle** is the portion of the wave that lies between two successive crests (or between any two successive points having the same phase). The **frequency** ν (nu) of the wave is the number of cycles passing a given point per unit time. If ν cycles pass a given point in unit time, the time it takes one cycle to pass a given point is $1/\nu$, and a crest has therefore traveled a distance λ in time $1/\nu$. If c is the speed of the wave, then since distance = rate $\times$ time, we have $\lambda = c(1/\nu)$ or

$$\lambda\nu = c \qquad (21.2)^*$$

The frequency is commonly given in the units s^{-1}. The SI system of units defines the frequency unit of one reciprocal second as one **hertz** (Hz): $1 \text{ Hz} \equiv 1 \text{ s}^{-1}$. Various multiples of the hertz are also used, for example, the kilohertz (kHz), megahertz (MHz), and gigahertz (GHz):

$$1 \text{ Hz} \equiv 1 \text{ s}^{-1}, \quad 1 \text{ kHz} = 10^3 \text{ Hz}, \quad 1 \text{ MHz} = 10^6 \text{ Hz}, \quad 1 \text{ GHz} = 10^9 \text{ Hz}$$

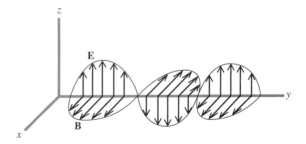

Figure 21.1

A portion of a plane-polarized electromagnetic wave. $1\frac{1}{2}$ cycles are shown.

Common units of λ include the **angstrom** (Å), the **micrometer** (μm), and the **nanometer** (nm):

$$1 \text{ Å} \equiv 10^{-8} \text{ cm} = 10^{-10} \text{ m}, \quad 1 \text{ } \mu\text{m} = 10^{-6} \text{ m}, \quad 1 \text{ nm} = 10^{-9} \text{ m} = 10 \text{ Å}$$

The **wavenumber** $\tilde{\nu}$ of a wave is the reciprocal of the wavelength:

$$\tilde{\nu} \equiv 1/\lambda \qquad \textbf{(21.3)}*$$

Most commonly, $\tilde{\nu}$ is expressed in cm^{-1}. The wavelength λ is the length of one cycle of the wave. The wavenumber $\tilde{\nu}$ expressed in cm^{-1} is the number of cycles of the wave in a 1-cm length.

The human eye is sensitive to electromagnetic radiation with λ in the range 400 nm (violet light) to 750 nm (red light), but there is no upper or lower limit to the values of λ and ν of an electromagnetic wave. Figure 21.2 shows the **electromagnetic spectrum,** the range of frequencies and wavelengths of electromagnetic waves. For convenience, the electromagnetic spectrum is divided into various regions, but there is no sharp boundary between adjacent regions.

All frequencies of electromagnetic radiation (light) travel at the same speed $c = 3 \times 10^8$ m/s in vacuum. The speed of light c_B in substance B depends on the nature of B and on the frequency of the light. The ratio c/c_B for a given frequency of light is the **refractive index** n_B of B for that frequency:

$$n_B \equiv c/c_B \qquad (21.4)$$

Some values of n_B at 25°C and 1 atm for yellow light of vacuum wavelength 589.3 nm ("sodium D light") follow:

Air	H$_2$O	C$_6$H$_6$	C$_2$H$_5$OH	CS$_2$	CH$_2$I$_2$	NaCl	Glass
1.0003	1.33	1.50	1.36	1.63	1.75	1.53	1.5–1.9

For quartz at 18°C and 1 atm, n decreases from 1.57 to 1.45 as the vacuum wavelength of the light increases from 185 to 800 nm. Organic chemists use n as a conveniently measured property to help characterize a liquid.

When a light beam passes obliquely from one substance to another, it is bent or *refracted,* because of the difference in speeds in the two substances. (Since $c = \lambda\nu$, if c changes, either λ or ν or both must change. It turns out that λ changes but ν stays the same as the wave goes from one medium to another.) The amount of refraction depends on the ratio of the speeds of the light in the two substances and so depends on the refractive indices of the substances. Because n of a substance depends on ν, light of different frequencies is refracted by different amounts. This allows us to separate or **disperse** an electromagnetic wave containing many frequencies into its component frequencies. An example is the dispersion of white light into the colors red, orange, yellow, green, blue, and violet by a glass prism.

So far in this section, we have presented the classical picture of light. However, in 1905, Einstein proposed that the interaction between light and matter could best be

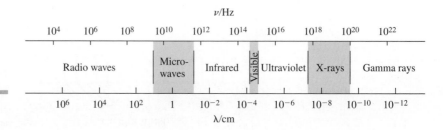

Figure 21.2

The electromagnetic spectrum.
The scale is logarithmic.

understood by postulating a particlelike aspect of light, each quantum (photon) of light having an energy $h\nu$ (Sec. 18.2). The direction of increasing frequency in Fig. 21.2 is thus the direction of increasing energy of the photons.

Although the classical picture of electromagnetic radiation as being produced by an accelerated charge is appropriate for the production of radio waves by electrons moving more or less freely in a metal wire (such electrons have a continuous range of allowed energies), the emission and absorption of radiation by atoms and molecules can generally be understood only by using quantum mechanics. The quantum theory of radiation pictures a photon as being produced or absorbed when an atom or molecule makes a transition between two allowed energy levels.

21.2 SPECTROSCOPY

Spectroscopy

In **spectroscopy,** one studies the absorption and emission of electromagnetic radiation (light) by matter. In a broader sense, spectroscopy deals with all interactions of electromagnetic radiation and matter and so also includes scattering of light (Sec. 21.10) and rotation of the plane of polarization of polarized light by optically active substances (Sec. 21.14).

The set of frequencies absorbed by a sample is its **absorption spectrum;** the frequencies emitted constitute the **emission spectrum.** A **line spectrum** contains only discrete frequencies. A **continuous spectrum** contains a continuous range of frequencies. A heated solid commonly gives a continuous emission spectrum. An example is the blackbody radiation spectrum (Fig. 18.1b). A heated gas that is not at high pressure gives a line spectrum, corresponding to transitions between the quantum-mechanically allowed energy levels of the individual molecules of the gas.

When a sample of molecules is exposed to electromagnetic radiation, the electric field of the radiation exerts a time-varying force on the electrical charges (electrons and nuclei) of each molecule. To treat the interaction of radiation and matter, one uses quantum mechanics, in particular, the time-dependent Schrödinger equation (18.10). Since the mathematics is complicated, we shall just quote the results and omit the derivations. [See *Levine* (1975), chap. 3.]

The quantum-mechanical treatment shows that when a molecule in the stationary state m is exposed to electromagnetic radiation, it may **absorb** a photon of frequency ν and make a transition to a higher-energy state n if the radiation's frequency satisfies $E_n - E_m = h\nu$ (Fig. 21.3a). This agrees with Eq. (18.7), given by Bohr. A molecule in stationary state n in the absence of radiation can spontaneously go to a lower stationary state m, emitting a photon whose frequency satisfies $E_n - E_m = h\nu$. This is **spontaneous emission** of radiation (Fig. 21.3b).

Exposing a molecule in state n to electromagnetic radiation whose frequency satisfies $E_n - E_m = h\nu$ will increase the probability that the molecule will undergo a transition to the lower state m with emission of a photon of frequency ν. Emission due to exposure to electromagnetic radiation is called **stimulated emission** (Fig. 21.3c).

We shall see in Sec. 21.3 that electronic states of a molecule are more widely spaced than vibrational states, which in turn are more widely spaced than rotational states. Transitions between molecular electronic states correspond to absorption in the ultraviolet (UV) and visible regions. Vibrational transitions correspond to absorption in the infrared (IR) region. Rotational transitions correspond to absorption in the microwave region. (See Fig. 21.2.)

The experimental techniques for absorption spectroscopy in the UV, visible, and IR regions are similar. Here, one passes a beam of light containing a continuous range of frequencies through the sample, disperses the radiation using a prism or diffraction

(a)
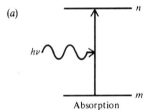
Absorption

(b)
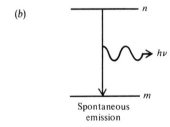
Spontaneous emission

(c)
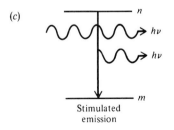
Stimulated emission

Figure 21.3

Absorption, spontaneous emission, and stimulated emission of radiation between states m and n.

grating, and at each frequency compares the intensity of the transmitted light with the intensity of a reference beam that did not pass through the sample.

The techniques of microwave spectroscopy are described in Sec. 21.7.

The radiation falling on the sample in a spectroscopy experiment not only causes absorption from the lower level of a transition but also produces stimulated emission from the upper level. This stimulated emission travels in the same direction as the incident radiation beam and so decreases the observed absorption signal. (Spontaneous emission is sent out in all directions and need not be considered.) Therefore the intensity of an absorption is proportional to the population difference between the lower and upper levels.

The energy of absorbed radiation is usually dissipated by intermolecular collisions to translational, rotational, and vibrational energies of the molecules, thereby increasing the temperature of the sample. Some of the absorbed energy may be radiated by the excited molecules (fluorescence and phosphorescence; Sec. 21.16). This occurs especially in low-pressure gases, where the average time between collisions is much greater than in liquids. The relative amount of emission depends on the average time between collisions compared with the average lifetimes of the various excited states. Sometimes the absorbed radiation leads to a chemical reaction (Sec. 21.16).

Selection Rules

The quantum-mechanical treatment of the interaction between radiation and matter shows that the probability of absorption or emission between the stationary states m and n is proportional to the square of the magnitude of the integral

$$\boldsymbol{\mu}_{mn} \equiv \int \psi_m^* \hat{\boldsymbol{\mu}} \psi_n \, d\tau \qquad \text{where } \hat{\boldsymbol{\mu}} = \sum_i Q_i \mathbf{r}_i \qquad (21.5)$$

where the integration is over the full range of electronic and nuclear coordinates and the sum goes over all the charged particles in the molecule. In the Born–Oppenheimer approximation (Sec. 20.2), the stationary-state wave functions ψ_m and ψ_n are each the product of electronic and nuclear wave functions. $\hat{\boldsymbol{\mu}}$ is the electric dipole-moment operator and occurs in Eq. (20.38). $\mathbf{r}_i$ is the displacement vector of charge Q_i from the origin. The integral $\boldsymbol{\mu}_{mn}$ is called the **transition (dipole) moment.** For pairs of states for which $\boldsymbol{\mu}_{mn} = 0$, the probability of a radiative transition is zero, and the transition is said to be **forbidden.** When $\boldsymbol{\mu}_{mn} \neq 0$, the transition is **allowed.** The allowed changes in quantum number(s) of a system constitute the **selection rule(s)** for the system. [The sum $\sum_i Q_i \mathbf{r}_i$ in (21.5) arises from the interaction of the electric field of the electromagnetic radiation with the charges of the molecule.]

EXAMPLE 21.1 Particle-in-a-box selection rule

Find the selection rule for a particle of charge Q in a one-dimensional box of length a.

To find the selection rule, we shall evaluate $\boldsymbol{\mu}_{mn}$ and see when it is nonzero. There is only one particle, so the sum in (21.5) has only one term. For this one-dimensional problem, the displacement from the origin equals the x coordinate. The wave functions are given by (18.35) as $(2/a)^{1/2} \sin(n\pi x/a)$. Therefore

$$\mu_{mn} = \frac{2Q}{a} \int_0^a x \sin \frac{m\pi x}{a} \sin \frac{n\pi x}{a} \, dx$$

Using the identities $\sin r \sin s = \frac{1}{2} \cos (r - s) - \frac{1}{2} \cos (r + s)$ and $\int x \cos cx \, dx = c^{-2} \cos cx + (x/c) \sin cx$, one finds (Prob. 21.6)

$$\mu_{mn} = \frac{Qa}{\pi^2} \left\{ \frac{\cos [(m - n)\pi] - 1}{(m - n)^2} - \frac{\cos [(m + n)\pi] - 1}{(m + n)^2} \right\} \quad (21.6)$$

We have $\cos \theta = 1$ for $\theta = 0, \pm 2\pi, \pm 4\pi, \pm 6\pi, \ldots$ and $\cos \theta = -1$ for $\theta = \pm \pi, \pm 3\pi, \pm 5\pi, \ldots$. If m and n are both even numbers or both odd numbers, then $m - n$ and $m + n$ are even numbers and μ_{mn} equals zero. If m is even and n is odd, or vice versa, then $m - n$ and $m + n$ are odd and μ_{mn} is nonzero. Hence, radiative transitions between particle-in-a-box states m and n are allowed only if the change $m - n$ in quantum numbers is an odd number. A particle in a box in the $n = 1$ ground state can absorb radiation of appropriate frequency and go to $n = 2$ or $n = 4$, etc., but cannot make a radiative transition to $n = 3$ or $n = 5$, etc. The particle-in-a-box selection rule is $\Delta n = \pm 1, \pm 3, \pm 5, \ldots$.

Exercise

For a hydrogen atom, evaluate the x, y, and z components of $\boldsymbol{\mu}_{mn}$ for m and n being the $1s$ and $2s$ states; take the origin at the nucleus. (*Answer:* 0, 0, 0.)

For the one-dimensional harmonic oscillator (Sec. 18.12), one finds that the transition moment is zero unless $\Delta v = \pm 1$, and this is the harmonic-oscillator selection rule. Thus, a harmonic oscillator in the $v = 2$ state can go only to $v = 3$ or $v = 1$ by absorption or emission of a photon. The selection rule for the two-particle rigid rotor (Sec. 18.14) is found to be $\Delta J = \pm 1$.

EXAMPLE 21.2 Rotational absorption frequencies

For a collection of identical two-particle rigid rotors, find the expression for the allowed rotational absorption frequencies, assuming that many rotational levels are populated.

Spectroscopy ordinarily deals with a collection of molecules distributed among states according to the Boltzmann distribution law (Sec. 15.9). The selection rule for absorption is $\Delta J = +1$. Some of the rigid rotors in the $J = 0$ level will absorb radiation and make a transition to the $J = 1$ level when exposed to radiation of the appropriate frequency. Some of the rotors in the $J = 1$ level will absorb radiation of appropriate frequency and go to $J = 2$; etc. See Fig. 21.13. To find the allowed transition frequencies, we set the photon energy $h\nu$ equal to the energy *difference* between the upper and lower rotational levels involved in the transition:

$$E_{\text{upper}} - E_{\text{lower}} = h\nu \quad (21.7)^*$$

Sometimes students fail to distinguish between the *energy* of a state and the *energy difference* between states and erroneously set the photon energy $h\nu$ equal to the energy of a stationary state. This error is similar to the failure to distinguish between the enthalpy H of a system in a thermodynamic state and the enthalpy change ΔH for a process.

Let J_1 and J_2 be the lower and upper rotational quantum numbers for the transition. The rotational levels are $E_{\text{rot}} = J(J + 1)\hbar^2/2I$ [Eq. (18.81)]. Equation (21.7) gives

$$h\nu = J_2(J_2 + 1)\hbar^2/2I - J_1(J_1 + 1)\hbar^2/2I$$

The rotational selection rule $\Delta J = 1$ (stated just before Example 21.2) gives $J_2 = J_1 + 1$, so

$$h\nu = (J_1 + 1)(J_1 + 2)\hbar^2/2I - J_1(J_1 + 1)\hbar^2/2I = (2J_1 + 2)\hbar^2/2I$$

$$\nu = (J_1 + 1)h/4\pi^2 I \quad \text{where } J_1 = 0, 1, 2, \ldots$$

The absorption frequencies are thus $h/4\pi^2 I$, $2h/4\pi^2 I$, $3h/4\pi^2 I$, ...

Exercise

A hypothetical one-dimensional quantum-mechanical system has the energy levels $E = (k + 2)b$, where b is a positive constant and the quantum number k has the values $k = 1, 2, 3, 4, \ldots$. For a collection of such systems with many energy levels populated and each system having the same value of b, find the allowed absorption frequencies of radiation if the selection rule is $\Delta k = \pm 1$. (*Answer: b/h.*)

The Beer–Lambert Law

The absorption of UV, visible, and IR light by a sample is often described by the Beer–Lambert law, which we now derive.

Consider a beam of light passing through a sample of pure substance B or of B dissolved in a solvent that neither absorbs radiation nor interacts strongly with B. The beam may contain a continuous range of wavelengths, but we shall focus attention on the radiation whose vacuum wavelength lies in the very narrow range from λ to $\lambda + d\lambda$.

Let $I_{\lambda,0}$ be the intensity of the radiation with wavelength in the range λ to $\lambda + d\lambda$ that is incident on the sample, and let I_λ be the intensity of this radiation after it has gone through a length x of the sample. The **intensity** is defined as the energy per unit time that falls on unit area perpendicular to the beam. The intensity is proportional to the number of photons incident on unit area in unit time. Let N_λ photons of wavelength between λ and $\lambda + d\lambda$ fall on the sample in unit time, and let dN_λ be the number of such photons absorbed by a thickness dx of the sample (Fig. 21.4). The probability that a given photon will be absorbed in the thickness dx is dN_λ/N_λ. This probability is proportional to the number of B molecules that a photon encounters as it passes through the layer of thickness dx. The number of B molecules encountered is proportional to the molar concentration c_B of B and to the layer thickness dx. Therefore, $dN_\lambda/N_\lambda \propto c_B \, dx$.

Let dI_λ be the change in light intensity at wavelength λ due to passage through the layer of thickness dx. Then $dI_\lambda \propto -dN_\lambda$. (The minus sign arises because dN_λ was defined as positive and dI_λ is negative.) Also, $I_\lambda \propto N_\lambda$. Hence, $dI_\lambda/I_\lambda \propto -dN_\lambda/N_\lambda$, and $dI_\lambda/I_\lambda \propto -c_B \, dx$. Letting α_λ be the proportionality constant and integrating along the length of the sample, we have

$$\frac{dI_\lambda}{I_\lambda} = -\alpha_\lambda c_B \, dx \quad \text{and} \quad \int_{I_{\lambda,0}}^{I_{\lambda,l}} \frac{dI_\lambda}{I_\lambda} = -\alpha_\lambda c_B \int_0^l dx \tag{21.8}$$

$$\ln \frac{I_{\lambda,l}}{I_{\lambda,0}} = 2.303 \log_{10} \frac{I_{\lambda,l}}{I_{\lambda,0}} = -\alpha_\lambda c_B l \tag{21.9}$$

where l is the sample's length and $I_{\lambda,l}$ is the intensity of the radiation transmitted by the sample. Letting $\varepsilon_\lambda \equiv \alpha_\lambda/2.303$ and defining the **absorbance** A_λ at wavelength λ as $\log_{10}(I_{\lambda,0}/I_{\lambda,l})$, we have

$$A_\lambda \equiv \log_{10}(I_{\lambda,0}/I_{\lambda,l}) = \varepsilon_\lambda c_B l \tag{21.10}$$

which is the **Beer–Lambert law.** Equations (21.9) and (21.10) can be written as

$$I_{\lambda,l} = I_{\lambda,0} e^{-\alpha_\lambda c_B l} = I_{\lambda,0} 10^{-\varepsilon_\lambda c_B l} \tag{21.11}$$

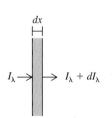

Figure 21.4

Radiation incident on and emerging from a thin slice of sample.

The intensity I_λ of the radiation decreases exponentially along the sample cell.

The fraction of incident radiation transmitted is the **transmittance** T_λ of the sample at wavelength λ. Thus $T_\lambda \equiv I_{\lambda,l}/I_{\lambda,0}$. Since $I_{\lambda,0}/I_{\lambda,l} = 10^{A_\lambda}$, we have

$$T_\lambda \equiv I_{\lambda,l}/I_{\lambda,0} = 10^{-A_\lambda}$$

For $A_\lambda = 1$, T_λ is $10^{-1} = 0.1$, and 90% of the radiation at λ is absorbed. For $A_\lambda = 2$, T_λ is 10^{-2}, and 99% of the radiation is absorbed. Figure 21.5 plots A_λ and T_λ versus l for $\varepsilon_\lambda = 10 \text{ dm}^3 \text{ mol}^{-1} \text{ cm}^{-1}$ and $c_B = 0.1 \text{ mol/dm}^3$.

The quantity ε_λ is the **molar absorption coefficient** or *molar absorptivity* (formerly called the molar extinction coefficient) of substance B at wavelength λ. Most commonly, the concentration c_B is expressed in mol/dm^3 and the path length l in centimeters, so ε is commonly given in dm^3 mol^{-1} cm^{-1}. Figure 21.37 shows ε as a function of λ for gas-phase benzene over a range of UV frequencies. Since the vertical scale in this figure is logarithmic, ε varies over an enormous range.

If several absorbing species B, C, . . . are present and there are no strong interactions between the species, then $dI_\lambda/I_\lambda = -(\alpha_B c_B + \alpha_C c_C + \cdots) \, dx$ and (21.10) becomes

$$A_\lambda \equiv \log_{10}(I_{\lambda,0}/I_{\lambda,l}) = (\varepsilon_{\lambda,B} c_B + \varepsilon_{\lambda,C} c_C + \cdots) l \qquad (21.12)$$

where $\varepsilon_{\lambda,B}$ is the molar absorption coefficient of B at wavelength λ. If the molar absorption coefficients are known for B, C, . . . at several wavelengths, measurement of $I_{\lambda,0}/I_{\lambda,l}$ at several wavelengths allows a mixture of unknown composition to be analyzed. Recall the use of spectroscopy to determine reaction rates—Sec. 17.2.

An application of (21.12) is the pulse oximeter, a device that continuously monitors the oxygenation of the blood of critically ill or anesthetized patients without drawing blood. Light of wavelengths 650 and 940 nm is sent through a fingertip of the patient, and the time-varying (pulsatile) components of the absorbances at these wavelengths are measured to give the absorption due to arterial blood. Reduced and oxygenated hemoglobin have different ε's at each of these wavelengths, and this allows the percentage of oxygenated hemoglobin to be found.

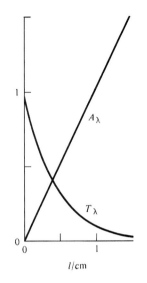

Figure 21.5

Absorbance and transmittance versus cell length for a sample with $\varepsilon_\lambda = 10 \text{ dm}^3 \text{ mol}^{-1} \text{ cm}^{-1}$ and $c_B = 0.1 \text{ mol/dm}^3$.

Lasers

The use of lasers has made possible many new types of spectroscopic experiments and has greatly improved the resolution and precision of spectroscopic investigations. The word *laser* is an acronym for "light amplification by stimulated emission of radiation."

To achieve laser action, one must first produce a **population inversion** in the system. This is a nonequilibrium situation with more molecules in an excited state than in a lower-lying state. Let the populations and energies of the two states involved be N_2 and N_1 and E_2 and E_1, with $E_2 > E_1$. Suppose we have a population inversion with $N_2 > N_1$. Photons of frequency $\nu_{12} = (E_2 - E_1)/h$ spontaneously emitted as molecules drop from state 2 to 1 will stimulate other molecules in state 2 to emit photons of frequency ν_{12} and fall to state 1 (Fig. 21.3c). Photons of frequency ν_{12} will also induce absorption from state 1 to 2, but because the system has $N_2 > N_1$, stimulated emission will predominate over absorption and we will get a net amplification of the radiation of frequency ν_{12}. A stimulated-emission photon is emitted in phase with the photon that produces its emission and travels in the same direction as this photon.

To see how a population inversion and laser action are produced, consider Fig. 21.6, in which the states 1, 2, and 3 are those involved in the laser action. Using either an electric discharge or light emitted from a flashlamp, one excites some molecules from the ground state, state 1, to state 3, a process called *pumping*. Suppose that the most probable fate of the molecules in state 3 is to rapidly give up energy to surrounding molecules and fall to state 2 without emitting radiation. Further suppose that the probability of spontaneous emission from state 2 to the ground state 1 is extremely

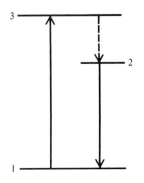

Figure 21.6

States involved in laser action.

small, so the population of state 2 builds up rapidly and eventually exceeds that of state 1, thereby giving a population inversion between states 2 and 1.

The laser system is contained in a cylindrical cavity whose ends have parallel mirrors. A few photons are spontaneously emitted as molecules go from state 2 to state 1. Those emitted at an angle to the cylinder's axis pass out of the system and play no part in the laser action. Those emitted along the laser axis travel back and forth between the end mirrors and stimulate emission of further photons of frequency $(E_2 - E_1)/h$. The presence of the end mirrors makes the laser a resonant optical cavity in which a standing-wave pattern is produced. If l is the distance between the mirrors, only light of wavelength λ such that $n\lambda/2$ is equal to l, where n is an integer, will resonate in the cavity. This makes the laser radiation nearly monochromatic (single-frequency). [Ordinarily, a given transition in a collection of molecules is spread over a range of frequencies, as a result of various effects; see *Levine* (1975), sec. 3.5.] One of the end mirrors is made partially transmitting to allow some of the laser radiation to leave the cavity.

The laser output is highly monochromatic, highly directional, intense, and coherent. *Coherent* means the phase of the radiation varies smoothly and nonrandomly along the beam. These properties make possible many applications in spectroscopy and kinetics. Thousands of different lasers exist. The material in which the laser action occurs may be a solid, liquid, or gas. The frequency emitted may lie in the infrared, visible, or ultraviolet region. The laser light may be emitted as a brief pulse (recall the use of lasers in flash photolysis—Sec. 17.14), or it may be continuously emitted, giving a CW (continuous wave) laser. Most lasers emit light of fixed frequency, but by using a tunable dye laser or tunable semiconductor laser, one can vary the frequency that will resonate in the cavity.

Kinds of lasers. A *solid-state metal-ion laser* contains a transparent crystal or glass to which a small amount of an ionic transition-metal or rare-earth compound has been added. For example, the *ruby laser* contains an Al_2O_3 crystal with a small amount of Cr_2O_3; the Cr^{3+} ions substitute for some of the Al^{3+} ions in the crystal structure. The Nd:YAG laser contains an yttrium aluminum garnet (YAG) crystal ($Y_3Al_5O_{12}$) with impurity Nd^{3+} ions substituting for some Y^{3+} ions. The laser action involves electronic energy levels of the Cr^{3+} or Nd^{3+} ions in the crystal (electric) field of the host crystal. These lasers are usually pumped by a surrounding flashlamp to achieve population inversion. Small Nd:YAG CW lasers are pumped by semiconductor-laser light.

In a *semiconductor laser* (also called a *diode laser*), the population inversion is achieved in a semiconducting solid. The electronic energy levels of solids occur in continuous bands rather than discrete energy levels (see Fig. 24.27 and the accompanying discussion). The highest occupied and lowest vacant energy bands in a semiconductor are called the valence band and the conduction band, respectively. A semiconductor laser contains a junction between an *n*-type semiconductor and a *p*-type semiconductor. An *n*-type (*n* is for negative) semiconductor contains impurity atoms that have more valence electrons than the atoms they replace, whereas a *p*-type (*p* is for positive) semiconductor contains impurity atoms with fewer valence electrons than the atoms they replace. For example, the semiconductor GaAs can be made an *n*-type semiconductor by adding excess As atoms, which replace some of the Ga atoms in the crystal structure; addition of excess Ga gives a *p*-type semiconductor. When a voltage is applied across the junction of a semiconductor laser, electrons drop from the conduction band of the *n*-type semiconductor to the valence band of the *p*-type semiconductor, emitting laser radiation, most commonly in the infrared. Semiconductor lasers can be tuned over a small frequency range by changing the temperature. Semiconductor lasers are tiny (a few millimeters in length) and are used in compact disk players and for high-resolution infrared spectroscopy.

A *gas laser* contains a gas at low pressure. An electric discharge through the gas leads to population inversion. In the *helium–neon laser*, an electric discharge is passed through a mixture of helium and neon with helium mole fraction 0.9. Collisions with electrons excite He atoms to the ^1S and ^3S terms of the $1s2s$ configuration. Collisions of excited He atoms with Ne atoms transfer energy and excite Ne atoms to a high electronic energy level, producing a population inversion in the Ne atoms. The He–Ne laser is a CW laser and is used in optical scanners such as supermarket bar-code readers. The CO_2 laser contains a mixture of CO_2, N_2, and He gases. An electric discharge raises N_2 molecules to the $v = 1$ first excited vibrational level. Collisions between N_2 and CO_2 molecules raise CO_2 molecules to excited vibration–rotation levels, producing a population inversion that gives laser emission.

In chemical lasers, a gas-phase exothermic chemical reaction produces products in excited vibrational levels; this population inversion leads to laser emission. The reactants flow continually through the cell of a chemical laser. See Sec. 23.3 for an example.

Another kind of gas-phase laser is an *exciplex laser*. An *exciplex* (excited complex) is a species that is formed from two atoms or molecules and that is stable (with respect to dissociation) in an excited electronic state but is unstable in its ground electronic state. When the two atoms or molecules that form such a species are identical, the species is called an *excimer* (excited dimer). (Very commonly, the word "excimer" is used whether or not the atoms or molecules forming the species are identical.) A simple excimer is He_2, which is unstable in its ground electronic state (no minimum in its potential-energy curve of electronic energy versus internuclear distance, except for a *very* slight minimum) but has bound excited electronic states that correspond to excitation of an electron from the antibonding $\sigma_u^* 1s$ MO to a higher MO that is bonding.

The excimers and exciplexes used in lasers are diatomic molecules. In a commonly used excimer laser, a continuous electric discharge is passed through a mixture of He, Kr, and F_2. The discharge produces Kr^+ and F^- ions according to $Kr + e^- \rightarrow Kr^+ + 2e^-$ and $F_2 + e^- \rightarrow F^- + F$. These ions combine to form the exciplex KrF* ($F^- + Kr^+ + He \rightarrow$ KrF* + He), where the star indicates the KrF species is formed in an excited electronic state (one with a minimum in its potential-energy curve), and the He is present as a third body to carry away energy and allow the formation of KrF* (Sec. 17.12). The KrF* (which has a 2-ns lifetime) rapidly drops to its ground electronic state emitting ultraviolet laser radiation. The unstable KrF ground state immediately dissociates to Kr and F. Recombination of F atoms forms the starting material F_2. This is a two-level laser in which population inversion is maintained because the lower level is unstable. Exciplex lasers are used in surgery.

In a *dye laser*, a solution of an organic dye flows through the laser cell. The dye is pumped to an excited electronic level S_1 by light from a flashlamp or from another laser. The lifetime for spontaneous emission from S_1 to vibrationally excited levels of the ground electronic state S_0 is of the order of 10^{-9} s. Intermolecular collisions in the liquid then cause extremely rapid (10^{-11} s) decay from the excited vibrational levels of S_0 to the ground vibrational level of S_0, thereby producing a population inversion between S_1 and the vibrationally excited levels of S_0, and we get laser emission. The vibration–rotation levels of S_0 are broadened by intermolecular collisions in the solution to give a continuous band of energy, so emission from S_1 to S_0 occurs over a continuous range of frequencies. By using a diffraction grating and changing the angle of the grating with respect to the laser beam, one can select the frequency of the laser radiation. The tunability of dye lasers makes them valuable in spectroscopy.

21.3 ROTATION AND VIBRATION OF DIATOMIC MOLECULES

We now consider nuclear motion in an isolated diatomic molecule. By "isolated" we mean that interactions with other molecules are slight enough to neglect. This condition is well met in a gas at low pressure.

Translation, Rotation, and Vibration

Recall from Sec. 20.2 that in the Born–Oppenheimer approximation, the potential energy for motion of the nuclei in a molecule is E_e, the electronic energy (including nuclear repulsion) as a function of the spatial configuration of the nuclei. Each electronic state of a diatomic molecule has its own E_e curve (for example, see Fig. 20.4). The Schrödinger equation (20.9) for nuclear motion in a particular electronic state is

$$(\hat{K}_N + E_e)\psi_N \equiv \hat{H}_N\psi_N = E\psi_N$$

where E is the total energy of the molecule, $\hat{K}_N$ is the operator for the kinetic energies of the nuclei, and ψ_N is the wave function for nuclear motion.

For a diatomic molecule composed of atoms A and B with masses m_A and m_B, we have $\hat{K}_N = -(\hbar^2/2m_A)\nabla_A^2 - (\hbar^2/2m_B)\nabla_B^2$ and $E_e = E_e(R)$, where R is the internuclear distance. The potential energy $E_e(R)$ in the Schrödinger equation for nuclear motion is a function of only the relative coordinates of the two nuclei. Therefore the results of Sec. 18.13 apply, and the center-of-mass motion and the internal motion can be dealt with separately. The total molecular energy is the sum of E_{tr}, the translational energy of the molecule as a whole, and E_{int}, the energy of the relative or internal motion of the two atoms:

$$E = E_{tr} + E_{int} \tag{21.13}$$

The allowed translational energies E_{tr} can be taken as those of a particle in a three-dimensional box, Eq. (18.47). The box is the container in which the gas molecules are confined.

From Sec. 18.13, the internal energy levels E_{int} are found from the Schrödinger equation for internal motion, which is

$$[-(\hbar^2/2\mu)\nabla^2 + E_e(R)]\psi_{int} = E_{int}\psi_{int} \tag{21.14}$$

The operator ∇^2 equals $\partial^2/\partial x^2 + \partial^2/\partial y^2 + \partial^2/\partial z^2$, where x, y, and z are the coordinates of one nucleus relative to the other; ψ_{int} is a function of x, y, and z. The reduced mass μ is given by Eq. (18.79) as $\mu = m_A m_B/(m_A + m_B)$. Instead of the relative cartesian coordinates x, y, and z, it is more convenient to use the spherical coordinates R, θ, and ϕ of one nucleus relative to the other (Fig. 21.7).

The kinetic energy of internal motion can be divided into kinetic energy of rotation and kinetic energy of vibration. Rotation changes the orientation of the molecular axis in space (that is, changes θ and ϕ) while the internuclear distance R remains fixed. Vibration changes the internuclear distance R. Since R is fixed for rotation, the potential energy $E_e(R)$ is associated with the vibration of the molecule. The rotational motion involves the coordinates θ and ϕ, whereas the vibrational motion involves R. It is therefore reasonable to treat the rotational and vibrational motions separately. This is actually an approximation, since there is interaction between rotation and vibration, as discussed later in this section.

The rotational energies of a rigid (fixed interparticle distance) two-particle rotor were given in Sec. 18.14. A diatomic molecule does not actually remain rigid while it rotates, because there is always some vibrational motion. Even the ground vibrational state has a zero-point vibrational energy. However, the nuclei vibrate about the equilibrium internuclear distance R_e, so to a good approximation, we can treat the molecule as a two-particle rigid rotor with separation R_e between the nuclei. From (18.81) and (18.82) the rotational energy of a *diatomic* molecule is (approximately)

$$E_{rot} \approx \frac{J(J + 1)\hbar^2}{2I_e}, \qquad I_e = \mu R_e^2, \qquad \mu \equiv \frac{m_A m_B}{m_A + m_B}, \qquad J = 0, 1, 2, \ldots \tag{21.15}*$$

$$E_{rot} \approx B_e h J(J + 1) \qquad \text{where } B_e \equiv h/8\pi^2 I_e \tag{21.16}$$

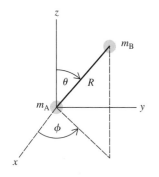

Figure 21.7

Coordinates R, θ, and ϕ for the internal nuclear motion in a diatomic molecule.

I_e is the **equilibrium moment of inertia.** B_e is the **equilibrium rotational constant.** Note that R_e differs for different electronic states of the same molecule (see, for example, Fig. 20.4). The rotational wave function ψ_{rot} depends on the angles θ and ϕ defining the orientation of the molecule in space (Fig. 21.7). The rotational levels (21.15) are $(2J + 1)$-fold degenerate, since the value of the quantum number M_J [Eq. (18.83)] does not affect E_{rot}.

Having dealt with rotation by using the rigid-rotor approximation, we now consider vibration of a diatomic molecule. The vibrational motion is along the internuclear separation R, so the vibrational part of the nuclear kinetic-energy operator $(-\hbar^2/2\mu)\nabla^2$ in (21.14) is $(-\hbar^2/2\mu)\, d^2/dR^2$. [This is an oversimplification. See *Levine* (1975), sec. 4.1, for a fuller discussion.] The potential energy for vibration is $E_e(R)$, as noted after (21.14). The vibrational wave function ψ_{vib} is a function of R. We have separated the rotational energy E_{rot} from the internal energy E_{int} in (21.14), so the energy that occurs in the vibrational Schrödinger equation is $E_{\text{int}} - E_{\text{rot}}$. The vibrational Schrödinger equation is therefore

$$\left[-\frac{\hbar^2}{2\mu}\frac{d^2}{dR^2} + E_e(R) \right]\psi_{\text{vib}}(R) = (E_{\text{int}} - E_{\text{rot}})\psi_{\text{vib}}(R) \qquad (21.17)$$

The potential-energy function $E_e(R)$ in (21.17) differs for each different electronic state. It is useful to expand $E_e(R)$ in a Taylor series about the equilibrium distance R_e. Equation (8.32) gives

$$E_e(R) = E_e(R_e) + E_e'(R_e)(R - R_e) + \tfrac{1}{2}E_e''(R_e)(R - R_e)^2$$
$$+ \tfrac{1}{6}E_e'''(R_e)(R - R_e)^3 + \cdots$$

where $E_e'(R_e)$ equals $dE_e(R)/dR$ evaluated at $R = R_e$, with similar definitions for $E_e''(R_e)$, etc. Since we are considering a bound state with a minimum in E_e at $R = R_e$ (Fig. 20.4), the first derivative $E_e'(R_e)$ equals 0. This eliminates the second term in the expansion. The nuclei vibrate about their equilibrium separation R_e. For the lower-energy vibrational levels, the vibrations will be confined to distances R reasonably close to R_e, so the terms involving $(R - R_e)^3$ and higher powers will be smaller than the term involving $(R - R_e)^2$; we shall neglect these terms. Therefore

$$E_e(R) \approx E_e(R_e) + \tfrac{1}{2}E_e''(R_e)(R - R_e)^2 \qquad (21.18)$$

We have found that in the region near R_e, the electronic energy curve $E_e(R)$ is approximately a parabolic (quadratic) function of $R - R_e$, the deviation of R from R_e. This is evident in Figs. 20.4 and 21.9. Figure 21.8 plots the approximation (21.18) and the exact E_e for the ground electronic state of H_2.

The quantity $E_e(R_e)$ in (21.18) is a constant for a given electronic state and will be called the **equilibrium electronic energy** E_{el} of the electronic state:

$$E_{\text{el}} \equiv E_e(R_e) \qquad (21.19)$$

For the H_2 ground electronic state, the equilibrium dissociation energy is $D_e = 4.75$ eV, so $E_e(R_e) = E_{\text{el}}$ is 4.75 eV below the energy of the dissociated molecule. The energy of two ground-state H atoms is given by (19.18) as $2(-13.60 \text{ eV}) = -27.20$ eV, so $E_{\text{el}} = -31.95$ eV for the H_2 ground electronic state (Figs. 21.8 and 20.4).

Let $x \equiv R - R_e$. Then $d^2/dx^2 = d^2/dR^2$. Substitution of (21.18) and (21.19) into (21.17) and rearrangement gives as the approximate vibrational Schrödinger equation for a diatomic molecule

$$\left[-\frac{\hbar^2}{2\mu}\frac{d^2}{dx^2} + \frac{1}{2}E_e''(R_e)x^2 \right]\psi_{\text{vib}} = (E_{\text{int}} - E_{\text{rot}} - E_{\text{el}})\psi_{\text{vib}} \qquad (21.20)$$

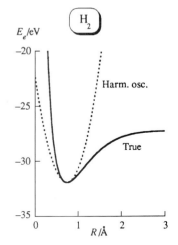

Figure 21.8

Harmonic-oscillator approximation to the potential-energy curve for nuclear motion in the H_2 ground electronic state compared with the true curve. The curve $E_e(R)$ is found by solving the electronic Schrödinger equation (20.7) at many internuclear distances R.

Since (21.20) has the same form as the Schrödinger equation (18.75) for a one-dimensional harmonic oscillator, the solutions to (21.20) are the harmonic-oscillator wave functions. The quantity $E_{int} - E_{rot} - E_{el}$ in (21.20) corresponds to the harmonic-oscillator energy E in (18.75). Therefore $E_{int} - E_{rot} - E_{el} = (v + \frac{1}{2})h\nu_e$. Combining this equation with $E = E_{tr} + E_{int}$ [Eq. (21.13)], we have as the molecular energy E:

$$E = E_{tr} + E_{rot} + E_{vib} + E_{el} \qquad \textbf{(21.21)}^*$$

$$E_{vib} \approx (v + \tfrac{1}{2})h\nu_e, \quad v = 0, 1, 2, \ldots \quad \text{diatomic molecule} \qquad \textbf{(21.22)}^*$$

The force constant k in (18.75) corresponds to $E_e''(R_e)$ in (21.20). Also, m in (18.75) corresponds to μ in (21.20). Therefore the **equilibrium** (or **harmonic**) **vibrational frequency** ν_e of the diatomic molecule is [Eq. (18.73)]

$$\nu_e = \frac{1}{2\pi}\left(\frac{k_e}{\mu}\right)^{1/2} = \frac{1}{2\pi}\left[\frac{E_e''(R_e)}{\mu}\right]^{1/2} \qquad (21.23)$$

From (18.71), the **equilibrium force constant** k_e has units of force over length.

In summary, we have seen that to a good approximation, the energy E of an isolated diatomic molecule is $E = E_{tr} + E_{rot} + E_{vib} + E_{el}$, where the translational energy levels E_{tr} are those of a particle in a box, the rotational energy levels E_{rot} can be approximated by the rigid-rotor energies $J(J + 1)\hbar^2/2I_e$, the vibrational levels E_{vib} can be approximated by the harmonic-oscillator levels $(v + \frac{1}{2})h\nu_e$, and the equilibrium electronic energy E_{el} is the energy at the minimum in the electronic energy curve $E_e(R)$. The translational, rotational, and vibrational energies represent energy over and above the energy at the minimum in the E_e curve. The molecule's internal energy [Eq. (21.13)] is $E_{rot} + E_{vib} + E_{el}$:

$$E_{int} \approx B_e h J(J + 1) + (v + \tfrac{1}{2})h\nu_e + E_{el} \qquad (21.24)$$

Anharmonicity

The approximation (21.18) for the potential energy of nuclear motion leads to harmonic-oscillator vibrational energy levels. Figure 21.8 shows that for $R \gg R_e$, the parabolic approximation (21.18) is poor. The potential energy $E_e(R)$ is not really a harmonic-oscillator potential energy, and this **anharmonicity** adds correction terms to the approximate vibrational-energy expression (21.22). One finds that the main correction term to (21.22) is

$$-h\nu_e x_e(v + \tfrac{1}{2})^2$$

where the *anharmonicity constant* $\nu_e x_e$ is nearly always positive and depends on $E_e'''(R_e)$ and $E_e^{(iv)}(R_e)$. The quantity x_e (which is a constant and not a coordinate) usually lies in the range 0.002 to 0.02 for diatomic molecules (see Table 21.1 in Sec. 21.4). Therefore, $\nu_e x_e \ll \nu_e$. As v increases, the magnitude of the term $-h\nu_e x_e(v + \frac{1}{2})^2$ becomes larger relative to $(v + \frac{1}{2})h\nu_e$.

With inclusion of the anharmonicity correction term, the spacing between the adjacent levels v and $v + 1$ becomes

$$(v + \tfrac{3}{2})h\nu_e - h\nu_e x_e(v + \tfrac{3}{2})^2 - \left[(v + \tfrac{1}{2})h\nu_e - h\nu_e x_e(v + \tfrac{1}{2})^2\right]$$

$$= h\nu_e - 2h\nu_e x_e(v + 1)$$

Thus the spacing between adjacent diatomic-molecule vibrational levels decreases as v increases. This is in contrast to the equally spaced levels of a harmonic oscillator. A harmonic oscillator has an infinite number of vibrational levels. However, a diatomic-molecule bound electronic state has only a finite number of vibrational levels, since once the vibrational energy exceeds the equilibrium dissociation energy D_e, the molecule dissociates. Figure 21.9 shows the vibrational levels of the H_2 ground electronic

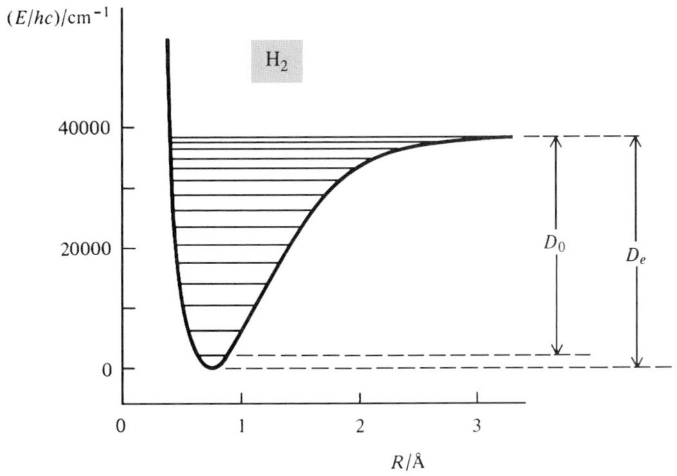

Figure 21.9

Ground-electronic-state vibrational
levels of H_2. [Data from S.
Weissman et al., *J. Chem. Phys.*,
39, 2226 (1963).]

state, which has 15 bound vibrational levels ($v = 0$ through 14). Note the decreasing spacing as v increases. The zero level of energy has been taken at the minimum of the potential-energy curve. The $v = 14$ level lies a mere 150 cm^{-1} below $E_e(\infty)/hc$.

Ground-Vibrational-State Dissociation Energy

The lowest molecular rotational energy is zero, since $E_{\text{rot}} = 0$ for $J = 0$ in (21.15). However, the lowest vibrational energy is nonzero. In the harmonic-oscillator approximation, the $v = 0$ ground vibrational state has the energy $\frac{1}{2}h\nu_e$. With inclusion of the anharmonicity term $-h\nu_e x_e (v + \frac{1}{2})^2$, the vibrational ground-state energy is $\frac{1}{2}h\nu_e - \frac{1}{4}h\nu_e x_e$. The dissociation energy measured from the lowest vibrational state is called the **ground-vibrational-state dissociation energy** D_0 (Fig. 21.9) and is somewhat less than the equilibrium dissociation energy D_e, because of the zero-point vibrational energy:

$$D_e = D_0 + \tfrac{1}{2}h\nu_e - \tfrac{1}{4}h\nu_e x_e \qquad (21.25)$$

For the H_2 ground electronic state, $D_0 = 4.48$ eV and $D_e = 4.75$ eV.

What is the relation between the molecular quantity D_0 and thermodynamic quantities? Consider the dissociation reaction $H_2(g) \to 2H(g)$. The standard-state thermodynamic properties refer to ideal gases, which have no intermolecular interactions. At the temperature absolute zero, the molecules will all be in the lowest available energy level. For H_2, this is the $v = 0$, $J = 0$ level of the ground electronic state. Therefore the thermodynamic quantity ΔU_0° for $H_2(g) \to 2H(g)$ will equal $N_A D_0$, where N_A is the Avogadro constant. Spectroscopic determination of D_0 for the H_2 ground electronic state gives $D_0 = 4.4781$ eV. Using (20.1), we get $\Delta U_0^\circ = 432.07$ kJ/mol. ΔU_{298}° differs from ΔU_0° because of the increase in translational energies of the 2 moles of H atoms and 1 mole of H_2 molecules and the increase in H_2 rotational energy on going from 0 to 298 K; see Prob. 22.69. (Excited vibrational and electronic levels of H_2 are not significantly occupied at 298 K.)

Vibration–Rotation Interaction

Equation (21.21) neglects the interaction between vibration and rotation. Because of the anharmonicity of the potential-energy curve $E_e(R)$, the average distance R_{av} between the nuclei increases as the vibrational quantum number increases (see Fig. 21.9). This increase in R_{av} increases the effective moment of inertia $I = \mu R_{\text{av}}^2$ and decreases the rotational energies, which are proportional to $1/I$. To allow for this effect, one adds the term $-h\alpha_e(v + \frac{1}{2})J(J + 1)$ to the energy, where the *vibration–rotation* *coupling constant α_e* is a positive constant that is much smaller than the rotational constant B_e in Eq. (21.16) (see Table 21.1).

Centrifugal Distortion

Since the molecule is not really a rigid rotor, the internuclear distance increases very slightly as the rotational energy increases, a phenomenon called *centrifugal distortion*. Centrifugal distortion increases the moment of inertia and hence decreases the rotational energy below that of a rigid rotor. One finds that the term $-hDJ^2(J + 1)^2$ must be added to the rotational energy, where the *centrifugal-distortion constant D* is a very small positive constant. Don't confuse D with the dissociation energy.

Internal Energy of a Diatomic Molecule

With inclusion of anharmonicity and vibration–rotation interaction, the expression (21.24) for the internal energy of a diatomic molecule becomes

$$E_{int} = E_{el} + h\nu_e(v + \tfrac{1}{2}) - h\nu_e x_e(v + \tfrac{1}{2})^2 + hB_e J(J + 1) - h\alpha_e(v + \tfrac{1}{2})J(J + 1)$$

$$(21.26)$$

Since the centrifugal-distortion term is tiny, it is omitted from E_{int}. The total energy equals $E_{int} + E_{tr}$. If vibration–rotation interaction is neglected, E_{int} is approximated as the sum of electronic, vibrational, and rotational energies:

$$E_{int} \approx E_{el} + h\nu_e(v + \tfrac{1}{2}) - h\nu_e x_e(v + \tfrac{1}{2})^2 + hB_e J(J + 1) \qquad (21.27)$$

Since ν_e [which depends on $E''_e(R_e)$] and B_e (which depends on R_e) each differ for different electronic states of the same molecule, each electronic state of a molecule has its own set of vibrational and rotational levels. Figure 21.10 shows some of the vibration–rotation levels of one electronic state of a diatomic molecule. The dots indicate higher rotational levels of each vibrational level. (Figure 21.9 shows only the energy levels with $J = 0$.)

Level Spacings

The translational energy levels for a particle in a cubic box of volume V are [Eq. (18.50)] $(n_x^2 + n_y^2 + n_z^2)h^2/8mV^{2/3}$. The spacing between adjacent translational levels with quantum numbers n_x, n_y, n_z and $n_x + 1$, n_y, n_z is $(2n_x + 1)h^2/8mV^{2/3}$, since $(n_x + 1)^2 - n_x^2 = 2n_x + 1$. A typical molecule has translational energy of roughly $\tfrac{1}{2}kT$ associated with motion in each direction (Chapter 15). The equation $\tfrac{1}{2}kT = n_x^2 h^2/8mV^{2/3}$ gives for a molecule of molecular weight 100 in a 1-cm³ box at room temperature: $n_x = 8 \times 10^8$; this gives a spacing of 5×10^{-30} J $= 3 \times 10^{-11}$ eV between adjacent translational levels. This energy gap is so tiny that, for all practical purposes, we can consider the translational energy levels of the gas molecules to be continuous rather than discrete. The very close spacing results from the macroscopic size of the container's volume V.

The spacing between the rotational levels with quantum numbers J and $J + 1$ is given by (21.15) as $(J + 1)\hbar^2/\mu R_e^2$, since $(J + 1)(J + 2) - J(J + 1) = 2(J + 1)$. For the CO ground electronic state, R_e is 1.1 Å. The reduced mass μ equals $m_A m_B/(m_A + m_B)$; the atomic masses m_A and m_B are found by dividing the molar masses by the Avogadro constant. Thus

$$\mu = \frac{[(12 \text{ g mol}^{-1})/N_A][(16 \text{ g mol}^{-1})/N_A]}{(28 \text{ g mol}^{-1})/N_A} = 1.1 \times 10^{-23} \text{ g}$$

This gives $\hbar^2/\mu R_e^2 = 8 \times 10^{-23}$ J $= 0.0005$ eV. The spacing between adjacent CO rotational levels is $J + 1$ times this number, where $J = 0, 1, 2, \ldots$.

The spacing between the harmonic-oscillator vibrational levels is $h\nu_e$. Observations on vibrational spectra show that ν_e is typically 10^{13} to 10^{14} s^{-1}, so the vibrational spacing is typically 7×10^{-21} to 7×10^{-20} J (0.04 to 0.4 eV).

Figure 21.10

Vibration–rotation energy levels for a diatomic molecule. For each vibrational state (defined by the quantum number v), there is a set of rotational levels (defined by the quantum number J). For the ground electronic state of H_2, v goes from 0 to 14 (Fig. 21.9).

Electronic absorption spectra show that the spacing $E_{el,2} - E_{el,1}$ between the ground and first excited electronic levels is typically 2 to 6 eV.

Thus, *electronic energy differences are substantially greater than vibrational energy differences, which in turn are much greater than rotational energy differences.* Figure 21.11 shows the typical ranges of energy differences between the lowest two rotational states, vibrational states, and electronic states of diatomic molecules.

The Boltzmann Distribution

In spectroscopy, one observes absorption or emission of radiation by a collection of many molecules in a gas, liquid, or solid. The molecules populate the various possible quantum states in accord with the Boltzmann distribution law. Absorption from a given quantum state will be observed only if there are a significant number of molecules in that state, so we want to examine the population of rotational, vibrational, and electronic states under typical conditions. The Boltzmann distribution law (15.74) gives the population ratio for states i and j as $N_i/N_j = e^{-(E_i - E_j)/kT}$. At room temperature, $kT = 0.026$ eV.

The typical molecular rotational-level spacing ΔE_{rot} found above is substantially less than kT at room temperature, so $e^{-\Delta E_{rot}/kT}$ is close to 1, and many excited rotational levels are populated at room temperature.

The vibrational frequency ν_e equals $(1/2\pi)(k_e/\mu)^{1/2}$. For heavy diatomic molecules (for example, Br_2 and I_2), the large value of μ gives a ν_e of roughly 10^{13} s^{-1} and a spacing $h\nu_e$ of roughly 0.04 eV, which is comparable to kT at room temperature. Hence, for heavy diatomic molecules, there is significant occupation of one or more excited vibrational levels at room temperature. However, relatively light diatomic molecules (for example, H_2, HCl, CO, and O_2) have ν_e values of roughly 10^{14} s^{-1} and vibrational spacings of roughly 0.4 eV. This is substantially greater than kT at room temperature, so $e^{-\Delta E_{vib}/kT}$ is very small and nearly all the molecules are in the ground vibrational level at room temperature. Figure 21.12 plots the fraction of molecules in the ground rotational state and in the ground vibrational state versus T for some diatomic molecules.

Since ΔE_{el} is substantially greater than kT at room temperature, excited electronic states are generally not occupied at room temperature.

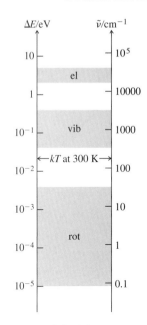

Figure 21.11

Typical ranges of spacings between electronic, vibrational, and rotational energy levels for diatomic molecules. $\tilde{\nu} = \Delta E/hc$. The scale is logarithmic.

21.4 ROTATIONAL AND VIBRATIONAL SPECTRA OF DIATOMIC MOLECULES

This section deals with radiative transitions between different vibration–rotation levels of the same electronic state of a diatomic molecule. Transitions between different electronic states are considered in Sec. 21.11.

Selection Rules

To find the spectrum line frequencies, we need the selection rules.

The Born–Oppenheimer approximation gives the molecular wave function as the product of electronic and nuclear wave functions: $\psi = \psi_e \psi_N$. For a transition between two states ψ and ψ' with no change in electronic state ($\psi_e = \psi_e'$), the transition-moment integral (21.5) determining the selection rules is

$$\int \int \psi_e^* \psi_N^* \hat{\boldsymbol{\mu}} \psi_e \psi_N' \, d\tau_e \, d\tau_N = \int \psi_N^* \psi_N' \left(\int \psi_e^* \hat{\boldsymbol{\mu}} \psi_e \, d\tau_e \right) d\tau_N = \int \psi_N^* \psi_N' \boldsymbol{\mu} \, d\tau_N$$

where (20.38) was used to introduce the electric dipole moment $\boldsymbol{\mu}$ of the electronic state. The magnitude μ of $\boldsymbol{\mu}$ depends on the internuclear distance. (The experimentally

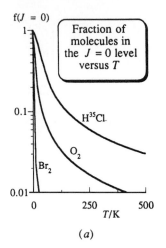

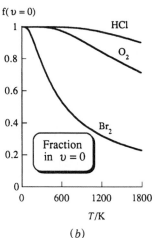

Figure 21.12

The fraction f of molecules in (a) the ground rotational state and (b) the ground vibrational state of some diatomic molecules plotted versus T. The vertical scale in (a) is logarithmic. The temperature scales in (a) and (b) differ.

observed value is an average over the molecular zero-point vibration.) One expands μ in a Taylor series about its value at R_e:

$$\mu(R) = \mu(R_e) + \mu'(R_e)(R - R_e) + \tfrac{1}{2}\mu''(R_e)(R - R_e)^2 + \cdots$$

where $\mu(R_e)$ is virtually the same as the experimentally observed dipole moment. One then substitutes this expansion and the expressions for ψ_N and ψ'_N into the above integral and evaluates this integral. The details are omitted.

The selection rules for diatomic-molecule transitions with no change in electronic state turn out to be

$$\Delta J = \pm 1 \qquad \qquad \text{(21.28)*}$$

$$\Delta v = 0, \pm 1 \ (\pm 2, \pm 3, \dots) \qquad \qquad \text{(21.29)*}$$

$$\Delta v = 0 \text{ not allowed if } \mu(R_e) = 0 \qquad \Delta v = \pm 1 \text{ not allowed if } \mu'(R_e) = 0$$
$$\text{(21.30)*}$$

where $\mu(R_e)$ is the molecule's electric dipole moment evaluated at the equilibrium bond distance and $\mu'(R_e)$ is $d\mu/dR$ evaluated at R_e. The parentheses in (21.29) indicate that the $\Delta v = \pm 2, \pm 3, \dots$ transitions are far less probable than the $\Delta v = 0$ and ± 1 transitions. If the molecule were a harmonic oscillator, and if terms after the $\mu'(R_e)(R - R_e)$ term in the $\mu(R)$ expansion were negligible, only $\Delta v = 0, \pm 1$ transitions would occur.

Rotational Spectra

Transitions with no change in electronic state and with $\Delta v = 0$ give the **pure-rotation spectrum** of the molecule. These transitions correspond to photons with energies in the microwave and far-IR regions. (The far-IR region is the portion of the IR region bordering the microwave region in Fig. 21.2.) Equation (21.30) shows that *a diatomic molecule has a pure-rotation spectrum only if it has a nonzero electric dipole moment.* A homonuclear diatomic molecule (for example, H_2, N_2, Cl_2) has no pure-rotation spectrum.

The pure-rotation spectrum is usually observed as an absorption spectrum. The absorption transitions all have $\Delta J = +1$ [Eq. (21.28)]. Because the rotational levels are not equally spaced, and because many excited rotational levels are populated at room temperature, there will be several lines in the pure-rotation spectrum. From $E_{\text{upper}} - E_{\text{lower}} = h\nu$, the absorption frequencies are $\nu = (E_{J+1} - E_J)/h$, where the energy levels are given by (21.26). For a pure-rotational transition, E_{el} is unchanged and v is unchanged, so the only terms in (21.26) that change are the last two terms, which involve the rotational quantum number J. The sum of these two terms is $hJ(J + 1) \cdot [B_e - \alpha_e(v + \tfrac{1}{2})]$, and the frequency of a diatomic-molecule pure-rotational transition between levels J and $J + 1$ is (recall Example 21.2 in Sec. 21.2)

$$\nu = (J + 1)(J + 2)[B_e - \alpha_e(v + \tfrac{1}{2})] - J(J + 1)[B_e - \alpha_e(v + \tfrac{1}{2})]$$

$$\nu = 2(J + 1)[B_e - \alpha_e(v + \tfrac{1}{2})] \equiv 2(J + 1)B_v \qquad \text{where } J = 0, 1, 2, \dots \qquad \text{(21.31)}$$

and where the *mean rotational constant* B_v for states with vibrational quantum number v is defined as

$$B_v \equiv B_e - \alpha_e(v + \tfrac{1}{2}) \qquad \qquad \text{(21.32)}$$

Tables sometimes list $B_0 = B_e - \tfrac{1}{2}\alpha_e$, instead of B_e. The rotational constant B_v allows for the increase in average internuclear distance and consequent decrease in rotational energies due to anharmonic vibration.

The wavenumbers of the pure-rotational transitions are $\tilde{\nu} = 1/\lambda = \nu/c = 2(J + 1)B_v/c$. We shall use a tilde to indicate division of a molecular constant by c. Thus, $\tilde{\nu} = 2(J + 1)\tilde{B}_v$, where $\tilde{B}_v \equiv B_v/c$. In particular [Eq. (21.16)],

$$\tilde{B}_0 \equiv B_0/c = h/8\pi^2 I_0 c \tag{21.33}$$

where I_0 is the moment of inertia averaged over the zero-point vibration. In the research literature, the tilde is often omitted.

For the majority of diatomic molecules, only the $v = 0$ vibrational level is significantly populated at room temperature, and the pure-rotation frequencies (21.31) become $2(J + 1)B_0$. The pure-rotation spectrum is a series of equally spaced lines at $2B_0, 4B_0, 6B_0, \ldots$ (if centrifugal distortion is neglected). The line at $2B_0$ is due to absorption by molecules in the $J = 0$ level; that at $4B_0$ is due to absorption by $J = 1$ molecules, etc. See Fig. 21.13.

If excited vibrational levels are appreciably populated, each rotational transition shows one or more nearby satellite lines due to transitions between rotational levels of the $v = 1$ vibrational level, between rotational levels of the $v = 2$ vibrational level, etc. These satellites are much weaker than the main line because the population of vibrational levels falls off rapidly as v increases.

Since the moments of inertia of different isotopic species of the same molecule differ, each isotopic species has its own pure-rotation spectrum. For $^{12}C^{16}O$, some observed pure-rotational transitions (all for $v = 0$) are (1 GHz = 10^9 Hz):

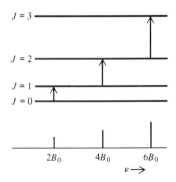

Figure 21.13

Diatomic-molecule pure-rotational absorption transitions. All the levels shown have $v = 0$.

$J \to J + 1$	$0 \to 1$	$1 \to 2$	$2 \to 3$	$3 \to 4$	$4 \to 5$
ν/GHz	115.271	230.538	345.796	461.041	576.268

The slight decreases in the successive spacings are due to centrifugal distortion. For $^{13}C^{16}O$ and $^{12}C^{18}O$, the $J = 0 \to 1$ pure-rotation transitions occur at 110.201 and 109.782 GHz, respectively.

From the observed frequency of a pure-rotational transition of a heteronuclear diatomic molecule, one can use (21.31) to calculate B_0. From B_0 one gets I_0, and from $I_0 = \mu R_0^2$ one gets R_0, the internuclear distance averaged over the zero-point vibration. If vibrational satellites are observed, they can be used to find α_e and then (21.32) allows calculation of B_e and hence of the equilibrium bond distance R_e.

EXAMPLE 21.3 Calculating a bond distance from the rotational spectrum

Calculate R_0 for $^{12}C^{16}O$ using the $J = 0 \to 1$ frequency of 115.271 GHz for $v = 0$.

From (21.31) with $J = 0$, we have

$$\nu = 2B_0 = 2h/8\pi^2 I_0 = h/4\pi^2\mu R_0^2 \quad \text{and} \quad R_0 = (h/4\pi^2\mu\nu)^{1/2}$$

This result also follows from $h\nu = E_{upper} - E_{lower} = 1(2)\hbar^2/2\mu R_0^2 - 0$, where the rotational levels (21.15) were used. An approximate value of μ for $^{12}C^{16}O$ was found near the end of Sec. 21.3, and use of the accurate atomic masses in the table inside the back cover gives the accurate value $\mu = 1.13850 \times 10^{-23}$ g. Substitution in $R_0 = (h/4\pi^2\mu\nu)^{1/2}$ gives

$$R_0 = \left[\frac{6.62607 \times 10^{-34}\,\text{J s}}{4\pi^2(1.13850 \times 10^{-26}\,\text{kg})(115.271 \times 10^9\,\text{s}^{-1})}\right]^{1/2} = 1.13089 \times 10^{-10}\,\text{m}$$

Exercise

The $J = 2 \to 3$ pure-rotational transition for the $v = 0$ state of $^{39}\text{K}^{37}\text{Cl}$ occurs at 22410 MHz. Neglecting centrifugal distortion, find R_0 for $^{39}\text{K}^{37}\text{Cl}$. (*Answer:* 2.6708 Å.)

Vibration–Rotation Spectra of Diatomics

Transitions with a change in vibrational state ($\Delta v \neq 0$) and with no change in electronic state give the **vibration–rotation spectrum** of the molecule. These transitions involve infrared photons. Equation (21.30) shows that a diatomic molecule has a vibration–rotation spectrum only if the change in dipole moment $d\mu/dR$ is nonzero at R_e. When a homonuclear diatomic molecule vibrates, its μ remains zero. For a heteronuclear diatomic molecule, vibration changes μ. Therefore, a *diatomic molecule shows an IR vibration–rotation spectrum only if it is heteronuclear.*

The vibration–rotation spectrum is usually observed as an absorption spectrum. From (21.29) the absorption transitions have $\Delta v = +1$ ($+2$, $+3$, . . .), where the transitions in parentheses are much less probable. Since $\Delta J = 0$ is not allowed by the selection rule (21.28), there is no pure-vibration spectrum for a diatomic molecule; when v changes, J also changes.

Vibrational levels are much more widely spaced than rotational levels, so the IR vibration–rotation spectrum consists of a series of bands. Each **band** corresponds to a transition between two particular vibrational levels v'' and v' and consists of a series of lines, each line corresponding to a different change in rotational state. (See Fig. 21.14.) We shall first ignore the rotational structure of the bands and shall calculate the frequency of a *hypothetical* transition where v changes but J is 0 in both the initial and final states; this gives the position of the **band origin.**

IR spectroscopists commonly work with wavenumbers rather than frequencies. The wavenumber $\tilde{\nu}_{\text{origin}}$ of the band origin for an absorption transition from vibrational level v'' to v' is $\tilde{\nu}_{\text{origin}} = 1/\lambda = \nu/c = (E_{v'} - E_{v''})/hc$. The use of the energy expression (21.26) with $J = 0$ gives

$$\tilde{\nu}_{\text{origin}} = \tilde{\nu}_e(v' - v'') - \tilde{\nu}_e x_e[v'(v' + 1) - v''(v'' + 1)] \qquad (21.34)$$

$$\tilde{\nu}_e \equiv \nu_e/c, \qquad \tilde{\nu}_e x_e \equiv \nu_e x_e/c \qquad (21.35)$$

where ν_e is the molecule's equilibrium vibrational frequency.

Throughout this chapter, *a double prime will be used on quantum numbers of the lower state and a single prime on quantum numbers of the upper state.*

Usually, most of the molecules are in the $v = 0$ vibrational level at room temperature, and the strongest band is the $v = 0 \to 1$ band, called the **fundamental band.** The $v = 0 \to 2$ band (the **first overtone**) is much weaker than the fundamental band. From (21.34), the fundamental, first overtone, second overtone, . . . bands occur at $\tilde{\nu}_e - 2\tilde{\nu}_e x_e$, $2\tilde{\nu}_e - 6\tilde{\nu}_e x_e$, $3\tilde{\nu}_e - 12\tilde{\nu}_e x_e$, . . . :

$v'' \to v'$	$0 \to 1$	$0 \to 2$	$0 \to 3$
$\tilde{\nu}_{\text{origin}}$	$\tilde{\nu}_e - 2\tilde{\nu}_e x_e$	$2\tilde{\nu}_e - 6\tilde{\nu}_e x_e$	$3\tilde{\nu}_e - 12\tilde{\nu}_e x_e$

Because the anharmonicity term $\nu_e x_e$ is much less than ν_e, the frequencies of the fundamental and overtone bands are *approximately* ν_e, $2\nu_e$, $3\nu_e$, etc. The wavenumber of the band origin of the fundamental band is called $\tilde{\nu}_0$:

$$\tilde{\nu}_0 = \tilde{\nu}_e - 2\tilde{\nu}_e x_e \quad \text{and} \quad \nu_0 = \nu_e - 2\nu_e x_e \qquad (21.36)$$

The symbols ω_e, ω_0, and $\omega_e x_e$ are commonly used for $\tilde{\nu}_e$, $\tilde{\nu}_0$, and $\tilde{\nu}_e x_e$, respectively.

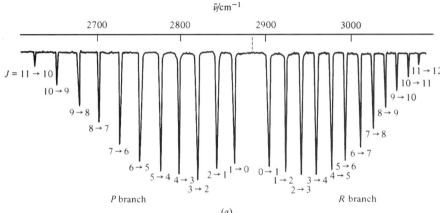

$\tilde{\nu}/\text{cm}^{-1}$

$J = 11 \rightarrow 10$
$10 \rightarrow 9$
$9 \rightarrow 8$
$8 \rightarrow 7$
$7 \rightarrow 6$
$6 \rightarrow 5$
$5 \rightarrow 4$ $4 \rightarrow 3$ $2 \rightarrow 1$ $1 \rightarrow 0$ $0 \rightarrow 1$
$3 \rightarrow 2$ $1 \rightarrow 2$ $3 \rightarrow 4$ $4 \rightarrow 5$
$2 \rightarrow 3$
$5 \rightarrow 6$
$6 \rightarrow 7$
$7 \rightarrow 8$
$8 \rightarrow 9$
$9 \rightarrow 10$
$10 \rightarrow 11$
$11 \rightarrow 12$

P branch

R branch

(*a*)

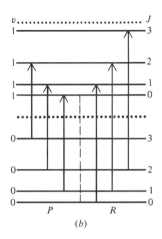

(*b*)

Figure 21.14

(*a*) The gas-phase low-pressure $v = 0 \rightarrow 1$ vibration band of $^1\text{H}^{35}\text{Cl}$ at room temperature. The dashed line indicates the position of the band origin ($\tilde{\nu}_{\text{origin}} = 2886$ cm^{-1}). (In a sample of naturally occurring HCl, each line is a doublet because of the presence of $^1\text{H}^{37}\text{Cl}$.) It is conventional to display infrared absorption spectra with absorption intensity increasing downward. (*b*) The first few *P*-branch and *R*-branch transitions on an energy-level diagram. Each arrow in (*b*) is positioned directly below the spectrum line it corresponds to in (*a*).

Some band origins of $^1\text{H}^{35}\text{Cl}$ IR bands are:

$v'' \rightarrow v'$	$0 \rightarrow 1$	$0 \rightarrow 2$	$0 \rightarrow 3$	$0 \rightarrow 4$	$0 \rightarrow 5$
$\tilde{\nu}_{\text{origin}}/\text{cm}^{-1}$	2886.0	5668.0	8346.8	10922.8	13396.2

From these data, $\tilde{\nu}_e$ and $\tilde{\nu}_e x_e$ can be found (Prob. 21.31).

If the vibrational frequency of the molecule is relatively low, or if the gas is heated, one observes absorption transitions originating from states with $v'' > 0$. These are called *hot bands*.

Having calculated the locations of IR band origins of diatomic molecules, we now deal with the rotational structure of each band. From the selection rule $\Delta J = \pm 1$ [Eq. (21.28)], each line in an IR band involves either $\Delta J = +1$ or $\Delta J = -1$. Vibration–rotation transitions with $\Delta J = +1$ give the **R branch** lines of the band; transitions with $\Delta J = -1$ give the **P branch**. Let J'' be the rotational quantum number of the lower vibration–rotation level and J' the rotational quantum number of the upper level (Fig. 21.15). The wavenumbers of the vibration–rotation lines are $\tilde{\nu} = (E_{J'v'} - E_{J''v''})/hc$. If we use the approximate energy expression (21.27) that neglects vibration–rotation interaction, then

$$\tilde{\nu} \approx \tilde{\nu}_e(v' - v'') - \tilde{\nu}_e x_e[v'(v' + 1) - v''(v'' + 1)] + \tilde{B}_e J'(J' + 1) - \tilde{B}_e J''(J'' + 1)$$

$$\tilde{\nu} \approx \tilde{\nu}_{\text{origin}} + \tilde{B}_e J'(J' + 1) - \tilde{B}_e J''(J'' + 1)$$

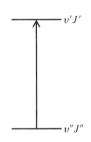

$v'J'$

$v''J''$

Figure 21.15

Energy levels of a vibration–rotation transition.

where (21.34) for $\tilde{\nu}_{\text{origin}}$ was used. For R branch lines, we have $J' = J'' + 1$. The use of this relation in the $\tilde{\nu}$ expression gives $\tilde{\nu}_R \approx \tilde{\nu}_{\text{origin}} + 2\tilde{B}_e(J'' + 1)$. Similarly, one finds (Prob. 21.28) $\tilde{\nu}_P \approx \tilde{\nu}_{\text{origin}} - 2\tilde{B}_e J''$. Thus

$$\tilde{\nu}_R \approx \tilde{\nu}_{\text{origin}} + 2\tilde{B}_e(J'' + 1) \qquad \text{where } J'' = 0, 1, 2, \ldots \qquad (21.37)$$

$$\tilde{\nu}_P \approx \tilde{\nu}_{\text{origin}} - 2\tilde{B}_e J'' \qquad \text{where } J'' = 1, 2, 3, \ldots \qquad (21.38)$$

$J'' = 0$ is excluded for the P branch lines because J' cannot be -1. The approximate equations (21.37) and (21.38) predict equally spaced lines on either side of $\tilde{\nu}_{\text{origin}}$ with no line at $\tilde{\nu}_{\text{origin}}$, since $J = 0 \rightarrow 0$ is forbidden for diatomic molecules.

The accurate expressions for $\tilde{\nu}_R$ and $\tilde{\nu}_P$ found using (21.26) (which includes vibration–rotation interaction) for E_{int} are (Prob. 21.29)

$$\tilde{\nu}_R = \tilde{\nu}_{\text{origin}} + [2\tilde{B}_e - \tilde{\alpha}_e(v' + v'' + 1)](J'' + 1) - \tilde{\alpha}_e(v' - v'')(J'' + 1)^2$$

$$(21.39)$$

$$\tilde{\nu}_P = \tilde{\nu}_{\text{origin}} - [2\tilde{B}_e - \tilde{\alpha}_e(v' + v'' + 1)]J'' - \tilde{\alpha}_e(v' - v'')J''^2 \qquad (21.40)$$

where $\tilde{\nu}_{\text{origin}}$ is given by (21.34) and the J'' values are given by (21.37) and (21.38). The vibration–rotation interaction constant α_e makes the R branch line spacings decrease as J'' increases and the P branch spacings increase as J'' increases.

Figure 21.14 shows the $v = 0 \rightarrow 1$ vibration–rotation band of $^1H^{35}Cl$.

Note that the line intensities in each branch in Fig. 21.14 first increase as J increases and then decrease at high J. The explanation for this is as follows. The Boltzmann distribution law $N_i/N_j = e^{-(E_i - E_j)/kT}$ gives the populations of the quantum-mechanical *states* i and j and not the populations of the quantum-mechanical energy levels. For each rotational energy level, there are $2J + 1$ rotational states, corresponding to the $2J + 1$ values of the M_J quantum number (Sec. 18.14), which does not affect the energy. Thus, the population of each rotational energy level J is $2J + 1$ times the population of the quantum-mechanical rotational state with quantum numbers J and M_J. For low J, this $2J + 1$ degeneracy factor outweighs the decreasing exponential in the Boltzmann distribution law, so the populations of rotational levels at first increase as J increases. For high J, the exponential dominates the $2J + 1$ factor, and the rotational populations decrease as J increases (Fig. 21.16).

Measurement of the wavenumbers of the lines of a vibration–rotation band allows $\tilde{\nu}_{\text{origin}}$, $\tilde{B}_e$, and $\tilde{\alpha}_e$ to be found from Eqs. (21.39) and (21.40). Knowing $\tilde{\nu}_{\text{origin}}$ for at least two bands, we can use (21.34) to find $\tilde{\nu}_e$ and $\tilde{\nu}_e x_e$. The rotational constant $\tilde{B}_e$ allows the moment of inertia and hence the internuclear distance R_e to be found. The vibrational frequency ν_e gives the force constant of the bond [Eq. (21.23)].

Vibrational and rotational constants for some diatomic molecules are listed in Table 21.1. Homonuclear diatomics have no pure-rotational (microwave) or vibration–rotation (IR) spectra, and their constants are found from electronic spectra (Sec. 21.11) or Raman spectra (Sec. 21.10). The values listed are for the ground electronic states, except that the CO* values are for one of the lower excited states of CO. Note that $D_e/hc > \tilde{\nu}_e \gg \tilde{B}_e$, in accord with the earlier conclusion that electronic energies are greater than vibrational energies, which are greater than rotational energies. Note also the large force constants for N_2 and CO, which have triple bonds. Other things being equal, *the stronger the bond, the larger the force constant and the shorter the bond length.*

The centrifugal-distortion constant D is of significant magnitude only for light molecules. Some values of $\tilde{D} \equiv D/c$ for ground electronic states are 0.05 cm^{-1} for $^1H_2$, 0.002 cm^{-1} for $^1H^{19}F$, 6×10^{-6} cm^{-1} for $^{14}N_2$, and 5×10^{-9} cm^{-1} for $^{127}I_2$.

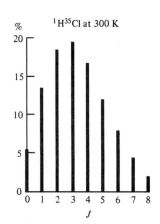

% $^1H^{35}Cl$ at 300 K

20

15

10

5

0
   0 1 2 3 4 5 6 7 8
                 J

Figure 21.16

Percentage populations of $^1H^{35}Cl$ $v = 0$ rotational energy levels at 300 K.

TABLE 21.1

Constants of Some Diatomic Molecules[a]

Molecule	D_e/hc cm^{-1}	R_e Å	k_e N/m	$\tilde{\nu}_e$ cm^{-1}	$\tilde{B}_e$ cm^{-1}	$\tilde{\alpha}_e$ cm^{-1}	$\tilde{\nu}_e x_e$ cm^{-1}
$^1\text{H}_2$	38297	0.741	576	4403.2	60.85	3.06	121.3
$^1\text{H}^{35}\text{Cl}$	37240	1.275	516	2990.9	10.593	0.31	52.8
$^{14}\text{N}_2$	79890	1.098	2295	2358.6	1.998	0.017	14.3
$^{12}\text{C}^{16}\text{O}$	90544	1.128	1902	2169.8	1.931	0.018	13.3
$^{12}\text{C}^{16}\text{O}*$	29424	1.370	555	1171.9	1.311	0.018	10.6
$^{127}\text{I}_2$	12550	2.666	172	214.5	0.0374	0.0001	0.6
$^{23}\text{Na}^{35}\text{Cl}$	34300	2.361	109	366	0.2181	0.0016	2.0
$^{12}\text{C}^1\text{H}$	29400	1.120	448	2858.5	14.457	0.53	63.0

[a]Data mainly from K. P. Huber and G. Herzberg, *Molecular Spectra and Molecular Structure,* vol. IV, *Constants of Diatomic Molecules,* Van Nostrand Reinhold, 1979.

21.5 MOLECULAR SYMMETRY

In Chapter 20, we used the symmetry of AOs to decide which AOs contribute to a given MO. We now discuss molecular symmetry elements as preparation for studying the rotational motion of polyatomic molecules. (The full application of symmetry to quantum chemistry requires group theory; see Sec. 21.17.)

Symmetry Elements

A molecule has an *n*-**fold axis of symmetry,** symbolized by C_n, if rotation by $360°/n$ (1/*n*th of a complete rotation) about this axis results in a nuclear configuration indistinguishable from the original one. For example, the bisector of the bond angle in HOH is a C_2 axis, since rotation by $360°/2 = 180°$ about this axis merely interchanges the two equivalent H atoms (Fig. 21.17). The hydrogens in the figure have been labeled H_a and H_b, but in reality they are physically indistinguishable from each other.

The NH_3 molecule has a C_3 axis passing through the N nucleus and the midpoint of the triangle formed by the three H nuclei (Fig. 21.17). Rotation by $360°/3 = 120°$ about this axis sends equivalent hydrogens into one another. The hexagonal molecule benzene (C_6H_6) has a C_6 axis perpendicular to the molecular plane and passing through the center of the molecule. Rotation by $360°/6$ about this axis sends equivalent nuclei into one another. This C_6 axis is also a C_3 axis and a C_2 axis, since $120°$ and $180°$ rotations about it send equivalent atoms into one another. Benzene has six other C_2 axes, each lying in the molecular plane; three of these go through two diagonally opposite carbons, and three bisect opposite pairs of carbon–carbon bonds.

A molecule has a **plane of symmetry,** symbolized by σ, if reflection of all nuclei through this plane sends the molecule into a configuration physically indistinguishable from the original one. Any planar molecule (for example, H_2O) has a plane of symmetry coinciding with the molecular plane, since reflection in the molecular plane leaves all nuclei unchanged in position. H_2O also has a second plane of symmetry; this lies perpendicular to the molecular plane (Fig. 21.18). Reflection through this second plane interchanges the equivalent hydrogens.

NH_3 has three planes of symmetry. Each symmetry plane passes through the nitrogen and one hydrogen and bisects the angle formed by the nitrogen and the other two hydrogens. Reflection in one of these planes leaves N and one H unchanged in position and interchanges the other two H's.

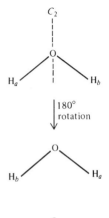

Figure 21.17

Symmetry axes (the dashed lines) in H_2O and NH_3.

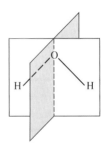

Figure 21.18

Symmetry planes in H_2O.

760

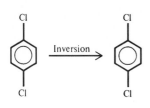

Figure 21.19

Inversion in *p*-dichlorobenzene and in *m*-dichlorobenzene. The coordinate origin is taken at the center of the ring.

A molecule has a **center of symmetry,** symbol *i,* if inversion of each nucleus through this center results in a configuration indistinguishable from the original one. By **inversion** through point P we mean moving a nucleus at *x, y, z* to the location $-x$, $-y, -z$ on the opposite side of P, where the origin is at P. Neither H_2O nor NH_3 has a center of symmetry; *p*-dichlorobenzene has a center of symmetry (at the center of the benzene ring), but *m*-dichlorobenzene does not (Fig. 21.19). A molecule with a center of symmetry cannot have a dipole moment.

A molecule has an *n*-**fold improper axis** (or **rotation–reflection axis**), symbol S_n, if rotation by 360°/*n* about the axis followed by reflection in a plane perpendicular to this axis sends the molecule into a configuration indistinguishable from the original one. Figure 21.20 shows a 90° rotation about an S_4 axis in CH_4 followed by a reflection in a plane perpendicular to this axis. The final configuration has hydrogens at the same locations in space as the original configuration. Note that the 90° rotation alone produces a configuration of the hydrogens that is distinguishable from the original one, so the S_4 axis in methane is not a C_4 axis. Methane has two other S_4 axes. Each S_4 axis goes through the centers of opposite faces of the cube in which the molecule has been inscribed. The C_6 axis in benzene is also an S_6 axis. (An S_2 axis is equivalent to a center of symmetry; see Prob. 21.39.)

Symmetry Operations

Associated with each symmetry element (C_n, S_n, σ, or *i*) of a molecule is a set of **symmetry operations.** For example, consider the C_4 axis of the square-planar molecule XeF_4. We can rotate the molecule by 90°, 180°, 270°, and 360° about this axis to give configurations indistinguishable from the original one (Fig. 21.21). The operation of rotating the molecule by 90° about the C_4 axis is symbolized by $\hat{C}_4$. Note the circumflex. Since a 180° rotation can be viewed as two successive 90° rotations, we symbolize a 180° rotation by $\hat{C}_4^2$, which equals $\hat{C}_4\hat{C}_4$. (Recall that multiplication of two operators means applying the operators in succession.) Similarly, 270° and 360° rotations are denoted by $\hat{C}_4^3$ and $\hat{C}_4^4$. Since a 360° rotation brings every nucleus back to its original location, we write $\hat{C}_4^4 = \hat{E}$, where $\hat{E}$ is the **identity operation,** defined as "do nothing." The operation $\hat{C}_4^5$, which is a 450° rotation about the C_4 axis, is the same as the 90° $\hat{C}_4$ rotation and is not counted as a new symmetry operation.

Since two successive reflections in the same plane bring the nuclei back to their original locations, we have $\hat{\sigma}^2 = \hat{E}$. Likewise, $\hat{i}^2 = \hat{E}$.

Since each symmetry operation must leave the location of the molecular center of mass unchanged, the symmetry elements of a molecule all intersect at the center of mass. Since a symmetry rotation must leave the orientation of the molecular dipole moment unchanged, the dipole moment of a molecule with a C_n axis must lie on this axis. A molecule with two or more noncoincident C_n axes has no dipole moment.

The set of symmetry operations of a molecule forms what mathematicians call a *group.*

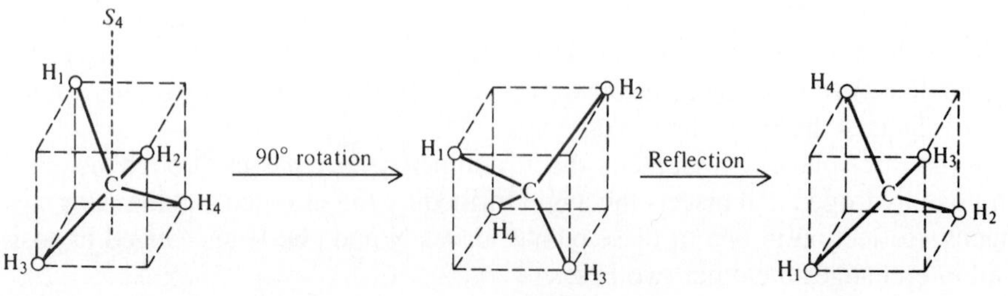

Figure 21.20

An S_4 axis in CH_4. The carbon atom is at the center of the cube.

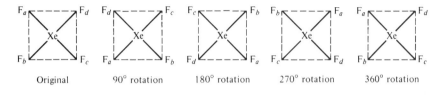

Original   90° rotation   180° rotation   270° rotation   360° rotation

Figure 21.21

Some of the symmetry rotations for XeF_4.

21.6   ROTATION OF POLYATOMIC MOLECULES

As with diatomic molecules, it is usually a good approximation to take the energy of a polyatomic molecule as the sum of translational, rotational, vibrational, and electronic energies. We now consider the rotational energy levels.

Classical Mechanics of Rotation

The classical-mechanical treatment of the rotation of a three-dimensional body is complicated, and this section gives the results of such a treatment without giving proofs.

The **moment of inertia** I_x of a system of mass points m_1, m_2, . . . about an arbitrary axis x is defined by

$$I_x \equiv \sum_i m_i r_{x,i}^2 \qquad (21.41)$$

where $r_{x,i}$ is the perpendicular distance from mass m_i to the x axis. Consider a set of three mutually perpendicular axes x, y, and z. The *products of inertia* I_{xy}, I_{xz}, and I_{yz} for the x, y, z system are defined by the sums $I_{xy} \equiv -\sum_i m_i x_i y_i$, etc., where x_i and y_i are the x and y coordinates of mass m_i. Any three-dimensional body possesses three mutually perpendicular axes a, b, and c passing through the center of mass and having the property that the products of inertia I_{ab}, I_{ac}, and I_{bc} are each zero for these axes; these three axes are called the **principal axes of inertia** of the body. The moments of inertia I_a, I_b, and I_c calculated with respect to the principal axes are the **principal moments of inertia** of the body.

Symmetry aids in locating the principal axes. One can show that: *A molecular symmetry axis coincides with one of the principal axes. A molecular symmetry plane contains two of the principal axes and is perpendicular to the third.*

EXAMPLE 21.4   Principal axes and principal moments

The Xe—F bond length in the square-planar molecule XeF_4 (Fig. 21.22) is 1.94 Å. Locate the principal axes of inertia, and calculate the principal moments of inertia of XeF_4.

The center of mass is at the Xe nucleus, and the three principal axes of inertia must intersect at this point. One of the principal axes coincides with the C_4 symmetry axis perpendicular to the molecular plane. The other two principal axes are perpendicular to the C_4 axis and lie in the molecular plane. They can be taken to coincide with the two C_2 axes that pass through the four F's, or they can be taken to coincide with the two C_2 axes that bisect the FXeF angles. (For highly symmetric molecules, the orientation of the principal axes may not be unique.) Let us take the two in-plane principal axes (which we label the a and b axes) to go through the F atoms. The perpendicular distances of the atoms from the a principal axis are then 0 for Xe, 0 for two F's, and 1.94 Å for two F's. Therefore Eq. (21.41) with x replaced by a gives

$$I_a = 2(19.0 \text{ amu})(1.94 \text{ Å})^2 = 143 \text{ amu Å}^2 = 2.37 \times 10^{-45} \text{ kg m}^2$$

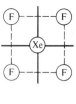

Figure 21.22

Two possible choices of in-plane principal axes in XeF_4.

since 1 amu = $(1 \text{ g/mol})/N_A$ (Sec. 1.4). Clearly, $I_b = I_a$. The c axis is perpendicular to the molecular plane and goes through the Xe atom. Therefore, r_c in (21.41) is 1.94 Å for each F, and $I_c = 4(19.0 \text{ amu})(1.94 \text{ Å})^2 = 286 \text{ amu Å}^2$.

Exercise

The octahedral molecule SF_6 has an S—F bond length of 1.56 Å. Find the principal moments of inertia of SF_6. (*Answers*: 185 amu Å^2, 185 amu Å^2, 185 amu Å^2.)

For molecules without symmetry, finding the principal axes is more complicated and is omitted.

The principal axes are important because the classical-mechanical rotational energy can be simply expressed in terms of the principal moments of inertia. The classical-mechanical rotational energy of a body turns out to be

$$E_{\text{rot}} = P_a^2/2I_a + P_b^2/2I_b + P_c^2/2I_c \qquad (21.42)$$

where P_a, P_b, and P_c are the components of the rotational angular momentum along the a, b, and c principal axes. Note the presence of three terms, each quadratic in a momentum, a fact mentioned in Sec. 15.10 on equipartition of energy.

If all the mass points lie on the same line (as in a linear molecule), this line is one of the principal axes, since it is a symmetry axis. The rotational angular momentum component of the body along this axis is zero, since the distance between all the masses and the axis is zero. Therefore, one of the three terms in (21.42) is zero, and a linear molecule has only two quadratic terms in its classical rotational energy.

The principal axes are labeled so that $I_a \leq I_b \leq I_c$.

A body is classified as a **spherical, symmetric,** or **asymmetric top,** according to whether three, two, or none of the principal moments I_a, I_b, I_c are equal:

Spherical top: $\qquad\qquad I_a = I_b = I_c$

Symmetric top: $\qquad\qquad I_a = I_b \neq I_c \quad$ or $\quad I_a \neq I_b = I_c$

Asymmetric top: $\qquad\qquad I_a \neq I_b \neq I_c$

It can be shown that *a molecule with one C_n or S_n axis with $n \geq 3$ is a symmetric top. A molecule with two or more noncoincident C_n or S_n axes with $n \geq 3$ is a spherical top.* NF_3, with one C_3 axis, is a symmetric top. XeF_4, with one C_4 axis, is a symmetric top. CCl_4, with four noncoincident C_3 axes (one through each C—Cl bond), is a spherical top. H_2O, with no C_3 or higher axis, is an asymmetric top.

Quantum-Mechanical Rotational Energy Levels

[For proofs of the following results, see *Levine* (1975), chap. 5.]

The **rotational constants** A, B, C of a polyatomic molecule are defined by [similar to (21.16)]

$$A \equiv h/8\pi^2 I_a, \quad B \equiv h/8\pi^2 I_b, \quad C \equiv h/8\pi^2 I_c \qquad (21.43)$$

The rotational energy levels of a spherical top are

$$E_{\text{rot}} = J(J + 1)\hbar^2/2I = BhJ(J + 1), \quad J = 0, 1, 2, \ldots \qquad (21.44)$$

where $I \equiv I_a = I_b = I_c$. The spherical-top rotational wave functions involve the quantum numbers J, K, and M, where K and M each range from $-J$ to J in integral steps. Each spherical-top rotational level is $(2J + 1)^2$-fold degenerate, corresponding to the $(2J + 1)^2$ different choices for K and M for a fixed J.

The rotational levels of a symmetric top with $I_b = I_c$ are

$$E_{rot} = \frac{J(J+1)\hbar^2}{2I_b} + K^2\hbar^2\left(\frac{1}{2I_a} - \frac{1}{2I_b}\right) = BhJ(J+1) + (A - B)hK^2$$

(21.45)

$$J = 0, 1, 2, \ldots; \qquad K = -J, -J + 1, \ldots, J - 1, J$$

There is also an M quantum number that does not affect E_{rot}. If $I_a = I_b$, then I_a and A in (21.45) are replaced by I_c and C. The quantity $K\hbar$ is the component of the rotational angular momentum along the molecular symmetry axis.

The axis of a linear molecule is a C_∞ axis, since rotations by $360°/n$, where $n = 2, 3, \ldots, \infty$, about this axis are symmetry operations. Hence, a linear molecule is a symmetric top. (This is also obvious from the fact that the moments of inertia about all axes that pass through the center of mass and are perpendicular to the molecular axis are equal.) Because all the nuclei lie on the C_∞ symmetry axis, there can be no rotational angular momentum along this axis and K must be zero for a linear molecule. Hence, Eq. (21.45) gives

$$E_{rot} = J(J + 1)\hbar^2/2I_b = BhJ(J + 1) \qquad \text{linear molecule}$$

where I_b is the moment of inertia about an axis through the center of mass and perpendicular to the molecular axis. A special case is a diatomic molecule, Eq. (21.15).

The rotational levels for an asymmetric top are extremely complicated and follow no simple pattern.

Selection Rules

The selection rules for the pure-rotational (microwave) spectra of polyatomic molecules are as follows.

Just as for diatomic molecules, *a polyatomic molecule must have a nonzero dipole moment to undergo a pure-rotational transition with absorption or emission of radiation.* Because of their high symmetry, all spherical tops (for example, CCl_4 and SF_6) and some symmetric tops (for example, C_6H_6 and XeF_4) have no dipole moment and exhibit no pure-rotation spectrum.

For a symmetric top with a dipole moment (for example, CH_3F), the pure-rotational selection rules are

$$\Delta J = \pm 1, \quad \Delta K = 0$$

(21.46)

The use of (21.46), (21.45), and $E_{upper} - E_{lower} = h\nu$ gives the frequency of the pure-rotational $J \rightarrow J + 1$ absorption transition as

$$\nu = 2B(J + 1), \quad J = 0, 1, 2, \ldots \qquad \text{symmetric top}$$

(21.47)

The microwave spectrum of a symmetric top consists of a series of equally spaced lines at $2B$, $4B$, $6B$, ... (provided centrifugal distortion is negligible).

The most populated vibrational state has all vibrational quantum numbers equal to zero, and the rotational constant B determined from the microwave spectrum is an average over the zero-point vibrations and is designated B_0. If excited vibrational levels are significantly populated at room temperature, vibrational satellites are observed, as discussed for diatomic molecules.

The selection rules for asymmetric tops are complicated, and are omitted.

21.7 MICROWAVE SPECTROSCOPY

Except for light diatomic molecules, pure-rotational spectra occur in the microwave portion of the electromagnetic spectrum. Figure 21.23 sketches a highly simplified version of a microwave spectrometer. A special electronic tube (either a klystron or a

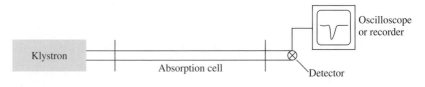

Figure 21.23

A microwave spectrometer.

backward-wave oscillator) generates virtually monochromatic microwave radiation, whose frequency can be readily varied over a wide range. The radiation is transmitted through a hollow metal pipe called a *waveguide*. A portion of waveguide sealed at both ends with mica windows is the absorption cell. The cell is filled with low-pressure (0.01 to 0.1 torr) vapor of the molecule to be investigated. (At medium and high pressures, intermolecular interactions broaden the rotational absorption lines, giving an essentially continuous rotational absorption spectrum.) The microwave radiation is detected with a metal-rod antenna mounted in the waveguide and connected to a semiconductor diode. When the klystron frequency coincides with one of the absorption frequencies of the gas, there is a dip in the microwave power reaching the detector.

Any substance with a dipole moment and with sufficient vapor pressure can be studied. The use of a waveguide heated to 1000 K enables the rotational spectra of the alkali halides to be observed. Very large molecules are hard to study. Such molecules have many low-energy vibrational levels that are significantly populated at room temperature; this produces a microwave spectrum with so many lines that it is extremely hard to figure out which rotational transitions the various lines correspond to. Some large molecules whose microwave spectra have been successfully investigated are azulene, β-fluoronaphthalene, and $C_6H_6Cr(CO)_3$.

The microwave spectra of molecular ions produced continuously by a glow discharge in the absorption cell and of free radicals produced by an electric discharge (or by reaction of products of an electric discharge with stable species) and continuously pumped through the absorption cell have been observed. Species studied include CO^+, HCO^+, CF_2, PH_2, and CH_3O. See E. Hirota, *Chem. Rev.*, **92,** 141 (1992). The microwave spectra of van der Waals molecules (Sec. 22.10) and hydrogen-bonded dimers such as $(H_2O)_2$ have been studied using molecular beams (Sec. 23.3). See *Gordy and Cook*, pp. 153–163; *Hollas,* sec. 4.10.2.

The microwave spectrum of a symmetric top is very simple [Eq. (21.47)] and yields the rotational constant B_0. The microwave spectrum of an asymmetric top is quite complicated, but once one has assigned several lines to transitions between specific rotational levels, one can calculate the three rotational constants A_0, B_0, C_0 and the corresponding principal moments of inertia.

The principal moments of inertia depend on the bond distances and angles and the conformation. Knowledge of one moment of inertia for a symmetric top or three moments of inertia for an asymmetric top is generally not enough information to determine the structure fully. One therefore prepares isotopically substituted species of the molecule and observes their microwave spectra to find their moments of inertia. The molecular geometry is determined by the nuclear configuration that minimizes the energy E_e in the electronic Schrödinger equation $\hat{H}_e \psi_e = E_e \psi_e$. The terms in the electronic Hamiltonian $\hat{H}_e$ are independent of the nuclear masses. Therefore, isotopic substitution does not affect the equilibrium geometry. (This isn't quite true, because of very slight deviations from the Born–Oppenheimer approximation.) When enough isotopically substituted species have been studied, the complete molecular structure can be determined. Some molecular structures found by microwave spectroscopy are:

CH$_2$F$_2$: R(CH) = 1.09 Å, R(CF) = 1.36 Å, ∠HCH = 113.7°, ∠FCF = 108.3°

CH$_3$OH: R(CH) = 1.09 Å, R(CO) = 1.43 Å, R(OH) = 0.94 Å,

   ∠HCH = 108.6°,   ∠COH = 108.5°

C$_6$H$_5$F: average R(CC) = 1.39 Å, R(CH) = 1.08 Å, R(CF) = 1.35 Å

H$_2$S: R(SH) = 1.34 Å, ∠HSH = 92.1°

SO$_2$: R(SO) = 1.43 Å, ∠OSO = 119.3°

O$_3$: R(OO) = 1.27 Å, ∠OOO = 116.8°

Molecular dipole moments can be found from microwave spectra. An insulated metal plate is inserted lengthwise in the waveguide. Application of a voltage to this plate subjects the gas molecules to an electric field, which shifts their rotational energies. (A shift in molecular energy levels due to an applied external electric field is called a *Stark effect.*) The magnitudes of these shifts depend on the components of the molecular electric dipole moment. Microwave spectroscopy gives quite accurate dipole moments. Moreover, it gives the components of the dipole-moment vector along the principal axes of inertia, so the orientation of the dipole-moment vector is known. Some dipole moments in debyes determined by microwave spectroscopy are

C$_3$H$_8$	HC(CH$_3$)$_3$	H$_2$O$_2$	H$_2$O	H$_2$S	azulene	NaCl	KCl	HCl	ClF	CH$_3$D
0.08	0.13	2.2	1.85	0.97	0.80	9.0	10.3	1.12	0.88	0.006

The nonzero dipole moments of the saturated hydrocarbons H$_2$C(CH$_3$)$_2$ and HC(CH$_3$)$_3$ are due to deviations from the $109\frac{1}{2}°$ tetrahedral bond angle and to polarization of the electron probability densities of the CH$_3$, CH$_2$, and CH groups by the unsymmetrical electric fields in the molecules [S. W. Benson and M. Luria, *J. Am. Chem. Soc.,* **97,** 704 (1975)]. The dipole moment of CH$_3$D is due to deviations from the Born–Oppenheimer approximation.

Besides molecular structures and dipole moments, microwave spectroscopy also yields values of barriers to internal rotation (Sec. 20.8) in polar molecules. Although internal rotation is a vibrational motion, it affects the pure-rotational spectrum, which allows barriers to be found. (See *Sugden and Kenney,* chap. 8.)

Many molecular species in interstellar space (in interstellar clouds of gas and dust or in circumstellar shells) have been detected mainly by observation of microwave emission lines using a radio telescope. Over 120 species have been found including OH, H$_3$O$^+$, H$_2$O, NH$_3$, SO$_2$, CH$_3$OH, CH$_3$CHO, C$_2$H$_5$OH, HCOOH, NH$_2$CHO, HC$_9$N, and NaCl; some evidence suggests the presence of the amino acid glycine, but this is not conclusive. See *Hollas,* sec. 4.7; D. Smith, *Chem. Rev.,* **92,** 1473 (1992); F. J. Lovas, *J. Phys. Chem. Ref. Data,* **21,** 181 (1992); www.cv.nrao.edu/~awootten/allmols.html.

Observed intensities of the microwave emission lines from ClO and O$_3$ in the Antarctic stratosphere show a strong correlation between increasing ClO and decreasing ozone, indicating that Antarctic ozone depletion is due to chlorofluorocarbons.

An impressive example of the interaction between theory and experiment is the study of the gas-phase microwave spectrum of the simplest amino acid, glycine, NH$_2$CH$_2$COOH. In 1978, the microwave spectrum of the glycine conformation labeled II in Fig. 21.24 was observed and rotational transitions were assigned, allowing its rotational constants to be found. However, ab initio SCF calculations predicted that conformation I in Fig. 21.24 would be 1 or 2 kcal/mol lower in energy than the observed conformation II. This prompted a further search of the microwave spectrum.

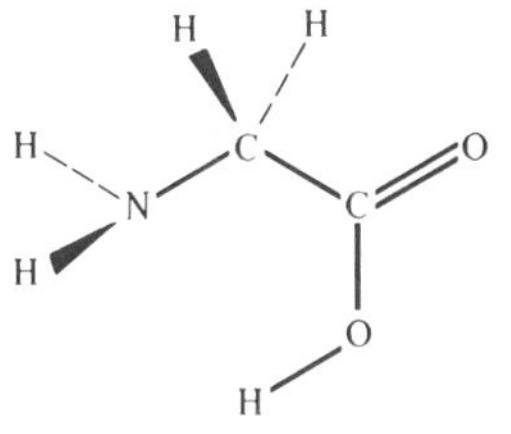

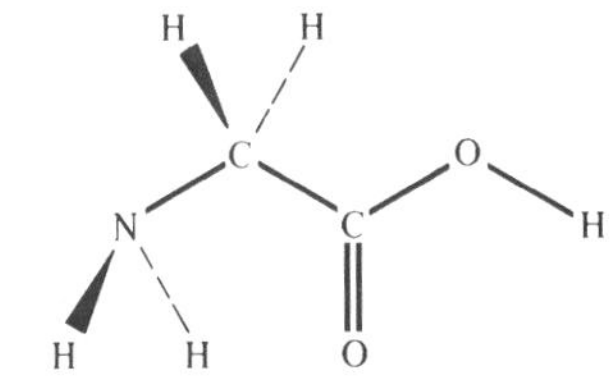

Figure 21.24

Two conformations of glycine. Conformer I has three intramolecular hydrogen bonds; II has one such bond.

Guided by the theoretically predicted rotational constants for conformation I, the experimentalists found transitions due to I and obtained its rotational constants, which are in excellent agreement with the theoretically predicted values. (The lowest-energy conformation I has a weaker microwave spectrum than II does because the dipole moment of I is much smaller.) The observed line intensities indicate that I is $1._4$ kcal/mol more stable than II. [See L. Schäfer et al., *J. Am. Chem. Soc.,* **102,** 6566 (1980); R. D. Suenram and F. J. Lovas, *ibid.,* **102,** 7180 (1980); D. T. Nguyen et al., *J. Comp. Chem.,* **18,** 1609 (1997).]

More than 1000 molecules have been studied by microwave spectroscopy.

21.8　VIBRATION OF POLYATOMIC MOLECULES

The nuclear wave function ψ_N of a molecule containing $\mathcal{N}$ atoms is a function of the $3\mathcal{N}$ spatial coordinates needed to specify the locations of the nuclei. The nuclear motions are translations, rotations, and vibrations. The translational wave function is a function of three coordinates, the x, y, and z coordinates of the center of mass. The rotational wave function of a linear molecule is a function of the two angles θ and ϕ (Fig. 21.7) needed to specify the spatial orientation of the molecular axis. To specify the orientation of a nonlinear molecule, one chooses some axis in the molecule, gives θ and ϕ for this axis, and gives the angle of rotation of the molecule itself about this axis. Thus, ψ_{rot} depends on three angles for a nonlinear molecule. The nuclear vibrational wave function therefore depends on $3\mathcal{N} - 3 - 2 = 3\mathcal{N} - 5$ coordinates for a linear molecule and $3\mathcal{N} - 3 - 3 = 3\mathcal{N} - 6$ coordinates for a nonlinear molecule. The number of independent coordinates needed to specify each kind of motion (translation, rotation, vibration) is called the number of **degrees of freedom** for that kind of motion (Fig. 21.25). Each such coordinate specifies a mode of motion, so the number of degrees of freedom is the number of modes of motion.

We first discuss the classical-mechanical treatment of molecular vibration and then the quantum-mechanical treatment.

DEGREES OF FREEDOM

	Lin.	Nonlin.
Trans.	3	3
Rot.	2	3
Vib.	$3\mathcal{N} - 5$	$3\mathcal{N} - 6$
Total	$3\mathcal{N}$	$3\mathcal{N}$

Figure 21.25

Degrees of freedom of a molecule with $\mathcal{N}$ atoms.

Classical Mechanics of Vibration

Consider a molecule with $\mathcal{N}$ nuclei vibrating about their equilibrium positions subject to the potential energy E_e, where E_e is a function of the nuclear coordinates and is found by solving the electronic Schrödinger equation (20.7). For small vibrations, we can expand E_e in a Taylor series and neglect terms higher than quadratic, as we did for a diatomic molecule [Eq. (21.18)]. Substituting this expansion into Newton's second law, one finds [*Levine* (1975), sec. 6.2] that any classical molecular vibration can be expressed as a linear combination of what are called normal modes of vibration.

In a given **normal mode,** all the nuclei vibrate about their equilibrium positions at the same frequency; the nuclei all vibrate in phase, meaning that each nucleus passes through its equilibrium position at the same time. However, the vibrational amplitudes of different nuclei may differ. A molecule has $3\mathcal{N} - 6$ or $3\mathcal{N} - 5$ normal modes, depending on whether it is nonlinear or linear, respectively.

For a diatomic molecule, $3\mathcal{N} - 5 = 1$. In the single normal mode, the two atoms vibrate along the internuclear axis. For a heteronuclear diatomic molecule, the vibrational amplitude of the heavier atom is less than that of the lighter atom.

Each normal mode has its own vibrational frequency. The forms and frequencies of the normal modes depend on the molecular geometry, the nuclear masses, and the force constants (the second derivatives of E_e with respect to the nuclear coordinates). If these quantities are known, one can find the normal modes and their frequencies.

(For each normal mode, the vibrational Hamiltonian has a kinetic-energy term that is quadratic in a momentum and a potential-energy term that is quadratic in a coordinate. This fact was used in the Sec. 15.10 discussion of equipartition.)

Figure 21.26

Normal vibrational modes of CO_2. The plus and minus signs indicate motion out of and into the plane of the paper.

Figure 21.26 shows the normal modes of the linear molecule CO_2, which has 3(3) − 5 = 4 normal modes. In the *symmetric stretching* vibration labeled ν_1, the carbon nucleus is motionless. In the *asymmetric stretch* ν_3, all the nuclei vibrate. The *bending* mode ν_{2b} is the same as ν_{2a} rotated by 90° about the molecular axis. Clearly, these two modes have the same frequency. Bond-bending frequencies are generally lower than bond-stretching frequencies. For CO_2, observation of IR and Raman spectra gives $\tilde{\nu}_1$ = 1340 cm^{-1}, $\tilde{\nu}_2$ = 667 cm^{-1}, $\tilde{\nu}_3$ = 2349 cm^{-1}, where $\tilde{\nu} = \nu/c$.

Figure 21.27 shows the three normal modes of the nonlinear molecule H_2O. Here, ν_1 and ν_3 are stretching vibrations, and ν_2 is a bending vibration.

In each of the diagrams of Figs. 21.26 and 21.27, the arrows show the directions of motion of the atoms at one moment during a normal mode of vibration. Since the atoms oscillate about their equilibrium positions, the directions of atomic motions change twice during each cycle of vibration. The relative magnitudes of the arrows indicate the relative amplitudes of the atomic vibrations. The amplitude of the heavy O atom is much less than that of an H atom.

Quantum Mechanics of Vibration

Neglecting terms higher than quadratic in the E_e expansion and expressing the vibrations in terms of normal modes, one finds [*Levine* (1975), sec. 6.4] that the Hamiltonian for vibration becomes the sum of harmonic-oscillator Hamiltonians, one for each normal mode. Hence, *the quantum-mechanical vibrational energy of a polyatomic molecule is approximately the sum of* $3\mathcal{N} - 6$ *or* $3\mathcal{N} - 5$ *harmonic-oscillator energies, one for each normal mode:*

$$E_{\text{vib}} \approx \sum_{i=1}^{3\mathcal{N}-6} (v_i + \tfrac{1}{2})h\nu_i \qquad (21.48)^*$$

$$v_1 = 0, 1, 2, \ldots, \quad v_2 = 0, 1, 2, \ldots, \quad \cdots \quad v_{3\mathcal{N}-6} = 0, 1, 2, \ldots$$

where ν_i is the frequency of the ith normal mode and v_i is its quantum number. For linear molecules, the upper limit is $3\mathcal{N} - 5$. The $3\mathcal{N} - 6$ or $3\mathcal{N} - 5$ vibrational quantum numbers vary independently of one another. The ground vibrational level has all v_i's equal to zero and has the zero-point energy $\frac{1}{2}\sum_i h\nu_i$ (anharmonicity neglected).

One finds that the most probable vibration–rotation absorption transitions are those for which one vibrational quantum number v_j changes by +1 with all others (v_k, $k \neq j$) unchanged. However, for the transition $v_j \to v_j + 1$ to occur, *the jth normal-mode vibration must change the molecular dipole moment*. Since the harmonic-oscillator levels are spaced by $h\nu_j$, the transition $v_j \to v_j + 1$ produces a band in the IR spectrum at the frequency ν_j [$\nu_{\text{light}} = (E_{\text{upper}} - E_{\text{lower}})/h = h\nu_j/h = \nu_j$].

Figure 21.27

The normal modes of H_2O. The heavy oxygen atom has a much smaller vibrational amplitude than the light hydrogen atoms.

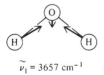

$\tilde{\nu}_1$ = 3657 cm^{-1}

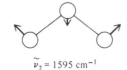

$\tilde{\nu}_2$ = 1595 cm^{-1}

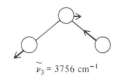

$\tilde{\nu}_3$ = 3756 cm^{-1}

For CO_2, the symmetric stretch ν_1 leaves the dipole moment unchanged, and no IR band is observed at ν_1. One says this vibration is **IR-inactive.** The CO_2 vibrations ν_2 and ν_3 each change the dipole moment and are **IR-active.**

In a molecule with no symmetry elements, all the normal modes change the dipole moment and all are IR-active. The only molecules with no IR-active modes are homonuclear diatomics.

The most populated vibrational level has all the v_i's equal to 0. A transition from this level to a level with $v_j = 1$ and all other vibrational quantum numbers equal to zero gives an **IR fundamental band.** Transitions where one v_j changes by 2 or more and all others are unchanged give **overtone bands.** Transitions where two vibrational quantum numbers change give **combination bands.**

> The selection rules determining the allowed overtone and combination bands are not so readily found as for the fundamental bands. For example, for the IR-active vibration ν_3 of CO_2, the first, third, fifth, . . . ($v_3' = 2, 4, 6 \ldots$) overtones are IR-inactive, and the second, fourth, sixth, . . . ($v_3' = 3, 5, 7 \ldots$) overtones are IR-active. An overtone of a forbidden fundamental may be allowed in certain cases. A combination band might involve a change in the quantum number of an IR-inactive vibration. The best way to determine which overtone and combination bands are allowed is to use group theory. (See *Herzberg,* vol. II, chap. III for the procedure.)

Overtone and combination bands are substantially weaker than fundamental bands.

The frequencies of the IR fundamental bands give (some of) the **fundamental vibration frequencies** of the molecule. Because of anharmonicity, the fundamental frequencies differ somewhat from the equilibrium vibrational frequencies [see Eq. (21.36)]. Usually, not enough data are available to determine the anharmonicity corrections, and so one works with the fundamental frequencies.

For H_2O, all three normal modes are IR-active. Wavenumbers of some IR band origins for H_2O vapor follow. Also listed are the band intensities (s = strong, m = medium, w = weak) and the quantum numbers $v_1'\, v_2'\, v_3'$ of the upper vibrational level. In all cases, the lower level is the 000 ground state.

$\tilde{\nu}_{\text{origin}}/\text{cm}^{-1}$	1595(s)	3152(m)	3657(s)	3756(s)	5331(m)	6872(w)
$v_1'\, v_2'\, v_3'$	010	020	100	001	011	021

The three strongest bands are the fundamentals and give the fundamental vibration wavenumbers as $\tilde{\nu}_1 = 3657 \text{ cm}^{-1}$, $\tilde{\nu}_2 = 1595 \text{ cm}^{-1}$, and $\tilde{\nu}_3 = 3756 \text{ cm}^{-1}$. Because of anharmonicity, the overtone transition $000 \rightarrow 020$ occurs at a bit less than twice the frequency of the $000 \rightarrow 010$ fundamental band. The combination band at 6872 cm^{-1} has $\tilde{\nu}_{\text{origin}} \approx 2\tilde{\nu}_2 + \tilde{\nu}_3 = 6946 \text{ cm}^{-1}$.

Molecules with more than, say, five atoms generally have one or more vibrational modes with frequencies low enough to have excited vibrational levels significantly populated at room temperature. This gives hot bands in the IR absorption spectrum. Figure 21.28 plots the fraction of molecules whose vibrational quantum number v_i is 0 versus the vibrational wavenumber $\tilde{\nu}_i$ for two temperatures. At room temperature, only vibrational modes with $\tilde{\nu}_i$ less than 700 cm^{-1} have excited vibrational levels significantly populated.

As with diatomic molecules, gas-phase IR bands of polyatomic molecules show a rotational structure. For certain vibrational transitions, the $\Delta J = 0$ transition is allowed, giving a line (called the Q **branch**) at the band origin.

The many modes of vibration of large molecules make it hard to correctly assign observed IR bands to various transitions. Rather accurate quantum-mechanical calculation of molecular vibrational frequencies is possible for molecules not too large; "it is virtually impossible to interpret and correctly assign the vibrational spectra of large

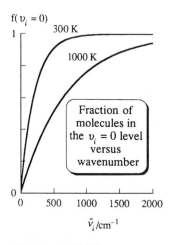

Figure 21.28

The fraction of molecules with vibrational quantum number $v_i = 0$ plotted versus vibrational wavenumber $\tilde{\nu}_i$ at two temperatures.

polyatomic molecules without quantum-mechanical calculations" (P. Pulay in D. R. Yarkony, ed., *Modern Electronic Structure Theory*, World Scientific, 1995, Part II, chap. 19).

The absorption of IR radiation by water vapor and CO_2 has a major effect on the earth's temperature. The sun's surface is at 6000 K, and most of the energy of its radiation (which is approximately that of a blackbody) lies in the visible region and the near-UV and near-IR regions bordering the visible. The earth's atmospheric gases do not have major absorptions in these regions, so most of the sun's radiation reaches the earth's surface. (Recall, however, the UV absorption by stratospheric O_3; Sec. 17.16.) The earth's surface is at 300 K, and most of the energy it radiates is in the mid-infrared. H_2O vapor and CO_2 in the atmosphere absorb a significant fraction of the IR radiation emitted by the earth and reradiate part of it back to the earth, making the earth warmer than if these gases were absent (the "greenhouse effect"). A *greenhouse gas* is one that absorbs in the IR region from 5 to 50 μm, where most of the earth's radiation is emitted. As noted earlier in this section, the most abundant atmospheric gases N_2, O_2, and Ar do not absorb IR radiation. H_2O and many minor atmospheric gases such as CO_2, CH_4, O_3, SF_6, and the chlorofluorocarbons (Sec. 17.16) are greenhouse gases.

The atmosphere's CO_2 content is steadily increasing because of the burning of fossil fuels. Computer models of the atmosphere indicate that if present trends continue, the 1990 to 2100 increase in the earth's average surface temperature will be in the range 1.4°C to 5.8°C (2.5°F to 10.4°F) [Third Assessment Report of the Intergovernmental Panel on Climate Change (IPCC) "Climate Change 2001: The Scientific Basis"; summary available at www.ipcc.ch/]. The most likely sea-level rise by 2100 is between 46 and 58 cm, due to melting of ice sheets and glaciers and to thermal expansion of the oceans (T. M. L. Wigley, "The Science of Climate Change," Pew Center on Global Climate Change, 1999; available at www.pewclimate.org/projects/). For more on global warming, see *Chem. Eng. News,* Aug. 9, 1999, pp. 16–23; www.ncdc.noaa.gov/ol/climate/globalwarming.html.

21.9 INFRARED SPECTROSCOPY

Figure 21.29 outlines a double-beam dispersion IR spectrometer. The radiation source is an electrically heated rod that emits continuous-frequency radiation. If the sample is in solution, the reference cell is filled with pure solvent; otherwise, it is empty. The chopper causes the sample beam and the reference beam to fall alternately on the prism. The prism disperses the radiation into its component frequencies. (The use of a diffraction grating instead of a prism gives higher resolution.) The mirror rotates, thereby changing the frequency of the radiation falling on the detector. The motor that drives this mirror also drives the chart paper on which the spectrum is recorded as a function of frequency. If the sample does not absorb at a given frequency, the sample and reference beams falling on the detector are equally intense and the ac amplifier

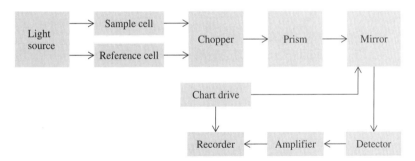

Figure 21.29

A double-beam infrared spectrometer.

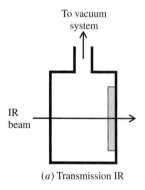

(*a*) Transmission IR

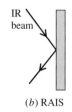

(*b*) RAIS

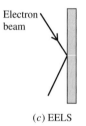

(*c*) EELS

Figure 21.30

Methods for studying the vibrations of a chemisorbed species.

receives no ac signal. At a frequency for which the sample absorbs, the light intensity reaching the detector varies at a frequency equal to the chopper frequency; the detector then puts out an ac signal whose strength depends on the intensity of absorption. IR detectors used include thermocouples, temperature-dependent resistors, and photoconductive materials. Dispersion IR spectrometers have been largely replaced by Fourier-transform IR spectrometers, which are discussed at the end of Sec. 21.9.

Applications of IR Spectroscopy

The sample may be gaseous, liquid, or solid. Solids are ground up with a hydrocarbon or fluorocarbon oil to form a paste, or they are ground up with KBr and squeezed in a die to form a transparent disk. Samples of biopolymers and synthetic polymers are often studied as very thin films formed by evaporating to dryness a solution of the polymer on the absorption-cell window. IR spectra of solids at high pressures can be obtained by using a diamond-anvil cell (Sec. 7.4).

One can study the IR spectrum of a species chemisorbed on a metal catalyst, by compressing tiny metal crystals supported on an inert oxide such as SiO_2 to form a thin disk. The disk is mounted inside an IR absorption cell, the cell is evacuated, and the adsorbate gas is admitted to the cell (Fig. 21.30*a*). The spectrum of the clean metal and support is subtracted from that of the metal, support, and chemisorbed gas. Instead of transmitting the IR radiation through the sample, one can get the IR spectrum of a chemisorbed species by reflecting the radiation off a single metal crystal containing the adsorbate (Fig. 21.30*b*); this is *reflection–absorption infrared spectroscopy* (RAIS or RAIRS; also called IRAS). Another way to obtain vibrational frequencies of chemisorbed species is by *electron-energy-loss spectroscopy* (EELS). Here, a monoenergetic beam of electrons is reflected off the metal-crystal surface (Fig. 21.30*c*), and the energies of the reflected electrons are measured. The losses in electron energy give the vibrational energy changes in the adsorbate, from which the adsorbate vibrational frequencies can be found.

The IR spectra of many short-lived free radicals and molecular ions have been studied. One technique (*matrix isolation*) is to cool a mixture of Ar and a relatively small amount of a stable species to a few kelvins. Photolysis (decomposition by light) of the frozen solid then produces long-lived radicals trapped in an inert solid matrix. Alternatively, the gas-phase mixture can be photolyzed at room temperature and then condensed on a very cold surface. Gas-phase spectra of transient species can be studied by producing the species in the IR absorption cell using an electric discharge or laser photolysis or by flowing the species (produced by a discharge or chemical reaction) through the absorption cell. (Recall the IR determination of the structure of the vinyl cation; Sec. 20.9.) By monitoring the IR absorption as a function of time, the reaction rates of radicals can be studied. For tabulations of species studied, see E. Hirota, *Chem. Rev.,* **92,** 141 (1992); P. F. Bernath, *Ann. Rev. Phys. Chem.,* **41,** 91 (1990).

Gas-phase IR bands under high resolution consist of closely spaced lines due to changes in the rotational quantum numbers. In liquids and solids, rotational structure is not observed, and each band appears as a broad absorption. In solids, the molecules do not rotate. In liquids (and in high-pressure gases), the high rate of intermolecular collisions shortens the lifetimes of vibrational–rotational states by inducing transitions. This lifetime shortening broadens the rotational fine-structure lines, since one form of the uncertainty principle is $\tau \, \Delta E \gtrsim h$, where ΔE is the uncertainty in the energy of a state whose lifetime is τ. The lines are also broadened due to the shifting of the rotational energy levels by the electric fields of intermolecular interactions. The net result is that the rotational lines are merged into a single broad absorption in liquids.

Use of a tunable laser as the IR source can dramatically improve the **resolution** (this is the minimum difference in wavenumber for which two nearby lines can be distinguished as separate lines) and the accuracy of line-frequency measurements. Use of a laser-source IR spectrometer with a resolution of 0.0005 cm^{-1} allowed the rotational structure of vibrational bands of gas-phase cubane, C_8H_8, to be studied [A. S. Pine et al., *J. Am. Chem. Soc.,* **106,** 891 (1984)].

The spacings between the rotational lines of an IR band depend on the moments of inertia, so analysis of the rotational structure of vibration–rotation bands enables molecular structures to be determined. Usually, isotopic species must also be studied to give enough information for a structural determination. Structures determined from IR spectra are not as accurate as those found from microwave spectra, but IR spectroscopy has the advantage of allowing structures of nonpolar molecules to be determined; nonpolar molecules (other than diatomics) have some IR-active vibrations.

Some nonpolar-molecule structures determined from IR spectra are

C_2H_6: $R(CH) = 1.10$ Å, $R(CC) = 1.54$ Å, $\angle HCC = 110°$

C_2H_4: $R(CH) = 1.09$ Å, $R(CC) = 1.34$ Å, $\angle HCC = 121°$

C_2H_2: $R(CH) = 1.06$ Å, $R(CC) = 1.20$ Å, $\angle HCC = 180°$

Analysis of the IR spectra of gas-phase H_3O^+ and D_3O^+ ions gave the structure of this pyramidal species as $R(OH) = 0.98$ Å and $\angle HOH = 111°$ [T. J. Sears et al., *J. Chem. Phys.,* **83,** 2676 (1985)].

Even if rotational fine structure is not resolved, information about the symmetry of the molecular structure can be obtained from observation of the IR spectrum and the Raman spectrum (Sec. 21.10). For example, the planar molecule oxalyl fluoride, FC(O)C(O)F, has $3(6) - 6 = 12$ normal modes of vibration. Group theory shows that the cis conformation will have 10 IR-active fundamentals and 12 Raman-active fundamentals, whereas the trans conformation will have 6 IR-active fundamentals and 6 Raman-active fundamentals. (The trans conformation has a center of symmetry, and group theory shows that in a molecule with a center of symmetry, no normal mode can be both IR-active and Raman-active.) Observation of IR and Raman spectra of oxalyl fluoride showed that in the solid only the trans conformer is present, but the liquid and gas contain a mixture of the two conformers [J. R. Durig et al., *J. Chem. Phys.,* **54,** 4428 (1971)].

For most molecules larger than benzene, the large number of vibrational modes, the presence in the spectrum of many hot bands, overtones, and combination bands, and the line broadening due to intermolecular collisions give an IR spectrum in which rotational structure is not observable and broad vibrational bands overlap one another, making it difficult to obtain structural information. However, the use of characteristic group frequencies (see below), supplementation of IR spectral information with Raman spectral information (Sec. 21.10), and comparison with spectra of similar compounds of known structure often allow useful structural information to be obtained from the vibrational spectra of large molecules.

Organic chemists use IR spectroscopy to help identify an unknown compound. Although most or all of the atoms are vibrating in each normal mode, certain normal modes may involve mainly motion of only a small group of atoms bonded together, with the other atoms vibrating only slightly. For example, aldehydes and ketones have a normal mode which is mainly a C=O stretching vibration, and its frequency is approximately the same in most aldehydes and ketones. Normal-mode analysis shows that two bonded atoms A and B in a compound will exhibit a characteristic vibrational frequency provided that either the force constant of the A—B bond differs greatly from the other force constants in the molecule or there is a large difference in mass between A and B. Double and triple bonds have much larger force constants than single bonds, so one observes characteristic vibrational frequencies for C=O, C=C, C≡C, and C≡N bonds. Since hydrogens are much lighter than carbons, one observes characteristic vibrational frequencies for OH, NH, and CH groups. In contrast, vibrations involving C—C, C—N, and C—O single bonds occur over a very wide range of frequencies.

Some characteristic IR wavenumbers in cm^{-1} for stretching vibrations are:

OH	NH	CH	C≡C	C=C	C=O
3200–3600	3100–3500	2700–3300	2100–2250	1620–1680	1650–1850

For a molecule with internal rotation about a single bond, the changes in electronic energy E_e produced by changes in the dihedral angle of internal rotation give rise to the potential-energy function for torsion (twisting) about that bond. Figure 20.33 shows this function for butane. Usually, one of the molecule's normal modes is essentially a torsional motion about the single bond. Because rotational barriers for single-bond torsion are low, the force constant and vibrational frequency for this normal mode are low. Typically $\tilde{\nu}$ is in the range 50 to 400 cm^{-1} for torsion about a single bond.

If one can observe and reliably assign the fundamental, overtones, and hot bands for the torsional mode of each conformer present, one can then attempt to derive a potential function for the torsion that will fit the torsional vibrational-level data. Additional information, such as the enthalpy difference(s) between conformers (which can be found from the temperature dependence of line-intensity ratios) can help determine the potential function. The torsional-potential function of butane in Fig. 20.33 was found from observed torsional-vibration transition frequencies of the gauche and trans conformers. For 1,3-butadiene, two different potential functions (the dashed curves in Fig. 20.35) were found to give good fits to observed torsional transition frequencies.

The most commonly studied portion of the IR spectrum is from 4000 to 400 cm^{-1} (2.5 to 25 μm).

The complete IR spectrum is a highly characteristic property of a compound and has been compared to a fingerprint.

Quantitative analysis of mixtures can be done by applying the Beer–Lambert law (Sec. 21.2) to IR absorption.

Molecular vibrational frequencies are affected by changes in bonding, conformation, and hydrogen bonding, so IR spectroscopy can be used to study such changes. Hydrogen bonding reduces the frequencies of OH and NH stretching vibrations. For example, water vapor shows symmetric and asymmetric OH stretches at 3657 and 3756 cm^{-1} (Fig. 21.27); in liquid water these are replaced by a broad absorption band centered at 3400 cm^{-1}. Conformations and hydrogen bonding of biological molecules and synthetic polymers have been studied using IR spectroscopy. [See E. G. Brame and J. G. Grasselli (eds.), *Infrared and Raman Spectroscopy*, pt. C, Dekker, 1977.]

IR spectroscopy and EELS are widely used to study the structure of species chemisorbed on solids. For CH$_2$=CH$_2$ chemisorbed on Pt, certain vibrational bands occur at about the same frequencies as IR bands of compounds containing n-alkyl groups bonded to metal atoms, thereby indicating the presence of σ-bonded ethylene (PtCH$_2$CH$_2$Pt), shown in Fig. 20.31a; other IR bands for C$_2$H$_4$ on Pt occur at about the same frequencies as IR bands of species known to contain C$_2$H$_4$ π bonded to a metal atom (for example, the complex ion [Pt(C$_2$H$_4$)Cl$_3$]$^-$), and they are evidence for the π complex shown in Fig. 20.31b. Still other bands occur at frequencies close to those of the compound CH$_3$CCo$_3$(CO)$_9$ and show the presence of the ethylidyne complex CH$_3$CPt$_3$ (Fig. 21.31). When adsorbed ethylidyne is formed, one H atom dissociates from C$_2$H$_4$ and is chemisorbed on the Pt surface and a second H atom shifts from one carbon to the other.

When H$_2(g)$ is chemisorbed on ZnO, one finds O—H and Zn—H IR bands at 3502 cm^{-1} and 1691 cm^{-1} due to dissociatively adsorbed hydrogen. In addition, there is a band at 4019 cm^{-1}, which is 142 cm^{-1} below the vibrational wavenumber $\tilde{\nu}_0 = 4161$ cm^{-1} of H$_2(g)$ [Table 21.1 and Eq. (21.36)]. For physically adsorbed diatomic molecules, one typically observes a downward frequency shift of about 20 cm^{-1} compared with the gas phase. The large shift for H$_2$ on ZnO is interpreted as evidence for nondissociatively chemisorbed H$_2$ (Fig. 20.31d); see C. C. Chang *et al.*, *J. Phys. Chem.*, **77,**

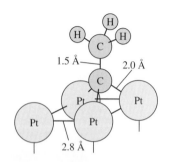

Figure 21.31

Structure of the ethylidyne complex formed when C$_2$H$_4$ (ethylene) is chemisorbed on the (111) surface of Pt at 300 K. (The notation is explained in Sec. 24.7.)

2634 (1973). (Although H_2 gas does not absorb IR radiation, chemisorbed H_2 molecules do absorb; the bond to the surface induces a dipole moment whose value changes as the H_2 internuclear distance changes.)

Fourier-Transform IR Spectroscopy

A **Fourier-transform IR (FT-IR) spectrometer** gives improved *sensitivity* (the ability to detect weak signals), speed, and accuracy of wavelength measurement as compared with the dispersion IR spectrometer of Fig. 21.29. An FT-IR spectrometer (Fig. 21.32) does not disperse the radiation (and so has no prism or grating) but uses a Michelson interferometer to form an interferogram, which is mathematically manipulated by a microcomputer to give the IR absorption spectrum. Continuous-frequency radiation from the source strikes the beam splitter, a flat, partially transparent plate that reflects half the incident light to the fixed mirror (beam *b*) and transmits half to the moving mirror (beam *c*). The moving mirror is driven by a motor and moves parallel to itself at constant speed. The beams *d* and *e* reflected from the two mirrors meet at the beam splitter, which transmits part of *d* and reflects part of *e* to give the combined beam *f*. Beam *f* passes through the sample, where absorption occurs to give beam *g*.

Temporarily let us suppose that the light from the source contains only a single wavelength λ, whose wavenumber is $\tilde{\nu} = 1/\lambda$. Beams *d* and *e* interfere with each other when they meet to form beam *f*, and by adding the electric fields of beams *d* and *e*, one finds that the intensity I_f of beam *f* is (for a derivation, see F. A. Jenkins and H. E. White, *Fundamentals of Optics*, 4th ed., McGraw-Hill, 1976, sec. 12.1)

$$I_f = \tfrac{1}{2}B_f[1 + \cos(2\pi\delta/\lambda)] = \tfrac{1}{2}B_f[1 + \cos(2\pi\delta\tilde{\nu})]$$

where δ is the difference in path lengths traveled by beams *c* and *e* as compared with beams *b* and *d*, and B_f is the intensity beam *f* would have if there were no path difference ($\delta = 0$). If x is the difference in distances of the moving mirror and the fixed mirror from the beam splitter, then $\delta = 2x$. Both x and δ change with time. When δ is a whole-number multiple of the wavelength ($\delta = n\lambda$), then $\cos(2\pi\delta/\lambda) = 1$ and $I_f = B_f$, because the beams meeting to form the combined beam are in phase. When $\delta = (n + \tfrac{1}{2})\lambda$, then $\cos(2\pi\delta/\lambda) = -1$ and $I_f = 0$, since the beams meet out of phase.

When beam *f* passes through the sample, the sample's absorption changes I_f and B_f in the above equation to I_g and B_g.

Since the radiation actually contains a continuous range of wavenumbers, we must integrate over all wavenumbers $\tilde{\nu}$ to get I_f and I_g. Also, B_g is a function of $\tilde{\nu}$ [$B_g = B_g(\tilde{\nu})$], since the sample's absorbance and the light emitted by the source are functions of $\tilde{\nu}$. Therefore

$$I_g(\delta) = \frac{1}{2}\int_0^\infty B_g(\tilde{\nu})[1 + \cos(2\pi\delta\tilde{\nu})]\,d\tilde{\nu}$$

$$= \frac{1}{2}\int_0^\infty B_g(\tilde{\nu})\,d\tilde{\nu} + \frac{1}{2}\int_0^\infty B_g(\tilde{\nu})\cos(2\pi\delta\tilde{\nu})\,d\tilde{\nu}$$

As the mirror moves, δ changes and I_g changes, so I_g is a function of the path-length difference δ. The detector records I_g as a function of time, and since δ is known at each time, the detector measures the function $I_g(\delta)$. At $\delta = 0$, light of all frequencies in beams *d* and *e* meets in phase and the intensity I_g is a maximum. The maximum intensity $I_g(0)$ is given by the above equation with $\delta = 0$ as $I_g(0) = \int_0^\infty B_g(\tilde{\nu})\,d\tilde{\nu}$. Hence the $I_g(\delta)$ equation becomes $I_g(\delta) = \tfrac{1}{2}I_g(0) + \tfrac{1}{2}\int_0^\infty B_g(\tilde{\nu})\cos(2\pi\delta\tilde{\nu})\,d\tilde{\nu}$. Defining $F(\delta) \equiv I_g(\delta) - \tfrac{1}{2}I_g(0)$, we have

$$F(\delta) \equiv I_g(\delta) - \tfrac{1}{2}I_g(0) = \frac{1}{2}\int_0^\infty B_g(\tilde{\nu})\cos(2\pi\delta\tilde{\nu})\,d\tilde{\nu}$$

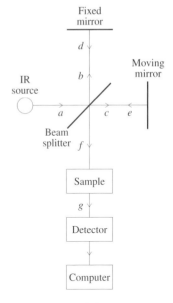

Figure 21.32

A Fourier-transform IR spectrometer.

The *interferogram function* $F(\delta)$ is known, since $I_g(\delta)$ and $I_g(0)$ are known. The function $B_g(\tilde{\nu})$ is the intensity of radiation reaching the detector as a function of wavenumber (with no path difference between the beams) and is exactly the spectrum we want to find. A mathematical theorem of Fourier analysis shows that if $F(\delta)$ and $B_g(\tilde{\nu})$ are related by the above equation and if $F(\delta)$ has zero slope at $\delta = 0$ [which it does because $I_g(\delta)$ is a maximum at $\delta = 0$ and $F(\delta)$ differs from $I_g(\delta)$ only by a constant], then $B_g(\tilde{\nu})$ is given by (see M. J. Lighthill, *Introduction to Fourier Analysis and Generalized Functions*, Cambridge, 1958, sec. 1.4)

$$B_g(\tilde{\nu}) = 8 \int_0^\infty F(\delta) \cos \left(2\pi\tilde{\nu}\delta \right) d\delta$$

For full accuracy in calculating $B_g(\tilde{\nu})$, one must make observations with δ extending to ∞. In practice, the path-length difference will have some maximum value δ_{max} (which is typically on the order of 1 to 20 cm), and the infinite limit in the integral (which is called the *Fourier cosine transform* of F) is approximated by δ_{max}. A computer, which is part of the spectrometer, calculates the spectrum $B_g(\tilde{\nu})$ from $F(\delta)$. Since only a single beam is used, one does a separate run with the sample cell empty and the computer combines the two spectra to give the sample's absorption spectrum. Because radiation of all wavenumbers reaches the detector at each instant and all this radiation is used to calculate the spectrum $B_g(\tilde{\nu})$, an FT-IR spectrometer gives much better signal-to-noise ratio than a dispersion instrument, where only a tiny portion of the radiation reaches the detector at each instant. Moreover, one can further improve the signal-to-noise ratio by making repeated runs and having the computer average the spectra. The noise then tends to cancel, since it is randomly positive and negative.

Currently, most commercially available IR spectrometers are Fourier-transform instruments.

21.10   RAMAN SPECTROSCOPY

Raman spectroscopy is quite different from absorption spectroscopy in that it studies light *scattered* by a sample rather than light absorbed or emitted. Suppose a photon collides with a molecule in state a. If the energy of the photon corresponds to the energy difference between state a and a higher level, the photon may be absorbed, the molecule making a transition to the higher level. No matter what the energy of the photon is, the photon–molecule collision may **scatter** the photon, meaning that the photon's direction of motion is changed. Although most of the scattered photons undergo no change in frequency and energy (*Rayleigh scattering*), a small fraction of the scattered photons exchange energy with the molecule during the collision. The resulting increase or decrease in energy of the scattered photons is the **Raman effect,** first observed by C. V. Raman in 1928.

Let ν_0 and ν_{scat} be the frequencies of the incident photon and the Raman-scattered photon, respectively, and let E_a and E_b be the energies of the molecule before and after it scatters the photon. Conservation of energy gives $h\nu_0 + E_a = h\nu_{scat} + E_b$, or

$$\Delta E \equiv E_b - E_a = h(\nu_0 - \nu_{scat}) \equiv h \, \Delta\nu \qquad (21.49)$$

The energy difference ΔE is the difference between two stationary-state energies of the molecule, so *measurement of the* **Raman shifts** $\Delta\nu \equiv \nu_0 - \nu_{scat}$ *gives molecular energy-level differences.*

In Raman spectroscopy, one exposes the sample (gas, liquid, or solid) to monochromatic radiation of any convenient frequency ν_0. Unlike absorption spectroscopy, ν_0 need have no relation to the difference between energy levels of the sample molecules.

Usually ν_0 lies in the visible or near-UV region. The Raman-effect lines are extremely weak (only about 0.001% of the incident radiation is scattered, and only about 1% of the scattered radiation is Raman-scattered). Therefore, the very intense light of a laser beam is used as the exciting radiation. Scattered light at right angles to the laser beam is focused on the entrance slit of a spectrometer (Fig. 21.33), which disperses the radiation using a diffraction grating and records light intensity versus frequency, giving the Raman spectrum.

Figure 21.34 shows the pattern of Raman lines for a gas of diatomic molecules. The strong central line at ν_0 is due to light scattered with no frequency change. On either side of ν_0 and close to it are lines corresponding to pure-rotational transitions in the molecule; the lines with $\nu_{scat} > \nu_0$ result from transitions where J decreases, and those with $\nu_{scat} < \nu_0$ result from increases in J. On the low-frequency side of ν_0 is a band of lines corresponding to $v = 0 \rightarrow 1$ vibration–rotation transitions in the molecule. If the $v = 1$ vibrational level is significantly populated, there is a weak band of lines on the high-frequency side, corresponding to $v = 1 \rightarrow 0$ vibration–rotation transitions. For historical reasons, the Raman lines with $\nu_{scat} < \nu_0$ are called *Stokes lines* and those with $\nu_{scat} > \nu_0$ are *anti-Stokes lines*.

Investigation of the Raman selection rules shows that spherical tops exhibit no pure-rotational Raman spectra, but all symmetric and asymmetric tops show pure-rotational Raman spectra. Raman scattering is a different process than absorption, and the rotational Raman selection rules differ from those for absorption. One finds that for linear molecules the pure-rotational Raman frequencies on either side of ν_0 correspond to $\Delta J = \pm 2$. The rotational Raman selection rules for nonlinear molecules are more complicated and are omitted. The pure-rotational Raman spectrum can yield the structure of a nonpolar molecule (for example, F_2, C_6H_6) that is not a spherical top. For benzene, the rotational Raman spectrum of *sym*-$C_6H_3D_3$, together with spectroscopic data for C_6H_6 and C_6D_6, gave $R_0(CC) = 1.397$ Å and $R_0(CH) = 1.08_6$ Å [H. G. M. Edwards, *J. Mol. Struct.*, **161**, 23 (1987)].

For a polyatomic molecule, the Raman spectrum shows several $\Delta v_j = 1$ fundamental bands, each corresponding to a Raman-active vibration. The kth normal mode is Raman-active if $(\partial \alpha / \partial Q_k)_e \neq 0$, where α is the molecular polarizability (Sec. 14.15), Q_k is the normal coordinate for the kth vibrational mode, and the derivative is evaluated at the equilibrium nuclear configuration. (The normal coordinate Q_k is a certain linear combination of the cartesian coordinates of the nuclei that vibrate in the kth normal mode; Q_k measures the extent of departure from the equilibrium configuration of the nuclei.) As noted in Sec. 14.15, α is different in different directions in the molecule, and one must examine the derivative of each component of α when deciding on Raman activity. It is hard to determine whether $(\partial \alpha / \partial Q_k)_e$ is zero by examining the atomic motions in the kth normal mode. The best way to determine which modes are Raman-active is to use group theory (see *Herzberg*, vol. II, chap. III).

Since the Raman selection rule differs from the IR selection rule, a given normal mode may be IR-active and Raman-inactive, IR-inactive and Raman-active, active in both the IR and Raman, or inactive in both the IR and Raman. For a molecule with no symmetry, all vibrational modes are Raman-active. Every molecule (including homonuclear diatomics) has at least one Raman-active normal mode.

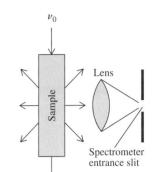

Figure 21.33

In Raman spectroscopy, one observes light scattered at right angles to the incident radiation.

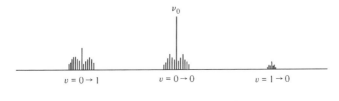

Figure 21.34

Raman spectrum of a gas of diatomic molecules.

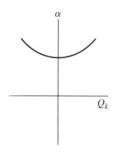

Figure 21.35

Behavior of each component of polarizability as a function of the coordinate describing the normal-mode vibrations ν_2 or ν_3 of CO_2.

The CO_2 symmetric stretching vibration ν_1 (Fig. 21.26), which is IR-inactive, is Raman-active. For the CO_2 vibrations ν_2 and ν_3, each component of α behaves as shown in Fig. 21.35, where $Q_k = 0$ corresponds to the equilibrium nuclear configuration. Although the ν_2 and ν_3 vibrations each change α, each of these vibrations is Raman-inactive because each has $(\partial\alpha/\partial Q_k)_e = 0$, as is evident from the zero slope at $Q_k = 0$.

Overtone and combination bands are usually too weak to be observed in Raman spectroscopy, so vibrational Raman spectra are simpler than IR spectra.

As in IR spectra, vibrational Raman spectra of gases show rotational structure, but rotational structure is absent in Raman spectra of liquids and solids.

It is hard (but not impossible) to obtain IR spectra below 100 cm^{-1}, but easy to obtain Raman spectra for the Raman shift $\Delta\tilde{\nu}$ in the range 10 to 100 cm^{-1}, so Raman spectroscopy is the preferred technique in this wavenumber range. For $\Delta\tilde{\nu}$ between 0 and 10 cm^{-1}, the Raman-shifted lines are hidden by the strong Rayleigh-scattered peak at $\tilde{\nu}_0$.

An important advantage of vibrational Raman spectroscopy over IR spectroscopy arises from the fact that liquid water shows only weak vibrational Raman scattering in the Raman-shift range 300–3000 cm^{-1} but has strong, broad IR absorptions in this range. Thus, vibrational Raman spectra of substances in aqueous solution can readily be studied without solvent interference. An aqueous solution of mercury(I) nitrate shows a strong peak at a Raman shift of 170 cm^{-1}, and this peak is absent from spectra of other nitrate salt solutions. The peak is due to vibration of the Hg—Hg bond and indicates that the mercury(I) ion is diatomic (Hg_2^{2+}). The vibrational Raman spectrum of CO_2 dissolved in water is almost the same as that of liquid CO_2 and shows no H_2CO_3 vibrational absorption bands, thereby confirming that CO_2 in water exists mainly as solvated CO_2 molecules, not as H_2CO_3. Vibrational Raman spectroscopy has been used to study ion pairing (Sec. 10.9) and solvation of ions. Vibrational Raman spectra of biological molecules in aqueous solution provide information on conformations and hydrogen bonding. (See P. R. Carey, *Biochemical Aspects of Raman and Resonance Raman Spectroscopy,* Academic Press, 1982; A. T. Wu, *Raman Spectroscopy and Biology,* Wiley, 1982.)

In **resonance Raman spectroscopy,** the exciting frequency ν_0 is chosen to coincide with an electronic absorption frequency (Sec. 21.11) of the species being studied. This dramatically increases the intensities of the Raman-scattered radiation for those vibrational modes that are localized in the portion of the molecule that is responsible for the electronic absorption at ν_0. Two important advantages of resonance Raman spectroscopy in the study of biological molecules are: (*a*) the increased scattering intensity allows study of solutions at the high dilutions (10^{-3} to 10^{-6} mol/dm^3) characteristic of biopolymers in organisms; (*b*) the selectivity of the intensity enhancement "samples" only the vibrations in one region of the molecule, simplifying the spectrum and allowing study of the bonding in that region. The resonance Raman spectra of the oxygen-carrying proteins hemoglobin and myoglobin give information on the bonding in the heme group; see J. M. Friedman et al., *Ann. Rev. Phys. Chem.,* **33,** 471 (1982).

When certain molecules or ions (mainly those containing O, N, or S atoms with lone pairs) are adsorbed on or are atomically close to a roughened surface or colloidal dispersion or vacuum-deposited thin film of certain metals (mainly Ag, Cu, or Au), the intensity of the molecules' Raman spectra is enormously increased, giving rise to *surface-enhanced Raman spectroscopy* (SERS). Interactions between the incident radiation and the electrons in the metal surface produce a great increase in the strength of the electric field of the electromagnetic radiation, thereby amplifying the Raman scattering. In some cases, chemical bonding between the molecule and the surface contributes to the effect.

21.11   ELECTRONIC SPECTROSCOPY

Electronic spectra involving transitions of valence electrons occur in the visible and UV regions and are studied in both absorption and emission. A photoelectric cell or photographic plate is the detector. One can study emission spectra of solids, liquids, and gases by raising the molecules to an excited electronic state using photons from a high-intensity lamp or a laser and then observing the emission at right angles to the incident beam. Spontaneous emission of light by excited-state atoms or molecules that rapidly follows absorption of light is called **fluorescence** (see also Sec. 21.16).

Atomic Spectra

For the hydrogen atom, the selection rule for n is found to be Δn = any value. The use of $E_a - E_b = h\nu$ and the energy-level formula (19.14) gives the spectral wavenumbers as

$$\tilde{\nu} = \frac{1}{\lambda} = \frac{\nu}{c} = \frac{\mu e^4}{8\varepsilon_0^2 c h^3}\left(\frac{1}{n_b^2} - \frac{1}{n_a^2}\right) \equiv R_{\rm H}\left(\frac{1}{n_b^2} - \frac{1}{n_a^2}\right) \qquad (21.50)$$

where the reduced mass depends on the electron and proton masses, $\mu = m_e m_p/(m_e + m_p)$ [Eq. (18.79)], and where the Rydberg constant for hydrogen is defined as $R_{\rm H} \equiv \mu e^4/8\varepsilon_0^2 c h^3 = 109678 \text{ cm}^{-1}$.

The H-atom spectrum consists of several series of lines, each ending in a continuous band. Transitions involving the n-value changes $2 \to 1, 3 \to 1, 4 \to 1, \ldots$ give the Lyman series in the UV (Fig. 21.36). As n_a in Eq. (21.50) goes to infinity, the Lyman-series lines converge to the limiting value $1/\lambda = R_{\rm H} = 109678 \text{ cm}^{-1}$, corresponding to $\lambda = 91.2$ nm. Beyond this limit is a continuous absorption or emission due to transitions between ionized H atoms and ground-state H atoms; the energy of an ionized atom takes on a continuous range of positive values. The position of the Lyman-series limit allows the ionization energy of H to be found.

The H-atom transitions $3 \to 2, 4 \to 2, 5 \to 2, \ldots$ give the Balmer series, which lies in the visible region. The transitions $4 \to 3, 5 \to 3, 6 \to 3, \ldots$ give the Paschen series in the IR.

Spectra of many-electron atoms are quite complicated, because of the many terms and levels arising from a given electron configuration. Once the spectrum has been unraveled, the atomic energy levels can be found.

Since each element has lines at frequencies characteristic of that element, atomic absorption and emission spectra are used to analyze for most chemical elements. For example, Ca, Mg, Na, K, and Pb in blood samples can be determined by atomic absorption spectroscopy.

For inner-shell electrons, the effective nuclear charge $Z_{\rm eff}$ is nearly equal to the atomic number; Eq. (19.52) shows that the differences between these inner-shell energies increase rapidly with increasing atomic number. For atoms beyond the second period, these energy differences correspond to x-ray photons. X-rays are produced when a beam of high-energy electrons penetrates a metal target. The deceleration of the electrons as they penetrate the target produces a continuous x-ray emission spectrum. In addition, an electron in the beam that collides with an inner-shell electron of

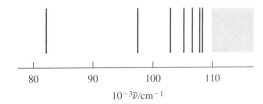

$10^{-3}\tilde{\nu}/\text{cm}^{-1}$

Figure 21.36

The Lyman series of H. Only the first seven lines are shown.

a target atom can knock this electron out of the atom. The spontaneous transition of a higher-level electron of the ionized atom into the vacancy thereby created will produce an x-ray photon of frequency corresponding to the energy difference, giving an x-ray emission line spectrum superimposed on the continuous emission.

Molecular Electronic Spectra

If ψ'' and ψ' are the lower and upper states in a molecular electronic transition, the transition frequencies are given by

$$h\nu = (E'_{\text{el}} - E''_{\text{el}}) + (E'_{\text{vib}} - E''_{\text{vib}}) + (E'_{\text{rot}} - E''_{\text{rot}}) \tag{21.51}$$

Each term in parentheses is substantially larger than the following term.

The electronic selection rules are rather complicated. Perhaps the most important one is that the transition-moment integral (21.5) vanishes unless

$$\Delta S = 0$$

where S is the total electronic spin quantum number. Actually, this selection rule does not hold rigorously, and weak transitions with $\Delta S \neq 0$ sometimes occur. The ground electronic state of most molecules has all electron spins paired and so has $S = 0$ (a **singlet state**). Some exceptions are O_2 (a **triplet** ground level with $S = 1$) and NO_2 (the odd number of electrons gives $S = \frac{1}{2}$ and a doublet ground level).

An electronic transition consists of a series of **bands,** each band corresponding to a transition between a given pair of vibrational levels. Gas-phase spectra under high resolution may show each band to consist of closely spaced lines arising from transitions between different rotational levels. For relatively heavy molecules, the close spacings between the rotational levels usually makes it impossible to resolve the rotational lines. In liquids, no rotational structure is observed and the vibrational bands are often broadened enough to merge them into a single broad absorption band for each electronic transition. Figure 21.37 shows the electronic absorption spectrum of gas-phase benzene.

Analysis of the rotational lines of an electronic absorption band allows the molecular vibrational and rotational constants to be obtained. Since all molecules show electronic absorption spectra, this allows one to obtain R_e and ν_e for homonuclear diatomics, which show no pure-rotational or vibration–rotation spectra. Excited electronic

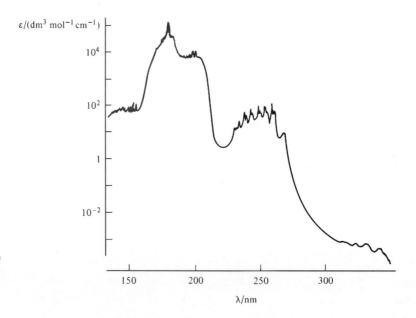

Figure 21.37

Sketch of the electronic absorption spectrum of gas-phase benzene. The vertical scale is logarithmic.

states often have geometries quite different from those of the ground electronic state. For example, HCN is nonlinear in some excited states.

Molecular dissociation energies and spacings between electronic energy levels are also obtained from electronic spectra. Figure 21.38 shows potential-energy curves and vibrational levels for the ground state and an excited electronic state of a diatomic molecule. Suppose we are able to observe transitions from the $v'' = 0$ vibrational level of the ground electronic state to each of the vibrational levels of the excited electronic state, as shown by the arrows in the figure. The bands corresponding to these transitions will come closer together as the excited-state vibrational quantum number v' increases, and these bands will be followed by a continuous absorption corresponding to transitions to states with energy above E'_{at}. Figure 21.38 gives

$$h\nu_{cont} = h\nu_{00} + D'_0 = D''_0 + E'_{at} - E''_{at} \tag{21.52}$$

where ν_{cont} is the frequency at which continuous absorption begins, ν_{00} is the frequency of the $v'' = 0$ to $v' = 0$ transition, D''_0 and D'_0 are the dissociation energies of the ground and excited electronic states, and E''_{at} and E'_{at} are the energies of the separated atoms into which the ground and excited electronic states dissociate. Analysis of the molecule's electronic spectrum will usually tell us which atomic states the two molecular electronic states dissociate into, so E''_{at} and E'_{at} can be found from known atomic energy levels (which are found from atomic spectra). Hence, Eq. (21.52) allows us to find the dissociation energies D''_0 and D'_0. Also, the quantity $h\nu_{00}$ gives the separation between the ground vibrational levels of the two electronic states.

Substances that absorb visible light are colored. Conjugated organic compounds often show electronic absorption in the visible region, due to excitation of a π electron to an antibonding π orbital. In the particle-in-a-box model of Prob. 20.51, the energy spacings are proportional to $1/a^2$, where a is the box length, so as the length of the conjugated chain increases, the lowest absorption frequency moves from the UV into the visible region. Transition-metal ions [for example, $Cu^{2+}(aq)$, $MnO_4^-(aq)$] frequently show visible absorption due to electronic transitions.

Certain groups called **chromophores** give characteristic electronic absorption bands. For example, the C=O group in the majority of aldehydes and ketones produces a weak electronic absorption in the region 270 to 295 nm with molar absorption coefficient $\varepsilon = 10$ to $30 \ dm^3 \ mol^{-1} \ cm^{-1}$ and a strong band at about 180 to 195 nm with $\varepsilon = 10^3$ to $10^4 \ dm^3 \ mol^{-1} \ cm^{-1}$. The weak band is due to an $n \rightarrow \pi^*$ transition, and the strong band is due to a $\pi \rightarrow \pi^*$ transition, where n signifies an electron in a nonbonding (lone-pair) orbital on oxygen, π is an electron in the π MO of the double bond, and π^* is an electron in the corresponding antibonding π MO. The longer-wavelength electronic absorptions of benzene in Fig. 21.37 are $\pi \rightarrow \pi^*$ transitions. Proteins absorb in the near-UV region (200 to 400 nm) as a result of $\pi \rightarrow \pi^*$ transitions

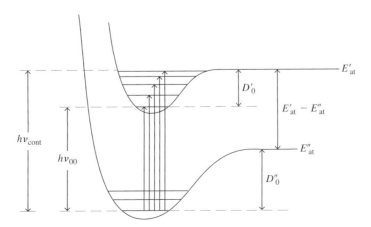

Figure 21.38

Determination of dissociation energies of electronic states of a diatomic molecule. E''_{at} and E'_{at} are the energies of the separated atoms produced by dissociation of the ground state and the first excited electronic state, respectively.

of electrons in the aromatic amino acid residues and $\pi \to \pi^*$ transitions in the amide groups. The proteins hemoglobin and myoglobin have strong visible absorptions because of $\pi \to \pi^*$ transitions in the heme group, which is a large conjugated chromophore.

Fluorescence Fluorescence spectroscopy is widely used in biochemistry, partly because it is capable of giving spectra from extremely small amounts of material. In absorption spectroscopy of a substance in solution, one compares the intensity of two beams, one that passed through the sample and one that went through pure solvent. If the absorber is present in a very small concentration, two nearly equal intensities are being compared, which is difficult. In contrast, in fluorescence spectroscopy, one compares the radiation emitted (at right angles to the exciting radiation beam) with darkness, which allows much greater sensitivity. Fluorescence spectroscopy allows quantitative analysis for substances present in trace amounts, for example, drugs, pesticides, and atmospheric pollutants. A molecule that shows fluorescence when exposed to the wavelength(s) used in the exciting radiation is called a **fluorophore.**

One can locate specific biomolecules in biological tissues and cells using *fluorescence microscopy*. Here, the specimen is illuminated with UV or visible light and is observed through a microscope using a filter that transmits only fluorescent light in a certain wavelength range. The biomolecules to be studied are either fluorescent or are bonded to a fluorescent labeling species.

In the technique of **laser-induced fluorescence** (LIF), one uses a laser to excite molecules to an excited electronic state and then observes the resulting fluorescence as the molecules drop to lower states. The near monochromaticity and tunability of the laser radiation allows one to control which vibrational state of the excited electronic state is populated, making it easier to study the fluorescence. LIF has been used to detect free radicals in kinetic studies of combustion.

Let S_0 denote the ground electronic state of a fluorophore, where the S stands for singlet. Let S_1 denote the lowest-lying electronically excited singlet state. Before absorption of the exciting radiation, most of the fluorophore molecules are in the ground vibrational state of S_0. Excited electronic states nearly always have equilibrium geometries that differ significantly from that of the ground electronic state. Figure 18.18 shows that the harmonic-oscillator ground-state wave function is concentrated near $x = 0$, which corresponds in Eqs. (21.18) and (21.20) to the equilibrium geometry. Because the ground vibrational wave functions of S_0 and S_1 have their probability densities concentrated in different regions of space, the magnitude of the transition dipole moment (21.5) is very small for this pair of vibrational states. Thus there is little probability for absorption of radiation (or emission of radiation) to occur between the ground vibrational states of S_0 and S_1. *The greatest probabilities are for excitation from the ground vibrational state of S_0 to highly excited vibrational states of S_1.*

Most of the molecules in these excited vibrational levels of S_1 will lose vibrational energy in collisions with neighboring molecules and drop to the ground vibrational level of S_1 before they fluoresce back to the ground state S_0. (This *vibrational relaxation* process is much faster than fluorescence.) Because the transition moment (21.5) is small for transitions between ground vibrational levels of S_0 and S_1, most of the fluorescence comes from molecules going from the ground vibrational level of S_1 to excited vibrational levels of S_0. The fluorescence emission transitions thus have smaller $|\Delta E|$ values than the excitation transitions. In fluorescence, the center of the fluorescence emission band almost always occurs at a longer wavelength (lower frequency and lower photon energy) than the exciting radiation absorbed. The difference in wavelength between the most strongly absorbed exciting wavelength and the most strongly emitted fluorescence wavelength is called the **Stokes shift.** The Stokes shift enables one to filter out scattered excitation radiation, making it easier to observe fluorescence.

DNA Sequencing Fluorescence plays a key role in automated DNA sequencing by capillary electrophoresis (Sec. 16.6). One mixes the single-stranded DNA to be sequenced with the following substances: (*a*) a *primer*, which is a short segment of nucleotide polymer whose base sequence is complementary to that of a known portion of the DNA to be sequenced (or is complementary to a known sequence that has been attached to the DNA); (*b*) DNA polymerase enzyme; (*c*) the four kinds of DNA nucleotide monomers, each consisting of deoxyribose, phosphate, and one of the four DNA bases bonded together; (*d*) small amounts of four kinds of modified DNA nucleotide monomers that each have dideoxyribose instead of deoxyribose and that have one of four dyes that fluoresce at different wavelengths. Which dye has been bonded to the modified nucleotide depends on which base the nucleotide contains. The primer molecules hydrogen bond to the single-stranded DNA molecules and then nucleotide monomers chemically bond to the primer molecules in an order that is complementary to the base sequence of the DNA template. When a modified nucleotide happens to bond to the lengthened primer, the structure of the modified nucleotide prevents further nucleotides bonding on. One thus ends up with a mixture of lengthened primers of all possible lengths, with the last nucleotide in each sequence having a base that corresponds to a particular fluorescent dye.

The lengthened primer molecules are separated from the DNA templates and the mixture undergoes electrophoresis, which separates the lengthened primers according to length. Near the end of the capillary, a laser excites fluorescence and the wavelength of the emitted fluorescent radiation reports on which base is at the end of the lengthened primer that is passing that point in the capillary. Each lengthened primer is one nucleotide longer than the one that preceded it past the observation point, so one has determined the base sequence that is complementary to that in the DNA.

Jet Cooling A technique used to simplify the electronic spectra of gas-phase radicals, ions, and large molecules is **jet cooling.** Here a mixture of a small amount of the species to be studied (gas B) and a large amount of He(g) or Ar(g) at high pressure (1 to 100 atm) expands through a small hole of diameter d_{hole} into a vacuum chamber. Conditions are such that $d_{\text{hole}} \gg \lambda$, where λ is the mean free path of gas molecules in the immediate vicinity of the hole. As explained after Eq. (15.58), when $\lambda \ll d_{\text{hole}}$, collisions in the region of the hole produce a bulk collective flow through the hole, and the expanding helium forms a flowing jet, which has substantial macroscopic kinetic energy [the term K in Eq. (2.39)]. The source of this macroscopic kinetic energy is the random molecular kinetic energy [the term $\frac{3}{2}RT$ in Eq. (2.89)] of the He atoms, so the He is cooled to a very low temperature, on the order of 1 K or less. Collisions of gas B molecules with the cold He gas in the region immediately beyond the hole cool the B molecules.

Translational energy of B molecules is transferred most easily, and the random translational motion of B molecules corresponds to a translational temperature T_{tr} of about 1 K. Rotational energy is transferred somewhat less easily than translational energy, and the rotation of the B molecules corresponds to a rotational temperature T_{rot} of perhaps 2 K. (T_{rot} is the temperature in the Boltzmann distribution law that corresponds to the observed distribution of molecules among rotational energy levels; see Prob. 21.61.) Vibrational cooling also occurs, but usually to a lesser extent. Typically T_{vib} is on the order of 100 K in the jet. The low value of T_{tr} in the jet makes the speed of sound in the jet [given in Sec. 15.5 as $(C_PRT_{\text{tr}}/C_VM)^{1/2}$] far less than the flow rate of the jet, so the jet is described as *supersonic*.

Because there is a reduction in the number of rotational and vibrational states that are populated, and because line broadening due to collisions is reduced in the low-density jet, the appearance of the spectrum is greatly simplified, making it easier to interpret. Vibrational bands that overlap to form a broad, featureless spectrum at room temperature are narrowed and become well separated. Rotational structure becomes more easily resolved. Molecules as large as zinc tetrabenzoporphine ($ZnN_4C_{36}H_{20}$) have been studied [U. Even, J. Jortner, and J. Friedman, *J. Phys. Chem.,* **86,** 2273 (1982)]. Because the B molecules are at a low density, a very sensitive technique such as LIF is used to study the electronic spectrum. By using an electric discharge or photolysis, radicals and ions (such as CH_2, CH_3, C_2H_5, C_5H_5, H_2O^+) can be produced and their spectra observed. For more on spectroscopy of jet-cooled species, see M. Ito et al., *Ann. Rev. Phys. Chem.,* **39,** 123 (1988); P. C. Engelking, *Chem. Rev.,* **91,** 399 (1991).

21.12   NUCLEAR-MAGNETIC-RESONANCE SPECTROSCOPY

Nuclear-magnetic-resonance (NMR) spectroscopy is "the most important spectroscopic technique in chemistry" [J. Jonas and H. S. Gutowsky, *Ann. Rev. Phys. Chem.,* **31,** 1 (1980)]. Before discussing NMR, we review the physics of magnetic fields.

The Magnetic Field

A magnetic field is produced by the motion of electric charge. Examples include the motion of electrons in a wire, the motion of electrons in free space, and the "spinning" of an electron about its own axis. The fundamental magnetic field vector **B** is called the **magnetic induction** or the **magnetic flux density.** There is a second magnetic field vector **H**, named the *magnetic field strength.* It used to be thought that **H** was the fundamental magnetic vector, but it is now known that **B** is. Really, **B** ought to be called the magnetic field strength, but it's too late to correct this injustice by giving **B** its proper name.

The definition of **B** is as follows. Imagine a positive test charge Q_t moving through point P in space with velocity **v**. If for arbitrary directions of **v** we find that a force $\mathbf{F}_\perp$ perpendicular to **v** acts on Q_t at P, we say that a magnetic field **B** is present at point P. One finds that there is one direction of **v** that makes $\mathbf{F}_\perp$ equal zero, and the direction of **B** is defined to coincide with this particular direction of **v**. The magnitude of **B** at point P is then defined by $B \equiv F_\perp/(Q_t v \sin \theta)$, where θ is the angle between **v** and **B**. (For $\theta = 0$, $F_\perp$ becomes 0.) Thus

$$F_\perp = Q_t vB \sin \theta \tag{21.53}$$

(In terms of vectors, $\mathbf{F}_\perp = Q_t \mathbf{v} \times \mathbf{B}$, where $\mathbf{v} \times \mathbf{B}$ is the vector cross-product. The magnetic force is perpendicular to both **v** and **B**.) If an electric field **E** is also present, it exerts a force $Q_t\mathbf{E}$ in addition to the magnetic force $\mathbf{F}_\perp$.

Equation (21.53) is written in SI units. The SI unit of B is the **tesla** (T), also called the Wb/m^2, where Wb stands for weber. From (21.53) and (2.7)

$$1 \text{ T} \equiv 1 \text{ N C}^{-1} \text{ m}^{-1} \text{ s} = 1 \text{ kg s}^{-1} \text{ C}^{-1} \tag{21.54}$$

(The strength of the earth's magnetic field at its surface varies from 26 to 60 μT, depending on location, and has declined about 10% in the last 100 years.)

Magnetic fields are produced by electric currents. Experiment shows that the magnetic field at a distance r from a very long straight wire in vacuum carrying a current I is proportional to I and inversely proportional to r; that is, $B = kI/r$, where k is a constant. The unit of electric current, the ampere (A), is defined (Sec. 16.5) so as to give k the value 2×10^{-7} T m A^{-1}. The constant k is also written as $\mu_0/2\pi$, where μ_0 is called the **permeability of vacuum.** Thus

$$\mu_0 \equiv 4\pi \times 10^{-7} \text{ T m A}^{-1} = 4\pi \times 10^{-7} \text{ N C}^{-2} \text{ s}^2 \tag{21.55}$$

where (21.54) and 1 A = 1 C/s were used.

Equation (21.1) for the speed of light in vacuum can be written as

$$c = (\varepsilon_0\mu_0)^{-1/2} \tag{21.56}$$

Consider a tiny loop of current I flowing in a circle of area A. At distances large compared with the radius of the loop, one finds (*Halliday and Resnick*, sec. 34-6) that the magnetic field produced by this current loop has the same mathematical form as the electric field of an electric dipole (Sec. 14.15) except that the electric dipole moment $\boldsymbol{\mu}$ is replaced by the **magnetic dipole moment m,** where **m** is a vector with magnitude $|\mathbf{m}| = IA$ and direction perpendicular to the plane of the current loop (Fig. 21.39); thus

$$|\mathbf{m}| \equiv IA \qquad (21.57)$$

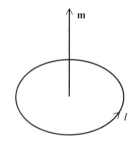

Figure 21.39

The magnetic dipole moment **m** is perpendicular to the plane of the current loop.

The tiny current loop is called a **magnetic dipole.** (The symbol $\boldsymbol{\mu}$ is often used for the magnetic dipole moment, but this can be confused with the symbol for electric dipole moment.)

A magnetic dipole acts like a tiny magnet with a north pole on one side of the current loop and a south pole on the other side. A bar magnet suspended in an external magnetic field has a preferred minimum-energy orientation in the field. Thus, a compass needle orients itself in the earth's magnetic field with one particular end of the needle pointing toward the earth's north magnetic pole. To turn the needle away from this orientation requires an input of energy. In general, a magnetic dipole **m** has a minimum-energy orientation in an externally applied magnetic field **B**. The potential energy V of interaction between **m** and the external field **B** can be shown to be (*Halliday and Resnick*, sec. 33-4)

$$V = -|\mathbf{m}|B \cos \theta \equiv -\mathbf{m} \cdot \mathbf{B} \qquad (21.58)$$

where θ is the angle between **m** and **B**. In (21.58), $\mathbf{m} \cdot \mathbf{B} \equiv |\mathbf{m}|B \cos \theta$ is the **dot product** of **m** and **B**. The minimum-energy orientation has **m** and **B** in the same direction, so that $\theta = 0$ and $V = -|\mathbf{m}|B$. The maximum-energy orientation has **m** and **B** in opposite directions ($\theta = 180°$) and $V = |\mathbf{m}|B$. The zero of potential energy has been arbitrarily chosen to make $V = 0$ at $\theta = 90°$.

> Chemists sometimes use gaussian units in dealing with magnetism. In gaussian units, the defining equation (21.53) is written as $F_{\perp} = (Q_i'vB' \sin \theta)/c$, where B' is the magnetic field in gaussian units and c is the speed of light. The gaussian unit of B' is the *gauss* (G). One finds that 1 T corresponds to 10^4 G.

Nuclear Spins and Magnetic Moments

Nuclei, like electrons, have a **spin angular momentum I.** A nucleus has two spin quantum numbers, I and M_I. These are analogous to s and m_s for an electron. The magnitude of the nuclear spin angular momentum is

$$|\mathbf{I}| = [I(I + 1)]^{1/2}\hbar \qquad (21.59)$$

and the possible values of the z component of **I** are [recall Eq. (19.29)]

$$I_z = M_I\hbar \qquad \text{where } M_I = -I, -I + 1, \ldots, I - 1, I \qquad (21.60)$$

The nuclear spin is the resultant of the spin and orbital angular momenta of the neutrons and protons that compose the nucleus. (See Prob. 19.47 for a discussion of how angular momenta combine in quantum mechanics.) The neutron and the proton each have a spin quantum number $\frac{1}{2}$. A nucleus with an odd mass number A has a half-integral I value ($\frac{1}{2}$ or $\frac{3}{2}$ or ...). A nucleus with A even has an integral value of I. A nucleus with A even and atomic number Z even has $I = 0$. Some values of I are listed in a table inside the back cover. For a nucleus with $I = \frac{1}{2}$ (for example, ^{1}H), the possible orientations of the spin vector **I** are the same as shown in Fig. 19.11 for an electron. For a nucleus with $I = 1$, the possible orientations of **I** are those shown in Fig. 19.10.

A moving charge produces a magnetic field. We can crudely picture spin as due to a particle rotating about one of its own axes. Hence, we expect a charged particle with spin to act as a tiny magnet. The magnetic properties of a particle with spin can be described in terms of the particle's magnetic dipole moment **m**.

Consider a particle of charge Q and mass m moving in a circle of radius r with speed v. The time for one complete revolution is $t = 2\pi r/v$, and the current flow is $I = Q/t = Qv/2\pi r$. From (21.57), the magnetic moment is $|\mathbf{m}| = \pi r^2 I = Qvr/2$. The particle's orbital angular momentum (Sec. 19.4) is $L = mvr$, and so the magnetic moment can be written as $|\mathbf{m}| = QL/2m$. The angular-momentum vector **L** is perpendicular to the circle, as is the magnetic-moment vector **m**. Therefore

$$\mathbf{m} = Q\mathbf{L}/2m \qquad (21.61)$$

A nucleus has a spin angular momentum **I**, and its magnetic dipole moment is given by an equation resembling (21.61). However, instead of using the charge and mass of the nucleus, it is more convenient to use the proton charge and mass e and m_p. Moreover, because of the composite structure of the nucleus, an extra numerical factor g_N must be included. Thus the **magnetic (dipole) moment m** of a nucleus is

$$\mathbf{m} = g_N \frac{e}{2m_p} \mathbf{I} \equiv \gamma \mathbf{I} \qquad (21.62)$$

where g_N is the **nuclear g factor** and the **magnetogyric** (or **gyromagnetic**) **ratio** γ of the nucleus is defined as

$$\gamma \equiv \frac{e}{2m_p} g_N = (4.78942 \times 10^7 \text{ Hz/T}) g_N \qquad (21.63)$$

where the table of physical constants and $1 \text{ C/kg} = 1 \text{ s}^{-1} \text{ T}^{-1} = 1 \text{ Hz/T}$ [see Eq. (21.54)] were used. Present theories of nuclear structure cannot predict g_N values. They must be determined experimentally. Values of g_N, I, atomic mass, and percent abundance are given for some isotopes in a table inside the back cover. The fact that g_N for ^{1}H is not a simple number indicates that the proton has an internal structure.

From (21.62) and (21.59), the magnitude of the magnetic moment of a nucleus is

$$|\mathbf{m}| = |g_N|(e/2m_p)[I(I + 1)]^{1/2}\hbar = |\gamma|[I(I + 1)]^{1/2}\hbar$$

This equation is often written as $|\mathbf{m}| = |g_N| \beta_N [I(I + 1)]^{1/2}$, where the **nuclear magneton** β_N is a physical constant defined as $\beta_N \equiv e\hbar/2m_p = 5.0508 \times 10^{-27}$ J/T. NMR spectroscopists prefer to write equations in terms of γ rather than β_N and g_N, so we won't use β_N.

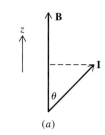

(a)

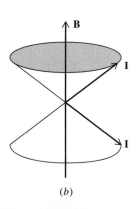

(b)

Figure 21.40

(a) Projection of **I** on the z axis. (b) The possible orientations of **I** for $I = \frac{1}{2}$ lie on the surfaces of two cones.

Nuclear-Magnetic-Resonance (NMR) Spectroscopy

In NMR spectroscopy, one applies an external magnetic field **B** to a sample containing nuclei with nonzero spin. For simplicity, we initially consider a single isolated nucleus with magnetic dipole moment **m**. The energy of the nuclear magnetic dipole in the applied field **B** depends on the orientation of **m** with respect to **B** and is given by (21.58) and (21.62) as

$$E = -\mathbf{m} \cdot \mathbf{B} = -\gamma \mathbf{I} \cdot \mathbf{B} = -\gamma |\mathbf{I}| B \cos \theta \qquad (21.64)$$

where $|\mathbf{I}|$ is the magnitude (length) of the spin-angular-momentum vector **I** and θ is the angle between **B** and **I**. Let the direction of the applied field be called the z direction. Figure 21.40a shows that $|\mathbf{I}| \cos \theta$ equals I_z, the z component of **I**. Hence, $E = -\gamma I_z B$. However, only certain orientations of **I** (and the associated magnetic moment **m**) in the field are allowed by quantum mechanics (Fig. 21.40b); I_z is quantized with the possible values $I_z = M_I \hbar$ [Eq. (21.60)] so $E = -\gamma M_I \hbar B$. The nuclear magnetic moment in the applied magnetic field therefore has the following set of quantized energy levels:

$$E = -\gamma \hbar B M_I, \qquad M_I = -I, \ldots, +I \qquad (21.65)$$

Figure 21.41 shows the allowed nuclear-spin energy levels of the nuclei ^{1}H (with $I = \frac{1}{2}$) and ^{2}H (with $I = 1$) as a function of the applied magnetic field. As B increases, the spacing between levels increases. In the absence of an external magnetic field, all orientations of the spin have the same energy.

By exposing the sample to electromagnetic radiation of appropriate frequency, one can observe transitions between these nuclear-spin energy levels. The selection rule is found to be

$$\Delta M_I = \pm 1 \tag{21.66}$$

The NMR absorption frequency satisfies $h\nu = |\Delta E| = |\gamma|\hbar B|\Delta M_I| = |\gamma|\hbar B$, where (21.65) was used. Hence

$$\nu = \frac{|\gamma|}{2\pi} B \tag{21.67}$$

[γ is negative for some nuclei since g_N in (21.63) is negative for some nuclei.] Although there are $2I + 1$ different energy levels for the nuclear magnetic dipole in the field, the selection rule (21.66) allows only transitions between adjacent levels, which are equally spaced. A collection of identical noninteracting nuclei therefore gives a single NMR absorption frequency.

Using g_N values from inside the back cover in (21.63), one finds the following $\gamma/2\pi$ values:

nucleus	^{1}H	^{13}C	^{15}N	^{19}F	^{31}P
$(\gamma/2\pi)$/(MHz/T)	42.577	10.708	-4.317	40.078	17.251

A typical magnetic field easily attained in the laboratory is 1 T (10000 G). For the ^{1}H nucleus (a proton) in this field, Eq. (21.67) and the $\gamma/2\pi$ table give the NMR absorption frequency as 42.577 MHz. This is in the radio-frequency (rf) portion of the electromagnetic spectrum. (FM radio stations broadcast from 88 to 108 MHz.)

NMR in bulk matter was first observed by Bloch and by Purcell in 1945. In his Nobel Prize acceptance speech, Purcell said: "I remember, in the winter of our first experiments, just seven years ago, looking on snow with new eyes. There the snow lay around my doorstep—great heaps of protons quietly precessing in the earth's magnetic field. To see the world for a moment as something rich and strange is the private reward of many a discovery." [*Science,* **118,** 431 (1953).]

Nuclei with $I = 0$ (for example, $^{12}_{6}$C, $^{16}_{8}$O, $^{32}_{16}$S) have no magnetic moment and no NMR spectrum. Nuclei with $I \geq 1$ have something called an electric quadrupole moment, which broadens the NMR absorption lines, tending to obscure the chemically

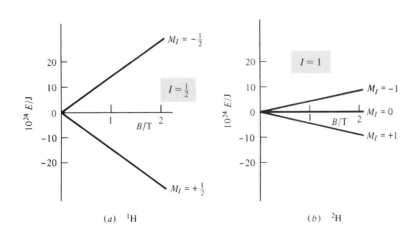

(a) ^{1}H

(b) ^{2}H

Figure 21.41

Nuclear-spin energy levels versus applied magnetic field for (a) ^{1}H; (b) ^{2}H.

Figure 21.42

An NMR spectrometer.

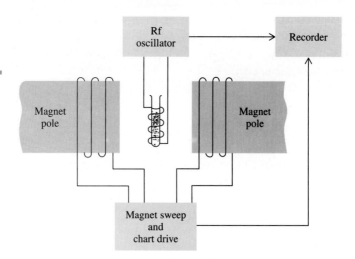

interesting details. Thus nuclei with $I = \frac{1}{2}$ are the most suitable for study. Some $I \neq \frac{1}{2}$ nuclei such as ^{2}H, ^{11}B, and ^{14}N have been studied. The most studied nuclei are ^{1}H and ^{13}C.

In spectroscopy, one usually varies the frequency ν of the incident electromagnetic radiation until absorption is observed. In NMR spectroscopy, one has the alternative of keeping ν fixed and varying the spacing between the levels by varying the magnitude B of the applied field until (21.67) is satisfied and absorption occurs. Both alternatives have been used in NMR.

Figure 21.42 shows a simplified version of an NMR spectrometer. The sample is usually a liquid. An electromagnet or permanent magnet applies a uniform magnetic field B_0. The value of B_0 is varied over a narrow range by varying the current through coils around the magnet poles. This current also drives the chart paper on which the spectrum is recorded. The sample tube is spun rapidly to average out any inhomogeneities in $\mathbf{B}_0$. A coil connected to an rf transmitter (oscillator) and wound around the sample exposes the sample to electromagnetic radiation of fixed frequency. The coil's inductance depends in part on what is inside the coil. The coil is part of a carefully tuned rf circuit in the transmitter. When the sample absorbs energy, its characteristics change, thereby detuning the transmitter circuit and decreasing the transmitter output. This decrease is recorded as the NMR signal. An alternative setup uses a separate detector coil at right angles to the transmitter coil. The flipping of the spins caused by absorption induces a current in the detector coil.

A spectrometer like that of Fig. 21.42 in which the sample is continuously exposed to rf radiation while one varies B_0 or ν in order to scan through the spectrum is called a *continuous-wave* (CW) spectrometer. CW NMR spectrometers are suitable for routine proton NMR studies, but for research work they have been superseded by Fourier-transform NMR spectrometers (these are discussed later in this section).

Recall that the intensity of an absorption line is proportional to the population difference between the levels involved (Sec. 21.2). In NMR spectroscopy, the separation ΔE between the energy levels (Fig. 21.41) is much less than kT at room temperature, and the Boltzmann distribution law $N_2/N_1 = e^{-(E_2-E_1)/kT}$ shows that there is only a slight difference between the populations of the two levels of the transition (Prob. 21.70). Hence the NMR absorption signal is quite weak, and NMR spectroscopy is hard to apply to very small samples.

Chemical Shifts

If Eq. (21.67) gave the NMR absorption frequencies for nuclei in molecules, NMR would be of no interest to chemists. However, the actual magnetic field experienced by a nucleus in a molecule differs very slightly from the applied field B_0, due to the magnetic field produced by the molecular electrons. Most ground-state molecules have all electrons paired, which makes the total electronic spin and orbital angular momenta equal to zero. With zero electronic angular momentum, the electrons produce no magnetic field. However, when an external magnetic field is applied to a molecule, this changes the electronic wave function slightly, thereby producing a slight contribution of the electronic motions to the magnetic field at each nucleus. (This effect is similar to the polarization of a molecule produced by an applied electric field.)

The magnetic field of the electrons usually opposes the applied magnetic field B_0 and is proportional to B_0. The electronic contribution to the magnetic field at a given nucleus i is $-\sigma_i B_0$, where the proportionality constant σ_i is called the **shielding constant** for nucleus i. The value of σ_i at a given nucleus depends on the electronic environment of the nucleus. For molecular protons, σ_i usually lies in the range 1×10^{-5} to 4×10^{-5}. For heavier nuclei (which have more electrons than H), σ_i may be 10^{-4} or 10^{-3}.

For each of the six protons in benzene (C_6H_6), σ_i is the same, since each proton has the same electronic environment. For chlorobenzene (C_6H_5Cl), there are three different values of σ_i for the protons, one value for the two ortho protons, one for the two meta protons, and one for the para proton. For CH_3CH_2Br, there is one value of σ_i for the CH_3 protons and a different value for the CH_2 protons. The low barrier to internal rotation about the carbon–carbon single bond makes the electronic environment of all three methyl protons the same (except at extremely low temperatures).

Addition of the electronic contribution $-\sigma_i B_0$ to the applied field B_0 gives the magnetic field B_i experienced by nucleus i as $B_i = B_0(1 - \sigma_i)$. Substitution in $\nu = |\gamma|B/2\pi$ [Eq. (21.67)] gives as the NMR frequencies of a molecule

$$\nu_i = (|\gamma_i|/2\pi)(1 - \sigma_i)B_0 \qquad (21.68)$$

where γ_i is the magnetogyric ratio of nucleus i. If one holds the frequency fixed and varies B_0, the values of the applied field B_0 at which absorption occurs are

$$B_{0,i} = \frac{2\pi\nu_{\text{spec}}}{|\gamma_i|(1 - \sigma_i)} \qquad (21.69)$$

where ν_{spec} is the fixed spectrometer frequency.

Different kinds of nuclei (^1H, ^{13}C, ^{19}F, etc.) have very different γ values, so their NMR absorption lines occur at very different frequencies. In a given NMR experiment, one examines the NMR spectrum of only one kind of nucleus, and γ_i in (21.69) has a single value. We now consider proton (^1H) NMR spectra.

Each chemically different kind of proton in a molecule has a different value of σ_i and a different NMR absorption frequency. Thus, C_6H_6 shows one NMR peak, C_6H_5Cl shows three NMR peaks, and CH_3CH_2Cl shows two NMR peaks (spin–spin coupling neglected; see the next subsection). The relative intensities of the peaks are proportional to the number of protons producing the absorption. For CH_3CH_2OH, the three peaks have a 3:2:1 ratio (Fig. 21.43). Nuclei in a molecule that have the same shielding constant as a result of either molecular symmetry (the protons in C_6H_6) or internal rotation about a single bond (the methyl protons in CH_3OH) are called **chemically equivalent.**

The variation in ν_i in (21.68) or $B_{0,i}$ in (21.69) due to variation in the chemical (that is, electronic) environment of the nucleus is called the chemical shift. The **chemical shift** δ_i of proton i is defined by

$$\delta_i \equiv (\sigma_{\text{ref}} - \sigma_i) \times 10^6 \qquad (21.70)$$

where σ_{ref} is the shielding constant for the protons of the reference compound tetramethylsilane (TMS), $(CH_3)_4Si$. All the TMS protons are equivalent, and TMS shows a single proton NMR peak. The factor 10^6 is included to give δ a convenient magnitude.

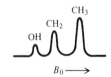

Figure 21.43

The low-resolution proton NMR spectrum of pure liquid CH_3CH_2OH.

Note that σ and δ are dimensionless. Also, the proportionality constants σ_i and σ_{ref} in (21.70) are molecular properties that are independent of the applied field B_0 and the spectrometer frequency ν_{spec}. Hence, δ_i *is independent of* B_0 *and* ν_{spec}. For a spectrometer in which B_0 is held fixed, the chemical shift δ can be expressed to a high degree of accuracy as (Prob. 21.80)

$$\delta_i = \frac{\nu_i - \nu_{ref}}{\nu_{ref}} \times 10^6 \tag{21.71}$$

where ν_{ref} and ν_i are the frequencies at which NMR absorption occurs for the reference nucleus and for nucleus i.

One finds that δ for the protons of a given kind of chemical group differs only slightly from compound to compound. Some typical proton δ values are

RCH_2R'	RCH_3	RNH_2	ROH	$CH_3C(O)R$	OCH_3	ArH	$RC(O)H$	$RCOOH$
1.1–1.5	0.8–1.2	1–4	1–6	2–3	3–4	6–9	9–11	10–13

where R and Ar are aliphatic and aromatic groups. Chemical shifts are affected by intermolecular interactions, so one usually observes the proton NMR spectrum of an organic compound in a dilute solution of an inert solvent, most commonly $CDCl_3$. The large δ range for alcohols is due to hydrogen bonding, the extent of which varies with the alcohol concentration.

NMR is an invaluable tool for structure determination.

Spin–Spin Coupling

Proton NMR spectra are more complex than we have so far indicated, because of the existence of nuclear **spin–spin coupling.** Each nucleus with spin $I \neq 0$ has a nuclear magnetic moment, and the magnetic field of this magnetic moment can affect the magnetic field experienced by a neighboring nucleus, thereby slightly changing the frequency at which the neighboring nucleus will undergo NMR absorption. Because of the rapid molecular rotation in liquids and gases, the direct nuclear spin–spin interaction averages to zero. However, there is an additional, indirect interaction between the nuclear spins that is transmitted through the bonding electrons. This interaction is unaffected by molecular rotation and causes splitting of the NMR peaks. The magnitude of the indirect spin–spin interaction depends on the number of bonds between the nuclei involved. *For protons separated by four or more bonds, this spin–spin interaction is usually negligible.* The magnitude of the spin–spin interaction between nuclei i and k is proportional to a quantity J_{ik}, called the **spin–spin coupling constant;** J_{ik} has units of frequency.

Some typical proton–proton J values in hertz are:

HC—CH	C=CH$_2$	*cis*-HC=CH	*trans*-HC=CH	HCOH	HCC(O)H
5 to 9	−3 to +3	5 to 12	12 to 19	5 to 10	1 to 3

The nonequivalent protons in CH_3CH_2Br are separated by three bonds (H—C_a, C_a—C_b, and C_b—H), and J for these protons is 7.2 Hz. The nonequivalent protons in CH_3OCH_2Br are separated by four bonds, and J is negligible for them.

The correct derivation of the NMR energy levels and frequencies allowing for spin–spin coupling requires a complicated quantum-mechanical treatment (often best done on a computer). Fortunately, for many compounds, a simple, approximate treatment called **first-order** analysis allows the spectrum to be accurately calculated. This approximation is valid provided both the following conditions hold:

1. The differences between NMR resonance frequencies of chemically different sets of protons are all much larger than the spin–spin coupling constants between nonequivalent protons.

2. There is only one coupling constant between any two sets of chemically equivalent spin-$\frac{1}{2}$ nuclei.

The table of δ values following (21.71) shows that in many cases the difference $\delta_i - \delta_j$ for chemically nonequivalent protons is equal to or greater than 1. Suppose this difference equals 1.5. From Eq. (21.68), the difference between the NMR absorption frequencies of protons i and j is $\nu_i - \nu_j = \gamma_p B_0(\sigma_j - \sigma_i)/2\pi$. But (21.70) gives $\delta_i - \delta_j = 10^6(\sigma_j - \sigma_i)$, so $\nu_i - \nu_j = \gamma_p B_0 10^{-6}(\delta_i - \delta_j)/2\pi$. Equation (21.69) gives $B_0 = 2\pi\nu_{\text{spec}}/\gamma_p$ (note that $\sigma_i \ll 1$), so

$$\nu_i - \nu_j = 10^{-6}\nu_{\text{spec}}(\delta_i - \delta_j) \tag{21.72}$$

Inexpensive commercially available proton NMR spectrometers typically have $\nu_{\text{spec}} = 60$ MHz. For $\delta_i - \delta_j = 1.5$ and $\nu_{\text{spec}} = 60$ MHz, Eq. (21.72) gives $\nu_i - \nu_j = 90$ Hz. This is substantially greater than J_{ij}, which is typically 10 Hz for protons. Hence condition 1 is met in many organic compounds. However, in many cases condition 1 is not met. For example, in CHR=CHR′, the protons have δ_i very close to δ_j, and the first-order treatment cannot be used for a 60-MHz spectrometer.

We shall illustrate the first-order treatment by applying it to CH_3CH_2OH. Consider first the CH_3 protons. They are separated by four bonds from the OH proton, and so the spin–spin interaction between these two groups is negligible. Since the methyl protons are separated by three bonds from the CH_2 protons, the spin–spin interaction between CH_2 and CH_3 protons splits the CH_3 peak. One can prove from quantum mechanics that, when the first-order treatment applies, *the spin–spin interactions between equivalent protons do not affect the spectrum.* Therefore we can ignore the spin–spin interactions between one methyl proton and another. Only the CH_2 protons affect the CH_3 peak.

Since $I = \frac{1}{2}$ for a proton, each CH_2 proton can have $M_I = +\frac{1}{2}$ or $-\frac{1}{2}$. Let up and down arrows symbolize these proton spin states. ($M_I\hbar$ is the proton spin-angular-momentum component in the z direction—the direction of the applied field B_0.) The two CH_2 protons can have the following possible nuclear-spin alignments:

$$\begin{array}{cccc} \uparrow\uparrow & \uparrow\downarrow & \downarrow\uparrow & \downarrow\downarrow \\ (a) & (b) & (c) & (d) \end{array} \tag{21.73}$$

Since the two CH_2 protons are indistinguishable, one actually takes symmetric and antisymmetric linear combinations of (b) and (c). States (a), (d), and the symmetric linear combination of (b) and (c) are analogous to the electron spin functions (19.37); the antisymmetric linear combination of (b) and (c) is analogous to (19.38). The two linear combinations of (b) and (c) each have a total M_I of 0. State (a) has a total M_I of 1. State (d) has a total M_I of -1.

In a sample of ethanol, 25% of the molecules will have the CH_2 proton spins aligned as in (a), 50% as in (b) or (c), and 25% as in (d). Alignments (b) and (c) do not affect the magnetic field experienced by the CH_3 protons, whereas alignments (a) and (d) either increase or decrease this field and so either increase or decrease the NMR absorption frequency of the CH_3 protons. The CH_2 protons therefore split the CH_3 NMR absorption peak into a triplet (Fig. 21.44). It turns out that *the frequency spacing between the lines of the triplet equals the coupling constant* $J_{CH_2CH_3}$ *between the* CH_2 *and* CH_3 *protons and is independent of the applied field* B_0. [Although one may vary the magnetic field and keep the frequency fixed, observed splittings in teslas are converted to hertz by multiplying by $\gamma_p/2\pi$; Eq. (21.67).] The preceding discussion shows that the intensity ratios of the members of the triplet are 1:2:1. These ratios deviate slightly from the experimental result in Fig. 21.44 because an approximate treatment is being used.

Figure 21.44

The high-resolution 60-MHz proton NMR spectrum of a dilute solution of CH_3CH_2OH in CCl_4 with a trace of acid. The different position of the OH peak compared with that in Fig. 21.43 is explained by hydrogen bonding in pure liquid ethanol. The relation between δ and the frequency-shift scale at the top is given by Eq. (21.72). J is the spin–spin coupling constant between CH_2 and CH_3 protons.

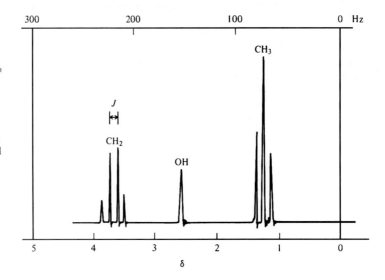

Now consider the CH_2 peak. The possible alignments of the CH_3 proton spins are

$$\uparrow\uparrow\uparrow \qquad \uparrow\uparrow\downarrow \qquad \uparrow\downarrow\uparrow \qquad \downarrow\uparrow\uparrow \qquad \uparrow\downarrow\downarrow \qquad \downarrow\uparrow\downarrow \qquad \downarrow\downarrow\uparrow \qquad \downarrow\downarrow\downarrow$$
$$(a) \qquad (b) \qquad (c) \qquad (d) \qquad (e) \qquad (f) \qquad (g) \qquad (h)$$

States (b), (c), and (d) have the same total M_I. States (e), (f), and (g) have the same total M_I. The CH_3 protons therefore act to split the CH_2 absorption into a quartet with 1:3:3:1 intensity ratios. The CH_2 protons are separated by three bonds from the OH proton, so we must also consider the effect of the OH proton. A trace of H_3O^+ or OH^- (including that coming from H_2O) will catalyze a rapid exchange of the OH protons between different ethanol molecules. This exchange eliminates the spin–spin interaction between the CH_2 protons and the OH proton, and the CH_2 peak remains a quartet with spacings equal to those in the CH_3 triplet. In pure ethanol, this exchange does not occur, and the OH proton acts to split each member of the CH_2 quartet into a doublet (corresponding to the OH proton spin states $\uparrow$ and $\downarrow$); the CH_2 absorption becomes an octet (eight lines) for pure ethanol. These eight lines are so closely spaced that it may be difficult to resolve all of them.

In ethanol containing a trace of acid or base, the OH proton NMR peak is a singlet. In pure ethanol, the OH absorption is split into a triplet by the CH_2 protons.

We have seen that two equivalent protons act to split the absorption peak of a set of adjacent protons into three lines, and three equivalent protons act to split such a peak into four lines. In general, one finds that *n equivalent protons act to split the absorption peak of a set of adjacent protons into n + 1 lines, provided the spectrum is first-order.*

What about spin–spin splittings from nuclei other than ^1H? Since ^{12}C, ^{16}O, and ^{32}S each have $I = 0$, these nuclei don't split proton NMR peaks. ^{14}N has $I = 1$; ^{35}Cl, ^{37}Cl, ^{79}Br, and ^{81}Br each have $I = \frac{3}{2}$; ^{127}I has $I = \frac{5}{2}$. It turns out that nuclei with $I > \frac{1}{2}$ generally don't split proton NMR peaks. ^{19}F has $I = \frac{1}{2}$ and does split proton NMR peaks.

For large organic molecules, the chances are good that two or more nonequivalent sets of protons will have similar δ values, making the first-order treatment invalid. The spectrum becomes very complicated and is hard to interpret. To overcome this difficulty, one can use a spectrometer with higher values of B_0 and ν_{spec}, which are proportional to each other [Eq. (21.69)]. Note from (21.72) that $\nu_i - \nu_j$ increases as ν_{spec} increases, so at sufficiently high ν_{spec} we have $\nu_i - \nu_j \gg J_{ij}$ (condition 1), and the first-order treatment applies. Values of B_0 for commercially available NMR spectrometers range from 1.4 to 21.1 T, corresponding to proton ν_{spec} values ranging from 60 to 900 MHz. The 60-MHz instrument is a CW spectrometer suitable only for routine work. Research NMR spectrometers are Fourier-transform instruments and have proton frequencies in the range 200 to 900 MHz. High-frequency, high-field NMR spectrometers use an electromagnet whose wires are made superconductors by being cooled to 4 K by liquid helium. In addition to simplifying the spectrum, an increase in B_0 increases the signal strength (Prob. 21.70), so smaller amounts of sample can be studied.

Fourier-Transform NMR Spectroscopy

The NMR spectrum of a molecule contains lines at several frequencies, and it takes about 10^3 s to scan through the spectrum using a CW spectrometer. ^{13}C NMR spectra of organic compounds provide information on the "backbone" of organic compounds. The isotope ^{13}C is present in only 1% natural abundance, which makes the ^{13}C NMR absorption signals very weak. One can scan the spectrum repeatedly and feed the results into a computer that adds the results of successive scans, thereby enhancing the signal. However, a sufficiently large number of scans takes several days and is impractical.

To overcome this difficulty, one uses **Fourier-transform** (FT) NMR spectroscopy. Here, instead of continuously exposing the sample to rf radiation while B_0 or ν is slowly varied, B_0 is kept fixed and the sample is irradiated with a very short pulse of high-power radiation from an rf transmitter whose frequency is ν_{trans}, where ν_{trans} is fixed at a value in the range of the NMR frequencies for the kind of nucleus being studied. For ^{13}C in a field of 10 T, the NMR absorption frequency is given by (21.67) as 107.1 MHz, so ν_{trans} is taken as this value. The pulse lasts for several microseconds, so it contains only a limited number of cycles of rf radiation. Because of this, one can show mathematically by a technique called Fourier analysis that the pulse of rf radiation is equivalent to a mixture of all frequencies of radiation. However, only frequencies reasonably close to ν_{trans} have significant amplitudes in the mixture, so the pulse briefly exposes the sample to a band of frequencies centered about ν_{trans}.

Chemically nonequivalent ^{13}C nuclei absorb at slightly different NMR frequencies, all in the region near 107.1 MHz, and the pulse will excite all the ^{13}C spins. After the pulse ends, one observes the signal in the detector coil for about 1 s, thereby obtaining the signal as a function of time. This signal-versus-time function is called the **free-induction decay** (FID).

For the simplest case where all the ^{13}C nuclei are chemically equivalent and ν_{trans} equals the NMR absorption frequency of these nuclei, absorption from the pulse occurs at only one frequency, and the FID function is a simple exponential decay with time, as the excited nuclei return to the ground state to reestablish the equilibrium distribution of spins (a process called **relaxation**). When several nonequivalent ^{13}C's are present in a molecule, absorption from the pulse occurs at several frequencies and the FID shows a very complicated appearance containing oscillations superimposed on an exponential decay. One can show that by taking the mathematical Fourier transform (which is a certain integral) of the FID of signal intensity versus time, one obtains the usual NMR spectrum of signal intensity versus absorption frequency. (Recall that a similar procedure is done in FT-IR; the path difference δ in Sec. 21.9 is a function of time, so one transforms from signal versus t to signal versus ν in FT-IR.) A computer built into the NMR spectrometer very rapidly does the Fourier transformation of the FID to give the absorption spectrum.

Because one needs to observe the FID for only about 1 s in order to obtain the spectrum, an FT-NMR spectrometer is much faster than a CW instrument. By adding the FIDs of many successive pulses and then performing the Fourier transformation, the spectrum's signal-to-noise ratio is increased and the ^{13}C NMR spectrum is readily observed in spite of the low abundance of ^{13}C.

Pulsed FT NMR is not restricted to ^{13}C studies, and virtually all high-quality commercial NMR spectrometers use this technique to increase sensitivity. Moreover, by using a complex sequence of pulses instead of a single pulse to produce the FID, one can obtain information that aids in assigning the signals in the spectra to the various proton and ^{13}C nuclei, and one can enhance the signals in the spectrum; see chap. 8 of *Friebolin*.

Double Resonance

In a double-resonance experiment, the sample is simultaneously exposed to rf radiation of two different frequencies, one frequency being used to observe radiation absorption and the second frequency to produce a perturbation that affects the spectrum. For example, in observing natural-abundance ^{13}C spectra in organic compounds, in addition to applying a pulse of rf radiation that covers the frequency range of the ^{13}C absorptions, one also usually applies continuous strong rf radiation whose frequencies cover the range of the proton absorption frequencies. The result is to remove the spin–spin coupling between the ^1H and ^{13}C nuclei (a process called **decoupling**), so the ^1H spins don't split the ^{13}C absorption lines. (Many other double-resonance techniques exist; see *Günther*.) Since the probability that two adjacent C nuclei are both ^{13}C is very small, there is no ^{13}C-^{13}C spin–spin splitting in natural-abundance ^{13}C NMR. *With no spin–spin splitting, the ^{13}C natural abundance spectrum contains one line for each set of nonequivalent carbons.* In ^{13}C NMR, the reference compound is $(CH_3)_4Si$ (TMS), and the ^{13}C chemical shifts in organic compounds usually lie in the range of δ values from -10 to 230. As with protons, the δ value is characteristic of the kind of carbon being observed. For example, δ for the C=O carbon in ketones is usually between 200 and 225. The combination of proton and ^{13}C NMR is an extremely powerful method of structure determination.

Dynamic NMR

Suppose we have a single nucleus moving back and forth at a frequency ν_{exch} between two environments 1 and 2 in which the magnetic field experienced by the nucleus differs; let ν_1 and ν_2 be the NMR absorption frequencies for the nucleus in environments 1 and 2. It can be shown that if the exchange frequency satisfies $\nu_{\text{exch}} \gg |\nu_1 - \nu_2|$, the NMR spectrum shows a single line at a frequency between ν_1 and ν_2, whereas if $\nu_{\text{exch}} \ll |\nu_1 - \nu_2|$, the spectrum shows a line at ν_1 and another line at ν_2. The rate of exchange varies with temperature, so by studying the temperature dependence of the NMR spectrum, one can obtain rate constants for the exchange reaction. First-order reactions with rate constants in the range 10^3 to 10^{-1} s^{-1} are most easily studied by NMR.

Figure 21.45

Dimethylformamide. Because of the partial double-bond character of the N—CO bond, the molecule is nearly planar except for the methyl hydrogens.

An example is dimethylformamide (Fig. 21.45). For this molecule, we can write a resonance structure with a double bond between C and N, and the partial double-bond character of this bond produces a substantial barrier to internal rotation. At room temperature, the 60-MHz proton NMR spectrum shows a line at $\delta = 8.0$ due to the CHO proton and shows two lines due to the methyl protons, one at $\delta = 2.79$ and one at 2.94. The presence of two methyl absorption lines shows that at 25°C the exchange rate of the two groups of methyl protons (labeled *a* and *b*) is far less than $|\nu_2 - \nu_1|$, and so the two sets of CH_3 protons are nonequivalent. (The *a* and *b* protons are separated by 4 bonds and do not split each other.) As the temperature is raised, the two methyl lines gradually coalesce, forming a single line at 120°C if a 60-MHz spectrometer is used. For 120°C and above, the internal rotation is so fast that $\nu_{\text{exch}} \gg |\nu_1 - \nu_2|$.

To obtain the rate constant at various temperatures, one must carry out a complicated quantum-mechanical analysis of the spectra for the intermediate temperatures, to find what value of k yields the observed spectrum at each T. Knowing k as a function of T, one can calculate the activation energy, which is the barrier to internal rotation, which is found to be 23 kcal/mol in dimethylformamide.

The temperature dependences of NMR spectra have been used to study the rates of internal rotation in substituted ethanes, proton exchange between alcohols and water, ring inversions, and inversion at nitrogen atoms; see *Friebolin*.

The Nuclear Overhauser Effect

Suppose the following double-resonance experiment is performed. The proton NMR spectrum of a molecule is recorded (using either a CW or FT spectrometer) while the sample is continuously irradiated with rf radiation of frequency ν_S that is the NMR absorption frequency of a specific set (which we call set S) of chemically equivalent protons in the molecule. One then finds that the intensities of all lines that are due to protons that are close to the set-S protons in the molecule are changed as compared with a spectrum taken without continuous radiation at ν_S. The radiation at ν_S changes the energy-level population distribution of the set-S protons, and the magnetic-dipole–magnetic-dipole interaction between the set-S protons and nearby protons changes the population distributions of the nearby protons, thereby changing the intensities of their NMR lines. This intensity change is the **nuclear Overhauser effect (NOE)**. The magnitude of the NOE is usually proportional to $1/r^6$, where r is the distance between the set-S protons and the protons producing the line whose intensity is changed. The NOE is negligible for $r > 4$ Å. The NOE can be used to help assign spectra and to find internuclear distances in a molecule.

For example, $(CH_3)_2NCHO$ (Fig. 21.45) at 25°C shows CH_3 absorption lines at $\delta = 2.79$ and 2.94. To determine which line goes with which set of CH_3 protons, one can use the NOE. When continuous rf radiation at the $\delta = 2.94$ frequency is applied, the intensity of the CHO proton line is increased by 18%, whereas rf radiation at $\delta = 2.79$ produces a 2% decrease in the CHO line. Hence the $\delta = 2.94$ protons must be closer to (that is, cis to) the CHO proton.

Two-Dimensional NMR Spectroscopy

In Fig. 21.46, the rectangles denote rf pulses applied to an NMR sample. The duration of each pulse is very short and is greatly exaggerated in the figure. Figure 21.46a shows a single rf pulse applied to the system, followed by observation of the FID as a function of time t. Fourier transformation of this function of t gives the NMR spectrum as a function of frequency ν. This is a one-dimensional (1D) NMR spectrum. In Fig. 21.46b, a pulse is applied, and then after a time t_1, a second pulse is applied, and then the FID is observed as a function of time t_2 up to a time $t_{2,\max}$. Fourier transformation of this function of t_2 gives the NMR spectrum as a function of frequency; the spectrum's appearance will be influenced by the first pulse. We have obtained the spectrum as a function of a single variable, so this is still a 1D spectrum.

Now suppose we repeat the experiment in Fig. 21.46b, except that we use a slightly longer time t_1 between the two pulses; we then repeat the experiment using successively larger values of the interval t_1 between the two pulses. By collecting all the FIDs from the successive experiments, we get an FID that is now a function $f(t_1, t_2)$ of two variables, t_1 and t_2. If we now do a Fourier transformation of $f(t_1, t_2)$ by integrating over both t_1 and t_2, we will get an NMR spectrum that will be a function of two frequencies ν_1 and ν_2. This spectrum is a **two-dimensional (2D) NMR spectrum**. The frequency ν_2 has the same meaning as the frequency in 1D NMR. The significance of the frequency ν_1 depends on the nature of the pulses used in the experiments. Over 100 different pulse patterns have been used in 2D NMR. Figures 21.46c and d show two other pulse patterns. In c, the time interval Δ remains fixed and t_1 is varied. In d, the second pulse is twice as long as the first.

The proton NMR spectra of large molecules such as steroids, oligosaccharides, proteins, and nucleic acids contain many overlapping lines, and the lines are extremely hard to assign to specific protons, even using very high-field spectrometers. By using 2D NMR, one can resolve very complicated spectra and determine the structure and

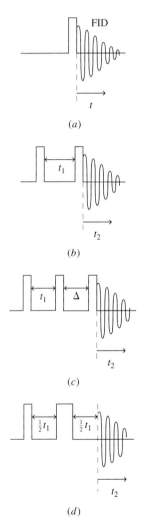

Figure 21.46

Some NMR pulse sequences.

conformation of compounds for which the 1D spectrum is hopelessly complex. The development of 2D NMR in the 1980s has produced a revolutionary increase in the power of NMR to deduce the structure of large molecules. See W. R. Croasmun and R. M. K. Carlson (eds.), *Two-Dimensional NMR Spectroscopy*, VCH, 1987; J. Schraml and J. M. Bellama, *Two-Dimensional NMR Spectroscopy*, Wiley, 1988.

For a protein whose sequence of amino acids is known, application of 2D (and 3D) NMR and the NOE enables one to determine the distances between various pairs of protons. From this information, one can deduce the three-dimensional structure (conformation) of the protein in aqueous solution. (See J. N. S. Evans, *Biomolecular NMR Spectroscopy*, Oxford, 1995; J. Cavanagh et al., *Protein NMR Spectroscopy*, Academic, 1996.) NMR structures for over 1000 proteins have been determined. [Protein structures are tabulated in the Protein Data Bank (www.rcsb.org/pdb/).] NMR spectroscopy enables structures of proteins with molecular weights up to 40000 to be determined. Methods to extend NMR structure determination to proteins with higher molecular weights are being developed (*Chem. Eng. News*, Feb. 15, 1999, p. 65).

NMR of Solids

As ordinarily observed, the NMR spectra of solids are of little value because the direct spin–spin interactions (which are averaged to zero in liquids, as noted earlier in this section), lead to broad, featureless absorptions with little information. However, by using special techniques, such as rapidly spinning the sample about an axis making a certain angle with B_0 [magic-angle spinning (MAS)], one can eliminate these spin–spin interactions and get high-resolution NMR spectra from solids. High-resolution solid-state NMR is used to study the structure and dynamics of solid polymers, biopolymers such as proteins, catalysts, molecules adsorbed on catalysts, coal, ceramics, glasses, etc. (see E. O. Stejskal and J. D. Memory, *High Resolution NMR in the Solid State*, Oxford, 1994).

Magnetic Resonance Imaging

NMR spectroscopy can be used to form cross-sectional images of body parts in living subjects by displaying the intensity of a particular NMR transition (for example, protons in water) as a function of the coordinates in the cross-sectional plane. This technique of magnetic-resonance imaging (MRI) is widely used to diagnose diseases such as cancer. See I. L. Pykett, *Scientific American,* May 1982, p. 78; E. R. Andrews, *Acc. Chem. Res.,* **16**, 114 (1983). For discussions on the controversy over the apportionment of credit for developing MRI, see B. H. Kevles, *Naked to the Bone,* Rutgers, 1997, chap. 8; *Physics Today,* Dec. 1997, pp. 100–102; *New York Times*, July 12, 1997, p. 33.

Classical Description of NMR Spectroscopy. NMR spectroscopy is a quantum-mechanical phenomenon, and a fully correct description of it must be a quantum-mechanical one. The quantum mechanics of many aspects of NMR is difficult. Hence, a classical description of NMR is widely used. Consider a nucleus with spin $I = \frac{1}{2}$ and with positive g_N. In the presence of an applied magnetic field $\mathbf{B}_0$ along the z axis, the angle θ the nuclear-spin vector $\mathbf{I}$ makes with the z axis satisfies $\cos \theta = I_z/|\mathbf{I}| = M_I \hbar/[I(I + 1)]^{1/2}\hbar = \pm\frac{1}{2}/(3/4)^{1/2} = \pm 1/3^{1/2}$, and $\theta = 54.7°$ or $125.3°$ (Fig. 21.40). As noted in Fig. 21.41a, the $M_I = +\frac{1}{2}$ orientation has lower energy. Since the magnetic moment $\mathbf{m}$ is proportional to $\mathbf{I}$, these are also the possible angles between $\mathbf{m}$ and $\mathbf{B}_0$. In classical mechanics, the interaction (21.58) between $\mathbf{m}$ and $\mathbf{B}_0$ produces a torque on $\mathbf{m}$ that makes $\mathbf{m}$ revolve around the field direction keeping its angle θ with $\mathbf{B}_0$ fixed, thereby sweeping out a conical surface. This motion is called **precession.** The precession frequency (called the *Larmor frequency*) equals $(\gamma/2\pi)\mathbf{B}_0$, where γ is the magnetogyric ratio of the nucleus.

For a collection of identical spin-$\frac{1}{2}$ nuclei in $\mathbf{B}_0$, slightly more than half the nuclei will have their $\mathbf{m}$ vectors on the cone with $\theta = 54.7°$, which has the lower energy. The magnetic moments $\mathbf{m}_i$ of the nuclei add vectorially to give a total magnetic moment $\Sigma_i \mathbf{m}_i$, and the quantity $\Sigma_i \mathbf{m}_i/V$, where V is the system's volume, is called the **magnetization M.**

Because the nuclear spins have random components in the x and y directions, the components of $\Sigma_i \mathbf{m}_i$ in the x and y directions are zero and $M_x = 0 = M_y$. Thus $\mathbf{M}$ lies along the z direction, the direction of $\mathbf{B}_0$. Note that $M_I = +\frac{1}{2}$ nuclei and $M_I = -\frac{1}{2}$ nuclei give opposite contributions to M_z, and M_z is nonzero because there is a slight excess of low-energy $M_I = +\frac{1}{2}$ nuclei at equilibrium. The magnitude of M_z is proportional to the population difference between the two states.

In pulse FT NMR spectroscopy, the magnetic field $\mathbf{B}_1$ of the pulse of electromagnetic radiation is perpendicular to the direction of $\mathbf{B}_0$ and lies in the xy plane. (In most branches of spectroscopy, the electric field of the electromagnetic radiation interacts with the electric charges of the molecule to produce the transition. In NMR, the interaction of the magnetic field of the radiation with the nuclear magnetic moments produces the transition.) The presence of the field $\mathbf{B}_1$ causes the magnetization vector $\mathbf{M}$ to precess about the direction of $\mathbf{B}_1$ at the Larmor frequency $(\gamma/2\pi)B_1$, thereby rotating $\mathbf{M}$ away from the z axis and toward the xy plane. The amount of rotation is measured by the angle θ between $\mathbf{M}$ and the z axis. The amount of rotation is proportional to the duration t_{pulse} of the pulse; $\theta = bt_{\text{pulse}}$, where b is a constant. The rotation frequency is $(\gamma/2\pi)B_1$ and the period of this rotation is the reciprocal of the frequency, namely, $2\pi/\gamma B_1$. After one period, θ will equal 2π. Hence $\theta = bt_{\text{pulse}}$ becomes $2\pi = b2\pi/\gamma B_1$ and $b = \gamma B_1$. Thus

$$\theta = \gamma B_1 t_{\text{pulse}}$$

In FT NMR, one usually chooses the pulse duration to make $\theta = 90°$ (a 90° pulse) so that the pulse moves $\mathbf{M}$ into the xy plane.

The situation $M_z = 0$ means that there are equal numbers of $M_I = +\frac{1}{2}$ and $M_I = -\frac{1}{2}$ spins. The pulse causes both absorption and stimulated emission of radiation. Because of a net absorption of radiation, the 90° pulse has excited enough low-energy $M_I = +\frac{1}{2}$ nuclei to the $M_I = -\frac{1}{2}$ state to equalize the populations of the states (a situation called *saturation*).

After the pulse, the spin system moves back toward equilibrium (a process called **relaxation**). The rate at which M_z moves back toward its equilibrium value is proportional to the quantity $1/T_1$, where T_1 is called the **spin–lattice relaxation time** or the **longitudinal relaxation time**. (The longitudinal direction is the z direction.) M_z moving back to its equilibrium value corresponds to the $M_I = \pm\frac{1}{2}$ populations moving back to equilibrium by having some high-energy spins make transitions to the low-energy $M_I = +\frac{1}{2}$ state. For the relatively low frequency of NMR transitions, it turns out that spontaneous emission of radiation is too slow to contribute significantly to this process of reestablishing the population equilibrium. Instead, spin-population relaxation occurs because interactions between the nuclear spins and their surroundings (called the "lattice" because this relaxation was first studied in crystalline solids) produce nonradiative transitions that transfer energy from the high-energy spins to molecular translational and rotational energies. The rate at which the component M_{xy} in the xy plane changes back to its equilibrium value of zero is proportional to the quantity $1/T_2$, where T_2 is the **spin–spin** (or **transverse**) **relaxation time**.

The classical vector model of NMR is useful for understanding many aspects of NMR. However, because NMR is a quantum-mechanical phenomenon, there are many NMR multiple-pulse experiments that cannot be correctly treated with the classical model and a complicated quantum-mechanical treatment is needed.

"There can be little doubt that the spectroscopic tool that has done the most for chemistry is NMR" [E. B. Wilson, *Ann. Rev. Phys. Chem.*, **30**, 1 (1979)].

21.13 ELECTRON-SPIN-RESONANCE SPECTROSCOPY

In **electron-spin-resonance** (ESR) **spectroscopy** [also called **electron paramagnetic resonance** (EPR) **spectroscopy**], one observes transitions between the quantum-mechanical energy levels of an unpaired electron spin magnetic moment in an external magnetic field. Most ground-state molecules have all electron spins paired, and such molecules show no ESR spectrum. One observes ESR spectra from free radicals such as H, CH_3, $(C_6H_5)_3C$, and $C_6H_6^-$, from transition-metal ions with unpaired electrons, and from excited triplet states of organic compounds. The sample may be solid, liquid, or gaseous.

An electron has spin quantum numbers $s = \frac{1}{2}$ and $m_s = +\frac{1}{2}$ and $-\frac{1}{2}$. Relativistic quantum mechanics and experiment show that the g value of a free electron is $g_e = 2.0023$. Analogous to the equation $\mathbf{m} = (g_N e/2m_p)\mathbf{I}$ [Eq. (21.62)] for a nuclear spin, the magnetic dipole moment of a free electron is

$$\mathbf{m}_e = (-g_e e/2m_e)\mathbf{S}, \qquad g_e \approx 2 \qquad (21.74)$$

where $-e$, m_e, and $\mathbf{S}$ are the electron charge, mass, and spin-angular-momentum vector. The magnitude of $\mathbf{S}$ is $[s(s+1)]^{1/2}\hbar = \frac{1}{2}\sqrt{3}\hbar$. An electron spin magnetic moment has two energy levels in an applied magnetic field, corresponding to the two orientations $m_s = +\frac{1}{2}$ and $m_s = -\frac{1}{2}$. The magnetic field B experienced by an unpaired electron in a molecular species differs somewhat from the applied field B_0, and we write $B = B_0(1 - \sigma)$, where σ is a shielding constant. The energy levels are [Eqs. (21.58) and (21.74)]

$$E = -\mathbf{m}_e \cdot \mathbf{B} = (g_e e/2m_e)\mathbf{S} \cdot \mathbf{B} = (g_e e/2m_e)B_0(1 - \sigma)S_z$$
$$= (g_e e/2m_e)B_0(1 - \sigma)(\pm\tfrac{1}{2}\hbar)$$
$$E = \pm\tfrac{1}{2}g_e\beta_e(1 - \sigma)B_0$$

where the **Bohr magneton** β_e is

$$\beta_e \equiv e\hbar/2m_e = 9.274 \times 10^{-24} \text{ J/T} \qquad (21.75)$$

(Don't confuse the electron mass m_e with the electron magnetic moment $\mathbf{m}_e$.) The energy-level separation is $g_e\beta_e(1 - \sigma)B_0$ and equating this to $h\nu$, we get the ESR transition frequency as

$$\nu = g_e\beta_e(1 - \sigma)B_0/h = g\beta_e B_0/h \qquad (21.76)$$

where the **g factor** for the molecular species is $g \equiv g_e(1 - \sigma)$. For organic radicals, g factors are close to the free-electron value $g_e = 2.0023$. For transition-metal ions, g can differ greatly from g_e. For $B_0 = 1$ T and $g = 2$, Eq. (21.76) gives the ESR frequency as 28 GHz, which is in the microwave region. ESR frequencies are much higher than NMR frequencies because the electron mass is 1/1836 the proton mass. Figure 21.47 is a block diagram of an ESR spectrometer. The klystron generates a fixed frequency of microwave radiation, which is propagated along the waveguide. The magnetic field applied to the sample is varied over a range to generate the spectrum.

Interaction between the electron spin magnetic moment and the nuclear spin magnetic moments splits the ESR absorption peak into several lines (*hyperfine splitting*). Both protons and electrons have spin $\frac{1}{2}$. Just as a proton NMR absorption peak is split into $n + 1$ peaks by n adjacent equivalent protons, the ESR absorption peak of a radical with a single unpaired electron is split into $n + 1$ peaks by n equivalent protons. For example, the ESR spectrum of the methyl radical $\cdot CH_3$ consists of 4 lines (Fig. 21.48). If the unpaired electron interacts with one set of m equivalent protons and a second set of n equivalent protons, the ESR spectrum has $(n + 1)(m + 1)$ lines. For example, the $\cdot CH_2CH_3$ ESR spectrum has 12 lines. The ^{12}C nuclei have $I = 0$ and don't split the ESR lines.

For the n-butyl radical (Fig. 21.49), the H's on the α carbon and the H's on the β carbon produce substantial ESR splitting; the H's on the γ carbon produce a very small splitting, which may or may not be resolved, depending on the quality of the spectrometer used. The H's on the δ carbon produce a negligible splitting. The n-butyl ESR spectrum will thus show either $3(3) = 9$ lines or $3(3)(3) = 27$ lines, depending

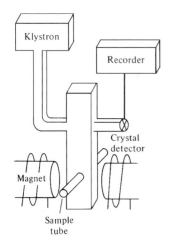

Figure 21.47

An ESR spectrometer.

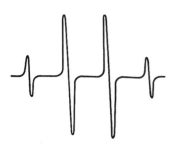

Figure 21.48

The ESR spectrum of the methyl radical. ESR spectrometers usually display the derivative (slope) of the absorption line. This slope is positive for the left half of the line and negative for the right half.

$$\overset{\delta}{C}H_3\overset{\gamma}{C}H_2\overset{\beta}{C}H_2\overset{\alpha}{C}H_2\cdot$$

Figure 21.49

The n-butyl radical.

on the spectrometer resolution. In radicals such as $C_6H_6^+$ and $C_6H_6^-$ formed from conjugated ring compounds, the odd electron is delocalized over the whole molecule and all the H's contribute to splitting the ESR lines.

ESR spectroscopy can detect free-radical reaction intermediates.

One can use ESR spectroscopy to obtain information on biological molecules by bonding an organic free radical to the macromolecule under study, a procedure called *spin labeling*. (See *Chang*, chap. 6; *Campbell and Dwek*, chap. 7.)

21.14 OPTICAL ROTATORY DISPERSION AND CIRCULAR DICHROISM

A molecule that rotates the plane of polarization of plane-polarized light (Fig. 21.1) is said to be **optically active.** Plane-polarized light is produced by passing unpolarized light through a polarizing crystal. A molecule that is not superimposable on its mirror image is optically active (provided the molecule and its mirror image cannot be interconverted by rotation about a bond with a low rotational barrier).

The **angle of optical rotation** α is the angle through which the light's electric-field vector has been rotated by passing through the sample. Most commonly, the optically active substance is in solution. α is proportional to the sample length l and to the concentration of optically active material in the solution. To get a quantity characteristic of the optically active substance, one defines the **specific rotation** $[\alpha]_\lambda$ of the optically active substance B as

$$[\alpha]_\lambda \equiv \frac{\alpha_\lambda}{[\rho_B/(g/cm^3)](l/dm)} \tag{21.77}$$

where ρ_B, the mass concentration of B, is the mass of B per unit volume in the solution [Eq. (9.2)] and l is the path length of the light through the sample. For a pure sample, ρ_B becomes the density of pure B. The optical rotation α and the specific rotation $[\alpha]$ depend on the radiation's wavelength (as indicated by the λ subscript) and depend on the temperature and solvent. If the polarization plane is rotated to the right (clockwise) as viewed looking toward the oncoming beam, the substance is *dextrorotatory* and $[\alpha]_\lambda$ is defined as positive. If the rotation is counterclockwise, the substance is *levorotatory* and $[\alpha]_\lambda$ is negative.

A plot of $[\alpha]_\lambda$ versus λ gives the **optical rotatory dispersion** (ORD) spectrum of the substance. $[\alpha]_\lambda$ changes very rapidly in a wavelength region where the optically active substance has an electronic absorption, and the ORD spectrum is of most interest in such regions—the UV and the visible.

Related to ORD is circular dichroism. For an optically active substance, the molar absorption coefficient ε_λ (Sec. 21.2) for left-circularly polarized light differs (typically by 0.01 to 0.1%) from ε_λ for right-circularly polarized light. **Circularly polarized** light is light in which the electric-field direction rotates about the direction of propagation as one moves along the wave, making one complete rotation in one wavelength of the light. The polarization is left or right depending on whether the field vector at a fixed position rotates counterclockwise or clockwise with time, respectively, when viewed from in front of the oncoming beam. Circularly polarized light can be produced by special devices.

The **circular dichroism** (CD) spectrum of a substance is a plot of $\Delta\varepsilon \equiv \varepsilon_L - \varepsilon_R$ versus light wavelength λ, where ε_L and ε_R are molar absorption coefficients [Eq. (21.10)] for left- and right-circularly polarized light. The circular dichroism $\Delta\varepsilon$ is nonzero only in regions of absorption and is most commonly measured in the UV and visible regions, the regions of electronic absorption. (Instead of $\Delta\varepsilon$, the *molar ellipticity* $[\theta] \equiv 3298\Delta\varepsilon$ is sometimes plotted.)

Many biological molecules are optically active. The ORD and CD spectra of proteins and nucleic acids are sensitive to the conformations of these macromolecules and can be used to follow changes in conformation as functions of temperature, binding of ligands, etc. Currently, CD spectra are used much more often than ORD.

By comparing the UV CD spectrum of a protein with CD spectra of proteins with known conformations one can (using one of many available computer programs) estimate with a fair degree of accuracy the percentages of α helix, parallel β sheet, antiparallel β sheet, etc., in the protein.

CD spectra are used to study the kinetics of protein folding and unfolding and have been used to study conformational changes in prions, infectious proteins believed responsible for such neurodegenerative diseases as mad-cow disease and Creutzfeldt–Jakob disease. The A, B, and Z forms of DNA can be distinguished from one another using their CD spectra.

CD spectroscopy is also used to measure $\Delta\varepsilon$ for IR absorption bands. By comparing a molecule's observed vibrational CD (VCD) spectrum with the spectrum predicted by a density-functional calculation (Sec. 20.10) for an assumed configuration, one can determine the absolute configuration of chiral molecules with up to 200 atoms.

For more on circular dichroism, see N. Berova and R. Woody (eds.), *Circular Dichroism*, 2nd ed., Wiley, 2000; G. D. Fasman (ed.), *Circular Dichroism and the Conformational Analysis of Biomolecules*, Plenum, 1996.

Measurement of the difference in Raman scattering intensity for right- and left-circularly polarized light as a function of Raman shift gives the branch of spectroscopy called *Raman optical activity* (ROA). ROA gives information about molecular configuration and conformation.

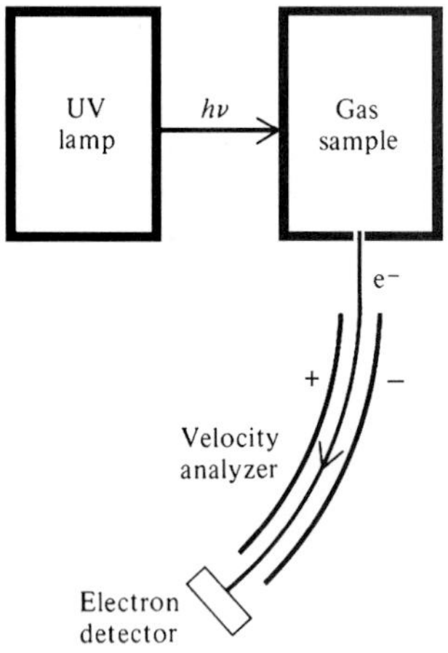

Figure 21.50

A UV photoelectron spectrometer. Electrons leave the sample chamber through a narrow slit. For a given value of the voltage difference between the charged plates of the velocity analyzer, only electrons with a particular velocity will reach the detector. Electrons with other speeds will hit one of the charged plates. By varying this voltage difference, one allows electrons with different speeds to reach the detector. [Velocity analyzers are used in EELS (Fig. 21.30) to produce a monoenergetic electron beam and to measure electron energies after reflection from the surface.]

21.15 PHOTOELECTRON SPECTROSCOPY

Conventional spectroscopy studies *photons* absorbed, emitted, or scattered by molecules. **Photoelectron spectroscopy** (PES), developed about 1960, studies the kinetic energies of *electrons* emitted when molecules of a sample are ionized by absorption of high-energy monochromatic radiation. In UV PES, the ionizing radiation (which is usually 58.4-nm UV radiation from a helium discharge tube) is passed through a gas sample and the energies of the emitted electrons (called *photoelectrons*) are found from their deflections in an electric or magnetic field (Fig. 21.50).

If $h\nu$ is the energy of the incident photon, $E(M)$ and $E(M^+)$ are the energies of the molecule M and the ion M^+ formed by the ionization, and $E(e^-)$ is the kinetic energy of the photoelectron, conservation of energy gives $h\nu + E(M) = E(e^-) + E(M^+)$, or

$$E(M^+) - E(M) = h\nu - E(e^-) \qquad (21.78)$$

Since $h\nu$ is known (it is 21.2 eV for 58.4-nm radiation), measurement of the electron's energy $E(e^-)$ gives the energy difference $E(M^+) - E(M)$ between the ion formed and the original molecule.

One finds that electrons with several different kinetic energies are emitted from a sample, and a plot of the rate of electron emission versus electron kinetic energy or versus ionization energy $E(M^+) - E(M)$ gives the photoelectron spectrum. Figure 21.51 sketches the UV photoelectron spectrum of $N_2(g)$ using 21.2 eV photons. Two horizontal scales are shown, one giving $E(e^-)$ and one giving the more interesting quantity $E(M^+) - E(M)$. Since $h\nu = 21.2$ eV, the sum of these scales is 21.2 eV at any point. Figure 21.51 shows three groups of lines, each group being called a *band*. From Fig. 20.14, the ground-state MO configuration of N_2 is

$$(\sigma_g 1s)^2(\sigma_u^* 1s)^2(\sigma_g 2s)^2(\sigma_u^* 2s)^2(\pi_u 2p)^4(\sigma_g 2p)^2$$

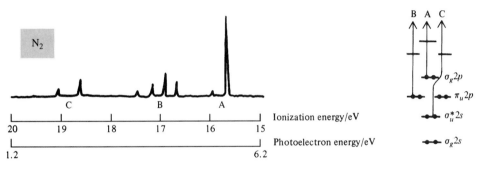

Figure 21.51

Sketch of the N_2 photoelectron spectrum produced by 21.2 eV photons. The MO diagram at the right shows which electrons are being ionized.

Each band results from removal of an electron from a different valence MO. The lowest ionization-energy band, the one at $15\frac{3}{4}$ eV, is produced by photoelectrons removed from the highest-energy MO, the $\sigma_g 2p$ MO, to give an N_2^+ ion with one $\sigma_g 2p$ electron (this gives the ground electronic state of N_2^+). The band at 17 eV results from the loss of a $\pi_u 2p$ electron and that at 19 eV from the loss of a $\sigma_u^* 2s$ electron; in each of these two cases, an excited electronic state of N_2^+ is produced. The 21.2-eV photons do not have enough energy to remove an electron from the $\sigma_g 2s$ MO or from the inner-shell orbitals.

Each band consists of a number of lines. Although the room-temperature un-ionized N_2 molecules are nearly all in their ground vibrational state, the N_2^+ ions are formed in various vibrational states. Each line in a band corresponds to a different vibrational state of the appropriate N_2^+ electronic state. The spacing between lines of a band is approximately equal to $h\nu_{vib}$, where ν_{vib} is the vibrational frequency of the N_2^+ electronic state.

By Koopmans' theorem (Sec. 20.8), the energy needed to remove an electron from a given MO is approximately equal to the Hartree–Fock orbital energy of the MO. The photoelectron spectrum provides a direct determination of MO energies and contributes to knowledge of molecular electronic structure.

Besides UV PES, another branch of photoelectron spectroscopy is x-ray PES, which uses x-ray photons to produce ionization of a gas or solid sample. The high-energy x-ray photons can remove an electron from an inner-shell orbital as well as from a valence MO, so one obtains binding energies for inner-shell electrons. The $1s$ energies of atoms vary rapidly with atomic number (Fig. 19.15), and measurement of the x-ray photoelectron spectrum allows both qualitative and quantitative analysis for elements present in a sample. For this reason, x-ray PES is often called *electron spectroscopy for chemical analysis* (**ESCA**). Typical $1s$ binding energies for second-row elements are shown in Fig. 21.52.

The binding energy of inner-shell electrons depends slightly on the electronic environment of the atom in the molecule. For example, the carbon $1s$ energies in CH_4 and CH_3F are 290.8 and 293.6 eV, respectively. Within a single compound, each nonequivalent carbon atom gives a $1s$ peak at a slightly different energy. For example, the ESCA spectrum of $CF_3COOCH_2CH_3$ has four carbon $1s$ peaks of equal intensity at ionization energies of 285, 286, 289, and 292 eV. Such ionization-energy differences for nonequivalent atoms of a given kind in a compound are called *chemical shifts* and (like NMR chemical shifts) provide structural information.

The observed ESCA electrons from a solid sample are emitted from a layer typically 20 to 100 Å deep. Electrons from greater depths are lost as a result of collisions

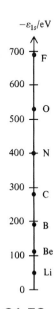

Figure 21.52

Typical $1s$ binding energies. (See also Fig. 19.15.)

with atoms of the solid. ESCA is used to study surface structures of solid catalysts and to study gases adsorbed on solids.

21.16   PHOTOCHEMISTRY

Photochemistry

Photochemistry is the study of chemical reactions produced by light. Absorption of a photon of light may raise a molecule to an excited electronic state, where it will be more likely to react than in the ground electronic state. In **photochemical reactions,** the activation energy is supplied by absorption of light. In contrast, the reactions studied in Chapter 17 are **thermal reactions,** in which the activation energy is supplied by intermolecular collisions.

The energy of a photon is $E_{\text{photon}} = h\nu = hc/\lambda$. The energy of one mole of photons is $N_A h\nu$. We find the following values of photon wavelength, energy, and molar energy:

λ/nm	200 (UV)	400 (violet)	700 (red)	1000 (IR)	
E_{photon}/eV	6.2	3.1	1.8	1.2	(21.79)
$N_A h\nu$/(kJ/mol)	598	299	171	120	

Since it usually takes at least $1\frac{1}{2}$ or 2 eV to put a molecule into an excited electronic state, photochemical reactions are initiated by UV or visible light.

Ordinarily, the number of photons absorbed equals the number of molecules making a transition to an excited electronic state. This is the **Stark–Einstein law** of photochemistry.

In exceptional circumstances, this law is violated. A high-power laser beam provides a very high density of photons (Prob. 21.91). There is some probability that a molecule will be hit almost simultaneously by two laser-beam photons, producing a transition in which a single molecule absorbs two photons at once. (For a discussion of two-photon absorption spectroscopy, see *Hollas,* sec. 8.2.14.) In rare cases, a single photon can excite two molecules in contact with each other. Liquid O_2 is light blue because of absorption of 630-nm (red) light; each photon absorbed excites two colliding O_2 molecules to the lowest-lying O_2 excited electronic state [E. A. Ogryzlo, *J. Chem. Educ.,* **42,** 647 (1965)].

Photochemical reactions are of tremendous biological importance. Most plant and animal life on earth depends on photosynthesis, a process in which green plants synthesize carbohydrates from CO_2 and water:

$$6CO_2 + 6H_2O \rightarrow C_6H_{12}O_6(\text{glucose}) + 6O_2 \qquad (21.80)$$

The reverse of this reaction provides energy for plants and animals. For (21.80), $\Delta G°$ is 688 kcal/mol, so the equilibrium lies far to the left in the absence of light. The presence of light and of the green pigment chlorophyll makes reaction (21.80) possible. Chlorophyll contains a conjugated ring system that allows it to absorb visible radiation. The main absorption peaks of chlorophyll are at 450 nm (blue) and 650 nm (red). Photosynthesis requires eight photons per molecule of CO_2 consumed.

The process of vision depends on photochemical reactions, such as the dissociation of the retinal pigment rhodopsin after it absorbs visible light. Other important photochemical reactions are the formation of ozone from O_2 in the earth's stratosphere, the formation of photochemical smog from automobile exhausts, the reactions in photography, and the formation of vitamin D and skin cancer by sunlight.

Photochemical reactions are more selective than thermal reactions. By using monochromatic light, we can excite one particular species in a mixture to a higher electronic state. (The monochromaticity, high power, and tunability of lasers make them ideal sources for photochemical studies.) In contrast, heating a sample increases the translational, rotational, and vibrational energies of all species. Organic chemists use photochemical reactions as a tool in syntheses.

Certain chemical reactions yield products in excited electronic states. Decay of these excited states may then produce emission of light, a process called **chemiluminescence.** Fireflies and many deep-sea fish show chemiluminescence. In a sense, chemiluminescence is the reverse of a photochemical reaction.

Consequences of Light Absorption

Let B* and B_0 denote a B molecule in an excited electronic state and in the ground electronic state, respectively. The initial absorption of radiation is $B_0 + h\nu \rightarrow B^*$. In most cases, the ground electronic state is a singlet with all electron spins paired. The selection rule $\Delta S = 0$ (Sec. 21.11) then shows that the excited electronic state B* is also a singlet.

Following light absorption, many things can happen.

The B* molecule is usually produced in an excited vibrational level. Intermolecular collisions (especially collisions with the solvent if the reaction is in solution) can transfer this extra vibrational energy to other molecules, causing B* to lose most of its vibrational energy and attain an equilibrium population of vibrational levels, a process called **vibrational relaxation.**

The B* molecule can lose its electronic energy by spontaneously emitting a photon, thereby falling to a lower singlet state, which may be the ground electronic state: $B^* \rightarrow B_0 + h\nu$. Spontaneous emission of radiation by an electronic transition in which the total electronic spin doesn't change ($\Delta S = 0$) is called **fluorescence.** Fluorescence is favored in low-pressure gases, where the time between collisions is relatively long. A typical lifetime of an excited singlet electronic state is 10^{-8} s in the absence of collisions.

The B* molecule can transfer its electronic excitation energy to another molecule during a collision, thereby returning to the ground electronic state, a process called **radiationless deactivation:** $B^* + C \rightarrow B_0 + C$, where B_0 and C on the right have extra translational, rotational, and vibrational energies.

The B* molecule (especially after undergoing vibrational relaxation) can make a radiationless transition to a different excited electronic state: $B^* \rightarrow B^{*\prime}$. Conservation of energy requires that B* and $B^{*\prime}$ have the same energy. Generally, the molecule $B^{*\prime}$ has a lower electronic energy and a higher vibrational energy than B*.

If B* and $B^{*\prime}$ are both singlet states (or both triplet states), then the radiationless process $B^* \rightarrow B^{*\prime}$ is called **internal conversion.** If B* is a singlet electronic state and $B^{*\prime}$ is a triplet (or vice versa), then $B^* \rightarrow B^{*\prime}$ is called **intersystem crossing.** Recall that a triplet state has two unpaired electrons and total electronic spin quantum number $S = 1$.

Suppose $B^{*\prime}$ is a triplet electronic state. It can lose its electronic excitation energy and return to the ground electronic state during an intermolecular collision or by intersystem crossing to form B_0 in a high vibrational energy level. In addition, $B^{*\prime}$ can emit a photon and fall to the singlet ground state B_0. Emission of radiation with $\Delta S \neq 0$ is called **phosphorescence.** Phosphorescence violates the selection rule $\Delta S = 0$ and has a very low probability of occurring. The lifetime of the lowest-lying excited triplet electronic state is typically 10^{-3} to 1 s in the absence of collisions.

[The term **luminescence** refers to any emission of light by electronically excited species, and includes fluorescence and phosphorescence (which are emissions that follow electronic excitation by absorption of light), chemiluminescence, luminescence following collisions with electrons (as in gas-discharge tubes or from television screens), etc.]

Figure 21.53

Photophysical processes. Dashed arrows indicate radiationless transitions. S_0 is the ground (singlet) electronic state. S_1 and S_2 are the lowest two excited singlet electronic states. T_1 is the lowest triplet electronic state. For simplicity, vibration–rotation levels are omitted.

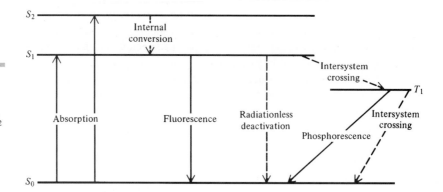

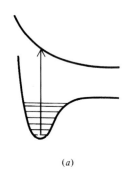

(a)

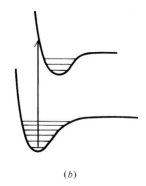

(b)

Figure 21.54

Electronic absorptions in a diatomic molecule that always lead to dissociation.

Figure 21.53 summarizes the preceding processes.

Besides the above physical processes, absorption of light can cause several kinds of chemical processes.

Since B* is often formed in a high vibrational level, the B* molecule may have enough vibrational energy to dissociate: B* → R + S. The decomposition products R and S may react further, especially if they are free radicals. If B* is a diatomic molecule with vibrational energy exceeding the dissociation energy D_e of the excited electronic state, then dissociation occurs in the time it takes one molecular vibration to occur, 10^{-3} s. For a polyatomic molecule with enough vibrational energy to break a bond, dissociation may take a while to occur. There are many vibrational modes, and it requires time for vibrational energy to flow into the bond to be broken. Excitation of a diatomic molecule to a repulsive electronic state [one with no minimum in the $E_e(R)$ curve] always causes dissociation. Excitation of a diatomic molecule to a bound excited electronic state with a minimum in the $E_e(R)$ curve causes dissociation if the vibrational energy of the excited molecule exceeds D_e. See Fig. 21.54. Sometimes a molecule undergoes internal conversion from a bound excited state to a repulsive excited electronic state, which then dissociates.

The vibrationally excited B* molecule may isomerize: B* → P. Many cis–trans isomerizations can be carried out photochemically.

The B* molecule may collide with a C molecule, the excitation energy of B* providing the activation energy for a bimolecular chemical reaction: B* + C → R + S.

The B* molecule may collide with an unexcited B or C molecule to form the excimer (BB)* or the exciplex (BC)* (Sec. 21.2), species stable only in an excited electronic state. This is especially common in solutions of aromatic hydrocarbons. The excimer or exciplex may then undergo fluorescence [(BB)* → 2B + $h\nu$ or (BC)* → $h\nu$ + B + C] or nonradiative decay to 2B or B + C.

The B* molecule may transfer its energy in a collision to another species D, which then undergoes a chemical reaction. Thus, B* + D → B + D*, followed by D* + E → products; alternatively, B* + D → B + P + R. This process is **photosensitization.** The species B functions as a photochemical catalyst. An example is photosynthesis, where the photosensitizer is chlorophyll.

All these chemical processes can be preceded by internal conversion or intersystem crossing, B* → B*′, so that it is B*′ that reacts.

The many possible chemical and physical processes make it hard to deduce the precise sequence of events in a photochemical reaction.

Photochemical Kinetics

A common setup for kinetic study of a photochemical reaction exposes the sample to a continuous beam of nearly monochromatic radiation. Of course, only radiation that is absorbed is effective in producing reaction. For example, exposing acetaldehyde to

400-nm radiation will have no effect, since radiation with a wavelength less than 350 nm is required to excite acetaldehyde to a higher electronic level. According to the Beer–Lambert law (21.11), the intensity I of radiation varies over the length of the reaction cell. Convection currents (and perhaps stirring) are usually sufficient to maintain a near-uniform concentration of reactants over the cell length, despite the variation in I.

As in any kinetics experiment, one follows the concentration of a reactant or product as a function of time. In addition, one measures the rate of absorption of light energy by comparing the energies reaching radiation detectors (such as photoelectric cells) after the beam passes through two side-by-side cells, one filled with the reaction mixture and one empty (or filled with solvent only).

The initial step in a photochemical reaction is

$$(1) \quad B + h\nu \rightarrow B^* \tag{21.81}$$

For the elementary process (21.81), the reaction rate is $r_1 \equiv d[B^*]/dt$, where $[B^*]$ is the molar concentration of B^*. From the Stark–Einstein law, r_1 equals $\mathcal{I}_a$, where $\mathcal{I}_a$ is defined as the number of moles of photons absorbed per second and per unit volume; $r_1 = \mathcal{I}_a$. We assume that B is the only species absorbing radiation. Let the reaction cell have length l, cross-sectional area $\mathcal{A}$, and volume $V = \mathcal{A}l$. Let I_0 and I_l be the intensities of the monochromatic beam as it enters the cell and as it leaves the cell, respectively. The intensity I is the energy that falls on unit cross-sectional area per unit time, so the radiation energy incident per second on the cell is $I_0 \mathcal{A}$ and the energy emerging per second is $I_l \mathcal{A}$. The energy absorbed per second in the cell is $I_0 \mathcal{A} - I_l \mathcal{A}$. Dividing by the energy $N_A h\nu$ per mole of photons and by the cell volume, we get $\mathcal{I}_a$, the moles of photons absorbed per unit volume per second:

$$r_1 = \mathcal{I}_a = \frac{I_0 \mathcal{A} - I_l \mathcal{A}}{V N_A h\nu} = \frac{I_0}{l N_A h\nu}(1 - e^{-\alpha[B]l}) \tag{21.82}$$

where the Beer–Lambert law (21.11) was used. In (21.82), $\alpha = 2.303\varepsilon$, where ε is the molar absorption coefficient of B at the wavelength used in the experiment.

The **quantum yield** Φ_X of a photochemical reaction is the number of moles of product X formed divided by the number of moles of photons absorbed. Division of numerator and denominator in this definition by volume and time gives

$$\Phi_X = \frac{d[X]/dt}{\mathcal{I}_a} \tag{21.83}$$

Quantum yields vary from 0 to 10^6. Quantum yields less than 1 are due to deactivation of B^* molecules by the various physical processes discussed above and to recombination of fragments of dissociation. The quantum yield of the photochemical reaction $H_2 + Cl_2 \rightarrow 2HCl$ with 400-nm radiation is typically 10^5. Absorption of light by Cl_2 puts it into an excited electronic state that "immediately" dissociates into Cl atoms. The Cl atoms then start a chain reaction (Sec. 17.13), yielding many, many HCl molecules for each Cl atom formed.

An example of photochemical kinetics is the dimerization of anthracene ($C_{14}H_{10}$), which occurs when a solution of anthracene in benzene is irradiated with UV light. A simplified version of the accepted mechanism is:

$$(1) \quad A + h\nu \rightarrow A^* \qquad r_1 = \mathcal{I}_a$$

$$(2) \quad A^* + A \rightarrow A_2 \qquad r_2 = k_2[A^*][A]$$

$$(3) \quad A^* \rightarrow A + h\nu' \qquad r_3 = k_3[A^*]$$

$$(4) \quad A_2 \rightarrow 2A \qquad r_4 = k_4[A_2]$$

where A is anthracene. Step (1) is absorption of a photon by anthracene to raise it to an excited electronic state; Eq. (21.82) gives $r_1 = \mathcal{I}_a$. Step (2) is dimerization. Step (3) is fluorescence. Step (4) is a unimolecular decomposition of the dimer.

The rate r for the overall reaction $2A \rightarrow A_2$ is

$$r = d[A_2]/dt = k_2[A^*][A] - k_4[A_2] \tag{21.84}$$

Use of the steady-state approximation for the reactive intermediate A* gives

$$d[A^*]/dt = 0 = \mathcal{I}_a - k_2[A][A^*] - k_3[A^*] \tag{21.85}$$

which gives $[A^*] = \mathcal{I}_a/(k_2[A] + k_3)$. Substitution in (21.84) gives

$$r = \frac{k_2[A]\mathcal{I}_a}{k_2[A] + k_3} - k_4[A_2] \tag{21.86}$$

Note that $\mathcal{I}_a$ depends on [A] in a complicated way [Eq. (21.82) with B replaced by A].

The quantum yield is given by (21.83) as

$$\Phi_{A_2} = \frac{d[A_2]/dt}{\mathcal{I}_a} = \frac{r}{\mathcal{I}_a} = \frac{k_2[A]}{k_2[A] + k_3} - \frac{k_4}{\mathcal{I}_a}[A_2] \tag{21.87}$$

If $k_4 = 0$ (no reverse reaction) and $k_3 = 0$ (no fluorescence), then Φ becomes 1. The first fraction on the right can be written as $k_2/(k_2 + k_3/[A])$; an increase in [A] increases Φ, since it increases r_2 (dimerization) compared with r_3 (fluorescence). This is the observed behavior. A typical Φ for this reaction is 0.2.

Instead of dealing with the individual rates of all the physical and chemical processes that follow absorption of radiation, one often adopts the simplifying approach of writing the initial step of the reaction as

$$\text{(I)} \quad B + h\nu \rightarrow R + S \qquad r_1 = \phi \mathcal{I}_a$$

Here, R and S are the first chemically different species formed following the absorption of radiation by B. Step I really summarizes several processes, namely, absorption of radiation by B to give B*, deactivation of B* by collisions and fluorescence, decomposition (or isomerization) of B* to R and S, and recombination of R and S immediately after their formation (recall the cage effect). The quantity ϕ, called the **primary quantum yield,** varies between 0 and 1. The greater the degree of collisional and fluorescent deactivation of B*, the smaller ϕ is. For absorption by gas-phase diatomic molecules that leads to dissociation, the dissociation is so rapid that deactivation is usually negligible and $\phi \approx 1$.

The Photostationary State

When a system containing a chemical reaction in equilibrium is placed in a beam of radiation that is absorbed by one of the reactants, the rate of the forward reaction is changed, thereby throwing the system out of equilibrium. Eventually a state will be reached in which the forward and reverse rates are again equal. This state will have a composition different from that of the original equilibrium state and is a **photostationary state.** It is a steady state (Sec. 1.2) rather than an equilibrium state, because removal of the system from its surroundings (the radiation beam) will alter the system's properties. An important photostationary state is the ozone layer in the earth's stratosphere (Sec. 17.16).

21.17 GROUP THEORY

Molecular symmetry elements and operations were discussed in Sec. 21.5. The full application of molecular symmetry uses the mathematics of group theory. This section gives an introduction to group theory and omits most proofs. For fuller details, see

Cotton or *Schonland*. Much of the material of this section is abstract, and is not as easy to read as a mystery thriller (but might well be mystifying the first time you read it).

Groups

Let $A, B, C, \ldots$ be a collection of entities, all of which are different from one another. The entities $A, B, C, \ldots$ might or might not be numbers. Let the symbol $*$ denote a specific rule for combining any two of the entities $A, B, C, \ldots$ to yield a third entity called the **product** of the two entities. For example, the equation $B*F = M$ says that M is the product of B and F. The rule for combining entities can be any well-defined rule, not necessarily ordinary multiplication. The entities $A, B, C, \ldots$ are said to form a **group** under the rule of combination $*$ if the following four conditions are satisfied: (*a*) closure, (*b*) associativity, (*c*) the existence of an identity entity, (*d*) the existence of an inverse for each entity. The meanings of these four conditions will be defined in the example that follows. The entities $A, B, C, \ldots$ that constitute the group are called the **elements** or **members** of the group.

Consider all the integers (whole numbers), positive, negative, and zero. Let the rule of combination be ordinary addition, so that $B*F$ becomes $B + F$.

The **closure** requirement means that if B and F are any two elements of the group (including the case where B and F are the same element), then the product $B*F$ is an element of the group. Since the sum of two integers is always an integer, the closure requirement is satisfied in this example.

The **associativity** requirement means that the rule of combination has the property that $(B*F)*J = B*(F*J)$, for all elements of the group. Since $(B + F) + J = B + (F + J)$, associativity holds. [Associativity should not be taken for granted. Is $(B/F)/J$ equal to $B/(F/J)$?]

The **identity element** I is a particular element of the group that has the property that $B*I = I*B = B$ for every element B in the group. For our example, the identity element is the integer zero. Since $B + 0 = 0 + B = B$, the requirement that an identity element exist is met.

The **inverse** B^{-1} of an element B has the property that $B*B^{-1} = B^{-1}*B = I$, where I is the identity element. If every element of the group has an inverse that is an element of the group, then the inverse requirement is met. For our example of integers, the inverse of B is $-B$ ($B^{-1} = -B$), since $B + (-B) = (-B) + B = 0$.

Since the four requirements are met, the set of all integers forms a group under the rule of combination of addition. The number of elements in a group is called its **order.** The group of integers under addition is of infinite order.

Note that there is no requirement that $B*D$ equal $D*B$. A group for which $B*D = D*B$ for all possible products is called **commutative** or **Abelian.**

From here on, the symbol $*$ for the rule of combination will be omitted and the product of the group elements B and D will be written as BD.

Symmetry Point Groups

We now show that the symmetry operations $\hat{A}, \hat{B}, \hat{C}, \ldots$ of a molecule (Sec. 21.5) form a group with the rule of combination for $\hat{B}$ and $\hat{F}$ being "take the product of the symmetry operations $\hat{B}$ and $\hat{F}$."

The **product** $\hat{B}\hat{F}$ of the symmetry operations $\hat{B}$ and $\hat{F}$ means we first apply the operation $\hat{F}$ to the molecule and we then apply $\hat{B}$ to the result found by applying $\hat{F}$. For example, consider the product $\hat{C}_2(z)\hat{\sigma}(xz)$ in the H_2O molecule. The $\hat{\sigma}(xz)$ operation (which is a reflection in the xz plane) interchanges the two hydrogens (Fig. 21.55). When the $\hat{C}_2(z)$ rotation is applied to the result, the two hydrogens are again interchanged. Since $\hat{C}_2(z)\hat{\sigma}(xz)$ returns all the atoms in H_2O to their original locations, one might think that $\hat{C}_2(z)\hat{\sigma}(xz)$ equals the identity operation $\hat{E}$, but this conclusion is too hasty. (Recall from Sec. 21.5 that the identity operation $\hat{E}$ does nothing.) The symmetry operation $\hat{\sigma}(yz)$,

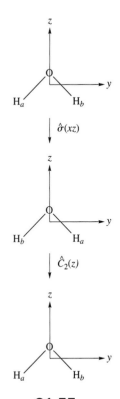

Figure 21.55

Effect of $\hat{C}_2(z)\hat{\sigma}(xz)$ on H_2O.

which is a reflection in the molecular plane, also leaves all atoms of H_2O unmoved. Symmetry operations are transformations of points in three-dimensional space. To see whether $\hat{C}_2(z)\hat{\sigma}(xz)$ equals $\hat{E}$ or $\hat{\sigma}(yz)$, we must consider what happens to a point at the three-dimensional location (x, y, z). The reflection $\hat{\sigma}(xz)$ in the xz plane leaves the x and z coordinates of any point unchanged and changes the y coordinate to its negative. The $\hat{C}_2(z)$ rotation about the z axis leaves the z coordinate of any point unchanged and sends the x and y coordinates to their respective negatives:

$$(x, y, z) \xrightarrow{\hat{\sigma}(xz)} (x, -y, z) \xrightarrow{\hat{C}_2(z)} (-x, y, z)$$

Thus the net effect of $\hat{C}_2(z)\hat{\sigma}(xz)$ is to send the x coordinate to its negative. This is what $\hat{\sigma}(yz)$ does, so $\hat{C}_2(z)\hat{\sigma}(xz) = \hat{\sigma}(yz)$. Note in Fig. 21.55, that, by convention, *the coordinate axes do not move when a symmetry operation is applied to the points in space.*

If $\hat{B}\hat{F} = \hat{F}\hat{B}$, the symmetry operations $\hat{B}$ and $\hat{F}$ are said to **commute.** Symmetry operations do not always commute (Prob. 21.99).

If $\hat{B}$ and $\hat{G}$ are symmetry operations of a molecule, each leaves the molecule in a position indistinguishable from the original position. Hence successive performance of these operations must leave the molecule in a position indistinguishable from the original, and the product $\hat{B}\hat{G}$ must be a symmetry operation. Thus the closure requirement is met.

Multiplication of symmetry operations can be shown to be associative, and the associativity requirement is met.

The identity element of a molecular symmetry group is the symmetry operation $\hat{E}$, the identity operation, which does nothing.

Clearly, each symmetry operation has an inverse, which undoes the effect of the operation. For example, the inverse of a $\hat{C}_4$ rotation is a $\hat{C}_4^3$ rotation about the same axis, since a 270° counterclockwise rotation $(\hat{C}_4^3)$ is the same as a 90° clockwise rotation. A reflection is its own inverse.

Since the four requirements are met, the set of symmetry operations of a molecule is a group. The symmetry groups of molecules are called **point groups,** since each symmetry operation leaves the point that is the molecular center of mass unmoved. A molecule can be classified as belonging to one of a number of possible point groups, depending on what symmetry elements are present.

Commonly Occurring Point Groups

A molecule whose symmetry elements are a C_n axis (where n can be 2 or 3 or 4 or . . .) and n planes of symmetry that each contain the C_n axis belongs to the point group C_{nv}. The v stands for "vertical." A **vertical** symmetry plane (symbol σ_v) is one that contains the highest-order axis of symmetry of the molecule. The H_2O molecule has as its symmetry elements a C_2 axis and two planes of symmetry that contain this axis (Figs. 21.17 and 21.18), so its point group is C_{2v}. The symmetry operations of H_2O are $\hat{E}$, $\hat{C}_2(z)$, $\hat{\sigma}(xz)$, and $\hat{\sigma}(yz)$; the order of the group C_{2v} is 4. The point group of NH_3 is C_{3v}, whose symmetry operations are $\hat{C}_3$, $\hat{C}_3^2$, $\hat{E}$, $\hat{\sigma}_a$, $\hat{\sigma}_b$, and $\hat{\sigma}_c$; the order is 6.

A molecule whose only symmetry element is a plane of symmetry belongs to the group C_s. Examples are the bent molecule HOCl, where the symmetry plane is the molecular plane, and the tetrahedral molecule $CHFBr_2$, where the symmetry plane contains the nuclei H, C, and F.

Molecules whose point group is D_{nh} have a C_n symmetry axis, n C_2 symmetry axes perpendicular to the C_n axis, a horizontal symmetry plane σ_h perpendicular to the C_n axis, n vertical symmetry planes containing the C_n axis, and a center of symmetry if n is even; the C_n axis is also an S_n axis. An example is benzene (Fig. 21.56), whose point group is D_{6h} (Prob. 21.103).

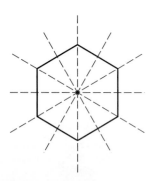

Figure 21.56

C_2 axes in C_6H_6.

The tetrahedral molecule CH_4 belongs to point group T_d (Prob. 21.104) and the octahedral molecule SF_6 belongs to O_h.

A molecule with no symmetry elements belongs to group C_1, since a $\hat{C}_1$ rotation equals the identity operation $\hat{E}$. The order of C_1 is 1.

In a linear molecule with no center of symmetry (for example, HF or OCS), the molecular axis is a C_∞ axis, since rotation about this axis by any angle is a symmetry operation. Also, any plane that contains the molecular axis is a symmetry plane, and there are an infinite number of such vertical symmetry planes. Hence linear molecules with no center of symmetry belong to group $C_{\infty v}$. In a linear molecule with a center of symmetry (for example, H_2 or CO_2), the molecular axis is a C_∞ axis, there are an infinite number of vertical symmetry planes, and also a horizontal symmetry plane perpendicular to the molecular axis. A centrosymmetric linear molecule has point group $D_{\infty h}$.

Some other point groups are considered in Probs. 21.106–21.108.

Multiplication Tables

The **multiplication table** of a group contains all possible products of members of the group. For example, the multiplication table of C_{2v}, the point group of H_2O, is shown in Table 21.2. The elements of the group are listed at the top of the table and at the far left of the table. Each of the 16 entries within the table is the product of the element at the far left of its row and at the top of its column. For example, the entry $\hat{\sigma}(xz)$ that lies in the column headed $\hat{\sigma}(yz)$ (the rightmost column) means that $\hat{C}_2(z)\hat{\sigma}(yz) = \hat{\sigma}(xz)$.

Each row (and each column) of a group multiplication table must contain each group element once and only once. To prove this, suppose the element F appeared twice in some row. Then we would have $DR = F$ and $DJ = F$, where R and J are two group elements. We have $DR = DJ$. Multiplication by D^{-1} on the left gives $D^{-1}DR = D^{-1}DJ$ and $IR = IJ$ (where I is the identity element), which becomes $R = J$. However, all elements of a group are different from one another, so R cannot equal J. Thus it is impossible for an element to appear twice in the same row of a group multiplication table.

The C_{2v} multiplication table is easily filled in as follows. The first row and first column of entries are easily found using the fact that $\hat{E}\hat{R} = \hat{R}$ and $\hat{S}\hat{E} = \hat{S}$, where $\hat{E}$ is the identity operation. The diagonal entries equal $\hat{E}$, since $\hat{C}_2^2 = \hat{E}$ and $\hat{\sigma}^2 = \hat{E}$. We saw (Fig. 21.55 and the associated discussion) that $\hat{C}_2(z)\hat{\sigma}(xz) = \hat{\sigma}(yz)$. All the remaining entries can then be filled in using the theorem that each element appears once in each row and once in each column. Note that C_{2v} is a commutative group.

Matrices

A **matrix** is a rectangular array of numbers (called the **elements** of the matrix). Matrices obey certain rules of combination. As well as being important in group theory, matrices play a key role in quantum-chemistry calculations. The most efficient

TABLE 21.2

Multiplication Table of C_{2v}

	$\hat{E}$	$\hat{C}_2(z)$	$\hat{\sigma}(xz)$	$\hat{\sigma}(yz)$
$\hat{E}$	$\hat{E}$	$\hat{C}_2(z)$	$\hat{\sigma}(xz)$	$\hat{\sigma}(yz)$
$\hat{C}_2(z)$	$\hat{C}_2(z)$	$\hat{E}$	$\hat{\sigma}(yz)$	$\hat{\sigma}(xz)$
$\hat{\sigma}(xz)$	$\hat{\sigma}(xz)$	$\hat{\sigma}(yz)$	$\hat{E}$	$\hat{C}_2(z)$
$\hat{\sigma}(yz)$	$\hat{\sigma}(yz)$	$\hat{\sigma}(xz)$	$\hat{C}_2(z)$	$\hat{E}$

way to solve the Hartree–Fock equations (19.54) and the Kohn–Sham equations (20.41) is by using matrices.

A matrix with m rows and n columns is called an m by n matrix and contains mn matrix elements. If $\mathbf{B}$ is a matrix, the notation b_{jk} symbolizes the element in row j and column k. Two matrices are said to be **equal** if they have the same number of rows, the same number of columns, and have all corresponding matrix elements equal to each other. The matrix equation $\mathbf{S} = \mathbf{T}$ is equivalent to the mn scalar equations $s_{jk} = t_{jk}$, where j goes from 1 to m and k goes from 1 to n.

Let $\mathbf{A}$ and $\mathbf{B}$ be 2 by 2 matrices. Let $\mathbf{C}$ denote the matrix product $\mathbf{AB}$. The product $\mathbf{AB}$ is defined as a 2 by 2 matrix whose elements are found as

$$\mathbf{AB} = \begin{pmatrix} a_{11} & a_{12} \\ a_{21} & a_{22} \end{pmatrix}\begin{pmatrix} b_{11} & b_{12} \\ b_{21} & b_{22} \end{pmatrix} = \begin{pmatrix} a_{11}b_{11} + a_{12}b_{21} & a_{11}b_{12} + a_{12}b_{22} \\ a_{21}b_{11} + a_{22}b_{21} & a_{21}b_{12} + a_{22}b_{22} \end{pmatrix}$$

$$= \begin{pmatrix} c_{11} & c_{12} \\ c_{21} & c_{22} \end{pmatrix} = \mathbf{C} \tag{21.88}$$

The element $c_{11} = a_{11}b_{11} + a_{12}b_{21}$ of $\mathbf{C}$ is found by adding the products of corresponding elements of row 1 of $\mathbf{A}$ and column 1 of $\mathbf{B}$. The element c_{12} is found from row 1 of $\mathbf{A}$ and column 2 of $\mathbf{B}$. The element c_{21} is found from row 2 of $\mathbf{A}$ and column 1 of $\mathbf{B}$.

EXAMPLE 21.5 Matrix multiplication

Find $\mathbf{AB}$ if

$$\mathbf{A} = \begin{pmatrix} 1 & 5 \\ 3 & 4 \end{pmatrix} \qquad \mathbf{B} = \begin{pmatrix} -1 & 6 \\ 3 & 2 \end{pmatrix} \tag{21.89}$$

We have

$$\mathbf{AB} = \begin{pmatrix} 1 & 5 \\ 3 & 4 \end{pmatrix}\begin{pmatrix} -1 & 6 \\ 3 & 2 \end{pmatrix} = \begin{pmatrix} 1(-1) + 5(3) & 1(6) + 5(2) \\ 3(-1) + 4(3) & 3(6) + 4(2) \end{pmatrix} = \begin{pmatrix} 14 & 16 \\ 9 & 26 \end{pmatrix}$$

Exercise

Find $\mathbf{BA}$. Does $\mathbf{BA}$ equal $\mathbf{AB}$? (*Answer:* The first-row elements are 17 and 19; the second-row elements are 9 and 23.)

As we saw in Example 21.5, matrix multiplication is not commutative; that is, $\mathbf{AB}$ and $\mathbf{BA}$ need not be equal.

If $\mathbf{T} = \mathbf{RS}$, where $\mathbf{R}$ and $\mathbf{S}$ are square matrices of the same size, the element in row j and column k of $\mathbf{T}$ is found by multiplying corresponding elements of row j of $\mathbf{R}$ and column k of $\mathbf{S}$ and adding the products (Prob. 21.112).

Addition of matrices is defined in Prob. 21.109.

A matrix that has one column is called a **column matrix** or a **column vector.** Let $\mathbf{V}$ be a 2 by 1 column vector. Multiplication of a square matrix by a column vector is illustrated by

$$\mathbf{AV} = \begin{pmatrix} a_{11} & a_{12} \\ a_{21} & a_{22} \end{pmatrix}\begin{pmatrix} v_{11} \\ v_{21} \end{pmatrix} = \begin{pmatrix} a_{11}v_{11} + a_{12}v_{21} \\ a_{21}v_{11} + a_{22}v_{21} \end{pmatrix} \tag{21.90}$$

$\mathbf{AV}$ is a column vector.

A **square** matrix has the same number of rows as columns. The **order** of a square matrix equals the number of rows. The elements $f_{11}, f_{22}, \ldots, f_{nn}$ (which go from the upper left corner to the lower right corner) of the square matrix $\mathbf{F}$ are said to lie on its **principal diagonal** and are called the **diagonal elements** of $\mathbf{F}$. Elements not on the principal diagonal of a square matrix are **off-diagonal elements.** A **diagonal matrix** is a square matrix all of whose off-diagonal elements are equal to zero.

A **unit matrix** is a diagonal matrix each of whose diagonal elements equals 1. For example, $\left(\begin{smallmatrix}1 & 0\\0 & 1\end{smallmatrix}\right)$ is a unit matrix of order 2. A unit matrix is denoted by the symbol $\mathbf{I}$. Multiplication of a square matrix $\mathbf{C}$ by a unit matrix of the same order as $\mathbf{C}$ does not change $\mathbf{C}$. Thus, $\mathbf{CI} = \mathbf{IC} = \mathbf{C}$.

The **inverse** of the square matrix $\mathbf{B}$ is a square matrix that satisfies $\mathbf{B}^{-1}\mathbf{B} = \mathbf{BB}^{-1} = \mathbf{I}$, where $\mathbf{B}^{-1}$ denotes the inverse and $\mathbf{I}$ is the unit matrix of the same order as $\mathbf{B}$. A square matrix has an inverse if and only if the determinant of its matrix elements is nonzero. When the determinant of the matrix is nonzero, the matrix is said to be **nonsingular.**

A **block-diagonal matrix** is a square matrix whose nonzero elements all lie on square blocks centered on the principal diagonal. For example, if

$$\mathbf{M} = \begin{pmatrix} -1 & 5 & 0 \\ 3 & 8 & 0 \\ 0 & 0 & 7 \end{pmatrix} \equiv \begin{pmatrix} \mathbf{M}_1 & \mathbf{0} \\ \mathbf{0} & \mathbf{M}_2 \end{pmatrix} \qquad \mathbf{N} = \begin{pmatrix} 2 & 9 & 0 \\ 4 & 1 & 0 \\ 0 & 0 & 6 \end{pmatrix} \equiv \begin{pmatrix} \mathbf{N}_1 & \mathbf{0} \\ \mathbf{0} & \mathbf{N}_2 \end{pmatrix} \qquad (21.91)$$

then $\mathbf{M}$ and $\mathbf{N}$ are block-diagonal matrices. $\mathbf{M}$ contains the 2 by 2 block $\mathbf{M}_1 = \left(\begin{smallmatrix}-1 & 5\\3 & 8\end{smallmatrix}\right)$ and the 1 by 1 block $\mathbf{M}_2 = (7)$. The product $\mathbf{MN}$ of the two block-diagonal matrices $\mathbf{M}$ and $\mathbf{N}$ is found to be a block-diagonal matrix whose nonzero blocks are the product of corresponding blocks of $\mathbf{M}$ and $\mathbf{N}$ (Prob. 21.113). That is,

$$\mathbf{MN} = \begin{pmatrix} \mathbf{M}_1\mathbf{N}_1 & \mathbf{0} \\ \mathbf{0} & \mathbf{M}_2\mathbf{N}_2 \end{pmatrix} \qquad (21.92)$$

If $\mathbf{P} = \mathbf{MN}$, then $\mathbf{P}$ is a block-diagonal matrix whose blocks are [Eq. (21.92)] $\mathbf{P}_1 = \mathbf{M}_1\mathbf{N}_1$ and $\mathbf{P}_2 = \mathbf{M}_2\mathbf{N}_2$. That is, *corresponding blocks of two block-diagonal matrices $\mathbf{M}$ and $\mathbf{N}$ multiply the same way as $\mathbf{M}$ and $\mathbf{N}$, if $\mathbf{M}$ and $\mathbf{N}$ have the same block-diagonal form.*

A diagonal matrix is a special case of a block-diagonal matrix that has its blocks all 1 by 1.

Representations

A symmetry operation moves each point in space to a new location. For example, the $\hat{\sigma}(xz)$ reflection of point group C_{2v} moves the point originally at (x, y, z) to $(x, -y, z)$. If we use a prime to denote the new location, we have $x' = x$, $y' = -y$, and $z' = z$. These three equations are equivalent to the matrix equation

$$\begin{pmatrix} x' \\ y' \\ z' \end{pmatrix} = \begin{pmatrix} 1 & 0 & 0 \\ 0 & -1 & 0 \\ 0 & 0 & 1 \end{pmatrix}\begin{pmatrix} x \\ y \\ z \end{pmatrix} \qquad (21.93)$$

The effect of $\hat{C}_2(z)$ is to move the point at (x, y, z) to $(-x, -y, z)$, and we can write a matrix equation to express this. Each of the symmetry operations of C_{2v} is thus described by a matrix. The matrices that correspond to the symmetry operations are

$$\hat{E}: \begin{pmatrix} 1 & 0 & 0 \\ 0 & 1 & 0 \\ 0 & 0 & 1 \end{pmatrix} \quad \hat{C}_2(z): \begin{pmatrix} -1 & 0 & 0 \\ 0 & -1 & 0 \\ 0 & 0 & 1 \end{pmatrix} \quad \hat{\sigma}(xz): \begin{pmatrix} 1 & 0 & 0 \\ 0 & -1 & 0 \\ 0 & 0 & 1 \end{pmatrix} \quad \hat{\sigma}(yz): \begin{pmatrix} -1 & 0 & 0 \\ 0 & 1 & 0 \\ 0 & 0 & 1 \end{pmatrix}$$
$$(21.94)$$

It isn't hard to show (Prob. 21.114) that these four matrices multiply the same way as the corresponding symmetry operations multiply. That is, if $\hat{R}\hat{T} = \hat{W}$, where

$\hat{R}$, $\hat{T}$, and $\hat{W}$ are symmetry operations of C_{2v} and if $\mathbf{R}$, $\mathbf{T}$, and $\mathbf{W}$ are the matrices in (21.94) that correspond to these symmetry operations, then $\mathbf{RT} = \mathbf{W}$.

A set of nonnull square matrices that multiply the same way the corresponding members of a point group multiply is said to be a **representation** of the group. (A *nonnull* matrix is one with at least one nonzero matrix element.) The matrices that constitute the representation are called the **matrix representatives** of the point-group operations. The order of the matrices in a representation is called the **dimension** of the representation. The matrices in (21.94) are a three-dimensional representation of C_{2v}.

Because the four matrices in the representation (21.94) are in diagonal form, which is a special case of block-diagonal form, the italicized statement after (21.92) shows that corresponding diagonal elements of these matrices must multiply in the same way as the 3 by 3 matrices in (21.94). Therefore, the diagonal elements give us three one-dimensional representations of C_{2v}. From (21.94), these one-dimensional representations are

$\hat{E}$	$\hat{C}_2(z)$	$\hat{\sigma}(xz)$	$\hat{\sigma}(yz)$	
(1)	(−1)	(1)	(−1)	
(1)	(−1)	(−1)	(1)	(21.95)
(1)	(1)	(1)	(1)	

For example, Table 21.2 gives $\hat{C}_2(z)\hat{\sigma}(xz) = \hat{\sigma}(yz)$. Taking the corresponding matrix representatives in the topmost representation in (21.95), we have $(-1)(1) = (-1)$, which is a valid matrix equation. (Matrices of order 1 multiply the same way ordinary numbers multiply.)

Whenever the matrices of a representation are in block-diagonal form, we can break these matrices into smaller matrices that give us representations of lower dimension, and the original representation is said to be a **reducible representation.**

If the matrices $\mathbf{A}$, $\mathbf{B}$, $\mathbf{C}$, . . . form a representation of a group, it can be proved (Prob. 21.116) that the matrices $\mathbf{P}^{-1}\mathbf{AP}$, $\mathbf{P}^{-1}\mathbf{BP}$, $\mathbf{P}^{-1}\mathbf{CP}$, . . . are a representation of the group; here, $\mathbf{P}$ is any nonsingular square matrix whose order is the same as that of $\mathbf{A}$, $\mathbf{B}$, $\mathbf{C}$, ... The transformation of $\mathbf{A}$ to $\mathbf{P}^{-1}\mathbf{AP}$ is called a **similarity transformation.** Two representations whose matrices are related by a similarity transformation are said to be **equivalent** to each other.

If the matrices $\mathbf{A}$, $\mathbf{B}$, $\mathbf{C}$, . . . are not in block-diagonal form but there exists some matrix $\mathbf{P}$ such that $\mathbf{P}^{-1}\mathbf{AP}$, $\mathbf{P}^{-1}\mathbf{BP}$, $\mathbf{P}^{-1}\mathbf{CP}$, . . . are all in the same block-diagonal form, then the representation $\mathbf{A}$, $\mathbf{B}$, $\mathbf{C}$, . . . is also called a **reducible representation.** If the matrix representatives are not in block-diagonal form and no similarity transformation exists that will put them into block-diagonal form, the representation is **irreducible.** All one-dimensional representations are considered to be irreducible.

The members of a group can be divided into **classes.** If $A, B, C, D, \ldots$ are the elements of a group, the elements that belong to the same class as B are found by taking $A^{-1}BA$, $B^{-1}BB$, $C^{-1}BC$, $D^{-1}BD$, ... One can show that an element of a group cannot belong to two different classes. For a commutative group, we have $A^{-1}BA = BA^{-1}A = BI = B$. Hence each element of a commutative group is in a class by itself. The group C_{2v} is commutative (Table 21.2) and has four members. Hence it has four classes.

A theorem of group theory states that *the number of nonequivalent irreducible representations of a group is equal to the number of classes of the group.* Since C_{2v} has four classes, it has four nonequivalent irreducible representations. We found three of them in (21.95).

Another theorem states that *if $d_1, d_2, \ldots, d_c$ are the dimensions of the nonequivalent irreducible representations of a group whose order is h, then*

$$d_1^2 + d_2^2 + \cdots + d_c^2 = h \tag{21.96}$$

(c is the number of classes.) The group C_{2v} has $h = 4$ and has four nonequivalent irreducible representations. We found three one-dimensional irreducible representations of C_{2v} in (21.95). Therefore, (21.96) gives us $1^2 + 1^2 + 1^2 + d_4^2 = 4$, and so $d_4 = 1$. The fourth irreducible representation of C_{2v} is found to be (Prob. 21.118)

$$
\begin{array}{cccc}
\hat{E} & \hat{C}_2(z) & \hat{\sigma}(xz) & \hat{\sigma}(yz) \\
\hline
(1) & (1) & (-1) & (-1)
\end{array}
\tag{21.97}
$$

Character Tables

The sum of the diagonal elements of a square matrix is called the **trace** of the matrix. For example, the trace of $\mathbf{A}$ in (21.89) is $1 + 4 = 5$. The traces of the matrices of a representation of a group are called the **characters** of the representation. For example, the characters of the reducible C_{2v} representation in (21.94) are 3, -1, 1, and 1. For most applications of group theory to quantum mechanics, one needs only the characters of representations and not the full matrices. A table of the sets of characters for the nonequivalent irreducible representations of a group gives the **character table** of the group. For C_{2v}, the nonequivalent irreducible representations are all one-dimensional, and the nonequivalent irreducible representations are (21.95) and (21.97). The C_{2v} character table is given in Table 21.3.

One-dimensional irreducible representations are labeled A or B, according to whether the character of the highest-order symmetry axis is $+1$ or -1, respectively. The numerical subscripts distinguish different irreducible representations. The one-dimensional representation whose characters are all equal to 1 exists for every point group and is called the **totally symmetric representation.** The letters x, y, and z will be explained later. [For molecules with a center of symmetry, each irreducible representation is labeled with a g or u subscript (gerade or ungerade) according to whether the character of $\hat{i}$ in that representation is positive or negative, respectively.]

The character table of group C_{3v}, the point group of NH_3 is also given in Table 21.3. The symmetry operations are $\hat{C}_3$, $\hat{C}_3^2$, $\hat{E}$, $\hat{\sigma}_a$, $\hat{\sigma}_b$, and $\hat{\sigma}_c$. One finds that C_{3v} has three classes. $\hat{E}$ is in a class by itself (this is always true; see Prob. 21.119). A second class consists of $\hat{C}_3$ and $\hat{C}_3^2$. The final class consists of $\hat{\sigma}_a$, $\hat{\sigma}_b$, and $\hat{\sigma}_c$. Note that the members of a class are closely related symmetry operations. Group theory shows that *symmetry operations in the same class have the same characters in a given representation.* Rather than listing each symmetry operation separately, the top row of a character table lists members of the same class together. Thus, $2\hat{C}_3(z)$ in the character table stands for $\hat{C}_3$ and $\hat{C}_3^2$, whose characters are equal; $3\hat{\sigma}_v$ in the character table stands for the three symmetry reflections.

Since C_{3v} has three classes, it has three nonequivalent irreducible representations. The letter E denotes a two-dimensional irreducible representation. Because the trace of a 2 by 2 unit matrix is $1 + 1 = 2$, the character of the identity operation $\hat{E}$ is 2 in a two-dimensional representation. (Some groups have three-dimensional irreducible

TABLE 21.3

Character Tables of C_{2v} and C_{3v}

C_{2v}	$\hat{E}$	$\hat{C}_2(z)$	$\hat{\sigma}(xz)$	$\hat{\sigma}(yz)$	
A_1	1	1	1	1	z
A_2	1	1	-1	-1	
B_1	1	-1	1	-1	x
B_2	1	-1	-1	1	y

C_{3v}	$\hat{E}$	$2\hat{C}_3(z)$	$3\hat{\sigma}_v$	
A_1	1	1	1	z
A_2	1	1	-1	
E	2	-1	0	(x, y)

representations and these are denoted by either T or F, according to the preference of the person making up the table. G and H denote four- and five-dimensional irreducible representations, respectively.) For C_{3v}, Eq. (21.96) becomes $1^2 + 1^2 + 2^2 = 6$.

Basis Functions

When a symmetry operation is applied to the points in three-dimensional space, the point originally at (x, y, z) is moved to the location (x', y', z'). One finds that $x' = d_{11}x + d_{12}y + d_{13}z$, where the d's are constants, with similar equations holding for y' and z'. Thus x' is a *linear combination* of the functions x, y and z [where this term is defined after Eq. (19.20)]. Let $\mathbf{r}$ and $\mathbf{r}'$ be column vectors whose elements are x, y, z and x', y', z', respectively. Because each of the primed coordinates is a linear combination of the unprimed coordinates, it follows from the rule for matrix multiplication that $\mathbf{r}' = \mathbf{D}\mathbf{r}$, where $\mathbf{D}$ is a square matrix of order three whose elements are d_{11}, d_{12}, etc. An example is (21.93). (The matrix $\mathbf{D}$ is not necessarily diagonal.) For each of the symmetry operations of a point group, we have a matrix that relates $\mathbf{r}'$ to $\mathbf{r}$. For example, for the C_{2v} operations, these matrices are given in (21.94).

We noted that the $\mathbf{D}$ matrices in (21.94) form a representation of C_{2v}. These $\mathbf{D}$ matrices consist of numbers that tell how the functions x, y, and z are transformed into linear combinations of x, y, and z by the group's symmetry operations. The functions x, y, and z are called *basis functions* for the representation that consists of the $\mathbf{D}$ matrices. In general, one can show that *whenever each of the set of functions $f_1, f_2, \ldots, f_n$ is transformed into a linear combination of $f_1, f_2, \ldots, f_n$ by the group's symmetry operations, then the matrices of coefficients of the linear combinations form a representation of the point group;* the functions $f_1, f_2, \ldots, f_n$ are called **basis functions** for this representation. (For this result to be valid, the functions $f_1, f_2, \ldots, f_n$ must be linearly independent, meaning none of them can be expressed as a linear combination of the other functions in the set.) Note that *the dimension of the representation whose basis functions are $f_1, f_2, \ldots, f_n$ equals n, the number of functions in the basis.*

For the group C_{2v}, the basis functions x, y, and z give matrices in (21.94) that are diagonal, so the function x alone is transformed into a linear combination of x (namely, $+x$ or $-x$), and x by itself is the basis for a representation of C_{2v}. From (21.94), the first representation in (21.95), and Table 21.3, this representation is B_1. Hence the function x is put on the same line as B_1 in the character table. Similarly for y and z. For the point group C_{3v}, one finds that the group operations transform z into itself and transform each of x and y into linear combinations of x and y. The italicized statements in the last paragraph thus show that z is the basis for a one-dimensional representation of C_{3v} and the set of functions x and y is the basis for a two-dimensional representation. This is the meaning of (x, y) on the E line of the C_{3v} character table.

One can consider the effects of symmetry operations on other functions besides x, y, and z. For example, consider a $2p_x$ AO, which has the form $2p_x = cxe^{-b(x^2+y^2+z^2)^{1/2}}$ [Eq. (19.23)], where c and b are constants. We saw that $\hat{C}_2(z)(x, y, z) \rightarrow (-x, -y, z)$, so $x' = -x$, $y' = -y$, and $z' = z$. Replacing x with $-x'$ and y with $-y'$ in $2p_x$, we get $-cx'e^{-b(x'^2+y'^2+z'^2)^{1/2}}$ so the transformed orbital is the negative of a $2p_x$ AO. Pictorially, a $2p_x$ AO has positive and negative lobes centered on the x axis, and a $180°$ $\hat{C}_2(z)$ rotation about the z axis interchanges the lobes and transforms $2p_x$ into $-2p_x$. A $90°$ counterclockwise $\hat{C}_4(z)$ rotation about the z axis transforms the $2p_x$ AO into the $2p_y$ AO (Fig. 21.57).

Group Theory and Quantum Mechanics

Let $\hat{H}$ be the electronic Hamiltonian operator in the molecular electronic Schrödinger equation. A symmetry operation of a molecule sends equivalent nuclei into one another. It therefore follows that each molecular symmetry operation commutes with $\hat{H}$ (commutation is defined at the end of Sec. 19.4). For a nondegenerate level, it fol-

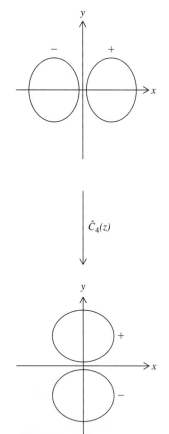

Figure 21.57

Effect of $\hat{C}_4(z)$ on $2p_x$.

lows from a certain theorem of quantum mechanics that the molecular electronic wave function must be an eigenfunction of each of the molecular symmetry operations. For a degenerate energy level, any linear combination of the wave functions of the level is an eigenfunction of $\hat{H}$ (as noted in Sec. 19.3), and one can show that when a symmetry operation commutes with $\hat{H}$, the application of that operation to the wave function of an n-fold degenerate level transforms the wave function into a linear combination of the n wave functions of the level. Therefore (by the italicized statements in the last subsection) the linearly independent wave functions of a degenerate energy level form a basis for a representation of the molecular point group. This representation might be reducible or irreducible. For reasons discussed elsewhere [*Levine* (1975), pp. 413–415], this representation is very unlikely to be reducible, and we shall assume it to be irreducible.

Thus, the wave functions of each electronic energy level form a basis for an irreducible representation of the molecular point group. This means that the symmetry operations transform each of the wave functions of a level into a linear combination of the wave functions of the level, where the coefficients in the linear combinations form matrices that give an irreducible representation of the point group. We can classify each electronic energy level according to one of the possible irreducible representations of the molecule's point group. Since the number of functions in the basis for a representation equals the dimension of the representation, the degeneracy of an energy level equals the dimension of the irreducible representation for which the wave functions of that level form a basis. In discussing the symmetry of wave functions, the term **symmetry species** is often used instead of irreducible representation.

For example, each electronic state of H_2O can be classified as belonging to one of the irreducible representations (symmetry species) A_1, A_2, B_1, or B_2 of the point group C_{2v}. Recall that atomic terms (^1S, ^3P, etc.) are specified by giving the spin multiplicity $2S + 1$ (where S is the total spin angular momentum quantum number) as a left superscript on the letter (S, P, D, . . .) specifying the total electronic orbital angular momentum. A molecular electronic term is specified by giving the spin multiplicity as a left superscript on the irreducible representation. Thus, for H_2O, one has electronic terms such as $^1A_1$, $^3A_1$, $^1B_1$, etc.

For most molecules, the irreducible representation (symmetry species) of the ground electronic state is the totally symmetric one and the electron spins are all paired to give a singlet state. The ground electronic state of H_2O is $^1A_1$.

One can show that each canonical MO of a molecule can be classified according to one of the irreducible representations (symmetry species) of the molecular point group. Lowercase letters are used for the symmetry species of MOs. The MOs belonging to a given symmetry species are numbered in order of increasing orbital energy. For example, the lowest a_1 MOs of H_2O are labeled $1a_1$, $2a_1$, $3a_1$, ... The **electron configuration** of a molecule is specified by giving the number of electrons in each shell, where a **shell** is a set of MOs with the same energy. (Recall that an atomic electron configuration such as $1s^2 2s^2 2p^4$ gives the number of electrons in each subshell, where a subshell is a set of AOs having the same energy.)

For H_2O, one finds the ground-electronic-state electron configuration to be $(1a_1)^2(2a_1)^2(1b_2)^2(3a_1)^2(1b_1)^2$. The lowest MO $1a_1$ is essentially the same as the $1s$ AO on the oxygen atom. This $1s$ AO is unaffected by each of the four molecular symmetry operations and so belongs to the totally symmetric symmetry species A_1. The $1b_1$ MO is a lone-pair $2p_x$ AO on oxygen, where the x axis is perpendicular to the molecular plane (Fig. 21.55). This AO is sent into its negative by $\hat{C}_2(z)$ and by $\hat{\sigma}(yz)$, and is unaffected by $\hat{\sigma}(xz)$ and $\hat{E}$, which corresponds to B_1 in the character table in Table 21.3. Other MOs are considered in Prob. 21.121.

Just as a given atomic electron configuration can give rise to several terms having different energies (Fig. 19.13), so can a given molecular electron configuration. For example, the excited electron configuration $(1a_1)^2(2a_1)^2(1b_2)^2(3a_1)^2(1b_1)(4a_1)$ of H_2O gives rise to the terms $^3B_1$ and $^1B_1$.

For NH_3 (point group C_{3v}), the ground electronic state is a $^1A_1$ state with the electron configuration $(1a_1)^2(2a_1)^2(1e)^4(3a_1)^2$. The E symmetry species of C_{3v} is two-dimensional, so there are two degenerate $1e$ MOs. These two degenerate MOs constitute a shell that holds four electrons, two in each MO. For CH_4, the ground electron configuration is $(1a_1)^2(2a_1)^2(1t_2)^6$, where the T_2 symmetry species has dimension 3, so there are three degenerate $1t_2$ MOs. The $1a_1$ MO is essentially a $1s$ AO on C and the other four occupied MOs are bonding delocalized canonical MOs (Sec. 20.6).

Since the Schrödinger equation omits electron spin, the degeneracy specified by the dimension of an irreducible representation does not include spin degeneracy, and this degeneracy is called *orbital degeneracy*.

Now consider molecular vibrations. One can show that each normal vibrational mode of a molecule belongs to one of the irreducible representations of the molecular point group. For example, application of each of the four C_{2v} symmetry operations to the H_2O normal mode ν_1 in Fig. 21.27 leaves the vectors of this mode unchanged. Hence the symmetry species of ν_1 is the totally symmetric species a_1. Likewise, ν_2 has species a_1. For ν_3, a $\hat{C}_2(z)$ rotation changes each vector to its negative and $\hat{\sigma}(xz)$ also changes each vector to its negative (the molecular plane is the yz plane). ν_3 is unchanged by $\hat{E}$ and by $\hat{\sigma}(yz)$. Thus the symmetry species of ν_3 is b_2.

Group theory enables one to do a lot more than just classify electronic states, MOs, and normal modes according to their symmetry species and degeneracies. Using group theory, one can predict the symmetry species of the normal modes of vibration of a molecule, and one can predict the symmetry species of the MOs formed from the basis AOs. Moreover, solution of the Hartree–Fock equation (19.54) is considerably simplified by group theory, since it enables one to deal separately with MOs of different symmetry species. Group theory is invaluable in deriving selection rules to find the allowed transitions in vibrational spectra and in electronic spectra.

Recall that a reducible representation can be put into block-diagonal form by a similarity transformation, with the blocks corresponding to irreducible representations of the point group. One says that the reducible representation is the **direct sum** of the irreducible representations that occur in the blocks. For example, if Γ is the reducible representation whose matrices are those in (21.94), then $\Gamma = B_1 \oplus B_2 \oplus A_1$, where $\oplus$ denotes the direct sum.

Many of the applications of group theory in quantum chemistry and spectroscopy depend on starting with a reducible representation and finding which irreducible representations it is the direct sum of. Let Γ be a reducible representation and let $a_{i\Gamma}$ denote the number of times the irreducible representation i occurs in Γ. For example, for the reducible representation (21.94), $a_{i\Gamma}$ is 1 for A_1, is 1 for B_1, is 1 for B_2, and is 0 for A_2. Let $\chi_\Gamma(\hat{R})$ and $\chi_i(\hat{R})$ denote the characters of the symmetry operation $\hat{R}$ in the representation Γ and in the irreducible representation i. For example, for (21.94), $\chi_\Gamma(\hat{E}) = 3$. Group theory shows that

$$a_{i\Gamma} = \frac{1}{h} \sum_{\hat{R}} \chi_i^*(\hat{R}) \chi_\Gamma(\hat{R}) \tag{21.98}$$

where h is the order of the group and the sum goes over the h different symmetry operations of the group. (The complex conjugate occurs because certain irreducible representations have complex characters.) For example, for Γ being the representation (21.94), the characters are 3, -1, 1, and 1, and Eq. (21.98) gives the number of times that B_1 occurs in Γ as $a_{B_1\Gamma} = \frac{1}{4}[(1)(3) + (-1)(-1) + (1)(1) + (-1)(1)] = 1$. Although this result is obvious by inspection of (21.94), it is not so obvious when the representation's matrices are not in block-diagonal form or when we know only the characters of the representation.

21.18 SUMMARY

Electromagnetic radiation (light) consists of oscillating electric and magnetic fields. Spectroscopy studies the interaction of light and matter, especially absorption and emission of radiation by matter. The energy $h\nu$ of the photon absorbed or emitted in a transition is equal to the energy difference between the stationary states m and n involved in the transition: $h\nu = |E_m - E_n|$. For a radiative transition to have significant probability of occurring, the transition-moment integral (21.5) must be nonzero. The selection rules give the allowed transitions for a given kind of system. The Beer–Lambert law (21.11) relates the fraction of radiation in a small wavelength range absorbed by a sample to the molar absorption coefficient ε.

The energy of a molecule is (to a good approximation) the sum of translational, rotational, vibrational, and electronic energies. The spacings between levels increase in the order: translational, rotational, vibrational, electronic.

For a diatomic molecule, the rotational energy is $E_{\text{rot}} \approx J(J + 1)\hbar^2/2I_e = B_e hJ(J + 1)$, where $J = 0, 1, 2, \ldots$, the equilibrium moment of inertia is $I_e = \mu R_e^2$ (where μ is the reduced mass of the two atoms), and B_e is the equilibrium rotational constant. Because vibration affects the average bond distances, B_e is replaced by B_v [Eq. (21.32)] when the molecule is in vibrational state v. For polyatomic molecules, the rotational-energy expression depends on whether the molecule is a spherical, symmetric, or asymmetric top.

The vibrational energy of a diatomic molecule is approximately that of a harmonic oscillator: $E_{\text{vib}} \approx (v + \frac{1}{2})h\nu_e$, where $v = 0, 1, 2, \ldots$ and the equilibrium vibration frequency is $\nu_e = (1/2\pi)(k_e/\mu)^{1/2}$. Anharmonicity adds a term proportional to $-(v + \frac{1}{2})^2$ to E_{vib} and causes the vibrational levels to converge as v increases (Fig. 21.9). For an $\mathcal{N}$-atom molecule, the vibrational energy is approximately the sum of $3\mathcal{N} - 6$ (or $3\mathcal{N} - 5$ if the molecule is linear) harmonic-oscillator energies, one for each normal vibrational mode [Eq. (21.48)].

The pure-rotational spectrum lies in the microwave (or far-IR) region and is studied in microwave spectroscopy. Molecular geometries are obtainable from the pure-rotational spectra of molecules with nonzero dipole moments. For a diatomic molecule, the radiative rotational selection rule is $\Delta J = \pm 1$.

The vibration–rotation spectrum lies in the infrared and consists of a series of bands. Each line in a band corresponds to a different rotational change. A normal mode is IR-active if it changes the dipole moment. Analysis of the IR spectrum yields the IR-active vibrational frequencies and (provided the rotational lines are resolved) the moments of inertia, from which the molecular structure can be found.

Pure-rotational and vibration–rotation transitions can also be observed in Raman spectroscopy. Here, rather than absorbing a photon, a molecule scatters a photon, exchanging energy with it in the process.

Transitions of valence electrons to higher levels produce absorption in the UV and visible regions.

In NMR spectroscopy, one observes transitions of nuclear spin magnetic moments in an applied magnetic field. The local field at a nucleus is influenced by the electronic environment, so nonequivalent nuclei in a molecule undergo NMR transitions at different frequencies. Moreover, because of interactions between nuclear spins within a molecule, the NMR absorption lines are split. Provided the spectrum is first-order, n equivalent protons act to split the absorption peak of a set of adjacent protons into $n + 1$ peaks. Spin–spin splitting is usually negligible for protons separated by more than three bonds.

In ESR spectroscopy, one observes transitions between the energy levels of an unpaired electron spin magnetic moment in an applied magnetic field.

Photochemistry studies chemical reactions produced by absorption of light.

The symmetry operations of a molecule form a mathematical group. Matrices that multiply the same way as the members of a group form a representation of the group. The character table of a group lists the characters of the nonequivalent irreducible representations of the group, where the characters are the traces of the matrices of the representations. Molecular wave functions, MOs, and normal modes can be classified according to their irreducible representations (symmetry species).

Important kinds of calculations discussed in this chapter include

- Calculation of absorption and emission frequencies and wavelengths from quantum-mechanical energy levels using $E_{upper} - E_{lower} = h\nu = hc/\lambda$ and the selection rules for the system.
- Evaluation of the transition-moment integral $\int \psi_m^*(\Sigma_i \, Q_i \mathbf{r}_i)\psi_n \, d\tau$ to find the selection rules for the allowed transitions.
- Use of the Beer–Lambert law $I = I_0 10^{-\varepsilon cl}$ to find the transmittance I/I_0 at a particular λ.
- Use of $E_{rot} = J(J+1)\hbar^2/2I$ and $I = \mu R_e^2$ to calculate rotational energy levels and bond distances in diatomic molecules.
- Use of $E_{vib} \approx (v + \frac{1}{2})h\nu$ and $\nu = (1/2\pi)(k/\mu)^{1/2}$ to find force constants of diatomic molecules.
- Calculation of relative populations of energy levels using the Boltzmann distribution law $N_i/N_j = (g_i/g_j)e^{-(E_i - E_j)/kT}$, where N_i, N_j and g_i, g_j are the populations and degeneracies of energy levels i and j.
- Calculation of the NMR transition frequency ν.
- Calculation of the specific optical rotation $[\alpha]$.
- Reduction of a reducible representation to the direct sum of irreducible representations.

FURTHER READING AND DATA SOURCES

Straughan and Walker; Harmony; Chang; Levine (1975); *Campbell and Dwek; Brand, Speakman, and Tyler; Herzberg; Gordy; Gordy and Cook; Mann and Akitt; Günther; Sugden and Kenney; Long; Hollas; Graybeal; Friebolin; Bovey.*

Molecular structures: *Landolt–Börnstein*, New Series, Group II, vols. 7, 15, and 21, *Structure Data of Free Polyatomic Molecules; Herzberg*, vol. III, app. VI.

Molecular spectroscopic constants: *Herzberg*, vol. III, app. VI; K. P. Huber and G. Herzberg, *Constants of Diatomic Molecules*, Van Nostrand Reinhold, 1979; *Landolt–Börnstein*, New Series, Group II, vols. 4, 6, 14a, 14b, 19, and 20.

Molecular dipole moments: R. D. Nelson et al., Selected Values of Electric Dipole Moments for Molecules in the Gas Phase, *Natl. Bur. Stand. U.S. Publ.* NSRDS-NBS 10, 1967; A. L. McClellan, *Tables of Experimental Dipole Moments*, vol. 1, Freeman, 1963, vol. 2, Rahara Enterprises, 1974, vol. 3, Rahara Enterprises, 1989.

Collections of spectra: W. M. Simons (ed.), *The Sadtler Handbook of Infrared Spectra*, Sadtler, 1978; C. J. Pouchert, *The Aldrich Library of Infrared Spectra*, 3d ed., Aldrich, 1981; W. Simons (ed.), *The Sadtler Handbook of Proton NMR Spectra*, Sadtler, 1978; C. J. Pouchert, *The Aldrich Library of NMR Spectra*, 2d ed., Aldrich, 1983. The Integrated Spectral Data Base System for Organic Compounds (SDBS) at www.aist.go.jp/RIODB/SDBS/menu-e.html contains thousands of ^{1}H and ^{13}C NMR, IR, Raman, and ESR spectra, and also mass spectra.

PROBLEMS

Section 21.1

21.1 True or false? (a) Electromagnetic radiation always travels at speed c. (b) In an electromagnetic wave traveling in the y direction, the electric field vector **E** always points in the same direction at each location along the wave. (c) The fields **E** and **B** are perpendicular to the direction of travel of an electromagnetic wave and are perpendicular to each other. (d) Microwaves have higher frequency than visible light. (e) An infrared photon has a lower energy than an ultraviolet photon.

21.2 Find the frequency, wavelength, and wavenumber of light with photons of energy 1.00 eV per photon.

21.3 Find the speed, frequency, and wavelength of sodium D light in water at 25°C. For data, see the material following Eq. (21.4).

Section 21.2

21.4 True or false? (a) When a molecule absorbs a photon and makes a transition to a stationary state of energy E_k, the absorption frequency ν satisfies $h\nu = E_k$. (b) When a molecule emits a photon of frequency ν, it undergoes an energy change given by $\Delta E = h\nu$. (c) When a molecule absorbs a photon of frequency ν, it undergoes an energy change given by $\Delta E = h\nu$. (d) The longer the wavelength of a transition, the smaller the energy difference between the two levels involved in the transition. (e) Exposing a molecule in state n to electromagnetic radiation of frequency $\nu = (E_n - E_m)/h$ will increase the probability that the molecule will make a transition to the lower state m with emission of a photon of frequency ν.

21.5 Give the SI units of (a) frequency; (b) wavenumber; (c) speed; (d) absorbance; (e) molar absorption coefficient.

21.6 Verify Eq. (21.6) for the particle-in-a-box transition moment.

21.7 Use the harmonic-oscillator selection rule $\Delta v = \pm 1$ to find the frequency or frequencies of light absorbed by a harmonic oscillator with vibrational frequency ν_{vib}.

21.8 Use Fig. 18.18 and Table 15.1 to evaluate the transition-moment integral (21.5) for each of the following pairs of states of a charged harmonic oscillator: (a) $v = 0$ and $v = 1$; (b) $v = 0$ and $v = 2$; (c) $v = 0$ and $v = 3$. Are the results consistent with the selection rule $\Delta v = \pm 1$?

21.9 For a certain quantum-mechanical system, the wavelength for an absorption transition from level A to level C is 485 nm and the wavelength for an absorption transition from level B to level C is 884 nm. Find λ for a transition between levels A and B.

21.10 A hypothetical quantum-mechanical system has the energy levels $E = bn(n + 2)$, where $n = 1, 2, 3, \ldots$ and b is a positive constant. The selection rule for radiative transitions is $\Delta n = \pm 2$. For a collection of such systems distributed among many energy levels, the lowest-frequency absorption transition is observed to be at 80 GHz. Find the next lowest absorption frequency.

21.11 (a) For an electron confined to a one-dimensional box of length 2.00 Å, calculate the three lowest possible absorption frequencies for transitions that start from the ground stationary state. (b) Repeat (a) without assuming that the initial state is the ground state.

21.12 A hypothetical quantum-mechanical system has the energy levels $E = an(n + 4)$, where $n = 0, 1, 2, \ldots$ and a is a positive constant. The selection rule for radiative transitions is $\Delta n = \pm 3$. Find the formula for the allowed absorption frequencies in terms of n_{lower}, a, and h.

21.13 A hypothetical quantum-mechanical system has the energy levels $E = AK(K + 3)$; $K = 1, 2, 3, \ldots$; $A > 0$. The radiative-transition selection rule is $\Delta K = \pm 1$. The $K = 2$ to 3 transition occurs at 60 GHz. A transition at 135 GHz is observed. What levels is this transition between?

21.14 For each of the absorbance values 0.1, 1, 2, and 10, calculate the transmittance and the percentage of radiation absorbed.

21.15 Ethylene has a UV absorption peak at 162 nm with $\varepsilon = 1.0 \times 10^4$ dm^3 mol^{-1} cm^{-1}. Calculate the transmittance of 162-nm radiation through a sample of ethylene gas at 25°C and 10 torr for a cell length of (a) 1.0 cm; (b) 10 cm.

21.16 Methanol has a UV absorption peak at 184 nm with $\varepsilon = 150$ dm^3 mol^{-1} cm^{-1}. Calculate the transmittance of 184-nm radiation through a 0.0010 mol dm^{-3} solution of methanol in a nonabsorbing solvent for a cell length of (a) 1.0 cm; (b) 10 cm.

21.17 A certain solution of the enzyme lysozyme (molecular weight 14600) in D_2O with a mass concentration of 80 mg/cm^3 in a 0.10-mm-long absorption cell is found to have a transmittance of 8.3% for infrared radiation of wavelength 6000 nm. Find the absorbance of the solution and the molar absorption coefficient of lysozyme at this wavelength.

21.18 A solution of 2.00 g of a compound transmits 60.0% of the 430-nm light incident on a 3.00-cm-long cell. What percent of 430-nm light will be transmitted by a solution of 4.00 g of this compound in the same cell?

21.19 At 330 nm, the ion $Fe(CN)_6^{3-}(aq)$ has $\varepsilon = 800$ dm^3 mol^{-1} cm^{-1}, and $Fe(CN)_6^{4-}(aq)$ has $\varepsilon = 320$ dm^3 mol^{-1} cm^{-1}. The reduction of $Fe(CN)_6^{3-}$ to $Fe(CN)_6^{4-}$ is being followed spectrophotometrically in 1.00-cm-long cell. The solution has an initial $Fe(CN)_6^{3-}$ concentration of 1.00×10^{-3} mol dm^{-3} and no $Fe(CN)_6^{4-}$. After 340 s, the absorbance is 0.701. Calculate the percent of $Fe(CN)_6^{3-}$ that has reacted.

Section 21.3

21.20 True or false? (a) The spacings between adjacent low-lying molecular translational, rotational, and vibrational levels satisfy $\Delta\varepsilon_{tr} < \Delta\varepsilon_{rot} < \Delta\varepsilon_{vib}$. (b) At room temperature, many rotational levels of gas-phase molecules are substantially populated. (c) At room temperature, many vibrational levels of $O_2(g)$ are substantially populated. (d) The vibrational energy levels of a diatomic molecule are accurately given by the harmonic-oscillator expression $(v + \frac{1}{2})h\nu$. (e) A bound electronic state of a diatomic molecule has a finite number of vibrational levels. (f) As the vibrational quantum number increases, the spacing between adjacent vibrational levels of a diatomic molecule decreases. (g) $D_0 > D_e$. (h) As the rotational quantum number J increases, the spacing between adjacent rotational levels of a diatomic molecule increases.

Section 21.4

21.21 True or false? (a) Diatomic-molecule vibration–rotation absorption bands always have $\Delta v = 1$. (b) For diatomic-molecule pure-rotational absorption spectra, only $\Delta J = 1$ lines occur. (c) Because only $\Delta J = 1$ is allowed in pure-rotational absorption spectra of diatomic molecules, a diatomic-molecule pure-rotational spectrum contains only one line.

21.22 Use data in Table 21.1 in Sec. 21.4 to calculate D_0 for the ground electronic state of (a) $^{14}N_2$; (b) $^{12}C^{16}O$.

21.23 (a) Explain why D_e and k_e for $D^{35}Cl$ are essentially the same as D_e and k_e for $H^{35}Cl$ but D_0 for these two species differs. (b) Use data in Table 21.1 in Sec. 21.4 to calculate D_0 for each of these two species. Neglect the difference in $\tilde{\nu}_e x_e$ for the two species.

21.24 As noted before Eq. (21.20), the vibrational coordinate x for a diatomic molecule equals $R - R_e$. We can estimate the typical departure A of a diatomic-molecule bond length from its equilibrium value due to zero-point vibration by setting the harmonic-oscillator ground-state vibrational energy equal to the classical-mechanical expression for the maximum potential energy of a harmonic oscillator. Show that this gives $A = (h/4\pi^2\nu_e\mu)^{1/2}$. Use Table 21.1 in Sec. 21.4 to calculate this typical departure for $H^{35}Cl$ and for $^{14}N_2$.

21.25 If the $J = 2$ to 3 rotational transition for a diatomic molecule occurs at $\lambda = 2.00$ cm, find λ for the $J = 6$ to 7 transition of this molecule.

21.26 The $J = 0 \rightarrow 1$, $v = 0 \rightarrow 0$ transition for $^1H^{79}Br$ occurs at 500.7216 GHz, and that for $^1H^{81}Br$ occurs at 500.5658 GHz. (a) Calculate the bond distance R_0 in each of these molecules. Use a table inside the back cover. Neglect centrifugal distortion. (b) Predict the $J = 1 \rightarrow 2$, $v = 0 \rightarrow 0$ transition frequency for $^1H^{79}Br$. (c) Predict the $J = 0 \rightarrow 1$, $v = 0 \rightarrow 0$ transition frequency for $^2H^{79}Br$. [Actually, each pure-rotational transition of these species is split into several lines because of the electric quadrupole moments (*Levine* (1975), p. 224) of the ^{79}Br and ^{81}Br nuclei. The frequencies given are for the centers of the $J = 0 \rightarrow 1$ transitions.]

21.27 The $J = 2 \rightarrow 3$ pure-rotational transition for the ground vibrational state of $^{39}K^{37}Cl$ occurs at 22410 MHz. Neglecting centrifugal distortion, predict the frequency of the $J = 0 \rightarrow 1$ pure-rotational transition of (a) $^{39}K^{37}Cl$; (b) $^{39}K^{35}Cl$.

21.28 Verify that neglect of vibration–rotation interaction gives Eq. (21.38) for $\tilde{\nu}_P$.

21.29 Verify Eqs. (21.39) and (21.40) for R- and P-branch wavenumbers.

21.30 For $^{16}O_2$, $\tilde{\nu}_e = 1580$ cm^{-1}. Find k_e for $^{16}O_2$.

21.31 (a) From the IR data following Eq. (21.36), calculate $\tilde{\nu}_e$ and $\tilde{\nu}_e x_e$ for $^1H^{35}Cl$. (b) Use the results of (a) to predict $\tilde{\nu}_{origin}$ for the $v = 0 \rightarrow 6$ transition of this molecule.

21.32 In the $v = 0 \rightarrow 1$ band of the $^{12}C^{16}O$ IR spectrum, the four lines closest to the band origin lie at 2150.858, 2147.084, 2139.427, and 2135.548 cm^{-1}, where the band origin is between the second and third of these lines. (a) Give the initial and final J values for each of these lines without looking at figures in the text. Give your reasoning. (b) Find $\tilde{\alpha}_e$, $\tilde{B}_e$, and $\tilde{\nu}_{origin}$. A good way to do this is to use the Excel Solver to minimize the sum of the squares of deviations of calculated from observed wavenumbers. (c) Find R_e for $^{12}C^{16}O$.

21.33 Use data in Table 21.1 to calculate the relative populations at 300 K of the $J = 0$ through $J = 6$ rotational levels of the $v = 0$ vibrational level of $^1H^{35}Cl$.

21.34 Match each of the symbols B_e, α_e, D, $\nu_e x_e$ with one of these terms: anharmonicity, vibration–rotation interaction, centrifugal distortion, rotational constant.

21.35 (a) For O_2, O_2^+, and O_2^-, which species has the largest k_e in the ground electronic state and which has the smallest k_e? (b) For N_2 and N_2^+, which species has the larger ν_e in the ground electronic state? (c) For N_2 and O_2, which molecule has the larger k_e in the ground electronic state? (d) For Li_2 and Na_2, which molecule has the greater rotational energy in the $J = 1$ level?

Section 21.5

21.36 True or false? (a) $\hat{C}_4^2 = \hat{C}_2$; (b) $\hat{C}_3^4 = \hat{C}_3$; (c) $\hat{\sigma}^3 = \sigma$; (d) $\hat{S}_3^3 = \hat{E}$.

21.37 List all the symmetry elements present in (a) H_2S; (b) CF_3Cl; (c) XeF_4; (d) PCl_5; (e) IF_5; (f) p-dibromobenzene; (g) HCl; (h) CO_2.

21.38 List all the symmetry operations for (a) H_2S; (b) CF_3Cl.

21.39 The inversion operation $\hat{i}$ moves a nucleus at x, y, z to $-x, -y, -z$. What effect does each of the following operations have on a nucleus at x, y, z? (a) A $\hat{C}_2$ rotation about the z axis; (b) a reflection in the xy plane; (c) an $\hat{S}_2$ rotation about the z axis. From the answer to (c), what statement can be made about the $\hat{S}_2$ operation?

Section 21.6

21.40 Without doing any calculations, describe as fully as you can the locations of the principal axes of inertia of (a) BF_3; (b) H_2O; (c) CO_2.

21.41 Classify each of the following as a spherical, symmetric, or asymmetric top: (a) SF_6; (b) IF_5; (c) H_2S; (d) PF_3; (e) benzene; (f) CO_2; (g) $^{35}Cl^{37}Cl_3$; (h) BF_3.

21.42 Each bond length in BF_3 is 1.313 Å. Calculate the principal moments of inertia of $^{11}B^{19}F_3$.

21.43 For PCl_5, the two axial bond lengths are 2.12 Å and the three equatorial bond lengths are 2.02 Å. Calculate I_a, I_b, and I_c for $^{31}P^{35}Cl_5$.

21.44 For CF_3I, the rotational constants are $\tilde{A} = 0.1910$ cm^{-1} and $\tilde{B} = 0.05081$ cm^{-1}. (a) Calculate E_{rot}/h for the $J = 0$ and J

= 1 rotational levels. (b) Calculate the two lowest microwave absorption frequencies.

21.45 The R_0 bond lengths in the linear molecule OCS are $R_{OC} = 1.160$ Å and $R_{CS} = 1.560$ Å. (a) The z coordinate of the center of mass (com) of a set of particles with masses m_i and z coordinates z_i is $z_{com} = (\Sigma_i \, m_i z_i)/(\Sigma_i \, m_i)$. Find the position of the center of mass of $^{16}O^{12}C^{32}S$. (b) Find the moment of inertia of $^{16}O^{12}C^{32}S$ about an axis through the center of mass and perpendicular to the molecular axis. (c) Find the three lowest microwave absorption frequencies of $^{16}O^{12}C^{32}S$.

21.46 Analysis of the infrared spectrum of $^{12}C^{16}O_2$ gives the rotational constant as $\tilde{B}_0 = 0.39021$ cm^{-1}. Find the CO bond length in CO_2.

Section 21.8

21.47 Give the number of normal vibrational modes of (a) SO_2; (b) C_2F_2; (c) CCl_4.

21.48 H_2O vapor has an IR absorption band at $\tilde{\nu}_{origin} = 7252$ cm^{-1}. The lower vibrational level for this band is 000. What are the possibilities for the upper vibrational level?

21.49 Use data in Sec. 21.8 to calculate the zero-point vibrational energy of (a) CO_2; (b) H_2O.

21.50 Two of the BF_3 normal modes of vibration are described below. State whether each mode is IR-active or IR-inactive. (a) Simultaneous stretching of each bond; (b) each atom moving perpendicular to the molecular plane, with B moving in the opposite direction as each F.

Section 21.9

21.51 State which of each of the following pairs of vibrations has the higher vibrational frequency. (a) C=C stretching, C≡C stretching; (b) C—H stretching, C—D stretching; (c) C—H stretching, CH_2 bending.

21.52 C—H stretching vibrations in organic compounds occur near 2900 cm^{-1}. Near what wavenumber would C—D stretching vibrations occur?

21.53 From the CH and C=O stretching frequencies listed in Sec. 21.9, estimate the force constants for stretching vibrations of these bonds.

Section 21.10

21.54 True or false? (a) The Raman shift of a given Raman spectral line does not change if the frequency ν_0 of the incident radiation is changed. (b) Stokes lines have positive Raman shifts. (c) The selection rules for pure-rotational Raman transitions are the same as for ordinary pure-rotational absorption transitions. (d) A molecule with no electric dipole moment will not show a pure-rotational absorption spectrum but may have pure-rotational lines in its Raman spectrum. (e) Every molecule has a vibrational Raman spectrum. (f) In vibrational Raman spectra, Stokes lines are more intense than anti-Stokes lines. (g) A normal mode that is IR-inactive must be Raman-active.

21.55 (a) Derive the formula for the Raman shifts of the pure-rotational Raman lines of a linear molecule. What is the spacing between adjacent pure-rotational Raman lines? (b) The

pure-rotational Raman spectrum of $^{14}N_2$ shows a spacing of 7.99 cm^{-1} between adjacent rotational lines. Find the bond distance in N_2. (c) What is the spacing between the unshifted line at ν_0 and each of the pure-rotational linear-molecule lines closest to ν_0? (d) If 540.8-nm radiation from an argon laser is used as the exciting radiation, find the wavelengths of the two pure-rotational Raman $^{14}N_2$ lines nearest the unshifted line.

Section 21.11

21.56 True or false? (a) Transitions between electronic states occuring with absorption or emission of radiation always have $\Delta S = 0$. (b) Most of the radiation emitted from a fluorophore is usually at longer wavelengths than the radiation exciting the fluorescence. (c) Excited electronic states of a molecule usually have equilibrium geometries quite close to that of the ground electronic state.

21.57 Calculate the wavelength of the series limit of the Balmer lines of the hydrogen-atom spectrum.

21.58 Calculate the wavelengths of the first three lines in the Paschen series of the hydrogen-atom spectrum.

21.59 Calculate the wavelength of the $n = 2 \rightarrow 1$ transition in Li^{2+}.

21.60 The Schumann–Runge bands of O_2 are due to a transition between the ground electronic state and an excited electronic state designated as the B state. The O_2 ground electronic state dissociates to two ground-state O atoms. The O_2 B state dissociates to one ground-state O atom and an O atom in an excited state 1.970 eV above the O ground state. The $v' = 0$ to $v'' = 0$ band of the Schumann–Runge bands is at 202.60 nm, and the bands converge to a continuous absorption beginning at 175.05 nm. Find D_0 of the O_2 ground state and of the B state.

21.61 For $^{12}C^{16}O$ molecules in a certain jet, the $J = 1$ to $J = 0$ population ratio is 0.181. Find the rotational temperature of the $^{12}C^{16}O$ molecules. See Table 21.1.

Section 21.12

21.62 True or false? (a) The proton spin state with $M_I = +\frac{1}{2}$ has lower energy in a magnetic field than that with $M_I = -\frac{1}{2}$. (b) The proton spin states $M_I = +\frac{1}{2}$ and $M_I = -\frac{1}{2}$ have the same energy in the absence of a magnetic field. (c) For all nuclei with $I = \frac{1}{2}$, the spin state with $M_I = +\frac{1}{2}$ has lower energy in a magnetic field than that with $M_I = -\frac{1}{2}$. (d) A nucleus such as ^{12}C or ^{16}O that has $I = 0$ has no NMR spectrum. (e) γ is a constant that has the same value for every nucleus. (f) g_N has the same value for every nucleus. (g) The nuclear magneton β_N has the same value for every nucleus. (h) The chemical shift δ_i does not change when B_0 and ν_{spec} change. (i) The frequency shift $\nu_i - \nu_{ref}$ does not change when B_0 changes. (j) J_{ij} does not change when B_0 changes. (k) The first-order treatment holds accurately for virtually all molecules studied in NMR spectroscopy. (l) A pulse FT-NMR spectrometer uses a fixed magnetic field B_0.

21.63 Calculate the force on an electron moving at 3.0×10^8 cm/s through a magnetic field of 1.5 T if the angle between the electron's velocity vector and the magnetic field is (a) 0°; (b) 45°; (c) 90°; (d) 180°.

21.64 Calculate the magnetic dipole moment of a particle with charge 2.0×10^{-16} C moving on a circle of radius 25 Å with speed 2.0×10^7 cm/s.

21.65 (a) Verify Eq. (21.63) for γ. (b) Verify the $\gamma/2\pi$ value for ^{1}H given after (21.67).

21.66 The nucleus ^{11}B has $I = \frac{3}{2}$ and $g_N = 1.792$. Calculate the energy levels of a ^{11}B nucleus in a magnetic field of (a) 1.50 T; (b) 15000 G.

21.67 Use data in Prob. 21.66 to find the NMR absorption frequency of ^{11}B in a magnetic field of (a) 1.50 T; (b) 2.00 T.

21.68 For an NMR spectrometer whose proton absorption frequency is 600 MHz, find the ^{13}C absorption frequency.

21.69 Calculate the value of B in a proton magnetic resonance spectrometer that has ν_{spec} equal to (a) 60 MHz; (b) 300 MHz.

21.70 (a) Calculate the ratio of the populations of the two nuclear-spin energy levels of a proton in a field of 1.41 T (the field in a 60-MHz spectrometer) at 25°C. (b) Explain why increasing the applied field B_0 increases the NMR absorption line strengths.

21.71 For an applied field of 1.41 T (the field used in a 60-MHz spectrometer), calculate the difference in NMR absorption frequencies for two protons whose δ values differ by 1.0.

21.72 (a) Sketch the proton NMR spectrum of acetaldehyde, CH_3CHO, for a 60-MHz spectrometer. Include both δ and ν scales. Estimate δ and J from tables in this chapter. (b) Repeat (a) for a 300-MHz spectrometer.

21.73 It is conventional to plot NMR spectra with the chemical shift δ_i increasing to the left. State whether each of the following increases to the left or to the right in a plotted NMR spectrum. (a) The absorption frequency ν_i for a fixed magnetic field B_0. (b) The frequency shift $\nu_i - \nu_{ref}$. (c) The shielding constant σ_i. (d) The magnetic field $B_{0,i}$ at which absorption occurs in a fixed-frequency NMR spectrometer.

21.74 For each of the following, state how many proton NMR peaks occur, the relative intensity of each peak, and whether each peak is a singlet, doublet, triplet, etc.: (a) benzene; (b) CH_3F; (c) C_2H_6; (d) $CH_3CH_2OCH_2CH_3$; (e) $(CH_3)_2CHBr$; (f) methyl acetate; (g) $CH_2=CHBr$ in a magnetic field large enough to produce a first-order spectrum; (h) C_2H_5CHO.

21.75 For each of the following molecules, state whether requirement 2 (only one coupling constant between any two sets of chemically equivalent spin-$\frac{1}{2}$ nuclei) is met: (a) $CH_2=CF_2$; (b) $CH_2=C=CF_2$.

21.76 Repeat Prob. 21.74 for the natural-abundance ^{13}C NMR spectra observed with proton–^{13}C spin–spin splittings eliminated by double resonance.

21.77 Give the number of lines in the spin-decoupled ^{13}C NMR spectrum of (a) o-xylene [$C_6H_4(CH_3)_2$]; (b) m-xylene; (c) p-xylene.

21.78 Use Fig. 21.44 to estimate J for the CH_2 and CH_3 protons in ethanol.

21.79 Suppose the proton NMR spectrum of CH_3CH_2OH is observed using a 60-MHz spectrometer and a 600-MHz spectrometer. State whether each of the following quantities is the same or different for the two spectrometers. If different, by what factor does the quantity change when one goes from 60 to 600 MHz? (a) $\delta_{CH_3} - \delta_{CH_2}$; (b) $\nu_{CH_3} - \nu_{CH_2}$; (c) J_{CH_2,CH_3}.

21.80 Verify Eq. (21.71). Use the fact that $\sigma \ll 1$.

21.81 (a) For a first-order process like the dimethylformamide internal rotation where there is no spin–spin coupling between the protons being interchanged and where the two singlet lines have equal intensities, one can show that the rate constant k_c at the temperature at which the lines coalesce satisfies $k_c = \pi|\nu_2 - \nu_1|/2^{1/2}$, where $\nu_2 - \nu_1$ is the frequency difference between the lines in the absence of interchange. Use data in the text to find k_c (at 120°C) for proton interchange in dimethylformamide. (b) If the spectrometer frequency ν_{spec} is increased, will the coalescence temperature for this process increase, decrease, or remain the same?

Section 21.13

21.82 Calculate the ESR frequency in a field of 2.500 T for $\cdot CH_3$, which has $g = 2.0026$.

21.83 How many lines will the ESR spectrum of the naphthalene negative ion $C_{10}H_8^-$ have?

21.84 Give the number of lines in the ESR spectrum of each of the following; assume that only protons on the α and β carbons cause splitting: (a) $(CH_3)_2CH\cdot$; (b) $CH_3CH_2CH_2CH_2\cdot$; (c) $(CH_3)_3C\cdot$; (d) $(CH_3)_2CHCH_2\cdot$.

Section 21.14

21.85 A solution of the amino acid L-lysine in water at 20°C containing 6.50 g of solute per 100 mL in a 2.00-dm-long tube has an observed optical rotation of $+1.90°$ for 589.3-nm light. Find the specific rotation at this wavelength.

21.86 For a freshly prepared aqueous solution of α-D-glucose, $[\alpha]_D^{20} = +112.2°$, where the subscript and superscript on the specific rotation indicate sodium D (589.3 nm) light and 20°C. For a freshly prepared aqueous solution of β-D-glucose, $[\alpha]_D^{20} = +17.5°$. As time goes by, $[\alpha]_D^{20}$ for each solution changes and reaches the limiting value 52.7°, which is for the equilibrium mixture of α- and β-D-glucose. Find the percentage of α-D-glucose present at equilibrium at 20°C in water.

Section 21.15

21.87 (a) When 30.4-nm radiation is used to produce the photoelectron spectrum of benzene, the highest-energy photoelectrons have kinetic energy of 31.5 eV. Find the ionization energy of the highest-energy MO in benzene (which is a π MO). (b) What would be the kinetic energy of the highest-energy photoelectrons emitted from benzene if 58.4-nm radiation were used?

21.88 The ground electronic state of N_2 has $\tilde{\nu}_0 = 2330$ cm^{-1}. The spacing between adjacent lines in the 17-eV band of the N_2 photoelectron spectrum is approximately 1800 cm^{-1}. Explain why this spacing is less than 2330 cm^{-1}.

21.89 For each of the following molecules, state the number and relative intensities of the carbon $1s$ ESCA peaks and do the same for the oxygen $1s$ ESCA peaks: (a) C_2H_5OH; (b) CH_3OCH_3; (c) CH_3COOCH_3; (d) C_6H_5OH (phenol).

Section 21.16

21.90 Verify the photon energy and molar-energy calculations in (21.79).

21.91 (a) Show that in a light beam whose intensity is I, the number of photons per unit volume equals $I/h\nu c'$, where c' is the speed of light in the medium through which the beam is passing. (*Hint:* Start with the definition of I.) (b) A pulsed argon laser operating at 488 nm can readily produce a focused beam with $I = 10^{15}$ W/m^2, where W stands for watts. If this beam passes through water with refractive index 1.34, find the number of photons per unit volume in the water and compare with the number of water molecules per unit volume.

21.92 In a certain photochemical reaction using 464-nm radiation, the incident-light power was 0.00155 W and the system absorbed 74.4% of the incident light; 6.80×10^{-6} mole of product was produced during an exposure of 110 s. Find the quantum yield.

21.93 The photochemical decomposition of HI proceeds by the mechanism

$$HI + h\nu \rightarrow H + I$$
$$H + HI \rightarrow H_2 + I$$
$$I + I + M \rightarrow I_2 + M$$

where the rate of the first step is $\phi \mathcal{I}_a$ with $\phi = 1$. (a) Show that $-d[HI]/dt = 2\mathcal{I}_a$. Hence the quantum yield with respect to HI is 2. (b) How many HI molecules will be decomposed when 1.00 kcal of 250-nm radiation is absorbed?

21.94 In the anthracene dimerization, the rate of the forward reaction is negligible in the absence of radiation, but let us assume a forward bimolecular reaction with rate constant k_5 in the absence of radiation:

$$(5) \quad 2A \rightarrow A_2 \qquad r_5 = k_5[A]^2$$

in addition to reactions (1) to (4) preceding Eq. (21.84). With inclusion of step 5, express the reaction rate r in terms of [A], [A$_2$], and $\mathcal{I}_a$. Set $r = 0$ to obtain the photostationary-state A$_2$ concentration. Compare with the equilibrium A$_2$ concentration in the absence of radiation.

Section 21.17

21.95 Which of the following are groups? (a) The numbers 1 and -1 with the rule of combination being ordinary multiplication. (b) The set of all integers, positive, negative, and zero, with multiplication as the rule of combination. (c) The numbers 1, 0, and -1 with the rule of combination being addition.

21.96 Which of these arithmetical operations are associative? (a) Addition. (b) Subtraction. (c) Multiplication. (d) Division.

21.97 (a) A group of order two has two elements A and I, where I is the identity element. Give the multiplication table of the group. (b) For a group of order three with the elements A, B, and I, find all possible forms of the multiplication table.

21.98 True or false? (a) The symmetry elements of a molecule are the members (elements) of the molecule's symmetry point group. (b) The symmetry operations of a molecule are the members (elements) of the molecule's symmetry point group.

21.99 Draw the octahedral molecule SF$_6$ with the z axis passing through two trans fluorines and with the x and y axes each bisecting FSF angles. Then number the F atoms. By applying operations to SF$_6$, find whether the operations in each of the following pairs commute with each other: (a) $\hat{C}_4(z)$ and $\hat{C}_2(x)$; (b) $\hat{C}_4(z)$ and $\hat{\sigma}(xz)$; (c) $\hat{i}$ and $\hat{\sigma}(xy)$.

21.100 Give the inverse of each of the following operations: (a) $\hat{E}$; (b) $\hat{\sigma}$; (c) $\hat{i}$; (d) $\hat{C}_5$; (e) $\hat{C}_5^2$; (f) $\hat{S}_3$.

21.101 True or false? (a) $\hat{C}_6^3 = \hat{C}_2$; (b) $\hat{S}_6^3 = \hat{S}_2$; (c) $\hat{S}_2 = \hat{i}$; (d) $\hat{S}_1 = \hat{\sigma}$.

21.102 For PCl$_3$, (a) list all the symmetry elements; (b) list all the symmetry operations.

21.103 For benzene, (a) list all the symmetry elements; (b) list all the symmetry operations.

21.104 For CH$_4$, list all the symmetry elements.

21.105 Find the point group of each of the following molecules: (a) CF$_4$; (b) HCN; (c) N$_2$; (d) BF$_3$; (e) SF$_6$; (f) IF$_5$; (g) XeF$_4$; (h) PCl$_3$; (i) PCl$_5$; (j) C$_2$H$_4$; (k) CHCl$_3$; (l) SF$_5$Br; (m) CHFClBr; (n) 1-fluoro-2-chlorobenzene.

21.106 The group C_n (where n can be 2, 3, 4, . . .) has a C_n axis as its only symmetry element. Give the symmetry operations of this group and give the order of this group.

21.107 The group C_{nh} (where n can be 2, 3, 4, . . .) contains a C_n axis, a σ_h plane of symmetry perpendicular to the axis, and a center of symmetry if n is even, but has no C_2 axes perpendicular to the C_n axis. For C_{nh}, (a) Is the C_n axis also an S_n axis? (b) When is the C_n axis also a C_2 axis? (c) Give the point group of each of these molecules: *cis*-dichloroethylene; *trans*-dichloroethylene; 1,1-dichloroethylene.

21.108 The group D_{nd} ($n = 2, 3, 4, . . .$) has a C_n axis, n C_2 axes perpendicular to the C_n axis, and n vertical planes of symmetry (called diagonal planes σ_d) that each contain the C_n axis and that bisect the angles between adjacent C_2 axes. The C_n axis is also an S_n axis. (A common error is to overlook the C_2 axes of a D_{nd} molecule.) Give the point group of each of these molecules: the staggered conformation of C$_2$H$_6$; allene. (When you see a molecule with two equal halves staggered with respect to each other, think D_{nd}.)

21.109 (a) The matrix sum $\mathbf{F} = \mathbf{A} + \mathbf{B}$ (which exists if $\mathbf{A}$ and $\mathbf{B}$ have the same number of rows and have the same number of columns) is the matrix each of whose elements is the sum of corresponding elements of $\mathbf{A}$ and $\mathbf{B}$; that is, $f_{ij} = a_{ij} + b_{ij}$. Find the sum of $\mathbf{A}$ and $\mathbf{B}$ in (21.89). (b) The matrix $k\mathbf{A}$, where k is a scalar, is the matrix each of whose elements is k times the corresponding element of $\mathbf{A}$. Find $3\mathbf{A}$, where $\mathbf{A}$ is given in (21.89).

21.110 If $\mathbf{A} = \begin{pmatrix} 0.2 & 4 \\ -1 & 3 \end{pmatrix}$ and $\mathbf{B} = \begin{pmatrix} 4 & 1 \\ 5 & 8 \end{pmatrix}$, find $\mathbf{AB}$ and $\mathbf{BA}$.

21.111 If $\mathbf{A} = \left(\begin{smallmatrix} 1 & 2 \\ 3 & 4 \end{smallmatrix}\right)$ and $\mathbf{W} = \left(\begin{smallmatrix} 2 \\ 6 \end{smallmatrix}\right)$, find $\mathbf{AW}$.

21.112 If $\mathbf{T} = \mathbf{RS}$, express the matrix element t_{mn} in terms of a certain sum of matrix elements of $\mathbf{R}$ and $\mathbf{S}$.

21.113 Verify the block-diagonal equation (21.92) for the matrices in (21.91).

21.114 Verify that the matrices in (21.94) multiply the same way as the corresponding symmetry operations multiply. To save time, note that the product of two diagonal matrices is a diagonal matrix whose elements are the products of corresponding elements of the matrices being multiplied.

21.115 Find the matrix that gives the effect of each of these operations on the point (x, y, z); (a) $\hat{C}_4(z)$ (assume the rotation is counterclockwise when viewed from the positive z axis); (b) $\hat{\imath}$; (c) $\hat{\sigma}(xy)$.

21.116 (a) If $\mathbf{AC} = \mathbf{F}$, prove that $(\mathbf{P}^{-1}\mathbf{AP})(\mathbf{P}^{-1}\mathbf{CP}) = \mathbf{P}^{-1}\mathbf{FP}$. (Matrix multiplication can be shown to be associative.) (b) Explain why if $\mathbf{A}, \mathbf{B}, \mathbf{C}, \ldots$ form a representation of a group then $\mathbf{P}^{-1}\mathbf{AP}, \mathbf{P}^{-1}\mathbf{BP}, \mathbf{P}^{-1}\mathbf{CP}, \ldots$ are a representation of the group.

21.117 Use (21.96) to prove that all the nonequivalent irreducible representations of a commutative group are one-dimensional.

21.118 Verify that (21.97) is a representation of C_{2v}.

21.119 Prove that the identity operation $\hat{E}$ is always in a class by itself.

21.120 Into what function is a $2p_y$ AO transformed by each of these operations (take all rotations as counterclockwise when viewed from the positive side of the rotation axis): (a) $\hat{C}_2(y)$; (b) $\hat{C}_2(z)$; (c) $\hat{C}_4(y)$; (d) $\hat{C}_4(z)$; (e) $\hat{\sigma}(yz)$; (f) $\hat{\sigma}(xz)$; (g) $\hat{\imath}$; (h) $\hat{S}_4(y)$.

21.121 Some occupied MOs of the ground electronic state of H_2O are described in this problem. For each MO, examine the symmetry behavior of the oxygen AOs and of the combination of hydrogen AOs, and then give the symmetry species (irreducible representation) of the MO. The z axis is the C_2 axis and the molecular plane is the yz plane. (a) This MO contains substantial contributions from the AOs $O2s$, $O2p_z$, and $H_11s + H_21s$. (b) This MO has substantial contributions from $O2p_y$ and from $H_11s - H_21s$.

21.122 Give the symmetry species of each of the following normal modes of vibration. (a) In H_2CO, a mode in which all atoms move perpendicular to the molecular yz plane with the carbon moving in the opposite direction as the other three atoms. (b) In NH_3, a mode in which the H atoms vibrate in the bond directions while the N atom vibrates along the C_3 axis (the z axis) with the vertical (z) component of motion of the three hydrogens being in the opposite direction as the z component of the N motion.

21.123 Use (21.98) to verify that the representation (21.94) is the direct sum of A_1, B_1, and B_2.

21.124 A certain representation of C_{2v} has the following characters:

$\hat{E}$	$\hat{C}_2(z)$	$\hat{\sigma}_v(xz)$	$\hat{\sigma}_v(yz)$
9	-1	1	3

(a) What is the dimension of this representation? (b) Express this representation as the direct sum of irreducible representations.

21.125 A certain representation of C_{3v} has the following characters:

$\hat{E}$	$2\hat{C}_3$	$3\hat{\sigma}_v$
293	-118	9

Express this representation as the direct sum of irreducible representations. [*Hint:* The sum in (21.98) is over the symmetry operations, not over the classes.]

21.126 Give the characters for a C_{3v} representation Γ for which $\Gamma = 4A_1 \oplus A_2 \oplus 6E$. (A similarity transformation does not change the trace of a matrix.)

General

21.127 Consider the molecules N_2, HBr, CO_2, H_2S, CH_4, CH_3Cl, and C_6H_6. (a) Which have pure-rotational absorption spectra? (b) Which have vibration–rotation absorption spectra? (c) Which have pure-rotational Raman spectra?

21.128 For H_2 and D_2, state which has the greater value of each of the following. In some cases, the value is the same for both. Neglect deviations from the Born–Oppenheimer approximation. (a) k_e; (b) ν_e; (c) I_e; (d) B_e; (e) D_e; (f) D_0; (g) number of bound vibrational levels; (h) fraction of molecules having $v = 0$ at 2000 K; (i) fraction of molecules having $J = 0$ at 300 K.

21.129 The abbreviation for a unit that is named after a person is capitalized. (a) Name at least 10 SI units named after persons. (b) Name 4 non-SI units named after persons.

21.130 True or false? (a) Linear molecules are symmetric tops. (b) A molecule with zero dipole moment cannot change its rotational state. (c) A molecule whose dipole moment is zero must have a center of symmetry. (d) Whenever a molecule goes from one energy level to another, it emits or absorbs a photon whose energy is equal to the energy difference between the levels. (e) An asymmetric top cannot have any axes of symmetry. (f) The Raman shift of a given Raman-spectrum line is independent of the value of the exciting frequency ν_0. (g) The rotational energy of any molecule is given by $BhJ(J + 1)$ provided centrifugal distortion is neglected. (h) The vibrational levels of a given electronic state of a diatomic molecule are unequally spaced. (i) For a system in thermal equilibrium, a state with a higher energy than another state always has a smaller population than the lower-energy state. (j) For a system in thermal equilibrium, an energy level with a higher energy always has a smaller population than the lower-energy level. (k) We all have moments of inertia.

Statistical
Mechanics

22.1   STATISTICAL MECHANICS

In Chapters 1–14 we used thermodynamics to study the macroscopic properties of matter. In Chapters 18–21 we used quantum mechanics to examine molecular properties. The link between quantum mechanics and thermodynamics is provided by **statistical mechanics,** whose aim is to deduce the macroscopic properties of matter from the properties of the molecules composing the system. Typical macroscopic properties are entropy, internal energy, heat capacity, surface tension, dielectric constant, viscosity, electrical conductivity, and chemical reaction rate. Molecular properties include molecular masses, molecular geometries, intramolecular forces (which determine the molecular vibration frequencies), and intermolecular forces. Because of the huge number of molecules in a macroscopic system, one uses statistical methods instead of attempting to consider the motion of each molecule in the system.

This chapter is restricted to **equilibrium statistical mechanics** (also called **statistical thermodynamics**), which deals with systems in thermodynamic equilibrium. *Nonequilibrium statistical mechanics* (whose theory is not as well developed as that of equilibrium statistical mechanics) deals with transport properties and chemical reaction rates. Very crude statistical-mechanical treatments of transport properties were given in Chapter 16. A statistical-mechanical theory of reaction rates is given in Chapter 23.

Statistical mechanics originated in the work of Maxwell and Boltzmann on the kinetic theory of gases (1860–1900). Major advances in the theory and the methods of calculation were made by Gibbs in his 1902 book *Elementary Principles in Statistical Mechanics* and by Einstein in a series of papers (1902–1904). Since quantum mechanics had not yet been discovered, these workers assumed that the system's molecules obeyed classical mechanics. This led to incorrect results in some cases. For example, calculated C_V's of gases of polyatomic molecules disagreed with experiment (Sec. 15.10). When quantum mechanics was discovered, the necessary modifications in statistical mechanics were easily made.

This chapter presents statistical mechanics using quantum mechanics and in a general form applicable to all forms of matter, not just gases. Section 22.2 derives the fundamental statistical-mechanical formulas that relate thermodynamic properties to the quantum-mechanical energy levels of the thermodynamic system. Although these formulas apply to all systems, it is easiest to apply these formulas when the system is an ideal gas. Most of this chapter (Secs. 22.3 to 22.8) is restricted to the statistical mechanics of ideal gases. Section 22.11 deals with nonideal gases and liquids. (Section 24.12 applies statistical mechanics to crystals.) As a preliminary to Sec. 22.11, Sec. 22.10 discusses intermolecular forces.

Statistical mechanics deals with both the microscopic (molecular) level and the macroscopic level, and it is important to define our terminology clearly. *The word* **system** *in this chapter refers only to a macroscopic thermodynamic system.* The fundamental microscopic entities that compose a (thermodynamic) system will be called **molecules** or **particles.** In some cases, these entities are not actually molecules. For example, one can apply statistical mechanics to the conduction electrons in a metal or to the photons in electromagnetic radiation.

The term **state of a system** has two meanings in statistical mechanics. The **thermodynamic state** of a system is specified by giving the values of enough macroscopic parameters to characterize the system. (For example, we might have 24.0 g of benzene plus 2.87 g of toluene at 52°C and 3.65 atm; or we might have 18.0 g of H_2O at 54°C in a volume of 17.2 cm^3.) On the other hand, we can talk about the **quantum state** of the system. By this we mean the following. Consider as an example the 18.0 g of H_2O at 54°C and 17.2 cm^3. We set up and solve the Schrödinger equation for a system composed of 6×10^{23} H_2O molecules to obtain a set of allowed wave functions and energy levels. (In practice, this is an impossible task, but we can imagine doing so in principle.) At any instant of time, the system will be in a definite quantum state j characterized by a certain wave function ψ_j (which is a function of a huge number of spatial and spin coordinates), an energy E_j, and a set of quantum numbers. (Actually, the system might well be in a nonstationary, time-dependent state, but to simplify the arguments we shall not consider this possibility.)

The term **macrostate** means the thermodynamic state of a system. The term **microstate** means the quantum state of a system. Because of the very large number of particles in a system, there are a huge number of different microstates that are compatible with a given macrostate. For example, suppose we have a fixed amount of an ideal monatomic gas whose internal energy and volume are fixed. The translational energy levels (18.47) lie extremely close together [see the discussion after (21.27)], and there are a huge number of different ways we can populate these levels and still end up with the same total energy. The macrostate is experimentally observable. The microstate is usually not observable.

If the molecules of the system do not interact with one another, we can also refer to the quantum states available to each molecule. We call these the **molecular states.** For example, consider a pure ideal gas of monatomic molecules. The macrostate (thermodynamic state) of the gas can be specified by giving the thermodynamic variables T, P, and n (or the variables T, V, and n, or some other set of three thermodynamic properties). The microstate (quantum state) of the gas is specified by saying how many molecules are in each of the available molecular translational quantum states (18.46), where n_x, n_y, and n_z each go from 1 on up. A molecular state is specified by giving the values of the three quantum numbers n_x, n_y, and n_z, since specification of these quantum numbers specifies the wave function of a molecule. In contrast, it takes a huge set of numbers to specify the system's microstate.

We now proceed to relate macrostates to microstates, in order to calculate macroscopic properties from molecular properties.

22.2 THE CANONICAL ENSEMBLE

This section derives some of the fundamental formulas of statistical mechanics. These derivations are abstract and mathematical. Don't be discouraged if you find it hard going. Later sections that apply the formulas to ideal gases are easier reading than this section.

Suppose we measure a macroscopic property of an equilibrium system, for example, the pressure. It takes time to make the measurement, and the observed pressure is a time average over the impacts of individual molecules on the walls. [Equation

(15.57) shows that for a gas at 1 atm and 25°C there are about 10^{17} impacts every microsecond on a 1-cm^2 wall area.] To calculate the value of the macroscopic property, we would have to take a time average over the changes in the microstate of the system. In some simple cases, the time-averaging calculation can be performed—recall the kinetic-theory calculation of ideal-gas pressure in Sec. 15.2. In general, however, it is not feasible to do a time-average calculation. Instead, one resorts to what is called an ensemble.

The Canonical Ensemble

An **ensemble** is a hypothetical collection of an infinite number of noninteracting systems, each of which is in the same macrostate (thermodynamic state) as the system of interest. Although the members of the ensemble are macroscopically identical, they show a wide variety of microstates, since many different microstates are compatible with a given macrostate. We postulate that

The measured time average of a macroscopic property in the system of interest is equal to the average value of that property in the ensemble.

This postulate enables us to replace the difficult calculation of a time average by the easier calculation of an average over systems in the ensemble.

Consider a single equilibrium thermodynamic system whose volume, temperature, and composition are held fixed. The system is enclosed in walls that are rigid (V constant), impermeable (composition constant), and thermally conducting, and the system is immersed in an extremely large constant-temperature bath. Because of thermal interactions with the bath, the system's microstate changes from moment to moment, and its pressure and quantum-mechanical energy fluctuate. Of course, these fluctuations are generally far too small to be detectable macroscopically, because of the large number of molecules in the system (Sec. 3.7).

We want to calculate such thermodynamic properties as pressure, internal energy, and entropy for this system. To do so, we imagine an ensemble in which each of the systems has the same temperature, volume, and composition as the system of interest. Each system of the ensemble sits in a very large constant-temperature bath (Fig. 22.1). An ensemble of systems each having fixed T, V, and composition is called a **canonical ensemble.** Canonical means standard, basic, authoritative. The name was chosen by Gibbs.

Consider an example. Suppose we were interested in calculating the thermodynamic properties of 1.00 millimole of H_2 at 0°C in a 2.00-L box:

$$1.00 \text{ mmol } H_2(273 \text{ K}, 2.00 \text{ L}) \qquad (22.1)$$

We would imagine an infinite number of macroscopic copies of this system, each copy containing 1.00 mmol of H_2 in a 2.00-L box that is immersed in a very large bath at 0°C.

We shall take an average at a fixed time over the microstates of the systems in the ensemble. The possible microstates are found by solving the Schrödinger equation

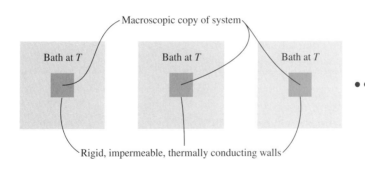

Macroscopic copy of system

Bath at T Bath at T Bath at T

• • •

Rigid, impermeable, thermally conducting walls

Figure 22.1

A canonical ensemble of macroscopically identical systems held at fixed T, V, and composition. The dots indicate the infinite number of systems and baths.

$\hat{H}\psi_j = E_j\psi_j$ for the (macroscopic) system. The possible wave functions ψ_j ($j = 1, 2,$ ...) and quantum energies E_j will depend on the composition of the system (since the number of molecules and the intermolecular and intramolecular forces depend on the composition) and will depend on the system's volume [recall from (18.50) that the translational energy of a particle in a cubic box depends on the volume of the box since $a^2 = V^{2/3}$]. We have

$$E_j = E_j(V, N_B, N_C, \ldots) \qquad (22.2)$$

where the system's composition is specified by giving $N_B, N_C, \ldots$, the number of molecules of each chemical species B, C, ... in the system. (Although our procedure is applicable to a multiphase system, for simplicity we shall usually consider only one-phase systems.) Note that E_j does not depend on the system's temperature. Temperature is a macroscopic nonmechanical property and does not appear in the quantum-mechanical Hamiltonian of the system or in the condition that the wave functions be well behaved.

We postulated that any macroscopic property of the system is to be calculated as an average over the ensemble at a fixed time. For example, the thermodynamic internal energy U will equal the average energy of the systems in the ensemble: $U = \langle E_j \rangle$. Because of the above-mentioned fluctuations, E_j will not be precisely the same for different systems in the ensemble. Similar to (15.40), we write

$$U = \langle E_j \rangle = \sum_j p_j E_j \qquad (22.3)$$

where p_j is a probability. This equation can be interpreted in two ways. If the sum is taken over the different possible energy values E_j of the system, then p_j is the probability that a system in the ensemble has energy E_j. Alternatively, if the sum is taken over the different possible quantum states of the system, then p_j is the probability that a system is in the microstate j (whose energy is E_j). We shall use the second interpretation. Thus, *the symbol $\sum_j$ indicates a sum over the allowed quantum states of the system* (and not a sum over energy levels); p_j *is the probability that the system is in the quantum state j* (and not the probability that the system has energy E_j).

A sum over states differs from a sum over energy levels because several different quantum states might have the same energy, a condition called degeneracy (Sec. 18.10). For the system (22.1), each system energy level is highly degenerate: there are 6×10^{20} molecules, each of which has translational energies in the x, y, and z directions, rotational energy, and vibrational energy; the number of ways a fixed amount of energy can be distributed among the various molecular quantum states is huge.

To find U in (22.3), we need the probabilities p_j (and the quantum-mechanical energies E_j). We postulate that

For a thermodynamic system of fixed volume, composition, and temperature, all quantum states that have equal energy have equal probability of occurring.

This postulate and the postulate of equality of ensemble averages and time averages are the two fundamental postulates of statistical mechanics. If microstates i and k of the (macroscopic) system have the same energy ($E_i = E_k$), then p_i equals p_k. Therefore p_j in (22.3) can depend only on the energy of state j (and not on how this energy is distributed among the molecules):

$$p_j = f(E_j) \qquad \text{fixed } T, V, \text{ composition} \qquad (22.4)$$

Evaluation of p_j

We now find how the probability p_j of being in microstate j depends on the energy E_j of the microstate. To find the function f, imagine that a second system of fixed volume, temperature, and composition is put into each bath of the ensemble (Fig. 22.2). Let I

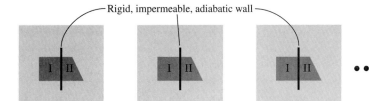

Rigid, impermeable, adiabatic wall

Figure 22.2

A canonical ensemble for a composite system composed of two noninteracting parts I and II.

and II denote the two systems in each bath. All the systems labeled I are macroscopically identical to one another. All the systems labeled II are macroscopically identical. However, systems I and II are not necessarily identical to each other; they can differ in volume and composition. We have

$$p_{\text{I},j} = f(E_{\text{I},j}) \qquad \text{and} \qquad p_{\text{II},k} = g(E_{\text{II},k}) \tag{22.5}$$

where $p_{\text{I},j}$ is the probability that system I is in the microstate j (whose energy is $E_{\text{I},j}$), $p_{\text{II},k}$ is the probability that system II is in the microstate k (whose energy is $E_{\text{II},k}$), and f and g are (at this point) unknown functions. Since systems I and II may differ, f and g are not necessarily the same function.

If we like, we can consider I and II to form a single composite system (I + II) of fixed volume, temperature, and composition. For this composite system,

$$p_{\text{I+II},i} = h(E_{\text{I+II},i}) \tag{22.6}$$

where $p_{\text{I+II},i}$ is the probability that system I + II is in the microstate i (whose energy is $E_{\text{I+II},i}$) and h is some function. The subsystems I and II are independent of each other, so $E_{\text{I+II},i} = E_{\text{I},j} + E_{\text{II},k}$. The microstate i of the composite system I + II is defined by the microstates j and k of the independent subsystems I and II. The probability that two independent events both happen is the product of the probabilities of each event, so the probability that system I + II is in microstate i is the product of the probabilities that system I is in microstate j and system II is in microstate k. Thus $p_{\text{I+II},i} = p_{\text{I},j}p_{\text{II},k}$. Substitution of (22.5) and (22.6) into $p_{\text{I+II},i} = p_{\text{I},j}p_{\text{II},k}$ gives

$$h(E_{\text{I},j} + E_{\text{II},k}) = f(E_{\text{I},j})g(E_{\text{II},k}) \tag{22.7}$$

Equation (22.7) has the form

$$h(x + y) = f(x)g(y) \qquad \text{where } x \equiv E_{\text{I},j}, \ y \equiv E_{\text{II},k} \tag{22.8}$$

$$h(z) = f(x)g(y) \qquad \text{where } z \equiv x + y \tag{22.9}$$

We now solve (22.9) for f. Taking $(\partial/\partial x)_y$ of (22.9), we have

$$[\partial h(z)/\partial x]_y = [df(x)/dx]\, g(y) = f'(x)g(y)$$

But $[\partial h(z)/\partial x]_y = [dh(z)/dz]\,(\partial z/\partial x)_y = dh/dz$, since $(\partial z/\partial x)_y = 1$. Therefore

$$dh/dz = f'(x)g(y)$$

Similarly, taking $(\partial/\partial y)_x$ of (22.9), we get $dh/dz = f(x)g'(y)$. Equating these two expressions for dh/dz, we have

$$f(x)g'(y) = f'(x)g(y)$$

$$\frac{g'(y)}{g(y)} = \frac{f'(x)}{f(x)} \equiv -\beta \tag{22.10}$$

where β is defined as $-f'(x)/f(x)$. Since the function $f'(x)/f(x)$ is independent of y, β is independent of y. The left side of (22.10) is independent of x, so β is independent of x. Therefore β is a constant. We have $df(x)/f(x) = -\beta\, dx$, which integrates to

$\ln f = -\beta x + \text{const.}$ Thus $f = e^{-\beta x} e^{\text{const}} = ae^{-\beta x}$, where $a \equiv e^{\text{const}}$ is a constant. Similarly, $g'(y)/g(y) = -\beta$ integrates to $g(y) = ce^{-\beta y}$, where c is a constant. We have

$$f(x) = ae^{-\beta x} \quad \text{and} \quad g(y) = ce^{-\beta y} \tag{22.11}$$

Since $f(x)g(y) = ace^{-\beta(x+y)}$, these functions satisfy (22.8).

Use of Eqs. (22.5) and (22.8) in (22.11) gives

$$p_{\text{I},j} = ae^{-\beta E_{\text{I},j}} \quad \text{and} \quad p_{\text{II},k} = ce^{-\beta E_{\text{II},k}} \tag{22.12}$$

To complete things, we must find a, β, and c. We know that these quantities are constants for a system of fixed temperature, volume, and composition. Therefore they might depend on one or more of T, V, and composition. However, they cannot depend on the microstate energies $E_{\text{I},j}$ (or $E_{\text{II},k}$), since β, a, and c are independent of x and y in the above derivation and x and y are $E_{\text{I},j}$ and $E_{\text{II},k}$.

Consider β. Equation (22.10) or (22.12) shows that any two systems I and II in the same constant-temperature bath have the same value of β. Two systems in the same bath have a common value of T but can differ in volume and in composition. We conclude that β can be a function of T only: $\beta = \phi(T)$, where ϕ is the same function for any two systems. In contrast, a can depend on all three temperature, volume, and composition.

The use of two systems was a temporary device, and we now go back to the ensemble with a single system in each bath. We thus write (22.12) as

$$p_j = ae^{-\beta E_j} \tag{22.13}$$

where β is a function of T and a is a function of T, V, and composition.

To evaluate a, we use the fact that the total probability must be 1. Thus $\sum_j p_j = 1 = a \sum_j e^{-\beta E_j}$, and

$$a = \frac{1}{\sum_j e^{-\beta E_j}} \tag{22.14}$$

Equation (22.13) becomes

$$p_j = \frac{e^{-\beta E_j}}{\sum_k e^{-\beta E_k}} \equiv \frac{e^{-\beta E_j}}{Z} \tag{22.15}$$

The summation index is a dummy variable (Sec. 1.8), so any letter can be used for it. To avoid later confusion, j in (22.14) was changed to k.

The sum in (22.15) plays a star role in statistical mechanics and is called the **canonical partition function** of the system:

$$Z \equiv \sum_j e^{-\beta E_j} \tag{22.16}*$$

where the sum goes over all possible quantum states of the system for a given composition and volume. E_j is the quantum-mechanical energy of the macroscopic system when it is in microstate j. From (22.15), the terms $e^{-\beta E_j}$ in the partition function govern how the systems of the ensemble are distributed or "partitioned" among the possible quantum states of the system. (The letter Z comes from the German word *Zustandssumme*, "sum over states.") The contribution $e^{-\beta E_j}$ of state j to Z is proportional to the probability p_j [Eq. (22.15)] that state j will occur.

Having evaluated a, we turn our attention to β in (22.15). We know that $\beta = \phi(T)$. We must find the function ϕ.

Evaluation of U and P

To help find $\phi(T)$, we shall get expressions for U and P, the internal energy and the pressure. Equations (22.3) and (22.15) give

$$U = \sum_j p_j E_j = \frac{\sum_j E_j e^{-\beta E_j}}{\sum_k e^{-\beta E_k}} = \frac{\sum_j E_j e^{-\beta E_j}}{Z} \qquad (22.17)$$

The canonical partition function Z is a function of β and the quantum-mechanical energy levels E_j, which depend on V and composition [Eq. (22.2)]. Therefore

$$Z = Z(\beta, V, N_B, N_C, \ldots) = Z(T, V, N_B, N_C, \ldots) \qquad (22.18)$$

since β is a function of T. Equation (22.18) shows that Z is a function of the thermodynamic state of the system. Partial differentiation of (22.16) gives

$$\left(\frac{\partial Z}{\partial \beta}\right)_{V,N_B} = \left(\frac{\partial}{\partial \beta}\right)_{V,N_B} \sum_j e^{-\beta E_j} = \sum_j \left(\frac{\partial e^{-\beta E_j}}{\partial \beta}\right)_{V,N_B} = -\sum_j E_j e^{-\beta E_j} \qquad (22.19)$$

since E_j is independent of T and hence of β. In (22.19) the subscript N_B indicates constant values of each of the composition variables $N_B, N_C, \ldots$. Equation (22.17) can thus be written as

$$U = -\frac{1}{Z}\left(\frac{\partial Z}{\partial \beta}\right)_{V,N_B} = -\left(\frac{\partial \ln Z}{\partial \beta}\right)_{V,N_B} \qquad (22.20)$$

which is the desired expression for U.

Now consider the system's pressure P. (Don't confuse P for pressure with p for probability.) Just as there are energy fluctuations among systems of the ensemble, there are also pressure fluctuations. Let P_j be the pressure in a system whose microstate is j. Our averaging postulate gives, similar to (22.3),

$$P = \sum_j p_j P_j \qquad (22.21)$$

What is P_j? Consider a single adiabatically enclosed system in quantum state j with pressure P_j and quantum energy E_j. Its thermodynamic energy is $U = E_j$. Imagine a reversible change in volume by dV while the system remains in state j. The quantum-mechanical expression for the energy change is $dU = (\partial E_j/\partial V)_{N_B} dV$. For example, consider a system consisting of two noninteracting distinguishable particles 1 and 2 having only translational energy and confined to a cubic box of volume V. The system's quantum energy E_j is the sum of the energies of the noninteracting particles. The particle-in-a-box energy-level formula (18.50) gives

$$E_j = (h^2/8m_1 V^{2/3})(n_{x,1}^2 + n_{y,1}^2 + n_{z,1}^2) + (h^2/8m_2 V^{2/3})(n_{x,2}^2 + n_{y,2}^2 + n_{z,2}^2)$$

$$\left(\frac{\partial E_j}{\partial V}\right)_{N_B} dV = -\frac{2}{3V^{5/3}}\left[\frac{h^2}{8m_1}(n_{x,1}^2 + n_{y,1}^2 + n_{z,1}^2) + \frac{h^2}{8m_2}(n_{x,2}^2 + n_{y,2}^2 + n_{z,2}^2)\right] dV$$

where, since the system's quantum state doesn't change, the quantum numbers $n_{x,1}$, $\ldots, n_{z,2}$ of the two particles stay fixed. Of course, actual thermodynamic systems have far more than two particles.

The thermodynamic expression for the reversible adiabatic energy change is $dU = dw_{rev} = -P_j dV$. (We restrict ourselves to systems capable of P-V work only.) Equating the thermodynamic and quantum-mechanical expressions for dU, we have $-P_j dV = (\partial E_j/\partial V)_{N_B} dV$, and

$$P_j = -(\partial E_j/\partial V)_{N_B} \qquad (22.22)$$

Substitution of (22.22) for P_j and (22.15) for p_j in (22.21) gives

$$P = \sum_j p_j P_j = -\frac{1}{Z} \sum_j e^{-\beta E_j} \left(\frac{\partial E_j}{\partial V} \right)_{N_B} \tag{22.23}$$

Partial differentiation of $Z = \sum_j e^{-\beta E_j}$ gives

$$\left(\frac{\partial Z}{\partial V} \right)_{T,N_B} = \sum_j \left(\frac{\partial e^{-\beta E_j}}{\partial V} \right)_{T,N_B} = \sum_j \frac{\partial e^{-\beta E_j}}{\partial E_j} \frac{\partial E_j}{\partial V} = -\sum_j \beta e^{-\beta E_j} \left(\frac{\partial E_j}{\partial V} \right)_{N_B}$$

(Recall that β is a function of T only.) Therefore Eq. (22.23) can be written as

$$P = \frac{1}{\beta Z} \left(\frac{\partial Z}{\partial V} \right)_{T,N_B} = \frac{1}{\beta} \left(\frac{\partial \ln Z}{\partial V} \right)_{T,N_B} \tag{22.24}$$

which is the desired expression for the pressure.

Evaluation of β

To find β, we evaluate $(\partial U/\partial V)_{T,N_B}$ using (22.20) for U and (22.24) for P:

$$\left(\frac{\partial U}{\partial V} \right)_T = -\left[\frac{\partial}{\partial V} \left(\frac{\partial \ln Z}{\partial \beta} \right)_V \right]_T = -\left[\frac{\partial}{\partial \beta} \left(\frac{\partial \ln Z}{\partial V} \right)_T \right]_V$$

$$= -\left[\frac{\partial}{\partial \beta} (\beta P) \right]_V = -P - \beta \left(\frac{\partial P}{\partial \beta} \right)_V \tag{22.25}$$

where we reversed the order of differentiation [Eq. (1.36)], and where constant composition is understood in (22.25) and the following few equations.

The thermodynamic identity (4.47) for $(\partial U/\partial V)_T$ gives at constant composition

$$\left(\frac{\partial U}{\partial V} \right)_T = T \left(\frac{\partial P}{\partial T} \right)_V - P = -\frac{1}{T} \left[\frac{\partial P}{\partial (1/T)} \right]_V - P \tag{22.26}$$

where we used $\partial P/\partial T = [\partial P/\partial(1/T)] [\partial(1/T)/\partial T] = -[\partial P/\partial(1/T)]/T^2$.

Equating (22.25) and (22.26), we get

$$-\beta \left(\frac{\partial P}{\partial \beta} \right)_V = -T^{-1} \left[\frac{\partial P}{\partial (1/T)} \right]_V$$

Let $Y \equiv 1/T$. Then $\beta (\partial P/\partial \beta)_V = Y (\partial P/\partial Y)_V$, and

$$\frac{\beta}{Y} = \left(\frac{\partial \beta}{\partial P} \right)_V \left(\frac{\partial P}{\partial Y} \right)_V = \left(\frac{\partial \beta}{\partial Y} \right)_V = \frac{d\beta}{dY}$$

since β and Y are each functions of T only. We have $dY/Y = d\beta/\beta$, which integrates to $\ln Y = \ln \beta + \text{const}$, so $Y = e^{\ln \beta} e^{\text{const}} = \beta k$, where $k \equiv e^{\text{const}}$ is a constant. So $\beta = Y/k$. Since $Y \equiv 1/T$, our final result for β is

$$\beta = \frac{1}{kT} \tag{22.27*}$$

We saw earlier that β is the same for any two systems in thermal equilibrium. Therefore k must be a universal constant. In Sec. 22.6, we shall find that k is Boltzmann's constant, Eq. (3.57).

Equations (22.20) and (22.24) for U and P become

$$U = -\frac{\partial \ln Z}{\partial \beta} = -\frac{\partial \ln Z}{\partial T} \frac{dT}{d\beta} = -\frac{\partial \ln Z}{\partial T} \left(-\frac{1}{k\beta^2} \right) = kT^2 \left(\frac{\partial \ln Z}{\partial T} \right)_{V,N_B} \tag{22.28}$$

$$P = kT \left(\frac{\partial \ln Z}{\partial V} \right)_{T,N_B} \tag{22.29}$$

Equation (22.29) allows the system's equation of state to be calculated from its canonical partition function.

The probability that a system of fixed volume, temperature, and composition is in quantum state j is given by (22.15) as $p_j = e^{-E_j/kT}/Z$, where Z is a constant at fixed T, V, and composition. Let $p(E_i)$ be the probability that this system has quantum energy E_i. As noted earlier, the quantum energy levels of a system with a large number of molecules are highly degenerate, with many different quantum states having the same energy. Let W_i be the number of quantum states that have energy E_i; that is, W_i is the degeneracy of the level E_i. The probability of being in any one of the states having energy E_i is $e^{-E_i/kT}/Z$. Hence, the probability $p(E_i)$ that the system has energy E_i is $W_i e^{-E_i/kT}/Z$:

$$p(E_i) = W_i e^{-E_i/kT}/Z \qquad (22.30)$$

The degeneracy W_i is a sharply increasing function of the system's energy E_i, since as E_i increases, the number of ways of distributing the energy among the molecules increases rapidly. The exponential function $e^{-E_i/kT}$ is a sharply decreasing function of E_i, especially since Boltzmann's constant k is a very small number. (The system energy E_i has the order of magnitude RT, and $-RT/kT = -N_A$.) The product of a sharply increasing function and a sharply decreasing function in (22.30) gives a probability $p(E_i)$ that is peaked extremely narrowly about a single value, which is the observed thermodynamic internal energy U (Fig. 22.3). The probabilities of significant fluctuations in the thermodynamic internal energy are *extremely* small.

Evaluation of S

We have found statistical-mechanical expressions for the thermodynamic internal energy U and pressure P [Eqs. (22.28) and (22.29)] in terms of the canonical partition function Z, where Z is a certain sum over the possible quantum states of the thermodynamic system [Eq. (22.16)]. There remains but one task—to find the statistical-mechanical expression for the entropy.

For a reversible process in a system of fixed composition and capable of P-V work only, we have $dU = T\,dS - P\,dV$. Solving for dS, we have (at constant composition)

$$dS = T^{-1}\,dU + PT^{-1}\,dV = d(T^{-1}U) + T^{-2}U\,dT + PT^{-1}\,dV$$

since $d(T^{-1}U) = -T^{-2}U\,dT + T^{-1}\,dU$. Substitution of (22.28) and (22.29) for U and P gives

$$dS = d(T^{-1}U) + k\left(\frac{\partial \ln Z}{\partial T}\right)_{V,N_B}\!\!dT + k\left(\frac{\partial \ln Z}{\partial V}\right)_{T,N_B}\!\!dV \qquad \text{const. } N_B \quad (22.31)$$

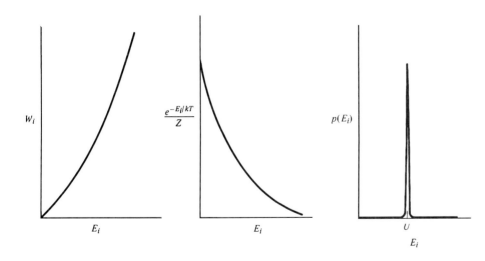

Figure 22.3

The probability $p(E_i) = W_i \times (e^{-E_i/kT}/Z)$ that the system has energy E_i is extremely sharply peaked about U.

From (22.18), $\ln Z$ is a function of T, V, and composition. At constant composition,

$$d \ln Z = \left(\frac{\partial \ln Z}{\partial T} \right)_{V,N_B} dT + \left(\frac{\partial \ln Z}{\partial V} \right)_{T,N_B} dV \qquad \text{const. } N_B$$

Therefore (22.31) becomes $dS = d(T^{-1}U) + k\, d \ln Z = d(T^{-1}U + k \ln Z)$. Integration gives

$$S = T^{-1}U + k \ln Z + C \qquad \text{const. } N_B \tag{22.32}$$

Since the derivation of (22.32) assumed constant composition, the integration "constant" C might be a function of composition: $C = f(N_B, N_C, \ldots)$. A detailed investigation (see E. Schrödinger, *Statistical Thermodynamics*, Cambridge University Press, 1952, pp. 16–17) shows that C will always cancel in calculating entropy changes for processes. Since only entropy *changes* are measurable, the value of C is of no significance and we can take C as zero. Thus

$$S = \frac{U}{T} + k \ln Z = kT \left(\frac{\partial \ln Z}{\partial T} \right)_{V,N_B} + k \ln Z \tag{22.33}$$

where (22.28) was used for U. Equation (22.33) is the desired expression for S.

The Helmholtz energy is $A = U - TS$. The use of (22.33) for S gives $A = U - T(U/T + k \ln Z) = -kT \ln Z$:

$$A = -kT \ln Z \tag{22.34}*$$

The Gibbs energy is $G = A + PV$ and can be found from (22.34) and (22.29).

The chemical potential of species B is given by (4.77) as $\mu_B = (\partial A/\partial n_B)_{T,V,n_{C \neq B}}$, where $n_B = N_B/N_A$ is the number of moles of B and N_A is the Avogadro constant. The use of (22.34) gives

$$\mu_B = -kT \left(\frac{\partial \ln Z}{\partial n_B} \right)_{T,V,n_{C \neq B}} = -RT \left(\frac{\partial \ln Z}{\partial N_B} \right)_{T,V,N_{C \neq B}} \tag{22.35}$$

since $dn_B = dN_B/N_A$ and we are anticipating the result (Sec. 22.6) that k is Boltzmann's constant: $N_A k = R$.

If the formula $A = -kT \ln Z$ is remembered, the equations for P, U, S, G, and μ_B can be quickly derived as follows. The Gibbs equation (4.77) reads $dA = -S\, dT - P\, dV + \Sigma_B \mu_B\, dn_B$. So, $S = -(\partial A/\partial T)_{V,N_B}$, $P = -(\partial A/\partial V)_{T,N_B}$, and $\mu_B = (\partial A/\partial n_B)_{T,V,n_{C \neq B}}$. Partial differentiation of (22.34) then gives (22.33) for S, (22.29) for P, and (22.35) for μ_B. Equation (22.28) for U is then derived from $U = A + TS$. The reason A is simply related to the canonical partition function is that the canonical ensemble consists of systems of constant T, V, and composition, and these are the "natural" variables for A.

Summary

We postulated that the thermodynamic properties of a system at fixed T, V, and composition can be found by averaging over the systems of a hypothetical ensemble, each system in the ensemble having the same T, V, and composition, but not necessarily the same quantum state. Using the postulate that quantum states of equal energy have equal probability of occurring in the ensemble, and using various thermodynamic relations, we found expressions for the average energy and pressure of the ensemble systems. Equating these averages to the thermodynamic U and P, we obtained the statistical-mechanical formulas (22.28) and (22.29) for U and P in terms of the canonical partition function Z, where Z is defined by (22.16). The expressions for U and P were used in $dU = T\, dS - P\, dV$ to find the statistical-mechanical expression (22.33) for S. The key statistical-mechanical formulas for calculating thermodynamic properties are

$$Z \equiv \sum_j e^{-E_j/kT} \tag{22.36}*$$

$$P = kT \left(\frac{\partial \ln Z}{\partial V} \right)_{T,N_B} \tag{22.37}$$

$$U = kT^2 \left(\frac{\partial \ln Z}{\partial T} \right)_{V,N_B} \tag{22.38}$$

$$S = U/T + k \ln Z \tag{22.39}$$

$$A = -kT \ln Z \tag{22.40}*$$

$$\mu_B = -RT \left(\frac{\partial \ln Z}{\partial N_B} \right)_{T,V,N_{C \neq B}} \tag{22.41}$$

where E_j is the energy of quantum state j of the thermodynamic system and the sum in (22.36) goes over all possible quantum states of the thermodynamic system. If species B is electrically charged, μ_B in (22.41) is to be replaced by the electrochemical potential $\tilde{\mu}_B$ (Sec. 14.3); this is true of all equations containing μ_B in this chapter. Equations (22.36) to (22.41) are valid for gases, liquids, and solids and for solutions as well as pure substances.

Having gone through the abstract reasoning of this section, what have we found? We have found how to relate macroscopic thermodynamic properties to molecular properties. This is done as follows.

1. We write down the Hamiltonian operator $\hat{H}$ for the thermodynamic system. (This requires knowledge of the kinds of intramolecular energies present and knowledge of the intermolecular forces.)
2. We solve the Schrödinger equation $\hat{H}\psi_j = E_j\psi_j$ for the entire thermodynamic system to obtain the quantum-mechanical energies E_j of the system's possible quantum states j.
3. We then evaluate the canonical partition function $Z \equiv \sum_j e^{-E_j/kT}$, where the sum is over the quantum states of the thermodynamic system.
4. We use $\ln Z$ to calculate the system's thermodynamic properties from Eqs. (22.37) to (22.41).

Provided we can solve the Schrödinger equation for the thermodynamic system, statistical mechanics allows us to calculate all the system's thermodynamic properties from its molecular properties.

22.3 CANONICAL PARTITION FUNCTION FOR A SYSTEM OF NONINTERACTING PARTICLES

Once a system's canonical partition function Z has been found by summation of $e^{-E_j/kT}$ over all possible quantum states of the system, all the system's thermodynamic properties (P, U, S, A, μ_B, μ_C, . . .) are readily found. However, the existence of forces between molecules makes Z extremely difficult to evaluate, since it is extremely hard to solve the Schrödinger equation for N interacting molecules to find the E_j's. For a system with no intermolecular forces, we can readily calculate Z.

Let the Hamiltonian operator $\hat{H}$ for the system be the sum of separate terms for the individual molecules, with no interaction terms between the molecules: $\hat{H} = \hat{H}_1$

$+ \hat{H}_2 + \cdots + \hat{H}_N$, where there are N molecules. Then the system's energy E_j is the sum of an energy for each molecule [Eq. (18.69)]:

$$E_j = \varepsilon_{1,r} + \varepsilon_{2,s} + \cdots + \varepsilon_{N,w} \qquad (22.42)$$

where $\varepsilon_{1,r}$ is the energy of molecule 1 when the system is in state j, and r denotes the quantum state of molecule 1. An epsilon is used to denote the energy of a single molecule, to avoid confusion with E_j, which is the quantum-mechanical energy of a thermodynamic system containing N molecules. From (18.70), the allowed energies for molecule 1 are found from the one-molecule Schrödinger equation $\hat{H}_1 \psi_{1,r} = \varepsilon_{1,r} \psi_{1,r}$. When the molecules exert forces on one another, Eq. (22.42) does not hold.

The canonical partition function (22.36) for a system of noninteracting molecules is [where $\beta = 1/kT$, Eq. (22.27)]

$$Z = \sum_j e^{-\beta E_j} = \sum_j e^{-\beta(\varepsilon_{1,r} + \varepsilon_{2,s} + \cdots + \varepsilon_{N,w})} \qquad (22.43)$$

First consider the case where the molecules are distinguishable from one another by being confined to different locations in space. This would occur in a crystal. (Of course, the molecules in a crystal do interact with each other, but we'll worry about that later; Sec. 24.12.) For distinguishable molecules, the state of the system is defined by giving the quantum state of each molecule; molecule 1 is in state r, molecule 2 in state s, etc. Therefore, to sum over all possible quantum states j of the system, we sum separately over all the possible states of each molecule, and (22.43) becomes

$$Z = \sum_r \sum_s \cdots \sum_w e^{-\beta \varepsilon_{1,r}} e^{-\beta \varepsilon_{2,s}} \cdots e^{-\beta \varepsilon_{N,w}} = \sum_r e^{-\beta \varepsilon_{1,r}} \sum_s e^{-\beta \varepsilon_{2,s}} \cdots \sum_w e^{-\beta \varepsilon_{N,w}}$$

$$(22.44)$$

where the sum identity (1.51) was used.

We define the **molecular partition functions** $z_1, z_2, \ldots$ as

$$z_1 \equiv \sum_r e^{-\beta \varepsilon_{1,r}}, \quad z_2 \equiv \sum_s e^{-\beta \varepsilon_{2,s}}, \quad \ldots$$

where $\varepsilon_{1,r}$ is the energy of molecule 1 in the molecular quantum state r and the first sum goes over the available quantum states of molecule 1. Equation (22.44) becomes

$$Z = z_1 z_2 \cdots z_N \qquad \text{distinguishable noninteracting molecules} \qquad (22.45)$$

If all the molecules happen to be the same kind, the set of molecular quantum states is the same for each molecule and $z_1 = z_2 = \cdots = z_N$. Thus

$$Z = z^N \qquad \text{identical disting. nonint. molecules} \qquad (22.46)$$

$$z \equiv \sum_s e^{-\beta \varepsilon_s} \qquad (22.47)$$

(To avoid confusion between Z and z in handwritten equations, use a script lowercase "zee" for the molecular partition function.)

If the molecules are not all alike but there are N_B molecules of species B, N_C molecules of species C, etc., (22.45) becomes

$$Z = (z_B)^{N_B} (z_C)^{N_C} \cdots \qquad \text{disting. nonint. molecules} \qquad (22.48)$$

$$z_B = \sum_r e^{-\beta \varepsilon_{B,r}}, \quad z_C = \sum_s e^{-\beta \varepsilon_{C,s}}, \quad \ldots$$

Now consider a pure ideal gas. The molecules move through the entire volume of the system and so are nonlocalized; moreover, they are all identical in a pure gas.

There is thus no way whatever of distinguishing one molecule from another. Therefore, a situation where molecule 1 is in state r and molecule 2 is in state s corresponds to the same quantum state as the situation where molecule 1 is in state s and molecule 2 is in state r (and the states of the other molecules are unchanged). Recall that for the helium atom [Eqs. (19.42) and (19.43)], we could not say that electron 1 was in the $1s$ orbital and electron 2 in the $2s$ orbital or that electron 1 was in the $2s$ orbital and electron 2 in the $1s$ orbital. Instead, a state of He with the electron configuration $1s2s$ corresponds to a wave function containing the terms $1s(1)2s(2)$ and $2s(1)1s(2)$; each electron is in both orbitals. Similarly, the wave function for a given quantum state of the gas of N identical nonlocalized molecules contains terms in which all the molecules are permuted among all the occupied molecular states. The microstate of the system depends not on which molecules are in which molecular states but on how many molecules are in each of the available molecular states.

How do we get Z when the noninteracting molecules are indistinguishable? Let us assume that the number of molecular states with significant probability of being occupied is much, much greater than the number of molecules in the gas. (This assumption will be justified below.) We then expect that the probability that two or more gas molecules are in the same molecular state will be very small, and we shall assume that no two molecules are in the same molecular state. If we were to use Eq. (22.44) or its equivalent (22.46) for Z, we would be counting each microstate of the system too many times. For example, suppose the system consists of three identical molecules, and let the system microstate j have one molecule in molecular state r, one molecule in t, and one in w. The system's wave function for state j would contain the term $\psi_r(1)\psi_t(2)\psi_w(3)$ and the five other terms that correspond to the $3! = 6$ permutations of molecules 1, 2, and 3 among the molecular states r, t, and w. [Thus, the Li-atom Slater determinant (19.51) when expanded contains six terms that involve permutations of the three electrons among the states $1s\alpha$, $1s\beta$, and $2s\alpha$.]

Equation (22.44), which sums separately over all molecular states for each molecule, counts $\psi_r(1)\psi_t(2)\psi_w(3)$ as a separate state from $\psi_t(1)\psi_r(2)\psi_w(3)$ and from the other four permutations, since it contains the separate terms

$$e^{-\beta\varepsilon_{1,r}}e^{-\beta\varepsilon_{2,t}}e^{-\beta\varepsilon_{3,w}}, \quad e^{-\beta\varepsilon_{1,t}}e^{-\beta\varepsilon_{2,r}}e^{-\beta\varepsilon_{3,w}}, \quad \text{etc.}$$

These six terms are numerically equal, since each contains the sum $\varepsilon_r + \varepsilon_t + \varepsilon_w$. Thus, Eq. (22.46) has $3! = 3 \cdot 2 \cdot 1 = 6$ equal terms where there should be only one term, and this situation holds for each quantum state of the system [*provided* there is a negligible probability that two molecules have the same molecular state; for $\psi_r(1)\psi_w(2)\psi_w(3)$ with molecules 2 and 3 both having state w, there are only two other possible permutations of the molecules among the occupied molecular states]. The correct value of Z for our hypothetical three-particle system can be obtained by dividing (22.46) by 3!. For a system of N indistinguishable noninteracting particles, the correct canonical partition function is obtained by dividing by $N!$ to give

$$Z = \frac{z^N}{N!} \qquad \text{for } \langle N_r \rangle \ll 1, \text{ pure ideal gas} \qquad \textbf{(22.49)}*$$

$$z \equiv \sum_r e^{-\beta\varepsilon_r} \qquad \textbf{(22.50)}*$$

where $\langle N_r \rangle$ is the average number of molecules in the molecular quantum state r and the inequality must hold for each available molecular state to ensure that there is negligible probability that two molecules will have the same molecular state.

For identical bosons (Sec. 19.6), any number of molecules can occupy a given molecular state and a system state like $\psi_r(1)\psi_w(2)\psi_w(3)$ can occur, but such states are incorrectly counted by the $1/N!$ factor. For identical fermions, no two molecules can be in the same molecular state, and a state like $\psi_r(1)\psi_w(2)\psi_w(3)$ is forbidden. However, evaluation of Z by independent summation over the molecular states as in (22.44) and subsequent division by $N!$ includes states with two or more molecules in the same molecular state and so gives an incorrect Z for fermions. For both fermions and bosons, the incorrect counting arises from terms in the sum (22.44) where two or more molecules have the same molecular state. When $\langle N_r \rangle \ll 1$, we expect terms with multiple occupancy of molecular states to make a negligible contribution to Z and so expect (22.49) to be accurate.

Now consider an ideal gas *mixture* with N_B molecules of species B, N_C of C, etc. Then we can distinguish B molecules from C molecules, but we cannot distinguish two B molecules from each other. We must correct (22.44) for permutations of B molecules with other B molecules by dividing by $N_B!$, for permutations of C molecules with other C molecules, etc. Instead of (22.48), we get

$$Z = \frac{(Z_B)^{N_B}}{N_B!} \frac{(Z_C)^{N_C}}{N_C!} \cdots \qquad \text{ideal gas mixture} \qquad (22.51)$$

provided $\langle N_{B,r} \rangle \ll 1$, $\langle N_{C,r} \rangle \ll 1$, etc.

Let us examine whether the assumption $\langle N_r \rangle \ll 1$ is justified. For an ideal gas, the molecular energy is the sum of translational, rotational, vibrational, and electronic energies. We shall consider only the translational states; inclusion of rotational, vibrational, and electronic states would only strengthen the result.

For a cubic box of volume V, Eq. (18.50) gives the translational energy of a molecule as $\varepsilon_{tr} = (h^2/8mV^{2/3})(n_x^2 + n_y^2 + n_z^2)$. We ask for the number of translational states whose energy is less than some maximum value ε_{max}. For $\varepsilon_{tr} \leq \varepsilon_{max}$, the translational quantum numbers must satisfy the inequality

$$n_x^2 + n_y^2 + n_z^2 \leq 8mV^{2/3}h^{-2}\varepsilon_{max} \qquad (22.52)$$

The average translational energy per molecule is $\frac{3}{2}kT$ [Eq. (15.15)], and most molecules have ε_{tr} within two or three times this average value (Fig. 15.11). We therefore take $\varepsilon_{max} = 3kT$ as the "maximum" ε_{tr} for a typical molecule. The translational states that have significant probability of being occupied have their energies between 0 and ε_{max}.

In Prob. 22.9, it is shown that the number of translational states that satisfy (22.52) and so have energy equal to or less than $3kT$ is

$$\frac{\pi}{6}(24mV^{2/3}h^{-2}kT)^{3/2} \approx 60\left(\frac{mkT}{h^2}\right)^{3/2}V$$

For $\langle N_r \rangle \ll 1$ to hold, we must have $60(mkT/h^2)^{3/2}V \gg N$, where N is the number of molecules. This inequality reads

$$\frac{1}{60}\left(\frac{h^2}{mkT}\right)^{3/2}\frac{N}{V} \ll 1 \qquad (22.53)$$

A typical system is 1 mole of O_2 at 0°C and 1 atm; here, $m = [32/(6 \times 10^{23})]$ g, $N = 6 \times 10^{23}$, and $V = 22400$ cm^3. One finds 4×10^{-8} for the left side of (22.53). Hence, $\langle N_r \rangle \ll 1$ is well satisfied. The translational levels are spaced extremely close together, so the number of available translational states is far greater than the number of gas molecules.

The inequality (22.53) does not hold for (*a*) extremely low temperatures, (*b*) extremely high densities, (*c*) extremely small particle masses. Physical examples of these conditions are (*a*) liquid helium at 2 or 3 K, (*b*) white-dwarf stars and neutron stars (pulsars), (*c*) the conduction electrons in metals.

When the condition $\langle N_r \rangle \ll 1$ is not met, Eqs. (22.49) and (22.51) for Z don't hold. The proper result for Z depends on whether the particles of the gas are bosons or fermions (Sec. 19.6). The boson and fermion expressions for Z in terms of molecular energy levels are given in Prob. 22.22.

It was formerly believed that having $\langle N_r \rangle \ll 1$ was sufficient to ensure the accuracy of $Z = z^N/N!$ [Eq. (22.49)]. Surprisingly, however, it has been found that under conditions of typical T and P, even though there are many more available molecular states than molecules, the dominant contribution to Z comes from terms with multiple occupancy of molecular states. This produces a huge error in the formula $Z = z^N/N!$. For example, in an ideal gas at 25°C and 1 atm, the true Z typically differs from $z^N/N!$ by a factor of $10^{(10^{15})}$ for bosons and $10^{(-10^{15})}$ for fermions. [F. Hynne, *Am. J. Phys.,* **49,** 125 (1981); H. Kroemer, ibid., **48,** 962 (1980); R. Baierlein, ibid., **65,** 314 (1997).] This monstrous error in Z is of no consequence because it is ln Z that determines all thermodynamic properties [Eqs. (22.37)–(22.41)], and the error in ln Z turns out to be negligible (Prob. 22.93).

Summary

For a system of N noninteracting molecules, we expressed the canonical partition function Z in terms of the molecular partition function z, where z is a sum over the molecular quantum states. For a pure ideal gas, $Z = z^N/N!$, where $z = \Sigma_r\, e^{-\beta\varepsilon_r}$, $\beta = 1/kT$, ε_r is the energy of molecular state r, and the sum goes over all the states of a molecule.

22.4   CANONICAL PARTITION FUNCTION OF A PURE IDEAL GAS

In Secs. 22.4, 22.6, and 22.7, we restrict ourselves to a pure ideal gas (all molecules alike). The thermodynamic functions of an ideal gas mixture are easily obtained as the sum of those for each gas in the mixture (Prob. 9.20).

In an ideal gas (no intermolecular forces) the system energy is the sum of molecular energies, and the canonical partition function is given by (22.49) as $Z = z^N/N!$, where z is the molecular partition function. Since ln Z (rather than Z) occurs in Eqs. (22.37) to (22.41) for the thermodynamic properties, we take ln Z:

$$\ln Z = \ln (z^N/N!) = N \ln z - \ln N! \qquad \text{pure ideal gas} \qquad (22.54)$$

It's a bit tiresome to evaluate a number like ln $(10^{23}!)$. Fortunately, an excellent approximation is available. **Stirling's formula,** valid for large N, is

$$N! = (2\pi)^{1/2} N^{N+1/2} e^{-N} \left(1 + \frac{1}{12N} + \frac{1}{288N^2} - \cdots \right) \qquad \text{for } N \text{ large} \qquad (22.55)$$

[For a derivation, see N. D. Mermin, *Am. J. Phys.,* **52,** 362 (1984).] Taking the natural log of (22.55), we have

$$\ln N! = \tfrac{1}{2} \ln 2\pi + (N + \tfrac{1}{2}) \ln N - N + \ln (1 + 1/12N + \cdots)$$

Since N is something like 10^{23}, we have $N + \tfrac{1}{2} \approx N$, $\tfrac{1}{2} \ln 2\pi - N \approx -N$, and $\ln (1 + 1/12N + \cdots) \approx \ln 1 = 0$. Therefore

$$\ln N! \approx N \ln N - N \qquad \text{for } N \text{ large} \qquad \textbf{(22.56)*}$$

To show how well (22.56) works, the following table compares ln $N!$ and $N \ln N - N$:

N	$\ln N!$	$N \ln N - N$	Error
10^3	5912.1	5907.8	−0.07%
10^4	82108.9	82103.4	−0.007%
10^6	12815518.4	12815510.6	−0.00006%

Each additional power of 10 in N decreases the percent error by roughly a factor of 10, so the percent error is entirely negligible for the usual numbers of molecules in a thermodynamic system.

Now consider the ideal-gas molecular partition function $z = \sum_r e^{-\varepsilon_r/kT}$ [Eq. (22.50)]. It is usually a good approximation to write the molecular energy as the sum of translational, rotational, vibrational, and electronic energies [Eq. (21.21)]

$$\varepsilon_r = \varepsilon_{tr,s} + \varepsilon_{rot,t} + \varepsilon_{vib,v} + \varepsilon_{el,u} \qquad (22.57)$$

where the subscripts s, t, v, u indicate the translational, rotational, vibrational, and electronic states. The molecular state r is defined by the translational, rotational, vibrational, and electronic states s, t, v, and u, so summation over r is accomplished by summing over s, t, v, and u. The quantum numbers of the four kinds of energies vary independently of one another. Using the sum identity (1.51), we have

$$z = \sum_r e^{-\beta\varepsilon_r} = \sum_s \sum_t \sum_v \sum_u e^{-\beta(\varepsilon_{tr,s} + \varepsilon_{rot,t} + \varepsilon_{vib,v} + \varepsilon_{el,u})}$$

$$= \sum_s e^{-\beta\varepsilon_{tr,s}} \sum_t e^{-\beta\varepsilon_{rot,t}} \sum_v e^{-\beta\varepsilon_{vib,v}} \sum_u e^{-\beta\varepsilon_{el,u}}$$

$$z = z_{tr}z_{rot}z_{vib}z_{el} \qquad (22.58)*$$

$$\ln z = \ln z_{tr} + \ln z_{rot} + \ln z_{vib} + \ln z_{el} \qquad (22.59)$$

$$z_{tr} \equiv \sum_s e^{-\beta\varepsilon_{tr,s}}, \qquad z_{rot} \equiv \sum_t e^{-\beta\varepsilon_{rot,t}}, \qquad z_{vib} \equiv \text{etc.} \qquad (22.60)$$

When the molecular energy is the sum of independent kinds of energies, the molecular partition function factors into a product of partition functions, one for each kind of energy. [Actually, Eq. (22.58) is a lie, but this will be cleared up in Sec. 22.6.]

Substitution of (22.59) and (22.56) into (22.54) gives for an ideal gas

$$\ln Z = N \ln z_{tr} + N \ln z_{rot} + N \ln z_{vib} + N \ln z_{el} - N(\ln N - 1) \qquad (22.61)$$

The thermodynamic internal energy is given by Eq. (22.38) as

$$U = kT^2(\partial \ln Z/\partial T)_{V,N}$$

Evaluation of $\partial \ln Z/\partial T$ from (22.61) gives for an ideal gas

$$U = NkT^2\left[\left(\frac{\partial \ln z_{tr}}{\partial T}\right)_V + \frac{d \ln z_{rot}}{dT} + \frac{d \ln z_{vib}}{dT} + \frac{d \ln z_{el}}{dT}\right]$$

$$U = U_{tr} + U_{rot} + U_{vib} + U_{el} \qquad (22.62)$$

$$U_{tr} \equiv NkT^2\left(\frac{\partial \ln z_{tr}}{\partial T}\right)_V, \qquad U_{rot} \equiv NkT^2\frac{d \ln z_{rot}}{dT}, \qquad U_{vib} \equiv \text{etc.} \quad (22.63)$$

(Only the translational energy depends on the volume.)

The entropy is given by (22.39) as $S = U/T + k \ln Z$. The use of (22.62) and (22.61) gives for a pure ideal gas

$$S = S_{tr} + S_{rot} + S_{vib} + S_{el} \qquad (22.64)$$

$$S_{tr} \equiv U_{tr}/T + Nk \ln z_{tr} - Nk(\ln N - 1) \qquad (22.65)$$

$$S_{rot} \equiv U_{rot}/T + Nk \ln z_{rot}, \qquad S_{vib} \equiv U_{vib}/T + Nk \ln z_{vib}, \qquad \text{etc.} \quad (22.66)$$

The translational contribution to S differs in form from the rotational, vibrational, and electronic contributions because it incorporates the $-\ln N!$ term. The $N!$ in Z is due to the indistinguishability of the molecules, which results from their being nonlocalized. Hence it is appropriate to include this term in S_{tr}.

These ideal-gas formulas are applied in Secs. 22.6 and 22.7 to give expressions for thermodynamic properties in terms of molecular properties.

22.5 THE BOLTZMANN DISTRIBUTION LAW FOR NONINTERACTING MOLECULES

We now derive the Boltzmann distribution law. Consider a thermodynamic system of noninteracting identical molecules, and let the possible quantum states available to each molecule be $r, s, t, u, \ldots$. Since there are no intermolecular interactions, the quantum energy of the system is

$$E_j = N_{r,j}\varepsilon_r + N_{s,j}\varepsilon_s + \cdots$$

where $N_{r,j}, N_{s,j}, \ldots$ are the numbers of molecules in the molecular states $r, s, \ldots$ when the system is in microstate j, and $\varepsilon_r, \varepsilon_s, \ldots$ are the energies of the molecular states $r, s, \ldots$. To avoid confusion, the quantum states of the thermodynamic system are labeled $i, j, k, \ldots$, whereas the quantum states of a single molecule are labeled $r, s, t, \ldots$.

Let us calculate $\langle N_s \rangle$, the average number of molecules in the molecular state s when the system is in a given macrostate. As we did with U and P, we average over the microstates in the canonical ensemble, and we have

$$\langle N_s \rangle = \sum_j p_j N_{s,j}$$

which is similar to $\langle E_j \rangle = \Sigma_j\, p_j E_j$ [Eq. (22.3)]. The probability p_j that the thermodynamic system is in microstate j is $p_j = e^{-\beta E_j}/Z$ [Eq. (22.15)]. Hence

$$\langle N_s \rangle = \frac{\sum\limits_j N_{s,j}\, \exp\left[-\beta(N_{r,j}\varepsilon_r + N_{s,j}\varepsilon_s + \cdots)\right]}{Z} \tag{22.67}$$

where the above expression for E_j was used.

Now consider the partial derivative $\partial Z/\partial\varepsilon_s$. We have

$$\frac{\partial Z}{\partial\varepsilon_s} = \frac{\partial}{\partial\varepsilon_s} \sum_j \exp\left[-\beta(N_{r,j}\varepsilon_r + N_{s,j}\varepsilon_s + \cdots)\right]$$

$$= -\beta \sum_j N_{s,j}\, \exp\left[-\beta(N_{r,j}\varepsilon_r + N_{s,j}\varepsilon_s + \cdots)\right]$$

Therefore the numerator in (22.67) equals $-(1/\beta)(\partial Z/\partial\varepsilon_s)$, and

$$\langle N_s \rangle = -\frac{1}{\beta}\frac{1}{Z}\frac{\partial Z}{\partial\varepsilon_s} = -\frac{1}{\beta}\frac{\partial \ln Z}{\partial\varepsilon_s} \tag{22.68}$$

where the partial derivative is at constant β (that is, constant T) and constant $\varepsilon_{r\neq s}$.

For a system of N noninteracting identical molecules, Eqs. (22.49) and (22.46) give (provided $\langle N_r \rangle \ll 1$ for all r) $Z = z^N/(N!)$ and $\ln Z = N \ln z - (\ln N!)$, where the $N!$ is present or absent according to whether the molecules are indistinguishable or distinguishable. In either case,

$$\frac{\partial \ln Z}{\partial\varepsilon_s} = N\frac{\partial \ln z}{\partial\varepsilon_s} = \frac{N}{z}\frac{\partial z}{\partial\varepsilon_s} = \frac{N}{z}\frac{\partial(e^{-\beta\varepsilon_r} + e^{-\beta\varepsilon_s} + \cdots)}{\partial\varepsilon_s} = -\beta\frac{N}{z}e^{-\beta\varepsilon_s}$$

where $z \equiv \Sigma_r\, e^{-\beta\varepsilon_r}$ [Eq. (22.50)] was used. Substitution in (22.68) gives

$$\frac{\langle N_s \rangle}{N} = \frac{e^{-\varepsilon_s/kT}}{z} = \frac{e^{-\varepsilon_s/kT}}{\displaystyle\sum_r e^{-\varepsilon_r/kT}} \qquad \text{for } \langle N_r \rangle \ll 1 \qquad \textbf{(22.69)*}$$

where $\langle N_s \rangle$ is the average number of molecules in molecular state s (and not the number of molecules in a particular energy level). Note that $\langle N_s \rangle/N$ is the probability that a molecule picked at random is in the molecular state s. This probability decreases exponentially as the energy ε_s of state s increases.

Equation (22.69) is the **Boltzmann** (or *Maxwell–Boltzmann*) **distribution law.** This equation is easily written down if it is remembered that $\langle N_s \rangle = \text{const} \times e^{-\varepsilon_s/kT}$ and $N = \Sigma_r\, \langle N_r \rangle = \Sigma_r\, (\text{const} \times e^{-\varepsilon_r/kT}) = \text{const}\,\Sigma_r\, e^{-\varepsilon_r/kT}$.

Equation (22.69) for state r is $\langle N_r \rangle/N = e^{-\varepsilon_r/kT}/z$, so

$$\frac{\langle N_r \rangle}{\langle N_s \rangle} = e^{-(\varepsilon_r - \varepsilon_s)/kT} \equiv e^{-\Delta\varepsilon/kT} \qquad \text{for } \langle N_r \rangle \ll 1 \qquad (22.70)$$

We proved (22.70) in Chapter 15 for a couple of specific cases, but we now have a general proof of its validity.

Often we want the average number of molecules having a given energy rather than the average number in a given molecular state. Several different molecular states may have the same energy (degeneracy, Sec. 18.10). If the molecular states s, t, and u all have the same energy ($\varepsilon_s = \varepsilon_t = \varepsilon_u$) and no other states have this energy, then the energy level ε_s is threefold degenerate and the average number of molecules having energy ε_s equals $\langle N_s \rangle + \langle N_t \rangle + \langle N_u \rangle = 3\langle N_s \rangle$, where (22.69) was used. In general, if the energy level ε_s is g_s-fold degenerate, (22.69) gives

$$\frac{\langle N(\varepsilon_s) \rangle}{N} = \frac{g_s e^{-\varepsilon_s/kT}}{z} \qquad \text{for } \langle N_r \rangle \ll 1 \qquad (22.71)$$

$$\frac{\langle N(\varepsilon_r) \rangle}{\langle N(\varepsilon_s) \rangle} = \frac{g_r}{g_s}\, e^{-(\varepsilon_r - \varepsilon_s)/kT} \qquad \text{for } \langle N_r \rangle \ll 1 \qquad \textbf{(22.72)*}$$

where $\langle N(\varepsilon_s) \rangle$ is the average number of molecules having energy ε_s. The degeneracy g_s is sometimes called the *statistical weight* of the molecular energy level ε_s. An example of (22.71) is Fig. 21.16, which plots populations of rotational levels; here, $g = 2J + 1$.

The molecular partition function $z = \Sigma_r\, e^{-\varepsilon_r/kT}$ is a sum over all one-molecule states. The term $e^{-\varepsilon_r/kT}$ has the same value for two or more states that have the same energy. Therefore, if we write z as a sum over molecular energy values (rather than as a sum over molecular states), we have

$$z = \sum_{m(\text{levels})} g_m e^{-\varepsilon_m/kT} \qquad (22.73)$$

where the sum is over the energy levels and g_m is the degeneracy of level m.

The above equations apply when the noninteracting molecules are all alike. When several species are present, each species has its own set of molecular energy levels and Z is given by (22.51) or (22.48). A derivation similar to the above gives

$$\frac{\langle N_{B,r} \rangle}{N_B} = \frac{e^{-\varepsilon_{B,r}/kT}}{z_B} \qquad \text{for } \langle N_{B,r} \rangle \ll 1 \qquad (22.74)$$

where $\langle N_{B,r} \rangle$ is the average number of species B molecules in the B molecular state r (whose energy is $\varepsilon_{B,r}$) and N_B is the number of B molecules in the system.

Let us show that *the Boltzmann distribution law can be applied to each kind of energy—translational, rotational, vibrational, and electronic*. Equation (22.69), with s changed to r, and Eqs. (22.57) and (22.58) give

$$\langle N_r \rangle = \frac{Ne^{-\varepsilon_r/kT}}{z} = \frac{Ne^{-\beta\varepsilon_{\text{tr},s}}e^{-\beta\varepsilon_{\text{rot},t}}e^{-\beta\varepsilon_{\text{vib},v}}e^{-\beta\varepsilon_{\text{el},u}}}{z_{\text{tr}}z_{\text{rot}}z_{\text{vib}}z_{\text{el}}}$$

Let $\langle N_{\text{vib},v} \rangle$ be the average number of molecules in the vibrational state v, without regard to what translational, rotational, and electronic states these molecules are in. To find $\langle N_{\text{vib},v} \rangle$ we must add up all the $\langle N_r \rangle$'s that correspond to different translational, rotational, and electronic states s, t, and u but the same vibrational state. Hence, we must sum $\langle N_r \rangle$ over s, t, and u, keeping the vibrational state v fixed:

$$\langle N_{\text{vib},v} \rangle = \frac{Ne^{-\beta\varepsilon_{\text{vib},v}}}{z_{\text{tr}}z_{\text{rot}}z_{\text{vib}}z_{\text{el}}} \sum_s e^{-\beta\varepsilon_{\text{tr},s}} \sum_t e^{-\beta\varepsilon_{\text{rot},t}} \sum_u e^{-\beta\varepsilon_{\text{el},u}}$$

$$\frac{\langle N_{\text{vib},v} \rangle}{N} = \frac{e^{-\varepsilon_{\text{vib},v}/kT}}{z_{\text{vib}}} = \frac{e^{-\varepsilon_{\text{vib},v}/kT}}{\displaystyle\sum_v e^{-\varepsilon_{\text{vib},v}/kT}} \qquad \text{for } \langle N_r \rangle \ll 1 \qquad (22.75)$$

where (22.60) was used.

Equation (22.75) has the same form as (22.69). Similar equations hold for translational, rotational, and electronic populations. The condition for the validity of (22.75) is $\langle N_r \rangle \ll 1$, but $\langle N_r \rangle \ll 1$ does not require that $\langle N_{\text{vib},v} \rangle$ be much less than 1. If two molecules are in the same vibrational state v but are in different translational states, the molecular state r differs for these molecules. The huge number of available translational states allows many molecules to have the same vibrational quantum number without violating $\langle N_r \rangle \ll 1$.

Some examples of the Boltzmann distribution are Eq. (15.52) for the Maxwell distribution of kinetic energy in gases and Eq. (15.72) for the distribution of potential energy of gas molecules in a gravitational field. The Boltzmann distribution plays a key role in the distribution of ions around a given ion in an electrolyte solution [note the kT in the Debye–Hückel equation (10.61)], in the degree of orientation polarization that occurs when a dielectric composed of polar molecules is placed in an electric field [note the kT in (14.87)], and in the distribution of double-layer ions and dipoles near a charged electrode (Sec. 14.14).

To get a feeling for the Boltzmann distribution, we shall apply it to the molecular vibrational energy levels of an ideal diatomic gas. In the harmonic-oscillator approximation, the vibrational energy of a diatomic molecule is given by (21.22) as $\varepsilon_{\text{vib}} = (v + \frac{1}{2})h\nu$, where $v = 0, 1, 2, \ldots$ and ν is the vibrational frequency. The zero level of energy is arbitrary, and we choose to measure the energy starting from the ground state $v = 0$; thus, we subtract $\frac{1}{2}h\nu$ from each level and write $\varepsilon_{\text{vib}} = vh\nu$. [If this seems disturbing, note that the numerator and each term in the denominator of (22.75) contain the factor $e^{-h\nu/2kT}$ and this factor cancels.] The levels are equally spaced, the spacing being $\Delta\varepsilon_v = h\nu = \varepsilon_{\text{vib}}/v$. Thus (22.75) can be written as

$$\frac{\langle N_v \rangle}{N} = \frac{e^{-v\Delta\varepsilon_v/kT}}{\displaystyle\sum_{v=0}^{\infty} e^{-v\Delta\varepsilon_v/kT}} = \frac{e^{-v\Delta\varepsilon_v/kT}}{z_{\text{vib}}} \qquad (22.76)$$

where $\langle N_v \rangle$ is the average number of molecules in vibrational state v.

The populations of vibrational levels thus depend on the ratio $\Delta\varepsilon_v/kT = h\nu/kT$ and on the quantum number v.

Figure 22.4

Fractional populations of the four lowest one-dimensional harmonic-oscillator vibrational levels for three different values of $kT/\Delta\varepsilon_v$.

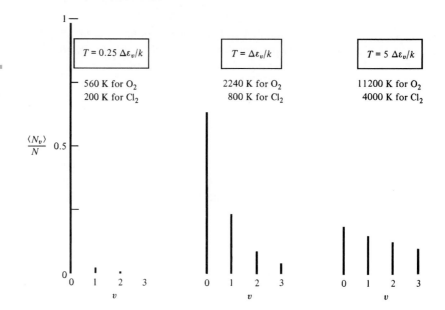

EXAMPLE 22.1 Boltzmann distribution law for harmonic oscillators

For a collection of one-dimensional harmonic oscillators, calculate the fractional populations of the lowest few vibrational levels for $h\nu/kT$ equal to (a) 4; (b) 1; (c) 0.2. Comment on the results.

For $\Delta\varepsilon_v/kT = 4$, the molecular partition function in (22.76) is $z_{vib} = e^0 + e^{-4} + e^{-8} + e^{-12} + \cdots = 1.01866$. [The terms decrease rapidly, so it is easy enough to sum the series on a calculator. However, time can be saved by noting that z_{vib} is a geometric series whose sum is $1/(1 - e^{-4})$; see Eq. (22.89).] Equation (22.76) with $\Delta\varepsilon_v/kT = 4$ gives the fractional populations $\langle N_v \rangle/N$ of the $v = 0, 1, 2, \ldots$ levels as e^0/z_{vib}, e^{-4}/z_{vib}, e^{-8}/z_{vib}, $\ldots$, where $z_{vib} = 1.01866$. The calculated percent populations $100\langle N_v \rangle/N$ are:

$$\Delta\varepsilon_v/kT = 4 \quad \begin{cases} \end{cases}$$

v	0	1	2	3
%	98.2	1.8	0.03	0.0006

Similarly, for $\Delta\varepsilon_v/kT = 1$ and $\Delta\varepsilon_v/kT = 0.2$, one finds:

$$\Delta\varepsilon_v/kT = 1 \quad \begin{cases} \end{cases}$$

v	0	1	2	3	4	5	6	7
%	63.2	23.3	8.6	3.1	1.2	0.4	0.2	0.06

$$\Delta\varepsilon_v/kT = 0.2 \quad \begin{cases} \end{cases}$$

v	0	1	2	3	4	5	$\cdots$	10	$\cdots$	15
%	18.1	14.8	12.2	9.9	8.1	6.7	$\cdots$	2.5	$\cdots$	0.9

When the vibrational energy spacing is substantially greater than kT, the molecules pile up in the ground vibrational level. This is the low-temperature behavior. When $\Delta\varepsilon_v \ll kT$ (high T), the distribution tends toward uniformity (Fig. 22.4). Figure 22.5 plots the fractional populations of the lowest three vibrational levels of Cl_2 versus T.

Exercise

For a harmonic oscillator with $\tilde{\nu} = 2000.0 \text{ cm}^{-1}$, find the fractional populations of the $v = 0$ and $v = 1$ levels at 1000.0 K. (*Answers:* 0.9437, 0.0531.)

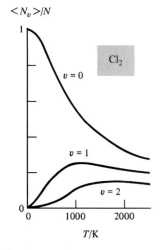

Figure 22.5

Populations of the $Cl_2(g)$ $v = 0, 1,$ and 2 vibrational levels (in the harmonic-oscillator approximation) versus T. The $v = 1$ and $v = 2$ populations are a maximum at 1150 K and 1970 K, respectively.

What is the physical reason behind the Boltzmann distribution? For a fixed system energy, any one given distribution of the total system energy among the molecules is as likely to occur as any other distribution. This follows from the postulate in Sec. 22.2 that the probability of a microstate is a function only of the system energy. The probability that a molecule is in a molecular state with a given energy is thus proportional to the number of ways of distributing the rest of the energy among the other molecules. As the amount of energy given one particular molecule increases, the number of ways of distributing the remaining energy decreases, so the probability $\langle N_s \rangle / N$ in (22.69) decreases with increasing ε_s.

When the condition $\langle N_r \rangle \ll 1$ is not obeyed, the Boltzmann distribution doesn't hold. The average populations of the one-particle states then depend on whether the particles are bosons or fermions. There is no limit on the number of bosons that can have the same molecular state, but for fermions, each molecular state can hold no more than one fermion (Sec. 19.8). For a system containing only one species of particle, the correct result for the average population of state r turns out to be [for a derivation, see *Andrews* (1975), chap. 9]

$$\langle N_r \rangle_{\text{BE}}^{\text{FD}} = \frac{1}{e^{-\mu/N_A kT} e^{\varepsilon_r/kT} \pm 1} \quad \text{indisting. ident. nonint. particles} \quad (22.77)$$

where μ is the chemical potential of the species and the upper sign is for fermions and the lower sign for bosons. For electrons and other charged species, μ is replaced by the electrochemical potential $\tilde{\mu}$. Note that with the plus sign, the denominator exceeds the numerator and $\langle N_r \rangle$ is less than 1, as it should be for fermions.

Bosons are said to obey **Bose–Einstein statistics.** Fermions obey **Fermi–Dirac statistics.** When $\langle N_r \rangle \ll 1$, Eq. (22.77) reduces to the Boltzmann distribution law (22.69) (see Prob. 22.21).

22.6 STATISTICAL THERMODYNAMICS OF IDEAL DIATOMIC AND MONATOMIC GASES

In Sec. 22.2, we expressed the thermodynamic properties of a system in terms of its canonical partition function Z. In Sec. 22.3 we found that for a pure ideal gas containing N molecules, $Z = z^N/N!$, where the molecular partition function z is the following sum over the molecular states: $z = \sum_r e^{-\beta \varepsilon_r}$. Because the energy ε_r of molecular state r is the sum of translational, rotational, vibrational, and electronic energies, we found in Sec. 22.4 that $z = z_{\text{tr}} z_{\text{rot}} z_{\text{vib}} z_{\text{el}}$ [Eq. (22.58)]. We used the relations $Z = z^N/N!$ and $z = z_{\text{tr}} z_{\text{rot}} z_{\text{vib}} z_{\text{el}}$ to express Z in terms of z_{tr}, z_{rot}, z_{vib}, and z_{el} [Eq. (22.61)] and thus obtained expressions for the thermodynamic properties U and S in terms of z_{tr}, z_{rot}, etc. [Eqs. (22.62) to (22.66)].

We now evaluate z_{tr}, z_{rot}, z_{vib}, and z_{el} for a pure ideal gas of diatomic or monatomic molecules, thereby expressing the gas's thermodynamic properties in terms of its molecular properties.

Translational Partition Function

Equation (22.60) gives $z_{\text{tr}} = \sum_s e^{-\beta \varepsilon_{\text{tr},s}}$. For the translational energies, we use the energies (18.47) of a particle in a rectangular box with sides a, b, c. Thus, $\varepsilon_{\text{tr}} = (h^2/8m) \cdot (n_x^2/a^2 + n_y^2/b^2 + n_z^2/c^2)$, where each quantum number goes from 1 to ∞, independently of the others, and a, b, and c are the dimensions of the container holding the gas. To sum over all translational quantum states, we sum over all the quantum numbers:

$$z_{\text{tr}} = \sum_{n_x=1}^{\infty} \sum_{n_y=1}^{\infty} \sum_{n_z=1}^{\infty} e^{-(\beta h^2/8m)(n_x^2/a^2 + n_y^2/b^2 + n_z^2/c^2)}$$

$$z_{\text{tr}} = \sum_{n_x=1}^{\infty} e^{-(\beta h^2/8ma^2)n_x^2} \sum_{n_y=1}^{\infty} e^{-(\beta h^2/8mb^2)n_y^2} \sum_{n_z=1}^{\infty} e^{-(\beta h^2/8mc^2)n_z^2} \quad (22.78)$$

The sums in (22.78) can be evaluated exactly, but this involves advanced mathematics. Instead, we shall use an approximate treatment that is simple and very accurate. The approximation we make is to replace each sum by an integral. This procedure is accurate when the terms in the sum change very little from one term to the next. Consider $\sum_{n=0}^{\infty} f(n)$, where $f(n) \approx f(n + 1)$ for all n. We have

$$\sum_{n=0}^{\infty} f(n) = f(0) + f(1) + \cdots = f(0) \int_0^1 dn + f(1) \int_1^2 dn + \cdots$$

$$= \int_0^1 f(0) \, dn + \int_1^2 f(1) \, dn + \cdots$$

Since the function $f(n)$ varies very slowly with n, we can set $f(0) \approx f(n)$ for n in the range 0 to 1 and $f(1) \approx f(n)$ for n in the range 1 to 2, etc. Therefore, we have $\sum_{n=0}^{\infty} f(n) \approx \int_0^1 f(n) \, dn + \int_1^2 f(n) \, dn + \cdots = \int_0^{\infty} f(n) \, dn$.

Because $f(n)$ varies slowly, we have $f(0) \approx f(1) \approx f(2) \approx \cdots$. Hence, many terms contribute substantially to the sum, and the relative contribution of any one term is small and can be neglected. If we like, we can start the sum at $n = 1$ instead of $n = 0$. Likewise, we can start the integration at $n = 1$ instead of $n = 0$. Therefore

$$\sum_{n=0}^{\infty} f(n) \approx \sum_{n=1}^{\infty} f(n) \approx \int_0^{\infty} f(n) \, dn \approx \int_1^{\infty} f(n) \, dn \qquad (22.79)$$

provided $\left| [f(n + 1) - f(n)]/f(n) \right| \ll 1$. Whether the lower limit is taken as 0 or 1 is simply a matter of which integral is easier to evaluate.

Let us check that the approximation (22.79) is applicable to the sums in (22.78). For the n_x sum we have

$$\frac{f(n_x + 1) - f(n_x)}{f(n_x)} = \frac{e^{-\beta \varepsilon(n_x + 1)} - e^{-\beta \varepsilon(n_x)}}{e^{-\beta \varepsilon(n_x)}} = e^{-\Delta \varepsilon/kT} - 1 \qquad (22.80)$$

where $\varepsilon(n_x) = (h^2/8ma^2)n_x^2$ and $\Delta \varepsilon$ is the spacing between adjacent levels of x translational energy. The translational levels are extremely close together, and for typical conditions one finds $\Delta \varepsilon/kT \approx 10^{-9}$ (Prob. 22.15). Therefore $e^{-\Delta \varepsilon/kT} - 1 = e^{-(10^{-9})} - 1 = (1 - 10^{-9} + \cdots) - 1 \approx -10^{-9}$, where the Taylor series (8.37) was used. Thus f decreases by 1 part in 10^9 for each increase of 1 in n_x, and replacement of the sum by an integral is eminently justified. More generally, we have shown that *the molecular-partition-function sum $\sum_s e^{-\varepsilon_s/kT}$ can be replaced by an integral whenever $\Delta \varepsilon_s \ll kT$,* where $\Delta \varepsilon_s$ is the spacing between adjacent levels of the kind of energy being considered (translational, rotational, etc.).

The use of (22.79) and integral 2 in Table 15.1 gives for the first sum in (22.78)

$$\sum_{n_x=1}^{\infty} e^{-(\beta h^2/8ma^2)n_x^2} = \int_0^{\infty} e^{-(\beta h^2/8ma^2)n_x^2} \, dn_x = \frac{1}{2}\left(\frac{8m\pi}{\beta h^2}\right)^{1/2} a$$

Similarly, the n_y and n_z sums equal $\frac{1}{2}(8m\pi/\beta h^2)^{1/2}b$ and $\frac{1}{2}(8m\pi/\beta h^2)^{1/2}c$. Thus $z_{\text{tr}} = \frac{1}{8}(8m\pi/\beta h^2)^{3/2}abc$. Since $\beta = 1/kT$ and $abc = V$ (the volume of the container holding the ideal gas), we have

$$z_{\text{tr}} = (2\pi mkT/h^2)^{3/2} V \qquad (22.81)$$

$$\ln z_{\text{tr}} = \tfrac{3}{2} \ln (2\pi mk/h^2) + \tfrac{3}{2} \ln T + \ln V \qquad (22.82)$$

where the logarithmic identities (1.67) and (1.68) were used.

Rotational Partition Function

From Secs. 21.3 and 18.14, the rotational quantum state of a diatomic molecule is defined by the quantum numbers J and M_J, and the energy levels (in the rigid-rotor approximation) are $\varepsilon_{rot} = (\hbar^2/2I)J(J+1)$, where I is the moment of inertia [Eq. (21.15)] and J goes from 0 to ∞. For J fixed, M_J takes on the $2J + 1$ integral values from $-J$ to $+J$. Since ε_{rot} is independent of M_J, each level is $(2J + 1)$-fold degenerate. To evaluate z_{rot} by summing over all rotational states as in Eq. (22.50), we must sum over both J and M_J. It's a bit easier to use a sum over energy levels instead of states. The energy levels are defined by J, and the degeneracy is $2J + 1$; therefore Eq. (22.73) gives

$$z_{rot} = \sum_{J=0}^{\infty} (2J + 1)e^{-(\hbar^2/2IkT)J(J+1)} = \sum_{J=0}^{\infty} (2J + 1)e^{-(\Theta_{rot}/T)J(J+1)}$$

where the **characteristic rotational temperature** Θ_{rot} was defined as

$$\Theta_{rot} \equiv \hbar^2/2Ik = \tilde{B}hc/k \qquad (22.83)$$

where (21.33) was used for the rotational constant $\tilde{B}$. The parameter Θ_{rot} has the dimensions of temperature but is not a temperature in the physical sense.

If Θ_{rot}/T is small, the rotational levels are closely spaced compared with kT and we can approximate the sum by an integral (as we did for z_{tr}). Thus

$$z_{rot} \approx \int_0^{\infty} (2J + 1)e^{-(\Theta_{rot}/T)J(J+1)}\, dJ = \int_0^{\infty} e^{-(\Theta_{rot}/T)w}\, dw$$

where we made the change of variable $w \equiv J(J+1) = J^2 + J$, $dw = (2J + 1)\, dJ$. Using $\int e^{-bw}\, dw = -b^{-1}e^{-bw}$, we get

$$z_{rot} \approx T/\Theta_{rot} \qquad \text{heteronuclear, } \Theta_{rot} \ll T \qquad (22.84)$$

Equation (22.84) is valid only for heteronuclear diatomic molecules. For homonuclear diatomics, the Pauli principle (Sec. 19.6) requires that ψ_N (the wave function for nuclear motion) be symmetric with respect to interchange of the identical nuclei if the nuclei are bosons or antisymmetric if they are fermions. This restriction on ψ_N cuts the number of available quantum states in half compared with a heteronuclear diatomic, for which there is no symmetry restriction. (For example, if the identical nuclei are bosons, a symmetric nuclear-spin function requires a symmetric rotational wave function and an antisymmetric nuclear-spin function requires an antisymmetric rotational function, but the combination of a symmetric spin function with an antisymmetric rotational function or an antisymmetric spin function with a symmetric rotational function is excluded.) Chemists customarily ignore the nuclear-spin quantum states in calculating z (see Sec. 22.9 for an explanation of why this is all right), but for consistency between homonuclear and heteronuclear partition functions, one must allow for the reduction in quantum states by dividing z_{rot} by 2 for a homonuclear diatomic. [For a fuller discussion see *McQuarrie* (1973), pp. 104–105.] Therefore, $z_{rot} \approx T/2\Theta_{rot}$ for a homonuclear diatomic.

To have a single formula for both homonuclear and heteronuclear diatomics, we define the **symmetry number** σ as 2 for a homonuclear diatomic and 1 for a heteronuclear diatomic, and we include a factor $1/\sigma$ in z_{rot}:

$$z_{rot} \approx \frac{T}{\sigma\Theta_{rot}} = \frac{2IkT}{\sigma\hbar^2} \qquad \text{for } T \gg \Theta_{rot} \qquad (22.85)$$

Some Θ_{rot} values and z_{rot} values at 300 and 1000 K for ground electronic states are

Molecule	H$_2$	H^{35}Cl	N$_2$	O$_2$	^{35}Cl$_2$	Na$_2$	I$_2$
Θ_{rot}/K	85.35	15.02	2.862	2.069	0.3500	0.2220	0.05369
$z_{rot,300}$	1.76	20.0	52	72	429	676	2794
$z_{rot,1000}$	5.86	66.6	175	242	1429	2252	9313

The light molecule H$_2$ has a small moment of inertia and hence a relatively high Θ_{rot}. Except for H$_2$ and its isotopic species D$_2$ and HD, Θ_{rot} for diatomics is much less than T at temperatures that chemists commonly deal with, so we can use the approximation (22.85).

Actually, the derivation of (22.85) was a bit phony. It is true that the rotational spacing is generally small compared with kT. However, the $2J + 1$ factor in the z_{rot} sum causes substantial term-to-term variation for the low-J terms, and we have not really justified replacing the sum by an integral. A more rigorous derivation is given in Prob. 22.34, where it is shown that

$$z_{rot} = \frac{T}{\sigma\Theta_{rot}}\left[1 + \frac{1}{3}\frac{\Theta_{rot}}{T} + \frac{1}{15}\left(\frac{\Theta_{rot}}{T}\right)^2 + \frac{4}{315}\left(\frac{\Theta_{rot}}{T}\right)^3 + \cdots\right] \quad (22.86)$$

For $T \gg \Theta_{rot}$, Eq. (22.86) reduces to (22.85). Equation (22.86) is accurate provided T is greater than Θ_{rot}. For $T < \Theta_{rot}$, one must evaluate the rotational partition function by direct term-by-term summation (Prob. 22.61).

Vibrational Partition Function

The harmonic-oscillator approximation gives the vibrational energies of a diatomic molecule as $(v + \frac{1}{2})h\nu$, where v goes from 0 to ∞ and there is no degeneracy (Sec. 21.3). The choice of zero level of energy is arbitrary, and *it is customary in molecular statistical mechanics to take the energy zero at the lowest available energy level of the molecule.* This level has zero rotational energy and $\frac{1}{2}h\nu$ vibrational energy. We therefore write $\varepsilon_{vib} = (v + \frac{1}{2})h\nu - \frac{1}{2}h\nu = vh\nu$, where ε_{vib} is measured from the $v = 0$ level. The thermodynamic internal energy obtained from the partition function will then be U relative to the lowest molecular energy level. (See also Prob. 22.5.) To be fully consistent, we should have subtracted the zero-point translational energy from ε_{tr}, but this is negligibly small.

We have

$$z_{vib} = \sum_v e^{-\beta\varepsilon_{vib,v}} = \sum_{v=0}^{\infty} e^{-vh\nu/kT} = \sum_{v=0}^{\infty} e^{-v\Theta_{vib}/T} \quad (22.87)$$

$$\Theta_{vib} \equiv h\nu/k = \tilde{\nu}hc/k \quad (22.88)$$

where Θ_{vib} is the **characteristic vibrational temperature** and (21.35) was used.

Some values of Θ_{vib} for ground electronic states are

Molecule	H$_2$	H^{35}Cl	N$_2$	O$_2$	^{35}Cl$_2$	I$_2$
Θ_{vib}/K	5990	4151	3352	2239	798	307

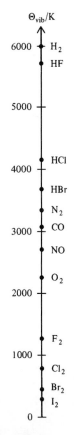

Figure 22.6

Characteristic vibrational temperatures for some diatomic molecules.

The generally high values of Θ_{vib} (Fig. 22.6) compared with ordinary temperatures show that the vibrational levels are not closely spaced compared with kT. (Recall from Sec. 21.3 that most diatomic molecules have very little occupancy of excited vibrational levels at room temperature.) Hence, the vibrational sum cannot be replaced by an integral. This is no cause for alarm, because the sum is easily evaluated exactly.

Recall from Eq. (8.8) the following formula for the sum of a geometric series:

$$1 + x + x^2 + x^3 + \cdots = \sum_{n=0}^{\infty} x^n = \frac{1}{1-x} \qquad \text{for } |x| < 1 \qquad (22.89)$$

[A derivation of (22.89) is given in Prob. 22.51.] The sum in z_{vib} in (22.87) corresponds to (22.89) with $x = e^{-h\nu/kT} < 1$ and $n = v$. Therefore

$$z_{vib} = \frac{1}{1 - e^{-h\nu/kT}} = \frac{1}{1 - e^{-\Theta_{vib}/T}} \qquad T \text{ not extremely high} \qquad (22.90)$$

The rigid-rotor and harmonic-oscillator expressions for ε_{rot} and ε_{vib} are only approximations, since they omit the effects of anharmonicity, centrifugal distortion, and vibration–rotation interaction (Sec. 21.3). Our expressions for z_{rot} and z_{vib} are therefore approximations. Fortunately, the corrections introduced by anharmonicity, etc., are usually small (except at high temperatures) and need only be considered in precise work. For details of these corrections, see *McQuarrie* (1973), prob. 6-24; *Davidson*, pp. 116–119.

The frequency ν in (22.90) is to be taken as ν_0 [Eq. (21.36)], the fundamental frequency corresponding to the $v = 0 \rightarrow 1$ transition (rather than as the equilibrium vibrational frequency ν_e). Likewise, I in (22.85) is I_0, the moment of inertia averaged over the zero-point vibrations, and is calculated from the rotational constant B_0 [Eq. (21.32) with $v = 0$].

There is another reason besides anharmonicity for (22.90) to fail at very high temperatures. z_{vib} in (22.90) is for a harmonic oscillator, which has an *infinite* number of vibrational levels. However, a diatomic molecule has only a *finite* number of vibrational levels (Fig. 21.9), and the sum in its z_{vib} should contain only a finite number of terms. This difference is unimportant at low and moderate temperatures where very high vibrational levels are not appreciably populated, but (22.90) becomes very inaccurate for temperatures where vibrational levels near the dissociation limit are significantly populated. One should use (22.90) only for $kT < 0.1D_e$, where D_e is the equilibrium dissociation energy (see *Knox*, sec. 6.3). For a typical D_e of 4 eV, this corresponds to $T < 4600$ K.

EXAMPLE 22.2 Calculation of Θ_{vib} and Θ_{rot}

Use Table 21.1 to find Θ_{vib} and Θ_{rot} for CO in its ground electronic state.
   We have [Eqs. (21.36) and (21.32)]

$$\tilde{\nu}_0 = \tilde{\nu}_e - 2\tilde{\nu}_e x_e = 2169.8 \text{ cm}^{-1} - 2(13.3 \text{ cm}^{-1}) = 2143.2 \text{ cm}^{-1}$$

$$\tilde{B}_0 = \tilde{B}_e - \tfrac{1}{2}\tilde{\alpha}_e = 1.931 \text{ cm}^{-1} - \tfrac{1}{2}(0.018 \text{ cm}^{-1}) = 1.922 \text{ cm}^{-1}$$

Use of (22.88) and (22.83) for Θ_{vib} and Θ_{rot} gives

$$\Theta_{vib} = \tilde{\nu}_0 hc/k, \qquad \Theta_{rot} = \tilde{B}_0 hc/k$$

$$hc/k = (6.6261 \times 10^{-34} \text{ J s})(2.9979 \times 10^8 \text{ m/s})/(1.38065 \times 10^{-23} \text{ J/K})$$

$$hc/k = 0.014388 \text{ m K} = 1.4388 \text{ cm K}$$

$$\Theta_{vib} = (2143.2 \text{ cm}^{-1})(1.4388 \text{ cm K}) = 3084 \text{ K}$$

$$\Theta_{rot} = (1.922 \text{ cm}^{-1})(1.4388 \text{ cm K}) = 2.765 \text{ K}$$

Exercise

Find Θ_{vib} and Θ_{rot} for the ground electronic state of $^{127}I_2$. (*Answers:* 307 K, 0.0537 K.)

fr($v = 0$)

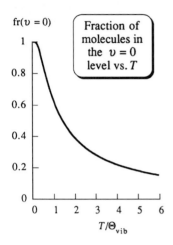

Fraction of molecules in the $v = 0$ level vs. T

Figure 22.7

Fraction of diatomic molecules in the $v = 0$ harmonic-oscillator level versus T/Θ_{vib}.

fr($J = 0$)

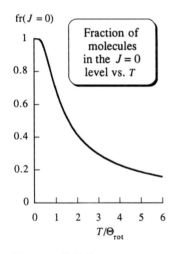

Fraction of molecules in the $J = 0$ level vs. T

Figure 22.8

Fraction of diatomic molecules in the $J = 0$ rigid-rotor level versus T/Θ_{rot}.

The physical significance of Θ_{vib} and Θ_{rot} can be seen from Figs. 22.7 and 22.8, which plot the fraction of molecules in the $v = 0$ vibrational level versus T/Θ_{vib} and the fraction in the $J = 0$ rotational level versus T/Θ_{rot}. At temperatures well below Θ_{vib} ($T < \frac{1}{3}\Theta_{vib}$) there is little occupation of excited vibrational levels. Similarly for Θ_{rot}.

Electronic Partition Function

We shall calculate z_{el} as a sum over electronic energy levels (rather than states), so we include the degeneracy g_{el} of each electronic level. Let the electronic energy levels be numbered 0, 1, 2, . . . in order of increasing energy, and let their degeneracies be $g_{el,0}$, $g_{el,1}$, $g_{el,2}$, ... As noted, we take the zero of energy at the lowest available level (the $J = 0$, $v = 0$ level of the ground electronic level). Hence the energy of the ground electronic level is taken as zero: $\varepsilon_{el,0} = 0$. The energy of each excited electronic level is then measured relative to the $v = 0$, $J = 0$ level of the ground electronic level. With $\varepsilon_{el,0} = 0$, we have

$$z_{el} = g_{el,0} + g_{el,1}e^{-\beta\varepsilon_{el,1}} + g_{el,2}e^{-\beta\varepsilon_{el,2}} + \cdots \qquad (22.91)$$

Since there is no general formula for the ε_{el}'s, the series is added term by term using spectroscopically observed electronic energies.

For nearly all common diatomic molecules, $\varepsilon_{el,1}$ is very much greater than kT at room temperature, and all terms in (22.91) after the first term contribute negligibly for all temperatures up to 5000 or 10000 K. Therefore

$$z_{el} = g_{el,0} \qquad T \text{ not extremely high} \qquad (22.92)$$

The main exception to (22.92) is NO, which has a very-low-lying excited electronic state (see *Kestin and Dorfman*, p. 261). O_2 has an excited electronic state that contributes significantly to z_{el} above 1500 K.

For most diatomics, the ground electronic level is nondegenerate: $g_{el,0} = 1$. An important exception is O_2, for which $g_{el,0} = 3$, as a result of spin degeneracy; recall that O_2 has a triplet ground level. Another exception is NO, which has an odd number of electrons; here, $g_{el,0} = 2$, as a result of the two possible orientations of the spin of the unpaired electron. The general rule for g_{el} of a diatomic molecule is given in *Hirschfelder, Curtiss, and Bird*, p. 119.

Monatomic molecules with filled subshells (Be, He, Ne, Ar, . . .) have $g_{el,0} = 1$. For H, Li, Na, K, . . . , the spin degeneracy of the odd electron gives $g_{el,0} = 2$. For F, Cl, Br, and I, $g_{el,0} = 4$ and there is a low-lying excited state that contributes to z_{el}. See *McQuarrie* (1973), sec. 5-2, for more details.

In deriving $z = z_{tr}z_{rot}z_{vib}z_{el}$ [Eq. (22.58)], we assumed the four kinds of energies to be independent of one another and summed separately over each kind of energy. Actually, this assumption is false. The bond distance and force constant of a diatomic molecule change from one electronic state to another, so each electronic state has a different vibrational frequency and a different moment of inertia. To allow for this, we must replace $z_{rot}z_{vib}z_{el}$ by

$$g_{el,0}z_{vib,0}z_{rot,0} + g_{el,1}e^{-\beta\varepsilon_{el,1}}z_{vib,1}z_{rot,1} + \cdots$$

where $z_{vib,0}$ and $z_{rot,0}$ are calculated using the vibrational frequency and moment of inertia of the ground electronic level, $z_{vib,1}$ and $z_{rot,1}$ use the parameters of the first excited electronic level, etc. Since the contributions of excited electronic levels are generally very small (except at very high T), it is usually an adequate approximation to ignore the change in z_{vib} and z_{rot} from one electronic state to another. The accurate expression then reduces to the form $z_{vib}z_{rot}z_{el}$ used earlier.

Pause for Refreshment and Review

We began by expressing all thermodynamic properties in terms of the system's canonical partition function Z. We then showed that for a pure ideal gas, $Z = z^N/N!$ and

$z = z_{tr}z_{rot}z_{vib}z_{el}$. We have now evaluated z_{tr}, z_{rot}, z_{vib}, and z_{el} for diatomic and monatomic molecules in terms of molecular properties (m, Θ_{rot}, Θ_{vib}, and $g_{el,0}$), so we are now ready to evaluate the thermodynamic properties of an ideal gas of such molecules in terms of molecular properties.

Equation of State

The first thermodynamic property we evaluate is the pressure P. Equation (22.37) reads $P = kT(\partial \ln Z/\partial V)_{T,N}$. For an ideal gas, $\ln Z$ is expressed in terms of z_{tr}, z_{rot}, z_{vib}, and z_{el} by Eq. (22.61). The molecular vibrational, rotational, and electronic energies depend on properties of the gas molecules but are independent of V. In contrast, ε_{tr} does depend on V. Therefore only z_{tr} is a function of V, as can be verified from (22.91), (22.90), (22.85), and (22.82). The use of (22.61) for $\ln Z$ gives

$$P = kT\left(\frac{\partial \ln Z}{\partial V}\right)_{T,N} = kT\left[\frac{\partial(N \ln z_{tr})}{\partial V}\right]_{T,N} = NkT\left(\frac{\partial \ln z_{tr}}{\partial V}\right)_T$$

Equation (22.82) for $\ln z_{tr}$ gives $(\partial \ln z_{tr}/\partial V)_T = 1/V$. Hence, $P = NkT/V$, or $PV = NkT$. Since $N = N_A n$ (where N_A is the Avogadro constant and n the number of moles of gas), we have $PV = nN_A kT$. The absolute temperature scale was defined in Sec. 1.5 to make $PV = nRT$ hold. Hence, $N_A k = R$, and $k = R/N_A$. This agrees with Eq. (3.57) and shows that k in (22.27) is Boltzmann's constant: $\beta = 1/kT = N_A/RT$. Also, $Nk = NR/N_A = nR$:

$$Nk = nR \qquad (22.93)$$

Internal Energy

We now calculate the internal energy of an ideal gas of diatomic molecules. Since we took the zero level of energy at the lowest available molecular energy level, we shall be finding $U - U_0$, where U_0 is the thermodynamic internal energy of a hypothetical ideal gas in which every molecule is in the lowest translational, rotational, vibrational, and electronic state. This would be the internal energy at absolute zero if the gas didn't condense and if the molecules were not fermions. Thus Eq. (22.62) reads $U - U_0 = U_{tr} + U_{rot} + U_{vib} + U_{el}$. From (22.63), (22.93), and (22.82), we have

$$U_{tr} = nRT^2\left(\frac{\partial \ln z_{tr}}{\partial T}\right)_V = nRT^2\frac{3}{2T} = \tfrac{3}{2}nRT$$

Similarly, U_{rot}, U_{vib}, and U_{el} are found from (22.63), (22.85), (22.90), and (22.92).

The results for an ideal gas of diatomic molecules at temperatures that are not extremely high or low are (Prob. 22.35)

$$U - U_0 = U_{tr} + U_{rot} + U_{vib} + U_{el} \qquad (22.94)$$

$$U_{tr} = \tfrac{3}{2}nRT \qquad (22.95)$$

$$U_{rot} = nRT \qquad (22.96)$$

$$U_{vib} = nR\frac{h\nu}{k}\frac{1}{e^{h\nu/kT} - 1} = nR\Theta_{vib}\frac{1}{e^{\Theta_{vib}/T} - 1} \qquad (22.97)$$

$$U_{el} = 0 \qquad (22.98)$$

Figure 22.9 plots the molar vibrational energy $U_{vib,m}$ versus T for $CO(g)$.

For a gas of monatomic molecules, there is no rotation or vibration, so $U_m - U_{m,0} = U_{tr,m} = \tfrac{3}{2}RT$, in agreement with the classical-kinetic-theory result (15.17).

Actually, Eq. (22.96) is not quite correct. When the high-temperature expression for U_{rot} is properly evaluated in the limit $T \gg \Theta_{rot}$ using (22.86) instead of (22.85) for z_{rot}, one finds (Prob. 22.63)

$$U_{rot} = nR(T - \Theta_{rot}/3)$$

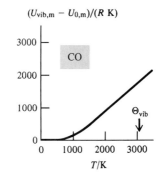

Figure 22.9

Vibrational contribution to the molar internal energy of $CO(g)$ versus temperature. $U_{0,m}$ is the molar zero-point vibrational energy.

Thus U_{rot} always deviates slightly at high T from the equipartition result nRT (Sec. 15.10). Since $\Theta_{\text{rot}}/3$ is generally quite small, Eq. (22.96) is only slightly in error. Similarly, because (22.95) for U_{tr} was found by approximating a sum by an integral, the true high-temperature expression for U_{tr} differs very slightly from $\frac{3}{2}nRT$, but the difference is utterly negligible [see G. Gutierrez and J. M. Yanez, *Am. J. Phys.*, **65**, 739 (1997)].

Heat Capacity

Differentiation of (22.94) with respect to T at constant V and n gives

$$C_V = C_{V,\text{tr}} + C_{V,\text{rot}} + C_{V,\text{vib}} + C_{V,\text{el}} \tag{22.99}$$

where $C_{V,\text{tr}} \equiv (\partial U_{\text{tr}}/\partial T)_{V,n}$, etc. Differentiation of (22.95) to (22.98) gives for an ideal gas of diatomic molecules at moderate temperatures:

$$C_{V,\text{tr}} = \tfrac{3}{2}nR \tag{22.100}$$

$$C_{V,\text{rot}} = nR \tag{22.101}$$

$$C_{V,\text{vib}} = nR\left(\frac{\Theta_{\text{vib}}}{T}\right)^2 \frac{e^{\Theta_{\text{vib}}/T}}{\left(e^{\Theta_{\text{vib}}/T} - 1\right)^2} \tag{22.102}$$

$$C_{V,\text{el}} = 0 \tag{22.103}$$

Note that U and C_V of the ideal gas are functions of T only, in agreement with (2.67) and (2.69).

Equations (22.100) and (22.101) for $C_{V,\text{tr,m}}$ and $C_{V,\text{rot,m}}$ agree with the classical-statistical-mechanical equipartition theorem of $\frac{1}{2}R$ for each quadratic term in the energy (Sec. 15.10), but $C_{V,\text{vib,m}}$ does not agree with the equipartition theorem. We obtained z_{tr} and z_{rot} by using the fact that the translational and rotational energy spacings are much less than kT, so the sums over discrete energy levels can be replaced by integrals over a continuous range of energy. Since continuous values for energy correspond to classical mechanics, we obtained the classical results for $C_{V,\text{tr}}$ and $C_{V,\text{rot}}$. However, vibrational levels are not closely spaced compared with kT, and the classical equipartition theorem fails for $C_{V,\text{vib}}$.

Figure 22.10 plots $C_{V,\text{vib,m}}$ of (22.102) versus T. At high T, the harmonic-oscillator $C_{V,\text{vib,m}}$ goes to the classical equipartition value R (see Prob. 22.36).

The result $C_{V,\text{rot}} = nR$ applies only at temperatures for which $T \gg \Theta_{\text{rot}}$. At low temperatures, we must use the series (22.86) for z_{rot} and calculate U_{rot} and $C_{V,\text{rot}}$ from (22.86). At very low temperatures, (22.86) fails, and z_{rot} must be calculated by direct

Figure 22.10

The vibrational contribution to $C_{V,\text{m}}$ of a gas of diatomic molecules in the harmonic-oscillator approximation. The dashed line is the classical equipartition result of R at all temperatures. (Because a real diatomic molecule has a finite number of vibrational levels, its $C_{V,\text{m,vib}}$ decreases at very high T.)

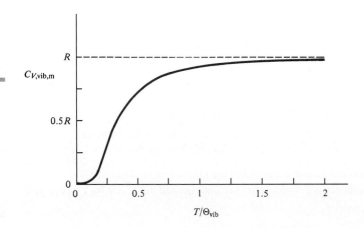

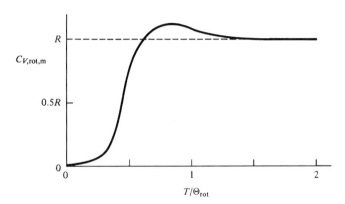

Figure 22.11

The rotational contribution to $C_{V,m}$ of a gas of diatomic molecules. The dashed line is the classical equipartition result.

term-by-term summation. We omit details but simply plot $C_{V,\text{rot},m}$ versus T in Fig. 22.11. The classical value R is reached at $T \approx 1.5\Theta_{\text{rot}}$.

For a gas of monatomic molecules with no low-lying electronic states, $C_{V,m} = C_{V,\text{tr},m} = \frac{3}{2}R$, in agreement with (15.18).

Entropy

Equations (22.64) to (22.66) allow S of an ideal gas to be calculated from z_{tr}, z_{rot}, etc. Using these equations, (22.95) to (22.98), (22.82), (22.85), (22.90), (22.92), and $PV = NkT$, we find for an ideal gas of diatomic molecules (Prob. 22.37)

$$S = S_{\text{tr}} + S_{\text{rot}} + S_{\text{vib}} + S_{\text{el}}$$

$$S_{\text{tr}} = \frac{5}{2}nR + nR \ln \left[\frac{(2\pi m)^{3/2}}{h^3} \frac{(kT)^{5/2}}{P} \right] \qquad (22.104)$$

$$S_{\text{rot}} = nR + nR \ln \frac{T}{\sigma\Theta_{\text{rot}}} \qquad (22.105)$$

$$S_{\text{vib}} = nR \frac{\Theta_{\text{vib}}}{T} \frac{1}{e^{\Theta_{\text{vib}}/T} - 1} - nR \ln \left(1 - e^{-\Theta_{\text{vib}}/T} \right) \qquad (22.106)$$

$$S_{\text{el}} = nR \ln g_{\text{el},0} \qquad (22.107)$$

where Θ_{rot} and Θ_{vib} are given by (22.83) and (22.88). S is the sum of (22.104) to (22.107), provided T is not extremely high or low. Figure 22.12 plots $S_{\text{vib},m}$ versus T for some gases.

Substitution of $P = (P/\text{bar}) (10^5 \text{ N/m}^2)$ [Eq. (1.11)], $m = (M_r/N_A) (10^{-3} \text{ kg/mol})$, where M_r is the (dimensionless) molecular weight, $T = (T/\text{K})$ K, where K is 1 kelvin, and the SI values of h, k, and N_A into (22.104) gives S_{tr} in a form convenient for calculations (Prob. 22.38):

$$S_{\text{tr},m} = R[1.5 \ln M_r + 2.5 \ln (T/\text{K}) - \ln (P/\text{bar}) - 1.1517] \qquad (22.108)$$

$$S_{\text{tr},m} = R[1.5 \ln M_r + 2.5 \ln (T/\text{K}) - \ln (P/\text{atm}) - 1.1649]$$

since $\ln(P/\text{bar}) = \ln (P/\text{atm}) + \ln (\text{atm/bar}) = \ln (P/\text{atm}) + \ln (760/750.06)$. From (22.108), $S_{\text{tr},m}$ increases as T increases, as $V (= nRT/P)$ increases, and as M_r increases (Prob. 22.60).

For a gas of monatomic molecules with a nondegenerate ground electronic state, $S_{\text{rot}} = S_{\text{vib}} = S_{\text{el}} = 0$, and $S = S_{\text{tr}}$.

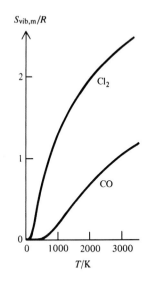

Figure 22.12

$S_{\text{vib},m}$ versus T for $Cl_2(g)$ and $CO(g)$. For Cl_2, $\Theta_{\text{vib}} = 3083$ K and $S_{\text{vib},m}$ is negligible for $CO(g)$ at room temperature.

EXAMPLE 22.3 Molar entropy of N_2

Calculate $S^\circ_{m,298}$ of $N_2(g)$.

Even though N_2 is slightly nonideal at 25°C and 1 bar, the standard state is the hypothetical ideal gas at 1 bar, so the above ideal-gas equations apply. The tables after (22.85) and (22.88) in this section give $\Theta_{rot} = 2.862$ K and $\Theta_{vib} = 3352$ K for N_2; also, $g_{el,0} = 1$. Substitution in (22.105) to (22.108) gives at 298.15 K

$$S^\circ_{tr,m} = (8.3145 \text{ J mol}^{-1}\text{K}^{-1})(1.5 \ln 28.013 + 2.5 \ln 298.15 - \ln 1 - 1.1517)$$
$$= 150.42 \text{ J mol}^{-1}\text{K}^{-1}$$

$$S^\circ_{rot,m} = (8.3145 \text{ J mol}^{-1}\text{K}^{-1})\left[1 + \ln \frac{298.15 \text{ K}}{2(2.862 \text{ K})}\right] = 41.18 \text{ J mol}^{-1}\text{K}^{-1}$$

$$\Theta_{vib}/T = (3352 \text{ K})/(298.15 \text{ K}) = 11.24$$

$$S^\circ_{vib,m} = (8.3145 \text{ J mol}^{-1}\text{K}^{-1})[11.24(e^{11.24} - 1)^{-1} - \ln(1 - e^{-11.24})]$$
$$= 0.0013 \text{ J mol}^{-1}\text{K}^{-1}$$

$$S^\circ_{el,m} = 0$$

$$S^\circ_{m,298} = S^\circ_{tr,m} + S^\circ_{rot,m} + S^\circ_{vib,m} + S^\circ_{el,m} = 191.60 \text{ J mol}^{-1}\text{K}^{-1}$$

For comparison, the experimental value (found by the methods of Sec. 5.7) is 192.1 J mol^{-1} K^{-1}.

The calculations of this example reflect an impressive synthesis of quantum mechanics, statistical mechanics, and thermodynamics.

Exercise

Calculate $S^\circ_{m,1200}$ for $N_2(g)$. (*Answer*: 234.16 J/mol-K.)

Exercise

Calculate $U^\circ_{m,298} - U^\circ_{m,0}$, $U^\circ_{m,1000} - U^\circ_{m,0}$, $C_{V,m,298}$, and $C_{V,m,1000}$ for $N_2(g)$. [*Answers*: 6.198 kJ/mol, 21.088 kJ/mol, 20.80 J/(mol K), 24.30 J/(mol K).]

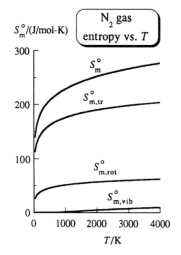

S°_m/(J/mol-K)

N$_2$ gas entropy vs. T

300

S°_m

200

$S^\circ_{m,tr}$

100

$S^\circ_{m,rot}$

$S^\circ_{m,vib}$

0

0   1000   2000   3000   4000

T/K

Figure 22.13

Translational, rotational, and vibrational contributions to S°_m of $N_2(g)$ plotted versus T.

Figure 22.13 plots the various contributions to S°_m of $N_2(g)$ versus T.

Physical Interpretation of z and S

Taking the zero level of molecular energy to coincide with the ground state (gs), we have $\varepsilon_{gs} = 0$, and the Boltzmann distribution law (22.70) gives $\langle N_r\rangle/\langle N_{gs}\rangle = e^{-\varepsilon_r/kT}$. If ε_r is less than kT or of the same magnitude as kT, then $e^{-\varepsilon_r/kT}$ will be reasonably close to 1 and the population $\langle N_r\rangle$ of state r will be a significant fraction of the ground-state population $\langle N_{gs}\rangle$. If ε_r is substantially greater than kT, then both $e^{-\varepsilon_r/kT}$ and $\langle N_r\rangle/\langle N_{gs}\rangle$ will be close to zero. Since the molecular partition function is $z \equiv \sum_r e^{-\varepsilon_r/kT}$, each state that is significantly populated at temperature T will contribute to z a term whose order of magnitude is 1, whereas each state not significantly populated will contribute negligibly to z. Therefore, *the numerical value of the molecular partition function z gives a very rough estimate of the number of molecular states that are significantly populated at T.*

EXAMPLE 22.4 Molecular partition functions and populations

In a previous example, we found $\Theta_{rot} = 2.765$ K and $\Theta_{vib} = 3084$ K for CO. Calculate z_{tr}, z_{rot}, z_{vib}, and z_{el} for 1.000 mol of CO(g) at 25°C and 1 atm, treating the gas as ideal. Relate the results to populations of molecular states. Are z_{tr}, z_{rot}, z_{vib}, and z_{el} extensive or intensive properties?

Equation (22.81) for z_{tr} gives $z_{tr} = (2\pi mkT/h^2)^{3/2}V$. We have $V = nRT/P = 24460$ cm^3 and $m = M/N_A$. Thus

$$z_{tr} = \left(\frac{2\pi MkT}{N_A h^2}\right)^{3/2} V$$

$$= \left[\frac{2\pi (0.028 \text{ kg/mol})(1.38 \times 10^{-23} \text{ J/K})(298 \text{ K})}{(6.02 \times 10^{23}/\text{mol})(6.63 \times 10^{-34} \text{ J s})^2}\right]^{3/2} (0.02446 \text{ m}^3)$$

$$z_{tr} = 3.5 \times 10^{30}$$

Since $T \gg \Theta_{rot}$ is satisfied, Eq. (22.85) for z_{rot} gives

$$z_{rot} = \frac{T}{\sigma \Theta_{rot}} = \frac{298 \text{ K}}{1(2.765 \text{ K})} = 108$$

Equations (22.90) and (22.92) for z_{vib} and z_{el} give

$$z_{vib} = \frac{1}{1 - e^{-\Theta_{vib}/T}} = \frac{1}{1 - e^{-(3084 \text{ K})/(298 \text{ K})}} = 1.00003$$

$$z_{el} = g_{el,0} = 1$$

The equation preceding (22.53) gives the number of translational states that are significantly populated in this system as

$$60(mkT/h^2)^{3/2}V = 60z_{tr}/(2\pi)^{3/2} = 60(3.5 \times 10^{30})/(2\pi)^{3/2} = 13 \times 10^{30}$$

which is the same order of magnitude as z_{tr}. Problem 22.54 uses the Boltzmann distribution law to show that 93% of the CO molecules at 25°C are in rotational states with $J \leq 16$. Since each rotational level is $(2J + 1)$-fold degenerate, there are $1 + 3 + 5 + \cdots + 33 = 289$ rotational states with $J \leq 16$, which is the same order of magnitude as z_{rot}. At room temperature, only one vibrational state and one electronic state are significantly populated, in accord with the z_{vib} and z_{el} values.

The equation $z_{tr} = (2\pi mkT/h^2)^{3/2}V$ shows that z_{tr} is a state function. Since z_{tr} is proportional to V, z_{tr} is extensive. Since z_{rot}, z_{vib}, and z_{el} are functions of T only, they are intensive properties. (Since these z's are not directly measurable, it is perhaps stretching things to call them "properties.")

Exercise

Consider the following gases: N_2, O_2, F_2, HF, $^{35}Cl_2$, $^{35}Cl^{37}Cl$. At 25°C and 1 bar, which one has the greatest z_{tr}? The greatest z_{rot}? The greatest z_{vib}? The greatest z_{el}? (*Answers*: $^{35}Cl^{37}Cl$, $^{35}Cl^{37}Cl$, $^{35}Cl^{37}Cl$, O_2.)

As T increases and more states become significantly populated, z increases and S [which has a term proportional to $\ln z$—Eqs. (22.65) and (22.66)] increases. Recall that entropy is related to the distribution of molecules over energy levels (Sec. 3.7).

Figures 22.14, 22.15, and 22.16 plot z, z_{rot}, and z_{vib} versus T for some gases.

The connection between entropy and distribution of molecules among molecular quantum states is shown directly by the fact that S can be expressed in terms of the mole-fraction populations of the molecular states; Prob. 22.62 shows that

$$S = -Nk \sum_r x_r \ln x_r - k \ln N!, \qquad x_r \equiv \langle N_r \rangle / N \qquad \text{pure ideal gas}$$

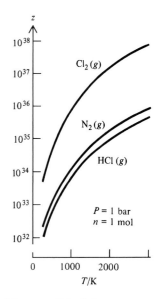

Figure 22.14

Molecular partition function z versus T for 1 mole of some gases at 1 bar. The vertical scale is logarithmic.

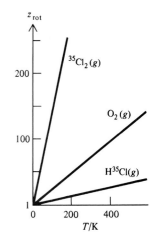

Figure 22.15

Rotational partition function versus T for some gases. As T increases, more rotational states become significantly populated and z_{rot} increases. At extremely low T, the z_{rot}-versus-T plot shows nonlinear behavior, which is not visible on the scale of this plot.

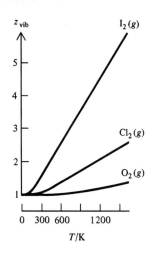

Figure 22.16

Plots of z_{vib} versus T for some gases. The Θ_{vib} values are 307 K for I_2, 798 K for Cl_2, and 2239 K for O_2.

where the sum goes over all molecular quantum states and x_r is the fractional population of state r. Note the close resemblance to the entropy-of-mixing formula (9.46). The function $-x_r \ln x_r$ goes to 0 as x_r goes to 0 or to 1 and is positive for $0 < x_r < 1$. Quantum states with negligible population or with fractional population extremely close to 1 contribute negligibly to S. As T increases and more quantum states become significantly populated, the sum $-\sum_r x_r \ln x_r$ increases and S increases.

Summary

This section found expressions for a diatomic molecule's translational, rotational, vibrational, and electronic partition functions z_{tr}, z_{rot}, z_{vib}, and z_{el}. For temperatures not extremely low, z_{tr} and z_{rot} were evaluated by replacing the sum over states by an integral. The harmonic-oscillator approximation was used for the vibrational energy levels and z_{vib} was found using the formula for the sum of a geometric series. From z_{tr}, z_{rot}, z_{vib}, and z_{el}, we found formulas for the translational, rotational, vibrational, and electronic contributions to $U - U_0$, C_V, and S of an ideal gas of diatomic molecules. These contributions depend on the following molecular properties: the molecular weight M_r, the moment of inertia I (which occurs in Θ_{rot}) and the symmetry number σ, the vibrational frequency ν, and the degeneracy $g_{el,0}$ of the ground electronic state.

22.7 STATISTICAL THERMODYNAMICS OF IDEAL POLYATOMIC GASES

The separation of the molecular energy ε into translational, rotational, vibrational, and electronic contributions holds well for the ground electronic states of most polyatomic molecules, so $z = z_{tr}z_{rot}z_{vib}z_{el}$ for an ideal gas of polyatomic molecules.

Translational Partition Function

Since ε_{tr} has the same form for polyatomics as for diatomics, Eq. (22.81) gives z_{tr} for polyatomic molecules.

Rotational Partition Function

For a linear polyatomic molecule, ε_{rot} and the rotational quantum numbers are the same as for a diatomic molecule, so (22.85) gives z_{rot} for a linear polyatomic molecule. For a linear molecule, $\sigma = 2$ if there is a center of symmetry (for example, HCCH, OCO), and $\sigma = 1$ if there is no center of symmetry (for example, HCCF, OCS).

For a nonlinear molecule, the exact rotational energies can be found if the molecule is a spherical or symmetric top (Sec. 21.6), but there is no simple algebraic formula for the quantum-mechanical rotational energies of an asymmetric top. Therefore, we have a problem in evaluating z_{rot}. We noted in Sec. 22.6 that replacing the sum in z by an integral amounts to treating the system classically. The classical-mechanical formula for the partition function will be given in Sec. 22.11. Using this formula and the known classical-mechanical expression for ε_{rot} of a polyatomic molecule, one finds for any nonlinear polyatomic molecule (see *McClelland*, sec. 11.6, for the derivation)

$$z_{rot} = \frac{\pi^{1/2}}{\sigma}\left(\frac{2kT}{\hbar^2}\right)^{3/2}(I_a I_b I_c)^{1/2} \qquad \text{nonlinear} \qquad (22.109)$$

where I_a, I_b, and I_c are the molecule's principal moments of inertia (Sec. 21.6). Equation (22.109) holds provided T is not extremely low. [For corrections to z_{rot} at very low T, see *Herzberg*, vol. II, pp. 505–506; K. F. Stripp and J. G. Kirkwood, *J. Chem. Phys.*, **19,** 1131 (1951).]

For a nonlinear molecule, the **symmetry number** σ in z_{rot} is the number of indistinguishable orientations obtainable from one another by rotations of the molecule.

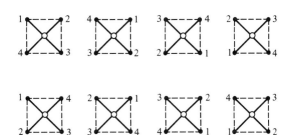

Figure 22.17

The eight indistinguishable
orientations of XeF_4 obtainable
from one another by rotations.

For example, $\sigma = 8$ for the square-planar molecule XeF_4 (Fig. 22.17). The second, third, and fourth orientations in Fig. 22.17 are obtained by 90°, 180°, and 270° rotations about the C_4 axis; the fifth orientation is obtained from the first by a 180° rotation about the C_2 axis passing through atoms 1 and 3; the sixth, seventh, and eighth orientations are obtained by 90°, 180°, and 270° rotations applied to the fifth orientation. For CH_3Cl, $\sigma = 3$. The factor $1/\sigma$ arises for the same reason as with diatomic molecules: The Pauli principle restricts the number of possible quantum states, so for a given nuclear-spin wave function, only $1/\sigma$ of the rotational wave functions are allowed.

Vibrational Partition Function

Equation (21.48) gives the harmonic-oscillator approximation to the vibrational energy of a polyatomic molecule containing $\mathcal{N}$ atoms as the sum of vibrational energies associated with the $3\mathcal{N} - 6$ (or $3\mathcal{N} - 5$) normal modes of vibration. The vibrational quantum numbers vary independently of one another. As usual, when the energy is the sum of independent energies, the partition function is the product of partition functions, one for each kind of energy [recall the derivation of Eq. (22.58)]:

$$z_{\text{vib}} = z_{\text{vib},1} z_{\text{vib},2} \cdots z_{\text{vib},3\mathcal{N}-6} = \prod_{s=1}^{3\mathcal{N}-6} z_{\text{vib},s} = \prod_{s=1}^{3\mathcal{N}-6} \frac{1}{1 - e^{-h\nu_s/kT}} \qquad (22.110)$$

where ν_s is the frequency of the sth normal mode and (22.90) was used for the partition function $z_{\text{vib},s}$ of a single vibrational mode. If the molecule is linear, change $3\mathcal{N} - 6$ to $3\mathcal{N} - 5$ here and below.

Internal Rotation For ethane, CH_3CH_3, one of the normal modes of vibration is a low-frequency torsional (twisting) motion of the two methyl groups about the C–C axis. The minimum energy of this vibration corresponds to staggered hydrogens and the maximum to eclipsed hydrogens (Fig. 22.18). Let b denote the energy difference between the eclipsed and staggered forms. The potential energy of Fig. 22.18 produces a set of torsional energy levels. Levels with energy substantially less than b resemble

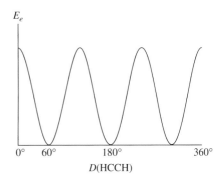

Figure 22.18

Electronic energy (including internuclear repulsions) in ethane versus HCCH dihedral angle.

the harmonic-oscillator energy-level pattern, since the potential energy in the immediate vicinity of 180° resembles the harmonic-oscillator potential energy. At temperatures for which $b \gg kT$, only the lowest levels are significantly populated, and the harmonic-oscillator approximation is accurate. The condition for the harmonic-oscillator approximation to be accurate turns out to be $b > 10kT$. From Sec. 20.8, the ethane barrier per mole is $B = N_A b = 2.9$ kcal/mol (and is less for many common molecules). At 300 K, $N_A kT = RT = 0.6$ kcal/mol, so the harmonic-oscillator approximation is not accurate for ethane. Instead, one must solve the Schrödinger equation numerically for the energy levels of the potential of Fig. 22.18 and evaluate the torsional-vibration-mode factor in the molecular partition function by summing over these levels. See *McClelland*, secs. 5.6, 10.5, and 10.6 for details. The situation becomes more complicated for molecules with torsional vibrations about several single bonds.

Electronic Partition Function

For nearly all stable polyatomic molecules, $g_{el,0}$ is 1, and there are no low-lying excited electronic states. Hence, z_{el} can usually be taken as 1. For species with an odd number of electrons (for example, NO_2 and the CH_3 radical), $g_{el,0}$ is 2, because of spin degeneracy.

Equation of State

As with diatomic molecules, only z_{tr} is a function of volume, and the equation of state for an ideal gas of polyatomic molecules is $PV = NkT$.

Internal Energy

The use of Eq. (22.63) to find U_{tr}, U_{rot}, etc., from z_{tr}, z_{rot}, etc., gives at temperatures neither extremely high nor extremely low (Prob. 22.68):

$$U_{tr} = \tfrac{3}{2}nRT, \qquad U_{el} = 0 \tag{22.111}$$

$$U_{rot} = \begin{cases} \tfrac{3}{2}nRT & \text{nonlinear} \\ nRT & \text{linear} \end{cases} \tag{22.112}$$

$$U_{vib} = nR \sum_{s=1}^{3\mathcal{N}-6} \frac{\Theta_{vib,s}}{e^{\Theta_{vib,s}/T} - 1} \qquad \text{where } \Theta_{vib,s} \equiv h\nu_s/k \tag{22.113}$$

The sum of these four energies is $U - U_0$.

Heat Capacity

Differentiation of the U's with respect to T gives

$$C_{V,tr} = \tfrac{3}{2}nR, \qquad C_{V,el} = 0 \tag{22.114}$$

$$C_{V,rot} = \begin{cases} \tfrac{3}{2}nR & \text{nonlinear} \\ nR & \text{linear} \end{cases} \tag{22.115}$$

$$C_{V,vib} = \sum_{s=1}^{3\mathcal{N}-6} C_{V,vib,s} \tag{22.116}$$

where $C_{V,vib,s}$ is given by (22.102) with Θ_{vib} replaced by $\Theta_{vib,s}$. Large polyatomic molecules often have low-frequency vibrations that contribute to C_V at room temperature. The high-T (classical) limit of each $C_{V,vib,s,m}$ is R (Fig. 22.10), and the high-T $C_{V,vib,m}$ is $(3\mathcal{N} - 6)R$ or $(3\mathcal{N} - 5)R$ (Sec. 15.10). [Because of vibrational anharmonicity (Sec. 21.3), the high-temperature $C_{V,vib,s,m}$ can exceed R (see Fig. 15.19).]

Entropy

Application of Eqs. (22.64) to (22.66) shows that S_{tr} is given by (22.104) and (22.108), that S_{rot} for a linear molecule is given by (22.105), and that

$$S_{rot} = \tfrac{3}{2} nR + nR \ln z_{rot} \qquad \text{nonlinear} \qquad (22.117)$$

$$S_{vib} = \sum_{s=1}^{3N-6} S_{vib,s}, \qquad S_{el} = nR \ln g_{el,0} \qquad (22.118)$$

where $S_{vib,s}$ is gotten by replacing Θ_{vib} by $\Theta_{vib,s}$ in (22.106) and z_{rot} is given by (22.109).

As a reminder, the equations of Secs. 22.6 and 22.7 apply only to ideal gases.

Molecular properties (molecular weights, vibrational frequencies, moments of inertia), partition functions, and thermodynamic properties differ slightly for different isotopic species. To calculate the thermodynamic properties of $CH_3Cl(g)$ using statistical mechanics, one does the calculations for $^{35}ClCH_3$ and for $^{37}ClCH_3$ and then takes a weighted average based on the percent abundances of ^{35}Cl and ^{37}Cl.

The calculation of thermodynamic properties of gases using statistical mechanics is surveyed in M. L. McGlashan (ed.), *Specialist Periodical Reports, Chemical Thermodynamics,* vol. 1, Chemical Society, 1973, pp. 268–316.

22.8 IDEAL-GAS THERMODYNAMIC PROPERTIES AND EQUILIBRIUM CONSTANTS

We have seen that $U - U_0$, S, and C_V of an ideal gas are readily calculated using statistical mechanics. The molecular properties needed for these calculations are: (*a*) the molecular weight (which occurs in z_{tr}), (*b*) the molecular geometry (which is needed to calculate the moments of inertia in z_{rot}), (*c*) the molecular vibration frequencies (which occur in z_{vib}), and (*d*) the degeneracy of the ground electronic level and the energies and degeneracies of any low-lying excited electronic levels (which occur in z_{el}). This information is obtained spectroscopically (Chapter 21). Hence, we can calculate the thermodynamic properties of an ideal gas from observations on the gas molecules' spectrum. For fairly small molecules, the gas-phase thermodynamic properties calculated by statistical mechanics are usually more accurate than the values determined from calorimetric measurements (Chapter 5), and many of the values listed in the Appendix are theoretical statistical-mechanical values.

If both calorimetric and spectroscopic data are lacking for a molecule (as is often true for reaction intermediates), one can use quantum-mechanical calculations to estimate the molecular properties and then calculate the gas-phase thermodynamic properties. Hartree–Fock and density-functional calculations (Chapter 20) usually give accurate molecular geometries. Calculated Hartree–Fock vibrational frequencies are typically 10% too high, so one multiplies them by a correction factor (called a *scale factor*) of 0.90 to improve the results. For density-functional B3LYP/6-31G* vibration frequencies, a scale factor of 0.96 is used. The main source of error in theoretical calculations of gas-phase thermodynamic properties is errors in the treatment of low-frequency torsional vibrations (internal rotation). A procedure that uses MP2/6-31G* geometries to calculate moments of inertia, scaled HF/6-31G* vibrational frequencies, and an MP2-calculated cosine potential for internal rotation gives $S_{m,298}^{\circ}$ values to an accuracy of 1 J/(mol K) for molecules with zero or one torsional mode and an accuracy of 2 J/(mol K) for molecules with two torsional modes [A. L. L. East and L. Radom, *J. Chem. Phys.,* **106,** 6655 (1997)].

We now examine the relations between the tabulated gas-phase thermodynamic quantities $\Delta_f H°$, $S_m°$, $C_{P,m}°$, $\Delta_f G°$, and $G_m° - H_{m,0}°$ and the calculated statistical-mechanical quantities $U_m - U_{m,0}$, $C_{V,m}°$, and $S_m°$. The standard state of any gas is the hypothetical ideal gas at 1 bar, so the ideal-gas formulas of Secs. 22.6 and 22.7 apply.

As a preliminary, consider the gas-phase reaction $0 \to \Sigma_i \nu_i B_i(g)$, where the stoichiometric coefficients ν_i are negative for reactants and positive for products. The reaction's standard change in internal energy at temperature T is $\Delta U_T° \equiv \Sigma_i \nu_i U_{m,T,i}°$. At $T = 0$, this equation becomes $\Delta U_0° = \Sigma_i \nu_i U_{m,0,i}°$, where $\Delta U_0°$ is the standard change in U for the gas-phase reaction at 0 K. Subtraction gives

$$\Delta U_T° - \Delta U_0° = \sum_i \nu_i U_{m,T,i}° - \sum_i \nu_i U_{m,0,i}°$$

$$\Delta U_T° = \Delta U_0° + \sum_i \nu_i (U_{m,T,i}° - U_{m,0,i}°) \qquad \text{gas-phase reaction} \qquad (22.119)$$

Each $U_{m,T,i}° - U_{m,0,i}°$ in (22.119) is calculated by statistical mechanics from Eqs. (22.94) to (22.98) and (22.111) to (22.113). Note that $U_{m,0,i}°$ is for *gas*-phase i.

To get $\Delta U_0°$ in (22.119), we use the following hypothetical path for the gas-phase reaction $bB(g) + cC(g) \to eE(g) + fF(g)$:

$$bB + cC \text{ at } 0 \text{ K} \xrightarrow{1} \text{gaseous atoms at } 0 \text{ K} \xrightarrow{2} eE + fF \text{ at } 0 \text{ K} \qquad (22.120)$$

At $T = 0$, all the molecules of each species are in their ground electronic, vibrational, and rotational states. Step 1 involves dissociation of each ground-state reactant into atoms. The energy needed to dissociate one molecule of ground-state B into atoms is the ground-state dissociation energy $D_{0,B}$ of B (shown in Fig. 21.9 for a diatomic molecule). The energy needed to dissociate b moles of B(g) at 0 K is $bN_A D_{0,B}$, where N_A is the Avogadro constant. Step 2 is the reverse of dissociation of the products, and $\Delta U_2 = -eN_A D_{0,E} - fN_A D_{0,F}$. We have $\Delta U_0° = \Delta U_1 + \Delta U_2 = bN_A D_{0,B} + cN_A D_{0,C} - eN_A D_{0,E} - fN_A D_{0,F}$ and

$$\Delta U_0° = -N_A \, \Delta D_0 \qquad \text{where } \Delta D_0 \equiv \sum_i \nu_i D_{0,i} \qquad \text{gas-phase reaction} \qquad (22.121)$$

For example, for $CH_4(g) + 2F_2(g) \to CH_2F_2(g) + 2HF(g)$, we have $\Delta U_0°/N_A = D_0(CH_4) + 2D_0(F_2) - D_0(CH_2F_2) - 2D_0(HF)$. To calculate $\Delta U_T°$ for a gas-phase reaction from statistical mechanics using (22.119), we need the dissociation energies D_0 of the species, which are found spectroscopically (Sec. 21.11).

Standard Enthalpies of Formation

$\Delta_f H_T°$ of a gaseous compound is readily calculated from $\Delta_f U_T°$ using $\Delta H° = \Delta U° + (\Delta n/\text{mol})RT$. If the elements forming the compound are all gaseous at T and 1 bar, then we can use statistical mechanics to calculate $U_{m,T}° - U_{m,0}°$ for each element and for the compound; we calculate $\Delta_f U_0°$ using (22.121) and then find $\Delta_f U_T°$ from (22.119).

If one or more of the elements are not gaseous, additional data are needed to find $\Delta_f U_T°$. For example, suppose we want $\Delta_f U_T°$ of $CH_4(g)$, according to C(graphite) + $2H_2(g) \to CH_4(g)$. For the solid graphite, $U_T° - U_0°$ is more accurately found from experimental heat-capacity data than from statistical-mechanical calculations. Since $(\partial H/\partial T)_P = C_P$, we have $H_{m,T}° - H_{m,0}° = \int_0^T C_{P,m}° \, dT$. Also, since step 1 in (22.120) involves vaporization of solid graphite to C(g), instead of $N_A D_0$ we use $\Delta U_{m,0}°$ of vaporization of graphite.

The quantity $H_{m,T}° - H_{m,0}°$, which is often given in thermodynamics tables, is readily calculated for a gas by statistical mechanics. We have $H_{m,T}° = U_{m,T}° + RT$ and $H_{m,0}° = U_{m,0}°$ for a gas-phase compound, so $H_{m,T}° - H_{m,0}° = U_{m,T}° - U_{m,0}° + RT$, where $H_{m,0}°$ and $U_{m,0}°$ are for the hypothetical gas-phase compound at 0 K.

Standard Entropies

In Sec. 22.9 we shall show that the convention of setting $C = 0$ in (22.32) for S agrees with the entropy convention used in thermodynamics (Sec. 5.7). Therefore, the thermodynamic $S°$ of any gas is found by setting $P = 1$ bar in S_{tr} and using (22.64).

Standard Heat Capacities

The equations in Secs. 22.6 and 22.7 give C_V for ideal gases. The ideal-gas equation $C_{P,m} = C_{V,m} + R$ then gives $C_{P,m}$.

Standard Gibbs Energies of Formation

We have $\Delta_f G_T° = \Delta_f H_T° - T \Delta_f S_T°$. Calculation of $\Delta_f H_T°$ is discussed above. $\Delta_f S_T°$ is found from the entropies of the compound and of its elements. If these are all gases, their entropies can all be calculated by statistical mechanics. For solid and liquid elements, the experimental S must be used.

The Quantity $G - H_0$

Thermodynamics tables often tabulate $(G_{m,T}° - H_{m,0}°)/T$ versus T (Sec. 5.9). We now relate this quantity to the molecular partition function of an ideal gas.

Since it is $U - U_0$ (and not U) that is calculated from the partition function, the relation $A \equiv U - TS = -kT \ln Z$ [Eq. (22.40)] must be modified to

$$A - U_0 = -kT \ln Z \tag{22.122}$$

to reflect our choice of zero energy at the lowest available molecular energy level. For a pure ideal gas, $Z = z^N/N!$ and

$$\ln Z = N \ln z - \ln N! = N \ln z - N \ln N + N \qquad \text{pure ideal gas}$$

Substitution for $\ln Z$ in (22.122) and use of $Nk = nR$ give

$$A - U_0 = -nRT \ln z + nRT \ln N - nRT$$

$$A_m - U_{m,0} = -RT \ln (z/N) - RT \qquad \text{pure ideal gas}$$

We have $G_m = A_m + PV_m = A_m + RT$ for an ideal gas, so

$$G_{m,T} - U_{m,0} = -RT \ln (z/N) \qquad \text{pure ideal gas} \tag{22.123}$$

Since $H_0 = U_0$ for an ideal gas, (22.123) allows us to calculate $(G_{m,T}° - H_{m,0}°)/T$ by setting $P = 1$ bar in z of the ideal gas.

Equilibrium Constants

For the ideal-gas reaction $0 \rightarrow \Sigma_i \nu_i B_i$, we have $\Delta G° = -RT \ln K_P°$. To calculate $K_P°$ using statistical mechanics, one calculates $G_{m,T}° - H_{m,0}°$ of each gas using (22.123), calculates $\Delta G_T°$ using $\Delta G_T° = \Sigma_i \nu_i (G_{m,T,i}° - H_{m,0,i}°) + \Delta H_0°$ [where $\Delta H_0°$ is found from (22.121)], and then calculates $K_P°$ from $\Delta G_T°$.

For theoretical discussions, it is useful to express the equilibrium constant in terms of the species' partition functions. From Sec. 6.1, the chemical potential of component i of an ideal gas mixture is given by $\mu_i = G_{m,i}^*(T, P_i)$, where P_i is the partial pressure of i in the mixture and the star denotes pure i. Since $P_i V = n_i RT$, if pure i is at temperature T and pressure P_i, its volume equals the mixture volume V. Thus, $\mu_i = G_{m,i}^*(T, V)$, where V is the volume of the gas mixture. For pure gas i, Eq. (22.123) gives $G_{m,T,i}^* = U_{m,0,i} - RT \ln (z_i/N_i)$, so

$$\mu_i = U_{m,0,i} - RT \ln (z_i/N_i) \qquad \text{ideal gas mixture} \tag{22.124}$$

where z_i is evaluated at the T and V of the mixture.

Consider the ideal-gas reaction $bB \rightleftharpoons eE$. Substitution of (22.124) for each μ into the equilibrium condition $\Sigma_i \nu_i \mu_i = 0$ gives

$$0 = -b\mu_B + e\mu_E = eU_{m,0,E} - bU_{m,0,B} - RT[e \ln (z_E/N_E) - b \ln (z_B/N_B)]$$

$$= \Delta U_0^{\circ} - RT \ln \frac{(z_E/N_E)^e}{(z_B/N_B)^b} \qquad (22.125)$$

We have

$$\frac{z_E}{N_E} = \frac{z_E/VN_A}{N_E/VN_A} = \frac{z_E/VN_A}{n_E/V} = \frac{z_E/VN_A}{c_E} \qquad (22.126)$$

where N_A is the Avogadro constant, n_E is the equilibrium number of moles of E, and $c_E = n_E/V$ is the equilibrium concentration of E. Using (22.126) and a similar equation for z_B/N_B in (22.125), we have

$$\frac{\Delta U_0^{\circ}}{RT} = \ln \left[\frac{(z_E/VN_A)^e}{(z_B/VN_A)^b} \frac{c_B^b}{c_E^e} \right] \quad \text{and} \quad \exp (\Delta U_0^{\circ}/RT) = \frac{(z_E/VN_A)^e}{(z_B/VN_A)^b} \frac{c_B^b}{c_E^e}$$

$$K_c \equiv \frac{c_E^e}{c_B^b} = e^{-\Delta U_0^{\circ}/RT} \frac{(z_E/VN_A)^e}{(z_B/VN_A)^b} \quad \text{ideal gases} \qquad (22.127)$$

where K_c is the concentration-scale equilibrium constant with units of $(\text{mol/L})^{e-b}$. The quantity ΔU_0° is calculated from (22.121).

Generalizing to the reaction $0 \rightarrow \Sigma_i \nu_i A_i$, one finds (Prob. 22.73)

$$K_c \equiv \prod_i (c_i)^{\nu_i} = \exp (-\Delta U_0^{\circ}/RT) \prod_i \left(\frac{z_i}{VN_A} \right)^{\nu_i} \quad \text{ideal gases} \qquad (22.128)$$

Equations (22.58) and (22.81) show that z_i is proportional to V. Therefore, z_i/V is independent of V and is a function of T only, as is K_c. The exponential factor in (22.128) corrects for the use of different zero levels of energy in z of each species. The VN_A converts from numbers of molecules N_i to concentrations c_i.

Equation (22.128) is actually an example of the Boltzmann distribution law. To see this, consider the special case of the isomerization reaction $B \rightleftharpoons D$. The Boltzmann distribution law (22.74) gives the average number of molecules in quantum state r of species B in an ideal gas mixture at equilibrium as $\langle N_{B,r} \rangle = N_B e^{-\beta \varepsilon_{B,r}}/z_B$, where N_B is the total number of B molecules. Let r be the ground state (state 0) of B. By convention, in calculating z_B we set the zero level of energy at the ground state of B, so that $\varepsilon_{B,0} = 0$. Therefore $\langle N_{B,0} \rangle = N_B/z_B$, and $N_B = \langle N_{B,0} \rangle z_B$. Similarly, $N_D = \langle N_{D,0} \rangle z_D$, where z_D is calculated taking $\varepsilon_{D,0} = 0$. Thus

$$\frac{N_D}{N_B} = \frac{\langle N_{D,0} \rangle z_D}{\langle N_{B,0} \rangle z_B}$$

But the Boltzmann distribution gives $\langle N_{D,0} \rangle / \langle N_{B,0} \rangle = \exp [-\beta(\varepsilon_{D,0} - \varepsilon_{B,0})] = \exp (-\Delta \varepsilon_0/kT) = \exp (-\Delta U_0^{\circ}/RT)$; therefore

$$\frac{N_D}{N_B} = \exp (-\Delta U_0^{\circ}/RT)\frac{z_D}{z_B} \quad \text{or} \quad \frac{c_D}{c_B} = \exp (-\Delta U_0^{\circ}/RT) \frac{z_D}{VN_A} \frac{VN_A}{z_B} \qquad (21.129)$$

which is (22.128). An extension of these arguments shows that for the general reaction $0 \rightleftharpoons \Sigma_i \nu_i B_i$, the Boltzmann distribution law yields (22.128). (The derivation for the $B + C \rightleftharpoons D$ case is given in *Knox*, sec. 11.2.)

The factor $\exp (-\Delta U_0^{\circ}/RT)$ in (22.129) is an energy (or enthalpy) factor that favors the species with the lower ground-state ε_0. The z_D/z_B partition-function factor in

(22.129) is an entropy factor (since z and S both increase as the number of significantly populated states increases) that favors the species with the greater number of thermally accessible states. For example, in Fig. 22.19, species B has a lower ε_0 and species D has a greater number of low-lying states. At low T, only the very lowest levels are significantly occupied, and the $\exp(-\Delta U_0^\circ/RT)$ factor makes $N_B > N_D$. At high T, the greater number of accessible states for D is more important than the energy factor, and we have $N_D > N_B$. (Recall a similar discussion in Sec. 6.2.)

22.9 ENTROPY AND THE THIRD LAW OF THERMODYNAMICS

We begin this section by deriving another statistical-mechanical formula for the entropy. Equation (22.30) reads $p(E_i) = W_i e^{-E_i/kT}/Z$, where $p(E_i)$ is the probability the thermodynamic system has quantum energy E_i and W_i is the degeneracy of level E_i (the number of system quantum states that have energy E_i). The total probability is 1, so $1 = \Sigma_{\text{levels}}\, p(E_i) = \Sigma_{\text{levels}} W_i e^{-E_i/kT}/Z$, where the sum goes over all the possible quantum energy levels E_i. As shown in Fig. 22.3, $p(E_i)$ is essentially zero unless E_i is extremely close to the thermodynamic internal energy U. Therefore we need include in the sum only those terms where E_i is very close to U. We have $1 \approx \Sigma'_{\text{levels}} W_i e^{-U/kT}/Z$ and $Z \approx e^{-U/kT} \Sigma'_{\text{levels}} W_i$, where the prime indicates that the sum goes only over levels for which $p(E_i)$ in Fig. 22.3 is significantly different from zero. The quantity $\Sigma'_{\text{levels}} W_i$ equals the total number of quantum states in the narrow band of energy levels around U for which there is a significant probability of finding the system. Let $W \equiv \Sigma'_{\text{levels}} W_i$. Then $Z \approx We^{-U/kT}$ and $\ln Z = \ln W - U/kT$. (A detailed investigation shows that this approximation is not accurate for Z but is extremely accurate for $\ln Z$, and it is $\ln Z$ which is used to calculate thermodynamic functions.) Equation (22.39) becomes $S = U/T + k \ln Z = U/T + k \ln W + k(-U/kT) = k \ln W$. This is the desired result:

$$S = k \ln W \qquad (22.130)$$

The entropy of a thermodynamic system is proportional to the log of W, where W is the total number of microstates for system energy levels that have a significant probability of being occupied. As the number of microstates available to the system increases, the system's entropy increases. (Recall the relation between entropy and spread over energy levels.) Equation (22.130) is **Boltzmann's principle** and is carved on his tombstone in Vienna. This equation is a more explicit formulation of Eq. (3.52). We were rather vague about the meaning of the probability in (3.52). Equation (22.130) is not as useful for practical calculations as the formula $S = U/T + k \ln Z$.

Let us estimate the magnitude of W. The entropy of a thermodynamic system containing 1 mole is on the order of R (where R is the gas constant). Equation (22.130) then gives the order of magnitude of W as

$$W = e^{S/k} \approx e^{R/k} = e^{N_A} \approx e^{(10^{23})} \approx 10^{(10^{22})} \approx 10^{10,000,000,000,000,000,000,000}$$

Recall from Fig. 22.3 that the degeneracy W is a very sharply increasing function of the system's quantum energy.

We now examine the conventions used for entropy. The statistical-mechanical result (22.32) reads $S = U/T + k \ln Z + C$, and we set the integration constant C equal to zero for all systems. In Sec. 5.7, we adopted the thermodynamic convention that $\lim_{T\to 0} S = 0$ for each element. This convention plus the third law of thermodynamics led to the result $\lim_{T\to 0} S = 0$ for every pure substance in internal equilibrium. We must now see whether the statistical-mechanical convention ($C = 0$) is consistent with the thermodynamic convention ($S_0 = 0$ for elements).

As $T \to 0$, all the systems in a canonical ensemble fall into the system's lowest available quantum energy level, so only this energy level is populated in the ensemble,

Figure 22.19

Low-energy states for two species, B and D. B has a lower-energy ground state. D has a greater number of low-energy states.

and W in Eq. (22.130) becomes W_0, where W_0 is the degeneracy of the lowest quantum energy level of the system. Equation (22.130) then gives

$$\lim_{T \to 0} S = k \ln W_0 \qquad (22.131)$$

For this statistical-mechanical result to be consistent with the thermodynamic result $\lim_{T \to 0} S = 0$ for a pure substance, the degeneracy W_0 of the ground level of a one-component system would have to be 1. Actually, W_0 is not 1 for a pure substance, for two reasons:

1. Each atomic nucleus has spin quantum numbers I and M_I (Sec. 21.12). I is a constant for a given nucleus, but M_I takes on the $2I + 1$ values from $-I$ to $+I$. In the absence of an external magnetic field, the $2I + 1$ states corresponding to different M_I values have the same energy [note that $E = 0$ for $B = 0$ in (21.65)]. Thus there is a nuclear-spin degeneracy of $2I + 1$ for each atom. The M_I quantum numbers of the atoms vary independently of one another, so the total ground-level nuclear-spin degeneracy is the product $\Pi_a (2I_a + 1)$, where I_a is the nuclear spin of atom a and the product goes over all atoms. For example, for a crystal containing N atoms each with $I = 1$, we get a nuclear-spin contribution to W_0 of 3^N and a contribution to S_0 of $k \ln 3^N = Nk \ln 3$.

2. When chemists talk of pure ClF, they mean a mixture of 75.8% $^{35}Cl^{19}F$ and 24.2% $^{37}Cl^{19}F$. These isotopic species are distinguishable from each other, and there is an entropy of mixing associated with mixing the pure isotopic species to get the naturally occurring isotopic mixture. Since isotopic species form very nearly ideal solutions, $\Delta_{mix}S$ is given by (9.46). In a crystal of naturally occurring ClF, there is a degeneracy associated with the various permutations of the ^{35}ClF and ^{37}ClF molecules among different locations in the crystal. Because there is a slight difference in intermolecular interactions for different isotopic species, at temperatures very close to absolute zero the true thermodynamic equilibrium state will be one with separate crystals of ^{35}ClF and ^{37}ClF, rather than a single solid solution of ^{35}ClF and ^{37}ClF. However, at such extremely low temperatures, the molecules do not have enough energy to overcome the energy barrier required for the unmixing and the crystal remains in a metastable state at low T with a frozen-in entropy of mixing.

To make the statistical-mechanical and thermodynamic entropies agree, chemists have adopted the convention of ignoring contributions to S made by nuclear spins and by isotopic mixing. Recall that we didn't sum over different nuclear-spin quantum states in calculating partition functions in Secs. 22.6 and 22.7, and we ignored the entropy of isotopic mixing. There is no harm in using these conventions because nuclear spins are not changed in chemical reactions and the amount of isotopic fractionation that occurs in chemical reactions is ordinarily negligible. With these conventions, we can expect W_0 to be 1 for a perfect crystal of a pure substance, so that $\lim_{T \to 0} S = k \ln 1 = 0$ for the statistical-mechanical entropy of a pure substance at absolute zero, in agreement with the thermodynamic result.

The traditional argument just given that uses (22.131) to justify the third law turns out to be incorrect. A successful theory of crystals is the Debye theory (Sec. 24.12). In the Debye theory, the spacings between vibrational levels of the solid are such that it is only for temperatures far below 10^{-6} K that (22.131) is a valid approximation to S. Moreover, the third law applies to gases, and it is only for temperatures far below 10^{-15} K that (22.131) is a valid approximation for gases. However, the experimental validity of the third law rests on data at temperatures on the order of 1 K, where (22.131) is irrelevant. A detailed analysis shows that the behavior of S at experimentally accessible low temperatures depends not at all on the degeneracy of the ground level of the system but on the density of states for low energies, where the *density of states* is the number of states per unit interval of energy. For most of the reasonable forms of the density of states, one finds that S approaches 0 as $T \to 0$. For details, see D. ter Haar, *Elements of Thermostatistics,* 2d ed., Holt, Rinehart, and Winston, 1966, chap. 9; R. B. Griffiths in E. B. Stuart et al. (eds), *A Critical Review of Thermodynamics,* Mono Book Corp., 1970, pp. 101–106; S. Mafé et al., *Am. J. Phys.,* **68,** 932 (2000).

Statistical-mechanical entropies are not absolute entropies, since they ignore the contributions of nuclear spins and isotopic mixing. Even if these contributions were included, we still would not have absolute entropies, since the statistical-mechanical formulas for S are based on the arbitrary convention of setting $C = 0$ in (22.32).

Comparison of gas-phase calorimetrically determined thermodynamic (td) entropies (Sec. 5.7) with statistical-mechanical (sm) entropies calculated from spectroscopic data (Secs. 22.6 and 22.7) shows that for most molecules, S_{td} and S_{sm} agree within the limits of experimental error. However, for several gases (for example, CO, N_2O, NO, H_2O, CH_3D, and H_2), S_{td} and S_{sm} at 298 K disagree by several J/mol-K.

To see the reason for these discrepancies, consider CO. There are two possible orientations for each CO molecule in the crystal (either CO or OC). The dipole moment of CO is very small (0.1 D), and so the difference in energy $\Delta\varepsilon$ for these two orientations is very small. When the crystal is formed at the CO normal melting point (66 K), $\Delta\varepsilon/kT$ is very small and the Boltzmann factor $e^{-\Delta\varepsilon/kT} \approx e^0 = 1$. Therefore the crystal is formed with approximately equal numbers of CO molecules with each orientation. As T is decreased toward zero, $\Delta\varepsilon/kT$ becomes large and $e^{-\Delta\varepsilon/kT} \to e^{-\infty} = 0$. Hence, if thermodynamic equilibrium were maintained, all the CO molecules would adopt the orientation with the lower energy. However, to have the incorrectly oriented CO molecules rotate 180° in the crystal requires a substantial activation energy, which is not available to the molecules at low T. The CO molecules remain locked into their nearly random orientations as T is lowered. The thermodynamic determination of S from observed C_P values is based on the third law, which applies only to systems in equilibrium. A CO crystal at 10 or 15 K is not in true thermodynamic equilibrium, and the calorimetrically measured entropy S_{td} is therefore in error. The correct entropy is that calculated by statistical mechanics; it is S_{sm}° that is listed in tables of thermodynamic properties.

The discrepancies for N_2O, NO, and CH_3D arise from the same reason as for CO. The discrepancy for H_2O is due to a randomness in the positions of H atoms in hydrogen bonds. H_2 is a special case; see *McClelland,* sec. 8.4.

Although the entropy of a thermodynamic system is calculable from molecular properties, entropy is not a molecular property. Entropy has meaning only for a large collection of molecules. Individual molecules do not have entropy.

22.10   INTERMOLECULAR FORCES

The statistical-mechanical formulas of Sec. 22.2 were applied to ideal gases (which have no intermolecular forces) in Secs. 22.3 to 22.8. Before treating systems with intermolecular forces, we discuss the nature of the forces between nonreacting molecules. Interactions between chemically reacting molecules are considered in Chapter 23.

In solids, liquids, and nonideal gases, the Hamiltonian of the thermodynamic system contains the potential energy $\mathcal{V}$ of interaction between the molecules. $\mathcal{V}$ is a function of the distances between the molecules and for polar molecules (and nonspherical nonpolar molecules) is also a function of the molecular orientations in space.

In most statistical-mechanical treatments, the force between molecules 1 and 2 is assumed to be unaffected by the nearby presence of a third molecule. This is an approximation because molecule 3 will polarize molecules 1 and 2 (Sec. 14.15), thereby changing the 1-2 intermolecular force. In gases not at high densities, there is little probability that three gas molecules will be simultaneously close together. Ignoring three-body forces is a very good approximation for low- and medium-density gases but not so good for solids, liquids, and high-density gases.

In dealing with gases and liquids, one often makes the simplifying approximation of averaging over the orientation dependence of $\mathcal{V}$, converting $\mathcal{V}$ into a function solely of the distances between molecules. This approximation is inapplicable to solids composed of polar molecules, since the molecules are held in fixed orientations in the solid.

With three-body forces neglected, $\mathcal{V}$ is a sum of pairwise potential energies of interaction v_{ij} between molecules i and j. With averaging over the orientation dependence, v_{ij} depends only on the distance r_{ij} between the centers of i and j; $v_{ij} = v_{ij}(r_{ij})$. For example, for a system with three molecules, $\mathcal{V} \approx v_{12}(r_{12}) + v_{13}(r_{13}) + v_{23}(r_{23})$. If the molecules are identical, v_{12}, v_{13}, and v_{23} are the same function. For a system of N molecules

$$\mathcal{V} \approx \sum_{i=1}^{N} \sum_{j>i} v_{ij}(r_{ij}) \tag{22.132}$$

The two-molecule potential energy v_{ij} could be calculated in principle by solving the Schrödinger equation for a system consisting of molecules i and j and averaging the result over all orientations. Such a calculation is extremely hard. Instead of quantum mechanics, one usually uses an approximate model of a molecule as an electric dipole having a certain electric polarizability (Sec. 14.15), and one uses classical electrostatics to calculate v_{ij}. Intermolecular forces between nonreacting molecules are a lot weaker than chemical-bonding forces and therefore have only small effects on the internal structures of the molecules. Thus we can get a fairly accurate representation of intermolecular interactions without dealing with the detailed structures of the molecules. An exception is hydrogen bonding.

The force $F(r)$ and potential energy $v(r)$ between two bodies are related by Eq. (2.21) as $F(r) = -dv(r)/dr$. For two ions, $F(r)$ is given by Coulomb's law (14.1), and $v(r) = Z_1 Z_2 e^2 / 4\pi\varepsilon_0 r$ [Eq. (19.1)]. Forces between ions occur in electrolyte solutions, ionic solids, and molten salts. This section considers only neutral molecules.

If molecules 1 and 2 have permanent electric dipole moments μ_1 and μ_2, the force one molecule exerts on the other will depend on μ_1, μ_2, the separation r, and the relative orientation of the two dipoles. Since we are ignoring the orientation dependence of v, we must average v over all orientations. If the two molecular dipoles were oriented completely randomly with respect to each other, their average interaction energy would be zero, because repulsive orientations would occur just as often as attractive orientations. However, the Boltzmann factor $e^{-v/kT}$ favors attractive orientations (which have lower energies) over repulsive ones. The orientation-averaged potential energy of two dipoles calculated with allowance for the Boltzmann distribution turns out to be

$$v_{\text{d-d}}(r) = -\frac{2}{3kT} \frac{\mu_1^2 \mu_2^2}{4\pi\varepsilon_0 r^6} \tag{22.133}$$

where μ_1 and μ_2 are in SI units (C m). As $T \to \infty$, the $e^{-\Delta\varepsilon/kT}$ Boltzmann factor goes to 1, all orientations become equally likely, and $v_{\text{d-d}} \to 0$. The term (22.133) is the **dipole–dipole** contribution to v. [For derivations of (22.133) to (22.135), see *Hirschfelder, Curtiss, and Bird*, secs. 13.3 and 13.5.]

Besides $v_{\text{d-d}}$ other interactions contribute to the intermolecular potential v. The permanent dipole moment of one molecule will induce a dipole moment in a second molecule (whether or not the second molecule has a permanent dipole moment); see Fig. 22.20. The (attractive) interaction between the permanent moment of one molecule and

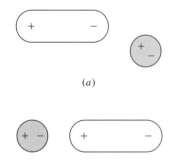

(a)

(b)

Figure 22.20

A molecule with a permanent dipole moment induces a dipole moment in a nearby molecule. Note that the interaction is attractive in both cases.

the induced moment of the second gives the **dipole–induced-dipole** contribution to v. This turns out to be

$$v_{\text{d-id}} = -\frac{\mu_1^2\alpha_2 + \mu_2^2\alpha_1}{(4\pi\varepsilon_0)^2 r^6} \tag{22.134}$$

where α_1 and α_2 are the polarizabilities of molecules 1 and 2 in SI units. Note the absence of kT in (22.134). The induced dipoles are born oriented, as noted in Sec. 14.15.

Even if neither molecule has a permanent dipole moment, there will still be an attractive force between the molecules. This must be so; otherwise, gases like He or N_2 would not condense to liquids. The electrons (and to a lesser extent the nuclei) are in continual motion within a molecule. The permanent dipole moment μ is calculated using the average locations of the charges. If μ is zero, the time-average charge distribution must be completely symmetric. However, the charge distribution at any instant in time need not be. For example, at a given instant, both electrons in a helium atom might be on the same side of the nucleus. The instantaneous dipole of a molecule induces a dipole in a nearby molecule. The interaction between the instantaneous dipole moment and the induced dipole moment produces a net attraction, whose form was calculated by London in 1930 using quantum mechanics. This **London** or **dispersion energy** is (approximately)

$$v_{\text{disp}} \approx -\frac{3I_1 I_2}{2(I_1 + I_2)}\frac{\alpha_1\alpha_2}{(4\pi\varepsilon_0)^2 r^6} \tag{22.135}$$

where I_1 and I_2 are the ionization energies of molecules 1 and 2. The order of magnitude of I is 10 eV for most molecules.

The net long-range attractive potential energy for two neutral molecules is the sum $v_{\text{d-d}} + v_{\text{d-id}} + v_{\text{disp}}$. Each term is proportional to $1/r^6$. The attractive force (called the **van der Waals force**) that results from these three effects is proportional to $1/r^7$ and falls off much faster with distance than a Coulomb's law force ($1/r^2$). Table 22.1 gives minus the coefficient of $1/r^6$ in (22.133) to (22.135) for several pure substances at 25°C.

Note that, *except for small, highly polar molecules* (for example, H_2O and HCN), *the dominant term in the van der Waals attraction is the dispersion energy*. Also, $v_{\text{d-id}}$ is always quite small. The molecular polarizability (which determines v_{disp}) is due

TABLE 22.1

Contributions to the Intermolecular Potential Energy of Like Molecules at 25°C

Molecule	μ/D	$(\alpha/4\pi\varepsilon_0)/\text{Å}^3$	I/eV	$-10^{79}vr^6/(\text{J m}^6)$		
				Dipole–dipole	Dipole–induced dipole	Dispersion
Ar	0	1.63	15.8	0	0	50
N_2	0	1.76	15.6	0	0	58
C_6H_6	0	9.89	9.2	0	0	1086
C_3H_8	0.08	6.29	11.1	0.0008	0.09	528
HCl	1.08	2.63	12.7	22	6	106
CH_2Cl_2	1.60	6.48	11.3	106	33	570
SO_2	1.63	3.72	12.3	114	20	205
H_2O	1.85	1.59	12.6	190	11	38
HCN	2.98	2.59	13.8	1277	46	111

mainly to the valence electrons and increases with increasing molecular size. Some $\alpha/4\pi\varepsilon_0$ values are:

Molecule	F_2	Cl_2	Br_2	I_2	CH_4	C_2H_6	C_3H_8
$(\alpha/4\pi\varepsilon_0)/\text{Å}^3$	1.3	4.6	6.7	10.2	2.6	4.5	6.3

The outer electrons in I_2 are farther from the nuclei than in Br_2 and so are more easily distorted. The increase in α increases v_{disp} in (22.135), so F_2 and Cl_2 are gases at room temperature, Br_2 is a liquid, and I_2 is a solid. The increase in boiling point as n increases in the series C_nH_{2n+2} is due to the increase in α and v_{disp}.

The molecular polarizability can be estimated as the sum of bond polarizabilities.

Besides the terms (22.133) to (22.135), another important intermolecular attraction is the **hydrogen bond.** This is an attraction between an electronegative atom in one molecule and a hydrogen atom bound to an electronegative atom in a second molecule. The electronegative atoms involved are F, O, N, and, to a lesser extent, Cl and S. Species showing substantial hydrogen bonding include HF, H_2O, NH_3, CH_3OH, CH_3NH_2, and CH_3COOH. In water, each OH bond is highly polar, and the small, positive H of one H_2O molecule is strongly attracted to the negative O of a nearby H_2O molecule, as symbolized by $HOH \cdots OH_2$. The magnitude of the hydrogen-bond energy (2 to 10 kcal/mol at the distance of lowest potential energy) is significantly greater than that of the energies (22.133) to (22.135) (which run 0.1 to 2 kcal/mol for molecules that are not large) but substantially less than the energy of an intramolecular covalent bond (30 to 230 kcal/mol, Sec. 20.1). The relatively high melting and boiling points of H_2O and NH_3 are due to hydrogen bonding. Hydrogen bonds hold together the paired bases in DNA, are responsible for binding most substrates to enzymes, and help determine the conformations of proteins.

Because of its occurrence in such key species as water, proteins, and DNA, the hydrogen bond has been intensively studied. A full quantum-mechanical understanding of the hydrogen bond has not yet been achieved. The hydrogen bond cannot be explained solely as an electrostatic attraction but has a significant amount of covalent character due to the sharing of lone-pair electrons on N, O, or F with H; see E. D. Isaacs et al., *Phys. Rev. Lett.,* **82,** 600 (1999). [For further discussion, see G. A. Jeffrey, *An Introduction to Hydrogen Bonding,* Oxford, 1997; S. Scheiner, *Hydrogen Bonding,* Oxford, 1997; D. Hadzi (ed.), *Theoretical Treatments of Hydrogen Bonding,* Wiley, 1997.] The hydrogen bond is a relatively strong, highly directional interaction, and we shall assume the absence of hydrogen bonding in the discussion below.

So far we have considered the forces that occur at relatively large separations between molecules. At small intermolecular distances, the molecules exert strong repulsions on each other, mainly because of the Pauli-exclusion-principle repulsion between the overlapping electron probability densities (Sec. 20.4). The very steep short-range repulsion can be crudely approximated by an inverse power of r; $v_{\text{rep}} \approx A/r^n$, where A is a positive constant and n is a large integer (8 to 18). The relative incompressibility of liquids and solids is due to v_{rep}. To experience v_{rep}, bang your fist on a table.

In addition, there are intermediate-range quantum-mechanical interactions whose calculation is extremely difficult.

If the intermolecular potential energy v is approximated as the sum of the short-range (v_{rep}) and long-range $(v_{\text{d-d}} + v_{\text{d-id}} + v_{\text{disp}})$ potentials, we get a function with the form $v = A/r^n - B/r^6$. Perhaps the most widely used intermolecular potential is the **Lennard-Jones 6-12 potential** (Prob. 22.81)

$$v = 4\varepsilon\left[\left(\frac{\sigma}{r}\right)^{12} - \left(\frac{\sigma}{r}\right)^6\right] \tag{22.136}$$

where ε is the depth of the minimum in the potential and σ is the intermolecular distance at which $v = 0$ (Fig. 22.21a). For $r < \sigma$, the $1/r^{12}$ term dominates, and v

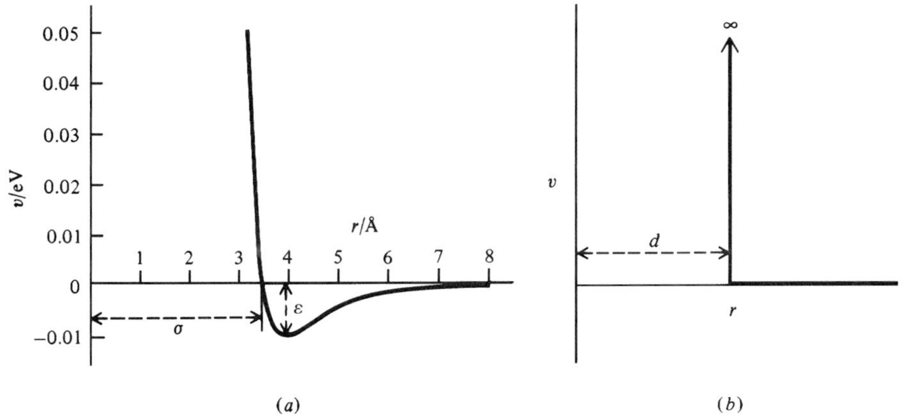

Figure 22.21

(a) The Lennard-Jones intermolecular potential for Ar. (b) The hard-sphere intermolecular potential.

increases steeply as r decreases. For $r > \sigma$, the $-1/r^6$ term dominates, and v decreases as r decreases. The parameter σ is an approximation to the sum of the average radii of the two molecules. Note the resemblance to Fig. 21.9, which is an *intra*molecular potential. The Lennard-Jones potential is not too bad but is not an accurate representation of intermolecular interactions, even for rare-gas atoms. (The physicist John E. Jones changed his surname to Lennard-Jones when he married Kathleen Mary Lennard in 1925.)

Note from Fig. 22.21 that intermolecular forces between neutral molecules are short-range. The attractive energy between two Ar atoms is nearly zero at a center-to-center distance of 8 Å, which is a bit more than twice the diameter of an Ar atom. At $r = 2.5\sigma$, Eq. (22.136) gives $v = -0.016\varepsilon$, and the intermolecular potential energy is a mere $1\frac{1}{2}$ percent of its value at the bottom of the curve.

The crudest approximation to intermolecular potentials is the **hard-sphere potential** function (Fig. 22.21b)

$$v = \begin{cases} 0 & \text{for } r \geq d \\ \infty & \text{for } r < d \end{cases} \tag{22.137}$$

where d is the sum of the radii of the colliding molecules, considered to be infinitely hard "billiard balls." The hard-sphere potential is often used in statistical mechanics, not because it accurately represents intermolecular forces, but because it greatly simplifies calculations.

We shall see in Sec. 22.11 that the second virial coefficient $B(T)$ in the virial equation of state (8.4) for gases can be expressed in terms of the intermolecular potential. If v is taken to be a Lennard-Jones function, the best fit to experimental $B(T)$ data for several gases is obtained with the ε and σ values in Table 22.2. The d_{vis} values are hard-sphere diameters calculated from gas viscosities using the kinetic-theory equation

TABLE 22.2

Lennard-Jones Parameters

Molecule	d_{vis}/Å	σ/Å	ε/eV	ε/kT_b
Ar	3.7	3.50	0.0101	1.3
Xe	4.9	4.10	0.0192	1.3
CH_4	4.1	3.78	0.0128	1.3
CO_2	4.6	4.33	0.0171	
C_6H_6	7.4	8.57	0.0209	0.7

(16.25) and are in rough agreement with the Lennard-Jones σ values. T_b is the normal boiling point. Interestingly, $\varepsilon \approx 1.3kT_b$ for the nonpolar, spherical molecules Ar, Xe, and CH_4, but this relation doesn't hold for C_6H_6, where v depends on the relative molecular orientation and the Lennard-Jones potential is a poor approximation. When the Lennard-Jones ε is equal to $\frac{3}{2}kT = 1.5kT$, the mean molecular translational energy equals the maximum intermolecular attraction energy, and so we should expect ε to correlate with the boiling point.

Note that ε is far less than the dissociation energies D_0 of diatomic molecules, which run 1 to 10 eV. The van der Waals attraction between two molecules is far weaker than chemical-bonding forces. Also, σ (which is 3 to 6 Å for molecules that are not very large) is substantially greater than chemical-bond distances (1 to 3 Å). Hydrogen bonds are 2 to 3 Å for the distance of minimum potential energy.

Equations (22.133) and (22.134) for $v_{\text{d-d}}$ and $v_{\text{d-id}}$ treat the molecule as an electric dipole, and their derivation is based on Eq. (14.81) for the electric potential of a dipole. Since (14.81) applies only at large distances from a system of charges, Eqs. (22.133) and (22.134) are not accurate for molecules that approach each other closely. These inaccuracies are partly compensated for by the use of empirically adjusted values of σ and ε in the Lennard-Jones potential. Note also that the orientation-averaged $v_{\text{d-d}}$ in (22.133) is temperature-dependent, but this is ignored in the Lennard-Jones potential.

For polar molecules, a substantial improvement on the Lennard-Jones potential is obtained by dropping the approximation that v depends only on the intermolecular distance r. A commonly used orientation-dependent potential is the *Stockmayer potential,* which includes a term depending on the angles defining the relative orientation of the two dipoles (see *Hirschfelder, Curtiss, and Bird,* pp. 35, 211–222).

The dispersion attractions in Ar_2 are strong enough for a small concentration of Ar_2 molecules to exist in argon gas at temperatures below 100 K. The Ar_2 dissociation energy and internuclear distance are $D_e = 0.012$ eV and $R_e = 3.76$ Å, and Ar_2 has several bound vibrational levels existing in the potential of Fig. 22.21a. Van der Waals molecules detected by spectroscopic methods include Ne_2, Ar_2, Kr_2, Xe_2, $Ar–N_2$, $(O_2)_2$, Mg_2, and $Ar–HCl$. One of the van der Waals molecules $NO–O_2$ and $(NO)_2$ might be a reaction intermediate in the $NO + O_2$ reaction; recall mechanisms (17.60) and (17.61).

The He_2 van der Waals molecule is the most weakly bound diatomic molecule. Because its zero-point vibrational energy is only slightly smaller than its equilibrium dissociation energy D_e, only the $v = 0$, $J = 0$ level exists. Ab initio calculations and experimental data show that $D_e = 0.00094$ eV for He_2 and that $D_0 \approx 0.0000001_0$ eV. Although the calculated minimum of the ground-state He_2 potential-energy curve is at $R_e = 2.97$ Å, the fact that the $v = 0$ level lies just below the dissociation limit makes the average internuclear distance in He_2 an astounding 55 Å [F. Luo et al., *J. Chem. Phys.,* **98**, 9687 (1993); **99**, 10084 (1993); J. B. Anderson et. al., *J. Chem. Phys.,* **99**, 345 (1993)]. He_2 was first detected in 1992 in a jet-cooled molecular beam at 0.001 K [see F. Luo et al., *J. Chem. Phys.,* **98**, 3564 (1993); **100**, 4023 (1994)].

For more on van der Waals molecules, see *Chem. Rev.,* **88**, 813–988 (1988); **94**, 1721–2160 (1994).

Experimental information on intermolecular forces is obtained from virial coefficients and transport properties of gases and from excess functions of liquid mixtures, since these macroscopic properties are related to intermolecular forces. More direct information on intermolecular forces is found from molecular-beam scattering experiments (Sec. 23.3) and study of van der Waals molecules. The very inadequate current knowledge of intermolecular forces is the major stumbling block in calculating macroscopic properties of liquids and high-pressure gases from molecular properties.

22.11   STATISTICAL MECHANICS OF FLUIDS

Ideal gases were treated in Secs. 22.3 to 22.8. We now consider nonideal gases and liquids. Section 24.12 treats the statistical mechanics of solids.

For liquids and nonideal gases, the potential energy $\mathcal{V}$ of intermolecular interactions makes it impossible to write the canonical partition function Z as a product of molecular partition functions. Moreover, it is a hopeless task to try and solve the Schrödinger equation for the entire system of 10^{23} molecules to obtain the system quantum energies E_j. One therefore resorts to the approximation of using classical statistical mechanics to evaluate part of Z.

In earlier sections we noted that when the spacing between the levels of a given kind of energy is much less than kT ($\Delta\varepsilon \ll kT$), this kind of energy can be treated classically, replacing the sum over states by an integral. Let H be the Hamiltonian (Sec. 18.11) for the entire system. Suppose that H can be written as

$$H = H_{cl} + H_{qu} \qquad (22.138)$$

where H_{cl} contains the energies that can be treated classically and H_{qu} contains the energies that must be treated quantum-mechanically. Further, suppose that the terms in H_{cl} are independent of those in H_{qu}, and vice versa. Because of the assumed independence of the terms in H_{cl} and H_{qu}, E_j in $Z = \sum_j e^{-\beta E_j}$ is the sum of classical and quantum energies and therefore

$$Z = Z_{scl} Z_{qu} \qquad (22.139)$$

where Z_{scl} is calculated classically using integration over states and Z_{qu} is calculated quantum-mechanically using summation over states. The subscript scl stands for semiclassical and will be explained later.

Except at very low T, the translational and rotational levels are closely spaced compared with kT, so H_{cl} contains the translational and rotational energies. The intermolecular potential $\mathcal{V}$ [Eq. (22.132)] is a function of the coordinates of the centers of mass of the molecules (since the distances between molecules depend on these coordinates) and is a function of the angles defining the spatial orientations of the molecules (since the force between polar molecules depends on their orientation). These are the same coordinates that describe molecular translations and rotations, so $\mathcal{V}$ is part of H_{cl}. The vibrational and electronic levels are not closely spaced compared with kT and are included in H_{qu} of the fluid. Thus

$$H_{cl} = H_{tr} + H_{rot} + \mathcal{V}, \qquad H_{qu} = H_{vib} + H_{el} \qquad (22.140)$$

The restoring forces in chemical bonds within a molecule are much stronger than intermolecular forces, so the molecular vibrations are not substantially affected by intermolecular forces. Also rotational and vibrational motions are approximately independent of each other. Although the intermolecular potential depends on the electronic states of the interacting molecules, at ordinary temperatures there is negligible occupation of excited electronic states. Therefore the separation (22.140) into noninteracting classical and quantum-mechanical motions is reasonably accurate.

We need the expression for Z_{scl} in (22.139). This is obtained by taking the limit of the quantum-mechanical canonical partition function for the relevant energies as $\Delta\varepsilon/kT \rightarrow 0$. The derivation is complicated [see *Hill*, sec. 22-6, or *McQuarrie* (1973), sec. 10-7] and is omitted. For a fluid containing N identical molecules, one obtains

$$Z_{scl} = \frac{1}{N!h^{fN}} \int \cdots \int \int \cdots \int e^{-H_{cl}/kT}\, dq_1 \cdots dq_{fN}\, dp_1 \cdots dp_{fN} \qquad (22.141)$$

where f is the number of coordinates of a molecule that are being treated classically and the p's and q's are explained below. We are treating the translational and rotational coordinates classically. Each molecule has three translational coordinates and two or three rotational coordinates, depending on whether the molecule is linear or nonlinear (Fig. 21.25). Thus, $f = 5$ for a linear molecule, and $f = 6$ for a nonlinear molecule.

In Eq. (22.141), q_1 to q_{fN} are the translational and rotational coordinates of the N molecules. These coordinates consist of the $3N$ translational cartesian coordinates x_1, $y_1, z_1, \ldots, x_N, y_N, z_N$ of the centers of mass of the N molecules, and the $2N$ or $3N$ angles of rotation for the N molecules. This gives a total of $5N$ or $6N$ coordinates: $fN = 5N$ or $6N$. The quantities $p_1, \ldots, p_{fN}$ are the momenta that correspond to these coordinates. For the translational coordinate x_1 of molecule 1, the corresponding momentum is the linear momentum $p_{x,1} \equiv mv_{x,1}$. For a rotational coordinate (that is, angle), the corresponding momentum is the angular momentum of rotation involving that angle.

H_{cl} is a function of the fN coordinates and the fN momenta. The integration in (22.141) consists of $2fN$ definite integrals over the full ranges of $q_1, \ldots, q_{fN}, p_1, \ldots, p_{fN}$. Each cartesian coordinate ranges from 0 to a or b or c, where a, b, and c are the x, y, and z dimensions of the container holding the fluid. Each linear momentum and each angular momentum ranges from $-\infty$ to ∞. The angular coordinates range from 0 to π or 0 to 2π, depending on the angle. For simplicity, the integration limits have been omitted in (22.141).

The state of a system in classical mechanics is defined by the coordinates and momenta of all the particles (Sec. 18.6). Hence, the quantum-mechanical summation of $e^{-\beta E_j}$ over states is replaced in (22.141) by a classical-mechanical integration of $e^{-\beta H_{cl}}$ over coordinates and momenta. The $1/N!$ arises from the indistinguishability of the molecules. The h^{-fN} factor cannot be understood classically, since Planck's constant h occurs only in quantum mechanics. Because of the h^{-fN} in (22.141), we use the designation semiclassical rather than classical.

Equation (22.141) is for a pure fluid. For a fluid mixture, $N!h^{fN}$ is replaced by $\prod_B N_B! h^{f_B N_B}$, where N_B is the number of molecules of species B, where f_B is the number of classical coordinates (five or six) of a B molecule, and the product goes over all species present. The rest of this section deals with a pure fluid.

H_{qu} in (22.140) contains the vibrational and electronic energies. These have been taken to be independent of intermolecular interactions. Therefore the system's vibrational and electronic energy levels are the sums of the vibrational and electronic energies of the isolated molecules, and the work of Secs. 22.6 and 22.7 applies. We have for a pure substance

$$Z_{qu} = z_{vib}^N z_{el}^N \qquad (22.142)$$

where z_{vib} is given by (22.110) and z_{el} by (22.92).

H_{cl} in (22.140) is

$$H_{cl} = H_{tr,1} + \cdots + H_{tr,N} + H_{rot,1} + \cdots + H_{rot,N} + \mathcal{V} \qquad (22.143)$$

$H_{tr,1}$ is the translational kinetic energy of molecule 1 and is given by Eq. (18.55) as $H_{tr,1} = (p_{x,1}^2 + p_{y,1}^2 + p_{z,1}^2)/2m$. $H_{rot,1}$ is the classical Hamiltonian for rotation of molecule 1 and is a function of the two or three rotational angular momenta and the rotational angles. The intermolecular potential energy $\mathcal{V}$ is a function of the center-of-mass coordinates of each molecule and of the rotational angles specifying the molecular orientations.

The next step is to substitute (22.143) for H_{cl} into the Z_{scl} expression (22.141) and integrate over the translational and rotational momenta. We omit the details (see *McClelland*, secs. 11.4 to 11.6) and just give the final result. One finds

$$Z_{scl} = \frac{1}{N!}\left(\frac{z_{tr}}{V}\right)^N z_{rot}^N Z_{con} \qquad (22.144)$$

where z_{tr} and z_{rot} are the ideal-gas molecular translational and rotational partition functions of Eqs. (22.81) and (22.85) or (22.109), and where the **configuration integral** (or *configurational partition function*) Z_{con} is

$$Z_{con} = \frac{1}{(4\pi \text{ or } 8\pi^2)^N} \int \cdots \int \int e^{-\mathcal{V}/kT} \sin \theta_1 \cdots \sin \theta_N \, dx_1 \cdots dz_N \, d(\text{angs}) \quad (22.145)$$

In (22.145) one uses 4π for linear molecules and $8\pi^2$ for nonlinear molecules. For a linear molecule, $d(\text{angs}) = d\theta_1 \, d\phi_1 \cdots d\theta_N \, d\phi_N$, where θ_1 and ϕ_1 are the spherical-coordinate angles that give the spatial orientation of the axis of molecule 1 (Fig. 21.7). For a nonlinear molecule, $d(\text{angs}) = d\theta_1 \, d\phi_1 \, d\chi_1 \cdots d\theta_N \, d\phi_N \, d\chi_N$, where θ_1 and ϕ_1 give the spatial orientation of some chosen axis in molecule 1 and χ_1 (which ranges from 0 to 2π) is the angle of rotation of the molecule about that axis. The coordinate integration limits in (22.145) are the same as in (22.141).

Substitution of (22.144) for Z_{scl} and (22.142) for Z_{qu} into $Z = Z_{scl}Z_{qu}$ [Eq. (22.139)] gives as the canonical partition function of a pure fluid:

$$Z = \frac{1}{N!}\left(\frac{z_{tr}}{V}\right)^N z_{rot}^N z_{vib}^N z_{el}^N Z_{con} \qquad \text{pure fluid} \qquad (22.146)$$

This differs from the ideal-gas Z of Eqs. (22.49) and (22.58) solely in the presence of the factor Z_{con}/V^N. We can write

$$\ln Z = \ln Z_{id} + \ln Z_{con} - N \ln V \qquad \text{pure fluid} \qquad (22.147)$$

where Z_{id} is for the corresponding ideal gas. *Equation (22.147) is the key result of this section.*

Once Z has been found, the thermodynamic properties of the fluid are readily calculated from Z by using Eqs. (22.37) to (22.41). For example, $A = -kT \ln Z$ and $P = kT(\partial \ln Z/\partial V)_{T,N}$. Since z_{rot}, z_{vib}, z_{el}, and z_{tr}/V [Eq. (22.81)] are independent of V, partial differentiation of (22.146) gives as the equation of state

$$P = kT\left(\frac{\partial \ln Z_{con}}{\partial V}\right)_{T,N} \qquad (22.148)$$

The obstacle to finding Z for a fluid is the difficulty in evaluating Z_{con}, which involves the intermolecular potential $\mathcal{V}$.

For a nonideal gas or a liquid, one approximates $\mathcal{V}$ as the sum of pairwise interactions, $\mathcal{V} \approx \Sigma\Sigma \, v_{ij}$ [Eq. (22.132)], chooses a form for v_{ij} (for example, the Lennard-Jones potential or the Stockmayer potential), and tries to evaluate Z_{con}.

Since direct evaluation of Z_{con} is extremely difficult, various mathematical gymnastics are used to simplify the problem. One approach leads to the following expression for Z_{con} as an infinite series (for a partial derivation, see *Kestin and Dorfman*, chap. 7, or *Jackson*, sec. 4.5):

$$\ln Z_{con} = N \ln V - N\left[\frac{N}{VN_A}B(T) + \frac{1}{2}\left(\frac{N}{VN_A}\right)^2 C(T) + \frac{1}{3}\left(\frac{N}{VN_A}\right)^3 D(T) + \cdots \right]$$

$$(22.149)$$

where $VN_A/N = V_m$ is the molar volume and B, C, ... are functions of temperature that can be expressed as certain integrals involving $e^{-\beta v_{ij}}$. [The general expressions for B and C are given in E. A. Mason and T. H. Spurling, *The Virial Equation of State*, Pergamon, 1969, eqs. (2.5.25) and (2.5.26).]

If v_{ij} depends only on the intermolecular distance r, it turns out that

$$B(T) = -2\pi N_A \int_0^\infty \left(e^{-v(r)/kT} - 1\right)r^2 \, dr \qquad \text{if } v = v(r) \qquad (22.150)$$

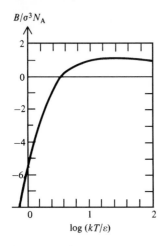

$B/\sigma^3 N_A$

Figure 22.22

The second virial coefficient $B(T)$ for the Lennard-Jones potential. ε and σ are the Lennard-Jones parameters.

where v is the intermolecular potential for two molecules separated by r. When the Lennard-Jones potential (22.136) is used in (22.150), one finds (see *Hirschfelder, Curtiss, and Bird,* pp. 163, 1114, and 1119) the results for $B(T)$ shown in Fig. 22.22. Note the resemblance to Fig. 8.2.

Substitution of (22.149) into (22.148) gives for the equation of state

$$P = \frac{RT}{V_m}\left[1 + \frac{B(T)}{V_m} + \frac{C(T)}{V_m^2} + \cdots\right] \qquad (22.151)$$

which is the virial equation (8.4).

Expressions for the thermodynamic functions $U - U_0$, S, $G - U_0$, etc., are readily calculated from (22.149) and (22.146); see Prob. 22.92 for the results.

If the intermolecular pair potential v is known (say, from a quantum-mechanical calculation), the virial coefficients B, C, ... can be calculated and the thermodynamic properties are then found from (22.151) and the equations of Prob. 22.92. Conversely, the experimentally measured virial coefficients found from P-V-T data can provide information on v. For gases of nonpolar, approximately spherical molecules (for example, Ar, N_2, CH_4), observed $B(T)$ data agree quite well with values calculated from the Lennard-Jones 6-12 potential.

Using reasonable potentials to calculate the virial coefficients, one finds that for gases at low or moderate densities, the successive terms in the virial expansion (22.149) rapidly decrease in magnitude and only the first few terms need be included. However, for gases at very high densities (low V_m) and for liquids, the successive terms do not decrease, and the series does not converge; the virial expansion fails, and one must use a different approach for liquids (see Sec. 24.14).

Let us evaluate the average translational energy for a molecule in the fluid. Using the log of the partition function (22.146) in Eq. (22.38), we get

$$U = NkT^2(\partial \ln z_{tr}/\partial T)_V + NkT^2(d \ln z_{rot}/dT) + NkT^2(d \ln z_{vib}/dT)$$
$$+ NkT^2(d \ln z_{el}/dT) + kT^2(\partial \ln Z_{con}/\partial T)_{V,N}$$

Clearly, the first term on the right is the total translational energy, the second term is the total rotational energy, etc. Thus, $U = U_{tr} + U_{rot} + U_{vib} + U_{el} + U_{intermol}$, where $U_{intermol}$ is the contribution of the intermolecular potential energies. Using (22.81) for z_{tr}, we get

$$U_{tr} = NkT^2(\partial \ln z_{tr}/\partial T)_V = \tfrac{3}{2}NkT \qquad (22.152)$$

Hence, the average translational energy per molecule is $\langle \varepsilon_{tr} \rangle = U_{tr}/N = \tfrac{3}{2}kT$. This holds for any fluid, not just an ideal gas [Eq. (15.16)].

Moreover, because the fluid's partition function (22.146) is the product of independent translational, rotational, vibrational, electronic, and intermolecular partition functions, the Boltzmann distribution law can be applied to each kind of molecular energy in the fluid. In particular, the distribution of translational energies is given by the Maxwell distribution (15.52). *The Maxwell distribution of speeds,* Eq. (15.44), *holds in any fluid,* not just in an ideal gas.

22.12   SUMMARY

By averaging over a canonical ensemble of thermodynamic systems of fixed volume, temperature, and composition, we found expressions for the thermodynamic state functions in terms of the system's canonical partition function Z, defined as $Z \equiv \sum_j e^{-\beta E_j}$, where $\beta = 1/kT$, E_j is the energy of the system's quantum state j, and the sum goes over all possible quantum states of the system. The quantum states and energies are found by solving the Schrödinger equation $\hat{H}\psi_j = E_j\psi_j$ for the entire thermodynamic system. The key formula relating Z and thermodynamic properties is $A = -kT \ln Z$.

For a system of N indistinguishable, identical, noninteracting molecules (a pure ideal gas), the canonical partition function is $Z = z^N/N!$, where the molecular partition function is $z \equiv \Sigma_r e^{-\beta \varepsilon_r}$, where ε_r is the energy of quantum state r of a molecule and the sum goes over all the possible molecular quantum states. The relation $Z = z^N/N!$ is valid provided the number of available molecular quantum states is much, much greater than the number of molecules. (If this condition doesn't hold, one must use either the Fermi–Dirac or Bose–Einstein expression for Z.) Writing $\varepsilon = \varepsilon_{tr} + \varepsilon_{rot} + \varepsilon_{vib} + \varepsilon_{el}$, we evaluated z for ideal-gas molecules as the product of translational, rotational, vibrational, and electronic molecular partition functions ($z = z_{tr} z_{rot} z_{vib} z_{el}$) and arrived at expressions for thermodynamic properties ($U - U_0$, S, C_V, etc.) of an ideal gas in terms of molecular properties (molecular weight, moments of inertia, vibrational frequencies, degeneracy of ground electronic state).

For a system of N noninteracting molecules, we derived the Boltzmann distribution law $\langle N_s \rangle / N = e^{-\varepsilon_s/kT}/z$ for the average number of molecules in quantum state s. This law holds provided $\langle N_r \rangle \ll 1$ for all molecular quantum states r. Otherwise, Bose–Einstein or Fermi–Dirac statistics must be used.

In evaluating z, we adopted the convention of taking the zero level of energy to coincide with the molecular ground state. With this convention, z gives a very rough estimate of the number of states that are significantly occupied.

The entropy of a thermodynamic system is $S = k \ln W$, where W is the total number of system quantum states that have a significant probability of being occupied.

The intermolecular potential energy for two nonreacting molecules can be approximated as the sum of an attractive term (which is the sum of dispersion, dipole–dipole, and dipole–induced-dipole interactions) and a repulsive term (which is due mainly to the Pauli repulsion between electrons).

For a fluid (liquid or nonideal gas) of N interacting molecules, one treats the translational, rotational, and intermolecular energies classically and treats the vibrational and electronic energies quantum-mechanically. This leads to the relation $\ln Z = \ln Z_{id} + \ln Z_{con} - N \ln V$, where Z_{id} is for the corresponding ideal gas and the configuration integral Z_{con} is a certain integral involving the intermolecular potential energy $\mathcal{V}$. If Z_{con} can be evaluated, then the fluid's canonical partition function Z is known and its thermodynamic properties can be calculated.

Important kinds of calculations in this chapter include:

- Use of Stirling's formula $\ln N! \approx N \ln N - N$ to find $\ln N!$ for large N.
- Calculation of populations of molecular states using the Boltzmann distribution law $\langle N_r \rangle / N = e^{-\varepsilon_r/kT}/z$ and of molecular energy levels using $\langle N(\varepsilon_s) \rangle / N = g_s e^{-\varepsilon_s/kT}/z$.
- Calculation of Θ_{rot} and Θ_{vib} from spectroscopic data using (22.83) and (22.88).
- Calculation of z_{tr}, z_{rot}, z_{vib}, and z_{el} for an ideal gas from Eqs. (22.81), (22.85), (22.90), and (22.92) for diatomic molecules and (22.81), (22.109), (22.110), and (22.92) for polyatomic molecules.
- Calculation of $U - U_0$ for an ideal gas from (22.94) to (22.98) for diatomic molecules and (22.111) to (22.113) for polyatomics.
- Calculation of C_V for an ideal gas from (22.99) to (22.103) for diatomic molecules or (22.114) to (22.116) for polyatomics.
- Calculation of S for an ideal gas from (22.105) to (22.108) or (22.108), (22.117), and (22.118).

FURTHER READING

Andrews (1975); *Davidson; Denbigh*, chaps. 11 and 12; *Hill; Jackson; Kestin and Dorfman; Knox; McClelland; McQuarrie* (1973); *Mandl; Reed and Gubbins*.

PROBLEMS

Section 22.2

22.1 (*a*) Which thermodynamic variables are held constant in each system of a canonical ensemble? (*b*) The canonical partition function Z of a one-phase system is a function of which thermodynamic variables? (*c*) Is the sum in $Z = \sum_j e^{-\beta E_j}$ a sum over the quantum states or the energy levels of the thermodynamic system?

22.2 What are the units of Z?

22.3 If system 1 is 10.0 g of water at 25°C and 1 atm and system 2 is 25.0 g of water at 25°C and 1 atm, and Z_1 and Z_2 are the canonical partition functions of these systems, what is the numerical value of $(\ln Z_2)/(\ln Z_1)$?

22.4 Verify that $G = kTV^2[\partial(V^{-1} \ln Z)/\partial V]_{T,N_B}$.

22.5 The choice of zero level of energy is arbitrary. Show that if a constant b is added to each of the system's possible energies E_j in (22.36), then (*a*) P in (22.37) is unchanged; (*b*) U is increased by b; (*c*) S is unchanged; (*d*) A is increased by b.

22.6 Verify that (22.33) for S can be written as $S = -k \sum_j p_j \ln p_j$.

Section 22.3

22.7 What are the units of z?

22.8 For neon gas in a 10-cm³ box at 300 K, calculate the number of available translational states with energy less than $3kT$.

22.9 This problem finds the number of translational quantum states with energy less than some maximum value. Let the axes of a cartesian coordinate system be labeled with the particle-in-a-box quantum numbers n_x, n_y, and n_z, and let a dot be placed at each point in space whose coordinates are integers. (*a*) Explain why the number of particle-in-a-box states with energy $\varepsilon_{tr} \leq \varepsilon_{max}$ equals the number of dots in one-eighth of a sphere whose radius is $r_{max} = (8mV^{2/3}h^{-2}\varepsilon_{max})^{1/2}$. [See Eq. (22.52).] (*b*) Let lines be drawn between the dots to form cubes, each cube having a dot at each of its 8 corners. Satisfy yourself that each dot is shared by 8 cubes—4 at the same altitude and 4 more above or below. The number of dots per cube is therefore 8/8 = 1, and the density of translational states is thus one per unit volume. Hence the number of translational states with $\varepsilon_{tr} \leq \varepsilon_{max}$ equals one-eighth the volume of a sphere of radius r_{max}. Show that for $\varepsilon_{max} = 3kT$, the number of translational states is $(24mV^{2/3}h^{-2}kT)^{3/2}\pi/6$.

Section 22.4

22.10 True or false? (*a*) For a pure ideal gas of a very large number of molecules, $\ln Z$ is directly proportional to the number of molecules present. (*b*) Because $\ln N! \approx N \ln N - N = \ln N^N - N$ is a good approximation for large N, the relation $N! \approx e^{\ln(N^N)-N} = N^N e^{-N}$ is a good approximation for large N.

22.11 Use Eq. (22.55) including the three terms in parentheses to estimate (*a*) 10!; (*b*) 100!; (*c*) 1000!. The accurate values are 3628800, $9.3326215444 \times 10^{157}$, and $4.0238726008 \times 10^{2567}$.

22.12 Use a computer or programmable calculator to accurately evaluate ln (300!). Compare the result with the approximation $\ln N! \approx N \ln N - N$.

22.13 Show that for a system of noninteracting, nonlocalized molecules, $\mu_B = -RT \ln (z_B/N_B)$ provided $\langle N_r \rangle \ll 1$.

Section 22.5

22.14 True or false? (*a*) For a system in thermodynamic equilibrium, the population of molecular energy levels always decreases as the energy of the levels increases. (*b*) It is impossible for a higher-energy molecular state to have a greater population of molecules than a lower-energy state. (*c*) For a thermodynamic system in equilibrium, molecular states that have the same energy must have the same population. (*d*) For a thermodynamic system in equilibrium, if we set the zero level of molecular energy at the ground state energy, then the fraction of molecules in the ground state equals $1/z$.

22.15 Most molecules in a gas have $\varepsilon_{tr} \leq 3kT$ and $\varepsilon_{tr,x} \leq kT$, where $\varepsilon_{tr,x}$ is the x component of translational energy. For N_2 in an 8-cm³ cubic box at 0°C, calculate $\Delta\varepsilon_x/kT$, where $\Delta\varepsilon_x$ is the spacing between adjacent x translational energies when $\varepsilon_{tr,x}$ equals kT.

22.16 The mean translational energy of molecules in a fluid is $\frac{3}{2}kT$. The relative populations of energy levels are determined by kT. Calculate kT in eV, kT/hc in cm⁻¹, and RT in kJ/mol and in kcal/mol at room temperature.

22.17 For N_2, the fundamental vibrational frequency is $\nu_0 = 6.985 \times 10^{13}$ s⁻¹. Calculate the ratio of the $v = 1$ to $v = 0$ populations at (*a*) 25°C; (*b*) 800°C; (*c*) 3000°C.

22.18 For N_2, the fundamental vibration wavenumber is $\tilde{\nu}_0 = 2329.8$ cm⁻¹. For 1.000 mol of $N_2(g)$, calculate the number of molecules in the $v = 0$ level and in the $v = 1$ level at (*a*) 25°C; (*b*) 800°C; (*c*) 3000°C.

22.19 For N_2, the rotational constant is $B_0 = 5.96 \times 10^{10}$ s⁻¹. Calculate the ratio of the $J = 1$ to $J = 0$ populations at (*a*) 200 K; (*b*) 600 K. (*c*) What is the limiting value of this ratio as $T \rightarrow \infty$?

22.20 In this problem, $\langle N_0 \rangle$, $\langle N_1 \rangle$, . . . are the average number of molecules in the $v = 0$, $v = 1$, . . . vibrational levels of a diatomic molecule. (*a*) Measurement of the intensities of spectroscopic absorption lines for a certain sample of $I_2(g)$ gives $\langle N_1 \rangle/\langle N_0 \rangle = 0.528$ and $\langle N_2 \rangle/\langle N_0 \rangle = 0.279$. Do these data indicate an equilibrium vibrational distribution? (*b*) For I_2, $\nu = 6.39 \times 10^{12}$ s⁻¹. Calculate the temperature of the I_2 sample. (*c*) For a certain gas of diatomic molecules (not I_2) in thermal equilibrium, one finds $\langle N_1 \rangle/\langle N_0 \rangle = 0.340$. Find $\langle N_3 \rangle/\langle N_0 \rangle$. Explain why your answer is only an approximation. [Spectroscopic observation of energy-level population ratios can be used to find temperatures of flames and stellar atmospheres; see C. S. McKee, *J. Chem. Educ.*, **58**, 605 (1981).]

22.21 When $\langle N_r \rangle \ll 1$, the denominator of the Fermi–Dirac/ Bose–Einstein distribution law (22.77) must be very large compared with the numerator, which is 1; hence, the ± 1 in the denominator can be neglected. Use the result of Prob. 22.13 to show that (22.77) reduces to the Boltzmann distribution when $\langle N_r \rangle \ll 1$.

22.22 The correct expression for the canonical partition function of a system of identical indistinguishable noninteracting particles is [*Mandl*, eqs. (11.15) and (11.31)]

$$Z_{BE}^{FD} = e^{-\beta \mu N/N_A} \prod_r \left(1 \pm e^{\beta(\mu/N_A - \varepsilon_r)}\right)^{\pm 1}$$

where $\beta = 1/kT$, μ is the chemical potential, the product goes over the molecular quantum states, the upper signs are for fermions, and the lower signs are for bosons. For $\langle N_r \rangle \ll 1$, we see from (22.77) that $e^{-\beta(\mu/N_A - \varepsilon_r)} \gg 1$. Show that Z_{BE}^{FD} reduces to (22.49) when $\langle N_r \rangle \ll 1$. [*Hint*: Take $\ln Z_{BE}^{FD}$ and use (8.36); then use (22.77) (with the ± 1 neglected) twice.]

22.23 For variables that are statistically independent of one another, the probability that the variables simultaneously have certain values is the product of the individual probabilities for each variable. Examine Eq. (22.77) to see whether the velocity components v_x, v_y, v_z in an ideal gas are statistically independent of one another (as assumed in the Chapter 15 derivation of the Maxwell distribution of speeds) when the condition $\langle N_r \rangle \ll 1$ does not hold.

Section 22.6
22.24 Consider the diatomic-molecule partition functions z_{tr}, z_{rot}, z_{vib}, z_{el}. (*a*) Which of them depend on T? (*b*) Which of them depend on V?

22.25 State which molecular properties each of these diatomic-molecule partition functions depends on: (*a*) z_{tr}; (*b*) z_{rot}; (*c*) z_{vib}.

22.26 For F_2 at 25°C, arrange these partition functions in order of increasing magnitude: z_{tr}; z_{rot}; z_{vib}; z_{el}.

22.27 Suppose a system consists of N noninteracting, indistinguishable, identical particles and that each particle has available to it only two quantum states, whose energies are $\varepsilon_1 = 0$ and $\varepsilon_2 = a$. (*a*) Find expressions for z, Z, U, C_V, and S. (See the note at the end of this problem.) (*b*) If $a = 1.0 \times 10^{-20}$ J and $N = 6.0 \times 10^{23}$, calculate z, S, and U at 400 K. (*c*) Find the limiting values of U and of C_V as $T \to \infty$. Give physical explanations for your results in terms of the populations of the levels. (*d*) Find the limiting value of S as $T \to \infty$. (*e*) Plot $C_{V,m}/R$ versus kT/a. (*f*) Explain why $C_{P,m}$ of O(g) decreases as T increases in the range 300 to 1500 K (see the Example 5.6 exercise). *Note*: The two quantum states referred to are the quantum states for the internal motion (Sec. 21.3) in each particle. In addition, each nonlocalized particle has a huge number of available translational states that allow $\langle N_r \rangle \ll 1$ to be satisfied. The problem therefore calculates the contributions of only the internal motions to the thermodynamic properties. The $1/N!$ should be omitted from Z, since the $1/N!$ belongs as part of the translational factor in Z, which is not being considered in this problem.

22.28 For a collection of harmonic oscillators, calculate the percent of oscillators in the $v = 0$ level when $h\nu/kT$ equals (*a*) 10; (*b*) 3; (*c*) 2; (*d*) 1; (*e*) 0.1.

22.29 (*a*) Show that if the harmonic-oscillator approximation is used, the fractional population of the diatomic-molecule vibrational level v is

$$\langle N_v \rangle / N = e^{-v\Theta_{vib}/T}\left(1 - e^{-\Theta_{vib}/T}\right)$$

(*b*) Show that $z_{vib} = N/\langle N_0 \rangle$, where $\langle N_0 \rangle$ is the number of molecules in the $v = 0$ level. Does a similar formula hold for z_{rot}? (*c*) Use the formula in (*a*) and Table 21.1 to calculate for one mole of $^{12}C^{16}O$ the number of molecules in the $v = 0$ level and in the $v = 1$ level at 25°C and at 1000°C. (*d*) For N_2, $\Theta_{vib} = 3352$ K. Plot the fractional population of the $v = 1$ vibrational level of $N_2(g)$ versus T from 0 to 15000 K, using the harmonic-oscillator approximation. Explain why the high-T part of the curve is not accurate.

22.30 Show that the temperature at which the fractional population of the harmonic-oscillator level v is a maximum is

$$T = \Theta_{vib}/\ln\left(1 + 1/v\right)$$

and that the fractional population of level v at this T is

$$\langle N_v \rangle / N = v^v/(v + 1)^{v+1}$$

[See Prob. 22.29(*a*).]

22.31 For a system of noninteracting molecules, let the molecular energy levels (not states) be $\varepsilon_1, \varepsilon_2, \varepsilon_3, \dots$. If $\varepsilon_2 > \varepsilon_1$, is it possible for a system in equilibrium to have more molecules in the level ε_2 than in ε_1? Explain.

22.32 A certain system is composed of 1 mole of identical, noninteracting, indistinguishable molecules. Each molecule has available to it only three energy levels, whose energies and degeneracies are $\varepsilon_1 = 0$, $g_1 = 1$; $\varepsilon_2/k = 100$ K, $g_2 = 3$; $\varepsilon_3/k = 300$ K, $g_3 = 5$. (k is Boltzmann's constant.) (*a*) Calculate z at 200 K. (*b*) Calculate the average number of molecules in each level at 200 K. (*c*) Calculate the average number of molecules in each level in the limit $T \to \infty$. (See the note at the end of Prob. 22.27.)

22.33 (*a*) Show that $\ln N! = \sum_{x=1}^{N} \ln x$. (*b*) Explain why this sum can be approximated by an integral if N is large. (*c*) Use (22.79) to show that $\ln N! \approx N \ln N - N$ for large N.

22.34 (*a*) The *Euler–Maclaurin summation formula* is

$$\sum_{n=a}^{\infty} f(n) = \int_a^{\infty} f(n)\, dn + \frac{f(a)}{2} - \frac{f'(a)}{12} + \frac{f'''(a)}{720} - \frac{f^{(v)}(a)}{30240} + \cdots$$

where $f^{(v)}$ is the fifth derivative of $f(n)$ evaluated at $n = a$. Use this formula to derive Eq. (22.86). (*b*) Calculate the percent error in z_{rot} of H_2 at 0°C when (22.85) is used. (*c*) Repeat (*b*) for N_2.

22.35 Verify Eqs. (22.96) to (22.98) for the contributions to U.

22.36 Show that $C_{V,vib}$ in (22.102) goes to nR as $T \to \infty$.

22.37 Verify Eqs. (22.104) to (22.107) for the contributions to S.

22.38 Verify (22.108) for $S_{tr,m}$.

22.39 Calculate $S^\circ_{m,298}$ for He(g), Ne(g), Ar(g), and Kr(g).

22.40 Calculate $S^\circ_{m,298}$ for H(g).

22.41 Calculate $C^\circ_{P,m,298}$ and $C^\circ_{V,m,298}$ for He(g) and Ne(g).

22.42 Calculate $U^\circ_{m,298} - U^\circ_{m,0}$ for He(g), Ne(g), and Ar(g).

22.43 For Ar(g), Probs. 22.39 and 22.42 give $S^\circ_{m,298} = 154.8$ J mol^{-1} K^{-1} and $U^\circ_{m,298} - U^\circ_{m,0} = 3718$ J/mol. Calculate ln Z, Z, and z for 1 mole of the ideal gas Ar at 25°C and 1 bar.

22.44 (a) Use data in Table 21.1 of Sec. 21.4 to calculate Θ_{vib} and Θ_{rot} of $^{14}N_2$. (b) For 1 mole of N_2, calculate z_{tr}, z_{rot}, z_{vib}, and z_{el} at 300 K and 2.00 atm. (c) Repeat (b) for 2500 K and 2.00 atm.

22.45 For HF, $\tilde{\nu}_0 = 3959$ cm^{-1}, and $\tilde{B}_0 = 20.56$ cm^{-1}. For HF(g), calculate (a) $S^\circ_{m,298}$; (b) $C^\circ_{V,m,298}$; (c) $C^\circ_{P,m,298}$.

22.46 For I_2, $\nu_0 = 6.395 \times 10^{12}$ s^{-1} and the internuclear distance is 2.67 Å. For I_2(g), calculate (a) $U^\circ_{m,500} - U^\circ_{m,0}$; (b) $H^\circ_{m,500} - U^\circ_{m,0}$; (c) $S^\circ_{m,500}$; (d) $G^\circ_{m,500} - U^\circ_{m,0}$.

22.47 The far-IR spectrum of HCl shows a series of lines with the approximately constant spacing 20.9 cm^{-1}. The near-IR spectrum shows a strong absorption band at 2885 cm^{-1}. Assume the data are for H^{35}Cl and calculate $S^\circ_{m,\,298}$ for H^{35}Cl(g).

22.48 (a) Without looking up data or doing elaborate calculations, estimate the room-temperature $C^\circ_{V,m}$ and $C^\circ_{P,m}$ of HF(g). Compare with the Appendix value. (b) Explain why several of the diatomic gases listed in the Appendix have the same $C^\circ_{P,m,298}$ value.

22.49 For NO, the ground electronic level and the first excited electronic level are each doubly degenerate. The separation between these two electronic levels is a mere 0.0149 eV. There are no other low-lying electronic levels. For NO(g), calculate z_{el} at 30 K, 150 K, and 300 K.

22.50 Use information in the preceding problem to calculate and plot $C_{V,m,el}$ versus T for NO(g) from 30 to 300 K. [See also Prob. 22.27(c).]

22.51 (a) Let $s = 1 + x + x^2 + \cdots$. Show that $s - xs = 1$. Then solve for s to get the geometric-series formula (22.89). (b) By considering the function $1/(1 - x)$, devise another derivation for (22.89).

22.52 (a) Find the expression for z_{vib} of a diatomic molecule if the zero of energy is taken at the bottom of the potential-energy curve, so that $\varepsilon_{vib} = (v + \frac{1}{2})h\nu$. (b) Then find U_{vib} and compare with (22.97). Does your result agree with Prob. 22.5(b)?

22.53 Sketch $C_{V,m}$ versus T for a typical ideal diatomic gas.

22.54 For CO (which has $\Theta_{rot} = 2.77$ K), find the percentage of molecules having $J \leq 16$ at 25°C (a) by using the approximation of replacing a sum by an integral; (b) by using a computer or programmable calculator to evaluate the required sum directly. In both (a) and (b), use (22.86) for z_{rot}. In working (b), also tabulate the percentages of molecules with $J = 0$, $J = 1$, $J = 2, \ldots, J = 16$.

22.55 In Example 22.3 in Sec. 22.6, we found that $S^\circ_{m,vib,298}$ is negligible for N_2. However, $S^\circ_{m,vib,298}$ is not completely negligible for F_2 or for FCl. Explain this in terms of the bonds in the molecules. Explain why $S^\circ_{m,vib,298}$ is much smaller for HF than for F_2.

22.56 (a) Replace the sum in (22.87) by an integral and perform the integration to get the semiclassical expression for z_{vib} of a diatomic molecule. (b) Verify that in the limit $T \rightarrow \infty$, Eq. (22.90) reduces to the result for (a).

22.57 Equation (22.104) predicts that $S_{tr} \rightarrow -\infty$ as $T \rightarrow 0$ at fixed n and P. But from the third law of thermodynamics we know that $S \rightarrow 0$ as $T \rightarrow 0$ for a substance. Explain this apparent contradiction.

22.58 (a) For the isoelectronic molecules CO(g) and N_2(g), state (without looking up any data) whether each of the following partition functions will be approximately equal or will differ substantially for temperatures in the range 100 to 1000 K: z_{tr}, z_{rot}, z_{vib}, z_{el}. (b) Estimate $S^\circ_{m,298}[CO(g)] - S^\circ_{m,298}[N_2(g)]$ by doing the *minimum* amount of calculation. Check your estimate with data in the Appendix.

22.59 Consider the species N_2(g), O_2(g), F_2(g), Br_2(g). Without looking up any data or formulas, answer the following questions. (a) Which one has the smallest Θ_{rot} and which the largest Θ_{rot}? (b) Which one has the smallest Θ_{vib} and which the largest Θ_{vib}? (c) Which one has the greatest z_{rot} at room temperature? (d) Which one has the greatest z_{vib} at room temperature? (e) Which one has the greatest $C_{V,m,rot}$ at room temperature? (f) Which one has the greatest $C_{V,m,vib}$ at room temperature?

22.60 According to (22.108), $S_{m,tr}$ increases as the molecular weight increases. Equations (22.104), (22.82), and (22.65) show that this increase is due to the increase in z_{tr} as M_r increases. Explain why z_{tr} increases as M_r increases at constant T.

22.61 For HF, $\Theta_{rot} = 29.577$ K. Evaluate z_{rot} for HF(g) at 20 K, 30 K, and 40 K, using (a) the high-T equation (22.85); (b) the expansion (22.86) with terms through $(\Theta_{rot}/T)^3$ included; (c) exact evaluation of z_{rot} in the equation preceding (22.83) by direct term-by-term addition. For (c), use a computer or programmable calculator; include a test to decide when to stop adding terms. Compare the approximate results of (a) and (b) with the exact result of (c).

22.62 Derive the formula $S = -Nk \sum_r x_r \ln x_r - k \ln N!$ (Sec. 22.6) as follows. Substitute (22.49) and (22.38) into (22.39). Then multiply the $Nk \ln z$ term by $\sum_s e^{-\varepsilon_s/kT}/z$ (which equals 1) and combine the sum in this term with the other sum in the S equation. Then use (22.69) and the log of (22.69). What condition (besides "ideal gas") must be satisfied for this formula for S to hold?

22.63 Obtain the correct high-T limit of $U_{m,rot}$ as follows. Use the Taylor series (8.36) to show that (22.86) gives

$$\ln z_{rot} = \ln T - \ln \sigma \Theta_{rot} + \Theta_{rot}/3T + \Theta^2_{rot}/90T^2 + \cdots$$

Then use (22.63) to show that the high-T limit of $U_{m,rot}$ is $RT - R\Theta_{rot}/3$. [For a more rigorous derivation, see I. N. Levine, *J. Chem. Educ.*, **62**, 53 (1985).]

Section 22.7

22.64 Give the symmetry number for (a) BF_3; (b) H_2O; (c) HCN; (d) CH_4; (e) C_2H_2; (f) C_2H_4; (g) NH_3.

22.65 For H_2S the fundamental vibrational wavenumbers are 2615, 1183, and 2628 cm^{-1}, and the rotational constants $\tilde{A}_0$, $\tilde{B}_0$, and $\tilde{C}_0$ are 10.37, 8.99, and 4.73 cm^{-1}. Calculate $S_{m,298}^\circ$ for $H_2S(g)$.

22.66 For CO_2 the fundamental vibrational wavenumbers are 1340, 667, 667, and 2349 cm^{-1}, and the rotational constant is $\tilde{B}_0 = 0.390$ cm^{-1}. Calculate $S_{m,298}^\circ$ for $CO_2(g)$.

22.67 Give the high-temperature values of $C_{P,m}^\circ$ and $C_{V,m}^\circ$ for (a) $CO_2(g)$; (b) $SO_2(g)$; (c) $C_2H_4(g)$. (Interpret "high temperature" to mean a temperature not high enough to give a significant population of excited electronic levels.)

22.68 Verify Eqs. (22.111) to (22.113) for the contributions to U.

Section 22.8

22.69 Given that $D_0 = 4.4780$ eV for the H_2 ground electronic state, (a) find ΔU_0° for $H_2(g) \to 2H(g)$; (b) without looking up data or doing elaborate calculations, give an accurate estimate of ΔH_{298}° of $H_2(g) \to 2H(g)$. Compare with the value obtained from Appendix data.

22.70 For Ar(g), calculate (a) $H_{m,298}^\circ - H_{m,0}^\circ$; (b) $H_{m,1000}^\circ - H_{m,0}^\circ$; (c) $G_{m,298}^\circ - H_{m,0}^\circ$; (d) $G_{m,1000}^\circ - H_{m,0}^\circ$.

22.71 A thermodynamics table gives $(G_{m,T}^\circ - H_{m,0}^\circ)/T = -257.7$ J/mol-K for $CH_3OH(g)$ at 1000 K. Find z for 1 mole of $CH_3OH(g)$ at 1000 K and 1 bar, assuming ideal-gas behavior.

22.72 For diamond, $H_{m,298}^\circ - H_{m,0}^\circ = 0.523$ kJ/mol and $S_{m,298}^\circ = 2.377$ J/mol-K. Calculate ln Z for 1 mole of diamond at 25°C and 1 bar.

22.73 Derive (22.128) for K_c.

Section 22.9

22.74 Calculate $W_{final}/W_{initial}$ for mixing 1 mole of benzene with 1 mole of toluene at 300 K and 1 atm (assume an ideal solution).

22.75 Calculate ln W for 1 mole of $N_2(g)$ at 25°C and 1 bar.

22.76 The formula $S = k \ln W$ seems to have a certain arbitrariness, since it was not precisely specified how wide a band of levels around U should be taken in Fig. 22.3 when computing S. Actually, this arbitrariness is of no consequence, since $k \ln W$ is extraordinarily insensitive to the value of W. Thus, suppose that Tom and Nan choose bands around U such that Tom's band includes $e^{(10^{12})}$ times as many states as Nan's. Calculate $S_{Tom} - S_{Nan}$. Is the difference significant?

Section 22.10

22.77 Which of these intermolecular interactions change with changing temperature in a gas? (a) dipole–dipole; (b) dipole–induced dipole; (c) London dispersion.

22.78 For each of the following pairs of molecules, state which one has the greater polarizability. (a) He or Ne; (b) H_2 or He; (c) F_2 or Cl_2; (d) H_2S or Ar; (e) CH_4 or C_2H_6.

22.79 Verify the values listed in Table 22.1 for v_{d-d}, v_{d-id}, and v_{disp} for CH_2Cl_2.

22.80 Use the Lennard-Jones parameters listed in Sec. 22.10 to estimate the force between two CH_4 molecules separated by (a) 8 Å; (b) 5 Å; (c) 3 Å.

22.81 With n in $v_{rep} \approx A/r^n$ set equal to 12, we get $v = A/r^{12} - B/r^6$. (a) Let σ be the value of r at which $v = 0$. Show that $v = B\sigma^6/r^{12} - B/r^6$. (b) Let r_{min} be the value of r at which v is a minimum. Show that $r_{min} = 2^{1/6}\sigma$. (c) Let $\varepsilon \equiv v(\infty) - v(r_{min})$ be the depth of the potential well. Show that $B = 4\sigma^6\varepsilon$. Hence v is given by the Lennard-Jones potential (22.136). (d) Show that (22.136) can be written as $v/\varepsilon = (r_{min}/r)^{12} - 2(r_{min}/r)^6$.

22.82 Use the result of Prob. 22.81(b) and data in Sec. 22.10 to calculate r_{min} in the Lennard-Jones potential for (a) Ar; (b) C_6H_6.

22.83 (a) The normal boiling point of Ne is 27.1 K. Estimate the Lennard-Jones ε for Ne. (The experimental value is 4.92×10^{-22} J.) (b) The critical temperature of C_2H_6 is 305 K. Estimate the Lennard-Jones ε for C_2H_6. (The experimental value is 3.18×10^{-21} J.)

22.84 (a) Sketch $F(r)$ versus r, where $F(r)$ is the force between two nonpolar molecules. (b) Let b be the value of r at which $F = 0$. Express b in terms of the Lennard-Jones σ.

22.85 An intermolecular potential intermediate in crudity between the hard-sphere and the Lennard-Jones potentials is the *square-well potential*, defined by $v = \infty$ for $r < \sigma$, $v = -\varepsilon$ for $\sigma \le r \le a$, and $v = 0$ for $r > a$. Sketch this potential.

22.86 Which of each of the following pairs has the greater normal boiling point? (a) Kr or Xe; (b) C_2H_5OH or $(CH_3)_2O$; (c) H_2O or H_2S.

22.87 Calculate the Lennard-Jones v/ε for each of these separations: (a) $2^{1/6}\sigma$; (b) 1.5σ; (c) 2σ; (d) 2.5σ; (e) 3σ.

Section 22.11

22.88 (a) Show that $Z_{con} = V^N$ for an ideal gas in a rectangular box with edges a, b, c. Consider separately the cases of linear and nonlinear molecules. (b) Show that (22.148) gives $P = NkT/V$ for an ideal gas.

22.89 Show that for the hard-sphere potential of Fig. 22.21b, $B = 4N_A V_{molec}$, where $V_{molec} = \frac{4}{3}\pi(d/2)^3$ is the volume of one molecule. (*Hint:* Break the integration range into two parts.)

22.90 The Lennard-Jones parameters for N_2 are $\varepsilon = 1.31 \times 10^{-21}$ J and $\sigma = 3.74$ Å. Use Fig. 22.22 to find the Lennard-Jones B of N_2 at 100, 300, and 500 K. The experimental values are -149, -4, and 17 cm^3/mol, respectively.

22.91 (a) For the square-well potential (defined in Prob. 22.85), find the expression for $B(T)$ in (22.150). (b) For N_2, the square-well-potential parameters (found by fitting B data) are $\varepsilon = 1.31 \times 10^{-21}$ J, $\sigma = 3.28$ Å, $a = 5.18$ Å. Calculate the square-well predictions for B of N_2 at 100, 300, and 500 K. The experimental values are -149, -4, and 17 cm^3/mol, respectively. (c) Use the result of (b) to estimate the compression factor $Z \equiv PV_m/RT$ for N_2 at 100 K and 3.0 atm.

22.92 Use Eqs. (22.149) and (22.147) to show that for a nonideal gas

$$U = U_{id} - nRT^2\left(\frac{1}{V_m}\frac{dB}{dT} + \frac{1}{2V_m^2}\frac{dC}{dT} + \cdots\right)$$

$$S = S_{id} - nR\left[\frac{1}{V_m}\left(B + T\frac{dB}{dT}\right) + \frac{1}{2V_m^2}\left(C + T\frac{dC}{dT}\right) + \cdots\right]$$

$$G = G_{id} + nRT\left(\frac{2}{1}\frac{B}{V_m} + \frac{3}{2}\frac{C}{V_m^2} + \cdots\right)$$

General

22.93 As noted at the end of Sec. 22.3, the formula $Z = z^N/N!$ gives a typical error in Z at 25°C and 1 atm of a factor of $10^{(\pm 10^{15})}$. Use (22.40) to find the error in A produced by such an error in Z at 25°C. Is this error in A significant?

22.94 In a certain gas at a certain temperature, $z_{rot} = 154.1$. What fraction of the molecules are in the ground rotational state?

22.95 Consider the quantities Z, $\ln Z$, z, $\ln z$, z_{tr}, z_{rot}, z_{vib}, and z_{el} for an ideal gas. (a) Which of these are proportional to the number of moles n? (b) Which are independent of n? (c) Which are independent of T? (d) Which are independent of P?

22.96 The formula $A = -kT \ln Z$ applies to which of the following kinds of thermodynamic systems? (a) Every thermodynamic system in equilibrium; (b) equilibrium thermodynamic systems where there are no intermolecular forces; (c) equilibrium thermodynamic systems where $\langle N_r \rangle \ll 1$; (d) equilibrium thermodynamic systems where there are no intermolecular forces and where $\langle N_r \rangle \ll 1$.

Theories of Reaction Rates

This chapter discusses the theoretical calculation of rate constants of elementary chemical reactions. If a reaction mechanism consists of several steps, the theories of this chapter can be applied to each elementary step to calculate its rate constant. Accurate theoretical prediction of reaction rates is a largely unsolved problem that is the subject of much current research.

Sections 23.1 to 23.7 deal with gas-phase reactions. Reactions in solution are considered in Sec. 23.8.

23.1 HARD-SPHERE COLLISION THEORY OF GAS-PHASE REACTIONS

The hard-sphere collision theory (developed about 1920) uses the following assumptions to arrive at an expression for the rate constant of an elementary bimolecular gas-phase reaction. (*a*) The molecules are hard spheres. (*b*) For a reaction to occur between molecules B and C, the two molecules must collide. (*c*) Not all collisions produce reaction. Instead, reaction occurs if and only if the relative translational kinetic energy along the line of centers of the colliding molecules exceeds a **threshold energy** ε_{thr} (Fig. 23.1). (*d*) The Maxwell–Boltzmann equilibrium distribution of molecular velocities is maintained during the reaction.

Assumption (*a*) is a crude approximation. Polyatomic molecules are not spherical but have a structure, and the hard-sphere potential function (Fig. 22.21*b*) is a grotesque caricature of intermolecular interactions. Assumption (*c*) seems reasonable, since a certain minimum energy is needed to initiate the breaking of the relevant bond(s) to cause reaction. We shall see in Sec. 23.3 that assumption (*c*) is somewhat inaccurate.

What about assumption (*d*)? According to assumption (*c*), it is the fast-moving molecules that react. Hence the gas mixture is continually being depleted of high-energy reactant molecules. The equilibrium distribution of molecular speeds is maintained by molecular collisions. In most reactions, the threshold energy (which we shall soon see is nearly the same as the experimental activation energy) is much greater than the mean molecular translational energy $\frac{3}{2}kT$. (Typical gas-phase bimolecular activation energies are 3 to 30 kcal/mol, compared with 0.9 kcal/mol for $\frac{3}{2}RT$ at room temperature.) Consequently, only a tiny fraction of collisions produces reaction. Since the collision rate is usually far greater than the rate of depletion of high-energy reactant molecules, the redistribution of energy by collisions is able to maintain the Maxwell distribution of speeds during the reaction. For the opposite extreme, where $\varepsilon_{\text{thr}} \approx 0$, virtually all B-C collisions lead to reaction, so the mixture is being depleted of low-energy as well as high-energy molecules. A careful study [B. M. Morris and R. D. Present, *J. Chem. Phys.*, **51**, 4862 (1969)] showed that for all values of $\varepsilon_{\text{thr}}/kT$, the corrections to the rate due to departure from the Maxwell distribution are slight.

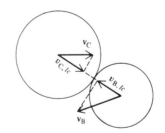

Figure 23.1

Line-of-centers components $v_{\text{C},lc}$ and $v_{\text{B},lc}$ of the velocities of two colliding molecules.

To calculate the gas-phase reaction rate, one simply calculates the rate of collisions for which the line-of-centers relative translational energy ε_{lc} exceeds ε_{thr}. Equations (18.77) to (18.80) show that $\varepsilon_{lc} = \frac{1}{2}\mu(v_{C,lc} - v_{B,lc})^2$, where $\mu = m_B m_C/(m_B + m_C)$ and $v_{C,lc}$ and $v_{B,lc}$ are shown in Fig. 23.1. Equation (15.62) gives Z_{BC}, the total number of B-C collisions per unit time per unit volume when the gas is in equilibrium. Z_{BC} was found from the Maxwell distribution of speeds, so, by assumption (d) above, we use (15.62) in the reacting system. The number of molecules reacting per unit time per unit volume equals Z_{BC} multiplied by the fraction ϕ of collisions for which $\varepsilon_{lc} \geq \varepsilon_{thr}$. Since the rigorous calculation of this fraction is complicated (see *Moore and Pearson,* chap. 4), we shall give only a plausible argument, rather than a derivation.

There are two components of velocity involved (namely, the component of velocity of each molecule along the line of centers), so we might suspect that ϕ equals the fraction of molecules in a hypothetical two-dimensional gas whose translational energy $\varepsilon = \varepsilon_x + \varepsilon_y = \frac{1}{2}mv_x^2 + \frac{1}{2}mv_y^2 = \frac{1}{2}mv^2$ exceeds ε_{thr}. A derivation similar to the one that gave the distribution (15.44) in a three-dimensional gas shows that the fraction of molecules with speed between v and $v + dv$ in a two-dimensional gas is (Prob. 15.17)

$$(m/2\pi kT)e^{-mv^2/2kT} \times 2\pi v\, dv = (m/kT)e^{-mv^2/2kT}v\, dv \qquad (23.1)$$

The use of $\varepsilon = \frac{1}{2}mv^2$ and $d\varepsilon = mv\, dv$ gives the fraction of molecules with energy between ε and $\varepsilon + d\varepsilon$ as $(1/kT)e^{-\varepsilon/kT}\, d\varepsilon$. Integration of this expression from ε_{thr} to ∞ gives the fraction ϕ with translational energy exceeding ε_{thr} as $e^{-\varepsilon_{thr}/kT}$. The rigorous derivation shows that ϕ does equal $e^{-\varepsilon_{thr}/kT}$.

Therefore, the number of B molecules reacting per unit volume per second in the elementary bimolecular reaction B + C → products is $Z_{BC}e^{-E_{thr}/RT}$, where $E_{thr} \equiv N_A\varepsilon_{thr}$ is the threshold energy on a per-mole basis, and N_A is the Avogadro constant. The reaction rate r in (17.3) is defined in terms of moles, so $r = Z_{BC}e^{-E_{thr}/RT}/N_A$. Since $r = k[B][C]$, the predicted rate constant is

$$k = \frac{Z_{BC}e^{-E_{thr}/RT}}{N_A[B][C]} \qquad (23.2)$$

The use of (15.62) for Z_{BC} with $N_B/V = N_A n_B/V = N_A[B]$ and $N_C/V = N_A[C]$ gives

$$k = N_A\pi(r_B + r_C)^2\left[\frac{8RT}{\pi}\left(\frac{1}{M_B} + \frac{1}{M_C}\right)\right]^{1/2}e^{-E_{thr}/RT} \qquad \text{for B} \neq \text{C} \quad (23.3)$$

For the bimolecular reaction 2B → products, the rate of reaction is given by (17.4) and (17.5) as $r \equiv -\frac{1}{2}d[B]/dt = k[B]^2$. The rate of disappearance of B is $-d[B]/dt = 2Z_{BB}e^{-E_{thr}/RT}/N_A$. The factor 2 appears because two B molecules disappear at each reactive collision. Therefore $k = -\frac{1}{2}(d[B]/dt)/[B]^2 = Z_{BB}e^{-E_{thr}/RT}/N_A[B]^2$. Substitution of (15.63) with $N_B/V = N_A[B]$ gives

$$k = \frac{1}{2^{1/2}}N_A\pi d_B^2\left(\frac{8RT}{\pi M_B}\right)^{1/2}e^{-E_{thr}/RT} \qquad \text{for B} = \text{C} \quad (23.4)$$

Equations (23.3) and (23.4) have the form $\ln k = \text{const} + \frac{1}{2}\ln T - E_{thr}/RT$. The definition (17.68) gives the activation energy as $E_a \equiv RT^2\, d\ln k/dT = RT^2(1/2T + E_{thr}/RT^2)$:

$$E_a = E_{thr} + \frac{1}{2}RT \qquad (23.5)$$

Substitution of (23.5) and (23.3) in $A \equiv ke^{E_a/RT}$ [Eq. (17.69)] gives the pre-exponential factor as

$$A = N_A\pi(r_B + r_C)^2\left[\frac{8RT}{\pi}\left(\frac{1}{M_B} + \frac{1}{M_C}\right)\right]^{1/2}e^{1/2} \qquad \text{for B} \neq \text{C} \quad (23.6)$$

Since $\frac{1}{2}RT$ is small, the hard-sphere threshold energy is nearly the same as the activation energy. The simple collision theory provides no means of calculating E_{thr} but gives only the pre-exponential factor A. Because of the crudities of the theory, the predicted $T^{1/2}$ dependence of A should not be taken seriously.

EXAMPLE 23.1 Hard-sphere collision-theory A

The bimolecular (elementary) reaction $CO + O_2 \rightarrow CO_2 + O$ has an observed activation energy of 51.0 kcal/mol and a pre-exponential factor 3.5×10^9 dm^3 s^{-1} mol^{-1} for the temperature range 2400 to 3000 K. Viscosity measurements and Eq. (16.25) give the hard-sphere diameters as $d(O_2) = 3.6$ Å and $d(CO) = 3.7$ Å. Calculate the hard-sphere collision-theory A factor and compare with the experimental value.

Equation (23.6) gives at the mean temperature 2700 K:

$$A = (6.0 \times 10^{23}\,\text{mol}^{-1})(3.14)(3.6_5 \times 10^{-10}\,\text{m})^2$$

$$\times \left[\frac{8(8.3\,\text{J/mol-K})(2700\,\text{K})}{3.14} \left(\frac{\text{mol}}{0.028\,\text{kg}} + \frac{\text{mol}}{0.032\,\text{kg}} \right) 2.72 \right]^{1/2}$$

$$A = 8.1 \times 10^8\,\text{m}^3\,\text{s}^{-1}\,\text{mol}^{-1} = 8.1 \times 10^{11}\,\text{dm}^3\,\text{s}^{-1}\,\text{mol}^{-1}$$

(The use of 2400 or 3000 K instead of 2700 K would change A only slightly.) The calculated A factor is 230 times the experimental A value. Since experimental A values are typically accurate to a factor of 3, the discrepancy between theory and experiment cannot be blamed on experimental error.

Exercise

The gas-phase reaction $F_2 + ClO_2 \rightarrow F + FClO_2$ has $A = 1.3 \times 10^7$ L mol^{-1} s^{-1} in the temperature range 230 to 250 K. What value of $r_B + r_C$ in the collision-theory equation is consistent with this value of A? Is this value of $r_B + r_C$ reasonable? (*Answer:* 0.030 Å.)

For most reactions, the calculated A values are much higher than the observed values. Hence, in the 1920s the hard-sphere collision theory was modified by adding a factor p to the right sides of Eqs. (23.2) to (23.4) and (23.6). The factor p is called the *steric* (or *probability*) *factor*. The argument is that the colliding molecules must be properly oriented for collision to produce a reaction; p (which lies between 0 and 1) supposedly represents the fraction of collisions in which the molecules have the right orientation. For example, for $CO + O_2 \rightarrow CO_2 + O$, we would expect the reaction to occur if the carbon end of the CO hits the O_2 but not if the oxygen end hits O_2. The hard-sphere collision theory provides no way to calculate p theoretically. Instead, p is found from the ratio of the experimental A to the collision-theory A. Thus, for $CO + O_2 \rightarrow CO_2 + O$, $p = 1/230 = 0.0043$. With this approach, p is simply an adjustable parameter that brings theory into agreement with experiment.

The idea of a proper orientation being needed for reaction is valid, but p also includes contributions that arise from using a hard-sphere potential and from ignoring molecular vibration and rotation. Typical p values range from 1 to 10^{-6}, tending to be smaller for reactions involving larger molecules, as would be expected if p were due at least in part to orientational requirements.

For trimolecular reactions, we need the three-body collision rate. Since the probability that three hard spheres collide at precisely the same instant is vanishingly small, a three-body collision is defined as one in which the three molecules are within a specified short distance from one another. With this definition, the trimolecular collision rate per unit volume is found to be proportional to [A][B][C]. The expression for k is omitted.

23.2 POTENTIAL-ENERGY SURFACES

The hard-sphere collision theory of chemical kinetics does not give accurate rate constants. A correct theory must use the true intermolecular forces between the reacting molecules and must take into account the internal structure of the molecules and their vibrations and rotations. In chemical reactions, bonds are being formed and broken, so we must consider the forces acting on the atoms in the molecules. During a molecular collision, the force on a given atom depends on both the intramolecular forces (which determine the vibrational motions in the molecule) and the intermolecular forces (Sec. 22.10). We cannot deal separately with each of the colliding molecules but must consider the two molecules to form a single quantum-mechanical entity, which we shall call a **supermolecule.** The supermolecule is not to be thought of as having any permanence or stability; it exists only during the collision process.

The force on a given atom in the supermolecule is determined by the potential energy V of the supermolecule; Eq. (2.21) gives $F_{x,a} = -\partial V/\partial x_a$, where $F_{x,a}$ is the x component of the force on atom a and V is the potential energy for atomic motion in the supermolecule. (The phrase "atomic motion" in the last sentence can be replaced by "nuclear motion," since the electrons follow the nuclear motion almost perfectly; Sec. 20.2.) The supermolecule's potential energy V is determined the same way the potential energy for nuclear vibrational motion in an ordinary molecule is calculated. Using the Born–Oppenheimer approximation, we solve the electronic Schrödinger equation $\hat{H}_e\psi_e = \varepsilon_e\psi_e$ [Eq. (20.7)] for a fixed nuclear configuration. The energy ε_e equals V at the chosen nuclear configuration. By varying the nuclear configuration, we get V as a function of the nuclear coordinates.

If the supermolecule has $\mathcal{N}$ atoms, there are $3\mathcal{N}$ nuclear coordinates. As with an ordinary molecule, there are three translational and three rotational coordinates. These leave V unchanged (since they leave all internuclear distances unchanged), so V is a function of $3\mathcal{N} - 6$ variables. If V were a function of two variables x and y, we could plot V in three-dimensional space; this plot is a surface (the *potential-energy surface*) whose distance above the xy plane at the point $x = a$, $y = b$ equals $V(a, b)$. Since V is usually a function of far more than two variables, such a plot usually cannot be made. Nevertheless, the function V is still called the **potential-energy surface,** no matter how many variables it depends on.

Consider some examples. The simplest bimolecular collision is that between two atoms. Here V is a function of only one variable, the interatomic distance R. (The supermolecule has $\mathcal{N} = 2$ and is linear. There are only two rotational coordinates, and $3\mathcal{N} - 3 - 2 = 1$.) The function $V(R)$ is the familiar potential-energy curve for the diatomic molecule formed by the two atoms. For example, for two colliding H atoms, the supermolecule is H_2. Its $V(R)$ is given by the second lowest curve in Fig. 20.4 if the electron spins are parallel, and by the lowest curve if the spins are antiparallel. Since there are three two-electron spin functions that have total electron-spin quantum number $S = 1$ and only one spin function that has $S = 0$ [Eqs. (19.37) and (19.38)], two H atoms will repel each other in 75% of collisions. Even when the spins are antiparallel, a stable molecule will not be formed, since a third body is needed to carry away some of the bond energy to prevent dissociation (Sec. 17.12).

Two colliding ground-state He atoms always repel each other (except for the very weak van der Waals attraction at relatively large R). A rough approximation to $V(R)$ is the Lennard-Jones potential.

Two colliding atoms are a very special case. Consider the general features of V when two stable, closed-shell polyatomic molecules collide. At relatively large distances, the weak van der Waals attraction (Sec. 22.10) causes V to decrease slightly as the molecules approach each other. When they are close enough for their electron probability densities to overlap substantially, the Pauli-principle repulsive force sets in, causing V to increase substantially. If the colliding molecules are not oriented properly to react, or if they don't have enough relative kinetic energy to overcome the intermolecular repulsion, the short-range repulsion causes them to bounce off each other without reacting. If, however, the molecules B and C are properly oriented and have enough relative kinetic energy to approach closely enough, a new chemical bond can be formed between them, usually accompanied by the simultaneous breaking of one or more bonds in the original molecules, thereby yielding the products D and E. The Pauli-principle repulsion between D and E then causes them to move away from each other, the potential energy V decreasing as this happens. During the course of the elementary reaction, the supermolecule's potential energy first decreases slightly, then rises to a maximum, and then falls off.

Recall from Sec. 22.10 that intermolecular interactions between nonreacting species can be reasonably estimated without considering the internal structures of the species. However, calculation of V in a chemical reaction requires detailed quantum-mechanical calculations on the supermolecule.

A much investigated potential-energy surface is that for the reaction $H + H_2 \rightarrow H_2 + H$. The reaction can be studied experimentally using isotopes ($D + H_2 \rightarrow DH + H$) or ortho and para H_2 ($H + $ para $H_2 \rightarrow$ ortho $H_2 + H$). [In ortho H_2, the nuclear (that is, proton) spins are parallel; in para H_2, they are antiparallel.] The supermolecule is H_3. The first quantum-mechanical calculation of the H_3 potential surface was made by Eyring and Polanyi in 1931, but reasonably accurate results were not obtained until the 1960s.

The first highly accurate ab initio calculation of the H_3 potential surface (including nonlinear geometries) was made by Liu and Siegbahn [B. Liu, *J. Chem. Phys.*, **58**, 1925 (1973); P. Siegbahn and B. Liu, ibid., **68**, 2457 (1978); B. Liu, ibid., **80**, 581 (1984).] The Liu–Siegbahn surface is estimated to be in error by less than 0.04 eV (1 kcal/mol) at every point. Recall from Sec. 20.8 that Hartree–Fock wave functions do not give correct values for the energy changes involved in breaking bonds. Therefore, accurate calculation of potential-energy surfaces in chemical reactions requires a procedure, such as configuration interaction (CI), that goes beyond the Hartree–Fock method; such procedures are much harder than SCF Hartree–Fock calculations. Liu and Siegbahn did CI calculations at 267 H_3 geometries. Once V has been calculated at many points, one can fit an algebraic function to the calculated V values. From this function, V and its partial derivatives can be found everywhere on the surface. For an H_3 surface of a few hundred points (geometries) accurate to 0.1 kcal/mol, see D. L. Diedrich and J. B. Anderson, *J. Chem. Phys.*, **100**, 8089 (1994). For an accurate analytic surface fitted to CI energies at 8701 geometries, see A. I. Boothroyd et al., *J. Chem. Phys.*, **104**, 7139 (1996).

For H_3, V is a function of $9 - 6 = 3$ variables. These can be taken as two interatomic distances R_{ab} and R_{bc} and an angle θ (Fig. 23.2a). The reaction is

$$H_a + H_b H_c \rightarrow H_a H_b + H_c \qquad (23.7)$$

The overlap between the electron probability densities of H_a and the H_2 molecule is a minimum when $\theta = 180°$. We therefore expect the Pauli repulsion to be minimized when the H_a atom approaches along the $H_b H_c$ axis. (Recall the steric factor of simple collision theory.) Accurate quantum-mechanical calculations bear this out.

Since V is a function of three variables, the potential-energy "surface" must be plotted in four dimensions. If we temporarily restrict ourselves to a fixed value of θ,

Figure 23.2

(a) Variables for the H + H$_2$ reaction. (b) The potential-energy surface for the H + H$_2$ reaction for $\theta = 180°$.

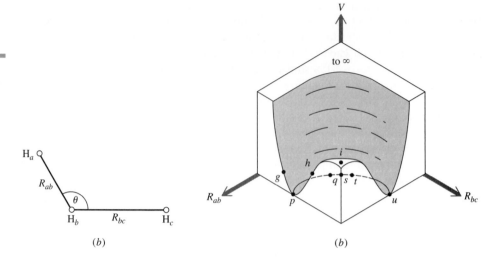

(b)

(b)

then V is a function of R_{ab} and R_{bc} only. We plot R_{ab} and R_{bc} on the two horizontal axes and $V(R_{ab}, R_{bc})$ for a fixed θ on the vertical axis. Figure 23.2b shows the $V(R_{ab}, R_{bc})$ surface for $\theta = 180°$. A contour map of this surface is shown in Fig. 23.3. Plots for other values of θ have the same general appearance. The solid lines in Fig. 23.3 are lines of constant V, each labeled with its V value.

At point p in Fig. 23.3, the distance R_{bc} equals the equilibrium bond length (0.74 Å) in H$_2$, and R_{ab} is large, indicating that atom H$_a$ is far from molecule H$_b$H$_c$. Point p corresponds to the reactants H$_a$ + H$_b$H$_c$. Point u, where R_{bc} is large and R_{ab} = 0.74 Å, corresponds to the products H$_a$H$_b$ + H$_c$. The energy of the reactants H$_a$ + H$_b$H$_c$ has been taken as zero, so points p and u each have zero energy. This is not true in general

Figure 23.3

Contour map of the H + H$_2$ potential-energy surface for $\theta = 180°$. Note that this diagram starts at $R_{ab} = 0.5$ Å $= R_{bc}$ rather than at zero. [Data from R. N. Porter and M. Karplus, *J. Chem. Phys.*, **40**, 1105 (1964).]

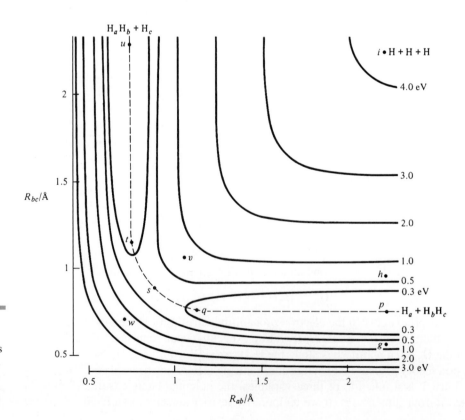

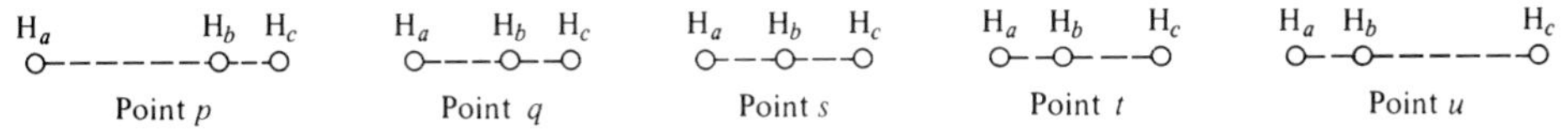

H_a $\quad$ H_b H_c $\quad\quad$ H_a $\quad$ H_b H_c $\quad\quad$ H_a $\quad$ H_b $\quad$ H_c $\quad\quad$ H_a $\quad$ H_b $\quad\quad$ H_c $\quad\quad$ H_a $\quad$ H_b $\quad\quad\quad$ H_c

o------o-o $\quad$ o---o-o $\quad$ o--o--o $\quad$ o-o--o $\quad$ o-o-----o

Point p $\quad\quad\quad$ Point q $\quad\quad\quad$ Point s $\quad\quad\quad$ Point t $\quad\quad\quad$ Point u

Figure 23.4

Configurations of the H_3 supermolecule for various points on the minimum-energy path.

for the reaction A + BC → AB + C, since the energy of AB will usually differ from that of BC.

Point i has both R_{ab} and R_{bc} very large and corresponds to three widely separated H atoms: $H_a + H_b + H_c$. The region around point i in Fig. 23.2b is a nearly flat plateau. The potential energy varies hardly at all here, since the atoms are so far apart that changing R_{ab} or R_{bc} doesn't affect V. The potential energy at i is 4.75 eV ($109\frac{1}{2}$ kcal/mol) above that at p, this being the equilibrium dissociation energy D_e of H_2. The region around i plays no part in the reaction (23.7). (This would be the reactant region for the trimolecular reaction H + H + H → H_2 + H.)

Along a line joining points g, p, and h, the distance R_{ab} between H_a and molecule H_bH_c is fixed, but R_{bc} in H_bH_c varies. This generates the ground-state diatomic potential-energy curve for H_2 (Fig. 21.9), which can be seen in Fig. 23.2b. V increases in going from p to g or from p to h, so point p lies at the bottom of a valley. Point u lies at the bottom of a second valley at right angles to the valley of p.

We now look for the path of minimum potential energy that connects reactants to products on the potential-energy surface. This is called the **minimum-energy path.** (It is often called the "reaction path" or the "reaction coordinate," but these names are open to objection. The term "reaction path" is misleading, in that the reacting molecules do not precisely follow the minimum-energy path; see Sec. 23.3. The term "reaction coordinate" has a meaning slightly different from that of "minimum-energy path"; see Sec. 23.4.) The minimum-energy path is the dotted line $pqstu$ in Figs. 23.2b and 23.3. Figure 23.4 shows the configurations of the H_3 supermolecule for some points on the minimum-energy path.

At point q, the atom H_a has approached fairly close to the molecule H_bH_c, and the bond distance R_{bc} has lengthened a bit, indicating a slight weakening of the bond. Point s has $R_{ab} = R_{bc}$, and here the H_a—H_b bond is half formed and the H_b—H_c bond is half broken. At point t, the atom H_c is retreating from the newly formed H_aH_b molecule. At point u, the reaction is over. Figures 23.2b and 23.3 show that the potential energy along the minimum-energy path increases from p to q to s, reaches a maximum at s, and then decreases from s to t to u (Fig. 23.5). Actually, there is an initial decrease

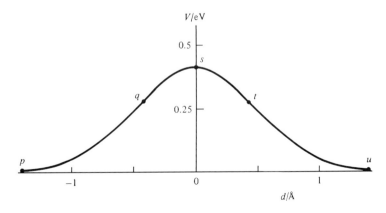

Figure 23.5

Potential energy V versus distance d along the minimum-energy path for the H + H_2 reaction.

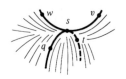

Figure 23.6

The H + H_2 potential-energy surface in the region of the saddle point s.

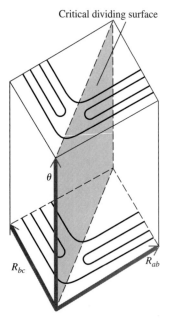

Critical dividing surface

Figure 23.7

Potential-energy contour plots for the H + H_2 reaction at various values of θ. There is a set of contours for each θ value, but to keep the figure simple, only two sets of contours are shown.

in V between points p and q due to the van der Waals attraction, but this is too slight to be visible in the figures.

Point s is the maximum point on the minimum-energy path and is what mathematicians call a **saddle point,** since the surface around s resembles a saddle (Fig. 23.6). We have $V_s < V_v$ and $V_s < V_w$, but $V_s > V_q$ and $V_s > V_t$. A hiker starting at point p and facing toward q is in a deep valley with walls rising to infinity on the left and to a high plateau on the right. As he walks from p to s, his elevation rises gradually from 0 to 0.4 eV (10 kcal/mol) [compared with 4.7 eV (110 kcal/mol) for the plateau height]. The region around s is a pass connecting the reactant valley to the product valley.

For the present, let us ignore the rotation and vibration of H_2 and use classical mechanics to consider a linear collision between an H atom and an H_2 molecule. The total energy of the colliding species in Fig. 23.4 remains fixed (conservation of energy). As H_a and H_bH_c approach each other, their potential energy increases (because of the Pauli-principle repulsion), so their kinetic energy decreases. If the relative translational kinetic energy of the colliding molecules is too small to allow the supermolecule to climb the potential hill to point s, the repulsion causes H_a to bounce off without reacting. If the relative kinetic energy is high enough, the supermolecule can go over point s and yield products (point u). An analogy is a ball rolled on the potential-energy surface from point p toward s. Whether or not it gets over the saddle point at s depends on its initial kinetic energy.

Since the colliding reactants need not have $\theta = 180°$, all values of θ from 0 to 180° must be considered. For each value of θ, one can do quantum-mechanical calculations of $V(R_{ab}, R_{bc})$. Contour plots of V for various values of θ resemble Fig. 23.3, and can be found in R. N. Porter and M. Karplus, *J. Chem. Phys.,* **40,** 1105 (1964), and D. G. Truhlar and C. J. Horowitz, ibid., **68,** 2466 (1978). Drawing a graph like Fig. 23.5 for each value of θ, one finds the potential-energy maximum E_{max} to be surmounted in going from reactants to products at various θ values to be 10 kcal/mol at $\theta = 180°$, 18 kcal/mol at 112°, 64 kcal/mol at 60°, 34 kcal/mol at 45°, and 10 kcal/mol at 0°. As noted earlier, reaction occurs most readily for a linear approach.

A plot of $V(R_{ab}, R_{bc}, \theta)$ requires four dimensions and is not readily made. Instead, we can set up coordinates with the R_{ab} and R_{bc} axes horizontal and the θ axis vertical and plot contours like Fig. 23.3 at various values of θ. Figure 23.7 indicates such a contour plot. (The meaning of the critical dividing surface is discussed in Sec. 23.4.) The angle θ may change during a given H_a-H_bH_c collision, and the full potential-energy function $V(R_{ab}, R_{bc}, \theta)$ is required to deal with an arbitrary collision process. Since a linear approach requires the least energy for reaction to occur, most collisions that lead to reaction will have θ reasonably close to 180°.

Since $\theta = 180°$ is the energetically most favored angle, point s in Fig. 23.3 (which is for $\theta = 180°$) lies at a minimum with respect to variations in θ, as well as with respect to variations along the line vsw. The configuration of the colliding molecules at the saddle point s is called the **transition state.** For H + H_2, the transition state is linear and symmetric, with each H—H distance equal to 0.93 Å (compared with $R_e = 0.74$ Å in H_2). The potential-energy difference between the transition state and the reactants (omitting zero-point vibrational energy) is the **(classical) barrier height** ε_b; that is, $\varepsilon_b \equiv V_s - V_p$. For H_2 + H, ab initio quantum-mechanical calculations give $\varepsilon_b = 0.42$ eV and $E_b \equiv N_A\varepsilon_b = 9.6$ kcal/mol. We shall see later that the barrier height is approximately (but not exactly) equal to the activation energy for the reaction.

The term "transition state" should not mislead one into thinking that the supermolecule at point s has any sort of stability. The transition state is just one particular point on the continuous path from reactants to products.

The 10-kcal/mol barrier height for $H_a + H_bH_c \rightarrow H_aH_b + H_c$ is much less than the 110 kcal/mol needed to break the bond in H_2 ($H_2 \rightarrow 2H$). In general, observed bimolecular activation energies are only a fraction of the energy needed to break the

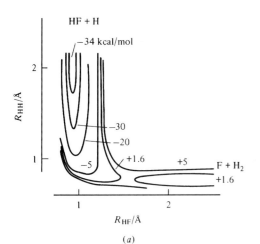

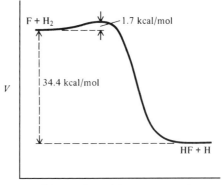

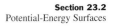

Figure 23.8

(*a*) Potential-energy contours for F + $H_2 \rightarrow$ HF + H for $\theta = 180°$. [Adapted from C. F. Bender, S. V. O'Neil, P. K. Pearson, and H. F. Schaefer, *Science, 176,* 1412–1414 (1972). Original figure copyright 1972 by the American Association for the Advancement of Science.] (*b*) Barrier height for this reaction.

relevant bond, because simultaneous formation of a new chemical bond largely compensates for the breaking of the old bond. As the H_b—H_c bond is breaking, the H_a—H_b bond is forming. When no new bonds are formed, E_a is quite high. Thus, E_a values for unimolecular decompositions are high and are approximately equal to the energy of the bond broken if the products have no bonds not present in the reactants.

Figure 23.8*a* shows potential energy contours found from Bender and coworkers' 1972 ab initio CI calculations for the F + $H_2 \rightarrow$ H + HF reaction (and for the reverse reaction H + HF $\rightarrow$ H_2 + F) for approach angle $\theta = 180°$. Energies are in kcal/mol, and the zero of energy is taken at the separated reactants. The figure is less symmetric than Fig. 23.3 for H_3, since the products differ from the reactants. The calculated barrier height is 1.7 kcal/mol for the forward reaction. (The experimental activation energy is 1.7 kcal/mol.) Since the calculated energy change for the reaction is -34.4 kcal/mol, the calculated barrier height for the reverse reaction is 36.1 kcal/mol (Fig. 23.8*b*).

The saddle-point (transition-state) geometry found by Bender and coworkers is linear with $R(HF) = 1.54$ Å and $R(HH) = 0.77$ Å, compared with $R(HH) = 0.74$ Å in the isolated H_2 molecule and $R(HF) = 0.93$ Å in HF. The transition state occurs early in the reaction, with the H—H bond distance only slightly elongated and the F atom relatively far from the H_2 molecule. The transition state for F + $H_2 \rightarrow$ H + HF resembles the reactants much more than the products. For the reverse reaction, the transition state (which is the same as for the forward reaction) resembles the products. (In 1955, Hammond postulated that in an exothermic reaction the transition state will be likely to resemble the reactants, while in an endothermic reaction it will resemble the products.)

Over a dozen calculations of the F + H_2 potential-energy surface have been done [see G. C. Schatz, *Science,* **262,** 1828 (1993)]. Although it was formerly believed that the minimum-energy path involved a linear geometry, the newer higher-level calculations give a nonlinear transition state that is bent about 60° from linearity and has a barrier that is 0.4 kcal/mol less than that for a linear approach angle. For experimental confirmation that the transition-state region of the potential-energy surface of this reaction is now accurately known, see D. E. Manolopoulos et al., *Science,* **262,** 1852 (1993).

The rate constant and the detailed course of an elementary chemical reaction depend on the shape of the entire potential-energy surface (see Sec. 23.3). However, the main features of an elementary reaction can be determined if only the barrier height and the structure of the transition state are known. The barrier height does not

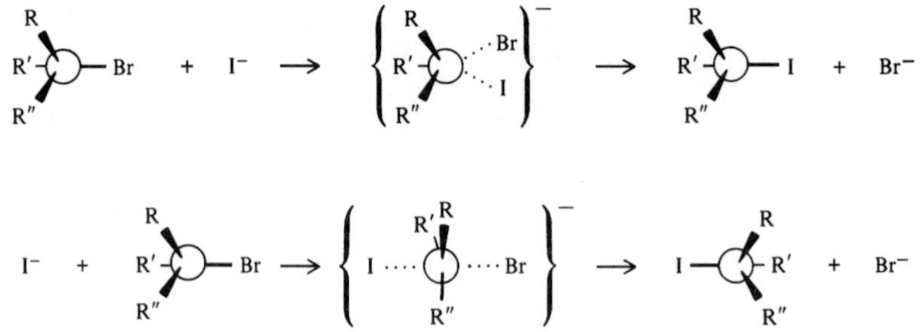

Figure 23.9

Two possible mechanisms for the reaction $I^- + RR'R''CBr \rightarrow RR'R''CI + Br^-$.

differ greatly from the activation energy, so a reaction with a low barrier height is rapid, whereas one with a high barrier is slow.

The transition state occurs at the maximum point on the minimum-energy path between reactants and products. The energy of the transition state relative to that of the reactants determines the barrier height. The geometrical structure of the transition state determines the stereochemistry of the reaction products.

An example is the elementary bimolecular reaction $I^- + CRR'R''Br \rightarrow CRR'R''I + Br^-$ (in the jargon of organic chemists, an S_N2 reaction, a bimolecular nucleophilic substitution). The I^- might attack the alkyl bromide on the same side of the molecule as the Br or on the opposite side (Fig. 23.9). The transition states are shown in braces. (These transition states are not reaction intermediates but are simply one point on the continuous path from reactants to products.) The product obtained in one case is the mirror image of that in the other case. For attack by I^- on the side opposite the Br, the transition state is expected to be of lower energy than that formed in attack on the same side as Br; this is because, for "backside" attack, the carbon can undergo gradual rehybridization to sp^2 AOs, which bond the three R groups, leaving a carbon p orbital to partially bond both the I and the Br in the transition state. The barrier should be lower for backside attack. This is borne out by the observation that the product is stereochemically inverted with respect to the reactant. The transition-state structure also explains why this reaction is very slow when the R groups are bulky. Large R groups mean a large Pauli-principle repulsion between I^- and the alkyl halide, and hence a greater barrier height and a slower reaction. This example shows why organic chemists spend so much time talking about transition-state structures. (Although S_N2 reactions in solution usually proceed by the second mechanism shown in Fig. 23.9, in the gas phase the reaction involves formation of two ion–dipole intermediates. See *Chem. Eng. News,* Jan. 13, 1992, p. 22.)

For a reaction between two five-atom molecules, the potential-energy surface is a function of 24 variables. Clearly the calculation of an accurate ab initio potential-energy surface for such a reaction is a fantastically difficult task, especially since a method (such as CI) that goes beyond the Hartree–Fock approximation is needed. Things can be simplified by the fact that only a small number of bond distances and angles change significantly during the reaction, so the number of variables can be substantially reduced. Even so, an accurate ab initio calculation of V is still extremely difficult and has been accomplished for only a few reactions.

Semiempirical calculations require much less computer time and storage capacity than ab initio calculations. The AM1 and PM3 methods (Sec. 20.11) have been used to calculate many potential-energy surfaces, barrier heights, and transition-state structures. Calculated semiempirical transition-state structures are usually in fairly good agreement with more accurate high-level ab initio transition-state structures, but are sometimes quite erroneous. Calculated semiempirical activation energies are often very inaccurate and the reliability of semiempirical potential-energy surfaces is often questionable; "there is no semiempirical method at present which can be used reliably in all situations [to study reactions]" (O. N. Ventura in S. Fraga, ed., *Computational Chemistry,* Part B, Elsevier, 1992).

The Woodward–Hoffmann rules use MO theory to decide which stereochemical path has the lower barrier height in various organic reactions. These rules are discussed in most current undergraduate organic chemistry texts. Pearson has used MO concepts to decide whether the activation energies of various elementary inorganic reactions are high or low. (See R. G. Pearson, *Symmetry Rules for Chemical Reactions,* Wiley-Interscience, 1976.)

Potential-energy surfaces for unimolecular and trimolecular reactions are discussed in Secs. 23.6 and 23.7.

23.3 MOLECULAR REACTION DYNAMICS

Molecular reaction dynamics studies what happens at the molecular level during an elementary chemical reaction. The kinetics methods discussed in Chapter 17 can tell us that the rate constant for the elementary reaction $Cl + H_2 \rightarrow HCl + H$ is given by $k = Ae^{-E_a/RT}$, where $A = 1.2 \times 10^{10}$ dm^3 mol^{-1} s^{-1} and $E_a = 4.3$ kcal/mol in the temperature range 250 to 450 K, but this statement gives little information about the details of the elementary reaction. Molecular dynamics considers such questions as: How does the probability for reaction vary with the angle the incoming Cl atom makes with the H—H line? How does this probability vary with the relative translational energy of the reactants? How is the product HCl distributed among its various translational, rotational, and vibrational states? How can quantum mechanics and statistical mechanics be used to calculate theoretically the rate constant at a given temperature?

The development of molecular reaction dynamics began in the 1930s, but it wasn't until the 1960s that new experimental techniques and the availability of electronic computers for theoretical calculations allowed reliable information to be obtained. With these developments, chemists are beginning to understand what happens in an elementary chemical reaction.

Trajectory Calculations

Suppose that the potential-energy surface for a given gas-phase elementary chemical reaction has been accurately calculated using quantum mechanics. How does one use this surface to calculate the rate constant at a given temperature?

The probability that two colliding gas molecules will react depends on the initial translational, rotational, and vibrational states of the molecules. One picks a pair of initial states for the reactant molecules and solves the time-dependent Schrödinger equation (18.10) to calculate the probability that reaction will occur for these initial states. The potential-energy V occurs in the Hamiltonian operator $\hat{H}$ in (18.9), so V is needed for this calculation. Since the gas molecules are distributed among many translational, rotational, and vibrational states, the process of solving the time-dependent Schrödinger equation must be repeated for sufficiently many different initial states to give a statistically representative sample of all possible initial states. Then one uses the Boltzmann distribution to calculate the relative number of collisions that occur for each set of initial states at temperature T, and a weighted average of the reaction probabilities is taken, thereby giving the rate constant at T. This procedure is the rigorously correct version of the simple collision theory of Sec. 23.1.

Solving the time-dependent Schrödinger equation is extremely difficult, and only a very few such calculations of rate constants have been made. Fortunately, there is an approximate approach that is much easier. Instead of using quantum mechanics to deal with the collision process, one uses classical mechanics. One chooses a pair of initial states for the reactant molecules (rotational and vibrational energies, relative translational energy, approach angle). The forces are obtained from the potential-energy surface ($F_{x,a} = -\partial V/\partial x_a$, etc.). Newton's second law $F = ma$ is numerically integrated on a computer to give the locations of the atoms as functions of time. The path of a particle is called its trajectory in classical mechanics, and such calculations are called **trajectory calculations.** After trajectories have been calculated for a representative set of initial conditions, suitable averaging using the Boltzmann distribution at the temperature of interest gives the rate constant.

Of course, the motions of atoms obey quantum mechanics, not classical mechanics. However, comparisons of the results of quantum-mechanical and classical calculations indicate that the classical-trajectory results are accurate except when very light species are involved. (As a particle's mass increases, its behavior approaches classical behavior.) Classical mechanics is deterministic, and the probability that a given initial state will lead to reaction is either 0 or 1. This is not so in quantum mechanics. In particular, there is some probability that molecules having relative kinetic energy less than the potential-energy barrier height will tunnel (Sec. 18.12) through the barrier and yield products. For reactions involving one of the species e^-, H^+, H, and H_2 (including reactions that transfer e^-, H^+, or H between two heavy molecules), tunneling is important and can easily multiply the rate constant by a factor of 3 or more. For heavier species, tunneling is unimportant. Methods for correcting classical-trajectory results to allow for tunneling have been developed. [See W. H. Miller, *Adv. Chem. Phys.,* **25,** 69 (1974); *Science,* **233,** 171 (1986).]

Consider the atom–diatomic-molecule reaction A + BC $\rightarrow$ AB + C. Figure 23.10 shows two typical classical trajectories for a linear collision plotted on the $\theta = 180°$ contour map. (The angle θ may well change during a collision, and the trajectory must be plotted on a figure like Fig. 23.7. Here we restrict ourselves to collisions in which θ stays constant at 180°.) In Fig. 23.10a, molecule BC is in the $v = 0$ vibrational level. Because of the zero-point vibration in BC, the trajectory of the supermolecule oscillates about the minimum-energy path (the dashed line), and the supermolecule is not likely to pass over the barrier precisely at the saddle point. Another reason the minimum-energy path is not precisely followed is that many reactive collisions have $\theta \neq 180°$. The wide vibrations of the product AB in Fig. 23.10a indicate that in this collision AB has been produced in an excited vibrational level. Figure 23.10b shows a nonreactive collision trajectory.

For the reaction H_2 + H, Karplus and coworkers calculated thousands of classical trajectories on a reasonably accurate semiempirical potential surface and calculated rate constants at several temperatures. Figure 23.11 shows the results of two such calculations.

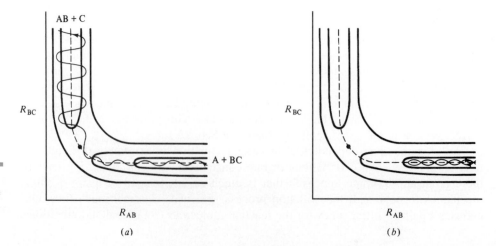

Figure 23.10

Classical trajectories for the reaction A + BC $\rightarrow$ AB + C. The saddle point is indicated by a dot. (*a*) A reactive collision. (*b*) A nonreactive collision.

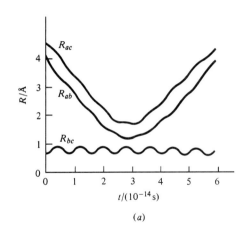

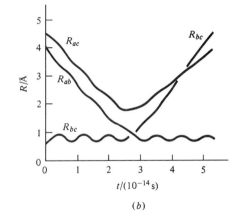

Figure 23.11

Two classical trajectories for the H + H$_2$ reaction. [M. Karplus, R. N. Porter, and R. D. Sharma, *J. Chem. Phys.*, **43**, 3259 (1965).]

In Fig. 23.11*a*, the distances R_{ab} and R_{ac} first decrease (indicating that atom H$_a$ is approaching molecule H$_b$H$_c$) and then increase (indicating that H$_a$ is moving away from H$_b$H$_c$); no reaction has occurred. The continual fluctuation in R_{bc} is due to zero-point vibration in molecule H$_b$H$_c$. In Fig. 23.11*b*, atom H$_a$ again approaches H$_b$H$_c$, colliding with it at 3×10^{-14} s. Now, however, R_{bc} goes to infinity after 3×10^{-14} s, indicating that the H$_b$—H$_c$ bond has broken, and R_{ab} fluctuates about the equilibrium H$_2$ bond distance 0.74 Å, indicating that atom H$_a$ has bonded to H$_b$. The reaction H$_a$ + H$_b$H$_c \rightarrow$ H$_a$H$_b$ + H$_c$ has occurred. Note that at 4×10^{-14} s atom H$_c$ is closer to H$_b$ in H$_a$H$_b$, but at 5×10^{-14} s it is closer to H$_a$. This indicates that the product molecule H$_a$H$_b$ is rotating. The initial state of the reactants had $J = 0$, and so part of the relative translational energy of the reactants has gone into rotational energy of a product. Trajectory calculations thus show how the energy of the products is distributed among rotational, vibrational, and relative translational energies.

Since most collisions are nonlinear and since θ changes during rotation, one needs to know V as a function of all three variables R_{ab}, R_{bc}, and θ in order to calculate k from H$_2$ + H trajectory calculations.

The H + H$_2$ calculations of Karplus, Porter, and Sharma showed that vibrational energy of the reactants can contribute to the energy needed to overcome the barrier, but that rotational energy does not contribute. Although $\theta = 180°$ is the energetically most favored approach angle, the large number of nonlinear approaches leads to an average approach angle of 160° for reactive collisions.

Rate constants calculated from classical trajectories agree reasonably well with experimental rate constants. The main sources of error in the calculated k values are inaccuracies in the potential-energy surface and tunneling.

In the hard-sphere collision theory (Sec 23.1), the probability p_r of a reactive collision varies with ε_{lc}, the *line-of-centers component* of the relative translational kinetic energy of the colliding molecules, according to Fig. 23.12*a*. When the hard-sphere p_r

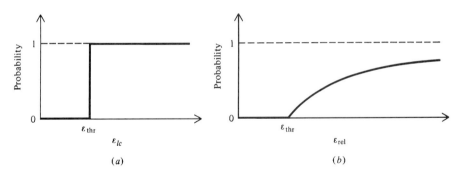

Figure 23.12

Hard-sphere collision-theory reaction probability as a function of (*a*) the line-of-centers kinetic energy and (*b*) the relative kinetic energy of the colliding molecules.

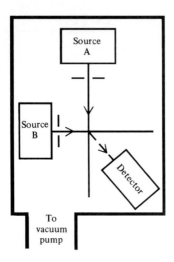

Figure 23.13

Schematic diagram of a crossed-molecular-beam apparatus. Each beam may be produced by effusion from a heated oven or by expansion through a nozzle from an unheated chamber containing gas at rather high pressure. Not shown are velocity selectors (rotating slotted disks), which may be placed in each beam before the collision region.

is calculated as a function of ε_{rel}, the *total* relative translational kinetic energy of the molecules, one obtains Fig. 23.12*b*. [See E. F. Greene and A. Kupperman, *J. Chem. Educ.*, **45**, 361 (1968) for the derivation.] The results of trajectory calculations on potential-energy surfaces and of molecular-beam experiments (see below) show that Fig. 23.12*b* is in error in two ways: (*a*) the true reaction probabilities are usually less than those of the hard-sphere theory (recall the steric factor), and (*b*) the true reaction probability reaches a maximum, and then at energies much greater than ε_{thr} falls off.

Molecular Beams

We now consider some experimental techniques in molecular reaction dynamics. In a crossed-molecular-beam experiment, a beam of A molecules crosses a beam of B molecules in a high-vacuum chamber (Fig. 23.13). The beams are usually produced by expansion from relatively high pressure chambers through a nozzle into vacuum, thereby giving jets whose translational and rotational temperatures are extremely low. (Recall the use of this technique as an aid to simplify molecular spectra—Sec. 21.11.) A narrow beam is produced from each jet by using skimmers to select only the central portion of each jet. Collisions in the region of intersection of the crossed beams can produce the chemical reaction A + B → C + D. A movable detector is used to detect the products. The most widely useful detector is a mass spectrometer. Product molecules that show visible or UV fluorescence can be detected using laser-induced fluorescence (LIF; Sec. 21.11). By varying the wavelength of the laser light used to excite the fluorescence and measuring the total (undispersed) fluorescence intensity versus this wavelength, one may be able to deduce the population distribution of product molecules among the vibration–rotation states of the ground electronic state of the product (see *Levine and Bernstein*, pp. 229–233).

The speeds of the reacting molecules in the beams can be controlled by velocity selectors (Fig. 15.8). Such experiments yield information on how the probability of reaction varies with the relative kinetic energy of the colliding molecules, on the angles at which the reaction products leave the collision site, and on the energy distribution of the products. Polar molecules can be oriented in a beam by applied electric fields (see *Chem. Eng. News*, Oct. 28, 1991, p. 19). This technique showed that in the reaction K + CH_3I → KI + CH_3, the reaction probability when the K atom collides with the I end of CH_3I is about twice that when K hits the CH_3 end. The steric factor p (Sec. 23.1) was determined to be 0.5 for this reaction. Reactions studied in molecular beams include D + H_2 → HD + H, F + H_2 → HF + H, K + Br_2 → KBr + Br, and Cl + CH_2=CHBr → CH_2=CHCl + Br. For details of the results, see *Moore and Pearson*, pp. 102–116.

Infrared Chemiluminescence

In the infrared-chemiluminescence technique, a gas-phase reaction is carried out at a pressure low enough for there to be a negligible probability that the products will lose vibrational and rotational energy by collisions. Instead, this energy is lost by emission of radiation (*chemiluminescence*, Sec. 21.16). For example, if the highly exothermic reaction H + F_2 → HF + F is studied, measurement of the intensities of the infrared vibration–rotation emission lines from HF tells us how the energy of the product HF is distributed among its various vibrational–rotational states, and thus it tells us the relative rates of formation of HF in these excited states. It turns out that a non-Boltzmann distribution is produced by the reaction, with a maximum population in the $v = 6$ HF vibrational level at 300 K (Fig. 23.14). (At ordinary pressures, molecular collisions would rapidly produce a Boltzmann distribution.) The energy-level distribution of product molecules is determined by the reaction's potential-energy surface [see *Laidler* (1987), sec. 12.3 for a discussion], so infrared-chemiluminescence studies give information about this surface.

Because the product HF vibrational levels show a population inversion, one can use the reaction $H + F_2 \rightarrow HF + F$ (which is an elementary step in the mechanism of the $H_2 + F_2 \rightarrow 2HF$ chain reaction) to produce a laser. A laser in which the population inversion is produced by a chemical reaction is called a *chemical laser*. The HF chemical laser was considered by the United States for use in its antimissile strategic defense ("star wars") system.

Laser Techniques

Lasers can be used to give information about molecular dynamics. For example, a laser can be used to excite a substantial fraction of one of the reactant species in a molecular beam to a specific vibrational level. We can then study the dependence of the reaction probability on the vibrational quantum state of this reactant.

A related idea is to use a tunable laser to selectively excite a particular normal vibrational mode that involves mainly vibration of a particular bond so as to preferentially break that bond, thereby controlling the outcome of the reaction with laser light. Because vibrational energy is transferred rapidly between different normal modes, this has proved to be a difficult task, but a few successes have been achieved. For example, in the competing gas-phase reactions $H + HOD \rightarrow OD + H_2$ and $H + HOD \rightarrow HO + HD$, by using a laser to excite the stretching normal mode in HOD that is mainly localized to an OH stretching vibration, one can preferentially break the OH bond in HOD; similarly, excitation of the OD stretching vibration in HOD leads to preferential breaking of the OD bond. [See *Science,* **255,** 1643 (1992); F. F. Crim, *J. Phys. Chem.,* **100,** 12725 (1996).] An alternative approach uses the coherence of laser light and certain quantum-mechanical effects in an attempt to produce desired products (P. Brumer and M. Shapiro, *Scientific American,* March 1995, p. 56). Whether control of chemical reactions by lasers will ever have commercial applications is unclear.

The breaking of a chemical bond in a molecule occurs over an extremely short time (10^{-13} to 10^{-12} s). By using very brief pulses of laser light, the bond-breaking process has been spectroscopically observed in several cases. This technique is called **femtosecond transition-state spectroscopy** (FTS) or *laser femtochemistry.* The laser light pulses used in FTS last on the order of 50 to 100 fs, where 1 femtosecond (fs) = 10^{-15} s. The experiments are done using either molecular beams or gaseous molecules in a chamber.

Unimolecular reactions are most easily studied by FTS. Figure 23.15 shows the potential-energy curves for stretching of the I—C bond for three electronic states of ICN. The ground electronic state is labeled A. The two excited states labeled B and C

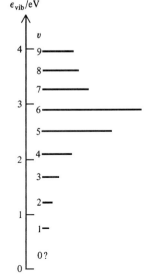

Figure 23.14

Distribution of vibrational energy in the HF formed in the gas-phase reaction $H + F_2 \rightarrow HF + F$ at 300 K as determined by infrared chemiluminescence using a low-pressure flow system. (Based on data of J. C. Polanyi and coworkers.) The length of each line is proportional to the number of HF molecules formed in vibrational level v. The $v = 0$ population cannot be found by this method because $v = 0$ molecules do not emit IR radiation. Note the unequal vibrational spacings.

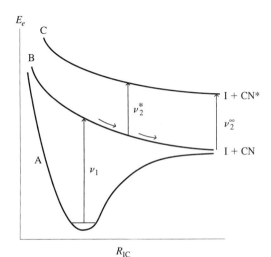

Figure 23.15

Potential-energy curves for changing the I—C bond length R_{IC} for three electronic states of ICN.

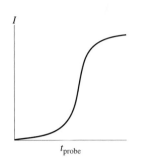

Figure 23.16

Intensity of LIF as a function of the time difference t_{probe} between pump and probe pulses in clocking experiments.

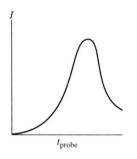

Figure 23.17

LIF intensity as a function of t_{probe} in experiments where the probe pulse has a frequency ν_2^* corresponding to the energy difference between states B and C in Fig. 23.15 at $R_{IC} = R^*$.

are repulsive, and each leads to dissociation. The dissociation products for state B are ground-state I and ground-state CN, and for state C are ground-state I and CN in an excited state symbolized by CN*. At time $t = 0$ a pulse of laser light (called the *pump pulse*) of frequency ν_1 excites ground-state ICN molecules to state B; the state-B molecules then dissociate over a short time to yield the fragments I + CN.

In one kind of experiment (called *clocking*), the pump pulse is followed at time t_{probe} by a laser pulse (called the *probe pulse*) whose frequency ν_2^∞ corresponds to the energy difference between CN and CN* (this is the energy difference between states B and C for I—C bond distance $R_{IC} = \infty$). The probe pulse therefore excites CN product molecules to the excited state CN*. One then observes the intensity I of the LIF (laser-induced fluorescence) from CN* as it returns to its ground electronic state. The experiment is then repeated many times with different values of t_{probe} (the time difference between pump and probe pulses). A smoothed plot of I versus t_{probe} looks like Fig. 23.16. As t_{probe} increases from zero, the LIF from CN* increases, thereby showing the increase in CN* concentration with time. It takes about 200 fs for the CN* concentration to reach half its maximum value, and this half-life time measures the time needed for the excited ICN molecule in state B to dissociate.

In the clocking experiment, the probe laser pulse is absorbed by the reaction product CN. In a second kind of experiment, the probe laser pulse is absorbed by state-B ICN molecules that are in the process of dissociating. Note from Fig. 23.15 that the spacing between states B and C decreases slightly as the distance R_{IC} decreases. Therefore, if the probe pulse frequency is set at a value ν_2^* that is less than ν_2^∞ and that corresponds to the vertical energy difference $E_{e,C}(R^*) - E_{e,B}(R^*)$ between the electronic energies of states B and C for I—C bond length equal to R^*, only those state-B ICN molecules whose value of R_{IC} is close to R^* will absorb from the probe pulse.

In this kind of experiment, one varies the time t_{probe} between pump and probe pulses and measures the LIF intensity (due to emission from state C) versus t_{probe}. A typical result is shown in Fig. 23.17. For times shortly after the pump pulse, the LIF intensity is weak because few ICN state-B molecules have R_{IC} close to R^*, and few state-C molecules are formed by absorption from the probe pulse. As t_{probe} increases, the LIF intensity goes through a maximum that occurs at a time when the greatest number of dissociating state-B ICN molecules have R_{IC} close to R^*. One can repeat this experiment with a different value of ν_2^*, corresponding to a different value of R^*.

It is hard to apply FTS to a bimolecular reaction, because in two intersecting molecular beams, collisions between reacting molecules occur at different times, so the reaction starts at different times for different pairs of reacting molecules. One way around this difficulty uses van der Waals molecules (Sec. 22.10), as follows. One prepares the van der Waals species IH $\cdots$ OCO in a jet-cooled mixture of HI and CO_2 in He and then dissociates the HI in the van der Waals molecules with a laser pulse. The H atom then reacts with the nearest O in OCO according to H + OCO $\rightarrow$ OH + CO. A probe laser pulse is used to detect the OH product.

One can use the results of FTS experiments to obtain information on the potential-energy surface for a reaction.

The 1999 Nobel Prize in chemistry was awarded to Ahmed Zewail for his work in femtochemistry. See A. H. Zewail, *J. Phys. Chem. A,* **104,** 5660 (2000); *Scientific American,* December 1990, p. 76.

23.4 TRANSITION-STATE THEORY FOR IDEAL-GAS REACTIONS

The rigorously correct way to calculate a reaction's rate constant theoretically is (*a*) to solve the electronic (time-independent) Schrödinger equation for a very large number of configurations of the nuclei, so as to generate the complete potential-energy surface for the reaction; (*b*) if light species are not involved, use this surface to perform classical trajectory calculations for a wide variety of initial reactant states and suitably average the results to obtain *k*. If light species are involved, corrections to the classical trajectory calculations to allow for tunneling are needed; alternatively, one can deal with the collisions using the time-dependent Schrödinger equation instead of classical mechanics.

The very great difficulties involved in this procedure make it highly desirable to have a simpler, approximate theory of rate constants. Such a theory is the **transition-state theory (TST),** also called the **activated-complex theory (ACT).** TST was developed in the 1930s by Pelzer and Wigner, Evans and M. Polanyi, and Eyring, and has been widely applied by Eyring and coworkers. For the history of the development of TST, see K. J. Laidler and M. C. King, *J. Phys. Chem.,* **87,** 2657 (1983). TST eliminates the need for trajectory calculations and requires that the potential-energy surface be known only in the region of the reactants and the region of the transition state. This section develops TST for ideal-gas reactions. TST for reactions in liquid solution is discussed in Sec. 23.8.

We saw in Sec. 23.2 that the potential-energy surface for a reaction has a reactant region and a product region that are separated by a barrier. TST chooses a boundary surface located between the reactant and product regions and assumes that all supermolecules that cross this boundary surface become products. The boundary surface, called the **(critical) dividing surface,** is taken to pass through the saddle point of the potential-energy surface. For the H_3 potential-energy contour map of Fig. 23.3, the critical dividing surface is a straight line that starts at the origin, passes through points *w*, *s*, and *v*, and extends out through the H + H + H region. (Most supermolecules cross the dividing line not too far from the saddle point *s*.) Figure 23.3 considers only collisions with $\theta = 180°$. The complete critical dividing surface for the H + H_2 reaction is shown in the more general Fig. 23.7.

TST assumes that *all supermolecules that cross the critical dividing surface from the reactant side become products.* This is reasonable, since once a supermolecule crosses the critical surface it is a downhill journey to products. A second assumption of TST is that *during the reaction, the Boltzmann distribution of energy is maintained for the reactant molecules.* This assumption was also used in the hard-sphere collision theory and, as noted in Sec. 23.1, is usually accurate. A third assumption is that *the supermolecules crossing the critical surface from the reactant side have a Boltzmann distribution of energy corresponding to the temperature of the reacting system.* These supermolecules originated from collisions of reactant molecules, and the reactant molecules have a Boltzmann distribution, so the third assumption is not unreasonable.

Let the elementary ideal-gas reaction under consideration be

$$B + C + \cdots \rightarrow E + F + \cdots$$

The reaction may have any molecularity, so "B + C + · · ·" is to be interpreted as "B," "B + C," or "B + C + D," according to whether the reaction is unimolecular, bimolecular, or trimolecular. Also, C might be the same as B.

As noted in Sec. 23.3, not all supermolecules cross the dividing surface at precisely the saddle point of the potential-energy surface. An **activated complex** is any supermolecule whose nuclear configuration corresponds to any point on the dividing surface or to any point within a short distance δ beyond the dividing surface. We expect that most activated complexes will have configurations reasonably close to the saddle-point configuration. The saddle point corresponds to the "equilibrium" structure of the activated complex, and points on the dividing surface near the saddle point correspond to "vibrations" about the "equilibrium" structure.

For H + H_2, the saddle-point geometry ("equilibrium" activated-complex structure) is the linear symmetric structure H · · · H · · · H. Points on the dividing line through *v*, *s*, *w* in Fig. 23.18 and lying to either side of *s* correspond to configurations in which R_{ab}

Figure 23.18

The region of existence of $\theta = 180°$ activated complexes for the $H + H_2$ reaction is between the parallel lines separated by δ.

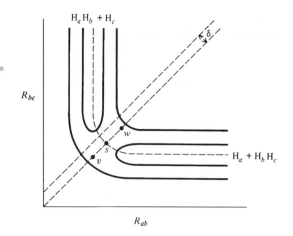

and R_{bc} are equal to each other but differ from the saddle-point R_{ab} distance (0.930 Å). Such points correspond to the symmetric stretching vibration ν_1 in Fig. 21.26. For linear $H + H_2$ collisions, the activated complexes have configurations lying between the two parallel dashed lines in Fig. 23.18. If we start at point s and move toward products on a line perpendicular to vsw, R_{ab} is decreasing and R_{bc} is increasing. This corresponds to the asymmetric stretching vibration ν_3 in Fig. 21.26. An arbitrary point in the region between the parallel lines in Fig. 23.18 corresponds to a superposition of the vibrations ν_1 and ν_3. When all $H + H_2$ collisions (including nonlinear ones) are considered, the activated complexes have configurations lying between the dividing surface in Fig. 23.7 and a second surface (not shown) that is a distance δ beyond the dividing surface. If we move up or down in Fig. 23.7, the angle θ varies, which corresponds to the degenerate bending vibration ν_2 in Fig. 21.26. An arbitrary activated complex corresponds to a superposition of the vibrations ν_1, ν_2, and ν_3.

A given activated complex exists only momentarily and does not actually undergo repeated bending and stretching vibrations. Instead, since supermolecules cross the dividing surface at various points, any given activated complex can be considered to be in a vibrational state that corresponds to the point at which the dividing surface is crossed. TST assumes these vibrational states to be populated in accord with the Boltzmann distribution.

The term **transition state** is used with a variety of slightly different meanings. One usage takes "transition state" as synonymous with "activated complex," and thus refers to all points lying on or between two parallel surfaces a distance δ apart. A very common usage by chemists defines the transition state as the saddle-point configuration, that is, the "equilibrium" configuration of the activated complex. A third usage is to take "transition state" as referring to any point on the (critical) dividing surface.

Denoting an activated complex by $X_f^\ddagger$, we write the elementary reaction $B + C + \cdots \rightarrow E + F + \cdots$ as

$$B + C + \cdots \rightarrow \{X_f^\ddagger\} \rightarrow E + F + \cdots \qquad (23.8)$$

The braces are a reminder that an activated complex is not a stable species or a reaction intermediate (Sec. 17.6) but is simply one stage in the smooth, continuous transformation of reactants to products in the elementary reaction. The subscript f indicates that we are talking about activated complexes that are crossing the dividing surface in the forward direction from reactants to products. If any reverse reaction $E + F + \cdots \rightarrow B + C + \cdots$ is occurring, there are also present activated complexes $\{X_b^\ddagger\}$ that cross the dividing surface in the reverse (back) direction; these are of no concern to us, since we are considering only the rate of the forward reaction.

TST postulates a Boltzmann distribution for the reactants B, C, . . . and for the activated complex $X_f^\ddagger$. The arguments preceding Eq. (22.129) show that when the species B, C, ... and the species $X_f^\ddagger$ (which is formed from B, C, ...) are present with each species distributed among its states according to the Boltzmann distribution law, then [see Eq. (22.129)]

$$\frac{N_f^\ddagger}{N_B N_C \cdots} = \frac{z_\ddagger}{z_B z_C \cdots} \exp\left(-\Delta\varepsilon_0^\ddagger/kT\right) \qquad (23.9)$$

where $N_f^\ddagger$, N_B, ... are the numbers of molecules of $X_f^\ddagger$, B, ..., where $z_\ddagger$, z_B, ... are the molecular partition functions of $X_f^\ddagger$, B, ..., and where

$$\Delta\varepsilon_0^\ddagger \equiv \varepsilon_0(X_f^\ddagger) - \varepsilon_0(B) - \varepsilon_0(C) - \cdots$$

is the difference between the energy of $X_f^\ddagger$ in its lowest state and the energies of the reactants B, C, ... in their lowest states. The quantity $\Delta\varepsilon_0^\ddagger$ differs somewhat from the classical barrier height ε_b, because of the zero-point vibrational energies of $X_f^\ddagger$, B, C, ...; see Fig. 23.19.

Division of each N in (23.9) by $N_A V$ converts it to a molar concentration. Therefore, (23.9) can be written as

$$K_f \equiv \frac{[X_f^\ddagger]}{[B][C]\cdots} = \frac{z_\ddagger/N_A V}{(z_B/N_A V)(z_C/N_A V)\cdots} \exp\left(-\Delta\varepsilon_0^\ddagger/kT\right) \qquad (23.10)$$

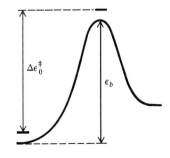

Figure 23.19

Relation between ε_b and $\Delta\varepsilon_0^\ddagger$.

Since (23.10) has the same form as (22.128), we defined K_f as $[X_f^\ddagger]/[B][C]\cdots$. The quantity K_f looks like an equilibrium constant, and it is often said that TST assumes that the activated complexes are in equilibrium with reactants. This statement is misleading and is best avoided. The word "equilibrium" suggests that molecules of $X_f^\ddagger$ sit around for a while, and then some of them go on to form products and some go back to reactants, but this is not what happens. The symbol $X_f^\ddagger$ denotes those supermolecules that cross the critical dividing surface from the reactant side. By hypothesis, such supermolecules always go on to form products. (It's true that supermolecules formed in low-energy collisions go only part way up the barrier and then "roll" back down to form separated reactants, but such supermolecules are not activated complexes, since they do not reach the critical dividing surface.) The activated complexes are not in a true chemical-reaction equilibrium with the reactants. Instead the activated complexes are assumed to be in *thermal* equilibrium with the reacting system, the activated-complex states being populated according to the Boltzmann distribution appropriate to the system's temperature.

A nonlinear activated complex with $\mathcal{N}$ atoms has three translational, three rotational, and $3\mathcal{N} - 6$ vibrational degrees of freedom (coordinates). The "equilibrium" structure of the activated complex lies at a saddle point. Since it lies on the minimum-energy path, the saddle point is a point of minimum potential energy for all coordinates but one, for which it is a maximum point. The first derivative vanishes at either a minimum or a maximum, so all the first derivatives of the potential energy V for nuclear motion are zero at the saddle point and V of the activated complex can be approximated as a quadratic function of the normal vibrational coordinates, just as for an ordinary molecule (Sec. 21.8).

The one vibrational normal coordinate of the activated complex for which V at the saddle point is a maximum is called the **reaction coordinate.** The normal "vibration" corresponding to the reaction coordinate is anomalous. Because the potential-energy surface slopes downhill along the reaction coordinate to either side of the saddle point, there is no restoring force for this "vibration," and a stable back-and-forth vibration along the reaction coordinate is impossible. Instead, nuclear motion along this coordinate breaks up the activated complex into products. For example, for the reaction $H_a + H_b H_c$, the anomalous vibrational mode of the H_3 activated complex is ν_3 (Fig. 21.26):

$$H_a \rightarrow \quad \leftarrow H_b \qquad H_c \rightarrow \qquad (23.11)$$

This mode leads to formation of the H_a—H_b bond and breaking of the H_b—H_c bond to give the products H_aH_b + H_c. As H_c moves away and H_a and H_b move toward each other, V decreases, so there is no restoring force to bring H_c back.

The reaction coordinate may be depicted as a line on the potential-energy contour map. In Fig. 23.18 for H + H_2, the reaction coordinate would be a line perpendicular to line vsw and going through point s. Along this line, R_{ab} decreases and R_{bc} increases. (The reaction coordinate is defined only in the region of the saddle point.) For H + H_2, the direction of the reaction coordinate is tangent to the minimum-energy path at the saddle point. This is not true for reactions with unsymmetrical activated complexes (for example, H + F_2). The TST critical dividing surface is defined to pass through the saddle point and to be oriented perpendicular to the reaction coordinate.

The partition function $z_{\ddagger}$ of the activated complex is given by Eq. (22.58) as $z_{\ddagger} = z_{tr}^{\ddagger} z_{rot}^{\ddagger} z_{vib}^{\ddagger} z_{el}^{\ddagger}$. From Eq. (22.110), $z_{vib}^{\ddagger}$ is the product of partition functions for the normal vibrational modes. Singling out the reaction coordinate for special attention, we write $z_{vib}^{\ddagger} = z_{rc} z_{vib}^{\ddagger\prime}$, where z_{rc} is the partition function for the anomalous motion along the reaction coordinate and $z_{vib}^{\ddagger\prime}$ is the partition function of the ordinary vibrational modes; $z_{vib}^{\ddagger\prime}$ is a product over $3\mathcal{N} - 7$ or $3\mathcal{N} - 6$ vibrational modes, according to whether the saddle-point geometry is nonlinear or linear. Therefore

$$z_{\ddagger} = z_{rc} z_{\ddagger}' \tag{23.12}$$

$$z_{\ddagger}' \equiv z_{tr}^{\ddagger} z_{rot}^{\ddagger} z_{vib}^{\ddagger\prime} z_{el}^{\ddagger} \tag{23.13}$$

We must decide how to treat motion along the reaction coordinate in order to calculate z_{rc}. Let Q_{rc} be the distance along the reaction coordinate. As noted, $\partial V/\partial Q_{rc} = 0$ at the saddle point. Hence, V is nearly constant along the reaction coordinate for the short distance δ (Fig. 23.18) that defines the region of existence of activated complexes. δ will cancel in the final expression for the rate constant, so we need not specify its value except to say that δ is short enough to prevent V from varying significantly along Q_{rc}. With $V \approx$ const along the reaction coordinate, the force component for motion along the reaction coordinate is approximately zero: $F_{rc} = -\partial V/\partial Q_{rc} \approx 0$. The motion of the activated complexes along the reaction coordinate is therefore treated as a one-dimensional translational motion of a free particle ("free" means no forces act) confined to a region of length δ, the region of existence of activated complexes. Translation along the reaction coordinate is an *internal* motion of activated-complex nuclei relative to one another [see, for example, Eq. (23.11)] and is to be distinguished from the ordinary translational motion of the activated complex through the three-dimensional space of the container. The former motion corresponds to z_{rc} in (23.12); the latter corresponds to $z_{tr}^{\ddagger}$ in (23.13).

The partition function for a particle undergoing translation in a one-dimensional box of length a is given by the first sum in Eq. (22.78), and this sum was found to equal $(2\pi mkT)^{1/2}a/h$. Replacing a by δ and m by m_{rc}, where m_{rc} is the effective mass for motion along the reaction coordinate, we get $(2\pi m_{rc}kT)^{1/2}\delta/h$. (The expression for m_{rc} can be worked out if desired; however, since m_{rc} will eventually cancel, we need not worry about it.) The first sum in (22.78) assumes motion in both positive and negative directions along the coordinate axis, but we are considering only activated complexes that are moving forward along Q_{rc}. The partition function is a sum over states, and half the possible states are missing when reverse motion is excluded. We therefore add a factor of $\frac{1}{2}$ to get z_{rc} for $X_f^{\ddagger}$; thus,

$$z_{rc} = \tfrac{1}{2}(2\pi m_{rc}kT)^{1/2}\delta/h \tag{23.14}$$

$$z_{\ddagger} = \tfrac{1}{2}(2\pi m_{rc}kT)^{1/2}\delta h^{-1} z_{\ddagger}' \tag{23.15}$$

where (23.12) was used.

Since TST assumes that all supermolecules crossing the dividing surface become products, we need only calculate the rate at which supermolecules cross the dividing surface to find the reaction rate. Let $v_{rc} \equiv dQ_{rc}/dt$ be the velocity component of a given activated complex along the reaction coordinate. At a given time t_0 let there be $N_f^{\ddagger}$ activated complexes in the system. Let τ be the average time needed for an activated complex to move a distance δ along Q_{rc}. At time $t_0 + \tau$ all $N_f^{\ddagger}$ complexes that were present at time t_0 will have passed through the surface located a distance δ beyond the critical dividing surface and become products. The reaction rate therefore equals $N_f^{\ddagger}/\tau$. But $\tau = \delta/\langle v_{rc} \rangle$ (where $\langle v_{rc} \rangle$ is the average value of v_{rc}), so the reaction rate is $N_f^{\ddagger} \langle v_{rc} \rangle / \delta$. This rate is in terms of number of molecules per unit time. In kinetics, chemists use r, the rate in terms of moles per unit volume per unit time. Dividing by N_A to convert to moles and by V to convert to rate per unit volume, we have $r = N_f^{\ddagger} \langle v_{rc} \rangle / N_A V \delta = [X_f^{\ddagger}] \langle v_{rc} \rangle / \delta$. The use of (23.10) for $[X_f^{\ddagger}]$ gives

$$r = \frac{\langle v_{rc} \rangle}{\delta} \frac{z_{\ddagger}/N_A V}{(z_B/N_A V)(z_C/N_A V) \cdots} \exp\left(-\Delta\varepsilon_0^{\ddagger}/kT\right)[B][C]\cdots \qquad (23.16)$$

To complete the calculation of r, we need $\langle v_{rc} \rangle$. Motion along Q_{rc} is being treated as a translation. As with the other degrees of freedom of the activated complex, we assume a Boltzmann distribution of energy for this translation. Hence the fraction of complexes with speeds along Q_{rc} in the range v_{rc} to $v_{rc} + dv_{rc}$ is $Be^{-m_{rc}v_{rc}^2/2kT} dv_{rc}$, where B is a constant [see Eq. (15.42)]. To find B, we integrate this expression from $v_{rc} = 0$ to ∞ (negative values of v_{rc} are excluded, as noted above) and set the total probability equal to one: $B \int_0^{\infty} e^{-m_{rc}v_{rc}^2/2kT} dv_{rc} = 1$. The use of integral 2 in Table 15.1 gives $B = 2(m_{rc}/2\pi kT)^{1/2}$. The probability density $g(v_{rc})$ for v_{rc} is therefore

$$g(v_{rc}) = 2(m_{rc}/2\pi kT)^{1/2}e^{-m_{rc}v_{rc}^2/2kT} \qquad (23.17)$$

The average value of v_{rc} is given by (15.38) as $\langle v_{rc} \rangle = \int_0^{\infty} v_{rc}g(v_{rc}) \, dv_{rc}$. Substitution of (23.17) and the use of integral 5 of Table 15.1 give (Prob. 23.4)

$$\langle v_{rc} \rangle = (2kT/\pi m_{rc})^{1/2} \qquad (23.18)$$

Substitution of (23.18) and (23.15) into (23.16) gives

$$r = \frac{1}{\delta}\left(\frac{2kT}{\pi m_{rc}}\right)^{1/2}\frac{(2\pi m_{rc}kT)^{1/2}\delta}{2h} \frac{z_{\ddagger}'/N_A V}{(z_B/N_A V)(z_C/N_A V) \cdots} \exp\left(-\Delta\varepsilon_0^{\ddagger}/kT\right)[B][C]\cdots$$

For the elementary reaction $B + C + \cdots \rightarrow$ products, the reaction rate is $r = k_r[B][C]\cdots$, where k_r is the rate constant. (The subscript r avoids confusion with the Boltzmann constant k.) We have $k_r = r/[B][C]\cdots$, so

$$k_r = \frac{kT}{h} \frac{z_{\ddagger}'/N_A V}{(z_B/N_A V)(z_C/N_A V) \cdots} \exp\left(-\Delta\varepsilon_0^{\ddagger}/kT\right) \qquad \text{ideal gas} \qquad (23.19)$$

which is the desired TST expression for the rate constant of an ideal-gas elementary reaction and is the key result of this section.

Sometimes a factor κ (kappa) is included on the right side of (23.19). The quantity κ (which lies between 0 and 1) is called the *transmission coefficient* and allows for the possibility that the shape of the potential-energy surface might be such that some of the supermolecules crossing the critical dividing surface are reflected back to give reactants. Since there is no simple way to calculate κ, and since it is likely to lie reasonably close to 1 in most cases, it will be omitted.

A weakness in the preceding derivation of (23.19) is the rather arbitrary use of the particle-in-a-box partition function for z_{rc}. Derivations that avoid this assumption exist (see *Bernasconi*, pt. I, p. 20; *Steinfeld, Francisco, and Hase*, sec. 10.3), and these derivations give the result (23.19).

To calculate $z_{\ddagger}' = z_{\mathrm{tr}}^{\ddagger} z_{\mathrm{rot}}^{\ddagger} z_{\mathrm{vib}}^{\ddagger'} z_{\mathrm{el}}^{\ddagger}$, we need the mass of the activated complex (to calculate $z_{\mathrm{tr}}^{\ddagger}$), its equilibrium structure (to calculate the moments of inertia in $z_{\mathrm{rot}}^{\ddagger}$), its vibrational frequencies, and the degeneracy of its ground electronic level. The equilibrium structure is given by the location of the saddle point. The normal-mode vibrational frequencies can be found if the potential-energy surface is known in the region of the activated complex; vibrational frequencies are related to force constants, and force constants are the second derivatives of V with respect to the normal vibrational coordinates [see Eq. (21.23)]. To calculate $\Delta\varepsilon_0^{\ddagger}$, we need the barrier height (and the vibrational frequencies, to correct for zero-point vibrations).

The quantity kT/h in (23.19) equals 0.6×10^{13} s^{-1} at 300 K and 2×10^{13} s^{-1} at 1000 K. At temperatures of 300 to 400 K, z_{tr} is typically 10^{24} V/cm^3 to 10^{27} V/cm^3, z_{rot} is typically 10 to 1000, and each normal vibrational mode typically contributes a factor 1 to 3 to z_{vib}.

The equation preceding (22.81), which was used to find z_{rc}, is a (semi)-classical statistical-mechanical result, since it was derived by assuming the levels to be closely spaced compared with kT, so that the sum could be replaced by an integral. Moreover, $\langle v_{\mathrm{rc}} \rangle$ was calculated using a classical-mechanical Boltzmann distribution. Thus, an additional assumption of TST is that *motion along the reaction coordinate can be treated classically.* For reactions involving light species, this assumption fails, and a correction for quantum-mechanical tunneling must be applied (Sec. 23.3). Several methods for calculating tunneling corrections for TST reaction rates have been proposed [B. C. Garrett and D. G. Truhlar, *J. Phys. Chem.,* **83**, 200 (1979); *J. Chem. Phys.,* **79**, 4931 (1983)]. Several quantum-mechanical versions of TST that may give results with a correct allowance for tunneling have been proposed [J. W. Tromp and W. H. Miller, *J. Phys. Chem.,* **90**, 3482 (1986)].

EXAMPLE 23.2 TST calculation of a rate constant

The most accurate calculation of the H_3 potential-energy surface in the region of the saddle point gives a classical barrier height of 9.61 kcal/mol and a linear transition state with bond distances $R_{ab}^{\ddagger} = R_{bc}^{\ddagger} = 0.930$ Å [D. L. Diedrich and J. B. Anderson, *Science,* **258**, 786 (1992)]. The vibrational wavenumbers of the DH_2 activated complex (calculated from the H_3 surface by allowing for the change in mass) are 1764 cm^{-1} (symmetric stretching) and 870 cm^{-1} (degenerate bending). The bond length and vibrational wavenumber of H_2 are $R_{bc} = 0.741$ Å and 4400 cm^{-1}. Use the transition-state theory to calculate the rate constant for $D + H_2 \rightarrow DH + H$ at 450 K. (The experimental value is 9×10^9 cm^3 mol^{-1} s^{-1}.)

Equation (23.19) with B = D and C = H_2 and Eq. (23.13) give

$$k_r = N_A kTh^{-1} \exp\left(-\Delta\varepsilon_0^{\ddagger}/kT\right) \frac{(z_{\mathrm{tr}}^{\ddagger}/V) z_{\mathrm{rot}}^{\ddagger} z_{\mathrm{vib}}^{\ddagger'} z_{\mathrm{el}}^{\ddagger}}{(z_{\mathrm{tr,D}}/V) z_{\mathrm{el,D}} (z_{\mathrm{tr,H_2}}/V) z_{\mathrm{rot,H_2}} z_{\mathrm{vib,H_2}} z_{\mathrm{el,H_2}}}$$

Since $\widetilde{\nu} = \nu/c$, the zero-point energy (ZPE) of the activated complex is

$$\tfrac{1}{2} hc(1764 \text{ cm}^{-1} + 870 \text{ cm}^{-1} + 870 \text{ cm}^{-1}) = 3.48 \times 10^{-13} \text{ erg}$$

The ZPE of D is zero, and that of H_2 is $\tfrac{1}{2} hc\widetilde{\nu} = 4.37 \times 10^{-13}$ erg. The change in ZPE is -0.89×10^{-13} erg. The classical barrier height ε_b on a per molecule basis is found by dividing 9.61 kcal/mol by N_A to give (Prob. 23.5) 6.68×10^{-13} erg. Hence (Fig. 23.19), $\Delta\varepsilon_0^{\ddagger} = 5.79 \times 10^{-13}$ erg.

We saw following Eq. (22.92) that $z_{el,D} = 2$. Since DH_2 also has one unpaired electron, we can expect that $z^{\ddagger}_{el} = 2$. The H_2 ground electronic level is nondegenerate, and $z_{el,H_2} = 1$.

Equation (22.85) for z_{rot} gives

$$\frac{z^{\ddagger}_{rot}}{z_{rot,H_2}} = \frac{I^{\ddagger}}{I_{H_2}}\frac{\sigma_{H_2}}{\sigma^{\ddagger}}$$

The symmetry numbers are $\sigma_{H_2} = 2$ and $\sigma^{\ddagger} = \sigma_{DH_2} = 1$. Using (18.82) and (21.41) for the moments of inertia, one finds (Prob. 23.6) $I_{H_2} = \frac{1}{2}m_H R_{bc}^2$ and $I^{\ddagger}/I_{H_2} = 8.66$. Therefore $z^{\ddagger}_{rot}/z_{rot,H_2} = 17.3$.

The use of $\tilde{\nu} = \nu/c$ and Eqs. (22.90) and (22.110) for z_{vib} gives $z_{vib,H_2} = 1.000$ and $z^{\ddagger\prime}_{vib} = 1.140$ at 450 K (Prob. 23.7).

Equation (22.81) for z_{tr} gives at 450 K (Prob. 23.7)

$$\frac{z^{\ddagger}_{tr}/V}{(z_{tr,D}/V)(z_{tr,H_2}/V)} = \left(\frac{m_{DH_2}}{m_D m_{H_2}}\right)^{3/2}\left(\frac{h^2}{2\pi kT}\right)^{3/2} = 5.51 \times 10^{-25}\ cm^3$$

Since z_{tr} is dimensionless, the above ratio must have units of volume.

Substitution in the expression for k_r gives at 450 K (Prob. 23.7)

$$k_r = 5.5 \times 10^9\ cm^3\ mol^{-1}\ s^{-1}$$

[The TST value is substantially lower than the experimental value $9 \times 10^9\ cm^3\ mol^{-1}\ s^{-1}$, mainly because of the neglect of tunneling. An approximate calculation of the tunneling correction gives a factor of 2.0 for this reaction at 450 K, which brings the TST rate into much better agreement with the experimental value; see B. C. Garrett and D. G. Truhlar, *J. Chem. Phys.*, **72**, 3460 (1980).]

The treatment of the symmetry of the transition state in TST has been the subject of controversy. See *Laidler* (1987), sec. 4.5.4.

Relation between TST and Hard-Sphere Collision Theory

For the bimolecular reaction B + C → products (with B ≠ C), suppose we ignore the internal structure of the colliding molecules and treat them as hard spheres of radii r_B and r_C. Then the reactants' partition functions are $z_B = z_{tr,B}$ and $z_C = z_{tr,C}$. Equation (22.81) gives

$$z_B/V = (2\pi m_B kT/h^2)^{3/2}, \qquad z_C/V = (2\pi m_C kT/h^2)^{3/2}$$

The most reasonable choice of transition state is the two hard spheres in contact. An ordinary diatomic molecule has one vibrational mode, so the "diatomic" activated complex has zero vibrational modes, since the reaction coordinate replaces the one vibration. The spheres of masses m_B and m_C are separated by a center-to-center distance of $r_B + r_C$ in the transition state, and (18.82) gives the moment of inertia as $I = \mu(r_B + r_C)^2$, where $\mu \equiv m_B m_C/(m_B + m_C)$ [Eq. (18.79)]. Equations (22.81) and (22.85) give the partition function of the activated complex as

$$\frac{z'_{\ddagger}}{V} = \frac{z^{\ddagger}_{tr}}{V}z^{\ddagger}_{rot} = \left[\frac{2\pi(m_B + m_C)kT}{h^2}\right]^{3/2}8\pi^2\frac{m_B m_C}{m_B + m_C}(r_B + r_C)^2\frac{kT}{h^2}$$

Substitution in the TST equation (23.19) gives for B ≠ C

$$k_r = N_A\pi(r_B + r_C)^2\left[\frac{8kT}{\pi}\left(\frac{m_B + m_C}{m_B m_C}\right)\right]^{1/2}\exp\left(-\Delta\varepsilon^{\ddagger}_0/kT\right) \qquad (23.20)$$

which is identical to the hard-sphere collision-theory result (23.3) if we take $\Delta\varepsilon_0^{\ddagger}$ to be the threshold energy $\varepsilon_{\text{thr}} = E_{\text{thr}}/N_A$. If B = C, then $\sigma^{\ddagger} = 2$, and TST reduces to (23.4). Thus, *TST reduces to the hard-sphere collision theory when the structure of the molecules is ignored.*

Temperature Dependence of the Rate Constant

To find the temperature dependence of the TST rate constant (23.19), we must examine the temperature dependences of the partition functions. Equations (22.81), (22.85), (22.109), and (22.92) give

$$z_{\text{tr}} \propto T^{3/2}, \quad z_{\text{rot,lin}} \propto T, \quad z_{\text{rot,nonlin}} \propto T^{3/2}, \quad z_{\text{el}} \propto T^0$$

The temperature dependence of z_{vib} is not so simple. At temperatures such that $kT \ll h\nu_s$ for all ν_s, Eq. (22.110) becomes $z_{\text{vib}} \approx 1 = T^0$. At temperatures such that $kT \gg h\nu_s$ for all ν_s, expansion of the exponentials in (22.110) gives (Prob. 23.10) $z_{\text{vib}} \propto T^{f_{\text{vib}}}$, where f_{vib} is the number of vibrational modes of the molecule. For an intermediate temperature, $z_{\text{vib}} \propto T^b$, where b is between 0 and f_{vib}. For most vibrations, ν_s is sufficiently high for the condition $kT \gg h\nu_s$ to be reached only at quite high temperatures. For a moderate temperature, we can expect that

$$z_{\text{vib}} \propto T^a \qquad \text{where } 0 \le a \le \tfrac{1}{2}f_{\text{vib}}$$

Over a restricted temperature range, each z_{vib} in (23.19) will have its a value approximately constant, and we can write

$$k_r \approx CT^m \exp\left(-\Delta E_0^{\ddagger}/RT\right) \tag{23.21}$$

where C and m are constants, and where $\Delta E_0^{\ddagger} \equiv N_A \Delta\varepsilon_0^{\ddagger}$.

Using the temperature dependences of the factors in the partition functions in (23.19), one can deduce the range of values of m. This is left as an exercise (Prob. 23.11). The results are as follows: (*a*) for a bimolecular gas-phase reaction between an atom and a molecule, m usually lies between -0.5 and 0.5; (*b*) for a bimolecular gas-phase reaction between two molecules, m usually lies between -2 and 0.5.

The activation energy is defined by (17.68) as $E_a \equiv RT^2\, d\ln k_r/dT$. Taking the log of (23.21) and differentiating, we get

$$E_a = \Delta E_0^{\ddagger} + mRT \tag{23.22}$$

Since m can be negative, zero, or positive, E_a can be less than, the same as, or greater than $\Delta E_0^{\ddagger}$. The quantity $\Delta E_0^{\ddagger}$ differs from the classical barrier height E_b by $\Delta\text{ZPE}^{\ddagger}$, the change in zero-point energy on formation of the activated complex (Fig. 23.19). Since $\Delta\text{ZPE}^{\ddagger}$ can be negative, zero, or positive, E_a can be less than, the same as, or greater than E_b.

The Arrhenius A factor is defined by (17.69) as $A = k_r e^{E_a/RT}$. The use of (23.22), (23.19), and (23.21) gives

$$A = \frac{kTe^m}{h} \frac{z'_{\ddagger}/N_A V}{(z_B/N_A V)(z_C/N_A V)\cdots} \approx Ce^m T^m \tag{23.23}$$

where m can be calculated if $z'_{\ddagger}$ is known. The accuracy of gas-kinetics data is often too poor to allow m to be found experimentally. The T^m dependence of the A factor is overwhelmed by the exponential function $\exp\left(-\Delta E_0^{\ddagger}/RT\right)$ in k_r, and kinetics data give only an A averaged over the temperature range of the data. Equation (23.21), which has three parameters (C, m, $\Delta E_0^{\ddagger}$), is commonly used to fit rate-constant data when accurate values over a wide temperature range have been measured.

For further discussion, see W. C. Gardiner, *Acc. Chem. Res.*, **10**, 326 (1977).

Isotope Effects

Consider a reaction whose rate-determining step involves breaking a C—H bond in a reactant. Suppose we substitute the hydrogen atom in this bond with deuterium. The frequency of the C—H stretching vibration is $\nu = (1/2\pi)(k/\mu)^{1/2}$. We have $\mu = m_1 m_2/(m_1 + m_2) \approx m_1 m_2/m_2 = m_1$, where $m_1 = m_H$ and m_2 is the mass of the rest of the molecule, and we used $m_2 \gg m_1$. Substitution of D for H does not change the force constant k and gives $\mu \approx m_1 = m_D \approx 2m_H$, so $\nu_{CD} \approx \nu_{CH}/2^{1/2}$. The zero-point vibrational energy of the reactant is lowered by $\frac{1}{2}h(\nu_{CH} - \nu_{CD}) = \frac{1}{2}h\nu_{CH}(1 - 1/2^{1/2}) = 0.146h\nu_{CH}$. In the activated complex, stretching the C—H or C—D bond corresponds to motion along the reaction coordinate that breaks up the activated complex to products (as discussed earlier in this section) and the CH or CD stretching "vibration" does not contribute at all to the zero-point energy of the activated complex. Of course, isotopic substitution has no effect on the potential-energy surface, which is found by solving the electronic Schrödinger equation.

Since the ZPE of the reactants is lowered by $0.146h\nu_{CH}$ and since the potential-energy surface and the ZPE of the activated complex are unaffected, substitution of D for H will increase $\Delta\varepsilon_0^{\ddagger}$ in Fig. 23.19 by $0.146h\nu_{CH}$. The factor $\exp(-\Delta\varepsilon_0^{\ddagger}/kT)$ in the TST equation (23.19) for k_r will then lower k_r substantially (by a factor of 8 at room temperature—Prob. 23.14). If tunneling is important, this will further lower the rate constant for the deuterated species, since the heavier D tunnels less readily than H. Isotopic substitution will also affect other vibrational frequencies, but these vibrations occur in both the reactant and the activated complex and their changes will result in only a small net change in rate. Likewise, the effects of isotopic substitution on z_{tr} and z_{rot} will affect k_r only slightly.

The preceding discussion assumes the H atom breaks away from the molecule without being transferred to another species. However, the more common situation is the transfer of an H atom, with a new bond being formed as the old one is broken: MH + N → M + HN, where the transition state is M $\cdots$ H $\cdots$ N. For this case, one finds that (in the absence of tunneling) the room-temperature factor-of-8 lowering on deuteration is the maximum effect and smaller effects occur if the transition state is unsymmetrical. (See *Moore and Pearson*, pp. 367–370; R. P. Bell, *The Tunnel Effect in Chemistry*, Chapman and Hall, 1980, chap. 4.)

If substitution of D for H in a C—H bond lowers the rate constant k_r substantially, the rate-determining step involves breaking that bond; if k_r changes only slightly on isotopic substitution, the rate-determining step may not involve breaking that bond. This is one technique used to determine reaction mechanisms. See T. H. Lowry and K. S. Richardson, *Mechanism and Theory in Organic Chemistry*, 2d ed., Harper and Row, 1981, pp. 205–212, 222–225.)

Tests of TST

Since accurate potential-energy surfaces are not known for most chemical reactions, one must usually guess at the structure of the activated complex and estimate its vibrational frequencies using approximate empirical rules that relate bond lengths and vibrational frequencies. Herschbach and coworkers used TST to calculate the A factors of 12 bimolecular gas-phase reactions. Experimental A values can be in error by a factor of 3 or 4. One might allow a factor of perhaps 2 or 3 as the error in $z_{\ddagger}'$, since it is found by guesswork. Thus, a TST A value that is no more than $10^{\pm 1}$ times the experimental value can be regarded as a confirmation of TST. Ten of the 12 calculated A values were within a factor of 10 of being correct. [D. Herschbach et al., *J. Chem. Phys.*, **25,** 736 (1956); *Knox*, pp. 237–239.]

Reactions for which the potential-energy surface has been accurately calculated involve light species, for which tunneling is important, so it is hard to test TST in these cases. However, a quantum-mechanical version of TST has given good agreement with experimental k_r values for the reaction $D + H_2 \rightarrow HD + H$. [See S. Chapman, B. C. Garrett, and W. H. Miller, *J. Chem. Phys.*, **63**, 2710 (1975).]

Another kind of test compares the predictions of TST with the results of trajectory calculations (Sec. 23.3). The majority of such comparisons have given pretty good agreement. However, in some cases, trajectory calculations show that the states of the activated complex are not populated according to the Boltzmann distribution. With certain shapes of potential-energy surface, trajectory calculations show that a significant fraction of supermolecules are reflected back to reactants after crossing the critical dividing surface. See K. Morokuma and M. Karplus, *J. Chem. Phys.*, **55**, 63 (1971); D. G. Truhlar, *J. Phys. Chem.*, **83**, 188 (1979).

Still another way to test TST is to make an isotopic substitution in a reactant and compare the observed effect on k_r with that calculated by TST. Kinetic-isotope tests tend to confirm the validity of TST. (See *Weston and Schwarz*, sec. 4.11.)

Although definitive results are not available, the weight of the evidence is that the transition-state theory works reasonably well in most (but not all) cases. Moreover, the theory is immensely helpful in providing a conceptual understanding of chemical reactions. A review of the theory concluded that "transition-state theory provides a useful conceptual framework for discussing chemical reactions under almost all conditions" [D. G. Truhlar, W. L. Hase, and J. T. Hynes, *J. Phys. Chem.*, **87**, 2664 (1983)].

For extensions to transition-state theory, see *Laidler* (1987), sec. 4.9; D. G. Truhlar et al., *J. Phys. Chem.*, **100**, 12771 (1996).

TST and Transport Properties

The applicability of TST is wider than the above derivation implies, and the theory can be applied to such rate processes as diffusion and viscous flow in liquids; see *Hirschfelder, Curtiss, and Bird*, sec. 9.2. (To diffuse from one location to another, a molecule in a liquid must squeeze past its neighbors; the variation in potential energy for this process resembles Fig. 23.5.)

23.5 THERMODYNAMIC FORMULATION OF TST FOR GAS-PHASE REACTIONS

Comparing (23.19) with (22.128), we see that the stuff that follows kT/h in (23.19) resembles an equilibrium constant, the only difference being that $z_{\ddagger}'$ is not the complete partition function of the activated complex but omits the contribution of z_{rc} [see Eq. (23.12)]. It is therefore customary to define $K_c^{\ddagger}$ by

$$K_c^{\ddagger} \equiv \frac{z_{\ddagger}'/N_A V}{(z_B/N_A V)(z_C/N_A V) \cdots} \exp\left(-\Delta\varepsilon_0^{\ddagger}/kT\right) \qquad (23.24)$$

Note from (23.10) that $K_c^{\ddagger} \neq [X_{\ddagger}']/[B][C] \cdots$. Instead Eqs. (23.10) and (23.15) give

$$K_c^{\ddagger} = K_f \left(\frac{2}{\pi m_{rc} kT}\right)^{1/2} \frac{h}{\delta} \qquad (23.25)$$

where $K_f \equiv [X_{\ddagger}']/[B][C] \cdots$. The TST equation (23.19) becomes

$$k_r = \frac{kT}{h} K_c^{\ddagger} \qquad (23.26)$$

In thermodynamics, the standard state used for ideal gases is the state at $P = P^\circ \equiv 1$ bar and temperature T. In kinetics, rate constants are usually expressed in terms of concentrations, and it is more convenient to use a standard state having unit concentration rather than unit pressure. Equation (6.4) and $P_i = n_i RT/V = c_i RT$ give $\mu_i = \mu_i^\circ + RT \ln P_i/P^\circ = \mu_i^\circ + RT \ln c_i RT/P^\circ = \mu_i^\circ + RT \ln RTc^\circ/P^\circ + RT \ln c_i/c^\circ$, where $c^\circ \equiv 1$ mol/dm^3. When $c_i = c^\circ$, the chemical potential becomes $\mu_i^\circ + RT \ln RTc^\circ/P^\circ$

$\equiv \mu_{c,i}^\circ$. Therefore, $\mu_i = \mu_{c,i}^\circ + RT \ln c_i/c^\circ$, where $\mu_{c,i}^\circ$ is the chemical potential of ideal gas i at 1 mol/L concentration and temperature T. Substituting into the equilibrium condition $\Sigma_i \nu_i\mu_i = 0$ and following the same steps that led to (6.14), we get

$$\Delta G_c^\circ = -RT \ln K_c^\circ = -RT \ln \left[K_c/(c^\circ)^{\Delta n/\text{mol}} \right] \tag{23.27}$$

where $\Delta G_c^\circ \equiv \Sigma_i \nu_i\mu_{c,i}^\circ$ and $K_c^\circ \equiv \Pi_i (c_i/c^\circ)^{\nu_i} = \Pi_i c_i^{\nu_i}/\Pi_i (c^\circ)^{\nu_i} = K_c/(c^\circ)^{\Delta n/\text{mol}}$, where $K_c \equiv \Pi_i c_i^{\nu_i}$.

For the process $B + C + \cdots \rightarrow \{X_f^\ddagger\}$, we have $\Delta n/\text{mol} = 1 - n$, where n is the molecularity of the reaction. By analogy to (23.27), we define $\Delta G_c^{\circ\ddagger}$, the **concentration-scale standard Gibbs energy of activation,** as

$$\Delta G_c^{\circ\ddagger} \equiv -RT \ln \left[K_c^\ddagger (c^\circ)^{n-1} \right] \tag{23.28}$$

The use of (23.28) in (23.26) gives

$$k_r = kTh^{-1}(c^\circ)^{1-n} e^{-\Delta G_c^{\circ\ddagger}/RT} \tag{23.29}$$

This is the thermodynamic version of the TST expression for the rate constant. The higher the value of $\Delta G_c^{\circ\ddagger}$, the slower the reaction.

In analogy with (6.25) and (6.36), we define $K_P^{\circ\ddagger}$ and $\Delta H^{\circ\ddagger}$ for gas-phase reactions as

$$K_P^{\circ\ddagger} \equiv K_c^{\circ\ddagger}(RTc^\circ/P^\circ)^{1-n} \tag{23.30}$$

$$\Delta H_c^{\circ\ddagger} = \Delta H^{\circ\ddagger} \equiv RT^2 \, d \ln K_P^{\circ\ddagger}/dT \tag{23.31}$$

Since ideal-gas enthalpies depend on T only, the standard enthalpy of activation is the same whether the standard state is $P = 1$ bar or $c = 1$ mol/L, that is, $H^{\circ\ddagger} = \Delta H_c^{\circ\ddagger}$.

The **concentration-scale standard entropy of activation** $\Delta S_c^{\circ\ddagger}$ is defined by $\Delta S_c^{\circ\ddagger} \equiv (\Delta H_c^{\circ\ddagger} - \Delta G_c^{\circ\ddagger})/T$, so

$$\Delta G_c^{\circ\ddagger} = \Delta H_c^{\circ\ddagger} - T \, \Delta S_c^{\circ\ddagger} \tag{23.32}$$

Substitution of (23.32) in (23.29) gives

$$k_r = kTh^{-1}(c^\circ)^{1-n} e^{\Delta S_c^{\circ\ddagger}/R} e^{-\Delta H_c^{\circ\ddagger}/RT} \tag{23.33}$$

The quantities $\Delta G_c^{\circ\ddagger}$, $\Delta H_c^{\circ\ddagger}$, and $\Delta S_c^{\circ\ddagger}$ are the changes in G, H, and S at temperature T when 1 mole of $X_f^\ddagger$ in its standard state (1 mol/L concentration) is formed from the pure, separated reactants in their 1 mol/L standard states, except that the contributions of motion along the reaction coordinate to the properties $G^\ddagger$, $H^\ddagger$, and $S^\ddagger$ of the activated complex $X_f^\ddagger$ are omitted. In other words, in calculating $G^\ddagger$, $H^\ddagger$, and $S^\ddagger$ from the statistical-mechanical equations of Chapter 22, we use $z_\ddagger'$ instead of $z_\ddagger$ (where $z_\ddagger' = z_\ddagger/z_{rc}$).

Equations (17.68) and (23.26) give for the activation energy $E_a \equiv RT^2 d \ln k_r/dT = RT + RT^2 d \ln K_c^{\circ\ddagger}/dT$. Equations (23.30) and (23.31) give $d \ln K_c^{\circ\ddagger}/dT = d \ln K_P^{\circ\ddagger}/dT + (n-1)/T = \Delta H^{\circ\ddagger}/RT^2 + (n-1)/T$. Therefore

$$E_a = \Delta H^{\circ\ddagger} + nRT \qquad \text{gas-phase reaction} \tag{23.34}$$

where n is the molecularity of the reaction.

Equation (5.10) gives $\Delta H^{\circ\ddagger} = \Delta U^{\circ\ddagger} + (1-n)RT$. Substitution in (23.34) gives $E_a = \Delta U^{\circ\ddagger} + RT$, where $\Delta U^{\circ\ddagger}$ equals U_m for $X_f^\ddagger$ at T (the contribution of motion along the reaction coordinate being omitted) minus $U_{m,B} + U_{m,C} + \cdots$ at T. This equation gives a simple physical interpretation to the activation energy E_a.

Equation (17.69) gives the pre-exponential factor as $A \equiv k_r e^{E_a/RT}$, and use of (23.33) and (23.34) gives as the relation between A and the activation entropy $\Delta S_c^{\circ\ddagger}$:

$$A = (kT/h)(c^\circ)^{1-n} e^n e^{\Delta S_c^{\circ\ddagger}/R} \qquad \text{gas-phase reaction} \tag{23.35}$$

From the experimental values of A and E_a, we can calculate $\Delta S_c^{\circ\ddagger}$ and $\Delta H_c^{\circ\ddagger}$ using (23.34) and (23.35). Equation (23.32) then gives $\Delta G_c^{\circ\ddagger}$.

For a bimolecular gas-phase reaction, the activated complex has fewer translational and rotational degrees of freedom and more vibrational degrees of freedom than the pair of reactant molecules. The larger spacing between vibrational levels compared with rotational and translational levels makes the contribution of a vibration to S less than that of a rotation or translation. The activation entropy $\Delta S_c^{\circ\ddagger}$ is therefore usually negative for a gas-phase bimolecular reaction.

23.6   UNIMOLECULAR REACTIONS

In a unimolecular decomposition or isomerization of a polyatomic molecule A, the elementary chemical reaction A* → products is preceded by the elementary physical reaction A + M → A* + M, which puts A into an excited vibrational level (Sec. 17.11).

For a gas-phase unimolecular reaction in the low-pressure falloff region, the rate of formation of vibrationally excited A molecules (symbolized by A*) falls below that needed to maintain the Boltzmann population of A*. Since the TST equation (23.19) assumes an equilibrium population of reactant states, we cannot apply (23.19) to the overall reaction A → products in the falloff region. In the high-pressure region, the Boltzmann population of A* is maintained, so TST can be used to calculate $k_{uni,P=\infty}$, the experimentally observed high-pressure rate constant.

For many unimolecular reactions, the potential-energy surface shows a saddle point, and we can identify the transition state with this saddle point.

For example, for the unimolecular isomerization reaction *cis*-CHF=CHF → *trans*-CHF=CHF, the minimum-energy path between the reactant and product involves a 180° rotation of one CHF group with respect to the other. As these groups rotate, the overlap between the carbon $2p$ AOs that form the π bond is gradually lost, becoming zero at a 90° rotation. The point of maximum potential energy on the minimum-energy path is at 90°, and this is the transition state. During the rotation from the cis to the trans compound, the bond distances will change, but these changes are small compared with the change in twist angle. Therefore, $V \approx V(\theta)$, where θ is the angle between the two CHF planes. With only one variable, the potential-energy surface becomes a line (Fig. 23.20). The carbon–carbon double- and single-bond energies are 615 and 344 kJ/mol (Table 20.1), so we expect the transition-state energy to lie about 271 kJ/mol above the reactant's energy. The observed E_a is 264 kJ/mol. The barrier is surmounted when enough vibrational energy flows into the vibrational normal mode that involves twisting about the CC axis.

For the unimolecular decomposition $CH_3CH_2Cl \rightarrow CH_2{=}CH_2 + HCl$, the likely transition state is shown in Fig. 23.20b. We can expect this transition state, with its partially broken HC and ClC bonds and partially formed CC π and HCl bonds, to lie at a maximum on the minimum-energy path.

There are, however, many unimolecular decompositions whose potential-energy surfaces have no saddle point. Figure 23.21 shows potential-energy contour maps for the molecules HCN and CO_2 with the bond angle θ restricted to its equilibrium value of 180°. The points at the centers of the 2-eV contours correspond to the HCN and CO_2 equilibrium geometries; each of these points lies at the bottom of a well. For the unimolecular reaction A → products, the supermolecule is A itself and the potential-energy surface for the reaction is the potential-energy surface of the A molecule. Thus, there is a well (rather than a saddle point as in triatomic bimolecular reactions) in the lower left corner of the linear contour maps of Fig. 23.21.

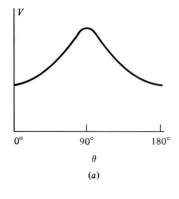

(a) Potential energy versus twist angle for *cis*-CHF=CHF → *trans*-CHF=CHF. (b) Transition state for the decomposition of CH_3CH_2Cl.

Figure 23.20

Figure 23.21

Potential-energy contour maps for linear configurations of (*a*) HCN; (*b*) CO_2.

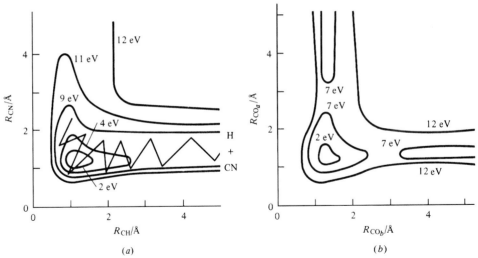

The zero of energy in Fig. 23.21 is taken at the bottom of the well (the equilibrium geometry). Figure 23.21*b* shows that the decomposition $CO_2 \rightarrow CO + O$ has a saddle point on the minimum-energy path (which involves stretching one CO bond until it breaks, the atoms remaining collinear); the saddle point lies between the 7-eV contours. However, the decomposition $HCN \rightarrow H + CN$ shows no saddle point; the potential energy in Fig. 23.21*a* increases continually as R_{CH} increases. The drop in V at large distances along the CO_2 decomposition path is due to formation of the third bond of the triple bond in carbon monoxide. We can expect that a unimolecular decomposition where no new bonds are formed in the products will show no saddle point. (The zigzag line in Fig. 23.21 is explained later.)

To apply TST where there is no saddle point, we must choose a dividing surface such that it is highly probable that a supermolecule crossing this surface will yield products. Clearly, the dividing surface must be much closer to the products than to the reactant molecule in this case. One speaks of a "loose" activated complex. For example, for the decomposition $C_2H_6 \rightarrow 2CH_3$, the dividing surface will correspond to a greatly elongated carbon–carbon bond. The activated complex will closely resemble two CH_3 radicals and will have more rotational degrees of freedom than C_2H_6. Two different methods have been proposed for choosing the dividing surface in a unimolecular reaction with no saddle point (for details, see *Forst*, chap. 11). The carbon–carbon distance in the loose $C_2H_6^{\ddagger}$ activated complex turns out to be 5 Å, compared with 1.5 Å in C_2H_6.

The zigzag line in Fig. 23.21 is a classical trajectory for the decomposition HCN* $\rightarrow$ H + CN. The trajectory starts well away from the equilibrium HCN geometry, corresponding to an excited vibrational level of HCN (as indicated by the star). The vibrations continue until the H—C bond has elongated sufficiently to break. This figure is incomplete, since it omits the bending vibrations.

For the unimolecular reaction A $\rightarrow$ products, the TST equation (23.19) gives the high-pressure rate constant as

$$k_{uni,P=\infty} = \frac{kT}{h} \frac{z'_{\ddagger}}{z_A} e^{-\Delta \varepsilon_0^{\ddagger}/kT} \tag{23.36}$$

What about the falloff region? A theory that allows calculation of k_{uni} in this region was developed by Marcus and Rice in 1951–1952. The Marcus–Rice theory is based in part on earlier work of Rice, Ramsperger, and Kassel and is therefore commonly called the *RRKM theory* of unimolecular reactions. For details of the RRKM theory, see *Holbrook, Pilling, and Robertson*; *Gardiner*, secs. 5-2 and 5-3. Comparison with experimental data shows that the RRKM theory works very well in nearly all cases. At high pressures, the RRKM result reduces to the TST equation (23.36).

23.7   TRIMOLECULAR REACTIONS

The best examples of trimolecular gas-phase reactions (Sec. 17.12) are recombinations of two atoms, for which a third species (M) is needed to carry away some of the bond energy to prevent dissociation. Figure 23.3 is the potential-energy contour map for collinear configurations of the recombination reaction $H + H + H \rightarrow H_2 + H$. A trajectory that starts at point i and is parallel to the R_{bc} axis corresponds to H_b and H_c approaching each other while H_a remains far away. Imagine a ball rolling on the potential-energy surface along this trajectory. The ball will roll down into the $H_a +$ H_bH_c valley, roll past the $R_{bc} = 0.74$ Å equilibrium internuclear distance, roll up the side of the valley with $R_{bc} < 0.74$ Å until it reaches an altitude of 110 kcal/mol (which was the initial potential energy of the system at point i), and then exactly reverse its path, ending up back at point i. Thus, if H_a does not participate, the newly formed H_bH_c molecule will dissociate back into atoms. A trajectory that leads to formation of a stable H_bH_c molecule must involve a decrease in both R_{ab} and R_{bc}.

The recombination reaction $A + B + M \rightarrow AB + M$ (where A and B are atoms) can proceed along a wide variety of trajectories, and no one configuration of A, B, and M can be picked out as *the* transition state. Thus TST does not apply to atomic recombination reactions. Several theories of trimolecular atomic recombination reactions have been proposed, but none is fully satisfactory [see *Gardiner,* pp. 141–147; H. O. Pritchard, *Acc. Chem. Res.,* **9,** 99 (1976)].

The reactions $2NO + X_2 \rightarrow 2NOX$, where X is Cl or O, are believed by some people to be trimolecular (Sec. 17.12). If these reactions are assumed to be trimolecular elementary reactions and TST is applied to them, a very good fit to the observed A values and the observed temperature dependence of k_r is obtained with plausible guesses for the transition-state structures [see *Laidler* (1987), sec. 5.2].

23.8   REACTIONS IN SOLUTION

Because of the strong intermolecular interactions in the liquid state, the theory of reactions in solution is far less developed than that of gas-phase reactions. In general, it is not currently possible to calculate reaction rates in solution from molecular properties. An exception is diffusion-controlled reactions; see Eq. (23.54).

Chemical reactions in solution can be divided into (*a*) **chemically controlled reactions,** in which the rate of chemical reaction between B and C molecules in a solvent cage is much less than the rate of diffusion of B and C toward each other through the solvent; (*b*) **diffusion-controlled reactions** (Sec. 17.15), in which the rate of diffusion is much less than the rate of chemical reaction in the cage; (*c*) **mixed-control reactions,** in which the rates of diffusion and of chemical reaction are of the same order of magnitude.

TST for Chemically Controlled Reactions in Solution

For the chemically controlled elementary reaction $B + C + \cdots \rightarrow \{X_f^{\ddagger}\} \rightarrow D + E + \cdots$, the transition-state theory gives r as the rate per unit volume at which supermolecules cross the critical dividing surface. From the paragraph preceding Eq. (23.16),

$$r = [X_f^{\ddagger}]\langle v_{rc}\rangle/\delta \tag{23.37}$$

where $[X_f^{\ddagger}]$ is the concentration of activated complexes (defined to exist for a length δ on the product side of the dividing surface). Whether the system is ideal or nonideal, it is the concentration $[X_f^‡]$ that appears in the expression for r. This is because r is always defined in terms of a concentration change: $r \equiv \nu_B^{-1}\, d[B]/dt$ [Eq. (17.4)].

To get $[X_f^‡]$, we modify Eq. (23.10) to allow for nonideality. Equation (23.10) expresses an apparent equilibrium between reactants and forward-moving activated complexes for an ideal-gas reaction. To allow for nonideality, we replace the concentrations in the apparent equilibrium constant $[X_f^‡]/[B][C] \cdots$ by activities, to give [see Eqs. (11.6) and (10.30)]

$$K_f^{\circ} = \frac{a_{‡}}{a_B a_C \cdots} = \frac{\gamma_{‡}}{\gamma_B \gamma_C \cdots} \frac{[X_f^‡]/c^{\circ}}{([B]/c^{\circ})([C]/c^{\circ}) \cdots} \qquad (23.38)$$

where $c^{\circ} \equiv 1$ mol/dm^3. (The c°'s are present because the activities and activity coefficients are dimensionless.) All activities and activity coefficients in this section are on the molar concentration scale (Sec. 10.4), so a c subscript on the a's and γ's is to be understood.

The use of (23.38) for $[X_f^‡]$ and (23.18) for $\langle v_{rc} \rangle$ in (23.37) gives

$$r = \frac{\gamma_B \gamma_C \cdots}{\gamma_{‡}} \left(\frac{2kT}{\pi m_{rc}} \right)^{1/2} \frac{1}{\delta} K_f^{\circ} (c^{\circ})^{1-n} [B][C] \cdots \qquad (23.39)$$

where n is the molecularity of the reaction. For a nonideal system, we define $K^{\circ ‡}$ by analogy to (23.25) as

$$K^{\circ ‡} \equiv K_f^{\circ} \left(\frac{2}{\pi m_{rc} kT} \right)^{1/2} \frac{h}{\delta} \qquad (23.40)$$

The rate constant is $k_r = r/[B][C] \cdots$, and (23.39) and (23.40) give

$$k_r = \frac{kT}{h} \frac{\gamma_B \gamma_C \cdots}{\gamma_{‡}} (c^{\circ})^{1-n} K^{\circ ‡} \qquad (23.41)$$

which differs from the ideal-gas expression (23.26) by the presence of the concentration-scale activity coefficients. The $(c^{\circ})^{1-n}$ is present in (23.41) [and absent from (23.45) below] because $K^{\circ ‡}$ is dimensionless whereas $K_c^‡$ in (23.26) has dimensions of concentration^{1-n}.

Because of the strong intermolecular interactions in the solution, we cannot express K_f° or $K^{\circ ‡}$ in terms of individual partition functions z_B, z_C, etc. We don't have an equation like (23.19), and calculation of k_r from molecular properties is not practical for reactions in solution.

Equation (23.41) can also be applied to nonideal-gas reactions, in which case the activity coefficients are replaced by fugacity coefficients.

The apparent equilibrium constant K_f° and the related constant $K^{\circ ‡}$ are functions of temperature, pressure, and solvent (since the concentration-scale standard states of the species B, C, . . . , and $X_f^‡$ depend on these things) but are independent of the solute concentrations. At infinite dilution, the γ's all become 1, and (23.41) becomes $k_r^{\infty} = (kT/h)(c^{\circ})^{1-n} K^{\circ ‡}$, where k_r^{∞} is the rate constant in the limit of infinite dilution. For a fixed T, P, and solvent, the TST equation (23.41) can thus be written as

$$k_r = (\gamma_B \gamma_C \cdots / \gamma_{‡}) k_r^{\infty} \qquad (23.42)$$

Comparison with Eq. (17.80) shows that Y in (17.80) equals $1/\gamma_{‡}$, where $\gamma_{‡}$ is the activity coefficient of the activated complex. Equation (23.42) (which is the **Brønsted–Bjerrum equation**) predicts that the rate constant should vary with reactant concentrations, since the γ's change with a change in solution composition. The low accuracy of most kinetics data makes this effect not worth worrying about, except for ionic reactions. Ionic solutions are highly nonideal, even at low concentrations.

Consider the bimolecular *elementary* ionic reaction $B^{z_B} + C^{z_C} \rightarrow \{X_f^{\ddagger(z_B+z_C)}\} \rightarrow$ products. The log of (23.42) reads

$$\log_{10} k_r = \log_{10} k_r^\infty + \log_{10} \gamma_B + \log_{10} \gamma_C - \log_{10} \gamma_\ddagger \qquad (23.43)$$

For ionic strengths up to 0.1 mol/dm³, the Davies equation (10.71) gives for aqueous solutions at 25°C and 1 atm

$$\log_{10} \gamma_B = -0.51 z_B^2 \left(\frac{I^{1/2}}{1 + I^{1/2}} - 0.30 I \right), \qquad I \equiv \frac{I_c}{c^\circ}$$

where $I_c \equiv \frac{1}{2}\sum_i z_i^2 c_i$ replaces I_m since we are using concentration-scale activity coefficients. Use of the Davies equation for γ_B, γ_C, and $\gamma_\ddagger$ in (23.43) gives [since the charge factor in $\log_{10} \gamma_B + \log_{10} \gamma_C - \log_{10} \gamma_\ddagger$ is $z_B^2 + z_C^2 - (z_B + z_C)^2 = -2z_B z_C$]

$$\log_{10} k_r = \log_{10} k_r^\infty + 1.02 z_B z_C \left(\frac{I^{1/2}}{1 + I^{1/2}} - 0.30 I \right) \qquad \text{dil. aq. soln., 25°C} \quad (23.44)$$

A plot of $\log_{10} k_r$ versus $I^{1/2}/(1 + I^{1/2}) - 0.30I$ should be linear, with slope $1.02 z_B z_C$, for ionic strengths up to 0.1 mol/dm³. This has been verified for many ionic reactions in solution. In calculating I, formation of ion pairs and complex ions must be taken into account.

The **primary kinetic salt effect** (23.44) is large, even at modest values of ionic strength. For example, for $z_B z_C = +2$, the values of k_r/k_r^∞ at $I = 10^{-3}$, 10^{-2}, and 10^{-1} are 1.15, 1.51, and 2.7, respectively. For $z_B z_C = -2$, the corresponding values are 0.87, 0.66, and 0.37. For $z_B z_C = 4$, one finds $k_r/k_r^\infty = 7.2$ at $I = 0.1$.

If the products have different charges than the reactants, the ionic strength can change markedly during a reaction, and k_r will change as the reaction proceeds. To avoid this, a large amount of inert salt is often added to ionic reaction mixtures to keep I essentially fixed during a given run. For a reaction of unknown mechanism, one can determine $z_B z_C$ for the rate-determining step by investigating the dependence of k_r on I (Prob. 23.20).

By analogy to (23.28), we define $\Delta G^{\circ\ddagger}$ for a reaction in solution by

$$\Delta G^{\circ\ddagger} \equiv -RT \ln K^{\circ\ddagger} \qquad (23.45)$$

where $K^{\circ\ddagger}$ is defined by (23.40). Equation (23.41) becomes

$$k_r = \frac{kT}{h} \frac{\gamma_B \gamma_C \cdots}{\gamma_\ddagger} (c^\circ)^{1-n} e^{-\Delta G^{\circ\ddagger}/RT} \qquad (23.46)$$

For nonionic reactions in dilute solutions, the γ's are reasonably close to 1 and (given the inaccuracy of most kinetics data) can be ignored:

$$k_r \approx kTh^{-1}(c^\circ)^{1-n} e^{-\Delta G^{\circ\ddagger}/RT} = kTh^{-1}(c^\circ)^{1-n} e^{\Delta S^{\circ\ddagger}/R} e^{-\Delta H^{\circ\ddagger}/RT} \qquad (23.47)$$

where we used $\Delta G^{\circ\ddagger} = \Delta H^{\circ\ddagger} - T\Delta S^{\circ\ddagger}$. Organic chemists are especially fond of (23.47), since it can be used to rationalize observed changes in rate constants in a series of reactions in terms of changes in standard-state activation entropies and enthalpies. The solvation of the reactants and the activated complex must be taken into account in discussing $\Delta S^{\circ\ddagger}$ and $\Delta H^{\circ\ddagger}$. For example, the elementary reaction $(CH_3)_3CCl \rightarrow (CH_3)_3C^+ + Cl^-$ in aqueous ethanol has $\Delta S^{\circ\ddagger}$ negative, even though a bond is partially broken in forming the transition state. The highly polar transition state $(CH_3)_3^{\delta+} \cdots Cl^{\delta-}$ produces a high degree of ordering in the surrounding solvent.

Equations (17.68) and (23.41) with the γ's omitted give $E_a \equiv RT^2 d \ln k_r/dT \approx RT^2(1/T + d \ln K^{\circ\ddagger}/dT)$. By analogy with equations following Eq. (17.71), we have $d \ln K^{\circ\ddagger}/dT \approx \Delta H^{\circ\ddagger}/RT^2 \approx \Delta U^{\circ\ddagger}/RT^2$. Therefore

$$E_a \approx \Delta H^{\circ\ddagger} + RT \qquad \text{nonionic reaction in dil. soln.} \qquad (23.48)$$

which differs from (23.34) for gases. Equation (17.69) gives $A \equiv k_r e^{E_a/RT}$, and the use of (23.48) and (23.47) gives

$$A \approx kTh^{-1}e(c^\circ)^{1-n}e^{\Delta S^{\circ\ddagger}/R} \qquad \text{nonionic reaction in dil. soln.} \qquad (23.49)$$

Equations (23.48) and (23.49) enable calculation of the activation enthalpy and entropy $\Delta H^{\circ\ddagger}$ and $\Delta S^{\circ\ddagger}$ (and hence $\Delta G^{\circ\ddagger}$) from experimental A and E_a values.

Because of the presence of the solvent, TST may well not be valid for certain conditions in solution; see *Steinfeld, Francisco, and Hase,* sec. 12.2; G. R. Fleming and P. G. Wolynes, *Physics Today,* May 1990, p. 36.

Diffusion-Controlled and Mixed-Control Reactions

The TST equation (23.41) does not apply to diffusion-controlled and mixed-control reactions, since their rate is governed at least in part by the rate at which reactants can diffuse through the solvent to encounter each other. Let the elementary reaction be $B + C \rightarrow$ products.

At the start of the reaction, species B and C are distributed randomly in the solution. Shortly after reaction begins, many pairs of B and C molecules that initially were close to each other will have encountered each other and reacted. The B molecules that remain will most likely be B molecules for which there were no C molecules in the immediate vicinity. If diffusion is not rapid enough to restore a random distribution, the region around a given B molecule will be somewhat depleted of C molecules (and vice versa). Let R be the distance from a given B molecule. As R increases, the average concentration of C molecules will increase, reaching its value in the bulk solution at some large value of R, which we shall call infinity. There is a concentration gradient of C in the region around a given B molecule. We shall assume that very shortly after the reaction begins a steady state is reached in which the rate of diffusion of C toward B equals the rate of reaction of C with B.

Fick's first law of diffusion, Eq. (16.30), gives the rate of diffusion of B molecules through the solvent A. To find the rate at which B and C molecules diffuse toward each other, we must use a diffusion coefficient that is the sum of the diffusion coefficients of B and C in the solvent A. We shall assume the solution is dilute, so the infinite-dilution diffusion coefficients can be used. Thus,

$$\frac{dn'_C}{dt} = (D_B^\infty + D_C^\infty)\mathcal{A}\frac{d[C]}{dR} \qquad (23.50)$$

where dn'_C is the net number of moles of C that in time dt cross a spherical surface of area $\mathcal{A} = 4\pi R^2$ surrounding a given B molecule and D_B^∞ and D_C^∞ are the infinite-dilution diffusion coefficients of B and C in the solvent A. The prime avoids confusion of the quantity (23.50), which refers to diffusion of C toward one particular B molecule, with the total rate of disappearance of C in the solution. We have $d[C]/dR > 0$, and we consider dn'_C/dt to be positive; hence a plus sign is used in (23.50).

Let us assume B and C are spherical molecules with radii r_B and r_C. For a steady state, dn'_C/dt is a constant. Integration of (23.50) between the limits $R = r_B + r_C$ (molecules B and C in contact) and $R = \infty$ (the bulk solution) gives

$$[C]_{R=\infty} - [C]_{R=r_B+r_C} = \frac{dn'_C}{dt}\frac{1}{D_B^\infty + D_C^\infty}\int_{r_B+r_C}^{\infty}\frac{1}{4\pi R^2}\,dR$$

$$[C]_{R=r_B+r_C} = [C] - \frac{dn'_C}{dt}\frac{1}{(D_B^\infty + D_C^\infty)4\pi(r_B + r_C)} \qquad (23.51)$$

where $[C] = [C]_{R=\infty}$ is the concentration of C in the bulk solution.

Let r be the observed reaction rate (per unit volume):

$$r = k[\text{B}][\text{C}] \tag{23.52}$$

where the observed rate constant k combines the effects of the diffusion rate and the chemical-reaction rate, and the concentrations are those in the bulk solution. The steady-state condition is that the rate at which C diffuses toward a given B molecule equal the reaction rate between C and that B molecule. Multiplication of r by the solution volume V gives the total reaction rate in the solution. The rate at one particular B molecule is then rV/N_B, where N_B is the number of B molecules in the solution. The steady-state condition then gives $dn'_\text{C}/dt = rV/N_\text{B} = rV/N_\text{A}n_\text{B} = r/N_\text{A}[\text{B}] = k[\text{C}]/N_\text{A}$, where (23.52) was used and N_A is the Avogadro constant. Substituting this expression for dn'_C/dt in (23.51), we get

$$[\text{C}]_{R=r_\text{B}+r_\text{C}} = [\text{C}]\left(1 - \frac{k}{4\pi N_\text{A}(D_\text{B}^\infty + D_\text{C}^\infty)(r_\text{B} + r_\text{C})}\right) = [\text{C}]\left(1 - \frac{k}{k_\text{diff}}\right) \tag{23.53}$$

$$k_\text{diff} \equiv 4\pi N_\text{A}(D_\text{B}^\infty + D_\text{C}^\infty)(r_\text{B} + r_\text{C}) \tag{23.54}$$

The significance of the defined quantity k_diff will be seen shortly.

Let k_chem be the rate constant for chemical reaction of B and C pairs in a solvent cage (Fig. 17.19). The steady-state hypothesis enables us to express r as

$$r = k_\text{chem}[\text{B}][\text{C}]_{R=r_\text{B}+r_\text{C}} \tag{23.55}$$

Equating (23.55) and (23.52), we get $k = k_\text{chem}[\text{C}]_{R=r_\text{B}+r_\text{C}}/[\text{C}]$. The use of (23.53) gives $k = k_\text{chem}(1 - k/k_\text{diff})$. Solving for k, we have as our final result

$$k = \frac{k_\text{diff}k_\text{chem}}{k_\text{diff} + k_\text{chem}} \quad \text{or} \quad \frac{1}{k} = \frac{1}{k_\text{diff}} + \frac{1}{k_\text{chem}} \tag{23.56}$$

For the limit $k_\text{chem} \to \infty$, the reaction is diffusion-controlled, and (23.56) becomes $1/k = 1/k_\text{diff}$ or $k = k_\text{diff}$. Thus, k_diff in (23.54) is the rate constant for the diffusion-controlled reaction. Equation (23.54) was given earlier as Eq. (17.111).

For the limit $k_\text{diff} \to \infty$ (or more precisely, $k_\text{diff} \gg k_\text{chem}$), Eq. (23.56) becomes $k = k_\text{chem}$. The reaction is chemically controlled. The rate constant k_chem is given by the TST equation (23.41).

When k_diff and k_chem are the same order of magnitude, the kinetics is mixed and k is given by (23.56). Figure 23.22 plots k in (23.56) versus k_chem for $k_\text{diff} = 0.7 \times 10^{10}$ L/mol-s, the typical k_diff value in water at 25°C (Sec. 17.15). For $k_\text{chem} > 10^{11.5}$ L/mol-s, we have $k \approx k_\text{diff}$, whereas for $k_\text{chem} < 10^{8.5}$ L/mol-s, we have $k \approx k_\text{chem}$.

For fast ionic reactions, the rate at which C and B diffuse toward each other is affected by the Coulombic attraction or repulsion. An extension of the above derivation (see *Weston and Schwarz*, secs. 6.2 and 6.3) gives Eq. (17.113).

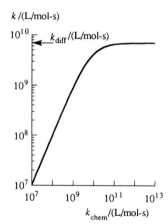

Figure 23.22

Plot of k in Eq. (23.56) versus k_chem for $k_\text{diff} = 0.7 \times 10^{10}$ L/mol-s.

23.9   SUMMARY

The hard-sphere collision theory of reaction rates equates the reaction rate to the rate of collisions of hard-sphere molecules in which the line-of-centers relative translational energy exceeds a threshold value. The theory fails.

The potential-energy surface for an elementary chemical reaction is a plot of the potential energy V of the supermolecule formed by the reacting molecules. The variables are the nuclear coordinates of the supermolecule (excluding translational and rotational coordinates). The minimum-energy path connecting reactants and products on the potential-energy surface for a bimolecular reaction usually goes through a maximum (which is a saddle point on the surface), and this maximum is the transition state.

Molecular reaction dynamics studies the motions (trajectories) of reacting species on the potential-energy surface. Suitable averaging of reaction probabilities for a representative set of trajectories allows the rate constant to be calculated from the potential-energy surface. Such calculations are extremely difficult. Molecular-beam experiments yield detailed information about reaction probabilities and the energy distribution of products.

The activated-complex or transition-state theory (ACT or TST) assumes that all supermolecules that cross the dividing surface located at the saddle point become products, that a Boltzmann distribution of energy holds for the activated complex and the reactants, and that motion along the reaction coordinate can be treated classically. TST leads to an expression [Eq. (23.19)] for a gas-phase rate constant in terms of the molecular partition functions of the activated complex and the reactants and the quantity $\Delta\varepsilon_0^{\ddagger}$, which is the difference in ground vibrational state energies of the transition state and the reactants (Fig. 23.19). TST explains kinetic isotope effects and kinetic salt effects (the dependence of rate constants on ionic strength).

Equation (23.54) relating the rate constant of a diffusion-controlled reaction to the diffusion coefficients and radii of the reactants was derived.

Important kinds of calculations dealt with in this chapter include:

- Calculation of the pre-exponential factor using hard-sphere collision theory.
- Calculation of the rate constant of a reaction using the transition-state theory.
- Calculation of $\Delta S_c^{\circ\ddagger}$, $\Delta H_c^{\circ\ddagger}$, and $\Delta G_c^{\circ\ddagger}$ from A and E_a for gas-phase reactions and reactions in solution.
- Calculation of the effect of ionic strength on the rate constant of an ionic reaction.

FURTHER READING

Levine and Bernstein; Moore and Pearson, chaps. 4, 5, and 7; *Knox*, chap. 12; *Denbigh*, secs. 15.7 to 15.9; *McClelland*, chap. 12; *Holbrook, Pilling, and Robertson; Bernasconi*, part I, pp. 1–96; *Gardiner*, chap. 4; *Weston and Schwarz*, chaps. 3–6; *Bamford and Tipper*, vols. 2 and 25; *Laidler* (1969); *Laidler* (1987), chaps. 4, 11, and 12; *Steinfeld, Francisco, and Hase*, chaps. 6–12.

PROBLEMS

Section 23.1

23.1 Give the equation that corresponds to (23.6) for the elementary reaction 2B → products.

23.2 (a) Use the hard-sphere collision theory to calculate the A factor for the elementary reaction $NO + O_3 \rightarrow NO_2 + O_2$. Reasonable values of molecular radii (calculated from the known molecular structures) are 1.4 Å for NO and 2.0 Å for O_3. Take $T = 500$ K. (b) The experimental A for this reaction is 8×10^{11} cm^3 mol^{-1} s^{-1}. Calculate the steric factor.

Section 23.3

23.3 For the ICN FTS experiments, $\lambda_2^{\infty} = 388.5$ nm. Which of the λ_2^{*} values 390 nm or 391 nm produces a maximum in the LIF intensity at an earlier time t_{probe}?

Section 23.4

23.4 Verify (23.18) for $\langle v_{\text{rc}} \rangle$.

23.5 Convert 9.61 kcal/mol to ergs per molecule.

23.6 (a) Locate the principal axes of the linear DH_2 transition state and express its moment of inertia I_b in terms of masses and bond distances. (*Hint:* Use the center-of-mass formula in Sec. 18.13.) (b) Verify the value of the ratio of rotational partition functions given in Example 23.2 in Sec. 23.4.

23.7 (a) Verify the numerical values of vibrational and translational partition functions in Example 23.2. (b) Verify the numerical value of k_r in this example.

23.8 Use TST and data in Example 23.2 to calculate k_r for D + $H_2 \rightarrow$ DH + H at 600 K. (The experimental value is $7_{.5} \times 10^{10}$ cm^3 mol^{-1} s^{-1}.) From this calculation and the example, is tunneling more or less important at higher T? Assume the discrepancy between theory and experiment is due mainly to tunneling.

23.9 Use TST and data in Example 23.2 to calculate k_r for H + $D_2 \rightarrow$ HD + D at 600 K. The HD_2 activated complex has vibrational wavenumbers 1762 cm^{-1} (symmetric stretching) and 694 cm^{-1} (degenerate bending). Begin by calculating the D$_2$ vibrational wavenumber from that of H$_2$. (The experimental k_r is 1.9×10^{10} cm^3 mol^{-1} s^{-1}.)

23.10 Show that in the high-T limit, z_{vib} in (22.110) is proportional to $T^{f_{vib}}$, where f_{vib} is the number of vibrational modes.

23.11 Work out the typical range of values for m in (23.21).

23.12 Use the TST equation (23.22) and data in Example 23.2 to calculate E_a for D + H$_2$ at 300 K. Assume that T is low enough for the vibrational partition functions to be neglected.

23.13 Use TST to derive the effusion equation (15.58). Take the critical dividing surface as coinciding with the hole and use (23.16).

23.14 Use data in Sec. 21.9 to estimate the factor by which k_r at 300 K is lowered by substitution of D for H in a CH bond broken in the rate-determining step. Neglect tunneling. Does this isotope effect increase or decrease as T increases?

23.15 By what factor is k_r at 300 K lowered by substitution of tritium (^{3}H) for H in a C—H bond broken in the rate-determining step? Neglect tunneling.

23.16 For the electrophilic aromatic substitution reaction ArH + X$^+$ → ArX + H$^+$ (where X$^+$ is an electrophile like NO$_2^+$), substitution of D for H causes only a small change in k_r. This observation rules out which of the following steps as the rate-determining step? (a) ArH + X$^+$ → ArX + H$^+$; (b) ArH + X$^+$ → ArHX$^+$; (c) ArHX$^+$ → ArX + H$^+$.

Section 23.5

23.17 (a) For the elementary gas-phase reaction O$_3$ + NO → NO$_2$ + O$_2$, one finds that E_a = 2.5 kcal/mol and A = 6×10^8 dm^3 mol^{-1} s^{-1} for the temperature range 220 to 320 K. Calculate $\Delta G_c^{\circ\ddagger}$, $\Delta H_c^{\circ\ddagger}$, and $\Delta S_c^{\circ\ddagger}$ for the midpoint of this temperature range. (b) The same as (a) for the gas-phase elementary reaction CO + O$_2$ → CO$_2$ + O; here, E_a = 51 kcal/mol and A = 3.5×10^9 dm^3 mol^{-1} s^{-1} for the range 2400 to 3000 K.

Section 23.8

23.18 (a) For an ionic elementary reaction between a +2 ion and a −3 ion, calculate k_r/k_r^∞ in water at 25°C for I = 10^{-3}, 10^{-2}, and 10^{-1}. (b) Do the same for reaction between a −2 ion and a −3 ion.

23.19 (a) For each of the following elementary reactions in aqueous solutions with very low ionic strength I, state whether the rate constant k_r will increase, decrease, or remain nearly constant as I increases: (i) CH$_3$Br + OH$^-$ → CH$_3$OH + Br$^-$; (ii) ClCH$_2$COO$^-$ + OH$^-$ → HOCH$_2$COO$^-$ + Cl$^-$; (iii) [Co(NH$_3$)$_5$Br]$^{2+}$ + NO$_2^-$ → [Co(NH$_3$)$_5$NO$_2$]$^{2+}$ + Br$^-$. (b) Does Eq. (23.44) predict that k_r will go through a maximum (or minimum) as I increases?

23.20 Measurement of the rate constant of the reaction S$_2$O$_8^{2-}$ + 2I$^-$ → 2SO$_4^{2-}$ + I$_2$ as a function of I = I_c/c° at 25°C in water yields the following data (where $k^\circ \equiv 1$ dm^3 mol^{-1} s^{-1}):

$10^3 I$	2.45	3.65	6.45	8.45	12.45
k/k°	0.105	0.112	0.118	0.126	0.140

Use a graph to find $z_B z_C$ for the rate-determining step.

23.21 For the elementary reaction CH$_3$Br + Cl$^-$ → CH$_3$Cl + Br$^-$ in acetone, one finds A = 2×10^9 dm^3 mol^{-1} s^{-1} and E_a = 15.7 kcal/mol. Calculate $\Delta H^{\circ\ddagger}$, $\Delta S^{\circ\ddagger}$, and $\Delta G^{\circ\ddagger}$ for this reaction at 300 K.

23.22 (a) Combine (23.53) and (23.56) to express the concentration of C in the immediate vicinity of B as a function of the bulk concentration of C and of k_{diff} and k_{chem}; then find the limiting values of this concentration for chemically controlled and for diffusion-controlled reactions. (b) Calculate the ratio of these two concentrations of C for k_{chem}/k_{diff} = 10, 1, and 0.1.

General

23.23 Name at least three kinds of reactions that involve departures from the Boltzmann distribution law for the reactants.

23.24 State whether $\Delta S_c^{\circ\ddagger}$ is expected to be quite positive, quite negative, or close to zero for the unimolecular decomposition of (a) C$_2$H$_5$Cl (Fig. 23.20b); (b) C$_2$H$_6$.

Solids and Liquids

24.1 SOLIDS AND LIQUIDS

A solid is classified as crystalline or amorphous. A **crystalline solid** shows a sharp melting point. [This is not quite true; recall surface melting (Sec. 7.4)]. Examination by the naked eye (or by a microscope if the sample is microcrystalline) shows crystals having well-developed faces and a characteristic shape. X-ray diffraction (Sec. 24.9) shows a crystalline solid to have a regular, ordered structure composed of identical repeating units having the same orientation throughout the crystal. The repeating unit is a group of one or more atoms, molecules, or ions.

An **amorphous solid** does not have a characteristic crystal shape. When heated, it softens and melts over a temperature range. X-ray diffraction shows a disordered structure. Polymers often form amorphous solids. As the liquid polymer is cooled, the polymer chains become twisted and tangled together in a random, irregular way. Some polymers do form crystalline solids. Others give solids that are partly crystalline and partly amorphous. A **glass** is an amorphous solid obtained by cooling a liquid. Amorphous solids also may form on deposition of a vapor on a cold surface and on evaporation of solvent from a solution. The most common kind of glass is prepared from molten SiO_2 with various amounts of dissolved metal oxides. It contains chains and rings involving Si—O bonds; the structure is disordered and irregular. Liquids that are viscous (for example, glycerol) tend to form glasses when cooled rapidly. The high viscosity makes it hard for the molecules to become ordered into a crystalline solid. A glass phase is thermodynamically metastable, having a higher G_m than does the crystalline form of the substance.

In solids, the structural units (atoms, molecules, or ions) are held more or less rigidly in place. In liquids, the structural units can move about by squeezing past their neighbors. The degree of order in a liquid is much less than in a crystalline solid. Liquids have a short-range order, in that molecules in the immediate environment of a given molecule tend to adopt a preferred orientation and intermolecular distance, but liquids have no long-range order, since there is no correlation between the orientations of two widely separated molecules and no restriction on the distance between them. Amorphous solids have the rigidity of solids but resemble liquids in having short-range order but not long-range order.

In 1984 a new kind of solid called a *quasicrystal* was discovered. Quasicrystals form when certain molten alloys are very rapidly cooled. Unlike amorphous solids, quasicrystals have both short-range and long-range order. However, despite their long-range order, the symmetry of the quasicrystal structure is incompatible with translational periodicity and so is of a type forbidden for ordinary crystals. The structure of quasicrystals is not fully understood. See P. W. Stephens and A. I. Goldman, *Scientific American,* April 1991, p. 44; *The New York Times,* Nov. 24, 1998, p. F1.

A **polymer** (or **macromolecule**) is a substance whose molecules consist of very many simple structural units. Polymer molecular weights usually lie in the range 10^4 to 10^7. **Biopolymers** are those occurring in living organisms. **Synthetic polymers** are manufactured.

Solid synthetic polymers are, in general, partly crystalline and partly amorphous. The degree of crystallinity depends on the polymer structure and on how the solid is prepared. Rapid cooling of the molten polymer favors formation of an amorphous solid. Regularity of structure in the polymer molecules favors formation of crystals.

At low temperatures, an amorphous solid polymer is hard and has a glassy appearance. When heated to a certain temperature, the amorphous solid becomes softer, rubberlike, and flexible, the polymer molecules now having enough energy to slide past one another; this temperature is called the *glass-transition temperature* T_g. Cooling the polymer below T_g locks the chains in fixed, random conformations to produce a hard, disordered, amorphous solid. A rubber ball cooled below T_g in liquid nitrogen will shatter when dropped on the floor.

A perfectly crystalline polymer would not show a glass transition but would melt at some temperature to a liquid. Perfect polymer crystals do not exist. A semicrystalline polymer shows both a glass-transition temperature T_g and a melting temperature T_m that lies above T_g. T_g is associated with the amorphous portions of the solid and T_m is associated with the crystalline portions. For temperatures between T_g and T_m, one has tiny crystallites embedded in a rubbery matrix. For nylon 66, T_g is 60°C and T_m is 265°C. For polyethylene, T_g is −125°C and T_m is 140°C. At the melting point of a perfect crystal, the volume changes discontinuously; $\Delta V \neq 0$. For polymers, one observes that $\Delta V = 0$ at T_g, but the slope of the V-versus-T curve changes. In the region of T_m, V changes rapidly, but not discontinuously, since the polymer is not a perfect crystal (Fig. 24.1).

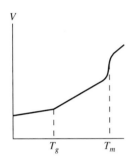

Figure 24.1

Typical plot of volume versus temperature for a semicrystalline polymer. T_g and T_m are the glass-transition temperature and the melting point.

The main kinds of biopolymers are proteins (whose structural units are amino acid residues), nucleic acids (DNA, RNA, whose structural units are nucleotides), and polysaccharides (cellulose, starch, glycogen, whose structural units are glucose residues).

Proteins can be classified as *fibrous* or *globular*. In a fibrous protein, the chain is coiled into a helix. The helix is stabilized by hydrogen bonds between one turn and the next. Hair and muscle proteins and collagen are fibrous. Fibrous proteins are generally water-insoluble.

In a globular protein, some portions of the chain are coiled into hydrogen-bond-stabilized helical segments. The helical content is from 0 to 75%, depending on the protein. Other portions of the chain are nearly fully extended and are hydrogen-bonded to adjacent parallel (or antiparallel—meaning adjacent chain portions run in opposite directions) portions of the chain to form what is called a β sheet, typically composed of two to five strands. The conformation of the remaining portions of the chain is not regular. The protein molecule folds on itself to give a roughly spherical or ellipsoidal overall shape. The folding is not random; rather, different portions of the chain are held together by S—S covalent bonds, hydrogen bonds, and van der Waals forces. Globular protein molecules contain many polar groups on their outer surface and are generally water-soluble. Most enzymes are globular proteins.

A crystal is classified as *ionic, covalent, metallic,* or *molecular,* according to the nature of the chemical bonding and the intermolecular forces in the crystal.

Ionic crystals consist of an array of positive and negative ions and are held together by the electrical attraction between oppositely charged ions. Examples are $NaCl$, MgO, $CaCl_2$, and KNO_3.

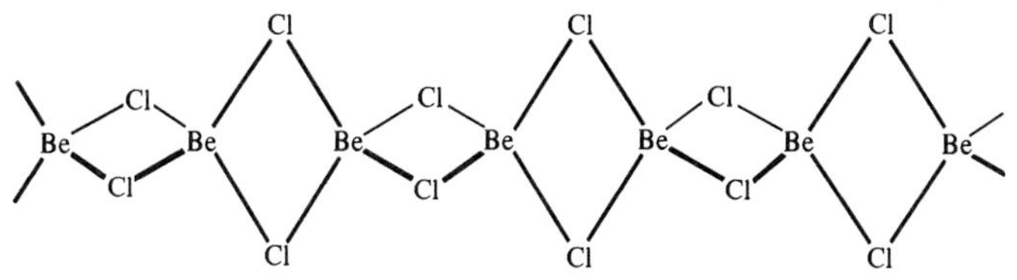

Figure 24.2

One-dimensional network structure of solid $BeCl_2$. The four Cl's around each Be are arranged approximately tetrahedrally.

Metallic crystals are composed of bonded metal atoms; some of the valence electrons are delocalized over the entire metal and hold the crystal together. Examples are Na, Cu, Fe, and various alloys.

Covalent (or **nonmetallic network**) **crystals** consist of an "infinite" network of atoms held together by polar or nonpolar covalent bonds, no individual molecules being present. Examples are carbon in the forms of diamond and graphite, Si, SiO_2, and SiC. In diamond (Fig. 24.19) each carbon is bonded to four others that surround it tetrahedrally, giving a three-dimensional network that extends throughout the crystal. Silicon has the same structure. SiO_2 has a three-dimensional network in which each Si is bonded to four O's at tetrahedral angles, and each O is bonded to two Si atoms. In many covalent crystals, the covalent bonds form a two-dimensional network. Examples are graphite and mica. Graphite consists of layers of fused hexagonal rings of bonded carbons; the bonds are intermediate between single and double bonds (as in benzene). Weak van der Waals forces hold the layers together. The covalent solids BeH_2 and $BeCl_2$ contain one-dimensional networks (Fig. 24.2).

Crystals of a long-chain polymer like polyethylene can be viewed as covalent solids with one-dimensional networks. In certain synthetic polymers, covalent chemical bonds between different chains join the whole piece of material into a single giant molecule. These polymers have three-dimensional networks and are called *cross-linked*. (In vulcanization of raw rubber, sulfur is added, which opens up double bonds in the polymer chains and cross-links adjacent molecules by bonds through one or more S atoms. This greatly increases the rubber's strength.)

In ionic, metallic, and three-dimensional covalent crystals, it is not possible to pick out individual molecules. The entire crystal is a single giant molecule.

The **coordination number** of an atom or ion in a solid is the number of nearest neighbors for that atom or ion. In NaCl (Fig. 24.16) each Na^+ has six Cl^- ions as nearest neighbors, and the coordination number of Na^+ is 6. The coordination number of carbon in diamond (Fig. 24.19) is 4. In SiO_2, the coordination number of Si is 4 and that of O is 2. The coordination number in a metal is usually 8 or 12.

Molecular crystals are composed of individual molecules. The atoms within each molecule are held together by covalent bonds. Relatively weak intermolecular forces hold the molecules together in the crystal. Molecular crystals are subdivided into **van der Waals crystals,** in which the intermolecular attractions are dipole–dipole, dipole–induced-dipole, and dispersion forces (Sec. 22.10), and **hydrogen-bonded crystals,** in which the main intermolecular attraction is due to hydrogen bonding. Some van der Waals crystals are Ar, CO_2, CO, O_2, HI, CH_3CH_2Br, $C_6H_5NO_2$, $HgCl_2$, C_{60}, and $SnCl_4$. Some hydrogen-bonded crystals are H_2O, HF, NH_3, and the amino acid $H_3^+NCH_2COO^-$.

The distinctions between the various kinds of crystals are not always clear-cut. For example, ZnS contains a three-dimensional network of Zn and S and is often classed as a covalent solid. However, each Zn—S bond has a substantial amount of ionic character, and one might consider the structure an ionic one in which the S^{2-} ions are substantially polarized (distorted) to give considerable covalent bonding.

24.4 COHESIVE ENERGIES OF SOLIDS

The **cohesive energy** (or *binding energy*) E_c of a crystal is the molar enthalpy change ΔH° for the isothermal conversion of the crystal into its ideal-gas-phase structural units. The structural units are isolated atoms for metallic and covalent crystals, molecules for molecular crystals, and ions for ionic crystals. E_c depends on temperature; the theoretically most significant value is that at 0 K. At 0 K, ΔH° differs negligibly from ΔU°. The NBS tables (Sec. 5.9) list $\Delta_f H^\circ$ values extrapolated to 0 K for solids and for gaseous atoms, ions, and molecules, and this allows easy calculation of E_c at 0 K.

Metals

The cohesive energy of the metal M is ΔH° for $M(c) \rightarrow M(g)$. This is the heat of sublimation of the solid to a monatomic gas at 1 bar (provided we neglect the slight difference between real-gas and ideal-gas enthalpies at 1 bar) and can be found from the solid's vapor pressure using the Clausius–Clapeyron equation (7.19). E_c values for metals are listed in thermodynamics tables as $\Delta_f H^\circ$ for $M(g)$. Some E_c values at 0 K in kJ/mol are

Na	K	Be	Mg	Cu	Ag	Cd	Al	Fe	W	Pt
108	90	320	146	337	284	112	324	414	848	564

The range for metals is from 80 to 850 kJ/mol (20 to 200 kcal/mol), which corresponds to 1 to 9 eV per atom. These energies are comparable to chemical-bond energies (Table 20.1). E_c values for transition metals tend to be high, because of covalent bonding involving d electrons. The melting points reflect the E_c values (Fig. 24.3). Some normal melting points are Na, 98°C; Mg, 650°C; Cu, 1083°C; Pt, 1770°C; W, 3400°C.

Covalent Solids

E_c for a covalent solid can be found from $\Delta_f H^\circ$ values of the solid and the gaseous atoms (Prob. 24.2). For example, E_c of SiC is ΔH° of $SiC(c) \rightarrow Si(g) + C(g)$. Some E_c values at 0 K in kJ/mol are

C(diamond)	C(graphite)	Si	SiC	SiO_2(quartz)
709	711	451	1227	1851

The cohesive energy of a covalent solid is due to covalent chemical bonds. The high E_c values are reflected in the hardness and the high melting points of covalent solids. Diamond sublimes (rather than melts) at 3500°C. Quartz melts at 1600°C. Diamond is the hardest known substance because of its high cohesive energy per unit volume and symmetrical bonding structure.

Ionic Solids

The cohesive energy of NaCl is ΔH° for the process $NaCl(c) \rightarrow Na^+(g) + Cl^-(g)$. Born and Haber pointed out that this process can be broken into the following isothermal steps:

$$NaCl(c) \xrightarrow{(a)} Na(c) + \tfrac{1}{2}Cl_2(g) \xrightarrow{(b)} Na(g) + Cl(g) \xrightarrow{(c)} Na^+(g) + Cl^-(g)$$

ΔH_a equals minus the enthalpy of formation $\Delta_f H^\circ$ of $NaCl(c)$ from its elements. ΔH_b equals $\Delta_f H^\circ$ of $Na(g)$ plus $\Delta_f H^\circ$ of $Cl(g)$. [$\Delta_f H^\circ$ of $Na(g)$ is the enthalpy of sublimation of solid Na. $\Delta_f H^\circ$ of $Cl(g)$ is closely related to the dissociation energy of Cl_2; see the discussion in Sec. 21.3 on H_2.] ΔH_c equals N_A times the ionization energy I of Na

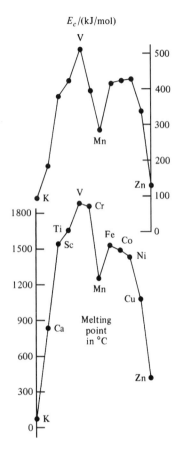

E_c/(kJ/mol)

Figure 24.3

Cohesive energies and melting points of the first 12 metals of the fourth row of the periodic table.

minus N_A times the electron affinity A of Cl (since excited electronic states of the atoms and ions are not populated at room temperature). Thus

$$E_c[\text{NaCl}(c)] = -\Delta_f H°[\text{NaCl}(c)] + \Delta_f H°[\text{Na}(g)]$$
$$+ \Delta_f H°[\text{Cl}(g)] + N_A I(\text{Na}) - N_A A(\text{Cl}) \qquad (24.1)$$

Using thermodynamic data in the Appendix, $I(\text{Na}) = 5.139$ eV, $A(\text{Cl}) = 3.614$ eV, and Eq. (20.1), we get $E_c = 787$ kJ/mol at 25°C (Prob. 24.3).

Some values of E_c in kJ/mol at 0 K for ionic solids (values in parentheses are approximate) are

NaCl	NaBr	LiF	CsCl	$ZnCl_2$	$MgCl_2$	MgO	CaO
786	751	1041	668	2728	2519	(3800)	(3400)

The factor-of-4 difference between NaCl and CaO is due to the ± 2 charges of Ca^{2+} and O^{2-}. Some normal melting points are NaCl, 801°C; NaBr, 755°C; MgO, 2800°C.

Molecular Solids

E_c for a molecular solid is the enthalpy of sublimation. Some values in kJ/mol at or near the melting point of the crystal are

Ar	Kr	CH_4	$n\text{-}C_{19}H_{40}$	H_2O	H_2S	Cl_2	I_2	
7.5	10.5	9.6	113	51	21	27	61	(24.2)

The weakness of intermolecular forces compared with chemical bonds means that E_c for a molecular crystal is typically an order of magnitude smaller than for a covalent, ionic, or metallic crystal. For molecules of comparable size, hydrogen-bonded crystals have higher E_c values than van der Waals crystals. E_c values and melting points increase as the molecular size increases, because of the increase in dispersion energy. Some normal melting points are Ar, -189°C; Kr, -157°C; CH_4, -182°C; $n\text{-}C_{19}H_{40}$, 32°C; H_2O, 0°C; H_2S, -86°C; Cl_2, -101°C; I_2, 114°C.

24.5 THEORETICAL CALCULATION OF COHESIVE ENERGIES

Covalent Solids

E_c for a covalent solid can be estimated as the sum of the bond energies for all the covalent bonds. Consider diamond, for example. In a mole, there are N_A atoms of carbon. Each carbon atom is bonded to four others. If we were to take $4N_A$ as the number of carbon–carbon single bonds per mole, we would be counting each C_a—C_b bond twice, once as one of the four bonds of carbon atom a and once as one of the four bonds of C_b. Hence, there are $2N_A$ bonds per mole. Table 20.1 gives the carbon–carbon single-bond energy as 344 kJ/mol. We therefore estimate E_c for diamond to be 688 kJ/mol. The actual value is 709 kJ/mol.

Use of an empirical C—C bond energy to estimate E_c of diamond is far from a true theoretical calculation of E_c. A quantum-mechanical calculation of E_c of diamond using a certain version of the density-functional method (Sec. 20.10) gave 730 kJ/mol [M. T. Yin and M. L. Cohen, *Phys. Rev. B.,* **24,** 6121 (1981)]. This method has been used to calculate cohesive energies, bond distances, and compressibilities of many solids with striking success; see M. L. Cohen et al., *Scientific American,* June 1982, p. 82; M. L. Cohen, *Int. J. Quantum Chem. Symp.,* **17,** 583 (1983).

Ionic Solids

E_c of an ionic crystal can be estimated by summing the energies of the interionic Coulombic attractions and repulsions and the Pauli repulsions arising from a slight overlap of the probability densities of ions in contact (recall Prob. 20.24). This approach is due to Born and Landé.

Consider NaCl as an example. In the NaCl crystal structure (Fig. 24.16) each Na^+ ion is surrounded by six nearby Cl^- ions at an equilibrium distance (averaged over zero-point vibrations) R_0. The next nearest neighbors to an Na^+ ion are 12 other Na^+ ions at a distance $\sqrt{2}R_0$. Then come 8 Cl^- ions at $\sqrt{3}R_0$, then 6 Na^+ ions at $\sqrt{4}R_0$, then 24 Cl^- ions at $\sqrt{5}R_0$, then 24 Na^+ ions at $\sqrt{6}R_0$, and so on. For a general separation R between Na^+ and Cl^- nearest neighbors, Eq. (19.1) gives as the potential energy of interaction between one Na^+ ion and all the other ions in the crystal

$$\frac{e^2}{4\pi\varepsilon_0 R}\left(-\frac{6}{1} + \frac{12}{\sqrt{2}} - \frac{8}{\sqrt{3}} + \frac{6}{\sqrt{4}} - \frac{24}{\sqrt{5}} + \frac{24}{\sqrt{6}} - \cdots\right) = -\frac{e^2}{4\pi\varepsilon_0 R}\mathcal{M} \quad (24.3)$$

where the **Madelung constant** $\mathcal{M}$ is the sum of the series. Because of the long range of interionic forces, the series converges very slowly, and special techniques must be used to evaluate it [see the references in W. B. Bridgman, *J. Chem. Educ.*, **46**, 592 (1969); E. L. Burrows and S. F. A. Kettle, ibid., **52**, 58 (1975)]. One finds $\mathcal{M} = 1.74756$ for NaCl. By symmetry, the potential energy of Coulombic interaction between a Cl^- ion and the other ions of the crystal is also $-e^2\mathcal{M}/4\pi\varepsilon_0 R$. Multiplication of $-2e^2\mathcal{M}/4\pi\varepsilon_0 R$ by N_A and division by 2 (to avoid counting each interionic interaction twice) gives the Coulomb contribution E_{Coul} to the NaCl cohesive energy as

$$E_{Coul} = -e^2\mathcal{M}N_A/4\pi\varepsilon_0 R \quad (24.4)$$

For NaCl, x-ray diffraction data give $R_0 = 2.820$ Å at 25°C and 2.798 Å at 0 K. Equation (24.4) gives (Prob. 24.6) $E_{Coul} = -868$ kJ/mol at 0 K.

The form of (24.4) and the Madelung constants for other crystal structures are discussed in D. Quane, *J. Chem. Educ.*, **47**, 396 (1970).

In addition to the Coulombic interactions between the ions considered as spheres, there is the Pauli-principle repulsion (Sec. 20.4) due to the slight overlap of neighboring-ion probability densities. A crude approximation to this repulsion energy between two ions is the function A/R^n, where n is a large power and A is a constant. (Recall that Lennard-Jones took $n = 12$; Sec. 22.10.) Each ion has six nearest neighbors, and there are $2N_A$ ions in 1 mole, so the total repulsion energy per mole is $6A(2N_A)/2R^n \equiv B/R^n$, where $B \equiv 6AN_A$ and we divided by 2 to avoid counting each repulsion twice. Thus, $E_{rep} = B/R^n$.

Let $E_p = E_{Coul} + E_{rep}$ be the potential energy of interaction between the ions in 1 mole of the crystal for the arbitrary nearest-neighbor separation R. The cohesive energy E_c is defined as a positive quantity, so $E_c = -E_p$, where E_p is evaluated at $R = R_0$. We have

$$-E_p = -(E_{Coul} + E_{rep}) = e^2\mathcal{M}N_A/4\pi\varepsilon_0 R - BR^{-n} \quad (24.5)$$

where B and n are as yet unknown.

To evaluate B, we use the condition that the equilibrium nearest-neighbor separation R_0 has the value that minimizes G at constant T and P. Thus $(\partial G/\partial R)_{T,P} = 0$ at $R = R_0$. The definition $G = U + PV - TS$ becomes $G = U + PV$ at 0 K. We have $(\partial G/\partial R)_{T,P} = (\partial U/\partial R)_{T,P} + P(\partial V/\partial R)_{T,P}$ at 0 K. At ordinary pressures, $P(\partial V/\partial R)_{T,P}$ is negligible compared with $(\partial U/\partial R)_{T,P}$ (see Prob. 24.9) and can be neglected to give $0 = (\partial G/\partial R)_{T,P} = (\partial U/\partial R)_{T,P}$ at 0 K. The ionic crystal's U at 0 K equals the energy E_p in (24.5) plus the zero-point vibrational energy E_{ZPE}. One finds $E_{ZPE} \ll E_p$, and E_{ZPE} will be neglected. Thus $0 = (\partial U/\partial R)_{T,P} = dE_p/dR$ at 0 K and for $R = R_0$.

Differentiation of (24.5) gives $-dE_p/dR = -e^2\mathcal{M}N_A/4\pi\varepsilon_0 R^2 + nBR^{-n-1}$. Setting $dE_p/dR = 0$ at $R = R_0$ at 0 K and solving for B, we get

$$B = (e^2/4\pi\varepsilon_0)\mathcal{M}N_A n^{-1}R_0^{n-1} \quad \text{for } T = 0 \quad (24.6)$$

Substitution of (24.6) in (24.5) gives

$$-E_p = (e^2/4\pi\varepsilon_0)\mathcal{M}N_A(R^{-1} - n^{-1}R_0^{n-1}R^{-n}) \tag{24.7}$$

$$-E_p = (e^2/4\pi\varepsilon_0)\mathcal{M}N_A R_0^{-1}[R_0/R - n^{-1}(R_0/R)^n] \tag{24.8}$$

$$E_c = -E_p(R_0) = \frac{e^2\mathcal{M}N_A}{4\pi\varepsilon_0 R_0}\left(1 - \frac{1}{n}\right) \qquad \text{for } T = 0 \tag{24.9}$$

The parameter n in the repulsive part of the potential can be found from compressibility data. Differentiation of $(\partial U_m/\partial V_m)_T = \alpha T/\kappa - P$ [Eq. (4.47)] with respect to V_m gives $(\partial^2 U_m/\partial V_m^2)_T$. Equating this at 0 K to $\partial^2 E_p/\partial V_m^2$, which is found by differentiation of (24.8), one finds (Prob. 24.10)

$$n = 1 + \frac{36\pi\varepsilon_0 V_{m,0} R_0}{\kappa e^2 \mathcal{M}N_A} \qquad \text{for } T = 0 \tag{24.10}$$

where $V_{m,0}$ and κ are the molar volume and compressibility at 0 K.

We shall see in Sec. 24.8 that the NaCl crystal is composed of cubic unit cells; each unit cell (Fig. 24.16) contains four Na^+–Cl^- ion pairs and has an edge length $2R_0$. A mole of NaCl(c) contains $N_A/4$ unit cells and has a volume $V_{m,0} = (N_A/4)(2R_0)^3 = 2N_A R_0^3$. Hence, for NaCl

$$n = 1 + 72\pi\varepsilon_0 R_0^4/\kappa e^2\mathcal{M} \qquad \text{for } T = 0 \tag{24.11}$$

Extrapolation of data to absolute zero gives $\kappa = 3.7 \times 10^{-12}$ cm^2 dyn^{-1} for NaCl at 1 atm. Using the 0-K R_0 value listed after Eq. (24.4), one finds $n = 8.4$ for NaCl (Prob. 24.7).

The theoretical NaCl cohesive energy is given by (24.9), (24.4), and the data following (24.4) as (868 kJ/mol)(1 − 1/8.4) = 765 kJ/mol at $T = 0$. The experimental E_c at 0 K listed after (24.1) is 786 kJ/mol. The agreement is good.

There are two corrections that should be included. The theoretical E_c should include the potential energy of the dispersion interaction (Sec. 22.10) between the ions. One finds that this adds 21 kJ/mol to the theoretical E_c, bringing it to 786 kJ/mol. The experimental E_c is $\Delta H°$ for NaCl(c) $\rightarrow$ Na$^+$(g) + Cl$^-$(g) and is less than the theoretical E_c at $T = 0$ because of the zero-point vibrational energy of NaCl(c). This zero-point energy is 6 kJ/mol and reduces the theoretical E_c to 780 kJ/mol, compared with the experimental 0-K value 786 kJ/mol.

Equation (24.9) shows that the binding energy E_c of an ionic solid is inversely proportional to the interionic distance R_0. An ionic solid that has a very large cation or anion may have E_c small enough to make its normal melting point lie below room temperature. For example, $[CH_3CH_2NH_3][NO_3]$ melts at 12°C and $[(CH_3(CH_2)_6)_4N]Cl$ melts at −9°C. *Room temperature ionic liquids* are good solvents for many organic and inorganic compounds and can serve as solvents for many reactions [T. Welton, *Chem. Rev.*, **99**, 2071 (1999)]. Their room-temperature vapor pressure is undetectably small and they are readily recycled, making them environmentally benign solvents.

Metals

Quantum-mechanical calculations of E_c of the first 50 metals in the periodic table using a density-functional method gave results generally within 15% of the true E_c; see V. L. Moruzzi et al., *Calculated Electronic Properties of Metals,* Pergamon, 1978; *Phys. Rev. B,* **15**, 2854 (1977).

Molecular Solids

E_c for a van der Waals crystal is found by summing the energies of the intermolecular van der Waals attractions and repulsions. For a hydrogen-bonded crystal, one also includes the energy of the hydrogen bonds.

Since the form of the intermolecular potential is not accurately known for most molecules, accurate calculation of E_c is difficult. For a simple crystal like Ar, a pretty accurate value of E_c can be calculated by using the Lennard-Jones 6-12 intermolecular potential and summing this over all pairs of Ar atoms in the crystal. This approach neglects three-molecule interactions. Each Ar atom in the crystal has 12 nearest neighbors (Sec. 24.8). Counting only nearest-neighbor interactions and using the Lennard-Jones potential (22.136), we have $E_c \approx -12(\frac{1}{2})N_A 4\varepsilon[(\sigma/R)^{12} - (\sigma/R)^6]$, where the factor $\frac{1}{2}$ avoids counting each interaction twice. Substitution of the ε and σ values in Sec. 22.10 and the experimental nearest-neighbor separation $R_0 = 3.75$ Å at 0 K gives (Prob. 24.13) 5.2 kJ/mol, compared with the experimental value 7.7 kJ/mol at 0 K. When non-nearest-neighbor interactions are included (Prob. 24.12), the Lennard-Jones potential predicts $E_c = 8.3$ kJ/mol. Thus, measurement of gas-phase properties of Ar allows the binding energy of solid Ar to be calculated.

For H_2O, experimental data and theoretical calculations for the gas-phase dimer $(H_2O)_2$ show that an $O \cdots HO$ hydrogen bond has an energy of roughly 5 kcal/mol (Sec. 20.9). Since each H atom in ice participates in one hydrogen bond, there are $2N_A$ hydrogen bonds per mole of ice. Let ΔE_0 be the energy difference at $T = 0$ between 1 mol of water vapor without molecular zero-point vibrational energy (ZPE) and 1 mol of ice without zero-point vibrational energy. ΔE_0 is the energy change for the hypothetical 0-K process

$$\text{ice(no ZPE)} \rightarrow \text{ice} \rightarrow H_2O(g) \rightarrow H_2O(g, \text{no ZPE})$$

For this process, $\Delta E_0 = \text{ZPE}_{\text{ice}} + \Delta_{\text{sub}}U - \text{ZPE}_{H_2O}$. The ΔU of sublimation of ice at 0 K can be accurately equated to $\Delta_{\text{sub}}H°$ for ice at 0 K, which is 11.3 kcal/mol. Also the ZPE of gas-phase water molecules is 13.2 kcal/mol and $\text{ZPE}_{\text{ice}} = 15.3$ kcal/mol. Therefore $\Delta E_0 = 13.4$ kcal/mol. Hydrogen bonds account for about 2(5 kcal/mol) = 10 kcal/mol of ΔE_0. The remainder is due to van der Waals forces.

In summary, although accurate calculation of the cohesive energy of a crystal is not always possible, the forces that determine the cohesive energy are well understood.

24.6 INTERATOMIC DISTANCES IN CRYSTALS

Ionic Crystals

Analysis of x-ray diffraction data of crystals (Sec. 24.9) enables one to find the electron probability density $\rho(x, y, z)$ (Sec. 20.8) in the crystal. Along the line joining the nuclei of adjacent Na^+ and Cl^- ions in an NaCl crystal, ρ is a local maximum at each nucleus (because of the s electrons) and decreases going away from each nucleus, reaching a minimum at some point between the nuclei. The distance from each nucleus to this minimum point gives the radius of each ion in the crystal. The crystal radius of a given ion is found to increase with increase in coordination number. Some ionic radii for coordination number 6 (based on room-temperature crystal structures) are

Li^+	Na^+	K^+	Rb^+	Cs^+	Be^{2+}	Mg^{2+}	Ca^{2+}	Al^{3+}	Cu^{2+}
0.90	1.16	1.52	1.66	1.81	0.41	0.86	1.14	0.68	0.87

O^{2-}	S^{2-}	F^-	Cl^-	Br^-	I^-
1.26	1.70	1.19	1.67	1.82	2.06

(See J. E. Huheey et al., *Inorganic Chemistry*, 4th ed., Addison-Wesley, 1993, Table 4.4 for more data.) Of course, cation radii are smaller than the corresponding atomic radii, and anion radii are larger than the corresponding atomic radii. These values can be used to estimate nearest-neighbor distances in ionic crystals.

The nearest-neighbor distance in NaCl(c) is 2.80 Å, compared with $R_e = 2.36$ Å in gas-phase NaCl. An ion in the crystal interacts with very many other ions, whereas an ion in the gas-phase molecule interacts with only one other ion. Hence, the potential energy is minimized at different values of R in the crystal and in the isolated molecule.

Molecular Crystals

Because of the weakness of intermolecular van der Waals forces compared with chemical-bonding forces, one usually finds bond distances within molecules to be almost unchanged on going from the gas phase to the solid phase. For example, the Raman spectrum of gas-phase benzene gives $R_0(CC) = 1.397$ Å, x-ray diffraction of solid benzene gives $R_0(CC) = 1.39_2$ Å, and neutron diffraction of solid benzene gives $R_0(CC) = 1.398$ Å.

The distance between molecules in contact in a crystal is determined by the intermolecular van der Waals attractive and repulsive forces. From intermolecular distances in crystals, one can deduce a set of **van der Waals radii** for atoms. For example, in solid Cl_2, the shortest Cl–Cl distance is 2.02 Å, and this is the bond distance in Cl_2; the single-bond covalent radius of Cl is 1.01 Å. The Cl_2 molecules in the crystal are arranged in layers. The closest Cl–Cl distance between neighboring molecules in the same layer is 3.34 Å, and between neighboring molecules in adjacent layers is 3.69 Å. The van der Waals radius of Cl thus lies between 3.34/2 and 3.69/2 Å and is usually taken as 1.8 Å.

Van der Waals radii are much larger than bond radii, since there are two pairs of valence electrons between the nonbonded atoms, compared with one pair between the bonded atoms. Van der Waals radii are found to be close to ionic radii; note that the outer part of a Cl atom in the molecule X:C̈l: resembles a :C̈l:⁻ ion. Some van der Waals radii in Å (L. Pauling, *The Nature of the Chemical Bond,* 3rd ed. Cornell University Press, 1960, p. 260) are

H	He	N	O	F	Ne	P	S	Cl
1.2	0.9	1.5	1.4	1.35	1.3	1.9	1.85	1.8

Van der Waals radii are used to fix the size of atoms in space-filling molecular models. One takes each atom as a truncated sphere (Fig. 24.4). [Detailed analysis of crystallographic data shows nonspherical shapes for most nonbonded atoms; see S. C. Nyburg et al., *Acta Cryst. B,* **41,** 274 (1985); **43,** 106 (1987).]

When hydrogen bonding exists, it may strongly influence the packing of the molecules in the crystal. Ice has a very open structure because of hydrogen bonding. When ice melts, the degree of hydrogen bonding decreases, and liquid water at 0°C is denser than ice.

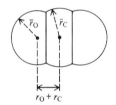

Figure 24.4

Model of CO_2. r_C and $\bar{r}_C$ are the double-bond radius and the van der Waals radius of carbon, respectively.

Covalent Crystals

Interatomic distances in covalent crystals are determined by essentially the same quantum-mechanical interactions as in isolated molecules. For example, the carbon–carbon bond distance in diamond at 18°C is 1.545 Å, virtually the same as in most saturated organic compounds (1.53 to 1.54 Å).

Metallic Crystals

The *metallic radius* of an atom is half the distance between adjacent atoms in a metallic crystal with coordination number 12. The metallic radius of an atom is a bit larger than its single-bond covalent radius. For example, the metallic radius of Cu is 1.28 Å, and its single-bond covalent radius is 1.17 Å.

Atoms at the surface of a metal experience different forces than those in the bulk, and low-energy-electron-diffraction experiments (Sec. 24.10) show that the distance between the surface layer and the second layer in a metal is typically 1 to 10% less than the distance between corresponding layers in the bulk.

24.7   CRYSTAL STRUCTURES

The Basis

A crystal contains a structural unit, called the **basis** (or **structural motif**), that is repeated in three dimensions to generate the crystal structure. The environment of each repeated unit is the same throughout the crystal (neglecting surface effects). The basis may be a single atom or molecule, or it may be a small group of atoms, molecules, or ions. Each repeated basis group has the same structure and the same spatial orientation as every other basis group in the crystal. Of course, the basis must have the same stoichiometric composition as the crystal.

For NaCl, the basis consists of one Na^+ ion and one Cl^- ion. For Cu, the basis is a single Cu atom. For Zn, the basis consists of two Zn atoms. For diamond, the basis is two C atoms; the two atoms of the basis are each surrounded tetrahedrally by four carbons, but the four bonds at one basis atom differ in orientation from those at the other atom; see Fig. 24.19 and the accompanying discussion. For CO_2, the basis is four CO_2 molecules. For benzene, it is four C_6H_6 molecules.

The Space Lattice

If we place a single point at the same location in each repeated basis group, the set of points obtained forms the **(space) lattice** of the crystal. Each point of the space lattice has the same environment. The space lattice is not the same as the crystal structure. Rather, the crystal structure is generated by placing an identical structural group (the basis) at each lattice point. The space lattice is a geometrical abstraction. Figure 24.5 shows a two-dimensional lattice and a hypothetical two-dimensional crystal structure formed by associating with each lattice point a basis consisting of an M atom and a W atom.

The W atoms in Fig. 24.5 do not lie at lattice points. Of course, the lattice points could have been chosen to coincide with the W atoms, in which case the M atoms would not lie at lattice points. In fact, no atom need lie at a lattice point. For example, in I_2, one chooses the lattice point at the center of one I_2 molecule of the basis, which consists of two I_2 molecules; see Fig. 24.20b.

The Unit Cell

The space lattice of a crystal can be divided into identical parallelepipeds by joining the lattice points with straight lines. (A parallelepiped is a six-sided geometrical solid whose faces are all parallelograms.) Each such parallelepiped is called a **unit cell.** The way in which a lattice is broken up into unit cells is not unique. Figure 24.6 shows two different ways of forming unit cells in the two-dimensional lattice of Fig. 24.5. The same kind of choice exists for three-dimensional lattices. In crystallography, one chooses the unit cell so that it has the maximum symmetry and has the smallest volume consistent with the maximum symmetry. The maximum-symmetry requirement implies the maximum number of perpendicular unit-cell edges.

In two dimensions, a unit cell is a parallelogram with sides of length a and b and angle γ between these sides. In three dimensions, a unit cell is a parallelepiped with

Figure 24.5

The crystal structure is generated by associating a basis group with each lattice point.

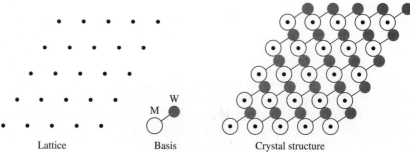

Lattice   Basis   Crystal structure

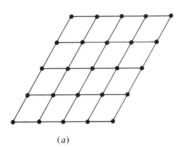 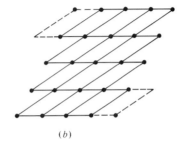

(*a*)   (*b*)

Figure 24.6

Two ways of breaking the two-dimensional lattice of Fig. 24.5 into unit cells.

edges of length a, b, c and angles α, β, γ, where α is the angle between edges b and c, etc.

In 1848, Bravais showed that there are 14 different kinds of lattices in three dimensions. The unit cells of the 14 Bravais lattices are shown in Fig. 24.7. The 14 Bravais lattices are grouped into seven **crystal systems,** based on unit-cell symmetry.

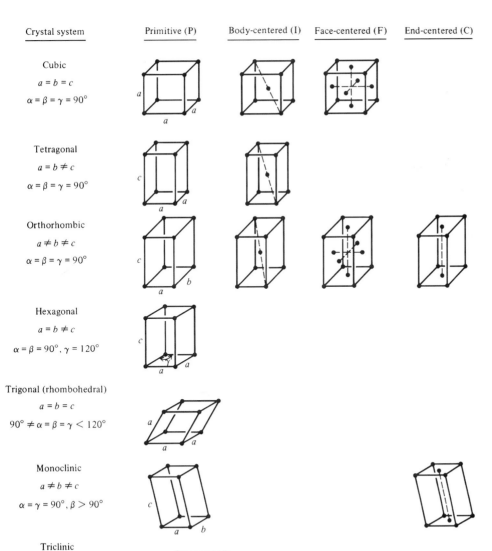

Figure 24.7

Unit cells of the 14 Bravais lattices.

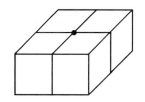

Figure 24.8

The lattice point shown is shared by four unit cells at the same level and four more unit cells (not shown) immediately above.

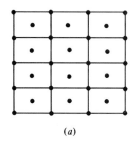

(a)

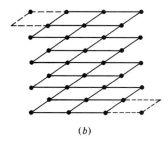

(b)

Figure 24.9

(a) A two-dimensional centered lattice. (b) The same lattice divided into primitive unit cells.

The relations between a, b, and c and between α, β, and γ for the seven systems are indicated in Fig. 24.7. (Some workers group the lattices into six systems; see *Buerger*, chap. 2.)

Unit cells that have lattice points only at their corners are called **primitive** (or **simple**) unit cells. Seven of the Bravais lattices have primitive (P) unit cells. A **body-centered** lattice (denoted by the letter I, from the German *innenzentrierte*) has a lattice point within the unit cell as well as at each corner of the unit cell. A **face-centered** (F) lattice has a lattice point on each of the six unit-cell faces as well as at the corners. The letter C denotes an **end-centered** lattice with a lattice point on each of the two faces bounded by edges of lengths a and b. The letters A and B have similar meanings.

Each point at a unit-cell corner is shared among eight adjacent unit cells in the lattice—four at the same level (Fig. 24.8) and four immediately above or below. Therefore a primitive unit cell has $8/8 = 1$ lattice point and 1 basis group per unit cell. Each point on a unit-cell face is shared between two unit cells, so an F unit cell has $8/8 + 6/2 = 4$ lattice points and 4 basis groups per unit cell.

One *could* use a primitive unit cell to describe any crystal structure, but since in many cases this cell would have less symmetry than a cell of a (nonprimitive) centered lattice, it is more convenient to use the centered lattice. For example, Fig. 24.9 shows a two-dimensional lattice broken into centered unit cells and into less symmetric primitive unit cells.

The structure of quasicrystals (Sec. 24.1) may involve unit cells that overlap neighboring unit cells [P. J. Steinhardt et al., *Nature,* **396,** 55 (1998); **403,** 267 (2000); feynman.princeton.edu/~steinh/quasi/].

Notation for Points and Planes

To designate the location of any point in the unit cell, we set up a coordinate system with origin at one corner of the unit cell and axes coinciding with the a, b, and c edges of the cell. Note that these axes are not necessarily mutually perpendicular. The position of a point in the cell is specified by giving its coordinates as fractions of the unit-cell lengths a, b, and c. Thus, the point at the origin is 000; the interior lattice point in an I lattice is at $\frac{1}{2}\frac{1}{2}\frac{1}{2}$; the point at the center of the face bounded by the b and c axes is $0\frac{1}{2}\frac{1}{2}$.

The orientation of a crystal plane is described by its **Miller indices** (hkl), which are obtained by the following steps: (1) find the intercepts of the plane on the a, b, c axes in terms of multiples of the unit-cell lengths a, b, c; (2) take the reciprocals of these numbers; (3) if fractions are obtained in step 2, multiply the three numbers by the smallest integer that will give whole numbers. If an intercept is negative, one indicates this by a bar over the corresponding Miller index.

One is usually interested only in planes densely populated by lattice points, since it is these planes which are important in x-ray diffraction (Sec. 24.9).

As an example, the shaded plane labeled r in Fig. 24.10 intercepts the a axis at $a/2$ and the b axis at $b/2$ and lies parallel to the c axis (intercept at ∞). Step 1 gives $\frac{1}{2}$, $\frac{1}{2}$, ∞. Step 2 gives 2, 2, 0. Hence the Miller indices are (220). The plane labeled s has Miller indices (110). The plane labeled t has intercepts $\frac{3}{2}a$, $\frac{3}{2}b$, ∞; step 2 gives $\frac{2}{3}$, $\frac{2}{3}$, 0, and the Miller indices are (220). Plane u has intercepts $2a$, $2b$, ∞, so step 2 gives $\frac{1}{2}$, $\frac{1}{2}$, 0, and step 3 gives (110). Also shown are a (111) plane and (100) planes. The higher the value of the Miller index h of a plane, the closer to the origin is the a intercept of the plane.

As well as giving the orientation of a single plane, one also uses Miller indices to denote a whole stack of parallel, equally spaced planes. The planes s, u, and an infinite number of planes parallel to them and separated by the same distance as s and u form the set of (110) planes. The set of (220) planes is considered to include the (110) planes plus the planes midway between the (110) planes. Planes r, s, t, u, . . . form the (220) set.

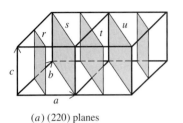
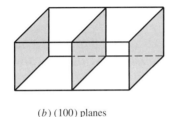

(a) (220) planes   (b) (100) planes   (c) (111) plane

Figure 24.10

(a) (220) planes. (b) (100) planes. (c) A (111) plane. Two unit cells are shown in (a) and in (b); one is shown in (c).

In determining the Miller indices of a set of parallel equally spaced planes, one looks at the intercepts of the plane closest to the origin but not containing the origin.

Each face of a macroscopic crystal contains a high density of lattice points. Examination of the macroscopic shape of a single crystal will generally tell which of the seven crystal systems the crystal belongs to (but will not tell what the Bravais lattice is) and will enable the a, b, and c axes to be located. One can therefore use Miller indices to specify the orientations of the macroscopic faces (surfaces) of the crystal, as well as to specify the orientations of planes within the crystal lattice.

Studies of chemisorption of gases on metals often use a single metal crystal cut to expose a particular face to the gas. One finds that the (100) surface of a metal has different adsorption properties (heats of adsorption, sticking probabilities) than other surfaces such as the (110) surface. The structure of an adsorbed species may differ on different surfaces of the same metal.

EXAMPLE 24.1 Miller indices

Find the Miller indices (hkl) of the surface s_2 and the set of planes p_1 in Fig. 24.11. All the planes and surfaces in this figure are parallel to the c axis. The lattice in Fig. 24.11 is primitive.

Since the planes and surfaces are parallel to the c axis, the c intercepts are all at ∞, and so the Miller l indices are all 0. We set up an a-b coordinate system with origin at point e, which is chosen as close as possible to the leftmost p_1 plane without being in this plane. For this origin, the leftmost p_1 plane intercepts the a axis at $1 \cdot a$ and the b axis at $1 \cdot b$, so the Miller indices are $h = 1/1 = 1$ and $k = 1/1 = 1$. The p_1 planes are the (110) planes. Similarly, with origin at r, we see that s_2 is a (110) surface.

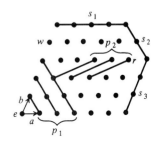

Figure 24.11

Planes and surfaces in a primitive lattice.

Exercise

Find the Miller indices of the surface s_1. [*Answer:* With origin at w, one finds (010).]

Anisotropy

The repeated structural group of a crystal has a fixed orientation in space. It follows that the properties of a crystal will, in general, be different in different directions. For example, we can see that the properties of the hypothetical two-dimensional crystal in Fig. 24.5 will differ in the a and b directions. A substance whose physical properties are the same in all directions is called **isotropic;** otherwise, it is **anisotropic.**

Solid

Liquid crystal

Liquid

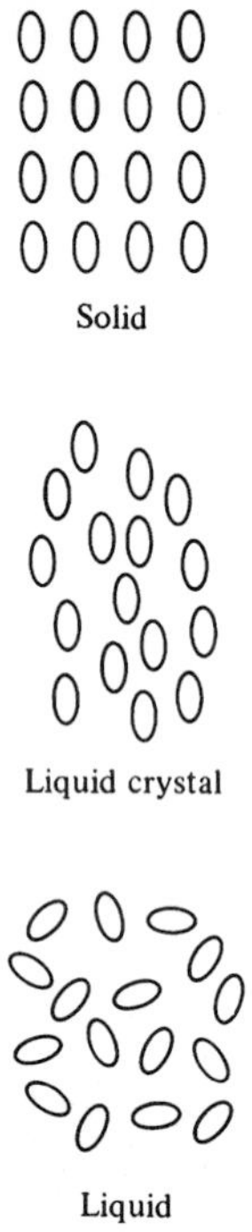

Figure 24.12

Orientation and spacing of
molecules in a solid, a liquid, and
a liquid crystal.

Gases and liquids are isotropic. An exception is liquid crystals. **Liquid crystals** flow like liquids but have much of the long-range order of solids. In a liquid crystal, the molecules can move about, and the intermolecular spacings are irregular. However, most of the molecules have the same spatial orientation (Fig. 24.12). (In certain liquid crystals, the molecules form layers; the molecules in a given layer all have the same orientation, but the orientation direction varies in a regular way from layer to layer. Such liquid crystals show bright temperature-dependent colors, and can be used to indicate temperature.) At a certain elevated temperature, a liquid crystal makes a transition to a true liquid state with random molecular orientations. Many liquid crystals are organic compounds having a long nonpolar chain and a polar group (for example, $C_6H_{13}C_6H_4COOH$). The liquid-crystal state may occur in biological-cell membranes.

Application of a weak electric field changes the orientation of molecules in a liquid crystal and thereby changes the appearance of the substance. This is how liquid-crystal displays for calculators, digital watches, and portable computers work.

A single crystal is, in general, anisotropic. A finely powdered solid is isotropic, since the random orientations of the tiny crystals produce isotropy. A single crystal has different refractive indices, coefficients of thermal expansion, electrical conductivities, speeds of sound, etc., in different directions. For example, the 20°C refractive indices of $AgNO_3$ along the a, b, and c axes are 1.73, 1.74, and 1.79 for sodium D light. The cubic crystal NaCl has the same refractive index along each axis and so is optically isotropic. However, NaCl is very anisotropic in its response to mechanical stress. For example, when crushed, an NaCl crystal cleaves only along planes containing unit-cell faces. Different faces of a crystal may show different catalytic activities and different rates of solution.

The Avogadro Constant

Let Z be the number of formula units per unit cell. The unit cell of NaCl (Fig. 24.16) has four Na^+ and four Cl^- ions, so $Z = 4$ for NaCl. The unit cell of CO_2 has four CO_2 molecules, so $Z = 4$ for $CO_2(c)$. The unit cell of diamond (Fig. 24.19) has eight carbon atoms, so $Z = 8$ for diamond.

One mole of a crystal contains N_A/Z unit cells. The volume of a right-angled unit cell is the product abc of its edges. (The volume formula for a nonrectangular cell is given in *Buerger*, p. 187.) The molar volume is therefore $V_m = abcN_A/Z$. The density is $\rho = M/V_m$, where M is the molar mass. Therefore

$$\rho = \frac{ZM}{abcN_A} \qquad \text{for } \alpha = \beta = \gamma = 90° \qquad (24.12)$$

X-ray diffraction (Sec. 24.9) allows a, b, c, and Z to be determined at temperature T. An accurate measurement of ρ at T then allows the Avogadro constant N_A to be found.

EXAMPLE 24.2 Density and unit-cell dimensions

Silicon crystallizes in the same face-centered cubic lattice as diamond, and so $a = b = c$ and $Z = 8$ for Si. For a single crystal of very pure Si, one finds at 22.5°C and 0 atm [P. Seyfried et al., *Z. Phys. B*, **87**, 289 (1992)]

$$\rho = 2.329032 \text{ g/cm}^3, \quad M = 28.08538 \text{ g/mol}, \quad a = 5.431020 \text{ Å}$$

Find N_A.

Substitution in (24.12) gives

$$N_A = \frac{ZM}{\rho a^3} = \frac{8(28.08538 \text{ g/mol})}{(2.329032 \text{ g/cm}^3)(5.431020 \times 10^{-8} \text{ cm})^3} = 6.02214 \times 10^{23} \text{ mol}^{-1}$$

Exercise

Potassium has a cubic lattice. At 25°C, the density of K is 0.856 g/cm³ and x-ray diffraction shows the unit-cell edge length is 5.33 Å. Find the number of formula units in a unit cell of K. What kind of cubic lattice does K have? (*Answer:* 2, body-centered cubic.)

Exercise

X-ray diffraction analysis of crystalline C_{60} (buckminsterfullerene) shows that the crystal structure at 300 K can be regarded as face-centered cubic with a one-molecule basis and a unit-cell edge length of 14.17 Å; the molecules are orientationally disordered due to rotation. Below 249 K, the molecules are orientationally ordered and the crystal structure is primitive cubic with a four-molecule basis and a unit-cell edge length of 14.04 Å at 11 K [P. A. Heiney et. al., *Phys. Rev. Lett.,* **66,** 2911 (1991)]. Find the density of $C_{60}(s)$ at 300 K and at 11 K. Find the center-to-center distance between nearest-neighbor C_{60} molecules in the solid at 300 K. (*Answers:* 1.682 g/cm³, 1.730 g/cm³, 10.02 Å.)

24.8   EXAMPLES OF CRYSTAL STRUCTURES

Metallic Crystals

Metal atoms are spherical, and the structures of metallic elements can be described in terms of the various ways of packing spheres.

Figure 24.13 shows a planar layer of spheres, with each sphere touching four others in the layer. For now, ignore the lines and the shaded spheres. Let successive layers of spheres be added with each sphere directly over a sphere in the layer beneath it. This gives a structure having a **simple cubic** space lattice (Fig. 24.7) with a basis of one atom at each lattice point. The coordination number (CN) is 6, since each atom touches four atoms in the same layer, one atom in the layer above it, and one atom in the layer below it. The structure is an open one, with only 52% of the volume occupied by the spheres (Prob. 24.21). The simple cubic lattice is very rare for metals; the only known example is Po.

Instead of adding the second layer directly over the first, let us add a second layer of spheres (shown shaded in Fig. 24.13) by placing a sphere in each of the hollows formed by the first-layer spheres. A third layer can then be placed in the hollows of the second layer, with each third-layer sphere lying directly above a first-layer sphere. A fourth layer is then added, with each fourth-layer sphere directly over a second-layer sphere; etc. This produces a **face-centered cubic** (fcc) space lattice (Fig. 24.7), the basis being one atom at each lattice point. The square base of a unit cell is outlined in Fig. 24.13. The four spheres with dots at their centers lie at the corners of the unit-cell base, and the sphere marked with a cross lies at the center of the unit-cell base. The four shaded second-layer spheres above the sphere with a cross lie at the centers of the four side faces of the unit cell. The third-layer sphere directly over the sphere with a cross lies at the center of the top face of the unit cell. If you have as much trouble as I do at three-dimensional visualization, construct models using coins or marbles held together with bits of clay.

The CN of the fcc structure is 12, since each atom touches four others in the same layer, four in the layer above, and four in the layer below. The high CN makes for a very close-packed structure with 74 percent of the volume occupied. [In 1611, Kepler conjectured that the fcc structure gives the densest possible packing of spheres. This fact wasn't proved until 1998. See *Science,* **281,** 1267 (1998).] The fcc structure is often called cubic close-packed (ccp). [An alternative description of the fcc structure

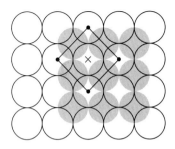

Figure 24.13

Formation of a simple cubic lattice and of a face-centered cubic lattice.

Figure 24.14

(a) Formation of a bcc lattice.
(b) Formation of an hcp structure.

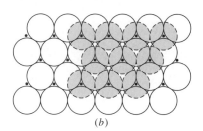

(a)   (b)

is often used. Drawing the (111) planes in the lattice, one finds that each atom touches six others in the same (111) plane, three others in the (111) plane above it, and three others in the (111) plane below it; Prob. 24.22.]

The fcc structure is very common for metals. Examples include Al, Cu, Au, Pb, Pt, Pd, Ni, and Ca.

Suppose we start with a layer of spheres arranged as in Fig. 24.14a with centers separated by $2/\sqrt{3} = 1.155$ times the spheres' diameter. We place a second layer (shaded spheres) in the hollows of the first layer, and a third layer with its spheres directly over the first-layer spheres, etc. This produces a **body-centered cubic** (bcc) space lattice (Fig. 24.7) with a basis of one atom at each lattice point. The CN is 8, since each atom touches four atoms in the layer above it, four in the layer below it, and none in its own layer. This structure fills 68% of the space and is quite common for metals. Examples include Cr, Mo, W, Ba, Li, Na, K, Rb, and Cs.

Finally, we can start with a layer in which each sphere touches six others (Fig. 24.14b). We place spheres in those hollows that are marked by dots to give a second layer (shaded spheres). We then place spheres in those second-layer hollows that lie directly over the centers of the first-layer spheres to give a third layer of spheres lying directly over the first-layer spheres. (An alternative choice exists for the third-layer spheres; this turns out to give the fcc lattice discussed above.) A fourth layer is then formed with its spheres lying directly above the second-layer spheres, etc. The CN is 12, since each atom touches six atoms in its own layer, three atoms in the layer above, and three atoms in the layer below. One finds that this structure has a (primitive) hexagonal space lattice; the basis consists of two atoms associated with each lattice point (Fig. 24.15). The second atom of the basis lies at the point $\frac{2}{3} \frac{1}{3} \frac{1}{2}$, which is not a lattice point. The unit cell is drawn with heavy lines in Fig. 24.15. This structure fills 74% of the volume (the same as the fcc structure) and is called **hexagonal close-packed** (hcp). Many metals have hcp structures, including Be, Mg, Cd, Co, Zn, Ti, and Tl.

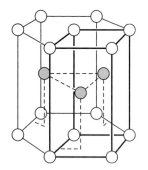

Figure 24.15

Atom positions in the hcp structure. The unit cell is rectangular and is indicated by the heavy lines. With the origin at the atom at the center of the hexagonal base, the shaded atoms lie at $\frac{2}{3} \frac{1}{3} \frac{1}{2}$.

Some metals undergo changes in structure as the temperature and pressure change. For example, Fe(c) at 1 atm is fcc between 906°C and 1401°C but is bcc both above and below this range.

In summary, most metal structures are hcp (CN 12), fcc (CN 12), or bcc (CN 8).

Ionic Crystals

Many M^+X^- ionic crystals have the NaCl-type structure. This is a face-centered cubic space lattice with a basis of one M^+ ion and one X^- ion associated with each lattice point. Figure 24.16a shows the NaCl unit cell. A Cl^- ion has been placed at each lattice point (the unit-cell corners and face centers), and an Na^+ ion has been placed a distance $\frac{1}{2}a$ directly above each Cl^- (where a is the cubic unit-cell edge length). The Na^+ ions do not lie at lattice points. (Of course, the lattice points could have been chosen to be at the Na^+ ions, in which case the Cl^- ions would not be at lattice points.) The four Na^+ ions on the bottom face of the unit cell are associated with lattice points in the unit cell below.

There are 8 Cl^- ions at the corners and 6 on the faces, so each unit cell has 8/8 + 6/2 = 4 Cl^- ions. There are 12 Na^+ ions on unit-cell edges, and these are shared with three other unit cells. There is 1 Na^+ at the unit-cell center. Hence there are 12/4 + 1

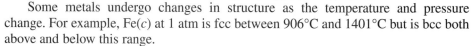

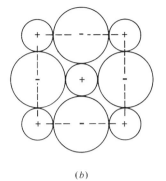

Figure 24.16

(a) The NaCl unit cell is outlined by the heavy lines. (b) Packing of ions in NaCl.

= 4 Na$^+$ ions per unit cell. The number of formula units per unit cell is $Z = 4$. The CN is 6.

The ions in Fig. 24.16a have been shrunk to allow the structure to be seen clearly. Figure 24.16b shows the actual arrangement of ions in a plane through the center of the unit cell. Note that oppositely charged ions are in contact. The center-to-center distance of nearest-neighbor oppositely charged ions is $\frac{1}{2}a$ [$\equiv R$ in Eq. (24.4)]; the centers of two nearest like-charged ions are $\frac{1}{2}a\sqrt{2}$ apart.

Compounds having the NaCl-type structure include many group 1 halides, hydrides, and cyanides (for example, LiH, KF, KH, KCN, LiCl, NaBr, NaCN), and many group 2 oxides and sulfides (for example, MgO, CaO, MgS). Further examples are AgBr, MnO, PbS, CeN, and FeO.

About 15 ionic compounds have the CsCl structure, whose unit cell is shown in Fig. 24.17. The CsCl space lattice is not body-centered cubic but is simple cubic. The basis consists of one Cs$^+$ ion and one Cl$^-$ ion associated with each lattice point. The Cl$^-$ ion at the center of the unit cell does not lie at a lattice point. The Cl$^-$ ions associated with the other seven lattice points in Fig. 24.17 lie in adjacent unit cells. Obviously, $Z = 1$. The CN is 8. Compounds having this structure include CsCl, CsBr, CsI, TlCl, TlBr, TlI, and NH$_4$Cl. Certain alloys have the CsCl structure, for example, CuZn (β-brass) and AgZn. (Recall the order–disorder transitions in NH$_4$Cl and β-brass; Sec. 7.5.)

Figure 24.18 shows the CaF$_2$ (fluorite) structure. The CaF$_2$ space lattice is face-centered cubic. The basis consists of one Ca^{2+} ion and two F$^-$ ions, the Ca^{2+} ions lying at the lattice points. For the Ca^{2+} ion at point 000, the associated F$^-$ ions lie at $\frac{1}{4}\frac{1}{4}\frac{1}{4}$ and $-\frac{1}{4}-\frac{1}{4}-\frac{1}{4}$. The CN of each Ca^{2+} is 8, and that of each F$^-$ is 4. Figure 24.18b

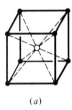

(a)

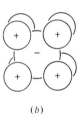

(b)

Figure 24.17

(a) The CsCl unit cell. (b) Packing of ions in CsCl.

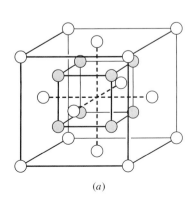

(a)

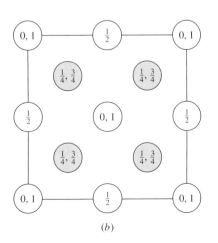

(b)

Figure 24.18

(a) The unit cell of CaF$_2$. The F atoms are shaded. (b) Projection of this unit cell on its base.

Figure 24.19

The unit cell of diamond. Shaded atoms lie within the unit cell and have c coordinate of $\frac{1}{4}$ or $\frac{3}{4}$. Dotted atoms lie on unit-cell faces.

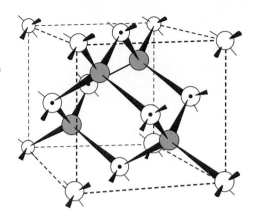

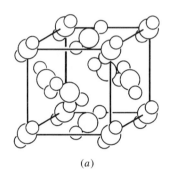

(a)

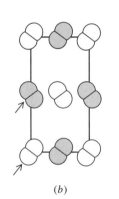

(b)

Figure 24.20

(a) The unit cell of CO_2. Each carbon atom lies at a corner or on a face of the unit cell. (b) The structure of $Br_2(c)$ and of $I_2(c)$. The c axis is perpendicular to the plane of the paper. Unshaded molecules lie on the base of the unit cell. Shaded molecules have a c coordinate of $\frac{1}{2}$. The top face of the unit cell contains molecules in the same positions and orientations as on the base. The arrows point to two molecules that constitute the basis. The atoms have been shrunk for clarity.

shows a projection of the unit-cell atom positions on the unit-cell base. The numbers in the circles give the c coordinate, the height above the base plane; when two numbers appear in a circle, there are two atoms directly above that position. Crystals with the CaF_2 structure include CaF_2, BaF_2, K_2O, and Na_2S.

Several other structures occur for ionic crystals, but discussion is omitted. (Detailed descriptions of many inorganic crystal structures are given in H. D. Megaw, *Crystal Structures,* Saunders, 1973.)

Covalent Crystals

The diamond space lattice is face-centered cubic. The basis consists of two C atoms. One atom of the basis occupies a lattice point, and the second atom is displaced in each direction by one-fourth the unit-cell edge. For example, there are atoms at 000 and at $\frac{1}{4}\frac{1}{4}\frac{1}{4}$ (see Fig. 24.19). Each of the two atoms of the basis is surrounded tetrahedrally by four carbons, but the bonds at each basis atom have different spatial orientations. The CN is 4. The unit cell contains $8/8 + 6/2 + 4 = 8$ carbon atoms, so $Z = 8$. Si and Ge have the diamond structure.

The ZnS space lattice is face-centered cubic, and the basis consists of one Zn atom and one S atom. The ZnS crystal structure is obtained from the diamond structure by replacing the two C atoms of each basis with one Zn atom and one S atom. The unit cell has four Zn atoms and four S atoms, so $Z = 4$. (Some people classify ZnS as an ionic, rather than covalent, crystal.) Crystals with the ZnS (zinc blende) structure include the following largely covalent crystals: SiC, AlP, CuCl, AgI, GaAs, ZnS, ZnSe, and CdS.

Molecular Crystals

Molecular crystals show a great variety of structures, and we shall consider only a few examples.

Crystals of essentially spherical molecules often have structures determined by the packing of spheres. For example, Ne, Ar, Kr, and Xe have fcc unit cells with a basis of one atom at each lattice point. He crystallizes in the hcp structure. CH_4 crystallizes in an fcc lattice with one CH_4 molecule per lattice point.

The CO_2 space lattice is simple cubic, with a basis of four CO_2 molecules. The molecules of one basis have their centers at the corner point 000 and at the centers of the three faces that meet at this corner. Each molecule is oriented with its axis parallel to one of the four cube diagonals. See Fig. 24.20a.

The Br_2 crystal has an end-centered orthorhombic space lattice with a basis of two Br_2 molecules associated with each lattice point; see Fig. 24.20b.

24.9   DETERMINATION OF CRYSTAL STRUCTURES

X-Ray Diffraction

Interatomic spacings in crystals are on the order of 1 Å. Electromagnetic radiation of 1-Å wavelength lies in the x-ray region. Hence, crystals act as diffraction gratings for x-rays. This was first realized by von Laue in 1912 and forms the basis for the determination of crystal structures.

Figure 24.21a shows a cross section of a primitive cubic lattice. The base of one of the unit cells is outlined. Dashed lines have been drawn through a particular set of equally spaced parallel planes, the (210) planes. Let a beam of monochromatic x-rays of wavelength λ fall on the crystal. Although a given lattice point will in general be associated with a group of atoms, let us temporarily assume that each lattice point is associated with a single atom. Figure 24.21b shows the x-ray beam incident at angle θ to one of the (210) planes. Most of the x-ray photons will pass through this plane with no change in direction, but a small fraction will collide with electrons in the atoms of this plane and will be *scattered,* that is, will undergo a change in direction. The x-ray photons are scattered in all directions by interaction with the electrons of the crystal.

Figure 24.21b shows waves scattered at angle β by two adjacent lattice points p and q. The observation point o for the scattered wave is essentially at infinity, so the path difference for waves scattered from p and q at angle β is $\overline{op} - (\overline{oq} + \overline{qr}) = (\overline{os} + \overline{sp}) - (\overline{oq} + \overline{qr}) = \overline{sp} - \overline{qr} = \overline{pq} \cos \beta - \overline{pq} \cos \theta = \overline{pq}(\cos \beta - \cos \theta)$. If this path difference is zero, scattered waves leaving points p and q at angle β will be in phase with each other and will give constructive interference; similarly, scattered waves leaving q and t at angle β will be in phase; etc. The zero-path-difference condition gives $0 = \overline{pq}(\cos \beta - \cos \theta)$, so $\beta = \theta$. Thus, waves scattered from a plane of lattice points at angle equal to the angle of incidence are in phase with one another. Waves scattered at other angles will generally be out of phase with one another and will give destructive interference. The single plane of lattice points acts as a "mirror" and "reflects" a small fraction of the incident x-rays. (It is true that, if the path difference between scattered waves leaving p and q equals λ or 2λ, etc., these waves are also in phase. However, it turns out that such waves can be regarded as being "reflected" from a set of differently oriented planes in Fig. 24.21a, so consideration of these waves will not add anything new. See C. Kittel, *Introduction to Solid State Physics,* 2d ed., Wiley, 1956, pp. 46–48.)

So far we have considered only waves scattered by one of the (210) planes. The x-ray beam will penetrate the crystal to a depth of tens of thousands of layers of planes, so we must consider the scattering from many, many (210) planes. Figure 24.21c shows the x-ray beam incident on successive (210) planes. At each plane there

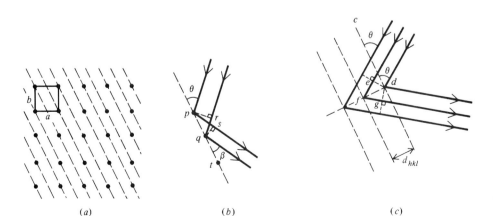

| (a) | (b) | (c) |

Figure 24.21

Derivation of the Bragg equation.

is some scattering with constructive interference at angle θ. For constructive interference between x-rays scattered by the entire set of (210) planes, the waves constructively scattered ("reflected") by two adjacent planes must have a path difference of an integral number of x-ray wavelengths. The path difference between adjacent layers is $\overline{ef} + \overline{fg} = 2\overline{ef}$. Since $\theta + \angle cde = 90° = \angle cde + \angle edf$, we have $\angle edf = \theta$. Therefore $\overline{ef} = \overline{fd} \sin\theta = d_{hkl} \sin\theta$ and the path difference between adjacent planes is $2d_{hkl} \sin\theta$, where d_{hkl} is the spacing between adjacent planes. (In this case, hkl is 210.) Therefore the condition for constructive interference is

$$2d_{hkl} \sin\theta = n\lambda, \qquad n = 1, 2, 3, \ldots \qquad (24.13)$$

The **Bragg equation** (24.13) is the fundamental equation of x-ray crystallography. Constructive interference between waves scattered by the lattice points produces a diffracted beam of x-rays only for the angles of incidence that satisfy (24.13).

For a set of planes to give a diffracted beam intense enough to be observed, each plane of the set must have a high density of electrons. This requires a high density of atoms, so each plane must have a high density of lattice points. Because the number of sets of such planes is limited, the number of values of d_{hkl} is limited and the Bragg condition (24.13) will be met only for certain values of θ.

X-ray crystallographers prefer to write (24.13) in the form

$$2d_{nh,nk,nl} \sin\theta = \lambda \qquad (24.14)$$

For example, for a primitive cubic lattice, the $n = 2$ diffracted beam from the (100) set of planes is considered to be the $n = 1$ diffracted beam from the (200) planes, whose spacing d_{200} is half that of the (100) planes. Similarly, the $n = 3$ diffracted beam for the (100) planes is considered to be the $n = 1$ diffracted beam for the (300) planes.

EXAMPLE 24.3 The Bragg equation

For x-rays with $\lambda = 3.00$ Å, what angles of incidence produce a diffracted beam from the (100) planes in a simple cubic lattice with $a = 5.00$ Å?

The (100) planes are spaced by a, so $\sin\theta_{100} = n\lambda/2d_{100} = n(3\text{ Å})/2(5\text{ Å}) = 0.300n = 0.300, 0.600,$ and 0.900. We find $\theta_{100} = 17.5°, 36.9°,$ and $64.2°$. These are the 100, 200, and 300 reflections.

Exercise

For x-rays with wavelength 4.00 Å, what angles of incidence produce a diffracted beam from the (110) planes in a simple cubic lattice with $a = 8.00$ Å? (*Answer:* 20.7°, 45.0°.)

For the Bragg equation to be applied, the x-ray wavelength must be accurately known. One way to measure λ is to calculate the unit-cell dimension a for a cubic crystal from its measured density using Eq. (24.12); one then diffracts the x-rays from this crystal and uses the Bragg equation (24.13) to find λ.

For some lattices, certain reflections that satisfy the Bragg equation do not give diffracted beams. For example, consider a face-centered lattice, not necessarily cubic (Fig. 24.22). For $n = 1$ and $hkl = 100$ in Eq. (24.13), the x-rays reflected from adjacent (100) planes have a path difference of one wavelength. Hence the path difference between x-rays reflected from the (200) plane labeled v in Fig. 24.22 and x-rays reflected from either of the (100) planes labeled t and w is one-half wavelength. These waves are out of phase and cancel. The cancellation is exact, because the number of lattice points in plane v of the unit cell is the same as the number of lattice points in

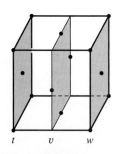

Figure 24.22

Unit cell of a face-centered lattice.

$t \qquad v \qquad w$

planes t and w of the unit cell. (There are $4/2 = 2$ lattice points in v, $4/8 + 1/2 = 1$ lattice point in t, and 1 lattice point in w.) For $n = 2$ in (24.13), the path difference between x-rays reflected from t and w is 2λ and between x-rays reflected from t and v is λ; hence, no cancellation occurs. Putting $n = 3, 4, \ldots$, we see that, in terms of the equation (24.14), a face-centered lattice gives 200, 400, 600, ... reflections, but not 100, 300, 500, ... reflections. These missing reflections are called **extinctions** or **systematic absences.**

A more general treatment (*Buerger,* chap. 5) shows that:

1. For a primitive lattice there are no extinctions.
2. For a face-centered lattice, the only reflections that occur are those whose three indices in (24.14) are either all even numbers or all odd numbers.
3. For a body-centered lattice, the only reflections that occur are those for which the sum of the indices is even.

The condition for end-centered lattices is omitted.

In an x-ray crystal-structure determination, one mounts a small single crystal on a glass fiber. A monochromatic (single-wavelength) x-ray beam impinges on the crystal. The crystal is slowly rotated, thereby bringing different sets of planes into proper orientation for the Bragg equation to be satisfied. The diffracted beams are recorded on photographic film to give a pattern of spots. Alternatively, an ionization counter or a scintillation counter is used to measure the intensities of the diffracted beams. In a scintillation counter, the x-rays fall on a material that fluoresces when illuminated with x-rays. A photoelectric cell measures the intensity of the fluorescent light.

Many modern x-ray diffractometers use a charge-coupled device (CCD) instead of a scintillation counter as the detector. The diffracted x-rays fall on a phosphor screen that fluoresces to produce visible light whose intensity at each location is proportional to the x-ray intensity at that location. The visible light falls on a CCD chip behind the phosphor screen. The CCD chip contains a rectangular array of 10^6 or 10^7 closely spaced semiconductor photodiodes. Light falling on a photodiode excites electrons from the valence band of the semiconductor to the conduction band (Sec. 24.11). The number of electrons excited is proportional to the light intensity. The CCD thus measures the light intensity (and, in effect, the x-ray intensity) at each location over a substantial area (and is an example of an area detector). (CCD light detectors are used in digital cameras and in scanners.) The output data of the CCD is stored in a computer. After a set of data is obtained at one crystal orientation, the crystal is rotated slightly and a new set of data is collected and stored; this process is repeated many times. CCD detectors have revolutionized x-ray crystallography, cutting the time for data collection and structure solution from days or weeks to hours in many cases (*Chem. Eng. News,* March 11, 1996, p. 30).

So far we have assumed one scatterer to be present at each lattice point. In reality, most crystals have several atoms associated with each lattice point. For simplicity, consider the crystal structure of Fig. 24.5 with two atoms, M and W, associated with each lattice point. The set of M atoms forms an array exactly like the array of lattice points, so x-ray diffraction from the M atoms occurs for those angles that satisfy the Bragg equation (24.13). Likewise the set of W atoms forms an array exactly like the array of lattice points, so x-ray diffraction from the W atoms occurs for those angles that satisfy the Bragg equation. However, the x-rays scattered by the W atoms are in general somewhat out of phase with those scattered by the M atoms. This diminishes the intensities of the observed diffraction spots on the film. The degree to which M-scattered and W-scattered rays are out of phase depends on the M-W interatomic distance and also on which set of planes (*hkl*) is doing the scattering. The intensity of a given spot on the diffraction pattern therefore depends on the M-W distance; this intensity also depends on the nature of M and W, since different atoms have different x-ray scattering powers.

The same argument holds when there are more than two atoms per lattice point. We conclude that the *locations* of the spots in the diffraction pattern depend only on the nature of the space lattice and the lengths and angles of the unit cell, and that the *intensities* of the spots depend on the kinds of atoms and the locations of atoms (interatomic distances) within the basis of the crystal structure.

Actually, we still have oversimplified things. Atoms are not points. Instead each atom contains a spatial distribution of electron probability density. Since it is the electrons that scatter the x-rays, the scattered-beam intensities depend on how the electron probability density varies from point to point in the basis. Analysis of the locations of the diffraction spots gives the space lattice and the lengths and angles of the unit cell. The unit-cell content Z is calculated from (24.12) using the known crystal density. Analysis of the intensities of the diffraction spots gives the electron probability density $\rho(x, y, z)$ as a function of position in the unit cell. From contours of ρ, the locations of the nuclei (other than H nuclei, which contribute little to ρ) are obvious (see, for example, Fig. 24.23), and bond lengths and angles can be found. Bond lengths determined by x-ray diffraction are generally less accurate than those found by spectroscopic methods.

Figure 24.23 shows a cross section of contours of constant ρ for benzene(*c*) as determined by x-ray-diffraction data.

Figure 24.23

Cross section of contours of constant electron-probability density in C_6H_6.

It should be emphasized that determination of ρ from the x-ray diffraction data is not generally a straightforward procedure. ρ could be directly determined if the relative phases of the diffracted beams were known, but measurement of intensities does not give these phases. To determine a structure, one may have to prepare derivatives with some atoms substituted by heavy atoms and examine the x-ray diffraction patterns of the derivatives. The difficulty of structure determination increases as the size of the unit cell increases.

For crystals with less than 100 independent nonhydrogen atoms per unit cell, the structure is most commonly solved for by the *direct method* (*Glusker and Trueblood,* chap. 8), which uses procedures of mathematical statistics and has been developed into computer programs that automatically solve for the structure. The 1985 Nobel Prize in chemistry was awarded to Hauptman and Karle for devising the direct method. Over 50000 crystal structures have been solved with this method.

The Hauptman–Karle direct method cannot be used with large molecules such as proteins. However, in the 1990s, Hauptman, Weeks, and Miller devised a direct method called *Shake and Bake* that enables crystal structures with as many as 1000 unique nonhydrogen atoms to be solved for; this method has solved many small-protein structures. [See R. Miller and C. M. Weeks in S. Fortier (ed.), *Direct Methods for Solving Macromolecular Structures,* Kluwer, 1998, pp 389–400; H. A. Hauptman, *J. Mol. Struct.,* **470,** 215 (1998).] Other methods, such as heavy-atom substitution, must be used for structures with more than 1000 nonhydrogen atoms.

For a crystal of a large molecule, obtaining an electron-probability-density map from the x-ray diffraction pattern is an extremely complicated problem and requires a huge number (millions or billions) of individual calculations. Nowadays, one uses an automated computer-controlled x-ray diffractometer to obtain the x-ray-diffraction positions and intensities, and the calculations to find the molecular structure are done on a computer. The Molecular Structure Corporation (www.msc.com/msc.html) uses x-ray diffraction to determine the structure of a crystal submitted to it; fees depend on the size of the molecule.

X-ray diffraction of crystals is the most widely used method of accurate molecular structure determination. The Cambridge Stuctural Database [www.ccdc.cam.ac.uk/prods/csd/csd.html; F. H. Allen et al., *J. Chem. Info. Comput. Sci.,* **31,** 187 (1991)] contains over 200000 structures of relatively small (most less than 60 atoms) organic and organometallic compounds found by x-ray and neutron diffraction. The Inorganic Crystal Structure Database (barns.ill.fr/dif/icsd/) has over 50000 structures.

Except for 2D NMR, the spectroscopic methods of structural determination discussed in Chapter 21 are inapplicable to large molecules. X-ray diffraction is therefore an invaluable tool in structural determinations of large molecules, including biologically important molecules such as proteins (which typically contain 10^3 nonhydrogen atoms per molecule). The three-dimensional structure of a protein whose amino acid sequence is known (from chemical analysis of fragments produced by partial hydrolysis) can frequently be determined by x-ray diffraction of a crystal of the protein. Protein crystals usually contain a substantial amount of solvent, so the environment of the protein molecule in a crystal is similar to its environment in solution, and the three-dimensional structures in the crystal and in solution are likely to be quite similar. The Protein Data Bank (PDB) at www.rcsb.org/pdb/ contains over 13000 structures of biological macromolecules (including about 1000 nucleic acids) found by x-ray crystallography and NMR. For more on protein crystallography, see D. E. McRee, *Practical Protein Crystallography,* 2nd ed., Academic, 1999.

X-ray diffraction has been used to study the crystal structure of many synthetic polymers. Crystalline polyethylene has a primitive orthorhombic space lattice with room-temperature unit-cell dimensions 7.39, 4.93, and 2.54 Å. A given polyethylene molecule in the crystal passes through many unit cells. Each unit cell contains four CH_2 groups. The basis is a fraction of one molecule and consists of two bonded CH_2 groups at one corner of the cell plus two bonded CH_2 groups in the center of the cell. Synthetic-polymer crystal structures are tabulated in J. Brandrup et al., (eds.) *Polymer Handbook,* 4th ed., Wiley, 1999, sec. VI.

In the 1950s, it was found that single crystals of many polymers could be grown by cooling a solution of the polymer. These crystals are plates about 100 Å thick. The polymer chains are oriented perpendicular to the top and bottom faces of the plate and fold over at the top and bottom faces.

Improvements in x-ray diffraction techniques in the 1970s made possible highly accurate determinations of the electron probability density ρ in molecular solids. Such densities are being used to study chemical bonding in transition-metal complexes and in organometallic compounds; see P. Coppens and M. B. Hall (eds.), *Electron Distributions and the Chemical Bond,* Plenum, 1982; P. Coppens, *X-Ray Charge Densities and Chemical Bonding,* Oxford, 1997.

Powder X-Ray Diffraction

For a cubic crystal, one can find the unit-cell edge length using a powdered (microcrystalline) sample. The tiny crystals in the sample have random orientations and, for any given set of planes [for example, the (220) planes], some of the tiny crystals will be oriented at the value of θ that satisfies the Bragg equation, (24.14). Hence we will get a Bragg reflection from each set of planes (except those where extinctions occur). The x-rays reflected from a plane make an angle 2θ with the direction of the incident beam (Fig. 24.24), so the diffraction angles are readily measured.

A little geometry (the derivation is omitted) shows that for a cubic crystal the perpendicular distance between adjacent planes with indices (*hkl*) is

$$d_{hkl} = a/(h^2 + k^2 + l^2)^{1/2} \qquad \text{cubic crystal} \qquad (24.15)$$

where a is the unit-cell edge length. If we relabel the indices nh, nk, and nl in (24.14) as *hkl* and substitute it into (24.15), we get

$$\sin \theta = (\lambda/2a)(h^2 + k^2 + l^2)^{1/2} \qquad (24.16)$$

As the indices *hkl* of a reflection increase, θ increases.

Figure 24.24

X-rays reflected from a crystal plane.

The extinction conditions stated earlier in this section show that the first several reflections observed for primitive, face-centered, and body-centered cubic lattices are

P	100, 110, 111, 200, 210, 211, 220, 221 and 300, 310, 311, . . .
F	111, 200, 220, 311, 222, 400, 331, 420, 422, 333 and 511, . . .
I	110, 200, 211, 220, 310, 222, 321, 400, 411 and 330, 420, . . .

where the reflections are listed in order of increasing $h^2 + k^2 + l^2$ and hence in order of increasing θ. The ratio of two $\sin^2 \theta$ values is

$$\sin^2 \theta_1 / \sin^2 \theta_2 = (h_1^2 + k_1^2 + l_1^2)/(h_2^2 + k_2^2 + l_2^2)$$

Comparison of the observed ratios with those expected for P, F, and I cubic lattices allows the lattice type and the *hkl* values of the reflections to be found. Hence *a* can be found from (24.15) if λ is known.

For example, suppose that for a crystal known to be cubic (by examination of its macroscopic appearance) the first several powder diffraction angles are 17.66°, 25.40°, 31.70°, 37.35°, 42.71°, 47.98°, and 59.08°. Calculation of $\sin^2 \theta$ gives 0.0920, 0.1840, 0.2761, 0.3681, 0.4601, 0.5519, and 0.7360. These numbers have the ratios 1:2:3:4:5:6:8. For P, F, and I cubic lattices, the expected $\sin^2 \theta$ ratios for the first several diffraction angles are found from the above-listed reflections to be

P	1:2:3:4:5:6:8
F	$1:1\frac{1}{3}:2\frac{2}{3}:3\frac{2}{3}:4:5\frac{1}{3}:6\frac{1}{3}$
I	1:2:3:4:5:6:7

Therefore the lattice is primitive cubic.

Except for crystals with simple structures, the powder method is usually unsuitable for structure determination, and one must use a single-crystal sample. However, by using a polystyrene matrix to orient the tiny crystals in a powder sample in approximately the same direction and by using a high-intensity x-ray beam, a method to determine complicated crystal structures from a powder sample has been devised [*Chem. Eng. News*, April 19, 1999, p. 11; T. Wessels et al., *Science*, **284,** 477 (1999)]. The high-intensity x-rays are produced by a *synchrotron*, a device in which electrons or positrons that have been accelerated to velocities near the speed of light are injected into a region containing a magnetic field, which causes the electrons to move on a circle, emitting radiation. Only a limited number of synchrotron x-ray sources exist.

Neutron Diffraction

A hydrogen atom contains but one electron and is therefore a very weak scatterer of x-rays. Consequently, H-atom positions are not accurately located by x-ray diffraction.

As noted in Chapter 18, microscopic particles like electrons, protons, and neutrons have wavelike properties and give diffraction effects. The wavelength of a neutron is given by the de Broglie relation $\lambda = h/mv$. Neutrons are scattered by atomic nuclei (not by electrons), and the neutron-scattering power of the H nucleus is comparable to that of other nuclei. This allows H atoms to be accurately located by neutron diffraction of crystals.

Neutron diffraction of crystals is not widely used since one must do the work at a nuclear-reactor facility.

24.10   DETERMINATION OF SURFACE STRUCTURES

Starting in the 1970s, extraordinary advances occurred in the determination of structures of crystal surfaces and of species adsorbed on crystal surfaces. A review article listed 26 techniques to determine surface structure, composition, and vibrational and electronic properties (S. Y. Tong, *Physics Today,* August 1984, p. 50).

LEED

Electrons give diffraction effects when scattered by a crystal surface (recall the Davisson–Germer experiment—Sec. 18.4). In **low-energy electron diffraction (LEED),** a monoenergetic beam of low-energy (10 to 500 eV) electrons is directed at one face of a single crystal. A few percent of the electrons are scattered elastically (that is, without a change in energy). Because of wave interference, these scattered electrons are concentrated along specific directions and form a diffraction pattern of spots on a fluorescent screen (Fig. 24.25). One can obtain additional information by varying the energy of the beam and its angle of incidence on the crystal.

Because of the low penetrating power of the electrons, the diffraction pattern is determined by the surface structure of the crystal. The locations of the spots on the screen tell us the size and orientation of the two-dimensional unit cell of the surface structure. As in x-ray diffraction, the intensities of the spots can be analyzed to determine the locations of the atoms. Analysis of LEED intensities is not straightforward, and bond lengths determined by LEED are typically in error by 0.05 to 0.1 Å.

LEED has revealed the bond shortening that occurs between the first and second layers of atoms in a metal (Sec. 24.6), a phenomenon called **surface relaxation.**

For certain crystals, LEED shows that the surface structure differs from the structure of corresponding planes of the bulk solid. Here, atoms on the surface change their locations and bonding to give a new structure, a process called **surface reconstruction.** For example, the (111) surface of annealed Si has a (7 × 7) reconstruction, meaning that the two-dimensional surface unit cell has sides each seven times longer than in the underlying bulk crystal. Surface reconstruction is much more common for semiconductors than for metals.

Most LEED work has been done on metallic and covalent crystals. Ionic and molecular crystals are harder to study.

LEED experiments (supplemented by other techniques such as EELS—Sec. 21.9) have determined the structures of several species adsorbed on metals. Recall the ethylidyne species of Fig. 21.31. For further results, see *Somorjai.*

Electron diffraction by gases can be used to determine molecular structures. Although the gas molecules have random orientations, the orientations of atoms with respect to one another in a given molecule are fixed, so one can extract structural information (bond distances and angles) from gas-phase electron diffraction. The method is not applicable to large molecules. For details, see *Brand, Speakman, and Tyler,* chap. 10; *Domenicano and Hargittai,* chap. 5.

The Scanning Tunneling Microscope

In the **scanning tunneling microscope (STM),** a precisely controlled, extremely sharp, metal needle tip is moved over the surface of an electrical conductor at a distance of about 5 Å. Application of a small voltage between the tip and the surface causes electrons to tunnel (Sec. 18.12) through the space between surface and tip. The magnitude of the tunneling current depends strongly on the tip–surface separation, and an electronic feedback loop maintains a constant current by adjusting the tip height. The plot of tip height versus surface position gives a contour map of the surface and can resolve single atoms.

Clean surfaces of single crystals contain terraces and steps (Fig. 24.26). The STM is used to study the role of such steps in heterogeneous catalysis. For example, when Pt is exposed to C_2H_4, this microscope shows that carbon atoms are deposited initially near the steps. See C. Quate, *Physics Today,* August 1986, p. 26.

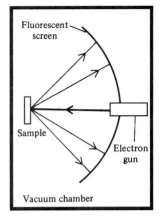

Figure 24.25

Schematic diagram of LEED apparatus. Not shown are grids placed in front of the fluorescent screen. The voltage on one of these grids is adjusted so that electrons scattered inelastically (that is, with loss of energy) do not reach the screen.

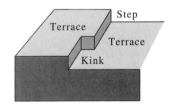

Figure 24.26

As a consequence of the way a crystal grows, the surface of a crystal contains a high concentration of terraces, monatomic steps, and kinks. Step and kink atoms have fewer neighbors than terrace atoms and so have different chemical properties than terrace atoms. A surface also contains a low concentration of point defects (Sec. 24.13), such as vacancies.

By using particular voltages in an STM, one can restrict the tunneling to specific orbitals of surface atoms, thereby producing images of bonding orbitals. See *Chem. Eng. News,* Sept. 1, 1986, p. 4 and *Physics Today,* January 1987, p. 19 for such images.

Certain biological specimens conduct well enough to be used in an STM, and the surfaces of viruses have been imaged with this microscope.

The Atomic Force Microscope

In an **atomic force microscope (AFM),** a probe with a very sharp tip is moved back and forth across the surface of a solid. The probe is attached to the underside of a flexible cantilever arm. The varying forces between the probe tip and the surface deflect the tip by varying amounts as the surface is scanned. The deflection of the tip is detected by monitoring a laser beam that is reflected off the top surface of the cantilever arm. As with the STM, one obtains an image of the surface, and in favorable cases, one can distinguish individual atoms if they are large enough (for example, Au). The surface need not be electrically conducting. In the most-common mode of operation, called contact mode, a constant loading force is applied to the cantilever, keeping the tip in contact with the surface. In tapping mode, the probe tip lies above the surface and the cantilever is rapidly oscillated vertically so that the probe tip makes contact with the surface for only a small fraction of each oscillation. As well as imaging solid surfaces, the AFM has been used to display the shape of biomolecules (such as DNA) in solution. Biological applications of the AFM are discussed in C. Bustamante and D. Keller, *Physics Today,* Dec. 1995, p. 32.

Field ion microscopy. In a field ion microscope (FIM), one places a single-crystal metal wire with a sharply pointed tip in a very strong electric field in high vacuum. Helium gas at low pressure is introduced. The electric field at atoms that protrude from the metal surface is greater than at nonprotruding atoms, and this electric field ionizes He atoms at these locations. The He^+ ions formed at the metal surface are attracted to a fluorescent screen maintained at a negative electric potential and form a highly magnified map of the ionization sites. The motions of single rhenium atoms adsorbed on tungsten have been followed using a FIM. However, the method is limited to metal surfaces and, moreover, only protruding atoms are imaged.

Electron microscopy. A light microscope uses a light beam to form an image. An electron microscope uses an electron beam to form an image. In a transmission electron microscope (TEM), the electron beam is passed through a thin sample of the material. Typically, a beam of 10^5-eV electrons is used (corresponding to a 0.04-Å de Broglie wavelength). The beam transmitted through the sample is focused and magnified using electric and magnetic fields to yield an image with magnification on the order of 5×10^5. To achieve high resolution, the sample must be very thin (10 to 100 Å). In a scanning transmission electron microscope (STEM), the incident electron beam is focused to a very small area and is scanned back and forth across the specimen. With the various forms of transmission electron microscopy, resolutions of $2\frac{1}{2}$ Å are obtainable. Isolated single heavy atoms of U and Th on 20-Å-thick carbon films have been observed, and some surface structures have been determined.

In a scanning electron microscope (SEM), a finely focused electron beam moves from point to point over the sample's surface. Electrons scattered by the surface and secondary electrons emitted by the surface are used to form an image of the surface. The resolution is about 100 Å.

24.11   BAND THEORY OF SOLIDS

A crystalline solid can be regarded as a single giant molecule. An approximate electronic wave function for this giant molecule can be constructed by the MO approach.

The electrons are fed into "crystal" orbitals that extend over the entire crystal. Each crystal orbital holds two electrons of opposite spin. Just as the MOs of a gas-phase molecule can be approximated as linear combinations of the orbitals of the atoms composing the molecule, the crystal orbitals can be approximated as linear combinations of orbitals of the species (atoms, ions, or molecules) composing the solid.

Consider Na(c) as an example. The electron configuration of an Na atom is $1s^2 2s^2 2p^6 3s$. Let the crystal contain N atoms, where N is on the order of 10^{23}. We construct crystal orbitals as linear combinations of the AOs of the N sodium atoms. Recall that when two Na atoms are brought together to form an Na$_2$ molecule, the two $1s$ AOs combine to give two MOs. In Na$_2$, the overlap between these inner-shell AOs is negligible, so their energies are nearly identical to the $1s$ AO energy of an isolated Na atom. In the Na(c) solid, which is being regarded as a single Na$_N$ molecule, the N $1s$ AOs of the isolated atoms form N crystal orbitals with energies nearly identical to that of a sodium $1s$ AO. These N $1s$ crystal orbitals have a capacity of $2N$ electrons and hold the $2N$ $1s$ electrons of the atoms. Similarly, the inner-shell $2s$ and $2p$ AOs form crystal orbitals with energies little changed from the AO energies, and these orbitals are filled in Na$_N$.

The N $3s$ sodium AOs form N delocalized crystal orbitals. The energy difference between the lowest and highest of these N crystal orbitals is of the same order of magnitude (a couple of eV or so) as the separation between the $\sigma_g 3s$ and $\sigma_g^* 3s$ MOs in Na$_2$. Because we have $N \approx 10^{23}$ crystal orbitals in an energy range of only a few eV, the energy spacing between adjacent crystal orbitals is extremely small, and for all practical purposes we can consider the crystal orbitals originating from the $3s$ AOs to form a continuous **band** of energy levels. This $3s$ band contains N orbitals and so has a capacity of $2N$ electrons. Since there are N $3s$ electrons in the N sodium atoms, the $3s$ band is only half filled (Fig. 24.27a). Hence, there are many vacant electronic energy levels in Na(c) that lie a negligible distance above the highest occupied electronic level. This makes Na(c) an excellent conductor of electricity, since electrons are easily excited to higher energy levels by an applied electric field. (Motion of electrons through the metal involves an increase in electronic energy.)

Direct evidence for the $3s$ band in Na(c) is obtained from the x-ray emission spectrum (Sec. 21.11). In Na(c) and in Na(g) one finds the $2p \rightarrow 1s$ x-ray line to be narrow, indicating that the $1s$ and $2p$ inner-shell bands in Na(c) are very narrow. In Na(g), the $3s \rightarrow 2p$ line is narrow. In Na(c), the $3s \rightarrow 2p$ line is centered at 407 Å and is broad (about 30 Å wide) because of the range of energies in the $3s$ band. These data give an energy range of 2.3 eV for the occupied part of the $3s$ band in Na(c) (Prob. 24.30).

An Mg atom has the electron configuration $1s^2 2s^2 2p^6 3s^2$. The $3s$ band in Mg(c) is exactly filled by the $2N$ electrons of the N magnesium atoms. One finds that the $3p$ band (formed by unoccupied $3p$ AOs of the Mg atoms) overlaps the $3s$ band in Mg(c);

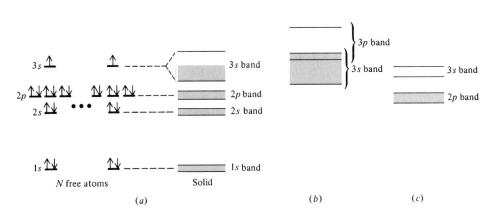

Figure 24.27

(a) Formation of crystal orbitals in sodium. (b) Crystal orbitals in Mg(c). (c) Crystal orbitals in Ne(c).

see Fig. 24.27b. There is no energy gap between the highest occupied and lowest vacant electronic energy levels, and Mg is an excellent conductor.

Consider Ne(c). The Ne electron configuration is $1s^2 2s^2 2p^6$. The $1s$ and $2s$ electrons give bands of very narrow widths. The interaction between the $2p$ AOs is not large (van der Waals forces are rather weak), and the $2p$ band is fairly narrow. Above the $2p$ band lies a fairly narrow $3s$ band (Fig. 24.27c). Because the $2p$ and $3s$ bands are fairly narrow, and because the $2p$ and $3s$ levels in an Ne atom are very widely separated, we get a substantial energy gap (several eV) between the top of the filled $2p$ band in Ne(c) and the bottom of the vacant $3s$ band. It takes a great deal of energy to excite electrons from the $2p$ band to the $3s$ band, and Ne(c) is therefore a nonconductor of electricity (an insulator).

In an ionic crystal such as NaCl, the bands are formed from the energy levels of the ions. The ions have rare-gas electron configurations, and arguments similar to those for Ne(c) show ionic crystals to be insulators. This conclusion is correct as far as conduction by electrons (electronic conduction) is concerned. However, at elevated temperatures, the ions can move fairly readily through an ionic crystal, and the crystal is an ionic conductor.

For a molecular crystal, we can construct crystal orbitals from the molecular orbitals of the molecules forming the crystal. Because of the weakness of the van der Waals interactions, the bands are rather narrow. The highest occupied band is filled, and there is a wide gap between this band and the lowest vacant band. As in Ne(c), the crystal is an insulator.

Consider diamond. The C atomic electron configuration is $1s^2 2s^2 2p^2$, and one might think that the band structure would involve overlapping $2s$ and $2p$ bands that have a total capacity of $8N$ electrons and are half filled with $4N$ electrons. However, a quantum-mechanical calculation shows that in diamond the $2s$ and $2p$ AOs combine to form two separate $2s$-$2p$ bands; the lower band holds $4N$ electrons and is completely filled; the upper band holds $4N$ electrons and is vacant. The separation between the two bands is several eV, so diamond is an insulator.

The energy gap E_g between the highest occupied band (called the **valence band**) and the lowest vacant band (the **conduction band**) in a solid can be found from the observed lowest frequency of visible or UV absorption by the solid. Results for the carbon-group elements are

Element	C(diamond)	Si	Ge	Sn(gray)	Pb
E_g/eV	6	1.2	0.7	0.08	0

With zero E_g, lead is a metal. With a large E_g, diamond is an insulator. (For NaCl, E_g is 7 eV; this gap is between the filled Cl^- $3p$ band and the higher-lying, vacant Na^+ $3s$ band.)

Si and Ge have fairly small band gaps and are classed as *intrinsic semiconductors*. Their electrical conductivity is substantially greater than that of insulators but far less than that of conductors. As the temperature is raised, the electrical conductivity of a semiconductor increases exponentially, as a result of excitation of electrons into the lowest vacant band. The electrons populate the bands according to the Boltzmann distribution law, with populations proportional to $e^{-E/kT}$. In contrast, the electrical conductivity of a metal decreases as T increases. The resistance of a metal arises from scattering of electrons by the positive ions. As T increases, the increase in thermal vibration of the ions increases the scattering.

Several organic polymers that are electrical conductors have been prepared (*Chem. Eng. News,* Oct. 16, 2000, p. 4).

One can use light to excite electrons into the lowest vacant band of a semiconductor. Semiconductors therefore exhibit *photoconductivity,* an increase in electrical conductivity when exposed to light.

24.12   STATISTICAL MECHANICS OF CRYSTALS

We shall regard the crystal as a single giant molecule with a set of allowed quantum-mechanical stationary states. This description is a natural one for metallic, covalent, and ionic crystals (as noted in Sec. 24.3). We shall also regard a molecular crystal as a single giant molecule containing weak "bonds" due to van der Waals forces and strong bonds due to ordinary chemical bonding. All the thermodynamic properties of the crystal are derivable from the crystal's canonical partition function $Z = \sum_j e^{-E_j/kT}$ [Eq. (22.16)], where E_j is the energy of stationary state j of the crystal and the sum goes over all the crystal's stationary states.

Consider a covalent or metallic crystal or an ionic crystal composed of monatomic ions. (Molecular crystals and ionic crystals with polyatomic ions will be dealt with later.) Let there be N atoms or ions present. For 1 mole of diamond, N is Avogadro's number. For 1 mole of NaCl, N is twice Avogadro's number.

What kinds of energy does the crystal have? The crystal considered as a giant molecule has $3N - 6$ vibrational modes of motion (*degrees of freedom*). There are also three translational and three rotational degrees of freedom of the crystal as a rigid body. These modes of motion can be activated by throwing the crystal around the laboratory. Since energy of bulk motion is not included in the thermodynamic internal energy U, there is no need to consider these six modes of motion. In addition to the $3N - 6$ vibrational-mode energies, there is electronic energy. Most solids are insulators or semiconductors and have a significant energy gap between the highest occupied and lowest vacant electronic energy levels (Sec. 24.11). Provided T is not extremely high, we can ignore electronic energy for these solids. For metals, excitation of electrons to higher levels occurs at all temperatures and must be allowed for. For now, we shall ignore the contribution of electronic energy to E_j for metals. This contribution will be added on at the end of the calculation. Thus we consider only vibrational energy in finding the energies E_j in the crystal's canonical partition function $Z = \sum_j e^{-E_j/kT}$.

As we did with molecules in Chapter 21, we can expand the potential energy E_p for vibrational motions of the N atoms or ions of the giant molecule in a Taylor series about the equilibrium configuration and neglect terms higher than quadratic. This gives the vibrational energy of the giant molecule as the sum of $3N - 6$ harmonic-oscillator energies, one for each normal mode of vibration [Eq. (21.48)]. Since N is something like 10^{23}, we omit the -6. In the harmonic-oscillator approximation, the quantum-mechanical vibrational energy levels of the giant molecule are

$$E_j = E_{p,\text{eq}} + \sum_{i=1}^{3N} (v_{i(j)} + \tfrac{1}{2}) h\nu_i = U_0 + \sum_{i=1}^{3N} v_{i(j)} h\nu_i \qquad (24.17)$$

$$U_0(V) \equiv E_{p,\text{eq}} + \tfrac{1}{2} h \sum_{i=1}^{3N} \nu_i \qquad (24.18)$$

where $v_{i(j)}$ is the vibrational quantum number of the ith normal mode when the crystal is in the quantum-mechanical state j, and ν_i is the vibrational frequency of the ith normal mode. The quantity $E_{p,\text{eq}}$ is the potential energy when all the atoms or ions are at their equilibrium (minimum-energy) separations. When the crystal's volume V is decreased by applying pressure, the equilibrium interatomic distances change and hence $E_{p,\text{eq}}$ changes. Also, the vibrational frequencies depend on V. Hence, U_0 depends on V. (U_0 is the crystal's energy at $T = 0$.) Because of the very large size of the giant molecule, we expect that many of the vibrational frequencies ν_i will be very low.

The sum in (24.17) is not over individual atoms or ions. The atoms or ions in a solid interact strongly with one another, and the energy cannot be taken as the sum of energies of noninteracting atoms or ions. The sum in (24.17) is over the crystal's normal modes (which are called **lattice vibrations**). In each normal mode, all the atoms or ions vibrate, and the energy of the ith normal mode is composed of kinetic and potential energies of all the atoms or ions.

The canonical partition function of the crystal is $Z = \Sigma_j e^{-E_j/kT}$. The crystal's Z will be like the vibrational partition function (22.110) of a polyatomic molecule. However, (22.110) takes the zero of energy at $U_0 = 0$, whereas in this chapter (to allow for the change in U_0 as V changes) we do not set $U_0 = 0$. Hence, we must add a factor $e^{-U_0/kT}$ to (22.110). Also, the product goes from 1 to $3N$. Thus, the crystal's canonical partition function is

$$Z = e^{-U_0/kT} \prod_{i=1}^{3N} \frac{1}{1 - e^{-h\nu_i/kT}} \qquad (24.19)$$

$$\ln Z = -\frac{U_0}{kT} - \sum_{i=1}^{3N} \ln\left(1 - e^{-h\nu_i/kT}\right) \qquad (24.20)$$

From $\ln Z$, all thermodynamic properties are readily calculated using (22.37) to (22.41). The only problem is that calculation of the $3N$ normal vibrational frequencies of a giant molecule of 10^{23} atoms ain't easy.

The Einstein Theory

The first application of quantum physics to the statistical mechanics of a material system was Einstein's 1907 treatment of solids. Einstein assumed (in effect) that every normal mode of the crystal has the same frequency ν_E (where E is for Einstein). Einstein knew this assumption was false, but he felt it would bring out the correct qualitative behavior of the thermodynamic properties. With each ν_i equal to ν_E, Eq. (24.20) becomes

$$\ln Z = -U_0(V)/kT - 3N \ln\left(1 - e^{-\Theta_E/T}\right) \qquad \text{where } \Theta_E \equiv h\nu_E/k \quad (24.21)$$

The use of $U = kT^2(\partial \ln Z/\partial T)_{V,N}$ [Eq. (22.38)] gives (Prob. 24.31)

$$U = U_0 + 3Nk\Theta_E/(e^{\Theta_E/T} - 1) \qquad (24.22)$$

The use of $C_V = (\partial U/\partial T)_V$ then gives for an Einstein crystal

$$C_V = 3Nk\left(\frac{\Theta_E}{T}\right)^2 \frac{e^{\Theta_E/T}}{(e^{\Theta_E/T} - 1)^2} \qquad (24.23)$$

which closely resembles (22.102) for a gas of diatomic molecules.

In the high-temperature limit, $e^{\Theta_E/T} - 1 = 1 + \Theta_E/T + \cdots - 1 \approx \Theta_E/T$, so (24.23) approaches $3Nk$ as T goes to infinity. This result agrees with experiment. (Recall the classical equipartition theorem.) For diamond, 1 mole has N_A atoms, so the high-T limit of $C_{V,m}$ is $3N_A k = 3R \approx 25$ J mol^{-1} K^{-1}. For NaCl, 1 mole has $2N_A$ ions, so the high-T limit of $C_{V,m}$ is $6N_A k = 6R$.

The theory gives C_V, but the experimentally measured quantity is C_P. From (4.53), $C_{P,m} = C_{V,m} + TV_m\alpha^2/\kappa$. Since $\alpha \to 0$ as $T \to 0$ (Prob. 5.60), the difference between $C_{P,m}$ and $C_{V,m}$ for a solid becomes negligible at low temperatures (below 100 K). At room temperature, this difference is moderately significant for solids (10^{-1} to 10^0 cal mol^{-1} K^{-1}). At high temperatures, C_V stays essentially constant at $3Nk$, but C_P increases essentially linearly with T, because of the $TV\alpha^2/\kappa$ term.

In the low-temperature limit, $e^{\Theta_E/T} \gg 1$, so C_V in (24.23) becomes proportional to $T^{-2}e^{-\Theta_E/T}$; as $T \to 0$, the exponential dominates the T^{-2} factor and C_V goes to zero. Again, this agrees with experiment.

For intermediate temperatures, the Einstein C_V function increases as T increases, approaching the $3Nk$ limit at high T. The overall appearance of the Einstein function resembles Fig. 24.28 for the Debye C_V function. Substitution of $T = \Theta_E$ in (24.23) shows that C_V has reached 92% of its high-T limit at Θ_E.

To apply the Einstein equation, one uses a single measured value of $C_{V,m}$ to evaluate the Einstein vibrational frequency ν_E and the Einstein characteristic temperature Θ_E. Equation (24.23) shows fairly good overall agreement with experiment. However, below about 40 K the agreement is poor. This is because of the very crude assumption that all normal-mode vibrations of the crystal have the same frequency. Θ_E values are 200 to 300 K for most elements, so most elements have nearly reached their high-T C_V values at room temperature. Some values of Θ_E are 67 K for Pb, 240 K for Al, and 1220 K for diamond.

The Debye Theory

In 1912, Debye presented a treatment that gives much better agreement with low-T C_V values than the Einstein theory. The large number of vibrational modes ($\approx 10^{24}$) makes it possible to treat the crystal's vibrational frequencies as continuously distributed over a range from a very low frequency (which can be taken as essentially zero) to a maximum frequency ν_m. Let $g(\nu)\,d\nu$ be the number of vibrational frequencies in the range between ν and $\nu + d\nu$. Changing the summation in (24.20) to a sum over vibrational frequencies (rather than over normal modes), we get $\ln Z = -U_0/kT - \Sigma_\nu \ln (1 - e^{-h\nu/kT})g(\nu)\,d\nu$. The sum over infinitesimals is a definite integral, and

$$\ln Z = -\frac{U_0}{kT} - \int_0^{\nu_m} \ln\left(1 - e^{-h\nu/kT}\right)g(\nu)\,d\nu \qquad (24.24)$$

Debye assumed that for frequencies in the range 0 to ν_m the distribution function $g(\nu)$ is the same as the distribution function of the elastic vibrations of a continuous solid calculated ignoring the solid's atomic structure. This latter quantity can be shown to be (*Kestin and Dorfman, sec. 9.3*)

$$g(\nu) = 12\pi V v_s^{-3}\nu^2 \qquad (24.25)$$

where v_s is essentially the average speed of sound waves in the solid. The total number of vibrational frequencies is $3N$, so $\int_0^{\nu_m} g(\nu)\,d\nu = 3N$. Substitution of (24.25) for $g(\nu)$ and integration gives

$$\nu_m = (3Nv_s^3/4\pi V)^{1/3} \qquad (24.26)$$

The use of (24.26) to eliminate v_s from (24.25) gives

$$g(\nu) = 9N\nu_m^{-3}\nu^2, \qquad 0 \le \nu \le \nu_m \qquad (24.27)$$

For $\nu > \nu_m$, the Debye $g(\nu)$ is 0.

Substitution of (24.27) in (24.24) gives

$$\ln Z = -\frac{U_0}{kT} - \frac{9N}{\nu_m^3} \int_0^{\nu_m} \nu^2 \ln\left(1 - e^{-h\nu/kT}\right)d\nu \qquad (24.28)$$

The use of $U = kT^2\,(\partial \ln Z/\partial T)_{V,N}$ and $C_V = (\partial U/\partial T)_V$ then gives (Prob. 24.37)

$$C_V = 9Nk\left(\frac{T}{\Theta_D}\right)^3 \int_0^{\Theta_D/T} \frac{x^4 e^x}{(e^x - 1)^2}\,dx \qquad (24.29)$$

where the **Debye characteristic temperature** Θ_D is defined as

$$\Theta_D \equiv h\nu_m/k \qquad (24.30)$$

(The quantity x in the definite integral is a dummy variable.) Expressions for other thermodynamic properties (such as S) are readily found from (24.24) and (24.27).

The integral in (24.29) cannot be expressed in terms of elementary functions but can be evaluated numerically by expanding the exponentials in infinite series and then integrating. Tabulations of the Debye $C_V/3Nk$ as a function of Θ_D/T are available. [See

Figure 24.28

C_V versus T for a solid according to the Debye theory.

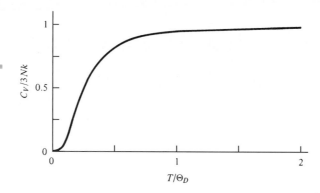

McQuarrie (1973), app. C.] Figure 24.28 plots the Debye C_V as a function of temperature.

At high T, the upper limit of the integral in (24.29) is close to zero, so the integration variable x is small. Hence, $e^x \approx 1$, and $e^x - 1 = 1 + x + \cdots -1 \approx x$. The integrand becomes x^2, and integration gives $C_V = 3Nk$, as found experimentally.

At low T, the upper limit in (24.29) can be taken as infinity. A table of definite integrals gives $\int_0^\infty x^4 e^x/(e^x - 1)^2 \, dx = 4\pi^4/15$, so the low-$T$ limit is

$$C_V = \frac{12\pi^4 Nk}{5}\left(\frac{T}{\Theta_D}\right)^3 \propto T^3 \qquad \text{low } T \qquad (24.31)$$

The **Debye T^3 law** (24.31) agrees well with experimental data, and is used to extrapolate heat-capacity data to absolute zero to evaluate entropies (Chapter 5).

The Debye characteristic temperature Θ_D can be found from (24.30) and (24.26). To achieve better agreement with experiment, one usually evaluates Θ_D empirically, so as to give a good fit to C_V data. The Debye theory works well at all temperatures but is far from perfect. The value of Θ_D for a given solid found by fitting C_V at one temperature differs somewhat from the value obtained using C_V at another temperature. Some Θ_D values (found from low-T C_V data) in kelvins are

Na	Cu	Ag	Be	Mg	diamond	Ge	SiO_2	NaCl	MgO	Ar	H_2O
160	343	225	1440	400	2230	370	470	320	950	93	192

Work subsequent to Debye's has allowed $g(\nu)$ to be calculated theoretically and measured for several solids. The Debye quadratic expression (24.27) with a cutoff at ν_m turns out to be a crude approximation to the actual $g(\nu)$; see Fig. 24.29. At low

Figure 24.29

Typical vibrational distribution function in a solid. The dashed line is the Debye expression (24.25).

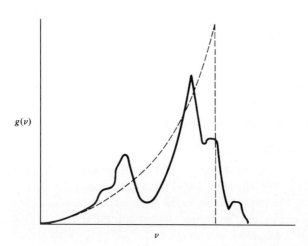

frequencies, the Debye $g(\nu)$ coincides with the true $g(\nu)$ curve. At T near zero, it is the low-frequency modes that are excited and that determine C_V, so the Debye theory gives the correct low-T behavior of C_V.

So far we have considered crystals of atoms or monatomic ions. Consider now a molecular crystal, for example, SO_3. There are N SO_3 molecules. Each SO_3 molecule has three translational degrees of freedom in which the molecule moves as a unit. The translational motions of one molecule are not independent of those of an adjacent molecule. Rather, intermolecular forces couple the $3N$ translational motions to form $3N$ lattice vibrations, whose frequencies are nearly continuous and can be treated by the Debye theory. (Each lattice vibration produces no change in bond distances or angles within each molecule.)

Each SO_3 molecule has three rotational degrees of freedom. As an SO_3 molecule rotates in a crystal, intermolecular interactions change its potential energy. The minimum potential energy is at the equilibrium position of the SO_3 molecule, so what was a free rotational motion in the gas becomes in the solid an oscillatory motion (vibration), which is called a *libration*. The frequencies of the three librations (corresponding to the three gas-phase rotational degrees of freedom) are determined by the intermolecular forces in the crystal.

Each SO_3 molecule has $3(4) - 6 = 6$ normal modes of intramolecular vibration, and these six vibrational frequencies are approximately the same as in gas-phase SO_3, since intermolecular forces are weaker than intramolecular chemical bonds.

The three librational frequencies and six intramolecular vibrational frequencies of each SO_3 are not continuous. Rather, each of these nine frequencies ν_i has its own characteristic temperature $h\nu_i/k$, and each frequency gives a contribution to C_V of the Einstein form (24.23), except that $3N$ in (24.23) is replaced by N, the number of SO_3 molecules. The crystal's C_V is then the sum of the Debye C_V (24.29) and nine more terms, each of these nine terms having the form of Eq. (22.102) with a different Θ_{vib}. At very low T, the contributions of the librations and intramolecular vibrations to C_V are negligible compared with those of the lattice vibrations (whose frequencies start at zero), so the Debye T^3 law holds.

A similar treatment holds for an ionic crystal with polyatomic ions. For example, consider a KNO_3 crystal with $N/2$ K^+ ions and $N/2$ NO_3^- ions. There are $3N$ lattice vibrations, which give a Debye-type contribution to C_V. Each NO_3^- ion has three librational motions (corresponding to gas-phase rotations) and has $3(4) - 6 = 6$ intraionic vibrations. These nine vibrations make Einstein-type contributions to C_V.

For metals, excitation of electrons must be considered. An approximate treatment (*Zemansky and Dittman,* sec. 12-4) shows that (provided T is not extremely high) electronic excitations contribute a term linear in T [recall Eq. (5.32)]:

$$C_{V,el,m} = bT \qquad (24.32)$$

Typically, b is 10^{-4} to 10^{-3} cal mol^{-1} K^{-2}. The room-temperature contribution of (24.32) to $C_{V,m}$ is 0.03 to 0.3 cal mol^{-1} K^{-1}, which is far less than $3R = 6$ cal mol^{-1} K^{-1}. At moderate temperatures, the electronic contribution (24.32) is thus only a small fraction of the total C_V. At very low T (below 4 K), the electronic contribution exceeds that of the lattice vibrations, since $(T/\Theta_D)^3$ goes to zero faster than bT.

24.13 DEFECTS IN SOLIDS

So far, we have assumed crystals to be perfect. In reality, all crystals show defects. These defects affect the crystal's density, heat capacity, and entropy only slightly, but they profoundly alter the mechanical strength, electrical conductivity, diffusion rate, and catalytic activity. Imperfections in solids are classified as point, line, or plane defects.

Point Defects

A *vacancy* is the absence of an atom, ion, or molecule from a site that would be occupied in a perfect crystal. A *substitutional impurity* is an impurity atom, molecule, or ion located at a place that is occupied by some other species in a perfect crystal; an *interstitial impurity* is located at a place (a *void*) that would be unoccupied in a perfect crystal. A *self-interstitial* is a nonimpurity atom, molecule, or ion located at a void. As the crystal's temperature is increased, the number of atoms, molecules, or ions having enough vibrational energy to break away from their perfect-crystal sites increases, thereby increasing the numbers of vacancies and self-interstitials. The presence of vacancies in Si limits the accuracy of the N_A value found in Example 24.2 (Sec. 24.7).

Catalytic sites on the surfaces of metal oxides are often due to anion or cation vacancies. Diffusion in solids and ionic conduction in solid salts involve vacancies and interstitials. Semiconductors used in transistors are generally *extrinsic* (or *impurity*) semiconductors (as contrasted to intrinsic semiconductors; Sec. 24.11), in which the electrical conductivity is due mainly to defects. For example, addition of a small amount of P as a substitutional impurity to Si greatly enhances the semiconductivity of the Si. The P atoms have five valence electrons compared with four for Si, and this produces extra electronic energy levels lying only slightly below the conduction band, allowing electrons to be more easily excited into the conduction band than in pure Si.

Line Defects

An *edge dislocation* (Fig. 24.30) is an extra plane of atoms that extends only part way through the crystal, thereby distorting its structure in the nearby planes and making the crystal mechanically weak. A more complicated kind of dislocation is a *screw dislocation;* see *Kittel,* chap. 20.

Figure 24.30

An edge dislocation in a solid.

Plane Defects

One kind of plane defect is a stacking error. For example, a hexagonal close-packed crystal might contain a few planes in which the packing is cubic close-packed.

Most crystalline solids do not consist of a single crystal but are composed of many tiny crystals held together. Neighboring crystals have random orientations, and the boundaries between the faces of neighboring crystals are plane defects.

24.14   LIQUIDS

Liquids have neither the long-range structural order of solids nor the small intermolecular interaction energy of gases and so are harder to deal with theoretically than solids or gases.

Liquids do have a short-range order that shows up in diffraction effects when x-rays are scattered by liquids. The scattered x-ray intensity as a function of angle shows broad maxima, in contrast to the sharp maxima obtained from solids. Analysis of these x-ray diffraction patterns allows one to obtain the **radial distribution function** (or **pair-correlation function**) $g(r)$ for the liquid. This function shows the variation in the average density of molecules with distance r from a given molecule. More explicitly, $g(r) = \rho(r)/\rho_{\text{bulk}}$, where ρ_{bulk} is the bulk molecular density ($\rho_{\text{bulk}} = N/V$, where N is the total number of molecules and V the total volume) and $\rho(r)$ is the local molecular density in the thin spherical shell from r to $r + dr$ around a given molecule. For nonspherical molecules, ρ/ρ_{bulk} depends on the directional angles θ and ϕ from the central molecule as well as on r. One obtains a radial pair-correlation function by averaging over the angles.

Figure 24.31 shows $g(r)$ calculated theoretically for a liquid of spherical molecules with a Lennard-Jones 6-12 intermolecular potential (22.136). The maxima at

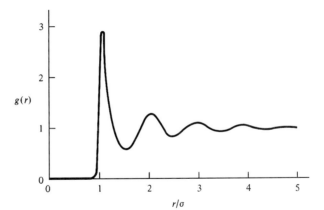

Figure 24.31

Radial distribution function for a
liquid obeying the Lennard-Jones
potential plotted at a typical
temperature and density. [Data
from the molecular-dynamics
calculations of L. Verlet, *Phys.
Rev.,* **165**, 201 (1968).]

$r = \sigma$, 2σ, and 3σ indicate three "shells" of molecules surrounding the central molecule. As r increases, the oscillations in $g(r)$ die out and $g(r)$ equals 1 for large r, indicating the lack of long-range order. For $r < \sigma$, $g(r)$ goes rapidly to zero, as a result of
the intermolecular Pauli repulsion. The experimentally observed $g(r)$ for Ar agrees
well with that calculated from the Lennard-Jones potential.

One finds that if the attractive part of the Lennard-Jones potential is omitted, the
calculated $g(r)$ for a liquid of spherical molecules is virtually unchanged from that
found with inclusion of the attractive term. This and other evidence [see D. Chandler,
Acc. Chem. Res., **7**, 246 (1974); *Ann. Rev. Phys. Chem.,* **29**, 441 (1978); D. Chandler,
J. D. Weeks, and H. C. Anderson, *Science,* **220**, 787 (1983)] indicates that the structure of most liquids is determined mainly by the intermolecular repulsive forces, with
intermolecular attractions playing only a relatively minor role in determining the
structure. The main exceptions are associated liquids with strong attractions due to
hydrogen bonding (H_2O, HF, HCl) or ionic forces (electrolyte solutions, molten salts).
The "structure" of a liquid refers to the spatial arrangement of molecules relative to
one another (the static structure) and the motions of the molecules in the liquid (the
dynamic structure). Although attractive forces usually have only a minor role in
determining liquid structure, they are important in determining thermodynamic properties such as internal energy and volume (recall Fig. 4.4).

The intermolecular repulsions are stronger, more rapidly varying, and shorter
range than the attractions. In a nonassociated liquid, the attractive forces (which are
vectors) roughly cancel, since each molecule is surrounded by others on all sides.
However, the attractive part of the potential energy (which is a scalar) does not cancel, but adds up. Repulsive forces don't cancel, since they are strong only when molecules are in contact.

The complexity of the intermolecular potential and the large number of degrees of
freedom make evaluation of the configuration integral (22.145) very difficult for a liquid. Many approximate statistical-mechanical theories of liquids have been developed.
Older theories such as Lennard-Jones' cell theory and Eyring's significant-structure
theory that are based on the resemblance between liquids and crystals have fallen out
of favor because an imperfect solid is a poor approximation to a liquid.

Modern theories usually express the liquid's thermodynamic properties in terms
of the pair-correlation function g rather than in terms of the configuration integral. The
most successful theories of liquids are perturbation theories, which initially calculate
g with the intermolecular attractions omitted, and then take them into account by treating them as a perturbation on the results obtained from the repulsive forces only (see
Rowlinson and Swinton, chaps. 7 and 8). Perturbation theories are quite successful in
calculating thermodynamic and transport properties of simple liquids and give pretty
good results for mixtures of simple liquids. Rowlinson and Swinton (op. cit., p. 312)

stated that, except for flexible-chain molecules, the "theory of the equilibrium properties of liquid mixtures, like that of pure liquids, is now an essentially solved problem." By this statement, they mean that *if* the intermolecular potentials are known, one can use present-day statistical-mechanical theories to calculate the thermodynamic properties with adequate accuracy. The hitch is that accurate intermolecular potentials are not known for most systems of interest.

Computer Simulation of Liquids

Liquids can be studied theoretically using the two *computer-simulation* approaches—the molecular dynamics and Monte Carlo methods. In the **molecular dynamics (MD) method** (developed by Alder and coworkers), one does calculations on a system containing 10^2 to 10^5 atoms. The molecules are in a box whose volume is fixed to correspond to a typical liquid density.

The system is viewed as a collection of atoms, sets of which are bonded together to form molecules. If one is dealing with monatomic molecules (for example, liquid Ar), the potential energy V for the motions of the atoms of the system is commonly taken as the sum of pairwise intermolecular potential energies v_{ij}, as in Eq. (22.132), where v_{ij} might be an empirical potential (such as the Lennard-Jones) or a quantum-mechanically calculated potential. For a system of molecules with more than one atom, one takes V as the sum of all the intramolecular and intermolecular interactions. Most commonly, one uses a molecular-mechanics potential energy of the form of Eq. (20.51): $V = V_{str} + V_{bend} + V_{tors} + V_{cross} + V_{es} + V_{vdW}$.

Suppose, for example, we are dealing with a system of 200 CH_3CH_2Cl molecules. Each molecule has 7 bonds, and V_{str} for the system has $7(200) = 1400$ terms. Similarly, for V_{bend}, V_{tors}, and V_{cross}. The term V_{vdW} contains contributions from van der Waals interactions between all pairs of atoms within each molecule that have a 1,4 or greater relation (Fig. 20.39) (that is, that are separated by three or more bonds); V_{vdW} also contains contributions involving pairs of atoms where the atoms of each pair lie in different molecules. For a CH_3CH_2Cl molecule, each of the three CH_3 hydrogens is separated by three bonds from the H, H, and Cl atoms on the second carbon, so there are nine contributions to V_{vdW} from interactions within a single molecule. For the 200 CH_3CH_2Cl molecules, intramolecular nonbonded interactions contribute $9(200) = 1800$ terms to V_{vdW}. An atom in one CH_3CH_2Cl molecule has a van der Waals interaction with the $8(199) = 1592$ atoms of the other 199 molecules. Multiplication by 8 gives $8(1592) = 12736$ terms for interactions between the atoms of one particular molecule and atoms in the other molecules. Multiplication by 200 and division by 2 (to avoid counting each interaction twice) gives $200(12736)/2 = 1.2736 \times 10^6$ terms in V_{vdW} due to intermolecular interactions. Similarly for V_{es}.

The atoms of the system are assigned initial positions and velocities. The velocities are chosen to be consistent with the Boltzmann distribution for the desired temperature. Using the expression for V of the system as a function of the coordinates of the atoms, the computer evaluates the force on each atom i from $F_{x,i} = -\partial V/\partial x_i$, etc. [Eq. (2.21)]. Using these forces and the initial positions and velocities, Newton's second law is used to find the positions and velocities at a short time interval δt later. (Of course, atoms obey quantum mechanics, not classical mechanics, so the use of classical mechanics introduces some error, which will be greatest for light atoms such as H.) Typically, δt is about 10^{-15} s, during which the forces are assumed to remain constant. At the new positions, the forces are re-evaluated and new positions and velocities calculated after another time interval δt. Typically, the system is followed for a total of 10^{-10} s. The MD method provides a "movie" of molecular motions.

Molecular dynamics calculations allow pair-correlation functions to be found by averaging over successive configurations and give information on structural details not accessible to experiment. For example, the patterns of hydrogen bonds in liquid water have been studied; see F. H. Stillinger, *Science,* **209,** 451 (1980). Furthermore, by taking the time average over the MD-calculated motions, one can obtain thermodynamic properties. For example, the time average of the intermolecular potential energy gives the thermodynamic internal energy U relative to U of the corresponding ideal gas (no intermolecular interactions). Moreover, the MD method is not restricted to equilibrium properties and can be used to calculate transport properties (for example, viscosity). [For details on how thermodynamic properties (including entropy and free energy relative to the corresponding ideal gas) and transport properties are calculated from molecular-dynamics simulations, see J. M. Haile, *Molecular Dynamics Simulation,* Wiley, 1992, chaps. 6 and 7.]

Besides the structure and properties of liquids, molecular dynamics is used to study phase transitions such as melting [D. Frenkel and J. P. McTague, *Ann. Rev. Phys. Chem.,* **31,** 491 (1980)], the structure of liquid–vapor interfaces, and intramolecular motions of protein molecules in solution (J. A. McCammon and M. Karplus, ibid., **31,** 29).

One of the major problems in molecular biology is to be able to understand how a globular protein folds into its native three-dimensional shape within seconds or less, starting from an unfolded random-coil state, given the astronomical number of possible protein conformations for a given amino acid sequence (Sec. 20.12). The ability to predict a protein's three-dimensional structure from its amino acid sequence would be of major help in such areas as drug design. Duan and Kollman used a molecular-dynamics simulation to study protein folding in aqueous solution. Their system consisted of one protein molecule (the villin headpiece subdomain) with 36 amino acid residues and about 3000 water molecules [Y. Duan and P. A. Kollman, *Science,* **282,** 740 (1998); www.psc.edu/science/Kollman98/kollman98.html]. They used a version of the AMBER force field. Time steps of 2×10^{-15} s were used and 5×10^8 integration steps were taken, so the system was followed for 1×10^{-6} s. The calculations required four months of supercomputer computing time. The small protein used is one of the fastest-folding proteins and is believed to require 10 to 100 microseconds to fold. Thus, the simulation followed the folding process for only a fraction of the required folding time. In the first 20 nanoseconds of the simulation, a burst of folding collapsed the structure to a quite compact form. During the first 200 nanoseconds, the protein oscillated between marginally stable compact forms and less-stable more-unfolded forms. From 250 to 400 nanoseconds, a "quiet" period occurred, during which the fluctuations between compact and unfolded states virtually ceased and the protein got closest to the stable native state. Eventually, it drifted away from this structure for the remainder of the simulation.

The **Monte Carlo (MC) method** (developed by Metropolis and coworkers and named after a European gambling resort, since the method involves randomness) is quite similar to the MD method. The main difference is that in the MC method, successive configurations of the system are not found by solving the equations of motion. Instead, the computer gives one molecule (picked at random) small random changes in position and orientation. If the potential energy $\mathcal{V}_2$ of the new configuration is less than the potential energy $\mathcal{V}_1$ of the original configuration, the new configuration is accepted. If $\mathcal{V}_2$ exceeds $\mathcal{V}_1$, the new configuration is rejected unless $\exp[-(\mathcal{V}_2 - \mathcal{V}_1)/kT]$ is greater than a randomly chosen number that lies between 0 and 1. It can be shown that this recipe produces a sequence of configurations such that the probability that a configuration with potential energy $\mathcal{V}$ appears in the sequence is proportional to the Boltzmann factor $e^{-\mathcal{V}/kT}$. One then averages over the sequence of configurations (typically, 10^6 to 10^7 configurations are used) to find the system's thermodynamic properties and radial distribution function. Unlike the MD method, the MC method is restricted to equilibrium properties.

In Sec. 22.2, we discussed thermodynamic properties as either time averages or ensemble averages over the system's microstates. The MD method uses time averages and the MC method uses ensemble averages over classical-mechanical microstates.

Beveridge and coworkers used a potential energy of interaction between two water molecules found from ab initio CI calculations and used quantum-mechanical H_2O–CH_4 and H_2O–ion potential-energy functions to do Monte Carlo calculations on liquid water, a dilute aqueous solution of CH_4, and dilute aqueous solutions of anions and cations [D. L. Beveridge et al. in J. M. Haile and G. A. Mansoori (eds.), *Molecular-Based Study of Fluids,* Amer. Chem. Soc. Adv. Chem. Ser. no. 204, p. 297 (1983); D. L. Beveridge et al. in P. Lykos (ed.), *Computer Modeling of Matter,* American Chemical Society, 1978, p. 191; D. L. Beveridge et al., *Annals N. Y. Acad. Sci.,* **367,** 108 (1981)]. They found an average coordination number of six water molecules around each Na^+ ion and found evidence for some increase in structure in the water molecules surrounding each CH_4 molecule (the *hydrophobic effect*).

Using a system of 125 water molecules at 25°C and runs of 0.5 to 4.4 million MC configurations, they calculated an oxygen–oxygen radial distribution function in good agreement with experiment. They found an internal energy of water (relative to the ideal gas) of −8.6 kcal/mol, compared with the experimental value −9.9 kcal/mol (Prob. 24.44) and a heat capacity of 20.1 cal/(mol K) compared with the true value 18.0 cal/(mol K). The error in U is probably due to neglect of three-body interactions (Sec. 22.10) and inaccuracies in the potential-energy function. The average coordination number was found to be 4.1, which is quite low, because of the open, hydrogen-bonded structure. (The average coordination number in liquid Ar is 10, compared with 12 in solid Ar. Solids have a well-defined coordination number. In liquids, we can talk about only an average coordination number.) The Helmholtz energy A of water was calculated from a 64-molecule system. [For the special techniques needed to calculate A, see D. L. Beveridge and F. M. DiCapua, *Ann. Rev. Biophys. Biophys. Chem.,* **18,** 431 (1989); P. Kollman, *Chem. Rev.,* **93,** 2395 (1993).] The result gave an entropy (relative to that of an ideal gas at the same density) of −15.6 cal/(mol K) compared with the experimental value −14.0 cal/(mol K) (Prob. 24.44). [Of course, S of $H_2O(g)$ can be found by the methods of Sec. 22.7.] Thus, a quantum-mechanically calculated potential-energy function for the interaction between two water molecules has been used to calculate the macroscopic thermodynamic properties U and S for liquid water, a truly impressive result.

24.15   SUMMARY

Crystalline solids are classified as ionic, covalent, metallic, or molecular. A solid's cohesive energy is $\Delta H°$ for sublimation of the solid to isolated atoms, molecules, or ions and is smallest for molecular crystals. From observed interatomic distances in crystals, one can estimate ionic radii and atomic van der Waals radii.

A crystal's structure is generated by placing an identical basis group at each point of the crystal's space lattice. The space lattice is divided into unit cells chosen to have the maximum symmetry and minimum volume. There are 14 kinds of space lattices, and their unit cells are classified as primitive, body-centered, face-centered, or end-centered. Miller indices (hkl) are used to denote crystal planes. Crystal structures are determined by x-ray diffraction. Structures of crystal surfaces and species adsorbed on surfaces can be determined by LEED.

The Debye statistical-mechanical theory of solids uses a quadratic expression with a high-frequency cutoff for the distribution function of normal-mode lattice vibrational frequencies of a crystal and yields expressions for the crystal's canonical partition function Z and thermodynamic properties. The Debye C_V is generally in good agreement with experiment and is proportional to T^3 near $T = 0$.

The structure of nonassociated liquids is determined mainly by the repulsive forces. Perturbation theories of liquids give good results for liquid properties, but their use is hampered by lack of knowledge of intermolecular forces. Computer simulation of liquids using the molecular dynamics or the Monte Carlo method provides valuable insights on liquid structure and enables calculation of liquid properties.

Important kinds of calculations include

- Calculation of E_c of a solid from thermodynamic data.
- Estimation of E_c of an ionic solid like NaCl from Eq. (24.9).
- Determination of Miller indices of crystal planes and faces.
- Calculation of the number of formula units in a unit cell from the density and unit-cell dimensions using (24.12).
- Calculation of the angles of incidence that produce a diffracted beam from a given set of planes using the Bragg equation.
- Calculation of C_V of solids using the Einstein or Debye theories.

FURTHER READING AND DATA SOURCES

Kittel; Buerger; Glusker and Trueblood; Sands; Somorjai; McClelland, secs. 6.3 to 6.5; *Denbigh,* chap. 13; *Kestin and Dorfman,* chap. 9.

Crystal-structure data: R. W. G. Wyckoff, *Crystal Structures,* 2d ed., vols. 1–6, Wiley-Interscience, 1963–1971; *Landolt–Börnstein,* 6th ed., vol. 1, pt. 4, 1955 and New Series, Group III, vols. 5–10b, 14, 1971–1988.

PROBLEMS

Section 24.3
24.1 Classify each of the following crystals as ionic, covalent, metallic, or molecular: (*a*) Co; (*b*) CO_2; (*c*) SiO_2; (*d*) BaO; (*e*) N_2; (*f*) $CsNO_3$; (*g*) Cs.

Section 24.4
24.2 Use thermodynamic data in the Appendix to find E_c at 25°C for (*a*) graphite; (*b*) SiC; (*c*) SiO_2.

24.3 Use (24.1) to calculate E_c for NaCl at 25°C.

24.4 Which of each of the following pairs of solids has the greater cohesive energy: (*a*) Br_2 or I_2; (*b*) NH_3 or PH_3; (*c*) SO_2 or SiO_2; (*d*) KF or MgO?

24.5 Use Appendix data to calculate the cohesive energy of *liquid* water at 25°C.

Section 24.5
24.6 Use (24.4) to calculate E_{Coul} for NaCl at 0 K.

24.7 Use (24.11) to calculate n for NaCl.

24.8 (*a*) KBr has the NaCl-type structure. For KBr at 1 atm, $R_0 = 3.299$ Å at 25°C and $\kappa \approx 5.5 \times 10^{-6}$ bar^{-1} at 0 K. Calculate E_c of KBr at 25°C and compare with the experimental value 718 kJ/mol. [Although Eq. (24.9) was derived for $T = 0$, its use at room temperature does not produce major error.] (*b*) CsCl (Fig. 24.17) has $\mathcal{M} = 1.762675$ and at 1 atm has unit-cell length $a = 4.123$ Å at 25°C and $\kappa \approx 6 \times 10^{-6}$ bar^{-1} at 0 K. Calculate E_c of CsCl at 25°C and compare with the experimental value 668 kJ/mol.

24.9 Use $U_m \approx E_p$, Eq. (24.8), and the expression for V_m that precedes Eq. (24.11) to calculate $(\partial U_m/\partial R)_{T,P}$ and $P(\partial V_m/\partial R)_{T,P}$ for R 0.01 Å longer than R_e for NaCl at 0 K and 1 atm and compare their magnitudes.

24.10 (*a*) Take $(\partial/\partial V_m)_T$ of $(\partial U_m/\partial V_m)_T = T(\partial P/\partial T)_V - P$ [Eq. (4.47)], and then take the limit of the result as $T \to 0$ to show that $(\partial^2 U_m/\partial V_m^2)_T = 1/\kappa V_m$ at $T = 0$. (*b*) U_m at $T = 0$ equals the potential energy E_p in (24.8) plus the zero-point vibrational energy ZPE. Neglecting the variation of the ZPE with V_m, we have $\partial^2 U_m/\partial V_m^2 = \partial^2 E_p/\partial V_m^2$ at $T = 0$. V_m is proportional to the cube of the nearest-neighbor distance, so $V_m = cR^3$, where c is a constant. Let $V_{m,0}$ and V_m be the volumes corresponding to R_0 and R. Show that (24.8) can be written as

$$-E_p = (e^2 \mathcal{M} N_A/4\pi\varepsilon_0 R_0)(V_{m,0}^{1/3} V_m^{-1/3} - n^{-1} V_{m,0}^{n/3} V_m^{-n/3})$$

Differentiate this equation twice with respect to V_m and then set $V_m = V_{m,0}$ to show that

$$-(\partial^2 E_p/\partial V_m^2)|_{R=R_0} = \tfrac{1}{9}(e^2 \mathcal{M} N_A/4\pi\varepsilon_0 R_0)V_{m,0}^{-2}(1 - n)$$

Substitute this result and the result of (*a*) into $\partial^2 U_m/\partial V_m^2 = \partial^2 E_p/\partial V_m^2$ at R_0 to derive (24.10) for n.

24.11 Evaluate the Madelung constant for a hypothetical one-dimensional crystal consisting of alternating $+1$ and -1 ions with equal spacing between ions. [*Hint:* Use (8.36).]

24.12 Using the Lennard-Jones intermolecular potential (22.136) for the interactions in Ar(*c*) and summing over the interactions between *all* pairs of atoms, one finds (*Kittel,* chap.

3) $-E_c = N_A\varepsilon[24.264(\sigma/R)^{12} - 28.908(\sigma/R)^6]$, where R is the nearest-neighbor distance. (a) Use the gas-phase Ar Lennard-Jones parameters $\varepsilon/k = 118$ K (where k is Boltzmann's constant) and $\sigma = 3.50$ Å and the experimental nearest-neighbor separation $R_0 = 3.75$ Å at 0 K to calculate E_c for Ar. (The experimental E_c is 1.85 kcal/mol at 0 K.) (b) Show that the expression for E_c in (a) predicts that $R_0/\sigma = 1.09$. Compare with the experimental value of this ratio.

24.13 Verify the nearest-neighbors-only value of E_c of Ar given near the end of Sec. 24.5.

Section 24.7

24.14 How many lattice points are there in a unit cell of (a) a body-centered lattice; (b) an end-centered lattice?

24.15 How many basis groups are there in a unit cell of a crystal with (a) a face-centered lattice; (b) a primitive lattice?

24.16 One form of $CaCO_3(c)$ is the mineral aragonite. Its room-temperature density is 2.93 g/cm^3. Its lattice is orthorhombic with $a = 4.94$ Å, $b = 7.94$ Å, and $c = 5.72$ Å at room temperature. Calculate the number of Ca^{2+} ions per unit cell of aragonite.

24.17 One form of TiO_2 is the mineral rutile, which has a tetragonal lattice with $a = 4.594$ Å and $c = 2.959$ Å at 25°C. There are two formula units per unit cell. Calculate the density of rutile at 25°C.

24.18 Find the Miller indices of the surface s_3 and the planes p_2 in Fig. 24.11.

24.19 Draw points of a two-dimensional lattice with square unit cells and then draw sets of the following lines (lines in a two-dimensional lattice are analogous to planes in a three-dimensional lattice; take the a direction as vertical): (a) (11); (b) (10); (c) (02); (d) (12); (e) $(1\bar{1})$.

24.20 The lattice of crystalline $COCl_2$ is body-centered tetragonal with 16 formula units per unit cell. How many molecules does the basis consist of?

Section 24.8

24.21 Calculate the percentage of empty space in the simple cubic lattice of Po.

24.22 Sketch a (111) plane of a metallic fcc structure and indicate which atoms touch an atom in the (111) plane.

24.23 The solid KF has the NaCl-type structure. The density of KF at 20°C is 2.48 g/cm^3. Calculate the unit-cell length and the nearest-neighbor distance in KF at 20°C.

24.24 CsBr has the CsCl-type structure. The density of CsBr at 20°C is 4.44 g/cm^3. Calculate the unit-cell a value and the nearest-neighbor distance in CsBr.

24.25 The density of CaF_2 at 20°C is 3.18 g/cm^3. Calculate the unit-cell length for CaF_2 at 20°C.

24.26 The density of diamond is 3.51 g/cm^3 at 25°C. Calculate the carbon–carbon bond distance in diamond.

24.27 Ar crystallizes in an fcc lattice with one atom at each lattice point. The unit-cell length is 5.311 Å at 0 K. Calculate the nearest-neighbor distance in Ar(c) at 0 K.

Section 24.9

24.28 A certain crystal has a simple cubic lattice with a unit-cell length of 4.70 Å. For x-rays with $\lambda = 1.54$ Å, calculate the diffraction angles from (a) the (100) planes; (b) the (110) planes.

24.29 Visual examination of crystals of Ag shows Ag(c) to belong to the cubic system. Using x-rays with $\lambda = 1.542$ Å, one finds the first few diffraction angles for a powder of Ag(c) to be 19.08°, 22.17°, 32.26°, 38.74°, 40.82°, 49.00°, and 55.35°. (a) Is the lattice P, F, or I? (b) Assign each of these diffraction angles to a set of planes. (c) Use each of these angles to calculate the unit-cell edge length.

Section 24.11

24.30 From data in Sec. 24.11 calculate the energy range for the occupied part of the 3s band in Na(c).

Section 24.12

24.31 (a) Verify the expression for U of an Einstein crystal. (b) Verify (24.23) for C_V of an Einstein crystal.

24.32 Find the expression for A of an Einstein crystal.

24.33 For an Einstein crystal, find the limiting form of U at (a) high T; (b) low T.

24.34 (a) Find the expression for S for an Einstein crystal. (b) Use the result for (a) and Θ_E values in Sec. 24.12 to calculate S_m for Al and diamond at 25°C; compare with the experimental values 6.77 and 0.568 cal/(mol K), respectively.

24.35 The Einstein temperature is 240 K for Al. What value of $C_{V,m}$ for Al does the Einstein model predict at (a) 50 K; (b) 100 K; (c) 240 K; (d) 400 K?

24.36 Given the following data for Cu(s) at 1 atm

T/K	50	100	150	200	300	500	800	1200
$C_{V,m}$/(J/mol-K)	6.24	16.0	20.3	22.4	23.8	24.5	25.4	26.0

use the Excel Solver to find the Θ_E value for which the Einstein theory gives the best least-squares fit to the data. Since this is a metal, you need to add a term bT [as in Eq. (5.32)] to the Einstein expression, where b is a positive constant. Take the initial guess of b as zero. Plot the Einstein-theory curve and the smoothed-data curve on the same graph.

24.37 (a) From (24.27), (24.24), (22.38), and the equation in Prob. 15.19, derive the expression for U for a Debye crystal. (To simplify the final result, multiply the numerator and denominator of the integrand in U by $e^{h\nu/kT}$.) (b) From the expression for U, find the Debye expression for C_V. Verify that the substitution $x \equiv h\nu/kT$ in the integrand gives (24.29).

24.38 The 1816 *law of Dulong and Petit* states that the product of the specific heat and the atomic weight of a metallic element is approximately 6 cal/(g °C). What is the statistical-mechanical basis for this law?

24.39 For I_2 one finds $C_{V,m} = 0.96$ cal mol^{-1} K^{-1} at 10 K. (*a*) Calculate Θ_D for I_2. (*b*) Calculate $C_{V,m}$ of I_2 at 12 K.

24.40 Use Fig. 24.28 and data in Sec. 24.12 to find the Debye $C_{V,m}$ at 298 K for (*a*) NaCl; (*b*) diamond. Compare the results with the $C_{P,m}$ data in the Appendix.

24.41 (*a*) Show that for a metal, a plot of $C_{V,m}/T$ versus T^2 at low T should give a straight line whose slope and intercept allow Θ_D and b to be calculated. (*b*) Use the following data for Ag to find Θ_D and b from a graph:

T/K	1.35	2.00	3.00	4.00
$10^3 C_{V,m}$/(cal mol^{-1} K^{-1})	0.254	0.626	1.57	3.03

24.42 (*a*) Use Eq. (24.31) and the fact that $C_V \approx C_P$ at very low T to show that for a nonmetallic solid at very low temperatures, $S_{\text{solid}} = 4\pi^4 NkT^3/5\Theta_D^3$. (*b*) At very low T, S for a gas has only the translational and electronic contributions (22.104) and (22.107). Use the relation $S_{\text{gas}} - S_{\text{solid}} = \Delta_{\text{sub}}H/T$ (where $\Delta_{\text{sub}}H$ is the enthalpy of sublimation) to obtain an expression for the vapor pressure P of a solid at low T in terms of $\Delta_{\text{sub}}H$. Show that the result has the form $P = aT^{5/2}e^{-c/T}$ and that $P \to 0$ as $T \to 0$.

Section 24.14

24.43 (*a*) Draw the radial distribution function $g(r)$ for a gas of molecules interacting with a hard-sphere potential (Fig. 22.21*b*). (*b*) Draw a rough sketch to show the general character of $g(r)$ for a solid.

24.44 Use Appendix data to find U_m and S_m for liquid water at 25°C and 1 bar relative to U_m and S_m for the corresponding ideal gas at the same temperature and density.

24.45 For an MD simulation using a system of 300 CH$_3$CHO molecules and a molecular-mechanics force field, how many terms are there in (*a*) V_{str}; (*b*) V_{vdW}?

24.46 For an MD simulation used to calculate equilibrium thermodynamic properties of a liquid, one usually discards the results of the first 10^5 integration steps. Explain why this is done.

General

24.47 State the contributions of each of the following scientists that were discussed in this book: (*a*) Einstein; (*b*) Born; (*c*) Debye.

24.48 If one treats Ar atoms as hard spheres, the fcc structure of solid Ar (Sec. 24.7) shows that the solid has 26% of empty space. At 1 atm and the normal melting point of 84 K, the density of solid Ar is 1.59 g/cm^3 and the density of liquid Ar is 1.42 g/cm^3. Find the percentage of empty space in (*a*) liquid Ar at 84 K and 1 atm; (*b*) gaseous Ar at 1 atm and 87 K (the normal boiling point).

24.49 True or false? (*a*) Each lattice point in a crystal has the same environment. (*b*) No atom need lie at a lattice point. (*c*) The basis must have the same stoichiometric composition as the entire crystal. (*d*) A crystal space lattice cannot have a positive ion at one lattice point and a negative ion at another lattice point.

BIBLIOGRAPHY

Adamson, A. W., and A. P. Gast: *The Physical Chemistry of Surfaces,* 6th ed., Wiley, 1997.

Allcock, H. R., and F. W. Lampe: *Contemporary Polymer Chemistry,* 2d ed., Prentice-Hall, 1990.

Andrews, F. C.: *Equilibrium Statistical Mechanics,* 2d ed., Wiley, 1975.

Andrews, F. C.: *Thermodynamics,* Wiley-Interscience, 1971.

Atkins, P. W., and R. S. Friedman: *Molecular Quantum Mechanics,* 3d ed., Oxford, 1999.

Aveyard, R., and D. A. Haydon: *An Introduction to the Principles of Surface Chemistry,* Cambridge, 1973.

Bamford, C. H., and C. F. H. Tipper (eds.): *Comprehensive Chemical Kinetics,* vols. 1–37, Elsevier, 1969–1999.

Bates, R. G.: *Determination of* pH, 2d ed., Wiley-Interscience, 1973.

Becker, E. D.: *High-Resolution NMR,* 3d ed., Academic, 1999.

Bernasconi, C. F. (ed.): *Investigation of Rates and Mechanisms of Reactions,* pts. 1 and 2, 4th ed. [vol. 6 of A. Weissberger (ed.), *Techniques of Chemistry*], Wiley, 1986.

Billmeyer, F. W.: *Textbook of Polymer Science,* 3d ed., Wiley, 1984.

Bird, R. B., W. E. Stewart, and E. N. Lightfoot: *Transport Phenomena,* Wiley, 1960.

Bockris, J. O.'M., and A. K. Reddy: *Modern Electrochemistry,* Plenum, 1970; 2d ed., vol. 1, Perseus, 1998.

Bovey, F. A.: *Nuclear Magnetic Resonance Spectroscopy,* 2d ed., Academic, 1988.

Brand, J. C. D., J. C. Speakman, and J. K. Tyler: *Molecular Structure,* 2d ed., Halsted, 1975.

Buerger, M. J.: *Contemporary Crystallography,* McGraw-Hill, 1970.

Campbell, I. D., and R. A. Dwek: *Biological Spectroscopy,* Benjamin/Cummings, 1984.

Chang, R.: *Basic Principles of Spectroscopy,* McGraw-Hill, 1971.

Chapra, S. C., and R. P. Canale: *Numerical Methods for Engineers,* 3d ed., McGraw-Hill, 1998.

Cotton, F. A.: *Chemical Applications of Group Theory,* 3rd ed., Wiley, 1990.

Craig, N.: *Entropy Analysis,* Wiley, 1992.

Davidson, N. R.: *Statistical Mechanics,* McGraw-Hill, 1962.

Davies, C. W.: *Ion Association,* Butterworth, 1962.

Debenedetti, P. G.: *Metastable Liquids,* Princeton, 1997.

Defay, R., I. Prigogine, A. Bellemans, and D. H. Everett: *Surface Tension and Adsorption,* Wiley, 1966.

de Heer, J.: *Phenomenological Thermodynamics with Applications to Chemistry,* Prentice-Hall, 1986.

DeKock, R. L., and H. B. Gray: *Chemical Bonding and Structure,* Benjamin/Cummings, 1980.

Denbigh, K.: *The Principles of Chemical Equilibrium,* 4th ed., Cambridge, 1981.

Dickerson, R. E.: *Molecular Thermodynamics,* Benjamin/Cummings, 1969.

Domenicano, A., and I. Hargittai: *Accurate Molecular Structures,* Oxford, 1992.

Espenson, J. H.: *Chemical Kinetics and Reaction Mechanisms,* 2d ed., McGraw-Hill, 1995.

Eyring, H., D. Henderson, and W. Jost (eds.): *Physical Chemistry: An Advanced Treatise,* Academic, 1967–1975.

Forst, W.: *Theory of Unimolecular Reactions,* Academic, 1973.

Franks, F. (ed.): *Water: A Comprehensive Treatise,* vols. 1–7, Plenum, 1972–1982.

Friebolin, H.: *Basic One- and Two-Dimensional NMR Spectroscopy,* 3d ed., Wiley, 1998.

Gardiner, W. C.: *Rates and Mechanisms of Chemical Reactions,* Benjamin, 1969.

Gasser, R. P. H.: *An Introduction to Chemisorption and Catalysis by Metals,* Oxford, 1985.

Glusker, J. P., and K. N. Trueblood: *Crystal Structure Analysis,* 2d ed., Oxford, 1985.

Gordy, W.: *Theory and Applications of Electron Spin Resonance,* Wiley, 1979.

Gordy, W., and R. L. Cook: *Microwave Molecular Spectra,* 3d ed., Wiley-Interscience, 1984.

Graybeal, J. D.: *Molecular Spectroscopy,* McGraw-Hill, 1988.

Guggenheim, E. A.: *Thermodynamics,* 5th ed., North-Holland, 1967.

Günther, H.: *NMR Spectroscopy,* 2d ed., Wiley, 1995.

Haase, R., and H. Schönert: *Solid–Liquid Equilibrium,* Pergamon, 1969.

Hague, D. N.: *Fast Reactions,* Wiley-Interscience, 1971.

Halliday, D., and R. Resnick: *Physics,* 3d ed., Wiley, 1978.

Hanna, M. W.: *Quantum Mechanics in Chemistry,* 3d ed., Benjamin-Cummings, 1981.

Harmony, M. D.: *Introduction to Molecular Energies and Spectra,* Holt, Rinehart and Winston, 1972.

Hehre, W. J., L. Radom, P. v. R. Schleyer, and J. A. Pople: *Ab Initio Molecular Orbital Theory,* Wiley, 1986.

Herzberg, G.: *Molecular Spectra and Molecular Structure,* vols. 1–3, Van Nostrand Reinhold, 1950, 1945, 1966.

Hill, T. L.: *An Introduction to Statistical Thermodynamics,* Addison-Wesley, 1960.

Hirschfelder, J. O., C. F. Curtiss, and R. B. Bird: *The Molecular Theory of Gases and Liquids,* Wiley, 1954.

Holbrook, K. A., M. J. Pilling, and S. H. Robertson: *Unimolecular Reactions,* 2d ed., Wiley, 1996.

Hollas, J. M.: *High Resolution Spectroscopy,* Butterworths, 1982.

Ives, D. J. G., and G. J. Janz (eds.): *Reference Electrodes,* Academic, 1961.

Jackson, E. A.: *Equilibrium Statistical Mechanics,* Prentice-Hall, 1968.

Jensen, F.: *An Introduction to Computational Chemistry,* Wiley, 1998.

Karplus, M., and R. N. Porter: *Atoms and Molecules,* Benjamin, 1970.

Kauzmann, W.: *Kinetic Theory of Gases,* Benjamin, 1966.

Kennard, E. H.: *Kinetic Theory of Gases,* McGraw-Hill, 1938.

Kestin, J.: *A Course in Thermodynamics,* vols. I and II, Blaisdell, 1966, 1968.

Kestin, J., and J. R. Dorfman: *A Course in Statistical Thermodynamics,* Academic, 1971.

Kirkwood, J. G., and I. Oppenheim: *Chemical Thermodynamics,* McGraw-Hill, 1961.

Kittel, C.: *Introduction to Solid State Physics,* 6th ed., Wiley, 1986.

Knox, J. H.: *Molecular Thermodynamics,* Wiley-Interscience, 1978.

Laidler, K. J.: *Theories of Chemical Reaction Rates,* McGraw-Hill, 1969.

Laidler, K. J.: *Chemical Kinetics,* 3d ed., Harper and Row, 1987.

Landolt-Börnstein: *Zahlenwerte und Funktionen* (*Numerical Data and Functional Relationships*), 6th ed., Springer-Verlag, 1950–1980; New Series, 1961–.

Leach, A. R.: *Molecular Modelling,* Addison-Wesley Longman, 1997.

Levine, I. N.: *Molecular Spectroscopy,* Wiley-Interscience, 1975.

Levine, I. N.: *Quantum Chemistry,* 5th ed., Prentice-Hall, 2000.

Levine, R. D., and R. B. Bernstein: *Molecular Reaction Dynamics and Chemical Reactivity,* Oxford, 1987.

Lewis, G. N., and M. Randall, revised by K. S. Pitzer and L. Brewer: *Thermodynamics,* 2d ed., McGraw-Hill, 1961.

Lide, D. R., and H. V. Kehiaian, *Handbook of Thermophysical and Thermochemical Data,* CRC Press, 1994.

Long, D. A.: *Raman Spectroscopy,* McGraw-Hill, 1977.

Lowe, J. P.: *Quantum Chemistry,* 2d ed., Academic, 1993.

McClelland, B. J.: *Statistical Thermodynamics,* Chapman and Hall, 1973.

McGlashan, M. L.: *Chemical Thermodynamics,* Academic, 1980.

McQuarrie, D. A.: *Statistical Thermodynamics,* Harper and Row, 1973.

McQuarrie, D. A.: *Quantum Chemistry,* University Science Books, 1983.

Mandl, F.: *Statistical Physics,* 2d ed., Wiley, 1988.

Mann, B. E., and J. W. Akitt: *NMR and Chemistry,* 4th ed., Stanley Thornes, 2000.

Moore, J. W., and R. G. Pearson: *Kinetics and Mechanism,* 3d ed., Wiley-Interscience, 1981.

Münster, A.: *Classical Thermodynamics,* Wiley-Interscience, 1970.

Nicholas, J. E.: *Chemical Kinetics,* Wiley, 1976.

Park, D.: *Introduction to the Quantum Theory,* 3d ed., McGraw-Hill, 1992.

Parr, R. G., and W. Yang, *Density-Functional Theory of Atoms and Molecules,* Oxford, 1989.

Pitzer, K. S. (ed.): *Activity Coefficients in Electrolyte Solutions,* CRC Press, 1991.

Pitzer, K. S.: *Thermodynamics,* 3d ed., McGraw-Hill, 1994.

Present, R. D.: *Kinetic Theory of Gases,* McGraw-Hill, 1958.

Press, W. H., et al.: *Numerical Recipes in FORTRAN,* 2d ed., Cambridge, 1993.

Prigogine, I., and R. Defay: *Chemical Thermodynamics,* Longmans Green, 1954.

Quinn, T. J.: *Temperature,* 2d ed., Academic, 1990.

Reed, T. R., and K. E. Gubbins: *Applied Statistical Mechanics,* McGraw-Hill, 1973.

Reid, R. C., J. M. Prausnitz, and B. E. Poling: *The Properties of Gases and Liquids,* 4th ed., McGraw-Hill, 1987.

Reynolds, W. C., and H. C. Perkins: *Engineering Thermodynamics,* McGraw-Hill, 1977.

Ricci, J. E.: *The Phase Rule and Heterogeneous Equilibrium,* Dover, 1966.

Robinson, R. A., and R. H. Stokes: *Electrolyte Solutions,* 2d ed., Butterworth, 1959.

Rosenberg, R. M., and I. M. Klotz: *Chemical Thermodynamics,* 6th ed., Wiley, 2000.

Rossiter, B. W., J. F. Hamilton, and R. C. Baetzold: *Physical Methods of Chemistry,* vols. I–X, 2d ed., Wiley, 1986–1993.

Rowlinson, J. S., and F. L. Swinton: *Liquids and Liquid Mixtures,* 3d ed., Butterworth, 1982.

Sands, D. E.: *Introduction to Crystallography,* Benjamin, 1969; Dover, 1993.

Schonland, D. S.: *Molecular Symmetry,* Van Nostrand Reinhold, 1965.

Shoemaker, D. P., C. W. Garland, and J. W. Nibler: *Experiments in Physical Chemistry,* 6th ed., McGraw-Hill, 1995.

Sokolnikoff, I. S., and R. M. Redheffer: *Mathematics of Physics and Modern Engineering,* 2d ed., McGraw-Hill, 1966.

Somorjai, G. A.: *Introduction to Surface Chemistry and Catalysis,* Wiley, 1994.

Steinfeld, J. I., J. S. Francisco, and W. L. Hase: *Chemical Kinetics and Dynamics,* 2d ed., Prentice-Hall, 1998.

Straughan, B. P., and S. Walker (eds.): *Spectroscopy,* vols. 1–3, 2d ed., Chapman and Hall, 1976.

Sugden, T. M., and C. N. Kenney: *Microwave Spectroscopy of Gases,* Van Nostrand, 1965.

Tabor, D.: *Gases, Liquids, and Solids,* 3d ed., Cambridge, 1991.

Tester, J., and M. Modell: *Thermodynamics and Its Applications,* 3d ed., Prentice-Hall, 1997.

Van Ness, H. C., and M. M. Abbott: *Classical Thermodynamics of Non-Electrolyte Solutions,* McGraw-Hill, 1982.

Van Wylen, G. S., R. E. Sonntag, and C. Borgnakke: *Fundamentals of Classical Thermodynamics,* 4th ed., Wiley, 1993.

Wayne, R. P.: *Principles and Applications of Photochemistry,* Oxford, 1988.

Weston, R., and H. Schwarz: *Chemical Kinetics,* Prentice-Hall, 1972.

Wilkinson, F.: *Chemical Kinetics and Reaction Mechanisms,* Van Nostrand Reinhold, 1980.

Young, D. A.: *Phase Diagrams of the Elements,* University of California, 1991.

Zemansky, M. W., and R. H. Dittman: *Heat and Thermodynamics,* 6th ed., McGraw-Hill, 1981.

APPENDIX

Standard-State Thermodynamic Properties at 25°C and 1 Bar[a]

Substance	$\Delta_f H^\circ_{298}$ kJ mol^{-1}	$\Delta_f G^\circ_{298}$ kJ mol^{-1}	$S^\circ_{m,298}$ J mol^{-1} K^{-1}	$C^\circ_{P,m,298}$ J mol^{-1} K^{-1}
$Ag^+(aq)$	105.56	77.09	72.8	
$Br(g)$	111.884	82.396	175.022	20.786
$Br^-(aq)$	−121.55	−103.97	82.4	−141.8
$Br_2(l)$	0	0	152.231	75.689
$Br_2(g)$	30.907	3.110	245.463	36.02
C(graphite)	0	0	5.740	8.527
C(diamond)	1.897	2.900	2.377	6.115
$C(g)$	716.682	671.257	158.096	20.838
$CF_4(g)$	−925.	−879.	261.61	61.09
$CH_4(g)$	−74.81	−50.72	186.264	35.309
$CO(g)$	−110.525	−137.168	197.674	29.116
$CO_2(g)$	−393.509	−394.359	213.74	37.11
$CO_3^{2-}(aq)$	−677.14	−527.81	−56.9	
$COF_2(g)$	−634.7	−619.2	258.60	46.82
$C_2H_2(g)$	226.73	209.20	200.94	43.93
$C_2H_4(g)$	52.26	68.15	219.56	43.56
$C_2H_6(g)$	−84.68	−32.82	229.60	52.63
$C_2H_5OH(l)$	−277.69	−174.78	160.7	111.46
$(CH_3)_2O(g)$	−184.05	−112.59	266.38	64.39
$C_3H_8(g)$	−103.85	−23.37	270.02	73.51
$C_6H_6(g)$	82.93	129.7	269.31	81.67
Cyclohexene(g) (C_6H_{10})	−5.36	107.0	310.86	105.02
α-D-Glucose(c) $(C_6H_{12}O_6)$	−1274.4	−910.1	212.1	218.8
Sucrose(c) $(C_{12}H_{22}O_{11})$	−2221.7	−1543.8	360.2	425.5
$CH_3(CH_2)_{14}COOH(c)$	−890.8	−314.5	455.2	460.7
$CaCO_3$(calcite)	−1206.92	−1128.79	92.9	81.88
$CaCO_3$(aragonite)	−1207.13	−1127.75	88.7	81.25
$CaO(c)$	−635.09	−604.03	39.75	42.80
$Cl(g)$	121.679	105.680	165.198	21.840
$Cl^-(aq)$	−167.159	−131.228	56.5	−136.4
$Cl_2(g)$	0	0	223.066	33.907
$Cu(c)$	0	0	33.150	24.435
$Cu^{2+}(aq)$	64.77	65.49	−99.6	
$F_2(g)$	0	0	202.78	31.30
$Fe(c)$	0	0	27.28	25.10
$Fe^{3+}(aq)$	−48.5	−4.7	−315.9	
$H(g)$	217.965	203.247	114.713	20.784
$H^+(aq)$	0	0	0	0
$H_2(g)$	0	0	130.684	28.824
$HD(g)$	0.318	−1.464	143.801	29.196
$D_2(g)$	0	0	144.960	29.196
$HBr(g)$	−36.40	−53.45	198.695	29.142
$HCl(g)$	−92.307	−95.299	186.908	29.12
$HF(g)$	−271.1	−273.2	173.779	29.133

Substance	$\Delta_f H^\circ_{298}$ kJ mol^{-1}	$\Delta_f G^\circ_{298}$ kJ mol^{-1}	$S^\circ_{m,298}$ J mol^{-1} K^{-1}	$C^\circ_{P,m,298}$ J mol^{-1} K^{-1}
$HN_3(g)$	294.1	328.1	238.97	43.68
$H_2O(l)$	-285.830	-237.129	69.91	75.291
$H_2O(g)$	-241.818	-228.572	188.825	33.577
$H_2O_2(l)$	-187.78	-120.35	109.6	89.1
$H_2S(g)$	-20.63	-33.56	205.79	34.23
$K^+(aq)$	-252.38	-283.27	102.5	21.8
$KCl(c)$	-436.747	-409.14	82.59	51.30
$Mg(c)$	0	0	32.68	24.89
$Mg(g)$	147.70	113.10	148.650	20.786
$MgO(c)$	-601.70	-569.44	26.94	37.15
$N(g)$	472.704	455.563	153.298	20.786
$N_2(g)$	0	0	191.61	29.125
$NH_3(g)$	-46.11	-16.45	192.45	35.06
$NH_2CH_2COOH(c)$	-528.10	-368.44	103.51	99.20
$NO(g)$	90.25	86.55	210.761	29.844
$NO_2(g)$	33.18	51.31	240.06	37.20
$NO_3^-(aq)$	-207.36	-111.25	146.4	-86.6
$N_2O_4(g)$	9.16	97.89	304.29	77.28
$Na(g)$	107.32	76.761	153.712	20.786
$Na^+(aq)$	-240.12	-261.905	59.0	46.4
$NaCl(c)$	-411.153	-384.138	72.13	50.50
$O(g)$	249.170	231.731	161.055	21.912
$O_2(g)$	0	0	205.138	29.355
$OH^-(aq)$	-229.994	-157.244	-10.75	-148.5
$PCl_3(g)$	-287.0	-267.8	311.78	71.84
$PCl_5(g)$	-374.9	-305.0	364.58	112.80
$SO_2(g)$	-296.830	-300.194	248.22	39.87
$Si(g)$	455.6	411.3	167.97	22.251
$SiC(\beta$, cubic$)$	-65.3	-62.8	16.61	26.86
$SiO_2(quartz)$	-910.94	-856.64	41.84	44.43
$Sn(gray)$	-2.09	0.13	44.14	25.77
$Sn(white)$	0	0	51.55	26.99
$SO_4^{2-}(aq)$	-909.27	-744.53	20.1	$-293.$

[a]Data obtained mainly by conversion from values in D. D. Wagman et al., Selected Values of Chemical Thermodynamic Properties, *Natl. Bur. Stand. Tech. Notes* 270-3, 270-4, 270-5, 270-6, 270-7, and 270-8, Washington, 1968–1981. The molality-scale standard state is used for solutes.

Answers to Selected Problems

1.1 (*a*) F; (*b*) T; (*c*) T; (*d*) F; (*e*) F. **1.2** (*a*) Closed nonisolated. **1.3** (*a*) 3; (*b*) 3. **1.5** (*a*) 19300 kg/m³. **1.6** (*a*) T; (*b*) F; (*c*) T; (*d*) T. **1.7** (*a*) 32.0; (*b*) 32.0 amu; (*c*) 32.0; (*d*) 32.0 g/mol. **1.8** x_{HCl} = 0.063. **1.9** (*a*) 1.99×10^{-23} g. **1.10** (*a*) T; (*b*) T; (*c*) F; (*d*) T; (*e*) F; (*f*) T. **1.11** (*a*) 5.5×10^6 cm³; (*b*) 1.0×10^4 bar; (*c*) 0.99×10^4 atm; (*d*) 1.5×10^3 kg/m³. **1.12** 652.4 torr. **1.13** (*a*) 33.9 ft; (*b*) 0.995 atm. **1.15** (*a*) 2.44 atm; (*b*) 16%. **1.17** 30.1 g/mol, 30.1. **1.18** 1.11 g/L. **1.19** 82.06_5 cm³ atm mol⁻¹ K⁻¹. **1.20** 31.0_7. **1.21** 0.133 mol N_2, 0.400 mol H_2, 1.33 mol NH_3. **1.24** 1800 kPa. **1.25** (*a*) 17.5 kPa; (*b*) 0.857 for H_2. **1.26** 32.3 cm³. **1.27** 0.60 mol and 0.40 mol. **1.28** (*a*) 2.5×10^{19}; (*b*) 3.2×10^{10}. **1.29** 0.0361 g and 0.619. **1.30** 5.3×10^{21} g. **1.31** (*a*) 6.4×10^{-5} bar; (*b*) 460 K; (*c*) 1.2×10^3 bar. **1.32** 0.247. **1.33** (*a*) P_{N_2} = 0.78 atm; (*b*) m_{N_2} = 75 kg. **1.35** 3. **1.36** (*a*) $6x^2e^{-3x} - 6x^3e^{-3x}$; (*c*) $1/x$; (*d*) $1/(1 - x)^2$; (*e*) $1/(x + 1) - x/(x + 1)^2 = 1/(x + 1)^2$; (*f*) $2e^{-2x}/(1 - e^{-2x})$; (*g*) 6 sin $3x$ cos $3x$. **1.40** (*a*) ax cos axy; (*c*) $-(x^2/y^2)e^{x/y}$; (*d*) 0. **1.41** (*a*) nR/P. **1.44** (*b*) 0.00137 atm. **1.45** 2.44×10^4 cm³/mol. **1.48** (*a*) 18.233 cm³/mol; (*b*) 18.15 cm³/mol. **1.51** 2.6×10^{-4} K⁻¹, 4.9×10^{-5} atm⁻¹, 5.3 atm/K. **1.54** 23 atm. **1.56** (*a*) 2000 atm. **1.57** (*a*) 25. **1.59** (*a*) $-190/3$; (*b*) 0.693; (*c*) 1/2; (*d*) -0.2233. **1.60** (*a*) $-a^{-1}$ cos $ax + C$. **1.63** (*a*) $x^4/2 + 3e^{5x}/5 + C$. **1.65** (*a*) 1750.62; (*b*) -458.73; (*c*) 5.43×10^{-139}. **1.68** 717.8 K. **1.69** (*a*) T; (*b*) F; (*c*) F; (*d*) T; (*e*) F; (*f*) T; (*g*) T; (*h*) T; (*i*) T; (*j*) F.

2.1 (*a*) T; (*b*) F. **2.2** (*a*) J; (*b*) J; (*c*) m³; (*d*) N; (*e*) m/s; (*f*) kg. **2.3** (*a*) 1 kg m² s⁻²; (*b*) 1 kg m⁻¹ s⁻²; (*c*) 10^{-3} m³; (*d*) 1 kg m s⁻². **2.4** (*a*) 15.2 J; (*c*) 14.0 m/s. **2.5** 1.00 Pa. **2.6** (*a*) F; (*b*) T; (*c*) T; (*d*) F; (*e*) F; (*f*) F; (*g*) F. **2.8** -18.0 J. **2.9** (*a*) -304 J. **2.11** (*a*) 0.107 cal/g-°C. **2.12** (*a*) T; (*b*) T; (*c*) F; (*d*) T; (*e*) F. **2.20** 18.001°C. **2.22** -200 J. **2.23** (*a*) T; (*b*) F. **2.26** No. **2.28** (*a*) T; (*b*) T. **2.31** (*a*) U; (*b*) H. **2.32** 15°C. **2.33** (*a*) 246 J/m³. **2.36** (*a*) T; (*b*) F; (*c*) F; (*d*) T; (*e*) T. **2.37** (*a*) 13.7 kJ, -13.7 kJ, 0, 0. **2.38** (*a*) 300 K, 0.500 atm; (*b*) 189 K, 0.315 atm. **2.39** (*a*) 0, 98.9 J, 98.9 J, 138.5 J. **2.41** (*a*) F; (*b*) T; (*c*) F; (*d*) F; (*e*) F. **2.42** (*a*) Process; (*b*) property; (*c*) process; (*d*) process. **2.43** (*a*) ∞; (*b*) $-∞$; (*c*) ∞; (*d*) 0. **2.44** (*a*) $q > 0$, $w < 0$, $\Delta U > 0$, $\Delta H > 0$; (*b*) $q > 0$, $w > 0$, $\Delta U > 0$, $\Delta H > 0$; (*c*) $q = 0$, $w < 0$, $\Delta U < 0$, $\Delta H < 0$; (*d*) $\Delta H = 0$, $\Delta U = 0$, $w < 0$, $q > 0$. **2.45** (*a*) $w = 0$, $q = 0$, $\Delta U = 0$. **2.47** (*a*) $q = 1447$ cal, $w = -397$ cal, $\Delta U = 1050$ cal, $\Delta H = 1447$ cal. **2.48** (*a*) 1436 cal, 0.039 cal, 1436 cal, 1436 cal; (*b*) 1801 cal, -0.019 cal, 1801 cal, 1801 cal; (*c*) 9717 cal, -741 cal, 8976 cal, 9717 cal. **2.49** (*a*) 6240 J, 10400 J. **2.52** (*a*) Kinetic; (*b*) kinetic; (*c*) both. **2.58** (*a*) 0.003637 K⁻¹. **2.64** Increase. **2.65** (*a*) Intensive, kg/m³; (*b*) extensive, J; (*c*) intensive, J/mol. **2.69** (*a*) F; (*b*) F; (*c*) F; (*d*) F; (*e*) F; (*f*) F; (*g*) F; (*h*) T; (*i*) F; (*j*) F; (*k*) F; (*l*) T; (*m*) F; (*n*) F; (*o*) F; (*p*) T.

3.1 (*a*) T; (*b*) T; (*c*) T; (*d*) F. **3.2** (*a*) 74.6%; (*b*) 746 J, 254 J. **3.3** 2830 K. **3.5** (*c*) 15 J. **3.8** (*a*) F; (*b*) T; (*c*) T; (*d*) T; (*e*) T; (*f*) F; (*g*) F; (*h*) F; (*i*) F; (*j*) T; (*k*) T. **3.9** (*a*) 17.9 cal/K; (*b*) -2.24 cal/K. **3.10** 4.16 cal/K. **3.11** 38.30 cal/K. **3.13** (*a*) 6.66 J/K; (*b*) 14.1 J/K. **3.15** -2.73 cal/K. **3.17** (*a*) 32.0°C; (*b*) -1.59 cal/K; (*c*) 1.87 cal/K; (*d*) 0.28 cal/K. **3.18** 8.14 J/K. **3.19** -2.03×10^{-5} cal/K. **3.21** (*a*) F; (*b*) F; (*c*) T; (*d*) T; (*e*) T; (*f*) F; (*g*) T; (*h*) F. **3.22** (*a*) positive; (*b*) positive; (*c*) zero; (*d*) positive. **3.25** (*a*) 373.2°M; 199.99°M. **3.28** (*b*) 7. **3.29** (*a*) q; (*b*) T. **3.32** (*a*) Reversible; (*b*) irreversible; (*c*) irreversible. **3.33** (*a*) The 10 g; the 10 g. **3.39** (*a*) 0; (*c*) 0.008 cal/K. **3.40** (*a*) J/K; (*e*) no units; (*f*) kg/mol. **3.41** (*c*), (*e*). **3.42** (*a*) F; (*b*) T; (*c*) F; (*d*) F; (*e*) T; (*f*) F.

4.1 (*a*) T; (*b*) F; (*c*) T; (*d*) F; (*e*) F; (*f*) T. **4.2** (*a*) $\Delta G = 0$, ΔA = 0.330 J. **4.3** (*a*) C_V. **4.6** (*a*) 2040 atm; (*b*) -0.0221 K/atm. **4.7** 72 J mol⁻¹ K⁻¹. **4.8** (*a*) 40 cal mol⁻¹ K⁻¹; (*b*) 0.8 cal mol⁻¹ atm⁻¹; (*c*) 70 cal/cm³; (*d*) 0.13 cal mol⁻¹ K⁻²; (*e*) -0.0012 cal mol⁻¹ K⁻¹ atm⁻¹. **4.18** (*a*) -1.66×10^{-6} cal mol⁻¹ K⁻¹ atm⁻¹; (*b*) 6.64 cal mol⁻¹ K⁻¹. **4.25** (*a*) T; (*b*) T. **4.26** -3220 J, -3220 J. **4.27** (*a*) $\Delta A < 0$, $\Delta G = 0$; (*b*) $\Delta A > 0$; $\Delta G = 0$; (*d*) $\Delta A < 0$, $\Delta G < 0$. **4.29** (*a*) 0; (*b*) -28 cal. **4.30** -840 J, -840 J. **4.32** 302 J. **4.34** T_{final} = 30.02°C. **4.36** (*a*) T; (*b*) T; (*c*) F; (*d*) T; (*e*) F; (*f*) T; (*g*) F. **4.39** (*a*) F; (*b*) F; (*c*) T; (*d*) F. **4.41** (*a*) Gas; (*b*) neither; (*c*) liquid; (*d*) solid. **4.43** -5 for O_2. **4.45** -0.45 mol. **4.46** (*a*) No; (*b*) no; (*c*) no; (*d*) yes; (*e*) yes; (*f*) no. **4.48** (*a*) All are 0; (*b*) $\Delta U = 0$. **4.52** (*a*) J. **4.58** (*a*) C_P; (*b*) μ_i^α and μ_i^β. **4.59** (*a*) F; (*b*) T; (*c*) F; (*d*) F; (*e*) T; (*f*) T; (*g*) F; (*h*) F; (*i*) T; (*j*) F; (*k*) T; (*l*) F; (*m*) F; (*n*) F; (*o*) F.

5.1 (*a*) F; (*b*) F; (*c*) F. **5.2** (*a*) F; (*b*) T; (*c*) T; (*d*) T. **5.4** (*a*) -638 kJ/mol; (*b*) -1276 kJ/mol; (*c*) 319 kJ/mol. **5.5** (*a*) F; (*b*) T; (*c*) T. **5.9** (*a*) T; (*b*) T; (*c*) T; (*d*) T; (*e*) F. **5.10** (*a*) -1124.06 kJ/mol; (*b*) -1036.04 kJ/mol; (*c*) -956.5 kJ/mol. **5.11** (*a*) -2801.6 kJ/mol, -2801.6 kJ/mol; (*b*) 24.755°C. **5.12** 24.755°C. **5.13** (*a*) 12.01 kJ/K; (*b*) -5186 kJ/mol, -5191 kJ/mol. **5.14** $\Delta U_{298} \approx -29.98$ kJ/mol, $\Delta H_{298} \approx -20.06$ kJ/mol. **5.15** (*a*) -3718.5 J/mol; (*b*) -3716.7 J/mol. **5.16** -248 kJ/mol, -239 kJ/mol. **5.17** -558 kJ/mol, -546 kJ/mol. **5.18** $-196\frac{1}{2}$ kcal/mol. **5.20** -85 kJ/mol. **5.21** (*b*) -4.6 J/mol. **5.22** (*a*) 0; (*b*) 288.2 J/mol. **5.23** (*a*) T; (*b*) T; (*c*) F; (*d*) F. **5.24** (*a*) -1118.7 kJ/mol; (*b*) -1036.7 kJ/mol. **5.30** F. **5.34** (*a*) 20.11_5 cal mol⁻¹ K⁻¹; (*b*) 20.11_9 cal mol⁻¹ K⁻¹. **5.37** (*a*) 69.91 J mol⁻¹ K⁻¹; (*b*) 81.59 J mol⁻¹ K⁻¹. **5.38** (*a*) -390.73 J mol⁻¹ K⁻¹; (*b*) -152.90 J mol⁻¹ K⁻¹. **5.39** (*a*) -374.76 J mol⁻¹ K⁻¹. **5.43** -197.35 kJ/mol. **5.44** (*b*) -1007.53 kJ/mol, -990.41 kJ/mol, -949.6 kJ/mol. **5.45** (*a*) -980.09 kJ/mol; (*b*) -979.39 kJ/mol. **5.46** (*a*)

−61.16 kJ/mol; (b) −306.67 kJ/mol. **5.49** 179.29 kJ/mol. **5.52** (a) 42 kJ/mol. **5.54** 61.8 cal mol^{-1} K^{-1}. **5.58** (a) Pa; (b) J. **5.61** (a) No; (b) no; (c) yes; (d) no; (e) yes; (f) yes; (g) yes. **5.62** 2400 K. **5.66** (a) T; (b) F; (c) F.

6.1 −6.9 kJ. **6.2** (a) T; (b) T; (c) T. **6.3** (a) 3.42, −10.2 kJ/mol. **6.4** 0.0709, 13.2 kJ/mol. **6.5** 24.0, −8.55 kJ/mol. **6.9** (a) T; (b) (b) F; (c) F; (d) F; (e) T; (f) F; (g) T; (h) T. **6.11** (a) $\Delta H° = 94._8$ kJ/mol, $\Delta G° = -3.05$ kJ/mol, $\Delta S° = 183$ J mol^{-1} K^{-1}. **6.21** (a) T; (b) F. **6.22** 22.2. **6.23** 1.50 mol CO_2, 0.50 mol CF_4, 0.0008 mol COF_2. **6.30** 0.633 mol, 2.633 mol, 2.367 mol. **6.31** 3.0×10^{-7}, 0.50; $x_{Cl_2} = 0.36_6$. **6.36** (c) 157 kJ/mol. **6.39** (a) No; (b) no; (c) no; (d) no; (e) no; (f) no; (g) yes; (h) no; (i) yes. **6.51** $\Delta S° = 34.5$ cal mol^{-1} K^{-1}, $\Delta C_P° = 0$. **6.54** (a) Yes; (b) no. **6.63** (a) F; (b) F; (c) T; (d) F; (e) F; (f) T; (g) T; (h) F; (i) F; (j) T; (k) T; (l) F; (m) F; (n) T.

7.1 (a) F; (b) F. **7.2** (a) 3; T, P, sucrose mole fraction; (b) 4; (c) 3; (d) 2; (e) 1. **7.4** (a) 3; (b) 4. **7.5** (a) 4; (b) 3; (c) 2; (e) 1. **7.7** (a) 2. **7.10** (a) T; (b) T; (c) F; (d) F; (e) T; (f) T; (g) T; (h) T; (i) T. **7.11** (a) Liquid; (b) gas. **7.12** (a) 1; (c) 0. **7.13** (a) 0.130 g liquid and 0.230 g vapor; (b) 0.360 g vapor. **7.14** (a) Gas; (b) solid; (c) gas. **7.16** (a) Ar; (b) H_2O. **7.21** (a) T; (b) T; (c) F; (d) T; (e) F; (f) F. **7.22** 545 torr. **7.23** (a) 134 atm. **7.24** (a) −38.4°C. **7.27** 42.7 kJ/mol. **7.28** (a) 1481 torr; (b) 85°C. **7.29** (c) 350°C. **7.34** (a) $15._4$ torr, 200 K; (b) 7.9 kJ/mol. **7.37** $7\frac{1}{2}$°C. **7.51** (b) 108 torr. **7.55** (a) T; (b) T; (c) F; (d) F; (e) T; (f) F; (g) T; (h) F; (i) F.

8.1 (a) Pa m^6/mol^2 and m^3/mol. **8.3** (a) 19.5 atm; 23.5 atm; (b) 1272 cm^3, 1465 cm^3. **8.7** (a) 317 atm; (b) 804 atm; (c) 172 atm. **8.10** (a) 1.34×10^6 cm^6 atm mol^{-2}, 32.0 cm^3/mol; (b) −131 cm^3/mol at 100 K. **8.20** (b) 0.375. **8.21** 132 atm. **8.27** 13 cal/mol, 0.028 cal mol^{-1} K^{-1}. **8.35** 49 atm, 565 K, 260 cm^3/mol. **8.40** (a) Ne, Ne, Ne, Ne. **8.41** 166 cm^3/mol, experimental value 176 cm^3/mol. **8.45** (a) F; (b) F.

9.1 (a) mol/m^3; (b) mol/kg; (c) no units. **9.2** c_i; c_i. **9.3** (a) 0.116 mol; (b) 0.398 mol; (c) 0.598 mol. **9.4** (a) 0.9518 g/cm^3; (b) 13.37 mol/kg. **9.5** 0.474 mol/kg, 0.00846. **9.10** (a) F; (b) F; (c) F; (d) T; (e) T; (f) F; (g) F; (h) T; (i) T. **9.11** 1011.9 cm^3. **9.13** 40.19 cm^3/mol. **9.14** 36.9 cm^3/mol, 18.1 cm^3/mol. **9.23** 16.9_5 cm^3/mol, 57.3 cm^3/mol. **9.24** (a) $14._0$ cm^3/mol, 40.7 cm^3/mol; (b) 16.5 cm^3/mol, 40.2 cm^3/mol. **9.31** (a) F; (b) T; (c) T; (d) F. **9.32** No. **9.33** (a) T; (b) T; (c) F; (d) F; (e) F; (f) T. **9.35** −3.98 kJ, 0, 13.6 J/K, 0. **9.36** (a) 40.4 torr, 10.2 torr; (b) 0.798, 0.202. **9.37** 0.211, 0.789. **9.38** 173 torr, 60.8 torr. **9.40** 0.8729 g/cm^3. **9.47** (a) 171.03 torr, 6.92 torr; (b) 0.9611_1 and 0.0388_9; (c) 692 torr; (d) 183.14 torr. **9.48** (b) 433.90 torr; (c) 903 torr. **9.52** (a) $6.8_2 \times 10^4$ atm; (b) 164 mg. **9.60** 2.40 atm, 0.83. **9.69** (a) T; (b) F; (c) F; (d) T; (e) F; (f) T.

10.1 (a) T; (b) T; (c) T; (d) T; (e) F. **10.2** (a) No; (b) yes; (c) yes; (d) yes. **10.5** (a) T; (b) T. **10.6** (a) 1.11, 2.04; (c) −702 J; (d)

−1.28 kJ. **10.8** (b) 0.843, 0.0337, 1.0036, 0.9635. **10.11** (a) 0.9823. **10.13** 1.327, 0.0349, 1.94. **10.25** (a) 1, 1, 1, −1; (b) 1, 2, 2, −1; (e) KCl. **10.28** (a) 1; (b) 1.587; (c) 1. **10.31** (a) 0.990; (b) 0.843. **10.32** (b) 2.85, 3.16. **10.35** 0.330 mol/kg. **10.39** 0.630. **10.47** (a) 0.214; (b) 0.553, 0.390. **10.50** (a) −79.885 kJ/mol, −55.84 kJ/mol, 80.66 J mol^{-1} K^{-1}. **10.56** (a) 16.8 kJ/mol; (b) −17.2 kJ/mol. **10.58** −131.23 kJ/mol, −167.16 kJ/mol, 56.3_6 J mol^{-1} K^{-1}. **10.62** (a) 0.997, 0.928; (b) $f_i = 2.32$ atm. **10.64** (a) 15.69 kJ; (b) 17.1 kJ. **10.67** (a) 0.9841, 0.9776, 0.9810, 0.9841. **10.74** (a) T; (b) F; (c) T; (d) T; (e) F; (f) F.

11.1 (a) T; (b) T. **11.2** (a) F; (b) F; (c) F; (d) F. **11.3** (b) **11.4** (a) 0.0064_2 mol/kg; (b) 0.0079 mol/kg; (c) 0.000169 mol/kg. **11.7** 1.27×10^{-7} mol/kg. **11.8** 1.34×10^{-7} mol/kg. **11.10** 1.05×10^{-7} mol/kg. **11.11** 4.28×10^{-4} mol/kg. **11.13** 2.2×10^{-9} mol/kg. **11.16** $\gamma_\pm^\dagger = 0.198$. **11.23** 1, 1.01, 1.11, 2.98. **11.24** 0.00805 mol/kg. **11.27** (a) 8.7; (b) 0.61. **11.28** 2.51×10^{-5} mol^2/kg^2. **11.29** 1.4×10^{-23}. **11.30** 2.72 mol Fe_3O_4. **11.31** (a) 0.039 mol $CaCO_3$; (b) 0 mol $CaCO_3$. **11.37** (b) 4.5 bar. **11.38** 79.9_0 kJ/mol, −75.3 mol^{-1} K^{-1}, 57.4_5 kJ/mol. **11.44** −3.47 kJ/mol. **11.47** (a) F; (b) T; (c) F; (d) F; (e) F; (f) F; (g) F.

12.1 (a) T; (b) T. **12.2** (a) T; (b) F. **12.3** 1073.4 torr. **12.4** (a) T; (b) F; (c) F; (d) T; (e) T; (f) T; (g) T. **12.5** 2.51°C. **12.6** 339. **12.9** (a) 180; (b) 10.8 kcal/mol. **12.12** 6.89 kcal/mol. **12.15** 0.027 mol naph., 0.014 mol anth. **12.17** F. **12.18** (a) 6.78 atm; (b) 0.99431, 0.99969. **12.19** 56000. **12.20** 736 cm. **12.21** 55500. **12.24** 29.0. **12.26** (b) 1.95 atm. **12.36** (a) $x_{B,v} = 0.75$; (b) $x_{B,l} = 0.04$. **12.38** 0.5 g nic. and 5.7 g water; 9.5 g nic. and 4.3 g water. **12.52** (a) 0.306; (b) 0.306; (c) 0.036. **12.68** Yes. **12.69** (a) 2; (b) 1; (c) 0. **12.70** 1.20. **12.73** (a) 100.15°C; (b) 23.61 torr. **12.75** (a) F; (b) T; (c) F; (d) F; (e) T; (f) T; (g) F; (h) T; (i) F.

13.1 (a) T; (b) T. **13.2** N/m. **13.3** (a) 4.8 cm^2; (b) 1.0×10^6 cm^2. **13.4** 220 ergs. **13.5** 20.2 mN/m. **13.6** 6.0%. **13.7** 4.9×10^{-6} atm. **13.8** (a) F; (b) T. **13.9** 762.7 torr. **13.10** 22.6 dyn/cm. **13.11** 3.22 cm. **13.12** 18.0 cm. **13.14** 53.0 dyn/cm. **13.16** 4.8×10^{-10} mol/cm^2. **13.18** -3.0×10^{-10} mol/cm^2. **13.22** 25 Å. **13.23** (a) 24 Å; (c) 2.4×10^{-10} mol/cm^2. **13.24** (a) 179 cm^2/g, 0.36 atm^{-1}. **13.30** (a) −210 kJ/mol. **13.35** (b) 17.726 torr.

14.1 (a) Yes; (b) no. **14.2** (a) T; (b) T; (c) T. **14.3** 4.6×10^{-8} N. **14.4** (a) 3.6×10^{10} V/m; (b) 0.90×10^{10} V/m. **14.5** −3.60 V. **14.6** (a) F; (b) F. **14.7** (a) 5.79×10^5 C; (b) -5.79×10^4 C. **14.8** T. **14.9** 1×10^4 J/mol. **14.10** (a) F; (b) T; (c) F. **14.12** (a) T; (b) T; (c) T; (d) F. **14.13** (a) 2; (b) 1; (c) 2; (d) 6; (e) 2. **14.14** −0.458 V. **14.17** (a) 0.1966 V; (b) 0.930, 0.786. **14.23** (a) F; (b) T. **14.24** (a) 0.355 V; (b) 0.38 V. **14.25** (a) −1.710 V. **14.26** (a) 2.1×10^{41}; (b) 5.0×10^{-27}. **14.28** (b) −0.164 V. **14.29** (b) 0.354 V; (c) 0.273 V. **14.30** 1.157 V. **14.31** −0.74 V. **14.32** 1.093 V. **14.35** (a) 0.0389 V. **14.36** −0.0205 V. **14.37** (a) T; (b) F. **14.38** (b) 0.0458 V; (c) 0.0458 V; (d) −8.84 kJ/mol, 10.6 kJ/mol, 65.2 J mol^{-1} K^{-1}. **14.39** 2. **14.41** 8×10^{-9}. **14.42** (a)

−54.4 kJ/mol, 3×10^9; (b) −27.2 kJ/mol, 6×10^4. **14.47** −0.152 V. **14.48** 0.84. **14.49** −131.2 kJ/mol, −131.2 kJ/mol. **14.51** (b) 1.75×10^{-5}. **14.53** 4.68. **14.54** −0.0104 V. **14.60** 1.25×10^{-39} C^2 N^{-1} m, 1.12×10^{-29} m^3. **14.61** (b) 1.0068_7. **14.62** 6.19×10^{-30} C m, 1.4×10^{-40} C^2 m N^{-1}. **14.70** (a) C; (b) m; (c) N/C = V/m; (h) J/mol.

15.1 (a) T; (b) F. **15.2** (a) T; (b) F; (c) T; (d) T. **15.3** (a) 3720 J. **15.4** (a) 1.18×10^{-20} J. **15.5** 1.366. **15.7** 18.5 K. **15.8** $1.3_7 \times 10^7$ J. **15.9** (a) 0 to ∞; (b) −∞ to ∞. **15.10** (a) T; (b) F; (c) T; (d) F. **15.11** (a) 1.1×10^{18}; (b) 7.45×10^{17}. **15.12** 1.24×10^{-6}. **15.15** (a) 4.50×10^4; (b) 1750 K. **15.20** (a) 5.32×10^4 cm/s; (b) 4.90×10^4 cm/s; (c) 4.35×10^4 cm/s. **15.30** C$_3$H$_6$. **15.31** 0.0152 torr. **15.32** 5.6 mg. **15.33** 8.9×10^{17}, 5.8×10^{-4} g. **15.35** (a) T; (b) F. **15.38** (a) 7.1×10^9 s^{-1}; (b) 8.7×10^{28} s^{-1} cm^{-3}. **15.40** (b) 5.1×10^5 Å. **15.41** 458 torr. **15.42** 0.80 torr. **15.43** 5060 m. **15.44** 0.064 torr. **15.50** 3.5, 15.17. **15.60** (a) F; (b) T; (c) T; (d) F; (e) F; (f) T; (g) F.

16.1 (a) T; (b) T; (c) F. **16.2** (a) 288 J; (b) 0.161 J/K. **16.3** 0.142 and 0.166 J K^{-1} m^{-1} s^{-1}. **16.5** 6.05 mJ K^{-1} cm^{-1} s^{-1}. **16.6** (a) F; (b) T; (c) T; (d) T. **16.8** (a) 5.66 cP; (b) 187. **16.9** (a) −35 Pa/m; (b) 17 cm/s; (c) 1100, 6400. **16.10** 0.58 mg/s. **16.11** 0.326 cP. **16.14** 420 cm/s, 0.37 cm/s. **16.15** 4.59 Å, 3.85 Å, 3.69 Å. **16.16** 1.20×10^{-4} P. **16.17** (a) 400 kg/mol, 500 kg/mol; (b) 300 kg/mol, 400 kg/mol. **16.18** 390000. **16.19** (a) F; (b) F; (c) T; (d) F; (e) T; (f) F. **16.20** (a) 2×10^{13} yr. **16.21** (a) 0.025 cm; (b) 0.19 cm; (c) 0.95 cm. **16.23** 6.6×10^{23}, 5.9×10^{23}, 7.8×10^{23}. **16.24** −0.083 μm, 2.5 μm. **16.25** (a) 0.16 cm^2/s; (b) 0.016 cm^2/s. **16.26** (b) 2.4×10^{-5} cm^2/s. **16.27** 2.0×10^{-5} cm^2/s. **16.28** (b) 0.40 cm^2/s. **16.29** 9.2×10^{-7} cm^2/s. **16.33** 6.3×10^4. **16.35** 6.2×10^{18}. **16.36** 0.0104 Ω. **16.37** 0.25 A. **16.38** 1.0 V/cm. **16.39** (a) T; (b) T; (c) F; (d) F; (e) T. **16.40** 1.19 g. **16.41** (c) 472.21 cm^2 Ω^{-1} mol^{-1}. **16.42** (a) 248.4 Ω^{-1} cm^2 mol^{-1}; (b) 124.2 Ω^{-1} cm^2 equiv^{-1}. **16.43** 4.668×10^{-4} cm^2 V^{-1} s^{-1}, 0.3894. **16.45** $t_+ = 0.4883$. **16.46** (a) 100.4 Ω^{-1} cm^2 mol^{-1}; (b) 391 Ω^{-1} cm^2 mol^{-1}. **16.48** (a) 6.99×10^{-4} cm^2 V^{-1} s^{-1}; (b) 0.017 cm/s; (c) 1.37 Å. **16.49** (b) 307 Ω^{-1} cm^2 mol^{-1}. **16.50** 0.426, 0.574. **16.52** 0.536 mmol. **16.56** (c) 89.9 Ω^{-1} cm^2 mol^{-1}. **16.57** (a) 140.7 cm^2 Ω^{-1} mol^{-1}, 0.000281 Ω^{-1} cm^{-1}; (b) 35.5 kΩ. **16.58** $9.9_7 \times 10^{-15}$ mol^2/L^2. **16.59** (a) 3.6×10^{-5} mol^2/L^2. **16.60** 1.74×10^{-5} mol/L.

17.1 (a) F; (b) F; (c) T; (d) F; (e) T; (f) F; (g) T; (h) F. **17.2** (a) s^{-1}; (b) L mol^{-1} s^{-1}. **17.4** −0.002 mol L^{-1} s^{-1}. **17.5** (a) 7.1×10^{-8} mol dm^{-3} s^{-1}, 8.5×10^{-7} mol/s; (b) −1.4 ×10^{-7} mol dm^{-3} s^{-1}; (c) 1.0×10^{18}; (d) 3.46×10^{-5} s^{-1}, 14×10^{-8} mol dm^{-3} s^{-1}, 17×10^{-7} mol/s. **17.13** (a) −k_1[B]; (b) −k_1[B], k_1[B] − k_2[E], k_2[E]. **17.14** (a) F; (b) F; (c) T. **17.15** (a) 0.00066 s^{-1}, 0.00133 s^{-1}; (b) 905 s, 1731 s. **17.16** (a) 2.00×10^4 s; (b) 5.0×10^{-4} mol/L. **17.18** (a) 0.030 mol NO$_2$, etc.; (b) 1.44×10^{-5} mol L^{-1} s^{-1}. **17.23** $n < 1$. **17.25** (a) 6.75; (b) 3. **17.28** (a) 3/2; (b) 0.008 dm$^{3/2}$ mol$^{-1/2}$ s^{-1}. **17.30** 0.00161 L mol^{-1} s^{-1}. **17.31** (b) 0.0071 L mol^{-1} s^{-1}. **17.32** (b) 0.33 dm^6 mol^{-2} s^{-1}. **17.33** 0.036 L^2 mol^{-2} s^{-1}. **17.35** 0.0188 L^3 mol^{-3} s^{-1}. **17.37** (a) T; (b) T. **17.38** 8.0×10^6 L mol^{-1} s^{-1}. **17.40** (a) T; (b) F. **17.44** (a) 60 s^{-1}. **17.59** (a) F; (b) F; (c) F. **17.60** 0.018 L mol^{-1} s^{-1}.

17.61 7×10^{10} L mol^{-1} s^{-1}, 162 kJ/mol. **17.62** 45.5 kcal/mol, 1.9×10^{11} L mol^{-1} s^{-1}. **17.64** (b) 77 min^{-1}. **17.65** (b) 3.87×10^{-7} s^{-1}; (c) 2.36×10^{10} s, 8.95×10^5 s, 795 s. **17.68** 342 kJ/mol. **17.70** (a) 145 kJ/mol; (b) 1.28. **17.72** (a) 1; (b) 83. **17.78** $k_1 = 0.256$ L mol^{-1} s^{-1}, $k_{-1}/k_2 = 2250$ L/mol. **17.85** (a) 1.0×10^{-7} mol/L, 3000, 0.0006 mol L^{-1} s^{-1}, 2×10^{-7} mol L^{-1} s^{-1}. **17.89** 1.3×10^{10} L mol^{-1} s^{-1}. **17.90** (b) 19 kJ/mol. **17.91** (a) F; (b) T; (c) F; (d) F. **17.92** $7_{.6} \times 10^4$ s^{-1}, $8_{.3} \times 10^{-3}$ mol/L. **17.94** 0. **17.97** (a) $1.3_8 \times 10^{15}$; (b) 0.080, 0.67. **17.98** (a) 5.6×10^{11} s. **17.99** 3.7×10^{-9} cm, 3.7×10^{-8} cm. **17.100** (b) 180 s. **17.102** 1.62×10^5 yr. **17.103** (a) 6.77×10^{11} s^{-1}; (b) 4.79×10^{11} s^{-1}. **17.105** (a) 1.08×10^{-10}%; (b) 0.0296 min^{-1} (g C)$^{-1}$. **17.108** (c) 6×10^{10}. **17.111** (a) T; (b) T; (c) T; (d) T; (e) T; (f) F; (g) T; (h) F; (i) T; (j) T; (k) F; (l) F; (m) F; (n) F.

18.1 (c) 1.76×10^{13} s^{-1}, 1.76×10^{14} s^{-1}; (d) 5900 K. **18.2** (b) 4.0×10^{26} J/s; (c) 1.4×10^{17} kg. **18.3** (a) 5.3×10^{14} s^{-1}, $1.2_1 \times 10^{15}$ s^{-1}, 5.7×10^{-5} cm, 2.5×10^{-5} cm. **18.4** 2.8×10^{-19} J. **18.5** 2.97×10^{20}. **18.7** (a) 6.6×10^{-10} cm. **18.9.** 7000 km/s. **18.11** (a) F; (b) F; (c) T; (d) T; (e) T; (f) T. **18.12** (a) 2; (b) 13$^{1/2}$; (c) 1. **18.17** (a) T; (b) F; (c) T; (d) T. **18.22** 1.4×10^{-5} cm. **18.23** (b) 0.0908, 0.2500, 0.3031. **18.24** (a) 3.45×10^{-5}; (b) 9.05×10^{-5}. **18.27** 1.2 nm. **18.32** (a) 17; (b) 6. **18.33** (a) T; (b) F; (c) F; (d) F; (e) F; (f) F; (g) T. **18.36** (a) $2f'(x)$. **18.37** (a) Three are linear, two nonlinear. **18.38** (a) Function; (b) operator. **18.40** (a) Three of these are eigenfunctions of d^2/dx^2; (b) −9, −16, 25. **18.44** (a) Nu, frequency; (b) vee, quantum number. **18.47** (a) 0. **18.50** (a) 0; (b) $(2\alpha)^{-1}$. **18.52** (a) 1.02×10^4 dyn/cm; (b) 5.1×10^{30}. **18.55** (a) 1.14×10^{-23} g; (b) 1.45×10^{-46} kg m^2; (d) 1.16×10^{11} s^{-1}, 2.32×10^{11} s^{-1}. **18.58** (a) 22%. **18.60** (a) T; (b) F; (c) T; (d) F; (e) T. **18.70** (a) T; (b) T; (c) T; (d) T; (e) F; (f) T; (g) F; (h) F; (i) F. **18.71** A good Internet search engine will find the answers to most of the questions. See also A. Pais, *Subtle Is the Lord,* Oxford, 1982; A. Pais, *Neil's Bohr's Times,* Oxford, 1993; W. Moore, *Schrödinger,* Cambridge, 1989.

19.1 (a) F; (b) T; (c) T. **19.4** (a) 7.7×10^{-19} J, 4.8 eV. **19.6** 1×10^{-12}. **19.7** 2.6 Å. **19.8** (a) T; (b) T; (c) F; (d) T; (e) F; (f) F; (g) T; (h) F; (i) F. **19.10** (a) T; (b) T; (c) F. **19.13** (a) 54.4 V; (b) 122.4 V. **19.14** 656.4 nm. **19.15** 0.5295 Å. **19.16** 6.8 V. **19.25** (b) $5a/Z$. **19.27** 0.981. **19.29** (a) T; (b) T; (c) T. **19.30** 45°, 90°, 135°. **19.32** (a) 0; (b) $h/2\pi$. **19.35** 54.7°, 125.3°. **19.38** (a) F; (b) F; (c) F; (d) T. **19.42** 3, 3/2. **19.47** (a) 3, 2, 1; 1, 0. **19.53** 4406 V. **19.58** (a) T; (b) T. **19.67** (a) 9.0×10^{-9}; (b) 1.4×10^{-9}; (c) 5.7×10^{-17}. **19.73** (a) F; (b) F; (c) F; (d) F; (e) F; (f) F; (g) F; (h) T; (i) F.

20.1 (a) 1.07 Å, 1.43 Å, 0.96 Å; (b) 1.07 Å, 1.15 Å. **20.3** (a) Bent with angle a bit less than 109.5°; (b) linear; (c) bent with angle a bit less than 120°. **20.6** (a) Trigonal bipyramidal; (b) square-based pyramid; (c) octahedral. **20.7** (a) Bent with angle close to 120°; (b) trigonal planar; (c) trigonal planar; (d) bent with angle close to 120°. **20.10** (a) −320 kJ/mol, −311.4 kJ/mol. **20.11** 1.4_5 D, 1.2 D. **20.12** (a) 1.9 D; (b) 1.9 D. **20.13** 4.3 D. **20.14** (a) 0.5; (b) 1.1; (c) 0.6. **20.19** −1. **20.21** (a) T; (b)

F; (c) F; (d) T; (e) T. **20.23** (a) 5.70 eV; (b) 10.4 D. **20.25** (a) T; (b) F; (c) T; (d) T. **20.26** (a) Even; (b) neither; (c) odd. **20.27** (a) 8.7×10^{-7}. **20.30** (a) 1/2; (b) 1; (c) 0; (d) 2. **20.39** (a) 241 kJ/mol. **20.42** (a) T; (b) T; (c) F. **20.43** (a) Yes; (b) no; (e) no. **20.45** (a) Yes; (b) no; (c) yes; (d) yes. **20.48** (a) T; (b) T; (c) F; (d) T; (e) T. **20.51** (a) 332 nm. **20.60** (a) 7; (b) 12; (e) 9. **20.62** (a) T; (b) F; (c) F; (d) F; (e) F; (f) F; (g) T; (h) T; (i) T.

21.1 (a) F; (b) F; (c) T; (d) F; (e) T. **21.2** 2.42×10^{14} Hz, 1.24×10^{-4} cm, 8060 cm^{-1}. **21.3** 2.25×10^{10} cm/s, 443 nm. **21.4** (a) F; (b) F; (c) T; (d) T; (e) T. **21.5** (a) s^{-1} or Hz; (b) m^{-1}; (c) m/s. **21.8** (b) 0. **21.10** 107 GHz. **21.11** (b) 6.82×10^{15} Hz, 1.14×10^{16} Hz, 1.59×10^{16} Hz. **21.14** Transmittances 0.79, 0.10, and 10^{-10}. **21.15** (a) 4.2×10^{-6}; (b) 1.6×10^{-54}. **21.17** 1.08, 2.0×10^{4} L mol^{-1} cm^{-1}. **21.18** 36%. **21.19** 20.6%. **21.20** (a) T; (b) T; (c) F; (d) F; (e) T; (f) T; (g) F; (h) T. **21.21** (a) F; (b) T; (c) F. **21.22** (a) 9.757 eV. **21.26** (a) 1.424258 Å, 1.424258 Å; (b) 1001.4432 GHz. **21.27** (a) 7470 MHz; (b) 7689 MHz. **21.30** 1.176×10^{6} dyn/cm. **21.31** (b) 15756 cm^{-1}. **21.33** 1, 2.714, 3.703, 3.839, 3.307, 2.450, 1.588 are the ratios to the $J = 0$ population. **21.36** (a) T; (b) T; (c) T; (d) F. **21.37** (a) A C_2 axis and two symmetry planes; (b) a C_3 axis through the CCl bond and three symmetry planes, each one containing C, Cl, and one F. **21.41** (a) Spherical; (b) symmetric; (c) asymmetric. **21.44** (a) 0, 3.046×10^{9} s^{-1}, 7.249×10^{9} s^{-1}; (b) 3046 MHz, 6093 MHz. **21.45** (c) 12.18 GHz, 24.36 GHz, 36.55 GHz. **21.46** 1.162 Å. **21.47** (a) 3; (b) 7. **21.49** (a) 0.311 eV. **21.52** 2050 cm^{-1}. **21.53** 490 N/m, 1240 N/m. **21.54** (a) T; (b) T; (c) F; (d) T; (e) T; (f) T; (g) F. **21.55** (a) 4B; (b) 1.098 Å. **21.56** (a) F; (b) T; (c) F. **21.57** 364.7 nm. **21.59** 13.50 nm. **21.62** (a) T; (b) T; (c) F; (d) T; (e) F; (f) F; (g) T; (h) T; (i) F; (j) T; (k) F; (l) T. **21.63** (a) 0; (b) 5.1×10^{-13} N. **21.64** 5.0×10^{-20} J/T. **21.66** (a) $E/(10^{-26}$ J$) = -2.04, -0.679, 0.679, 2.04$. **21.67** (a) 20.49 MHz. **21.69** (a) 1.41 T; (b) 7.05 T. **21.70** (a) 0.9999903. **21.71** 60 Hz. **21.74** (a) One singlet peak; (d) A quartet of relative intensity 2 and a triplet of relative intensity 3. **21.82** 70.1 GHz. **21.83** 25. **21.85** 14.6°. **21.86** 37.2%. **21.87** (a) 9.3 eV. **21.92** 13.8. **21.95** (a) Yes; (b) no; (c) no. **21.96** (a) Yes; (b) no; (c) yes; (d) no. **21.98** (a) F; (b) T. **21.100** (a) $\hat{E}$; (b) $\hat{\sigma}$; (c) $\hat{\imath}$. **21.101** (a) T; (b) T; (c) T; (d) T. **21.102** (a) C_3 and three σ_v. **21.105** (a) T_d; (b) $C_{\infty v}$; (c) $D_{\infty h}$; (d) D_{3h}; (e) O_h; (f) C_{4v}; (g) D_{4h}. **21.109** (a) $\begin{pmatrix} 0 & 11 \\ 6 & 6 \end{pmatrix}$; (b) $\begin{pmatrix} 3 & 15 \\ 9 & 12 \end{pmatrix}$. **21.110** $\begin{pmatrix} 20.8 & 32.2 \\ 11 & 23 \end{pmatrix}$ and $\begin{pmatrix} -0.2 & 19 \\ -7 & 44 \end{pmatrix}$. **21.111** $\begin{pmatrix} 14 \\ 30 \end{pmatrix}$. **21.120** (a) $2p_y$; (b) $-2p_y$; (c) $2p_y$; (d) $-2p_x$. **21.122** (b) A_1. **21.124** (a) 9. **21.126** 17, -1, 3. **21.127** (a) HBr, H$_2$S, CH$_3$Cl. **21.130** (a) T; (b) F; (c) F; (d) F; (e) F; (f) T; (g) F; (h) T; (i) T; (j) F.

22.1 (c) states. **22.3** 2.50. **22.8** 3×10^{27}. **22.10** (a) T; (b) F. **22.11** (a) 3628810; (b) $9.332621569 \times 10^{157}$. **22.12** 1414.905850, 1411.134742. **22.14** (a) F; (b) F; (c) T; (d) T. **22.15** 1.8×10^{-9}. **22.17** (a) 1.3×10^{-5}; (b) 0.044. **22.19** (a) 2.92. **22.20** (b) 480 K; (c) 0.0393. **22.24** (a) All. **22.25** (a) m; (b) I and σ; (c) ν. **22.27** (b) $z = 1.163$, $U = 844$ J, $S = 3.36$ J/K. **22.28** (a) 0.999955; (b) 0.950; (c) 0.865. **22.32** (a) 3.935; (b) 1.53×10^{23}, 2.78×10^{23}, 1.71×10^{23}; (c) 0.669×10^{23}, 2.01×10^{23}, 3.35×10^{23}. **22.39** 126.15 J mol^{-1} K^{-1}, etc. **22.42** 3718 J/mol for each. **22.44** (a) 3352.3 K, 2.862 K. **22.45** (a) 173.75 J mol^{-1} K^{-1}; (b) 20.79 J mol^{-1} K^{-1}; (c) 29.10 J mol^{-1} K^{-1}. **22.46** (a) 13404 J/mol; (b) 17561 J/mol; (c) 279.70 J mol^{-1} K^{-1}. **22.47** 186.7 J mol^{-1} K^{-1}. **22.49** 2.006, 2.63, 3.12. **22.54** (a) 93.9%; (b) 93.2%. **22.64** (a) 6; (b) 2; (c) 1; (d) 12. **22.65** 205.73 J mol^{-1} K^{-1}. **22.66** 213.79 J mol^{-1} K^{-1}. **22.67** (a) 62.36 and 54.04 J mol^{-1} K^{-1}. **22.75** 1.388×10^{25}. **22.77** (a) Yes; (b) no; (c) no. **22.78** (c) Cl$_2$. **22.80** (a) -7.1×10^{-8} dyn. **22.82** (a) 3.9 Å. **22.83** (a) 4.86×10^{-15} erg. **22.90** -170, -3, and 16 cm^3/mol. **22.91** (b) -163, -4, 17 cm^3/mol. **22.94** 0.00649. **22.96** (a).

23.2 (a) 3×10^{14} cm^3 mol^{-1} s^{-1}; (b) 0.003. **23.8** 5.6×10^{10} cm^3 mol^{-1} s^{-1}. **23.9** 1.5×10^{10} cm^3 mol^{-1} s^{-1}. **23.12** 8.0 kcal/mol. **23.14** k is multiplied by 0.122. **23.17** (a) 7.4 kcal/mol, 1.4_3 kcal/mol, -22.1 cal mol^{-1} K^{-1}. **23.18** (a) 0.65, 0.29, 0.052. **23.20** 2. **23.21** 15.1 kcal/mol, -18.0 cal mol^{-1} K^{-1}, 20.5 kcal/mol.

24.1 (a) Metallic; (b) molecular; (c) covalent; (d) ionic; (e) molecular. **24.2** (a) 716.7 kJ/mol; (b) 1237.6 kJ/mol. **24.3** 787.3 kJ/mol. **24.5** 44.0 kJ/mol. **24.6** -867.7 kJ/mol. **24.7** 8.4. **24.8** (a) 666 kJ/mol; (b) 620 kJ/mol. **24.11** 2 ln 2. **24.12** (a) 8.33 kJ/mol. **24.14** (a) 2; (b) 2. **24.16** 4. **24.17** 4.249 g/cm^3. **24.20** 8. **24.21** 47.6%. **24.23** 5.38 Å, 2.69 Å. **24.24** 4.30 Å, 3.72 Å. **24.25** 5.46 Å. **24.26** 1.54_5 Å. **24.27** 3.755 Å. **24.28** (a) 9.4°, 19.1°, 29.4°, 40.9°, 55.0°, 79.4°; (b) 13.4°, 27.6°, 44.0°, 67.9°. **24.29** (c) 4.085 and 4.086 Å. **24.34** (b) 7.41 and 0.514 cal mol^{-1} K^{-1}. **24.35** (a) 4.8 J mol^{-1} K^{-1}; (b) 15.8 J mol^{-1} K^{-1}. **24.39** (a) 78 K; (b) 1.66 cal mol^{-1} K^{-1}. **24.40** (a) 47 J mol^{-1} K^{-1}. **24.48** (a) 34%; (b) 99.74%. **24.49** (a) T; (b) T; (c) T; (d) T.

Index

Fundamental Constants[a]

Constant	Symbol	SI value	Non-SI value
Gas constant	R	8.3145 J mol^{-1} K^{-1} 8.3145 m^3 Pa mol^{-1} K^{-1}	8.3145×10^7 erg mol^{-1} K^{-1} 83.145 cm^3 bar mol^{-1} K^{-1} 82.057_5 cm^3 atm mol^{-1} K^{-1} 1.9872 cal mol^{-1} K^{-1}
Avogadro constant	N_A	6.022142×10^{23} mol^{-1}	
Faraday constant	F	96485.34 C mol^{-1}	
Speed of light in vacuum	c	2.99792458×10^8 m s^{-1}	$2.99792458 \times 10^{10}$ cm s^{-1}
Planck constant	h	6.62607×10^{-34} J s	6.62607×10^{-27} erg s
Boltzmann constant	k	1.38065×10^{-23} J K^{-1}	1.38065×10^{-16} erg K^{-1}
Proton charge	e	1.602176×10^{-19} C	
Electron rest mass	m_e	9.10938×10^{-31} kg	9.10938×10^{-28} g
Proton rest mass	m_p	1.672622×10^{-27} kg	1.672622×10^{-24} g
Permittivity of vacuum	ε_0	$8.8541878 \times 10^{-12}$ C^2 N^{-1} m^{-2}	
	$4\pi\varepsilon_0$	$1.112650056 \times 10^{-10}$ C^2 N^{-1} m^{-2}	
	$1/4\pi\varepsilon_0$	8.98755179×10^9 N m^2 C^{-2}	
Permeability of vacuum	μ_0	$4\pi \times 10^{-7}$ N C^{-2} s^2	
Gravitational constant	G	6.673×10^{-11} m^3 s^{-2} kg^{-1}	6.673×10^{-8} cm^3 s^{-2} g^{-1}

[a]Adapted from P. J. Mohr and B. N. Taylor, *Rev. Mod. Phys.*, **72**, 351 (2000).

Defined Constants

Standard gravitational acceleration $g_n \equiv 9.80665$ m/s^2
Zero of the Celsius scale $\equiv 273.15$ K

Greek Alphabet

Alpha	A	α	Iota	I	ι	Rho	P	ρ
Beta	B	β	Kappa	K	κ	Sigma	Σ	σ
Gamma	Γ	γ	Lambda	Λ	λ	Tau	T	τ
Delta	Δ	δ	Mu	M	μ	Upsilon	Y	υ
Epsilon	E	ε	Nu	N	ν	Phi	Φ	ϕ
Zeta	Z	ζ	Xi	Ξ	ξ	Chi	X	χ
Eta	H	η	Omicron	O	o	Psi	Ψ	ψ
Theta	Θ	θ	Pi	Π	π	Omega	Ω	ω